1s		
	1 H Hydrogen	2 He Helium

	13	14	15	16	17	18
p	5 B Boron	6 C Carbon	7 N Nitrogen	8 O Oxygen	9 F Fluorine	10 Ne Neon
	13 Al Aluminum	14 Si Silicon	15 P Phosphorus	16 S Sulfur	17 Cl Chlorine	18 Ar Argon
	31 Ga Gallium	32 Ge Germanium	33 As Arsenic	34 Se Selenium	35 Br Bromine	36 Kr Krypton
	49 In Indium	50 Sn Tin	51 Sb Antimony	52 Te Tellurium	53 I Iodine	54 Xe Xenon
	81 Tl Thallium	82 Pb Lead	83 Bi Bismuth	84 Po Polonium	85 At Astatine	86 Rn Radon

8	9	10	11	12
26 Fe Iron	27 Co Cobalt	28 Ni Nickel	29 Cu Copper	30 Zn Zinc
44 Ru Ruthenium	45 Rh Rhodium	46 Pd Palladium	47 Ag Silver	48 Cd Cadmium
76 Os Osmium	77 Ir Iridium	78 Pt Platinum	79 Au Gold	80 Hg Mercury
108 Hs Hassium	109 Mt Meitnerium			

67 Ho Holmium	68 Er Erbium	69 Tm Thulium	70 Yb Ytterbium
99 Es Einsteinium	100 Fm Fermium	101 Md Mendelevium	102 No Nobelium

LANGE'S HANDBOOK OF CHEMISTRY

LANGE'S HANDBOOK OF CHEMISTRY

James G. Speight, Ph.D.

CD&W Inc., Laramie, Wyoming

Sixteenth Edition

McGRAW-HILL

New York Chicago San Francisco Lisbon London Madrid
Mexico City Milan New Delhi San Juan Seoul
Singapore Sydney Toronto

The McGraw-Hill Companies

Library of Congress Catalog Card Number 84-643191
ISSN 0748-4585

2 3 4 5 6 7 8 9 0 DOC/DOC 0 1 0 9 8 7 6

ISBN 0-07-143220-5

The sponsoring editor for this book was Kenneth P. McCombs and the production supervisor was Sherri Souffrance. It was set in Times Roman by International Typesetting and Composition. The art director for the cover was Anthony Landi.

Printed and bound by RR Donnelley.

This book is printed on acid-free paper.

CONTENTS

PREFACE TO THE SIXTEENTH EDITION

This Sixteenth Edition of *Lange's Handbook of Chemistry* takes on a new format under a new editor. Nevertheless, the Handbook remains the one-volume source of factual information for chemists and chemical engineers, both professionals and students. The aim of the Handbook remains to provide sufficient data to satisfy the general needs of the user without recourse to other reference sources. The many tables of numerical data that have been compiled, as well as additional tables, will provide the user with a valuable time-saver.

The new format involves division of the Handbook into four major sections, instead of the 11 sections that were part of previous editions. Section 1, Inorganic Chemistry, contains a group of tables relating to the physical properties of the elements (including recently discovered elements) and several thousand compounds. Likewise, Section 2, Organic Chemistry, contains a group of tables relating to the physical properties of the elements and several thousand compounds. Following these two sections, Section 3, Spectroscopy, presents the user with the fundamentals of the various spectroscopic techniques. This section also contains tables that are relevant to the spectroscopic properties of elements, inorganic compounds, and organic compounds. Section 4, General Information and Conversion Tables, contains all of the general information and conversion tables that were previously found in different sections of the Handbook.

In Sections 1 and 2, the data for each compound include (where available) name, structural formula, formula weight, density, refractive index, melting point, boiling point, flash point, dielectric constant, dipole moment, solubility (if known) in water and relevant organic solvents, thermal conductivity, and electrical conductivity. The presentation of alternative names, as well as trivial names of long-standing use, has been retained. Section 2 also contains expanded information relating to the names and properties of condensed polynuclear aromatic compounds.

Enthalpies and Gibbs Energies of Formation, Entropies, and Heat Capacities of Organic and Inorganic Compounds, and Heats of Melting, Vaporization, and Sublimation and Specific Heat at Various Temperatures, are also presented in Sections 1 and 2 for organic and inorganic compounds, as well as information on the critical properties (critical temperature, critical pressure, and critical volume).

As in the previous edition, Section 3, Spectroscopy, retains subsections on infrared spectroscopy, Raman spectroscopy, fluorescence spectroscopy, mass spectrometry, and X-ray spectrometry. The section on Practical Laboratory Information (now Section 4), has been retained as it offers valuable information and procedures for laboratory methods.

As stated in the prefaces of earlier editions, every effort has been made to select the most useful and reliable information and to record it with accuracy. It is hoped that users of this Handbook will continue to offer suggestions of material that might be included in, or even excluded from, future editions and call attention to errors. These communications should be directed to the editor through the publisher, McGraw-Hill.

JAMES G. SPEIGHT, PH.D.
Laramie, Wyoming

PREFACE TO THE FIFTEENTH EDITION

This new edition, the fifth under the aegis of the present editor, remains the one-volume source of factual information for chemists, both professionals and students—the first place in which to "look it up" on the spot. The aim is to provide sufficient data to satisfy all one's general needs without recourse to other reference sources. A user will find this volume of value as a time-saver because of the many tables of numerical data that have been especially compiled.

Descriptive properties for a basic group of approximately 4300 organic compounds are compiled in Section 1, an increase of 300 entries. All entries are listed alphabetically according to the senior prefix of the name. The data for each organic compound include (where available) name, structural formula, formula weight, Beilstein reference (or if un- available, the entry to the *Merck Index*, 12th ed.), density, refractive index, melting point, boiling point, flash point, and solubility (citing numerical values if known) in water and various common organic solvents. Structural formulas either too complex or too ambiguous to be rendered as line formulas are grouped at the bottom of each facing double page on which the entries appear. Alternative names, as well as trivial names of long-standing usage, are listed in their respective alphabetical order at the bottom of each double page in the regular alphabetical sequence. Another feature that assists the user in locating a desired entry is the empirical formula index.

Section 2 on General Information, Conversion Tables, and Mathematics has had the table on general conversion factors thoroughly reworked. Similarly the material on Statistics in Chemical Analysis has had its contents more than doubled.

Descriptive properties for a basic group of inorganic compounds are compiled in Section 3, which has undergone a small increase in the number of entries. Many entries under the column "Solubility" supply the reader with precise quantities dissolved in a stated solvent and at a given temperature. Several portions of Section 4, Properties of Atoms, Radicals, and Bonds, have been significantly enlarged. For example, the entries under "Ionization Energy of Molecular and Radical Species" now number 740 and have an additional column with the enthalpy of formation of the ions. Likewise, the table on "Electron Affinities of the Elements, Molecules, and Radicals" now contains about 225 entries. The Table of Nuclides has material on additional radionuclides, their radiations, and the neutron capture cross sections.

Revised material for Section 5 includes the material on surface tension, viscosity, dielectric constant, and dipole moment for organic compounds. In order to include more data at several temperatures, the material has been divided into two separate tables. Material on surface tension and viscosity constitute the first table with 715 entries; included is the temperature range of the liquid phase. Material on dielectric constant and dipole moment constitute another table of 1220 entries. The additional data at two or more temperatures permit interpolation for intermediate temperatures and also permit limited extrapolation of the data. The Properties of Combustible Mixtures in Air has been revised and expanded to include over 450 compounds. Flash points are to be found in Section 1. Completely revised are the tables on Thermal Conductivity for gases, liquids, and solids. Van der Waals' constants for gases have been brought up to date and expanded to over 500 substances.

Section 6, which includes Enthalpies and Gibbs Energies of Formation, Entropies, and Heat Capacities of Organic and Inorganic Compounds, and Heats of Melting, Vaporization, and Sublimation and Specific Heat at Various Temperatures for organic and inorganic compounds, has expanded by

11 pages, but the major additions have involved data in columns where it previously was absent. More material has also been included for critical temperature, critical pressure, and critical volume.

The section on Spectroscopy has been retained but with some revisions and expansion. The section includes ultraviolet-visible spectroscopy, fluorescence, infrared and Raman spectroscopy, and X-ray spectrometry. Detection limits are listed for the elements when using flame emission, flame atomic absorption, electrothermal atomic absorption, argon induction coupled plasma, and flame atomic fluorescence. Nuclear magnetic resonance embraces tables for the nuclear properties of the elements, proton chemical shifts and coupling constants, and similar material for carbon-13, boron-11, nitrogen-15, fluorine-19, silicon-29, and phosphorus-31.

In Section 8, the material on solubility constants has been doubled to 550 entries. Sections on proton transfer reactions, including some at various temperatures, formation constants of metal complexes with organic and inorganic ligands, buffer solutions of all types, reference electrodes, indicators, and electrode potentials are retained with some revisions. The material on conductance has been revised and expanded, particularly in the table on limiting equivalent ionic conductance.

Everything in Sections 9 and 10 on physiochemical relationships, and on polymers, rubbers, fats, oils, and waxes, respectively, has been retained.

Section 11, Practical Laboratory Information, has undergone significant changes and expansion. Entries in the table on "Molecular Elevation of the Boiling Point" have been increased. McReynolds' constants for stationary phases in gas chromatography have been reorganized and expanded. The guide to ion-exchange resins and discussion is new and embraces all types of column packing and membrane materials. Gravimetric factors have been altered to reflect the changes in atomic weights for several elements. Newly added are tables listing elements precipitated by general analytical reagents, and giving equations for the redox determination of the elements with their equivalent weights. Discussion on the topics of precipitation and complexometric titration include primary standards and indicators for each analytical technique. A new topic of masking and demasking agents includes discussion and tables of masking agents for various elements, for anions and neutral molecules, and common demasking agents. A table has been added listing the common amino acids with their pI and pK_a values and their 3-letter and I-letter abbreviations. Lastly a 9-page table lists the threshold limit value (TL V) for gases and vapors.

As stated in earlier prefaces, every effort has been made to select the most useful and reliable information and to record it with accuracy. However, the editor's 50 years of involvement with textbooks and handbooks bring a realization of the opportunities for gremlins to exert their inevitable mischief. It is hoped that users of this handbook will continue to offer suggestions of material that might be included in, or even excluded from, future editions and call attention to errors. These communications should be directed to the editor. The street address will change early in 1999, as will the telephone number.

JOHN A. DEAN
Knoxville, Tennessee

PREFACE TO THE FIRST EDITION

This book is the result of a number of years' experience in the compiling and editing of data useful to chemists. In it an effort has been made to select material to meet the needs of chemists who cannot command the unlimited time available to the research specialist, or who lack the facilities of a large technical library which so often is not conveniently located at many manufacturing centers. If the information contained herein serves this purpose, the compiler will feel that he has accomplished a worthy task. Even the worker with the facilities of a comprehensive library may find this volume of value as a time-saver because of the many tables of numerical data which have been especially computed for this purpose.

Every effort has been made to select the most reliable information and to record it with accuracy. Many years of occupation with this type of work bring a realization of the opportunities for the occurrence of errors, and while every endeavor has been made to prevent them, yet it would be remarkable if the attempts towards this end had always been successful. In this connection it is desired to express appreciation to those who in the past have called attention to errors, and it will be appreciated if this be done again with the present compilation for the publishers have given their assurance that no expense will be spared in making the necessary changes in subsequent printings.

It has been aimed to produce a compilation complete within the limits set by the economy of available space. One difficulty always at hand to the compiler of such a book is that he must decide what data are to be excluded in order to keep the volume from becoming unwieldy because of its size. He can hardly be expected to have an expert's knowledge of all branches of the science nor the intuition necessary to decide in all cases which particular value to record, especially when many differing values are given in the literature for the same constant. If the expert in a particular field will judge the usefulness of this book by the data which it supplies to him from fields other than his specialty and not by the lack of highly specialized information in which only he and his co-workers are interested (and with which he is familiar and for which he would never have occasion to consult this compilation), then an estimate of its value to him will be apparent. However, if such specialists will call attention to missing data with which they are familiar and which they believe others less specialized will also need, then works of this type can be improved in succeeding editions.

Many of the gaps in this volume are caused by the lack of such information in the literature. It is hoped that to one of the most important classes of workers in chemistry, namely the teachers, the book will be of value not only as an aid in answering the most varied questions with which they are confronted by interested students, but also as an inspiration through what it suggests by the gaps and inconsistencies, challenging as they do the incentive to engage in the creative and experimental work necessary to supply the missing information.

While the principal value of the book is for the professional chemist or student of chemistry, it should also be of value to many people not especially educated as chemists. Workers in the natural sciences—physicists, mineralogists, biologists, pharmacists, engineers, patent attorneys, and librarians—are often called upon to solve problems dealing with the properties of chemical products or materials of construction. For such needs this compilation supplies helpful information and will serve not only as an economical substitute for the costly accumulation of a large library of monographs on specialized subjects, but also as a means of conserving the time required to search for information so widely scattered throughout the literature. For this reason especial care has been taken in compiling a comprehensive index and in furnishing cross references with many of the tables.

It is hoped that this book will be of the same usefulness to the worker in science as is the dictionary to the worker in literature, and that its resting place will be on the desk rather than on the bookshelf.

N. A. LANGE
Cleveland, Ohio
May 2, 1934

LANGE'S HANDBOOK OF CHEMISTRY

SECTION 1

INORGANIC CHEMISTRY

SECTION 1
INORGANIC CHEMISTRY

1.1 NOMENCLATURE OF INORGANIC COMPOUNDS

The following synopsis of rules for naming inorganic compounds and the examples given in explanation are not intended to cover all the possible cases.

1.1.1 Writing Formulas

1.1.1.1 Mass Number, Atomic Number, Number of Atoms, and Ionic Charge. The mass number, atomic number, number of atoms, and ionic charge of an element are indicated by means of four indices placed around the symbol:

mass number atomic number	**SYMBOL**	ionic charge number of atoms	${}^{15}_{7}N^{3-}_{2}$

Ionic charge should be indicated by an Arabic superscript numeral preceding the plus or minus sign: Mg^{2+}, PO_4^{3-}

1.1.1.2 Placement of Atoms in a Formula. The electropositive constituent (cation) is placed first in a formula. If the compound contains more than one electropositive or more than one electronegative constituent, the sequence within each class should be in alphabetical order of their symbols. The alphabetical order may be different in formulas and names; for example, $NaNH_4HPO_4$, ammonium sodium hydrogen phosphate.

Acids are treated as hydrogen salts. Hydrogen is cited last among the cations.

When there are several types of ligands, anionic ligands are cited before the neutral ligands.

1.1.1.3 Binary Compounds between Nonmetals. For binary compounds between nonmetals, that constituent should be placed first which appears earlier in the sequence:

Rn, Xe, Kr, Ar, Ne, He, B, Si, C, Sb, As, P, N, H, Te, Se, S, At, I, Br, Cl, O, F

Examples: $AsCl_3$, SbH_3, H_3Te, BrF_3, OF_2, and N_4S_4.

1.1.1.4 Chain Compounds. For chain compounds containing three or more elements, the sequence should be in accordance with the order in which the atoms are actually bound in the molecule or ion.

Examples: SCN^- (thiocyanate), HSCN (hydrogen thiocyanate or thiocyanic acid), HNCO (hydrogen isocyanate), HONC (hydrogen fulminate), and HPH_2O_2 (hydrogen phosphinate).

1.1.1.5 Use of Centered Period. A centered period is used to denote water of hydration, other solvates, and addition compounds; for example, $CuSO_4 \cdot 5H_2O$, copper(II) sulfate 5-water (or pentahydrate).

1.1.1.6 Free Radicals. In the formula of a polyatomic radical an unpaired electron(s) is (are) indicated by a dot placed as a right superscript to the parentheses (or square bracket for coordination compounds). In radical ions the dot precedes the charge. In structural formulas, the dot may be placed to indicate the location of the unpaired electron(s).

Examples: $(HO)^{\cdot}$ $(O_2)^{2\cdot}$ $(\dot{N}H^{+}{}_{3})$

1.1.1.7 Enclosing Marks. Where it is necessary in an inorganic formula, enclosing marks (parentheses, braces, and brackets) are nested within square brackets as follows:

[( )], [{ ( ) }], [{ [( )] }], [{ { [( )] } }]

1.1.1.8 Molecular Formula. For compounds consisting of discrete molecules, a formula in accordance with the correct molecular weight of the compound should be used.

Examples: S_2Cl_2, S_8, N_2O_4, and $H_4P_2O_6$; not SCl, S, NO_2, and H_2PO_3.

1.1.1.9 Structural Formula and Prefixes. In the structural formula the sequence and spatial arrangement of the atoms in a molecule are indicated.

Examples: NaO(O=C)H (sodium formate), Cl—S—S—Cl (disulfur dichloride).

Structural prefixes should be italicized and connected with the chemical formula by a hyphen: *cis-*, *trans-*, *anti-*, *syn-*, *cyclo-*, *catena-*, *o-* or *ortho-*, *m-* or *meta-*, *p-* or *para-*, *sec-* (secondary), *tert-* (tertiary), *v-* (vicinal), *meso-*, *as-* for asymmetrical, and *s-* for symmetrical.

The sign of optical rotation is placed in parentheses, (+) for dextrorotary, (–) for levorotary, and (±) for racemic, and placed before the formula. The wavelength (in nanometers is indicated by a right subscript; unless indicated otherwise, it refers to the sodium D-line.

The italicized symbols *d-* (for deuterium) and *t-* (for tritium) are placed after the formula and connected to it by a hyphen. The number of deuterium or tritium atoms is indicated by a subscript to the symbol.

Examples:	*cis*-$[PtCl_2(NH_3)_2]$	methan-d_3-ol
	di-*tert*-butyl sulfate	$(+)_{589}$ $[Co(en)_3]Cl_2$
	methan-ol-*d*	

1.1.2 Naming Compounds

1.1.2.1 Names and Symbols for Elements. Names and symbols for the elements are given in Table 1.3. Wolfram is preferred to tungsten but the latter is used in the United States. In forming a complete name of a compound, the name of the electropositive constituent is left unmodified except when it is necessary to indicate the valency (see oxidation number and charge number, (formerly the Stock and Ewens-Bassett systems). The order of citation follows the alphabetic listing of the names of the cations followed by the alphabetical listing of the anions and ligands. The alphabetical citation is maintained regardless of the number of each ligand.

Example: $K[AuS(S_2)]$ is potassium (disulfido)thioaurate (1–).

1.1.2.2 Electronegative Constituents. The name of a monatomic electronegative constituent is obtained from the element name with its ending (-en, -ese, -ic, -ine, -ium, -ogen, -on, -orus, -um, -ur, -y, or -ygen) replaced by -ide. The elements bismuth, cobalt, nickel, zinc, and the noble gases are used unchanged with the ending -ide. Homopolyatomic ligands will carry the appropriate prefix. A few Latin names are used with affixes: cupr- (copper), aur- (gold), ferr- (iron), plumb- (lead), argent- (silver), and stann- (tin).

For binary compounds the name of the element standing later in the sequence in Sec. 1.1.1.3 is modified to end in -ide. Elements other than those in the sequence of Sec. 1.1.1.3 are taken in the reverse order of the following sequence, and the name of the element occurring last is modified to end in -ide; e.g., calcium stannide.

ELEMENT SEQUENCE

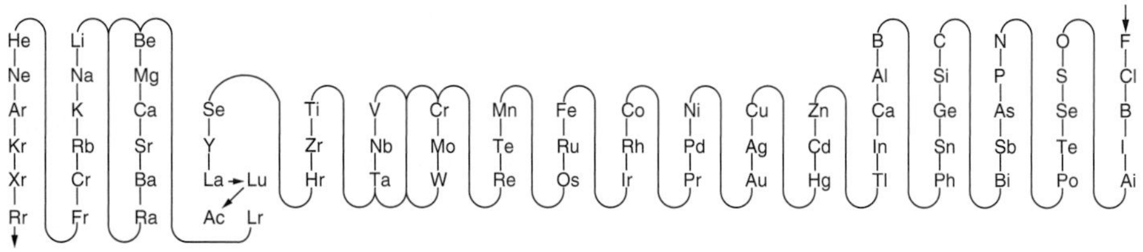

1.1.2.3 Stoichiometric Proportions. The stoichiometric proportions of the constituents in a formula may be denoted by Greek numerical prefixes: mono-, di-, tri-, tetra-, penta-, hexa-, hepta-, octa-, nona- (Latin), deca-, undeca- (Latin), dodeca-, …, icosa- (20), henicosa- (21), …, triconta- (30), tetraconta- (40), …, hecta- (100), and so on, preceding without a hyphen the names of the elements to which they refer. The prefix mono can usually be omitted; occasionally hemi- (1/2) and sesqui- (3/2) are used. No elisions are made when using numerical prefixes except in the case of icosa- when the letter "i" is elided in docosa- and tricosa-. Beyond 10, prefixes may be replaced by Arabic numerals.

When it is required to indicate the number of entire groups of atoms, the multiplicative numerals bis-, tris-, tetrakis-, pentakis-, and so on, are used (i.e., -kis is added starting from tetra-). The entity to which they refer is placed in parentheses.

Examples: $Ca[PF_6]_2$, calcium bis(hexafluorophosphate); and $(C_{10}H_{21})_3PO_4$, tris(decyl) phosphate instead of tridecyl which is $(C_{13}H_{27}–)$.

Composite numeral prefixes are built up by citing units first, then tens, then hundreds, and so on. For example, 43 is written tritetraconta- (or tritetracontakis-).

In indexing it may be convenient to italicize a numerical prefix at the beginning of the name and connect it to the rest of the name with a hyphen; e.g., *di*-nitrogen pentaoxide (indexed under the letter "n").

1.1.2.4 Oxidation and Charge Numbers. The *oxidation number* (Stock system) of an element is indicated by a Roman numeral placed in parentheses immediately following the name of the element. For zero, the cipher 0 is used. When used in conjunction with symbols, the Roman numeral may be placed above and to the right. The *charge number* of an ion (Ewens-Bassett system) rather than the oxidation state is indicated by an Arabic numeral followed by the sign of the charge cited and is placed in parentheses immediately following the name of the ion.

Examples: P_2O_5, diphosphorus pentaoxide or phosphorus(V) oxide; Hg_2^{2+}. mercury(I) ion or dimercury (2+) ion; $K_2[Fe(CN)_6]$, potassium hexacyanoferrate(II) or potassium hexacyanoferrate(4–); $Pb_2^{II}Pb^{IV}O_4$, dilead(II) lead(IV) oxide or trilead tetraoxide.

Where it is not feasible to define an oxidation state for each individual member of a group, the overall oxidation level of the group is defined by a formal ionic charge to avoid the use of fractional oxidation states; for example, O_2^-.

1.1.2.5 Collective Names. Collective names include:

Halogens (F, Cl, Br, I, At)

Chalcogens (O, S, Se, Te, Po)

Alkali metals (Li, Na, K, Rb, Cs, Fr)

Alkaline-earth metals (Ca, Sr, Ba, Ra)

Lanthanoids or lanthanides (La to Lu)

Rare-earth metals (Sc, Y, and La to Lu inclusive)

Actinoids or actinides (Ac to Lr, those whose 5*f* shell is being filled)

Noble gases (He to Rn)

A transition element is an element whose atom has an incomplete *d* subshell, or which gives rise to a cation or cations with an incomplete *d* subshell.

1.1.2.6 Isotopically Labeled Compounds. The hydrogen isotopes are given special names: ^1H (protium), ^2H or D (deuterium), and ^3H or T (tritium). The superscript designation is preferred because D and T disturb the alphabetical ordering in formulas.

Other isotopes are designated by mass numbers: ^{10}B (boron-10).

Isotopically labeled compounds may be described by inserting the italic symbol of the isotope in brackets into the name of the compound; for example, $H^{36}Cl$ is hydrogen chloride[^{36}*Cl*] or hydrogen chloride-36, and $^2H^{38}Cl$ is hydrogen [^{2}*H*] chloride[^{38}*Cl*] or hydrogen-2 chloride-38.

1.1.2.7 Allotropes. Systematic names for gaseous and liquid modifications of elements are sometimes needed. Allotropic modifications of an element bear the name of the atom together with the descriptor to specify the modification. The following are a few common examples:

Symbol	Trivial name	Systematic name
H	Atomic hydrogen	Monohydrogen
O_2	(Common oxygen)	Dioxygen
O_3	Ozone	Trioxygen
P_4	White phosphorus	Tetraphosphorus
S_8	α-Sulfur, β-Sulfur	Octasulfur
S_n	μ-Sulfur (plastic sulfur)	Polysulfur

Trivial (customary) names are used for the amorphous modification of an element.

1.1.2.8 Heteroatomic and Other Anions. A few heteroatomic anions have names ending in -ide. These are

—OH, hydroxide ion (not hydroxyl)

—CN, cyanide ion

—NH_2^- hydrogen difluoride ion

—NH_2, amide ion

—NH—, imide ion

—NH—NH_2, hydrazide ion

—NHOH, hydroxylamide ion

—HS^-, hydrogen sulfide ion

Added to these anions are

—triiodide ion

—N_3, axide ion

—O_3, ozonide ion

—O—O—, peroxide ion

—S—S—, disulfide ion

1.1.2.9 Binary Compounds of Hydrogen. Binary compounds of hydrogen with the more electropositive elements are designated hydrides (NaH, sodium hydride).

Volatile hydrides, except those of Periodic Group VII and of oxygen and nitrogen, are named by citing the root name of the element (penultimate consonant and Latin affixes, Sec. 1.1.2.2) followed by the suffix -ane. Exceptions are water, ammonia, hydrazine, phosphine, arsine, stibine, and bismuthine.

Examples: B_2H_6, diborane; $B_{10}H_{14}$, decaborane (14); $B_{10}H_{16}$, decaborane (16); P_2H_4, diphosphane; Sn_2H_6, distannane; H_2Se_2, diselane; H_2Te_2, ditellane; H_2S_5, pentasulfane; and pbH_4, plumbane.

1.1.2.10 Neutral Radicals. Certain neutral radicals have special names ending in -yl:

HO	hydroxyl	PO	phosphoryl
CO	carbonyl	SO	sulfinyl (thionyl)
ClO	chlorosyl*	SO_2	sulfonyl (sulfuryl)
ClO_2	chloryl*	S_2O_5	disulfuryl
ClO_3	perchloryl*	SeO	seleninyl
CrO_2	chromyl	SeO_2	selenoyl
NO	nitrosyl	UO_2	uranyl
NO_2	nitryl (nitroyl)	NpO_2	neptunyl†

Radicals analogous to the above containing other chalcogens in place of oxygen are named by adding the prefixes thio-, seleno-, and so on; for example, PS, thiophosphoryl; CS, thiocarbonyl.

*Similarly for the other halogens.
†Similarly for the other actinide elements.

1.1.3 Cations

1.1.3.1 Monatomic Cations. Monatomic cations are named as the corresponding element; for example, Fe^{2+}, iron(II) ion; Fe^{3+}, iron(III) ion.

This principle also applies to polyatomic cations corresponding to radicals with special names ending in -yl (Sec. 1.1.2.10); for example, PO^+, phosphoryl cation; NO^+, nitrosyl cation; NO_2^{2+}, nitryl cation; O_2^{2+} oxygenyl cation.

Use of the oxidation number and charge number extends the range for radicals; for example, UO_2^{2+} uranyl(VI) or uranyl(2+) cation; UO_2^+, uranyl(V) or uranyl(1+) cation.

1.1.3.2 Polyatomic Cations. Polyatomic cations derived by addition of more protons than required to give a neutral unit to polyatomic anions are named by adding the ending -onium to the root of the name of the anion element; for example, PH_4^+phosphonium ion; H_2I^+, iodonium ion; H_3O^+, oxonium ion; $CH_3OH_2^+$methyl oxonium ion.

Exception: The name ammonium is retained for the NH_4^+ ion; similarly for substituted ammonium ions; for example, NF_4^+ tetrafluoroammonium ion.

Substituted ammonium ions derived from nitrogen bases with names ending in -amine receive names formed by changing -amine into -ammonium. When known by a name not ending in -amine, the cation name is formed by adding the ending -ium to the name of the base (eliding the final vowel); e.g., anilinium, hydrazinium, imidazolium, acetonium, dioxanium.

Exceptions are the names uronium and thiouronium derived from urea and thiourea, respectively.

1.1.3.3 Multiple Ions from One Base. Where more than one ion is derived from one base, the ionic charges are indicated in their names: $N_2H_5^+$, hydrazinium(1+) ion; $N_2H_6^{2+}$, hydrazinium(2+) ion.

1.1.4 Anions

See Secs. 1.1.2.2 and 1.1.2.8 for naming monatomic and certain polyatomic anions. When an organic group occurs in an inorganic compound, organic nomenclature (*q.v.*) is followed to name the organic part.

1.1.4.1 Protonated Anions. Ions such as HSO_4^- are recommended to be named hydrogensulfate with the two words written as one following the usual practice for polyatomic anions.

1.1.4.2 Other Polyatomic Anions. Names for other polyatomic anions consist of the root name of the central atom with the ending -ate and followed by the valence of the central atom expressed by its oxidation number. Atoms and groups attached to the central atom are treated as ligands in a complex.

Examples: $[Sb(OH)_6^-]$, hexahydroxoantimonate(V); $[Fe(CN_6]^{3-}$, hexacyanoferrate(III); $[Co(NO_2)_6]^{3-}$, hexanitritocobaltate(III); $[TiO(C_2O_4)_2(H_2O)_2]^{2-}$, oxobisoxalatodiaquatitanate(IV); $[PCl_6]^-$, hexachlorophosphate(V).

Exceptions to the use of the root name of the central atom are antimonate, bismuthate, carbonate, cobaltate, nickelate (or niccolate), nitrate, phosphate, tungstate (or wolframate), and zincate.

1.1.4.3 Anions of Oxygen. Oxygen is treated in the same manner as other ligands with the number of -oxo groups indicated by a suffix; for example, SO_3^{2-}, trioxosulfate.

The ending -ite, formerly used to denote a lower state of oxidation, may be retained in trivial names in these cases (note Sec. 1.1.5.3 also):

---

†Similarly for the other actinoid elements.

AsO_3^{3-}	arsenite	NOO_2^-	peroxonitrite
BrO^-	hypobromite	PO_3^{3-}	phosphite*
ClO^-	hypochlorite	SO_3^{2-}	sulfite
ClO_2^-	chlorite	$S_2O_5^{2-}$	disulfite
IO^-	hypoiodite	$S_2O_4^{2-}$	dithionite
NO_2^-	nitrite	$S_2O_2^{2-}$	thiosulfite
$N_2O_2^{2-}$	hyponitrite	SeO_3^{2-}	selenite

However, compounds known to be double oxides in the solid state are named as such; for example, Cr_2CuO_4 (actually $Cr_2O_3 \cdot CuO$) is chromium(III) copper(II) oxide (and not copper chromite).

1.1.4.4 Isopolyanions. Isopolyanions are named by indicating with numerical prefixes the number of atoms of the characteristic element. It is not necessary to give the number of oxygen atoms when the charge of the anion or the number of cations is indicated.

Examples: $Ca_3Mo_7O_{24}$, tricalcium 24-oxoheptamolybdate, may be shortened to tricalcium heptamolybdate; the anion, $Mo_7O_{24}^{6-}$, is heptamolybdate(6–); $S_2O_7^{2-}$, disulfate(2–); $P_2O_7^{4-}$, diphosphate(V)(4-).

When the characteristic element is partially or wholly present in a lower oxidation state than corresponds to its Periodic Group number, oxidation numbers are used; for example, $[O_2HP—O—PO_3H]^{2-}$, dihydrogendiphosphate(III, V)(2–).

A bridging group should be indicated by adding the Greek letter μ immediately before its name and separating this from the rest of the complex by a hyphen. The atom or atoms of the characteristic element to which the bridging atom is bonded, is indicated by numbers.

Examples: $[O_3P—S—PO_2—O—PO_3]^{5-}$, 1, 2-$\mu$-thiotriphosphate(5–)
$[S_3P—O—PS_2—O—PS_3]^{5-}$, di-$\mu$-oxo-octathiotriphosphate(5–)

1.1.5 Acids

1.1.5.1 Acids and -ide Anions. Acids giving rise to the -ide anions (Sec. 1.1.2.2) should be named as hydrogen … -ide; for example, HCl, hydrogen chloride; HN_3, hydrogen azide.

Names such as hydrobromic acid refer to an aqueous solution, and percentages such as 48% HBr denote the weight/volume of hydrogen bromide in the solution.

1.1.5.2 Acids and -ate Anions. Acids giving rise to anions bearing names ending in -ate are treated as in Sec. 1.1.5.1; for example, H_2GeO_4, hydrogen germanate; $H_4[Fe(CN)_6]$, hydrogen hexacyanoferrate(II).

1.1.5.3 Trivial Names. Acids given in Table 1.1 retain their trivial names due to long-established usage. Anions may be formed from these trivial names by changing -ous acid to -ite, and -ic acid to -ate. The prefix hypo- is used to denote a lower oxidation state and the prefix per- designates a higher oxidation state. The prefixes ortho- and meta- distinguish acids of differing water content; for example, H_4SiO_4 is orthosilicic acid and H_2SiO_3 is metasilicic acid. The anions would be named silicate (4–) and silicate(2–), respectively.

1.1.5.4 Peroxo- Group. When used in conjunction with the trivial names of acids, the prefix peroxo- indicates substitution of —O—by—O—O—.

*Named for esters formed from the hypothetical acid $P(OH)_3$.

TABLE 1.1 Trivial Names for Acids

H_3AsO_4	arsenic acid
H_3AsO_3	arsenious acid
H_3BO_3	orthoboric acid (or boric acid)
HBO_2	metaboric acid
$HBrO_3$	bromic acid
$HBrO_2$	bromous acid
HBrO	hypobromous acid
H_2CO_3	carbonic acid
HOCN	cyanic acid
HNCO	isocyanic acid
HONC	fulminic acid
$HClO_4$	perchloric acid
$HClO_3$	chloric acid
$HClO_2$	chlorous acid
HClO	hypochlorous acid
H_2CrO_4	chromic acid
$H_2Cr_2O_7$	dichromic acid
H_5IO_6	orthoperiodic acid
HIO_4	periodic acid
HIO_3	iodic acid
HIO	hypoiodous acid
$HMnO_4$	permanganic acid
H_2MnO_4	manganic acid
HNO_4	peroxonitric acid
HNO_3	nitric acid
HNO_2	nitrous acid
H_2NO_2	nitroxylic acid
$H_2N_2O_2$	hyponitrous acid
HOONO	peroxonitrous acid
H_3PO_4	orthophosphoric acid (or phosphoric acid)
HPO_3	metaphosphoric acid
H_3PO_5	peroxomonophosphoric acid
$H_4P_2O_7$	diphosphoric acid (or pyrophosphoric acid)
$H_4P_2O_8$	peroxodiphosphoric acid
$(HO)_2OP$—$PO(OH)_2$	diphosphoric(IV) acid or hypophosphoric acid
$(HO)_2P$—O—$P(O)(OH)_2$	diphosphoric(III,V) acid
H_2PHO_3	phosphonic acid
$H_2P_2H_2O_5$	diphosphonic acid
HPH_2O_2	phosphinic acid (formerly hypophosphorous acid)
$HReO_4$	perrhenic acid
H_2ReO_4	rhenic acid
H_2SO_4	sulfuric acid
$H_2S_2O_7$	disulfuric acid
H_2SO_5	peroxomonosulfuric acid
$H_2S_2O_3$	thiosulfuric acid
$H_2S_2S_6$	dithionic acid
H_2SO_3	sulfurous acid
$H_2S_2O_5$	disulfurous acid
$H_2S_2O_2$	thiosulfurous acid
$H_2S_2O_4$	dithionous acid
$H_2S_xO_6$ ($x = 3, 4, \ldots$)	polythionic acid (tri-, tetra-, . . .)
H_2SO_2	sulfoxylic acid
$HSb(OH)_6$	hexahydrooxoantimonic acid
H_2SeO_4	selenic acid
H_2SeO_3	selenious acid
H_4SiO_4	orthosilicic acid
H_2SiO_3	metasilicic acid
$HTcO_4$	pertechnetic acid
H_2TcO_4	technetic acid
H_6TeO_6	orthotelluric acid

1.1.5.5 Replacement of Oxygen by Other Chalcogens. Acids derived from oxoacids by replacement of oxygen by sulfur are called thioacids, and the number of replacements are given by prefixes di-, tri-, and so on. The affixes seleno- and telluro- are used analogously.

Examples: HOO—C=S, thiocarbonic acid; HSS—C=S, trithiocarbonic acid.

1.1.5.6 Ligands Other than Oxygen and Sulfur. See Sec. 1.1.7, Coordination Compounds, for acids containing ligands other than oxygen and sulfur (selenium and tellurium).

1.1.5.7 Differences between Organic and Inorganic Nomenclature. Organic nomenclature is largely built upon the scheme of substitution, that is, the replacement of hydrogen atoms by other atoms or groups. Although rare in inorganic nomenclature: NH_2Cl is called chloramine and $NHCl_2$ dichloroamine. Other substitutive names are fluorosulfonic acid and chlorosulfonic acid derived from HSO_3H. These and the names aminosulfonic acid (sulfamic acid), iminodisulfonic acid, and nitrilotrisulfonic acid should be replaced by the following based on the concept that these names are formed by adding hydroxyl, amide, imide, and so on, groups together with oxygen atoms to a sulfur atom:

HSO_3F	fluorosulfuric acid	$NH(SO_3H)_2$	imidobis(sulfuric) acid
HSO_3Cl	chlorosulfuric acid	$N(SO_3H)_3$	nitridotris(sulfuric) acid
NH_2SO_3H	amidosulfuric acid		

1.1.6 Salts and Functional Derivatives of Acids

1.1.6.1 Acid Halogenides. For acid halogenides the name is formed from the corresponding acid radical if this has a special name (Sec. 1.1.2.10); for example, NOCl, nitrosyl chloride. In other cases these compounds are named as halogenide oxides with the ligands listed alphabetically; for example, BiClO, bismuth chloride oxide; VCl_2O, vanadium(IV) dichloride oxide.

1.1.6.2 Anhydrides. Anhydrides of inorganic acids are named as oxides; for example, N_2O_5, dinitrogen pentaoxide.

1.1.6.3 Esters. Esters of inorganic acids are named as the salts; for example, $(CH_3)_2SO_4$, dimethyl sulfate. However, if it is desired to specify the constitution of the compound, the nomenclature for coordination compounds should be used.

1.1.6.4 Amides. Names for amides are derived from the names of the acid radicals (or from the names of acids by replacing acid by amide); for example, $SO_2(NH_2)_2$, sulfonyl diamide (or sulfuric diamide); NH_2SO_3H, sulfamidic acid (or amidosulfuric acid).

1.1.6.5 Salts. Salts containing acid hydrogen are named by adding the word hydrogen before the name of the anion (however, see Sec. 1.1.4.1), for example, KH_2PO_4, potassium dihydrogen phosphate; $NaHCO_3$, sodium hydrogen carbonate (not bicarbonate); $NaHPHO_3$, sodium hydrogen phosphonate (only one acid hydrogen remaining).

Salts containing O^{2-} and HO^- anions are named oxide and hydroxide, respectively. Anions are cited in alphabetical order which may be different in formulas and names.

Examples: FeO(OH), iron(III) hydroxide oxide; $VO(SO_4)$, vanadium(IV) oxide sulfate.

1.1.6.6 Multiplicative Prefixes. The multiplicative prefixes bis, tris, etc., are used with certain anions for indicating stoichiometric proportions when di, tri, etc., have been preempted to designate condensed anions; for example, $AlK(SO_4)_2 \cdot 12H_2O$, aluminum potassium bis(sulfate) 12-water (recall that disulfate refers to the anion $S_2O_7^{2-}$).

1.1.6.7 Crystal Structure. The structure type of crystals may be added in parentheses and in italics after the name; the latter should be in accordance with the structure. When the typename is also the mineral name of the substance itself, italics are not used.

Examples: $MgTiO_3$, magnesium titanium trioxide (*ilmenite* type); $FeTiO_3$, iron(II) titanium trioxide (ilmenite).

1.1.7 Coordination Compounds

1.1.7.1 Naming a Coordination Compound. To name a coordination compound, the names of the ligands are attached directly in front of the name of the central atom. The ligands are listed in alphabetical order regardless of the number of each and with the name of a ligand treated as a unit. Thus "diammine" is listed under "a" and "dimethylamine" under "d." The oxidation number of the central atom is stated last by either the oxidation number or charge number.

1.1.7.2 Anionic Ligands. Whether inorganic or organic, the names for anionic ligands end in -o (eliding the final -e, if present, in the anion name). Enclosing marks are required for inorganic anionic ligands containing numerical prefixes, and for thio, seleno, and telluro analogs of oxo anions containing more than one atom.

If the coordination entity is negatively charged, the cations paired with the complex anion (with -ate ending) are listed first. If the entity is positively charged, the anions paired with the complex cation are listed immediately afterward.

The following anions do not follow the nomenclature rules:

F^-	fluoro	HO_2^-	hydrogen peroxo
Cl^-	chloro	S^{2-}	thio (only for single sulfur)
Br^-	bromo	S_2^{2-}	disulfido
I^-	iodo	HS^-	mercapto
O^{2-}	oxo	CN^-	cyano
H^-	hydrido (or hydro)	CH_3O^-	methoxo or methanolato
OH^-	hydroxo	CH_3S^-	methylthio or methanethiolato
O_2^{2-}	peroxo		

1.1.7.3 ***Neutral and Cationic Ligands.*** Neutral and cationic ligands are used without change in name and are set off with enclosing marks. Water and ammonia, as neutral ligands, are called "aqua" and "ammine," respectively. The groups NO and CO, when linked directly to a metal atom, are called nitrosyl and carbonyl, respectively.

1.1.7.4 ***Attachment Points of Ligands.*** The different points of attachment of a ligand are denoted by adding italicized symbol(s) for the atom or atoms through which the attachment occurs at the end of the name of the ligand; e.g., glycine-*N* or glycinato-*O*, *N*. If the same element is involved in different possible coordination sites, the position in the chain or ring to which the element is attached is indicated by numerical superscripts: e.g., tartrato(3–)-O^1, O^2, or tartrato(4–)-O^2, O^3 or tartrato(2–) O^1, O^4

1.1.7.5 ***Abbreviations for Ligand Names.*** Except for certain hydrocarbon radicals, for ligand (L) and metal (M), and a few with H, all abbreviations are in lowercase letters and do not involve hyphens. In formulas, the ligand abbreviation is set off with parentheses. Some common abbreviations are

Ac	acetyl	en	ethylenediamine
acac	acetylacetonato	Him	imidazole
Hacac	acetylacetone	H_2ida	iminodiacetic acid
Hba	benzoylacetone	Me	methyl
Bzl	benzyl	H_3nta	nitrilotriacetic acid
Hbg	biguanide	nbd	norbornadiene
bpy	2, 2′-bipyridine	ox	oxalato(2–) from parent H_2ox
Bu	Butyl	phen	1, 10-phenanthroline
Cy	cyclohexyl	Ph	phenyl
D_2dea	diethanolamine	pip	piperidine
dien	diethylenetriamine	Pr	propyl
dmf	dimethylformamide	pn	propylenediamine
H_2dmg	dimethylglyoxime	Hpz	pyrazole
dmg	dimethylglyoximato(2–)	py	pyridine
Hdmg	dimethylglyoximato(1–)	thf	tetrahydrofuran
dmso	dimethylsulfoxide	tu	thiourea
Et	ethyl	H_3tea	triethanolamine
H_4edta	ethylenediaminetetraacetic acid	tren	2, 2′, 2″-triaminotriethylamine
Hedta, edta	coordinated ions derived from H_4edta	trien	triethylenetetraamine
		tn	trimethylenediamine
Hea	ethanolamine	ur	urea

Examples: $Li[B(NH_2)_4]$, lithium tetraamidoborate(1−) or lithium tetraamidoborate(III); $[Co(NH_3)_5Cl]Cl_3$, pentaamminechlorocobalt(III) chloride or pentaamminechlorocobalt(2+) chloride; $K_3[Fe(CN)_5CO]$, potassium carbonylpentacyanoferrate(II) or potassium carbonylpentacyanoferrate(3–); $[Mn\{C_6H_4(O)(COO)\}_2(H_2O)_4]^-$, tetraaquabis[salicylato(2–)]manganate(III) ion; $[Ni(C_4H_7N_2O_2)_2]$ or [Ni(dmg)] which can be named bis-(2, 3-butanedione dioximate)nickel(II) or bis[dimethylglyoximato(2–)]nickel(II).

1.1.8 Addition Compounds

The names of addition compounds are formed by connecting the names of individual compounds by a dash (—) and indicating the numbers of molecules in the name by Arabic numerals separated by the solidus (diagonal slash). All molecules are cited in order of increasing number; those having the same number are cited in alphabetic order. However, boron compounds and water are always cited last and in that order.

Examples: $3CdSO_4 \cdot 8H_2O$, cadmium sulfate—water (3/8); $Al_2(SO_4)_3 \cdot K_2SO_4 \cdot 24H_2O$, aluminum sulfate—potassium sulfate—water (1/1/24); $AlCl_3 \cdot 4C_2H_5OH$, aluminum chloride—ethanol (1/4).

1.1.9 Synonyms and Mineral Names

TABLE 1.2 Synonyms and Mineral Names

Acanthite, *see* Silver sulfide
Alabandite, *see* Manganese sulfide
Alamosite, *see* Lead(II) silicate(2−)
Altaite, *see* Lead telluride
Alumina, *see* Aluminum oxide
Alundum, *see* Aluminum oxide
Alunogenite, *see* Aluminum sulfate 18-water
Amphibole, *see* Magnesium silicate(2−)
Andalusite, *see* Aluminum silicon oxide (1/1)
Anglesite, *see* Lead sulfate
Anhydrite, *see* Calcium sulfate
Anhydrone, *see* Magnesium perchlorate
Aragonite, *see* Calcium carbonate
Arcanite, *see* Potassium sulfate
Argentite, *see* Silver sulfide
Argol, *see* Potassium hydrogen tartrate
Arkansite, *see* Titanium(IV) oxide
Arsenolite, *see* Arsenic(III) oxide dimer
Arsine, *see* Arsenic hydride
Auric and aurous, *see* under Gold
Azoimide, *see* Hydrogen azide
Azurite, *see* Copper(II) carbonate—dihydroxide (2/1)
Baddeleyite, *see* Zirconium(IV) oxide
Baking soda, *see* Sodium hydrogen carbonate
Barite (barytes), *see* Barium sulfate
Bieberite, *see* Cobalt sulfate 7-water
Bismuthine, *see* Bismuth hydride
Bismuthinite, *see* Bismuth sulfide
Bleaching powder, *see* Calcium hydrochlorite
Bleaching solution, *see* Sodium hydrochlorite
Blue copperas, *see* Copper(II) sulfate 7-water
Boracic acid, *see* Hydrogen borate
Borax, *see* Sodium tetraborate 10-water
Braunite, *see* Manganese(III) oxide
Brimstone, *see* Sulfur
Bromellite, *see* Beryllium oxide
Bromosulfonic acid, *see* Hydrogen bromosulfate
Bromyrite, *see* Silver bromide
Brookite, *see* Titanium(IV) oxide
Brucite, *see* Magnesium hydroxide
Bunsenite, *see* Nickel oxide
Cacodylate, *see* Sodium dimethylarsonate 3-water
Caesium, *see* under Cesium
Calamine, *see* Zinc carbonate
Calcia, *see* Calcium oxide
Calcite, *see* Calcium carbonate
Calomel, *see* Mercury(I) chloride
Caro's acid, *see* Hydrogen peroxosulfate
Cassiopeium, *see* Lutetium
Cassiterite, *see* Tin(IV) oxide
Caustic potash, *see* Potassium hydroxide
Caustic soda, *see* Sodium hydroxide
Celestite, *see* Strontium sulfate
Cementite, *see* tri-Iron carbide
Cerargyrite, *see* Silver chloride
Cerussite, *see* Lead carbonate
Chalcanthite, *see* Copper(II) sulfate 5-water
Chalcocite, *see* Copper(I) sulfide
Chalk, *see* Calcium carbonate
Chile nitre, *see* Sodium nitrate
Chile saltpeter, *see* Sodium nitrate
Chloromagnesite, *see* Magnesium chloride
Chlorosulfonic acid, *see* Hydrogen chlorosulfate
Cinnabar, *see* Mercury(II) sulfide
Claudetite, *see* Arsenic(III) oxide dimer

(*Continued*)

TABLE 1.2 Synonyms and Mineral Names (*Continued*)

Clausthalite, *see* Lead selenide
Clinoenstatite, *see* Magnesium silicate(2−)
Columbium, *see* under Niobium
Corrosive sublimate, *see* Mercury(II) chloride
Corundum, *see* Aluminum oxide
Cotunite, *see* Lead chloride
Covellite, *see* Copper(II) sulfide
Cream of tartar, *see* Potassium hydrogen tartrate
Crocoite, *see* Lead chromate(VI)(2−)
Cryolite, *see* Sodium hexafluoroaluminate
Cryptohalite, *see* Ammonium hexafluorosilicate
Cupric and cuprous, *see* under Copper
Cuprite, *see* Copper(I) oxide

Dakin's solution, *see* Sodium hypochlorite
Dehydrite, *see* Magnesium perchlorate
Dental gas, *see* Nitrogen(I) oxide
Diamond, *see* Carbon
Dichlorodisulfane, *see* di-Sulfur dichloride
Diuretic salt, *see* Potassium acetate
Dolomite, *see* Calcium magnesium carbonate (1/1)
Dry ice, *see* Carbon dioxide (solid)

Enstatite, *see* Magnesium silicate(2−)
Epsom salts, *see* Magnesium sulfate 7-water
Epsomite, *see* Magnesium sulfate 7-water
Eriochalcite, *see* Copper(II) chloride

Fayalite, *see* Iron(II) silicate(4−)
Ferric and ferrous, *see* under Iron
Fluorine oxide, *see* Oxygen difluoride
Fluoristan, *see* Tin(II) fluoride
Fluorite, *see* Calcium fluoride
Fluorosulfonic acid, *see* Hydrogen fluorosulfate
Fluorspar, *see* Calcium fluoride
Forsterite, *see* Magnesium silicate(4−)
Freezing salt, *see* Sodium chloride
Fulminating mercury, *see* Mercury fulminate

Galena, *see* Lead sulfite
Glauber's salt, *see* Sodium sulfate 10-water
Goethite, *see* Iron(II) hydroxide oxide
Goslarite, *see* Zinc sulfate 7-water
Graham's salt, *see* Sodium phosphate(1−)
Graphite, *see* Carbon
Greenockite, *see* Cadmium sulfide
Gruenerite, *see* Iron(II) silicate(2−)
Guanajuatite, *see* Bismuth selenide
Gypsum, *see* Calcium sulfate 2-water

Halite, *see* Sodium chloride
Hausmannite, *see* Manganese(II,IV) oxide
Heavy hydrogen, *see* Hydrogen[^{2}H] or name followed by *-d*
Heavy water, *see* Hydrogen[^{2}H] oxide
Heazlewoodite, *see* *tri*-Nickel disulfide
Hematite, *see* Iron(III) oxide
Hermannite, *see* Manganese silicate
Hessite, *see* Silver telluride
Hieratite, *see* Potassium hexafluorosilicate
Hydroazoic acid, *see* Hydrogen azide
Hydrophilite, *see* Calcium chloride
Hydrosulfite, *see* Sodium dithionate(III)
Hypo (photographic), *see* Sodium thiosulfate 5-water
Hypophosphite, *see* under Phosphinate

Ice, *see* Hydrogen oxide (solid)
Iceland spar, *see* Calcium carbonate
Iodyrite, *see* Silver iodide

Jeweler's borax, *see* Sodium tetraborate 10-water
Jeweler's rouge, *see* Iron(III) oxide

Kalinite, *see* Aluminum potassium bis(sulfate)
Kernite, *see* Sodium tetraborate
Kyanite, *see* Aluminum silicon oxide (1/1)

Laughing gas, *see* Nitrogen(I) oxide
Lautarite, *see* Calcium iodate
Lawrencite, *see* Iron(II) chloride
Lechatelierite, *see* Silicon dioxide
Lime, *see* Calcium oxide
Litharge, *see* Lead(II) oxide
Lithium aluminum hydride, *see* Lithium tetrahydridoaluminate
Lodestone, *see* Iron(II,III) oxide
Lunar caustic, *see* Silver nitrate
Lye, *see* Sodium hydroxide

Magnesia, *see* Magnesium oxide
Magnesite, *see* Magnesium carbonate
Magnetite, *see* Iron(II,III) oxide
Malachite, *see* Copper carbonate dihydroxide
Manganosite, *see* Manganese(II) oxide
Marcasite, *see* Iron disulfide
Marshite, *see* Copper(I) iodide
Mascagnite, *see* Ammonium sulfate
Massicotite, *see* Lead oxide
Mercuric and mercurous, *see* under Mercury
Metacinnabar, *see* Mercury(II) sulfide
Millerite, *see* Nickel sulfide
Mirabilite, *see* Sodium sulfate
Mohr's salt, *see* Ammonium iron(II) sulfate 6-water
Moissanite, *see* Silicon carbide
Molybdenite, *see* Molybdenum disulfide
Molybdite, *see* Molybdenum(VI) oxide
Molysite, *see* Iron(III) chloride
Montroydite, *see* Mercury(II) oxide
Morenosite, *see* Nickel sulfate 7-water
Mosaic gold, *see* Tin disulfide
Muriatic acid, *see* Hydrogen chloride, aqueous solutions

Nantokite, *see* Copper(I) chloride
Natron, *see* Sodium carbonate
Naumannite, *see* Silver selenide
Neutral verdigris, *see* Copper(II) acetate
Nitre (niter), *see* Potassium nitrate

TABLE 1.2 Synonyms and Mineral Names (*Continued*)

Nitric oxide, *see* Nitrogen(II) oxide
Nitrobarite, *see* Barium nitrate
Nitromagnesite, *see* Magnesium nitrate 6-water
Nitroprusside, *see* Sodium pentacyanonitrosylferrate(II) 2-water

Oldhamite, *see* Calcium sulfide
Opal, *see* Silicon dioxide
Orpiment, *see* Arsenic trisulfide
Oxygen powder, *see* Sodium peroxide

Paris green, *see* Copper acetate arsenate(III) (1/3)
Pawellite, *see* Calcium molybdate(VI)(2−)
Pearl ash, *see* Potassium carbonate
Perborax, *see* Sodium peroxoborate
Periclase, *see* Magnesium oxide
Persulfate, *see* Peroxodisulfate
Phosgene, *see* Carbonyl chloride
Phosphine, *see* Hydrogen phosphide
Pickling acid, *see* Hydrogen sulfate
Pitchblende, *see* Uranium(IV) oxide
Plaster of Paris, *see* Calcium sulfate hemihydrate
Plattnerite, *see* Lead(IV) oxide
Polianite, *see* Manganese(IV) oxide
Polishing powder, *see* Silicon dioxide
Potash, *see* Potassium carbonate
Potassium acid phthalate, *see* Potassium hydrogen phthalate
Prussic acid, *see* Hydrogen cyanide
Pyrite, *see* Iron disulfide
Pyrochroite, *see* Manganese(II) hydroxide
Pyrohytpophosphite, *see* diphosphate(IV)
Pyrolusite, *see* Manganese(IV) oxide
Pyrophanite, *see* Manganese titanate(IV)(2−)
Pyrophosphate, *see* Diphosphate(V)
Pyrosulfuric acid, *see* Hydrogen disulfate

Quartz, *see* Silicon dioxide
Quicksilver, *see* Mercury

Realgar, *see* di-Arsenic disulfide
Red lead, *see* Lead(II,IV) oxide
Rhodochrosite, *see* Manganese carbonate
Rhodonite, *see* Manganese silicate(1−)
Rochelle salt, *see* Potassium sodium tartrate 4-water
Rock crystal, *see* Silicon dioxide
Rutile, *see* Titanium(IV) oxide

Sal soda, *see* Sodium carbonate 10-water
Saltpeter, *see* Potassium nitrate
Scacchite, *see* Manganese chloride
Scheelite, *see* Calcium tungstate(VI)(2−)
Sellaite, *see* Magnesium fluoride
Senarmontite, *see* Antimony(III) oxide
Siderite, *see* Iron(II) carbonate
Siderotil, *see* Iron(II) sulfate 5-water
Silica, *see* Silicon dioxide
Silicotungstic acid, *see* Silicon oxide—tungsten oxide—water (l/12/26)
Sillimanite, *see* Aluminum silicon oxide (l/1)
Smithsonite, *see* Zinc carbonate
Soda ash, *see* Sodium carbonate
Spelter, *see* Zinc metal
Sphalerite, *see* Zinc sulfide
Spherocobaltite, *see* Cobalt(II) carbonate
Spinel, *see* Magnesium aluminate(2−)
Stannic and stannous, *see* under Tin
Stibine, *see* Antimony hydride
Stibnite, *see* Antimony(III) sulfide
Stolzite, *see* Lead tungstate(VI)(2−)
Strengite, *see* Iron(III) phosphate
Strontianite, *see* Strontium carbonate
Sugar of lead, *see* Lead acetate
Sulfamate, *see* Amidosulfate
Sulphate, *see* Sulfate
Sulfurated lime, *see* Calcium sulfide
Sulfuretted hydrogen, *see* Hydrogen sulfide
Sulphur, *see* Sulfur
Sulfuryl, *see* Sulfonyl
Sycoporite, *see* Cobalt sulfide
Sylvite, *see* Potassium chloride
Szmikite, *see* Manganese(II) sulfate hydrate

Tarapacaite, *see* Potassium chromate(VI)
Tellurite, *see* Tellurium dioxide
Tenorite, *see* Copper(II) oxide
Tephroite, *see* Manganese silicate(1−)
Thenardite, *see* Sodium sulfate
Thionyl, *see* Sulfinyl
Thorianite, *see* Thorium dioxide
Topaz, *see* Aluminum hexafluorosilicate
Tridymite, *see* Silicon dioxide
Troilite, *see* Iron(II) sulfide
Trona, *see* Sodium carbonate—hydrogen carbonate dihydrate
Tschermigite, *see* Aluminum ammonium bis(sulfate)
Tungstenite, *see* Tungsten disulfide
Tungstite, *see* Hydrogen tungstate

Uraninite, *see* Uranium(IV) oxide

Valentinite, *see* Antimony(III) oxide
Verdigris, *see* Copper acetate hydrate
Vermillion, *see* Mercury(II) sulfide
Villiaumite, *see* Sodium fluoride
Vitamin B_3, *see* Calcium (+)pantothenate

Washing soda, *see* Sodium carbonate 10-water
Whitlockite, *see* Calcium phosphate
Willemite, *see* Zinc silicate(4−)
Wolfram, *see* Tungsten
Wuestite, *see* Iron(II) oxide
Wulfenite, *see* Lead molybdate(VI)(2−)
Wurtzite, *see* Zinc sulfide

Zincite, *see* Zinc oxide
Zincosite, *see* Zinc sulfate
Zincspar, *see* Zinc carbonate
Zirconia, *see* Zirconium oxide

1.2 PHYSICAL PROPERTIES OF INORGANIC COMPOUNDS

Names follow the IUPAC Nomenclature. Solvates are listed under the entry for the anhydrous salt. Acids are entered under hydrogen and acid salts are entered as a subentry under hydrogen.

Formula weights are based upon the International Atomic Weights and are computed to the nearest hundredth when justified. The actual significant figures are given in the atomic weights of the individual elements. Each element that has neither a stable isotope nor a characteristic natural isotopic composition is represented in this table by one of that element's commonly known radioisotopes identified by mass number and relative atomic mass.

1.2.1 Density

Density is the mass of a substance contained in a unit volume. In the SI system of units, the ratio of the density of a substance to the density of water at 15°C is known as the *specific gravity* (*relative density*). Various units of density, such as kg/m^3, lb-mass/ft^3, and g/cm^3, are commonly used. In addition, molar densities or the density divided by the molecular weight is often specified.

Density values are given at room temperature unless otherwise indicated by the superscript figure; for example, 2.487^{15} indicates a density of 2.487 g/cm^3 for the substance at 15°C. A superscript 20 over a subscript 4 indicates a density at 20°C relative to that of water at 4°C. For gases the values are given as grams per liter (g/L).

1.2.2 Melting Point (Freezing Temperature)

The *melting point* of a solid is the temperature at which the vapor pressure of the solid and the liquid are the same and the pressure totals one atmosphere and the solid and liquid phases are in equilibrium. For a pure substance, the *melting point* is equal to the *freezing point*. Thus, the *freezing point* is the temperature at which a liquid becomes a solid at normal atmospheric pressure.

The *triple point* of a material occurs when the vapor, liquid, and solid phases are all in equilibrium. This is the point on a *phase diagram* where the solid-vapor, solid-liquid, and liquid-vapor equilibrium lines all meet. A *phase diagram* is a diagram that shows the state of a substance at different temperatures and pressures.

Melting point is recorded in a certain case as 250 d and in some other cases as d 250, the distinction being made in this manner to indicate that the former is a melting point with decomposition at 250°C while in the latter decomposition only occurs at 250°C and higher temperatures. Where a value such as $-6H_2O$, 150 is given it indicates a loss of 6 moles of water per formula weight of the compound at a temperature of 150°C. For hydrates the temperature stated represents the compound melting in its water of hydration.

1.2.3 Boiling Point

The normal boiling point (boiling temperature) of a substance is the temperature at which the vapor pressure of the substance is equal to atmospheric pressure.

At the boiling point, a substance changes its state from liquid to gas. A stricter definition of boiling point is the temperature at which the liquid and vapor (gas) phases of a substance can exist in equilibrium. When heat is applied to a liquid, the temperature of the liquid rises until the *vapor pressure* of the liquid equals the pressure of the surrounding atmosphere (gases). At this point there is no further rise in temperature, and the additional heat energy supplied is absorbed as *latent heat* of vaporization to transform the liquid into gas. This transformation occurs not only at the surface of the liquid (as in the case of *evaporation*) but also throughout the volume of the liquid, where bubbles of gas are formed. The boiling point of a liquid is lowered if the pressure of the surrounding atmosphere (gases) is decreased. On the other hand, if the pressure of the surrounding atmosphere (gases) is increased, the boiling point is raised. For this reason, it is customary when the boiling point of a substance is given to include the pressure at which it is observed, if that pressure is other than standard, i.e., 760 mm of mercury or 1 atmosphere (STP, Standard Temperature and Pressure). The boiling point of a solution is usually higher than that of the pure solvent; this boiling-point elevation is one of the colligative properties common to all solutions.

Boiling point is given at atmospheric pressure (760 mm of mercury or 101 325 Pa) unless otherwise indicated; thus 82^{15mm} indicates that the boiling point is 82°C when the pressure is 15 mm of mercury. Also, subl 550 indicates that the compound sublimes at 550°C. Occasionally decomposition products are mentioned.

1.2.4 Refractive Index

The refractive index n is the ratio of the velocity of light in a particular substance to the velocity of light in vacuum. Values reported refer to the ratio of the velocity in air to that in the substance saturated with air. Usually the yellow sodium doublet lines are used; they have a weighted mean of 589.26 nm and are symbolized by D. When only a single refractive index is available, approximate values over a small temperature range may be calculated using a mean value of 0.000 45 per degree for dn/dt, and remembering that n_D decreases with an increase in temperature. If a transition point lies within the temperature range, extrapolation is not reliable.

The *specific refraction* r_D is given by the Lorentz and Lorenz equation,

$$r_D = \frac{n_D^2 - 1}{n_D^2 + 2} \cdot \frac{1}{\rho}$$

where ρ is the density at the same temperature as the refractive index, and is independent of temperature and pressure. The molar refraction is equal to the specific refraction multiplied by the molecular weight. It is a more or less additive property of the groups or elements comprising the compound. An extensive discussion will be found in Bauer, Fajans, and Lewin, in *Physical Methods of Organic Chemistry*, 3d ed., A. Weissberger (ed.), vol. 1, part II, chap. 28, Wiley-Interscience, New York, 1960.

The empirical Eykman equation

$$\frac{n_D^2 - 1}{n_D + 0.4} \cdot \frac{1}{\rho} = \text{constant}$$

offers a more accurate means for checking the accuracy of experimental densities and refractive indices, and for calculating one from the other, than does the Lorentz and Lorenz equation.

The refractive index of moist air can be calculated from the expression

$$(n-1) \times 10^6 = \frac{103.49}{T} p_1 + \frac{177.4}{T} p_2 + \frac{86.26}{T}\left(1 + \frac{5748}{T}\right) p_3$$

where p_1 is the partial pressure of dry air (in mmHg), p_2 is the partial pressure of carbon dioxide (in mmHg), p_3 is the partial pressure of water vapor (in mmHg), and T is the temperature (in kelvins).

Example: 1-Propynyl acetate has $n_D = 1.4187$ and density = 0.9982 at 20°C; the molecular weight is 98.102. From the Lorentz and Lorenz equation,

$$r_D = \frac{(1.4187)^2 + 1}{(1.4187)^2 + 2} \cdot \frac{1}{0.9982} = 0.2528$$

The molar refraction is

$$Mr_D = (98.102)(0.2528) = 24.80$$

From the atomic and group refractions, the molar refraction is computed as follows:

6 H	6.600
5 C	12.090
1 C≡C	2.398
1 O(ether)	1.643
1 O(carbonyl)	2.211
	$Mr_D = 24.942$

TABLE 1.3 Physical Constants of Inorganic Compounds

Abbreviations Used in the Table

a, acid
abs, absolute
abs ale, anhydrous ethanol
acet, acetone
alk, alkali (aq NaOH or KOH)
anhyd, anhydrous
aq, aqueous
aq reg, aqua regia
atm, atmosphere
BuOH, butanol
bz, benzene
c, solid state
ca., approximately
chl, chloroform
cone, concentrated
cub, cubic
d, decomposes
dil, dilute
disprop, disproportionates
EtOAc, ethyl acetate
eth, diethyl ether
EtOH, 95% ethanol
expl, explodes
fcc, face-centered cubic
fctetr, face-centered tetragonal
FP, flash point
fum, fuming
fus, fusion, fuses
g, gas, gram
glyc, glycerol
h, hot
hex, hexagonal
HOAc, acetic acid
i, insoluble
ign, ignites
L, liter
lq, liquid
MeOH, methanol
min, mineral
mL, milliliter
org, organic
oxid, oxidizing
PE, petroleum ether
pyr, pyridine
s, soluble
satd, saturated
sl, slightly
soln, solution
solv, solvent (s)
subl, sublimes
sulf, sulfides
tart, tartrate
THF, tetrahydrofuran
v, very
vac, vacuum
viol, violently
volat, volatilizes
<, less than
>, greater than

Name	Formula	Formula weight	Density	Melting point, °C	Boiling point, °C	Solubility in 100 parts solvent
Actinium-227	Ac	227.0278	10.07	1050(50)	ca. 3200	d aq; s acids
bromide	$AcBr_3$	466.74	5.85	subl 800		s aq
Aluminum	Al	26.981539	2.70	660.323	2518	s HCl, H_2SO_4, alk
acetylacetonate	$Al(C_5H_7O_2)_3$	324.31	1.27	190–193	315	i aq; v s alc; s bz, eth
ammonium bis(sulfate) 12-water	$AlNH_4(SO_4)_2 \cdot 12H_2O$	453.33	1.65	anhyd >280		14.3 g/100 mL aq; s glyc; i alc
antimonide	AlSb	148.74	4.26	1060		
arsenide	AlAs	101.90	3.76	1740		
bis(acetylsalicylate)	$Al(OOCC_6H_4OCOCH_3)_2OH$	402.30				v sl s aq, alc, eth
borate (2/1)	$2Al_2O_3 \cdot B_2O_3$	273.54		ca. 1050		i aq
bromide	$AlBr_3$	266.69	3.205_0^{18}	97.5	subl 253	d (viol) aq; s alc, acet, bz, CS_2
butoxide, *sec*-	$Al(C_4H_9O)_3$	246.33	0.967		$200–206^{30mm}$	FP 27; v s org solv
butoxide, *tert*-	$Al(C_4H_9O)_3$	246.33	1.025_0^{20}		subl 180	v s org solv
carbide (4/3)	Al_4C_3	143.96	2.360	2100	d $>2200^{400mm}$	d aq; fire hazard
chlorate	$Al(ClO_3)_3$	277.35				v s aq; s alc
chloride	$AlCl_3$	133.34	2.440^{25}	192.6	subl 181.1	g/100 mL: 70 aq (viol), 100^{12} abs alc; s CCl_4, eth; sl s bz
ethoxide	$Al(C_2H_5O)_3$	162.16	1.142_0^{20}	140	205^{14mm}	s hot aq d; v sl s alc, eth
fluoride	AlF_3	83.98	2.882_4^{25}	1090	subl 1272	0.56 aq; i a, alk, alc, acet
hydroxide	$Al(OH)_3$	78.01	2.42	to Al_2O_3, 300		i aq; s acids, alkalis
iodide	AlI_3	407.69	3.98^{17}	191.0	382	d aq; s alc, eth, CS_2
isopropoxide	$Al(C_3H_7O)_3$	204.25	1.0346_0^{20}	118.5	135^{10mm}	d aq; s alc, bz, chl, PE
methoxide	$Al(CH_3O)_3$	72.07		0	130	

nitrate 9-water	$Al(NO_3)_3 \cdot 9H_2O$	375.13	1.72	73	d 135	g/100 mL: 64 aq, 100 alc; s acet
nitride	AlN	40.99	3.05	d 2517		d aq, acid, alkali
oxide (alpha-)	AlO_3	101.96	3.97	2054(6)	2980	i aq; v sl s a, alk
perchlorate 6-water	$Al(ClO_4)_3 \cdot 6H_2O$	433.43	2.020	120.8	anhyd 178	133 g/100 mL^{20} aq
phenoxide	$Al(C_6H_5O)_3$	306.27	1.23	d 265		d aq; s alc, chl, eth
phosphate	$AlPO_4$	121.95	2.56	>1460		i aq; sl s a
phosphide	AlP	57.96	2.85^{15}_{4}	2550		d aq
phosphinate (hypophosphite)	$Al(H_2PO_2)_3$	221.94		d to PH_3, 220		i aq; s HCl, warm alkali
potassium bis(sulfate) 12-water	$AlK(SO_4)_2 \cdot 12H_2O$	474.39	1.757^{20}	$-9H_2O$, 92	anhyd, 200	11.4 g/100 mL aq; v s glyc; i alc
propoxide	$Al(C_3H_7O)_3$	204.25	1.0578^{20}_{0}	106	248^{14mm}	d aq; s alc
selenide	Al_2Se_3	290.84	3.437^{20}_{4}	947		d aq, acid
silicon oxide (1/1)	$Al_2O_3 \cdot SiO_2$	162.05	3.247			i aq; d HF; s fused alkali
sodium bis(sulfate) 12-water	$AlNa(SO_4)_2 \cdot 12H_2O$	458.28	1.675^{20}	61		110 g/100 mL^{15} aq; i alc
stearate	$Al(C_{18}H_{35}O_2)_3$	877.41	1.070	117–120		i aq, alc; s bz, alk
sulfate	$Al_2(SO_4)_3$	342.15	1.61	770 d		36.4 g/100 mL^{20} aq; sl s alc
sulfate 18-water	$Al_2(SO_4)_3 \cdot 18H_2O$	666.46	1.69^{17}	d 86.5		87 g/100 mL^{0} aq; i alc
sulfide	Al_2S_3	150.16	2.20^{13}	1097	subl 1500	hyd aq; s acid
tetrahydridoborate	$Al(BH_4)_3$	71.53		−64.5	44.5	d aq; ign air; expl in O_2, 20
Americium	Am	243	12	1176	2011	s a
Ammonia	NH_3	17.03	lq: 0.6818 at bp g: $0.6175^{15,\ 7.2atm}$	−77.75	−33.35	g/100 mL: 34 aq; 13.2 alc; s eth, organic solvents
Ammonium acetate	$NH_4C_2H_3O_2$	77.08	1.17^{20}	114	d	g/100 mL: 148^{4} aq, 7.9^{15} MeOH; s alc
amidosulfate	$NH_4SO_3NH_2$	114.13		131	d 160	v s aq; sl s alc
benzoate	$NH_4C_6H_5O_2$	139.15	1.260	198	subl 160	g/100 mL: 20^{15} aq, 2.8 alc; s glyc
bromide	NH_4Br	97.94	2.429	452 (subl under pressure)	d 397 vacuo	76 g/100 mL^{20} aq; v s acet, alc, eth
calcium arsenate 6-water	$NH_4CaAsO_4 \cdot 6H_2O$	305.13	1.905^{15}	d 140		0.02 aq; s NH_4Cl
carbamate	NH_4COONH_2	78.07		subl 60		v s aq; sl s alc; i eth
carbonate 1-water	$(NH_4)_2CO_3 \cdot H_2O$	114.10		volatilizes 60		v s aq; i alc
chloride	NH_4Cl	53.49	1.5274^{25}	237.8	520	g/100 mL: 26^{15} aq, 0.6^{19} abs alc; i acet, eth
chromate(VI)	$(NH_4)_2CrO_4$	152.07	1.91^{12}	d 185		34 g/100 mL^{20} aq; sl s MeOH
chromium(III) bissulfate 12-water	$NH_4Cr(SO_4)_2 \cdot 12H_2O$	478.34	1.72	94 d		7.2 g/100 mL^{0} aq
copper(II) tetrachloride 2-water	$(NH_4)_2CuCl_4 \cdot 2H_2O$	277.46	1.993	anhyd, 110	d >120	40.3 g/100 mL^{20} aq; s alc

(*Continued*)

TABLE 1.3 Physical Constants of Inorganic Compounds (*Continued*)

Name	Formula	Formula weight	Density	Melting point, °C	Boiling point, °C	Solubility in 100 parts solvent
cyanide	NH_4CN	44.06	1.10	d 36		v s aq, alc
dichromate(VI)	$(NH_4)_2Cr_2O_7$	252.07	2.155	d 180 to Cr_2O_3		35.6 g/100 mL^{20} aq; s alc; flammable
dihydrogen arsenate	$NH_4H_2AsO_4$	158.97	2.311	d 300		v s aq
dihydrogen phosphate	$NH_4H_2PO_4$	115.03	1.803^{19}	d 190		37 g/100 mL^{20} aq; sl s alc; i acet
disulfatocobatate(II) 6-water	$(NH_4)_2[Co(SO_4)_2] \cdot 6H_2O$	395.23	1.902			18 g/100 mL^{20} aq; v sl s alc
disulfatoferrate(II) 6-water	$(NH_4)_2[Fe(SO_4)_2] \cdot 6H_2O$	392.14	1.864	d 100		36.4 g/100 mL^{20} aq; i alc
disulfatoferrate(III) 12-water	$NH_4[Fe(SO_4)_2] \cdot 12H_2O$	482.19	1.71	39–41	d 230	124 g/100 mL aq
disulfatonickelate(II) 6-water	$(NH_4)_2[Ni(SO_4)_2] \cdot 6H_2O$	395.00	1.923			8.95 g/100 mL^{20} aq
dithiocarbamate	$NH_4S(C{=}S)NH_2$	110.20	1.451^{20}_{4}	99 d		v s aq; s alc; sl s eth
diuranate(VI)	$(NH_4)_2U_2O_7$	624.22				v sl s aq, alk; s acids
fluoride	NH_4F	37.04	1.009^{25}	d to NH_3 + HF		100 g/100 mL^{0} aq; s alc
formate	NH_4OOCH	63.06	1.27	116	d 180	143 g/100 mL^{20} aq; s alc, eth
heptamolybdate(VI)(6−) 4-water	$(NH_4)_2Mo_7O_{24} \cdot 4H_2O$	1235.86	2.498	anhyd 90	d 190	43 g/100 mL aq; s acids; i alc
hexachloropalladate(IV)	$(NH_4)_2[PdCl_6]$	355.20	2.418	d		sl s aq
hexachloroplatinate(IV)	$(NH_4)_2[PtCl_6]$	443.87	3.065	d 380		0.5 aq
hexadecanoate	$NH_4OOC(CH_2)_{14}CH_3$	273.45		21–22		s aq; sl s bz; i alc, acet
hexafluoroaluminate(3−)	$(NH_4)_3[AlF_6]$	195.09	1.78	d >100		v s aq
hexafluorogallate	$(NH_4)_3GaF_6$	237.83	2.10	d 200		
hexafluorogermanate	$(NH_4)_2GeF_6$	222.68	2.564	380	subl	s aq; i eth
hexafluorophosphate	$NH_4[PF_6]$	163.00	2.180^{18}_{4}	d 68		74.8 g/100 mL^{20} aq; s alc, acet
hexafluorosilicate	$(NH_4)_2[SiF_6]$	178.15	2.011	d		18.6 g/100 mL^{20} aq; i alc, acet
hexanitratocerate(IV)	$(NH_4)_2[Ce(NO_3)_6]$	548.22				135 g/100 mL^{20} aq; s alc, HNO
hydrogen carbonate	NH_4HCO_3	79.06	1.586	107 (rapid heating)		g/100 mL: 17.4^{20} aq, 10 glyc
hydrogen citrate	$(NH_4)_2HC_6H_5O_7$	226.19	1.48			100 g/100 mL aq; sl s alc
hydrogen difluoride	NH_4HF_2	57.04	1.51	124.6	240 d	v s aq; sl s alc
hydrogen oxalate hydrate	$NH_4HC_2O_4 \cdot H_2O$	125.08	1.556	anhyd, 170		s aq, alc; i bz, eth
hydrogen phosphate	$(NH_4)_2HPO_4$	132.06	1.619	d 155		69 g/100 mL^{20} aq; i alc, acet
hydrogen sulfate	NH_4HSO_4	115.11	1.78	146.9	d 350	100 g/100 mL aq; i alc, acet

hydrogen sulfide	NH_4HS	51.11	1.17	d 25 to NH_3 + H_2S		128 g/100 mL^0 aq; s glyc; i alc, acet
hydrogen sulfite	NH_4HSO_3	99.11	2.03	subl 150 in N_2		267 g/100 mL^{10} aq
hydrogen (±)tartrate	$NH_4HC_4H_4O_6$	167.12	1.68	d 200		2.2^{15} aq; i alc
hydroxide	NH_4OH	35.05		−77		49% dissolved NH_3
hypophosphite	$NH_4H_2PO_2$	83.03		d		v s aq; sl s alc; i acet
iodate	NH_4IO_3	192.94	3.309	d 150		2.6^{15} aq
iodide	NH_4I	144.94	2.514^{25}	subl 551	220 vacuo	167 g/100 mL^{20} aq; v s alc, acet
lactate	$NH_4C_3H_5O_3$	107.11	1.2^{15}	92		v s aq, alc, glyc; i acet, eth
magnesium arsenate 6-water	$NH_4MgAsO_4 \cdot 6H_2O$	289.36	1.923	d		0.038^{20} aq
molybdate(VI)(2−)	$(NH_4)_2MoO_4$	196.04	2.276^{25}_{4}	d		s acids
nitrate	NH_4NO_3	80.04	1.725^{25}	169.6	d 210	g/100 mL: 192^{20} aq; 3.8^{20} alc; 17^{20} MeOH; s acet
octadecanoate	$NH_4OOC(CH_2)_{16}CH_3$	301.50		21–22		sl s aq; s alc; i acet
octanoate	$NH_4OOC(CH_2)_6CH_3$	161.24		d on standing		v s aq, alc, acet; sl s eth
oxalate hydrate	$(NH_4)_2C_2O_4 \cdot H_2O$	142.11	1.50	d 70		5.1^{20} aq; s alc
oxodioxalatotitanate(IV)	$(NH_4)_2TiO(C_2O_4)_2$	276.02				v s aq
perchlorate	NH_4ClO_4	117.49	1.95	d 240		g/100 mL^{25}: 21.9 aq, 1.49 EtOH, 0.014 BuOH, 0.029 EtOAc
permanganate	NH_4MnO_4	136.97	2.208^{10}	explodes, 110		0.8^{15} aq
peroxodisulfate	$(NH_4)_2S_2O_8$	228.20	1.982	d 120	expl 180	58 g/100 mL^0 aq
phosphinate	$NH_4PH_2O_2$	83.04	1.634	200	d 240	g/100 mL: 100 aq, 5 alc; i acet
phosphomolybdate hydrate	$(NH_4)_3PO_4 \cdot 12MoO_3 \cdot H_2O$	1894.36		d		sl s aq
picrate	$NH_4C_6H_2N_3O_7$	246.14	1.719	d	expl 423	1.1^{20} aq; sl s alc
selenate(VI)	$(NH_4)_2SeO_4$	179.04	2.193^{20}_{4}	d		117 g/100 mL^7 aq; s HOAC; i alc
stearate	$NH_4C_{18}H_{35}O_2$	301.51	0.89	22		sl s aq, bz; s alc; i acet
sulfamate	$NH_4NH_2SO_3$	114.13		131	d 160	v s aq; sl s alc
sulfate	$(NH_4)_2SO_4$	132.14	1.769^{20}	d >280		43.5 g/100 mL^{20} aq; i alc, acet
sulfide	$(NH_4)_2S$	68.14		d ≈0		v s aq; s alc, alk
sulfite hydrate	$(NH_4)_2SO_3 \cdot H_2O$	134.16	1.41	d 60		75 g/100 mL^{20} aq; i alc, acet
(±)tartrate	$(NH_4)_2C_4H_4O_6$	184.15	1.601	d		58 g/100 mL^{15} aq; sl s alc
tetraborate 4-water	$(NH_4)_2B_4O_7 \cdot 4H_2O$	263.44				s aq; i alc

(Continued)

TABLE 1.3 Physical Constants of Inorganic Compounds (*Continued*)

Name	Formula	Formula weight	Density	Melting point, °C	Boiling point, °C	Solubility in 100 parts solvent
tetrachloroaluminate	$NH_4[AlCl_4]$	186.83		304		s aq, eth
tetrachloropalladate(II)	$(NH_4)_2[PdCl_4]$	284.29	2.170	d		v s aq; i abs alc
tetrachloroplatinate(II)	$(NH_4)_2[PtCl_4]$	372.97	2.936	140 d		s aq; i alc
tetrachlorozincate	$(NH_4)_2[ZnCl_4]$	243.28	1.879	150 d	subl 341	v s aq
tetrafluoroborate	$NH_4[BF_4]$	104.84	1.871	subl		25 g/100 mL^{16} aq
thiocyanate	NH_4SCN	76.12	1.305	149.6	d 170	128 g/100 mL^{0} aq; v s alc; s acet
thiosulfate	$(NH_4)_2S_2O_3$	148.21	1.679	d 150		2.15^{15} aq; i alc, eth
vanadate(V)(1−)	NH_4VO_3	116.98	2.326	d 200		0.48^{20} aq
Antimony	Sb	121.760(1)	6.697^{25}	630.7	1587	s hot conc H_2SO_4, aqua regia
arsenide	SbAs	196.68	6.0	≈680		
(III) bromide	$SbBr_3$	361.47	4.35	96.6	280	s acet, bz, chl
(III) chloride	$SbCl_3$	228.12	3.14^{20}_{4}	73.4	220.3	10 g/100 mL^{20} aq; s alc, bz, chl
(V) chloride	SCl_5	299.02	2.336^{20}_{4}	3.5	79^{22mm}	d aq; s HCl, chl, CCl_4
(III) fluoride	SbF_3	178.75	4.379^{20}_{20}	292	376	444 g/100 mL^{20} aq
(V) fluoride	SbF_3	216.75	2.99^{23}	8.3	141	d viol aq; s HOAc; forms solids with alc, bz, CS_2, eth
hydride (stibine)	SbH_3	124.78	5.475 g/L	−91.5	−18.4	20 mL/100 mL^{20} aq; s CS_2, alc
(III) iodide	SbI_3	502.47	4.92	168	401	g/100 g^{25}: 1.16 bz, 1.24 tol, 0.16 chl
(III) oxide (valentinite)	Sb_2O_3	291.52	5.7	655	1425	v sl s aq; s HCl, KOH
(V) oxide	Sb_2O_3	323.52	3.78	$-O_2$, >300		v sl s aq; sl s warm KOH, eth
(III) selenide	Sb_2Se_3	480.40	5.81	612		v sl s aq; s conc HCl
(III) sulfate	$SB_2(SO_4)_3$	531.71	3.62	d		sl s aq
(III) sulfide	Sb_2S_3	339.72	4.56	546		0.002^{20} aq (d); s H_2SO_4
(V) sulfide	Sb_2S_5	403.85	4.120	75 d		i aq; s HCl (d), NaOH
(III) telluride	Sb_2Te_3	626.32	6.52	620		i aq; s HNO_3
triethyl	$Sb(C_2H_5)_3$	209.0	1.324^{14}	−29	159.5	i aq
trimethyl	$Sb(CH_3)_3$	166.9	1.523^{15}		80.6	sl s aq
Argon	Ar	39.948(1)	1.7824 g/L^{0}	−189.38	−185.87	3.36 mL/100 mL^{20} aq
Arsenic	As	74.92159(2)	5.727^{25}_{4}	817	subl 615	i aq; s HNO_3
(III) bromide	$AsBr_3$	314.63	3.3972^{25}_{4}	31.1	220.0	hyd aq; s HCl, CS_2, PE
(III) chloride	$AsCl_3$	181.28	2.1497^{25}_{4}	−16.2	130.2	misc chl, CCl_4, eth; s HCl
(*di-*) disulfide	As_2S_2	213.97	3.254^{19}	320	565	s alkali; v sl s bz
(III) fluoride	AsF_3	131.92	2.73^{15}_{15}	−5.95	57.8	s alc, bz, eth, HF
(V) fluoride	AsF_5	169.91	7.46 g/L	−79.8	−52.8	hyd aq; s alc, bz, eth

(III) hydride (arsine)	AsH_3	77.95	3.420 g/L	−116.9	−62.5	28 mL/100 mL^{20} aq; s bz, chl
(III) iodide	AsI_3	455.63	4.73	140.9	424	s bz, tol; sl s aq, alc, eth
(III) oxide (arsenolite)	As_2O_3	197.84	3.86	274	460	1.8^{20} aq; s alc
(III) oxide (claudetite)	As_2O_3	197.84	3.74	313	460	sl s aq; s dil acid, alk
(V) oxide	As_2O_5	229.84	4.32	315	d 800	66 g/100 mL^{20} aq; s alc
(III) selenide	As_2Se_3	386.72	4.75	260		s alkali, HNO_3
(III) sulfide	As_2S_3	246.04	3.460	310	707	i aq; s alk, slowly s hot HCl
(V) sulfide	As_2S_5	310.17		subl 500		0.0003 aq; s alkali, HNO_3
(III) telluride	As_2Te_3	532.64	6.50	621		
Astatine	At	210		302		
Barium	Ba	137.33	3.51^{20}	726.9	1845	d aq to Ba(OH)
acetate hydrate	$Ba(C_2H_3O_2)_2 \cdot H_2O$	273.43	2.19	anhyd 110	d 150	58.8 g/100 mL^{0} aq; 0.014 alc
benzenesulfonate	$Ba(O_3SC_6H_5)_2$	451.70				s aq; sl s alc
bromate hydrate	$Ba(BrO_3)_2 \cdot H_2O$	411.14	3.99^{18}	d 260		0.96^{30} aq; s acet; i alc
bromide	$BaBr_2$	297.14	4.781	856	1835	92 g/100 mL^{0} aq; s MeOH, acet
carbonate	$BaCO_3$	197.34	4.2865	d 1300 to BaO + CO_2		0.0024 aq; s acids
chlorate hydrate	$Ba(ClO_3)_2 \cdot H_2O$	322.24	3.179	anhyd 120	$-O_2$, 250	34 g/100 mL^{20} aq; sl s alc, acet
chloride	$BaCl_2$	208.24	3.856^{24}	962	1560	36 g/100 mL^{20} aq; s MeOH; i acet, EtAc
chloride dihydrate	$BaCl_2 \cdot 2H_2O$	244.26	3.097	anhyd 113		31.7 g/100 mL^{0} aq
chromate(VI)	$BaCrO_4$	253.33	4.498^{20}	d		0.001^{20} aq; s mineral acids
cyanide	$Ba(CN)_2$	189.36				80 g/100 mL^{14} aq; s alc
fluoride	BaF_2	175.32	4.89	1368	2260	0.161^{20} aq; s acids
hexafluorosilicate	$Ba[SiF_6]$	279.40	4.29$^{21}_{4}$	d 300		0.0235^{25} aq; s NH_4Cl soln; i alc
hydrogen phosphate	$BaHPO_4$	233.31	4.165^{15}	d 410		0.01 aq; s HCl, HNO_3
hydroxide 8-water	$Ba(OH)_2 \cdot 8H_2O$	315.48	2.18^{16}	78		3.9^{20} aq
iodate	$Ba(IO_3)_2$	487.13	5.23^{20}	d 476		0.033^{20} aq; s HCl
iodide	BaI_2	391.14	5.15	711	2027	169 g/100 mL^{20} aq; s alc, acet
manganate(VI)(2−)	$BaMnO_4$	256.26	4.85			disprop to $Ba(MnO_4)_2$ + MnO_2
molybdate	$BaMoO_4$	297.27	4.975	1450		0.0058^{25} aq
niobate	$Ba(NbO_3)_2$	419.14	5.44	1455		i aq
nitrate	$Ba(NO_3)_2$	261.34	3.24^{23}	592	d	5.0 aq; v sl s alc, acet
nitrite hydrate	$Ba(NO_2)_2 \cdot H_2O$	247.35	3.173^{30}	d 115		54.8 g/100 mL^{0} aq; i alc

(Continued)

TABLE 1.3 Physical Constants of Inorganic Compounds (*Continued*)

Name	Formula	Formula weight	Density	Melting point, °C	Boiling point, °C	Solubility in 100 parts solvent
oxalate	BaC_2O_4	225.35	2.658	400 d		i aq
oxide	BaO	153.33	5.72	1973	3088	3.5^{20} aq; s acids, EtOH
perchlorate	$Ba(ClO_4)_2$	336.23	3.20	505		g/100 mL^{25}: 129 aq, 78 EtOH, 42 BuOH, 81 EtOAc; i eth
perchlorate 3-water	$Ba(ClO_4)_2 \cdot 3H_2O$	390.27	2.74	d 400		198 g/100 mL^{25} aq; s MeOH; sl s acet
permanganate	$Ba(MnO_4)_2$	375.20	3.77	d 200		v s aq
peroxide	BaO_2	169.33	4.96	450 d	$-O_2$, 800	1.5^{0} aq
selenide	BaSe	216.29	5.02	1780		d aq
stearate	$Ba(C_{18}H_{35}O_2)_2$	704.28	1.145	160		i aq
sulfate	$BaSO_4$	233.39	4.50^{15}	1580	d >1600	0.00285 aq
sulfide	BaS	169.39	4.25^{15}	2230		7.9^{20} aq; dec in acids
sulfite	$BaSO_3$	217.39	4.44	d		0.02^{0} aq; i alc
tetracyanoplatinate(II)-4-water	$Ba[Pt(CN)_4] \cdot 4H_2O$	508.54	2.076			2.86 aq; i alc
thiocyanate 2-water	$Ba(SCN)_2 \cdot 2H_2O$	289.53	2.286^{18}	d 160		170 g/100 mL^{20} aq; s alc, acet
thiosulfate hydrate	$BaS_2O_3 \cdot H_2O$	267.47	3.5^{18}	d 220		0.21^{20} aq; i alc, acet, eth, CS
titanate(IV)(2−)	$BaTiO_3$	233.19	6.02	1625		i aq
vanadate	$Ba_3(VO_4)_2$	641.86	5.14	707		
zirconate	$BaZrO_3$	276.55	5.52	2500		i aq, alk; sl s acids
Berkelium (α form)	Bk	247	14.78	1050		
(β form)	Bk	247	13.25	986		
Beryllium	Be	9.012	1.8477^{20}	1287	2467	i aq; s acid, alk
bromide	$BeBr_2$	168.82	3.465^{25}	508	521	v s aq; s alc; 18.6 pyr
carbide	Be_2C	30.04	1.90^{15}	d >2127		d aq; s acids, alkali giving CH_4
chloride	$BeCl_2$	79.92	1.899^{25}	415 (alpha)	482.3	42 g/100 mL aq; s alc, eth, pyr, CS_2
fluoride	BeF_2	47.01	1.986	555	subl 1036^{1mm}	v s aq (slowly)
hydride	BeH_2	11.03	0.65	$-H_2$, 220		d aq (slowly), acids (rapidly)
hydroxide	$Be(OH)_2$	43.03	1.909	93		s hot conc acids and alkali (viol)
iodide	BeI_2	262.82	4.32	480	487	hyd aq violently; s alc, eth, CS_2
nitrate 3-water	$Be(NO_3)_2 \cdot 3H_2O$	187.07	1.557	60.5	d 125	166 g/100 mL^{20} aq
nitride	Be_3N_2	55.05	2.71	2200		d hot aq, alkali
oxide	BeO	25.01	3.025	2578 (alpha)	3787	s conc H_2SO_4
selenate 4-water	$BeSeO_4 \cdot 4H_2O$	224.03	2.03	anhyd 300	d 560	49 g/100 mL^{25} aq

silicate	Be_2SiO_4	110.11	3.0	1560		i aq
sulfate 4-water	$BeSO_4 \cdot 4H_2O$	177.14	1.713	anhyd 270	d 580	39 g/100 mL^{20} aq; i alc
sulfide	BeS	41.08	2.36	d		i aq; s HNO_3
Bismuth	Bi	208.9804	9.78	271.5	1564	i aq; s hot H_2SO_4
(III) bromide	$BiBr_3$	448.69	5.72	218	453	d aq; s dil acids, acet
bromide oxide	BiBrO	304.88	8.082^{15}	d		i aq; s acids
(III) chloride	$BiCl_3$	315.34	4.75	233.5	447	d aq; s HCl, alc, eth, acet
chloride oxide	BiClO	260.43	7.72^{15}	d		i aq; s HCl
(III) fluoride	BiF_3	265.98	8.32	727	900	i aq; s HF
(V) fluoride	BiF_5	303.97	5.55^{25}	154.4	subl 550	d (viol) aq giving O_3 + BiF_3
hydride	BiH_3	212.00	9.303 g/L	−67	16.8	very unstable liquid
(III) hydroxide	$Bi(OH)_3$	260.00	4.962^{15}	−water, 100		d aq; s HCl
(III) iodide	BiI_3	589.69	5.778^{20}_{4}	408.6	subl 439	i aq; s HCl, alc
iodide oxide	BiIO	351.88	7.922	d red heat		i aq; s HCl
(III) nitrate 5-water	$Bi(NO_3)_3 \cdot 5H_2O$	485.07	2.83	anhyd 80		d aq; s HNO_3, acet, glyc
(III) oxide	Bi_2O_3	465.96	8.76	817	1890	i aq; s HCl, HNO_3
(V) oxide	Bi_2O_5	497.96	5.10	d 150		i aq; s KOH
(III) phosphate	$BiPO_4$	303.95	6.323^{15}	d		s conc HCl, HNO_3
(III) selenide	Bi_2Se_3	654.84	7.70^{20}_{4}	710 d	d	i aq; d aq reg
(III) sulfate	$Bi_2(SO_4)_3$	706.14	5.08	d 405		d aq, alc; s HCl
(III) sulfide	Bi_2S_3	514.16	6.78	850		i aq, EtAc; s HNO_3, HCl
(III) telluride	Bi_2Te_3	800.76	7.74	588.5		i aq; s alc
Boranes						
diborane(6)	B_2H_6	27.67	1.214 g/L	−165.5	−92.5	FP −68; s NH_4OH, conc H_2SO_4
tetraborane(10)	B_4H_{10}	53.32	2.340 g/L	−120	18	sl s aq; s bz
pentaborane(9)	B_5H_9	63.13	0.60	−46.81	60.0	hyd aq
pentaborane(11)	B_5H_{11}	65.14	0.745	−123	63	d aq
hexaborane(10)	B_6H_{10}	74.95	0.67	−62.3	108 d	d hot aq
decaborane(14)	$B_{10}H_{14}$	122.22	0.948	99.5	213	sl s aq; s bz, CS_2, eth
Borazine	$B_3H_6N_3$	80.50	lq: 0.81^{bp}	−58	55	sl s aq (d)
Boric acids, *see* under Hydrogen						
Boron	B	10.811	2.34	2076	3864	i aq
carbide	B_4C	55.25	2.510^{25}_{4}	2350	>3500	s fused alkalis
tribromide	BBr_3	250.52	2.6	−46.0	91.3	d aq, alc
trichloride	BCl_3	117.17	5.141 g/L	−107	12.7	d aq, alc

(Continued)

TABLE 1.3 Physical Constants of Inorganic Compounds (*Continued*)

Name	Formula	Formula weight	Density	Melting point, °C	Boiling point, °C	Solubility in 100 parts solvent
trifluoride	BF_3	67.81	3.077 g/L^{STP}	−127.1	−100.4	332 g/100 mL^{0} aq; s bz, chl, CCl_4
trifluoride 1-diethyl ether	$BF_3 \cdot O(C_2H_5)_2$	141.94	1.125	−60.4	125.7	d aq
trifluoride 1-methanol	$BF_3 \cdot HOCH_3$	131.89	1.203		59^{4mm}	
nitride	BN	24.82	2.18	2967		sl s hot acids
oxide	B_2O_3	69.62	2.55	450.0	2065	3.3 aq (slowly); s alc, glyc
Bromine	Br_2	159.808	3.1023$^{25}_{4}$	−7.25	58.8	3.4 g/100 mL^{20} aq; v s alc, chl, eth
pentafluoride	BF_5	174.90	2.460	−60.5	40.76	explodes with water; s HF
trifluoride	BF_3	136.90	2.803^{25}	8.77	125.74	d viol aq; d alk; smokes in air
Cadmium	Cd	112.411	8.65^{25}	321	765	i aq, alk; s HNO_3, hot HCl
acetate	$Cd(C_2H_3O_2)_2$	230.50	2.341	255	d	v s aq; s alc
bromide	$CdBr_2$	272.22	5.192	566	963	99 g/100 mL^{20} aq; s acet; sl s eth
carbonate	$CdCO_3$	172.42	4.258^{4}	d 500		s acids, NH_4OH
chloride	$CdCl_2$	183.32	4.05^{25}	568	960	120 g/100 mL^{25} aq
cyanide	$Cd(CN)_2$	164.44	2.226	d 200		1.71 g/100 mL^{15} aq; sl s alc
fluoride	CdF_2	150.41	6.33	1110	1748	4.3 g/100 mL^{25} aq
hydroxide	$Cd(OH)_2$	146.43	4.79	$-H_2O$, 130	CaO, 200	0.00026^{20} aq; s acids
iodide	CdI_2	366.22	5.670	388	742	84.7 g/100 mL^{20} aq; s alc, acet, eth
nitrate 4-water	$Cd(NO_3)_2 \cdot 4H_2O$	308.48	2.455	59.4		167 g/100 mL^{25} aq; s alc, acet
oxide	CdO	128.41	8.15 cubic	1540		i aq; s acids
phosphide	Cd_3P_2	399.18	5.96	700		s dil acid
selenide	CdSe	191.37	5.81^{15}	1350		i aq; d acids
sulfate-water (3/8)	$3CdSO_4 \cdot 8H_2O$	769.56	3.08	monohydrate, 80		94.4 g/100 mL^{25} aq; i alc, EtAc
sulfide	CdS	144.48	4.83	1750		0.13^{18} aq; s acids
telluride	CdTe	240.01	6.20$^{15}_{4}$	1041		i aq; d HNO_3
tungstate(VI)	$CdWO_4$	360.25	8.0			i aq, dil acids; s alkali CN's
Calcium	Ca	40.078(4)	1.55	842	1484	d aq; s acids
acetate	$Ca(C_2H_3O_2)_2$	158.17	1.50	d > 160		37.4 g/100 mL^{0} aq; i alc, bz, acet
arsenate	$Ca_3(AsO_4)_2$	398.07	3.620			0.013^{25} aq
bromide	$CaBr_2$	199.89	3.38	742	1815	143 g/100 mL^{20} aq; v s alc, acet
carbide	CaC_2	64.10	2.222	2300		reacts with aq giving C_2H_2
carbonate (aragonite)	$CaCO_3$	100.09	2.83	d 825 to CaO		s dil acids
carbonate (calcite)	$CaCO_3$	100.09	2.711	d 825 to CaO		0.0013 g/100 mL^{20}; s acids
chlorate 2-water	$Ca(ClO_3)_2 \cdot 2H_2O$	243.01	2.711	anhyd 100		167 g/100 mL^{20} aq; s alc

chloride	$CaCl_2$	110.98	2.16^{25}_{4}	775	ca. 1940	42 g/100 mL^{20} aq; s alc, acet
chloride 6-water	$CaCl_2 \cdot 6H_2O$	219.07	1.71	anhyd 200		74.5 g/100 mL20 aq; v s alc
chlorite	$Ca(ClO_2)_2$	174.99	2.71^{25}	100		167 g/100 mL aq; s alc
chromate(VI) 2-water	$CaCrO_4 \cdot 2H_2O$	192.10	2.50	anhyd 200		sl s aq; s dil acids
citrate 4-water	$CaC_6H_6O_7 \cdot 4H_2O$	570.51		anhyd 120		0.10 aq; i alc
cyanamide	$CaCN_2$	80.10	2.29	ca. 1340	subl	no known solv without dec
cyanide	$Ca(CN)_2$	92.11		s >350		s aq
dichromate(VI)	$CaCr_2O_7$	256.10	2.370^{30}_{4}	d >100		v s aq; i eth; d alc
dihydrogen phosphate hydrate	$Ca(H_2PO_4)_2 \cdot H_2O$	252.07	2.220^{18}_{4}	anhyd 100	d 200	1.8^{30} aq
diphosphate (pyrophosphate)	$Ca_2P_2O_7$	254.10	3.09	1353		i aq; s HCl, HNO_3
fluoride	CaF_2	78.08	3.180	1418	2533	0.0015^{20} aq; s conc mineral acids
formate	$Ca(CHO_2)_2$	130.11	2.015	300 d		16.6 g/100 mL^{20} aq; i alc
(+)gluconate	$Ca[OOC(CHOH)_2CH_4OH]_2$	430.38				3.72^{20} aq
glycerophosphate	$Ca[C_3H_5(OH)_3]PO_4$	210.16		d >170		1.66^{20} aq; i alc
hexafluorosilicate	$Ca[SiF_6]$	182.17	2.662			i aq, acet
hydride	CaH_2	42.09	1.70	1000		d aq, alc
hydroxide	$Ca(OH)_2$	74.09	2.343	$-H_2O$, 580		0.17^{10} aq; s acids
hypochlorite	$Ca(OCl)_2$	142.99	2.35	100 d		d aq evolving Cl_2; i alc
iodate	$Ca(IO_3)_2$	389.88	4.519^{15}_{4}	d >540		0.10^{0} aq; i alc
iodide	CaI_2	293.89	3.956	783	1755	68 g/100 mL^{20} aq; v s alc, acet; i eth
lactate 5-water	$Ca(C_3H_5O_3)_2 \cdot 5H_2O$	308.30		$-3H_2O$, 100	anhyd 120	5.4^{15} aq; v sl s alc
magnesium carbonate	$Ca[Mg(CO_3)_2]$	184.41	2.872	d 730		0.032^{18} aq; s HCl
molybdate(VI)(2−)	$CaMoO_4$	200.02	4.35			s conc mineral acids
nitrate	$Ca(NO_3)_2$	164.09	2.504	561		152 g/100 mL^{30} aq
nitride	Ca_3N_2	148.25	2.67	1195		d aq; s dilute acids (d)
nitrite 4-water	$Ca(NO_2)_2 \cdot 4H_2O$	204.15	1.674	d		84.5 g/100 mL^{18} aq; sl s alc
oleate	$Ca(C_{18}H_{33}O_2)_2$	603.01		83–84	d >400	0.04 aq; s chl, bz; v sl s alc, eth
oxalate hydrate	$CaC_2O_4 \cdot H_2O$	146.11	2.2	anhyd 200		0.0006 aq; s acids
oxide	CaO	56.08	3.34	2900	3500	0.13^{25} aq; s acids
palmitate	$Ca(C_{16}H_{31}O_2)_2$	550.93		d >155		0.003 aq; sl s bz, chl, HOAc
(+)panthothenate (vitamin B_3)	$Ca[O_2CH_2CH_2NHO\text{-}CH(OH)C(CH_3)_2CH_2OH]_2$	476.55		d 195–196		36 g/100 mL aq; sl s alc, acet
perchlorate	$Ca(ClO_4)_2$	238.98	2.65	d 270		g/100 mL^{25}: 112 aq, 89.5 EtOH, 68 BuOH, 57 EtOAc, 43 acet

(Continued)

TABLE 1.3 Physical Constants of Inorganic Compounds (*Continued*)

Name	Formula	Formula weight	Density	Melting point, °C	Boiling point, °C	Solubility in 100 parts solvent
permanganate 5-water	$Ca(MnO_4)_2 \cdot 5H_2O$	368.03	2.4	d		338 g/100 mL aq
peroxide	CaO_2	72.08	2.92	explodes 275		sl s aq; s acids
phenoxide	$Ca(OC_6H_5)_2$	226.28	d in air			sl s aq, alc
phosphate	$Ca_3(PO_4)_2$	310.18	3.14	1670		0.03^{25} aq; s HCl, HNO_3; i alc
phosphide	Ca_3P_2	182.18	2.51	ca. 1600		d aq; s acids; i alc, eth
phosphinate	$Ca(PH_2O_2)_2$	170.06		d >300		15.4 g/100 mL aq; sl s glyc
propanoate	$Ca(OOCC_3H_5)_2$	186.22				s aq; sl s alc; i acet, bz
salicylate 2-water	$Ca(C_7H_5O_3)_2 \cdot 2H_2O$	350.34		anhyd 200	d 240	2.8^{15} aq; 0.015^{16} EtOH
selenate 2-water	$CaSeO_4 \cdot 2H_2O$	219.07	2.75	anhyd 200	d 698	9.2 g/100 mL^{25} aq
selenide	CaSe	119.04	3.82			
silicate	Ca_2SiO_4	172.24	3.27	2130		i aq
stearate	$Ca(C_{18}H_{35}O_2)_2$	607.04		179–180		0.004^{15} aq; s hot pyr; i acet, chl
succinate 3-water	$CaC_4H_6O_4 \cdot 3H_2O$	212.22				1.28^{20} aq; s acids; i alc
sulfate	$CaSO_4$	136.14	2.960	1460		0.20 aq; s acids
sulfate hemihydrate	$CaSO_4 \cdot 0.5H_2O$	145.15		anhyd 163		0.3^{20} aq; s acids, glyc
sulfate 2-water	$CaSO_4 \cdot 2H_2O$	172.17	2.32	−1.5 H_2O, 128	anhyd 163	0.26^{20} aq; s acid, glyc
sulfide	CaS	72.14	2.59	2525		0.02 (d) aq; d acids
sulfite 2-water	$CaSO_3 \cdot 2H_2O$	156.17		anhyd 100		0.004 aq; s acids d; sl s alc
(±)tartrate 4-water	$CaC_4H_4O_6 \cdot 4H_2O$	260.21		anhyd 200		0.0045^{25} aq; s acids; sl s alc
telluride	CaTe	167.68	4.873			
tetraborate	CaB_4O_7	195.36				s dil acids
tetrahydridoaluminate	$Ca[AlH_4]_2$	102.10		ign moist air		d viol aq, alc; i bz, eth
thiocyanate 3-water	$Ca(SCN)_2 \cdot 3H_2O$	210.29		d >160		150 g/100 mL aq; v s alc
thioglycollate 3-water	$Ca(\text{-}OOCCH_2S\text{-}) \cdot 3H_2O$	184.24		−H_2O, >95	d >220	s aq; v sl s alc, chl; i bz, eth
thiosulfate 6-water	$CaS_2O_3 \cdot 6H_2O$	260.30	1.872	d >45		92 g/100 mL^{25} aq; i alc
titanate	$CaTiO_3$	135.84	3.98	1980		
tungstate(VI)(2−)	$CaWO_4$	287.93	6.062^{20}			0.0032 aq; d hot acids
Californium-252	Cf	252.1		900		
chloride	$CfCl_3$	358.5	5.88			
Carbon (diamond)	C	12.011	3.513	$3500^{63.5atm}$	3930	i aq, alc
(graphite)	C		2.267	subl 3915–4020		
dioxide	CO_2	44.01	c: 1.56^{-79} g: 1.975 g/L^0	−78.44 subl		88 mL/100 mL^{20} aq
diselenide	CSe_2	169.93	2.6626^{25}_{4}	−45.5	125.1	i aq; s acet, eth; misc CCl_4; d alc
disulfide	CS_2	76.14	1.2555	−111.6	46.56	FP −30; 0.29^{20} aq; s alc, eth

hydride (methane)	CH_4	16.04	0.415^{-164}	−182.48	−161.49	s bz
monoxide	CO	28.01	lq: 0.814^{-195} g: 1.250 g/L^0	−205.05	−191.49	2.3 mL/100 mL^{20} aq; 16 mL/100 ml alc; s HOAc, EtAc
suboxide	C_3O_2	68.03	1.114^{0}_{4} 2.985 g/L	−111.3	6.8	d aq to malonic acid; sl s CS_2
tetrabromide	CBr_4	331.65	3.42	90.1	190	i aq; s alc, chl, eth
tetrachloride	CCl_4	153.82	1.589^{25}_{25}	−22.9	76.7	0.05 mL/100 mL aq; s alc, chl, eth
tetrafluoride	CF_4	88.00	1.96^{-184}	−183.6	−127.8	sl s aq
tetraiodide	CI_4	519.63	4.34^{20}_{4}	171	subl 130	slowly hyd aq; s bz, chl, eth
Carbonyl bromide	$COBr_2$	187.82	2.5		64.5	hyd aq
chloride	$COCl_2$	98.92	4.340 g/L	−127.9	8.2	hyd aq; s bz, HOAc
fluoride	COF_2	66.01	lq: 1.139 g: 2.896 g/L	−114.0	−83.1	hyd aq
sulfide	COS	60.07	2.636 g/L	−138.81	−50.23	54 mL/100 mL^{20} aq; s alc, CS_2
Cerium	Ce	140.11	6.773	795	3440	i aq; s acids
(III) bromide	$CeBr_3$	379.83	5.18	733	1460	s aq, alc
(III) chloride	$CeCl_3$	246.47	3.97^{25}	817	1730	s aq, alc
(III) fluoride	CeF_3	197.11	6.157	1430	2327	i but slowly hyd aq; s H_2SO_4
(IV) fluoride	CeF_4	216.11	4.77	d >550		i aq
(III) iodide	CeI_3	520.83		766	1400	s aq
(III) nitrate 3-water	$Ce(NO_3)_3\ 3H_2O$	380.17		anhyd 150	d 200	234 g/100 mL^{20} aq
(IV) oxide	CeO_2	172.11	7.65	2400		i aq; s acids
(III) sulfate	$Ce_2(SO_4)_3$	568.42	3.912	d 1000		9.72 g/100 mL^{21} aq
(IV) sulfate	$Ce(SO_4)_2$	332.24	3.91	d 195		hyd aq; s dil H_2SO_4
Cesium	Cs	132.9054	1.8785^{15}	28.44	668.2	d aq; s acids
bromide	CsBr	212.81	4.44	636	≈1300	107 g/100 mL^{18} aq; s alc; i acet
carbonate	Cs_2CO_3	325.82	4.24	792		v s aq; 11 g/100 mL^{20} alc; s eth
chloride	CsCl	168.36	3.99	646	1300	g/100 mL: 187^{20} aq; 34^{25} MeOH; v s alc
fluoride	CsF	151.90	4.115	703	1231	322 g/100 mL^{18} aq
hydroxide	CsOH	149.91	3.68	272	990	386 g/100 mL^{15} aq; s alc
iodate	$CsIO_3$	307.81	4.934^{20}	565		2.6^{25} aq
iodide	CsI	259.81	4.510	621	≈1280	76.5 g/100 mL^{20} aq; s EtOH; i acet
nitrate	$CsNO_3$	194.91	3.66	414	d 849	23 g/100 mL^{20} aq; s acet; v sl s alc

(Continued)

TABLE 1.3 Physical Constants of Inorganic Compounds (*Continued*)

Name	Formula	Formula weight	Density	Melting point, °C	Boiling point, °C	Solubility in 100 parts solvent
oxide	Cs_2O	281.81	4.65	490		v s aq
perchlorate	$CsClO_4$	232.36	3.327	250		g/100 mL^{25}: 1.96, 0.0086 EtOH, 0.118 acet, 0.0048 BuOH; i EtOAc, eth
selenate	Cs_2SeO_4	408.77	4.453			244 g/100 mL^{12} aq
sulfate	Cs_2SO_4	361.87	4.243	1005		179 g/100 mL^{20} aq; i alc, acet, pyr
Chlorine	Cl_2	70.905	g: 2.98^{20} g/L lq: 1.5649^{-35}	−101.5	−34.04	199 mL/100 mL^{25} aq
dioxide	ClO_2	67.45	2.960 g/L	−59.6	10.9	11.2 g/100 mL^{10} aq
fluoride	ClF	54.45	4.057 g/L	−155.6	−100.1	d viol aq; organics burst into flame
heptoxide	Cl_2O_7	182.90	1.805^{25}	−91.5	82	hyd aq slowly; explodes on concussion or on contact with flame or I_2
monoxide	Cl_2O	86.90	3.813 g/L	−120.6	2.2	v s aq (forms HClO); s CCl_4
pentafluoride	ClF_5	130.44	5.724 g/L	−103	−13.1	
trifluoride	ClF_3	92.45	g: 4.057 g/L lq: 1.825^{bp}_{20}	−76.3	11.75	hyd viol aq; organic matter and glass wool burst into flame
trioxide (dimer)	$(ClO_3)_2$	166.90	1.92^{20}	3.5	≈200	reacts with aq
Chromium	Cr	51.996	7.15	1907	2679	s dil HCl
(II) acetate	$Cr(C_2H_3O_2)_2$	170.09	1.79			sl s aq, alc; s a; i eth
(III) acetate	$Cr(C_2H_3O_2)_3$	229.13				s aq
(II) bromide	$CrBr_2$	211.80	4.236	842		s aq, alc
(III) bromide	$CrBr_3$	291.71	4.68			s hot aq; v s alc
(II) chloride	$CrCl_2$	122.90	2.88^{25}	814	subl 1300	v s aq
(III) chloride	$CrCl_3$	158.35	2.87	1152	d >1300	s aq, alc (slow); i acet
(II) fluoride	CrF_2	89.99	3.79	894		sl s aq; s hot HCl
(III) fluoride	CrF_3	108.99	3.8	1400		aq, alc; s HF, HCl
(III) formate 6-water	$Cr(CHO_2)_3 \cdot 6H_2O$	295.15		d >300		s aq
hexacarbonyl	$Cr(CO)_6$	220.06	1.77	d 130	explodes 210	i aq, alc; s eth, chl
(III) hydroxide	$Cr(OH)_3$	101.02		d		i aq; s acids
(III) nitrate 9-water	$Cr(NO_3)_3 \cdot 9H_2O$	400.15	1.80	66	d >100	208 g/100 mL^{15} aq; s alc
(III) oxide	Cr_2O_3	151.99	5.21	2330	≈3000	i aq, alc; sl s acids, alkalis
(IV) oxide	CrO_2	84.00	4.89	197	$-O_2$, 250	i aq; s HNO_3
(VI) oxide	CrO_3	99.99	2.70^{25}	198	d 250	61.7 g/100 mL aq; may ign organics
(III) phosphate	$CrPO_4$	146.97	4.6	>1800		i aq, acids, aq reg

potassium bissulfate 12-water	$CrK(SO_4)_2 \cdot 12H_2O$	499.41	1.826^{25}	89	anhyd 400	22 g/100 mL^{25} aq; i alc
(II) sulfate 7-water	$CrSO_4 \cdot 7H_2O$	274.17				22.9 g/100 mL^{0} aq; sl s alc
(III) sulfate 18-water	$Cr_2(SO_4)_3 \cdot 18H_2O$	716.45	1.7	d 100		220 g/100 mL^{20} aq
Chromyl chloride	CrO_2Cl_2	154.90	1.9145^{25}_{4}	−96.5	117	d aq; s bz, chl, eth, CCl_4
fluoride	CrO_2F_2	121.99		31.6^{885mm}	subl 29.6	
Cobalt	Co	58.9332	8.90	1494	2927	i aq; s dil HNO_3
(II) acetate 4-water	$Co(C_2H_3O_2)_2 \cdot 4H_2O$	249.08	1.705^{19}	anhyd 140		s aq; 2.1 g/100 mL^{15} MeOH
(III) acetate	$Co(C_2H_3O_2)_3$	236.07		d >100		s aq, HOAc, alc
(II) bromide	$CoBr_2$	218.74	4.909^{25}_{4}	678 (in N_2)		112 g/100 mL^{20} aq; s alc, acet
(II) carbonate	$CoCO_3$	118.94	4.13	d		0.18^{15} aq; s hot acids
(II) chloride	$CoCl_2$	129.84	3.367^{25}_{4}	735	1049	53 g/100 mL^{20} aq; s alc, acet, eth, glyc, pyr
(II) chloride 6-water	$CoCl_2 \cdot 6H_2O$	237.93	1.924	anhyd 110		97 g/100 mL^{20} aq
(II) chromate	$CoCrO_4$	174.93	≈4.0	d		i aq; s acids
(II) cyanide	$Co(CN)_2$	110.97	1.872^{25}_{4}	d 300		0.0042^{18} aq; s KCN
(II) fluoride	CoF_3	96.93	4.46	1127	≈1400	1.36^{20} aq; s warm mineral acids
(III) fluoride	CoF_3	115.93	3.88	926		d aq
(II) formate 2-water	$Co(CHO_2)_2 \cdot 2H_2O$	185.00	2.129^{22}_{4}	anhyd 140	d 175	5.03 g/100 mL^{30} aq; i alc
(II) hydroxide	$Co(OH)_2$	92.95	3.37	168 (vacuo)		0.00018 aq; v s acids
(III) hydroxide	$Co(OH)_3$	109.96	4.46	$-H_2O$, 100	d	0.00032 aq; s acids
(II) iodide (alpha, black)	CoI_2	312.74	5.584^{25}_{4}	515 (vacuo)	570 (vacuo)	203 aq
(II) nitrate 6-water	$Co(NO_3)_3 \cdot 6H_2O$	291.03	1.88	55	d >74	155 g/100 mL^{30} aq; v s alc
(II) oxalate	CoC_2O_4	146.95	3.021	d 250		0.002^{18} aq
(II) oxide	CoO	74.93	6.44	−s1935		i aq; s acids, alkalis
(II,III) oxide	Co_3O_4	240.80	6.07	d >900		i aq; s acids, alkalis
(II) phosphate 8-water	$Co_3(PO_4)_2 \cdot 8H_2O$	510.87	2.769	anhyd 200		v sl s aq; s mineral acids
(II) sulfate 7-water	$CoSO_4 \cdot 7H_2O$	281.10	2.03	anhyd 420	d 1140	65 g/100 mL^{20} aq; sl s alc
(II) sulfide	CoS	91.00	5.45^{18}	1180		i aq; s acids
(II) thiocyanate 3-water	$Co(SCN)_2 \cdot 3H_2O$	229.14		anhyd 105		7.8^{18} aq; s alc, eth
Copper	Cu	63.546	8.96^{20}	1084.62	2561.5	i; s HNO_3, hot H_2SO_4
(II) acetate 1-water	$Cu(C_2H_3O_2) \cdot H_2O$	199.65	1.882	115	d 240	8 g/100 mL aq; 0.48 MeOH; sl s eth
acetate *meta*-arsenate (1/3)	$Cu(C_2H_3O_2)_2 \cdot 3Cu(AsO_2)_2$	1013.80				unstable in acids, bases; s NH_4OH
(II) borate(1−)	$Cu(BO_2)_2$	149.17	3.859			s a; i aq
(I) bromide	CuBr	143.45	4.98	497	1345	v sl s aq; s HCl, HBr, NH_4OH
(II) bromide	$CuBr_2$	223.35	4.71	498	900	126 g/100 mL aq; s alc, acet, pyr; i

(Continued)

TABLE 1.3 Physical Constants of Inorganic Compounds (*Continued*)

Name	Formula	Formula weight	Density	Melting point, °C	Boiling point, °C	Solubility in 100 parts solvent
(II) carbonate hydroxide (1/1) (malachite)	$CuCO_3 \cdot Cu(OH)_2$	221.12	4.0	d 200		i aq; s acids
(II) chlorate 6-water	$Cu(ClO_3)_2 \cdot 6H_2O$	338.54		65	d 100	242 g/100 mL^{18} aq; v s alc; s acet
(I) chloride	$CuCl$	99.00	4.14	430	≈1400	0.024 aq; s conc HCl, conc NH_4OH
(II) chloride	$CuCl_2$	134.45	3.386	300 d		73 g/100 mL^{20} aq; s alc, acet
(II) chloride 2-water	$CuCl_2 \cdot 2H_2O$	170.48	2.51	anhyd 200	d >300	76.4 g/100 mL^{25} aq; v s alc; s acet
(I) chromium(III) oxide (1/1)	$Cr_2O_3 \cdot Cu_2O$	295.07	5.24^{20}	d >900		i aq; s HNO_3
(II) citrate 2.5-water	$Cu_2C_6H_4O_7 \cdot 2.5H_2O$	360.22		anhyd 100		0.17 aq; s acids
(I) cyanide	$CuCN$	89.56	2.92	473 (in N_2)	d	i aq; s NH_4OH, KCN; d hot dil HCl
(II) fluoride	CuF_2	101.54	4.23	836	1676	4.75 g/100 mL^{20} aq; s acids
(II) formate	$Cu(CHO_2)_2$	153.58	1.831			12.5 aq
(II) hexafluorosilicate 4-water	$Cu[SiF_6] \cdot 4H_2O$	277.60	2.56	d		124 g/100 mL^{20} aq
(II) hydroxide	$Cu(OH)_2$	97.56	3.368	d 160		i aq; s acids
(I) iodide	CuI	190.45	5.67	606	≈1290	i aq; s KCN, NH_4OH, KI
(II) nitrate 3-water	$Cu(NO_3)_2 \cdot 3H_2O$	241.60	2.32	114.5	170 d	138 g/100 mL^0 aq; v s alc
(II) oleate	$Cu(OOCC_{17}H_{33})_2$	626.46				i aq; sl s alc; s eth
(II) oxalate hemihydrate	$CuC_2O_4 \cdot 0.5H_2O$	160.57		anhydr >200	d 310	0.002 aq; s NH_4OH
(I) oxide	Cu_2O	143.09	6.0_4^{25}	1235	$-O_2$, 1800	i aq; s HCl
(II) oxide	CuO	79.54	6.315_4^{14}	1450		i aq, alc; s acids, KCN
(II) perchlorate	$Cu(ClO_4)_2$	262.45	2.225^{23}	d >130		146 g/100 mL^{30} aq; s eth, EtAc; i bz
(II) phosphate 3-water	$Cu_3(PO_4)_2 \cdot 3H_2O$	434.63		d		i aq; s acids
(II) salicylate 4-water	$Cu(C_7H_5O_3)_2 \cdot 4H_2O$	409.83		dehyd in air		v s aq; s alc
(II) selenate 5-water	$CuSeO_4 \cdot 5H_2O$	296.58	2.559	anhyd 265	d ca. 480	25 g/100 mL^{20} aq; v sl s acet
(I) selenide	Cu_2Se	206.05	6.84_4^{21}	1113		d HCl
(II) selenide	$CuSe$	142.51	6.0	d 550		s acids
(II) stearate	$Cu(OOCC_{17}H_{35})_2$	630.50		≈250		i aq, alc, eth; s hot bz, pyr
(II) sulfate	$CuSO_4$	159.61	3.603	d >560		14.3 g/100 mL^0 aq; i alc
(II) sulfate 5-water	$CuSO_4 \cdot 5H_2O$	249.69	2.284_4^{16}	anhyd 200		32 g/100 mL^{20} aq; s MeOH, glyc
(I) sulfide	Cu_2S	159.16	5.6_4^{20}	1130		i aq; d HNO_3, s KCN
(II) sulfide	CuS	95.61	4.76			i aq; s hot HNO_3, KCN

(I) sulfite hydrate	$Cu_2SO_3 \cdot H_2O$	225.16	3.83^{15}	d		sl s aq; s HCl
(II) tartrate 3-water	$CuC_4H_4O_6 \cdot 3H_2O$	265.66				0.4220 aq; s acids, alkalis
(I) thiocyanate	CuSCN	121.62	2.85	1084		0.00044 aq; s NH_4OH, eth, alkali SCN
(II) tungstate(VI)(2-)	$CuWO_4 \cdot 2H_2O$	347.41				0.1^{15} aq; d acids; s NH_4OH
Curium-244	Cm	244.063	13.51	1340	≈3110	s acids
Cyanogen	NC—CN	52.03	2.335 g/L	−27.84	−21.15	mL/100 mL: 450^{20} aq, 230 alc;
azide	$NC—N_3$	68.04				s acetonitrile; pure azide detonates upon shock. Handle only in solvents.
bromide	NCBr	105.92	2.005	52	61.5	v s aq, alc, eth
chloride	NCCl	61.47	2.697 g/L	−6.5	13.8	s aq, alc, eth
fluoride	NCF	45.02	1.975 g/L	−82	−46	
Deuterium	D_2 or $^2H_2$	4.03	0.169^{mp} lq	−252.89	−249.49	sl s aq
oxide	D_2O	20.03	1.1056^{20}	3.82	101.43	misc aq
Dysprosium	Dy	162.50	8.540^{25}	1412	2567	s acids
bromide	$DyBr_3$	402.21	4.78	880	1480	s aq
chloride	$DyCl_3$	268.86	3.67	680	1530	s aq
fluoride	DyF_3	219.50	7.465	1154	2230	i aq
oxide	Dy_2O_3	373.00	7.81^{27}	2408		s aq
Einsteinium	Es	252.083	8.84	860		
Erbium	Er	167.26	9.066	1529	2868	s acid
chloride	$ErCl_3$	273.62	4.1	776	1500	s aq; sl s alc
oxide	Er_2O_3	382.52	8.640	2418		0.0005^{25} aq; s acids
sulfate 8-water	$Er_2(SO_4)_3 \cdot 8H_2O$	766.83	3.205	anhyd 110	d 630	16.0 g/100 mL^{20} aq
Europium	Eu	151.965	5.244	822	1527	s acids
(III) chloride	$EuCl_3$	258.32	4.89	623 d		s aq
(III) oxide	Eu_2O_3	351.93	7.42	2350		i aq; s acids
(III) sulfate 8-water	$Eu_2(SO_4)_3 \cdot 8H_2O$	736.24	$-8H_2O$, 375			2.56^{20} aq
Fermium-257	Fm	257.0951		1527		
Fluorine	F_2	38.00	1.513^{bp} lq 1.667 g/L	−219.61	−188.13	d aq viol; ignites organics and silicates
nitrate	$FONO_2$	81.00	1.507^{bp} lq	−175	−45.9	hyd aq; s acet; ignites alc, eth; liquid explodes on slight concussion
perchlorate	$FOClO_3$	118.45	5.20 g/L	−167.3	−15.9	explodes on slightest provocation
Francium-223	Fr	223.02				

(Continued)

TABLE 1.3 Physical Constants of Inorganic Compounds (*Continued*)

Name	Formula	Formula weight	Density	Melting point, °C	Boiling point, °C	Solubility in 100 parts solvent
Gadolinium	Gd	157.25	7.90	1312	3273	s acids
chloride	$GdCl_3$	263.61	4.52^{0}	~609	1580	s aq
fluoride	GdF_3	214.25	7.047	1231	2277	i aq
nitrate 6-water	$Gd(NO_3)_3 \cdot 6H_2O$	451.36	2.332	91		s aq, alc
oxide	Gd_2O_3	362.50	7.407^{15}	2340		s acids
sulfate 8-water	$Gd_2(SO_4)_2 \cdot 8H_2O$	746.81	3.010^{18}	anhyd 400	d 500	4.08 aq
Gallium	Ga	69.723	$5.904^{29.6}$ (c) $6.095^{29.8}$ (lq)	29.7646	2203	s conc HCl, halogens, alkalis
antimonide	GaSb	191.48	5.614	712		s HCl
arsenide	GaAs	144.65	5.318^{25}_{4}	1238		s HCl
chloride	$GaCl_3$	176.08	2.47	77.9	201.2	d aq; s bz, CCl_4, CS_2
fluoride	GaF_3	126.72	4.47	>1000	subl 950	0.004^{25} aq; s HF
nitrate	$Ga(NO_3)_3$	255.74		d 110	→ Ga_2O_3, 200	v s aq
phosphide	GaP	100.70		1465		
selenide	GaSe	148.68	5.03^{25}	960	d	
triethyl	$Ga(C_2H_5)_3$	146.90	1.058^{30}	−82.3	142.8	
trimethyl	$Ga(CH_3)_3$	114.84	1.151^{15}	−15.7	55.8	
Germanium	Ge	72.61	5.323	937.3	2830	i aq; s hot H_2SO_4
(IV) bromide	$GeBr_4$	392.23	3.132	26.1	186.4	hyd aq; s bz, eth
IV) chloride	$GeCl_4$	214.42	1.879	−49.5	86.5	hyd aq; s bz, eth; sl s dil HCl
(IV) fluoride	GeF_4	148.60	6.521 g/L	−15	d >1000	hyd aq; s dil HCl
hydride (germane)	GeH_4	76.64	3.363 g/L	−164.8	−88.1	sl s hot HCl
(IV) oxide	GeO_2	104.61	4.25	1115	1200	0.43^{20} aq; s acids, alkalis
sulfide	GeS_2	136.74	3.01	530		
Gold	Au	196.967	19.3	1064.18	2856	s aq reg, KCN, hot H_2SO_4
(I) chloride	AuCl	232.42	7.57	289		s HCl, HBr, KCN
(III) chloride	$AuCl_3$	303.33	4.7	d >160	subl 180	68 g/100 mL^{20} aq; s EtOH
(I) cyanide	AuCN	222.99	7.14^{20}_{4}	d		s aq reg, KCN, NH_4OH
(III) cyanide 3-water	$Au(CN)_3 \cdot 3H_2O$	329.07		d 50		v s aq; sl s alc
diantimonide	$AuSb_2$	440.47		460		
(III) fluoride	AuF_3	253.96	6.75	subl 300	d 500	
(III) oxide	Au_2O_3	441.93		d 150		s HCl, KCN

(I) sodium thiosulfate 2-water	$AuNa_3(S_2O_3)_2 \cdot 2H_2O$	526.24	3.09	anhyd 160		50 g/100 mL aq; i alc
stannide	AuSn	315.66		418		
(III) sulfide	Au_2S_3	490.13	8.754	d 197		i aq; s Na_2S
Hafnium	Hf	178.49	13.31	2227	4450	s HF
chloride	$HfCl_4$	320.30		432	subl 317	hyd aq; s acet, MeOH
oxide	HfO_2	210.49	9.68^{20}	2774		i aq
Helium	He	4.00260	0.176 g/L 0.1249 (lq)	-272.15^{25atm}	−268.935	0.861 mL/100 mL^{20} aq
Holmium	Ho	164.9304	8.79	1474	2720	s acids; oxidizes in moist air
bromide	$HoBr_3$	404.64	4.86	914	1470	s aq
chloride	$HoCl_3$	271.29	3.7	718	1510	s aq
Hydrazine	$H_2N—NH_2$	32.05	1.0036^{25}_{4}	2.0	113.5	FP 52; misc aq, alc
hydrate	$H_2N—NH_2 \cdot H_2O$	50.06	1.030	−51.7 & −65	118–119	misc aq, alc; i chl, eth
Hydrazinium(1+) chloride	$H_2N—NH_3Cl$	68.51	1.5	89	d 240	v s aq; i org solv
(2+) chloride	$ClH_3N—NH_3Cl$	104.97	1.423	198	d 200	v s aq; sl s alc
(1+) iodide	$H_2N—NH_3I$	159.96		125		s aq
(+1) perchlorate	$H_2N—NH_3ClO_4$	132.51	1.939^{15}	137	d 145	d aq; s alc
(2+) sulfate	$(H_3NNH_3)SO_4$	130.13	1.378	254	d	3.4^{20} aq; i alc
(1+) tartrate	$(H_2N—NH_3)_2C_4H_4O_6$	182.13		183		6.0 g/100 mL^{0} aq
Hydrogen	H_2	2.0159 0.07099^{bp} (lq)	0.088 g/L	−259.35	−252.88	1.9 mL aq
amidosulfate (sulfamate)	H_2NSO_3H	97.09	2.126	205	d	14.7 g/100 mL aq; sl s alc, acet
azide	HN_3	43.03	1.126^{0}	−80	37	v s aq; (very explosive)
borate(1−) (cubic)	HBO_2	43.83	2.486	236		v sl s aq
borate(3−) (ortho)	H_3BO_3	61.83	1.435^{15}	171.0	d 357	5.56 g/100 mL^{30} aq
bromide	HBr	80.91	3.388 g/L^{20}	−86.87	−66.71	193 g/100 mL^{25} aq; misc alc
bromide (constant boiling)	48% HBr + H_2O		1.49	−11	126	v s aq
bromide-*d*	^{2}HBr	81.91	3.39 g/L^{20}	−87.46	−66.5	v s aq
bromosulfate	$HOSO_2Br$	240.90		−6 to −8	d	hyd aq
chlorate (40% solution)	$HClO_3$	84.46	1.282^{20}_{4}			
chloride	HCl	36.46	1.526 g/L^{20}	−114.18	−85.05	72 g/100 mL^{20} aq
chloride (constant boiling)	20.24% HCl + H_2O		1.097		110	v s aq
chloride-*d*	^{2}HCl	37.47	1.49 g/L^{25}	−114.64	−84.72	v s aq
chlorosulfate	HSO_3Cl	116.52	1.753	−80	152	hyd viol → HCl + H_2SO_4
cyanate	HOCN	43.03	1.140^{-20}_{4}	−86	23.5	s aq d; s bz, eth

(Continued)

TABLE 1.3 Physical Constants of Inorganic Compounds (*Continued*)

Name	Formula	Formula weight	Density	Melting point, °C	Boiling point, °C	Solubility in 100 parts solvent
cyanide	HCN	27.03	0.687	−13.4	25.6	misc aq
deuteride	$^1H^2H$ or HD	3.02		−256.56	−251.03	
diphosphate(IV)	$(HO)_2OP—PO(OH)_2$	162.01	70	d 100	d	aq
diphosphate(V)	$H_4P_2O_7$	177.98		61		709 g/100 mL^{23} aq
fluoride	HF	20.01	0.922 g/L^0	−83.57	19.52	v s aq, alc; 2.54 g/100 g^5 bz
fluoride (constant boiling)	35.35% HF + H_2O				120	v s aq
fluoride-*d*	^2HF	21.02		−83.6	18.65	s aq
fluoroborate	$H[BF_4]$	87.81		d 130		v s aq
fluorophosphate	H_2PO_3F	99.99	1.818	−80		
fluorosulfate	$HOSO_2F$	100.07	1.726^{25}_{4}	−87.3	165.5	s aq
hexafluorosilicate 2-water	$H_2[SiF_6] \cdot 2H_2O$	180.11	1.463	19		60–70% aq solution
iodate	HIO_3	175.91	4.629^{0}_{4}	110 →H_5IO_6	220 →I_2O_5	269 g/100 mL^{20} aq; s alc; i eth
iodide	HI	127.91	5.37 g/L^{20}	−50.8	−35.1	234 g/100 mL^{10} aq; misc alc
iodide (constant boiling)	57% HI + H_2O		1.70		127	v s aq
iodide-*d*	HI	128.91		−51.87	−35.7	v s aq
molybdate hydrate	$H_2MoO_4 \cdot H_2O$	179.97	3.124^{15}	−H O, 70		0.133^{18} aq; s alk
nitrate	HNO_3	63.02	1.5492^0 lq	−41.59	83	v s
nitrate (constant boiling)	69% HNO_3 + H_2O		1.41^{20}		120.5	misc aq
oxide (water)	H_2O	18.02	1.000	0.00	100.00	
oxide-d_2	D_2O or ^2H_2O	20.03	1.1044^{25}	3.81	101.42	misc aq
perchlorate 2-water	$HClO_4 \cdot 2H_2O$	136.49	1.67^{20}	−17.8	203	v s aq (commercial 72% acid)
periodate(1−) (meta)	HIO_4	191.91		subl 110	d 138	440 g/100 mL^{25} aq
periodate(5−)	H_5IO_6	227.94		122	d 130–140	misc aq; s alc
peroxide	H_2O_2	34.01	1.463^0	−0.43	152	misc aq; s alc, eth
peroxodisulfate	$HO_3S—O—OSO_3H$	194.14		d 60		v s aq
phosphate(V)(1−) (meta)	HPO_3	79.98	2.2–2.5	subl	red heat	slowly s aq → H_3PO_4; s alc
phosphate(V)(3−) (ortho)	H_3PO_4	98.00	1.868^{25}	42.35	d 213	v s aq
commercial 85% acid			1.685	anhyd 150	$H_4P_2O_7$, 200	→ HPO_3, >300
phosphate(V)(3−)-d_3	$^2H_3PO_4$	101.03	1.908^{25}	46.0		v s aq
phosphide, *see* Phosphine						
phosphinate	HPH_2O_2	66.0	1.493^{19}	26.5	d 50	s aq
phosphonate (phosphorous acid)	H_2PHO_3	82.00	1.651^{25}_{4}	≈73	d >180	v s aq, alc

selenate	H_2SeO_4	144.98	2.9508_4^{15}	58	260	vs aq (viol)
selenide	H_2Se	80.98	2.12_4^{-bp}	−65.73	−41.4	9.5 mL/100 mL^{20} aq; s CS_2
sulfate	H_2SO_4	98.08	1.8318^{20}	10.38	335.5	misc aq
sulfate-d_2	$^2H_2SO_4$ or D_2SO_4	100.09	1.8620	14.35		misc aq
sulfide	H_2S	34.08	1.5392 g/L^0	−85.49	−60.33	0.334 mL^{25} aq
tellurate(IV)	H_2TeO_3	177.63	3.0	d to TeO_2		0.0007 aq; s acid, alkali
tellurate(VI) (monoclinic)	H_6TeO_6	229.66	3.068	$-2H_2O$, 120	320 → TeO	30 g/100 mL^{18} aq
telluride	H_2Te	129.62	5.687 g/L	−49	−2	s aq d
trithiocarbonate	$(HS)_2CS$	110.21	1.483_4^{20}	−26.9	57.8	d aq, alc
tungstate(VI)(2−)	H_2WO_4	249.86	5.5	anhyd 100		i aq; s HF, alkalis
Hydroxylamine	$HONH_2$	33.03	1.204_4^{40}	33	58^{22mm}	v s aq, MeOH; sl s bz, eth
Hydroxylammonium chloride	$HONH_3Cl$	69.49	1.680^{20}	150.5	d	g/100 mL: 83^{17} aq, 12.5^{20} MeOH, 5.1^{20} EtOH; s glyc
sulfate	$(HONH_3)_2SO_4$	164.14		170		69 g/100 mL^{20} aq
Indium	In	114.82	7.31	156.60	2072	s acids
antimonide	InSb	236.58	5.77	525		i aq
arsenide	InAs	189.74	5.67	942		
chloride	$InCl_3$	221.18	4.0	583	subl 500	212 g/100 mL^{25} aq
fluoride	InF_3	171.82	4.39	1170		0.040^{25} aq; s dilute acids
oxide	In_2O_3	277.63	7.179		850	s hot mineral acids
phosphide	InP	145.79	4.81	1062		v sl s acids
telluride	In_2Te_3	612.44	5.75	667		
trimethyl	$In(CH_3)_3$	159.93	1.568	88.4	135.8	d aq; s acet, bz
Iodine	I_2	253.809	4.63^{25}	113.60	185.24	g/100 mL^{25}: 0.029 aq, 14.1 bz, 16.5 CS_2, 21.4 EtOH, 25.2 eth, 2.6 CCl_4; s chl, HOAc
heptafluoride	**IF_7**	259.89	lq: 2.8^6	6.45	4.77 subl	s aq (d), s NaOH
monobromide	**IBr**	206.81	4.416	40	116 d	s aq, alc, eth, CS_2
monochloride	**ICl**	162.36	3.10_4^{29}	27.2 α-form	97 d	d aq; s alc, eth, HOAc
pentafluoride	**IF_5**	221.90	3.19^{25}	9.43	100.5	d aq viol
pentoxide	**I_2O_5**	333.81	4.98	d 275		187 g/100 mL^{13} aq
trichloride	**ICl_3**	233.26	3.202^{-4}	~33	64 subl	d aq; s alc, bz, HCl
Iridium	Ir	192.217	22.65_4^{20}	2447	~2550	s K_2SO_4 fusion, KOH + KNO_3 fusion
hexafluoride	IrF_6	306.21	4.82	44.4	53.6	d aq
(III) oxide	Ir_2O_3	432.43		d ~ 1000 to Ir + O_2		s boiling HCl
(IV) oxide	**IrO_2**	224.22	11.7	d 1100		0.0002^{20} aq; s HCl
trichloride	**$IrCl_3$**	298.58	5.30	d 763		i acids, alkalis

(Continued)

TABLE 1.3 Physical Constants of Inorganic Compounds (*Continued*)

Name	Formula	Formula weight	Density	Melting point, °C	Boiling point, °C	Solubility in 100 parts solvent
Iron	Fe	55.845	7.86	1535	2861	i aq; s acids
(III) arsenate 2-water	$FeAsO_4 \cdot 2H_2O$	230.79	3.18	1020		v sl s aq; s acids
(II) bromide	$FeBr_2$	126.75	3.16	677	1023	117 g/100 mL^{20} aq; v s alc
(III) bromide	$FeBr_3$	295.67	4.5	d		s aq, alc, eth, HOAc
(*tri-*) carbide	Fe_3C	179.55	7.694	1227		s acids
(II) carbonate	$FeCO_3$	115.85	3.9	d		0.072^{18} aq; s acids
(II) chloride	$FeCl_2$	126.75	3.16	677	1024	62.5 g/100 mL^{20} aq; v s alc, acet
(III) chloride	$FeCl_3$	162.20	2.898	304	≈316	74 g/100 mL^{0} aq; s alc, acet, eth
disulfide (pyrite)	FeS_2	119.98	5.02	d 602		s acids d
(II) fluoride	FeF_2	93.84	4.09	1100	1837	sl s aq; s dil HF; i alc, bz, eth
(III) fluoride	FeF_3	112.84	3.87	subl 1000		0.091^{25} aq; s HF
(III) hexacyanoferrate(II)	$Fe_4[Fe(CN)_6]_3$	859.23	1.80	250 d		i aq; s HCl
(II) hydroxide	$Fe(OH)_2$	89.86	3.4			0.006 aq; s acids
(III) hydroxide oxide	$FeO(OH)$	88.85	4.26	anhyd 136		i aq, alc; s HCl
(II) iodide	FeI_2	309.65	5.315	587	1093	s aq
(III) nitrate 9-water	$Fe(NO_3)_3 \cdot 9H_2O$	404.00	1.684	47	d 100	138 g/100 mL^{20} aq
(*di-*) nitride	Fe_2N	125.70	6.35	d 200		s HCl
(II) oxalate 2-water	$FeC_2O_4 \cdot 2H_2O$	179.89	2.28	d 150		0.044^{18} aq; s mineral acids
(II) oxide	FeO	71.84	6.0	1377	d 3414	i aq; s acids
(II,III) oxide	Fe_3O_4	231.53	5.17	1597		i aq; s acids
(III) oxide	Fe_2O_3	159.69	5.25	1565		i aq; s HCl
pentacarbonyl	$Fe(CO)_5$	195.90	1.49	−20.0	103.9	FP −20; i aq; s alc, bz, eth
(II) phosphate 8-water	$Fe_3(PO_4)_2 \cdot 8H_2O$	501.60	2.58			i aq; s acids
phosphide	Fe_2P	142.66	6.85	1370		s hot mineral acids
(II) selenide	$FeSe$	134.81	6.78	d		s HCl
(II) silicate(2−)	$FeSiO_3$	131.93	3.5	1140		
(II) silicate(4−)	Fe_2SiO_4	203.77	4.30	1220		d HCl
(II) sulfate 7-water	$FeSO_4 \cdot 7H_2O$	278.01	1.89	anhyd 300	d 671	48 g/100 mL^{20} aq
(III) sulfate	$Fe_2(SO_4)_3$	399.88	3.097^{18}	d 1178		slowly s aq (hyd); sl s alc
(II) sulfide	FeS	87.92	4.7	1190	d	0.0006^{18} aq; s acid
(III) thiocyanate	$Fe(SCN)_3$	230.09				v s aq
Krypton	Kr	83.80	3.7493 g/L	−157.36	−153.22	5.94 mL/100 mL^{20} aq
difluoride	KrF_2	121.80	3.24	subl−60		s anhyd HF
Lanthanum	La	138.9055	6.162	920	3464	i aq; s HCl
chloride	$LaCl_3$	245.26	3.84	852	1812	v s aq
chloride 7-water	$LaCl_3 \cdot 7H_2O$	371.37		anhyd 852 (in HCl atm)		v s aq; s alc

fluoride	LaF_3	195.90	5.9	1493	2327	
nitrate 6-water	$La(NO_3)_3 \cdot 6H_2O$	433.01		40	d 126	181 g/100 mL^{20} aq; v s alc
oxide	La_2O_3	325.81	6.51	2305	4200	s acids
sulfate	$La_2(SO_4)_3$	566.00	3.60	d white heat		2.33 g/100 mL^{20} aq; i alc
sulfate 9-water	$La_2(SO_4)_3 \cdot 9H_2O$	728.14	2.821	anhyd 400		2.92 g/100 mL^{20} aq; i alc
Lawrencium	Lr	262		1627		
Lead	Pb	207.2	11.34_4^{20} (fcc)	327.43	1749	s hot conc HNO_3, HCl, H_2SO_4
(II) acètate 3-water	$Pb(C_2H_3O_2)_2 \cdot 3H_2O$	427.3	2.55	75	d >200	g/100 mL: 63^{15} aq, 3.3 alc
(IV) acetate	$Pb(C_2H_3O_2)_4$	443.4	2.228	≈75–180		s hot HOAc, bz, chl, conc HX acids
(II) azide	$Pb(N_3)_2$	291.2	4.7	expl 350 or when shocked		0.023^{18} aq; v s HOAc
(II) borate(1−) hydrate	$Pb(BO_2)_2 \cdot H_2O$	310.8	5.598 anhyd	anhyd 160	mp 500	s acids
(II) bromide	$PbBr_2$	367.0	6.69	371	912	0.450^{0} aq; s acids; i alc
(II) carbonate	$PbCO_3$	267.2	6.61	d 340 → PbO		i aq; s acids, alkalis
(II) chlorate	$Pb(ClO_3)_2$	374.1	3.89	d 230		140 g/100 mL^{18} aq; v s alc
(II) chloride	$PbCl_2$	278.1	5.98	501	950	0.99^{20} aq
(II) chloride fluoride	PbClF	261.7	7.05			
(II) chromate(VI)(2−)	$PbCrO_4$	323.2	6.12	844	d	i aq; s dil HNO_3, alkalis
(II) fluoride	PbF_2	245.2	8.445	830	1297	0.064^{20} aq
(IV) fluoride	PbF_4	283.2	6.7	≈ 600		hyd aq
(II) formate	$Pb(CHO_2)_2$	297.2	4.63	d 190		1.6 g/100 mL^{20} aq
(II) hydrogen arsenate	$PbHAsO_4$	347.1	5.94	d 280 to $Pb_2As_2O_7$		s HNO_3, alkalis
(II) hydroxide	$Pb(OH)_2$	241.2	7.59	d 145		0.016^{20} aq; s acids, alkalis
(II) iodide	PbI_2	461.0	6.16	410	872	0.063^{20} aq; s KI, $Na_2S_2O_3$, alkalis
(II) molybdate(VI)(2−)	PbMoO	367.1	6.7	1065		s acids, alkalis
(II) nitrate	$Pb(NO_3)_2$	331.2	4.53	470		g/100 mL: 56^{20} aq, 1.3 MeOH
(II) oleate	$Pb(C_{18}H_{33}O_2)_2$	770.1				s alc, bz, eth
(II) oxalate	PbC_2O_4	295.2	5.28	d 300		s acids, alkalis
(II) oxide (litharge)	PbO	223.2	9.35 (red)	886	1472 d	0.0017^{20} aq; s HNO_3
(IV) oxide	PbO_2	239.2	9.64	d 290, Pb_3O_4	d 595, PbO	s HCl, dil HNO_3 + H_2O_{23}, $H_2C_2O_4$
(II,IV) oxide (red lead)	Pb_3O_4	685.6	8.92	d 595 → PbO		s HNO_3, hot HCl
(II) phosphate	$Pb_3(PO_4)_2$	811.5	7.0	1014		s HNO_3, alkalis
(II) selenide	PbSe	286.2	8.15	1078		s HNO_3
(II) silicate(2−)	$PbSiO_3$	283.3	6.5	764		s acids
(II) silicate(4−)	Pb_2SiO_4	506.5	7.60	743		
(II) stearate	$Pb(C_{18}H_{35}O_2)_2$	774.2	1.4	≈125		0.05^{35} aq; s hot alc
(II) sulfate	$PbSO_4$	303.3	6.29	1170		0.00425 aq; s NaOH
(II) sulfide	PbS	239.3	7.60	1118	1300 subl	0.0006^{18} aq; s HNO_3, hot dil HCl

(*Continued*)

TABLE 1.3 Physical Constants of Inorganic Compounds (*Continued*)

Name	Formula	Formula weight	Density	Melting point, °C	Boiling point, °C	Solubility in 100 parts solvent
(II) telluride	PbTe	334.8	8.16	924		i acids and alkalis
tetraethyl	$Pb(C_2H_5)_4$	323.45	1.653	−137	≈200	i aq; s bz, hydrocarbons
tetramethyl	$Pb(CH_3)_4$	267.35	1.995	−30.2	110	s hydrocarbons
(II) thiocyanate	$Pb(SCN)_2$	323.4	3.82	d 190		0.44^{18} aq, s HNO_3, NaOH
Lithium	Li	6.941	0.534^{20}	180.54	1341	d aq to LiOH
acetate 2-water	$LiC_2H_3O_2 \cdot 2H_2O$	102.02	1.3	58	d	63 g/100 mL^{20} aq; v s alc
aluminate(1−)	$LiAlO_2$	65.92	2.554	1700		
amide	$LiNH_2$	22.96	1.178	380	d 450 vacuo	d aq (→ LiOH + NH_3); i bz, eth
benzoate	$LiC_7H_5O_2$	128.06		>300		g/100 mL: 33 aq; 7.7 alc
borate(1−)	$LiBO_2$	49.75	2.18	849	1719	2.7 g/100 mL^{20} aq; i alc
borohydride	$Li[BH_4]$	21.78	0.66	268	d 380	s aq, eth, THF, aliphatic amines
bromate	$LiBrO_3$	134.85	3.62			179 g/100 mL^{20} aq
bromide	LiBr	86.84	3.464	552	1289	164 g/100 mL aq; s alc, eth
carbonate	Li_2CO_3	73.89	2.11	720	d 1300	1.3 g/100 mL^{20} aq; i alc; s acids
chloride	LiCl	42.39	2.07	613	1360	77 g/100 mL^{20} aq; s alc, acet
chromate(VI)(2−) 2-water	$Li_2CrO_4 \cdot 2H_2O$	165.91	2.15	anhyd 75		142 g/100 mL^{18} aq; s EtOH
citrate 4-water	$Li_3C_6H_5O_7 \cdot 4H_2O$	281.98		anhyd 105		61 g/100 mL^{15} aq; sl s alc
fluoride	LiF	25.94	2.640	848	1681	0.13^{25} aq; s acids
hexafluoroaluminate(3−)	$Li_3[AlF_6]$	161.79		1012		
hydride	LiH	7.95	0.76–0.77	680	d 950	no solvent known; flammable
hydride-*d*	Li^2H or LiD	8.96	0.881	686		
hydroxide	LiOH	23.95	1.45	471.2	1626	12.4 g/100 mL^{20} aq; sl s alc
iodate	$LiIO_3$	181.84	4.502	450		66 g/100 mL aq; in alc
iodide	LiI	133.84	4.061	469	1174	165 g/100 mL^{20} aq & alc; v s acet
nitrate	$LiNO_3$	68.95	2.38	~255		50 g/100 mL^{20} aq; s alc
nitride	Li_3N	34.83	1.27	813		d aq
oxide	Li_2O	29.88	2.013	1570	2563	forms LiOH in aq
perchlorate	$LiClO_4$	106.39	2.43	236	d ~ 400 LiCl + O_2	47.4 g/100 mL^{25} aq; v s organic solv
peroxide	Li_2O_2	45.88	2.31	d > 195 to Li_2O		
silicate(2−)	Li_2SiO_3	89.97	2.52^{25}_{4}	1201		d dil HCl
sulfate	Li_2SO_4	109.95	2.22	859		34.5 g/100 mL^{20} aq; i alc
tetraborate(2−)	$Li_2B_4O_7$	169.12		917		sl s aq
tetrahydridoaluminate	$Li[AlH_4]$	37.95	0.917	d 137		d aq, alc; g/100 mL: 30 eth, 13 THF; flammable
tetrahydridoborate	$LiBH_4$	21.79	0.666	268	d 380	s aq pH >7; s eth, THF
Lutetium	Lu	174.967	9.841	1663	3402	s acids
chloride	$LuCl_3$	281.33	3.98	892	subl >750	s aq
sulfate 8-water	$Lu_2(SO_4)_3 \cdot 8H_2O$	782.25				42.3 g/100 mL^{20} aq

Magnesium	Mg	24.305	1.738[20]	651	1100	i aq; s dilute acids
acetate	$Mg(C_2H_3O_2)_2$	142.00	1.42	323 d		53.4 g/100 mL[20] aq; v s alc
aluminate(2−)	$MgAl_2O_4$	142.25	3.6	2135		v sl s HCl
amide	$Mg(NH_2)_2$	56.35	1.39^{25}_{4}	ign in air		d viol water giving NH_3
borate(1−) 8-water	$Mg(BO_2)_2 \cdot 8H_2O$	254.04	2.30			sl s aq; s acids
bromide	$MgBr_2$	184.11	3.722	711 d	1158	101 g/100 mL[20] aq
carbonate	$MgCO_3$	84.31	3.05	990		0.01 aq; s acids
chloride	$MgCl_2$	95.21	2.33	714	1412	54.6 g/100 mL[20] aq
fluoride	MgF_2	62.30	3.148	1263	2270	0.013[25] aq; s HNO_3
(*di-*) germanide	Mg_2Ge	121.22	3.09	1115		
hexafluorosilicate 6-water	$Mg[SiF_6] \cdot 6H_2O$	274.47	1.788	$-SiF_4$, 120		51 g/100 mL[20] aq; i alc
hydride	MgH_2	26.32	1.45	d 200 vacuo	ign in air	d aq and alc violently
hydrogen phosphate 3-water	$MgHPO_4 \cdot 3H_2O$	174.33	2.13[15]	anhyd 205	d 550	sl s aq; s acids
hydroxide	$Mg(OH)_2$	58.32	2.36	350 d		0.00125 aq; s acids
iodide	MgI_2	278.12	4.43	634	0	140 g/100 mL[20] aq; s alc
lactate 3-water	$MgC_6H_{10}O_6 \cdot 3H_2O$	256.51				4 g/100 mL aq; sl s alc
mandelate	$MgC_{16}H_{14}O_6$	326.59				0.004[100] aq; i alc
nitrate	$Mg(NO_3)_2 \cdot 6H_2O$	256.41	1.464	95	d 129	120 g/100 mL[20] aq; v s alc
nitride	Mg_3N_2	100.93	2.712	d 270		d aq; s acids
oleate	$Mg(C_{18}H_{33}O_2)_2$	587.22				sl s alc, eth, PE
oxide	MgO	40.30	3.65–3.75	2800	3600	i aq, alc; s acids
perchlorate	$Mg(ClO_4)_2$	223.21	2.21	d >251		g/100 mL[25]: 73 aq, 18 EtOH, 44.6 BuOH, 54 EtOAc, 32 acet
permanganate	$Mg(MnO_4)_2$	262.19				v s aq
peroxide	MgO_2	56.30	≈3.0	d 100		s acids
peroxoborate 7-water	$Mg(BO_3)_2 \cdot 7H_2O$	268.09				sl s aq d; s dilute acids
phosphate 5-water	$Mg_3(PO_4)_2 \cdot 5H_2O$	352.96	1.64[15]	anhyd ~400		0.02 aq; s acids
silicate(2−)	$MgSiO_3$	100.39	3.192^{25}_{4}	d 1557		i aq; v sl s HF
silicate(4−)	Mg_2SiO_4	140.69	3.21	1898		i aq; d hot HCl
(*di-*) silicide	Mg_2Si	76.70	2.0	1100		d aq, HCl
(*di-*) stannide	Mg_2Sn	167.32	3.60	778		s aq, HCl
sulfate 7-water	$MgSO_4 \cdot 7H_2O$	246.47	1.67	anhyd 250		27.2 g/100 mL aq; sl s alc
sulfite 6-water	$MgSO_3 \cdot 6H_2O$	212.46	1.725	anhyd 200	mp: 2227	0.66[25] aq
tungstate(VI)(2−)	$MgWO_4$	272.14	6.89			i aq; d acids
Manganese	Mn	54.9380	7.21[20]	1244 fctetr	2095	d aq; s acids
acetate 4-water	$Mn(C_2H_3O_2)_2 \cdot 4H_2O$	245.09	1.589	80		38 g/100 mL[50] aq; v s alc

(Continued)

TABLE 1.3 Physical Constants of Inorganic Compounds (*Continued*)

Name	Formula	Formula weight	Density	Melting point, °C	Boiling point, °C	Solubility in 100 parts solvent
bromide	$MnBr_2$	214.75	4.39	698	1027	147 g/100 mL^{20} aq; s alc
(*tri*-) carbide	Mn_3C	176.83	6.89	1520		d aq; s acid
carbonate	$MnCO_3$	114.95	3.125	d >200		0.0065^{25} aq; s acids
chloride	$MnCl_2$	125.84	2.977	650	1210	74 g/100 mL^{20} aq; s alc, pyr; i eth
chloride 4-water	$MnCl_2 \cdot 4H_2O$	187.91	2.01	97.5	anhyd 198	143 g/100 mL aq; s alc; i eth
decacarbonyl	$Mn_2(CO)_{10}$	389.98	1.75	d 110		i aq; s organic solvents
diphosphate	$Mn_2P_2O_7$	283.82	3.707	1196		i aq; s acid
(II) fluoride	MnF_2	92.93	3.98	930	1820	0.66^{40} aq; s HF, conc HCl
(III) fluoride	MnF_3	111.93	3.54	d >600		hyd aq; s acid
hydroxide	$Mn(OH)_2$	88.95	3.258	d		0.002^{18} aq; s acids
iodide	MnI_2	308.75	5.04	638	1017	s aq
nitrate 6-water	$Mn(NO_3)_2 \cdot 6H_2O$	287.04	1.8	25.8		v s aq, alc
(II) oxide	MnO	70.94	5.37	1840		i aq; s acids
(III) oxide	Mn_2O_3	157.87	4.89	877 d		i aq; s HCl giving off Cl_2
(IV) oxide	MnO_2	86.94	5.08	$-O_2$, 530		s HCl; i HNO_3, cold H_2SO_4
(II,IV) oxide	Mn_3O_4	228.81	4.84	1567		i aq; s HCl
(VII) oxide	Mn_2O_7	221.87	2.396	ca. −20	ca. 25	explodes 85; v s aq
phosphinate hydrate	$Mn(PH_2O_2)_2 \cdot H_2O$	202.93		d to PH_3		15 g/100 mL aq; i alc
silicate, meta-	$MnSiO_3$	131.02	3.48	1290		i aq, HCl
sulfate	$MnSO_4$	151.00	3.25	700	d 850	52 g/100 mL aq; i alc
sulfate hydrate	$MnSO_4 \cdot H_2O$	169.02	2.95	anhyd 400–450		70 g/100 mL^{20} aq
sulfate 7-water	$MnSO_4 \cdot 7H_2O$	277.11	2.09	anhyd 280		115 g/100 mL^{20} aq
sulfide	MnS	87.00	3.99	1610		0.0006^{18} aq; s acids
titanate(IV)(2−)	Mn_2TiO_4	150.84	4.54	1360		
Mercury	Hg	200.59	13.534	−38.83	356.7	i aq; s HNO_3, hot conc H_2SO_4
(II) acetate	$Hg(C_2H_3O_2)_2$	318.68	3.28	178–180 d		g/100 mL: 40^{10} aq, 7.5^{15} MeOH
(II) benzoate	$Hg(C_7H_5O_2)_2$	424.83		165		v s NaCl soln; sl s alc
(I) bromide	Hg_2Br_2	560.99	7.307	subl 393 d		i aq, alc, eth; d hot HCl
(II) bromide	$HgBr_2$	360.40	6.05	237	322 subl	g/100 mL: 0.56^{20} aq; 20^{25} alc; v s HCl, HBr
(I) chloride	Hg_2Cl_2	472.09	7.16	subl 382	d without melting	s aqua regia; i aq, alc, eth
(II) chloride	$HgCl_2$	271.50	5.4	277	304	g/100 mL^{20}: 7.15 aq, 26 alc, 4 eth 8.3 glyc, 0.5 bz; s HOAc, EtAc
(II) cyanide	$Hg(CN)_2$	252.63	4.00	d 320		g/100 mL^{20}: 9.3 aq, 25 MeOH, 8 EtOH
(I) fluoride	Hg_2F_2	439.18	8.73	>570 d		hydrolyses in water

(II) fluoride	HgF_2	238.59	8.95	d 645	d >650	hyd aq; s HF
(II) fulminate	$Hg(ONC)_2$	284.62	4.42	explodes		sl s aq; s alc; dangerously flammable
(I) iodide	Hg_2I_2	654.99	7.70	290 d	subl 140	i aq, alc, eth; s KI
(II) iodide	HgI_2	454.40	6.28	259	350 subl	g/100 mL: 0.006^{25} aq, 0.8 alc, 0.8 eth, 1.7 acet
(I) nitrate 2-water	$Hg_2(NO_3)_2 \cdot 2H_2O$	561.22	4.79	70 d		hyd aq; s HNO_3
(II) nitrate	$Hg(NO_3)_2$	324.60	4.3	79	d	v s aq; s acet
(I) oxide	Hg_2O	417.18	9.8	d 100		i aq; s HNO_3
(II) oxide	HgO	216.59	11.14	d 500		0.005^{25} aq; s dil HCl, HNO , I^-, CN^-
(I) sulfate	Hg_2SO_4	497.24	7.56	d		0.06^{25} aq; s HNO_3
(II) sulfate	$HgSO_4$	296.65	6.47	d		d aq; s acid
(II) sulfide (cinnabar)	HgS	232.66	8.17	subl 583	→ blk HgO, 386	i aq; s aqua regia
(II) thiocyanate	$Hg(SCN)_2$	316.76	3.71	d 165		0.063^{25} aq; s HCl
Molybdenum	Mo	95.94	10.28	2622	4825	s hot H_2SO_4, HNO_3, fused KNO_3
(III) bromide	$MoBr_3$	335.65	4.89	subl 977		d alkalis
(IV) chloride	$MoCl_4$	237.75		317	407	s conc acids
(V) chloride	$MoCl_5$	273.19	2.928	194	268	s conc acids, dry eth, dry alc
(VI) fluoride	MoF_6	209.93	2.54	17.6	35.0	hyd aq; s alkalis; 31 g/100 g HF
hexacarbonyl	$Mo(CO)_6$	264.00	1.96	150 d	subl	s bz
(IV) oxide	MoO_2	127.94	6.47	d ≈1100		i aq
(VI) oxide	MoO_3	143.94	4.696^{26}_{4}	801	1155	0.05^{28} aq; s conc mineral acids, alk
(III) sulfide	Mo_2S_3	288.07	5.91^{15}	1807	d 1867	d hot HNO_3
(IV) sulfide	MoS_2	160.07	5.06^{15}_{15}	2375	subl 450	s aqua regia
Neodymium	Nd	144.24	7.01	1024	3074	s hot aq, acids
chloride	$NdCl_3$	250.60	4.134	760	1600	98 g/100 mL^{20} aq; s alc
oxide	Nd_2O_3	336.48	7.28	1900		s dilute acids
sulfate 8-water	$Nd_2(SO_4)_3 \cdot 8H_2O$	720.79	2.85	d 700–800		8.87 g/100 mL^{20} aq
Neon	Ne	20.180	0.8999 g/L^0	−248.67	−246.05	1.05 mL^{20} aq
Neptunium	Np	237.0482	20.2	644	>3900	s HCl
(IV) oxide	NpO_2	269	11.1	2547		
Nickel	Ni	58.69	8.908^{20}	1453	2884	i aq; s HNO_3
acetate 4-water	$Ni(C_2H_3O_2)_2 \cdot 4H_2O$	248.86	1.744	d		16 g/100 mL aq; s alc

(Continued)

TABLE 1.3 Physical Constants of Inorganic Compounds (*Continued*)

Name	Formula	Formula weight	Density	Melting point, °C	Boiling point, °C	Solubility in 100 parts solvent
acetylacetonate	$Ni(C_5H_7O_2)_2$	256.91	1.455^{17}	230	235^{11atm}	s aq, alc, bz, chl; i eth
bromide	$NiBr_2$	218.50	5.098	963	subl	100 g/100 mL^{20} aq
carbonate hydroxide (1/2)	$NiCO_3 \cdot 2Ni(OH)_2$	304.12	2.6			s dilute acids
carbonyl	$Ni(CO)_4$	170.73	1.31	−19.3	43 (expl 60)	s EtOH, bz, acet
chloride	NiCl	129.60	3.51	1009	subl 973	61 g/100 mL^{20} aq
chloride 6-water	$NiCl_2 \cdot 6H_2O$	237.69				100 g/100 mL^{20} aq; s alc
cyanide 4-water	$Ni(CN)_2 \cdot 4H_2O$	182.79		anhyd 400		0.006^{18} aq; s KCN, NH_4OH
dimethylglyoxime	$Ni(HC_2H_6N_2O_2)_2$	288.92		subl 250		i aq; s abs alc, dilute acids
(*tri-*) disulfide	Ni_3S_2	240.21	5.87	790	d 2967	s HNO_3
fluoride	NiF_2	96.69	4.72	1450	1740	4 g/100 mL^{20} aq; i alc, eth
formate 2-water	$Ni(CHO_2)_2 \cdot 2H_2O$	184.78	2.154^{20}	anhyd 130	d 180–200	s aq; i alc
nitrate 6-water	$Ni(NO_3)_2 \cdot 6H_2O$	290.81	2.05	56.7	136.7	150 g/100 mL^{20} aq
(II) oxide	NiO	74.71	7.45	2000		s acids
(III) oxide	Ni_2O_3	165.42	4.83	$-O_2$, 600		s hot HCl, HNO_3, H_2SO_4
sulfate	$NiSO_4$	154.78	3.68	$-SO_3$, 840		29 g/100 mL^0 aq
sulfate 6-water	$NiSO_4 \cdot 6H_2O$	262.86	2.07	anhyd 280		40 g/100 mL^{20} aq
sulfide	NiS	90.77	5.3–5.6	976	d 2047	s HNO_3, KHS
tetracarbonyl	$Ni(CO)_4$	170.74	1.3185^{17}	−19.3	42.3	explodes 63; FP −4; s organic solvents
Niobium	Nb	92.9064	8.57^{20}	2468	4860	s fused alkali hydroxides
(V) chloride	$NbCl_5$	270.20	2.75	206	247.0	s HCl, CCl_4
(V) fluoride	NbF_5	187.91	2.696^{80}_{4}	80.0	234.9	hyd aq, alc; sl s CS_2, CCl_4
(V) oxide	Nb_2O_5	265.82	4.55	1512		s HF, hot H_2SO_4
Nitrogen	N_2	28.0341	1.165 g/L^{20}	−210.01	−195.79	mL/100 mL: 1.6^{20} aq, 0.112 alc
	$^{15}N_2$	30.01	1.25 g/L^{20}	−209.952	−195.73	
(I) oxide	N_2O	44.02	1.843 g/L^{20}	−90.81	−88.46	130^0 mL aq; s alc, eth
(II) oxide	NO	30.01	1.249 g/L^{20}	−163.64	−151.76	4.6 mL/100 mL^{20} aq
(III) oxide	N_2O_3	76.02	1.447 g/L^2	−100.7	2	s eth
(IV) oxide dimer	N_2O_4	92.02	1.448^{20}_{4}	−9.3	21.15 d	s conc HNO_3, conc H_2SO_4, chl
(V) oxide	N_2O_5	108.01	2.05	30	47.0	v s chl; s CCl_4
selenide	N_4Se_4	371.87	4.2	explosive		sl s bz, CS_2
sulfide	N_4S_4	184.28	2.24^{18}	180	185	s organic solvents
trichloride	NCl_3	120.37	1.653^{20}	−27	71	i aq; s bz, CS_2, CCl_4
trifluoride	NF_3	70.01	2.96 g/L^{20}	−208.5	−129.06	
Nitrosyl chloride	NOCl	65.47	1.592^{-5}	−61.5	−5.5	hyd aq; s fuming H_2SO_4
fluoride	NOF	49.01	2.788 g/L^{20}	−132.5	−59.9	hyd aq
hydrogen sulfate	$NOHSO_4$	127.08		d 73.5		d aq; s H_2SO_4
tetrafluoroborate	$NO[BF_4]$	116.83	2.185^{25}_{4}	subl $250^{0.01mm}$		d aq

Nitryl chloride	NO_2Cl	81.46	2.81 g/L^{100}	−145	−14.3	d aq
fluoride	NO_2F	65.00	2.7 g/L^{20}	−166.0	−72.4	d aq
Osmium	Os	190.2	22.61^{20}	3045	5225	s molten alkali or oxidizing fluxes
hexafluoride	OsF_6	304.2		32.1	45.9	hyd aq
tetrachloride	$OsCl_4$	332.0	4.38$^{20}_{4}$	subl 450		slow hyd aq
tetraoxide	OsO_4	254.20	4.91	40.6	130.0	g/100 mL: 7.24^{25} aq; 375^{25} CCl_4; s bz, eth, alc
Oxygen	O_2	31.9988	1.331 g/L^{20}	−218.4	−182.96	mL/100 mL^{20}: 3.13 aq, 14.3 alc
difluoride	OF_2	54.00	2.26 g/L^{20}	−223.8	−145.3	6.8 mL/100 mL^{0} aq
(*di-*) difluoride	O_2F_2	70.00	1.45^{bp} (lq)	−154	d − 100	
Ozone	O_3	48.00	1.998 g/L^{20}	−192.5	−111.9	49.4 mL/100 mL^{0} aq
Palladium	Pd	106.42	12.023^{20}	1555	3167	s hot HNO_3, H_2SO_4
acetate	$Pd(C_2H_3O_2)_2$	224.49		205 d		i aq, alc; s acet, chl, eth
chloride	$PdCl_2$	177.30	4.0^{18}	680	d >680	s alc, acet, HCl
nitrate	$Pd(NO_3)_2$	230.42		d		s dil HNO_3
oxide	PdO	122.40	8.70^{20}	879 d		s 48% HBr; sl s aqua regia
Perchloryl fluoride	ClO_3F	102.46	0.637 g/L	−147.74	−46.67	
Phosphorus (white)	P_4 molecules	123.8950	1.823^{25}	44.15	280.3	g/100 mL: 2.86 bz, 2.50 chl, 1.25 CS_2; 0.025 abs alc, 1.0 eth
(red)	P_4	123.8950	2.34	597	subl 416	i aq; ignites in air, 260
hydride, *see* Phosphine						
pentabromide	PBr_5	430.56	3.46^{20}	106 d		d aq; s CCl_4, CS_2
pentachloride	PCl_5	208.27	2.119^{20}	subl 100	166 d	hyd aq; s CCl_4, CS_2
pentafluoride	PF_5	125.98	5.805 g/L	−93.8	−84.6	hyd aq
pentoxide (dimer)	P_4O_{10}	283.88	2.30	340	subl 360	d aq; s H_2SO_4
pentasulfide	P_2S_5	222.29	2.09	288	514	hyd aq; s alkali; 0.222^{17} CS_2
tribromide	PBr_3	270.73	2.85^{15}	−41.5	173.2	d aq, alc; s acet, CS_2
trichloride	PCl_3	137.35	1.575$^{20}_{4}$	−93.6	76.1	d aq, alc; s bz, chl
trifluoride	PF_3	87.98	3.907 g/L	−151.30	−101.38	hyd aq
trioxide (dimer)	P_4O_6	219.90	2.136$^{20}_{4}$	23.8	173 (N_2 atm)	hyd aq; s bz, CS_2
(*tetra-*) triselenide	P_4Se_3	360.80	1.31	245–246	360–400	flammable in air; s bz, acet, chl, CS_2
(*tetra-*) trisulfide	P_4S_3	220.09	2.03^{17}	167	407	100 g/100 mL^{17} CS_2; s tolune

(Continued)

TABLE 1.3 Physical Constants of Inorganic Compounds (*Continued*)

Name	Formula	Formula weight	Density	Melting point, °C	Boiling point, °C	Solubility in 100 parts solvent
Phosphine	PH_3	34.00	1.529 g/L	−133.81	−87.78	mL/100 mL^{17}: 1025 CS_2, 726 bz, 319 HOAc, 26 aq; s alc, eth
Phosphonium iodide	PH_4I	161.91	2.86	18.5	subl 62.5	d aq
Phosphoryl chloride difluoride	$POClF_2$	120.43	1.656^{0}	−96.4	3.1	
dichloride fluoride	$POCl_2F$	136.89	1.5497^{20}	−80.1	52.90	
tribromide	$POBr_3$	286.72	2.822	56	191.7 d	s bz, CS_2, eth
trichloride	$POCl_3$	153.35	1.645^{25}	1.25	105	d aq, alc
Platinum	Pt	195.08	21.09^{20}	1769	3824	s aqua regia, fused alkali
(II) chloride	$PtCl_2$	266.00	5.87	d 581		i aq, alc; s HCl, NH_4OH
(IV) chloride	$PtCl_4$	336.90	4.303^{25}	d 370		143 g/100 mL^{25} aq
(VI) fluoride	PtF_6	309.08	3.826 (lq)	61.3	69.14	
(II) oxide	PtO	211.09	14.9^{15}	d 550		i aq; s HCl
(IV) oxide	PtO_2	227.09	10.2	450		i aqua regia
(IV) sulfide	PtS_2	259.22	7.66	d 225		s HCl, HNO_3
Plutonium	Pu	239.052	19.816^{20}_{4}	639.5	3230	i aq; s acids
(III) bromide	$PuBr_3$	478.79	6.69	681	d > 1300	s aq
(III) chloride	$PuCl_3$	345.42	5.70	760	1767	i aq; v s acids
(III) fluoride	PuF_3	296.06	9.32	1425	d 2000	hyd aq
(IV) fluoride	PuF_4	315.05	7.00	1037 d		i aq
(VI) fluoride	PuF_6	353.05	4.86	51.59	62.16	
(II) hydride	PuH_2	241.08	10.40	ca. 727		
(III) hydride	PuH_3	242.08	9.61	ca. 327		
(II) oxide	PuO	255.05	13.9	1900		
(III) oxide	Pu_2O_3	526.12	10.2	2085 (in He)		
(IV) oxide	PuO_2	271.05	11.46	2390 (in He)	d 2800	
(III) sulfide	Pu_2S_3	574.30	9.95	1727		
Polonium	Po	208.9824	9.196 alpha; 9.398 beta	254	962	sl s aq; s acids
(IV) chloride	$PoCl_4$	350.79		300 (in Cl_2)	390 (in Cl_2)	sl hyd aq; v s HCl; s alc, acet
(IV) oxide	PoO	240.98	d 550			v s dilute HCl

Potassium	K	39.0983	0.89	63.38	759	d aq to KOH; s acids
acetate	$KC_2H_3O_2$	98.14	1.57	292		g/100 mL: 200 aq, 34 alc
arsenate	K_3AsO_4	256.21	2.8	1310		19 g/100 mL aq; slowly s glyc; s alc
borate(1−)	KBO_2	81.91		947	1401	71 g/100 mL^{30} aq
bromate	$KBrO_3$	167.00	3.27	≈350	d 370	6.9 g/100 mL^{20} aq
bromide	KBr	119.00	2.75	734	1435	g/100 mL: 65^{20} aq, 22 glyc, 0.4 alc
carbonate	K_2CO_3	138.21	2.29	901	d to K_2O	90 g/100 mL^{20} aq; i alc
chlorate	$KClO_3$	122.55	2.32	368	d >400	g/100 mL: 7.3^{20} aq, 2 glyc
chloride	KCl	74.55	1.988	771	1437	g/100 mL: 34^{20} aq, 7 glyc, 0.4 alc
chromate(VI)	K_2CrO_4	194.19	2.732	975		64 g/100 mL^{20} aq; i alc
citrate hydrate	$K_3C_6H_5O_7 \cdot H_2O$	324.42	1.98	anhyd 180	d 230	g/100 mL: 154 aq; 40 glyc
cyanate	KOCN	81.11	2.05	d ≈700		s aq; sl s alc
cyanide	KCN	65.12	1.55	634	1625	g/100 mL: 50 aq, 50 glyc, 4 MeOH
dichromate(VI)	$K_2Cr_2O_7$	294.19	2.676_4^{25}	398	d 500	11.7 g/100 mL^{20} aq
dicyanoargentate(I)	$K[Ag(CN)_2]$	199.01	2.36			25 g/100 mL^{30} aq
dihydrogen arsenate	KH_2AsO_4	180.03	2.867	288		g/100 mL: 19^6 aq, 63 glyc; i alc
dihydrogen phosphate	KH_2PO_4	136.09	2.338	d 400 (KPO_3)		22.6 g/100 mL^{20} aq; i alc
dioxide	KO_2	71.10	2.14	509	d	v s aq with decomposition
diphosphate(V) 3-water	$K_4P_2O_7 \cdot 3H_2O$	384.38	2.33	anhyd 300	mp: 1090	s aq; i alc
disulfate(IV)	$K_2S_2O_5$	222.32				s aq; flammable if ground
disulfate(VI) (pyrosulfate)	$K_2S_2O_7$	254.32	2.28	≈325		s aq
ethyldithiocarbonate	$KOCSSC_2H_5$	160.30	1.558	d 200		v s aq
fluoride	KF	58.10	2.48	859.9	1505	95 g/100 mL^{20} aq
formate	KCHO	84.12	1.91	167.5	d >mp	250 g/100 mL aq
gluconate	$KC_6H_{11}O_7$	234.25		d 180		v s aq; i alc, bz, chl
heptaiodobis-muthate(III)(4−)	$K_4[BiI_7]$	1253.82				d aq; s alkali iodide solutions
hexachloroplatinate(IV)	$K_2[PtCl_6]$	485.99	3.50	d 250		0.48^{20} aq
hexacyanoferrate(II) 3-water	$K_4[Fe(CN)_6] \cdot 3H_2O$	422.39	1.85	anhyd 100	d	28 g/100 mL^{20} aq
hexacyanoferrate(III)	$K_3[Fe(CN)_6]$	329.25	1.89	d		40 g/100 mL^{20} aq (slow); sl s alc
hexafluorosilicate	$K_2[SiF_6]$	220.27	2.27	d		sl s aq; i alc
hexafluorozirconate	$K_2[ZrF_6]$	283.41	3.58			2.7 g/100 mL^{20} aq
hexanitritocobaltate(III) 1.5-water	$K_3[Co(NO_2)_6] \cdot 1.5H_2O$	479.30		d 200		0.089^{18} aq; s HOAc; v sl s alc
hydride	KH	40.11	1.43	417 d		d aq

(Continued)

TABLE 1.3 Physical Constants of Inorganic Compounds (*Continued*)

Name	Formula	Formula weight	Density	Melting point, °C	Boiling point, °C	Solubility in 100 parts solvent
hydrogen carbonate	$KHCO_3$	100.11	2.17	d >100		34 g/100 mL^{20} aq; i alc
hydrogen difluoride	KHF	78.10	2.37	238.80	d 477	39 g/100 mL^{20} aq; s alc
hydrogen phosphate	K_2HPO_4	174.18		d to $K_2P_2O_7$		150 g/100 mL aq
hydrogen phthalate	$KHC_8H_4O_4$	204.22	1.636	d		8.3 g/100 mL aq; sl s alc
hydrogen sulfate	$KHSO_4$	136.17	2.24	197	d to $K_2S_2O_7$	48 g/100 mL^{20} aq
hydrogen sulfide	KHS	72.17	1.70	≈455		s aq, alc
hydrogen tartrate	$KHC_4H_4O_6$	188.18	1.956			0.5^{20} aq; s acids; v sl s alc
hydroxide	KOH	56.11	2.044	406	1323	g/100 mL: 112^{20} aq, 33 alc, 40 glyc
iodate	KIO_3	214.00	3.89	560 d		8.1 g/100 mL^{20} aq; i alc
iodide	KI	166.00	3.12	681	1345	g/100 mL: 144^{20} aq, 4.5 alc, 50 glyc
manganate(VI)	K_2MnO_4	197.13		190 d		s aq; stable in KOH
molybdate(VI)	K_2MoO_4	238.14	2.3	919	d 1400	160 g/100 mL aq
nitrate	KNO_3	101.10	2.11	333	d 400	g/100 mL: 32^{20} aq, 0.16 alc, s glyc
nitrite	KNO_2	85.10	1.915	441	d 350	306 g/100 mL^{20} aq; sl s alc
oxalate hydrate	$K_2C_2O_4 \cdot H_2O$	184.23	2.13	anhyd 160	d to K_2CO_3	36 g/100 mL^{20} aq
oxide	K_2O	94.20	2.35	350 d		d aq to KOH, s alc
oxobisoxalatodiaquatitanate(IV)	$K_2[TiO(C_2O_4)_2(H_2O)_2]$	354.18				v s aq
perchlorate	$KClO_4$	138.55	2.52	d 400		2.04^{25} aq; 0.0036^{25} BuOH; 0.0013 EtOAc
periodate	KIO_4	230.010	3.618	582		0.42^{20} aq, sl s KOH
permanganate	$KMnO_4$	158.03	2.7	d 240 → O_2		6.34 g/100 mL^{20} aq; d HCl
peroxide	K_2O_2	110.20		490		d aq
peroxodicarbonate hydrate	$K_2C_2O_6 \cdot H_2O$	216.24				6.5 g/100 mL aq; d hot aq
peroxodisulfate	$K_2S_2O_8$	270.32	2.48	d 100		2.5 g/100 mL^{20} aq; i alc
perrhenate	$KReO_4$	289.30	4.38	555	1370	0.99^{20} aq
phenolsulfonate hydrate	$KC_6H_4(OH)SO_3 \cdot H_2O$	240.28	1.87			s aq, alc
phosphate	K_3PO_4	212.27	2.564^{17}_{4}	1340		50.8 g/100 mL^{20} aq; i alc
selenocyanate	KSeCN	144.08		d 100		s aq
silicate(2−)	K_2SiO_3	154.29		976		s aq
sodium hexanitritocobaltate(III) hydrate	$K_2Na[Co(NO_2)_6] \cdot H_2O$	454.18	1.633	d 135		0.07 aq
sodium tartrate 4-water	$KNaC_4H_4O_6 \cdot 4H_2O$	282.23	1.790	70–80	anhyd 130–140	54 g/100 mL^{15} aq
sorbate	$KC_6H_7O_2$	150.22	1.363^{25}_{20}	d >270		g/100 mL: 58.2^{20} aq, 6.5 alc
stannate(IV) 3-water	$K_2SnO_3 \cdot 3H_2O$	298.94	3.197	anhyd 140		100 g/100 mL^{20} aq; i alc

stearate	$KOOCC_{17}H_{35}$	322.57				readily soluble hot aq or alc
sulfate	K_2SO_4	174.26	2.66	1069	1670	g/100 mL: 11^{20} aq, 1.3 glyc, i alc
sulfide	K_2S	110.26	1.74	948		
sulfite 2-water	$K_2SO_3 \cdot 2H_2O$	194.29		d		28.6 g/100 mL^{20} aq
tartrate hemihydrate	$K_2C_4H_4O_6 \cdot 0.5H_2O$	235.28	1.98	anhyd 155	d 200	138 g/100 mL^{20} aq
tellurate(IV)	K_2TeO_3	253.79				s aq
tetrachloroaurate(III)	$K[AuCl_4]$	377.88	3.75	d 357		61.8 g/100 mL^{20} aq
tetrafluoroborate	$K[BF_4]$	125.90	2.505_4^{20}	530		0.45^{20} aq
tetrahydridoborate	$K[BH_4]$	53.94	1.11	d 497		g/100 mL: 21^{25} aq, 3.5^{20} MeOH
tetraiodocadmate 2-water	$K_4[CdI_4] \cdot 2H_2O$	698.21	3.359_4^{21}			g/100 mL: 137^{15} aq, 71^{15} alc, 4 eth
tetraiodomercurate(II)	$K_2[HgI_4]$	786.48				v s aq; s alc, acet, eth
thiocyanate	KSCN	97.18	1.89	173	d 500	g/100 mL: 217^{20} aq, 200 acet, 8 alc
thiosulfate	$K_2S_2O_3$	190.33		d 400		155 g/100 mL^{20} aq; i alc
trihydrogen bisoxalate 2-water	$KH_3(C_2O_4) \cdot 2H_2O$	254.20	1.836	d		1.8 aq
trisoxalatoantimonate(III)	$K_3[Sb(C_2O_4)_3]$	503.12				a aq
trithiocarbonate	K_2CS_3	186.41		d		v s aq
uranyl(VI) acetate hydrate	$K(UO_2)(C_2H_3O_2)_2 \cdot H_2O$	504.28	3.296^{15}	anhyd 275		s aq
Praseodymium	Pr	140.9077	6.475 α-form	935	3520	s hot water and acids
chloride	$PrCl_3$	247.27	4.0	769 to 782	1710	104 g/100 mL^{13} aq; s alc
(III) oxide	Pr_2O_3	329.81	7.07	oxidizes to Pr_6O_{11}		i aq; s acids
(IV)	PrO_2	172.91	6.82	tr 350 to Pr_6O_{11}		
Promethium-147	Pm	146.915	7.22	1080	3000 est	
bromide	$PmBr_3$	386.7	5.38	727	1667	s aq
chloride	$PmCl_3$	153.4		737	1670	s aq
Protoactinium	Pa	231.0359	15.37	1568(8)	4227	
(IV) chloride	$PaCl_4$	372.85	4.72	subl 400		i aq; s HCl
(V) chloride	$PaCl_5$	408.31	3.74	301	420	hyd aq; s THF, CH_3CN
Radium	Ra	226.03	5.5	700.1	1737	d aq; s acids
bromide	$RaBr_2$	385.88	5.79	728	subl 900	s aq
chloride	$RaCl_2$	296.93	4.91	1000		s aq
Radon	Rn	222.0	9.73 g/L	−71	−62	23 mL/100 mL^{20} aq; s org solv
Rhenium	Re	186.207	21.02	3180	5678	s HNO_3

(*Continued*)

TABLE 1.3 Physical Constants of Inorganic Compounds (*Continued*)

Name	Formula	Formula weight	Density	Melting point, °C	Boiling point, °C	Solubility in 100 parts solvent
chloride trioxide	$ReClO_3$	269.66		4.5	128	hyd in water to $HReO_4$; s CCl_4
(IV) fluoride	ReF_4	262.20	5.38	124.5	795	hyd aq
(VI) fluoride	ReF_6	300.20	3.58	18.5	33.8	52.5 g/100 mL anhyd HF; s HNO_3
(VII) fluoride	ReF_7	319.20	3.65	48.3	73.7	hyd aq
(VI) oxide	ReO_3	234.20	6.9–7.4	disprop 400	750	s HNO_3
(VII) oxide	Re_2O_7	484.41	6.1	300.3	360.3	v s aq, org solv
(VII) sulfide	Re_2S_7	596.88	4.866	d 460		i aq; s HNO_3
(VI) tetrachloride oxide	$ReCl_4O$	344.02	3.309	29.3	225	hyd aq; s CCl_4
Rhodium	Rh	102.9055	12.41^{20}	1963	3727	s fused $KHSO_4$
(III) chloride	$RhCl_3$	209.26	5.38	d 450		i aq; s KOH, KCN
(III) fluoride	RhF_3	159.90	5.4	subl 600		i acids, alkalis
(III) oxide	Rh_2O_3	253.81	8.20	d 1100		i aq reg, KOH
tetracarbonyldi-μ-chloro-dichloride	$Rh_2(CO)_4Cl_2$	388.76		124–125		s org solv except hydrocarbons
Rubidium	Rb	85.4678	1.532	39.31	691	d aq to RbOH
acetate	$RbC_2H_3O_2$	144.52		246		86 g/100 mL^{45} aq
bromide	RbBr	165.37	3.35	682	1346	108 g/100 mL^{20} aq
carbonate	Rb_2CO_3	230.95		837	d 900	g/100 mL: 450^{20} aq, 0.74_{19} alc
chlorate	$RbClO_3$	168.94	3.184	342		5.4 g/100 mL^{20} aq
chloride	RbCl	120.92	2.76	715	1390	g/100 mL: 91^{20} aq, 1.1 MeOH
dihydrogen phosphate	RbH_2PO_4	182.47		840		s aq
fluoride	RbF	104.47	3.2	833	1410	131 g/100 mL^{18} aq
hexachloroplatinate(IV)	$Rb_2[PtCl_6]$	578.75	3.94	d		0.028^{20} aq
hydroxide	RbOH	102.47	3.20	301		180 g/100 mL^{18} aq; s alc
iodide	RbI	212.37	3.55	642	1304	163 g/100 mL^{25} aq; s alc
nitrate	$RbNO_3$	147.47	3.11	305		19.5 g/100 mL^{20} aq
oxide	Rb_2O	186.93	4.0	400 d		s aq → RbOH
sulfate	Rb_2SO_4	267.00	3.5	1050		48 g/100 mL^{20} aq
Ruthenium	Ru	101.07	12.45_4^{20}	2334	4150	s fused alkali, oxidizing fluxes
(III) chloride (hexagonal)	$RuCl_3$	207.43	3.11	d >500		i aq; s HCl, alc
(V) fluoride	RuF_5	196.06	3.90	86.5	227	d aq
(IV) oxide	RuO_2	133.07	6.97	d		i aq; s fused alkali
Samarium	Sm	150.36	7.52	1074	1794	s acids
(II) chloride	$SmCl_2$	221.27	3.687	855	2030	s aq dec; i alc
(III) chloride	$SmCl_3$	256.72	4.46	682	d	93.4 g/100 mL^{20} aq
(III) fluoride	SmF_3	207.36	6.643	1306	2427	i aq; s H_2SO_4
(III) oxide	Sm_2O_3	348.72	8.347	2335		s acids
(III) sulfate 8-water	$Sm_2(SO_4)_3 \cdot 8H_2O$	733.03	2.93	anhyd 450		2.7 g/100 mL^{20} aq

Scandium	Sc	44.956	2.985 hex	1541	2836	d aq
chloride	$ScCl_3$	151.31	2.39	967	967	v s aq; i alc
oxide	Sc_2O_3	137.91	3.864	2485		s hot or conc acids
sulfate 5-water	$Sc_2(SO_4)_3 \cdot 5H_2O$	468.17	2.519	anhyd 250	d 550	54.6 g/100 mL^{25} aq
Selenium (hexagonal)	Se	78.96	4.81_4^{20}	217	685	s eth, KOH, KCN; i aq, alc
(IV) bromide	$SeBr_4$	398.58	4.029	123		d aq; s HBr, chl, CS_2
(IV) chloride	$SeCl_4$	220.77	2.6	305	subl 196	d aq
(*di-*) dibromide	Se_2Br_2	317.73	3.604_4^{15}		225 d	d aq; s chl, CS_2
dibromide oxide	$SeBr_2O$	254.77	3.38^{50}	41.6	217 d	d aq
(*di-*) dichloride	Se_2Cl_2	228.83	2.774_4^{25}	−85	127 dec	d aq; s bz, chl, CS_2
dichloride oxide	$SeCl_2O$	165.867	2.44	8.5	177.2	d aq; misc bz, chl, CCl_4, CS_2
difluoride oxide	SeF_2O	132.96	2.8	15	125	d aq
(IV) fluoride	SeF_4	154.95	2.75	−10	106	reacts aq viol; misc alc, eth; s chl
(VI) fluoride	SeF_6	192.95	8.467 g/L	−34.6		
(*di-*) hexasulfide	Se_2S_6	350.32	2.44	121.5		s CS_2; 1.2 g/100 mL^{20} bz
(IV) oxide	SeO_2	110.96	3.95	340	subl 315	w/w %: 38^{14} aq, 10^{12} MeOH, 4.35 acet, 6.7^{14} EtOH, 1.1^{12} HOAc; s H_2SO_4
(*tetra-*) tetrasulfide	Se_4S_4	444.10	3.20	113 d		i aq; 0.04 g/100 mL^{20} bz; s CS_2
Silane	SiH_4	32.12	1.409 g/L	−185	−111.9	d aq slowly; i alc, bz, chl, eth
chloro-	SiH_3Cl	66.56	2.921 g/L	−118	−30.4	
dichloro-	SiH_2Cl_2	101.01	4.432 g/L	−122	8.3	d aq
iodo-	SiH_3I	158.01	2.035	−57	45.5	d aq
trichloro-	$SiHCl_3$	135.45	1.331	−128	33	d aq; s bz, chl
Silicon	Si	28.0855	2.33	1412	3265	s HF + HNO_3, fused alkali oxides
carbide (beta)	SiC	40.10	3.16	2830		s fused alkali oxides
dioxide (α quartz)	SiO_2	60.08	2.648	573 tr β quartz	2950	i aq; s HF
dioxide - tungsten trioxide - water (silicotungstic acid)	$SiO_2 \cdot 12WO_3 \cdot 26H_2O$	3310.66				v s aq, alc
disulfide	SiS_2	92.22	2.04	1090		s d aq, alc; i bz

(*Continued*)

TABLE 1.3 Physical Constants of Inorganic Compounds (*Continued*)

Name	Formula	Formula weight	Density	Melting point, °C	Boiling point, °C	Solubility in 100 parts solvent
tetrabromide	$SiBr_4$	347.70	2.81	5.2	154	hyd aq viol
tetrachloride	$SiCl_4$	169.90	1.5	−68.8	57.6	hyd aq; s bz, CCl_4, eth
tetrafluoride	SiF_4	104.08	4.567 g/L	−90.3	−86	hyd aq; s HF
tetraiodide	SiI_4	535.70	4.1	120.5	287.3	d aq; 2.2 g/100 mL^{27} CS_2
(*tri-*) tetranitride	Si_3N_4	140.28	3.17	1878		i aq; s HF
Silver	Ag	107.8682	10.49	961.78	2164	s HNO_3
acetate	$AgC_2H_3O_2$	166.91	3.259	d		1.04^{20} aq; s dil HNO_3
antimonide	Ag_3Sb	445.35		559		
azide	AgN_3	149.89	4.9	exp ~252		i aq; s KCN, HNO_3 (explosive)
bromide	AgBr	187.77	6.473	432	1500	i aq; s KCN
carbonate	Ag_2CO_3	275.75	6.077	218		0.003^{20} aq; s KCN, HNO_3, NH_4OH
chlorate	$AgClO_3$	191.32	4.430_4^{20}	231	d 270	10 g/100 mL^{15} aq
chloride	AgCl	143.32	5.56	455	1547	i aq; 7.7 g/100 mL NH_4OH, KCN, $Na_2S_2O_3$
chromate(VI)	Ag_2CrO_4	331.73	5.625^{25}			0.002^{20} aq; s HNO_3, NH_4OH
cyanide	AgCN	133.89	3.95	320 d		i aq; s KCN
fluoride	AgF	126.87	5.852	435	≈1150	182 g/100 mL^{20} aq; s HF, CH_3CN
(II) fluoride	AgF_2	145.87	4.57	690	d 700	hyd viol aq
iodate	$AgIO_3$	282.77	5.525^{20}	>200	d	0.053^{25} aq; 40 g/100 mL 10% NH_4OH
iodide (alpha)	AgI	234.77	5.683^{30}	558	1505	i aq; s KCN, KI, $(NH_4)_2CO_3$
nitrate	$AgNO_3$	169.87	4.352^{19}	212	d 440	g/100 mL: 216^{20} aq, 3.3 alc, 0.4 acet
nitrite	$AgNO_2$	153.87	4.453	d > 140		0.33^{25} aq; d dilute acids
oxalate	$Ag_2C_2O_4$	303.76	5.03^4	explodes 140		0.004^{20} aq; s HNO_3, NH_4OH
oxide	Ag_2O	231.73	7.22_4^{25}	d 200 (d light)		0.002^{25} aq; s dil HNO_3, NH_4OH
(II) oxide	AgO	123.87	7.483_4^{25}	d > 100		i aq; d alk and acids
perchlorate	$AgClO_4$	207.32	2.806^{25}	d 486		557 g/100 mL^{20} aq; s bz, glyc, pyr
permanganate	$AgMnO_4$	226.80	4.49	d by light		0.9 aq; d alc
phosphate	Ag_3PO_4	418.62	6.37	849		0.006 aq; v s dil HNO_3, KCN, $(NH_4)_2CO_3$
selenate(IV)	Ag_2SeO_3	342.69	5.93	530	d > 530	sl s aq; s HNO_3
sulfate	Ag_2SO_4	311.80	5.45	660	d 1085	0.80^{20} aq (slow); s HNO_3, NH_4OH, H_2SO_4
sulfide (agentite)	Ag_2S	247.80	7.234_4^{20}	845	d	i aq; s HNO_3, alk CN's
Sodium	Na	22.98977	0.968^{20}	97.82	881.4	d aq to NaOH
acetate	$NaC_2H_3O_2$	82.03	1.528	324		75 g/100 mL^{20} aq
acetate 3-water	$NaC_2H_3O_2 \cdot 3H_2O$	136.08	1.45	anhyd 120	d > 120	g/100 mL: 125^{20} aq, 5.1 alc
aluminate(1−)	$NaAlO_2$	81.97	4.63	1650		v s aq; i alc

aluminum sulfate 12-water	$NaAl(SO_4)_2 \cdot 12H_2O$	458.28	1.61	−60		110 g/100 mL[15] aq; i alc
amide	$NaNH_2$	39.01	1.39	210	subl 400	d >500, reacts aq viol
ammonium phosphate 4-water	$NaNH_4HPO_4 \cdot 4H_2O$	209.07	1.54	≈80	anhyd >280	14.3 g/100 mL aq
arsenate(III)(1−)	$NaAsO_2$	129.91	1.87			v s aq; sl s alc
ascorbate	$NaC_6H_7O_6$	198.11		d 218		62 g/100 mL[20] aq
azide	NaN_3	65.01	1.846^{20}	d to Na + N_2		41 g/100 mL[20] aq; 0.3 alc
benzoate	$NaO_2C_6H_5$	144.11				g/100 mL: 63[25] aq; 1.3 alc
bismuthate(V)(1−)	$NaBiO_3$	279.96		d		i cold aq; dec by hot aq & acids
bismuthide	Na_3Bi	277.95		766		d aq
bromate	$NaBrO_3$	150.89	3.34	381 d		40 g/100 mL[20] aq; i alc
bromide	NaBr	102.89	3.200^{20}_{4}	755	1390	g/100 mL: 90[20] aq, 6 alc; 16 MeOH
carbonate	Na_2CO_3	105.99	2.533^{20}	858.1	d	29 g/100 mL[20] aq; s glyc; i alc
carbonate hydrate	$Na_2CO_3 \cdot H_2O$	124.00	2.25	anhyd 100		g/100 mL: 33 aq, 14 glyc; i alc
carbonate 10-water	$Na_2CO_3 \cdot 10H_2O$	286.14	1.46	34 d		50 g/100 mL aq; s glyc
carbonate - hydrogen carbonate 2-water (trona)	$Na_2CO_3 \cdot NaHCO_3 \cdot 2H_2O$	226.02	2.112			13 g/100 mL[0] aq
chlorate(V)	$NaClO_3$	106.44	2.5	248	d >300 → O_2	g/100 mL: 96[20] aq, 0.77 alc, 25 glyc
chloride	NaCl	58.44	2.17	800.8	1465	g/100 mL: 36[20] aq, 10 glyc
chlorite	$NaClO_2$	90.44		d 180–200		34 g/100 mL[17] aq
chromate(VI)	Na_2CrO_4	161.97	2.72	792		84 g/100 mL[20] aq
citrate 2-water	$Na_3C_6H_5O_7 \cdot 2H_2O$	294.10		anhyd 150		77 g/100 mL[25] aq; i alc
cyanate	NaOCN	65.01	1.89	550		s aq d; 0.22[0] alc
cyanide	NaCN	49.01	1.6	563		58.7 g/100 mL[20] aq
cyanohydridoborate	$Na[BH_3CN]$	62.84	1.12	>240 d		g/100 mL: 212 aq, 37.2 THF; v s
dichromate 2-water	$Na_2Cr_2O_7 \cdot 2H_2O$	298.00	2.348^{25}_{4}	anhyd 100; mp 356	d 400	73.1 g/100 mL[20] aq
diethyldithiocarbamate	$NaS_2CN(C_2H_5)_2 \cdot 3H_2O$	225.31		anhyd 94–96		s aq, alc
dihydrogen arsenate(V) hydrate	$NaH_2AsO_4 \cdot H_2O$	181.94	2.53	anhyd 130	d 200	s aq
dihydrogen diphosphate(V)	$Na_2H_2P_2O_7$	221.94	1.9	d 220		4.5 g/100 mL[0] aq
dihydrogen phosphate(V) dihydrate	$NaH_2PO_4 \cdot 2H_2O$	156.01	1.91	anhyd 100	d $NaPO_3$, 200	71 g/100 mL[0] aq; i alc

(Continued)

TABLE 1.3 Physical Constants of Inorganic Compounds (*Continued*)

Name	Formula	Formula weight	Density	Melting point, °C	Boiling point, °C	Solubility in 100 parts solvent
dimethylarsonate 3-water (cacodylate)	$NaO_2As(CH_3)_2$	214.03		anhyd 120		g/100 mL: 200 aq, 40 alc
dioxide	NaO_2	54.99		552		
diphosphate(V)	$Na_4P_2O_7$	265.90	2.53	988		2.26^0 aq
dithionate(V) 2-water	$Na_2S_2O_6 \cdot 2H_2O$	242.14	2.19	anhyd 110	d 267 to Na_2SO_4 + SO_2	13.4 g/100 mL^{20} aq; i alc
dithionate(III)	$Na_2S_2O_4$	174.11		d		22 g/100 mL^{20} aq; sl s alc
diuranate(VI)	$Na_2U_2O_7$	634.03				i aq; s acids
dodecylbenzenesulfonate	$NaO_3SC_6H_4C_{12}H_{25}$	348.49				
dodecylsulfate	$NaO_3SOC_{12}H_{25}$	288.38				10 g/100 mL aq
ethoxide	$NaOC_2H_5$	68.06		>300		d aq; s abs alc
ethylenebis(iminodiacetate) (EDTA)	$(NaOOCCH_2)_2NC_2H_4$-$N(CH_2COONa)_2$	380.20				103 g/100 mL aq
ethylsulfate	$NaO_3SOC_2H_5$	148.12				140 g/100 mL aq; s alc
fluoride	NaF	41.99	2.78	996	1704	4 g/100 mL^{15} aq; i alc
formate	$NaHCO_2$	68.01	1.92	253	d >253	81 g/100 mL^{20} aq; s glyc; sl s alc
gluconate	$NaC_6H_{11}O_7$	218.14				59 g/100 mL^{25} aq; sl s alc; i eth
glycerophosphate	$Na_2C_3H_5(OH)_2PO_4$	216.04		d >130		67 g/100 mL aq; i alc
hexachloroplatinate(IV) 6-water	$Na_2[PtCl_6] \cdot 6H_2O$	561.88	2.50	$-6H_2O$, 110		v s aq; s alc
hexacyanoferrate(II) 10-water	$Na_4[Fe(CN)_6] \cdot 10H_2O$	484.06	1.46	anhyd 82	d 435	28 g/100 mL^{20} aq
hexacyanoferrate(III) hydrate	$Na_3[Fe(CN)_6] \cdot H_2O$	298.93				18.9 g/100 mL^0 aq
hexafluoroaluminate	$Na_3[AlF_6]$	209.94	2.97	1009		s aq
hexanitritocobaltate(III)	$Na_3[Co(NO_2)_6]$	403.98				v s aq; sl s alc
hydride	NaH	24.00	1.39	425 d		ign spontaneously moisture; d alc viol
hydrogen arsenate(V) 7-water	$Na_2HAsO_4 \cdot 7H_2O$	312.01	1.87	anhyd 130	d 150	61 g/100 mL^{15} aq; s glyc; sl s alc
hydrogen carbonate	$NaHCO_3$	84.01	2.20	to Na_2CO_3	270	8 g/100 mL^{20} aq; i alc
hydrogen difluoride	$NaHF_2$	62.00	2.08	d >160		3.7 g/100 mL^{20} aq
hydrogen phosphate 7-water	$Na_2HPO_4 \cdot 7H_2O$	268.07	1.7	d		25 g/100 mL^{40} aq; v sl s alc
hydrogen sulfate	$NaHSO_4$	120.06	2.435	315	d	50 g/100 mL^{20} aq; d alc
hydrogen sulfide	NaHS	56.06	1.79	350		s aq, alc, eth
hydrogen sulfite	$NaHSO_3$	104.06	1.48	d		g/100 mL: 29 aq, 1.4 alc

hydroxide	$NaOH$	40.00	2.130	323	1388	g/100 mL: 108^{20} aq, 14 abs alc, 24 MeOH; s glyc
hydroxymethanesulfinate dihydrate	$Na[HOCH_2SO_2] \cdot 2H_2O$	154.12		63–64	d >64	v s aq; i abs alc, bz, eth
hypochlorite 5-water	$NaClO \cdot 5H_2O$	164.52	1.6	18	d by CO_2 from air	29 g/100 mL^0 aq
iodate	$NaIO_3$	197.89	4.28	d		8.1 g/100 mL^{20} aq
iodide	NaI	149.89	3.67	660	1304	g/100 mL: 200^{20} aq, 100 glyc, 50 alc; s acet
lactate	$NaOOCCHOHCH_3$	112.06		d		misc aq, alc
methoxide	$NaOCH_3$	54.02		>300		d aq; s alc
molybdate(VI) 2-water	$Na_2MoO_4 \cdot 2H_2O$	241.95	≈3.5	anhyd 100	mp 687	65 g/100 mL^{20} aq
nitrate	$NaNO_3$	85.00	2.26	307	d ≈500	g/100 mL: 88^{20} aq, 0.8 alc
nitrite	$NaNO_2$	69.00	2.17	271	d >320	67 g/100 mL^{20} aq
oxalate	$Na_2C_2O_4$	134.00	2.34	d ≈250		3.4 g/100 mL^{20} aq; i alc
oxide	Na_2O	61.98	2.27	dull red heat	d >400	d aq to NaOH violently
pentacyanonitrosylferrate(III) 2-water (nitroprusside)	$Na_2[Fe(CN)_5NO] \cdot 2H_2O$	297.65	1.72			40 g/100 mL^{16} aq
perchlorate	$NaClO_4$	122.44	2.52	480 d		g/100 mL^{25}; 114 aq, 1.5 BuOH, 8.4 EtOAc
periodate	KIO_4	213.89	3.865	d ≈300		10.3 g/100 mL^{20} aq
peroxide	Na_2O_2	77.98	2.805	675	d	v s aq (dec)
peroxoborate 4-water	$NaBO_3 \cdot 4H_2O$	153.88		d >60		2.5 g/100 mL aq
peroxodisulfate(VI)	$Na_2S_2O_8$	238.11		d		55 g/100 mL aq; d by alc
perrhenate	$NaReO_4$	273.19	5.24	300		33 g/100 mL^{20} aq
phosphate	Na_3PO_4	163.94	2.537	1340		12.1 g/100 mL^{20} aq
phosphate 12-water	$Na_3PO_4 \cdot 12H_2O$	380.12	1.62	73.4	$-11H_2O$, 100	28.3 g/100 mL^{20} aq; i alc
phosphinate hydrate	$NaPH_2O_2 \cdot H_2O$	105.99		anhyd 200	d to PH_3	100 g/100 mL^{20} aq; s glyc, alc
propanoate	$NaOOCC_2H_5$	96.06				g/100 mL^{25}: 100 aq, 4.1 alc
salicylate	$NaOOCC_6H_4OH$	160.10				g/100 mL: 110^{20} aq, 11 alc, 25 glyc
selenate(VI)	Na_2SeO_4	188.94	3.098			27 g/100 mL^{20} aq
silicate(2−) meta-	Na_2SiO_3	122.06	2.614	1089		s aq; hyd by hot aq; i alc
silicate(2−) 5-water	$Na_2SiO_3 \cdot 5H_2O$	212.14	1.749	72.2	anhyd 100	v s aq
silicate(4−)	Na_4SiO_4	184.04		1018		s aq

(*Continued*)

TABLE 1.3 Physical Constants of Inorganic Compounds (*Continued*)

Name	Formula	Formula weight	Density	Melting point, °C	Boiling point, °C	Solubility in 100 parts solvent
stannate(IV) 3-water	$Na_2SnO_3 \cdot 3H_2O$	266.71		d 140 (slow)		59 g/100 mL^{20} aq; i alc
stearate	$NaOOCC_{17}H_{35}$	306.47		d		sl s aq
sulfate	Na_2SO_4	142.04	2.7	8800	d 2227	28 g/100 mL^{20} aq
sulfate 10-water	$Na_2SO_4 \cdot 10H_2O$	322.20	1.46	32.4	anhyd 100	67 g/100 mL^{25} aq; s glyc; i alc
sulfide	Na_2S	78.05	1.856	1172 vacuo		18.6 g/100 mL^{20} aq; sl s alc
sulfide 9-water	$Na_2S \cdot 9H_2O$	240.18	1.43	d ≈50		200 g/100 mL aq; sl s alc
sulfite	Na_2SO_3	126.04	2.63	d		31 g/100 mL^{20} aq; s glyc; i alc
tartrate dihydrate	$Na_2C_4H_4O_6 \cdot 2H_2O$	230.08	1.82	anhyd ~120		29 g/100 mL^{6} aq; i alc
tetraborate	$Na_2B_4O_7$	201.22	2.4	742.5		2.6^{20} aq
tetraborate 10-water (borax)	$Na_2B_4O_7 \cdot 10H_2O$	381.37	1.73	75 d	anhyd 320	g/100 mL: 6.3 aq, 100 glyc
tetrachloroaluminate	$Na[AlCl_4]$	191.78	2.01	151		s aq
tetrachloroaurate	$Na[AuCl_4] \cdot 2H_2O$	397.80		d >100		166 g/100 mL^{27} aq; s alc, chl
tetrafluoroborate	$Na[BF_4]$	109.82	2.47	384	d	108 g/100 mL^{27} aq
tetrahydridoborate	$Na[BH_4]$	37.83	1.074	497	d 315	18^{25} DMF; 16.4^{20} MeOH (reacts)
thiocyanate	NaSCN	81.07		287		134 g/100 mL^{20} aq
thiosulfate	$Na_2S_2O_3$	158.11	2.345			s aq; i alc
thiosulfate 5-water	$Na_2S_2O_3 \cdot 5H_2O$	248.19	1.69	anhyd 100	d >100	70 g/100 mL^{20} aq (dec slowly)
trimetaphosphate 6-water	$(NaPO_3)_3 \cdot 6H_2O$	414.04	1.786	53	anhyd 100	22 g/100 mL aq; i alc
tungstate(VI) dihydrate	$Na_2WO_4 \cdot 2H_2O$	329.85	3.25	anhyd 100	mp: 695.6	88 g/100 mL^{0} aq; i alc
vanadate(V)	$NaVO_3$	121.93				s hot aq
Strontium	Sr	87.62	2.64	757	1366	d to $Sr(OH)_2$ in water
bromide	$SrBr_2$	247.43	4.216	657	2045	100 g/100 mL^{20} aq
carbonate	$SrCO_3$	147.63	3.5	d 1100 to SrO + CO_2		i aq; s acids
chlorate	$Sr(ClO_3)_2$	254.52	3.152	120 d → O_2		167 g/100 mL^{20} aq
chloride	$SrCl_2$	158.53	3.052	874	1250	52.9 g/100 mL^{20} aq
chromate(VI)	$SrCrO_4$	203.61	3.89	d		0.12^{20} aq; s HCl
fluoride	SrF_2	125.62	4.24	1477	2460	0.011^{20} aq; s hot HCl
hydrogen phosphate	$SrHPO_4$	183.60	3.544			i aq; s acids
hydroxide	$Sr(OH)_2$	121.64	3.625	535	$-H_2O$, 744	0.8^{20} aq
iodate	$Sr(IO_2)_2$	437.43	5.045^{15}			0.03^{15} aq
iodide	SrI_2	341.43	4.42	402	1773 d	178 g/100 mL^{20} aq; s alc
lactate 3-water	$Sr(OOCCHOHCH_3)_2 \cdot 3H_2O$	319.81		anhyd 150		33 g/100 mL aq
nitrate	$Sr(NO_3)_2$	211.63	2.99	570	645	69.5 g/100 mL^{20} aq; sl s alc, acet
oxide	SrO	103.62	4.7	2430		0.69^{20} aq
perchlorate	$Sr(ClO_4)_2$	286.52	3.00^{25}			g/100 mL^{25}: 157 aq, 71 BuOH, 77

						EtOAc, 90 acet
peroxide	SrO_2	119.62	4.78	215 d		0.018[20] aq; d hot aq
sulfate	$SrSO_4$	183.68	3.96	1607		0.013[20] aq; sl s acid
sulfide	SrS	119.69	3.70	2227		sl s aq; s acid (dec)
Sulfinyl bromide (Thionyl)	$SOBr_2$	207.87	2.688_4^{20}	−52	140	hyd aq (slow); misc bz, chl, CCl_4
chloride	$SOCl_2$	118.97	1.638	−104.5	76	hyd aq; misc bz, chl, CCl_4
fluoride	SOF_2	86.06	3.776 g/L	−129.5	−43.8	hyd aq; s bz, chl, eth
Sulfonyl chloride (Sulfuryl)	SO_2Cl_2	134.97	1.6674_4^{20}	−54.1	69.3	hyd aq; misc bz, eth, HOAc
diamide	$SO_2(NH_2)_2$	96.11	1.807	93	d 250	s aq, hot EtOH, acet
fluoride	SO_2F_2	102.06	4.478 g/L	−135.8	−55.38	mL gas/100 mL: 4 aq, 24 alc, 136 CCl_4, 210 toluene
Sulfur (gamma)	S	32.066	1.92	106.8	444.72	23 g/100 mL[0] CS_2; s alc, bz
(alpha) orthorhombic	S_8	256.53	2.08[20]	tr 94.5 to beta form	444.6	i aq; s organic solvents
(beta) monoclinic tr slowly to rhombic	S_8	256.53	1.96	115.21	444.6	23 g/100 mL[0] CS ; s alc, bz
(*di-*) decafluoride	S_2F_{10}	254.11	2.08	−52.7	30	d fusion with KOH
(*di-*) dichloride	ClSSCl	135.04	1.688	−77	137	hyd aq; s alc, bz, eth, CS_2, CCl_4
dichloride	SCl_2	102.97	1.622	−122	59.5	hyd aq
dioxide	SO_2	64.07	2.811 g/L	−75.47	−10	mL/100 mL: 3937[20] aq, 25 alc, 32 MeOH; s chl, eth
hexafluoride	SF_6	146.06	6.409 g/L	−50.8	subl −63.8	sl s aq; s alc, KOH
tetrafluoride	SF_4	108.06	4.742 g/L	−121.0	−38	d aq viol; v s bz
trioxide (alpha)	SO_3	80.06		62.3	vp 73mm at 25	stable modification
(beta)	SO_3	80.06		32.5	vp 344mm at 25	
(gamma)	SO_3	80.06	1.92	16.8	44.8	v s aq (slow)
Sulfuryl, *see* Sulfonyl						
Tantalum	Ta	180.9479	16.69	2996	5429	s HF, fused alkali (slowly)
(V) bromide	$TaBr_5$	580.47	4.99	265	349	hyd aq; s abs alc, eth
carbide	TaC	192.96	14.3	3880	4780	sl s HF
(*di-*) carbide	Ta_2C	373.91	15.1	3327		
(V) chloride	$TaCl_5$	358.21	3.68	216	239.3	hyd aq; s abs alc
diboride	TaB_2	202.57	11.2	3140		
(V) fluoride	TaF_5	275.94	4.74[20]	96.8	229.5	s aq, eth, conc HNO_3
(V) iodide	TaI	815.47	5.80	496	543	hyd aq; s eth
nitride	TaN	194.95	13.7	3090		sl s aq reg; reacts alkalis
(V) oxide	Ta_2O_5	441.89	8.2	1785		s HF; d fused $KHSO_4$ or KOH
Technetium-98	Tc	97.9072	11	2157	4265	s HNO_3, aq reg, conc H_2SO_4
(VI) fluoride	TcF_6	212.91	3.0	37.4	55.3	s HCl
(IV) oxide	TcO_2	130.91	6.9	subl 1000		s acid, alkali
(VII) oxide	Tc_2O_7	309.81		119.5	310.6	s aq

(*Continued*)

TABLE 1.3 Physical Constants of Inorganic Compounds (*Continued*)

Name	Formula	Formula weight	Density	Melting point, °C	Boiling point, °C	Solubility in 100 parts solvent
Tellurium	Te	127.60	6.24	449.8	989.9	s HNO_3, KOH, conc H_2SO_4
(IV) bromide	TeBr	447.22	4.3	380	≈20 d	s HBr, eth, HOAc
(II) chloride	$TeCl_2$	198.51	6.9	208	328	disprop with eth, diox; s acid
(IV) chloride	$TeCl_4$	269.41	3.0	225	380	hyd aq; s HCl, abs alc, bz
(IV) fluoride	TeF_4	203.59		129	d > 195	d aq
(VI) fluoride	TeF_6	241.59	10.601 g/L	−37.68	subl −38.9	hyd aq, KOH
(IV) iodide	TeI_4	635.22	5.05	280		hyd aq; s HI, alkali; sl s acet
(IV) oxide	TeO_2	159.60	5.9	733	1245	s HCl, HF, NaOH
Terbium	Tb	158.9254	8.23	1356	3230	s acids
chloride	$TbCl_3$	265.28	4.35	588	1550	v s aq
nitrate 6-water	$Tb(NO_3)_3 \cdot 6H_2O$	453.03		89.3		s aq
Thallium	Tl	204.383	11.85	303.5	1457	i aq; s HNO_3
(I) bromide	TlBr	284.29	7.5	460	820	0.05^{20} aq; s alc
(I) carbonate	Tl_2CO_3	468.78	7.11	272		4.1 g/100 mL^{20} aq; i alc
(I) chloride	TlCl	239.84	7.00	430	720	0.33^{20} aq; i alc
(I) cyanide	TlCN	230.40	6.523	d		16.8 g/100 mL^{28} aq; s alc, acid
(I) ethoxide	$TlOC_2H_5$	249.44	3.49	−3	d 130	s eth; sl s alc; d aq
(I) fluoride	TlF	223.38	8.36	326	826	$78.6\%^{15}$ aq
(III) fluoride	TlF_3	261.38	8.65	550 d		d aq
(I) iodide (rhombic)	TlI	331.29	7.1	442	823	i aq, alc; s KI
(I) nitrate	$TlNO_3$	266.39	5.55	206	d 450	9.55 g/100 mL^{20} aq; i alc
(I) oxide	Tl_2O	424.77	9.52	579	1080	v s aq; s acid, alc
(III) oxide (hexagonal)	Tl_2O_3	456.77	10.2	834	$-O_2$, 875	i aq; d by HCl, H_2SO_4
(I) selenate(VI)	Tl_2SeO_4	551.73	6.875	>400		2.8 g/100 mL^{20} aq; i alc, eth
(I) selenide	Tl_2Se	487.73	9.05	340		i aq, acid
(I) sulfate	Tl_2SO_4	504.83	6.77	632	d	4.87 g/100 mL^{20} aq
(I) sulfide	Tl_2S	440.83	8.39	448	1367	0.02^{20} aq; s mineral acids
Thiocarbonyl chloride	S=CCl	114.98	1.509^{15}		73.5	d aq; s eth
Thiocyanogen	$(SCN)_2$	116.16		ca. −2		d aq; s alc, CS_2, eth
Thionyl, *see* Sulfinyl						
Thiophosphoryl tribromide	$PSBr_3$	302.78	2.85^{17}	38.0	209 d	s aq, eth, CS_2
trichloride (alpha)	$PSCl_3$	169.41	1.635	−40.8	125	hyd aq; s bz, chl, CS_2
trifluoride	PSF_3	120.03		−148.8	−52.2	
Thiosulfinyl difluoride	$S{=}SF_2$	102.13		−165	−10.6	hyd aq
Thorium	Th	232.038	11.7	1750	4788	s acids
chloride	$ThCl_4$	373.85	4.59	770	921	s aq, alc
fluoride	ThF_4	308.03	6.1	1110	1680	s acids
iodide	ThI_4	739.66	6.00	570	837	hyd aq
nitrate	$Th(NO_3)_4$	400.06		d 630, ThO_2		191 g/100 mL^{20} aq; v s alc

oxide	ThO_2	264.04	10.0	3390	4400	s hot H_2SO_4
sulfate 9-water	$Th(SO_4)_2 \cdot 9H_2O$	586.30	2.77	anhyd 400		1.57 g/100 mL^{25} aq
Thullium	Tm	168.9342	9.32	1545	1950	s acids
chloride	$TmCl_3$	275.29		824	1490	s aq, alc
fluoride	TmF_3	225.93	7.971	1158	2230	s H_2SO_4
Tin (white)	Sn	118.710	7.265	231.928	2602	s conc HCl, hot H_2SO_4
(II) acetate	$Sn(C_2H_3O_2)_2$	236.80	2.31	182.5	240	d aq; s dilute HCl
(II) bromide	$SnBr_2$	278.52	5.12	215	639	85 g/100 mL^0 aq; s alc, eth
(IV) bromide	$SnBr_4$	438.33	3.34	31	205	v a (hyd) aq; s acet, alc
(II) chloride	$SnCl_2$	189.61	3.90	246.9	623	84 g/100 mL^0 aq; s acet, alc, eth
(IV) chloride	$SnCl_4$	260.52	2.234	−3.3	114.1	s aq (hyd), alc, acet, bz, eth
(II) fluoride	SnF_2	156.71	4.57	213	850	30% aq
(IV) fluoride	SnF_4	194.70	4.78		subl 705	hyd aq
hexafluorozirconate	$Sn[ZrF_6]$	323.92	4.21			s aq
(II) iodide	SnI_2	372.52	5.285	320	714	0.98^{20} aq (d); s bz, chl, alk Cl^- or I^-
(IV) iodide	SnI_4	626.33	4.46	143	364	hyd aq; s alc, bz, chl, eth, CCl_4, CS_2
(II) oxalate	SnC_2O_4	206.73	3.56	280 d		s dilute HCl
(II) oxide	SnO	134.71	6.45	to SnO_2, 300		s acids, conc KOH
(IV) oxide	SnO_2	150.71	6.95	1630		s hot conc KOH (slow)
(II) selenide	SnSe	197.67	6.179	861		s aqua regia, alkali sulfides
(II) sulfate	$SnSO_4$	214.77	4.15	to SnO_2, 378		18.9 g/100 mL^{20} aq; s dilute H_2SO_4
(II) sulfide	SnS	150.78	5.08	880	1210	s conc HCl, hot conc H_2SO_4
(IV) sulfide	SnS_2	182.84	4.5	d 600		s aq reg, alkali hydroxides & sulfides
(II) telluride	SnTe	246.31	6.5	790		i aq
Titanium (hexagonal)	Ti	47.867	4.506	1668	3287	s hot acid, HF
(III) bromide	$TiBr_3$	287.58	4.24		subl 794	
(IV) bromide	$TiBr_4$	367.48	3.37	39	230	hyd aq; 187 g/100 mL abs alc
(II) chloride	$TiCl_2$	118.77	3.13	1035	1500	d aq; s alc
(III) chloride	$TiCl_3$	154.23	2.64	425 d		s aq (heat evolved), alc
(IV) chloride	$TiCl_4$	189.68	1.73	−25	136.4	s cold aq, alc
dihydride	TiH_2	49.88	3.752	d 450		
(IV) fluoride	TiF_4	123.86	2.798	>400	subl 285.5	s aq (slow hyd); s alc, pyr
(IV) iodide	TiI_4	555.49	4.3	150	377	s dry nonpolar solvents
(IV) isopropoxide	$Ti[OCH(CH_3)_2]_4$	284.22	0.9711_4^{20}	~20	220	d aq; s bz, chl, eth
(II) oxide	TiO	63.87	4.95	1750	3660	s H_2SO_4

(Continued)

TABLE 1.3 Physical Constants of Inorganic Compounds (*Continued*)

Name	Formula	Formula weight	Density	Melting point, °C	Boiling point, °C	Solubility in 100 parts solvent
(III) oxide	Ti_2O_3	143.73	4.486	1842		s H_2SO_4, hot HF
(IV) oxide (rutile)	TiO_2	79.87	4.23	1843		s HF, hot conc H_2SO_4
oxide sulfate	$TiOSO_4$	159.94				d aq
(III) sulfate	$Ti_2(SO_4)_3$	383.93				s dilute HCl, dilute H_2SO_4
Tungsten	W	183.84	19.25	3387	5900	s HNO_3 + HF, fusion NaOH + $NaNO_3$
(V) bromide	WBr_5	583.36		286	333	hyd aq; s chl, eth
(VI) bromide	WBr_6	663.26	6.9	309	subl 327	hyd aq; s eth CS_2
(V) chloride	WCl_5	361.10	3.875	242	286	hyd aq
(VI) chloride	WCl_6	396.56	3.52	279	347	hyd aq; s CS_2, CCl_4
dichloride dioxide	WCl_2O_2	286.74	4.67	265	d 369	hyd aq; s HCl
(VI) fluoride	WF_6	297.83	3.441	2.3	17.5	hyd aq; s anhyd HF
(IV) oxide	WO_2	215.84	10.8	1550	d 1724	s acids, KOH
(VI) oxide	WO_3	231.84	7.16	1472	1837	i aq; s hot alkali
(IV) sulfide	WS_2	247.97	7.6	d 1250		s HNO_3 + HF
tetrachloride oxide	WCl_4O	341.65	11.92	211	227	hyd aq
tetrafluoride oxide	WF_4O	275.83	5.07	106	186	
Uranium	U	238.0289	19.1	1135	4131	s acid
(IV) bromide	UBr_4	557.65	5.55	519	777	v s aq
(III) chloride	UCl_3	344.39	5.51	837	1657	v s aq
(IV) chloride	UCl_4	379.84	4.725	590	790	v s aq (d); s polar org solvents
(V) chloride	UCl_5	415.29		287	527	d aq; s CS_2
(VI) chloride	UCl_6	450.75	3.6	177	392	hyd aq; s chl
(IV) fluoride	UF_4	314.02	6.70	1036	1417	s conc acids (d); alk (d)
(VI) fluoride	UF_6	352.02	5.09	64.0	subl 56.5	hyd aq; s chl, CCl_4
(III) hydride	UH_3	241.05	11.1			i aq
(IV) iodide	UI_4	745.65	5.6	506	757	s aq
(IV) oxide (pitchblende)	UO_2	270.03	10.97	2827		s conc HNO_3
(VI) oxide	UO_3	286.03	7.29	d 1300		i aq; s HCl, HNO_3
octaoxide [(V,VI) oxide]	U_3O_8	842.08	8.38	d 1300 to UO_2		s HNO_3
peroxide 2-water	$UO_4 \cdot 2H_2O$	338.06		d 90–195 to U_2O_7 (slow)	d >200 to UO_2	d by HCl
Uranyl(VI) acetate 2-water	$UO_2(C_2H_3O_2)_2 \cdot 2H_2O$	422.13	2.893	anhyd 110	d 275	7.7 g/100 mL[15] aq; sl s alc
chloride	UO_2Cl_2	340.93	5.43	577		320 g/100 mL[18] aq; s acet, alc
fluoride	UO_2F_2	308.03	6.37	d 300		v s aq

nitrate 6-water	$UO_2(NO_3)_2 \cdot 6H_2O$	502.13	2.807	60	d 118	155 g/100 mL^{20} aq; v s alc, eth
sulfate 3-water	$UO_2SO_4 \cdot 3H_2O$	420.14	3.28	d 100		g/100 mL: 21 aq, 4 alc
Vanadium	V	50.9415	6.11^{19}	1917	3421	s HF, HNO_3, hot H_2SO_4, aq reg
(IV) chloride	VCl_4	192.75	1.82	−25.7	148	hyd aq; s nonpolar solvents
dichloride oxide	VCl_2O	137.86	2.88	disprop 384		hyd (slow) aq; s abs alc, HOAc
(III) fluoride	VF_3	107.94	3.363	≈1400	subl 800	i almost all organic solvents
(IV) fluoride	VF_4	126.94	3.15	subl 120 (vac) & disprop		s aq, acet, HOAc
(V) fluoride	VF_5	145.93	2.50	19.5	48	hyd aq; v s anhyd HF, acet, alc
(II) oxide	VO	66.94	5.76	1790		s HCl
(III) oxide	V_2O_3	149.88	4.87	1940		sl s acids
(IV) oxide	VO_2	82.94	4.34	1967		s acids, alkalis
(V) oxide	V_2O_5	181.88	3.35	670	d 1800	0.07 aq; s conc acids, alkalis
(IV) oxide sulfate	$VOSO_4$	163.00				s aq
(III) sulfate	$V_2(SO_4)_3$	390.07		410 (vac)		s (slow) aq, HNO_3
(III) sulfide	V_2S_3	198.08	4.72	d 600		s hot acids, alkali sulfides
Xenon	Xe	131.29	5.761 g/L	−111.8	−108.04	10.8 mL/100 mL^{20} aq
difluoride	XeF	169.29	4.32	129.0	subl 114.3	2.5 g/100 mL^{0} aq
hexafluoride	XeF_6	245.28	3.56	49.5	75.6	hyd aq
tetrafluoride	XeF_4	207.28	4.04	117.1	subl 115.7	hyd aq; s F_3CCOOH
trioxide	XeO_3	179.29	4.55	explodes 25		s aq giving xenic acid
Ytterbium	Yb	173.04	6.90	819	1196	s acids
(II) chloride	$YbCl_2$	243.95	5.27	721	1930	s aq
(III) chloride 6-water	$YbCl_3 \cdot 6H_2O$	387.49	2.57	anhyd 180	mp: 865	v s aq
(III) fluoride	YbF_3	230.04	8.17	1157	2230	s H_2SO_4
(III) nitrate 4-water	$Yb(NO_3)_3 \cdot 4H_2O$	431.12				s aq
(III) oxide	Yb_2O_3	394.08	9.18	2435		s dilute acids
(III) sulfate 8-water	$Yb_2(SO_4)_3 \cdot 8H_2O$	778.39	3.3			34.8 g/100 mL^{20} aq
Yttrium	Y	88.9059	4.472	1522	3345	s hot water (d)
chloride	YCl_3	195.26	2.61	721	1510	79 g/100 mL^{20} aq; s alc
fluoride	YF_3	145.90	4.0	1152	2230	s conc acids (d)
nitrate 6-water	$Y(NO_3)_3 \cdot 6H_2O$	383.01	2.68	$-3H_2O$, 100		171 g/100 mL^{20} aq
oxide	Y_2O_3	225.81	5.03	2440	4300	s acids
sulfate 8-water	$Y_2(SO_4)_3 \cdot 8H_2O$	610.12	2.56	anhyd 400	d > 1000	9.6 g/100 mL^{20} aq
Zinc	Zn	65.39	7.14	419.527	907	i aq; s acids, alkalis (slow)
acetate dihydrate	$Zn(C_2H_3O_2)_2 \cdot 2H_2O$	219.51	1.735	237 d		g/100 mL: 41.6^{20} aq, 3.3 alc
arsenate(III)(1−)	$Zn(AsO_2)_2$	279.23				s acids

(Continued)

TABLE 1.3 Physical Constants of Inorganic Compounds (*Continued*)

Name	Formula	Formula weight	Density	Melting point, °C	Boiling point, °C	Solubility in 100 parts solvent
arsenate(V)(3−) 8-water	$Zn_3(AsO_4)_2 \cdot 8H_2O$	618.13	3.33			s acids and alkalis
bromide	$ZnBr_2$	225.20	4.5	394	697	g/100 mL: 471^{25} aq, 200 alc; s KOH, eth
carbonate	$ZnCO_3$	125.40	4.4	$-CO_2$, 300		0.02^{25} aq; s acids, KOH, NH_4 salts
chloride	$ZnCl_2$	136.29	2.907	290	732	g/100 ml: 395^{20} aq, 77 alc, 50 glyc; v s acet
chromate(VI)	$ZnCrO_4$	181.39	3.40			s acids
cyanide	$Zn(CN)_2$	117.43	1.852	d 800		0.058^{18} aq; s acids, KCN, KOH
fluoride	ZnF_2	103.39	4.9	872	1500	s HNO_3, HCl, NH_4OH
hexafluorosilicate 6-water	$Zn[SiF_6] \cdot 6H_2O$	315.56	2.104	d 100		v s aq
iodate	$Zn(IO_3)_2$	415.20	5.063	d		0.87^{20} aq; s HNO_3, KOH
iodide	ZnI_2	319.20	4.74	446	625 d	g/100 mL: 332^{20} aq, 50 glyc; v s alc
nitrate 6-water	$Zn(NO_3)_2 \cdot 6H_2O$	297.49	2.067	$-6H_2O$, 131		146 g/100 mL^0 aq; v s alc
oxide	ZnO	81.39	5.60	1975		i aq; s acids, KOH, NH_4OH
peroxide	ZnO_2	97.39	1.57	d >150	explodes 212	d (slow) aq; s dilute acids (d)
1,4-phenolsulfonate 8-water	$Zn[C_6H_4(OH)SO_3]_2 \cdot 8H_2O$	555.84		anhyd 120		g/100 mL: 63 aq, 56 alc
phosphate(V)	$Zn_3(PO_4)_2$	386.11	3.998	900		s acids, NH_4OH
phosphide	Zn_3P_2	258.12	4.55	420	1100	d aq, HCl (viol); s bz, CS_2
propionate	$Zn(C_3H_5O_2)_2$	211.53				$32\%^{15}$ aq; $2.8\%^{15}$ alc
selenide	ZnSe	144.35	5.65	>1100		d dilute HNO_3
silicate(2−)	Zn_2SiO_4	222.86	4.10	1512		i aq or dilute acids
stearate	$Zn(C_{18}H_{35}O_2)_2$	632.34	1.095	130		d dil acids; s bz; i aq, alc, eth
sulfate	$ZnSO_4$	161.45	3.8	680 d		$53.8\%^{20}$ aq
sulfate 7-water	$ZnSO_4 \cdot 7H_2O$	287.56	1.97	anhyd 280	d >500	g/100 mL: 167 aq, 40 glyc; i alc
sulfide (wirzite)	ZnS	97.46	4.09	1722		i aq; s dilute mineral acids
telluride	ZnTe	192.99	6.34	1239		d (slow) aq or dilute HCl
thiocyanate	$Zn(SCN)_2$	181.56				0.14 aq; s alc
Zirconium	Zr	91.224	6.52	1852	3577	s aq reg, HF, hot H_3PO_4, fusion with KOH + KNO_3
(IV) bromide	$ZrBr_4$	410.84	3.98	450	subl 357	
carbide	ZrC	103.23	6.73	3532	5100	sl s conc H_2SO_4
(II) chloride	$ZrCl_2$	162.13	3.6	727	1292	d aq

(IV) chloride	$ZrCl_4$	233.03	2.80	437 (25 atm)	subl 334	hyd aq to $ZrCl_2O$; s alc, eth
diboride	ZrB_2	112.85	6.17	3245	d 4193	
dichloride oxide 8-water	$ZrCl_2O \cdot 8H_2O$	322.25	1.91	anhyd 210	d 410	v s aq, alc
dihydride	ZrH_2	93.24	5.61			i aq
(IV) fluoride	ZrF_4	167.22	4.436	932^{tp}	subl 912	1.32 g/100 mL^{20} aq
(IV) hydroxide	$Zr(OH)_4$	159.25	3.25	to ZrO_2, 500		s mineral acids
(IV) iodide	ZrI_4	598.84		499 (sealed tube)	subl 432.5	s aq (d), eth
(IV) nitrate 5-water	$Zr(NO_3)_4 \cdot 5H_2O$	429.32		d 100		v s aq; s alc
(IV) oxide	ZrO_2	123.22	5.68	2678	4300	s hot H_2SO_4, HF (slow)
(IV) silicate(4−)	$ZrSiO_4$	183.31	4.56	d 1540 to $ZrO_2 + SiO_2$		unaffected by aqueous reagents
sulfate 4-water	$Zr(SO_4)_2 \cdot 4H_2O$	355.41	2.80	anhyd 380		52.5 g/100 g aqueous solution

TABLE 1.4 Color, Crystal Symmetry and Refractive Index of Inorganic Compounds

Abbreviations Used in the Table

Color				Crystal Symmetry	
B	*brown*	*R*	*red*	*C*	*cubic*
BE	*blue*	*SL*	*silver*	*H*	*hexagonal*
BK	*black*	*V*	*violet*	*M*	*monoclinic*
CL	*colorless*	*W*	*white*	*R*	*rhombic*
G	*gray*	*Y*	*yellow*	*RH*	*Rhombohedral*
GN	*green*			*T*	*tetragonal*
O	*orange*			*TG*	*trigonal*
P	*purple*			*TR*	*triclinic*

Compound	Formula	Molecular weight	Color	Crystal symmetry	Refractive index n_D
Actinium					
Bromide	$AcBr_3$	466.7	W	H	
Chloride	$AcCl_3$	333.4	W	H	
Fluoride	AcF_3	284.0	W	H	
Oxide	Ac_2O_3	502.0	W	H	
Aluminum					
Bromide	$AlBr_3$	266.7	CL	R	
Carbide	Al_4C_3	143.9	Y	H	2.70
Chloride	ACl_3	133.3	W	H	1.56
Fluoride	AlF_3	84.0	CL	TR	1.38
Hydroxide	$Al(OH)_3$	78.0	W	M	
Iodide	AlI_3	407.7	W		
Nitrate	$Al(NO_3)_3 \cdot 9H_2O$	375.1	CL	R	1.54
Nitride	AlN	41.0	W	H	
Oxide	Al_2O_3	102.0	CL	H	1.68
Phosphate	$AlPO_4$	122.0	W	R	1.56
Silicate	Al_2SiO_5	162.0	W	R	1.66
Sulfate	$Al_2(SO_4)_3$	342.2	W	R	1.47
Sulfide	Al_2S_3	150.2	Y	H	
Americium					
Oxide IV	AmO_2	275.1	B	C	
Ammonium					
Bromide	NH_4Br	98.0	W	C	1.711
Carbonate	$(NH_4)_2CO_3 \cdot H_2O$	114.1	W	C	
Chlorate	NH_4ClO_3	101.5	W	M	
Chloride	NH_4Cl	53.5	W	C	1.642
Chromate	$(NH_4)_2CrO_4$	152.1	Y	M	
Fluoride	NH_4F	37.0	W	H	1.315
Iodate	NH_4IO_3	192.9	W	R	
Iodide	NH_4I	144.9	W	C	1.703
Nitrate	NH_4NO_3	80.0	W	R	1.413
Nitrite	NH_4NO_2	64.0	Y		
Oxalate	$(NH_4)_2C_2O_4 \cdot H_2O$	142.1	CL	R	1.44–1.59
Perchlorate	NH_4ClO_4	117.5	W	R	1.49
Hydrogen Phosphate	$(NH_4)_2HPO_4$	132.1	W	M	1.53
Dihydrogen Phosphate	$NH_4H_2PO_4$	115.0	W	T	1.48–1.53
Sulfate	$(NH_4)_2SO_4$	132.1	W	R	1.53
Hydrogen sulfide	NH_4HS	51.1	W	R	1.74
Thiocyanate	NH_4SCN	76.1	CL	M	1.61–1

TABLE 1.4 Color, Crystal Symmetry and Refractive Index of Inorganic Compounds (*Continued*)

Compound	Formula	Molecular weight	Color	Crystal symmetry	Refractive Index n_D
Antimony					
Bromide III	$SbBr_3$	361.5	CL	R	1.74
Chloride III	$SbCl_3$	228.1	CL	R	1.74
Chloride V	$SbCl_5$	299.0	W	LIQ	1.601[1]
Fluoride III	SbF_3	178.8	CL	R	
Fluoride V	SbF_5	216.7	CL	LIQ	
Hydride III	SbH_3	124.8	CL	GAS	
Iodide III	SbI_3	502.5	RD	H	
Iodide V	SbI_5	756.3	B		
Oxide III	Sb_2O_3	291.5	CL	R	2.35
Oxide V	Sb_2O_5	323.5	Y	C	
Oxychloride III	$SbOCl$	173.2	W	M	
Sulfate III	$Sb_2(SO_4)_3$	531.7	W		
Sulfide III	Sb_2S_3	339.7	BK	R	4.064
Sulfide V	Sb_2S_5	403.8	Y		
Arsenic					
Acid, ortho	$H_3AsO_4 \cdot {}^1/_2H_2O$	151.0	CL		
Bromide III	$AsBr_3$	314.7	CL	R	
Chloride III	$AsCl_3$	181.3	CL	LIQ	1.598
Chloride V	$AsCl_5$	252.2	CL		
Fluoride III	AsF_3	131.9	CL	LIQ	
Fluoride V	AsF_5	169.9	CL	GAS	
Hydride III	AsH_3	77.9	CL	GAS	
Iodide III	AsI_3	455.6	R	H	
Iodide V	AsI_5	709.5	B	M	
Oxide III	As_2O_3	197.2	CL	C	
Oxide V	As_2O_5	229.9	W		
Sulfide II	As_2S_2	214.0	R	M	2.46–2.52
Sulfide III	As_2S_3	246.0	Y	M	2.4–2.6
Sulfide V	As_2S_5	310.2	Y	M	
Barium					
Bromate	$Ba(BrO_3)_2 \cdot H_2O$	411.2	CL	M	
Bromide	$BaBr_2$	297.2	CL	R	1.75
Carbide	BaC_2	161.4	G	T	
Carbonate	$BaCO_3$	197.4	W	R	1.676
Chlorate	$Ba(ClO_3)_2 \cdot H_2O$	322.3	CL	M	1.56–1
Chloride	$BaCl_2$	208.3	CL	M	1.736
Chromate	$BaCrO_4$	253.3	Y	R	
Fluoride	BaF_2	175.3	CL	C	1.474
Hydride	BaH_2	139.4	G		
Hydroxide	$Ba(OH)_2 \cdot 8H_2O$	315.5	CL	M	1.502
Iodide	BaI_2	391.2	CL	M	
Nitrate	$Ba(NO_3)_2$	261.4	CL	C	1.572
Oxalate	BaC_2O_4	225.4	W		
Oxide	BaO	153.3	CL	C	1.98
Perchlorate	$Ba(ClO_4)_2$	336.2	CL	H	
Sulfate	$BaSO_4$	233.4	W	R	1.636
Sulfide	BaS	169.4	CL	C	2.155
Titanate	$BaTiO_3$	233.3		T/H	2.40

(Continued)

TABLE 1.4 Color, Crystal Symmetry and Refractive Index of Inorganic Compounds (*Continued*)

Compound	Formula	Molecular weight	Color	Crystal symmetry	Refractive index n_D
Beryllium					
Bromide	$BeBr_2$	168.8	W	OR	
Carbide	Be_2C	30.0	Y	H	
Chloride	$BeCl_2$	79.9	W	OR	
Fluoride	BeF_2	47.0	CL	T	
Hydroxide	$Be(OH)_2$	43.0	W	R	
Iodide	BeI_2	262.8	CL	RH	
Nitrate	$Be(NO_3)_2 \cdot 3H_2O$	187.1	W		
Nitride	Be_3N_2	55.1	CL	C	
Oxide	BeO	25.0	W	H	1.72
Sulfate	$BeSO_4$	105.1	CL	T	
Sulfate	$BeSO_4 \cdot 4H_2O$	177.1	CL	T	1.44–1.47
Bismuth					
Bromide III	$BiBr_3$	448.7	Y		
Chloride III	$BiCl_3$	315.4	W		
Fluoride III	BiF_3	266.0	G	C	1.74
Hydroxide III	$Bi(OH)_3$	260.0	W		
Iodide III	BiI_3	589.7	RD	H	
Nitrate III	$Bi(NO_3)_3 \cdot 5H_2O$	485.1	CL	TR	
Nitrate, Basic III	$BiO(NO_3) \cdot H_2O$	305.0	W	H	
Oxide III	Bi_2O_3	466.0	Y	R	1.91
Oxide IV	$Bi_2O_4 \cdot 2H_2O$	518.0	B		
Oxide V	Bi_2O_5	498.0	B		
Oxychloride III	BiOCl	260.5	W	T	2.15
Phosphate III	$BiPO_4$	304.0	W	M	
Sulfate III	$Bi_2(SO_4)_3$	706.1	W		
Sulfide III	Bi_2S_3	514.2	B	R	1.34–1.46
Boron					
Arsenate	$BAsO_4$	149.7	W	T	1.68
Boric Acid	H_3BO_3	61.8	W	TR	
Bromide	BBr_3	250.5	CL	LIQ	1.5312^{16}
Carbide	B_4C	55.3	BK	RH	
Chloride	BCl_3	117.2	CL	LIQ	
Diborane	B_2H_6	27.7	CL	GAS	
Fluoride	BF_3	67.8	CL	GAS	
Iodide	BI_3	391.6	W		
Nitride	BN	24.8	W	H	
Oxide	B_2O_3	69.6	W	C	
Sulfide	B_2S_3	117.8	W		
Bromine					
Chloride I	BrCl	115.4	R	GAS	
Fluoride I	BrF	98.9	B	GAS	
Fluoride III	BrF_3	136.9	CL	LIQ	1.4536^{25}
Fluoride V	BrF_5	174.9	CL	LIQ	1.3529^{25}
Hydride I	H Br	80.9	CL	GAS	1.325^{10}
Cadmium					
Bromide	$CdBr_2$	272.2	W	H	
Carbonate	$CdCO_3$	172.4	W	TG	
Chloride	$CdCl_2$	228.4	W	H	

TABLE 1.4 Color, Crystal Symmetry and Refractive Index of Inorganic Compounds (*Continued*)

Compound	Formula	Molecular weight	Color	Crystal wymmetry	Refractive index n_D
Cadmium (*Continued*)					
Fluoride	CdF_2	150.4	W	C	1.56
Hydroxide	$Cd(OH)_2$	146.4	W	TR	
Iodide	CdI_2	366.2	B	H	
Nitrate	$Cd(NO_3)_2 \cdot 4H_2O$	308.5	W		
Oxide	CdO	128.4	B	C	
Sulfate	$CdSO_4$	208.5	W	R	
Sulfate	$3CdSO_4 \cdot 8H_2O$	769.6	CL	M	1.565
Sulfide	CdS	144.5	Y	H	2.51
Calcium					
Bromate	$CaBrO_3 \cdot H_2O$	313.9		M	
Bromide	$CaBr_2 \cdot 6H_2O$	308.0	CL	H	
Carbide	CaC_2	64.1	CL	T	1.75
Carbonate	$CaCO_3$	100.1	CL	R	1.681
Chloride	$CaCl_2$	111.0	CL	C	1.52
Chloride	$CaCl_2 \cdot 6H_2O$	219.1	C	T	1.417
Chromate	$CaCrO_4 \cdot 2H_2O$	192.1	Y	M	
Fluoride	CaF_2	78.1	CL	C	1.434
Hydride	CaH_2	42.1	W	R	
Hydroxide	$Ca(OH)_2$	74.1	CL	H	1.574
Iodide	CaI_2	293.9	W	H	
Nitrate	$Ca(NO_3)_2$	164.1	CL	C	
Nitrate	$Ca(NO_3)_2 \cdot 4H_2O$	236.2	CL	M	1.498
Nitride	Ca_3N_2	148.3	B	H	
Oxalate	CaC_2O_4	128.1	CL	C	
Oxide	CaO	56.1	CL	C	1.838
Perchlorate	$Ca(ClO_4)_2$	239.0	CL		
Peroxide	CaO_2	72.1	W	T	
Sulfate	$CaSO_4$	136.1	CL	M	1.576
Sulfate	$CaSO_4 \cdot 2H_2O$	172.2	CL	M	1.5226
Sulfide	CaS	72.1	CL	C	2.137
Carbon					
Dioxide	CO_2	44.0	CL	GAS	
Disulfide	CS_2	76.1	CL	LIQ	1.6290
Monoxide	CO	28.0	CL	GAS	
Oxybromide	$COBr_2$	187.8	CL	LIQ	
Oxychloride	$COCl_2$ (Phosgene)	98.9	CL	GAS	
Oxysulfide	COS	60.1	CL	GAS	
Cerium					
Bromide III	$CeBr_3$	380.0		H	
Chloride III	$CeCl_3$	246.5	CL	H	
Fluoride III	CeF_3	197.1	W	H	
Iodate IV	$Ce(IO_3)_4$	839.7	Y		
Iodide III	CeI_3	520.8	Y	R	
Molybdate III	$Ce_2(MoO_4)_3$	760.0	Y	T	2.01
Nitrate III	$Ce(NO_3)_3 \cdot 6H_2O$	434.2	CL		
Oxide III	Ce_2O_3	328.2	GN	H	
Oxide IV	CeO_2	172.1	W	C	
Sulfate III	$Ce_2(SO_4)_3$	568.4	CL	M/R	
Sulfide	Ce_2S_3	376.4	Y	C	

(*Continued*)

TABLE 1.4 Color, Crystal Symmetry and Refractive Index of Inorganic Compounds (*Continued*)

Compound	Formula	Molecular weight	Color	Crystal symmetry	Refractive index n_D
Cesium					
Bromide	$CsBr$	212.8	CL	C	1.642
Carbonate	Cs_2CO_3	325.8	CL		
Chloride	$CsCl$	168.4	CL	C	1.534
Fluoride	CsF	151.9	CL	C	1.481
Hydroxide	$CsOH$	149.9	W		
Iodide	CsI	259.8		C	1.661; 1.669
Iodide III	CsI_3	513.7	BK	R	
Nitrate	$CsNO_3$	194.9	W	H	1.55
Oxide	Cs_2O	281.8	R		
Perchlorate	$CsClO_4$	232.4	CL	R	1.479
Periodate	$CsIO_4$	323.8	W	R	
Peroxide	Cs_2O_2	297.8	Y	R	
Sulfate	Cs_2SO_4	361.9	CL	R	1.564
Superoxide	CsO_2	164.9	Y		
Trioxide	Cs_2O_3	313.8	B	C	
Chlorine					
Dioxide	ClO_2	67.5	Y	GAS	
Fluoride	ClF	54.5	CL	GAS	
Trifluoride	ClF_3	92.5	CL	GAs	
Monoxide	Cl_2O	86.9	B	GAS	
Hydrochloric Acid	HCl	36.5	CL	GAS	1.254^{10}
Perchloric Acid	$HClO_4$	100.5	CL	LIQ	
Chromium					
Bromide II	$CrBr_2$	211.8	W	M	
Carbide III	Cr_3C_2	180.0	G	R	
Chloride II	$CrCl_2$	122.9	W	R	
Chloride III	$CrCl_3$	158.4	V	R	
Fluoride II	CrF_2	90.0	GN	M	
Fluoride III	CrF_3	109.0	GN	R	
Iodide II	CrI_2	305.8	B	M	
Nitrate III	$Cr(NO_3)_3$	238.0	GN		
Nitrate III	CrN	66.0		C	
Oxide II	CrO	68.0	BK	H	
Oxide III	Cr_2O_3	152.0	GN	H	2.551
Oxide IV	CrO_2	84.0	B		
Oxide VI	CrO_3	100.0	RD	R	
Phosphate III	$CrPO_4 \cdot 6H_2O$	255.1	V	TR	
Sulfate III	$Cr_2(SO_4) \cdot 18H_2O$	716.5	V	C	1.564
Sulfide II	CrS	84.1	BK	M	
Sulfide III	Cr_2S_3	200.2	B	TG	
Cobalt					
Bromide II	$CoBr_2$	218.8	GN	H	
Chlorate II	$Co(ClO_3)_2 \cdot 6H_2O$	333.9	R	C	1.55
Chloride II	$CoCl_2$	129.8	BE	H	
Fluoride II	CoF_2	96.9	R	M	
Fluoride III	CoF_3	115.9	B	H	
Hydroxide II	$Co(OH)_2$	92.9	R	R	
Iodate II	$Co(IO_3)_2$	408.7	V		

TABLE 1.4 Color, Crystal Symmetry and Refractive Index of Inorganic Compounds (*Continued*)

Compound	Formula	Molecular weight	Color	Crystal symmetry	Refractive index n_D
Cobalt (*Continued*)					
Iodide II	CoI_2	312.7	BK	H	
Nitrate II	$Co(NO_3)_2 \cdot 6H_2O$	291.0	R	M	
Oxide II	CoO	74.9	GN	C	
Oxide III	Co_2O_3	165.9	B	R	
Oxide II–III	Co_3O_4	240.8	BK	C	
Perchlorate II	$Co(ClO_4)_2$	257.8	R		1.50
Sulfate II	$CoSO_4$	155.0	BE	C	
Sulfate II	$CoSO_4 \cdot 7H_2O$	281.1	R	M	1.48
Sulfide II	CoS	91.0	R	H	
Sulfide III	Co_2S_3	214.1	BK		
Copper					
Bromide I	CuBr	143.5	W	C	
Bromide II	$CuBr_2$	223.4	BK	M	
Carbonate, Basic II	$2CuCO_3 \cdot Cu(OH)_2$	344.7	BE	M	1.731
Chloride I	CuCl	99.0	W	C	
Chloride II	$CuCl_2$	134.5	Y	M	
Chloride II	$CuCl_2 \cdot 2H_2O$	170.5	Y	R	
Fluoride II	$CuF_2 \cdot 2H_2O$	137.6	W	M	
Hydroxide I	CuOH	80.6	Y		
Hydroxide II	$Cu(OH)_2$	97.6	BE		
Iodide I	CuI	190.5	W	C	2.346
Nitrate II	$Cu(NO_3)_2 \cdot 3H_2O$	241.6	BE		
Oxide I	Cu_2O	143.1	R	C	2.705
Oxide II	CuO	79.5	BK	TR	2.63
Sulfate II	$CuSO_4$	159.6	W	R	
Sulfate II	$CuSO_4 \cdot 5H_2O$	249.7	BE	TR	1.52
Sulfide I	Cu_2S	159.1	BK	C	
Sulfide II	CuS	95.6	BK	H	
Thiocyanate I	CuSCN	121.6	W		
Curium					
Bromide III	$CmBr_3$	488		R	
Chloride III	$CmCl_3$	353	W	H	
Fluoride III	CmF_3	304	W	H	
Fluoride IV	CmF_4	323	B	M	
Iodide III	CmI_3	628	W	H	
Dysprosium					
Bromide	$DyBr_3$	402.3	CL	R	
Chloride	$DyCl_3$	268.9	Y	M	
Fluoride	DyF_3	219.5	CL	H	
Iodide	DyI_3	543.2	GN	H	
Nitrate	$Dy(NO_3)_3 \cdot 5H_2O$	438.6	Y	TR	
Oxide	Dy_2O_3	373.0	W	C	
Sulfate	$Dy_2(SO_4)_3 \cdot 8H_2O$	757.3	Y	M	
Erbium					
Bromide	$ErBr_3$	407.1	V	R	
Chloride	$ErCl_3$	273.6	V	M	
Fluoride	ErF_3	224.3	RD	R	

(*Continued*)

TABLE 1.4 Color, Crystal Symmetry and Refractive Index of Inorganic Compounds (*Continued*)

Compound	Formula	Molecular weight	Color	Crystal symmetry	Refractive index n_D
Erbium (*Continued*)					
Iodide	ErI_3	548.0	V	H	
Oxide	Er_2O_3	382.6	R	C	
Sulfate	$Er_2(SO_4)_3$	622.7	W		
Sulfide	Er_2S_3	263.5	R	M	
Europium					
Bromide II	$EuBr_2$	311.8		R	
Bromide III	$EuBr_3$	391.7	G	R	
Chloride II	$EuCl_2$	222.9	W	R	
Chloride III	$EuCl_3$	258.3	Y	H	
Fluoride II	EuF_2	190.0	Y	C	
Fluoride III	EuF_3	209.0	W	R	
Iodide II	EuI_2	405.8	GN	M	
Iodide III	EuI_3	532.7			
Oxide III	Eu_2O_3	351.9	R	C	
Sulfate III	$Eu_2(SO_4)_3 \cdot 8H_2O$	736.2	R	M	
Fluorine					
Dioxide	F_2O_2	70.0	B	GAS	
Hydride	HF	20.0	CL	GAS	
Oxide	F_2O	54.0	CL	GAS	
Cadolinium					
Bromide	$GdBr_3$	397.0	W	H	
Chloride	$GdCl_3$	263.6	W	H	
Fluoride	GdF_3	214.3	W	R	
Iodide	GdI_3	538.0	Y	H	
Nitrate	$Gd(NO_3)_3 \cdot 6H_2O$	451.4		T	
Oxide	Gd_2O_3	362.5	W	C	
Sulfate	$Gd_2(SO_4)_3$	602.7	CL		
Sulfide	Gd_2S_3	410.7	Y	C	
Gallium					
Arsenide III	GaAs	144.6	G	C	
Bromide III	$GaBr_3$	309.5	CL		
Chloride II	Ga_2Cl_4	281.3	W		
Chloride III	$GaCl_3$	176.0	CL	TR	
Fluoride III	GaF_3	126.7	W	RH	
Iodide III	GaI_3	450.4	Y		
Oxide I	Ga_2O	155.4	G		
Oxide III	Ga_2O_3	187.4	G	M (β)	1.95
Sulfide I	Ga_2S	171.5	G		
Sulfide II	Ga_2S_3	235.6	Y	H	
Germanium					
Bromide IV	$GeBr_4$	392.2	G		1.627
Chloride IV	$GeCl_4$	214.4	CL	LIQ	1.464
Fluoride IV	GeF_4	148.6	CL	GAS	
Hydride IV	GeH_4 (Germane)	76.6	CL	GAS	1.00089
Iodide IV	GeI_4	580.2	R	C	
Oxide II	GeO	88.6	G		1.607

TABLE 1.4 Color, Crystal Symmetry and Refractive Index of Inorganic Compounds (*Continued*)

Compound	Formula	Molecular weight	Color	Crystal symmetry	Refractive index n_D
Germanium (*Continued*)					
Oxide IV	GeO_2	104.6	CL	H	
Sulfide II	GeS	104.7	Y	R	
Sulfide IV	GeS_2	136.7	W	R	
Gold					
Bromide I	AuBr	276.9	G		
Bromide III	$AuBr_3$	436.7	B		
Chloride I	AuCl	232.4	Y	R	
Chloride III	$AuCl_3$	303.3	R		
Hydroxide III	$Au(OH)_3$	248.0	B		
Iodide	AuI	323.9	Y	TR	
Iodide III	AuI_3	577.7	G		
Sulfate III	$Au_2(SO_4)_3 \cdot H_2O$	490.5	B		
Sulfide I	Au_2S	426.0	B		
Sulfide III	Au_2S_3	490.1	B		
Hafnium					
Bromide	$HfBr_4$	498.1	W		
Carbide	HfC	190.5		C	
Chloride	$HfCl_4$	320.3	W		
Fluoride	HfF_4	254.5	CL	M	1.56
Iodide	HfI_4	686.1			
Nitride	HfN	192.5	Y	C	
Oxide	HfO_2	210.5	W	T	
Sulfide	HfS_2	242.6		H	
Holmium					
Bromide	$HoBr_3$	404.7	Y	R	
Chloride	$HoCl_3$	271.3	Y	M	
Fluoride	HoF_3	221.9	B	H	
Iodide	HoI_3	545.6	Y		
Oxide	Ho_2O_3	377.9		C	
Hydrogen					
Bromide	HBr	80.9	CL	GAS	2.77^{-67}
Chloride	HCl	36.5	CL	GAS	
Fluoride	HF	20.0	CL	GAS	
Iodide	HI	127.9	CL	GAS	1.466
Oxide	H_2O	18.0	CL	LIQ	1.3333
Oxide-Deutero	$2H_2O$	20.0	CL	LIQ	1.3284
Peroxide	H_2O_2	34.0	CL	LIQ	1.414^{22}
Selenide	H_2Se	81.0	CL	GAS	
Sulfide	H_2S	34.1	CL	GAS	1.374
Telluride	H_2Te	129.9	CL	GAS	
Indium					
Bromide I	InBr	194.7	B		
Bromide III	$InBr_3$	354.5	CL		
Chloride I	InCl	150.3	R	C	
Chloride III	$InCl_3$	221.2	CL	M	
Fluoride III	InF_3	171.8	CL	H	

(*Continued*)

TABLE 1.4 Color, Crystal Symmetry and Refractive Index of Inorganic Compounds (*Continued*)

Compound	Formula	Molecular weight	Color	Crystal symmetry	Refractive index n_D
Indium (*Continued*)					
Iodide I	InI	241.7	B		
Iodide III	InI_3	495.5	Y	M	
Oxide III	In_2O_3	277.6	Y	C	
Sulfate III	$In_2(SO_4)_3$	517.8	W	M	
Sulfide III	In_2S_3	325.8	R (β)	C	
Iodine					
Bromide I	IBr	206.8	BK	OR	
Chloride I, α	ICl	162.4	R	C	
Chloride I, β	ICl	162.4	R	LIQ	
Chloride III	ICl_3	233.3	Y	R	
Fluoride V	IF_5	221.9	CL	LIQ	
Fluoride VII	IF_7	259.9	CL	GAS	
Oxide IV	I_2O_4	317.8	Y		
Oxide V	I_2O_5	333.8	CL		
Iodic Acid	HIO_3	175.9	W	R	
Hydrogen Iodide	HI	127.9	CL	GAS	1.466
Iridium					
Bromide II	$IrBr_3 \cdot 4H_2O$	504.0	GN		
Bromide IV	$IrBr_4$	511.8	BK		
Chloride III	$IrCl_3$	298.6	GN	H	
Chloride IV	$IrCl_4$	334.0	R	C	
Fluoride VI	IrF_6	306.2	Y	T	
Iodide III	IrI_3	572.9	GN		
Iodide IV	IrI_4	699.8	BK		
Oxide IV	IrO_2	224.2	BK		
Sulfide IV	IrS_2	256.3	BK		
Iron					
Arsenide	FeAs	130.8	W	R	
Arsenide, di–	$FeAs_2$	205.7	G	R	
Bromide II	$FeBr_2$	215.7	GN	H	
Bromide III	$FeBr_3 \cdot 6H_2O$	403.7	R		
Carbide	Fe_3C	179.6	G	C	
Carbonate II	$FeCO_3$	115.9	G		
Chloride II	$FeCl_2$	126.8	G	H	
Chloride III	$FeCl_3$	162.2	GN	H	
Fluoride III	FeF_3	112.9	W	R	
Hydroxide II	$Fe(OH)_2$	89.9	GN	H	
Hydroxide III	$Fe(OH)_3$	106.9	B		
Iodide II	FeI_2	309.7	BK	H	
Nitrate II	$Fe(NO_3)_2 \cdot 6H_2O$	288.0	GN	R	
Nitrate III	$Fe(NO_3)_3 \cdot 9H_2O$	404.0	CL	M	
Nitride	Fe_2N	125.7	G		
Oxide II	FeO	71.9	BK	C	2.32
Oxide III	Fe_2O_3	159.7	B	TG	3.04
Oxide II-III	Fe_3O_4	231.6	BK	C	2.42
Phosphate III	$FePO_4 \cdot 2H_2O$	186.9	W	M	1.35
Phosphide	Fe_2P	142.7	G	H	
Sulfate II	$FeSO_4 \cdot 7H_2O$	278.0	GN	M	1.48

TABLE 1.4 Color, Crystal Symmetry and Refractive Index of Inorganic Compounds (*Continued*)

Compound	Formula	Molecular weight	Color	Crystal symmetry	Refractive index n_D
Iron (***Continued***)					
Sulfate III	$Fe_2(SO_4)_3$	399.9	Y	R	1.81
Sulfate II, Ammonium	$(NH_4)_2\ Fe(SO_4) \cdot 6H_2O$	392.2	GN	M	1.49
Sulfide II	FeS	87.9	BK	H	
Sulfide III	Fe_2S_3	207.9	BK	H	
Sulfide, di	FeS_2	120.0	Y	C	
Lanthanum					
Bromate	$La(BrO_3)_3 \cdot 9H_2O$	684.8		H	
Bromide	$LaBr_3$	378.6	W	H	
Chloride	$LaCl_3$	245.3	W	H	
Fluoride	LaF_3	195.9	W	H	
Iodide	LaI_3	519.6	G	R	
Molybdate	$La_2(MoO_4)_3$	757.6		T	
Oxide	La_2O_3	325.8	W	R	
Sulfate	$La_2(SO_4)_3$	566.0	W		
Sulfide	La_2S_3	374.0	Y	H	
Lead					
Acetate II	$Pb(C_2H_3O_2)_2$	325.3	W		
Acetate IV	$Pb(C_2H_3O_2)_4$	443.4	CL	M	
Arsenate II	$Pb_3(AsO_4)_2$	899.4	W		
Bromide II	$PbBr_2$	367.0	W	R	
Carbonate II	$PbCO_3$	267.2	CL	R	1.80–2.08
Chloride II	$PbCl_2$	278.1	W	R	2.22
Chloride IV	$PbCl_4$	349.0	Y	LIQ	
Chromate II	$PbCrO_4$	323.2	Y	M	2.33
Fluoride II	PbF_2	245.2	CL	R	
Hydroxide II	$Pb(OH)_2$	241.2	W	H	
Iodate II	$Pb(IO_3)_2$	557.0	W		
Iodide II	PbI_2	461.0	Y	H	
Molybdate II	$PbMoO_4$	367.2	CL	T	2.30
Nitrate II	$Pb(NO_3)_2$	331.2	CL	C	1.782
Oxide II	PbO	223.2	R	T	
Oxide IV	PbO_2	239.2	B	T	
Oxide II–IV	Pb_3O_4	685.6	R	T	
Phosphate, III	$Pb_3(PO_4)_2$	811.6	W	H	1.95
Sulfate II	$PbSO_4$	303.3	W	R	1.85
Sulfide II	PbS	239.3	BK	C	3.911
Tungstate II	$PbWO_4$	455.1	CL	M	
Lithium					
Aluminum Hydride	$LiAlH_4$	37.9	W		
Bromide	$LiBr$	86.9	W	C	1.784
Carbonate	Li_2CO_3	73.9	W	M	1.43; 1.5
Chloride	$LiCl$	42.4	W	C	1.662
Fluoride	LiF	25.9	W	C	1.391
Hydride	LiH	8.0	CL	C	
Hydroxide	$LiOH$	24.0	W	T	1.46
Iodide	LiI	133.9	W	C	1.955
Nitrate	$LiNO_3$	68.9	W	TG	1.435;1.439
Oxide	Li_2O	29.9	W	C	1.644

(*Continued*)

TABLE 1.4 Color, Crystal Symmetry and Refractive Index of Inorganic Compounds (*Continued*)

Compound	Formula	Molecular weight	Color	Crystal symmetry	Refractive index n_D
Lithium (*Continued*)					
Peroxide	Li_2O_2	45.9		H	
Perchlorate	$LiClO_4$	160.4	W	H	
Phosphate	Li_3PO_4	115.8	CL	R	
Sulfate,	Li_2SO_4	109.9	CL	M	1.465
Sulfide	Li_2S	45.9	W	C	
Lutetium					
Bromide	$LuBr_3$	414.7	W	TG	
Chloride	$LuCl_3$	281.3	W	M	
Fluoride	LuF_3	232.0	W	R	
Iodide	LuI_3	555.7	B	H	
Oxide	Lu_2O_3	397.9		C	
Magnesium					
Aluminate	$MgO \cdot Al_2O_3$	142.3	CL	C	1.723
Bromide	$MgBr_2$	184.1	W	H	
Carbonate	$MgCO_3$	84.3	W	TG	1.51; 1.70
Chloride	$MgCl_2$	95.2	W	H	1.59; 1.67
Fluoride	MgF_2	62.3	CL	T	1.38
Hydroxide	$Mg(OH)_2$	58.3	CL	H	1.57
Iodide	MgI_2	278.2	W	H	
Nitrate	$Mg(NO_3)_2 \cdot 6H_2O$	256.4	CL	M	
Oxide	MgO	40.3	CL	C	1.736
Silicide	Mg_2Si	76.7	BE	C	
Silicate, m	$MgSiO_3$	100.4	W	M	1.66
Silicate, o	Mg_2SiO_4	140.7	W	R	1.65
Sulfate	$MgSO_4$	120.4	CL	R	
Sulfide	MgS	56.4	R	C	2.271
Manganese					
Bromide II	$MnBr_2$	214.8	W	H	
Carbonate II	$MnCO_3$	114.9	W	R	1.817
Chloride II	$MnCl_2$	125.9	W	H	
Fluoride II	MnF_2	92.9	R	T	
Iodide II	MnI_2	308.8	W	H	
Oxide II	MnO	70.9	GN	C	2.16
Oxide III	Mn_2O_3	157.9	BK	C	
Oxide IV	MnO_2	86.9	BK	R	
Oxide II–IV	Mn_3O_4	228.8	BK	R	
Potassium Permanganate	$KMnO_4$	158.0	P	R	1.59
Silicide	$MnSi$	83.0		C	
Sulfate II	$MnSO_4$	151.0	R		
Sulfide II	MnS	87.0	GN	C	
Mercury					
Bromide I	Hg_2Br_2	561.1	W	T	
Bromide II	$HgBr_2$	360.4	CL	R	
Chloride I	Hg_2Cl_2	472.1	W	T	1.97; 2.66
Chloride II	$HgCl_2$	271.5	CL	R	1.72; 1.97
Cyanide II	$Hg(CN)_2$	252.7	CL	T	1.645
Fluoride I	Hg_2F_2	439.2	Y	C	

TABLE 1.4 Color, Crystal Symmetry and Refractive Index of Inorganic Compounds (*Continued*)

Compound	Formula	Molecular weight	Color	Crystal symmetry	Refractive index n_D
Mercury (*Continued*)					
Fluoride II	HgF_2	238.6	CL	C	
Iodide I	Hg_2I_2	655.0	Y	T	
Iodide II	HgI_2	454.4	R/Y	T/R	2.45; 2.7
Nitrate I	$Hg_2(NO_3)_2 \cdot 2H_2O$	561.2	CL	M	
Nitrate II	$Hg(NO_3)_2 \cdot {}^1/_2H_2O$	333.6	W		
Oxide I	Hg_2O	417.2	BK		
Oxide II	HgO	216.6	Y/R	R	2.37; 2.6
Sulfate I	Hg_2SO_4	497.3	CL	M	
Sulfate II	$HgSO_4$	296.7	CL	R	
Sulfide III	HgS	232.7	R	H	2.85; 3.2
Molybdenum					
Carbide II	Mo_2C	203.9	W	H	
Carbide IV	MoC	108.0	G	H	
Chloride II	$MoCl_2$	166.9	Y		
Chloride III	$MoCl_3$	202.3	R		
Chloride V	$MoCl_5$	273.2	BK	M	
Fluoride VI	MoF_6	202.9	Cl		
Iodide II	MoI_2	349.8	B		
Molybdic Acid	$H_2MoO_4 \cdot 4H_2O$	180.0	Y	M	
Oxide IV	MoO_2	127.9	G	T	
Oxide VI	MoO_3	143.9	CL	R	
Silicide IV	$MoSi_2$	152.1	G	T	
Sulfide IV	MoS_2	160.1	BK	H	4.7
Neodymium					
Bromide	$NdBr_3$	384.0	V	R	
Chloride	$NdCl_3$	250.6	V	H	
Fluoride	NdF_3	201.2	V	H	
Iodide	NdI_3	524.9	G	R	
Oxide	Nd_2O_3	336.5	BE	H	
Sulfide	Nd_2S_3	384.7	GN		
Neptunium					
Bromide II	$NpBr_3$	476.7	GN	R	
Chloride III	$NpCl_3$	343.4	GN	H	
Chloride IV	$NpCl_4$	378.8	BN	T	
Fluoride III	NpF_3	294.0	P	H	
Fluoride VI	NpF_6	351.0	O	R	
Iodide III	NpI_3	617.7	B	R	
Oxide IV	NpO_2	269.0	GN	C	
Nickel					
Arsenide	NiAs	133.6	W	H	
Bromide II	$NiBr_2$	218.5	Y		
Carbonyl	$Ni(CO)_4$	170.7	CL	LIQ	1.458^{10}
Chloride II	$NiCl_2$	129.6	Y	H	
Fluoride II	NiF_2	96.7	Y	T	
Hydroxide II	$Ni(OH)_2$	92.7	GN		
Iodide II	NiI_2	312.5	BK	H	
Nitrate II	$Ni(NO_3)_2 \cdot 6H_2O$	290.8	GN	M	
Oxide II	NiO	74.7	G	C	2.37

(*Continued*)

TABLE 1.4 Color, Crystal Symmetry and Refractive Index of Inorganic Compounds (*Continued*)

Compound	Formula	Molecular weight	Color	Crystal symmetry	Refractive index n_D
Nickel (***Continued***)					
Phosphide	Ni_2P	148.4	G		
Sulfate II	$NiSO_4$	154.8	Y	C	
Sulfide II	NiS	90.8	BK	TR	
Niobium					
Bromide	$NbBr_5$	492.5	R	R	
Carbide	NbC	104.9	BK	C	
Chloride	$NbCl_5$	270.2	W	M	
Fluoride	NbF_5	187.9	CL	M	
Iodide	NbI_5	727.4	BRASS	M	
Oxide	Nb_2O_5	265.8	W	R	
Nitrogen					
Ammonia	NH_3	17.0	CL	GAS	1.325
Hydrazine	N_2H_4	32.0	CL	LIQ	1.4707
Hydrazoic Acid	NH_3	43.0	CL	LIQ	
Hydroxylamine	NH_2OH	33.0	W	R	$1.440^{23.5}$
Nitric Acid	HNO_3	63.0	CL	LIQ	1.397^{16}
Chloride	NCl_3	120.4	Y	LIQ	
Fluoride	NF_3	71.0	CL	GAS	
Iodide	NI_3	394.7	BK		
Oxide I (nitrous-)	N_2O	44.0	CL	GAS	
Oxide II (nitric-)	NO	30.0	CL	GAS	1.193^{16}
Oxide III (tri-)	N_2O_3	76.0	B	GAS	
Oxide IV (per-)	NO_2	46.0	B	GAS	
Oxide V (penta-)	N_2O_5	108.0	W	R	
Sulfide II	N_4S_4	184.3	O	M	2.046
Nitrosyl Chloride	NOCl	65.5	O	GAS	
Nitrosyl Fluoride	NOF	49.0	CL	GAS	
Nitryl Chloride	NO_2Cl	81.5	CL	GAS	
Osmium					
Chloride IV	$OsCl_4$	332.0	R		
Fluoride V	OsF_5	285.2	G	M	
Fluoride VI	OsF_6	304.2	GN	C	
Fluoride VIII	OsF_8	342.2	Y		
Iodide IV	OsI_4	697.8	BK		
Oxide IV	OsO_2	222.2	BK	T	
Oxide VIII	OsO_4	254.1	CL	M	
Sulfide IV	OsS_2	254.3	BK	C	
Oxygen					
Fluoride	OF_2	54.0	B	GAS	
Ozone	O_3	48.0	CL	GAS	
Palladium					
Bromide II	$PdBr_2$	266.6	B		
Chloride II	$PdCl_2$	177.3	R	C	
Fluoride II	PdF_2	144.4	B	T	
Iodide II	PdI_2	360.2	BK		
Oxide II	PdO	122.4	G	T	
Sulfide II	PdS	138.5	BK	T	

TABLE 1.4 Color, Crystal Symmetry and Refractive Index of Inorganic Compounds (*Continued*)

Compound	Formula	Molecular weight	Color	Crystal symmetry	Refractive index n_D
Phosphorus					
Hypophosphorous Acid	H_3PO_2	66.0	CL		
Phosphoric Acid	H_3PO_4	98.0	CL	R	
Phosphorous Acid	H_3PO_3	82.0	CL		
Bromide III	PBr_3	270.7	CL	LIQ	1.6945^{19}
Bromide V	PBr_5	430.5	Y	R	
Chloride III	PCl_3	137.3	CL	LIQ	
Chloride V	PCl_5	208.3	W	T	
Fluoride III	PF_3	88.0	CL	GAS	
Fluoride V	PF_5	126.0	CL	GAS	
Hydride (Phosphine)	PH_3	34.0	CL	GAS	
Iodide III	PI_3	411.7	R	H	
Oxide III	P_4O_6	219.9	W	M	
Oxide IV	PO_2	63.0	CL	R	
Oxide V	P_2O_5	142.0	W	H	
Oxybromide V	$POBr_3$	286.7	CL		
Oxychloride	$POCl_3$	153.4	CL	LIQ	
Oxyfluoride	POF_3	104.0	CL	GAS	
Sulfide	P_4S_7	348.4	Y		
Sulfide V	P_2S_5	222.3	Y		
Thiobromide V	$PSBr_3$	302.8	Y	C	
Thiochloride V	$PSCl_3$	169.4	CL	LIQ	1.635^{25}
Platinum					
Bromide II	$PtBr_2$	354.9	B	C	
Bromide IV	$PtBr_4$	514.8	B		
Chloride II	$PtCl_2$	260.0	GN	H	
Chloride IV	$PtCl_4$	336.9	B		
Fluoride IV	PtF_4	271.2	R		
Fluoride VI	PtF_6	309.1	R		
Hydroxide II	$Pt(OH)_2$	229.1	BK		
Hydroxide IV	$Pt(OH)_4$	263.1	B		
Iodide II	PtI_2	448.9	BK		
Oxide II	PtO	211.1	G	T	
Oxide IV	PtO_2	227.1	BK		
Sulfate IV	$Pt(SO_4)_2 \cdot 4H_2O$	459.4	Y		
Sulfide II	PtS	227.2	BK	T	
Sulfide III	Pt_2S_3	486.6	G		
Sulfide IV	PtS_2	259.2	G		
Plutonium					
Bromide III	$PuBr_3$	481.7	GN	R	
Carbide IV	PuC	256.0	SL	C	
Chloride III	$PuCl_3$	346.4	GN	H	
Fluoride III	PuF_3	299.0	P	H	
Fluoride IV	PuF_4	318.0	B	M	
Fluoride VI	PuF_6	356.0	B	R	
Iodide III	PuI_3	622.7	GN	R	
Nitride III	PuN	256.0	BK	C	
Oxide IV	PuO_2	274.0	GN	C	2.4

(*Continued*)

TABLE 1.4 Color, Crystal Symmetry and Refractive Index of Inorganic Compounds (*Continued*)

Compound	Formula	Molecular weight	Color	Crystal symmetry	Refractive index n_D
Polonium (*Continued*)					
Bromide IV	$PoBr_4$	529.7	R	C	
Chloride II	$PoCl_2$	281.0	R	R	
Chloride IV	$PoCl_4$	351.9	Y	M	
Oxide IV	PoO_2	242.0	R/Y	T/C	
Potassium					
Bromate	$KBrO_3$	167.0	CL	TR	
Bromide	KBr	119.0	CL	C	1.559
Carbonate	K_2CO_3	138.2	CL	M	1.426; 1.431
Chlorate	$KClO_3$	122.6	CL	M	1.409; 1.423
Chloride	KCl	74.6	CL	C	1.490
Cyanide	KCN	65.1	CL	C	1.410
Dichromate	$K_2Cr_2O_7$	294.2	O	M/TR	1.738 TR
Ferrocyanide	$K_4[Fe(CN)_6] \cdot 3H_2O$	422.4	Y	M/T	1.577
Fluoride	KF	58.1	CL	C	1.35
Hydroxide	KOH	56.1	W	C/R	
Iodate	KIO_3	214.0	CL	M	
Iodide	KI	166.0	W	C	1.677
Nitrate	KNO_3	101.1	CL	R/TR	1.335; 1.?
Oxide	K_2O	94.2	CL	C	
Perchlorate	$KClO_4$	138.6	CL	R	1.47
Periodate	KIO_4	230.0	CL	T	1.63
Permanganate	$KMnO_4$	158.0	P	R	1.59
Peroxide	K_2O_2	110.2	Y	R	
Phosphate, o	K_3PO_4	212.3	CL	TR	
Sulfate	K_2SO_4	174.3	CL	R/H	1.495
Sulfide	K_2S	110.3	B	C	
Superoxide	KO_2	71.1	Y	T	
Thiocyanate	KSCN	97.2	CL	R	
Praseodymium					
Bromide	$PrBr_3$	380.6	GN	H	
Chloride	$PrCl_3$	247.3	GN	H	
Fluoride	PrF_3	197.9	GN	H	
Iodide	PrI_3	521.6	G	R	
Oxide	Pr_2O_3	329.8	Y	H	
Sulfate	$Pr_2(SO_4)_3 \cdot 8H_2O$	714.1	GN	M	1.55
Sulfide	Pr_2S_3	378.0	B		
Protactinium					
Bromide IV	$PaBr_4$	470.9	R	T	
Chloride IV	$PaCl_4$	372.9	GN	T	
Fluoride IV	PaF_4	307.1	B	M	
Iodide III	PaI_3	611.8	BK	R	
Oxide IV	PaO_2	263.1	BK	C	
Radium					
Bromide	$RaBr_2$	385.8	Y	M	
Chloride	$RaCl_2$	296.1	Y	M	
Sulfate	$RaSO_4$	322.1	CL	R	

TABLE 1.4 Color, Crystal Symmetry and Refractive Index of Inorganic Compounds (*Continued*)

Compound	Formula	Molecular weight	Color	Crystal symmetry	Refractive index n_D
Rhenium					
Bromide III	$ReBr_3$	425.9	B		
Chloride III	$ReCl_3$	292.6	R		
Chloride V	$ReCl_5$	363.5	B		
Fluoride IV	ReF_4	262.5	GN	T	
Flouride VI	ReF_6	300.2	Y	LIQ	
Flouride VII	ReF_7	319.2	O	C	
Oxide IV	ReO_2	218.2	BK	M	
Oxide VI	ReO_3	234.2	R	C	
Oxide VII	Re_2O_7	484.4	Y	H	
Oxybromide VII	ReO_3Br	314.1	W		
Oxychloride VII	ReO_3Cl	269.7	CL	LIQ	
Sulfide IV	ReS_2	250.4	BK	H	
Sulfide VII	Re_2S_7	596.9	BK	T	
Rhodium					
Chloride III	$RhCl_3$	209.3	R		
Fluoride III	RhF_3	159.9	R	R	
Hydroxide III	$Rh(OH)_3$	155.9	Y		
Oxide III	Rh_2O_3	253.8	G		
Oxide IV	RhO_2	134.9	B		
Sulfide III	Rh_2S_3	302.0	BK		
Rubidium					
Bromate	$RbBrO_3$	213.4	CL	C	
Bromide	$RbBr$	165.4	CL	C	1.5530
Carbonate	Rb_2CO_3	231.0	CL		
Chloride	$RbCl$	120.9	CL	C	1.493
Fluoride	RbF	104.5	CL	C	1.398
Hydroxide	$RbOH$	102.5	W	R	
Iodide	RbI	212.4	CL	C	1.6474
Nitrate	$RbNO_3$	147.5	CL		1.52
Oxide	Rb_2O	187.0	Y	C	
Perchlorate	$RbClO_4$	189.4		C/R	1.4701
Peroxide	Rb_2O_2	202.9	Y	C	
Sulfate	Rb_2SO_4	267.0	CL	R	1.513
Sulfide	Rb_2S	203.0	Y		
Superoxide	RbO_2	117.5	Y	T	
Ruthenium					
Chloride III	$RuCl_3$	207.4	R	TR/H	
Fluoride V	RuF_5	196.1	GN	M	
Oxide IV	RuO_2	133.1	BE	T	
Oxide VIII	RuO_4	165.1	Y	R	
Sulfide IV	RuS_2	165.2	BK	C	
Samarium					
Bromate III	$Sm(BrO_3)_3 \cdot 9H_2O$	696.2	Y	H	
Bromide II	$SmBr_2$	310.2	B		
Bromide III	$SmBr_3$	390.1	Y	R	
Chloride II	$SmCl_2$	221.3	B	R	

(*Continued*)

TABLE 1.4 Color, Crystal Symmetry and Refractive Index of Inorganic Compounds (*Continued*)

Compound	Formula	Molecular weight	Color	Crystal symmetry	Refractive index n_D
Samarium (*Continued*)					
Chloride III	$SmCl_3$	256.7	Y	H	
Fluoride II	SmF_2	188.4	Y	C	
Fluoride III	SmF_3	207.4	W	R	
Iodide II	SmI_2	404.2	Y	M	
Iodide III	SmI_3	531.1	Y	H	
Nitrate III	$Sm(NO_3)_3 \cdot 6H_2O$	444.5	Y	TR	
Oxide III	Sm_2O_3	348.7	Y	M	
Sulfate III	$Sm_2(SO_4)_3 \cdot 8H_2O$	733.0	Y	M	1.55
Sulfide III	Sm_2S_3	396.9	Y	C	
Scandium					
Bromide	$ScBr_3$	284.7	W		
Chloride	$ScCl_3$	151.3	CL	RH	
Fluoride	ScF_3	102.0		RH	
Iodide	ScI_3	425.7	W	H	
Nitrate	$Sc(NO_3)_3$	231.0	CL		
Oxide	Sc_2O_3	137.9	W	C	
Sulfate	$Sc_2(SO_4)_3$	378.1	CL		
Selenium					
Bromide I	Se_2Br_2	317.7	R	LIQ	
Bromide IV	$SeBr_4$	398.6	B		
Chloride I	Se_2Cl_2	228.8	B	LIQ	
Chloride IV	$SeCl_4$	220.8	CL	C	1.807
Fluoride IV	SeF_4	154.9	CL	LIQ	
Fluoride VI	SeF_6	192.9	CL	GAS	1.895
Hydride II	H_2Se	81.0	CL	GAS	
Oxide IV	SeO_2	111.0	CL	T	>1.76
Oxide VI	SeO_3	127.0	W	T	
Oxybromide	$SeOBr_2$	254.8	O	LIQ	
Oxychloride	$SeOCl_2$	165.9	Y	LIQ	1.651
Oxyfluoride	$SeOF_2$	133.0	CL	LIQ	
Selenic Acid	H_2SeO_4	145.0	W	R	
Selenous Acid	H_2SeO_3	129.0	CL	H	
Silicon					
Bromide	$SiBr_4$	347.7	CL	LIQ	1.5797[1]
Carbide	SiC	40.1	BK	C/H	2.67
Chloride	$SiCl_4$	169.9	CL	LIQ	
Fluoride	SiF_4	104.1	CL	GAS	
Hydride (silane)	SiH_4	32.1	CL	GAS	
Hydride (disilane)	Si_2H_6	62.2	CL	GAS	
Hydride (trisilane)	Si_3H_8	92.3	CL	LIQ	
Iodide	SiI_4	535.7	CL	C	
Nitride	Si_3N_4	140.3	G	H	
Oxide II	SiO	44.1	W	C	
Oxide IV (amorph)	SiO_2	60.1	CL		1.4588
Oxychloride	Si_2OCl_6	284.9	CL	LIQ	
Sulfide	SiS_2	92.2	W	R	

TABLE 1.4 Color, Crystal Symmetry and Refractive Index of Inorganic Compounds (*Continued*)

Compound	Formula	Molecular weight	Color	Crystal symmetry	Refractive index n_D
Silver					
Bromate	$AgBrO_3$	235.8	CL	T	1.874,1.904
Bromide	AgBr	187.8	Y	C	2.253
Carbonate	Ag_2CO_3	257.8	Y		
Chlorate	$AgClO_3$	191.3	W	T	
Chloride	AgCl	143.3	W	C	2.071
Cyanide	AgCN	133.9	W	H	1.685,1.9
Fluoride	AgF	126.9	Y	C	
Iodate	$AgIO_3$	282.8	CL	R	
Iodide	AgI	234.8	Y	H/C	2.21
Nitrate	$AgNO_3$	169.9	CL	R	1.74
Nitrite	$AgNO_2$	153.9	Y	R	
Oxide	Ag_2O	231.8	B	C	
Perchlorate	$AgClO_4$	207.4	W	C	
Phosphate, o	Ag_3PO_4	418.6	Y	C	
Sulfate	Ag_3SO_4	311.8	W	R	
Sulfide	Ag_2S	247.8	BK	C/R	
Telluride	Ag_2Te	343.4	G	M	
Thiocyanate	AgSCN	166.0	CI		
Sodium					
Bicarbonate	$NaHCO_3$	84.0	W	M	1.500
Bromate	$NaBrO_3$	150.9	CL	C	1.594
Bromide	NaBr	102.9	Cl	C	1.6412
Carbonate	Na_2CO_3	106.0	W		1.535
Chlorate	$NaClO_3$	106.4	CL	C	1.513
Chloride	NaCl	58.4	CL	C	1.544
Cyanide	NaCN	49.0	CL	C	1.452
Fluoride	NaF	42.0	CL	C	1.336
Hydride	NaH	24.0	SL	C	1.470
Hydroxide	NaOH	40.0	W	R/C	1.358
Iodate	$NaIO_3$	197.9	W	R	
Iodide	NaI	149.9	CL	C	1.775
Nitrate	$NaNO_3$	85.0	CL	TR	1.34;1
Nitrite	$NaNO_2$	69.0	Y	R	
Oxide	Na_2O	62.0	G	C	
Perchlorate	$NaClO_4$	122.4	W	C/R	1.46
Periodate	$NaIO_4$	213.9	CL	T	
Peroxide	Na_2O_2	78.0	Y	H	
Phosphate, o	Na_3PO_4	163.9	W		
Silicate, m	Na_2SiO_3	122.1	CL	M	1.52
Sulfate	Na_2SO_4	142.1	CL	R	1.48
Sulfide	Na_2S	78.1	W	C	
Sulfite	Na_2SO_3	126.1	W	H	1.5
Thiosulfate	$Na_2S_2O_3$	158.1	CL	M	
Strontium					
Bromide	$SrBr_2$	247.5	W	R	1.575
Carbonate	$SrCO_3$	147.6	CL	R	1.521
Chloride	$SrCl_2$	158.5	CL	C	1.650
Fluoride	SrF_2	125.6	CL	C	1.442
Hydride	SrH_2	89.6	W	R	

(*Continued*)

TABLE 1.4 Color, Crystal Symmetry and Refractive Index of Inorganic Compounds (*Continued*)

Compound	Formula	Molecular weight	Color	Crystal symmetry	Refractive index n_D
Strontium (*Continued*)					
Hydroxide	$Sr(OH)_2$	121.7	W		
Iodate	$Sr(IO_3)_2$	437.4		TR	
Iodide	SrI_2	341.4	CL	—	
Nitrate	$Sr(NO_3)_2$	211.7	CL	C	1.567
Oxide	SrO	103.6	W	C	1.870
Peroxide	SrO_2	119.6	CL	T	
Sulfate	$SrSO_4$	183.7	CL	R	1.62
Sulfide	SrS	119.7	CL	C	2.107
Sulfur					
Bromide I	S_2Br_2	224.0	R	LIQ	1.736
Chloride I	S_2Cl_2	135.0	Y	LIQ	1.666^{14}
Chloride II	SCl_2	103.0	R	LIQ	1.557
Chloride IV	SCl_4	173.9	R	LIQ	
Fluoride I	S_2F_2	102.1	CL	GAS	
Fluoride VI	SF_6	146.0	CL	GAS	
Hydride	H_2S	34.1	CL	GAS	1.374
Oxide IV	SO_2	64.1	CL	GAS	
Oxide VI	SO_3	80.1	CL	LIQ	
Pyrosulfuric Acid	$H_2S_2O_7$	178.1	CL	LIQ	
Sulfuric Acid	H_2SO_4	98.1	CL	LIQ	1.429^{23}
Sulfuryl Chloride	SO_2Cl_2	135.0	CL	LIQ	1.444^{12}
Thionyl Bromide	$SOBr_2$	207.9	Y	LIQ	
Thionyl Chloride	$SOCl_2$	119.0	CL	LIQ	1.527^{10}
Tantalum					
Bromide	$TaBr_5$	580.5	Y	R	
Carbide	TaC	193.0	BK	C	
Chloride	$TaCl_5$	358.2	Y	M	
Fluoride	TaF_5	275.9	CL	M	
Iodide	TaI_5	815.4	BK	R	
Nitride	TaN	194.9	BK	H	
Oxide	Ta_2O_5	441.9	CL	R	
Sulfide	Ta_2S_4	490.1	BK	H	
Tellurium					
Bromide II	$TeBr_2$	287.4	GN		
Bromide V	$TeBr_4$	447.3	Y		
Chloride II	$TeCl_2$	198.5	GN		
Chloride IV	$TeCl_4$	269.4	W	M	
Fluoride VI	TeF_6	241.6	CL	GAS	
Hydride	H_2Te	129.6	CL	GAS	
Iodide IV	TeI_4	635.2	BK	R	
Oxide IV	TeO_2	159.6	W	T/R	2.00–2.35
Oxide VI	TeO_3	175.6	Y		
Telluric Acid, o	H_2TeO_6	229.7	W	C	
Terbium					
Bromide	$TbBr_3$	398.6	W		
Chloride	$TbCl_3$	265.3	W		

TABLE 1.4 Color, Crystal Symmetry and Refractive Index of Inorganic Compounds (*Continued*)

Compound	Formula	Molecular weight	Color	Crystal symmetry	Refractive index n_D
Terbium (*Continued*)					
Fluoride	TBF_3	215.9	W	R	
Iodide	TbI_3	539.6		H	
Nitrate	$Tb(NO_3)_3 \cdot 6H_2O$	453.0	CL	M	
Oxide	Tb_2O_3	365.8	W	C	
Thalliun					
Bromide I	$TlBr$	284.3	W	C	2.4–2.8
Carbonate I	Tl_2CO_3	468.8	CL	M	
Chloride I	$TlCl$	239.8	W	C	2.247
Chloride III	$TlCl_3$	310.8	W	H	
Fluoride	TlF	223.4	CL	R	
Hydroxide I	$TlOH$	221.4	Y	R	
Iodide I	TlI	331.3	Y/R	R/C	2.78
Nitrate I	$TlNO_3$	266.4	W	C/TR	
Oxide I	Tl_2O	424.7	BK	RH	
Oxide III	Tl_2O_3	456.7	CL	C	
Sulfate I	Tl_2SO_4	504.8	CL	R	1.87
Sulfide I	Tl_2S	440.8	BK	T	
Thorium					
Bromide	$ThBr_4$	551.7	W	T	
Carbide	ThC_2	256.1	Y	T	
Chloride	$ThCl_4$	373.9	W	T	
Fluoride	ThF_4	308.0	W	M	
Iodide	ThI_4	739.7	Y	M	
Oxide	ThO_2	264.0	W	C	
Sulfate	$Th(SO_4)_2$	424.2	W	M	
Sulfide	ThS_2	296.2	BK	R	
Thulium					
Bromide	$TmBr_3$	408.7	W	H	
Chloride	$TmCl_3$	275.2	Y	M	
Fluoride	TmF_3	225.9	W	R	
Iodide	TmI_3	549.6	Y	H	
Oxide	Tm_2O_3	385.9	Y	C	
Tin					
Bromide II	$SnBr_2$	278.5	Y	R	
Bromide IV	$SnBr_4$	438.4	CL	R	
Chloride II	$SnCl_2$	189.6	W	R	
Chloride IV	$SnCl_4$	260.5	CL	LIQ	1.512
Fluoride II	SnF_2	156.7	W	M	
Fluoride IV	SnF_4	194.7	W	M	
Hydride	SnH_4	122.7		GAS	
Iodide II	SnI_2	372.5	R	R	
Iodide IV	SnI_4	626.3	R	C	2.106
Oxide II	SnO	143.7	BK	T	
Oxide IV	SnO_2	150.7	W	T	1.996
Sulfide II	SnS	150.8	BK	R	
Sulfide IV	SnS_2	182.8	Y	H	

(*Continued*)

TABLE 1.4 Color, Crystal Symmetry and Refractive Index of Inorganic Compounds (*Continued*)

Compound	Formula	Molecular weight	Color	Crystal symmetry	Refractive index n_D
Titanium					
Bromide IV	$TiBr_4$	367.6	O	M	
Carbide IV	TiC	59.9	G	C	
Chloride II	$TiCl_2$	118.8	BK	H	
Chloride III	$TiCl_3$	154.3	V	H	
Chloride IV	$TiCl_4$	189.7	Y	LIQ	1.61
Fluoride IV	TiF_4	123.9	W		
Iodide IV	TiI_4	555.5	B	C	
Nitride	TiN	61.9	Y	C	
Oxide II	TiO	63.9	BK	C	
Oxide IV	TiO_2	79.9	BK	T	2.55
Sulfide IV	TiS_2	112.0	Y	H	
Tungsten					
Bromide V	WBr_5	583.4	B		
Carbide II	W_2C	379.7	G	H	
Carbide IV	WC	195.9	G	C	
Chloride V	WCl_5	361.1	GN		
Chloride VI	WCl_6	396.6	BE	C	
Fluoride VI	WF_6	297.8	CL	GAS	
Oxide IV	WO_2	215.9	B	T	
Oxide VI	WO_3	231.9	Y	M	
Sulfide IV	WS_2	248.0	BK	H	
Tungstic Acid	H_2WO_4	250.0	Y	R	2.24
Uranium					
Bromide III	UBr_3	477.8	R	H	
Bromide IV	UBr_4	557.7	B	M	
Carbide	UC	250.0	BK	C	
Carbide	UC_2	262.0	BK	T	
Chloride III	UCl_3	344.4	R	H	
Chloride IV	UCl_4	379.9	GN	T	
Fluoride IV	UF_4	314.1	GN	M	
Fluoride VI	UF_6	352.1	Y	R	1.38
Nitride	UN	252.0	B	C	
Oxide IV	UO_2	270.1	BK	C	
Oxide VI	UO_3	286.1	R	H	
Oxide IV–VI	U_3O_8	842.2	BK	R	
Uranyl Acetate	$UO_2(C_2H_3O_2)_2 \cdot 6H_2O$	422.1	Y	R	
Uranyl Nitrate	$UO_2(NO_3)_2 \cdot 6H_2O$	502.1	Y	R	1.49
Vanadium					
Carbide IV	VC	62.9	BK	C	
Chloride IV	VCl_4	192.7	R	LIQ	1
Fluoride III	VF_3	107.9	GN	R	
Fluoride V	VF_5	145.9	CL	R	
Iodide II	VI_2	304.7	V	H	
Oxide III	V_2O_3	149.9	BK	RH	
Oxide IV	VO_2	82.9	BE	T	
Oxide V	V_2O_5	181.9	R	R	
Oxychloride V	$VOCl_3$	173.3	Y	LIQ	
Sulfide II	VS	83.0	BK	H	

TABLE 1.4 Color, Crystal Symmetry and Refractive Index of Inorganic Compounds (*Continued*)

Compound	Formula	Molecular weight	Color	Crystal symmetry	Refractive index n_D
Xenon					
Fluoride II	XeF_2	169.3	CL	T	
Fluoride IV	XeF_4	207.3	CL	M	
Fluoride VI	XeF_6	245.3	CL	M	
Oxide VI	XeO_3	179.3	CL	R	1.79
Yttebium					
Bromide III	$YbBr_3$	412.8	CL		
Chloride II	$YbCl_2$	244.0	GN	R	
Chloride III	$YbCl_3$	279.3	W	M	
Fluoride III	YbF_3	230.0	W	R	
Iodide II	YbI_2	426.9	BK	H	
Iodide III	YbI_3	553.8	Y	H	
Oxide III	Yb_2O_3	394.1	CL	C	
Sulfate III	$Yb_2(SO_4)_3$	634.3	CL		
Yttrium					
Bromide	YBr_3	328.6	W		
Chloride	YCl_3	195.3	W	M	
Fluoride	YF_3	145.9	W		
Iodide	YI_3	469.6	W	H	
Oxide	Y_2O_3	225.8	W	C	
Sulfate	$Y_2(SO_4)_3$	466.0	W		
Zinc					
Acetate	$Zn(C_2H_3O_2)_2$	183.5	CL	M	
Bromide	$ZnBr_2$	225.2	CL	R	1.5452
Calbonate	$ZnCO_3$	125.4	CL	TR	1.168
Chloride	$ZnCl_2$	136.3	W	H	1.687
Fluoride	ZnF_2	103.4	CL	M	
Hydroxide	$Zn(OH)_2$	99.4	CL	R	
Iodide	ZnI_2	319.2	CL	C	
Nitrate	$Zn(NO_3)_2 \cdot 6H_2O$	297.5	CL	T	
Oxide	ZnO	81.4	W	H	2.01
Sulfate	$ZnSO_4$	161.4	CL	R	1.669
Sulfide	ZnS	97.5	CL	C/H	2.36
Zirconium					
Bromide	$ZrBr_4$	410.9	W		
Carbide	ZrC	103.2	G	C	
Chloride	$ZrCI_4$	233.1	W	C	
Fluoride	ZrF_4	167.2	W	M	1.59
Iodide	ZrI_4	598.8	W		
Nitride	ZrN	105.2	B		
Oxide	ZrO_2	123.2	W	M	

TABLE 1.5 Refractive Index of Minerals

Mineral name	Refractive index	Mineral name	Refractive index
Actinolite	1.618–1.641	Crocoite	2.31–2.66
Adularia moonstone	1.525	Cuprite	2.85
Adventurine feldspar	1.532–1.542		
Adventurine quartz	1.544–1.533	Danburite	1.633
Agalmatoite	1.55	Demantoid garnet	1.88
Agate	1.544–1.553	Diamond	2.417–2.419
Albite feldspar	1.525–1.536	Diopsite	1.68–1.71
Albite moonstone	1.535	Dolomite	1.503–1.682
Alexandrite	1.745–1.759	Dumortierite	1.686–1.723
Almandine garnet	1.76–1.83		
Almandite garnet	1.79	Ekanite	1.60
Amazonite feldspar	1.525	Elaeolite	1.532–1.549
Amber	1.540	Emerald	1.576–1.582
Amblygonite	1.611–1.637	Enstatite	1.663–1.673
Amethyst	1.544–1.553	Epidote	1.733–1.768
Anatase	2.49–2.55	Euclase	1.652–1.672
Andalusite	1.634–1.643		
Andradite garnet	1.82–1.89	Fibrolite	1.659–1.680
Anhydrite	1.571–1.614	Fluorite	1.434
Apatite	1.632–1.648		
Apophyllite	1.536	Gaylussite	1.517
Aquamarine	1.577–1.583	Glass	1.44–1.90
Aragonite	1.530–1.685	Grossular garnet	1.738–1.745
Augelite	1.574–1.588		
Axinite	1.675–1.685	Hambergite	1.559–1.631
Azurite	1.73–1.838	Hauynite	1.502
		Hematite	2.94–3.22
Barite	1.636–1.648	Hemimorphite	1.614–1.636
Barytocalcite	1.684	Hessonite garnet	1.745
Benitoite	1.757–1.8	Hiddenite	1.655–1.68
Beryl	1.577–1.60	Howlite	1.586–1.609
Beryllonite	1.553–1.562	Hypersthene	1.67–1.73
Brazilianite	1.603–1.623		
Brownite	1.567–1.576	Idocrase	1.713–1.72
		Iolite	1.548
Calcite	1.486–1.658	Ivory	1.54
Cancrinite	1.491–1.524		
Cassiterite	1.997–2.093	Jadeite	1.66–1.68
Celestite	1.622–1.631	Jasper	1.54
Cerussite	1.804–2.078	Jet	1.66
Ceylanite	1.77–1.80		
Chalcedony	1.53–1.539	Kornerupine	1.665–1.682
Chalybite	1.63–1.87	Kunzite	1.655–1.68
Chromite	2.1	Kyanite	1.715–1.732
Chrysoberyl	1.745		
Chrysocolla	1.50	Labradorite feldspar	1.565
Chrysoprase	1.534	Lapis gem	1.50
Citrine	1.55	Lazulite	1.615–1.645
Clinozoisite	1.724–1.734	Leucite	1.5085
Colemanite	1.586–1.614		
Coral	1.486–1.658	Magnesite	1.515–1.717
Cordierite	1.541	Malachite	1.655–1.909
Corundum	1.766–1.774	Meerschaum	1.53.... none

TABLE 1.5 Refractive Index of Minerals (*Continued*)

Mineral name	Refractive index	Mineral name	Refractive index
Microcline feldspar	1.525	Serpentine	1.555
Moldavite	1.50	Shell	1.53–1.686
Moss agate	1.54–1.55	Sillimanite	1.658–1.678
		Sinhalite	1.699–1.707
Natrolite	1.48–1.493	Smaragdite	1.608–1.63
Nephrite	1.60–1.63	Smithsonite	1.621–1.849
Nephrite jade	1.600–1.627	Sodalite	1.483
		Spessartite garnet	1.81
Obsidian	1.48–1.51	Spinel	1.712–1.736
Oligoclase feldspar	1.539–1.547	Sphalerite	2.368–2.371
Olivine	1.672	Sphene	1.885–2.05
Onyx	1.486–1.658	Spodumene	1.65–1.68
Opal	1.45	Staurolite	1.739–1.762
Orthoclase feldspar	1.525	Steatite	1.539–1.589
		Stichtite	1.52–1.55
Painite	1.787–1.816	Sulfur	1.96–2.248
Pearl	1.52–1.69		
Periclase	1.74	Taaffeite	1.72
Peridot	1.654–1.69	Tantalite	2.24–2.41
Peristerite	1.525–1.536	Tanzanite	1.691–1.70
Petalite	1.502–1.52	Thomsonite	1.531
Phenakite	1.65–1.67	Tiger eye	1.544–1.553
Phosgenite	2.117–2.145	Topaz (white)	1.638
Prase	1.54–1.533	Topaz (blue)	1.611
Prasiolite	1.54–1.553	Topaz (pink, yellow)	1.621
Prehnite	1.61–1.64	Tourmaline	1.616–1.652
Proustite	2.79–3.088	Tremolite	1.60–1.62
Purpurite	1.84–1.92	Tugtupite	1.496–1.50
Pyrite	1.81	Turquoise	1.61–1.65
Pyrope	1.74	Turquoise gem	1.61
Quartz	1.55	Ulexite	1.49–1.52
		Uvarovite	1.87
Rhodizite	1.69		
Rhodochrisite	1.60–1.82	Variscite	1.55–1.59
Rhodolite garnet	1.76	Vivianite	1.580–1.627
Rhodonite	1.73–1.74		
Rock crystal	1.544–1.553	Wardite	1.59–1.599
Ruby	1.76–1.77	Willemite	1.69–1.72
Rutile	2.61–2.90	Witherite	1.532–1.68
		Wulfenite	2.300–2.40
Sanidine	1.522		
Sapphire	1.76–1.77	Zincite	2.01–1.03
Scapolite	1.54–1.56	Zircon	1.801–2.01
Scapolite (yellow)	1.555	Zirconia (cubic)	2.17
Scheelite	1.92–1.934	Zoisite	1.695

TABLE 1.6 Properties of Molten Salts

Material	Melting point Tm (°K)	Boiling point (°K)	Density at melting point ($g \cdot cm^{-3}$)	Critical temperature (°K)	Volume change on melting $\Delta V_f/\Delta V_s$ 100	Surface tension at melting point ($dynes \cdot cm^{-1}$)	Viscosity at melting point (centipoise)	Sound velocity at melting point ($m \cdot cm^{-1}$)	Cryoscopic constant (°K/mole · kg)
LiF	1121	1954	1.83	4140	29.4	252		2546	2.77
NaF	1268	1977	1.96	4270	27.4	185		2080	16.6
KF	1131	1775	1.91	3460	17.2	141		1827	21.8
RbF	1048	1681	—	3280	—	167			38.4
LiCl	883	1655	1.60	3080	26.2	137	1.73	2038	13.7
NaCl	1073	1738	1.55	3400	25.0	116	1.43	1743	20.0
KCl	1043	1680	1.50	3200	17.3	99	1.38	1595	25.4
LiBr	823	1583	2.53	3020	24.3	—		1470	27.6
NaBr	1020	1665	2.36	3200	22.4	100		1325	34.0
KBr	1007	1656	2.133	3170	16.6	90		1256	55.9
$NaNO_2$	544	*d* > 593	1.81		—	120			
KNO_2	692	d623	—		—	109			
$LiNO_3$	527	—	1.78		21.4	116	5.46	1853	5.93
$NaNO_3$	583	d653	1.90		10.7	116	2.89	1808	15.4
KNO_3	610	d > 613	1.87		3.32	110	2.93	1754	30.8
$RbNO_3$	589	—	2.48		−0.23	109			89.0
$AgNO_3$	483	d > 485	3.97			148	4.25	1607	25.9
$TlNO_3$	480	706	4.90			94			58
Li_2SO_4	1132	—	2.00			225			142
Na_2SO_4	1157	—	2.07			192			66.3
K_2SO_4	1347	—	1.88			144			68.7
$ZnCl_2$	548	1005	2.39			53		1002	
$HgCl_2$	550	577	4.37			—			39.3
$PbCl_2$	771	1227	3.77			137	4.25	4952	
Na_2WO_4	969	—	3.85			202			
Na_3AlF_6	1273	—	1.84			135			
KCNS	450	—	1.60			101			12.7

Notes: (a) 5893 Å; (b) 5890 Å.

Material	Heat capacity, Cp (cal./°K · mole)	Heat of fusion at melting point (kcal · mole^{-1})	Entropy of fusion at melting point (entropy units)	Equivalent conductance at 1.1 Tm [(ohm)$^{-1}$cm^2 (equiv)$^{-1}$]	Decom-position potential of melt (volts)	Measurement temperature for decomposition potential (°K)	Molar refractivity at 5461 Å (cm^3 · mole^{-1})	Refractive index at 5461 Å	Measurement temperature for refractive index, (°K)
LiF	15.50	6.47	5.77	151	2.20	1273	2.89	1.32	1223
NaF	16.40	8.03	6.33	120	2.76	1273	3.41	1.25	1273
KF	16.00	6.75	5.97	148	2.54	1273	5.43	1.28	1173
RbF		6.15	5.76						
LiCl	15.0	4.76	5.39	178.5	3.30	1073	8.32	1.501	883
NaCl	16.0	6.69	6.23	152.3	3.25	1073	9.65	1.320	1173
KCl	16.0	6.34	6.08	122.4	3.37	1073	11.75	1.329	1173
LiBr		4.22	5.13	181	2.95	1073	11.81	1.60	843
NaBr		6.24	6.12	149	2.83	1073	13.19	1.486	1173
KBr		6.10	6.06	108	2.97	1073	15.40	1.436	1173
$NaNO_2$				58			9.63[a]	1.416[a]	573
KNO_2				~87			11.67	1.356[a]	873
$LiNO_3$	26.6	5.961	11.66	44			10.74	1.467	573
$NaNO_3$	37.0	3.696	6.1	58			11.54	1.431	573
KNO_3	29.5	2.413	4.58	46			13.57	1.426	573
$RbNO_3$		1.105	1.91	35			15.31[b]	1.431[b]	573
$AgNO_3$	30.6	2.886		38			16.20[a]	1.660[a]	573
$TlNO_3$		2.264		27			21.38	1.688[b]	573
Li_2SO_4		1.975		123			14.87	1.452	1173
Na_2SO_4		5.67		90			16.53	1.395	1173
K_2SO_4	47.8	9.06		157			20.93	1.388	1173
$ZnCl_2$	24.1	2.45		~0.08	1.43	973	18.2	1.588	593
$HgCl_2$	25.0	4.15		0.00096	0.86	973	22.9	1.661	563
$PbCl_2$		4.40		52.3	1.12	973	26.1	2.024	873
Na_2WO_4				46			24.58	1.542	1173
Na_3AlF_6		27.64					17.2	1.290	1273
KCNS		3.07		17.3			19.65	1.537	573

TABLE 1.7 Triple Points of Various Materials

Substance	Triplet point, oK	Pressure, mmHg
Ammonia	195.46	45.58
Argon	83.78	516
Boron tribromide	226.67	
Bromine	280.4	44.1
Carbon dioxide	216.65	
Cyclopropane	145.59	
Deuterium oxide	276.97	
1-Hexene	133.39	
Hydrogen, normal	13.95	54
Hydrogen, para	13.81	
Hydrogen bromide	186.1	~232
Hydrogen chloride	158.8	
Iodine heptafluoride	279.6	
Krypton	115.95	548
Methane	90.67	87.60
Methane-d_1	90.40	84.52
Methane-d_2	90.14	81.80
Methane-d_3	89.94	80.12
Methane-d_4	89.79	79.13
Molybdenum oxide tetrafluoride	370.3	
Molybdenum pentafluoride	340	
Neon	24.55	324
Neptunium hexafluoride	328.25	758.0
Niobium pentabromide	540.6	
Niobium pentachloride	476.5	
Nitrogen	63.15	94
1-Octene	171.45	
Oxygen	54.34	
Phosphorus, white	863	32 760
Plutonium hexafluoride	324.74	533.0
Propene	103.95	
Radon	202	~500
Rhenium dioxide trifluoride	363	
Rhenium heptafluoride	321.4	
Rhenium oxide pentafluoride	313.9	
Rhenium pentafluoride	321	
Succinonitrile (NIST standard)	331.23	
Sulfur dioxide	197.68	1.256
Tantalum pentabromide	553	
Tantalum pentachloride	489.0	
Tungsten oxide tetrafluoride	377.8	
Uranium hexafluoride	337.20	1 139.6
Water	273.16	
Xenon	161.37	612

TABLE 1.8 Density of Mercury and Water

The density of mercury and pure air-free water under a pressure of 101, 325 Pa(1 atm) is given in units of grams per cubic centimeter ($g \cdot cm^{-3}$). For mercury, the values are based on the density at 20°C being 13.545 884 $g \cdot cm^{-3}$. Water attains its maximum density of 0.999 973 $g \cdot cm^{-3}$ at 3.98°C. For water, the temperature (t_m, °C) of maximum density at different pressures (p) in atmospheres is given by

$$t_m = 3.98 - 0.0225(p - 1)$$

Density of water	Temp., °C	Density of mercury	Density of water	Temp., °C	Density of mercury
	−20	13.644 59	0.987 12	52	13.467 68
	−18	13.639 62	0.986 18	54	13.462 82
	−16	13.634 66	0.985 21	56	13.457 96
	−14	13.629 70	0.984 22	58	13.453 09
	−12	13.624 75	0.983 20	60	13.448 23
	−10	13.619 79	0.982 16	62	13.443 37
	−8	13.614 85	0.981 09	64	13.438 52
	−6	13.609 90	0.980 01	66	13.433 67
	−4	13.604 96	0.978 90	68	13.428 82
	−2	13.600 02	0.977 77	70	13.423 97
0.999 84	0	13.595 08	0.976 61	72	13.419 13
0.999 94	2	13.590 15	0.975 44	74	13.414 28
0.999 97	4	13.585 22	0.974 24	76	13.409 43
0.999 94	6	13.580 29	0.973 03	78	13.404 60
0.999 85	8	13.575 36	0.971 79	80	13.399 77
0.999 70	10	13.570 44	0.970 53	82	13.394 92
0.999 50	12	13.565 52	0.969 26	84	13.390 09
0.999 24	14	13.560 60	0.967 96	86	13.385 26
0.998 94	16	13.555 70	0.966 65	88	13.380 42
0.998 60	18	13.550 79	0.965 31	90	13.375 60
0.998 20	20	13.545 88	0.963 96	92	13.370 77
0.997 77	22	13.540 97	0.962 59	94	13.365 94
0.997 30	24	13.536 06	0.961 20	96	13.361 12
0.996 78	26	13.531 17	0.959 79	98	13.356 30
0.996 23	28	13.526 26	0.958 36	100	13.351 48
0.995 65	30	13.521 37		120	13.303 4
0.995 03	32	13.516 47		140	13.255 4
0.994 37	34	13.511 58		160	13.207 6
0.993 69	36	13.506 70		180	13.159 8
0.992 97	38	13.501 82		200	13.112 0
0.992 22	40	13.496 93		220	13.064 5
0.991 44	42	13.492 07		240	13.016 9
0.990 63	44	13.487 18		260	12.969 2
0.989 79	46	13.482 29		280	12.921 5
0.988 93	48	13.477 42		300	12.873 7
0.988 04	50	13.472 56			

TABLE 1.9 Specific Gravity of Air at Various Temperatures

The table below gives the weight in grams · 10^4 of 1 mL of air at 760 mm of mercury pressure and at the temperature indicated. Density in grams per milliliter is the same as the specific gravity referred to water at 4°C as unity. To convert to density referred to air at 70°F as unity, divide the values below by 12.00.

t°C.	Sp.Gr. × 10^4	t°C.	Sp.Gr. × 10^4	t°C.	Sp.Gr. × 10^4	t°C.	Sp.Gr. × 10^4
−25	14.240	15	12.255	60	10.596	140	8.541
−24	14.182	16	12.213	62	10.532	142	8.500
−23	14.125	17	12.170	64	10.470	144	8.459
−22	14.069	18	12.129	66	10.408	146	8.419
−21	14.013	19	12.087	68	10.347	148	8.379
−20	13.957	20	12.046	70	10.286	150	8.339
−19	13.902	21	12.004	72	10.227	155	8.242
−18	13.847	22	11.964	74	10.168	160	8.147
−17	13.793	23	11.923	76	10.109	165	8.054
−16	13.739	24	11.883	78	10.052	170	7.963
−15	13.685	25	11.843	80	9.995	175	7.874
−14	13.632	26	11.803	82	9.938	180	7.787
−13	13.580	27	11.764	84	9.882	185	7.702
−12	13.527	28	11.725	86	9.828	190	7.619
−11	13.476	29	11.686	88	9.773	195	7.537
−10	13.424	30	11.647	90	9.719	200	7.457
−9	13.373	31	11.609	92	9.666	205	7.379
−8	13.322	32	11.570	94	9.613	210	7.303
−7	13.272	33	11.533	96	9.561	215	7.228
−6	13.222	34	11.495	98	9.509	220	7.155
−5	13.173	35	11.458	100	9.458	230	7.013
−4	13.124	36	11.420	102	9.408	240	6.881
−3	13.075	37	11.383	104	9.358	250	6.753
−2	13.026	38	11.347	106	9.308	260	6.624
−1	12.978	39	11.310	108	9.259	270	6.504
0	12.931	40	11.274	110	9.211	280	6.389
+1	12.883	41	11.238	112	9.163	290	6.277
2	12.836	42	11.202	114	9.116	300	6.166
3	12.790	43	11.167	116	9.069	310	6.062
4	12.743	44	11.132	118	9.022	320	5.942
5	12.697	45	11.097	120	8.976	330	5.847
6	12.652	46	11.062	122	8.931	340	5.755
7	12.606	47	11.027	124	8.886	350	5.664
8	12.561	48	10.993	126	8.841	360	5.578
9	12.517	49	10.958	128	8.797	370	5.493
10	12.472	50	10.924	130	8.753	380	5.407
11	12.428	52	10.857	132	8.710	400	5.248
12	12.385	54	10.791	134	8.667	420	5.101
13	12.341	56	10.725	136	8.625	440	4.952
14	12.298	58	10.660	138	8.583	460	4.812

TABLE 1.10 Boiling Points of Water

psi	Boiling point, °F	psi	Boiling point, °F	psi	Boiling point, °F
0.5	79.6	44	273.1	150	358.5
1	101.7	46	275.8	175	371.8
2	126.0	48	278.5	200	381.9
3	141.4	50	281.0	225	391.9
4	125.9	52	283.5	250	401.0
5	162.2	54	285.9	275	409.5
6	170.0	56	288.3	300	417.4
7	176.8	58	290.5	325	424.8
8	182.8	60	292.7	350	431.8
9	188.3	62	294.9	375	438.4
10	193.2	64	297.0	400	444.7
11	197.7	66	299.0	425	450.7
12	201.9	68	301.0	450	456.4
13	205.9	70	303.0	475	461.9
14	209.6	72	304.9	500	467.1
14.69	212.0	74	306.7	525	472.2
15	213.0	76	308.5	550	477.1
16	216.3	78	310.3	575	481.8
17	219.4	80	312.1	600	486.3
18	222.4	82	313.8	625	490.7
19	225.2	84	315.5	650	495.0
20	228.0	86	317.1	675	499.2
22	233.0	88	318.7	700	503.2
24	237.8	90	320.3	725	507.2
26	242.3	92	321.9	750	511.0
28	246.4	94	323.4	775	514.7
30	250.3	96	324.9	800	518.4
32	254.1	98	326.4	825	521.9
34	257.6	100	327.9	850	525.4
36	261.0	105	331.4	875	528.8
38	264.2	110	334.8	900	532.1
40	267.3	115	338.1	950	538.6
42	270.2	120	341.3	1000	544.8

TABLE 1.11 Boiling Points of Water

A. Barometric Pressures at Various Temperatures

Temp. °C.	0.0°	0.2°	0.4°	0.6°	0.8°
	mm of Hg	mm of Hg	mm of Hg	mm of Hg	mm of Hg
80	355.40	358.28	361.19	364.11	367.06
81	370.03	373.01	376.02	379.05	382.09
82	385.16	388.25	391.36	394.49	397.64
83	400.81	404.00	407.22	410.45	413.71
84	416.99	420.29	423.61	426.95	430.32
85	433.71	437.12	440.55	444.01	447.49
86	450.99	454.51	458.06	461.63	465.22
87	468.84	472.48	476.14	479.83	483.54
88	487.28	491.04	494.82	498.63	502.46
89	506.32	510.20	514.11	518.04	521.99
90	525.97	529.98	534.01	538.07	542.15
91	546.26	550.40	554.56	558.75	562.96
92	567.20	571.47	575.76	580.08	584.43
93	588.80	593.20	597.63	602.09	606.57
94	611.08	615.62	620.19	624.79	629.41
95	634.06	638.74	643.45	648.19	652.96
96	657.75	662.58	667.43	672.32	677.23
97	682.18	687.15	692.15	697.19	702.25
98	707.35	712.47	717.63	722.81	728.03
99	733.28	738.56	743.87	749.22	754.59
100	760.00	765.44	770.91	776.42	781.95

B. Boiling Points of Water at Various Pressures

Pressure, atm.	Boiling Point, °C.	Pressure, atm.	Boiling Point, °C.	Pressure, atm.	Boiling Point, °C.	Pressure, atm.	Boiling Point, °C.
0.5	80.9	7	164.2	14	194.1	21	213.9
1	100.0	8	169.6	15	197.4	22	216.2
2	119.6	9	174.5	16	200.4	23	218.5
3	132.9	10	179.0	17	203.4	24	220.8
4	142.9	11	183.2	18	206.1	25	222.9
5	151.1	12	187.1	19	208.8	26	225.0
6	158.1	13	190.7	20	211.4	27	227.0

TABLE 1.12 Refractive Index, Viscosity, Dielectric Constant, and Surface Tension of Water at Various Temperatures

Temp., °C	Refractive index, n_D	Viscosity mN · s · m^{-2}	Dielectric constant, ε	Surface tension mN · s · m^{-2}
0	1.333 95	1.793	87.90	75.83
5	1.333 88	1.521	85.84	75.09
10	1.333 69	1.307	83.96	74.36
15	1.333 39	1.135	82.00	73.62
20	1.333 00	1.002	80.20	72.88
25	1.332 50	0.890 3	78.35	72.14
30	1.331 94	0.797 7	76.60	71.40
35	1.331 31	0.719 0	74.83	70.66
40	1.330 61	0.653 2	73.17	69.92
50	1.329 04	0.547 0	69.58	68.45
60	1.327 25	0.466 5	66.73	66.97
70	1.325 11	0.404 0	63.73	65.49
80		0.354 4	60.86	64.01
90		0.314 5	58.12	62.54
100		0.281 8	55.51	61.07

TABLE 1.13 Compressibility of Water

In the table below are given the relative volumes of water at various temperatures and pressures. The volume at 0°C and one normal atmosphere (760 mm of Hg) is taken as unity.

P, atm	− 10°C.	0°C.	10°C.	20°C.	40°C.	60°C.	80°C.
1	1.0017	1.0000	1.0001	1.0016	1.0076	1.0168	1.0287
500	0.9788	0.9767	0.9778	0.9804	0.9867	0.9967	1.0071
1000	0.9581	0.9566	0.9591	0.9619	0.9689	0.9780	0.9884
1500	0.9399	0.9394	0.9424	0.9456	0.9529	0.9617	0.9717
2000	0.9223	0.9241	0.9277	0.9312	0.9386	0.9472	0.9568
2500	0.9083	0.9112	0.9147	0.9183	0.9257	0.9343	0.9437
3000	0.8962	0.8993	0.9028	0.9065	0.9139	0.9225	0.9315
3500	0.8852	0.8884	0.8919	0.8956	0.9030	0.9115	0.9203
4000	0.8751	0.8783	0.8818	0.8855	0.8931	0.9012	0.9097
4500	0.8658	0.8692	0.8725	0.8762	0.8838	0.8919	0.9001
5000	0.8573	0.8606	0.8639	0.8675	0.8752	0.8832	0.8913
6000	......	0.8452	0.8481	0.8517	0.8595	0.8674	0.8752
7000	......	......	0.8340	0.8374	0.8456	0.8534	0.8610
8000	......	......	......	0.8244	0.8330	0.8408	0.8483
9000	......	......	......	0.8128	0.8219	0.8297	0.8371
10000	......	......	......	0.8027	0.8119	0.8196	0.8268
11000	......	......	......	......	0.8023	0.8101	0.8172
12000	......	......	......	......	0.7931	0.8009	0.8080

TABLE 1.14 Flammability Limits of Inorganic Compounds in Air

Compound	Limits of Flammability	
	Lower volume %	Upper volume %
Ammonia	15.50	27.00
Carbon monoxide	12.50	74.20
Carbonyl sulfide	11.90	28.50
Cyanogen	6.60	42.60
Hydrocyanic acid	5.60	40.00
Hydrogen	4.00	74.20
Hydrogen sulfide	4.30	45.50

1.3 THE ELEMENTS

The chemical elements are the fundamental materials of which all matter is composed. From the modern viewpoint a substance that cannot be broken down or reduced further is, by definition, an element.

The Periodic Table presents organized information about the chemical elements. The elements are grouped into eight classes according to their properties.

The electronic configuration for an element's ground state is a shorthand representation giving the number of electrons (superscript) found in each of the allowed sublevels (*s, p, d, f*) above a noble gas core (indicated by brackets). In addition, values for the thermal conductivity, the electrical resistance, and the coefficient of linear thermal expansion are included.

Hund's Rule states that for a set of equal-energy orbitals, each orbital is occupied by one electron before any oribital has two. Therefore, the first electrons to occupy orbitals within a sublevel have parallel spins.

TABLE 1.15 Subdivision of Main Energy Levels

Main energy level	1	2		3			4			
Number of sublevels(n)	1	2		3			4			
Number of orbitals(n^2)	1	4		9			16			
Kind and no. of orbitals per sublevel	s 1	s 1	p 3	s 1	p 3	d 5	s 1	p 3	d 5	f 7
Maximum no. of electrons per sublevel	2	2	6	2	6	10	2	6	10	14
Maximum no. of electrons per main level ($2n^2$)	2	8		18			32			

TABLE 1.16 Chemical Symbols, Atomic Numbers, and Electron Arrangements of the Elements

Element name	Chemical symbol	Atomic number
Actinium	Ac	89
Aluminum	Al	13
Americium	Am	95
Antimony	Sb	51
Argon	Ar	18
Arsenic	As	33
Astatine	At	85
Barium	Ba	56
Berkelium	Bk	97
Beryllium	Be	4
Bismuth	Bi	83
Bohrium	Bh	107
Boron	B	5
Bromine	Br	35
Cadmium	Cd	48
Calcium	Ca	20
Californium	Cf	98
Carbon	C	6
Cerium	Ce	58
Cesium	Cs	55
Chlorine	Cl	17
Chromium	Cr	24
Cobalt	Co	27
Copper	Cu	29
Curium	Cm	96
Dubnium	Db	105
Dysprosium	Dy	66
Einsteinium	Es	99
Erbium	Er	68
Europium	Eu	63
Fermium	Fm	100
Fluorine	F	9
Francium	Fr	87
Gadolinium	Gd	64
Gallium	Ga	31
Germanium	Ge	32
Gold	Au	79
Hafnium	Hf	72
Hassium	Hs	108
Helium	He	2
Holmium	Ho	67
Hydrogen	H	1
Indium	In	49
Iodine	I	53
Iridium	Ir	77
Iron	Fe	26
Krypton	Kr	36
Lanthanum	La	57
Lawrencium	Lr or Lw	103
Lead	Pb	82
Lithium	Li	3
Lutetium	Lu	71
Magnesium	Mg	12
Manganese	Mn	25

(Continued)

TABLE 1.16 Chemical Symbols, Atomic Numbers, and Electron Arrangements of the Elements (*Continued*)

Element name	Chemical symbol	Atomic number
Meitnerium	Mt	109
Mendelevium	Md	101
Mercury	Hg	80
Molybdenum	Mo	42
Neodymium	Nd	60
Neon	Ne	10
Neptunium	Np	93
Nickel	Ni	28
Niobium	Nb	41
Nitrogen	N	7
Nobelium	No	102
Osmium	Os	76
Oxygen	O	8
Palladium	Pd	46
Phosphorus	P	15
Platinum	Pt	78
Plutonium	Pu	94
Polonium	Po	84
Potassium	K	19
Praseodymium	Pr	59
Promethium	Pm	61
Protactinium	Pa	91
Radium	Ra	88
Radon	Rn	86
Rhenium	Re	75
Rhodium	Rh	45
Rubidium	Rb	37
Ruthenium	Ru	44
Rutherfordium	Rf	104
Samarium	Sm	62
Scandium	Sc	21
Seaborgium	Sg	106
Selenium	Se	34
Silicon	Si	14
Silver	Ag	47
Sodium	Na	11
Strontium	Sr	38
Sulfur	S	16
Tantalum	Ta	73
Technetium	Tc	43
Tellurium	Te	52
Terbium	Tb	65
Thallium	Tl	81
Thorium	Th	90
Thulium	Tm	69
Tin	Sn	50
Titanium	Ti	22
Tungsten	W	74
Ununbium	Uub	112
Ununhexium	Uuh	116
Ununnilium	Uun	110
Ununoctium	Uuo	118
Ununquadium	Unq	114
Unununium	Uuu	111
Uranium	U	92

TABLE 1.16 Chemical Symbols, Atomic Numbers, and Electron Arrangements of the Elements (*Continued*)

Element name	Chemical symbol	Atomic number
Vanadium	V	23
Xenon	Xe	54
Ytterbium	Yb	70
Yttrium	Y	39
Zinc	Zn	30
Zirconium	Zr	40

*As of the time of writing, there were no known elements with atomic numbers 113, 115, or 117.

Hydrogen (1) Symbol, H. A colorless, odorless gas at room temperature. The most common isotope has atomic weight 1.00794. The lightest and most abundant element in the universe.

- Electrons in first energy level: 1

Helium (2) Symbol, He. A colorless, odorless gas at room temperature. The most common isotope has atomic weight 4.0026. The second lightest and second most abundant element in the universe.

- Electrons in first energy level: 2

Lithium (3) Symbol, Li. Classified as an alkali metal. In pure form it is silver-colored. The lightest elemental metal. The most common isotope has atomic weight 6.941.

- Electrons in first energy level: 2
- Electrons in second energy level: 1

Beryllium (4) Symbol, Be. Classified as an alkaline earth. In pure form it has a grayish color similar to that of steel. Has a relatively high melting point. The most common isotope has atomic weight 9.01218.

- Electrons in first energy level: 2
- Electrons in second energy level: 2

Boron (5) Symbol, B. Classified as a metalloid. The most common isotope has atomic weight 10.82. Can exist as a powder or as a black, hard metalloid. Boron is not found free in nature.

- Electrons in first energy level: 2
- Electrons in second energy level: 3

Carbon (6) Symbol, C. A nonmetallic element that is a solid at room temperature. Has a characteristic hexagonal crystal structure. Known as the basis of life on Earth. The most common isotope has atomic weight 12.011. Exists in three well-known forms: *graphite* (a black powder) which is common, *diamond* (a clear solid) which is rare, and *amorphous*.

Another form of carbon is graphite. Used in electrochemical cells, air-cleaning filters, thermocouples, and noninductive electrical resistors. Also used in medicine to absorb poisons and toxins in the stomach and intestines. Abundant in mineral rocks such as

- Electrons in first energy level: 2
- Electrons in second energy level: 4

Nitrogen (7) Symbol, N. A nonmetallic element that is a colorless, odorless gas at room temperature. The most common isotope has atomic weight 14.007. The most abundant component of the earth's atmosphere (approximately 78 percent at the surface). Reacts to some extent with certain combinations of other elements.

- Electrons in first energy level: 2
- Electrons in second energy level: 5

Oxygen (8) Symbol, O. A nonmetallic element that is a colorless, odorless gas at room temperature. The most common isotope has atomic weight 15.999. The second most abundant component of the earth's atmosphere (approximately 21 percent at the surface).

Combines readily with many other elements, particularly metals. One of the oxides of iron, for example, is known as common rust. Normally, two atoms of oxygen combine to form a molecule (O_2). In this form, oxygen is essential for the sustenance of many forms of life on Earth. When three oxygen atoms form a molecule (O_3), the element is called *ozone*. This form of the element is beneficial in the upper atmosphere because it reduces the amount of ultraviolet radiation reaching the earth's surface. Ozone is, ironically, also known as an irritant and pollutant in the surface air over heavily populated areas.

- Electrons in first energy level: 2
- Electrons in second energy level: 6

Fluorine (9) Symbol, F. The most common isotope has atomic weight 18.998. A gaseous element of the halogen family. Has a characteristic greenish or yellowish color. Reacts readily with many other elements.

- Electrons in first energy level: 2
- Electrons in second energy level: 7

Neon (10) Symbol, Ne. The most common isotope has atomic weight 20.179. A noble gas present in trace amounts in the atmosphere.

- Electrons in first energy level: 2
- Electrons in second energy level: 8

Sodium (11) Symbol, Na. The most common isotope has atomic weight 22.9898. An element of the alkali-metal group. A solid at room temperature.

- Electrons in first energy level: 2
- Electrons in second energy level: 8
- Electrons in third energy level: 1

Magnesium (12) Symbol, Mg. The most common isotope has atomic weight 24.305. A member of the alkaline earth group. At room temperature it is a whitish metal.

- Electrons in first energy level: 2
- Electrons in second energy level: 8
- Electrons in third energy level: 2

Aluminum (13) Symbol, Al. The most common isotope has atomic weight 26.98. A metallic element and a good electrical conductor. Has many of the same characteristics as magnesium, except it reacts less easily with oxygen in the atmosphere.

- Electrons in first energy level: 2
- Electrons in second energy level: 8
- Electrons in third energy level: 3

Silicon (14) Symbol, Si. The most common isotope has atomic weight 28.086. A metalloid abundant in the earth's crust. Especially common in rocks such as granite, and in many types of sand.

- Electrons in first energy level: 2
- Electrons in second energy level: 8
- Electrons in third energy level: 4

Phosphorus (15) Symbol, P. The most common isotope has atomic weight 30.974. A nonmetallic element of the nitrogen family. Found in certain types of rock.

- Electrons in first energy level: 2
- Electrons in second energy level: 8
- Electrons in third energy level: 5

Sulfur (16) Symbol, S. Also spelled *sulphur*. The most common isotope has atomic weight 32.06. A nonmetallic element. Reacts with some other elements.

- Electrons in first energy level: 2
- Electrons in second energy level: 8
- Electrons in third energy level: 6

Chlorine (17) Symbol, Cl. The most common isotope has atomic weight 35.453. A gas at room temperature and a member of the halogen family. Reacts readily with various other elements.

- Electrons in first energy level: 2
- Electrons in second energy level: 8
- Electrons in third energy level: 7

Argon (18) Symbol, A or Ar. The most common isotope has atomic weight 39.94. A gas at room temperature; classified as a noble gas. Present in small amounts in the atmosphere.

- Electrons in first energy level: 2
- Electrons in second energy level: 8
- Electrons in third energy level: 8

Potassium (19) Symbol, K. The most common isotope has atomic weight 39.098. A member of the alkali metal group.

- Electrons in first energy level: 2
- Electrons in second energy level: 8
- Electrons in third energy level: 8
- Electrons in fourth energy level: 1

Calcium (20) Symbol, Ca. The most common isotope has atomic weight 40.08. A metallic element of the alkaline-earth group. Calcium carbonate, or calcite, is abundant in the earth's crust, especially in limestone

- Electrons in first energy level: 2
- Electrons in second energy level: 8
- Electrons in third energy level: 8
- Electrons in fourth energy level: 2

Scandium (21) Symbol, Sc. The most common isotope has atomic weight 44.956. In the pure form it is a soft metal. Classified as a transition metal.

- Electrons in first energy level: 2
- Electrons in second energy level: 8
- Electrons in third energy level: 9
- Electrons in fourth energy level: 2

Titanium (22) Symbol, Ti. The most common isotope has atomic weight 47.88. Classified as a transition metal.

- Electrons in first energy level: 2
- Electrons in second energy level: 8
- Electrons in third energy level: 10
- Electrons in fourth energy level: 2

Vanadium (23) Symbol, V. The most common isotope has atomic weight 50.94. Classified as a transition metal. In its pure form it is whitish in color.

- Electrons in first energy level: 2
- Electrons in second energy level: 8
- Electrons in third energy level: 11
- Electrons in fourth energy level: 2

Chromium (24) Symbol, Cr. The most common isotope has atomic weight 51.996. Classified as a transition metal. In its pure form it is grayish in color.

- Electrons in first energy level: 2
- Electrons in second energy level: 8
- Electrons in third energy level: 13
- Electrons in fourth energy level: 1

Manganese (25) Symbol, Mn. The most common isotope has atomic weight 54.938. Classified as a transition metal. In its pure form it is grayish in color.

- Electrons in first energy level: 2
- Electrons in second energy level: 8
- Electrons in third energy level: 13
- Electrons in fourth energy level: 2

Iron (26) Symbol, Fe. The most common isotope has atomic weight 55.847. In its pure form it is a dull gray metal.

- Electrons in first energy level: 2
- Electrons in second energy level: 8
- Electrons in third energy level: 14
- Electrons in fourth energy level: 2

Cobalt (27) Symbol, Co. The most common isotope has atomic weight 58.94. Classified as a transition metal. In the pure form it is silvery in color.

- Electrons in first energy level: 2
- Electrons in second energy level: 8

- Electrons in third energy level: 15
- Electrons in fourth energy level: 2

Nickel (28) Symbol, Ni. The most common isotope has atomic weight 58.69. Classified as a transition metal. In its pure form it is light gray to white.

- Electrons in first energy level: 2
- Electrons in second energy level: 8
- Electrons in third energy level: 16
- Electrons in fourth energy level: 2

Copper (29) Symbol, Cu. The most common isotope has atomic weight 63.546. Classified as a transition metal. In its pure form it has a characteristic red or wine color.

- Electrons in first energy level: 2
- Electrons in second energy level: 8
- Electrons in third energy level: 18
- Electrons in fourth energy level: 1

Zinc (30) Symbol, Zn. The most common isotope has atomic weight 65.39. Classified as a transition metal. In pure form, it is a dull blue-gray color.

- Electrons in first energy level: 2
- Electrons in second energy level: 8
- Electrons in third energy level: 18
- Electrons in fourth energy level: 2

Gallium (31) Symbol, Ga. The most common isotope has atomic weight 69.72. A semiconducting metal. In pure form it is light gray to white.

- Electrons in first energy level: 2
- Electrons in second energy level: 8
- Electrons in third energy level: 18
- Electrons in fourth energy level: 3

Germanium (32) Symbol, Ge. The most common isotope has atomic weight 72.59. A semiconducting metalloid.

- Electrons in first energy level: 2
- Electrons in second energy level: 8
- Electrons in third energy level: 18
- Electrons in fourth energy level: 4

Arsenic (33) Symbol, As. The most common isotope has atomic weight 74.91. A metalloid used as a dopant in the manufacture of semiconductors. In its pure form it is gray in color.

- Electrons in first energy level: 2
- Electrons in second energy level: 8
- Electrons in third energy level: 18
- Electrons in fourth energy level: 5

Selenium (34) Symbol, Se. The most common isotope has atomic weight 78.96. Classified as a nonmetal. In its pure form it is gray in color.

- Electrons in first energy level: 2
- Electrons in second energy level: 8
- Electrons in third energy level: 18
- Electrons in fourth energy level: 6

Bromine (35) Symbol, Br. The most common isotope has atomic weight 79.90. A nonmetallic element of the halogen family. A reddish-brown liquid at room temperature. Has a characteristic unpleasant odor. Reacts readily with various other elements.

- Electrons in first energy level: 2
- Electrons in second energy level: 8
- Electrons in third energy level: 18
- Electrons in fourth energy level: 7

Krypton (36) Symbol, Kr. The most common isotope has atomic weight 83.80. Classified as a noble gas. Colorless and odorless. Present in trace amounts in the earth's atmosphere. Some common isotopes of this element are radioactive.

- Electrons in first energy level: 2
- Electrons in second energy level: 8
- Electrons in third energy level: 18
- Electrons in fourth energy level: 8

Rubidium (37) Symbol, Rb. The most common isotope has atomic weight 85.468. Classified as an alkali metal. In its pure form it is silver-colored. Reacts easily with oxygen and chlorine.

- Electrons in first energy level: 2
- Electrons in second energy level: 8
- Electrons in third energy level: 18
- Electrons in fourth energy level: 8
- Electrons in fifth energy level: 1

Strontium (38) Symbol, Sr. The most common isotope has atomic weight 87.62. A metallic element of the alkaline-earth group. In pure form it is gold-colored.

- Electrons in first energy level: 2
- Electrons in second energy level: 8
- Electrons in third energy level: 18
- Electrons in fourth energy level: 8
- Electrons in fifth energy level: 2

Yttrium (39) Symbol, Y. The most common isotope has atomic weight 88.906. Classified as a transition metal. In its pure form it is silver-colored.

- Electrons in first energy level: 2
- Electrons in second energy level: 8
- Electrons in third energy level: 18
- Electrons in fourth energy level: 9
- Electrons in fifth energy level: 2

Zirconium (40) Symbol, Zr. The most common isotope has atomic weight 91.22. Classified as a transition metal. In its pure form it is grayish in color.

- Electrons in first energy level: 2
- Electrons in second energy level: 8
- Electrons in third energy level: 18
- Electrons in fourth energy level: 10
- Electrons in fifth energy level: 2

Niobium (41) Symbol, Nb. The most common isotope has atomic weight 92.91. Classified as a transition metal. This element is sometimes called *columbium.* In pure form it is shiny, and is light gray to white in color.

- Electrons in first energy level: 2
- Electrons in second energy level: 8
- Electrons in third energy level: 18
- Electrons in fourth energy level: 12
- Electrons in fifth energy level: 1

Molybdenum (42) Symbol, Mo. The most common isotope has atomic weight 95.94. Classified as a transition metal. In its pure form, it is hard and silver-white.

Used as a catalyst, as a component of hard alloys for the aeronautical and aerospace industries, and in steel-hardening processes. It is known for high thermal conductivity, low thermal-expansion coefficient, high melting point, and resistance to corrosion. Most molybdenum compounds are relatively nontoxic.

- Electrons in first energy level: 2
- Electrons in second energy level: 8
- Electrons in third energy level: 18
- Electrons in fourth energy level: 13
- Electrons in fifth energy level: 1

Technetium (43) Symbol, Tc. Formerly called *masurium.* The most common isotope has atomic weight 98. Classified as a transition metal. In its pure form, it is grayish in color. This element is not found in nature; it occurs when the uranium atom is split by nuclear fission. It also occurs when molybdenum is bombarded by high-speed deuterium nuclei (particles consisting of one proton and one neutron). This element is radioactive.

- Electrons in first energy level: 2
- Electrons in second energy level: 8
- Electrons in third energy level: 18
- Electrons in fourth energy level: 14
- Electrons in fifth energy level: 1

Ruthenium (44) Symbol, Ru. The most common isotope has atomic weight 101.07. A rare element, classified as a transition metal. In pure form it is silver-colored.

- Electrons in first energy level: 2
- Electrons in second energy level: 8
- Electrons in third energy level: 18
- Electrons in fourth energy level: 15
- Electrons in fifth energy level: 1

Rhodium (45) Symbol, Rh. The most common isotope has atomic weight 102.906. Classified as a transition metal. In its pure form it is silver-colored. Occurs in nature along with platinum and nickel.

- Electrons in first energy level: 2
- Electrons in second energy level: 8
- Electrons in third energy level: 18
- Electrons in fourth energy level: 16
- Electrons in fifth energy level: 1

Palladium (46) Symbol, Pd. The most common isotope has atomic weight 106.42. Classified as a transition metal. In its pure form it is light gray to white. In nature, palladium is found with copper ore.

- Electrons in first energy level: 2
- Electrons in second energy level: 8
- Electrons in third energy level: 18
- Electrons in fourth energy level: 18
- Electrons in fifth energy level: 0

Silver (47) Symbol, Ag. The most common isotope has atomic weight 107.87. Classified as a transition metal. In its pure form it is a bright, shiny, and silverish-white colored metal.

- Electrons in first energy level: 2
- Electrons in second energy level: 8
- Electrons in third energy level: 18
- Electrons in fourth energy level: 18
- Electrons in fifth energy level: 1

Cadmium (48) Symbol, Cd. The most common isotope has atomic weight 112.41. Classified as a transition metal. In its pure form it is silver-colored.

- Electrons in first energy level: 2
- Electrons in second energy level: 8
- Electrons in third energy level: 18
- Electrons in fourth energy level: 18
- Electrons in fifth energy level: 2

Indium (49) Symbol, In. The most common isotope has atomic weight 114.82. A metallic element used as a dopant in semiconductor processing. In pure form it is silver-colored. In nature, it is often found along with zinc.

- Electrons in first energy level: 2
- Electrons in second energy level: 8
- Electrons in third energy level: 18
- Electrons in fourth energy level: 18
- Electrons in fifth energy level: 3

Tin (50) Symbol, Sn. The most common isotope has atomic weight 118.71. In pure form it is a white or grayish metal. It changes color (from white to gray) when it is cooled through a certain temperature range. It is ductile and malleable.

- Electrons in first energy level: 2
- Electrons in second energy level: 8

- Electrons in third energy level: 18
- Electrons in fourth energy level: 18
- Electrons in fifth energy level: 4

Antimony (51) Symbol, Sb. The most common isotope has atomic weight 121.76. Classified as a metalloid. In pure form, it is blue-white or blue-gray in color. Has a characteristic flakiness and brittleness.

- Electrons in first energy level: 2
- Electrons in second energy level: 8
- Electrons in third energy level: 18
- Electrons in fourth energy level: 18
- Electrons in fifth energy level: 5

Tellurium (52) Symbol, Te. The most common isotope has atomic weight 127.60. A rare metalloid element related to selenium. In pure form, it is silverish-white and has high luster. In nature it is found along with other metals such as copper. It has a characteristic brittleness.

- Electrons in first energy level: 2
- Electrons in second energy level: 8
- Electrons in third energy level: 18
- Electrons in fourth energy level: 18
- Electrons in fifth energy level: 6

Iodine (53) Symbol, I. The most common isotope has atomic weight 126.905. A member of the halogen family. In pure form it has a black or purple-black color.

- Electrons in first energy level: 2
- Electrons in second energy level: 8
- Electrons in third energy level: 18
- Electrons in fourth energy level: 18
- Electrons in fifth energy level: 7

Xenon (54) Symbol, Xe. The most common isotope has atomic weight 131.29. Classified as a noble gas. Colorless and odorless; present in trace amounts in the earth's atmosphere.

- Electrons in first energy level: 2
- Electrons in second energy level: 8
- Electrons in third energy level: 18
- Electrons in fourth energy level: 18
- Electrons in fifth energy level: 8

Cesium (55) Symbol, Cs. Also spelled *caesium* (in Britain). The most common isotope has atomic weight 132.91. Classified as an alkali metal. In pure form, it is silver-white in color, is ductile, and is malleable.

- Electrons in first energy level: 2
- Electrons in second energy level: 8
- Electrons in third energy level: 18
- Electrons in fourth energy level: 18
- Electrons in fifth energy level: 8
- Electrons in sixth energy level: 1

Barium (56) Symbol, Ba. The most common isotope has atomic weight 137.36. Classified as an alkaline earth. In pure form it is silver-white in color, and is relatively soft; it is sometimes mistaken for lead.

- Electrons in first energy level: 2
- Electrons in second energy level: 8
- Electrons in third energy level: 18
- Electrons in fourth energy level: 18
- Electrons in fifth energy level: 8
- Electrons in sixth energy level: 2

Lanthanum (57) Symbol, La. The most common isotope has atomic weight 138.906. Classified as a rare earth. In pure form it is white in color, malleable, and soft.

- Electrons in first energy level: 2
- Electrons in second energy level: 8
- Electrons in third energy level: 18
- Electrons in fourth energy level: 18
- Electrons in fifth energy level: 9
- Electrons in sixth energy level: 2

Cerium (58) Symbol, Ce. The most common isotope has atomic weight 140.13. Classified as a rare earth. In pure form it is light silvery-gray. It reacts readily with various other elements and is malleable and ductile.

- Electrons in first energy level: 2
- Electrons in second energy level: 8
- Electrons in third energy level: 18
- Electrons in fourth energy level: 20
- Electrons in fifth energy level: 8
- Electrons in sixth energy level: 2

Praseodymium (59) Symbol, Pr. The most common isotope has atomic weight 140.908. Classified as a rare earth. In pure form it is silver-gray, soft, malleable, and ductile.

- Electrons in first energy level: 2
- Electrons in second energy level: 8
- Electrons in third energy level: 18
- Electrons in fourth energy level: 21
- Electrons in fifth energy level: 8
- Electrons in sixth energy level: 2

Neodymium (60) Symbol, Nd. The most common isotope has atomic weight 144.24. Classified as a rare earth. In pure form it is shiny and is silvery in color.

- Electrons in first energy level: 2
- Electrons in second energy level: 8
- Electrons in third energy level: 18
- Electrons in fourth energy level: 22

- Electrons in fifth energy level: 8
- Electrons in sixth energy level: 2

Promethium (61) Symbol, Pm. Formerly called *illinium*. The most common isotope has atomic weight 145. Classified as a rare earth. In pure form it is gray in color, and is highly radioactive.

- Electrons in first energy level: 2
- Electrons in second energy level: 8
- Electrons in third energy level: 18
- Electrons in fourth energy level: 23
- Electrons in fifth energy level: 8
- Electrons in sixth energy level: 2

Samarium (62) Symbol, Sm. The most common isotope has atomic weight 150.36. Classified as a rare earth. In pure form it is silvery-white in color with high luster.

- Electrons in first energy level: 2
- Electrons in second energy level: 8
- Electrons in third energy level: 18
- Electrons in fourth energy level: 24
- Electrons in fifth energy level: 8
- Electrons in sixth energy level: 2

Europium (63) Symbol, Eu. The most common isotope has atomic weight 151.96. Classified as a rare earth. In pure form it is silver-gray in color, and has ductility similar to that of lead.

- Electrons in first energy level: 2
- Electrons in second energy level: 8
- Electrons in third energy level: 18
- Electrons in fourth energy level: 25
- Electrons in fifth energy level: 8
- Electrons in sixth energy level: 2

Gadolinium (64) Symbol, Gd. The most common isotope has atomic weight 157.25. Classified as a rare earth. In pure form it is silver in color, is ductile, and is malleable.

- Electrons in first energy level: 2
- Electrons in second energy level: 8
- Electrons in third energy level: 18
- Electrons in fourth energy level: 25
- Electrons in fifth energy level: 9
- Electrons in sixth energy level: 2

Terbium (65) Symbol, Tb. The most common isotope has atomic weight 158.93. Classified as a rare earth. In pure form it is silver-gray, soft, malleable, and ductile.

- Electrons in first energy level: 2
- Electrons in second energy level: 8
- Electrons in third energy level: 18

- Electrons in fourth energy level: 27
- Electrons in fifth energy level: 8
- Electrons in sixth energy level: 2

Dysprosium (66) Symbol, Dy. The most common isotope has atomic weight 162.5. Classified as a rare earth. In pure form it has a bright, shiny silver color. It is soft and malleable, but it has a relatively high melting point.

- Electrons in first energy level: 2
- Electrons in second energy level: 8
- Electrons in third energy level: 18
- Electrons in fourth energy level: 28
- Electrons in fifth energy level: 8
- Electrons in sixth energy level: 2

Holmium (67) Symbol, Ho. The most common isotope has atomic weight 164.93. Classified as a rare earth. In pure form it is silver in color. It is soft and malleable.

- Electrons in first energy level: 2
- Electrons in second energy level: 8
- Electrons in third energy level: 18
- Electrons in fourth energy level: 29
- Electrons in fifth energy level: 8
- Electrons in sixth energy level: 2

Erbium (68) Symbol, Er. The most common isotope has atomic weight 167.26. Classified as a rare earth. In pure form it is silverish, soft, malleable, and ductile.

- Electrons in first energy level: 2
- Electrons in second energy level: 8
- Electrons in third energy level: 18
- Electrons in fourth energy level: 30
- Electrons in fifth energy level: 8
- Electrons in sixth energy level: 2

Thulium (69) Symbol, Tm. The most common isotope has atomic weight 168.93. Classified as a rare earth. In pure form this element is grayish in color, soft, malleable, and ductile.

- Electrons in first energy level: 2
- Electrons in second energy level: 8
- Electrons in third energy level: 18
- Electrons in fourth energy level: 31
- Electrons in fifth energy level: 8
- Electrons in sixth energy level: 2

Ytterbium (70) Symbol, Yb. The most common isotope has atomic weight 173.04. Classified as a rare earth. In pure form it is silver-white in color, soft, malleable, and ductile.

- Electrons in first energy level: 2
- Electrons in second energy level: 8

- Electrons in third energy level: 18
- Electrons in fourth energy level: 32
- Electrons in fifth energy level: 8
- Electrons in sixth energy level: 2

Lutetium (71) Symbol, Lu. The most common isotope has atomic weight 174.967. Classified as a rare earth. In its pure form, it is silver-white and radioactive, with a half-life on the order of thousands of millions of years.

- Electrons in first energy level: 2
- Electrons in second energy level: 8
- Electrons in third energy level: 18
- Electrons in fourth energy level: 32
- Electrons in fifth energy level: 9
- Electrons in sixth energy level: 2

Hafnium (72) Symbol, Hf. The most common isotope has atomic weight 178.49. Classified as a transition metal. In pure form, it is silver-colored, shiny, and ductile.

- Electrons in first energy level: 2
- Electrons in second energy level: 8
- Electrons in third energy level: 18
- Electrons in fourth energy level: 32
- Electrons in fifth energy level: 10
- Electrons in sixth energy level: 2

Tantalum (73) Symbol, Ta. The most common isotope has atomic weight 180.95. Classified as a transition metal; an element of the vanadium family. In pure form it is grayish-silver in color, ductile, and hard, with a high melting point.

- Electrons in first energy level: 2
- Electrons in second energy level: 8
- Electrons in third energy level: 18
- Electrons in fourth energy level: 32
- Electrons in fifth energy level: 11
- Electrons in sixth energy level: 2

Tungsten (74) Symbol, W. Also known as *wolfram.* The most common isotope has atomic weight 183.85. Classified as a transition metal. In pure form it is silver-colored. It has an extremely high melting point.

- Electrons in first energy level: 2
- Electrons in second energy level: 8
- Electrons in third energy level: 18
- Electrons in fourth energy level: 32
- Electrons in fifth energy level: 12
- Electrons in sixth energy level: 2

Rhenium (75) Symbol, Re. The most common isotope has atomic weight 186.207. Classified as a transition metal. In pure form it is silver-white, has high density, and has a high melting point.

- Electrons in first energy level: 2
- Electrons in second energy level: 8
- Electrons in third energy level: 18
- Electrons in fourth energy level: 32
- Electrons in fifth energy level: 13
- Electrons in sixth energy level: 2

Osmium (76) Symbol, Os. The most common isotope has atomic weight 190.2. A transition metal of the platinum group. In pure form it is bluish-silver in color, dense, hard, and brittle.

- Electrons in first energy level: 2
- Electrons in second energy level: 8
- Electrons in third energy level: 18
- Electrons in fourth energy level: 32
- Electrons in fifth energy level: 14
- Electrons in sixth energy level: 2

Iridium (77) Symbol, Ir. The most common isotope has atomic weight 192.22. A transition metal of the platinum group. In pure form it is yellowish-white in color with high luster; it is hard, brittle, and has high density.

- Electrons in first energy level: 2
- Electrons in second energy level: 8
- Electrons in third energy level: 18
- Electrons in fourth energy level: 32
- Electrons in fifth energy level: 15
- Electrons in sixth energy level: 2

Platinum (78) Symbol, Pt. The most common isotope has atomic weight 195.08. Classified as a transition metal. In pure form it has a brilliant, shiny, white luster. It is malleable and ductile.

- Electrons in first energy level: 2
- Electrons in second energy level: 8
- Electrons in third energy level: 18
- Electrons in fourth energy level: 32
- Electrons in fifth energy level: 17
- Electrons in sixth energy level: 1

Gold (79) Symbol, Au. The most common isotope has atomic weight 196.967. A transition metal. In pure form it is shiny, yellowish, ductile, malleable, and comparatively soft.

- Electrons in first energy level: 2
- Electrons in second energy level: 8
- Electrons in third energy level: 18
- Electrons in fourth energy level: 32

- Electrons in fifth energy level: 18
- Electrons in sixth energy level: 1

Mercury (80) Symbol, Hg. The most common isotope has atomic weight 200.59. Classified as a transition metal. In pure form it is silver-colored and liquid at room temperature.

- Electrons in first energy level: 2
- Electrons in second energy level: 8
- Electrons in third energy level: 18
- Electrons in fourth energy level: 32
- Electrons in fifth energy level: 18
- Electrons in sixth energy level: 2

Thallium (81) Symbol, Tl. The most common isotope has atomic weight 204.38. A metallic element. In pure form it is bluish-gray or dull gray, soft, malleable, and ductile.

- Electrons in first energy level: 2
- Electrons in second energy level: 8
- Electrons in third energy level: 18
- Electrons in fourth energy level: 32
- Electrons in fifth energy level: 18
- Electrons in sixth energy level: 3

Lead (82) Symbol, Pb. The most common isotope has atomic weight 207.2. A metallic element. In pure form it is dull gray or blue-gray, soft, and malleable; relatively low melting temperature.

- Electrons in first energy level: 2
- Electrons in second energy level: 8
- Electrons in third energy level: 18
- Electrons in fourth energy level: 32
- Electrons in fifth energy level: 18
- Electrons in sixth energy level: 4

Bismuth (83) Symbol, Bi. The most common isotope has atomic weight 208.98. A metallic element. In pure form it is pinkish-white and brittle.

- Electrons in first energy level: 2
- Electrons in second energy level: 8
- Electrons in third energy level: 18
- Electrons in fourth energy level: 32
- Electrons in fifth energy level: 18
- Electrons in sixth energy level: 5

Polonium (84) Symbol, Po. The most common isotope has atomic weight 209. Classified as a metalloid. It is produced from the decay of radium and is sometimes called radium-F. Polonium is radioactive; it emits primarily alpha particles.

- Electrons in first energy level: 2
- Electrons in second energy level: 8

- Electrons in third energy level: 18
- Electrons in fourth energy level: 32
- Electrons in fifth energy level: 18
- Electrons in sixth energy level: 6

Astatine (85) Symbol, At. The most common isotope has atomic weight 210. Formerly called *alabamine*. Classified as a halogen. The element is radioactive.

- Electrons in first energy level: 2
- Electrons in second energy level: 8
- Electrons in third energy level: 18
- Electrons in fourth energy level: 32
- Electrons in fifth energy level: 18
- Electrons in sixth energy level: 7

Radon (86) Symbol, Rn. The most common isotope has atomic weight 222. Classified as a noble gas. It is radioactive, emitting primarily alpha particles, and has a short half-life. Radon is a colorless gas that results from the disintegration of radium.

- Electrons in first energy level: 2
- Electrons in second energy level: 8
- Electrons in third energy level: 18
- Electrons in fourth energy level: 32
- Electrons in fifth energy level: 18
- Electrons in sixth energy level: 8

Francium (87) Symbol, Fr. The most common isotope has atomic weight 223. Classified as an alkali metal. This element is radioactive, and all isotopes decay rapidly. Produced as a result of the radioactive disintegration of actinium.

- Electrons in first energy level: 2
- Electrons in second energy level: 8
- Electrons in third energy level: 18
- Electrons in fourth energy level: 32
- Electrons in fifth energy level: 18
- Electrons in sixth energy level: 8
- Electrons in seventh energy level: 1

Radium (88) Symbol, Ra. The most common isotope has atomic weight 226. Classified as an alkaline earth. In pure form it is silver-gray, but darkens quickly when exposed to air. This element is radioactive, emitting alpha particles, beta particles, and gamma rays. It has a moderately long half-life.

- Electrons in first energy level: 2
- Electrons in second energy level: 8
- Electrons in third energy level: 18
- Electrons in fourth energy level: 32
- Electrons in fifth energy level: 18
- Electrons in sixth energy level: 8
- Electrons in seventh energy level: 2

Actinium (89) Symbol, Ac. The most common isotope has atomic weight 227. Classified as a rare earth. In pure form it is silver-gray in color. This element is radioactive, emitting beta particles. The most common isotope has a half-life of 21.6 years.

- Electrons in first energy level: 2
- Electrons in second energy level: 8
- Electrons in third energy level: 18
- Electrons in fourth energy level: 32
- Electrons in fifth energy level: 18
- Electrons in sixth energy level: 9
- Electrons in seventh energy level: 2

Thorium (90) Symbol, Th. The most common isotope has atomic weight 232.038. Classified as a rare earth and a member of the actinide series. In pure form it is silver-colored, soft, ductile, and malleable.

- Electrons in first energy level: 2
- Electrons in second energy level: 8
- Electrons in third energy level: 18
- Electrons in fourth energy level: 32
- Electrons in fifth energy level: 18
- Electrons in sixth energy level: 10
- Electrons in seventh energy level: 2

Protactinium (91) Symbol, Pa. Formerly called *protoactinium*. The most common isotope has atomic weight 231.036. Classified as a rare earth. In pure form it is silver-colored.

- Electrons in first energy level: 2
- Electrons in second energy level: 8
- Electrons in third energy level: 18
- Electrons in fourth energy level: 32
- Electrons in fifth energy level: 20
- Electrons in sixth energy level: 9
- Electrons in seventh energy level: 2

Uranium (92) Symbol, U. The most common isotope has atomic weight 238.029. Classified as a rare earth. In pure form it is silver-colored, malleable, and ductile.

- Electrons in first energy level: 2
- Electrons in second energy level: 8
- Electrons in third energy level: 18
- Electrons in fourth energy level: 32
- Electrons in fifth energy level: 21
- Electrons in sixth energy level: 9
- Electrons in seventh energy level: 2

Neptunium (93) Symbol, Np. The most common isotope has atomic weight 237. Classified as a rare earth. In pure form it is silver-colored, and reacts with various other elements to form compounds.

- Electrons in first energy level: 2
- Electrons in second energy level: 8

- Electrons in third energy level: 18
- Electrons in fourth energy level: 32
- Electrons in fifth energy level: 23
- Electrons in sixth energy level: 8
- Electrons in seventh energy level: 2

Plutonium (94) Symbol, Pu. The most common isotope has atomic weight 244. Classified as a rare earth. In pure form it is silver-colored; when it is exposed to air, a yellow oxide layer forms. Plutonium reacts with various other elements to form compounds.

- Electrons in first energy level: 2
- Electrons in second energy level: 8
- Electrons in third energy level: 18
- Electrons in fourth energy level: 32
- Electrons in fifth energy level: 24
- Electrons in sixth energy level: 8
- Electrons in seventh energy level: 2

Americium (95) Symbol, Am. The most common isotope has atomic weight 243. Classified as a rare earth. In pure form it is silver-white and malleable.

- Electrons in first energy level: 2
- Electrons in second energy level: 8
- Electrons in third energy level: 18
- Electrons in fourth energy level: 32
- Electrons in fifth energy level: 25
- Electrons in sixth energy level: 8
- Electrons in seventh energy level: 2

Curium (96) Symbol, Cm. The most common isotope has atomic weight 247. Classified as a rare earth. In pure form it is silvery in color, and it reacts readily with various other elements. This element, like most transuranic elements, is dangerously radioactive.

- Electrons in first energy level: 2
- Electrons in second energy level: 8
- Electrons in third energy level: 18
- Electrons in fourth energy level: 32
- Electrons in fifth energy level: 25
- Electrons in sixth energy level: 9
- Electrons in seventh energy level: 2

Berkelium (97) Symbol, Bk. The most common isotope has atomic weight 247. Classified as a rare earth. It is radioactive with a short half-life. Berkelium is a human-made element and is not known to occur in nature.

- Electrons in first energy level: 2
- Electrons in second energy level: 8
- Electrons in third energy level: 18
- Electrons in fourth energy level: 32

- Electrons in fifth energy level: 26
- Electrons in sixth energy level: 9
- Electrons in seventh energy level: 2

Californium (98) Symbol, Cf. The most common isotope has atomic weight 251. Classified as a rare earth. It is radioactive, emitting neutrons in large quantities. It is human-made element, not known to occur in nature.

- Electrons in first energy level: 2
- Electrons in second energy level: 8
- Electrons in third energy level: 18
- Electrons in fourth energy level: 32
- Electrons in fifth energy level: 28
- Electrons in sixth energy level: 8
- Electrons in seventh energy level: 2

Einsteinium (99) Symbol, E or Es. The most common isotope has atomic weight 252. Classified as a rare earth. It is radioactive with a short half-life. Einsteinium is a human-made element and is not known to occur in nature.

- Electrons in first energy level: 2
- Electrons in second energy level: 8
- Electrons in third energy level: 18
- Electrons in fourth energy level: 32
- Electrons in fifth energy level: 29
- Electrons in sixth energy level: 8
- Electrons in seventh energy level: 2

Fermium (100) Symbol, Fm. The most common isotope has atomic weight 257. Classified as a rare earth. It has a short half-life, is human-made, and is not known to occur in nature.

- Electrons in first energy level: 2
- Electrons in second energy level: 8
- Electrons in third energy level: 18
- Electrons in fourth energy level: 32
- Electrons in fifth energy level: 30
- Electrons in sixth energy level: 8
- Electrons in seventh energy level: 2

Mendelevium (101) Symbol, Md or Mv. The most common isotope has atomic weight 258. Classified as a rare earth. It has a short half-life, is human-made, and is not known to occur in nature.

- Electrons in first energy level: 2
- Electrons in second energy level: 8
- Electrons in third energy level: 18
- Electrons in fourth energy level: 32
- Electrons in fifth energy level: 31
- Electrons in sixth energy level: 8
- Electrons in seventh energy level: 2

Nobelium (102) Symbol, No. The most common isotope has atomic weight 259. Classified as a rare earth. It has a short half-life (seconds or minutes, depending on the isotope), is human-made, and is not known to occur in nature.

- Electrons in first energy level: 2
- Electrons in second energy level: 8
- Electrons in third energy level: 18
- Electrons in fourth energy level: 32
- Electrons in fifth energy level: 32
- Electrons in sixth energy level: 8
- Electrons in seventh energy level: 2

Lawrencium (103) Symbol, Lr or Lw. The most common isotope has atomic weight 262. Classified as a rare earth. It has a half-life less than one minute, is human-made, and is not known to occur in nature.

- Electrons in first energy level: 2
- Electrons in second energy level: 8
- Electrons in third energy level: 18
- Electrons in fourth energy level: 32
- Electrons in fifth energy level: 32
- Electrons in sixth energy level: 9
- Electrons in seventh energy level: 2

Rutherfordium (104) Symbol, Rf. Also called *unnilquadium* (Unq) and *Kurchatovium* (Ku). The most common isotope has atomic weight 261. Classified as a transition metal. It has a half-life on the order of a few seconds to a few tenths of a second (depending on the isotope), is human-made, and is not known to occur in nature.

- Electrons in first energy level: 2
- Electrons in second energy level: 8
- Electrons in third energy level: 18
- Electrons in fourth energy level: 32
- Electrons in fifth energy level: 32
- Electrons in sixth energy level: 10
- Electrons in seventh energy level: 2

Dubnium (105) Symbol, Db. Also called *unnilpentium* (Unp) and *Hahnium* (Ha). The most common isotope has atomic weight 262. Classified as a transition metal. It has a half-life on the order of a few seconds to a few tenths of a second (depending on the isotope), is human-made, and is not known to occur in nature.

- Electrons in first energy level: 2
- Electrons in second energy level: 8
- Electrons in third energy level: 18
- Electrons in fourth energy level: 32
- Electrons in fifth energy level: 32
- Electrons in sixth energy level: 11
- Electrons in seventh energy level: 2

Seaborgium (106) Symbol, Sg. Also called *unnilhexium* (Unh). The most common isotope has atomic weight 263. Classified as a transition metal. It has a half-life on the order of one second or less, is human-made, and is not known to occur in nature.

- Electrons in first energy level: 2
- Electrons in second energy level: 8
- Electrons in third energy level: 18
- Electrons in fourth energy level: 32
- Electrons in fifth energy level: 32
- Electrons in sixth energy level: 12
- Electrons in seventh energy level: 2

Bohrium (107) Symbol, Bh. Also called *unnilseptium* (Uns). The most common isotope has atomic weight 262. Classified as a transition metal. It is human-made and is not known to occur in nature.

- Electrons in first energy level: 2
- Electrons in second energy level: 8
- Electrons in third energy level: 18
- Electrons in fourth energy level: 32
- Electrons in fifth energy level: 32
- Electrons in sixth energy level: 13
- Electrons in seventh energy level: 2

Hassium (108) Symbol, Hs. also called *unniloctium* (Uno). The most common isotope has atomic weight 265. Classified as a transition metal. It is human-made and not known to occur in nature.

- Electrons in first energy level: 2
- Electrons in second energy level: 8
- Electrons in third energy level: 18
- Electrons in fourth energy level: 32
- Electrons in fifth energy level: 32
- Electrons in sixth energy level: 14
- Electrons in seventh energy level: 2

Meitnerium (109) Symbol, Mt. Also called *unnilenium* (Une). The most common isotope has atomic weight 266. Classified as a transition metal. It is human-made and not known to occur in nature.

- Electrons in first energy level: 2
- Electrons in second energy level: 8
- Electrons in third energy level: 18
- Electrons in fourth energy level: 32
- Electrons in fifth energy level: 32
- Electrons in sixth energy level: 15
- Electrons in seventh energy level: 2

Ununnilium (110) Symbol, Uun. The most common isotope has atomic weight 269. Classified as a transition metal. It is human-made and not known to occur in nature.

- Electrons in first energy level: 2
- Electrons in second energy level: 8

- Electrons in third energy level: 18
- Electrons in fourth energy level: 32
- Electrons in fifth energy level: 32
- Electrons in sixth energy level: 17
- Electrons in seventh energy level: 1

Unununium (111) Symbol, Uuu. The most common isotope has atomic weight 272. Classified as a transition metal. It is human-made and not known to occur in nature.

- Electrons in first energy level: 2
- Electrons in second energy level: 8
- Electrons in third energy level: 18
- Electrons in fourth energy level: 32
- Electrons in fifth energy level: 32
- Electrons in sixth energy level: 18
- Electrons in seventh energy level: 1

Ununbium (112) Symbol, Uub. The most common isotope has atomic weight 277. Classified as a transition metal. It is human-made and not known to occur in nature.

- Electrons in first energy level: 2
- Electrons in second energy level: 8
- Electrons in third energy level: 18
- Electrons in fourth energy level: 32
- Electrons in fifth energy level: 32
- Electrons in sixth energy level: 18
- Electrons in seventh energy level: 2

(113) As of this writing, no identifiable atoms of an element with atomic number 113 have been reported. The synthesis of or appearance of such an atom is believed possible because of the observation of ununqadium (Uuq, element 114) in the laboratory.

Ununquadium (114) Symbol, Uuq. The most common isotope has atomic weight 285. First reported in January 1999. It is human-made and not known to occur in nature.

(115) As of this writing, no identifiable atoms of an element with atomic number 115 have been reported. The synthesis or appearance of such an atom is believed possible because of the observation of ununhexium (Uuh, element 116) in the laboratory.

Ununhexium (116) Symbol, Uuh. The most common isotope has atomic weight 289. First reported in January 1999. It is a decomposition product of ununoctium, and it in turn decomposes into ununquadium. It is not known to occur in nature.

(117) As of this writing, no identifiable atoms of an element with atomic number 117 have been reported. The synthesis or appearance of such an atom is believed possible because of the observation of ununoctium (Uuo, element 118) in the laboratory.

Ununoctium (118) Symbol, Uuo. The most common isotope has atomic weight 293. It is the result of the fusion of krypton and lead and decomposes into ununhexium. It is not known to occur in nature.

TABLE 1.17 Atomic Numbers, Periods, and Groups of the Elements (The Periodic Table)

Group Period	1	2		3	4	5	6	7	8	9	10	11	12	13	14	15	16	17	18
1	1																	1	2
	H																	H	He
2	3	4												5	6	7	8	9	10
	Li	Be												B	C	N	O	F	Ne
3	11	12												13	14	15	16	17	18
	Na	Mg												Al	Si	P	S	Cl	Ar
4	19	20		21	22	23	24	25	26	27	28	29	30	31	32	33	34	35	36
	K	Ca		Sc	Ti	V	Cr	Mn	Fe	Co	Ni	Cu	Zn	Ga	Ge	As	Se	Br	Kr
5	37	38		39	40	41	42	43	44	45	46	47	48	49	50	51	52	53	54
	Rb	Sr		Y	Zr	Nb	Mo	Tc	Ru	Rh	Pd	Ag	Cd	In	Sn	Sb	Te	I	Xe
6	55	56	*	71	72	73	74	75	76	77	78	79	80	81	82	83	84	85	86
	Cs	Ba		Lu	Hf	Ta	W	Re	Os	Ir	Pt	Au	Hg	Tl	Pb	Bi	Po	At	Rn
7	87	88	**	103	104	105	106	107	108	109	110	111	112	113	114	115	116	117	118
	Fr	Ra		Lr	Unq	Unp	Unh	Uns	Uno	Mt	Uun	Uuu	Uub	Uut	Uuq	Uup	Uuh	Uus	Uuo
*Lanthanides			*	57	58	59	60	61	62	63	64	65	66	67	68	69	70		
				La	Ce	Pr	Nd	Pm	Sm	Eu	Gd	Tb	Dy	Ho	Er	Tm	Yb		
†Actinides			**	89	90	91	92	93	94	95	96	97	98	99	100	101	102		
				Ac	Th	Pa	U	Np	Pu	Am	Cm	Bk	Cf	Es	Fm	Md	No		

TABLE 1.18 Atomic Weights of the Elements

Name	Atomic number	Symbol	Atomic weight
Actinium	89	Ac	[227]
Aluminium	13	Al	26.981538
Americium	95	Am	[243]
Antimony	51	Sb	121.76
Argon	18	Ar	39.948
Arsenic	33	As	74.9216
Astatine	85	At	[210]
Barium	56	Ba	137.327
Berkelium	97	Bk	[247]
Beryllium	4	Be	9.012182
Bismuth	83	Bi	8.98038
Bohrium	107	Bh	[264]
Boron	5	B	10.811
Bromine	35	Br	79.904
Cadmium	48	Cd	112.411
Caesium	55	Cs	132.90545
Calcium	20	Ca	40.078
Californium	98	Cf	[251]
Carbon	6	C	12.0107
Cerium	58	Ce	140.116
Chlorine	17	Cl	35.4527
Chromium	24	Cr	51.9961
Cobalt	27	Co	8.9332
Copper	29	Cu	63.546
Curium	96	Cm	[247]
Dubnium	105	Db	[262]
Dysprosium	66	Dy	162.5
Einsteinium	99	Es	[252]
Erbium	68	Er	167.26
Europium	63	Eu	151.964
Fermium	100	Fm	[257]
Fluorine	9	F	18.9984032
Francium	87	Fr	[223]
Gadolinium	64	Gd	157.25
Gallium	31	Ga	69.723
Germanium	32	Ge	72.61
Gold	79	Au	196.96655
Hafnium	72	Hf	178.49
Hassium	108	Hs	[265]
Helium	2	He	4.002602
Holmium	67	Ho	164.93032
Hydrogen	1	H	1.00794
Indium	49	In	114.818
Iodine	53	I	126.90447
Iridium	77	Ir	192.217
Iron	26	Fe	55.845
Krypton	36	Kr	83.8
Lanthanum	57	La	138.9055
Lawrencium	103	Lr	[262]
Lead	82	Pb	207.2
Lithium	3	Li	6.941
Lutetium	71	Lu	174.967
Magnesium	12	Mg	24.305
Manganese	25	Mn	54.938049
Meitnerium	109	Mt	[268]
Mendelevium	101	Md	[258]

TABLE 1.18 Atomic Weights of the Elements (*Continued*)

Name	Atomic number	Symbol	Atomic weight
Mercury	80	Hg	200.59
Molybdenum	42	Mo	95.94
Neodymium	60	Nd	144.24
Neon	10	Ne	20.1797
Neptunium	93	Np	[237]
Nickel	28	Ni	58.6934
Niobium	41	Nb	92.90638
Nitrogen	7	N	14.00674
Nobelium	102	No	[259]
Osmium	76	Os	190.23
Oxygen	8	O	15.9994
Palladium	46	Pd	106.42
Phosphorus	15	P	30.973761
Platinum	78	Pt	195.078
Plutonium	94	Pu	[244]
Polonium	84	Po	[209]
Potassium	19	K	39.0983
Praseodymium	59	Pr	140.90765
Promethium	61	Pm	[145]
Protactinium	91	Pa	231.03588
Radium	88	Ra	[226]
Radon	86	Rn	[222]
Rhenium	75	Re	186.207
Rhodium	45	Rh	102.9055
Rubidium	37	Rb	85.4678
Ruthenium	44	Ru	101.07
Rutherfordium	104	Rf	[261]
Samarium	62	Sm	150.36
Scandium	21	Sc	44.95591
Seaborgium	106	Sg	[263]
Selenium	34	Se	78.96
Silicon	14	Si	28.0855
Silver	47	Ag	107.8682
Sodium	11	Na	22.98977
Strontium	38	Sr	87.62
Sulfur	16	S	32.066(6)
Tantalum	73	Ta	180.9479
Technetium	43	Tc	[98]
Tellurium	52	Te	127.6
Terbium	65	Tb	158.92534
Thallium	81	Tl	204.3833
Thorium	90	Th	232.0381
Thulium	69	Tm	168.93421
Tin	50	Sn	118.71
Titanium	22	Ti	47.867
Tungsten	74	W	183.84
Ununbium	112	Uub	[277]
Ununnilium	110	Uun	[269]
Unununium	111	Uuu	[272]
Uranium	92	U	238.0289
Vanadium	23	V	50.9415
Xenon	54	Xe	131.29
Ytterbium	70	Yb	173.04
Yttrium	39	Y	88.90585
Zinc	30	Zn	65.39
Zirconium	40	Zr	91.224

TABLE 1.19 Physical Properties of the Elements

The relative atomic masses in the following table are based on the $^{12}C = 12$ scale; a value in brackets denotes the mass number of the most stable isotope. The data are based on the most recent values adopted by IUPAC, with a maximum of six significant figures.
ρ denotes density, $\theta_{C,m}$ denotes melting temperature, $\theta_{C,b}$ denotes boiling temperature, and c_p denotes specific heat capacity.
subl. denotes sublimes

Element	Symbol	Atomic number	Relative atomic mass	ρ/g cm^{-3}	$\theta_{C,m}$/°C	$\theta_{C,b}$/°C	c_p/J kg^{-1} K^{-1}	Oxidation states
Actinium	Ac	89	227.028	10.1	1050	3200		3
Aluminium	Al	13	26.9815	2.70	660	2470	900	3
Americium	Am	95	(243)	11.7	(1200)	(2600)	140	3, 4, 5, 6
Antimony	Sb	51	121.75	6.62	630	1380	209	3, 5
Argon	Ar	18	39.948	1.40 (87 K)	−189	−186	519	
Arsenic (α, grey)	As	33	74.9216	5.72		613 subl.	326	3, 5
Astatine	At	85	(210)		(302)	(380)	(140)	
Barium	Ba	56	137.33	3.51	714	1640	192	2
Berkelium	Bk	97	(247)					3, 4
Beryllium	Be	4	9.01218	1.85	1280	2477	1.82×10^3	2
Bismuth	Bi	83	208.980	9.80	271	1560	121	3,5
Boron	B	5	10.81	2.34	2300	3930	1.03×10^3	3
Bromine	Br	35	79.904	3.12	−7.2	58.8	448	1, 3, 4, 5, 6
Cadmium	Cd	48	112.41	8.64	321	765	230	2
Caesium	Cs	55	132.905	1.90	28.7	690	234	1
Calcium	Ca	20	40.08	1.54	850	1487	653	2
Californium	Cf	98	(251)					3
Carbon	C	6	12.011	2.25 (graphite) 3.51 (diamond)	3730 subl.	4830	711 (graphite) 519 (diamond)	2,4
Cerium	Ce	58	140.12	6.78	795	3470	184	3,4
Chlorine	Cl	17	35.453	1.56 (238 K)	−101	−34.7	477	1, 3, 4, 5, 6, 7
Chromium	Cr	24	51.996	7.19	1890	2482	448	2,3,6
Cobalt	Co	27	58.9332	8.90	1492	2900	435	2,3
Copper	Cu	29	63.546	8.92	1083	2595	385	1,2
Curium	Cm	96	(247)					3
Dysprosium	Dy	66	162.50	8.56	1410	2600	172	3
Einsteinium	Es	99	(252)					3

Erbium	Er	68	167.26	9.16	1500	2900	167	3
Europium	Eu	63	151.96	5.24	826	1440	138	2, 3
Fermium	Fm	100	(257)					3
Fluorine	F	9	18.9984	1.11 (85 K)	−220	−188	824	1
Francium	Fr	87	(223)		(27)	(680)	(140)	1
Gadolinium	Gd	64	157.25	7.95	1310	3000	234	3
Gallium	Ga	31	69.72	5.91	29.8	2400	381	3
Germanium	Ge	32	72.59	5.35	937	2830	322	4
Gold	Au	79	196.967	19.3	1063	2970	130	1, 3
Hafnium	Hf	72	178.49	13.3	2220	5400	146	4
Helium	He	2	4.00260	0.147 (4 K)	−270	−269	5.19×10^3	
Holmium	Ho	67	164.930	8.80	1460	2600	163	3
Hydrogen	H	1	1.0079	0.070 (20 K)	−259	−252	1.43×10^4	1
Indium	In	49	114.82	7.30	157	2000	238	1, 3
Iodine	I	53	126.905	4.93	114	184	218	1, 3, 5, 7
Iridium	Ir	77	192.22	22.5	2440	5300	134	2, 3, 4, 6
Iron	Fe	26	55.847	7.86	1535	3000	448	2, 3, 6
Krypton	Kr	36	83.80	2.16 (121 K)	−157	−152	247	2
Lanthanum	La	57	138.906	6.19	920	3470	201	3
Lawrencium	Lr	103	(260)					
Lead	Pb	82	207.2	11.3	327	1744	130	2, 4
Lithium	Li	3	6.941	0.53	180	1330	3.39×10^3	1
Lutetium	Lu	71	174.967	9.84	1650	3330	155	3
Magnesium	Mg	12	24.305	1.74	650	1110	1.03×10^3	2
Manganese	Mn	25	54.9380	7.20	1240	2100	477	2, 3, 4, 6, 7
Mendelevium	Md	101	(258)					3
Mercury	Hg	80	200.59	13.6	−38.9	357	138	1, 2
Molybdenum	Mo	42	95.94	10.2	2610	5560	251	2, 3, 4, 5, 6
Neodymium	Nd	60	144.24	7.00	1020	3030	188	3
Neon	Ne	10	20.179	1.20 (27 K)	−249	−246	1.03×10^3	
Neptunium	Np	93	237.048	20.4	640			3, 4, 5, 6
Nickel	Ni	28	58.69	8.90	1453	2730	439	2, 3
Niobium	Nb	41	92.9064	8.57	2470	3300	264	3, 5

(Continued)

TABLE 1.19 Physical Properties of the Elements (*Continued*)

Element	Symbol	Atomic number	Relative atomic mass	ρ/g cm^{-3}	$\theta_{C,m}$/°C	$\theta_{C,b}$/°C	c_p/J kg^{-1} K^{-1}	Oxidation states
Nitrogen	N	7	14.0067	0.808 (77 K)	−210	−196	1.04×10^3	1, 2, 3, 4, 5
Nobelium	No	102	(259)					
Osmium	Os	76	190.2	22.5	3000	5000	130	2, 3, 4, 6, 8
Oxygen	O	8	15.9994	1.15 (90 K)	−218	−183	916	2
Palladium	Pd	46	106.42	12.0	1550	3980	243	2, 4
Phosphorus	P	15	30.9738	1.82 (white) 2.34 (red)	44.2 (white) 590 (red)	280 (white)	757 (white) 670 (red)	3, 5
Platinum	Pt	78	195.08	21.4	1769	4530	134	2, 4, 6
Plutonium	Pu	94	(244)	19.8	640	3240		3, 4, 5, 6
Polonium	Po	84	(209)	9.4	254	960	126	2, 4
Potassium	K	19	39.0983	0.86	63.7	774	753	1
Praseodymium	Pr	59	140.908	6.78	935	3130	192	3, 4
Promethium	Pm	61	(145)		1030	2730	184	3
Protoactinium	Pa	91	231.036	15.4	1230		121	4, 5
Radium	Ra	88	226.025	5.0	700	1140	121	2
Radon	Rn	86	(222)	4.4 (211 K)	−71	−61.8	92	
Rhenium	Re	75	186.207	20.5	3180	5630	138	2, 4, 5, 6, 7
Rhodium	Rh	45	102.906	12.4	1970	4500	243	2, 3, 4
Rubidium	Rb	37	85.4678	1.53	38.9	688	360	1
Ruthenium	Ru	44	101.07	12.3	2500	4900	238	3, 4, 5, 6, 8
Samarium	Sm	62	150.36	7.54	1070	1900	197	2, 3
Scandium	Sc	21	44.9559	2.99	1540	2730	556	3
Selenium	Se	34	78.96	4.81	217	685	322	2, 4, 6
Silicon	Si	14	28.0855	2.33	1410	2360	711	4
Silver	Ag	47	107.868	10.5	961	2210	234	1
Sodium	Na	11	22.9898	0.97	97.8	890	1.23×103	1
Strontium	Sr	38	87.62	2.62	768	1380	284	2
Sulphur (α, rhombic)	S	16	32.06	2.07 (α) 1.96 (β)	113 (α) 119 (β)	445	732	2, 4, 6
Tantalum	Ta	73	180.948	16.6	3000	5420	138	5
Technetium	Tc	43	(98)	11.5	2200	3500	243	7
Tellurium	Te	52	127.60	6.25	450	990	201	2, 4, 6

Terbium	Tb	65	158.925	8.27	1360	2800	184	3, 4
Thallium	Tl	81	204.383	11.8	304	1460	130	1, 3
Thorium	Th	90	232.038	11.7	1750	3850	113	3, 4
Thulium	Tm	69	168.934	9.33	1540	1730	159	2, 3
Tin (white)	Sn	50	118.71	7.28 (white) 5.75 (grey)	232	2270	218	2, 4
Titanium	Ti	22	47.88	4.54	1675	3260	523	2, 3, 4
Tungsten	W	74	183.85	19.4	3410	5930	134	2, 4, 5, 6
Uranium	U	92	238.029	19.1	1130	3820	117	3, 4, 5, 6
Vanadium	V	23	50.9415	5.96	1900	3000	481	2, 3, 4, 5
Xenon	Xe	54	131.29	3.52 (165 K)	−112	−108	159	2, 4, 6, 8
Ytterbium	Yb	70	173.04	6.98	824	1430	146	2, 3
Yttrium	Y	39	88.9059	4.34	1500	2930	297	3
Zinc	Zn	30	65.39	7.14	420	907	385	2
Zirconium	Zr	40	91.224	6.49	1850	3580	276	2, 3, 4

TABLE 1.20 Conductivity and Resistivity of the Elements

Name	Symbol	Atomic number	Electronic configuration	Thermal conductivity, $W \cdot (m \cdot K)^{-1}$ at 25°C	Electrical resistivity, $\mu\Omega \cdot cm$ at 20°C	Coefficient of linear thermal expansion (25°C), $m \cdot m^{-1}(\times 10^6)$
Actinium	Ac	89	[Rn] $6d^2\ 7s$	12		
Aluminum	Al	13	[Ne] $3s^2\ 3p$	237	2.6548	23.1
Americium	Am	95	[Rn] $5f^7\ 7s^2$	10		
Antimony	Sb	51	[Kr] $4d^{10}\ 5s^2\ 5p^3$	24.4	41.7	11.0
Argon	Ar	18	[Ne] $3s^2\ 3p^6$	0.017 72		
Arsenic	As	33	[Ar] $3d^{10}\ 4s^2\ 4p^3$	50.2	33.3	
Astatine	At	85	[Xe] $4f^{14}\ 5d^{10}\ 6s^2\ 6p^5$	1.7		
Barium	Ba	56	[Xe] $6s^2$	18.4	33.2	20.6
Berkelium	Bk	97	[Rn] $5f^8\ 6d\ 7s^2$	10		
Beryllium	Be	4	[He] $2s^2$	200	3.56	11.3
Bismuth	Bi	83	[Xe] $4f^{14}\ 5d^{10}\ 6s^2\ 6p^3$	7.97	129	13.4
Boron	B	5	[He] $2s^2\ 2p$	27.4	1.5×10^{12}	5–7
Bromine	Br	35	[Ar] $3d^{10}\ 4s^2\ 4p^5$	0.122	7.8×10^{18}	
Cadmium	Cd	48	[Kr] $4d^{10}\ 5s^2$	96.6	7.27 (22°C)	30.8
Calcium	Ca	20	[Ar] $4s^2$	201	3.36	22.3
Californium	Cf	98	[Rn] $5f^{10}\ 7s^2$			
Carbon	C	6	[He] $2s^2\ 2p^2$			
(amorphous)				1.59		
(diamond)				900–2320	0.8	
(graphite)				119–165	1375	
Cerium	Ce	58	[Xe] $4f\ 5d\ 6s^2$	11.3	82.8 (β, hex)	6.3
Cesium	Cs	55	[Xe] $6s$	35.9	20.5	
Chlorine	Cl	17	[Ne] $3s^2\ 3p^5$	0.0089	$>10^9$	
Chromium	Cr	24	[Ar] $3d^5\ 4s$	93.9	12.5	4.9
Cobalt	Co	27	[Ar] $3d^7\ 4s^2$	100	6.24	13.0
Copper	Cu	29	[Ar] $3d^{10}\ 4s$	401	1.678	16.5
Curium	Cm	96	[Rn] $5f^7\ 6d\ 7s^2$			
Dysprosium	Dy	66	[Xe] $4f^{10}\ 6s^2$	10.7	92.6	9.9
Einsteinium	Es	99	[Rn] $5f^{11}\ 7s^2$			
Erbium	Er	68	[Xe] $4f^{14}\ 6s^2$	14.5	86.0	12.2
Europium	Eu	63	[Xe] $4f^7\ 6s^2$	13.9	90.0	35.0

Fermium	Fm	100	[Rn] $5f^{12}\ 7s^2$			
Fluorine	F	9	[He] $2s^2\ 2p^5$	0.0277		
Francium	Fr	87	[Rn] $7s$			
Gadolinium	Gd	64	[Xe] $4f^7\ 5d\ 6s^2$	10.5	131	9.4 (100°C)
Gallium	Ga	31	[Ar] $3d^{10}\ 4s^2\ 4p$	29.4(lq) 40.6(c)	25.795 (30°C)	120
Germanium	Ge	32	[Ar] $3d^{10}\ 4s^2\ 4p^2$	60.2	53 000	6.0
Gold (aurum)	Au	79	[Xe] $4f^{14}\ 5d^{10}\ 6s$	318	2.214	14.2
Hafnium	Hf	72	[Xe] $4f^{14}\ 5d^2\ 6s^2$	23.0	33.1	5.9
Helium	He	2	$1s^2$	0.1513		
Holmium	Ho	67	[Xe] $4f^{11}\ 6s^2$	16.2	81.4	11.2
Hydrogen	H	1	$1s$	0.1805		
Indium	In	49	[Kr] $4d^{10}\ 5s^2\ 5p$	81.8	8.37	32.1
Iodine	I	53	[Kr] $4d^{10}\ 5s^2\ 5p^5$	449	1.3×10^{15} (0°C)	
Iridium	Ir	77	[Xe] $4f^{14}\ 5d^7\ 6s^2$	147	4.71	6.4
Iron	Fe	26	[Ar] $3d^6\ 4s^2$	80.4	9.61	11.8
Krypton	Kr	36	[Ar] $3d^{10}\ 4s^2\ 4p^6$	9.43		
Lanthanum	La	57	[Xe] $5d\ 6s^2$	13.4	61.5	12.1
Lawrencium	Lr	103	[Rn] $4f^{14}\ 6d\ 7s^2$			
Lead	Pb	82	[Xe] $4f^{14}\ 5d^{10}\ 6s^2\ 6p^2$	35.3	20.8	28.9
Lithium	Li	3	$1s^2\ 2s$	84.8	9.28	46
Lutetium	Lu	71	[Xe] $4f^{14}\ 5d\ 6s^2$	16.4	58.2	9.9
Magnesium	Mg	12	[Ne] $3s^2$	156	4.39	24.8
Manganese	Mn	25	[Ar] $3d^5\ 4s^2$	7.81	144	21.7
Mendelevium	Md	101	[Rn] $5f^{13}\ 7s^2$			
Mercury	Hg	80	[Xe] $4f^{14}\ 5d^{10}\ 6s^2$	8.30	95.8(lq); 21(c)	
Molybdenum	Mo	42	[Kr] $4d^5\ 5s$	138	5.34	4.8
Neodymium	Nd	60	[Xe] $4f^4\ 6s^2$	16.5	64.3	9.6
Neon	Ne	10	$1s^2\ 2s^2\ 2p^6$	0.0491		
Neptunium	Np	93	[Rn] $5f^4\ 6d\ 7s^2$	6.3	122.0 (22°C)	
Nickel	Ni	28	[Ar] $3d^8\ 4s^2$	90.9	6.93	13.4
Niobium	Nb	41	[Kr] $4d^4\ 5s$	53.7	15.2 (0°C)	7.3
Nitrogen	N	7	$1s^2\ 2s^2\ 2p^3$	0.025 83		
Nobelium	No	102	[Rn] $5f^{14}\ 7s^2$			
Osmium	Os	76	[Xe] $4f^{14}\ 5d^6\ 6s^2$	87.6	8.12 (0°C)	5.1
Oxygen	O	8	$1s^2\ 2s^2\ 2p^4$	0.026 58 (g) 0.149 (lq)		
Palladium	Pd	46	[Kr] $4d^{10}$	71.8	10.54	11.8

(*Continued*)

TABLE 1.20 Conductivity and Resistivity of the Elements (*Continued*)

Name	Symbol	Atomic number	Electronic configuration	Thermal conductivity, $W \cdot (m \cdot K)^{-1}$ at 25°C	Electrical resistivity, $\mu\Omega \cdot cm$ at 20°C	Coefficient of linear thermal expansion (25°C), $m \cdot m^{-1} (\times 10^6)$
Phosphorus	P	15	[Ne] $3s^2\ 3p^3$	0.236 17	10	
Platinum	Pt	78	[Xe] $4f^{14}\ 5d^9\ 6s$	71.6	10.6	8.8
Plutonium	Pu	94	[Rn] $5f^6\ 7s^2$	6.74	146.0 (0°C)	46.7
Polonium	Po	84	[Xe] $4f^{14}\ 5d^{10}\ 6s^2\ 6p^4$	0.2	40.0 (0°C) alpha	
Potassium	K	19	[Ar] $4s$	102.5	7.2	
Praseodymium	Pr	59	[Xe] $4f^3\ 6s^2$	12.5	70.0	6.7
Promethium	Pm	61	[Xe] $4f^5\ 6s^2$	17.9	64.0 (25°C)	est [11.]
Protactinium	Pa	91	[Rn] $5f^2\ 6d\ 7s^2$	47	19.1 (22°C)	
Radium	Ra	88	[Rn] $7s^2$	18.6	100	
Radon	Rn	86	[Xe] $4f^{14}\ 5d^{10}\ 6s^2\ 6p^6$	0.003 61		
Rhenium	Re	75	[Xe] $5f^{14}\ 5d^5\ 6s^2$	48.0	19.3	6.2
Rhodium	Rh	45	[Kr] $4d^8\ 5s$	150	4.33 (0°C)	8.2
Rubidium	Rb	37	[Kr] $5s$	58.2	12.8	
Ruthenium	Ru	44	[Kr] $4d^7\ 5s$	117	7.1 (0°C)	6.4
Samarium	Sm	62	[Xe] $4f^6\ 6s^2$	13.3	94.0	12.7
Scandium	Sc	21	[Ar] $3d\ 4s^2$	15.8	56.2	10.2
Selenium	Se	34	[Ar] $3d^{10}\ 4s^2\ 4p^4$	0.519	1.2 (0°C)	37
Silicon	Si	14	[Ne] $3s^2\ 3p^2$	149	10^5	
Silver	Ag	47	[Kr] $4d^{10}\ 5s$	429	1.587	18.9
Sodium	Na	11	[Ne] $3s$	142	4.77	71
Strontium	Sr	38	[Kr] $5s^2$	35.4	13.2	22.5
Sulfur	S	16	[Ne] $3s^2\ 3p^4$	0.205	2×10^{23}	
Tantalum	Ta	73	[Xe] $4f^{14}\ 5d^3\ 6s^2$	57.5	13.5	6.3
Technetium	Tc	43	[Kr] $4d^5\ 5s^2$	50.6	22.6 (100°C)	
Tellurium	Te	52	[Kr] $4d^{10}\ 5s^2\ 5p^4$	1.97–3.38	$(5.8–33) \times 10^3$	
Terbium	Tb	65	[Xe] $4f^9\ 6s^2$	11.1	115	10.3
Thallium	Tl	78	[Xe] $4f^{14}\ 5d^{10}\ 6s^2\ 6p$	46.1	18	29.9
Thorium	Th	90	[Rn] $6d^2\ 7s^2$	54.0	15.4 (22°C)	11.1
Thullium	Tm	69	[Xe] $4f^{13}\ 6s^2$	16.9	67.6	13.3
Tin (stannum)	Sn	50	[Kr] $4d^{10}\ 5s^2\ 5p^2$	66.8	11.5 (0°C)	22.0

Titanium	Ti	22	[Ar] $3d^2\ 4s^2$	21.9	42.0	8.6
Tungsten (wolframium)	W	74	[Xe] $4f^{14}\ 5d^4\ 6s^2$	173	5.28	4.5
Uranium	U	92	[Rn] $5f^3\ 6d\ 7s^2$	27.5	28.0 (0°C)	13.9
Vanadium	V	23	[Ar] $3d^3\ 4s^2$	30.7	19.7	8.4
Xenon	Xe	54	[Kr] $4d^{10}\ 5s^2\ 5p^6$	0.005 65		
Ytterbium	Yb	70	[Xe] $4f^{14}\ 6s^2$	38.5	25	26.3
Yttrium	Y	39	[Kr] $4d\ 5s^2$	17.2	59.6	10.6
Zinc	Zn	30	[Ar] $3d^{10}\ 4s^2$	116	5.9	30.2
Zirconium	Zr	40	[Kr] $4d^2\ 5s^2$	22.6	42.1	5.7

TABLE 1.21 Work Functions of the Elements

The work function ϕ is the energy necessary to just remove an electron from the metal surface in thermoelectric or photoelectric emission. Values are dependent upon the experimental technique (vacua of 10^{-9} or 10^{-10} torr, clean surfaces, and surface conditions including the crystal face identification).

Element	ϕ, eV	Element	ϕ, eV	Element	ϕ,eV
Ag	4.64	Hg	4.50	Ru	4.80
Al	4.19	In	4.08	Sb	4.56
As	(3.75)	Ir	5.6	Sc	3.5
Au	5.32	K	2.30	Se	5.9
B	(4.75)	La	3.40	Si	4.85
Ba	2.35	Li	3.10	Sm	2.95
Be	5.08	Mg	3.66	Sn	4.35
Bi	4.36	Mn	3.90	Sr	2.76
C	(5.0)	Mo	4.30	Ta	4.22
Ca	2.71	Na	2.70	Tb	3.0
Cd	4.12	Nb	4.20	Te	4.70
Ce	2.80	Nd	3.1	Th	3.71
Co	4.70	Ni	5.15	Ti	4.10
Cr	4.40	Os	4.83	Tl	4.02
Cs	1.90	Pb	4.18	U	3.70
Cu	4.70	Pd	5.00	V	4.44
Eu	2.50	Po	4.6	W	4.55
Fe	4.65	Pr	2.7	Y	3.1
Ga	4.25	Pt	5.40	Zn	4.30
Ge	5.0	Rb	2.20	Zr	4.00
Gd	3.1	Re	4.95		
Hf	3.65	Rh	4.98		

TABLE 1.22 Relative Abundances of Naturally Occurring Isotopes

Element	Mass number	Percent	Element	Mass number	Percent
Aluminum	27	100	Cadmium	106	1.25(4)
Antimony	121	57.21(5)		108	0.89(2)
	123	42.79(5)		110	12.49(12)
Argon	36	0.337(3)		111	12.80(8)
	38	0.063(1)		112	24.13(14)
	40	99.600(3)		113	12.22(8)
Arsenic	75	100		114	28.7(3)
Barium	130	0.106(2)		116	7.49(9)
	132	0.101(2)	Calcium	40	96.941(18)
	134	2.42(3)		42	0.647(9)
	135	6.59(2)		43	0.135(6)
	136	7.85(4)		44	2.088(12)
	137	11.23(4)		46	0.004(3)
	138	71.70(7)		48	0.187(4)
Beryllium	9	100	Carbon	12	98.89(1)
Bismuth	209	100		13	1.11(1)
Boron	10	19.9(2)	Cerium	136	0.19(1)
	11	80.1(2)		138	0.25(1)
Bromine	79	50.69(7)		140	88.43(10)
	81	49.31(7)		142	11.13(10)

TABLE 1.22 Relative Abundances of Naturally Occurring Isotopes (*Continued*)

Element	Mass number	Percent	Element	Mass number	Percent
Cesium	133	100	Iodine	127	100
Chlorine	35	75.77(7)	Iridium	191	37.27(9)
	37	24.23(7)		193	62.73(9)
Chromium	50	4.345(13)	Iron	54	5.85(4)
	52	83.79(2)		56	91.75(4)
	53	9.50(2)		57	2.12(1)
	54	2.365(7)		58	0.26(1)
Cobalt	59	100	Krypton	78	0.35(2)
Copper	63	69.17(3)		80	2.25(2)
	65	30.83(3)		82	11.6(1)
Dysprosium	156	0.06(1)		83	11.5(1)
	158	0.10(1)		84	57.0(3)
	160	2.34(6)		86	17.3(2)
	161	18.9(2)	Lanthanum	138	0.0902(2)
	162	25.5(2)		139	99.9098(2)
	163	24.9(2)	Lead	204	1.4(1)
	164	28.2(2)		206	24.1(1)
Erbium	162	0.14(1)		207	22.1(1)
	164	1.61(2)		208	52.4(1)
	166	33.6(2)	Lithium	6	7.5(2)
	167	22.95(15)		7	92.5(2)
	168	26.8(2)	Lutetium	175	97.41(2)
	170	14.9(2)		176	2.59(2)
Europium	151	47.8(5)	Magnesium	24	78.99(3)
	153	52.2(5)		25	10.00(1)
Fluorine	19	100		26	11.01(2)
Gadolinium	152	0.20(1)	Manganese	55	100
	154	2.18(3)	Mercury	196	0.15(1)
	155	14.80(5)		198	9.97(8)
	156	20.47(4)		199	16.87(10)
	157	15.65(3)		200	23.10(16)
	158	24.84(12)		201	13.18(8)
	160	21.86(4)		202	29.86(20)
Gallium	69	60.108(9)		204	6.87(4)
	71	39.892(9)	Molybdenum	92	14.84(4)
Germanium	70	21.23(4)		94	9.25(3)
	72	27.66(3)		95	15.92(5)
	73	7.73(1)		96	16.68(5)
	74	35.94(2)		97	9.55(3)
	76	7.44(2)		98	24.13(7)
Gold	197	100		100	9.63(3)
Hafnium	174	0.162(3)	Neodymium	142	27.13(12)
	176	5.206(5)		143	12.18(6)
	177	18.606(13)		144	23.80(12)
	178	27.297(4)		145	8.30(6)
	179	13.629(6)		146	17.19(9)
	180	35.100(7)		148	5.76(3)
Helium	4	100		150	5.64(3)
Holmium	165	100	Neon	20	90.48(3)
Hydrogen	1	99.985(1)		21	0.27(1)
	2	0.015(1)		22	9.25(3)
Indium	113	4.29(2)	Nickel	58	68.077(9)
	115	95.71(2)		60	26.223(8)

(*Continued*)

TABLE 1.22 Relative Abundances of Naturally Occurring Isotopes (*Continued*)

Element	Mass number	Percent	Element	Mass number	Percent
	61	1.140(1)		154	22.7(2)
	62	3.634(2)	Scandium	45	100
	64	0.926(1)	Selenium	74	0.89(2)
Niobium	93	100		76	9.36(11)
Nitrogen	14	99.634(9)		77	6.63(6)
	15	0.366(9)		78	23.78(9)
Osmium	184	0.020(3)		80	49.61(10)
	186	1.58(2)		82	8.73(6)
	187	1.6(4)	Silicon	28	92.23(2)
	188	13.3(1)		29	4.67(2)
	189	16.1(1)		30	3.10(1)
	190	26.4(2)	Silver	107	51.839(7)
	192	41.0(3)		109	48.161(7)
Oxygen	16	99.76(1)	Sodium	23	100
	17	0.04	Strontium	84	0.56(1)
	18	0.20(1)		86	9.86(1)
Palladium	102	1.02(1)		87	7.00(1)
	104	11.14(8)		88	82.58(1)
	105	22.33(8)	Sulfur	32	95.02(9)
	106	27.33(3)		33	0.75(4)
	108	26.46(9)		34	4.21(8)
	110	11.72(9)		36	0.02(1)
Phosphorus	31	100	Tantalum	180	0.012(2)
Platinum	190	0.01(1)		181	99.988(2)
	192	0.79(6)	Tellurium	120	0.096(2)
	194	32.9(6)		122	2.603(4)
	195	33.8(6)		123	0.908(2)
	196	25.3(6)		124	4.816(6)
	198	7.2(2)		125	7.139(6)
Potassium	39	93.258(4)		126	18.952(11)
	40	0.0117(1)		128	31.687(11)
	41	6.730(3)		130	33.799(10)
Praseodymium	141	100	Terbium	159	100
Protoactinium	230	100	Thallium	203	29.52(1)
Rhenium	185	37.40(2)		205	70.48(1)
	187	62.60(2)	Thorium	228	100
Rhodium	103	100	Thullium	169	100
Rubidium	85	72.17(2)	Tin	112	0.97(1)
	87	27.83(2)		114	0.65(1)
Ruthenium	96	5.52(6)		115	0.34(1)
	98	1.88(6)		116	14.53(11)
	99	12.7(1)		117	7.68(7)
	100	12.6(1)		118	24.23(11)
	101	17.0(1)		119	8.59(4)
	102	31.6(2)		120	32.59(10)
	104	18.7(2)		122	4.63(3)
Samarium	144	3.1(1)		124	5.79(5)
	147	15.0(2)	Titanium	46	8.25(3)
	148	11.3(1)		47	7.44(2)
	149	13.8(1)		48	73.72(3)
	150	7.4(1)		49	5.41(2)
	152	26.7(2)		50	5.4(1)

TABLE 1.22 Relative Abundances of Naturally Occurring Isotopes (*Continued*)

Element	Mass number	Percent
Tungsten	180	0.12(1)
	182	26.50(3)
	183	14.31(1)
	184	30.64(1)
	186	28.43(4)
Uranium	234	0.0055(5)
	235	0.720(1)
	238	99.275(2)
Vanadium	50	0.250(2)
	51	99.750(2)
Xenon	124	0.10(1)
	126	0.09(1)
	128	1.91(3)
	129	26.4(6)
	130	4.1(1)
	131	21.2(4)
	132	26.9(5)
	134	10.4(2)
	136	8.9(1)
Ytterbium	168	0.13(1)
	170	3.05(6)
	171	14.3(2)
	172	21.9(3)
	173	16.12(2)
	174	31.8(4)
	176	12.7(2)
Yttrium	89	100
Zinc	64	48.6(3)
	66	27.9(2)
	67	4.1(1)
	68	18.8(4)
	70	0.6(1)
Zirconium	90	51.45(3)
	91	11.22(4)
	92	17.15(2)
	94	17.38(4)
	96	2.80(2)

TABLE 1.23 Radioactivity of the Elements (Neptunium Series)

Element	Symbol	Radiation	Half-life
Plutonium ↓	^{241}Pu	β	13.2 years
Americium ↓	^{241}Am	α	462 years
Neptunium ↓	^{237}Np	α	2.20×10^6 years
Protactinium ↓	^{233}Pa	β	27.4 days
Uranium ↓	^{233}U	α	1.62×10^5 years
Thorium ↓	^{229}Th	α	7.34×10^3 years
Radium ↓	^{225}Ra	β	14.8 days
Actinium ↓	^{225}Ac	α	10.0 days
Francium ↓	^{221}Fr	α	4.8 min
Astatine ↓	^{217}At	α	1.8×10^{-2} sec

(*Continued*)

TABLE 1.23 Radioactivity of the Elements (Neptunium Series) (*Continued*)

Element	Symbol	Radiation	Half-life
Bismuth 98% ∣ 2% (branches to Polonium and Thallium)	^{213}Bi	β and α	47 min
↓ Polonium	^{213}Po	α	4.2×10^{-6} sec
↓ Thallium	^{209}Tl	β	2.2 min
↓ Lead	^{209}Pb	β	3.32 hr
↓ Bismuth (End Product)	^{209}Bi	Stable	—

TABLE 1.24 Radioactivity of the Elements (Thorium Series)

Radioelement	Corresponding element	Symbol	Radiation	Half-life
Thorium	Thorium	^{232}Th	α	1.39×10^{10} years
↓ Mesothorium I	Radium	^{228}Ra	β	6.7 years
↓ Mesothorium II	Actinium	^{228}Ac	β	6.13 hr
↓ Radiothorium	Thorium	^{228}Th	α	1.91 years
↓ Thorium X	Radium	^{224}Ra	α	3.64 days
↓ Th Emanation	Radon	^{220}Rn	α	52 sec
↓ Thorium A	Polonium	^{216}Po	α	0.16 sec
↓ Thorium B	Lead	^{212}Pb	β	10.6 hr
↓ Thorium C 66.3% ∣ 33.7% (branches to Thorium C′ and Thorium C″)	Bismuth	^{212}Bi	β and α	60.5 min
↓ Thorium C′	Polonium	^{212}Po	α	3×10^{-7} sec
↓ Thorium C″	Thallium	^{208}Tl	β	3.1 min
↓ Thorium D (End Product)	Lead	^{208}Pb	Stable	—

TABLE 1.25 Radioactivity of the Elements (Actinium Series)

Radioelement	Corresponding element	Symbol	Radiation	Half-life
Actinouranium ↓	Uranium	^{235}U	α	7.13×10^8 years
Uranium Y ↓	Thorium	^{231}Th	β	25.6 hr
Protactinium ↓	Protactinium	^{231}Pa	α	3.43×10^4 years
Actinium 98.8% | 1.2%	Actinium	^{227}Ac	β and α	21.8 years
Radioactinium	Thorium	^{227}Th	α	18.4 days
Actinium K ↓	Francium	^{223}Fr	β	21 min
Actinium X ↓	Radium	^{223}Ra	α	11.7 days
Ac Emanation ↓	Radon	^{219}Rn	α	3.92 sec
Actinium A ~100% | ~5 × 10⁻⁴%	Polonium	^{215}Po	α and β	1.83×10^{-3} s
Actinium B	Lead	^{211}Pb	β	36.1 min
Astatine-215 ↓	Astatine	^{215}At	α	$\sim 10^{-4}$ sec
Actinium C 99.7% | 0.3%	Bismuth	^{211}Bi	α and β	2.16 min
Actinium C′	Polonium	^{211}Po	α	0.52 sec
Actinium C″ ↓	Thallium	^{207}Tl	β	4.8 min
Actinium D (End Product)	Lead	^{207}Pb	Stable	—

TABLE 1.26 Radioactivity of the Elements (Uranium Series)

Radioelement	Corresponding element	Symbol	Radiation	Half-life
Uranium I ↓	Uranium	^{238}U	α	4.51×10^9 years
Uranium X_1 ↓	Thorium	^{234}Th	β	24.1 days
Uranium X_2* ↓	Protactinium	^{234}Pa	β	1.18 min
Uranium II ↓	Uranium	^{234}U	α	2.48×10^5 years
Ionium ↓	Thorium	^{230}Th	α	8.0×10^4 years
Radium ↓	Radium	^{226}Ra	α	1.62×10^3 years

(Continued)

TABLE 1.26 Radioactivity of the Elements (Uranium Series) (*Continued*)

Radioelement	Corresponding element	Symbol	Radiation	Half-life
Ra Emanation ↓	Radon	^{222}Rn	α	3.82 days
Radium A 99.98% ∣ 0.02%	Polonium	^{218}Po	α and β	3.05 min
Radium B	Lead	^{214}Pb	β	26.8 min
Astatine-218 ↓	Astatine	^{218}At	α	2 sec
Radium C 99.96% ∣ 0.04%	Bismuth	^{214}Bi	β and α	19.7 min
Radium C′	Polonium	^{214}Po	α	1.6×10^{-4} sec
Radium C″ ↓	Thallium	^{210}Tl	β	1.32 min
Radium D ↓	Lead	^{210}Pb	β	19.4 years
Radium E ~100% ∣ 2×10^{-4}%	Bismuth	^{210}Bi	β and α	5.0 days
Radium F	Polonium	^{210}Po	α	138.4 days
Thallium-206 ↓	Thallium	^{206}Tl	β	4.20 min
Radium G (End Product)	Lead	^{206}Pb	Stable	—

*Uranium X_2 is an excited state of ^{234}Pa and undergoes isomeric transition to a small extent to form uranium Z (^{234}Pa in its ground state); the latter has a half-life of 6.7 h, emitting beta radiation and forming uranium II (^{234}U).

1.4 IONIZATION ENERGY

TABLE 1.27 Ionization Energy of the Elements

The minimum amount of energy required to remove the least strongly bound electron from a gaseous atom (or ion) is called the ionization energy and is expressed in $MJ \cdot mol^{-1}$.

At. no.	Element	Spectrum (in $MJ \cdot mol^{-1}$) I	II	III	IV	V	VI
1	H	1.312					
2	He	2.372	5.251				
3	Li	0.520	7.298	11.815			
4	Be	0.899	1.757	14.849	21.007		
5	B	0.801	2.427	3.660	25.027	32.828	
6	C	1.086	2.353	4.620	6.223	37.832	47.191
7	N	1.402	2.856	4.578	7.475	9.445	53.268
8	O	1.314	3.388	5.300	7.469	10.989	13.326
9	F	1.681	3.374	6.147	8.408	11.022	15.164

TABLE 1.27 Ionization Energy of the Elements (*Continued*)

At. no.	Element	Spectrum (in MJ · mol^{-1}) I	II	III	IV	V	VI
10	Ne	2.081	3.952	6.122	9.370	12.177	15.238
11	Na	0.496	4.562	6.912	9.543	13.353	16.610
12	Mg	0.738	1.451	7.733	10.540	13.629	17.994
13	Al	0.578	1.817	2.745	11.577	14.831	18.377
14	Si	0.786	1.577	3.231	4.355	16.091	19.784
15	P	1.012	1.903	2.912	4.956	6.274	21.268
16	S	1.000	2.251	3.361	4.564	7.004	8.495
17	Cl	1.251	2.297	3.822	5.158	6.54	9.362
18	Ar	1.521	2.666	3.931	5.771	7.238	8.787
19	K	0.419	3.051	4.411	5.877	7.976	9.649
20	Ca	0.590	1.145	4.912	6.474	8.144	10.496
21	Sc	0.631	1.235	2.389	7.089	8.844	10.719
22	Ti	0.658	1.310	2.652	4.175	9.573	11.516
23	V	0.650	1.414	2.828	4.507	6.299	12.362
24	Cr	0.653	1.592	2.987	4.743	6.70	8.738
25	Mn	0.717	1.509	3.248	4.94	6.99	9.22
26	Fe	0.759	1.561	2.957	5.63	7.24	9.56
27	Co	0.758	1.646	3.232	4.95	7.67	9.84
28	Ni	0.737	1.753	3.393	5.30	7.34	10.4
29	Cu	0.745	1.958	3.555	5.536	7.70	9.9
30	Zn	0.906	1.733	3.833	5.73	7.95	10.4
31	Ga	0.579	1.979	2.963	6.2		
32	Ge	0.762	1.537	3.302	4.410	9.022	
33	As	0.947	1.798	2.735	4.837	6.043	12.31
34	Sc	0.941	2.045	2.974	4.143	6.99	7.883
35	Br	1.140	2.10	3.47	4.56	5.76	8.55
36	Kr	1.351	2.350	3.565	5.07	6.24	7.57
37	Rb	0.403	2.632	3.9	5.08	6.85	8.14
38	Sr	0.549	1.064	4.138	5.5	6.91	8.76
39	Y	0.616	1.181	1.980	5.96	7.43	8.97
40	Zr	0.660	1.267	2.218	3.313	7.75	
41	Nb	0.664	1.382	2.416	3.695	4.877	9.847
42	Mo	0.685	1.558	2.621	4.477	5.91	6.641
43	Tc	0.702	1.472	2.850			
44	Ru	0.711	1.617	2.747			
45	Rh	0.720	1.744	2.997			
46	Pd	0.805	1.875	3.177			
47	Ag	0.731	2.073	3.361			
48	Cd	0.868	1.631	3.616			
49	In	0.558	1.821	2.704	5.2		
50	Sn	0.709	1.412	2.943	3.930	6.974	
51	Sb	0.834	1.595	2.44	4.26	5.4	10.4
52	Te	0.869	1.795	2.698	3.610	5.668	6.82
53	I	1.008	1.846	3.2			
54	Xe	1.170	2.046	3.099			
55	Cs	0.376	2.234				
56	Ba	0.503	0.965				
57	La	0.538	1.067	1.850	4.820	5.94	
58	Ce	0.528	1.047	1.949	3.547	6.325	7.487
59	Pr	0.523	1.018	2.086	3.761	5.551	
60	Nd	0.530	1.035	2.13	3.90		

(*Continued*)

TABLE 1.27 Ionization Energy of the Elements (*Continued*)

At. no.	Element	Spectrum (in MJ · mol^{-1}) I	II	III	IV	V	VI
61	Pm	0.535	1.052	2.15	3.97		
62	Sm	0.543	1.068	2.26	3.99		
63	Eu	0.547	1.085	2.40	4.12		
64	Gd	0.592	1.167	1.99	4.26		
65	Tb	0.564	1.112	2.114	3.839		
66	Dy	0.572	1.126	2.20	3.99		
67	Ho	0.581	1.139	2.204	4.10		
68	Er	0.589	1.151	2.194	4.13		
69	Tm	0.596	1.163	2.285	4.13		
70	Yb	0.603	1.174	2.417	4.203		
71	Lu	0.524	1.34	2.022	4.366		
72	Hf	0.68	1.44	2.25	3.216		
73	Ta	0.761					
74	W	0.770					
75	Re	0.760					
76	Os	0.84					
77	Ir	0.88					
78	Pt	0.87	1.791				
79	Au	0.890	1.98				
80	Hg	1.007	1.810	3.30			
81	Tl	0.589	1.971	2.878			
82	Pb	0.716	1.450	3.081	4.083	6.64	
83	Bi	0.703	1.610	2.466	4.371	5.40	8.52
84	Po	0.812					
85	At						
86	Rn	1.037					
87	Fr						
88	Ra	0.509	0.979				
89	Ac	0.67	1.17				
90	Th	0.587	1.11	1.93	2.78		
91	Pa	0.568					
92	U	0.598					
93	Np	0.605					
94	Pu	0.585					
95	Am	0.578					
96	Cm	0.581					
97	Bk	0.601					
98	Cf	0.608					
99	Es	0.619					
100	Fm	0.627					
101	Md	0.635					
102	No	0.642					

TABLE 1.28 Ionization Energy of Molecular and Radical Species

Species	Ionization energy		$\Delta_f H$ (ion) in $kJ \cdot mol^{-1}$
	In $MJ \cdot mol^{-1}$	In electron volts	
Aluminum tribromide	1.00	10.4	593
Aluminum trichloride	1.159	12.01	573
Aluminum trifluoride	1.394	14.45	282
Aluminum triiodide	0.88	9.1	673
Amidogen (NH_2)	1.075(1)	11.14(1)	1264
Ammonia	0.980(1)	10.16(1)	934
Antimony trichloride	0.97(1)	10.1(1)	661
Arsenic trichloride	1.018(3)	10.55(3)	754
Arsenic trifluoride	1.239(5)	12.84(5)	452
Arsine	0.954	9.89	1021
Barium oxide	0.667(6)	6.91(6)	543
Bismuth trichloride	1.00	10.4	736
Borane (BH_3)	1.19(1)	12.3(1)	1287
Boron dioxide (BO_2)	1.30(3)	13.5(3)	1001
Boron oxide (B_2O_3)	1.303(14)	13.50(15)	460
Boron tribromide	1.014(2)	10.51(2)	809
Boron trichloride	1.119(2)	11.60(2)	718
Boron trifluoride	1.501(3)	15.56(3)	365
Boron triodide	0.893(3)	9.25(3)	964
Bromine (Br_2)	1.0146(5)	10.515(5)	1046
Bromine chloride (BrCl)	1.062	11.01	1079
Bromine fluoride (BrF)	1.136(1)	11.77(1)	1077
Bromine pentafluoride	1.271(1)	13.17(1)	840
Bromosilane ($BrSiH_3$)	1.02	10.6	943
Calcium oxide	0.67	6.9	691
Cesium chloride	0.756(5)	7.84(5)	510
Cesium fluoride	1.221(1)	12.65(1)	1170
Cesium fluoride	0.849(10)	8.80(10)	489
Chlorine (Cl_2)	1.1424(5)	11.840(5)	1108
Chlorine difluoride	1.232(5)	12.77(5)	1128
Chlorine dioxide	1.000(2)	10.36(2)	1096
Chlorine oxide	1.057	10.95	1159
Chlorine trifluoride	1.221(5)	12.65(5)	1057
Chlorosilane ($ClSiH_3$)	1.10	11.4	899
Chromyl chloride (CrO_2Cl_2)	1.12	11.6	580
Diborane (B_2H_6)	1.098(3)	11.38(3)	1134
Dichlorosilane (Cl_2SiH_2)	1.10	11.4	765
Difluoramine (HNF_2)	1.112(8)	11.53(8)	1046
Difluoroamidogen (NF_2)	1.122(1)	11.628(1)	1155
Difluorosilane (F_2SiH_2)	1.18	12.2	386
Dioxygen fluoride	1.22(2)	12.6(2)	1228
Disilane	0.94	9.7	1015
Disulfur oxide	1.017(4)	10.54(4)	967
Fluorine (F_2)	1.5146(3)	15.697(3)	1515
Fluorosilane ($FSiH_3$)	1.13	11.7	752
Gallium bromide	1.003	10.40	711
Gallium chloride	1.112	11.52	648
Gallium triiodide	0.907	9.40	765
Gallium(I) fluoride	0.93(5)	9.6(5)	700

(*Continued*)

TABLE 1.28 Ionization Energy of Molecular and Radical Species (*Continued*)

Species	Ionization energy		$\Delta_f H$ (ion) in kJ · mol^{-1}
	In MJ · mol^{-1}	In electron volts	
Germane (GeH_4)	1.093	11.33	1185
Germanium oxide (GeO)	1.085(1)	11.25(1)	1044
Germanium sulfide (GeS)	0.963(2)	9.98(2)	1055
Germanium tetrachloride	1.1270(5)	11.68(5)	629
Germanium tetrafluoride	1.50	15.5	307
Germanium tetraiodide	0.909	9.42	850
Hafnium bromide	1.05	10.9	366
Hafnium chloride	1.13	11.7	246
Hexaborane (B_6H_{10})	0.87	9.0	965
Hydrazine	7.82(14)	8.10(15)	877
Hydrazoic acid (HN_3)	1.0344(24)	10.720(25)	1328
Hydrogen (H_2)	1.488413(5)	15.42589(5)	1488
Hydrogen bromide	1.125(3)	11.66(3)	1087
Hydrogen chloride	1.2299	12.747	1137
Hydrogen fluoride	1.5481(3)	16.044(3)	1276
Hydrogen iodide	1.0004(1)	10.368(1)	1028
Hydrogen peroxide	1.017	10.54	881
Hydrogen selenide	0.9535(1)	9.882(1)	983
Hydrogen sulfide	1.0085(8)	10.453(8)	988
Hydroperoxy (HOO)	1.095(1)	11.35(1)	1106
Hydroxyl (OH)	1.254	13.00	1293
Hydroxylamine (NH_2OH)	0.947	10.00	923
Hypochlorous acid (HOCl)	1.073(1)	11.12(1)	993
Hypofluorous acid (HOF)	1.226(1)	12.71(1)	1130
Imidogen (NH)	1.302(1)	13.49(1)	1678
Iodine (I_2)	0.90694(12)	9.3995(12)	969
Iodine bromide	0.9446(4)	9.790(4)	986
Iodine chloride	0.9734(10)	10.088(10)	991
Iodine fluoride	1.025	10.62	930
Iodine pentafluoride	1.2488(5)	12.943(5)	408
Lead oxide (PbO)	0.976(10)	9.08(10)	939
Lead(II) chloride	0.96	10.0	789
Lead(II) fluoride	1.11	11.5	679
Lead(II) sulfide	0.825	8.5(5)	954
Lithium bromide	0.84	8.7	685
Lithium chloride	0.923	9.57	727
Lithium hydride	0.74	7.7	882
Lithium iodide	0.72	7.5	633
Lithium oxide	0.815	8.45(20)	895
Magnesium fluoride	1.29	13.4	569
Magnesium oxide	0.93	9.7	992
Mercapto (SH)	1.001	10.37	1140
Mercury(II) bromide	1.019(3)	10.560(3)	935
Mercury(II) chloride	1.0988(3)	11.380(3)	952
Mercury(II) iodide	0.91748(22)	9.5088(22)	900
Molybdenum hexafluoride	1.40(1)	14.5(1)	−159
Molybdenum(V) chloride	0.84	8.7	392
Niobium(V) chloride	1.058	10.97	656
Nitric acid	1.153(1)	11.95(1)	1019

TABLE 1.28 Ionization Energy of Molecular and Radical Species (*Continued*)

Species	Ionization energy		$\Delta_f H$ (ion) in kJ · mol^{-1}
	In Mj · mol^{-1}	In electron volts	
Nitric oxide	0.893900(6)	9.26436(6)	985
Nitrogen (N_2)	1.59336	15.5808	1503
Nitrogen dioxide	0.941(1)	9.75(1)	974
Nitrogen pentoxide	1.15	11.9	1161
Nitrogen tetroxide	1.04(2)	10.8(2)	1050
Nitrogen trichloride	0.9765(10)	10.12(10)	1244
Nitrogen trifluoride	1.254(2)	13.00(2)	1125
Nitrosyl bromide	0.981(3)	10.17(3)	1065
Nitrosyl chloride (NOCl)	1.049(1)	10.87(1)	1099
Nitrosyl fluoride (NOF)	1.219(3)	12.63(3)	1152
Nitrous acid (HONO)	1.09	11.3	977
Nitrous oxide (N_2O)	1.2433	12.886	1325
Nitryl chloride (NO_2Cl)	1.142	11.84	1155
Nitryl fluoride (NO_2F)	1.263	13.09	1154
Osmium tetroxide	1.1895	12.320	850
Oxygen (O_2)	1.1647(1)	12.071(1)	1165
Oxygen dichloride	1.056	10.94	1135
Oxygen difluoride (OF_2)	1.265(1)	13.11(1)	1290
Oxygen fluoride	1.232	12.77	1341
Ozone (O_3)	1.199	12.43	1342
Pentaborane (B_5H_9)	0.955(4)	9.90(4)	1028
Perchloryl fluoride (ClO_3F)	1.2490(5)	12.945(5)	1224
Phosphine (PH_3)	0.9522(2)	9.869(2)	958
Phosphorus (P_2)	1.016	10.53	1160
Phosphorus nitride	1.143	11.85	1248
Phosphorus pentachloride	1.03	10.7	656
Phosphorus pentafluoride	1.46	15.1	−137
Phosphorus sulfur trichloride ($PSCl_3$)	0.956	9.91	668
Phosphorus tribromide	0.94	9.7	798
Phosphorus trichloride	0.956	9.91	668
Phosphorus trifluoride	1.104	11.44	146
Phosphoryl chloride ($POCl_3$)	1.096(2)	11.36(2)	540
Phosphoryl trifluoride (POF_3)	1.231(1)	12.76(1)	−24
Potassium bromide	0.757(10)	7.85(10)	578
Potassium chloride	0.77(4)	8.0(4)	557
Potassium iodide	0.696(29)	7.21(30)	570
Rhenium(VII) oxide	1.23(2)	12.7(2)	125
Rubidium bromide	0.766(3)	7.94(3)	583
Rubidium chloride	0.820(3)	8.50(3)	590
Ruthenium tetroxide	1.172(3)	12.15(3)	988
Silane	1.124	11.65	1158
Silicon oxide (SiO)	1.103	11.43	1002
Silicon tetrachloride	1.136(1)	11.79(1)	527
Silicon tetrafluoride	1.51	15.7	−100
Silver chloride	0.973	10.08	1065
Silver fluoride	1.06(3)	11.0(3)	1071
Sodium bromide	0.802(10)	8.31(10)	660
Sodium chloride	0.861(6)	8.92(6)	681
Sodium iodide	0.737(2)	7.64(2)	659
Stibine (SbH_3)	0.920(3)	9.54(3)	1067

(*Continued*)

TABLE 1.28 Ionization Energy of Molecular and Radical Species (*Continued*)

Species	Ionization energy		$\Delta_f H$ (ion) in kJ · mol^{-1}
	In Mj · mol^{-1}	In electron volts	
Strontium oxide	0.675(14)	7.00(15)	662
Sulfur (S_2)	0.9027(2)	9.356(2)	1031
Sulfur chloride pentafluoride	1.1921(5)	12.335(5)	144
Sulfur dichloride	0.912(3)	9.45(3)	895
Sulfur difluoride	0.973	10.08	676
Sulfur dioxide	1.189(2)	12.32(2)	892
Sulfur hexafluoride	1.479(3)	15.33(3)	259
Sulfur oxide (SO)	0.996(2)	10.32(2)	1001
Sulfur pentafluoride	1.01(1)	10.5(1)	97
Sulfur trioxide	1.235(4)	12.80(4)	839
Sulfuryl chloride (SO_2Cl_2)	1.163	12.05	807
Sulfuryl fluoride (SO_2F_2)	1.110	11.5	679
Tantalum(V) chloride	1.069	11.08	348
Tetraborane (B_4H_{10})	1.038(4)	10.76(4)	1105
Tetrafluorohydrazine (gauche)	1.152(3)	11.94(3)	1119
Thallium(I) bromide	0.882(2)	9.14(2)	844
Thallium(I) chloride	0.936(3)	9.70(3)	869
Thallium(I) fluoride	1.015	10.52	835
Thionitrosyl fluoride (NSF)	1.111(4)	11.51(4)	1090
Thionyl chloride	1.058	10.96	844
Thionyl fluoride	1.182	12.25	688
Thiophosphoryl trifluoride (PSF_3)	1.066(4)	11.05(4)	58
Thorium(IV) oxide	0.847(14)	8.70(15)	342
Tin(II) bromide	0.87	9.0	830
Tin(II) chloride	0.965	10.0	760
Tin(II) fluoride	1.07	11.1	586
Tin(II) oxide	0.926(2)	9.60(2)	944
Tin(II) sulfide	0.85	8.8	966
Tin(IV) bromide	1.02	10.6	709
Tin(IV) chloride	1.146(5)	11.88(5)	673
Tin(IV) hydride	1.037	10.75	1200
Titanium(IV) bromide	0.99	10.3	375
Titanium(IV) chloride	1.124(14)	11.65(15)	363
Titanium(IV) oxide	0.920(10)	9.54(10)	623
trans-Difluorodiazine	1.24	12.8	1315
Trifluoramine oxide (NOF_3)	1.279(1)	13.26(1)	1116
Trifluorosilane (F_3SiH)	1.35	14.0	150
Trisilane	0.89	9.2	1009
Tungsten(VI) chloride	0.92	9.5	348
Uranium hexafluoride	1.350(10)	14.00(10)	−796
Uranium(IV) oxide	5.2(1)	5.4(1)	57
Uranium(VI) oxide	1.01(5)	10.5(5)	214
Vanadium(IV) chloride	0.89	9.2	210
Vanadium(V) oxychloride ($VOCl_3$)	1.120	11.61	425
Water	1.2170(10)	12.612(10)	975
Xenon difluoride	1.192(1)	12.35(1)	1083
Xenon tetrafluoride	1.221(10)	12.65(10)	1016
Zirconium bromide	1.03	10.7	388
Zirconium chloride	1.08	11.2	392

Source: Sharon, G., et al., *J. Phys. Chem. Ref. Data*, **17**:Suppl. No 1 (1988).

1.5 ELECTRONEGATIVITY

Electronegativity χ is the relative attraction of an atom for the valence electrons in a covalent bond. It is proportional to the effective nuclear charge and inversely proportional to the covalent radius:

$$\chi = \frac{0.31(n+1\pm c)}{r} + 0.50$$

where n is the number of valence electrons, c is any formal valence charge on the atom and the sign before it corresponds to the sign of this charge, and r is the covalent radius. Originally the element fluorine, whose atoms have the greatest attraction for electrons, was given an arbitrary electronegativity of 4.0. A revision of Pauling's values based on newer data assigns –3.90 to fluorine. Values in Table 1.29 refer to the common oxidation states of the elements.

TABLE 1.29 Electronegativity Values of the Elements

H																
2.20																
Li	Be										B	C	N	O	F	
0.98	1.57										2.04	2.55	3.04	3.44	3.90	
Na	Mg										Al	Si	P	S	Cl	
0.93	1.31										1.61	1.90	2.19	2.58	3.16	
K	Ca	Sc	Ti	V	Cr	Mn	Fe	Co	Ni	Cu	Zn	Ga	Ge	As	Se	Br
0.82	1.00	1.36	1.54	1.63	1.66	1.55	1.83	1.88	1.91	1.90	1.65	1.81	2.01	2.18	2.55	2.96
Rb	Sr	Y	Zr	Nb	Mo	Tc	Ru	Rh	Pd	Ag	Cd	In	Sn	Sb	Te	I
0.82	0.95	1.22	1.33	1.6	2.16	2.10	2.2	2.28	2.20	1.93	1.69	1.78	1.96	2.05	2.1	2.66
Cs	Ba	La	Hf	Ta	W	Re	Os	Ir	Pt	Au	Hg	Tl	Pb	Bi	Po	At
0.79	0.89	1.10	1.3	1.5	1.7	1.9	2.2	2.2	2.2	2.4	1.9	1.8	1.8	1.9	2.0	2.2
Fr	Ra	Ac														
0.7	0.9	1.1														
			Ce	Pr	Nd		Sm		Gd		Dy	Ho	Er	Tm		Lu
Lanthanides			1.12	1.13	1.14		1.17		1.20		1.22	1.23	1.24	1.25		1.0
			Th	Pa	U	Np	Pu	Am	Cm	Bk	Cf	Es	Fm	Md	No	
Actinides			1.3	1.5	1.7	1.3	1.3	1.3	1.3	1.3	1.3	1.3	1.3	1.3	1.3	

The greater the difference is electronegativity, the greater is the ionic character of the bond. The amount of ionic character I is given by:

$$I = 0.46\,|\,\chi_A - \chi_B\,| + 0.035(\chi_A - \chi_B)^2$$

The bond is fully covalent when $(\chi_A - \chi_B) < 0.5$ (and $I < 6\%$).

1.6 ELECTRON AFFINITY

TABLE 1.30 Electron Affinities of Elements, Molecules, and Radicals

Electron affinity of an atom (molecule or radical) is defined as the energy difference between the lowest (ground) state of the neutral and the lowest state of the corresponding negative ion in the gas phase.

$$A(g) + e^- = A^-(g)$$

Data are limited to those negative ions which, by virtue of their positive electron affinity, are stable. Uncertainty in the final data figures is given in parentheses. Calculated values are enclosed in brackets.

A. Atoms

Atom	Electron affinity, in eV	Electron affinity, in kJ · mol^{-1}
Aluminum	0.441(10)	42.5(10)
Antimony	1.046(5)	100.9(5)
Arsenic	0.81(3)	78.(3)
Astatine	[2.8(3)]	[270.(30)]
Barium	[0.15]	[14.]
Bismuth	0.946(10)	91.3(10)
Boron	0.277(10)	26.7(10)
Bromine	3.363590(3)	324.5367(3)
Calcium	0.0185(25)	1.78(24)
Carbon	1.2629(3)	121.85(3)
Cesium	0.471626(25)	45.5048(24)
Chlorine	3.61269	348.570
Chromium	0.666(12)	64.3(12)
Cobalt	0.662(3)	63.9(3)
Copper	1.235(5)	119.2(5)
Fluorine	3.401190(4)	328.1638(4)
Francium	[0.46]	[44]
Gallium	0.30(15)	29.(15)
Germanium	1.233(3)	119.0(3)
Gold	2.30863(3)	222.748(3)
Hafnium	[≈0.]	[≈0.]
Hydrogen	0.75195(19)	72.552(18)
Hydrogen-d_1 deuterium	0.75459(7)	72.807(7)
Indium	0.3(2)	29.(2)
Iodine	3.05904(1)	295.151(1)
Iridium	1.565(8)	151.0(8)
Iron	0.151(3)	14.6(3)
Lanthanum	[0.5(3)]	[48.(30)]
Lead	0.364(8)	35.1(8)
Lithium	0.6180(5)	59.63(5)
Molybdenum	0.748(2)	72.2(2)
Nickel	1.156(10)	111.5(10)
Niobium	0.893(25)	86.2(24)
Osmium	[0.2(1)]	[19.(10)]
Oxygen	1.4611103(7)	140.97523(7)
Palladium	0.562(5)	54.2(5)
Phosphorus	0.7465(3)	72.03(3)
Platinum	2.128(2)	205.3(2)
Polonium	[1.9(3)]	[183.(30)]

TABLE 1.30 Electron Affinities of Elements, Molecules, and Radicals (*Continued*)

A. Atoms

Atom	Electron affinity, in eV	Electron affinity, in kJ · mol^{-1}
Potassium	0.50147(10)	48.384(10)
Rhenium	[0.15(15)]	[14.(14)]
Rubidium	0.48592(2)	46.884(2)
Ruthenium	[1.05(15)]	[101.(14)]
Scandium	0.188(20)	18.1(19)
Selenium	2.020670(25)	194.9643(24)
Silver	1.302(7)	125.6(7)
Sodium	0.547926(25)	52.86666(24)
Strontium	0.048(6)	4.6(6)
Sulfur	2.077104(1)	200.4094(1)
Tantalum	0.322(12)	31.1(12)
Technetium	[0.55(20)]	[53.(19)]
Tellurium	1.9708(3)	190.15(3)
Thallium	0.2(2)	19.(19)
Tin	1.112(4)	107.3(4)
Titanium	0.079(14)	7.6(14)
Tungsten	0.815(2)	78.6(2)
Vanadium	0.525(12)	50.7(12)
Yttrium	0.307(12)	29.6(12)
Zirconium	0.426(14)	41.1(14)

B. Molecules

Molecule	Electron affinity, in eV	Electron affinity, in kJ · mol^{-1}
BF_3	2.65	256
BH_3	0.038(15)	3.7(15)
1,4-Benzoquinone	1.91(10)	184.(10)
Br_2	2.55(10)	246.(10)
$CBrF_3$	0.91(20)	89.(19)
CF_3I	1.57(20)	151.(19)
COS	0.46(20)	44.(19)
CS_2	0.895(20)	86.3(19)
C_6F_6 hexafluorobenzene	0.52(10)	50.(10)
1,2-$C_6H_4(NO_3)_2$ (also 1,3-)	1.65(10)	159.(10)
1,4-$C_6H_4(NO_3)_2$	2.00(10)	193.(10)
C_6H_5Br bromobenzene	1.15(11)	111.(11)
C_6H_5Cl chlorobenzene	0.82(11)	79.(11)
C_6H_5I iodobenzene	1.41(11)	136.(11)
$C_6H_5NO_2$ nitrobenzene	1.01(10)	97.(10)
1,4-$C_6H_4(CN)NO_2$	1.72(10)	166.(10)
Cl_2	2.38(10)	229.(10)
CoH_2	1.450(14)	139.9(13)
CsCl	0.455(10)	43.9(10)
CuO	1.777(6)	171.5(6)
F_2	3.08(10)	297.(10)
FeO	1.493(5)	144.1(5)
I_2	2.55(5)	246.(5)

(*Continued*)

TABLE 1.30 Electron Affinities of Elements, Molecules, and Radicals (*Continued*)

B. Molecules (*continued*)

Molecule	Electron affinity, in eV	Electron affinity, in $kJ \cdot mol^{-1}$
IBr	2.55(10)	246.(10)
IrF_6	6.5(4)	627.(40)
KBr	0.642(10)	61.9(10)
KCl	0.582(10)	56.1(10)
KI	0.728(10)	70.2(10)
LiCl	0.593(10)	54.3(10)
LiH	0.342(12)	33.0(12)
MoO_3	2.9(2)	280.(20)
NO	0.026(5)	2.5(5)
NO_2	2.273(5)	219.3(5)
N_2O	0.22(10)	21.(10)
NaBr	0.788(10)	76.0(10)
NaCl	0.727(10)	70.1(10)
NaI	0.865(10)	83.5(10)
NaK	0.465(30)	44.9(30)
O_2	0.451(7)	43.5(7)
O_3	2.103(3)	202.9(9)
OsF_6	6.0(3)	579.(29)
PBr_3	1.59(15)	153.(14)
PCl_3	0.82(10)	79.(10)
PF_5	0.75(15)	72.(14)
$POCl_3$	1.41(2)	136.(2)
PbO	0.722(6)	69.7(6)
PtF_6	7.0(4)	675.(40)
RbCl	0.544(10)	52.5(10)
RuF_6	7.5(3)	724.(28)
SF_4	1.5(2)	145.(19)
SF_6	1.05(10)	101.(10)
SO_2	1.107(8)	106.8(8)
SeF_6	2.9(2)	280.(19)
SeO	1.456(20)	140.5(19)
SeO_2	1.823(50)	175.9(48)
TeF_6	3.34(17)	322.(16)
TeO	1.695(22)	163.5(21)
UF_6	5.1(2)	492.(19)
V_4O_{10}	4.2(6)	405.(60)
WO_3	3.9(2)	376.(19)

C. Radicals

Radical	Electron affinity in eV	Electron affinity in $kJ \cdot mol^{-1}$
AsH_2	1.27(3)	123.(3)
CCl_2	1.591(10)	153.5(10)
CF_2	0.165(10)	15.9(10)
CH	1.238(8)	119.4(8)
CHBr	1.454(5)	140.3(5)
CHCl	1.210(5)	117.5(5)
CHF	0.542(5)	52.3(5)

TABLE 1.30 Electron Affinities of Elements, Molecules, and Radicals (*Continued*)

C. Radical

Radical	Electron affinity, in eV	Electron affinity, in $kJ \cdot mol^{-1}$
CHI	1.42(17)	137.(17)
CHO_2	3.498(5)	337.5(5)
CH_2	0.652(6)	62.9(6)
CH_2S	0.465(23)	44.9(22)
$CH_2{=}SiH$	2.010(10)	193.9(10)
CH_3	0.08(3)	7.7(3)
CH_3CH_2O ethoxide	1.726(33)	166.5(32)
CH_3O	1.570(22)	151.5(21)
CH_3S	1.861(4)	179.6(4)
CH_3SCH_2	0.868(51)	83.7(49)
CH_3Si	0.852(10)	82.2(10)
CH_3SiH_2	1.19(4)	115.(4)
C_2F_2 difluorovinylidene	2.255(6)	217.6(6)
C_2H_2 vinylidene	0.490(6)	47.3(6)
$CH_2{=}CH$ vinyl	0.667(24)	64.3(23)
C_2H_3O acetaldehyde enolate	1.82476(12)	176.062(12)
CH_3CH_2S	1.953(6)	188.4(6)
$HC{\equiv}C{-}CH_2$	0.893(25)	86.2(24)
CH_3CHCN	1.247(12)	120.3(12)
C_2H_5O ethoxide	1.726(33)	166.5(31)
C_2H_5S ethyl sulfide	1.953(6)	188.4(6)
C_3H_3 propargyl radical	0.893(25)	86.2(24)
$CH_3CH{-}CN$	1.247(12)	120.3(12)
C_3H_5 allyl	0.362(19)	34.9(18)
C_3H_5O acetone enolate	1.758(19)	169.2(18)
propionaldehyde enolate	1.621(6)	156.4(6)
$C_3H_5O_2$ methyl acetate enolate	1.80(6)	174.(6)
C_3H_7O propoxide	1.789(33)	172.6(31)
isopropyl oxide	1.839(29)	177.4(28)
C_3H_7S propyl sulfide	2.00(2)	193.(2)
isopropyl sulfide	2.02(2)	195.(2)
C_4H_5O cyclobutanone enolate	1.801(8)	173.8(8)
C_4H_7O butyraldehyde enolate	1.67(5)	161.(5)
C_4H_9O *tert*-butoxyl	1.912(54)	184.5(52)
C_4H_9S butyl sulfide	2.03(2)	196.(2)
tert-butyl sulfide	2.07(2)	200.(2)
C_5H_5 cyclopentadienyl	1.804(7)	174.1(7)
C_5H_7 pentadienyl	0.91(3)	88.(3)
C_5H_7O cyclopentanone enolate	1.598(7)	154.2(7)
C_5H_9O 3-pentanone enolate	1.69(5)	163.(5)
$C_5H_{11}S$ pentyl sulfide	2.09(2)	202.(2)
C_6H_5 phenyl	1.096(6)	105.7(6)
C_6H_5NH anilide	1.70(3)	164.(3)
C_6H_5O phenoxyl	2.253(6)	217.4(6)
C_6H_5S thiophenoxide	$\leq$2.47(6)	$\leq$238.(6)
$C_6H_5CH_2$ benzyl	0.912(6)	88.0(6)
$C_6H_5CH_2O$ benzyl oxide	2.14(2)	206.(2)
C_6H_9O cyclohexanone enolate	1.526(10)	147.2(10)
$H_2C{=}CH{-}CH{=}CH{-}CH{=}CH{-}CH_2$ heptatrienyl	1.27(3)	122.(3)
CN	3.862(4)	372.6(4)

(*Continued*)

TABLE 1.30 Electron Affinities of Elements, Molecules, and Radicals (*Continued*)

C. Radical

Radical	Electron affinity, in eV	Electron affinity, in kJ · mol^{-1}
$CNCH_2$ cyanomethyl	1.543(14)	148.9(14)
CO_3	2.69(14)	259.(14)
CS	0.205(21)	19.8(20)
ClO	2.275(6)	219.5(6)
HCO	0.313(5)	30.2(5)
HNO	0.338(15)	32.6(14)
HO_2	1.078(17)	104.0(6)
FO	2.272(6)	219.2(6)
N_3	2.70(12)	260.(12)
NCO	3.609(5)	348.2(5)
NCS	3.537(5)	341.3(5)
NH	0.370(4)	35.7(4)
NO_3	3.937(14)	379.9(14)
NS	1.194(11)	115.2(11)
O_2Aryl	0.52(2)	50.(2)
OClO	2.140(8)	206.5(8)
OH	1.82767(2)	176.343(2)
OIO	2.577(8)	248.6(8)
PH	1.028(10)	99.2(10)
PH_2	1.27(1)	123.(1)
PO	1.092(10)	105.4(10)
PO_2	3.42(1)	330.(1)
SF	2.285(6)	220.5(6)
SH	2.314344(4)	223.300(4)
SO	1.125(5)	108.5(5)
SeḢ	2.21252(3)	213.475(3)
SiF_3	≤2.95(10)	285.(10)
SiH	1.277(9)	123.2(9)
SiH_2	1.124(20)	108.4(19)
SiH_3	1.406(14)	106.7(14)

Source: H. Hotop and W. C. Lineberger, *J. Phys. Chem. Reference Data* **14**:731 (1985).

1.7 BOND LENGTHS AND STRENGTHS

Distances between centers of bonded atoms are called *bond lengths*, or *bond distances*. Bond lengths vary depending on many factors, but in general, they are very consistent. Of course the bond orders affect bond length, but bond lengths of the same order for the same pair of atoms in various molecules are very consistent.

The *bond order* is the number of electron pairs shared between two atoms in the formation of the bond. Bond order for C=C and O=O is 2. The amount of energy required to break a bond is called *bond dissociation energy* or simply *bond energy*. Since bond lengths are consistent, bond energies of similar bonds are also consistent.

Bonds between the same type of atom are *covalent bonds*, and bonds between atoms when their electronegativity differs slightly are also predominant covalent in character. Theoretically, even ionic bonds have some covalent character. Thus, the boundary between ionic and covalent bonds is not a clear line of demarcation.

For covalent bonds, bond energies and bond lengths depend on many factors: electron affinities, sizes of atoms involved in the bond, differences in their electronegativity, and the overall structure of the molecule. There is a general trend in that *the shorter the bond length, the higher the bond energy* but there is no formula to show this relationship, because of the widespread variation in bond character.

1.7.1 Atom Radius

The *atom radius* of an element is the shortest distance between like atoms. It is the distance of the centers of the atoms from one another in metallic crystals and for these materials the atom radius is often called the metal radius. Except for the lanthanides (CN = 6), CN = 12 for the elements.

1.7.2 Ionic Radii

One of the major factors in determining the structures of the substances that can be thought of as made up of cations and anions packed together is ionic size. It is obvious from the nature of wave functions that no ion has a precisely defined radius. However, with the insight afforded by electron density maps and with a large base of data, new efforts to establish tables of ionic radii have been made.

Effective ionic radii are based on the assumption that the ionic radius of O^{2-} (CN 6) is 140 pm and that of F^- (CN 6) is 133 pm. Also taken into consideration is the coordination number (CN) and electronic spin state (HS and LS, high spin and low spin) of first-row transition metal ions. These radii are empirical and include effects of covalence in specific metal-oxygen or metal-fluorine bonds. Older "crystal ionic radii" were based on the radius of F^- (CN 6) equal to 119 pm; these radii are 14–18 percent larger than the effective ionic radii.

1.7.3 Covalent Radii

Covalent radii are the distance between two kinds of atoms connected by a covalent bond of a given type (single, double, etc.).

TABLE 1.31 Atom Radii and Effective Ionic Radii of Elements

Element	Atom radius, pm	Effective ionic radii, pm				
		Ion charge	Coordinator number 4	6	8	12
Actinium	187.8	3+		111		
Aluminum	143.1	3+	39	53.5		
Americium	173	2+			126	
		3+		97.5	109	
		4+		89	95	
		5+		86		
		6+		80		
Antimony	145	3−		245		
		1+		89		
		3+	76	76		
		5+		60		

(*Continued*)

TABLE 1.31 Atom Radii and Effective Ionic Radii of Elements (*Continued*)

Element	Atom radius, pm	Effective ionic radii, pm				
		Ion charge	Coordinator number			
			4	6	8	12
Arsenic	124.8	3−		222		
		3+		58		
		5+	33.5	46		
Astatine		1−		227		
		5+		57		
		7+		62		
Barium	217.3	2+		136	142	160
Berkelium		2+		118		
		3+		98		
		4+		87	93	
Beryllium	111.3	1−	195			
		2+	27	45		
Bismuth	154.7	3−		213		
		3+		103	111	
		5+		76		
Boron	86	1+	35			
		3+	11	27		
Bromine		1−		196		
		3+	59			
		5+	31*	47		
		7+		25		
Cadmium	148.9	2+	78	95	110	131
Calcium	197	2+		100	112	135
Californium	186(2)	2+		117		
		3+		95		
		4+		82.1		
Carbon		4−	260			
		4+	15	16		
Cerium	181.8	3+		102	114.3	134
		4+		87	97	114
Cesium	265	1+		167	174	188
Chlorine		1−		181		
		5+	34			
		7+	8	27		
Chromium	128	1+	81			
		2+		73 LS		
				80 HS		
		3+		61.5		
		4+	41	55		
		5+	34.5	49	57	
		6+	26	44		
Cobalt	125	2+	38	65 LS	90	
				74.5 HS		
		3+		54.5 LS		
				61 HS		
		4+	40	53 HS		
Copper	128	1+	60	77		
		2+	57	73		
		3+		54 LS		

*CN = 3

TABLE 1.31 Atom Radii and Effective Ionic Radii of Elements (*Continued*)

Element	Atom radius, pm	Ion charge	Effective ionic radii, pm — Coordinator number 4	6	8	12
Curium	174	3+		97		
		4+		85	95	
Dysprosium	178.1	2+		107	119	
		3+		91.2	102.7	
Einsteinium	186(2)	3+		98		
Erbium	176.1	3+		89.0	100.4	
Europium	208.4	2+		117	125	135
		3+		94.7	106.6	
Fluorine	71.7	1−	131	133		
		7+		8		
Francium	270	1+		180		
Gadolinium	180.4	3+		93.8	105.3	
Gallium	135	2+		120		
		3+	47	62.0		
Germanium	128	2+		73		
		4+	39.0	53.0		
Gold	144	1+		137		
		3+	68	85		
Hafnium	159	4+	58	71	83	
Holmium	176.2	3+		90.1	101.5*	112
Hydrogen		1−		154		
Indium	167	1+		140		
		3+	62	80.0	92	
Iodine		1−		220		
		5+		95		
		7+	42	53		
Iridium	135.5	3+		68		
		4+		62.5		
		5+		57		
Iron	126	2+		61 LS		
			63 HS	78 HS	92 HS	
		3+		55 LS		
			49 HS	64.5 HS	78 HS	
		4+		58.5		
		6+	25			
Lanthanum	183	3+		103.2	116.0	136
Lead	175	2+	98	119	129	149
		4+		78	94	
Lithium	152	1+	59	76		
Lutetium	173.8	3+		86.1	97.7	
Magnesium	160	2+	57	72.0	89	
Manganese	127	2+	66 HS	67 LS	96	
				83 HS		
		3+		58 LS		
				64.5 HS		
		4+	39	53		
		5+	33			
		6+	25.5			
		7+	25	46		

*CN = 10

(*Continued*)

TABLE 1.31 Atom Radii and Effective Ionic Radii of Elements (*Continued*)

Element	Atom radius, pm	Effective ionic radii, pm				
		Ion charge	Coordinator number			
			4	6	8	12
Mercury	151	1+	111*	119		
		2+	96	102	114	
Molybdenum	139	3+		69		
		4+		65.0		
		5+	46	61		
		6+	41	59	73†	
Neodymium	181.4	2+			129	
		3+		98.3	110.9	127
Neptunium	155	2+		110		
		3+		101		
		4+		87	98	
		5+		75		
		6+		72		
		7+		71		
Nickel	124	2+	55	69.0		
		3+		56 LS		
				60 HS		
		4+		48 LS		
Niobium	146	3+		72		
		4+		68	79	
		5+	48	64	74	
Nitrogen		3−	146			
		1+	25			
		3+		16		
		5+		13		
Nobelium		2+		110		
Osmium	135	4+		63.0		
		5+		57.5		
		6+		54.5		
		7+		52.5		
		8+	39			
Oxygen		2−	138	140	142	
Palladium	137	2+	64	86		
		3+		76		
		4+		61.5		
Phosphorus	108	3−		212		
		3+		44		
		5+	17	38		
Platinum	138.5	2+		80		
		4+		62.5		
		5+		57		
Plutonium	159	3+		100		
		4+		86	96	
		5+		74		
		6+		71		

*CN = 3
†CN = 7

TABLE 1.31 Atom Radii and Effective Ionic Radii of Elements (*Continued*)

Element	Atom radius, pm	Effective ionic radii, pm				
		Ion charge	Coordinator number			
			4	6	8	12
Polonium	164	2−		(230)		
		4+		94	108	
		6+		67		
Potassium	232	1+	137	138	151	164
Praseodymium	182.4	3+		99	112.6	
		4+		85	96	
Promethium	183.4	3+		97	109.3	
Protoactinium	163	3+		104		
		4+		90	101	
		5+		78	91	
Radium	(220)	2+			148	170
Rhenium	137	4+		63		
		5+		58		
		6+		55		
		7+	38	53		
Rhodium	134	3+		66.5		
		4+		60		
		5+		55		
Rubidium	248	1+		152	161	172
Ruthenium	134	3+		68		
		4+		62.0		
		5+		56.5		
		7+	38			
		8+	36			
Samarium	180.4	2+			127	
		3+		95.8	107.9	124
Scandium	162	3+		74.5	87.0	
Selenium	116	2−		198		
		4+		50		
		6+		42		
Silicon	118	4+	26	40.0		
Silver	144	1+	100	115	130	
		2+	79	94		
		3+	67	75		
Sodium	186	1+	99	102	118	139
Strontium	215	2+		118	126	144
Sulfur	106	2−		184		
		4+		37		
		6+	12	29		
Tantalum	146	3+		72		
		4+		68		
		5+		64	74	
Technetium	136	4+		64.5		
		5+		60		
		7+	37	56		
Tellurium	142	2−		221		
		4+	66	97		
		6+	43	56		

(*Continued*)

TABLE 1.31 Atom Radii and Effective Ionic Radii of Elements (*Continued*)

Element	Atom radius, pm	Effective ionic radii, pm				
		Ion Charge	Coordinator number			
			4	6	8	12
Terbium	177.3	3+		92.3	104.0	
		4+		76	88	
Thallium	170	1+		150	159	170
		3+	75	88.5	98	
Thorium	179	4+		94	105	121
Thullium	175.9	2+		103		
		3+		88.0	99.4	105*
Tin	151	2+		118		
		4+	55	69.0	81	
Titanium	147	2+		86		
		3+		67.0		
		4+	42	60.5	74	
Tungsten	139	4+		66		
		5+		62		
		6+	42	60		
Uranium	156	3+		102.5		
		4+		89	·100	117
		5+		76		
		6+	52	73	86	
Vanadium	134	2+		79		
		3+		64.0		
		4+		58	72	
		5+	35.5	54		
Xenon		8+	40	48		
Ytterbium	193.3	2+		102	114	
		3+		86.8	98.5	104*
Yttrium	180	3+		90.0	101.9	108*
Zinc	134	2+	60	74.0	90	
Zirconium	160	4+	59	72	84	89*

*CN = 11

TABLE 1.32 Approximate Effective Ionic Radii in Aqueous Solutions at 25°C

å (in Å)	Inorganic ions	å (in Å)	Organic ions
2.5	Rb^-, Cs^+, NH_4^+, Tl^+, Ag^+	3.5	$HCOO^-$, H_2Cit^-, $CH_3NH_3^+$, $(CH_3)_2NH_2^+$
3	K^+, Cl^-, Br^-, I^-, CN^-, NO_2^-, NO_3^-	4	$H_3N^+CH_2COOH$, $(CH_3)_3NH^+$, $C_2H_5NH_3^+$
3.5	OH^-, F^-, SCN^-, OCN^-, HS^-, ClO_3^-, ClO_4^-, BrO_3^-, IO_4^-, MnO_4^-	4.5	CH_3COO^-, $ClCH_2COO^-$, $(CH_3)_4N^+$, $(C_2H_5)_2NH_2^+$, $H_2NCH_2COO^-$, oxalate^{2-}, $HCit^{2-}$
4	Na^-, $CdCl^+$, Hg_2^{2+}, ClO_2^-, IO_3^-, HCO_3^-, $H_2PO_4^-$, HSO_3^-, $H_2AsO_4^-$, SO_4^{2-}, $S_2O_3^{2-}$, $S_2O_8^{2-}$, SeO_4^{2-}, CrO_4^{2-}, HPO_4^{2-}, $S_2O_6^{2-}$, PO_4^{3-}, $Fe(CN)_6^{3-}$, $Cr(NH_3)_6^{3+}$, $Co(NH_3)_6^{3+}$, $Co(NH_3)_5H_2O^{3+}$	5	Cl_2CHCOO^-, Cl_3COO^-, $(C_2H_5)_3NH^+$, $C_3H_7NH_3^+$, Cit^{3-}, succinate^{2-}, malonate^{2-}, tartrate^{2-}
4.5	Pb^{2-}, CO_3^{2-}, SO_3^{2-}, MoO_4^{2-}, $Co(NH_3)_5Cl^{2+}$, $Fe(CN)_5NO^{2-}$	6	benzoate$^-$, hydroxybenzoate$^-$, chlorobenzoate$^-$, phenylacetate$^-$, vinylacetate$^-$, $(CH_3)_2C{=}CHCOO^-$, $(C_2H_5)_4N^+$, $(C_3H_7)_2NH_2^+$, phthalate^{2-}, glutarate^{2-}, adipate^{2-}
5	Sr^{2-}, Ba^{2+}, Ra^{2+}, Cd^{2+}, Hg^{2+}, S^{2-}, $S_2O_4^{2-}$, WO_4^{2-}, $Fe(CN)_6^{4-}$	7	trinitrophenolate$^-$, $(C_3H_7)_3NH^+$, methoxybenzoate$^-$, pimelate^{2-}, suberate^{2-}, Congo red anion^{2-}
6	Li^+, Ca^{2+}, Cu^{2+}, Zn^{2+}, Sn^{2+}, Mn^{2+}, Fe^{2+}, Ni^{2+}, Co^{2+}, $Co(en)_3^{3+}$, $Co(S_2O_3)(CN)_5^{4-}$	8	$(C_6H_5)_2CHCOO^-$, $(C_3H_7)_4N^+$
8	Mg^{2+}, Be^{2+}		
9	H^+, Al^{3+}, Fe^{3+}, Cr^{3+}, Sc^{3+}, Y^{3+}, La^{3+}, In^{3+}, Ce^{3+}, Pr^{3+}, Nd^{3+}, Sm^{3+}, $Co(SO_3)_2(CN)_4^{5-}$		
11	Th^{4+}, Zr^{4+}, Ce^{4+}, Sn^{4+}		

TABLE 1.33 Covalent Radii for Atoms

Element	Single-bond radius, pm*	Double-bond radius, pm	Triple-bond radius, pm
Aluminum	126		
Antimony	141	131	
Arsenic	121	111	
Beryllium	106		
Boron	88		
Bromine	114	104	
Cadmium	148		
Carbon	77.2	66.7	60.3
Chlorine	99	89	
Copper	135		
Fluorine	64	54	
Gallium	126		
Germanium	122	112	
Hydrogen	30		
Indium	144		
Iodine	133	123	
Magnesium	140		
Mercury	148		
Nitrogen	70	60	55
Oxygen	66	55	
Phosphorus	110	100	93
Silicon	117	107	100
Selenium	117	107	
Silver	152		
Sulfur	104	94	87
Tellurium	137	127	
Tin	140	130	
Zinc	131		

* Single-bond radii are for a tetrahedral (CN = 4) structure.

TABLE 1.34 Octahedral Covalent Radii for CN = 6

Atom	Octahedral covalent radius, pm	Atom	Octahedral covalent radius, pm
Cobalt(II)	132	Nickel(III)	130
Cobalt(III)	122	Nickel(IV)	121
Gold(IV)	140	Osmium(II)	133
Iridium(III)	132	Palladium(IV)	131
Iron(II)	123	Platinum(IV)	131
Iron(IV)	120	Rhodium(III)	132
Nickel(II)	139	Ruthenium(II)	133

TABLE 1.35 Bond Lengths between Elements

Elements	Bond type	Bond Length, pm
	Boron	
B-B	B_2H_6	177(1)
B-Br	BBr_3	187(2)
B-Cl	BCl_3	172(1)
B-F	BF_3, R_2BF	129(1)
B-H	Boranes	121(2)
	Bridge	139(2)
B-N	Borazoles	142(1)
B-O	$B(OH)_3$, $(RO)_3B$	136(5)
	Hydrogen	
H-Al	AlH	164.6
H-As	AsH_3	151.9
H-Be	BeH	134.3
H-Br	HBr	140.8
H-Ca	CaH	200.2
H-Cl	HCl	127.4
H-F	HF	91.7
H-Ge	GeH_4	153
H-I	HI	160.9
H-K	KH	224.4
H-Li	LiH	159.5
H-Mg	MgH	173.1
H-Na	NaH	188.7
H-Sb	H_3Sb	170.7
H-Se	H_2Se	146.0
H-Sn	SnH_4	170.1
D-Br	DBr (^{2}HBr)	141.44
D-C1	DCl	127.46
D-I	DI	161.65
T-Br	TBr (^{3}HBr)	141.44
T-Cl	TCl	127.40
	Nitrogen	
N-Cl	NO_2Cl	179(2)
N-F	NF_3	136(2)
N-H	NH^+_4	103.4(3)
	NH_3, RNH_2	101.2
	H_2NNH_2	103.8
	$R—CO—NH_2$	99(3)
	HN=C=S	101.3(3)
N-D	ND (N^2H)	104.1
N-N	HN_3	102(1)
	R_2NNH_2	145.1(5)
	N_2O	112.6(2)
	N_2^+	111.6
N-O	NO_2Cl	124(1)
	$RO—NO_2$	136(2)
	NO_2	118.8(5)
N=O	N_2O	118.6(2)
	RNO_2	122(I)
	NO^+	106.19
N-Si	SiN	157.2
	Oxygen	
O-H	H_2O	95.8
	ROH	97(1)
	OH^+	102.89
	HOOH	96.0(5)
	D_2O (^2H_2O)	95.75
	OD	96.99
O-O	HO—OH	148(1)
	O_2^+	122.7
	O_2^-	126(2)
	O_3^{2-}	149(2)
	O_3	127.8(5)
O-Al	AlO	161.8
O-As	As_2O_6 bridges	179
O-Ba	BaO	190.0
O-Cl	ClO_2	148.4
	OCl_2	168
O-Mg	MgO	174.9
O-Os	OsO_4	166
O-Pb	PbO	193.4
	Phosphorus	
P-Br	PBr_3	223(1)
P-Cl	PCl_3	200(2)
P-F	$PFCl_2$	155(3)
P-H	PH_3, PH_4^+	142.4(5)
P-I	PI_3	252(1)
P-N	Single bond	149.1
P-O	Single bond	144.7
	p^3 bonding	167
	sp^3 bonding	154(4)
P-S	p^3 bonding	212(5)
	sp^3 bonding	208(2)
	In rings	220(3)
P-C	Single bond	156.2
	p^3 bonding	187(2)
	Silicon	
Si-Br	$SiBr_4$, R_3SiBr	216(1)
Si-Cl	$SiCl_4$, R_3SiCl	201.9(5)
Si-F	SiF_4, R_3SiF	156.1(3)
	SiF_6	158
Si-H	SiH_4	148.0(5)
	R_3SiH	147.6(5)
Si-I	SiI_4	234
	R_3SiI	246(2)
Si-O	R_3SiOR	153.3(5)
Si-Si	H_3SiSiH_3	230(2)
	Sulfur	
S-Br	$SOBr_2$	227(2)
S-Cl	S_2Cl_2	158.5(5)
S-F	SOF_2	158.5(5)
S-H	H_2S	133.3
	RSH	132.9(5)
	D_2S	134.5
S-O	SO_2	143.21
	$SOCl_2$	145(2)
S-S	RSSR	205(1)

TABLE 1.36 Bond Dissociation Energies

The bond dissociation energy (enthalpy change) for a bond A—B which is broken through the reaction

$$AB \rightarrow A + B$$

is defined as the standard-state enthalpy change for the reaction at a specified temperature, here at 298 K. That is,

$$\Delta Hf_{298} = \Delta Hf_{298}(A) + \Delta Hf_{298}(B) - \Delta Hf_{298}(AB)$$

All values refer to the gaseous state and are given at 298 K. Values of 0 K are obtained by subtracting $RT from the value at 298 K.

To convert the tabulated values to kcal/mol, divide by 4.184.

Bond	ΔHf_{298}, kJ/mol
Aluminum	
Al—Al	186(9)
Al—As	180
Al—Au	326(6)
Al—Br	439(8)
Al—C	255
Al—Cl	494(13)
AlCl—Cl	402(8)
$AlCl_2$—Cl	372(8)
AlO—Cl	515(84)
Al—Cu	216(10)
Al—D	291
Al—F	664(6)
AlF—F	546(42)
AlF_2—F	544(46)
AlO—F	761(42)
Al—H	285(6)
Al—I	368(4)
Al—Li	176(15)
Al—N	297(96)
Al—O	512(4)
AlCl—O	540(41)
AlF—O	582
Al—P	213(13)
Al—Pd	259(12)
Al—S	374(8)
Al—Se	334(10)
Al—Si	251(3)
Al—Te	268(10)
Al—U	326(29)
Antimony	
Sb—Sb	299(6)
Sb—Br	314(59)
Sb—Cl	360(50)
Sb—F	439(96)
Sb—N	301(50)
Antimony (*continued*)	
Sb—O	372(84)
Sb—P	357
Sb—S	379
Sb—Te	277.4(38)
Arsenic	
As—As	382(11)
As—Cl	448
As—Ga	209.6(12)
As—H	272(12)
As—N	582(126)
As—O	481(8)
As—P	534(13)
As—S	(478)
As—Se	96
As—Tl	198(15)
Astatine	
At—At	(115.9)
Barium	
Ba—Br	370(8)
Ba—Cl	444(13)
Ba—F	487(7)
Ba—I	>431(4)
Ba—O	563(42)
Ba—OH	477(42)
Ba—S	400(19)
Beryllium	
Be—Be	59
Be—Br	381(84)
Be—Cl	388(9)

TABLE 1.36 Bond Dissociation Energies (*Continued*)

Bond	ΔHf_{298}, kJ/mol
Beryllium (*continued*)	
BeCl—Cl	540(63)
Be—F	577(42)
Be—H	226(21)
Be—O	448(21)
Be—S	372(59)
Bismuth	
Bi—Bi	197(4)
Bi—Br	267(4)
Bi—Cl	305(8)
Bi—D	284
Bi—F	259(29)
Bi—Ga	159(17)
Bi—H	279
Bi—O	343(6)
Bi—P	280(13)
Bi—Pb	142(15)
Bi—S	316(5)
Bi—Sb	251(4)
Bi—Se	280(6)
Bi—Te	232(11)
Bi—Tl	121(13)
Boron	
B—B	297(21)
H_3B—BH_3	146
OB—BO	506(84)
B—Br	435(21)
B—C	448(29)
B—Cl	536(29)
BO—Cl	460(42)
B—D	341(6)
B—F	766(13)
BF—F	523(63)
BF_2—F	557(84)
B—H	330(4)
B—I	384(21)
B—N	389(21)
B—O	806(5)
BCl—O	715(41)
B—P	347(17)
B—S	581(9)
B—Se	462(15)
B—Si	289(29)
B—Te	354(20)
Bromine	
Br—Br	193.870(4)
Br—C	280(21)
Br—CH_3	284(8)
Br—CH_2Br	255(13)
Br—$CHBr_2$	259(17)
Br—CBr_3	209(13)
Br—CCl_3	218(13)
Br—CF_3	285(13)
Br—CF_2CF_3	287.4(63)
Br—$CF_2CF_2CF_3$	278.2(63)
Br—CHF_2	289
Br—Cl	218.84(4)
Br—CN	381
Br—CO—C_6H_5	268
Br—F	233.8(2)
Br—N	276(21)
Br—NF_2	222
Br—NO	120.1(63)
Br—O	235.1(4)
Cadmium	
Cd—Cd	11.3(8)
Cd—Br	159(96)
Cd—Cl	206.7(34)
Cd—F	305(21)
Cd—H	69.0(4)
Cd—I	138(21)
Cd—In	138
Cd—O	142(42)
Cd—S	196
Cd—Se	310
Calcium	
Ca—Ca	14.98(46)
Ca—Br	321(23)
Ca—Cl	398(13)
Ca—F	527(21)
Ca—H	167.8
Ca—I	285(63)
Ca—O	464(84)
Ca—S	314(19)
Cerium	
Ce—Ce	243(21)
Ce—F	582(42)
Ce—N	519(21)
Ce—O	795(13)
Ce—S	573(13)
Ce—Se	495(15)
Ce—Te	389(42)

(*Continued*)

TABLE 1.36 Bond Dissociation Energies (*Continued*)

Bond	ΔHf_{298}, kJ/mol	Bond	ΔHf_{298}, kJ/mol
Cesium		Chromium (*continued*)	
Cs—Cs	41.75(93)	Cr—Cu	155(21)
Cs—Br	397.5(42)	Cr—F	437(20)
Cs—Cl	439(21)	Cr—Ge	170(29)
Cs—F	514(8)	Cr—H	280(50)
Cs—H	178.1(38)	Cr—I	287(24)
Cs—I	339(4)	Cr—N	378(19)
Cs—O	297(25)	Cr—O	427(29)
Cs—OH	385(13)	OCr—O	531(63)
		O_2Cr—O	477(84)
Chlorine		Cr—S	339(21)
Cl—Cl	242.580(16)	Cobalt	
Cl—C	338(42)		
Cl—CH_3	339(21)	Co—Co	167(25)
Cl—CH_3^+	213	Co—Br	331(42)
Cl—$C(CH_3)_3$	328.4	Co—Cl	398(8)
Cl—CH_2Cl	310(13)	Co—Cu	162(17)
Cl—CCl_3	293(21)	Co—F	435(63)
Cl—CF_3	360(33)	Co—Ge	239(25)
Cl—CCl_2F	305(8)	Co—I	235(81)
Cl—$CClF_2$	318(8)	Co—O	368(21)
Cl—CF_2CF_2	346.0(71)	Co—S	343(21)
Cl—CH=CH_2	351		
Cl—CN	439	Copper	
Cl—COCl	328		
Cl—$COCH_3$	349.4	Cu—Cu	202(4)
Cl—COC_6H_5	310(13)	Cu—Br	331(25)
Cl—Cl^+	393	Cu—Cl	383(21)
Cl—ClO	143.3(42)	Cu—F	431(13)
O_3Cl—ClO_4	243	Cu—Ga	216(15)
Cl—F	250.54(8)	Cu—Ge	209(21)
O_3Cl—F	255	Cu—H	280(8)
Cl—N	389(50)	Cu—I	197(21)
Cl—NCl	280	Cu—Ni	206(17)
Cl—NCl_2	381	Cu—O	343(63)
Cl—NF_2	*ca.* 134	Cu—S	285(17)
Cl—NH_2	251(25)	Cu—Se	293(38)
Cl—NO	159(6)	Cu—Sn	177(17)
Cl—NO_2	142(4)	Cu—Te	176(38)
Cl—O	272(4)		
OCl—O	243(13)	Curium	
O_2Cl—O	201(4)		
Cl—P	289(42)	Cm—O	736
Cl—$SiCl_3$	464		
		Dysprosium	
Chromium			
		Dy—F	527(21)
Cr—Cr	155(21)	Dy—O	611(42)
Cr—Br	328(24)	Dy—Se	322(42)
Cr—Cl	366(24)	Dy—Te	234(42)

TABLE 1.36 Bond Dissociation Energies (*Continued*)

Bond	ΔHf_{298}, kJ/mol
Erbium	
Er—F	565(17)
Er—O	611(13)
Er—S	418(42)
Er—Se	326(42)
Er—Te	239(42)
Europium	
Eu—Eu	33.5(165)
Eu—Cl	*ca.* 326
Eu—F	528(18)
Eu—O	557(13)
Eu—S	364(15)
Eu—Se	301(15)
Eu—Te	243(15)
Fluorine	
F—F	156.9(96)
F—F^+	>251
F—CH_3	452(21)
F—$C(CH_3)_3$	439
F—C_6H_5	485
F—CCl_3	444(21)
F—CCl_2F	460(25)
F—$CClF_2$	490(25)
F—CF_3	523(17)
F—$COCH_3$	498
F—FO	272(13)
F—FO_2	81.0
F—N	301(42)
F—NF	318(25)
F—NF_2	243(8)
F—NO	235.6(42)
F—NO_2	197(25)
Gadolinium	
Gd—F	590(27)
Gd—O	716(17)
Gd—S	525(15)
Gd—Se	431(15)
Gallium	
Ga—Ga	138(21)
Ga—Br	444(17)
$(CH_3)_3Ga$—CH_3	253
Ga—Cl	481(13)
Ga—F	577(15)
Ga—H	<274
Ga—I	339(10)
Gallium (*continued*)	
Ga—O	285(63)
Ga—P	230(13)
Ga—Sb	209(13)
Ga—Te	251(25)
Germanium	
Ge—Ge	274(21)
Ge—Br	255(29)
Ge—Cl	431.8(4)
Ge—F	485(21)
Ge—H	321.3(8)
Ge—O	662(13)
Ge—S	551.0(25)
Ge—Se	490(21)
Ge—Si	301(21)
Ge—Te	402(8)
Gold	
Au—Au	221.3(21)
Au—B	368(11)
Au—Be	285(8)
Au—Bi	293(84)
Au—Cl	343(10)
Au—Co	215(13)
Au—Cr	215(6)
Au—Cu	232(9)
Au—Fe	187(17)
Au—Ga	294(15)
Au—Ge	277(15)
Au—H	314(10)
Au—La	80(5)
Au—Li	68.0(16)
Au—Mg	243(42)
Au—Mn	185(13)
Au—Ni	274(21)
Au—Pb	130(42)
Au—Pd	143(21)
Au—Rh	231(29)
Au—S	418(25)
Au—Si	312(12)
Au—Sn	244(17)
Au—Te	247(67)
Au—U	318(29)
Hafnium	
Hf—C	548(63)
Hf—N	534(29)
Hf—O	791(8)

(*Continued*)

TABLE 1.36 Bond Dissociation Energies (*Continued*)

Bond	ΔHf_{298}, kJ/mol	Bond	ΔHf_{298}, kJ/mol
Hydrogen		Hydrogen (*continued*)	
H—H	436.002(4)	H—$CHCl_2$	414.2
H—^{2}H or H—D	439.446(4)	H—CCl_3	377(8)
^{2}H—^{2}H or D—D	443.546(4)	H—CBr_3	377(8)
H—Br	365.7(21)	H—CCl_2CHCl_2	393(8)
H—C	337.2(8)	H—CH_2F	423(8)
H—CH	452(33)	H—CHF_2	423(8)
H—CH_2	473(4)	H—CF_3	444(13)
H—CH_3	431(8)	H—CF_2Cl	435(4)
^{2}H—C^2H_3 or D—CD_3	442.75(25)	H—CH_2CF_3	446(45)
H—C≡CH	523(4)	H—CF_2CH_3	416(4)
H—CH=CH_2	427	H—CF_2CF_3	431(63)
H—CH_2CH_3	410(4)	H—CH_2I	431(8)
H—CH_2C≡CH	392.9(50)	H—CHI_2	431(8)
H—CH_2CH=CH_2	356	H—CN	540(25)
H—cyclopropyl	423(13)	H—CH_2CN	*ca.* 389
H—$CH_2CH_2CH_3$	410(8)	H—$CH(CH_3)CN$	377(8)
H—$CH(CH_3)_2$	395.4	H—$C(CH_3)_2CN$	364(8)
H—cyclobutyl	397(13)	H—CH_2NH_2	397(8)
H—$CH_2CH(CH_3)_2$	360	H—$CH_2Si(CH_3)_3$	414(4)
H—$CH(CH_3)CH_2CH_3$	397(4)	H—CH_2COCH_3	393(75)
H—$C(CH_3)_3$	381	H—Cl	431.8(4)
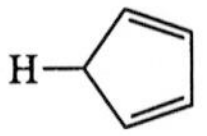	339(4)	H—CO	126(8)
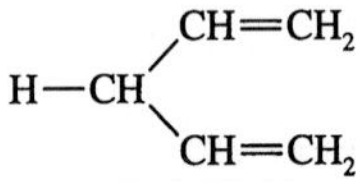	335(4)	H—CHO	364(4)
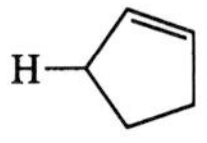	343(4)	H—COOH	377
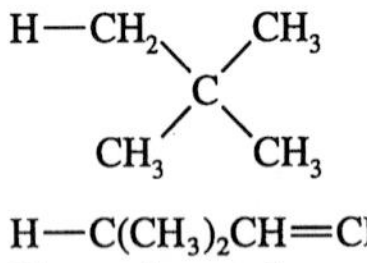	414(4)	H—$COCH_3$	364(4)
H—$C(CH_3)_2CH$=CH_2	331	H—$COCH_2CH_3$	364(4)
H—cyclopentyl	395(42)	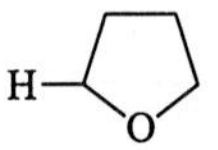	385
H—$CH_2C(CH_3)_3$	418(4)	H—COC_6H_5	364(4)
H—C_6H_5	431	H—$COCF_3$	381(8)
H—$CH_2C_6H_5$	356(4)	H—F	568.6(13)
H—$C(C_6H_5)_3$	314	H—I	298.7(8)
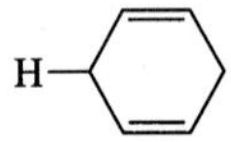	310	H—N	314(17)
H—cyclohexyl	399.6(42)	H—NH	377(8)
H—cycloheptyl	387.0(42)	H—NH_2	435(8)
H—norbornyl	406(13)	H—$NHCH_3$	431(8)
H—CH_2Br	410(25)	H—$N(CH_3)_2$	397(8)
H—$CHBr_2$	435	H—NHC_6H_5	335(13)
H—CH_2Cl	423	H—$N(CH_3)C_6H_5$	310(13)
		HNF_2	318(13)
		H—N_3	356
		H—NO	<205
		H—O	428.0(21)
		H—OH	498.7(8)
		H—OCH_3	436.8(42)
		H—OCH_2CH_3	436.0
		H—$OC(CH_3)_3$	439(4)
		H—OC_6H_5	368(25)
		H—ONO	327.6(25)

TABLE 1.36 Bond Dissociation Energies (*Continued*)

Bond	ΔHf_{298}, kJ/mol	Bond	ΔHf_{298}, kJ/mol
Hydrogen (*continued*)		**Iridium**	
H—ONO_2	423.4(25)	Ir—O	352(21)
H—OOH	374(8)	Ir—Si	463(21)
H—$OOCCH_3$	469(17)	**Iron**	
H—$OOCCH_2CH_3$	460(17)		
H—$OOCC_3H_7$	431(17)		
H—P	343(29)	Fe—Fe	100(21)
H—S	344(12)	Fe—Br	247(96)
H—SH	381(4)	Fe—Cl	*ca.* 352
H—SCH_3	*ca.* 368	Fe—O	409(13)
H—Se	305(2)	Fe—S	339(21)
H—Si	298.49(46)	Fe—Si	297(25)
H—SiH_3	393(13)	**Krypton**	
H—$Si(CH_3)_3$	377(13)		
H—Te	268(2)	Kr—Kr	5.4(8)
Indium		Kr—F	54
In—In	100(8)	**Lanthanum**	
In—Br	418(21)		
In—Cl	439(8)	La—La	247(21)
In—F	506(15)	La—C	506(63)
In—O	360(21)	La—F	598(42)
In—P	197.9(85)	La—N	519(42)
In—S	289(17)	La—O	799(13)
In—Sb	152(11)	La—S	577(25)
In—Se	247(17)	**Lead**	
In—Te	218(17)		
Iodine		Pb—Pb	339(25)
		Pb—Br	247(38)
I—I	152.549(8)	$Pb(CH_3)_3$—CH_3	207(42)
I—Br	179.1(4)	Pb—Cl	301(29)
I—CH_3	232(13)	Pb—F	356(8)
I—C_2H_5	223.8	Pb—H	176(21)
I—$CH(CH_3)_2$	222	Pb—I	197(38)
I—$C(CH_3)_3$	207.1	Pb—O	378(4)
I—CH_2CF_3	234(4)	Pb—S	346.0(17)
I—CF_2CH_3	216(4)	Pb—Se	303(4)
I—C_3F_7	209(4)	Pb—Te	251(13)
I—CH=$CHCH_3$	172	**Lithium**	
I—C_6H_5	268(4)		
I—C_6F_5	276	Li—Li	106(4)
I—Cl	213.3(4)	Li—Br	423(21)
I—$COCH_3$	219.7	Li—Cl	469(13)
I—CN	305(4)	Li—F	577(21)
I—F	280(4)	Li—H	247
I—N	159(17)	Li—I	352(13)
I—NO	71(4)	Li—Na	88
I—NO_2	75(4)	Li—O	341(6)
I—O	184(21)	Li—OH	427(21)

(*Continued*)

TABLE 1.36 Bond Dissociation Energies (*Continued*)

Bond	ΔHf_{298}, kJ/mol
Lutetium	
Lu—Lu	142(34)
Lu—F	569(42)
Lu—O	695(13)
Lu—S	507(15)
Lu—Te	326(17)
Magnesium	
Mg—Mg	8.522(4)
Mg—Br	297(63)
Mg—Cl	318(13)
Mg—F	462(21)
MgF—F	569(42)
Mg—H	197(50)
Mg—I	*ca.* 285
Mg—O	394(35)
Mg—OH	238(21)
Mg—S	310(75)
Manganese	
Mn—Mn	42(29)
Mn—Br	314(10)
Mn—Cl	361(10)
Mn—F	423(15)
Mn—I	283(10)
Mn—Cu	159(17)
Mn—O	402(34)
Mn—S	301(17)
Mn—Se	201(13)
Mercury	
Hg—Hg	17.2(21)
Hg—Br	72.8(42)
CH_3—$HgCH_3$	240.6
C_2H_5—HgC_2H_5	182.8(42)
C_3H_7—HgC_3H_7	197.1
Isopropyl—Hgisopropyl	170.3
C_6H_5—HgC_6H_5	285
Hg—Cl	100(8)
Hg—F	130(38)
Hg—H	39.8
Hg—I	38
Hg—K	8.24(21)
Hg—Na	>6.7
Hg—S	213
Hg—Se	(167)
Hg—Te	(142)
Molybdenum	
Mo—I	372
Mo—O	607(34)
MoO—O	678(84)
MoO_2—O	565(84)
Neodymium	
Nd—F	545(13)
Nd—O	703(34)
Nd—S	474(15)
Nd—Se	385(17)
Nd—Te	305(17)
Neon	
Ne—Ne	3.93
Neptunium	
Np—O	720(29)
Nickel	
Ni—Ni	261.9(25)
Ni—Br	360(13)
Ni—Cl	372(21)
Ni—F	435
Ni—H	289(13)
Ni—I	293(21)
Ni—O	391.6(38)
Ni—S	360(21)
Ni—Si	318(17)
Niobium	
Nb—O	753(13)
Nitrogen	
N—N	945.33(59)
N—Br	276(21)
ON—Br	28.7(15)
N—Cl	389(50)
ON—Cl	159(6)
O_2N—Cl	142(4)
N—F	301(42)
FN—F	318(21)
F_2F—N	243(8)
ON—F	236(4)
O_2N—F	188(21)

TABLE 1.36 Bond Dissociation Energies (*Continued*)

Bond	ΔHf_{298}, kJ/mol	Bond	ΔHf_{298}, kJ/mol
Nitrogen (*continued*)		Oxygen (*continued*)	
N—I	159(17)	$C_2H_5O—OC_2H_5$	159
$F_2N—NF_2$	88(4)	$C_3H_7O—OC_3H_7$	155
$H_2N—NH_2$	297(8)	Palladium	
$H_2N—NHCH_3$	271	Pd—O	234(29)
$H_2N—N(CH_3)_2$	264	Phosphorus	
$H_2N—NHC_6H_5$	213	P—P	490(11)
$HN—N_2$	38	P—Br	266.5
ON—N	480.7(42)	P—C	513(8)
$ON—NO_2$	39.8(8)	P—Cl	289(42)
$O_2N—NO_2$	57.3(21)	P—F	439(96)
HN═NH	456(42)	P—H	343(29)
N≡N	946	P—N	617(21)
N—O	630.57(13)	P—O	596.6
HN═O	481	$Br_3P═O$	498(21)
NN—O	167	$Cl_3P═O$	510(21)
ON—O	305	$F_3P═O$	544(21)
N—P	617(21)	P—S	346.0(17)
N—S	464(21)	P═S	347
Osmium		P—Se	363(10)
$O_3Os—O$	301(21)	P—Te	298(10)
Oxygen		Platinum	
O—O	498.34(20)	Pt—B	478(17)
O—Br	235.1(4)	Pt—H	352(38)
$HO—CH_3$	377(13)	Pt—O	347(34)
$HO—CH═CH_2$	364	Pt—P	417(17)
$HO—CH_2CH═CH_2$	456	Pt—Si	501(18)
$HO—C_6H_5$	431	Potassium	
$HO—CH_2C_6H_5$	322	K—K	57.3(42)
HO—CHO	402(13)	K—Br	383(8)
$HO—COCH_3$	452(21)	K—Cl	427(8)
$HO—COC_2H_5$	180	K—F	497.5(25)
O—Cl	272(4)	K—H	183(15)
HO—Cl	251(13)	K—I	331(13)
O—F	222(17)	K—Na	63.6(29)
O—FO	467	K—O	239(34)
FO—OF	261(84)	K—OH	343(8)
O—I	184(21)	Praseodymium	
HO—I	234(13)	Pr—F	582(46)
O—N	630.57(13)	Pr—O	753(17)
$HO—NCH_3$	209	Pr—S	492.5(46)
$HO—OC(CH_3)_3$	192(8)		
HO—OH	213.8(21)		
O—OH	268(4)		
$CF_3O—OCF_3$	192		
$CH_3O—OCH_3$	157.3(8)		

(*Continued*)

TABLE 1.36 Bond Dissociation Energies (*Continued*)

Bond	ΔHf_{298}, kJ/mol	Bond	ΔHf_{298}, kJ/mol
Praseodymium (*continued*)		**Scandium**	
Pr—Se	446(23)	Sc—Sc	163(21)
Pr—Te	326(42)	Sc—Br	444(63)
		Sc—C	393(63)
Promethium		Sc—Cl	318
		Sc—F	589(13)
Pm—F	540(42)	Sc—N	469(84)
Pm—O	674(63)	Sc—O	674(13)
Pm—S	423(63)	Sc—S	478(13)
Pm—Se	339(63)	Sc—Se	385(17)
Pm—Te	255(63)	Sc—Te	289(17)
Radium		**Selenium**	
Ra—Cl	343(75)	Se—Se	332.6(4)
		Se—Br	297(84)
Rhodium		Se—C	582(96)
		Se—Cl	322
Rh—Rh	285(21)	Se—F	339(42)
Rh—B	476(21)	Se—H	305(2)
Rh—C	583.7(63)	Se—N	381(63)
Rh—O	377(63)	Se—O	423(13)
Rh—Si	395(18)	Se—P	364(10)
Rh—Ti	391(15)	Se—S	381(21)
		Se—Si	531(25)
Rubidium		Se—Te	268(8)
Rb—Rb	45.6(21)	**Silicon**	
Rb—Br	389(13)		
Rb—Cl	448(21)	Si—Si	327(10)
Rb—F	494(21)	Si—Br	343(50)
Rb—H	167(21)	Si—C	435(21)
Rb—I	335(13)	Si—Cl	456(42)
Rb—O	255(84)	Si—F	540(13)
Rb—OH	351(8)	Si—H	298.49(46)
		Si—I	339(84)
Ruthenium		Si—N	439(38)
		Si—O	798(8)
Ru—O	481(63)	Si—S	619(13)
O_3Ru—O	439	Si—Se	531(25)
Ru—Si	397(21)	H_3Si—SiH_3	339(17)
Ru—Th	592(42)	$(CH_3)_3$Si—Si$(CH_3)_3$	339
		(Aryl)$_3$Si—Si(aryl)$_3$	368(31)
Samarium		Si—Te	506(38)
		Silver	
Sm—Cl	423(13)		
Sm—F	531(18)	Ag—Ag	163(8)
Sm—O	619(13)	Ag—Au	203(9)
Sm—S	389	Ag—Bi	193(42)
Sm—Se	331(15)		
Sm—Te	272(15)		

TABLE 1.36 Bond Dissociation Energies (*Continued*)

Bond	ΔHf_{298}, kJ/mol	Bond	ΔHf_{298}, kJ/mol
Silver (*continued*)		Tantalum	
Ag—Br	293(29)	Ta—N	611(84)
Ag—Cl	341.4	Ta—O	805(13)
Ag—Cu	176(8)	Tellurium	
Ag—F	354(16)		
Ag—Ga	180(15)		
Ag—Ge	175(21)	Te—B	354(20)
Ag—H	226(8)	Te—H	268(2)
Ag—I	234(29)	Te—I	193(42)
Ag—In	176(17)	Te—O	391(8)
Ag—O	213(84)	Te—P	298(10)
Ag—Sn	136(21)	Te—S	339(21)
Ag—Te	293(96)	Te—Se	268(8)
Sodium		Terbium	
Na—Na	77.0	Tb—F	561(42)
Na—Br	370(13)	Tb—O	707(13)
Na—Cl	410(8)	Tb—S	515(42)
Na—F	481(8)	Tb—Te	339(42)
Na—H	201(21)	Thallium	
Na—I	301(8)		
Na—K	63.6(29)		
Na—O	257(17)	Tl—Tl	63
Na—OH	381(13)	Tl—Br	333.9(17)
Na—Rb	59(4)	Tl—Cl	372.8(21)
		Tl—F	445(19)
Strontium		Tl—H	188(8)
		Tl—I	272(8)
Sr—Br	332(19)		
Sr—Cl	406(13)	Thorium	
Sr—F	542(7)		
Sr—H	163(8)	Th—Th	289
Sr—I	263(42)	Th—C	484(25)
Sr—O	454(15)	Th—N	577.4(21)
Sr—OH	381(42)	Th—O	854(13)
Sr—S	314(21)	Th—P	377
Sulfur		Thullium	
S—S	429(6)	Tm—F	569(42)
S—Cl	255	Tm—O	557(13)
S—F	343(5)	Tm—S	368(42)
O_2S—F	71	Tm—Se	276(42)
S—N	464(21)	Tm—Te	276(42)
S—O	521.70(13)		
OS—O	551.4(84)	Tin	
O_2S—O	348.1(42)		
HS—SH	272(21)	Sn—Sn	195(17)
		Sn—Br	339(4)

(*Continued*)

TABLE 1.36 Bond Dissociation Energies (*Continued*)

Bond	ΔHf_{298}, kJ/mol	Bond	ΔHf_{298}, kJ/mol
Tin (*continued*)		Vanadium (*continued*)	
BrSn—Br	326	V—Cl	477(63)
Br_3Sn—Br	272	V—F	590(63)
$(C_2H_5)_3Sn$—C_2H_5	*ca.* 238	V—N	477(8)
Sn—Cl	406(13)	V—O	644(21)
Sn—F	467(13)	V—S	490(16)
Sn—H	267(17)	V—Se	347(21)
Sn—I	234(42)	Xenon	
Sn—O	548(21)	Xe—Xe	6.53(30)
Sn—S	464(3)	Xe—F	13.0(4)
Sn—Se	401.3(59)	Xe—O	36.4
Sn—Te	319.2(8)	Ytterbium	
Titanium		Yb—Cl	322
Ti—Ti	141(21)	Yb—F	521(10)
Ti—Br	439	Yb—H	159(38)
Ti—C	435(25)	Yb—O	397.9(63)
Ti—Cl	494	Yb—S	167
Ti—F	569(34)	Yttrium	
Ti—H	*ca.* 159	Y—Y	159(21)
Ti—I	310(42)	Y—Br	485(84)
Ti—N	464	Y—C	418(63)
Ti—O	662(16)	Y—Cl	527(42)
Ti—S	426(8)	Y—F	605(21)
Ti—Se	381(42)	Y—N	481(63)
Ti—Te	289(17)	Y—O	715.1(30)
Tungsten		Y—S	528(11)
W—Cl	423(42)	Y—Se	435(13)
W—F	548(63)	Y—Te	339(13)
W—O	653(25)	Zinc	
OW—O	632(84)	Zn—Zn	29
O_2W—O	598(42)	Zn—Br	142(29)
W—P	305(4)	C_2H_5C—C_2H_5	*ca.* 201
Uranium		Zn—Cl	229(20)
U—O	761(17)	Zn—F	368(63)
OU—O	678(59)	Zn—H	85.8(21)
O_2U—O	644(88)	Zn—I	138(29)
U—S	523(10)	Zn—O	284.1
Vanadium		Zn—S	205(13)
V—V	242(21)	Zn—Se	136(13)
V—Br	439(42)	Zn—Te	205
V—C	469(63)		

TABLE 1.36 Bond Dissociation Energies (*Continued*)

Zirconium		Zirconium (*continued*)	
Zr—C	561(25)	Zr—O	760(8)
Zr—F	623(63)	Zr—S	575(17)
Zr—N	565(25)		

1.8 DIPOLE MOMENTS

The dipole moment is the mathematical product of the distance between the centers of charge of two atoms multiplied by the magnitude of that charge. Thus, the dipole moment (μ) of a compound or molecule is:

$$\mu = Q \times r$$

where Q is the magnitude of the electrical charge(s) that are separated by the distance r; the unit of measurement is the Debye (D)

All bonds between equal atoms are given zero values. Because of their symmetry, methane and ethane molecules are nonpolar. The principle of bond moments thus requires that the CH_3 group moment equal one H—C moment. Hence the substitution of any aliphatic H by CH_3 does not alter the dipole moment, and all saturated hydrocarbons have zero moments as long as the tetrahedral angles are maintained.

TABLE 1.37 Bond Dipole Moments

Bond	Moment, D*	Bond	Moment, D*
H—C		C—N, aliphatic	0.45
Aliphatic	0.3	C=N	1.4
Aromatic	0.0	C≡N (nitrile)	3.6
C—C	0.0	NC (isonitrile)	3.0
C≡C	0.0	N—H	1.31
C—O		N—O	0.3
Ether, aliphatic	0.74	N=O	2.0
Alcohol, aliphatic	0.7	N (lone pair on sp^3 N)	1.0
C=O		C—P, aliphatic	0.8
Aliphatic	2.4	P—O	(0.3)
Aromatic	2.65	P=O	2.7
O—H	1.51	P—S	0.5
C—S	0.9	P=S	2.9
C=S	2.0	B—C, aliphatic	0.7
S—H	0.65	B—O	0.25
S—O	(0.2)	Se—C	0.7
S=O		Si—C	1.2
Aliphatic	2.8	Si—H	1.0
Aromatic	3.3	Si—N	1.55

*To convert debye units D into coulomb-meters, multiply by 3.33564×10^{-30}.

TABLE 1.38 Group Dipole Moments

Bond	Moment, D*	Bond	Moment, D*
H—Sb	−0.08	Br—F	1.3
H—As	−0.10	Cl—F	0.88
H—P	0.36	Li—C	1.4
H—I	0.38	K—Cl	10.6
H—Br	0.78	K—F	7.3
H—Cl	1.08	Cs—Cl	10.5
H—F	1.94	Cs—F	7.9
C—Te	0.6		
N—F	0.17	Dative (coordination) bonds	
P—I	0.3		
P—Br	0.36	N → B	2.6
P—Cl	0.81	O → B	3.6
As—I	0.78	S → B	3.8
As—Br	1.27	P → B	4.4
As—Cl	1.64	N → O	4.3
As—F	2.03	P → O	2.9
Sb—I	0.8	S → O	3.0
Sb—Br	1.9	As → O	4.2
Sb—Cl	2.6	Se → O	3.1
S—Cl	0.7	Te → O	2.3
Cl—O	0.7	P → S	3.1
I—Br	1.2	P → Se	3.2
I—Cl	1	Sb → S	4.5
Br—Cl	0.57		

*To convert debye units D into coulomb-meters, multiply by 3.33564×10^{-30}.

The group moment always includes the C—X bond. When the group is attached to an aromatic system, the moment contains the contributions through resonance of those polar structures postulated as arising through charge shifts around the ring.

1.8.1 Dielectric Constant

The *dielectric constant* (also referred to as the *relative permittivity*, *K*) is the ratio of the permittivity of the material to the permittivity of free space and is the property of a material that determines the relative speed with which an electrical signal will travel in that material.

$$K = \epsilon_T/\epsilon_0$$

Signal speed is roughly inversely proportional to the square root of the dielectric constant. A low dielectric constant will result in a high signal propagation speed and a high dielectric constant will result in a much slower signal propagation speed.

The *dielectric loss factor* is the tangent of the loss angle and the *loss tangent* (tan Δ) is defined by the relationship:

$$\tan \Delta = 2\sigma/\varepsilon \upsilon$$

σ is the electrical conductivity, ε is the dielectric constant, and υ is the frequency. The loss tangent is roughly wavelength independent.

TABLE 1.39 Dipole Moments and Dielectric Constants

Substance	Dielectric constant, ε	Dipole moment, D
Air	1.000 536 4	
$AlBr_3$	3.38^{100}	5.2
Ar		
(g)	1.000 517 2	
(lq)	1.538^{-191}, 1.325^{-132}	0
$AsBr_3$	8.83^{35}	1.61
$AsCl_3$	12.6^{20}	1.59
AsH_3 (arsine)	2.40^{-72}, 2.05^{20}	0.20
BBr_3	2.58^{0}	0
BCl_3		0
BF_3		0
B_2H_6 (diborane)	$1.872^{-92.5}$	0
B_4H_{10}		0.486
B_5H_9	21.1^{25}	2.13
B_6H_{10}		2.50
$B_3H_6N_3$		0
Br_2 (g)	1.0128^{20}	
(lq)	3.1484^{25}	0
BrF_3	106.8^{25}	1.1
BrF_5	$7.91^{24.5}$	1.51
Cl_2 (g)		0
(lq)	2.147^{-65}, 1.91^{14}	
ClF_3	4.394^{20}, 4.29^{25}	0.554
ClF_5	4.28^{-80}	
ClO_3F	2.194^{-123}	0.023
CO (g)	$1.000\ 70^{0}$	0.112
(lq)		
CO_2 (g)	1.000 922	0
(lq)	$1.60^{°C,\ 50\ atm}$, 1.449^{23}	
$COCl_2$	4.34^{22}	1.17
COF_2		0.95
COS	4.47^{-88}	0.712
COSe	3.47^{10}	0.73
CS		1.98
CS_2 (g)	1.0029^{0}	0
(lq)	2.632^{20}	
CrO_2Cl_2	2.6^{20}	0.47
D_2 (deuterium)	1.290^{-255}, 1.277^{-253}	
DH	$1.269^{16.78\ K}$	
D_2O	79.75^{20}, 78.25^{25}	1.87
F_2	1.491^{-220}, 1.54^{-202}	
$GaCl_3$		0.85
$GeBr_4$		
$GeBr_4$	2.955^{26}	
$GeCl_4$	2.463^{0}, 2.430^{25}	0
$GeClH_3$		2.13
H_2(g)	1.000 253 8	0
t		
(lq)	$1.279^{13.5\ K}$, $1.228^{20.4\ K}$	
HBr(g)	$1.003\ 13^{0}$	0.827
(lq)	8.23^{-86}, 3.82^{25}	
He (g)	1.000 0565 0	0
(lq) (II)	$1.055^{2.055\ K}$	
(III)		
(IV)		
HCl (g)	1.0046^{0}	1.109
(lq)	14.3^{-114}, 4.60^{28}	
HClO		1.3
HCN	114.9^{20}	2.98
HCNO (isocyanate)		1.6
HCNS		1.7
HF	83.6^{0}	1.826
HFO		2.23
HI (g)	$1.002\ 34^{0}$	0.448
(lq)	3.87^{-53}, 2.90^{22}	
HN_3 (azide)		1.70
H_2O (see Table 1.12)		
H_2O_2	84.2^{0}, 74.6^{17}	1.573
HNO_3		2.17
H_2S (g)	1.0040^{0}	0.97
(lq)	5.93^{10}	
H_2Se		0.24
HSO_3Cl	60^{60}	
HSO_3F	ca. 120^{25}	
H_2SO_4	100^{25}	
H_2Te		<0.2
Hg		0
I_2	11.1^{118}	0
IBr		0.726
IF		1.95
IF_5	37.13^{20}	2.18
IF_7	1.97^{23}	
IOF_5	1.75^{25}	
Kr (g)		<0.05
(lq)	$1.644^{-153.4}$	
Mn_2O_7	3.28^{20}	
Ne (g)	$1.000\ 063\ 9^{20}$	0
(lq)	$1.1907^{-247.1}$	
N_2 (g)	$1.000\ 548\ 0^{20}$	0
(lq)	1.468^{-210}, 1.454^{-203}	
NH_3 (g)	1.0072^{0}	1.471
(lq)	$22.4^{-33.5}$ 16.61^{20}	

(*Continued*)

TABLE 1.39 Dipole Moments and Dielectric Constants (*Continued*)

Substance	Dielectric constant, ε	Dipole moment, D
N_2H_4 (hydrazine)	52.9^{20}, 51.7^{25}	1.75
$Ni(CO)_4$		
NO		0.159
N_2O (g)	$1.001\ 13^{0}$	0.161
(lq)	1.52^{15}	
NO_2		0.316
N_2O_4	2.56^{25}, 2.44^{20}	0.5
N_2O_3		2.122
NOBr	13.4^{15}	1.8
NOCl	18.2^{12}	1.9
NO_2Cl		0.53
NOF		1.73
NO_2F		0.47
NO_3	31.13^{-70}	
O_2 (g)	$1.000\ 494\ 7^{20}$	0
(lq)	$1.568^{-218.7}$, 1.507^{-193}	
O_3	4.75^{-183}	0.534
OF_2		0.297
O_2F_2 (FOOF)		1.44
OsO_4		0
P (lq)	4.096^{34}	
PBr_3	3.9^{20}	0.56
PCl_3	3.43^{25}, 3.50^{17}	0.78
PCl_5	2.85^{160}, 2.7^{165}	0.9
PCl_2F_3	2.813^{-45}	
PCl_3F_2	2.375^{-5}	
PCl_4F	$2.65^{0.5}$	
PF_3		1.03
PF_5		0
PH_3	2.9^{15}	0.574
PI_3	4.12^{65}	0
PO_3		
$POCl_3$	13.7^{25}	2.54
POF_3		1.868
$PSCl_3$	5.8^{22}	1.42
PSF_3		0.64
$PbCl_4$	2.78^{20}	
ReO_2Cl_3		
ReO_3Cl		
S	3.499^{134}	
SCl_2	2.915^{25}	0.36
S_2Cl_2 dimer	4.79^{15}	1.0
S_2F_2		
FSSF isomer		1.45
S = SF_2 isomer		1.03
SF_4		0.632
SF_6	1.81^{-50}	0
S_2F_{10}	2.020^{20}	0
SO_2 (g)	1.0093^{0}	1.63
(lq)	16.3^{25}	
SO_3	3.11^{18}	0
$SOBr_2$	9.06^{20}	9.11
$SOCl_2$	9.25^{20}, 8.675^{25}	1.45
SOF_2		1.63
SO_2Cl_2	9.15^{20}	1.81
SO_2F_2		1.12
$SbCl_3$	33.2^{75}	3.93
$SbCl_5$	3.22^{20}	0
SbF_5		
SbH_3		0.12
Se (lq)	$5.44^{237.5}$	
SeF_4		1.78
SeF_6		0
$SeOCl_2$	46.2^{20}	2.64
SeO_2		2.62
$SiCl_4$	2.248^{0}	0
SiF_4		0
SiH_4		0
$SiHCl_3$		0.86
SiH_3Cl		1.31
$SnBr_4$	3.169^{30}	0
$SnCl_4$	3.014^{0}, 2.89^{20}	0
TeF_6		0
$TiCl_4$	2.843^{14}, 2.80^{20}	0
UF_6 (g)	$1.002\ 92^{67}$	0
(lq)	2.18^{65}	
VCl_4	3.05^{25}	0
$VOBr_3$	3.6^{25}	
$VOCl_3$	3.4^{25}	0.3
Xe (g)	1.001 23	0
(lq, II)	$1.880^{-111.9}$	
XeF_6	4.10^{125}	

1.9 MOLECULAR GEOMETRY

Molecular geometry is the specific three-dimensional arrangement of atoms and the positions of the atomic nuclei in a molecule.

Various instrumental techniques such as x-ray crystallography and other experimental techniques can be used to derive information about the locations of atoms in a molecule.

Thus, molecular geometry is associated with the specific orientation of bonding atoms. A careful analysis of electron distribution in various orbitals will usually result in correct determination of the molecular geometry.

TABLE 1.40 Spatial Orientation of Common Hybrid Bonds

On the assumption that the pairs of electrons in the valency shell of a bonded atom in a molecule are arranged in a definite way which depends on the number of electron pairs (coordination number), the geometrical arrangement or shape of molecules may be predicted. A multiple bond is regarded as equivalent to a single bond as far as molecular shape is concerned.

Coordination number	Orbitals hybridized	Geometrical arrangement	Minimum radius ratio
2	sp dp	Linear	
	p^2 ds d^2	Bent (angular)	
3	sp^2 ds^2	Trigonal planar	0.155
	p^3 d^2p	Trigonal pyramidal	
4	sp^2d p^2d^2	Square planar	
	sp^3 d^3s	Tetrahedral	0.225
	d^4	Tetragonal pyramidal	
5	sp^3d d^3sp	Trigonal bipyramidal	0.155
6	d^2sp^3	Octahedral	0.414
	d^4sp	Trigonal prism	
7		One atom above the face of an octahedron, which is distorted chiefly by separating the atoms at the corners of this face.	0.592
8	d^4sp^3	Square antiprism (dodecahedral)	0.645
		Cube	0.732
9		Formed by adding atoms beyond each of the vertical faces of a right triangular prism.	0.732
12		Cube-octahedron	1.000

TABLE 1.41 Crystal Lattice Types

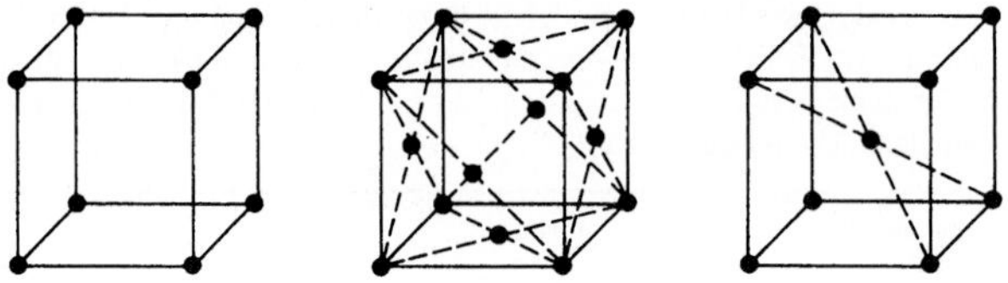

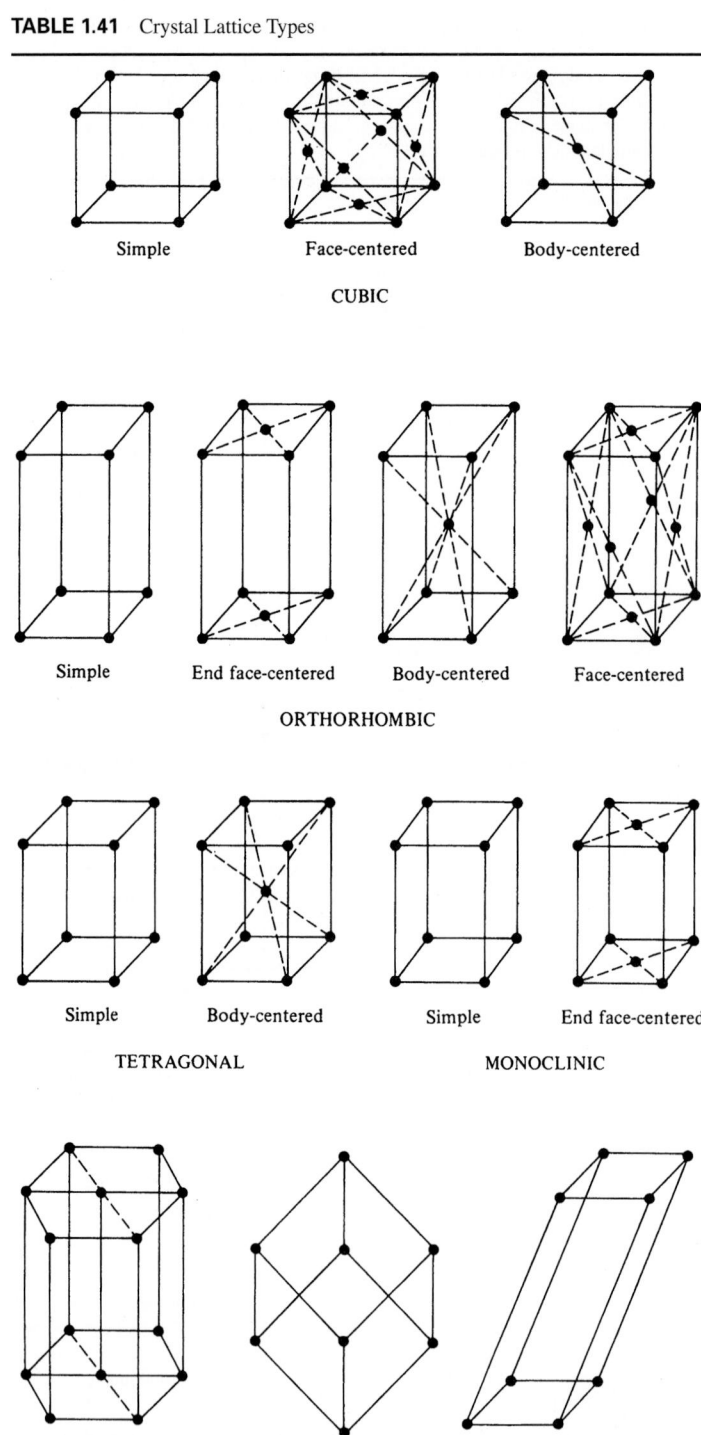

Simple | Face-centered | Body-centered

CUBIC

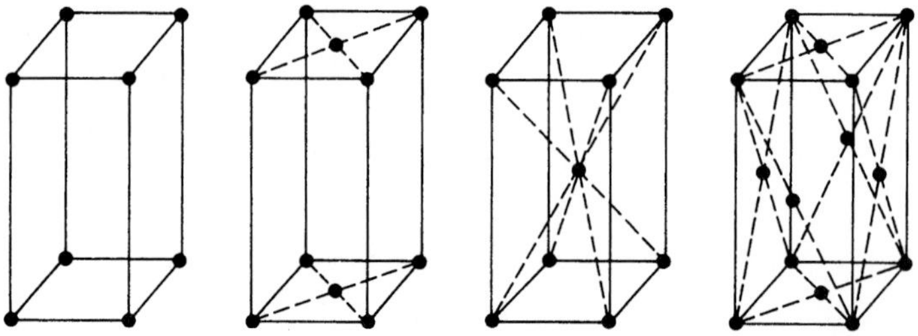

Simple | End face-centered | Body-centered | Face-centered

ORTHORHOMBIC

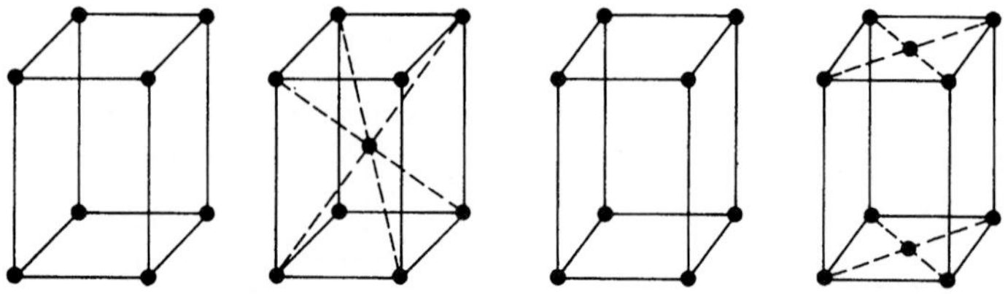

Simple | Body-centered | Simple | End face-centered

TETRAGONAL | MONOCLINIC

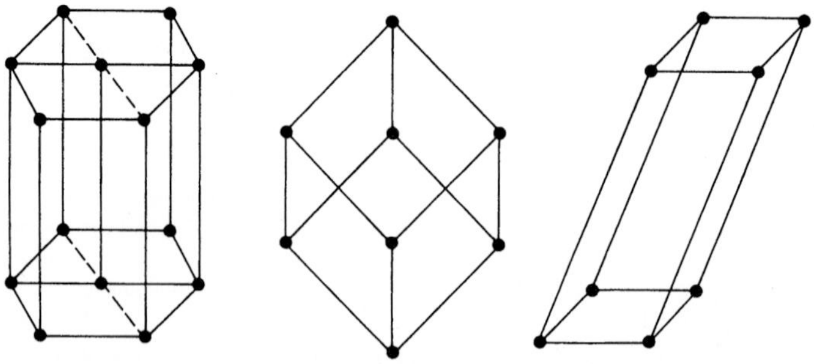

HEXAGONAL | RHOMBOHEDRAL | TRICLINIC

TABLE 1.42 Crystal Structure

Unit cells of the different lattice types in each system are illustrated in Table 1.41

System	Characteristics	Essential symmetry	Axes in unit cell	Angles in unit cell
Cubic	Three axes equal and mutually perpendicular	Four threefold axes	$a = b = c$	$\alpha = \beta = \gamma = 90°$
Tetragonal	Two equal axes and one unequal axis mutually perpendicular	One fourfold axis	$a = b \neq c$	$\alpha = \beta = \gamma = 90°$
Orthorhombic (or rhombic)	Three unequal axes mutually perpendicular	Three mutually perpendicular twofold axes, or two planes intersecting in a twofold axis	$a \neq b \neq c$	$\alpha = \beta = \gamma = 90°$
Hexagonal or trigonal	Three equal axes inclined at 120° with a fourth axis unequal and perpendicular to the other three	One sixfold axis or one threefold axis	$a = b \neq c$ $a = b = c$	$\alpha = \beta = 90°$; $\gamma = 120°$ $\alpha = \beta = \gamma \neq 90°$
Monoclinic	Two axes at an oblique angle with a third perpendicular to the other two	One twofold axis or one plane	$a \neq b \neq c$	$\alpha = \beta = 90°$; $\gamma \neq 90°$
Triclinic	Three unequal axes intersecting obliquely	No planes or axes of symmetry	$a \neq b \neq c$	$\alpha \neq \beta \neq \gamma \neq 90°$
Rhombohedral	Two equal axes making equal angle with each other			

1.10 NUCLIDES

The nuclide is the nucleus of a particular isotope.

TABLE 1.43 Table of Nuclides

Explanation of Column Headings

Nuclide. Each nuclide is identified by element name and the mass number *A*, equal to the sum of the numbers of protons *Z* and neutrons *N* in the nucleus. The *m* following the mass number (for example, ^{69m}Zn) indicates a metastable isotope. An asterisk preceding the mass number indicates that the radionuclide occurs in nature.

Half-life. The following abbreviations for time units are employed: y = years, d = days, h = hours, min = minutes, s = seconds, ms = milliseconds, and ns = nanoseconds.

Natural abundance. The natural abundances listed are on an "atom percent" basis for the stable nuclides present in naturally occurring elements in the earth's crust.

Thermal neutron absorption cross section. Simply designated "cross section," it represents the ease with which a given nuclide can absorb a thermal neutron (energy less than or equal to 0.025 eV) and become a different nuclide. The cross section is given here in units of barns (1 barn = 10^{-24} cm^2). If the mode of reaction is other than (n, γ), it is so indicated.

Major radiations. In the last column are the principal modes of disintegration and energies of the radiations in million electronvolts (MeV). Symbols used to represent the various modes of decay are:

α, alpha particle emission	K, electron capture
β^-, beta particle, negatron	IT, isomeric transition
β^+, positron	x, X-rays of indicated element (e.g., O-x, oxygen X-rays, and the type, K or L)
γ, gamma radiation	

For β^- and β^+, values of E_{max} are listed. Radiation types and energies of minor importance are omitted unless useful for identification purposes.

(Continued)

TABLE 1.43 Table of Nuclides (*Continued*)

Element	A	Half-life	Natural abundance, %	Cross section, barns	Radiation (MeV)
Hydrogen	1		99.985(1)	0.332(2)	
	2		0.015(1)	0.000 52(1)	
	3	12.32 y			β^-(0.0186)
Beryllium	7	53.28 d			K, γ(0.478)
	9		100	0.008(1)	
	10	1.52×10^6 y			β^-(0.555)
Boron	10		19.9(2)	3837(10)(n, α)	
	11		80.1(6)	0.005(3)	
Carbon	11	20.3 min			β^+(0.961)
	12		98.89(1)	0.0035(1)	
	14	5715 y			β^-(0.156)
Nitrogen	13	9.965 min			β^+(1.190)
	14		99.634(9)	1.8(1)(n, p)	
Oxygen	15	122.2 s			β^+(2.754)
	19	26.9 s			β^-(4.82); γ(0.197, 1.357)
Fluorine	18	1.8295 h			β^+(0.635); K, O-x
	19		100	0.0095(7)	β^+(2.754)
	20	11.00 s			β^-(5.40); γ(1.63)
Sodium	22	2.605 y		2800.(300)(n, p)	β^+(0.545, 1.83); K, Ne-x, γ(1.275)
	23		100	0.53	
	24	14.659 h			β^-(1.39); γ(2.75, 1.37)
Magnesium	24		78.89(3)	0.053(6)	
	25		10.00(1)	0.17(5)	
	27	9.45 min		0.07(2)	β^-(1.75, 1.59); γ(0.844, 1.014)
	28	20.90 h			β^-(0.459); γ(1.342, 0.942, 0.401, 0.031)
Aluminum	26	7.1×10^5 y			β^+(1.16); K, Mg-x; γ(1.809)
	27		100	0.230(2)	
	28	2.25 min			β^-(2.865); γ(1.778)
Silicon	28		92.23(2)	0.17(1)	
	29		4.67(2)	0.12(1)	
	30		3.10(1)	0.107(4)	
	31	2.62 h		0.073(6)	β^-(1.471); γ(1.266)
	32	1.6×10^2 y			β^-(0.213)
Phosphorus	30	2.50 min			β^+(3.245)
	31		100	0.16(2)	
	32	14.28 d			β^-(1.710)
	33	25.3 d			β^-(0.249)
Sulfur	32		95.02(9)	0.55(2)	
	34		4.21(8)	0.29(6)	
	35	87.51 d			β^-(0.167)
	37	5.05 min			β^-(4.75, 1.64); γ(3.103, 0.908)
	38	2.84 h			β^-(1.00, 3.0); γ(1.942, 0.196)

TABLE 1.43 Table of Nuclides (*Continued*)

Element	A	Half-life	Natural abundance, %	Cross section, barns	Radiation (MeV)
Chlorine	35		75.77(5)	43.7(4)	
	36	3.01×10^5 y		46.(2)	β^-(0.709); K, S-x
	37		24.23(5)	0.4	
	38	37.24 min			β^-(4.91, 1.11, 2.77); γ(2.168, 1.642)
	39	55.6 min			β^-(1.91, 2.18, 3.45); γ(1.267, 0.250, 1.52)
Argon	37	35.0 d			K, Cl-x
	39	268 y			β^-(0.565)
	40		99.600(3)	0.64(3)	
	41	1.82 h		0.5(1)	β^-(1.20, 2.49); γ(1.29)
	42	33 y			β^-(0.60)
Potassium	39		93.258(4)	2.1(2)	
	*40	1.26×10^9 y	0.0117(1)	30.(8)	β^-(1.312); K, Ar-x; γ(1.461)
	41		6.730(4)	1.46(3)	
	42	12.360 h			β^-(3.523, 1.97); γ(1.525)
	43	22.3 h			β^-(0.825, 0.45, 1.24, 1.814); γ(0.618, 0.373, 0.39, 0.221)
Calcium	40		96.941(18)	0.41(3)	
	42	1.02×10^5 y		≈4	
	43		0.135(6)	6.(1)	
	44		2.086(12)	0.8(2)	
	45	162.7 d		≈15	β^-(0.257)
	47	4.536 d			β^-(1.98, 0.684); γ(1.297)
	49	8.72 min			β^-(1.95, 0.89); γ(3.084, 4.07)
Scandium	42*m*	61.6 s			β^+(2.82); γ(0.438, 1.227, 1.524)
	43	3.89 h			β^+(1.22)
	44*m*	2.442 d			IT, Sc-x; γ(0.271)
	44	3.927 h			β^+(1.47); K, γ(1.16)
	45		100	27	
	46*m*	19.5 s			γ(0.142)
	46	83.81 d		8.(1)	β^-(0.357); γ(1.12, 0.889); Ti-x
	47	3.341 d			β^-(0.439, 0.60); γ(0.159)
	48	1.821 d			β^-(0.65); γ(1.31, 1.04, 0.984)
Titanium	44	47.3 y			K, γ(0.68, 0.078)
	45	3.08 h			β^+(1.044); K, Sc-x
	48		73.72(3)	7.9(9)	
	49		5.41(2)	1.9(5)	
	50		5.18(2)	0.179(3)	
	51	5.76 min			β^-(2.14, 1.50); γ(0.320, 0.928)
Vanadium	48	16.0 d			β^+(0.698); γ(0.511, 0.945, 0.983, 1.312, 2.24)

(*Continued*)

TABLE 1.43 Table of Nuclides (*Continued*)

Element	A	Half-life	Natural abundance, %	Cross section, barns	Radiation (MeV)
Vanadium (*cont.*)	49	330 d			K, Ti-x
	50	>1.4 × 10^{17} y	0.250(2)	40.(20)	
	51		99.750(2)	4.9(1)	
	52	3.75 min			β^-(2.47); γ(1.434)
Chromium	48	21.6 h			K, V-x; γ(0.116, 0.305)
	50		4.345(13)	15.(1)	
	51	27.70 d			K, V-x; γ(0.320)
	52		83.79(2)	0.8(1)	
	53		9.50(2)	18.(2)	
Manganese	51	46.2 min			β^+(2.2); γ(0.749, 1.15)
	52	5.60 d			β^+(0.575); γ(0.511, 0.744, 1.434)
	53	3.7 × 10^6 y		70.(10)	
	54	312.2 d		<10	γ(0.834)
	55		100	13.3(1)	
	56	2.5785 h			β^-(1.028, 1.03, 0.718); γ(0.847, 1.81, 2.11)
Iron	52	8.275 h			β^+(0.804); K, Mn-x; γ(0.169)
	54		5.85(4)	2.7(5)	
	55	2.73 y		13.(2)	K, Mn-x
	56		91.75(4)	2.6(2)	
	57		2.12(1)	2.5(5)	
	59	44.51 d		13.(3)	β^-(0.273, 0.475); γ(1.10, 1.29)
Cobalt	55	17.53 h			β^+(1.04, 1.50); K, Fe-x; γ(0.932, 0.480, 1.41)
	56	77.3 d			β^+(1.46); K, Fe-x; γ(0.847, 1.04, 1.24, 1.77, 2.60, 3.26, 2.02)
	57	271.77 d			K, Fe-x; γ(0.136, 0.122)
	58*m*	9.1 h		1.4(1) × 10^5	γ(0.025)
	58	70.88 d		1.9(2) × 10^3	K, β^+(0.474); Fe-x; γ(0.811)
	59		100	19	
	60*m*	10.47 min		58.(8)	β^-(1.55)
	60	5.271 y		2.0(2)	β^-(0.318); γ(1.173, 1.332)
	61	1.650 h			β^-(1.22); γ(0.842–0.909)
Nickel	56	6.08 d			K, Co-x; γ(0.158, 0.270, 0.480, 0.75, 0.812, 1.56)
	57	35.6 h			K, β^+(0.849, 0.712); Co-x, γ(1.378, 0.0127, 1.76)
	58		68.077(9)	4.6(4)	
	60		26.22(1)	2.9(3)	
	63	100 y		24.(3)	β^-(0.067)

TABLE 1.43 Table of Nuclides (*Continued*)

Element	A	Half-life	Natural abundance, %	Cross section, barns	Radiation (MeV)
Nickel (*cont.*)	64		0.926(1)	1.8(1)	
	65	2.517 h		22.(2)	β^-(2.14, 0.65, 1.020); γ(1.48, 0.366, 1.116)
	66	2.275 d			β^-(0.23)
Copper	61	3.408 h			β^+(1.220); K, Ni-x; γ(0.283, 0.656)
	63		69.17(3)	4.5(2)	
	64	12.701 h		≈270	β^-(0.578); β^+(0.65); Ni-x; γ(1.346)
	65		30.83(3)	2.17(3)	
	66	5.07 min		$1.4(1) \times 10^2$	β^-(2.74); γ(1.039)
	67	2.580 d			β^-(0.395, 0.484, 0.577); γ(0.185, 0.092)
Zinc	62	9.26 h			K, β^+(0.66); Cu-x; γ(0.041, 0.597)
	64		48.6(3)	0.46	
	65	243.8 d		66.(8)	K, β^+(0.325), Cu-x; γ(1.116)
	66		27.9(2)	1.0(2)	
	67		4.1(1)	6.9(1)	
	68		18.8(4)	0.87	
	69*m*	13.76 h			IT, Zn-x, γ(0.439)
	69	56 min			β^-(0.905)
	71*m*	3.97 h			β^-(1.45); γ(0.386, 0.487, 0.620)
	72	46.5 h			β^-(0.30, 0.25); γ(0.145, 0.191)
Gallium	66	9.5 h			β^+(1.84, 4.153); γ(1.039, 2.752)
	67	3.260 d			K, Zn-x; γ(0.093, 0.184, 0.300)
	68	1.130 h			β^+(1.83); K, Zn-x; γ(1.077)
	69		60.108(9)	1.68(7)	
	70	21.1 min			β^-(1.65); γ(0.175, 1.042)
	71		39.892(9)	4.7(2)	
	72	14.10 h			β^-(0.64, 1.51, 2.52, 3.15); γ(0.63, 2.20, 2.50)
	73	3.120 d			β^-(1.59); γ(0.053, 0.297)
Germanium	66	2.66 h			K, β^+(1.02); Ga-x; γ(0.044, 0.382)
	68	270.8 d			Ga, K-x
	69	1.63 d			β^+(0.70, 1.22); γ(1.107, 0.574)
	71	11.2 d			Ga-x
	72		27.66(3)	0.9(2)	
	73		7.73(1)	15.(1)	
	74		35.94(2)	0.3	
	75	1.380 h			β^-(1.19); γ(0.265, 0.419)

(*Continued*)

TABLE 1.43 Table of Nuclides (*Continued*)

Element	A	Half-life	Natural abundance, %	Cross section, barns	Radiation (MeV)
Germanium (*cont.*)	77	11.30 h			β^-(0.71, 1.38, 2.19); γ(0.211, 0.215, 0.264)
	78	1.45 h			β^-(0.95); γ(0.277, 0.294)
Arsenic	71	2.70 d			K, β^+(0.81); Ge-x; γ(0.175, 1.096)
	72	1.083 d			β^+(3.339, 2.498, 1.884); K, Ge-x; γ(0.834, 1.051)
	73	80.30 d			K, γ(0.0534, 0.0133)
	74	17.78 d			β^+(0.94); β^-(0.71, 1.35); γ(0.596, 0.635)
	75		100	4.0(4)	
	76	1.096 d			β^-(2.97, 2.41, 1.79); γ(0.559, 0.657)
	77	38.8 h			β^-(0.683); γ(0.239, 0.250, 0.521)
	78	91 min			β^-(4.21); γ(0.614, 0.70, 1.31)
Selenium	72	8.40 d			K, As-x; γ(0.046)
	73	7.1 h			β^+(1.32); γ(0.361, 0.067)
	74		0.89(2)	50.(4)	
	75	119.78 d			K, γ(0.265, 0.136); As-x
	77*m*	17.5 s			γ(0.162)
	77		7.63(6)	42.(4)	
	80		49.61(10)	0.5	
	81	18.5 min			β^-(1.58); γ(0.276, 0.290, 0.828)
Bromine	75	1.62 h			β^+(3.03); γ(0.287)
	76	16.2 h		224.(42)	β^+(1.9, 3.68); K, Se-x; γ(0.559, 1.86)
	77	2.376 d			γ(0.239, 0.521)
	79		50.69(7)	10.8	
	80*m*	4.42 h			IT, Br-x; γ(0.037, 0.049)
	80	17.66 min			β^-(1.997, 1.38); K, β^+(0.85), Se-x; γ(0.617)
	81		49.31(7)	2.6	
	82	1.4708 d			β^-(0.444); γ(0.554, 0.619, 0.776)
Krypton	76	14.8 h			K, γ(0.252)
	77	1.24 h			β^+(1.875, 1.700, 1.550); K, Br-x; γ(0.130, 0.147)
	79	1.455 d			β^+(1.626); γ(0.261, 0.398, 0.606)
	81*m*	13 s			IT, Kr-x; γ(0.190)
	81	2.10×10^5 y			K, Br-x; γ(0.276)
	83		11.5(1)	183.(30)	
	84		57.0(3)	0.10	
	85*m*	4.48 h			β^-(0.83); γ(0.151, 0.305)

TABLE 1.43 Table of Nuclides (*Continued*)

Element	A	Half-life	Natural abundance, %	Cross section, barns	Radiation (MeV)
Krypton (*cont.*)	85	10.72 y			β^-(0.67); γ(0.517)
	87	1.27 h			β^-(3.49, 0.389, 1.38); γ(0.403, 2.55)
	88	2.84 h			β^-(2.91); γ(0.196, 2.392)
Rubidium	84	32.9 d			β^-(0.894); β^+(2.681); γ(0.882)
	85		72.17(2)	0.5	
	86	18.65 d		< 20	β^-(1.775); γ(1.08)
	87	4.88×10^{10} y	27.83(2)	0.10(1)	β^-(0.283)
	88	17.7 min		1.2(3)	β^-(5.31); γ(1.836, 0.898)
	89	15.4 min			β^-(1.26, 2.2, 4.49); γ(1.032, 1.248, 2.196)
Strontium	82	25.36 d			K, Rb-x
	85*m*	1.126 h			K, Rb-x, Sr-x; γ(0.150, 0.231)
	85	64.84 d			K, Rb-x; γ(0.514)
	87*m*	2.795 h			IT, γ(0.388)
	88		82.58(1)	0.0058(4)	
	89	50.52 d		0.42(4)	β^-(1.497); γ(0.909)
	90	29.1 y		0.0097(7)	β^-(0.546)
	91	9.5 h			β^-(1.09, 1.36, 2.66); γ(0.556, 0.750, 1.024)
	92	2.71 h			β^-(0.55, 1.5); γ(1.383)
Yttrium	85*m*	4.86 h			β^+(2.24); K, Sr-x; γ(0.767, 0.232, 2.124)
	85	2.68 h			β^+(1.58, 1.15); K, Sr-x; γ(0.504, 0.232)
	86	14.74 h			β^+(5.24); γ(0.307, 0.628, 1.077, 1.153, 1.921)
	87*m*	12.9 h			Y-x; γ(0.381)
	88	106.6 d			β^-(0.76); γ(0.898, 1.836, 2.734, 3.219)
	90	2.67 d		<7	β^-(2.28); γ(2.186)
	91*m*	49.71 min			Y-x; IT; γ(0.556)
	91	58.5 d		1.4(3)	β^-(1.545); γ(1.21)
	92	3.54 h			β^-(3.64); γ(0.448, 0.561, 0.934, 1.405)
	93	10.2 h			β^-(2.88); γ(0.267, 0.947, 1.918)
Zirconium	86	16.5 h			K, Y-x; γ(0.243, 0.612)
	87	1.73 h			β^+(2.260); K, Y-x; γ(0.381, 1.228)
	88	83.4 d			K, Y-x; γ(0.393)
	89	3.27 d			K, β^+(0.897); Y-x; γ(0.909)
	91		11.22(4)	1.2(3)	
	93	1.5×10^6 y			β^-(0.091)
	95	64.02 d			β^-(0.366, 0.400); γ(0.724, 0.757)
	97	16.90 h			β^-(1.91); γ(0.743)

(*Continued*)

TABLE 1.43 Table of Nuclides (*Continued*)

Element	A	Half-life	Natural abundance, %	Cross section, barns	Radiation (MeV)
Niobium	89	2.03 h			β^+(3.320); γ(1.627)
	90	14.60 h			β^+(1.50); K, Zr-x; γ(0.141, 1.129, 2.186, 2.319)
	91*m*	62 d			IT, Nb-x; γ(0.1045, 1.205)
	91	700 y			Mo-x
	92*m*	10.15 d			K, γ(0.913, 0.934, 1.848)
	93*m*	16.1 y			Nb-x
	93		100	1.1	
	94*m*	6.26 min			γ(0.871)
	94	2.4×10^4 y			β^-(0.473); γ(0.703, 0.871)
	95*m*	3.61 d			γ(0.204, 0.236)
	95	35.0 d		<7	β^-(0.160); γ(0.765)
	96	23.4 h			β^-(0.748, 0.500); γ(0.778, 1.091)
	97*m*	58.1 s			IT; γ(0.766)
	97	1.23 h			β^-(1.267); γ(0.481, 0.658)
Molybdenum	90	5.67 h			K, β^+(1.085); Nb-x; γ(0.122, 0.257)
	93*m*	6.85 h			IT, Mo-x; γ(0.264, 0.685, 1.477)
	95		15.92(5)	13.4(5)	
	97		9.55(3)	2.5(3)	
	98		24.13(7)	0.14(1)	
	99	2.75 d			β^-(1.357); Tc-x; γ(0.181, 0.366, 0.739)
	101	14.6 min			β^-(2.23, 0.7); γ(0.192, 0.591)
Technetium	93	2.73 h			β^+(0.81); γ(1.363, 1.477, 1.520)
	94	4.88 h			β^+(4.256); γ(0.449, 0.703, 0.850, 0.871)
	95*m*	61 d			β^+(0.71); γ(0.204, 0.582, 0.835)
	95	20.0 h			K, Mo-x; γ(0.766, 1.074)
	96	4.3 d			K, Mo-x; γ(0.778, 0.813, 0.850, 1.122)
	97*m*	90 d			K, Tc-x; γ(0.0965)
	97	2.6×10^6 y			K, Mo-x
	98	4.2×10^6 y			β^-(0.40); γ(0.652, 0.745)
	99*m*	6.012 h			IT, Tc-x; γ(0.141, 0.143)
	99	2.13×10^5 y		20	β^-(0.292)
Ruthenium	95	1.64 h			β^+(1.20, 0.91); γ(0.290, 0.336, 0.627)
	97	2.88 d			K, Tc-x; γ(0.216, 0.324, 0.461)
	100		12.6(1)	5.8(6)	

TABLE 1.43 Table of Nuclides (*Continued*)

Element	A	Half-life	Natural abundance, %	Cross section, barns	Radiation (MeV)
Ruthenium (*cont.*)	101		17.0(1)	5.(1)	
	102		31.6(2)	1.2(1)	
	103	39.27 d		<20	β^-(0.12, 0.223); γ(0.295, 0.4444, 0.497, 0.557, 0.610)
	105	4.44 h			β^-(1.187, 0.11, 1.134); γ(0.149, 0.263, 0.317, 0.469, 0.676, 0.724)
	106	1.020 y			β^-(0.0394)
Rhodium	99*m*	4.7 h			β^+(0.74); γ(0.277, 0.341, 0.618, 1.261)
	99	16 d			β^+(0.54, 0.68); γ(0.089, 0.353, 0.528)
	100	20.8 h			β^+(2.62, 2.07); γ(0.446, 0.540, 0.588, 0.823, 1.553, 2.376)
	101*m*	4.35 d			K, IT, Ru-x, Rh-x; γ(0.127, 0.307, 0.545)
	101	3.3 y			K, Ru-x; γ(0.127, 0.198, 0.325)
	102*m*	207 d			β^-(1.15); β^+(1.29, 0.82); γ(0.469, 0.475, 0.557, 0.628, 1.103)
	102	2.9 y			K, Ru-x; γ(0.475, 0.631, 0.697, 0.767, 1.047, 1.103)
	103*m*	56.12 min			IT, Rh-x, γ(0.0.040)
	103		100	145	
	104*m*	4.36 min		800.(100)	γ(0.051, 0.097, 0.556)
	104	42.3 s		40.(30)	β^-(2.44), γ(0.358, 0.556, 1.237)
	105*m*	40 s			IT, Rh-x; γ(0.130)
	105	35.4 h		$1.1(3) \times 10^4$	β^-(0.567, 0.247); γ(0.280, 0.306, 0.319)
	106*m*	2.18 h			β^-(0.92); γ(0.222, 0.451, 0.512, 0.616, 0.717, 0.784, 1.046, 1.528)
	106	29.80 s			β^-(3.54, 3.0, 2.4); γ(0.512, 0.622)
Palladium	100	3.63 d			K, Rh-x; γ(0.0748, 0.0840, 0.0327)
	101	8.47 h			K, Rh-x; β^+(0.776); γ(0.296, 0.590)
	103	16.99 d			K, Rh-x; γ(0.357, 0.497)
	105		22.33(8)	22.(2)	
	107	6.5×10^6 y		1.8(2)	β^-(0.03)
	108		26.46(9)	8.7	
	109	13.5 h			β^-(1.028); Ag-x; γ(0.088, 0.311, 0.636)

(*Continued*)

TABLE 1.43 Table of Nuclides (*Continued*)

Element	A	Half-life	Natural abundance, %	Cross section, barns	Radiation (MeV)
Palladium (*cont.*)	111*m*	5.5 h			β^-(0.35, 0.77); γ(0.070, 0.172, 0.391)
	111	23.4 min			β^-(2.2); γ(0.060, 0.245, 0.580, 0.650, 1.389, 1.459)
	112	21.4 h			β^-(0.28); γ(0.018)
Silver	103	1.10 h			β^+(1.7, 1.3); γ(0.119, 0.148)
	104	69 min			β^+(0.99); γ(0.556, 0.926, 0.942)
	105	41.29 d			K, Pd-x; γ(0.064, 0.280, 0.344, 0.443)
	106*m*	8.4 d			K, Pd-x; γ(0.451, 0.512, 0.717, 1.046)
	107*m*	44.2 s			K, Ag-x; γ(0.093)
	107		51.839(7)	35	
	108*m*	130 y			γ(0.434, 0.614, 0.723)
	108	2.42 min			β^-(1.65); β^+(0.90); γ(0.434, 0.619, 0.633)
	109		48.161(7)	91	
	110*m*	249.8 d		82.(11)	β^-(0.087, 0.530); IT, γ(0.658, 0.764, 0.885, 0.937, 1.384)
	111*m*	1.08 min			K, Ag-x; γ(0.060, 0.245)
	111	7.47 d		3.(2)	β^-(1.04); γ(0.245, 0.342)
	112	3.13 h			β^-(3.94, 3.4); γ(0.607, 0.617, 1.39)
Cadmium	107	6.52 h			β^+(0.302); K, Ag-x; γ(0.093, 0.829)
	109	462 d			K, Ag-x; γ(0.088)
	111*m*	48.5 min			K, Cd-x; γ(0.151, 0.245)
	111		12.80(8)	24.(3)	
	113*m*	14.1 y			β^-(0.59); γ(0.264)
	113	9×10^{15} y	12.22(6)	20 060.(40)	
	115*m*	44.6 d			β^-(1.62); γ(0.934, 1.29, 0.485)
	115	2.228 d			β^-(1.11, 0.593); In-x; γ(0.231, 0.260, 0.336, 0.492, 0.528)
	117*m*	3.4 h			β^-(0.72); γ(0.159, 0.553); In-x
	117	2.49 h			β^-(0.67, 2.2); γ(0.221, 0.273, 0.345, 1.303)
Indium	109	4.2 h			K, Cd-x; β^+(0.79); γ(0.203, 0.623)
	110*m*	4.9 h			γ(0.658, 0.885, 0.937)
	110	1.15 h			β^+(2.22); K, Cd-x; γ(0.658)
	111	2.805 d			K, Cd-x; γ(0.171, 0.245)

TABLE 1.43 Table of Nuclides (*Continued*)

Element	A	Half-life	Natural abundance, %	Cross section, barns	Radiation (MeV)
Indium (*cont.*)	113*m*	1.658 h			IT, In-x; γ(0.392)
	114*m*	49.51 d			IT, K, In-x; γ(0.190)
	114	1.1983 min			β^-(1.99); K, Cd-x, β^+(0.40); γ(0.558, 0.573, 1.30)
	115*m*	4.486 h			β^-(0.83); K, In-x; γ(0.336, 0.497)
	*115	4.4×10^{14} y	95.71(2)	205	β^-(0.495)
	116*m*	54.1 min			β^-(1.00); γ(0.138, 0.417, 1.09, 1.293)
	117*m*	1.94 h			β^-(1.77); γ(0.159, 0.315, 0.553)
	117	44 min			β^-(0.74); γ(0.159, 0.397, 0.553)
Tin	110	4.1 h			K, In-x; γ(0.283)
	113	115.1 d		≈9	K, In-x, γ(0.392, 0.255)
	116		14.53(11)	1.1(1)	
	117*m*	13.60 d			K, Sn-x; γ(0.159)
	119*m*	293 d			K, Se-x; γ(0.239)
	119		8.59(4)	2.(1)	
	121*m*	≈55 y			β^-(0.354); K, In-x; γ(0.0372)
	121	1.128 d			β^-(0.383)
	123	129.2 d			β^-(1.42); γ(0.160, 1.030, 1.089)
	125	9.63 d			β^-(2.35); γ(1.067)
	127	2.10 h			β^-(2.42, 3.2); γ(0.823, 1.096)
Antimony	115	32.1 min			β^+(1.51); γ(0.499)
	116*m*	1.00 h			β^+(1.16); γ(0.407, 0.543, 0.973, 1.293)
	117	2.80 h			β^+(0.57); γ(0.159)
	118*m*	5.00 h			γ(0.254, 1.051, 1.280)
	118	3.6 min			β^+(2.65); γ(1.230)
	119	38.1 h			γ(0.0239)
	120	15.89 min			β^+(1.72); γ(0.704, 1.171)
	121		57.21(5)	6	
	122	2.72 d			β^-(1.414); β^+(1.980); γ(0.564, 0.693, 1.141, 1.257)
	123		42.7(9)	3.3	
	124	60.20 d			β^-(0.61, 2.301); γ(0.603, 0.646, 1.69, 0.723)
	126	12.4 d			β^-(1.9); γ(0.279, 0.415, 0.666, 0.695, 0.720)
	127	3.84 d			β^-(0.89, 1.10, 1.50); γ(0.252, 0.291, 0.412, 0.437, 0.686, 0.784)
	128	9.1 h			β^-(2.3); γ(0.215, 0.314, 0.527, 0.743, 0.754)

(*Continued*)

TABLE 1.43 Table of Nuclides (*Continued*)

Element	A	Half-life	Natural abundance, %	Cross section, barns	Radiation (MeV)
Antimony (*cont.*)	129	4.40 h			β^-(0.65); γ(0.181, 0.359, 0.460, 0.545, 0.813, 0.915, 1.030)
Tellurium	116	2.49 h			γ(0.0937)
	117	1.03 h			β^+(1.78); γ(0.920, 1.716, 2.300)
	119*m*	4.69 d			γ(0.154, 0.271, 1.213)
	119	16.0 h			β^+(0.627; γ(0.644, 0.700)
	121*m*	≈154 d			γ(0.212)
	121	16.8 d			γ(0.508, 0.573)
	123*m*	119.7 d			γ(0.159)
	125		7.139(6)	1.6(2)	
	127*m*	109 d			β^-(0.77); γ(0.088)
	127	9.35 h			β^-(0.696); γ(0.360)
	129*m*	33.6 d			β^-(1.60); γ(0.460, 0.696)
	129	1.160 h			β^-(1.453, 0.989); I-x, γ(0.460, 0.487)
	131*m*	1.35 d			β^-(0.42); IT, Te-x, I-x; γ(0.150, 0.774, 0.794, 0.852)
	131	25.0 min			β^-(2.14, 1.69, 1.35); I-x; γ(0.150, 0.453, 0.493)
	132	25.0 min			β^-(0.215); γ(0.050, 0.112, 0.228)
Iodine	121	2.12 h			β^+(1.2); γ(2.12)
	122	3.6 min			β^+(3.1); γ(0.564)
	123	13.2 h			K, Te-x; γ(0.159)
	124	4.18 d			β^+(1.54, 2.14, 0.75); γ(0.603, 0.723, 1.691)
	125	59.4 d		$9.(1) \times 10^2$	K, Te-x; γ(0.035)
	126	13.0 d			β^+(1.13); β^-(0.87, 1.25); γ(0.389, 0.662)
	127		100	6.15(10)	
	128	24.99 min		22.(4)	β^-(2.13); γ(0.443, 0.527)
	129	1.7×10^7 y			β^-(0.15); γ(0.040)
	130	12.36 h		18.(3)	β^+(1.13); β^-(0.87, 1.25); γ(0.389, 0.662)
	131	8.040 d		≈0.7	β^-(0.606); γ(0.284, 0.364, 0.637)
	132	208 h			β^-(0.80, 1.03, 1.2, 1.6, 2.16); γ(0.098, 0.506, 0.523, 0.630, 0.651, 0.667, 0.723, 0.955)
	133	20.8 h			β^-(1.24); γ(0.511, 0.530, 0.875)
	135	6.57 h			β^-(0.9, 1.3); γ(0.418, 0.527, 1.132, 1.260)
Xenon	123	2.00 h			β^+(1.51); γ(0.149, 0.178)
	125	17.1 h			γ(0.188, 0.243)

TABLE 1.43 Table of Nuclides (*Continued*)

Element	A	Half-life	Natural abundance, %	Cross section, barns	Radiation (MeV)
Xenon (*cont.*)	127*m*	1.15 min			γ(0.127, 0.173)
	127	36.4 d			γ(0.172, 0.203, 0.375)
	129*m*	8.89 d			γ(0.040, 0.197)
	129		26.4(6)	22.(5)	
	131*m*	11.9 d			γ(0.164)
	131		21.2(4)	90.(10)	
	133*m*	2.19 d			γ(0.233)
	133	5.243 d		190.(90)	β^-(0.346); Cs-x; γ(0.081)
	135*m*	15.3 min			γ(0.527)
	135	9.1 h			β^-(0.91); γ(0.250, 0.608)
Cesium	126	1.64 min			β^+(3.4, 3.7); γ(0.0389, 0.491, 0.925)
	127	6.2 h			β^+(0.65, 1.06); γ(0.125, 0.412)
	128	3.62 min			β^+(2.44, 2.88); γ(0.443)
	129	1.336 d			γ(0.372, 0.412)
	132	6.48 d			γ(0.465, 0.630, 0.668)
	133		100	28	
	134*m*	2.91 h			IT, K, Cs-x; γ(0.127)
	134	2.065 y		140.(10)	β^-(0.658, 0.089); γ(0.563, 0.569, 0.605, 0.796)
	135	2.3×10^6 y		8.9(5)	β^-(0.205)
	136	13.16 d			β^-(0.341); γ(0.341, 0.819, 1.048)
	137	30.2 y			β^-(0.514); K, Ba-x; γ(0.662)
Barium	126	1.65 h			γ(0.218, 0.234, 0.258)
	128	2.43 d			γ(0.273); K, Cs-x
	129*m*	2.17 h			γ(0.177, 0.182, 0.202, 1.459)
	129	2.2 h			β^+(1.42); γ(0.129, 0.214, 0.221)
	131	11.7 d			γ(0.124, 0.216, 0.496)
	133*m*	1.621 d			γ(0.276)
	133	10.53 y		4.(1)	γ(0.081, 0.356)
	135*m*	1.196 d			IT, Ba-x; γ(0.268)
	135		6.59(2)	5.8	
	137		11.23(4)	5.(1)	
	137*m*	2.552 min			IT, K, Ba-x; γ(0.662)
	138		71.70(7)	0.41(2)	
	139	1.396 h		5.1	β^-(2.27, 2.14); K, La-x; γ(0.166, 1.254, 1.421)
	140	12.75 d			β^-(0.48, 1.02); γ(0.163, 0.305, 0.537)
	142	10.7 min			β^-(1.0, 1.1); γ(0.231, 0.255, 0.309, 1.204)

(*Continued*)

TABLE 1.43 Table of Nuclides (*Continued*)

Element	A	Half-life	Natural abundance, %	Cross section, barns	Radiation (MeV)
Lanthanum	131	59 min			β^+(1.42, 1.94); γ(0.526, 0.109, 0.366)
	132	4.8 h			β^+(2.6, 3.2, 3.7); γ(0.465, 0.567)
	133	3.91 h			β^+(1.2); γ(0.279, 0.290, 0.302)
	134	6.5 min			β^+(2.67); γ(0.605)
	135	19.5 h			γ(0.481)
	136	8.87 min			β^+(1.8); γ(0.816)
	*138	1.06×10^{11} y		57.(6)	
	139		99.9098(2)	9.2(2)	
	140	1.68 d		2.7(3)	β^-(1.670, 1.35)
	141	3.90 h			β^-(2.43)
	142	1.54 h			β^-(2.11, 2.98, 4.52)
Cerium	132	3.5 h			γ(0.154, 0.182)
	133	5.4 h			β^+(1.3); γ(0.058, 0.131, 0.472, 0.510)
	135	17.7 h			β^+(0.8); γ(0.266, 0.300, 0.607)
	137*m*	1.43 d			IT K, Ce-x; γ(0.169, 0.254)
	137	9.0 h			γ(0.447)
	139	137.6 d			γ(0.166)
	140		88.43(10)	0.58(4)	
	141	32.50 d			β^-(0.436, 0.581); K, Pr-x; γ(0.145)
	142		11.13(10)	0.97(3)	
	143	1.38 d		6.1(7)	β^-(1.404, 1.110); K, Pr-x; γ(0.293)
	144	284.6 d		1.0(1)	β^-(0.318, 0.185); K, Pr-x; γ(0.080, 0.134)
Praseodymium	136	13.1 min			β^+(2.98); γ(0.540, 0.552)
	137	1.28 h			β^+(1.68); γ(0.434, 0.514, 0.837)
	138*m*	2.1 h			β^+(1.65); γ(0.304, 0.789, 1.038)
	139	4.41 h			β^+(1.09); γ(0.255, 1.347, 1.631)
	141		100	11.5	
	142	19.12 h		20.(3)	β^-(2.164); γ(1.576)
	143	13.57 d		90.(10)	β^-(0.933); γ(0.742)
	145	5.98 h			β^-(1.80); γ(0.073, 0.676, 0.748)
Neodymium	139*m*	5.5 h			β^+(1.17); γ(0.114, 0.738)
	141	2.49 h			β^+(0.802)
	142		27.13(2)	19.(1)	
	143		12.18(6)	220.(10)	
	*144	2.1×10^{15} y	23.8(1)	3.6(3)	
	145		8.3(6)	47.(6)	

TABLE 1.43 Table of Nuclides (*Continued*)

Element	A	Half-life	Natural abundance, %	Cross section, barns	Radiation (MeV)
Neodymium (*cont.*)	146		17.19(9)	1.5(2)	
	147	10.98 d		440.(150)	β^-(0.805); γ(0.091, 0.531)
	149	1.73 h			β^-(1.03, 1.13); γ(0.211, 0.114)
Promethium	143	265 d			K, Nd-x; γ(0.742)
	144	360 d			K, Nd-x; γ(0.618, 0.696)
	146	5.53 y		8.4(2) × 10^3	K, β^-(0.795); Nd-x; γ(0.453, 0.75)
	147	2.6234 y		180	β^-(0.224); γ(0.122, 0.197)
	148*m*	41.29 d		106.(8) × 10^2	β^-(0.69, 0.50, 0.40); IT, Pm-x, Sm-x; γ(0.550, 0.630, 0.726)
	148	5.37 d		≈1000	β^-(1.02, 2.47); γ(0.550, 0.915, 1.465)
	149	2.212 d		14.(2) × 10^2	β^-(1.072, 0.78); γ(0.286, 0.591, 0.859)
	150	2.68 h			β^-(1.6, 2.3, 1.8); γ(0.334, 1.166, 0.132)
	151	1.183 d		≈150	β^-(0.84); γ(0.168, 0.275, 0.340)
Samarium	142	1.208 h			β^+(1.0); K, Pr-x
	144		3.1(1)	1.6(1)	
	145	340 d		280.(20)	γ(0.061, 0.492); K, Pm-x
	146	1.03 × 10^8 y			α(2.50)
	*147	1.06 × 10^{11} y	15.0(2)	56.(4)	α(2.23)
	148	7 × 10^{15} y	11.3(1)	2.4(6)	α(1.96)
	149	10^{16} y	13.8(1)	401.(6) × 10^2	
	150		7.4(1)	102.(5)	
	151	90 y			β^-(0.076)
	152		26.7(2)	206.(15)	
	153	1.929 d		420.(180)	β^-(0.64, 0.69); γ(0.103)
	154		22.7(2)	7.5(3)	
	155	22.2 min			β^-(1.52); γ(0.104)
	156	9.4 h			β^-(0.43, 0.71); γ(0.166, 0.204)
Europium	148	54.5 d			β^+(0.92); γ(0.550, 0.630)
	149	93.1 d			K, Sm-x; γ(0.277, 0.328)
	150*m*	12.8 h			β^-(1.013); γ(0.334, 0.407)
	150	36 y			γ(0.334, 0.439, 0.584)
	151		47.8(5)	9000	
	152*m*	9.30 h			β^-(1.85); γ(0.122, 0.841, 0.963)
	152	13.48 y		11.(2) × 10^3	K, β^-(1.47, 0.690); K, Gd-x, K, Sm-x; γ(0.122, 0.344, 1.408)
	153		52.2(5)	320.(20)	

(*Continued*)

TABLE 1.43 Table of Nuclides (*Continued*)

Element	A	Half-life	Natural abundance, %	Cross section, barns	Radiation (MeV)
Europium (*cont.*)	154	8.59 y		1.5(3) × 10^3	β^-(0.27, 0.58, 0.843, 1.87); γ(0.123, 0.723, 1.274)
	155	4.76 y		3.9(2) × 10^3	β^-(0.15); γ(0.087, 0.105)
	156	15.2 d			β^-(0.30, 0.49, 1.2, 2.45); γ(0.089, 0.646, 0.723, 0.812)
	157	15.13 h			β^-(1.30); γ(0.064, 0.371, 0.411)
	158	45.9 min			β^-(2.5); γ(0.898, 0.944, 0.977)
Gadolinium	146	48.3 d			β^+(0.35); γ(0.115, 0.155)
	147	1.588 d			β^+(0.93); γ(0.229, 0.370, 0.396, 0.929)
	151	124 d			α(2.73); γ(0.154, 0.243)
	153	241.6 d			γ(0.94, 0.103)
	155		14.80(5)	61.(1) × 10^3	
	157		15.65(3)	2.54(3) × 10^5	
	158		24.84(12)	2.3(5)	
	159	18.56 h			β^-(0.971); Tb-x; γ(0.363)
	160		21.86(4)	1.5(7)	
Terbium	158	180 y			γ(0.944, 0.962)
	159		100	23.2(5)	
	160	72.3 d		5.7(11) × 10^2	β^-(0.57, 0.86); γ(0.299, 0.879, 0.966)
Dysprosium	159	144 d		8.(2) × 10^3	K, Tb-x; γ(0.326)
	161		18.9(2)	600.(150)	
	162		25.5(2)	170.(20)	
	163		24.9(2)	120.(10)	
	164		28.2(2)	2000	
	165	2.33 h		3.5(3) × 10^3	β^-(1.29); Ho-x; γ(0.095)
	165*m*	1.26 min			γ(0.108, 0.515)
Holmium	156	56 min			γ(0.138, 0.267)
	159	33.0 min			γ(0.121, 0.132, 0.253, 0.310)
	167	3.1 h			β^-(0.31, 0.62, 0.96); γ(0.238, 0.321, 0.347)
	165		100	61	
	166*m*	1.2 × 10^3 y		9.14(65) × 10^3	Er-x; γ(0.810, 0.712, 0.184)
	166	1.117 d			β^-(1.855, 1.776); γ(1.379)
Erbium	166		33.6(2)	20	
	167		22.95(15)	7.(2) × 10^2	
	168		26.8(2)	2.0(6)	
	169	9.40 d			β^-(0.35)
	170		14.9(2)	6.2(2)	

TABLE 1.43 Table of Nuclides (*Continued*)

Element	A	Half-life	Natural abundance, %	Cross section, barns	Radiation (MeV)
Erbium (*cont.*)	171	7.52 h		370.(40)	β^-(1.49); Tm-x; γ(0.112, 0.296, 0.308)
	172	2.05 d			β^-(0.28, 0.36); γ(0.407, 0.610)
Thullium	166	7.70 h			γ(0.184, 0.779, 1.273, 2.052)
	169		100	106	
	170	128.6 d		100.(20)	β^-(0.968, 0.884)
	171	1.92 y		≈160	β^-(0.096); γ(0.067)
	172	2.65 d			β^-(1.79, 1.86); γ(1.387, 1.466, 1.530, 1.609)
	173	8.2 h			β^-(0.80, 0.86); γ(0.399, 0.461)
Ytterbium	165	9.9 min			β^+(1.58); γ(1.090)
	166	2.363 d			γ(0.184, 0.779, 1.273, 2.052)
	169	32.03 d		$3.6(3) \times 10^3$	γ(0.110, 0.177, 0.198)
	171		14.3(2)	50.(10)	
	173		16.12(21)	16.(2)	
	174		31.8(4)	120	
	175	4.19 d			β^-(0.466); Lu-x; γ(0.396)
	176		12.7(2)	3.1(2)	
	177	1.9 h			β^-(1.40); K, Lu-x; γ(0.150)
	178	1.23 h			β^-(0.25); γ(0.141, 0.325, 0.352, 0.381, 0.613)
Lutetium	164	3.14 min			β^+(1.6, 3.8); γ(0.124, 0.262, 0.740, 0.864, 0.880)
	165	16.7 min			β^+(2.06); γ(0.121, 0.132, 0.174, 0.204)
	175		97.41(2)	24	
	176*m*	3.66 h			β^-(1.229, 1.317); Hf-x; γ(0.0884)
	176	3.8×10^{16} y		2100	γ(0.202, 0.307)
	177	6.75 d		$10.(3) \times 10^2$	β^-(0.497), Hf-x; γ(0.113, 0.208)
Hafnium	178		27.297(4)	85	
	179		13.629(6)	46	
	†179m_1	18.7 s			γ(0.161, 0.214)
	†179m_2	25.1 d			γ(0.123, 0.146, 0.363, 0.454)
	180		35.100(7)	13.(1)	
	180*m*	5.519 h			IT, Hf-x; γ(0.215, 0.332, 0.443)
	181	42.4 d		30.(25)	β^-(0.408); Ta-x; γ(0.133, 0.346, 0.482)

†Two different metastable states possessing the same mass number but different half-lives.

(*Continued*)

TABLE 1.43 Table of Nuclides (*Continued*)

Element	A	Half-life	Natural abundance, %	Cross section, barns	Radiation (MeV)
Hafnium (*cont.*)	183	1.07 h			β^-(1.18, 1.54); γ(0.459, 0.784)
	184	4.1 h			β^-(0.74, 0.85, 1.10); γ(0.139, 0.345)
Tantalum	181		99.988(2)	20	
	182*m*	16.5 min			γ(0.147, 0.172, 0.184)
	182	114.43 d		$8.2(6) \times 10^3$	β^-(0.25, 0.44, 0.52); γ(0.068, 1.121)
	183	5.1 d			β^-(0.62); γ(0.108, 0.246, 0.304)
	184	8.7 h			β^-(1.17); γ(0.253, 0.414)
Tungsten	182		26.50(3)	20.(1)	
	183		14.31(1)	10.5(3)	
	184		30.64(1)	2	
	185	74.8 d		≈3.3	β^-(0.433); γ(0.125)
	186		28.43(4)	37.(2)	
	187	23.9 h		70.(10)	β^-(1.315, 0.624; K, Re-x; γ(0.072, 0.480, 0.686)
	188	69.4 d			β^-(0.349); γ(0.227, 0.291)
Rhenium	182*m*	12.7 h			β^+(0.55, 1.74); γ(1.121, 1.221)
	184	38 d			γ(0.790, 0.903)
	185		37.40(2)	110	
	186	3.718 d			β^-(1.07, 0.933); K, W-x, Os-x; γ(0.123, 0.137, 0.632, 0.768)
	*187	4.2×10^{10}	62.60(2)	74	
	188	16.94 h			β^-(2.12, 1.96); Os-x; γ(0.155)
	189	24 h			β^-(1.01); γ(0.147, 0.22, 0.245)
Osmium	186	2×10^{15} y	1.58(2)	≈80	
	188		13.3(1)	≈5	
	190*m*	9.9 min			IT, Os-x; γ(0.187, 0.361, 0.503, 0.616)
	190		26.4(2)	13	
	191	15.4 d		$3.8(6) \times 10^2$	β^-(0.143); Os-x; γ(0.129)
	192		41.0(3)	3.(1)	
	193	30.5 h			β^-(1.04); Ir-x; γ(0.139, 0.460)
	196	34.9 min			β^-(0.84); γ(0.126, 0.408)
Iridium	184	3.0 h			β^+(2.3, 2.9); γ(0.120, 0.264, 0.390)
	185	14 h			γ(0.254, 1.829)
	186	15.7 h			γ(0.137, 0.296, 0.435)
	188	1.72 d			γ(0.155, 0.478, 0.633, 2.215)

TABLE 1.43 Table of Nuclides (*Continued*)

Element	A	Half-life	Natural abundance, %	Cross section, barns	Radiation (MeV)
Iridium (*cont.*)	189	13.2 d			K, Os-x; γ(0.245)
	190	11.8 d			γ(0.187, 0.407, 0.519, 0.558, 0.605)
	191		37.27(9)	920	
	192	73.83 d			β^-(0.672); K, Pt-x; γ(0.316, 0.468)
	193		62.73(9)	116	
	194	19.3 h		1.5(3) × 10^3	β^-(2.25); γ(0.294, 0.328, 0.645)
	195*m*	3.9 h			β^-(0.41, 0.97); γ(0.320, 0.365, 0.433, 0.685)
Platinum	187	2.35 h			γ(0.105, 0.110, 0.201, 0.285, 0.709)
	188	10.2 d			γ(0.188, 0.195)
	189	10.89 h			K, Ir-x; γ(0.094, 0.608, 0.721)
	194		32.9(6)	1.2	
	195*m*	4.02 d			IT, Pt-x; γ(0.099)
	195		33.8(6)	28.(1)	
	196		25.3(6)	55	
	197*m*	1.573 h			IT, Pt-x; γ(0.053, 0.346)
	197	18.3 h			β^-(0.719); K, Au-x; γ(0.191, 0.269)
	199*m*	14.1 s			γ(0.392)
	199	30.8 min		≈16	β^-(0.90, 1.14); γ(0.186, 0.317, 0.494, 0.549)
	200	12.5 h			γ(0.136, 0.227, 0.244)
Gold	197		100	98.7(1)	
	197*m*	7.8 s			IT, K, Au-x; γ(0.130, 0.279)
	198	2.694 d		26.5(15) × 10^3	β^-(0.961); K, Hg-x; γ(0.412)
	199	3.139 d			β^-(0.292, 0.250); K, Hg-x; γ(0.158, 0.208)
	200*m*	18.7 h			β^-(0.56); γ(0.111, 0.368, 0.498, 0.597, 0.760)
	200	48.4 min			β^-(2.2); γ(0.368, 1.225)
Mercury	196		0.15(1)	3150	
	197*m*	23.8 h			IT, K, Hg-x; γ(0.134)
	197	2.6725 d			K, Au-x; γ(0.077)
	199*m*	42.6 min			γ(0.158)
	199		16.87(10)	2.1(2) × 10^3	
	200		23.10(16)	<60	
	202		29.86(20)	4.9(5)	
	203	46.61 d			β^-(0.213); γ(0.279)
Thallium	201	3.040 d			K, Hg-x; γ(0.135, 0.167)
	202	12.23 d			K, Hg-x; γ(0.440)
	203		29.52(1)	11.(1)	
	204	3.78 y		22.(2)	β^-(0.763); K, Hg-x

(*Continued*)

TABLE 1.43 Table of Nuclides (*Continued*)

Element	A	Half-life	Natural abundance, %	Cross section, barns	Radiation (MeV)
Thallium	205		70.48(1)	0.11(2)	
(*cont.*)	*206	4.20 min			β^-(1.53); K, Pb-x; γ(0.803)
	*207	4.77 min			β^-(1.43); γ(0.897)
	208	3.053 min			β^-(1.796, 1.28, 1.52); γ(0.277, 0.511, 0.583, 0.614)
	209	2.16 min			β^-(1.8); γ(1.567, 0.465)
	210	1.30 min			β^-(1.9, 1.3); γ(0.298, 0.798)
Lead	201	9.33 h			γ(0.331, 0.361)
	203	2.1615 d			γ(0.279)
	204*m*	1.120 h			IT, Pb-x; γ(0.375, 0.899, 0.912)
	207		22.1(1)	0.70(1)	
	209	3.253 h			β^-(0.645)
	*210	22.6 y			α(3.72)
	*211	36.1 min			β^-(1.36); γ(0.405, 0.427, 0.832)
	*212	10.64 h			β^-(0.569, 0.28); Bi-x; γ(0.239)
	*214	26.9 min			β^-(0.67, 0.73); γ(0.24, 0.30, 0.352)
Bismuth	205	15.31 d			γ(0.703, 1.764)
	206	6.243 d			γ(0.516, 0.803, 0.881)
	209		100	0.034	
	*210	5.013 d			β^-(1.16); γ(0.266, 0.352)
	212	1.0092 h			β^-(2.25); γ(0.288, 0.727, 0.786, 1.621); Tl-x; α(6.05, 6.09)
	*214	19.7 min			β^-(3.26); γ(0.609, 1.120, 1.764)
Polonium	204	3.53 h			γ(0.270, 0.884, 1.016)
	205	1.7 h			γ(0.837, 0.850, 0.872, 1.001)
	206	8.8 d			α(5.233); γ(0.286, 0.312, 0.807)
	208	2.898 y			α(5.116)
	209	102 y			α(4.88); IT, K, Bi-x; γ(0.260, 0.896)
	210	138.38 d			α(5.304); γ(0.803)
	212	298 ns			α(8.784)
	214	0.1637 ms			α(7.686)
	216	145 ms			α(6.778)
	218	3.04 min			α(5.18)
Astatine	207	1.81 h			α(5.76); γ(0.168, 0.588, 0.814)
	208	1.63 h			α(5.641); K, Po-x, γ(0.177, 0.660, 0.685, 0.845, 1.028)

TABLE 1.43 Table of Nuclides (*Continued*)

Element	A	Half-life	Natural abundance, %	Cross section, barns	Radiation (MeV)
Astatine (*cont.*)	209	5.41 h			α(5.65), K, Po-x; γ(0.545, 0.782, 0.790)
	210	8.1 h			K, Po-x; γ(0.245, 0.528, 1.181, 1.437, 1.483)
	211	7.214 h			α(5.87); K, Po-x; γ(0.669, 0.742)
Radon	210	2.4 h			α(6.039); γ(0.196, 0.458, 0.571, 0.649)
	211	14.68 h			α(5.784, 5.851); γ(0.169, 0.250, 0.370, 0.674, 0.678, 1.363)
	212	24 min			α(6.260)
	220	55.6 s			α(6.288)
	222	2.8235 d		0.74(5)	α(5.49); γ(0.510)
Francium	212	20 min			α(6.41, 6.26); γ(1.186, 1.275)
	220	27.4 s			α(6.686, 0.641, 6.582); γ(0.106, 0.154, 0.162)
	221	4.8 min			α(6.341); γ(0.218, 0.409)
	222	14.3 min			β^-(0.178)
	223	22.0 min			β^-(0.117)
Radium	*224	3.66 d		12.0(5)	α(5.685, 5.45); K, Rn-x; γ(0.241, 0.409, 0.650)
	*226	1599 y		≈13	α(4.78, 4.60); K, Rn-x; γ(0.186, 0.262)
	*228	5.76 y		36.(5)	γ(0.0135)
Actinium	*227	21.77 y		$8.8(7) \times 10^2$	β^-(0.045); α(4.95, 4.94); K, Th-x; γ(0.084, 0.160, 0.270)
	*228	6.15 h			β^-(2.18, 1.85, 1.11); K, Th-x; γ(0.339, 0.911, 0.969)
Thorium	226	30.6 min			α(6.337, 6.228); γ(0.206, 0.242)
	228	1.913 y		$1.2(2) \times 10^2$	α(5.42, 5.34, 5.18); K, Ra-x
	*230	7.54×10^4 y		23.4(5)	α(4.68, 4.62); K, Ra-x; γ(0.068)
	231	1.063 d			β^-(0.305, 0.218, 0.138)
	*232	1.405×10^{10} y		7.37(4)	α(4.01, 3.95); γ(0.059)
	233	22.3 min		$1.5(1) \times 10^3$	β^-(1.245); γ(0.459)
	*234	24.10 d		1.8(5)	β^-(0.198, 0.102); K, Pa-x
Protactinium	230	17.4 d		$1.5(3) \times 10^3$	β^-(0.51); γ(0.444, 0.455, 0.899, 0.952)
	*231	3.25×10^4 y		$2.0(1) \times 10^2$	α(5.06, 5.03, 5.01, 4.95, 4.73); K, Ac-x; γ(0.260, 0.284, 0.300, 0.330)

(*Continued*)

TABLE 1.43 Table of Nuclides (*Continued*)

Element	A	Half-life	Natural abundance, %	Cross section, barns	Radiation (MeV)
Protactinium (*cont.*)	232	1.31 d		4.6(10) × 10^2	β^-(1.34); γ(0.109, 0.150, 0.894, 0.969)
	233	27.0 d			β^-(0.256, 0.15, 0.568); K,L U-x; γ(0.300, 0.312, 0.341)
	234*m*	1.17 min			β^-(2.29); IT, K, U-x
	235	24.4 min			β^-(1.4)
Uranium	230	20.8 d			α(5.89, 5.82)
	232	68.9 y		73.(2)	α(5.320, 5.263)
	233	1.592 × 10^5 y		47.(2)	α(4.825, 4.783); L, Th-x; γ(0.029, 0.042, 0.055, 0.097, 0.119, 0.146, 0.164, 0.22, 0.291, 0.32)
	*234	2.454 × 10^5 y	0.0055(5)	96.(2)	α(4.776, 4.723); L, Th-x; γ(0.121)
	*235	7.037 × 10^8 y	0.720(1)	95.(5)	α(4.40, 4.37, 4.22); K,L Th-x; γ(0.14, 0.16, 0.186, 0.20)
	237	6.75 d		≈100	
	*238	4.46 × 10^9 y	99.2745(15)	2.7(1)	α(4.196, 4.147)
	239	23.47 min		22.(2)	β^-(1.21, 1.29)
Neptunium	236	1.55 × 10^5 y			β^-(0.49), γ(0.104, 0.160)
	237	2.14 × 10^6 y		180	α(4.79, 4.77); K,L Pa-x
	238	2.117 d		51	β^-(1.2); γ(0.984, 1.029)
	239	2.355 d		5.1(2) × 10^2	β^-(0.438, 0.341); γ(0.228, 0.278)
Plutonium	237	45.7 d			K,L Np-x
	238	87.74 y			α(5.50, 5.46); K, U-x; γ(0.0435)
	239	2.411 × 10^4 y		2.7(1) × 10^2	α(5.16, 5.14, 5.11); K, U-x; γ(0.375, 0.414, 0.129)
	240	6.537 × 10^3 y		2.9(1) × 10^2	α(5.168, 5.124); L, U-x
	242	3.763 × 10^5 y		19.(1)	α(4.90, 4.86); γ(0.045, 0.103)
	244	8.2 × 10^7 y		1.7(1)	α(4.59, 4.55); L, U-x
	246	10.85 d			β^-(0.150, 0.35); γ(0.224)
Americium	241	432.2 y		600	α(5.49, 5.44); γ(0.12, 0.14)
	243	7370 y		80	α(5.277, 5.234); γ(0.075)
Curium	242	162.8 d		≈20	α(6.113, 6.069); L, Pu-x
	243	28.5 y		1.3(1) × 10^2	α(5.786, 5.742)
	244	18.11 y		15.(1)	α(5.805, 5.753); γ(0.099, 1.526)
Berkelium	247	1.4 × 10^3 y			α(5.532, 5.678, 5.712)
	249	320 d		7.(1) × 10^2	α(5.42); β^-(0.125)
	250	3.217 h			β^-(0.74); γ(0.989, 1.032)

TABLE 1.43 Table of Nuclides (*Continued*)

Element	A	Half-life	Natural abundance, %	Cross section, barns	Radiation (MeV)
Californium	251	900 y		$2.9(2) \times 10^2$	α(5.677, 5.851, 6.014)
	252	2.645 y		20.(2)	α(6.118, 6.076); L, Cm-x; γ(0.043, 0.100)
Einsteinium	253	20.47 d		186	α(6.64); γ(0.389)
	254	275.7 d		28.(3)	α(6.43)
	255	40 d		≈55	β^-(0.29); α(6.26)
Fermium	255	20.1 h		26.(3)	α(7.023)
	257	100.5 d			α(6.519); L, Cf-x; γ(0.179, 0.241)
Mendelevium	258	51.5 d			α(6.718, 6.763); γ(0.368)
	260	32 d			
Nobelium	255	3.1 min			α(8.12, 7.93); γ(0.187)
	259	58 min			α(7.52, 7.55)
Lawrencium	260	3 min			
	261	40 min			
	262	3.6 h			

1.11 VAPOR PRESSURE

Vapor pressure is the pressure exerted by a pure component at equilibrium, at any temperature, when both liquid and vapor phases exist and thus extends from a minimum at the triple point temperature to a maximum at the critical temperature (the critical pressure), and is the most important of the basic thermodynamic properties affecting liquids and vapors.

Except at very high total pressures (above about 10 MPa), there is no effect of total pressure on vapor pressure. If such an effect is present, a correction can be applied. The pressure exerted above a solid-vapor mixture may also be called vapor pressure but is normally only available as experimental data for common compounds that sublime.

1.11.1 Vapor Pressure Equations

Numerous mathematical formulas relating the temperature and pressure of the gas phase in equilibrium with the condensed phase have been proposed. The Antoine equation (Eq. 1) gives good correlation with experimental values. Equation 2 is simpler and is often suitable over restricted temperature ranges. In these equations, and the derived differential coefficients for use in the Haggenmacher and Clausius-Clapeyron equations, the p term is the vapor pressure of the compound in pounds per square inch (psi), the t term is the temperature in degrees Celsius, and the T term is the absolute temperature in kelvins (t°C + 273.15).

Eq.	Vapor-pressure equation	dp/dT	$-[d(\ln p)/d(1/T)]$
1	$\log p = A - \dfrac{B}{t+C}$	$\dfrac{2.303\,pB}{(t+C)^2}$	$\dfrac{2.303\,BT^2}{(t+C)^2}$
2	$\log p = A - \dfrac{B}{T}$	$\dfrac{2.303\,pB}{T^2}$	$2.303B$
3	$\log p = A - \dfrac{B}{T} - C\,\log T$	$p\left(\dfrac{2.303\,B}{T^2} - \dfrac{C}{T}\right)$	$2.303B - CT$

Equations 1 and 2 are easily rearranged to calculate the temperature of the normal boiling point:

$$t = \frac{B}{A - \log p} - C \tag{5.1}$$

$$T = \frac{B}{A - \log p} \tag{5.2}$$

The constants in the Antoine equation may be estimated by selecting three widely spaced data points and substituting in the following equations in sequence:

$$\left(\frac{y_3 - y_2}{y_2 - y_1}\right)\left(\frac{t_2 - t_1}{t_3 - t_2}\right) = 1 - \left(\frac{t_3 - t_1}{t_3 + C}\right)$$

$$B = \left(\frac{y_3 - y_1}{t_2 + t_1}\right)(t_1 + C)(t_3 + C)$$

$$A = y_2 + \left(\frac{B}{t_2 + C}\right)$$

In these equations, $y_i = \log p_i$.

TABLE 1.44 Vapor Pressures of Selected Elements at Different Temperatures

Element	Atomic number	Atomic symbol	Boiling point, °C	Vapor pressure temperature, °C								
				E-08	E-07	E-06	E-05	E-04	E-03	E-02	E-01	1
Aluminum	13	Al	2467	685	742	812	887	972	1082	1217	1367	1557
Antimony	52	Sb	1750	279	309	345	383	425	475	533	612	757
Arsenic	33	As	613	104	127	150	174	204	237	277	317	372
Barium	56	Ba	1140	272	310	354	402	462	527	610	711	852
Beryllium	4	Be	2970	707	762	832	907	997	1097	1227	1377	1557
Bismuth	83	Bi	1560	347	367	409	459	517	587	672	777	897
Boron	5	B	2550	1282	1367	1467	1582	1707	1867	2027	2247	2507
Cadmium	48	Cd	765	74	95	119	146	177	217	265	320	392
Calcium	20	Ca	1484	282	317	357	405	459	522	597	689	802
Carbon	6	C	4827	1657	1757	1867	1987	2137	2287	2457	2657	2897
Cobalt	27	Co	2870	922	992	1067	1157	1257	1382	1517	1687	1907
Chromium	24	Cr	2672	837	902	977	1062	1157	1267	1397	1552	1737
Copper	29	Cu	2567	722	787	852	937	1027	1132	1257	1417	1617
Dysprosium	66	Dy	2562	625	682	747	817	897	997	1117	1262	1437
Erbium	68	Er	2510	649	708	777	852	947	1052	1177	1332	1527
Europium	63	Eu	1597	283	319	361	409	466	532	611	708	827
Gallium	31	Ga	2403	619	677	742	817	907	1007	1132	1282	1472
Germanium	32	Ge	2830	812	877	947	1037	1137	1257	1397	1557	1777
Gold	79	Au	2807	807	877	947	1032	1132	1252	1397	1567	1767
Indium	77	In	2000	488	539	597	664	742	837	947	1082	1247
Iron	26	Fe	2750	892	957	1032	1127	1227	1342	1477	1647	1857
Lanthanum	57	La	3469	1022	1102	1192	1297	1422	1562	1727	1927	2177
Lead	82	Pb	1740	342	383	429	485	547	625	715	832	977
Lithium	49	Li	1347	235	268	306	350	404	467	537	627	747
Magnesium	12	Mg	1107	185	214	246	282	327	377	439	509	605
Manganese	25	Mn	1962	505	554	611	675	747	837	937	1082	1217
Mercury	80	Hg	357	−72	−59	−44	−27	7	16	46	80	125
Molybdenum	42	Mo	4612	1592	1702	1822	1957	2117	2307	2527	2787	3117
Nickel	28	Ni	2732	927	997	1072	1157	1262	1382	1527	1697	1907
Niobium	41	Nb	4927	1762	1867	1987	2127	2277	2447	2657	2897	3177
Palladium	46	Pd	2927	842	912	992	1082	1192	1317	1462	1647	1877
Phosphorus	15	P	2804	54	69	88	108	129	157	185	222	261

(*Continued*)

TABLE 1.44 Vapor Pressures of Selected Elements at Different Temperatures (*Continued*)

Element	Atomic number	Atomic symbol	Boiling point, °C	Vapor pressure temperature, °C								
				E-08	E-07	E-06	E-05	E-04	E-03	E-02	E-01	1
Platinum	78	Pt	3827	1292	1382	1492	1612	1747	1907	2097	2317	2587
Potassium	19	K	774	21	42	65	91	123	161	208	267	345
Praseodymium	59	Pr	3127	797	867	947	1042	1147	1277	1427	1617	1847
Rhenium	75	Re	5627	1947	2077	2217	2387	2587	2807	3067	3407	3807
Rhodium	45	Rh	3727	1277	767	1472	1582	1707	1857	2037	2247	2507
Scandium	21	Sc	2832	772	837	917	1007	1107	1232	1377	1567	1797
Selenium	34	Se	685	63	83	107	133	164	199	243	297	363
Silicon	14	Si	4827	992	1067	1147	1237	1337	1472	1632	1817	2057
Silver	47	Ag	2212	574	626	685	752	832	922	1027	1162	1322
Sodium	11	Na	553	74	97	123	155	193	235	289	357	441
Strontium	38	Sr	1384	241	273	309	353	394	465	537	627	732
Sulfur	16	S	45	−10	3	17	37	55	80	109	147	189
Tantalum	73	Ta	5425	1957	2097	2237	2407	2587	2807	3057	3357	3707
Tellurium	52	Te	990	155	181	209	242	280	323	374	433	518
Thallium	81	TI	1457	283	319	359	407	463	530	609	706	827
Tin	50	Sn	2270	682	747	807	897	997	1107	1247	1412	1612
Titanium	22	Ti	3287	1062	1137	1227	1327	1442	1577	1737	1937	2177
Tungsten	74	W	5660	2117	2247	2407	2567	2757	2977	3227	3537	3917
Ytterbium	70	Yb	1466	247	279	317	365	417	482	557	647	787
Yttrium	39	Y	3337	957	1032	1117	1217	1332	1467	1632	1832	2082
Zinc	30	Zn	907	123	147	177	209	247	292	344	408	487

TABLE 1.45 Vapor Pressures of Inorganic Compounds up to 1 Atmosphere

Compound name	Formula	Pressure, mm Hg 1	5	10	20	40	60	100	200	400	760	Melting point, °C
		Temperature, °C										
Aluminum	Al	1284	1421	1487	1555	1635	1684	1749	1844	1947	2056	660
borohydride	$Al(BH_4)_3$		−52.2	−42.9	−32.5	−20.9	−13.4	−3.9	+11.2	28.1	45.9	−64
bromide	$AlBr_3$	81.3	103.8	118.0	134.0	150.6	161.7	176.1	199.8	227.0	256.3	97
chloride	Al_2Cl_6	100.0	116.4	123.8	131.8	139.9	145.4	152.0	161.8	171.6	180.2	192.4
fluoride	AlF_3	1238	1298	1324	1350	1378	1398	1422	1457	1496	1537	1040
iodide	AlI_3	178.0	207.7	225.8	244.2	265.0	277.8	294.5	322.0	354.0	385.5	
oxide	Al_2O_3	2148	2306	2385	2465	2549	2599	2665	2766	2874	2977	2050
Ammonia	NH_3	−109.1	−97.5	−91.9	−85.8	−79.2	−74.3	−68.4	−57.0	−45.4	−33.6	−77.7
heavy	ND_3						−74.0	−67.4	−57.0	−45.4	−33.4	−74.0
Ammonium bromide	NH_4Br	198.3	234.5	252.0	270.6	290.0	303.8	320.0	345.3	370.8	396.0	
carbamate	$N_2H_6CO_2$	−26.1	−10.4	−2.9	+5.3	14.0	19.6	26.7	37.2	48.0	58.3	
chloride	NH_4Cl	160.4	193.8	209.8	226.1	245.0	256.2	271.5	293.2	316.5	337.8	520
cyanide	NH_4CN	−50.6	−35.7	−28.6	−20.9	−12.6	−7.4	−0.5	+9.6	20.5	31.7	36
hydrogen sulfide	NH_4HS	−51.1	−36.0	−28.7	−20.8	−12.3	−7.0	0.0	+10.5	21.8	33.3	
iodide	NH_4I	210.9	247.0	263.5	282.8	302.8	316.0	331.8	355.8	381.0	404.9	
Antimony	Sb	886	984	1033	1084	1141	1176	1223	1288	1364	1440	630.5
tribromide	$SbBr_3$	93.9	126.0	142.7	158.3	177.4	188.1	203.5	225.7	250.2	275.0	96.6
trichloride	$SbCl_3$	49.2	71.4	85.2	100.6	117.8	128.3	143.3	165.9	192.2	219.0	73.4
pentachloride	$SbCl_5$	22.7	48.6	61.8	75.8	91.0	101.0	114.1				2.8
triiodide	SbI_3	163.6	203.8	223.5	244.8	267.8	282.5	303.5	333.8	368.5	401.0	167
trioxide	Sb_4O_6	574	626	666	729	812	873	957	1085	1242	1425	656
Argon	A	−218.2	−213.9	−210.9	−207.9	−204.9	−202.9	−200.5	−195.6	−190.6	−185.6	−189.2
Arsenic	As	372	416	437	459	483	498	518	548	579	610	814
Arsenic tribromide	$AsBr_3$	41.8	70.6	85.2	101.3	118.7	130.0	145.2	167.7	193.6	220.0	
trichloride	$AsCl_3$	−11.4	+11.7	+23.5	36.0	50.0	58.7	70.9	89.2	109.7	130.4	−18
trifluoride	AsF_3					−2.5	+4.2	13.2	26.7	41.4	56.3	−5.9
pentafluoride	AsF_5	−117.9	−108.0	−103.1	−98.0	−92.4	−88.5	−84.3	−75.5	−64.0	−52.8	−79.8
trioxide	As_2O_3	212.5	242.6	259.7	279.2	299.2	310.3	332.5	370.0	412.2	457.2	312.8
Arsine	AsH_3	−142.6	−130.8	−124.7	−117.7	−110.2	−104.8	−98.0	−87.2	−75.2	−62.1	−116.3
Barium	Ba		984	1049	1120	1195	1240	1301	1403	1518	1638	850

(*Continued*)

TABLE 1.45 Vapor Pressures of Inorganic Compounds up to 1 Atmosphere (*Continued*)

Compound name	Formula	Pressure, mm Hg: 1	5	10	20	40	60	100	200	400	760	Melting point, °C
		Temperature, °C										
Beryllium borohydride	$Be(BH_4)_2$	+1.0	19.8	28.1	36.8	46.2	51.7	58.6	69.0	79.7	90.0	123
bromide	$BeBr_2$	289	325	342	361	379	390	405	427	451	474	490
chloride	$BeCl_2$	291	328	346	365	384	395	411	435	461	487	405
iodide	BeI_2	283	322	341	361	382	394	411	435	461	487	488
Bismuth	Bi	1021	1099	1136	1177	1217	1240	1271	1319	1370	1420	271
tribromide	$BiBr_3$		261	282	305	327	340	360	392	425	461	218
trichloride	$BiCl_3$		242	264	287	311	324	343	372	405	441	230
Diborane hydrobromide	B_2H_5Br	−93.3	−75.3	−66.3	−56.4	−45.4	−38.2	−29.0	−15.4	0.0	+16.3	−104.2
Borine carbonyl	BH_3CO	−139.2	−127.3	−121.1	−114.1	−106.6	−101.9	−95.3	−85.5	−74.8	−64.0	−137.0
triamine	$B_3N_3H_6$	−63.0	−45.0	−35.3	−25.0	−13.2	−5.8	+4.0	18.5	34.3	50.6	−58.2
Boron hydrides												
dihydrodecaborane	$B_{10}H_{14}$	60.0	80.8	90.2	100.0	117.4	127.8	142.3	163.8			99.6
dihydrodiborane	B_2H_6	−159.7	−149.5	−144.3	−138.5	−131.6	−127.2	−120.9	−111.2	−99.6	−86.5	−169
dihydropentaborane	B_5H_9		−40.4	−30.7	−20.0	−8.0	−0.4	+9.6	24.6	40.8	58.1	−47.0
tetrahydropentaborane	B_5H_{11}	−50.2	−29.9	−19.9	−9.2	+2.7	10.2	20.1	34.8	51.2	67.0	
tetrahydrotetraborane	B_4H_{10}	−90.9	−73.1	−64.3	−54.8	−44.3	−37.4	−28.1	−14.0	+0.8	16.1	−119.9
Boron tribromide	BBr_3	−41.4	−20.4	−10.1	+1.5	14.0	22.1	33.5	50.3	70.0	91.7	−45
trichloride	BCl_3	−91.5	−75.2	−66.9	−57.9	−47.8	−41.2	−32.4	−18.9	−3.6	+12.7	−107
trifluoride	BF_3	−154.6	−145.4	−141.3	−136.4	−131.0	−127.6	−123.0	−115.9	−108.3	−100.7	−126.8
Bromine	Br_2	−48.7	−32.8	−25.0	−16.8	−8.0	−0.6	+9.3	24.3	41.0	58.2	−7.3
pentafluoride	BrF_5	−69.3	−51.0	−41.9	−32.0	−21.0	−14.0	−4.5	+9.9	25.7	40.4	−61.4
Cadmium	Cd	394	455	484	516	553	578	611	658	711	765	320.9
chloride	$CdCl_2$		618	656	695	736	762	797	847	908	967	568
fluoride	CdF_2	1112	1231	1286	1344	1400	1436	1486	1561	1651	1751	520
iodide	CdI_2	416	481	512	546	584	608	640	688	742	796	385
oxide	CdO	1000	1100	1149	1200	1257	1295	1341	1409	1484	1559	
Calcium	Ca		926	983	1046	1111	1152	1207	1288	1388	1487	851
Carbon (graphite)	C	3586	3828	3946	4069	4196	4273	4373	4516	4660	4827	
dioxide	CO_2	−134.3	−124.4	−119.5	−114.4	−108.6	−104.8	−100.2	−93.0	−85.7	−78.2	−57.5
disulfide	CS_2	−73.8	−54.3	−44.7	−34.3	−22.5	−15.3	−5.1	+10.4	28.0	46.5	−110.8
monoxide	CO	−222.0	−217.2	−215.0	−212.8	−210.0	−208.1	−205.7	−201.3	−196.3	−191.3	−205.0

oxyselenide	COSe	−117.1	−102.3	−95.0	−86.3	−76.4	−70.2	−61.7	−49.8	−35.6	−21.9	
oxysulfide	COS	−132.4	−119.8	−113.3	−106.0	−98.3	−93.0	−85.9	−75.0	−62.7	−49.9	−138.8
selenosulfide	CSeS	−47.3	−26.5	−16.0	−4.4	+8.6	17.0	28.3	45.7	65.2	85.6	−75.2
subsulfide	C_3S_2	14.0	41.2	54.9	69.3	85.6	96.0	109.9	130.8			+0.4
tetrabromide	CBr_4					96.3	106.3	119.7	139.7	163.5	189.5	90.1
tetrachloride	CCl_4	−50.0	−30.0	−19.6	−8.2	+4.3	12.3	23.0	38.3	57.8	76.7	−22.6
tetrafluoride	CF_4	−184.6	−174.1	−169.3	−164.3	−158.8	−155.4	−150.7	−143.6	−135.5	−127.7	−183.7
Cesium	Cs	279	341	375	409	449	474	509	561	624	690	28.5
bromide	CsBr	748	838	887	938	993	1026	1072	1140	1221	1300	636
chloride	CsCl	744	837	884	934	989	1023	1069	1139	1217	1300	646
fluoride	CsF	712	798	844	893	947	980	1025	1092	1170	1251	683
iodide	CsI	738	828	873	923	976	1009	1055	1124	1200	1280	621
Chlorine	Cl_2	−118.0	−106.7	−101.6	−93.3	−84.5	−79.0	−71.7	−60.2	−47.3	−33.8	−100.7
fluoride	ClF		−143.4	−139.0	−134.3	−128.8	−125.3	−120.8	−114.4	−107.0	−100.5	−145
trifluoride	ClF_3		−80.4	−71.8	−62.3	−51.3	−44.1	−34.7	−20.7	−4.9	+11.5	−83
monoxide	Cl_2O	−98.5	−81.6	−73.1	−64.3	−54.3	−48.0	−39.4	−26.5	−12.5	+2.2	−116
dioxide	ClO_2			−59.0	−51.2	−42.8	−37.2	−29.4	−17.8	−4.0	+11.1	−59
heptoxide	Cl_2O_7	−45.3	−23.8	−13.2	−2.1	+10.2	+18.3	29.1	44.6	62.2	78.8	−91
Chlorosulfonic acid	HSO_3Cl	32.0	53.5	64.0	75.3	87.6	95.2	105.3	120.0	136.1	151.0	−80
Chromium	Cr	1616	1768	1845	1928	2013	2067	2139	2243	2361	2482	1615
carbonyl	$Cr(CO)_6$	36.0	58.0	68.3	79.5	91.2	98.3	108.0	121.8	137.2	151.0	
oxychloride	CrO_2Cl_2	−18.4	+3.2	13.8	25.7	38.5	46.7	58.0	75.2	95.2	117.1	
Cobalt chloride	$CoCl_2$					770	801	843	904	974	1050	735
nitrosyl tricarbonyl	$Co(CO)_3NO$				−1.3	+11.0	18.5	29.0	44.4	62.0	80.0	−11
Columbium fluoride	CbF_3			86.3	103.0	121.5	133.2	148.5	172.2	198.0	225.0	75.5
Copper	Cu	1628	1795	1879	1970	2067	2127	2207	2325	2465	2595	1083
Cuprous bromide	Cu_2Br_2	572	666	718	777	844	887	951	1052	1189	1355	504
chloride	Cu_2Cl_2	546	645	702	766	838	886	960	1077	1249	1490	422
iodide	Cu_2I_2		610	656	716	786	836	907	1018	1158	1336	605
Cyanogen	C_2N_2	−95.8	−83.2	−76.8	−70.1	−62.7	−57.9	−51.8	−42.6	−33.0	−21.0	−34.4
bromide	CNBr	−35.7	−18.3	−10.0	−1.0	+8.6	14.7	22.6	33.8	46.0	61.5	58
chloride	CNCl	−76.7	−61.4	−53.8	−46.1	−37.5	−32.1	−24.9	−14.1	−2.3	+13.1	−6.5
fluoride	CNF	−134.4	−123.8	−118.5	−112.8	−106.4	−102.3	−97.0	−89.2	−80.5	−72.6	
Deuterium cyanide	DCN	−68.9	−54.0	−46.7	−38.8	−30.1	−24.7	−17.5	−5.4	+10.0	26.2	−12
Fluorine	F_2	−223.0	−216.9	−214.1	−211.0	−207.7	−205.6	−202.7	−198.3	−193.2	−187.9	−223
oxide	F_2O	−196.1	−186.6	−182.3	−177.8	−173.0	−170.0	−165.8	−159.0	−151.9	−144.6	−223.9

(Continued)

TABLE 1.45 Vapor Pressures of Inorganic Compounds up to 1 Atmosphere (*Continued*)

Compound name	Formula	Pressure, mm Hg										Melting point, °C
		1	5	10	20	40	60	100	200	400	760	
		Temperature, °C										
Germanium bromide	$GeBr_4$		43.3	56.8	71.8	88.1	98.8	113.2	135.4	161.6	189.0	26.1
chloride	$GeCl_4$	−45.0	−24.9	−15.0	−4.1	+8.0	16.2	27.5	44.4	63.8	84.0	−49.5
hydride	GeH_4	−163.0	−151.0	−145.3	−139.2	−131.6	−126.7	−120.3	−111.2	−100.2	−88.9	−165
Trichlorogermane	$GeHCl_3$	−41.3	−22.3	−13.0	−3.0	+8.8	16.2	26.5	41.6	58.3	75.0	−71.1
Tetramethylgermane	$Ge(CH_3)_4$	−73.2	−54.6	−45.2	−35.0	−23.4	−16.2	−6.3	+8.8	26.0	44.0	−88
Digermane	Ge_2H_6	−88.7	−69.8	−60.1	−49.9	−38.2	−30.7	−20.3	−4.7	+3.3	31.5	−109
Trigermane	Ge_3H_6	−36.9	−12.8	−0.9	+11.8	26.3	35.5	47.9	67.0	88.6	110.8	−105.6
Gold	Au	1869	2059	2154	2256	2363	2431	2521	2657	2807	2966	1063
Helium	He	−271.7	−271.5	−271.3	−271.1	−270.7	−270.6	−270.3	−269.8	−269.3	−268.6	
para–Hydrogen	H_2	−263.3	−261.9	−261.3	−260.4	−259.6	−258.9	−257.9	−256.3	−254.5	−252.5	−259.1
Hydrogen bromide	HBr	−138.8	−127.4	−121.8	−115.4	−108.3	−103.8	−97.7	−88.1	−78.0	−66.5	−87.0
chloride	HCl	−150.8	−140.7	−135.6	−130.0	−123.8	−119.6	−114.0	−105.2	−95.3	−84.8	−114.3
cyanide	HCN	−71.0	−55.3	−47.7	−39.7	−30.9	−25.1	−17.8	−5.3	+10.2	25.9	−13.2
fluoride	H_2F_2		−74.7	−65.8	−56.0	−45.0	−37.9	−28.2	−13.2	+2.5	19.7	−83.7
iodide	HI	−123.3	−109.6	−102.3	−94.5	−85.6	−79.8	−72.1	−60.3	−48.3	−35.1	−50.9
oxide(water)	H_2O	−17.3	+1.2	11.2	22.1	34.0	41.5	51.6	66.5	83.0	100.0	0.0
sulfide	H_2S	−134.3	−122.4	−116.3	−109.7	−102.3	−97.9	−91.6	−82.3	−71.8	−60.4	−85.5
disulfide	HSSH	−43.2	−24.4	−15.2	−5.1	+6.0	12.8	22.0	35.3	49.6	64.0	−89.7
selenide	H_2Se	−115.3	−103.4	−97.9	−91.8	−84.7	−80.2	−74.2	−65.2	−53.6	−41.1	−64
telluride	H_2Te	−96.4	−82.4	−75.4	−67.8	−59.1	−53.7	−45.7	−32.4	−17.2	−2.0	−49.0
Iodine	I_2	38.7	62.2	73.2	84.7	97.5	105.4	116.5	137.3	159.8	183.0	112.9
heptafluoride	IF	−87.0	−70.7	−63.0	−54.5	−45.3	−39.4	−31.9	−20.7	−8.3	+4.0	5.5
Iron	Fe	1787	1957	2039	2128	2224	2283	2360	2475	2605	2735	1535
pentacarbonyl	$Fe(CO)_5$		−6.5	+4.6	16.7	30.3	39.1	50.3	68.0	86.1	105.0	−21
Ferric chloride	Fe_2Cl_6	194.0	221.8	235.5	246.0	256.8	263.7	272.5	285.0	298.0	319.0	304
Ferrous chloride	$FeCl_2$			700	737	779	805	842	897	961	1026	
Krypton	Kr	−199.3	−191.3	−187.2	−182.9	−178.4	−175.7	−171.8	−165.9	−159.0	−152.0	−156.7
Lead	Pb	973	1099	1162	1234	1309	1358	1421	1519	1630	1744	327.5
bromide	$PbBr_2$	513	578	610	646	686	711	745	796	856	914	373
chloride	$PbCl_2$	547	615	648	684	725	750	784	833	893	954	501
fluoride	PbF_2		861	904	950	1003	1036	1080	1144	1219	1293	855

iodide	PbI_2	479	540	571	605	644	668	701	750	807	872	402
oxide	PbO	943	1039	1085	1134	1189	1222	1265	1330	1402	1472	890
sulfide	PbS	852	928	975	1005	1048	1074	1108	1160	1221	1281	1114
Lithium	Li	723	838	881	940	1003	1042	1097	1178	1273	1372	186
bromide	LiBr	748	840	888	939	994	1028	1076	1147	1126	1310	547
chloride	LiCl	783	880	932	987	1045	1081	1129	1203	1290	1382	614
fluoride	LiF	1047	1156	1211	1270	1333	1372	1425	1503	1591	1681	870
iodide	LiI	723	802	841	883	927	955	993	1049	1110	1171	446
Magnesium	Mg	621	702	743	789	838	868	909	967	1034	1107	651
chloride	$MgCl_2$	778	877	930	968	1050	1088	1142	1223	1316	1418	712
Manganese	Mn	1292	1434	1505	1583	1666	1720	1792	1900	2029	2151	1260
chloride	$MnCl_2$		736	778	825	879	913	960	1028	1108	1190	650
Mercury	Hg	126.2	164.8	184.0	204.6	228.8	242.0	261.7	290.7	323.0	357.0	−38.9
Mercuric bromide	$HgBr_2$	136.5	165.3	179.8	194.3	211.5	221.0	237.8	262.7	290.0	319.0	237
chloride	$HgCl_2$	136.2	166.0	180.2	195.8	212.5	222.2	237.0	256.5	275.5	304.0	277
iodide	HgI_2	157.5	189.2	204.5	220.0	238.2	249.0	261.8	291.0	324.2	354.0	259
Molybdenum	Mo	3102	3393	3535	3690	3859	3964	4109	4322	4553	4804	2622
hexafluoride	MoF_6	−65.5	−49.0	−40.8	−32.0	−22.1	−16.2	−8.0	+4.1	17.2	36.0	17
oxide	MoO_3	734	785	814	851	892	917	955	1014	1082	1151	795
Neon	Ne	−257.3	−255.5	−254.6	−253.7	−252.6	−251.9	−251.0	−249.7	−248.1	−246.0	−248.7
Nickel	Ni	1810	1979	2057	2143	2234	2289	2364	2473	2603	2732	1452
carbonyl	$Ni(CO)_4$					−23.0	−15.9	−6.0	+8.8	25.8	42.5	−25
chloride	$NiCl_2$	671	731	759	789	821	840	866	904	945	987	1001
Nitrogen	N_2	−226.1	−221.3	−219.1	−216.8	−214.0	−212.3	−209.7	−205.6	−200.9	−195.8	−210.0
Nitric oxide	NO	−184.5	−180.6	−178.2	−175.3	−171.7	−168.9	−166.0	−162.3	−156.8	−151.7	−161
Nitrogen dioxide	NO_2	−55.6	−42.7	−36.7	−30.4	−23.9	−19.9	−14.7	−5.0	+8.0	21.0	−9.3
Nitrogen pentoxide	N_2O_5	−36.8	−23.0	−16.7	−10.0	−2.9	+1.8	7.4	15.6	24.4	32.4	30
Nitrous oxide	N_2O	−143.4	−133.4	−128.7	−124.0	−118.3	−114.9	−110.3	−103.6	−96.2	−85.5	−90.9
Nitrosyl chloride	NOCl					−60.2	−54.2	−46.3	−34.0	−20.3	−6.4	−64.5
fluoride	NOF	−132.0	−120.3	−114.3	−107.8	−100.3	−95.7	−88.8	−79.2	−68.2	−56.0	−134
Osmium tetroxide (yellow)	OsO_4	3.2	22.0	31.3	41.0	51.7	59.4	71.5	89.5	109.3	130.0	56
(white)	OsO_4	−5.6	+15.6	26.0	37.4	50.5	59.4	71.5	89.5	109.3	130.0	42
Oxygen	O_2	−219.1	−213.4	−210.6	−207.5	−204.1	−201.9	−198.8	−194.0	−188.8	−183.1	−218.7
Ozone	O_3	−180.4	−168.6	−163.2	−157.2	−150.7	−146.7	−141.0	−132.6	−122.5	−111.1	−251
Phosgene	$COCl_2$	−92.9	−77.0	−69.3	−60.3	−50.3	−44.0	−35.6	−22.3	−7.6	+8.3	−104
Phosphorus (yellow)	P	76.6	111.2	128.0	146.2	166.7	179.8	197.3	222.7	251.0	280.0	44.1
(violet)	P	237	271	287	306	323	334	349	370	391	417	590
tribromide	PBr_3	7.8	34.4	47.8	62.4	79.0	89.8	103.6	125.2	149.7	175.3	−40

(*Continued*)

TABLE 1.45 Vapor Pressures of Inorganic Compounds up to 1 Atmosphere (*Continued*)

Compound name	Formula	Pressure, mm Hg 1	5	10	20	40	60	100	200	400	760	Melting point, °C
		Temperature, °C										
trichloride	PCl_3	−51.6	−31.5	−21.3	−10.2	+2.3	10.2	21.0	37.6	56.9	74.2	−111.8
pentachloride	PCl_5	55.5	74.0	83.2	92.5	102.5	108.3	117.0	131.3	147.2	162.0	
Phosphine	PH_3					−129.4	−125.0	−118.8	−109.4	−98.3	−87.5	−132.5
Phosphonium bromide	PH_4Br	−43.7	−28.5	−21.2	−13.3	−5.0	+0.3	7.4	17.6	28.0	38.3	
chloride	PH_4Cl	−91.0	−79.6	−74.0	−68.0	−61.5	−57.3	−52.0	−44.0	−35.4	−27.0	−28.5
iodide	PH_4I	−25.2	−9.0	−1.1	+7.3	16.1	21.9	29.3	39.9	51.6	62.3	
Phosphorus trioxide	P_4O_6		39.7	53.0	67.8	84.0	94.2	108.3	129.0	150.3	173.1	22.5
pentoxide	P_4O_{10}	384	424	442	462	481	493	510	532	556	591	569
oxychloride	$POCl_3$			2.0	13.6	27.3	35.8	47.4	65.0	84.3	105.1	2
thiobromide	$PSBr_3$	50.0	72.4	83.6	95.5	108.0	116.0	126.3	141.8	157.8	175.0	38
thiochloride	$PSCl_3$	−18.3	+4.6	16.1	29.0	42.7	51.8	63.8	82.0	102.3	124.0	−36.2
Platinum	Pt	2730	3007	3146	3302	3469	3574	3714	3923	4169	4407	1755
Potassium	K	341	408	443	483	524	550	586	643	708	774	62.3
bromide	KBr	795	892	940	994	1050	1087	1137	1212	1297	1383	730
chloride	KCl	821	919	968	1020	1078	1115	1164	1239	1322	1407	790
fluoride	KF	885	988	1039	1096	1156	1193	1245	1323	1411	1502	880
hydroxide	KOH	719	814	863	918	976	1013	1064	1142	1233	1327	380
iodide	KI	745	840	887	938	995	1030	1080	1152	1238	1324	723
Radon	Rn	−144.2	−132.4	−126.3	−119.2	−111.3	−106.2	−99.0	−87.7	−75.0	−61.8	−71
Rhenium heptoxide	Re_2O_7	212.5	237.5	248.0	261.0	272.0	280.0	289.0	307.0	336.0	362.4	296
Rubidium	Rb	297	358	389	422	459	482	514	563	620	679	38.5
bromide	RbBr	781	876	923	975	1031	1066	1114	1186	1267	1352	682
chloride	RbCl	792	887	937	990	1047	1084	1133	1207	1294	1381	715
fluoride	RbF	921	982	1016	1052	1096	1123	1168	1239	1322	1408	760
iodide	RbI	748	839	884	935	991	1026	1072	1141	1223	1304	642
Selenium	Se	356	413	442	473	506	527	554	594	637	680	217
dioxide	SeO_2	157.0	187.7	202.5	217.5	234.1	244.6	258.0	277.0	297.7	317.0	340
hexafluoride	SeF_6	−118.6	−105.2	−98.9	−92.3	−84.7	−80.0	−73.9	−64.8	−55.2	−45.8	−34.7
oxychloride	$SeOCl_2$	34.8	59.8	71.9	84.2	98.0	106.5	118.0	134.6	151.7	168.0	8.5
tetrachloride	$SeCl_4$	74.0	96.3	107.4	118.1	130.1	137.8	147.5	161.0	176.4	191.5	

Silicon	Si	1724	1835	1888	1942	2000	2036	2083	2151	2220	2287	1420
dioxide	SiO_2			1732	1798	1867	1911	1969	2053	2141	2227	1710
tetrachloride	$SiCl_4$	−63.4	−44.1	−34.4	−24.0	−12.1	−4.8	+5.4	21.0	38.4	56.8	−68.8
tetrafluoride	SiF_4	−144.0	−134.8	−130.4	−125.9	−120.8	−117.5	−113.3	−170.2	−100.7	−94.8	−90
Trichlorofluorosilane	$SiFCl_3$	−92.6	−76.4	−68.3	−59.0	−48.8	−42.2	−33.2	−19.3	−4.0	+12.2	−120.8
Iodosilane	SiH_3I		−53.0	−47.7	−33.4	−21.8	−14.3	−4.4	+10.7	27.9	45.4	−57.0
Diiodosilane	SiH_2I_2		3.8	18.0	34.1	52.6	64.0	79.4	101.8	125.5	149.5	−1.0
Disiloxan	$(SiH_3)_2O$	−112.5	−95.8	−88.2	−79.8	−70.4	−64.2	−55.9	−43.5	−29.3	−15.4	−144.2
Trisilane	Si_3H_8	−68.9	−49.7	−40.0	−29.0	−16.9	−9.0	+1.6	17.8	35.5	53.1	−117.2
Trisilazane	$(SiH_3)_3N$	−68.7	−49.9	−40.4	−30.0	−18.5	−11.0	−1.1	+14.0	31.0	48.7	−105.7
Tetrasilane	Si_4H_{10}	−27.7	−6.2	+4.3	15.8	28.4	36.6	47.4	63.6	81.7	100.0	−93.6
Octachlorotrisilane	Si_3Cl_3	46.3	74.7	89.3	104.2	121.5	132.0	146.0	166.2	189.5	211.4	
Hexachlorodisiloxane	$(SiCl_3)_2O$	−5.0	17.8	29.4	41.5	55.2	63.8	75.4	92.5	113.6	135.6	−33.2
Hexachlorodisilane	Si_2Cl_6	+4.0	27.4	38.8	51.5	65.3	73.9	85.4	102.2	120.6	139.0	−1.2
Tribromosilane	$SiHBr_3$	−30.5	−8.0	+3.4	16.0	30.0	39.2	51.6	70.2	90.2	111.8	−73.5
Trichlorosilane	$SiHCl_3$	−80.7	−62.6	−53.4	−43.8	−32.9	−25.8	−16.4	−1.8	+14.5	31.8	−126.6
Trifluorosilane	$SiHF_3$	−152.0	−142.7	−138.2	−132.9	−127.3	−123.7	−118.7	−111.3	−102.8	−95.0	−131.4
Dibromosilane	SiH_2Br_2	−60.9	−40.0	−29.4	−18.0	−5.2	+3.2	14.1	31.6	50.7	70.5	−70.2
Difluorosilane	SiH_2F_2	−146.7	−136.0	−130.4	−124.3	−117.6	−113.3	−107.3	−98.3	−87.6	−77.8	
Monobromosilane	SiH_3Br		−85.7	−77.3	−68.3	−57.8	−51.1	−42.3	−28.6	−13.3	+2.4	−93.9
Monochlorosilane	SiH_3Cl	−117.8	−104.3	−97.7	−90.1	−81.8	−76.0	−68.5	−57.0	−44.5	−30.4	
Monofluorosilane	SiH_3F	−153.0	−145.5	−141.2	−136.3	−130.8	−127.2	−122.4	−115.2	−106.8	−98.0	
Tribromofluorosilane	$SiFBr_3$	−46.1	−25.4	−15.1	−3.7	+9.2	17.4	28.6	45.7	64.6	83.8	−82.5
Dichlorodifluorosilane	SiF_2Cl_2	−124.7	−110.5	−102.9	−94.5	−85.0	−78.6	−70.3	−58.0	−45.0	−31.8	−139.7
Trifluorobromosilane	SiF_3Br								−69.8	−55.9	−41.7	−70.5
Trifluorochlorosilane	SiF_3Cl	−144.0	−133.0	−127.0	−120.5	−112.8	−108.2	−101.7	−91.7	−81.0	−70.0	−142
Hexafluorodisilane	Si_2F_6	−81.0	−68.8	−63.1	−57.0	−50.6	−46.7	−41.7	−34.2	−26.4	−18.9	−18.6
Dichlorofluorobromosilane	$SiFCl_2Br$	−86.5	−68.4	−59.0	−48.8	−37.0	−29.0	−19.5	−3.2	+15.4	35.4	−112.3
Dibromochlorofluorosilane	$SiFClBr_2$	−65.2	−45.5	−35.6	−24.5	−12.0	−4.7	+6.3	23.0	43.0	59.5	−99.3
Silane	SiH_4	−179.3	−168.6	−163.0	−156.9	−150.3	−146.3	−140.5	−131.6	−122.0	−111.5	−185
Disilane	Si_2H_6	−114.8	−99.3	−91.4	−82.7	−72.8	−66.4	−57.5	−44.6	−29.0	−14.3	−132.6
Silver	Ag	1357	1500	1575	1658	1743	1795	1865	1971	2090	2212	960.5
chloride	AgCl	912	1019	1074	1134	1200	1242	1297	1379	1467	1564	455
iodide	AgI	820	927	983	1045	1111	1152	1210	1297	1400	1506	552
Sodium	Na	439	511	549	589	633	662	701	758	823	892	97.5
bromide	NaBr	806	903	952	1005	1063	1099	1148	1220	1304	1392	755
chloride	NaCl	865	967	1017	1072	1131	1169	1220	1296	1379	1465	800

(Continued)

TABLE 1.45 Vapor Pressures of Inorganic Compounds up to 1 Atmosphere (*Continued*)

Compound name	Formula	Pressure, mm Hg 1	5	10	20	40	60	100	200	400	760	Melting point, °C
		Temperature, °C										
cyanide	NaCN	817	928	983	1046	1115	1156	1214	1302	1401	1497	564
fluoride	NaF	1077	1186	1240	1300	1363	1403	1455	1531	1617	1704	992
hydroxide	NaOH	739	843	897	953	1017	1057	1111	1192	1286	1378	318
iodide	NaI	767	857	903	952	1005	1039	1083	1150	1225	1304	651
Strontium	Sr		847	898	953	1018	1057	1111	1192	1285	1384	800
Strontium oxide	SrO	2068	2198	2262	2333	2410						2430
Sulfur	S	183.8	223.0	243.8	264.7	288.3	305.5	327.2	359.7	399.6	444.6	112.8
monochloride	S_2Cl_2	−7.4	+15.7	27.5	40.0	54.1	63.2	75.3	93.5	115.4	138.0	−80
hexafluoride	SF_5	−132.7	−120.6	−114.7	−108.4	−101.5	−96.8	−90.9	−82.3	−72.6	−63.5	−50.2
Sulfuryl chloride	SO_2Cl_2		−35.1	−24.8	−13.4	−1.0	+7.2	17.8	33.7	51.3	69.2	−54.1
Sulfur dioxide	SO_2	−95.5	−83.0	−76.8	−69.7	−60.5	−54.6	−46.9	−35.4	−23.0	−10.0	−73.2
trioxide (α)	SO_3	−39.0	−23.7	−16.5	−9.1	−1.0	+4.0	10.5	20.5	32.6	44.8	16.8
trioxide (β)	SO_3	−34.0	−19.2	−12.3	−4.9	+3.2	8.0	14.3	23.7	32.6	44.8	32.3
trioxide (γ)	SO_3	−15.3	−2.0	+4.3	11.1	17.9	21.4	28.0	35.8	44.0	51.6	62.1
Tellurium	Te	520	605	650	697	753	789	838	910	997	1087	452
chloride	$TeCl_4$			233	253	273	287	304	330	360	392	224
fluoride	TeF_5	−111.3	−98.8	−92.4	−83.0	−78.4	−73.8	−67.9	−57.3	−48.2	−38.6	−37.8
Thallium	Tl	825	931	983	1040	1103	1143	1196	1274	1364	1457	3035
Thallous bromide	TlBr		490	522	559	598	621	653	703	759	819	460
chloride	TlCl		487	517	550	589	612	645	694	748	807	430
iodide	TlI	440	502	531	567	607	631	663	712	763	823	440
Thionyl bromide	$SOBr_2$	−6.7	+18.4	31.0	44.1	58.8	68.3	80.6	99.0	119.2	139.5	−52.2
Thionyl chloride	$SOCl_2$	−52.9	−32.4	−21.9	−10.5	+2.2	10.4	21.4	37.9	56.5	75.4	−104.5
Tin	Sn	1492	1634	1703	1777	1855	1903	1968	2063	2169	2270	231.9
Stannic bromide	$SnBr_4$		58.3	72.7	88.1	105.5	116.2	131.0	152.8	177.7	204.7	31.0
Stannous chloride	$SnCl_2$	316	366	391	420	450	467	493	533	577	623	246.8
Stannic chloride	$SnCl_4$	−22.7	−1.0	+10.0	22.0	35.2	43.5	54.7	72.0	92.1	113.0	−30.2
iodide	SnI_4		156.0	175.8	196.2	218.8	234.2	254.2	283.5	315.5	348.0	144.5
hydride	SnH_4	−140.0	−125.8	−118.5	−111.2	−102.3	−96.6	−89.2	−78.0	−65.2	−52.3	−149.9

Tin tetramethyl	$Sn(CH_3)_4$	−51.3	−31.0	−20.6	−9.3	+3.5	11.7	22.8	39.8	58.5	78.0	
trimethyl-ethyl	$Sn(CH_3)_3 \cdot C_2H_5$	−30.0	−7.6	+3.8	16.1	30.0	38.4	50.0	67.3	87.6	108.8	
trimethyl-propyl	$Sn(CH_3)_3 \cdot C_3H_7$	−12.0	+10.7	21.8	34.0	48.5	57.5	69.8	88.0	109.6	131.7	
Titanium chloride	$TiCl_4$	−13.9	+9.4	21.3	34.2	48.4	58.0	71.0	90.5	112.7	136.0	−30
Tungsten	W	3990	4337	4507	4690	4886	5007	5168	5403	5666	5927	3370
Tungsten hexafluoride	WF_6	−71.4	−56.5	−49.2	−41.5	−33.0	−27.5	−20.3	−10.0	+1.2	17.3	−0.5
Uranium hexafluoride	UF_6	−38.8	−22.0	−13.8	−5.2	+4.4	10.4	18.2	30.0	42.7	55.7	69.2
Vanadyl trichloride	$VOCl_3$	−23.2	+0.2	12.2	26.6	40.0	49.8	62.5	82.0	103.5	127.2	
Xenon	Xe	−168.5	−158.2	−152.8	−147.1	−141.2	−137.7	−132.8	−125.4	−117.1	−108.0	−111.6
Zinc	Zn	487	558	593	632	673	700	736	788	844	907	419.4
chloride	$ZnCl_2$	428	481	508	536	566	584	610	648	689	732	365
fluoride	ZnF_2	970	1055	1086	1129	1175	1207	1254	1329	1417	1497	872
diethyl	$Zn(C_2H_5)_2$	−22.4	0.0	+11.7	24.2	38.0	47.2	59.1	77.0	97.3	118.0	−28
Ziroconium bromide	$ZrBr_4$	207	237	250	266	281	289	301	318	337	357	450
chloride	$ZrCl_4$	190	217	230	243	259	268	279	295	312	331	437
iodide	ZrI_4	264	297	311	329	344	355	369	389	409	431	499

TABLE 1.46 Vapor Pressures of Various Inorganic Compounds

Substance	State	Eq.	Range,°C	*A*	*B*	*C*
Aluminum						
$AlCl_3$		2	70–190	16.24	6 006	
Al_2O_3		2	1840–2000	14.22	28 200	
Ammonium						
NH_3	c*	1		9.963 82	1 617.907	272.55
	liq	1		7.360 50	926.132	240.17
NH_4Br	subl c	1		9.220 0	3 947	227.0
NH_4Cl	subl c	1		9.355 7	3 703.7	232.0
NH_4I	subl c	1		9.147 0	3 858	226.0
NH_4N_3	c	1		10.433 4	2 821.0	240.0
Antimony						
Sb	c	2	1070–1325	9.051	9 871	
$SbBr_3$		2	235–324	8.005	2 873	
$SbCl_3$		2	170–253	8.090	2 582.3	
SbI_3		2	330–445	7.831	3 350.55	
Sb_2Se_3	subl c	2		8.790 6	6 432.3	
Argon						
Ar	c	1		7.505 81	399.085	272.63
	liq	1		6.616 51	304.227	267.32
Arsenic						
As		2	440–815	10.800	6 947	
		2	800–860	6.692	2 460	
$AsCl_3$		2	50–100	7.953	2 042.7	
As_2O_3		2	100–310	12.127	5 815.81	
		2	315–490	6.513	2 722.2	
Barium						
Ba		2	930–1130	15.765	18 280	
BaH_2 [97% pure]		2	500–1000	6.86	4 000	
Bismuth						
Bi		2	1210–1420	8.876	10 446	
$BiCl_3$		2	91–213	2.681	685.519	
Boron						
BBr_3		2	−40 to 90	7.655	1 740.3	
BCl_3		1		6.188 11	756.89	214.0
$B(CH_3)_3$		2	−118 to −20	7.459 5	1 157.99	
B_2H_6	liq	1		6.366 38	521.490	241.98
B_5H_{11}	liq	2	−43 to 8.4	7.901	1 690.3	
Bromine						
Br_2	c	1		9.7209	2 041.3	260.1
	liq	1		6.877 80	1 119.68	221.38
BrF_3	liq	1		7.729 74	1 673.95	219.48
BrF_5	liq	1		7.273 68	1 219.28	236.40
BrO_2F	liq	1		7.436 51	1 195.8	260.1
Cadmium						
Cd		2	150–321	8.564	5 693	
		2	500–840	7.897	5 218	
CdI_2		2	385–450	9.269	6 383	
Calcium						
Ca		2	500–700	9.697	10 185	
		2	960–1100	16.240	19 325	

*Crystalline solid.

TABLE 1.46 Vapor Pressures of Various Inorganic Compounds (*Continued*)

Substance	State	Eq.	Range,°C	*A*	*B*	*C*
Carbon						
C [as C(g)]	liq	1		11.042 8	37 736	302.2
[as C_2(g)]	liq	1		12.583 2	43 281	318.3
[all species]	liq	1		9.381 3	27 240	264.0
Carbon						
CNBr	subl c	1		9.488 9	2 041.8	251.70
CNF		1	−76 to −47	6.778 9	697.61	224.95
CO	c I	1		7.414 8	342.50	269.0
	liq	1		6.694 22	291.743	267.99
CO_2	c	1		9.810 66	1 347.786	273.00
C_3O_2	liq	1	−71 to 7	7.188 99	1 100.94	249.15
$COCl_2$	liq	1		6.971 33	998.770	236.68
COF_2		1	−109 to −84	6.885 5	576.70	228.58
COS		1	−111 to −49	6.907 23	804.48	250.0
CS_2		1	3–80	6.942 79	1 169.11	241.59
CSe_2		1	0–50	6.776 73	1 353.20	219.95
CSeS		1	−16 to 84	6.699 6	1 161.97	219.59
Cesium						
Cs		2	200–350	6.949	3 833.7	
CsBr		2	978–1305	7.990	8 022.53	
CsCl		2	986–1295	8.340	8 523.94	
CsF		2	1033–1255	7.703	7 359.21	
CsH		2	245–378	11.79	5 900	
		2	340–440	9.25	4 410	
CsI		2	1052–1280	9.124	9 699.11	
Chlorine						
Cl_2	c	1		9.705 12	1 444.19	267.13
	liq	1		6.937 90	861.34	246.33
ClF	liq	1		6.989	682.1	256
ClF_3	liq	1		7.366 85	1 096.28	232.63
ClF_5		1		6.269 33	653.06	206.6
ClO_2	liq	1		6.036 11	590.09	176.15
Cl_2O	liq	1		7.132 68	1 021.56	238.16
$ClOClO_3$	liq	1		7.538 67	1 404.18	257.00
Cl_2O_7	liq	1		6.869 29	1 214.00	220.79
ClO_2F	liq	1		6.677 15	809.78	218.96
ClO_3F	liq	1		6.895 19	791.73	243.88
Copper						
CuBr		2	997–1351	5.460	4 173.2	
CuCl		2	878–1369	5.454	4 215.0	
CuI		2	991–1154	5.570	4 215.0	
Fluorine						
F_2	liq	1		6.765 88	304.35	266.54
FNO_3	liq	1		6.658 6	769.5	248.0
Germanium						
$GeCl_4$		2	10.4–86	7.340	2 010.9	
Helium						
^3He	liq	1	−271.13 to −270.86	4.272 7	5.594	273.840
	liq	1	−271.13 to −269.92	5.100 0	11.062	274.950
^4He		1	−271.4 to −270.1	4.558 7	8.1548	273.710
		1	−271.4 to −268.9	5.320 75	14.6515	274.950
		1	−271.4 to −268.1	6.004 60	24.0668	276.650

(*Continued*)

TABLE 1.46 Vapor Pressures of Various Inorganic Compounds (*Continued*)

Substance	State	Eq.	Range,°C	*A*	*B*	*C*
Hydrogen						
$^1H_2$ normal, 25% para	c	1		6.043 86	66.507	274.630
	liq	1		5.824 38	67.5078	275.700
equilibrium	c	1		6.042 07	65.961	274.60
	liq	1		5.814 64	66.7945	275.650
$^1H^2H$ (DH)	c	1		6.960 08	99.968	276.590
	liq	1		6.016 12	77.1349	275.620
$^2H_2$ (D_2) normal, 66.7% ortho	c	1		7.726 05	135.461	278.550
	liq	1		6.128 25	83.5251	275.216
$^2H_2$ equilibrium, 97.8% ortho	c	1		7.751 10	135.58	278.50
	liq	1		6.044 68	79.5888	274.680
$^3H_2$ (T_2) normal, 25% para	c	1		6.184 03	76.7445	271.850
	liq	1		6.089 21	81.8971	273.650
^{1}HBr	c	1		7.667 61	878.57	253.2
	liq	1		6.287 53	540.82	225.44
^{2}HBr (DBr)	c	1		7.500 93	820.68	247.3
	liq	1		6.162 38	505.68	220.6
^{1}HCl	c	1		8.134 73	941.57	268.06
	liq	1		7.170 00	745.80	258.88
^{2}HCl (DCl)	c	1		7.850 47	843.32	258.32
	liq	1		6.935 96	668.20	249.50
HCN	liq	1	−16 to 46	7.528 2	1329.5	260.4
^{1}HF	liq	1		7.680 98	1475.60	287.88
^{2}HF (DF)	liq	1		7.217 04	1268.37	273.87
^{1}HI	c	1		7.315 6	894.32	239.6
	liq	1		5.608 9	416.04	188.1
^{2}HI (DI)	c	1		7.314 9	889.52	238.8
	liq	1		5.601 8	413.98	187.8
HN_3	liq	1		6.857	1 066	232
HNO_3	liq	1		7.511 9	1 406	221.0
^1H_2O			[See Tables 5.4 and 5.6]			
^2H_2O (D_2O)			[See Table 5.7]			
$H_2{}^{18}O$		1	0–60	8.133 2	1 762.39	235.660
		1	60–120	7.972 08	1 668.84	227.700
H_2O_2	liq	1		7.969 17	1 886.76	220.6
HPO_2F	liq	1		6.735 3	1 342.9	232.0
H_2S	c	1		7.614 18	885.319	250.25
	liq	1		6.993 92	768.130	249.09
H_2S_2	liq	1		6.974	1 232	225
H_2S_3	liq	1		6.807	1 488	209
H_2S_4	liq	1		6.945	1 772	196
H_2S_5	liq	1		7.320	2 104	189
HSO_3Cl	liq	1		7.049	1 480	201
HSO_3F	liq	1		7.399 5	1 521	174.0
H_2Se	c	1		7.635 4	927.6	240.0
	liq	1		6.966 0	787.67	235.0
H_2Te	liq	1		7.000	935	229
Iodine						
I_2	c	1		9.810 9	2 901.0	256.00
	liq	1		7.018 1	1 610.9	205.0
ICl	liq	1		7.702 1	1 517.9	217.0
IF_5	c	1		10.964	2 538	245
	liq	1		7.464 8	1 460	216.0
IF_7	c	1		7.998	1 340	256

TABLE 1.46 Vapor Pressures of Various Inorganic Compounds (*Continued*)

Substance	State	Eq.	Range,°C	*A*	*B*	*C*
Iridium						
IrF_6	c	2	0.4–44	8.618	1 868	
	liq	2	44–54	7.952	1 657	
Iron						
$FeCl_2$	liq	2	708–834	9.794	7 455	
	liq	2	700–930	8.33	7 061	
$FeCl_3$	c	2	160–304	15.11	7 142	
FeI_2		2	517–577	13.183	10 778	
		2	601–686	9.674	7 716	
Krypton						
Kr	c	1		7.539 55	539.48	269.8
	liq	1		6.630 70	416.38	264.45
Lead						
Pb		2	525–1325	7.827	9 845.4	
$PbBr_2$		2	735–918	8.064	6 163.1	
$PbCl_2$		2	500–950	8.961	7 411.4	
PbF_2		2	1078–1289	8.391	8 623.2	
Lithium						
LiBr		2	1010–1265	8.068	7 975.5	
LiCl		2	1045–1325	7.939	8 142.7	
LiF		2	1398–1666	8.753	11 407	
LiH		2	500–650	11.227	9 600	
		2	700–800	9.926	8 204	
LiI		2	940–1140	8.011	7 500	
Magnesium						
Mg		2	900–1070	12.993	13 579.8	
MgH_2		2	337–415	9.78	3 857	
Mercury						
Hg				[See Table 5.3]		
$HgBr_2$		2	130–270	10.094	4 168.0	
$HgCl_2$		2	130–270	10.094	4 118.34	
		2	275–309	8.409	3 187.1	
Hg_2Cl_2		1		8.521 51	3 110.96	168.0
HgI_2		2	266–360	8.115	3 278.5	
Neon						
Ne	c	1		7.065 16	110.61	272.00
	liq	1		6.084 44	78.380	270.550
Neptunium						
NpF_6	liq	3	55.1–76.8	0.010 23	1 191.1	−2.582 5
Nickel						
$Ni(CO)_4$		2	2–40	7.780	1 556.5	
Niobium						
$NbBr_5$	liq	2		8.92	3 850	
$NbCl_5$	liq	2	210–254	8.37	2 827	
NbF_5	liq	2		8.439	2 824	
Nitrogen						
N_2 natural	c	1		7.345 12	322.222	269.980
	liq	1		6.494 57	255.680	266.550
$^{15}N_2$	c	1		7.363 96	323.17	269.88
	liq	1		6.494 14	255.535	266.451
NCl_3		1		6.956	1 190	221
NF_3	liq	1		6.779 66	501.913	257.79
NH_3				[See Table 1.49]		

(*Continued*)

TABLE 1.46 Vapor Pressures of Various Inorganic Compounds (*Continued*)

Substance	State	Eq.	Range,°C	*A*	*B*	*C*
Nitrogen (*cont.*)						
N_2H_4	liq	1		7.801 9	1 679.07	227.7
NO natural	c	1		9.628 26	758.736	266.00
	liq	1		8.743 00	682.938	268.27
N_2O	c	1		9.437 00	1 174.020	268.22
	liq	1		7.003 94	654.260	247.16
N_2O_4 equilibrium	c	1		10.736 31	2 075.53	252.80
mixture	liq	1		8.917 12	1 798.54	276.80
N_2O_5	c	1		11.644 5	2 510	253.0
NOCl	c	1		8.540 8	1 397.3	261.0
	liq	1		7.361 54	1 094.73	249.70
N_2O_3		2	−25 to 0	10.30	2 057.9	
NOF	liq	1		6.443 5	556.13	216.0
NO_2Cl	liq	1		5.372 3	395.40	174.0
NO_2F	liq	1		6.833 4	654.55	238.0
Osmium						
OsF_5		2	75–180	9.75	3 429	
OsF_6		2	34–48	7.470	1 473	
OsF_8		2	38–47	7.650	1 525	
OsO_4		2	−38 to 40	10.710 0	2 951.00	
OsO_3F_2		2	59–105	7.994	1 911	
Oxygen						
O_2	liq	1		6.691 44	319.013	266.697
O_3	liq	1		6.837	552.5	251.0
OF_2	liq	1		7.236 19	545.05	269.91
O_2F_2	liq	1		6.779 02	756.39	250.16
O_3F_2		2	79–114	6.134 3	675.57	
Palladium						
$PdCl_2$		2	680–857	6.32	5 032	
Phosphorus						
P red, V	subl c	1		11.060	5 323	220
white	subl c	1		6.936 9	1 907.6	190.0
P_4 black, o-rh		1		12.405	6 671	247
PBr_3	liq	1	−40 to 173	6.915 5	1 590.5	221.0
PBr_5	liq	1	to 104	6.948	1 320	214
$PBrF_2$	liq	1	−133 to −16	6.904 2	885.12	236.0
PBr_2F	liq	1	−115 to 78	6.858 0	1 210.3	226.0
PCl_3	liq	1	−92 to 76	6.826 7	1 196	227.0
PCl_5	c	1	to 160	10.206 8	2 903.1	237.0
	liq	1		7.033	1 490	200.0
$PClF_2$	liq	1	−165 to −47	6.639 6	780.88	255.0
PCl_2F	liq	1	−144 to 14	6.796 56	982.332	237.00
$P(OCN)_3$	liq	2	−2 to 169	8.745 5	2 595	
PF_3	liq	1	−152 to −101	6.860 4	620.22	257.0
PF_5	liq	1	−93.8 to −84.5	6.914 4	647.21	245.0
PH_3	c	1		7.482 35	794.496	265.20
	liq	1		6.715 59	645.512	256.066
P_2H_4	liq	1		6.862 8	1 137	227.0
P_4O_6	liq	1	24–175	6.716 37	1 412.8	193.0
P_4O_{10}	c III	1		9.707 0	3 822	201.0
	c I	1		10.843 2	6 424	213
	liq	1		6.935 2	3 069	152
$POBr_3$	liq	1	51–192	7.007 8	1 609.2	198.0
$POBrCl_2$	liq	1	31–165	6.924	1 411	213
POBrClF	liq	1		6.914	1 214	222

TABLE 1.46 Vapor Pressures of Various Inorganic Compounds (*Continued*)

Substance	State	Eq.	Range,°C	*A*	*B*	*C*
Phosphorus (*continued*)						
$POBrF_2$	liq	1	−85 to 32	7.101 9	1 118.9	233.0
$POBr_2F$	liq	1	−117 to 110	6.721 2	1 328.9	236.0
$POCl_3$	liq	1	1.2–105	6.865 8	1 297.2	220.0
$POClF_2$	liq	1	−96 to 3	6.926 6	946.96	231.0
$POCl_2F$	liq	1	−80 to 53	7.084 65	1 201.86	233.00
POF_3	c	1		10.930 5	1 783	261.0
	liq	1		7.115 5	810.1	231.0
$PO(OCN)_3$		2	5–193	9.168 2	2 931	
$PO(SCN)_3$		2	14–300	8.533 0	3 240	
P_4S_{10}		2		9.17	4 940	
$PSBr_3$	c	2		10.105	3 196.2	
	liq	2		8.338 3	2 641.9	
$PS(OCN)_3$		2		10.032	3 492	
Platinum						
Pt		2	1425–1765	7.786	25 384	
PtF_6	liq	1	61.3–81.7	89.15	5 686	27.49
Polonium						
Po	liq	1		7.041 4	5 017.6	241.0
$PoCl_4$	liq	1		7.554	2 360	115
Potassium						
K		2	260–760	7.183	4 434.33	
KBr		2	1095–1375	7.936	8 555.3	
KCl		2	1116–1418	8.130	8 863.4	
KF		2	1278–1500	9.000	10 838	
KOH		2	1170–1327	7.330	7 103.3	
KI		2	1063–1333	7.949	8 132.2	
Protactinium	liq	2		17.27	7 377	
Radon						
Rn	c	1		7.495 5	884.41	255.0
	liq	1		6.701 5	718.25	250.0
Rhenium						
ReF_5	c	2		9.024	3 037	
ReF_6	c	3	−3.45 to 18.5	9.123 0	1 765.4	0.1790
	liq	3	18.5–48	18.208 1	1 956.7	3.599
ReF_7	c	3	−14.5 to 48.3	13.043 2	2 205.8	1.470 3
	liq	3	48.3–74.6	−21.583 5	244.28	−9.908 3
ReO_2	c	2	650–785	11.65	14 437	
	liq	2	480–660	5.345	4 742	
ReO_3	c	2	325–420	15.16	10 882	
	liq	2	300–480	7.745	4 966	
Re_2O_7	liq	2	230–360	8.98	3 868	
$ReOF_4$	liq	2	108–172	10.09	3 206	
$ReOF_5$	liq	2	41–73	7.727	1 679	
ReS_2	c	2	500–700	3.214	4 976	
Re_2S_7	c	2	260–410	8.86	4 800	
Rubidium						
Rb		2	250–370	6.976	3 969.5	
RbCl		2	1142–1395	9.111	10 373	
RbF		2	1142–1400	8.570	9 568.4	
Ruthenium						
$RuOF_4$		2	120–160	8.60	2 616	
Selenium						
Se	liq	1		7.631 6	4 213.0	202.0
$SeCl_4$	c	1		10.250 9	3 068.8	225.0

(*Continued*)

TABLE 1.46 Vapor Pressures of Various Inorganic Compounds (*Continued*)

Substance	State	Eq.	Range,°C	*A*	*B*	*C*
Selenium (*Continued*)						
SeF_4	liq	1		7.888 7	1 603.0	215.0
SeF_6	c	1		8.385 4	1 121.4	250.0
SeO_2		1		6.577 81	1 879.81	179.0
$SeOCl_2$	liq	1		6.257 3	970.87	112.0
$SeOF_2$	liq	1		7.420	1 380	178
Silicon						
$SiCl_4$	liq	1	0–53	6.857 26	1 138.92	228.88
SiH_4		2	−160 to −112	6.881	645.9	
Si_2H_6		2	−115 to −14.6	7.258	1 133.4	
Si_3H_8		2	−70 to 52	7.676	1 559.1	
Silver						
AgCl		2	1255–1442	8.179	9 688.7	
Sodium						
Na		2	180–883	7.553	5 395.4	
NaCl		2	976–1155	8.329 7	9 417.07	
NaCI		2	1156–1430	8.548	9 704.3	
NaCN		2	800–1360	7.472	8 122.81	
NaF		2	1562–1701	8.640	11 396.6	
NaI		2	1063–1307	8.371	8 623.2	
NaOH		2	1010–1402	7.030	6 894	
Strontium						
Sr		2	940–1140	16.056	18 802.8	
Sulfur						
S equilibrium	liq	1		6.843 59	2 500.12	186.30
S_2Br_2	liq	1		7.177	1 660	185
SCl_2	liq	1		8.454	1 594	227
S_2Cl_2	liq	1		6.783 6	1 341	206.0
S_2F_2	liq	1		6.684	628	256
SF_4	liq	1		6.839 5	823.4	248.0
SF_6	c	1		8.416 0	1 096.5	262.0
S_2F_{10}	liq	1		7.067 6	1 100.6	234.0
SO_2	c	1		9.754 3	1 553.8	225.0
	liq	1		7.282 28	999.900	237.190
SO_3 "icelike"	c III	1		10.565 7	2 273.8	255.0
"woollike"	c II	1		11.590 1	2 665.6	264.0
	c I	1		14.255 9	3 692.1	273.0
	liq	1		9.050 85	1 735.31	236.50
$SOBr_2$	liq	1		7.056	1 445	206
$SOCl_2$	liq	1		7.287 45	1 446.7	252.7
SOClF	liq	1		7.173 1	1 100.1	244.00
SOF_2	liq	1		6.959 06	775.48	234.00
SOF_4	liq	1		7.071 8	840.3	249.0
$S_2O_2F_{10}$	liq	1		6.874	1 110	229
$S_2O_5Cl_2$	liq	1		7.019	1 460	202
S_2O_5ClF	liq	1		7.015 6	1 257.4	204.0
$S_2O_5F_2$	liq	1		6.881	1 120	229
$S_2O_5F_4$	liq	1		6.885	1 140	227
SO_2BrF	liq	1		7.142 8	1 155	231.0
SO_2Cl_2	liq	1		7.001 7	1 209	224.0
SO_2ClF	liq	1		6.521 5	793.73	210.70
SO_2F_2	liq	1		6.907 0	784.3	250
Tantalum						
$TaBr_5$	liq	2		8.11	3 260	
$TaCl_5$	liq	2	220–240	8.68	2 970	

TABLE 1.46 Vapor Pressures of Various Inorganic Compounds (*Continued*)

Substance	State	Eq.	Range,°C	*A*	*B*	*C*
Tantalum (*Continued*)						
TaF_5	liq	2		8.524	2 834	
TaI_5	liq	2		7.67	3 950	
Technetium						
TcF_6	liq	3	37.4–51.7	24.808 7	2 405	5.803 6
TcO_3F	liq	2	18.3–51.8	8.417	2 065	
Tc_2O_7	c	2		18.279	7 205	
	liq	2		8.999	3 571	
Tellurium						
Te	liq	1		7.301 0	5 370.6	221
$TeCl_4$	liq	1		7.558 6	2 355	115
TeF_6	liq	1		6.748 8	807.0	247.0
Te_2F_{10}	liq	1		6.901 8	1 150	227.0
TeO_2		2	450–733	12.328 4	13 222	
Thallium						
Tl		2	950–1200	6.1240	6 268	
TlF		2	282–298	12.52	5 484	
Thorium						
ThF_4	liq	2		10.821	15 270	
ThH_2		2	up to 883	9.50	7 650	
Tin						
$SnCl_4$		2	−52 to −38	9.824	2 441.23	
SnH_4		2	−148 to −49	7.400	999.68	
Titanium						
$TiCl_2$	subl c	2		9.30	8 500	
$TiCl_3$	subl c	2	455–550	10.401	8 296	
$TiCl_4$	liq	2	−23 to 136	7.683	1 964	
TiI_4	liq	2	160–360	7.577	3 054	
Tungsten						
W		2	2230–2770	9.920	46 850	
Uranium						
UF_6	liq	1	64–116	6.994 64	1 126.288	221.963
	liq	1	116–230	7.690 69	1 683.165	302.148
UH_3 dissociation		2	200–430	9.39	4 590	
U^2H_3 (UD_3)		2		9.43	4 500	
U^3H_3 (UT_3)		2		9.46	4 471	
Vanadium						
VBr_2	c	2	541–716	9.08	10 460	
	subl c	2	800–905	5.9	9 830	
VBr_3		2	314–427	11.12	7 470	
VCl_2	subl c	2	910–1100	5.725	9 721	
VCl_3		2	352–567	11.20	9 777	
VCl_4	liq	2	30–153	7.62	2 020	
VF_3	subl c	2	650–920	12.357	15 603	
VF_5	subl c	2	−20 to 19.5	8.168	2 608	
	liq	2	19.5–45.5	7.549	2 423	
VI_2	subl c	2	850–1016	2.56	5 600	
$VOCl_3$	liq	2	15.4–125	7.69	1 920	
Xenon						
Xe	c	1		7.484 5	714.896	264.0
	liq	1		6.642 89	566.282	258.660
XeF_2	subl c	1		10.019 47	2 683.96	261.68
XeF_4	subl c	1		10.913 87	3 095.06	269.56
Zinc						
Zn	c	2	250–419	9.200	6 946.6	

TABLE 1.47 Vapor Pressure of Mercury

Temp. °C	mm of Hg	Temp. °C	mm of Hg	Temp. °C	mm of Hg
0	0.000 185	92	0.1769	184	10.116
2	0.000 228	94	0.1976	186	10.839
4	0.000 276	96	0.2202	188	11.607
6	0.000 335	98	0.2453	190	12.423
8	0.000 406	100	0.2729	192	13.287
10	0.000 490	102	0.3032	194	14.203
12	0.000 588	104	0.3366	196	15.173
14	0.000 706	106	0.3731	198	16.200
16	0.000 846	108	0.4132	200	17.287
18	0.001 009	110	0.4572	202	18.437
20	0.001 201	112	0.5052	204	19.652
22	0.001 426	114	0.5576	206	20.936
24	0.001 691	116	0.6150	208	22.292
26	0.002 000	118	0.6776	210	23.723
28	0.002 359	120	0.7457	212	25.233
30	0.002 777	122	0.8198	214	26.826
32	0.003 261	124	0.9004	216	28.504
34	0.003 823	126	0.9882	218	30.271
36	0.004 471	128	1.084	220	32.133
38	0.005 219	130	1.186	222	34.092
40	0.006 079	132	1.298	224	36.153
42	0.007 067	134	1.419	226	38.318
44	0.008 200	136	1.551	228	40.595
46	0.009 497	138	1.692	230	42.989
48	0.010 98	140	1.845	232	45.503
50	0.012 67	142	2.010	234	48.141
52	0.014 59	144	2.188	236	50.909
54	0.016 77	146	2.379	238	53.812
56	0.019 25	148	2.585	240	56.855
58	0.022 06	150	2.807	242	60.044
60	0.025 24	152	3.046	244	63.384
62	0.028 83	154	3.303	246	66.882
64	0.032 87	156	3.578	248	70.543
66	0.037 40	158	3.873	250	74.375
68	0.042 51	160	4.189	252	78.381
70	0.048 25	162	4.528	254	82.568
72	0.054 69	164	4.890	256	86.944
74	0.061 89	166	5.277	258	91.518
76	0.069 93	168	5.689	260	96.296
78	0.078 89	170	6.128	262	101.28
80	0.088 80	172	6.596	264	106.48
82	0.100 0	174	7.095	266	111.91
84	0.112 4	176	7.626	268	117.57
86	0.126 1	178	8.193	270	123.47
88	0.1413	180	8.796	272	129.62
90	0.1582	182	9.436	274	136.02

TABLE 1.47 Vapor Pressure of Mercury (*Continued*)

Temp. °C	mm of Hg	Temp. °C	mm of Hg	Temp. °C	mm of Hg
276	142.69	332	478.13	388	1299.1
278	149.64	334	497.12	390	1341.9
280	156.87	336	516.74	392	1386.1
282	164.39	338	537.00	394	1431.3
284	172.21	340	557.90	396	1477.7
286	180.34	342	579.45	398	1525.2
288	188.79	344	601.69	400	1574.1
290	197.57	346	624.64	430	2464
292	206.70	348	648.30	460	3715
294	216.17	350	672.69	490	5420
296	226.00	352	697.83	520	7691
298	236.21	354	723.73	550	10650
300	246.80	356	750.43	600	22.87 atm
302	257.78	358	777.92	650	35.49 atm
304	269.17	360	806.23	700	52.51 atm
306	280.98	362	835.38	750	74.86 atm
308	293.21	364	865.36	800	103.31 atm
310	305.89	366	896.23	850	138.42 atm
312	319.02	368	928.02	900*	180.92 atm
314	332.62	370	960.66	950	226.58 atm
316	346.70	372	994.34	1000	290.5 atm
318	361.26	374	1028.9	1050	358.1 atm
320	376.33	376	1064.4	1100	437.3 atm
322	391.92	378	1100.9	1150	521.3 atm
324	408.04	380	1138.4	1200	616.8 atm
326	424.71	382	1177.0	1250	721.4 atm
328	441.94	384	1216.6	1300	835.9 atm
330	459.74	386	1257.3		

*Critical point.

TABLE 1.48 Vapor Pressure of Ice in Millimeters of Mercury

For temperatures from –99 to 0°C.

The values in the table are for ice in contact with its own vapor. Where the ice is in contact with air at a temperature t°C, this correction must be added: Correction = $20p/(100)(t + 273)$.

t, °C	p, mm Hg	t, °C	p, mm Hg	t, °C	p, mm Hg
−99	0.000 012	−51	0.026 1	−16.5	1.080
−98	0.000 015	−50	0.029 6	−16.0	1.132
−97	0.000 018	−49	0.033 4	−15.5	1.186
−96	0.000 022	−48	0.037 8	−15.0	0.241
−95	0.000 027	−47	0.042 6	−14.5	1.300
−94	0.000 033	−46	0.048 1	−14.0	1.361
−93	0.000 040	−45	0.054 1	−13.5	1.424
−92	0.000 048	−44	0.060 9	−13.0	1.490
−91	0.000 058	−43	0.068 4	−12.5	1.559
−90	0.000 070	−42	0.076 8	−12.0	1.632
−89	0.000 084	−41	0.086 2	−11.5	1.707
−88	0.000 10	−40	0.096 6	−11.0	1.785
−87	0.000 12	−39	0.108 1	−10.5	1.866
−86	0.000 14	−38	0.120 9	−10.0	1.950
−85	0.000 17	−37	0.135 1	−9.8	1.985
−84	0.000 20	−36	0.150 7	−9.6	2.021
−83	0.000 24	−35	0.168 1	−9.4	2.057
−82	0.000 29	−34	0.187 3	−9.2	2.093
−81	0.000 34	−33	0.208 4	−9.0	2.131
−80	0.000 40	−32	0.231 8	−8.8	2.168
−79	0.000 47	−31	0.257 5	−8.6	2.207
−78	0.000 56	−30.0	0.285 9	−8.4	2.246
−77	0.000 66	−29.5	0.301	−8.2	2.285
−76	0.000 77	−29.0	0.317	−8.0	2.326
−75	0.000 90	−28.5	0.334	−7.8	2.367
−74	0.001 05	−28.0	0.351	−7.6	2.408
−73	0.001 23	−27.5	0.370	−7.4	2.450
−72	0.001 43	−27.0	0.389	−7.2	2.493
−71	0.001 67	−26.5	0.409	−7.0	2.537
−70	0.001 94	−26.0	0.430	−6.8	2.581
−69	0.002 25	−25.5	0.453	−6.6	2.626
−68	0.002 61	−25.0	0.476	−6.4	2.672
−67	0.003 02	−24.5	0.500	−6.2	2.718
−66	0.003 49	−24.0	0.526	−6.0	2.765
−65	0.004 03	−23.5	0.552	−5.8	2.813
−64	0.004 64	−23.0	0.580	−5.6	2.862
−63	0.005 34	−22.5	0.609	−5.4	2.912
−62	0.006 14	−22.0	0.640	−5.2	2.962
−61	0.007 03	−21.5	0.672	−5.0	3.013
−60	0.008 08	−21.0	0.705	−4.8	3.065
−59	0.009 25	−20.5	0.740	−4.6	3.117
−58	0.010 6	−20.0	0.776	−4.4	3.171
−57	0.012 1	−19.5	0.814	−4.2	3.225
−56	0.013 8	−19.0	0.854	−4.0	3.280
−55	0.015 7	−18.5	0.895	−3.8	3.336
−54	0.017 8	−18.0	0.939	−3.6	3.393
−53	0.020 3	−17.5	0.984	−3.4	3.451
−52	0.023 0	−17.0	1.031	−3.2	3.509

TABLE 1.48 Vapor Pressure of Ice in Millimeters of Mercury (*Continued*)

t, °C	p, mm Hg	t, °C	p, mm Hg	t, °C	p, mm Hg
−3.0	3.568	−1.8	3.946	−0.8	4.287
−2.8	3.360	−1.6	4.012	−0.6	4.359
−2.6	3.691	−1.4	4.079	−0.4	4.431
−2.4	3.753	−1.2	4.147	−0.2	4.504
−2.2	3.816	−1.0	4.217	0.0	4.579
−2.0	3.880				

TABLE 1.49 Vapor Pressure of Liquid Ammonia, NH_3

t°C.	p in atm	t°C.	p in atm	t°C.	p in atm
–78	0.0582	–6	3.3677	66	29.784
–76	0.0683	–4	3.6405	68	31.211
–74	0.0797	–2	3.9303	70	32.687
–72	0.0929	0	4.2380	72	34.227
–70	0.1078	+2	4.5640	74	35.813
–68	0.1246	4	4.9090	76	37.453
–66	0.1437	6	5.2750	78	39.149
–64	0.1651	8	5.6610	80	40.902
–62	0.1891	10	6.0685	82	42.712
–60	0.2161	12	6.4985	84	44.582
–58	0.2461	14	6.9520	86	46.511
–56	0.2796	16	7.4290	88	48.503
–54	0.3167	18	7.9310	90	50.558
–52	0.3578	20	8.4585	92	52.677
–50	0.4034	22	9.0125	94	54.860
–48	0.4536	24	9.5940	96	57.111
–46	0.5087	26	10.2040	98	59.429
–44	0.5693	28	10.8430	100	61.816
–42	0.6357	30	11.512	102	64.274
–40	0.7083	32	12.212	104	66.804
–38	0.7875	34	12.943	106	69.406
–36	0.8738	36	13.708	108	72.084
–34	0.9676	38	14.507	110	74.837
–32	1.0695	40	15.339	112	77.668
–30	1.1799	42	16.209	114	80.578
–28	1.2992	44	17.113	116	83.570
–26	1.4281	46	18.056	118	86.644
–24	1.5671	48	19.038	120	89.802
–22	1.7166	50	20.059	122	93.045
–20	1.8774	52	21.121	124	96.376
–18	2.0499	54	22.224	126	99.796
–16	2.2349	56	23.372	128	103.309
–14	2.4328	58	24.562	130	106.913
–12	2.6443	60	25.797	132	110.613
–10	2.8703	62	27.079	132.3	111.3(c.p.)
-8	3.1112	64	28.407		

TABLE 1.50 Vapor Pressure of Water

For temperatures from –10 to 120°C.

The values in the table are for water in contact with its own vapor. Where the water is in contact with air at a temperature t in degrees. Celsius, the following correction must be added: Correction (when $t \leq 40°C$) = $p(0.775 - 0.000\ 313t)/100$; correction (when $t > 50°C$) = $p(0.0652 - 0.000\ 087\ 5t)/100$.

t, °C	p, mm Hg	t, °C	p, mm Hg	t, °C	p, mm Hg	t, °C	p, mm Hg
−10.0	2.149	13.0	11.231	23.4	21.583	32.6	36.891
−9.5	2.236	13.5	11.604	23.6	21.845	32.8	37.308
−9.0	2.326	14.0	11.987	23.8	22.110	33.0	37.729
−8.5	2.418	14.5	12.382	24.0	22.387	33.2	38.155
−8.0	2.514	15.0	12.788	24.2	22.648	33.4	38.584
−7.5	2.613	15.2	12.953	24.4	22.922	33.6	39.018
−7.0	2.715	15.4	13.121	24.6	23.198	33.8	39.457
−6.5	2.822	15.6	13.290	24.8	23.476	34.0	39.898
−6.0	2.931	15.8	13.461	25.0	23.756	34.2	40.344
−5.5	3.046	16.0	13.634	25.2	24.039	34.4	40.796
−5.0	3.163	16.2	13.809	25.4	24.326	34.6	41.251
−4.5	3.284	16.4	13.987	25.6	24.617	34.8	41.710
−4.0	3.410	16.6	14.166	25.8	24.912	35.0	42.175
−3.5	3.540	16.8	13.347	26.0	25.209	35.2	42.644
−3.0	3.673	17.0	14.530	26.2	25.509	35.4	43.117
−2.5	3.813	17.2	14.715	26.4	25.812	35.6	43.595
−2.0	3.956	17.4	14.903	26.6	26.117	35.8	44.078
−1.5	4.105	17.6	15.092	26.8	26.426	36.0	44.563
−1.0	4.258	17.8	15.284	27.0	26.739	36.2	45.054
−0.5	4.416	18.0	15.477	27.2	27.055	36.4	45.549
0.0	4.579	18.2	15.673	27.4	27.374	36.6	46.050
0.5	4.750	18.4	15.871	27.6	27.696	36.8	46.556
1.0	4.926	18.6	16.071	27.8	28.021	37.0	47.067
1.5	5.107	18.8	16.272	28.0	28.349	37.2	47.582
2.0	5.294	19.0	16.477	28.2	28.680	37.4	48.102
2.5	5.486	19.2	16.685	28.4	29.015	37.6	48.627
3.0	5.685	19.4	16.894	28.6	29.354	37.8	49.157
3.5	5.889	19.6	17.105	28.8	29.697	38.0	49.692
4.0	6.101	19.8	17.319	29.0	30.043	38.2	50.231
4.5	6.318	20.0	17.535	29.2	30.392	38.4	50.774
5.0	6.543	20.2	17.753	29.4	30.745	38.6	51.323
5.5	6.775	20.4	17.974	29.6	31.102	38.8	51.879
6.0	7.013	20.6	18.197	29.8	31.461	39.0	52.442
6.5	7.259	20.8	18.422	30.0	31.824	39.2	53.009
7.0	7.513	21.0	18.650	30.2	32.191	39.4	54.580
7.5	7.775	21.2	18.880	30.4	32.561	39.6	54.156
8.0	8.045	21.4	19.113	30.6	32.934	39.8	54.737
8.5	8.323	21.6	19.349	30.8	33.312	40.0	55.324
9.0	8.609	21.8	19.587	31.0	33.695	40.5	56.81
9.5	8.905	22.0	19.827	31.2	34.082	41.0	58.34
10.0	9.209	22.2	20.070	31.4	34.471	41.5	59.90
10.5	9.521	22.4	20.316	31.6	34.864	42.0	61.50
11.0	9.844	22.6	20.565	31.8	35.261	42.5	63.13
11.5	10.176	22.8	20.815	32.0	35.663	43.0	64.80
12.0	10.518	23.0	21.068	32.2	36.068	43.5	66.51
12.5	10.870	23.2	21.324	32.4	36.477	44.0	68.26

TABLE 1.50 Vapor Pressure of Water (*Continued*)

t, °C	p, mm Hg	t, °C	p, mm Hg	t, °C	p, mm Hg	t, °C	p, mm Hg
44.5	70.05	63.0	171.38	81.5	377.3	97.0	682.07
45.0	71.88	63.5	175.35	82.0	384.9	97.2	687.04
45.5	73.74	64.0	179.31	82.5	392.8	97.4	692.05
46.0	75.65	64.5	183.43	83.0	400.6	97.6	697.10
46.5	77.61	65.0	187.54	83.5	408.7	97.8	702.17
47.0	79.60	65.5	191.82	84.0	416.8	98.0	707.27
47.5	81.64	66.0	196.09	84.5	425.2	98.2	712.40
48.0	83.71	66.5	200.53	85.0	433.6	98.4	717.56
48.5	85.85	67.0	204.96	85.5	442.3	98.6	722.75
49.0	88.02	67.5	209.57	86.0	450.9	98.8	727.98
49.5	90.24	68.0	214.17	86.5	459.8	99.0	733.24
50.0	92.51	68.5	218.95	87.0	468.7	99.2	738.53
50.5	94.86	69.0	223.73	87.5	477.9	99.4	743.85
51.0	97.20	69.5	228.72	88.0	487.1	99.6	749.20
51.5	99.65	70.0	233.7	88.5	496.6	99.8	754.58
52.0	102.09	70.5	238.8	89.0	506.1	100.0	760.00
52.5	104.65	71.0	243.9	89.5	515.9	101.0	787.57
53.0	107.20	71.5	249.3	90.0	525.76	102.0	815.86
53.5	109.86	72.0	254.6	90.5	535.83	103.0	845.12
54.0	112.51	72.5	260.2	91.0	546.05	104.0	875.06
54.5	115.28	73.0	265.7	91.5	556.44	105.0	906.07
55.0	118.04	73.5	271.5	92.0	566.99	106.0	937.92
55.5	120.92	74.0	277.2	92.5	577.71	107.0	970.60
56.0	123.80	74.5	283.2	93.0	588.60	108.0	1004.42
56.5	126.81	75.0	289.1	93.5	599.66	109.0	1038.92
57.0	129.82	75.5	295.3	94.0	610.90	110.0	1074.56
57.5	132.95	76.0	301.4	94.5	622.31	111.0	1111.20
58.0	136.08	76.5	307.7	95.0	633.90	112.0	1148.74
58.5	139.34	77.0	314.1	95.2	638.59	113.0	1187.42
59.0	142.60	77.5	320.7	95.4	643.30	114.0	1227.25
59.5	145.99	78.0	327.3	95.6	648.05	115.0	1267.98
60.0	149.38	78.5	334.2	95.8	652.82	116.0	1309.94
60.5	152.91	79.0	341.0	96.0	657.62	117.0	1352.95
61.0	156.43	79.5	348.1	96.2	662.45	118.0	1397.18
61.5	160.10	80.0	355.1	96.4	667.31	119.0	1442.63
62.0	163.77	80.5	362.4	96.6	672.20	120.0	1489.14
62.5	167.58	81.0	369.7	96.8	677.12		

TABLE 1.51 Vapor Pressure of Deuterium Oxide

t, °C	p, mm Hg	t, °C	p, mm Hg	t, °C	p, mm Hg
0	3.65	20	15.2	80	331.6
1	3.93	30	28.0	90	495.5
2	4.29	40	49.3	100	722.2
3	4.65	50	83.6	101.43	760.0
3.8	5.05	60	136.6		
10	7.79	70	216.1		

1.12 VISCOSITY AND SURFACE TENSION

Viscosity is the shear stress per unit area at any point in a confined fluid divided by the velocity gradient in the direction perpendicular to the direction of flow. If this ratio is constant with time at a given temperature and pressure for any species, the fluid is called a Newtonian fluid.

The *absolute viscosity* (μ) is the sheer stress at a point divided by the velocity gradient at that point. The most common unit is the poise (1 kg/m sec) and the SI unit is the Pa.sec (1 kg/m sec). As many common fluids have viscosities in the hundredths of a poise the centipoise (cp) is often used. One centipoise is then equal to one mPa sec.

The *kinematic viscosity* (v) is ratio of the absolute viscosity to density at the same temperature and pressure. The most common unit corresponding to the poise is the stoke (1 cm^2/sec) and the SI unit is m^2/sec.

The molecules in a gas-liquid interface are in tension and tend to contract to a minimum surface area. This tension may be quantified by the surface tension (σ), which is the force in the plane of the surface per unit length.

TABLE 1.52 Viscosity and Surface Tension of Inorganic Substances

For the majority of compounds the dependence of the surface tension γ on the temperature can be given as:

$$\gamma = a - bt$$

where a and b are constants and t is the temperature in degrees Celsius. The values of the dipole moment are for the gas phase.

Substance	Viscosity, $mN \cdot s \cdot m^{-2}$	Surface tension $mN \cdot m^{-1}$	
		a	b
Air	0.0182^{20}, 0.0231^{127}		
$AlBr_3$			
Ar			
(g)	0.0233^{20}, 0.0288^{127}		
(lq)		34.28	0.2493
$AsBr_3$		54.41	0.1043
$AsCl_3$		41.67	0.097 81
AsH_3 (arsine)			
BBr_3		31.90	0.1280
BCl_3			
BF_3	0.0171^{27}, 0.0217^{127}	−2.92	0.2030
B_2H_6 (diborane)		−3.13	0.1783
B_4H_{10}			
B_5H_9			
B_6H_{10}			
$B_3H_6N_3$			
Br_2 (g)			
(lq)	1.252^{0}, 1.03^{16}, 0.744^{25}	45.5	0.1820
BrF_3	2.22^{20}	38.30	0.0999
BrF_5	0.62^{24}	25.24	0.1098

TABLE 1.52 Viscosity and Surface Tension of Inorganic Substances (*Continued*)

Substance	Viscosity, $mN \cdot s \cdot m^{-2}$	Surface tension $mN \cdot m^{-1}$	
		a	*b*
Cl_2 (g)	0.0132^{20}		
(lq)		19.87	0.1897
ClF_3	0.48^{12}	26.9	0.1660
ClF_5			
ClO_3F		12.24	0.1576
CO (g)	0.0175^{20}, 0.0221^{127}		
(lq)		−30.20	0.2073
CO_2 (g)	0.0147^{20}, 0.0197^{127}		
(lq)	0.071^{20}	6.14^{-10}	2.67^{10}
$COCl_2$		22.59	0.1456
COF_2			
COS		12.12	0.1779
COSe			
CS			
CS_2 (g)			
(lq)	0.429^{0}, 0.375^{20}, 0.352^{25}	35.29	0.1484
CrO_2Cl_2			
D_2 (deuterium)	0.0126^{27}, 0.0154^{127}		
DH		6.537	0.1883
D_2O	0.0111^{25} (g), 1.098^{25} (lq)	71.72^{20}	68.38^{40}
F_2		−16.10	0.1646
$GaCl_3$		35.0	0.1000
$GeBr_4$		35.51^{30}	33.70^{50}
$GeBr_4$		35.51^{30}	33.70^{50}
$GeCl_4$		22.44^{30}	
$GeClH_3$			
H_2 (g)	0.0088^{20},		
t	0.109^{127}		
(lq)		2.80^{-258}	2.12^{-254}
HBr (g)			
(lq)	0.83^{-67}	13.10	0.2079
He (g)	0.0196^{27}, 0.0244^{27}		
(lq) (II)		$0.351^{0.50\ K}$	$0.317^{2.00\ K}$
(III)		$0.151^{3.61\ K}$	$0.131^{1.13\ K}$
(IV)		$0.372^{0.50\ K}$	$0.354^{1.40\ K}$
HCl (g)	0.0146^{27}, 0.0197^{127}		
(lq)	0.51^{-95}		

(*Continued*)

TABLE 1.52 Viscosity and Surface Tension of Inorganic Substances (*Continued*)

Substance	Viscosity, $mN \cdot s \cdot m^{-2}$	Surface tension $mN \cdot m^{-1}$	
		a	*b*
HClO			
HCN	0.235^{0}, 0.206^{18}, 0.183^{25}	19.45^{10}	18.33^{20}
HCNO (iso-cyanate)			
HCNS			
HF	0.256^{0}	10.41	0.078 67
HFO			
HI (g)			
(lq)			
HN_3 (azide)			
H_2O (see Table 5.19)			
H_2O_2	1.25^{20}	78.97	0.1549
HNO_3			
H_2S (g)			
(lq)	0.412^{0}	48.95	0.1758
H_2Se		22.32	0.1482
HSO_3Cl	2.43^{20}		
HSO_3F	1.56^{25}		
H_2SO_4	24.54^{25}		
H_2Te		29.03	0.2619
Hg	1.552^{20}, 1.526^{25}, 1.402^{50}	490.6	0.2049
I_2	1.98^{116}		
IBr			
IF			
IF_5		33.16	0.1318
IF_7			
IOF_5			
Kr (g)	0.0250^{20}, 0.0331^{127}		
(lq)		40.576 (in K)	0.2890 (in K)
Mn_2O_7			
Ne (g)	0.0303^{20}, 0.0389^{127}		
(lq)			
N_2 (g)	0.0176^{20}, 0.0222^{127}		
(lq)		26.42 (in K)	0.2265 (in K)
NH_3 (g)			
(lq)	$0.254^{-33.5}$	37.91^{-50}	35.38^{-40}
N_2H_4 (hydra-zine)	0.97^{20}, 0.876^{25}, 0.628^{50}	72.41	0.2407
$Ni(CO)_4$		18.11	0.1117
NO	0.0192^{27}, 0.0238^{127}	−67.48	0.5853

TABLE 1.52 Viscosity and Surface Tension of Inorganic Substances (*Continued*)

Substance	Viscosity, $mN \cdot s \cdot m^{-2}$	Surface tension $mN \cdot m^{-1}$	
		a	*b*
N_2O (g)	0.0146^{20}, 0.0194^{127}		
(lq)		5.09	0.2032
NO_2	0.532^{0}, 0.402^{25}		
N_2O_4			
N_2O_3			
NOBr			
NOCl		29.49	0.1493
NO_2Cl			
NOF		14.00	0.1165
NO_2F		8.26	0.1854
NO_3			
O_2 (g)	0.0204^{20}, 0.0261^{127}		
(lq)		−33.72	0.2561
O_3		38.1^{-183}	
OF_2			
O_2F_2 (FOOF)			
OsO_4			
P (lq)			
PBr_3		45.34	0.1283
PCl_3	0.662^{0}, 0.529^{25}, 0.439^{50}	31.14	0.1266
PCl_5			
PCl_2F_3			
PCl_3F_2			
PCl_4F			
PF_3			
PF_5			
PH_3			
PI_3		61.66	0.067 71
PO_3		40.44	0.1158
$POCl_3$	1.065^{25}	35.22	0.1275
POF_3			
$PSCl_3$		37.00	0.1272
PSF_3			
$PbCl_4$			
ReO_2Cl_3		57.00	0.2485
ReO_3Cl		54.05	0.1979
S			
SCl_2			
S_2Cl_2 dimer		46.23	0.1464
S_2F_2			
FSSF isomer			
$S{=}SF_2$ isomer			
SF_4		12.87	0.1734
SF_6	0.0153^{27}, 0.0198^{127}	5.66	0.1190

(*Continued*)

TABLE 1.52 Viscosity and Surface Tension of Inorganic Substances (*Continued*)

Substance	Viscosity, mN · s · m^{-2}	Surface tension mN · m^{-1}	
		a	*b*
S_2F_{10}			
SO_2 (g)	0.0129^{27}, 0.0175^{127}		
(lq)		26.58	0.1948
SO_3			
$SOBr_2$		46.28	0.0750
$SOCl_2$		36.10	0.1416
SOF_2			
SO_2Cl_2		32.10	0.1328
SO_2F_2			
$SbCl_3$		47.87	0.1238
$SbCl_5$			
SbF_5		49.07	0.1937
SbH_3			
Se (lq)			
SeF_4		38.61	0.1274
SeF_6			
$SeOCl_2$			
SeO_2			
$SiCl_4$	99.4^{25}, 96.2^{50}	20.78	0.099 62
SiF_4			
SiH_4			
$SiHCl_3$	0.415^{0}, 0.326^{25}	20.43	0.1076
SiH_3Cl			
$SnBr_4$			
$SnCl_4$		29.92	0.1134
TeF_6			
$TiCl_4$		33.54^{20}	31.06^{40}
UF_6 (g)			
(lq)		25.5	0.1240
VCl_4			
$VOBr_3$			
$VOCl_3$		36.36^{20}	33.60^{40}
Xe (g)	0.0228^{20}, 0.030^{127}		
(lq, II)		$0.345^{1.00\ K}$	$0.317^{2.00\ K}$
XeF_6			

1.13 THERMAL CONDUCTIVITY

The thermal conductivity is a measure of the effectiveness of a material as a thermal insulator. The energy transfer rate through a body is proportional to the temperature gradient across the body and the cross sectional area of the body. In the limit of infinitesimal thickness and temperature difference, the fundamental law of heat conduction is:

$$Q = \lambda A dT/dx$$

where Q is the heat flow, A is the cross-sectional area, dT/dx is the temperature/thickness gradient, and λ is the thermal conductivity.

A substance with a large thermal conductivity value is a good conductor of heat; one with a small thermal conductivity value is a poor heat conductor i.e. a good insulator.

TABLE 1.53 Thermal Conductivity of the Elements

Element number	Element symbol	Thermal conductivity (W/m)/K 27°C, 81°F	Element number	Element symbol	Thermal conductivity (W/m)/K 27°C, 81°F
1	H	0.1815	2	He	0.152
3	Li	84.7	4	Be	200
5	B	27	6	C	155
7	N	0.02598	8	O	0.02674
9	F	0.0279	10	Ne	0.0493
11	Na	141	12	Mg	156
13	Al	237	14	Si	148
15	P	0.235	16	S	0.269
17	Cl	0.0089	18	Ar	0.0177
19	K	102.5	20	Ca	200
21	Sc	15.8	22	Ti	21.9
23	V	30.7	24	Cr	93.7
25	Mn	7.82	26	Fe	80.2
27	Co	100	28	Ni	90.7
29	Cu	401	30	Zn	116
31	Ga	40.6	32	Ge	59.9
33	As	50	34	Se	2.04
35	Br	0.122	36	Kr	0.00949
37	Rb	58.2	38	Sr	35.3
39	Y	17.2	40	Zr	22.7
41	Nb	53.7	42	Mo	138
43	Tc	50.6	44	Ru	117
45	Rh	150	46	Pd	71.8
47	Ag	429	48	Cd	96.8
49	In	81.6	50	Sn	66.6
51	Sb	24.3	52	Te	2.35
53	I	0.449	54	Xe	0.00569
55	Cs	35.9	56	Ba	18.4
57	La	13.5	58	Ce	11.4
59	Pr	12.5	60	Nd	16.5
61	Pm	17.9	62	Sm	13.3
63	Eu	13.9	64	Gd	10.6
65	Tb	11.1	66	Dy	10.7
67	Ho	16.2	68	Er	14.3

TABLE 1.54 Thermal Conductivity of Various Solids

All values of thermal conductivity, *k*, are in millijoules $cm^{-1} \cdot s^{-1} \cdot K^{-1}$. To convert to $mW \cdot m^{-1} \cdot K^{-1}m$, divide values by 10. For values in millicalories, divide by 4.184.

Substance	t, °C	*k*
Asphalt	20	7.447
Basalt	20	21.76
Bauxite	600	5.56
Boiler scale	66	13.1
Brick, common	20	6.3
Blotting paper	20	0.628
Cardboard	20	2.1
Cement, Portland	90	2.97
Chalk	20	9.2
Chemical elements, *see* Table 4.1		
Coal	0	1.69
Concrete	20	9.2
Cork, sp. grav. = 0.2	30	0.54
Cork meal	100	0.556
Cotton, sp. grav. = 0.081	0	0.569
Diatomaceous earth	20	0.54
Ebonite	0	1.58
Eiderdown	20	0.046
Feathers (with air)	9	0.238
Feldspar	20	23.4
Felt (dark gray)	40	0.623
Fire brick	20	4.6
Flannel	60	0.148
Flint	20	10.0
Glass, crown	12.5	6.82
flint	12.5	5.98
Jena	22	9.50
quartz	0	13.89
	100	19.12
soda	20	7.1
	100	7.5
Granite	20	34.2
Graphite, sp. grav. = 1.58	50	441.4
Graphite powder, sp. grav. = 0.7	40	11.92
Gypsum	0	13.0
Horse hair, sp. grav. = 0.172	20	0.510
Ice		23.8
Leather, cowhide	84	1.76
Linen	20	0.879
Magnesia brick	20	11.3
	1130	30.1
Marble, white		32.6
Mica	41	3.60
Naphthalene	0	3.77
Paper	20	1.3
Paraffin	0	2.88
Plaster of Paris	20	2.93
Porcelain	95	10.38
Quartz, parallel to axis	0	136.0
	100	90.0

TABLE 1.54 Thermal Conductivity of Various Solids (*Continued*)

Substance	t, °C	*k*
Quartz, perpendicular to axis	0	72.43
	100	55.77
Plastics, *see* Section 10		
Roofing paper	0	1.90
Rubber, natural and synthetic, *see* Section 10		
Sand, dry	20	3.89
Sandstone, sp. grav. = 2.259	40	18.37
Silk, sp. grav. = 0.101	0	0.510
Slate	20	19.66
Soil, dry	20	1.38
Wax, bees	20	0.866
Wood, maple, parallel to face	20	4.25
perpendicular to face	50	1.82
Wood, oak, parallel to face	15	3.49
perpendicular to face	15	2.09
Wood, pine, parallel to face	20	3.49
perpendicular to face	15	1.51

1.14 CRITICAL PROPERTIES

Critical temperature (T^c), critical pressure (P_c), and critical volume (V_c) represent three widely used pure component constants. These critical constants are very important properties in chemical engineering field because almost all other thermo chemical properties are predictable from boiling point and critical constants with using corresponding state theory. Therefore, precise prediction of critical constants is very necessary.

1.14.1 Critical Temperature

The critical temperature of a compound is the temperature above which a liquid phase cannot be formed, no matter what the pressure on the system. The critical temperature is important in determining the phase boundaries of any compound and is a required input parameter for most phase equilibrium thermal property or volumetric property calculations using analytic equations of state or the theorem of corresponding states. Critical temperatures are predicted by various empirical methods according to the type of compound or mixture being considered.

Another somewhat simpler method for estimating the critical temperature of pure compounds requires the normal boiling point, the relative density, and the compound family.

$$\log Tc = A + B \log_{10} (\text{relative density}) + C \log T_b$$

where T_c and T_b are the critical and normal boiling temperatures, respectively, expressed in degrees Kelvin. The relative density of the liquid at 15°C is 0.1 MPa. The regression constants *A*, *B*, and C are available by family (Table 2-384).

For pure inorganic compounds, the method only requires the normal boiling point as input.

$$T_c = 1.64T_b$$

1.14.2 Critical Pressure

The critical pressure of a compound is the vapor pressure of that compound at the critical temperature. Below the critical temperature, any compound above its vapor pressure will be a liquid.

1.14.3 Critical Volume

The critical volume of a compound is the volume occupied by a specified mass of a compound at its critical temperature and critical pressure.

1.14.4 Critical Compressibility Factor

The critical compressibility factor of a compound is calculated from the experimental or predicted values of the critical properties.

$$Z_c = (P_c V_c)/(RT_c)$$

Critical compressibility factors are used as characterization parameters in corresponding states methods to predict volumetric and thermal properties. The factor varies from approximately 0.23 for water to 0.26-0.28 for most hydrocarbons to above 0.30 for light gases.

TABLE 1.55 Critical Properties

Substance	T_c, °C	P_c, atm	P_c, MPa	V_c, $cm^3 \cdot mol^{-1}$	ρ_c, $g \cdot cm^{-3}$
Air	−140.6	37.2	3.77	92.7	0.313
Aluminum tribromide	490	28.5	2.89	310	0.860
Aluminum trichloride	356	26	2.63	261	0.510
Ammonia	132.4	111.3	11.28	72.5	0.235
Antimony tribromide	631.4	56	5.67		
Antimony trichloride	521			270	0.84
Argon	−122.3	48.1	4.87	74.6	0.536
Arsenic	1400				
Arsenic trichloride	318	58.4	5.91	252	0.720
Arsine	99.9	63.3	6.41	133	0.588
Arsine-d_3	98.9				
Bismuth tribromide	946			301	1.49
Bismuth trichloride	906	118	11.96	261	1.21
Boron pentafluoride	205				
Boron tribromide	308	48.1	4.87	272	0.921
Boron trichloride	178.8	38.2	3.87	266	0.441
Boron trifluoride	−12.3	49.2	4.98	124	0.549
Bromine	315	102	10.3	135	1.184
Antimony tribromide	631.4	56	5.67		
Antimony trichloride	521			270	0.84
Argon	−122.3	48.1	4.87	74.6	0.536
Arsenic	1400				
Arsenic trichloride	318	58.4	5.91	252	0.720
Arsine	99.9	63.3	6.41	133	0.588
Arsine-d_3	98.9				
Benzaldehyde	422	45.9	4.65	324	0.327
Benzene	288.90	48.31	4.895	255	0.306
Benzoic acid	479	41.55	4.21	341	0.358
Benzonitrile	426.3	41.55	4.21	339	0.304
Benzyl alcohol	422	42.4	4.3	334	0.324
Biphenyl	516	38.0	3.85	502	0.307
Bismuth tribromide	946			301	1.49
Bismuth trichloride	906	118	11.96	261	1.21
Boron pentafluoride	205				
Boron tribromide	308	48.1	4.87	272	0.921
Boron trichloride	178.8	38.2	3.87	266	0.441

TABLE 1.55 Critical Properties (*Continued*)

Substance	T_c, °C	P_c, atm	P_c, MPa	V_c, $cm^3 \cdot mol^{-1}$	ρ_c, $g \cdot cm^{-3}$
Carbon dioxide	31.1	72.8	7.38	94.0	0.468
Carbon disulfide	279	78.0	7.90	173	0.41
Carbon monoxide	−140.2	34.5	3.50	93.1	0.301
Carbonyl chloride	182	56	5.67	190	0.52
Carbonyl sulfide	102	58	5.88	140	0.44
Cesium	1806			300	0.44
Chlorine	143.8	76.1	7.71	124	0.573
Chlorine pentafluroide	142.6	51.9	5.26	230.9	0.565
Chlorine trifluoride	153.5				
Deuterium (equilibrium)	−234.8	16.28	1.650	60.4	0.0668
Deuterium (normal)	−234.7	16.43	1.665	60.3	0.0669
Deuterium bromide	88.8				
Deuterium chloride	50.3				
Deuterium hydride (DH)	−237.3	14.64	1.483	62.8	0.0481
Deuterium iodide	148.6				
Deuterium oxide	370.9	213.8	21.66	55.6	0.360
Diborane	166	39.5	4.00		
Dihydrogen disulfide	299	58.3	5.91		
Dihydrogen heptasulfide	742	33	3.34		
Dihydrogen hexasulfide	707	36	3.65		
Dihydrogen octasulfide	767	32	3.24		
Dihydrogen pentasulfide	657	38.4	3.89		
Dihydrogen tetrasulfide	582	43.1	4.37		
Dihydrogen trisulfide	465	50.6	5.13		
Flurorine	−129.0	51.47	5.215	66.2	0.574
Germanium tetrachloride	276.9	38	3.85	330	0.650
Hafnium tetrabromide	473			415	1.20
Hafnium tetrachloride	450	57.0	5.86	304	1.05
Hafnium tetraiodide	643			528	1.30
Helium (equilibrium)	−267.96	2.261	0.2289		0.06930
Helium-3	−269.85	1.13	0.1182	72.5	0.0414
Helium-4	−267.96	2.24	0.227	57.3	0.0698
Hydrazine	380	14.5	1.47	96.1	0.333
Hydrogen (equilibrium)	−240.17	12.77	1.294	65.4	0.0308
Hydrogen (normal)	−239.91	12.8	1.297	65.0	0.0310
Hydrogen bromide	89.8	84.4	8.55	100.0	0.809
Hydrogen chloride	51.40	82.0	8.31	81.0	0.45
Hydrogen cyanide	183.5	53.2	5.39	139	0.195
Hydrogen deuteride	−237.25	14.64	1.483	62.8	0.048
Hydrogen fluoride	188	64	6.5	69	0.29
Hydrogen iodide	150.7	82.0	8.31	131	0.976
Hydrogen selenide	137	88	8.9		
Hydrogen sulfide	100.4	88.2	8.94	98.5	0.31
Iodine	546	115	11.7	155	0.164
Krypton	−63.75	54.3	5.50	91.2	0.9085
Mercury	1477	1587	160.8		
Mercury(II) bromide	789				
Mercury(II)chloride	700				
Mercury(II) iodide	799				
Neon	−228.71	27.2	2.77	41.7	0.4835
Niobium pentabromide	737			469	1.05
Niobium pentachloride	534			400	0.68
Niobium pentafluoride	464	62	6.28	155	1.21

(*Continued*)

TABLE 1.55 Critical Properties (*Continued*)

Substance	T_c, °C	P_c, atm	P_c, MPa	V_c, $cm^3 \cdot mol^{-1}$	ρ_c, $g \cdot cm^{-3}$
Nitric oxide	−92.9	64.6	6.55	58	0.52
Nitrogen-14	146.94	33.5	3.39	89.5	0.313
Nitrogen-15	146.8	33.5	3.39	90.4	0.332
Nitrogen chloride difluoride	64.3	50.8	5.15		
Nitrogen dioxide (equilibrium)	158.2	100	10.1	170	0.557
Nitrogen trideuteride (ND_3)	132.4				
Nitrogen trifluoride	−39.3	44.7	4.53		
Nitrous oxide	36.434	71.596	7.2545	97.4	0.4525
Nitrosyl chloride	167	90	9.12	139	0.471
Nitryl fluoride	76.3				
Osmium tetroxide	132	170	17.2		
Oxygen	−118.56	49.77	5.043	73.4	0.436
Oxygen difluoride	−58.0	48.9	4.95	97.7	0.553
Ozone	−12.10	53.8	5.45	88.9	0.540
Phosgene	182	56	5.67	190	0.52
Phosphine	51.3	64.5	6.54		
Phosphine-d_3	50.4				
Phosphonium chloride	49.1	72.7	7.37		
Phosphorus	721				
Phosphorus bromide difluoride	113				
Phosphorus chloride difluoride	89.2	44.6	4.52		
Phosphorus dibromide fluoride	254				
Phosphorus dichloride fluoride	189.9	49.3	5.00		
Phosphorus pentachloride	372				
Phosphorus trichloride	290			260	0.528
Phosphorus trifluoride	−1.9	42.7	4.33		
Phosphoryl chloride difluoride	150.7	43.4	4.40		
Phosphoryl trichloride	329				
Phosphoryl trifluoride	73.4	41.8	4.24		
Radon	104	62	6.28	139	1.6
Rhenium(VII) oxide	669			334	
Rhenium(VI) oxide tetrachloride	508			161	0.95
Rubidium	1832			250	0.34
Selenium	1493				
Silane	−3.5	47.8	4.84		
Silicon chloride trifluoride	34.5	34.2	3.47		
Silicon tetrabromide	390				
Silicon tetrachloride	234	37	3.75	326	0.521
Silicon tetrafluoride	−14.0	36.7	3.72		
Silicon trichloride fluoride	165.4	35.3	3.57		
Sulfur	1041	116	11.7		
Sulfur dioxide	157.7	77.8	7.88	122	0.5240
Sulfur hexafluoride	45.6	37.1	3.76	198	0.734
Sulfur tetrafluoride	91.7				
Sulfur trioxide	217.9	81	8.2	130	0.633
Tantalum pentabromide	701			461	1.26
Tantalum pentachloride	494			400	0.89
Tin(IV) chloride	318.7	37.0	3.75	351	0.742
Titanium tetrachloride	365	46	4.66	340	0.558
Tungsten (VI) oxide tetrachloride	509			338	1.01
Uranium hexafluoride	232.7	45.5	4.61	250	1.41
Water	374.2	217.6	22.04	56.0	0.325

TABLE 1.55 Critical Properties (*Continued*)

Substance	T_c, °C	P_c, atm	P_c, MPa	V_c, $cm^3 \cdot mol^{-1}$	ρ_c, $g \cdot cm^{-3}$
Xenon	16.583	57.64	5.84	118	1.105
Zirconium tetrabromide	532			415	0.99
Zirconium tetrachloride	505	56.9	5.77	319	0.730
Zirconium tetraiodide	687			528	1.13

1.15 THERMODYNAMIC FUNCTIONS (CHANGE OF STATE)

All substances can exist in one of three forms (also called *states* or *phases*) that basically depend on the temperature of the substance. These states or phases are (1) solid, (2) liquid, and (3) gas.

The solid-to-liquid transition is a melting process, and the heat required is the heat of melting. The liquid-to-solid transition is the reverse process, and the heat liberated is the heat of freezing. The solid-to-gas transition is a sublimation process, and the heat required is the heat of sublimation. The liquid-to-gas transition is a vaporization process, and the heat required is the heat of vaporization (heat of boiling). Both the gas-to-solid and the gas-to-liquid processes are condensation processes and have an associated heat of condensation.

Each change of state is accompanied by a change in the energy of the system. Wherever the change involves the disruption of intermolecular forces, energy must be supplied. The disruption of intermolecular forces accompanies the state going toward a less ordered state. As the strengths of the intermolecular forces increase, greater amounts of energy are required to overcome them during a change in state. The melting process for a solid is also referred to as fusion, and the enthalpy-change associated with melting a solid is often called the heat of fusion (AH_{fus}). The heat needed for the vaporization of a liquid is called the heat of vaporization (AH_{vap}).

The specific heat is the amount of heat per unit mass required to raise the temperature by one degree Celsius. The relationship between heat and temperature change is usually expressed in the form shown below where c is the specific heat. The relationship does not apply if a phase change is encountered, because the heat added or removed during a phase change does not change the temperature.

$$Q = cm\Delta T$$

i.e., heat added is equal to the specific heat multiplied by the mass (weight) multiplied by the temperature difference ($\Delta T = t_{final} - t_{initial}$)

TABLE 1.56 Enthalpies and Gibbs Energies of Formation, Entropies, and Heat Capacities of the Elements and Inorganic Compounds

Substance	Physical state	$\Delta_f H°$ $kJ \cdot mol^{-1}$	$\Delta_f G°$ $kJ \cdot mol^{-1}$	$S°$ $J \cdot deg^{-1} \cdot mol^{-1}$	C_p° $J \cdot deg^{-1} \cdot mol^{-1}$
Ac Actinium	c	0	0	56.5	27.2
Al Aluminum	c	0	0	28.30(10)	24.4
	g	330.0(40)	289.4	164.554(4)	21.4
Al^{3+} std. state	aq	−538.4(15)	−485.3	−325.(10)	
Al_6BeO_{10}	c	−5624	−5317	175.6	265.19
$Al(BH_4)_3$	lq	−16.3	145.0	289.1	194.6
$AlBr_3$	c	−527.2	−488.5	180.2	100.58
std. state	aq	−895	−799	−74.5	
Al_4C_3	c	−216	−203	89	
$Al(CH_3)_3$	lq	136.4	−10.0	209.4	155.6
$Al(OAc)_3$	c	−1892.4			
$AlCl_3$	c	−704.2	−628.8	109.29	91.13
std. state	aq	−1033	−878	−152.3	
$AlCl_3 \cdot 6H_2O$	c	−2692	−2269	377	
AlF_3	c	−1510.4(13)	−1431.1	66.5(5)	75.13
std. state	aq	−1531.0	−1322	−363.2	

(*Continued*)

TABLE 1.56 Enthalpies and Gibbs Energies of Formation, Entropies, and Heat Capacities of the Elements and Inorganic Compounds (*Continued*)

Substance	Physical state	$\Delta_f H°$ kJ · mol^{-1}	$\Delta_f G°$ kJ · mol^{-1}	$S°$ J · deg^{-1} · mol^{-1}	$C_p°$ J · deg^{-1} · mol^{-1}
$AlF_3 \cdot H_2O$	c	−2297	−2052	209	
AlH_3	c	−46.0		30.0	40.2
AlI_3	c	−313.8	−300.8	159.0	98.7
std. state	aq	−699	−640	12.1	
$AlK(SO_4)_2 \cdot 12H_2O$	c	−6061.8	−5141.7	687.4	651.0
AlN	c	−318.1	−287.0	20.14	30.10
$Al(NO_3)_3$ std. state	aq	−1155	−820	117.6	
$Al(NO_3)_3 \cdot 6H_2O$	c	−2850.5	−2203.9	467.8	433.0
$Al(NO_3)_3 \cdot 9H_2O$	c	−3757.1	−2929.6	569	
AlO_2^- std. state	aq	−930.9	−830.9	−36.8	
Al_2O_3 corundum	c	−1675.7(13)	−1582.3	50.92(10)	79.15
$Al(OH)_3$	c	−1284	−1306	71	93.1
$Al(OH)_4^-$ std. state	aq	−1502.5	−1305.3	102.9	
AlP	c	−166.5			
$AlPO_4$ berlinite	c	−1733.8	−1618.0	90.79	93.18
Al_2S_3	c	−724.0	−640	116.85	105.06
Al_2Se_3	c	−565			
Al_2SiO_5 andalusite	c	−2592.0	−2444.8	93.2	122.76
$Al_2(SO_4)_3$	c	−3435	−3507	239.3	259.4
std. state	aq	−3790	−3205	−583.3	
Al_2Te_3	c	−326			
Americium					
Am	c	0	0	62.7	
Am^{3+}	aq	−682.8	−671.5	−159.0	
Am^{4+}	aq	−511.7	−461.1	−372	
Am_2O_3	c	−1757	−1678	154.7	
AmO_2	c	−1005.0	950.2	96.2	
Ammonium					
NH_3	g	−45.94(35)	−16.4	192.776(5)	35.65
undissoc; std. state	aq	−80.29	−26.57	111.3	
ND_3	g	−58.6	−26.0	203.9	38.23
NH_4^+ std. state	aq	−133.26(25)	−79.37	111.17(40)	79.9
NH_4OH undissoc; std. state	aq	−361.2	−254.0	165.5	
ionized; std. state	aq	−362.50	−236.65	102.5	−68.6
NH_4OAc	c	−616.14			
std. state	aq	−618.52	−448.78	200.0	73.6
$NH_4Al(SO_4)_2$	c	−2352.2	−2038.4	216.3	226.44
std. state	aq	−2481	−2054	−168.2	
NH_4AsO_2 std. state	aq	−561.54	−429.41	154.8	
$NH_4H_2AsO_3$ std. state	c	−847.30	−666.60	223.8	
$NH_4H_2AsO_4$	c	−1059.8	−833.0	172.05	151.17
std. state	aq	−1042.07	−832.66	230.5	
$(NH_4)_2HAsO_4$ std. state	aq	−1171.1	−873.20	225.1	
$(NH_4)_3AsO_4$ std. state	aq	−1286.7	−886.63	177.4	
NH_4Br	c	−271.8	−175.2	113.0	96.0
std. state	aq	−254.05	−183.34	194.97	−61.9
NH_4BrO_3	aq	−199.58	−60.84	275.10	
NH_4 carbamate	c	−657.60	−448.07	133.5	
NH_4Cl	c	−314.5	−202.9	94.6	84.1
std. state	aq	−299.66	−210.62	169.9	−56.5
NH_4ClO_3 std. state	aq	−236.48	−87.40	275.7	

TABLE 1.56 Enthalpies and Gibbs Energies of Formation, Entropies, and Heat Capacities of the Elements and Inorganic Compounds (*Continued*)

Substance	Physical state	$\Delta_f H°$ kJ · mol^{-1}	$\Delta_f G°$ kJ · mol^{-1}	$S°$ J · deg^{-1} · mol^{-1}	$C_p°$ J · deg^{-1} · mol^{-1}
NH_4ClO_4	c	−295.3	−88.8	186.2	128.1
std. state	aq	−261.84	−87.99	295.4	
NH_4CN	c	0.4			134.0
std. state	aq	18.0	92.9	207.5	
NH_4CNO cyanate std. state	aq	−278.7	−177.0	220.1	
$(NH_4)_2CO_3$ std. state	aq	−942.15	−686.64	169.9	
$(NH_4)_2C_2O_4$ oxalate	c	−1123.0			226.0
$(NH_4)_2CrO_4$	c	−1167.3			
std. state	aq	−1144.3	−886.59	277.0	
$(NH_4)_2Cr_2O_7$	aq	−1755.2	−1459.5	488.7	
NH_4 dithiocarbonate	c	−126.8			
NH_4F	c	−463.96	−348.78	71.97	65.27
std. state	aq	−465.14	−358.19	99.6	−26.8
NH_4 formate std. state	aq	−558.06	−430.5	205.0	−7.9
NH_4HCO_3	c	−849.4	−665.9	120.9	
	aq	−824.5	−666.1	204.6	
NH_4I	c	−201.4	−112.5	117.0	81.8
std. state	aq	−187.69	−130.96	224.7	−62.3
NH_4IO_3	c	−385.8			
std. state	aq	−354.0	−207.5	231.8	
NH_4N_3 azide	c	115.5	274.1	112.6	
	aq	142.7	268.6	221.3	
NH_4NO_2	aq	−237.2	−111.6	236.4	−17.6
NH_4NO_3	c	−365.56	−184.01	151.08	139.3
std. state	aq	−339.87	−190.71	259.8	−6.7
$NH_4H_2PO_4$	c	−1145.07	−1210.56	151.96	142.26
std. state	aq	−1428.79	−1209.76	203.8	
$(NH_4)_2HPO_4$	c	−1556.91		188.0	
std. state	aq	−1557.16	−1248.00	193.3	
$NH_4H_3P_2O_7$	aq	−2409.1	−2102.6	326.0	
NH_4HS	c	−156.9	−50.6	97.5	
	aq	−150.2	−67.2	176.1	
NH_4HSO_3	aq	−758.7	−607.0	253.1	
NH_4HSO_4	c	−1026.96			
std. state	aq	−1019.85	−835.38	245.2	−3.8
$(NH_4)_3PO_4$	c	−1671.9			
std. state	aq	−1674.9	−1256.9	117	
$(NH_4)_4P_2O_7$ std. state	aq	−2801.2	−2236.8	335	
$(NH_4)_2PtCl_6$	c	−803.3			237.7
NH_4ReO_4	c	−945.6	−774.9	232.6	
$(NH_4)_2S$	aq	−231.8	−72.8	212.1	
NH_4SCN	aq	−56.1	13.4	257.7	39.7
NH_4HSeO_4 std. state	aq	−714.2	−531.6	262.8	
$(NH_4)_2SeO_4$	aq	−864.0	−599.8	280.7	
$(NH_4)_2SiF_6$	c	−2681.69	−2365.3	280.24	228.11
$(NH_4)_2SO_3$	aq	−900.4	−645.0	197.5	
$(NH_4)_2SO_4$	c	−1180.9	−901.70	220.1	187.49
std. state	aq	−1174.28	−903.37	246.9	−133.1
$(NH_4)_2S_2O_8$	c	−1648.08			
std. state	aq	−1610.0	−1273.6	471.1	
NH_4VO_3	c	−1053.1	−888.3	140.6	129.33

(*Continued*)

TABLE 1.56 Enthalpies and Gibbs Energies of Formation, Entropies, and Heat Capacities of the Elements and Inorganic Compounds (*Continued*)

Substance	Physical state	$\Delta_f H°$ kJ · mol^{-1}	$\Delta_f G°$ kJ · mol^{-1}	$S°$ J · deg^{-1} · mol^{-1}	$C_p°$ J · deg^{-1} · mol^{-1}
Antimony					
Sb	c	0	0	45.7	25.2
	g	262.3	222.1	180.3	20.8
$SbBr_3$	c	−259.4	−239.3	207.1	
	g	−194.6	−223.9	372.9	80.2
$SbCl_3$	c	−382.0	−323.7	184.1	107.9
$SbCl_5$	lq	−440.16	−350.2	301	
SbF_3	c	−915.5			
SbH_3	g	145.11	147.74	232.8	41.05
SbI_3	c	−100.4		215.5	97.57
Sb_2O_3	c	−708.8		123.01	101.25
Sb_2O_5	c	−971.9	−829.2	125.1	117.61
Sb_2S_3	c	−174.9		182.0	117.74
Sb_2Te_3	c	−56.5	−55.2	234	
Argon					
Ar	g	0	0	154.846(3)	20.79
Arsenic					
As gray	c	0	0	35.1	24.64
$AsBr_3$	g	−130.0	−159.0	363.9	79.16
$AsCl_3$	lq	−305.0	−259.4	216.3	133.5
	g	−261.5	−248.9	327.06	75.73
AsF_3	lq	−821.3	−774.2	181.2	126.2
	g	−785.8	−770.8	289.1	65.6
AsH_3	g	66.44	68.91	222.8	38.07
AsI_3	c	−58.2	−59.4	213.05	105.77
AsO_2^-	aq	−429.0	−350.0	40.6	
AsO_4^{3-}	aq	−888.1	−648.4	−162.8	
As_2O_5	c	−924.87	−782.3	105.4	116.5
As_4O_6 octahedral	c	−1313.94	−1152.52	214.2	191.29
As_2S_3	c	−169.0	−168.6	163.6	116.3
Astatine					
At	c	0	0	121.3	
Barium					
Ba	c	0	0	62.48	28.10
Ba^{2+} std. state	aq	−537.64	−560.74	9.6	
$Ba(OAc)_2$ acetate	c	−1484.5			
std. state	aq	−1509.67	−1299.55	182.8	
$BaBr_2$	c	−757.3	−736.8	146.0	77.0
std. state	aq	−780.73	−768.68	174.5	
$BaBr_2 \cdot 2H_2O$	c	−1366.1	−1230.5	226	
$Ba(BrO_3)_2$	c	−752.66	−577.4	243	
BaC_2O_4 oxalate	c	−1368.6			
$BaCl_2$	c	−855.0	−806.7	123.67	75.14
$BaCl_2 \cdot 2H_2O$	c	−1456.9	−1293.2	202.9	161.96
$Ba(ClO_3)_2$	c	−762.7			
$Ba(ClO_3)_2 \cdot H_2O$	c	−1691.6	−1270.7	393	
$BaCO_3$ witherite	c	−1213.0	−1134.4	112.1	86.0
$BaCrO_4$	c	−1446.0	−1345.3	158.6	
BaF_2	c	−1207.1	−1156.8	96.4	71.20
std. state	aq	−1202.90	−1118.38	−17.0	
$Ba(HCO_3)_2$ std. state	aq	−1921.63	−1734.4	192.1	

TABLE 1.56 Enthalpies and Gibbs Energies of Formation, Entropies, and Heat Capacities of the Elements and Inorganic Compounds (*Continued*)

Substance	Physical state	$\Delta_f H°$ kJ · mol^{-1}	$\Delta_f G°$ kJ · mol^{-1}	$S°$ J · deg^{-1} · mol^{-1}	$C_p°$ J · deg^{-1} · mol^{-1}
$Ba(H_2PO_2)_2$	c	−1762.3			
BaI_2	c	−602.1	−601.4	165.1	77.49
std. state	aq	−648.02	−663.92	232.2	
$Ba(IO_3)_2$	c	−1027.2	−864.8	249.4	187.4
std. state	aq	−980.3	−816.7	246.4	
$BaMnO_4$	c	−1548	−1439.7	138	140.6
$BaMoO_4$	c	−1507.5	−1439.7	144.3	114.7
$Ba(NO_2)_2$	c	−768.2			
$Ba(NO_3)_2$	c	−988.0	−792.6	213.8	151.38
std. state	aq	−952.36	−783.41	302.5	
BaO	c	−548.0	−520.4	72.07	47.28
BaO_2	c	−634.3			
$Ba(OH)_2$	c	−944.7	−859.5	107	101.6
$Ba(OH)_2 \cdot H_2O$	c	−3342.2	−2793.2	427	
BaS	c	−460.0	−456.0	78.2	49.37
BaSe	c	−372			
$BaSeO_3$	c	−1040.6	−968.2	167	
$BaSiF_6$	c	−1952.2	−2794.1	163	
$BaSO_3$	c	−1179.5			
$BaSO_4$	c	−1473.19	−1362.2	132.2	101.75
$BaTiO_3$	c	−1659.8	−1572.4	108.0	102.47
Beryllium					
Be	c	0	0	9.50(8)	16.38
	g	324.(5)		136.275(3)	
Be^{2+} std. state	aq	−382.8	−379.7	−129.7	
$BeAl_2O_4$ chrysoberyl	c	−2301.0	−2178.5	66.29	105.38
$BeBr_2$	c	−353.5	−337	108.0	69.4
Be_2C	c	91	−88	16.3	43.2
$BeCl_2$ β form	c	−490.4	−445.6	75.81	62.43
$BeCO_3$	c	1025.0		52.0	65.0
BeF_2 α form	c	−1026.8	−979.4	53.35	51.82
BeI_2	c	−192.5	−187	121.0	71.1
Be_3N_2 cubic	c	−588.3	−532.9	34.13	64.36
BeO α form	c	−609.4(25)	−580.1	13.77(4)	25.56
BeO_2^{2-}	aq	−790.8	−640.1	−159.0	
$3BeO \cdot B_2O_3$	c	−3105	−2939	100	139.7
$Be(OH)_2$ β form	c	−902.5	−815.0	45.5	62.1
BeS	c	−234.3	−233.0	34.0	34.0
$BeSeO_4$	c	−1205.2	−1093.8	77.9	85.7
std. state	aq	−982.0	−820.9	−75.7	
Be_2SiO_4	c	−2117	−2003	64.19	95.6
$BeSO_4$	c	−1200.8	−1089.4	77.97	85.70
std. state	aq	−1290.0	−1124.3	−109.6	
$BeSO_4 \cdot H_2O$	c	−2423.75	−2080.66	232.97	216.61
$BeWO_4$	c	−1513	−1405	88.4	97.3
Bismuth					
Bi	c	0	0	56.7	25.5
	g	207.1	168.2	187.0	20.8
$BiBr_3$	c	264	234	226	109
$BiCl_3$	c	−379.1	−315.1	177.0	105.0
BiH_3	g	277.8			

(*Continued*)

TABLE 1.56 Enthalpies and Gibbs Energies of Formation, Entropies, and Heat Capacities of the Elements and Inorganic Compounds (*Continued*)

Substance	Physical state	$\Delta_f H°$ kJ · mol^{-1}	$\Delta_f G°$ kJ · mol^{-1}	$S°$ J · deg^{-1} · mol^{-1}	$C_p°$ J · deg^{-1} · mol^{-1}
BiI_3	c	−100.4	−175.3		
Bi_2O_3	c	−574.0	−493.7	151.5	113.5
BiOCl	c	−366.9	−322.2	120.5	
Bi_2S_3	c	−143.1	−140.6	200.4	122.2
$Bi_2(SO_4)_3$	c	−2544.3			
Bi_2Te_3	c	−78.24		260.91	152.21
Boron					
B	c	0	0	5.90(8)	11.1
	g	565.(5)		153.436(15)	
BBr_3	lq	−239.7	−238.5	229.7	128.03
B_4C	c	−62.7	−62.1	27.18	53.76
BCl_3	g	−403.8	−388.7	290.1	62.7
BF_3	g	−1136.0(8)	−1119.4	254.42(20)	50.45
BF_4^- std. state	aq	−1574.9	−1487.0	179.9	
BH_3	g	100.0	111	187.9	36.22
BH_4^- std. state	aq	48.16	114.27	110.5	
B_2H_6 diborane(6)	g	35.6	86.7	232.1	56.9
B_5H_9 pentaborane(9)	lq	42.7	171.8	184.2	151.13
$B_{10}H_{14}$ decaborane(14)	c	−29.83	212.9	234.9	221.2
BN	c	−254.4	−228.4	14.80	19.72
$B_3N_3H_6$ borazine	lq	−541.0	−392.7	199.6	
	g	−510	−389	288.61	96.94
BO_2^- std. state	aq	−772.37	−678.94	−37.24	
B_2O_3	c	−1273.5(14)	−1194.3	53.97(30)	62.8
$B(OH)_4^-$ std. state	aq	−1344.03	−1153.32	102.5	
$B_3O_3H_3$ boroxin	c	−1262	−11.56	167	98.3
B_2S_3	c	−240.6		100.0	111.7
Bromine					
Br atomic	g	111.87(12)	82.4	175.018(4)	20.8
Br^- std. state	aq	−121.41(15)	−103.97	82.55(20)	−141.8
Br_2	lq	0	0	152.21(30)	75.67
	g	30.91(11)		245.468(5)	
Br_3^- std. state	aq	−130.42	−107.07	215.5	
BrCl	g	14.6	−0.96	239.91	34.98
BrF	g	−93.8	−109.2	229.0	32.97
BrF_3	lq	−300.8	−240.5	178.2	124.6
	g	−255.6	229.4	292.5	66.6
BrF_5	lq	−458.6	−351.9	225.1	
	g	−428.9	−351.6	323.2	99.6
BrO^- std. state	aq	−94.1	−33.5	42.0	
BrO_3^- std. state	aq	−67.07	18.6	161.71	
BrO_4^-	aq	13.0	118.1	199.6	
Cadmium					
Cd	c	0	0	51.80(15)	25.9
	g	111.80(20)		167.749(4)	20.8
Cd^{2+}	aq	−75.92(60)		−72.8(15)	
$CdBr_2$	c	−316.18	−296.31	137.2	76.7
std. state	aq	−318.99	−285.52	91.6	
$CdCl_2$	c	−391.6	−343.9	115.3	74.7
std. state	aq	−410.20	−340.12	39.8	
$CdCl_2 \cdot 5/2H_2O$	c	−1131.94	−944.08	227.2	

TABLE 1.56 Enthalpies and Gibbs Energies of Formation, Entropies, and Heat Capacities of the Elements and Inorganic Compounds (*Continued*)

Substance	Physical state	$\Delta_f H°$ kJ · mol^{-1}	$\Delta_f G°$ kJ · mol^{-1}	$S°$ J · deg^{-1} · mol^{-1}	$C_p°$ J · deg^{-1} · mol^{-1}
$Cd(CN)_2$	c	162.3			
std. state	aq	225.5	267.4	115.1	
$CdCO_3$	c	−750.6	−669.4	92.5	
$Cd(OAc)_2$ std. state	aq	−1047.9	−816.4	100	
CdF_2	c	−700.4	−647.7	77.4	
std. state	aq	−741.15	−635.21	−100.8	
CdI_2	c	−203.3	−201.4	161.1	80.0
std. state	aq	−186.3	−180.8	149.4	
CdI_4^- std. state	aq	−341.8	−315.9	326	
$Cd(NH_3)_4^{2+}$ std. state	aq	−450.2	−226.4	336.4	
$Cd(NO_3)_2$	c	−456.3			
std. state	aq	−490.6	−300.2	219.7	
CdO	c	−258.35(40)	−228.7	54.8(15)	43.4
$Cd(OH)_2$	c	−560.7	−473.6	96.0	
CdS	c	−161.9	−156.5	64.9	55.5
$CdSO_4$	c	−933.4	−822.7	123.0	99.6
std. state	aq	−985.2	−822.2	−53.1	
$CdSO_4 \cdot 8/3H_2O$	c	−1729.30(80)	−1465.3	229.65(40)	213.3
$CdSeO_4$	c	−633.0	−531.8	164.4	
std. state	aq	−674.9	−518.8	−19.3	
CdTe	c	−92.5	−92.0	100.0	
Calcium					
Ca	c	0	0	41.59(40)	25.9
	g	177.8(8)		154.887(4)	
Ca^{2+} std. state	aq	−543.0(10)	−553.54	−56.2(10)	
$Ca(OAc)_2$	c	−1479.5			
std. state	aq	−1514.73	−1292.35	120.1	
$Ca_3(AsO_4)_2$	c	−3298.7	−3063.1	226	
$Ca(BO_2)_2$	c	−2030.9	−1924.1	104.85	103.98
CaB_4O_7	c	−3360.3	−3167.1	134.7	157.9
$CaBr_2$	c	−682.8	−663.6	130.0	75.04
std. state	aq	−785.9	−761.5	111.7	
CaC_2	c	−59.8	−64.9	69.96	62.72
$CaCl_2$	c	−795.4	−748.8	108.4	72.9
std. state	aq	−877.13	−816.05	59.8	
$CaCl_2 \cdot 2H_2O$	c	−1402.9			738
$CaCN_2$ cyanamide	c	−350.6			
$Ca(CN)_2$	c	−184.5			
$CaCO_3$ calcite	c	−1207.6	−1129.1	91.7	83.5
aragonite	c	−1207.8	−1128.2	88.0	82.3
	aq	−1220.0	−1081.4	−110.0	
CaC_2O_4	c	−1360.6			
$CaC_2O_4 \cdot H_2O$	c	−1674.9	−1514.0	156.5	152.8
$CaCrO_4$	c	−1379.1	−1277.4	134	
CaF_2	c	−1228.0	−1175.6	68.6	67.0
	aq	−1208.1	−1111.2	−80.8	
$Ca(formate)_2$	c	1386.6			
CaH_2	c	−181.5	−142.5	41.4	41.0
$CaHPO_4 \cdot 2H_2O$	c	−2403.58	−2154.75	189.45	197.07
$Ca(H_2PO_2)_2$ hypophosphite	c	−1752.7			
$Ca(H_2PO_4)_2$ std. state	aq	−3135.41	−2814.33	127.6	

(*Continued*)

TABLE 1.56 Enthalpies and Gibbs Energies of Formation, Entropies, and Heat Capacities of the Elements and Inorganic Compounds (*Continued*)

Substance	Physical state	$\Delta_f H°$ kJ · mol^{-1}	$\Delta_f G°$ kJ · mol^{-1}	$S°$ J · deg^{-1} · mol^{-1}	$C_p°$ J · deg^{-1} · mol^{-1}
$Ca(H_2PO_4)_2 \cdot H_2O$	c	−3409.67	−3058.42	259.8	258.82
CaI_2	c	−533.5	−528.9	142.0	77.16
std. state	aq	−653.2	−656.7	169.5	
$Ca(IO_3)_2$	c	−1002.5	−839.3	230	
$Ca[Mg(CO_3)_2]$ dolomite	c	−2326.3	−2163.6	155.18	157.53
$CaMoO_4$	c	−1541.4	−1434.7	122.6	114.3
Ca_3N_2	c	−439.3		105.0	113.0
$Ca(NO_2)_2$	c	−741.4			
$Ca(NO_3)_2$	c	−938.2	−742.8	193.3	149.37
std. state	aq	−957.55	−776.22	239.7	
CaO	c	−634.92(90)	−603.3	38.1(4)	42.0
$Ca(OH)_2$	c	−985.2	−897.5	83.4	87.5
Ca_3P_2	c	−506			
$Ca_3(PO_4)_2$	c	−4120.8	−3884.8	236.0	227.8
$Ca_2P_2O_7$	c	−3338.8	−3132.1	189.24	187.8
$Ca_{10}(PO_4)_6F_2$ fluoroapatite	c	−13,744	−12,983	775.7	751.9
CaS	c	−482.4	−477.4	56.5	47.4
CaSe	c	−368.2	−363.2	67	
$CaSiO_3$	c	−1634.9	−1549.7	81.92	85.27
Ca_2SiO_4	c	−2307.5	−2192.8	127.7	128.8
$3CaO \cdot SiO_2$	c	−2929.2	−2784.0	168.6	171.9
$CaSO_3 \cdot 2H_2O$	c	−1752.7	−1555.2	184	178.7
$CaSO_4$	c	−1425.2	−1309.1	108.4	99.0
	aq	−1451.1	−1298.1	−33.1	
$CaSO_4 \cdot ½H_2O$	c	−1576.7	−1436.8	130.5	119.4
$CaSO_4 \cdot 2H_2O$	c	−2022.6	−1797.5	194.1	186.0
$Ca(VO_3)_2$	c	−2329.3	−2169.7	179.1	166.8
$CaWO_4$	c	−1645.15	−1538.50	126.40	114.14
Carbon					
C graphite	c	0	0	5.74(10)	8.517
	g	716.68(45)		158.100(3)	
diamond	c	1.897	2.900	2.377	6.116
CN^-	aq	150.6	172.4	94.1	
$(CN)_2$ cyanogen	g	306.7	297.2	241.9	56.9
CNBr	g	186.2	165.3	248.36	46.9
CNCl	g	137.95	131.02	236.2	45.0
CNF	g			224.7	41.8
CNI	c	166.2	185.0	96.2	
	g	225.5	196.6	256.8	48.3
CNN_3 cyanogen azide	c	387.4			
OCN^-	aq	−146.0	−97.4	106.7	
CO	g	−110.53(17)	−137.16	197.660(4)	29.14
CO_2	g	−393.51(13)	394.39	213.785(10)	37.13
undissoc; std. state	aq	−413.26(20)	−386.0	119.36(60)	
CO_3^{2-}	aq	−675.23(25)	−527.9	−50.0(10)	
C_3O_2 suboxide	g	−93.7	−109.8	276.4	67.0
$COBr_2$	g	−96.2	−110.9	309.1	61.8
$COCl_2$ phosgene	g	−219.1	−204.9	283.50	57.70
COClF	g			276.7	52.4
COF_2	g	−639.8	−623.33	258.89	46.8

TABLE 1.56 Enthalpies and Gibbs Energies of Formation, Entropies, and Heat Capacities of the Elements and Inorganic Compounds (*Continued*)

Substance	Physical state	$\Delta_f H^\circ$ kJ · mol^{-1}	$\Delta_f G^\circ$ kJ · mol^{-1}	S° J · deg^{-1} · mol^{-1}	C_p° J · deg^{-1} · mol^{-1}
COS carbonyl sulfide	g	−142.0	−166.9	231.56	41.50
CS_2	lq	89.0			74.6
	g	117.7	67.1	237.8	45.4
CTe_2	lq	164.8			
Cerium					
Ce γ, fcc	c	0	0	72.0	26.9
Ce^{3+} std. state	aq	−696.2	−672.0	−205.0	
Ce^{4+} std. state	aq	−537.2	−503.8	−301.0	
$CeCl_3$	c	−1060.5	−984.8	151.0	87.4
std. state	aq	−1197.5	−1065.7	−38.0	
CeF_3	c	−1635.9	−1556	115.1	99.3
CeI_3	c	−669.3	−674	209	
$Ce(NO_3)_3$	c	−1225.9			
CeO_2	c	−1088.7	−1024.7	62.30	61.63
Ce_2O_3	c	−1796.2	−1706.2	150.6	114.6
CeS	c	−459.4	−451.5	78.2	50.0
$Ce_2(SO_4)_3$	c	−3954.3			
std. state	aq	−4176.9	−3652.6	−318	
$Ce_2(SO_4)_3 \cdot 8H_2O$	c	−5522.9	−5607.4		
Cesium					
Cs	c	0	0	85.23(40)	32.20
	lq	2.087	0.025	92.1	32.4
	g	76.5(10)		175.601(3)	
Cs^+ std. state	aq	−258.00(50)	−292.0	132.1(5)	−10.5
Cs acetate	aq	−744.3	−661.3	219.7	
$CsBO_2$	c	−972.0	−915.0	104.4	80.6
CsBr	c	−405.8	−391.4	113.05	52.93
std. state	aq	−379.8	−396.0	215.5	
CsCl	c	−442.8	414.4	101.18	52.44
std. state	aq	−425.4	−423.3	189.4	−146.9
$CsClO_4$	c	−443.1	−314.3	175.1	108.3
Cs_2CO_3	c	−1139.7	−1054.4	204.5	123.9
std. state	aq	−1193.7	−1111.9	209.2	
CsF	c	−553.5	−525.5	92.8	51.1
std. state	aq	−590.9	−570.8	119.2	
Cs formate	aq	−683.8	−643.0	226.0	
$CsHCO_3$	c	−966.1			
CsHF	c	−923.8	−858.9	135.2	87.3
$CsHSO_4$	c	−1158.1			
	aq	−1145.6	−1047.9	264.8	
CsI	c	−346.6	−340.6	123.1	52.8
std. state	aq	−313.5	−343.6	244.4	−152.7
$CsIO_3$	c	−525.9	−433.9		167
$CsNO_3$	c	−506.0	−406.6	155.2	
std. state	aq	−465.6	−403.3	279.5	−99.2
Cs_2O	c	−345.8	−308.2	146.9	76.0
CsOH	c	−417.2	370.7	98.7	67.9
std. state	aq	−488.3	−449.3	122.3	
Cs_2PtCl_6 std. state	aq	−1184.9	−1066.9	485.8	
Cs_2S	aq	−483.7	−498.3	251.0	
Cs_2Se	aq		454.8		

(*Continued*)

TABLE 1.56 Enthalpies and Gibbs Energies of Formation, Entropies, and Heat Capacities of the Elements and Inorganic Compounds (*Continued*)

Substance	Physical state	$\Delta_f H^\circ$ kJ · mol^{-1}	$\Delta_f G^\circ$ kJ · mol^{-1}	S° J · deg^{-1} · mol^{-1}	C_p° J · deg^{-1} · mol^{-1}
Cs_2SO_4	c	−1443.0	−1323.6	211.9	134.9
std. state	aq	−1425.8	−1328.6	286.2	
Chlorine					
Cl atomic	g	121.301(8)		165.190(4)	
Cl^- std. state	aq	−167.08(10)	−131.3	56.60(20)	−136.4
Cl_2	g	0	0	233.08(10)	33.95
ClF	g	−50.3	−51.84	217.9	32.08
ClF_3	g	−163.2	−123.0	281.6	63.85
ClF_5	g	−239	−147	310.74	97.17
ClO	g	101.8	98.1	226.6	31.5
ClO^- std. state	aq	−107.1	−36.8	41.8	
ClO_2	g	102.5	120.5	256.8	42.00
ClO_2^- std. state	aq	−66.5	17.2	101.3	
ClO_3^- std. state	aq	−104.0	−8.0	162.3	
ClO_3F perchloryl fluoride	g	−23.8	48.2	279.0	64.9
ClO_4^- std. state	aq	−128.10(40)	−8.62	184.0(15)	
Cl_2O	g	80.3	97.9	266.2	45.4
Cl_2O_7	lq	238.1			
	g	1138			
Chromium					
Cr	c	0	0	23.8	23.43
Cr^{2+} std. state	aq	−143.5			
$CrBr_2$	c	−302.1			
$CrCl_2$	c	−395.4	−356.0	115.3	71.2
$CrCl_3$	c	−556.5	−486.1	123.0	91.8
$Cr(CO)_6$ hexacarbonyl	c	−1077.8		293.01	226.23
CrF_2	c	−778.0			
CrF_3	c	−1159	−1088	93.9	78.7
Cr_2FeO_4	c	−1444.7	−1343.8	146.0	133.6
CrI_2	c	−156.9			
CrI_3	c	−205.0			
CrN	c	−117	−93	38	52.7
CrO_2	c	−598.0			
Cr_2O_3	c	−1140	−1058.1	81.2	118.7
Cr_3O_4	c	−1131.0			
CrO_2Cl_2	g	−538.1	−501.6	329.8	84.5
CrO_4^{2-} std. state	aq	−881.15	−727.85	50.21	
$HCrO_4^-$ std. state	aq	−878.22	−764.84	184.1	
$Cr_2O_7^{2-}$ std. state	aq	−1490.3	−1301.2	261.9	
$Cr_2(SO_4)_3$	c	−609.6		269.9	302.6
Cobalt					
Co	c	0	0	30.0	24.8
Co^{2+} std. state	aq	−58.2	−54.4	−113	
Co^{3+} std. state	aq	92	134	−305	
$CoBr_2$	c	−220.9			79.5
std. state	aq	−301.3	−262.3	50	
$CoCl_2$	c	−312.5	−269.8	109.2	78.49
std. state	aq	−392.5	−316.7	0	
$CoCO_3$	c	−713.0			
CoF_2	c	−692	−647	82.4	68.9
CoF_3	c	−790	−719	95	92

TABLE 1.56 Enthalpies and Gibbs Energies of Formation, Entropies, and Heat Capacities of the Elements and Inorganic Compounds (*Continued*)

Substance	Physical state	$\Delta_f H°$ kJ · mol^{-1}	$\Delta_f G°$ kJ · mol^{-1}	$S°$ J · deg^{-1} · mol^{-1}	$C_p°$ J · deg^{-1} · mol^{-1}
CoI_2	c	−88.7			
	aq	−168.6	−157.7	109.0	
$Co(NH_3)_6^{2+}$ std. state	aq	−584.9	−157.3	146	
$Co(NH_3)_6^{3+}$ std. state	aq		−189.5		
$Co(NO_3)_2$	c	−420.5			
std. state	aq	−472.8	−277.0	180	
CoO	c	−237.7	−214.0	53.0	55.3
Co_3O_4	c	−891	−774	102.5	123.4
$Co(OH)_2$	c	−539.7	−454.4	79.0	
CoS	c	−82.8			
Co_2S_3	c	−147.3			
$CoSO_4$	c	−888.3	−782.4	118.0	103
std. state	aq	−967.3	−799.1	−92.0	
$CoSO_4 \cdot 7H_2O$	c	−2979.93	−2473.83	406.06	390.49
Copper					
Cu	c	0	0	33.15(8)	24.44
	g	337.4(12)		166.398(4)	
Cu^+ std. state	aq	71.67	50.00	40.6	
Cu^{2+} std. state	aq	64.9(10)	65.52	−98.(4)	
$Cu(OAc)_2$ acetate	c	−893.3			
std. state	aq	−907.25	−673.29	73.6	
$Cu_3(AsO_4)_2$ std. state	aq	−1581.97	−1100.48	−804.2	
CuBr	c	−104.6	−100.8	96.2	54.7
$CuBr_2$	c	−141.84			
CuCl	c	−137.2	−119.9	86.2	48.5
$CuCl_2$	c	−220.1	−175.7	108.09	71.88
$Cu(ClO_4)_2$ std. state	aq	−193.89	48.28	264.4	
CuCN	c	95.0	108.4	90.00	61.04
CuCNS std. state	aq	138.11	142.67	184.93	
$Cu(CNS)_2$ std. state	aq	217.65	250.87	189.1	
CuF	c	−280	−260	64.9	51.9
CuF_2	c	−542.7	−492	77.45	65.55
$Cu(formate)_2$	aq	−786.34	−636.4	84	
CuI	c	67.8	−69.5	96.7	54.1
$Cu(NH_3)_4^{2+}$ std. state	aq	−348.5	−111.3	273.6	
$Cu(NO_3)_2$	c	−302.9			
std. state	aq	−349.95	−157.15	193.3	
CuO	c	−157.3	−129.7	42.6	42.2
Cu_2O	c	−168.6	−149.0	93.1	63.6
$Cu(OH)_2$	c	−450	−373	108.4	95.19
CuS	c	−53.1	−53.7	66.5	47.8
Cu_2S	c	−79.5	−86.2	120.9	76.3
CuSe	c	−39.5			
Cu_2Se	c	−59.4		157.3	88.70
$CuSO_4$	c	−771.4(12)	−662.2	109.2(4)	98.87
std. state	aq	−844.50	−679.11	−79.5	
$CuSO_4 \cdot 5H_2O$	c	−2279.65	−1880.04	300.4	280
$CuWO_4$	c	−1105.0			
Dysprosium					
Dy	c	0	0	75.6	27.7
Dy^{3+} std. state	aq	−699.0	−665.0	−231.0	21.0

(*Continued*)

TABLE 1.56 Enthalpies and Gibbs Energies of Formation, Entropies, and Heat Capacities of the Elements and Inorganic Compounds (*Continued*)

Substance	Physical state	$\Delta_f H°$ kJ · mol^{-1}	$\Delta_f G°$ kJ · mol^{-1}	$S°$ J · deg^{-1} · mol^{-1}	$C_p°$ J · deg^{-1} · mol^{-1}
$DyCl_3$	c	−1000			100.0
	aq	−1197.0	−1059.0	−61.9	−389.0
DyF_3	c	−1711.0			
Dy_2O_3	c	−1863.1	−1771.5	149.8	116.27
Erbium					
Er	c	0	0	73.18	28.12
Er^{3+} std. state	aq	−705.4	−669.1	−244.3	21.0
$ErCl_3$	c	−998.7			100.0
	aq	−1207.1	−1062.7	−75.3	−389.0
Er_2O_3	c	−1897.9	−1808.7	155.6	108.49
Europium					
Eu	c	0	0	77.78	27.66
Eu^{2+} std. state	aq	−527.0	540.2	−8.0	
Eu^{3+}	aq	−605.0	−574.0	−222.0	8.0
$EuCl_2$	aq	−862.0			
$EuCl_3$	c	−936.0	−856	144.1	
	aq	−1106.2	−967.7	−54.0	−402.0
EuF_3	c	−1571			
Eu_2O_3 monoclinic	c	−1651.4	−1556.9	146	122.2
Eu_3O_4	c	−2272.0	−2142.0	205.0	
$Eu(OH)_3$	c	−1332	−1195	119.9	
Fluorine					
F atomic	g	79.38(30)	62.3	158.751(4)	22.7
F^-	aq	−335.35(65)	−278.8	−13.8(8)	−106.7
F_2	g	0	0	202.791(5)	31.30
FNO_3	g	10.5	73.7	292.9	65.22
FO	g	109.0	105.0	216.8	30.5
F_2O	g	24.7	41.9	247.4	43.3
F_2O_2	g	18.0			
Francium					
Fr	c	0	0	95.40	31.80
FrCl	c	−439		113.0	53.56
Fr_2O	c	−338	299.2	156.9	
Gadolinium					
Gd	c	0	0	68.07	37.03
Gd^{3+} std. state	aq	−686.0	−661.0	−205.9	
$GdCl_3$	c	−1008.0	−933	151.4	88.0
std. state	aq	−1188.0	−1059.0	−36.8	−410.0
GdF_3	lq	−1297			
Gd_2O_3 monoclinic	c	−1819.6	−1730	150.6	106.7
Gallium					
Ga	c	0	0	40.8	26.06
	lq	5.6			
	g	272.0	233.7	169.0	25.3
Ga^{3+}	aq	−211.7	−159.0	−331.0	
GaAs	c	−71.0	−67.8	64.2	46.2
$GaBr_3$	c	−386.6	−359.8	180.0	
$GaCl_3$	c	−524.7	−454.8	142.0	
GaF_3	c	−1163.0	−1085.3	84	
GaI_3	c	−238.9		205.0	100
Ga_2O_3 rhombic	c	−1089.1	−998.3	84.98	92.1

TABLE 1.56 Enthalpies and Gibbs Energies of Formation, Entropies, and Heat Capacities of the Elements and Inorganic Compounds (*Continued*)

Substance	Physical state	$\Delta_f H°$ kJ · mol^{-1}	$\Delta_f G°$ kJ · mol^{-1}	$S°$ J · deg^{-1} · mol^{-1}	$C_p°$ J · deg^{-1} · mol^{-1}
$Ga(OH)_3$	c	−964.4	−831.3	100.0	
GaSb	c	−41.8	−38.9	76.07	48.53
Germanium					
Ge	c	0	0	31.09(15)	23.3
	g	372.0(30)	331.2	167.904(5)	30.7
$GeBr_4$	lq	−347.7	−331.4	280.8	
	g	−300.0	−318.0	396.2	101.8
$GeCl_4$	lq	−531.8	−462.8	245.6	
	g	−495.8	−457.3	347.7	96.1
GeF_4	g	−1190.20(50)	−1150.0	301.9(10)	81.84
GeH_4	g	90.8	113.4	217.02	45.02
GeI_4	c	−141.8	−144.4	271.1	
	g	−56.9	−106.3	428.9	104.1
GeO_2 tetragonal	c	−580.0(10)	−521.4	39.71(15)	52.1
GeP	c	−21.0	−17.0	63.0	
GeS	c	−69.0	−71.6	71	
Gold					
Au	c	0	0	47.4	25.36
AuBr	c	−14.0			
$AuBr_3$	c	−53.3			
AuCl	c	−34.7		92.9	48.74
$AuCl_3$	c	−117.6		148.1	94.81
$AuCl_4^-$ std. state	aq	−322.2	−237.32	266.9	
$Au(CN)_2^-$ std. state	aq	242.3	285.8	172	
AuF_3	c	−363.6		114.2	91.29
$AuSb_2$	c	−19.46		119.2	77.40
AuSn	c	−30.5		93.7	49.41
Hafnium					
Hf hexagonal	c	0	0	43.56	25.69
HfC	c	−230.1		41.21	34.43
$HfCl_4$	c	−990.4	−901.3	190.8	120.46
HfF_4 monoclinic	c	−1930.5	−1830.5	113	
HfO_2	c	−1144.7	−1088.2	59.3	60.25
Helium					
He	g	0	0	126.153(2)	20.786
Holmium					
Ho	c	0	0	75.3	27.15
Ho^{3+} std. state	aq	−705.0	−673.7	226.8	17.0
$HoCl_3$	c	−1005.4			88
std. state	aq	−1206.7	−1067.3	−57.7	−393.0
HoF_3	c	−1707.0			
Ho_2O_3	c	−1880.7	−1791.2	158.2	115.0
Hydrogen					
H atomic	g	217.998(6)	203.3	114.717(2)	20.8
H^+ std. state	aq	0	0	0	0
H_2	g	0	0	130.680(3)	28.84
H^2H	g	0.321	−1.463	143.80	29.20
$^2H_2$ (D_2) deuterium	g	0	0	144.96	29.19
$HAsO_2^-$ undissoc; std. state	aq	−456.5	−402.71	125.9	

(*Continued*)

TABLE 1.56 Enthalpies and Gibbs Energies of Formation, Entropies, and Heat Capacities of the Elements and Inorganic Compounds (*Continued*)

Substance	Physical state	$\Delta_f H°$ kJ · mol^{-1}	$\Delta_f G°$ kJ · mol^{-1}	$S°$ J · deg^{-1} · mol^{-1}	$C_p°$ J · deg^{-1} · mol^{-1}
$H_2AsO_3^-$ undissoc; std. state	aq	−714.79	−587.22	110.5	
H_3AsO_3 undissoc; std. state	aq	−742.2	−639.90	195.0	
$HAsO_4^{2-}$ undissoc; std. state	aq	−906.34	−714.70	−1.7	
$H_2AsO_4^-$ undissoc; std. state	aq	−909.56	−753.29	117	
H_3AsO_3	c	−906.30			
undissoc; std. state	aq	−902.5	−766.1	184	
HBO_2	c	−794.3	−723.4	38	54.4
H_3BO_3	c	−1094.8(8)	−968.9	89.95(60)	86.1
undissoc	aq	−1072.8(8)		162.4(6)	
HBr	g	−36.29(16)	−53.4	198.700(4)	29.1
std. state	aq	−121.55	−103.97	82.4	−141.8
HBrO undissoc; std. state	aq	−113.0	−82.4	142	
$HBrO_3$ std. state	aq	−67.07	18.54	161.71	
HCl	g	−92.31(10)	−95.30	186.902(5)	29.12
std. state	aq	−167.15	−131.25	56.5	−136.4
^{2}HCl deuterium chloride	g	−93.35	−95.94	192.63	29.17
HClO	g	−78.7	−66.1	236.7	37.15
undissoc; std. state	aq	−120.9	−79.9	142	
$HClO_2$ undissoc; std. state	aq	−51.9	5.9	188.3	
$HClO_3$ std. state	aq	−103.97	−8.03	162.3	
$HClO_4$	lq	−40.58			
std. state	aq	−129.33	−8.62	182.0	
$HClO_4 \cdot H_2O$	c	−302.21			
$HClO_4 \cdot 2H_2O$	lq	−677.98			
HCN	lq	108.87	124.93	112.84	70.63
	g	135.1	124.7	201.81	35.86
ionized; std. state	aq	150.6	172.4	94.1	
undissoc; std. state	aq	107.11	119.66	124.7	
HCNO ionized; std. state	aq	−146.0	−97.5	106.7	
undissoc; std. state	aq	−154.39	−117.2	144.8	
HCNS ionized; std. state	aq	76.44	92.68	144.4	−40.2
$HCOO^-$ formate	aq	−425.6	−351.0	92.0	−87.9
CH_3COO^- acetate	aq	−486.0	−369.3	86.6	−6.3
HCO_3^- std. state	aq	−689.93(20)	−586.85	98.4(5)	
H_2CO_3 std. state	aq	−699.65	−623.16	187.4	
$HC_2O_4^-$	aq	−818.4	−698.3	149.4	
$H_2C_2O_4$	c	−821.7	−723.7	109.8	91.0
$C_2O_4^{2-}$	aq	−825.1	−673.9	45.6	
H_2CS_3 trithiocarbonic acid	lq	25.1	27.82	233.0	149.8
HF	g	−273.30(70)	−275.4	173.779(3)	29.14
	lq	−299.78	75.40	51.67	
undissoc; std. state	aq	−320.08	−296.86	88.7	
F^-	aq	−332.63	−278.8	−13.8	−106.7
^{2}HF	g	−275.5	−277.27	179.70	29.14
HF_2^- std. state	aq	−649.94	−578.15	92.5	
H_2F_2 dimer	g	−572.66	−544.51	238	44.89
$H_2Fe(CN)_6^{2-}$ std. state	aq	455.6	658.44	218	

TABLE 1.56 Enthalpies and Gibbs Energies of Formation, Entropies, and Heat Capacities of the Elements and Inorganic Compounds (*Continued*)

Substance	Physical state	$\Delta_f H°$ kJ · mol^{-1}	$\Delta_f G°$ kJ · mol^{-1}	$S°$ J · deg^{-1} · mol^{-1}	$C_p°$ J · deg^{-1} · mol^{-1}
HFO	g	98	−86	226.8	35.93
HI	g	26.50(10)	1.7	206.590(4)	29.16
std. state	aq	−55.19	−51.59	111.3	−142.3
HIO undissoc; std. state	aq	−138.1	−99.2	95.4	
HIO_3	c	−230.1			
H_2MoO_4	c	−1046.0			
HN	g	351.5	345.6	181.2	29.2
HN_3	lq	264.0	327.2	140.6	
	g	294.1	328.1	239.0	43.7
H_2N	g	184.9	194.6	195.0	33.9
$^2H_2N_2$ *cis*-diazine	g	207	241	224.09	39.02
HNCO isocyanic acid	g	−116.73	−107.36	238.11	44.85
HNCS isothiocyanic acid	g	127.61	112.88	248.03	46.40
HNO_2	g	−79.5	−46.0	254.1	45.5
HNO_3	lq	−174.1	−80.7	155.60	109.9
	g	−133.9	−73.54	266.9	54.1
std. state	aq	−207.36	−111.34	146.4	−86.6
$H_2N_2O_2$ hyponitrous acid	aq	−57.3	36.0	218	
HO hydroxyl	g	39.0	34.2	183.64	30.00
HO^-	aq	−230.015	−157.28	−10.90	−148.5
HO_2	g	10.5	22.6	229.0	34.9
HO_2^- std. state	aq	−160.33	67.4	23.9	
H_2O	c	−292.72			37.11
	lq	−285.830(40)	−237.14	69.95(3)	75.35
	g	−241.826(40)	−228.61	188.835(10)	33.60
$^1H^2HO$	g	−245.37	−233.18	199.51	33.79
^2H_2O deuterium oxide	g	−249.20	−234.54	198.33	34.25
H_2O_2 hydrogen peroxide	lq	−187.78	−120.42	109.6	89.1
	g	−136.3	−105.6	232.7	43.14
undissoc; std. state	aq	−191.17	−134.10	143.9	
HOCN undissoc; std. state	aq	−154.39	−117.2	144.8	
OCN^- cyanate std. state	aq	−146.02	−97.5	106.7	
HPO_3	c	−948.51			
HPO_4^{2-} std. state	aq	−1299.0(15)	−1089.26	−33.5(15)	
$H_2PO_4^-$ std. state	aq	−1302.6(15)	−1130.39	92.5(15)	
HPH_2O_2 hypophosphorous acid	c	−604.6			
H_3PO_3	c	−964.4			
H_3PO_4	c	−1284.4	−1124.3	110.5	106.1
	lq	−1271.7	−1123.6	150.8	145.06
ionized; std. state	aq	−1277.4	−1018.8	222	
undissoc; std. state	aq	−1288.34	−1142.65	158.2	
$HP_2O_7^{3-}$	aq	−2274.8	−1972.2	46.0	
$H_2P_{27}^{2-}$	aq	−2278.6	−2010.2	163.0	
$H_4P_2O_7$	c	−2241.0			
undissoc; std. state	aq	−2268.6	−2032.2	268	
$HReO_4$	c	−762.3	−656.4	158.2	
HS	g	142.7	113.3	195.7	32.3
HS^- std. state	aq	−16.3(15)	12.05	67.(5)	
H_2S	g	−20.6(5)	−33.4	205.81(5)	34.19
undissoc; std. state	aq	−38.6(15)	−27.87	126.(5)	

(*Continued*)

TABLE 1.56 Enthalpies and Gibbs Energies of Formation, Entropies, and Heat Capacities of the Elements and Inorganic Compounds (*Continued*)

Substance	Physical state	$\Delta_f H°$ kJ · mol^{-1}	$\Delta_f G°$ kJ · mol^{-1}	$S°$ J · deg^{-1} · mol^{-1}	$C_p°$ J · deg^{-1} · mol^{-1}
^2H_2S	g	−23.9	−35.3	215.3	35.76
H_2S_2	g	15.5			51.5
$HSbO_2$ undissoc; std. state	aq	−487.9	−407.5	46.6	
HSCN undissoc; std. state	aq	76.4	97.7	144.3	−40.2
SCN^- std. state	aq	76.44	92.68	144.5	−40.2
HSe^- std. state	aq	15.9	43.9	79.0	
H_2Se	g	29.7	15.9	219.0	34.7
$HSeO_3^-$ std. state	aq	−514.55	−411.54	135.1	
H_2SeO_3	c	−524.46			
undissoc; std. state	aq	−507.48	−426.22	207.9	
$HSeO_4^-$ std. state	aq	−581.6	−452.3	149.4	
H_2SeO_4	c	−530.1			
H_2SiO_3	c	−1188.67	−1092.4	134.0	
undissoc; std. state	aq	−1182.8	−1079.5	109	
H_4SiO_4	c	−1481.1	−1333.0	192	
undissoc; std. state	aq	−1468.6	−1316.7	180	
HSO_3^- std. state	aq	−626.22	−527.8	139.8	
HSO_4^-	aq	−886.9(10)	−755.9	131.7(30)	−84.0
HSO_3Cl	lq	−601.2			
HSO_3F	lq	−795.0			
	g	−753	−691	297	75.24
H_2SO_3 undissoc; std. state	aq	−608.81	−537.90	232.2	
H_2SO_4	lq	−814.0	−689.9	156.90	138.9
std. state	aq	−909.27	−744.63	20.1	293
$H_2SO_4 \cdot H_2O$	lq	−1127.6	−950.3	211.5	214.3
$H_2SO_4 \cdot 2H_2O$	lq	−1427.1	−1199.6	276.4	261.5
$H_2SO_4 \cdot 3H_2O$	lq	−1720.4	−1443.9	345.4	319.1
$H_2SO_4 \cdot 4H_2O$	lq	−2011.2	−1685.8	414.5	386.4
$H_2S_2O_7$	c	−1273.6			
H_2Te	g	99.6		228.9	35.56
H_2WO_4	c	−1131.8	−1003.9	145	113
Indium					
In	c	0	0	57.8	26.7
In^{3+}	aq	−105.0	−98.0	−151.0	
InAs	c	−58.6	−53.6	75.7	47.78
$InBr_3$	c	−428.9			
$InCl_3$	c	−537.2			
InF	g	−203.4			
InH	g	215.5	190.3	207.53	29.58
InI	c	−116.3	−120.5	130.0	
InI_3	c	−238.0			
$InOH^{2+}$	aq	−370.3	−313.0	−88.0	
$In(OH)_2^+$	aq	−619.0	−525.0	25.0	
In_2O_3	c	−925.27	−830.73	104.2	92
InP	c	−88.7	−77.0	59.8	45.44
InS	c	−138.1	−131.8	67	
In_2S_3	c	−427	−412.5	163.6	118.0
In_2Se_3	c	−343			
InSb	c	−30.5	−25.5	86.2	49.5
Iodine					
I atomic	g	106.76(4)	70.2	180.787(4)	20.8

TABLE 1.56 Enthalpies and Gibbs Energies of Formation, Entropies, and Heat Capacities of the Elements and Inorganic Compounds (*Continued*)

Substance	Physical state	$\Delta_f H°$ kJ · mol^{-1}	$\Delta_f G°$ kJ · mol^{-1}	$S°$ J · deg^{-1} · mol^{-1}	$C_p°$ J · deg^{-1} · mol^{-1}
I^- std. state	aq	−56.78(5)	−51.59	106.45(30)	−142.3
I_2	c	0	0	116.14(30)	54.44
	g	62.42(8)	19.37	260.687(5)	36.86
std. state	aq	22.6	16.40	137.2	
I_3^- std. state	aq	−51.5	−51.5	239.3	
IBr	c	−10.5			
	g	40.8	3.7	258.8	36.4
ICl	c	−35.4	−14.05	97.93	55.23
	lq	−23.93	−13.6	135.1	
	g	17.8	−5.5	247.6	35.6
ICl_3	c	−89.5	−22.34	167.4	
IF	g	−95.7	−118.5	236.3	33.4
IF_5	lq	−864.8			
	g	−822.5	−751.5	327.7	99.2
IF_7	g	−961.1	−835.8	347.7	134.5
IO	g	175.1	149.8	245.5	32.9
IO^- std. state	aq	−107.5	−38.5	−5.4	
IO_3^- std. state	aq	−221.3	−128.0	118.4	
IO_4^- std. state	aq	−151.5	−58.6	222	
I_2O_5	c	−158.07			
Iridium					
Ir	c	0	0	35.48	25.06
$IrCl_3$	c	−245.6	180	113	
IrF_6	c	−579.65	−461.66	247.7	
IrO_2	c	−274.1		57.3	57.32
IrS_2	c	−138.0			
Iron					
Fe alpha	c	0	0	27.32	25.09
Fe^{2+} std. state	aq	−89.1	−78.87	−137.7	
Fe^{3+} std. state	aq	−48.5	−4.7	−315.9	
$FeBr_2$	c	−249.8	−238.1	140.7	80.2
std. state	aq	−332.2	−286.81	27.2	
$FeBr_3$	c	−286.2			
	aq	−413.4	−316.7	−68.6	
Fe_3C α-cementite	c	25.1	20.1	104.6	105.9
$FeCl_2$	c	−341.8	−302.3	118.0	76.7
	aq	−423.4	−341.3	−24.7	
$FeCl_3$	c	−399.4	−333.9	142.34	96.65
std. state	aq	−550.2	−398.3	−146.4	
$Fe(CN)_6^{3-}$ std. state	aq	561.9	729.3	270.3	
$Fe(CN)_6^{4-}$ std. state	aq	455.6	694.9	95.0	
$FeCNS^{2+}$ std. state	aq	23.4	71.1	−130	
$FeCO_3$	c	−740.6	−666.7	92.9	82.1
$Fe(CO)_5$	lq	−774.0	−705.3	338.1	240.6
$FeCr_2O_4$	c	−1446.0	−1343.9	146.2	133.8
FeF_2	c	−711.3	−668.6	86.99	68.12
std. state	aq	−754.4	−636.5	−165.3	
FeF_3	c	−1042	−972	98	91.0
	aq	−1046.4	−840.9	−357.3	
FeI_2	c	−113.0	−111.7	167.4	83.7
std. state	aq	−199.6	−182.1	84.9	

(*Continued*)

TABLE 1.56 Enthalpies and Gibbs Energies of Formation, Entropies, and Heat Capacities of the Elements and Inorganic Compounds (*Continued*)

Substance	Physical state	$\Delta_f H^\circ$ kJ · mol^{-1}	$\Delta_f G^\circ$ kJ · mol^{-1}	S° J · deg^{-1} · mol^{-1}	C_p° J · deg^{-1} · mol^{-1}
FeI_3	aq	−214.2	−159.4	18.0	
$FeMoO_4$	c	−1075.0	−975.0	129.3	118.5
Fe_2N	c	−3.8		101.3	70.0
$Fe(NO_3)_3$ std. state	aq	−670.7	−338.5	123.4	
FeO	c	−272.0	−251.4	60.75	49.91
Fe_2O_3 hematite	c	−824.2	−742.2	87.40	103.9
Fe_3O_4 magnetite	c	−1118.4	−1015.4	145.27	143.4
$FeOH^+$ std. state	aq	−324.7	−277.4	−29	
$Fe(OH)^{2+}$ std. state	aq	−290.8	−229.4	−142	
$Fe(OH)_2$	c	−574.0	−490.0	87.9	97.1
$Fe(OH)_3$	c	−833	−705	104.6	101.7
FeS	c	−100.0	−100.4	60.32	50.52
FeS_2 marcasite	c	−167.4	−156.1	53.87	62.39
FeS_2 pyrite	c	−178.2	−166.9	52.92	62.12
$FeSiO_3$	c	−1155		87.5	89.4
Fe_2SiO_4	c	−1479.9	−1379.0	145.18	132.9
$FeSO_4$	c	−928.4	−820.8	107.5	100.6
std. state	aq	−998.3	−823.4	−117.6	
$Fe_2(SO_4)_3$	c	−2583.0	−2262.7	307.5	264.8
std. state	aq	−2825.0	−2243.0	−571.5	
$FeTiO_3$	c	−1246.4		105.9	99.5
$FeWO_4$	c	−1155.0	−1054.0	131.8	114.4
Krypton					
Kr	g	0	0	164.085(3)	20.786
Lanthanum					
La	c	0	0	56.9	27.11
La^{3+}	aq	−707.1	683.7	−217.6	−13.0
$LaCl_3$	c	−1072.2		144.4	108.8
std. state	aq	−1208.8	−1077.4	−50.0	−423.0
$LaCl_3 \cdot 7H_2O$	c	−3178.6	−2713.3	462.8	431.0
LaI_3	c	−668.9			
$La(NO_3)_3$	c	−1254.4			
std. state	aq	−1329.3			
La_2O_3	c	−1793.7	−1705.8	127.32	108.78
$La_2(SO_4)_3$	c	−3941.3		280	
La_2Te_3	c	−724	−714.6	231.63	132.13
Lead					
Pb	c	0	0	64.80(30)	26.84
	g	195.2(8)	162.2	175.375(5)	20.8
Pb^{2+}	aq	0.92(25)	−24.4	18.5(10)	
$Pb(OAc)_2$	c	−964.4			
$Pb(BO_2)_2$	c	−1556	−1450	131	107.1
PbB_4O_7	c	−2858	−2667	167	168
$PbBr_2$	c	−278.7	−261.9	161.5	80.1
	aq	−244.8	−232.3	175.3	
$Pb(CH_3)_4$	lq	97.9			
$Pb(C_2H_5)_4$	lq	52.7		464.6	307.4
$PbCl_2$	c	−359.4	−314.1	136	77.1
	aq	−336.0	−286.9	123.4	
$PbCl_4$	lq	−329.3			
PbClF	c	−534.7	−488.3	121.8	

TABLE 1.56 Enthalpies and Gibbs Energies of Formation, Entropies, and Heat Capacities of the Elements and Inorganic Compounds (*Continued*)

Substance	Physical state	$\Delta_f H°$ kJ · mol^{-1}	$\Delta_f G°$ kJ · mol^{-1}	$S°$ J · deg^{-1} · mol^{-1}	C_p° J · deg^{-1} · mol^{-1}
$PbCO_3$	c	−699.2	−625.5	131.0	87.40
PbC_2O_4	c	−851.4	−750.2	146.0	105.4
$PbCrO_4$	c	−930.9			
PbF_2	c	−664	−617.1	110.5	72.3
	aq	−666.9	−582.0	−17.2	
PbF_4	c	−941.8			
PbI_2	c	−175.5	−173.58	174.9	77.4
	aq	−112.1	−127.6	233.0	
$PbMoO_4$	c	−1051.9	−951.4	166.1	119.70
$Pb(N_3)_2$ monoclinic	c	478.2	624.7	148.1	
$Pb(NO_3)_2$	c	−451.9			
	aq	−416.3	−246.9	303.3	
PbO litharge	c	−219.0	−188.9	66.5	45.8
PbO_2	c	−277.4	−217.3	68.60	64.6
Pb_3O_4	c	−718.4	−601.2	211.3	146.9
$Pb_3(PO_4)_2$	c	−2595.3	−2432.6	353.1	256.3
PbS	c	−100.4	−98.7	91.3	49.4
PbSe	c	−102.9	−101.7	102.5	50.2
$PbSeO_4$	c	−609.2	505.0	167.8	
$PbSiO_3$	c	−1145.7	−1062.1	109.6	90.04
$PbSiO_4$	c	−2023.8	−1909.6	84.01	98.66
Pb_2SiO_4	c	−1363.1	−1252.6	186.6	137.2
$PbSO_3$	c	−669.9			
$PbSO_4$	c	−919.97(40)	−813.0	148.50(60)	103.2
$PbSO_4 \cdot PbO$	c	−1182.0		225.06	150.16
PbTe	c	−70.7	−69.5	110.0	50.5
Lithium					
Li	c	0	0	29.12(20)	24.8
	g	159.3(10)		138.782(10)	
Li^+ std. state	aq	−278.47(8)	−293.30	12.24(15)	68.6
Li_3AlF_6 cryolite	c	−3317	−3152	238.5	215.7
$LiAlH_4$	c	−116.3	−44.7	78.7	83.2
$LiAlO_2$	c	−1188.7	−1126.3	53.3	67.78
$LiBeF_3$	c	−1651.8	−1576.3	89.2	91.8
$LiBH_4$	c	−190.8	−125.0	75.9	82.6
$LiBH_4 \cdot$ tetrahydrofuran	c	−415.5	−220.5	289	
Li_2BeF_4	c	−2274	−2171	130.6	135.3
$LiBO_2$	c	−1032.2	−976.1	51.5	59.8
$Li_2B_4O_7$	c	−3362	−3170	156	183.0
LiBr	c	−351.2	−342.00	74.27	48.91
std. state	aq	−400.03	−397.27	95.81	−73.2
$LiBrO_3$	c	−346.98			
std. state	aq	−345.56	−274.89	174.9	
LiCl	c	−408.6	−384.4	59.3	48.03
	aq	−445.6	−424.6	69.9	−67.8
$LiClO_4$	c	−381.0	−254	126	105
std. state	aq	−407.81	−302.1	195.4	−7.5
Li_2CO_3	c	−1215.9	−1132.12	90.4	99.1
	aq	−1234.1	−1114.6	−29.7	
LiF	c	−616.0	−587.7	35.66	41.6
std. state	aq	−611.12	−571.9	−0.4	−38.1

(*Continued*)

TABLE 1.56 Enthalpies and Gibbs Energies of Formation, Entropies, and Heat Capacities of the Elements and Inorganic Compounds (*Continued*)

Substance	Physical state	$\Delta_f H°$ kJ · mol^{-1}	$\Delta_f G°$ kJ · mol^{-1}	$S°$ J · deg^{-1} · mol^{-1}	$C_p°$ J · deg^{-1} · mol^{-1}
LiH	c	−90.5	−68.45	20.04	27.96
LiI	c	−270.4	−270.3	86.8	51.0
std. state	aq	−333.67	−344.8	124.7	−73.6
$LiIO_3$	c	−503.38			
std. state	aq	−499.82	−421.33	131.4	−55.2
Li_3N	c	−164.6	−128.6	62.59	75.27
$LiNO_2$	c	−372.4	−302.0	96.0	
$LiNO_3$	c	−483.1	−381.1	90.0	
std. state	aq	−485.9	−404.5	160.2	−18.0
Li_2O	c	−597.9	−561.2	37.6	
Li_2O_2	c	−634.3	−578.9	56.5	70.6
LiOH	c	−484.9	−439	42.82	49.7
std. state	aq	−508.40	−451.9	7.1	
Li_3PO_4	c	−2095.8			
Li_2SiO_3	c	−1648.1	−1557.2	79.8	99.1
$Li_2Si_2O_5$	c	−2561	−2417	125.5	138.1
Li_2SO_4	c	−1436.4	−1321.7	115.1	117.6
std. state	aq	−1466.2	−1331.2	7.3	−155.6
Li_2TiO_3	c	−1670.7	−1579.8	91.8	109.9
Lutetium					
Lu	c	0	0	50.96	26.86
Lu^{3+}	aq	−665.0	−628.0	−264.0	25.0
$LuCl_3$	c	−945.6			
std. state	aq	−1167.0	−1021.0	−96.0	−385.0
LuI_3	c	−548.0			
Lu_2O_3	c	−1878.2	−1789.1	109.96	101.75
Magnesium					
Mg	c	0	0	32.67(10)	24.87
	g	147.1(8)		148.648(3)	
Mg^{2+} std. state	aq	−467.0(6)	−454.8	−137.(4)	
$MgAl_2O_4$	c	−2299	−2177	89.0	116.20
$MgBr_2$	c	−524.3	−503.8	117.2	73.16
std. state	aq	−709.94	−662.8	26.8	
$MgBr_2 \cdot 6H_2O$	c	−2410.0	−2056.0	397	
$MgCl_2$	c	−641.3	−591.8	89.63	71.38
std. state	aq	−801.15	−717.1	−25.1	
$MgCl_2 \cdot 6H_2O$	c	−2499.0	−2115.0	315.1	
$Mg(ClO_4)_2$	c	−568.90			
std. state	aq	−725.51	−472.0	225.4	
$Mg(ClO_4)_2 \cdot 6H_2O$	c	−2445.6	−1863.1	520.1	
$MgCO_3$	c	−1095.8	−1012.1	65.7	75.51
MgC_2O_4	c	−1269.0			
std. state	aq	−1292.0	−1128.8	−92.5	
MgF_2	c	−1124.2(12)	1071.1	57.2(5)	61.5
Mg_2Ge	c	−108.8	−105.9	86.48	69.54
MgH_2	c	−75.3	−35.9	31.1	35.4
MgI_2	c	−364.0	−358.2	129.7	74.8
std. state	aq	−577.22	−558.1	84.5	
Mg_3N_2	c	−461.1	−400.9	87.9	104.5
$MgNH_4PO_4 \cdot 6H_2O$	c	−3681.9			
$Mg(NO_3)_2$	c	−790.65	−589.5	164.0	141.9
std. state	aq	−881.6	−677.4	154.8	

TABLE 1.56 Enthalpies and Gibbs Energies of Formation, Entropies, and Heat Capacities of the Elements and Inorganic Compounds (*Continued*)

Substance	Physical state	$\Delta_f H°$ kJ · mol^{-1}	$\Delta_f G°$ kJ · mol^{-1}	$S°$ J · deg^{-1} · mol^{-1}	$C_p°$ J · deg^{-1} · mol^{-1}
$Mg(NO_3)_2 \cdot 6H_2O$	c	−2613.3	−2080.7	452	
MgO microcrystal	c	−601.6(3)	−569.3	26.95(15)	37.2
$Mg(OH)_2$	c	−924.7	−833.7	63.24	77.25
std. state	aq	−926.8	−769.4	−149.0	
$Mg_3(PO_4)_2$	c	−3780.7	−3538.8	189.20	213.47
MgS	c	−346.0	−341.8	50.3	45.6
$MgSeO_4$	c	−968.51			
std. state	aq	−1066.1	−896.2	−84.1	
Mg_2Si	c	−77.8	−77.1	81.6	67.9
$MgSiO_3$ clinoenstatite	c	−1548.9	−1462.0	67.8	81.9
Mg_2SiO_4 forsterite	c	−2174.0	−2055.1	95.1	118.5
$Mg_3Si_4O_{10}(OH)_2$ talc	c	−5922.5	−5543.0	260.7	321.8
$MgSO_3 \cdot 3H_2O$	c	−1931.8			
$MgSO_3 \cdot 6H_2O$	c	−2817.5			
$MgSO_4$	c	−1284.9	−1170.6	91.6	96.5
std. state	aq	−1376.1	−1199.5	−118.01	
$MgSO_4 \cdot H_2O$ kieserite	c	−1602.1	−1428.8	126.4	
$MgSO_4 \cdot 7H_2O$ epsomite	c	−3388.71	−2871.9	372	
$MgTiO_3$	c	−1497.6	−1420.1	111.08	91.88
Mg_2TiO_4	c	−2164.0	−2048	115.0	129
$MgTi_2O_5$	c	−2509	−2369	135.6	146.9
$Mg_2V_2O_7$ triclinic	c	−2835.9	−2645.29	200.4	203.47
$MgWO_4$	c	−1516	−1404	101.2	109.1
Manganese					
Mn	c	0	0	32.01	26.30
Mn^{2+} std. state	aq	−220.75	−228.1	−73.6	50
$MnBr_2$	c	−384.9	−372	138.1	75.31
std. state	aq	−464.0	−409.2		
Mn_3C	c	−4.6	5.4	98.7	93.51
$MnCl_2$	c	−481.3	−440.5	118.20	72.9
std. state	aq	−555.05	−490.8	38.9	−222
$MnCO_3$	c	−894.1	−816.7	85.8	81.5
$Mn_2(CO)_{10}$	c	−1677.4			
MnF_2	c	−795.0	−749	92.26	67.99
MnI_2	c	−242.7		150.6	75.35
	aq	−331.0			
$Mn(NO_3)_2$	c	−576.26			
std. state	aq	−635.6	−451.0	218.0	−121.0
MnO	c	−385.2	−362.9	59.8	45.4
MnO_2	c	−520.1	−465.2	53.1	54.1
Mn_2O_3	c	−959.0	−881.2	110.5	107.7
MnO_4^-	aq	−541.4	−447.3	191.2	−82.0
MnO_4^{2-}	aq	−653.0	−500.8	59	
Mn_3O_4	c	−1387.8	−1283.2	155.6	139.7
$Mn_3(PO_4)_2$	c	−3116.7			
MnS	c	−214.2	−218.4	78.2	50.0
MnSe	c	−106.7	−111.7	90.8	51.0
$MnSiO_3$	c	−1320.9	−1240.6	89.1	86.4
$MnSiO_4$	c	−1730.5	−1632.1	163.2	129.9
$MnSO_4$	c	−1065.3	−957.42	112.1	100.4
std. state	aq	−1130.1	−972.8	−53.6	−243
$MnTiO_3$	c	−1355.6		105.9	99.8

(*Continued*)

TABLE 1.56 Enthalpies and Gibbs Energies of Formation, Entropies, and Heat Capacities of the Elements and Inorganic Compounds (*Continued*)

Substance	Physical state	$\Delta_f H°$ kJ · mol^{-1}	$\Delta_f G°$ kJ · mol^{-1}	$S°$ J · deg^{-1} · mol^{-1}	$C_p°$ J · deg^{-1} · mol^{-1}
Mercury					
Hg	lq	0	0	75.90(12)	28.00
	g	61.38(4)	31.8	174.971(5)	20.8
Hg^{2+}	aq	170.21(20)		−36.19(80)	
Hg^{+}	aq	166.87(50)		65.74(80)	
$HgBr_2$	c	−170.7	−153.1	172.0	75.3
Hg_2Br_2	c	−206.9	−181.1	218.0	104.6
$Hg(CH_3)_2$	lq	59.8	140.2	209	
$Hg(C_2H_5)_2$	lq	30.1			
$HgCl_2$	c	−224.3	−178.6	146.0	73.9
Hg_2Cl_2	c	−265.37(40)	−210.7	191.6(8)	102.0
$Hg(CN)_2$	c	263.6			
Hg_2CO_3	c	−553.5	−468.1	180.0	
HgC_2O_4	c	−678.2			
HgF_2	c	−405	−362	134.3	74.86
Hg_2F_2	c	−485	−469	161	100.4
HgI_2	c	−105.4	−101.7	180.0	77.75
Hg_2I_2	c	−121.3	−111.1	233.5	105.9
$Hg_2(N_3)_2$	c	594.1	746.4	205	
HgO	c	−90.79(12)	−58.49	70.25(30)	44.06
HgS	c	−58.2	−50.6	82.4	48.4
$HgSO_4$	c	−707.5	−594		
Hg_2SO_4	c	−743.09(40)	−625.8	200.70(20)	131.96
HgTe	c	−42.0			
Molybdenum					
Mo	c	0	0	28.71	24.13
$MoBr_3$	c	−284	−259	175	105.4
$MoCl_4$	c	−477	−402	224	128
$MoCl_5$	c	−527	−423	238	155.6
$MoCl_6$	c	−523	−391	255	175
$Mo(CO)_6$	c	−982.8	−877.8	325.9	242.3
MoF_6	lq	−1585.66	−1473.17	259.69	169.8
MoO_2	c	−588.9	−533.0	46.3	56.0
MoO_3	c	−745.2	−668.1	77.8	75.0
MoO_4^{2-} std. state	aq	−997.9	−836.4	27.2	
MoS_2	c	−235.1	−225.9	62.57	63.56
Mo_2S_3	c	−270.3	−278.6	181.2	109.3
Neodymium					
Nd	c	0	0	71.6	27.5
Nd^{3+} std. state	aq	−696.2	−671.5	−206.7	−21
$NdCl_3$	c	−1041.0			113
std. state	aq	−1197.9	−1065.7	−37.7	−431
NdF_3	c	−1657.0			
$Nd(NO_3)_3$	c	−1230.9			
Nd_2O_3	c	−1807.9	−1720.9	158.6	111.3
Neon					
Ne	g	0	0	146.328(3)	20.786
Neptunium					
Np	c	0	0		29.46
NpF_6	c	−1937			
NpO_2	c	−1029	−979	80.3	66.1

TABLE 1.56 Enthalpies and Gibbs Energies of Formation, Entropies, and Heat Capacities of the Elements and Inorganic Compounds (*Continued*)

Substance	Physical state	$\Delta_f H^\circ$ kJ · mol^{-1}	$\Delta_f G^\circ$ kJ · mol^{-1}	S° J · deg^{-1} · mol^{-1}	C_p° J · deg^{-1} · mol^{-1}
Nickel					
Ni	c	0	0	29.87	26.1
Ni^{2+} std. state	aq	−54.0	−45.6	−128.9	
$Ni(OAc)_2$ std. state	aq	−1025.9	−784.5	44.4	
$NiBr_2$	c	−212.1			
	aq	−297.1	−253.6	36.0	
$NiCl_2$	c	−305.3	−259.0	97.7	71.66
std. state	aq	−388.3	−307.9	−15.1	
$Ni(CN)_4^{2-}$ std. state	aq	367.8	472.0	218	
$Ni(CO)_4$	lq	−633.0	−588.2	313	404.6
	g	−602.9	−587.2	410.6	145.2
NiC_2O_4	c	−856.9			
NiF_2	c	−651.5	−604.2	73.6	64.1
	aq	−719.2	−603.3	−156.5	
NiI_2	c	−78.8			
	aq	−164.4	−149.0	93.7	
$Ni(NO_3)_2$	c	−415.1			
std. state	aq	−468.6	−268.6	164.0	
NiO	c	−240.6	−211.7	38.00	44.31
Ni_2O_3	c	−489.5			
$NiOH^+$	aq	−287.9	−227.6	−71.0	
$Ni(OH)_2$	c	−529.7	−447.3	88.0	
NiS	c	−82.0	−79.5	53.0	47.1
Ni_3S_2	c	−216.0	−210	133.9	117.7
NiS_2	c	−131.4	−124.7	72	70.6
$NiSO_4$	c	−872.9	−759.8	92.0	138.0
std. state	aq	−963.2	−790.3	−108.8	327.9
$NiSO_4 \cdot 7H_2O$	c	−2976.3	−2462.2	378.94	364.59
$NiWO_4$	c	−1128.4		118.0	136.0
Niobium					
Nb	c	0	0	36.4	24.67
$NbBr_5$	c	−556	−508	258.8	147.9
NbC	c	−138.9	−136.8	34.98	36.23
$NbCl_5$	c	−797.5	−683.3	210.5	148.1
NbF_5	c	−1813.8	−1699.0	160.3	134.7
NbI_5	c	−268.6		343	155.6
NbN	c	−236.4	−205.9	34.5	39.0
NbO	c	−405.8	−392.6	48.1	41.3
NbO_2	c	−796.2	−740.5	54.5	57.45
Nb_2O_5	c	−1899.5	−1765.8	137.3	132.0
$NbOCl_3$	c	−879.5	−782	159	120.0
Nitrogen					
N atomic	g	472.68(40)		153.301(3)	
N_2	g	0	0	191.609(4)	29.124
N_3^-	aq	275.1	348.2	107.9	
NCl_3	lq	230.0			
NF_2	g	43.1	57.8	249.9	41.0
NF_3	g	−132.1	−90.6	260.8	53.37
H_2NOH	c	−114.2			
N_2F_2 *cis*	g	69.5	109	259.8	49.96
trans	g	82.0	120.5	262.6	53.47

(*Continued*)

TABLE 1.56 Enthalpies and Gibbs Energies of Formation, Entropies, and Heat Capacities of the Elements and Inorganic Compounds (*Continued*)

Substance	Physical state	$\Delta_f H°$ kJ · mol^{-1}	$\Delta_f G°$ kJ · mol^{-1}	$S°$ J · deg^{-1} · mol^{-1}	$C_p°$ J · deg^{-1} · mol^{-1}
N_2F_4	g	−8.4	79.9	301.2	79.2
N_2H_4 hydrazine	lq	50.6	149.3	121.2	98.84
$N_2^2H_4$ hydrazine-d_4	g	81.6	150.9	248.86	55.52
$N_2H_5^+$ std. state	aq	−7.5	82.4	151	70.3
N_2H_5Br	c	−155.6			
std. state	aq	−128.9	−21.8	233.1	−71.6
N_2H_5Cl	c	−197.1			
std. state	aq	−174.9	−49.0	207.1	−66.1
$N_2H_5Cl \cdot HCl$	c	−367.4			
N_2H_5OH	lq	−242.7			
undissoc; std. state	aq	−251.50	−109.2	207.9	73.2
$N_2H_5NO_3$	c	−251.58			
std. state	aq	−215.10	−28.91	297	
$(N_2H_5)_2SO_4$	c	−959.0			
std. state	aq	−924.7	−579.9	322	−151
NO	g	91.29	87.60	210.76	29.85
NOBr	g	82.23	82.42	273.7	45.48
NOCl	g	51.71	66.10	261.68	44.7
NOF	g	−66.5	−51.0	248.02	41.3
NOF_3	g	−163	−96	278.40	67.86
NO_2	g	33.1	51.3	240.1	37.2
NO_2^-	aq	−104.6	−32.2	123.0	−97.5
NO_2Cl	g	12.6	54.4	272.19	53.19
NO_2F	g	−109	−66	260.2	49.8
NO_3	g	69.41	114.35	252.5	46.9
NO_3^-	aq	−206.85(40)	−111.3	146.70(40)	−86.6
N_2O	g	81.6	103.7	220.0	38.62
N_2O_2	g	170.37	202.88	287.52	63.51
$N_2O_2^{2-}$ hyponitrite	aq	−17.2	138.9	27.6	
N_2O_3	g	86.6	142.4	314.7	72.72
N_2O_4	lq	−19.5	97.5	209.20	142.71
	g	11.1	99.8	304.38	79.2
N_2O_5	g	11.3	117.1	355.7	95.30
NSF	g			259.8	44.1
Osmium					
Os	c	0	0	32.6	24.7
$OsCl_3$	c	−190.4	−121	130	
$OsCl_4$	c	−254.8	−159	155	
OsF_6	g			358.1	120.8
OsO_4	c	−394.1	−305.0	143.9	
	g	−337.2	−292.8	293.8	74.1
Oxygen					
O atomic	g	249.18(10)	231.7	161.059(3)	21.9
O_2	g	0	0	205.152(5)	29.4
O_3	g		142.7	163.2	238.92
OF_2	g	24.5	41.8	247.5	57.11
O_2F_2	g	18.0	61.42	268.11	54.06
OH^-	aq	−230.015(40)	−157.28	−10.90(20)	−148.5
Palladium					
Pd	c	0	0	37.61	25.94
Pd^{2+} std. state	aq	149.0	176.6	−184.0	

TABLE 1.56 Enthalpies and Gibbs Energies of Formation, Entropies, and Heat Capacities of the Elements and Inorganic Compounds (*Continued*)

Substance	Physical state	$\Delta_f H°$ kJ · mol^{-1}	$\Delta_f G°$ kJ · mol^{-1}	$S°$ J · deg^{-1} · mol^{-1}	$C_p°$ J · deg^{-1} · mol^{-1}
$PdBr_2$	c	−104.2			
$PdBr_4^{2-}$ std. state	aq	−384.9	−318.0	247	
$PdCl_2$	c	−171.5	−125.1	105	
$PdCl_4^{2-}$ std. state	aq	−550.2	−416.7	167	
Pd_2H	c	−19.7	−5.0	91.6	
PdO	c	−85.4		56.1	31.5
PdS	c	−75	−67	46	
PdS_2	c	−81.2	−74.5	80	
Phosphorus					
P white	c	0	0	41.09(25)	23.83
	g	316.5(10)	280.1	163.1199(3)	20.8
red, V	c	−17.46	−12.46	22.85	21.19
P_2	g	144.0(20)		218.123(4)	
P_4	g	58.9(3)	24.4	280.01(50)	67.16
PBr_3	lq	−184.5	−175.5	240.2	
	g	−139.3	−162.8	348.15	76.02
PBr_5	c	−269.9			
PCl_3	lq	−319.7	−272.4	217.2	
	g	−227.1	−267.8	311.8	71.8
PCl_5	c	−443.5			
	g	−374.9	−305.0	364.6	112.8
PF_3	g	−958	−937	273.1	58.69
PF_5	g	−1594.4	−1520.7	300.8	84.8
PH_3	g	5.4	13.4	210.24	37.10
std. state	aq	−9.50	25.31	120.1	
PH_4Br	c	−127.6	−47.7	110.0	
PH_4Cl	c	−145.2			
PH_4I	c	−69.9	0.8	123.0	109.6
PH_4OH undissoc; std. state	aq	−295.35	−211.88	190.0	
PI_3	c	−45.6			
PO_2	g	−279.9	−281.6	252.1	39.5
PO_3^-	aq	−977.0			
PO_4^{3-} std. state	aq	−1277.4	−1018.8	−220.5	
$P_2O_7^{4-}$ std. state	aq	−2271.1	−1919.2	−117.0	
$(P_2O_3)_2$ dimer	c	−1640.1			
P_4O_{10}	c	−3009.9	−2723.3	228.78	211.71
$POBr_3$	c	−458.6			
	g	−389.11	−390.91	−359.84	89.87
$POCl_3$	lq	−597.1	−520.9	222.46	138.82
	g	−558.5	−512.9	325.5	84.94
$POClF_2$	g	−970.7	−924.1	301.68	68.83
$POCl_2F$	g	−765.7	−721.6	320.38	79.32
POF_3	g	−1254.0	−1206	285.4	68.82
$PSCl_3$	g	−363.2	−347.7	337.23	89.83
PSF_3	g	−1009	−985	298.1	74.55
P_4S_3	c	−155	−159	201	146
Platinum					
Pt	c	0	41.63	25.87	
$PtBr_2$	c	−82.0			
$PtBr_3$	c	−120.9			
$PtBr_4$	c	−156.5			

(*Continued*)

TABLE 1.56 Enthalpies and Gibbs Energies of Formation, Entropies, and Heat Capacities of the Elements and Inorganic Compounds (*Continued*)

Substance	Physical state	$\Delta_f H°$ kJ · mol^{-1}	$\Delta_f G°$ kJ · mol^{-1}	$S°$ J · deg^{-1} · mol^{-1}	$C_p°$ J · deg^{-1} · mol^{-1}
$PtCl_2$	c	−123.4		117	
$PtCl_3$	c	−182.0	−134	151	
$PtCl_4$	c	−3218			
$PtCl_4^{2-}$	c	−231.8	−172	176	
$PtCl_4^{2-}$ std. state	aq	−499.2	−361.5	155	
$PtCl_6^{2-}$ std. state	aq	−668.2	−482.8	220.1	
PtF_6	g			348.3	122.8
PtI_4	c	−72.8			
PtS	c	−81.6	−76.2	55.06	43.39
PtS_2	c	−108.8	−99.6	74.68	65.90
Plutonium					
Pu	c	0	0	51.5	35.5
Pu^{3+}	aq	−579.9	−587.9	−163	
Pu^{4+}	aq	−579.9	−1490		
$PuBr_3$	c	−831.8	−804.6	192.88	107.86
$PuCl_3$	c	−961.5	−892.7	159.00	102.84
$PuCl_4$	c	−1381			
PuF_3	c	−1552	−1478.8	112.97	96.82
PuF_4	c	−1732	−1644.7	161.9	120.8
PuF_6	c	25.48	27.2	222.59	167.36
PuH_2	c	−139.3	−101.7	59.8	39.0
PuH_3	c	−138	−82.4	64.9	43.2
PuI_3	c	−648.5	−643.9	214.2	111.8
PuO	c	−565	−538.9	70.7	51.3
PuO_2	c	−1058.1	−1005.8	82.4	68.6
Pu_2O_3 beta	c	−1715.4	−1632.3	152.3	131.0
$Pu(SO_4)_2$	c	−2200.8	−1969.5	163.18	181.96
PuS	c	−439.3	−436.7	78.24	53.97
Pu_2S_3	c	−989.5	−985.5	192.46	129.66
Polonium					
Po	c	0	0	62.8	26.4
PoO_2	c	−251	−197	71	61.5
Potassium					
K	c	0	0	64.68(20)	29.60
	lq	2.284	0.264	71.46	32.72
	g	89.0(8)		160.341(3)	
K^+ std. state	aq	−252.14(8)	−283.26	101.20(20)	21.8
KOAc acetate	c	−723.0			
	aq	−738.39	−652.66	189.1	15.5
$KAg(CN)_2$	aq	18.0	22.2	297	
$KAgCl_2$	aq	−497.4	−498.7	333.9	
K_2AgI_3	aq	−686.6	−720.5	458.1	
$KAlCl_4$	c	97	−1094	197	156.4
K_3AlCl_6	c	−2092.0	−1938	377	248.9
K_3AlF_6	c	−3358.1		284.5	221.1
$KAl(SO_4)_2$	c	−2470.2	−2240.1	204.47	192.92
K_3AsO_4 std. state	aq	−1645.27	−1498.29	144.8	
KBF_4	c	−1887	−1785	133.9	114.48
std. state	aq	−1827.2	−1770.3	285	
KBH_4	c	−227.4	−160.2	106.31	96.57
std. state	aq	−204.22	−168.99	212.97	

TABLE 1.56 Enthalpies and Gibbs Energies of Formation, Entropies, and Heat Capacities of the Elements and Inorganic Compounds (*Continued*)

Substance	Physical state	$\Delta_f H^\circ$ kJ · mol⁻¹	$\Delta_f G^\circ$ kJ · mol⁻¹	S° J · deg⁻¹ · mol⁻¹	C_p° J · deg⁻¹ · mol⁻¹
KBO_2	c	−981.6	−923.4	79.98	66.7
std. state	aq	−1024.75	−962.19	65.3	
$K_2B_4O_7$	c	−3334.2	−3136.8	208	170.5
KBr	c	−393.8	−380.7	95.9	52.3
std. state	aq	−373.92	−387.23	184.9	−120.1
$KBrO_3$	c	−360.2	−271.2	149.2	105.2
	aq	−319.45	−264.72	264.22	
$KBrO_4$	c	−287.86	−174.47	170.01	120.2
KCl	c	−436.5	−408.5	82.55	51.29
std. state	aq	−419.53	−414.51	159.0	−114.6
KClO std. state	aq	−359.4	−320.1	146	
$KClO_2$ std. state	aq	−318.8	−266.1	203.8	
$KClO_3$	c	−397.73	−296.31	143.1	100.3
std. state	aq	−356.35	−291.29	264.9	
$KClO_4$	c	−432.8	−303.1	151.0	112.41
std. state	aq	−381.71	−291.88	284.5	
KCN	c	−113.1	−101.9	128.52	66.3
std. state	aq	−101.7	−110.9	196.7	
K_2CO_3	c	−1151.0	−1063.5	155.5	114.44
std. state	aq	−1181.90	−1094.41	148.1	
$K_2C_2O_4$	c	−1346.0			
	aq	−1329.72			
K_2CrO_4	c	−1403.7	−1295.8	200.12	145.98
std. state	aq	−1385.91	−1294.36	255.2	
$K_2Cr_2O_7$	c	−2061.5	−1882.0	291.2	219.2
$K_2CuCl_4 \cdot 2H_2O$	c	−1707.1	−1492.9	355.43	253.22
KF	c	−567.2	−537.8	66.5	48.98
std. state	aq	−585.01	−562.08	88.7	−84.9
$K_3Fe(CN)_6$	c	−249.8	−129.7	426.06	
std. state	aq	−139.4	−120.5	577.8	
$K_4Fe(CN)_6$	c	−594.1	−453.1	418.8	322.2
std. state	aq	−554.0	−438.11	505.0	
K formate	c	−679.73			
std. state	aq	−677.93	−634.3	192	−66.1
K glycinate	aq	−722.16	−598.23	221.8	
KH	c	−57.72	−53.01	50.21	37.91
K_2HAsO_4 std. state	aq	−1411.10	−1281.22	203.3	
KH_2AsO_4	c	−1180.7	−1036.0	155.02	126.73
std. state	aq	−1161.94	−1036.54	218	
$KHCrO_4$ std. state	aq	−1130.5	−1048.1	286.6	
$KHCO_3$	c	−963.2	−863.6	115.5	
std. state	aq	−944.33	−870.10	193.7	
KHC_2O_4 std. state	aq	−1070.7	−981.7	251.9	
KHF_2	c	−927.7	−859.7	104.3	76.94
	aq	−902.32	−861.40	195.0	
$KHgBr_3$	c	−550.20			
std. state	aq	−545.6	−542.7	360	
K_2HgBr_4	c	−963.6			
std. state	aq	−935.5	−937.6	515	
$KHgCl_3$	c	−671.1			
std. state	aq	−641.0	−592.5	314	

(*Continued*)

TABLE 1.56 Enthalpies and Gibbs Energies of Formation, Entropies, and Heat Capacities of the Elements and Inorganic Compounds (*Continued*)

Substance	Physical state	$\Delta_f H^\circ$ kJ · mol^{-1}	$\Delta_f G^\circ$ kJ · mol^{-1}	S° J · deg^{-1} · mol^{-1}	C_p° J · deg^{-1} · mol^{-1}
$K_2Hg(CN)_4$	c	−32.2			
std. state	aq	21.8	51.9	510	
K_2HgI_4	c	−775.0			
std. state	aq	−739.7	−778.2	565	
KH_2PO_4	c	−1568.33	−1415.95	134.85	116.57
std. state	aq	−1548.67	−1622.85	192.9	
K_2HPO_4 std. state	aq	−1796.90	−1655.78	171.5	
$K_2H_2P_2O_7$	c	−2815.8			
	aq	−2783.2	−2576.9	368	
$K_3HP_2O_7$	aq	−3032.1	−2822.1	351	
KHS	c	−265.10			75.3
std. state	aq	−269.9	−271.21	165.3	
$KHSO_3$	aq	−878.60	−811.07	242.3	
$KHSO_4$	c	−1160.6	−1131.4	138.1	
std. state	aq	−1139.72	−1039.26	234.3	−63.0
KI	c	−327.9	−324.9	106.3	52.9
	aq	−307.57	−334.85	213.8	−120.5
KIO_3	c	−510.43	−418.4	151.46	106.48
	aq	−473.6	−411.3	220.9	
KIO_4	c	−467.23	−361.41	175.7	
	aq	−403.8	−341.8	322	
$KMnO_4$	c	−837.2	−737.6	171.71	117.6
K_2MoO_4	c	−1498.71			
std. state	aq	−1502.5	−1402.9	232.2	
KNH_2 amide	c	−128.9			
KNO_2	c	−369.82	−306.60	152.09	107.40
std. state	aq	−356.9	−315.5	225.5	
KNO_3	c	−494.63	−394.93	133.05	96.4
std. state	aq	−459.74	−394.59	249.0	−64.9
$K_2Ni(CN)_4$ std. state	aq	−136.8	−94.6	423	
K_2O	c	−361.5	−322.1	94.1	83.7
KO_2	c	−284.9	−239.4	122.5	77.53
K_2O_2	c	−494.1	−425.1	102.0	110
KOCN cyanate	c	−418.65			
std. state	aq	−398.3	−380.7	209.2	
KOH	c	−424.7	−378.7	78.9	64.9
std. state	aq	−482.37	−440.53	91.6	−126.8
K_2PdBr_4	c	−938.1			
std. state	aq	−889.5	−884.5	452	
K_3PO_4	c	−1950.2			
std. state	aq	−2034.7	−1868.6	87.9	
$K_4P_2O_7$	aq	−3280.7	−3052.2	293	
K_2PtBr_4	c	−915.0			
std. state	aq	−872.8	−828.4	326.4	
K_2PtBr_6	c	−1021.3			
std. state	aq	−975.3	−898.7	368	
K_2PtCl_4	c	−1054.4			180.2
std. state	aq	−1003.7	−928.0	360	
K_2PtCl_6	c	−1229.3	−1078.6	333.9	205.60
std. state	aq	−1171.8	−1049.4	424.7	
K_2ReCl_6	c	−1310.4	−1172.8	371.71	214.68
std. state	aq	−1266.92	−1156.0	460	

TABLE 1.56 Enthalpies and Gibbs Energies of Formation, Entropies, and Heat Capacities of the Elements and Inorganic Compounds (*Continued*)

Substance	Physical state	$\Delta_f H°$ kJ · mol^{-1}	$\Delta_f G°$ kJ · mol^{-1}	$S°$ J · deg^{-1} · mol^{-1}	$C_p°$ J · deg^{-1} · mol^{-1}
$KReO_4$	c	−1097.0	−994.5	167.82	122.55
std. state	aq	−1039.7	−977.8	303.8	8.4
K_2S	c	−380.7	−364.0	105.0	74.7
std. state	aq	−471.5	−480.7	190.4	
K_2S_2	c	−432.2			
	aq	−474.5	−487.0	233.5	
KSCN	c	−200.16	−178.32	124.26	88.53
std. state	aq	−175.94	−190.58	246.9	−18.4
K_2SeO_3	c	−979.5			
std. state	aq	−1013.8	−936.4	218.0	
K_2SeO_4	c	−1110.02	−1002.9	222	
std. state	aq	−1103.7	−1007.9	259.0	
K_2SiF_6	c	−2956.0	−2798.7	225.9	
std. state	aq	−2893.7	−2766.0	327.2	
K_2SiO_3	c	−1548.1	−1455.7	146.1	118.4
K_2SnBr_6	c	−1218.0	−1160.2	443.1	246.0
K_2SnCl_6	c	−1477.0	−1333.0	366.5	246.0
K_2SO_3	c	−1125.5			
std. state	aq	−1140.1	−1053.1	176	
K_2SO_4	c	−1437.8	−1321.4	175.6	131.5
	aq	−1414.0	−1311.1	225.1	−251.0
K_2SO_6	c	−1437.7	−1319.6	175.5	131.3
std. state	aq	−1414.02	−1311.14	225.1	−251
$K_2S_2O_3$	c	−1173.6			
std. state	aq	−1156.9	−1089.1	272	
$K_2S_2O_4$	aq	−1258.1	−1166.9	297	
$K_2S_2O_7$	c	−1986.6	−1791.6	255	
$K_2S_2O_8$	c	−1916.10	−1697.41	278.7	213.2
std. state	aq	−1849.3	−1681.6	449.4	
$K_2S_4O_6$	c	−1780.7	−1613.43	309.66	230.79
std. state	aq	−1728.8	−1607.1	462.3	−24.3
KSO_3F	c	−1159.0			
K_2UO_4	c	−1921.3			
KVO_4	c	−1154.8			
std. state	aq	−1140.6	−1066.9	155	
$K_2Zn(CN)_4$	c	−100.0			
std. state	aq	−162.3	−119.7	431	
Praseodymium					
Pr	c	0	0	73.2	27.20
Pr^{3+} std. state	aq	−704.6	−679.1	−209.0	−29.0
$Pr(OAc)_3$ std. state	aq	−2147.52	−1805.56	164.9	
$PrCl_3$	c	−1056.9			100.0
std. state	aq	−1206.3	−1072.8	−42.0	−439.0
$Pr(NO_3)_3$	c	−1229.3			
Pr_2O_3	c	−1809.6			117.40
Promethium					
$PmCl_3$	c	−1054.0			
Protactinium					
Pa	c	0	0	51.8	
Pa^{4+}	aq	−619.2			
$PaBr_4$	c	−824.0	−787.9	234.0	
$PaBr_5$	c	−862	−820	289	

(*Continued*)

TABLE 1.56 Enthalpies and Gibbs Energies of Formation, Entropies, and Heat Capacities of the Elements and Inorganic Compounds (*Continued*)

Substance	Physical state	$\Delta_f H°$ kJ · mol^{-1}	$\Delta_f G°$ kJ · mol^{-1}	$S°$ J · deg^{-1} · mol^{-1}	$C_p°$ J · deg^{-1} · mol^{-1}
$PaCl_4$	c	−1043.1	−953.0	192.0	
$PaCl_5$	c	−1144.7	−1034.3	238.0	
Radium					
Ra	c	0	0	71	
Ra^{2+}	aq	−527.6	−561.5	54.0	
$RaCl_2$ std. state	aq	−861.9	−823.8	167.0	
$Ra(NO_3)_2$	c	−992	−796.2	222	
std. state	aq	−942.2	−784.1	347.0	
$RaSO_4$	c	−1471.1	−1365.7	138	
std. state	aq	−1436.8	−1306.2	75.0	
Radon					
Rn	g	0	0	176.235	20.79
Rhenium					
Re	c	0	0	36.9	25.5
	g	769.9	724.6	188.9	20.8
Re^- std. state	aq	46.0	10.1	230.0	
$ReBr_3$	c	−167.0			
$ReCl_3$	c	−264	−188	123.9	92.4
$ReCl_6^{2-}$ std. state	aq	−761	−590	251	
ReO_2	c	−423	−368	172	
ReO_3	c	−605.0	−531	257.3	
Re_2O_7	c	−1240.1	−1066.1	207.1	166.1
	g	−1100.0	−994.0	452.0	
Rhodium					
Rh	c	0	0	31.51	24.98
$RhCl_3$	c	−299.2			
Rh_2O_3	c	−343.0		110.9	104.0
Rubidium					
Rb	c	0	0	76.78(30)	31.06
	g	80.9(8)	53.1	170.094(3)	20.8
Rb^+ std. state	aq	−251.12(10)	−283.97	121.75(25)	
Rb acetate	aq	−737.2	−653.3	207.9	
$RbBO_2$	c	−971.0	−913.0	94.3	74.1
RbBr	c	−394.59	−381.79	109.96	52.84
std. state	aq	−372.71	−387.94	203.93	
$RbBrO_3$	c	−367.27	−278.11	161.1	
Rb_2CO_3	c	−1136.0	−1051.0	181.33	117.61
std. state	aq	−1179.5	−1095.8	186.2	
RbCl	c	−435.35	−407.81	95.90	52.41
std. state	aq	−418.32	−415.22	178.0	
$RbClO_3$	c	−402.9	−300.4	151.9	103.2
std. state	aq	−355.14	−291.9	283.68	
$RbClO_4$	c	−437.19	−306.9	161.1	
std. state	aq	−380.49	−292.59	303.3	
RbF	c	−557.7		75.3	50.5
std. state	aq	−583.79	−562.79	107.53	
Rb formate	aq	−676.7	−635.1	213.0	
$RbHCO_3$	c	−963.2	−863.6	121.3	
std. state	aq	−943.16	−870.82	212.71	
$RbHF_2$	c	−922.6	−855.6	120.08	79.37
std. state	aq	−901.11	−862.11	213.8	

TABLE 1.56 Enthalpies and Gibbs Energies of Formation, Entropies, and Heat Capacities of the Elements and Inorganic Compounds (*Continued*)

Substance	Physical state	$\Delta_f H°$ kJ · mol^{-1}	$\Delta_f G°$ kJ · mol^{-1}	$S°$ J · deg^{-1} · mol^{-1}	$C_p°$ J · deg^{-1} · mol^{-1}
$RbHSO_4$	c	−1159.0			
std. state	aq	−1138.51	−1039.98	253.1	
RbI	c	−333.8	−328.9	118.4	53.18
std. state	aq	−306.35	−335.56	232.6	
$RbNO_2$	c	−367.4	−306.2	172.0	
$RbNO_3$	c	−495.05	−395.85	147.3	102.1
std. state	aq	−458.52	−395.30	267.8	
Rb_2O	c	−339			
Rb_2O_2	c	−472.0			
RbOH	c	−418.19			
std. state	aq	−481.16	−441.24	110.75	
Rb_2PtCl_6	c	−1245.6	−1109.6	406	
std. state	aq	−1170.7	−1056.6	464	
$RbReO_4$	c	−1102.9	−996.2	167	
std. state	aq	−1038.5	−978.6	322.6	
Rb_2S	aq	−469.4	−482.0	228.4	
Rb_2SeO_4	c	−1114.2			
std. state	aq	−1101.7	−1009.2	297.1	
Rb_2SO_4	c	−1435.61	−1316.96	197.44	134.06
std. state	aq	−1411.60	−1312.56	263.2	
Ruthenium					
Ru	c	0	0	28.53	24.1
$RuBr_3$	c	−138.0			
$RuCl_3$	c	−205.0			
RuI_3	c	−65.7			
RuO_2	c	−305.0			
RuO_4	c	−239.3	−152.3	146.4	
	lq	−228.5	−152.3	183.3	
Samarium					
Sm	c	0	0	69.58	29.54
Sm^{3+} std. state	aq	−691.6	−666.5	−211.7	−21
$SmCl_2$	c	−815.5			
$SmCl_3$	c	−1025.9			
std. state	aq	−1193.3	−1060.2	−42.7	−431
SmF_3	c	−1778.0			
$SmF_3 \cdot ½H_2O$	c	−1825.1			
SmI_3	c	−620.1			
$Sm(IO_3)_3$	c	−1381			
$Sm(NO_3)_2$	c	−1212.1			
Sm_2O_3	c	−1823.0	−1734.7	151.0	114.5
$Sm_2(SO_4)_3$	c	−3899.1			
Scandium					
Sc	c	0	0	34.64	25.52
Sc^{3+} std. state	aq	−614.2	−586.6	−255.0	
$ScBr_3$	c	−743.1			
$ScCl_3$	c	−925.1		121.3	93.64
ScF_3	c	−1629.2	−1555.6	92	
$ScOH^{2+}$	aq	−861.5	−801.2	−134.0	
Sc_2O_3	c	−1908.8	−1819.41	76.99	94.2

(*Continued*)

TABLE 1.56 Enthalpies and Gibbs Energies of Formation, Entropies, and Heat Capacities of the Elements and Inorganic Compounds (*Continued*)

Substance	Physical state	$\Delta_f H°$ kJ · mol^{-1}	$\Delta_f G°$ kJ · mol^{-1}	$S°$ J · deg^{-1} · mol^{-1}	$C_p°$ J · deg^{-1} · mol^{-1}
Selenium					
Se	c	0	0	41.97	24.98
	g	227.1	187.0	174.8	22.1
$SeBr_2$	g	−21.0			
$SeCl_4$	c	−188.3			
SeF_6	g	−1117.0	−1017.0	313.8	110.5
SeO	g	53.4	26.8	234.0	31.3
SeO_2	c	−225.4			
SeO_3	c	−166.9			
SeO_3^{2-} std. state	aq	−509.2	−369.9	13	
SeO_4^{2-}	aq	−599.2	−441.4	54.0	
Silicon					
Si	c	0	0	18.81(8)	20.00
	g	450.(8)		167.981(4)	
$SiBr_4$	lq	−457.3	−433.9	277.5	146.4
	g	−415.5	−431.8	377.9	97.1
$SiBrCl_3$	g			350.1	90.9
SiC alpha	c	−62.8	−60.2	16.49	26.76
beta	c	−65.3	−62.8	16.61	26.9
$SiCl_4$	lq	−686.93	−620.0	239.7	145.3
	g	−657.0	−617.0	330.7	90.26
$SiClBr_3$	g			377.1	95.3
$SiClF_3$	g	−1318	−1280	309	79.4
SiF_4	g	−1615.0(8)	−1572.7	282.76(50)	73.62
SiH_4	g	34.3	56.8	204.65	42.83
$SiHBr_3$	g	−317.6	−328.5	348.6	80.8
$SiHCl_3$	lq	−539.3	−482.5	227.6	
	g	−513.0	−482.0	313.7	75.8
$SiHF_3$	g			271.9	60.5
SiH_2Cl_2	g	−320.5	−295.0	285.7	60.5
SiH_3Cl	g	−142	−119	250.8	51.10
SiH_3F	g	−377	−353	238.4	47.20
Si_2H_6	g	80.3	127.2	272.7	80.79
SiI_4	c	−189.5	−191.6	258.1	108.0
	lq	−174.60	−187.49	294.30	159.79
Si_3N_4	c	−743.5	−642.1	101.3	99.5
SiO	g	−99.6	−126.4	211.6	29.9
SiO_2 quartz	c	−910.7(10)	−856.4	41.46(20)	44.4
high cristobalite	c	−905.5	−853.6	50.05	26.58
$SiOF_2$	g	−967	−951	271.3	53.69
SiS_2	c	−213.4	−212.6	80.3	77.5
Silver					
Ag	c	0	0	42.55(20)	25.4
	g	284.9(8)		172.997(4)	
Ag^+ std. state	aq	105.79(8)	77.12	73.45(40)	21.8
Ag^{2+} in 4*M* $HClO_4$	aq	268.6	269.0	−88	
AgAt	c	−45.2		133.1	55.7
AgBr	c	−100.37	−96.90	107.11	52.38
$AgBrO_3$	c	−10.5	71.3	151.9	
AgCl	c	−127.01(5)	−109.8	96.25(20)	50.79
$AgClO_2$	c	8.79	75.7	134.56	87.32

TABLE 1.56 Enthalpies and Gibbs Energies of Formation, Entropies, and Heat Capacities of the Elements and Inorganic Compounds (*Continued*)

Substance	Physical state	$\Delta_f H°$ kJ · mol^{-1}	$\Delta_f G°$ kJ · mol^{-1}	$S°$ J · deg^{-1} · mol^{-1}	$C_p°$ J · deg^{-1} · mol^{-1}
$AgClO_3$	c	−30.3	64.5	142.0	
$AgClO_4$	c	−31.13		162.3	
std. state	aq	−23.77	68.49	254.8	
AgCN	c	146.0	156.9	107.19	66.73
$Ag(CN)_2^-$ std. state	aq	270.3	305.4	192	
Ag_2CrO_4	c	−731.74	−641.83	217.6	142.26
Ag_2CO_3	c	−505.9	−436.8	167.4	112.26
$Ag_2C_2O_4$	c	−673.2	−584.1	209	
AgF	c	−204.6		83.7	51.92
AgF_2	c	−360.0			
AgI	c	−61.84	−66.19	115.5	56.82
$AgIO_3$	c	−171.1	−93.7	149.4	102.93
AgN_3	c	308.8	376.1	104.2	
$Ag(NH_3)_2^+$ std. state	aq	−111.29	−17.24	245.2	
$AgNO_3$	c	−124.4	−33.47	140.92	93.05
std. state	aq	−101.80	−34.23	219.2	−64.9
AgO	c	−12.15	13.83	58.5	44.0
Ag_2O	c	−31.1	−11.21	121.3	65.86
Ag_2O_3	c	33.9	121.4	100.0	
Ag_2S argentite	c	−32.59	−40.67	143.9	76.53
Ag_3Sb	c	−23.0		171.5	101.7
AgSCN	c	87.9	101.38	131.0	63
Ag_2Se	c	−38	−44.4	150.71	81.76
Ag_2SO_4	c	−715.9	−618.4	200.4	131.4
std. state	aq	−698.10	−590.36	165.7	−251
Ag_2Te	c	−37.2	−43.1	154.8	87.5
Sodium					
Na	c	0	0	51.30(20)	28.15
	g	107.5(7)		153.718(3)	
Na^+ std. state	aq	−240.34(6)	−261.88	58.45(15)	46.4
$NaAg(CN)_2$ std. state	aq	30.12	43.5	251	
NaOAc	c	−708.81	−607.27	123.0	79.9
std. state	aq	−726.13	−631.28	145.6	40.2
$NaAlCl_4$	c	−1142.0	−996.4	188.3	154.98
Na_3AlCl_6	c	−1979.0	−1829	347.0	244.1
$NaAlF_4$	g	−1869.0	−1827.5	345.7	105.9
Na_3AlF_6	c	−3361.2	−3136.7	239.5	215.89
$NaAlH_4$	c	−115.5			
$NaAlO_2$	c	−1137.3	−1069.2	70.40	73.64
$NaAl(SO_4)_2$ std. state	aq	−2590	−2238	−222.6	
$NaAlSiO_4$	c	−2092.8	−1978.2	124.3	
$NaAsO_2$	c	−660.53			
std. state	aq	−669.15	−611.91	99.6	
Na_3AsO_4	c	−1540			
std. state	aq	−1608.50	−1434.19	14.2	
$NaAu(CN)_2$	aq	2.1	23.9	230	
$NaBF_4$	c	−1844.7	−1750.1	145.31	120.3
std. state	aq	−1812.1	−1748.9	243	
$NaBH_4$	c	−188.6	−123.9	101.3	86.8
std. state	aq	−199.60	−147.61	169.5	

(*Continued*)

TABLE 1.56 Enthalpies and Gibbs Energies of Formation, Entropies, and Heat Capacities of the Elements and Inorganic Compounds (*Continued*)

Substance	Physical state	$\Delta_f H°$ kJ · mol^{-1}	$\Delta_f G°$ kJ · mol^{-1}	$S°$ J · deg^{-1} · mol^{-1}	$C_p°$ J · deg^{-1} · mol^{-1}
$NaBO_2$	c	−977.0	−920.7	73.54	65.94
std. state	aq	−1012.49	−940.81	21.8	
$NaBO_3 \cdot 4H_2O$	c	−2114.2			
$Na_2B_4O_7$	c	−3291.1	−3096.0	189.0	186.8
std. state	aq	−3271.1	−3076.9	192.9	
$Na_2B_4O_7 \cdot 10H_2O$	c	−6298.6	−5516.6	586	614.5
NaBr	c	−361.08	−349.00	86.82	51.38
std. state	aq	−361.66	−365.85	141.4	−95.4
$NaBr_3$ std. state	aq	−370.54	−368.95	274.5	
NaBrO std. state	aq	−384.3	−295.4	100	
$NaBrO_3$	c	−334.09	−242.6	128.9	
std. state	aq	−307.19	−243.34	220.9	
$NaBrO_4$ std. state	aq	−227.19	−143.93	−258.57	
$Na_2[Cd(CN)_4]$	aq	−52.3	−16.3	439	
NaCl	c	−411.2	−384.1	72.1	50.51
std. state	aq	−407.27	−393.17	115.5	−90.0
NaClO std. state	aq	−347.3	−298.7	100	
$NaClO_2$	c	−307.02		115.9	
std. state	aq	−306.7	−244.8	160.3	
$NaClO_3$	c	−365.77	−262.34	123.4	
std. state	aq	−344.09	−269.91	221.3	
$NaClO_4$	c	−383.3	−254.9	142.3	111.3
std. state	aq	−369.45	−270.50	241.0	
NaCN	c	−87.5	−76.4	115.6	70.4
std. state	aq	−89.5	−89.5	153.1	
$Na_3[Co(NO_2)_6]$	c	−1423.0			
Na_2CO_3	c	−1130.7	−1044.4	135.0	112.3
	aq	−1157.4	−1051.6	61.6	
$Na_2CO_3 \cdot H_2O$	c	−1431.26	−1285.41	168.11	145.60
$Na_2CO_3 \cdot 10H_2O$	c	−4081.32	−3428.20	564.0	550.32
$Na_2C_2O_4$	c	−1318.0			142
std. state	aq	−1305.4	−1197.9	163.6	
Na_2CrO_4	c	−1342.2	−1235.0	176.61	142.13
std. state	aq	−1361.39	−1251.64	168.2	
$Na_2Cr_2O_7$	c	−1978.6			
std. state	aq	−1970.7	−1825.1	379.9	
Na ethoxide	c	−413.80			
NaF	c	−576.6	−546.3	51.11	46.85
std. state	aq	−572.75	−540.70	45.2	−60.3
$Na_3[Fe(CN)_6]$ std. state	aq	−158.6	−56.5	447.3	
$Na_4[Fe(CN)_6]$ std. state	aq	−505.0	−352.63	231.0	
Na formate	c	−666.5	−600.00	103.76	82.68
std. state	aq	−666.67	−613.0	151	−41.4
NaH	c	−56.34	−33.55	40.02	36.39
Na_2HAsO_4 std. state	aq	−1386.58	−1238.51	116.3	
NaH_2AsO_4 std. state	aq	−1149.68	−1015.16	176	
$NaHCO_3$	c	−950.81	−851.0	101.7	87.61
std. state	aq	−932.11	−848.72	150.2	
$NaHCrO_4$ std. state	aq	−1118.4	−1026.8	243.1	
$NaHF_2$	c	−920.27	−852.20	90.92	75.02
std. state	aq	−890.06	−840.02	151.5	

TABLE 1.56 Enthalpies and Gibbs Energies of Formation, Entropies, and Heat Capacities of the Elements and Inorganic Compounds (*Continued*)

Substance	Physical state	$\Delta_f H°$ kJ · mol^{-1}	$\Delta_f G°$ kJ · mol^{-1}	$S°$ J · deg^{-1} · mol^{-1}	$C_p^°$ J · deg^{-1} · mol^{-1}
$Na_2H_2[Fe(CN)_6]$	aq	−24.7	134.64	335	
NaH_2PO_4	c	−1536.8	−1386.2	127.49	116.86
std. state	aq	−1536.4	−1392.27	149.4	
Na_2HPO_4	c	−1748.1	−1608.3	150.50	135.31
std. state	aq	−1772.38	−1613.06	84.5	
$Na_2H_2P_2O_7$	c	−2764.8	−2522.5	220.20	198.15
NaHS	c	−237.23			
std. state	aq	−257.73	−249.83	121.8	
$NaHSeO_3$	c	−759.23			
std. state	aq	−754.67	−673.41	194.1	
$NaHSeO_4$	c	−821.40			
std. state	aq	−821.74	−714.2	208.4	
$NaHSO_4$	c	−1125.5	−992.9	113.0	
std. state	aq	−1127.46	−1017.88	190.8	−38
NaI	c	−287.9	−286.1	98.50	52.1
std. state	aq	−295.31	−313.47	170.3	−95.8
NaI_3	aq	−291.6	−313.4	298.3	
$NaIO_3$	c	−481.79		135.1	92.1
std. state	aq	−461.50	−389.95	177.4	
$NaIO_4$	c	−429.28	−323.09	163.0	
std. state	aq	−391.62	−320.49	280	
Na methoxide	c	−367.8	−294.80	110.58	69.45
std. state	aq	−433.59	−332.46	17.6	
$NaMnO_4$ std. state	aq	−781.6	−709.2	250.2	
Na_2MnO_4	c	−1156.0			
std. state	aq	−1134	−1024.7	176	
Na_2MoO_4	c	−1468.12	−1354.30	159.70	141.71
std. state	aq	−1478.2	−1360.2	145.2	
$Na_2Mo_2O_7$	c	−2245.05	−2058.19	250.6	217.15
NaN_3	c	21.71	93.76	96.86	76.61
std. state	aq	35.02	86.2	166.9	
$NaNH_2$	c	−123.9	−64.0	76.90	66.15
$NaNbO_3$	c	−1315.9	1233.0	117	
std. state	aq	−1265.7	−1194.1	155	
$NaNO_2$	c	−358.65	−284.60	103.8	
std. state	aq	−344.8	−294.1	182.0	−51.0
$NaNO_3$	c	−467.85	−367.06	116.52	92.88
std. state	aq	−447.48	−373.21	205.4	−40.2
$Na_2[Ni(CN)_4]$	aq	−112.6	−51.9	335	
NaO_2	c	−260.2	−218.4	115.9	72.14
Na_2O	c	−414.2	−375.5	75.04	69.10
Na_2O_2	c	−510.9	−449.6	94.8	89.3
NaOCN cyanate	c	−405.39	−358.2	96.7	86.6
std. state	aq	−386.2	−359.4	165.7	
NaOH	c	−425.6	−379.4	64.4	59.5
std. state	aq	−469.15	−419.20	48.1	−102.1
Na_3PO_4	c	−1917.40	−1788.87	173.80	153.47
std. state	aq	−1997.9	−1804.6	−46	
$Na_4P_2O_7$	c	−3188	−2969.4	270.29	241.12
std. state	aq	−3231.7	−2966.9	117	

(*Continued*)

TABLE 1.56 Enthalpies and Gibbs Energies of Formation, Entropies, and Heat Capacities of the Elements and Inorganic Compounds (*Continued*)

Substance	Physical state	$\Delta_f H°$ kJ · mol^{-1}	$\Delta_f G°$ kJ · mol^{-1}	$S°$ J · deg^{-1} · mol^{-1}	$C_p°$ J · deg^{-1} · mol^{-1}
$NaReO_4$	c	−1057.09	−953.74	151.5	133.89
std. state	aq	−1027.6	−956.5	260.2	
Na_2S	c	−364.8	−349.8	83.7	82.8
std. state	aq	−443.3	−438.1	103.3	
Na_2S_2	c	−397.0	−392	151	
std. state	aq	−450.2	−444.3	146.4	
NaSCN	c	−170.50			
std. state	aq	−163.68	−169.20	203.84	6.3
Na_2Se	c	−341.4			
Na_2SeO_3	c	−958.6			
std. state	aq	−989.5	−893.7	130	
Na_2SeO_4	c	−1069.0			
Na_2SiF_6	c	−2909.6	−2754.2	207.1	187.1
Na_2SiO_3	c	−1554.9	−1462.8	113.8	111.9
$Na_2Si_2O_5$	c	−2470.1	−2324.1	164.1	157.0
$NaSnBr_3$	aq	−615.1	−608.8	310	
$NaSnCl_3$	aq	−727.2	−692.0	318	
Na_2SO_3	c	−1100.8	−1012.5	145.94	120.25
std. state	aq	−1115.87	−1010.44	87.9	
Na_2SO_4	c	−1387.1	−1270.2	149.6	128.2
std. state	aq	−1389.51	−1268.40	138.1	−201
$Na_2SO_4 \cdot 10H_2O$	c	−4327.26	−3647.40	592.0	
$Na_2S_2O_3$	c	−1123.0	−1028.0	155	
std. state	aq	−1132.40	−1046.0	184.1	
$Na_2S_2O_3 \cdot 5H_2O$	c	−2607.93	−2230.1		
$Na_2S_2O_4$ dithionate	c	−1232.2			
std. state	aq	−1233.9	−1124.2	209.2	
$Na_2S_2O_7$	c	−1925.1	−1722.1	202.1	
$Na_2S_2O_8$	aq	−1825.1	−1638.9	362.3	
Na_2Te	c	−349.4			
Na_2TeO_4	c	−1270.7			
Na_2TiO_3	c	−1591.2	−1496.2	121.67	125.65
Na_2UO_4 beta	c	−1893.3	−1777.78	166.02	146.65
Na_3UO_4	c	−2025.1	−1901.2	198.20	173.01
$NaVO_3$	c	−1145.79	−1064.12	113.68	97.57
std. state	aq	−1128.4	−1045.6	109	
Na_3VO_4	c	−1757.87	−1637.83	190.0	164.85
$Na_4V_2O_7$	c	−2918.84	−2712.52	318.4	269.74
Na_2WO_4	c	−1544.7	−1429.8	160.3	139.8
$Na_2[Zn(CN)_4]$	aq	−138.1	−77.0	343	
Strontium					
Sr	c	0	0	55.0	26.79
Sr^{2+} std. state	aq	−545.8	−559.44	−32.6	
$Sr(OAc)_2$	c	−1487.4			
$Sr_3(AsO_4)_2$	c	−3317.1	−3080.3	255	
$SrBr_2$	c	−717.6	−697.1	135.1	75.3
	aq	−788.89	−767.39	132.2	
$SrCl_2$	c	−828.9	−781.1	114.9	75.59
std. state	aq	−880.10	−821.95	80.3	
$Sr(ClO_4)_2$	c	−762.69			
std. state	aq	−804.46	−576.68	331.4	

TABLE 1.56 Enthalpies and Gibbs Energies of Formation, Entropies, and Heat Capacities of the Elements and Inorganic Compounds (*Continued*)

Substance	Physical state	$\Delta_f H°$ kJ · mol^{-1}	$\Delta_f G°$ kJ · mol^{-1}	$S°$ J · deg^{-1} · mol^{-1}	$C_p°$ J · deg^{-1} · mol^{-1}
$SrCO_3$	c	−1220.1	−1140.1	97.1	81.42
	aq	−1222.9	−1087.3	−89.5	
SrC_2O_4	c	−1370.7			
SrF_2	c	−1216.3	−1164	82.1	70.0
Sr formate	c	−1393.3			
$SrHPO_4$	c	−1821.7	−1688.7	121	
$Sr(H_2PO_4)_2$	c	−3134.7			
SrI_2	c	−558.1	−557.7	159.1	77.95
std. state	aq	−656.18	−662.62	190.0	
$Sr(IO_3)_2$	c	−1019.2	−855.2	234	
$SrMoO_4$	c	−1561.1		128.9	117.07
$Sr(NO_2)_2$	c	−762.3			
$Sr(NO_3)_2$	c	−978.22	−780.0	194.56	149.87
std. state	aq	−960.52	−782.12	260.2	
SrO	c	−592.0	−561.9	54.4	45.0
SrO_2	c	−654.4		54	79.45
$Sr(OH)_2$	c	−959	−881	97	74.9
$Sr_3(PO_4)_2$	c	−4122.9			
SrS	c	−472.4	−467	68.2	48.7
SrSe	c	−385.8			
$SrSeO_3$	c	−1047.7			
$SrSeO_4$	c	−1142.7			
$SrSiO_3$	c	−1633.9	−1549.8	96.7	88.53
Sr_2SiO_4	c	−2304.6	−2191.2	153.1	134.26
$SrSO_3$	c	−1177.0			
$SrSO_4$	c	−1453.1	−1341.0	117.0	107.78
	aq	−1455.1	−1304.0	−12.6	
Sr_2TiO_4	c	−2287.4	−2178.6	159.0	143.68
Sulfur					
S rhombic	c	0	0	32.054(50)	22.60
monoclinic	c	0.360	−0.070	33.03	23.23
	g	277.17(15)		167.829(6)	
S_2^{2-}	aq	33.1	85.8	−14.6	
S_2	g	128.60(30)		228.167(10)	
S_8	g	101.25	49.16	430.20	156.06
S_2Br_2	lq	−13.0			
SCl_2	lq	−50.0	−28.5	184	91.0
$SClF_5$	lq	−1065.7			
S_2Cl_2	lq	−59.4	−39	224	124.3
SCN^-	aq	76.4	92.7	144.3	−40.2
SF_4	g	−763.2	−722.0	299.6	77.60
SF_6	g	−1220.5	−1116.5	291.5	96.96
S_2F_{10}	g	−2064	−1861	397	176.7
SO	g	6.3	−19.9	222.0	30.2
SO_2	g	−296.81(20)	−300.13	248.223(50)	39.88
SO_3	g	−395.7	−371.02	256.77	50.66
$SOCl_2$	g	−212.50	−198.3	309.8	66.5
SOF_2	g	−544	−502	278.7	56.81
SO_2Cl_2	g	−364.0	−320.0	311.9	77.01
SO_2ClF	g	−556	−513	303	71.6
SO_2F_2	g	−759	−712	284.0	66.0

(*Continued*)

TABLE 1.56 Enthalpies and Gibbs Energies of Formation, Entropies, and Heat Capacities of the Elements and Inorganic Compounds (*Continued*)

Substance	Physical state	$\Delta_f H°$ kJ · mol^{-1}	$\Delta_f G°$ kJ · mol^{-1}	$S°$ J · deg^{-1} · mol^{-1}	$C_p°$ J · deg^{-1} · mol^{-1}
SO_3^{2-}	aq	−635.5	−486.5	−29.0	
SO_4^{2-}	aq	−909.34(40)	−744.5	18.50(40)	−293.0
$S_2O_3^{2-}$	aq	−652.3	−522.5	67.0	
$S_2O_4^{2-}$	aq	−753.5	−600.3	92.0	
$S_2O_8^{2-}$	aq	−1344.7	−1114.9	244.3	
Tantalum					
Ta	c	0	0	41.47	25.40
TaB_2	c	−209.2		44.4	48.12
$TaBr_5$	c	−598.3		305.4	155.73
TaC	c	−144.1	−142.7	42.37	36.79
Ta_2C	c	−197.5		83.7	60.96
$TaCl_5$	c	−859.0	−746	222	148
TaF_5	c	−1903.6		195.0	130.46
Ta_2H	c	−32.6	−69.0	79.1	90.8
TaI_5	c	−490		343	155.6
TaN	c	−251		50.6	42.1
TaO_2	g	−201	−209	280	44.0
Ta_2O_5	c	−2046	−1911.0	143.1	135.0
$TaOCl_3$	g	−780.7		361.5	98.53
Technetium					
Tc	c	0	0	33.47	24.27
Tc_2O_7	c	−1113			
Tellurium					
Te	c	0	0	49.70	25.70
$TeBr_4$	c	−190.4			
$TeCl_4$	c	−326.4		209	138.5
TeF_6	g	−1318.0		335.77	116.90
TeO_2	c	−322.6	−270.3	79.5	63.89
$Te(OH)_3^+$	aq	−322.6	−496.1	111.7	
Terbium					
Tb	c	0	0	73.22	28.91
Tb^{3+} std. state	aq	−682.8	−651.9	−226.0	17.0
$TbCl_3$	c	−997.1			
std. state	aq	−1184.1	−1045.6	−59.0	−393.0
TbO_2	c	−971.5			
Tb_2O_3	c	−1865.2			115.9
$Tb_2(SO_4)_3$ std. state	aq	−4131.7	−3597.4		
Thallium					
Tl	c	0	0	64.18	26.32
Tl^+ std. state	aq	5.36	−32.38	125.5	
Tl^{3+} std. state	aq	196.6	214.6	−192.0	
TlBr	c	−173.2	−167.36	120.5	50.50
std. state	aq	−116.19	−136.36	207.9	
$TlBr_3$	aq	−168.2	−97.1	54.0	
$TlBrO_3$	c	−136.4	−53.14	168.6	
std. state	aq	−78.2	−30.5	288.7	
TlCl	c	−204.10	−184.93	111.30	50.92
std. state	aq	−161.80	−163.64	182.00	
$TlCl_3$	c	−315.1			
std. state	aq	−305.0	−179.1	−23.0	
$TlClO_3$	aq	−93.7	−35.6	287.9	

TABLE 1.56 Enthalpies and Gibbs Energies of Formation, Entropies, and Heat Capacities of the Elements and Inorganic Compounds (*Continued*)

Substance	Physical state	$\Delta_f H^\circ$ kJ · mol^{-1}	$\Delta_f G^\circ$ kJ · mol^{-1}	S° J · deg^{-1} · mol^{-1}	C_p° J · deg^{-1} · mol^{-1}
Tl_2CO_3	c	−700	−614.6	155.2	
TlF	c	−324.6		83.3	54.77
std. state	aq	−327.27	−311.21	111.7	
TlI	c	−123.9	−125.39	127.6	52.51
std. state	aq	−49.83	−83.97	236.8	
$TlNO_3$	c	−243.93	−152.46	160.7	99.50
	aq	−202.0	−143.7	272.0	
Tl_2O	c	−178.7	−147.3	126	
TlOH	c	−238.9	−195.8	88	
std. state	aq	−224.64	−189.66	114.6	
Tl_2S	c	−97.1	−93.7	151.0	
Tl_2Se	c	−59.0	−59.0	172.0	
Tl_2SO_4	c	−931.8	−830.48	230.5	
std. state	aq	−898.56	−809.40	271.1	
Thorium					
Th	c	0	0	51.8(5)	27.32
	g	602.(6)		190.17(5)	
Th^{4+} std. state	aq	−769.0	−705.1	−422.6	
$ThBr_4$	c	−965.3	−927.2	230	
$ThC_{1.94}$	c	−146	−147.7	68.49	56.69
$ThCl_4$	c	−1186.2	−1094.1	190.4	120.3
ThF_3	g	−1166.1	−1160.6	339.2	73.3
ThF_4	c	−2097.8	−2003.4	142.05	110.7
undissoc; std. state	aq	−2115.0	−1947.2	−105	
ThH_2	c	−139.8	−100.0	50.71	36.69
ThI_4	c	−664.8	−655.2	255	
ThN	c	−391.2	−363.6	56.07	45.2
Th_3N_4	c	−1315.0	−1212.9	201	155.90
$Th(NO_3)_4$	c	−1441.4			
ThO_2	c	−1226.4(35)	−1169.20	65.23(20)	61.76
$ThOCl_2$	c	−1232.2	−1156.0	123.4	91.25
$ThOF_2$	c	−1665.2	−1589.5	105	
$Th(OH)^{3+}$	aq	−1030.1	−920.5	−343.0	
$Th(OH)_2^{2+}$	aq	−1282.4	−1140.9	−218.0	
Th_3P_4	c	−1140.2	−1112.9	221.8	
ThS_2	c	−626.3	−620.1	96.2	
Th_2S_3	c	−1083.7	−1077.0	180	
$Th(SO_4)_2$	c	−2542.6	−2310.4	159.0	173.47
Thullium					
Tm	c	0	0	74.01	27.03
Tm^{3+} std. state	aq	−697.9	−661.9	−243.0	25.0
$TmCl_3$	c	−986.6			
std. state	aq	−1199.1	−1055.6	−75.0	−385.0
Tm_2O_3	c	−1888.7	−1794.5	139.8	116.7
Tin					
Sn white	c	0	0	51.08(8)	26.99
	aq	301.2(15)		168.492(4)	
gray	c	−2.09	0.13	44.14	25.77
Sn^{2+} in aqueous HCl	aq	−8.9(10)	−27.2	−16.7(40)	
Sn^{4+} in aqueous HCl	aq	30.5	2.5	−117	
$SnBr_2$	c	−243.5			

(*Continued*)

TABLE 1.56 Enthalpies and Gibbs Energies of Formation, Entropies, and Heat Capacities of the Elements and Inorganic Compounds (*Continued*)

Substance	Physical state	$\Delta_f H°$ kJ · mol^{-1}	$\Delta_f G°$ kJ · mol^{-1}	$S°$ J · deg^{-1} · mol^{-1}	$C_p°$ J · deg^{-1} · mol^{-1}
$SnBr_4$	c	−377.4	−350.2	264.4	136.44
	g	−314.6	−331.4	411.9	103.4
$SnCl_2$	c	−325.1		130	79.33
std. state	aq	−329.7	−299.6	172	
$SnCl_4$	lq	−511.3	−440.2	258.6	165.3
	g	−471.5	−432.2	365.8	98.3
SnH_4	g	162.8	188.3	227.7	48.95
SnI_2	c	−143.5			
SnI_4	g			446.1	105.4
SnO tetragonal	c	−280.71(20)	−251.9	57.17(30)	44.31
SnO_2 tetragonal	c	−577.63(20)	−515.8	49.04(10)	52.59
$Sn(OH)^+$	aq	−286.2	−254.8	50.0	
$Sn(OH)_2$	c	−561.1	−491.6	155.0	
SnS	c	−100	−98.3	77.0	49.25
SnS_2	c	−167.4		87.4	70.12
Titanium					
Ti	c	0	0	30.72(10)	25.0
	g	473.(3)		180.298(10)	
TiB	c	−160	−160	35	29.7
TiB_2	c	−280	−275	28.5	44.3
$TiBr_2$	c	−402	−383	108	78.7
$TiBr_3$	c	−548.5	−523.8	176.6	101.7
$TiBr_4$	c	−616.7	−589.5	243.5	131.5
TiC	c	−184	−180	24.2	33.81
$TiCl_2$	c	−513.8	−464.4	87.4	69.8
$TiCl_3$	c	−720.9	−653.5	139.7	97.2
$TiCl_4$	lq	−804.2	−737.2	252.3	145.2
	g	−763.2(30)	−726.3	353.2(40)	95.4
TiF_3	c	−1435	−1362	88	92
TiF_4	c	−1649	−1559	133.96	114.27
TiH_2	c	−144	−105.1	29.71	30.09
TiI_4	c	−375	−371.5	249.4	125.6
TiN	c	−265.8	−243.8	52.73	37.08
TiO	c	−519.7	−495.0	50.0	39.9
TiO_2	c	−944.0(8)	−888.8	50.62(30)	55.0
Ti_2O_3	c	−1520.9	−1434.2	78.8	97.4
Ti_3O_5	c	−2459.4	−2317.4	129.3	154.8
Tungsten					
W	c	0	0	32.6	24.3
WBr_5	c	−312	−270	272	155
WBr_6	c	−348.5	−290.8	314	181.4
$W(CO)_6$	c	−953.5		331.8	242.5
WCl_4	c	−443	−360	198.3	129.7
WCl_5	c	−515	−402	217.6	155.6
WCl_6	c	−602.5	−456	238.5	175.4
WF_6	lq	−1747.7	−1631.4	251.5	
	g	−1721.7	−1631.4	341.1	119.0
WO_2	c	−589.9	−533.86	50.5	56.1
WO_3	c	−842.9	−764.1	75.9	73.8
WO_4^{2-}	aq	−1075.7			
$WOCl_4$	c	−671	−549	173	146

TABLE 1.56 Enthalpies and Gibbs Energies of Formation, Entropies, and Heat Capacities of the Elements and Inorganic Compounds (*Continued*)

Substance	Physical state	$\Delta_f H°$ kJ · mol^{-1}	$\Delta_f G°$ kJ · mol^{-1}	$S°$ J · deg^{-1} · mol^{-1}	$C_p°$ J · deg^{-1} · mol^{-1}
WOF_4	c	−1407	−1298	176.0	133.6
WO_2Cl_2	c	−780	−703	200.8	104.4
Uranium					
U	c	0	0	50.20(20)	27.66
	g	533.(8)		199.79(10)	
U^{3+}	aq	−489.1	−476.2	−188.0	
U^{4+}	aq	−591.2	−531.9	−410.0	
UB_2	c	−161.6	−159.4	55.52	55.77
UBr_3	c	−699.2	−673.6	192	108.8
UBr_4	c	−802.5	−767.8	238.0	128.0
UBr_5	c	−810.9	−769.9	293	160.7
UC	c	−98.3	−99.2	59.20	50.12
UCl_3	c	−866.5	−799.1	159.0	102.5
UCl_4	c	−1019.2	−930.1	197.1	122.0
	aq	−1259.8	−1056.8	−184.0	
UCl_5	c	−1058	−950	242.7	144.6
UCl_6	c	−1092	−962	285.8	175.7
UF_3	c	−1502.1	−1433.4	123.43	95.10
UF_4	c	−1921.2	−1823.3	151.67	116.02
UF_5	c	−2075.3	−1958.6	199.6	132.3
UF_6	c	−2197.0	−2068.6	227.6	166.8
UH_3	c	−127.2	−72.8	63.68	49.29
UI_3	c	−460.7	−459.8	222	112.1
UI_4	c	−512.1	−506.7	264	134.3
UN	c	−290.8	−265.7	62.43	47.57
UO_2	c	−1085.0(10)	−1031.8	77.03(20)	63.60
UO_2^{2+} std. state	aq	−1019.0(15)	−953.5	−98.2(30)	
UO_3 gamma	c	−1223.8(12)	−1145.7	96.11(40)	81.67
U_3O_7	c	−3427.1	−3242.9	250.5	215.5
U_3O_8	c	−3574.8(25)	−3369.8	282.55(50)	238.36
U_4O_9	c	−4510.4	−4275.1	334.1	293.3
$UOBr_2$	c	−973.6	−929.7	158.00	98.00
$UOCl_2$	c	−1066.9	−996.2	138.32	95.06
UOF_2	c	−1499.1	−1428.8	119.2	
$UO_2(OAc)_2$	c	−1963.55			
UO_2Br_2	c	−1137.6	−1066.5	169.5	
UO_2Cl_2	c	−1243.9	−1146.4	150.5	107.86
std. state	aq	−1353.9	−1215.9	15.5	
UO_2CO_3	c	−1691.2	−1562.7	138	
std. state	aq	−1696.6	−1481.6	−154.4	
$UO_2C_2O_4$	c	−1796.94			
UO_2F_2	c	−1653.5	−1557.4	135.56	103.22
std. state	aq	−1684.0	−1551.3	−125.1	
$UO_2(NO_3)_2$	c	−1349.3	−1105.0	243	
std. state	aq	−1434.3	−1176.1	195.4	
$UO_2(OH)_2$ std. state	aq	−1479.5	−1267.8	−118.8	
UO_2SO_4	c	−1845.1	−1683.6	154.8	145.2
std. state	aq	−1928.8	−1698.3	−77.4	
US_2	c	−527	−526.4	110.42	74.64
US_3	c	−549.4	−547.3	138.49	95.60

(*Continued*)

TABLE 1.56 Enthalpies and Gibbs Energies of Formation, Entropies, and Heat Capacities of the Elements and Inorganic Compounds (*Continued*)

Substance	Physical state	$\Delta_f H°$ kJ · mol^{-1}	$\Delta_f G°$ kJ · mol^{-1}	$S°$ J · deg^{-1} · mol^{-1}	$C_p°$ J · deg^{-1} · mol^{-1}
Vanadium					
V	c	0	0	28.94	24.90
VBr_4	g	−336.8			
VCl_2	c	−452	−406	97.1	72.22
VCl_3	c	−580.7	−511.3	131.0	93.18
VCl_4	lq	−569.4	−503.8	255.0	161.7
VF_5	lq	−1480.3	−1373.2	175.7	
	g	−1433.9	−1369.8	320.9	98.58
VN	c	−217.15	−191.08	37.28	38.00
VO	c	−431.8	−404.2	39.0	45.5
VO_2	c	−717.6		51.5	62.59
VO_2^+ std. state	aq	−649.8	−587.0	−42.3	
VO_2^{2+} std. state	aq	−486.6	−446.4	−133.9	
VO_3^- std. state	aq	−888.3	−783.7	50.2	
V_2O_3	c	−1218.8	−1139.3	98.3	103.2
V_2O_4	c	−1427	−1318.4	103	115.4
V_2O_5	c	−1550	−1419.3	130	130.6
V_3O_5	c	−1933	−1803	163	
$VOCl_3$	lq	−734.7	−668.6	244.4	150.62
	g	−695.6	−659.3	344.4	89.9
$VOSO_4$	c	−1309.2	−1169.9	108.8	
Xenon					
Xe	g	0	0	169.685(3)	20.786
XeF_2	c	−164.0			
XeF_4	c	−261.5	−123.0		
XeF_6	c	−360			
	g	−297			
XeO_3	c	402			
$XeOF_4$	lq	146			
Ytterbium					
Yb	c	0	0	59.87	26.74
Yb^{2+} std. state	aq		−527.0		
Yb^{3+} std. state	aq	−674.5	−643.9	238.0	25.0
$Yb(OAc)_3$ undissoc; std. state	aq	−2105.0	−1772.84	183.3	
$YbCl_2$	c	−799.6			
$YbCl_3$	c	−959.8			
std. state	aq	−1176.1	−1037.6	−71.0	−385.0
$Yb(NO_3)_3$ std. state	aq	−1296.6			
Yb_2O_3	c	−1814.6	−1726.7	133.1	115.35
Yttrium					
Y	c	0	0	44.4	26.51
Y^{3+} std. state	aq	−723.4	−693.7	−251.0	
YCl_3	c	−1000		136.8	75.0
YF_3	c	−1718.8	−1644.7	100	
Y_2O_3	c	−1905.31	−1816.65	99.08	102.51
$Y(OH)_3$	c	−1435	−1291	99.2	
Zinc					
Zn	c	0	0	41.63(15)	25.40
	g	130.40(40)		160.990(4)	
Zn^{2+} std. state	aq	−153.39(20)	−147.1	−109.8(5)	46.0

TABLE 1.56 Enthalpies and Gibbs Energies of Formation, Entropies, and Heat Capacities of the Elements and Inorganic Compounds (*Continued*)

Substance	Physical state	$\Delta_f H°$ kJ · mol^{-1}	$\Delta_f G°$ kJ · mol^{-1}	$S°$ J · deg^{-1} · mol^{-1}	$C_p°$ J · deg^{-1} · mol^{-1}
$ZnBr_2$	c	−328.65	−312.13	138.5	65.7
std. state	aq	−396.98	−354.97	52.72	−238.0
$ZnCl_2$	c	−415.05	−369.45	111.46	71.34
std. state	aq	−488.19	−409.53	0.84	−226.0
$Zn(CN)_4^{2-}$ std. state	aq	342.3	446.9	226	
$ZnCO_3$	c	−812.78	−731.57	82.4	79.71
ZnF_2	c	−764.4	−713.3	73.68	65.7
std. state	aq	−819.14	−704.67	−139.8	−167.0
ZnI_2	c	−208.03	−208.95	161.1	65.69
	aq	−264.3	−250.2	110.5	−238.0
$Zn(NO_3)_2$	c	−483.7			
	aq	−568.6	−369.6	180.7	−126.0
ZnO	c	−350.46(27)	−320.52	43.65(40)	40.25
$Zn(OH)_2$	c	−641.91	−553.59	81.2	
std. state	aq	−613.88	−461.62	−133.5	−251
ZnS sphalerite	c	−205.98	−201.29	57.7	46.02
wurtzite	c	−192.6			
ZnSe	c	−163	−163	84.0	
$ZnSO_4$	c	−982.84	−871.5	110.5	99.2
	aq	−1063.2	−891.6	−92.0	−247.0
Zn_2SiO_4	c	−1636.7	−1523.2	131.42	123.3
Zirconium					
Zr	c	0	0	39.0	25.40
ZrB	c	−322	−318.2	35.94	48.24
$ZrBr_2$	c	−405	−382	116	86.7
$ZrBr_4$	c	−760.7	−725.3	224	124.8
ZrC	c	197	−193	33.32	37.90
$ZrCl_2$	c	−502.0	−386	110	72.6
$ZrCl_3$	c	−714	−646	146	96
$ZrCl_4$	c	−981	−890	181.4	119.8
ZrF_2	c	−962	−913	75	66
ZrF_4	c	−1911.3	−1810.0	104.7	103.6
ZnH_2	c	−169.0	−128.8	35.0	31.0
ZrI_2	c	−259	−258	150.2	94.1
ZrI_3	c	−397.5	−394.9	204.6	103.8
ZrI_4	c	−488	−485.4	260	127.8
ZrN	c	−365	−336.7	38.86	40.44
ZrO_2	c	−1100.6	−1042.8	50.36	56.19
$ZrSiO_4$	c	−2033.4	−1919.1	84.1	98.7
$ZrSO_4$	c	−2217.1			172.0

TABLE 1.57 Heats of Fusion, Vaporization, and Sublimation and Specific Heat at Various Temperatures of the Elements and Inorganic Compounds

Abbreviation Used in the Table

Hm, enthalpy of melting (at the melting point) in $kJ \cdot mol^{-1}$
Hv, enthalpy of vaporization (at the boiling point) in $kJ \cdot mol^{-1}$
Hs, enthalpy of sublimation (or vaporization at 298 K) in $kJ \cdot mol^{-1}$
C_p, specific heat (at temperature specified on the Kelvin scale) for the physical state in existence (or specified: c, lq, g) at that temperature in $J \cdot K^{-1}$ (mol^{-1}
Ht, enthalpy of transition (at temperature specified, superscript, measured in degrees Celsius) in $kJ \cdot mol^{-1}$

				C_p			
Substance	ΔHm	ΔHv	ΔHs	400 K	600 K	800 K	1000 K
Aluminum							
Al	10.71	294.0	326.4	25.8	27.9	30.6	34.9(lq)
$Al(BH_4)_3$		30					
Al_6BeO_{10}	402			324.3	380.6	407.8	425.2
$AlBr_3$	11.25	23.5		125.0	125.0	125.0	125.0
Al_4C_3				138.5	159.2	169.7	176.1
$AlCl_3$	35.4		116	100.1	117.7	135.2	152.8
AlF_3, $\Delta Ht = 0.56^{455}$	98			86.3	97.3	98.5	100.8
AlI_3	15.9	32.2	112	108.5	121.3		
AlN				36.7	43.5	46.8	48.5
Al_2O_3 corundum	111.4			96.1	112.5	120.1	124.8
AlOCl				64.3	72.6	76.9	79.3
Al_2SiO_5 andalusite				149.6	174.5	186.1	194.0
kyanite				148.3	176.2	188.3	196.2
sillimanite				147.5	173.0	185.0	193.5
$Al_6Si_2O_{13}$ mullite				390.7	459.8	494.1	513.4
Al_2S_3	55			115.0	124.1	129.7	134.0
Al_2TiO_5				162.0	182.8	192.9	200.0
Americium							
Am	14.39						
Ammonium							
NH_3	5.66	23.35	19.86	38.7	45.3	51.1	56.2
ND_3 ammonia-d_3				42.9	51.5	58.6	64.3
NH_4Br, $\Delta Ht = 3.22^{138}$							
NH_4Cl, $\Delta Ht = 1.046^{-30.6}$				103			
$\Delta Ht = 3.950^{184.6}$							
NH_4ClO_4				148.7			
NH_4I, $\Delta Ht = 2.93^{-13}$	20.9		168.5^{525}	89.0	103.3	117.7	
NH_4NO_3	6.40						
Antimony							
Sb	19.87	193.43		25.9	27.7	29.5	31.4
$SbBr_3$	14.6	59		125.5(lq)	81.6(g)	82.2	82.5
$SbCl_3$	12.7	45.2		123.4(lq)	81.6(g)	82.2	82.5
$SbCl_5$	10.0	48.4					
SbH_3		21.3					
SbI_3	22.8	68.6		106.6(lq)	143.5(lq)	82.2(g)	82.5(g)
Sb_2O_3, $\Delta Ht = 7.1^{573}$	54.4	74.6		108.5	122.8	137.1	150.6
Sb_2S_3				123.3	134.4	145.4	
Argon							
Ar	1.12	6.43		20.8	20.8	20.8	20.8

TABLE 1.57 Heats of Fusion, Vaporization, and Sublimation and Specific Heat at Various Temperatures of the Elements and Inorganic Compounds (*Continued*)

				C_p			
Substance	ΔHm	ΔHv	ΔHs	400 K	600 K	800 K	1000 K
Arsenic							
As	24.44			25.6	27.5	29.3	
$AsBr_3$	11.7	41.8					
$AsCl_3$	10.1	35.0		133.5(lq)	88.3(g)	88.3	
AsF_3	10.4	29.7					
AsF_5		20.8					
AsH_3		16.7		45.4	53.2	58.8	63.9
AsI_3		59.3					
As_2O_3	18.4			116.4			
Barium							
Ba	7.12	140.3		33.2	33.9(c)		39.1(lq)
$BaBr_2$	32.2			79.2	83.5	87.9	92.2
$BaCl_2$, $\Delta Ht = 16.9^{925}$	15.85	246.4		77.3	80.4	84.3	89.5
$BaCO_3$, $\Delta Ht = 18.8^{806}$	40			99.0	113.0	124.2	134.6
BaF_2, $\Delta Ht = 2.67^{1207}$	17.8	285.4	405.1	75.9	80.3	84.9	94.6
BaH_2	25						
BaI_2	26.5	43.9	302.5	79.5	83.5	87.5(c)	113.0(lq)
$BaMoO_4$				129.5	143.5	152.2	159.3
BaO	46	330.6	424.3	49.9	53.2	55.4	57.1
$Ba(OH)_2$	16			112.6	122.7(c)	141.0(lq)	
BaS	63						
$BaSO_4$	40			119.4	131.6	135.9	137.9
$BaTiO_3$, $\Delta Ht = 0.067^{75}$				111.5	121.8	126.1	128.7
Beryllium							
Be	7.895	297	291	20.0	23.3	25.5	27.3
$BeAl_2O_4$, chrysoberyl	170.0			130.3	155.0	166.8	174.2
$BeBr_2$	18	100.0	515	70.6	77.6(c)	113.0(lq)	113.0
Be_2C	75.3			47.6	51.9	64.7	73.2
$BeCl_2$, $\Delta Ht = 6.8^{403}$	8.66	105	136.0	68.7	75.8(c)	121.4(lq)	121.4
BeF_2, $\Delta Ht = 0.92^{227}$	4.77	199.4		62.5	67.5	74.1(c)	85.6(lq)
BeI_2	18	70.5	125	76.9	84.2		
Be_3N_2	129.3			84.4	106.5	117.6	123.6
BeO, $\Delta Ht = 6.7^{2100}$	86			33.8	42.4	46.7	49.3
BeS				120.8	149.2	166.0	174.1
Be_2SiO_4				103.9	126.8	149.8	174.4
$BeSO_4$, $\Delta Ht = 1.113^{590}$	6			103.9	126.8	149.8	174.4
$\Delta Ht = 19.55^{635}$							
$BeWO_4$				113.0	131.3	142.9	153.0
Bismuth							
Bi	11.30	151		27.0(c)	31.8(lq)	31.8	31.8
$BiBr_3$	21.7	75.4					
$BiCl_3$	10.9	72.6					
BiI_3		20.9					
Bi_2O_3, $\Delta Ht = 116.7^{717}$	28.5			116.9	123.6	130.3	137.0
Bi_2S_3				131.1	136.2	141.3	146.4
Bi_2Te_3	120.5			164.3	179.7	192.3	
Boron							
B	50.2	480	552	15.7	20.8	23.4	25.0
BBr_3		30.5		72.6(g)	77.6	79.8	81.1
B_4C	105			76.4	98.4	107.7	114.3

(*Continued*)

TABLE 1.57 Heats of Fusion, Vaporization, and Sublimation and Specific Heat at Various Temperatures of the Elements and Inorganic Compounds (*Continued*)

Substance	ΔHm	ΔHv	ΔHs	C_p 400 K	C_p 600 K	C_p 800 K	C_p 1000 K
BCl_3	2.10	23.8	23.1	68.4(g)	75.0	78.2	79.8
BF_3	4.20	19.3	57.5	67.1	72.6	75.8	
$F_2B\text{-}BF_2$		28					
BH_3				38.9	45.4	52.3	58.4
B_2H_6	4.44	14.3		74.3	101.3	121.7	136.4
B_4H_9	6.13	28.4		130.2(g)	187.6	227.4	254.4
B_4H_{10}		27.1					
B_5H_{11}		31.8					
$B_{10}H_{14}$	32.5	48.5	76.7	250.0(lq)	351.6(g)	417.2	460.4
BI_3		40.5					
BN	81		728	26.3	35.2	40.5	44.3
$B_3N_3H_6$ borazine		32.1		126.9	169.4	197.2	216.6
B_2O_3	24.56	390.4		77.9	98.1(c)	129.7(lq)	129.7
$B_3O_3H_3$ boroxin			44.8	120.1	162.8	194.6	214.2
Bromine							
Br_2	10.57	29.96	30.9	36.7(g)	37.3	37.6	37.8
BrCl	10.4	34.7					
BrF		25.1					
BrF_3	12.05	47.6		72.6	78.0	80.1	81.2
BrF_5	5.67	30.6		113.0	123.2	127.3	129.3
Cadmium							
Cd	6.19	99.9		27.1(c)	29.7(lq)	29.7	29.7
$CdBr_2$	20.9	115					
$CdCl_2$	48.58	124.3		79.8	86.3	92.7	104.6
CdF_2	22.6	214					
CdI_2	15.3	115					
$Cd(NO_3)_2 \cdot 4H_2O$	32.6						
CdO			225.1	43.8	45.6	47.3	49.1
CdS			209.6	55.5	56.2	57.0	57.7
$CdSO_4$				108.3	123.8	139.2	154.7
Calcium							
Ca, $\Delta Ht = 0.93^{4}$	8.54	154.7		26.9	30.0	33.8	39.7
$Ca(BO_2)_2$	74.1			125.0	144.9	157.2	176.2
CaB_4O_7	113.4			202.0	243.0	267.7	287.8
$CaBr_2$	29.1	200	298.3	78.0	80.5	83.5	88.6
CaC_2 carbide	32						
$CaCl_2$	28.05	235		75.6	78.2	80.9	85.8
$CaCN_2$ cyanamide	0.432						
$CaCO_3$	36						
CaF_2, $\Delta Ht = 4.8^{1151}$	29.3	308.9	441	73.9	78.5	83.9	90.1
CaH_2	6.7						
CaI_2	41.8	179.4	243	79.2	83.1	87.1	91.0
$Ca[Mg(CO_3)_2]$ dolomite				143.3	163.3	176.8	188.3
$CaMoO_4$				131.3	144.9	153.5	150.6
Ca_3N_2				122.2	140.8	159.2	
$Ca(NO_3)_2$	21.4			173.7	210.5	243.4	
CaO	79.5			46.6	50.5	52.4	53.7
$Ca(OH)_2$, $\Delta Hdec = 99.2$				98.4	107.4		
$Ca_3(PO_4)_2$, $\Delta Ht = 15.5^{1100}$				255.1	295.6	331.3	365.7
CaS	70			49.2	51.5	53.0	54.1

TABLE 1.57 Heats of Fusion, Vaporization, and Sublimation and Specific Heat at Various Temperatures of the Elements and Inorganic Compounds (*Continued*)

				C_p			
Substance	ΔHm	ΔHv	ΔHs	400 K	600 K	800 K	1000 K
$CaSiO_3$, $\Delta Ht = 7.1^{1190}$	56.1			100.4	113.0	119.2	123.8
Ca_2SiO_4, $\Delta Ht = 4.44^{675}$ $\Delta Ht = 3.26^{1420}$				146.4	162.8	179.2	184.0
$3CaO \cdot SiO_2$				196.4	218.4	230.8	240.4
$CaSO_4$	28.0			109.7	129.5	149.2	169.0
$CaSO_4 \cdot \frac{1}{2}H_2O$				147.4	167.2	186.9	206.7
$CaSO_4 \cdot 2H_2O$				260.7	280.3	300.0	319.8
$CaTiO_3$, $\Delta Ht = 2.30^{1257}$				112.3	123.1	127.7	130.4
$Ca(VO_2)_2$				182.9	206.7	230.5	254.4
$CaWO_4$				127.6	140.2	147.3	152.8
Carbon							
C graphite	117			12.0	16.6	19.7	21.7
$(CN)_2$ cyanogen	8.1	23.3	19.7	61.9(g)	68.2	72.9	76.4
CNBr			45.4	50.19(g)	53.7	56.2	58.1
CNCl	11.4			48.7	52.8	55.7	57.7
CNI			59.4	50.8	53.7	55.8	57.4
CO, $\Delta Ht = 0.632^{-211.6}$	0.837	6.04		29.3	30.4	31.9	33.2
CO_2	9.02	15.8	25.2	41.3	47.3	51.4	54.3
C_2O_3	5.40	$26.9^{43.5}$		75.0	85.5	92.7	97.7
$COCl_2$	5.74	24.4		63.9	71.1	75.0	77.4
COF_2		16.1		54.8	64.9	70.8	74.4
COS	7.73	18.6		45.9	51.3	54.7	57.0
CS_2	4.40	26.7	27.5	49.7	54.6	57.4	59.3
Cerium							
Ce, $\Delta Ht = 3.01^{730}$	5.46	398	419	30.6	30.8	32.1	33.8
$CeCl_3$	54.4	170.1	326				
CeI_3	51.9						
CeO_2				66.9	69.0	71.1	73.2
Cesium							
Cs	2.09	63.9	76.6	31.5	31.0	30.9(lq)	20.8(g)
CsBr	23.6	151		52.9	55.0	57.2(c)	77.4(lq)
CsCl, $\Delta Ht = 3.77^{470}$	15.9	115.1		54.7	59.1	63.7(c)	77.4(lq)
CsF	21.7	115.5		53.8	57.4	60.9(c)	74.1(lq)
CsI	23.9	150.2		51.9	57.8(c)	65.5(lq)	67.8
$CsIO_3$	13.0						
CsOH, $\Delta Ht = 1.30^{137}$ $\Delta Ht = 6.1^{220}$	4.56	120		74.4(c)	81.6(lq)	81.6	81.6
Cs_2SO_4, $\Delta Ht = 4.3^{667}$	35.7		76.5	112.1	132.2	163.2	194.2
Chlorine							
Cl_2	6.406	20.41	17.65	35.3	36.6	37.1	37.4
ClF		24		33.8	35.6	36.5	37.0
ClF_3	7.61	27.5		70.6(g)	76.8	79.4	80.7
ClF_5		22.9		110.0	121.6	126.3	128.6
ClO				33.2	35.3	36.3	36.9
ClO_2		30		46.1	51.4	54.2	55.8
ClO_3F	3.83	19.33		75.9	89.2	96.1	100.0
Cl_2O		25.9		51.4	54.7	56.2	56.9
Cl_2O_7		34.69					
Chromium							
Cr, $\Delta Ht = 0.0008^{38.5}$	21.0	339.5	397	25.2	27.7	29.4	31.9

(*Continued*)

TABLE 1.57 Heats of Fusion, Vaporization, and Sublimation and Specific Heat at Various Temperatures of the Elements and Inorganic Compounds (*Continued*)

				C_p			
Substance	ΔHm	ΔHv	ΔHs	400 K	600 K	800 K	1000 K
$CrCl_2$	32.2	196.7		72.6	77.0	81.5	85.9
$CrCl_3$			237.7	93.1	99.0	104.9	110.7
$Cr(CO)_6$			72.0	233.9			
CrN, $\Delta Hdec = 112$			49.1	50.4	51.7	53.0	
CrO_2Cl_2		35.1					
CrO_2F_2	23.4	34.3					
CrO_3	15.77			63.9	72.5	76.7	78.8
Cr_2O_3	129.7			112.7	120.5	124.3	127.0
$Cr_2(SO_4)_3$				316.9	345.2	373.5	401.8
Cobalt							
Co, $\Delta Ht = 0.452^{427}$	16.2	377	424	26.5	29.7	32.4	37.0
$CoCl_2$	45	146	219	81.7	84.6	86.8	88.2
CoF_2	59	202	315	75.7	80.8	82.9	84.2
CoF_3				97	100	102	104
CoO				52.9	54.3	54.8	56.0
Co_3O_4				143	163	185	210
$CoSO_4$, $\Delta Ht = 2.1^{691}$				119	141	152	158
Copper							
Cu	13.26	300.4	337.7	25.3	26.5	27.4	28.7
CuBr, $\Delta Ht = 5.86^{380}$	9.6			56.5	59.8(c)	66.9(lq)	66.9
$\Delta Ht = 2.9^{465}$							
CuCl	10.2	54	241.8	56.9	61.5(c)	66.9(lq)	66.9
$CuCl_2$, $\Delta Ht = 0.700^{402}$	20.4			76.3	80.2(c)	82.4(lq)	100.0
$\Delta Ht = 15.001^{598}$							
CuCN		12			66.7	73.1	78.0
CuF			268	55.5	59.6		
CuF_2	55	156	261	72.4	81.9	87.0	90.4
CuI	10.9			55.4	57.8	60.2	66.9
CuO	11.8			46.8	50.8	53.2	55.0
Cu_2O	64.8			67.6	73.3	77.6	81.5
CuS				48.8	51.0	53.2	55.4
Cu_2S, $\Delta Ht = 3.85^{103}$	10.9			97.3	97.3	85.0	85.0
$\Delta Ht = 0.84^{350}$							
Cu_2Se, $\Delta Ht = 4.85^{110}$				90.9	91.7	92.5	93.4
$CuSO_4$				114.9	136.3	147.7	153.8
Dysprosium							
Dy	11.06	280	290.4				
Erbium							
Er	19.90	280	317.2				
Europium							
Eu	9.21	176	178				
Fluorine							
F_2, $\Delta Ht = 0.728^{-227.6}$	0.510	6.62		33.0	35.2	36.3	37.1
FNO_3				75.1	87.8	94.8	98.9
Gadolinium							
Gd	10.05	301.3		36.6	35.5	34.5	33.5
Gd_2O_3				113.4	120.1	124.4	127.9
Gallium							
Ga	5.59	254		27.1(lq)	26.7	26.6	26.6
$GaBr_3$	12.1	38.9					

TABLE 1.57 Heats of Fusion, Vaporization, and Sublimation and Specific Heat at Various Temperatures of the Elements and Inorganic Compounds (*Continued*)

Substance	ΔHm	ΔHv	ΔHs	C_p 400 K	600 K	800 K	1000 K
$GaCl_3$	11.13	23.9					
GaI_3	12.9	56.5					
Ga_2O_3	100			91.4	112.5	133.5	
GaSb	25.1						
Germanium							
Ge, $\Delta Ht = 37.03^{938.3}$	36.94	334		24.3	25.4	26.2	26.9
$GeBr_4$		41.4					
$GeCl_4$		27.9		100.7	104.6	106.1	106.8
GeH_4		14.1					
Ge_2H_6		25.1					
Ge_3H_8		32.2					
GeO_2	43.9			61.39	69.1	72.4	75.0
Gold							
Au	12.55	324		25.8	26.8	27.8	28.8
AuSn	25.6			54.1	63.3(c)	60.6(lq)	
Hafnium							
Hf, $\Delta Ht = 5.9^{1750}$	27.2	571	618.4	26.7	28.6	30.3	31.9
$HfCl_4$	75		99.6	125.4	105.8	106.7	107.1
HfO_2, $\Delta Ht = 10.5^{1700}$	104.6			67.7	73.9	77.3	79.9
Helium							
He	0.0138	0.0829		20.79	20.79	20.79	20.79
Holmium							
Ho	16.8	71		280	317		
Hydrogen							
H_2	0.117	0.904		29.2	29.3	29.6	30.2
$^1H^2H$				29.2	29.4	29.9	30.7
$^2H_2$				29.2	29.6	30.5	31.6
HBO_2	14.3		242.1	61.5(c)			
H_3BO_3	22.3						
HBr	2.406	17.61	12.7	29.2	29.8	31.1	32.3
HCl, $\Delta Ht = 1.188^{-174.77}$	1.992	16.14	9.1	19.2	29.2	29.6	31.6
^2HCl				29.4	30.6	32.1	33.5
HClO				40.0	44.0	46.6	48.5
HCN	8.406	25.22		39.4	44.2	47.9	51.0
HF	4.58			29.1	29.2	29.5	30.2
^2HF				29.2	29.5	30.5	31.6
H_2F_2 dimer				49.7	56.5	61.0	64.4
HFO				38.6	42.8	45.7	47.9
HI	2.87	19.77	17.4	29.3	30.3	31.8	33.1
HNCO isocyanic acid				50.6	58.3	63.5	67.5
HNCS isothiocyanic acid				53.2	61.0	65.9	69.3
HNO_2 *cis*				51.4	59.9	65.4	69.2
trans				52.1	60.3	65.6	69.3
HNO_3	10.47	39.46	39.1	63.1	76.8	85.0	90.4
HN_3		30.5					
H_2O	6.009	40.66	44.0	34.3(g)	36.4	38.8	41.4
$^1H^2HO$				34.8	37.5	40.4	43.3
^2H_2O				35.6	38.8	42.2	45.4
H_2O_2	12.50		51.63	48.5	55.7	59.8	66.7
$^2H_2O_2$	12.68		52.4				

(*Continued*)

TABLE 1.57 Heats of Fusion, Vaporization, and Sublimation and Specific Heat at Various Temperatures of the Elements and Inorganic Compounds (*Continued*)

				C_p			
Substance	ΔHm	ΔHv	ΔHs	400 K	600 K	800 K	1000 K
HPH_2O_2	9.67						
H_3PO_3	12.84						
H_3PO_4	13.4			175.7	236.0	296.2	365.5
H_2S, $\Delta Ht = 1.531^{-169.61}$	23.8	18.67	14.1	38.9	42.5	45.8	
H_2S_2		33.8					
H_2Se		19.7					
HSO_3F				87.5	102.6	111.0	116.3
H_2SO_4	10.71	50.2		158.2	197.0(lq)	125.9(g)	132.7
$H_2SO_4 \cdot H_2O$	19.46			228.5			
$H_2SO_4 \cdot 2H_2O$	18.24			294.6			
$H_2SO_4 \cdot 3H_2O$	24.0			347.8			
$H_2SO_4 \cdot 4H_2O$	30.64			410.3			
H_2Te		19.2					
Indium							
In	3.28	231.8	243.1	28.5(c)	30.1(lq)	30.1	30.1
InBr	15	92					
$InBr_3$	26						
InCl	21.3						
$InCl_3$	27						
InF_3	64						
InI	17.3	90.8					
InI_3	18.5						
In_2O_3	105						
InSb	25.5						
Iodine							
I_2	150.66	41.6	62.4	79.6(lq)	37.6(g)	37.9	38.1
ICl	11.60		52.9	98.3(lq)	90.0	81.6	73.2
IF				35.1	36.6	37.3	37.7
IF_5		41.3		476.1(g)	516.7	533.0	541.4
IF_7				152.0(g)	167.6	173.9	177.0
Iridium							
Ir	41.12	231.8	243.1	28.5(c)	30.1(lq)	30.1	30.1
IrF_6	8.40	36					
IrO_2				63.8	76.5	89.2	102.0
Iron							
Fe, $\Delta Ht = 0.90^{911}$ $\Delta Ht = 0.837^{1392}$	13.81	340	415.5	27.4	32.1	38.0	54.4
$FeBr_2$	50.2						
$FeBr_3$, $\Delta Ht = 0.418^{377}$	50.2		207.5	83.0	87.0	91.4	95.9
Fe_2C, $\Delta Ht = 0.75^{190}$	51.5			115.7	114.7	117.2	119.8
$FeCl_2$	43.01	26.3		79.7	83.1	85.5	101.2
$FeCl_3$	43.1	43.76		106.7(c)	133.9(lq)	82.3(g)	81.5
$FeCO_3$				93.5	115.9	138.3	
$Fe(CO)_5$	13.23	33.72		189.0	209.8	223.1	232.2
$FeCr_2O_4$				152.0	167.7	175.9	182.2
FeF_2	51.9	224.4	316	72.0	77.1	80.3	82.1
FeF_3			274	96.4	96.8	99.3	101.8
FeI_2, $\Delta Ht = 0.8^{377}$	45	104.6	192	83.9	84.4	110.9	113.0(lq)
Fe_3N				72.6	77.7	82.8	87.9
FeO	24.06			51.8	54.9	57.3	59.4

TABLE 1.57 Heats of Fusion, Vaporization, and Sublimation and Specific Heat at Various Temperatures of the Elements and Inorganic Compounds (*Continued*)

				C_p			
Substance	ΔHm	ΔHv	ΔHs	400 K	600 K	800 K	1000 K
Fe_2O_3, $\Delta Ht = 0.67^{677}$				120.1	141.2	158.2	150.6
Fe_3O_4	138.1			171.1	212.5	252.9	
$Fe(OH)_2$			243.5	102.1	111.3	118.9	123.4
$Fe(OH)_3$				118.0	140.6	154.8	164.9
FeS, $\Delta Ht = 0.40^{138}$	31.5			89.2	62.0	58.6	59.0
$\Delta Ht = 0.095^{325}$							
FeS_2 marcasite				69.2	74.6	78.7	82.8
pyrite				68.9	74.3	78.3	82.5
$FeSiO_3$				100.8	114.3	124.5	133.9
Fe_2SiO_4	92			150.9	168.5	179.7	189.1
$FeSO_4$				116.7	138.0	149.4	
$Fe_2(SO_4)_3$				307.0	363.3	393.3	409.2
$FeTiO_3$ ilminite	90.8	111.4	122.0	128.1	132.8		
Krypton							
Kr	1.37	9.08					
Lanthanum							
La, $\Delta Ht = 2.85^{868}$	6.20	402.1		28.5	29.8	31.2	32.5
$LaCl_3$	43.1	192.1		105.8	110.1	114.3	118.7
La_2O_3				117.3	124.7	128.9	132.3
Lead							
Pb	4.77	179.5	195.2	27.7	29.4	30.0	29.4
$Pb(BO_2)_2$				129.7	162.3		
PbB_4O_7				207	265	305	330
$PbBr_2$	16.44	133	173	81.3	88.8	112.1(lq)	112.1
$Pb(CH_3)_4$	10.86						
$Pb(C_2H_5)_4$	8.80						
$PbCl_2$	21.9	127	185.3	80.1	85.9	111.5(lq)	111.5
$PbCO_3$				99.7	123.6	147.6	
PbF_2, $\Delta Ht = 1.46^{310}$	14.7	157		76.1	82.5	89.1	95.6
PbI_2	23.4	104	172	78.9	83.7(c)	108.6(lq)	108.6
$PbMoO_4$				135.3	148.9	159.0	168.2
PbO, $\Delta Ht = 0.17^{488}$	25.5	207		50.4	55.4	55.0	57.8
PbO_2				67.6			
Pb_3O_4				173.1	190.8	199.2	
PbS	18.8	230		50.5	52.4	54.3	56.2
$PbSiO_3$	26.0			101.5	113.5	125.6	138.4
Pb_2SiO_4	51.0			152.0	173.3	184.2	189.1
$PbSO_4$, $\Delta Ht = 17.2^{866}$	40.2			108.7	128.6	152.4	177.3
$PbSO_4 \cdot PbO$				157.3	182.5	211.7	242.0
Lithium							
Li	3.00	147.1	159.3	27.6(c)	29.5(lq)	28.9	28.8
Li_3AlF_6, $\Delta Ht = 9.5^{562}$	110.5			236.4	262.8	290.8	318.6
$LiAlO_2$	87			81.5	92.7	98.2	102.0
$LiBH_4$				91.0			
$LiBeF_3$	27.2			104.6	129.7(c)	159.0(lq)	159.0
Li_2BeF_4	44.0			150.5	180.2(c)	232.1(lq)	232.1
$LiBO_2$	33.8	265		81.1	85.1	96.9	108.3
$Li_2B_4O_7$	121			197.6	241.1	274.4	300.2
LiBr	17.6	107.1		51.3	56.1	64.5(c)	65.3(lq)
LiCl	19.9			51.0	55.6	65.8	

(*Continued*)

TABLE 1.57 Heats of Fusion, Vaporization, and Sublimation and Specific Heat at Various Temperatures of the Elements and Inorganic Compounds (*Continued*)

				C_p			
Substance	Δ*Hm*	Δ*Hv*	Δ*Hs*	400 K	600 K	800 K	1000 K
$LiClO_4$	29			130.0(c)	161.0(lq)	161	161
Li_2CO_3, $\Delta Ht = 0.561^{350}$	41			112.2	149.4	159.0	
$\Delta Ht = 2.238^{410}$							
LiF	27.09	146.8	276.1	46.5	51.6	55.7	59.6
LiH	22.6		231.3	34.8	46.4	57.3	
LiI	14.6						
$LiIO_3$, $\Delta Ht = 2.22^{260}$							
Li_3N				87.1	106.4	124.4	141.0
$LiNO_3$	24.9						
Li_2O	58.6			64.0	73.8	80.6	86.2
Li_2O_2				82.7(c)	80.2(g)	81.4	82.1
LiOH	20.88	187.9	250.6	58.0	68.2(c)	87.1(lq)	87.1
Li_2SiO_3	28.0			118.8	134.3	144.4	152.3
$Li_2Si_2O_5$, $\Delta Ht = 0.941^{936}$	53.8			174.9	205.7	222.6	235.4
Li_2SO_4, $\Delta Ht = 28.5^{575}$	7.50			139.2	168.5	196.1	223.4
Li_2TiO_3, $\Delta Ht = 11.51^{1212}$	110.7			127.4	141.5	149.0	153.9
Lutetium							
Lu	(22)	414					
Magnesium							
Mg	8.48	128	147	26.1	28.2	30.5	
$MgAl_2O_4$	192			138.0	157.9	169.5	178.7
$MgBr_2$	39.3	149	222	77.3	81.4	84.5	
$MgCl_2$	43.1	156.2	249.2	75.7	79.9	82.5	
$MgCO_3$	59			89.9	109.0	122.3	131.8
MgF_2	58.5	274.1	399.5	68.5	75.3	78.6	80.5
MgH_2	14						
MgI_2	26		206	78.4	83.0	96.3(c)	100.4(lq)
Mg_3N_2, $\Delta Ht = 0.46^{550}$			107.6	113.8	119.9	123.8	
$\Delta Ht = 0.92^{788}$							
$Mg(NO_3)_2$				168.5	225.5		
MgO	77			42.6	47.4	49.7	51.2
$Mg(OH)_2$				91.7			
$Mg_3(PO_4)_2$	121			240.2	282.2	320.6	351.5
MgS	63						
Mg_2Si	85.8			73.8	79.8	83.9	87.4
$MgSiO_3$, $\Delta Ht = 0.67^{630}$	71			94.2	107.0	115.8	120.3
$\Delta Ht = 1.63^{985}$							
Mg_2SiO_4				137.6	156.4	167.1	174.6
$MgSO_4$	14.6			110.0	127.6	140.5	151.7
$MgTiO_3$				105.2	118.5	125.4	129.9
Mg_2TiO_4				146	164	175	184
$MgWO_4$				123.4	137.0	146.1	154.8
Manganese							
Mn, $\Delta Ht = 2.23^{727}$	12.9	221		28.5	31.9	34.9	37.5
$\Delta Ht = 2.12^{1101}$							
$\Delta Ht = 1.88^{1137}$							
$MnBr_2$	33	113		77.8	82.8	87.7	
Mn_3C, $\Delta Ht = 14.94^{1037}$				104.4	115.0	121.7	127.4
$MnCl_2$	30.7	149.0		77.2	81.8	85.1	96.2(lq)
$Mn_2(CO)_{10}$			62.8				

TABLE 1.57 Heats of Fusion, Vaporization, and Sublimation and Specific Heat at Various Temperatures of the Elements and Inorganic Compounds (*Continued*)

				C_p			
Substance	ΔHm	ΔHv	ΔHs	400 K	600 K	800 K	1000 K
MnF_2	23.0			70.6	75.7	80.7	85.9
MnI_2	42			78.1	83.6	89.0	108.8
MnO	54.4			47.5	50.3	52.4	54.2
MnO_2				63.4	71.1	75.1	
Mn_2O_3				109.0	120.8	129.4	137.2
Mn_3O_4, $\Delta Ht = 20.79^{1172}$				157.3	169.5	179.7	189.3
MnS	26.4			50.7	52.2	53.7	55.2
$MnSiO_3$	66.9			100.9	113.1	119.5	124.2
$MnSO_4$				119.0	136.7	147.7	
$MnTiO_3$				111.7	121.2	125.7	128.8
Mercury							
Hg	2.29	59.1	61.4	27.4	27.1(lq)	20.8(g)	20.8
$HgBr_2$	17.9	58.9		78.3	102.1(lq)	102.1	102.1
Hg_2Br_2				109.6	115.6		
$HgCl_2$	19.41	58.9		77.0(c)	102.9(lq)		
Hg_2Cl_2				106.0	112.1		
HgF_2	23.0	92		77.0	81.2	85.4(c)	102.9(lq)
Hg_2F_2				104.7	111.7	116.9	
HgI_2, $\Delta Ht = 2.52^{129}$	18.9	59.2		82.0(c)	84.1(lq)	62.2(g)	62.2
Hg_2I_2	27.8			110.4(c)	136.4(lq)		
HgO				48.3	54.1		
HgS, $\Delta Ht = 4.2^{386}$				48.0	51.0	54.1	
Molybdenum							
Mo	37.48	617	664	25.1	26.5	27.4	28.4
$MoBr_3$				106.9	109.8	112.7	
$MoCl_4$	17	61.5		135.0(c)	146.4(lq)		
$MoCl_5$	18.8	62.8		167.4(c)	175.7(lq)	175.7	175.7
$Mo(CO)_6$		72.5	69.9				
MoF_6, $\Delta Ht = 8.17^{-9.65}$	4.33	27.2	28.0	133.1	145.3	150.4	153.0
MoO_2				63.5	71.2	76.5	81.4
MoO_3	48	138		83.1	91.8	100.0	109.0
MoS_2				68.9	73.6	76.2	78.2
Mo_2S_3	130			117.5	127.4	135.2	142.3
Neodymium							
Nd, $\Delta Ht = 2.98^{862}$	7.14	289		28.2	32.1	36.9	42.0
Nd_2O_3				120.3	130.0	137.7	144.4
Neon							
Ne	0.335	1.71					
Neptunium							
Np, $\Delta Ht = 8.37^{280}$	3.20	336		34.8			
Nickel							
Ni	17.48	377.5		28.5	30.0	31.0	32.2
$NiCl_2$	71.2		231.0	76.3	79.9	80.9	
$Ni(CO)_4$	13.8	29.3		160.4(g)	173.2	182.1	188.6
NiF_2				76.4	78.5	82.6	
NiO				52.2	51.8	53.6	55.2
NiS, $\Delta Ht = 6.4^{379}$	30.1			12.1	13.2	13.7	15.1
Ni_3S_2, $\Delta Ht = 56.2^{556}$	19.7			127.1	139.9	150.7	188.6
NiS_2	65.7			72.8	70.0	81.0	85.2
$NiSO_4$				142.6	150.8	159.2	167.4

(*Continued*)

TABLE 1.57 Heats of Fusion, Vaporization, and Sublimation and Specific Heat at Various Temperatures of the Elements and Inorganic Compounds (*Continued*)

				C_p			
Substance	ΔHm	ΔHv	ΔHs	400 K	600 K	800 K	1000 K
$NiWO_4$				138.9	144.6	150.3	155.9
Niobium							
Nb	30	689.9	726	25.4	26.3	27.2	28.0
$NbBr_5$	24.0	50.2	112.5	147.9(c)	147.9(lq)		
$NbCl_5$	38.3	52.7		170.7(c)	127.9(g)	129.8	130.7
NbF_5	12.2	52.3		43.5(lq)			
NbI_5	37.7	58.6		182.0(c)			
NbN, $\Delta Ht = 4.2^{1370}$	46.0			45.4	49.9	51.6	53.2
NbO	85	618		44.0	47.2	49.5	51.5
NbO_2, $\Delta Ht = 3.42^{817}$	92		598.0	63.5	71.7	70.5	87.5
Nb_2O_5	104.3			145.0	160.7	170.0	175.5
Nitrogen							
N_2, $\Delta Ht = 0.230^{-237.53}$	0.720	5.577		29.2	30.1	31.4	32.7
NF_3		11.6		61.9	71.4	76.0	78.4
N_2F_2 *cis*	15.4	91.6		58.2	68.3	73.6	76.6
trans	14.2	87.9		60.2	68.9	73.8	76.7
N_2F_4		13.3					
NH_3 (*see* Ammonium)							
N_2H_4	12.66	41.8	44.7	61.7(g)	77.6	88.2	96.4
NO	2.30	13.83		29.9	31.2	32.8	34.0
NOCl		25.8		47.1	50.7	53.2	54.9
NOF		19.3		44.6	48.9	51.7	53.5
NOF_3				78.7	90.9	97.0	100.5
NO_2				40.5	46.4	50.4	53.0
NO_2Cl		25.7		59.6	68.1	73.1	76.1
NO_2F		18.0		57.0	66.4	71.9	75.3
NO_3				55.9	67.4	73.3	76.5
N_2O	6.54	16.53		42.7	48.4	52.2	54.9
N_2O_4	14.65	38.12		88.5	104.0	113.4	119.2
N_2O_5			62.3	110.9	128.4	137.0	141.4
NSF		22.2					
Osmium							
Os	57.85	738		25.1	25.9	26.7	27.4
OsF_6		28.62					
OsO_4	9.8	39.54					
Oxygen							
O_2, $\Delta Ht = 0.092^{-249.49}$	0.444	6.820	8.204	30.11	32.09	33.74	34.88
$\Delta Ht = 0.745^{-229.38}$							
O_3		10.84		43.74	49.86	53.15	55.02
OF_2		11.09		64.3	72.4	76.4	78.6
O_2F_2		19.1					
Palladium							
Pd	16.74	362		26.5	27.7	28.8	30.0
$PdCl_2$	40.1						
PdO				37.6	49.5	61.3	
Phosphorus							
P		0.66	12.4	14.2			
P_4, $\Delta Ht = 0.521^{-77.8}$	0.659	56.5	58.9	73.3(g)	78.4	80.4	81.4
PBr_3		38.8		78.9	81.2	82.0	82.4
$PClF_2$		17.6					

TABLE 1.57 Heats of Fusion, Vaporization, and Sublimation and Specific Heat at Various Temperatures of the Elements and Inorganic Compounds (*Continued*)

Substance	ΔHm	ΔHv	ΔHs	C_p 400 K	C_p 600 K	C_p 800 K	C_p 1000 K
$PClF_3$		17.6					
PCl_2F		24.9					
PCl_3	7.10	30.5	32.1	76.0(g)	79.7	81.2	81.9
PCl_5			64.9	120.1(g)	126.8	129.5	130.7
PF_3		16.5		66.3(g)	74.0	77.6	79.5
PF_5		17.2		99.2(g)	114.7	121.9	125.6
PH_3	1.130	14.60		41.8	50.9	58.5	64.3
P_2H_4		28.8					
PI_3		43.9					
P_4O_6	14.06	43.43		172.1	200.8	213.5	220.0
P_4O_{10}	27.2		106.0	260.3	336.0(c)		
$POBr_3$	38						
$POCl_3$	13.1	34.3	38.6	92.0(g)	99.1	102.5	108.5
$POClF_2$		25.4		79.3	91.6	97.7	101.1
$POCl_2F$		30.96		87.7	96.6	100.9	103.2
POF_3	15.06	23.22	21.1	79.1	91.2	97.4	100.9
$PSCl_3$				96.5	102.4	104.8	105.9
PSF_3		19.58		84.5	95.3	100.3	102.9
P_4S_3	9.2	59.8		184.1	184.1(lq)	155.0(g)	155.0
Platinum							
Pt	22.17	469	545	26.4	27.5	28.5	29.6
PtS				51.4	53.8	56.2	58.6
PtS_2				69.9	75.9	81.9	87.9
Plutonium							
Pu, $\Delta Ht = 13.4^{122}$	2.82	333.5		39.5	46.9	40.6	40.6
$\Delta Ht = 2.9^{206}$							
$\Delta Ht = 3.3^{319}$							
$\Delta Ht = 66.9^{480}$							
$PuBr_3$	55.2	236.4	292.5				
$PuCl_3$	63.6	241.0	304.6				
PuF_3	59.8		374.9				
PuF_4	65.3		299.6				
PuF_6	17.6	29.9	48.5				
PuI_3	50.2						
PuO_2		559.8					
Polonium							
Po		102.91					
Potassium							
K	2.321	76.90	88.8	31.5(lq)	30.1	29.8	30.7
$KAlCl_4$				165.5	183.2	196.6	202.1
K_3AlCl_6				259.2	279.5	295.8	
K_3AlF_6				244.5	269.4	286.8	302.0
KBF_4, $\Delta Ht = 14.06^{283}$	17.7			130.8	142.1	150.9	167.2
KBH_4				100.9	106.0	118.4	
KBO_2	31	238.9		76.7	89.8	98.5	
$K_2B_4O_7$	104			206.3	250.5	271.1	283.3
KBr	25.5	149.2		53.8	56.4	60.4	68.0
KCl	26.53	124.3		53.0	55.9	59.2	64.0
$KClO_4$, $\Delta Ht = 13.77^{299.6}$				138.5	165.3		
KCN, $\Delta Ht = 1.167^{-104.9}$	14.6	157.1		66.3	66.4	66.5(c)	66.5(lq)

(*Continued*)

TABLE 1.57 Heats of Fusion, Vaporization, and Sublimation and Specific Heat at Various Temperatures of the Elements and Inorganic Compounds (*Continued*)

Substance	ΔHm	ΔHv	ΔHs	C_p			
				400 K	600 K	800 K	1000 K
K_2CO_3	27.6			128.1	150.7	170.0	189.0
K_2CrO_4	29.0						
$K_2Cr_2O_7$	36.7						
KF	27.2	141.8	231.8	51.0	54.3	57.4	61.2
KH				44.1	51.9		
KHF_2, $\Delta Ht = 11.22^{196.7}$	6.62			86.1(c)	104.6(lq)		
KI	24.0	190.9	202.4	53.9	57.3	62.6(c)	72.4(lq)
KNO_3, $\Delta Ht = 5.10^{128}$	10.1			108.4	120.5		
K_2O, $\Delta Ht = 6.20^{372}$				79.1	100.0	100.0	100.0
KO_2, $\Delta Ht = 0.302^{-79.7}$				83.9	90.2		
$\Delta Ht = 0.157^{-42.3}$							
K_2O_2				107	121		
KOH, $\Delta Ht = 6.4^{243}$	8.60	142.7	192	72.5	79.0(c)	83.0(lq)	83.0
KPO_3	8.8						
K_3PO_4	37.2						
$K_2P_2O_7$	58.6						
$KReO_4$	85.4						
K_2S	16.15	77.3	82.5	87.7			
K_2SiO_3	50			135.6	157.7	170.7	179.1
K_2SO_4, $\Delta Ht = 8.45^{584}$	34.39			147.6	172.5	199.6	226.1
K_2WO_4	19.5						
K_2ZrCl_6	23.0						
Praseodymium							
Pr	6.89	331	356				
Promethium							
Pm	7.13	289	328				
Protactinium							
Pa	12.34	481					
$PaCl_3$	92.9	61.3					
Radium							
Ra	8.5	113					
Radon							
Rn	3.247	18.10					
Rhenium							
Re	60.43	704	779	26.0	26.9	28.0	29.1
ReF_5		58.1					
ReF_6	4.6	28.7					
ReF_7	7.5	38.3					
ReO_2			274.6				
ReO_3	21.8		208.4				
Re_2O_7	64.2	74.1					
$ReOCl_4$		45.6					
$ReOF_4$	13.5	61.0					
$ReOF_5$		32.0	37.4				
Rhodium							
Rh	26.59	494	556	26.0	28.0	30.0	32.0
Rh_2O_3				109.9	121.4	133.0	144.5
Rubidium							
Rb	2.19	75.77		31.7	30.9	30.7	
RbBr	15.5	154.8		52.8	54.9	57.1(c)	66.9(lq)

TABLE 1.57 Heats of Fusion, Vaporization, and Sublimation and Specific Heat at Various Temperatures of the Elements and Inorganic Compounds (*Continued*)

				C_p			
Substance	ΔHm	ΔHv	ΔHs	400 K	600 K	800 K	1000 K
RbCl	18.4	165.7		52.3	54.3	56.4(c)	64.0(lq)
$RbClO_4$, $\Delta Ht = 12.59^{284}$							
RbF	17.3	177.8		51.9	57.9	64.9	72.3
RbI	12.5	150.6			55.1	57.3(c)	66.9(lq)
$RbNO_3$	5.61						
RbOH	6.78						
Ruthenium							
Run, $\Delta Ht = 0.13^{1035}$	38.59	591.6		24.5	25.7	27.0	28.2
$\Delta Ht = 0.96^{1500}$							
Samarium							
Sm, $\Delta Ht = 3.11^{917}$	8.62	165	207	33.3	39.1	44.3	49.3
Sm_2O_3, $\Delta Ht = 1.05^{922}$				125.2	135.3	141.4	146.3
Scandium							
Sc	14.1	332.7	376				
$ScCl_3$				96.7	102.7	108.7	114.6
Sc_2O_3				106.4	111.1	115.8	120.5
Selenium							
Se, $\Delta Ht = 0.75^{150}$	6.69	95.48		28.1(c)	35.2(lq)	35.1	
SeF_4		47.2					
SeF_6	8.4		26.8	127.9	141.3	147.1	150.7
SeO_2		94.5					
$SeOCl_2$	4.23	42.7					
Silicon							
Si	50.21	359	450	22.3	24.5	25.7	26.5
$SiBr_4$		37.9		146.4(lq)	104.9(g)	106.2	106.2
SiC beta				34.1	41.8	45.9	48.4
$SiCl_4$	7.60	28.7	29.7	96.9(g)	102.6	104.8	106.0
$SiClF_3$		18.7		88.3	97.5	101.7	103.8
$SiCl_2F_2$		21.2					
SiF_4			25.7	83.1	94.1	99.4	102.3
SiH_4	0.67	12.1		51.5	65.9	76.7	84.5
Si_2H_6		21.2					
Si_3H_8		28.5					
SiH_3Br		24.4					
SiH_2Br_2		31					
$SiHBr_3$		34.8					
SiH_3Cl		21		60.7	74.0	83.1	89.4
SiH_2Cl_2		25.2	24.2	71.5	82.9	90.0	94.6
$SiHCl_3$		26.6	25.7	83.7	92.5	97.2	100.2
SiH_3F		18.8		57.2	71.8	81.7	88.3
SiH_2F_2		16.3					
$SiHF_3$		16.2					
SiI_4	19.7	56.9	79	164.0(lq)	106.0(g)	106.9	107.3
Si_3N_4				110.7	129.7	145.8	158.2
SiO_2 cristobalite	8.51						
SiO_2 quartz	7.7		600	53.5	64.4	76.2	68.94
$\Delta Ht = 0.73^{574}$							
$\Delta Ht = 2.0^{806}$							
$SiOF_2$				61.3	70.4	75.0	77.6
SiS_2	20.9			78.6	81.7	83.4	85.4

(*Continued*)

TABLE 1.57 Heats of Fusion, Vaporization, and Sublimation and Specific Heat at Various Temperatures of the Elements and Inorganic Compounds (*Continued*)

Substance	ΔHm	ΔHv	ΔHs	C_p 400 K	C_p 600 K	C_p 800 K	C_p 1000 K
Silver							
Ag	11.95	258		25.7	26.8	28.4	30.0
AgBr	9.12	198		59.0	71.8(c)	62.3(lq)	62.3
AgCl	13.2	199		56.9	54.4	54.4	54.4
Ag_2CO_3					122.6		
AgF	16.7	179.1		54.1(c)	58.4		
AgI, $\Delta Ht = 6.15^{147}$	9.41	143.9		64.7	56.5	56.5	58.6(lq)
$AgNO_3$, $\Delta Ht = 2.5^{160}$	11.5			112.5	128.0		
Ag_2O				73.0			
Ag_2S, $\Delta Ht = 5.86^{176}$	14.1			86.6	90.5	90.5	90.5
$\Delta Ht = 5.86^{586}$							
Sodium							
Na	2.60	97.42	107.5	31.5(lq)	29.3	29.9	29.0
$NaAlCl_4$				164.8(c)			
Na_3AlCl_6				254.4	273.0		
Na_3AlF_6, $\Delta Ht = 8.37^{565}$	107.28			234.6	261.8	196.8	282.8
$\Delta Ht = 0.42^{880}$							
$NaAlO_2$, $\Delta Ht = 1.297^{467}$				83.4	94.3	98.7	102.3
$NaBH_4$, $\Delta Ht = 0.999^{-83.3}$				94.6	108.6		
$NaBO_2$	36.2	239.7	322.2	75.4	88.6	97.2	103.2
$Na_2B_4O_7$	76.9			221.7	268.6	444.9(lq)	
NaBr	26.11	160.7	217.5	53.5	56.1	58.6	61.1
$NaBrO_3$	28.11						
NaCl	28.16			52.3	55.5	59.3	72.5
$NaClO_3$	22.1						
$NaClO_4$, $\Delta Ht = 13.98^{308}$				136.0(c)			
NaCN	8.79	148.1	172.8	68.7	68.8	69.0	
Na_2CO_3, $\Delta Ht = 0.690^{450}$	29.64			125.1	163.3	153.3	179.8
NaF	33.35	176.1	284.9	49.6	52.7	55.7	59.5
NaH				42.5	50.7		
NaI	23.60			53.8	56.2	58.5(c)	64.9(lq)
$NaIO_3$, $\Delta Ht = 35.1^{422}$							
$NaNO_3$	15						
NaO_2, $\Delta Ht = 1.464^{-76.7}$				76.3	84.5	92.6	
$\Delta Ht = 1.548^{-49.9}$							
Na_2O, $\Delta Ht = 1.76^{750.1}$	47.7			75.8	85.7	91.3	94.9
$\Delta Ht = 11.92^{970.1}$							
Na_2O_2, $\Delta Ht = 5.73^{512}$				97.7	108.4	113.6	
NaOH, $\Delta Ht = 72^{299.6}$	6.60	175.3	228.2	64.9(c)	86.1(lq)	84.9	83.7
Na_2S	19.3			20.1	20.9	21.5	22.0
Na_2S_2				104.3	115.4(c)	124.7(lq)	124.7
Na_2SiO_3	51.8			127.8	147.1	159.7	169.4
$Na_2Si_2O_5$, $\Delta Ht = 0.42^{678}$	35.6			183.4	217.6	235.2	292.9
Na_2SO_4, $\Delta Ht = 10.91^{241}$	23.6			145.1	175.3	187.3	200.3
Na_2TiO_3	70.3						
Na_2WO_4, $\Delta Ht = 30.85^{587.7}$	23.80			155.3	178.2	198.7	
$\Delta Ht = 4.113^{588.9}$							
Strontium							
Sr, $\Delta Ht = 0.84^{547}$	7.43	136.9	164.0	27.8	29.8	31.9	34.1
$SrBr_2$, $\Delta Ht = 12.2^{645}$	10.1	194.1	310	79.0	82.7	87.6(c)	116.4(lq)

TABLE 1.57 Heats of Fusion, Vaporization, and Sublimation and Specific Heat at Various Temperatures of the Elements and Inorganic Compounds (*Continued*)

Substance	ΔHm	ΔHv	ΔHs	C_p 400 K	600 K	800 K	1000 K
$SrCl_2$, $\Delta Ht = 6.0^{727}$	17.5	248.1	356	78.9	83.7	90.8	105.8
$SrCO_3$, $\Delta Ht = 19.7^{924}$	40			95.1	107.1	116.1	124.0
SrF_2, $\Delta Ht = 0.04^{1148}$ $\Delta Ht = 0.04^{1211}$	28.5	320	451.0	74.7	79.8	81.0	85.8
SrI_2	19.67	189.7	286.6	80.7	86.3	91.8(c)	110.0(lq)
SrH_2	23						
$SrMoO_4$				131.5	145.4	154.0	161.2
SrO	81			48.5	52.0	54.3	56.1
SrO_2				81.3	85.0		
$Sr(OH)_2$	23			88.5	115.0(c)	157.8(lq)	157.8
SrS	63			50.2	53.2	54.9	56.2
$SrSO_4$	36			113.5	124.6	135.7	146.9
Sulfur							
S monoclinic $\Delta Ht = 0.400^{95.2}$	1.727	45	62.2	23.2	23.3(lq)	21.8(g)	21.5
S_8				167.1	177.9	186.7	193.6
SCl_2		32.4		53.6	56.0	56.9	57.4
S_2Cl_2		36.0		124.3(lq)	80.8(g)	82.6	83.5
SF_4		26.4		87.5	97.3	101.7	103.8
SF_6	5.02	17.1	9.0	116.4	136.1	144.8	149.3
S_2F_{10}				211.4	246.4	261.8	269.2
SO_2	7.40	24.94	22.92	43.43	48.9	52.3	54.3
SO_3	8.60	40.7	43.14	57.7	67.3	72.8	76.0
$SOCl_2$		31.7	31	71.3	76.4	78.9	80.3
SOF_2		21.8		64.3	72.4	76.4	78.6
SO_2Cl_2		31.38	30.1	85.2	94.5	99.4	102.1
SO_2ClF				81.1	92.1	97.9	101.1
SO_2F_2		20.0		76.5	89.3	96.1	99.9
Tantalum							
Ta	36.57	732.8	778	25.8	26.8	27.5	27.9
TaB_2	83.7			57.6	66.6	72.2	83.3
$TaBr_5$	45.6	62.3		168.2			
TaC	105			41.7	46.5	49.1	51.1
Ta_2C				66.7	72.4	76.2	79.5
$TaCl_5$	41.6	54.8	94.1	148.(c)	129.(g)	131	132
TaF_5	18.8	56.9		182.0(lq)			
TaI_5	41.8	64.9		164.6	182.0(c)	120.0(g)	120.6
TaN	67			45.4	51.9	58.5	65.0
TaO_2				47.7	52.3	54.6	55.7
Ta_2O_5	120			147.5	164.4	175.2	182.8
Technetium							
Tc	33.29	585.2		25.1	26.8	28.5	30.1
TcF_6	4.72	31.1					
TcO_3F	22.5	39.5					
Tellurium							
Te	17.49	114.1		28.0	32.3(c)	37.7(lq)	37.7
$TeCl_4$	18.8	77		138.9(c)	222.6(lq)	108.8(g)	108.8
TeF_4		34.3					
TeF_6			28.2	132.2	143.8	148.7	151.7
Te_2F_{10}		39.5					

(*Continued*)

TABLE 1.57 Heats of Fusion, Vaporization, and Sublimation and Specific Heat at Various Temperatures of the Elements and Inorganic Compounds (*Continued*)

				C_p			
Substance	Δ*Hm*	Δ*Hv*	Δ*Hs*	400 K	600 K	800 K	1000 K
TeH_2		23.9					
TeO_2	29.1			67.9	72.5	76.1	79.2
Terbium							
Tb	10.15	293	389				
Thallium							
Tl, $\Delta Ht = 0.38^{234}$	4.14	165	181	27.5(c)	30.1(lq)	30.1	30.1
TlBr	16.4	99.6		53.5	59.5(c)	75.5(lq)	67.8
TlCl	15.56	102.2		53.6	55.2(c)	59.4(lq)	59.4
Tl_2CO_3	18.4						
TlF	13.87	115.9			66.8(lq)	67.3	
TlI	14.73	104.7		53.9	60.6(c)	72.0(lq)	72.0
$TlNO_3$	9.56						
Tl_2O	30.3						
Tl_2O_3	53						
Tl_2S	12	154					
Tl_2SO_4	23.0						
Thorium							
Th, $\Delta Ht = 2.73^{1360}$	13.81	514		28.4	30.5	32.7	34.4
$ThBr_4$	66.9						
$ThCl_4$, $\Delta Ht = 5.0^{406}$	40.2	146.4		126.7	132.7	136.4	139.6
ThF_4	44.0	258					
ThI_4	61.4	56.9					
Th_3N_4				169.5	196.5	222.7	
ThO_2	1218.0			67.4	72.4	75.3	77.7
$ThOCl_2$				97.0	102.5	105.9	108.6
$Th(SO_4)_2$				197.0	243.2	289.4	
Thullium							
Tm	16.84	247	232.2				
Tin							
Sn white, $\Delta Ht = 2.09^{13}$	7.03	296.1		28.9	28.9(c)	28.7(lq)	28.7
$SnBr_2$	7.2	102					
$SnBr_4$	11.9	43.5		158.0(lq)	106.8(g)	107.3	107.5
$SnCl_2$	12.8	86.8		83.3(c)	92.1(lq)	92.1	92.1
$SnCl_4$	9.20	34.9					
SnH_4		19.1					
SnI_2		105					
SnO				45.8	48.7	51.7	54.6
SnO_2, $\Delta Ht = 1.88^{410}$				64.4	73.9	78.5	81.8
$\Delta Ht = 1.26^{540}$							
SnS, $\Delta Ht = 0.67^{602}$				50.5	55.5	61.3	
SnS_2				71.9	75.4	79.0	82.5
Titanium							
Ti, $\Delta Ht = 4.2^{893}$	14.15	425	469	26.9	28.6	29.5	32.1
TiB				40.3	48.6	50.9	51.9
TiB_2	100.4			54.9	66.2	72.1	76.9
$TiBr_2$			206.2	79.9	82.1	84.4	86.7
$TiBr_3$			138.8	105.8	125.5	147.3	156.7
$TiBr_4$	12.9	44.4		151.9(lq)	106.1(g)	106.9	107.3
TiC	71			40.7	47.7	49.9	51.2
$TiCl_2$		232	212	73.4	78.4	82.2	85.9

TABLE 1.57 Heats of Fusion, Vaporization, and Sublimation and Specific Heat at Various Temperatures of the Elements and Inorganic Compounds (*Continued*)

Substance	ΔHm	ΔHv	ΔHs	C_p			
				400 K	600 K	800 K	1000 K
$TiCl_3$		124	166.3	98.6	102.0	104.4	106.7
$TiCl_4$	9.97	36.2		146.2(lq)	104.4(g)	106.0	106.7
TiF_3			222	93	98	103	109
TiF_4			97.9	126.7(c)	100.2(g)	103.3	104.9
TiH_2				39.3	53.8	63.1	68.5
TiI_2			217	87.0	88.4	89.9	91.3
TiI_3				117.5	119.0	120.4(c)	20.6(g)
TiI_4, $\Delta Ht = 9.9^{106}$	19.8	58.4		148.1(c)	156.6(lq)	25.7(g)	27.8
TiN	66.9			43.8	48.7	50.6	52.1
TiO, $\Delta Ht = 4.2^{992}$	41.8			45.0	50.8	55.2	59.1
TiO_2 rutile	58.0		673	63.6	70.9	73.9	75.3
Ti_2O_3, $\Delta Ht = 1.138^{197}$	105			117.5	136.4	143.0	146.4
Tungsten							
W	52.31	806.7	851	24.9	25.9	26.7	27.6
WBr_5	17.1	81.5		166.(c)	182.(lq)	132.2(g)	132.5
WBr_6				192.5(c)	156.3(g)	157.0	157.4
WCl_4				135.3	146.2(c)	106.7(g)	107.2
WCl_5	20.5	68.1	100	167.4(c)	129.5(g)	131.0	131.8
WCl_6, $\Delta Ht = 4.1^{177}$	6.60	52.7	79.2	192.5(c)	200.8(lq)	155.8(g)	156.6
$W(CO)_6$			72.0				
WF_6, $\Delta Ht = 2.067^{-8.5}$	4.10	27.05	26.65	132.4(g)	145.0	150.3	153.0
WO_2			666.3	63.4	71.3	75.5	78.2
WO_3, $\Delta Ht = 1.49^{777}$	73.4	76.6	550.2	82.2	93.1	98.2	101.7
$WOCl_4$	45	67.8		157.(c)	123.2(g)	127.0	129.1
WOF_4	5.0	56		107.8	119.8	125.0	127.8
WO_2Cl_2				115.1	135.6(c)		
Uranium							
U, $\Delta Ht = 2.93^{672}$	9.14	417.1	525	29.0	34.8	41.6	41.8
$\Delta Ht = 4.791^{772}$							
UBr_3	43.9						
UBr_4	55.2	119.2		131.4	140.1(c)	163.2(lq)	163.2
UC				64.6	58.3	60.3	62.2
UCl_3	46.4	193.0		102.8	107.7	113.6	119.9
UCl_4	44.8	141.4		126.1	134.4	142.0	162.5
UCl_5	35.6	75.3		150.9	159.8(c)	186.7(lq)	134.5(g)
UCl_6	20.9	50.2		182.8	214.0	158.8	168.0
UF_3				99.0	104.9	111.0	117.2
UF_4	42.7	221.8		119.1	125.0	130.9	136.8
UF_5	33.5			136.4	143.1(c)	166.6(lq)	
UF_6	19.19	28.90	48.20	140.5(g)	148.7	152.2	154.4
UH_3				50.9	57.4	66.1	
UI_4	70.7	130.6		140.6	149.5(c)	165.7(lq)	165.7
UN				52.2	56.3	58.3	59.8
UO_2				72.7	79.8	83.2	85.5
UO_3				88.9	95.3	99.0	
U_3O_8				266.0	290.7	304.2	
$UOCl_2$				101.9	109.6	115.1	
UO_2Cl_2				118.1	126.2	130.0	
UO_2F_2				113.9	122.5	126.7	129.5

(*Continued*)

TABLE 1.57 Heats of Fusion, Vaporization, and Sublimation and Specific Heat at Various Temperatures of the Elements and Inorganic Compounds (*Continued*)

Substance	ΔHm	ΔHv	ΔHs	C_p 400 K	C_p 600 K	C_p 800 K	C_p 1000 K
Vanadium							
V	21.5	459	516	26.2	27.5	28.7	30.1
VCl_4	2.30	41.4	42.5	161.7(lq)	100.1(g)	102.6	104.7
VF_5	50.0	44.5					
VN, $\Delta Hdec = 227.6^{2346}$			741	43.3	48.2	51.2	53.7
VO	63			49.6	53.5	57.1	60.5
VO_2, $\Delta Ht = 4.21^{72}$	56.9			67.2	74.3	77.8	80.2
V_2O_3, $\Delta Ht = 1.623^{-104.3}$	117.2			117.5	127.3	132.6	138.0
V_2O_4, $\Delta Ht = 9.0^{67}$	112.1			135.3	148.4	155.5	160.7
V_2O_5	64.5	263.6		151.0	168.3	177.3	183.7
$VOCl_3$		36.8					
Xenon							
Xe	1.81	12.64		20.79(g)	20.79	20.79	20.79
Ytterbium							
Yb	7.66	159					
Yttrium							
Y, $\Delta Ht = 4.97^{1485}$	11.42	365	425	27.3	28.5	29.9	31.5
Y_2O_3, $\Delta Ht = 1.30^{1057}$	105			113.3	121.3	124.7	126.9
Zinc							
Zn	7.32	123.6		26.3	28.6(c)	31.4(lq)	31.4
$ZnBr_2$	16.7	118		70.1(c)	78.8(lq)	113.8	61.5(g)
$ZnCl_2$	10.25	126		69.9(c)	100.8(lq)	100.8	100.8
ZnF_2		190.1		66.9	69.1	71.4	73.7
ZnO, $\Delta Ht = 13.4^{1020}$	52.3			49.4	52.4	54.1	55.5
Zn_2SiO_4				129.4	141.4	153.4	165.4
$ZnSO_4$, $\Delta Ht = 20.3^{740}$				116.0	137.4	139.7	142.0
Zirconium							
Zr, $\Delta Ht = 4.02^{862}$	21.00	573	610.0	25.9	27.3	29.0	31.1
ZrB_2	104.6			57.5	65.8	69.7	72.1
$ZrBr_2$	63	131.5	230	87.9	90.2	92.5	94.8
$ZrBr_4$				129.3	133.3(c)	107.2(g)	107.6
ZrC	79.5			43.6	49.4	52.3	53.4
$ZrCl_2$	27	45.0		76.0	80.0	83.1	85.9
$ZrCl_3$			190	101	106	109	112
$ZrCl_4$	50		110.5	125.4	131.1(c)	106.5(g)	107.1
ZrF_2	33	289	404	70	76	81	84
ZrF_4	64.2		237.7	113.5	124.0	129.4	134.1
ZrI_2	25.1	113		95.0	96.6	106.1	123.6
ZrI_3			176	105.9	106.7	107.1(c)	82.9(g)
ZrI_4			126.4	131.0	134.6(c)	107.6(g)	107.6
ZrN	67.4			44.8	48.7	50.9	52.7
ZrO_2, $\Delta Ht = 5.02^{1205}$	87.0	624		63.9	70.2	73.5	75.7
$ZrSiO_4$				114.6	133.7	142.7	147.3

1.16 ACTIVITY COEFFICIENTS

The activity coefficient is the ratio of the chemical activity of any substance to its molar concentration. The measured concentration of a substance may not be an accurate indicator of its chemical effectiveness, as represented by the equation for a particular reaction, in which case an activity coefficient is arbitrarily established and used instead of the concentration…

Although it is not possible to measure an individual ionic activity coefficient, f_i, it may be estimated from the following equation of the Debye-Hückel theory:

$$-\log f_i = \frac{Az_i^2 \sqrt{I}}{I + B\mathring{a} \sqrt{I}}$$

where I is the ionic strength of the medium, and $\mathring{a}$ is the ion-size parameter—the effective ionic radius (Table 1.32). The values of A and B vary with the temperature and dielectric constant of the solvent; values from 0 to 100C for aqueous medium ($\mathring{a}$ in angstrom units) are listed in Table 1.59. Corresponding values of A and B for unit weight of solvent (when employing molality) can be obtained by multiplying the corresponding values for unit volume (molarity units) by the square root of the density of water at the appropriate temperature.

The ionic strength can be estimated from the summation of the product molarity times ionic charge squared for all the ionic species present in the solution, i.e., $I = 0.5\ (c_1 z_1^2 + c_2 z_2^2 + \cdots + c_i z_i^2)$.

Values for the activity coefficients of ions in water at 25°C are given in Table 8.1 in terms of their effective ionic radii.

At moderate ionic strengths a considerable improvement is effected by subtracting a term bI from the Debye-Hückel expression; b is an adjustable parameter which is 0.2 for water at 25°C. Table 1.58 gives the values of the ionic activity coefficients (for z_i from 1 to 6) with $\mathring{a}$ taken to be 4.6Å.

In general, the mean ionic activity coefficient is given by

$$f_\pm = \sqrt[(x+y)]{f_+^x f_-^y}$$

where f_+, f_- are the individual ionic activity coefficients, and x,y are the charge numbers (z_+, z_-) of the respective ions. In binary electrolyte solution.

$$f_\pm = \sqrt{f_+ f_-}$$

In ternary electrolytes, e.g., $BaCl_2$ or K_2SO_4,

$$f_\pm = \sqrt[3]{f_+ f_-^2} \quad \text{or} \quad f_\pm = \sqrt[3]{f_+^2 f_-}$$

In quaternary electrolytes, e.g., $LaCl_3$ or $K_3[Fe(CN)_6]$,

$$f_\pm = \sqrt[4]{f_+ f_-^3} \quad \text{or} \quad f_\pm = \sqrt[4]{f_+^3 f_-}$$

TABLE 1.58 Individual Activity Coefficients of Ions in Water at 25°C

Effective Ionic Radii å (in Å)	f_i at Ionic Strength of 0.001	0.005	0.01	0.05	0.1
Univalent Ions					
9	0.967	0.933	0.914	0.86	0.83
8	0.966	0.931	0.912	0.85	0.82
7	0.965	0.930	0.909	0.845	0.81
6	0.965	0.929	0.907	0.835	0.80
5	0.964	0.928	0.904	0.83	0.79
4	0.964	0.928	0.902	0.82	0.775
3.5	0.964	0.926	0.900	0.81	0.76
3	0.964	0.925	0.899	0.805	0.755
2.5	0.964	0.924	0.898	0.80	0.75
Divalent Ions					
8	0.872	0.755	0.69	0.52	0.45
7	0.872	0.755	0.685	0.50	0.425
6	0.870	0.749	0.675	0.485	0.405
5	0.868	0.744	0.67	0.465	0.38
4.5	0.868	0.741	0.663	0.45	0.36
4	0.867	0.740	0.660	0.445	0.355
Trivalent Ions					
6	0.731	0.52	0.415	0.195	0.13
5	0.728	0.51	0.405	0.18	0.115
4	0.725	0.505	0.395	0.16	0.095
Tetravalent Ions					
11	0.588	0.35	0.255	0.10	0.065
5	0.57	0.31	0.20	0.048	0.021
Pentavalent Ions					
9	0.43	0.18	0.105	0.020	0.009

TABLE 1.59 Constants of the Debye-Hückel Equation from 0 to 100°C

$$-\log f_i = \frac{A z_i^2 \sqrt{I}}{I + B\mathring{a}\sqrt{I}}$$

Temp., °C	Unit Volume of Solvent		Temp., °C	Unit Volume of Solvent	
	A	B		A	B
0	0.4918	0.3248	55	0.5432	0.3358
5	0.4952	0.3256	60	0.5494	0.3371
10	0.4989	0.3264	65	0.5558	0.3384
15	0.5028	0.3273	70	0.5625	0.3397
20	0.5070	0.3282	75	0.5695	0.3411
25	0.5115	0.3291	80	0.5767	0.3426
30	0.5161	0.3301	85	0.5842	0.3440
35	0.5211	0.3312	90	0.5920	0.3456
40	0.5262	0.3323	95	0.6001	0.3471
45	0.5317	0.3334	100	0.6086	0.3488
50	0.5373	0.3346			

The values for unit weight of solvent (molality scale) can be obtained by multiplying the corresponding values for unit volume by the square root of the density of water at the appropriate temperature.

TABLE 1.60 Individual Ionic Activity Coefficients at Higher Ionic Strengths at 25°C

The values were calculated from the modified Debye-Hückel equation utilizing the modifications proposed by Robinson and by Guggenheim and Bates:

$$-\frac{\log f_i}{z_i^2} = \frac{0.511I}{1+1.5I} - 0.2I$$

where I is the ionic strength and $\mathring{a}$ is assumed to be 4.6 Å.

I	$-\frac{\log_{10} f_i}{z_i^2}$	f_i for z_i = 1	2	3	4	5	6
0.05	0.0756	0.840	0.498	0.209	0.0617	0.0129	0.00190
0.1	0.0896	0.814	0.438	0.156	0.0369	0.00576	0.000595
0.2	0.0968	0.800	0.410	0.138	0.0283	0.00380	0.000328
0.3	0.0936	0.806	0.422	0.144	0.0318	0.00457	0.000427
0.4	0.0858	0.821	0.454	0.169	0.0424	0.00716	0.000815
0.5	0.0753	0.841	0.500	0.210	0.0624	0.0131	0.00195
0.6	0.0631	0.865	0.559	0.270_5	0.0978	0.0265	0.00535
0.7	0.0496	0.892	0.633	0.358	0.161	0.0575_5	0.0164
0.8	0.0352	0.922	0.723	0.482	0.273	0.132	0.0541
0.9	0.0201	0.955	0.831	0.659	0.477	0.314	0.189
1.0	0.0044	0.900	0.960	0.913	0.850	0.776	0.694

1.17 BUFFER SOLUTIONS

A buffer solution is a solution that resists changes in pH when small quantities of an acid or an alkali are added.

An acidic buffer solution is a solution that has a pH less than 7. Acidic buffer solutions are commonly made from a weak acid and one of its salts. A common example is a mixture of ethanoic acid and sodium ethanoate in solution. In this case, if the solution contained equal molar concentrations of both the acid and the salt, the pH would be 4.76. The pH of the buffer solution can be changed by changing the ratio of acid to salt, or by choosing a different acid and one of its salts.

An alkaline buffer solution has a pH greater than 7. Alkaline buffer solutions are commonly made from a weak base and one of its salts. An example is a mixture of ammonia solution and ammonium chloride solution. If these were mixed in equal molar proportions, the solution would have a pH of 9.25.

To prepare the standard pH buffer solutions recommended by the National Bureau of Standards (U.S.), the indicated weights of the pure materials should be dissolved in water of specific conductivity not greater than 5 micromhos. The tartrate, phthalate, and phosphates can be dried for 2 h at 100°C before use. Potassium tetroxalate and calcium hydroxide need not be dried. Fresh-looking crystals of borax should be used. Before use, excess solid potassium hydrogen tartrate and calcium hydroxide must be removed. Buffer solutions pH 6 or above should be stored in plastic containers and should be protected from carbon doxide with soda-lime traps. The solutions should be replaced within 2 to 3 weeks, or sooner if formation of mold is noticed. A crystal of thymol may be added as a preservative.

1.17.1 Standards for pH Measurement of Blood and Biological Media

Blood is a well-buffered medium. In addition to the NBS phosphate standard of 0.025 M (pH_s = 6.480 at 38°C), another reference solution containing the same salts, but in the molal ratio 1:4, has an ionic strength of 0.13. It is prepared by dissolving 1.360 g of KH_2PO_4 and 5.677 g of Na_2HPO_4 (air weights) in carbon dioxide-free water to make 1 liter of solution. The pH_s is 7.416 ± 0.004 at 37.5 and 38°C.

The compositions and pH_s values of *tris*(hydroxymethyl)aminomethane, covering the pH range 7.0 to 8.9, are listed in Table 1.63.

When there are two or more acid groups per molecule, or a mixture is composed of several overlapping acids, the useful range is larger. Universal buffer solutions consist of a mixture of acid groups which overlap such that successive pK_a values differ by 2 pH units or less. The Prideaux-Ward mixture comprises phosphate, phenyl acetate, and borate plus HCl and covers the range from 2 to 12 pH units. The McIlvaine buffer is a mixture of citric acid and Na_2HPO_4 that covers the range from pH 2.2 to 8.0. The Britton-Robinson system consists of acetic acid, phosphoric acid, and boric acid plus NaOH and covers the range from pH 4.0 to 11.5. A mixture composed of Na_2CO_3, NaH_2PO_4, citric acid, and 2-amino-2-methyl-1,3-propanediol covers the range from pH 2.2 to 11.0.

General directions for the preparation of buffer solutions of varying pH but fixed ionic strength are given by Bates.* Preparation of McIlvaine buffered solutions at ionic strengths of 0.5 and 1.0 and Britton-Robinson solutions of constant ionic strength have been described by Elving et al.† and Frugoni,‡ respectively.

*Bates, *Determination of pH, Theory and Practice*, Wiley, New York, 1964, pp. 121–122.
†Elving, Markowitz, and Rosenthal, *Anal. Chem.*, **28:**1179 (1956).
‡Frugoni, *Gazz. Chim. Ital.*, **87:**L403 (1957).

TABLE 1.61 National Bureau of Standards (U.S.) Reference pH Buffer Solutions

Temperature °C	Secondary standard 0.05 *M* K tetraoxalate	KH tartrate (saturated at 25°C)	0.05 *M* KH_2 citrate	0.05 *M* KH phthalate	0.025 *M* KH_2PO_4, 0.025 *M* Na_2HPO_4	0.0087 *M* KH_2PO_4, 0.0302 *M* Na_2HPO_4	0.01 *M* $Na_2B_4O_7$	0.025 *M* $NaHCO_3$, 0.025 *M* Na_2CO_3	Secondary standard $Ca(OH)_2$ (saturated at 25°C)
0	1.666		3.860	4.003	6.984	7.534	9.464	10.317	13.423
5	1.668		3.840	3.999	6.951	7.500	9.395	10.245	13.207
10	1.638		3.820	3.997	6.923	7.472	9.332	10.179	13.003
15	1.642		3.802	3.998	6.900	7.448	9.276	10.118	12.810
20	1.644		3.788	4.002	6.881	7.429	9.225	10.062	12.627
25	1.646	3.557	3.776	4.005	6.865	7.413	9.180	10.012	12.454
30	1.648	3.552	3.766	4.011	6.853	7.400	9.139	9.966	12.289
35		3.549	3.759	4.018	6.844	7.389	9.102	9.925	12.133
38	1.649	3.548	3.756	4.030	6.840	7.384	9.088	9.910	12.043
40	1.650	3.547	3.753	4.035	6.838	7.380	9.068	9.889	11.984
45		3.547		4.047	6.834	7.373	9.038		11.841
50	1.653	3.549	3.749	4.050	6.833	7.367	9.011	9.828	11.705
55		3.554		4.075	6.834		8.985		11.574
60	1.660	3.560		4.081	6.836		8.962		11.449
70	1.671	3.580		4.116	6.845		8.921		
80	1.689	3.609		4.164	6.859		8.885		
90	1.72	3.650		4.205	6.877		8.850		
95	1.73	3.674		4.227	6.886		8.833		
Dilution value $\Delta pH_{1/2}$	+0.186	+0.049	0.024	+0.052	+0.080	+0.070	+0.01	0.079	−0.28

Source: R. G. Bates, *J. Res. Natl. Bur. Stand. (U.S.)*, **66A:**179(1962) and B. R. Staples and R. G. Bates, *J. Res. Natl. Bur. Stand. (U.S.)*, **73A:**37 (1969).
Note: The uncertainty is ±0.003 in pH in the range 0–50°C, rising to ±0.02 above 70°C.

TABLE 1.62 Compositions of Standard pH Buffer Solutions [National Bureau of Standards (U.S.)]

Standard	Weight, g
$KH_3(C_2O_4)_2 \cdot 2H_2O$, 0.05*M*	12.61
Potassium hydrogen tartrate, about 0.034*M*	Saturated at 25°C
Potassium hydrogen phthalate, 0.05*M*	10.12
Phosphate:	
KH_2PO_4, 0.025*M*	3.39
Na_2HPO_4, 0.025*M*	3.53
Phosphate:	
KH_2PO_4, 0.008665*M*	1.179
Na_2HPO_4, 0.03032*M*	4.30
$Na_2B_4O_7 \cdot 10H_2O$, 0.01*M*	3.80
Carbonate:	
$NaHCO_3$, 0.025*M*	2.10
Na_2CO_3, 0.025*M*	2.65
$Ca(OH)_2$, about 0.0203*M*	Saturated at 25°C

TABLE 1.63 Composition and pH Values of Buffer Solutions 8.107

Values based on the conventional activity pH scale as defined by the National Bureau of Standards (U.S.) and pertain to a temperature of 25°C [Ref: Bower and Bates, *J. Research Natl. Bur. Standards (U.S.)*, **55:**197 (1955) and Bates and Bower, *Anal. Chem.*, **28:**1322 (1956)]. Buffer value is denoted by column headed β.

25 ml 0.2*M* KCl + *x* ml 0.2*M* HCl, Diluted to 100 ml			50 ml 0.1*M* KH Phthalate + *x* ml 0.1*M* HCl, Diluted to 100 ml			50 ml 0.1*M* KH Phthalate + *x* ml 0.1*M* NaOH, Diluted to 100 ml		
pH	x	β	pH	x	β	pH	x	β
1.00	67.0	0.31	2.20	49.5		4.20	3.0	0.017
1.20	42.5	0.34	2.40	42.2	0.036	4.40	6.6	0.020
1.40	26.6	0.19	2.60	35.4	0.033	4.60	11.1	0.025
1.60	16.2	0.077	2.80	28.9	0.032	4.80	16.5	0.029
1.80	10.2	0.049	3.00	22.3	0.030	5.00	22.6	0.031
2.00	6.5	0.030	3.20	15.7	0.026	5.20	28.8	0.030
2.20	3.9	0.022	3.40	10.4	0.023	5.40	34.1	0.025
			3.60	6.3	0.018	5.60	38.8	0.020
			3.80	2.9	0.015	5.80	42.3	0.015

TABLE 1.63 Composition and pH Values of Buffer Solutions 8.107 (*Continued*)

50 ml 0.1*M* KH_2PO_4 + *x* ml 0.1*M* NaOH, Diluted to 100 ml			50 ml 0.1*M* *Tris*(hydroxymethyl)aminomethane + *x* ml 0.1*M* HCl, Diluted to 100 ml $\Delta pH/\Delta t \simeq -0.028$ $I = 0.001x$			50 ml of a Mixture 0.1*M* with Respect to Both KCl and H_3BO_3 + *x* ml 0.1*M* NaOH, Diluted to 100 ml		
pH	*x*	*β*	pH	*x*	*β*	pH	*x*	*β*
5.80	3.6		7.00	46.6		8.00	3.9	
6.00	5.6	0.010	7.20	44.7	0.012	8.20	6.0	0.011
6.20	8.1	0.015	7.40	42.0	0.015	8.40	8.6	0.015
6.40	11.6	0.021	7.60	38.5	0.018	8.60	11.8	0.018
6.60	16.4	0.027	7.80	34.5	0.023	8.80	15.8	0.022
6.80	22.4	0.033	8.00	29.2	0.029	9.00	20.8	0.027
7.00	29.1	0.031	8.20	22.9	0.031	9.20	26.4	0.029
7.20	34.7	0.025	8.40	17.2	0.026	9.40	32.1	0.027
7.40	39.1	0.020	8.60	12.4	0.022	9.60	36.9	0.022
7.60	42.4	0.013	8.80	8.5	0.016	9.80	40.6	0.016
7.80	44.5	0.009	9.00	5.7		10.00	43.7	0.014
8.00	46.1					10.20	46.2	

50 ml 0.025*M* Borax + *x* ml 0.1*M* HCl, Diluted to 100 ml $\Delta pH/\Delta t \simeq -0.008$ $I = 0.025$			50 ml 0.025*M* Borax + *x* ml 0.1*M* NaOH, Diluted to 100 ml $\Delta pH/\Delta t \simeq -0.008$ $I = 0.001(25 + x)$			50 ml 0.05*M* $NaHCO_3$ + *x* ml 0.1*M* NaOH, Diluted to 100 ml $\Delta pH/\Delta t \simeq -0.009$ $I = 0.001(25 + 2x)$		
pH	*x*	*β*	pH	*x*	*β*	pH	*x*	*β*
8.00	20.5		9.20	0.9		9.60	5.0	
8.20	19.7	0.010	9.40	3.6	0.026	9.80	6.2	0.014
8.40	16.6	0.012	9.60	11.1	0.022	10.00	10.7	0.016
8.60	13.5	0.018	9.80	15.0	0.018	10.20	13.8	0.015
8.80	9.4	0.023	10.00	18.3	0.014	10.40	16.5	0.013

50 ml 0.025*M* Borax + *x* ml 0.1*M* HCl, Diluted to 100 ml $\Delta pH/\Delta t \simeq -0.008$ $I = 0.025$			50 ml 0.025*M* Borax + *x* ml 0.1*M* NaOH, Diluted to 100 ml $\Delta pH/\Delta t \simeq -0.008$ $I = 0.001(25 + x)$			50 ml 0.05*M* $NaHCO_3$ + *x* ml 0.1*M* NaOH, Diluted to 100 ml $\Delta pH/\Delta t \simeq -0.009$ $I = 0.001(25 + 2x)$		
pH	*x*	*β*	pH	*x*	*β*	pH	*x*	*β*
9.00	4.6	0.026	10.20	20.5	0.009	10.60	19.1	0.012
9.10	2.0		10.40	22.1	0.007	10.80	21.2	0.009
			10.60	23.3	0.005	11.00	22.7	

(*Continued*)

TABLE 1.63 Composition and pH Values of Buffer Solutions 8.107 (*Continued*)

50 ml 0.05M Na_2HPO_4 + x ml 0.1M NaOH, Diluted to 100 ml $\Delta pH/\Delta t \simeq -0.025$ $I = 0.001(77 + 2x)$			25 ml 0.2M KCl + x ml 0.2M NaOH, Diluted to 100 ml $\Delta pH/\Delta t \simeq -0.033$ $I = 0.001(50 + 2x)$		
pH	x	β	pH	x	β
11.00	4.1	0.009	12.00	6.0	0.028
11.20	6.3	0.012	12.20	10.2	0.048
11.40	9.1	0.017	12.40	16.2	0.076
11.60	13.5	0.026	12.60	25.6	0.12
11.80	19.4	0.034	12.80	41.2	0.21
11.90	23.0	0.037	13.00	66.0	0.30

The phosphate-succinate system gives the values of pH_s

Molality KH_2PO_4 = Molality $Na_2HC_6H_5O_7$	pH_s	$\Delta(pH_s/\Delta t)$
0.005	6.251	−0.000 86 deg^{-1}
0.010	6.197	−0.000 71
0.015	6.162	
0.020	6.131	
0.025	6.109	−0.004

TABLE 1.64 Standard Reference Values pH for the Measurement of Acidity in 50 Weight Percent Methanol-Water

Temperature, °C	0.02m HOAc, 0.02m NaOAc, 0.02m NaCl	0.02m NaHSuc, 0.02m NaCl	0.02m KH_2PO_4, 0.02m Na_2HPO_4, 0.02m NaCl
10	5.560	5.806	7.937
15	5.549	5.786	7.916
20	5.543	5.770	7.898
25	5.540	5.757	7.884
30	5.540	5.748	7.872
35	5.543	5.743	7.863
40	5.550	5.741	7.858

OAc = acetate Suc = succinate

Reference: R. G. Bates, *Anal Chem.*, **40**(6):35A (1968).

TABLE 1.65 pH Values for Buffer Solutions in Alcohol-Water Solvents at 25°C

Liquid-junction potential not included.

Solvent composition (weight per cent alcohol)	0.01*M* $H_2C_2O_4$, 0.01*M* $NH_4HC_2O_4$	0.01*M* H_2Suc, 0.01*M* LiHSuc	0.01*M* HSal, 0.01*M* NaSal
	Methanol-Water Solvents		
0	2.15	4.12	
10	2.19	4.30	
20	2.25	4.48	
30	2.30	4.67	
40	2.38	4.87	
50	2.47	5.07	
60	2.58	5.30	
70	2.76	5.57	
80	3.13	6.01	
90	3.73	6.73	
92	3.90	6.92	
94	4.10	7.13	
96	4.39	7.43	
98	4.84	7.89	
99	5.20	8.23	
100	5.79	8.75	7.53
	Ethanol-Water Solvents		
0	2.15	4.12	
30	2.32	4.70	
50	2.51	5.07	
71.9	2.98	5.71	
100			8.32

Suc = succinate Sal = salicylate

1.17.2 Buffer Solutions Other Than Standards

The range of the buffering effect of a single weak acid group is approximately one pH unit on either side of the pK_a. The ranges of some useful buffer systems are collected in Table 1.66. After all the components have been brought together, the pH of the resulting solution should be determined at the temperature to be employed with reference to standard reference solutions. Buffer components should be compatible with other components in the system under study; this is particularly significant for buffers employed in biological studies. Check tables of formation constants to ascertain whether metal-binding character exists.

TABLE 1.66 pH Values of Biological and Other Buffers for Control Purposes

Materials	Acronym	pK_a	pH range
p-Toluenesulfonate and *p*-toluenesulfonic acid		1.7	1.1–3.3
Glycine and HCl		2.35	1.0–3.7
Citrate and HCl		3.13	1.3–4.7
Formate and HCl		3.71	2.8–4.6
Succinate and borax		4.21, 5.64	3.0–5.8
Phenyl acetate and HCl		4.31	3.5–5.0
Acetate and acetic acid		4.76	3.7–5.6
Succinate and succinic acid		4.21, 5.64	4.8–6.3
2-(*N*-Morpholino)ethanesulfonic acid	MES	6.1	5.5–6.7
Bis(2-hydroxyethyl)iminotris(hydroxymethyl)methane	BIS-TRIS	6.5	5.8–7.2
KH_2PO_4 and borax		2.2, 7.2; 9	5.8–9.2
N-(2-Acetamido)-2-iminodiacetic acid	ADA	6.6	6.0–7.2
2-[(2-Amino-2-oxoethyl)amino]ethanesulfonic acid	ACES	6.8	6.1–7.5
Piperazine-*N,N'*-bis(2-ethanesulfonic acid)	PIPES	6.8	6.1–7.5
3-(*N*-Morpholino)-2-hydroxypropanesulfonic acid	MOPSO	6.9	6.2–7.6
1,3-Bis[tris(hydroxymethyl)methylamino]propane	BIS-TRIS PROPANE	6.8, 9.0	6.3–9.5
KH_2PO_4 and Na_2HPO_4		7.2	6.1–7.5
N,N-Bis(2-hydroxyethyl)-2-aminoethanesulfonic acid	BES	7.1	6.4–7.8
3-(*N*-Morpholino)propanesulfonic acid	MOPS	7.2	6.5–7.9
N-(2-Hydroxyethyl)piperazine-*N'*-(2-ethanesulfonic acid)	HEPES	7.5	6.8–8.2
N-Tris(hydroxymethyl)methyl-2-aminoethanesulfonic acid	TES	7.5	6.8–8.2
3-[*N,N*-Bis(2-hydroxyethyl)amino]-2-hydroxypropanesulfonic acid	DIPSO	7.6	7.0–8.2
3-[*N*-tris(hydroxymethyl)methylamino]-2-hydroxypropanesulfonic acid	TAPSO	7.6	7.0–8.2
5,5-Diethylbarbiturate (veronal) and HCl		8.0	7.0–8.5
Tris(hydroxymethyl)aminoethane	TRIZMA	8.1	7.0–9.1
N-(2-hydroxyethyl)piperazine-*N'*-(2-hydroxypropanesulfonic acid)	HEPPSO	7.8	7.1–8.5
Piperazine-*N,N'*-bis(2-hydroxypropanesulfonic acid)	POPSO	7.8	7.2–8.5
Triethanolamine	TEA	7.8	6.9–8.5
N-Tris(hydroxymethyl)methylglycine	TRICINE	8.1	7.4–8.8
Borax and HCl			7.6–8.9
N,N-Bis(2-hydroxyethyl)glycine	BICINE	8.3	7.6–9.0
N-Tris(hydroxymethyl)methyl-3-aminopropanesulfonic acid	TAPS	8.4	7.7–9.1
3-[(1,1-Dimethyl-2-hydroxyethyl)-2-hydroxypropanesulfonic acid	AMPSO	9.0	8.3–9.7
Ammonia (aqueous) and NH_4Cl		9.2	8.3–9.2
2-(*N*-Cyclohexylamino)-2-hydroxy-1-propanesulfonic acid	CHES	9.3	8.6–10.0
Glycine and NaOH		9.7	8.2–10.1
Ethanolamine (2-aminoethanol) and HCl		9.5	8.6–10.4
3-(Cyclohexylamino)-2-hydroxy-1-propanesulfonic acid	CAPSO	9.6	8.9–10.3
2-Amino-2-methyl-1-propanol	AMP	9.7	9.0–10.5
Carbonate and hydrogen carbonate		10.3	9.2–11.0
Borax and NaOH			9.4–11.1
3-(Cyclohexylamino)-1-propanesulfonic acid	CAPS	10.4	9.7–11.1
Na_2HPO_4 and NaOH		11.9	11.0–12.0

TABLE 1.66 pH Values of Biological and Other Buffers for Control Purposes (*Continued*)

x mL of 0.2*M* Sodium Acetate (27.199 g NaOAc · $3H_2O$ per liter) plus *y* mL of 0.2*M* Acetic Acid

pH	NaOAc, mL	Acetic Acid, mL
3.60	7.5	92.5
3.80	12.0	88.0
4.00	18.0	82.0
4.20	26.5	73.5
4.40	37.0	63.0
4.60	49.0	51.0
4.80	60.0	40.0
5.00	70.5	29.5
5.20	79.0	21.0
5.40	85.5	14.5
5.60	90.5	9.5

x mL of 0.1*M* KH_2PO_4 (13.617 g · L^{-1}) plus *y* mL of 0.05*M* Borax Solution (19.404 g $Na_2B_4O_7 \cdot 10H_2O$ per Liter)

pH	KH_2PO_4, mL	Borax, mL	pH	KH_2PO_4, mL	Borax, mL
5.80	92.1	7.9	7.60	51.7	48.3
6.00	87.7	12.3	7.80	49.2	50.8
6.200	83.0	17.0	8.00	46.5	53.5
6.40	77.8	22.2	8.20	43.0	57.0
6.60	72.2	27.8	8.40	38.7	61.3
6.80	66.7	33.3	8.60	34.0	66.0
7.00	62.3	37.7	8.80	27.6	72.4
7.20	58.1	41.9	9.00	17.5	82.5
7.40	55.0	45.0	9.20	5.0	95.0

x mL of Veronal (20.6 g Na Diethylbarbiturate per Liter) plus *y* mL of 0.1*M* HCl

pH	Veronal, mL	HCl, mL
7.00	53.6	46.4
7.20	55.4	44.6
7.40	58.1	41.9
7.60	61.5	38.5
7.80	66.2	33.8
8.00	71.6	28.4
8.20	76.9	23.1
8.40	82.3	17.7
8.60	87.1	12.9
8.80	90.8	9.2
9.00	93.6	6.4

x mL of 0.2*M* Aqueous NH_3 Solution plus *y* mL of 0.2*M* NH_4Cl (10.699 g · L^{-1})

pH	Aq NH_3, mL	NH_4Cl, mL
8.00	5.5	94.5
8.20	8.5	91.5
8.40	12.5	87.5
8.60	18.5	81.5
8.80	26.0	74.0
9.00	36.0	64.0
9.25	50.0	50.0
9.40	58.5	41.5
9.60	69.0	31.0
9.80	78.0	22.0
10.00	85.0	15.0

x mL of 0.1*M* Citrate (21.0 g Citric Acid Monohydrate + 200 mL 1*M* NaOH per Liter) plus *y* mL of 0.1*M* NaOH

pH	Citrate, mL	NaOH mL
5.10	90.0	10.0
5.30	80.0	20.0
5.50	71.0	29.0
5.70	67.0	33.0
5.90	62.0	38.0

x mL of 0.2*M* NaOH Added to 100 mL of Stock Solution (0.04*M* Acetic Acid, 0.04*M* H_3PO_4, and 0.04*M* Boric Acid)

pH	NaOH, mL	pH	NaOH, mL	pH	NaOH, mL	pH	NaOH, mL
1.81	0.0	4.10	25.0	6.80	50.0	9.62	75.0
1.89	2.5	4.35	27.5	7.00	52.5	9.91	77.5
1.98	5.0	4.56	30.0	7.24	55.0	10.38	80.0
2.09	7.5	4.78	32.5	7.54	57.5	10.88	82.5
2.21	10.0	5.02	35.0	7.96	60.0	11.20	85.0
2.36	12.5	5.33	37.5	8.36	62.5	11.40	87.5
2.56	15.0	5.72	40.0	8.69	65.0	11.58	90.0
2.87	17.5	6.09	42.5	8.95	67.5	11.70	92.5
3.29	20.0	6.37	45.0	9.15	70.0	11.82	95.0
3.78	22.5	6.59	47.5	9.37	72.5	11.92	97.5

(*Continued*)

TABLE 1.66 pH Values of Biological and Other Buffers for Control Purposes (*Continued*)

x mL of 0.1*M* HCl plus y mL of 0.1*M* Glycine (7.505 g Glycine + 5.85 g NaCl per Liter)			x mL of 0.1*M* HCl plus y mL of 0.1*M* Citrate (21.008 g Citric Acid Monohydrate + 200 ml 1*M* NaOH per Liter)			x mL of 0.05*M* Succinic Acid (5.90 g · L^{-1}) plus y mL of Borax Solution (19.404 g $Na_2B_4O_7 \cdot 10H_2O$ per Liter)		
pH	HCl, mL	Glycine, mL	pH	HCl, mL	Citrate, mL	pH	Succinic Acid, mL	Borax, mL
1.20	84.0	16.0	3.50	52.8	47.2	3.60	90.5	9.5
1.40	71.0	29.0	3.60	51.3	48.7	3.80	86.3	13.7
1.60	61.8	38.2	3.80	48.6	51.4	4.00	82.2	17.8
1.80	55.2	44.8	4.00	43.8	56.2	4.20	77.8	22.2
2.00	49.1	50.9	4.20	38.6	61.4	4.40	73.8	26.2
2.20	42.7	57.3	4.40	34.6	65.4	4.60	70.0	30.0
2.40	36.5	63.5	4.60	24.3	75.7	4.80	66.5	33.5
2.60	30.3	69.7	4.80	11.0	89.0	5.00	63.2	36.8
2.80	24.0	76.0				5.20	60.5	39.5
3.00	17.8	82.2				5.40	57.9	42.1
3.30	10.8	89.2				5.60	55.7	44.3
3.60	6.0	94.0				5.80	54.0	46.0

x mL of 0.2*M* $Na_2HPO_4 \cdot 2H_2O$ (35.599 g · L^{-1}) plus y mL of 0.1*M* Citric Acid (19.213 g · L^{-1})

pH	Na_2HPO_4, mL	Citric Acid, mL	pH	Na_2HPO_4, mL	Citric Acid, mL	pH	Na_2HPO_4, mL	Citric Acid, mL
2.20	2.00	98.00	4.20	41.40	58.60	6.20	66.10	33.90
2.40	6.20	93.80	4.40	44.10	55.90	6.40	69.25	30.75
2.60	10.90	89.10	4.60	46.75	53.25	6.60	72.75	27.25
2.80	15.85	84.15	4.80	49.30	50.70	6.80	77.25	22.75
3.00	20.55	79.45	5.00	51.50	48.50	7.00	82.35	17.65
3.20	24.70	75.30	5.20	53.60	46.40	7.20	86.95	13.05
3.40	28.50	71.50	5.40	55.75	44.25	7.40	90.85	9.15
3.60	32.20	67.80	5.60	58.00	42.00	7.60	93.65	6.35
3.80	35.50	64.50	5.80	60.45	39.55	7.80	95.75	4.25
4.00	38.55	61.45	6.00	63.15	36.85	8.00	97.25	2.75

1.18 SOLUBILITY AND EQUILIBRIUM CONSTANT

The equilibrium constant is the value of the reaction quotient for a system at equilibrium. The reaction quotient is the ratio of molar concentrations of the reactants to those of the products, each concentration being raised to the power equal to the coefficient in the equation.

For the hypothetical chemical reaction

$$A + B \leftrightarrow C + D$$

the equilibrium constant, K, is:

$$K = [C][D]/[A][B]$$

The notation [A] signifies the molar concentration of species A. An alternative expression for the equilibrium constant can involve the use of partial pressures.

The equilibrium constant can be determined by allowing a reaction to reach equilibrium, measuring the concentrations of the various solution-phase or gas-phase reactants and products, and substituting these values into the relevant equation.

TABLE 1.67 Solubility of Gases in Water

The column (or line entry) headed "α" gives the volume of gas (in milliliters) measured at standard conditions (0°C and 760 mm or 101.325 kN · m^{-2}) dissolved in 1 mL of water at the temperature stated (in degrees Celsius) and when the pressure of the gas without that of the water vapor is 760 mm. The line entry "A" indicates the same quantity except that the gas itself is at the uniform pressure of 760 mm when in equilibrium with water.

The column headed "l" gives the volume of the gas (in milliliters) dissolved in 1 mL of water when the pressure of the gas plus that of the water vapor is 760 mm.

The column headed "q" gives the weight of gas (in grams) dissolved in 100 g of water when the pressure of the gas plus that of the water vapor is 760 mm.

Temp., °C	Acetylene		Air*		Ammonia		Bromine	
	α	q	$\alpha(\times 10^3)$	% oxygen in air	α	q	α	q
0	1.73	0.200	29.18	34.91	1130	89.5	60.5	42.9
1	1.68	0.194	28.42	34.87	—	—	—	—
2	1.63	0.188	27.69	34.82	—	—	54.1	38.3
3	1.58	0.182	26.99	34.78	—	—	—	—
4	1.53	0.176	26.32	34.74	1047	79.6	48.3	34.2
5	1.49	0.171	25.68	34.69	—	—	—	—
6	1.45	0.167	25.06	34.65	—	—	43.3	30.6
7	1.41	0.162	24.47	34.60	—	—	—	—
8	1.37	0.157	23.90	34.56	947	72.0	38.9	27.5
9	1.34	0.154	23.36	34.52	—	—	—	—
10	1.31	0.150	22.84	34.47	870	68.4	35.1	24.8
11	1.27	0.146	22.34	34.43	—	—	—	—
12	1.24	0.142	21.87	34.38	857	65.1	31.5	22.2
13	1.21	0.138	21.41	34.34	837	63.6	—	—
14	1.18	0.135	20.97	34.30	—	—	28.4	20.0
15	1.15	0.131	20.55	34.25	770	—	—	—
16	1.13	0.129	20.14	34.21	775	58.7	25.7	18.0
17	1.10	0.125	19.75	34.17	—	—	—	—
18	1.08	0.123	19.38	34.12	—	—	23.4	16.4
19	1.05	0.119	19.02	34.08	—	—	—	—
20	1.03	0.117	18.68	34.03	680	52.9	21.3	14.9
21	1.01	0.115	18.34	33.99	—	—	—	—
22	0.99	0.112	18.01	33.95	—	—	19.4	13.5
23	0.97	0.110	17.69	33.90	—	—	—	—
24	0.95	0.107	17.38	33.86	639	48.2	17.7	12.3
25	0.93	0.105	17.08	33.82	—	—	—	—
26	0.91	0.102	16.79	33.77	—	—	16.3	11.3
27	0.89	0.100	16.50	33.73	—	—	—	—
28	0.87	0.098	16.21	33.68	586	44.0	15.0	10.3
29	0.85	0.095	15.92	33.64	—	—	—	—
30	0.84	0.094	15.64	33.60	530	41.0	13.8	9.5
35	—	—	—	—	—	—	—	—
40	—	—	14.18	—	400	31.6	9.4	6.3
45	—	—	—	—	—	—	—	—
50	—	—	12.97	—	290	23.5	6.5	4.1
60	—	—	12.16	—	200	16.8	4.9	2.9
70	—	—	—	—	—	11.1	3.8	1.9
80	—	—	11.26	—	—	6.5	3.0	1.2
90	—	—	—	—	—	3.0	—	—
100	—	—	11.05	—	—	0.0	—	—

*Free from NH_3 and CO_2; total pressure of air + water vapor is 760 mm.

TABLE 1.67 Solubility of Gases in Water

Temp. °C	Carbon dioxide		Carbon monoxide		Chlorine		Ethane		Ethylene		Hydrogen	
	α	q	α	q	l	q	α	q	α	q	α	q
0	1.713	0.334 6	0.035 37	0.004 397	—	—	0.098 74	0.013 17	0.226	0.028 1	0.021 48	0.000 192 2
1	1.646	0.321 3	0.034 55	0.004 293	—	—	0.094 76	0.012 63	0.219	0.027 2	0.021 26	0.000 190 1
2	1.584	0.309 1	0.033 75	0.004 191	—	—	0.090 93	0.012 12	0.211	0.026 2	0.021 05	0.000 188 1
3	1.527	0.297 8	0.032 97	0.004 092	—	—	0.087 25	0.011 62	0.204	0.025 3	0.020 84	0.000 186 2
4	1.473	0.287 1	0.032 22	0.003 996	—	—	0.083 72	0.011 14	0.197	0.024 4	0.020 64	0.000 184 3
5	1.424	0.277 4	0.031 49	0.003 903	—	—	0.080 33	0.010 69	0.191	0.023 7	0.020 44	0.000 182 4
6	1.377	0.268 1	0.030 78	0.003 813	—	—	0.077 09	0.010 25	0.184	0.022 8	0.020 25	0.000 180 6
7	1.331	0.258 9	0.030 09	0.003 725	—	—	0.074 00	0.009 83	0.178	0.022 0	0.020 07	0.000 178 9
8	1.282	0.249 2	0.029 42	0.003 640	—	—	0.071 06	0.009 43	0.173	0.021 4	0.019 89	0.000 177 2
9	1.237	0.240 3	0.028 78	0.003 559	—	—	0.068 26	0.009 06	0.167	0.020 7	0.019 72	0.000 175 6
10	1.194	0.231 8	0.028 16	0.003 479	3.148	0.997 2	0.065 61	0.008 70	0.162	0.020 0	0.019 55	0.000 174 0
11	1.154	0.223 9	0.027 57	0.003 405	3.047	0.965 4	0.063 28	0.008 38	0.157	0.019 4	0.019 40	0.000 172 5
12	1.117	0.216 5	0.027 01	0.003 332	2.950	0.934 6	0.061 06	0.008 08	0.152	0.018 8	0.019 25	0.000 171 0
13	1.083	0.209 8	0.026 46	0.003 261	2.856	0.905 0	0.058 94	0.007 80	0.148	0.018 3	0.019 11	0.000 169 6
14	1.050	0.203 2	0.025 93	0.003 194	2.767	0.876 8	0.056 94	0.007 53	0.143	0.017 6	0.018 97	0.000 168 2
15	1.019	0.197 0	0.025 43	0.003 130	2.680	0.849 5	0.055 04	0.007 27	0.139	0.017 1	0.018 83	0.000 166 8
16	0.985	0.190 3	0.024 94	0.003 066	2.597	0.823 2	0.053 26	0.007 03	0.136	0.016 7	0.018 69	0.000 165 4
17	0.956	0.184 5	0.024 48	0.003 007	2.517	0.797 9	0.051 59	0.006 80	0.132	0.016 2	0.018 56	0.000 164 1
18	0.928	0.178 9	0.024 02	0.002 947	2.440	0.773 8	0.050 03	0.006 59	0.129	0.015 8	0.018 44	0.000 162 8
19	0.902	0.173 7	0.023 60	0.002 891	2.368	0.751 0	0.048 58	0.006 39	0.125	0.015 3	0.018 31	0.000 161 6
20	0.878	0.168 8	0.023 19	0.002 838	2.299	0.729 3	0.047 24	0.006 20	0.122	0.014 9	0.018 19	0.000 160 3
21	0.854	0.164 0	0.022 81	0.002 789	2.238	0.710 0	0.045 89	0.006 02	0.119	0.014 6	0.018 05	0.000 158 8
22	0.829	0.159 0	0.022 44	0.002 739	2.180	0.691 8	0.044 59	0.005 84	0.116	0.014 2	0.017 92	0.000 157 5
23	0.804	0.154 0	0.022 08	0.002 691	2.123	0.673 9	0.043 35	0.005 67	0.114	0.013 9	0.017 79	0.000 156 1
24	0.781	0.149 3	0.021 74	0.002 646	2.070	0.657 2	0.042 17	0.005 51	0.111	0.013 5	0.017 66	0.000 154 8
25	0.759	0.144 9	0.021 42	0.002 603	2.019	0.641 3	0.041 04	0.005 35	0.108	0.013 1	0.017 54	0.000 153 5
26	0.738	0.140 6	0.021 10	0.002 560	1.970	0.625 9	0.039 97	0.005 20	0.106	0.012 9	0.017 42	0.000 152 2
27	0.718	0.136 6	0.020 80	0.002 519	1.923	0.611 2	0.038 95	0.005 06	0.104	0.012 6	0.017 31	0.000 150 9
28	0.699	0.132 7	0.020 51	0.002 479	1.880	0.597 5	0.037 99	0.004 93	0.102	0.012 3	0.017 20	0.000 149 6
29	0.682	0.129 2	0.020 24	0.002 442	1.839	0.584 7	0.037 09	0.004 80	0.100	0.012 1	0.017 09	0.000 148 4
30	0.665	0.125 7	0.019 98	0.002 405	1.799	0.572 3	0.036 24	0.004 68	0.098	0.011 8	0.016 99	0.000 147 4

35	0.592	0.110 5	0.018 77	0.002 231	1.602	0.510 4	0.032 30	0.004 12	—	—	0.016 66	0.000 142 5
40	0.530	0.097 3	0.017 75	0.002 075	1.438	0.459 0	0.029 15	0.003 66	—	—	0.016 44	0.000 138 4
45	0.479	0.086 0	0.016 90	0.001 933	1.322	0.422 8	0.026 60	0.003 27	—	—	0.016 24	0.000 134 1
50	0.436	0.076 1	0.016 15	0.001 797	1.225	0.392 5	0.024 59	0.002 94	—	—	0.016 08	0.000 128 7
60	0.359	0.057 6	0.014 88	0.001 522	1.023	0.329 5	0.021 77	0.002 39	—	—	0.016 00	0.000 117 8
70	—	—	0.014 40	0.001 276	0.862	0.279 3	0.019 48	0.001 85	—	—	0.016 0	0.000 102
80	—	—	0.014 30	0.000 980	0.683	0.222 7	0.018 26	0.001 34	—	—	0.016 0	0.000 079
90	—	—	0.014 2	0.000 57	0.39	0.127	0.017 6	0.000 8	—	—	0.016 0	0.000 046
100	—	—	0.014 1	0.000 00	0.00	0.000	0.017 2	0.000 0	—	—	0.016 0	0.000 000
0	4.670	0.706 6	0.055 63	0.003 959	0.073 81	0.009 833	0.023 54	0.002 942	0.048 89	0.006 945	79.789	22.83
1	4.522	0.683 9	0.054 01	0.003 842	0.071 84	0.009 564	0.022 97	0.002 869	0.047 58	0.006 756	77.210	22.09
2	4.379	0.661 9	0.052 44	0.003 728	0.069 93	0.009 305	0.022 41	0.002 798	0.046 33	0.006 574	74.691	21.37
3	4.241	0.640 7	0.050 93	0.003 619	0.068 09	0.009 057	0.021 87	0.002 730	0.045 12	0.006 400	72.230	20.66
4	4.107	0.620 1	0.049 46	0.003 513	0.066 32	0.008 816	0.021 35	0.002 663	0.043 97	0.006 232	69.828	19.98
5	3.977	0.600 1	0.048 05	0.003 410	0.064 61	0.008 584	0.020 86	0.002 600	0.042 87	0.006 072	67.485	19.31
6	3.852	0.580 9	0.046 69	0.003 312	0.062 98	0.008 361	0.020 37	0.002 537	0.041 80	0.005 918	65.200	18.65
7	3.732	0.562 4	0.045 39	0.003 217	0.061 40	0.008 147	0.019 90	0.002 477	0.040 80	0.005 773	62.973	18.02
8	3.616	0.544 6	0.044 13	0.003 127	0.059 90	0.007 943	0.019 45	0.002 419	0.039 83	0.005 632	60.805	17.40
9	3.505	0.527 6	0.042 92	0.003 039	0.058 46	0.007 747	0.019 02	0.002 365	0.038 91	0.005 498	58.697	16.80
10	3.399	0.511 2	0.041 77	0.002 955	0.057 09	0.007 560	0.018 61	0.002 312	0.038 02	0.005 368	56.647	16.21
11	3.300	0.496 0	0.040 72	0.002 879	0.055 87	0.007 393	0.018 23	0.002 263	0.037 18	0.005 246	54.655	15.64
12	3.206	0.481 4	0.039 70	0.002 805	0.054 70	0.007 233	0.017 86	0.002 216	0.036 37	0.005 128	52.723	15.09
13	3.115	0.467 4	0.038 72	0.002 733	0.053 57	0.007 078	0.017 50	0.002 170	0.035 59	0.005 014	50.849	14.56
14	3.028	0.454 0	0.037 79	0.002 665	0.052 50	0.006 930	0.017 17	0.002 126	0.034 86	0.004 906	49.033	14.04
15	2.945	0.441 1	0.036 90	0.002 599	0.051 47	0.006 788	0.016 85	0.002 085	0.034 15	0.004 802	47.276	13.54
16	2.865	0.428 7	0.036 06	0.002 538	0.050 49	0.006 652	0.016 54	0.002 045	0.033 48	0.004 703	45.578	13.05
17	2.789	0.416 9	0.035 25	0.002 478	0.049 56	0.006 524	0.016 25	0.002 006	0.032 83	0.004 606	43.939	12.59
18	2.717	0.405 6	0.034 48	0.002 422	0.048 68	0.006 400	0.015 97	0.001 970	0.032 20	0.004 514	42.360	12.14
19	2.647	0.394 8	0.033 76	0.002 369	0.047 85	0.006 283	0.015 70	0.001 935	0.031 61	0.004 426	40.838	11.70
20	2.582	0.384 6	0.033 08	0.002 319	0.047 06	0.006 173	0.015 45	0.001 901	0.031 02	0.004 339	39.374	11.28
21	2.517	0.374 5	0.032 43	0.002 270	0.046 25	0.006 059	0.015 22	0.001 869	0.030 44	0.004 252	37.970	10.88
22	2.456	0.364 8	0.031 80	0.002 222	0.045 45	0.005 947	0.014 98	0.001 838	0.029 88	0.004 169	36.617	10.50
23	2.396	0.355 4	0.031 19	0.002 177	0.044 69	0.005 838	0.014 75	0.001 809	0.029 34	0.004 087	35.302	10.12
24	2.338	0.346 3	0.030 61	0.002 133	0.043 95	0.005 733	0.014 54	0.001 780	0.028 81	0.004 007	34.026	9.76
25	2.282	0.337 5	0.030 06	0.002 091	0.043 23	0.005 630	0.014 34	0.001 751	0.028 31	0.003 931	32.786	9.41
26	2.229	0.329 0	0.029 52	0.002 050	0.042 54	0.005 530	0.014 13	0.001 724	0.027 83	0.003 857	31.584	9.06

(Continued)

TABLE 1.67 Solubility of Gases in Water (*Continued*)

Temp. °C	Carbon dioxide		Carbon monoxide		Chlorine		Ethane		Ethylene		Hydrogen	
	α	q	α	q	l	q	α	q	α	q	α	q
26	2.229	0.329 0	0.029 52	0.002 050	0.042 54	0.005 530	0.014 13	0.001 724	0.027 83	0.003 857	31.584	9.06
27	2.177	0.320 8	0.029 01	0.002 011	0.041 88	0.005 435	0.013 94	0.001 698	0.027 36	0.003 787	30.422	8.73
28	2.128	0.313 0	0.028 52	0.001 974	0.041 24	0.005 342	0.013 76	0.001 672	0.026 91	0.003 718	29.314	8.42
29	2.081	0.305 5	0.028 06	0.001 938	0.040 63	0.005 252	0.013 58	0.001 647	0.026 49	0.003 651	28.210	8.10
30	2.037	0.298 3	0.027 62	0.001 904	0.040 04	0.005 165	0.013 42	0.001 624	0.026 08	0.003 588	27.161	7.80
35	1.831	0.264 8	0.025 46	0.001 733	0.037 34	0.004 757	0.012 56	0.001 501	0.024 40	0.003 315	22.489	6.47
40	1.660	0.236 1	0.023 69	0.001 586	0.035 07	0.004 394	0.011 84	0.001 391	0.023 06	0.003 082	18.766	5.41
45	1.516	0.211 0	0.022 38	0.001 466	0.033 11	0.004 059	0.011 30	0.001 300	0.021 87	0.002 858	—	—
50	1.392	0.188 3	0.021 34	0.001 359	0.031 52	0.003 758	0.010 88	0.001 216	0.020 90	0.002 657	—	—
60	1.190	0.148 0	0.019 54	0.001 144	0.029 54	0.003 237	0.010 23	0.001 052	0.019 46	0.002 274	—	—
70	1.022	0.110 1	0.018 25	0.000 926	0.028 10	0.002 668	0.009 77	0.000 851	0.018 33	0.001 856	—	—
80	0.917	0.076 5	0.017 70	0.000 695	0.027 00	0.001 984	0.009 58	0.000 660	0.017 61	0.001 381	—	—
90	0.84	0.041	0.017 35	0.000 40	0.026 5	0.001 13	0.009 5	0.000 38	0.017 2	0.000 79	—	
100	0.81	0.000	0.017 0	0.000 00	0.026 3	0.000 00	0.009 5	0.000 00	0.017 0	0.000 00	—	—

*Atmospheric nitrogen containing 98.815% N_2 by volume + 1.185% inert gases.

TABLE 1.67 Solubility of Gases in Water

Substance		0°	10°	20°	30°	40°	60°	80°
Argon	α	0.052 8	0.041 3	0.033 7	0.028 8	0.025 1	0.020 9	0.018 4
Helium	A	0.009 8	0.009 11	0.008 6	0.008 39	0.008 41	0.009 02	$0.009\ 42^{70^\circ}$
Hydrogen bromide	1	612	582		533^{25°		469^{50°	406^{75°
Hydrogen chloride	α	512	475	442	412	385	339	
Krypton	α	0.110 5	0.081 0	0.062 6	0.051 1	0.043 3	0.035 7	
Neon	A		$0.011\ 7^{9^\circ}$	0.010 6	0.010 0	$0.009\ 48^{42^\circ}$		$0.009\ 84^{73^\circ}$
Nitrous oxide	A		0.88	0.63				
Ozone	$g \cdot L^{-1}$	0.039 4	$0.029\ 9^{12^\circ}$	$0.021\ 0^{19^\circ}$	0.0139^{27°	0.004 2	0	
Radon	α	0.510	0.326	0.222	0.162	0.126	0.085	
Xenon	α	0.242	0.174	0.123	0.098	0.082		

TABLE 1.68 Solubility of Inorganic Compounds and Metal Salts of Organic Acids in Water at Various Temperatures

Solubilities are expressed as the number of grams of substance of stated molecular formula which when dissolved in 100 g of water make a saturated solution at the temperature stated (°C).

Substance	Formula	0°	10°	20°	30°	40°	60°	80°	90°	100°
Aluminum chloride	$AlCl_3$	43.9	44.9	45.8	46.6	47.3	48.1	48.6		49.0
fluoride	AlF_3	0.56	0.56	0.67	0.78	0.91	1.1	1.32		1.72
nitrate	$Al(NO_3)_3$	60.0	66.7	73.9	81.8	88.7	106	132	153	160
perchlorate	$Al(ClO_4)_3$	122	128	133						182
sulfate	$Al_2(SO_4)_3$	31.2	33.5	36.4	40.4	45.8	59.2	73.0	80.8	89.0
thallium(I) sulfate	$Al_2Tl_2(SO_4)_4$	3.15	4.60	6.39	9.37	14.39	35.35			
Ammonium aluminum										
sulfate	$NH_4Al(SO_4)_2$	2.10	5.00	7.74	10.9	14.9	26.7			
azide	NH_4N_3	16.0		25.3		37.1				
bromide	NH_4Br	60.5	68.1	76.4	83.2	91.2	108	125	135	145
chloride	NH_4Cl	29.4	33.2	37.2	41.4	45.8	55.3	65.6	71.2	77.3
chloroiridate(IV)	$(NH_4)_2TrCl_6$	0.56	0.71	0.95	1.20	1.56	2.45	4.38		
chloroplatinate(IV)	$(NH_4)_2PtCl_6$	0.289	0.374	0.499	0.637	0.815	1.44	2.16	2.61	3.36
chromate	$(NH_4)_2CrO_4$	25.0	29.2	34.0	39.3	45.3	59.0	76.1		
chromium(III) sulfate	$(NH_4)Cr(SO_4)_2$	3.95			18.8	32.6				
cobalt(II) sulfate	$(NH_4)_2Co(SO_4)_2$	6.0	9.5	13.0	17.0	22.0	33.5	49.0	58.0	75.1
dichromate	$(NH_4)_2Cr_2O_7$	18.2	25.5	35.6	46.5	58.5	86.0	115		156
dihydrogen arsenate	$NH_4H_2AsO_4$	33.7		48.7		63.8	83.0	107	122	
dihydrogen phosphate	$NH_4H_2PO_4$	22.7	29.5	37.4	46.4	56.7	82.5	118		173
dithionate	$(NH_4)_2S_2O_6$	133	151	166	179					
formate	NH_4CHO_2	102		143		204	311	533		
hydrogen carbonate	NH_4HCO_3	11.9	16.1	21.7	28.4	36.6	59.2	109	170	354
hydrogen phosphate	$(NH_4)_2HPO_4$	42.9	62.9	68.9	75.1	81.8	97.2			
hydrogen tartrate	$NH_4C_4H_5O_6$	1.00	1.88	2.70						
iodide	NH_4I	155	163	172	182	191	209	229		250
iron(II) sulfate	$(NH_4)_2Fe(SO_4)_2$	12.5	17.2	26.4	33	46				
Ammonium magnesium										
sulfate	$(NH_4)_2Mg(SO_4)_2$	11.8	14.6	18.0	21.7	25.8	35.1	48.3		65.7
nickel sulfate	$(NH_4)_2Ni(SO_4)_2$	1.00	4.00	6.50	9.20	12.0	17.0			
nitrate	NH_4NO_3	118	150	192	242	297	421	580	740	871
oxalate	$(NH_4)_2C_2O_4$	2.2	3.21	4.45	6.09	8.18	14.0	22.4	27.9	34.7
perchlorate	NH_4ClO_4	12.0	16.4	21.7	27.7	34.6	49.9	68.9		
selenite	$(NH_4)_2SeO_3$	96	105	115	126	143	192			

sulfate	$(NH_4)_2SO_4$	70.6	73.0	75.4	78.0	81	88	95		103
sulfite	$(NH_4)_2SO_3$	47.9	54.0	60.8	68.8	78.4	104	144	150	153
tartrate	$(NH_4)_2C_4H_4O_6$	45.0	55.0	63.0	70.5	76.5	86.9			
thioantimonate(V)	$(NH_4)_3SbS_4$	71.2		91.2	120					
thiocyanate	NH_4SCN	120	144	170	208	234	346			
vanadate	NH_4VO_3			0.48	0.84	1.32	2.42			
zinc sulfate	$(NH_4)_2Zn(SO_4)_2$	7.0	9.5	12.5	16.0	20.0	30.0	46.6	58.0	72.4
Antimony(III) chloride	$SbCl_3$	602		910	1087	1368	[completely miscible at 72°]			
fluoride	SbF_3	385		444	562					
Arsenic hydride										
(760 mm), cc	AsH_3	42	30	28						
oxide (pent-)	As_2O_5	59.5	62.1	65.8	69.8	71.2	73.0	75.1		76.7
oxide (tri-)	As_2O_3	1.20	1.49	1.82	2.31	2.93	4.31	6.11		8.2
Barium acetate	$Ba(C_2H_3O_2)_2 \cdot 3H_2O$	58.8	62	72	75	78.5	75.0	74.0		74.8
azide	$Ba(N_3)_2$	12.5	16.1	$17.4^{17°}$						
bromate	$Ba(BrO_3)_2 \cdot H_2O$	0.29	0.44	0.65	0.95	1.31	2.27	3.52	4.26	5.39
bromide	$BaBr_2 \cdot 2H_2O$	98	101	104	109	114	123	135		149
n-butyrate	$Ba(C_4H_7O_2)_2$	37.0	36.1	35.4	34.9	35.2	37.2	41.7	45.5	$48.1^{95°}$
caproate	$Ba(C_6H_{11}O_2)_2 \cdot 3.5H_2O$	11.71	8.38	6.89	5.87	5.79	8.39	14.71	19.28	
chlorate	$Ba(ClO_3)_2 \cdot H_2O$	20.3	26.9	33.9	41.6	49.7	66.7	84.8		105
chloride	$BaCl_2 \cdot 2H_2O$	31.2	33.5	35.8	38.1	40.8	46.2	52.5	55.8	59.4
chlorite	$Ba(ClO_2)_2$	43.9	44.6	45.4		47.9	53.8	66.6		80.8
fluoride	BaF_2		0.159	0.160	0.162					
formate	$Ba(CHO_2)_2$	26.2	28.0	29.9	31.9	34.0	38.6	44.2	47.6	51.3
hydroxide	$Ba(OH)_2$	1.67	2.48	3.89	5.59	8.22	20.94	101.4		
iodate	$Ba(IO_3)_2$			0.035	0.046	0.057				
iodide	$BaI_2 \cdot 2H_2O$	182	201	223	250		264		291	301
nitrate	$Ba(NO_3)_2$	4.95	6.67	9.02	11.48	14.1	20.4	27.2		34.4
nitrite	$Ba(NO_2)_2 \cdot H_2O$	50.3	60	72.8		102	151	222	261	325
perchlorate	$Ba(ClO_4)_2 \cdot 3H_2O$	239		336		416	495	575		653
propionate	$Ba(C_3H_5O_2)_2 \cdot H_2O$	57.2	56.8		57.5	59.0	62.0	67.8	73.0	82.7
*iso*succinate	$BaC_4H_4O_4$	0.421	0.432	0.418	0.393	0.366	0.306	0.237		
sulfamate	$Ba(SO_3NH_2)_2$	18.3	22.3	26.8	32.5	38.5	49.6	61.5		73.5
sulfide	BaS	2.88	4.89	7.86	10.38	14.89	27.69	49.91	67.34	60.29
tartrate	$Ba(C_2H_2O_3)_2$	0.021	0.024	0.028	0.032	0.035	0.044	0.053		
Beryllium nitrate	$Be(NO_3)_2$	97	102	108	113	125	178			
sulfate	$BeSO_4$	37.0	37.6	39.1	41.4	45.8	53.1	67.2		82.8
Boric acid	H_3BO_3	2.67	3.73	5.04	6.72	8.72	14.81	23.62	30.38	40.25
Cadmium bromide	$CdBr_2$	56.3	75.4	98.8	129	152	153	156		160

(Continued)

TABLE 1.68 Solubility of Inorganic Compounds and Metal Salts of Organic Acids in Water at Various Temperatures (*Continued*)

Substance	Formula	0°	10°	20°	30°	40°	60°	80°	90°	100°
chlorate	$Cd(ClO_3)_2$	299	308	322	348	376	455			
chloride	$CdCl_2 \cdot 2.5H_2O$	90	100	113	132					
	$CdCl_2 \cdot H_2O$		135	135	135	135	136	140		147
formate	$Cd(CHO_2)_2$	8.3	11.1	14.4	18.6	25.3	59.5	80.5	85.2	94.6
iodide	CdI_2	78.7		84.7	87.9	92.1	100	111		125
nitrate	$Cd(NO_3)_2$	122	136	150	167	194	310	713		
perchlorate	$Cd(ClO_4)_2 \cdot 6H_2O$		180	188	195	203	221	243		272
selenate	$CdSeO_4$	72.5	68.4	64.0	58.9	55.0	44.2	32.5	27.2	22.0
sulfate	$CdSO_4$	75.4	76.0	76.6		78.5	81.8	66.7	63.1	60.8
Calcium acetate	$Ca(OAc)_2 \cdot 2H_2O$	37.4	36.0	34.7	33.8	33.2	32.7	33.5	31.1	29.7
benzoate	$Ca(OBz)_2 \cdot 3H_2O$	2.32	2.45	2.72	3.02	3.42	4.71	6.87	8.55	8.70
bromide	$CaBr_2 \cdot 6H_2O$	125	132	143	$185^{34°}$	213	278	295		$312^{105°}$
butyrate	$Ca(C_4H_7O_2)_2$	20.31	19.15	18.20	17.25	16.40	15.15	14.95		15.85
cacodylate	$Ca(C_2H_6AsO_2)_2 \cdot 9H_2O$	48	52	59	71					
chloride	$CaCl_2 \cdot 6H_2O$	59.5	64.7	74.5	100	128	137	147	154	159
chromate	$CaCrO_4$	4.5		2.25	1.83	1.49	0.83			
(mn)	$CaCrO_4 \cdot 2H_2O$	17.3		16.6	16.1					
formate	$Ca(CHO_2)_2$	16.15		16.60		17.05	17.50	17.95		18.40
gluconate	$Ca(C_6H_{11}O_7)_2 \cdot H_2O$			3.72		5.29		12.11	36.80	$57.2^{96°}$
hydrogen carbonate	$Ca(HCO_3)_2$	16.15		16.60		17.05	17.50	17.95		18.40
hydroxide	$Ca(OH)_2$	0.189	0.182	0.173	0.160	0.141	0.121		0.086	0.076
Calcium iodate	$Ca(IO_3)_2 \cdot 6H_2O$	0.090		0.24	0.38	0.52	0.65	0.66	0.67	
iodide	CaI_2	64.6	66.0	67.6	69.0	70.8	74	78		81
lactate	$Ca(C_3H_5O_3)_2 \cdot 5H_2O$	3.1		$5.4^{15°}$	7.9					
levulinate	$Ca(C_{10}H_{14}O_6) \cdot 2H_2O$	38.1		$45.1^{16°}$	55.0	$70.3^{45°}$	$88.7^{55°}$			
malonate	$Ca(C_3H_2O_4)$	0.29	0.33	0.36	0.40	0.42	0.46	0.48		
nitrate	$Ca(NO_3)_2 \cdot 4H_2O$	102	115	129	152	191		358		363
nitrite	$Ca(NO_2)_2 \cdot 4H_2O$	63.9		$84.5^{18°}$	104		134	151	166	178
propionate	$Ca(C_3H_5O_2)_2 \cdot H_2O$	42.80		39.85			38.25	39.85	42.15	48.44
selenate	$CaSeO_4 \cdot 2H_2O$	9.73	9.77	9.22	8.79	7.14				
succinate	$Ca(C_3H_2O_2)_2 \cdot 3H_2O$	1.127	1.22	1.28		1.18	0.89	0.68		0.66
sulfamate	$Ca(SO_3NH_2)_2$	56.5	62.8	72.3	84.5	100.1	150.0	215.2	$242^{95°}$	
sulfate	$CaSO_4 \cdot \frac{1}{2}H_2O$			0.32	$0.29^{25°}$	$0.26^{35°}$	$0.21^{45°}$	$0.145^{65°}$	$0.12^{75°}$	0.071
	$CaSO_4 \cdot 2H_2O$	0.223	0.244	$0.255^{18°}$	0.264	0.265	$0.244^{65°}$	$0.234^{75°}$		0.205
tartrate	$CaC_4H_4O_6 \cdot 4H_2O$	0.026	0.029	0.034	0.046	0.063	0.091	0.130		
uranyl carbonate	$Ca_2UO_2(CO_3)_3 \cdot 10H_2O$	0.1		0.4^{23}		0.8	$1.5^{55°}$			

valerate	$Ca(C_5H_9O_2)_2$	9.82	9.25	8.80	8.40	8.05	7.78	7.95	8.20	8.78
*iso*valerate	$Ca(C_5H_9O_2)_2 \cdot 3H_2O$	26.05	22.70	21.80	21.68	22.00	18.38	16.88	16.65	16.55
Carbon disulfide	CS_2	0.204	0.194	0.179	0.155	0.111				
oxide sulfide (STP) mL/100 mL	COS	133.3	83.6	56.1	40.3					
tetrafluoride (STP) mL/100 g	CF_4		0.595	0.490	0.415	0.366				
Cerium(III) ammonium nitrate	$Ce(NH_4)_2(NO_3)_5$		242	276	318	376	681			
(IV) ammonium nitrate	$Ce(NH_4)_2(NO_3)_6$			135	150	169	213			
(III) ammonium sulfate	$Ce(NH_4)(SO_4)_2$			5.53	4.49	3.48	2.02	1.33		
(III) selenate	$Ce_2(SeO_3)_3$	39.5	37.2	35.2	33.2	32.6	13.7	4.6	2.1	
(III) sulfate	$Ce_2(SO_4)_3 \cdot 9H_2O$	21.4		9.84	7.24	5.63	3.87			
	$Ce_2(SO_4)_3 \cdot 8H_2O$			9.43	7.10	5.70	4.04			
Cesium aluminum sulfate	$Cs_2Al_2(SO_4)_4$	18.8	0.30	0.40	0.61	0.85	2.00	5.40	10.5	22.7
bromate	$CsBrO_3$	0.21		3.66$^{25°}$	4.53	5.30$^{35°}$				
chlorate	$CsClO_3$		3.8	6.2	9.5	13.8	26.2	45.0	58.0	79.0
chloride	CsCl	2.46	175	187	197	208	230	250	260	271
chloroaurate(III)	$CsAuCl_4$	161	0.5	0.8	1.7	3.3	8.9	19.5	27.7	37.9
chloroplatinate(IV)	Cs_2PtCl_6	0.0047	0.0064	0.0087	0.0119	0.0158	0.0290	0.0525	0.0675	0.0914
formate	$CsCHO_2$	335	381	450	533	694				
iodide	CsI	44.1	58.5	76.5	96	124$^{45°}$	150	190	205	
nitrate	$CsNO_3$	9.33	14.9	23.0	33.9	47.2	83.8	134	163	197
perchlorate	$CsClO_4$	0.8	1.0	1.6	2.6	4.0	7.3	14.4	20.5	30.0
sulfate	Cs_2SO_4	167	173	179	184	190	200	210	215	220
Chlorine dioxide	ClO_2	2.76	6.00	8.70$^{15°}$						
Chromium(III) nitrate	$Cr(NO_3)_3$	108$^{5°}$	124$^{15°}$	130$^{25°}$	152$^{35°}$					
(VI) oxide	CrO_3	164.8		167.2		172.5	183.9	191.6		206.8
(III) perchlorate	$Cr(ClO_4)_3$	104	123	130						
Cobalt(II) bromide	$CoBr_2$	91.9		112	128	163	227	241		257
chlorate	$Co(ClO_3)_2$	135	162	180	195	214	316			
chloride	$CoCl_2$	43.5	47.7	52.9	59.7	69.5	93.8	97.6	101	106
iodate	$Co(IO_3)_2$			1.02	0.90	0.88	0.82	0.73		0.70
nitrate	$Co(NO_3)_2$	84.0	89.6	97.4	111	125	174	204	300	
nitrite	$Co(NO_2)_2$	0.076	0.24	0.40	0.61	0.85				
sulfate	$CoSO_4$	25.5	30.5	36.1	42.0	48.8	55.0	53.8	45.3	38.9
	$CoSO_4 \cdot 7H_2O$	44.8	56.3	65.4	73.0	88.1	101			

(*Continued*)

TABLE 1.68 Solubility of Inorganic Compounds and Metal Salts of Organic Acids in Water at Various Temperatures (*Continued*)

Substance	Formula	0°	10°	20°	30°	40°	60°	80°	90°	100°
Copper(II) ammonium chloride	$CuCl_2 \cdot 2NH_4Cl$	28.2	$32.0^{12°}$	35.0	38.3	43.8	56.6	76.5	76.5	
ammonium sulfate	$CuSO_4 \cdot (NH_4)_2SO_4$	11.5	15.1	19.4	24.4	30.5	46.3	69.7	86.1	107
bromide	$CuBr_2$	107	116	126	128	$131^{50°}$				
chloride	$CuCl_2$	68.6	70.9	73.0	77.3	87.6	96.5	104	108	120
fluorosilicate	$CuSiF_6$	73.5	76.5	81.6	$84.1^{25°}$	$91.2^{50°}$		$93.2^{75°}$		
nitrate	$Cu(NO_3)_2$	83.5	100	125	156	163	182	208	222	247
potassium sulfate	$CuSO_4 \cdot K_2SO_4$	5.1	7.2	10.0	13.6	18.2				
selenate	$CuSeO_4$	12.04	14.53	17.51	21.04	25.22	36.50	53.68		
sulfate	$CuSO_4 \cdot 5H_2O$	23.1	27.5	32.0	37.8	44.6	61.8	83.8		114
tartrate	$CuC_4H_4O_6 \cdot 3H_2O$		$0.020^{15°}$	0.042	0.089	0.142	0.197	0.144		
Gadolinium bromate	$Gd(BrO_3)_3 \cdot 9H_2O$	50.2	70.1	95.6	126	166				
sulfate	$Gd_2(SO_4)_3$	3.98	3.30	2.60	2.32					
Germanium(IV) oxide	GeO_2		0.49	0.43	0.50	0.61				
Holmium sulfate	$Ho_2(SO_4)_3 \cdot 8H_2O$			8.18	$6.71^{25°}$	4.52				
Hydrazinium (1+) nitrate	$N_2H_5NO_3$		175	266	402	607	2127			
(2+) sulfate	$N_2H_6SO_4$			2.87	3.89	4.15	9.08	14.39		
(1+) sulfate	$(N_2H_5)_2SO_4$				221	300	554			
Hydrogen bromide	HBr	221.2	210.3	$204.0^{15°}$		$171.5^{50°}$		$150.5^{75°}$		130.0
chloride	HCl	82.3	77.2	72.1	67.3	63.3	56.1			
selenide, mL at STP	H_2Se	386	351	289						
Iodine	I_2	0.014	0.020	0.029	0.039	0.052	0.100	0.225	0.315	0.445
Iridium(IV) ammonium chloride	$(NH_4)_2IrCl_6$	0.556	0.706	0.77	1.21	1.57	2.46	4.38	dec	
sodium chloride	Na_2IrCl_6		$34.46^{15°}$		56.17	96.00	191.2	279.3		
Iron(II) ammonium sulfate	$FeSO_4 \cdot (NH_4)_2SO_4 \cdot 6H_2O$	17.23	31.0	36.47	45.0					
(II) bromide	$FeBr_2$	101	109	117	124	133	144	168	176	184
(II) chloride	$FeCl_2$	49.7	59.0	62.5	66.7	70.0	78.3	88.7	92.3	94.9
(III) chloride	$FeCl_3 \cdot 6H_2O$	74.4		91.8	106.8					
(II) fluorosilicate	$FeSiF_6 \cdot 6H_2O$	72.1	74.4		$77.0^{25°}$		$83.7^{50°}$	$88.1^{75°}$		$100.1^{106°}$
(II) nitrate	$Fe(NO_3)_2 \cdot 6H_2O$	113	134				266			
(III) nitrate	$Fe(NO_3)_3 \cdot 9H_2O$	112.0		137.7		175.0				
(III) perchlorate	$Fe(ClO_4)_3$	289		368	422	478	772			
(II) sulfate	$FeSO_4 \cdot 7H_2O$	28.8	40.0	48.0	60.0	73.3	100.7	79.9	68.3	57.8
Lanthanum bromate	$La(BrO_3)_3$	98	120	149	200					
nitrate	$La(NO_3)_3$	100		136		168	247			

selenate	$La_2(SeO_3)_3$	50.5	45	45	45	45	18.5	5.4	2.2	
sulfate	$La_2(SO_4)_3$	3.00	2.72	2.33	1.90	1.67	1.26	0.91	0.79	0.68
Lead(II) acetate	$Pb(C_2H_3O_2)_2$	19.8	29.5	44.3	69.8	116				
bromide	$PbBr_2$	0.45	0.63	0.86	1.12	1.50	2.29	3.23	3.86	4.55
chloride	$PbCl_2$	0.67	0.82	1.00	1.20	1.42	1.94	2.54	2.88	3.20
fluorosilicate	$PbSiF_6$	190		222			403	428		463
Germanium(IV) oxide	GeO_2		0.49	0.43	0.50	0.61				
Holmium sulfate	$Ho_2(SO_4)_3 \cdot 8H_2O$			8.18	$6.71^{25°}$	4.52				
Hydrazinium (1+) nitrate	$N_2H_5NO_3$		175	266	402	607	2127			
(2+) sulfate	$N_2H_6SO_4$			2.87	3.89	4.15	9.08	14.39		
(1+) sulfate	$(N_2H_5)_2SO_4$				221	300	554			
Hydrogen bromide	HBr	221.2	210.3	$204.0^{15°}$		$171.5^{50°}$		$150.5^{75°}$		130.0
chloride	HCl	82.3	77.2	72.1	67.3	63.3	56.1			
selenide, mL at STP	H_2Se	386	351	289						
Iodine	I_2	0.014	0.020	0.029	0.039	0.052	0.100	0.225	0.315	0.445
Iridium(IV) ammonium chloride	$(NH_4)_2IrCl_6$	0.556	0.706	0.77	1.21	1.57	2.46	4.38	dec	
sodium chloride	Na_2IrCl_6		$34.46^{15°}$		56.17	96.00	191.2	279.3		
Iron(II) ammonium sulfate	$FeSO_4 \cdot (NH_4)_2SO_4 \cdot 6H_2O$	17.23	31.0	36.47	45.0					
(II) bromide	$FeBr_2$	101	109	117	124	133	144	168	176	184
(II) chloride	$FeCl_2$	49.7	59.0	62.5	66.7	70.0	78.3	88.7	92.3	94.9
(III) chloride	$FeCl_3 \cdot 6H_2O$	74.4		91.8	106.8					
(II) fluorosilicate	$FeSiF_6 \cdot 6H_2O$	72.1	74.4		$77.0^{25°}$		$83.7^{50°}$	$88.1^{75°}$		$100.1^{106°}$
(II) nitrate	$Fe(NO_3)_2 \cdot 6H_2O$	113	134				266			
(III) nitrate	$Fe(NO_3)_3 \cdot 9H_2O$	112.0		137.7		175.0				
(III) perchlorate	$Fe(ClO_4)_3$	289		368	422	478	772			
(II) sulfate	$FeSO_4 \cdot 7H_2O$	28.8	40.0	48.0	60.0	73.3	100.7	79.9	68.3	57.8
Lanthanum bromate	$La(BrO_3)_3$	98	120	149	200					
nitrate	$La(NO_3)_3$	100		136		168	247			
selenate	$La_2(SeO_3)_3$	50.5	45	45	45	45	18.5	5.4	2.2	
sulfate	$La_2(SO_4)_3$	3.00	2.72	2.33	1.90	1.67	1.26	0.91	0.79	0.68
Lead(II) acetate	$Pb(C_2H_3O_2)_2$	19.8	29.5	44.3	69.8	116				
bromide	$PbBr_2$	0.45	0.63	0.86	1.12	1.50	2.29	3.23	3.86	4.55
chloride	$PbCl_2$	0.67	0.82	1.00	1.20	1.42	1.94	2.54	2.88	3.20
fluorosilicate	$PbSiF_6$	190		222			403	428		463

(Continued)

TABLE 1.68 Solubility of Inorganic Compounds and Metal Salts of Organic Acids in Water at Various Temperatures (*Continued*)

Substance	Formula	0°	10°	20°	30°	40°	60°	80°	90°	100°
iodide	PbI_2	0.044	0.056	0.069	0.090	0.124	0.193	0.294		0.42
nitrate	$Pb(NO_3)_2$	37.5	46.2	54.3	63.4	72.1	91.6	111		133
Lithium acetate	$LiC_2H_3O_2$	31.2	35.1	40.8	50.6	68.6				
ammonium sulfate	$LiNH_4SO_4$		55.2		55.9	56.1	56.5			
azide	LiN_3	61.3	64.2	67.2	71.2	75.4	86.6			100
benzoate	$LiC_7H_5O_2$	38.9	41.6	44.7	53.8					
borate (meta-)	$LiBO_2$	0.90	1.3	2.7	5.7	10.9				
bromate	$LiBrO_3$	154	166	179	198	221	269	308	329	355
bromide	$LiBr$	143	147	160	183	211	223	245		266
carbonate	Li_2CO_3	1.54	1.43	1.33	1.26	1.17	1.01	0.85		0.72
chlorate	$LiClO_3$	241	283	372	488	604	777			
chloride	$LiCl$	69.2	74.5	83.5	86.2	89.8	98.4	112	121	128
chloroaurate(III)	$LiAuCl_4$		113	136	167	206	324	599		
cyanoplatinate(II)	$Li_2Pt(CN)_4$	105		141	153	160	178	216	239	
formate	$LiCHO_2$	32.3	35.7	39.3	44.1	49.5	64.7	92.7	116	138
hydrogen phosphite	Li_2HPO_3	9.97			7.61	7.11	6.03			4.43
hydroxide	$LiOH$	11.91	12.11	12.35	12.70	13.22	14.63	16.56		19.12
iodide	LiI	151	157	165	171	179	202	435	440	481
molybdate	Li_2MoO_4	82.6		79.5	79.4	78.0				73.9
nitrate	$LiNO_3$	53.4	60.8	70.1	138	152	175			
nitrite	$LiNO_2$	70.9	82.5	96.8	114	133	177	233	272	324
perchlorate	$LiClO_4$	42.7	49.0	56.1	63.6	72.3	92.3	128	151	
phosphate (meta-)	$LiPO_3$	0.101		$0.058^{25°}$		0.048				
selenite	Li_2SeO_3	25.0	23.3	21.5	19.6	17.9	14.7	11.9	11.1	9.9
sulfate	Li_2SO_4	36.1	35.5	34.8	34.2	33.7	32.6	31.4	30.9	
tartrate (*d*-)	$Li_2C_4H_4O_6$	42.0	31.8	27.1	26.6	27.2	29.5			
thiocyanate	$LiSCN$			114	131	153				
vanadate	Li_3VO_4	2.50		4.82	6.28	4.38	2.67			
Magnesium acetate	$Mg(C_2H_3O_2)_2$	56.7	59.7	53.4	68.6	75.7	118			
bromide	$MgBr_2$	98	99	101	104	106	112			125
chlorate	$Mg(ClO_3)_2$	114	123	135	155	178	242		268	
chloride	$MgCl_2$	52.9	53.6	54.6	55.8	57.5	61.0	66.1	69.5	73.3
fluorosilicate	$MgSiF_6$	26.3		30.8		34.9	44.4			
formate	$Mg(CHO_2)_2$	14.0	14.2	14.4	14.9	15.9	17.9	20.5	22.2	23.9
iodate	$Mg(IO_3)_2$		7.2	8.6	10.0	11.7	15.2	15.5	15.6	
iodide	MgI_2	120		140		173		186		

Magnesium nitrate	$Mg(NO_3)_2$	62.1	66.0	69.5	73.6	78.9	78.9	91.6	106	
selenate	$MgSeO_4$	20.0	30.4	38.3	44.3	48.6	55.8			
sulfate	$MgSO_4$	22.0	28.2	33.7	38.9	44.5	54.6	55.8	52.9	50.4
sulfite	$MgSO_3$	0.339	0.446	0.573	0.751	0.959	0.779	0.642	0.622	
tartrate	$MgC_4H_4O_6$	0.54	0.78	1.06		1.02				
Manganese bromide	$MnBr_2$	127	136	147	157	169	197	225	226	228
chloride	$MnCl_2$	63.4	68.1	73.9	80.8	88.5	109	113	114	115
fluoride	MnF_2			1.06		0.67	0.44			0.48
nitrate	$Mn(NO_3)_2$	102	118	139	206					
oxalate	MnC_2O_4	0.020	0.024	0.028	0.033					
sulfate	$MnSO_4$	52.9	59.7	62.9	62.9	60.0	53.6	45.6	40.9	35.3
Mercury(II) bromide	$HgBr_2$	0.30	0.40	0.56	0.66	0.91	1.68	2.77		4.9
(II) chloride	$HgCl_2$	3.63	4.82	6.57	8.34	10.2	16.3	30.0		61.3
(I) perchlorate	$Hg_2(ClO_4)_2$	282	325	367	407	455	499	541		580
Molybdenum trioxide	MoO_3			0.134	0.285	0.454	1.08	1.74		
Neodymium bromate	$Nd(BrO_3)_3$	43.9	59.2	75.6	95.2	116				
chloride	$NdCl_3$		96.7	98.0	99.6	102	105			
nitrate	$Nd(NO_3)_3$	127	133	142	145	159	211			
selenate	$Nd_2(SeO_3)_3$	46.2	44.6	41.8	39.9	39.9	43.9	7.0	3.3	
sulfate	$Nd_2(SO_4)_3$	13.0	9.7	7.1	5.3	4.1	2.8	2.2	1.2	
Nickel bromide	$NiBr_2$	113	122	131	138	144	153	154		155
chlorate	$Ni(ClO_3)_2$	111	120	133	155	181	221	308		
chloride	$NiCl_2$	53.4	56.3	60.8	70.6	73.2	81.2	86.6		87.6
fluoride	NiF_2		2.55	2.56			2.56		2.59	
iodate	$Ni(IO_3)_2$				1.15		1.06		1.00	
	$Ni(IO_3)_2 \cdot 4H_2O$	0.74		1.09	1.43					
iodide	NiI_2	124	135	148	161	174	184	187	188	
nitrate	$Ni(NO_3)_2$	79.2		94.2	105	119	158	187	188	
perchlorate	$Ni(ClO_4)_2$	105	107	110	113	117				
Nickel sulfate	$NiSO_4 \cdot 6H_2O$	(pale blue)		40.1	43.6	47.6				
		(green)		44.4	46.6	49.2	55.6	64.5	70.1	76.7
	$NiSO_4 \cdot 7H_2O$	26.2	32.4	37.7	43.4	50.4				
Osmium tetroxide	OsO_4	5.26	5.75	6.43						
Oxalic acid	$H_2C_2O_4$	3.54	6.08	9.52	14.23	21.52	44.32	84.5	120	
Potassium acetate	$KC_2H_3O_2$	216	233	256	283	324	350	381	398	
aluminum sulfate	$KAl(SO_4)_2$	3.00	3.99	5.90	8.39	11.7	24.8	71.0	109	
azide	KN_3	41.4	46.2	50.8	55.8	61.0				106
benzoate	$KC_7H_5O_2$		65.8	70.7	76.7	82.1				

(Continued)

TABLE 1.68 Solubility of Inorganic Compounds and Metal Salts of Organic Acids in Water at Various Temperatures (*Continued*)

Substance	Formula	0°	10°	20°	30°	40°	60°	80°	90°	100°
bromate	$KBrO_3$	3.09	4.72	6.91	9.64	13.1	22.7	34.1		49.9
bromide	KBr	53.6	59.5	65.3	70.7	75.4	85.5	94.9	99.2	104
cadmium bromide	$KCdBr_3$	116	133	150	170	191	233	276	298	325
cadmium chloride	$KCdCl_3$	26.6	32.3	38.9	45.6	53.1	67.5	83.5		101
carbonate	K_2CO_3	105	108	111	114	117	127	140	148	156
chlorate	$KClO_3$	3.3	5.2	7.3	10.1	13.9	23.8	37.6	46.0	56.3
chloride	KCl	28.0	31.2	34.2	37.2	40.1	45.8	51.3	53.9	56.3
chloroaurate(III)	$KAuCl_4$		38.3	61.8	94.9	145	405			
chloroplatinate(IV)	K_2PtCl_6	0.48	0.60	0.78	1.00	1.36	2.45	3.71		5.03
chromate	K_2CrO_4	56.3	60.0	63.7	66.7	67.8	70.1		74.5	
citrate	$K_3C_6H_5O_7$		153	172	194					
cobalt(II) sulfate	$K_2Co(SO_4)_2$	8.5	11.7	15.5	19.3	23.3	32.5	47.7		
copper(II) sulfate	$K_2Cu(SO_4)_2$	5.1	7.2	10.0	13.6	18.2				
cyanoplatinate(II)	$K_2Pt(CN)_4$	11.6	19.8	33.9	52.0	78.3	139	177	194	
dichromate	$K_2Cr_2O_7$	4.7	7.0	12.3	18.1	26.3	45.6	73.0		
dihydrogen phosphate	KH_2PO_4	14.8	18.3	22.6	28.0	33.5	50.2	70.4	83.5	
dithionate	$K_2S_2O_6$	2.6	4.2	6.6	9.3					
ferricyanide	$K_3Fe(CN)_6$	30.2	38	46	53	59.3	70			91
ferrocyanide	$K_4Fe(CN)_6$	14.3	21.1	28.2	35.1	41.4	54.8	66.9	71.5	74.2
fluoride	KF	44.7	53.5	94.9	108	138	142	150		
fluorogermanate(IV)	K_2GeF_6	0.25	0.36	0.50	0.66	0.96				
fluorosilicate	K_2SiF_6	0.077	0.102	0.151	0.202	0.253				
fluorotitanate(IV)	K_2TiF_6	0.55	0.91	1.28						
formate	$KCHO_2$		313	337	361	398	471	580	658	
hydrogen carbonate	$KHCO_3$	22.5	27.4	33.7	39.9	47.5	65.6			
Potassium hydrogen fluoride	KHF_2	24.5	30.1	39.2	46.8	56.5	78.8	114		
hydrogen selenite	$KH_3(SeO_3)_2$	115	162	215	300	408	900			
hydrogen sulfate	$KHSO_4$	36.2		48.6	54.3	61.0	76.4	96.1		122
hydrogen tartrate	$KC_4H_5O_6$	0.231	0.358	0.523	0.762					
hydroxide	KOH	95.7	103	112	126	134	154			178
iodate	KIO_3	4.60	6.27	8.08	10.3	12.6	18.3	24.8		32.3
iodide	KI	128	136	144	153	162	176	192	198	206
iron(II) sulfate	$K_2Fe(SO_4)_2$	19.6	24.5	32.1	39.1	44.9	57.2			
magnesium sulfate	$K_2Mg(SO_4)_2$	14.0	19.5	25.0	30.4	36.6	50.2	63.4		

nickel sulfate	$K_2Ni(SO_4)_2$	3.37	4.50	5.94	7.72	9.85	15.4	23.0	27.8	33.4
nitrate	KNO_3	13.9	21.2	31.6	45.3	61.3	106	167	203	245
nitrite	KNO_2	279	292	306	320	329	348	376	390	410
oxalate	$K_2C_2O_4$	25.5	31.9	36.4	39.9	43.8	53.2	63.6	69.2	75.3
perchlorate	$KClO_4$	0.76	1.06	1.68	2.56	3.73	7.3	13.4	17.7	22.3
periodate	KIO_4	0.17	0.28	0.42	0.65	1.0	2.1	4.4	5.9	
permanganate	$KMnO_4$	2.83	4.31	6.34	9.03	12.6	22.1			
peroxodisulfate	$K_2S_2O_8$	1.65	2.67	4.70	7.75	11.0				
perrhenate	$KReO_4$	0.34	0.63	0.99	1.47	2.2	4.58	8.7		
phosphate	K_3PO_4		81.5	92.3	108	133				
salicylate	$KC_7H_5O_3$	21.2	32.4	47.1	61.3	78.6	116	156		
selenate	K_2SeO_4	107	109	111	113	115	119	121		122
selenite	K_2SeO_3	169	186	203	217	217	220			217
sulfate	K_2SO_4	7.4	9.3	11.1	13.0	14.8	18.2	21.4	22.9	24.1
sulfite	K_2SO_3	106		106	107	107	108			112
tellurate	K_2TeO_4	8.8		27.5	50.4					
thioantimonate(V)	K_3SbS_4	306	320		302	315		381		
thiocyanate	KSCN	177	198	224	255	289	372	492	571	675
thiosulfate	$K_2S_2O_3$	96		155	175	205	238	293	312	
zinc sulfate	$K_2Zn(SO_4)_2 \cdot 6H_2O$	13.0	18.9	25.9	35.0	44.9	72.1			
Praseodymium bromate	$Pr(BrO_3)_3$	55.9	73.0	91.8	114	144				
nitrate	$Pr(NO_3)_3$			112	162	178				
selenate	$Pr_2(SeO_3)_3$	36.2			32.4	31.2	30.4	5.43	3.6	
sulfate	$Pr_2(SO_4)_3$	19.8	15.6	12.6	9.89	2.56	5.04	3.5	1.1	0.91
Rubidium aluminum sulfate	$Rb_2Al_2(SO_4)_4$	0.72	1.05	1.50	2.20	3.25	7.40	21.6		
bromate	$RbBrO_3$				3.6	5.1				
bromide	RbBr	90	99	108	119	132	158			
chlorate	$RbClO_3$	2.1	3.4	5.4	8.0	11.6	22	38	49	63
chloride	RbCl	77	84	91	98	104	115	127	133	143
chloroaurate(III)	$RbAuCl_4$		4.8	9.9	15.5	21.5	36.2	54.6	65.8	79.2
chloroplatinate(IV)	Rb_2PtCl_6	0.014	0.020	0.028	0.040	0.056	0.090	0.182	0.247	0.33
chromate	Rb_2CrO_4	62.0	67.5	73.6	78.9	85.6	95.7			
cobalt sulfate	$Rb_2Co(SO_4)_2$	5.10	7.47	10.8	14.5	18.2	30.2	44.9	55.0	70.1
dichromate (mn)	$Rb_2Cr_2O_7$			5.9	10.0	15.2	32.3			
(tric)				5.8	9.5	14.8	32.4			
formate	$RbCHO_2$		443	554	614	694	900			
iron(III) sulfate	$RbFe(SO_4)_2 \cdot 12H_2O$		8.0	20	35	52				
nitrate	$RbNO_3$	19.5	33.0	52.9	81.2	117	200	310	374	452

(*Continued*)

TABLE 1.68 Solubility of Inorganic Compounds and Metal Salts of Organic Acids in Water at Various Temperatures (*Continued*)

Substance	Formula	0°	10°	20°	30°	40°	60°	80°	90°	100°
perchlorate	$RbClO_4$	1.09	1.19	1.55	2.20	3.26	6.27	11.0	15.5	22.0
salicylate	$RbC_7H_5O_3$		187	212	238	268	324			
sulfate	Rb_2SO_4	37.5	42.6	48.1	53.6	58.5	67.5	75.1	78.6	81.8
Samarium bromate	$Sm(BrO_3)_3$	34.2	47.6	62.5	79.0	98.5				
chloride	$SmCl_3$		92.4	93.4	94.6	96.9				
Selenic acid	H_2SeO_4	426		567	1328					
Selenious acid	H_2SeO_3	90.1	122.2	166.7	235.6	344.4	383.1	383.1	385.4	
Selenium dioxide	SeO_2		222	257	291	335	440			
Silver acetate	$AgC_2H_3O_2$	0.73	0.89	1.05	1.23	1.43	1.93	2.59		
bromate	$AgBrO_3$		0.11	0.16	0.23	0.32	0.57	0.94	1.33	
chlorate	$AgClO_3$		10.4	15.3	20.9	26.8				
fluoride	AgF	85.9	120	172	190	203				
nitrate	$AgNO_3$	122	167	216	265	311	440	585	652	733
nitrite	$AgNO_2$	0.16	0.22	0.34	0.51	0.73	1.39			
perchlorate	$AgClO_4$	455	484	525	594	635				793
sulfamate	$AgNH_2SO_3$	2.30	4.82	7.53	10.3	15.3	28.5			
sulfate	Ag_2SO_4	0.57	0.70	0.80	0.89	0.98	1.15	1.30	1.36	1.41
Sodium acetate	$NaC_2H_3O_2$	36.2	40.8	46.4	54.6	65.6	139	153	161	170
aluminum sulfate	$Na_2Al_2(SO_4)_4$	37.4	39.3	39.7	41.7	43.8				
azide	NaN_3	38.9	39.9	40.8						55.3
benzoate	$NaC_7H_5O_2$	62.6	62.8	62.8	62.9	63.1	64.5	68.6	70.6	73.3
borate (penta-)	$Na_2B_{10}O_{16}$	6.4	8.6	12.0	16.4	22.0	37.9	63.4	83.5	108
borate (tetra-)	$Na_2B_4O_7$	1.11	1.60	2.56	3.86	6.67	19.0	31.4	41.0	52.5
bromate	$NaBrO_3$	24.2	30.3	36.4	42.6	48.8	62.6	75.7		90.8
bromide	$NaBr$	80.2	85.2	90.8	98.4	107	118	120	121	121
carbonate	Na_2CO_3	7.00	12.5	21.5	39.7	49.0	46.0	43.9	43.9	
chlorate	$NaClO_3$	79.6	87.6	95.9	105	115	137	167	184	204
chloride	$NaCl$	35.7	35.8	35.9	36.1	36.4	37.1	38.0	38.5	39.2
chloroaurate(III)	$NaAuCl_4$		139	151	178	227	900			
chloroiridate(IV)	Na_2IrCl_6		31.6	39.3	56.2	96.1	192	279		
chromate	Na_2CrO_4	31.7	50.1	84.0	88.0	96.0	115	125		126
cyanide	$NaCN$	40.8	48.1	58.7	71.2					
dichromate	$Na_2Cr_2O_7$	163	172	183	198	215	269	376	405	415
diethyl barbiturate	$NaC_8H_{11}N_2O_3$		12.7	21.5	24.7				48.0	
dihydrogen phosphate (ortho-)	NaH_2PO_4	56.5	69.8	86.9	107	133	172	211	234	

dihydrogen phosphate (pyro-)	$Na_2H_2P_2O_7$	4.47	6.95	12.0	17.1	18.4				
dithionate	$Na_2S_2O_6$	6.3	11.1	15.1	19.6	24.7	36.1	49.3	56.3	64.7
dodecanesulfonate	$NaC_{12}H_{25}SO_3$			0.13	0.25	6.54				
dodecanoate	$NaC_{12}H_{23}O_2$				4.58	22.7	105	170		
EDTA (Y)*	$Na_2H_2Y \cdot 2H_2O$	10.6		11.1	12.8	14.2	17.0	22.2	24.3	$27.0^{98°}$
ferrocyanide	$Na_4Fe(CN)_6$	11.2	14.8	18.8	23.8	29.9	43.7	62.1		
fluoride	NaF	3.66		4.06	4.22	4.40	4.68	4.89		5.08
fluoroberyllate	Na_2BeF_4	1.33		1.44		1.92	2.24	2.62	2.73	
fluorogermanate	Na_2GeF	1.52	1.68		2.25	2.83		3.36		
fluorosilicate	Na_2SiF_6	4.35	5.7	7.2	8.6	10.3	14.3	18.7	21.5	24.5
formate	$NaCHO_2$	43.9	62.5	81.2	102	108	122	138	147	160
germanate	Na_2GeO_3	14.4	18.8	23.8	28.7	37.2	65.0	116		
hydrogen arsenate	Na_2HAsO_4	5.9	13.0	33.9	49.3	69.5	144	186	188	198
hydrogen carbonate	$NaHCO_3$	7.0	8.1	9.6	11.1	12.7	16.0			
hydrogen phosphate	Na_2HPO_4	1.68	3.53	7.83	22.0	55.3	82.8	92.3	102	104
hydrogen phosphite	Na_2HPO_3	418	424	429	566					
hydrogen succinate	$NaC_4H_5O_4$	17.5	25.3	34.8	47.7	61.6	74.5	90.1		
hydroxide	NaOH		98	109	119	129	174			
hydroxostannate(IV)	$Na_2Sn(OH)_6$	46.0		43.7	42.7	38.9				
hypochlorite	NaClO	29.4	36.4	53.4	100	110				
iodate	$NaIO_3$	2.48	4.59	8.08	10.7	13.3	19.8	26.6	29.5	33.0
iodide	NaI	159	167	178	191	205	257	295		302
molybdate	Na_2MoO_4	44.1	64.7	65.3	66.9	68.6	71.8			
nitrate	$NaNO_3$	73.0	80.8	87.6	94.9	102	122	148		180
nitrite	$NaNO_2$	71.2	75.1	80.8	87.6	94.9	111	133		160
oxalate	$Na_2C_2O_4$	2.69	3.05	3.41	3.81	4.18	4.93	5.71		6.50
perchlorate	$NaClO_4$	167	183	201	222	245	288	306		329
periodate	$NaIO_4$	1.83	5.6	10.3	19.9	30.4				
phosphate	Na_3PO_4	4.5	8.2	12.1	16.3	20.2	29.9	60.0	68.1	77.0
potassium tartrate	$NaKC_4H_4O_6$	31.9	46.6	67.8	102					
salicylate	$NaC_7H_5O_3$		44.7	95.3	111	117	130	144		
selenate	Na_2SeO_4	13.3	25.2	26.9	77.0	81.8	78.6	74.8	73.0	72.7
selenite	Na_2SeO_3	78.6	81.2	86.2	94.2	96.5	91.6	86.6	84.5	82.5
sulfate	Na_2SO_4	4.9	9.1	19.5	40.8	48.8	45.3	43.7	42.7	42.5
	$Na_2SO_4 \cdot 7H_2O$	19.5	30.0	44.1						
sulfide	Na_2S	9.6	12.1	15.7	20.5	26.6	39.1	55.0	65.3	
sulfite	Na_2SO_3	14.4	19.5	26.3	35.5	37.2	32.6	29.4	27.9	
thioantimonate(V)	Na_3SbS_4	13.4	20.0	27.9	37.2	49.3	53.8	88.3		
thiocyanate	NaSCN		111	134	164	176	192	210	218	

(Continued)

TABLE 1.68 Solubility of Inorganic Compounds and Metal Salts of Organic Acids in Water at Various Temperatures (*Continued*)

Substance	Formula	0°	10°	20°	30°	40°	60°	80°	90°	100°
thiosulfate	$Na_2S_2O_3 \cdot 5H_2O$	50.2	59.7	70.1	83.2	104				
tungstate	Na_2WO_4	71.5		73.0		77.6		90.8		97.2
vanadate	$NaVO_3$			19.3	22.5	26.3	33.0	40.8		
Strontium acetate	$Sr(C_2H_3O_2)_2$	37.0	42.9	41.1	39.5	38.3	36.8	36.1	36.2	36.4
bromide	$SrBr_2$	85.2	93.4	102	112	123	150	182		223
chloride	$SrCl_2$	43.5	47.7	52.9	58.7	65.3	81.8	90.5		101
chromate	$SrCrO_4$		0.085	0.090				0.058		
Strontium fluoride	SrF_2	0.0113		0.0117	0.0119					
formate	$Sr(CHO_2)_2$	9.1	10.6	12.7	15.2	17.8	25.0	31.9	32.9	34.4
hydroxide	$Sr(OH)_2$	0.91	1.25	1.77	2.64	3.95	8.42	20.2	44.5	91.2
iodide	SrI_2	165		178		192	218	270	365	383
nitrate	$Sr(NO_3)_2$	39.5	52.9	69.5	88.7	89.4	93.4	96.9	98.4	
nitrite	$Sr(NO_2)_2$			65	72	79	97	130	134	
oxide	SrO				1.03	1.05	3.40	9.15	13.13	12.15
sulfate	$SrSO_4$	0.0113	0.0129	0.0132	0.0138	0.0141	0.0131	0.0116	0.0115	
Sulfamic acid	H_2NSO_3H	14.7	18.6	21.3	26.1	29.5	37.1	47.1		
Telluric acid	H_2TeO_4	16.2	33.8	41.6	50.0	57.2	77.5	106		155
Terbium bromate	$Tb(BrO_3)_3 \cdot 9H_2O$	66.4	89.7	117	152	198				
Thallium(I) azide	TlN_3	0.171	0.236	0.364						
bromide	TlBr	0.022	0.032	0.048	0.068	0.097	0.177			
carbonate	Tl_2CO_3			5.3			12.2			27.2
chlorate	$TlClO_3$	2.00		3.92		$12.7^{50°}$		36.6		57.3
chloride	TlCl	0.21	0.25	0.33	0.42	0.52	0.80	1.20		1.80
hydroxide	TlOH	25.4	29.6	35.0	40.4	49.4	73.3	106	126	150
iodide	TlI	0.002		0.006		0.015	0.035	0.070		0.120
nitrate	$TlNO_3$	3.90	6.22	9.55	14.3	21.0	46.1	110	200	414
nitrite	$TlNO_2$	17.9	28.9	40.3	53.2	83.6	216	1150	750	
perchlorate	$TlClO_4$	6.00	8.04	13.1	19.7	28.3	50.8	81.5		
picrate	$TlOC_6H_2(NO_2)_3$	0.135		0.40	0.57	0.83	1.73			
selenate	Tl_2SeO_4		2.17	2.80				8.50		10.8
sulfate	Tl_2SO_4	2.73	3.70	4.87	6.16	7.53	11.0	14.6	16.5	18.4
Thorium nitrate	$Th(NO_3)_4$	186	187	191						
sulfate	$Th(SO_4)_2 \cdot 4H_2O$					4.04	1.63			
	$Th(SO_4)_2 \cdot 9H_2O$	0.74	0.99	1.38	1.99	3.00				
Tin(II) iodide	SnI_2			0.99	1.17	1.42	2.11	3.04	3.58	4.20
Uranium(IV) sulfate	$U(SO_4)_2 \cdot 4H_2O$				10.1	9.0	7.7			
	$U(SO_4)_2 \cdot 8H_2O$			11.9	17.9	29.2	55.8			

Uranyl nitrate	$UO_2(NO_3)_2$	98	107	122	141	167	317	388	426	474
oxalate	$UO_2C_2O_4$		0.45	0.50	0.61	0.80	1.22	1.94		3.16
Ytterbium sulfate	$Yb_2(SO_4)_3$	44.2	37.5		22.2	17.2	10.4	6.4	5.8	4.7
Yttrium bromide	YBr_3	63.9		75.1		87.3	101	116	123	
chloride	YCl_3	77.3	78.1	78.8	79.6	80.8				
nitrate	$Y(NO_3)_3$	93.1	106	123	143	163	200			
sulfate	$Y_2(SO_4)_3$	8.05	7.67	7.30	6.78	6.09	4.44	2.89	2.2	
Zinc bromide	$ZnBr_2$	389		446	528	591	618	645		672
chlorate	$Zn(ClO_3)_2$	145	152	200	209	223				
chloride	$ZnCl_2$	342	363	395	437	452	488	541		614
formate	$Zn(CHO_2)_2$	3.70	4.30	5.20	6.10	7.40	11.8	21.2	28.8	38.0
iodide	ZnI_2	430		432		445	467	490		510
nitrate	$Zn(NO_3)_2$	98			138	211				
sulfate (rh)	$ZnSO_4$	41.6	47.2	53.8	61.3	70.5	75.4	71.1		60.5
sulfate (mn)			54.4	60.0	65.5					
tartrate	$ZnC_4H_4O_6$			0.022	0.041	0.060	0.104	0.059		

*Properly called dihydrogen ethylenediaminetetraacetate (Na_2H_2 EDTA · $2H_2O$).

TABLE 1.69 Dissociation Constants of Inorganic Acids

The *dissociation constant* of an acid K_a may conveniently be expressed in terms of the pK_a value where p$K_a = -\log_{10} (K_a/\text{mol dm}^{-3})$. The values given in the following table are for aqueous solutions at 298 K: the pK_1, pK_2, and pK_3 values refer to the first, second, and third ionizations respectively.

Name	Formula	pK$_a$
Aluminium ion (hydrated)	$[Al(H_2O)_6]^{3+}$	4.9 (pK_1)
Ammonium ion	NH_4^+	9.25
Arsenic(III) acid	H_3AsO_3	9.22 (pK_1)
Arsenic(V) acid	H_3AsO_4	2.30 (pK_1)
Boric acid	H_3BO_3	9.24 (pK_1)
Bromic(1) acid	HOBr	8.70
Carbonic acid	H_2CO_3	6.38[a] (pK_1) 10.32 (pK_2)
Chloric(I) acid	HOCl	7.43
Chloric(III) acid	$HClO_2$	2.0
Chromium(III) ion (hydrated)	$[Cr(H_2O)_6]^{3+}$	3.9 (pK_1)
Hydrazinium ion	$N_2H_5^+$	7.93
Hydrocyanic acid	HCN	9.40
Hydrofluoric acid	HF	3.25
Hydrogen peroxide	H_2O_2	11.62 (pK_1)
Hydrogen sulphide	H_2S	7.05 (pK_1) 12.92 (pK_2)
Hydroxyammonium ion	NH_3OH^+	5.82
Iodic(I) acid	HOI	10.52
Iodic(V) acid	HIO_3	0.8
Iron(III) ion (hydrated)	$[Fe(H_2O)_6]^{3+}$	2.22 (pK_1)
Lead(II) ion (hydrated)	$[Pb(H_2O)_n]^{2+}$	7.8 (pK_1)
Nitrous acid	HNO_2	3.34
Phosphinic acid	H_3PO_2	2.0
Phosphoric(V) acid	H_3PO_4	2.15 (pK_1) 7.21 (pK_2) 12.36 (pK_3)
Phosphonic acid	H_3PO_3	2.00 (pK_1) 6.58 (pK_2)
Silicic acid	H_2SiO_3	9.9 (pK_1) 11.9 (pK_2)
Sulphuric acid	H_2SO_4	1.92 (pK_2)
Sulphurous acid	H_2SO_3	1.92 (pK_1) 7.21 (pK_2)

[a]Some of the unionized acid exists as dissolved CO_2 molecules rather than H_2CO_3: pK_1 for the molecular species H_2CO_3 is approximately 3.7.

TABLE 1.70 Ionic Product Constant of Water

This table gives values of p*Kw* on a modal scale, where *Kw* is the ionic activity product constant of water. Values are from W. L. Marshall and E. U. Franck, *J. Phys. Chem. Ref. Data*, **10:**295 (1981).

Temp., °C	p*Kw*	Temp., °C	p*Kw*	Temp., °C	p*Kw*
0	14.938	45	13.405	95	12.345
5	14.727	50	13.275	100	12.264
10	14.528	55	13.152	125	11.911
15	14.340	60	13.034	150	11.637
18	14.233	65	12.921	175	11.431
20	14.163	70	12.814	200	11.288
25	13.995	75	12.711	225	11.207
30	13.836	80	12.613	250	11.192
35	13.685	85	12.520	275	11.251
40	13.542	90	12.431	300	11.406

TABLE 1.71 Solubility Product Constants

The data refer to various temperatures between 18 and 25°C, and were complied from values cited by Bjerrum, Schwarzenbach, and Sillen, *Stability Constants of Metal Complexes*, Part II, Chemical Society, London, 1958, and values taken from publications of the IUPAC Solubility Data Project: *Solubility Data Series*, international Union of Pure and Applied Chemistry, Pergamon Press, Oxford, 1979–1992; H. L. Clever, and F. J. Johnston, *J. Phys Chem. Ref. Data*, **9:**751 (1980); Y. Marcus, *Ibid.* **9:**1307 (1980); H. L. Clever, S. A. Johnson, and M. E. Derrick, *Ibid.* **14:**631 (1985), and **21:**941 (1992).

In the table, "L" is the abbreviation of the organic ligand.

Compound	Formula	pK_{sp}	K_{sp}
Actinium			
hydroxide	$Ac(OH)_3$	15	1×10^{-15}
Aluminum			
arsonate	$AlAsO_4$	15.80	1.6×10^{-16}
cupferrate	AlL_3	18.64	2.3×10^{-19}
hydroxide	$Al(OH)_3$	32.89	1.3×10^{-33}
phosphate	$AlPO_4$	20.01	9.84×10^{-21}
8-quinolinolate	AlL_3	29.00	1.00×10^{-29}
selenide	Al_2Se_3	24.4	4×10^{-25}
sulfide	Al_2S_3	6.7	2×10^{-7}
Americium			
(III) hydroxide	$Am(OH)_3$	19.57	2.7×10^{-20}
(IV) hydroxide	$Am(OH)_4$	56	1×10^{-56}
Ammonium			
uranyl arsenate	$NH_4UO_2AsO_4$	23.77	1.7×10^{-24}
Arsenic			
(III) sulfide	As_2S_3	21.68	2.1×10^{-22}

(Continued)

TABLE 1.71 Solubility Product Constants (*Continued*)

Compound	Formula	pK_{sp}	K_{sp}
Barium			
arsenate	$Ba_3(AsO_4)_2$	50.11	8.0×10^{-51}
bromate	$Ba(BrO_3)_2$	5.50	2.43×10^{-4}
carbonate	$BaCO_3$	8.59	2.58×10^{-9}
chromate	$BaCrO_4$	9.93	1.17×10^{-10}
ferricyanide 6-hydrate	$Ba_2[Fe(CN)_6]\cdot 6H_2O$	7.49	3.2×10^{-8}
fluoride	BaF_2	6.74	1.84×10^{-7}
hexafluorosilicate	$BaSiF_6$	6	1×10^{-6}
hydrogen phosphate	$BaHPO_4$	6.49	3.2×10^{-7}
hydroxide 8-hydrate	$Ba(OH)_2\cdot 8H_2O$	3.59	2.55×10^{-4}
iodate hydrate	$Ba(IO_3)_2\cdot H_2O$	8.40	4.01×10^{-9}
molybdate	$BaMoO_4$	7.45	3.54×10^{-8}
niobate	$Ba(NbO_3)_2$	16.50	3.2×10^{-17}
nitrate	$Ba(NO_3)_2$	2.33	4.64×10^{-3}
oxalate	BaC_2O_4	6.79	1.6×10^{-7}
oxalate hydrate	$BaC_2O_4\cdot H_2O$	7.64	2.3×10^{-8}
permanganate	$Ba(MnO_4)_2$	9.61	2.5×10^{-10}
perrhenate	$Ba(ReO_4)_2$	1.28	5.2×10^{-2}
phosphate	$Ba_3(PO_4)_2$	22.47	3.4×10^{-23}
pyrophosphate	$Ba_2P_2O_7$	10.50	3.2×10^{-11}
8-quinolinolate	BaL_2	8.30	5.0×10^{-9}
selenate	$BaSeO_4$	7.47	3.40×10^{-8}
sulfate	$BaSO_4$	9.97	1.08×10^{-10}
sulfite	$BaSO_3$	9.30	5.0×10^{-10}
thiosulfate	BaS_2O_3	4.79	1.6×10^{-5}
Beryllium			
carbonate 4-hydrate	$BeCO_3\cdot 4H_2O$	3	1×10^{-3}
hydroxide (amorphous)	$Be(OH)_2$	21.16	6.92×10^{-22}
molybdate	$BeMoO_4$	1.49	3.2×10^{-2}
niobate	$Be(NbO_3)_2$	15.92	1.2×10^{-16}
Bismuth			
arsenate	$BiAsO_4$	9.35	4.43×10^{-10}
cupferrate	BiL_3	27.22	6.0×10^{-28}
hydroxide	$Bi(OH)_3$	30.4	6.0×10^{-31}
iodide	BiI_3	18.11	7.71×10^{-19}
oxide bromide	$BiOBr$	6.52	3.0×10^{-7}
oxide chloride	$BiOCl$	30.75	1.8×10^{-31}
oxide hydroxide	$BiO(OH)$	9.4	4×10^{-10}
oxide nitrate	$BiO(NO_3)$	2.55	2.82×10^{-3}
oxide nitrite	$BiO(NO_2)$	6.31	4.9×10^{-7}
oxide thiocyanate	$BiO(SCN)$	6.80	1.6×10^{-7}
phosphate	$BiPO_4$	22.89	1.3×10^{-23}
sulfide	Bi_2S_3	97	1×10^{-97}
Cadmium			
anthranilate	CdL_2	8.27	5.4×10^{-9}
arsenate	$Cd_3(AsO_4)_2$	32.66	2.2×10^{-33}
benzoate 2-hydrate	$CdL_2\cdot 2H_2O$	2.7	2×10^{-3}
borate, *meta*	$Cd(BO_2)_2$	8.64	2.3×10^{-9}
carbonate	$CdCO_3$	12.0	1.0×10^{-12}
cyanide	$Cd(CN)_2$	8.0	1.0×10^{-8}
ferrocyanide	$Cd_2[Fe(CN)_6]$	16.49	3.2×10^{-17}
fluoride	CdF_2	2.19	6.44×10^{-3}

TABLE 1.71 Solubility Product Constants (*Continued*)

Compound	Formula	pK_{sp}	K_{sp}
hydroxide	$Cd(OH)_2$ fresh	14.14	7.2×10^{-15}
iodate	$Cd(IO_3)_2$	7.60	2.5×10^{-8}
oxalate 3-water	$CdC_2O_4 \cdot 3H_2O$	7.85	1.42×10^{-8}
phosphate	$Cd_3(PO_4)_2$	32.60	2.53×10^{-33}
quinaldate	CdL_2	12.30	5.0×10^{-13}
sulfide	CdS	26.10	8.0×10^{-27}
tungstate	$CdWO_4$	5.7	2×10^{-6}
Calcium			
acetate 3-water	$Ca(OAc)_2 \cdot 3H_2O$	2.4	4×10^{-3}
arsenate	$Ca_3(AsO_4)_2$	18.17	6.8×10^{-19}
benzoate 3-water	$CaL_2 \cdot 3H_2O$	2.4	4×10^{-3}
carbonate	$CaCO_3$	8.54	2.8×10^{-9}
carbonate (calcite)	$CaCO_3$	8.47	3.36×10^{-9}
carbonate (aragonite)	$CaCO_3$	8.22	6.0×10^{-9}
carbonatomagnesium	$Ca[Mg(CO_3)_2]$ dolomite	11	1×10^{-11}
chromate	$CaCrO_4$	3.15	7.1×10^{-4}
fluoride	CaF_2	8.28	5.3×10^{-9}
hexafluorosilicate	$Ca[SiF_6]$	3.09	8.1×10^{-4}
hydrogen phosphate	$CaHPO_4$	7.0	1.0×10^{-7}
hydroxide	$Ca(OH)_2$	5.26	5.5×10^{-6}
iodate 6-water	$Ca(IO_3)_2 \cdot 6H_2O$	6.15	7.10×10^{-7}
molybdate	$CaMoO_4$	7.84	1.46×10^{-8}
niobate	$Ca(NbO_3)_2$	17.06	8.7×10^{-18}
oxalate hydrate	$CaC_2O_4 \cdot H_2O$	8.63	2.32×10^{-9}
phosphate	$Ca_3(PO_4)_2$	28.68	2.07×10^{-29}
8-quinolinolate	CaL_2	11.12	7.6×10^{-12}
selenate	$CaSeO_4$	3.09	8.1×10^{-4}
selenite	$CaSeO_3$	5.53	8.0×10^{-6}
silicate, *meta*	$CaSiO_3$	7.60	2.5×10^{-8}
sulfate	$CaSO_4$	4.31	4.93×10^{-5}
sulfate dihydrate	$CaSO_4 \cdot 2H_2O$	4.50	3.14×10^{-5}
sulfite	$CaSO_3$	7.17	6.8×10^{-8}
sulfite 0.5-water	$CaSO_3 \cdot 0.5H_2O$	6.51	3.1×10^{-7}
tartrate dihydrate	$CaL \cdot 2H_2O$	6.11	7.7×10^{-7}
tungstate	$CaWO_4$	8.06	8.7×10^{-9}
Cerium			
(III) fluoride	CeF_3	15.1	8×10^{-16}
(III) hydroxide	$Ce(OH)_3$	19.80	1.6×10^{-20}
(IV) hydroxide	$Ce(OH)_4$	47.7	2×10^{-48}
(III) iodate	$Ce(IO_3)_3$	9.50	3.2×10^{-10}
(IV) iodate	$Ce(IO_3)_4$	16.3	5×10^{-17}
(III) oxalate 9-water	$Ce_2(C_2O_4)_3 \cdot 9H_2O$	25.50	3.2×10^{-26}
(III) phosphate	$CePO_4$	23	1×10^{-23}
(III) selenite	$Ce_2(SeO_3)_3$	24.43	3.7×10^{-25}
(III) sulfide	Ce_2S_3	10.22	6.0×10^{-11}
(III) tartrate	Ce_2L_3	19.0	1.0×10^{-19}
Cesium			
bromate	$CsBrO_3$	1.7	5×10^{-2}
chlorate	$CsClO_3$	1.4	4×10^{-2}
cobaltihexanitrite	$Cs_3[Co(NO_2)_6]$	15.24	5.7×10^{-16}
hexachloroplatinate(IV)	$Cs_2[PtCl_6]$	7.50	3.2×10^{-8}
hexafluoroplatinate(IV)	$Cs_2[PtF_6]$	5.62	2.4×10^{-6}
hexafluorosilicate	$Cs_2[SiF_6]$	4.90	1.3×10^{-5}

(*Continued*)

TABLE 1.71 Solubility Product Constants (*Continued*)

Compound	Formula	pK_{sp}	K_{sp}
perchlorate	$CsClO_4$	2.40	3.95×10^{-3}
periodate	$CsIO_4$	5.29	5.16×10^{-6}
permanganate	$CsMnO_4$	4.08	8.2×10^{-5}
perrhanate	$CsReO_4$	3.40	4.0×10^{-4}
tetrafluoroborate	$Cs[BF_4]$	4.7	5×10^{-5}
Chromium(II)			
hydroxide	$Cr(OH)_2$	15.7	2×10^{-16}
Chromium(III)			
arsenate	$CrAsO_4$	20.11	7.7×10^{-21}
fluoride	CrF_3	10.18	6.6×10^{-11}
hydroxide	$Cr(OH)_3$	30.20	6.3×10^{-31}
phosphate 4-water	$CrPO_4 \cdot 4H_2O$ green	22.62	2.4×10^{-23}
	violet	17.00	1.0×10^{-17}
Cobalt			
anthranilate	CoL_2	9.68	2.1×10^{-10}
arsenate	$Co_3(AsO_4)_2$	28.17	6.80×10^{-29}
carbonate	$CoCO_3$	12.84	1.4×10^{-13}
ferrocyanide	$Co_2[Fe(CN)_6]$	14.74	1.8×10^{-15}
hydrogen phosphate	$CoHPO_4$	6.7	2×10^{-7}
(II) hydroxide	$Co(OH)_2$ fresh	14.23	5.92×10^{-15}
(III) hydroxide	$Co(OH)_3$	43.80	1.6×10^{-44}
iodate	$Co(IO_3)_2$	4.0	1.0×10^{-4}
phosphate	$Co_3(PO_4)_2$	34.69	2.05×10^{-35}
selenite	$CoSeO_3$	6.80	1.6×10^{-7}
quinaldate	CoL_2	10.80	1.6×10^{-11}
8-quinolinolate	CoL_2	24.80	1.6×10^{-25}
sulfide	α-CoS	20.40	4.0×10^{-21}
	β-CoS	24.70	2.0×10^{-25}
Copper(I)			
azide	CuN_3	8.31	4.9×10^{-9}
bromide	CuBr	8.20	6.27×10^{-9}
chloride	CuCl	6.76	1.72×10^{-7}
cyanide	CuCN	19.46	3.47×10^{-20}
hydroxide	CuOH	14	1×10^{-14}
iodide	CuI	11.90	1.27×10^{-12}
sulfide	Cu_2S	47.60	2.5×10^{-48}
tetraphenylborate	CuL	8.0	1.0×10^{-8}
thiocyanate	CuSCN	12.75	1.77×10^{-13}
Copper(II)			
anthranilate	CuL_2	13.22	6.0×10^{-14}
arsenate	$Cu_3(AsO_4)_2$	35.10	7.95×10^{-36}
azide	$Cu(N_3)_2$	9.20	6.3×10^{-10}
carbonate	$CuCO_3$	9.86	1.4×10^{-10}
chromate	$CuCrO_4$	5.44	3.6×10^{-6}
dithiooxamide	CuL	15.12	7.67×10^{-16}
ferrocyanide	$Cu_2[Fe(CN)_6]$	15.89	1.3×10^{-16}
hydroxide	$Cu(OH)_2$	19.66	2.2×10^{-20}
iodate	$Cu(IO_3)_2$	7.16	6.94×10^{-8}
oxalate	CuC_2O_4	9.35	4.43×10^{-10}
phosphate	$Cu_3(PO_4)_2$	36.85	1.40×10^{-37}
pyrophosphate	$Cu_2P_2O_7$	15.08	8.3×10^{-16}
quinaldate	CuL_2	16.80	1.6×10^{-17}
8-quinolinolate	CuL_2	29.70	2.0×10^{-30}

TABLE 1.71 Solubility Product Constants (*Continued*)

Compound	Formula	pK_{sp}	K_{sp}
selenite	$CuSeO_3$	7.68	2.1×10^{-8}
sulfide	CuS	35.20	6.3×10^{-36}
Dysprosium			
chromate 10-water	$Dy_2(CrO_4)_3 \cdot 10H_2O$	8	1×10^{-8}
hydroxide	$Dy(OH)_3$	21.85	1.4×10^{-22}
Erbium			
hydroxide	$Er(OH)_3$	23.39	4.1×10^{-24}
Europium			
hydroxide	$Eu(OH)_3$	23.03	9.38×10^{-24}
Gadolinium			
hydrogen carbonate	$Gd(HCO_3)_3$	1.7	2×10^{-2}
hydroxide	$Gd(OH)_3$	22.74	1.8×10^{-23}
Gallium			
ferrocyanide	$Ga_4[Fe(CN)_6]_3$	33.82	1.5×10^{-34}
hydroxide	$Ga(OH)_3$	35.14	7.28×10^{-36}
8-quinolinolate	GaL_3	40.80	1.6×10^{-41}
Germanium			
oxide	GeO_2	57.0	1.0×10^{-57}
Gold(I)			
chloride	AuCl	12.70	2.0×10^{-13}
iodide	AuI	22.80	1.6×10^{-23}
Gold(III)			
chloride	$AuCl_3$	24.50	3.2×10^{-25}
hydroxide	$Au(OH)_3$	45.26	5.5×10^{-46}
iodide	AuI_3	46	1×10^{-46}
oxalate	$Au_2(C_2O_4)_3$	10	1×10^{-10}
Hafnium			
hydroxide	$Hf(OH)_3$	25.40	4.0×10^{-26}
Holmium			
hydroxide	$Ho(OH)_3$	22.3	5.0×10^{-23}
Indium			
ferrocyanide	$In_4[Fe(CN)_6]_3$	43.72	1.9×10^{-44}
hydroxide	$In(OH)_3$	33.2	6.3×10^{-34}
quinolinolate	InL_3	31.34	4.6×10^{-32}
selenite	$In_2(SeO_3)_3$	32.60	4.0×10^{-33}
sulfide	In_2S_3	73.24	5.7×10^{-74}
Iron(II)			
carbonate	$FeCO_3$	10.50	3.13×10^{-11}
fluoride	FeF_2	5.63	2.36×10^{-6}
hydroxide	$Fe(OH)_2$	16.31	4.87×10^{-17}
oxalate dihydrate	$FeC_2O_4 \cdot 2H_2O$	6.50	3.2×10^{-7}
sulfide	FeS	17.20	6.3×10^{-18}
Iron(III)			
arsenate	$FeAsO_4$	20.24	5.7×10^{-21}
ferrocyanide	$Fe_4[Fe(CN)_6]_3$	40.52	3.3×10^{-41}
hydroxide	$Fe(OH)_3$	38.55	2.79×10^{-39}
phosphate dihydrate	$FePO_4 \cdot 2H_2O$	15.00	9.91×10^{-16}
quinaldate	FeL_3	16.89	1.3×10^{-17}
selenite	$Fe_2(SeO_3)_3$	30.70	2.0×10^{-31}
Lanthanum			
bromate 9-water	$La(BrO_3)_3 \cdot 9H_2O$	2.50	3.2×10^{-3}
fluoride	LaF_3	16.2	7×10^{-17}

(*Continued*)

TABLE 1.71 Solubility Product Constants (*Continued*)

Compound	Formula	pK_{sp}	K_{sp}
hydroxide	$La(OH)_3$	18.70	2.0×10^{-19}
iodate	$La(IO_3)_3$	11.12	7.50×10^{-12}
molybdate	$La_2(MoO_4)_3$	20.4	4×10^{-21}
oxalate 9-water	$La_2(C_2O_4)_3$	26.60	2.5×10^{-27}
phosphate	$LaPO_4$	22.43	3.7×10^{-23}
sulfide	La_2S_3	12.70	2.0×10^{-13}
tungstate trihydrate	$La_2(WO_4)_3 \cdot 3H_2O$	3.90	1.3×10^{-4}
Lead			
acetate	$Pb(OAc)_2$	2.75	1.8×10^{-3}
anthranilate	PbL_2	9.81	1.6×10^{-10}
arsenate	$Pb_3(AsO_4)_3$	35.39	4.0×10^{-36}
azide	$Pb(N_3)_2$	8.59	2.5×10^{-9}
borate, *meta*	$Pb(BO_2)_3$	10.78	1.6×10^{-11}
bromate	$Pb(BrO_3)_2$	1.70	2.0×10^{-2}
bromide	$PbBr_2$	6.82	6.60×10^{-6}
carbonate	$PbCO_3$	13.13	7.4×10^{-14}
chloride	$PbCl_2$	4.77	1.70×10^{-5}
chloride fluoride	$PbClF$	8.62	2.4×10^{-9}
chlorite	$Pb(ClO_2)_2$	8.4	4×10^{-9}
chromate	$PbCrO_4$	12.55	2.8×10^{-13}
ferrocyanide	$Pb_2[Fe(CN)_6]$	14.46	3.5×10^{-15}
fluoride	PbF_2	7.48	3.3×10^{-8}
fluoride iodide	$PbFI$	8.07	8.5×10^{-9}
hydrogen phosphate	$PbHPO_4$	9.90	1.3×10^{-10}
hydrogen phosphite	$PbHPO_3$	6.24	5.8×10^{-7}
hydroxide	$Pb(OH)_2$	14.84	1.43×10^{-15}
hydroxide bromide	$PbOHBr$	14.70	2.0×10^{-15}
hydroxide chloride	$PbOHCl$	13.7	2×10^{-14}
hydroxide nitrate	$PbOHNO_3$	3.55	2.8×10^{-4}
iodate	$Pb(IO_3)_2$	12.43	3.69×10^{-13}
iodide	PbI_2	8.01	9.8×10^{-9}
molybdate	$PbMoO_4$	13.00	1.0×10^{-13}
niobate	$Pb(NbO_3)_2$	16.62	2.4×10^{-17}
oxalate	PbC_2O_4	9.32	4.8×10^{-10}
phosphate	$Pb_3(PO_4)_2$	42.10	8.0×10^{-43}
quinaldate	PbL_2	10.60	2.5×10^{-11}
selenate	$PbSeO_4$	6.84	1.37×10^{-7}
selenite	$PbSeO_3$	11.50	3.2×10^{-12}
sulfate	$PbSO_4$	7.60	2.53×10^{-8}
sulfide	PbS	27.10	8.0×10^{-28}
thiocyanate	$Pb(SCN)_2$	4.70	2.0×10^{-5}
thiosulfate	PbS_2O_3	6.40	4.0×10^{-7}
tungstate	$PbWO_4$	6.35	4.5×10^{-7}
Lead(IV)			
hydroxide	$Pb(OH)_4$	65.50	3.2×10^{-66}
Lithium			
carbonate	Li_2CO_3	1.60	2.5×10^{-2}
fluoride	LiF	2.74	1.84×10^{-3}
phosphate	Li_3PO_4	10.63	2.37×10^{-11}
uranylarsenate	$LiUO_2AsO_4$	18.82	1.5×10^{-19}
Lutetium			
hydroxide	$Lu(OH)_3$	23.72	1.9×10^{-24}

TABLE 1.71 Solubility Product Constants (*Continued*)

Compound	Formula	pK_{sp}	K_{sp}
Magnesium			
ammonium phosphate	$MgNH_4PO_4$	12.60	2.5×10^{-13}
arsenate	$Mg_3(AsO_4)_2$	19.68	2.1×10^{-20}
carbonate	$MgCO_3$	5.17	6.82×10^{-6}
carbonate trihydrate	$MgCO_3 \cdot 3H_2O$	5.62	2.38×10^{-6}
fluoride	MgF_2	10.29	5.16×10^{-11}
hydroxide	$Mg(OH)_2$	11.25	5.61×10^{-12}
iodate 4-water	$Mg(IO_3)_2 \cdot 4H_2O$	2.50	3.2×10^{-3}
niobate	$Mg(NbO_3)_2$	16.64	2.3×10^{-17}
oxalate dihydrate	$MgC_2O_4 \cdot 2H_2O$	5.32	4.83×10^{-6}
phosphate	$Mg_3(PO_4)_2$	23.98	1.04×10^{-24}
8-quinolinolate	MgL_2	15.40	4.0×10^{-16}
selenite	$MgSeO_3$	4.89	1.3×10^{-5}
sulfite	$MgSO_3$	2.50	3.2×10^{-3}
Manganese			
anthranilate	MnL_2	6.75	1.8×10^{-3}
arsenate	$Mn_3(AsO_4)_2$	28.72	1.9×10^{-29}
carbonate	$MnCO_3$	10.63	2.34×10^{-11}
ferrocyanide	$Mn_2[Fe(CN)_6]$	12.10	8.0×10^{-13}
iodate	$Mn(IO_3)_2$	6.36	4.37×10^{-7}
hydroxide	$Mn(OH)_2$	12.72	1.9×10^{-13}
oxalate dihydrate	$MnC_2O_4 \cdot 2H_2O$	6.77	1.70×10^{-7}
8-quinolinolate	MnL_2	21.70	2.0×10^{-22}
selenite	$MnSeO_3$	6.90	1.3×10^{-7}
sulfide	MnS amorphous	9.60	2.5×10^{-10}
	MnS crystalline	12.60	2.5×10^{-13}
Mercury(I)			
azide	$Hg_2(N_3)_2$	9.15	7.1×10^{-10}
bromide	Hg_2Br_2	22.19	6.40×10^{-23}
carbonate	Hg_2CO_3	16.44	3.6×10^{-17}
chloride	Hg_2Cl_2	17.84	1.43×10^{-18}
cyanide	$Hg_2(CN)_2$	39.3	5×10^{-40}
chromate	Hg_2CrO_4	8.70	2.0×10^{-9}
ferricyanide	$(Hg_2)_3[Fe(CN)_6]_2$	20.07	8.5×10^{-21}
fluoride	Hg_2F_2	5.51	3.10×10^{-6}
hydrogen phosphate	Hg_2HPO_4	12.40	4.0×10^{-13}
hydroxide	$Hg_2(OH)_2$	23.70	2.0×10^{-24}
iodate	$Hg_2(IO_3)_2$	13.71	2.0×10^{-14}
iodide	Hg_2I_2	28.72	5.2×10^{-29}
oxalate	$Hg_2C_2O_4$	12.76	1.75×10^{-13}
quinaldate	Hg_2L_2	17.90	1.3×10^{-18}
selenite	Hg_2SeO_3	14.20	8.4×10^{-15}
sulfate	Hg_2SO_4	6.19	6.5×10^{-7}
sulfite	Hg_2SO_3	27.0	1.0×10^{-27}
sulfide	Hg_2S	47.0	1.0×10^{-47}
thiocyanate	$Hg_2(SCN)_2$	19.49	3.2×10^{-20}
tungstate	Hg_2WO_4	16.96	1.1×10^{-17}
Mercury(II)			
bromide	$HgBr_2$	19.21	6.2×10^{-20}
hydroxide	$Hg(OH)_2$	25.52	3.2×10^{-26}
iodate	$Hg(IO_3)_2$	12.49	3.2×10^{-13}
iodide	HgI_2	28.54	2.9×10^{-29}
1,10-phenanthroline	HgL_2	24.70	2.0×10^{-25}

(*Continued*)

TABLE 1.71 Solubility Product Constants (*Continued*)

Compound	Formula	pK_{sp}	K_{sp}
quinaldate	HgL_2	16.80	1.6×10^{-17}
selenite	$HgSeO_3$	13.82	1.5×10^{-14}
sulfide	HgS red	52.4	4×10^{-53}
	HgS black	51.80	1.6×10^{-52}
Neodymium			
carbonate	$Nd_2(CO_3)_3$	32.97	1.08×10^{-33}
hydroxide	$Nd(OH)_3$	21.49	3.2×10^{-22}
Neptunyl(VI)			
hydroxide	$NpO_2(OH)_2$	21.60	2.5×10^{-22}
Nickel			
ammine perrhenate	$[Ni(NH_3)_6][ReO_4]_2$	3.29	5.1×10^{-4}
anthranilate	NiL_2	9.09	8.1×10^{-10}
arsenate	$Ni_3(AsO_4)_2$	25.51	3.1×10^{-26}
carbonate	$NiCO_3$	6.85	1.42×10^{-7}
ferrocyanide	$Ni_2[Fe(CN)_6]$	14.89	1.3×10^{-15}
hydrazine sulfate	$[Ni(N_2H_4)_3]SO_4$	13.15	7.1×10^{-15}
hydroxide	$Ni(OH)_2$ fresh	15.26	5.48×10^{-16}
iodate	$Ni(IO_3)_2$	4.33	4.71×10^{-5}
oxalate	NiC_2O_4	9.4	4×10^{-10}
phosphate	$Ni_3(PO_4)_2$	31.32	4.74×10^{-32}
pyrophosphate	$Ni_2P_2O_7$	12.77	1.7×10^{-13}
quinaldate	NiL_2	10.1	8×10^{-11}
8-quinolinolate	NiL_2	26.1	8×10^{-27}
selenite	$NiSeO_3$	5.0	1.0×10^{-5}
α-sulfide	α-NiS	18.50	3.2×10^{-19}
β-sulfide	β-NiS	24.0	1.0×10^{-24}
γ-sulfide	γ-NiS	25.70	2.0×10^{-26}
Palladium			
(II) hydroxide	$Pd(OH)_2$	31.0	1.0×10^{-31}
(IV) hydroxide	$Pd(OH)_4$	70.20	6.3×10^{-71}
quinaldate	PdL_2	12.90	1.3×10^{-13}
thiocyanate	$Pd(SCN)_2$	22.36	4.39×10^{-23}
Platinum			
(IV) bromide	$PtBr_4$	40.50	3.2×10^{-41}
(II) hydroxide	$Pt(OH)_2$	35	1×10^{-35}
Plutonium			
(III) fluoride	PuF_3	15.60	2.5×10^{-16}
(IV) fluoride	PuF_4	19.20	6.3×10^{-20}
(IV) hydrogen phosphate	$Pu(HPO_4)_2 \cdot xH_2O$	27.7	2×10^{-28}
(III) hydroxide	$Pu(OH)_3$	19.70	2.0×10^{-20}
(IV) hydroxide	$Pu(OH)_4$	55	1×10^{-55}
(IV) iodate	$Pu(IO_3)_4$	12.3	5×10^{-13}
(VI) carbonate	PuO_2CO_3	12.77	1.7×10^{-13}
(V) hydroxide	$PuO_2(OH)$	9.3	5×10^{-10}
(VI) hydroxide	$PuO_2(OH)_2$	24.7	2×10^{-25}
Polonium			
sulfide	PoS	28.26	5.6×10^{-29}
Potassium			
hexabromoplatinate	$K_2[PtBr_6]$	4.20	6.3×10^{-5}
hexachloropalladinate	$K_2[PdCl_6]$	5.22	6.0×10^{-6}
hexachloroplatinate	$K_2[PtCl_6]$	5.13	7.48×10^{-6}
hexafluoroplatinate	$K_2[PtF_6]$	4.54	2.9×10^{-5}

TABLE 1.71 Solubility Product Constants (*Continued*)

Compound	Formula	pK_{sp}	K_{sp}
hexafluorosilicate	$K_2[SiF_6]$	6.06	8.7×10^{-7}
hexafluorozirconate	$K_2[ZrF_6]$	3.3	5×10^{-4}
iodate	KIO_4	3.43	3.74×10^{-4}
perchlorate	$KClO_4$	1.98	1.05×10^{-2}
sodium cobaltinitrite hydrate	$K_2Na[Co(NO_2)_6] \cdot H_2O$	10.66	2.2×10^{-11}
tetraphenylborate	$K[B(C_6H_5)_4]$	7.66	2.2×10^{-8}
uranyl arsenate	$K[UO_2AsO_4]$	22.60	2.5×10^{-23}
uranyl carbonate	$K_4[UO_2(CO_3)_3]$	4.20	6.3×10^{-5}
Praseodymium			
hydroxide	$Pr(OH)_3$	23.45	3.39×10^{-24}
Promethium			
hydroxide	$Pm(OH)_3$	21	1×10^{-21}
Radium			
iodate	$Ra(IO_3)_2$	8.94	1.16×10^{-9}
sulfate	$RaSO_4$	10.44	3.66×10^{-11}
Rhodium			
hydroxide	$Rh(OH)_3$	23	1×10^{-23}
Rubidium			
cobaltinitrite	$Rb_3[Co(NO_2)_6]$	14.83	1.5×10^{-15}
hexachloroplatinate	$Rb_2[PtCl_6]$	7.20	6.3×10^{-8}
hexafluoroplatinate	$Rb_2[PtF_6]$	6.12	7.7×10^{-7}
hexafluorosilicate	$Rb_2[SiF_6]$	6.30	5.0×10^{-7}
perchlorate	$RbClO_4$	2.52	3.0×10^{-3}
periodate	$RbIO_4$	3.26	5.5×10^{-4}
Ruthenium			
hydroxide	$Ru(OH)_3$	36	1×10^{-36}
Samarium			
hydroxide	$Sm(OH)_3$	22.08	8.3×10^{-23}
Scandium			
fluoride	ScF_3	23.24	5.81×10^{-24}
hydroxide	$Sc(OH)_3$	30.65	2.22×10^{-31}
Silver			
acetate	AgOAc	2.71	1.94×10^{-3}
arsenate	Ag_3AsO_4	21.99	1.03×10^{-22}
azide	AgN_3	8.54	2.8×10^{-9}
bromate	$AgBrO_3$	4.27	5.38×10^{-5}
bromide	AgBr	12.27	5.35×10^{-13}
carbonate	Ag_2CO_3	11.07	8.46×10^{-12}
chloride	AgCl	9.75	1.77×10^{-10}
chlorite	$AgClO_2$	3.70	2.0×10^{-4}
chromate	Ag_2CrO_4	11.95	1.12×10^{-12}
cobaltinitrite	$Ag_3[Co(NO_2)_6]$	20.07	8.5×10^{-21}
cyanamide	Ag_2CN_2	10.14	7.2×10^{-11}
cyanate	AgOCN	6.64	2.3×10^{-7}
cyanide	AgCN	16.22	5.97×10^{-17}
dichromate	$Ag_2Cr_2O_7$	6.70	2.0×10^{-7}
dicyanimide	$AgN(CN)_2$	8.85	1.4×10^{-9}
ferrocyanide	$Ag_4[Fe(CN)_6]$	40.81	1.6×10^{-41}
hydroxide	AgOH	7.71	2.0×10^{-8}
hyponitrite	$Ag_2N_2O_2$	18.89	1.3×10^{-19}
iodate	$AgIO_3$	7.50	3.17×10^{-8}

(*Continued*)

TABLE 1.71 Solubility Product Constants (*Continued*)

Compound	Formula	pK_{sp}	K_{sp}
iodide	AgI	16.07	8.52×10^{-17}
molybdate	Ag_2MoO_4	11.55	2.8×10^{-12}
nitrite	$AgNO_2$	3.22	6.0×10^{-4}
oxalate	$Ag_2C_2O_4$	11.27	5.40×10^{-12}
phosphate	Ag_3PO_4	16.05	8.89×10^{-17}
quinaldate	AgL	16.89	1.3×10^{-17}
perrhenate	$AgReO_4$	4.10	8.0×10^{-5}
selenate	Ag_2SeO_4	7.25	5.7×10^{-8}
selenite	Ag_2SeO_3	15.00	1.0×10^{-15}
selenocyanate	AgSeCN	15.40	4.0×10^{-16}
sulfate	Ag_2SO_4	4.92	1.20×10^{-5}
sulfite	Ag_2SO_3	13.82	1.50×10^{-14}
sulfide	Ag_2S	49.20	6.3×10^{-50}
thiocyanate	AgSCN	11.99	1.03×10^{-12}
vanadate	$AgVO_3$	6.3	5×10^{-7}
tungstate	Ag_2WO_4	11.26	5.5×10^{-12}
Sodium			
ammonium cobaltinitrite	$Na(NH_4)_2[Co(NO_2)_6]$	10.66	2.2×10^{-11}
antimonate	$Na[Sb(OH)_6]$	7.4	4×10^{-8}
hexafluoroaluminate	$Na_2[AlF_6]$	9.39	4.0×10^{-10}
uranyl arsenate	$NaUO_2AsO_4$	21.87	1.3×10^{-22}
Strontium			
arsenate	$Sr_3(AsO_4)_2$	18.37	4.29×10^{-19}
carbonate	$SrCO_3$	9.25	5.60×10^{-10}
chromate	$SrCrO_4$	4.65	2.2×10^{-5}
fluoride	SrF_2	8.36	4.33×10^{-9}
iodate	$Sr(IO_3)_2$	6.94	1.14×10^{-7}
iodate hydrate	$Sr(IO_3)_2 \cdot H_2O$	6.42	3.77×10^{-7}
molybdate	$SrMoO_4$	6.7	2×10^{-7}
niobate	$Sr(NbO_3)_2$	17.38	4.2×10^{-18}
oxalate hydrate	$SrC_2O_4 \cdot H_2O$	6.80	1.6×10^{-7}
phosphate	$Sr_3(PO_4)_2$	27.39	4.0×10^{-28}
8-quinolinolate	SrL_2	9.3	5×10^{-10}
selenate	$SrSeO_4$	3.09	8.1×10^{-4}
selenite	$SrSeO_3$	5.74	1.8×10^{-6}
sulfate	$SrSO_4$	6.46	3.44×10^{-7}
sulfite	$SrSO_3$	7.4	4×10^{-8}
tungstate	$SrWO_4$	9.77	1.7×10^{-10}
Terbium			
hydroxide	$Tb(OH)_3$	21.70	2.0×10^{-22}
Tellurium			
hydroxide	$Te(OH)_4$	53.52	3.0×10^{-54}
Thallium(I)			
azide	TlN_3	3.66	2.2×10^{-4}
bromate	$TlBrO_3$	4.96	1.10×10^{-5}
bromide	TlBr	5.43	3.71×10^{-6}
chloride	TlCl	3.73	1.86×10^{-4}
chromate	Tl_2CrO_4	12.06	8.67×10^{-13}
ferrocyanide dihydrate	$Tl_4[Fe(CN)_6] \cdot 2H_2O$	9.3	5×10^{-10}
hexachloroplatinate	$Tl_2[PtCl_6]$	11.40	4.0×10^{-12}
iodate	$TlIO_3$	5.51	3.12×10^{-6}
iodide	TlI	7.26	5.54×10^{-8}

TABLE 1.71 Solubility Product Constants (*Continued*)

Compound	Formula	pK_{sp}	K_{sp}
oxalate	$Tl_2C_2O_4$	3.7	2×10^{-4}
selenate	Tl_2SeO_4	4.00	1.0×10^{-4}
selenite	Tl_2SeO_3	38.7	2×10^{-39}
sulfide	Tl_2S	20.30	5.0×10^{-21}
thiocyanate	TlSCN	3.80	1.57×10^{-4}
Thallium(III)			
hydroxide	$Tl(OH)_3$	43.77	1.68×10^{-44}
8-quinolinolate	TlL_3	32.40	4.0×10^{-33}
Thorium			
hydrogen phosphate	$Th(HPO_4)_2$	20	1×10^{-20}
hydroxide	$Th(OH)_4$	44.40	4.0×10^{-45}
iodate	$Th(IO_3)_4$	14.60	2.5×10^{-15}
oxalate	$Th(C_2O_4)_2$	22	1×10^{-22}
phosphate	$Th_3(PO_4)_4$	78.60	2.5×10^{-79}
Thullium			
hydroxide	$Tm(OH)_3$	23.48	3.3×10^{-24}
Tin			
(II) hydroxide	$Sn(OH)_2$	27.26	5.45×10^{-28}
(IV) hydroxide	$Sn(OH)_4$	56	1×10^{-56}
(II) sulfide	SnS	25.00	1.0×10^{-25}
Titanium			
(III) hydroxide	$Ti(OH)_3$	40	1×10^{-40}
(IV) oxide hydroxide	$TiO(OH)_2$	29	1×10^{-29}
Uranium(IV)			
fluoride 2.5-water	$UF_4 \cdot 2.5H_2O$	21.24	5.7×10^{-22}
Uranyl(VI)(2+)			
carbonate	UO_2CO_3	11.73	1.8×10^{-12}
ferrocyanide	$UO_2[Fe(CN)_6]$	13.15	7.1×10^{-14}
hydrogen arsenate	UO_2HAsO_4	10.50	3.2×10^{-11}
hydrogen phosphate	UO_2HPO_4	10.67	2.1×10^{-11}
hydroxide	$UO_2(OH)_2$	21.95	1.1×10^{-22}
iodate hydrate	$UO_2(IO_3)_2 \cdot H_2O$	7.50	3.2×10^{-8}
oxalate trihydrate	$UO_2C_2O_4 \cdot 3H_2O$	3.7	2×10^{-4}
phosphate	$(UO_2)_3(PO_4)_2$	46.7	2×10^{-47}
sulfite	UO_2SO_3	8.58	2.6×10^{-9}
thiocyanate	$(UO_2)(SCN)_2$	3.4	4×10^{-4}
Vanadium			
(IV) hydroxide	$VO(OH)_2$	22.13	5.9×10^{-23}
(III) phosphate	$(VO_2)_3PO_4$	24.1	8×10^{-25}
Ytterbium			
hydroxide	$Yt(OH)_3$	23.60	2.5×10^{-24}
Yttrium			
carbonate	$Y_2(CO_3)_3$	2.99	1.03×10^{-3}
fluoride	YF_3	20.06	8.62×10^{-21}
hydroxide	$Y(OH)_3$	22.00	1.00×10^{-22}
iodate	$Y(IO_3)_3$	9.95	1.12×10^{-10}
oxalate	$Y_2(C_2O_4)_3$	28.28	5.3×10^{-29}
Zinc			
anthranilate	ZnL_2	9.23	5.9×10^{-10}
arsenate	$Zn_3(AsO_4)_2$	27.55	2.8×10^{-28}
borate hydrate	$Zn(BO_2)_2 \cdot H_2O$	10.18	6.6×10^{-11}
carbonate	$ZnCO_3$	9.94	1.46×10^{-10}
ferrocyanide	$Zn_2[Fe(CN)_6]$	15.40	4.0×10^{-15}

(*Continued*)

TABLE 1.71 Solubility Product Constants (*Continued*)

Compound	Formula	pK_{sp}	K_{sp}
fluoride	ZnF_2	1.52	3.04×10^{-2}
hydroxide	$Zn(OH)_2$	16.5	3×10^{-17}
iodate dihydrate	$Zn(IO_3)_2 \cdot 2H_2O$	5.37	4.1×10^{-6}
oxalate dihydrate	$ZnC_2O_4 \cdot 2H_2O$	8.86	1.38×10^{-9}
phosphate	$Zn_3(PO_4)_2$	32.04	9.0×10^{-33}
quinaldate	ZnL_2	13.80	1.6×10^{-14}
8-quinolinolate	ZnL_2	24.30	5.0×10^{-25}
selenide	ZnSe	25.44	3.6×10^{-26}
selenite hydrate	$ZnSeO_3 \cdot H_2O$	6.80	1.57×10^{-7}
sulfide	α-ZnS	23.80	1.6×10^{-24}
	β-ZnS	21.60	2.5×10^{-22}
Zirconium			
oxide hydroxide	$ZrO(OH)_2$	48.20	6.3×10^{-49}
phosphate	$Zr_3(PO_4)_4$	132	1×10^{-132}

TABLE 1.72 Stability Constants of Complex Ions

The stability constant of a complex ion is a measure of its stability with respect to dissociation into its constituent species at a given temperature, e.g. the formation of the tetra-amminecopper(II) ion may be represented by the equation

$$Cu^{2+} + 4NH_3 = [Cu(NH_3)_4]^{2+}$$

and the stability constant is given by

$$K_{stab} = \frac{[Cu(NH_3)_4^{2+}]}{[Cu^{2+}][NH_3]^4}$$

The higher the stability constant the more stable the complex ion. v denotes the stoichiometric number of a molecule, atom or ion, and is positive for a product and negative for a reactant.

Equilibrium	$\frac{K_{stab}}{(mol \cdot dm^{-3})^{\Sigma v}}$	$\log_{10}\left\{\frac{K_{stab}}{(mol \cdot dm^{-3})^{\Sigma v}}\right\}$
$Ag^+ + 2CN^- = [Ag(CH)_2]^-$	$1{\cdot}0 \times 10^{21}$	21·0
$Ag^+ + NH_3 = [Ag(NH_3)]^+$	$2{\cdot}5 \times 10^{3}$	3·4
$[Ag(NH_3)]^+ + NH_3 = [Ag(NH_3)_2]^+$	$6{\cdot}3 \times 10^{3}$	3·8
$Ag^+ + 2NH_3 = [Ag(NH_3)_2]^+$	$1{\cdot}7 \times 10^{7}$	7·2
$Ag^+ + 2S_2O_3^{2-} = [Ag(S_2O_3)_2]^{3-}$	$1{\cdot}0 \times 10^{13}$	13·0
$Al^{3+} + 6F^- = [AlF_6]^{3-}$	6×10^{19}	19·8
$Al(OH)_3 + OH^- = [Al(OH)_4]^-$	40	1·6
$Cd^{2+} + 4CN^- = [Cd(CN)_4]^{2-}$	$7{\cdot}1 \times 10^{16}$	16·9
$Cd^{2+} + 4I^- = [CdI_4]^{2-}$	2×10^{6}	6·3
$Cd^{2+} + 4NH_3 = [Cd(NH_3)_4]^{2+}$	$4{\cdot}0 \times 10^{6}$	6·6
$Co^{2+} + 6NH_3 = [Co(NH_3)_6]^{2+}$	$7{\cdot}7 \times 10^{4}$	4·9
$Co^{3+} + 6NH_3 = [Co(NH_3)_6]^{3+}$	$4{\cdot}5 \times 10^{33}$	33·7
$Cr(OH)_3 + OH^- = [Cr(OH)_4]^-$	1×10^{-2}	−2
$Cu^+ + 4CN^- = [Cu(CN)_4]^{3-}$	$2{\cdot}0 \times 10^{27}$	27·3
$Cu^{2+} + 4Cl^- = [CuCl_4]^{2-}$	$4{\cdot}0 \times 10^{5}$	5·6
$Cu^+ + 2NH_3 = [Cu(NH_3)_2]^+$	1×10^{11}	11

TABLE 1.72 Stability Constants of Complex Ions (*Continued*)

Equilibrium	$\frac{K_{stab}}{(mol \cdot dm^{-3})^{\Sigma\nu}}$	$\log_{10}\left\{\frac{K_{stab}}{(mol \cdot dm^{-3})^{\Sigma\nu}}\right\}$
$Cu^{2+} + NH_3 = [Cu(NH_3)]^{2+}$	$2{\cdot}0 \times 10^4$ (K_1)	4·3
$[Cu(NH_3)]^{2+} + NH_3 = [Cu(NH_3)_2]^{2+}$	$4{\cdot}2 \times 10^3$ (K_2)	3·6
$[Cu(NH_3)_2]^{2+} + NH_3 = [Cu(NH_3)_3]^{2+}$	$1{\cdot}0 \times 10^3$ (K_3)	3·0
$[Cu(NH_3)_3]^{2+} + NH_3 = [Cu(NH_3)_4]^{2+}$	$1{\cdot}7 \times 10^2$ (K_4)	2·2
$Cu^{2+} + 4NH_3 = [Cu(NH_3)_4]^{2+}$	$1{\cdot}4 \times 10^{13}$ ($K = K_1K_2K_3K_4$)	13·1
$Fe^{2+} + 6CN^- = [Fe(CN)_6]^{4-}$	ca. 10^{24}	ca. 24
$Fe^{3+} + 6CN^- = [Fe(CN)_6]^{3-}$	ca. 10^{31}	ca. 31
$Fe^{3+} + 4Cl^- = [FeCl_4]^-$	8×10^{-2}	−1·1
$Fe^{3+} + SCN^- = [Fe(SCN)]^{2+}$	$1{\cdot}4 \times 10^2$	2·1
$[Fe(SCN)]^{2+} + SCN^- = [Fe(SCN)_2]^+$	16	1·2
$[Fe(SCN)_2]^+ + SCN^- = Fe(SCN)_3$	1	0
$Hg^{2+} + 4CN^- = [Hg(CN)_4]^{2-}$	$2{\cdot}5 \times 10^{41}$	41·4
$Hg^{2+} + 4Cl^- = [HgCl_4]^{2-}$	$1{\cdot}7 \times 10^{16}$	16·2
$Hg^{2+} + 4I^- = [HgI_4]^{2-}$	$2{\cdot}0 \times 10^{30}$	30·3
$I^- + I_2 = I_3^-$	$7{\cdot}1 \times 10^2$	2·9
$Ni^{2+} + 6NH_3 = [Ni(NH_3)_6]^{2+}$	$4{\cdot}8 \times 10^7$	7·7
$Pb(OH)_2 + OH^- = [Pb(OH)_3]^-$	50	1·7
$Sn(OH)_4 + 2OH^- = [Sn(OH)_6]^{2-}$	5×10^3	3·7
$Zn^{2+} + 4CN^- = [Zn(CN)_4]^{2-}$	5×10^{16}	16·7
$Zn^{2+} + 4NH_3 = [Zn(NH_3)_4]^{2+}$	$3{\cdot}8 \times 10^9$	9·6
$Zn(OH)_2 + 2OH^- = [Zn(OH)_4]^{2-}$	10	1·0

TABLE 1.73 Saturated Solutions

The following table provides the data for making saturated solutions of the substances listed at the temperature designated. Data are provided for making saturated solutions by weight (g of substance per 100 g of saturated solution) and by volume (g of substance per 100 ml of saturated solution and the ml of water required to make such a solution).

To make one *fluid ounce* of a saturated solution: multiply the grams of substance per 100 ml of saturated solution by 4.55 to obtain the number of grains required, by 0.01039 to obtain the number of avoirdupois ounces, by 0.00947 to obtain the number of apothecaries (Troy) ounces; also multiply the ml of water by 16.23 to obtain the number of minims, or divide by 100 to obtain the number of fluid ounces.

To make one *fluid dram*: multiply the grams of substance per 100 ml of saturated solution by 0.5682 to obtain the number of grains required; also multiply the ml of water by 0.60 to obtain the number of minims required.

Substance	Formula	Temp, °C	g/100 g satd soln	g/100 ml satd soln	ml water/ 100 ml satd soln	Specific gravity
acetanilide	$C_6H_5NHCOCH_3$	25	0.54	0.54	99.2	0.997
p-acetophenetidin	$C_6H_4(OC_2H_5)NHCH_3CO$	25	0.0766	0.0766	99.92	1.00
p-acetotoluide	$CH_3CONHC_6H_4CH_3$	25	0.12	0.12	99.7	0.9979
alanine	$CH_3CH(NH_2)COOH$	25	14.1	14.7	89.5	1.042
aluminum ammonium sulfate	$Al_2(SO_4)_3(NH_4)_2SO_4 \cdot 24H_2O$	25	12.4	13	92	1.05
aluminum chloride hydrated	$AlCl_3 \cdot 6H_2O$	25	55.5	75	60	1.35
aluminum fluoride	$Al_2F_6 \cdot 5H_2O$	20	0.499	0.5015	100.0	1.0051
aluminum potassium sulfate	$AlK(SO_4)_2$	25	6.62	7.02	99.1	1.061
aluminum sulfate	$Al_2(SO_4)_3 \cdot 18H_2O$	25	48.8	63	66	1.29

(*Continued*)

TABLE 1.73 Saturated Solutions (*Continued*)

Substance	Formula	Temp, °C	g/100 g satd soln	g/100 ml satd soln	ml water/ 100 ml satd soln	Specific gravity
o-aminobenzoic acid	$C_6H_4NH_2COOH$	25	0.52	0.519	99.4	0.999
DL-α-amino-*n*-butyric acid	$CH_3CH_2CH(NH_2)COOH$	25	17.8	18.6	86.2	1.046
DL-α-aminoisobutyric acid	$(CH_3)_2C(NH_2)COOH$	25	13.3	13.7	89.5	1.031
ammonium arsenate	$NH_4H_2AsO_4$	20	32.7	40.2	83.0	1.228
ammonium benzoate	$NH_4C_7H_5O_2$	25	18.6	19.4	84.7	1.040
ammonium bromide	NH_4Br	15	41.7	53.8	75.2	1.290
ammonium carbnonate		25	20	22	88	1.10
ammonium chloride	NH_4Cl	15	26.3	28.3	79.3	1.075
ammonium citrate, dibasic	$(NH_4)_2HC_6H_5O_7$	25	48.7	60.5	61.5	1.22
ammonium dichromate	$(NH_4)_2Cr_2O_7$	25	27.9	33	85	1.18
ammonium iodide	NH_4I	25	64.5	106.2	58.3	1.646
ammonium molybdate	$(NH_4)_6Mo_7O_{24} \cdot 4H_2O$	25	30.6	39	88	1.27
ammonium nitrate	NH_4NO_3	25	68.3	90.2	41.8	1.320
ammonium oxalate	$(NH_4)_2C_2O_4 \cdot H_2O$	25	4.95	5.06	97.0	1.019
ammonium perchlorate	NH_4ClO_4	25	21.1	23.7	88.7	1.123
ammonium periodate	NH_4IO_4	16	2.63	2.68	99.2	1.018
ammonium persulfate	$(NH_4)_2S_2O_8$	25	42.7	53	71	1.24
ammonium phosphate, dibasic	$(NH_4)_2 \cdot HPO_4$	14.5	56.2	75.5	58.8	1.343
ammonium phosphate, monobasic	$NH_4H_2PO_4$	25	28.4	33	83	1.16
ammonium salicylate	$NH_4C_7H_5O_3$	25	50.8	58.2	56.4	1.145
ammonium silicofluoride	$(NH_4)_2SiF_6$	17.5	15.7	17.2	92.3	1.095
ammonium sulfate	$(NH_4)_2SO_4$	20	42.6	53.1	71.7	1.248
ammonium sulfite	$(NH_4)_2SO_3.H_2O$	25	39.3	47.3	73.2	1.204
ammonium thiocyanate	NH_4CNS	25	62.2	71	43	1.14
amyl alcohol	$C_5H_{11}OH$	25	2.61	2.60	96.9	0.995
aniline	$C_6H_5NH_2$	22	3.61	3.61	96.2	0.998
aniline hydrochloride	$C_6H_5NH_2 \cdot HCl$	25	49	54	56	1.10
aniline sulfate	$(C_6H_5NH_2)_2 \cdot H_2SO_4$	25	5.88	6	96	1.02
L-asparagine	$NH_2COCH_2CH(NH_2)COOH$	25	2.44	2.46	98.2	1.007
barium bromide	$BaBr_2$	20	51	87.2	83.8	1.710
barium chlorate	$Ba(ClO_3)_2$	25	28.5	36.8	92.6	1.294
barium chloride	$BaCl_2$	20	26.3	33.4	93.8	1.27
barium iodide	$BaI_2 \cdot 7\frac{1}{2}H_2O$	25	68.8	157.0	71.1	2.277
barium nitrate	$Ba(NO_3)_2$	25	9.4	10.2	97.9	1.080
barium nitrite	$Ba(NO_2)_2$	17	40	59.6	89.4	1.490
barium perchlorate	$Ba(ClO_4)_2$	25	75.3	145.8	47.8	1.936
benzamide	$C_6H_5CONH_2$	25	1.33	1.33	98.6	0.999
benzoic acid	$C_7H_6O_2$	25	0.367	0.367	99.63	1.00
beryllium sulfate	$BeSO_4 \cdot 4H_2O$	25	28.7	37.3	93.0	1.301
boric acid	H_3BO_3	25	4.99	5.1	97	1.02
n-butyl alcohol	$CH_3(CH_2)_2CH_2OH$	25	79.7	67.3	17.1	0.845
cadmium bromide	$CdBr_2 \cdot 4H_2O$	25	52.9	94.0	83.9	1.775
cadmium chlorate	$Cd(ClO_3)_2 \cdot 12H_2O$	18	76.4	174.5	54.0	2.284
cadmium chloride	$CdCl_2 \cdot 2\frac{1}{2}H_2O$	25	54.7	97.2	80.8	1.778
cadmium iodide	CdI_2	20	45.9	73.0	86.3	1.590
cadmium sulfate	$3(CdSO_4) \cdot 8H_2O$	25	43.4	70.3	91.8	1.619
calcium bromide	$CaBr_2$	20	58.8	107.2	75.0	1.82

TABLE 1.73 Saturated Solutions (*Continued*)

Substance	Formula	Temp, °C	g/100 g satd soln	g/100 ml satd soln	ml water/ 100 ml satd soln	Specific gravity
calcium chlorate	$Ca(ClO_3)_2 \cdot 2H_2O$	18	64.0	110.7	62.3	1.729
calcium chloride	$CaCl_2 \cdot 6H_2O$	25	46.1	67.8	79.2	1.47
calcium chromate	$CaCrO_4 \cdot 2H_2O$	18	14.3	16.4	98.7	1.149
calcium ferrocyanide	$Ca_2Fe(CN)_6$	25	36.5	49.6	86.2	1.357
calcium iodide	CaI_2	20	67.6	143.8	69.0	2.125
calcium lactate	$Ca(C_3H_5O_3)_2 \cdot 5H_2O$	25	4.95	5	96	1.01
calcium nitrite	$Ca(NO_2)_2 \cdot 4H_2O$	18	45.8	65.7	77.8	1.427
calcium sulfate	$CaSO_4 \cdot 2H_2O$	25	0.208	0.208	99.70	0.999
camphoric acid	$C_8H_{14}(COOH)_2$	25	0.754	0.754	99.246	1.00
carbon disulfide	CS_2	22	0.173	0.173	99.63	0.998
cerium nitrate	$Ce(NO_3)_3 \cdot 6H_2O$	25	63.7	119.9	68.2	1.880
cesium bromide	CsBr	21.4	53.1	89.8	79.5	1.693
cesium chloride	CsCl	25	65.7	126.3	65.9	1.923
cesium iodide	CsI	22.8	48.0	74.1	80.5	1.545
cesium nitrate	$CsNO_3$	25	21.9	26.1	92.9	1.187
cesium perchlorate	$CsClO_4$	25	2.01	2.03	99.0	1.010
cesium periodate	$CsIO_4$	15	2.10	2.13	99.5	1.017
cesium sulfate	Cs_2SO_4	25	64.5	129.8	71.7	2.013
chloral hydrate	$CCl_3CHO \cdot H_2O$	25	79.4	120	31	1.51
chloroform	$CHCl_3$	29.4	0.703	0.705	99.57	1.0028
chromic oxide	CrO_3	18	62.5	106.3	64.0	1.703
chromium potassium sulfate	$Cr_2K_2(SO_4)_4 \cdot 24H_2O$	25	19.6	22	90	1.12
citric acid	$(CH_2)_2COH(COOH)_3 \cdot H_2O$	25	67.5	88.6	42.7	1.311
cobalt chlorate	$Co(ClO_3)_2$	18	64.2	119.3	66.5	1.857
cobalt nitrate	$Co(NO_3)_2$	18	49.7	78.2	79.1	1.572
cobalt perchlorate	$Co(ClO_4)_2$	26	71.8	113.5	44.7	1.581
cupric ammonium chloride	$CuCl_2 \cdot 2NH_4Cl \cdot 2H_2O$	25	30.3	35.5	82	1.17
cupric ammonium sulfate	$CuSO_4 \cdot (NH_4)_2SO_4$	19	15.3	17.3	96.0	1.131
cupric bromide	$CuBr_2$	25	55.8	102.5	81.2	1.84
cupric chlorate	$Cu(ClO_3)_2$	18	62.2	105.2	64.1	1.692
cupric chloride	$CuCl_2 \cdot 2H_2O$	25	53.3	80	70	1.50
cupric nitrate	$Cu(NO_3)_2 \cdot 6H_2O$	20	56.0	94.5	74.3	1.688
cupric selenate	$CuSeO_4$	21.2	14.7	17.2	99.4	1.165
cupric sulfate	$CuSO_4 \cdot 5H_2O$	25	18.5	22.3	98.7	1.211
dextrose	$C_6H_{12}O_6 \cdot H_2O$	25	49.5	59	60	1.19
ether	$(C_2H_5)_2O$	22	5.45	5.34	93.0	0.985
ethyl acetate	$CH_3COOC_2H_5$	25	7.47	7.44	92.1	0.996
ferric ammonium citrate		25	67.7	97	46	1.43
ferric ammonium oxalate	$Fe(NH_4)_3(C_2O_4)_3 \cdot 3H_2O$	25	51.5	65	61	1.26
ferric ammonium sulfate	$FeSO_4 \cdot (NH_4)_2SO_4$	16.5	19.1	22.4	94.3	1.165
ferric chloride	$FeCl_3$	25	73.1	131.1	48.3	1.793
ferric nitrate	$Fe(NO_3)_3$	25	46.8	70.2	79.8	1.50
ferric perchlorate	$Fe(ClO_4)_3 \cdot 10H_2O$	25	79.9	132.1	33.2	1.656
ferrous sulfate	$FeSO_4 \cdot 7H_2O$	25	42.1	52.8	72.7	1.255
gallic acid	$C_6H_2(OH)_3COOH \cdot H_2O$	25	1.15	1.15	99.05	1.002
D-glutamic acid	$C_5H_9O_4N$	25	0.86	0.86	99.15	1.0002
glycine	NH_2CH_2COOH	25	20.0	21.7	86.8	1.083
hydroquinone	$C_6H_4(OH)_2$	20	6.7	6.78	94.4	1.012
m-hydroxybenzoic acid	$C_6H_4OHCOOH$	25	0.975	0.975	99.03	1.000

(*Continued*)

TABLE 1.73 Saturated Solutions (*Continued*)

Substance	Formula	Temp, °C	g/100 g satd soln	g/100 ml satd soln	ml water/ 100 ml satd soln	Specific gravity
lactose	$C_{12}H_{22}O_{11} \cdot H_2O$	25	15.9	17	90	1.07
lead acetate	$Pb(C_2H_3O_2)_2$	25	36.5	49.0	85.1	1.340
lead bromide	$PbBr_2$	25	0.97	0.98	99.6	1.006
lead chlorate	$Pb(ClO_3)_2$	18	60.2	117.0	77.3	1.944
lead chloride	$PbCl_2$	25	1.07	1.08	99.6	1.007
lead iodide	PbI_2	25	0.08	0.08	99.7	0.998
lead nitrate	$Pb(NO_3)_2$	25	37.1	53.6	91.0	1.445
DL-leucine	$C_6H_{13}O_2N$	25	0.976	0.975	98.9	0.999
L-leucine	$C_6H_{13}O_2N$	25	2.24	2.24	97.85	1.0012
lithium benzoate	$LiC_7H_5O_2$	25	27.7	30.4	79.6	1.100
lithium bromate	$LiBrO_3$	18	60.4	110.5	72.5	1.830
lithium carbonate	Li_2CO_3	15	1.36	1.38	100.0	1.014
lithium chloride	$LiCl \cdot H_2O$	25	45.9	59.5	70.2	1.296
lithium citrate	$Li_3C_6H_5O_7$	25	31.8	38.6	82.8	1.213
lithium dichromate	$Li_2Cr_2O_7 \cdot H_2O$	18	52.6	82.9	74.8	1.574
lithium fluoride	LiF	18	0.27	0.27	99.9	1.002
lithium formate	$LiCHO_2$	18	27.9	31.8	80.4	1.140
lithium iodate	$LiIO_3$	18	44.6	69.9	86.8	1.566
lithium nitrate	$LiNO_3$	19	48.9	64.5	67.5	1.318
lithium perchlorate	$LiClO_4 \cdot 3H_2O$	25	37.5	47.6	79.5	1.269
lithium salicylate	$LiC_7H_5O_3$	25	52.7	63.6	57.1	1.206
lithium sulfate	$Li_2SO_4 \cdot H_2O$	25	27.2	33	88.5	1.21
magnesium bromide	$MgBr_2 \cdot 6H_2O$	18	50.1	83.1	82.8	1.655
magnesium chlorate	$Mg(ClO_3)_2$	18	56.3	90.0	69.7	1.594
magnesium chloride	$MgCl_2 \cdot 6H_2O$	25	62.5	79	47.5	1.26
magnesium chromate	$MgCr_2O_4 \cdot 7H_2O$	18	42.0	59.7	82.5	1.422
magnesium dichromate	$MgCrO_7 \cdot 5H_2O$	25	81.0	138.8	32.6	1.712
magnesium iodate	$Mg(IO_3)_2 \cdot 4H_2O$	18	6.44	6.95	100.8	1.078
magnesium iodide	$MgI_2.8H_2O$	18	59.7	114.0	77.1	1.909
magnesium molybdate	$MgMoO_4$	25	15.9	18.4	97.4	1.159
magnesium nitrate	$Mg(NO_3)_2 \cdot 6H_2O$	25	42.1	58.6	80.5	1.388
magnesium perchlorate	$Mg(ClO_4)_2 \cdot 6H_2O$	25	49.9	73.6	73.9	1.472
magnesium selenate	$MgSeO_4$	20	35.3	50.8	93.0	1.440
magnesium sulfate	$MgSO_4 \cdot 7H_2O$	25	55.3	72	58.5	1.30
manganese chloride	$MnCl_2$	25	43.6	63.2	82.0	1.449
manganese nitrate	$Mn(NO_3)_2 \cdot 6H_2O$	18	57.3	93.2	69.2	1.624
manganese silicofluoride	$MnSiF_6$	17.5	37.7	54.5	90.1	1.446
manganese sulfate	$MnSO_4$	25	39.4	59.1	90.8	1.499
mercuric acetate	$Hg(C_2H_3O_2)_2$	25	30.2	38	88	1.26
mercuric bromide	$HgBr_2$	25	0.609	0.610	99.6	1.0023
mercury bichloride	$HgCl_2$	25	6.6	6.96	98.5	1.054
methylene blue	$C_{16}H_{18}N_3ClS \cdot 3H_2O$	25	4.25	4.3	97	1.01
methyl salicylate	$C_6H_4OHCOOCH_3$	25	0.12	0.12	99.88	1.00
monochloracetic acid	$CH_2ClCOOH$	25	78.8	105	28	1.33
β-naphthalenesulfonic acid	$C_{10}H_7SO_3H$	30	56.9	67.9	51.4	1.193
nickel ammonium sulfate	$NiSO_4(NH_4)_2SO_4 \cdot 6H_2O$	25	9.0	9.5	96	1.05
nickel chlorate	$Ni(ClO_3)_2$	18	56.7	94.2	72.0	1.658
nickel chlorate	$Ni(ClO_3)_2 \cdot 6H_2O$	18	64.5	107.2	59.1	1.661
nickel nitrate	$Ni(NO_3)_2 \cdot 6H_2O$	25	77	122	36	1.58

TABLE 1.73 Saturated Solutions (*Continued*)

Substance	Formula	Temp, °C	g/100 g satd soln	g/100 ml satd soln	ml water/ 100 ml satd soln	Specific gravity
nickel perchlorate	$Ni(ClO_4)_2$	26	70.8	112.2	46.4	1.584
nickel perchlorate	$Ni(ClO_4)_2 \cdot 9H_2O$	18	52.4	82.7	75.1	1.576
nickel sulfate	$NiSO_4 \cdot 6H_2O$	25	47.3	64	71	1.35
DL-norleucine	$C_6H_{13}NO_2$	25	1.13	1.13	98.97	0.999
oxalic acid	$H_2C_2O_4 \cdot 2H_2O$	25	9.81	10.3	94.2	1.044
phenol	C_6H_5OH	20	6.1	6.14	94.5	1.0057
β-phenylalanine	$C_6H_5CH_2CH(NH_2)COOH$	25	2.88	2.89	97.5	1.0035
m-phenylenediamine	$C_6H_8N_2$	20	23.1	23.8	79.3	1.032
p-phenylenediamine	$C_6H_8N_2$	20	3.69	3.70	96.67	1.0038
phenyl salicylate	$C_6H_4OHCOOC_6H_5$	25	0.015	0.015	99.84	0.999
phenyl thiourea	$CS(NH_2)NHC_6H_5$	25	0.24	0.24	99.6	0.998
phosphomolybdic acid	$20MoO_3 \cdot 2H_3PO_4 \cdot 48H_2O$	25	74.3	135	46	1.81
phosphotungstic acid	Approx. $20WO_3 \cdot 2H_3PO_4 \cdot 25H_2O$	25	71.4	160	64	2.24
potassium acetate	$KC_2H_3O_2$	25	68.7	97.1	44.3	1.413
potassium antimony tartrate	$KSbOC_4H_4O_6$	25	7.64	8.02	96.9	1.049
potassium bicarbonate	$KHCO_3$	25	26.6	31.6	87.5	1.188
potassium bitartrate	$KC_4H_5O_6$	25	0.65	0.65	99.3	0.999
potassium bromate	$KBrO_3$	25	7.53	7.89	97.5	1.054
potassium bromide	KBr	25	40.6	56.0	82.0	1.380
potassium carbonate	$K_2CO_3 \cdot 1\frac{1}{2}H_2O$	25	52.9	82.2	73.5	1.559
potassium chlorate	$KClO_3$	25	8.0	8.41	96.6	1.051
potassium chloride	KCl	25	26.5	31.2	86.8	1.178
potassium chromate	K_2CrO_4	25	39.4	54.1	83.7	1.381
potassium citrate	$K_3C_6H_5O_7$	25	60.91	92.1	59.2	1.514
potassium dichromate	$K_2Cr_2O_7$	25	13.0	14.2	95.0	1.092
potassium ferricyanide	$K_3Fe(CN)_6$	22	32.1	38.1	80.8	1.187
potassium ferrocyanide	$K_4Fe(CN)_6$	25	24.0	28.2	89.2	1.173
potassium fluoride	$KF \cdot 2H_2O$	18	48.0	72.0	78.0	1.500
potassium formate	$KCHO_2$	18	76.8	120.6	36.4	1.571
potassium hydroxide	KOH	15	51.7	79.2	74.2	1.536
potassium iodate	KIO_3	25	8.40	8.99	98.0	1.071
potassium iodide	KI	25	59.8	103.2	69.1	1.721
potassium meta-antimonate	$KSbO_3$	18	2.73	2.81	99.7	1.025
potassium nitrate	KNO_3	25	28.0	33.4	86.0	1.193
potassium nitrite	KNO_2	20	74.3	121.5	42.3	1.649
potassium oxalate	$K_2C_2O_4 \cdot H_2O$	25	28.3	34	86	1.20
potassium perchlorate	$KClO_4$	25	2.68	2.72	99.0	1.014
potassium periodate	KIO_4	13	0.658	0.661	99.83	1.005
potassium permanganate	$KMnO_4$	25	7.10	7.43	97.3	1.046
potassium sodium tartrate	$KNaC_4H_4O_6 \cdot 4H_2O$	25	39.71	51.9	78.8	1.308
potassium stannate	K_2SnO_3	15.5	42.7	69.2	92.9	1.620
potassium sulfate	K_2SO_4	25	10.83	11.8	96.9	1.086
quinine salicylate	$C_{20}H_{24}N_2O_2 \cdot C_6H_4(OH)COOH.2H_2O$	25	0.065	0.065	99.84	0.999
resorcinol	$C_6H_4(OH)_2$	25	58.8	67.2	47.2	1.142
rubidium bromate	$RbBrO_3$	16	2.15	2.18	99.4	1.016
rubidium bromide	RbBr	25	52.7	85.6	76.9	1.625
rubidium chloride	RbCl	25	48.6	72.8	77.1	1.050
rubidium iodate	$RbIO_3$	15.6	2.72	2.78	99.5	1.022
rubidium iodide	RbI	24.3	63.6	117.7	67.3	1.850
rubidium nitrate	$RbNO_3$	25	40.1	55.0	82.4	1.375

(*Continued*)

TABLE 1.73 Saturated Solutions (*Continued*)

Substance	Formula	Temp, °C	g/100 g satd soln	g/100 ml satd soln	ml water/ 100 ml satd soln	Specific gravity
rubidium perchlorate	$RbClO_4$	25	1.88	1.90	99.3	1.012
rubidium periodate	$RbIO_4$	16	0.645	0.648	99.85	1.0052
rubidium sulfate	Rb_2SO_4	25	33.8	45.6	89.7	1.354
silicotungstic acid	$H_4SiW_{12}O_{40}$	18	90.6	258	26.8	2.843
silver acetate	$Ag(C_2H_3O_2)$	25	1.10	1.11	99.40	1.0047
silver bromate	$AgBrO_3$	25	0.204	0.2037	99.65	0.9985
silver fluoride	$AgF \cdot 2H_2O$	15.8	64.5	168.4	92.7	2.61
silver nitrate	$AgNO_3$	25	71.5	164	65.5	2.29
silver perchlorate	$AgClO_4 \cdot H_2O$	25	84.5	237.1	43.5	2.806
sodium acetate	$NaC_2H_3O_2$	25	33.6	40.5	80.0	1.205
sodium ammonium sulfate	$NaNH_4SO_4$	15	25.2	29.6	87.9	1.174
sodium arsenate	$Na_3AsO_4 \cdot 12H_2O$	17	21.1	23.5	88.0	1.119
sodium benzenesulfonate	$NaC_6H_5SO_3$	25	16.4	17.6	90.1	1.076
sodium benzoate	$NaC_7H_5O_2$	25	36.0	41.5	73.9	1.152
sodium bicarbonate	$NaHCO_3$	15	8.28	8.80	97.6	1.061
sodium bisulfate	$NaHSO_4 \cdot H_2O$	25	59	87	60	1.47
sodium bromide	$NaBr \cdot 2H_2O$	25	48.6	75.0	79.4	1.542
sodium carbonate	$Na_2CO_3 \cdot 10H_2O$	25	22.6	28.1	96.5	1.242
sodium chlorate	$NaClO_3$	25	51.7	74.3	69.6	1.440
sodium chloride	$NaCl$	25	26.5	31.7	88.1	1.198
sodium chromate	Na_2CrO_4	18	40.1	57.4	85.7	1.430
sodium citrate	$Na_3C_6H_5O_7 \cdot 5H_2O$	25	48.1	61.2	66.0	1.272
sodium dichromate	$Na_2Cr_2O_7$	18	63.9	111.4	63.0	1.743
sodium ferrocyanide	$Na_4Fe(CN)_6$	25	17.1	19.4	93.9	1.131
sodium fluoride	NaF	25	3.98	4.14	99.7	1.038
sodium formate	$NaCHO_2$	18	44.7	58.9	73.0	1.316
sodium hydroxide	$NaOH$	25	50.8	77	74	1.51
sodium hypophosphite	NaH_2PO_2	16	52.1	72.4	66.6	1.386
sodium iodate	$NaIO_3 \cdot H_2O$	25	8.57	9.21	98.5	1.075
sodium iodide	NaI	25	64.8	124.3	67.7	1.919
sodium molybdate	Na_2MoO_4	18	39.4	56.6	87.0	1.435
sodium nitrate	$NaNO_3$	25	47.9	66.7	72.5	1.391
sodium nitrite	$NaNO_2$	20	45.8	62.3	73.8	1.359
sodium oxalate	$Na_2(CO_2)_2$	25	3.48	3.58	99.1	1.025
sodium paratungstate	$(Na_2O)_3(WO_3)_7 \cdot 16H_2O$	0	26.7	35.2	96.5	1.316
sodium perchlorate	$NaClO_4$	25	67.8	114.1	54.1	1.683
sodium periodate	$NaIO_4 \cdot 3H_2O$	25	12.6	13.9	96.2	1.103
sodium phenolsulfonate	$C_6H_4(OH)SO_3Na$	25	16.1	17.4	90.5	1.079
sodium phosphate dibasic	Na_2HPO_4	17	4.2	4.4	99.9	1.043
sodium phosphate tribasic	Na_3PO_4	14	9.5	10.5	99.8	1.103
sodium pyrophosphate	$Na_2H_2P_2O_7 \cdot 6H_2O$	25	13.0	14.4	95.8	1.104
sodium salicylate	$NaC_7H_5O_3$	25	53.6	67.0	58.0	1.248
sodium selenate	Na_2SeO_4	18	29.0	38.1	93.4	1.313
sodium silicofluoride	$NaSiF_6$	20	0.773	0.737	99.76	1.0054
sodium sulfate	Na_2SO_4	25	21.8	26.4	94.5	1.208
sodium sulfate	$Na_2SO_4 \cdot 10H_2O$	25	27.7	33.3	87.0	1.207
sodium sulfide	$Na_2S \cdot 9H_2O$	25	52.3	63	57	1.20
sodium sulfite, anhydrous	Na_2SO_3	25	23	28.5	95.5	1.24
sodium thiocyanate	$NaCNS$	25	62.9	87	51	1.38

TABLE 1.73 Saturated Solutions (*Continued*)

Substance	Formula	Temp, °C	g/100 g satd soln	g/100 ml satd soln	ml water/ 100 ml satd soln	Specific gravity
sodium thiosulfate	$Na_2S_2O_3 \cdot 5H_2O$	25	66.8	93	46	1.39
sodium tungstate	$Na_2WO_4 \cdot 10H_2O$	18	42.0	66.1	91.3	1.573
stannous chloride	$SnCl_2$	15	72.9	133.1	49.5	1.827
strontium chlorate	$Sr(ClO_3)_2$	18	63.6	117.0	67.0	1.839
strontium chloride	$SrCl_2 \cdot 6H_2O$	15	33.4	45.5	90.7	1.36
strontium iodide	$SrI_2 \cdot 6H_2O$	20	64.0	137.8	77.5	2.15
strontium nitrate	$Sr(NO_3)_2$	25	44.2	65.3	82.5	1.477
strontium nitrite	$Sr(NO_2)_2$	19	39.3	56.8	87.8	1.445
strontium perchlorate	$Sr(ClO_4)_2$	25	75.6	158.5	50.8	2.084
strontium salicylate	$Sr(C_7H_5O_3)_2$	25	4.58	4.68	97.5	1.019
succinic acid	$(CH_2)_2(COOH)_2$	25	7.67	7.82	94.5	1.021
succinimide	$(CH_2CO)_2NH \cdot H_2O$	25	30.6	32.7	74.2	1.067
sucrose	$C_{12}H_{22}O_{11}$	25	67.89	90.9	43.0	1.340
tartaric acid	$C_2H_2(OH)_2(COOH)_2$	15	58.5	76.9	54.7	1.31
tetraethyl ammonium iodide	$N(C_2H_5)_4I$	25	32.9	36.2	74.0	1.102
tetramethyl ammonium iodide	$N(CH_3)_4I$	25	5.51	5.60	96.1	1.016
thallium chloride	TlCl	25	0.40	0.40	99.6	1.0005
thallium nitrate	$TlNO_3$	25	10.4	11.4	98.0	1.093
thallium nitrite	$TlNO_2$	25	32.1	43.7	92.5	1.360
thallium perchlorate	$TlClO_4$	25	13.5	15.2	97.1	1.122
thallium sulfate	Tl_2SO_4	25	5.48	5.74	99.0	1.047
trichloroacetic acid	CCl_3COOH	25	92.3	149.6	12.41	1.615
uranyl chloride	UO_2Cl_2	18	76.2	208.5	65.2	2.736
uranyl nitrate	$UO_2(NO_3)_2 \cdot 6H_2O$	25	68.9	120	54.5	1.74
urea	$(NH_2)_2CO$	25	53.8	62	53.5	1.15
urea phosphate	$CO(NH_2)_2 \cdot H_3PO_4$	24.5	52.4	66.1	60.1	1.26
urethan	$NH_2CO_2C_2H_5$	25	82.8	88.8	18.5	1.073
D-valine	$(CH_3)_2CHCH(NH_2)COOH$	25	8.14	8.26	93.3	1.015
DL-valine	$(CH_3)_2CHCH(NH_2)COOH$	25	6.61	6.68	94.5	1.012
zinc acetate	$Zn(C_2H_3O_2)_2$	25	25.7	30.0	86.5	1.165
zinc benzenesulfonate	$Zn(C_6H_5SO_3)_2$	25	29.5	34.9	83.4	1.182
zinc chlorate	$Zn(ClO_3)_2$	18	65.0	124.4	67.0	1.914
zinc chloride	$ZnCl_2$	25	67.5	128	61	1.89
zinc iodide	ZnI_2	18	81.2	221.3	51.2	2.725
zinc phenolsulfonate	$(C_6H_5OSO_3)_2Zn \cdot 8H_2O$	25	39.8	47.3	71.5	1.185
zinc selenate	$ZnSeO_4$	22	37.8	58.9	97.0	1.559
zinc silicofluoride	$ZnSiF_6 \cdot 6H_2O$	20	32.9	47.2	96.3	1.434
zinc sulfate	$ZnSO_4 \cdot 7H_2O$	25	36.7	54.6	94.7	1.492
zinc valerate	$Zn(C_5H_9O_2)_2$	25	1.27	1.27	98.8	1.001

1.19 PROTON TRANSFER REACTIONS

A proton transfer reaction is a reaction in which the main feature is the intermolecular or intramolecular transfer of a proton from one binding site to another.

In the detailed description of proton transfer reactions, especially of rapid proton transfers between electronegative atoms, it should always be specified whether the term is used to refer to the overall process, including the more-or-less *encounter-controlled* formation of a hydrogen bonded complex and the separation of the products or, alternatively, the proton transfer event (including solvent rearrangement) by itself.

For the general proton transfer reaction:

$$HB = H^+ + B$$

the acidic dissociation constant is formulated as follows:

$$K_a = \frac{[H^+][B]}{[HB]}$$

The most common charge types for the acid HB and its conjugate base B are

$$CH_3COOH = H^+ + CH_3COO\text{–(acetic acid, acetate ion)}$$
$$HSO_4^- = H^+ + SO_4^{2-} \text{ (hydrogen sulfate ion, sulfate ion)}$$
$$NH_4^+ = H^+ + NH_3 \text{ (ammonium ion, ammonia)}$$

Acids which have more than one acidic hydrogen ionize in steps, as shown for phosphoric acid:

$H_3PO_4 = H^+ + H_2PO_4^-$	$pK_1 = 2.148$	$K_1 = 7.11 \times 10^{-3}$
$H_2PO_4^- = H^+ + HPO_4^{2-}$	$pK_2 = 7.198$	$K_2 = 6.34 \times 10^{-8}$
$HPO_4^{2-} = H^+ + PO_4^{3-}$	$pK_3 = 11.90$	$K_3 = 1.26 \times 10^{-12}$

If the basic dissociation constant K_b for the equilibrium such as

$$NH_3 + H_2O = NH_4 + OH$$

is required, pK_b may be calculated from the relationship

$$pK_b = pK_w - pK_a$$

I_a general, for an organic acid, a useful estimate of its pK_a value can sometimes be obtained by making a comparison with recognizably similar compounds for which pK_a values are known: (1) alkyl chains, alicyclic rings, or saturated carbocyclic rings fused to aromatic or heterocyclic rings can be replaced by methyl or ethyl groups; (2) acid-strengthening inductive and mesomeric effects of a nitro group attached to an aromatic ring are very similar to those of a nitrogen atom located at the same position in a heteroaromatic ring (e.g., 3-hydroxypyridine and 3-nitrophenol).

1.19.1 Calculation of the Approximate pH Value of Solutions

Strong acid:	$pH = -\log [\text{acid}]$
Strong base:	$pH = 14.00 + \log [\text{base}]$
Weak acid:	$pH = 1/2pK_a - 1/2 \log [\text{acid}]$
Weak base:	$pH = 14.00 - 1/2pK_b + 1/2 \log [\text{base}]$

Salt formed by a weak acid and a strong base:

$$\mathrm{pH} = 7.00 + 1/2\mathrm{p}K_a + 1/2\ \log[\mathrm{salt}]$$

Acid salts of a dibasic acid:

$$\mathrm{pH} = 1/2\mathrm{p}K_1 + 1/2\mathrm{p}K_2 - 1/2\ \log\ [\mathrm{salt}] + 1/2\ \log(K_1 + [\mathrm{salt}])$$

Buffer solution consisting of a mixture of a weak acid and its salt:

$$\mathrm{pH} = \mathrm{p}K_a + \log\left(\frac{[\mathrm{salt}]+[\mathrm{H_3O^+}]-[\mathrm{OH^-}]}{[\mathrm{acid}]+[\mathrm{H_3O^+}]-[\mathrm{OH^-}]}\right)$$

1.19.2 Calculation of Concentrations of Species Present at a Given pH

$$\alpha_0 = \frac{[\mathrm{H^+}]^n}{[\mathrm{H^+}]^n + K_1[\mathrm{H^+}]^{n-1} + K_1K_2[\mathrm{H^+}]^{n-2} + \cdots + K_1K_2\cdots K_n} = \frac{[\mathrm{H}_n\mathrm{A}]}{C_{\mathrm{acid}}}$$

$$\alpha_1 = \frac{K_1[\mathrm{H^+}]^{n-1}}{[\mathrm{H^+}]^n + K_1[\mathrm{H^+}]^{n-1} + K_1K_2[\mathrm{H^+}]^{n-2} + \cdots + K_1K_2\cdots K_n} = \frac{[\mathrm{H}_{n-1}\mathrm{A^-}]}{C_{\mathrm{acid}}}$$

$$\alpha_2 = \frac{K_1K_2[\mathrm{H^+}]^{n-2}}{[\mathrm{H^+}]^n + K_1[\mathrm{H^+}]^{n-1} + K_1K_2[\mathrm{H^+}]^{n-2} + \cdots + K_1K_2\cdots K_n} = \frac{[\mathrm{H}_{n-2}\mathrm{A^{2-}}]}{C_{\mathrm{acid}}}$$

$$\vdots$$

$$\alpha_n = \frac{K_1K_2\cdots K_n}{[\mathrm{H^+}]^n + K_1[\mathrm{H^+}]^{n-1} + K_1K_2[\mathrm{H^+}]^{n-2} + \cdots + K_1K_2\cdots K_n} = \frac{[\mathrm{A}^{n-}]}{C_{\mathrm{acid}}}$$

TABLE 1.74 Proton Transfer Reactions of Inorganic Materials in Water at 25°C

Substance	Formula or remarks	pK_1	pK_2
Aluminic acid	H_3AlO_3	11.2	
Aluminum ion (aquo)	Al^{3+} (aquo)	4.98(4)	
Americium(III) ion	Am^{3+} (aquo) $\mu = 0.1$	5.92	
Ammonium ion	NH_4^+	9.246(2)	
Ammonium-d_3	ND_3H^+	9.757	
Antimonic acid	$HSb(OH)_6 = Sb(OH)_6^- + H^+$ $\mu = 0.5$	2.55	
Antimony(III) ion	$SbO^+ + H_2O = Sb(OH)_3 + H^+$ $\mu = 1.0$	1.42	
Barium ion	pK_b of $Ba(OH)^+$ $\mu = 0.1$	0.64	
Berkelium(III) ion	pK for hydrolysis of Bk^{3+} $\mu = 0.1$	5.66	
Beryllium(II) ion	Be^{2+} (aquo) $= BeOH^+ + H^+$ $\mu = 1.0$	6.5	
Bismuth(III) ion	$Bi^{3+} = BiOH^{2+} + H^+$ $\mu = 3.0$	1.58	
Boric acid, tetra-	$H_2B_4O_7$	4	9
Bromine	$Br_2 + H_2O = HBrO + H^+ + Br^-$	7.92	
Cadmium ion	Cd^{2+} (aquo) hydrolysis	9.2(1)	
Calcium ion	Ca^{2+} (aquo) hydrolysis	12.67(3)	
Californium(III) ion	Cf^{3+} (aquo) hydrolysis $\mu = 0.1$	5.62	
Carbon dioxide	CO_2 (aquo)	6.352(1)	10.329
	CO_2 in D_2O	6.77	10.93
Cerium(III) ion	Ce^{3+} (aquo) hydrolysis	*ca.* 9.3	
Cerium(IV) ion	Hydrolysis to $Ce(OH)^{3+}$ and $Ce(OH)_2^{2+}$	−1.15	0.82
Chromium(III) ion	Cr^{3+} (aquo) hydrolysis	3.95	
Cobalt(II) ion	Co^{2+} (aquo) hydrolysis	8.9	
Cobalt(III) ion	Co^{3+} (aquo) hydrolysis $m = 1$	1.75	
Copper(II) ion	Cu^{2+} (aquo) hydrolysis	7.34	
Curium(III) ion	Cm^{3+} (aquo) hydrolysis $m = 0.1$	6.00(5)	
Deuterium oxide	D_2O (molal scale)	14.956(1)	
Dysprosium(III) ion	Dy^{3+} (aquo) hydrolysis	8.10	
Erbium(III) ion	Er^{3+} (aquo) hydrolysis $\mu = 3$	9.0	
Europium(III) ion	Eu^{3+} (aquo) hydrolysis	8.03	
Fermium(III) ion	Fm^{3+} hydrolysis $\mu = 0.1$	3.8	
Gadolinium(III) ion	Gd^{3+} hydrolysis	8.27	
Gallium(III) ion	Ga^{3+} (successive values for hydrolysis)	2.92	3.77
		pK_3 4.75	
Gold(III) hydroxide	H_3AuO_3	<11.7	13.36
Hafnium(IV) ion	Hf^{4+} hydrolysis $\mu = 1$	−0.12	0.23
Hexaminotriphosphazene	$N_3P_3(NH_2)_6$	<3.2	7.68(3)
Holmium(III) ion	Ho^{3+} hydrolysis $\mu = 0.3$	8.04	

Hydrazinium(2+) ion	$^{+}H_3N—NH_3^{+}$	0.27	7.94(3)
Hydrogen amidodisulfonate	$HNSO(OH)_2$	pK_3 8.50	
Hydrogen amidophosphate	$H_2NPO(OH)_2$ (26°C)	2.739	8.102
Hydrogen arsenate	H_3AsO_4	2.223	6.760
Hydrogen-d_3 arsenate	D_3AsO_4	2.596	
Hydrogen arsenite	$HAsO_2$	9.28(10)	
Hydrogen azide	HN_3	4.62	
Hydrogen-*d* azide	DN_3 (in D_2O)	5.115	
Hydrogen borate (3−)	H_3BO_3	9.236	
Hydrogen bromate	$HBrO_3$ (in formamide)	1.02	
Hydrogen bromide	HBr	−8.72(15)	
Hydrogen chlorate	$HClO_3$ (theoretical prediction)	−2.7	
Hydrogen chloride	HCl	−6.2(1)	
Hydrogen-*d* chloride	DCl (in dimethylformamide)	3.58	
Hydrogen chlorite	$HClO_2$	1.94	
Hydrogen chromate	H_2CrO_4	0.74	6.488
Hydrogen cyanate	HOCN	3.46	
Hydrogen cyanide	HCN	9.21	
Hydrogen-*d* cyanide	DCN (in D_2O) $\mu = 0.11$	8.97	
Hydrogen diamidophosphate	$(NH_2)PO(OH)$ (30°C)	1.279(+1)	4.889
Hydrogen diamidothiophosphate	$(NH_2)PO(SH)$ (20°C)	2.0(+1)	4.3
Hydrogen diimidotriphosphate	$(HO)_2PO(NH)PO(OH)(NH)PO(OH)_2$ $\mu = 0.1$	~1	~2
		pK_3 3.03	pK_4 6.61
		pK_5 9.84	
Hydrogen diphosphate	$H_4P_2O_7$	0.91	2.10
		pK_3 6.70	pK_4 9.35
Hydrogen disulfate	$H_2S_2O_7$ (theoretical prediction)	−12	−8
Hydrogen dithionate	$H_2S_2O_6$	−3.4	−0.2
Hydrogen dithionite	$H_2S_2O_4$	0.35	2.45
Hydrogen fluoride	H_2F_2	3.20(4)	
Hydrogen germanate	H_2GeO_4	9.01	12.30
Hydrogen hexafluorosilicate	H_2SiF_6		1.92
Hydrogen hydrosulfite	$H_2S_2O_4$	0.35	2.50
Hydrogen hypobromite	HBrO	8.55	
Hydrogen hypochlorite	HClO	7.537	
Hydrogen hypoiodite	HIO	10.5(5)	
Hydrogen hyponitrite	$H_2N_2O_2$	7.21	11.45(10)
Hydrogen iodate	HIO_3	0.804	

(Continued)

TABLE 1.74 Proton Transfer Reactions of Inorganic Materials in Water at 25°C (*Continued*)

Substance	Formula or remarks	pK_1	pK_2
Hydrogen-*d* iodate	DIO_3 (in D_2O)	1.15	
Hydrogen iodide	HI	−8.56	
Hydrogen manganate(VI)	H_2MnO_4 (35°C) $\mu = 0.1$		10.15
Hydrogen nitrate	HNO_3	−1.37(7)	
Hydrogen nitrite	HNO_2	3.14(1)	
Hydrogen perchlorate	$HClO_4$	−1.6	
Hydrogen periodate	HIO_4	1.64	
Hydrogen peroxide	H_2O_2	11.64(2)	
Hydrogen peroxophosphate	H_3PO_5 $\mu = 0.2$	1.1	5.5
		pK_3 12.8	
Hydrogen peroxosulfate	H_2SO_5	1.0	9.86
Hydrogen perrhenate	$HReO_4$	−1.25	
Hydrogen pertechnetate	$HTcO_4$	0.3	
Hydrogen perthiocarbonate	H_2CS_4	3.54	7.24
Hydrogen perxenate	H_4XeO_6	pK_3 10.5	
Hydrogen phosphate(3−)	H_3PO_4	2.148(20)	7.198(10)
		pK_3 12.32(6)	
Hydrogen-d_2 phosphate	D_2PO_4 (in D_2O)	7.780	
Hydrogen phosphinate	H_2PHO_2	1.23	
Hydrogen phosphonate	H_2PHO_3	1.43	6.68(14)
Hydrogen selenate	H_2SeO_4		1.66
Hydrogen selenide	H_2Se $\mu = 0.03$	3.89	11.0
Hydrogen selenite	H_2SeO_3	2.62	8.30(15)
Hydrogen silicate(4−)	H_4SiO_4	9.60(10)	11.8(1)
Hydrogen sulfamate	H_2NSO_3H	0.99	
Hydrogen sulfate	H_2SO_4		1.99(1)
Hydrogen sulfide	H_2S	6.97	12.90
Hydrogen sulfite	$SO_2 + H_2O = HSO_3^- = H^+$	1.89	7.205
Hydrogen tellurate	H_6TeO_6	7.65(5)	11.00(5)
Hydrogen telluride	H_2Te (18°C)	2.64	11−12
Hydrogen tellurite	H_2TeO_3 (20°C)	6.27	8.43
Hydrogen tetrafluoroborate	HBF_4	0.5	
Hydrogen tetracyanonickelate	$H_2Ni(CN)_4$	4.69	6.59
Hydrogen tetraperoxochromate	H_3CrO_8 (30°C) $\mu = 3$	7.16	
Hydrogen tetrapolyphosphate	$H_4P_4O_{13}$ $\mu = 0.034$	1.99	2.64
		pK_3 6.62	pK_4 8.2

Hydrogen tetrathiophosphate	H_3PS_4	1.5	3.5
		pK_3 6.6	
Hydrogen thiocyanate	$HSCN$ $\mu = 3$	−1.8	
Hydrogen thiophosphate	H_3PO_3S	1.788	5.427
		pK_3 10.08	
Hydrogen thiosulfate	$H_2S_2O_3$	0.6	1.74
Hydrogen tripolyphosphate	$H_3P_3O_9$	~1	1.7
		pK_3 2.00(10)	
		pK_4 5.83(7)	
		pK_5 8.51(6)	
Hydrogen triselenocarbonate	H_2CSe_3	1.16	7.70
Hydrogen trithiocarbonate	H_2CS_3 (20°C)	2.68	8.18
Hydrogen tungstate	H_2WO_4	2.20	3.70
Hydrogen vanadate(−1)	HVO_3	3.80	
Hydrogen vanadate(3−)	H_3VO_4	3.78	7.78(4)
Hydroxylamine-*N*,*N*-disulfonic acid	$HON(SO_3H)_2$ $\mu = 1.6$	pK_3 11.85	
Hydroxylamine *O*-sulfonate	$^+H_3NOSO_3^-$ $\mu = 1$	1.48	
Imidodiphosphoric acid	$(HO)_2PO(NH)PO(OH)_2$ $\mu = 0.2$	~2	2.85
		pK_3 7.08	pK_4 9.72
Indium(III) ion	In^{3+} hydrolysis	3.54	4.28
Iridium(III) ion	Ir^{3+} hydrolysis $\mu = 1$	4.37	5.20
Iron(II) ion	Fe^{2+} hydrolysis $\mu = 1$	6.8	
Iron(III) ion	Fe^{3+} hydrolysis	2.19	
Lanthanum(III) ion	La^{3+} hydrolysis	9.06	
Lead(II) ion	Pb^{2+} hydrolysis $\mu = 0.3$	7.8	
Lead(IV) ion	Pb^{4+} hydrolysis	1.8	3.2
Lithium(I) ion	Li^+	13.8	
Lutetium(III) ion	Lu^{3+} hydrolysis	7.94	
Magnesium(II) ion	Mg^{2+} hydrolysis	11.41	
Manganese(II) ion	Mn^{2+} hydrolysis	10.59	
Manganese(III) ion	Mn^{3+} hydrolysis	0.4	
Mercury(I) ion	Hg_2^{2+} hydrolysis $\mu = 0.5$	5.0	
Mercury(II) ion	Hg^{2+} hydrolysis $\mu = 0.5$	3.70	2.65
Neodymium(III) ion	Nd^{3+} hydrolysis $\mu = 3$	9.0(5)	
Neptunium(III) ion	Np^{3+} hydrolysis $\mu = 0.3$	7.43	
Neptunium(IV) ion	Np^{4+} hydrolysis $\mu = 2$	2.30	
Neptunium(V) ion	NpO_2^+ hydrolysis	8.90(2)	
Nickel(II) ion	Ni^{2+} hydrolysis	9.86	
Osmium tetroxide	OsO_4 hydrolysis $\mu = 1$	12.1	
Palladium(II) ion	Pd^{2+} (stepwise pK_b values)	13.0	12.8
Pentacyanoaquoferrate(II) ion	$Fe(CN)_5(H_2O)^{3-}$ $\mu = 0.1$	2.63	

(Continued)

TABLE 1.74 Proton Transfer Reactions of Inorganic Materials in Water at 25°C (*Continued*)

Substance	Formula or remarks	pK_1	pK_2
Plutonium(III) ion	Pu^{3+} hydrolysis $\mu = 0.07$	7.2(2)	
Plutonium(IV) ion	Pu^{4+} hydrolysis $\mu = 2$	1.26	
Plutonium(V) ion	PuO_2^+ hydrolysis $\mu = 0.003$	9.7	
Plutonium(VI) ion	PuO_2^{2+} hydrolysis	3.33	4.05
Polonium(IV) ion	Po^{4+} hydrolysis	0.48	2.74
		pK_3 5.58	
Praseodymium(III) ion	Pr^{3+} hydrolysis $\mu = 0.3$	8.55	
Protoactinium(IV) ion	Pa^{4+} hydrolysis $\mu = 3$	0.14	0.38
Protoactinium(V) ion	Pa^{5+} hydrolysis $\mu = 3$	1.05	
Scandium(III) ion	Sc^{3+} hydrolysis $\mu = 0.05$	4.58(3)	
Silver(I) ion	Ag^+ hydrolysis	>11.1	
Sodium ion	Na^+ (aquo)	14.67(10)	
Strontium ion	Sr^{2+} (aquo)	13.18	
Terbium(III) ion	Tb^{3+} hydrolysis $\mu = 0.3$	8.16	
Thallium(I) ion	Tl^+	13.36(15)	
Thallium(III) ion	Tl^{3+} hydrolysis $\mu = 3$	1.14	
Thorium(IV) ion	Th^{4+} hydrolysis $\mu = 0.5$	3.89	4.20
Tin(II) ion	Sn^{2+} hydrolysis $\mu = 3$	3.81(10)	
Titanium(III)	Ti^{3+} hydrolysis $\mu = 3$	2.55	
Titanium(IV)	$TiO^{2+} + H_2O = TiO(OH)^+ + H^+$	1.3	
Tritium oxide	pK_w for $T_2O = T^+ + OH^-$	15.21	
Uranium(IV) ion	U^{4+} hydrolysis	0.68	
Uranyl(VI) ion	UO_2^{2+} $\mu = 0.035$	5.82	
Vanadium(II) ion	V^{2+} hydrolysis	6.85	
Vanadium(III) ion	V^{3+} hydrolysis	2.92	3.5
Vanadyl(IV) ion	VO^{2+} hydrolysis	6.86(10)	
Vanadyl(V) ion	VO_2^+(20°C) $\mu = 0.1$	1.83	
Xenon trioxide	$XeO_3 + H_2O = HXeO_4^- + H^+$	10.5	
Ytterbium(III) ion	Yb^{3+} hydrolysis	7.99(6)	
Yttrium(III) ion	Y^{3+} hydrolysis $\mu = 0.3$	8.34	
Zinc ion	Zn^{2+} hydrolysis	8.96	
Zirconium(IV) ion	Zr^{4+} hydrolysis $\mu = 1$	−0.32	0.06
		pK_3 0.35	

Source: J. J. Christensen, L. D. Hansen, and R. M. Izatt, *Handbook of Proton Ionization Heats and Related Thermodynamic Quantities,* Wiley-Interscience, New York, 1976; D. D. Perrin, *Ionisation Constants of Inorganic Acids and Bases in Aqueous Solution*, 2d ed., Pergamon Press, 1982.

1.20 FORMATION CONSTANTS

The formation constant of a metal complex is the equilibrium constant for the formation of a complex ion from its components in solution.

Each value listed is the logarithm of the overall formation constant for the cumulative binding of a ligand L to the central metal cation M, viz.:

	Comulative formation constant	Stepwise stability constants
$M + L = ML$	K_1	k_1
$M + 2L = ML_2$	K_2	k_1k_2
....		
$M + nL = ML_n$	K_n	$k_1k_2 \cdots k_n$

As an example, the entries in Table 1.75 for the zinc ammine complexes represent these equilibria:

$$Zn^{2+} + NH_3 = Zn(NH_3)^{2+} \qquad K_1 = \frac{[Zn(NH_3)^{2+}]}{[Zn^{2+}][NH_3]}$$

$$Zn^{2+} + 2NH_3 = Zn(NH_3)_2^{2+} \qquad K_2 = \frac{[Zn(NH_3)_2^{2+}]}{[Zn^{2+}][NH_3]^2}$$

$$Zn^{2+} + 3NH_3 = Zn(NH_3)_3^{2+} \qquad K_3 = \frac{[Zn(NH_3)_3^{2+}]}{[Zn^{2+}][NH_3]^3}$$

$$Zn^{2+} + 4NH_3 = Zn(NH_3)_4^{2+} \qquad K_4 = \frac{[Zn(NH_3)_4^{2+}]}{[Zn^{2+}][NH_3]^4}$$

If the stepwise stability or formation constants of the reactions are desired, for the first step $\log K_1 = \log k_1 = 2.37$. For the second and succeeding steps the equilibria and corresponding constants are as follows:

$$Zn(NH_3)^{2+} + NH_3 = Zn(NH_3)_2^{2+} \qquad \log k_2 = \log k_2 - \log k_1 = 2.44$$

$$Zn(NH_3)_2^{2+} + NH_3 = Zn(NH_3)_3^{2+} \qquad \log k_3 = \log k_2 - \log k_1 = 3.50$$

$$Zn(NH_3)_3^{2+} + NH_3 = Zn(NH_3)_4^{2+} \qquad \log k_4 = \log k_4 - \log k_3 = 2.15$$

The reverse of the association or formation reactions would represent the dissociation or instability constant for the systems, i.e., $-\log K_f = \log K_{\text{instab}}$.

The data in the tables generally refer to temperatures of about 20 to 25°C. Most of the values in Table 1.75 refer to zero ionic strength, but those in Table 1.76 often refer to a finite ionic strength.

TABLE 1.75 Cumulative Formation Constants for Metal Complexes with Inorganic Ligands

	log K_1	log K_2	log K_3	log K_4	log K_5	log K_6
Ammonia						
Cadmium	2.65	4.75	6.19	7.12	6.80	5.14
Cobalt(II)	2.11	3.74	4.79	5.55	5.73	5.11
Cobalt(III)	6.7	14.0	20.1	25.7	30.8	35.2
Copper(I)	5.93	10.86				
Copper(II)	4.31	7.98	11.02	13.32	12.86	
Iron(II)	1.4	2.2				
Manganese(II)	0.8	1.3				
Mercury(II)	8.8	17.5	18.5	19.28		
Nickel	2.80	5.04	6.77	7.96	8.71	8.74
Platinum(II)						35.3
Silver(I)	3.24	7.05				
Zinc	2.37	4.81	7.31	9.46		
Bromide						
Astatine	2.51 [AtBr]					
Bismuth(III)	4.30	5.55	5.89	7.82		9.70
Bromine	1.24 [Br_3^-]					
Cadmium	1.75	2.34	3.32	3.70		
Cerium(III)	0.42					
Copper(I)		5.89				
Copper(II)	0.30					
Gold(I)		12.46				
Indium	1.30	1.88	2.48			
Iodine	2.64 [IBr]					
Iron(III)	−0.30	−0.50				
Lead	1.2	1.9		1.1		
Mercury(II)	9.05	17.32	19.74	21.00		
Palladium(II)				13.1		
Platinum(II)				20.5		
Rhodium(III)		14.3	16.3	17.6	18.4	17.2
Scandium	2.08	3.08				
Silver(I)	4.38	7.33	8.00	8.73		
Thallium(I)	0.93					
Thallium(III)	9.7	16.6	21.2	23.9	29.2	31.6
Tin(II)	1.11	1.81	1.46			
Uranium(IV)	0.18					
Yttrium	1.32					
Chloride						
Americium(III)	1.17					
Antimony(III)	2.26	3.49	4.18	4.72		
Bismuth(III)	2.44	4.7	5.0	5.6		
Cadmium	1.95	2.50	2.60	2.80		
Cerium(III)	0.48					
Copper(I)		5.5	5.7			
Copper(II)	0.1	−0.6				
Curium(III)	1.17					
Gold(III)		9.8				
Indium	1.42	2.23	3.23			
Iron(II)	0.36					
Iron(III)	1.48	2.13	1.99	0.01		
Lead	1.62	2.44	1.70	1.60		
Manganese(II)	0.96					
Mercury(II)	6.74	13.22	14.07	15.07		

TABLE 1.75 Cumulative Formation Constants for Metal Complexes with Inorganic Ligands (*Continued*)

	log K_1	log K_2	log K_3	log K_4	log K_5	log K_6
Palladium(II)	6.1	10.7	13.1	15.7		
Platinum(II)		11.5	14.5	16.0		
Plutonium(III)	1.17					
Silver(I)	3.04	5.04		5.30		
Thallium(I)	0.52					
Thallium(III)	8.14	13.60	15.78	18.00		
Thorium	1.38	0.38				
Tin(II)	1.51	2.24	2.03	1.48		
Tin(IV)						4
Uranium(IV)	0.8					
Uranium(VI)	0.22					
Zinc	0.43	0.61	0.53	0.20		
Zirconium	0.9	1.3	1.5	1.2		
Cyanide						
Cadmium	5.48	10.60	15.23	18.78		
Copper(I)		24.0	28.59	30.30		
Gold(I)		38.3				
Iron(II)						35
Iron(III)						42
Mercury(II)				41.4		
Nickel				31.3		
Silver(I)		21.1	21.7	20.6		
Zinc				16.7		
Fluoride						
Aluminum	6.10	11.15	15.00	17.75	19.37	19.84
Beryllium	5.1	8.8	12.6			
Cerium(III)	3.20					
Chromium(III)	4.41	7.81	10.29			
Gadolinium	3.46					
Gallium	5.08					
Indium	3.70	6.25	8.60	9.70		
Iron(III)	5.28	9.30	12.06			
Lanthanum	2.77					
Magnesium	1.30					
Manganese(II)	5.48					
Plutonium(III)	6.77					
Scandium						17.3
Thallium(I)	0.1					
Thallium(III) [TlO^+]	6.44					
Thorium	7.65	13.46	17.97			
Titanium(IV) [TiO^{2+}]	5.4	9.8	13.7	18.0		
Uranium(VI)	4.59	7.93	10.47	11.84		
Yttrium	4.81	8.54	12.14			
Zirconium	8.80	16.12	21.94			
Hydroxide						
Aluminum	9.27			33.03		
Antimony(III)		24.3	36.7	38.3		
Arsenic [as AsO^+]	14.33	18.73	20.60	21.20		
Beryllium	9.7	14.0	15.2			
Bismuth(III)	12.7	15.8		35.2		
Cadmium	4.17	8.33	9.02	8.62		
Cerium(III)	14.6					
Cerium(IV)	13.28	26.46				

(*Continued*)

TABLE 1.75 Cumulative Formation Constants for Metal Complexes with Inorganic Ligands (*Continued*)

	log K_1	log K_2	log K_3	log K_4	log K_5	log K_6
Chromium(III)	10.1	17.8		29.9		
Copper(II)	7.0	13.68	17.00	18.5		
Dysprosium	5.2					
Erbium(III)	5.4					
Gadolinium	4.6					
Gallium	11.0	21.7		34.3	38.0	40.3
Indium	9.9	19.8		28.7		
Iodine	9.49	11.24				
Iron(II)	5.56	9.77	9.67	8.58		
Iron(III)	11.87	21.17	29.67			
Lanthanum	3.3					
Lead(II)	7.82	10.85	14.58			61.0
Lutetium	6.6					
Magnesium	2.58					
Manganese(II)	3.90		8.3			
Neodymium	5.5					
Nickel	4.97	8.55	11.33			
Praseodymium	4.30					
Plutonium(III)	7.0					
Plutonium(IV)	12.39					
Plutonium [as PuO_2^{2+}]	8.3	16.6	20.9			
Samarium(III)	4.8					
Scandium	8.9					
Tellurium(IV)			41.6	53.0	64.8	72.0
Thallium(III)	12.86	25.37				
Titanium(III)	12.71					
Uranium(IV)	13.3				41.2	
Uranium(VI) [as UO_2^{2+}]	9.5	22.80		32.4		
Vanadium(III)	11.1	21.6				
Vanadium(IV) [as VO^{2+}]	8.6		[25.8 for $V_2O_4(OH)^-$]			
Vanadium(V) [as VO^{3+}]		25.2		46.2	58.5	
Yttrium	5.0					
Zinc	4.40	11.30	14.14	17.66		
Zirconium	14.3	28.3	41.9	55.3		
Iodide						
Bismuth	3.63			14.95	16.80	18.80
Cadmium	2.10	3.43	4.49	5.41		
Copper(I)		8.85				
Indium	1.00	2.26				
Iodine	2.89	5.79				
Iron(III)	1.88					
Lead	2.00	3.15	3.92	4.47		
Mercury(II)	12.87	23.82	27.60	29.83		
Silver	6.58	11.74	13.68			
Thallium(I)	0.72	0.90	1.08			
Thallium(III)	11.41	20.88	27.60	31.82		
Iodate						
Barium	1.05					
Calcium	0.89					
Magnesium	0.72					
Strontium	1.00					
Thorium	2.88	4.79	7.15			

TABLE 1.75 Cumulative Formation Constants for Metal Complexes with Inorganic Ligands (*Continued*)

	log K_1	log K_2	log K_3	log K_4	log K_5	log K_6
Nitrate						
Barium	0.92					
Beryllium	1.62					
Bismuth(III)	1.26					
Cadmium	0.40					
Calcium	0.28					
Cerium(III)	1.04	2.55				
Curium(III)	0.57					
Hafnium	0.92	2.43	4.32	6.40	8.48	10.29
Iron(III)	1.0					
Lanthanum	0.26	0.69	1.27			
Lead	1.18					
Mercury(II)	0.35					
Neodymium	0.52	1.18				
Neptunium(IV)	0.38					
Plutonium(III)	0.77	1.93	3.09			
Plutonium(IV)	0.54					
Strontium	0.82					
Thallium(I)	0.33					
Thallium(III)	0.92					
Thorium	0.78	1.89	2.89	3.63		
Uranium(IV)	0.20	0.37				
Uranium(VI)	0.34	0.45				
Ytterbium	0.45	1.30	2.42			
Zirconium [as ZrO^{2+}]		1.91		3.54		
Pyrophosphate						
Barium	4.6					
Calcium	4.6					
Cadmium	5.6					
Copper(II)	6.7	9.0				
Lead		5.3				
Magnesium	5.7					
Nickel	5.8	7.4				
Strontium	4.7					
Yttrium		9.7				
Zirconium		6.5				
Sulfate						
Cerium(III)	3.40					
Erbium	3.58					
Gadolinium	3.66					
Holmium	3.58					
Indium	1.78	1.88	2.36			
Iron(III)	2.03	2.98				
Lanthanum	3.64					
Neodymium	3.64					
Nickel	2.4					
Plutonium(IV)	3.66					
Praseodymium	3.62					
Samarium	3.66					
Thorium	3.32	5.50				
Uranium(IV)	3.24	5.42				
Uranium(VI)	1.70	2.45	3.30			

(*Continued*)

TABLE 1.75 Cumulative Formation Constants for Metal Complexes with Inorganic Ligands (*Continued*)

	log K_1	log K_2	log K_3	log K_4	log K_5	log K_6
Yttrium	3.47					
Ytterbium	3.58					
Zirconium	3.79	6.64	7.77			
Sulfite						
Copper(I)	7.5	8.5	9.2			
Mercury(II)		22.66				
Silver	5.30	7.35				
Thiocyanate						
Bismuth	1.15	2.26	3.41	4.23		
Cadmium	1.39	1.98	2.58	3.6		
Chromium(III)	1.87	2.98				
Cobalt(II)	−0.04	−0.70	0	3.00		
Copper(I)	12.11	5.18				
Gold(I)		23		42		
Indium	2.58	3.00	4.63			
Iron(III)	2.95	3.36				
Mercury(II)		17.47		21.23		
Nickel	1.18	1.64	1.81			
Ruthenium(III)	1.78					
Silver		7.57	9.08	10.08		
Thallium(I)	0.80					
Uranium(IV)	1.49	2.11				
Uranium(VI)	0.76	0.74	1.18			
Vanadium(III)	2.0					
Vanadium(IV)	0.92					
Zinc	1.62					
Thiosulfate						
Cadmium	3.92	6.44				
Copper(I)	10.27	12.22	13.84			
Iron(III)	2.10					
Lead		5.13	6.35			
Mercury(II)		29.44	31.90	33.24		
Silver	8.82	13.46				

TABLE 1.76 Cumulative Formation Constants for Metal Complexes with Organic Ligands

Temperature is 25°C and ionic strengths are approaching zero unless indicated otherwise: (*a*) At 20°C, (*b*) at 30°C, (*c*) 0.1 *M* uni-univalent salt, (*d*) 1.0 *M* uni-univalent salt, (*e*) 2.0 *M* uni-univalent salt present.

	log K_1	log K_2	log K_3	log K_4
Acetate				
Ag(I)	0.73	0.64		
Ba(II)	0.41			
Ca(II)	0.6			
Cd(II)	1.5	2.3	2.4	
Ce(III)	1.68	2.69	3.13	3.18
Co(II)	1.5	1.9		
Cr(III)	1.80	4.72		
Cu(II) *a*	2.16	3.20		
Fe(II) *c*	3.2	6.1	8.3	
Fe(III) *a,d*	3.2			
In(III)	3.50	5.95	7.90	9.08
Hg(II)		8.43		
La(III) *a,e*	1.56	2.48	2.98	2.95
Mg(II)	0.8			
Mn(II)	9.84	2.06		
Ni(II)	1.12	1.81		
Pb(II)	2.52	4.0	6.4	8.5
Rare earths *a,e*	1.6–1.9	2.8–3.0	3.3–3.7	
Sr(II)	0.44			
Tl(III)				15.4
UO_2(II) *a,e*	2.38	4.36	6.34	
Y(III) *a,e*	1.53	2.65	3.38	
Zn(II)	1.5			
Acetylacetone				
Al(III) *b*	8.6	15.5		
Be(II)	7.8	14.5		
Cd(II)	3.84	6.66		
Ce(III)	5.30	9.27	12.65	
Cr(II)	5.9	11.7		
Co(II)	5.40	9.54		
Cu(II)	8.27	16.34		
Dy(III) *b*	6.03	10.70	14.04	
Er(III) *b*	5.99	10.67	14.09	
Eu(III) *b*	5.87	10.35	13.64	
Fe(II)	5.07	8.67		
Fe(III)	11.4	22.1	26.7	
Ga(III)	9.5	17.9	23.6	
Gd(III) *b*	5.90	10.38	13.79	
Hf(IV)	8.7	15.4	21.8	28.1
Ho(III)	6.05	10.73	14.13	
In(III)	8.0	15.1		
La(III) *b*	5.1	8.90	11.90	
Lu(III) *b*	6.23	11.00	13.63	
Mg(II)	3.65	6.27		
Mn(II)	4.24	7.35		
Mn(III)			3.86	
Nd(III)	5.6	9.9	13.1	
Ni(II) *a*	6.06	10.77	13.09	

(*Continued*)

TABLE 1.76 Cumulative Formation Constants for Metal Complexes with Organic Ligands (*Continued*)

	log K_1	log K_2	log K_3	log K_4
Pd(II) *b*	16.2	27.1		
Pr(III) *b*	5.4	9.5	12.5	
Pu(IV) *c*	10.5	19.7	28.1	34.1
Sc(III) *b*	8.0	15.2		
Sm(III) *b*	5.9	10.4		
Tb(III) *b*	6.02	10.63	14.04	
Th(IV)	8.8	16.2	22.5	26.7
Tm(IV) *b*	6.09	10.85	14.33	
U(IV) *a,c*	8.6	17.0	23.4	29.5
UO_2(II) *b*	7.74	14.19		
VO(II)	8.68	15.79		
V(II)	5.4	10.2	14.7	
Y(III) *b*	6.4	11.1	13.9	
Yb(III) *b*	6.18	11.04	13.64	
Zn(II) *b*	4.98	8.81		
Zr(IV)	8.4	16.0	23.2	30.1
Alizarin red				
Cr(VI)	4.7			
Cu(II)	4.1			
Hf(IV)		10.4		
Mo(VI)		9.6		
Pb(II)	6.0			
Th(IV)		8.24		
UO_2(II)	4.22			
V(V)		8.6		
W(VI)		7.8		
Arsenazo				
Hf(IV)	10.07			
Zr(IV)	12.95			
Aurintricarboxylic acid				
Be(II)	4.54			
Cu(II)	4.1	8.81		
Fe(III)	4.68			
Th(IV)	5.04			
UO_2(II)	4.77			
Benzoylacetone (75% dioxane)				
Ba(II)		9.4		
Be(II)	12.59	24.01		
Cd(II)	7.79	14.36		
Ce(III)	10.09	19.42	27.04	
Co(II)	9.42	17.83		
Cu(II)	12.05	23.01		
La(III)	6.33	11.66	16.78	
Mg(II)	7.69	14.09		
Mn(II)	8.66	15.78		
Ni(II)	9.58	18.00		
Pb(II)	8.84	16.35		
Pr(III)	7.02	13.62	18.74	
UO_2(II)	12.15	23.27		
Y(III)	8.24	14.98	20.57	
Zn(II)	9.62	17.90		

TABLE 1.76 Cumulative Formation Constants for Metal Complexes with Organic Ligands (*Continued*)

	log K_1	log K_2	log K_3	log K_4
Calmagite				
Ca	6.05			
Mg	8.05			

	Complex of HL^{2-} Anion		Complex of L^{3-} Anion		Complex of H_2L^-
	log K_1	log K_2	log K_1	log K_2	log K_3
Citric acid					
Ag	7.1				
Al	7.0		20.0		
Ba	2.98				
Be	4.52				
Ca	4.68				
Cd	3.98		11.3		
Ce(III)		6.18		9.65	3.2
Co(II)	4.8		12.5		
Cu(II)	4.35		14.2		
Eu(III)		6.46		9.80	
Fe(II)	3.08		15.5		
Fe(III)	12.5		25.0		
La		6.97		9.45	6.22
Mg	3.29				
Mn(II)	3.67				
Nd(III)		6.32		9.70	
Ni	5.11		14.3		
Pb	6.50				
Pr					3.4
Ra	2.36				
Sr	2.8				
Tl(I)	1.04				
UO_2	8.5	10.8			
Y					3.6
Yb				8	
Zn	4.71		11.4		

	log K_1	log K_2	log K_3	
1,2-Diaminocyclohexane-*N,N,N′,N′*-tetraacetic acid				
Al *c*	17.63			
Ba *c*	8.64			
Ca *c*	12.3			
Cd *c*	19.88			
Ce(III) *c*	16.76			
Co(II) *c*	19.57			
Cu(II) *c*	21.95			
Dy(III) *c*	19.69			
Er(III) *c*	20.20			
Eu(III) *c*	18.77			
Fe(III) *c*	27.48			
Ga *c*	22.91			

(*Continued*)

TABLE 1.76 Cumulative Formation Constants for Metal Complexes with Organic Ligands (*Continued*)

	log K_1	log K_2	log K_3	log K_4
Gd *c*	18.80			
Hg(II) *c*	24.4			
Ho *c*	19.89			
La *c*	16.35			
Lu *c*	21.51			
Mg *c*	10.41			
Mn(II) *c*	17.43			
Nd *c*	17.69			
Ni *c*	19.4			
Pb *c*	20.33			
Pr *c*	17.23			
Sm(III) *c*	18.63			
Sr *c*	8.92			
Tb *c*	19.30			
Tm *c*	20.46			
VO(II) *c*	19.40			
Y *c*	19.41			
Yb *c*	20.80			
Zn *c*	18.6			
Dibenzoylmethane (75% dioxane)				
Ba	6.10	11.50		
Be	13.62	26.03		
Ca	7.17	13.55		
Cd	8.67	16.63		
Ce(III)	10.99	21.53	30.38	
Co(II)	10.35	20.05		
Cu(II)	12.98	24.98		
Cs	3.42			
Fe(II)	11.15	21.50		
K	3.67			
Li	5.95			
Mg	8.54	16.21		
Mn(II)	9.32	17.79		
Na	4.18			
Ni	10.83	20.72		
Pb	9.75	18.79		
Rb	3.52			
Sr	6.40	12.10		
Zn	10.23	19.65		
	log K_1	log K_2	log K_3	log K_f [MHL]
4,5-Dihydroxybenzene-1,3-disulfonic acid (Tiron)				
Al	19.02	31.10	33.5	
Ba	4.10			14.6
Ca	5.80			14.8
Cd *d*	7.69	13.29		
Ce(III)		3.75		
Co(II) *d*	8.19	14.41		15.7
Cu(II) *d*	12.76	23.73		18.1

TABLE 1.76 Cumulative Formation Constants for Metal Complexes with Organic Ligands (*Continued*)

	log K_1	log K_2	log K_3	log K_f [MHL]
Fe(III) *a,c*	20.7	35.9	46.9	22.6
La	12.9			18.6 [La(OH)L]
Mg *a,c*	6.86			14.6
Mn(II) *c*	8.6			
Ni *a,c*	8.56	14.90		15.6
Pb *d*	11.95	18.28		
Sr *c*	4.55			
UO_2(II) *c*	15.90			
VO(II)	15.88			
Zn *d*	9.00	16.91		15.9

	log K_1	log K_2	log K_f [M_2L_3]
2,3-Dimercaptopropan-1-of (BAL)			
Fe(II)	15.8		
Fe(III)	30.6 [Fe(OH)L]		28
Mn(II)	5.23	10.43	
Ni		22.78	
Zn	13.48	23.3	40.6

	log K_1	log K_2	log K_3	log K_4
Dimethylglyoxime (50% dioxane)				
Cd	5.7	10.7		
Co(II)	9.80	18.94		
Cu(II)	12.00	33.44		
Fe(II)		7.25		
La	6.6	12.5		
Ni	11.16			
Pb	7.3			
Zn	7.7	13.9		
2,2′-Dipyridyl				
Ag	3.65	7.15		
Cd	4.26	7.81	10.47	
Co(II)	5.73	11.57	17.59	
Cr(II)	4.5	10.5	14.0	
Cu(I)		14.2		
Cu(II)	8.0	13.60	17.08	
Fe(II)	4.36	8.0	17.45	
Hg(II)	9.64	16.74	19.54	
Mg	0.5			
Mn(II) *d*	4.06	7.84	11.47	
Ni	6.80	13.26	18.46	
Pb	3.0			
Ti(III)			25.28	
V(II)	4.9	9.6	13.1	
Zn	5.30	9.83	13.63	
Eriochrome Black T				
Ca	5.4			
Mg	7.0			
Zn	13.5	20.6		

(*Continued*)

TABLE 1.76 Cumulative Formation Constants for Metal Complexes with Organic Ligands (*Continued*)

				log K_4
Ethanolamine				
Ag	3.29	6.92		
Cu(II)		6.68		16.48
Hg(II)	8.51	17.32		
Ethylenediamine				
Ag	4.70	7.70		
Cd *a*	5.47	10.09	12.09	
Co(II)	5.91	10.64	13.94	
Co(III)	18.7	34.9	48.69	
Cr(II)	5.15	9.19		
Cu(I)		10.8		
Cu(II)	10.67	20.00	21.0	
Fe(II)	4.34	7.65	9.70	
Hg(II)	14.3	23.3		
Mg	0.37			
Mn(II)	2.73	4.79	5.67	
Ni	7.52	13.84	18.33	
Pd(II)		26.90		
V(II)	4.6	7.5	8.8	
Zn	5.77	10.83	14.11	
Ethylenediamine-*N, N, N′, N′*-tetraacetic acid				
Ag	7.32			
Al	16.11			
Am(III)	18.18			
Ba	7.78			
Be	9.3			
Bi	22.8			
Ca	11.0			
Cd	16.4			
Ce(III)	16.80			
Cf(III)	19.09			
Cm(III)	18.45			
Co(II)	16.31			
Co(III)	36			
Cr(II)	13.6			
Cr(III)	23			
Cu(II)	18.7			
Dy	18.0			
Er	18.15			
Eu(III)	17.99			
Fe(II)	14.33			
Fe(III)	24.23			
Ga	20.25			
Gd	17.2			
Hg(II)	21.80			
Ho	18.1			
In	24.95			
La	16.34			
Li	2.79			
Lu	19.83			
Mg	8.64			
Mn(II)	13.8			
Mo(V)	6.36			

TABLE 1.76 Cumulative Formation Constants for Metal Complexes with Organic Ligands (*Continued*)

				log K_4
Na	1.66			
Nd	16.6			
Ni	18.56			
Pb	18.3			
Pd(II)	18.5			
Pm(III)	17.45			
Pr	16.55			
Pu(III)	18.12			
Pu(IV)	17.66			
Pu(VI)	17.66			
Ra	7.4			
Sc	23.1			
Sm	16.43			
Sn(II)	22.1			
Sr	8.80			
Tb	17.6			
Th	23.2			
Ti(III)	21.3			
TiO(II)	17.3			
Tl(III)	22.5			
Tm	19.49			
U(IV)	17.50			
V(II)	12.70			
V(III)	25.9			
VO(II)	18.0			
V(V)	18.05			
Y	18.32			
Yb	18.70			
Zn	16.4			
Zr	19.40			
Glycine				
Ag	3.41	6.89		
Ba	0.77			
Be		4.95		
Ca	1.38			
Cd	4.74	8.60		
Co(II)	5.23	9.25	10.76	
Cu(II)	8.60	15.54	16.27	
Dy		12.2		
Er		12.7		
Fe(II) *a*	4.3	7.8		
Fe(III) *a,d*	10.0			
Gd		11.9		
Hg(II)	10.3	19.2		
La		11.2		
Mg	3.44	6.46		
Mn(II)	3.6	6.6		
Ni	6.18	11.14	15	
Pb	5.47	8.92		
Pd(II)	9.12	17.55		
Pr		11.5		
Sm		11.7		

(*Continued*)

TABLE 1.76 Cumulative Formation Constants for Metal Complexes with Organic Ligands (*Continued*)

				log K_4
Sr	0.91			
Y		12.5		
Yb		13.0		
Zn	5.52	9.96		
***N'*-(2-Hydroxyethyl)ethylenediamine-*N,N,N'*-triacetic acid**				
Ba *c*	5.54			
Ca *c*	8.43			
Cd *c*	13.0			
Ce(III) *c*	14.11			
Co(II) *c*	14.4			
Cu(II) *c*	17.40			
Dy *c*	15.30			
Er *c*	15.42			
Eu(III) *c*	15.35			
Fe(II) *c*	11.6			
Fe(III) *c*	19.8			
Gd *c*	15.22			
Hg(II) *c*	20.1			
Ho *c*	15.32			
La *c*	13.46			
Lu *c*	15.88			
Mg *c*	5.78			
Mn(II) *c*	10.7			
Nd *c*	14.86			
Ni *c*	17.0			
Pb *c*	15.5			
Pr *c*	14.61			
Sm *c*	15.28			
Sr *c*	6.92			
Tb *c*	15.32			
Th *c*	18.5			
Tm *c*	15.59			
Y *c*	14.65			
Yb *c*	15.88			
Zn *c*	14.5			
8-Hydroxy-2-methylquinoline (50% dioxane)				
Cd	9.00	9.00	16.60	
Ce(III)	7.71			
Co(II)	9.63	18.50		
Cu(II)	12.48	24.00		
Fe(II)	8.75	17.10		
Mg	5.24	9.64		
Mn(II)	7.44	13.99		
Ni	9.41	17.76		
Pb	10.30	18.50		
UO_2(II)	9.4	17		
Zn	9.82	18.72		
8-Hydroxyquinoline-5-sulfonic acid				
Ba	2.31			
Ca	3.52			
Cd	7.70	14.20		
Ce(III)	6.05	11.05	14.95	

TABLE 1.76 Cumulative Formation Constants for Metal Complexes with Organic Ligands (*Continued*)

				log K_4
Co(II)	8.11	15.05	20.41	
Cu(II)	11.92	21.87		
Er	7.16	13.34	18.56	
Fe(II)	8.4	15.7	21.75	
Fe(III)	11.6	22.8	35.65	
Gd	6.64	12.37	17.27	
La	5.63	10.13	13.83	
Mg	4.79	8.19		
Mn(II)	5.67	10.72		
Nd	6.3	11.6	16.0	
Ni	9.57	18.27	22.9	
Pb	8.53	16.13		
Pr	6.17	11.37	15.67	
Sm	6.58	12.28	17.04	
Sr	2.75			
Th	9.56	18.29	25.92	32.04
UO_2(II)	8.52	15.67		
Zn	8.65	16.15		
Lactic acid				
Ba	0.64			
Ca	1.42			
Cd	1.70			
Ce(III) *a,c*	2.76	4.73	5.96	
Co(II)	1.90			
Cu(II)	3.02	4.85		
Er	2.77	5.11	6.70	
Eu(III)	2.53	4.60	5.88	
Fe(III)	7.1			
Gd	2.53	4.63	5.91	
Ho	2.71	4.97	6.55	
La *a,c*	2.60	4.34	5.64	
Li	0.20			
Mg	1.37			
Mn(II)	1.43			
Nd	2.47	4.37	5.60	
Ni	2.22			
Pb	2.40	3.80		
Pr *a,c*	2.85	4.90	6.10	
Rare earths *a,c*	2.8–3.0	4.9–5.4	6.1–7.8	
Sm	2.56	4.58	5.90	
Sr	0.98			
Tb	2.61	4.73	6.01	
Y	2.53	4.70	6.12	
Yb	2.85	5.27	7.96	
Zn	2.20	3.75		
Nitrilotriacetic acid				
Al	>10			
Ba *a*	5.88			
Ca	7.60	11.61		
Cd *c*	9.80	15.2		
Ce(III) *c*	10.83	18.67		

(*Continued*)

TABLE 1.76 Cumulative Formation Constants for Metal Complexes with Organic Ligands (*Continued*)

				log K_4
Co(II) *c*	10.38	14.5		
Cr(III)	>10			
Cu(II) *c*	13.10			
Dy *c*	11.74	21.15		
Er *c*	12.03	21.29		
Eu(III) *c*	11.52	20.70		
Fe(II) *c*	8.84			
Fe(III) *c*	15.87	24.32		
Gd *c*	11.54	20.80		
Hg(II)	12.7			
Ho *c*	11.90	21.25		
In	15			
La *c*	10.36	17.60		
Li *a*	3.28			
Lu *c*	12.49	21.91		
Mg *c*	5.36	10.2		
Mn(II)	8.60	11.1		
Na	2.15			
Nd *c*	11.26	19.73		
Ni	11.26	16.0		
Pb *a,c*	11.8			
Pr *c*	11.07	19.25		
Sm(III) *c*	11.53	20.53		
Sr	6.73			
Tb *c*	11.59	20.97		
Tl(I)	3.44			
Th *c*	12.4			
Tm *c*	12.22	21.45		
Y *c*	11.48	20.43		
Yb *c*	12.40	21.69		
Zn *c*	10.45	13.45		
Zr *c*	20.8			
1-Nitroso-2-naphthol (75% dioxane)				
Ag	7.74			
Cd	6.18	11.38		
Co(II)	10.67	22.81		
Cu(II)	12.52	23.37		
Mg	6.2	10.60		
Nd	9.5	17.7	25.6	
Ni	10.75	21.29	28.09	
Pb	9.73	17.31		
Pr	9.04	17.06	23.85	
Th *c*	8.50	16.13	24.03	30.29
Y	9.02	17.74	25.04	
Zn	9.32	17.02		
Zr	3.6			
Oxalate				
Ag	2.41			
Al	7.26	13.0	16.3	
Am(III)		9.8		[$Am(HL)_4^-$ 11.0]
Ba	2.31			

TABLE 1.76 Cumulative Formation Constants for Metal Complexes with Organic Ligands (*Continued*)

				log K_4
Be	4.90			
Ca	3.0			
Cd	3.52	5.77		
Ce(III)	6.52	10.5	11.3	
Co(II)	4.79	6.7	9.7	
Co(III)			~20	
Cu(II)	6.16	8.5		
Er	4.82	8.21	10.03	
Fe(II)	2.9	4.52	5.22	
Fe(III)	9.4	16.2	20.2	
Gd	7.04			
Hg(II)		6.98		
Mg	3.43	4.38		
Mn(II)	3.97	5.80		
Mn(III) *e*	9.98	16.57	19.42	
Mo(III)	3.38			
Mo(VI)				[$MoO_3(L)^{2-}$ 13.0]
Nd	7.21	11.5	>14	
Ni	5.3	7.64	~8.5	
NpO_2(II)	3.30	7.07		
Pb		6.54		
Pu(III)	9.31	18.70	28	
Pu(IV)	8.74	16.91	23.39	27.50
PuO_2(II)		11.4		
Sr	2.54			
Th				24.48
TiO(II)	2.67			
Tl(I)	2.03			
UO_2(II)		10.57		
VO(II)		9.80		
V(II)	~2.7			
Y	6.52	10.10	11.47	
Yb	7.30	11.7	>14	
Zn	4.89	7.60	8.15	
Zr	9.80	17.14	20.86	21.15
1,10-Phenanthroline				
Ag	5.02	12.07		
Ca	0.7			
Cd	5.93	10.53	14.31	
Co(II)	7.25	13.95	19.90	
Cu(II)	9.08	15.76	20.94	
Fe(II)	5.85	11.45	21.3	
Fe(III)	6.5	11.4	23.5	
Hg(II)		19.65	23.35	
Mg	1.2			
Mn(II)	3.88	7.04	10.11	
Ni	8.80	17.10	24.80	
Pb	4.65	7.5	9	
VO(II)	5.47	9.69		
Zn	6.55	12.35	17.55	

(*Continued*)

TABLE 1.76 Cumulative Formation Constants for Metal Complexes with Organic Ligands (*Continued*)

				log K_4
Phthalic acid				
Ba	2.33			
Ca	2.43			
Cd	2.5			
Co(II)	1.81	4.51		
Cu(II)	3.46	4.83		
La		7.74		
Ni	2.14			
Pb *d*	3.4			
UO_2(II)	4.38			
Zn	2.2			
Piperidine				
Ag	3.30	6.48		
Hg(II)	8.70	17.44		
Pt(II)			log K_5 5.7	log K_6 8.2
Propylene-1,2-diamine				
Cd *b,c*		9.97	12.12	
Co(II) *d*	5.42	11.47	14.72	
Cu(II) *c*	6.41	20.06		
Hg(II) *c*	10.78	23.53	23.25	
Ni *d*	7.43	13.62	17.89	
Zn *b,c*	5.89	10.87	12.57	
Pyridine				
Ag	1.97	4.35		
Cd	1.40	1.95	2.27	2.50
Co(II)	1.14	1.54		
Cu(I)		3.34	4.51	5.44
				log K_6 6.89
Cu(II)	2.59	4.33	5.93	6.54
			log K_5 7.00	log K_6 10.2
Fe(II)	0.71			
Hg(II)	5.1	10.0	10.4	
Mn(II)	1.92	2.77	3.37	3.50
VO(II)	−1.70			
Zn	1.41	1.11	1.61	1.93
Pyridine-2,6-dicarboxylic acid				
Ba *a,d*	3.46			
Ca *a,d*	4.6	7.2		
Cd *a,d*	5.7	10.0		
Ce(III) *a,d*	8.34	14.42	18.80	
Co(II) *a,d*	7.0	12.5		
Cu(II) *a,d*	9.14	16.52		
Dy *a,d*	8.69	16.19	22.14	
Er *a,d*	8.77	16.39	22.14	
Eu(III) *a,d*	8.84	15.98	21.00	
Fe(II) *a,d*	5.71	10.36		
Fe(III) *a,d*	10.91	17.13		
Gd *a,d*	8.74	16.06	21.83	
Ho *a,d*	8.72	16.23	22.08	
La *a,d*	7.98	13.79	18.06	
Lu *a,d*	9.03	16.80	21.48	

TABLE 1.76 Cumulative Formation Constants for Metal Complexes with Organic Ligands (*Continued*)

				log K_4
Hg(II) *a,d*	20.28			
Mg *a,d*	2.7			
Mn(II) *a,d*	5.01	8.49		
Nd *a,d*	8.78	15.60	20.66	
Ni *a,d*	6.95	13.50		
Pb *a,d*	8.70	10.60		
Pr *a,d*	8.63	15.10	19.94	
Sm *a,d*	8.86	15.88	21.23	
Sr *a,d*	3.89			
Tb *a,d*	8.68	16.11	22.03	
Tm *a,d*	8.83	16.54	22.04	
Y *a,d*	8.46	15.73	21.34	
Yb *a,d*	8.85	16.61	21.83	
Zn *a,d*	6.35	11.88		
1-(2-Pyridylazo)-2-naphthol (PAN)				
Co(II)	>12			
Cu(II)	16			
Mn(II)	8.5	16.4		
Ni	12.7	25.3		
Tl(III)	2.29			
Zn	11.2	21.7		

	log K_f [ML]	log K_f [MHL]	log K_f [M(HL)$_2$]
4-(2-Pyridylazo)resorcinal (PAR)			
Co(II)		>12	
Cu(II)	10.3		
Mn(II)		9.7	18.9
Ni		13.2	26.0
Sc	4.8		
Tl(III)	4.23		
Zn		12.4	23.5

	log K_f [ML]	log K_f [M$_2$L]	log K_f [MHL]
Pyrocatechol-3,5-disulfonate (Pyrocatechol Violet)			
Al	19.13	4.95	
Bi	27.07	5.25	
Cd	8.13		5.86
Co(II)	9.01		6.53
Cu(II)	16.47		11.18
Ga	22.18	4.65	
In	18.10	4.81	
Mg	4.42	4.6	3.66
Mn(II)	7.13		5.36
Ni	9.35	4.38	6.85
Pb	13.25		10.19
Th	23.36	4.42	
Zn	10.41	6.21	7.21
Zr	27.40	4.18	

(*Continued*)

TABLE 1.76 Cumulative Formation Constants for Metal Complexes with Organic Ligands (*Continued*)

	log K_1	log K_2	log K_3	log K_4
8-Quinolinol				
Ba	2.07			
Be	3.36			
Ca (75% dioxane)	7.3	13.2		
Cd	7.2	13.4		
Ce(III) (50% dioxane)	9.15	17.13		
Co(II)	9.1	17.2		
Cu(II)	12.2	23.4		
Fe(II)	8.58	16.93	22.23	
Fe(III)	12.3	23.6	33.9	
La	5.85	16.95		
Mg (50% dioxane)	6.38	11.81		
Mn(II) (50% dioxane)	8.28	15.45		
Ni (50% dioxane)	11.44	21.38		
Pb (50% dioxane)	10.61	18.70		
Sm	6.84		19.50	
Sr	2.89	6.08		
Th	10.45	20.40	29.85	38.80
UO_2(II) (50% dioxane)	11.25	20.89		
V(II)	12.8	23.6		
VO(II)	10.97	20.19		
Y	8.15	14.90	20.25	
Zn (50% dioxane)	9.96	18.86		

	log K_f [MHL^+]	log K_f [$M(HL)_2$]
Salicylaldoxime		
Ba	0.53	3.72
Be	<7	
Ca	0.92	3.72
Cd	<4.4	
Co(II)		8.13
Cu(II)		8.13
Mg	0.64	4.10
Ni		3.77
Sr		3.77
Zn	<5.2	

	log K_1	log K_2	log K_3	log K_4
Salicylic acid				
Al	14.11			
Be	17.4			
Cd	5.55			
Ce(III)	2.66			
Co(II)	6.72	11.42		
Cr(II)	8.4	15.3		
Cu(II)	10.60	18.45		
Fe(II)	6.55	11.25		
Fe(III) *a,c*	16.48	28.12	36.80	
La	2.64			

TABLE 1.76 Cumulative Formation Constants for Metal Complexes with Organic Ligands (*Continued*)

	log K_1	log K_2	log K_3	log K_4
Mg (75% dioxane)	4.7			
Mn(II)	5.90	9.80		
Nd	2.70			
Ni	6.95	11.75		
Pr	2.68			
Th	4.25	7.60	10.05	11.60
TiO(II)	6.09			
UO_2(II)	13.4			
V(II)	6.3			
Zn	6.85			
Succinic acid				
Ba	2.08			
Be	3.08			
Ca	2.0			
Cd	2.2			
Co(II)	2.22			
Cu(II)	3.33			
Fe(III)	7.49			
Hg(II)		7.28		
La	3.96			
Mg	1.20			
Mn(II)	2.26			
Nd	8.1			
Ni	2.36			
Pb	2.8			
Ra	1.0			
Sr	1.06			
Zn	1.6			
5-Sulfosalicylic acid				
Al *c*	13.20	22.83	28.89	
Be *c*	11.71	20.81		
Cd *c*	16.68	29.08		
Co(II) *c*	6.13	9.82		
Cr(II) *c*	7.1	12.9		
Cr(III) *c*	9.56			
Cu(II) *c*	9.52	16.45		
Fe(II) *c*	5.90			
Fe(III) *c*	14.64	25.18	32.12	
La *c*	9.11			
Mn(II) *c*	5.24	8.24		
NbO(III) *c*	4.0	7.7		
Ni *c*	6.42	10.24		
UO_2(II) *c*	11.14	19.20		
Zn *c*	6.05	10.65		
Tartaric acid				
Ba		1.62		
Bi			8.30	
Ca	2.98	9.01		
Cd	2.8			
Co(II)	2.1			
Cu(II)	3.2	5.11	4.78	6.51 log K_f 19.14 [$Cu(OH)_2L^{2-}$]

(*Continued*)

TABLE 1.76 Cumulative Formation Constants for Metal Complexes with Organic Ligands (*Continued*)

	log K_1	log K_2	log K_3	log K_4
Eu(III)	4.98	8.11		
Fe(III)	7.49			
La	3.06			
Mg		1.36		
Nd	9.0			
Pb	3.78		4.7	log K_f 14.1 [$Pb(OH)_2L^{2-}$]
Ra	1.24			
Sr	1.60			
Zn	2.68	8.32		
Thioglycolic acid				
Ce(III) *a,c*	1.99	3.03		
Co(II)	5.84	12.15		
Fe(II)		10.92		
Hg(II)		43.82		
La *a,c*	1.98	2.98		
Mn(II)	4.38	7.56		
Pb	8.5			
Ni	6.98	13.53		
Rare earths *a,c*	1.9–2.1	3.0–3.3		
Y *a,c*	1.91	3.19		
Zn	7.86	15.04		
Thiourea				
Ag	7.4	13.1		
Bi				log K_6 11.9
Cd	0.6	1.6	2.6	4.6
Cu(I)			13	15.4
Hg(II)		22.1	24.7	26.8
Pb	1.4	3.1	4.7	8.3
Ru(III)	1.21		0.72	
Thoron				
Th		10.15		
Triethanolamine				
Ag	2.30	3.64		
Co(II)	1.73			
Cu(II)	4.30			
Hg(II)	6.90	13.08		
Ni	2.7			
Zn	2.00			
Triethylenetetramine (Trien)				
Ag	7.7			
Cd	10.75	13.9		
Co(II)	11.0			
Cu(II)	20.4			
Fe(II)	7.8			
Fe(III)	21.9			
Hg(II)	25.26			
Mn(II)	4.9			
Ni	14.0			
Pb	10.4			
Zn	11.9			

TABLE 1.76 Cumulative Formation Constants for Metal Complexes with Organic Ligands (*Continued*)

	log K_1	log K_2	log K_3	log K_4
1,1,1-Trifluoro-3-2′-Thenoylacetone (TTA)				
Ba		10.6		
Cu(II)	6.55	13.0		
Fe(III)	6.9			
Ni	10.0			
Pr	9.53			
Pu(III)	9.53			
Pu(IV)	8.0			
Th	8.1			
U(IV)	7.2			
Zr	3.03 [as ZrL^{3+}]			
Xylenol orange				
Bi	5.52			
Fe(III)	5.70			
Hf	6.50			
Tl(III)	4.90			
Zn	6.15			
Zr	7.60			
Zincon				
Zn	13.1			

1.21 ELECTRODE POTENTIALS

The electrode potential is the difference between the charge on an electrode and the charge in the solution.

The electrode potential is denoted as the electromotive force (EMF) and the electromotive force of any electrolytic cell is the sum of the potentials produced at two electrodes.

TABLE 1.77 Potentials of the Elements and Their Compounds at 25°C

Standard potentials are tabulated except when a solution composition is stated; the latter are formal potentials and the concentrations are in mol/liter.

Half-reaction	Standard or formal potential	Solution composition
Actinium		
$Ac^{3+} + 3e^- = Ac$	−2.13	
Aluminum		
$Al^{3+} + 3e^- = Al$	−1.676	
$AlF_6^{3-} + 3e^- = Al + 6F^-$	−2.07	
$Al(OH)_4^- + 3e^- = Al + 4OH^-$	−2.310	
Americium		
$AmO_2^{2+} + 4H^+ + 2e^- = Am^{4+} + 2H_2O$	1.20	
$AmO_2^{2+} + e^- = AmO_2^+$	1.59	
$AmO_2^+ + 4H^+ + e^- = Am^{4+} + 2H_2O$	0.82	
$AmO_2^+ + 4H^+ + 2e^- = Am^{3+} + 2H_2O$	1.72	
$Am^{4+} + e^- = Am^{3+}$	2.62	
$Am^{4+} + 4e^- = Am$	−0.90	
$Am^{3+} + 3e^- = Am$	−2.07	
Antimony		
$Sb(OH)_4^- + 2e^- = SbO_2^- + 2OH^- + 2H_2O$	−0.465	1 NaOH
$SbO_2^- + 2H_2O + 3e^- = Sb + 4OH^-$	0.639	1 NaOH
$Sb + 3H_2O + 3e^- = SbH_3 + 3OH^-$	−1.338	1 NaOH
$Sb_2O_5 + 6H^+ + 4e^- = 2SbO^+ + 3H_2O$	0.605	
$Sb_2O_5 + 4H^+ + 4e^- = Sb_2O_3 + 2H_2O$	0.699	
$Sb_2O_5 + 2H^+ + 2e^- = Sb_2O_4 + H_2O$	1.055	
$Sb_2O_4 + 2H^+ + 2e^- = Sb_2O_3 + H_2$	0.342	
$SbO^+ + 2H^+ + 3e^- = Sb + H_2O$	0.204	
$Sb + 3H^+ + 3e^- = SbH_3$	−0.510	
Arsenic		
$H_3AsO_4 + 2H^+ + 2e^- = HAsO_2 + 2H_2O$	0.560	
$HAsO_2 + 3H^+ + 3e^- = As + 2H_2O$	0.240	
$As + 3H^+ + 3e^- = AsH_3$	−0.225	
$AsO_4^{3-} + 2H^+ + 2e^- = AsO_2^- + 4OH^-$	−0.67	
$AsO_2^- + 2H_2O + 3e^- = As + 4OH^-$	−0.68	
$As + 3H_2O + 3e^- = AsH_3 + 3OH^-$	−1.37	
Astatine		
$HAtO_3 + 4H^+ + 4e^- = HAtO + 2H_2$	*ca.* 1.4	
$2HAtO + 2H^+ + 2e^- = At_2 + 2H_2O$	*ca.* 0.7	
$At_2 + 2e^- = 2At^-$	0.20	
Barium		
$BaO_2 + 4H^+ + 2e^- = Ba^{2+} + 2H_2O$	2.365	
$Ba^{2+} + 2e^- = Ba$	−2.92	

TABLE 1.77 Potentials of the Elements and Their Compounds at 25°C (*Continued*)

Half-reaction	Standard or formal potential	Solution composition
Berkelium		
$Bk^{4+} + 4e^- = Bk$	−1.05	
$Bk^{4+} + e^- = Bk^{3+}$	1.67	
$Bk^{3+} + 3e^- = Bk$	−2.01	
Beryllium		
$Be^{2+} + 2e^- = Be$	−1.99	
Bismuth		
Bi_2O_4 (bismuthate) $+ 4H^+ + 2e^- = 2BiO^+ + 2H_2O$	1.59	
$Bi^{3+} + 3e^- = Bi$	0.317	
$Bi + 3H^+ + 3e^- = BiH_3$	−0.97	
$BiCl_4^- + 3e^- = Bi + 4Cl^-$	0.199	
$BiBr_4^- + 3e^- = Bi + 4Br^-$	0.168	
$BiOCl + 2H^+ + 3e^- = Bi + H_2O + Cl^-$	0.170	
Boron		
$B(OH)_3 + 3H^+ + 3e^- = B + 3H_2O$	−0.890	
$BO_2^- + 6H_2O + 8e^- = BH_3^- + 8OH^-$	−1.241	
$B(OH)_4^- + 3e^- = B + 4OH^-$	−1.811	
Bromine		
$BrO_4^- + 2H^+ + 2e^- = BrO_3^- + H_2O$	1.853	
$BrO_3^- + 6H^+ + 6e^- = Br^- + 3H_2O$	1.478	
$BrO_3^- + 5H^+ + 4e^- = HBrO + 2H_2O$	1.444	
$2BrO_3^- + 12H^+ + 10e^- = Br_2 + 6H_2O$	1.5	
$2HBrO + 2H^+ + 2e^- = Br_2 + 2H_2O$	1.604	
$HBrO + H^+ + 2e^- = Br^- + H_2O$	1.341	
$BrO^- + H_2O + 2e^- = Br^- + 2OH^-$	0.76	1 NaOH
$Br_3^- + 2e^- = 3Br^-$	1.050	
$Br_2(aq) + 2e^- = 2Br^-$	1.087	
Cadmium		
$Cd^{2+} + 2e^- = Cd$	−0.403	
$Cd^{2+} + Hg + 2e^- = Cd(Hg)$	−0.352	
$CdCl_4^{2-} + 2e^- = Cd + 4Cl^-$	−0.453	
$Cd(CN)_4^{2-} + 2e^- = Cd + 4CN^-$	−0.943	
$Cd(NH_3)_4^{2+} + 2e^- = Cd + 4NH_3$	−0.622	
$Cd(OH)_4^{2-} + 2e^- = Cd + 4OH^-$	−0.670	
Calcium		
$CaO_2 + 4H^+ + 2e^- = Ca^{2+} + H_2O$	2.224	
$Ca^{2+} + 2e^- = Ca$	−2.84	
$Ca + 2H^+ + 2e^- = CaH_2$	0.776	
Californium		
$Cf^{3+} + 3e^- = Cf$	−1.93	
$Cf^{3+} + e^- = Cf^{2+}$	−1.6	
$Cf^{2+} + 2e^- = Cf$	−2.1	
Carbon		
$CO_2 + 2H^+ + 2e^- = CO + H_2O$	−0.106	
$CO_2 + 2H^+ + 2e^- = HCOOH$	−0.20	
$2CO_2 + 2H^+ + 2e^- = H_2C_2O_4$	−0.481	
$C_2O_4^{2-} + 2H^+ + 2e^- = 2HCOO^-$	0.145	
$HCOOH + 2H^+ + 2e^- = HCHO + H_2O$	0.034	

(*Continued*)

TABLE 1.77 Potentials of the Elements and Their Compounds at 25°C (*Continued*)

Half-reaction	Standard or formal potential	Solution composition
$C_2N_2 + 2H^+ + 2e^- = 2HCN$	0.373	
$HCNO + 2H^+ + 2e^- = CO + H_2O$	0.330	
$HCHO + 2H^+ + 2e^- = CH_3OH$	0.2323	
$CNO^- + H_2O + 2e^- = CN^- + 2OH^-$	−0.97	
Cerium		
$Ce(IV) + e^- = Ce(III)$	1.70	1 $HClO_4$
	1.61	1 HNO_3
	1.44	0.5 H_2SO_4
	1.28	1 HCl
$Ce^{3+} + 3e^- = Ce$	−2.34	
Cesium		
$Cs^+ + e^- = Cs$	−2.923	
$Cs^+ + Hg + e^- = Cs(Hg)$	−1.78	
Chlorine		
$ClO_4^- + 2H^+ + 2e^- = ClO_3^- + H_2O$	1.201	
$2ClO_4^- + 16H^+ + 14e^- = Cl_2 + 8H_2O$	1.392	
$ClO_4^- + 8H^+ + 8e^- = Cl^- + 4H_2O$	1.388	
$ClO_3^- + 2H^+ + e^- = ClO_2(g) + H_2O$	1.175	
$ClO_3^- + 3H^+ + 2e^- = HClO_2 + H_2O$	1.181	
$2ClO_3^- + 12H^+ + 10e^- = Cl_2 + 6H_2O$	1.468	
$ClO_3^- + 6H^+ + 6e^- = Cl^- + 3H_2O$	1.45	
$ClO_2(g) + H^+ + e^- = HClO_2$	1.188	
$HClO_2 + 2H^+ + 2e^- = HClO + H_2O$	1.64	
$HClO_2 + 3H^+ + 4e^- = Cl^- + 2H_2O$	1.584	
$2HClO_2 + 6H^+ + 6e^- = Cl_2(g) + 4H_2O$	1.659	
$2ClO^- + 2H_2O + 2e^- = Cl_2(g) + 4OH^-$	0.421	1 NaOH
$ClO^- + H_2O + 2e^- = Cl^- + 2OH^-$	0.890	1 NaOH
$Cl_3^- + 2e^- = 3Cl^-$	1.415	
$Cl_2(aq) + 2e^- = 2Cl^-$	1.396	
Chromium		
$Cr_2O_7^{2-} + 14H^+ + 6e^- = 2Cr^{3+} + 7H_2O$	1.36	
	1.15	0.1 H_2SO_4
	1.03	1 $HClO_4$
$CrO_4^{2-} + 4H_2O + 3e^- = Cr(OH)_4^- + 4OH^-$	−0.13	1 NaOH
$Cr^{3+} + e^- = Cr^{2+}$	−0.424	
$Cr^{3+} + 3e^- = Cr$	−0.74	
$Cr^{2+} + 2e^- = Cr$	0.90	
Cobalt		
$CoO_2 + 4H^+ + e^- = Co^{3+} + 2H_2O$	1.416	
$Co(H_2O)_6^{3+} + e^- = Co(H_2O)_6^{2+}$	1.92	
$Co(NH_3)_6^{3+} + e^- = Co(NH_3)_6^{2+}$	0.058	7 NH_3
$Co(OH)_3 + e^- = Co(OH)_2 + OH^-$	0.17	
$Co(en)_3^{3+} + e^- = Co(en)_3^{2+}$ [en = ethylenediamine]	−0.2	0.1 en
$Co(CN)_6^{3-} + e^- = Co(CN)_5^{3-} + CN^-$	−0.8	0.8 KOH
$Co^{2+} + 2e^- = Co$	−0.277	
$Co(NH_3)_6^{2+} + 2e^- = Co + 6NH_3$	−0.422	
$[Co(CO)_4]_2 + 2e^- = 2Co(CO)_4^-$	−0.40	

TABLE 1.77 Potentials of the Elements and Their Compounds at 25°C (*Continued*)

Half-reaction	Standard or formal potential	Solution composition
Copper		
$Cu^{2+} + 2e^- = Cu$	0.340	
$Cu^{2+} + e^- = Cu^+$	0.159	
$Cu^+ + e^- = Cu$	0.520	
$Cu^{2+} + Cl^- + e^- = CuCl$	0.559	
$Cu^{2+} + 2Br^- + e^- = CuBr_2^-$	0.52	1 KBr
$Cu^{2+} + I^- + e^- + CuI$	0.86	
$Cu^{2+} + 2CN^- + e^- = Cu(CN)_2^-$	1.12	
$Cu(NH_3)_4^{2+} + e^- = Cu(NH_3)_2^+ + 2NH_3$	0.10	1 NH_3
$Cu(en)_2^{2+} + e^- = Cu(en)^+ + en$	−0.35	
$Cu(CN)_2^- + e^- = Cu + 2CN^-$	−0.44	
$CuCl_3^{2-} + e^- = Cu + 3Cl^-$	0.178	1 HCl
$Cu(NH_3)_2^+ + e^- = Cu + 2NH_3$	−0.100	
Curium		
$Cm^{4+} + e^- = Cm^{3+}$	3.2	1 $HClO_4$
$Cm^{3+} + 3e^- = Cm$	−2.06	
Dysprosium		
$Dy^{3+} + 3e^- = Dy$	−2.29	
$Dy^{3+} + e^- = Dy^{2+}$	−2.5	
$Dy^{2+} + 2e^- = Dy$	−2.2	
Einsteinium		
$Es^{3+} + 3e^- = Es$	−2.0	
$Es^{3+} + e^- = Es^{2+}$	−1.5	
$Es^{2+} + 2e^- = Es$	−2.2	
Erbium		
$Er^{3+} + 3e^- = Er$	−2.32	
Europium		
$Eu^{3+} + 3e^- = Eu$	−1.99	
$Eu^{3+} + e^- = Eu^{2+}$	−0.35	
$Eu^{2+} + 2e^- = Eu$	−2.80	
Fermium		
$Fm^{3+} + 3e^- = Fm$	−1.96	
$Fm^{3+} + e^- = Fm^{2+}$	−1.15	
$Fm^{2+} + 2e^- = Fm$	−2.37	
Fluorine		
$F_2 + 2H^+ + 2e^- = 2HF$	3.053	
$F_2 + H^+ + 2e^- = HF_2^-$	2.979	
$F_2 + 2e^- = 2F^-$	2.87	
$OF_2 + 3H^+ + 4e^- = HF_2^- + H_2O$	2.209	
Francium		
$Fr^+ + e^- = Fr$	*ca.* −2.9	
Gadolinium		
$Gd^{3+} + 3e^- = Gd$	−2.28	
Gallium		
$Ga^{3+} + 3e^- = Ga$	−0.529	
$Ga^{3+} + e^- = Ga^{2+}$	−0.65	
$Ga^{2+} + 2e^- = Ga$	−0.45	

(*Continued*)

TABLE 1.77 Potentials of the Elements and Their Compounds at 25°C (*Continued*)

Half-reaction	Standard or formal potential	Solution composition
Germanium		
GeO_2(tetr) + $2H^+$ + $2e^-$ = GeO(yellow) + H_2O	−0.255	
GeO_2(tetr) + $4H^+$ + $2e^-$ = Ge^{2+} + $2H_2O$	−0.210	
GeO_2(hex) + $4H^+$ + $2e^-$ = Ge^{2+} + $2H_2O$	−0.132	
H_2GeO_3 + $4H^+$ + $4e^-$ = Ge + $3H_2O$	0.012	
Ge^{4+} + $2e^-$ = Ge^{2+}	0.0	
Ge^{2+} + $2e^-$ = Ge	0.247	
GeO + $2H^+$ + $2e^-$ = Ge + H_2O	−0.255	
Ge + $4H^+$ + $4e^-$ = GeH_4	−0.29	
Gold		
Au^{3+} + $3e^-$ = Au	1.52	
Au^{3+} + $2e^-$ = Au^+	1.36	
Au^+ + e^- = Au	1.83	
$AuCl_4^-$ + $2e^-$ = $AuCl_2^-$ + $2Cl^-$	0.926	
$AuBr_4^-$ + $2e^-$ = $AuBr_2^-$ + $2Br^-$	0.802	
$Au(SCN)_4^-$ + $2e^-$ = $Au(SCN)_2^-$ + $2SCN^-$	0.623	
$AuBr_4^-$ + $3e^-$ = Au + $4Br^-$	0.854	
$AuCl_4^-$ + $3e^-$ = Au + $4Cl^-$	1.002	
$Au(SCN)_4^-$ + $3e^-$ = Au + $4SCN^-$	0.662	
$Au(OH)_3$ + $3H^+$ + $3e^-$ = Au + $3H_2O$	1.45	
$AuBr_2^-$ + e^- = Au + $2Br^-$	0.960	
$AuCl_2^-$ + e^- = Au + $2Cl^-$	1.15	
AuI_2^- + e^- = Au + $2I^-$	0.576	
$Au(CN)_2^-$ + e^- = Au + $2CN^-$	−0.596	
$Au(SCN)_2$ + e^- = Au + $2SCN^-$	0.69	
Hafnium		
Hf^{4+} + $4e^-$ = Hf	−1.70	
HfO_2 + $4H^+$ + $4e^-$ = Hf + $2H_2O$	−1.57	
Holmium		
Ho^{3+} + $3e^-$ = Ho	−2.23	
Hydrogen		
$2H^+$ + $2e^-$ = H_2	0.0000	
$2D^+$ + $2e^-$ = D_2	0.029	
$2H_2O$ + $2e^-$ = H_2 + $2OH^-$	−0.828	
Indium		
In^{3+} + $3e^-$ = In	−0.338	
In^{3+} + $2e^-$ = In^+	−0.444	
In^+ + e^- = In	−0.126	
Iodine		
H_5IO_6 + H^+ + $2e^-$ = IO_3^- + $3H_2O$	1.603	
IO_3^- + $5H^+$ + $4e^-$ = HIO + $2H_2O$	1.14	
HIO_3 + $5H^+$ + $2Cl^-$ + $4e^-$ = ICl_2^- + $3H_2O$	1.214	
$2IO_3^-$ + $12H^+$ + $10e^-$ = I_2(c) + $3H_2O$	1.195	
IO_3^- + $3H_2O$ + $6e^-$ = I^- + $6OH^-$	0.257	
$2IBr_2^-$ + $2e^-$ = I_2Br^- + $3Br^-$	0.821	
$2IBr_2^-$ + $2e^-$ = I_2(c) + $4Br^-$	0.874	
2IBr + $2e^-$ = I_2Br^- + Br^-	0.973	
2IBr + $2e^-$ = I_2 + $2Br^-$	1.02	
2ICl + $2e^-$ = I_2(c) + $2Cl^-$	1.20	

TABLE 1.77 Potentials of the Elements and Their Compounds at 25°C (*Continued*)

Half-reaction	Standard or formal potential	Solution composition
$2ICl_2^- + 2e^- = I_2(c) + 4Cl^-$	1.07	
$2ICN + 2H^+ + 2e^- = I_2(c) + 2HCN$	0.695	
$2ICN + 2H^+ + 2e^- = I_2(aq) + 2HCN$	0.609	
$2HIO + 2H^+ + 2e^- = I_2 + 2H_2O$	1.45	
$HIO + H^+ + 2e^- = I^- + H_2O$	0.985	
$I_3^- + 2e^- = 3I^-$	0.536	
$I_2(aq) + 2e^- = 2I^-$	0.621	
$I_2(c) + 2e^- = 2I^-$	0.5355	
Iridium		
$IrBr_6^{2-} + e^- = IrBr_6^{3-}$	0.805	
$IrCl_6^{2-} + e^- = IrCl_6^{3-}$	0.867	
$IrI_6^{2-} + e^- = IrI_6^{3-}$	0.49	
$IrO_2 + 4H^+ + e^- = Ir^{3+} + 2H_2O$	0.223	
$IrO_2 + 4H^+ + 4e^- = Ir + 2H_2O$	0.935	1 H_2SO_4
$Ir^{3+} + 3e^- = Ir$	1.156	
$IrCl_6^{2-} + 4e^- = Ir + 6Cl^-$	0.835	
$IrCl_6^{3-} + 3e^- = Ir + 6Cl^-$	0.77	
Iron		
$FeO_4^{2-} + 8H^+ + 3e^- = Fe^{3+} + 4H_2O$	2.2	
$FeO_4^{2-} + 2H_2O + 3e^- = FeO_2^- + 4OH^-$	0.55	10 NaOH
$Fe^{3+} + e^- = Fe^{2+}$	0.771	
	0.70	1 HCl
	0.67	0.5 H_2SO_4
	0.44	0.3 H_3PO_4
$Fe(CN)_6^{3-} + e^- = Fe(CN)_6^{4-}$	0.361	
	0.71	1 HCl
$Fe(EDTA)^- + e^- = Fe(EDTA)^{2-}$	0.12	0.1 EDTA, pH 4–6
$Fe(OH)_4^- + e^- = Fe(OH)_4^{2-}$	−0.73	1 NaOH
$Fe^{2+} + 2e^- = Fe$	−0.44	
$[Fe(CO)_4]_3 + 6e^- = 3Fe(CO)_4^{2-}$	−0.70	
Lanthanum		
$La^{3+} + 3e^- = La$	−2.38	
Lawrencium		
$Lr^{3+} + 3e^- = Lr$	−2.0	
Lead		
$Pb^{4+} + 2e^- = Pb^{2+}$	1.65	
$PbO_2(alpha) + SO_4^{2-} + 4H^+ + 2e^- = PbSO_4 + 2H_2O$	1.690	
$PbO_2 + 4H^+ + 2e^- = Pb^{2+} + 2H_2O$	1.46	
$PbO_2 + 2H^+ + 2e^- = PbO + H_2O$	0.28	
$PbO^{2-} + H_2O + 2e^- = HPbO_2^- + 3OH^-$	0.3	2 NaOH
$Pb^{2+} + 2e^- = Pb$	−0.126	
$HPbO_2^- + H_2O + 2e^- = Pb + 3OH^-$	−0.54	
$PbHPO_4 + 2e^- = Pb + HPO_4^{2-}$	−0.465	
$PbSO_4 + 2e^- = Pb + SO_4^{2-}$	−0.356	
$PbF_2 + 2e^- = Pb + 2F^-$	−0.344	
$PbCl_2 + 2e^- = Pb + 2Cl^-$	−0.268	
$PbBr_2 + 2e^- = Pb + 2Br^-$	−0.280	
$PbI_2 + 2e^- = Pb + 2I^-$	−0.365	
$Pb + 2H^+ + 2e^- = PbH_2$	−1.507	

(*Continued*)

TABLE 1.77 Potentials of the Elements and Their Compounds at 25°C (*Continued*)

Half-reaction	Standard or formal potential	Solution composition
Lithium		
$Li^+ + e^- = Li$	−3.040	
$Li^+ + Hg + e^- = Li(Hg)$	−2.00	
Lutetium		
$Lu^{3+} + 3e^- = Lu$	−2.30	
Magnesium		
$Mg^{2+} + 2e^- = Mg$	−2.356	
$Mg(OH)_2 + 2e^- = Mg + 2OH^-$	−2.687	
Manganese		
$MnO_4^- + e^- = MnO_4^{2-}$	0.56	
$MnO_4^- + 4H^+ + 3e^- = MnO_2(beta) + 2H_2O$	1.70	
$MnO_4^- + 2H_2O + 3e^- = MnO_2 + 4OH^-$	0.60	
$MnO_4^- + 8H^+ + 5e^- = Mn^{2+} + 4H_2O$	1.51	
$MnO_4^{2-} + e^- = MnO_4^{3-}$	0.27	
$MnO_4^{2-} + 2H_2O + 2e^- = MnO_2 + 4OH^-$	0.62	
$MnO_4^{3-} + 2H_2O + e^- = MnO_2 + 4OH^-$	0.96	
$MnO_2 + 4H^+ + e^- = Mn^{3+} + 2H_2O$	0.95	
$MnO_2(beta) + 4H^+ + 2e^- = Mn^{2+} + 2H_2O$	1.23	
$Mn^{3+} + e^- = Mn^{2+}$	1.5	
$Mn(H_2P_2O_7)_3^{3-} + 2H^+ + e^- = Mn(H_2P_2O_7)_2^{2-} + H_4P_2O_7$	1.15	0.4 $H_2P_2O_7^{2-}$
$Mn(CN)_6^{3-} + e^- = Mn(CN)_6^{4-}$	−0.24	1.5 NaCN
$Mn^{2+} + 2e^- = Mn$	−1.17	
Mendelevium		
$Md^{3+} + 3e^- = Md$	−1.7	
$Md^{3+} + e^- = Md^{2+}$	−0.15	
$Md^{2+} + 2e^- = Md$	−2.4	
Mercury		
$2Hg^{2+} + 2e^- = Hg_2^{2+}$	0.911	
$2HgCl_2 + 2e^- = Hg_2Cl_2 + 2Cl^-$	0.63	
$Hg^{2+} + 2e^- = Hg(lq)$	0.8535	
$HgO(c,red) + 2H^+ + 2e^- = Hg + H_2O$	0.926	
$Hg_2^{2+} + 2e^- = 2Hg$	0.7960	
$Hg_2F_2 + 2e^- = 2Hg + 2F^-$	0.656	
$Hg_2Cl_2 + 2e^- = 2Hg + 2Cl^-$	0.2682	
$Hg_2Br_2 + 2e^- = 2Hg + 2Br^-$	0.1392	
$Hg_2I_2 + 2e^- = 2Hg + 2I^-$	−0.0405	
$Hg_2SO_4 + 2e^- = 2Hg + SO_4^{2-}$	0.614	
Molybdenum		
$MoO_4^{2-} + 4H_2O + 6e^- = Mo + 8OH^-$	−0.913	
$H_2MoO_4 + 6H^+ + 6e^- = Mo + 4H_2O$	0.114	
$H_2MoO_4 + 2H^+ + 2e^- = MoO_2 + 2H_2O$	0.646	
$MoO_2 + 4H^+ + 4e^- = Mo + 2H_2O$	−0.152	
$H_2MoO_4 + 6H^+ + 3e^- = Mo^{3+} + 4H_2O$	0.428	
$Mo(CN)_8^{3-} + e^- = Mo(CN)_8^{4-}$	0.725	
$Mo^{3+} + 3e^- = Mo$	−0.2	
Neodymium		
$Nd^{3+} + 3e^- = Nd$	−2.32	
$Nd^{3+} + e^- = Nd^{2+}$	−2.6	
$Nd^{2+} + 2e^- = Nd$	−2.2	

TABLE 1.77 Potentials of the Elements and Their Compounds at 25°C (*Continued*)

Half-reaction	Standard or formal potential	Solution composition
Neptunium		
$NpO_3^+ + 2H^+ + e^- = NpO_2^{2+} + H_2O$	2.04	
$NpO_2^{2+} + e^- = NpO_2^+$	1.34	
$NpO_2^{2+} + 4H^+ + 2e^- = Np^{4+} + 2H_2O$	0.95	
$Np^{4+} + e^- = Np^{3+}$	0.18	
$Np^{4+} + 4e^- = Np$	−1.30	
$Np^{3+} + 3e^- = Np$	−1.79	
Nickel		
$NiO_4^{2-} + 4H^+ + 2e^- = NiO_2 + 2H_2O$	1.8	
$NiO_2 + 4H^+ + 2e^- = Ni^{2+} + 2H_2O$	1.593	
$NiO_2 + 2H_2O + 2e^- = Ni(OH)_2 + 2OH^-$	0.490	
$Ni(CN)_4^{2-} + e^- = Ni(CN)_3^{2-} + CN^-$	−0.401	
$Ni^{2+} + 2e^- = Ni$	−0.257	
$Ni(OH)_2 + 2e^- = Ni + 2OH^-$	−0.72	
$Ni(NH_3)_6^{2+} + 2e^- = Ni + 6NH_3$	−0.49	
Niobium		
$Nb_2O_5 + 10H^+ + 4e^- = 2Nb^{3+} + 5H_2O$	−0.1	
$Nb_2O_5 + 10H^+ + 10e^- = 2Nb + 5H_2O$	−0.65	
$Nb^{3+} + 3e^- = Nb$	−1.1	
Nitrogen		
$2NO_3^- + 4H^+ + 2e^- = N_2O_4 + 2H_2O$	0.803	
$NO_3^- + 3H^+ + 2e^- = HNO_2 + H_2O$	0.94	
$N_2O_4 + 2H^+ + 2e^- = 2HNO_2$	1.07	
$HNO_2 + H^+ + e^- = NO + H_2O$	0.996	
$2HNO_2 + 4H^+ + 4e^- = N_2O(g) + 3H_2O$	1.297	
$2HNO_2 + 4H^+ + 4e^- = H_2N_2O_2 + 2H_2O$	0.86	
$2NO + 2H^+ + 2e^- = H_2N_2O_2$	0.71	
$2NO + 2H^+ + 2e^- = N_2O + H_2O$	1.59	
$H_2N_2O_2 + 6H^+ + 4e^- = 2HONH_3^+$	0.496	
$N_2O + 2H^+ + 2e^- = N_2 + H_2O$	1.77	
$N_2O + 6H^+ + H_2O + 4e^- = 2HONH_3^+$	−0.05	
$N_2 + 2H_2O + 4H^+ + 2e^- = 2HONH_3^+$	−1.87	
$N_2 + 5H^+ + 4e^- = N_2H_5^+$	−0.23	
$HONH_3^+ + 2H^+ + 2e^- = NH_4^+ + H_2O$	1.35	
$2HONH_3^+ + H^+ + 2e^- = N_2H_5^+ + 2H_2O$	1.41	
$N_2H_5^+ + 3H^+ + 2e^- = 2NH_4^+$	1.275	
$3N_2 + 2H^+ + 2e^- = 2HN_3$	−3.40	
Nobelium		
$No^{3+} + 3e^- = No$	−1.2	
$No^{3+} + e^- = No^{2+}$	1.4	
$No^{2+} + 2e^- = No$	−2.5	
Osmium		
$OsO_4(aq) + 4H^+ + 4e^- = OsO_2 \cdot 2H_2O + 2H_2O$	0.964	
$OsO_4(c, yellow) + 8H^+ + 8e^- = Os + 4H_2O$	0.85	
$OsO_2 + 4H^+ + 4e^- = Os + 2H_2O$	0.687	
$OsCl_6^{2-} + e^- = OsCl_6^{3-}$	0.45	
$OsBr_6^{2-} + e^- = OsBr_6^{3-}$	0.35	
Oxygen		
$O_3 + 2H^+ + 2e^- = O_2 + H_2O$	2.075	
$O_3 + H_2O + 2e^- = O_2 + 2OH^-$	1.240	1 NaOH

(*Continued*)

TABLE 1.77 Potentials of the Elements and Their Compounds at 25°C (*Continued*)

Half-reaction	Standard or formal potential	Solution composition
$O_2 + 4H^+ + 4e^- = 2H_2O$	1.229	
$O_2 + 2H^+ + 2e^- = H_2O$	0.695	
$O_2 + H_2O + 2e^- = HO_2^- + OH^-$	−0.076	
$H_2O_2 + 2H^+ + 2e^- = 2H_2O$	1.763	
$HO_2^- + H_2O + 2e^- = 3OH^-$	0.867	1 NaOH
$O_2 + 2H_2O + 4e^- = 4OH^-$	0.401	
Palladium		
$PdO_3 + 2H^+ + 2e^- = PdO_2 + H_2O$	2.030	
$PdCl_6^{2-} + 2e^- = PdCl_4^{2-} + 2Cl^-$	1.470	
$PdBr_6^{2-} + 2e^- = PdBr_4^{2-} + 2Br^-$	0.99	
$PdI_6^{2-} + 2e^- = PdI_4^{2-} + 2I^-$	0.48	
$Pd^{2+} + 2e^- = Pd$	0.915	
$PdCl_4^{2-} + 2e^- = Pd + 4Cl^-$	0.62	1 HCl
$PdBr_4^{2-} + 2e^- = Pd + 4Br^-$	0.49	
$Pd(NH_3)_4^{2+} + 2e^- = Pd + 4NH_3$	0.0	1 NH_3
$Pd(CN)_4^{2-} + 2e^- = Pd + 4CN^-$	−1.35	1 KCN
Phosphorus		
$H_3PO_4 + 2H^+ + 2e^- = H_3PO_3 + H_2O$	−0.276	
$2H_3PO_4 + 2H^+ + 2e^- = H_4P_2O_6 + 2H_2O$	−0.933	
$H_4P_2O_6 + 2H^+ + 2e^- = 2H_3PO_3$	0.380	
$H_3PO_3 + 2H^+ + 2e^- = HPH_2O_2 + H_2O$	−0.499	
$HPH_2O_2 + H^+ + e^- = P + 2H_2O$	−0.365	
$H_3PO_3 + 3H^+ + 3e^- = P + 3H_2O$	−0.502	
$2P(white) + 4H^+ + 4e^- = P_2H_4$	−0.100	
$P_2H_4 + 2H^+ + 2e^- = 2PH_3$	−0.006	
$P(white) + 3H^+ + 3e^- = PH_3$	−0.063	
Platinum		
$PtO_3 + 2H^+ + 2e^- = PtO_2 + H_2O$	2.0	
$PtO_2 + 2H^+ + 2e^- = PtO + H_2O$	1.045	
$PtCl_6^{2-} + 2e^- = PtCl_4^{2-} + 2Cl^-$	0.726	
$PtBr_6^{2-} + 2e^- = PtBr_4^{2-} + 2Br^-$	0.613	1 KBr
$PtI_6^{2-} + 2e^- = PtI_4^{2-} + 2I^-$	0.321	1 KI
$Pt^{2+} + 2e^- = Pt$	1.188	
$PtCl_4^{2-} + 2e^- = Pt + 4Cl^-$	0.758	
$PtBr_4^{2-} + 2e^- = Pt + 4Br^-$	0.698	
Plutonium		
$PuO_2^{2+} + e^- = PuO_2^+$	1.02	
$PuO_2^{2+} + 4H^+ + 2e^- = Pu^{4+} + 2H_2O$	1.04	
$Pu^{4+} + e^- = Pu^{3+}$	1.01	
	0.80	1 H_3PO_4
	0.50	1 HF
$Pu^{4+} + 4e^- = Pu$	−1.25	
$Pu^{3+} + 3e^- = Pu$	−2.00	
Polonium		
$PoO_2 + 4H^+ + 2e^- = Po^{2+} + 2H_2O$	1.1	
$Po^{4+} + 4e^- = Po$	0.73	
$Po^{2+} + 2e^- = Po$	0.37	
$Po + 2H^+ + 2e^- = H_2Po$	*ca.* −1.0	

TABLE 1.77 Potentials of the Elements and Their Compounds at 25°C (*Continued*)

Half-reaction	Standard or formal potential	Solution composition
Potassium		
$K^+ + e^- = K$	−2.924	
$K^+ + Hg + e^- = K(Hg)$	*ca.* −1.9	
Praseodymium		
$Pr^{4+} + e^- = Pr^{3+}$	3.2	
$Pr^{3+} + e^- = Pr$	−2.35	
Promethium		
$Pm^{3+} + 3e^- = Pm$	−2.42	
Protoactinium		
$PaOOH^{2+} + 3H^+ + e^- = Pa^{4+} + 2H_2O$	−0.10	
$PaOOH^{2+} + 3H^+ + 5e^- = Pa + 2H_2O$	−1.19	
$Pa^{4+} + 4e^- = Pa$	−1.46	
Radium		
$Ra^{2+} + 2e^- = Ra$	−2.916	
Rhenium		
$ReO_4^- + 2H^+ + e^- = ReO_3 + H_2O$	0.768	
$ReO_4^- + 4H^+ + 3e^- = ReO_2 + 2H_2O$	0.51	
$ReO_4^- + 2H_2O + 3e^- = ReO_2 + 4OH^-$	−0.594	
$ReO_4^- + 6Cl^- + 8H^+ + 3e^- = ReCl_6^{2-} + 4H_2O$	0.12	
$2ReO_4^- + 10H^+ + 8e^- = Re_2O_3 + 5H_2O$	−0.808	
$ReO_3 + 2H^+ + 2e^- = ReO_2 + H_2O$	0.63	
$ReO_2 + 4H^+ + 4e^- = Re + 2H_2O$	0.22	
$ReCl_6^{2-} + 4e^- = Re + 6Cl^-$	0.51	
$Re + e^- = Re^-$	−0.10	
Rhodium		
$RhO_2 + 4H^+ + e^- = Rh^{3+} + 2H_2O$	1.881	
$Rh^{3+} + 3e^- = Rh$	0.76	
$RhCl_6^{3-} + 3e^- = Rh + 6Cl^-$	0.5	
Rubidium		
$Rb^+ + e^- = Rb$	−2.924	
$Rb^+ + Hg + e^- = Rb(Hg)$	−1.81	
Ruthenium		
$RuO_4 + e^- = RuO_4^-$	0.89	
$RuO_4 + 4H^+ + 4e^- = RuO_2 + 2H_2O$	1.4	
$RuO_4 + 8H^+ + 8e^- = Ru + 4H_2O$	1.04	
$RuO_4^- + e^- = RuO_4^{2-}$	0.593	
$RuO_4^{2-} + 4H^+ + 2e^- = RuO_2 + 2H_2O$	2.0	
$RuO_2 + 4H^+ + 4e^- = Ru + 2H_2O$	0.68	
$Ru(H_2O)_6^{3+} + e^- = Ru(H_2O)_6^{2+}$	0.249	
$Ru(NH_3)_6^{3+} + e^- = Ru(NH_3)_6^{2+}$	0.10	
$Ru(CN)_6^{3-} + e^- = Ru(CN)_6^{4-}$	0.86	
$Ru^{3+} + e^- = Ru^{2+}$	0.249	
Samarium		
$Sm^{3+} + 3e^- = Sm$	−2.30	
$Sm^{3+} + e^- = Sm^{2+}$	−1.55	
$Sm^{2+} + 2e^- = Sm$	−2.67	
Scandium		
$Sc^{3+} + 3e^- = Sc$	−2.03	

(*Continued*)

TABLE 1.77 Potentials of the Elements and Their Compounds at 25°C (*Continued*)

Half-reaction	Standard or formal potential	Solution composition
Selenium		
$SeO_4^{2-} + 4H^+ + 2e^- = H_2SeO_3 + H_2O$	1.151	
$H_2SeO_3 + 4H^+ + 4e^- = Se + 3H_2O$	0.74	
$Se(c) + 2H^+ + 2e^- = H_2Se(aq)$	−0.115	
$Se + H^+ + 2e^- = HSe^-$	−0.227	
$Se + 2e^- = Se^{2-}$	−0.670	1 NaOH
Silicon		
$SiO_2(quartz) + 4H^+ + 4e^- = Si + 2H_2O$	−0.909	
$SiO_2 + 2H^+ + 2e^- = SiO + H_2O$	−0.967	
$SiO_2 + 8H^+ + 8e^- = SiH_4 + 2H_2O$	−0.516	
$SiF_6^{2-} + 4e^- = Si + 6F^-$	−1.37	
$SiO + 2H^+ + 2e^- = Si + H_2O$	−0.808	
$Si + 4H^+ + 4e^- = SiH_4(g)$	−0.143	
Silver		
$AgO^+ + 2H^+ + e^- = Ag^{2+} + H_2O$	1.360	
$Ag_2O_3 + 2H^+ + 2e^- = 2AgO + H_2O$	1.569	
$Ag_2O_3 + H_2O + 2e^- = 2AgO + 2OH^-$	0.739	1 NaOH
$Ag_2O_3 + 6H^+ + 4e^- = 2Ag^+ + 3H_2O$	1.670	
$Ag^{2+} + e^- = Ag^+$	1.980	
$AgO + 2H^+ + e^- = Ag^+ + H_2O$	1.772	
$Ag^+ + e^- = Ag$	0.7991	
$Ag_2SO_4 + 2e^- = 2Ag + SO_4^{2-}$	0.653	
$Ag_2C_2O_4 + 2e^- = 2Ag + C_2O_4^{2-}$	0.47	
$Ag_2CrO_4 + 2e^- = 2Ag + CrO_4^{2-}$	0.447	
$Ag(NH_3)_2^+ + e^- = Ag + 2NH_3$	0.373	
$AgCl + e^- = Ag + Cl^-$	0.2223	
$AgBr + e^- = Ag + Br^-$	0.071	
$AgCN + e^- = Ag + CN^-$	−0.017	
$AgI + e^- = Ag + I^-$	−0.152	
$Ag(CN) + e^- = Ag + 2CN^-$	−0.31	
$AgSCN + e^- = Ag + SCN^-$	0.09	
$Ag_2S + 2e^- = 2Ag + S^{2-}$	−0.71	
Sodium		
$Na^+ + e^- = Na$	−2.713	
$Na^+ + Hg + e^- = Na(Hg)$	−1.84	
Strontium		
$SrO_2 + 4H^+ + 2e^- = Sr^{2+}$	2.33	
$Sr^{2+} + 2e^- = Sr$	−2.89	
Sulfur		
$S_2O_8^{2-} + 2e^- = 2SO_4^{2-}$	1.96	
$S_2O_8^{2-} + 2H^+ + 2e^- = 2HSO_4^-$	2.08	
$2SO_4^{2-} + 4H^+ + 2e^- = S_2O_6^{2-} + 2H_2O$	−0.25	
$SO_4^{2-} + 4H^+ + 2e^- = SO_2(aq) + H_2O$	0.158	
$SO_4^{2-} + H_2O + 2e^- = SO_3^{2-} + 2OH^-$	−0.936	
$S_2O_6^{2-} + 4H^+ + 2e^- = 2H_2SO_3$	0.569	
$S_2O_6^{2-} + 2e^- = 2SO_3^{2-}$	0.037	
$2HSO_3^- + 2H^+ + 2e^- = S_2O_4^{2-} + 2H_2O$	0.099	
$2SO_3^{2-} + 2H_2O + 2e^- = S_2O_4^{2-} + 4OH^-$	−1.13	
$4H_2SO_3 + 4H^+ + 6e^- = S_4O_6^{2-} + 6H_2O$	0.507	

TABLE 1.77 Potentials of the Elements and Their Compounds at 25°C (*Continued*)

Half-reaction	Standard or formal potential	Solution composition
$4HSO_3^- + 8H^+ + 6e^- = S_4O_6^{2-} + 6H_2O$	0.577	
$2SO_2(aq) + 2H^+ + 4e^- = S_2O_3^{2-} + H_2O$	0.400	
$2SO_3^{2-} + 3H_2O + 4e^- = S_2O_3^{2-} + 6OH^-$	−0.576	1 NaOH
$SO_3^{2-} + 3H_2O + 4e^- = S + 6OH^-$	−0.59	1 NaOH
$S_4O_6^{2-} + 2e^- = 2S_2O_3^{2-}$	0.080	
$S_2O_3^{2-} + 6H^+ + 4e^- = 2S + 3H_2O$	0.5	
$SF_4(g) + 4e^- = S + 4F^-$	0.97	
$S_2Cl_2(g) + 2e^- = 2S + 2Cl^-$	1.19	
$S + H^+ + 2e^- = HS^-$	0.287	
$S + 2H^+ + 2e^- = H_2S(aq)$	0.144	
$S + 2H^+ + 2e^- = H_2S(g)$	0.174	
$S + 2e^- = S^{2-}$	−0.407	
Tantalum		
$Ta_2O_5 + 10H^+ + 10e^- = 2Ta + 5H_2O$	−0.81	
$TaF_7^{2-} + 5e^- = Ta + 7F^-$	−0.45	
Technetium		
$TcO_4^- + 4H^+ + 3e^- = TcO_2 + 2H_2O$	0.738	
$TcO_4^- + 2H^+ + e^- = TcO_3 + H_2O$	0.700	
$TcO_4^- + e^- = TcO_4^{2-}$	0.569	
$TcO_4^- + 8H^+ + 7e^- = Tc + 4H_2O$	0.472	
$TcO_4^{2-} + 4H^+ + 2e^- = TcO_2 + 2H_2O$	1.39	
$TcO_2 + 4H^+ + 4e^- = Tc + 2H_2O$	0.272	
$Tc + e^- = Tc^-$	*ca.* −0.5	
Tellurium		
$H_2TeO_4 + 6H^+ + 2e^- = Te^{4+} + 4H_2O$	0.929	
$H_2TeO_4 + 2H^+ + 2e^- = TeO_2(c) + 2H_2O$	1.02	
$TeO_4^{2-} + 2H^+ + 2e^- = TeO_3^{2-} + H_2O$	0.897	
$TeOOH^+ + 3H^+ + 4e^- = Te + 2H_2O$	0.559	
$H_2TeO_3 + 4H^+ + 4e^- = Te + 3H_2O$	0.589	
$TeO_3^{2-} + 6H^+ + 4e^- = Te + 3H_2O$	0.827	
$TeO_3^{2-} + 3H_2O + 4e^- = Te + 6OH^-$	−0.415	
$TeO_2(c) + 4H^+ + 4e^- = Te + 2H_2O$	0.521	
$Te + 2H^+ + 2e^- = H_2Te(aq)$	−0.740	
$Te + H^+ + 2e^- = HTe^-$	−0.817	
$Te^{2-} + 2H^+ + 2e^- = 2HTe^-$	−0.794	
Terbium		
$Tb^{3+} + 3e^- = Tb$	−2.31	
Thallium		
$Tl^{3+} + 2e^- = Tl^+$	1.25	1 $HClO_4$
	0.77	1 HCl
$Tl^{3+} + 3e^- = Tl$	0.72	
$Tl^+ + e^- = Tl$	−0.336	
$TlCl + e^- = Tl + Cl^-$	−0.557	
$TlBr + e^- = Tl + Br^-$	−0.658	
$TlI + e^- = Tl + I^-$	−0.752	
Thorium		
$Th^{4+} + 4e^- = Th$	−1.83	

(*Continued*)

TABLE 1.77 Potentials of the Elements and Their Compounds at 25°C (*Continued*)

Half-reaction	Standard or formal potential	Solution composition
Thullium		
$Tm^{3+} + 3e^- = Tm$	−2.32	
Tin		
$Sn^{4+} + 2e^- = Sn^{2+}$	0.154	
$SnCl_6^{2-} + 2e^- = SnCl_4^{2-} + 2Cl^-$	0.14	
$SnO_3^{2-} + 6H^+ + 2e^- = Sn^{2+} + 3H_2O$	0.849	
$SnF_6^{2-} + 4e^- = Sn + 6F^-$	−0.200	
$Sn^{2+} + 2e^- = Sn$	−0.1375	
$SnCl_4^{2-} + 2e^- = Sn + 4Cl^-$	−0.19	1 HCl
$HSnO_2^- + H_2O + 2e^- = Sn + 3OH^-$	−0.91	
$Sn + 4H^+ + 4e^- = SnH_4$	−1.07	
Titanium		
$TiO^{2+} + 2H^+ + e^- = Ti^{3+} + H_2O$	−0.10	
$TiO^{2+} + 2H^+ + 4e^- = Ti + H_2O$	−0.86	
$Ti^{3+} + e^- = Ti^{2+}$	−0.37	
$Ti^{3+} + 3e^- = Ti$	−1.21	
$Ti^{2+} + 2e^- = Ti$	−1.63	
Tungsten		
$2WO_3 + 2H^+ + 2e^- = W_2O_5 + H_2O$	−0.029	
$WO_3 + 6H^+ + 6e^- = W + 3H_2O$	−0.090	
$WO_4^{2-} + 4H_2O + 6e^- = W + 8OH^-$	−1.074	
$WO_4^{2-} + 2H_2O + 2e^- = WO_2 + 4OH^-$	−1.259	
$W_2O_5 + 2H^+ + 2e^- = 2WO_2 + H_2O$	−0.031	
$W(CN)_8^{3-} + e^- = W(CN)_8^{4-}$	0.457	
$WO_2 + 4H^+ + 4e^- = W + 2H_2O$	−0.119	
$WO_2 + 2H_2O + 4e^- = W + 4OH^-$	−0.982	
Uranium		
$UO_2^{2+} + e^- = UO_2^+$	0.16	
$UO_2^{2+} + 4H^+ + 2e^- = U^{4+} + 2H_2O$	0.27	
$UO_2^+ + 4H^+ + e^- = U^{4+} + 2H_2O$	0.38	
$U^{4+} + e^- = U^{3+}$	−0.52	
$U^{4+} + 4e^- = U$	−1.38	
$U^{3+} + 3e^- = U$	−1.66	
Vanadium		
$VO_2^+ + 2H^+ + e^- = VO^{2+} + H_2O$	1.000	
$VO_2^+ + 4H^+ + 2e^- = V^{3+} + 2H_2O$	0.668	
$VO_2^+ + 4H^+ + 3e^- = V^{2+} + 2H_2O$	0.361	
$VO_2^+ + 4H^+ + 5e^- = V + 4H_2O$	−0.236	
$VO^{2+} + 2H^+ + e^- = V^{3+} + H_2O$	0.337	
$V^{3+} + e^- = V^{2+}$	−0.255	
$V^{2+} + 2e^- = V$	−1.13	
Xenon		
$H_4XeO_6 + 2H^+ + 2e^- = XeO_3 + 3H_2O$	2.42	
$HXeO_6^{3-} + 2H_2O + e^- = HXeO_4 + 4OH^-$	0.9	
$XeO_3 + 6H^+ + 2F^- + 4e^- = XeF_2 + 3H_2O$	1.6	
$XeO_3 + 6H^+ + 6e^- = Xe(g) + 3H_2O$	2.10	
$XeF_2 + e^- = XeF + F^-$	0.9	
$XeF_2 + 2H^+ + 2e^- = Xe(g) + 2HF$	2.64	
$XeF + e^- = Xe(g) + F^-$	3.4	

TABLE 1.77 Potentials of the Elements and Their Compounds at 25°C (*Continued*)

Half-reaction	Standard or formal potential	Solution composition
Ytterbium		
$Yb^{3+} + e^- = Yb^{2+}$	−1.05	
$Yb^{2+} + 2e^- = Yb$	−2.8	
$Yb^{3+} + 3e^- = Yb$	−2.22	
Yttrium		
$Y^{3+} + 3e^- = Y$	−2.37	
Zinc		
$Zn^{2+} + 2e^- = Zn$	−0.7626	
$Zn(NH_3)_4^{2+} + 2e^- = Zn + 4NH_3$	−1.04	
$Zn(CN)_4^{2-} + 2e^- = Zn + 4CN^-$	−1.34	
$Zn(tartrate)_4^{6-} + 2e^- = Zn + 4(tartrate)^{2-}$	−1.15	
$Zn(OH)_4^{2-} + 2e^- = Zn + 4OH^-$	−1.285	
Zirconium		
$Zr^{4+} + 4e^- = Zr$	−1.55	
$ZrO_2 + 4H^+ + 4e^- = Zr + 2H_2O$	−1.45	

Source: A. J. Bard, R. Parsons, and J. Jordan (eds.), *Standard Potentials in Aqueous Solution* (prepared under the auspices of the International Union of Pure and Applied Chemistry), Marcel Dekker, New York, 1985; G. Charlot et al., *Selected Constants: Oxidation-Reduction Potentials of Inorganic Substances in Aqueous Solution*, Butterworths, London, 1971.

TABLE 1.78 Potentials of Selected Half-Reactions at 25°C

A summary of oxidation-reduction half-reactions arranged in order of decreasing oxidation strength and useful for selecting reagent systems.

Half-reaction	$E°$, volts
$F_2(g) + 2H^+ + 2e^- = 2HF$	3.053
$O_3 + H_2O + 2e^- = O_2 + 2OH^-$	1.246
$O_3 + 2H^+ + 2e^- = O_2 + H_2O$	2.075
$Ag^{2+} + e^- = Ag^+$	1.980
$S_2O_8^{2-} + 2e^- = 2SO_4^{2-}$	1.96
$HN_3 + 3H^+ + 2e^- = NH_4^+ + N_2$	1.96
$H_2O_2 + 2H^+ + 2e^- = 2H_2O$	1.763
$Ce^{4+} + e^- = Ce^{3+}$	1.72
$MnO_4^- + 4H^+ + 3e^- = MnO_2(c) + 2H_2O$	1.70
$2HClO + 2H^+ + 2e^- = Cl_2 + H_2O$	1.630
$2HBrO + 2H^+ + 2e^- = Br_2 + H_2O$	1.604
$H_5IO_6 + H^+ + 2e^- = IO_3 + 3H_2O$	1.603
$NiO_2 + 4H^+ + 2e^- = Ni^{2+} + 2H_2O$	1.593
$Bi_2O_4(bismuthate) + 4H^+ + 2e^- = 2BiO^+ + 2H_2O$	1.59
$MnO_4^- + 8H^+ + 5e^- = Mn^{2+} + 4H_2O$	1.51
$2BrO_3^- + 12H^+ + 10e^- = Br_2 + 6H_2O$	1.478
$PbO_2 + 4H^+ + 2e^- = Pb^{2+} + 2H_2O$	1.468
$Cr_2O_7^{2-} + 14H^+ + 6e^- = 2Cr^{3+} + 7H_2O$	1.36
$Cl_2 + 2e^- = 2Cl^-$	1.3583
$2HNO_2 + 4H^+ + 4e^- = N_2O + 3H_2O$	1.297
$N_2H_5^+ + 3H^+ + 2e^- = 2NH_4^+$	1.275
$MnO_2 + 4H^+ + 2e^- = Mn^{2+} + 2H_2O$	1.23
$O_2 + 4H^+ + 4e^- = 2H_2O$	1.229
$ClO_4^- + 2H^+ + 2e^- = ClO_3^- + H_2O$	1.201

(*Continued*)

TABLE 1.78 Potentials of Selected Half-Reactions at 25°C (*Continued*)

Half-reaction	$E°$, volts
$2IO_3^- + 12H^+ + 10e^- = I_2 + 3H_2O$	1.195
$N_2O_4 + 2H^+ + 2e^- = 2HNO_3$	1.07
$2ICl_2^- + 2e^- = 4Cl^- + I_2$	1.07
$Br_2(1q) + 2e^- = 2Br^-$	1.065
$N_2O_4 + 4H^+ + 4e^- = 2NO + 2H_2O$	1.039
$HNO_2 + H^+ + e^- = NO + H_2O$	0.996
$NO_3^- + 4H^+ + 3e^- = NO + 2H_2O$	0.957
$NO_3^- + 3H^+ + 2e^- = HNO_2 + H_2O$	0.94
$2Hg^{2+} + 2e^- = Hg_2^{2+}$	0.911
$Cu^{2+} + I^- + e^- = CuI$	0.861
$OsO_4(c) + 8H^+ + 8e^- = Os + 4H_2O$	0.84
$Ag^+ + e^- = Ag$	0.7991
$Hg_2^{2+} + 2e^- = 2Hg$	0.7960
$Fe^{3+} + e^- = Fe^{2+}$	0.771
$H_2SeO_3 + 4H^+ + 4e^- = Se + 3H_2O$	0.739
$HN_3 + 11H^+ + 8e^- = 2NH_4^+$	0.695
$O_2 + 2H^+ + 2e^- = H_2O_2$	0.695
$Ag_2SO_4 + 2e^- = 2Ag + SO_4^{2-}$	0.654
$Cu^{2+} + Br^- + e^- = CuBr(c)$	0.654
$Au(SCN)_4^- + 3e^- = Au + 4SCN^-$	0.636
$2HgCl_2 + 2e^- = Hg_2Cl_2(c) + 2Cl^-$	0.63
$Sb_2O_5 + 6H^+ + 4e^- = 2SbO^+ + 3H_2O$	0.605
$H_3AsO_4 + 2H^+ + 2e^- = HAsO_2 + 2H_2O$	0.560
$TeOOH^+ + 3H^+ + 4e^- = Te + 2H_2O$	0.559
$Cu^{2+} + Cl^- + e^- = CuCl(c)$	0.559
$I_3^- + 2e^- = 3I^-$	0.536
$I_2 + 2e^- = 2I^-$	0.536
$Cu^+ + e^- = Cu$	0.53
$4H_2SO_3 + 4H^+ + 6e^- = S_4O_6^{2-} + 6H_2O$	0.507
$Ag_2CrO_4 + 2e^- = 2Ag + CrO_4^{2-}$	0.449
$2H_2SO_3 + 2H^+ + 4e^- = S_2O_3^{2-} + 3H_2O$	0.400
$UO_2^+ + 4H^+ + e^- = U^{4+} + 2H_2O$	0.38
$Fe(CN)_6^{3-} + e^- = Fe(CN)_6^{4-}$	0.361
$Cu^{2+} + 2e^- = Cu$	0.340
$VO^{2+} + 2H^+ + e^- = V^{3+} + H_2O$	0.337
$BiO^+ + 2H^+ + 3e^- = Bi + H_2O$	0.32
$UO_2^{2+} + 4H^+ + 2e^- = U^{4+} + 2H_2O$	0.27
$Hg_2Cl_2(c) + 2e^- = 2Hg + 2Cl^-$	0.2676
$AgCl + e^- = Ag + Cl^-$	0.2223
$SbO^+ + 2H^+ + 3e^- = Sb + H_2O$	0.212
$CuCl_3^{2-} + e^- = Cu + 3Cl^-$	0.178
$SO_4^{2-} + 4H^+ + 2e^- = H_2SO_3 + H_2O$	0.158
$Sn^{4+} + 2e^- = Sn^{2+}$	0.15
$S + 2H^+ + 2e^- = H_2S$	0.144
$Hg_2Br_2(c) + 2e^- = 2Hg + 2Br^-$	0.1392
$CuCl + e^- = Cu + Cl^-$	0.121
$TiO^{2+} + 2H^+ + e^- = Ti^{3+} + H_2O$	0.100
$S_4O_6^{2-} + 2e^- = 2S_2O_3^{2-}$	0.08
$AgBr + e^- = Ag + Br^-$	0.0711
$HCOOH + 2H^+ + 2e^- = HCHO + H_2O$	0.056
$CuBr + e^- = Cu + Br^-$	0.033
$2H^+ + 2e^- = H_2$	0.0000
$Hg_2I_2 + 2e^- = 2Hg + 2I^-$	−0.0405

TABLE 1.78 Potentials of Selected Half-Reactions at 25°C (*Continued*)

Half-reaction	$E°$, volts
$Pb^{2+} + 2e^- = Pb$	−0.125
$Sn^{2+} + 2e^- = Sn$	−0.136
$AgI + e^- = Ag + I^-$	−0.1522
$N_2 + 5H^+ + 4e^- = N_2H_5^+$	−0.225
$V^{3+} + e^- = V^{2+}$	−0.255
$Ni^{2+} + 2e^- = Ni$	−0.257
$Co^{2+} + 2e^- = Co$	−0.277
$Ag(CN)_2^- + e^- = Ag + 2CN^-$	−0.31
$PbSO_4 + 2e^- = Pb + SO_4^{2-}$	−0.3505
$Cd^{2+} + 2e^- = Cd$	−0.4025
$Cr^{3+} + e^- = Cr^{2+}$	−0.424
$Fe^{2+} + 2e^- = Fe$	−0.44
$H_3PO_3 + 2H^+ + 2e^- = HPH_2O_2 + H_2O$	−0.499
$2CO_2 + 2H^+ + 2e^- = H_2C_2O_4$	−0.49
$U^{4+} + e^- = U^{3+}$	−0.52
$Zn^{2+} + 2e^- = Zn$	−0.7626
$Mn^{2+} + 2e^- = Mn$	−1.18
$Al^{3+} + 3e^- = Al$	−1.67
$Mg^{2+} + 2e^- = Mg$	−2.356
$Na^+ + e^- = Na$	−2.714
$K^+ + e^- = K$	−2.925
$Li^+ + e^- = Li$	−3.045
$3N_2 + 2H^+ + 2e^- = 2HN_3$	−3.10

TABLE 1.79 Overpotentials for Common Electrode Reactions at 25°C

The overpotential is defined as the difference between the actual potential of an electrode at a given current density and the reversible electrode potential for the reaction.

	Current Density, A/cm²					
	0.001	0.01	0.1	0.5	1.0	5.0
Electrode	Overpotential, volts					
Liberation of H_2 from 1*M* H_2SO_4						
Ag	0.097	0.13	0.3		0.48	0.69
Al	0.3	0.83	1.00		1.29	
Au	0.017		0.1		0.24	0.33
Bi	0.39	0.4			0.78	0.98
Cd		1.13	1.22		1.25	
Co		0.2				
Cr		0.4				
Cu			0.35		0.48	0.55
Fe		0.56	0.82		1.29	
Graphite	0.002		0.32		0.60	0.73
Hg	0.8	0.93	1.03		1.07	
Ir	0.0026	0.2				
Ni	0.14	0.3			0.56	0.71
Pb	0.40	0.4			0.52	1.06
Pd	0	0.04				
Pt (smooth)	0.0000	0.16	0.29		0.68	
Pt (platinized)	0.0000	0.030	0.041		0.048	0.051
Sb		0.4				
Sn		0.5	1.2			
Ta		0.39	0.4			
Zn	0.48	0.75	1.06		1.23	
Liberation of O_2 from 1*M* KOH						
Ag	0.58	0.73	0.98		1.13	
Au	0.67	0.96	1.24		1.63	
Cu	0.42	0.58	0.66		0.79	
Graphite	0.53	0.90	1.09		1.24	
Ni	0.35	0.52	0.73		0.85	
Pt (smooth)	0.72	0.85	1.28		1.49	
Pt (platinized)	0.40	0.52	0.64		0.77	
Liberation of Cl_2 from saturated NaCl solution						
Graphite			0.25	0.42	0.53	
Platinized Pt	0.006		0.026	0.05		
Smooth Pt	0.008	0.03	0.054	0.161	0.236	
Liberation of Br_2 from saturated NaBr solution						
Graphite		0.002	0.027	0.16	0.33	
Platinized Pt		0.002	0.012	0.069	0.21	
Smooth Pt		0.002	0.006*	0.26	0.38†	
Liberation of I_2 from saturated NaI solution						
Graphite	0.002	0.014	0.097			
Platinized Pt		0.006	0.032		0.196	
Smooth Pt		0.003	0.03	0.12	0.22	

* At 0.23 A/cm².† At 0.72 A/cm².

The overpotential required for the evolution of O_2 from dilute solutions of $HClO_4$, HNO_3, H_3PO_4 or H_2SO_4 onto smooth platinum electrodes is approximately 0.5 V.

TABLE 1.80 Half-Wave Potentials of Inorganic Materials

All values are in volts vs. the saturated calomel electrode.

Element	$E_{1/2}$, volts	Solvent system
Aluminum		
3+	−0.5	0.2*M* acetate, pH 4.5–4.7, plus 0.07% azo dye Pontochrome Violet SW; reduction wave of complexed dye is 0.2 V more negative than that of the free dye.
Antimony		
3+ to 0	−0.15	1*M* HCl
	−0.31(1)	1*M* HNO_3 (or 0.5*M* H_2SO_4)
	−0.8	0.5*M* tartrate, pH 4.5
	−1.0; −1.2	0.5*M* tartrate, pH 9 (waves not distinct)
	−1.26	1*M* NaOH; also anodic wave (3+ to 5+) at −0.45
	−1.32	0.5*M* tartrate plus 0.1*M* NaOH
5+	0.0; −0.257	6*M* HCl. First wave (5+ to 3+) starts at the oxidation potential of Hg; second wave is 3+ to 0.
5+ to 0	−0.35	1*M* HCl plus 4*M* KBr
Arsenic		
3+ to 5+	−0.26	0.5*M* KOH (anodic wave); only suitable wave
3+	−0.8; −1.0	0.1*M* HCl; ill-defined waves
	−0.7; −1.0	0.5*M* H_2SO_4 (or 1*M* HNO_3)
Barium		
2+ to 0	−1.94	0.1*M* $(C_2H_5)_4NI$
Bismuth		
3+ to 0	−0.025(15)	1*M* HNO_3 (or 0.5*M* H_2SO_4)
	−0.09	1*M* HCl
	−0.29	0.5*M* tartrate, pH 4.5
	−0.7	0.5*M* tartrate (pH 9), wave not well-developed
	−1.0	0.5*M* tartrate plus 0.1*M* NaOH, poor wave
Bromine		
5+ to 1−	−1.75	0.1*M* alkali chlorides (or 0.1*M* NaOH)
	0.13	0.05*M* H_2SO_4
0 to 1−	0.0	Wave (anodic) starts at zero; Hg_2Br_2 forms
Br^-	0.1	Oxidation of Hg to form mercury(I) bromide
Cadmium		
2+ to 0	−0.60	0.1*M* KCl, or 0.5*M* H_2SO_4, or 1*M* HNO_3
	−0.64	0.5*M* tartrate at pH 4.5 or 9
	−0.81	1*M* NH_4Cl plus 1*M* NH_3
Calcium		
2+ to 0	−2.22	0.1*M* $(C_2H_5)_4NCl$
	−2.13	0.1*M* $(C_2H_5)_4NCl$ in 80% ethanol
Cerium		
3+ to 0	−1.97	0.02*M* alkali sulfate
Cesium		
1+ to 0	−2.05	0.1*M* $(C_2H_5)_4NOH$ in 50% ethanol
Chlorine		
Cl^-	0.25	Oxidation of Hg to form Hg_2Cl_2
Chromium		
6+ to 3+	−0.85	CrO_4^{2-} to CrO_2^- in 0.1 to 1*M* NaOH
3+ to 0	−0.35; −1.70	1*M* NH_4Cl—NH_3 buffer (pH 8–9); 3+ to 2+ to 0
3+ to 2+	−0.95	0.1*M* pyridine–0.1*M* pyridinium chloride

(*Continued*)

TABLE 1.80 Half-Wave Potentials of Inorganic Materials (*Continued*)

Element	$E_{1/2}$, volts	Solvent system
2+ to 0	−1.54	1*M* KCl
2+ to 3+	−0.40	1*M* KCl (anodic wave)
Cobalt		
3+ to 0	−0.5; −1.3	1*M* NH_4Cl plus 1*M* NH_3; 3+ to 2+ to 0
2+ to 0	−1.07	0.1*M* pyridine plus pyridinium chloride
	−1.03	Neutral 1*M* potassium thiocyanate
	−1.4	$Co(H_2O)_6^{2+}$ in noncomplexing systems
3+ to 2+	0.0	1*M* sodium oxalate in acetate buffer (pH 5); diffusion current measured between 0 and −0.1 V
Copper		
2+ to 0	0.04	0.1*M* KNO_3, 0.1*M* NH_4ClO_4, or 1*M* Na_2SO_4
	−0.085	0.1*M* $Na_4P_2O_7$ plus 0.2*M* Na acetate, pH 4.5
	−0.09	0.5*M* Na tartrate, pH 4.5
	−0.20	0.1*M* potassium oxalate, pH 5.7 to 10
	−0.22	0.5*M* potassium citrate, pH 7.5
	−0.4	0.5*M* Na tartrate plus 0.1*M* NaOH (pH 12)
	−0.568	0.1*M* KNO_3 plus 1*M* ethylenediamine
2+	0.04; −0.22	1*M* KCl; consecutive waves: 2+ to 1+ to 0
	−0.02; −0.39	0.1*M* KSCN; consecutive waves: 2+ to 1+ to 0
	0.05; −0.25	0.1*M* pyridine plus 0.1*M* pyridinium chloride; consecutive waves: 2+ to 1+ to 0
	−0.24; −0.50	1*M* NH_4Cl plus 1*M* NH_3; consecutive waves
Gallium		
3+ to 0	−1.1	Not more than 0.001*M* HCl or wave masked by hydrogen wave which immediately follows
Germanium		
2+ to 0	−0.45	6*M* HCl; prior reduction with HPH_2O_2 to 2+
Gold		
3+ to 1+	0	1*M* KCN; wave starts at 0 V
1+ to 0	−1.4	$Au(CN)_2^-$ wave best for analytical purposes
Indium		
3+ to 0	−0.60	1*M* KCl
		In Na acetate, pH 3.9 to 4.2
Iodine		
IO_4^-	0.36	First wave at pH 0 (shifts to −0.08 at pH 12); second wave corresponds to iodate reduction
IO_3^-	−0.075	0.2*M* KNO_3 (shifts −0.13 V/pH unit increase)
	−0.305	0.1*M* hydrogen phthalate, pH 3.2
	−0.500	0.1*M* acetate plus 0.1*M* KCl, pH 4.9
	−0.650	0.1*M* citrate, pH 5.95
	−1.050	0.2*M* phosphate, pH 7.10
	−1.20	0.05*M* borax + 0.1*M* KCl, pH 9.2; or NaOH plus 0.1*M* KCl, pH 13.0
0 to 1−	0.0	Wave starts from zero in acid media; Hg_2I_2 formed
1−	−0.1	Oxidation of Hg to form Hg_2I_2
Iron		
3+	−0.44; −1.52	1*M* $(NH_4)_2CO_3$; two waves; 3+ to 2+ to 0
	−0.17; −1.50	0.5*M* Na tartrate, pH 5.8; two waves; 3+ to 2+ to 0
	−0.9; −1.5	0.1 to 5*M* KOH plus 8% mannitol; 3+ to 2+ to 0

TABLE 1.80 Half-Wave Potentials of Inorganic Materials (*Continued*)

Element	$E_{1/2}$, volts	Solvent system
3+ to 2+	−0.13	0.1*M* EDTA plus 2*M* Na acetate, pH 6–7
	−0.27	0.2*M* Na oxalate, pH 7.9 or less
	−0.28	0.5*M* Na citrate, pH 6.5
	−1.46(2)	1*M* NH_4ClO_4
	−1.36	0.1*M* KHF_2, pH 4 or less
2+ to 3+	−0.28	0.5*M* Na citrate, pH 6.5
	−0.27	0.2*M* Na oxalate, pH 7.9 or less
	−0.17	0.5*M* Na tartrate, pH 5.8
	−1.36	0.1*M* KHF_2, pH 4 or less
Lead		
2+ to 0	−0.405	1*M* HNO_3
	−0.435	1*M* KCl (or HCl)
	−0.49(1)	0.5*M* Na tartrate, pH 4.5 or 9
	−0.72	1*M* KCN
	−0.75	1*M* KOH or 0.5*M* Na tartrate plus 0.1*M* NaOH
Lithium		
1+ to 0	−2.31	0.1*M* $(C_2H_5)_4$ NOH in 50% ethanol
Magnesium		
2+ to 0	−2.2	0.1*M* $(C_2H_5)_4NCl$ (poorly defined wave)
Manganese		
2+ to 0	−1.65	1*M* NH_4Cl plus 1*M* NH_3
	−1.55	1*M* KCNS
	−1.33	1.5*M* KCN
Molybdenum		
6+	−0.26; −0.63	0.3*M* HCl, two waves: 6+ to 5+ to 3+
Nickel		
2+ to 0	−0.70	1*M* KSCN
	−0.78	1*M* KCl plus 0.5*M* pyridine
	−1.09	1*M* NH_4Cl plus 1*M* NH_3
	−1.1	$Ni(H_2O)_6^{2+}$ in NH_4ClO_4 or KNO_3
	−1.36	$Ni(CN)_4^{2-}$ in 1*M* KCN (alkaline media)
Niobium		
5+ to 3+	−0.80(4)	1*M* HNO_3
Nitrogen		
Nitrate	−1.45	0.017*M* $LaCl_3$ (reduced to hydroxylamine)
HNO_2	−0.77	0.1*M* HCl
C_2N_2	−1.2; −1.55	0.1*M* Na acetate, two waves
Oxamic acid	−1.55	0.1*M* Na acetate
Cyanide	−0.45	0.1*M* NaOH; anodic wave starts at −0.45
Thiocyanate	0.18	Anodic wave; neutral or weakly alkaline medium
Osmium		
OsO_4	0.0; −0.41; −1.16	Sat'd $Ca(OH)_2$. Three waves: first starts at 0; second wave is OsO_4^{2-} to Os(V); and third wave is Os(V) to Os(III)
Oxygen		
O_2	−0.05; −0.9	Buffer solutions of pH 1 to 10. Two waves: O_2 to H_2O_2, and H_2O_2 to H_2O. Second wave extends from −0.5 to −1.3
H_2O_2	−0.9	Very extended wave (see above); sharper in presence of Aerosol OT

(*Continued*)

TABLE 1.80 Half-Wave Potentials of Inorganic Materials (*Continued*)

Element	$E_{1/2}$, volts	Solvent system
Palladium		
2+ to 0	−0.31	1*M* pyridine plus 1*M* KCl
	−0.64	0.1*M* ethylenediamine plus 1*M* KCl
	−0.72	1*M* NH_4Cl plus 1*M* NH_3
Potassium		
1+ to 0	−2.10	0.1*M* $(C_2H_5)_4NOH$ in 50% ethanol
Rhenium		
7+ to 4+	−0.44	2*M* HCl or (better) 4*M* $HClO_4$
4+ to 3+	−0.51	$ReCl_6^{2-}$ ion in 1*M* HCl
Rhodium		
3+ to 2+	−0.41	1*M* pyridine plus 1*M* KCl
Rubidium		
1+ to 0	−1.99	0.1*M* $(C_2H_5)_4NOH$ in 50% ethanol
Scandium		
3+ to 0	−1.80	0.1*M* LiCl, KCl, or $BaCl_2$
Selenium		
4+ to 2−	−1.44	1*M* NH_4Cl plus NH_3, pH 8.0
	−1.54	Same system adjusted to pH 9.5
2−	−0.49	Anodic wave at pH 0 due to HgSe
	−0.94	Anodic wave at pH 12 (0.01*M* NaOH)
Silver		
1+ to 0		Wave starts at oxidation potential of Hg
1+ to 0	−0.3	0.0014*M* $KAg(CN)_2$ without excess cyanide
Sodium		
1+ to 0	−2.07	0.1*M* $(C_2H_5)_4NOH$ in 50% ethanol
Strontium		
2+ to 0	−2.11	0.1*M* $(C_2H_5)_4NI$, water or 80% ethanol
Sulfur		
SO_2	−0.38	1*M* HNO_3 (or other strong acid); 4+ to 2+
$S_2O_4^{2-}$	−0.43	0.5*M* $(NH_4)_2HPO_4$ plus 1*M* NH_3 (anodic wave)
$S_2O_3^{2-}$	−0.15	1*M* strong acid; anodic mercury wave
0 to 2−	−0.50	90% methanol, 9.5% pyridine, 0.5% HCl (pH 6)
HS^-	−0.76	0.1*M* NaOH (anodic mercury wave)
Tellurium		
4+ to 0	−0.4	Citrate buffer, pH 1.6 (second of two waves)
	−0.63	Ammoniacal buffer, pH 9.4
4+ to 2−	−1.22	0.1*M* NaOH
2− to 0	−0.72	1*M* HCl (true anodic reversible wave)
	−0.08	1*M* NaOH (same as above; intermediate values at pH 1 to 13)
Thallium		
3+ to 0	−0.48	1*M* KCl, KNO_3, K_2SO_4, KOH, or NH_3
Tin		
4+ to 2+	−0.25; −0.52	4*M* NH_4Cl + 1*M* HCl; two waves: 4+ to 2+ to 0
2+ to 0	−0.59	0.5*M* tartrate, pH 4.3
	−1.22	1*M* NaOH (stannite ion to tin)
2+ to 4+	−0.28	0.5*M* Na tartrate, pH 4.3 (anodic wave)
	−0.73	1*M* NaOH (stannite ion to stannate ion)

TABLE 1.80 Half-Wave Potentials of Inorganic Materials (*Continued*)

Element	$E_{1/2}$, volts	Solvent system
Titanium		
4+ to 3+	−0.173	0.1*M* $K_2C_2O_4$ plus 1*M* H_2SO_4
	−1.22	0.4*M* tartrate, pH 6.5
Tungsten		
6+	0.0; −0.64	6*M* HCl; two waves: first wave starts at zero and is W(VI) to W(V), the second wave is W(V) to W(III)
Uranium		
6+	−0.180; −0.92	UO_2^{2+} to UO_2^+, then U^{3+} in 0.02*M* HCL
Vanadium		
5+ to 4+ to 2+	−0.97; −1.26	1*M* NH_4Cl plus 1*M* NH_3 and 0.08*M* Na_2SO_3
4+ to 2+	−0.98	0.05*M* H_2SO_4
3+ to 2+	−0.55	0.5*M* H_2SO_4
4+ to 5+	−0.32	1*M* NH_4Cl, 1*M* NH_3, and 0.08*M* Na_2SO_3
4+ to 5+	0.76	0.05*M* H_2SO_4; anodic wave starting from zero
2+ to 3+	−0.55	0.5*M* H_2SO_4; anodic wave
Zinc		
2+ to 0	−0.995	0.1*M* KCl
	−1.01	0.1*M* KSCN
	−1.15	0.5*M* tartrate, pH 9
	−1.23	0.5*M* tartrate, pH 4.5
	−1.33	1*M* NH_4Cl plus 1*M* NH_3
	−1.53	1*M* NaOH

TABLE 1.81 Standard Electrode Potentials for Aqueous Solutions

Acidic solutions ($[H^+]$ = 1.0 mol kg^{-1})	
Half-reaction	$E°(V)$
$Li^+ + e^- \rightleftharpoons Li$	−3.045
$K^+ + e^- \rightleftharpoons K$	−2.925
$Na^+ + e^- \rightleftharpoons Na$	−2.714
$La^{3+} + 3e^- \rightleftharpoons La$	−2.37
$Mg^{2+} + 2e^- \rightleftharpoons Mg$	−2.356
$\frac{1}{2}H_2 + e^- \rightleftharpoons H^-$	−2.25
$Be^{2+} + 2e^- \rightleftharpoons Be$	−1.97
$Zr^{4+} + 4e^- \rightleftharpoons Zr$	−1.70
$Al^{3+} + 3e^- \rightleftharpoons Al$	−1.67
$Ti^{3+} + 3e^- \rightleftharpoons Ti$	−1.21
$Mn^{2+} + 2e^- \rightleftharpoons Mn$	−1.18
$V^{2+} + 2e^- \rightleftharpoons V$	−1.13
$SiO_2(glass) + 4H^+ + 4e^- \rightleftharpoons Si + 2H_2O$	−0.888
$Zn^{2+} + 2e^- \rightleftharpoons Zn$	−0.763
$U^{4+} + e^- \rightleftharpoons U^{3+}$	−0.52
$Fe^{2+} + 2e^- \rightleftharpoons Fe$	−0.44
$Cr^{3+} + e^- \rightleftharpoons Cr^{2+}$	−0.424
$Cd^{2+} + 2e^- \rightleftharpoons Cd$	−0.403
$PbSO_4 + 2e^- \rightleftharpoons Pb + SO_4^{2-}$	−0.351
$Eu^{3+} + e^- \rightleftharpoons Eu^{2+}$	−0.35

(*Continued*)

TABLE 1.81 Standard Electrode Potentials for Aqueous Solutions (*Continued*)

Acidic solutions ($[H^+] = 1.0$ mol kg^{-1})	
Half-reaction	$E°(V)$
$Co^{2+} + 2e^- \rightleftharpoons Co$	−0.277
$H_3PO_4 + 2H^+ + 2e^- \rightleftharpoons H_3PO_3 + H_2O$	−0.276
$Ni^{2+} + 2e^- \rightleftharpoons Ni$	−0.257
$V^{3+} + e^- \rightleftharpoons V^{2+}$	−0.255
$2SO_4^{2-} + 4H^+ + 2e^- \rightleftharpoons S_2O_6^{2-} + 2H_2O$	−0.253
$N_2 + 5H^+ + 4e^- \rightleftharpoons N_2H_5^+$	−0.23
$CO_2 + 2H^+ + 2e^- \rightleftharpoons HCOOH$	−0.16
$AgI + e^- \rightleftharpoons Ag + I^-$	−0.152
$Sn^{2+} + 2e^- \rightleftharpoons Sn$	−0.136
$Pb^{2+} + 2e^- \rightleftharpoons Pb$	−0.125
$2H^+ + 2e^- \rightleftharpoons H_2$	0.000
$HCOOH + 2H^+ + 2e^- \rightleftharpoons HCHO + H_2O$	+0.056
$AgBr + e^- \rightleftharpoons Ag + Br^-$	+0.071
$TiO^{2+} + 2H^+ \rightleftharpoons + e^- Ti^{3+} + H_2O$	+0.100
$S + 2H^+ + 2e^- \rightleftharpoons H_2S$	+0.144
$Sn^{4+} + 2e^- \rightleftharpoons Sn^{2+}$	+0.15
$SO_4^{2-} + 4H^+ + 2e^- \rightleftharpoons H_2SO_3 + H_2O$	+0.158
$Cu^{2+} + e^- \rightleftharpoons Cu^+$	+0.159
$AgCl + e^- \rightleftharpoons Ag + Cl^-$	+0.222
$HCHO + 2H^+ + 2e^- \rightleftharpoons CH_3OH$	+0.232
$UO_2^{2+} + 4H^+ + 2e^- \rightleftharpoons U^{4+} + 2H_2O$	+0.27
$VO^{2+} + 2H^+ + e^- \rightleftharpoons V^{3+} + H_2O$	+0.337
$Cu^{2+} + 2e^- \rightleftharpoons Cu$	+0.340
$Fe(CN)_6^{3-} + e^- \rightleftharpoons Fe(CN)_6^{4-}$	+0.361
$2H_2SO_3 + 2H^+ + 4e^- \rightleftharpoons S_2O_3^{2-} + 3H_2O$	+0.400
$H_2SO_3 + 4H^+ + 4e^- \rightleftharpoons S + 3H_2O$	+0.500
$4H_2SO_3 + 4H^+ + 6e^- \rightleftharpoons S_4O_6^{2-} + 6H_2O$	+0.507
$Cu^+ + e^- \rightleftharpoons Cu$	+0.520
$I_2 + 2e^- \rightleftharpoons 2I^-$	+0.5355
$I_3^- + 2e^- \rightleftharpoons 3I^-$	+0.536
$MnO_4^- + e^- \rightleftharpoons MnO_4^{2-}$	+0.56
$S_2O_6^{2-} + 4H^+ + 2e^- \rightleftharpoons 2H_2SO_3$	+0.569
$CH_3OH + 2H^+ + 2e^- \rightleftharpoons CH_4 + H_2O$	+0.59
$HN_3 + 11H^+ + 8e^- \rightleftharpoons 3NH_4^+$	+0.695
$O_2 + 2H^+ + 2e^- \rightleftharpoons H_2O_2$	+0.695
$Rh^{3+} + 3e^- \rightleftharpoons Rh$	+0.76
$(NCS)_2 + 2e^- \rightleftharpoons 2NCS^-$	+0.77
$Fe^{3+} + e^- \rightleftharpoons Fe^{2+}$	+0.771
$Hg_2^{2+} + 2e^- \rightleftharpoons 2Hg$	+0.796
$Ag^+ + e^- \rightleftharpoons Ag$	+0.799
$2NO_3^- + 4H^+ + 2e^- \rightleftharpoons N_2O_4 + 2H_2O$	+0.803
$Hg^{2+} + 2e^- \rightleftharpoons Hg$	+0.911
$NO_3^- + 3H^+ + 2e^- \rightleftharpoons HNO_2 + H_2O$	+0.94
$NO_3^- + 4H^+ + 3e^- \rightleftharpoons NO + 2H_2O$	+0.957
$NHO_2 + H^+ + e^- \rightleftharpoons NO + H_2O$	+0.996
$N_2O_4 + 4H^+ + 4e^- \rightleftharpoons 2NO + 2H_2O$	+1.039
$Br_2 + 2e^- \rightleftharpoons 2Br^-$	+1.065
$N_2O_4 + 2H^+ + 2e^- \rightleftharpoons 2HNO_2$	+1.07
$H_2O_2 + H^+ + e^- \rightleftharpoons \cdot OH + H_2O$	+1.14
$ClO_4^- + 2H^+ + 2e^- \rightleftharpoons ClO_3^- + H_2O$	+1.201
$O_2 + 4H^+ + 4e^- \rightleftharpoons 2H_2O$	+1.229
$MnO_2 + 4H^+ + 2e^- \rightleftharpoons Mn^{2+} + 2H_2O$	+1.23

TABLE 1.81 Standard Electrode Potentials for Aqueous Solutions (*Continued*)

Acidic Solutions ($[H^+] = 1.0$ mol kg^{-1})	
Half-reaction	$E°(V)$
$N_2H_5^+ + 3H^+ + 2e^- \rightleftharpoons 2NH_4^+$	+1.275
$Cl_2 + 2e^- \rightleftharpoons 2Cl^-$	+1.358
$Cr_2O_7^{2-} + 14H^+ + 6e^- \rightleftharpoons 2Cr^{3+} + 7H_2O$	+1.36
$PbO_2 + 4H^+ + 2e^- \rightleftharpoons Pb^{2+} + 2H_2O$	+1.468
$2BrO_3^- + 12H^+ + 10e^- \rightleftharpoons Br_2 + 6H_2O$	+1.478
$Mn^{3+} + e^- \rightleftharpoons Mn^{2+}$	+1.51
$Au^{3+} + 3e^- \rightleftharpoons Au$	+1.52
$NiO_2 + 4H^+ + 2e^- \rightleftharpoons Ni^{2+} + 2H_2O$	+1.593
$2HBrO + 2H^+ + 2e^- \rightleftharpoons Br_2 + 2H_2O$	+1.604
$2HClO + 2H^+ + 2e^- \rightleftharpoons Cl_2 + 2H_2O$	+1.630
$PbO_2 + SO_4^{2-} + 4H^+ + 2e^- \rightleftharpoons PbSO_4 + 2H_2O$	+1.698
$MNO_4^- + 4H^+ + 3e^- \rightleftharpoons MnO_2 + 2H_2O$	+1.70
$Ce^{4+} + e^- \rightleftharpoons Ce^{3+}$	+1.72
$H_2O_2 + 2H^+ + 2e^- \rightleftharpoons 2H_2O$	+1.763
$Au^+ + e^- \rightleftharpoons Au$	+1.83
$Co^{3+} + e^- \rightleftharpoons Co^{2+}$	+1.92
$HN_3 + 3H^+ + 2e^- \rightleftharpoons NH_4^+ + N_2$	+1.96
$S_2O_8^{2-} + 2e^- \rightleftharpoons 2SO_4^{2-}$	+1.96
$O_3 + 2H^+ + 2e^- \rightleftharpoons O_2 + H_2O$	+2.075
$(OH + H^+ + e^- \rightleftharpoons H_2O$	+2.38
$F_2 + 2H^+ + 2e^- \rightleftharpoons 2HF$	+3.053

Basic Solutions ($[OH^-] = 1.0$ mol kg^{-1})	
Half-reaction	$E°(V)$
$Ca(OH)_2 + 2e^- \rightleftharpoons Ca + 2OH^-$	−3.026
$Mg(OH)_2 + 2e^- \rightleftharpoons Mg + 2OH^-$	−2.687
$Al(OH)_4^- + 3e^- \rightleftharpoons Al + 4OH^-$	−2.310
$SiO_3^{2-} + 3H_2O + 4e^- \rightleftharpoons Si + 6OH^-$	−1.7
$Mn(OH)_2 + 2e^- \rightleftharpoons Mn + 2OH^-$	−1.56
$2TiO_2 + H_2O + 2e^- \rightleftharpoons Ti_2O_3 + 2OH^-$	−1.38
$Cr(OH)_3 + 3e^- \rightleftharpoons Cr + 3OH^-$	−1.33
$Zn(OH)_4^{2-} + 2e^- \rightleftharpoons Zn + 4OH^-$	−1.285
$Zn(NH_3)_4^{2+} + 2e^- \rightleftharpoons Zn + 4NH_3$	−1.04
$MnO_2 + 2H_2O + 4e^- \rightleftharpoons Mn + 4OH^-$	−0.980
$Cd(CN)_4^{2-} + 2e^- \rightleftharpoons Cd + 4CN^-$	−0.943
$SO_4^{2-} + H_2O + 2e^- \rightleftharpoons SO_3^{2-} + 2OH^-$	−0.94
$2H_2O + 2e^- \rightleftharpoons H_2 + 2OH^-$	−0.828
$HFeO_2^- + H_2O + 2e^- \rightleftharpoons Fe + 3OH^-$	−0.8
$Co(OH)_2 + 2e^- \rightleftharpoons Co + 2OH^-$	−0.733
$CrO_4^{2-} + 4H_2O + 3e^- \rightleftharpoons Cr(OH)_4^- + 4OH^-$	−0.72
$Ni(OH)_2 + 2e^- \rightleftharpoons Ni + 2OH^-$	−0.72
$FeO_2^- + H_2O + e^- \rightleftharpoons HFeO_2^- + OH^-$	−0.69
$2SO_3^{2-} + 3H_2O + 4e^- \rightleftharpoons S_2O_3^{2-} + 6OH^-$	−0.58
$Ni(NH_3)_6^{2+} + 2e^- \rightleftharpoons Ni + 6NH_3$	−0.476
$S + 2e^- \rightleftharpoons S^{2-}$	−0.45
$O_2 + e^- \rightleftharpoons O_2^-$	−0.33
$CuO + H_2O + 2e^- \rightleftharpoons Cu + 2OH^-$	−0.29
$Mn_2O_3 + 3H_2O + 2e^- \rightleftharpoons 2Mn(OH)_2 + 2OH^-$	−0.25
$2CuO + H_2O + 2e^- \rightleftharpoons Cu_2O + 2OH^-$	−0.22
$O_2 + H_2O + 2e^- \rightleftharpoons HO_2^- + OH^-$	−0.065
$MnO_2 + 2H_2O + 2e^- \rightleftharpoons Mn(OH)_2 + 2OH^-$	−0.05

(*Continued*)

TABLE 1.81 Standard Electrode Potentials for Aqueous Solutions (*Continued*)

Basic solutions ($[OH^-] = 1.0$ mol kg^{-1})	
Half-reaction	$E°(V)$
$NO_3^- + H_2O + 2e^- \rightleftharpoons NO_2^- + 2OH^-$	+0.01
$Co(NH_3)_6^{3+} + e^- \rightleftharpoons Co(NH_3)_6^{2+}$	+0.058
HgO (red form) $+ H_2O + 2e^- \rightleftharpoons Hg + 2OH^-$	+0.098
$N_2H_4 + 2H_2O + 2e^- \rightleftharpoons 2NH_3 + 2OH^-$	+0.1
$Co(OH)_3 + e^- \rightleftharpoons Co(OH)_2 + OH^-$	+0.17
$HO_2^- + H_2O + e^- \rightleftharpoons {}^-OH + 2OH^-$	+0.184
$O_2^- + H_2O + e^- \rightleftharpoons HO_2^- + OH^-$	+0.20
$ClO_3^- + H_2O + 2e^- \rightleftharpoons ClO_2^- + 2OH^-$	+0.295
$Ag_2O + H_2O + 2e^- \rightleftharpoons 2Ag + 2OH^-$	+0.342
$Ag(NH_3)_2^+ + e- \rightleftharpoons Ag + 2NH_3$	+0.373
$ClO_4^- + H_2O + 2e^- \rightleftharpoons ClO_3^- + 2OH^-$	+0.374
$O_2 + 2H_2O + e^- \rightleftharpoons 4OH^-$	+0.401
$NiO_2 + 2H_2O + 2e^- \rightleftharpoons Ni(OH)_2 + 2OH^-$	+0.490
$FeO_4^{2-} + 2H_2O + 3e^- \rightleftharpoons FeO_2^- + 4OH^-$	+0.55
$BrO_3^- + 3H_2O + 6e^- \rightleftharpoons Br^- + 6OH^-$	+0.584
$MnO_4^{2-} + 2H_2O + 2e^- \rightleftharpoons MnO_2 + 4OH^-$	+0.62
$ClO_2^- + H_2O + 2e^- \rightleftharpoons ClO^- + 2OH^-$	+0.681
$BrO^- + H_2O + 2e^- \rightleftharpoons Br^- + 2OH^-$	+0.766
$HO_2^- + H_2O + 2e^- \rightleftharpoons 3OH^-$	+0.867
$ClO^- + H_2O + 2e^- \rightleftharpoons Cl^- + 2OH^-$	+0.890
$ClO_2 + e^- \rightleftharpoons ClO_2^-$	+1.041
$O_3 + H_2O + 2e^- \rightleftharpoons O_2 + 2OH^-$	+1.246
$OH + e^- \rightleftharpoons OH^-$	+1.985

TABLE 1.82 Potentials of Reference Electrodes in Volts as a Function of Temperature

Liquid-junction potential included.

Temp., °C	0.1*M* KCl Calomel*	1.0*M* KCl Calomel*	3.5*M* KCl Calomel*	Satd. KCl Calomel*	1.0*M* KCl Ag/AgCl†	1.0*M* KBr Ag/AgBr‡	1.0*M* KI Ag/AgI§
0	0.3367	0.2883		0.25918	0.23655	0.08128	−0.14637
5					0.23413	0.07961	−0.14719
10	0.3362	0.2868	0.2556	0.25387	0.23142	0.07773	−0.14822
15	0.3361			0.2511	0.22857	0.07572	−0.14942
20	0.3358	0.2844	0.2520	0.24775	0.22557	0.07349	−0.15081
25	0.3356	0.2830	0.2501	0.24453	0.22234	0.07106	−0.15244
30	0.3354	0.2815	0.2481	0.24118	0.21904	0.06856	−0.15405
35	0.3351			0.2376	0.21565	0.06585	−0.15590
38	0.3350		0.2448	0.2355			
40	0.3345	0.2782	0.2439	0.23449	0.21208	0.06310	−0.15788
45					0.20835	0.06012	−0.15998
50	0.3315	0.2745		0.22737	0.20449	0.05704	−0.16219
55					0.20056		
60	0.3248	0.2702		0.2235	0.19649		
70					0.18782		
80				0.2083	0.1787		
90					0.1695	0.0251	

* Bates et al., *J. Research Natl. Bur. Standards*, **45**, 418 (1950).
† Bates and Bower, *J. Research Natl. Bur. Standards*, **53**, 283 (1954).
‡ Hetzer, Robinson and Bates, *J. Phys. Chem.*, **66**, 1423 (1962).
§ Hetzer, Robinson and Bates, *J. Phys. Chem.*, **68**, 1929 (1964).

TABLE 1.83 Potentials of Reference Electrodes (in Volts) at 25°C for Water-Organic Solvent Mixtures

Solvent, wt %	Methanol, Ag/AgCl	Ethanol, Ag/AgCl	2-Propanol, Ag/AgCl	Acetone, Ag/AgCl	Dioxane, Ag/AgCl	Ethylene glycol, Ag/AgCl	Methanol, calomel	Dioxane, calomel
5			0.2180	0.2190		0.2190		
10	0.2153	0.2146	0.2138	0.2156		0.2160		
20	0.2090	0.2075	0.2063	0.2079	0.2031	0.2101	0.255	0.2501
30		0.2003				0.2036		
40	0.1968	0.1945		0.1859		0.1972	0.243	
45					0.1635			0.2104
50		0.1859		0.158				
60	0.1818	0.173				0.1807		
70		0.158			0.0659		0.216	0.1126
80	0.1492	0.136						
82					−0.0614			−0.0014
90	0.1135	0.096		−0.034				
94.2	0.0841							
98		0.0215						
99							0.103	
100	−0.0099	−0.0081		−0.53				

1.22 CONDUCTANCE

Conductivity. The standard unit of conductance is electrolytic conductivity (formerly called specific conductance) κ, which is defined as the reciprocal of the resistance [Ω^{-1}] of a 1-m cube of liquid at a specified temperature [$\Omega^{-1} \cdot m^{-1}$]. See Table 1.86 and the definition of the cell constant.

In accurate work at low concentrations it is necessary to subtract the conductivity of the pure solvent (Table 2.69) from that of the solution to obtain the conductivity due to the electrolyte.

Resistivity (Specific Resistance)

$$\rho = \frac{1}{\mathrm{k}} \qquad [\Omega \cdot m]$$

Conductance of an Electrolyte Solution

$$\frac{1}{R} = \mathrm{k}\frac{S}{d} \qquad [\Omega^{-1}]$$

where S is the surface area of the electrode, or the mean cross-sectional area of the solution [m^2], and d is the mean distance between the electrodes [m].

Equivalent Conductivity

$$\Lambda = \frac{k}{C} \qquad [\Omega^{-1} \cdot m^2 \cdot \text{equiv}^{-1}]$$

In the older literature, C is the concentration in equivalents per liter. The volume of the solution in cubic centimeters per equivalent is equal to $1000/C$, and $\Lambda = 1000\ \kappa/C$, the units employed in Table 8.32 $[\Omega^{-1} \cdot cm^2 \cdot \text{equiv}^{-1}]$. The formula unit used in expressing the concentration must be specified; for example, NaCl, $^1/_2K_2SO_4$, $^1/_3LaCl_3$.

The equivalent conductivity of an electrolyte is the sum of contributions of the individual ions. At infinite dilution: $\Lambda^\circ = \lambda_c^\circ + \lambda_a^\circ$, where λ_c° and λ_a° are the ionic conductances of cations and anions, respectively, at infinite dilution (Table 1.87).

Ionic Mobility and Ionic Equivalent Conductivity

$$\lambda_c = Fu_c \qquad \text{and} \qquad \lambda_a = Fu_a \qquad [\Omega^{-1} \cdot m^2 \cdot \text{equiv}^{-1}]$$

where F is the Faraday constant, and u_c, u_a are the ionic mobilities $[m^2 \cdot s^{-1} \cdot V^{-1}]$.

$$\Lambda = \alpha F(u_c + u_a) = \alpha(\lambda_c + \lambda_a)$$

where α is the degree of electrolytic dissociation, Λ/Λ°. The electric mobility u of a species is the magnitude of the velocity in an electric field $[m \cdot s^{-1}]$ divided by the magnitude of the strength of the electric field $E[V \cdot m^{-1}]$.

Ostwald Dilution Law

$$K_d = \frac{\alpha^2 C}{1-\alpha}$$

where K_d is the dissociation constant of the weak electrolyte. In general for an electrolyte which yields n ions:

$$K_d = \frac{C^{(n-1)}\Lambda^n}{\Lambda^{\circ(n-1)}(\Lambda^\circ - \Lambda)}$$

Transference Numbers or Hittorf Transport Numbers

$$T_c = \frac{\lambda_c}{\lambda_c + \lambda_a} \qquad T_a = \frac{\lambda_a}{\lambda_c + \lambda_a} \qquad T_c + T_a = 1$$

$$\frac{T_c}{T_a} = \frac{u_c}{u_a} = \frac{\lambda_c}{\lambda_a}$$

$$\lambda_c = T_c\Lambda \qquad \lambda_a = T_a\Lambda$$

TABLE 1.84 Properties of liquid Semi-conductors

Material	Melting point (°K)	Density at °K* (g cm^{-3}) Solid	Density at °K* (g cm^{-3}) Liquid	Electrical conductivity at °K* $\Omega^{-1} \cdot cm^{-1}$ Solid	Electrical conductivity at °K* $\Omega^{-1} \cdot cm^{-1}$ Liquid	Atomization energy (kcal/mole)	Heat of fusion (kcal/mole)	Entropy of fusion (e.u.)	Thermo-electric power (μV per °K) Solid	Thermo-electric power (μV per °K) Liquid	Activation energy for viscous flow (kcal per mole)	Entropy of viscous flow (e.u.)	Viscosity of liquid at °K* (centipoises)
Si	1693	2.30	2.53	580	12000	204	12.1	7.1			8.63	2.1	0.348
Ge	1210	5.26	5.51	1250	14000	178	8.35	6.9	−90	0	2.74	2.85	0.135
AlSb	1353	4.18	4.72	160	9900	160	14.2	5.2	−160	−60	10	2.2	0.250
GaSb	985	5.60	6.06	280	10600	134	12.0	6.1	−60	0	2.7	5.0	0.368
InSb	809	5.76	6.48	2900	10000	121	11.6	7.2	−120	−20	2.0	9.4	0.363
GaAs	1511	5.16	5.71	300	7900	146	23.2	7.7	—	—	6.5	7.8	0.320
InAs	1215	5.5	5.89	3600	6800	130	12.6	5.2	—	—	6.2	3	0.174
ZnTe	1512					109					9.0	6.5	0.868
CdTe	1365					99					5.75	7.7	0.435
CuI	875	5.36	4.84				2.6		550	490			0.432
Ga_2Te_3	1063	5.35	5.086						−290	−85	11	7	0.546
In_2Te_3	940	5.77	5.54						−50	30	13	7.5	0.323
Mg_2Si	1375	1.84	2.27	1120	9800		20.4	5.0			13.9	2.3	0.299
Mg_2Ge	1388		3.20	1140	8400						9.5	3.8	0.311
Mg_2Sn	1051	3.45	3.52	2040	10600		11.4	3.6			9.5	5.8	0.520
Mg_2Pb	823	5.00	5.20	3530	8600		9.3	3.8			9.6	6.2	0.560
GeTe	998	5.97	5.57	2400	2600		11.3	5.7	130	21	4.70	5.8	0.375
SnTe	1063	6.15	5.85	1440	1800		8.0	3.7	140	28	4.90	6.1	0.348
PbTe	1190	7.69	7.45	420	1520		7.5	3.1	−60	−10	6.85	6.0	0.243
PbSe	1361	7.57	7.10	300	450		8.5	3.1	−120	−60	6.85	7.5	0.240
PbS	1392	7.07	6.45	250	220		8.7	3.1	−220	−220	9.80	7.5	0.319
Bi_2Se_3	979	7.27	6.97	450	900				−90	−35	9.7	5.05	0.540
Bi_2Te_3	858	7.5	7.26	1250	2580		28.35	6.6	−45	−3	2.7	7.80	0.198
Sb_2Te_3	895	6.29	6.09	900	1850		23.65	5.3	90	11	6.1	7.35	0.513
Se (hex)	493	4.69	3.975				1.5	3			3.94	6.7	6.63
Te	725	6.1	5.775				4.17	5.7			1.18	6.7	0.357

*At melting point.

TABLE 1.85 Limiting Equivalent Ionic Conductances in Aqueous Solutions

In 10^{-4} $m^2 \cdot S \cdot equiv^{-1}$ *or* $mho \cdot cm^2 \cdot equiv^{-1}$.

Ion	Temperature, °C 0	18	25
Inorganic cations			
Ag^+	33	54.5	61.9
Al^{3+}	29		61
Ba^{2+}	33.6	54.3	63.9
Be^{2+}			45
Ca^{2+}	30.8	51	59.5
Cd^{2+}	28	45.1	54
Ce^{3+}			70
Co^{2+}	28	45	53
$Co(NH_3)_6^{3+}$			100
$Co(ethylenediamine)_3^{3+}$			74.7
Cr^{3+}			67
Cs^+	44	68	77.3
Cu^{2+}	28	45.3	56.6
D^+ (deuterium)		213.7	
Dy^{3+}			65.7
Er^{3+}			66.0
Eu^{3+}			67.9
Fe^{2+}	28	45.3	53.5
Fe^{3+}			69
Gd^{3+}			67.4
H^+	224.1	315.8	350.1
Hg_2^{2+}			68.7
Hg^{2+}			63.6
Ho^{3+}			66.3
K^+	40.3	64.6	73.5
La^{3+}	35.0	59.2	69.6
Li^+	19.1	33.4	38.69
Mg^{2+}	28.5	46	53.06
Mn^{2+}	27	44.5	53.5
NH_4^+	40.3	64	73.7
$N_2H_5^+$ (hydrazinium 1+)			59
Na^+	25.85	43.5	50.11
Nd^{3+}			69.6
Ni^{2+}	28	45	50
Pb^{2+}	37.5	60.5	71
Pr^{3+}			69.6
Ra^{2+}	33	56.6	66.8
Rb^+	43.5	67.5	77.8
Sc^{3+}			64.7
Sm^{3+}			68.5
Sr^{2+}	31	51	59.46
Tl^+	43.3	66	74.9
Tm^{3+}			65.5
UO_2^{2+}			32
Y^{3+}			62
Yb^{3+}			65.2
Zn^{2+}	28	45.0	52.8

TABLE 1.85 Limiting Equivalent Ionic Conductances in Aqueous Solutions (*Continued*)

	Temperature, °C		
Ion	0	18	25
Inorganic anions			
$Au(CN)_2^-$			50
$Au(CN)_4^-$			36
$B(C_6H_5)_4^-$			21
Br^-	43.1	67.6	78.1
Br_3^-			43
BrO_3^-	31.0	49.0	55.7
Cl^-	41.4	65.5	76.31
ClO_2^-			52
ClO_3^-	36	55.0	64.6
ClO_4^-	37.3	59.1	67.3
CN^-			78
CO_3^{2+}	36	60.5	69.3
$Co(CN)_6^{3-}$			98.9
CrO_4^{2-}	42	72	85
F^-		46.6	55.4
$Fe(CN)_6^{4-}$			110.4
$Fe(CN)_6^{3-}$			100.9
$H_2AsO_4^-$			34
HCO_3^-			44.5
HF_2^-			75
HPO_4^{2-}			33
$H_2PO_4^-$		28	33
HS^-	40	57	65
HSO_3^-	27		50
HSO_4^-			50
$H_2SbO_4^-$			31
I^-	42.0	66.5	76.9
IO_3^-	21.0	33.9	40.5
IO_4^-		49	54.5
MnO_4^-	36	53	61.3
MoO_4^{2-}			74.5
N_3^-			69.5
$N(CN)_2^-$			54.5
NO_2^-	44	59	71.8
NO_3^-	40.2	61.7	71.42
$NH_2SO_3^-$ (sulfamate)			48.6
OCN^- (cyanate)		54.8	64.6
OH^-	117.8	175.8	198
PF_6^-			56.9
PO_3F^{2-}			63.3
PO_4^{3-}			69.0
$P_2O_7^{4-}$			96
$P_3O_9^{3-}$			83.6
$P_3O_{10}^{5-}$			109
ReO_4^-		46.5	54.9
SCN^- (thiocyanate)	41.7	56.6	66.5
$SeCN^-$			64.7
SeO_4^{2-}		65	75.7
SO_3^{2-}			79.9

(*Continued*)

TABLE 1.85 Limiting Equivalent Ionic Conductances in Aqueous Solutions (*Continued*)

Ion	Temperature, °C		
	0	18	25
SO_4^{2-}	41	68.3	80.0
$S_2O_3^{2-}$			85.0
$S_2O_4^{2-}$	34		66.5
$S_2O_6^{2-}$			93
$S_2O_8^{2-}$			86
WO_4^{2-}	35	59	69.4
Organic cations			
Decylpyridinium$^+$			29.5
Diethylammonium$^+$			42.0
Dimethylammonium$^+$			51.5
Dipropylammonium$^+$			30.1
Dodecylammonium$^+$			23.8
Ethylammonium$^+$			47.2
Ethyltrimethylammonium$^+$			40.5
Isobutylammonium$^+$			38.0
Methylammonium$^+$			58.3
Piperidinium$^+$			37.2
Propylammonium$^+$			40.8
Tetrabutylammonium$^+$			19.5
Tetraethylammonium$^+$			32.6
Tetramethylammonium$^+$			44.9
Tetrapropylammonium$^+$			23.5
Triethylsulfonium$^+$			36.1
Trimethylammonium$^+$			47.2
Trimethylsulfonium$^+$			51.4
Tripropylammonium$^+$			26.1
Organic anions			
Acetate$^-$	20	34	41
Benzoate$^-$			32.4
Bromoacetate$^-$			39.2
Bromobenzoate$^-$			30
Butanoate$^-$			32.6
Chloroacetate$^-$			42.2
m-Chlorobenzoate$^-$			31
o-Chlorobenzoate$^-$			30.5
Citrate(3−)			70.2
Crotonate$^-$			33.2
Cyanoacetate$^-$			43.4
Cyclohexanecarboxylate$^-$			28.7
Cyclopropane-1,3-dicarboxylate^{2-}			53.4
Decylsulfonate$^-$			26
Dichloroacetate$^-$			38.3
Diethylbarbiturate(2−)			26.3
Dihydrogencitrate$^-$			30
Dimethylmalonate(2−)			49.4
3,5-Dinitrobenzoate$^-$			28.3
Dodecylsulfonate$^-$			24
Ethylmalonate$^-$			49.3
Ethylsulfonate$^-$			39.6

TABLE 1.86 Standard Solutions for Calibrating Conductivity Vessels

The values of conductivity κ are corrected for the conductivity of the water used. The cell constant θ of a conductivity cell can be obtained from the equation

$$\theta = \frac{KRR_{\text{solv}}}{R_{\text{solv}} - R}$$

where R is the resistance measured when the cell is filled with a solution of the composition stated in the table below, and R_{solv} is the resistance when the cell is filled with solvent at the same temperature.

Grams KCI per kilogram solution (in vacuo)	Conductivity in $\text{ohm}^{-1} \cdot \text{cm}^{-1}$ at		
	0°C	18°C	25°C
71.135 2	$0.065\ 14_4$	$0.097\ 79_0$	$0.111\ 28_7$
7.419 13	$0.007\ 134_4$	$0.011\ 161_2$	$0.012\ 849_7$
0.745 263*	$0.000\ 773\ 2_6$	$0.001\ 219\ 9_2$	$0.001\ 408\ 0_8$

*Virtually 0.0100 *M*.

From the data of Jones and Bradshaw, *J. Am. Chem. Soc.*, **55,** 1780 (1933). The original data have been converted from (int. ohm)$^{-1}$cm^{-1}.

TABLE 1.87 Equivalent Conductivities of Electrolytes in Aqueous Solutions at 18°C

The unit of Λ in the table is $\Omega^{-1} \cdot cm^{-2} \cdot equiv^{-1}$. The entities to which the equivalent relates are given in the first column.

Electrolyte	Concentration, *N*										
	0.001	0.005	0.01	0.05	0.1	0.5	1.0	2.0	3.0	4.0	5.0
Acetic acid	41	20.0	14.3	6.48	4.60	2.01	1.32		0.54		0.29
$AgNO_3$	113.2	110.0	107.8	99.5	94.3	77.8	67.8	56.0	48.2	42.1	37.2
½Ag_2SO_4	116.3	108.4	102.9								
⅓$AlBr_3$ (25°)	132	124	119	103	97						
⅓$AlCl_3$	121.1	105.0	93.8			65.0	56.2	44.2	34.7	27.2	
⅓AlI_3 (25°)	131	124	119	108							
⅓$Al(NO_3)_3$ (25°)	123	115	110	94	88						
⅙$Al_2(SO_4)_3$ (25°)	107.2	76.8	60.6								
½$Ba(OAc)_2$	85.0	80.4	77.1	65.7	60.2	43.8	34.3				
½$Ba(BrO_3)_2$ (25°)	113.6	106.8	102.7								
½$BaCl_2$	115.6	112.3	106.7	96.0	90.8	77.3	70.1	60.3	52.3		
½$Ba(NO_3)_2$	111.7	105.3	101.0	86.8	78.9	56.6	48.4		29.8	23.4	
½$Ba(OH)_2$	216	213	207	191	180						
Butyric acid						1.66	0.98	0.46	0.26	0.18	0.11
½$Ca(OAc)_2$	79.6	75.0	71.9	60.3	54.0	36.3	26.3				
½$CaCl_2$	112.0	106.7	103.4	93.3	88.2	74.9	67.5	58.3	49.7	42.4	35.6
½$Ca(NO_3)_2$	108.5	103.0	99.5	88.4	82.5	65.7	55.9	43.5	35.5	26.0	21.5
½$Ca(OH)_2$		233	226								
½$CaSO_4$	104.3	86.3	77.4								
½$CdBr_2$		86.5	76.3	53.2	44.6	25.3	18.3	12.5	9.1	6.8	5.3
½$CdCl_2$		91	83	59	50	30.8	22.4	14.4	9.9	7.1	5.4
½CdI_2		76.7	65.6	40.1	31.0	18.3	15.4	12.3	9.7	8.0	
½$Cd(NO_3)_2$		100	96	86.4	80.8	63.9	54.5	41.0	31.4	23.7	17.6
½$CdSO_4$	97.7	79.7	70.3	49.6	42.2	28.7	23.6	17.7	14.0	11.0	8.35
⅓$CeCl_3$ (25°)	137.4		122.1		99.0						
⅙$Ce_2(C_2O_4)_3$ (25°)	85.5	54	45.8	29							
Chloroacetic acid (25°)					42.9	20.2	13.6	8.1	5.6	4.2	3.3
Citric acid	88.4	54	42.5	22.0	16.1	7.3	5.4				
½$CoCl_2$		99.3	95.6	82.3	75.0	51.5	45.3	40.3	35.4	30.5	26.4
⅓$CrCl_3$						68.6	56.8	44.8	35.2		

½CrO_3(H_2CrO_4) (25°)	201	195	193	191	186						
CsCl	130.7	127.5	125.2		113.5	104.3	100.3	95.7	85.1		
½Cu$(OAc)_2$ (25°)	55.7	50.6	47.2	34.9	28.4						
½$CuCl_2$								41.2	31.5	24.5	19.1
½$Cu(NO_3)_2$ (15°)	107.9	97.1	93.7	83.7	78.2	67.5	56.8	45.4	35.3	27.8	21.4
½$CuSO_4$	98.5	81.0	71.7	53.6	43.8	30.5	25.6	19.7	16.5		
Dichloroacetic acid (25°)					207.5	119	82	44.6	26.5	16.3	9.6
½$FeCl_2$ (25°)	131	125	120	103	93						
⅓$FeCl_3$						66.5	52.9	37.6	28.1	20.5	15.9
½$FeSO_4$	82	75	70	54	44.5	30.8	25.8	19.5	15.37		
Formic acid	125.6						5.18	3.68	2.93	2.39	1.92
H_3AsO_4 (1 *M*) (25°)	308.2	230.0	187.0	103.4	80.4						
H_3BO_3	13.5										
HBr					356	306	282	243	214	179	
$HBrO_3$ (25°)	401	387	373	272	156						
HCl	377	373	370	360	351	327	301		215		152.2
$HClO_3$					343	317	292	247	207		
$HClO_4$ (25°)	413	406	402	392	386	358					
HF		90	60	35.9	31.3	27.0	25.7		24.2		24.0
HI					347	322	297	255	215	179	
HIO_3	343.3	332.8	323.9		253	175	141	106	87	71	
HNO_3	375	371	368	357	350	324	310		220		156
H_3PO_4 (1 *M*)	318	279	255				66		53.1		51.3
HSCN (25°)	399	394	390	377	370						
½H_2SO_4	361	330	308	253	225	205	198		166.8		135.0
½$HgCl_2$				1.85	1.23						
⅓$InBr_3$					53.9	37.0	28.7	19.8	14.4	10.1	
KOAc	98.3	95.7	94.0	87.7	83.8	71.6	63.4	50.0	40.7	31.4	24.5
KBr	129.4	126.4	124.4	117.8	114.2	105.4	102.5	98.0	93.3	87.9	
$KBrO_3$	109.9	106.9	104.7	97.3	93.0						
⅓K_3citrate		109.9	103	87.8	80.8						
KCl	127.3	124.4	122.4	115.8	112.0	102.4	98.3	92.0	88.9		
$KClO_3$	116.9	113.6	111.6	103.7	99.2	85.3					
$KClO_4$ (25°)	137.9	134.2	131.5	121.6	115.2						

(*Continued*)

TABLE 1.87 Equivalent Conductivities of Electrolytes in Aqueous Solutions at 18°C (*Continued*)

The unit of Λ in the table is $\Omega^{-1} \cdot cm^{-2} \cdot equiv^{-1}$. The entities to which the equivalent relates are given in the first column.

	Concentration, *N*										
Electrolyte	0.001	0.005	0.01	0.05	0.1	0.5	1.0	2.0	3.0	4.0	5.0
KCN (15°)						104.2	99.7				
½K_2CO_3	133.0	121.6	115.5	100.7	94.1	77.8	70.7	65.0	55.6	49.2	42.9
½$K_2C_2O_4$	122.4	116.7	112.5	100.8	94.9	80.4	73.7				
½K_2CrO_4					100.5	86.4	79.5	72.0	59.9		
½$K_2Cr_2O_7$					98.2	85.4					
KF	108.9	106.2	104.3	97.7	94.0	82.6	76.0	63.4	56.5	51.7	46.5
⅓$K_3[Fe(CN)_6]$	163.1	150.7									
¼$K_4[Fe(CN)_6]$	167.2	146.1	134.8	107.7	97.9						
$KHCO_3$ (25°)	115.3	112.2	110.1			86.5	78.9				
KH phthalate	119.3	103.7	99.9	89.3	83.8						
KHS						92.5	91.7	86.4	80.7		69.3
$KHSO_4$						21.0	18.4	15.2			
KH_2PO_4 (1 *M*) (25°)	107.1	100.8	98.0	90.7	85.6	60.0[18]	45.8[18]				
KI	128.2	125.3	123.4	117.3	114.0	106.2	103.6	101.3	96.4	89.0	81.2
KIO_3	96.0	93.2	91.2	84.1	79.7						
KIO_4 (25°)	124.9	121.2	118.5	106.7	98.1						
$KMnO_4$ (25°)	133.3		126.5		113						
KNO_3	123.6	120.5	118.2	109.9	104.8	89.2	80.5	69.4	61.3		
KOH	234	230	228	219	213	197	184		140.6		105.8
$KReO_4$ (25°)	125.1	121.3	118.5	106.4	97.4						
½K_2S							135.6	119.7	108.3	97.2	86.1
KSCN	118.6	115.8	113.9	107.7	104.3	95.7	91.6	86.8	74.6		
½K_2SO_4	126.9	120.3	115.8	101.9	94.9	78.5	71.6				
⅓$LaCl_3$ (25°)	137.0	127.5	121.8	106.2	99.1						
⅓$La(NO_3)_3$				86.1	72.1	65.4	54.0	39.1	28.5	19.9	
⅙$La_2(SO_4)_3$				25.7	21.5						
Lactic acid	108.9	53.5	39	18.1	13.2						
LiOAc					51.3	37.7	28.9	18.2	11.9	7.2	
LiBr				87.9	84.4	73.9	67.2	57.7		44.2	
LiCl	96.5	93.9	92.1	86.1	82.4	70.7	63.4	53.1	45.3		33.3
$LiClO_4$ (25°)	103.4	100.6	98.6	92.2	88.6						

½Li_2CO_3				64.2	59.1						
LiI						75.3	69.2	61.0			
$LiIO_3$	65.3	62.9	61.2	55.3	51.5	39.0	31.2	21.4	14.6		
$LiNO_3$	92.9	90.3	88.6	82.7	79.2	68.0	60.8	50.3	34.9	27.3	
LiOH						149.0	134.5	113.5	95.7		
½Li_2SO_4	96.4		86.9	74.7	68.2	50.5	41.3	30.7	23.3	18.1	13.9
½$MgCl_2$	106.4	101.3	98.1	88.5	83.4	69.6	61.5	52.3	43.3	35.0	28.0
½$Mg(NO_3)_2$	102.6	97.7	94.7	85.3	80.5	67.0	59.0	47.0	39.8		
½$MgSO_4$	99.8	84.5	76.2	56.9	49.7	35.4	28.9	23.0	17.3	12.9	9.3
½$MnCl_2$					86.0	68.5	61.0	48.5	38.8	30.2	23.0
½$MnSO_4$						27.6	24.4	18.3	14.0	10.5	7.3
NH_3(aq)	28.0	13.2	9.6	4.6	3.3	1.35	0.89		0.36		0.20
NH_4OAc		92.9	91.4	84.9		60.5	54.7	42.9	34.0	26.5	
NH_4Cl	127.3	124.3	122.1	115.2	110.7	101.4	97.0	92.1	88.2	85.0	80.7
NH_4F					90.1	74.5	65.7	55.3	47.9	42.2	
NH_4I				118.0	115.0	106.0	103.1	100.0		91.4	84.5
NH_4NO_3	124.5		118.0	110.0	106.6	94.5	88.8	85.1		71.9	47.6
NH_4SCN					104.3	94.0	89.9	84.7	79.2	74.0	
½$(NH_4)_2SO_4$		120.0	116.5		89.0	79.5	73.0	65.0		55.2	
NaOAc	75.2	72.4	70.2	64.2	61.1	49.4	41.2	29.8	21.5	15.3	10.5
NaBr				99.1	96.0	84.6	78.1	69.1		53.0	
$NaBrO_3$						61.8	54.5	44.1			
Na *n*-butyrate (25°)	80.3	77.6	75.8	69.3	65.3						
NaCl	106.5	103.8	102.0	95.7	92.0	80.9	74.3	64.8	56.5	49.4	42.7
$NaClO_4$	114.9[25]	111.7[25]	109.6[25]	102.4[25]	98.4[25]	71.7	65.0	55.1	46.0	38.8	
½Na_2CO_3	112	102.5	96.2	80.3	72.9	54.5	45.5	34.5	27.2		
½Na_2CrO_4					82.5	66.4	57.7	46.6	38.3	31.1	
½$Na_2Cr_2O_7$ (25°)		103		98.3	94.9						
NaF	87.8	85.2	83.5	77.0	73.1	60.0	51.9				
¼$Na_4[Fe(CN)_6]$ (25°)		129.6	120.0	97.0	88.2						
Na formate	88.6					61.4	53.7	43.1	34.8	28.2	
$NaHCO_3$ (25°)	93.5	90.5	88.4	80.6	76.0						
⅓Na_2HPO_4	58.4		54.0		44.0	33.5	28.0				
NaH_2PO_4	67.9	65.8	64.4	57.8	54.1						
¼$Na_2H_2P_2O_7$	41.1	39.4	38.2	34.6	32.5	25.4					
NaI	124.2	121.2	119.2	112.8	108.8	97.5	89.9	78.6	69.9	62.2	

(*Continued*)

TABLE 1.87 Equivalent Conductivities of Electrolytes in Aqueous Solutions at 18°C (*Continued*)

The unit of Λ in the table is $\Omega^{-1} \cdot cm^{-2} \cdot equiv^{-1}$. The entities to which the equivalent relates are given in the first column.

	Concentration, *N*										
Electrolyte	0.001	0.005	0.01	0.05	0.1	0.5	1.0	2.0	3.0	4.0	5.0
$NaIO_3$	75.2	72.6	70.9	64.4	60.5						
½Na_2MoO_4	120.8	113	110								
NaN_3 (25°)	117.1	113.8	110.5	101.3	95.7		68.0				
$NaNO_2$ (25°)							75.9	63.1	53.6		39.7
$NaNO_3$	102.9	100.1	98.2	91.4	87.2	74.1	65.9	54.5	46.0	39.0	
NaOH	208	203	200	190	183	172	160		108.0		69.0
Na picrate (25°)	78.6	75.7	73.7	66.3	61.8						
⅓Na_3PO_4	125	122	119	91							
Na propionate (25°)	83.5	80.9	79.1								
½Na_2S						117.0	104.3	85.0	71.0	59.0	47.2
NaSCN						74.3	68.9	59.8	50.9	43.7	
½Na_2SiO_3	144	139	136	124	116	88	72	51	38	27	19
½Na_2SO_4	106.7	100.8	96.8	83.9	78.4	59.7	50.8	40.0	33.5		
(mono) Na tartrate	120	81.5	74.8	64.3	60.4						
½Na_2WO_4 (25°)	116.1	109.2	104.8	92.2	85.8						
½$NiSO_4$	96.3	79.5	70.8	51.0	43.8	30.4	25.1	19.3	15.1		
½Oxalic acid	180.7		158.2	132.9	116.9	75.9	59.4				
½$Pb(NO_3)_2$	116.1	108.6	103.5	86.3	77.3	53.2	42.0	31.0			
Propionic acid						1.57	1.00	0.54		0.20	
RbCl	130.3	127.4	125.3	117.8	113.9		101.9	97.1	92.7	87.2	
RbOH					220.6	204.8	192.0	170.0	148.3		
¼$SnCl_4$						216.8	121.7	66.9	47.9	32.7	
½$SrCl_2$	114.5	108.9	105.4	94.4	90.2	75.7	68.5	58.7	49.9	42.2	
½$Sr(NO_3)_2$	108.3	102.7	99.0	87.3	80.9	62.7	52.1	38.0	29.3	29.3	16.4
Tartaric acid (15°)							7.03	4.58	3.32	2.48	1.83
¼$ThCl_4$						61.0	54.0	44.3	36.3	29.8	
TlCl	128.2	123.7	120.2								
TlF	113.3	108.2	105.4	97.4	92.6	78.8	71.5	62.7			
$TlNO_3$	124.7	121.1	118.4	107.9	101.2						
½Tl_2SO_4	127.4	118.4	112.3	92.7	83.1						
Trichloroacetic acid (25°)						273	207	127	79	44	19
½UO_2F_2 (25°)	26.10	12.31	9.17	5.43	4.74	3.75	3.22				
½UO_2SO_4 (25°)	106.5	63.2	49.2	27.6	22.2	14.4	11.6				2.7
⅓YCl_3 (25°)	129	122	118	109							
½$Zn(OAc)_2$ (25°)	83	77	73	58	49						
½$ZnCl_2$	107	101	98	87	82	65	55	39.6	29.6	23.2	18.5
½$Zn(NO_3)_2$	120	114	111	100							
½$ZnSO_4$	98.4	82.1	73.2	53.0	45.6	32.3	26.6	20.0	15.9	12.0	9.0

TABLE 1.88 Conductivity of Very Pure Water at Various Temperatures and the Equivalent Conductances of Hydrogen and Hydroxyl Ions

Temp., °C	Conductivity, μS · cm^{-1}	Resistivity, MΩ · cm	Equivalent conductance, cm^2 · ohm^{-1} · equivalent^{-1}	
			λ^0, H^+	λ^0, OH^-
0	0.011 61	86.14	224.1	117.8
5	0.016 61	60.21	250.0	133.6
10	0.023 15	43.21	275.6	149.6
15	0.031 53	31.71	300.9	165.9
18	0.037 54	26.64	315.8	491.6
20	0.042 05	23.78	325.7	182.5
25	0.055 08	18.15	350.1	199.2
30	0.070 96	14.09	374.0	216.1
35	0.090 05	11.10	397.4	233.0
40	0.112 7	8.88	420.0	267.2
45	0.139 3	7.18	442.0	267.2
50	0.170 2	5.88	463.3	284.3
55	0.205 5	4.86	483.8	301.4
60	0.245 7	4.06	503.4	318.5
65	0.291 2	3.43	522.0	335.4
70	0.341 6	2.93	539.7	352.2
75	0.397 8	2.51	556.4	368.8
80	0.459 3	2.18	572.0	385.2
85	0.525 8	1.90	586.4	401.4
90	0.597 7	1.67	599.6	417.3
95	0.675 3	1.48	611.6	432.8
100	0.756 9	1.32	622.2	448.1
150	1.84	0.543		
200	2.99	0.334	824	701
250	3.31	0.302		
300	2.42	0.413	894	821

Source: Data from T. S Light and S.L. Licht. *Anal Chem.*, **59**: 2327–2330(1987).

1.23 THERMAL PROPERTIES

TABLE 1.89 Eutectic Mixtures

The *eutectic temperature* $\theta_{C,E}$ is the lowest temperature at which both the solid components of a mixture are in equilibrium with the liquid phase. $\theta_{C,m}$ denotes melting temperature.

Component 1	$\theta_{c,m}/°C$	Component 2	$\theta_{C,m}/°C$	$\theta_{C,E}/°C$	Composition of eutectic mixture (per cent by mass)	
Sn	232	Pb	327	183	Sn, 63·0	Pb, 37·0
Sn	232	Zn	420	198	Sn, 91·0	Zn, 9·0
Sn	232	Ag	961	221	Sn, 96·5	Ag, 3·5
Sn	232	Cu	1083	227	Sn, 99·2	Cu, 0·8
Sn	232	Bi	271	140	Sn, 42·0	Bi, 58·0
Sb	630	Pb	327	246	Sb, 12·0	Pb, 88·0
Bi	271	Pb	327	124	Bi, 55·5	Pb, 44·5
Bi	271	Cd	321	146	Bi, 60·0	Cd, 40·0
Cd	321	Zn	420	270	Cd, 83·0	Zn, 17·0

TABLE 1.90 Transition Temperatures

$\theta_{C,t}$ *denotes transition temperature*

Substance	System	$\theta_{C,t}/°C$
sulphur	Rhombic (α) $\rightleftharpoons$ Monoclinic (β)	95.6
Tin	Grey (α) White (β)	
Iron	α (body-centered cubic) $\rightleftharpoons$ γ (face-centered cubic)	906
	γ (body-centered cubic) $\rightleftharpoons$ δ (face-centered cubic)	1401
Sodium sulphate	Na_2So_4 $10H_2O \rightleftharpoons Na_2SO_4 + 10H_2O$	32.4
Mercury(II) iodide	Tetragonal (red) $\rightleftharpoons$ Orthorhombic (yellow)	126
Ammonium chloride	α (CsCl structure) $\rightleftharpoons$ β (NaCl structure)	184
Caesium chloride	CsCl structure $\rightleftharpoons$ NaCl structure	445
Copper(I) mercury(II) Iodide	Tetragonal (red) $\rightleftharpoons$ Cubic (dark brown)	69

SECTION 2
ORGANIC CHEMISTRY

SECTION 2
ORGANIC CHEMISTRY

2.1 NOMENCLATURE OF ORGANIC COMPOUNDS

The following synopsis of rules for naming organic compounds and the examples given in explanation are not intended to cover all the possible cases.

2.1.1 Nonfunctional Compounds

2.1.1.1 Alkanes. The saturated open-chain (acyclic) hydrocarbons (C_nH_{2n+2}) have names ending in -ane. The first four members have the trivial names *methane* (CH_4), *ethane* (CH_3CH_3 or C_2H_6), *propane* (C_3H_8), and *butane* (C_4H_{10}). For the remainder of the alkanes, the first portion of the name is derived from the Greek prefix that cites the number of carbons in the alkane followed by -ane with elision of the terminal -a from the prefix.

TABLE 2.1 Straight-Chain Alkanes

*n**	Name	*n**	Name	*n**	Name	*n**	Name
1	Methane	11	Undecane‡	21	Henicosane	60	Hexacontane
2	Ethane	12	Dodecane	22	Docosane	70	Heptacontane
3	Propane	13	Tridecane	23	Tricosane	80	Octacontane
4	Butane	14	Tetradecane			90	Nonacontane
5	Pentane	15	Pentadecane	30	Triacontane	100	Hectane
6	Hexane	16	Hexadecane	31	Hentriacontane	110	Decahectane
7	Heptane	17	Heptadecane	32	Dotriacontane	120	Icosahectane
8	Octane	18	Octadecane			121	Henicosahectane
9	Nonane†	19	Nonadecane	40	Tetracontane		
10	Decane	20	Icosane§	50	Pentacontane		

* n = total number of carbon atoms.
† Formerly called enneane.
‡ Formerly called hendecane.
§ Formerly called eicosane.

For branching compounds, the parent structure is the longest continuous chain present in the compound. Consider the compound to have been derived from this structure by replacement of hydrogen by various alkyl groups. Arabic number prefixes indicate the carbon to which the alkyl group is attached. Start numbering at whichever end of the parent structure that results in the lowest-numbered locants. The arabic prefixes are listed in numerical sequence, separated from each other by commas and from the remainder of the name by a hyphen.

If the same alkyl group occurs more than once as a side chain, this is indicated by the prefixes di-, tri-, tetra-, etc. Side chains are cited in alphabetical order (before insertion of any multiplying prefix). The name of a complex radical (side chain) is considered to begin with the first letter of its complete name. Where names of complex radicals are composed of identical words, priority for citation is given to that radical which contains the lowest-numbered locant at the first cited point of difference in the radical. If two or more side chains are in equivalent positions, the one to be assigned the lowest-numbered locant is that cited first in the name. The complete expression for the side chain may be enclosed in parentheses for clarity or the carbon atoms in side chains may be indicated by primed locants.

If hydrocarbon chains of equal length are competing for selection as the parent, the choice goes in descending order to (1) the chain that has the greatest number of side chains, (2) the chain whose side chains have the lowest-numbered locants, (3) the chain having the greatest number of carbon atoms in the smaller side chains, or (4) the chain having the least-branched side chains.

These trivial names may be used for the unsubstituted hydrocarbon only:

Isobutane	$(CH_3)_2CHCH_3$	Neopentane	$(CH_3)_4C$
Isopentane	$(CH_3)_2CHCH_2CH_3$	Isohexane	$(CH_3)_2CHCH_2CH_2CH_3$

Univalent radicals derived from saturated unbranched alkanes by removal of hydrogen from a terminal carbon atom are named by adding -yl in place of -ane to the stem name. Thus the alkane *ethane* becomes the radical *ethyl*. These exceptions are permitted for unsubstituted radicals only:

Isopropyl	$(CH_3)_2CH$—	Isopentyl	$(CH_3)_2CHCH_2CH_2$—
Isobutyl	$(CH_3)_2CHCH_2$—	Neopentyl	$(CH_3)_3CCH_2$—
sec-Butyl	$CH_3CH_2CH(CH_3)$—	*tert*-Pentyl	$CH_3CH_2C(CH_3)_2$—
tert-Butyl	$(CH_3)_3C$—	Isohexyl	$(CH_3)_2CHCH_2CH_2CH_2$—

Note the usage of the prefixes iso-, neo-, *sec*-, and *tert*-, and note when italics are employed. Italicized prefixes are never involved in alphabetization, except among themselves; thus *sec*-butyl would precede isobutyl, isohexyl would precede isopropyl, and *sec*-butyl would precede *tert*-butyl.

Examples of alkane nomenclature are

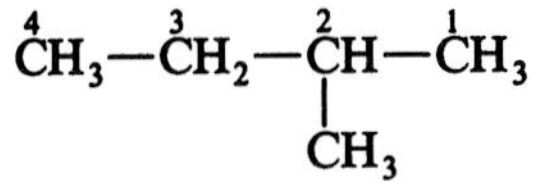

2-Methylbutane (or the trivial name, isopentane)

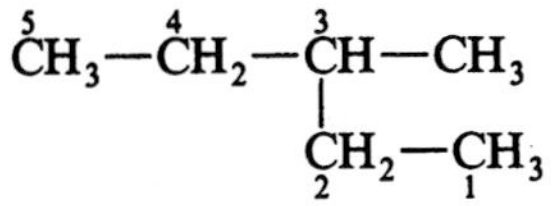

3-Methylpentane (not 2-ethylbutane)

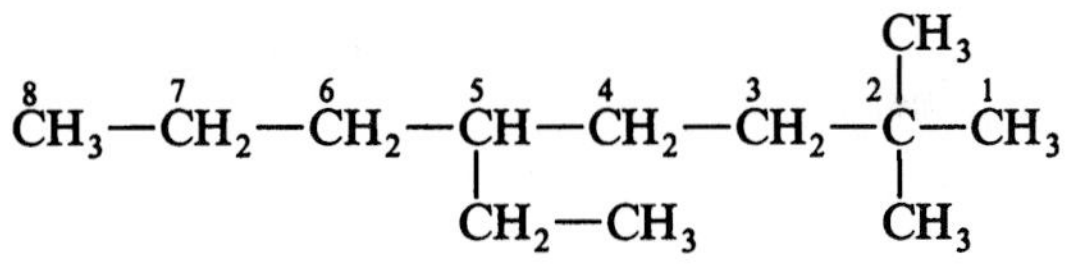

5-Ethyl-2,2-dimethyloctane (note cited order)

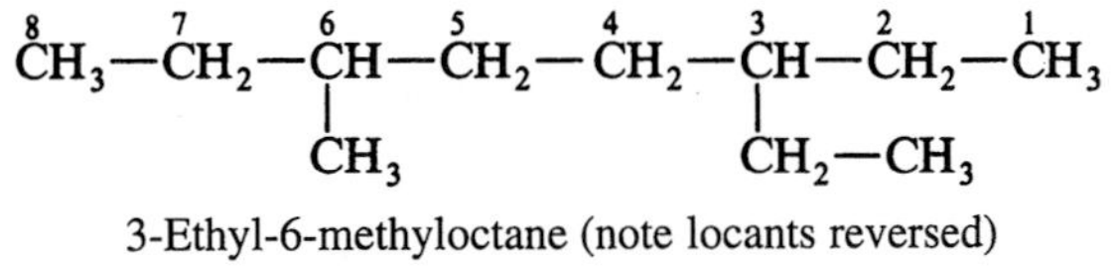

3-Ethyl-6-methyloctane (note locants reversed)

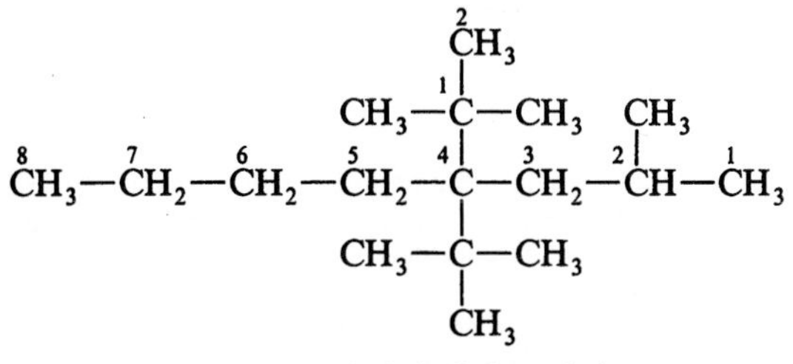

4,4-Bis(1,1-dimethylethyl)-2-methyloctane
4,4-Bis-1′,1′-dimethylethyl-2-methyloctane
4,4-Bis(*tert*-butyl)-2-methyloctane

Bivalent radicals derived from saturated unbranched alkanes by removal of two hydrogen atoms are named as follows: (1) If both free bonds are on the same carbon atom, the ending -ane of the hydrocarbon is replaced with -ylidene. However, for the first member of the alkanes it is methylene rather than methylidene. Isopropylidene, *sec*-Butylidene, and neopentylidene may be used for the unsubstituted group only. (2) If the two free bonds are on different carbon atoms, the straight-chain group terminating in these two carbon atoms is named by citing the number of methylene groups comprising the chain. Other carbon groups are named as substituents. Ethylene is used rather than dimethylene for the first member of the series, and propylene is retained for $CH_3—CH—CH_2—$ (but trimethylene is $—CH_2—CH_2—CH_2—$).

Trivalent groups derived by the removal of three hydrogen atoms from the same carbon are named by replacing the ending -ane of the parent hydrocarbon with -ylidyne.

2.1.1.2 Alkenes and Alkynes. Each name of the corresponding saturated hydrocarbon is converted to the corresponding alkene by changing the ending -ane to -ene. For alkynes the ending is -yne. With more than one double (or triple) bond, the endings are -adiene, -atriene, etc. (or -adiyne, -atriyne, etc.). The position of the double (or triple) bond in the parent chain is indicated by a locant obtained by numbering from the end of the chain nearest the double (or triple) bond; thus $CH_3CH_2CH{=}CH_2$ is 1-butene and $CH_3C{\equiv}CCH_3$ is 2-butyne.

For multiple unsaturated bonds, the chain is so numbered as to give the lowest possible locants to the unsaturated bonds. When there is a choice in numbering, the double bonds are given the lowest locants, and the alkene is cited before the alkyne where both occur in the name. Examples:

$CH_3CH_2CH_2CH_2CH{=}CH—CH{=}CH_2$ 1,3-Octadiene

$CH_2{=}CHC{\equiv}CCH{=}CH_2$ 1,5-Hexadiene-3-yne

$CH_3CH{=}CHCH_2C{\equiv}CH$ 4-Hexen-1-yne

$CH{\equiv}CCH_2CH{=}CH_2$ 1-Penten-4-yne

Unsaturated branched acyclic hydrocarbons are named as derivatives of the chain that contains the maximum number of double and/or triple bonds. When a choice exists, priority goes in sequence to (1) the chain with the greatest number of carbon atoms and (2) the chain containing the maximum number of double bonds.

These nonsystematic names are retained.

Ethylene $CH_2{=}CH_2$

Allene $CH_2{=}C{=}CH_2$

Acetylene $HC{\equiv}CH$

An example of nomenclature for alkenes and alkynes is

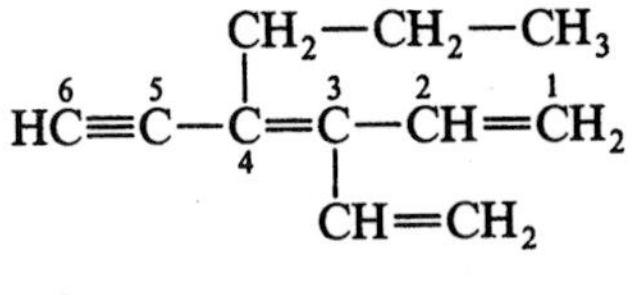

4-Propyl-3-vinyl-1,3-hexadien-5-yne

Univalent radicals have the endings -enyl, -ynyl, -dienyl, -diynyl, etc. When necessary, the positions of the double and triple bonds are indicated by locants, with the carbon atom with the free valence numbered as 1. Examples:

$CH_2{=}CH—CH_2—$ 2-Propenyl

$CH_3—C{\equiv}C—$ 1-Propynyl

$CH_3—C{\equiv}C—CH_2CH{=}CH_2—$ 1-Hexen-4-ynyl

These names are retained:

Vinyl (for ethenyl) $CH_2{=}CH{-}$

Allyl (for 2-propenyl) $CH_2{=}CH{-}CH_2{-}$

Isopropenyl (for 1-methylvinyl but for unsubstituted radical only) $CH_2{=}C(CH_3){-}$

Should there be a choice for the fundamental straight chain of a radical, that chain is selected which contains (1) the maximum number of double and triple bonds, (2) the largest number of carbon atoms, and (3) the largest number of double bonds. These are in descending priority.

Bivalent radicals derived from unbranched alkenes, alkadienes, and alkynes by removing a hydrogen atom from each of the terminal carbon atoms are named by replacing the endings -ene, -diene, and -yne by -enylene, -dienylene, and -ynylene, respectively. Positions of double and triple bonds are indicated by numbers when necessary. The name *vinylene* instead of ethenylene is retained for $-CH{=}CH-$.

2.1.1.3 Monocyclic Aliphatic Hydrocarbons. Monocyclic aliphatic hydrocarbons (with no side chains) are named by prefixing cyclo- to the name of the corresponding open-chain hydrocarbon having the same number of carbon atoms as the ring. Radicals are formed as with the alkanes, alkenes, and alkynes. Examples:

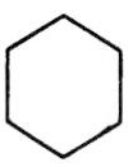

Cyclohexane — Cyclohexyl- (for the radical)

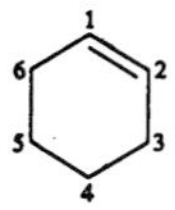

Cyclohexene — 1-Cyclohexenyl- (for the radical with the free valence at carbon 1)

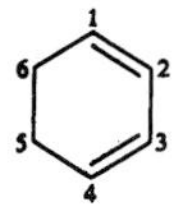

1,3-Cyclohexandiene — Cyclohexadienyl- (the unsaturated carbons are given numbers as low as possible, numbering from the carbon atom with the free valence given the number 1)

For convenience, aliphatic rings are often represented by simple geometric figures: a triangle for cyclopropane, a square for cyclobutane, a pentagon for cyclopentane, a hexagon (as illustrated) for cyclohexane, etc. It is understood that two hydrogen atoms are located at each corner of the figure unless some other group is indicated for one or both.

2.1.1.3 Monocyclic Aromatic Compounds. Except for six retained names, all monocyclic substituted aromatic hydrocarbons are named systematically as derivatives of benzene. Moreover, if the substituent introduced into a compound with a retained trivial name is identical with one already present in that compound, the compound is named as a derivative of benzene. These names are retained:

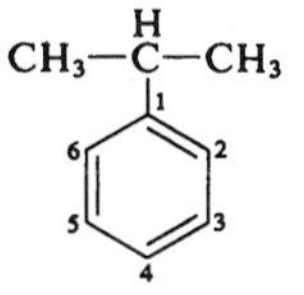

Cumene

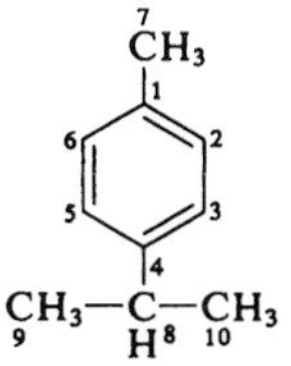

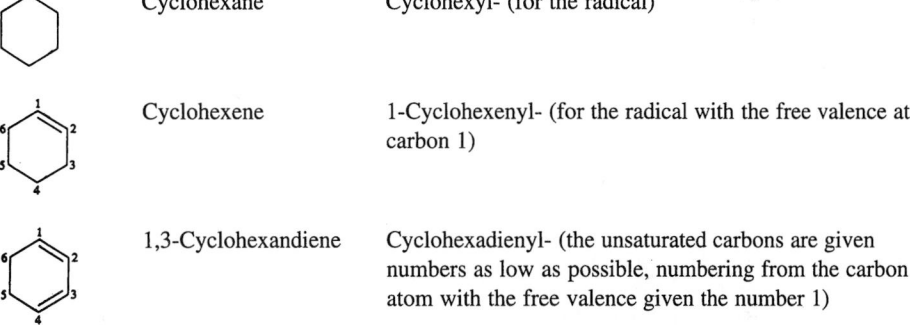

Cymene (all three forms; *para-* shown)

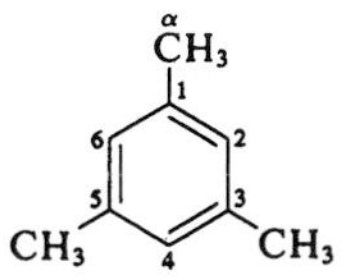

Mesitylene

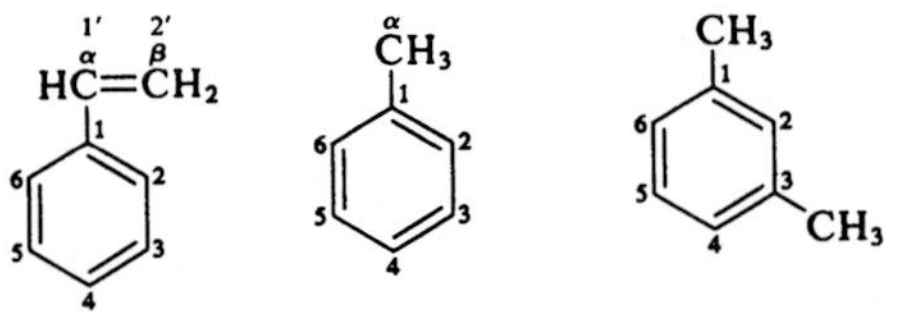

Styrene Toluene Xylene (all three forms; *meta-* shown)

The position of substituents is indicated by numbers, with the lowest locant possible given to substituents. When a name is based on a recognized trivial name, priority for lowest-numbered locants is given to substituents implied by the trivial name. When only two substituents are present on a benzene ring, their position may be indicated by *o-* (*ortho-*), *m-* (*meta-*), and *p-* (*para-*) (and alphabetized in the order given) used in place of 1,2-, 1,3-, and 1,4-, respectively.

Radicals derived from monocyclic substituted aromatic hydrocarbons and having the free valence at a ring atom (numbered 1) are named phenyl (for benzene as parent, since benzyl is used for the radical $C_6H_5CH_2—$), cumenyl, mesityl, tolyl, and xylyl. All other radicals are named as substituted phenyl radicals. For radicals having a single free valence in the side chain, these trivial names are retained:

Benzyl $C_6H_5CH_2—$

Benzhydryl (alternative to diphenylmethyl) $(C_6H_5)_2CH—$

Cinnamyl $C_6H_5CH{=}CH—CH_2—$

Phenethyl $C_6H_5CH_2CH_2—$

Styryl $C_6H_5CH{=}CH—$

Trityl $(C_6H_5)_3C—$

Otherwise, radicals having the free valence(s) in the side chain are named in accordance with the rules for alkanes, alkenes, or alkynes.

The name *phenylene* (*o-*, *m-*, or *p-*) is retained for the radical $—C_6H_4—$. Bivalent radicals formed from substituted benzene derivatives and having the free valences at ring atoms are named as substituted phenylene radicals, with the carbon atoms having the free valences being numbered 1,2-, 1,3-, or 1,4-, as appropriate.

Radicals having three or more free valences are named by adding the suffixes -triyl, -tetrayl, etc. to the systematic name of the corresponding hydrocarbon.

2.1.1.4 Fused Polycyclic Hydrocarbons. The names of polycyclic hydrocarbons containing the maximum number of conjugated double bonds end in -ene. Here the ending does not denote one double bond. Names of hydrocarbons containing five or more fixed benzene rings in a linear arrangement are formed from a numerical prefix followed by -acene.

Numbering of each ring system is fixed but it follows a systematic pattern. The individual rings of each system is oriented so that the greatest number of rings are (1) in a horizontal row and (2) the maximum number of rings is above and to the right (upper-right quadrant) of the horizontal row. When two orientations meet these requirements, the one is chosen that has the fewest rings in the lower-left quadrant. Numbering proceeds in a clockwise direction, commencing with the carbon atom not engaged in ring fusion that lies in the most counterclockwise position of the uppermost ring (upper-right quadrant); omit atoms common to two or more rings. Atoms common to two or more rings are designated by adding lowercase roman letters to the number of the position immediately preceding. Interior atoms follow the highest number, taking a clockwise sequence wherever there is a choice. Anthracene and phenanthrene are two exceptions to the rule on numbering. Two examples of numbering follow:

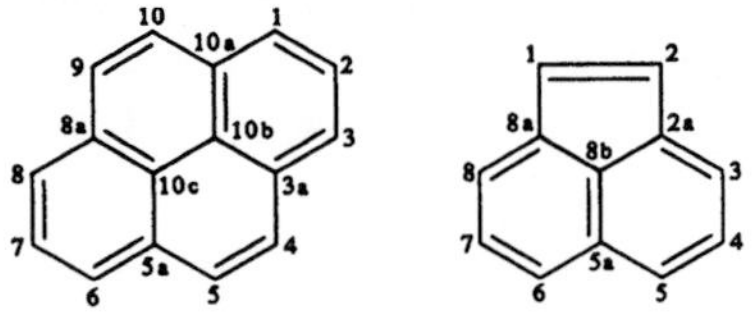

When a ring system with the maximum number of conjugated double bonds can exist in two or more forms differing only in the position of an "extra" hydrogen atom, the name can be made specific by indicating the position of the extra hydrogen(s). The compound name is modified with a locant followed by an italic capital *H* for each of these hydrogen atoms. Carbon atoms that carry an indicated hydrogen atom are numbered as low as possible. For example, 1*H*-indene is illustrated in Table 2.2; 2*H*-indene would be

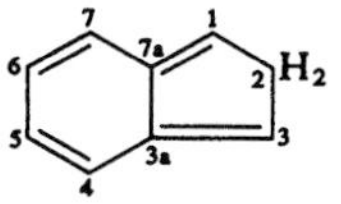

Names of polycyclic hydrocarbons with less than the maximum number of noncumulative double bonds are formed from a prefix dihydro-, tetrahydro-, etc., followed by the name of the corresponding unreduced hydrocarbon. The prefix perhydro- signifies full hydrogenation. For example, 1,2-dihydronaphthalene is

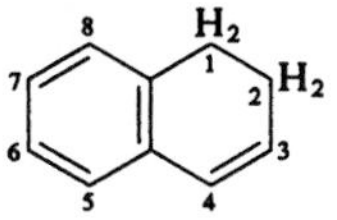

Examples of retained names and their structures are as follows:

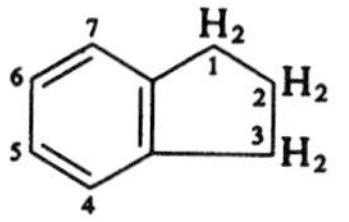

Indan

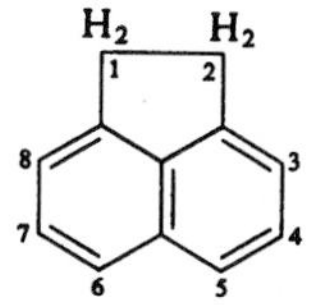

Acenaphthene

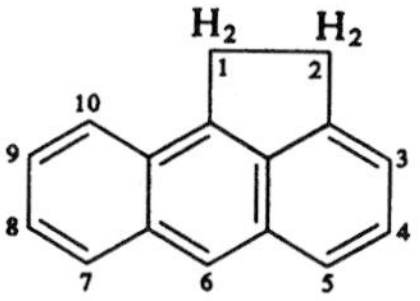

Aceanthrene

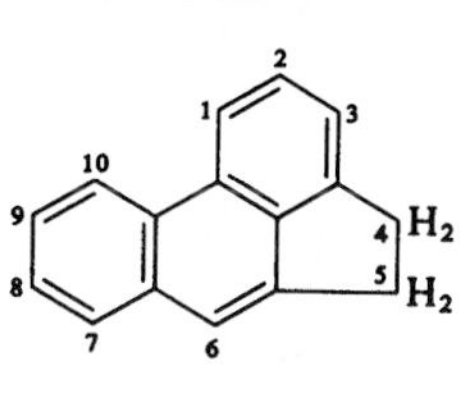

Acephenanthrene

Polycyclic compounds in which two rings have two atoms in common or in which one ring contains two atoms in common with each of two or more rings of a contiguous series of rings and which contain at least two rings of five or more members with the maximum number of noncumulative double bonds and which have no accepted trivial name are named by prefixing to the name of the parent ring or ring system designations of the other components. The parent name should contain as many rings as possible (provided it has a trivial name). Furthermore, the attached component(s) should be as simple as possible. For example, one writes dibenzophenanthrene and not naphthophenanthrene because the attached component benzo- is simpler than napththo-. Prefixes designating attached components are formed by changing the ending -ene into -eno-; for example, indeno- from indene. Multiple prefixes are arranged in alphabetical order. Several abbreviated prefixes are recognized; the parent is given in parentheses:

Acenaphtho-	(acenaphthylene)	Naphtho-	(naphthalene)
Anthra-	(anthracene)	Perylo-	(perylene)
Benzo-	(benzene)	Phenanthro-	(phenanthrene)

TABLE 2.2 Fused Polycyclic Hydrocarbons

Listed in order of increasing priority for selection as parent compound.

1. Pentalene

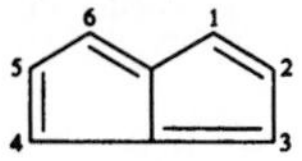

2. Indene

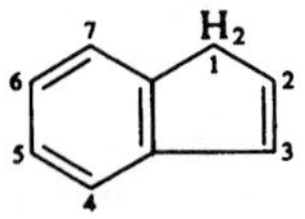

3. Naphthalene

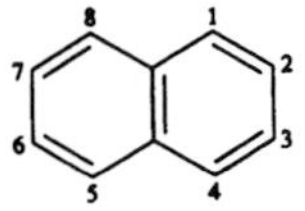

4. Azulene

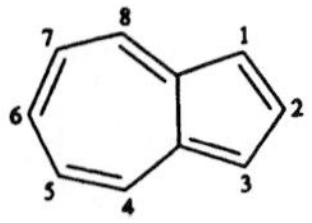

5. Heptalene

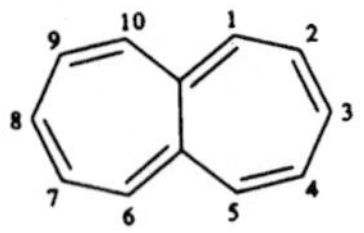

6. Biphenylene

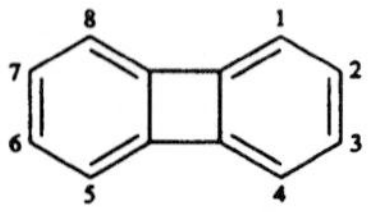

7. *asym*-Indacene

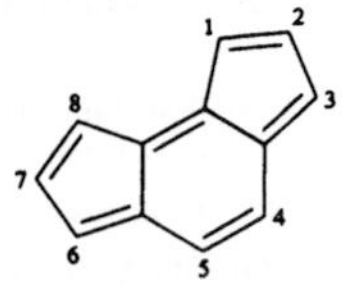

8. *sym*-Indacene

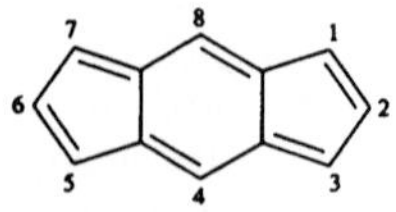

9. Acenaphthylene

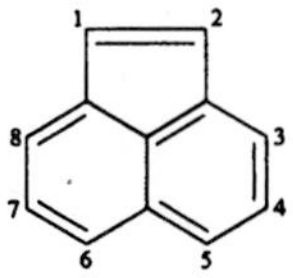

10. Fluorene

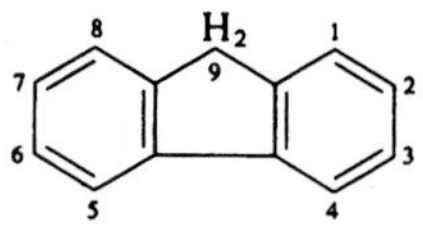

11. Phenalene

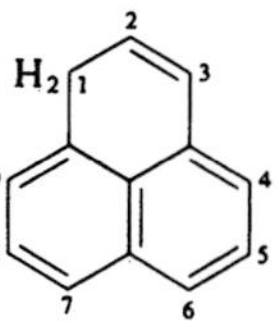

12. Phenanthrene*

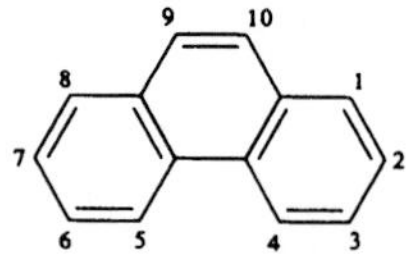

13. Anthracene*

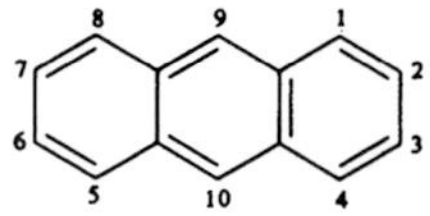

14. Fluoranthene

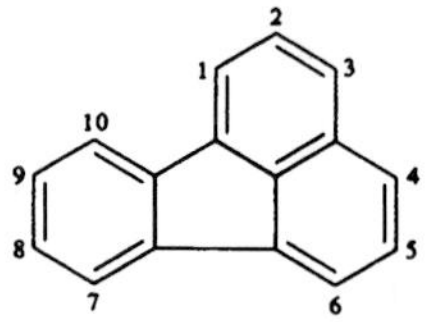

15. Acephenanthrylene

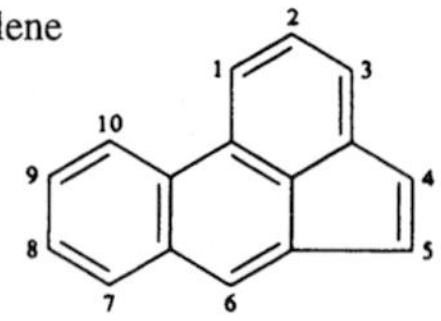

16. Aceanthrylene

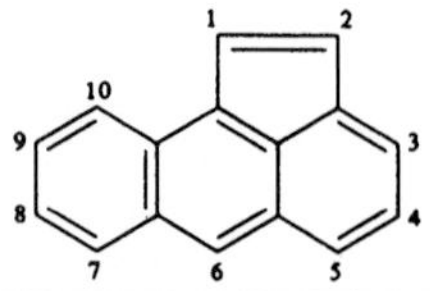

*Asterisk after a compound denotes exception to systematic numbering.

TABLE 2.2 Fused Polycyclic Hydrocarbons (*Continued*)

17. Triphenylene 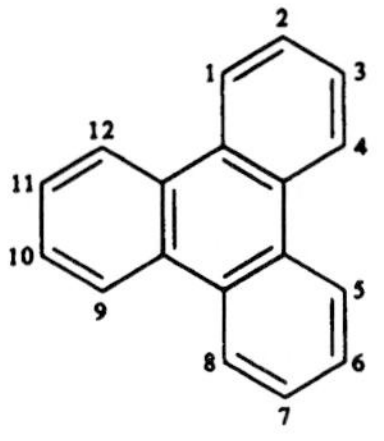	19. Chrysene 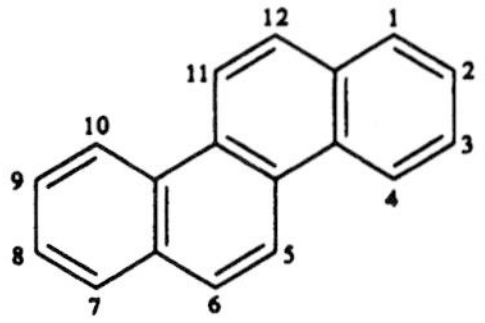
18. Pyrene 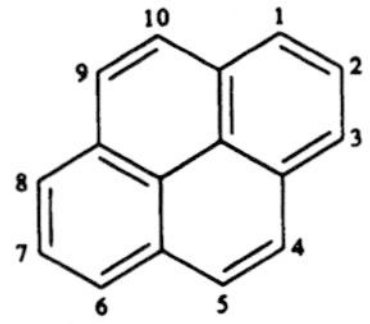	20. Naphthacene 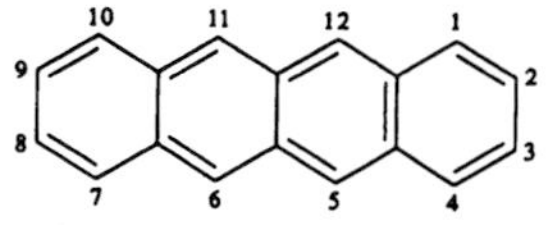

For monocyclic prefixes other than benzo-, the following names are recognized, each to represent the form with the maximum number of noncumulative double bonds: cyclopenta-, cyclohepta-, cycloocta-, etc.

Isomers are distinguished by lettering the peripheral sides of the parent beginning with *a* for the side 1,2, and so on, lettering every side around the periphery. If necessary for clarity, the numbers of the attached position (1,2, for example) of the substituent ring are also denoted. The prefixes are cited in alphabetical order. The numbers and letters are enclosed in square brackets and placed immediately after the designation of the attached component. Examples are

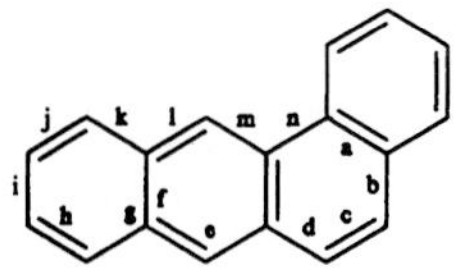

Benz[α]anthracene

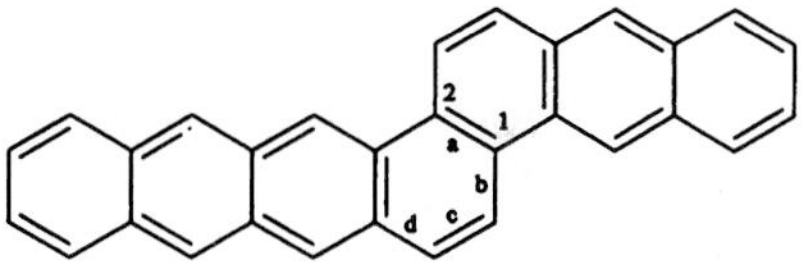

Anthra[2,1-α]naphthacene

2.1.1.5 Bridged Hydrocarbons. Saturated alicyclic hydrocarbon systems consisting of two rings that have two or more atoms in common take the name of the open-chain hydrocarbon containing the same total number of carbon atoms and are preceded by the prefix bicyclo-. The system is numbered commencing with one of the bridgeheads, numbering proceeding by the longest possible path to the second bridgehead. Numbering is then continued from this atom by the longer remaining unnumbered path back to the first bridgehead and is completed by the shortest path from the atom next to the first bridgehead. When a choice in numbering exists, unsaturation is given the lowest numbers. The number of carbon atoms in each of the bridges connecting the bridgeheads is indicated in brackets in descending order. Examples are

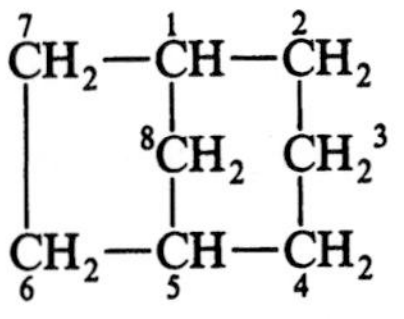

Bicyclo[3.2.1]octane

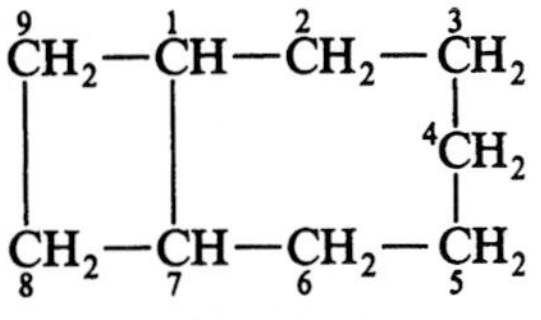

Bicyclo[5.2.0]nonane

2.1.1.6 Hydrocarbon Ring Assemblies. Assemblies are two or more cyclic systems, either single rings or fused systems, that are joined directly to each other by double or single bonds. For identical systems naming may proceed (1) by placing the prefix bi- before the name of the corresponding radical or (2), for systems joined through a single bond, by placing the prefix bi- before the name of the corresponding hydrocarbon. In each case, the numbering of the assembly is that of the corresponding radical or hydrocarbon, one system being assigned unprimed numbers and the other primed numbers. The points of attachment are indicated by placing the appropriate locants before the name; an unprimed number is considered lower than the same number primed. The name *biphenyl* is used for the assembly consisting of two benzene rings. Examples are

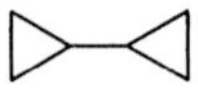

1,1′-Bicyclopropyl or 1,1′-bicyclopropane

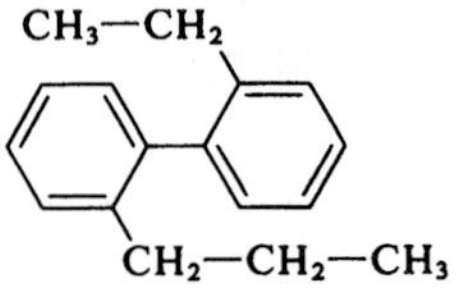

2-Ethyl-2′-propylbiphenyl

For nonidentical ring systems, one ring system is selected as the parent and the other systems are considered as substituents and are arranged in alphabetical order. The parent ring system is assigned unprimed numbers. The parent is chosen by considering the following characteristics in turn until a decision is reached: (1) the system containing the larger number of rings, (2) the system containing the larger ring, (3) the system in the lowest state hydrogenation, and (4) the highest-order number of ring systems. Examples are given, with the deciding priority given in parentheses preceding the name:

(1) 2-Phenylnaphthalene
(2) and (4) 2-(2′-Naphthyl)azulene
(3) Cyclohexylbenzene

2.1.1.7 Radicals from Ring Systems. Univalent substituent groups derived from polycyclic hydrocarbons are named by changing the final *e* of the hydrocarbon name to -yl. The carbon atoms having free valences are given locants as low as possible consistent with the fixed numbering of the hydrocarbon. Exceptions are naphthyl (instead of naphthalenyl), anthryl (for anthracenyl), and phenanthryl (for phenanthrenyl). However, these abbreviated forms are used only for the simple ring systems. Substituting groups derived from fused derivatives of these ring systems are named systematically.

2.1.1.8 Cyclic Hydrocarbons with Side Chains. Hydrocarbons composed of cyclic and aliphatic chains are named in a manner that is the simplest permissible or the most appropriate for the chemical intent. Hydrocarbons containing several chains attached to one cyclic nucleus are generally named as derivatives of the cyclic compound, and compounds containing several side chains and/or cyclic radicals attached to one chain are named as derivatives of the acyclic compound. Examples are

2-Ethyl-l-methylnaphthalene	Diphenylmethane
1,5-Diphenylpentane	2,3-Dimethyl-l-phenyl-l-hexene

Recognized trivial names for composite radicals are used if they lead to simplifications in naming. Examples are

1-Benzylnaphthalene 1,2,4-Tris(3-*p*-tolylpropyl)benzene

Fulvene, for methylenecyclopentadiene, and stilbene, for 1,2-diphenylethylene, are trivial names that are retained.

2.1.1.9 Heterocyclic Systems. Heterocyclic compounds can be named by relating them to the corresponding carbocyclic ring systems by using replacement nomenclature. Heteroatoms are denoted by prefixes ending in *a*. If two or more replacement prefixes are required in a single name, they are cited in the order of their listing in the table. The lowest possible numbers consistent with the numbering of

TABLE 2.3 Heterocyclic Systems

Heterocyclic atoms are listed in decreasing order of priority.

Element	Valence	Prefix	Element	Valence	Prefix
Oxygen	2	Oxa-	Antimony	3	Stiba-*
Sulfur	2	Thia-	Bismuth	3	Bisma-
Selenium	2	Selena-	Silicon	4	Sila-
Tellurium	2	Tellura-	Germanium	4	Germa-
Nitrogen	3	Aza-	Tin	4	Stanna-
Phosphorus	3	Phospha-*	Lead	4	Plumba-
Arsenic	3	Arsa-*	Boron	3	Bora-
			Mercury	2	Mercura-

*When immediately followed by -in or -ine, phospha- should be replaced by phosphor-, arsa- by arsen-, and stiba- by antimon-. The saturated six-membered rings corresponding to phosphorin and arsenin are named *phosphorinane* and *arsenane*. A further exception is the replacement of borin by borinane.

the corresponding carbocyclic system are assigned to the heteroatoms and then to carbon atoms bearing double or triple bonds. Locants are cited immediately preceding the prefixes or suffixes to which they refer. Multiplicity of the same heteroatom is indicated by the appropriate prefix in the series: di-, tri-, tetra-, penta-, hexa-, etc.

If the corresponding carbocyclic system is partially or completely hydrogenated, the additional hydrogen is cited using the appropriate *H*- or hydro- prefixes. A trivial name along with the state of hydrogenation may be used. In the specialist nomenclature for heterocyclic systems, the prefix or prefixes (Table 2.3) are combined with the appropriate stem from Table 2.4, ending in an *a* where necessary. Examples of acceptable usage, including (1) replacement and (2) specialist nomenclature, are

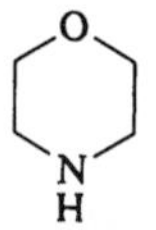

(1) 1-Oxa-4-azacyclohexane
(2) 1,4-Oxazoline
Morpholine

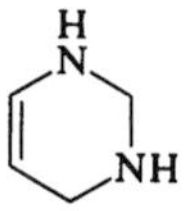

(1) 1,3-Diazacyclohex-5-ene
(2) 1,2,3,4-Tetrahydro-1,3-diazine

(1) Thiacyclopropane
(2) Thiirane
Ethylene sulfide

TABLE 2.4 Suffixes for Heterocyclic Systems

Number of ring members	Rings Containing Nitrogen		Rings Containing Nitrogen	
	Unsaturation*	Saturation	Unsaturation*	Saturation
3	-irine	-iridine	-irene	-irane
4	-ete	-etidine	-ete	-etane
5	-ole	-olidine	-ole	-olane
6	-ine†	‡	-in	-ane§
7	-epine	‡	-epin	-epane
8	-ocine	‡	-ocin	-ocane
9	-onine	‡	-onin	-onane
10	-ecine	‡	-ecin	-ecane

*Unsaturation corresponding to the maximum number of noncumulative double bonds. Heteroatoms have the normal valences.

† For phosphorus, arsenic, antimony, and boron, there are special provisions (Table 2.3).

‡ Expressed by prefixing perhydro- to the name of the corresponding unsaturated compound.

§ Not applicable to silicon, germanium, tin, and lead; perhydro- is prefixed to the name of the corresponding unsaturated compound.

TABLE 2.5 Trivial Names of Heterocyclic Systems Suitable for Use in Fusion Names

Listed in order of increasing priority as senior ring system.

Structure	Parent name	Radical name	Structure	Parent name	Radical name
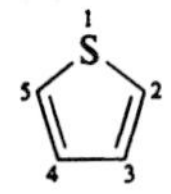	Thiophene	Thienyl	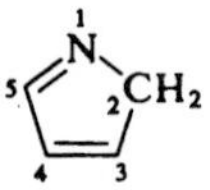	2*H*-Pyrrole	2*H*-Pyrrolyl
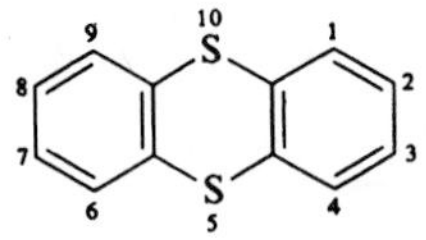	Thianthrene	Thianthrenyl	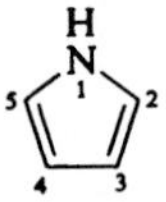	Pyrrole	Pyrrolyl
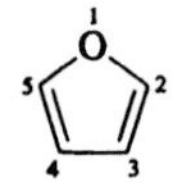	Furan	Furyl	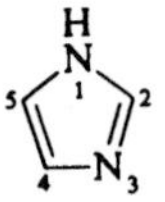	Imidazole	Imidazolyl
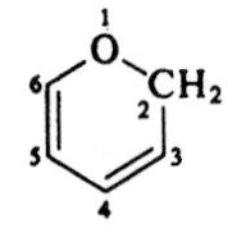	Pyran (2*H*-shown)	Pyranyl	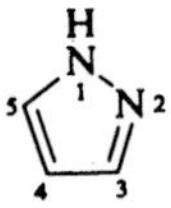	Pyrazole	Pyrazolyl
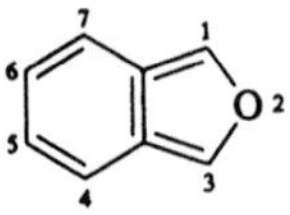	Isobenzofuran	Isobenzo-furanyl	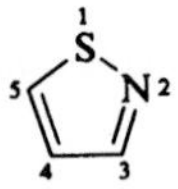	Isothiazole	Isothiazolyl
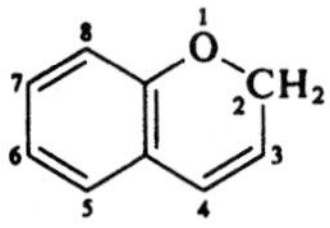	Chromene (2*H*-shown)	Chromenyl	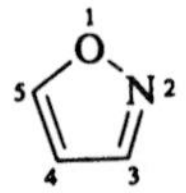	Isoxazole	Isoxazolyl
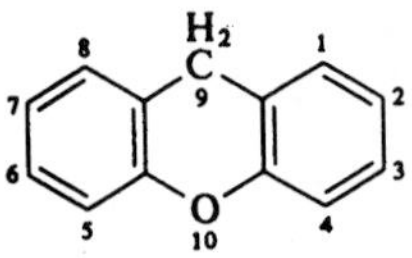	Xanthene*	Xanthenyl	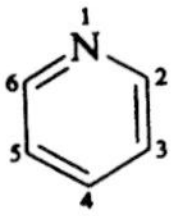	Pyridine	Pyridyl
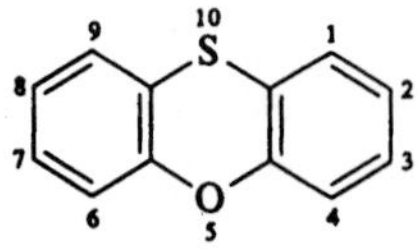	Phenoxathiin	Phenoxa-thiinyl	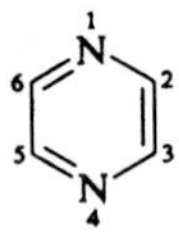	Pyrazine	Pyrazinyl
			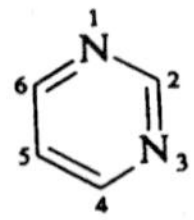	Pyrimidine	Pyrimidinyl
			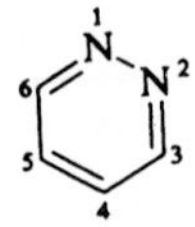	Pyridazine	Pyridazinyl

* Asterisk after a compound denotes exception to systematic numbering.

TABLE 2.5 Trivial Names of Heterocyclic Systems Suitable for Use in Fusion Names (*Continued*)

Structure	Parent name	Radical name	Structure	Parent name	Radical name
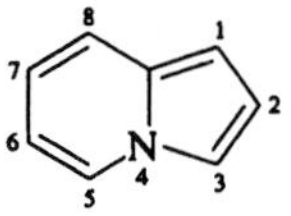	Indolizine	Indolizinyl	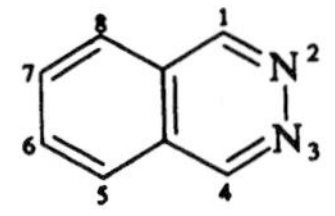	Phthalazine	Phthalazinyl
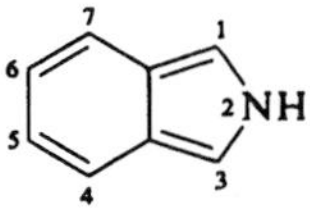	Isoindole	Isoindolyl	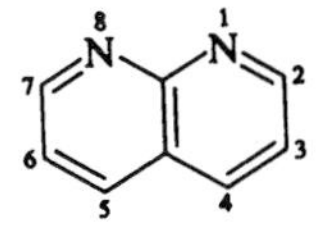	Naphthyridine (1,8-shown)	Naphthyridinyl
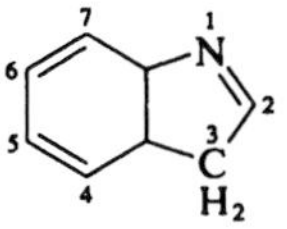	3*H*-Indole	3*H*-Indolyl	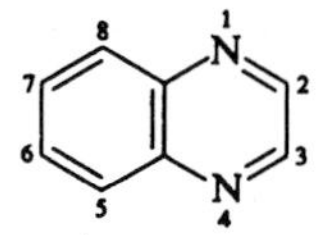	Quinoxaline	Quinoxalinyl
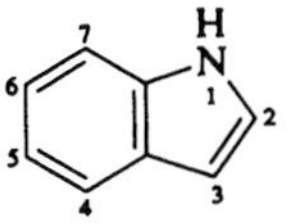	Indole	Indolyl	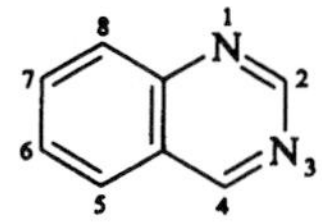	Quinazoline	Quinazolinyl
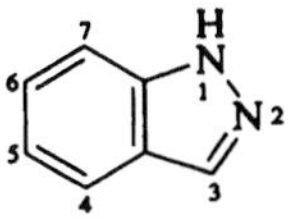	1*H*-Indazole	1*H*-Indazolyl	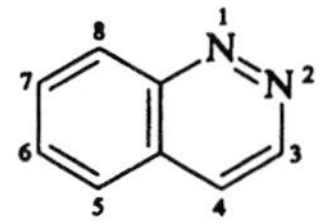	Cinnoline	Cinnolinyl
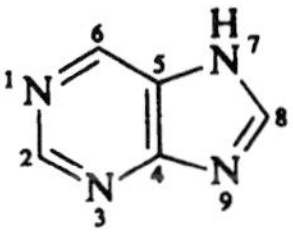	Purine*	Purinyl	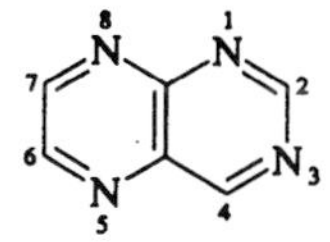	Pteridine	Pteridinyl
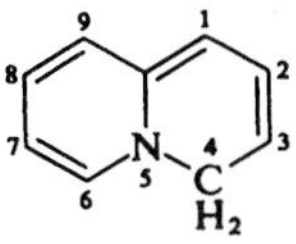	4*H*-Quinolizine	4*H*-Quinolizinyl	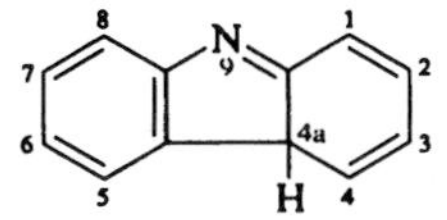	4α*H*-Carbazole*	4α*H*-Carbazolyl
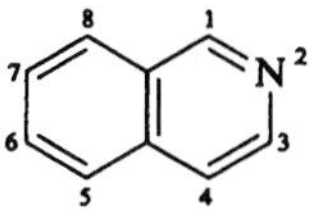	Isoquinoline	Isoquinolyl	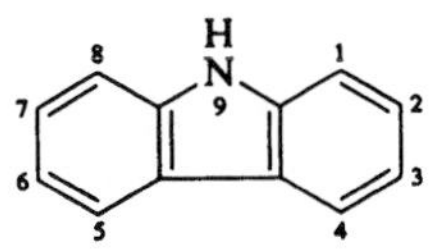	Carbazole*	Carbazolyl
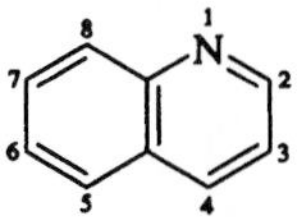	Quinolone	Quinolyl			

* Asterisk after a compound denotes exception to systematic numbering.

Radicals derived from heterocyclic compounds by removal of hydrogen from a ring are named by adding -yl to the names of the parent compounds (with elision of the final *e*, if present). These exceptions are retained:

Furyl (from furan)	Furfuryl (for 2-furylmethyl)
Pyridyl (from pyridine)	Furfurylidene (for 2-furylmethylene)
Piperidyl (from piperidine)	Thienyl (from thiophene)
Quinolyl (from quinoline)	Thenylidyne (for thienylmethylidyne)
Isoquinolyl	Furfurylidyne (for 2-furylmethylidyne)
Thenylidene (for thienylmethylene)	Thenyl (for thienylmethyl)

Also, piperidino- and morpholino- are preferred to 1-piperidyl- and 4-morpholinyl-, respectively.

TABLE 2.5 Trivial Names of Heterocyclic Systems Suitable for Use in Fusion Names (*Continued*)

Structure	Parent name	Radical name	Structure	Parent name	Radical name
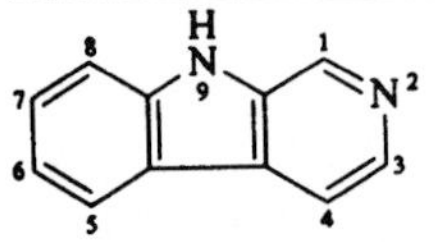	β-Carboline	β-Carbolinyl	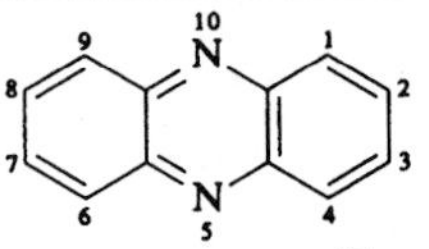	Phenazine	Phenazinyl
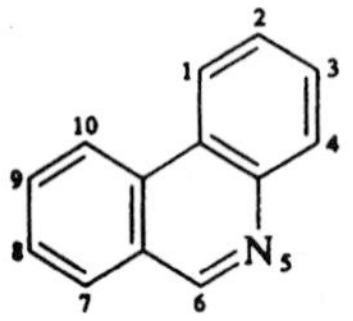	Phenanthri-dine	Phenanthri-dinyl	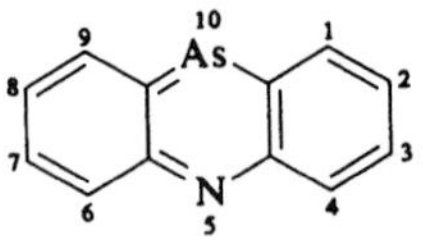	Phenarsazine	Phenarsazinyl
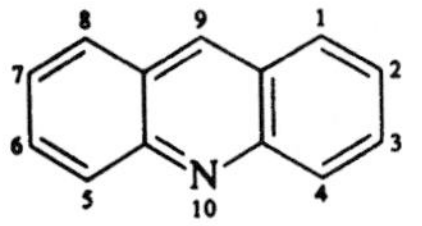	Acridine*	Acridinyl	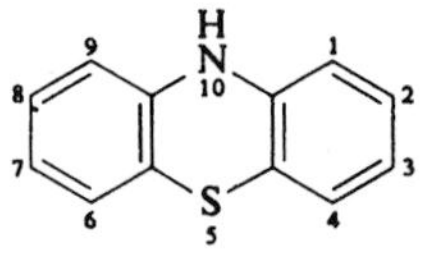	Phenothiazine	Phenothiazinyl
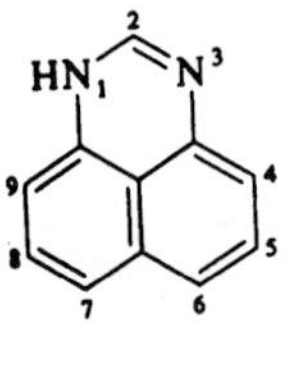	Perimidine	Perimidinyl		Furazan	Furazanyl
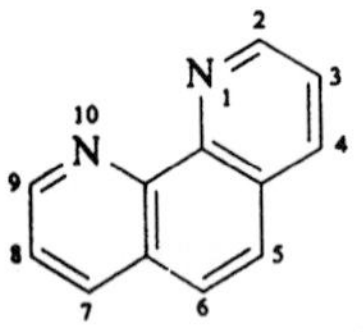	Phenanthroline (1,10-shown)	Phenanthrolinyl	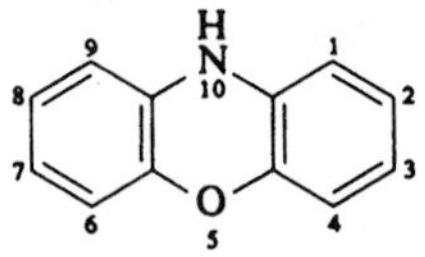	Phenoxazine	Phenoxazinyl

* Asterisk after a compound denotes exception to systematic numbering.

TABLE 2.6 Trivial Names for Heterocyclic Systems That Are Not Recommended for Use in Fusion Names

Listed in order of increasing priority.

Structure	Parent name	Radical name	Structure	Parent name	Radical name
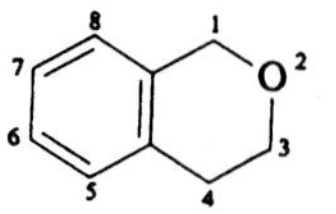	Isochroman	Isochromanyl	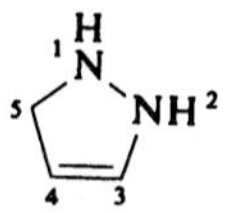	Pyrazoline (3-shown*)	Pyrazolinyl
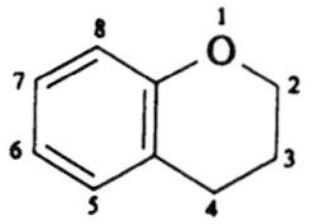	Chroman	Chromanyl	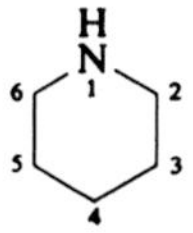	Piperidine	Piperidyl†
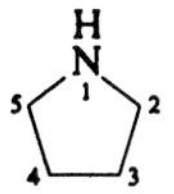	Pyrrolidine	Pyrrolinyl	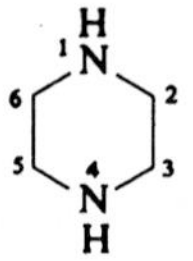	Piperazine	Piperazinyl
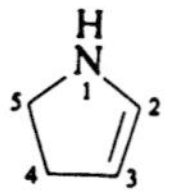	Pyrroline (2-shown*)	Pyrrolinyl	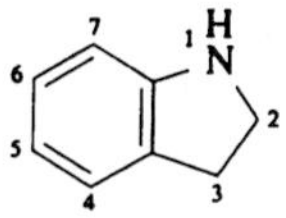	Indoline	Indolinyl
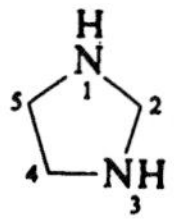	Imidazolidine	Imidazolidinyl	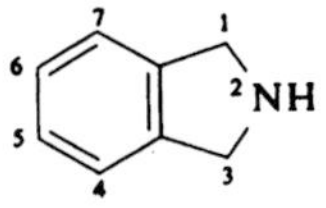	Isoindoline	Isoindolinyl
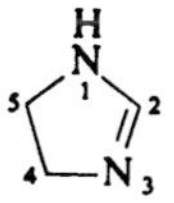	Imidazoline (2-shown*)	Imidazolinyl	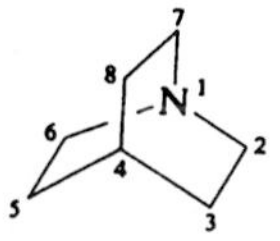	Quinuclidine	Quinuclidinyl
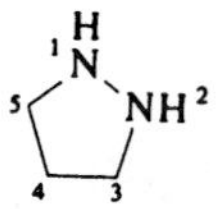	Pyrazolidine	Pyrazolidinyl	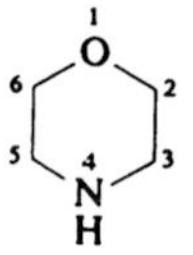	Morpholine	Morpholinyl‡

* Denotes position of double bond.
† For 1-piperidyl, use piperidino.
‡ For 4-morpholinyl, use morpholino.

If there is a choice among heterocyclic systems, the parent compound is decided in the following order of preference:

1. A nitrogen-containing component
2. A component containing a heteroatom, in the absence of nitrogen, as high as possible (Table 2.3).
3. A component containing the greater number of rings

4. A component containing the largest possible individual ring
5. A component containing the greatest number of heteroatoms of any kind
6. A component containing the greatest variety of heteroatoms
7. A component containing the greatest number of heteroatoms first listed in Table 2.3

If there is a choice between components of the same size containing the same number and kind of heteroatoms, choose as the base component that one with the lower numbers for the heteroatoms before fusion. When a fusion position is occupied by a heteroatom, the names of the component rings to be fused are selected to contain the heteroatom.

2.1.2 Functional Compounds

There are several types of nomenclature systems that are recognized. Which type to use is sometimes obvious from the nature of the compound. Substitutive nomenclature, in general, is preferred because of its broad applicability, but radicofunctional, additive, and replacement nomenclature systems are convenient in certain situations.

2.1.2.1 Substitutive Nomenclature. The first step is to determine the kind of characteristic (functional) group for use as the principal group of the parent compound. A characteristic group is a recognized combination of atoms that confers characteristic chemical properties on the molecule in which it occurs. Carbon-to-carbon unsaturation and heteroatoms in rings are considered nonfunctional for nomenclature purposes.

Substitution means the replacement of one or more hydrogen atoms in a given compound by some other kind of atom or group of atoms, functional or nonfunctional. In substitutive nomenclature, each substituent is cited as either a prefix or a suffix to the name of the parent (or substituting radical) to which it is attached; the latter is denoted the parent compound (or parent group if a radical).

When oxygen is replaced by sulfur, selenium, or tellurium, the priority for these elements is in the descending order listed. The higher valence states of each element are listed before considering the successive lower valence states. Derivative groups have priority for citation as principal group after the respective parents of their general class.

Systematic names formed by applying the principles of substitutive nomenclature are single words except for compounds named as acids. First, select the parent compound, and thus the suffix, from the characteristic group (Table 2.7). All remaining functional groups are handled as prefixes that precede, in alphabetical order, the parent name. Two examples are:

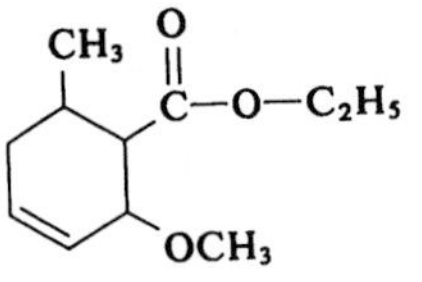

Structure I

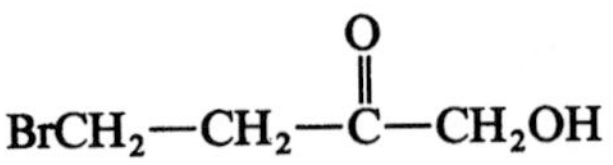

Structure II

Structure I contains an ester group and an ether group. Since the ester group has higher priority, the name is ethyl 2-methoxy-6-methyl-3-cyclohexene-1-carboxylate. Structure II contains a carbonyl group, a hydroxy group, and a bromo group. The latter is never a suffix. Between the other two, the carbonyl group has higher priority, the parent has -one as suffix, and the name is 4-bromo-l-hydroxy-2-butanone.

Selection of the principal alicyclic chain or ring system is governed by these selection rules:

1. For purely alicyclic compounds, the selection process proceeds successively until a decision is reached: (a) the maximum number of substituents corresponding to the characteristic group

(Table 2.7) (b) the maximum number of double and triple bonds considered together, (c) the maximum length of the chain, and (d) the maximum number of double bonds.

2. If the characteristic group occurs only in a chain that carries a cyclic substituent, the compound is named as an aliphatic compound into which the cyclic component is substituted; a radical prefix is used to denote the cyclic component. This chain need not be the longest chain.
3. If the characteristic group occurs in more than one carbon chain and the chains are not directly attached to one another, then the chain chosen as parent should carry the largest number of the characteristic group. If necessary, the selection is continued as in rule 1.
4. If the characteristic group occurs only in one cyclic system, that system is chosen as the parent.
5. If the characteristic group occurs in more than one cyclic system, that system is chosen as parent which (a) carries the largest number of the principal group or, failing to reach a decision, (b) is the senior ring system.
6. If the characteristic group occurs both in a chain and in a cyclic system, the parent is that portion in which the principal group occurs in largest number. If the numbers are the same, that portion is chosen which is considered to be the most important or is the senior ring system.

TABLE 2.7 Characteristic Groups for Substitutive Nomenclature

Listed in order of decreasing priority for citation as principal group or parent name.

Class	Formula*	Prefix	Suffix
1. Cations:		-onio-	-onium
	H_4N^+	Ammonio-	-ammonium
	H_3O^+	Oxonio-	-oxonium
	H_3S^+	Sulfonio-	-sulfonium
	H_3Se^+	Selenonio-	-selenonium
	H_2Cl^+	Chloronio-	-chloronium
	H_2Br^+	Bromonio-	-bromonium
	H_2I^+	Iodonio-	-iodonium
2. Acids:			
Carboxylic	—COOH	Carboxy-	-carboxylic acid
	—(C)OOH		-oic acid
	—C(=O)OOH		-peroxy···carboxylic acid
	—(C=O)OOH		-peroxy···oic acid
Sulfonic	$—SO_3H$	Sulfo-	-sulfonic acid
Sulfinic	$—SO_2H$	Sulfino-	-sulfinic acid
Sulfenic	—SOH	Sulfeno-	-sulfenic acid
Salts	—COOM		Metal···carboxylate
	—(C)OOM		Metal···oate
	$—SO_3M$		Metal···sulfonate
	$—SO_2M$		Metal···sulfinate
	—SOM		Metal···sulfenate
3. Derivatives of acids:			
Anhydrides	—C(=O)OC(=O)—		-carboxylic anhydride
	—(C=O)O(C=O)—		-oic anhydride
Esters	—COOR	R-oxycarbonyl-	R···carboxylate
	—C(OOR)		R···oate
Acid halides	—CO—halogen	Haloformyl	-carbonyl halide
Amides	$—CO—NH_2$	Carbamoyl-	-carboxamide
	$(C)O—NH_2$		-amide

(Continued)

TABLE 2.7 Characteristic Groups for Substitutive Nomenclature (*Continued*)

Class	Formula*	Prefix	Suffix
Hydrazides	—CO—NHNH$_2$	Carbonyl-hydrazino-	-carbohydrazide
	—(CO)—NHNH$_2$		-ohydrazide
Imides	—CO—NH—CO—	R-imido-	-carboximide
Amidines	—C(=NH)—NH$_2$	Amidino-	-carboxamidine
	—(C=NH)—NH$_2$		-amidine
4. Nitrile (cyanide)	—CN	Cyano-	-carbonitrile
	—(C)N		-nitrile
5. Aldehydes	—CHO	Formyl-	-carbaldehyde
	—(C=O)H	Oxo-	-al
	(then their analogs and derivatives)		
6. Ketones	>(C=O)	Oxo-	-one
	(then their analogs and derivatives)		
7. Alcohols (and phenols)	—OH	Hydroxy-	-ol
Thiols	—SH	Mercapto-	-thiol
8. Hydroperoxides	—O—OH	Hydroperoxy-	
9. Amines	—NH$_2$	Amino-	-amine
Imines	>NH	Imino-	-imine
Hydrazines	—NHNH$_2$	Hydrazino-	-hydrazine
10. Ethers	—OR	R-oxy-	
Sulfides	—SR	R-thio-	
11. Peroxides	—O—OR	R-dioxy-	

*Carbon atoms enclosed in parentheses are included in the name of the parent compound and not in the suffix or prefix.

TABLE 2.8 Characteristic Groups Cited Only as Prefixes in Substitutive Nomenclature

Characteristic group	Prefix	Characteristic group	Prefix
—Br	Bromo-	—IX$_2$	X may be halogen or a radical; dihalogenoiodo- or diacetoxyiodo-, e.g., —ICl$_2$ is dichloroido-
—Cl	Chloro-		
—ClO	Chlorosyl-		
—ClO$_2$	Chloryl-	>N$_2$	Diazo-
—ClO$_3$	Perchloryl-	—N$_3$	Azido-
—F	Fluoro-	—NO	Nitroso-
—I	Iodo-	—NO$_2$	Nitro-
—IO	Iodosyl-	>N(=O)OH	*aci*-Nitro-
—IO$_2$	Iodyl*	—OR	R-oxy-
—I(OH)$_2$	Dihydroxyiodo-	—SR	R-thio-
		—SeR (—TeR)	R-seleno- (R-telluro-)

*Formerly iodoxy.

7. When a substituent is itself substituted, all the subsidiary substituents are named as prefixes and the entire assembly is regarded as a parent radical.
8. The seniority of ring systems is ascertained by applying the following rules successively until a decision is reached: (a) all heterocycles are senior to all carbocycles, (b) for heterocycles, the preference follows the decision process described under Heterocyclic Systems (p. 1.11) (c) the largest number of rings, (d) the largest individual ring at the first point of difference, (e) the largest number of atoms in common among rings, (f) the lowest letters in the expression for ring functions, (g) the lowest numbers at the first point of difference in the expression for ring junctions, (h) the lowest state of hydrogenation, (i) the lowest-numbered locant for indicated hydrogen, (j) the lowest-numbered locant for point of attachment (if a radical), (k) the lowest-numbered locant for an attached group expressed as a suffix, (l) the maximum number of substituents cited as prefixes, (m) the lowest-numbered locant for substituents named as prefixes, hydro prefixes, -ene, and -yne, all considered together in one series in ascending numerical order independent of their nature, and (n) the lowest-numbered locant for the substituent named as prefix which is cited first in the name.

2.1.2.2 Numbering of Compounds. If the rules for aliphatic chains and ring systems leave a choice, the starting point and direction of numbering of a compound are chosen so as to give lowest-numbered locants to these structural factors, if present, considered successively in the order listed below until a decision is reached. Characteristic groups take precedence over multiple bonds.

1. Indicated hydrogen, whether cited in the name or omitted as being conventional
2. Characteristic groups named as suffix following ranking order (Table 2.7)
3. Multiple bonds in acyclic compounds; in bicycloalkanes, tricycloalkanes, and polycycloalkanes, double bonds having priority over triple bonds; and in heterocyclic systems whose names end in -etine, -oline, or -olene
4. The lowest-numbered locant for substituents named as prefixes, hydro prefixes, -ene, and -yne, all considered together in one series in ascending numerical order
5. The lowest locant for that substituent named as prefix which is cited first in the name

For cyclic radicals, indicated hydrogen and thereafter the point of attachment (free valency) have priority for the lowest available number.

2.1.2.3 Prefixes and Affixes. Prefixes are arranged alphabetically and placed before the parent name; multiplying affixes, if necessary, are inserted and *do not* alter the alphabetical order already attained. The parent name includes any syllables denoting a change of ring number or relating to the structure of a carbon chain. Nondetachable parts of parent names include

1. Forming rings; cyclo-, bicyclo-, spiro-
2. Fusing two or more rings: benzo-, naphtho-, imidazo-
3. Substituting one ring or chain member atom for another: oxa-, aza-, thia-
4. Changing positions of ring or chain members: iso-, *sec*-, *tert*-, neo-
5. Showing indicated hydrogen
6. Forming bridges: ethano-, epoxy-
7. Hydro-

Prefixes that represent complete terminal characteristic groups are preferred to those representing only a portion of a given group. For example, for the prefix $—C(=O)CH_3$, the name (formylmethyl) is preferred to (oxoethyl).

The multiplying affixes di-, tri-, tetra-, penta-, hexa-, hepta-, octa-, nona-, deca-, undeca-, and so on are used to indicate a set of *identical* unsubstituted radicals or parent compounds. The forms bis-, tris-, tetrakis-, pentakis-, and so on are used to indicate a set of identical radicals or parent compounds *each substituted in the same way*. The affixes bi-, ter-, quater-, quinque-, sexi-, septi-, octi-, novi-, deci-, and so on are used to indicate the number of identical rings joined together by a single or double bond.

Although multiplying affixes may be omitted for very common compounds when no ambiguity is caused thereby, such affixes are generally included throughout this handbook in alphabetical listings. An example would be ethyl ether for diethyl ether.

*2.1.2.4 **Conjunctive Nomenclature.*** Conjunctive nomenclature may be applied when a principal group is attached to an acyclic component that is directly attached by a carbon-carbon bond to a cyclic component. The name of the cyclic component is attached directly in front of the name of the acyclic component carrying the principal group. This nomenclature is not used when an unsaturated side chain is named systematically. When necessary, the position of the side chain is indicated by a locant placed before the name of the cyclic component. For substituents on the acyclic chain, carbon atoms of the side chain are indicated by Greek letters proceeding from the principal group to the cyclic component. The terminal carbon atom of acids, aldehydes, and nitriles is omitted when allocating Greek positional letters. Conjunctive nomenclature is not used when the side chain carries more than one of the principal group, except in the case of malonic and succinic acids.

The side chain is considered to extend only from the principal group to the cyclic component. Any other chain members are named as substituents, with appropriate prefixes placed before the name of the cyclic component.

When a cyclic component carries more than one identical side chain, the name of the cyclic component is followed by di-, tri-, etc., and then by the name of the acyclic component, and it is preceded by the locants for the side chains. Examples are

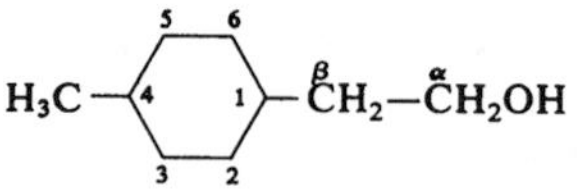

4-Methyl-1-cyclohexaneethanol

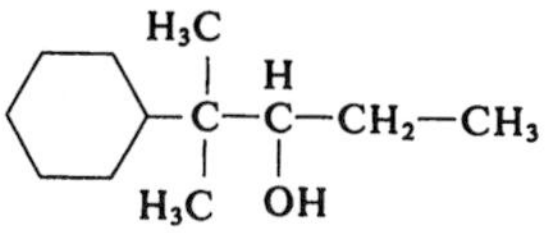

α-Ethyl-β,β-dimethylcyclohexaneethanol

When side chains of two or more different kinds are attached to a cyclic component, only the senior side chain is named by the conjunctive method. The remaining side chains are named as prefixes. Likewise, when there is a choice of cyclic component, the senior is chosen. Benzene derivatives may be named by the conjunctive method only when two or more identical side chains are present. Trivial names for oxo carboxylic acids may be used for the acyclic component. If the cyclic and acyclic components are joined by a double bond, the locants of this bond are placed as superscripts to a Greek capital delta that is inserted between the two names. The locant for the cyclic component precedes that for the acyclic component, e.g., indene-$\Delta^{1,\alpha}$-acetic acid.

*2.1.2.5 **Radicofunctional Nomenclature.*** The procedures of radicofunctional nomenclature are identical with those of substitutive nomenclature except that suffixes are never used. Instead, the functional class name (Table 2.9) of the compound is expressed as one word and the remainder of the molecule as another that precedes the class name. When the functional class name refers to a characteristic group that is bivalent, the two radicals attached to it are each named, and when different, they are written as separate words arranged in alphabetical order. When a compound contains more than one kind of group, that kind is cited as the functional group or class name that occurs higher in the table, all others being expressed as prefixes.

Radicofunctional nomenclature finds some use in naming ethers, sulfides, sulfoxides, sulfones, selenium analogs of the preceding three sulfur compounds, and azides.

TABLE 2.9 Radicofunctional Nomenclature

Groups are listed in order of decreasing priority.

Group	Functional class names
X in acid derivatives	Name of X (in priority order: fluoride, chloride, bromide, iodide, cyanide, azide; then the sulfur and selenium analogs)
—CN, —NC	Cyanide, isocyanide
>CO	Ketone; then S and Se analogs
—OH	Alcohol; then S and Se analogs
—O—OH	Hydroperoxide
>O	Ether or oxide
>S, >SO, >SO_2	Sulfide, sulfoxide, sulfone
>Se, >SeO, >SeO_2	Selenide, selenoxide, selenone
—F, —Cl, —Br, —I	Fluoride, chloride, bromide, iodide
—N_3	Azide

2.1.2.5 Replacement Nomenclature. Replacement nomenclature is intended for use only when other nomenclature systems are difficult to apply in the naming of chains containing heteroatoms. When no group is present that can be named as a principal group, the longest chain of carbon and heteroatoms terminating with carbon is chosen and named as though the entire chain were that of an acyclic hydrocarbon. The heteroatoms within this chain are identified by means of prefixes aza-, oxa-, thia-, etc. Locants indicate the positions of the heteroatoms in the chain. Lowest-numbered locants are assigned to the principal group when such is present. Otherwise, lowest-numbered locants are assigned to the heteroatoms considered together and, if there is a choice, to the heteroatoms cited earliest in Table 2.3. An example is

$$HO-\overset{13}{C}H_2-\overset{12}{O}-\overset{11}{C}H_2-\overset{10}{C}H_2-\overset{9}{O}-\overset{8}{C}H_2-\overset{7}{C}H_2-\underset{H}{\overset{6}{N}}-\overset{5}{C}H_2-\overset{4}{C}H_2-\underset{H}{\overset{3}{N}}-\overset{2}{C}H_2-\overset{1}{C}OOH$$

13-Hydroxy-9,12-dioxa-3,6-diazatridecanoic acid

2.1.3 Specific Functional Groups

2.1.3.1 Acetals and Acylals. Acetals, which contain the group $>C(OR)_2$, where R may be different, are named (1) as dialkoxy compounds or (2) by the name of the corresponding aldehyde or ketone followed by the name of the hydrocarbon radical(s) followed by the word *acetal*. For example, $CH_3-CH(OCH_3)_2$ is named either (1) 1,1-dimethoxyethane or (2) acetaldehyde dimethyl acetal.

A cyclic acetal in which the two acetal oxygen atoms form part of a ring may be named (1) as a heterocyclic compound or (2) by use of the prefix methylenedioxy for the group $-O-CH_2-O-$ as a substituent in the remainder of the molecule. For example,

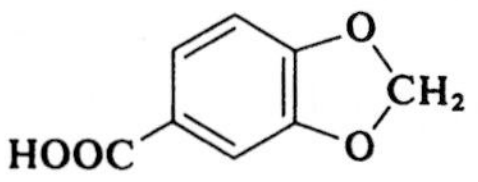

(1) 1,3-Benzo[*d*]dioxole-5-carboxylic acid

(2) 3,4-Methylenedioxybenzoic acid

Acylals, $R^1R^2C(OCOR^3)_2$, are named as acid esters;

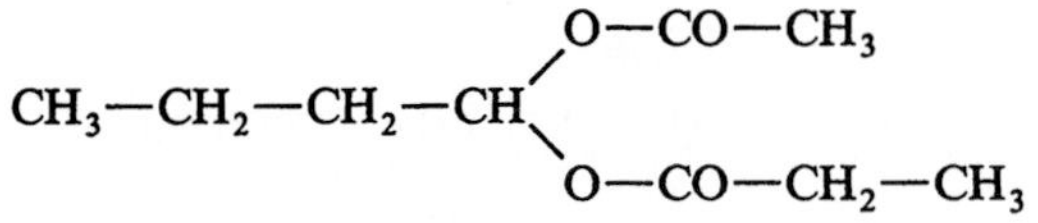

Butylidene acetate propionate

α-Hydroxy ketones, formerly called acyloins, had been named by changing the ending -ic acid or -oic acid of the corresponding acid to -oin. They are preferably named by substitutive nomenclature; thus

$CH_3—CH(OH)—CO—CH_3$ 3-Hydroxy-2-butanone (formerly acetoin)

2.1.3.2 Acid Anhydrides. Symmetrical anhydrides of monocarboxylic acids, when unsubstituted, are named by replacing the word *acid* by *anhydride*. Anhydrides of substituted monocarboxylic acids, if symmetrically substituted, are named by prefixing bis- to the name of the acid and replacing the word *acid* by *anhydride*. Mixed anhydrides are named by giving in alphabetical order the first part of the names of the two acids followed by the word *anhydride*, e.g., acetic propionic anhydride or acetic propanoic anhydride. Cyclic anhydrides of polycarboxylic acids, although possessing a heterocyclic structure, are preferably named as acid anhydrides. For example,

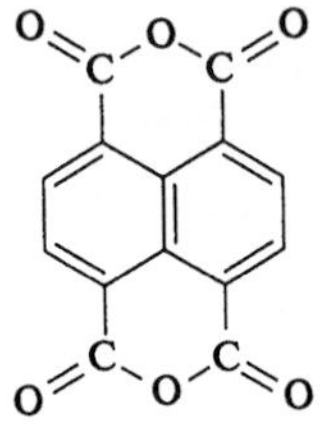

1,8;4,5-Napthalenetetracarboxylic dianhydride (note the use of a semicolon to distinguish the pairs of locants)

2.1.3.3 Acyl Halides. Acyl halides, in which the hydroxyl portion of a carboxyl group is replaced by a halogen, are named by placing the name of the corresponding halide after that of the acyl radical. When another group is present that has priority for citation as principal group or when the acyl halide is attached to a side chain, the prefix haloformyl- is used as, for example, in fluoroformyl-.

2.1.3.4 Alcohols and Phenols. The hydroxyl group is indicated by a suffix -ol when it is the principal group attached to the parent compound and by the prefix hydroxy- when another group with higher priority for citation is present or when the hydroxy group is present in a side chain. When confusion may arise in employing the suffix -ol, the hydroxy group is indicated as a prefix; this terminology is also used when the hydroxyl group is attached to a heterocycle, as, for example, in the name 3-hydroxythiophene to avoid confusion with thiophenol (C_6H_5SH). Designations such as isopropanol, *sec*-butanol, and *tert*-butanol are incorrect because no hydrocarbon exists to which the suffix can be added. Many trivial names are retained. (Table 2.10).

TABLE 2.10 Alcohols and Phenols

Ally alcohol	$CH_2{=}CHCH_2OH$
tert-Butyl alcohol	$(CH_3)_3COH$
Benzyl alcohol	$C_6H_5CH_2OH$
Phenethyl alcohol	$C_6H_5CH_2CH_2OH$
Ethylene glycol	$HOCH_2CH_2OH$
1,2-Propylene glycol	$CH_3CHOHCH_2OH$
Glycerol	$HOCH_2CHOHCH_2OH$
Pentaerythritol	$C(CH_2OH)_4$
Pinacol	$(CH_3)_2COHCOH(CH_3)_2$
Phenol	C_6H_5OH
Xylitol	$HOCH_2CH(OH)—CH(OH)—CH(OH)—CH_2OH$
Geraniol	$(CH_3)_2C{=}CHCH_2CH_2C(CH_3){=}CHCH_2OH$

TABLE 2.10 Alcohols and Phenols (*Continued*)

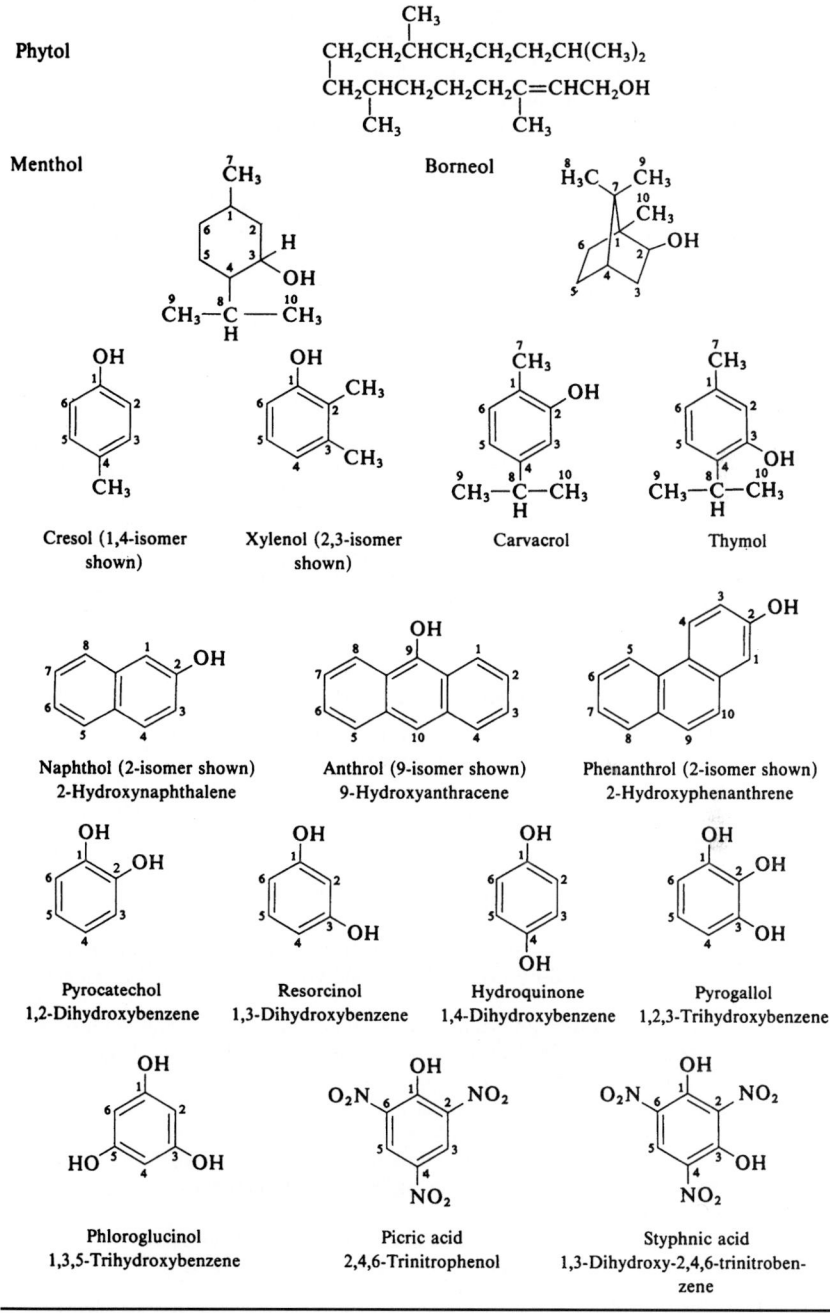

Phytol

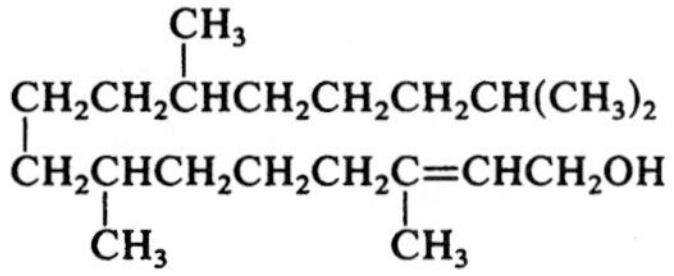

Menthol

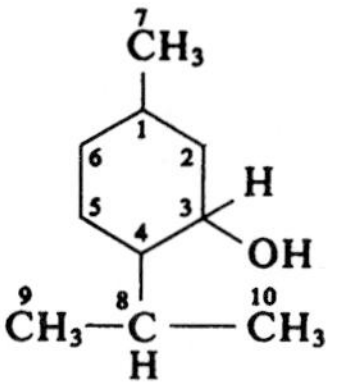

Borneol

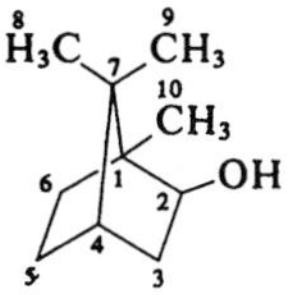

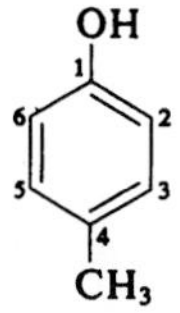

Cresol (1,4-isomer shown)

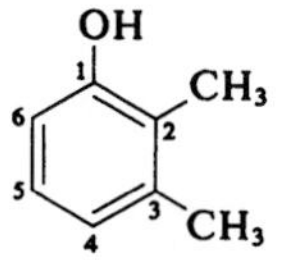

Xylenol (2,3-isomer shown)

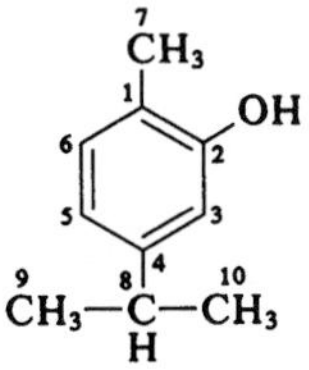

Carvacrol

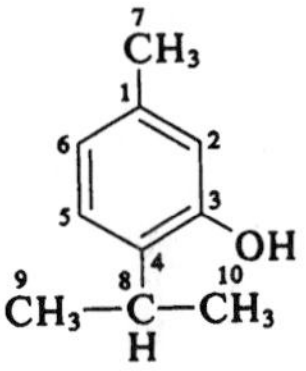

Thymol

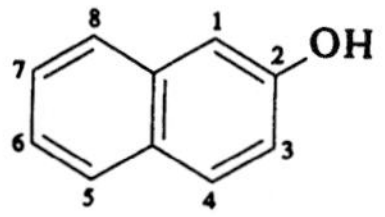

Naphthol (2-isomer shown)
2-Hydroxynaphthalene

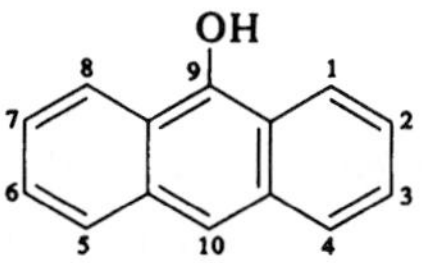

Anthrol (9-isomer shown)
9-Hydroxyanthracene

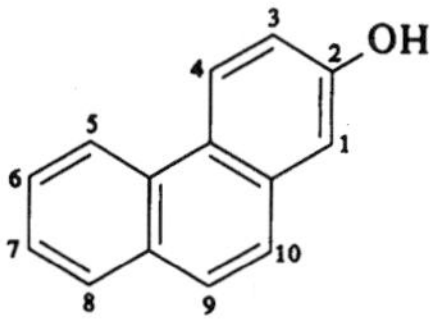

Phenanthrol (2-isomer shown)
2-Hydroxyphenanthrene

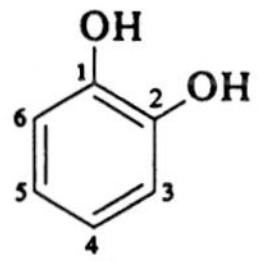

Pyrocatechol
1,2-Dihydroxybenzene

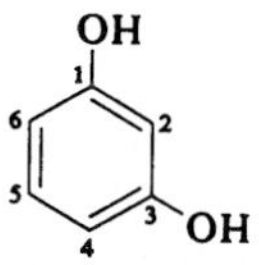

Resorcinol
1,3-Dihydroxybenzene

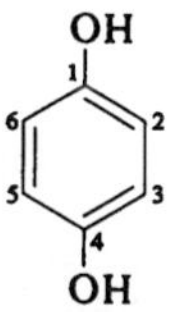

Hydroquinone
1,4-Dihydroxybenzene

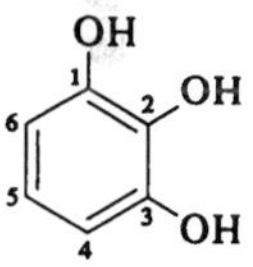

Pyrogallol
1,2,3-Trihydroxybenzene

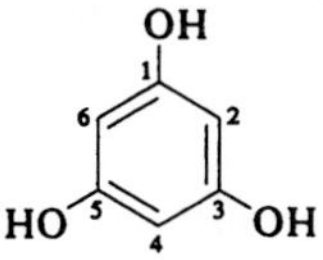

Phloroglucinol
1,3,5-Trihydroxybenzene

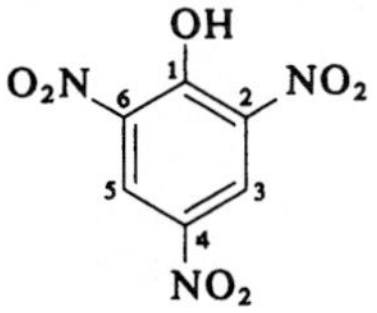

Picric acid
2,4,6-Trinitrophenol

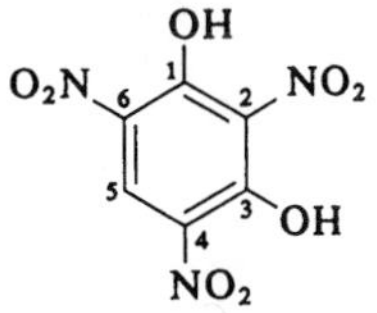

Styphnic acid
1,3-Dihydroxy-2,4,6-trinitrobenzene

The radicals (RO—) are named by adding -oxy as a suffix to the name of the R radical, e.g., pentyloxy for $CH_3CH_2CH_2CH_2CH_2O$—. These contractions are exceptions: methoxy (CH_3O—), ethoxy (C_2H_5O—), propoxy (C_3H_7O—), butoxy (C_4H_9O—), and phenoxy (C_6H_5O—). For unsubstituted radicals only, one may use isopropoxy [$(CH_3)_2CH$—O—], isobutoxy [$(CH_3)_2CH_2CH$—O—], *sec*-butoxy [$CH_3CH_2CH(CH_3)$—O—], and *tert*-botoxy [$(CH_3)_3C$—O—].

Bivalent radicals of the form O—Y—O are named by adding -dioxy to the name of the bivalent radicals except when forming part of a ring system. Examples are —O—CH_2—O— (methylenedioxy), —O—CO—O— (carbonyldioxy), and —O—SO_2—O— (sulfonyldioxy). Anions derived from alcohols or phenols are named by changing the final -ol to -olae.

Salts composed of an anion, RO—, and a cation, usually a metal, can be named by citing first the cation and then the RO anion (with its ending changed to -yl oxide), e.g., sodium benzyl oxide for $C_6H_5CH_2ONa$. However, when the radical has an abbreviated name, such as methoxy, the ending -oxy is changed to -oxide. For example, CH_3ONa is named sodium methoxide (not sodium methylate).

*2.1.3.5 **Aldehydes.*** When the group —C(=O)H, usually written —CHO, is attached to carbon at one (or both) end(s) of a linear acyclic chain the name is formed by adding the suffix -al (or -dial) to the name of the hydrocarbon containing the same number of carbon atoms. Examples are butanal for $CH_3CH_2CH_2CHO$ and propanedial for, $OHCCH_2CHO$.

Naming an acyclic polyaldehyde can be handled in two ways. First, when more than two aldehyde groups are attached to an unbranched chain, the proper affix is added to -carbaldehyde, which becomes the suffix to the name of the longest chain carrying the maximum number of aldehyde groups. The name and numbering of the main chain do not include the carbon atoms of the aldehyde groups. Second, the name is formed by adding the prefix formyl- to the name of the -dial that incorporates the principal chain. Any other chains carrying aldehyde groups are named by the use of formylalkyl- prefixes. Examples are

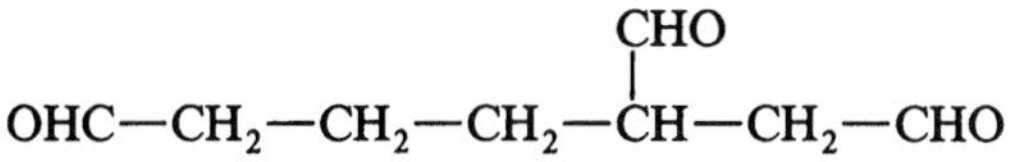

(1) 1,2,5-Pentanetricarbaldehyde
(2) 3-Formylheptanedial

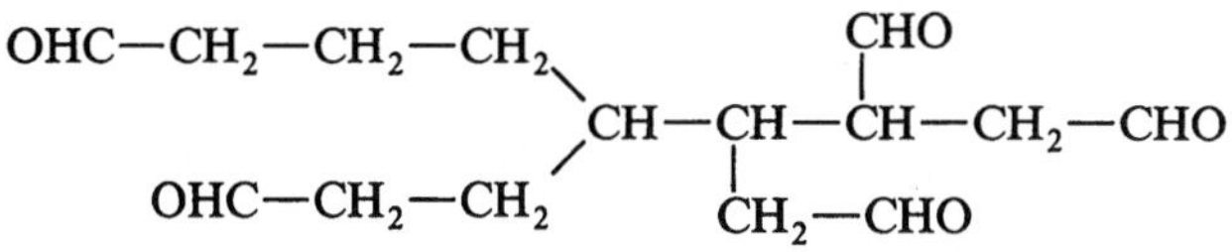

(1) 4-(2-Formylethyl)-3-(formylmethyl)-1,2,7-heptanetricarbaldehyde
(2) 3-Formyl-5-(2-formylethyl)-4-(formylmethyl)nonanedial

When the aldehyde group is directly attached to a carbon atom of a ring system, the suffix-carbaldehyde is added to the name of the ring system, e.g., 2-naphthalenecarbaldehyde. When the aldehyde group is separated from the ring by a chain of carbon atoms, the compound is named (1) as a derivative of the acyclic system or (2) by conjunctive nomenclature, for example, (1) (2-naphthyl)propionaldehyde or (2) 2-naphthalenepropionaldehyde.

An aldehyde group is denoted by the prefix formyl- when it is attached to a nitrogen atom in a ring system or when a group having priority for citation as principal group is present and part of a cyclic system.

When the corresponding monobasic acid has a trivial name, the name of the aldehyde may be formed by changing the ending -ic acid or -oic acid to -aldehyde. Examples are

Formaldehyde	Acrylaldehyde (not acrolein)
Acetaldehyde	Benzaldehyde
Propionaldehyde	Cinnamaldehyde
Butyraldehyde	2-Furaldehyde (not furfural)

The same is true for polybasic acids, with the proviso that all the carboxyl groups must be changed to aldehyde; then it is not necessary to introduce affixes. Examples are

Glyceraldehyde
Glycolaldehyde
Malonaldehyde
Succinaldehyde
Phthalaldehyde (*o*-, *m*-, *p*-)

These trivial names may be retained: citral (3,7-dimethyl-2,6-octadienal), vanillin (4-hydroxy-3-methoxybenzaldehyde), and piperonal (3,4-methylenedioxybenzaldehyde).

2.1.3.6 Amides. For primary amides the suffix -amide is added to the systematic name of the parent acid. For example, CH_3—CO—NH_2 is acetamide. Oxamide is retained for H_2N—CO—CO—NH_2. The name -carboxylic acid is replaced by -carboxamide.

For amino acids having trivial names ending in -ine, the suffix -amide is added after the name of the acid (with elision of *e* for monomides). For example, H_2N—CH_2—CO—NH_2 is glycinamide.

In naming the radical R—CO—NH—, either (1) the -yl ending of RCO— is changed to -amido or (2) the radicals are named as acylamino radicals. For example,

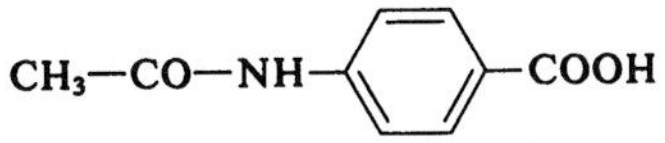

(1) 4-Acetamidobenzoic acid
(2) 4-Acetylaminobenzoic acid

The latter nomenclature is always used for amino acids with trivial names.

N-substituted primary amides are named either (1) by citing the substitutents as *N* prefixes or (2) by naming the acyl group as an *N* substituent of the parent compound. For example,

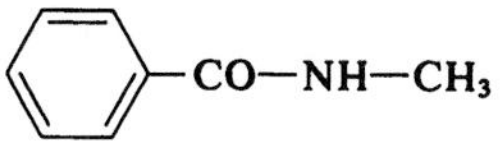

(1) *N*-Methylbenzamide
(2) Benzoylaminomethane

2.1.3.7 Amines. Amines are preferably named by adding the suffix -amine (and any multiplying affix) to the name of the parent radical. Examples are

$CH_3CH_2CH_2CH_2CH_2NH_2$ Pentylamine
$H_2NCH_2CH_2CH_2CH_2CH_2NH_2$ 1,5-Pentyldiamine or pentamethylenediamine

Locants of substituents of symmetrically substituted derivatives of symmetrical amines are distinguished by primes or else the names of the complete substituted radicals are enclosed in parentheses. Unsymmetrically substituted derivatives are named similarly or as *N*-substituted products of a primary amine (after choosing the most senior of the radicals to be the parent amine). For example,

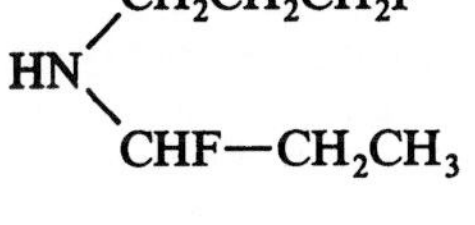

(1) 1,3′-Difluorodipropylamine
(2) 1-Fluoro-*N*-(3-fluoropropyl)propylamine
(3) (1-Fluoropropyl)(3-fluoropropyl)amine

Complex cyclic compounds may be named by adding the suffix -amine or the prefix amino- (or aminoalkyl-) to the name of the parent compound. Thus three names are permissible for

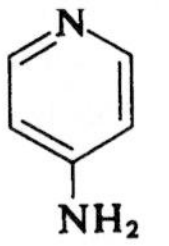

(1) 4-Pyridylamine
(2) 4-Pyridinamine
(3) 4-Aminopyridine

Complex linear polyamines are best designated by replacement nomenclature. These trivial names are retained: aniline, benzidene, phenetidine, toluidine, and xylidine.

The bivalent radical —NH— linked to two identical radicals can be denoted by the prefix imino-, as well as when it forms a bridge between two carbon ring atoms. A trivalent nitrogen atom linked to three identical radicals is denoted by the prefix nitrilo-. Thus ethylenediaminetetraacetic acid (an allowed exception) should be named ethylenedinitrilotetraacetic acid.

2.1.3.8 Ammonium Compounds. Salts and hydroxides containing quadricovalent nitrogen are named as a substituted ammonium salt or hydroxide. The names of the substituting radicals precede the word *ammonium*, and then the name of the anion is added as a separate word. For example, $(CH_3)_4N^+I^-$ is tetramethylammonium iodide.

When the compound can be considered as derived from a base whose name does not end in -amine, its quaternary nature is denoted by adding ium to the name of that base (with elision of *e*), substituent groups are cited as prefixes, and the name of the anion is added separately at the end. Examples are

$C_6H_5NH_3^+HSO_4^-$ Anilinium hydrogen sulfate

$[(C_6H_5NH_3)^+]_2PtCl_6^{2-}$ Dianilinium hexachloroplatinate

The names *choline* and *betaine* are retained for unsubstituted compounds.

In complex cases, the prefixes amino- and imino- may be changed to ammonio- and iminio- and are followed by the name of the molecule representing the most complex group attached to this nitrogen atom and are preceded by the names of the other radicals attached to this nitrogen. Finally the name of the anion is added separately. For example, the name might be 1-trimethylammonio-acridine chloride or 1-acridinyltrimethylammonium chloride.

When the preceding rules lead to inconvenient names, then (1) the unaltered name of the base may be used followed by the name of the anion or (2) for salts of hydrohalogen acids only the unaltered name of the base is used followed by the name of the hydrohalide. An example of the latter would be 2-ethyl-*p*-phenylenediamine monohydrochloride.

2.1.3.9 Azo Compounds. When the azo group (—N=N—) connects radicals derived from identical unsubstituted molecules, the name is formed by adding the prefix azo- to the name of the parent unsubstituted molecules. Substituents are denoted by prefixes and suffixes. The azo group has priority for lowest-numbered locant. Examples are azobenzene for C_6H_5—N=N—C_6H_5, azobenzene-4-sulfonic acid for C_6H_5—N=N—$C_6H_5SO_3H$, and 2′,4-dichloroazobenzene-4′-sulfonic acid for ClC_6H_4—N=N—$C_6H_3ClSO_3H$.

When the parent molecules connected by the azo group are different, azo is placed between the complete names of the parent molecules, substituted or unsubstituted. Locants are placed between the affix azo and the names of the molecules to which each refers. Preference is given to the more complex parent molecule for citation as the first component, e.g., 2-aminonaphthalene-l-azo-(4′-chloro-2′-methylbenzene).

In an alternative method, the senior component is regarded as substituted by RN=N-, this group R being named as a radical. Thus 2-(7-phenylazo-2-naphthylazo)anthracene is the name by this alternative method for the compound named anthracene-2-azo-2′-naphthalene-7′-azobenzene.

2.1.3.10 Azoxy Compounds. Where the position of the azoxy oxygen atom is unknown or immaterial, the compound is named in accordance with azo rules, with the affix azo replaced by azoxy. When the position of the azoxy oxygen atom in an unsymmetrical compound is designated, a prefix *NNO*- or *ONN*- is used. When both the groups attached to the azoxy radical are cited in the name of the compound, the prefix *NNO*- specifies that the second of these two groups is attached directly to —N(O)—; the prefix *ONN*- specifies that the first of these two groups is attached directly to —N(O)—. When only one parent compound is cited in the name, the prefixed *ONN*- and *NNO*- specify that the group carrying the primed and unprimed substituents is connected, respectively, to the —N(O)— group. The prefix *NON*- signifies that the position of the oxygen atom is unknown; the azoxy group is then written as —N_2O—. For example,

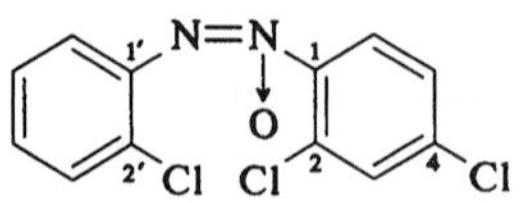

2,2′,4-Trichloro-*NNO*-azoxybenzene

2.1.3.11 Boron Compounds. Molecular hydrides of boron are called boranes. They are named by using a multiplying affix to designate the number of boron atoms and adding an Arabic numeral within parentheses as a suffix to denote the number of hydrogen atoms present. Examples are pentaborane(9) for B_5H_9 and pentaborane(11) for B_5H_{11}.

Organic ring systems are named by replacement nomenclature. Three- to ten-membered monocyclic ring systems containing uncharged boron atoms may be named by the specialist nomenclature for heterocyclic systems. Organic derivatives are named as outlined for substitutive nomenclature.

2.1.3.12 Carboxylic Acids. Carboxylic acids may be named in several ways. First, —COOH groups replacing CH_3— at the end of the main chain of an acyclic hydrocarbon are denoted by adding -oic acid to the name of the hydrocarbon. Second, when the —COOH group is the principal group, the suffix -carboxylic acid can be added to the name of the parent chain whose name and chain numbering *does not include* the carbon atom of the —COOH group. The former nomenclature is preferred unless use of the ending -carboxylic acid leads to citation of a larger number of carboxyl groups as suffix. Third, carboxyl groups are designated by the prefix carboxy- when attached to a group named as a substituent or when another group is present that has higher priority for citation as principal group. In all cases, the principal chain should be linked to as many carboxyl groups as possible even though it might not be the longest chain present. Examples are

$CH_3CH_2CH_2CH_2CH_2CH_2COOH$	(1) Heptanoic acid (2) 1-Hexanecarboxylic acid
$C_6H_{11}COOH$	(2) Cyclohexanecarboxylic acid

$$\begin{array}{ccccccc} & & COOH & & CH_2COOH & & \\ & & | & & | & & \\ CH_3—CH_2— & & CH & —CH_2— & CH & —CH_2—COOH \end{array}$$

(3) 2-(Carboxymethyl)-1,4-hexanedicarboxylic acid

Removal of the OH from the —COOH group to form the acyl radical results in changing the ending -oic acid to -oyl or the ending -carboxylic acid to -carbonyl. Thus the radical $CH_3CH_2CH_2CH_2CO$— is named either pentanoyl or butanecarbonyl. When the hydroxyl has not been removed from all carboxyl groups present in an acid, the remaining carboxyl groups are denoted by the prefix carboxy-. For example, $HOOCCH_2CH_2CH_2CH_2CH_2CO$— is named 6-carboxyhexanoyl.

Many trivial names exist for acids (Table 2.11). Generally, radicals are formed by replacing -ic acid by -oyl.* When a trivial name is given to an acyclic monoacid or diacid, the numeral 1 is always given as locant to the carbon atom of a carboxyl group in the acid or to the carbon atom with a free valence in the radical RCO—.

2.1.3.13 Ethers (R^1—O—R^2). In substitutive nomenclature, one of the possible radicals, R—O—, is stated as the prefix to the parent compound that is senior from among R^1 or R^2. Examples are methoxyethane for $CH_3OCH_2CH_3$ and butoxyethanol for $C_4H_9OCH_2CH_2OH$.

When another principal group has precedence and oxygen is linking two identical parent compounds, the prefix oxy- may be used, as with 2,2′-oxydiethanol for $HOCH_2CH_2OCH_2CH_2OH$.

Compounds of the type RO—Y—OR, where the two parent compounds are identical and contain a group having priority over ethers for citation as suffix, are named as assemblies of identical units. For example, HOOC—CH_2—O—CH_2CH_2—O—CH_2—COOH is named 2,2′-(ethylenedioxy) diacetic acid.

*Exceptions: formyl, acetyl, propionyl, butyryl, isobutyryl, valeryl, isovaleryl, oxalyl, malonyl, succinyl, glutaryl, furoyl, and thenoyl.

TABLE 2.11 Names of Some Carboxylic Acids

Systematic name	Trivial name	Systematic name	Trivial name
Methanoic	Formic	*trans*-Methylbutenedioic	Mesaconic*
Ethanoic	Acetic		
Propanoic	Propionic	1,2,2-Trimethyl-1,3-cyclopentanedicarboxylic acid	Camphoric
Butanoic	Butyric		
2-Methylpropanoic	Isobutyric*		
Pentanoic	Valeric	Benzenecarboxylic	Benzoic
3-Methylbutanoic	Isovaleric*	1,2-Benzenedicarboxylic	Phthalic
2,2-Dimethylpropanoic	Pivalic*	1,3-Benzenedicarboxylic	Isophthalic
Hexanoic	(Caproic)	1,4-Benzenedicarboxylic	Terephthalic
Heptanoic	(Enanthic)	Naphthalenecarboxylic	Naphthoic
Octanoic	(Caprylic)	Methylbenzenecarboxylic	Toluic
Decanoic	(Capric)	2-Phenylpropanoic	Hydratropic
Dodecanoic	Lauric*	2-Phenylpropenoic	Atropic
Tetradecanoic	Myristic*	*trans*-3-Phenylpropenoic	Cinnamic
Hexadecanoic	Palmitic*	Furancarboxylic	Furoic
Octadecanoic	Stearic*	Thiophenecarboxylic	Thenoic
		3-Pyridinecarboxylic	Nicotinic
Ethanedioic	Oxalic	4-Pyridinecarboxylic	Isonicotinic
Propanedioic	Malonic		
Butanedioic	Succinic	Hydroxyethanoic	Glycolic
Pentanedioic	Glutaric	2-Hydroxypropanoic	Lactic
Hexanedioic	Adipic	2,3-Dihydroxypropanoic	Glyceric
Heptanedioic	Pimelic*	Hydroxypropanedioic	Tartronic
Octanedioic	Suberic*	Hydroxybutanedioic	Malic
Nonanedioic	Azelaic*	2,3-Dihydroxybutanedioic	Tartaric
Decanedioic	Sebacic*	3-Hydroxy-2-phenylpropanoic	Tropic
Propenoic	Acrylic	2-Hydroxy-2,2-diphenylethanoic	Benzilic
Propynoic	Propiolic		
2-Methylpropenoic	Methacrylic	2-Hydroxybenzoic	Salicylic
trans-2-Butenoic	Crotonic	Methoxybenzoic	Anisic
cis-2-Butenoic	Isocrotonic	4-Hydroxy-3-methoxybenzoic	Vanillic
cis-9-Octadecenoic	Oleic		
trans-9-Octadecenoic	Elaidic	3,4-Dimethoxybenzoic	Veratric
cis-Butenedioic	Maleic	3,4-Methylenedioxybenzoic	Piperonylic
trans-Butenedioic	Fumaric	3,4-Dihydroxybenzoic	Protocatechuic
cis-Methylbutenedioic	Citraconic*	3,4,5-Trihydroxybenzoic	Gallic

* Systematic names should be used in derivatives formed by substitution on a carbon atom.
Note: The names in parentheses have been discontinued.

Linear polyethers derived from three or more molecules of aliphatic dihydroxy compounds, particularly when the chain length exceeds ten units, are most conveniently named by open-chain replacement nomenclature. For example, $CH_3CH_2—O—CH_2CH_2—O—CH_2CH_3$ could be 3,6-dioxaoctane or (2-ethoxy)ethoxyethane.

An oxygen atom directly attached to two carbon atoms already forming part of a ring system or to two carbon atoms of a chain may be indicated by the prefix epoxy-. For example, $CH_2—CH—CH_2Cl$ (with O bridging the first two carbons) is named 1-chloro-2,3-epoxypropane.

Symmetrical linear polyethers may be named (1) in terms of the central oxygen atom when there is an odd number of ether oxygen atoms or (2) in terms of the central hydrocarbon group when there is an even number of ether oxygen atoms. For example, $C_2H_5—O—C_4H_8—O—C_4H_8—O—C_2H_5$ is bis-(4-ethoxybutyl)ether, and 3,6-dioxaoctane (earlier example) could be named 1,2-bis(ethoxy)ethane.

Partial ethers of polyhydroxy compounds may be named (1) by substitutive nomenclature or (2) by stating the name of the polyhydroxy compound followed by the name of the etherifying radical(s) followed by the word *ether*. For example,

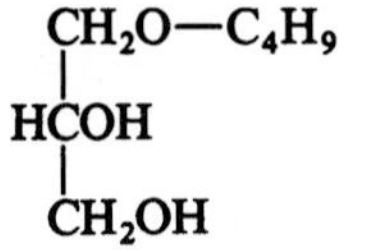

(1) 3-Butoxy-1,2-propanediol
(2) Glycerol 1-butyl ether; also, 1-*O*-butylglycerol

Cyclic ethers are named either as heterocyclic compounds or by specialist rules of heterocyclic nomenclature. Radicofunctional names are formed by citing the names of the radicals R^1 and R^2 followed by the word *ether*. Thus methoxyethane becomes ethyl methyl ether and ethoxyethane becomes diethyl ether.

2.1.3.14 Halogen Derivatives. Using substitutive nomenclature, names are formed by adding prefixes listed in Table 2.8 to the name of the parent compound. The prefix perhalo- implies the replacement of all hydrogen atoms by the particular halogen atoms.

Cations of the type $R^1R^2X^+$ are given names derived from the halonium ion, H_2X^+, by substitution, e.g., diethyliodonium chloride for $(C_2H_5)_2I^+Cl^-$.

Retained are these trivial names; bromoform ($CHBr_3$), chloroform ($CHCl_3$), fluoroform (CHF_3), iodoform (CHI_3), phosgene ($COCl_2$), thiophosgene ($CSCl_2$), and dichlorocarbene radical ($>CCl_2$). Inorganic nomenclature leads to such names as carbonyl and thiocarbonyl halides (COX_2 and CSX_2) and carbon tetrahalides (CX_4).

2.1.3.15 Hydroxylamines and Oximes. For RNH—OH compounds, prefix the name of the radical R to hydroxylamine. If another substituent has priority as principal group, attach the prefix hydroxyamino- to the parent name. For example, C_6H_5NHOH would be named *N*-phenylhydroxylamine, but HOC_6H_4NHOH would be (hydroxyamino)phenol, with the point of attachment indicated by a locant preceding the parentheses.

Compounds of the type $R^1NH—OR^2$ are named (1) as alkoxyamino derivatives of compound R^1H, (2) as *N,O*-substituted hydroxylamines. (3) as alkoxyamines (even if R^1 is hydrogen), or (4) by the prefix aminooxy- when another substituent has priority for parent name. Examples of each type are

1. 2-(Methoxyamino)-8-naphthalenecarboxylic acid for $CH_3ONH—C_{10}H_6COOH$
2. *O*-Phenylhydroxylamine for $H_2N—O—C_6H_5$ or *N*-phenylhydroxylamine for $C_6H_5NH—OH$
3. Phenoxyamine for $H_2N—O—C_6H_5$ (not preferred to *O*-phenylhydroxylamine)
4. Ethyl (aminooxy)acetate for $H_2N—O—CH_2CO—OC_2H_5$

Acyl derivatives, RCO—NH—OH and $H_2N—O—CO—R$, are named as *N*-hydroxy derivatives of amides and as *O*-acylhydroxylamines, respectively. The former may also be named as hydroxamic acids. Examples are *N*-hydroxyacetamide for $CH_3CO—NH—OH$ and *O*-acetylhydroxylamine for $H_2N—O—CO—CH_3$. Further substituents are denoted by prefixes with *O*- and/or *N*-locants. For example, $C_6H_5NH—O—C_2H_5$ would be *O*-ethyl-*N*-phenylhydroxylamine or *N*-ethoxylaniline.

For oximes, the word *oxime* is placed after the name of the aldehyde or ketone. If the carbonyl group is not the principal group, use the prefix hydroxyimino-. Compounds with the group $>$N—OR are named by a prefix alkyloxyimino- oxime *O*-ethers or as *O*-substituted oximes. Compounds with the group $>$C=N(O)R are named by adding *N*-oxide after the name of the alkylideneaminc compound. For amine oxides, add the word *oxide* after the name of the base, with locants. For example, $C_5H_5N—O$ is named pyridine *N*-oxide or pyridine 1-oxide.

2.1.3.16 Imines. The group $>$C=NH is named either by the suffix -imine or by citing the name of the bivalent radical $R^1R^2C<$ as a prefix to amine. For example, $CH_3CH_2CH_2CH=NH$ could be named 1-butanimine or butylideneamine. When the nitrogen is substituted, as in $CH_2=N—CH_2CH_3$, the name is *N*-(methylidene)ethylamine.

Quinones are exceptions. When one or more atoms of quinonoid oxygen have been replaced by >NH or >NR, they are named by using the name of the quinone followed by the word *imine* (and preceded by proper affixes). Substituents on the nitrogen atom are named as prefixes. Examples are

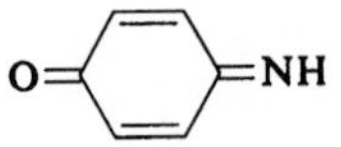

p-Benzoquinone monoimine

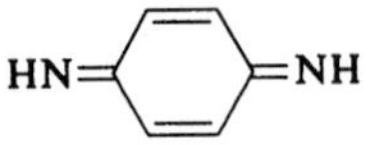

p-Benzoquinone diimine

2.1.3.17 Ketenes. Derivatives of the compound ketene, $CH_2{=}C{=}O$, are named by substitutive nomenclature. For example, $C_4H_9CH{=}C{=}O$ is butyl ketene. An acyl derivative, such as $CH_3CH_2{-}CO{-}CH_2CH{=}C{=}O$, may be named as a polyketone, 1-hexene-1,4-dione. Bisketene is used for two to avoid ambiguity with diketene (dimeric ketene).

2.1.3.18 Ketones. Acyclic ketones are named (1) by adding the suffix -one to the name of the hydrocarbon forming the principal chain or (2) by citing the names of the radicals R^1 and R^2 followed by the word *ketone*. In addition to the preceding nomenclature, acyclic monoacyl derivatives of cyclic compounds may be named (3) by prefixing the name of the acyl group to the name of the cyclic compound. For example, the three possible names of

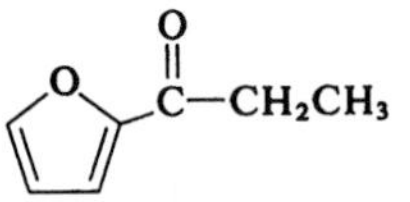

(1) 1-(2-Furyl)-1-propanone
(2) Ethyl 2-furyl ketone
(3) 2-Propionylfuran

When the cyclic component is benzene or naphthalene, the -ic acid or -oic acid of the acid corresponding to the acyl group is changed to -ophenone or -onaphthone, respectively. For example, $C_6H_5{-}CO{-}CH_2CH_2CH_3$ can be named either butyrophenone (or butanophenone) or phenyl propyl ketone.

Radicofunctional nomenclature can be used when a carbonyl group is attached directly to carbon atoms in two ring systems and no other substituent is present having priority for citation.

When the methylene group in polycarbocyclic and heterocyclic ketones is replaced by a keto group, the change may be denoted by attaching the suffix -one to the name of the ring system. However, when ≥CH in an unsaturated or aromatic system is replaced by a keto group, two alternative names become possible. First, the maximum number of noncumulative double bonds is added after introduction of the carbonyl group(s), and any hydrogen that remains to be added is denoted as indicated hydrogen with the carbonyl group having priority over the indicated hydrogen for lower-numbered locant. Second, the prefix oxo- is used, with the hydrogenation indicated by hydro prefixes; hydrogenation is considered to have occurred before the introduction of the carbonyl group. For example,

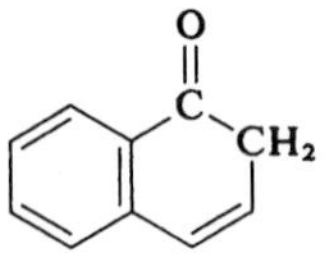

(1) 1-(2*H*)-Naphthalenone
(2) 1-Oxo-1,2-dihydronaphthalene

When another group having higher priority for citation as principal group is also present, the ketonic oxygen may be expressed by the prefix oxo-, or one can use the name of the carbonyl-containing radical, as, for example, acyl radicals and oxo-substituted radicals. Examples are

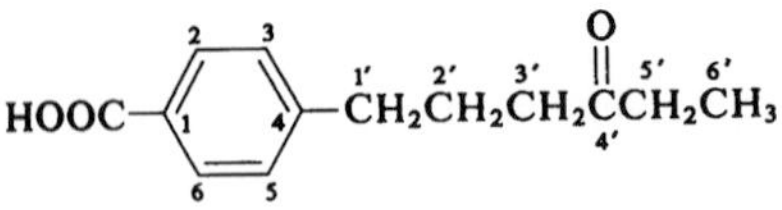

4-(4′-Oxohexyl)-1-benzoic acid

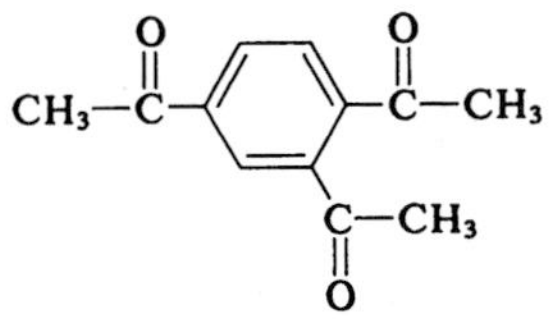

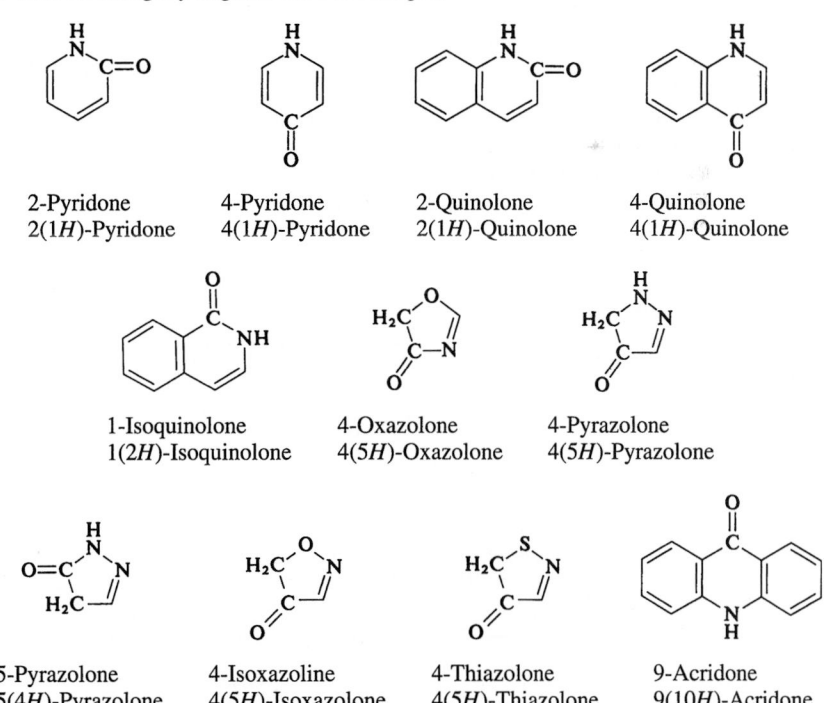

1,2,4-Triacetylbenzene

Diketones and tetraketones derived from aromatic compounds by conversion of two or four >CH groups into keto groups, with any necessary rearrangement of double bonds to a quinonoid structure, are named by adding the suffix -quinone and any necessary affixes.

Polyketones in which two or more contiguous carbonyl groups have rings attached at each end may be named (1) by the radicofunctional method or (2) by substitutive nomenclature. For example,

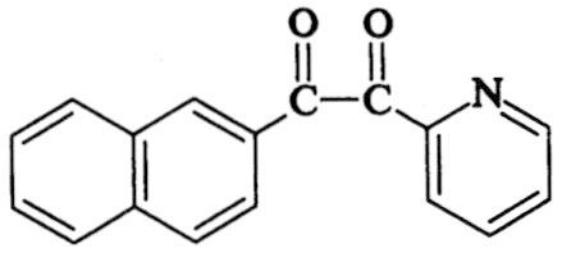

(1) 2-Naphthyl 2-pyridyl diketone
(2) 1-(2-Naphthyl)-2-(2-pyridyl)ethanedione

Some trivial names are retained: acetone (2-propanone), biacetyl (2,3-butanedione), propiophenone ($C_6H_5-CO-CH_2CH_3$), chalcone ($C_6H_5-CH{=}CH-CO-C_6H_5$), and deoxybenzoin ($C_6H_5-CH_2-CO-C_6H_5$).

These contracted names of heterocyclic nitrogen compounds are retained as alternatives for systematic names, sometimes with indicated hydrogen. In addition, names of oxo derivatives of fully saturated nitrogen heterocycles that systematically end in -idinone are often contracted to end in -idone when no ambiguity might result. For example,

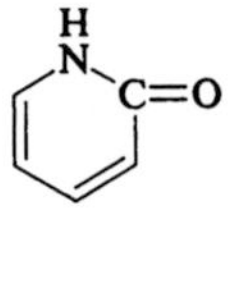

2-Pyridone
2(1*H*)-Pyridone

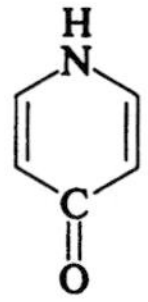

4-Pyridone
4(1*H*)-Pyridone

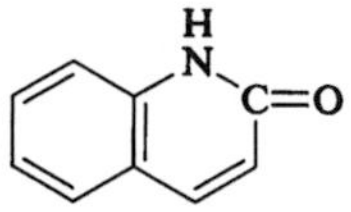

2-Quinolone
2(1*H*)-Quinolone

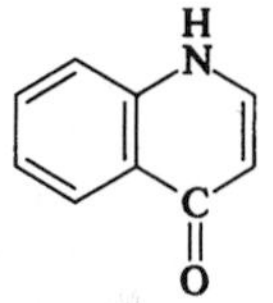

4-Quinolone
4(1*H*)-Quinolone

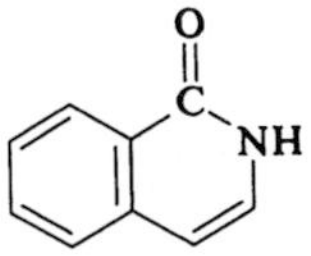

1-Isoquinolone
1(2*H*)-Isoquinolone

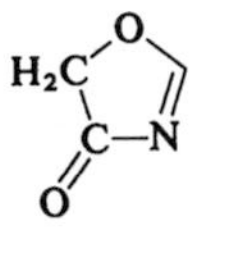

4-Oxazolone
4(5*H*)-Oxazolone

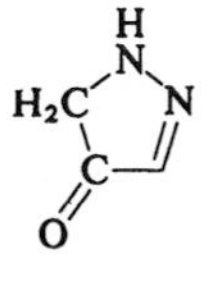

4-Pyrazolone
4(5*H*)-Pyrazolone

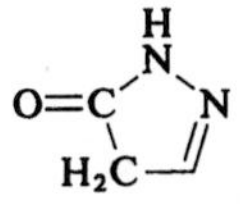

5-Pyrazolone
5(4*H*)-Pyrazolone

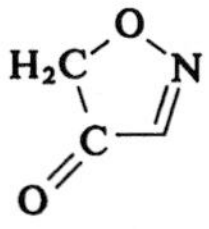

4-Isoxazoline
4(5*H*)-Isoxazolone

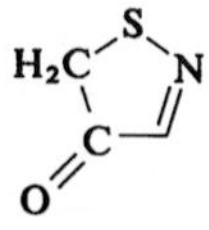

4-Thiazolone
4(5*H*)-Thiazolone

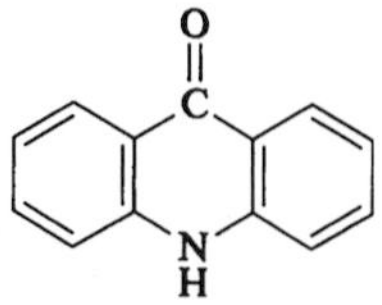

9-Acridone
9(10*H*)-Acridone

2.1.3.19 Lactones, Lactides, Lactams, and Lactims. When the hydroxy acid from which water may be considered to have been eliminated has a trivial name, the lactone is designated by substituting -olactone for -ic acid. Locants for a carbonyl group are numbered as low as possible, even before that of a hydroxyl group.

Lactones formed from aliphatic acids are named by adding -olide to the name of the nonhydroxylated hydrocarbon with the same number of carbon atoms. The suffix -olide signifies the change of $>CH\cdots CH_3$ into $>C\cdots C{=}O$ (with the carbonyl carbon and the first carbon joined through —O—).

Structures in which one or more (but not all) rings of an aggregate are lactone rings are named by placing -carbolactone (denoting the —O—CO— bridge) after the names of the structures that remain when each bridge is replaced by two hydrogen atoms. The locant for —CO— is cited before that for the ester oxygen atom. An additional carbon atom is incorporated into this structure as compared to the -olide.

These trivial names are permitted: γ-butyrolactone, γ-valerolactone, and δ-valerolactone. Names based on heterocycles may be used for all lactones. Thus, γ-butyrolactone is also tetrahydro-2-furanone or dihydro-2(3*H*)-furanone.

Lactides, intermolecular cyclic esters, are named as heterocycles. *Lactams* and *lactims*, containing a —CO—NH— and —C(OH) =N—group, respectively, are named as heterocycles, but they may also be named with -lactam or -lactim in place of -olide. For example,

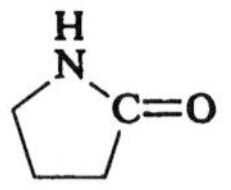

(1) 2-Pyrrolidinone
(2) 4-Butanelactam

2.1.3.20 Nitriles and Related Compounds. For acids whose systematic names end in -carboxylic acid, nitriles are named by adding the suffix -carbonitrile when the —CN group replaces the —COOH group. The carbon atom of the —CN group is excluded from the numbering of a chain to which it is attached. However, when the triple-bonded nitrogen atom is considered to replace three hydrogen atoms at the end of the main chain of an acyclic hydrocarbon, the suffix -nitrile is added to the name of the hydrocarbon. Numbering begins with the carbon attached to the nitrogen. For example, $CH_3CH_2CH_2CH_2CH_2CN$ is named (1) pentanecarbonitrile or (2) hexanenitrile.

Trivial acid names are formed by changing the endings -oic acid or -ic acid to -onitrile. For example, CH_3CN is acetonitrile. When the —CN group is not the highest priority group, the —CN group is denoted by the prefix cyano-.

In order of decreasing priority for citation of a functional class name, and the prefix for substitutive nomenclature, are the following related compounds:

Functional group	Prefix	Radicofunctional ending
—NC	Isocyano-	Isocyanide
—OCN	Cyanato-	Cyanate
—NCO	Isocyanato-	Isocyanate
—ONC	—	Fulminate
—SCN	Thiocyanato-	Thiocyanate
—NCS	Isothiocyanato-	Isothiocyanate
—SeCN	Selenocyanato-	Selenocyanate
—NCSe	Isoselenocyanato-	Isoselenocyanate

2.1.3.21 Peroxides. Compounds of the type R—O—OH are named (1) by placing the name of the radical R before the word *hydroperoxide* or (2) by use of the prefix hydroperoxy- when another parent name has higher priority. For example, C_2H_5OOH is ethyl hydroperoxide.

Compounds of the type $R^1O—OR^2$ are named (1) by placing the names of the radicals in alphabetical order before the word *peroxide* when the group —O—O— links two chains, two rings, or a ring and a chain, (2) by use of the affix dioxy to denote the bivalent group —O—O— for naming assemblies of identical units or to form part of a prefix, or (3) by use of the prefix epidioxy- when the peroxide group forms a bridge between two carbon atoms, a ring, or a ring system. Examples are methyl propyl peroxide for CH_3—O—O—C_3H_7 and 2,2′-dioxydiacetic acid for HOOC—CH_2—O—O—CH_2—COOH.

2.1.3.21 Phosphorus Compounds. Acyclic phosphorus compounds containing only one phosphorus atom, as well as compounds in which only a single phosphorus atom is in each of several functional groups, are named as derivatives of the parent structures (Table 2.12). Often these are purely hypothetical parent structures. When hydrogen attached to phosphorus is replaced by a hydrocarbon group, the derivative is named by substitution nomenclature. When hydrogen of an —OH group is replaced, the derivative is named by radicofunctional nomenclature. For example, $C_2H_5PH_2$ is ethylphosphine; $(C_2H_5)_2PH$, diethylphosphine; $CH_3P(OH)_2$, dihydroxy-methyl-phosphine or methylphosphonous acid; C_2H_5—PO(Cl)(OH), ethylchlorophosphonic acid or ethylphosphonochloridic acid or hydrogen chlorodioxoethylphosphate(V); $CH_3CH(PH_2)COOH$, 2-phosphinopropionic acid; $HP(CH_2COOH)_2$, phosphinediyldiacetic acid; $(CH_3)HP(O)OH$, methylphosphinic acid or hydrogen hydridomethyldioxophosphate(V); $(CH_3O)_3PO$, trimethyl phosphate; and $(CH_3O)_3P$, trimethyl phosphite.

2.1.3.22 Salts and Esters of Acids. Neutral salts of acids are named by citing the cation(s) and then the anion, whose ending is changed from -oic to -oate or from -ic to -ate. When different acidic residues are present in one structure, prefixes are formed by changing the anion ending -ate to -ato- or -ide to -ido-. The prefix carboxylato- denotes the ionic group —COO^-. The phrase (metal) salt of (the acid) is permissible when the carboxyl groups are not all named as affixes.

Acid salts include the word *hydrogen* (with affixes, if appropriate) inserted between the name of the cation and the name of the anion (or word *salt*).

Esters are named similarly, with the name of the alkyl or aryl radical replacing the name of the cation. Acid esters of acids and their salts are named as neutral esters, but the components are cited

TABLE 2.12 Phosphorus-Containing Compounds

Formula	Parent name	Substitutive prefix	Radicofunctional ending
H_3P	Phosphine	H_2P— Phosphino-	Phosphide
H_5P	Phosphorane	H_4P— Phosphoranyl- H_3P< Phosphoroanediyl- H_2P≤ Phosphoranetriyl-	
H_3PO	Phosphine oxide		
H_3PS	Phosphine sulfide		
H_3PNH	Phosphine imide		
$P(OH)_3$	Phosphorous acid		Phosphite
$HP(OH)_2$	Phosphonous acid		Phosphonite
H_2POH	Phosphinous acid		Phosphinite
$P(O)(OH)_3$	Phosphoric acid	P(O)≤ Phosphoryl-	Phosphate(V)
$HP(O)(OH)_2$	Phosphonic acid	HP(O)< Phosphonoyl- —$P(O)OH_2$ Phosphono-	Phosphonate
$H_2P(O)OH$	Phosphinic acid	$H_2P(O)$— Phosphinoyl- >P(O)OH Phosphinoco- Phosphinato-	Phosphinate

in the order: cation, alkyl or aryl radical, hydrogen, and anion. Locants are added if necessary. For example,

$$\begin{array}{l} CH_2{-}CO{-}OC_2H_5 \\ |\\ HOC{-}COO^- \quad K^+ \; H^+ \\ | \\ CH_2{-}COO^- \end{array}$$

Potassium 1-ethyl hydrogen citrate

Ester groups in R^1—CO—OR^2 compounds are named (1) by the prefix alkoxycarbonyl- or aryloxycarbonyl- for —CO—OR^2 when the radical R^1 contains a substituent with priority for citation as principal group or (2) by the prefix acyloxy- for R^1—CO—O— when the radical R^2 contains a substituent with priority for citation as principal group. Examples are

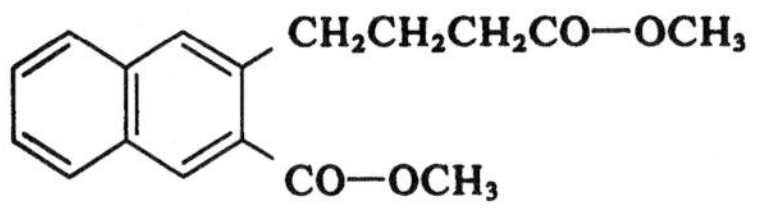

Methyl 3-methoxycarbonyl-2-naphthalenebutyrate

$[CH_3O{-}CO{-}CH_2CH_2\overset{+}{N}(CH_3)_3]Cl^-$ [(2-Methoxycarbonyl)ethyl]trimethylammonium chloride

$C_6H_5{-}CO{-}OCH_2CH_2COOH$ 3-Benzoyloxypropionic acid

The trivial name *acetoxy* is retained for the CH_3—CO—O— group. Compounds of the type $R^2C(OR^2)_3$ are named as R^2 esters of the hypothetical ortho acids. For example, $CH_3C(OCH_3)_3$ is trimethyl orthoacetate.

2.1.3.22 Silicon Compounds. SiH_4 is called silane; its acyclic homologs are called disilane, trisilane, and so on, according to the number of silicon atoms present. The chain is numbered from one end to the other so as to give the lowest-numbered locant in radicals to the free valence or to substitutents on a chain. The abbreviated form silyl is used for the radical SiH_3—. Numbering and citation of side chains proceed according to the principles set forth for hydrocarbon chains. Cyclic nonaromatic structures are designated by the prefix cyclo-.

When a chain or ring system is composed entirely of alternating silicon and oxygen atoms, the parent name *siloxane* is used with a multiplying affix to denote the number of silicon atoms present. The parent name *silazane* implies alternating silicon and nitrogen atoms; multiplying affixes denote the number of silicon atoms present.

The prefix sila- designates replacement of carbon by silicon in replacement nomenclature. Prefix names for radicals are formed analogously to those for the corresponding carbon-containing compounds. Thus silyl is used for SiH_3—, silyene for —SiH_2—, silylidyne for —SiH<, as well as trily, tetrayl, and so on for free valences(s) on ring structures.

2.1.3.23 Sulfur Compounds *Bivalent Sulfur.* The prefix thio, placed before an affix that denotes the oxygen-containing group or an oxygen atom, implies the replacement of that oxygen by sulfur. Thus the suffix -thiol denotes —SH, -thione denotes —(C)═S and implies the presence of an ═S at a nonterminal carbon atom, -thioic acid denotes [(C)═S]OH ⇌ [(C)═O]SH (that is, the *O*-substituted acid and the *S*-substituted acid, respectively), -dithioc acid denotes [—C(S)SH, and -thial denotes —(C)HS (or -carbothialdehyde denotes —CHS). When -carboxylic acid has been used for acids, the sulfur analog is named -carbothioic acid or -carbodithioic acid.

Prefixes for the groups HS— and RS— are mercapto- and alkylthio-, respectively; this latter name may require parentheses for distinction from the use of thio- for replacement of oxygen in a trivially named acid. Examples of this problem are 4-C_2H_5—C_6H_4—CSOH named *p*-ethyl(thio)benzoic acid and 4-C_2H_5—S—C_6H_4—COOH named *p*-(ethylthio)benzoic acid. When —SH is not the principal group, the prefix mercapto- is placed before the name of the parent compound to denote an unsubstituted —SH group.

The prefix thioxo- is used for naming =S in a thioketone. Sulfur analogs of acetals are named as alkylthio- or arylthio-. For example, $CH_3CH(SCH_3)OCH_3$ is 1-methoxy-1-(methylthio)ethane. Prefix forms for -carbothioic acids are hydroxy(thiocarbonyl)- when referring to the *O*-substituted acid and mercapto(carbonyl)- for the *S*-substituted acid.

Salts are formed as with oxygen-containing compounds. For example, C_2H_5—S—Na is named either sodium ethanethiolate or sodium ethyl sulfide. If mercapto- has been used as a prefix, the salt is named by use of the prefix sulfido- for —S^-.

Compounds of the type R^1—S—R^2 are named alkylthio- (or arylthio-) as a prefix to the name of R^1 or R^2, whichever is the senior.

2.1.3.24 Sulfonium Compounds. Sulfonium compounds of the type $R^1R^2R^3S^+X^-$ are named by citing in alphabetical order the radical names followed by -sulfonium and the name of the anion. For heterocyclic compounds, -ium is added to the name of the ring system. Replacement of >CH by sulfonium sulfur is denoted by the prefix thionia-, and the name of the anion is added at the end.

2.1.3.25 Organosulfur Halides. When sulfur is directly linked only to an organic radical and to a halogen atom, the radical name is attached to the word *sulfur* and the name(s) and number of the halide(s) are stated as a separate word. Alternatively, the name can be formed from R—SOH, a sulfenic acid whose radical prefix is sulfenyl-. For example, CH_3CH_2—S—Br would be named either ethylsulfur monobromide or ethanesulfenyl bromide. When another principal group is present, a composite prefix is formed from the number and substitutive name(s) of the halogen atoms in front of the syllable thio. For example, BrS—COOH is (bromothio)formic acid.

2.1.3.26 Sulfoxides. Sulfoxides, R^1—SO—R^2, are named by placing the names of the radicals in alphabetical order before the word *sulfoxide*. Alternatively, the less senior radical is named followed by sulfinyl- and concluded by the name of the senior group. For example, CH_3CH_2—SO—$CH_2CH_2CH_3$ is named either ethyl propyl sulfoxide or 1-(ethylsulfinyl)propane.

When an >SO group is incorporated in a ring, the compound is named an oxide.

2.1.3.27 Sulfones. Sulfones, R^1—SO_2—R^2, are named in an analogous manner to sulfoxides, using the word *sulfone* in place of *sulfoxide*. In prefixes, the less senior radical is followed by -sulfonyl-. When the >SO_2 group is incorporated in a ring, the compound is named as a dioxide.

2.1.3.28 Sulfur Acids. Organic oxy acids of sulfur, that is, —SO_3H, —SO_2H, and —SOH, are named sulfonic acid, sulfinic acid, and sulfenic acid, respectively. In subordinate use, the respective prefixes are sulfo-, sulfino, and sulfeno-. The grouping —SO_2—O—SO_2— or —SO—O—SO is named sulfonic or sulfinic anhydride, respectively.

Inorganic nomenclature is employed in naming sulfur acids and their derivatives in which sulfur is linked only through oxygen to the organic radical. For example, $(C_2H_5O)_2SO_2$ is diethyl sulfate and C_2H_5O—SO_2—OH is ethyl hydrogen sulfate. Prefixes *O*- and *S*- are used where necessary to denote attachment to oxygen and to sulfur, respectively, in sulfur replacement compounds. For example, CH_3—S—SO_2—ONa is sodium *S*-methyl thiosulfate.

When sulfur is linked only through nitrogen, or through nitrogen and oxygen, to the organic radical, naming is as follows: (1) *N*-substituted amides are designated as *N*-substituted derivatives of the sulfur amides and (2) compounds of the type *R*—NH—SO_3H may be named as *N*-substituted sulfamic acids or by the prefix sulfoamino- to denote the group HO_3S—NH—. The groups —N=SO and —N=SO_2 are named sulfinylamines and sulfonylamines, respectively.

2.1.3.29 Sultones and Sultams. Compounds containing the group —SO_2—O— as part of the ring are called -sultone. The —SO_2— group has priority over the —O— group for lowest-numbered locant.

Similarly, the —SO_2—N= group as part of a ring is named by adding -sultam to the name of the hydrocarbon with the same number of carbon atoms. The —SO_2— has priority over —N= for lowest-numbered locant.

2.1.4 Stereochemistry

Concepts in stereochemistry, that is, chemistry in three-dimensional space, are in the process of rapid expansion. This section will deal with only the main principles. The compounds discussed will be those that have identical molecular formulas but differ in the arrangement of their atoms in space. *Stereoisomers* is the name applied to these compounds.

Stereoisomers can be grouped into three categories: (1) Conformational isomers differ from each other only in the way their atoms are oriented in space, but can be converted into one another by rotation about sigma bonds. (2) Geometric isomers are compounds in which rotation about a double bond is restricted. (3) Configurational isomers differ from one another only in configuration about a chiral center, axis, or plane. In subsequent structural representations, a broken line denotes a bond projecting behind the plane of the paper and a wedge denotes a bond projecting in front of the plane of the paper. A line of normal thickness denotes a bond lying essentially in the plane of the paper.

2.1.4.1 Conformational Isomers. A molecule in a conformation into which its atoms return spontaneously after small displacements is termed a *conformer*. Different arrangements of atoms that can be converted into one another by rotation about single bonds are called *conformational isomers* (see Fig. 2.1). A pair of conformational isomers can be but do not have to be mirror images of each other. When they are not mirror images, they are called *diastereomers*.

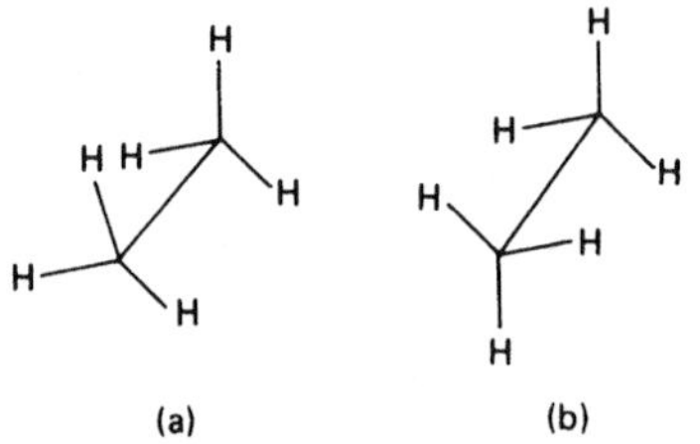

FIGURE 2.1 Conformations of ethane. (*a*) Eclipsed; (*b*) staggered.

2.1.4.2 Acyclic Compounds. Different conformations of acyclic compounds are best viewed by construction of ball-and-stick molecules or by use of Newman projections (see Fig. 2.2). Both types of representations are shown for ethane. Atoms or groups that are attached at opposite ends of a single bond should be viewed along the bond axis. If two atoms or groups attached at opposite ends of the bond appear one directly behind the other, these atoms or groups are described as eclipsed. That portion of the molecule is described as being in the eclipsed conformation. If not eclipsed, the atoms or groups and the conformation may be described as staggered. Newman projections show these conformations clearly.

Certain physical properties show that rotation about the single bond is not quite free. For ethane there is an energy barrier of about 3 kcal · mol^{-1} (12 kJ · mol^{-1}). The potential energy of the molecule is at a minimum for the staggered conformation, increases with rotation, and reaches a maximum at the eclipsed conformation. The energy required to rotate the atoms or groups about the carbon-carbon bond is called *torsional energy*. Torsional strain is the cause of the relative instability of the eclipsed conformation or any intermediate skew conformations.

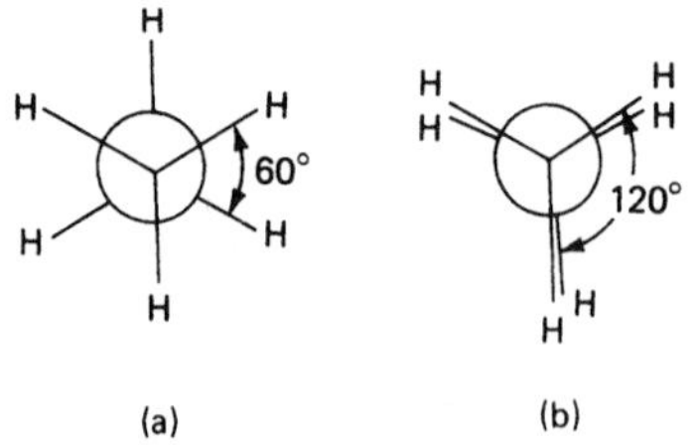

FIGURE 2.2 Newman projections for ethane. (*a*) Staggered; (*b*) eclipsed.

In butane, with a methyl group replacing one hydrogen on each carbon of ethane, there are several different staggered conformations (see Fig. 2.3). There is the *anti*-conformation in which the methyl groups are as far apart as they can be (dihedral angle of 180°). There are two *gauche* conformations in which the methyl groups are only 60° apart; these are two nonsuperimposable mirror images of each other. The *anti*-conformation is more stable than the *gauche* by about 0.9 kcal · mol^{-1} (4 kJ · mol^{-1}). Both are free of torsional strain. However, in a *gauche* conformation the methyl groups are closer together than the sum of their van der Waals' radii. Under these conditions van der Waals' forces are repulsive and raise the energy of conformation. This strain can affect not only the relative stabilities of

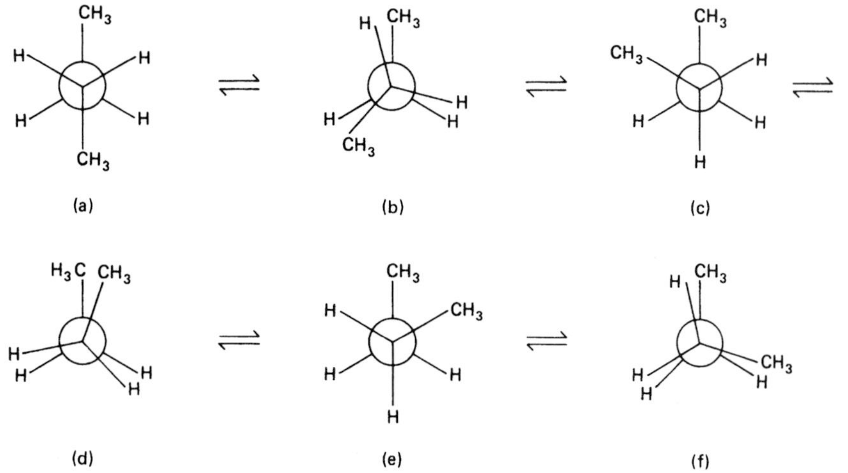

FIGURE 2.3 Conformations of butane. (*a*) Anti-staggered; (*b*) eclipsed; (*c*) gauche-staggered; (*d*) eclipsed; (*e*) gauche-staggered; (*f*) eclipsed. (Eclipsed conformations are slightly staggered for convenience in drawing; actually they are superimposed.)

various staggered conformations but also the heights of the energy barriers between them. The energy maximum (estimated at 4.8 to 6.1 kcal · mol^{-1} or 20 to 25 kJ · mol^{-1}) is reached when two methyl groups swing past each other (the eclipsed conformation) rather than past hydrogen atoms.

2.1.4.3 Cyclic Compounds. Although cyclic aliphatic compounds are often drawn as if they were planar geometric figures (a triangle for cyclopropane, a square for cyclobutane, and so on), their structures are not that simple. Cyclopropane does possess the maximum angle strain if one considers the difference between a tetrahedral angle (109.5°) and the 60° angle of the cyclopropane structure. Nevertheless the cyclopropane structure is thermally quite stable. The highest electron density of the carbon-carbon bonds does not lie along the lines connecting the carbon-carbon bonds does not lie along the lines connecting the carbon atoms. Bonding electrons lie principally outside the triangular internuclear lines and result in what is known as *bent bonds* (see Fig. 2.4).

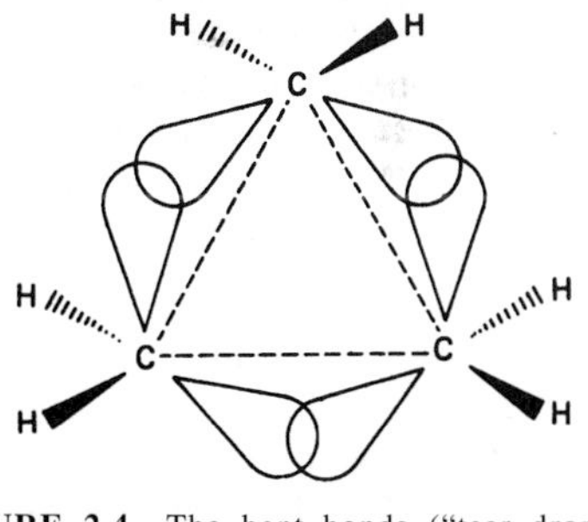

FIGURE 2.4 The bent bonds ("tear drops") of cyclopropane.

Cyclobutane has less angle strain than cyclopropane (only 19.5°). It is also believed to have some bent-bond character associated with the carbon-carbon bonds. The molecule exists in a nonplanar conformation in order to minimize hydrogen-hydrogen eclipsing strain.

Cyclopentane is nonplanar, with a structure that resembles an envelope (see Fig. 2.5). Four of the carbon atoms are in one plane, and the fifth is out of that plane. The molecule is in continual motion so that the out-of-plane carbon moves rapidly around the ring.

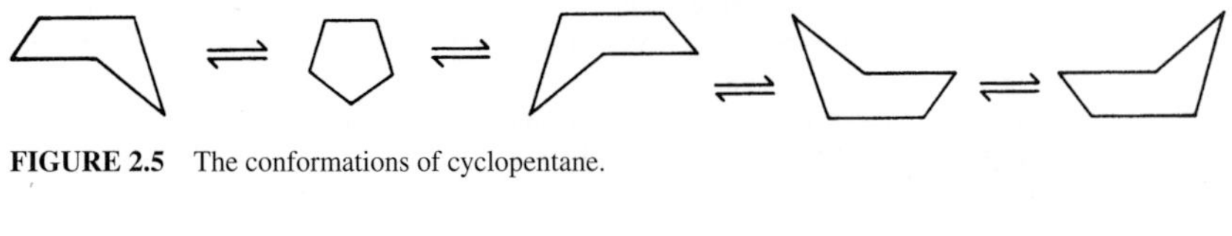

FIGURE 2.5 The conformations of cyclopentane.

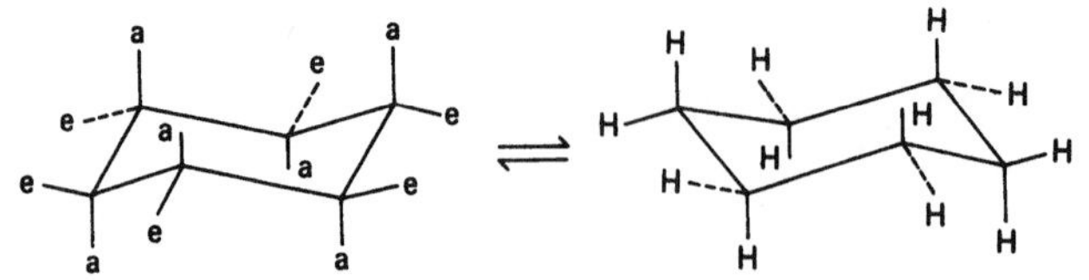

FIGURE 2.6 The two chair conformations of cyclohexane; *a* = axial hydrogen atom and *e* = equatorial hydrogen atom.

The 12 hydrogen atoms of cyclohexane do not occupy equivalent positions. In the chair conformation six hydrogen atoms are perpendicular to the average plane of the molecule and six are directed outward from the ring, slightly above or below the molecular plane (see Fig. 2.6). Bonds which are perpendicular to the molecular plane are known as *axial bonds*, and those which extend outward from the ring are known as *equatorial bonds*. The three axial bonds directed upward originate from alternate carbon atoms and are parallel with each other; a similar situation exists for the three axial bonds directed downward. Each equatorial bond is drawn so as to be parallel with the ring carbon-carbon bond once removed from the point of attachment to that equatorial bond. At room temperature, cyclohexane is interconverting rapidly between two chair conformations. As one chair form converts to the other, all the equatorial hydrogen atoms become axial and all the axial hydrogens become equatorial. The interconversion is so rapid that all hydrogen atoms on cyclohexane can be considered equivalent. Interconversion is believed to take place by movement of one side of the chair structure to produce the twist boat, and then movement of the other side of the twist boat to give the other chair form. The chair conformation is the most favored structure for cyclohexane. No angle strain is encountered since all bond angles remain tetrahedral. Torsional strain is minimal because all groups are staggered.

In the boat conformation of cyclohexane (see Fig 2.7) eclipsing torsional strain is significant, although no angle strain is encountered. Nonbonded interaction between the two hydrogen atoms across the ring from each other (the "flagpole" hydrogens) is unfavorable. The boat conformation is about 6.5 kcal · mol^{-1} (27 kJ · mol^{-1}) higher in energy than the chair form at 25°C.

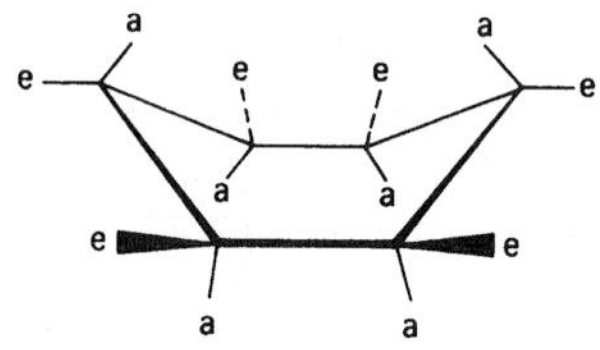

FIGURE 2.7 The boat conformation of cyclohexane. *a* = axial hydrogen atom and *e* = equatorial hydrogen atom.

A modified boat conformation of cyclohexane, known as the twist boat (see Fig. 2.8), or skew boat, has been suggested to minimize torsional and nonbounded interactions. This particular conformation is estimated to be about 1.5 kcal · mol^{-1} · (6 kJ · mol^{-1}) lower in energy than the boat form at room temperature.

The medium-size rings (7 to 12 ring atoms) are relatively free of angle strain and can easily take a variety of spatial arrangements. They are not large enough to avoid all nonbonded interactions between atoms.

FIGURE 2.8 Twist-boat conformation of cyclohexane.

Disubstituted cyclohexanes can exist as *cis-trans* isomers as well as axial-equatorial conformers. Two isomers are predicted for 1,4-dimethylcyclohexane (see Fig. 2.9). For the *trans* isomer the diequatorial conformer is the energetically favorable form. Only one *cis* isomer is observed, since the two conformers of the *cis* compound are identical. Interconversion takes place between the conformational (equatorial-axial isomers) but not configurational (*cis-trans*) isomers.

The bicyclic compound decahydronaphthalene, or bicyclo[4.4.0]decane, has two fused six-membered rings. It exists in *cis* and *trans* forms (see Fig. 2.10), as determined by the configurations at the

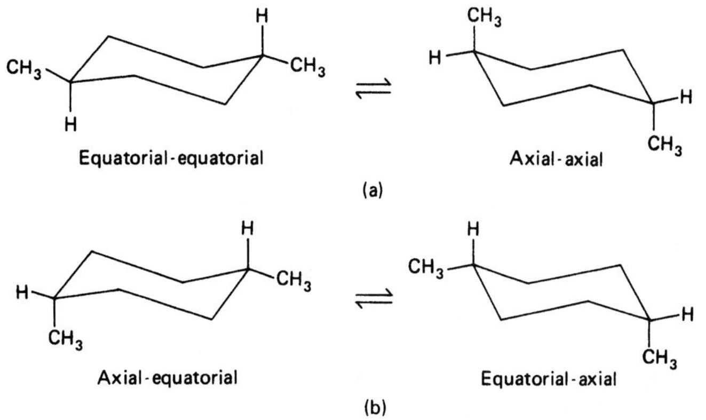

FIGURE 2.9 Two isomers of 1,4-dimethylcyclohexane. (*a*) *Trans* isomer; (*b*) *cis* isomer.

bridgehead carbon atoms. Both *cis-* and *trans*-decahydronaphthalene can be constructed with two chair conformations.

2.1.4.4 Geometrical Isomerism. Rotation about a carbon-carbon double bond is restricted because of interaction between the *p* orbitals which make up to pi bond. Isomerism due to such restricted rotation about a bond is known as *geometric isomerism*. Parallel overlap of the *p* orbitals of each carbon atom of the double bond forms the molecular orbital of the pi bond. The relatively large barrier to rotation about the pi bond is estimated to be nearly 63 kcal · mol^{-1} (263 kJ · mol^{-1}).

When two different substituents are attached to each carbon atom of the double bond, *cis-trans* isomers can exist. In the case of *cis*-2-butene (see Fig. 2.11a), both methyl groups are on the same side of the double bond. The other isomer has the methyl groups on opposite sides and is designated as *trans*-2-butene (see Fig. 2.11b). Their physical properties are quite different. Geometric isomerism can also exist in ring systems; examples were cited in the previous discussion on conformational isomers.

For compounds containing only double-bonded atoms, the reference plane contains the double bonded atoms and is perpendicular to the plane containing these atoms and those directly attached to them. It is customary to draw the formulas so that the reference plane is perpendicular to that of the paper. For cyclic compounds the reference plane is that in which the ring skeleton lies or to which it approximates. Cyclic structures are commonly drawn with the ring atoms in the plane of the paper.

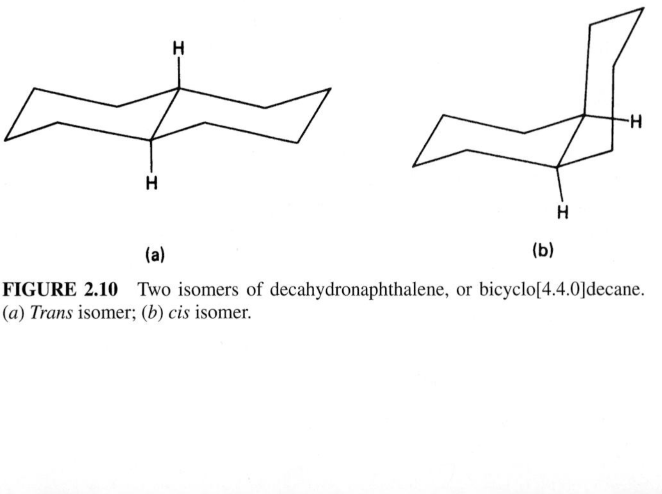

FIGURE 2.10 Two isomers of decahydronaphthalene, or bicyclo[4.4.0]decane. (*a*) *Trans* isomer; (*b*) *cis* isomer.

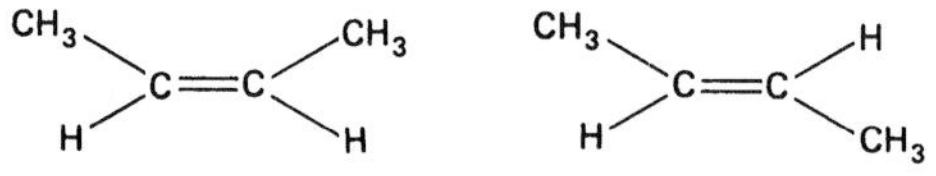

(a) (b)

FIGURE 2.11 Two isomers of 2-butene. (*a*) *Cis* isomer, bp 3.8°C, mp −138.9°C, dipole moment 0.33 D; (*b*) *trans* isomer, bp 0.88°C, mp −105.6°C, dipole moment 0 D.

2.1.4.5 Sequence Rules for Geometric Isomers and Chiral Compounds. Although *cis* and *trans* designations have been used for many years, this approach becomes useless in complex systems. To eliminate confusion when each carbon of a double bond or a chiral center is connected to different groups, the Cahn, Ingold, and Prelog system for designating configuration about a double bond or a chiral center has been adopted by IUPAC. Groups on each carbon atom of the double bond are assigned a first (1) or second (2) priority. Priority is then compared at one carbon relative to the other. When both first priority groups are on the *same side* of the double bond, the configuration is designated as Z (from the German *zusammen*, "together"), which was formerly *cis*. If the first priority groups are on *opposite sides* of the double bond, the designation is *E* (from the German *entgegen*, "in opposition to"), which was formerly *trans*. (See Fig. 2.12).

When a molecule contains more than one double bond, each *E* or *Z* prefix has associated with it the lower-numbered locant of the double bond concerned. Thus (see also the rules that follow)

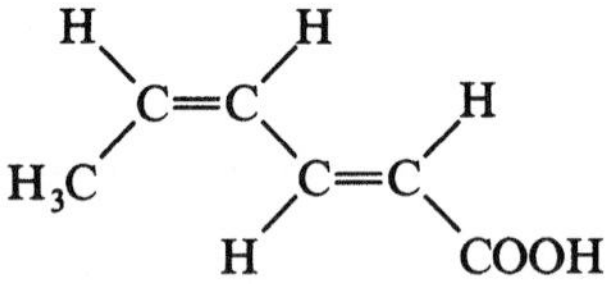

(2*E*,4*Z*)-2,4-Hexadienoic acid

When the sequence rules permit alternatives, preference for lower-numbered locants and for inclusion in the principal chain is allotted as follows in the order stated: *Z* over *E* groups and *cis* over *trans* cyclic groups. If a choice is still not attained, then the lower-numbered locant for such a preferred group at the first point of difference is the determining factor. For example,

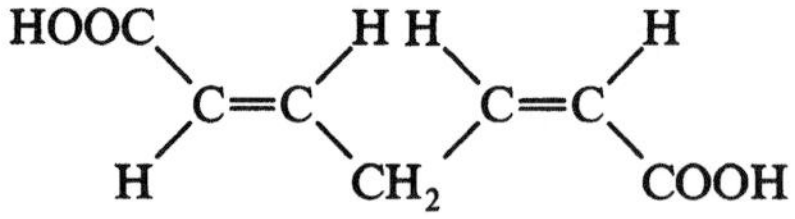

(2*Z*,5*E*)-2,5-Heptadienedioic acid

Rule 1. Priority is assigned to atoms on the basis of atomic number. Higher priority is assigned to atoms of higher atomic number. If two atoms are isotopes of the same element, the atom of higher mass number has the higher priority. For example, in 2-butene, the carbon atom of each methyl group receives first priority over the hydrogen atom connected to the same carbon atom. Around the asymmetric carbon atom in chloroiodomethanesulfonic acid, the priority sequence is I, Cl, S, H. In 1-bromo-1-deuteroethane, the priority sequence is Cl, C, D, H.

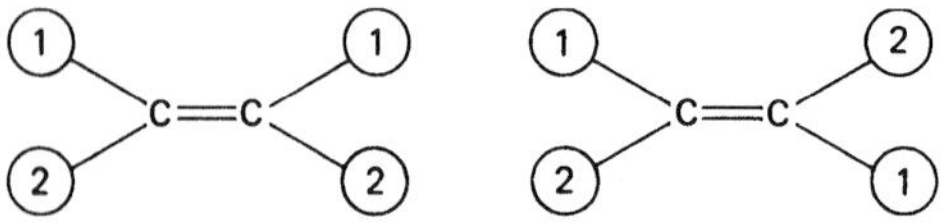

(a) (b)

FIGURE 2.12 Configurations designated by priority groups. (*a*) *Z* (*cis*); (*b*) *E* (*trans*).

Rule 2. When atoms attached directly to a double-bonded carbon have the same priority, the second atoms are considered and so on, if necessary, working outward once again from the double bond or chiral center. For example, in 1-chloro-2-methylbutene, in CH_3 the second atoms are H, H, H and in CH_2CH_3 they are C, H, H. Since carbon has a higher atomic number than hydrogen, the ethyl group has the next highest priority after the chlorine atom.

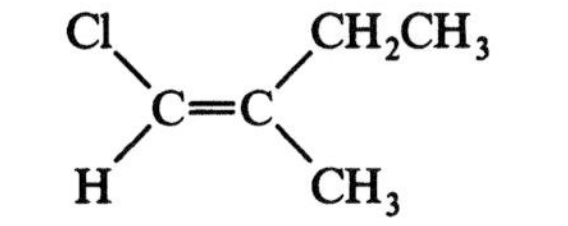

(*Z*)-1-Chloro-2-methylbutene

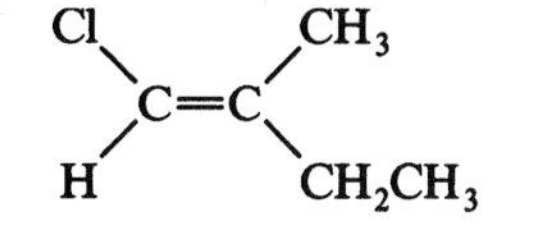

(*E*)-1-Chloro-2-methylbutene

Rule 3. When groups under consideration have double or triple bonds, the multiple-bonded atom is replaced conceptually by two or three single bonds to that same kind of atom. Thus, $=$A is considered to be equivalent to two A's, $\langle^{A}_{A}$ or and $\equiv$A equals $\overset{A}{\underset{A}{\leftarrow}}$A. However, a real $\langle^{A}_{A}$ has priority over $=$A; likewise a real $\overset{A}{\underset{A}{\leftarrow}}$A has priority over $\equiv$A. Actually, both atoms of a multiple bond are duplicated, or triplicated, so that C $=$ O is treated as $\begin{matrix} C-O \\ | \quad | \\ O \quad C \end{matrix}$, that is $\begin{matrix} C-O \\ | \\ (O) \end{matrix}$ and $\begin{matrix} C-O \\ | \\ (C) \end{matrix}$, and C $\equiv$ N is treated as (N)/C——N\\(C) with (N) on C and (C) on N, i.e. $\underset{(N)\ \ (N)}{C}$——$\underset{(C)\ \ (C)}{N}$. A phenyl carbon becomes —C(CH)(C)(CH).
Only the double-bonded atoms themselves are duplicated, not the atoms or groups attached to them. The duplicated atoms (or phantom atoms) may be considered as carrying atomic number zero. For example, among the groups OH, CHO, CH_2OH, and H, the OH group has the highest priority, and the C(O, O, H) of CHO takes priority over the C(O, H, H) of CH_2OH.

2.1.4.6 Chirality and Optical Activity. A compound is chiral (the term *dissymmetric* was formerly used) if it is not superimposable on its mirror image. A chiral compound does not have a plane of symmetry. Each chiral compound possesses one (or more) of three types of chiral element, namely, a chiral center, a chiral axis, or a chiral plane.

2.1.4.7 Chiral Center. The chiral center, which is the chiral element most commonly met, is exemplified by an asymmetric carbon with a tetrahedral arrangement of ligands about the carbon. The ligands comprise four different atoms or groups. One "ligand" may be a lone pair of electrons; another, a phantom atom of atomic number zero. This situation is encountered in sulfoxides or with a nitrogen atom. Lactic acid is an example of a molecule with an asymmetric (chiral) carbon. (See Fig. 2.13.)

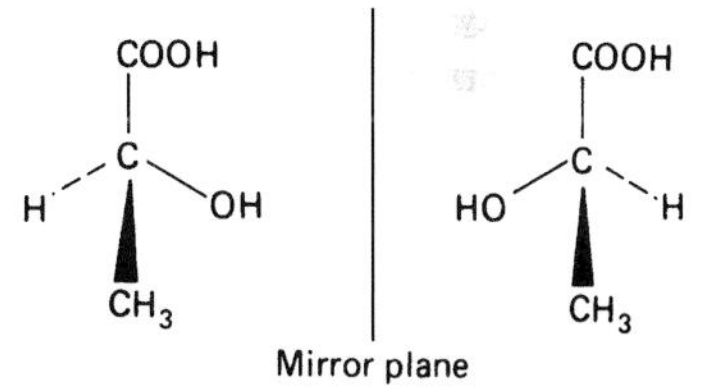

FIGURE 2.13 Asymmetric (chiral) carbon in the lactic acid molecule.

A simpler representation of molecules containing asymmetric carbon atoms is the Fischer projection, which is shown here for the same lactic acid configurations. A Fischer projection involves

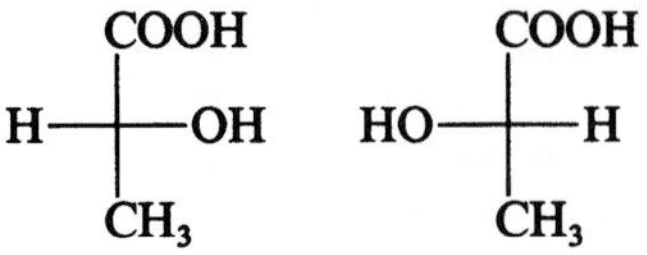

drawing a cross and attaching to the four ends the four groups that are attached to the asymmetric carbon atom. The asymmetric carbon atom is understood to be located where the lines cross. The horizontal lines are understood to represent bonds coming toward the viewer out of the plane of the paper. The vertical lines represent bonds going away from the viewer behind the plane of the paper as if the vertical line were the side of a circle. The principal chain is depicted in the vertical direction;

the lowest-numbered (locant) chain member is placed at the top position. These formulas may be moved sideways or rotated through 180° in the plane of the paper, but they may not be removed from the plane of the paper (i.e., rotated through 90°). In the latter orientation it is essential to use thickened lines (for bonds coming toward the viewer) and dashed lines (for bonds receding from the viewer) to avoid confusion.

2.1.4.8 Enantiomers. Two nonsuperimposable structures that are mirror images of each other are known as *enantiomers*. Enantiomers are related to each other in the same way that a right hand is related to a left hand. Except for the direction in which they rotate the plane of polarized light, enantiomers are identical in all physical properties. Enantiomers have identical chemical properties except in their reactivity toward optically active reagents.

Enantiomers rotate the plane of polarized light in opposite directions but with equal magnitude. If the light is rotated in a clockwise direction, the sample is said to be dextrorotatory and is designed as (+). When a sample rotates the plane of polarized light in a counterclockwise direction, it is said to be levorotatory and is designed as (−). Use of the designations *d* and *l* is discouraged.

2.1.4.9 Specific Rotation. Optical rotation is caused by individual molecules of the optically active compound. The amount of rotation depends upon how many molecules the light beam encounters in passing through the tube. When allowances are made for the length of the tube that contains the sample and the sample concentration, it is found that the amount of rotation, as well as its direction, is a characteristic of each individual optically active compound.

Specific rotation is the number of degrees of rotation observed if a 1-dm tube is used and the compound being examined is present to the extent of 1 g per 100 mL. The density for a pure liquid replaces the solution concentration.

$$\text{Specific rotation} = [\alpha] = \frac{\text{observed rotation (degrees)}}{\text{length (dm)} \times \text{(g/100 ml)}}$$

The temperature of the measurement is indicated by a superscript and the wavelength of the light employed by a subscript written after the bracket; for example, $[\alpha]^{20}_{590}$ implies that the measurement was made at 20°C using 590-nm radiation.

2.1.4.10 Optically Inactive Chiral Compounds. Although chirality is a necessary prerequisite for optical activity, chiral compounds are not necessarily optically active. With an equal mixture of two enantiomers, no net optical rotation is observed. Such a mixture of enantiomers is said to be *racemic* and is designated as (±) and not as *dl*. Racemic mixtures usually have melting points higher than the melting point of either pure enantiomer.

A second type of optically inactive chiral compounds, *meso* compounds, will be discussed in the next section.

2.1.4.11 Multiple Chiral Centers. The number of stereoisomers increases rapidly with an increase in the number of chiral centers in a molecule. A molecule possessing two chiral atoms should have four optical isomers, that is, four structures consisting of two pairs of enantiomers. However, if a compound has two chiral centers but both centers have the same four substituents attached, the total number of isomers is three rather than four. One isomer of such a compound is not chiral because it is identical with its mirror image; it has an internal mirror plane. This is an example of a diastereomer. The achiral structure is denoted as a *meso* compound. Diastereomers have different physical and chemical properties from the optically active enantiomers. Recognition of a plane of symmetry is usually the easiest way to detect a *meso* compound. The stereoisomers of tartaric acid are examples of compounds with multiple chiral centers (see Fig. 2.14), and one of its isomers is a *meso* compound.

When the asymmetric carbon atoms in a chiral compound are part of a ring, the isomerism is more complex than in acyclic compounds. A cyclic compound which has two different asymmetric carbons with different sets of substituent groups attached has a total of $2^2 = 4$ optical isomers: an enantiometric pair of *cis* isomers and an enantiometric pair of *trans* isomers. However, when the two

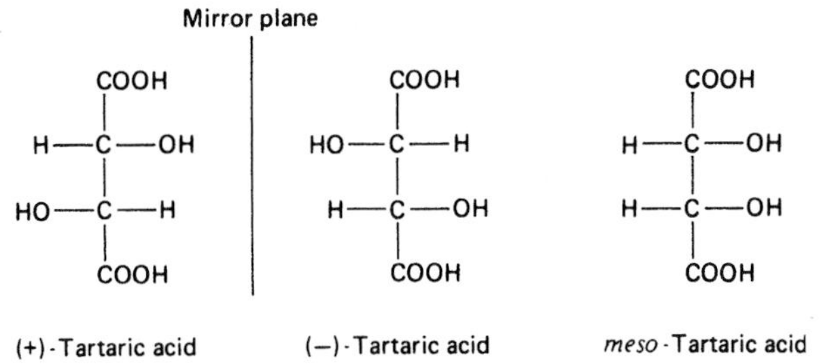

FIGURE 2.14 Isomers of tartaric acid.

asymmetric centers have the same set of substituent groups attached, the *cis* isomer is a *meso* compound and only the *trans* isomer is chiral. (See Fig. 2.15).

2.1.4.12 Torsional Asymmetry. Rotation about single bonds of most acyclic compounds is relatively free at ordinary temperatures. There are, however, some examples of compounds in which nonbonded interactions between large substitutent groups inhibit free rotation about a sigma bond. In some cases these compounds can be separated into pairs of enantiomers.

A *chiral axis* is present in chiral biaryl derivatives. When bulky groups are located at the *ortho* positions of each aromatic ring in biphenyl, free rotation about the single bond connecting the two rings is inhibited because of torsional strain associated with twisting rotation about the central single bond. Interconversion of enantiomers is prevented (see Fig. 2.16).

For compounds possessing a chiral axis, the structure can be regarded as an elongated tetrahedron to be viewed along the axis. In deciding upon the absolute configuration it does not matter from which end it is viewed; the nearer pair of ligands receives the first two positions in the order of precedence (see Fig. 2.17).

A *chiral plane* is exemplified by the plane containing the benzene ring and the bromine and oxygen atoms in the chiral compound (see Fig. 2.18). Rotation of the benzene ring around the oxygen-to-ring single bonds is inhibited when x is small (although no critical size can be reasonably established).

2.1.4.13 Absolute Configuration. The terms absolute stereochemistry and absolute configuration are used to describe the three-dimensional arrangement of substituents around a chiral element. A general system for designating absolute configuration is based upon the priority system and sequence rules. Each group attached to a chiral center is assigned a number, with number one the highest-priority group. For example, the groups attached to the chiral center of 2-butanol (see Fig. 2.19) are assigned

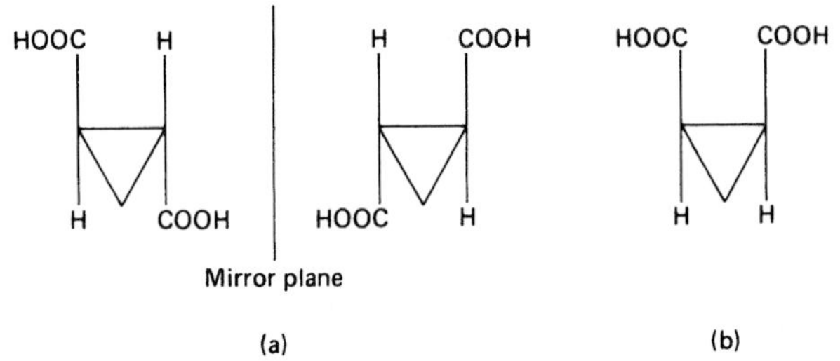

FIGURE 2.15 Isomers of cyclopropane-1,2-dicarboxylic acid. (*a*) *Trans* isomer; (*b*) *meso* isomer.

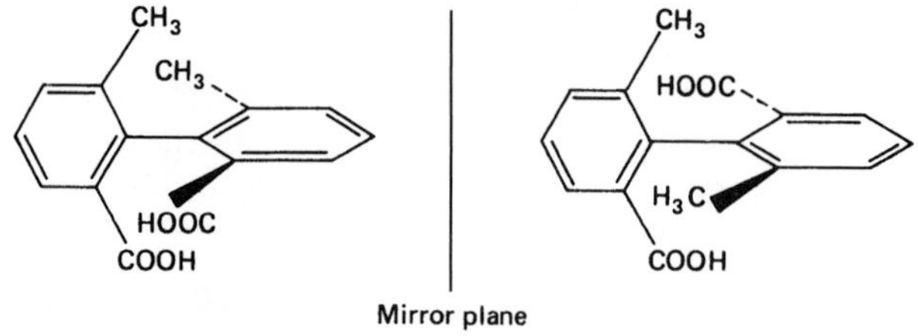

FIGURE 2.16 Isomers of biphenyl compounds with bulky groups attached at the *ortho* positions.

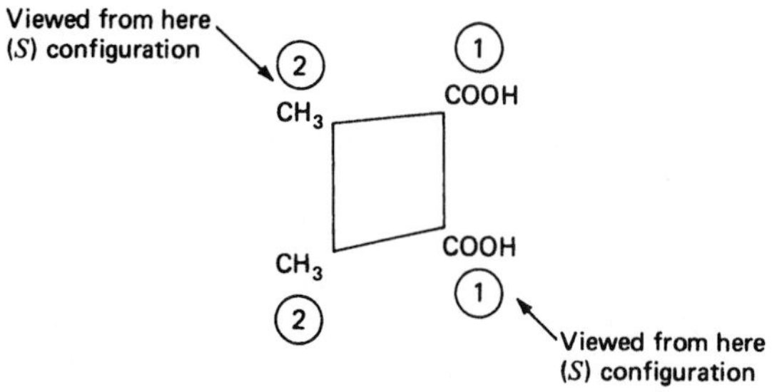

FIGURE 2.17 Example of a chiral axis.

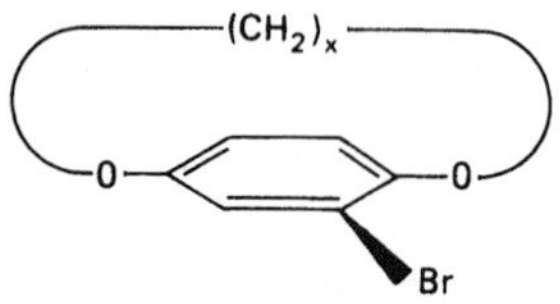

FIGURE 2.18 Example of a chiral plane.

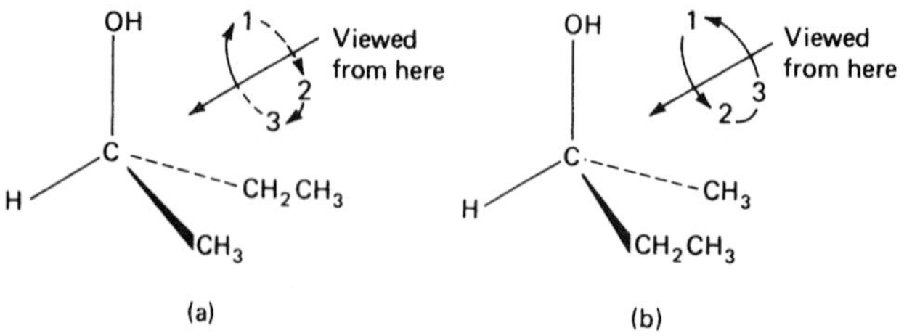

FIGURE 2.19 Viewing angle as a means of designating the absolute configuration of compounds with a chiral axis. (*a*) (*R*)-2-Butanol (sequence clockwise); (*b*) (*S*)-2-butanol (sequence counterclockwise).

these priorities: 1 for OH, 2 for CH_2CH_3, 3 for CH_3, and 4 for H. The molecule is then viewed from the side opposite the group of lowest priority (the hydrogen atom), and the arrangement of the remaining groups is noted. If, in proceeding from the group of highest priority to the group of second priority and thence to the third, the eye travels in a clockwise direction, the configuration is specified *R* (from the Latin *rectus*, "right"); if the eye travels in a counterclockwise direction, the configuration is specified *S* (from the Latin *sinister*, "left"). The complete name includes both configuration and direction of optical rotation, as for example, (*S*)-(+)-2-butanol.

The relative configurations around the chiral centers of many compounds have been established. One optically active compound is converted to another by a sequence of chemical reactions which are stereospecific; that is, each reaction is known to proceed spatially in a specific way. The configuration of one chiral compound can then be related to the configuration of the next in sequence. In order to establish absolute configuration, one must carry out sufficient stereospecific reactions to relate a new compound to another of known absolute configuration. Historically the configuration of D-(+)-2,3-dihydroxypropanal has served as the standard to which all configuration has been compared. The absolute configuration assigned to this compound has been confirmed by an X-ray crystallographic technique.

2.1.5 Amino Acids

An *amino acid* is an organic compound containing an amine group ($-NH_2$) and a carboxylic acid group ($-CO_2H$) in the same molecule. While there are many forms of amino acids, all of the important amino acids found in living organisms are alpha-amino acids. Alpha amino acids have the carboxylic acid group and the amino group attached to the same carbon atom.

The simplest amino acid is glycine (H_2NCH_2COOH) and contains no asymmetric carbon atoms (tetrahedral carbon atoms with four different groups attached). All of the other amino acids contain an asymmetric carbon atom and are therefore optically active. Under physiological aqueous conditions a proton transfer from the acid to the base occurs, forming a dipolar ion or zwitterion, because

TABLE 2.13 Formula and Nomenclature of Amino Acids

Name	Abbr.	Linear structural formula
Alanine	ala	$CH_3—CH(NH_2)—COOH$
Arginine	arg	$HN{=}C(NH_2)—NH—(CH_2)_3\text{-}CH(NH_2)\text{-}COOH$
Asparagine	asn	$H_2N—CO—CH_2—CH(NH_2)—COOH$
Aspartic acid	asp	$HOOC—CH_2—CH(NH_2)—COOH$
Cysteine	cys	$HS—CH_2—CH(NH_2)—COOH$
Glutamine	gln	$H_2N—CO—(CH_2)_2—CH(NH_2)—COOH$
Glutamic acid	glu	$HOOC—(CH_2)_2—CH(NH_2)—COOH$
Glycine	gly	$NH_2—CH_2—COOH$
Histidine	his	$NH\text{-}CH{=}N—CH{=}C—CH_2CH(NH_2)—COOH$ (ring: NH joined to C)
Isoleucine	ile	$CH_3\text{-}CH_2\text{-}CH(CH_3)—CH(NH_2)—COOH$
Leucine	leu	$(CH_3)_2—CH—CH_2—CH(NH_2)—COOH$
Lysine	lys	$H_2N—(CH_2)_4—CH(NH_2)—COOH$
Methionine	met	$CH_3—S—(CH_2)_2—CH(NH_2)—COOH$
Phenylalanine	phe	$C_6H_5—CH_2—CH(NH_2)—COOH$
Proline	pro	$NH—(CH_2)_3—CH\text{-}COOH$ (ring: NH joined to CH)
Serine	ser	$HO—CH_2—CH(NH_2)—COOH$
Threonine	thr	$CH_3—CH(OH)—CH(NH_2)—COOH$
Tryptophan	trp	$C_6H_4—NH—CH{=}C—CH_2—CH(NH_2)—COOH$ (ring: C_6H_4 joined to C)
Tyrosine	tyr	$HO—p—C_6H_4—CH_2—CH(NH_2)—COOH$
Valine	val	$(CH_3)_2—CH—CH(NH_2)—COOH$

TABLE 2.14 Acid-Base Properties of Amino Acids

Amino acid	pK_{a1}*	pK_{a2}*	pl
Glycine	2.34	9.60	5.97
Alanine	2.34	9.69	6.00
Valine	2.32	9.62	5.96
Leucine	2.36	9.60	5.98
Isoleucine	2.36	9.60	6.02
Methionine	2.28	9.21	5.74
Proline	1.99	10.60	6.30
Phenylalanine	1.83	9.13	5.48
Tryptophan	2.83	9.39	5.89
Asparagine	2.02	8.80	5.41
Glutamine	2.17	9.13	5.65
Serine	2.21	9.15	5.68
Threonine	2.09	9.10	5.60
Tyrosine	2.20	9.11	5.66

*In all cases pK_{a1} corresponds to ionization of the carboxyl group; pK_{a2} corresponds to deprotonation of the ammonium ion.

the carboxylic acid is a much stronger acid than is the ammonium ion. The actual structure of glycine in solution, for example, is $^{+}H_3NCH_2COO^{-}$ at pH 7 rather than H_2NCH_2COOH. At very low pH the acid group can be protonated and at very high pH the ammonium group can be deprotonated, but the forms of amino acids relevant to living organisms are the zwitterions.

TABLE 2.15 Acid-Base Properties of Amino Acids with Ionizable Side Chains

Amino acid	pK_{a1}*	pK_{a2}	pK_a of side chain	pl
Aspartic acid	1.88	9.60	3.65	2.77
Glutamic acid	2.19	9.67	4.25	3.22
Lysine	2.18	8.95	10.53	9.74
Arginine	2.17	9.04	12.48	10.76
Histidine	1.82	9.17	6.00	7.59

*In all cases pK_{a1} corresponds to ionization of the carboxyl group of $RCH(\overset{+}{N}H_3)CO_2H$, and pK_{a2} to ionization of the ammonium ion.

2.1.6 Carbohydrates

Carbohydrates consist of the elements carbon, hydrogen, and oxygen. In their basic form, carbohydrates are simple sugars or *monosaccharides*. These simple sugars can combine with each other to form more complex carbohydrates. The combination of two simple sugars is a *disaccharide*. Carbohydrates consisting of two to ten simple sugars are called *oligosaccharides*, and those with a larger number are called *polysaccharides*.

2.1.6.1 ***Sugars.*** Sugars are white crystalline carbohydrates that are soluble in water and generally have a sweet taste. Monosaccharides are simple sugars

The classification system of monosaccharides is based on the number of carbons in the sugar:

Number of carbon atoms	Category name	Examples
4	Tetrose	Erythrose, Threose
5	Pentose	Arabinose, Ribose, Ribulose, Xylose, Xylulose, Lyxose
6	Hexose	Allose, Altrose, Fructose, Galactose, Glucose, Gulose, Idose, Mannose, Sorbose, Talose
7	Heptose	Sedoheptulose

Many saccharide structures differ only in the orientation of the hydroxyl groups (-OH). This slight structural difference makes a big difference in the biochemical properties, organoleptic properties (e.g., taste), and in the physical properties such as melting point and Specific Rotation (how polarized light is distorted). A chain-form monosaccharide that has a carbonyl group (C═O) on an end carbon forming an aldehyde group (-CHO) is classified as an *aldose*. When the carbonyl group is on an inner atom forming a ketone, it is classified as a *ketose*.

2.1.6.1.1 Tetroses

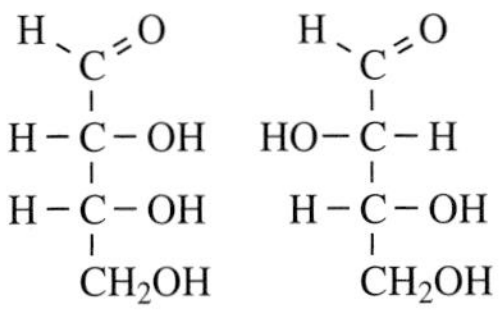

D-Erythrose D-Threose

2.1.6.1.2 Pentoses The ribose structure is a component of deoxyribonucleic acid (DNA) and ribonucleic acids (RNA).

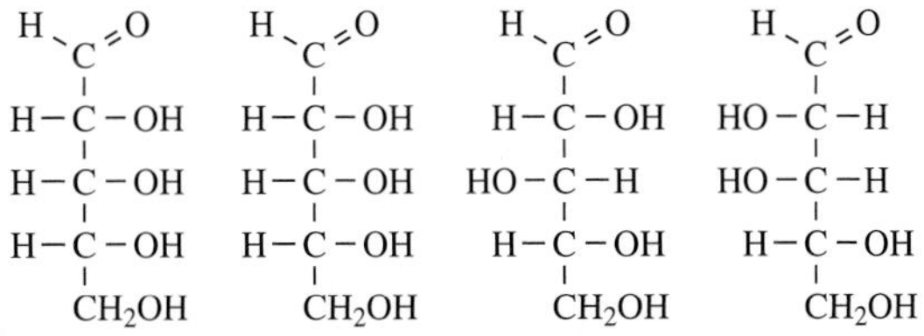

D-Ribose D-Arabinose D-Xylose D-Lyxose

2.1.6.1.3 Hexoses. Hexoses, such as the ones illustrated here, have the molecular formula $C_6H_{12}O_6$.

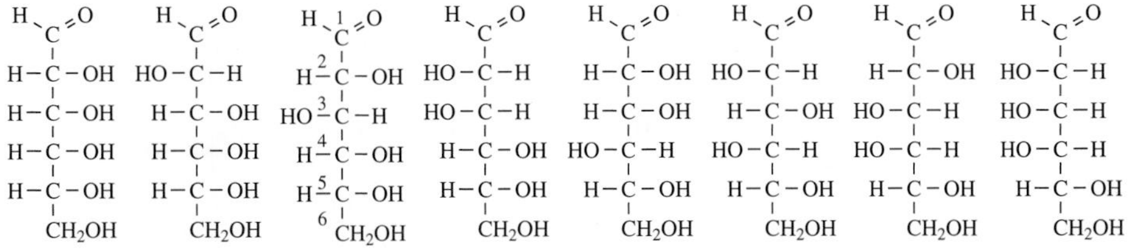

D-Allose D-Altrose D-Glucose D-Mannose D-Gulose D-Idose D-Galactose D-Talose

Structures that have opposite configurations of a hydroxyl group at only one position, such as glucose and mannose, are called *epimers*.

Glucose, also called dextrose, is the most widely distributed sugar in the plant and animal kingdoms and it is the sugar present in blood as "blood sugar". The chain form of glucose is a polyhydric aldehyde, meaning that it has multiple hydroxyl groups and an aldehyde group. Fructose, also called levulose, is shown here in the chain and ring forms.

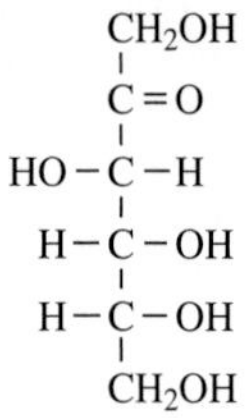

D-Fructose (a ketose)

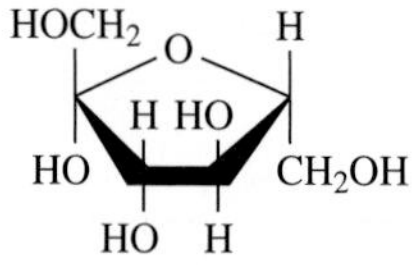

Fructose

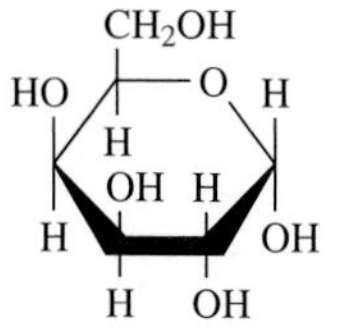

Galactose

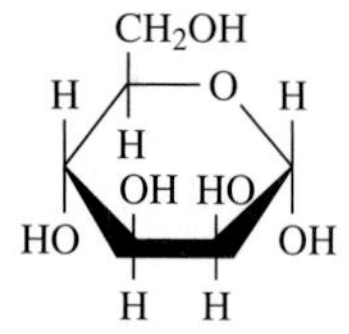

Mannose

2.1.6.1.4 Heptoses Sedoheptulose has the same structure as fructose, but it has one extra carbon.

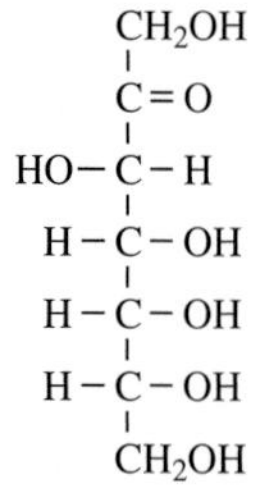

D-Sedoheptulose

2.1.6.1.5 Chain and Ring Structure. Many simple sugars can exist in a chain form or a ring form, as illustrated by the hexoses above. The ring form is favored in aqueous solutions, and the mechanism of ring formation is similar for most sugars. The glucose ring form is created when the oxygen on carbon number 5 links with the carbon comprising the carbonyl group (carbon number 1) and transfers its hydrogen to the carbonyl oxygen to create a hydroxyl group. The rearrangement produces *alpha*-glucose when the hydroxyl group is on the opposite side of the -CH_2OH group, or *beta*-glucose when the hydroxyl group is on the same side as the -CH_2OH group. Isomers that differ only in their configuration about their carbonyl carbon atom are called *anomers*.

The symbol 'd' (or 'D') is used to indicate that the shows that a sugar is *dextrorotary*, i.e., it rotates polarized light to the right, but can also denote a specific configuration. On the other hand, the symbol 'l' (or 'L') indicates that the sugar is *laevorotatory*, i.e., it rotates polarized light to the left. Again the symbol may be used to indicate a specific configuration.

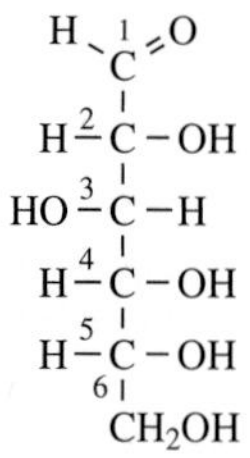

d-Glucose (an aldose)

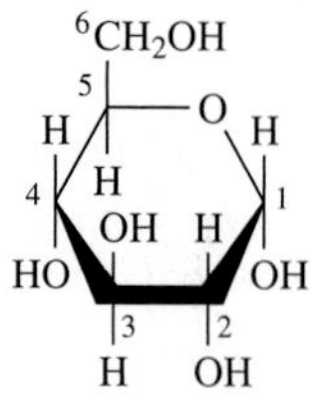

α-d-Glucose

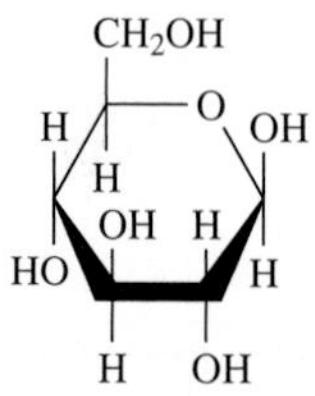

β-d-Glucose

2.1.6.2 *Stereochemistry.* Saccharides with identical functional groups but with different spatial configurations have different chemical and biological properties. Stereochemistry is the study of the arrangement of atoms in three-dimensional space. Stereoisomers are compounds in which the atoms are linked in the same order but differ in their spatial arrangement. Compounds that are mirror images of each other but are not identical are called *enantiomers*. The following structures illustrate the difference between *β*-D-glucose and *β*-L-glucose. Identical molecules can be made to correspond to each other by flipping and rotating. However, enantiomers cannot be made to correspond to their mirror images by flipping and rotating. Glucose is sometimes illustrated as a "chair form" because it is a more accurate representation of the bond angles of the molecule.

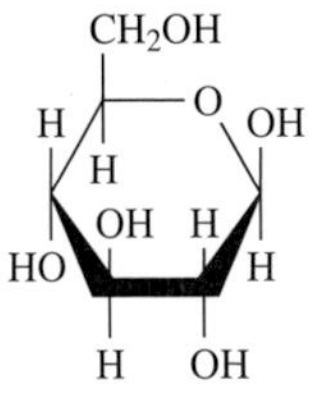

β-d-Glucose

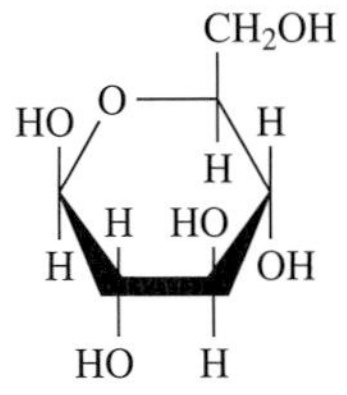

β-l-Glucose

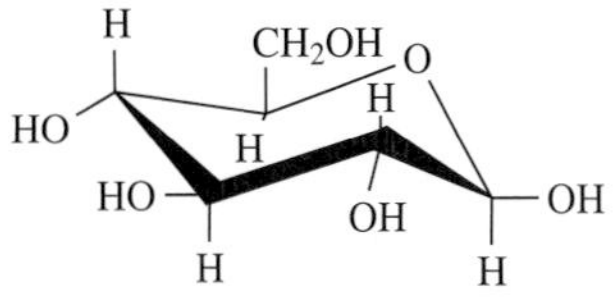

β-d-Glucose (chair form)

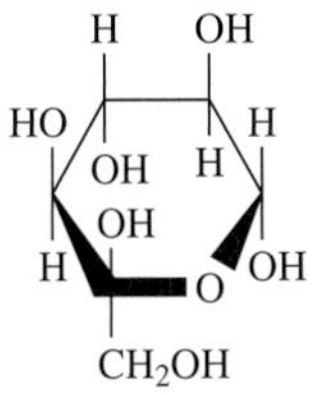

β-d-Glucose

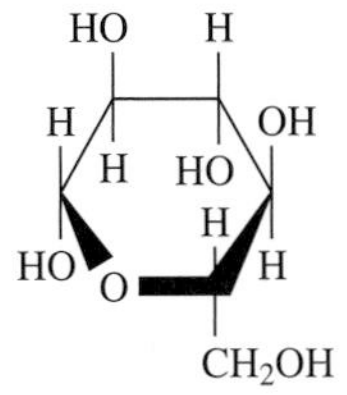

β-l-Glucose

2.1.6.3 *Sugar Alcohols, Amino Sugars, and Uronic Acids.* Sugars may be modified by natural or laboratory processes into compounds that retain the basic configuration of saccharides, but have different functional groups. *Sugar alcohols*, also known as polyols, polyhydric alcohols, or polyalcohols, are the hydrogenated forms of the aldoses or ketoses. For example, glucitol, also known as sorbitol, has the same linear structure as the chain form of glucose, but the aldehyde (-CHO) group is replaced with a -CH_2OH group. Other common sugar alcohols include the monosaccharides erythritol and xylitol and the disaccharides lactitol and maltitol. Sugar alcohols have about half the calories of sugars and are frequently used in low-calorie or "sugar-free" products.

Amino sugars or aminosaccharides replace a hydroxyl group with an amino (-NH_2) group. Glucosamine is an amino sugar used to treat cartilage damage and reduce the pain and progression of arthritis.

Uronic acids have a carboxyl group (-COOH) on carbon number six.

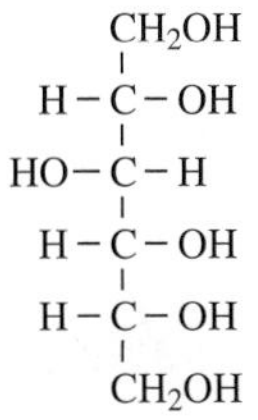

Glucitol
Sorbitol
(a sugar alcohol)

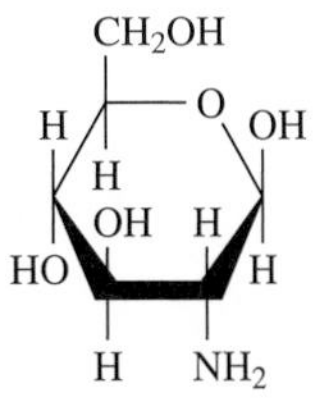

or Glucosamine
(an amino sugar)

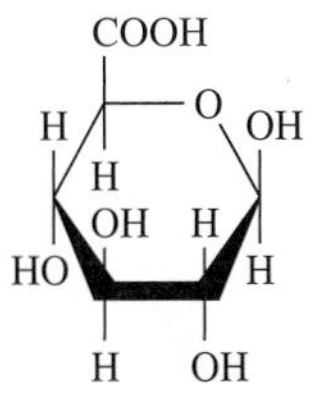

Glucuronic acid
(a uronic acid)

2.1.6.3 Disaccharides. Disaccharides consist of two simple sugars and the common disaccharides are sucrose, lactose, and maltose.

Disaccharide	Description	Component monosaccharides
Sucrose	common table sugar	Glucose + fructose
Lactose	main sugar in milk	galactose + glucose
Maltose	product of starch hydrolysis	glucose + glucose

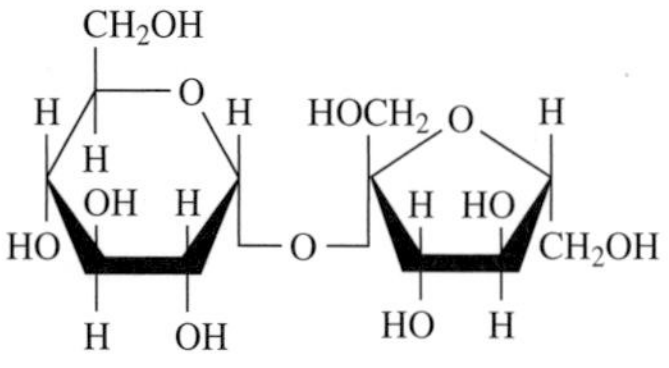

Sucrose

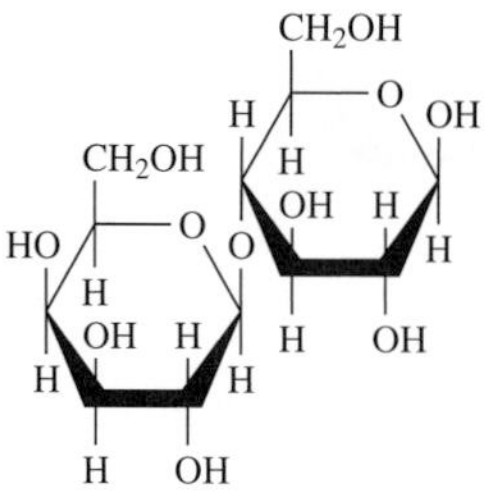

Lactose

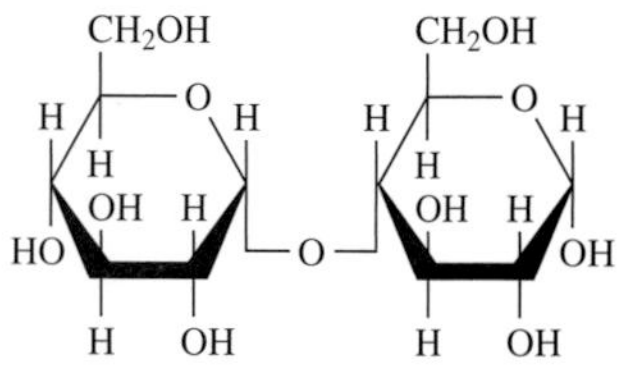

Maltose

Lactose has a molecular structure consisting of galactose and glucose. It is of interest because it is associated with lactose intolerance, which is the intestinal distress caused by a deficiency of lactase, an intestinal enzyme needed to absorb and digest lactose in milk. Undigested lactose ferments in the colon and causes abdominal pain, bloating, gas, and diarrhea. Yogurt does not cause these problems because lactose is consumed by the bacteria that transform milk into yogurt.

Maltose consists of two α-D-glucose molecules with the alpha bond at carbon 1 of one molecule attached to the oxygen at carbon 4 of the second molecule. This is called a $1\alpha \rightarrow 4$ linkage.

Cellobiose is a disaccharide consisting of two β-D-glucose molecules that have a $1\beta \rightarrow 4$ linkage. Cellobiose has no taste, whereas maltose is about one-third as sweet as sucrose.

2.1.6.4 **Polysaccharides.** Polysaccharides are polymers of simple sugars but, unlike sugars, polysaccharides are insoluble in water.

2.1.6.4.1 Starch. Starch is the major form of stored carbohydrate in plants. Starch is composed of a mixture of two substances: *amylose*, an essentially linear polysaccharide, and *amylopectin*, a highly branched polysaccharide. Both forms of starch are polymers of α-d-glucose. Natural starch contains 10–20% amylose and 80–90% amylopectin.

Amylose molecules consist typically of 200 to 20,000 glucose units that form a helix as a result of the bond angles between the glucose units.

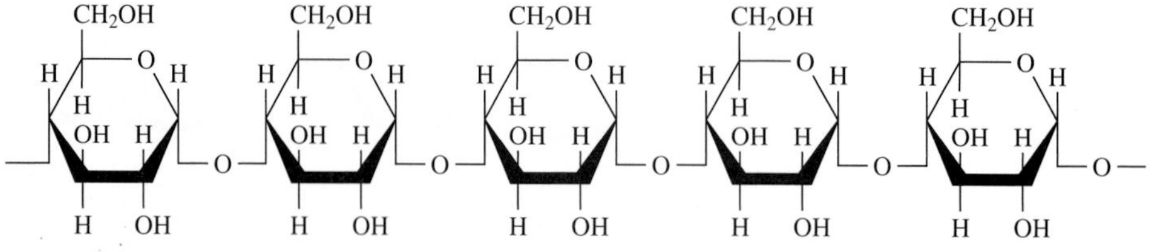

Amylose

Amylopectin differs from amylose in being highly branched. Short side chain of about 30 glucose units are attached approximately every twenty to thirty glucose units along the chain. Amylopectin molecules may contain up to two million glucose units.

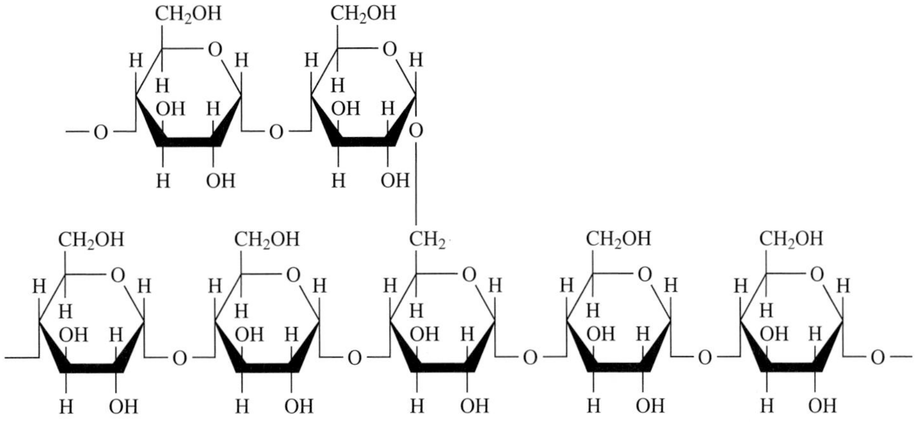

Amylopectin

Starches are transformed into many commercial products by hydrolysis with acids or enzymes. The resulting products are assigned a Dextrose Equivalent (DE) value that is related to the degree of hydrolysis. A DE value of 100 corresponds to completely hydrolyzed starch, which is pure glucose (dextrose). Maltodextrins are not sweet and have DE values less than 20. Syrups, such as corn syrup, have DE values from 20 to 95. "High fructose corn syrup," commonly used to sweeten soft drinks, is made by enzymatically isomerizing a portion of the glucose into fructose, which is about twice as sweet as glucose.

2.1.6.4.2 Glycogen. Glucose is stored as glycogen in animal tissues by the process of glycogenesis. When glucose cannot be stored as glycogen or used immediately for energy, it is converted to fat. Glycogen is a polymer of α-d-glucose identical to amylopectin, but the branches in glycogen tend to be shorter (about 13 glucose units) and more frequent. The glucose chains are organized globularly, like the branches of a tree, surrounding a pair of molecules of glycogenin, a protein with a molecular weight of 38,000 that acts as a primer at the core of the structure. Glycogen is easily converted back to glucose to provide energy.

2.1.6.4.2 Cellulose. Cellulose is a polymer of β-d-glucose, which in contrast to starch, is oriented with -CH_2OH groups alternating above and below the plane of the cellulose molecule thus producing long, unbranched chains. The absence of side chains allows cellulose molecules to lie close together and form rigid structures. Cellulose is the major structural material of plants. Wood is largely cellulose, and cotton is almost pure cellulose. Cellulose can be hydrolyzed to its constituent glucose units by microorganisms that inhabit the digestive tract of termites and ruminants. Cellulose may be modified in the laboratory by treating it with nitric acid (HNO_3) to replace all the hydroxyl groups with nitrate groups (-ONO_2) to produce cellulose nitrate that is an explosive component of smokeless powder.

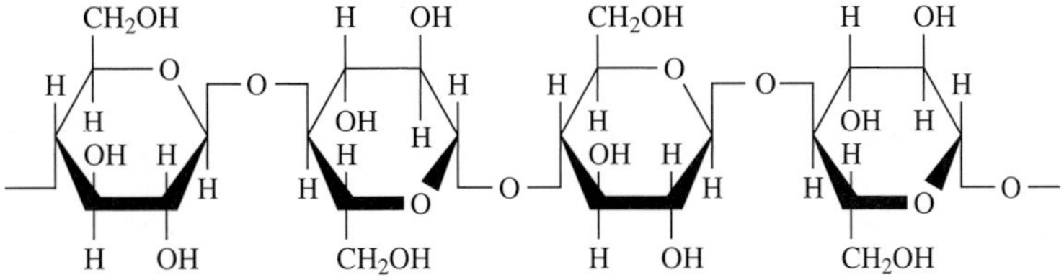

Cellulose

2.1.7 Miscellaneous Compounds

TABLE 2.16 Representative Terpenes

Monoterpenes

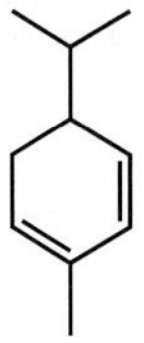

α-Phellandrene
(eucalyptus)

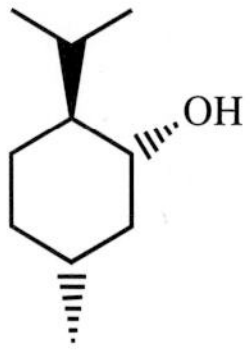

Menthol
(peppermint)

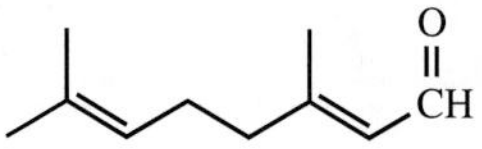

Citral
(lemon grass)

Sesquiterpenes

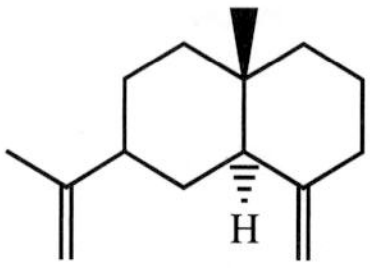

α-Selinene
(celery)

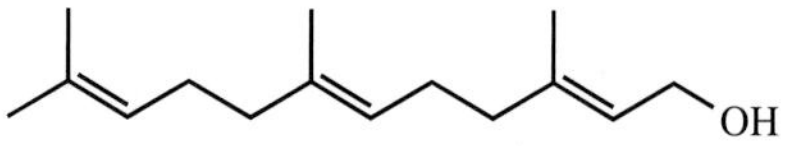

Farnesol
(ambrette)

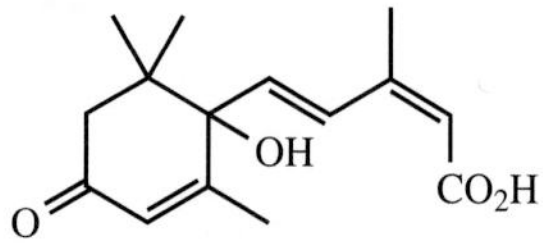

Abscisic acid
(a plant hormone)

Diterpenes

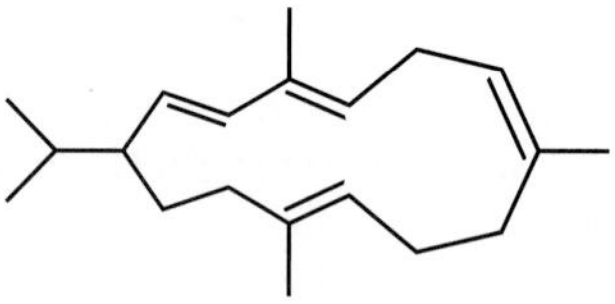

Cembrene
(pine)

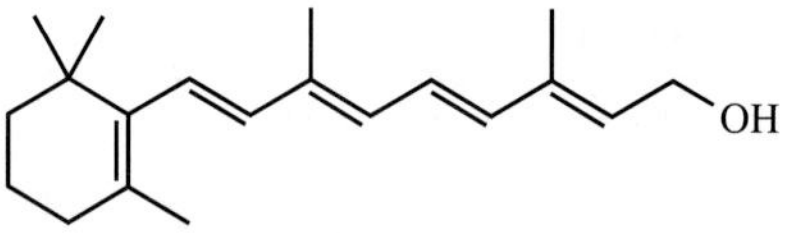

Vitamin A
(present in mammalian tissue and fish oil;
important substance in the chemistry of vision)

Triterpenes

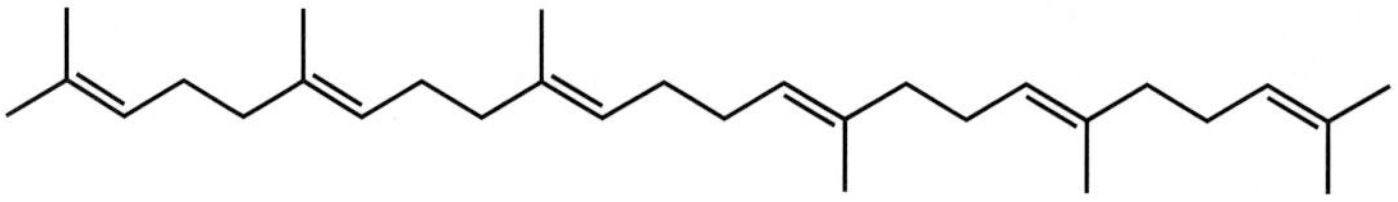

Squalene
(shark liver oil)

Tetraterpenes

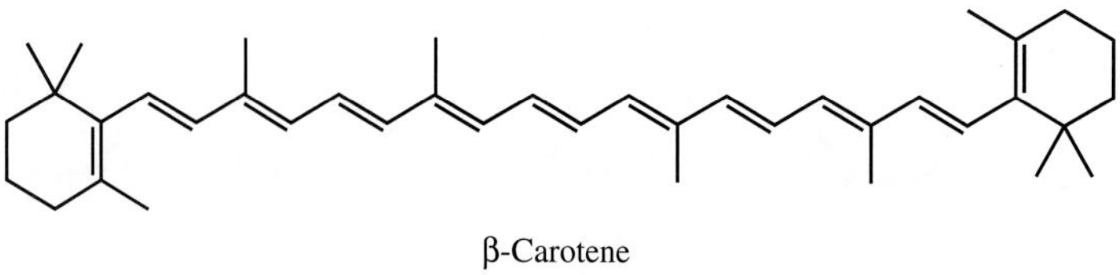

β-Carotene
(present in carrots and other vegetables;
enzymes in the body cleave β-carotene to vitamin A)

TABLE 2.17 Representative Fatty Acids

Number of Carbons	Common name	Systematic name	Structural formula	Melting point °C
Saturated fatty acids				
12	Lauric acid	Dodecanoic acid	$CH_3(CH_2)_{10}CO_2H$	44
14	Myristic acid	Tetradecanoic acid	$CH_3(CH_2)_{12}CO_2H$	58
16	Palmitic acid	Hexadecanoic acid	$CH_3(CH_2)_{14}CO_2H$	63
18	Stearic acid	Octadecanoic acid	$CH_3(CH_2)_{16}CO_2H$	69
20	Arachidic acid	Icosanoic acid	$CH_3(CH_2)_{18}CO_2H$	75
Unsaturated fatty acids				
18	Oleic acid	*cis*-9-Octadecenoic acid	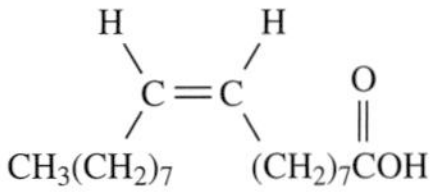	4
18	Linoleic acid	*cis,cis*-9, 12-Octadecadienoic acid	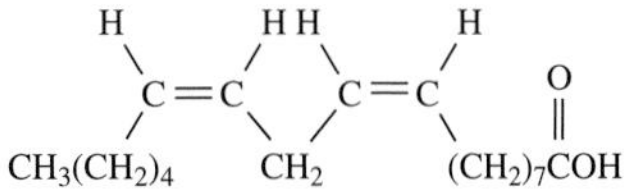	−12
18	Linoleimic acid	*cis,cis,cis*-9, 12, 15-Octadecatrienoic acid	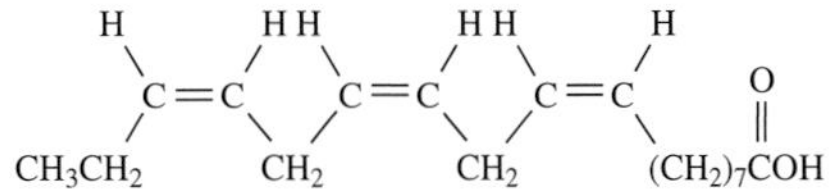	—
20	Arachidonic acid	*cis,cis,cis,cis*-5, 8, 11, 14-Icosatetraenoic acid	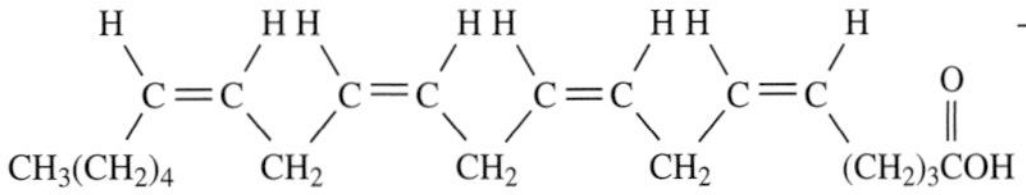	−49

TABLE 2.18 Pyrimidines and Purines That Occur in DNA and RNA

	Name	Structure	Occurrence
Pyrimidines			
	Cytosine	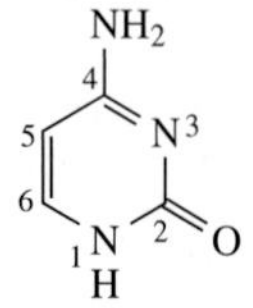	DNA and RNA
	Thymine	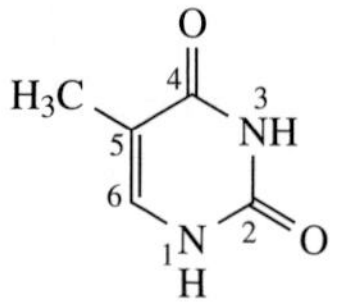	DNA
	Uracil	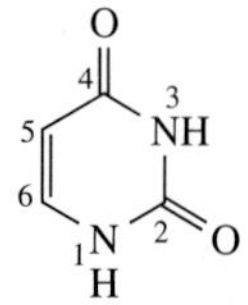	RNA
Purines			
	Adenine	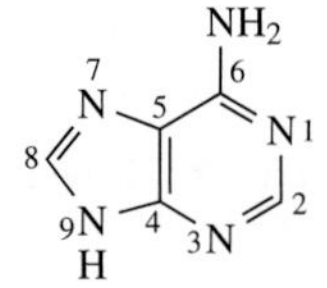	DNA and RNA
	Guanine	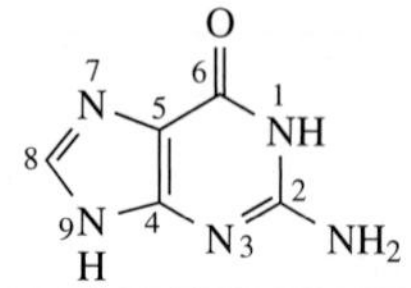	DNA and RNA

TABLE 2.19 Organic Radicals

For more comprehensive lists, see the various lists of radicals given in the subject indexes of the annual and decennial indexes of Chemical Abstracts.

Name	Formula	Name	Formula
Acenaphthenyl	$C_{12}H_9$—	Azido	N_3—
Acenaphthenylene	—$C_{12}H_8$—	Azino	=N—N=
Acenaphthenylidene	$C_{12}H_8$=	Azo	—N=N—
Acetamido	CH_3—CO—NH—	Azoxy	—N(O)—N—
Acetimidoyl	CH_3C(=NH)—	Azulenyl	$C_{10}H_7$—
Acetoacetyl	CH_3—CO—CH_2—CO—	Benzamido	C_6H_5—CO—NH—
Acetohydrazonoyl	CH_3—C(=NNH_2)—	Benzeneazo	C_6H_5—N=N—
Acetohydroximoyl	CH_3—C(=NOH)—	Benzeneazoxy	C_6H_5—N_2O—
Acetonyl	CH_3—CO—CH_2—	1,2-Benzenedicarbonyl,	
Acetonylidene	CH_3—CO—CH=	*see* Phthaloyl	
Acetoxy	CH_3—CO—O—	1,3-Benzenedicarbonyl (*or*	—CO—C_6H_4—CO— (*m*-)
Acetyl (*not ethanoyl*)	CH_3—CO—	*isophthaloyl*)	
Acetylamino	CH_3—CO—NH—	1,4-Benzenedicarbonyl (*or*	—CO—C_6H_4—CO— (*p*-)
Acetylhydrazino	CH_3—CO—NH—NH—	*terephthaloyl*)	
Acetylimino	CH_3—CO—N=	Benzenesulfinyl	C_6H_5—SO—
Acridinyl (*from acridine*)	$NC_{13}H_8$—	Benzenesulfonamido	C_6H_5—SO_2—NH—
Acroyloyl (*or propenoyl*)	CH_2=CH—CO—	Benzenesulfonyl	C_6H_5—SO_2—
Adipoyl (*or hexanedioyl*)	—CO—$[CH_2]_4$—CO—	Benzenesulfonylamino	C_6H_5—SO_2—NH—
Alanyl	CH_3—CH(NH_2)—CO—	Benzenetriyl	C_6H_3—
β-Alanyl	H_2N—CH_2—CH_2—CO—	Benzhydryl (*or diphenyl-*	$(C_6H_5)_2$CH—
Allyl (*or 2-propenyl*)	CH_2=CH—CH_2—	*methyl*)	
Allylidene	CH_2=CH—CH=	Benzidino	*p*-H_2N—C_6H_4—C_6H_4—
Allyloxy	CH_2=CH—CH_2—O—		NH—
Amidino	H_2N—C(=NH)—	Benziloyl (*or 2-hydroxy-*	$(C_6H_5)_2$C(OH)—CO—
Amino	H_2N—	*2,2-diphenylethanoyl*)	
Aminomethyleneamino	H_2N—CH=N—	Benzimidazolyl	$N_2C_7H_5$—
Aminooxy	H_2N—O—	Benzimidoyl	C_6H_5—C(=NH)—
Ammonio	$^+H_3$N—	Benzofuranyl	OC_8H_5—
Amyl, *see* Pentyl		Benzopyranyl	OC_9H_7—
Anilino	C_6H_5—NH—	Benzoquinonyl (1,2- or	$(O=)_2C_6H_3$—
Anisidino (*o-, m-, or*	CH_3O—C_6H_4—NH—	1,4-)	
p-)		Benzo[*b*]thienyl	SC_8H_5—
Anisoyl (*o-, m-, or*	CH_3O—C_6H_4—CO—	Benzoyl	C_6H_5—CO—
p-; or methoxyben-		Benzoylamino	C_6H_5—CO—NH—
zoyl)		Benzoylhydrazino	C_6H_5—CO—NH—NH—
Anthraniloyl	*o*-NH_2—C_6H_4—CO—	Benzoylimino	C_6H_5—CO—N=
Anthryl (*from anthracene*)	$C_{14}H_9$—	Benzoyloxy	C_6H_5—CO—O—
Anthrylene	—$C_{14}H_8$—	Benzyl	C_6H_5—CH_2—
Arginyl	H_2N—C(=NH)—NH—	Benzylidene	C_6H_5—CH=
	$[CH_2]_3$—CH(NH)—	Benzylidyne	C_6H_5—C≡
	CO—	Benzyloxy	C_6H_5—CH_2—O—
Asparaginyl	H_2N—CO—CH_2—	Benzyloxycarbonyl	C_6H_5—CH_2—O—CO—
	CH(NH_2)—CO—	Benzylthio	C_6H_5—CH_2—S—
Aspartoyl	—CO—CH_2—	Biphenylenyl	$C_{12}H_7$—
	CH(NH_2)—CO—	Biphenylyl	C_6H_5—C_6H_4—
α-Aspartyl	HO_2C—CH_2CH(NH_2)—	Bornenyl	$C_{10}H_{15}$—
Atropoyl (*or 2-phenylpro-*	C_6H_5—C(=CH_2)—CO—	Bornyl (*not camphyl or*	$C_{10}H_{17}$—
penoyl)		*bornylyl*)	
Azelaoyl, *see* Nonane-		Bromo	Br—
dioyl		Bromoformyl	Br—CO—

(*Continued*)

TABLE 2.19 Names and Formulas of Organic Radicals (*Continued*)

Name	Formula
Bromonio	$^+$HBr—
Butadienyl (1,3- shown)	CH_2=CH—CH=CH—
Butanedioyl, *see* Succinyl	
Butanediylidene	=CH—CH_2—CH_2—CH=
Butanediylidyne	≡C—CH_2—CH_2—C≡
Butanoyl, *see* Butyryl	
cis-Butenedioyl, *see* Maleoyl	
trans-Butenedioyl, *see* Fumaroyl Butenoyl, *see* Crotonoyl and Isocrotonoyl	
1-Butenyl	CH_3—CH_2—CH=CH—
2-Butenyl (*not crotyl*)	CH_3—CH=CH—CH_2—
2-Butenylene	—CH_2—CH=CH—CH_2—
Butenylidene (2- *shown*)	CH_3CH=CH—CH=
Butenylidyne (2- *shown*)	CH_3—CH=CH—C≡
Butoxy	CH_3—$[CH_2]_3$—O—
sec-Butoxy (*unsubstituted only*)	C_2H_5—CH(CH_3)—O—
tert-Butoxy (*unsubstituted only*)	$(CH_3)_3$C—O—
Butyl	CH_3—$[CH_2]_3$— or C_4H_9—
sec-Butyl (*unsubstituted only*)	C_2H_5—CH(CH_3)—
tert-Butyl (*unsubstituted only*)	$(CH_3)_3$C—
Butylidene	CH_3—CH_2—CH_2—CH=
sec-Butylidene (*unsubstituted only*)	$C_2H_5C(CH_3)$=
Butylidyne	CH_3—$[CH_2]_2$—C≡
Butyryl (*or butanoyl*)	CH_3—CH_2—CH_2—CO—
Camphoroyl	$C_{10}H_{14}O_2$—
Carbamoyl	H_2N—CO—
Carbazolyl	$NC_{12}H_8$—
Carbazoyl	H_2N—NH—CO—
Carbonimidoyl	—C(=NH)—
Carbonohydrazido (*preferred to carbohydazido or carbazido*)	H_2N—NH—CO—NH—NH—
Carbonyl	—CO— or =C(O)
Carbonyldioxy	—O—CO—O—
Carboxy	HO_2C—
Carboxylato	—O_2C—
Chloro	Cl—
Chlorocarbonyl, *see* Chloroformyl	
Chloroformyl	Cl—C(O)—
Chlorosyl	OCl—
Chlorothio	ClS—
Chloryl	O_2Cl—
Cinnamoyl (*or 3-phenylpropenoyl*)	C_6H_5—CH=CH—CO
Cinnamyl	C_6H_5—CH=CH—CH_2—
Cinnamylidene	C_6H_5—CH=CH—CH=
Citraconoyl (*unsubstituted only*)	HC—CO— ‖ CH_3—C—CO—
Crotonoyl	CH_3—CH=CH—CO— (*trans*)
Crotyl, *see* 2-Butenyl	
Cumenyl (*o*-, *m*-, *or p*-)	$(CH_3)_2$CH—C_6H_4—
Cyanato	NCO—
Cyano	NC—
Cyclobutyl	C_4H_7—
Cycloheptyl	C_7H_{13}—
	CH=CH—CH— CH=CH—CH_2
Cyclohexadienylidene (2,4- shown)	CH—CH_2—C< CH—CH=CH
Cyclohexanecarbonyl	C_6H_{11}—CO—
Cyclohexanecarbothioyl	C_6H_{11}—CS—
Cyclohexanecarboxamido	C_6H_{11}—CO—NH—
Cyclohexanecarboximidoyl	C_6H_{11}—C(=NH)—
Cyclohexenyl	C_6H_9—
2-Cyclohexenylidene	CH=CH—C< H_2C—CH_2—CH_2
Cyclohexyl	C_6H_{11}—
Cyclohexylcarbonyl	C_6H_{11}—CO—
Cyclohexylene	—C_6H_{10}—
Cyclohexylidene	CH_2—CH_2—C< CH_2—CH_2—CH_2
Cyclohexylthiocarbonyl	C_6H_{11}—CS—
Cyclopentadienyl	C_5H_5—
Cyclopentadienylidene	CH=CH—CH=CH—C=
Cyclopenta[*a*]phenanthryl	$C_{17}H_{17}$—
1,2-Cyclopentenophenanthryl	$C_{17}H_{11}$—
Cyclopentenyl	C_5H_7—
Cyclopentyl	C_5H_9—
Cyclopentylene	—C_5H_8—
Cyclopropyl	C_3H_5—
Cysteinyl	HS—CH_2—CH(NH_2)—CO—
Cystyl	—CO—CH(NH_2)—CH_2—S—S—CH_2—CH(NH_2)—CO—
Decanedioyl	—CO—$[CH_2]_8$—CO—
Decanoyl	CH_3—$[CH_2]_8$—CO—
Decyl	CH_3—$[CH_2]_9$—
Diacetoxyiodo	(CH_3—CO—O$)_2$I—
Diacetylamino	(CH_3—CO$)_2$N—
Diaminomethyleneamino	$(NH_2)_2$C=N—

TABLE 2.19 Names and Formulas of Organic Radicals (*Continued*)

Name	Formula
Diazo	=N$_2$
Diazoamino	—N=N—NH—
Dibenzoylamino	(C$_6$H$_5$—CO)$_2$N—
Dichloroiodo	Cl$_2$I—
Diethylamino	(C$_2$H$_5$)$_2$N—
3,4-Dihydroxybenzoyl, *see* Protocatechuoyl	
2,3-Dihydroxybutanedioyl, *see* Tartaroyl	
Dihydroxyiodo	(HO)$_2$I—
2,3-Dihydroxypropanoyl, *see* Glyceroyl	
3,4-Dimethoxybenzoyl, *see* Veratroyl	
3,4-Dimethoxyphenethyl	3,4-(CH$_3$O)$_2$C$_6$H$_3$CH$_2$CH—
3,4-Dimethoxyphenylacetyl	3,4-(CH$_3$O)$_2$C$_6$H$_3$CH$_2$CO—
Dimethylamino	(CH$_3$)$_2$N—
Dimethylbenzoyl	(CH$_3$)$_2$C$_6$H$_6$H$_3$—CO—
Dioxy	—O—O—
Diphenylamino	(C$_6$H$_5$)$_2$N—
Diphenylmethylene	(C$_6$H$_5$)$_2$C=
Dithio	—S—S—
Diethiocarboxy	HSSC—
Dithiosulfo	HOS$_2$—
Dodecanoyl	CH$_3$[CH$_2$]$_{10}$—CO—
Dodecyl	CH$_3$[CH$_2$]$_{11}$—
Elaidoyl (*or* trans-*9-octadecenoyl*)	CH$_3$[CH$_2$]$_7$CH=CH—[CH$_2$]$_7$—CO—
Epidioxy (as a bridge)	—O—O—
Epidiseleno (as bridge)	—Se—Se—
Epidithio (as a bridge)	—S—S—
Epimino (as a bridge)	—NH—
Episeleno (as a bridge)	—Se—
Epithio (as a bridge)	—S—
Epoxy (as a bridge)	—O—
Ethanesulfonamide	C$_2$H$_5$—SO$_2$—NH—
Ethanoyl, *see* Acetyl	
Ethenyl, *see* Vinyl	
Ethoxalyl	C$_2$H$_5$—OOC—CO—
Ethoxy	C$_2$H$_5$—O—
Ethoxycarbonyl	C$_2$H$_5$—O—CO—
Ethyl	C$_2$H$_5$— or CH$_3$—CH$_2$—
Ethylamino	C$_2$H$_5$—NH—
Ethylene	—CH$_2$—CH$_2$—
Ethylenedioxy	—O—CH$_2$—CH$_2$—O—
Ethylidene	CH$_3$—CH=
Ethylidyne	CH$_3$—C≡
Ethylsulfonylamino	C$_2$H$_5$—SO$_2$—NH—
Ethylthio	C$_2$H$_5$—S—
Ethynyl	HC≡C—
Ethynylene	—C≡C—
Fluoranthenyl	C$_{16}$H$_9$—

Name	Formula
Fluorenyl	C$_{13}$H$_9$—
Fluoro	F—
Fluoroformyl	F—CO—
Formamido	OCH—NH—
Formimidoyl	CH(=NH)—
Formyl (*not methanoyl*)	OCH— or —C(O)H
Formylamino	H—CO—NH—
Formylimino	H—CO—N=
Formyloxy	H—CO—O—
Fumaroyl (*or* trans-*butenedioyl*)	—CO—CH=CH—CO— (*trans*)
Furancarbonyl, *see* Furoyl	
Furfuryl (*2- only; preferred to 2-furylmethyl*)	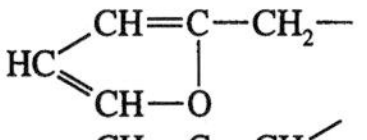
Furfurylidene (*2- only*)	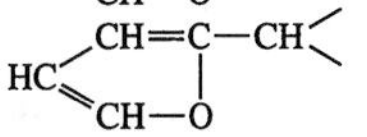
Furoyl (*3- shown; preferred to furancarbonyl*)	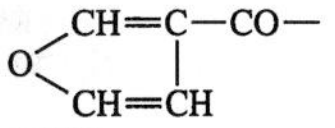
Furyl	OC$_4$H$_3$—
3-Furylmethyl	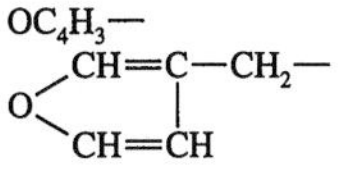
Galloyl (*or 3,4,5-trihydroxybenzoyl*)	3,4,5-(HO)$_3$C$_6$H$_2$—CO—
Geranyl (*from geraniol*)	C$_{10}$H$_{17}$—
Glutaminyl	H$_2$N—CO—CH$_2$—CH$_2$—CH(NH$_2$)—CO—
Glutamoyl	—CO—CH$_2$—CH$_2$—CH(NH$_2$)—CO—
α-Glutamyl	HOOC[CH$_2$]$_2$CH(NH$_2$)—CO—
γ-Glutamyl	HOOC—CH(NH$_2$)—[CH$_2$]$_2$—CO—
Glutaryl (*or pentanedioyl*)	—CO—[CH$_2$]$_3$—CO—
Glyceroyl (*or 2,3- dihydroxypropanoyl*)	HO—CH$_2$—CH(OH)—CO—
Glycoloyl (*or hydroxyethanoyl*)	HO—CH$_2$—CO—
Glycyl	H$_2$N—CH$_2$—CO—
Glycylamino	H$_2$N—CH$_2$—CO—NH—
Glyoxyloyl	OHC—CO—
Guanidino	H$_2$N—C(=NH)—NH—
Guanyl, *see* Amidino	
Heptanamido	CH$_3$—[CH$_2$]$_5$—CO—NH—
Heptanedioyl	—CO—[CH$_2$]$_5$—CO—
Heptanoyl	CH$_3$—[CH$_2$]$_5$—CO—
Heptyl	CH$_3$—[CH$_2$]$_5$—CH$_2$—
Hexadecanoyl	CH$_3$—[CH$_2$]$_{14}$—CO—
Hexadecyl	CH$_3$—[CH$_2$]$_{14}$—CH$_2$—
Hexamethylene	—[CH$_2$]$_6$—
Hexanamido	CH$_3$—[CH$_2$]$_4$—CO—NH—
Hexanedioyl (*or adipoyl*)	—CO—[CH$_2$]$_4$—CO—

(*Continued*)

TABLE 2.19 Names and Formulas of Organic Radicals (*Continued*)

Name	Formula	Name	Formula
Hexanimidoyl	$CH_3—[CH_2]_4—C(=NH)—$	Iodonio	$^+HI—$
Hexanoyl	$CH_3—[CH_2]_4—CO—$	Iodosyl	OI—
Hexanoylamino	$CH_3—[CH_2]_4—CO—NH—$	Iodyl	$O_2I—$
Hexyl	$CH_3—[CH_2]_4—CH_2—$	Isobutoxy (*unsubstituted only*)	$(CH_3)_2CH—CH_2—O—$
Hexylidene	$CH_3—[CH_2]_4—CH=$	Isobutyl (*unsubstituted only*)	$(CH_3)_2CH—CH_2—$
Hexyloxy	$CH_3[CH_2]_5—O—$	Isobutylidene (*unsubstituted only*)	$(CH_3)_2CH—CH=$
Hippuroyl	$C_6H_5—CO—NH—CH_2—CO—$	Isobutylidyne (*unsubstituted only*)	$(CH_3)_2CH—C\equiv$
Histidyl	$N_2C_3H_3—CH_2—CH(NH_2)—CO—$	Isobutyryl (*unsubstituted only; or 2-methylpropanoyl*)	$(CH_3)_2CH—CO—$
Homocysteinyl	$HS—CH_2—CH_2—CH(NH_2)—CO—$	Isocarbonohydrazido	$H_2N—N=C(OH)—NH—NH—$
Homoseryl	$HO—CH_2—CH_2—CH(NH_2)—CO—$	Isocrotonoyl	$CH_3—CH=CH—CO—$ (*cis*)
Hydantoyl	$H_2N—CO—NH—CH_2—CO—$	Isocyanato	OCN—
Hydratropoyl (*or 2-phenylpropanoyl*)	$C_6H_5—CH(CH_3)—CO—$	Isocyano	CN—
Hydrazi	—NH—NH— (to single atom)	Isohexyl (*unsubstituted only*)	$(CH_3)_2CH—[CH_2]_3—$
Hydrazino	$H_2N—NH—$	Isoleucyl	$C_2H_5—CH(CH_3)—CH(NH_2)—CO$
Hydrazo	—NH—NH— (to different atoms)	Isonicotinoyl (*or 4-pyridinecarbonyl*)	$NC_5H_4—CO—$ (4-)
Hydrazono	$H_2N—N=$	Isopentyl (*unsubstituted only*)	$(CH_3)_2CH—CH_2—CH_2—$
Hydroperoxy	HO—O—	Isophthaloyl (*or 1,3-benzenedicarbonyl*)	$—CO—C_6H_4—CO—$ (*m*-)
Hydroseleno	HSe—	Isopropenyl (*unsubstituted only; or 1-methylvinyl*)	$CH_2=C(CH_3)—$
Hydroxy	HO—	Isopropoxy (*unsubstituted only*)	$(CH_3)_2CH—O—$
Hydroxyamino	HO—NH—	Isopropyl (*unsubstituted only*)	$(CH_3)_2CH—$
o-Hydroxybenzoyl (*or salicyloyl*)	o-$HO—C_6H_4—CO—$	*p*-Isopropylbenzoyl	p-$(CH_3)_2CH—C_6H_4—CO—$
m-Hydroxybenzoyl	m-$HO—C_6H_4—CO—$	Isopropylbenzyl	$(CH_3)_2CH—C_6H_4—CH_2—$
p-Hydroxybenzoyl	p-$HO—C_6H_4—CO—$	Isopropylidene	$(CH_3)_2C=$
Hydroxybutanedioyl, *see* Maloyl		Isoselenocyanato	SeCN—
2-Hydroxy-2,2-diphenyl ethanoyl, *see* Benziloyl		Isosemicarbazido	$H_2N—NH—C(OH)=N—$
Hydroxyethanoyl, *see* Glycoloyl		Isothiocyanato	SCN—
Hydroxyimino	HO—N=	Isothioureido	HN=C(SH)—NH—, $H_2N—C(SH)=N—$
4-Hydroxy-3-methoxybenzoyl (*or vanilloyl*)	4-$HO,3$-$CH_3O—C_6H_3—CO—$	Isoureido	HN=C(OH)—NH—, $H_2N—C(OH)=N—$
3-Hydroxy-2-phenylpropanoyl (*or tropoyl*)	$C_6H_5—CH(CH_2OH)—CO—$	Isovaleryl (*unsubstituted only; or 3-methylbutanoyl*)	$(CH_3)_2CH—CH_2—CO—$
Hydroxypropanedioyl (*or tartronoyl*)	—CO—CH(OH)—CO—	Lactoyl	$CH_3—CH(OH)—CO—$
2-Hydroxypropanoyl (*or lactoyl*)	$CH_3—CH(OH)—CO—$	Lauroyl (*unsubstituted only*)	$CH_3—[CH_2]_{10}—CO—$
Icosyl	$CH_3—[CH_2]_{18}—CH_2—$		
Imino	—NH—, HN=		
Iminomethylamino	HN=CH—NH—		
Iodo	I—		
Iodoformyl	I—CO—		

TABLE 2.19 Names and Formulas of Organic Radicals (*Continued*)

Name	Formula
Leucyl	$(CH_3)_2CH—CH_2—CH(NH_2)—CO—$
Lysyl	$H_2N—[CH_2]_4—CH(NH_2)—CO—$
Maleoyl	$—CO—CH{=}CH—CO—$
Malonyl	$—CO—CH_2—CO—$
Maloyl	$—CO—CH(OH)—CH_2—CO—$
Mercapto-	$HS—$
Mesaconoyl (*unsubstituted only*)	$—CO—CH$ ‖ $CH_3—C—CO—$
Mesityl	$2,4,6\text{-}(CH_3)_3C_6H_2—$
Mesoxalo	$HOOC—CO—CO—$
Mesoxalyl	$—CO—CO—CO—$
Mesyl	$CH_3—SO_2—$
Methacryloyl (*or 2-methylpropenoyl*)	$CH_2{=}C(CH_3)—CO—$
Methaneazo	$CH_3—N{=}N—$
Methaneazoxy	$CH_3—N_2O—$
Methanesulfinamido	$CH_3—SO—NH—$
Methanesulfinyl	$CH_3—SO—$
Methanesulfonamido	$CH_3—SO_2—NH—$
Methanesulfonyl, *see* Mesyl	
Methanoyl, *see* Formyl	
Methionyl	$CH_3—S—CH_2—CH_2—CH(NH_2)—CO—$
Methoxalyl	$CH_3OOC—CO—$
Methoxy	$CH_3O—$
Methoxybenzoyl (*o-, m-, or p-*)	$CH_3O—C_6H_4—CO—$
Methoxycarbonyl	$CH_3O—CO—$
Methoxyimino	$CH_3O—N{=}$
Methoxyphenyl	$CH_3O—C_6H_4—$
Methoxysulfinyl	$CH_3O—SO—$
Methoxysulfonyl	$CH_3O—SO_2—$
Methoxy(thiosulfonyl)	$CH_3O—S_2O—$
Methyl	$CH_3—$
Methylallyl	$CH_2{=}C(CH_3)—CH_2—$
Methylamino	$CH_3—NH—$
Methylazo	$CH_3—N{=}N—$
Methylazoxy	$CH_3—N_2O—$
α-Methylbenzyl	$C_6H_5—CH(CH_3)—$
Methylbenzyl	$CH_3—C_6H_4—CH_2—$
3-Methylbutanoyl	$(CH_3)_2CH—CH_2—CO—$
cis-Methylbutenedioyl	$HC—CO—$ ‖ $CH_3—C—CO—$
trans-Methylbutenedioyl	$—CO—CH$ ‖ $CH_3—C—CO—$
Methyldithio	$CH_3—S—S—$
Methylene	$—CH_2—$, $H_2C{=}$
Methylenedioxy	$—O—CH_2—O—$
3,4-Methylenedioxybenzoyl	$3,4\text{-}CH_2O_2{:}C_6H_3—CO—$
5-Methylhexyl	$(CH_3)_2CH—[CH_2]_4—$
Methylidyne	$HC{\equiv}$
Methylsulfinimidoyl	$CH_3—S({=}NH)—$
Methylsulfinohydrazonoyl	$CH_3—S({=}NNH_2)—$
Methylsulfinohydroximoyl	$CH_3—S({=}N—OH)—$
Methylsulfinyl	$CH_3—SO—$
Methylsulfinylamino	$CH_3—SO—NH—$
Methylsulfonohydrazonoyl	$CH_3—S(O)(NNH_2)—$
Methylsulfonimidoyl	$CH_3—S(O)({=}NH)—$
Methylsulfonohydroxamoyl	$CH_3—S(O)(N—OH)—$
Methylsulfonyl	$CH_3—SO_2—$
Methylthio	$CH_3S—$
(Methylthio)sulfonyl	$CH_3S—SO_2—$
1-Methylvinyl, *see* Isopropenyl	
Morpholino (*4- only*)	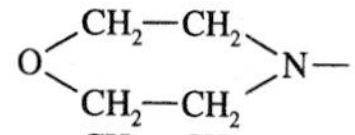
Morpholinyl (*3- shown*)	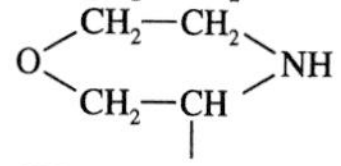
Myristoyl (*unsubstituted only*)	$CH_3—[CH_2]_{12}—CO—$
Naphthalenazo	$C_{10}H_7—N{=}N—$
Naphthalenecarbonyl, *see* Naphthoyl	
Naphthoyl	$C_{10}H_7—CO—$
Naphthoyloxy	$C_{10}H_7—CO—O—$
Naphthyl	$C_{10}H_7—$
Naphthylazo	$C_{10}H_7—N{=}N—$
Naphthylene	$—C_{10}H_6—$
Naphthylenebisazo	$—N{=}N—C_{10}H_6—N{=}N—$
Naphthyloxy	$C_{10}H_7—O—$
Neopentyl (*unsubstituted only*)	$(CH_3)_3C—CH_2—$
Nicotinoyl	$NC_5H_4—CO—$ (3-)
Nitrilo	$N{\equiv}$
Nitro	$O_2N—$
aci-Nitro	$HO—(O{=})N{=}$
Nitroso	$ON—$
Nonanedioyl	$—CO—[CH_2]_7—CO—$
Nonanoyl	$CH_3—[CH_2]_7—CO—$
Nonyl	$CH_3—[CH_2]_7—CH_2—$
Norbornyl	$C_7H_{11}—$
Norbornylyl, *see* Norbornyl	
Norcamphyl, *see* Norbornyl	
Norleucyl	$CH_3—[CH_2]_3—CH(NH_2)—CO—$
Norvalyl	$CH_3—CH_2—CH_2—CH(NH_2)—CO—$
Octadecanoyl	$CH_3—[CH_2]_{16}—CO—$

(*Continued*)

TABLE 2.19 Names and Formulas of Organic Radicals (*Continued*)

Name	Formula	Name	Formula
cis-9-Octadecenoyl	H[CH_2]$_8$—CH=CH—[CH_2]$_7$—CO—	Phenylsulfamoyl	C_6H_5—NH—SO_2
Octadecyl	CH_3—[CH_2]$_{16}$—CH_2—	Phenylsulfinyl	C_6H_5—SO—
Octanedioyl	—CO—[CH_2]$_6$—CO—	Phenylsulfonyl	C_6H_5—SO_2—
Octanoyl	CH_3—[CH_2]$_6$—CO—	Phenylsulfonylamino	C_6H_5—SO_2—NH—
Octyl	CH_3—[CH_2]$_6$—CH_2—	Phenylthio	C_6H_5—S—
Oleoyl	H[CH_2]$_8$—CH=CH—[CH_2]$_7$—CO—	3-Phenylureido	C_6H_5—NH—CO—NH—
Ornithyl	H_2N—[CH_2]$_3$—CH(NH_2)—CO—	Phthalamoyl	H_2N—CO—C_6H_4—CO— (*o*-)
Oxalacetyl	—CO—CH_2—CO—CO—	Phthalidyl	C_6H_4—CO—O—CH— (ring)
Oxalaceto	HOOC—CO—CH_2—CO—	Phthalimido	CO—C_6H_4—CO—N— (ring)
Oxalo	HOOC—CO—	Phthaloyl	—CO—C_6H_4—CO— (*o*-)
Oxalyl	—CO—CO—	Picryl	2,4,6-(NO_2)$_3C_6H_2$—
Oxamoyl	H_2N—CO—CO—	Pimeloyl (*unsubstituted only*)	—CO—[CH_2]$_5$—CO—
Oxido	$^-$O— (ion)	Piperidino (*1- only*)	$C_5H_{10}N$—
Oxo	O=	Piperidyl (*2-, 3-, 4-*)	NC_5H_{10}—
Oxonio	^+H_2O—	Piperonyl	3,4-CH_2O_2:C_6H_3—CH_2—
Oxy	—O—	Pivaloyl (*unsubstituted only*)	(CH_3)$_3$C—CO—
Palmitoyl (*unsubstituted only*)	CH_3—[CH_2]$_{14}$—CO—	Polythio	—S_4—
Pentafluorothio	F_5S—	Propanedioyl, *see* Malonyl	
Pentamethylene	—CH_2—CH_2—CH_2—CH_2—CH_2—	Propanoyl, *see* Propionyl	
Pentanedioyl, *see* Glutaryl		Propargyl, *see* 2-Propynyl	
Pentanoyl, *see* Valeryl		Propenoyl, *see* Acryloyl	
Pentenyl (*2- shown*)	CH_3—CH_2—CH=CH—CH_2—	1-Propenyl	CH_3—CH=CH—
Pentyl	CH_3—CH_2—CH_2—CH_2—CH_2—	2-Propenyl, *see* Allyl	
Pentyloxy	CH_3—[CH_2]$_4$—O—	Propenylene	—CH_2—CH=CH—
Perchloryl	O_3Cl—	Propioloyl	CH≡C—CO—
Phenacyl	C_6H_5—CO—CH_2—	Propionamido	CH_3—CH_2—CO—NH—
Phenacylidene	C_6H_5—CO—CH=	Propionyl	CH_3—CH_2—CO—
Phenanthryl	$C_{14}H_9$—	Propionylamino	CH_3—CH_2—CO—NH—
Phenethyl	C_6H_5—CH_2—CH_2—	Propionyloxy	CH_3—CH_2—CO—O—
Phenetidino (*o-, m-, or p-*)	C_2H_5O—C_6H_4—NH—	Propoxy	CH_3—CH_2—CH_2—O
Phenoxy	C_6H_5—O—	Propyl	CH_3—CH_2—CH_2—
Phenyl	C_6H_5—	Propylene	—CH(CH_3)—CH_2—
Phenylacetyl	C_6H_5—CH_2—CO—	Propylidene	CH_3—CH_2—CH=
Phenylazo	C_6H_5—N=N—	Propylidyne	CH_3—CH_2—C≡
Phenylazoxy	C_6H_5—N_2O—	Propynoyl, *see* Propiolyl	
Phenylcarbamoyl	C_6H_5—NH—CO	1-Propynyl	CH_3—C≡C—
Phenylene	—C_6H_4—	2-Propynyl	HC≡C—CH_2—
Phenylenebisazo	—N=N—C_6H_4—N=N—	Protocatechuoyl	3,4-(HO)$_2C_6H_3$—CO—
Phenylimino	C_6H_5—N=	3-Pyridinecarbonyl	NC_5H_4—CO— (3-)
2-Phenylpropanoyl	C_6H_5—CH(CH_3)—CO—	4-Pyridinecarbonyl	NC_5H_4—CO— (4-)
3-Phenylpropenoyl, *see* Cinnamoyl		Pyridinio	$^+NC_5H_5$— (ion)
3-Phenylpropyl	C_6H_5—CH_2—CH_2—CH_2—	Pyridyl	NC_5H_4—
		2-Pyridylcarbonyl	NC_5H_4—CO— (2-)
		Pyridyloxy	NC_5H_4—O—
		Pyruvoyl	CH_3—CO—CO—
		Salicyl	*o*-HO—C_6H_4—CH_2—
		Salicylidene	*o*-HO—C_6H_4—CH=
		Salicyloyl	*o*-HO—C_6H_4—CO—
		Sarcosyl	CH_3—NH—CH_2—CO—

TABLE 2.19 Names and Formulas of Organic Radicals (*Continued*)

Name	Formula
Sebacoyl (*unsubstituted only*)	—CO—[CH$_2$]$_8$—CO—
Seleneno	HOSe—
Selenino	HO$_2$Se—
Seleninyl	OSe═
Seleno	—Se—
Selenocyanato	NC—Se—
Selenoformyl	HSeC—
Selenonio	$^+$H$_2$Se— (ion)
Selenono	HO$_3$Se—
Selenonyl	O$_2$Se—
Selenoureido	H$_2$N—CSe—NH—
Selenoxo	(C)═Se
Semicarbazido	H$_2$N—CO—NH—NH—
Semicarbazono	H$_2$N—CO—NH—N═
Seryl	HO—CH$_2$—CH(NH$_2$)—CO—
Stearoyl (*unsubstituted only*)	CH$_3$—[CH$_2$]$_{16}$—CO—
Styryl	C$_6$H$_5$—CH═CH—
Suberoyl (*unsubstituted only*)	—CO—[CH$_2$]$_6$—CO—
Succinamoyl	H$_2$N—CO—CH$_2$—CH$_2$—CO—
Succinimido	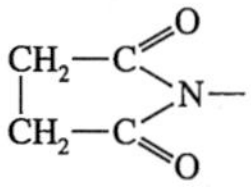
Succinimidoyl	—C(═NH)—CH$_2$—CH$_2$C(═NH)—
Succinyl	—CO—CH$_2$—CH$_2$—CO—
Sulfamoyl	H$_2$N—SO$_2$—
Sulfanilamido	*p*-H$_2$N—C$_6$H$_4$—SO$_2$—NH—
Sulfanilyl	*p*-H$_2$N—C$_6$H$_4$—SO$_2$—
Sulfenamoyl	H$_2$N—S—
Sulfeno	HO—S—
Sulfido	$^-$S— (ion)
Sulfinamoyl	H$_2$N—SO—
Sulfino	HO$_2$S—
Sulfinyl	—SO—
Sulfo	HO—SO$_2$—
Sulfoamino	HO$_2$S—NH—
Sulfonato	$^-$O$_3$S— (ion)
Sulfonio	$^+$H$_2$S— (ion)
Sulfonyl	—SO$_2$—
Sulfonyldioxy	—O—SO$_2$—O—
Tartaroyl	—CO—CH(OH)—CH(OH)—CO—
Tartronoyl	—CO—CH(OH)—CO—
Tauryl	H$_2$N—CH$_2$—CH$_2$—SO$_2$—
Telluro	Te replacing O
Terephthaloyl	—CO—C$_6$H$_4$—CO— (*p*-)
Terphenylyl	C$_6$H$_5$—C$_6$H$_4$—C$_6$H$_4$—

Name	Formula
(Terthiophen)yl	SC$_4$H$_3$—SC$_4$H$_2$—SC$_4$H$_2$—
Tetradecanoyl	CH$_3$—[CH$_2$]$_{12}$—CO—
Tetradecyl	CH$_3$—[CH$_2$]$_{12}$—CH$_2$—
Tetramethylene	—CH$_2$—CH$_2$—CH$_2$—CH$_2$—
Thenoyl (*2- shown*)	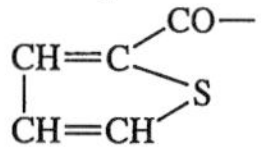
Thenyl	SC$_4$H$_3$—CH$_2$—
Thienyl	SC$_4$H$_3$—
Thio	—S—
Thioacetyl	CH$_3$—CS—
Thiobenzoyl	C$_6$H$_5$—CS—
Thiocarbamoyl	H$_2$N—CS—
Thiocarbazono	HN═N—CS—NH—NH—
Thiocarbodiazono	HN═N—CS—N═N—
Thiocarbonohydrazido	H$_2$N—NH—CS—NH—NH—
Thiocarbonyl	—CS—, SC═
Thiocarboxy	HSOC—, HS—CO—
Thiocyanato	NCS—
Thioformyl	SHC—, HCS—
Thiophenecarbonyl, *see* Thenoyl	
Thiosemicarbazido	H$_2$N—CS—NH—NH—
Thiosulfino	HOS$_2$—
Thiosulfo	HO$_2$S$_2$—
Thioreido	H$_2$N—CS—NH—
Thioxo	S═
Threonyl	CH$_3$—CH(OH)—CH(NH$_2$)—CO—
Toluenesulfonyl (*o-*, *m-*)	CH$_3$—C$_6$H$_4$—SO$_2$—
Toluidino (*o-*, *m-*, *or p-*)	CH$_3$—C$_6$H$_4$—NH—
Toluoyl (*o-*, *m-*, *or p-*)	CH$_3$—C$_6$H$_4$—CO—
Tolyl (*o-*, *m-*, *or p-*)	CH$_3$—C$_6$H$_4$—
Tolylsulfonyl	CH$_3$—C$_6$H$_4$—SO$_2$—
Tosyl (*p- only*)	*p*-CH$_3$—C$_6$H$_4$—SO$_2$—
Triazano	H$_2$N—NH—NH—
Triazeno	H$_2$N—N═N—
Trichlorothio	Cl$_3$S—
Tridecanoyl	CH$_3$—[CH$_2$]$_{11}$—CO—
Tridecyl	CH$_3$—[CH$_2$]$_{12}$—
Trifluorothio	F$_3$S—
3,4,5-Trihydroxybenzoyl	3,4,5-(HO)$_3$C$_6$H$_2$—CO—
Trimethylammonio	(CH$_3$)$_3$N$^+$— (ion)
Trimethylanilino (*all isomers*)	(CH$_3$)$_3$C$_6$H$_2$—NH—
Trimethylene	—CH$_2$—CH$_2$—CH$_2$—
Trimethylenedioxy	—O—CH$_2$—CH$_2$—CH$_2$—O—
Triphenylmethyl	(C$_6$H$_5$)$_3$C—
Trithio	—S$_3$—
Trithiosulfo	HS—S$_3$—

(*Continued*)

TABLE 2.19 Names and Formulas of Organic Radicals (*Continued*)

Name	Formula	Name	Formula
Trityl	$(C_6H_5)_3C—$	Vanilloyl	3,4-$CH_3O(HO)C_6H_3—$ $CO—$
Tropoyl	$C_6H_5—CH(CH_2OH)—$ $CO—$	Vanillyl	3,4-$CH_3O(HO)C_6H_3—$ $CH_2—$
Tyrosyl	*p*-$HO—C_6H_4—CH_2—$ $CH(NH_2)—CO—$	Veratroyl	3,4-$(CH_3O)_2C_6H_3—$ $CO—$
Undecanoyl	$CH_3—[CH_2]_9—CO—$	Veratryl	3,4-$(CH_3O)_2C_6H_2—$ $CH_2—$
Undecyl	$CH_3—[CH_2]_9—CH_2—$	Vinyl	$CH_2{=}CH—$
Ureido	$H_2N—CO—NH—$	Vinylene	$—CH{=}CH—$
Ureylene	$—NH—CO—NH—$	Xylidino (*all isomers*)	$(CH_3)_2C_6H_3—NH—$
Valeryl	$CH_3—[CH_2]_3—CO—$	Xylyl (*all isomers*)	$(CH_3)_2C_6H_3—$
Valyl	$(CH_3)_2CH—CH(NH_2)—$ $CO—$		

2.2 PHYSICAL PROPERTIES OF ORGANIC COMPOUNDS

Names of the compounds (Table 2.20) are arranged alphabetically. Usually substitutive nomenclature is employed; exceptions generally involve ethers, sulfides, sulfones, and sulfoxides. Each compound is given a number within its letter classification; thus compound c209 is 3-chlorophenol.

Formula Weights are based on the International Atomic Weights of 1993 and are computed to the nearest hundredth when justified. The actual significant figures are given in the atomic weights of the individual elements; see Table 3.2.

Density values are given at room temperature unless otherwise indicated by the superscript figure; thus 0.9711^{112} indicates a density of 0.9711 for the substance at 112°C. A density of 0.899^{16}_{14} indicates a density of 0.899 for the substance at 16°C relative to water at 4°C.

Refractive Index, unless otherwise specified, is given for the sodium line at 589.6 nm. The temperature at which the measurement was made is indicated by the superscript figure; otherwise it is assumed to be room temperature.

Melting Point is recorded in certain cases as 250 d and in some other cases as d 250, the distinction being made in this manner to indicate that the former is a melting point with decomposition at 250°C, while the latter decomposition occurs only at 250°C and higher temperatures. Where a value such as $—2H_2O$, 120 is given, it indicates a loss of 2 moles of water per formula weight of the compound at a temperature of 120°C.

Boiling Point is given at atmospheric pressure (760 mmHg) unless otherwise indicated; thus 82^{15mm} indicates that the boiling point is 82°C when the pressure is 15 mm Hg. Also, subl 550 indicates that the compound sublimes at 550°C.

Flash Point is given in degrees Celsius, usually using a closed cup. When the method is known, the acronym appears in parentheses after the value: closed cup (CC), Cleveland closed cup (CCC), open cup (OC), Tag closed cup (TCC), and Tag open cup (TOC). Because values will vary with the specific procedure employed, and many times the method was not stated, the values listed for the flash point should be considered only as indicative.

Solubility is given in parts by weight (of the formula weight) per 100 parts by weight of the solvent and at room temperature. Other temperatures are indicated by the superscript. Another way in which solubility is explicitly stated is in weight (in grams) per 100 mL of the solvent. In the case of gases, the solubility is often expressed as 5 mL^{10}, which indicates that at 10°C, 5 mL of the gas is soluble in 100 g (or 100 mL, if explicitly stated) of the solvent.

TABLE 2.20 Physical Constants of Organic Compounds

Abbreviations Used in the Table

abs, absolute
acet, acetone
alc, alcohol (ethanol usually)
alk, alkali (aqueous NaOH or KOH)
anhyd, anhydrous
aq, aqueous, water
as, asymmetrical
atm, atmosphere
BuOH, 1-butanol
bz, benzene
c, cold
chl, chloroform
conc, concentrated
d, decomposes or decomposed
D, dextrorotatory
deliq, deliquescent
dil, dilute
diox, 1,4-dioxane
DL, inactive (50% D and 50% L)
DMF, dimethylformamide
E, trans (German "entgegen")
EtOAc, ethyl acetate
eth, diethyl ether
EtOH, ethanol, 95%
expl, explodes
glyc, glycerol
h, hot
HOAc, acetic acid
hyd, hydrolysis
hygr, hygroscopic
i, insoluble
ign, ignites
i-PrOH, isopropyl alcohol, 2-propanol
L, levorotatory
m, meta configuration
Me, methyl
MeOH methanol
misc, miscible; soluble in all proportions
NaOH, aqueous sodium hydroxide
o, ortho configuration
org, organic
p, para configuration
PE, petroleum ether
pyr, pyridine
s, soluble
sec, secondary
sl, slight, slightly
soln, solution
solv, solvent
subl, sublimes
s, symmetrical
sym, symmetrical
tert, tertiary
v, very
v sl s, very slightly soluble
v, s very soluble
vac, vacuo or vacuum
vols, volumes
Z, cis (German "zusamman")
>, greater than
<, less than
~, approximately
±, inactive [50% (+) and 50% (−)]
α, alpha (first) position
β, beta (second) position
γ, gamma (third) position
δ, delta (fourth) position
ω, omega position (farthest from parent functional group)

No.	Name	Formula	Formula weight	Beilstein reference	Density, g/mL	Refractive index	Melting point,°C	Boiling point,°C	Flash point,°C	Solubility in 100 parts solvent
a1	(−)-Abietic acid		302.46	9^2, 424			172–175			i aq; s acet, alc, bz, chl, CS_2, eth, dil alk
a2	Acenaphthene		154.21	5, 586	1.189	1.6048^{95}	93.4	279		i aq; 3.2 alc; 20 bz; 10 chl; 1.8 MeOH; 3.2 g in 100 mL HOAc
a3	Acenaphthylene		152.20	5, 625	0.899_4^{16}		88–91	280		i aq; v s alc, eth
a4	Acetaldehyde	CH_3CHO	44.05	1, 594	0.788_6^{16}	1.3316^{20}	−123	21	−38(CC)	misc aq, alc, eth
a5	Acetaldoxime	$CH_3CH{=}NOH$	59.07	1, 608	0.966	1.415^{20}	46.5(α) 12(β)	114.5	40	v s aq, alc, eth
a6	Acetamide	CH_3CONH_2	59.07	2^2, 177	0.999^{78}	1.4158^{110}	81	222		70 aq; 50 alc; 16 pyr; s chl, glyc, hot bz

(Continued)

TABLE 2.20 Physical Constants of Organic Compounds (*Continued*)

No.	Name	Formula	Formula weight	Beilstein reference	Density, g/mL	Refractive index	Melting point, °C	Boiling point, °C	Flash point, °C	Solubility in 100 parts solvent
a7	Acetamidine HCl	$CH_3C(=NH)NH_2 \cdot HCl$	94.54	2, 185			164–166			v s aq; s alc; i acet, eth
a8	*N*-(2-Acetamido)-2-aminoethanesulfonic acid	$H_2N(CO)CH_2NHCH_2CH_2SO_3H$	182.20				>220 dec			
a9	4-Acetamidobenzaldehyde	$CH_3CONHC_6H_4CHO$	163.18	14, 38			156–158			s aq, bz; sl s alc
a10	4-Acetamidobenzenesulfonyl chloride	$CH_3CONHC_6H_4SO_3Cl$	233.67	14, 702			148 dec			d aq; v s alc, bz, eth, acet
a11	2-Acetamidobenzoic acid	$CH_3CONHC_6H_4CO_2H$	179.18	14, 337			185–187			sl s aq; v s alc, bz, eth, acet
a12	4-Acetamidobenzoic acid	$CH_3CONHC_6H_4CO_2H$	179.18	14, 432			262 dec			i aq; s alc; sl s eth
a13	2-Acetamidofluorene		223.28	12, 1331			192–196			i aq; s alc, glycols
a14	*N*-(2-Acetamido)iminodiacetic acid	$H_2NCOCH_2N(CH_2CO_2H)_2$	190.16				219 d			
a15	2-Acetamidophenol	$CH_3CONHC_6H_4OH$	151.17	13, 370			207–210			
a16	3-Acetamidophenol	$CH_3CONHC_6H_4OH$	151.17	13, 415			146–149			
a17	4-Acetamidophenol	$CH_3CONHC_6H_4OH$	151.17	13, 460	1.293_4^{21}		170–172			s alc, acet
a18	Acetanilide	$CH_3CONHC_6H_5$	135.17	12, 237	1.219_4^{15}		114	304–305	173	0.56 aq^{25}; 25 acet; 29 alc; 2 bz; 27 chl; 5 eth
a19	Acetic acid	CH_3CO_2H	60.05	2, 96	1.0492_4^{20}	1.3718^{20}	16.7	118	39 (CC)	misc aq, alc, eth, CCl_4
a20	Acetic acid-*d*	CH_3CO_2D	61.06	2^3, 202	1.059	1.2715^{20}		115.5	40	misc aq, alc, eth, CCl_4
a21	Acetic-d_3 acid-*d*	CD_3CO_2D	64.08	2^3, 203	1.137	1.3687^{20}		114.4	40	misc aq, alc, eth, CCl_4
a22	Acetic anhydride	$(CH_3CO)_2O$	102.09	2, 166	1.080_4^{15}	1.3904^{20}	−73	139	54 (CC)	s chl, eth; slowly s aq forming HOAc, alc forming EtOAc
a23	Acetic anhydride-d_6	$(CD_3CO)_2O$	108.14			1.3875^{20}		65^{65mm}	54	see acetic anhydride
a24	Acetoacetanilide	$CH_3COCH_2CONHC_6H_5$	177.20	12, 518	1.260^{20}		85	dec	185	s alc, hot bz, acids, alkalis, chl, eth
a25	Acetoacetic acid	CH_3COCH_2COOH	102.09	3, 630			36–37	d viol 100		misc aq, alc
a26	Acetone	CH_3COCH_3	58.08	1, 635	0.7908_4^{20}	1.3591^{20}	−94	56	−20	misc aq, alc, chl, DMF
a27	Acetone-d_6	CD_3COCD_3	64.13		0.872	1.3554^{20}	−93.8	55.5	−17	see acetone

a28	Acetone oxime	$(CH_3)_2C{=}NOH$	73.10	1, 649	0.911^{62}_{2}		60	135		v s aq, alc, eth
a29	Acetonitrile	CH_3CN	41.05	2, 183	0.7875^{15}_{4}	1.3460^{15}	−44	81.6	6	misc aq, acet, alc, chl, eth, EtOAc
a30	Acetonitrile-d_3	CD_3CN	44.08	2^4, 428	0.844	1.3406^{20}		80.7	5	misc aq, alc, chl
a31	Acetophenone	$C_6H_5COCH_3$	120.15	7, 271	1.026^{20}_{4}	1.5372^{20}	20	202	77	0.55 aq; s alc, chl, eth, glyc
a32	Acetophenone-methyl-d_3	$C_6H_5COCD_3$	123.18	7^4, 626	1.055	1.5325^{20}		201–202	82	
a33	4-Acetylbenzenesulfonic acid, sodium salt	$CH_3COC_6H_5SO_3^- Na^+$	222.20	11^2, 186			>300			
a34	Acetylbiphenyl	$C_6H_5C_6H_4COCH_3$	196.25	7^2, 337			116–118	325–327		i aq; v s alc, acet
a35	Acetyl bromide	CH_3COBr	122.95	2, 174	1.663^{16}_{4}	1.4486^{20}	−96	76	>110	dec viol by aq or alc; misc bz, chl, eth
a36	2-Acetylbutyrolactone		128.13	17^3, 5837	1.1846^{20}_{4}	1.4585^{20}		107^{5mm}	>110	20% v/v aq
a37	Acetyl chloride	CH_3COCl	78.50	2, 173	1.104^{20}_{4}	1.3896^{20}	−113	51	4 (CC)	dec viol aq or alc; misc bz, chl, eth, HOAc, PE
a38	Acetylcholine bromide	$(CH_3)_3N(Br)CH_2CH_2$-O_2CCH_3	226.11	4^1, 428			144–146			v s aq (dec by hot aq or alkalis); s alc; i eth
a39	Acetylcholine chloride	$(CH_3)_3N(Cl)CH_2CH_2$-O_2CCH_3	181.66	4, 281			150–152			v s aq, alc; dec by hot aq or alkalis; i eth
a40	2-Acetylcyclopentanone		126.16	7, 558	1.043	1.4905^{20}		75^{8mm}	72	
a41	Acetylene	$HC{\equiv}CH$	26.04	1, 228	0.90(*g*)		−85(subl)		−18	1 vol in 1 vol aq, in 6 vol HOAc or alc; s bz, eth; acet dissolves 25 vol[15] but 300 vols at 12 atm
a42	Acetylenedicarboxylic acid	$HO_2CC{\equiv}CCO_2H$	114.06	2, 801			180 d			v s aq, alc, eth
a43	Acetyl fluoride	CH_3OF	62.04	2, 172	1.002^{15}_{4}		<−60	20.8		5 aq(dec); sl s acet, alc, bz, eth
a44	2-Acetylfuran		110.11	17, 286	1.098	1.5065^{20}	29–30	67^{10mm}	71	

(*Continued*)

TABLE 2.20 Physical Constants of Organic Compounds (*Continued*)

No.	Name	Formula	Formula weight	Beilstein reference	Density, g/mL	Refractive index	Melting point, °C	Boiling point, °C	Flash point, °C	Solubility in 100 parts solvent
a45	*N*-Acetyl-(−)-glutamic acid	$HO_2CCH_2CH_2CH(NHCOCH_3)CO_2CH_3$	189.17	4^2, 908			200–201			
a46	*N*-Acetylglycine	$CH_3CONHCH_2CO_2H$	117.10	4, 354			206–208			2.7 aq^{15}; s alc; sl s acet, chl, HOAc; i bz, eth
a47	1-Acetylimidazole		110.12				103–105			
a48	Acetyl iodide	CH_3COI	169.95	2, 174	2.0674^{20}_{4}	1.5491^{20}		108		dec aq, alc; s bz, eth
a49	Acetyl-2-methylcholine chloride	$CH_3CO_2CH(CH_3)CH_2$–$N(Br)(CH_3)_3$	195.69	Merck: 12, 6003			172–173			v s aq, alc, chl; i eth; dec by alkalis, eth
a50	2-Acetylphenothiazine		241.31				180–185			
a51	2-Acetylphenylacetonitrile	$C_6H_5CH(CN)COCH_3$	159.19	10, 699				92–94		
a52	1-Acetyl-4-pipidone		141.17		1.146	1.5026^{20}		218	>110	
a53	2-Acetylpyridine	$(C_5H_4N)COCH_3$	121.14	21, 279	1.080	1.5203^{20}		188–189	73	v s alc, eth
a54	3-Acetylpyridine	$(C_5H_4N)COCH_3$	121.14	21, 279	1.102	1.5336^{20}		220	150	v s acids, alc, eth; s aq
a55	4-Acetylpyridine	$(C_5H_4N)COCH_3$	121.14	21, 279	1.095	1.5350^{20}		212	>110	v s alc, eth
a56	Acetylsalicylic acid	$HO_2C_6H_4$-2-O_2CCH_3	180.16	10, 67	1.35		135			0.33 aq^{25}; 29 acet; 20 alc; 5.9 chl; 5 eth; s bz
a57	2-Acetylthiophene	$(C_4H_3S)COCH_3$	126.18	17, 287	1.168^{22}_{4}	1.5564^{20}	10–11	214		sl s aq; misc alc, eth
a58	1-Acetyl-2-thiourea	$CH_3C(O)NHC(S)NH_3$	118.16	3. 191			167			s hot aq, alc; sl s eth
a59	*N*-Acetyl-(±)-tryptophan		246.27	22^2, 469			206			s aq, alc; v s eth
a60	Acridine		179.22	20, 459	1.005^{20}_{4}		106–110 subl 100	346		s alc, eth, CS_2, PE; sl s hot aq
a61	Acrylamide	$H_2C{=}CHCONH_2$	71.08	2, 400	1.222^{30}_{4}		84.5	192.6		at 30°, g/100 mL: 215 aq, 155 MeOH, 86 EtOH, 63 acet, 12.6 EtOAc, 2.7 chl, 0.3 bz
a62	Acrylic acid	$H_2C{=}CHCO_2H$	72.06	2, 397	1.0511^{20}	1.4224^{20}	12–14	141	50	misc aq, alc, bz, eth, chl, acet
a63	Acrylonitrile	$H_2C{=}CHCN$	53.06	2, 400	0.8060^{20}_{4}	1.3911^{20}	−83.5	77.3	0	7.3 aq; misc org solv

a63a	Acryloyl chloride	$H_2C{=}CHCOCl$	90.51	2, 400	1.114	1.4350^{20}		72–76	15	d aq; v s chl
a64	1-Adamantanamine		151.25	Merck: 12, 389			160–190			sl s aq
a65	Adamantane		136.24	Merck: 12, 149	1.09	1.568	270 (sealed tube)	205 subl		s acet
a66	Adenine		135.13	26, 420			360 dec	subl 220		0.005 aq; sl s alc; i chl, eth
a67	(−)-Adenosine		267.24	31, 27			235			s aq; i alc
a68	(±)-*α*-Alanine	$CH_3CH(NH_2)CO_2H$	89.09	4, 387	1.424		264–269 (depends on heating rate)	subl >200		16.7 aq^{25}; 0.009 alc^{25}; i eth
a69	(−)-*α*-Alanine	$CH_3CH(NH_2)CO_2H$	89.09	4, 381	1.401		dec 297			16.7 aq^{25}; 0.2 alc^{25}; i eth
a70	*β*-Alanine	$H_2NCH_2CH_2CO_2H$	89.09	4, 401	1.437^{-5}		197 dec			v s aq; sl s alc; i eth
a71	Allantoin		158.12	25, 474			238			0.45 aq; 0.2 alc; i eth
a72	Allene	$H_2C{=}C{=}CH_2$	40.06	1, 248	1.787	1.4168	−136	−34		
a73	Alloxan monohydrate		160.09	24, 500			anhyd: 256 dec			s aq, alc, acet, HOAc; sl s chl, EtOAc, PE
a74	Allyl acetate	$H_2C{=}CHCH_2OCOCH_3$	100.12	2, 136	0.977^{20}_{4}	1.4040^{20}		104	22	i aq; misc alc, eth
a75	Allyl alcohol	$H_2C{=}CHCH_2OH$	58.08	2, 436	0.8540^{20}_{4}	1.4134^{20}	−129	97	21	misc aq, alc, chl, eth
a76	Allylamine	$H_2C{=}CHCH_2NH_2$	57.10	4, 205	0.761^{20}_{4}	1.4185^{20}	−88.2	53–55	−29	misc aq, alc, chl, eth
a77	*N*-Allylaniline	$C_6H_5NHCH_2CH{=}CH_2$	133.19	12, 170	0.982^{25}	1.5630^{20}		220	89	i aq; s alc, eth
a78	Allylbenzene	$C_6H_5CH_2CH{=}CH_2$	118.18	5, 484	0.892^{20}_{0}	1.5122^{20}		157	33	i aq; s alc, eth
a79	Allyl bromide	$H_2C{=}CHCH_2Br$	120.98	1, 201	1.398^{20}_{4}	1.4654^{20}	−119	70	−2	sl s aq; misc org solv
a80	Allyl butanoate	$CH_3CH_2CH_2COOCH_2$-$CH{=}CH_2$	128.17	2, 272	0.902	1.4142^{20}	44^{15mm}	41		
a81	Allyl chloride	$H_2C{=}CHCH_2Cl$	76.53	1, 198	0.938^{20}_{4}	1.4154^{20}	−134.5	44–46	−31 (CC)	sl s aq; misc alc, chl, eth, PE
a82	Allyl chloroformate	$H_2C{=}CHCH_2OOCCl$	120.54	3, 12	1.136	1.4223	110	27	31	
a83	Allylcyclohexylamine	$(C_6H_{11})NHCH_2CH{=}CH_2$	139.24		0.962	1.4664^{20}		66^{12mm}	53	
a84	4-Allyl-1,2-dimethoxybenzene	$H_2C{=}CHCH_2C_6H_3(OCH_3)_2$	178.23	6, 963	1.036	1.5344^{20}	−4	255		
a85	*N*-Allyl-*N,N*-dimethylamine	$H_2C{=}CHCH_2N(CH_3)_2$	85.0			1.4010^{20}		64		
a86	Allyl ethyl ether	$H_2C{=}CHCH_2OCH_2CH_3$	86.13	1, 438	0.765^{20}_{4}	1.3881^{20}		68	−20	i aq; misc alc, eth
a87	Allyl iodide	$H_2C{=}CHCH_2I$	167.98	1, 202	1.825^{20}_{4}	$1,5540^{21}$	−99	103		i aq; misc alc, eth
a88	Allyl isothiocyanate	$H_2C{=}CHCH_2NCS$	99.16	4, 214	1.013^{25}_{4}	1.5248^{25}	−80	152	46	0.2 aq; misc org solv

(Continued)

TABLE 2.20 Physical Constants of Organic Compounds (*Continued*)

No.	Name	Formula	Formula weight	Beilstein reference	Density, g/mL	Refractive index	Melting point, °C	Boiling point, °C	Flash point, °C	Solubility in 100 parts solvent
a89	Allyl methacrylate	$H_2C{=}C(CH_3)COOCH_2$-$CH{=}CH_2$	126.16	2^3, 1290	0.938	1.4360		61^{43mm}	33	
a90	Allyl methyl sulfide	$H_2C{=}CHCH_2SCH_3$	88.17	1, 440	0.803	1.4714^{20}		91–93	18	
a91	1-Alloxy-2,3-epoxypropane	$\underset{\diagdown O \diagup}{H_2C{-}CH}CH_2OCH_2$-$CH{=}CH_2$	114.14		0.962	1.4332^{20}		154	57	
a92	3-Alloxy-1,2-propanediol	$H_2C{=}CHCH_2$-$CH_2CH(OH)CH_2OH$	132.16	1, 513	1.068	1.4620^{20}		142^{28mm}	>110	
a93	Allyloxytrimethylsilane	$H_2C{=}CHCH_2OSi(CH_3)_3$	130.26		0.7830	1.4075^{25}		102	0	
a94	2-Allylphenol	$H_2C{=}CHCH_2C_6H_4OH$	134.18	6, 572	1.033^{15}_{4}	1.5450^{20}	10	220	88	s alc, eth
a95	Allyl phenyl ether	$H_2C{=}CHCH_2OC_6H_5$	134.18	6, 144	0.983^{15}_{4}	1.5200^{20}		192	62	i aq; s alc, misc eth
a96	Allyl propyl ether	$H_2C{=}CHCH_2OC_3H_7$	100.16	1, 438	0.767^{20}_{4}	1.3990^{20}		90–92	−5	s alc; misc eth
a97	1-Allyl-2-thiourea	$H_2C{=}CHCH_2NHC(S)NH_2$	116.19	4, 211	1.219^{20}_{20}		70–72			3.3 aq; s alc; i bz; v sl s eth
a98	Allyltrichlorosilane	$H_2C{=}CHCH_2SiCl_3$	175.52	4^3, 1909	1.2011^{20}_{4}	1.4550^{20}		117.5	31	
a99	Allyltriethoxysilane	$H_2C{=}CHCH_2Si(OC_2H_5)_3$	204.34	4^3, 1909	0.9030^{20}	1.4062^{20}		176^{740mm}	21	
a100	Allyl trifluoroacetate	$CF_3COOCH_2CH{=}CH_2$	154.09	2^4, 464	1.183	1.3350^{20}		66–67	−1	
a101	Allyltrimethylsilane	$H_2C{=}CHCH_2Si(CH_3)_3$	114.27		0.7193^{20}_{4}	1.4080^{20}		84–88	7	
a102	Allylurea	$H_2C{=}CHCH_2NHCONH_2$	100.12	4, 209			85			v s aq, alc; i chl, CS_2 eth, toluene
a103	Aminoacetonitrile	H_2NCH_2CN	56.07	4, 344				58^{15mm} d		s acids, alc
a104	Aminoacetonitrile hydrogen sulfate	$H_2NCH_2CN \cdot H_2SO_4$	154.14	4, 344			121	d 165		v s aq; sl s alc; i eth
a105	2′-Aminoacetophenone	$H_2NC_6H_4COCH_3$	135.17	14, 41				70^{3mm}	>110	v sl s aq; s alc, eth
a106	3′-Aminoacetophenone	$H_2NC_6H_4COCH_3$	135.17	14, 45			99	290		
a107	4′Aminoacetophenone	$H_2NC_6H_4COCH_3$	135.17	14, 46			106	293–295		s hot aq, alc, eth, HOAc; sl s bz
a108	1-Aminoanthraquinone		223.23	14, 177			ca. 250	subl		i aq; v s alc, bz, chl, eth, HOAc, HCl
a109	2-Aminoanthraquinone		223.23	14, 191			295 d	subl		i aq, eth; s alc, bz
a110	4-Aminoantipyrine		203.25	24, 273			109			s aq, alc, bz; sl s eth
a111	*p*-Aminoazobenzene	$C_6H_5N{=}NC_6H_4NH_2$	197.24				128	>360		sl a aq; v s alc, bz, chl, eth
a112	2-Aminobenzamide	$H_2NC_6H_4CONH_2$	136.15	14, 320			110	300 sl d		v s hot aq, alc; i bz; sl s eth
a113	4-Aminobenzene-	$H_2NC_6H_4AsO(OH)_2$	217.06	16, 878			232			s hot aq; alk CO_3,

	arsonic acid									conc'd mineral acids; i acet, bz, chl, eth
a114	5-Aminobenzene-1,3-dicarboxylic acid	$H_2NC_6H_3(COOH)_2$	181.15	14^1, 636			>300			
a115	2-Aminobenzene-sulfonic acid	$H_2NC_6H_4SO_3H$	173.19	14, 681			ca. d 325			1.5 aq^{15}; v sl s alc, eth
a116	3-Aminobenzene-sulfonic acid	$H_2NC_6H_4SO_3H$	173.19	14, 688	1.69		>300			2 aq^{15}; sl s alc, MeOH
a117	4-Aminobenzene-sulfonic acid	$H_2NC_6H_4SO_3H$	173.19	14, 695			d 288			1 aq^{20}; sl s hot MeOH; i alc, bz, eth
a118	2-Aminobenzoic acid	$H_2NC_6H_4COOH$	137.14	14, 310			144–146	subl		v s hot aq, alc, eth
a119	3-Aminobenzoic acid	$H_2NC_6H_4COOH$	137.14	14, 383	1.511^4		172–174			v s hot aq, alc; s eth
a120	4-Aminobenzoic acid	$H_2NC_6H_4COOH$	137.14	14, 418	1.374		187			0.59 aq; 12 alc; 2 eth; s EtOAc, HOAc
a121	2-Aminobenzonitrile	$H_2NC_6H_4CN$	118.14	14, 322			49	268	>110	s alc, eth
a122	3-Aminobenzonitrile	$H_2NC_6H_4CN$	118.14	14, 391			53	288–290	>110	s hot aq; v s alc, eth
a123	4-Aminobenzonitrile	$H_2BC_6H_4CN$	118.14	14, 425			85	dec		v s hot aq, alc, eth
a124	2-Aminobenzophenone	$H_2NC_6H_4COC_6H_5$	197.24	14, 76			108	223–226		sl s aq; s alc, eth
a125	2-Aminobenzothiazole		150.20	27, 182			132	dec		v s conc'd acids, alc, chl, eth
a126	2-Aminobenzotri-fluoride	$H_2NC_6H_4CF_3$	161.13	12^{12}, 453	1.290^{25}	1.4785^{25}	34	175	55	
a127	3-Aminobenzotri-fluoride	$H_2NC_6H_4CF_3$	161.13	12, 870	1.290	1.4800^{20}	6	187	85	
a128	4-Aminobenzotri-fluoride	$H_2NC_6H_4CF_3$	161.13	12^3, 2151	1.283^{27}	1.4815^{25}	38	83^{12mm}	86	
a129	*N*-(4-Aminobenzoyl)-glycine	$H_2NC_6H_4CONHCH_2COOH$	194.19	14^2, 258			198–199			i aq; s alc, bz, chi
a130	2-Aminobiphenyl	$H_2NC_6H_4C_6H_5$	169.23	12, 1317			50–53	299	>110	sl s aq; s alc
a131	4-Aminobiphenyl	$H_2NC_6H_4C_6H_5$	169.23	12, 1318			52–54	191^{15mm}	>110	s hot aq, alc, eth
a132	2-Amino-5-bromo-benzoic acid	$Br(NH_2)C_6H_3COOH$	216.03	14, 370			218–219			s alc, bz, chl, eth, HOAc; v s acet
a133	(±)-2-Aminobutanoic acid	$CH_3CH_2CH(NH_2)COOH$	103.12	4, 408			304 d	subl >300		21 aq^{25}, 0.18 hot alc; i eth
a133a	3-Aminobutanoic acid	$H_3CCH_2CH(NH_2)COOH$	103.12	4, 412			193–194			125 aq; i alc, eth
a134	4-Aminobutanoic acid	$H_2NCH_2CH_2CH_2COOH$	103.12	4, 413			195 d			v s aq; i org solv
a135	2-Amino-1-butanol	$CH_2CH_2CH(NH_2)CH_2OH$	89.14	4, 291	0.944^{20}_{20}	1.4521^{20}	−2	176–178	74 (OC)	misc aq; s alc

(*Continued*)

TABLE 2.20 Physical Constants of Organic Compounds (*Continued*)

No.	Name	Formula	Formula weight	Beilstein reference	Density, g/mL	Refractive index	Melting point, °C	Boiling point, °C	Flash point, °C	Solubility in 100 parts solvent
a136	3-(4-Aminobutyl)-piperidine	$(HNC_5H_9)(CH_2)_4NH_2$	156.27	22^3, 3788	0.910		39–42	148^{10mm}	>110	
a137	4-Amino-6-chloro-1,3-benzenedisulfonamide	$H_2NC_6H_2(Cl)(SO_2NH_2)_2$	285.73	14^4, 2810			257–261			
a138	2-Amino-4-chlorobenzoic acid	$H_2N(Cl)C_6H_3COOH$	171.58	14, 365			231–233			
a139	5-Amino-2-chlorobenzoic acid	$H_2N(Cl)C_6H_3COOH$	171.58	14, 412			188 d			
a140	2-Amino-4′-chlorobenzophenone	$H_2NC_6H_4COC_6H_4Cl$	231.68	14^1, 389			104			
a141	2-Amino-5-chlorobenzophenone	$H_2N(Cl)C_6H_3COC_6H_5$	231.68	14, 79			98–100			
a142	2-Amino-5-chlorobenzotrifluoride	$H_2N(Cl)C_6H_3CF_3$	195.57	12^3, 1921	1.386	1.5069^{20}		67^{3mm}	none	
a143	5-Amino-2-chlorobenzotrifluoride	$H_2N(Cl)C_6H_3CF_3$	195.57				36–38		>110	
a144	2-(3-Amino-4-chlorobenzoyl)benzoic acid	$H_2N(Cl)C_6H_3COC_6H_4COOH$	275.69	14, 661			171–173			
a145	4-Amino-4′-chlorobiphenyl	$H_2NC_6H_4$-C_6H_4Cl	203.67				128–134			i aq; s alc, acet, bz, chl, HOAc
a146	4-Amino-5-chloro-2-methoxybenzoic acid	$H_2NC_6H_2(Cl)(OCH_3)COOH$	201.61				206 d			
a147	2-Amino-4-chlorophenol	$H_2N(Cl)C_6H_3OH$	143.57	13, 383			139–143			
a148	2-Amino-5-chloropyridine	$H_2N(Cl)(C_5H_3N)$	129.56	22^2, 332			135–138	128^{11mm}		
a149	3-Aminocrotononitrile	$CH_3C(NH_2){=}CHCN$	82.11	3, 660						
a150	1-[(2-Aminoethyl)-amino]-2-propanol	$CH_3CH(OH)CH_2NHCH_2$-CH_2NH_2	118.18		0.9837_4^{25}	1.4788^{25}		112^{10mm}		
a151	5-Amino-2,3-dihydro-1,4-phthalazinedione		177.16	25^1, 698			319–320			
a152	2-Amino-4,6-dihydroxypyrimidine		127.10	24, 468			>300			
a153	4-Amino-2,6-dihydroxypyrimidine		127.10	24, 469			>300			
a154	2-Amino-3,3-dimethylbutane	$(CH_3)_3CCH(NH_2)CH_3$	101.19	4, 193	0.755	1.4130^{20}	−20	102–103	1	

a155	2-Amino-4,6-dimethyl-pyridine	$(CH_3)_2(NH_2)(C_5H_2N)$	122.17	22, 435			63–64	235		
a156	4-Amino-2,6-dimethyl-pyrimidine		123.16	24^2, 45			184–186			156 aq; 18.9 alc
a157	6-Amino-1,3-dimethyl-uracil		155.16	24, 471			295 d			
a158	5-Amino-2,6-dioxo-1,2,3,6-tetrahydro-4-pyrimidinecarboxylic acid		171.11	25, 264			>300			
a159	α-Aminodiphenyl-methane	$(C_6H_5)_2CHNH_2$	183.25	12, 1323	1.0635^{22}_{4}	1.5950^{20}	34	304	>110	sl s aq; s acids
a160	2-Aminoethanesulfonic acid	$H_2NCH_2CH_2SO_3H$	125.15	4, 528			d ca. 300			5.45 aq^{12}; 0.004 alc^{17}
a161	2-Aminoethanethiol	$HSCH_2CH_2NH_2$	77.14	4, 286			97–99			v s aq; s alc
a162	1-Aminoethanol	$CH_2CH(OH)NH_2$	61.08				97	110 d		s aq; sl s eth
a163	2-Aminoethanol	$H_2NCH_2CH_2OH$	61.08	4, 274	1.0117^{25}_{4}	1.4539^{20}	10.3	171	93	misc aq, org solv
a164	2-(2-Aminoethoxy)-ethanol	$H_2NCH_2CH_2OCH_2CH_2OH$	105.14	4^3, 642	1.048			218–224		
a165	2-(2-Aminoethyl-amino)ethanol	$H_2NCH_2CH_2NHCH_2CH_2OH$	104.15	4, 286	1.030	1.4861^{20}		240^{753mm}	>110	v s aq, alc; sl s eth
a166	1-[(2-Aminoethyl)-amino]-2-propanol	$CH_3CH(OH)CH_2NHCH_2$-CH_2NH_2	118.18	Merck: 12, 458	0.9837^{25}_{4}	1.4738^{25}		112^{10mm}		s acids
a167	3-(2-Aminoethyl-amino)propyltri-methoxysilane	$H_2NCH_2CH_2NHCH_2CH_2$-$Si(OCH_3)_3$	222.1		1.01^{25}_{4}	1.4418^{25}		140^{15mm}	150	
a168	3-Amino-9-ethylcarba-zole		210.28	22^1, 642			98–100			
a169	2-Aminoethyl hydro-gen sulfate	$H_2NCH_2CH_2OSO_3H$	141.15	4, 276			277 d			
a170	3-(2-Aminoethyl)-indole		160.22	22^1, 636			118	$137^{0.15mm}$		i aq, bz, chl, eth; s alc, acet, HCl
a171	S-2-Aminoethyliso-thiouronium bro-mide HBr		281.01	Merck: 12, 176			194–195			
a172	*N*-(2-Aminoethyl)-morpholine		130.19	27^3, 370	0.992	1.4755^{20}	25.6	205	175	s aq, alc, bz, acet, acids
a173	4-(2-Aminoethyl)-phenol	$HOC_6H_4CH_2CH_2NH_2$	137.18	13, 625			164–165	166^{2mm}		1 aq^{15}; 10 boiling alc; s HCl

(Continued)

TABLE 2.20 Physical Constants of Organic Compounds (*Continued*)

No.	Name	Formula	Formula weight	Beilstein reference	Density, g/mL	Refractive index	Melting point, °C	Boiling point, °C	Flash point, °C	Solubility in 100 parts solvent
a174	*N*-(2-Aminoethyl)-piperazine		129.21		0.985^{20}_{20}	1.4983^{20}	−26	218–222	93 (OC)	
a175	*N*-(2-Aminoethyl)-1,3-propanediamine	$H_2NCH_2CH_2CH_2NHCH_2CH_2NH_2$	117.20		0.928	1.4815^{20}			96	
a176	2-Amino-2-ethyl-1,3-propanediol	$HOCH_2C(NH_2)(C_2H_5)CH_2OH$	119.16	4,3,850	1.099^{20}_{20}	1.490^{20}	38	152^{10mm}	>110	misc aq; s alc
a177	2-(2-Aminoethyl)-pyridine	$H_2NCH_2CH_2(C_5H_4N)$	122.17	22, 434	1.021	1.5360^{20}		93^{12mm}	100	
a178	4-(2-Aminoethyl)-pyridine	$H_2NCH_2CH_2(C_5H_4N)$	122.17		1.012	1.5403^{20}		104^{9mm}		
a179	2-Amino-5-fluorobenzotrifluoride	$H_2N(F)C_6H_3CF_3$	179.12	12^3, 1991	1.3781	1.4608^{20}		81^{20mm}	70	
a180	Aminoguanidine hydrogen carbonate	$H_2NNHC({=}NH)NH_2 \cdot H_2CO_3$	136.11	3, 117			172 d			i aq; d hot aq
a181	*N*-Aminohexamethyleneimine	$(C_6H_{12}N)NH_2$	114.19		0.984	1.4850^{20}		165	56	
a182	(±)-2-Aminohexanoic acid	$CH_3(CH_2)_3CH(NH_2)COOH$	131.17	4, 433	1.172		301			1.15 aq^{25}; 0.42 alc^{25}; s acids
a183	6-Aminohexanoic acid	$H_2N(CH_2)_5COOH$	131.17	4, 434			204–206			v s aq; i alc, s acids
a184	6-Amino-1-hexanol	$H_2N(CH_2)_5CH_2OH$	117.19	4^2, 748			56–58	135^{30mm}		
a185	(−)-2-Amino-3-hydroxybutanoic acid	$CH_3CH(OH)CH(NH_2)COOH$	119.12	4, 514			d 255			v s aq; i alc, chl, eth
a186	(±)-4-Amino-3-hydroxybutanoic acid	$H_2NCH_2CH(OH)CH_2COOH$	119.12	4^2, 938			218 d			s aq; sl s alc, chl, eth, EtOAc
a187	4-Amino-6-hydroxy-2-mercaptopyrimidine hydrate		161.18	24, 476			>300			
a188	2-Amino-4-hydroxy-6-methylpyrimidine		125.13	24, 343			>300			
a189	4-Amino-3-hydroxy-1-naphthalenesulfonic acid		239.25	14, 846			295 d			i aq, alc, bz, eth
a190	4-Amino-5-hydroxy-1-naphthalenesulfonic acid		239.25	14, 835						sl s aq; i alc, eth
a191	5-Amino-6-hydroxy-2-naphthalenesulfonic acid		239.25							sl s hot aq; i eth
a192	6-Amino-7-hydroxy-2-naphthalenesulfonic acid		239.25	14, 849			>300			

a193	2-Amino-3-hydroxy-pyridine	$H_2N(HO)(C_5H_3N)$	110.12	12^2, 408			172–174			
a194	4-Amino-2-hydroxy-pyrimidine		111.10	24, 314			>300			0.77 aq; sl s alc
a195	1-Aminoindane		133.19	12, 1191	1.038^{15}_{4}	1.5613^{20}	1.5	97^{8mm}	94	sl s aq
a196	5-Aminoindane		133.19	12^1, 511			36	249^{745mm}	>110	sl s aq
a197	5-Aminoindazole		133.15	25^2, 308			175–178			
a198	6-Aminoindazole		133.15	25, 317			206 d			
a199	2-Amino-5-iodobenzoic acid	$H_2N(I)C_6H_3COOH$	263.03	14, 373			221 d			sl s aq, PE; s alc
a200	(±)-2-Amino-4-mercaptobutanoic acid	$HSCH_2CH_2CH(NH_2)COOH$	135.19	4^3, 1647			232–233			
a201	Aminomethanesulfonic acid	$H_2NCH_2SO_3H$	111.12	1, 583			185 d			v s aq
a202	3-Amino-4-methoxy-benzoic acid	$CH_3O(NH_2)C_6H_3COOH$	167.16	14^1, 657			210			
a203	2-Amino-1-methoxy-propane	$CH_3OCH_2CH(CH_3)NH_2$	84.14	4^4, 1615	0.845	1.4065^{20}		93	8	
a204	5-Amino-2-methoxy-pyridine	$CH_3O(NH_2)(C_5H_3N)$	124.14	22^2, 408		1.5745^{20}	31	90^{1mm}	>110	
a205	4′-Amino-*N*-methyl-acetanilide	$CH_3ON(CH_3)C_6H_4NH_2$	164.21	13^1, 30			90–92			
a206	4-Amino-3-methyl-benzenesulfonic acid	$H_2NC_6H_3(CH_3)SO_3H$	187.22	14, 726			>300			
a207	2-Amino-5-methyl-benzoic acid	$H_2N(CH_3)C_6H_3COOH$	151.17	14, 481			175 d			sl s aq; s alc, eth
a208	3-Amino-4-methyl-benzoic acid	$H_2N(CH_3)C_6H_3COOH$	151.17	14, 487			167–169			s aq
a209	2-Amino-3-methyl-1-butanol	$(CH_3)_2CHCH(NH_2)CH_2OH$	103.17	4^3, 805	0.906	1.4543^{20}	35–36	80^{8mm}	90	
a210	2-(Aminomethyl)-1-ethylpyrrolidine		128.22		0.887	1.4665^{20}		60^{16mm}	60	
a211	2-Amino-3-methyl-1-pentanol	$CH_2CH_2CH(CH_3)CH(NH_2)$-CH_2OH	117.19			1.4589^{20}	30	97^{14mm}	100	
a212	2-Amino-4-methyl-1-pentanol	$CH_3CH(CH_3)CH_2CH(NH_2)$-CH_2OH	117.19	4, 298	0.917	1.4496^{20}		200	90	
a213	4-Amino-3-methyl-phenol	$H_2N(CH_3)C_3H_3OH$	123.16				179			
a214	4-(Aminomethyl)-piperidine		114.19			1.4900^{20}	25	200	78	

(*Continued*)

TABLE 2.20 Physical Constants of Organic Compounds (*Continued*)

No.	Name	Formula	Formula weight	Beilstein reference	Density, g/mL	Refractive index	Melting point, °C	Boiling point, °C	Flash point, °C	Solubility in 100 parts solvent
a215	2-Amino-2-methyl-1,3-propanediol	$HOCH_2C(CH_3)(NH_2)CH_2OH$	105.14				108–110	151^{10mm}		250 aq^{20}; s alc
a216	2-Amino-2-methyl-1-propanol	$(CH_3)_2C(NH_2)CH_2OH$	89.14	4^3, 783	0.934_{20}^{20}	1.4480^{20}	25	165	67	misc aq; s alc, org solv
a217	2-Amino-2-methyl-propionic acid	$(CH_3)_2C(NH_2)COOH$	103.12	4, 414			335 (sealed tube)	280 subl		v s aq
a218	2-(Aminomethyl)-pyridine	$H_2NCH_2(C_5H_4N)$	108.14		1.049	1.5440^{20}		85^{12mm}	90	
a219	3-(Aminomethyl)-pyridine	$H_2NCH_2(C_5H_4N)$	108.14		1.062	1.5510^{20}	−21	74^{1mm}	100	
a220	4-(Aminomethyl)-pyridine	$H_2NCH_2(C_5H_4N)$	108.14	22^3, 4181	1.065	1.5515^{20}	−8	230	108	
a221	2-Amino-3-methyl-pyridine	$H_2N(CH_3)(C_5H_3N)$	108.14	22^2, 342	1.073	1.5823^{20}	32–34	222	111	
a222	2-Amino-4-methyl-pyridine	$H_2N(CH_3)(C_4H_3N)$	108.14	22^2, 342			98–100	230		v s aq, alc, DMF
a223	2-Amino-6-methyl-pyridine	$H_2N(CH_3)(C_4H_3N)$	108.14	22^1, 633			42–45	209	103	v s aq
a224	2-Amino-4-methyl-pyrimidine		109.13	24, 84			160	subl		s hot aq; s alc
a225	2-Amino-4-methyl-thiazole		114.17	27, 159			44–46	232	>110	v s aq, alc, eth
a226	2-Aminomethyl-3,5,5-trimethylcyclo-hexanol		171.29		0.969	1.4904^{20}	43–48	265	>110	
a227	*N*-Aminomorpholine		102.14	27, 8	1.059	1.4772^{20}		168	58	
a228	1-Aminonaphthalene	$(C_{10}H_7)NH_2$	143.18	12, 1212	1.13		48–50	301	157	0.17 aq; v s alc, eth
a229	2-Aminonaphthalene	$(C_{10}H_7)NH_2$	143.18	12, 1212			111–113	306		s hot aq, alc, eth
a230	2-Amino-1-naphtha-lenesulfonic acid	$H_2N(C_{10}H_6)SO_3H$	223.25	14, 736			dec			0.031 aq; sl s hot aq; s dil alkali
a231	5-Amino-2-naphtha-lenesulfonic acid	$H_2N(C_{10}H_6)SO_3H$	223.25	14, 758			180			sl s aq; s hot aq
a232	8-Amino-2-naphthol	$H_2NC_{10}H_6OH$	159.19	13, 685			207			
a233	2-Amino-4-nitro-benzoic acid	$H_2N(NO_2)C_6H_3COOH$	182.14	14, 374			270 d			i aq; v s alc, eth
a234	2-Amino-5-nitro-benzonitrile	$H_2N(NO_2)C_6H_3CN$	163.14	14^2, 234			200–207			
a235	5-Amino-5-nitro-benzophenone	$C_6H_5COC_6H_4(NH_2)NO_2$	242.23	14, 79			166–168			

a236	2-Amino-6-nitrobenzothiazole		195.20	27^2, 232			247–249			
a237	4-Amino-3-nitrobenzotrifluoride	$H_2N(NO_2)C_6H_3CF_3$	206.12				105–106			
a238	2-Amino-4-nitrophenol	$O_2N(NH_2)C_6H_3OH$	154.13	13^2, 192			143–145			
a239	2-Amino-5-nitrophenol	$O_2N(NH_2)C_6H_3OH$	154.13	13, 390			202 d			
a240	4-Amino-2-nitrophenol	$O_2N(NH_2)C_6H_3OH$	154.13	13, 520			125–127			
a241	D-(−)-*threo*-2-Amino-1-(4-nitrophenyl)-1,3-propanediol	$HOCH_2C(NH_2)C(OH)$-$C_6H_4NO_2$	212.21				163–165			
a242	2-Amino-5-(4-nitrophenylsulfonyl-thiazole		285.30				222–226			
a243	2-Amino-5-nitropyridine	$H_2N(C_5H_3N)NO_2$	139.11	22^1, 631			186–188			sl s aq, bz, eth
a244	2-Amino-5-nitrothiazole		145.14	Merck: 12, 477			d 202			s sl s aq; 0.7 alc; 0.4 ether; s dil acids
a245	*exo*-2-Aminonorbornane		111.19	12^3, 160	0.938	1.4807^{20}		49^{10mm}	35	
a246	2-Aminopentane	$H(CH_2)_3CH(NH_2)CH_3$	87.17	4, 177	0.739^{20}	1.4047^{20}		91–92		s aq, alc, eth, PE
a247	3-Aminopentane	$C_2H_5CH(NH_2)C_2H_5$	87.17	4, 179	0.749^{20}_{4}	1.4055^{20}		91	1	misc aq, alc, eth
a248	DL-2-Aminopentanoic acid	$H(CH_2)_3CH(NH_2)COOH$	117.15	4, 416			303	320 subl		5.5 aq^{18}; v sl s alc, chl, eth, PE
a249	5-Aminopentanoic acid	$H_2N(CH_2)_4COOH$	117.15	4, 418			158–161			v s aq; sl s alc; i eth
a250	5-Amino-1-pentanol	$H_2N(CH_2)_5OH$	103.17	4^1, 441	0.949	1.4615^{20}	35–37	122^{16mm}	65	
a251	2-Aminophenethyl alcohol	$H_2NC_6H_4CH_2CH_2OH$	137.18	13^3, 1679	1.045	1.5849^{20}		148^{4mm}	>112	
a252	2-Aminophenol	$H_2NC_6H_4OH$	109.13	13, 354			170–174			2 aq; 4.3 alc; v s eth
a253	3-Aminophenol	$H_2NC_6H_4OH$	109.13	13, 401			122–123	164^{11mm}		2.5 aq; v s hot aq, alc, eth
a254	4-Aminophenol	$H_2NC_6H_4OH$	109.13	13, 427			190	150^{3mm}		0.65 aq; 4.5 alc; 9.3 $EtMeKetone^{58}$; s eth
a255	4′-Aminophenylacetonitrile	$H_2NC_6H_4CH_2CN$	132.17				45–48	312	>110	sl s hot aq; s alc
a256	1-(3-Aminophenyl)-ethanol	$H_2NC_6H_4CH(CH_3)OH$	137.18	13^3, 1654			68–71			
a257	2-Amino-1-phenylethanol	$H_2NCH_2CH(C_6H_5)OH$	137.18	13^2, 361			56–58	160^{17mm}		v s aq; s alc

(*Continued*)

TABLE 2.20 Physical Constants of Organic Compounds (*Continued*)

No.	Name	Formula	Formula weight	Beilstein reference	Density, g/mL	Refractive index	Melting point, °C	Boiling point, °C	Flash point, °C	Solubility in 100 parts solvent
a258	1*S*,2*S*-(+)-2-Amino-1-phenyl-1,3-propanediol	$C_6H_5CH(OH)CH(NH_2)$-CH_2OH	167.21	13,*4*, 2968			109–113			
a259	L-2-Amino-3-phenyl-1-propanol	$C_6H_5CH_2(NH_2)CH_2OH$	151.21	13^3, 1757			92–94			
a260	3-Amino-1-phenyl-2-pyrazolin-5-one		175.19				210 d			
a261	*N*-Aminopiperidine		100.17	20, 89	0.928	1.4750^{20}		146^{730mm}	36	
a262	3-Amino-1,2-propanediol	$H_2NCH_2CH(OH)CH_2OH$	91.11	4, 301	1.175	1.4920^{20}		265^{739mm}	>110	
a263	DL-1-Amino-2-propanol	$CH_3CH(OH)CH_2NH_2$	75.11	4, 289	0.973	1.4483^{20}	−2	160	76	v s aq, alc; i eth
a264	DL-2-Amino-1-propanol	$CH_3CH(NH_2)CH_2OH$	75.11	4^1, 432	0.943	1.4495^{20}		173–176	83	v s aq, alc, eth
a265	*S*-(+)-2-Amino-1-propanol	$CH_3CH(NH_2)CH_2OH$	75.11	4^3, 735	0.965	1.4498^{20}		176	62	v s aq, alc, eth
a266	3-Amino-1-propanol	$H_2NCH_2CH_2CH_2OH$	75.11	4, 288	0.982	1.4610^{20}	10–12	188	79 (TOC)	s aq, alc
a267	2-Amino-1-propene-1,1,3-tricarbonitrile	$NCC(CN){=}C(NH_2)CH_2CN$	132.13	Merck: 11, 495			171–173			s aq
a268	3-Aminopropyl-(diethoxy)methylsilane	$H_2N(CH_2)_3Si(CH_3)$-$(OCH_2CH_3)_2$	191.4		0.916^{20}_{4}	1.427^{20}		88^{8mm}		
a269	1-(3-Aminopropyl)-imidazole		125.18	23^3, 577	1.049	1.5190^{20}			>110	
a270	*N*-(3-Aminopropyl)-iminodiethanol	$H_2N(CH_3)N(CH_2CH_2OH)_2$	162.23		0.1071	1.4980^{20}		170^{2mm}	137	
a271	*N*-(3-Aminopropyl)-morpholine		144.22		0.9872^{20}_{20}	1.4761^{20}	−15	224	98	misc aq, alc, bz
a272	*N*-(3-Aminopropyl)-2-pyrolidinone		142.20		1.014	1.500^{20}		123^{1mm}	>110	
a273	3-Aminopropyltriethoxysilane	$H_2N(CH_2)_3Si(OC_2H_5)_3$	221.37	0.9506^{20}_{4}	1.4225^{20}			217	104	
a274	3-Aminopropyltrimethoxysilane	$H_2N(CH_2)_3Si(OCH_3)_3$	179.29		1.01^{25}_{4}	1.420^{25}		80^{8mm}	83	
a275	2-Aminopyridine	$(C_5H_4N)NH_2$	94.12	22, 428			58.1	210.6	92	s aq, alc, bz, eth
a276	3-Aminopyridine	$(C_5H_4N)NH_2$	94.12	22, 431			64	250–252		s aq, alc, bz, eth
a277	4-Aminopyridine	$(C_5H_4N)NH_2$	94.12	22, 433			160–162	273		s aq, alc; sl s bz, eth
a278	2-Aminopyrimidine		95.11	24, 80			125–127	subl		v s aq
a279	4-Aminoquinaldine		158.20	22, 453			167–169	333		sl s aq; v s alc, eth, acet; s hot bz

a280	4-Aminosalicylic acid	$H_2NC_6H_3(OH)CO_2H$	153.14	14, 579			150–151			0.2 aq; 4.8 alc; s dil acids, alk; sl s eth
a281	5-Aminosalicylic acid	$H_2NC_6H_3(OH)CO_2H$	153.14	14, 579			280 d			
a282	2-Aminoterephthalic acid	$H_2NC_6H_3(CO_2H)_2$	181.15	14, 558			324 d			
a283	5-Amino-1,2,3,4-tetrazole hydrate		103.08	26, 403			204 d			
a284	2-Amino-1,3,4-thiadiazole		101.13	27, 624			190–192			
a285	2-Aminothiazole		100.14	27, 155			93			sl s aq, alc, eth; s hot aq, HCl
a286	2-Amino-2-thiazoline		100.14	27, 136			79–82			s HCl
a287	2-Aminothiophenol	$H_2NC_6H_4SH$	125.19	13, 397	1.170	1.6420^{20}	19–21	$72^{0.1mm}$	79	
a288	2-Aminotoluene-5-sulfonic acid	$H_2NC_6H_3(CH_3)SO_3H$	187.22	14, 726			>300			i aq^{12}; v s hot aq
a289	3-Amino-1,2,4-triazole		84.08	26, 137			150–153			s aq, alc, chl
a290	5-Amino-1,3,3-trimethylcyclohexanemethylamine	$H_2N(C_6H_7)(CH_3)_3CH_2NH_2$	170.30		0.922	1.4880^{20}	10	247	>110	
a291	5-Amino-2,2,4-trimethylcyclopentanemethylamine		156.27		0.901	1.4733^{20}		221	97	
a292	11-Aminoundecanoic acid	$H_2N(CH_2)_{10}CO_2H$	201.31				190–192			
a293	Aniline	$C_6H_5NH_2$	93.12	12, 59	1.027^{20}_{20}	1.5863^{20}	−6	184–186	70	3.5 aq^{25}; s acids; misc most org solv
a294	Aniline hydrochloride	$C_6H_5NH_2 \cdot HCl$	129.59	Merck: 12, 696	1.222		198	245	193 (CC)	100 aq; v s alc
a295	2-Anilinoethanol	$C_6H_5NHCH_2CH_2OH$	137.18	12, 182	1.085	1.5793^{20}		152^{10mm}	153	sl s aq; v s alc, chl, eth
a296	3-Anilinopropionitrile	$C_6H_5NHCH_2CH_2CN$	146.19				52–53		>110	
a297	Anthracene		178.23	5, 657	1.25^{27}_{4}		215-218	339–342	121 (CC)	1.5 alc; 1.6 bz; 1.2 chl; 3.1 CS_2; 0.5 eth; i aq
a298	9,10-Anthraquinone		208.20	7, 781	1.43^{20}_{4}		286	377	185 (CC)	0.44 alc^{25}; 0.6 chl^{20}; 0.2 bz^{20}; 0.11 eth^{25}
a299	Antipyrine		188.23	24, 27	1.088^{113}_{4}		111–114	319		100 aq; 77 alc; 100 chl; 2.3 eth
a300	L-(+)-Arabinose		150.13	31, 32			157–160			100 aq; 0.4 alc
a301	L-(+)-Arginine	$H_2NC(=NH)NH(CH_2)_3CH(NH_2)CO_2H$	174.20	4, 420			d 240			15 aq^{21}; sl s alc

(*Continued*)

TABLE 2.20 Physical Constants of Organic Compounds (*Continued*)

No.	Name	Formula	Formula weight	Beilstein reference	Density, g/mL	Refractive index	Melting point, °C	Boiling point, °C	Flash point, °C	Solubility in 100 parts solvent
a302	L-(+)-Ascorbic acid		176.12	18^3, 3038	1.65^{25}		190–192			33 aq; 3.3 alc; 1 glyc; i bz, chl, eth, PE
a303	L-(+)-Asparagine	$H_2NCOCH_2CH(NH_2)CO_2H$	132.12	4, 476			235			3.5 aq^{28}; s alkalis, acids; i alc, bz, eth
a304	L-(+)-Aspartic acid	$HO_2CCH_2CH(NH_2)CO_2H$	133.10	4, 472	$1.661^{12.5}$		270–272			0.45 aq; s alkalis, acids; i alc, eth
a305	Atropine		289.38	21, 27			114–116	subl 110 high vac		0.22 aq; 50 alc; 4 eth; 100 chl; 3.9 glyc; s bz, dil acids
a306	Aurintricarboxylic acid, triammonium salt		473.44	10^2, 775			225 d			v s aq
a307	2-Azacyclooctanone		127.19	21, 242			35–38	148^{10mm}	>110	
a308	2-Azacyclotridecanone		197.32				150–153			
a309	Azidotrimethylsilane	$(CH_3)_3SiN_3$	115.21		0.868	1.4140^{20}	−95	95–96	23	
a310	Azidotriphenylsilane	$(C_6H_5)_3SiN_3$	301.4				83–84	$100^{0.01mm}$		
a311	1-Aziridineethanol	$(C_2H_4N)CH_2CH_2OH$	87.12		1.088	1.4560^{20}		168	67	
a312	Azobenzene	$C_6H_5N{=}NC_6H_5$	182.23	16, 8	1.203_4^{20}		67–68	293		4.2 alc^{20}; s eth, HOAc
a313	2,2′-Azobis(2-methylpropionitrile)	$(CH_3)_2C(CN)N{=}N{-}C(CN)(CH_3)_2$	164.21	4, 563				107 d		2 $EtOH^{20}$; 5 $MeOH^{20}$; can explode in acetone
a314	Azodicarbonamide	$H_2NCON{=}NCONH_2$	116.08	3, 123			225 d			i aq, alc; s hot aq
a315	4,4′-Azoxydianisole	$H_3OC_6H_4N{=}N(\rightarrow O)C_6H_4{-}OCH_3$	258.28	16, 637			120			
a316	Azulene		128.17	5^2, 432			99–100	242		i aq; s org solvents
b1	Barbituric acid		128.09	24, 467			252 d			s hot aq, dil acids
b2	Basic fuchsin		337.86	13, 765	1.22		250 d			0.3 aq; s alc, acids
b3	Benzaldehyde	C_6H_5CHO	106.12	7, 174	1.050_4^{15}	1.5456^{20}	−26	179	63	0.3 aq; misc alc, eth
b4	Benzamide	$C_6H_5CONH_2$	121.13	9, 195	1.341^4		129–130	288–290		1.3 aq; 17 alc; 30 pyr
b5	Benzanilide	$C_6H_5CONHC_6H_5$	197.24	12, 262	1.315		163	117^{10mm}		i aq; 1.7 alc; sl s eth
b6	1,2-Benzanthracene		228.29	5, 718			155–157	437.6		sl s hot aq; s org solv
b7	2,3-Benzanthracene		228.29	5^2, 628	1.35		357 (Cu block)	subl		sl s most org solv
b8	Benzene	C_6H_6	78.11	5, 179	0.8787_4^{15}	1.5011^{20}	5.5	80.0	−11 (CC)	0.17 aq; misc most org solv
b9	Benzene-*1,3,5-d*$_3$	$C_6H_3D_3$	81.14	5^3, 518	0.908	1.4990^{20}		80	−11 (CC)	similar to ordinary benzene
b10	Benzene-$^{13}C_6$	$^{13}C_6H_6$	84.07		0.949	1.5010^{20}	5.5	80	−11 (CC)	similar to ordinary benzene

b11	Benzene-d_6	C_6D_6	84.16	5^3, 519	0.950	1.4986^{20}	6.8	79.1	−11 (CC)	similar to ordinary benzene
b12	Benzenearsonic acid	$C_6H_5AsO(OH)_2$	202.03	16, 868	1.760^{25}		162			2.5 aq; 2 alc; i chl
b13	Benzeneboronic acid	$C_6H_5B(OH)_2$	121.94	16, 920			216			2.5 aq; 1.8 bz; 30 eth; 178 MeOH
b14	1,4-Benzenedicarbaldehyde	$C_6H_4(CHO)_2$	134.13	7, 675			113	248		i aq; 6 bz; 17 acet; 2 eth; 14 diox; 46 MeOH
b15	1,2-Benzenedicarbonyl dichloride	$C_6H_4(COCl)_2$	203.02	9, 834	1.409^{20}		15–16	280–282		d aq, alc; s eth
b16	1,4-Benzenedicarbonyl dichloride	$C_6H_4(COCl)_2$	203.02	9, 844			81	266	180	37 bz; 9 CCl_4
b17	1,3-Benzenedicarboxylic acid	$C_6H_4(COOH)_2$	166.13	9, 832			345–348	subl		0.012 aq; v s alc, HOAc; i bz, PE
b18	1,4-Benzenedicarboxylic acid	$C_6H_4(COOH)_2$	166.13	9, 841			subl 402			sl s alc; s alkalis; v sl s aq, chl, eth
b19	1,4-Benzenedimethanol	$C_6H_4(CH_2OH)_2$	138.17	6, 919	1.100^{117}		117–119	143^{1mm}	188	
b20	Benzenehexacarboxylic acid	$C_6(COOH)_6$	342.17	9, 1008			286 d			v s aq, alc
b21	Benzenesulfinic acid	$C_6H_5S(=O)OH$	142.16	11, 2			85	100 d		sl s aq; s alc, bz, eth
b22	Benzenesulfonamide	$C_6H_5SO_2NH_2$	157.19	11, 39			150–152			i aq; sl s alc; s eth
b23	Benzenesulfonic acid	$C_6H_5SO_2OH$	158.18	11, 26			50–51			v s aq, alc; sl s bz; i CS_2, eth
b24	Benzenesulfonyl chloride	$C_6H_5SO_2Cl$	176.62	11, 34	1.3842^{15}_{15}	1.5518^{20}	14.5	120^{10mm}	>110	i aq; s alc, eth
b25	Benzenesulfonyl fluoride	$C_6H_5SO_2F$	160.17	11^2, 23	1.3286^{20}_{4}	1.4920^{20}	−5	207–208	87	s alc, eth
b26	Benzenesulfonyl hydrazide	$C_6H_5SO_2NHNH_2$	172.21	11, 52			d 104			flammable solid
b27	1,2,4,5-Benzenetetracarboxylic acid	$C_6H_2(COOH)_4$	254.15	9, 997			276			1.5 aq; v s alc
b28	1,2,4,5-Benzenetetracarboxyl dianhydride		218.12	19, 196			283–286	397–400		
b29	1,2,3-Benzenetricarboxylic acid dihyrate	$C_6H_3(COOH)_3\cdot 2H_2O$	246.18	9, 976			192 d			sl s aq; v s eth

(Continued)

TABLE 2.20 Physical Constants of Organic Compounds (*Continued*)

No.	Name	Formula	Formula weight	Beilstein reference	Density, g/mL	Refractive index	Melting point, °C	Boiling point, °C	Flash point, °C	Solubility in 100 parts solvent
b30	1,2,4-Benzenetricarboxylic acid	$C_6H_3(COOH)_3$	210.14	9, 997			231 d			2.1 aq; 25.3 alc; 7.9 acet; v s eth
b31	1,3,5-Benzenetricarboxylic acid	$C_6H_3(COOH)_3$	210.14	9, 978			>330			sl s aq; v s alc; s eth
b32	1,2,4-Benzenetricarboxylic anhydride		192.13	18, 468			161–163	245^{14mm}		50 acet; 22 EtOAc; 15 DMF
b33	1,3,5-Benzenetricarboxylic trichloride	$C_6H_3(COCl)_3$	265.48				35–36	180^{16mm}	>110	
b34	1,2,4-Benzenetriol	$C_6H_3(OH)_3$	126.11	6, 1087			141			v s aq, alc, eth, EtOAc
b35	Benzil	$C_6H_5CO—COC_6H_5$	210.23	7, 747	1.23_4^{15}		95	346–348		i aq; s alc, bz, chl, EtOAc, eth
b36	Benzil dioxime	$C_6H_5C(=NOH)-C(=NOH)C_6H_5$	240.25	7³, 3816			(α) 240 (β) 214			i aq, HOAc, eth; sl s alc; s NaOH
b37	Benzilic acid	$(C_6H_5)_2C(OH)COOH$	228.24	10, 342			150			sl s aq; v s alc, eth hot aq
b38	Benzil monohydrazone	$C_6H_5C(=NNH_2)COC_6H_5$	224.26	7¹, 394			150–152			
b39	Benzimidazole		118.13	23, 131			170.5	>360		sl s aq, eth; v s alc
b40	7,8-Benzo-1,3-diazaspiro[4,5]decane-2,4-dione		216.23	Merck: 12, 9372			268			s alc, HOAc
b41	1,4-Benzodioxan		136.15		1.142	1.5490^{20}		103^{6mm}	87	
b42	2,3-Benzofuran		118.13	17, 54	1.072	1.5660^{20}	<−18	173–175	56	i aq; misc alc, bz, eth, PE
b43	Benzofurazan-1-oxide		136.11	27¹, 740			69–71			
b44	Benzoic acid	C_6H_5COOH	122.12	9, 92	1.321		122.4	249	121 (CC)	0.29 aq²⁵; 43 alc; 10 bz; 22 chl; 33 eth; 33 acet; 30 CS_2
b45	Benzoic anhydride	$(C_6H_5CO)_2O$	226.22	9, 164	1.1989_4^{15}		42	360	110	i aq; s alc, acet, chl bz, HOAc, EtOAc
b46	DL-Benzoin	$C_6H_5COCH(OH)C_6H_5$	212.25	8, 167	1.3100_4^{20}		137	194^{12mm}		s hot alc, acet; 20 pyr; sl s eth
b47	Benzoin ethyl ether	$C_6H_5CH(C_2H_5)COC_6H_5$	240.30	8, 174	1.1016_4^{17}	1.5727^{17}	62	195^{20mm}		s alc, bz, eth
b48	Benzoin isobutyl ether	$C_6H_5CH[OCH_2CH(CH_3)_2]-COC_6H_5$	268.36		0.985	1.5485^{20}		$133^{0.5mm}$	85	
b49	Benzoin methyl ether	$C_6H_5CH(OCH_3)COC_6H_5$	226.28	8, 174	1.1278_4^{14}		48	189^{15mm}	>110	v s alc, bz, eth
b50	α-Benzoinoxime	$C_6H_5CH(OH)C(=NOH)-C_6H_5$	227.26	8, 175			152–156			sl s aq; s alc, NH_4OH
b51	Benzonitrile	C_6H_5CN	103.12	9, 275	1.010	1.5289^{20}	−12.7	191	71	0.2 aq; misc org solv
b52	1,2-Benzophenanthrene		202.26	5, 718	1.274_4^{20}		258	448		i aq; s alc, eth

b53	Benzophenone	$C_6H_5COC_6H_5$	182.22	7, 411	1.1108_4^{18}	1.5975^{45}	48	305	>110	13.3 alc; 17 eth; s chl
b54	Benzophenone hydrazone	$C_6H_5C(=NNH_2)C_6H_5$	196.25	7, 417			95–98	230^{55mm}		
b55	1-Benzopyran-4(4*H*)-one		146.15	17, 327			55–60			
b56	1,2-Benzo[*a*]pyrene		252.32	Merck: 12, 1134			179	312^{10mm}		s bz; sl s alc
b57	4,5-Benzo[*e*]pyrene		252.32	Merck: 12, 1105			179			s bz
b58	1,4-Benzoquinone	$C_6H_4(=O)_2$	108.10	7, 609	1.318_4^{20}		116			sl s aq; s alc, hot bz, eth, hot PE; alkalis with dec
b59	Benzothiazole		135.19	Merck: 12, 1139	1.2460_4^{20}	1.6379^{20}	2	131^{34mm}	>110	sl s aq; v s alc, CS_2
b60	Benzo[*b*]thiophene		134.20	17, 59	1.1937^{40}	1.6302^{40}	32	221	>110	s alc, bz, chl, eth
b61	1,2,3-Benzotriazole		119.13	26, 38	1.238	1.6420^{20}	98.5	204 may explode		sl s aq; s alc, bz, chl, DMF
b62	Benzoxazole		119.12	27, 42		1.5594	30	182	58	sl s aq
b63	1-Benzoylacetone	$C_6H_5COCH_2COCH_3$	162.19	7, 680	1.090_{60}^{60}		60	260 sl d		sl s aq; v s alc, eth
b64	2-Benzoylbenzoic acid	$C_6H_5COC_6H_4COOH$	226.23	10, 747			129	265		sl s aq; v s alc, eth
b65	Benzoyl bromide	C_6H_5COBr	185.03	9, 195	1.5467^{20}	1.5883^{20}	−24	219	90	d aq, alc; misc eth
b66	Benzoyl chloride	C_6H_5COCl	140.57	9, 182	1.211_4^{20}	1.5537^{20}	−1.0	197.2	88 (CC)	d aq, alc; misc bz, eth CS_2
b67	Benzoyl cyanide	C_6H_5COCN	131.13	10, 659	1.106		32	206		i aq
b68	Benzoyl fluoride	C_6H_5COF	124.11	9, 181	1.140	1.4960^{20}	−28	161	48	d hot aq; v s alc, eth
b69	Benzoylformic acid	$C_6H_5COCOOH$	150.13	10, 654			67–69			
b70	*N*-Benzoylglycine	$C_6H_5CONHCH_2COOH$	179.18	9, 225			179			0.4 aq; 0.1 chl; 0.25 eth; sl s alc; i bz, PE
b71	Benzoylhydrazine	$C_6H_5CONHNH_2$	136.15	9, 319			117			
b71a	Benzoyl peroxide	$(C_6H_5CO)_2O_2$	242.23	9, 179			103–106	explodes		2.5 CS_2; s bz, chl, eth
b72	3-Benzoylpropanoic acid	$C_6H_5COCH_2CH_2COOH$	178.19	10, 696			117–119			sl s aq; s alc
b73	2-Benzoylpyridine	$C_6H_5CO(C_5H_4N)$	183.21	21, 330			44	317	150	
b74	3-Benzoylpyridine	$C_6H_5CO(C_5H_4N)$	183.21	21, 331			40	397	150	s alc, bz, eth
b75	4-Benzoylpyridine	$C_6H_5CO(C_5H_4N)$	183.21	21, 331			71	315	150	s alc, bz, eth
b76	Benzyl acetate	$CH_3CO_2CH_2C_6H_5$	150.18	6, 435	1.050_4^{25}	1.4998^{25}	−51.5	213.5	102 (CC)	i aq; misc alc, eth
b77	Benzyl acetoacetate	$CH_3COCH_2CO_2CH_2C_6H_5$	192.21	6, 438	1.112	1.5121^{20}		159^{10mm}	>110	

(Continued)

TABLE 2.20 Physical Constants of Organic Compounds (*Continued*)

No.	Name	Formula	Formula weight	Beilstein reference	Density, g/mL	Refractive index	Melting point, °C	Boiling point, °C	Flash point, °C	Solubility in 100 parts solvent
b77a	Benzylacetone	$C_6H_5CH_2CH_2COCH_3$	148.21	7, 314	0.989	1.5122^{20}		235	98	
b78	Benzyl alcohol	$C_6H_5CH_2OH$	108.14	6, 428	1.0453^{20}_{4}	1.5403^{20}	−15.2	205	93 (CC)	0.08 aq; misc alc, chl, eth
b79	Benzylamine	$C_6H_5CH_2NH_2$	107.16	12, 1013	0.9831^{19}_{4}	1.5401^{20}	10	185	60	misc aq, alc, eth
b80	*N*-Benzylaminoethanol	$C_6H_5CH_2NHCH_2CH_2OH$	151.21	12, 1040	1.065	1.5435^{20}		156^{12mm}	>110	
b81	3-(Benzylamino)-propanonitrile	$C_6H_5CH_2NHCH_2CH_2CN$	160.22		1.024	1.5308^{20}			>110	
b82	*N*-Benzylbenzamide	$C_6H_5CH_2NHCH_2C_6H_5$	211.26				106			
b83	Benzyl benzoate	$C_6H_5CO_2CH_2C_6H_5$	212.25	9, 121	1.118^{25}_{4}	1.5681^{21}	21	323	148	misc alc, chl, eth
b84	2-Benzylbenzoic acid	$C_6H_5CH_2C_6H_4COOH$	212.24	9², 471			110–113			sl s aq; s alc, bz, chl, eth
b85	Benzyl bromide	$C_6H_5CH_2Br$	171.04	5, 306	1.4380^{22}_{0}	1.5752^{20}	−3.9	199	86	slowly dec aq
b86	Benzyl 2-bromoacetate	$BrCH_2CO_2CH_2C_6H_5$	229.08	6¹, 220	1.446	1.5440^{20}		170^{22mm}	>110	
b87	Benzyl-*tert*-butanol	$C_6H_5CH_2CH_2C(CH_3)_2OH$	164.25	6, 548		1.5090^{20}	31–33	144^{85mm}	>110	
b88	Benzyl butyl 1,2-phthalate	$C_6H_5CH_2O_2C_6H_4CO_2C_4H_9$	312.37	9², 594	1.119^{25}_{25}	1.5400^{20}			199	
b89	Benzyl carbamate	$C_6H_5CH_2OCONH_2$	151.17	6, 437			87–89	220 d		v s alc; sl s eth
b90	Benzyl chloride	$C_6H_5CH_2Cl$	126.59	5, 292	1.100^{20}_{20}	1.5381^{20}	−43 to −49	179	67	misc alc, chl, eth
b91	Benzyl chloroformate	$C_6H_5CH_2OCOCl$	170.60	6, 437	1.195	1.5190^{20}		103^{20mm}	91	dec aq; s eth
b92	Benzyl chlorothiol-formate	$C_6H_5CH_2SCOCl$	186.5		1.237^{30}_{4}	1.5711^{30}		$80^{0.13mm}$	118	
b93	Benzyl cinnamate	$C_6H_5CH{=}CHCO_2CH_2C_6H_5$	238.29	9, 584			39	200^{5mm}	>110	s alc, eth; i aq, glyc
b94	*S*-Benzyl-L-cysteine	$C_6H_5CH_2SCH_2CH(NH_2)$-COOH	211.28	6, 465			214 d			
b95	Benzyl *N*,*N*-dimethyl-dithiocarbamate	$(CH_3)_2NCS_2CH_2C_6H_5$	211.35				41		>110	
b96	Benzyldimethylstearyl-ammonium chloride hydrate	$C_6H_5CH_2N[(CH_2)_{17}CH_3]$-$(CH_3)_2Cl \cdot H_2O$	442.18	12³, 2212			67–69			
b96a	*N*-Butyl-*N*-ethylaniline	$C_6H_5N(CH_2C_6H_5)C_2H_5$	211.31	12, 1026	1.029	1.5950^{20}		164^{6mm}	>110	
b97	Benzyl ethyl ether	$C_6H_5CH_2OC_2H_5$	136.20	Merck: 12, 1168	0.9478^{20}	1.4955^{20}		186		misc alc, eth; i aq
b98	*N*-Benzylformamide	$C_6H_5CH_2NHCHO$	135.17	12, 1043			61			
b99	Benzyl formate	$C_6H_5CH_2O_2CH$	136.15	Merck: 12, 1169	1.081^{20}_{4}			203		i aq; s alc
b100	Benzyl 4-hydroxy-benzoate	$HOC_6H_4CO_2CH_2C_6H_5$	228.25	10,3, 311			110–112			
b101	*O*-Benzylhydroxyl-amine hydrochlo-ride	$C_6H_5CH_2ONH_2 \cdot HCl$	159.62	6, 440				238 subl	>110	

b102	Benzylidineaniline	$C_6H_5N{=}CHC_6H_5$	181.24	12, 195	1.045^{50}_{4}		56	300	>110	s alc, chl, CS_2
b103	Benzylidenemalononitrile	$C_6H_5CH{=}C(CN)_2$	154.17	9, 895			83–85			
b104	*N*-Benzylidenemethylamine	$C_6H_5CH{=}NCH_3$	119.17	7, 213	0.967	1.5520^{20}		80^{18mm}	>112	
b105	3-Benzylidenephthalide		124.21	17, 376			99–102			
b106	Benzyl mercaptan	$C_6H_5CH_2SH$	222.24	6, 453	1.058^{20}	1.5751^{20}		206^{30mm}	>110	
b107	Benzyl methacrylate	$H_2C{=}C(CH_3)CO_2CH_2C_6H_5$	176.22	6^3, 1481	1.040	1.5120^{20}		98^{4mm}	77	
b108	*N*-Benzylmethylamine	$C_6H_5CH_2NHCH_3$	138.23	12, 1019	0.939	1.5230^{20}		184–189	77	
b109	3-(*N*-Benzyl-*N*-methylamino)-1,2-propanediol	$C_6H_5CH_2N(CH_3)CH_2CH(OH)CH_2OH$	195.26		1.084	1.5341^{20}		206^{30mm}	>110	
b110	Benzyl methyl sulfide	$C_6H_5CH_2SCH_3$	138.23	6, 453	1.015	1.5620^{20}		195–198	73	
b111	1-Benzyl-3-methyl-2-thiourea	$C_6H_5CH_2NHC({=}S)NHCH_3$	180.27	12, 1052			74–76			
b112	Benzyl nicotinate	$(C_5H_4N)CO_2CH_2C_6H_5$	213.24	22,*3*, 366	1.165	1.5700^{20}	21–23	189^{12mm}	>110	
b113	4-Benzyloxybenzaldehyde	$C_6H_5CH_2OC_6H_4CHO$	212.25	8, 73			73–74			
b114	4-Benzyloxybenzyl alcohol	$C_6H_5CH_2OC_6H_4CH_2OH$	214.26				86–87			
b115	2-Benzyloxyethanol	$C_6H_5CH_2OCH_2CH_2OH$	152.20	6^2, 413	1.07^{20}_{20}	1.5210^{20}		265	129	0.4 aq
b116	4-Benzyloxy-3-methoxybenzaldehyde	$C_6H_5CH_2OC_6H_4(OCH_3)CHO$	242.29				63–65			
b117	4-(Benzyloxymethyl)-2,2-dimethyl-1,3-dioxolane		222.28	19^2, 73	1.051	1.4940^{20}		$91^{0.1mm}$	>110	
b118	Benzyl phenyl sulfide	$C_6H_5CH_2SC_6H_5$	200.30	6, 454			41–44	197^{27mm}	>110	i aq; sl s alc; s eth
b119	1-Benzylpiperazine		176.26		1.014	1.5467^{20}			>110	s aq, alc, eth
b120	4-Benzylpiperidine		175.28	20, 296	0.997	1.5379^{20}	6–7	279	>110	
b121	1-Benzyl-4-piperidone		189.26		1.021	1.5399^{20}		134^{7mm}	>110	
b122	2-Benzylpyridine	$C_6H_5CH_2(C_5H_4N)$	169.23	20, 425	1.054	1.5790^{20}	8–10	276	125	i aq; v s alc, eth
b123	4-Benzylpyridine	$C_6H_5CH_2(C_5H_4N)$	169.23	20, 426	1.061^{20}_{0}	1.5818^{20}		287	115	s alc; v s eth
b124	1-Benzyl-2-pyrrolidinone		175.23		1.095	1.5525^{20}			>110	
b125	Benzyl salicylate	$HOC_6H_4CO_2CH_2C_6H_5$	228.25	Merck: 12, 1181	1.175^{20}			208^{25mm}		sl s aq; misc alc, eth
b126	Benzyl thiocyanate	$C_6H_5CH_2SCN$	149.22	6, 460			43	235	>110	i aq; s alc; v s eth
b127	Benzyltributylammonium chloride	$C_6H_5CH_2N(C_4H_9)_3^+Cl^-$	312.94				164 d			

(Continued)

TABLE 2.20 Physical Constants of Organic Compounds (*Continued*)

No.	Name	Formula	Formula weight	Beilstein reference	Density, g/mL	Refractive index	Melting point, °C	Boiling point, °C	Flash point, °C	Solubility in 100 parts solvent
b128	Benzyltrichlorosilane	$C_6H_5CH_2SiCl_3$	225.28	16, 912	1.288_4^{20}	1.5250^{20}		142^{100mm}	93	
b129	Benzyltriethoxysilane	$C_6H_5CH_2Si(OC_2H_5)_3$	254.40		0.986_4^{20}			175^{70mm}		
b130	Benzyltriethylammonium chloride	$C_6H_5CH_2N(C_2H_5)_3^+Cl^-$	227.78	12, 1021			185 d			
b131	Benzyltrimethylammonium chloride	$C_6H_5CH_2N(CH_3)_3^+Cl^-$	185.70	12, 1021			239 d		none	
b132	Benzyltrimethylsilane	$C_6H_5CH_2Si(CH_3)_3$	164.32	16, *1*, 526	0.8933^{20}	1.4941^{20}		190	57	
b133	Betaine	$(CH_3)_3N^+CH_2COO^-$	117.15	4, 347			dec 310			160 aq; 55 MeOH; 8.7 EtOH
b134	Bicyclo[2.2.1]hepta-2,5-diene		92.14		0.909^{20}	1.4707^{20}	−20	89	−11	i aq; s PE
b135	Bicyclo[2.2.1]-2-heptene		94.16				44–46	96	−15	s eth
b136	Bicyclo[2.2.1]-2-heptene-2-carbaldehyde		122.16		1.108	1.4883^{20}		70^{12mm}	51	
b137	Biguanide	$H_2NC(=NH)NH-C(=NH)NH_2$	101.11	3, 93			130	dec 142		s aq, alc; i bz, chl, eth
b138	Biphenyl	$C_6H_5-C_6H_5$	154.20	5, 578	0.991_4^{75}	1.588^{77}	69–71	256	113 (CC)	i aq; s alc, eth
b139	4-Biphenylcarboxylic acid	$C_6H_5-C_6H_4COOH$	198.22	9, 671			226	subl		v s alc, eth; s bz; i aq
b140	4,4′-Biphenyldiamine	$H_2NC_6H_4-C_6H_4NH_2$	184.24	13, 214			120	*ca.* 400		s alc; 2 eth; 20 hot alc
b141	2,2′-Biphenyldicarboxylic acid	$HOOCC_6H_4-C_6H_4COOH$	242.23	9, 922			228–229			0.06 aq; s org solvents
b142	4-Biphenylsulfonic acid	$C_6H_5-C_6H_4SO_3H$	234.26				138			
b143	2-Biphenylyl glycidyl ether		226.28				30–32	$120^{0.1mm}$		
b144	2,2-Bis[4-(allyloxy)-phenyl]-propane	$H_2C=CHCH_2OC_6H_4-C(CH_3)_2C_6H_4-OCH_2CH=CH_2$	308.42		1.022	1.5636^{20}			>110	
b145	*N,N′*-Bis(3-aminopropyl)ethylenediamine	$H_2N(CH_2)_3NHCH_2-CH_2NH(CH_2)_3NH_2$	174.29		0.952	1.4910^{20}		160^{5mm}	>110	
b146	*N,N′*-Bis(3-aminopropyl)piperazine		200.33	23^2, 12	0.973	1.5015^{20}	15	152^{2mm}	162	
b147	*N,N′*-Bis(3-aminopropyl)-1,3-propanediamine	$H_2N(CH_2)_3NHCH_2CH_2CH_2-NH(CH_2)_3NH_2$	188.32	4^4, 1278	0.920	1.4915^{20}		103^{1mm}		
b148	Bis(2-bromoethyl) ether	$BrCH_2CH_2OCH_2CH_2Br$	231.92					107^{20mm}		

b149	1,3-Bis(bromoethyl)-tetramethyldisiloxane	$[BrCH_2Si(CH_3)_2]_2O$	320.17		1.3918^{20}_{4}	1.4719^{20}		104^{15mm}		
b150	2,2-Bis(bromomethyl)-1,3-propanediol	$HOCH_2CH(CH_2Br)_2CH_2OH$	261.95	1^1, 251			114			
b151	Bis(2-butoxyethyl)-ether	$(C_4H_9OCH_2CH_2)_2O$	218.34		0.8853^{20}_{20}	1.4240^{20}	−60.2	256	118	0.3 aq; misc alc, esters, eth, CCl_4 ketones
b152	Bis[2-(2-butoxyethoxy)-ethyl] adipate	$[-CH_2CH_2CO_2(CH_2CH_2O)_2-(CH_2)_3CH_3]_2$	434.58	2^3, 1718	1.010	1.4480^{20}	−11		110	
b153	2,5-Bis(5-*tert*-butyl-2-2′-benzoxazolyl)-thiophene		430.57				201			
b154	Bis(*sec*-butyl) disulfide	$[CH_3CH_2CH(CH_3)]_2S_2$	178.36	1^3, 1549	0.957	1.4920^{20}		164^{739mm}	112	
b155	Bis(*tert*-butyl) disulfide	$(CH_3)_3CSSC(CH_3)_3$	178.36	1, 379	0.909	1.4930^{20}		204	79	
b156	1,1-Bis(*tert*-butylperoxy)cyclohexane	$C_6H_{10}[OOC(CH_3)_3]_2$	260.38		0.970	1.4570^{20}		54^{15mm}	90	
b157	2,5-Bis(*tert*-butylperoxy)-2,5-dimethylhexane	$[(CH_3)_3COOC(CH_3)_2CH_2-]_2$	290.45		0.877	1.4230^{20}		57^{7mm}	41	
b158	2,5-Bis(*tert*-butylperoxy)-2,5-dimethyl-3-hexyne	$(CH_3)_3COOC(CH_3)_2C{\equiv}C-C(CH_3)_2OOC(CH_3)_3$	286.41	1^4, 2701	0.881	1.4320^{20}		67^{2mm}	85	
b159	Bis[1-(*tert*-butylperoxy)-1-methylethyl-benzene	$C_6H_4[C(CH_3)_2OOC(CH_3)_3]_2$	338.49				44–48			flammable solid oxidizer
b160	1,1-Bis(*tert*-butylperoxy)-3,3,5-trimethyl-cyclohexane	$[(CH_3)_3COO]_2C_6H_7(CH_3)_3$	302.46		0.906	1.4410^{20}			87	
b161	1,2-Bis(2-chloroethoxy)ethane	$(ClCH_2CH_2OCH_2-)_2$	187.07	1^3, 2079	1.197^{20}_{4}	1.4610^{20}		235	121	
b162	Bis(2-chloroethoxy)-methylsilane	$H(CH_3)Si(OCH_2CH_2Cl)_2$	203.1		1.1643^{20}_{4}	1.4431^{20}		97^{18mm}		
b163	Bis(2-chloroethyl) ether	$ClCH_2CH_2OCH_2CH_2Cl$	143.01	1^2, 335	1.2220^{20}_{20}	1.4575^{20}	−50 to −52	178.5	55	s most org solvents
b164	Bis(2-chloroethyl)-*N*-methylamine	$CH_3N(CH_2CH_2Cl)_2$	156.07		1.118^{25}_{4}		−60	75^{10mm}		v sl s aq; misc most org solvents

(*Continued*)

TABLE 2.20 Physical Constants of Organic Compounds (*Continued*)

No.	Name	Formula	Formula weight	Beilstein reference	Density, g/mL	Refractive index	Melting point, °C	Boiling point, °C	Flash point, °C	Solubility in 100 parts solvent
b165	Bis(chloromethyl)dimethylsilane	$(CH_3)_2Si(CH_2Cl)_2$	157.12	4^3, 1845	1.975^{20}_{4}	1.4600^{20}		160	46	
b165a	Bis(chloromethyl) ether	$ClCH_2OCH_2Cl$	114.96	Merck: 12, 3119	1.315^{20}_{4}	1.4346	−41.5	106		dec aq
b166	Bis(2-chloro-1-methyl)ethyl ether	$ClCH_2CH(CH_3)OCH(CH_3)$-CH_2Cl	171.07		1.1122^{20}_{20}			187.3	85	
b167	1,3-Bis(chloromethyl)-tetramethyldisiloxane	$[ClCH_2Si(CH_3)_2]_2O$	231.3	4^3, 1864	1.050	1.4405^{20}		205	73	
b168	Bis(4-chlorophenoxy)-acetic acid	$(ClC_6H_4O)_2CHCOOH$	313.14				140–142			
b169	2,2-Bis(4-chloro-phenyl)-1,1-dichloroethane	$(ClC_6H_4)_2CHCHCl_2$	320.05	5^3, 1830			110			similar to b168
b170	1,1-Bis(4′-chloro-phenyl)ethanol	$(ClC_6H_4)_2C(OH)CH_3$	267.16	6^3, 3396			69			s org solvents
b171	Bis(4-chlorophenyl) sulfone	$ClC_6H_4SO_2C_6H_4Cl$	287.16	6, 327			145–148	250^{10mm}		
b172	Bis(4-chlorophenyl) sulfoxide	$ClC_6H_4S(O)C_6H_4Cl$	271.17	6^1, 149			141–144			
b173	1,1-Bis(4-chloro-phenyl)-2,2,2-trichloroethane	$(ClC_6H_4)_2CHCCl_3$	354.49	5^3, 1833			109–111			58 acet; 78 bz; 45 chl; v s pyr, 1,4-dioxane
b174	1,2-Bis(dichlorometh-ylsilyl)ethane	$[-CH_2Si(CH_3)Cl_2]_2$	256.11	4^4, 192	1.263	1.4760^{20}	33–35	210	90	
b175	1,3-Bis(dichloro-methyl)tetramethyl-disiloxane	$[ClCH(CH_3)_2Si]_2O$	300.16		1.2213^{20}_{4}	1.4660^{20}		149^{40mm}		
b176	*N,N*-Bis(2,2-diethoxy-ethyl)methylamine	$[(C_2H_5O)_2CHCH_2]_2NCH_3$	263.38	4, 311	0.945	1.4259^{20}		222^{244mm}	60	
b177	4,4′-Bis(diethyl-amino)benzo-phenone	$[(C_2H_5)_2NC_6H_4]_2C{=}O$	324.47	14, 98			95			
b178	4,4′-Bis(dimethyl-amino)benzophe-none	$(CH_3)_2NC_6H_4]_2C{=}O$	268.35	14, 89			172	>360 d		s alc, warm bz; v sl s eth; i aq
b179	Bis(dimethylamino)di-methylsilane	$[(CH_3)_2N]Si(CH_3)_2$	146.31	4^4, 4143	0.810^{22}	1.4170^{20}	−98	128–129	−7	
b180	1,3-Bis(dimethyl-amino)-2-propanol	$[(CH_3)_2NCH_2]_2CHOH$	146.23	4, 290	0.897	1.4422^{20}			>110	

b181	2,4-Bis(α,α-dimethylbenzyl)phenol	$[C_6H_5C(CH_3)_2]_2C_6H_3OH$	330.47	6^4, 5076			63–65	206^{15mm}		
b182	1,1-Bis(3,4-dimethylphenyl)ethane	$[(CH_3)_2C_6H_3]_2CHCH_3$	238.38	5^3, 1908	0.982	1.5640^{20}		174^{5mm}	>110	
b183	Bis(dimethylthiocarbamyl) disulfide	$[(CH_3)_2NC(=S)S-]_2$	240.43	4, 76	1.29		155–156			s alc, eth; sl s bz, acet; i aq
b184	Bis(3,4-epoxycyclohexylmethyl) adipate		366.46		1.149	1.4930			>110	
b185	1,4-Bis(2,3-epoxypropoxy)butane	$[\underset{O}{\underbrace{H_2C—CH}}CH_2OCH_2CH_2-]_2$	202.25		1.049	1.4530^{20}		160^{11mm}	>110	
b186	Bis(2-ethoxyethyl) ether	$(C_2H_5OCH_2CH_2)_2O$	162.23	1^2, 519	0.907_4^{20}	1.4110^{20}	−45	188	82	v s aq, alc, org solvents
b187	Bis(2-ethylhexyl) adipate	$[-CH_2CH_2CO_2CH(C_2H_5)(CH_2)_3CH_3]_2$	370.58	2^3, 1715	0.990	1.4425^{20}		167^{1mm}	>110	
b188	Bis(2-ethylhexyl)amine	$[CH_3(CH_2)_3CH(C_2H_5)(CH_2)_3CH_3]_2$	241.46	4^3, 388	0.805	1.4425^{20}		123^{5mm}	>110	
b189	Bis(2-ethylhexyl) chlorendate		613.28		1.240	1.500^{20}		$233^{0.3mm}$	>110	
b190	Bis(2-ethylhexyl) decanedioate	$CH_3(CH_2)_3CH(C_2H_5)CH_2OOC(CH_2)_8COOCH_2CH(C_2H_5)(CH_2)_3CH_3$	426.66		0.9119_{25}^{25}	1.4496^{25}				
b191	Bis(2-ethylhexyl) hydrogen phosphate	$[CH_3(CH_2)_3CH(C_2H_5)CH_2O]_2P(O)OH$	322.43	1^4, 1786	0.965	1.4430^{20}	−60	209^{10mm}	>110	
b192	Bis(2-ethylhexyl) hydrogen phosphite	$[CH_3(CH_2)_3CH(C_2H_5)CH_2O]_2POH$	306.43		0.916	1.4420^{20}			>110	
b193	Bis(2-ethylhexyl) *o*-phthalate	$[CH_3(CH_2)_3CH(C_2H_5)CH_2OOC]_2C_6H_4$	390.56	Merck: 12, 1291	0.9843^{20}	1.4859^{20}	−50 to −55	384	218	0.01 aq
b194	Bis(2-ethylhexyl) 1,4-phthalate	$[CH_3(CH_2)_3CH(C_2H_5)CH_2OOC]_2C_6H_4$	390.56	9,*4*, 3306	0.980	1.4900^{20}	30–34	400	>110	
b195	Bis(4-fluorophenyl)methane	$(FC_6H_4)_2CH_2$	204.22	5^3, 1789	1.145	1.5362^{20}	29–30	260^{742mm}	>110	
b196	Bis(hexamethylene)triamine	$[H_2N(CH_2)_6]_2NH$	215.39				33–36	165^{4mm}	>110	
b197	1,4-Bis(2-hydroxyethoxy)-2-butyne	$HOCH_2CH_2OCH_2C{\equiv}CCH_2OCH_2CH_2OH$	174.20		1.144	1.4850^{20}			>110	

(*Continued*)

TABLE 2.20 Physical Constants of Organic Compounds (*Continued*)

No.	Name	Formula	Formula weight	Beilstein reference	Density, g/mL	Refractive index	Melting point, °C	Boiling point, °C	Flash point, °C	Solubility in 100 parts solvent
b198	Bis(2-hydroxyethyl) ether	$HOCH_2CH_2OCH_2CH_2OH$	106.12	1, 468	1.1184^{20}_{20}	1.4460^{20}	−10.4	246	118	misc aq, alc, acet, eth
b199	*N,N*-Bis(2-hydroxyethyl)glycine	$(HOCH_2CH_2)_2NCH_2COOH$	163.17	Merck: 12, 1248			193–195			17.9 aq^0
b200	2,6-Bis(hydroxymethyl)-*p*-cresol	$CH_3C_6H_2(CH_2OH)_2OH$	168.19	6, 1127			128–130			
b201	2,2-Bis(hydroxymethyl)propanoic acid	$(HOCH_2)_2C(CH_3)COOH$	134.13	3, 401			181–185			s aq, MeOH; sl s acet; i bz
b202	4,8-Bis(hydroxymethyl)tricyclo[5.2.1.0^{2,6}]decane		196.29	6^4, 5538		1.5280^{20}			110	
b203	4,4-Bis(4-hydroxyphenyl)pentanoic acid	$CH_3C(C_6H_4OH)_2CH_2$-CH_2COOH	286.33	Merck: 12, 3370			171–172 higher melting form			s hot aq, acet, alc, HOAc, MeEtKe
b204	Bis(2-hydroxypropyl) ether	$HO(CH_2)_3O(CH_2)_3OH$	134.18	1^2, 537	1.0252^{20}_{20}	1.4410^{20}		231.8	137	misc aq, alc
b205	1,3-Bis(isocyanatomethyl)benzene	$C_6H_4(CH_2NCO)_2$	188.19	13^3, 334	1.202	1.5910^{20}	−7	130^{2mm}	>110	
b206	1,3-Bis(isocyanatomethyl)cyclohexane	$C_6H_{10}(CH_2NCO)_2$	194.24		1.101	1.4850^{20}			>110	
b207	1,3-Bis(1-isocyanato-1-methylethyl)benzene	$C_6H_4[C(CH_3)_2NCO]_2$	244.30		1.060	1.5110^{20}		$106^{0.9mm}$	153	
b208	Bis(2-mercaptoethyl) ether	$(HSCH_2CH_2)_2O$	138.25		1.114		−80	217	98	
b209	Bis(2-mercaptoethyl) sulfide	$(HSCH_2CH_2)_2S$	154.32		1.183	1.5961^{20}		136^{10mm}	90	
b210	1,4-Bis(methanesulfonoxy)butane	$(CH_3SO_2OCH_2CH_2$-$)_2$	246.30				115–117			sl hyd aq; 0.1 alc; 1.4 acet
b211	1,2-Bis(methoxyethoxy)ethane	$(CH_3OCH_2CH_2OCH_2$-$)_2$	178.23		0.990^{20}_4	1.4224^{20}	−45	216	110	misc aq
b212	Bis[2-(2-methoxyethoxy)ethyl] ether	$(CH_3OCH_2CH_2OCH_2CH_2)_2O$	228.28	1^3, 2107	1.0087^{20}_4	1.4330^{20}	−27	275	140	s aq
b213	Bis(2-methoxyethyl)amine	$(CH_3OCH_2CH_2)_2NH$	133.19	4^3, 691	0.902	1.4190^{20}		172	58	

b214	Bis(2-methoxyethyl) ether	$(CH_3OCH_2CH_2)_2O$	134.18	1^2, 520	0.9440^{25}	1.4043^{25}	−64 to −68	162	67	misc aq
b214a	2,2-Bis(4-methoxy-phenyl)-1,1,1-trichloroethane	$(CH_3OC_6H_4)_2CHCCl_3$	345.66	6, 1007			86–88			v sl s aq; s alc
b215	Bis(2-methylallyl) carbonate	$[H_2C{=}C(CH_3)CH_2O]_2C{=}O$	170.21		0.943	1.4370^{20}		202	72	
b216	Bis(3-nitrophenyl) disulfide	$O_2NC_6H_4SSC_6H_4NO_2$	308.33	6, 339			83			i aq; s alc; v s eth
b217	Bis(octadecyl)penta-erythritol diphos-phite	$[C_{18}H_{37}OP(OCH_2)_2\text{-}]_2$	721.01		0.925	1.457	40		261	
b218	1,4-Bis(5-phenyloxa-zol-2-yl)benzene		364.40				244			
b219	*N,N′*-Bis(salicylidene)-1,4-butanediamine	$HOC_6H_4CH{=}N(CH_2)_4\text{-}N{=}CHC_6H_4OH$	296.37	8^3, 163			88–90			
b220	*N,N′*-Bis(salicylidene)-ethylenediamine	$(\text{-}CH_2N{=}CHC_6H_4OH)_2$	268.32	8, 48			128			
b221	*N,N′*-Bis(salicylidene)-1,6-hexanediamine	$HOC_6H_4CH{=}N(CH_2)_6\text{-}N{=}CHC_6H_4OH$	324.44	8^3, 165			69			
b222	Bis(*p*-tolyl) disulfide	$CH_3C_6H_4SSC_6H_4CH_3$	246.39	6, 425			43–46			i aq; s alc; v s eth
b223	Bis(*p*-tolyl) sulfoxide	$CH_3C_6H_4S(\rightarrow O)C_6H_4CH_3$	230.33	6, 419			94–96			v s alc, bz, chl, eth
b224	Bis(tributyltin) oxide	$(C_4H_9)_3SnOSn(C_4H_9)_3$	596.08		1.170	1.4860^{20}		180^{2mm}	>110	
b225	1,4-Bis(trichloro-methyl)benzene	$Cl_3CC_6H_4CCl_3$	312.84	5, 385			108–110			i aq; 26 acet; 38 bz
b226	Bis(2,4,5-trichloro-phenyl) disulfide	$Cl_3C_6H_2SSC_6H_2Cl_3$	425.01				140–144			
b227	1,2-Bis(trichlorosilyl)-ethane	$Cl_3SiCH_2CH_2SiCl_3$	296.94	4^4, 4266	1.483^{20}_{4}	1.4750^{20}	24.5	202	65	
b228	3,5-Bis(trifluoro-methyl)aniline	$(F_3C)_2C_6H_3NH_2$	229.13		1.467	1.4340^{20}		85^{15mm}	83	
b229	1,3-Bis(trifluoro-methyl)benzene	$(F_3C)_2C_6H_4$	214.11	5^3, 834	1.3790^{25}	1.3916^{25}		116	26	
b230	*N,O*-Bis(trimethyl-silyl)acetamide	$CH_3{-}C(O{-}Si(CH_3)_3){=}N{-}Si(CH_3)_3$	203.43		0.832^{20}_{4}	1.4170^{20}		73^{35mm}	11	
b231	Bis(trimethylsilyl)-acetylene	$(CH_3)_3SiC{\equiv}CSi(CH_3)_3$	170.41		0.770^{20}_{4}	1.4270^{20}		137	2	

(*Continued*)

TABLE 2.20 Physical Constants of Organic Compounds (*Continued*)

No.	Name	Formula	Formula weight	Beilstein reference	Density, g/mL	Refractive index	Melting point, °C	Boiling point, °C	Flash point, °C	Solubility in 100 parts solvent
b232	Bis(trimethylsilyl)formamide	$HC(=NSi(CH_3)_3)OSi(CH_3)_3$	189.41		0.885	1.4381^{20}		55^{13mm}		
b233	*N,O*-Bis(trimethylsilyl)hydroxylamine	$(CH_3)_3SiONHSi(CH_3)_3$	177.40		0.830	1.4112^{20}		80^{100mm}	28	
b234	1,2-Bis(trimethylsilyloxy)ethane	$(CH_3)_3SiOCH_2CH_2OSi(CH_3)_3$	206.43		0.842	1.4034^{20}		166	46	
b235	*N,O*-Bis(trimethylsilyl)trifluoroacetamide	$F_3C[=NSi(CH_3)_3]OSi(CH_3)_3$	257.40		0.969	1.3839^{20}	−10	50^{14mm}	23	
b236	1,3-Bis(trimethylsilyl)urea	$(CH_3)_3SiNHCONHSi(CH_3)_3$	204.42				232 dec			
b237	1,3-Bis[tris(hydroxymethyl)methylamino]propane	$CH_2[CH_2NHC(CH_2OH)_3]_2$	282.34	4^3, 859			170			s aq
b238	Biuret	$H_2NC(=O)NHC(=O)NH_2$	103.08	3, 70	1.467_4^{-5}		anhyd 110	dec 190		v s alc; 2 aq^{25}
b239	Borane-*tert*-butylamine	$(CH_3)_3CNH_2 \cdot BH_3$	86.97				100 dec			
b240	Borane-*N,N*-diethylaniline	$C_6H_5N(C_2H_5)_2 \cdot BH_3$	163.07				−30		21	
b241	Borane-*N,N*-diisopropylethylamine	$[(CH_3)_2CH]_2C_2H_5 \cdot BH_3$	143.08		0.822	1.4600^{20}	15–17		40	
b242	Borane-dimethylamine	$(CH_3)_2NH \cdot BH_3$	58.92				36		43	
b243	Borane-dimethyl sulfide	$(CH_3)_2S \cdot BH_3$	75.97		0.801				18	
b244	Borane-pyridine	$C_5H_4N \cdot BH_3$	92.93		0.920	1.5320^{20}	10–11		21	
b245	(1*S*-*endo*)-(−)-Borneol		154.25	6, 72	1.011_4^{20}		204	210^{779mm}	65	i aq; 176 alc; s eth
b246	(−)-1-Bornyl acetate		196.29	6, 82	0.982	1.4626	27	224	84	v sl s aq; s alc, eth
b247	*N*-Bromoacetamide	$CH_3CON(Br)H$	137.96	2, 181			102–105			sl s aq; v s eth
b248	*p*-Bromoacetanilide	$BrC_6H_4NHCOCH_3$	214.06	12, 642	1.717		168			s alc, bz, chl, EtOAc
b249	Bromoacetic acid	$BrCH_2COOH$	138.95	2, 213	1.934_4^{50}	1.4804^{50}	50	208	>110	v s aq, alc
b250	Bromoacetonitrile	$BrCH_2CN$	119.95	2, 216	1.722	1.4800^{20}		62^{24mm}	>110	
b251	2-Bromoacetophenone	$C_6H_5COCH_2Br$	199.05	7, 283	1.647_4^{20}		50	135^{18mm}	>110	v s alc, bz, chl, eth
b253	*p*-Bromoacetophenone	$BrC_6H_4COCH_3$	199.05	7, 283	1.647		54	255	>110	s alc, bz, CS_2, HOAc PE
b254	Bromoacetyl bromide	$BrCH_2COBr$	201.86	2, 215	2.317_{22}^{22}	1.5480^{20}		150	none	dec aq, alc
b255	Bromoacetyl chloride	$BrCH_2COCl$	157.40	2, 215	1.908	1.4960^{20}		128	none	dec aq, alc

b256	2-Bromoaniline	$BrC_6H_4NH_2$	172.03	12, 631	1.578^{20}_{4}	1.6223^{20}	31	229	>110	i aq; s alc, eth
b257	3-Bromoaniline	$BrC_6H_4NH_2$	172.03	12, 633	1.580^{20}_{4}	1.6250^{20}	16.8	251	>110	sl s aq; s alc, eth
b258	4-Bromoaniline	$BrC_6H_4NH_2$	172.03	12, 636	1.4970^{100}_{4}		66.3			i aq; v s alc, eth
b259	2-Bromoanisole	$BrC_6H_4OCH_3$	187.04	6, 197	1.502	1.5740^{20}	2	223	96	
b260	4-Bromoanisole	$BrC_6H_4OCH_3$	187.04	6, 199	1.494	1.5640^{20}	9–10	223	94	
b261	3-Bromobenzaldehyde	BrC_6H_4CHO	185.03	7, 238	1.587	1.5935^{20}		230	96	i aq; v s alc, eth
b262	Bromobenzene	C_6H_5Br	157.01	5, 206	1.4952^{20}_{4}	1.5602^{20}	−30.6	156	51	0.045 aq^{30}; 10.4 alc^{25}; 71.6 eth^{25}; misc bz, chl, PE
b263	Bromobenzene-d_5	C_6D_5Br	162.06		1.539	1.5585^{20}		53^{23mm}	51	
b264	4-Bromobenzene-sulfonyl chloride	$BrC_6H_4SO_2Cl$	255.52	11, 57			74.5	153^{15mm}		i aq; s alc (dec); v s eth
b265	2-Bromobenzoic acid	BrC_6H_4COOH	201.02	9, 347			148–150			
b266	4-Bromobenzoic acid	BrC_6H_4COOH	201.02	9, 351	1.929^{25}_{4}		251–253			0.18 aq^{25}; s alc, eth
b267	4-Bromobenzo-phenone	$BrC_6H_4COC_6H_5$	261.12	7, 422				82	350	i alc; sl s bz, eth
b268	2-Bromobenzotri-fluoride	$BrC_6H_4CF_3$	225.01		1.652^{20}	1.4820^{20}		168	51	
b269	3-Bromobenzotri-fluoride	$BrC_6H_4CF_3$	225.01		1.613	1.4730^{20}		152	43	
b270	3-Bromobenzoyl chloride	BrC_6H_4COCl	219.47	9, 350	1.662	1.5965^{20}		$75^{0.5mm}$	107	
b271	4-Bromobenzyl bromide	$BrC_6H_4CH_2Br$	249.94	5, 308		1.6193^{20}	61	124^{12mm}	>110	s aq, alc, bz, eth, CS_2, HOAc
b272	α-Bromobenzyl cya-nide	$C_6H_5CH(Br)CN$	196.05		1.539^{29}_{4}	1.5696^{20}	29	242 dec	>110	sl s aq; v s alc, acet, eth. A war gas.
b273	4-Bromobiphenyl	$BrC_6H_4C_6H_5$	233.11	5, 580	0.9327^{25}_{4}		90–92	310		i aq; s alc, bz, eth
b274	1-Bromobutane	$CH_3CH_2CH_2CH_2Br$	137.02	1, 119	1.2686^{25}_{4}	1.4374^{25}	−112.4	101.6	18	i aq; s alc, bz, eth
b275	2-Bromobutane	$CH_3CH_2CHBrCH_3$	137.02	1, 119	1.2585^{20}	1.4360^{20}	−112.7	91.4	21	<0.1 aq; v s alc, eth
b276	1-Bromo-2-butene	$CH_3CH{=}CHCH_2Br$	135.01	1, 205	1.312	1.4765^{20}		99	11	
b277	2-Bromo-2-butene	$CH_3CH{=}C(Br)CH_3$	135.01	1, 205	1.328	1.4590^{20}		90^{740mm}	1	Mixture of *cis*, *trans*
b278	4-Bromo-1-butene	$BrCH_2CH_2CH{=}CH_2$	135.01	1^1, 84	1.3230^{20}_{4}	1.4608^{20}		100	9	i aq; s alc, eth
b279	4-Bromobutyl acetate	$CH_3CO_2(CH_2)_4Br$	195.06	2^3, 39	1.348	1.4600^{20}		93^{12mm}	109	
b280	1-Bromo-4-*tert*-butyl-benzene	$(CH_3)3CC_6H_4Br$	213.12	5, 416	1.229	1.5330^{20}	15–16	81^{2mm}	97	
b281	4-Bromobutyl phenyl ether	$C_6H_5O(CH_2)_4Br$	229.12	6^2, 82			41–43	156^{18mm}	>110	

(*Continued*)

TABLE 2.20 Physical Constants of Organic Compounds (*Continued*)

No.	Name	Formula	Formula weight	Beilstein reference	Density, g/mL	Refractive index	Melting point, °C	Boiling point, °C	Flash point, °C	Solubility in 100 parts solvent
b282	2-Bromobutyric acid	$CH_3CH_2CH(Br)COOH$	167.00	2, 281	1.5669^{20}_{20}	1.4720^{20}	−4	103^{10mm}	>110	6.7 aq; s alc, eth
b283	α-Bromo-γ-butyrolactone		164.99		1.990^{20}	1.5080^{20}		138^{6mm}	>110	
b284	[1*R*-*endo*]-(+)-3-Bromocamphor		231.14	7, 120	1.449		75–78	244		15 alc; 200 chl; 62 eth; s olive oil
b285	1-Bromocarbonyl-1-methylethyl acetate	$CH_3CO_2C(CH_3)_2COBr$	209.05		1.431	1.4570^{20}		77^{12mm}	110	
b286	2-Bromo-4′-chloroacetophenone	$ClC_6H_4COCH_2Br$	233.50							
b287	2-Bromochlorobenzene	BrC_6H_4Cl	191.46	5, 209	1.6382^{25}_{4}	1.5789^{25}		204	79	i aq; v s bz
b288	3-Bromochlorobenzene	BrC_6H_4Cl	191.46	5, 209	1.6302^{20}_{4}	1.5770^{20}	−21	196	80	i aq; v s alc, bz, eth
b296	4-Bromochlorobenzene	BrC_6H_4Cl	191.46	5, 209	1.576^{71}_{4}	1.5531^{70}	66	196		0.1 aq; misc MeOH, eth
b297	3-Bromo-4-chlorobenzotrifluoride	$Br(Cl)C_6H_3CF_3$	259.46	5^3, 715	1.726	1.4990^{20}	−22	190	94	
b298	1-Bromo-4-chlorobutane	$ClCH_2CH_2CH_2CH_2Br$	171.47	5^3, 294	1.488	1.4875^{20}		82^{30mm}	60	i aq; s alc, chl, eth
b299	4′-Bromo-4-chlorobutyrophenone	$BrC_6H_4CO(CH_2)_3Cl$	261.55				36–38		>110	
b300	4-Bromo-6-chloro-*o*-cresol	$Br(Cl)C_6H_2(OH)CH_3$	221.49	6, 360			45–47		>110	
b301	Bromochlorodifluoromethane	$Br(Cl)CF_2$	165.36		6.579 g/L		−160	−3.7		
b302	3-Bromo-1-chloro-5,5-dimethyl-hydantoin		241.48				160–164			
b303	1-Bromo-2-chloroethane	$ClCH_2CH_2Br$	143.41	1, 89	1.7392^{20}_{4}	1.4917^{20}	−18.4	106.6		0.7 aq; misc org solv
b303a	Bromochlorofluoromethane	$Br(Cl)CHF$	149.37		1.9771^{0}	1.4144^{55}	−115	36		
b304	7-Bromo-5-chloro-8-hydroxyquinoline		258.51	21^1, 222			177–179			
b305	Bromochloromethane	$ClCH_2Br$	129.38	1, 67	1.923^{25}_{4}	1.480^{25}	−88	68		0.9 aq; misc MeOH, eth
b306	1-Bromo-3-chloro-2-methylpropane	$ClCH_2CH(CH_3)CH_2Br$	171.47	1^3, 324	1.467	1.4809^{20}		154	>110	

b307	1-Bromo-3-chloro-propane	$ClCH_2CH_2CH_2Br$	157.44	1, 109	1.492	1.4851^{20}	< −50	143.5		0.1 aq; misc org solv
b308	2-Bromo-2-chloro-1,1,1-trifluoroethane	$BrCH(Cl)CF_3$	197.39	1^4 156	1.8636^{25}	1.3691^{20}		50.2	none	
b309	2-Bromocinnam-aldehyde	$C_6H_4CH{=}C(Br)CHO$	211.06	7, 358			66–68			
b310	Bromocycloheptane	$Br(C_7H_{13})$	177.09	5, 29	1.2887^{22}_{4}	1.5052^{20}		72^{10mm}	68	i aq; v s chl, eth
b311	Bromocyclohexane	$Br(C_6H_{11})$	163.06	5, 24	1.3264^{15}_{4}	1.4956^{15}		165.8	62	0.1 aq; 10 MeOH; 71 eth
b312	3-Bromocyclohexene		161.04	5^2, 40	1.3890^{20}_{4}	1.5292^{20}		65^{15mm}	54	
b313	Bromocyclopentane	$Br(C_5H_9)$	149.04	5, 19	1.3900^{20}_{4}	1.4881^{20}		137–139	35	
b314	Bromocyclopropane	$Br(C_3H_5)$	120.98		1.510	1.4605^{29}		69	−6	
b315	1-Bromodecane	$CH_3(CH_2)_9Br$	221.18	1^2, 130	1.0658^{20}_{4}	1.4560^{20}	−30	238–240	94	i aq; v s chl, eth
b316	Bromodichloro-methane	$BrCHCl_2$	163.83	1, 67	1.980^{20}	1.4967^{20}	−55	87	none	sl s aq; misc org solv
b317	2-Bromo-1,1-di-ethoxyethane	$BrCH_2CH(OC_2H_5)_2$	197.08	1, 625	1.310	1.4385^{20}		67^{18mm}	51	s hot alc
b318	4-Bromo-1,2-di-methoxybenzene	$BrC_6H_3(OCH_3)_2$	217.07	6, 784	1.702	1.5743^{20}	256	109		
b319	2-Bromo-1,1-di-methoxyethane	$BrCH_2CH(OCH_3)_2$	169.02	1, 624	1.430	1.4450^{20}		150	53	
b320	1-Bromo-2,2-di-methoxypropane	$CH_3C(OCH_3)_2CH_2Br$	185.05		1.355	1.4475^{20}		87^{80mm}	40	
b321	4-Bromo-2,6-di-methylphenol	$BrC_6H_2(CH_3)_2OH$	201.07	6, 485			79–81			
b322	3-Bromo-2,2-dimethyl-1-propanol	$BrCH_2C(CH_3)_2CH_2OH$	167.05	1^1, 201	1.358	1.4794^{20}		184–187	75	
b323	2-Bromo-4,6-dinitro-aniline	$BrC_6H_2(NO_2)_2NH_2$	262.03	12, 761			154	subl		v s hot alc, hot acet
b324	1-Bromo-2,4-dinitro-benzene	$BrC_6H_3(NO_2)_2$	247.01				71–73			
b325	4-Bromodiphenyl ether	$BrC_6H_4OC_6H_5$	249.11	6^1, 105	1.423	1.6070^{20}	18	305	>110	
b326	1-Bromodiphenyl-methane	$C_6H_5CH(Br)C_6H_5$	247.14	5, 592			40–42	184^{20mm}	>110	
b327	1-Bromododecane	$CH_3(CH_2)_{11}Br$	249.24	1^2, 133	1.038	1.4580^{20}	−11	135^{6mm}	>110	0.1 aq; s alc, eth
b328	1-Bromo-2,3-epoxy-propane	$\underset{\backslash O /}{H_2C{-}CHCH_2Br}$	136.98	17, 9	1.601^{20}	1.4820^{20}	−40	134–136	56	i aq; sl s alc; s eth

(*Continued*)

TABLE 2.20 Physical Constants of Organic Compounds (*Continued*)

No.	Name	Formula	Formula weight	Beilstein reference	Density, g/mL	Refractive index	Melting point, °C	Boiling point, °C	Flash point, °C	Solubility in 100 parts solvent
b329	Bromoethane	CH_3CH_2Br	108.97	1, 88	1.4612^{20}_{4}	1.4242^{20}	−119	38.2	−23	0.91 aq^{20}; misc alc, chl, eth
b330	2-Bromoethanesulfonic acid, sodium salt	$BrCH_2CH_2SO_3^- Na^+$	211.02	4, 7			283 dec			
b331	2-Bromoethanol	$BrCH_2CH_2OH$	124.98	1, 338	1.7629^{20}_{4}	1.4936^{20}		57^{20mm}	>110	misc aq; s org solv except PE
b332	2-Bromoethyl acetate	$CH_3CO_2CH_2CH_2Br$	167.01	2^1, 57	1.514^{20}_{4}	1.4547^{20}	−13.8	159	71	v s aq; misc alc, eth
b333	2-Bromoethylamine HBr	$BrCH_2CH_2NH_2 \cdot HBr$	204.90	4, 134			172–174			v s aq, alc
b334	(1-Bromoethyl)-benzene	$C_6H_5CH(CH_3)Br$	185.07	5, 355	1.356	1.5600^{20}		94^{16mm}	81	
b334a	(2-Bromoethyl)-benzene	$C_6H_5CH_2CH_2Br$	185.07	5, 355	1.355	1.5560^{20}		221	89	
b335	1-Bromo-2-ethyl-benzene	$BrC_6H_4CH_2CH_3$	185.07	5, 355	1.338	1.5490^{20}		194^{16mm}	71	
b336	Bromoethylene	$H_2C{=}CHBr$	106.95	1, 188	1.493^{20}	1.4380^{20}	−139	15.8	none	i aq; misc alc, eth
b337	2-Bromoethyl ethyl ether	$BrCH_2CH_2OCH_2CH_3$	153.02	1, 338	1.3572^{20}_{4}	1.4450^{20}		150	21	sl s aq; misc alc, eth
b338	2-Bromoethyl phenyl ether	$BrCH_2CH_2OC_6H_5$	201.07	6, 142			34	144^{40mm}	65	i sq; v s alc, eth
b339	*N*-(2-Bromoethyl)-phthalimide		254.09	21, 461			81–84			s hot aq; v s eth
b340	1-Bromo-2-fluoro-benzene	BrC_6H_4F	175.01		1.601	1.5337^{20}		156	43	
b341	1-Bromo-3-fluoro-benzene	BrC_6H_4F	175.01		1.567	1.5257^{20}		150	38	
b342	1-Bromo-4-fluoro-benzene	BrC_6H_4F	175.01	5, 209	1.593^{15}	1.5310^{15}	−17.4	152	60	
b343	1-Bromoheptane	$H(CH_2)_7Br$	179.11	1, 155	1.1384^{20}_{4}	1.4505^{20}	−58	180	60	i aq; v s alc, eth
b344	2-Bromoheptane	$H(CH_2)_5CH(Br)CH_3$	179.11	1, 155	1.142	1.4470^{20}		66^{21mm}	47	
b345	1-Bromohexadecane	$H(CH_2)_{16}Br$	305.35	1^2, 138	0.9991	1.4618^{20}	17.8	336	177	i aq; misc org solv
b346	1-Bromohexane	$H(CH_2)_6Br$	165.08	1, 144	1.1763^{20}_{4}	1.4475^{20}	−85	154–158	57	i aq; misc alc, eth
b347	DL-2-Bromohexanoic acid	$CH_3(CH_2)_3CH(Br)COOH$	195.06	2, 325	1.370	1.4720^{20}		138^{18mm}	>110	s alc, eth
b348	5-Bromoisatin		226.03	21, 453			251–253			
b350	(2-Bromoisopropyl)-benzene	$C_6H_5CH(CH_3)CH_2Br$	199.10	5^1, 191	1.316	1.5480^{20}		108^{18mm}	91	

b351	2-Bromo-4-isopropyl-1-methylbenzene	$CH_3(Br)C_6H_3CH(CH_3)_2$	213.0		1.253_{25}^{25}	1.535^{25}	−20	120		i aq; 50 MeOH; misc org solvents
b352	Bromomaleic anhydride		176.96	17, 435	1.905	1.5400^{20}		215	>110	
b353	2-Bromomesitylene	$1,3,5-(CH_3)_3C_6H_2Br$	199.10	5, 408	1.301	1.5520^{20}	2	255	96	
b354	Bromomethane	CH_3Br	94.94	1, 67	1.732_{0}^{0}	1.4234^{10}	−94	3.56	none	0.1 aq; s alc, chl, eth
b355	4-Bromomandelic acid	$BrC_6H_4CH(OH)COOH$	231.05	10, 210			117–118			sl s aq
b356	5-Bromo-2-methoxybenzaldehyde	$BrC_6H_3(OCH_3)CHO$	215.05	8, 55			116–119			
b357	2-Bromo-1-methoxybenzene	$BrC_6H_4OCH_3$	187.04	6, 197	1.5018_{4}^{20}	1.5737^{20}	2	223	96	i aq; v s alc, eth
b358	3-Bromo-1-methoxybenzene	$BrC_6H_4OCH_3$	187.04	6, 198	1.477	1.5635^{20}	211	93		i aq; s alc, eth
b359	4-Bromo-1-methoxybenzene	$BrC_6H_4OCH_3$	187.04	6, 199	1.4564_{4}^{20}	1.5630^{20}	10	223	94	sl s aq; v s alc, eth
b360	4-Bromo-2-methylaniline	$CH_3(Br)C_6H_3NH_2$	186.06	12, 838			57–59	240	>110	sl s aq; v s alc
b361	1-Bromo-3-methylbenzyl alcohol	$BrC_6H_4CH(CH_3)OH$	201.07	6^2, 447	1.460		36–38	121^{7mm}	63	
b362	1-Bromo-3-methylbutane	$(CH_3)_2CHCH_2CH_2Br$	151.05	1, 136	1.210_{4}^{15}	1.4409^{20}	−112	119.7	32	0.02 aq; misc alc, eth
b363	2-Bromo-2-methylbutane	$C_2H_5C(CH_3)_2Br$	151.05	1, 136	1.182	1.4423^{20}		107^{735mm}	5	
b364	2-Bromo-3-methylbutanoic acid	$(CH_3)_2CHCH(Br)COOH$	181.04	2, 317			44	126^{20mm}	107	sl s aq; s alc, eth
b365	4-Bromo-2-methyl-2-butene	$BrCH_2C{=}C(CH_3)_2$	149.04	1^2, 189	1.293	1.4898^{20}		60^{60mm}	32	
b366	(Bromomethyl)chlorodimethylsilane	$BrCH_2Si(CH_3)_2Cl$	187.5	4^4, 4024	1.375	1.4650^{20}		130^{740mm}	41	
b367	(Bromomethyl)cyclohexane	$(C_6H_{11})CH_2Br$	177.09	5^2, 18	1.269	1.4907^{20}		77^{26mm}	57	
b368	2-Bromomethyl-1,3-dioxalane		167.01	19^2, 8	1.613	1.4817^{20}		82^{27mm}	62	
b369	Bromomethyl methyl ether	$BrCH_2OCH_3$	124.97	1, 582	1.531	1.4550^{20}		87	26	
b370	1-Bromo-2-methylnaphthalene	$Br(C_{10}H_6)CH_3$	221.10	5, 568	1.418	1.6486^{20}		296	>110	

(Continued)

TABLE 2.20 Physical Constants of Organic Compounds (*Continued*)

No.	Name	Formula	Formula weight	Beilstein reference	Density, g/mL	Refractive index	Melting point, °C	Boiling point, °C	Flash point, °C	Solubility in 100 parts solvent
b371	1-Bromo-2-methylpropane	$(CH_3)_2CHCH_2Br$	137.03	1, 126	1.2641^{20}	1.4362^{20}	−119	91.5	18	0.06 aq; misc alc, eth
b372	2-Bromo-2-methylpropane	$(CH_3)_3CBr$	137.03	1, 127	1.2125^{25}_{4}	1.425^{25}	−16.2	73.1	18	i aq; misc org solv
b373	2-Bromo-2-methylpropanoic acid	$BrC(CH_3)_2COOH$	167.01	2, 295	1.52		48–49	200	>110	sl s aq; s alc, eth; dec by hot aq
b374	2-Bromo-2-methylpropionyl bromide	$(CH_3)_2C(Br)COBr$	229.91	2, 297	1.860	1.5064^{24}		164	110	
b375	2-Bromo-2-methylpropiophenone	$C_6H_5CO(CH_3)_2Br$	227.11	7, 316	1.350	1.5561^{20}		148^{30mm}	>112	
b376	1-Bromonaphthalene	$(C_{10}H_7)Br$	207.07	5, 547	1.4834^{20}_{4}	1.6580^{20}	−1.8	281	>110	misc alc, bz, chl, eth
b377	1-Bromo-1-naphthol	$BrC_{10}H_6OH$	233.07	6, 650			78	130 dec		i aq; s alc, bz, eth
b378	1-Bromo-2-naphthol	$BrC_{10}H_6OH$	223.07	6, 650			78–81			
b379	1-Bromo-2-nitrobenzene	$BrC_6H_4NO_2$	202.01	5^1, 247	1.6245^{80}_{4}		43	261	110	v s alc; s bz, eth
b380	5-Bromo-2-nitrobenzotrifluoride	$O_2N(Br)C_6H_3CF_3$	270.02	5^3, 755	1.7992^{25}	1.5180^{25}	33–35	100^{5mm}	>110	
b381	2-Bromo-2-nitro-1,3-propanediol	$(HOCH_2)_2C(Br)NO_2$	199.99	1, 476			120–122			s aq, alc, EtOAc; sl s bz, acet, chl, eth
b382	1-Bromononane	$H(CH_2)_9Br$	207.16	1^1, 63	1.084	1.4540^{20}		201	90	i aq; s chl, eth
b383	*exo*-2-Bromo-norbornane		175.07		1.363	1.5148^{20}		82^{29mm}	60	
b384	1-Bromooctadecane	$H(CH_2)_{18}Br$	333.41	1^1, 69	0.976		23	216^{12mm}	>110	i aq; s alc, eth
b385	1-Bromooctane	$H(CH_2)_8Br$	193.13	1, 160	1.108^{25}_{4}	1.4518^{25}	−55	201	78	i aq; misc alc, eth
b386	Bromopentafluorobenzene	BrC_6F_5	246.97		1.947^{20}	1.4490^{20}	−31	137	87	
b387	1-Bromopentane	$H(CH_2)_5Br$	151.05	1, 131	1.2237^{15}_{4}	1.4444^{20}	−88	129.6	31	i aq; s alc; misc eth
b388	2-Bromopentane	$CH_3CH_2CH_2CH(Br)CH_3$	151.05	1, 131	1.2039^{20}_{4}	1.4403^{20}		117	20	
b389	3-Bromopentane	$C_2H_5CH(Br)C_2H_5$	151.05	1^1, 43	1.216	1.4445^{20}		119	18	
b390	5-Bromopentyl acetate	$CH_3CO_2(CH_2)_5Br$	209.09	2^3, 249	1.255	1.4620^{20}		110^{15mm}	>110	
b391	9-Bromophenanthrene		257.14	5, 671	1.409^{101}_{4}		54–58	190^{2mm}	>110	i aq; s alc, eth
b392	2-Bromophenol	BrC_6H_4OH	173.01	6, 197	1.492	1.5892^{20}	6	194	42	s aq; misc chl, eth
b393	3-Bromophenol	BrC_6H_4OH	173.01	6, 198			32	236	>110	
b394	4-Bromophenol	BrC_6H_4OH	173.01	6, 198	1.5875^{80}		64	238		14 aq; v s alc, chl
b395	1-(4-Bromophenoxy)-1-ethoxyethane	$CH_3CH(OC_6H_4Br)OC_2H_5$	245.12		1.348	1.5229^{20}		125^{8mm}	106	

b396	4-Bromophenylacetic acid	$BrC_6H_4CH_2COOH$	215.05	9, 451			119			sl s aq; v s alc, eth
b397	4-Bromophenylacetonitrile	$BrC_6H_4CH_2CN$	196.05	9, 451			47–49		>110	i aq; sl s alc; v s bz
b398	4-Bromophenyl phenyl ether	$BrC_6H_4OC_6H_5$	249.11	6^1, 105	1.423	1.6070^{20}	18	305	>110	
b399	1-Bromo-3-phenylpropane	$BrC_6H_4CH_2CH_2CH_2Br$	199.10	5, 391	1.310	1.5450^{20}		220	101	
b400	1-Bromopropane	$CH_3CH_2CH_2Br$	122.99	1, 108	1.3597^{15}	1.4370^{15}	−110.1	71.0		0.23 aq^{30}; misc alc
b401	2-Bromopropane	$CH_3CH(Br)CH_3$	123.99	1, 108	1.3222^{15}	1.4285^{15}	−89.0	59.5	19	0.3 aq^{18}; misc alc, bz, chl, eth
b402	3-Bromo-1-propanol	$BrCH_2CH_2CH_2OH$	139.00	1, 356	1.5374_4^{20}	1.4858^{20}		62^{5mm}	93	s aq; misc alc, eth
b403	1-Bromo-2-propanone	CH_3OCH_2Br	136.98	Merck: 12, 1422	1.634^{23}	1.4697^{15}	−36.5	137		v sl s aq; s alc, acet
b404	1-Bromo-1-propene	$CH_3CH{=}CHBr$	120.98	1, 200	1.4133_4^{20}	1.4538^{20}	−116	70	−6	i aq
b405	2-Bromo-2-propene	$CH_3C(Br){=}CH_2$	120.98	1, 200	1.362_4^{20}	1.4425^{20}	−125	47–49	4	
b406	2-Bromopropionic acid	$CH_3CH(Br)COOH$	152.98	2, 254	1.7000^{20}	1.4750^{20}	25.7	203	100	v s aq, alc, bz, chl, eth
b407	3-Bromopropionic acid	$BrCH_2CH_2COOH$	152.98	2, 256	1.480		62.5		65	s aq, alc, bz, chl, eth
b408	3-Bromopropionitrile	$BrCH_2CH_2CN$	133.98	2^2, 231	1.6152_4^{20}	1.4800^{20}		78^{10mm}	98	v s alc, eth
b409	2-Bromopropionyl bromide	$CH_3CH(Br)COBr$	215.88	2, 256	2.061	1.5182^{20}		50^{10mm}	>110	
b410	2-Bromopropionyl chloride	$CH_3CH(Br)COCl$	171.43	2, 256	1.700^{11}	1.4800^{20}		133	51	d aq; s chl, eth
b411	3-Bromopropionyl chloride	$CH_3CH(Br)COCl$	171.43	2^2, 231	1.701	1.4968^{20}		57^{17mm}	79	
b412	2-Bromopropiophenone	$C_6H_5COCH(Br)CH_3$	213.08	7, 302	1.430_4^{20}	1.5715^{20}		250	>110	s alc, bz, eth, acet
b413	3-Bromopropyl phenyl ether	$C_6H_5OCH_2CH_2CH_2Br$	215.10	6, 142	1.365	1.5464^{20}	10–11	134^{14mm}	96	
b414	3-Bromopropyltrichlorosilane	$Br(CH_2)_3SiCl_3$	256.44		1.605	1.4900^{20}		202–204	76	
b415	3-Bromopropyne	$BrCH_2C{\equiv}CH$	118.97	1, 248	1.335	1.4905^{20}		88–90	18	
b416	2-Bromopyridine	$Br(C_5H_4N)$	158.00	20, 233	1.657^{18}	1.5720^{20}		194	54	i aq; s org solv
b417	3-Bromopyridine	$Br(C_5H_4N)$	158.00	20, 233	1.645_4^0	1.5695^{20}	142–143	173	51	s aq; v s alc, eth
b418	3-Bromoquinoline		208.06	20, 363	1.533	1.6640^{20}	15	276	>110	s HOAc
b419	5-Bromosalicylic acid	$Br(HO)C_6H_3COOH$	217.02	10, 107			166			0.3 aq^{80}; 85 alc^{25}; 70 eth^{25}
b420	β-Bromostyrene	$C_6H_5CH{=}CHBr$	183.05	5, 477	1.422_4^{20}	1.6066^{20}	7	112^{20mm}	79	i aq; misc alc, eth

(Continued)

TABLE 2.20 Physical Constants of Organic Compounds (*Continued*)

No.	Name	Formula	Formula weight	Beilstein reference	Density, g/mL	Refractive index	Melting point, °C	Boiling point, °C	Flash point, °C	Solubility in 100 parts solvent
b421	(±)-Bromosuccinic acid	$HOOCH_2CH(Br)COOH$	196.99	2, 621	2.073		161			18 aq; s alc, acet, eth
b422	*N*-Bromosuccinimide		177.99	21, 380	2.098		173 sl dec			1.5 aq^{25}; 14.4 acet^{25}; 3.1 HOAc^{25}
b423	1-Bromotetradecane	$H(CH_2)_{14}Br$	277.30	1^2, 136	1.0124^{25}_{4}	1.4600^{20}	6	178^{20mm}	>110	s alc; v s chl; misc bz, acet
b424	3-Bromotetrahydro-2-methyl-2*H*-pyran		179.06	17^3, 75	1.366	1.4830^{20}		61^{17mm}	57	
b425	3-Bromothioanisole	$BrC_6H_4SCH_3$	203.11	6, 330			38–40		>110	
b426	2-Bromothiophene	$Br(C_5H_3S)$	163.04	17, 33	1.684^{20}_{4}	1.5860^{20}		151	60	v s acet, eth
b427	3-Bromothiophene	$Br(C_5H_3S)$	163.04		1.740	1.5910^{20}		150	56	
b428	4-Bromothiophenol	BrC_6H_4SH	189.08	6, 330			76	239		
b429	2-Bromotoluene	$BrC_6H_4CH_3$	171.04	5, 304	1.422^{25}_{25}	1.552^{25}	−26	181	78	0.1 aq; misc alc, bz, chl, eth
b430	3-Bromotoluene	$BrC_6H_4CH_3$	171.04	5, 305	1.4099^{20}	1.5517^{20}	−39.8	183.7	60	s alc, bz, eth
b431	4-Bromotoluene	$BrC_6H_4CH_3$	171.04	5, 305	1.3959^{35}_{35}	1.5490	28.5	184.5	85	s alc, bz, eth
b432	Bromotrichloromethane	$BrCCl_3$	198.28	1, 67	1.997^{25}_{25}	1.5063^{20}	−6	104–105		misc org solv
b433	1-Bromotridecane	$H(CH_2)_{13}Br$	263.27	1^2, 134	1.0262^{20}_{4}	1.4592^{20}	7	150^{10mm}	>110	v s chl
b434	Bromotrifluoromethane	$BrCF_3$	148.91	1^3, 83	6.087 g/L		−168 to −172	−57.8		v s chl
b435	5-Bromo-1,2,4-trimethylbenzene	$BrC_6H_2(CH_3)_3$	199.10	5, 403			73	235		i aq; s alc
b436	2-Bromo-1,3,5-trimethylbenzene	$BrC_6H_2(CH_3)_3$	199.10	5, 408	1.301	1.5511^{20}	2	225	96	i aq; s bz; v s eth
b437	Bromotrimethylgermane	$(CH_3)_3GeBr$	197.60		1.544^{18}	1.4705^{20}	−25	113.7	37	
b438	Bromotrimethylsilane	$(CH_3)_3SiBr$	153.10		1.160	1.4140^{20}		79	32	
b439	Bromotriphenylethylene	$(C_6H_5)_2C{=}C(Br)C_6H_5$	335.22	5, 722			115–117			
b440	Bromotriphenylmethane	$(C_6H_5)_3CBr$	323.24	5, 704			152–154	230^{15mm}		
b441	1-Bromoundecane	$CH_3(CH_2)_{10}Br$	235.22	1^2, 132	1.954	1.4563^{20}	−9	138^{18mm}	>110	
b442	11-Bromoundecanoic acid	$Br(CH_2)_{10}COOH$	265.20	2^2, 315			51	174^{2mm}	>110	i aq; v s alc
b443	α-Bromo-1,2-xylene	$BrCH_2C_6H_3CH_3$	185.07	5, 365	1.381^{23}	1.381^{20}	21	224	82	s alc, eth
b444	α-Bromo-1,3-xylene	$BrCH_2C_6H_3CH_3$	185.07	5, 374	1.370^{23}	1.5560^{20}		185^{340mm}	82	s alc, eth
b445	2-Bromo-1,4-xylene	$BrCH_2C_6H_3CH_3$	185.07	5, 385	1.340	1.5505^{20}	9–10	199–201	79	v s chl, hot ether

b446	4-Bromo-1,2-xylene	$BrCH_2C_6H_3CH_3$	185.07	5, 365	1.370_{15}^{15}	1.5560^{20}		215	80	v s alc, eth
b447	Brucine		394.45	27^2, 797			178			77 alc; 1 bz; 20 chl; 4 EtOAc
b448	1,2-Butadiene	$CH_3CH{=}C{=}CH_2$	54.09	1, 249	0.676^{10}	1.4205^{1}	−136.2	10.9		misc alc, eth
b449	1,3-Butadiene	$H_2C{=}CHCH{=}CH_2$	54.09	1, 249	2.211 g/L	1.4293^{-25}	−108.9	−4.4	−76	misc alc, eth
b450	Butadiene sulfone		118.15	17^3, 144			66		>110	
b451	1,3-Butadienyl acetate	$CH_3CO_2CH{=}CHCH{=}CH_2$	112.13	2^3, 295	0.945	1.4690^{20}		60^{40mm}	33	
b452	1,3-Butadiyne	$HC{\equiv}CC{\equiv}CH$	50.06	1^3, 1056	0.7364_4^0	1.4189^5	−36	10.3		v s eth; s acet, bz
b453	2-Butanamine	$CH_3CH_2CH(NH_2)CH_3$	73.14	4, 160	0.7308_4^{15}	1.3963^{15}	−104.5	66	−19	misc aq, alc
b454	Butane	$CH_3CH_2CH_2CH_3$	58.12	1, 118	0.6011^0	1.3562^{-13}	−138.3	−0.50	−60	1 vol aq dissolves 0.15 vol and 1 vol alc 18 vols at 17° and 770 mm; 1 vol ether or $CHCl_3$ dissolves 25 or 30 vols, resp.
b455	1,4-Butanediamine	$H_2NCH_2CH_2CH_2CH_2NH_2$	88.15	4, 264	0.877_4^{25}	1.4569^{20}	28	158–160	51	s aq
b456	Butanedinitrile	$NCCH_2CH_2CN$	80.09	2, 615	0.9867_4^{60}	1.4173^{60}	54.5	266	132	11.5 aq; s acet, chl, 1,4-dioxane; sl s bz
b457	1,2-Butanediol	$CH_3CH_2CH(OH)CH_2OH$	90.12	1, 477	1.006_0^{18}	1.4380^{20}		207.5	93	s aq, alc, acet
b457a	1,3-Butanediol	$CH_3CH(OH)CH_2CH_2OH$	90.12	1, 477	1.0053_{20}^{20}	1.441^{20}	<−50	207.5	121	s aq, alc, acet; 9 eth
b457b	1,4-Butanediol	$HOCH_2CH_2CH_2CH_2OH$	90.12	1, 478	1.016_4^{25}	1.4452^{20}	20	235	121	misc aq, alc, acet; 0.3 bz; 3.1 eth; 0.9 PE
b458	*meso*-2,3-Butanediol	$CH_3CH(OH)CH(OH)CH_3$	90.12	1, 479	0.9939_4^{25}	1.4324^{35}	25	182	85	misc aq, alc
b459	1,4-Butanediol dimethanesulfonate	$CH_3SO_2O(CH_2)_4OSO_2CH_3$	246.30	4^4, 19			114–117			2.4 acet[25]; 0.1 alc[25]
b460	1,3-Butanediol diacetate	$CH_3CO_2CH_2CH_2CH(CH_3)$-O_2CCH_3	174.20	2, 143	1.028	1.4199^{20}		99^{8mm}	85	
b461	1,4-Butanediol diacrylate	$(H_2C{=}CHCO_2CH_2CH_2\text{-})_2$	198.22	2^4, 170	1.051	1.4560^{20}		$83^{0.3mm}$	>110	
b462	1,3-Butanediol dimethacrylate	$H_2C{=}C(CH_3)CO_2CH_2CH_2$-$CH(CH_3)O_2CC(CH_3){=}CH_2$	226.28		1.010	1.4520^{20}		290	>110	
b463	1,4-Butanediol dimethacrylate	$[H_2C{=}C(CH_3)CO_2CH_2CH_2\text{-}]_2$	226.28	2^4, 1534	1.010	1.4560^{20}		134^{4mm}	>110	
b464	1,4-Butanediol divinyl ether	$(\text{-}CH_2CH_2OCH{=}CH_2)_2$	142.20	1^4, 2518	0.898	1.444^{20}	−8	64^{10mm}	62	
b465	1,4-Butanediol vinyl ether	$H_2C{=}CHO(CH_2)_4OH$	116.16	1^4, 2518	0.939	1.4440^{20}		95^{20}	85	

(*Continued*)

TABLE 2.20 Physical Constants of Organic Compounds (*Continued*)

No.	Name	Formula	Formula weight	Beilstein reference	Density, g/mL	Refractive index	Melting point, °C	Boiling point, °C	Flash point, °C	Solubility in 100 parts solvent
b466	2,3-Butanedione	$CH_3C(=O)C(=O)CH_3$	86.09	1, 769	0.990_{15}^{15}	1.3951^{20}		86	7	25 aq; misc alc, eth
b467	2,3-Butanedione monoxide	$CH_3C(=NOH)C(=O)CH_3$	101.11	1, 772			75–78	186		
b468	1,4-Butanedithiol	$HSCH_2CH_2CH_2CH_2SH$	122.25	1, 479	1.042	1.5290^{20}		106^{30mm}	70	i aq; v s alc
b468a	Butanenitrile	$CH_3CH_2CH_2CN$	69.11	2^2, 252	0.7936	1.4440^{20}	−112	117.6	24	3.3 aq; misc alc, eth
b469	1,2,3,4-Butanetetracarboxylic acid	$[\text{-}CH(COOH)CH_2COOH]_2$	234.16	2, 863			196			
b470	1-Butanethiol	$CH_3CH_2CH_2CH_2SH$	90.19	1, 370	0.8367_4^{25}	1.4430^{25}	−116	98.5	2	0.06 aq; v s alc, eth
b471	2-Butanethiol	$CH_3CH_2CH(SH)CH_3$	90.19	1, 373	0.8246_4^{25}	1.4338^{25}	−165	85.0	21	sl s aq; v s alc, eth
b472	1,2,4-Butanetriol	$HOCH_2CH_2CH(OH)CH_2OH$	106.12	1, 519	1.190^{20}	1.4748^{20}		191^{18mm}	167	v s aq, alc
b473	1-Butanol	$CH_3CH_2CH_2CH_2OH$	74.12	1, 367	0.8097_4^{20}	1.3993^{20}	−89.5	117.7	37	7.4 aq; misc alc, eth
b474	2-Butanol	$CH_3CH_2CH(OH)CH_3$	74.12	1, 371	0.8069_4^{20}	1.3972^{20}	−114.7	99.5	24	12.5 aq; misc alc, eth
b475	2-Butanone	$CH_3CH_2COCH_3$	72.11	1, 666	0.8054_4^{20}	1.3788^{20}	−86.7	79.6	−9	24 aq; misc alc, bz, eth
b476	2-Butanone oxime	$CH_3CH_2C(=NOH)CH_3$	87.12	1, 668	0.924	1.4420^{20}		60^{15mm}	60	
b477	1-Butene	$CH_3CH_2CH=CH_2$	56.11	1, 203	0.6255_4^{mp}	1.3962^{20}	−185.3	−6.5	−80	i aq; v s alc, eth
b478	*cis*-2-Butene	$CH_3CH=CHCH_3$	56.11	1^3, 728	0.6213	1.3931^{-25}	−139.3	3.7	−73	i aq; v s alc, eth
b479	*trans*-2-Butene	$CH_3CH=CHCH_3$	56.11	1, 205	0.6041	1.3848^{-25}	−105.8	0.9	−73	i aq; v s alc, eth
b480	*cis*-2-Butene-1,4-diol	$HOCH_2CH=CHCH_2OH$	88.11	1^2, 567	1.0700_4^{20}	1.4780^{20}	2	234	128	s aq; v s alc
b481	*trans*-2-Butene-1,4-diol	$HOCH_2CH=CHCH_2OH$	88.11	1^3, 2252	1.070_4^{20}	1.4755^{20}	25	132		v s aq, alc
b482	3-Butenenitrile	$H_2C=CHCH_2CN$	67.09	2, 408	0.8341_4^{20}	1.4060^{20}	−87	119	21	sl s aq; misc alc, eth
b483	*cis*-2-Butenoic acid	$CH_3CH=CHCOOH$	86.09	2, 412	1.0267_4^{20}	1.4483^{14}	14–15	168–169		v s aq; s alc
b484	*trans*-2-Butenoic acid	$CH_3CH=CHCOOH$	86.09	2, 408	0.9604_4^{80}	1.4248^{77}	72	185	87	55 aq; 52 EtOH; 53 acet; 37 toluene
b485	3-Butenoic acid	$H_2C=CHCH_2COOH$	86.09	2, 407	1.0091_4^{20}	1.4249^{20}	−39	163	65	s aq; misc alc, eth
b486	*cis*-2-Buten-1-ol	$CH_3CH=CHCH_2OH$	72.11	1, 442	0.8662_4^{20}	1.4342^{20}	−89.4	123.6	56	16.6 aq; misc alc
b487	*trans*-2-Buten-1-ol	$CH_3CH=CHCH_2OH$	72.11	1, 442	0.8524_4^{20}	1.4289^{20}	<−30	121.2	56	16.6 aq; misc alc
b488	3-Buten-2-one	$H_2C=CHCOCH_3$	70.09	1, 728	0.8636_4^{20}	1.4086^{20}		81.4	−6	v s aq, alc, acet, eth
b489	1-Buten-3-yne	$HC\equiv CCH=CH_2$	52.07	1^3, 1032	0.7095_4^1	1.4161		5.1		
b490	4-Butoxyaniline	$CH_3(CH_2)_3OC_6H_4NH_2$	165.24	13^2, 226	0.992	1.5543^{20}		149^{13mm}	>110	
b491	4-Butoxybenzoic acid	$CH_3(CH_2)_3OC_6H_4COOH$	194.23	10^2, 93			150			
b492	Butoxycarbonylmethyl butyl phthalate	$2\text{-}[CH_3(CH_2)_3O_2CCH_2O_2C]\text{-}C_6H_4CO_2(CH_2)_3CH_3$	336.39	9,3, 4187	1.100	1.4900^{20}		219^{5mm}	>110	
b493	2-Butoxyethanol	$CH_3(CH_2)_3OCH_2CH_2OH$	118.18	1^2, 519	0.9012_4^{20}	1.4198^{20}	−75	168	69	5 aq; s most org solv
b494	1-*tert*-Butoxy-2-ethoxyethane	$(CH_3)_3COCH_2CH_2OC_2H_5$	146.23	1^3, 2085	0.834	1.4015^{20}		148	33	

b495	2-(2-Butoxyethoxy)-ethanol	$HOCH_2CH_2OCH_2CH_2OC_4H_9$	162.23	1^2, 521	0.9536^{20}_{20}	1.4306^{20}	−68.1	230.4	100	misc aq, alc, bz, acet, CCl_4, PE
b496	2-(2-Butoxyethoxy)-ethyl acetate	$CH_3CO_2(CH_2CH_2O)_2CH_2$-$CH_2CH_2CH_3$	204.27	2^3, 308	0.978	1.4260^{20}		245	>110	
b497	2-Butoxyethyl acetate	$CH_3CO_2CH_2CH_2O(CH_2)_3CH_3$	160.22	2^3, 307	0.942	1.4136^{20}		192	76	
b498	2-*tert*-Butoxy-2-methoxyethane	$(CH_3)_3CO_2CH_2CH_2OCH_3$	132.20	1^3, 2084	0.840	1.3985^{20}		132	25	
b499	1-*tert*-Butoxy-2-propanol	$(CH_3)_3COCH_2CH(OH)CH_3$	132.10	1^3, 2148	0.874	1.4130^{20}		143–145	44	
b500	3-Butoxypropylamine	$CH_3(CH_2)_3O(CH_2)_3NH_2$	73.14	4^3, 739	0.853	1.4260^{20}		170	63	
b501	Butyl acetate	$C_4H_9O_2CH_3$	116.16	2, 130	0.8813^{20}_{4}	1.3941^{20}	−77/−78	126	22	0.43 aq; misc alc, eth; s most org solvents
b502	DL-*sec*-Butyl acetate	$CH_3CO_2CH(CH_3)C_2H_5$	116.16	2^2, 131	0.8748^{20}	1.3888^{20}	−99	112	31	0.62 aq; s alc, eth
b503	*tert*-Butyl acetate	$(CH_3)_3CO_2CCH_3$	116.16	2, 131	0.8665^{20}_{4}	1.3870^{20}		95.1	16	i aq; misc alc, eth
b504	*tert*-Butylacetic acid	$(CH_3)CCH_2COOH$	116.16	2, 337	0.912	1.4115^{20}	6–7	190		
b505	*tert*-Butyl acetoacetate	$(CH_3)_3COC({=}O)CH_2$-$C({=}O)CH_3$	158.20		0.954	1.4180^{20}			60	
b506	2-Butylacrolein	$(CH_3(CH_2)_3C({=}CH_2)CHO$	112.17	1^4, 3482	0.843	1.4348^{20}		139	33	
b507	*N*-*tert*-Butylacrylamide	$H_2C{=}CHCONHC(CH_3)_3$	127.19	4^4, 664			128–129			
b507a	Butyl acrylate	$H_2{=}CHCO_2(CH_2)_3CH_3$	128.17	2^2, 388	0.894	1.4180^{20}	−64	145	39	0.14 aq[20]
b508	*tert*-Butyl acrylate	$H_2C{=}CHCO_2C(CH_3)_3$	128.17	2^3, 1228	0.875	1.4108^{20}		63^{60mm}	17	
b509	Butylamine	$CH_3CH_2CH_2CH_2NH_2$	73.14	4, 156	0.7327^{25}_{4}	1.3992^{25}	−50/−49	77	−12	misc aq, alc, eth
b510	(±)-*sec*-Butylamine	$C_2H_5CH(NH_2)CH_3$	73.14	4, 160	0.724^{20}_{4}	1.3928^{20}	−104	63	−9	misc aq, alc
b511	*tert*-Butylamine	$(CH_3)_3CNH_2$	73.14	4, 173	0.6951^{20}_{4}	1.3788^{20}	−66	44	−9	misc aq, alc
b512	Butyl-4-aminobenzoate	$H_2NC_6H_4CO_2(CH_2)_3CH_3$	193.25	14^2, 249			57–59	174^{8mm}		v sl s aq; s dil acids, alc, chl, eth
b513	2-(*tert*-Butylamino)-ethanol	$(CH_3)_3CNHCH_2CH_2OH$	117.19				42–45	92^{25mm}	68	
b514	2-(*tert*-Butylamino)-ethyl methacrylate	$H_2C{=}C(CH_3)CO_2CH_2$-$CH_2NC(CH_3)_3$	185.27	4^4, 1509	0.914	1.4420^{20}		82^{10mm}	71	
b515	3-(*tert*-Butylamino)-1,2-propanediol	$(CH_3)_3CNHCH_2CH(OH)$-CH_2OH	147.22				70	92^{1mm}		
b516	2-Butylaniline	$CH_3(CH_2)_3C_6H_4NH_2$	149.24	12^2, 633	0.953	1.5380^{20}		123^{12mm}	108	
b517	2-*sec*-Butylaniline	$C_2H_5CH(CH_3)C_6H_4NH_2$	149.24	12^3, 2721	0.957	1.5410^{20}		122^{16mm}	>110	
b518	4-Butylaniline	$CH_3(CH_2)_3C_6H_4NH_2$	149.24	12^1, 503	0.945	1.5350^{20}		120^{15mm}	101	
b519	4-*sec*-Butylaniline	$C_2H_5CH(CH_3)C_6H_4NH_2$	149.24	12^2, 635	0.977	1.5370^{20}		245^{727mm}	107	

(*Continued*)

TABLE 2.20 Physical Constants of Organic Compounds (*Continued*)

No.	Name	Formula	Formula weight	Beilstein reference	Density, g/mL	Refractive index	Melting point, °C	Boiling point, °C	Flash point, °C	Solubility in 100 parts solvent
b520	2-*tert*-Butylanthra-quinone		264.32				98–100			
b521	Butylbenzene	$CH_3CH_2CH_2CH_2C_6H_5$	134.22	5, 413	0.8604_{4}^{20}	1.4898^{20}	−88	183	71	misc alc, bz, eth
b522	*sec*-Butylbenzene	$C_2H_5CH(CH_3)C_6H_5$	134.22	5, 414	0.8608_{4}^{20}	1.4890^{20}	−82.7	173	52	misc alc, bz, eth
b523	*tert*-Butylbenzene	$(CH_3)_3CC_6H_5$	134.22	5, 415	0.8669_{4}^{20}	1.4923^{20}	−58.1	168.5	60	misc alc, bz, eth
b524	Butyl benzoate	$C_6H_5CO_2C_4H_9$	178.23	9, 112	1.0000^{20}	1.496	−22	250	106	i aq; s alc, eth
b525	2-Butylbenzofuran		174.25		0.987	1.5330^{20}			101	
b526	4-*tert*-Butylbenzoic acid	$(CH_3)_3CC_6H_4COOH$	178.23	9, 560	1.142_{4}^{20}		166.3			i aq; v s alc, bz
b527	4-*tert*-Butylbenzoyl chloride	$(CH_3)_3CC_6H_4COCl$	196.68		1.007	1.5364^{20}		135^{20mm}	87	
b528	*N*-(*tert*-Butyl)benzyl-amine	$C_6H_5CH_2NHC(CH_3)_3$	163.27	12, 1022	0.881	1.4968^{20}		80^{5mm}	80	
b529	*tert*-Butyl bromo-acetate	$BrCH_2CO_2C(CH_3)_3$	195.06	2^1, 96	1.321	1.4450^{20}		50^{10mm}	49	
b530	Butyl 2-butoxy-2-hydroxyacetate	$CH_3(CH_2)_3OCH(OH)CO_2$-$(CH_2)_3CH_3$	204.27	3^4, 1497	0.996	1.4291^{20}		90^{40mm}	74	
b531	Butyl butyrate	$CH_3CH_2CH_2CCO_2C_4H_9$	144.22	2, 271	0.8692_{4}^{20}	1.4064^{20}	−91.5	166	49	i aq; misc alc, eth
b532	Butyl carbamate	$H_2NCO_2(CH_2)_3CH_3$	117.15				53–55		108	
b533	Butyl carbazate	$H_2NNHCO_2C(CH_3)_3$	132.16				39–42	$65^{0.03mm}$	91	
b534	4-*tert*-Butylcatechol	$(CH_3)_3C_6H_3$-1,2-$(OH)_2$	166.22		1.049_{25}^{60}		52–55	285	151	0.2 aq;[80] 240 eth;[25] s alc; v s acet
b535	*tert*-Butyl chloro-acetate	$ClCH_2CO_2C(CH_3)_3$	150.61	2^3, 444	1.053	1.4230^{20}		49^{11mm}	46	
b536	4-*tert*-Butyl-1-chloro-benzene	$(CH_3)_3C_6H_4Cl$	158.67	5, 416	1.006	1.5108^{20}	23–25	217		
b537	*tert*-Butylchlorodi-phenylsilane	$(CH_3)_3CSi(C_6H_5)_2Cl$	274.87		1.057	1.5675^{20}		$90^{0.02mm}$	>110	
b538	Butyl chloroformate	$ClCO_2C_4H_9$	136.58	3^2, 11	1.074_{4}^{25}	1.4114^{20}		142	25	d aq, alc; misc eth
b539	Butyl cyanoacetate	$NCCH_2CO_2C_4H_9$	141.17	2^1, 255	0.993	1.4254^{20}		115^{15mm}	87	
b540	*tert*-Butyl cyanoacetate	$NCCH_2CO_2C(CH_3)_3$	141.17		0.972	1.4200^{20}		108	91	
b541	Butylcyclohexane	$(C_6H_{11})_6C_4H_9$	140.27	5^1, 20	0.818	1.4400^{20}	−78	178–180	41	
b542	*tert*-Butylcyclohexane	$(C_6H_{11})C(CH_3)_3$	140.27	5^1, 20	0.831	1.4470^{20}		167	42	
b543	2-*tert*-Butylcyclo-hexanol	$(CH_3)_3C(C_6H_{10})OH$	145.27	6^3, 126	0.902		43–46		79	i aq
b544	4-*tert*-Butylcyclo-hexanol	$(CH_3)_3C(C_6H_{10})OH$	156.27	6^1, 18			62–70	115^{15mm}	105	i aq

b545	2-*tert*-Butylcyclohexanone	$(CH_3)_3C(C_6H_9)({=}O)$	154.25	7^3, 143	0.896	1.4565^{20}		63^{4mm}	72	
b546	4-*tert*-Butylcyclohexanone	$(CH_3)_3C(C_6H_9)({=}O)$	154.25	7^1, 29			47–50	116^{20mm}	96	i aq
b547	Butyl decyl *o*-phthalate	$C_4H_9O_2C_6H_4CO_2C_{10}H_{21}$	362.51		0.994^{25}_{25}				202	
b548	4-*sec*-Butyl-2,6-di-*tert*-butylphenol	$C_2H_5CH(CH_3)C_6H_2(OH)$-$[C(CH_3)_3]$	262.44	6,*3*, 2094	0.902		25	142^{10mm}	>110	
b549	*N*-Butyldiethanolamine	$C_4H_9N(CH_2CH_2OH)_2$	161.25	4, 285	0.986^{20}_{20}	1.4625^{20}	−70	276	126	
b550	Butyl 3,4-dihydro-2,2-dimethyl-4-oxo-2*H*-pyran-6-carboxylate		226.27		1.054^{25}_{25}	1.4767^{20}		256–270	>110	
b551	*tert*-Butyldimethylchlorosilane	$(CH_3)_3CSi(CH_3)_2Cl$	150.73	4,*4*, 4076			89	124–126	22	
b552	6-*tert*-Butyl-2,4-dimethylphenol	$(CH_3)_3CC_6H_2(CH_3)_2OH$	178.28	6^3, 2020		1.5178^{20}	23	249	111	
b553	*N*-Butylethanolamine	$HOCH_2CH_2NHC_4H_9$	117.19		0.89^{20}	1.444^{20}	−3.5	192	77	
b554	Butyl ethyl ether	$C_4H_9OC_2H_5$	102.18	1, 369	0.7495^{20}_{4}	1.3818^{20}	−124	92	4	i aq; misc alc, eth
b555	2-Butyl-2-ethyl-1,5-pentanediamine	$H_2N(CH_2)_3C[(CH_2)_3CH_3]$-$(C_2H_5)CH_2NH_2$	186.34		0.876	1.4700^{20}		269^{750mm}	>110	
b556	2-Butyl-2-ethyl-1,3-propanediol	$HOCH_2C(C_2H_5)(C_4H_9)CH_2OH$	160.25	1^3, 2228	0.931^{50}_{20}	1.4587^{25}	41–44	178^{50mm}	>110	0.8 aq
b557	Butyl ethyl sulfide	$C_4H_9SC_2H_5$	118.24	1^3, 1522	0.8376^{20}_{4}	1.4491^{20}	−95.1	144.2		s chl
b558	*N*-*tert*-Butylformamide	$HCONHC(CH_3)_3$	101.15	4^3, 324	0.903	1.4330^{20}	16	202	95	
b559	Butyl formate	$HCO_2C_4H_9$	102.13	2, 21	0.892	1.3889^{20}	−91.5	106	18	
b560	Butyl glycidyl ether	$H_2C{-}CHCH_2OC_4H_9$ (epoxide, ring O)								
b561	*tert*-Butyl glycidyl ether	$H_2C{-}CHCH_2OC(CH_3)_3$ (epoxide, ring O)	130.19	17^3, 988	0.917	1.4166^{20}			43	
b562	*tert*-Butylhydrazine HCl	$(CH_3)_3CNHNH_2\cdot HCl$	124.61	4^3, 1734			194			
b563	*tert*-Butyl hydroperoxide	$(CH_3)_3C{-}O{-}OH$	90.12	1^3, 1579	0.896^{20}_{4}	1.4007^{20}	−8	34^{17mm}	37	s aq, alc, chl, eth
b564	1-Butylimidazole		124.19	23^2, 36	0.945	1.4800^{20}		116^{12mm}	>110	
b565	Butyl isocyanate	C_4H_9NCO	99.13		0.880	1.4061^{20}		115	17	

(*Continued*)

TABLE 2.20 Physical Constants of Organic Compounds (*Continued*)

No.	Name	Formula	Formula weight	Beilstein reference	Density, g/mL	Refractive index	Melting point, °C	Boiling point, °C	Flash point, °C	Solubility in 100 parts solvent
b566	*tert*-Butyl isocyanate	$(CH_3)_3CNCO$	99.13	4, 175	0.868	1.3865^{20}		86	−4	
b567	Butyl lactate	$CH_3CH(OH)CO_2C_4H_9$	148.19	3^2, 207	0.984	1.4210^{20}	−28	185–187	69	
b568	Butyl levulinate	$CH_3COCH_2CH_2CO_2C_4H_9$	172.22		0.974	1.4270^{20}		$108^{5.5mm}$	91	
b569	Butyl 3-mercapto-propionate	$HSCH_2CH_2CO_2C_4H_9$	162.25		0.795	1.4100^{20}		101^{12mm}	93	
b570	Butyl methacrylate	$H_2C{=}C(CH_3)CO_2C_4H_9$	142.19	2^3, 1286	0.889^{25}_{15}	1.4230^{25}		170	50	i aq; misc alc, eth
b571	*sec*-Butyl-2-methyl-2-butenoate	$CH_3CH{=}C(CH_3)CO_2\text{-}CH(CH_3)C_2H_5$	156.23		0.889	1.4350^{20}		85^{27mm}	66	
b572	*tert*-Butyl methyl ether	$(CH_3)_3C{-}O{-}CH_3$	88.15	1, 381	0.7404^{20}_{4}	1.3689^{20}	−109	52	−28	4.8 aq; v s alc, eth; unstable acid solns
b573	2-*tert*-Butyl-4-methyl-phenol	$(CH_3)_3CC_6H_3(CH_3)OH$	164.25		0.9247^{75}_{4}	1.4969^{75}	51.7	237	100	i aq; s org solv
b574	2-*tert*-Butyl-5-methyl-phenol	$(CH_3)_3CC_6H_3(CH_3)OH$	164.25	6^2, 507	0.964	1.5192^{20}		118^{12mm}	105	
b575	2-*tert*-Butyl-6-methyl-phenol	$(CH_3)_3CC_6H_3(CH_3)OH$	164.25			1.5190^{20}	30–32	230	107	
b576	*tert*-Butyl-1-methyl-2-propynyl ether	$(CH_3)_3COCH(CH_3)C{\equiv}CH$	126.20		0.795	1.4100^{20}		41^{25mm}	10	
b577	*tert*-Butyl methyl sulfide	$(CH_3)_3CSCH_3$	104.21	1^3, 1591	0.826^{20}_{4}	1.441^{20}	−97.8	102	−3	v s alc
b578	Butyl nitrite	C_4H_9ONO	103.12	1, 369	0.9114^{0}_{4}	1.3768		78	−13	misc alc, eth
b579	*tert*-Butyl nitrite	$(CH_3)_3CONO$	103.12	1, 382	0.8671^{20}_{4}	1.3687^{20}		63	−13	sl s aq; v s alc, chl, eth, CS_2
b580	Butyl 4-nitrobenzoate	$O_2NC_6H_4CO_2C_4H_9$	223.23	9^2, 259			35–39	160^{8mm}	>110	
b581	Butyl octadecanoate	$CH_3(CH_2)_{16}CO_2C_4H_9$	340.60	2^2, 352	0.8551^{20}_{4}	1.4422^{25}	26.3	343	160	s alc; v s acet
b581a	Butyl *cis*-9-octa-decenoate	$CH_3(CH_2)_8CH{=}CH(CH_2)_7\text{-}CO_2C_4H_7$	338.57		0.8704^{15}	1.4480^{25}	−26		180	s eth
b582	Butyl 4-oxopentanoate	$CH_3C({=}O)CH_2CH_2CO_2C_4H_9$	172.22		0.9735^{20}_{4}	1.4270^{20}		107^{6mm}	91	s alc, acet, eth
b583	4-(1-Butylpentyl)-pyridine	$C_4H_9CHC(CH_2)_3CH_3$ with C_5H_4N on CH	205.35	20^3, 2872	0.887	1.4877^{20}		267	>110	
b584	*tert*-Butyl peroxo-benzoate	$C_6H_5C({=}O)O{-}O{-}C(CH_3)_3$	194.23		1.021	1.4990^{20}		$76^{0.2mm}$	93	
b585	2-*sec*-Butylphenol	$CH_3CH_2CH(CH_3)C_6H_4OH$	150.22		0.982	1.5222^{20}	12	228	112	i aq; s alc; v s eth
b586	2-*tert*-Butylphenol	$(CH_3)_3CC_6H_4OH$	150.22	6^2, 489	0.9783^{20}_{4}	1.5228^{20}	−7	221–224	>110	
b587	4-*sec*-Butylphenol	$CH_3CH_2CH(CH_3)C_6H_4OH$	150.22	6, 522	0.969^{20}_{4}	1.5150	62	136^{25mm}	115	s hot aq, alc, eth
b588	4-*tert*-Butylphenol	$(CH_3)_3CC_6H_4OH$	150.22	6, 524	0.908^{114}_{4}	1.4787^{114}	98	237		i aq; s alc, eth

b589	*tert*-Butyl 4-phenoxyphenol ketone	$C_6H_5OC_6H_4C(=O)C(CH_3)_3$	254.33	8^3, 491			52–54	175^{3mm}	>110	
b590	*tert*-Butyl phenyl carbonate	$C_6H_5OC(=O)OC(CH_3)_3$	194.23		1.047	1.4805^{20}		$79^{0.8mm}$	101	
b591	Butyl phenyl ether	$CH_3CH_2CH_2CH_2OC_6H_5$	150.22	6, 143	0.9351^{20}_4	1.4970^{20}	−19	210.3	82 (OC)	
b592	4-*tert*-Butylphenyl salicylate	$HOC_6H_4CO_2CH_6H_4C(CH_3)_3$	270.31				62–64			<0.1 aq; 79 alc; 153 EtOAc; 158 toluene
b593	Butyl propionate	$CH_3CH_2CO_2C_4H_9$	130.19	2, 241	0.8818^{15}	1.3982^{25}	−89	146.8	38	v s alc, eth; v sl s aq
b594	*tert*-Butyl propionate	$CH_3CH_2CO_2C(CH_3)_3$	130.19	2^3, 528	0.865	1.3930^{20}		118	20	
b595	4-*tert*-Butyl pyridine	$(CH_3)_3C(C_5H_4N)$	135.21	20, 252	0.915	1.4952^{20}		197	63	
b596	*tert*-Butyl 1-pyrrolecarboxylate	$(C_4H_4N)CO_2C(CH_3)_3$	167.21		1.000	1.4685^{20}		92^{20mm}	75	
b597	1-Butylpyrrolidine	$(C_4H_8N)C_4H_9$	127.23	20^2, 4	0.814	1.4440^{20}		157	36	
b598	4-*tert*-Butylstyrene	$(CH_3)_3CC_6H_4CH=CH_2$	160.26	5^3, 1254	0.875	1.5260^{20}	−37	92^{9mm}	80	
b599	1-Butyl-3-sulfanilylurea	$4-(H_2N)C_6H_4SO_2NH-CONHC_4H_9$	271.34	14,*4*, 2667			143–145			
b600	Butyltin trichloride	$C_4H_9SnCl_3$	282.17	4^4, 4346	1.693	1.5229^{20}		93^{10mm}	81	
b601	Butyltin tris(2-ethylhexanoate)	$[CH_3(CH_2)_3CH(C_2H_5)CO_2]_3-SnC_4H_9$	605.43		1.105	1.4650^{20}			>110	
b602	4-*tert*-Butyltoluene	$(CH_3)_3CC_6H_4CH_3$	148.25	5, 439	0.8612^{20}	1.4918^{20}	−52	190	68	
b603	Butyltrichlorosilane	$C_4H_9SiCl_3$	191.56	4,*1*, 582	1.160	1.4370^{20}		149	45	
b604	*tert*-Butyltrichlorosilane	$(CH_3)_3CSiCl_3$	191.56	4^3, 1905			97–100	132–134	40	
b605	Butyl trifluoroacetate	$CF_3CO_2C_4H_9$	170.1		1.0268^{22}	1.353^{22}		100.1		
b606	Butyltrimethoxysilane	$C_4H_9Si(OCH_3)_3$	178.3		0.9312^{20}_4	1.3979^{20}		164–165		
b607	*tert*-Butyl trimethylsilyl peroxide	$(CH_3)_3C—O—O—Si(CH_3)_3$	162.3		0.8219^{20}_4	1.3935^{20}	dec 135	41^{41mm}		
b608	Butylurea	$C_4H_9NHCONH_2$	116.16	4^1, 371			96–98			s aq, alc, eth
b609	Butyl vinyl ether	$C_4H_9OCH=CH_2$	100.16		0.7792^{20}	1.4007^{20}	−92	94.2	−9	0.3 aq
b610	5-*tert*-Butyl-*m*-xylene	$(CH_3)_3CC_6H_3(CH_3)_2$	162.28	5, 447	0.867	1.4946^{20}		205–206	72	
b610a	1-Butyne	$CH_3CH_2C\equiv CH$	54.09		2.211 g/L		−126	8.1		
b610b	2-Butyne	$CH_3C\equiv C—CH_3$	54.09		0.688		−32	27		
b611	2-Butyne-1,4-diol	$HOCH_2C\equiv CCH_2OH$	86.09	1^1, 261		1.450^{25}	56–58	238	152	374 aq; 83 als; 0.04 bz; 2.6 eth; 70 acet
b612	Butyraldehyde	$CH_3CH_2CH_2CHO$	72.11	1, 662	0.8016^{20}_4	1.3843^{20}	−96/−99	74.8	−22	7.1 aq; misc alc, acet, eth, EtOAc

(*Continued*)

TABLE 2.20 Physical Constants of Organic Compounds (*Continued*)

No.	Name	Formula	Formula weight	Beilstein reference	Density, g/mL	Refractive index	Melting point, °C	Boiling point, °C	Flash point, °C	Solubility in 100 parts solvent
b613	Butyramide	$CH_3CH_2CH_2CONH_2$	87.12	2, 275			116	216		16 aq; s alc
b614	Butyric acid	$CH_3CH_2CH_2COOH$	88.11	2, 264	0.9582^{20}_{4}	1.3991^{20}	−5.3/−5.7	163.5	72	misc aq, alc, eth
b615	Butyric anhydride	$[CH_3CH_2CH_2C(=O)]_2O$	158.20	2, 274	0.9668^{20}_{4}	1.4070^{20}	−75/−66	199.5	54	s aq (dec); alc (dec), eth
b616	β-Butyrolactone		86.09	17^1, 130	1.056	1.4109^{20}	−43.5	204	60	
b617	γ-Butyrolactone		86.09	17, 234	1.124^{25}_{4}	1.4348^{25}	−43.5	204	98	misc aq; s alc, acet, bz, eth
b618	Butyronitrile	$CH_3CH_2CH_2CN$	69.11	2^2, 252	0.7954^{15}_{4}	1.4440^{20}	−112	117.6	24	3.3 aq; misc alc, eth
b619	Butyrophenone	$C_6H_5C(=O)C_3H_7$	148.21	7, 313	1.021	1.5195^{20}	11–13	230	88	
b620	Butyryl chloride	$CH_3CH_2CH_2COCl$	106.55	2, 274	1.0263^{21}_{4}	1.412^{20}	−89	102	21	s aq (dec), alc (dec); misc eth
c1	Caffeine		194.19	26, 461	1.23^{18}_{4}		238	subl 178		2.1 aq; 1.5 alc; 18 chl; 0.19 eth; 1 bz; 2 acet
c2	(±)-Camphene		136.24	5, 156	0.8422^{54}_{4}	1.4551^{54}	51–52	159	36	i aq; s alc, chl, eth
c3	(1*R*)-(+)-Camphor		152.24	7, 101	0.992^{25}_{4}	1.5462	179	207	66	100 alc; 100 eth; 200 chl; 250 acet
c4	(1*R*,3*S*)-Camphoric acid		200.23	9, 745	1.186^{20}_{4}		186–188			at 25°C: 0.8 aq, 100 alc, 250 acet, 200 eth, 200 HOAc; s chl
c5	(±)-10-Camphor-sulfonic acid		232.30	11, 314			194 dec			deliq moist air; sl s HOAc, EtOAc; i eth
c6	Carbazole		167.21	20, 433	1.10^{18}_{4}		245	355		16 pyr; 11 acet; 3 eth; 0.8 bz; sl s HOAc, PE
c7	4-Carbethoxy-2-methyl-3-cyclo-hexen-1-one		182.22	10, 631	1.078	1.4880^{20}		268–272	>110	
c8	Carbobenzyloxy-glycine	$C_6H_5CH_2OC(=O)NH-CH_2COOH$	209.20				122			
c9	Carbohydrazide	$H_2NNHC(=O)NHNH_2$	90.08	3, 121			157–158			v s aq; i alc, bz, eth; forms salts with acids
c10	Carbon disulfide	CS_2	76.14	3, 197	1.2632^{20}_{4}	1.6270^{20}	−111.6	46.5	−30	0.3 aq; misc bz, chl, eth, CCl_4

c11	Carbon monoxide	CO	28.01	Merck: 12, 1861	1.145 g/L		−205	−191.5		2.3 aq; 16 alc; s chl, EtOAc, HOAc
c12	Carbon oxide sulfide	COS	60.07		2.456 g/L		−138.8	−50		
c13	Carbon tetrabromide	CBr_4	331.65	1, 68	3.42		90	190	none	
c14	Carbon tetrachloride	CCl_4	153.82	1, 64	1.589^{25}_{25}	1.4607^{20}	−23	76.7	none	0.05 aq; misc alc, bz, chl, eth, CS_2, PE
c15	Carbon tetrafluoride	CF_4	88.01	1, 59	1.89^{-183} liq		−183.6	−127.8		
c16	Carbon tetraiodide	CI_4	519.63	1, 74	4.32^{20}_{4}		171			s bz, chl; dec hot alc
c17	4-Carboxybenzene-sulfonamide	$HOOCC_6H_4SO_2NH_2$	201.20	11, 390			dec 280			v s alc; s alkalis; i aq, bz, eth
c18	(4-Carboxybutyl)tri-phenylphosphonium bromide	$HOOC(CH_2)_4(C_6H_5)_3Br$	443.33				205–207			
c19	1-(Carboxymethyl)-pyridinium chloride		173.60				189 dec			
c20	*R*-(−)-Carvone		150.22	7, 157	0.965^{20}_{4}	1.4989^{20}	<15	230	88	i aq; misc alc
c21	Catechol	C_6H_4-1,2-$(OH)_2$	110.11		1.344		104–106	245	137	43 aq; v s alkalis, pyr; s alc, bz, chl, eth
c22	Catecholborane		119.92		1.125	1.5070^{20}	12	50^{50mm}	2	
c23	Chalcone	$C_6H_5CH{=}CHCOC_6H_5$	208.26	7, 478	1.0712^{62}_{4}		55–57	208^{25mm}	>110	v s bz, chl, CS_2, eth; sl s alc
c23a	Chloroacetaldehyde	$ClCH_2CHO$	78.50	1, 610			−16	85–86		s aq, alc, eth
c24	2-Chloroacetamide	$ClCH_2CONH_2$	93.51	2, 199			119	225 dec		10 aq; 10 alc; sl s eth
c25	2′-Chloroacetanilide	$ClC_6H_4NHCOCH_3$	169.61	12, 559			88–90			s alc
c26	3′-Chloroacetanilide	$ClC_6H_4NHCOCH_3$	169.61	12, 604			79–81			v s alc, bz, CS_2
c26a	4′-Chloroacetanilide	$ClC_6H_4NHCOCH_3$	169.61	12, 611	1.385^{20}_{4}		179			i aq; v s alc, eth, CS_2
c27	Chloroacetic acid	$ClCH_2COOH$	94.50	2, 194	1.580 (*c*)	1.4297^{65}	61	189	126	v s aq; s alc, bz, eth
c28	Chloroacetic anhydride	$[ClCH_2C({=}O)]_2O$	170.98	2, 199	1.5494^{20}_{4}		46	203		v s chl, eth; sl s bz; dec by aq, alc
c29	4′-Chloroacetoacet-anilide	$CH_3COCH_2CH_2CO$-NHC_6H_4Cl	211.65				134	dec	160 (CC)	
c30	Chloroacetonitrile	$ClCH_2CN$	75.50	2, 201	1.193	1.4225^{20}		126	47	
c31	2-Chloroacetophenone	$C_6H_5COCH_2Cl$	154.60	7, 282	1.324^{15}		54–56	245		i aq; v s alc, bz, eth
c32	*o*-Chloroacetophenone	$ClC_6H_4COCH_3$	154.60	7^1, 151	1.188	1.5438^{20}		228^{738mm}	88	sl s aq; s eth

(*Continued*)

TABLE 2.20 Physical Constants of Organic Compounds (*Continued*)

No.	Name	Formula	Formula weight	Beilstein reference	Density, g/mL	Refractive index	Melting point, °C	Boiling point, °C	Flash point, °C	Solubility in 100 parts solvent
c33	*p*-Chloroacetophenone	$ClC_6H_4COCH_3$	154.60	7, 281	1.192_4^{20}	1.555^{20}	20–21	237	90	i aq; misc alc, eth
c34	Chloroacetyl chloride	$ClCH_2COCl$	112.94	2, 199	1.420_4^{20}	1.4541^{20}	−21.8	106	none	dec by aq, MeOH
c36	2-Chloroacrylonitrile	$H_2C{=}C(Cl)CN$	87.51		1.096	1.4290^{20}	−65	89	6	
c37	2-Chloro-4-amino-toluene	$ClC_6H_3(CH_3)NH_2$	141.60	12, 988	1.1671	1.5840^{20}	24–25	238	100	
c38	2-Chloroaniline	$ClC_6H_4NH_2$	127.57	12, 597	1.2125_4^{20}	1.5895^{20}	−14	208.8	97	0.88 aq; s acids, most common org solvents
c39	3-Chloroaniline	$ClC_6H_4NH_2$	127.57	12, 602	1.2150_4^{22}	1.5931^{20}	−10.4	230.5	123	i aq; s most common org solvents
c40	4-Chloroaniline	$ClC_6H_4NH_2$	127.57	12, 607	1.169_4^{77}	1.5546^{85}	72.5	232		s hot aq; v s alc, acet, eth, CS_2
c41	1-Chloroanthra-quinone		242.66	7, 787			160	sublimes		sl s alc; s hot bz; misc eth
c42	2-Chloroanthra-quinone		242.66	7, 787			211	sublimes		sl s alc, bz; i eth
c43	2-Chlorobenzaldehyde	ClC_6H_4CHO	140.57	7, 233	1.2483_4^{20}	1.5658	11	215	87	sl s aq; s alc, bz, eth
c44	3-Chlorobenzaldehyde	ClC_6H_4CHO	140.57	7, 234	1.241	1.5545^{20}	18	214	88	
c45	4-Chlorobenzaldehyde	ClC_6H_4CHO	140.57	7, 235	1.196_4^{61}	1.552^{61}	47	214	87	s aq; v s alc, bz, eth
c46	2-Chlorobenzamide	$ClC_6H_4CONH_2$	155.58	9, 336			142–144			
c47	Chlorobenzene	C_6H_5Cl	112.56	5, 199	1.1063^{20}	1.5248^{20}	−45.3	131.7	28	0.049 aq^{30}; v s alc, bz, chl, eth
c48	4-Chlorobenzene-sulfonamide	$ClC_6H_4SO_2NH_2$	191.64	11, 55			146			s hot aq, hot alc, hot eth
c49	4-Chlorobenzene-sulfonic acid	$ClC_6H_4SO_3H$	192.62	11, 54				149^{22mm}	107	
c50	4-Chlorobenzene-sulfonyl chloride	$ClC_6H_4SO_2Cl$	211.07	11, 55			55	141^{15mm}	107	dec aq, alc; v s bz, eth
c51	2-Chlorobenzoic acid	ClC_6H_4COOH	156.57	9, 334	1.544_4^{20}		140			0.11 aq; v s alc, eth
c52	3-Chlorobenzoic acid	ClC_6H_4COOH	156.57	9, 337	1.496_4^{25}		158			0.04 aq; v s alc, eth
c53	4-Chlorobenzoic acid	ClC_6H_4COOH	156.57	9, 340			241–243			0.02 aq; v s alc, eth
c54	2-Chlorobenzonitrile	ClC_6H_4CN	137.57	9, 336			46	232	108	s alc, eth
c55	4-Chlorobenzonitrile	ClC_6H_4CN	137.57	9, 341			93	223		s alc, bz, chl, eth
c56	2-Chlorobenzo-phenone	$ClC_6H_4COC_6H_5$	216.67	7, 419			44–47	300	>110	
c57	4-Chlorobenzo-phenone	$ClC_6H_4COC_6H_5$	216.67	7, 419			77	196^{17mm}		s alc, acet, bz, eth

c58	2-Chlorobenzotri-chloride	$ClC_6H_4CCl_3$	229.92	5, 302	1.508	1.5817^{20}	29	264	98	
c59	4-Chlorobenzotri-chloride	$ClC_6H_4CCl_3$	229.92	5, 303	1.495	1.5722^{20}		245	>110	
c60	2-Chlorobenzotri-fluoride	$ClC_6H_4CF_3$	180.56	5^3, 692	1.3540^{25}	1.4513^{25}	−6.4	152	58	
c61	3-Chlorobenzotri-fluoride	$ClC_6H_4CF_3$	180.56	5^3, 692	1.3311^{25}	1.4438^{25}	−56.7	137.7	38	
c62	4-Chlorobenzotri-fluoride	$ClC_6H_4CF_3$	180.56		1.353^{20}	1.4463	−36	138.7	47	
c63	2-(4-Chlorobenzoyl)-benzoic acid	$ClC_6H_4COC_6H_4COOH$	260.68	10, 750			150			s alc, bz, eth
c64	2-Chlorobenzoyl chloride	ClC_6H_4COCl	175.01	9, 336	1.382	1.5718^{20}	−3	238	>110	dec by aq & alc
c65	4-Chlorobenzoyl chloride	ClC_6H_4COCl	175.01	9, 341	1.377	1.5780^{20}	14	222	105	dec by aq & alc
c66	4-Chlorobenzyl alcohol	$ClC_6H_4CH_2OH$	142.59	6, 444			72	234		v s alc, eth
c67	2-Chlorobenzylamine	$ClC_6H_4CH_2NH_2$	141.60	12, 1073	1.173	1.5630^{20}		104^{11mm}	88	
c68	4-Chlorobenzylamine	$ClC_6H_4CH_2NH_2$	141.60	12, 1074	1.164	1.5586^{20}		215	90	
c69	2-Chlorobenzyl chloride	$ClC_6H_4CH_2Cl$	161.03	5, 297	1.274	1.5591^{20}	−17	214	82	
c70	4-Chlorobenzyl chloride	$ClC_6H_4CH_2Cl$	161.03	5, 308			30	222	97	s alc, v s eth
c71	2-Chlorobenzyl cyanide	$ClC_6H_4CH_2CN$	151.60	9, 448		1.5540^{20}	24	242	>110	
c72	4-Chlorobenzyl cyanide	$ClC_6H_4CH_2CN$	151.60	9, 448			30.3	267	>110	
c73	4-Chlorobenzyl mercaptan	$ClC_6H_4CH_2SH$	158.65	6, 466	1.202	1.5893^{20}	20		76	
c74	1-Chloro-1,3-butadiene	$H_2C{=}CHCH{=}CHCl$	88.54	1^3, 949	0.9601^{20}_4	1.4712^{20}		68	−20	v s chl
c74a	2-Chloro-1,3-butadiene	$H_2C{=}CHC(Cl){=}CH_2$	88.54		0.952			59		
c75	1-Chlorobutane	$CH_3CH_2CH_2CH_2Cl$	92.57	1, 118	0.8864^{20}_4	1.4021^{20}	−123.1	78.4	−9	0.11 aq; misc alc, eth
c76	2-Chlorobutane	$CH_3CH_2CH(Cl)CH_3$	92.57	1, 119	0.8732^{20}_4	1.3971^{20}	−131.3	68.2	−15	0.1 aq; misc alc, eth
c77	4-Chloro-1-butanol	$ClCH_2CH_2CH_2CH_2OH$	108.56	1^2, 398	1.0883^{20}_4	1.4518^{20}		89^{20mm}	32	s alc, eth
c78	3-Chloro-2-butanone	$CH_3CH(Cl)C({=}O)CH_3$	106.55	1, 669	1.055	1.4172^{20}		117	21	v s alc, eth
c79	*cis*-1-Chloro-2-butene	$CH_3CH{=}CHCH_2Cl$	90.55	1^2, 176	0.9426^{20}	1.4390^{20}		84.1	−15	s alc, acet

(Continued)

TABLE 2.20 Physical Constants of Organic Compounds (*Continued*)

No.	Name	Formula	Formula weight	Beilstein reference	Density, g/mL	Refractive index	Melting point, °C	Boiling point, °C	Flash point, °C	Solubility in 100 parts solvent
c80	*trans*-1-Chloro-2-butene	$CH_3CH{=}CHCH_2Cl$	90.55	1^{2}, 176	0.929	1.4390^{20}		85	−5	s alc, acet
c81	3-Chloro-1-butene	$CH_3CH(Cl)CH{=}CH_2$	90.55	1^{2}, 174	0.9001_{4}^{20}	1.4155^{20}		65	−20	v s acet
c82	4-Chlorobutyl acetate	$CH_3CO_2CH_2CH_2CH_2CH_2Cl$	150.61	2^{2}, 141	1.072	1.4338^{20}		92^{22mm}	64	
c83	3-Chloro-1-butyne	$CH_3CH(Cl)C{\equiv}CH$	88.54	1^{4}, 970	0.961	1.4280^{20}		68–70	1	
c84	3-Chlorobutyric acid	$CH_3CH(Cl)CH_2COOH$	122.55	2, 277	1.186_{4}^{20}	1.4421^{20}	16.3	109^{17mm}	>110	s alc, eth
c85	4-Chlorobutyric acid	$ClCH_2CH_2CH_2COOH$	122.55	2, 278	1.2236_{4}^{20}	1.4521^{20}	12–16	196^{22mm}	>110	sl s aq; v s eth
c86	4-Chlorobutyronitrile	$ClCH_2CH_2CH_2CN$	103.55	2, 278	1.158	1.4413^{20}		197	85	s alc, eth
c87	4-Chlorobutyryl chloride	$ClCH_2CH_2CH_2COCl$	141.00	2, 278	1.258	1.4609^{20}		174	72	dec by aq, alc; s eth
c88	Chloro(chloromethyl)-dimethylsilane	$ClCH_2Si(CH_3)_2Cl$	143.09		1.086	1.4373^{20}		114^{752mm}	21	
c89	3-Chloro-2-chloro-methyl-1-propene	$H_2C{=}C(CH_2Cl)_2$	125.00	1^{2}, 181	1.080	1.4753^{20}	−14	138	36	
c90	*trans*-2-Chloro-cinnamic acid	$ClC_6H_4CH{=}CHCO_2H$	182.61	9, 594			208–210			
c91	Chlorocyclohexane	ClC_6H_{11}	118.61	5, 21	1.000_{4}^{20}	1.4620^{20}	−44	142	28	i aq; s alc, eth
c92	1-Chloro-3-cyclo-hexylpropane	$C_6H_{11}(CH_2)_3Cl$	160.69	5^{2}, 23	0.997	1.4662^{20}		79^{5mm}	78	
c93	Chlorocyclopentane	C_5H_9Cl	104.58	5, 19	1.0051_{4}^{20}	1.4512^{20}		114	15	i aq
c94	1-Chlorodecane	$CH_3(CH_2)_9Cl$	176.73	1, 168	0.868	1.4362^{20}	−34	223	83	i aq
c95	Chlorodicyclohexyl-borane	$(C_6H_{11})_2BCl$	212.57	16^{4}, 1637	0.970			101^{1mm}		
c96	2-Chloro-1,1-diethoxy-ethane	$ClCH_2CH(OC_2H_5)_2$	152.62	1, 611	1.018	1.4157^{20}		157	29	
c97	3-Chloro-1,1-diethoxy-propane	$ClCH_2CH_2CH(OC_2H_5)_2$	166.65	1, 632	0.995	1.4240^{20}		84^{25mm}	36	
c98	Chlorodifluoroacetic acid	$F_2C(Cl)COOH$	130.48	2, 201	1.540	1.3559^{20}	24–26	122	none	
c99	1-Chloro-2,4-difluoro-benzene	$ClC_6H_3F_2$	148.54	5^{4}, 653	1.353	1.4750^{20}		127	32	
c100	1-Chloro-1,1-difluoro-ethane	$CH_3C(Cl)F_2$	100.50	1^{3}, 138	4.108 g/L		−131	−10		0.19 aq
c100a	1-Chloro-2,2-difluoro-ethylene	$ClCH{=}CF_2$	98.48		4.025 g/L		−138.5	−18.5		
c101	Chlorodifluoro-methane	$HCClF_2$	86.47	1^{3}, 41	1.4909^{-69}		−157	−40.8		0.30 aq

c102	1-Chloro-2,4-dihydroxybenzene	$ClC_6H_3(OH)_2$	144.56	6^2, 818			107	147^{18mm}		v s aq, alc, chl, eth
c103	2-Chloro-1,4-dihydroxybenzene	$ClC_6H_3(OH)_2$	144.56	6, 849			101–102	263		v s aq; i alc, s eth
c104	2-Chloro-1,4-dimethoxybenzene	$ClC_6H_3(OCH_3)_2$	172.61	6^3, 4432	1.211	1.5467^{20}		234	110	
c105	2-Chloro-1,1-dimethoxyethane	$ClCH_2CH(OCH_3)_2$	124.57		1.094^{20}_{20}	1.4148^{20}		130	28	
c107	2-Chloro-4,6-dimethylaniline	$ClC_6H_2(CH_3)_2NH_2$	155.63	12, 1125	1.110		38–40		>110	
c108	4-Chloro-3,5-dimethylphenol	$ClC_6H_2(CH_3)_2OH$	156.61	6^2, 463			115.5	246		0.03 aq; 100 alc; s bz, eth, alkalis
c109	1-Chloro-2,2-dimethylpropane	$(CH_3)_3CCH_2Cl$	106.59	1, 141	0.866^{20}_4	1.4042^{20}	−20	84.4	32	
c110	3-Chloro-2,2-dimethyl-1-propanol	$ClCH_2C(CH_3)_2CH_2OH$	122.60			1.4504^{20}	34–36	87^{35mm}	71	
c111	Chlorodimethylsilane	$(CH_3)_2Si(Cl)H$	94.62		0.852^{20}_4	1.3827^{20}	−111	36	−28	
c112	Chlorodimethylvinylsilane	$(CH_3)_2Si(Cl)CH{=}CH_2$	120.7	4,*4*, 4080	0.884^{25}_4	1.414^{25}		82.5	−5	
c113	6-Chloro-2,4-dinitroaniline	$ClC_6H_2(NO_2)_2NH_2$	217.57	12^1, 367			159			
c114	1-Chloro-2,4-dinitrobenzene	$ClC_6H_3(NO_2)_2$	202.55	5, 263	1.4982^{75}_4	1.5857^{60}	52–54	315	186	sl s alc; s hot alc, bz, eth
c115	2-Chloro-3,5-dinitrobenzoic acid	$ClC_6H_2(NO_2)_2COOH$	246.56	9, 415			198	241 explodes		0.3 aq
c116	Chlorodiphenylmethane	$C_6H_5CH(Cl)C_6H_5$	202.68	5^2, 500	1.140^{20}_4	1.5951^{20}	17	140^{3mm}	>110	
c117	Chlorodiphenylmethylsilane	$(C_6H_5)_2Si(Cl)CH_3$	232.8	16^2, 606	1.1277^{20}_4	1.5742^{20}		295	>110	
c118	Chlorodiphenylphosphine	$(C_6H_5)_2PCl$	220.64	16, 763	1.229	1.6338^{20}		320	>110	
c119	1-Chlorododecane	$CH_3(CH_2)_{11}Cl$	204.79		0.8673^{20}_4	1.4426	−9	116	93	v s alc; s bz
c120	1-Chloro-2,3-epoxypropane	$\underset{\backslash O /}{H_2C—CH}CH_2Cl$	92.53	17, 6	1.1812^{20}_4	1.4358^{20}	−57.2	116.1	31	5.9 aq; misc alc, chl
c121	Chloroethane	CH_3CH_2Cl	64.52	1, 82	0.9214^0_4	1.3742^{10}	−139	12.3	−50	0.45 aq^0; 48 alc; misc eth
c122	2-Chloroethanol	$ClCH_2CH_2OH$	80.52	1, 337	1.2019^{20}	1.4422^{20}	−67.5	128.6	60	misc aq, alc

(Continued)

TABLE 2.20 Physical Constants of Organic Compounds (*Continued*)

No.	Name	Formula	Formula weight	Beilstein reference	Density, g/mL	Refractive index	Melting point, °C	Boiling point, °C	Flash point, °C	Solubility in 100 parts solvent
c123	2-(2-Chloroethoxy)-ethanol	$ClCH_2CH_2OCH_2CH_2OH$	124.57	1, 467	1.180	1.4529^{20}		81^{5mm}	90	
c124	2-[2-(2-Chloroethoxy)-ethoxy]ethanol	$ClCH_2CH_2OCH_2CH_2O$-CH_2CH_2OH	168.62	1, 468	1.160	1.4580^{20}		120^{5mm}	107	
c125	2-Chloroethoxytri-methylsilane	$ClCH_2CH_2OSi(CH_3)_3$	152.70	4^3, 1856	0.944	1.4140^{20}		134	30	
c126	2-Chloroethylamine hydrochloride	$ClCH_2CH_2NH_2 \cdot HCl$	115.99	4, 133			146			
c127	1-Chloro-2-ethyl-benzene	$ClC_6H_4C_2H_5$	140.61		1.055^{25}_{25}		−81	179.2	66	i aq; misc alc, eth
c128	(2-Chloroethyl)-benzene	$C_6H_5CH_2CH_2Cl$	140.61	5, 354	1.069	1.5300^{20}		84^{16mm}	66	s alc, bz, eth
c129	Chloroethylene	$H_2C{=}CHCl$	62.50	1, 186	0.97^{-14}		−154	−13.4	−78	sl s aq; s alc
c130	*N*-(2-Chloroethyl)-*N*-ethylamine	$C_6H_5N(C_2H_5)CH_2CH_2Cl$	183.68	12^3, 263	1.075	1.5584^{20}		164^{42mm}	>110	
c131	2-Chloroethyl ethyl ether	$ClCH_2CH_2OCH_2CH_3$	108.57	1, 337	0.989	1.4120^{20}		107	15	
c132	2-Chloroethyl methyl ether	$ClCH_2CH_2OCH_3$	94.54	1, 337	1.035	1.4090^{20}		90	15	
c133	*N*-(2-Chloroethyl)-morpholine HCl		186.08				186			
c133a	2-Chloroethyl phenyl ether	$C_6H_5OCH_2CH_2Cl$	156.61	6^3, 675	1.129	1.5340^{20}		98^{15mm}	100	
c134	*N*-(2-Chloroethyl)-piperidine HCl		184.11	20, 17			236			
c135	2-Chloroethyl *p*-toluenesulfonate	$CH_3C_6H_4SO_3CH_2CH_2Cl$	234.70	11^2, 45	1.294	1.5290^{20}		$153^{0.3mm}$	>110	
c136	2-Chloroethyl vinyl ether	$H_2C{=}CHOCH_2CH_2Cl$	106.55	1^2, 473	1.0525^{15}_{15}	1.4370^{20}	−69.7	110	16	0.6 aq
c137	1-Chloro-2-fluoro-benzene	ClC_6H_4F	130.55	5^1, 110	1.244	1.5010^{20}	−42.4	138.5	31	s alc, eth
c138	1-Chloro-3-fluoro-benzene	ClC_6H_4F	130.55		1.219	1.4944^{20}		126	20	s alc, eth
c139	2-Chloro-6-fluoro-benzyl chloride	$Cl(F)C_6H_3CH_2Cl$	179.02		1.401	1.5372^{20}			93	
c140	4-Chloro-4′-fluoro-butyrophenone	$FC_6H_4C({=}O)CH_2CH_2CH_2Cl$	200.64		1.220	1.5255^{20}			>110	

c141	3-Chloro-4-fluoro-nitrobenzene	$Cl(F)C_6H_3NO_2$	175.55	5^1, 130	1.6028^{17}	1.5674^{17}	41.5	127^{17mm}		
c142	2-Chloro-4-fluoro-phenol	$Cl(F)C_6H_3OH$	146.55	6^4, 880	1.344	1.5300	23	88^{4mm}	75	
c143	2-Chloro-6-fluoro-toluene	$Cl(F)C_6H_3CH_3$	144.58		1.191	1.5026^{20}		156	46	
c144	4-Chloro-2-fluoro-toluene	$Cl(F)C_6H_3CH_3$	144.58	5^4, 813	1.186	1.4998^{20}		158^{743mm}	51	
c145	Chloroform	$CHCl_3$	119.39	1, 61	1.4832^{20}	1.4459^{20}	−63.6	61.1		0.50 aq[25]; misc alc, bz, eth, PE, CCl_4
c146	Chloroform-*d*	$CDCl_3$	120.39	1^3, 63	1.500	1.4445^{20}	−64	60.9		see under chloroform
c147	1-Chloroheptane	$CH_3(CH_2)_6Cl$	134.65	1, 154	0.881^{16}_{0}	1.4250^{20}	−69	159−161	41	misc alc, eth
c148	1-Chlorohexadecane	$CH_3(CH_2)_{15}Cl$	260.89	1, 172	0.865	1.4490^{20}		149^{1mm}	>110	
c149	1-Chlorohexane	$CH_3(CH_2)_5Cl$	120.62	1, 143	0.8780^{20}_{4}	1.4195^{20}	−94	134	26	i aq
c150	6-Chloro-1-hexanol	$Cl(CH_2)_6OH$	136.62		1.204	1.4560^{20}		110^{14mm}	98	sl s aq; v s alc, eth
c151	4-Chloro-4′-hydroxy-benzophenone	$ClC_6H_4C(=O)C_6H_4OH$	232.67	8^2, 187			175−178	257^{13mm}		
c152	5-Chloro-8-hydroxy-7-iodoquinoline		305.50	21, 98			172			i alc, eth; 0.8 chl; 0.6 HOAc
c153	5-Chloro-8-hydroxy-quinoline		179.61	21, 95			130			sl s aq HCl
c154	1-Chloro-4-iodo-benzene	ClC_6H_4I	238.46	5, 221	1.186^{57}_{4}		53−54	227	108	s alc
c155	1-Chloro-3-iodo-propane	$Cl(CH_2)_3I$	204.44	1, 114	1.904	1.5463^{20}		170−172	>110	
c156	1-Chloro-3-mercapto-2-propanol	$HSCH_2CH(OH)CH_2Cl$	126.61	1^3, 2156	1.277	1.5276^{20}		$57^{1.3mm}$	97	
c157	Chloromethane	CH_3Cl	50.49	1, 59	2.064 g/L	1.3712^{-24}	−97.7	−24.2	<0	0.48 aq;[25] s alc,; misc chl, eth, HOAc
c158	3-Chloro-4-methoxy-aniline	$ClC_6H_3(OCH_3)NH_2$	157.60	13, 511			50−55		110	
c159	5-Chloro-2-methoxy-aniline	$ClC_6H_3(OCH_3)NH_2$	157.60	13, 383			83−85			
c160	1-Chloro-2-methoxy-benzene	$ClC_6H_4OCH_3$	142.59	6, 184	1.123	1.5445^{20}		196	76	i aq; s alc, eth
c161	5-Chloro-2-methoxy-benzoic acid	$ClC_6H_3(OCH_3)COOH$	186.59	10, 103			98−100			

(*Continued*)

TABLE 2.20 Physical Constants of Organic Compounds (*Continued*)

No.	Name	Formula	Formula weight	Beilstein reference	Density, g/mL	Refractive index	Melting point, °C	Boiling point, °C	Flash point, °C	Solubility in 100 parts solvent
c162	2-Chloro-6-methoxy-pyridine	$CH_3O(Cl)(C_5H_3N)$	143.57		1.207	1.5263^{20}		186		
c163	2-Chloro-6-methyl-aniline	$CH_3O(Cl)C_6H_3NH_2$	141.60	12^1, 388	1.152	1.5761^{20}	2	215	98	s alc
c164	3-Chloro-2-methyl-aniline	$CH_3O(Cl)C_6H_3NH_2$	141.60	12, 836	1.185	1.5874^{20}	2	117^{10mm}	>110	
c165	3-Chloro-4-methyl-aniline	$CH_3O(Cl)C_6H_3NH_2$	141.60	12, 988		1.5830^{20}	25	238	100	
c166	4-Chloro-2-methyl-aniline	$CH_3O(Cl)C_6H_3NH_2$	141.60	12, 835		1.5848^{20}	27	241	99	s hot alc
c167	5-Chloro-2-methyl-aniline	$CH_3O(Cl)C_6H_3NH_2$	141.60	12, 835		1.5840^{20}	22	237	160	
c168	3-(Chloromethyl)-benzoyl chloride	$ClCH_2C_6H_4COCl$	189.04	9^2, 325	1.330	1.5748^{20}		150^{20mm}	>110	
c169	DL-4-Chloro-2-(α-methylben-zyl)phenol	$C_6H_5CH(CH_3)C_6H_3(Cl)OH$	232.71	6^4, 4710	1.238	1.5994^{20}		155^{2mm}	>110	
c169a	1-Chloro-3-methyl-butane	$ClCH_2CH_2CH(CH_3)CH_3$	106.60		0.8750^{20}	1.4084^{20}	−104	99	<21	sl s aq; misc alc, eth
c170	2-Chloro-2-methyl-butane	$CH_3CH_2CCl(CH_3)_2$	106.59	1, 134	0.8650^{20}_{4}	1.4052^{20}	−73.7	85	−9	i aq; s alc, eth
c171	Chloromethyldichloro-methylsilane	$ClCH_2Si(Cl)_2CH_3$	163.5	4^3, 1888	1.286	1.4494^{20}		121	110	
c172	Chloromethyl ethyl ether	$ClCH_2OCH_2CH_3$	94.54	1^2, 645	1.04^{20}_{4}	1.4040^{20}		79–83	19	s alc; v s eth
c172a	3-(Chloromethyl)-heptane	$CH_3CH_2CH_2CH_2CH(CH_2Cl)CH_2CH_3$	148.68		0.8769^{20}	1.4319^{20}		172	60	
c173	Chloromethyl methyl ether	$ClCH_2OCH_3$	80.51	1, 580	1.0703^{20}_{4}	1.3961^{20}	−103.5	57–59	15	dec by aq; s acet, CS_2
c174	Chloromethyl methyl sulfide	$ClCH_2SCH_3$	95.48		1.153	1.4963^{20}		105	17	
c175	1-(Chloromethyl)-naphthalene	$C_{10}H_7CH_2Cl$	176.65	5, 566	1.180	1.6380^{20}	32	169^{25mm}	>110	
c176	4-Chloro-2-methyl-phenol	$CH_3(Cl)C_6H_3OH$	142.59	6, 359			45–48	220–225	>110	sl s aq

c177	4-Chloro-3-methylphenol	$CH_3(Cl)C_6H_3OH$	142.59	6, 381			65–68	235		i aq; s alc, bz, chl, eth, acet
c178	1-Chloro-2-methyl-2-phenylpropane	$C_6H_5(CH_3)_2CH_2Cl$	168.67	5^2, 320	1.047	1.5240^{20}		96^{10mm}	92	
c179	1-Chloro-2-methylpropane	$(CH_3)_2CHCH_2Cl$	92.57	1, 124	0.8829^{15}	1.4010^{15}	−130.3	68.9	<21	0.09 aq; misc alc, eth
c180	2-Chloro-2-methylpropane	$(CH_3)_3CCl$	92.57	1, 125	0.8420^{20}	1.3856^{20}	−26	50.8	<0	sl s aq; misc alc, eth
c181	1-Chloro-2-methylpropene	$(CH_3)_2C{=}CHCl$	90.55	1, 209	0.9186^{20}_{4}	1.4225^{20}		68.1	−1	misc alc, eth
c182	3-Chloro-2-methylpropene	$ClCH_2C(CH_3){=}CH_2$	90.55	1, 209	0.9210^{15}_{4}	1.4272^{20}	−80	72	−12	misc alc, eth
c183	Chloromethyltrimethylsilane	$ClCH_2Si(CH_3)_3$	122.67	4^3, 1844	0.8861^{20}_{4}	1.4180^{20}		99	−2	
c184	6-(Chloromethyl)uracil		160.56	23^1, 328			257 dec			
c185	1-Chloronaphthalene	$C_{10}H_7Cl$	162.62	5, 541	1.1938^{20}_{4}	1.6326^{20}	−2.3	259	121	s alc, bz, PE
c186	2-Chloronaphthalene	$C_{10}H_7Cl$	162.62		1.1377^{71}	1.6079^{71}	60	256		s alc, bz, chl, eth
c187	4′-Chloro-3′-nitroacetophenone	$ClC_6H_3(NO_2)C({=}O)CH_3$	199.60	7^3, 995			101			
c188	2-Chloro-4-nitroaniline	$ClC_6H_3(NO_2)NH_2$	172.57	12, 733			107–109			sl s aq; v s alc, eth
c189	2-Chloro-5-nitroaniline	$ClC_6H_3(NO_2)NH_2$	172.57	12, 732			119–121			
c190	4-Chloro-2-nitroaniline	$ClC_6H_3(NO_2)NH_2$	172.57	12, 729			117–119			v s alc, eth
c191	4-Chloro-3-nitroaniline	$ClC_6H_3(NO_2)NH_2$	172.57	12, 731			99–101			v s alc; s eth
c192	1-Chloro-2-nitrobenzene	$ClC_6H_4NO_2$	157.56	5, 241	1.348		33	246	123	s alc, bz, eth
c193	1-Chloro-3-nitrobenzene	$ClC_6H_4NO_2$	157.56	5, 243	1.534^{20}_{4}		44	236	103	sl s alc; v s chl, eth
c194	1-Chloro-4-nitrobenzene	$ClC_6H_4NO_2$	157.56	5, 243	1.520		83–84	242	>110	sl s alc; v s eth, CS_2
c195	2-Chloro-4-nitrobenzoic acid	$ClC_6H_3(NO_2)COOH$	201.57	9, 404			139–141			s hot aq, hot bz
c196	2-Chloro-5-nitrobenzoic acid	$ClC_6H_3(NO_2)COOH$	201.57	9, 403	1.608^{18}		166–168			sl s aq; s alc, bz, eth
c197	4-Chloro-3-nitrobenzoic acid	$ClC_6H_3(NO_2)COOH$	201.57	9, 402	1.645^{18}		180–183			sl s alc; s hot aq
c198	4-Chloro-3-nitrobenzophenone	$ClC_6H_3(NO_2)C({=}O)C_6H_5$	261.66	7^1, 230			104–106	235^{13mm}		

(Continued)

TABLE 2.20 Physical Constants of Organic Compounds (*Continued*)

No.	Name	Formula	Formula weight	Beilstein reference	Density, g/mL	Refractive index	Melting point, °C	Boiling point, °C	Flash point, °C	Solubility in 100 parts solvent
c199	2-Chloro-5-nitro-benzotrifluoride	$ClC_6H_3(NO_2)CF_3$	225.55		1.527	1.5083^{20}		231	98	
c200	4-Chloro-3-nitro-benzotrifluoride	$ClC_6H_3(NO_2)CF_3$	225.55		1.511	1.4893^{20}	−2.5	222	101	
c201	4-Chloro-2-nitrophenol	$ClC_6H_3(NO_2)OH$	173.56	6, 238			85–87			
c202	2-Chloro-6-nitro-toluene	$ClC_6H_3(NO_2)CH_3$	171.58	5, 327		1.5377^{70}	36	238	125	i aq
c203	4-Chloro-2-nitro-toluene	$ClC_6H_3(NO_2)CH_3$	171.58	5, 327			39	240^{718mm}	>110	i aq
c203a	1-Chlorooctadecane	$CH_3(CH_2)_{17}Cl$	288.95	1^3, 566	0.849	1.4516^{20}		$158^{1.5mm}$	>110	
c204	1-Chlorooctane	$CH_3(CH_2)_7Cl$	148.68	1, 159	0.875	1.4298^{20}	−58	182	70	0.02 aq; misc alc, eth
c204a	1-Chloropentane	$CH_3(CH_2)_4Cl$	106.60	1, 130	0.8820^{20}	1.4115^{20}	−99	107–108	13	
c205	3-Chloro-2,4-pentane-dione	$CH_3COCH(Cl)COCH_3$	134.56	1, 785	1.129	1.4830^{20}		52^{18mm}	12	
c206	5-Chloro-2-pentanone	$ClCH_2CH_2CH_2COCH_3$	120.58	1^2, 738	1.0571^{18}_4	1.4390^{20}		72^{20mm}	35	s acet, eth
c207	3-Chloroperoxy-benzoic acid	$ClC_6H_5C(O)OOH$	172.57	9^4, 972			69–71			
c208	2-Chlorophenol	ClC_6H_4OH	128.56	6, 183	1.2573^{23}_4	1.5565^{20}	9.8	175	63	sl s aq; v s alc, eth, caustic alkali
c209	3-Chlorophenol	ClC_6H_4OH	128.56	6, 185	1.245^{45}_4	1.5565^{40}	33	214	>110	sl s aq; s alc, eth
c210	4-Chlorophenol	ClC_6H_4OH	128.56	6, 186	1.2238^{78}_4	1.5479^{40}	43	220	115	sl s aq; v s alc, chl, eth, $CHCl_3$, glyc
c211	4-Chlorophenoxyacetic acid	$ClC_6H_4OCH_2COOH$	186.59	6, 187			157–159			s aq; MeOH
c212	2-(4-Chlorophenoxy)-2-methylpropanoic acid	$ClC_6H_4OC(CH_3)_2COOH$	214.65	Merck: 12, 2437			118–119			
c213	(±)-2-(4-Chlorophen-oxy)propanoic acid	$ClC_6H_4OCH(CH_3)COOH$	200.62	6^3, 695			117			
c214	4-Chlorophenylacetic acid	$ClC_6H_4CH_2COOH$	170.60	9, 448			108			v s aq, alc, eth; s bz
c215	(4-Chlorophenyl)-acetonitrile	$ClC_6H_4CH_2CN$	151.60	9, 448			30.5	265–267	>110	
c216	2-Chloro-1,4-phenyl-enediamine sulfate	$H_2NC_6H_3(Cl)NH_2 \cdot H_2SO_4$	240.67	13, 117			251–253			s aq
c217	4-Chloro-1,2-phenyl-enediamine	$ClC_6H_3(NH_2)_2$	142.59	13, 25			70–73			s mineral acids
c218	1-(4-Chlorophenyl)-ethanol	$ClC_6H_4CH(CH_3)OH$	156.61	6^1, 236	1.171	1.5410^{20}		119^{10mm}	>110	

c219	3-Chlorophenyl isocyanate	ClC_6H_4NCO	153.57	12, 606	1.260	1.5576^{20}	−4.4	114^{43mm}	86	
c220	4-Chlorophenyl isocyanate	ClC_6H_4NCO	153.57	12, 616	1.200	1.5618^{20}	29–31	204	>110	
c221	4-Chlorophenyl phenyl sulfone	$ClC_6H_4SO_2C_6H_5$	252.72	6^1, 149			94			at 20°C: 74 acet; 44 bz; 5 CCl_4; 65 diox; 21 i-PrOH
c222	1-Chloro-3-phenylpropane	$C_6H_5(CH_2)_3Cl$	154.64	5, 391	1.080	1.5207^{20}		219	87	
c223	4-Chlorophenyl sulfone	$(ClC_6H_5)_2SO_2$	287.17	6, 327			145–148	250^{10mm}		
c224	3-Chlorophthalide		168.58	17^1, 162			58	150^{10mm}		
c225	1-Chloropropane	$CH_3CH_2CH_2Cl$	78.54	1, 104	0.8899^{20}	1.3886^{20}	−122.8	46–47	−31	0.27 aq; misc alc, eth
c226	2-Chloropropane	$CH_3CHClCH_3$	78.54	1, 105	0.8563^{20}	1.3777^{20}	−117	35–36	−35	0.2 aq^{20}; misc alc, bz, chl, eth
c227	3-Chloro-1,2-propanediol	$ClCH_2CH(OH)CH_2OH$	110.54	1, 473	1.3218^{20}_4	1.4805^{20}		213	>110	s aq, alc, eth
c228	2-Chloropropanoic acid	$CH_3CH(Cl)COOH$	108.52	2, 248	1.182	1.4345^{20}		170–190	101	misc aq, alc, eth
c229	3-Chloropropanoic acid	$ClCH_2CH_2COOH$	108.52	2, 249			41	200^{765mm}	>110	v s aq, alc, chl; s eth
c230	1-Chloro-2-propanol	$CH_3CH(OH)CH_2Cl$	94.54	1, 363	1.115^{20}	1.4375^{20}_4		126–127	51	misc aq; s alc
c231	3-Chloro-1-propanol	$ClCH_2CH_2CH_2OH$	94.54	1, 356	1.1309^{20}_4	1.4450^{20}		160–162	73	
c232	Chloro-2-propanone	$ClCH_2COCH_3$	92.53	1, 653	1.135^{15}	1.4320^{20}	−44.5	119.7	27	10 aq; misc alc, chl, eth
c233	3-Chloropropanonitrile	$ClCH_2CH_2CN$	89.53	2, 250	1.1443^{18}	1.4341^{20}	−51	95^{50mm}	75	
c234	3′-Chloropropanophenone	$ClC_6H_4C(=O)CH_2CH_3$	168.62	7^3, 1028			45–47	d > 130 124^{14mm}	>110	
c235	2-Chloropropanyl chloride	$CH_3CH(Cl)COCl$	126.97	2, 248	1.308	1.4400^{20}		109–111	31	dec aq, alc
c236	3-Chloropropanyl chloride	$ClCH_2CH_2COCl$	126.97	2, 250	1.3307^{13}	1.4570^{20}		143–145	61	i aq; d hot aq, hot alc; s alc; v s eth
c236a	3-Chloro-1-propene	$ClCH_2CH{=}CH_2$	76.53	1, 198	0.938^{20}_4	1.4154^{20}	−134.5	45	−32	0.36 aq; misc alc, PE
c237	3-Chloropropylacetate	$CH_3CO_2(CH_3)_3Cl$	130.02	4, 148			148–150			
c238	3-Chloropropyl thiolactate	$CH_3C(=O)SCH_2CH_2Cl$	152.64	2^3, 493	1.159	1.4946^{20}		84^{10mm}	77	

(*Continued*)

TABLE 2.20 Physical Constants of Organic Compounds (*Continued*)

No.	Name	Formula	Formula weight	Beilstein reference	Density, g/mL	Refractive index	Melting point, °C	Boiling point, °C	Flash point, °C	Solubility in 100 parts solvent
c239	(3-Chloropropyl)triethoxysilane	$Cl(CH_2)_3Si(OC_2H_5)_3$	240.81		1.009^{20}_{4}	1.420^{20}		102^{10mm}		
c240	(3-Chloropropyl)trimethoxysilane	$Cl(CH_2)_3Si(OCH_3)_3$	198.72		1.077^{25}_{4}	1.4183^{25}		195^{750mm}	78	
c241	3-Chloropropyne	$ClCH_2C{\equiv}CH$	74.51	1, 248	1.0306^{25}_{4}	1.4560^{20}	−78	57	−13	misc alc, bz, eth, EtOAc
c242	2-Chloropyridine	$Cl(C_5H_4N)$	113.55	20, 230	1.205^{15}	1.5320^{20}		166^{714mm}	65	sl s aq; s alc, eth
c243	3-Chloropyridine	$Cl(C_5H_4N)$	113.55	20, 230	1.194	1.5300^{20}		148	65	
c244	4-Chlororesorcinol	ClC_6H_3-1,3$(OH)_2$	144.56	6^2, 818			106–108	147^{18mm}		
c245	4-Chlorosalicylic acid	$ClC_6H_3(2\text{-}OH)COOH$	172.57	10, 101			210–212			
c246	5-Chlorosalicylic acid	$ClC_6H_3(2\text{-}OH)COOH$	172.57	10, 102			172			
c247	*N*-Chlorosuccinimide		133.53	21, 380	1.65		150–151			1.4 aq; 0.67 alc; 2 bz; sl s chl, CCl_4, eth
c248	Chlorosulfonic acid	$ClHO_3S$	116.52	Merck: 12, 2218	1.753^{20}_{4}	1.437^{14}	−80	152^{755mm}	none	s pyr, dichloroethane; aq dec with violence
c249	Chlorosulfonyl isocyanate	$ClSO_2NCO$	141.53		1.626	1.4470^{20}	−44	107	none	
c250	1-Chlorotetradecane	$CH_3(CH_2)_{13}Cl$	232.84	1^2, 135	0.859	1.4460^{20}		142^{4mm}	>110	
c251	2-Chlorothiophene	$Cl(C_5H_3S)$	118.59	17, 32	1.286	1.5483^{20}	−72	127–129	22	i aq; misc alc, eth
c252	4-Chlorothiophenol	ClC_6H_4SH	144.62	6, 326			49–52	205–207	>110	
c253	8-Chlorothiophylline		214.61	26, 473			dec 290			s alkali
c254	Chlorotitanium triisopropoxide	$[(CH_3)_2CHO]_3TiCl$	260.62		1.091				22	
c255	2-Chlorotoluene	$ClC_6H_4CH_3$	126.59	5, 290	1.0826^{20}_{4}	1.5268^{20}	−35.6	159.0	47	sl s aq; v s alc, bz, chl, eth
c256	3-Chlorotoluene	$ClC_6H_4CH_3$	126.59	5, 291	1.0760^{19}_{4}	1.5218^{20}	−47.8	161.8	50	s alc, bz, chl; misc eth
c257	4-Chlorotoluene	$ClC_6H_4CH_3$	126.59	5, 292	1.0697^{20}_{4}	1.5150^{20}	7.5	162.4	49	sl s aq; s alc, bz, eth
c258	*N*-Chloro-*p*-toluene sulfonamide, sodium salt	$CH_3C_6H_4SO_2NCl^- Na^+$	227.67				167 dec			s aq; i bz, chl, eth
c259	4-(4-Chloro-*o*-tolyloxy)butyric acid	$ClC_6H_3(CH_3)O(CH_2)_3COOH$	228.68				99–100			
c260	Chlorotriethylgermane	$(C_2H_5)_3GeCl$	195.23	4^3, 1912	1.175	1.4590^{20}			>110	
c261	Chlorotriethylsilane	$(C_2H_5)_3SiCl$	150.73	4, 624	0.898	1.4300^{20}		142–144	29	
c262	Chloro-2,2,2-trifluoroethane	CF_3CH_2Cl	118.5	1,3, 138	1.389^{0}	1.3090^{0}	−105	6.9		
c263	Chlorotrifluoroethylene	$CF_2{=}CFCl$	116.47	1^3, 646	1.315		−158.2	−28		

c264	Chlorotrifluoromethane	$ClCF_3$	104.46	1^3, 42	4.270 g/L		−181	−81		
c265	Chlorotrimethylgermane	$(CH_3)_3GeCl$	153.16		1.2382^{22}	1.4283^{20}	−13	102	1	
c266	Chlorotrimethylsilane	$(CH_3)_3SiCl$	108.64	4,*3*, 1857	0.8580^{20}_{4}	1.3870^{20}	−40	57	−27	
c267	Chlorotriphenylmethane	$(C_6H_5)_3CCl$	278.78	5, 700			110–112	235^{20mm}		v s bz, chl, eth
c268	Chlorotriphenyltin	$(C_6H_5)_3SnCl$	385.46	12, 914			108 dec	240^{14mm}		
c268a	Chloro-tris(dimethylamino)silane	$[(CH_3)_2N]_3SiCl$	195.8		0.975^{20}_{4}	1.442^{20}		63^{12mm}		
c269	*α*-Chloro-*o*-xylene	$CH_3C_6H_4CH_2Cl$	140.61	5, 364	1.063	1.5391^{20}		96^{25mm}	73	i aq; misc alc, eth
c270	*α*-Chloro-*m*-xylene	$CH_3C_6H_4CH_2Cl$	140.61	5, 373	1.064^{20}	1.5350^{20}		195–196	75	i aq; misc alc, eth
c271	*α*-Chloro-*p*-xylene	$CH_3C_6H_4CH_2Cl$	140.61	5, 384		1.5330^{20}	4.5	200	75	misc alc, bz, eth, acet
c272	2-Chloro-*p*-xylene	$ClC_6H_3(CH_3)_2$	140.61	5, 384	1.049	1.5240^{20}	2	186	57	
c273	4-Chloro-*p*-xylene	$ClC_6H_3(CH_3)_2$	140.61	5, 363	1.047	1.5280^{20}		221–223	66	misc alc, bz, eth, acet
c274	Cholesterol		386.66	6,*3*, 2607	1.052^{19}_{19}		148.5	$203^{0.5mm}$		1.3 alc; 35 eth; 22 chl; s bz, PE
c275	Cholic acid		408.58	10^3, 2162			198			(15°): 0.03 aq; 3.1 alc; 2.8 acet; 15.2 HOAc; 0.5 chl; 0.036 bz
c276	Cinchonine		194.40	23^2, 369			ca. 260			1.6 alc; 0.9 chl; 0.2 eth
c277	1,8-Cineole		154.25	17, 23	0.921^{25}_{25}	$1,4572^{20}$	1	176.4	48	misc alc, chl, eth
c278	*trans*-Cinnamaldehyde	$C_6H_5CH{=}CHCHO$	132.16	7, 348	1.050^{25}_{25}	1.6219^{20}	−7.5	136^{20mm}	71	0.014 aq; misc alc, chl, eth
c279	*trans*-Cinnamic acid	$C_6H_5CH{=}CHCOOH$	148.16	9, 573	1.2475^{4}_{4}		133	300		0.05 aq; 16 alc; 8 chl
c280	*trans*-Cinnamoyl chloride	$C_6H_5CH{=}CHCOCl$	166.61	9^2, 390	1.1617^{25}_{4}	1.614^{43}	35–36	258	>110	s hot alc, CCl_4
c281	Cinnamyl acetate	$CH_3CO_2CH_2CH{=}CHC_6H_5$	176.22	6^2, 527	1.0571	1.5421^{20}		265	>110	
c282	Cinnamyl alcohol	$C_6H_5CH{=}CHCH_2OH$	134.18	6, 570	1.0397^{35}_{35}	1.5758^{33}	33	250.0	>110	s aq; v s common organic solvents
c283	Cinnamyl chloride	$C_6H_5CH{=}CHCH_2Cl$	159.62	5, 482	1.096	1.5840^{20}	−19	108^{12mm}	79	
c284	Citraconic acid	$CH_3C(COOH){=}CHCOOH$	130.10	2, 768	1.62		92 dec			v s aq, alc, eth; sl s chl; i bz, PE
c285	Citraconic anhydride		112.08	17, 440	1.247	1.4712^{20}	8	214	101	
c286	Citral (geranial plus neral, *cis* and *trans* forms, resp.)	$(CH_3)_2C{=}CHCH_2CH_2$-$C(CH_3){=}CHCHO$	152.24		0.888	1.4876^{20}		229	101	

(*Continued*)

TABLE 2.20 Physical Constants of Organic Compounds (*Continued*)

No.	Name	Formula	Formula weight	Beilstein reference	Density, g/mL	Refractive index	Melting point, °C	Boiling point, °C	Flash point, °C	Solubility in 100 parts solvent
c287	Citral dimethyl acetal	$(CH_3)_2C{=}CHCH_2CH_2$-$C(CH_3){=}CH(OCH_3)_2$	198.31	1^4, 3570	0.890	1.4540^{20}		106^{10mm}	92	
c288	Citrazinic acid		155.11	22, 254			carbonizes without melting >300			i aq; s alkali
c289	Citric acid	$HOOCCH_2C(OH)(COOH)$-CH_2COOH	192.12	3, 556	1.665		154			59 aq
c290	*β*-Citronellol	$(CH_3)_2C{=}CHCH_2CH_2$-$CH(CH_3)CH_2CH_2OH$	156.27	1^1, 232	0.8570^{20}_4	1.4560^{20}		222	98	v sl s aq; misc alc, eth
c299	Cocaine		303.35	22^2, 150		1.5022^{98}	98	$187^{0.1mm}$		0.17 aq; 15 alc; 140 chl; 28 eth; s acet; EtOAc, CS_2
c300	Coumarin		146.15	17, 328	0.935^{20}_4		68–70	298		0.25 aq; v s alc, chl, eth; s alkali
c301	Creatine	$HOOCCH_2N(CH_3)$-$C({=}NH)NH_2$	131.14	4, 363			dec 303			1.3 aq; 0.11 alc; i eth
c302	Creatinine		113.12	24, 245			255 dec			8 aq; sl s alc; i eth
c303	*o*-Cresol	$CH_3C_6H_4OH$	108.14	6, 349	1.0273^{41}	1.5361^{41}	30	191	81	3.1 aq^{40}; misc alc, chl, eth; s alkali
c304	*m*-Cresol	$CH_3C_6H_4OH$	108.14	6, 373	1.034^{20}_4	1.5438^{20}	12	202.2	86	2.5 aq^{40}; misc alc, chl, eth; s alkali
c305	*p*-Cresol	$CH_3C_6H_4OH$	108.14	6, 389	1.0179^{41}	1.5312^{41}	34.8	201.9	86	2.3 aq^{40}; misc alc, chl, eth; s alkali
c306	*trans*-Crotonaldehyde	$CH_3CH{=}CHCHO$	70.09	1, 728	0.8516^{20}	1.4373^{20}	−76	102–104	13	18.1 aq^{20}
c307	Crotonic acid	$CH_3CH{=}CHCOOH$	86.19	2, 408	0.964^8_4	1.4228^{80}	71.6	185	87	54.6 aq^{20}; 52.5 $EtOH^{25}$; 53 acet; 37.5 toluene
c308	Crotonic anhydride	$(CH_3CH{=}CHO)_2O$	154.17	2, 411	1.040	1.4740^{20}		248	110	
c309	Crotononitrile	$CH_3CH{=}CHCN$	67.09	2, 412	1.4190^{20}	1.4190^{20}		121	20	
c310	Crotonyl chloride	$CH_3CH{=}CHCOCl$	104.54	2, 411	1.091	1.4600^{20}		120–123	35	
c311	Crotyl alcohol	$CH_3CH{=}CHCH_2OH$	72.11	1, 442	0.845	1.4270^{20}		122	37	17 aq; misc alc
c312	Crotyl chloride	$CH_3CH{=}CHCH_2Cl$	90.55	1^2, 176	0.929	1.4360^{20}		85	−5	
c313	12-Crown-4		176.21		1.089	1.4630^{20}		$70^{0.5mm}$	>110	specific for Li^+
c314	18-Crown-6		264.32				42–45		>110	
c315	Crystal Violet		407.99	13, 756			215 dec			
c316	Cumene hydroperoxide	$C_6H_5C(CH_3)_2OH$	152.20	6^3, 1814	1.030	1.5210^{20}		101^{8mm}	56	

c316a	Cumylphenol	$C_6H_5C(CH_3)_2C_6H_4OH$	212.29				74–76	335		
c317	Cupferron	$C_6H_5N(NO)O^- NH_4^+$	155.16	16^1, 395			163–164			v s aq, alc
c318	Cyanamide	H_2NCN	42.04	3^2, 63	1.282_4^{20}		46	83^{380mm}	>110	78 aq; 29 BuOH; 42 EtOAc; s alc, eth
c319	2-Cyanoacetamide	$NCCH_2CONH_2$	84.08	2, 589			119.5		215	25 aq; 3.1 alc
c320	Cyanoacetic acid	$NCCH_2COOH$	85.06	2, 583			66	108^{15mm}	107	s aq, alc, eth; sl s bz
c321	Cyanoacetohydrazide	$NCCH_2C(=O)NHNH_2$	99.09	Merck: 11, 2688			115	dec		v s aq; s alc; i eth
c322	Cyanoacetylurea	$NCCH_2C(=O)NH-C(=O)NH_2$	127.10	3, 66			214 dec			
c323	2-Cyanoethanol	$NCCH_2CH_2OH$	71.08	3^2, 213	1.0588^0			108^{11mm}		misc aq, alc; sl s eth
c324	2-Cyanoethyl acrylate	$H_2C=CHCO_2CH_2CH_2CN$	125.13	3^3, 543	1.052	1.4470^{20}		108^{12mm}	103	
c325	Cyanogen bromide	BrCN	105.93	3, 39	2.015_4^{20}		52	61–62	5	v s aq, alc, eth
c326	1-Cyano-3-methylisothiourea, sodium salt	$CH_2NH(=NCN)S^- Na^+$	137.14	4, 71			290 dec			
c327	1-Cyanonaphthalene	$C_{10}H_7CN$	153.18	9, 649	1.1113_{25}^{25}	1.6298^{18}	38	299		i aq; v s alc, eth
c328	2-Cyanopyridine	$NC(C_5H_4N)$	104.11	22, 36	1.081	1.5288^{20}	26–28	215	89	s aq; v s alc, bz, eth
c329	3-Cyanopyridine	$NC(C_5H_4N)$	104.11	22, 41			50–52	201	84	v s aq, alc, bz, eth
c330	4-Cyanopyridine	$NC(C_5H_4N)$	104.11	22, 46			78–80			s aq, alc, bz, eth
c331	Cyanotrimethylsilane	$(CH_3)_3SiCN$	99.21	4^4, 3893	0.783_4^{20}	1.3924^{20}	11–12	118–119	1	
c332	Cyanuric acid		129.08	26, 239	1.768^0		>360; dec to HOCN			0.5 aq; s hot alc, pyr; i acet, bz, chl, eth
c333	Cyclobutane	C_4H_8	56.10	5, 17	0.7038^0	1.3752^0	−91	13		i aq; v s alc, acet
c334	Cyclobutanecarboxylic acid	$(C_4H_7)COOH$	100.12	9, 5	1.047	1.4433^{20}	−20 to −7.5	195	83	
c335	Cyclodecane	$C_{10}H_{20}$	140.27		0.871	1.4707^{20}		201	65	
c336	Cyclododecanol	$C_{12}H_{23}OH$	184.32				77			
c337	Cyclododecanone	$C_{12}H_{22}(=O)$	182.31	7^2, 48	0.906^{62}		59–61	85^{1mm}		
c338	*trans,trans,cis*-1,5,9-cyclododecatriene		162.28	5^4, 1115	0.8925_4^{20}	1.5070^{20}	−18	231	87	
c339	Cyclododecene		166.31		0.863	1.4822^{20}		232–245	93	
c340	Cyclododecylamine	$(C_{12}H_{23})NH_2$	183.34				28–30	124^{7mm}	121	
c341	Cycloheptane	C_7H_{14}	98.18	5, 29	0.811_4^{20}	1.4455^{20}	−8.0	118	6	v s alc, eth
c342	Cycloheptanol	$C_7H_{13}OH$	114.19	6, 10	0.948_4^{20}	1.4760^{20}	2	185	71	sl s aq; v s alc, eth
c343	Cycloheptanone	$C_7H_{12}(=O)$	112.17	7, 13	0.9490_4^{20}	1.4611^{20}		179–181	55	i aq; v s alc; s eth

(*Continued*)

TABLE 2.20 Physical Constants of Organic Compounds (*Continued*)

No.	Name	Formula	Formula weight	Beilstein reference	Density, g/mL	Refractive index	Melting point, °C	Boiling point, °C	Flash point, °C	Solubility in 100 parts solvent
c344	1,3,5-Cyclohepta-triene		92.13	5, 280	0.888	1.5211^{20}	−75.3	115.5	26	s alc, eth; v s bz, chl
c345	Cycloheptene	C_7H_{12}	96.17	5, 65	0.824	1.4585^{20}		114.7	−6	s alc, eth
c346	8-Cyclohexadecene-1-one		236.40	7^3, 521		1.4890^{20}		195^{19mm}	>110	
c347	Cyclohexane	C_6H_{12}	84.16	5, 20	0.7786^{20}_{4}	1.4262^{20}	6.6	80.7	−20	0.01 aq; misc acet, alc, bz, CCl_4, eth
c348	Cyclohexane-d_{12}	C_6D_{12}	92.26	5^3, 36	0.893	1.4210^{20}		78	−18	
c349	1,3-Cyclohexanebis-(methylamine)	$C_{10}H_{10}(NHCH_3)_2$	142.25		0.945	1.4930^{20}			106	
c350	1,3-Cyclohexane-carbonitrile	$C_6H_{11}CN$	109.17	9, 9	0.919	1.4505^{20}		76^{16mm}	65	
c351	Cyclohexanecarbonyl chloride	$C_6H_{11}COCl$	146.62	9, 9	1.096	1.4700^{20}		184	66	
c352	Cyclohexanecarbox-aldehyde	$C_6H_{11}CHO$	112.17	7, 19	0.926	1.4500^{20}		163	40	
c353	Cyclohexanecarboxylic acid	$C_6H_{11}COOH$	128.17	9, 7	1.0480^{15}_{4}	1.4530^{20}	29	232.5	>110	0.21 aq; s alc, bz, eth
c354	*trans*-1,2-Cyclo-hexanediamine	$C_6H_{10}(NH_2)_2$	114.19	13^3, 8	0.951	1.4884^{20}	14–15	92^{18mm}	68	
c355	1,3-Cyclohexanedi-carboxylic acid	$C_6H_{10}(COOH)_2$	172.18	9, 732			132–141			
c356	*cis*-1,2-Cyclohexanedi-carboxylic anhydride		154.17	17, 452			32–34	158^{17mm}	>110	
c357	1,4-Cyclohexanedi-methanol	$C_6H_{10}(CH_2OH)_2$	144.21		0.978^{100}_{4}	1.4893^{20}	43	283	161	misc aq, alc; 2.5 eth
c358	1,4-Cyclohexane-divinyl ether	$C_6H_{10}(OCH{=}CH_2)_2$	196.29		0.919	1.4720^{20}		126^{14mm}	>110	
c359	1,4-Cyclohexanediol	$C_6H_{10}(OH)_2$	116.16	6, 741			98–100	150^{20mm}	65	
c360	1,3-Cyclohexanedione	$C_6H_8({=}O)_2$	112.13	7, 554	1.0861^{91}	1.4576^{102}	103–105			s aq, alc, acet, chl
c361	1,2-Cyclohexanedione dioxime	$C_6H_8({=}NOH)_2$	142.16	7^2, 526			185–188			s aq
c362	Cyclohexanemethyl-amine	$C_6H_{11}CH_2NH_2$	113.20	12, 12	0.870	1.4630^{20}		145–147	43	
c363	Cyclohexanepropionic acid	$C_6H_{11}CH_2CH_2COOH$	156.23	9, 82	0.912	1.4636^{20}	14–17	275.8	>110	
c364	Cyclohexanethiol	$C_6H_{11}SH$	116.23	6, 8	0.950	1.4921^{20}		158–160	43	
c365	Cyclohexanol	$C_6H_{11}OH$	100.16	6, 5	0.9416^{30}	1.4629^{30}	25.4	161	68	3.8 aq^{25}; misc alc, bz

c366	Cyclohexanone	$C_6H_{10}({=}O)$	98.15	7, 8	0.9478^{20}_{4}	1.4510^{20}	−31	155.7	44	15 aq^{10}; s alc, eth
c367	Cyclohexanone oxime	$C_6H_{10}({=}NOH)$	113.16	7, 10			89–91	206–210		s aq, eth; sl s alc
c368	Cyclohexene	C_6H_{10}	82.15	5, 63	0.8094^{20}_{4}	1.4464^{20}	−103.5	83.0	−12	0.02 aq; misc alc, bz, acet, eth
c369	3-Cyclohexene-1-methanol	$C_6H_9CH_2OH$	112.17	6^3, 215	0.961	1.4853^{20}		85^{18mm}	76	
c370	Cyclohexene oxide		98.15	17, 21	0.970	1.4520^{20}		130	27	
c371	2-Cyclohexene-1-one	$C_6H_8({=}O)$	96.13	7^2, 55	0.993	1.4885^{20}	−53	168	56	v s alc
c372	4-(3-Cyclohexene-1-yl)pyridine		159.23	20^3, 3239	1.021	1.5480^{20}		141^{20mm}	>110	
c373	Cyclohexyl acetate	$CH_3CO_2C_6H_{11}$	142.20	6, 7	0.966	1.4395^{20}		173	57	sl s aq; s org solv
c374	Cyclohexylacetic acid	$C_6H_{11}CH_2COOH$	142.20	9^2, 9	1.007	1.4630^{20}	31–33	242–244	>110	
c375	Cyclohexylamine	$C_6H_{11}NH_2$	99.18	12, 5	0.8671^{20}	1.4593^{20}	−18	134	31	misc aq, alc, chl, eth
c376	Cyclohexylbenzene	$C_6H_{11}C_6H_5$	160.26	5, 503	0.9502^{20}_{4}	1.5258^{20}	7	240	98	i aq; v s alc, eth
c377	Cyclohexyldimethoxy-methylsilane	$C_6H_{11}Si(OCH_3)_2CH_3$	188.35		0.940	1.4390^{20}		201.2	73	
c378	2-Cyclohexylethanol	$C_6H_{11}CH_2CH_2OH$	128.22	6, 17	0.919	1.4647^{20}		207^{745mm}	86	
c379	Cyclohexylethyl acetate	$CH_3CO_2CH_2CH_2C_6H_{11}$	170.25		0.949	1.4461		98^{15mm}	81	
c380	*N*-Cyclohexyl-formamide	$C_6H_{11}NHCHO$	127.18	12^2, 11			38–40	113^{10mm}	>110	
c381	Cyclohexyl isocyanate	$C_6H_{11}NCO$	125.17	12^2, 12	0.980	1.4551^{20}		168–170	48	
c382	Cyclohexyl isothio-cyanate	$C_6H_{11}NCS$	141.24	12^2, 12	0.996	1.5350^{20}		219	95	
c383	Cyclohexyl meth-acrylate	$H_2C{=}C(CH_3)CO_2C_6H_{11}$	168.24	6^3, 25	0.964	1.4580^{20}		70^{4mm}	82	
c384	Cyclohexylmethanol	$C_6H_{11}CH_2OH$	114.19	6, 14	0.9215^{25}_{4}	1.4640^{25}		181	71	s alc, eth
c385	3-Cyclohexyl-1-propanol	$C_6H_{11}CH_2CH_2CH_2OH$	142.24	6^1, 15	1.007	1.4975^{20}		218	101	
c386	*N*-Cyclohexyl-2-pyrrolidinone		167.25	21^3, 3149	1.026	1.495	12	284	>110	
c387	*cis,cis*-1,3-Cyclo-octadiene		108.18	5^4, 401	0.869	1.4928^{20}	−53 to −51	55^{34mm}	24	
c388	1,5-Cyclooctadiene		108.18	5, 116	0.8818^{25}_{4}	1.4905^{25}	−69	149–150	31	s CCl_4
c389	Cyclooctane	C_8H_{16}	112.22	5, 35	0.834	1.4574^{20}	14.8	151.1	30	
c390	*trans*-1,2-Cyclo-octanediol	$C_8H_{14}(OH)_2$	144.21	6^3, 4094	1.080	1.4980^{20}	32	$94^{0.5mm}$	>110	

(Continued)

TABLE 2.20 Physical Constants of Organic Compounds (*Continued*)

No.	Name	Formula	Formula weight	Beilstein reference	Density, g/mL	Refractive index	Melting point, °C	Boiling point, °C	Flash point, °C	Solubility in 100 parts solvent
c391	Cyclooctanol	$C_8H_{15}OH$	128.22	6^2, 25	0.9740_4^{20}	1.4850^{20}	14–15	108^{22mm}	86	
c392	Cyclooctanone	$C_8H_{14}(=O)$	126.20	7, 21	0.9584_4^{20}	1.6494^{20}	41–43	195–197	72	
c393	*cis*-Cyclooctene	C_8H_{14}	110.20	5^1, 35	0.846	1.4698^{20}	−16	145–146	25	
c394	Cyclooctylamine	$C_8H_{15}NH_2$	127.23		0.928	1.4804^{20}	−48	190	62	
c395	Cyclopentadiene		66.10	Merck: 12, 2807	0.8021_4^{20}	1.4463^{16}	−85	41–42		misc alc, bz, CCl_4, eth; s aniline, HOAc, CS_2
c396	Cyclopentane	C_5H_{10}	70.13	5, 19	0.7460_4^{20}	1.4068^{20}	−94	49.3	−37	i aq; misc alc, eth
c397	Cyclopentane-carboxylic acid	C_5H_9COOH	114.14	9, 6	1.053_4^{20}	1.4540^{20}	4	216	93	sl s aq; s MeOH
c398	Cyclopentanol	C_5H_9OH	86.13	6, 5	0.9488_4^{20}	1.4521^{20}	−19	140	51	sl s aq; s alc
c399	Cyclopentanone	$C_5H_8(=O)$	84.12	7, 5	0.9509_4^{18}	1.4366^{20}	−51	130.6	26	sl s aq; misc alc, eth
c400	Cyclopentanone oxime	$C_5H_8(=NOH)$	99.13	7, 7			53–55	196	92	s aq, alc, bz, chl, eth
c401	Cyclopentene	C_5H_8	68.11	5, 61	0.7720^{20}	1.4228^{20}	−135.1	44.2	−29	
c402	2-Cyclopentene-1-acetic acid	$C_5H_7CH_2COOH$	126.16	9, 42	1.047	1.4675^{20}	19	$94^{2.5mm}$	>110	
c403	*N*-(1-Cyclopenten-1-yl)morpholine		153.23		0.957	1.5105^{20}		106^{12mm}	60	
c404	Cyclopentylamine	$C_5H_9NH_2$	85.15	12, 4	0.863	1.4482^{20}		106–108	17	
c405	3-Cyclopentyl-propanoic acid	$C_5H_9CH_2CH_2COOH$	142.20		0.996	1.4570^{20}		130^{12mm}	46	
c406	Cyclopropane	C_3H_6	42.08	5, 15	0.720_4^{-79}		−127	−32.8		37 mL/100 mL aq^{15}; v s alc, eth
c407	Cyclopropanecarbo-nitrile	C_3H_5CN	67.09	9, 4	0.911^{16}	1.4207^{20}		135	32	s eth
c408	Cyclopropanecarbonyl chloride	C_3H_5COCl	104.54	9, 4	1.152	1.4522^{20}		119	23	
c409	Cyclopropane-carboxylic acid	C_3H_5COOH	86.09	9, 4	1.088	1.4380^{20}	17–19	182–184	71	sl s hot aq; s alc, eth
c410	Cyclopropyl methyl ketone	$C_3H_5COCH_3$	84.12	7, 7	0.8993_4^{20}	1.4240^{20}		114	21	s aq, alc, eth
c411	L-Cysteine	$HSCH_2CH(NH_2)COOH$	121.16	4, 506			220 dec			v s aq, alc; i bz, eth
c412	L-Cystine	$HOOCCH(NH_2)SSCH_2$-$CH(NH_2)COOH$	240.30	4, 507			dec 240			0.01 aq; s acid, alkali; i alc
d1	1,9-Decalene	$H_2C=CH(CH_2)_6CH=CH_2$	138.25	1^1, 123	0.750	1.4320^{20}		169	41	
d2	*cis*-Decahydro-naphthalene	$C_{10}H_{18}$	138.25	5, 92	0.8963_4^{20}	1.4810^{20}	−43	195.8	58 (CC)	v s alc, chl, eth; misc most ketones, esters

d3	*trans*-Decahydro-naphthalene	$C_{10}H_{18}$	138.25	5², 56	0.8700^{20}_{4}	1.4690^{20}	−30.4	187.3	54	see under *cis*
d4	Decahydro-2-naphthol	$C_{10}H_{17}OH$	154.25	6, 67	0.996	1.500^{20}		109^{14mm}	>110	
d5	Decamethylcyclo-pentasiloxane	$[\text{—}Si(CH_3)_2O\text{—}]_5$	370.78	4⁴, 4128	0.9593^{20}_{4}	1.3982^{20}	−38	101^{20mm}	72	i aq
d6	Decamethyltetra-siloxane	$(CH_3)_3SiO[Si(CH_3)_2O]_2$-$Si(CH_3)_3$	310.69	4³, 1879	0.8536^{20}_{4}	1.3895^{20}	−68	194	62	sl s alc; s bz, PE
d7	Decanal	$H(CH_2)_9CHO$	156.27	1, 711	0.830^{15}_{4}	1.4280^{20}	−5	208–209	85	i aq; s alc, eth
d8	Decane	$CH_3(CH_2)_8CH_3$	142.29	1, 168	0.7301^{20}_{4}	1.4110^{20}	−29.7	174.1	46	0.07 aq
d9	1,10-Decanediamine	$H_2N(CH_2)_{10}NH_2$	172.32	4, 273			62–63	140^{12mm}		
d10	Decanedioic acid	$HOOC(CH_2)_8COOH$	202.25	2, 718	1.207^{20}_{4}	1.422^{134}	134.5	232^{10mm}		0.1 aq^{20}, eth^{17}; v s alc, esters, ketones
d11	1,2-Decanediol	$CH_3(CH_2)_7CH(OH)CH_2OH$	174.28	1, 494			48–50	255	>110	
d12	1,10-Decanediol	$HO(CH_2)_{10}OH$	174.28	1², 560			74	170^{8mm}	>110	sl s aq, eth; v s alc
d13	Decanedioyl dichloride	$ClC(\text{=}O)(CH_2)_8COCl$	239.14	2, 719	1.1212^{20}_{4}	1,4678^{20}		220^{75mm}	>110	dec aq, alc
d13a	Decanenitrile	$CH_3(CH_2)_8CN$	153.27	2, 356	0.8295^{15}_{4}	1.4295^{20}	−15	235–237		misc alc, chl, eth
d14	1-Decanethiol	$CH_3(CH_2)_9SH$	174.35	1², 459	0.841	1.4565^{20}	−26	114^{13mm}	98	
d15	Decanoic acid	$CH_3(CH_2)_8COOH$	172.27	2², 309	0.8752^{50}_{4}	1.4288^{40}	32	270	>110	0.015 aq; s alc, bz, chl, CS_2
d16	1-Decanol	$CH_3(CH_2)_9OH$	158.29	1, 425	0.8297^{20}_{4}	1.4359^{20}	6.9	232	82	i aq; s alc, eth
d17	δ-Decanolactone		170.25	17⁵,9, 91	0.954	1.4580^{20}		$120^{0.02mm}$	>110	
d18	2-Decanone	$CH_3(CH_2)_7COCH_3$	156.27	1, 711	0.825	1.4250^{20}	3.5	211	71	
d19	3-Decanone	$CH_3(CH_2)_6COC_2H_5$	156.27	1¹, 367	0.825	1.4241^{20}	−3.8	205	25	
d20	4-Decanone	$CH_3(CH_2)_6C(\text{=}O)(CH_2)_2CH_3$	156.27	1, 711	0.824^{20}_{0}	1.4237^{20}		207	71	i aq; misc alc, eth
d21	Decanoyl chloride	$CH_3(CH_2)_8COCl$	190.71	2, 356	0.919	1.4410^{20}	−34.5	96^{5mm}	106	dec aq. alc; s eth
d22	1-Decene	$H(CH_2)_8CH\text{=}CH_2$	140.27	1³, 858	0.7408^{20}	1.4210^{20}	−66	170.6	47	i aq; misc alc, eth
d23	Decylamine	$H(CH_2)_{10}NH_2$	157.30	4, 199	0.787	1.4360^{20}	12–14	216–218	85	sl s aq; misc alc, bz, eth, acet
d24	Dehydroabeitylamine		285.48	12⁴, 3005		1.5460^{20}			>110	
d25	Dehydroacetic acid		168.15	17, 559			111–113	270		at 25°: 22 acet; 18 bz; 5 eth; 3 EtOH; 5 MeOH
d26	Deoxybenzoin	$C_6H_5CH_2COC_6H_5$	196.25	7², 368	1.201^{0}_{4}		55–56	320	110	i aq; v s alc, eth
d27	Diacetoxydimethyl-silane	$(CH_3)_2Si(OOCCH_3)_2$	176.3		1.054^{20}_{4}	1.4030^{20}		164–166		
d28	*trans*-1,1-Diacetoxy-2-butene	$(CH_3CO_2)_2CHCH\text{=}CHCH_3$	172.18	2, 154	1.057	1.4290^{20}		106^{20mm}	87	
d29	1,1-Diacetoxy-2-propene	$(CH_3CO_2)_2CHCH\text{=}CH_2$	158.16	2, 154	1.078	1.4190^{20}		184	78	

(*Continued*)

TABLE 2.20 Physical Constants of Organic Compounds (*Continued*)

No.	Name	Formula	Formula weight	Beilstein reference	Density, g/mL	Refractive index	Melting point, °C	Boiling point, °C	Flash point, °C	Solubility in 100 parts solvent
d30	Diallylamine	$(H_2C{=}CHCH_2)_2NH$	97.16	4, 208	0.787	1.4405^{20}	−88	112	15	
d31	Diallyl ether	$(H_2C{=}CHCH_2)_2O$	98.15	1, 438	0.805^{18}_{0}	1.4160^{20}		94–95	−6 (OC)	i aq; misc alc, eth
d32	Diallyl maleate	$H_2C{=}CHCH_2O_2CCH{=}CH\text{-}CO_2CH_2CH{=}CH_2$	196.20	2^3, 1926	1.073	1.4702^{20}	−47	116^{4mm}	>110	
d33	Diallyl 1,2-phthalate	$C_6H_4(CO_2CH_2CH{=}CH_2)_2$	246.27	9^3, 4120	1.121	1.5187^{20}		167^{5mm}	>110	
d34	Diallyl sulfide	$(H_2C{=}CHCH_2)_2S$	114.21	1, 440	0.8877^{27}_{4}	1.4889^{20}	−85	138	46	sl s aq; misc alc, eth
d35	(+)-*N,N*-Diallyl-tartardiamide	$[\text{-}CH(OH)CONHCH_2\text{-}CH{=}CH_2]_2$	228.25	4, 218			186–188			
d36	1,2-Diaminoanthra-quinone		238.25	14^1, 459			289–291			sl s alc, eth
d37	1,4-Diaminoanthra-quinone		238.25	14, 197			265–269			sl s aq, alc; v s bz
d38	1,5-Diaminoanthra-quinone		238.25	14, 203			308 dec			
d39	2,6-Diaminoanthra-quinone		238.25	14, 215			>325			sl s hot aq, pyr
d40	3,5-Diaminobenzoic acid	$(H_2N)_2C_6H_3COOH$	152.15	14, 453			228	$-H_2O$, 110		sl s aq; s alc, eth
d41	1,4-Diaminobutane	$H_2N(CH_2)_4NH_2$	88.15	4, 264	0.877	1.4569^{20}	27.3	158–160	51	
d42	4,4′-Diaminodiphenyl-amine sulfate	$H_2NC_6H_4NHC_6H_4NH_2 \cdot H_2SO_4$	297.33	13, 110			300			s aq
d43	*trans*-1,2-Diamino-cyclohexane	$C_6H_{10}(NH_2)_2$	114.19	13^3, 8	0.951	1.2886^{20}	14–15	81^{15mm}	68	
d44	*trans*-1,4-Diamino-cyclohexane	$C_6H_{10}(NH_2)_2$	114.19	13^1, 3			69–72	197	71	
d45	*trans*-1,2-Diamino-cyclohexane-*N,N,N′,N′*-tetra-acetic acid hydrate	$C_6H_{10}[N(CH_2COOH)_2]_2 \cdot H_2O$	364.36	13^3, 10			213–216			v s aq
d46	4,4′-Diaminodiphenyl-methane	$H_2NC_6H_4CH_2C_6H_4NH_2$	198.27	13, 238			91–92	398	221	sl s aq; v s alc, bz, eth
d47	3,3′-Diaminodiphenyl sulfone	$H_2NC_6H_4SO_2C_6H_4NH_2$	248.30	13, 426			170–173			i aq; s alc, bz
d48	4,4′-Diaminodiphenyl sulfone	$H_2NC_6H_4SO_2C_6H_4NH_2$	248.30	13, 536			175–176			i aq; s alc, acet, dil HCl
d49	2,4-Diamino-6-hydroxypyrimidine		126.12	24, 469			285 dec			s aq
d50	Diaminomaleonitrile	$NCC(NH_2){=}C(NH_2)CN$	108.10	4^2, 949			178–179			

d51	1,8-Diamino-*p*-menthane		170.30	13, 4	0.914	1.4805^{20}	−45	125^{10mm}	93	
d52	3,3′-Diamino-*N*-methyldipropylamine	$CH_3N[(CH_2)_3NH_2]_2$	145.25	4^4, 1279	0.901	1.4725^{20}		112^{6mm}	102	
d53	2,4-Diamino-6-phenyl-1,3,5-triazine		187.21	26^1, 69	1.40_4^{25}		227–228			0.06 aq; s alc, eth, dil HCl; sl s DMF
d54	1,2-Diaminopropane	$CH_3CH(NH_2)CH_2NH_2$	74.13	4, 257	0.878	1.4460^{20}		119–120	33	v s aq
d55	1,3-Diaminopropane	$H_2N(CH_2)_3NH_2$	74.13	4, 261	0.888	1.4570^{20}	−12	140	48	v s aq
d56	1,3-Diamino-2-propanol	$H_2NCH_2CH(OH)CH_2NH_2$	90.13	4, 290			40–45	235	>110	
d58	2,6-Diaminopyridine	$(H_2N)_2C_5H_3N$	109.13	22^1, 647			120–122			s aq, alc
d59	2,4-Diaminotoluene	$(H_2N)_2C_6H_3CH_3$	122.17	13, 124			97–99	283–285		
d60	3,4-Diaminotoluene	$(H_2N)_2C_6H_3CH_3$	122.17	13, 148			91–93	156^{18mm}		
d61	1,4-Diazabicyclo[2.2.2]-octane		112.18	23^3, 484			158–160	174	62	45 aq; 77 EtOH; 51 bz; 13 acet; 26 MeEtKe
d62	1,8-Diazabicyclo[5.4.0]-undec-7-ene		152.24		1.018	1.5219^{20}		$80^{0.6mm}$	>110	
d63	Diazomethane	$CH_2{=}N{=}N$	42.04	23, 25			−145	−23		VERY EXPLOSIVE; s eth, dioxane
d64	1-Diazo-2-naphthol-4-sulfonic acid		272.22	16, 595			160 dec			
d65	1,2,5,6-Dibenzanthracene		278.33	5^1, 369			266 subl	524		s bz, PE; sl s alc, eth
d66	Dibenzofuran		168.20	17, 70	1.0886_4^{99}	1.6079^{99}	81–83	285		s alc, bz, eth; i aq
d67	Dibenzothiophene		184.26	17, 72			97–100	332–333		s aq; v s alc, bz
d68	Dibenzoylmethane	$C_6H_5COCH_2COC_6H_5$	224.26	7, 769			78–79	220^{18mm}		4.4 alc; s eth, aq NaOH
d69	Dibenzoyl peroxide	$C_6H_5C({=}O)OOC({=}O)C_6H_5$	242.23	9, 179			103–106	may explode when heated		sl s aq, alc; s bz, chl, eth
d70	(−)-Dibenzoyl-L-tartaric acid hydrate	$[(C_6H_5COOCH(COOH)\text{-}]_2 \cdot H_2O$	376.34	9, 170			90–92			
d71	Dibenzylamine	$C_6H_5CH_2NHCH_2C_6H_5$	197.28	12, 1035	1.026	1.5731^{20}	−26	300	143	i aq; s alc, eth
d72	Dibenzyldisulfide	$C_6H_5CH_2SSCH_2C_6H_5$	246.39	6, 465			69	d > 270		s hot alc, bz, eth
d73	Dibenzyl ether	$C_6H_5CH_2OCH_2C_6H_5$	198.27	6, 434	1.0014_4^{20}	1.5168^{20}	2	298	135 (CC)	misc alc, acet, chl, eth

(*Continued*)

TABLE 2.20 Physical Constants of Organic Compounds (*Continued*)

No.	Name	Formula	Formula weight	Beilstein reference	Density, g/mL	Refractive index	Melting point, °C	Boiling point, °C	Flash point, °C	Solubility in 100 parts solvent
d74	*N,N'*-Dibenzyl-ethylenediamine	$(C_6H_5CH_2NHCH_2\text{-})_2$	240.35	12, 1067	1.024^{20}_{4}	1.5624^{20}	26	195^{4mm}	>110	v s alc, bz, chl, eth
d75	Dibenzyl malonate	$CH_2[CO_2CH_2C_6H_5]_2$	284.31	6, 436	1.137	1.5447^{20}		$188^{0.2mm}$	>110	
d76	Dibromoacetic acid	$Br_2CHCOOH$	217.86	2, 218			39–41	130^{16mm}	>110	
d77	Dibromoacetonitrile	Br_2CHCN	198.86	2, 219	2.296	1.5393^{20}		69^{24mm}		
d78	2,4'-Dibromoaceto-phenone	$BrC_6H_4C(=O)CH_2Br$	277.96	7, 285			108–110			v s warm alc; s eth
d79	1,4-Dibromobenzene	$C_6H_4Br_2$	235.92	5, 211	0.9641^{100}	1.5743^{100}	87.3	220		1.4 alc; v s eth; s bz
d80	4,4'-Dibromobiphenyl	$BrC_6H_4C_6H_4Br$	312.00	5, 580			167–170	355–360		s bz; sl s hot alc
d81	1,2-Dibromobutane	$CH_3CH_2CH(Br)CH_2Br$	215.93	1, 120	1.789	1.5141^{20}		60^{20mm}	>110	
d82	1,3-Dibromobutane	$CH_3CH(Br)CH_2CH_2Br$	215.93	1, 120	1.800^{20}	1.5085^{20}		175		s chl, eth
d83	1,4-Dibromobutane	$BrCH_2CH_2CH_2CH_2Br$	215.93	1, 120	1.8080^{20}_{4}	1.5186^{20}	−20	198	110	s chl
d84	*meso*-2,3-Dibromo-butane	$CH_3CH(Br)CH(Br)CH_3$	215.93	1.121	1.767	1.5100^{20}		74^{47mm}	>110	
d85	2,3-Dibromo-1,4-butanediol	$HOCH_2CH(Br)CH(Br)\text{-}CH_2OH$	247.93	1^3, 2176			88–90	$150^{1.5mm}$		
d86	1,4-Dibromo-2,3-butanediol	$BrCH_2C(=O)C(=O)CH_2Br$	243.89	1, 774			117–119			
d87	*trans*-2,3-Dibromo-2-butene-1,4-diol	$HOCH_2C(Br)=C(Br)CH_2OH$	245.91	1^1, 260			112–114			
d88	Dibromochloro-methane	$HCClBr_2$	208.29	1, 67	2,451	1.5465^{20}	−22	120^{748mm}	none	misc alc, bz, eth
d89	*trans*-1,2-Dibromo-cyclohexane	$C_6H_{10}Br_2$	241.96	5, 24	1.784	1.5515^{20}		146^{10mm}	>110	
d90	1,2-Dibromo-2-chloro-1,1,2-trifluoroethane	$FCCl(Br)C(Br)F_2$	276.5		2.2478^{20}	1.4275^{20}		93–94	none	
d91	1,10-Dibromodecane	$Br(CH_2)_{10}Br$	300.09	1^1, 64	1.335^{30}	1.4912^{20}	27	160^{15mm}	>110	sl s alc; s eth
d92	1,2-Dibromo-1,1-difluoroethane	$CH_2BrC(Br)F_2$	223.87	1, 92	2.2238^{20}	1.4456^{20}	−61.3	92.4	none	i aq
d93	Dibromodifluoro-methane	Br_2CF_2	209.81	1^1, 16	2.288^{15}_{4}	1.4016^{20}	−110	25	none	0.1 aq; misc alc, bz, chl, eth
d94	1,2-Dibromo-3,3-di-methylbutane	$(CH_3)_3CCH(Br)CH_2Br$	243.98	1, 151	1.610	1.5053^{20}		73^{3mm}	83	
d95	1,3-Dibromo-5,5-di-methylhydantoin		185.93				197 dec			
d96	1,1-Dibromoethane	CH_3CHBr_2	187.86	1, 90	2.055^{20}_{4}	1.5379^{20}		113	none	i aq; v s alc, eth
d97	1,2-Dibromoethane	$BrCH_2CH_2Br$	187.86	1, 90	2.1802^{20}_{4}	1.5387^{20}	10.0	131.7	none	0.43 aq; misc alc, eth
d98	(1,2-Dibromoethyl)-benzene	$C_6H_5CH(Br)CH_2Br$	263.97	5, 356			70–74	140^{15mm}		

d99	*cis*-1,2-Dibromo-ethylene	BrCH=CHBr	185.86	1, 190	2.21_4^{17}	1.5431^{18}	−53	112.5	none	s alc, bz, chl, eth
d100	*trans*-1,2-Dibromo-ethylene	BrCH=CHBr	185.86	1, 190	2.246	1.5505^{18}	−6.5	108	none	
d101	1,2-Dibromoethyltri-chlorosilane	$BrCH_2CH(Br)SiCl_3$	321.3		2.046_4^{20}	1.537^{20}		90^{11mm}		
d102	4′5′-Dibromo-fluorescein		490.12	19, 228			270–273			s hot alc, HOAc
d103	1,4-Dibromo-2-fluoro-benzene	$Br_2C_6H_3F$	253.91	5^4, 684			33–36	216	101	
d104	2,4-Dibromo-1-fluoro-benzene	$Br_2C_6H_3F$	253.91		2.047^{20}	1.5840^{20}		105^{22mm}	92	
d104a	Dibromofluoro-methane	Br_2CHF	191.83				−78	65		
d105	1,2-Dibromohexa-fluoropropane	$CF_3CF(Br)C(Br)F_2$	309.84	1^4, 218	2.169	1.3605^{20}	−95	72^{734mm}	none	
d106	1,6-Dibromohexane	$Br(CH_2)_6Br$	243.98	1, 145	1.586_4^{18}	1.5066^{20}		243	>110	misc eth
d107	2,5-Dibromo-3,4-hexanedione	$CH_3CHBrC(=O)C(=O)$-$CH(Br)CH_3$	271.95	1^3, 3132	1.766	1.5120^{20}		103^{10mm}	>110	
d108	5,7-Dibromo-8-hydroxyquinoline		302.96	21, 97			200–201	subl		s alc, bz; v s eth
d109	2,4-Dibromomesitylene	1,3,5-$(CH_3)_3$-C_6HBr_2	278.00	5, 408			61–63	278–279		
d110	Dibromomethane	CH_2Br_2	173.85	1, 67	2.4956_4^{20}	1.5419^{20}	−52.7	96–97	none	1.15 aq; misc alc, bz, acet, chl, eth
d111	2,6-Dibromo-4-methyl-phenol	$Br_2C_6H_2(CH_3)OH$	265.94	6, 406			49–50		>110	
d112	5,7-Dibromo-2-methyl-8-quinolinol		316.99	21^3, 1240			126–130			
d113	1,6-Dibromo-2-naphthol	$Br_2C_{10}H_5OH$	301.98	6, 652			105–107			
d114	2,6-Dibromo-4-nitro-aniline	$Br_2C_6H_2(NO_2)NH_2$	295.93	12, 743			206–208			sl s aq; s HOAc
d115	2,5-Dibromonitro-benzene	$Br_2C_6H_3NO_2$	280.91	5, 250	2.374		82–84			s bz, hot alc
d116	1,8-Dibromooctane	$Br(CH_2)_8Br$	272.03	1, 160	1.477	1.4981^{20}	15–16	272	>110	
d117	1,4-Dibromopentane	$CH_3CH(Br)CH_2CH_2CH_2Br$	229.95	1, 131	1.687	1.5085^{20}	−34	99^{25mm}	>110	
d118	1,5-Dibromopentane	$Br(CH_2)_5Br$	229.95	1, 131	1.6879_4^{15}	1.5092^{20}	−34	110^{15mm}	>110	
d119	2,4-Dibromophenol	$Br_2C_6H_3OH$	251.92	6, 202			40–42	154^{11mm}	>110	

(*Continued*)

TABLE 2.20 Physical Constants of Organic Compounds (*Continued*)

No.	Name	Formula	Formula weight	Beilstein reference	Density, g/mL	Refractive index	Melting point, °C	Boiling point, °C	Flash point, °C	Solubility in 100 parts solvent
d120	1,2-Dibromopropane	$CH_3CH(Br)CH_2Br$	201.90	1, 109	1.933^{20}	1.5203^{20}	−55.5	142	none	0.2 aq; misc alc, bz, chl, eth
d121	1,3-Dibromopropane	$BrCH_2CH_2CH_2Br$	201.90	1, 110	1.9712^{25}_{4}	1.5233^{20}	−36	166.8	54	0.17 aq; s alc, eth
d122	1,3-Dibromo-2-propanol	$BrCH_2CH(OH)CH_2Br$	217.90	1, 365	2.136	1.5514^{20}		83^{7mm}	46	
d123	2,3-Dibromo-1-propanol	$BrCH_2CH(Br)CH_2OH$	217.90	1, 357	2.120^{20}_{4}	1.5599^{20}		97^{10mm}	>110	sl s aq; misc alc, bz, acet, eth
d124	2,3-Dibromopropene	$BrCH_2C(Br){=}CH_2$	199.88	1, 201	1.9336^{20}_{4}	1.5470^{20}		140–143	81	
d125	2,3-Dibromopropionic acid	$BrCH_2CH(Br)COOH$	231.88	2, 258			64–66	160^{20mm}		s aq, alc, bz
d126	2,3-Dibromopropionitrile	$BrCH_2CH(Br)CN$	212.88	2, 259	2.140	1.5450^{20}		173		
d127	2,6-Dibromopyridine	BrC_5H_3N	236.91	20², 153			118–119	255		
d128	*meso*-2,3-Dibromosuccinic acid	$HOOCCH(Br)CH(Br)COOH$	275.89	2, 625			275 subl			v s aq, alc
d129	1,2-Dibromotetrachloroethane	$BrCCl_2CCl_2Br$	325.65	1, 93	2.713		222 dec		none	
d130	1,2-Dibromotetrafluoroethane	$BrCF_2CF_2Br$	259.83		2.149^{25}	1.367^{25}	−110.5	47	none	
d131	2,5-Dibromothiophene	$Br_2C_4H_2S$	241.94	17, 33	2.147^{23}_{23}	1.6289^{20}	−6	211	99	i aq; v s alc, eth
d132	α,α-Dibromotoluene	$C_6H_5CHBr_2$	249.94	5, 308	1.510^{15}	1.6147^{20}		156^{23mm}	>110	i aq; misc alc, eth
d133	1,2-Dibromo-1,1,2-trifluoroethane	$HC(Br)FC(Br)F_2$	241.8	1, 92	2.274^{27}	1.4191^{24}		76.5		
d134	α,α-Dibromo-*o*-xylene	$C_6H_4(CH_2Br)_2$	263.97	5, 366	1.960		92–94			sl s alc, chl, eth
d135	α,α-Dibromo-*p*-xylene	$C_6H_4(CH_2Br)_2$	263.97	5, 386	1.012^{0}		72–74	261		v s alc, chl; s eth
d136	Dibutoxydibutyltin	$[CH_3(CH_2)_3O]_2Sn[(CH_2)_3CH_3]_2$	379.15		1.110	1.4740^{20}		$138^{0.05mm}$	40	
d137	1,2-Dibutoxyethane	$C_4H_9OCH_2CH_2OC_4H_9$	174.28		0.8374^{20}_{20}	1.4131^{20}	−69.1	203.6	85	0.2 aq; misc alc, acet
d138	Dibutyl adipate	$[\text{-}CH_2CH_2CO_2(CH_2)_3CH_3]_2$	258.36	2², 575	0.962	1.4360^{20}		305	>110	
d139	Dibutylamine	$(C_4H_9)_2NH$	129.25	4, 157	0.7670^{20}	1.4177^{20}	−62	159.6	47	0.47 aq; s alc, acet, eth EtOAc, PE
d140	Di-*sec*-butylamine	$[C_2H_5CH(CH_3)]_2NH$	129.25	4, 162	0.753	1.4100^{20}		135	20	
d141	*N,N*-Dibutylaminoethanol	$(C_4H_9)_2NCH_2CH_2OH$	173.29	4³, 682	0.860^{20}_{20}	1.444^{20}	< −70	229–230	91	
d142	*N,N*-Dibutylaniline	$C_6H_5N(C_4H_9)_2$	205.34	12³, 95	0.904^{20}	1.5297^{20}		267–275	>110	i aq, MeOH; s acet, bz, EtOH, EtOAc, eth
d143	Dibutyl decanedioate	$C_4H_9O_2C(CH_2)_8CO_2C_4H_9$	214.45	2, 719	0.9366^{20}	1.4415^{20}	−10	344–345	178	0.004 aq

d144	Di-*tert*-butyl dicarbonate	$(CH_3)_3COC(=O)OC(CH_3)_3$	218.25		0.950	1.4103^{20}	23	$56^{0.5mm}$	37	
d145	2,5-Di-*tert*-butyl-1,4-dihydroxybenzene	$[(CH_3)_3C]_2C_6H_2(OH)_2$	222.33				217–219			
d146	Dibutyl disulfide	$C_4H_9SSC_4H_9$	178.36	1^2, 400	0.9383^{20}_{4}	1.4920^{20}	−71	231.2	93	i aq; misc alc, eth
d147	Di-*tert*-butyl disulfide	$(CH_3)_3CSSC(CH_3)_3$	178.36		0.935	1.4920		229–233	93	
d148	Dibutyl ether	$C_4H_9OC_4H_9$	130.22	1, 369	0.7689^{20}_{4}	1.3992^{20}	−95	140	25	0.03 aq; misc alc, eth
d149	2,6-Di-*tert*-butyl-4-(dimethylaminomethyl)phenol	$(CH_3)_2NCH_2C_6H_2$-$[C(CH_3)_3]_2OH$	263.43	13^4, 2014			93–94	172^{30mm}		
d150	*N,N*-Dibutylethylenediamine	$[CH_3(CH_2)_3]_2NCH_2CH_2NH_2$	172.32	4^4, 1182	0.823	1.4430^{20}		117^{24mm}	87	
d151	*N,N*-Dibutylformamide	$HC(=O)N(C_4H_9)_2$	157.26		0.864	1.4429^{20}		120^{15mm}	100	
d152	Dibutyl hexanedioate	$[-CH_2CH_2CO_2(CH_2)_3CH_3]_2$	258.36	2^2, 575	0.962	1.4358^{20}		305	>110	
d153	2,5-Di-*tert*-butylhydroquinone	$[(CH_3)_3C]_2C_6H_2$-1,4-$(OH)_2$	222.33	6, *3*, 4741			217–219			
d154	Dibutyl maleate	$C_4H_9O_2CCH=CHCO_2C_4H_9$	228.29	2^3, 1925	0.9950^{20}	1.4454^{20}	< −80	281	141	0.05 aq
d155	Di-*tert*-butyl malonate	$CH_2CO_2C(CH_3)_3$ | $CO_2C(CH_3)_3$	216.27	2^3, 1621		1.4184^{20}	−6.0	93^{10mm}	88	
d156	2,6-Di-*tert*-butyl-4-methylphenol	$[(CH_3)_3C]_2C_6H_2(CH_3)OH$	220.36	6^3, 2073	1.048^{20}_{4}	1.4859^{75}	70	265	127	s alc, bz, acet, PE
d157	Dibutyl octanedioate	$[-(CH_2)_3CO_2(CH_2)_3CH_3]_2$	286.41	2^3, 1767	0.948	1.4390^{20}		$176^{4.5mm}$	>110	
d158	Dibutyl oxalate	$C_4H_9O_2CCO_2C_4H_9$	202.25	2, 540	0.986^{20}_{20}	1.4232^{20}	−30.0	239–240	108	misc alc, ketones, PE
d159	Di-*tert*-butyl peroxide	$(CH_3)_3CO—OC(CH_3)_3$	146.23	1^3, 1580	0.794^{20}	1.3890^{20}	−40	110	1	misc acet, octane
d160	2,4-Di-*tert*-butylphenol	$[(CH_3)_3C]_2C_6H_3OH$	206.33				56.5	263.5	115	s hot alc; i alk
d161	2,6-Di-*sec*-butylphenol	$[CH_3CH_2CH(CH_3)]_2C_6H_3OH$	206.23		0.918	1.5100^{20}	−42	255–260	127	
d162	2,6-Di-*tert*-butylphenol	$[(CH_3)_3C]_2C_6H_3OH$	206.23	6^3, 2061			35–38	253	118	s hot alc; i alk
d163	3,5-Di-*tert*-butylphenol	$[(CH_3)_3C]_2C_6H_3OH$	206.23				87–89			
d164	Dibutyl phosphite	$(C_4H_9O)_2P(O)H$	194.21	1^1, 187	0.995	1.4239^{20}		119^{11mm}	121	
d165	Dibutyl 1,2-phthalate	C_6H_4 1,2-$[CO_2C_4H_9]_2$	278.35	9^2, 586	1.0465^{20}_{4}	1.4911^{20}	−35	340	157	0.01 aq; v s alc, bz, acet, eth
d166	*N,N*-Dibutyl-1,3-propanediamine	$C_4H_9NH(CH_2)_3NHC_4H_9$	186.34		0.827	1.4463^{20}		205	103	
d167	Dibutyl suberate	$CH_3(CH_2)_3O_2C(CH_2)_6CO_2$-$(CH_2)_3CH_3$	286.41	2^3, 1767	0.948	1.4390^{20}		$175.5^{4.5mm}$	>110	
d168	Dibutyl succinate	$[C_4H_9O_2CCH_2-]_2$	230.30	2^2, 551	0.9768^{20}_{4}	1.4299^{20}	−29.0	274.5		i aq; s alc, eth
d169	Dibutyl sulfate	$C_4H_9OSO_2OC_4H_9$	210.29		1.059^{25}_{4}	1.4213^{20}		132^{11mm}		
d170	Dibutyl sulfide	$C_4H_9SC_4H_9$	146.30	1, 370	0.8386^{20}	1.4530^{20}	−80	185	76	i aq; v s alc, eth

(Continued)

TABLE 2.20 Physical Constants of Organic Compounds (*Continued*)

No.	Name	Formula	Formula weight	Beilstein reference	Density, g/mL	Refractive index	Melting point, °C	Boiling point, °C	Flash point, °C	Solubility in 100 parts solvent
d171	Di-*tert*-butyl sulfide	$(CH_3)_3CSC(CH_3)_3$	146.30		0.815	1.4506^{20}		151	48	
d172	Dibutyl sulfite	$(C_4H_9O)_2S(=O)$	194.29	1^2, 397	0.9944^{22}_{4}	1.4310^{20}		108^{15mm}		
d173	Dibutyl sulfone	$(C_4H_9)_2SO_2$	178.29	1, 371			46	295	143	i aq; s alc, eth
d174	Dibutyl L-tartrate	$[-CH(OH)CO_2(CH_2)_3CH_3]_2$	262.31	3, 518	1.091	1.4465^{20}	22	175^{5mm}	>110	
d175	*N,N*-Dibutyl-2-thiourea	$C_4H_9NC(=S)NHC_4H_9$	188.34				63–65			i aq; s alc; sl s eth
d176	Dibutyltin diacetate	$(CH_3CO_2)_2Sn(C_4H_9)_2$	351.01		1.320	1.4700^{20}		145^{10mm}	>110	
d177	Dibutyltin dichloride	$(C_4H_9)SnCl_2$	303.83				39–41	135^{10mm}	>110	
d178	Dibutyltin dilaurate	$[CH_3(CH_2)_{10}CO_2]_2Sn(C_4H_9)_2$	631.56	Merck: 12, 3089	1.066	1.4683^{20}	22–24		>110	s PE, bz, acet, eth, org esters
d179	Dibutyltin maleate		346.98				135–140			
d180	Dibutyltin oxide	$(C_4H_9)_2SnO$	248.92	4^1, 588			> 300			
d181	Dicaprolactone 2-(acryloxy)ethyl ester	$HO(CH_2)_5CO_2(CH_2)_5CO_2$-$CH_2CH_2O_2CCH=CH_2$	344.41		1.100	1.4660^{20}			>110	
d182	Dichloroacetic acid	$Cl_2CHCOOH$	128.94	2, 202	1.563^{20}_{4}	1.4462^{20}	9–11	193–194	>110	misc aq, alc, eth
d183	1,1-Dichloroacetone	$CH_3C(=O)CHCl_2$	126.97	1, 654	1.305^{18}_{15}	1.4455^{20}		120	24	s sl aq; s alc, eth
d184	1,3-Dichloroacetone	$ClCH_2C(=O)CH_2Cl$	126.97	1, 655	1.383		39–41	173	89	
d185	2′,4′-Dichloroacetophenone	$Cl_2C_6H_3C(=O)CH_3$	189.04	7, 282		1.5635^{20}	33–34	145^{15mm}	>110	i aq
d186	Dichloroacetyl chloride	$Cl_2CHC(=O)Cl$	147.39	2, 204	1.5315^{16}_{4}	1.4603^{20}		107–108	none	dec aq, alc; misc eth
d187	2,3-Dichloroaniline	$Cl_2C_6H_3NH_2$	162.02	12, 621		1.5969^{20}	23–24	252	>110	s alc; v s eth
d188	2,4-Dichloroaniline	$Cl_2C_6H_3NH_2$	162.02	12, 621	1.567^{20}		59–62	245		sl s aq; s alc, eth
d189	2,5-Dichloroaniline	$Cl_2C_6H_3NH_2$	162.02	12, 625			49–51	251	>110	s alc, bz, eth
d190	2,6-Dichloroaniline	$Cl_2C_6H_3NH_2$	162.02	12, 626			38–41		>110	
d191	3,4-Dichloroaniline	$Cl_2C_6H_3NH_2$	162.02	12, 626			70–72	272		s alc, eth; sl s bz
d192	3,5-Dichloroaniline	$Cl_2C_6H_3NH_2$	162.02	12, 626			51–53	259^{741mm}	>110	i aq; s alc, eth
d193	1,5-Dichloroanthraquinone		277.11	7, 787			245–247			sl s alc, bz, acet
d194	2,3-Dichlorobenzaldehyde	$Cl_2C_6H_3CHO$	175.01	7^3, 878			64–67			
d195	2,4-Dichlorobenzaldehyde	$Cl_2C_6H_3CHO$	175.01	7, 236			69–73	233		i aq; s alc
d196	2,4-Dichlorobenzamide	$Cl_2C_6H_3CONH_2$	190.03	9^3, 1376			191–194			
d197	2,6-Dichlorobenzamide	$Cl_2C_6H_3CONH_2$	190.03	9^1, 149			196–199			
d198	1,2-Dichlorobenzene	$C_6H_4Cl_2$	147.00	5, 201	1.3059^{20}_{4}	1.5510^{20}	−17.0	180.4	66	misc alc, bz, eth
d199	1,3-Dichlorobenzene	$C_6H_4Cl_2$	147.00	5, 202	1.2884^{20}_{4}	1.5460^{20}	−24.8	173.1	72	0.01 aq; s alc, eth
d200	1,4-Dichlorobenzene	$C_6H_4Cl_2$	147.00	5, 203	1.2417^{6}	1.5285^{20}	53	174.1	66	s alc, bz, chl, eth

d201	2,5-Dichlorobenzene-sulfonyl chloride	$Cl_2C_6H_3SO_2Cl$	245.51	11[1], 15			36–37		>110	d hot alc, hot aq
d202	2,4-Dichlorobenzoic acid	$Cl_2C_6H_3COOH$	191.01	9, 342			157–160			s hot aq, alc, bz, chl
d203	2,5-Dichlorobenzoic acid	$Cl_2C_6H_3COOH$	191.01	9, 342			154–157	301		sl s aq; s alc, eth
d204	3,4-Dichlorobenzoic acid	$Cl_2C_6H_3COOH$	191.01	9, 343			207–209			s hot aq, eth; v s alc
d205	4,4′-Dichlorobenzo-phenone	$(ClC_6H_4)_2C{=}O$	251.11	7, 420			144–146	353		s hot alc, v s chl, eth
d206	2,4-Dichlorobenzo-trifluoride	$Cl_2C_6H_3CF_3$	215.00	5[3], 698	1.484	1.4810^{20}		117–118	72	
d207	3,4-Dichlorobenzo-trifluoride	$Cl_2C_6H_3CF_3$	215.00	5[3], 698	1.478	1.4750^{20}	−12	173–174	65	
d208	2,4-Dichlorobenzoyl chloride	$Cl_2C_6H_3C({=}O)Cl$	209.46	9, 342	1.494	1.5297^{20}	16–18	150^{34mm}	137	dec aq, alc
d209	3,4-Dichlorobenzoyl chloride	$Cl_2C_6H_3C({=}O)Cl$	209.46	9, 344			30–33	242	142	dec aq, alc
d210	1,4-Dichlorobutane	$ClCH_2CH_2CH_2CH_2Cl$	127.01	1, 119	1.1598^{20}_{4}	1.4566^{20}	−38	161–163	40	i aq; s chl
d211	*cis*-1,4-Dichloro-2-butene	$ClCH_2CH{=}CHCH_2Cl$	125.00	1[3], 743	1.188^{25}_{4}	1.4887^{25}	−48	152	55	i aq; s org solvents
d212	3,4-Dichloro-1-butene	$ClCH_2CH(Cl)CH{=}CH_2$	125.00	1[3], 725	1.150	1.4658^{20}	−61	123	28	
d213	1,4-Dichloro-2-butyne	$ClCH_2C{\equiv}CCH_2Cl$	122.98	1[3], 927	1.258^{20}_{4}	1.5048^{20}		165–168	160	
d214	Dichloro(2-chloro-ethyl)methylsilane	$ClCH_2CH_2SiCl_2(CH_3)$	177.53	4[3], 1892	1.261	1.4580^{20}		157^{744mm}	32	
d215	Dichloro(3-chloro-propyl)methylsilane	$Cl(CH_2)_3Si(CH_3)Cl_2$	191.56	4[4], 4170	1.227	1.4620^{20}		80^{18mm}	59	
d216	1,10-Dichlorodecane	$Cl(CH_2)_{10}Cl$	211.18	1[3], 522	0.999	1.4605^{20}	15.6	168^{28mm}	>110	
d217	1,1-Dichloro-2,2-di-ethoxyethane	$Cl_2CHCH(OC_2H_5)_2$	187.07	1, 614	1.138	1.4360^{20}		183–184	60	
d218	Dichlorodifluoro-methane	Cl_2CF_2	120.91	1, 61	1.486^{-30}		−158	−29.8		0.01 aq; 9 bz; 5.5 chl; 6 diox; s alc, eth
d219	1,1-Dichloro-3,3-dimethylbutane	$(CH_3)_3CCH_2CHCl_2$	155.07	1[3], 409	1.027	1.4388^{20}	−56	148	36	
d220	1,3-Dichloro-3,5-dimethylhydantoin		197.02	24[2], 158			134–136			
d221	Dichlorodiphenyl-methane	$(C_6H_5)_2CCl_2$	237.13	5, 590	1.235	1.6040^{20}		305	>110	

(*Continued*)

TABLE 2.20 Physical Constants of Organic Compounds (*Continued*)

No.	Name	Formula	Formula weight	Beilstein reference	Density, g/mL	Refractive index	Melting point, °C	Boiling point, °C	Flash point, °C	Solubility in 100 parts solvent
d222	Dichlorodimethylsilane	$(CH_3)_2SiCl_2$	129.06		1.064_4^{20}	1.4038^{20}	−16	70	−16	
d223	Dichlorodiphenylsilane	$(C_6H_5)_2SiCl_2$	253.20	16, 910	1.222^{20}		308–309	157	dec aq, alc	
d224	1,12-Dichlorododecane	$Cl(CH_2)_{12}Cl$	239.23	1^1, 67			28–30	172^{10mm}	>110	
d225	1,1-Dichloroethane	CH_3CHCl_2	98.96	1, 83	1.1757_4^{20}	1.4164^{20}	−97	57.3	−17	0.51 aq; misc alc
d226	1,2-Dichloroethane	$ClCH_2CH_2Cl$	98.96	1, 84	1.2351_4^{20}	1.4448^{20}	−35.7	83.5	13	0.8 aq; misc alc, chl, eth
d227	1,1-Dichloroethylene	$H_2C{=}CCl_2$	96.94	1, 186	1.2129_4^{20}	1.4247^{20}	−122.6	31.6	−28	0.01 aq; s alc, bz, chl, eth
d228	*cis*-1,2-Dichloroethylene	$ClCH{=}CHCl$	96.94	1, 188	1.2838_4^{20}	1.4490^{20}	−80.1	60	2	0.7 aq; s alc, eth
d229	*trans*-1,2-Dichloroethylene	$ClCH{=}CHCl$	96.94	1, 188	1.2565^{20}	1.4452^{20}	−49.8	48.7	2	0.6 aq; s alc, eth
d230	2,2′-Dichloroethyl ether	$ClCH_2CH_2OCH_2CH_2Cl$	143.01	1^2, 335	1.2220_{20}^{20}	1.457^{20}		178.5	55	1.1 aq; s alc, bz, eth
d231	2,2-Dichloroethyl methyl ether	$Cl_2CHCH_2OCH_3$	128.99		1.226	1.4375^{20}			33	
d232	Dichloroethylmethylsilane	$(C_2H_5)Si(CH_3)Cl_2$	143.09		1.063	1.4190^{20}		100	43	
d233	Dichlorofluoromethane	$FCHCl_2$	102.92	1, 61	1.405^9	1.3724^9	−135	8.9		69 HOAc; 108 diox; s alc, eth; i aq
d234	1,6-Dichlorohexane	$Cl(CH_2)_6Cl$	155.07	1, 144	1.068	1.4568^{20}		87^{15mm}	73	s chl
d235	Dichloromethane	CH_2Cl_2	84.93	1, 60	1.3265^{20}	1.4246^{20}	−95	40	none	1.3 aq; misc alc, eth
d236	Dichloromethane-d_2	CD_2Cl_2	86.95	1^4, 39	1.3621	1.4218^{20}		40	none	
d237	α,α-Dichloromethyl methyl ether	Cl_2CHOCH_3	114.96		1.271	1.4300^{20}		85	42	
d238	Dichloro(methyl)octylsilane	$CH_3(CH_2)_7Si(CH_3)Cl_2$	227.25	4, *4*, 4182	0.973	1.4440^{20}		94^{6mm}	98	
d239	Dichloro(methyl)phenylsilane	$C_6H_5Si(CH_3)Cl_2$	191.13		1.176	1.5190^{20}		205	82	
d240	Dichloro(methyl)silane	$HSi(CH_3)Cl_2$	115.04	4^1, 581	1.105	1.398^{20}	−93	41	−32	
d241	Dichloro(methyl)vinylsilane	$H_2C{=}CHSi(CH_3)Cl_2$	141.07		1.087_4^{20}	1.4300^{20}		92	4	
d242	2,4-Dichloro-1-naphthol	$Cl_2C_{10}H_5OH$	213.06	6, 612			108			

d243	2,3-Dichloro-1,4-naphthoquinone		227.05	7, 729			190–192			sl s alc, bz, eth
d244	2,6-Dichloro-4-nitroaniline	$Cl_2C_6H_2(NO_2)NH_2$	207.02	12, 735			190–192			
d245	2,3-Dichloronitrobenzene	$Cl_2C_6H_3NO_2$	192.00	5, 245	1.721^{14}		61–62	257–258	123	s PE
d246	2,4-Dichloronitrobenzene	$Cl_2C_6H_3NO_2$	192.00	5, 245	1.439^{80}		29–32	258	>110	s hot alc; misc eth
d247	2,5-Dichloronitrobenzene	$Cl_2C_6H_3NO_2$	192.00	5, 245			54–57	266–269	>110	
d248	3,4-Dichloronitrobenzene	$Cl_2C_6H_3NO_2$	192.00	5, 246	1.456^{75}		41–44	256	123	
d249	2,4-Dichloro-6-nitrophenol	$Cl_2C_6H_2(NO_2)OH$	208.00	6, 241			118–120			
d250	1,7-Dichlorooctamethyltetrasiloxane	$[Cl(CH_3)_2SiOSi(CH_3)_2\text{-}]_2$	351.53	4^3, 1884	1.011^{20}_{4}	1.403^{20}	−62	222		
d251	1,5-Dichloropentane	$Cl(CH_2)_5Cl$	141.04	1, 131	1.1058^{20}_{4}	1.4553^{20}	−72	66^{10mm}	26	i aq; s alc, eth
d252	2,3-Dichlorophenol	$Cl_2C_6H_3OH$	163.00	6^1, 102			58–60	206		s alc, eth
d253	2,4-Dichlorophenol	$Cl_2C_6H_3OH$	163.00	6, 189			42–43	210	113	v s alc, bz, chl, eth
d254	2,5-Dichlorophenol	$Cl_2C_6H_3OH$	163.00	6, 189			56–58	211		v s alc, bz, eth
d255	2,6-Dichlorophenol	$Cl_2C_6H_3OH$	163.00	6, 190			65–68	218–220		v s alc, eth
d256	2,4-Dichlorophenoxyacetic acid	$Cl_2C_6H_3OCH_2COOH$	221.04				136–140	$160^{0.4mm}$		s alc, bz, chl, eth
d257	4-(2,4-Dichlorophenoxy)butanoic acid	$Cl_2C_6H_3O(CH_2)_3CO_2H$	249.10	6^3, 708			117–119			46 ppm aq^{25}; s acet, alc, eth; sl s bz
d258	2-(2,4-Dichlorophenoxy)propanoic acid	$Cl_2C_6H_3OCH(CH_3)CO_2H$	235.07	6, 189			110–112			350 ppm aq^{20}; v s org solvents
d259	3,4-Dichlorophenyl isocyanate	$Cl_2C_6H_3NCO$	188.01	12^3, 1405			42–44	120^{18mm}	>110	
d260	Dichlorophenylphosphine	$C_6H_5PCl_2$	178.99	16, 763	1.319	1.5980^{20}	−51	222	>112	
d261	4,5-Dichloro-*o*-phthalic acid	$Cl_2C_6H_2(CO_2H)_2$	235.02	9^1, 366			201–203			s aq; v s eth
d262	1,2-Dichloropropane	$CH_3CH(Cl)CH_2Cl$	112.99	1, 105	1.1558^{20}	1.4390^{20}	−100	96	4	0.26 aq; misc alc, bz, chl, eth
d263	1,3-Dichloropropane	$ClCH_2CH_2CH_2Cl$	112.99	1, 105	1.1878^{20}_{4}	1.4487^{20}	−99.5	120–122	32	v s alc, eth
d264	1,3-Dichloro-2-propanol	$ClCH_2CH(OH)CH_2Cl$	128.99	1, 364	1.198	1.4835^{20}	−4	174.3	85	9.1 aq; misc alc, eth

(*Continued*)

TABLE 2.20 Physical Constants of Organic Compounds (*Continued*)

No.	Name	Formula	Formula weight	Beilstein reference	Density, g/mL	Refractive index	Melting point, °C	Boiling point, °C	Flash point, °C	Solubility in 100 parts solvent
d265	1,3-Dichloropropene	$ClCH_2CH{=}CHCl$	110.97	1, 199	1.217_4^{20}	1.470^{20}		97–112	25	i aq; s chl, eth
d266	2,3-Dichloro-1-propene	$ClCH_2C(Cl){=}CH_2$	110.97	1, 199	1.204_{25}^{25}	1.4611^{20}		94	10	misc alc; s eth
d267	3,6-Dichloropyridazine		148.98				66–69			
d268	2,6-Dichloropyridine	$Cl_2C_5H_3N$	147.99	20, 231			86–88			
d269	3,5-Dichloropyridine	$Cl_2C_5H_3N$	147.99	20, 231			65–67			
d270	4,7-Dichloroquinoline		198.05	20^3, 3384			84–86	148^{10mm}		
d270a	Dichlorosilane	Cl_2SiH_2	101.01				−122	8.3		
d270b	1,1-Dichlorotetra-fluoroethane	F_3CCFCl_2	170.92		1.455^{25} satd pressure	1.3092^{0}	−57	4		
d271	1,2-Dichloro-1,1,2,2-tetrafluoroethane	$ClCF_2CF_2Cl$	170.93	1^3, 152	1.470_4^{20} satd pressure	1.3092^{20}	−94	3.6		s alc, eth
d272	2,5-Dichlorothiophene	$Cl_2(C_4H_2S)$	153.03	17, 33	1.442	1.5621^{20}	−40.5	162	59	i aq; misc alc, eth
d273	α,α-Dichlorotoluene	$C_6H_5CHCl_2$	161.03	5, 297	1.254	1.5500^{20}	−16/−17	205	92	v s alc, eth
d274	2,4-Dichlorotoluene	$Cl_2C_6H_3CH_3$	161.03	5, 295	1.2460_{20}^{20}	1.5511^{20}	−13	200.5	79	i aq
d275	2,6-Dichlorotoluene	$Cl_2C_6H_3CH_3$	161.03	5, 296	1.254	1.5507^{20}		196–203	82	i aq; s chl
d276	3,4-Dichlorotoluene	$Cl_2C_6H_3CH_3$	161.03	5, 296	1.251_{25}^{25}	1.5472^{20}	−15	209	85	i aq
d277	α,α-Dichloro-*o*-xylene	$C_6H_4(CH_2Cl)_2$	175.06	5, 364			55–57	239–241	107	
d278	α,α-Dichloro-*p*-xylene	$C_6H_4(CH_2Cl)_2$	175.06	5, 384			99–101	254		22.5 acet; 20 bz; 4.5 CCl_4; 11 eth; 18 EtOAc
d279	2,5-Dichloro-*p*-xylene	$Cl_2C_6H_2(CH_3)_2$	175.06	5, 384			71	222		27 acet; 44 bz; 39 eth; 32 EtOAc; 5 MeOH
d280	Dicumyl peroxide	$[C_6H_5C(CH_3)_2]_2O_2$	270.37				39–41		>110	
d281	Dicyandiamide	$H_2NC({=}NH)NHCN$	84.08	3, 91	1.400_4^{25}		208–211			2.3 aq; 1.3 alc; i bz
d282	1,2-Dicyanobenzene	$C_6H_4(CN)_2$	128.13	9, 815			139–141			v s bz, alc; s hot eth
d283	1,3-Dicyanobenzene	$C_6H_4(CN)_2$	128.13	9, 836			158–160			s alc, bz, chl, eth
d284	1,4-Dicyanobutane	$NC(CH_2)_4CN$	108.14	2, 653	0.951	1.4380^{20}	1–3	295	93	
d285	1,6-Dicyanohexane	$NC(CH_2)_6CN$	136.20	2, 694	0.954	1.4436^{20}	−3.5	185^{15mm}	>110	

d286	2,4-Dicyano-3-methyl-glutaramide	$CH_3CH[CH(CN)CONH_2]_2$	194.19	2^2, 704			159–160			
d287	1,5-Dicyanopentane	$NC(CH_2)_5CN$	122.17	2, 671	0.951	1.4410^{20}		176^{14mm}	>110	
d288	Dicyclohexyl	$C_6H_{11}C_6H_{11}$	166.31	5, 108	0.864	1.4782^{20}	3–4	227	92	7 MeOH; misc bz, acet, eth
d289	Dicyclohexylamine	$(C_6H_{11})_2NH$	181.32	12, 6	0.910	1.4842^{20}	−2	255.8	96	misc alc, bz, chl, eth
d290	*N,N′*-Dicyclohexyl-carbodiimide	$C_6H_{11}N{=}C{=}NC_6H_{11}$	206.33	Merck: 12, 3146			35–36	124^{6mm}	110	
d291	Dicyclohexyl *o*-phthalate	C_6H_4-1,2-$(CO_2C_6H_{11})_2$	330.43	9, 799			64–66			
d292	Dicyclopentadiene		132.21	5, 495	0.930^{25}_4	1.5050^{25}	−1	170	26	s alc, eth
d293	Dicyclopentenyl methacrylate		218.30	6^3, 1942	1.050	1.5080^{20}		137^{13mm}	>110	
d294	Dicyclopropyl ketone	$(C_3H_5)_2C{=}O$	110.16		0.977	1.4670^{20}		160–162	39	
d295	Didodecyl 3,3′-thiodi-propionate	$S[CH_2CH_2CO_2(CH_2)_{11}CH_3]_2$	514.86	3^3, 556	0.915		40–42		>110	
d296	Dieldrin		380.92	17^3, 526			176–177			i aq; s common org solvents except PE
d297	Diethanolamine	$HOCH_2CH_2NHCH_2CH_2OH$	105.14	4, 283	1.0881^{30}_4	1.4747^{30}	28.0	269	172	96 aq; 4 bz; 0.8 eth; misc MeOH, acet
d298	2,2-Diethoxyacet-ophenone	$C_6H_5C({=}O)CH(OC_2H_5)_2$	208.26	7^1, 361	1.034	1.4995^{20}		134^{10mm}	>110	
d299	4,4-Diethoxybutyl-amine	$H_2N(CH_2)_3CH(OC_2H_5)_2$	161.25	4, 319	0.933	1.4275^{20}		196	62	
d300	2,2-Diethoxy-*N,N*-di-methylethylamine	$(C_2H_5O)_2CHCH_2N(CH_3)_2$	161.25	4, 308	0.883	1.4129^{20}		170	45	
d301	Diethoxydimethyl-silane	$(C_2H_5O)_2Si(CH_3)_2$	148.28		0.840^{20}_4	1.3811^{20}	−87	114	11	
d302	Diethoxydiphenyl-silane	$(C_2H_5O)_2Si(C_6H_5)_2$	272.42	16^2, 608	1.0329^{20}_4	1.5269^{20}		139^{2mm}	>110	
d303	1,1-Diethoxyethane	$CH_3CH(OC_2H_5)_2$	118.18	1, 603	0.8254^{20}_4	1.3819^{20}	−100	102.2	−21	5 aq; misc alc, eth
d304	1,2-Diethoxyethane	$C_2H_5OCH_2CH_2OC_2H_5$	118.18	1, 468	0.842	1.3922^{20}	−74	121.4	27	21 aq
d305	2,2-Diethoxyethanol	$(C_2H_5O)_2CHCH_2OH$	134.18	1, 818	0.888^{24}_4	1.4160^{20}		167	67	s alc, eth
d306	2,2-Diethoxyethyl-amine	$(C_2H_5O)_2CHCH_2NH_2$	133.19	4, 308	0.916	1.4170		162–163	45	
d307	Diethoxymethane	$(C_2H_5O)_2CH_2$	104.15		0.839	1.3732^{20}		87–88	−5	

(*Continued*)

TABLE 2.20 Physical Constants of Organic Compounds (*Continued*)

No.	Name	Formula	Formula weight	Beilstein reference	Density, g/mL	Refractive index	Melting point, °C	Boiling point, °C	Flash point, °C	Solubility in 100 parts solvent
d308	3-(Diethoxymethylsilyl)propylamine	$CH_3Si(OC_2H_5)_2(CH_2)_3NH_2$	191.35	4, *4*, 4201	0.916	1.4260^{20}		88^{8mm}	75	
d309	2,5-Diethoxynitrobenzene	$(C_2H_5O)_2C_6H_3NO_2$	211.22	6, 857			48–51	169^{13mm}	>110	
d310	Diethoxymethylvinylsilane	$(C_2H_5O)_2Si(CH_3)CH{=}CH_2$	160.29	4^4, 4183	0.858^{20}_{4}	1.400^{20}		133–134	17	
d311	1,1-Diethoxypropane	$CH_3CH_2CH(OC_2H_5)_2$	132.20	1, 630	0.8232^{20}_{4}	1.3884^{20}		122.8	7	v s alc, eth
d312	3,3-Diethoxy-1-propene	$(C_2H_5O)_2CHCH{=}CH_2$	130.19	1, 727	0.854	1.4000^{20}		125	4	
d313	2,2-Diethoxytriethylamine	$(C_2H_5O)_2CHCH_2N(C_2H_5)_2$	189.30	4, 309	0.850	1.4189^{20}		194–195	65	
d314	*N,N*-Diethylacetamide	$CH_3C({=}O)N(C_2H_5)_2$	115.18	4, 110	0.925	1.4401^{20}		182–186	70	
d315	Diethyl 1,3-acetonedicarboxylate	$C_2H_5OOCCH_2C({=}O)CH_2$-$CO_2C_2H_5$	202.21	3, 791	1.113	1.4385^{20}		250	86	
d316	Diethyl 2-acetylglutarate	$C_2H_5O_2CCH_2CH_2CH$-$[C({=}O)CH_3]CO_2C_2H_5$	230.26	3, 809	1.071	1.4386^{20}		154^{11mm}	>110	
d317	Diethyl acetylsuccinate	$C_2H_5O_2CCH_2CH[C({=}O)$-$CH_3]CO_2C_2H_5$	216.23	3, 801	1.081	1.4346^{20}		183^{50mm}	>110	
d318	Diethyl adipate	$C_2H_5O_2C(CH_2)_4CO_2C_2H_5$	202.25	2, 652	1.009	1.4270^{20}	−18	251	110	
d319	Diethyl allylmalonate	$C_2H_5O_2CCH(CH_2CH{=}CH_2)$-$CO_2C_2H_5$	200.23	2, 776	1.015	1.4304^{20}		222–223	71	
d320	Diethylaluminum chloride	$(C_2H_5)_2AlCl$	120.56	4^3, 1972	0.961		−50	126^{50mm}	−18	
d321	Diethylaluminum ethoxide	$(C_2H_5)_2AlOC_2H_5$	130.17	4^3, 1972	0.850		2.5–4.5	109^{10mm}	−18	
d322	Diethylaluminum iodide	$(C_2H_5)_2AlI$	212.01	4^2, 1024	1.609			120^{4mm}	−18	
d323	Diethylamine	$(C_2H_5)_2NH$	73.14	4, 95	0.7074^{20}_{4}	1.3864^{10}	−50.0	55.5	−23	misc aq, alc
d324	Diethylamine HCl	$(C_2H_5)_2NH \cdot HCl$	109.60	4, 95	1.048^{21}_{4}		227–230	320–330		s aq, alc, chl; i eth
d325	2-(Diethylamino)acetonitrile	$(C_2H_5)_2NCH_2CN$	112.18	4, 350	0.866	1.4260^{20}		170	53	
d326	4-(Diethylamino)benzaldehyde	$(C_2H_5)_2NC_6H_4CHO$	177.25	14^2, 25			39–41	174^{7mm}	>110	
d327	2-Diethylaminoethanol	$(C_2H_5)_2NCH_2CH_2OH$	117.19	4, 282	0.8800^{25}	1.4389^{20}	−70	163	48	s aq, alc, bz, eth
d328	2-Diethylaminoethylchloride HCl	$ClCH_2CH_2N(C_2H_5)_2 \cdot HCl$	172.10	4^2, 618			108–210			
d329	2-(Diethylamino)ethyl methacrylate	$H_2C{=}C(CH_3)CO_2CH_2CH_2$-$N(C_2H_5)_2$	185.27	4^3, 676	0.922	1.4440^{20}		80^{10mm}	76	

d330	3-(Diethylamino)-phenol	$(C_2H_5)_2NC_6H_4OH$	165.24	13, 408			65–69	170^{15mm}		s aq, alc, eth
d331	3-Diethylamino-1,2-propanediol	$(C_2H_5)_2NCH_2CH(OH)CH_2OH$	147.22	4, 302	0.973^{20}_{20}	1.4602^{20}		233–235	107	s aq, alc, chl, eth
d332	1-Diethylamino-2-propanol	$(C_2H_5)_2NCH_2CH(OH)CH_3$	131.22	4^2, 737	0.889	1.4255^{20}	13.5	59^{13mm}	33	s alc
d333	3-Diethylamino-1-propanol	$(C_2H_5)_2NCH_2CH_2CH_2OH$	131.22	4, 288	0.884	1.4435		83^{15mm}	65	
d334	3-Diethylaminopropylamine	$(C_2H_5)_2NCH_2CH_2CH_2NH_2$	130.24		0.826	1.4416^{20}		159	58	
d335	*N,N*-Diethylaniline	$C_6H_5N(C_2H_5)_2$	149.24	12, 164	0.9302^{25}_{4}	1.5394^{25}	−38	216	97	1 aq; sl s alc, eth
d336	2,6-Diethylaniline	$(C_2H_5)_2C_6H_3NH_2$	149.24		0.906	1.5452^{20}	3	243	123	
d337	Diethyl azelate	$C_2H_5O_2C(CH_2)_7CO_2C_2H_5$	244.33	2, 709	0.973	1.4350^{20}	−16	172^{18mm}	>110	
d338	Diethyl azodi-carboxylate	$C_2H_5O_2CN{=}NCO_2C_2H_5$	174.16	3, 123	1.106	1.4280^{20}		106^{13mm}	>110	
d339	5,5-Diethylbarbituric acid		184.19	24^2, 279	1.220		188–192			0.7 aq; 7 alc; 1.3 chl; 3.2 eth; s acet, HOAc
d340	Diethyl benzalmalonate	$C_6H_5CH{=}C(CO_2C_2H_5)_2$	248.28	9, 892	1.107	1.5365^{20}		215^{30mm}	>110	
d340a	1,2-Diethylbenzene	$C_6H_4(C_2H_5)_2$	134.22	5, 426	0.880	1.5020^{20}	−31	184	49	
d341	1,3-Diethylbenzene	$C_6H_4(C_2H_5)_2$	134.22	5, 426	0.8640^{20}_{4}	1.4950^{20}	−83.9	181.1	50	s alc, eth
d342	1,4-Diethylbenzene	$C_6H_4(C_2H_5)_2$	134.22	5, 426	0.8620^{20}_{4}	1.4940^{20}	−42.8	183.8	56	s alc, eth
d343	Diethyl benzyl-malonate	$C_6H_5CH_2CH(CO_2C_2H_5)_2$	250.29	9, 869	1.064	1.4868^{20}		162^{10mm}	>110	
d344	Diethyl benzo-phosphonate	$C_6H_5CH_2P(O)(OC_2H_5)_2$	228.23	12, 164	1.095	1.4970^{20}		108^{1mm}	>110	
d345	Diethyl bis(hydroxy-methyl)malonate	$(HOCH_2)_2C(CO_2C_2H_5)_2$	220.22				49–51		>110	
d346	Diethyl bromomalonate	$BrCH(CO_2C_2H_5)_2$	239.07	2, 594	1.4022^{25}_{4}	1.4550^{20}	−54	235 dec	>110	i aq; misc alc, eth
d347	Diethyl butylmalonate	$C_4H_9CH(CO_2C_2H_5)_2$	216.28	2^1, 282	0.983	1.4220		235–240	93	v s alc, eth
d348	Diethylcarbamoyl chloride	$(C_2H_5)_2N(O)Cl$	135.59	4, 120	1.070	1.4515^{20}	−32	187–190	75	d hot aq, hot alc
d349	Diethyl carbonate	$(C_2H_5O)_2C{=}O$	118.13	3, 5	0.9764^{20}_{4}	1.3843^{20}	−43.0	126	25	69 aq; misc alc, bz, eth, esters
d350	Diethyl chloro-phosphate	$(C_2H_5O)_2P(O)Cl$	172.55	1, 332	1.194	1.4165^{20}		60^{2mm}	61	

(*Continued*)

TABLE 2.20 Physical Constants of Organic Compounds (*Continued*)

No.	Name	Formula	Formula weight	Beilstein reference	Density, g/mL	Refractive index	Melting point, °C	Boiling point, °C	Flash point, °C	Solubility in 100 parts solvent
d351	Diethyl chlorothiophosphate	$(C_2H_5O)_2P(S)Cl$	188.61	1^3, 1332	1.200	1.4715^{20}		45^{3mm}	>110	
d352	Diethyl cyanophosphate	$(C_2H_5O)_2P(O)CN$	163.11		1.075	1.4012^{20}		105^{19mm}	80	
d353	*N,N*-Diethylcyclohexylamine	$C_6H_{11}N(C_2H_5)_2$	155.29	12, 6	0.850	1.4562^{20}		194–195	57	
d354	Diethyl diethylmalonate	$(C_2H_5)_2C(CO_2C_2H_5)_2$	216.28	2, 686	0.990	1.4230^{20}		228–230	94	
d355	1,3-Diethyl-1,3-diphenylurea	$[C_6H_5N(C_2H_5)]_2C{=}O$	268.36	12, 422			73–75			
d356	Diethyl disulfide	$C_2H_5SSC_2H_5$	122.25	1, 347	0.998^{20}_{4}	1.5063^{20}	−101.5	154.0	40	sl s aq; misc alc, eth
d357	Diethyldithiocarbamic acid, sodium salt	$(C_2H_5)_2NC({=}S)S^- Na^+ \cdot 3H_2O$	225.31	4^2, 613			95–99			
d358	Diethyl dithiophosphate	$(C_2H_5O)_2P(S)SH$	186.23	1, 333	1.111	1.5120^{20}		60^{1mm}	82	
d359	*N,N*-Diethyldodecanamide	$CH_3(CH_2)_{10}C({=}O)N(C_2H_5)_2$	255.45		0.847	1.4545^{20}		166^{2mm}	>110	
d360	Diethyl dodecanedioate	$C_2H_5O_2C(CH_2)_{10}CO_2C_2H_5$	186.41	2^2, 616	0.951	1.4402^{20}	15	193^{14mm}	>110	
d361	Diethylene glycol	$(HOCH_2CH_2)_2O$	106.12	1, 468	1.1197^{15}	1.4460^{20}	−10	246	124	
d362	Diethylenetriamine	$(H_2NCH_2CH_2)NH$	103.17	4, 255	0.9542^{20}_{20}	1.4826^{20}	−35/−39	207	98	misc aq, alc, bz, eth
d363	Diethylenetriaminepentaacetic acid	$[(HO_2CCH_2)_2NCH_2CH_2]_2N$-$(CH_2CO_2H)N(CH_2CO_2H)_2$	393.35	4^4, 2454			219–220			
d364	*N,N*-Diethylethanolamine	$HOCH_2CH_2N(C_2H_5)_2$	117.19	4, 282	0.884	1.4410^{20}		161	48	
d365	Diethyl ether	$C_2H_5OC_2H_5$	74.12	1, 314	0.7134^{20}_{4}	1.3527^{20}	−116.3	34.6	−45	6 aq; misc alc, bz, chl
d366	Diethyl ethoxymethylenemalonate	$(C_2H_5O_2C)_2C{=}CHOC_2H_5$	216.23	3, 469	1.070	1.4620^{20}		279–281	155	
d367	*N,N*-Diethylethylenediamine	$(C_2H_5)_2NCH_2CH_2NH_2$	116.21	4, 251	0.827	1.4360^{20}		145–147	30	
d368	Diethyl ethylmalonate	$C_2H_5CH(CO_2C_2H_5)_2$	188.22	2, 644	1.004^{20}_{20}	1.4158^{20}		77^{5mm}	88	sl s aq; v s alc, eth
d369	*N,N*-Diethylformamide	$(C_2H_5)_2NCHO$	101.15	4, 109	0.908	1.4340^{20}		176–177	60	misc aq; v s alc, eth
d370	Diethyl fumarate	$C_2H_5O_2CCH{=}CHCO_2C_2H_5$	172.18	2, 742	1.052^{20}_{4}	1.4406^{20}	1–2	218–219	91	
d371	Diethyl glutarate	$C_2H_5O_2CCH_2CH_2CH_2$-$CO_2C_2H_5$	188.22	2, 633	1.022	1.4240^{20}	−23.8	237	96	0.9 aq; v s alc; s eth
d372	2,4-Diethyl-2,6-heptadienal	$H_2C{=}CHCH_2CH(C_2H_5)$-$CH{=}C(C_2H_5)CHO$	166.27		0.862	1.4676^{20}		91^{12mm}	86	
d373	Diethyl heptanedioate	$C_2H_5O_2C(CH_2)_5CO_2C_2H_5$	216.28	2, 671	0.9945^{20}	1.4280^{20}	−24	192^{100mm}	>110	i aq; s alc, eth

d374	Di-(2-ethylhexyl)-*o*-phthalate	$C_6H_4[CO_2CH_2CH(C_2H_5)-C_4H_9]_2$	390.56	10, 1248	0.981^{25}_{25}	1.4853^{20}	−50	384	207	
d375	Diethyl hydrogen phosphonate	$(C_2H_5O)_2P(O)H$	138.10	1, 330	1.079^{20}_{4}	1.4076^{20}		51^{2mm}	90	hyd aq; s alc, eth
d376	*N,N*-Diethylhydroxylamine	$(C_2H_5)_2NOH$	89.14	4, 536	1.867	1.4195^{20}	−25	125–130	45	
d377	Diethyl maleate	$C_2H_5O_2CCH{=}CHCO_2C_2H_5$	172.18	2, 751	1.0687^{20}	1.4400^{20}	−8.8	225.3	93	1.4 aq; s alc, eth
d378	Diethyl malonate	$C_2H_5O_2CCH_2CO_2C_2H_5$	160.17	2, 573	1.0550	1.4136^{20}	−49.9	199.3	93	2.7 aq; misc alc, eth
d379	Diethylmalonic acid	$HO_2CC(C_2H_5)_2CO_2H$	160.17	2, 686			127	170–180		v s aq, alc, eth
d380	*N,N*-Diethylmethylamine	$(C_2H_5)_2NH_3$	87.17	4, 99	0.720	1.3887^{20}		63–65	−23	
d381	Diethyl methylmalonate	$C_2H_5O_2CCH(CH_3)CO_2C_2H_5$	174.20	2, 629	1.018^{20}_{4}	1.4130^{20}		198	76	
d382	Diethyl 2-methyl-2′-oxosuccinate	$C_2H_5O_2CCH(CH_3)C({=}O)-CO_2C_2H_5$	202.21	3, 794	1.073	1.4313^{20}		138^{23mm}	>110	
d383	*N,N*-Diethyl-4-nitrosoaniline	$C_6H_4(NO)N(C_2H_5)_2$	178.24	12, 684			82–84			
d384	Diethyl octanedioate	$C_2H_5O_2C(CH_2)_6CO_2C_2H_5$	230.30	2, 693	0.9822^{20}_{4}	1.4323^{20}	5.9	282	>112	i aq; s alc, eth
d385	Diethyl oxalate	$C_2H_5O_2CCO_2C_2H_5$	146.14	2, 535	1.0785^{20}_{4}	1.4102^{20}	−40.6	185.4	76	3.6 aq (gradual dec); misc alc, eth
d386	Diethyl oxydiformate	$[C_2H_5OC({=}O)]_2O$	162.14	Merck: 12, 8182	1.12^{20}_{4}	1.3980^{20}		93^{18mm}	69	50 alc; s esters, ketones; s aq
d386a	3,3-diethylpentane	$C(C_2H_5)_4$	128.26		0.7536^{20}	1.4206^{20}	−33	146		
d387	*N*[1],*N*[1]-Diethyl-1,4-pentanediamine	$CH_3CH(NH_2)(CH_2)_3N(C_2H_5)_2$	158.29	Merck: 12, 6819	0.817	1.4429^{20}		200	68	s aq, alc, eth
d388	*N*[1],*N*[1]-Diethyl-1,4-phenylenediamine	$(C_2H_5)_2NC_6H_4NH_2$	164.25	13, 75	0.988	1.5710^{20}		116^{5mm}	>110	
d389	Diethyl phenylmalonate	$C_6H_5CH(CO_2C_2H_5)_2$	236.27	9, 854	1.0950^{20}_{4}	1.4913^{20}	16	170^{14mm}	>110	i aq; s alc
d390	Diethyl phosphite	$(C_2H_5)_2P(O)H$	138.10	1, 330	1.079^{20}_{4}	1.4079^{20}		51^{2mm}	90	hyd aq; s alc, eth
d391	Diethyl *o*-phthalate	$C_6H_4(CO_2C_2H_5)_2$	222.24	9, 798	1.232^{14}_{4}	1.5049^{14}	−40	295	160	i aq; misc alc, eth
d392	*N,N*-Diethyl-1,3-propanediamine	$(C_2H_5)_2NCH_2CH_2CH_2NH_2$	130.24		0.826	1.4416^{20}		159	58	
d393	2,2-Diethyl-1,3-propanediol	$(C_2H_5)_2C(CH_2OH)_2$	132.20		1.052^{20}	1.4574^{25}	61.3	125^{10mm}		25 aq; v s alc, eth

(*Continued*)

TABLE 2.20 Physical Constants of Organic Compounds (*Continued*)

No.	Name	Formula	Formula weight	Beilstein reference	Density, g/mL	Refractive index	Melting point, °C	Boiling point, °C	Flash point, °C	Solubility in 100 parts solvent
d394	Diethyl propylmalonate	$C_2H_5O_2CCH(C_3H_7)CO_2C_2H_5$	202.25	2, 657	0.987	1.4185^{20}		221–222	91	
d395	Diethyl sebacate	$C_2H_5O_2C(CH_2)_8CO_2C_2H_5$	258.36	2, 717	0.963	1.4360^{20}	1–2	312	>110	0.14 aq; misc alc, eth
d396	Diethyl succinate	$C_2H_5O_2C(CH_2)_2CO_2C_2H_5$	174.20	2, 609	1.040_4^{20}	1.4200^{20}	−21	217.7	100	i aq; misc alc, eth
d397	Diethyl sulfate	$(C_2H_5O)_2SO_2$	154.18	1, 327	1.172_4^{25}	1.4004^{20}	−25	208	78	i aq; misc alc, eth
d398	Diethyl sulfide	$(C_2H_5)_2S$	90.19	1, 344	0.8367_4^{20}	1.4430^{20}	−103.9	92.1	−9	i aq; misc alc, eth
d399	Diethyl sulfite	$(C_2H_5O)_2SO$	138.19	1, 325	1.883	1.450^{20}		158	53	s aq(dec), alc
d400	(+)-Diethyl-L-tartrate	$[-CH(OH)CO_2C_2H_5]_2$	206.19	3, 512	1.205_4^{20}	1.4460^{20}	17	280	93	sl s aq; misc alc, eth
d401	(−)-Diethyl-D-tartrate	$[-CH(OH)CO_2C_2H_5]_2$	206.19	3^1, 181	1.205	1.4460^{20}		162^{19mm}	93	sl s aq; misc alc, eth
d402	*N,N*-Diethyl-*m*-toluamide	$CH_3C_6H_4C(=O)N(C_2H_5)_2$	191.27	9^2, 325	0.996_4^{20}	1.5212^{20}		111^{1mm}	>110	i aq; v s alc, bz, eth
d403	*N,N*-Diethyl-*m*-toluidine	$CH_3C_6H_4CN(C_2H_5)_2$	163.26	12, 857	0.922	1.5360^{20}		231–232	100	
d404	*N,N*-Diethyl-1,1,1-trimethylsilylamine	$(C_2H_5)_2NSi(CH_3)_3$	145.32	4^3, 1861	0.767	1.4110^{20}		125–126	10	
d405	Diethylzinc	$(C_2H_5)_2Zn$	123.49	6, 672	1.2065_4^{20}	1.4983^{20}	−28	118	−23	
d406	1,2-Difluorobenzene	$C_6H_4F_2$	114.09	5^2, 147	1.158	1.4430^{20}	−34	92	2	
d406a	1,4-Difluorobenzene	$C_6H_4F_2$	114.09	5, 199	1.1701^{20}	1.4410^{20}	−13	89	2	
d407	1,1-Difluoroethane	CH_3CHF_2	66.05	1^3, 130	0.909^{21}	1.3011^{-72}	−117	−24.7		0.32 aq
d408	1,1-Difluoroethylene	$CH_2=CF_2$	64.04	1, 186			−144	−86		
d409	Difluoromethane	CH_2F_2	52.02	1, 59	2.126 g/L		−136	−51.6		FLAMMABLE GAS
d410	2,4-Difluoronitrobenzene	$F_2C_6H_3NO_2$	159.09	5^1, 129	1.451	1.5110^{20}	9–10	203–204	90	
d411	1,1-Difluorotetrachloroethane	ClF_2CCCl_3	203.83	1, 86	1.649	1.413	41	91	none	sl s alc; v s eth
d412	1,2-Difluorotetrachloroethane	FCl_2CCCl_2F	203.83	1^3, 365	1.6447_4^{25}	1.413^{25}	23.8	203.8		i aq; s alc, eth
d413	Dihexylamine	$(C_6H_{13})_2NH$	185.36	4^1, 384	0.795	1.4320^{20}		192–195	95	s alc, eth
d414	Dihexyl ether	$(C_6H_{13})_2O$	186.34	1^3, 1656	0.7936_4^{20}	1.4204^{20}		226.2	77	i aq; s ethers
d415	9,10-Dihydroanthracene		180.25	5, 641	0.880		108–110	312		i aq; s alc, bz, eth
d416	(+)-Dihydrocarvone		152.24	7^3, 337	0.929^{19}	1.4718^{20}		221–222	81	
d417	Dihydrocoumarin		148.16	17, 315	1.169^{18}	1.5563^{20}	25	272	>110	sl s alc, eth; s chl
d418	2,5-Dihydro-2,5-dimethoxyfurfurylamine		159.19	18^3, 7426	1.102	1.4600^{20}		96^{12mm}	96	
d419	2,3-Dihydro-2,2-dimethyl-7-benzofuranol		164.21	17^5, 4, 47	1.101	1.5410^{20}			110	

d420	3,4-Dihydro-2-ethoxy-2*H*-pyran		128.17		0.957	1.4394^{20}		42^{16mm}	24	
d421	2,3-Dihydrofuran		70.09	17^3, 141	0.927	1.4239^{20}		54–55	−24	
d422	3,4-Dihydro-2-methoxy-2*H*-pyran		114.14			1.4425^{20}			16	
d423	3,4-Dihydro-1(2*H*)-naphthalenone		146.19	7, 370	1.099	1.5685^{20}	5–6	116^{6mm}	>110	
d424	3,4-Dihydro-2*H*-pyran		84.12		0.922^{19}_{15}	1.4410^{20}	−70	86	−15	s aq, alc
d425	2′,4′-Dihydroacetophenone	$(HO)_2C_6H_3C(=O)CH_3$	152.15	8, 266	1.180		145–147			s warm alc, HOAc, pyr; i bz, eth
d426	1,8-Dihydroxyanthraquinone		240.21	8, 458			193–197	subl		0.005 alc; 0.2 eth; s chl
d427	2,4-Dihydroxybenzaldehyde	$(HO)_2C_6H_3CHO$	138.12	8, 241			135–136	226^{22mm}		v s aq, alc, chl, eth
d428	1,2-Dihydroxybenzene	$C_6H_4(OH)_2$	110.11	6, 759	1.344^4		104–106	245.5	137	43 aq; s alc, bz, chl, eth; v s pyr, alkalis
d429	1,3-Dihydroxybenzene	$C_6H_4(OH)_2$	110.11	6^2, 802	1.272^{15}		109–110	276	171	110 aq; 110 alc; v s eth, glyc; sl s chl
d430	1,4-Dihydroxybenzene	$C_6H_4(OH)_2$	110.11	6, 836	1.332^{15}		170–171	285–287		7 aq; v s alc, eth
d431	2,4-Dihydroxybenzoic acid	$(HO)_2C_6H_3CO_2H$	154.12	10, 377			213 rapid heating			s hot aq, alc, eth
d432	2,5-Dihydroxybenzoic acid	$(HO)_2C_6H_3CO_2H$	154.12	10, 384			199–200			0.5 aq; s alc, eth
d433	3,4-Dihydroxybenzoic acid	$(HO)_2C_6H_3CO_2H$	154.12	10, 389	1.54		200–202			2 aq; s alc, eth
d434	3,5-Dihydroxybenzoic acid	$(HO)_2C_6H_3CO_2H$	154.12	10, 404			236 dec			sl s aq; s alc, eth
d435	2,4-Dihydroxybenzophenone	$(HO)_2C_6H_3C(=O)C_6H_5$	214.22	8, 312			144–145			v s alc, eth, HOAc
d436	2,2′-Dihydroxybiphenyl	$HOC_6H_4C_6H_4OH$	186.21	6, 989			110	315		s alc, bz, eth; sl s aq
d437	4,6-Dihydroxy-2-mercaptopyrimidine		144.15	24, 476			236			
d438	1,2-Dihydroxy-4-methylbenzene	$(HO)_2C_6H_3CH_3$	124.14	6, 878	1.129^{74}_4	1.5425^{74}	67–69	251		v s aq, alc, eth
d439	1,5-Dihydroxynaphthalene	$C_{10}H_6(OH)_2$	160.17	6, 980			259 dec			sl s aq; s alc; v s eth

(*Continued*)

TABLE 2.20 Physical Constants of Organic Compounds (*Continued*)

No.	Name	Formula	Formula weight	Beilstein reference	Density, g/mL	Refractive index	Melting point, °C	Boiling point, °C	Flash point, °C	Solubility in 100 parts solvent
d440	1,6-Dihydroxynaphthalene	$C_{10}H_6(OH)_2$	160.17	6, 981			138–140			v s alc, eth
d441	2,3-Dihydroxynaphthalene	$C_{10}H_6(OH)_2$	160.17	6, 982			162–164			v s alc, eth
d442	2,7-Dihydroxynaphthalene	$C_{10}H_6(OH)_2$	160.17	6, 985			187 dec			sl s aq; v s alc, eth
d443	1,4-Dihydroxy-2-naphthoic acid	$(HO)_2C_{10}H_5CO_2H$	204.19	10, 442			220 dec			
d444	3,5-Dihydroxy-2-naphthoic acid	$(HO)_2C_{10}H_5CO_2H$	204.19	10, 444			277 dec			
d445	1,3-Dihydroxy-2-propanone	$HOCH_2C(=O)CH_2OH$	90.08	1, 846			65–71			v s aq, alc, acet, eth
d446	7-(2,3-Dihydroxypropyl)theophylline		254.25				158			33 aq; 2 alc; 1 chl
d447	3,6-Dihydroxypyridazine		112.09	24, 312			306–308			sl s ahot alc; s hot aq
d448	2,3-Dihydroxypyridine	$(HO)_2C_5H_3N$	111.10	21^2, 107			245 dec			
d449	1,4-Diiodobenzene	$C_6H_4I_2$	329.91	5, 227			131–133	285		sl s alc; v s eth
d450	1,4-Diiodobutane	$I(CH_2)_4I$	309.92	1, 123	2.350	1.6212^{20}	6	152^{26mm}	none	
d451	1,2-Diiodoethane	ICH_2CH_2I	281.86	1, 99	2.132^{10}		81–84	200		sl s aq; s alc, eth
d452	Diiodomethane	CH_2I_2	267.84	1, 71	3.325^{20}_{4}	1.7425^{20}	6	181	>110	0.12 aq; misc alc, bz, eth, PE
d453	1,5-Diiodopentane	$I(CH_2)_5I$	323.94	1, 133	2.177	1.6002^{20}		102^{3mm}	>110	
d454	1,3-Diiodopropane	$I(CH_2)_3I$	295.88	1, 115	2.5755^{20}_{4}	1.6423^{20}	−13	222	>110	i aq; s chl, eth
d455	Diisobutylaluminum chloride	$[(CH_3)_2CHCH_2]_2AlCl$	176.67	4^4, 4403	0.905	1.4506^{20}	−40	152^{10mm}	−18	
d456	Diisobutylaluminum hydride	$[(CH_3)_2CHCH_2]_2AlH$	142.22	4^4, 4400	0.798			118^{1mm}	−18	
d457	Diisobutylamine	$[(CH_3)_2CHCH_2]_2NH$	129.25	4, 166	0.740	1.4081^{20}	−77	137–139	29	s alc, acet, eth, chl
d458	Diisobutyl ether	$[(CH_3)_2CHCH_2]_2O$	130.22		0.761^{15}			122–124	8	i aq; misc alc, eth
d459	Diisobutyl hexanedioate	$[(CH_3)_2CHCH_2O_2CCH_2CH_2\text{-}]_2$	258.36		0.950^{25}_{25}				160	
d460	Diisobutyl *o*-phthalate	$C_6H_4[CO_2CH_2CH(CH_3)_2]_2$	278.35	9^2, 587	1.038^{25}_{25}	1.4900^{20}			174	
d461	1,6-Diisocyanatohexane	$OCN(CH_2)_6NCO$	168.20	4^2, 711	1.040	1.4525^{20}		255	140	
d462	Diisodecyl phenyl phosphite	$(C_{10}H_{21}O)_2P(O)C_6H_5$	438.64		0.940	1.4800^{20}		176^{5mm}		
d463	Diisoheptyl *o*-phthalate	$C_6H_4(CO_2C_7H_{15})_2$			0.990	1.4860^{20}			>110	

d464	Diisononyl *o*-phthalate	$C_6H_4(CO_2C_9H_{19})_2$			0.972	1.4850^{20}			>110	
d465	Diisooctyl nonanedioate	$C_8H_{17}O_2C(CH_2)_7CO_2C_8H_{17}$	412.66		0.905	1.4510^{10}		210^{2mm}	>110	
d466	Diisooctyl *o*-phthalate	$C_6H_4(CO_2C_8H_{17})_2$	390.56		0.983	1.4860^{20}			>110	
d466a	Diisopentyl ether	$[(CH_3)_2CHCH_2CH_2]_2O$	158.28		0.7777^{20}	1.4085^{20}		172.5		
d467	1,3-Diisopropenylbenzene	$C_6H_4[C(CH_3){=}CH_2]_2$	158.25		0.925	1.5571^{20}		231	91	
d468	Diisopropylamine	$[(CH_3)_2CH]_2NH$	101.19	4, 154	0.7153^{20}	1.3924^{20}	−61	83.5	−1	11 aq; s alc
d469	2-(Diisopropylamino)ethanol	$[(CH_3)_2CH]_2NCH_2CH_2OH$	145.25	4^1, 430	0.826	1.4417^{20}		187–192	57	
d470	3-Diisopropylamino-1,2-propanediol	$[(CH_3)_2CH]_2NCH_2CH(OH)CH_2OH$	175.27		0.962	1.4583^{20}		131^{10mm}	>110	
d471	2,6-Diisopropylaniline	$[(CH_3)_2CH]_2C_6H_3NH_2$	177.29	12, 168	0.940	1.5332^{20}	−45	257	123	
d472	Diisopropyl azodicarboxylate	$(CH_3)_2CHO_2CNCO_2CH(CH_3)_2$	202.21		1.027	1.4200^{20}		$75^{0.25mm}$	106	
d473	1,3-Diisopropylbenzene	$C_6H_4[CH(CH_3)_2]_2$	162.28	5, 447	0.856_4^{20}	1.4890^{20}	−63	203	76	misc alc, bz, eth, acet
d474	1,4-Diisopropylbenzene	$C_6H_4[CH(CH_3)_2]_2$	162.28	5^2, 339	0.857_4^{20}	1.4889^{20}	−17	204	76	misc alc, bz, acet, eth
d475	Diisopropylcyanamide	$[(CH_3)_2CH]_2NCN$	126.20	4^3, 279	0.839	1.4270^{20}		93^{25mm}	78	
d476	Diisopropyl ether	$[(CH_3)_2CH]_2O$	102.17	1, 362	0.7258_4^{20}	1.3679^{20}	−86.9	68.4	−28	1.2 aq; misc alc, bz, chl, eth
d477	*N,N*-Diisopropylethylamine	$[(CH_3)_2CH]_2NC_2H_5$	129.25	4, *4*, 511	0.742	1.4133^{20}	< −50	127	10	
d478	Diisopropyl malonate	$(CH_3)_2CHO_2CCH_2CO_2CH(CH_3)_2$	188.22	2^3, 1620	0.991	1.4120^{20}		95^{12mm}	88	
d479	2,6-Diisopropylphenol	$[(CH_3)_2CH]_2C_6H_3OH$	178.28	6^1, 272	0.962	1.5140^{20}	18	256	110	
d480	Diisopropyl phosphite	$[(CH_3)_2CHO]_2P(O)H$	166.16	1, 363	0.997	1.4070^{20}		$72–75^{20}$	>110	
d481	(+)-Diisopropyl L-tartrate	$[-CH(OH)CO_2CH(CH_3)_2]_2$	234.25	3, 517	1.114	1.4387^{20}		152^{12mm}	109	
d482	1,3-Diisopropyl-2-thiourea	$(CH_3)_2CHNHCSNHCH(CH_3)_2$	160.28	4, 155			143–145			
d483	Diketene		84.07	17^3, 4297	1.090	1.4330^{20}		127	34	
d484	*threo*-1,4-Dimercapto-2,3-butanediol	$HSCH_2CH(OH)CH(OH)CH_2SH$	154.25				42.43			v s aq, alc, chl, eth
d485	2,3-Dimercapto-1-propanol	$HSCH_2CH(SH)CH_2OH$	124.22		1.2385_4^{25}	1.5270^{25}		120^{15mm}	>110	8 aq(dec); s alc, eth

(Continued)

TABLE 2.20 Physical Constants of Organic Compounds (*Continued*)

No.	Name	Formula	Formula weight	Beilstein reference	Density, g/mL	Refractive index	Melting point, °C	Boiling point, °C	Flash point, °C	Solubility in 100 parts solvent
d486	2,5-Dimercapto-1,3,4-thiadiazole		150.24	27, 677			162 dec			
d487	3′4′-Dimethoxyacetophenone	$(CH_3O)_2C_6H_3COCH_3$	180.20	8^2, 298			49–51	286–288	>110	sl s aq, alc, eth
d488	2,4-Dimethoxyaniline	$(CH_3O)_2C_6H_3NH_2$	153.18	13, 784	1.075		34–37		>110	s alc, bz, eth
d489	2,5-Dimethoxyaniline	$(CH_3O)_2C_6H_3NH_2$	153.18	13, 788			80–82	270		s aq, alc
d490	3,4-Dimethoxyaniline	$(CH_3O)_2C_6H_3NH_2$	153.18	13, 780			88	176^{22mm}		s hot eth
d491	2,5-Dimethoxybenzaldehyde	$(CH_3O)_2C_6H_3CHO$	166.18	8, 245			49–52	146^{10mm}	>110	
d492	3,4-Dimethoxybenzaldehyde	$(CH_3O)_2C_6H_3CHO$	166.18	8, 255			42–43	281	>110	v s alc, eth
d493	1,2-Dimethoxybenzene	$C_6H_4(OCH_3)_2$	138.17	6, 771	1.0819^{25}	1.5232^{25}	22.5	206.3	87	sl s aq; s alc, eth
d494	1,3-Dimethoxybenzene	$C_6H_4(OCH_3)_2$	138.17	6, 813	1.055	1.5240	−55	87^{7mm}	87	s alc, bz, eth
d495	1,4-Dimethoxybenzene	$C_6H_4(OCH_3)_2$	138.17	6, 843	1.036^{65}_{8}		55–60	213		s alc; v s bz, eth
d496	3,4-Dimethoxybenzoic acid	$(CH_3O)_2C_6H_3CO_2H$	182.18	10^1, 188			180–181			0.05 aq; v s alc, eth
d497	3,5-Dimethoxybenzoic acid	$(CH_3O)_2C_6H_3CO_2H$	182.18	10, 405			182–184			
d498	2,6-Dimethoxybenzoyl chloride	$(CH_3O)_2C_6H_3COCl$	200.62	10^3, 1402			64–66			
d499	3,4-Dimethoxybenzyl alcohol	$(CH_3O)_2C_6H_3CH_2OH$	168.19	5, 1113	1.157	1.5520^{20}		297^{732mm}	>110	
d500	2,2-Dimethoxycyclohexanol	$(CH_3O)_2C_6H_9OH$	160.22		1.072	1.4620^{20}		90^{9mm}	40	
d501	2,5-Dimethoxy-2,5-dihydrofuran		130.14		1.073	1.4339^{20}		160–162	47	
d502	Dimethoxydimethylsilane	$(CH_3O)_2Si(CH_3)_2$	120.23		0.880	1.3690^{20}		81.4	10	
d503	Dimethoxydiphenylsilane	$(C_6H_5)_2Si(OCH_3)_2$	244.4		1.0771^{20}_{4}	1.5447^{20}		161^{15mm}		
d504	1,1-Dimethoxyethane	$CH_2CH(OCH_3)_2$	90.12	1, 603	0.8502^{20}	1.3668^{20}	−113	64.5	−17	s aq, alc, chl, eth
d505	1,2-Dimethoxyethane	$CH_3OCH_2CH_2OCH_3$	90.12	1, 467	0.8620^{20}_{4}	1.3796^{20}	−68	85.2	1	misc aq, alc; s PE
d506	(2,2′-Dimethoxy)ethylamine	$H_2NCH_2CH(OCH_3)_2$	105.14	4^2, 758	0.965	1.4170^{20}		135^{95mm}	53	
d507	Dimethoxymethane	$CH_2(OCH_3)_2$	76.10	1, 574	0.8601^{20}_{20}	1.3514^{20}	−104.8	42.3	−32	32 aq
d508	1,1-Dimethoxy-2-methylaminoethane	$CH_3NHCH_2CH(OCH_3)_2$	119.16	4^2, 759	0.928	1.4115^{20}		140	29	

d509	Dimethoxymethylvinylsilane	$CH_3Si(OCH_3)_2CH{=}CH_2$	132.24		0.884	1.3950^{20}		106	3	
d510	Dimethoxymethylphenylsilane	$(CH_3O)_2Si(CH_3)C_6H_5$	182.3		0.993^{20}_{4}	1.469^{20}		199–200		
d511	1,2-Dimethoxy-4-nitrobenzene	$(CH_3O)_2C_6H_3NO_2$	183.16	6, 789	1.1888^{133}_{4}		95–98	230^{17mm}		v s alc, eth; s chl
d512	2,6-Dimethoxyphenol	$(CH_3O)_2C_6H_3OH$	154.17	6, 1081			53–56	261	>110	s alc, alk; v s eth
d513	3,4-Dimethoxyphenylacetic acid	$(CH_3O)_2C_6H_3CO_2H$	196.20	10, 409			96–98			s aq; v s alc, eth
d514	3,4-Dimethoxyphenylacetonitrile	$(CH_3O)_2C_6H_3CN$	177.20	10^1, 198			62–63	178^{10mm}		
d515	2,2-Dimethoxy-2-phenylacetophenone	$C_6H_5C(O)C(OCH_3)_2C_6H_5$	256.30				67–70			
d516	1,1-Dimethoxy-2-phenylethane	$C_6H_5CH_2CH(OCH_3)_2$	166.22	7, 293	1.004	1.4950^{20}		221	83	
d517	2-(3,4-Dimethoxyphenyl)ethylamine	$(CH_3O)_2C_6H_3CH_2CH_2NH_2$	181.24	13, 800	1.074	1.5464^{20}		188^{15mm}	>110	
d518	1,2-Dimethoxypropane	$CH_3CH(OCH_3)CH_2OCH_3$	104.15	1^4, 2471	0.855	1.3835^{20}		96	0	
d519	2,2-Dimethoxypropane	$(CH_3)_2C(OCH_3)_2$	104.15	1, 648	0.847	1.3780^{20}		83	−11	
d520	1,1-Dimethoxy-2-propanone	$CH_3C(O)CH(OCH_3)_2$	118.13	1^1, 395	0.976	1.3978^{20}		143–147	37	
d521	3,3-Dimethoxy-1-propene	$(CH_3O)_2CHCH{=}CH_2$	102.13	1^1, 378	0.862	1.3954^{20}		89–90	−2	
d522	1,2-Dimethoxy-4-propenylbenzene	$CH_3CH{=}CHC_6H_5(OCH_3)_2$	178.23	6, 956	1.055	1.5680^{20}		262–264	>110	
d523	3,3-Dimethoxypropionitrile	$(CH_3O)_2CHCH_2CN$	115.13	3^4, 521	1.026	1.4130^{20}		92^{30mm}	86	
d524	2,6-Dimethoxypyridine	$(CH_3O)_2C_5H_3N$	139.15		1.053	1.5129^{20}		178–180	61	
d525	2,5-Dimethoxytetrahydrofuran	$(CH_3O)_2C_4H_6O$	132.16		1.020	1.4180^{20}		145–147	35	
d526	*N,N*-Dimethylacetamide	$CH_3C(O)N(CH_3)_2$	87.12	4, 59	0.9366^{25}	1.4376^{20}	−20	165.5	70	misc aq, alc, bz, eth
d527	2′,6′-Dimethylacetanilide	$CH_3C(O)NHC_6H_3(CH_3)_2$	163.22	12, 1109			182–184			
d528	Dimethyl 1,3-acetonedicarboxylate	$[CH_3O_2CCH_2]_2C{=}O$	174.15	3, 790	1.185	1.4434^{20}		150^{25mm}	>110	

(Continued)

TABLE 2.20 Physical Constants of Organic Compounds (*Continued*)

No.	Name	Formula	Formula weight	Beilstein reference	Density, g/mL	Refractive index	Melting point, °C	Boiling point, °C	Flash point, °C	Solubility in 100 parts solvent
d529	Dimethyl acetylenedicarboxylate	$CH_3O_2CC{\equiv}CCO_2CH_3$	142.11	2, 803	1.156	1.4470^{20}		98^{19mm}	86	
d530	Dimethyl acetylsuccinate	$CH_3O_2CC_2CH(COCH_3)$-CO_2CH_3	188.18	3^4, 1825	1.160		33	134^{12mm}	>110	
d531	*N,N*-Dimethylacrylamide	$H_2C{=}CHC(O)N(CH_3)_2$	99.13	4^3, 130	0.962	1.4730^{20}		81^{20mm}	71	
d532	3,3-Dimethylacrylic acid	$(CH_3)_2C{=}CHCO_2H$	100.12	2, 432			69	195		
d533	Dimethylaluminum chloride	$(CH_3)_2AlCl$	92.51	4^3, 1971	0.996		−21	126–127	−18	
d534	Dimethylamine	$(CH_3)_2NH_2$	45.08	4, 39	0.680^0_4	1.350^{17}	−92.2	6.9	20	v s aq; s alc, eth
d535	Dimethylaminoacetonitrile	$(CH_3)_2NCH_2CN$	84.12	4, 346	0.863	1.4101^{20}		138	36	
d536	4-(Dimethylamino)benzaldehyde	$(CH_3)_2NC_6H_4CHO$	149.19	14, 31			74	176^{17mm}		s alc, chl, eth, HOAc
d537	3-Dimethylaminobenzoic acid	$(CH_3)_2NC_6H_4CO_2H$	165.19	14, 392			148–152			
d538	4-Dimethylaminobenzoic acid	$(CH_3)_2NC_6H_4CO_2H$	165.19	14, 426			241 dec			s alc; sl s eth
d539	2-(Dimethylamino)ethanol	$(CH_3)_2NCH_2CH_2OH$	89.14	4, 276	0.8876^{20}_4	1.4294^{20}		135	40	misc aq, alc, eth
d540	2-[2-(Dimethylamino)ethoxy]ethanol	$(CH_3)_2NCH_2CH_2OCH_2$-CH_2OH	133.19	4^2, 719	0.954	1.4420^{20}		95^{15mm}	92	
d541	2-(Dimethylamino)ethyl acrylate	$H_2C{=}CHCO_2CH_2CH_2$-$N(CH_3)_2$	143.19	4^3, 649	0.943	1.4280^{20}		64^{12mm}	58	
d542	2-(Dimethylamino)ethyl benzoate	$C_6H_5CO_2CH_2CH_2N(CH_3)_2$	193.26		1.014	1.5077^{20}		159^{20mm}	>110	
d543	2-(Dimethylamino)ethyl methacrylate	$H_2C{=}C(CH_3)CO_2CH_2CH_2$-$N(CH_3)_2$	157.22	4^3, 649	0.933	1.4400^{20}		182–192	70	
d544	3-Dimethylaminophenol	$(CH_3)_2NC_6H_4OH$	137.18	13, 405	1.5895^{25}		82–84	265–268		v s alc, bz, eth, acet
d545	3-Dimethylamino-1,2-propanediol	$(CH_3)_2NCH_2CH(OH)CH_2OH$	119.16	4, 302	1.004	1.4609^{20}		216–217	105	s aq, alc, chl, eth
d546	1-Dimethylamino-2-propanol	$CH_3CH(OH)CH_2N(CH_3)_2$	103.17		0.837	1.4193^{20}		121–127	35	
d547	3-Dimethylamino-1-propanol	$(CH_3)_2NCH_2CH_2OH$	103.17	4^1, 433	0.872	1.4360^{20}		163–164	36	

d548	3-(Dimethylamino)-propionitrile	$(CH_3)_2NCH_2CH_2CN$	98.15	4^3, 1265	0.870	1.4258^{20}	−43	171^{750mm}	62	
d549	3-Dimethylamino-propylamine	$(CH_3)_2N(CH_2)_3NH_2$	102.18	4^3, 554	0.812	1.4350		133	15	
d550	*N*-[3-(Dimethylamino)-propyl]methacryl-amide	$H_2C{=}C(CH_3)CONH(CH_2)_3N(CH_3)_2$	170.26		0.940	1.4790^{20}		134^{2mm}	>110	
d551	4-(Dimethylamino)-pyridine	$(CH_3)_2N(C_5H_4N)$	122.17	22^2, 341			112–114			v s aq, alc, bz, chl
d552	Dimethyl 2-amino-1,4-phthalate	$H_2NC_6H_3(CO_2CH_3)_2$	209.20	14, 559			127–130			
d553	*N,N*-Dimethylaniline	$C_6H_5N(CH_3)_2$	121.18	12, 141	0.9559^{20}_4	1.5584^{20}	2.5	194.2	63	v s alc, chl, eth
d554	2,3-Dimethylaniline	$(CH_3)_2C_6H_3NH_2$	121.18	12, 1101	0.9933^{20}	1.5685^{20}	< −15	221–222	97	sl s aq; s alc, eth
d555	2,4-Dimethylaniline	$(CH_3)_2C_6H_3NH_2$	121.18	12, 1111	0.9723^{20}	1.55686^{20}	−14.3	214	90	s alc, bz, eth
d556	2,5-Dimethylaniline	$(CH_3)_2C_6H_3NH_2$	121.18	12, 1135	0.9790^{21}_4	1.5592^{20}	15.5	214	93	sl s aq; s alc, eth
d557	2,6-Dimethylaniline	$(CH_3)_2C_6H_3NH_2$	121.18	12, 1107	0.9842^{20}	1.5601^{20}	11.2	215	96	sl s aq; s alc, eth
d558	3,4-Dimethylaniline	$(CH_3)_2C_6H_3NH_2$	121.18	12, 1103	1.076^{18}		51	228	98	sl s aq; s alc
d559	3,5-Dimethylaniline	$(CH_3)_2C_6H_3NH_2$	121.18	12, 1131	0.9706^{20}	1.5578^{20}	9.8	220.5	93	sl s aq; s alc
d560	Dimethylarsinic acid	$(CH_3)_2As(O)OH$	138.00	4, 610			195–196			v s alc; 200 aq; i eth
d561	1,3-Dimethylbarbituric acid		156.14	24, 471			124–126			
d562	*N,N*-Dimethylbenz-amide	$C_6H_5CON(CH_3)_2$	149.19	9, 201			43–45	133^{15mm}	>110	
d563	3,4-Dimethylbenzoic acid	$(CH_3)_2C_6H_3CO_2H$	150.18	9^2, 353			165–167	subl		s alc, bz
d564	2,5-Dimethylbenzo-nitrile	$(CH_3)_2C_6H_3CN$	131.18	9, 535	0.957	1.5284^{20}	13–14	223^{730mm}	92	
d565	*N,N*-Dimethylbenzyl-amine	$C_6H_5CH_2N(CH_3)_2$	135.21	12, 1019	0.900	1.5011^{20}	−75	183	54	
d566	2,3-Dimethyl-1,3-butadiene	$H_2C{=}C(CH_3)C(CH_3){=}CH_2$	82.15	1^3, 991	0.7222^{25}_4	1.4362^{25}	−76.0	69.2	−22	
d567	2,2-Dimethylbutane	$CH_3CH_2C(CH_3)_3$	86.18	1, 150	0.6492^{20}	1.3688^{20}	−99.9	49.7	−48	
d568	2,3-Dimethylbutane	$(CH_3)_2CHCH(CH_3)_2$	86.18	1, 151	0.6616^{20}	1.3750^{20}	−128.5	58.0	−29	
d569	2,3-Dimethyl-2,3-butanediol	$(CH_3)_2C(OH)C(OH)(CH_3)_2$	86.18	1, 487			41.1	174.4	77	v s hot aq, alc, eth
d570	2,3-Dimethyl-2-butanol	$(CH_3)_2CHC(CH_3)_2OH$	102.18	1, 413	0.8236^{20}_4	1.4176^{20}	−14	118	29	s aq; misc alc, eth
d570a	3,3-Dimethyl-1-butanol	$(CH_3)_3CCH_2CH_2OH$	102.18	1^3, 1677	0.824^{20}	1.4176^{20}	−60	143	47	

(Continued)

TABLE 2.20 Physical Constants of Organic Compounds (*Continued*)

No.	Name	Formula	Formula weight	Beilstein reference	Density, g/mL	Refractive index	Melting point, °C	Boiling point, °C	Flash point, °C	Solubility in 100 parts solvent
d571	3,3-Dimethyl-2-butanol	$(CH_3)_3CCH(OH)CH_3$	102.18	1, 412	0.8185^{20}_{4}	1.4151^{20}	5.6	120	28	s alc; misc eth
d572	3,3-Dimethyl-2-butanone	$(CH_3)_3CCOCH_3$	100.16	1, 694	0.7250^{25}_{25}	1.3939^{25}	−52.5	106	23	2.5 aq; s alc, eth
d572a	2,3-Dimethyl-1-butene	$(CH_3)_2CHC(CH_3){=}CH_2$	84.16	1^3, 816	0.680	1.3890^{20}	−157	55.6	−18	
d573	2,3-Dimethyl-2-butene	$(CH_3)_2C{=}C(CH_3)_2$	84.16	1, 218	0.7081^{20}_{4}	1.4124^{20}	−75	73	−16	s alc, eth
d574	3,3-Dimethyl-1-butene	$(CH_3)_3CCH{=}CH_2$	84.16	1.217	0.6531^{20}_{4}	1.3762^{20}	−115	41	−28	
d575	*N,N*-Dimethylbutylamine	$CH_3(CH_2)_3N(CH_3)_2$	101.19	4, *1*, 371	0.721	1.3980^{20}		93^{750mm}	−3	
d576	2,2-Dimethylbutyric acid	$C_2H_5C(CH_3)_2CO_2H$	116.16	2, 335	0.928	1.4154^{20}		96^{5mm}	79	
d577	3,3-Dimethylbutyric acid	$(CH_3)_3CCH_2CO_2H$	116.16	2, 337	0.9124^{20}_{4}	1.4100^{20}	6–7	190	88	s alc, eth
d578	Dimethylcadmium	$(CH_3)_2Cd$	142.48		1.9846^{17}_{4}	1.5488	−4.5	105.5	>150 explodes	dec aq; s PE
d579	Dimethylcarbamyl chloride	$(CH_3)_2NCOCl$	107.54	4, 73	1.168	1.4540^{20}	−33	168	68	
d580	Dimethyl carbonate	$(CH_3O)_2C{=}O$	90.08	3, 4	1.065^{17}_{4}	1.3682^{20}	0.5	90–91	18	i aq; misc alc, eth
d581	Dimethyl chloromalonate	$ClCH(CO_2CH_3)_2$	166.56	2, 592	1.305	1.4370^{20}		106^{19mm}	106	
d582	Dimethyl chlorothiophosphate	$(CH_3O)_2P(S)Cl$	160.56	1^1, 143	1.322	1.4819^{20}		67^{16mm}	105	
d583	Dimethylcyanamide	$(CH_3)_2NCN$	70.09	4, 74	0.867	1.4100^{20}		161–163	58	
d584	Dimethyl *N*-cyanothioiminocarbonate	$(CH_3S)_2C{=}NCN$	146.23	3, 220			46–50		110	
d584a	1,1-Dimethylcyclohexane	$(CH_3)_2C_6H_{10}$	112.22	5, 35	0.777	1.4280^{20}	−33	120	7	
d585	*cis*-1,2-Dimethylcyclohexane	$(CH_3)_2C_6H_{10}$	112.22	5, 36	0.7963^{20}	1.4335^{20}	−49.9	129.7	16	i aq; s alc, bz
d586	*trans*-1,2-Dimethylcyclohexane	$(CH_3)_2C_6H_{10}$	112.22	5, 36	0.7760^{20}	1.4273^{20}	−90	123.4	11	i aq; s alc, bz
d587	*cis*-1,3-Dimethylcyclohexane	$(CH_3)_2C_6H_{10}$	112.22	5, 36	0.784	1.4230^{20}	−76	120	5	
d587a	*trans*-1,3-Dimethylcyclohexane	$(CH_3)_2C_6H_{10}$	112.22	5^2, 21	0.780	1.4305^{20}	−90	124.5	7	
d588	*cis*-1,4-Dimethylcyclohexane	$(CH_3)_2C_6H_{10}$	112.22	5^2, 22	0.783	1.4297^{20}	−88	125	6	
d589	5,5-Dimethyl-1,3-cyclohexanedione		140.18	7, 559			dec 149			0.4 aq; s alc, bz

d590	2,3-Dimethylcyclohexanol	$(CH_3)_2C_6H_9OH$	128.22		0.934	1.4653^{20}			65	
d591	3,5-Dimethylcyclohexanol	$(CH_3)_2C_6H_9OH$	128.22	6, 18	0.892	1.4552	11–12	186	73	
d592	2,6-Dimethylcyclohexanone	$(CH_3)_2C_6H_8({=}O)$	126.20	7, 23	0.925	1.4460^{20}		175	51	i aq; s alc, eth
d593	*N,N*-Dimethylcyclohexylamine	$C_6H_{11}N(CH_3)_2$	127.23		0.849	1.4535^{20}		159	42	
d594	2,3-Dimethylcyclohexylamine	$(CH_3)_2C_6H_9NH_2$	127.23		0.835	1.4595^{20}		160	51	
d595	1,5-Dimethyl-1,5-cyclooctadiene		136.24		0.867	1.4896^{20}		74^{16mm}	55	
d596	Dimethyl 1,1-cyclopropanedicarboxylate	$C_3H_4(CO_2CH_3)_2$	158.16	9^1, 314	1.147	1.4410^{20}		196–198	95	
d597	Dimethyl decanedioate	$CH_3O_2C(CH_2)_8CO_2CH_3$	230.30	2, 719	0.983^{30}_{20}	1.4335^{28}	23	144^{5mm}	145	i aq; s alc, eth
d598	2,2-Dimethyl-1,3-dioxane-4,6-dione		144.13				94–96			s aq, acet
d599	2,2-Dimethyl-1,3-dioxolane-4-methanol		132.16	19, 65	1.063	1.4340^{20}		188–189	80	misc aq, alc, bz, esters, eth, PE, acetals
d600	Dimethyl disulfide	CH_3SSCH_3	94.20	1, 291	1.0625^{20}	1.5289^{20}	−84.7	109.8	24	i aq; misc alc, eth
d601	Dimethyldithiocarbamic acid, Zn salt	$[(CH_3)_2NCS_2]_2Zn$	305.80	4^3, 149	1.66		250–252			< 0.2 alc, eth; < 0.5 acet, bz; 0.5 naphtha
d602	*N,N*-Dimethyldodecylamine	$CH_3(CH_2)_{11}N(CH_3)_2$	213.41	4^3, 409	0.775	1.4375^{20}	−20	112^{3mm}	>110	
d603	Dimethyl ether	$(CH_3)_2O$	46.07	1, 281	0.661^{20}		−141.5	−24.9	−41	35 aq(5 atm); 15 bz; 11.8 acet
d604	*N,N*-Dimethylethylamine	$C_2H_5N(CH_3)_2$	73.14	4, 94	0.675	1.3720^{20}	−140	36–38	−36	
d605	*N,N*-Dimethylethylenediamine	$C_2H_5NCH_2CH_2NH_2$	88.15	4^2, 690	0.803	1.4260^{20}		106	23	
d606	*N,N*-Dimethylformamide	$(CH_3)_2NCHO$	73.10	4, 58	0.9445^{25}_{4}	1.4305^{20}	−60.4	153.0	57	misc aq, alc, bz, eth
d607	*N,N*-Dimethylformamide dimethyl acetal	$(CH_3)_2NCH(OCH_3)_2$	119.16		0.897	1.3972^{20}		103^{720mm}	7	

(Continued)

TABLE 2.20 Physical Constants of Organic Compounds (*Continued*)

No.	Name	Formula	Formula weight	Beilstein reference	Density, g/mL	Refractive index	Melting point, °C	Boiling point, °C	Flash point, °C	Solubility in 100 parts solvent
d608	Dimethyl fumarate	$CH_3O_2CCH{=}CHCO_2CH_3$	144.13	2, 741	1.045^{106}		105	193		sl s alc, eth
d609	2,5-Dimethylfuran	$(CH_3)_2(C_4H_2O)$	96.13	17, 41	0.9000^{20}_{4}	1.4414^{20}	−62	93	−1	i aq; misc alc, eth
d610	Dimethylglyoxime	$CH_3C({=}NOH)$-$C({=}NOH)CH_3$	116.12	1, 772			240			s alc, acet, eth, pyr
d611	2,4-Dimethyl-1,6-heptadienal	$H_2C{=}CHCH_2CH(CH_3)$-$CH{=}C(CH_3)CHO$	138.21		0.870	1.4664^{20}		47^{2mm}	64	
d612	2,4-Dimethyl-2,6-heptadien-1-ol	$H_2C{=}CHCH_2CH(CH_3)$-$CH{=}C(CH_3)CH_2OH$	140.23		1.351	1.4640^{20}		86^{10mm}	78	
d613	2,6-Dimethyl-2,5-heptadien-4-one	$(CH_3)_2C{=}CHC({=}O)$-$CH{=}(CH_3)_2$	138.21	1, 751	0.885^{20}_{4}	1.4968^{21}	28	198–199	79	sl s aq; s alc, eth
d613a	2,2-Dimethylheptane	$(CH_3)_3C(CH_2)_4CH_3$	128.26		0.7105^{20}	1.4016^{20}	−113	132.7		
d614	Dimethyl heptane-dioate	$CH_3O_2C(CH_2)_5CO_2CH_3$	188.22	2^1, 281	1.0625^{20}_{4}	1.4314^{20}	−21	122^{11mm}	>110	s alc
d615	2,6-Dimethyl-4-heptanol	$(CH_3)_2CHCH_2CH(OH)$-$CH_2CH(CH_3)_2$	144.26	1, 425	0.809	1.4236^{20}		178	66	
d616	2,6-Dimethyl-4-heptanone	$[(CH_3)_2CHCH_2]_2C{=}O$	142.24	1, 710	0.806^{20}_{20}	1.4114^{20}	−41.5	169.4	49	0.06 aq; misc alc, bz, chl, eth
d616a	2,4-Dimethylhexane	$C_2H_5CH(CH_3)CH_2CH(CH_3)_2$	114.23	1, 162	0.6962^{25}	1.3929^{25}		109.5	10	
d617	Dimethyl hexanedioate	$CH_3O_2C(CH_2)_4CO_2CH_3$	174.20	1, 652	1.0600^{20}_{4}	1.4285^{20}	8	112^{10mm}	107	i aq; s alc, eth
d618	2,5-Dimethyl-2,5-hexanediol	$[(CH_3)_2C(OH)CH_2\text{-}]_2$	146.23	1, 492			86–90	214–215	126	
d619	1,5-Dimethylhexyl-amine	$(CH_3)_2CH(CH_2)_3$-$CH(NH_2)CH_3$	129.25	Merck: 11, 6678	0.767	1.4209^{20}		154–156	48	
d620	2,5-Dimethyl-3-hexyne-2,5-diol	$(CH_3)_2C(OH)C{\equiv}C$-$C(OH)(CH_3)_2$	142.20	1, 501			94–95	205–206		
d621	3,5-Dimethyl-1-hexyn-3-ol	$(CH_3)_2CHCH_2C(CH_3)(OH)C{\equiv}CH$	126.20	1^2, 507	0.859	1.4335^{20}		151	44	
d622	5,5-Dimethylhydantoin		128.13	24, 289			176–178			v s aq, alc, bz, chl, eth, acet
d623	1,1-Dimethylhydrazine	$(CH_3)_2NNH_2$	60.10	4, 547	0.791^{22}_{4}	1.4075^{20}	−58	63.9	1	misc aq, alc, eth, PE
d624	1,2-Dimethylhydrazine	$CH_3NHNHCH_3$	60.10	4, 547	0.8274^{20}_{4}	1.4209^{20}		81	flam-mable	misc aq, alc, eth, PE
d625	Dimethyl hydrogen phosphonate	$(CH_3O)_2P(O)H$	110.05	1, 285	1.200^{20}_{4}	1.4009^{20}		170–171	29	s aq(hyd); misc alc, acet, eth
d626	1,2-Dimethylimidazole		96.13	23, 66	1.084		29–30	204	92	
d627	1,3-Dimethyl-2-imidazolidinone		114.15		1.044	1.4720^{20}		108^{17mm}	80	

d628	*N,N*-Dimethyliso-propylamine	$(CH_3)_2CHN(CH_3)_2$	87.17	4^2, 630	0.715	1.3905^{20}		66	−9	
d629	Dimethyl maleate	$CH_3O_2CCH{=}CHCO_2CH_3$	144.13	2, 751	1.1606^{20}	1.4422^{20}	−19	202	113	8.7 aq
d630	Dimethyl malonate	$CH_3O_2CCH_2CO_2CH_3$	132.12	2, 572	1.154^{20}_{4}	1.4135^{20}	−62	180–181	90	sl s aq; misc alc, eth
d631	Dimethylmercury	$(CH_3)_2Hg$	230.66	4, 678	3.1874^{20}	1.5452^{20}	−43	92–94	5	i aq; s alc, eth
d632	3,4-Dimethyl-1-methoxybenzene	$(CH_3)_2C_6H_3OCH_3$	136.19	6, 481	0.9744^{14}_{4}	1.5198^{14}		200		i aq; s alc, bz, eth
d633	3,5-Dimethyl-1-methoxybenzene	$(CH_3)_2C_6H_3OCH_3$	136.19	6, 493	0.9627^{15}_{4}	1.5107^{15}		193	65	i aq; s alc, bz, eth
d634	Dimethyl methyl-malonate	$CH_3CH(CO_2CH_3)_2$	146.14	2, 628	1.098	1.4140^{20}		176–177	76	
d635	Dimethyl methyl-phosphonate	$(CH_3O)_2P(O)CH_3$	124.08	4^1, 572	1.145	1.4130^{20}		181	68	
d636	Dimethyl methyl-succinate	$CH_3O_2CCH_2CH(CH_3)$-CO_2CH_3	160.17	2^3, 1696	1.076	1.4200^{20}		196	83	
d637	2,6-Dimethyl-morpholine		115.18		0.9346^{20}	1.4470^{20}	−85	147	48	misc aq, alc, bz
d637a	1,2-Dimethylnaphtha-lene	$C_{10}H_6(CH_3)_2$	156.23	5^1, 267	1.0179^{20}	1.6166^{20}	0.8	266.5	>110	
d638	1,2-Dimethyl-3-nitro-benzene	$(CH_3)_2C_6H_3NO_2$	151.17	5, 367	1.129	1.5434^{20}	7–9	245	107	i aq; s alc
d639	1,2-Dimethyl-4-nitro-benzene	$(CH_3)_2C_6H_3NO_2$	151.17	5, 368	1.139		29–31	143^{20mm}	>110	i aq; s alc
d640	1,3-Dimethyl-2-nitro-benzene	$(CH_3)_2C_6H_3NO_2$	151.17	5, 378	1.112	1.5220^{20}	14–16	225^{744mm}	87	i aq; s alc
d641	1,3-Dimethyl-4-nitro-benzene	$(CH_3)_2C_6H_3NO_2$	151.17	5, 378	1.117	1.5497^{20}	2	237–239	107	s alc, bz, chl, eth
d642	*N,N*-Dimethyl-4-nitrosoaniline	$(CH_3)_2NC_6H_4NO$	150.18	12, 677			86	flammable solid		i aq; s alc, eth
d643	Dimethyl 2-nitro-1,4-phthalate	$O_2NC_6H_3$-1,4-$(CO_2CH_3)_2$	239.18	9, 826			72–75			
d644	*cis*-3,7-Dimethyl-2,6-octadienal		152.24		0.8888^{20}_{4}	1.4898^{20}		229	101	misc alc, eth, glyc
d645	*trans*-3,7-Dimethyl-2,6-octadienal		152.24		0.8869^{20}_{4}	1.4869^{20}		229	101	misc alc, eth, glyc
d646	3,7-Dimethyl-1-octanol	$(CH_3)_2CH(CH_2)_3CH(CH_3)$-$CH_2CH_2OH$	158.29	1, 426	0.840	1.4355^{20}		96^{9mm}	95	
d647	3,7-Dimethyl-3-octanol	$(CH_3)_2CH(CH_2)_3$-$C(OH)(CH_3)C_2H_5$	158.29	1, 426	0.826	1.4336^{20}		73^{6mm}	76	

(Continued)

TABLE 2.20 Physical Constants of Organic Compounds (*Continued*)

No.	Name	Formula	Formula weight	Beilstein reference	Density, g/mL	Refractive index	Melting point, °C	Boiling point, °C	Flash point, °C	Solubility in 100 parts solvent
d648	2,6-Dimethyl-2,4,6-octatriene	$CH_3CH{=}C(CH_3)CH{=}CH\text{-}CH{=}C(CH_3)_2$	136.24	1^{3}, 1050	0.811	1.5429^{20}		75^{14mm}	68	
d649	*N,N*-Dimethyloctylamine	$CH_3(CH_2)_7N(CH_3)_2$	157.30	4^{1}, 386	0.765	1.4243^{20}	−57	195	65	
d650	3,6-Dimethyl-4-octyne-3,6-diol	$C_2H_5C(CH_3)(OH)C{\equiv}C\text{-}C(CH_3)(OH)C_2H_5$	170.35	1^{1}, 263			53–55	214^{680mm}	>110	
d651	Dimethyl octanedioate	$CH_3O_2C(CH_2)_6CO_2CH_3$	202.25	2, 693	1.0210^{20}_{4}	1.4325^{20}	−4.8	268		i aq; s alc
d652	Dimethyl oxalate	$CH_3O_2CCO_2CH_3$	118.09	2, 534	1.148^{54}	1.379^{80}	50–54	163.5	75	6 aq; s alc, eth
d653	3,3-Dimethyloxetane		86.13	17^{2}, 21	0.835	1.3990		81	−9	
d654	2,3-Dimethylpentane	$C_2H_5CH(CH_3)CH(CH_3)_2$	100.21	1^{2}, 120	0.6951^{20}_{4}	1.3920^{20}		89.8	<−7	i aq; s alc, eth
d655	2,4-Dimethylpentane	$(CH_3)_2CHCH_2CH(CH_3)_2$	100.21	1, 158	0.6727^{20}	$1{,}3815^{20}$	−120	80.4	−12	
d656	Dimethyl pentanedioate	$CH_3O_2C(CH_2)_3CO_2CH_3$	160.17	2, 633	1.0876^{20}	1.4244^{20}	−42.5	214	102	v s alc, eth
d657	2,4-Dimethyl-3-pentanol	$(CH_3)_2CHCH(OH)CH(CH_3)_2$	116.20	1, 417	0.829^{20}_{4}	1.4254^{20}		140	37	sl s aq; s alc, eth
d658	2,4-Dimethyl-3-pentanone	$(CH_3)_2CHC(O)CH(CH_3)_2$	114.19	1, 703	0.8062^{20}_{4}	1.3986^{20}	−69	125	15	
d659	2,3-Dimethylphenol	$(CH_3)_2C_6H_3OH$	122.17	6, 480		1.5420^{20}	72.8	217		v s alc, bz, chl, eth
d660	2,4-Dimethylphenol	$(CH_3)_2C_6H_3OH$	122.17	6, 486	1.0276^{14}_{4}	1.5420^{14}	24.5	211	>110	v s alc, bz, chl, eth
d661	2,5-Dimethylphenol	$(CH_3)_2C_6H_3OH$	122.17	6, 494	0.965^{80}		74.5	211.5		v s alc, bz, chl, eth
d662	2,6-Dimethylphenol	$(CH_3)_2C_6H_3OH$	122.17	6, 485			45.7	201	73	v s alc, bz, chl, eth
d663	3,4-Dimethylphenol	$(CH_3)_2C_6H_3OH$	122.17	6, 480	0.9830^{20}		60.8	227		v s alc, bz, chl, eth
d664	3,5-Dimethylphenol	$(CH_3)_2C_6H_3OH$	122.17	6, 492	0.9680^{20}		64	222		v s alc, bz, chl, eth
d665	*N,N*-Dimethyl-1,4-phenylenediamine	$(CH_3)_2NC_6H_4NH_2$	136.20	13, 72			36	262	90	v s aq; s alc
d666	4,4-Dimethyl-2-phenyl-2-oxazoline		175.23	27^{4}, 1114	1.025	1.5322^{20}	20–24	124^{20mm}	102	
d667	2,2-Dimethyl-3-phenyl-1-propanol	$C_6H_5CH_2C(CH_3)_2CH_2OH$	164.25				35	126^{15mm}	109	
d668	Dimethyl 1,2-phthalate	$C_6H_4(CO_2CH_3)_2$	194.19	9, 797	1.1905^{20}	1.5138^{20}	5.5	283.7	146	0.4 aq; misc alc, chl, eth; i PE
d669	Dimethyl 1,3-phthalate	$C_6H_4(CO_2CH_3)_2$	194.19	9, 834	1.194^{20}_{4}	1.5168^{20}	67–68	282		i aq
d670	Dimethyl 1,4-phthalate	$C_6H_4(CO_2CH_3)_2$	194.19	9, 843			140–142	288		0.3 hot aq; s hot alc; s eth
d671	1,4-Dimethylpiperazine		114.19	23, 7	0.844	1.4463^{20}		132^{750mm}	18	
d672	*cis*-2,6-Dimethylpiperidine		113.20	20, 108	0.840	1.4394^{20}		127	11	

d673	2,2-Dimethylpropane	$(CH_3)_4C$	72.15	Merck: 12, 6545	0.613^{0}	1.3476^{6}	−16.6	9.5	−65	
d674	2,2-Dimethyl-1,3-propanediamine	$H_2NCH_2C(CH_3)_2CH_2NH_2$	102.18	4^3, 595	0.851	1.4566^{20}	31	154	47	
d675	2,2-Dimethyl-1,3-propanediol	$(CH_3)_2C(CH_2OH)_2$	104.15	1, 483	1.11^{25}		127–128	208–210	107	180 aq; 12 bz; 60 acet; v s alc, eth
d676	2,2-Dimethyl-1-propanol	$(CH_3)_3CCH_2OH$	88.15	1, 406	0.812^{20}_{4}		52.5	113.1	36	3.6 aq; misc alc, eth
d677	2,2-Dimethylpropionaldehyde	$(CH_3)_3CCHO$	186.25		0.793	1.3794^{20}	6	74^{730mm}	<1	
d678	*N,N*-Dimethylpropionamide	$C_2H_5C(O)N(CH_3)_2$	101.15	4^3, 126	0.920	1.4400^{20}	−45	175	62	
d679	2,2-Dimethylpropionic acid	$(CH_3)_3CCO_2H$	102.13	2, 319	0.905^{50}	1.3931^{37}	35.5	163.8	63	2.5 aq; v s alc, eth
d680	2,2-Dimethylpropionic anhydride	$[(CH_3)_3DD(O)]_2O$	186.25	2, 320	0.918	1.4092^{20}		193	57	
d681	2,2-Dimethylpropionyl chloride	$(CH_3)_3CC(O)Cl$	120.58	2, 320	0.979	1.4120^{20}		105–106	<1	dec aq, alc; v s eth
d682	1,1-Dimethylpropylamine	$CH_3CH_2C(CH_3)_2NH_2$	87.17	4, 179	0.731^{25}_{4}	1.3996^{20}	−105	77	65	misc aq, alc, eth
d683	1,1-Dimethyl-2-propynylamine	$HC{\equiv}CC(CH_3)_2NH_2$	83.13		0.790	1.4235^{20}		79–80	2	
d684	3,5-Dimethylpyrazole		96.13	23, 74			108	218		s aq; v s bz, eth
d685	2,3-Dimethylpyridine	$(CH_3)_2(C_5H_3N)$	107.16	20, 243	0.945	1.5080	−15	163	50	
d686	2,4-Dimethylpyridine	$(CH_3)_2(C_5H_3N)$	107.16	20, 244	0.9309^{20}	1.5010^{20}	< −64	158.3	37	17 aq; v s alc, bz, eth
d687	2,6-Dimethylpyridine	$(CH_3)_2(C_5H_3N)$	107.16	20, 244	0.9226^{20}	1.4956^{20}	−6.0	144	33	43 aq^{45}; s alc, eth
d688	3,4-Dimethylpyridine	$(CH_3)_2(C_5H_3N)$	107.16	20, 246	0.954^{25}_{4}	1.5100^{25}	−12	164	53	sl s aq; s alc, eth
d689	3,5-Dimethylpyridine	$(CH_3)_2(C_5H_3N)$	107.16	20, 246	0.939^{25}_{4}	1.5033^{25}	−9	170	53	s aq, alc, eth
d690	Dimethyl pyrocarbonate	$O(CO_2CH_3)_2$	134.09	3^4, 17	1.250	1.3933^{20}		46^{5mm}	80	
d691	Dimethyl succinate	$CH_3O_2CCH_2CH_2CO_2CH_3$	146.14	2, 609	1.1198^{20}	1.4190^{20}	19	196.4	85	0.83 aq; 2.9 alc
d692	Dimethylsulfamoyl chloride	$(CH_3)_2NSO_2Cl$	143.59	4, 84	1.337	1.4518^{20}		114^{75mm}	94	
d693	Dimethyl sulfate	$(CH_3O)_2SO_2$	126.13	1, 283	1.3322^{20}_{4}	1.3874^{20}	−31.8	188 dec	83	2.8 aq(hyd); s acet, bz, dioxane, eth
d694	Dimethyl sulfide	$(CH_3)_2S$	62.13	1, 288	0.8483^{20}	1.4438^{20}	−98.3	37.3	−36	2 aq; s alc, eth
d695	Dimethyl sulfite	$(CH_3O)_2SO$	110.13	1, 282	1.294	1.4083^{20}		126–127	30	
d696	Dimethyl sulfone	$(CH_3)_2SO_2$	94.13	1, 289			109	238	143	v s aq, alc, acet
d697	Dimethyl sulfoxide	$(CH_3)_2SO$	78.13	1, 289	1.101^{20}_{4}	1.4170^{20}	18.5	189.0	95	s alc, acet, bz, chl

(Continued)

TABLE 2.20 Physical Constants of Organic Compounds (*Continued*)

No.	Name	Formula	Formula weight	Beilstein reference	Density, g/mL	Refractive index	Melting point, °C	Boiling point, °C	Flash point, °C	Solubility in 100 parts solvent
d698	Dimethyl-d_6 sulfoxide	$(CD_3)_2SO$	84.18	1^4, 1279	1.190	1.4758^{20}		55^{5mm}	95	
d699	(+)-Dimethyl L-tartrate	$CH_3O_2CH(OH)CH(OH)-CO_2CH_3$	178.14	3, 510	1.328^{20}_{4}		48–50	163^{23mm}	>110	s aq; 200 alc[15]; v s bz
d700	Dimethyltelluride	$(CH_3)_2Te$	157.68	1, 291			−10	91–92		dec aq; v s alc; i eth
d701	2,5-Dimethyltetrahydrofuran	$(CH_3)_2(C_4H_2O)$	100.16	17, 14	0.833	1.4041		90–92	26	
d702	1,3-Dimethyl-3,4,5,6-tetrahydro-2(1*H*)-pyrimidinone		128.18	24^3, 32	1.060	1.4880^{20}		146^{44mm}	>110	
d703	Dimethyl 3,3′-dithiopropionate	$(CH_3O_2CCH_2CH_2)_2S$	206.26		1.198	1.4740^{20}		148^{18mm}	>110	
d704	*N,N*-Dimethylthioformamide	$(CH_3)_2NC(S)H$	89.16	4, 70	1.047	1.5757^{20}		58^{1mm}	99	
d705	*N,N′*-Dimethylthiourea	$(CH_3NH)_2C{=}S$	104.18	4, 70			60–62			v s aq, alc, acet
d706	*N,N*-Dimethyl-*p*-toluidine	$CH_3C_6H_4N(CH_3)_2$	135.21	12, 902	0.937	1.5458^{20}		211	83	
d707	*N,N*-Dimethyltrimethylsilylamine	$(CH_3)_3SiN(CH_3)_2$	117.27		0.732	1.3970^{20}		84	−19	
d708	1,3-Dimethylurea	$(CH_3NH)_2C{=}O$	88.11	4, 65			101–104	268–270		v s aq, alc; i eth
d709	Dimethylzinc	$(CH_3)_2Zn$	95.45	Merck: 12, 3312	0.724		−40	46	−1	misc bz, PE; s eth
d710	2,4-Dinitroaniline	$(O_2N)_2C_6H_3NH_2$	183.12	12, 747	1.615^{14}		176–178			i aq; 0.75 alc
d711	1,3-Dinitrobenzene	$C_6H_4(NO_2)_2$	168.11	5, 258	1.368		89–90	297		0.05 aq; 2.7 alc; v s bz, chl, EtOAc
d712	2,4-Dinitrobenzenesulfenyl chloride	$(O_2N)_2C_6H_3SCl$	234.62	6^2, 316			96			s bz, HOAc; dec alc
d713	3,5-Dinitrobenzoic acid	$(O_2N)_2C_6H_3CO_2H$	212.12	9, 413			205–207			1.9 hot aq; v s alc; sl s bz, eth
d714	3,5-Dinitrobenzoyl chloride	$(O_2N)_2C_6H_3COCl$	230.56	9, 414			69–71	196^{11mm}		dec aq, alc; s eth
d715	2,6-Dinitro-*p*-cresol	$(O_2N)_2C_6H_2(OH)CH_3$	198.13	6, 414			77–79			
d716	4,6-Dinitro-*o*-cresol	$(O_2N)_2C_6H_2(OH)CH_3$	198.13	6, 368			83–87			v s alc, acet, eth, alk
d717	2,4-Dinitrodiphenylamine	$(O_2N)_2C_6H_3NHC_6H_5$	259.22	12, 751			159–161			
d718	2,4-Dinitro-1-fluorobenzene	$FC_6H_3(NO_2)_2$	186.10	5, 262	1.482	1.5690^{20}	27–30	178^{25mm}	>110	s bz, eth, glyc
d719	1,5-Dinitronaphthalene	$C_{10}H_6(NO_2)_2$	218.17	5, 558			216–217	subl		s bz; v s eth; sl s alc

d720	2,4-Dinitrophenol	$(O_2N)_2C_6H_3OH$	184.11	6, 251	1.683		106–108			s alc, bz; 16 EtOAc; 36 acet; 5 chl; 20 pyr
d721	2,4-Dinitrophenylhydrazine	$(O_2N)_2C_6H_3NHNH_2$	198.14	15, 489			*ca.* 200			sl s aq, alc; s acid
d722	3,5-Dinitrosalicylic acid	$(O_2N)_2C_6H_2(OH)CO_2H$	228.12	10, 122			169–172			s aq; v s alc, eth
d723	2,4-Dinitrotoluene	$CH_3C_6H_3(NO_2)_2$	182.14	5, 339	1.321^{71}	1.442	67–70	300 sl d		1.2 alc; 9 eth
d724	2,6-Dinitrotoluene	$CH_3C_6H_3(NO_2)_2$	182.14	5, 341	1.2833^{111}	1.479	64–66			s alc
d725	Dinonyl hexanedioate	$C_9H_{19}O_2C(CH_2)_4CO_2C_9H_{19}$	398.63		0.917^{25}_{25}				218	
d726	Dioctadecyl phosphite	$(C_{18}H_{37}O)_2P(O)H$	586.97				57–59			
d727	Dioctadecyl 3,3′-thiopropionate	$S[CH_2CH_2CO_2(CH_2)_{17}CH_3]_2$	683.18				65–67			
d728	Dioctylamine	$(C_8H_{17})_2NH$	241.46	4, 196	0.799	1.4432^{20}	14–16	298	>110	i aq; v s alc, eth
d729	Dioctyl ether	$(C_8H_{17})_2O$	242.45	1, 419	0.806	1.4318^{20}	−7.6	287	>110	
d730	Dioctyl sulfide	$(C_8H_{17})_2S$	258.51	1, 419	0.842	1.4610^{20}		180^{10mm}	>110	
d731	4,9-Dioxa-1,12-dodecanediamine	$H_2N(CH_2)_3O(CH_2)_4O(CH_2)_3NH_2$	204.32		0.962	1.4609^{20}		136^{4mm}	>110	
d732	1,3-Dioxane		88.11	19, 2	1.032	1.4180^{20}	−45	106	15	
d733	1,4-Dioxane		88.11	19, 3	1.0329^{20}_{4}	1.4224^{20}	11.8	101.2	12	misc aq, alc, bz, chl, eth, PE
d734	1,3-Dioxolane		74.08	19^2, 2	1.060^{20}_{4}	1.4000^{20}	−95	78	2	misc aq; s alc, eth
d735	Dipentaerythritol	$(HOCH_2)_3CCH_2OCH_2C(CH_2OH)_3$	254.28				215–218			
d736	Dipentene		136.24	5, 137	0.8402^{21}_{4}	1.4739^{20}	−95.5	178	45	i aq; misc alc
d737	Dipentylamine	$(C_5H_{11})_2NH$	157.29	4^1, 378	0.777	1.4272		195–202	52	v s alc, eth
d738	Dipentyl ether	$(C_5H_{11})_2O$	158.29	1^1, 193	0.7833^{20}_{4}	1.4120^{20}	−69.4	190	57	misc alc, eth; s acet
d739	*N,N*-Diphenylacetamide	$CH_3CON(C_6H_5)_2$	211.26	12, 247			103	$130^{0.02mm}$		sl s aq; s alc, eth
d740	Diphenylacetic acid	$(C_6H_5)_2CHCO_2H$	212.25	9, 673	1.258^{15}_{15}		148	195^{5mm}		s hot aq, alc, chl, eth
d741	Diphenylacetonitrile	$(C_6H_5)_2CHCN$	193.25	9, 674			71–73	181^{12mm}		
d742	Diphenylacetylene	$C_6H_5C{\equiv}CC_6H_5$	178.23	5, 656	0.990		62.5	300		v s eth, hot alc
d743	Diphenylamine	$(C_6H_5)_2NH$	169.23	12, 174	1.160		53	302	152	45 alc; v s bz, eth
d744	*cis,trans*-1,4-Diphenyl-1,3-butadiene	$C_6H_5CH{=}CHCH{=}CHC_6H_5$	206.29	5, 676	0.9974^{22}_{4}	1.0653^{22}	149.7	350^{720mm}		s alc; sl s eth
d745	Diphenylcarbamoyl chloride	$(C_6H_5)_2NC(O)Cl$	231.68				82–84			

(Continued)

TABLE 2.20 Physical Constants of Organic Compounds (*Continued*)

No.	Name	Formula	Formula weight	Beilstein reference	Density, g/mL	Refractive index	Melting point, °C	Boiling point, °C	Flash point, °C	Solubility in 100 parts solvent
d746	1,5-Diphenylcarbohydrazide	$(C_6H_5NHNH)_2C{=}O$	242.28	15, 292			168–171			s hot alc, acet, HOAc
d747	Diphenyl carbonate	$(C_6H_5O)_2C{=}O$	214.22	6, 158			80–81	301–302		s hot alc, bz, eth
d748	Diphenyl chlorophosphate	$(C_6H_5O)_2P(O)Cl$	268.64	6, 179	1.296	1.5500^{20}		316^{272mm}	>110	
d749	Diphenyl diselenide	$C_6H_5SeSeC_6H_5$	312.13	6, 346	1.557_4^{80}		61–63			s hot alc
d750	Diphenyl disulfide	$C_6H_5SSC_6H_5$	218.34	6, 323	1.353_4^{20}		58–60	310		s alc, bz, eth; i aq
d751	Diphenylenimine		167.21	20, 433	1.10_4^{18}		246	355		0.8 bz; 3 eth; 16 pyr; 11 acet; i aq
d752	1,2-Diphenylethane	$C_6H_5CH_2CH_2C_6H_5$	182.27	5, 598	0.995_4^{20}	1.5338	52.5	284	>110	s alc; v s chl, eth
d753	Diphenyl ether	$C_6H_5OC_6H_5$	170.21	6, 146	1.0661_4^{30}	1.5763^{30}	26.9	258	112	s alc, bz, eth, HOAc
d754	*N,N'*-Diphenylformamidine	$C_6H_5N{=}CHNHC_6H_5$	196.25	12, 236			138–141			s eth; v s chl
d755	1,3-Diphenylguanidine	$C_6H_5NHC({=}NH)NHC_6H_5$	211.27	12, 369	1.13		148–150	dec 170		s alc, hot bz, chl
d756	5,5-Diphenylhydantoin		252.27	24, 410			294–297			i aq; 1.7 alc; 3.3 acet
d757	1,2-Diphenylhydrazine	$C_6H_5NHNHC_6H_5$	184.24	15, 123	1.158_4^{16}		123–126			v s alc; sl s bz
d758	Diphenylmercury	$(C_6H_5)_2Hg$	354.81	16, 946	2.318^4		128–129	dec >306		s chl; sl s hot alc
d759	Diphenylmethane	$C_6H_5CH_2C_6H_5$	168.24	5^2, 498	1.006	1.5768^{20}	25	265	>110	v s alc, bz, chl, eth
d760	Diphenylmethanol	$C_6H_5CH(OH)C_6H_5$	184.24	6, 678			66.7	298		0.05 aq; v s alc, chl, eth
d761	1,1-Diphenylmethylamine	$C_6H_5CH(NH_2)C_6H_5$	183.25	12, 1323	1.0635_4^{22}	1.5956^{99}	34	295	>112	sl s aq
d762	2,5-Diphenyloxazole		221.26	27, 78			72–74	360		
d763	Diphenyl phosphite	$(C_6H_5O)_2P(O)H$	234.19	6^1, 94	1.223	1.5575^{20}	12	219^{26mm}	176	
d764	Diphenylphosphoryl azide	$(C_6H_5O)_2P(O)N_3$	275.20		1.277	1.5518^{20}		$157^{0.17mm}$	>110	
d765	Diphenyl *o*-phthalate	$C_6H_4(CO_2C_6H_5)_2$	318.33	9, 801			74–76			
d766	2,2-Diphenyl-1-picrylhydrazyl		394.32	16^2, 363			127 dec			
d767	1,3-Diphenyl-2-propanone	$C_6H_5CH_2(C{=}O)CH_2C_6H_5$	210.28	7, 445	1.2		32–34	330		i aq; v s alc, eth
d768	2,2-Diphenylpropionic acid	$CH_3C(C_6H_5)_2CO_2H$	226.28	9^2, 474			175–177	300		s alc; v s bz, eth
d769	Diphenylsilanediol	$(C_6H_5)_2Si(OH)_2$	216.31	16, 909			140 dec		53	
d770	Diphenyl sulfide	$(C_6H_5)_2S$	186.28	6, 299	1.118_{15}^{15}	1.6327^{20}	−40	296	>110	misc bz, eth, CS_2
d771	Diphenyl sulfone	$(C_6H_5)_2SO_2$	218.27	6, 300			128–129	379		i aq; s hot alc, bz
d772	Diphenyl sulfoxide	$(C_6H_5)_2SO$	202.28	6, 300			69–71	207^{13mm}		
d773	Diphenylthiocarbazone	$C_6H_5N{=}NC(S)NHNHC_6H_5$	256.33	16, 26			168 dec			i aq; v s chl, CCl_4

d774	1,3-Diphenyl-2-thiourea	$C_6H_5NHC(S)NHC_6H_5$	228.32	12, 394	1.32		154			i aq; v s alc, eth
d775	1,3-Diphenylurea	$C_6H_5NHC(O)NHC_6H_5$	212.35	12, 352	1.239		238	260 dec		0.015 aq; s eth, HOAc
d776	Dipiperidinomethane		182.31		0.915	1.4820^{20}		123^{15mm}	91	
d777	Dipropylamine	$(C_3H_7)_2NH$	101.19	4, 138	0.7375^{20}_{4}	1.4043^{20}	−63	109.2	17	4 aq; v s alc, eth, PE
d778	3-Dipropylamino-1,2-propanediol	$(C_3H_7)_2NCH_2CH(OH)CH_2OH$	175.27	4[3], 841	0.949	1.4554^{20}		143^{9mm}	>110	
d779	Dipropylene glycol	$HO(CH_2)_3O(CH_2)_3OH$	134.18	1[2], 537	1.023	1.4410^{20}			137	
d780	Dipropylene glycol butyl ether	$CH_3CH(OH)CH_2OCH_2CH(OC_4H_9)CH_3$	190.29	1, *4*, 2474	0.917^{25}_{25}	1.425^{25}		229	96	
d781	Dipropylene glycol *tert*-butyl ether	$(CH_3)_3CO(CH_2)_3O(CH_2)_3OH$	190.29		0.900	1.4240^{20}		220–222	87	
d782	Dipropylene glycol dibenzoate	$[C_6H_5CO_2(CH_2)_3]_2O$	342.40	9[2], 108	1.120	1.5280^{20}		232^{5mm}	>110	
d783	Dipropylene glycol isopropyl ether	$CH_3CH(OH)CH_2OCH_2CH[OCH(CH_3)_2]CH_3$	176.2		0.878^{25}_{25}	1.421^{25}		80.1	90	
d784	Dipropylene glycol methyl ether	$CH_3CH(OH)CH_2OCH_2CH(OCH_3)CH_3$	148.2		0.951^{20}_{20}	1.419^{20}	−117	188.3	74	
d785	Dipropylene glycol acetate	$CH_3CO_2(CH_2)_3O(CH_2)_3OCH_3$	190.24		0.970	1.4180^{20}		200	85	
d786	Dipropyl ether	$(C_3H_7)_2O$	102.18	1, 354	0.7466^{20}	1.3803^{20}	−126.2	89.6	21	0.4 aq
d787	Dipropyl hexanedioate	$C_3H_7O_2C(CH_2)_4CO_2C_3H_7$	230.30	2[2], 574	0.9790^{20}_{4}	1.4314^{20}	−20	144^{10mm}		i aq; s alc, eth v s PE
d788	Dipropyl sulfate	$(C_3H_7O)_2SO_2$	182.24	1, 354	1.106^{20}_{4}		dec 140	120^{20mm}		
d789	Dipropyl sulfone	$(C_3H_7)_2SO_2$	150.24	1, 359	1.028^{50}_{4}		28–30	270	126	
d790	2,2′-Dipyridyl		156.19	23, 199			70–73	273		0.5 aq; v s alc, bz, chl, eth, PE
d791	Disilane	H_3SiSiH_3	62.22	Merck: 12, 3419	0.686^{-25}_{4}		−132	−14.3	ignites in air	s alc, bz, CS_2
d792	1,3-Dithiane		120.24				53–55		90	
d793	4,4′-Dithiobutyric acid	$HO_2C(CH_2)_3SS(CH_2)_3CO_2H$	238.32	3, 312			110			
d794	3,3′-Dithiodipropionic acid	$HO_2C(CH_2)_2SS(CH_2)_2CO_2H$	210.27				157–159			
d795	Dithiooxamide	$H_2NC(S)C(S)NH_2$	120.20	2, 565			245			
d796	2,2′-Dithiosalicylic acid	$S_2(C_6H_4CO_2H)_2$	306.36	10, 129			287–290			sl s aq; s alc; i eth
d797	1,3-Di-*o*-tolylguanidine	$(CH_3C_6H_4NH)_2C{=}NH$	239.32	12, 803	1.10^{20}_{4}		176–178			s hot alc, eth
d798	Divinyl ether	$H_2C{=}CHOCH{=}CH_2$	70.09	Merck: 12, 10133	0.773^{20}_{4}	1.3989^{20}	−101	28.3	< −30	0.53 aq; misc alc, eth

(*Continued*)

TABLE 2.20 Physical Constants of Organic Compounds (*Continued*)

No.	Name	Formula	Formula weight	Beilstein reference	Density, g/mL	Refractive index	Melting point, °C	Boiling point, °C	Flash point, °C	Solubility in 100 parts solvent
d799	1,3-Divinyltetra-methyldisiloxane	$[CH_2{=}CHSi(CH_3)_2]_2O$	186.39	4,4,4080	0.811_4^{20}	1.4110^{20}	−99	139	24	
d800	3,9-Divinyl-2,4,8,10-tetraoxaspiro[5.5]-undecane		212.25	19^3, 5679	1.251		43–46	110^{2mm}	110	
d801	Docasane	$CH_3(CH_2)_{20}CH_3$	310.61	1, 174	0.7782^{45}	1.4358^{45}	43–45	369	>110	i aq; sl s alc; v s eth
d802	1-Docosanol	$CH_3(CH_2)_{21}OH$	326.61	1, 431			65–72	180^{22mm}		sl s eth; s alc, chl
d803	Dodecane	$CH_3(CH_2)_{10}CH_3$	170.34	1, 171	0.7490_4^{20}	1.4216^{20}	−10	216.2	74	
d804	1,12-Dodecanediamine	$H_2N(CH_2)_{12}NH_2$	200.37	4, 273			71		155	
d805	Dodecanedioic acid	$HO_2C(CH_2)_{10}CO_2H$	230.30	2, 729			128–130	245^{10mm}		
d806	1,2-Dodecanediol	$CH_3(CH_2)_9CH(OH)CH_2OH$	202.34	1^3, 2237			58–60			
d807	1,12-Dodecanediol	$HOCH_2(CH_2)_{10}CH_2OH$	202.34	1^2, 562			81–84	189^{12mm}		
d808	1-Dodecanethiol	$CH_3(CH_2)_{11}SH$	202.40		0.845_{20}^{20}	1.4587^{20}		266–283	87	i aq; s alc, eth
d809	Dodecanoic acid	$CH_3(CH_2)_{10}CO_2H$	200.32	2, 359	0.869^{14}	1.4183^{82}	43	225^{100mm}	>110	i aq; 100 alc; v s bz, eth; 40 PrOH
d810	1-Dodecanol	$CH_3(CH_2)_{11}OH$	186.34	1, 428	0.8308_4^{25}	1.4413^{25}	24	259	>110	i aq; s alc, eth
d811	δ-Dodecanolactone		198.31	17^5,9,100	0.942	1.4602^{20}	−12	126^{1mm}	>110	
d812	Dodecanoyl peroxide	$[CH_3(CH_2)_{10}CO]_2O_2$	398.63	2^3, 893			55–57			
d813	1-Dodecene	$CH_3(CH_2)_9CH{=}CH_2$	168.32	1, 225	0.7584_4^{20}	1.4294^{20}	−35.2	213.4	79	s alc, eth, PE
d814	2-Dodecen-1-ylsuccinic anhydride		266.38				41–43	180^{5mm}	177	
d815	Dodecyl acetate	$CH_3CO_2(CH_2)_{11}CH_3$	228.38	2, 136	0.865	1.4318^{20}		150^{15mm}	>110	
d816	Dodecyl acrylate	$H_2C{=}CHCO_2(CH_2)_{11}CH_3$	240.39	2^3, 1230	0.884	1.4450^{20}			>110	
d817	Dodecyl aldehyde	$CH_3(CH_2)_{10}CHO$	184.32	1, 714	0.835	1.4344^{20}		185^{100mm}	101	
d818	Dodecylamine	$CH_3(CH_2)_{11}NH_2$	185.36	4, 200	0.808		30–32	247–249	>110	misc alc, bz, chl, eth
d819	Dodecyl methacrylate	$H_2C{=}C(CH_3)CO_2(CH_2)_{11}CH_3$	254.42	2^3, 1290	0.868	1.4460^{20}	−7	142^{4mm}	>110	
d820	Dodecyl sulfate, sodium salt	$CH_3(CH_2)_{11}SO_3^-\ Na^+$	288.38	1^3, 1786			204–207			10 aq
d821	Dodecyltrichlorosilane	$CH_3(CH_2)_{11}SiCl_3$	303.8	4^3, 1907	1.020	1.458^{20}		294	>110	
d822	Dodecyl vinyl ether	$CH_3(CH_2)_{11}OCH{=}CH_2$	212.38		0.817	1.4382^{20}		117–120	>110	
d823	Dotriacontane	$CH_3(CH_2)_{30}CH_3$	450.88	1,177	0.8124_4^{20}	1.4364^{70}	68–70	467		sl s alc, bz, eth
d824	Dulcitol		182.17	1, 544	1.47^{20}		188–191	280^{1mm}		3.3 aq; sl s alc
e1	Eicosane	$CH_3(CH_2)_{18}CH_3$	282.56	1, 174	0.7823 (s)		37	343	>110	
e2	1*R*,2*S*-(−)-Ephedrine	$CH_3NHCH(CH_3)CH(OH)C_6H_5$	165.24	13, 636	1.124		39	255	85	s aq, alc, chl, eth
e3	1,2-Epoxybutane	$H_2C{-}CHCH_2CH_3$ O	72.11	17^2, 17	0.8297^{20}	1.3850^{20}	−150	63	−22	6 aq; misc alc, bz, chl, eth

e4	1,2-Epoxy-5,9-cyclo-dodecadiene		178.28		0.980	1.5045^{20}		83^{1mm}	>110	
e5	1,2-Epoxycyclo-dodecane		182.31		0.939	1.4773^{20}			>110	
e6	1,2-Epoxycyclopentane		84.12	17, 21	0.964	1.4336^{20}		102	10	
e7	1,2-Epoxydecane	$\underbrace{H_2C—CH}_{O}CH_2(CH_2)_6CH_3$	156.27	17, 18	0.840	1.4290^{20}		94^{15mm}	78	
e8	1,2-Epoxydodecane	$\underbrace{H_2C—CH}_{O}CH_2(CH_2)_8CH_3$	184.32	17^3, 136	0.844	1.4355^{20}		125^{15mm}	105	
e9	1,2-Epoxyethylbenzene	$\underbrace{H_2C—CH}_{O}C_6H_5$	120.15	17, 49	1.0523^{16}_{4}	1.5338^{20}	−37	194	79	i aq; s alc, eth
e10	1,2-Epoxyhexadecane	$\underbrace{H_2C—CH}_{O}CH_2(CH_2)_{12}CH_3$	240.43	17, 20	0.846	1.4452^{20}	21–22	180^{12mm}	93	
e11	1,2-Epoxyhexane	$\underbrace{H_2C—CH}_{O}CH_2CH_2CH_2CH_3$	100.16	17^4, 86	0.831	1.4056^{20}		118–120	15	
e12	1,2-Epoxy-5-hexene	$\underbrace{H_2C—CH}_{O}CH_2CH_2CH{=}CH_2$	98.15	17^3, 163	0.870	1.4252^{20}		121	15	
e13	1,2-Epoxyoctadecane	$\underbrace{H_2C—CH}_{O}CH_2(CH_2)_{14}CH_3$	268.49	17^3, 140			33–35	$137^{0.5mm}$	>110	
e14	1,2-Epoxy-3-phenoxy-propane	$\underbrace{H_2C—CH}_{O}CH_2OC_6H_5$	150.18	17, 105	1.109	1.530^{20}	3.5	245	>110	
e15	1,2-Epoxypropane	$\underbrace{H_2C—CH}_{O}CH_3$	58.08	17, 6	0.859^{0}_{4}	1.3660^{20}	−112	35	−37	41 aq; misc alc, eth
e16	2,3-Epoxy-1-propanol	$\underbrace{H_2C—CH}_{O}CH_2OH$	74.08	17, 104	1.1143^{25}_{4}	1.4315^{20}		$66^{2.5mm}$	81	misc aq
e17	2,3-Epoxypropyl-methacrylate	$\underbrace{H_2C—CH}_{O}CH_2O_2C(CH_3)$-$CH{=}CH_2$	142.16		1.042	1.4494^{20}		189	76	
e18	1,2-Epoxy-3,3,3-tri-chloropropane	$\underbrace{H_2C—CH}_{O}CCl_3$	161.42	17^2, 14	1.495	1.4778^{20}		151^{745mm}	66	
e19	*meso*-Erythritol	$HOCH_2[CH(OH)]_2CH_2OH$	122.12	1, 525			120–123	329–331		
e20	Ethane	CH_3CH_3	30.07	1, 80	1.356^{0} g/L		−182.8	−88	−135	4.7 mL aq; 46 mL alc[4]
e21	1,2-Ethanediamine	$H_2NCH_2CH_2NH_2$	60.10	4, 230	0.8977^{20}_{4}	1.4568^{20}	11	117.3	33	misc aq, alc; i bz

(Continued)

TABLE 2.20 Physical Constants of Organic Compounds (*Continued*)

No.	Name	Formula	Formula weight	Beilstein reference	Density, g/mL	Refractive index	Melting point, °C	Boiling point, °C	Flash point, °C	Solubility in 100 parts solvent
e21a	1,2-Ethanediol	$HOCH_2CH_2OH$	62.07	1, 465	1.1135^{20}_{4}	1.4318^{20}	−12.6	197.3	110	misc aq, alc, glyc, pyr
e22	1,2-Ethanediol diacetate	$CH_3CO_2CH_2CH_2O_2CCH_3$	146.14	2, 142	1.1043^{20}	1.4150^{20}	−31	190.2	82	misc alc, eth
e23	1,2-Ethanediol dimethacrylate	$[H_2C{=}C(CH_3)CO_2CH_2{-}]_2$	198.22	2^3, 1292	1.051	1.4549^{20}		100^{5mm}	>110	
e24	1,2-Ethanedithiol	$HSCH_2CH_2SH$	94.20	1, 471	1.123^{24}	1.5580^{20}		146	50	v s alc, alk
e25	Ethanesulfonic acid	$C_2H_5SO_3H$	110.13	4, 5	1.350	1.4340^{20}	−17	$123^{0.01mm}$	>110	
e26	Ethanesulfonyl chloride	$CH_3CH_2SO_2Cl$	128.57	4, 6	1.357^{22}	1.4330^{20}		177	83	dec aq, alc; v s eth
e26a	Ethanethiol	CH_3CH_2SH	62.13	1, 340	0.8315^{25}	1.420^{25}	−147.9	35.0	−17	0.7 aq; s alc, eth
e27	Ethanol	CH_3CH_2OH	46.07	1, 292	0.7894^{20}_{4}	1.3611^{20}	−114	78.3	13	misc aq, alc, chl, eth
e28	Ethanol-*d*	CH_3CH_2OD	47.08	1^3, 1287	0.801	1.3595^{20}		78.8	12	misc aq, alc, eth
e29	Ethanolamine	$H_2NCH_2CH_2OH$	61.08	Merck: 12, 3712	1.0180^{20}	1.4539^{20}	10.5	170.8	86	misc aq, alc, acet
e30	Ethoxyacetic acid	$CH_3CH_2OCH_2CO_2H$	104.11	3, 233	1.1021^{20}_{4}	1.4190^{20}		97^{11mm}	97	s aq, alc, eth
e31	3-Ethoxyacrylonitrile	$C_2H_5OCH{=}CHCN$	97.12	3^3, 681	0.944	1.4545^{20}		91^{19mm}	81	
e32	4-Ethoxyaniline	$CH_3CH_2OC_6H_5NH_2$	137.18	13, 436	1.0652^{16}_{4}	1.5609^{20}	4	250	115	i aq; s alc
e33	2-Ethoxybenzaldehyde	$CH_3CH_2OC_6H_4CHO$	150.18	8, 43	1.074	1.5422	20	136^{24mm}	107	misc alc, eth
e34	4-Ethoxybenzaldehyde	$CH_3CH_2OC_6H_4CHO$	150.18	8, 73	1.080^{25}_{25}	1.5584^{20}	13–14	255	>110	v s alc, bz, eth
e35	2-Ethoxybenzamide	$CH_3CH_2OC_6H_4CONH_2$	165.19	10, 93			132–134			sl s aq; s alc, eth
e36	Ethoxybenzene	$CH_3CH_2OC_6H_5$	122.17	6, 140	0.967^{20}_{4}	1.5074^{20}	−29.5	169.8	63	v s alc, eth
e37	2-Ethoxybenzoic acid	$CH_3CH_2OC_6H_4CO_2H$	166.18	10, 64	1.105	1.5400^{20}	19.4	174^{15mm}	>110	sl s aq
e38	4-Ethoxybenzoic acid	$CH_3CH_2OC_6H_4CO_2H$	166.18	10, 156			197–199			sl s hot aq
e39	Ethoxycarbonyl isothiocyanate	$CH_3CH_2OC({=}O)NCS$	131.15	3^3, 279	1.112	1.5000^{20}		56^{18mm}	50	
e40	2-Ethoxyethanol	$CH_3CH_2OCH_2CH_2OH$	90.12	1, 467	0.9295^{20}	1.4075^{20}	−70	134.8	43	misc aq, alc, acet, eth
e41	2-(2-Ethoxyethoxy)-ethanol	$C_2H_5OCH_2CH_2OCH_2CH_2OH$	134.18	1^2, 520	0.9841^{25}_{4}	1.4254^{25}	−76	196	96	misc aq, alc, bz, chl, acet, pyr
e41a	2-(2-Ethoxyethoxy)-ethanol acetate	$CH_3CO_2CH_2CH_2OCH_2CH_2$-$OCH_2CH_3$	176.21		1.0096^{20}	1.4213^{20}	−25	218.5	110	
e42	2-Ethoxyethyl acetate	$CH_3CO_2CH_2CH_2OCH_2CH_3$	132.16	2^2, 155	0.9749^{20}_{4}	1.4023^{20}	−61.7	156.3	57	29 aq; misc alc, eth
e43	2-Ethoxyethyl acrylate	$H_2C{=}CHCO_2CH_2CH_2OC_2H_5$	144.17	2^3, 1232	0.982	1.4270^{20}		78^{23mm}	65	
e44	2-Ethoxyethylamine	$CH_3CH_2OCH_2CH_2NH_2$	89.14	4^2, 718	0.8512^{20}_{4}	1.4101^{20}		107	21	misc aq, alc, eth
e45	2-Ethoxyethyl methacrylate	$H_2C{=}C(CH_3)CO_2CH_2CH_2OC_2H_5$	158.20	2^3, 1291	0.964	1.4285^{20}		93^{35mm}	71	
e46	3-Ethoxy-4-hydroxy-benzaldehyde	$C_2H_5OC_6H_3(OH)CHO$	166.18	8, 256			76–78			s eth, glycols; 50 alc

e47	3-Ethoxy-4-methoxy-benzaldehyde	$C_2H_5OC_6H_3(OCH_3)CHO$	180.2	8, 256			51–53		>110	s alc, bz, chl, eth
e48	1-Ethoxy-2-methoxy-benzene	$C_2H_5OC_6H_4OCH_3$	152.19	6, 771	1.044	1.5240^{20}		217–218	90	
e49	Ethoxymethylene-malononitrile	$CH_3CH_2OCH{=}C(CN)_2$	122.13	3^1, 162			64–66	160^{12mm}		
e50	1-Ethoxynaphthalene	$C_{10}H_7OCH_2CH_3$	172.23	6, 606	1.060^{20}_4	1.6040^{20}	5.5	280	>110	i aq; v s alc, eth
e51	2-Ethoxyphenol	$C_2H_5OC_6H_4OH$	138.17	6, 771	1.090	1.5288^{20}	29	217	91	
e52	*trans*-2-Ethoxy-5-(1-propenyl)phenyl	$C_2H_5OC_6H_3(CH{=}CHCH_3)OH$	178.23	6^2, 918			86–88			
e53	3-Ethoxypropionitrile	$C_2H_5OCH_2CH_2CN$	99.14	3, 298	0.911	1.4065^{20}		171–172	63	
e54	3-Ethoxypropylamine	$C_2H_5OCH_2CH_2CH_2NH_2$	103.17	4^3, 739	0.861	1.4178^{20}		136–138	32	
e55	3-Ethoxysalicyl-aldehyde	$C_2H_5OC_6H_3(OH)CHO$	166.18	8^2, 267			66–68	264		
e56	Ethoxytrimethylsilane	$(CH_3)_3SiOC_2H_5$	118.3	4^3, 1856	0.7573^{20}_4	1.3742^{20}		75–76	−18	
e57	Ethyl acetate	$CH_3CO_2C_2H_5$	88.11	2, 125	0.9006^{20}_4	1.3724^{20}	−84	77	−4	9.7 aq; misc alc, acet, chl, eth
e58	Ethyl acetoacetate	$CH_3COCH_2CO_2C_2H_5$	130.15	3, 632	1.0213^{25}_4	1.4174^{20}	−45	180.8	57	2.9 aq; misc alc, chl
e59	*p*-Ethylacetophenone	$C_2H_5C_6H_4COCH_3$	148.21	7^4, 1101	0.993	1.5293^{20}	−20.6	114^{11mm}	90	
e60	Ethyl acrylate	$H_2C{=}CHCO_2C_2H_5$	100.12	2, 399	0.9234^{20}	1.4060^{20}	−71	99	10	1.5 aq; s alc, eth
e61	Ethylaluminum dichloride	$C_2H_5AlCl_2$	126.95	4^3, 1973	1.207^{50}		32	113^{50mm}	−18	
e62	Ethylaluminum sesquichloride	$C_2H_5AlCl_2 \cdot ClAl(C_2H_5)_2$	247.51		1.092		−50	204	−18	
e63	Ethylamine	$C_2H_5NH_2$	45.09	4, 87	0.689^{15}_{15}	1.3663^{20}	−81	16.6	< −18	misc aq, alc, eth
e64	Ethyl 2-aminobenzoate	$H_2NC_6H_4CO_2C_2H_5$	165.19	14, 319	1.088^{15}	1.5640^{20}	13–15	266–268	>110	i aq; s alc, eth
e65	Ethyl 4-aminobenzoate	$H_2NC_6H_4CO_2C_2H_5$	165.19	14, 422			88–90	310		0.04 aq; 20 alc; 50 chl, 25 eth; s dil acid
e66	Ethyl 3-amino-crotonate	$CH_3C(NH_2){=}CHCO_2C_2H_5$	129.16	3, 654	1.021^{20}_4		33–35	210–215	97	i aq; s alc, bz, eth
e67	2-(Ethylamino)ethanol	$CH_3CH_2NHCH_2CH_2OH$	89.14	4, 282	0.914^{20}_4	1.4402^{20}	−90	170	71	v s aq, alc, eth
e68	*N*-Ethylaniline	$C_6H_5NHC_2H_5$	121.18	12, 159	0.958^{25}_{25}	1.5559^{20}	−63.5	203	85	i aq; misc alc, eth
e69	2-Ethylaniline	$CH_3CH_2C_6H_4NH_2$	121.18	12^2, 584	0.983	1.5590^{20}	−44	210	91	sl s aq; v s alc, eth
e70	4-Ethylaniline	$CH_3CH_2C_6H_4NH_2$	121.18	12, 1090	0.975	1.5542^{20}	−5	216	85	sl s aq; v s alc, eth
e71	2-Ethylanthraquinone		236.27	7^1, 425			108–111			
e72	4-Ethylbenzaldehyde	$C_2H_5C_6H_4CHO$	134.18	7, 307	0.979	1.5390^{20}		221	92	
e73	Ethylbenzene-d_{10}	$C_6D_5CD_2CD_3$	116.25		0.949	1.4920^{20}		134.6	31	

(*Continued*)

TABLE 2.20 Physical Constants of Organic Compounds (*Continued*)

No.	Name	Formula	Formula weight	Beilstein reference	Density, g/mL	Refractive index	Melting point, °C	Boiling point, °C	Flash point, °C	Solubility in 100 parts solvent
e74	Ethylbenzene	$C_6H_5CH_2CH_3$	106.17	5^2, 274	0.8670^{20}_{4}	1.4959^{20}	−95.0	136.2	22	0.01 aq; misc alc, bz, chl, eth
e75	4-Ethylbenzene-sulfonic acid	$C_2H_5C_6H_4SO_3H$	186.23	11, 120	1.229	1.5331			>110	
e76	Ethyl benzoate	$C_6H_5CO_2C_2H_5$	150.18	9, 110	1.051^{15}	1.5000^{20}	−34.7	212.4	84	0.05 aq; misc alc, chl, bz, eth, PE
e77	Ethyl benzoylacetate	$C_6H_5(C{=}O)CH_2CO_2C_2H_5$	192.21	10, 674	1.110	1.5338^{20}		265–270	63	i aq; misc alc, eth
e78	Ethyl 3-benzoyl-acrylate	$C_6H_5(C{=}O)CH{=}CHCO_2C_2H_5$	204.23	10^2, 501	1.112	1.5435^{20}		185^{25mm}	>110	
e79	Ethyl 2-benzylaceto-acetate	$CH_3C({=}O)CH(CH_2C_6H_5)CO_2C_2H_5$	220.27	10, 710	1.036	1.4996^{20}		276	>110	
e80	*N*-Ethylbenzylamine	$C_6H_5CH_2NHC_2H_5$	135.21	12, 1020	0.909	1.5117^{20}		194	66	
e81	Ethyl (2-benzyl)-benzoylacetate	$C_6H_5C({=}O)CH(CH_2C_6H_5)$-$CO_2C_2H_5$	282.34	10, 764	1.110	1.5567^{20}		270^{80mm}	>110	
e82	Ethyl *N*-benzyl-*N*-cyclopropylcarba-mate	$C_6H_5CH_2N(C_3H_5)CO_2C_2H_5$	219.28		0.997	1.5104^{20}			>110	
e83	Ethyl bromoacetate	$BrCH_2CO_2CH_2CH_3$	167.01	2, 214	1.506^{20}_{20}	1.4510^{20}	< −20	159	47	i aq; misc alc, eth
e84	Ethyl 4-bromo-benzoate	$BrC_6H_4CO_2C_2H_5$	229.08	9, 352	1.403	1.5440^{20}		131^{14mm}	>110	
e85	Ethyl 2-bromobutyrate	$CH_3CH_2CH(Br)CO_2C_2H_5$	195.06	2^2, 255	1.329^{20}_{20}	1.4470^{20}		177 dec	58	i aq; misc alc, eth
e86	Ethyl 4-bromobutyrate	$BrCH_2CH_2CH_2CO_2C_2H_5$	195.06	2, 283	1.363	1.4559^{20}		82^{10mm}	90	
e87	Ethyl 2-bromo-heptanoate	$CH_3(CH_2)_3CH(Br)CO_2C_2H_5$	237.14	2, 341	1.211	1.4524^{20}		109^{10mm}	104	
e88	Ethyl 6-bromo-hexanoate	$Br(CH_2)_5CO_2C_2H_5$	223.12	2^3, 737	1.254	1.4590^{20}		130^{16mm}	>110	
e89	Ethyl 2-bromoiso-butyrate	$(CH_3)_2C(Br)CO_2C_2H_5$	195.06	2, 296	1.329^{20}_{4}	1.4446^{20}		67^{11mm}	60	i aq; misc alc, eth
e90	Ethyl 2-bromo-octanoate	$CH_3(CH_2)_5CH(Br)CO_2C_2H_5$	251.17	2, 349	1.167	1.4520^{20}			106	
e91	Ethyl 3-bromo-2-oxo-propionate	$BrCH_2C({=}O)CO_2C_2H_5$	195.02	3^2, 409	1.554	1.4695^{20}		100^{10mm}	98	
e92	Ethyl 2-bromo-pentanoate	$CH_3(CH_2)_2CH(Br)CO_2C_2H_5$	209.09	2, 302	1.116	1.4486^{20}		190–192	77	i aq; misc alc, eth
e93	Ethyl 2-bromo-propionate	$CH_3CH(Br)CO_2C_2H_5$	181.03	2, 255	1.394	1.4460^{20}		156–160	51	i aq; misc alc, eth
e94	Ethyl 3-bromo-propionate	$BrCH_2CH_2CO_2C_2H_5$	181.03	2, 256	1.4123^{18}_{4}	1.4569^{18}		136^{50mm}	79	i aq; misc alc, eth

e95	2-Ethyl-1-butanol	$(C_2H_5)_2CHCH_2OH$	102.18	1, 412	0.8330^{20}	1.4224^{20}	< −15	146	58	0.63 aq
e95a	2-Ethyl-1-butene	$(C_2H_5)_2C{=}CH_2$	84.16	1^2, 95	0.689	1.3960^{20}	−131	65	−26	
e96	2-Ethylbutyl acetate	$CH_3CO_2CH_2CH(C_2H_5)_2$	144.21	2^3, 257	0.876	1.4100^{20}		160^{740mm}	52	
e97	*N*-Ethylbutylamine	$CH_3(CH_2)_3NHC_2H_5$	101.19	4, 157	0.740^{20}_{4}	1.4050^{20}		108	18	
e98	2-Ethylbutyraldehyde	$(C_2H_5)_2CHCHO$	100.16	1, 693	0.8162^{20}_{20}	1.4018^{20}	−89	116.7	21	0.31 aq
e99	Ethyl butyrate	$CH_3CH_2CH_2CO_2C_2H_5$	116.16	2, 270	0.879^{20}_{4}	1.3998^{20}	−98	121	24	0.49 aq; misc alc, eth
e100	2-Ethylbutyric acid	$(C_2H_5)_2CHCO_2H$	116.16	2, 333	0.9225^{20}_{20}	1.4133^{20}	−14	194	87	
e101	Ethyl butyrylacetate	$CH_3(CH_2)_2C({=}O)CH_2$-$CO_2C_2H_5$	158.20	3, 684	1.001	1.4270^{20}		104^{22mm}	78	
e102	Ethyl carbamate	$H_2NCO_2C_2H_5$	89.09	3, 22	1.056		49–50	182–184	92	200 aq; 125 alc; 111 chl; 67 eth
e103	Ethyl carbazate	$H_2NNHCO_2C_2H_5$	104.11	3, 98			44–47	110^{22mm}	86	
e104	*N*-Ethylcarbazole		195.27	20, 436			68–70			
e105	Ethyl chloroacetate	$ClCH_2CO_2C_2H_5$	122.55	2, 197	1.1498^{20}_{4}	1.4227^{20}	−21	144	65	i aq; misc alc, eth
e106	Ethyl 2-chloro-acetoacetate	$CH_3C({=}O)CH(Cl)CO_2C_2H_5$	164.59	3, 662	1.190	1.4430^{20}		107^{14mm}	50	i aq; s alc, eth
e107	Ethyl 4-chloro-acetoacetate	$ClCH_2C({=}O)CH_2CO_2C_2H_5$	164.59	3, 663	1.218^{17}_{4}	1.4520^{20}		115^{14mm}	96	i aq; misc alc, eth
e108	Ethyl 4-chlorobutyrate	$ClCH_2CH_2CH_2CO_2C_2H_5$	150.61	2, 278	1.0754^{20}_{4}	1.4306^{20}		186	51	s alc, acet, eth
e109	Ethyl chloroformate	$ClCO_2C_2H_5$	108.52	3, 10	1.1403^{20}_{4}	1.3941^{20}	−81	93	13	misc alc, bz, chl, eth
e110	Ethyl 2-chloro-propionate	$CH_3CH(Cl)CO_2C_2H_5$	136.58	2, 248	1.087^{20}_{4}	1.4185^{20}	146–149	38		
e111	Ethyl 3-chloro-propionate	$ClCH_2CH_2CO_2C_2H_5$	136.58	2, 250	1.1086^{20}_{4}	1.4249^{20}		162–163	54	misc alc, eth
e112	Ethyl chrysanthemumate		196.29	9^2, 45	0.906	1.4600^{20}		112^{10mm}	84	
e113	Ethyl *trans*-cinnamate	$C_6H_5CH{=}CHCO_2C_2H_5$	176.22	9^2, 385	1.0495^{20}_{4}	1.5598^{20}	10	271	>110	misc alc, eth; i aq
e114	Ethyl crotonate	$CH_3CH{=}CHCO_2C_2H_5$	114.14	2, 411	0.9175^{20}_{4}	1.4240^{20}		138	28	i aq; s alc, eth
e115	Ethyl cyanoacetate	$NCCH_2CO_2C_2H_5$	113.12	2, 585	1.0564^{25}_{4}	1.4176^{20}	−22	206	110	i aq; misc alc, eth
e116	Ethyl 2-cyano-3,3-diphenylacrylate	$(C_6H_5)_2C{=}C(CN)CO_2C_2H_5$	277.33	9^3, 4601			97–99	$174^{0.2mm}$		
e117	Ethylcyclohexane	$C_6H_{11}CH_2CH_3$	112.22	5, 35	0.7879^{20}	1.4330^{20}	−111	131.8	35	
e118	4-Ethylcyclohexanol	$CH_3CH_2C_6H_{10}OH$	128.22	6^2, 26	0.889	1.4625^{20}		84^{10mm}	77	
e118a	Ethylcyclopentane	$C_2H_5(C_5H_9)$	98.19	5^2, 19	0.763	1.4190^{20}	−138	103	15	
e119	Ethyl cyclopropane-carboxylate	$C_3H_5CO_2CH_2CH_3$	114.14	9, 4	0.960	1.4197^{20}		129–133	18	
e120	Ethyl decanoate	$CH_3(CH_2)_8CO_2C_2H_5$	200.32	2, 356	0.862^{20}	1.4248^{20}		245	102	misc alc, chl, eth
e121	Ethyl diazoacetate	$N_2CH_2CO_2C_2H_5$	114.10	3^1, 211	1.0852^{18}_{4}	1.4588^{18}	−22	141^{710mm}	26	misc alc, bz, eth

(Continued)

TABLE 2.20 Physical Constants of Organic Compounds (*Continued*)

No.	Name	Formula	Formula weight	Beilstein reference	Density, g/mL	Refractive index	Melting point, °C	Boiling point, °C	Flash point, °C	Solubility in 100 parts solvent
e122	Ethyl 2,3-dibromo-propionate	$BrCH_2CH(Br)CO_2C_2H_5$	259.94	2, 259	1.788_4^{16}	1.4986^{20}		214	91	s alc, eth
e123	Ethyl dichloro-phosphate	$CH_3CH_2OP(O)Cl_2$	162.94	1, 332	1.373	1.4338^{20}		65^{10mm}	>110	
e124	Ethyl dichlorothio-phosphate	$CH_3CH_2OP(S)Cl_2$	179.01	1, 353	1.353	1.5040^{20}		68^{10mm}	>110	
e125	*N*-Ethyldiethanolamine	$CH_3CH_2N(CH_2CH_2OH)_2$	133.19	4, 284	1.014	1.4665^{20}	−50	246–252	123	
e126	Ethyl 3,3-dimethyl-acrylate	$(CH_3)_2C{=}CHCO_2C_2H_5$	128.17	2, 433	0.9247_4^{20}	1.4350^{20}		155	33	
e127	Ethyl 4-dimethyl-aminobenzoate	$(CH_3)_2NC_6H_4CO_2C_2H_5$	193.25	14^1, 571			64–66			
e128	Ethyl 2,2-dimethyl-propionate	$(CH_3)_3CCO_2C_2H_5$	130.19	2, 320	0.8584_4^{18}	1.3922^{18}		118.2	16	s alc, eth
e129	Ethyl 3,5-dinitro-benzoate	$(O_2N)_2C_6H_3CO_2C_2H_5$	240.17	9, 414			94–95			
e130	5-Ethyl-1,3-dioxane-5-methanol		146.19	19^5,2,382	1.090	1.4630^{20}		105^{5mm}	>110	
e131	Ethylene	$H_2C{=}CH_2$	28.05	1, 180	1.147 g/L		−169.4	−104		11 mL aq^{25}; 200 alc^{25}; v s eth; s acet, bz
e132	Ethylene carbonate		88.06	19, 100	1.3214^{39}	1.4199^{40}	36.4	248	143	misc aq
e133	Ethylenediamine	$H_2NCH_2CH_2NH_2$	60.10	4, 230	0.879^{20}	1.4566^{20}	11	117	40	
e134	Ethylenediamine-*N,N,N′,N′*-tetra-acetic acid	$(HO_2CCH_2)_2NCH_2CH_2$-$N(CH_2CO_2H)_2$	292.24	4^3, 1187			250 dec			0.05 aq
e135	Ethylene glycol	$HOCH_2CH_2OH$	62.07	1, 465	1.113	1.4310^{20}		196–198	>110	
e136	Ethylene glycol bis-(mercaptoacetate)	$(HSCH_2CO_2CH_2{-})_2$	210.27		1.313	1.5211^{20}		139^{2mm}	>110	
e137	Ethylene glycol diacetate	$CH_3CO_2CH_2CH_2O_2CCH_3$	146.14	2, 142	1.1043^{20}	1.4159^{20}	−31	190	88	
e138	Ethylene glycol diethyl ether	$C_2H_5OCH_2CH_2OC_2H_5$	118.18	1, 468	0.8484^{20}	1.3860^{20}	−74	119	35	
e139	Ethylene glycol diglycidyl ether	$(H_2C{-}CHCH_2OCH_2{-})_2$ (epoxide, bridging O)	174.20	1, 468	0.842	1.3923^{20}	−74	121	20	
e140	Ethylene glycol dimethacrylate	$[H_2C{=}C(CH_3)CO_2CH_2{-}]_2$	198.22	2^3, 1292	1.051	1.4549^{20}		100^{5mm}	>110	

e141	Ethylene glycol dimethyl ether	$CH_3OCH_2CH_2OCH_3$	90.12	1, 467	0.8691^{20}	1.3796^{20}	−58	85	−2	
e142	Ethylene glycol divinyl ether	$H_2C{=}CHOCH_2CH_2OCH{=}CH_2$	114.14	1^3, 2807	0.914	1.4350^{20}		125–127	27	
e143	Ethylene glycol methyl ether acrylate	$H_2C{=}CHCO_2CH_2CH_2OCH_3$	130.14	2^3, 1232	1.012	1.4270^{20}		56^{12mm}	60	
e144	Ethylene glycol methyl ether methacrylate	$H_2C{=}C(CH_3)CO_2CH_2CH_2OCH_3$	144.17	2^3, 1291	0.993	1.4310^{20}		65^{12mm}	60	
e145	Ethylene glycol phenyl ether acrylate	$H_2C{=}CHCO_2CH_2CH_2OC_6H_5$	192.21	6^3, 572	1.104	1.5180^{20}		$84^{0.2mm}$	>110	
e146	Ethyleneimine	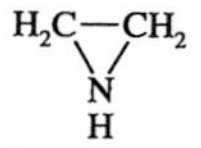	43.07		0.8321^{25}_{4}	1.4123^{25}	−78	56	−11	misc aq; s alc
e147	Ethylene oxide	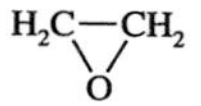	44.05	17, 4	0.891^{0}_{4}	1.3597^{7}	−111	10.6	−18	misc aq; s alc, eth
e148	Ethylene sulfide	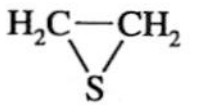	60.12	17^2, 12	1.010	1.4935^{20}		55–56	10	sl s alc, eth
e149	Ethyl 2-ethoxy-2-hydroxyacetate	$HOCH(OC_2H_5)CO_2C_2H_5$	148.16	3, 601	1.079	1.4200^{20}		137	49	
e150	Ethyl (ethoxymethylene)cyanoacetate	$C_2H_5OCH{=}C(CN)CO_2C_2H_5$	169.18	3, 470			51–53	190^{30mm}	>110	
e151	Ethyl 3-ethoxypropionate	$C_2H_5OCH_2CH_2CO_2C_2H_5$	146.19	3, 298	0.949	1.4050^{20}		166	52	
e152	Ethyl 4-{[(ethylphenylamino)methylene]amino}-benzoate	$C_6H_5N(C_2H_5)CH{=}N{-}C_6H_4{-}CO_2C_2H_5$	296.37				62–65	215^{2mm}		
e153	Ethyl fluoroacetate	$FCH_2CO_2C_2H_5$	106.10	2, 193	1.0926^{21}	1.3755^{20}		119	30	s aq
e154	Ethyl formate	$HCO_2C_2H_5$	74.08	2, 19	0.917^{20}_{4}	1.3590^{20}	−80	54	−20	10 aq; misc alc, eth
e155	Ethyl 2-furoate		140.14	18, 275	1.117^{20}_{4}		35–37	196	70	i aq; s alc, eth
e156	Ethyl heptanoate	$CH_3(CH_2)_5CO_2C_2H_5$	158.24	2^2, 295	0.8685^{20}_{4}	1.4144^{15}	−66	189	66	s alc, eth
e157	Ethyl hexadecanoate	$CH_3(CH_2)_{14}CO_2C_2H_5$	284.48	2^2, 336	0.8577^{25}_{4}	1.4347^{34}	22	191^{10mm}		s alc, eth
e158	2-Ethylhexanaldehyde	$CH_3(CH_2)_3CH(C_2H_5)CHO$	128.22	1, 707	0.822	1.4155		$55^{13.5mm}$	42	
e158a	3-Ethylhexane	$(C_2H_5)CHCH_2CH_2CH_3$	114.23	1^4, 431	0.7136^{20}	1.4018^{20}		118.6		s alc, eth
e159	2-Ethyl-1,3-hexanediol	$CH_3(CH_2)_2CH(OH){-}CH(C_2H_5)CH_2OH$	146.23	Merck: 12, 3790	0.9325^{22}_{4}	1.4530^{22}	−40	244	127	0.6% (w/w) aq; s alc, propylene glycol

(Continued)

TABLE 2.20 Physical Constants of Organic Compounds (*Continued*)

No.	Name	Formula	Formula weight	Beilstein reference	Density, g/mL	Refractive index	Melting point, °C	Boiling point, °C	Flash point, °C	Solubility in 100 parts solvent
e160	Ethyl hexanoate	$CH_3(CH_2)_4CO_2C_2H_5$	144.21	2, 323	0.871_4^{20}	1.4075^{20}	−67	166–168	49	i aq; misc alc, eth
e161	2-Ethylhexanoic acid	$CH_3(CH_2)_3CH(C_2H_5)CO_2H$	144.21	2, 349	0.9077	1.4241^{20}	−118.4	228	127	0.25 aq
e162	2-Ethyl-1-hexanol	$CH_3(CH_2)_3CH(C_2H_5)CH_2OH$	130.23	Merck: 12, 3854	0.8319^{25}	1.4300^{20}	−70	184.6	73	0.07 aq; s alc, bz, chl
e163	2-Ethylhexanoyl chloride	$CH_3(CH_2)_3CH(C_2H_5)COCl$	162.66	2^2, 304	0.939	1.4335^{20}		68^{11mm}	69	
e164	2-Ethylhexyl acetate	$CH_3(CH_2)_3CH(C_2H_5)$-$CH_2O_2CCH_3$	172.27	Merck: 12, 6860	0.8718	1.4204^{20}	−80	199	71	0.03 aq; misc alc, oils, org liquids
e165	2-Ethylhexyl acrylate	$H_2C{=}CCO_2CH(C_2H_5)(CH_2)_3CH_3$	184.28	2^3, 1229	0.885	1.4358		214–219	79	
e166	2-Ethylhexylamine	$CH_3(CH_2)_3CH(C_2H_5)CH_2NH_2$	129.31	4^3, 388	0.789	1.4300^{20}	−76	169	60	i aq; s alc, acet, eth
e167	2-Ethylhexyl chloroformate	$CH_3(CH_2)_3CH(C_2H_5)CH_2O_2CCl$	192.69	3^4, 28	0.981	1.4312^{20}		107^{30mm}	81	
e168	2-Ethylhexyl cyanoacetate	$NCCH_2CO_2CH_2CH(C_2H_5)$-$(CH_2)_3CH_3$	197.28		0.975	1.4380^{20}		150^{11mm}	>110	
e169	2-Ethylhexyl 2-cyano-3,3-diphenylacrylate	$(C_6H_5)_2C{=}C(CN)CO_2CH_2$-$CH(C_2H_5)(CH_2)_3CH_3$	361.49		1.051	1.5670^{20}	−10	$218^{1.5mm}$	>110	
e170	2-Ethylhexyl 4-(dimethylamino)benzoate	$(CH_3)_2NC_6H_4CO_2CH_2$-$CH(C_2H_5)(CH_2)_3CH_3$	277.41		0.995	1.5420^{20}		325	>110	
e171	2-Ethylhexyl glycidyl ether	$CH_3(CH_2)_3CH(C_2H_5)CH_2$-$OCH_2\underbrace{CH{-}CH_2}_{O}$	186.30		0.891	1.4340^{20}		$61^{0.3mm}$	96	
e172	2-Ethylhexyl methacrylate	$H_2C{=}C(CH_3)CO_2CH_2$-$CH(C_2H_5)(CH_2)_3CH_3$	198.31	2^3, 1289	0.885	1.4381^{20}		120^{18mm}	92	
e173	2-Ethylhexyl nitrate	$CH_3(CH_2)_3CH(C_2H_5)CH_2ONO_2$	175.23		0.963	1.4320^{20}			75	explodes when heated
e174	2-Ethylhexyl salicylate	$2\text{-}(HO)C_6H_4CO_2CH_2$-$CH(C_2H_5)(CH_2)_3CH_3$	250.34	10^3, 124	1.014	1.5020^{20}		190^{21mm}	>110	
e175	2-Ethylhexyl vinyl ether	$CH_3(CH_2)_3CH(C_2H_5)$-$CH_2OCH{=}CH_2$	156.26		0.8102	1.4273^{20}	−85	177–178	52	0.01 aq
e176	Ethyl hydrocinnamate	$C_6H_5CH_2CH_2CO_2C_2H_5$	178.23	9, 511	1.010	1.4940^{20}		247–248	107	
e177	Ethyl hydrogen hexanedioate	$HO_2C(CH_2)_4CO_2C_2H_5$	174.20	2^1, 277		1.4387^{20}	28–29	180^{18mm}	>110	
e178	Ethyl 4-hydroxybenzoate	$HOC_6H_4CO_2C_2H_5$	166.18	10, 159			116–118	297–298		0.07 aq; v s alc, eth
e179	Ethyl 3-hydroxybutyrate	$CH_3CH(OH)CH_2CO_2C_2H_5$	132.16	3, 309	1.017_4^{20}	1.4205^{20}		170	64	s aq, alc
e180	Ethyl 2-hydroxyethyl sulfide	$HOCH_2CH_2SCH_2CH_3$	106.19	1^2, 525	1.020	1.4869^{20}		180–184	>110	s eth

e181	Ethyl 6-hydroxyhexanoate	$HO(CH_2)_5CO_2C_2H_5$	160.22	3, 3,628	0.985	1.4370^{20}		128^{12mm}	>110	
e182	Ethyl 2-hydroxyisobutyrate	$(CH_3)_2C(OH)CO_2C_2H_5$	132.16	3, 315	0.965	1.4078^{20}		150	44	dec by hot aq
e183	2-Ethyl-2-(hydroxymethyl)-1,3-propanediol	$C_2H_5C(CH_2OH)_3$	134.18	1^3, 2349			60–62	161^{2mm}		
e184	2-Ethyl-2-(hydroxymethyl)-1,3-propanedioltriacrylate	$(H_2C{=}CHCO_2CH_2)_3CC_2H_5$	296.32		1.100	1.4736^{20}		157	>110	
e185	2-Ethyl-2-(hydroxymethyl)-1,3-propanedioltrimethacrylate	$[H_2C{=}C(CH_3)CO_2CH_2]_3CC_2H_5$	338.40		1.060	1.4724^{20}			>110	
e186	*N*-Ethyl-3-hydroxypiperidine		129.20	Merck: 12, 3890	0.970	1.4754^{20}		95^{15mm}	47	
e187	2,2′-Ethylidenebis(4,6-di-*tert*-butylphenol)	$CH_3CH\{C_6H_2[C(CH_3)_3]_2OH\}_2$	438.70				162–164			
e188	2,2′-Ethylidenebis(4,6-di-*tert*-butylphenyl) fluorophosphite		486.66				201–203			
e189	4,4′-Ethylidenebisphenol	$CH_3CH_2CH(C_6H_4OH)_2$	214.26	6, 1006			123–127			
e190	5-Ethylidene-2-norborene		120.20		0.893	1.4895			38	
e191	2-Ethylimidazole		96.13	23, 78			86	268		
e192	Ethyl isobutyrate	$(CH_3)_2CHCO_2C_2H_5$	116.16	2, 291	0.870^{20}	1.3903^{20}	−88	110	13	misc alc, eth; sl s aq
e193	Ethyl isothiocyanate	CH_3CH_2NCS	87.14	4, 123	1.003^{18}_{4}	1.5142^{18}	−6	130–132	32	i aq; misc alc, eth
e194	Ethyl (−)-lactate	$CH_3CH(OH)CO_2C_2H_5$	118.13	3, 264	1.0328^{20}	1.4124^{20}	−26	154–155	46	misc aq, alc, eth, esters, PE
e195	Ethyl (±)-mandelate	$C_6H_5CH(OH)CO_2C_2H_5$	180.21	10, 202	1.115	1.5120^{20}	33–34	253–255	>110	
e196	Ethyl 2-mercaptoacetate	$HSCH_2CO_2C_2H_5$	120.17	3, 255	1.0964	1.4571^{20}		54^{12mm}	47	s alc, eth
e197	Ethyl 3-mercaptopropionate	$HSCH_2CH_2CO_2C_2H_5$	134.20	3^3, 555	1.039	1.4570^{20}		76^{10mm}	72	

(*Continued*)

TABLE 2.20 Physical Constants of Organic Compounds (*Continued*)

No.	Name	Formula	Formula weight	Beilstein reference	Density, g/mL	Refractive index	Melting point, °C	Boiling point, °C	Flash point, °C	Solubility in 100 parts solvent
e198	Ethylmercury chloride	CH_3CH_2HgCl	165.13		3.5		192	sublimes		0.78 eth; 2.6 chl
e199	Ethyl methacrylate	$H_2C{=}C(CH_3)CO_2C_2H_5$	114.14	2, 423	0.917	1.4116^{25}		118	15	i aq; s alc, eth
e200	Ethyl 4-methoxy-phenylacetate	$CH_3OC_6H_4CO_2C_2H_5$	194.23	10^1, 83	1.097	1.5075^{20}		138^{7mm}	46	
e201	Ethyl 2-methylaceto-acetate	$CH_3C({=}O)CH(CH_3)CO_2C_2H_5$	144.17	3, 679	1.019	1.4280^{20}		187	62	i aq; s alc, eth
e202	*N*-Ethyl-2-methyl-allylamine	$H_2C{=}C(CH_3)CH_2NHC_2H_5$	99.18	4^4, 1104	0.753	1.4221^{20}		105	7	
e203	*N*-Ethyl-*N*-methyl-aniline	$C_6H_5N(CH_3)C_2H_5$	135.21	12, 162	0.947	1.5470^{20}		203–205	74	i aq; misc alc, eth
e204	Ethyl 2-methyl-benzoate	$CH_3C_6H_4CO_2C_2H_5$	164.21	9, 463	1.032	1.5070^{20}		221^{731mm}	91	
e205	Ethyl 3-methyl-benzoate	$CH_3C_6H_4CO_2C_2H_5$	164.21	9, 476	1.030	1.5054^{20}		110^{20mm}	101	
e206	Ethyl 4-methyl-benzoate	$CH_3C_6H_4CO_2C_2H_5$	164.21	9, 484	1.025	1.5085^{20}		235	99	
e207	Ethyl 2-methylbutyrate	$CH_3CH_2CH(CH_3)CO_2C_2H_5$	130.19	2, 305	0.869	1.3969^{20}		133	26	
e208	Ethyl 3-methylbutyrate	$(CH_3)_2CHCH_2CO_2C_2H_5$	130.19	2^2, 275	0.8656^{20}	1.3962^{20}	−99	135	26	0.2 aq; misc alc, bz
e209	2-Ethyl-2-methyl-1,3-dioxolane		116.16	19^2, 11	0.929	1.4090^{20}		116–117	12	
e210	Ethyl methyl ether	$C_2H_5OCH_3$	60.10	1, 314	2.456 g/L		−113	7.4		s aq; misc alc, eth
e210a	3-Ethyl-4-methyl-hexane	$(C_2H_5)_2CHCH(CH_3)C_2H_5$	128.26		0.7420^{20}	1.4134^{20}		140	24	
e211	2-Ethyl-4-methyl-imidazole		110.16	23^2, 72	0.975	1.5000^{20}	47–54	292–295	137	
e212	Ethyl 4-methyl-5-imidazolecarboxy-late		154.17	25^1, 534			204–206			
e213	4-Ethyl-2-methyl-2-(3-methylbutyl)-oxazolidine		185.3		0.877	1.4420^{20}		194	82	
e214	3-Ethyl-2-methyl-pentane	$(C_2H_5)_2CHCH(CH_3)_2$	114.24	1^3, 489	0.7193^{20}_4	1.4040^{20}	−115.0	115.7	<21	i aq; sl s alc; s eth
e215	3-Ethyl-3-methyl-pentane	$(C_2H_5)_3CCH_3$	114.24		0.7274^{20}	1.4078^{20}	−90.9	118.3		i aq; s eth
e216	Ethyl 1-methyl-2-piperidinecarboxy-late		171.24	22^1, 485	0.975	1.4519^{20}		96^{11mm}	73	

e217	Ethyl 1-methyl-3-piperidinecarboxylate		171.24		0.954	1.4510^{20}		89^{11mm}	68	
e218	Ethyl 3-methyl-1-piperidine propionate		199.30	2^2, 59	0.945	1.4530^{20}		112^{13mm}	99	
e219	2-Ethyl-2-methyl-1,3-propanediol	$HOCH_2C(C_2H_5)(CH_3)CH_2OH$	118.18	1, 487			41–44	226	>110	
e220	5-Ethyl-2-methylpyridine	$C_2H_5(CH_3)C_5H_3N$	121.18	20, 248	0.919	1.4970^{20}		178	66	s alc, bz, eth, acid
e221	Ethyl methyl sulfide	$CH_3CH_2SCH_3$	76.15	1, 343	0.842	1.4392^{20}	−106	66.7	−15	i aq; misc alc, eth
e222	Ethyl (methylthio)-acetate	$CH_3SCH_2CO_2C_2H_5$	134.20		1.043	1.4587^{20}		72^{25mm}	59	
e223	*N*-Ethylmorpholine		115.18	27^1, 203	0.905	1.4410^{20}	−63	139	27	misc aq, alc, eth
e224	Ethyl nitrate	$CH_3CH_2ONO_2$	91.13	1, 329	1.100^{25}_{4}	1.3849^{22}	−94.6	87.7	10 (CC)	1 aq; misc alc, eth
e225	Ethyl nitrite	CH_3CH_2ONO	75.07	1, 329	0.90^{15}_{15}			17	−35	misc alc, eth
e226	4-Ethylnitrobenzene	$C_2H_5C_6H_4NO_2$	151.17	5, 358	1.118	1.5445^{20}	−32	245–246	>110	v s alc, eth
e227	Ethyl 4-nitrobenzoate	$O_2NC_6H_4CO_2C_2H_5$	195.17	9, 390			55–59			
e228	Ethyl nonanoate	$CH_3(CH_2)_7CO_2C_2H_5$	186.30	2, 353	0.866	1.4219^{20}	−37	227	94	i aq; misc alc, eth
e229	Ethyl *cis,cis*-9,12-octadecadienoic acid	$H(CH_2)_5CH{=}CHCH_2$-$CH{=}CH(CH_2)_7CO_2C_2H_5$	308.51	2^2, 461	0.8846	1.4675^{20}		193^{6mm}	>110	misc DMF, oils
e230	Ethyl *cis*-9-octadecenoate	$CH_3(CH_2)_7CH{=}CH(CH_2)_7$-$CO_2C_2H_5$	310.53	2, 467	0.869	1.4500^{20}	−32	216^{15mm}	>110	i aq; misc alc, eth
e231	Ethyl octanoate	$CH_3(CH_2)_7CO_2C_2H_5$	172.27	2, 348	0.878	1.4166	−43	208	75	i aq; misc alc, eth
e232	Ethyl oxalyl chloride	$CH_3CH_2OC({=}O)C({=}O)Cl$	136.53	2, 541	1.2223	1.4164^{20}		135	41	d aq, alc; s bz, eth
e233	Ethyl oxamate	$CH_3CH_2OC({=}O)C({=}O)NH_2$	117.10	2, 544			114–116			s aq, eth; i bz
e234	2-Ethyl-2-oxazoline		99.13		0.982	1.4370^{20}	−62	128	29	
e235	Ethyl 2-oxocyclopentanecarboxylate	$(O{=})(C_5H_7)CO_2C_2H_5$	156.18	10, 597	1.054	1.4485^{20}		102^{11mm}	77	
e236	Ethyl 4-oxopentanoate	$CH_3C({=}O)CH_2CH_2CO_2C_2H_5$	144.17	3, 675	1.012	1.4222^{20}		205–206		v s aq; misc alc
e237	Ethyl 2-oxopropionate	$CH_3C({=}O)CO_2C_2H_5$	116.12	3, 616	1.060^{16}_{4}	1.408^{16}		144	45	sl s aq; misc alc, eth
e238	3-Ethylpentane	$(C_2H_5)_3CH$	100.20	1^3, 441	0.6982^{20}_{4}	1.3934^{20}	−118.6	93.5		i aq; s alc, eth
e239	Ethyl pentanoate	$CH_3(CH_2)_3CO_2C_2H_5$	130.19	2, 301	0.877^{20}_{4}	1.3732^{20}	−91.3	145.5		0.2 aq; misc alc, eth
e240	2-Ethylphenol	$C_2H_5C_6H_4OH$	122.17	5, 470	1.037	1.5372^{20}	−18	204	78	
e241	3-Ethylphenol	$C_2H_5C_6H_4OH$	122.17	6, 471	1.001	1.5330^{20}	−4	110^{15mm}	94	
e242	4-Ethylphenol	$C_2H_5C_6H_4OH$	122.17	6, 472	1.011	1.5239	45	218	100	i aq; misc alc, eth
e243	Ethyl phenylacetate	$C_6H_5CH_2CO_2C_2H_5$	164.20	9, 434	1.031	1.4980^{20}		229	77	i aq; misc alc, eth

(Continued)

TABLE 2.20 Physical Constants of Organic Compounds (*Continued*)

No.	Name	Formula	Formula weight	Beilstein reference	Density, g/mL	Refractive index	Melting point, °C	Boiling point, °C	Flash point, °C	Solubility in 100 parts solvent
e244	Ethyl 3-phenylglycidate		192.21		1.102	1.5180^{20}		$96^{0.5mm}$	>110	
e245	1-Ethylpiperazine		114.19	23^2, 5	0.899	1.4690^{20}		157	43	
e246	Ethyl *N*-piperazinocarboxylate		158.20	23^2, 9	1.080	1.4765^{20}		273	>110	
e247	1-Ethylpiperidine		113.20	20, 17	0.834	1.4440^{20}		131	18	
e248	2-Ethylpiperidine		113.20	20, 104	0.858	1.4510^{20}		143	31	s aq
e249	Ethyl 3-piperidinecarboxylate		157.21		1.012	1.4601^{20}		104^{7mm}	90	
e250	Ethyl 4-piperidinecarboxylate		157.21		1.010	1.4591^{20}		204	80	s aq, alc, bz, eth
e251	Ethyl *N*-piperidinepropionate		185.27	20, 62	0.927	1.4545^{20}		217–219	87	
e252	Ethyl 1-propenyl ether	$CH_3CH{=}CHOC_2H_5$	86.13	1, 435	0.778	1.3980^{20}		67–76	−18	
e253	Ethyl propionate	$CH_3CH_2CO_2C_2H_5$	102.13	2, 240	0.8917^{20}	1.3839^{20}	−73.9	99	12	1.7 aq; misc alc, eth
e254	Ethyl propyl ether	$CH_3CH_2OCH_2CH_2CH_3$	88.15	1, 354	0.739	1.3695^{20}	−79	62–63	32	sl s aq; misc alc, eth
e255	Ethyl propyl sulfide	$CH_3CH_2SCH_2CH_2CH_3$	104.21	1^3, 1432	0.8270	1.4462^{20}	−117.0	118.5		s alc
e256	2-Ethylpyridine	$CH_3CH_2(C_5H_4N)$	107.16	20, 241	0.937	1.4964^{20}		149	29	sl s aq; s alc, eth
e257	3-Ethylpyridine	$CH_3CH_2(C_5H_4N)$	107.16	20, 242	0.954	1.5015^{20}		162–165	48	v s alc, eth; sl s aq
e258	4-Ethylpyridine	$CH_3CH_2(C_5H_4N)$	107.16	20, 243	0.942	1.5009^{20}		168	47	sl s aq; s alc, eth
e259	Ethyl 2-pyridinecarboxylate		151.17	22, 35	1.1194	1.5088^{20}	2	240–241	107	misc aq, alc, eth
e260	1-Ethyl-2-pyrrolidinone		113.16		0.992	1.4652^{20}		97^{20mm}	76	
e261	Ethyl salicylate	$C_6H_4(OH)CO_2C_2H_5$	166.18	10, 73	1.131	1.5219^{20}	2–3	232–234	107	misc alc, eth; sl s aq
e262	Ethyl sorbate	$CH_3CH{=}CHCH{=}CHCO_2C_2H_5$	140.18	2, 484	0.956	1.4942^{20}		195.5	69	
e262a	2-Ethyltoluene	$CH_3C_6H_4C_2H_5$	120.19	5^1, 192	0.865	1.5040^{20}	−81	165	39	
e262b	3-Ethyltoluene	$CH_3C_6H_4C_2H_5$	120.19	5, 398	0.865	1.4960^{20}	−95	161	38	
e262c	4-Ethyltoluene	$CH_3C_6H_4C_2H_5$	120.19	5, 397	0.861	1.4950^{20}	−62	162	36	
e263	Ethyl 4-toluenesulfonate	$CH_3C_6H_4SO_2OC_2H_5$	200.26	11, 99	1.166^{45}_{4}	1.5110^{20}	33	173^{15mm}	157	i aq; s alc, eth
e264	*N*-Ethyl-*m*-toluidine	$CH_3C_6H_4NHC_2H_5$	135.21	12, 857	0.957	1.5451^{20}		221	89	
e265	*N*-Ethyl-*o*-toluidine	$CH_3C_6H_4NHC_2H_5$	135.21		0.938	1.5470^{20}		218	88	
e266	6-Ethyl-*o*-toluidine	$C_2H_5C_6H_3(CH_3)NH_2$	135.21		0.968	1.5525^{20}	−33	231	89	
e267	2-(*N*-Ethyl-*m*-toluidino)ethanol	$CH_3C_6H_4N(C_2H_5)CH_2CH_2OH$	179.26		1.019	1.5540^{20}		115^{1mm}	>110	
e268	Ethyl trichloroacetate	$Cl_3CCO_2C_2H_5$	191.44	2, 209	1.383^{20}_{4}	1.4447^{20}		168	65	i aq; s alc, eth
e269	Ethyltrichlorosilane	$C_2H_5SiCl_3$	163.51	4, 630	1.238	1.4252^{20}	−106	99	13	

e270	Ethyltriethoxysilane	$C_2H_5Si(OC_2H_5)_3$	192.33	4^{4}, 4223	0.895	1.3920^{20}		158–166	38	
e271	Ethyltriphenylphosphonium iodide	$C_2H_5P(C_6H_5)_3I$	418.26	16, 760			169–171			
e272	Ethyl undecanoate	$CH_3(CH_2)_{10}CO_2C_2H_5$	214.35	2, 358	0.859	1.4280^{20}		105^{4mm}	>110	i aq; s org solvents
e273	Ethyl 10-undecenoate	$H_2C{=}CH(CH_2)_8CO_2C_2H_5$	212.34	2, 459	0.879	1.4390^{20}		258–259	>110	
e274	Ethylurea	$CH_3CH_2NHC({=}O)NH_2$	88.11	4, 115	1.213^{18}		93–96			v s aq; 80 alc; i eth
e275	*N*-Ethylurethane	$CH_3CH_2NHCO_2C_2H_5$	117.15	4, 114	0.981^{20}_{4}	1.4211^{20}		85^{20mm}	75	63 aq
e276	Ethyl vinyl ether	$CH_3CH_2OCH{=}CH_2$	72.11	1, 433	0.7589^{20}	1.3767^{20}	−116	35	< −45	0.9 aq; s alc, eth
e277	*N*-Ethyl-2,3-xylidine	$(CH_3)_2C_6H_3NHC_2H_5$	149.24	12, 1101	0.917	1.5468^{20}		228	71	
e278	1-Ethynyl-1-cyclohexanol	$HOC_6H_{10}C{\equiv}CH$	124.18	6^{2}, 100	0.967		31–33	180	62	2.4 aq; misc alc, bz, acet, ketones, PE
e279	Eugenol	4-$(H_2C{=}CHCH_2)C_6H_3$-2-$(OCH_3)OH$	164.20	6, 961	1.066	1.5410^{20}	−12/−10	254	>110	
f1	Fluoranthene		202.26	5, 685	1.252^{0}_{4}		108	384		sl s alc; s bz, eth
f2	Fluorene		166.22	5, 625	1.203^{0}_{4}		115	295		v s HOAc; s bz, eth
f3	Fluorenone		180.21	7, 465	1.1300^{99}_{4}	1.6369^{99}	82–85	342		s alc, bz; v s eth
f4	Fluorescein		332.31	19, 222			320			s hot alc, hot HOAc
f5	Fluoroacetic acid	FCH_2CO_2H	78.04	2, 193			33	165		sl s aq, alc
f6	4-Fluoroacetophenone	$FC_6H_4COCH_3$	138.14		1.138	1.5110^{20}		196	71	
f7	2-Fluoroaniline	$FC_6H_4NH_2$	111.12	12^{1}, 296	1.151	1.5420^{20}	−29	183	60	
f8	4-Fluoroaniline	$FC_6H_4NH_2$	111.12	12, 597	1.1725	1.5395^{20}	−2	187	73	sl s aq; s alc, eth
f9	2-Fluorobenzaldehyde	FC_6H_4CHO	124.11	7^{1}, 132	1.178	1.5220^{20}	−44.5	91^{46mm}	55	
f10	4-Fluorobenzaldehyde	FC_6H_4CHO	124.11	7^{1}, 132	1.157	1.5200^{20}	−10	181	56	
f11	Fluorobenzene	C_6H_5F	96.11	5, 198	1.0240^{20}_{4}	1.4657^{20}	−42.2	84.7	−15	0.15 aq; misc alc
f12	2-Fluorobenzoic acid	$FC_6H_5CO_2H$	140.11	9, 333	1.460^{25}_{4}		123–125			sl s aq; s alc, eth
f13	4-Fluorobenzoic acid	$FC_6H_5CO_2H$	140.11	9, 333	1.479^{25}_{4}		184–187			0.1 aq; s alc, eth
f14	2-Fluorobenzoyl chloride	FC_6H_5COCl	158.56	9^{1}, 136	1.328	1.5365^{20}	4	92^{15mm}	82	
f15	4-Fluorobenzoyl chloride	FC_6H_5COCl	158.56	9^{1}, 137	1.342	1.5296^{20}	9	82^{20mm}	82	
f16	4-Fluorobenzyl chloride	$FC_6H_5CH_2Cl$	144.58		1.207	1.5130^{20}		82^{26mm}	60	
f17	Fluoroethane	CH_3CH_2F	48.06	1, 82			−143.2	−37.7		198 mL aq; v s alc, eth
f18	Fluoromethane	CH_3F	34.04	1, 59	1.195 g/L		−141.8	−78.4		166 mL aq; v s alc, eth
f19	3-Fluoro-1-methoxybenzene	$FC_6H_4OCH_3$	126.13		1.104	1.4880^{20}		158^{743mm}	43	
f20	4-Fluoro-1-methoxybenzene	$FC_6H_4OCH_3$	126.13	6^{1}, 98	1.114	1.4877^{20}	−45	157	43	s eth

(*Continued*)

TABLE 2.20 Physical Constants of Organic Compounds (*Continued*)

No.	Name	Formula	Formula weight	Beilstein reference	Density, g/mL	Refractive index	Melting point, °C	Boiling point, °C	Flash point, °C	Solubility in 100 parts solvent
f21	2-Fluoro-2-methylpropane	$(CH_3)_3CF$	76.11	1^4, 286			−77	12	−12	
f22	4-Fluoro-3-nitroaniline	$FC_6H_3(NO_2)NH_2$	156.12	12, 729			96–98		91	
f23	1-Fluoro-4-nitrobenzene	$FC_6H_4NO_2$	141.10	5, 241	1.3300^{20}_{4}	1.5312^{20}	21	205	83	i aq; s alc, eth
f24	4-Fluoro-3-nitrotoluene	$CH_3C_6H_3(NO_2)F$	155.13		1.262	1.5240^{20}	28–30	241	>110	
f25	4-Fluorophenol	FC_6H_4OH	112.10	6, 183			46–48	185	68	
f26	2-Fluoropyridine	$F(C_5H_4N)$	97.09	20^1, 80	1.128	1.4680^{20}		126	28	
f27	2-Fluorotoluene	$FC_6H_4CH_3$	110.13	5, 290	1.0014^{17}	1.4716^{17}	−62	115	12	v s alc, eth
f28	3-Fluorotoluene	$FC_6H_4CH_3$	110.13	5, 290	0.9974^{20}	1.4691^{20}	−87	115	9	s alc, eth
f29	4-Fluorotoluene	$FC_6H_4CH_3$	110.13	5, 290	0.9975^{20}	1.4698^{20}	−56	117	17	s alc, eth
f30	Fluorotrichloromethane	$FCCl_3$	137.37	1, 64	1.494	1.3821^{20}	−110	24	none	
f31	Formaldehyde	$H_2C{=}O$	30.03	1, 558	0.815^{-20}_{4}	0.8153^{-20}	−92	−19.5	56	122 aq; s alc, eth
f32	Formamide	$HC({=}O)NH_2$	45.04	2, 26	1.1334^{20}_{4}	1.4475^{20}	2.6	220	154	misc aq, alc, acet
f33	Formamidine acetate	$HC({=}NH)NH_2 \cdot HO_2CCH_3$	104.11				158 dec			
f34	Formamidinesulfinic acid	$H_2NC({=}NH)S(O)OH$	108.12	3^1, 36			126 dec			
f35	Formanilide	C_6H_5NHCHO	121.14	12, 230	1.144		47	271	>110	2.5 aq
f36	Formic acid	HCO_2H	46.03	2, 8	1.220^{20}_{4}	1.3704^{20}	8.3	100.8	68	misc aq, alc, eth
f37	2-Formylbenzoic acid	$HO_2CC_6H_4HCO$	150.13	10, 666	1.404		96–98			s aq; v s alc, eth
f38	Formylhydrazine	$HC({=}O)NHNH_2$	60.06	2, 93			54–56		>110	v s alc, chl, eth; s bz
f39	4-Formylmorpholine		115.13	27^3, 274	1.145	1.4848^{20}		236–237	>110	
f40	*N*-Formylpiperidine		113.16	20, 45	1.019	1.4780^{20}		222	91	
f41	D-(−)-Fructose		180.16	31, 321				122 dec		v s aq; 6.7 alc; s pyr
f42	Fumaric acid	$HO_2CCH{=}CHCO_2H$	116.07	2, 737	1.635^{20}_{4}		287	subl 300		0.6 aq; 9 alc; 0.7 eth
f43	Fumaroyl dichloride	$ClC({=}O)CH{=}CHC({=}O)Cl$	152.96	2, 743	1.408^{20}	1.4988^{20}		161–164	73	dec aq, alc
f44	2-Furaldehyde		96.09	17^2, 305	1.1598^{20}_{4}	1.5262^{20}	−36.5	161.8	60	8 aq; misc alc, eth
f45	Furan		68.07	17, 27	0.9514^{20}	1.4214^{20}	−85.6	31.4	−35	1 aq; misc alc, eth
f46	2-Furanacrylic acid		138.12	18, 300			142–144	286		0.2 aq; 1.1 bz; s alc, eth, HOAc
f47	2,5-Furandimethanol		128.13	17^1, 90			74–76			
f48	2-Furanmethanethiol		114.17	17^2, 116	1.132	1.5304^{20}		155	45	
f49	Furfuryl acetate		140.14	17^2, 115	1.1175^{20}_{4}	1.4618^{20}		175–177	65	i aq; s alc, eth
f50	Furfuryl alcohol		98.10	17, 112	1.1295^{20}_{4}	1.4868^{20}	−31	171	75	misc aq(dec); v s alc, eth
f51	Furfurylamine		97.12	18, 584	1.0995^{20}_{4}	1.4900^{20}	−70	145–146	46	misc aq; s alc, eth

f52	Furfuryl methacrylate		166.18	17^3, 1248	1.078	1.4820^{20}		82^{5mm}	90	
f53	α-Furildioxime		220.18	19, 166			166–168			v s alc, eth; sl s bz
f54	2-Furoic acid		112.08	18, 272			133–134	230–232		4 aq; s alc; v s eth
f55	2-Furoyl chloride		130.53	18, 276	1.324	1.5310^{20}	−2	170	85	dec aq, alc; s eth
g1	D-(+)-Galactose		180.16	31, 295			167			200 aq; s pyr; sl s alc
g2	Geraniol	$(CH_3)_2C{=}CHCH_2CH_2$-$C(CH_3){=}CHCH_2OH$	154.25	1, 457	0.8894^{20}_4	1.4766^{20}		230	76	i aq; misc alc, eth
g3	Geranyl acetate	$(CH_3)_2C{=}CHCH_2CH_2$-$C(CH_3){=}CHCH_2O_2CCH_3$	196.29	2, 140	0.9174^{15}_{15}	1.4628^{15}		138^{25mm}	104	v s alc; misc eth
g4	Gerard reagent P	$[(C_5H_5N)CH_2C({=}O)NHNH_2]^+ Cl^-$	187.63	Merck: 12, 4436			dec 200			less soluble in polar solvents than T
g5	Gerard reagent T	$[(CH_3)_3NCH_2C({=}O)NHNH_2]^+ Cl^-$	167.64	Merck: 12, 4436			192			v s aq, HOAc, glyc, ethylene glycol
g6	D-Gluconic acid		196.16	3, 542			131			v s aq; sl s alc; i eth
g7	δ-Gluconolactone		178.14	18^1, 405			153			50 aq; 1 alc; i eth
g8	α-D-(+)-Glucose		180.16	31, 83	1.5620^{18}_4		153–156			91 aq; 0.83 MeOH; s pyr
g9	α-D-Glucose pentaacetate		390.34	31, 119			109–111			0.15 aq; 1.3 alc; 3 eth
g11	D-Glucurono-3,6-lactone		176.12	Merck: 11, 4362			176–178			27 aq; 2.8 MeOH
g12	(*S*)-(+)-Glutamic acid	$HO_2CCH_2CH_2CH(NH_2)CO_2H$	147.13	4, 488	1.538^{20}_4		d 247	subl 200		0.8 aq; i alc, eth
g13	(*S*)-(+)-Glutamine	$H_2NC({=}O)CH_2CH_2$-$CH(NH_2)CO_2H$	146.15	4, 491			185 dec			5 aq; 0.0035 MeOH; i bz, chl, eth, acet
g14	Glutaric acid	$HO_2CCH_2CH_2CH_2CO_2H$	132.12	2, 631	1.429^{20}_4	1.4188^{106}	98	303		43 aq^{20}; v s alc, eth; s bz, chl; sl s PE
g15	Glutaric anhydride		114.10	17, 411			55–57	150^{10mm}	>110	
g16	Glutaric dialdehyde	$OCHCH_2CH_2CH_2CHO$	100.12	1, 776		1.4338^{25}		187–189	none	s aq, alc
g17	Glutaronitrile	$NCCH_2CH_2CH_2CN$	94.12	2, 635	0.9888^{23}	1.4345^{20}	−29	286	>110	s aq, alc, chl; i eth
g18	Glutaryl dichloride	$ClC({=}O)(CH_2)_3C({=}O)Cl$	169.01	2, 634	1.324	1.4720^{20}		216–218	106	dec aq, alc; s eth
g19	Glycerol	$HOCH_2CH(OH)CH_2OH$	92.09	1, 502	1.2613^{20}	1.4746^{20}	18	290	199	misc aq, alc; 0.2 eth
g20	Glyceryl tris(butyrate)	$(CH_3CH_2CH_2CO_2CH_2)_2CH$-$O_2CCH_2CH_2CH_3$	302.37	2, 273	1.032^{20}_4	1.4359^{20}	−75	287–288	173	i aq; v s alc, eth
g21	Glyceryl tris-(dodecanoate)	$[CH_3(CH_2)_{10}CO_2CH_2]_2CH$-$O_2C(CH_2)_{10}CH_3$	639.02	2, 362	0.894^{60}_4	1.4404^{60}	46			v s bz, eth; sl s alc
g22	Glyceryl tris(nitrate)	$O_2NOCH_2CH(ONO_2)CH_2ONO_2$	227.09	1, 516	1.594^{20}_4	1.4786^{12}	13.3	160^{5mm}	explodes 270	0.18 aq; 54 alc; misc eth

(Continued)

TABLE 2.20 Physical Constants of Organic Compounds (*Continued*)

No.	Name	Formula	Formula weight	Beilstein reference	Density, g/mL	Refractive index	Melting point, °C	Boiling point, °C	Flash point, °C	Solubility in 100 parts solvent
g23	Glyceryl tris(oleate)	$[CH_3(CH_2)_7CH{=}CH(CH_2)_7CO_2CH_2]_2CHO_2C(CH_2)_7CH{=}CH(CH_2)_7CH_3$	885.46	4, 468	0.915^{15}_{4}	1.4621^{40}	−4/−5	235^{15mm}		s chl, eth, CCl_4
g24	Glyceryl tris(palmitate)	$[CH_3(CH_2)_{14}CO_2CH_2]_2CHO_2C(CH_2)_{14}CH_3$	807.35	2, 373	0.8663^{80}_{4}	1.4381^{80}	65–66	310–320		v s bz, chl, eth
g25	Glyceryl tris-(tridecanoate)	$[CH_3(CH_2)_{11}CO_2CH_2]_2CHO_2C(CH_2)_{11}CH_3$	723.18	2, 367	0.885^{60}_{4}	1.4428^{60}	57			v s alc, bz, chl
g26	Glycine	$H_2NCH_2CO_2H$	75.07	4, 333	1.1607		dec 240			25 aq; 0.6 pyr; i eth
g27	*N*-Glycylglycine	$H_2NCH_2C({=}O)NHCH_2CO_2H$	132.12	4, 371			260 dec			s hot aq; sl s alc
g28	Glyoxal	$HC({=}O)CHO$	58.04	1, 759	1.14	1.3826^{20}	15	50.4		viol rxn aq; s anhyd solvents; mixtures with air may explode
g29	Glyoxylic acid	$HC({=}O)CO_2H$	74.04	3, 594			98			v s aq; sl s alc, eth
g30	Guanidine	$H_2NC({=}NH)NH_2$	59.07	3, 82			*ca.* 50	dec 160		v s aq, alc
g31	Guanine		151.13	26, 449			>300			s alk soln, dil acids; sl s alc, eth
h1	Heptadecane	$CH_3(CH_2)_{15}CH_3$	140.41	1, 173	0.7767^{22}	1.4360^{25}	22.0	302.2	148	s eth; sl s alc
h1a	1-Heptadecanol	$CH_3(CH_2)_{16}OH$	256.48	1^1, 220			53.8	333	>110	
h2	Heptafluorobutyric acid	$CF_3CF_2CF_2CO_2H$	214.04		1.625	$<1.300^{20}$		120	none	
h3	Heptaldehyde	$CH_3(CH_2)_5CHO$	114.19	1^2, 750	0.8216^{15}_{4}	1.4285^{20}	−43	153	35	misc alc, eth; sl s aq
h4	2,2,4,4,6,8,8-Heptamethylnonane	$(CH_3)_3CCH_2C(CH_3)_2CH_2CH(CH_3)CH_2C(CH_3)_3$	226.45		0.793	1.4391^{20}		240	95	
h5	1,1,1,3,5,5,5-Heptamethyltrisiloxane	$[(CH_3)_3SiO]_2SiHCH_3$	222.51	4^3, 1874	0.819	1.3820^{20}		142	27	
h6	Heptane	$CH_3(CH_2)_5CH_3$	100.21	1, 154	0.6838^{20}_{4}	1.3877^{20}	−90.6	98.4	−4 (CC)	s alc, chl, eth
h7	Heptanedioic acid	$HO_2C(CH_2)_5CO_2H$	160.17	2, 670	1.329^{15}		105.8	212^{10mm}		5 aq; v s alc, eth
h8	1-Heptanethiol	$CH_3(CH_2)_6SH$	132.27	1, 415			−43.2	176.9	46	i aq
h9	Heptanoic acid	$CH_3(CH_2)_5CO_2H$	130.19	2, 338	0.9181^{20}_{4}	1.4221^{20}	−8	222	>110	0.25 aq; s alc, eth
h10	Heptanoic anhydride	$[CH_3(CH_2)_5CO]_2O$	242.36	2, 340	0.923	1.4332^{20}	−12.4	268	>110	i aq; s alc, eth
h11	1-Heptanol	$CH_3(CH_2)_6OH$	116.20	1, 414	0.8219^{20}_{4}	1.4242^{20}	−34	176.4	73	misc alc, eth
h12	2-Heptanol	$CH_3(CH_2)_5CH(OH)CH_3$	116.20	1, 415	0.8167^{20}	1.4210^{10}		159	71	0.35 aq; s alc, bz, eth
h13	3-Heptanol	$CH_3(CH_2)_3CH(OH)CH_2CH_3$	116.20	1^1, 205	0.8227^{20}	1.4214^{20}	−70	157	60	sl s aq
h14	2-Heptanone	$HC_3(CH_2)_4C({=}O)CH_3$	114.19	1, 699	0.8197^{15}_{4}	1.4116^{15}	−35	151	39	s alc, eth
h15	3-Heptanone	$CH_3(CH_2)_3C({=}O)CH_2CH_3$	114.19	1, 699	0.8197^{20}_{20}	1.4055^{20}	−39	147	46	0.43 aq; s alc, eth
h16	4-Heptanone	$CH_3(CH_2)_2C({=}O)(CH_2)_2CH_3$	114.19	1, 699	0.817	1.4068^{20}	−32.1	143.7	48 (CC)	0.53 aq; misc alc, eth
h17	Heptanoyl chloride	$CH_3(CH_2)_5C({=}O)Cl$	148.63	2, 340	0.960	1.4300^{20}		173	58	dec aq, alc; s eth

h18	1-Heptene	$CH_3(CH_2)_4CH{=}CH_2$	98.90	1, 219	0.6970^{20}	1.3999^{20}	−120	93.6	−8	0.1 aq; s alc, eth
h18a	*cis*-2-Heptene	$CH_3(CH_2)_3CH{=}CHCH_3$	98.19	1^3, 825	0.708^{20}	1.406^{20}		98.4	−6	
h18b	*trans*-2-Heptene	$CH_3(CH_2)_3CH{=}CHCH_3$	98.19	1, 219	0.7012^{20}	1.4045^{20}	−109.5	98	−1	
h19	1-Heptylamine	$CH_3(CH_2)_6NH_2$	115.22	4, 193	0.777	1.4243^{20}	−23	154–56	35	s alc, acet, eth, PE
h20	1-Heptyne	$CH_3(CH_2)_4C{\equiv}CH$	96.17	1, 256	0.733	1.4075^{20}	−81	99–100	−2	
h21	Hexachloroacetone	$Cl_3CC({=}O)CCl_3$	264.75	1, 657	1.743	1.5112^{20}	−30	66^{6mm}	none	sl s aq; s acet
h22	Hexachlorobenzene	CCl_6	284.78	5, 205	2.044^{24}		232	325	242	s bz, chl, eth
h23	Hexachloro-1,3-butadiene	$CL_2C{=}CClCCl{=}CCl_2$	260.76	1, 250	1.655	1.5550^{20}	−21	215	none	s alc, eth
h24	1,2,3,4,5,6-Hexachlorocyclohexane, γ-isomer	$C_6H_6Cl_6$	290.83	5^1, 8	1.87^{20}		113–115			s bz, chl
h25	Hexachlorocyclo-1,3-pentadiene		272.77		1.701^{25}_4	1.5644^{20}	−10	239	none	
h27	Hexachloroethane	Cl_3CCCl_3	236.74	1, 87	2.091		187	sublimes	none	s alc, bz, chl, eth
h28	1,4,5,6,7,7-Hexachloro-5-norbornene-2,3-dicarboxylic anhydride		370.83	9^3, 4049			239–242			
h29	Hexachlorophene	$CH_3[C_6H(Cl)_3OH]_2$	406.91	6^3, 5407			163–165		none	
h30	Hexachloropropene	$Cl_3CC(Cl){=}CCl_2$	248.75	1, 200	1.765	1.5480^{20}		210	none	
h31	Hexadecane	$CH_3(CH_2)_{14}CH_3$	226.45	1, 172	0.7733^{20}_4	1.4345^{20}	18.2	286.8	135	misc eth
h32	1,2-Hexadecanediol	$CH_3(CH_2)_{13}CH(OH)CH_2OH$	258.45	1^3, 2244			72–74			
h33	1-Hexadecanethiol	$CH_3(CH_2)_{15}SH$	258.51	1, 430	0.840	1.4720^{20}	18–20	184^{7mm}	101	sl s alc, s eth
h34	Hexadecanoic acid	$CH_3(CH_2)_{14}CO_2H$	256.43	2, 370	0.852^{62}_4	1.4273^{80}	62	351		s hot: chl, eth
h35	1-Hexadecanol	$CH_3(CH_2)_{15}OH$	242.45	1, 429	0.8116^{60}	1.4355^{60}	49.3	334	135	s alc, chl, eth
h36	1-Hexadecene	$CH_3(CH_2)_{14}CH{=}CH_2$	224.43	1, 226	0.783^{20}_4	1.4401	4.1	284	132	s alc, eth, PE
h37	1-Hexadecylamine	$CH_3(CH_2)_{15}NH_2$	241.46	4, 202			45–48	330	140	v s alc, eth; s bz, chl
h38	2,4-Hexadienal	$CH_3CH{=}CHCH{=}CHCHO$	96.13	1^2, 809	0.871	1.5386^{20}		76^{30mm}	67	
h39	1,5-Hexadiene	$H_2C{=}CHCH_2CH_2CH{=}CH_2$	82.15	1, 253	0.6923^{20}_4	1.4042^{20}	−140.7	59.5	−27	s alc, eth
h40	2,4-Hexadienoic acid	$CH_3CH{=}CHCH{=}CO_2H$	112.13	2, 483			134.5	119^{10mm}	127	0.2 aq; 13 alc; 9 acet; 2.3 bz; 11 diox; 1 CCl_4
h41	Hexafluorobenzene	C_6F_6	186.05	5^3, 523	1.6182^{20}	1.3781^{20}	5.1	80.3	10	
h42	Hexafluoroethane	F_3CCF_3	138.01	1^3, 132	1.590^{-78}		−100.7	−78.3		sl s alc, eth
h43	1,1,1,3,3,3-Hexafluoro-2-propanol	$(CF_3)_2CHOH$	168.04		1.596^{25}	1.2750^{20}	−3	58.2	none	s aq, bz, CCl_4
h44	Hexafluoropropene	$CF_3CF{=}CF_2$	150.02	1^3, 697			−153	−28		

(*Continued*)

TABLE 2.20 Physical Constants of Organic Compounds (*Continued*)

No.	Name	Formula	Formula weight	Beilstein reference	Density, g/mL	Refractive index	Melting point, °C	Boiling point, °C	Flash point, °C	Solubility in 100 parts solvent
h45	Hexamethylcyclotri-siloxane	[-Si$(CH_3)_2$O-]$_3$	222.48	4^3, 1884			64–66	133–135	35	
h46	1,1,1,3,3,3-Hexamethyl-disilazane	$(CH_3)_3SiNHSi(CH_3)_3$	161.40	4,3, 1861	0.774^{20}_{4}	1.4071^{20}		126	8	
h47	Hexamethyldisiloxane	$(CH_3)_3SiOSi(CH_3)_3$	162.38	4^3, 1859	0.764^{20}_{4}	1.3775^{20}	−67	101	−2	
h48	Hexamethyleneimine		99.18	20, 94	0.880	1.4631^{20}		138^{749mm}	18	
h49	Hexamethylene-tetramine		140.19	1, 583	1.331^{-5}		280 subl		250	67 aq; 8 alc; 10 chl
h50	Hexamethyl-phosphoramide	$[(CH_3)_2N]_3P({=}O)$	179.20		1.027^{20}	1.4588^{20}	7	232^{740mm}	105	misc aq
h51	Hexanaldehyde	$CH_3(CH_2)_4CHO$	100.16	1^2, 745	0.8335^{20}_{4}	1.4035^{20}	−56	131	32	v s alc, eth; sl s aq
h52	Hexane	$CH_3(CH_2)_4CH_3$	86.18	1, 142	0.6594^{20}_{4}	1.3749^{20}	−95.4	68.7	−22	misc alc, chl, eth
h53	1,6-Hexanediamine	$H_2N(CH_2)_6NH_2$	116.21	4, 269			42	205	81	v s aq; sl s alc, bz
h54	1,6-Hexanedioic acid	$HO_2C(CH_2)_4CO_2H$	146.14	2, 649	1.360^{25}_{4}		152–154	337.5	196	1.4 aq; v s alc; s acet
h55	DL-1,2-Hexanediol	$CH_3(CH_2)_3CH(OH)CH_2OH$	118.18	1^1, 251	0.951	1.4425^{20}		223–224	>110	
h56	1,6-Hexanediol	$HO(CH_2)_6OH$	118.18	1, 484	0.958	1.4579^{25}	42.8	208	101	v s aq, alc
h57	2,5-Hexanediol	$CH_3CH(OH)CH_2CH_2CH(OH)CH_3$	118.18	1, 485	0.9617^{45}_{16}	1.4465^{20}	−50	220.8	101	s aq, alc, eth
h58	1,6-Hexanediol diacrylate	$[H_2C{=}CHCO_2(CH_2)_3\text{-}]_2$	226.28		1.010	1.4562^{20}			>110	
h59	1,6-Hexanediol dimethacrylate	$[H_2C{=}C(CH_3)CO_2(CH_2)_3\text{-}]_2$	254.33		0.995	1.4580^{20}		> 350	>110	
h60	2,5-Hexanedione	$CH_3C({=}O)CH_2CH_2C({=}O)CH_3$	114.14	1, 788	0.973^{20}_{4}	1.4260^{20}	−9	188	78	misc aq, alc, eth
h61	Hexanenitrile	$CH_3(CH_2)_4CN$	97.16	2, 324	0.8052^{20}	1.4069^{20}	−80.3	163.6	43	i aq; s alc, eth
h62	1-Hexanethiol	$CH_3(CH_2)_5SH$	118.24	1^3, 1659	0.8424^{20}_{4}	1.4496^{20}	−80.5	152.7	20	i aq; v s alc, eth
h63	1,2,6-Hexanetriol	$HOCH_2CH(OH)(CH_2)_3CH_2OH$	134.17	1^4, 2784	1.1063^{20}_{20}	1.58^{20}	−32.8	178^{5mm}	191	misc alc, acet; i bz
h64	Hexanoic acid	$CH_3(CH_2)_4CO_2H$	116.16	2, 321	0.9265^{20}_{4}	1.4168^{20}	−3	205	102	1.1 aq; v s alc, eth
h65	Hexanoic anhydride	$[CH_3(CH_2)_4C({=}O)]_2O$	214.31	2, 324	0.926	1.4280^{20}	−41	246–248	>110	s alc
h66	1-Hexanol	$CH_3(CH_2)_5OH$	102.18	1, 407	0.8136^{20}	1.4182^{20}	−44.6	157.5	63	8 aq; misc bz, eth; s alc
h67	2-Hexanol	$CH_3(CH_2)_3CH(OH)CH_3$	102.18	1, 408	0.8108^{25}_{4}	1.4128^{25}	−47	139.9	41	sl s aq; s alc, eth
h68	3-Hexanol	$CH_3CH_2CH_2CH(OH)CH_2CH_3$	102.18	1, 408	0.8193^{20}_{4}	1.4160^{20}		135	41	
h69	6-Hexanolactone		114.14	17^2, 290	1.030	1.4630^{20}	−18	215	109	
h70	2-Hexanone	$CH_3(CH_2)_3C({=}O)CH_3$	100.16	1, 689	0.8113^{20}	1.4007^{20}	−55.5	127.6	25	v s alc, eth
h71	3-Hexanone	$CH_3CH_2CH_2C({=}O)CH_2CH_3$	100.16	1, 690	0.815	1.4002^{20}		123	35	
h72	Hexanoyl chloride	$CH_3(CH_2)_4C({=}O)Cl$	134.61	2, 324	0.9754^{20}_{4}	1.4263^{20}	−87	153	50	dec aq, alc; s eth
h73	1-Hexene	$CH_3(CH_2)_3CH{=}CH_2$	84.16	1, 215	0.6732^{20}	1.3879^{20}	−139.8	63.5	−9	0.005 aq
h74	*trans*-2-Hexenoic acid	$CH_3(CH_2)_2CH{=}CHCO_2H$	114.14	2^4, 1563	0.965	1.4885^{20}	33–35	217	>110	
h75	*trans*-3-Hexenoic acid	$CH_3CH_2CH{=}CHCH_2CO_2H$	114.14	2, 435	0.963	1.4398^{20}	11–12	119^{22mm}	>110	

h76	*trans*-2-Hexen-1-ol	$CH_3CH_2CH_2CH{=}CHCH_2OH$	100.16	1^2, 486	0.849	1.4343^{20}		158–160	54	
h77	5-Hexen-2-one	$H_2C{=}CHCH_2CH_2C({=}O)CH_3$	98.15	1, 734	0.847	1.4197^{20}		128–129	23	
h78	*trans*-2-Hexenyl acetate	$CH_3C({=}O)CH_2CH{=}CHCH_2$-$CH_2CH_3$	142.20	2^2, 151	0.898	1.4275^{20}		166	58	
h79	Hexyl acetate	$CH_3(CH_2)_5O_2CCH_3$	144.21	2, 132	0.860^{20}_{20}	1.4090^{20}	−81	171	45	0.13 aq; v s alc, eth
h80	Hexyl acrylate	$H_2C{=}CHCO_2(CH_2)_5CH_3$	156.23	2^3, 1228	0.888	1.4280^{20}		90^{24mm}	68	
h81	Hexylamine	$CH_3(CH_2)_5NH_2$	101.19	4, 188	0.763^{25}_{4}	1.4180^{20}	−23	133	8	sl s aq; misc alc, eth
h82	1-Hexyne	$CH_3(CH_2)_3C{\equiv}CH$	82.14	1, 253	0.7152^{20}_{4}	1.3989^{20}	−131.9	71.3	−21	i aq; s alc, eth
h83	L-Histidine		155.16	25, 513			282 dec			41 aq; v sl s alc
h84	Hydantoin		100.08	24, 242			221–223			s alc, alk; sl s eth
h85	Hydrazine	H_2NNH_2	32.05	Merck: 12, 4809	1.0036^{25}_{4}	1.4700^{20}	1.4	113.5	52	misc aq, alc
h86	1,4-Hydroquinone	C_6H_4-1,4-$(OH)_2$	110.11	6, 836	1.332^{15}		172	286		7 aq; v s alc, eth; sl s bz
h87	Hydroxyacetaldehyde	$HOCH_2CHO$	60.05	1, 817	1.366^{100}		93–94	110^{12mm}		v s aq, alc; sl s eth
h88	Hydroxyacetic acid	$HOCH_2CO_2H$	76.05	3, 228			80	100		s aq, alc, acet, eth
h89	1′-Hydroxy-2′-acetonaphthone	$C_{10}H_6(OH)C({=}O)CH_3$	186.21	8, 149			98–100	325 sl d		i aq; v s bz; s HOAc
h90	Hydroxyacetone	$HOCH_2C({=}O)CH_3$	74.08	1^1, 84	1.082	1.4315^{20}	−17	146	56	misc aq, alc, eth
h91	2′-Hydroxyacetophenone	$HOC_6H_4C({=}O)CH_3$	136.15	8, 85	1.131^{21}_{4}	1.5584^{20}	4–6	213^{717mm}	>110	misc alc, eth; sl s aq
h92	3′-Hydroxyacetophenone	$HOC_6H_4C({=}O)CH_3$	136.15	8, 86	1.100^{100}	1.535^{100}	87–89	296		s aq; v s alc, bz, eth
h93	4′-Hydroxyacetophenone	$HOC_6H_4C({=}O)CH_3$	136.15	8, 87	1.109^{100}		109–111	148^{3mm}		v s alc, eth; sl s aq
h94	2-Hydroxybenzaldehyde	$C_6H_4(OH)CHO$	122.12	8, 31	1.1674^{20}	1.5740^{20}	−7	196.7	78	1.7 aq^{86}; s alc, eth
h95	3-Hydroxybenzaldehyde	$C_6H_4(OH)CHO$	122.12	8, 58			103–105	191^{50mm}		s alc, bz, eth; sl s aq
h96	4-Hydroxybenzaldehyde	$C_6H_4(OH)CHO$	122.12	8, 64	1.129^{130}_{4}		117–119			1 aq; 70 acet; 4 bz^{65}; v s alc, eth
h97	2-Hydroxybenzaldehyde oxime	$C_6H_4(OH)CH{=}NOH$	137.14	8, 49			57	dec		v s alc, bz, eth, acids
h98	2-Hydroxybenzamide	$C_6H_4(OH)C({=}O)NH_2$	137.14	10, 87			140	dec 270		0.2 aq; s alc, chl, eth
h99	2-Hydroxybenzoic acid	$C_6H_4(OH)CO_2H$	138.12	10, 43	1.443^{20}_{4}		157–159	211^{20mm}		0.2 aq; 37 alc; 33 eth; 33 acet; 2 chl; 0.7 bz

(*Continued*)

TABLE 2.20 Physical Constants of Organic Compounds (*Continued*)

No.	Name	Formula	Formula weight	Beilstein reference	Density, g/mL	Refractive index	Melting point, °C	Boiling point, °C	Flash point, °C	Solubility in 100 parts solvent
h100	3-Hydroxybenzoic acid	$C_6H_4(OH)CO_2H$	138.12	10, 134	1.473		201–203			0.8 aq; 10 eth
h101	4-Hydroxybenzoic acid	$C_6H_4(OH)CO_2H$	138.12	10, 149	1.468^{4}		215–217			0.2 aq; v s alc; 23 eth
h102	4-Hydroxybenzoic hydrazide	$HOC_6H_4C(=O)NHNH_2$	152.15	10, 174			266 dec			
h103	4-Hydroxybenzophenone	$HOC_6H_4C(=O)C_6H_5$	198.22	8^{2}, 184			132–135			v s alc, eth; sl s aq
h104	1-Hydroxybenzotriazole		135.13	26, 41			155–158			
h105	6-Hydroxy-1,3-benzoxathiol-2-one		168.17	19^{4}, 2508			158–160			
h106	2-Hydroxybenzyl alcohol	$HOC_6H_4CH_2OH$	124.13	6, 891	1.161^{25}		83–85	subl 100		6.6 aq; v s alc, chl, eth; s bz
h107	1-Hydroxy-2-butanone	$CH_3CH_2C(=O)CH_2OH$	88.11	1, 826	1.026	1.4282^{20}		78^{60mm}	60	
h108	3-Hydroxy-2-butanone	$CH_3C(=O)CH(OH)CH_3$	88.11	1, 827	0.9972^{17}_{4}	1.4171^{20}	15	148	50	misc aq, alc; sl s eth
h109	4-Hydroxycinnamic acid	$HOC_6H_4CH=CHCO_2H$	164.16	10, 297			210–213			s alc, eth; sl s aq
h111	7-Hydroxycoumarin		162.14	18, 27			226–228			v s alc, chl, alk, HOAc
h112	1-Hydroxy-1-cyclohexanecarbonitrile	$C_6H_{10}(OH)CN$	125.17	10, 5	1.031	1.4576^{20}	29		60	
h113	2-Hydroxy-3,5-diiodobenzoic acid	$I_2C_6H_2(OH)CO_2H$	389.91	10, 113			232–235			v s alc, eth; i bz, chl
h114	4-Hydroxy-3,5-dinitrobenzoic acid	$HOC_6H_2(NO_2)CO_2H$	228.12	1, 183			245 dec			
h115	3-Hydroxydiphenylamine	$HOC_6H_4NHC_6H_5$	185.23	13, 410			80–82	340		
h116	(2-Hydroxydiphenyl)methane	$HOC_6H_4CH_2C_6H_5$	184.24	6, 675		1.5994^{20}	54	312	>110	s organic solvents, alk
h117	(4-Hydroxydiphenyl)methane	$HOC_6H_4CH_2C_6H_5$	184.24	6, 675			84	322		s hot aq, org solvents, HOAc, alkalis
h118	2-(2-Hydroxyethoxy)phenol	$HOCH_2CH_2C_6H_4OH$	154.17	6^{2}, 782			99–100	$128^{0.7mm}$		
h119	*N*-(2-Hydroxyethyl)acetamide	$HOCH_2CH_2NHC(=O)CH_3$	103.12	4^{1}, 430	1.1233^{20}_{20}	1.4575^{20}	63–65	155^{5mm}	176	misc aq; sl s bz
h120	2-Hydroxyethyl acetate	$CH_3CO_2CH_2CH_2OH$	104.11	2, 141	1.108^{15}	1.4201^{20}		188	88	misc aq, alc, chl, eth
h121	2-Hydroxyethyl acrylate	$H_2C=CHCO_2CH_2CH_2OH$	116.12	2^{4}, 1469	1.011	1.4500^{20}		92^{12mm}	98	

h122	3-(1-Hydroxyethyl)-aniline	$CH_3CH(OH)C_6H_4NH_2$	137.18	13^3, 1654			66–69			
h123	2-Hydroxyethyl disulfide	$HOCH_2CH_2SSCH_2CH_2OH$	154.25	1, 471	1.261	1.5655^{20}	25–27	$158^{3.5mm}$	>110	
h124	*N*-(2-Hydroxyethyl)-ethylenediamine-*N,N,N′*-triacetic acid	$HO_2CCH_2N(CH_2CH_2OH)$-$CH_2CH_2N(CH_2CO_2H)_2$	278.26				212 dec			
h125	2-Hydroxyethyl-hydrazine	$HOCH_2CH_2NHNH_2$	76.10	4^1, 562	1.123	1.4961^{20}	−70	220	73	misc aq; s alc
h126	2-Hydroxyethyl methacrylate	$HOCH_2CH_2O_2CC(CH_3){=}CH_2$	130.14		1.073	1.4520^{20}		$67^{3.5mm}$	97	
h127	*N*-(2-Hydroxyethyl)-morpholine		131.18	27, 7	1.083	1.4760^{20}		227	99	misc aq
h128	*N*-(2-Hydroxyethyl)-phthalimide		191.19	21, 469			126–128			
h129	1-(2-Hydroxyethyl)-piperazine		130.19	23^2, 6	1.061	1.5065^{20}		246	>110	
h130	*N*-(2-Hydroxyethyl)-piperazine-*N′*-ethane-sulfonic acid		238.31	Merck: 12, 4687			234 dec			sat'd aq: 2.25M^0
h131	*N*-(2-Hydroxyethyl)-piperidine		129.20	20, 25	1.0059^{15}_{4}	1.4804^{20}		199–202	68	
h132	*N*-(2-Hydroxyethyl)-pyridine	$HOCH_2CH_2NC_5H_4$	123.16	21, 50	1.093	1.5368^{20}		116^{9mm}	92	v s aq, alc, chl
h133	*N*-(2-Hydroxyethyl)-pyrrolidine	$HOCH_2CH_2NC_4H_8$	115.8	20^2, 5	0.985	1.4713^{20}		81^{13mm}	56	
h134	*N*-(2-Hydroxyethyl)-2-pyrrolidinone		129.16	21^4, 3142	1.143	1.4960^{20}		142^{2mm}	>110	
h135	2-Hydroxyethyl salicylate	$(HO)C_6H_4CO_2CH_2CH_2OH$	182.18	10, 81	1.224	1.5480^{20}		166^{13mm}	>110	
h136	(2-Hydroxyethyl)tri-phenylphosphonium bromide	$HOCH_2CH_2P(C_6H_5)_3Br$	387.26	16, 761			217–219			
h137	8-Hydroxy-7-iodo-5-quinolinesulfonic acid		351.12	22, 408			269–270 dec			

(*Continued*)

TABLE 2.20 Physical Constants of Organic Compounds (*Continued*)

No.	Name	Formula	Formula weight	Beilstein reference	Density, g/mL	Refractive index	Melting point, °C	Boiling point, °C	Flash point, °C	Solubility in 100 parts solvent
h138	2-Hydroxyisobutyric acid	$(CH_3)_2C(OH)CO_2H$	104.11	3, 313			82	$84^{1.5mm}$		v s aq, alc, eth
h138a	2-Hydroxyisobutyronitrile	$(CH_3)_2C(OH)CN$	85.11	3, 316	0.932	1.3990^{20}	−19	82^{23mm}	63	
h139	Hydroxylamine HCl	$H_2NOH \cdot HCl$	69.49		1.670		159 dec			
h140	4-Hydroxy-2-mercapto-6-methylpyrimidine		142.18	24, 351			330 dec			v s aq NH_3, alkalis; sl s alc, acet
h141	4-Hydroxy-2-mercapto-6-propylpyrimidine		170.23				219–221			0.1 aq; 1.7 alc; 1.7 acet; v s alkalis
h142	4-Hydroxy-3-methoxybenzaldehyde	$CH_3OC_6H_3(OH)CHO$	152.15	8, 247	1.056		80–81	285		1 aq; s alc, chl, pyr
h143	4-Hydroxy-3-methoxybenzoic acid	$CH_3OC_6H_3(OH)CO_2H$	168.15	10, 392			210–213			0.12 aq; v s alc
h144	2-Hydroxy-4-methoxybenzophenone	$CH_3OC_6H_3(OH)C(=O)C_6H_5$	228.25	8, 312			63–66	160^{5mm}		v s alc, chl, eth
h145	4-Hydroxy-3-methoxybenzyl alcohol	$CH_3OC_6H_3(OH)CH_2OH$	154.17	6, 1113			113–115			
h146	*N*-(Hydroxymethyl)acrylamide	$H_2C{=}CHC(=O)NHCH_2OH$	101.11	2^4, 1472	1.074	1.430^{20}			none	
h147	4-Hydroxy-3-methyl-2-butanone	$HOCH_2CH(CH_3)C(=O)CH_3$	102.13	1^1, 422	0.993	1.4340^{20}		92^{15mm}	81	
h148	7-Hydroxy-4-methylcoumarin		176.17	18, 31			190–192			s alc, HOAc; sl s eth
h149	*N*-(Hydroxymethyl)nicotinamide	$(C_5H_4N)C(=O)NHCH_2OH$	152.15	10, 4750			152–154			
h150	4-Hydroxy-4-methyl-2-pentanone	$(CH_3)_2C(OH)CH_2C(=O)CH_3$	116.16	Merck: 12, 3008	0.9306^{25}_{4}	1.4235^{20}	−44	167.91	58	misc aq
h151	*N*-(Hydroxymethyl)phthalimide		177.16	21, 475			147–149			sl s aq, alc, bz
h152	4-Hydroxy-*N*-methylpiperidine		115.18	21^1, 188		1.4775^{20}	29–31	200		
h153	2-Hydroxy-2-methylpropionitrile	$(CH_3)_2C(OH)CN$	85.10	3, 316	0.9267^{25}_{4}	1.3992^{20}	−19	95	63	s aq, alc, chl, eth
h154	2-Hydroxy-2-methylpropiophenone	$C_6H_5C(=O)C(CH_3)_2OH$	164.20	8^1, 553	1.077	1.5330^{20}		103^{4mm}	>110	
h155	5-Hydroxy-2-methylpyridine	$HO(C_5H_2N)CH_3$	109.13	21^3, 480			168–170			

h156	3-Hydroxy-2-methyl-4-pyrone		126.11				161–162			1.2 aq; v s hot aq; s alc, alk; sl s bz, eth
h157	2-Hydroxy-1-naphthaldehyde	$C_{10}H_6(OH)CHO$	172.18	8, 143			82–85	192^{27mm}		
h158	1-Hydroxy-2-naphthoic acid	$C_{10}H_6(OH)CO_2H$	188.18	10, 331			191–192			v s alc, bz, eth, alk
h159	2-Hydroxy-1-naphthoic acid	$C_{10}H_6(OH)CO_2H$	188.18	10, 328			167 dec			
h160	3-Hydroxy-2-naphthoic acid	$C_{10}H_6(OH)CO_2H$	188.18	10, 333			222–223			v s alc, eth; s bz, chl
h161	2-Hydroxy-1,4-naphthoquinone		174.16	8, 300			dec >191			s HOAc
h162	4-Hydroxy-3-nitrobenzenearsonic acid	$HOC_6H_3(NO_2)AsO(OH)_2$	263.04	16^1, 456			>300			v s alc, acet, HOAc, alk; sl s aq; i eth
h163	4-Hydroxy-3-nitrobenzoic acid	$HOC_6H_3(NO_2)CO_2H$	183.12	10, 181			184–185			
h164	5-Hydroxy-2-pentanone	$CH_3C(=O)CH_2CH_2CH_2OH$	102.13	1, 831	1.007^{20}_{4}	1.4372^{20}		144^{100mm}	93	misc aq; s alc, eth
h165	4-Hydroxyphenylacetic acid	$HOC_6H_4CH_2CO_2H$	152.15	10, 190			149–151			v s alc, eth; sl s aq
h166	4-(4-Hydroxyphenyl)-2-butanone	$HOC_6H_4CH_2CH_2C(=O)CH_3$	164.20	8^2, 117			82–83			
h167	4-Hydroxyphenylglycine	$HOC_6H_4CH(NH_2)CO_2H$	167.16	14^1, 659			240 dec			sl s aq, alc, bz, acet
h168	*N*-(4-Hydroxyphenyl)glycine	$HOC_6H_4NHCH_2CO_2H$	167.16	13, 488			244 dec			s alk, acid; v sl s aq, alc, acet, bz, eth
h169	2′-Hydroxy-3-phenylpropiophenone	$HOC_6H_4C(=O)CH_2CH_2C_6H_5$	226.28	8^2, 202		1.5968^{20}	36–37		>110	
h170	1-(3-Hydroxyphenyl)urea	$HOC_6H_4NHC(=O)NH_2$	152.15	13, 417			182–184			
h171	*N*-Hydroxyphthalimide		163.13	21, 500			233 dec			
h172	2-Hydroxypropionitrile	$CH_3CH(OH)CN$	71.08	3^2, 209	0.9834^{25}	1.4027^{25}	−40	103^{50mm}	76	misc aq, alc; s eth
h173	3-Hydroxypropionitrile	$HOCH_2CH_2CN$	71.08	3, 298	1.0404^{25}_{4}	1.4248^{20}	−46	221	129	misc aq, alc, acet; 2,3 eth; i bz, PE
h174	2′-Hydroxypropiophenone	$HOC_6H_4C(=O)CH_2CH_3$	150.18	8, 102	1.094	1.5480^{20}		115^{15mm}	>110	v s alc, eth; sl s aq
h175	4′-Hydroxypropiophenone	$HOC_6H_4C(=O)CH_2CH_3$	150.18	8, 102			148			v s alc, eth; sl s aq

(Continued)

TABLE 2.20 Physical Constants of Organic Compounds (*Continued*)

No.	Name	Formula	Formula weight	Beilstein reference	Density, g/mL	Refractive index	Melting point, °C	Boiling point, °C	Flash point, °C	Solubility in 100 parts solvent
h176	1-(2-Hydroxy-1-propoxy)-2-propanol	$CH_3CH(OH)CH_2OCH_2CH(OH)CH_3$	134.18		1.0252^{20}_{20}	1.4440^{20}		231.8	138	misc aq, alc
h177	Hydroxypropyl acrylate	$H_2C{=}CHCO_2(CH_2)_3OH$	130.14	2^4, 1469	1.044	1.4450^{20}		77^{5mm}	89	
h178	Hydroxypropyl methacrylate	$H_2C{=}C(CH_3)CO_2(CH_2)_3OH$	144.17	2^4, 1532	1.066	1.4470^{20}		$57^{0.5mm}$	96	
h179	2-Hydroxypyridine	HOC_5H_4N	95.10	21, 43			105–107	280–281		aq, alc, bz, sl s eth
h180	3-Hydroxypyridine	HOC_5H_4N	95.10			126–129		151^{3mm}		v s aq, alc; sl s eth
h181	4-Hydroxypyridine	HOC_5H_4N	95.18					230^{12mm}		v s aq; i alc, bz, eth
h182	2-Hydroxypyridine-5-carboxylic acid	$HO(C_5H_3N)CO_2H$	139.11	22, 215			>300			sl s aq, alc, eth
h183	3-Hydroxypyridine-*N*-oxide	$(HO)C_5H_4N{=}O$	111.10				190–192			
h184	8-Hydroxyquinoline		145.16	21, 91			72–74	267^{742mm}		v s alc, acet, bz, chl
h185	8-Hydroxyquinoline-5-sulfonic acid		225.22	22, 407			>300			v s aq; sl s alc, eth
h186	DL-Hydroxysuccinic acid	$HO_2CCH(OH)CH_2CO_2H$	134.09	3, 435			131–133			56 aq; 45 EtOH; 18 acet; 0.8 eth; 23 diox
h187	(−)-Hydroxysuccinic acid	$HO_2CCH(OH)CH_2CO_2H$	134.09	3, 419			100			36 aq; 87 EtOH; 61 acet; 2.7 eth; 75 diox
h188	*N*-Hydroxysuccinimide		115.09	21, 380			95–98			v s aq
i1	Icosane	$CH_3(CH_2)_{18}CH_3$	282.56	1, 174	0.7777^{37}	1.4346^{40}	36.4	343.8	>112	
i2	1-Icosene	$CH_3(CH_2)_{17}CH{=}CH_2$	280.54	1^3, 881			28.7	342.4		
i3	1*H*-Imidazole		68.08	23, 45			90–91	257	145	v s aq, alc, chl, eth
i4	2-Imidazolidinethione		102.16	24,4			203–204			2 aq; s alc, pyr; i bz, acet, chl, eth
i5	2-Imidazolidone		86.09	24, 16			133–135			v s aq, hot alc
i6	3,3′-Iminobis(*N,N*-dimethyl)propylamine	$HN[(CH_2)_3N(CH_3)_2]_2$	187.33	4^3, 565	0.841	1.4490^{20}	−78	131^{20mm}	98	
i7	Iminodiacetic acid	$HO_2CCH_2NHCH_2CO_2H$	133.10	4, 365			243 dec			2 aq; v sl s bz, eth
i8	Iminodiacetonitrile	$NCCH_2NHCH_2CN$	95.11	4, 367			77			s aq, alc; sl s eth
i9	Iminodibenzyl		195.27				105–108			
i10	Indane		118.18	Merck: 12, 4966	0.9639^{20}_{4}	1.5383^{20}	−51.4	178	50	s alc, chl, eth; i aq
i11	5-Indanol		134.18	6, 575			51–53	255	>110	v s alc, eth; sl s aq
i12	1-Indanone		132.16	7, 360	1.1090^{45}_{4}	1.561^{45}	40–42	243–245	111	s alc, eth; sl s aq

i13	1,2,3-Indantrione hydrate		178.14	Merck: 12, 6645			dec 241			v s aq; s alc
i14	Indene		116.16	5, 515	0.9968_4^{20}	1.5762^{20}	−1.8	181.6	58	misc alc, bz, chl, eth
i15	Indole		117.15	20, 304	1.0643	1.609^{60}	52.54	253–254	>110	s hot aq, bz, eth
i16	Indole-3-acetic acid		175.19	22, 66			168–170			v s alc; s acet, eth
i17	Indole-2,3-dione		147.13	21, 432			203.5 dec			s hot aq, hot alc, alk
i18	Indoline		119.17	20, 257	1.063	1.5906^{20}		221	92	sl s aq
i19	Inositol		180.16	6^2, 1157	1.752		225			14 aq; sl s alc; i eth
i20	Iodoacetamide	ICH_2CONH_2	184.96	2, 223			93–96			s hot aq
i21	Iodoacetic acid	ICH_2CO_2H	185.95	2, 222			79–82			s aq, alc; v sl s eth
i22	3-Iodoaniline	$IC_6H_4NH_2$	219.03	12, 670	1.821	1.6820^{20}	25	146^{15mm}	>110	i aq; s alc, eth
i23	Iodobenzene	C_6H_5I	204.01	5, 215	1.8308^{20}	1.6200^{20}	−31	188	74	misc alc, chl, eth
i24	Iodobenzene diacetate	$C_6H_5I(O_2CCH_3)_2$	322.10	5, 218			163–165			
i25	2-Iodobenzoic acid	$IC_6H_4CO_2H$	248.02	9, 363	2.249_4^{25}		162–164			s alc, eth; sl s aq
i26	1-Iodobutane	$HC_3CH_2CH_2CH_2I$	184.02	1, 123	1.6154^{20}	1.4999^{20}	−103.5	130–131	33	i aq; s alc, eth
i27	2-Iodobutane	$CH_3CH_2CH(I)CH_3$	184.02	1, 123	1.5920^{20}	1.4991^{20}	−104.0	120	23	i aq; s alc, eth
i28	Iodocyclohexane	$C_6H_{11}I$	210.06	5^2, 13	1.626_{15}^{15}	1.5472^{20}		180		i aq; s eth
i29	1-Iododecane	$CH_3(CH_2)_9I$	268.19	1, 168	1.257_4^{20}	1.4850^{20}		132^{15mm}	>110	i aq; s alc, eth
i30	2-Iodododecane	$CH_3(CH_2)_{11}I$	296.24	1^1, 67	1.201	1.4844	−3	160^{15mm}	>110	
i31	Iodoethane	CH_3CH_2I	155.97	1, 96	1.9358^{20}	1.5130^{20}	−111	72.4	none	0.4 aq; misc alc, bz, chl, eth
i32	2-Iodoethanol	ICH_2CH_2OH	171.97	1, 339	2.2197_4^{20}	1.5694^{20}		75^{5mm}	65	s aq; v s alc, eth
i33	Iodoform	CHI_3	393.73	1, 73	4.008		120–123		none	1.4 alc; 10 chl; 13 eth; v s bz, acet
i34	1-Iodoheptane	$CH_3(CH_2)_6I$	226.10	1, 155	1.373_4^{20}	1.4900^{20}	−48	204	78	i aq; s alc, eth
i35	1-Iodohexadecane	$CH_3(CH_2)_{15}I$	352.35	1, 172	1.121	1.4806^{20}	23	207^{10mm}	>110	
i36	1-Iodohexane	$CH_3(CH_2)_5I$	212.08	1, 146	1.437_4^{20}	1.4920^{20}		179–180	61	i aq
i37	1-Iodomethane	CH_3I	141.94	1, 69	2.2789_4^{20}	1.5308^{20}	−66.5	42.5	none	1.4 aq; misc alc, eth
i38	1-Iodo-2-methyl-propane	$(CH_3)_2CHCH_2I$	184.02	1, 128	1.6035^{20}	1.4960^{20}	−93.5	121	12	i aq; misc alc, eth
i39	2-Iodo-2-methyl-propane	$(CH_3)_3CI$	184.02	1^3, 326	1.571_0^0	1.4918^{20}	−38	100	7	dec aq; misc alc, eth
i40	1-Iodo-3-nitrobenzene	$IC_6H_4NO_2$	249.01	5, 253	1.9477_4^{50}		36–38	280	71	i aq; s alc, eth
i41	1-Iodo-4-nitrobenzene	$IC_6H_4NO_2$	249.01	5, 252			175–177	289^{772mm}	>110	
i42	1-Iodononane	$CH_3(CH_2)_8I$	254.18	1, 166	1.288	1.4870^{20}		108^{8mm}	85	
i43	1-Iodooctadecane	$CH_3(CH_2)_{17}I$	380.40	1, 173			33–35	197^{2mm}	>110	
i44	1-Iodooctane	$CH_3(CH_2)_7I$	240.13	1, 160	1.330_4^{20}	1.4889^{20}	−46	226	95	s alc, eth
i47	1-Iodopentane	$CH_3(CH_2)_4I$	198.06	1, 133	1.512_4^{20}	1.4954^{20}	−85	155	51	sl s aq; s alc, eth

(Continued)

TABLE 2.20 Physical Constants of Organic Compounds (*Continued*)

No.	Name	Formula	Formula weight	Beilstein reference	Density, g/mL	Refractive index	Melting point, °C	Boiling point, °C	Flash point, °C	Solubility in 100 parts solvent
i48	1-Iodopropane	$CH_3CH_2CH_2I$	169.99	1, 113	1.7489^{20}	1.5058^{20}	−101	102	44	0.1 aq; misc alc, eth
i49	2-Iodopropane	$(CH_3)_2CHI$	169.99	1, 114	1.7042^{20}	1.4992^{20}	−90	89.5	42	0.14 aq; misc alc, eth
i50	3-Iodo-1-propene	$ICH_2CH{=}CH_2$	167.97	1, 202	1.845^{22}_{4}	1.5540^{21}	−99	103	18	misc alc, chl, eth
i51	5-Iodosalicylic acid	$IC_6H_3(OH)CO_2H$	264.02	10, 112			189–191			v s alc; i bz, chl
i52	2-Iodothiophene		210.04	17, 34	1.902	1.6520^{20}	−40	73^{15mm}	71	v s eth
i53	2-Iodotoluene	$IC_6H_4CH_3$	218.04	5, 310	1.713	1.6079^{20}		211	90	i aq; s alc, eth
i54	3-Iodotoluene	$IC_6H_4CH_3$	218.04	5, 311	1.698	1.6040^{20}		82^{10mm}	82	i aq; misc alc, eth
i55	4-Iodotoluene	$IC_6H_4CH_3$	218.04	5, 312			34–36	211	90	
i56	Iodotrimethylsilane	$(CH_3)_3SiI$	200.10		1.406^{20}_{4}	1.4710^{20}		106	−31	
i57	1-Iodoundecane	$CH_3(CH_2)_{10}I$	282.21	1^1, 66	1.220	1.4849^{20}		130^{5mm}	>110	
i58	α-Ionone		192.30	7, 168	0.932^{20}	1.4980^{20}		124^{11mm}	104	s alc, bz, chl, eth
i59	β-Ionone		192.30	7, 167	0.946^{17}	1.521^{17}		128^{12mm}	>110	s alc, bz, chl, eth
i60	Isatoic anhydride		163.13	27, 264			233 dec			sl s aq, hot alc, acet
i61	D-(−)-Isoascorbic acid		176.12				169 dec			s aq, alc, acet, pyr
i62	DL-Isoborneol		154.25	6^2, 80			214 subl			v s alc, chl, eth
i63	Isobutyl acetate	$(CH_3)_2CHCH_2O_2CCH_3$	116.16	2, 131	0.8712^{20}	1.3902^{20}	−99	116.5	18	0.7 aq; v s alc
i64	Isobutyl acetoacetate	$CH_3COCH_2CO_2CH_2CH(CH_3)_2$	158.20		0.980	1.4240^{20}		100^{22mm}	78	
i65	Isobutyl acrylate	$H_2C{=}CHCO_2CH_2CH(CH_3)_2$	128.19	2^3, 1227	0.890	1.4140		132	32	
i66	Isobutylamine	$(CH_3)_2CHCH_2NH_2$	73.14	4, 163	0.724^{20}_{4}	1.3972^{20}	−86.6	68	−9	misc aq, alc, acet, eth
i67	Isobutylbenzene	$C_6H_5CH_2CH(CH_3)_2$	134.22	5, 414	0.8532^{20}	1.4866^{20}	−51.5	172.8	55	misc alc, eth
i68	Isobutyl chloroformate	$ClCO_2CH_2CH(CH_3)_2$	136.58	3, 12	1.053	1.4070^{20}		128.8	27	misc bz, chl, eth
i69	Isobutyl formate	$HCO_2CH_2CH(CH_3)_2$	102.13	2, 21	0.8776^{20}	1.3855^{20}	−95.5	98.4	10	1 aq; misc alc, eth
i70	Isobutyl isobutyrate	$(CH_3)_2CHCH_2O_2CH(CH_3)_2$	144.22	2, 291	0.8542^{20}	1.3999^{20}	−80.7	148.5	38	0.5 aq; misc alc
i71	Isobutyl methacrylate	$H_2C{=}C(CH_3)CO_2CH_2CH(CH_3)_2$	142.19	2^3, 1287	0.882^{25}_{15}	1.4170^{25}		155	41	misc alc, eth
i72	Isobutyl nitrate	$(CH_3)_2CHCH_2ONO_2$	119.12	1, 377	1.015^{20}_{4}	1.4028^{20}		123	21	i aq; misc alc, eth
i73	Isobutyl nitrite	$(CH_3)_2CHCH_2ONO$	103.12	1, 377	0.870^{22}_{4}	1.3715^{22}		67	−21	misc alc; sl s aq (dec)
i74	Isobutyl propionate	$C_2H_5CO_2CH_2CH(CH_3)_2$	130.19	2, 241	0.888^{0}_{4}	1.3974^{20}	−71	137	26	i aq; misc alc
i75	Isobutyl stearate	$CH_3(CH_2)_{16}CO_2CH_2CH(CH_3)_2$	340.57				*ca.* 20			
i76	Isobutyltriethoxy-silane	$(CH_3)_2CHCH_2Si(OC_2H_5)_3$	220.39		0.880	1.400^{20}		190–191	60	
i77	Isobutyltrimethoxy-silane	$(CH_3)_2CHCH_2Si(OCH_3)_3$	178.30		0.930	1.3960^{20}		137	39	
i78	Isobutyl vinyl ether	$(CH_3)_2CHCH_2OCH{=}CH_2$	100.16	1^3, 1862	0.7702^{20}_{20}	1.3950^{20}	−112	83.4	−13	0.2 aq
i79	Isobutyraldehyde	$(CH_3)_2CHCHO$	72.11	1, 671	0.7988^{20}_{4}	1.3723^{20}	−65.9	64.5	−18 (CC)	11 aq; misc alc, bz, acet, chl, eth
i80	Isobutyramide	$(CH_3)_2CHCONH_2$	87.12	2, 293	1.013		127–129	216–220		
i81	Isobutyric acid	$(CH_3)_2CHCO_2H$	88.11	2, 288	0.9681^{20}	1.3925^{20}	−46	154	56	17 aq; misc alc, chl, eth

i82	Isobutyric anhydride	$[(CH_3)_2CHCO]_2O$	158.20	2, 292	0.954	1.4062^{20}	−56	182	59	
i83	Isobutyronitrile	$(CH_3)_2CHCN$	69.11	2, 294	0.7704^{20}	1.3720^{20}	−71.5	104	8	v s alc, eth; sl s aq
i84	Isobutyrophenone	$C_6H_5COCH(CH_3)_2$	148.21	7, 316	0.988^{20}	1.5172		217	84	
i85	Isobutyryl chloride	$(CH_3)_2CHCOCl$	106.55	2, 293	1.017	1.4073^{20}	−90	91–93	1	dec aq, dec alc; s eth
i86	Isodecyl acrylate	$H_2C{=}CHCO_2C_{10}H_{21}$	212.34		0.875	1.4420^{20}		121^{10mm}	106	
i87	Isodecyl methacrylate	$H_2C{=}C(CH_3)CO_2C_{10}H_{21}$	226.36		0.878	1.4430^{20}		126^{10mm}	>110	
i88	L-Isoleucine	$C_2H_5CH(CH_3)CH(NH_2)CO_2H$	131.18	4, 454			288 dec	subl 168		4 aq; sl s hot alc
i89	Isooctyl acrylate	$H_2C{=}CHCO_2C_8H_{17}$	184.25		0.880	1.4370^{20}		125^{20mm}	80	
i90	Isooctyl diphenyl phosphite	$(C_6H_5O)_2POC_8H_{17}$	346.41		1.045	1.5220^{20}		188		
i91	Isopentyl acetate	$CH_3CO_2CH_2CH_2CH(CH_3)_2$	130.19	2, 132	0.876^{15}_{4}	1.4007^{20}	−78.5	142	25	0.25 aq; misc alc, eth
i92	Isopentyl nitrite	$ONOCH_2CH_2CH(CH_3)_2$	117.15	1, 402	0.872	1.3860^{20}		99	10	misc alc, eth; sl s aq
i93	Isophorone		138.21	7, 65	0.955^{20}	1.4759^{20}	−8.1	215.2	84	1.2 aq
i94	Isophorone diisocyanate		222.29		1.049	1.4841^{20}		159^{15mm}	>110	
i95	Isopropenyl acetate	$CH_3CO_2C(CH_3){=}CH_2$	100.12	2^2, 278	0.909	1.4005^{20}		94	18	
i96	3-Isopropenyl-α,α-dimethylbenzyl isocyanate	$H_2C{=}C(CH_3)C_6H_4C(CH_3)_2NCO$	201.27		1.108	1.5300^{20}		268–271	>110	
i97	2-Isopropoxyethanol	$(CH_3)_2CHOCH_2CH_2OH$	104.15	1^2, 519	0.903	1.4104^{20}		44^{13mm}	45	
i98	3-Isopropoxypropylamine	$(CH_3)_2CHO(CH_2)_3NH_2$	117.19	4^3, 739	0.845	1.4195^{20}		79^{85mm}	39	
i99	Isopropyl acetate	$(CH_3)_2CHO_2CCH_3$	102.13	2, 130	0.8718^{20}	1.3770^{20}	−73	89	2	3 aq; misc alc, eth
i100	Isopropylamine	$(CH_3)_2CHNH_2$	59.11	4, 152	0.686^{25}_{4}	1.3711^{25}	−95	31.7	−37	misc aq, alc, eth
i101	2-Isopropylaniline	$(CH_3)_2CHC_6H_4NH_2$	135.2	12, 1147	0.955	1.5477^{20}		222	95	
i102	4-Isopropylbenzaldehyde	$(CH_3)_2CHC_6H_4CHO$	148.21	7, 318	0.977	1.5298^{20}		236	93	
i103	Isopropylbenzene	$(CH_3)_2CHC_6H_5$	120.20	5, 393	0.864^{20}_{4}	1.4915^{20}	−96	152–154	36	s alc, bz, eth
i104	4-Isopropylbenzyl alcohol	$(CH_3)_2CHC_6H_4CH_2OH$	150.22	6, 543	0.982^{15}	1.5206^{20}	28	248.4	>110	misc alc, eth; i aq
i105	*N*-Isopropylbenzylamine	$C_6H_5CH_2NHCH(CH_3)_2$	149.24		0.892	1.5025^{20}		200	87	
i106	Isopropyl butyrate	$CH_3CH_2CH_2CO_2CH(CH_3)_2$	130.19	2, 271	0.859	1.3932^{20}		131	30	
i107	Isopropyl chloroacetate	$ClCH_2CO_2CH(CH_3)_2$	136.58	2, 198	1.096	1.4190^{20}		149–150	70	
i108	Isopropylcyclohexane	$C_6H_{11}CH(CH_3)_2$	126.24	5, 41	0.8023^{20}_{4}	1.4399^{20}	−90	155	35	v s alc, eth

(*Continued*)

TABLE 2.20 Physical Constants of Organic Compounds (*Continued*)

No.	Name	Formula	Formula weight	Beilstein reference	Density, g/mL	Refractive index	Melting point, °C	Boiling point, °C	Flash point, °C	Solubility in 100 parts solvent
i109	Isopropyl hexadecanoate	$CH_3(CH_2)_{14}CO_2CH(CH_3)_2$	298.51	2^2, 336	0.862	1.4385^{20}			>110	
i110	4,4′-Isopropylidene-bis(2,6-dibromo-phenoxy)ethanol	$(CH_3)_2C[C_6H_2(Br)_2OCH_2CH_2OH]_2$	632.01				107			
i111	4,4′-Isopropylidene-bis(diisodecyl phenyl phosphite)	$[(C_{10}H_{21}O)_2POC_6H_4]_2C(CH_3)_2$	917.34		0.964	1.4980^{20}		336	>110	
i112	4,4′-Isopropylidene-dicyclohexanol	$(CH_3)_2C(C_6H_{10}OH)_2$	240.39	6^2, 761				234^{14mm}	>110	
i113	4,4′-Isopropylidene-diphenol	$(CH_3)_2C(C_6H_4OH)_2$	228.29	6, 1011			137–140	220^{4mm}		
i114	2-Isopropylimidazole		110.16	23, 83			129–131	256–260		
i115	Isopropyl isocyanate	$(CH_3)_2CHCNO$	85.11	4, 155	0.866	1.3825^{20}		74–75	−2	
i116	Isopropyl *S*-(−)-lactate	$(CH_3)_2CHO_2CCH(OH)CH_3$	132.16	3, 282	0.998^{20}_{20}	1.4082^{25}		166–168	57	s aq, alc, eth
i117	2-Isopropyl-6-methyl-aniline	$(CH_3)_2CHC_6H_3(CH_3)NH_2$	149.24		0.957	1.5440^{20}			41	
i118	2-Isopropyl-1-methyl-benzene	$(CH_3)_2CHC_6H_4CH_3$	134.21	5, 419	0.8766^{20}_{4}	1.5006^{20}	−71.5	178.2		misc alc, eth
i119	3-Isopropyl-1-methyl-benzene	$(CH_3)_2CHC_6H_4CH_3$	134.21	5, 419	0.8610^{20}_{4}	1.4930^{20}	−63.8	175.1		misc alc, eth
i120	4-Isopropyl-1-methyl-benzene	$(CH_3)_2CHC_6H_4CH_3$	134.21	5, 420	0.8573^{20}_{4}	1.4909^{20}	−68.9	177.1	47	misc alc, eth
i121	2-Isopropyl-5-methyl-phenol	$(CH_3)_2CHC_6H_3(CH_3)OH$	150.22	6, 532	0.925^{80}_{4}		51.5	232.5		i aq; v s alc, chl, eth
i122	4-Isopropyl-3-methyl-phenol	$(CH_3)_2CHC_6H_3(CH_3)OH$	150.22	6^2, 491			111–114			
i123	5-Isopropyl-3-methyl-phenol	$(CH_3)_2CHC_6H_3(CH_3)OH$	150.22	6, 526			51		>110	
i124	Isopropyl nitrate	$(CH_3)_2CHONO_2$	105.09	1, 363	1.036^{19}_{19}	1.391^{20}		102	12	
i125	Isopropyl nitrite	$(CH_3)_2CHONO$	89.09	Merck: 12, 5235	0.844^{25}_{4}	1.3520^{20}		39^{752mm}		
i126	1-Isopropyl-4-nitro-benzene	$(CH_3)_2CHC_6H_4NO_2$	165.19	5^2, 308	1.090	1.5380^{20}		107^{11mm}	>110	
i127	2-Isopropylphenol	$(CH_3)_2CHC_6H_4OH$	136.19	6, 504	1.012^{20}	1.5259^{20}	15–16	212–213	88	misc alc, eth
i128	3-Isopropylphenol	$(CH_3)_2CHC_6H_4OH$	136.19	6, 505	0.994	1.5250^{20}	25	228	104	
i129	4-Isopropylphenol	$(CH_3)_2CHC_6H_4OH$	136.19	6, 505	0.990^{20}		59–61	212		316 alc; 350 eth

i130	4-Isopropylpyridine	$(CH_3)_2CH(C_5H_4N)$	121.18	20, 248	0.938	1.4980^{20}		173	66	
i131	Isopropyl tetradecanoate	$(CH_3)_2CHO_2C(CH_2)_{12}CH_3$	270.46	2^3, 923	0.850	1.4350^{20}	*ca.* 3	193^{20mm}	>110	s caster oil, cottonseed oil, acet, EtOAc, EtOH, toluene, mineral oil
i132	Isopulegol		154.25	6, 65	0.912	1.4725^{20}		91^{12mm}	78	v sl s aq
i133	Isoquinoline		129.16	20, 380	1.0910^{30}_{4}	1.6208^{30}	26.5	243.5	107	sl s aq; s acid
k1	Ketene	$H_2C{=}C{=}O$	42.04	1, 724			−151	−49.8		s acet, eth; dec aq
k2	8-Ketotricyclo-[5.2.1.0^{2,6}]decane		150.22	7^2, 133	1.063	1.5020^{20}		132^{30mm}	101	
L1	DL-Lactic acid	$CH_3CH(OH)CO_2H$	90.08	3, 268	1.249^{15}_{4}		16.8	122^{14mm}	>110	s aq, alc; i chl, PE
L2	L-(+)-Lactic acid	$CH_3CH(OH)CO_2H$	90.08	3, 261	1.2060^{25}_{4}	1.4270^{20}	53	119^{12mm}	>110	v s aq, alc, eth
L3	α-Lactose		342.32	31, 408			202			20 aq; v sl s alc
L4	β-Lactose		342.32	31, 408	1.525^{20}		202			45 aq; i alc, eth
L5	DL-Leucine	$(CH_3)_2CHCH_2CH(NH_2)CO_2H$	131.18	4, 447			dec 332	subl 293		1 aq; 0.13 alc; i eth
L6	L-Leucine	$(CH_3)_2CHCH_2CH(NH_2)CO_2H$	131.18	4, 437	1.293^{18}		293 dec	subl 145		2.4 aq^{25}; 0.07 alc; 1 HOAc; i eth
L7	*R*-(+)-Limonene		136.24	5, 133	0.8411^{20}_{4}	1.4730	−96.5	178	49	misc alc, eth
L8	*S*-(−)-Limonene		136.24	5, 136	0.841^{20}_{4}	1.4746^{20}	−96.5	178	48	misc alc, eth
L9	(+)-Limonene oxide		152.24	17, 44	0.929	1.4661^{20}		114^{50mm}	65	
L10	Linalool		154.25	1, 462	0.865^{15}	1.4615^{20}		197^{720mm}	76	misc alc, eth
L11	Linalyl acetate		196.29	2, 141	0.895^{20}_{4}	1.4460^{20}		220	90	misc alc, eth
L12	*S*-(+)-Lysine	$H_2N(CH_2)_4CH(NH_2)CO_2H$	146.19	4, 435			212 dec			v s aq; sl s alc; i eth
m1	Maleic acid	$HO_2CH{=}CHCO_2H$	116.07	2, 748	1.590		130.5			70 aq; 70 alc; s acet, HOAc; sl s eth
m2	Maleic anhydride		98.06	17, 432	1.48		52.8	202	103	s aq (to acid), alc (to ester); 227 acet; 53 chl; 50 bz; 112 EtOAc
m3	Malonic acid	$HO_2CCH_2CO_2H$	104.06	2, 566	1.63		135–137			154 aq; 42 alc; 8 eth; 14 pyr
m4	Malonodiamide	$H_2NCOCH_2CONH_2$	102.09	2, 582			172–175			9 aq; i alc, eth
m5	Malononitrile	$NCCH_2CN$	66.06	2, 589	1.1910^{20}_{4}	1.4146^{34}	32–34	220	112	13 aq, 40 alc; 20 eth
m6	Malonyl dichloride	$ClCOCH_2COCl$	140.95	2^1, 252	1.4486^{19}_{4}	1.4620^{20}		55^{19mm}	47	dec hot aq; s eth
m7	D-(+)-Maltose hydrate		360.32	31, 386	1.540^{17}		119–121	dec 130		v s aq; sl s alc; i eth
m8	DL-Mandelic acid	$C_6H_5CH(OH)CO_2H$	152.15	10, 192	1.300^{20}_{4}		120–122			16 aq; 100 alc; s eth
m9	Mandelonitrile	$C_6H_5CH(OH)CN$	133.15	10, 193	1.117	1.5315^{20}	−10	170	97	v s alc, cho, eth; i aq
m10	Mannitol		182.17	1, 534	1.52^{20}		166–168	$290^{3.5mm}$		18 aq; 1.2 alc; i eth

(*Continued*)

TABLE 2.20 Physical Constants of Organic Compounds (*Continued*)

No.	Name	Formula	Formula weight	Beilstein reference	Density, g/mL	Refractive index	Melting point, °C	Boiling point, °C	Flash point, °C	Solubility in 100 parts solvent
m11	D-(+)-Mannose		180.16	31, 284	1.54^{20}		128–130			250 aq; 28 pyr; 0.8 alc
m12	(−)-Menthol		156.27	6, 28	0.890^{15}_{15}	1.458^{25}	41–43	212	93	v s alc, chl, eth, PE
m13	(−)-Menthone		154.25	7, 38	0.895^{20}_{4}	1.4510^{20}	−6	207	72	misc alc, eth; sl s aq
m14	*S*-(+)-Menthyl acetate		198.31	6, 32	1.4480^{20}			229–230	77	
m15	Menthyl anthranilate		275.40	14^3, 885	1.040	1.5420^{20}		179^{3mm}	>110	
m16	Mercaptoacetic acid	$HSCH_2CO_2H$	92.12	3, 245	1.325	1.5030^{20}	−16.5	96^{5mm}	>110	misc aq, alc, bz, eth
m17	2-Mercaptobenz-imidazole		150.20	24, 119			301–305			sl s aq; s alc
m18	2-Mercaptobenzoic acid	$HSC_6H_4CO_2H$	154.19	10, 125			165–168			v s alc, HOAc
m19	2-Mercaptobenzo-thiazole		167.25	27, 185	1.42^{20}_{4}		180–181	dec		2 alc; 1 eth; 10 acet; 1 bz; s alk; i aq
m20	2-Mercaptoethanol	$HSCH_2CH_2OH$	78.13	1, 470	1.1143^{20}_{4}	1.5006^{20}		156.9	73	misc aq, alc, bz, eth
m21	3-Mercapto-1,2-propanediol	$HSCH_2CH(OH)CH_2OH$	108.16	1, 519	1.295^{14}_{14}	1.5243^{20}		118^{5mm}	>110	misc alc; v s acet
m22	2-Mercaptopropionic acid	$CH_3CH(SH)CO_2H$	106.14	3, 289	1.220^{15}_{4}	1.4809^{20}	10–14	102^{16mm}	87	misc aq, alc, eth, acet
m23	3-Mercaptopropionic acid	$HSCH_2CH_2CO_2H$	106.14	3, 299	1.218	1.4911^{20}	17–19	111^{15mm}	93	
m24	(3-Mercaptopropyl)-trimethoxysilane	$HS(CH_2)_3Si(OCH_3)_3$	196.34		1.039^{20}_{4}	1.4440^{20}		198	48	
m25	Mercaptosuccinic acid	$HO_2CCH_2CH(SH)CO_2H$	150.15	3, 439			5–7			50 aq; 50 alc; s eth
m26	2-Mercaptothiazoline		119.21	27, 140			105–107			
m27	Methacrylaldehyde	$H_2C{=}C(CH_3)CHO$	70.09	1, 731	0.847	1.4160^{20}	−81	69	−15	6 aq; misc alc, eth
m28	Methacrylamide	$H_2C{=}C(CH_3)CONH_2$	85.11	2^2, 399			109–111			s alc; sl s eth
m29	Methacrylic acid	$H_2C{=}C(CH_3)CO_2H$	86.09	2, 421	1.0153^{20}_{4}	1.4314^{20}	16	163	77	9 aq; misc alc, eth
m30	Methacrylic anhydride	$[H_2C{=}C(CH_3)CO]_2O$	154.17	2^3, 1293	1.035	1.4530^{20}		87^{13mm}	84	
m30a	Methacrylonitrile	$H_2C{=}C(CH_3)CN$	67.91	2, 423	0.8001^{20}_{4}	1.4007^{20}	−35.8	90.3	1.1	2.6 aq; misc acet, bz
m31	Methacryloyl chloride	$H_2C{=}C(CH_3)COCl$	104.54	2^2, 394	1.070	1.4420^{20}		95–96	2	
m32	Methallylidene diacetate	$(CH_3CO_2)_2CHC(CH_3){=}CH_2$	172.18	2^4, 292	1.039	1.4245^{20}	−15	191	83	
m33	Methane	CH_4	16.04	1, 56	0.7168 g/L 0.4240^{bp}		−182.5	−161.5		3.3 mL aq; 47 mL alc
m34	Methanesulfonic acid	CH_3SO_3H	96.10	4, 4	1.4812^{18}_{4}	1.4303^{20}	20	167^{10mm}	>110	1.5 bz; misc aq
m35	Methanesulfonic anhydride	$(CH_3SO_2)_2O$	174.19	4, 5			71	138^{10mm}		v s aq (dec)
m36	Methanesulfonyl chloride	CH_3SO_2Cl	114.55	4, 5	1.4805^{18}_{4}	1.4518^{20}	−32	161	>110	s alc, eth

m37	Methanethiol	CH_3SH	48.11	1, 288	1.966 g/L		−123	6.0		2.3 aq; v s alc, eth
m38	Methanol	CH_3OH	32.04	1, 273	0.7913^{20}_{4}	1.3284^{20}	−97.7	64.7	11	misc aq, alc, bz, chl, eth
m39	Methanol-*d*	CH_3OD	33.05	1^3, 1186	0.8127^{20}_{4}	1.3270^{20}	−110	65.5	11	misc aq, alc, eth
m40	Methanol-d_4	CD_3OD_1	36.07	1^3, 1187	0.888	1.3256^{20}		65.4	11	misc aq, alc, eth
m41	Methanol-^{13}C	$^{13}CH_3OH$	33.03	1^3, 1187	0.815	1.3290^{20}	−97.8	64	12	
m42	DL-Methionine	$CH_3SCH_2CH_2CH(NH_2)CO_2H$	149.21	4^2, 938	1.340		281 dec			3 aq; i eth; v sl s alc
m43	Methoxyacetic acid	$CH_3OCH_2CO_2H$	90.08	3, 232	1.174	1.4158^{20}		202–204	>110	misc aq, alc, eth
m44	2′-Methoxyacetophenone	$CH_3OC_6H_4COCH_3$	150.18	8, 85	1.090^{20}_{4}	1.5393^{20}		131^{18mm}	108	
m45	3′-Methoxyacetophenone	$CH_3OC_6H_4COCH_3$	150.18	8, 86	1.094	1.5410^{20}		239–241	>110	s aq
m46	4′-Methoxyacetophenone	$CH_3OC_6H_4COCH_3$	150.18	8, 87	1.082^{41}_{4}	1.5335	36–38	154^{26mm}	>110	v s alc, eth
m47	3-Methoxyacrylonitrile	$CH_3OCH{=}CHCN$	83.09		0.990	1.4550^{20}			76	
m48	2-Methoxyaniline	$CH_3OC_6H_4NH_2$	123.16	13, 358	1.098^{15}_{15}	1.5730^{20}	5–6	225	98	i aq; misc alc, eth
m49	3-Methoxyaniline	$CH_3OC_6H_4NH_2$	123.16	13, 404	1.096	1.5794^{20}	−10	251	>110	s alc, acid; sl s aq
m50	4-Methoxyaniline	$CH_3OC_6H_4NH_2$	123.16	13, 435	1.087		57–60	240–243		v s alc; sl s aq
m51	2-Methoxybenzaldehyde	$CH_3OC_6H_4CHO$	136.15	8, 43	1.127	1.560^{20}	37–39	238	117	sl s alc, bz; i eth
m52	3-Methoxybenzaldehyde	$CH_3OC_6H_4CHO$	136.15	8, 59	1.119	1.5533^{20}		143^{50mm}	>110	
m53	4-Methoxybenzaldehyde	$CH_3OC_6H_4CHO$	136.15	8, 67	1.119	1.5713^{20}	−1	248	108	misc alc
m54	4-Methoxybenzamide	$CH_3OC_6H_4CONH_2$	151.17	10^2, 100			164–167	295	108	s aq; v s alc; sl s eth
m55	Methoxybenzene	$CH_3OC_6H_5$	108.14	6, 138	0.9942^{20}	1.5170^{20}	−37.5	153.8	51	1 aq; misc alc, eth
m56	4-Methoxybenzenesulfonyl chloride	$CH_3OC_6H_4SO_2Cl$	206.65	11, 243			40–43		>110	dec aq; s alc, eth
m57	2-Methoxybenzoic acid	$CH_3OC_6H_4CO_2H$	152.15	10, 64	1.180		100	200		0.5 aq; v s alc, eth
m58	3-Methoxybenzoic acid	$CH_3OC_6H_4CO_2H$	152.15	10, 137			104	172^{10mm}		s hot aq, alc, eth
m59	4-Methoxybenzoic acid	$CH_3OC_6H_4CO_2H$	152.15	10, 154	1.385^4		185	275–280		0.04 aq; v s alc, chl
m60	4-Methoxybenzoyl chloride	$CH_3OC_6H_4COCl$	170.60	10, 163		1.5810^{20}	22	145^{14mm}	87	i aq (dec); s alc (dec); s acet, bz

(Continued)

TABLE 2.20 Physical Constants of Organic Compounds (*Continued*)

No.	Name	Formula	Formula weight	Beilstein reference	Density, g/mL	Refractive index	Melting point, °C	Boiling point, °C	Flash point, °C	Solubility in 100 parts solvent
m61	4-Methoxybenzyl alcohol	$CH_3OC_6H_4CH_2OH$	138.17	6, 897	1.109_4^{25}	1.5442^{20}	23–25	259	>110	i aq; s alc, eth
m62	4-Methoxybenzylamine	$CH_3OC_6H_4CH_2NH_2$	137.18	13, 606	1.050^{15}	1.5462^{20}		236–237	>110	v s aq, alc, eth
m63	2-Methoxybiphenyl	$CH_3OC_6H_4C_6H_5$	184.24	6, 672	1.023	1.6105^{20}	30–33	274	>110	
m64	3-Methoxy-1-butanol	$CH_3OCH(CH_3)CH_2CH_2OH$	104.15		0.9229_{20}^{20}	1.4145^{20}	−85	161.1	46	misc aq
m65	4-Methoxy-3-buten-2-one	$CH_3OCH{=}CHCOCH_3$	100.12		0.982	1.4680^{20}		200	63	
m66	2-Methoxycinnamaldehyde	$CH_3OC_6H_4CH{=}CHCHO$	162.19				44–48	$130^{0.6mm}$	>110	
m67	1-Methoxy-1,4-cyclohexadiene		110.16	6^3, 367	0.940	1.4819^{20}		148–150	36	
m68	2-Methoxydibenzofuran		198.22	17^3, 1590			42–45		>110	
m69	7-Methoxy-3,7-dimethyloctanal	$(CH_3)_2C(OCH_3)(CH_2)_3$-$CH(CH_3)CH_2CHO$	186.30		0.877	1.4374^{20}		$60^{0.45mm}$	98	
m70	2-Methoxy-1,3-dioxolane		104.11	19^4, 617	1.092	1.4091^{20}		129–130	31	
m71	2-Methoxyethanol	$CH_3OCH_2CH_2OH$	76.10	1, 467	0.9646^{20}	1.4021^{20}	−85.1	124	39	misc aq
m72	2-(2-Methoxyethoxy)-acetic acid	$CH_3OCH_2CH_2OCH_2CO_2H$	134.13	3^3, 374	1.180	1.4380^{20}		245–250	>110	
m73	2-(2-Methoxyethoxy)-ethanol	$CH_3OCH_2CH_2OCH_2CH_2OH$	120.15		1.035_4^{20}	1.4264^{20}	−50	194	96	misc aq, alc, bz, eth, ketones
m74	2-Methoxyethoxymethyl chloride	$CH_3OCH_2CH_2OCH_2Cl$	124.57		1.091	1.4270^{20}		50^{13mm}	>110	
m75	2-Methoxyethyl acetate	$CH_3CO_2CH_2CH_2OCH_3$	118.13	2, 141	1.0049^{20}	1.4002^{20}	−70	144	49	misc aq
m76	2-Methoxyethyl acetoacetate	$CH_3COCH_2CO_2CH_2CH_2OCH_3$	160.17		1.090	1.4339^{20}		120^{20mm}	103	
m77	2-Methoxyethylamine	$CH_3OCH_2CH_2NH_2$	75.11	4^2, 718	0.864	1.4054^{20}		95	9	v s aq, alc
m78	2-Methoxyethyl cyanoacetate	$CH_3OCH_2CH_2O_2CCH_2CN$	143.14	2^4, 1891	1.127	1.4340^{20}		100^{1mm}	>110	
m79	1-Methoxy-2-indanol		164.20	6, 970	1.128	1.5482^{20}		146^{11mm}	>110	
m80	2-Methoxy-5-methylaniline	$CH_3OC_6H_3(CH_3)NH_2$	137.18	13^2, 388			52–54	235	>110	s aq; v s alc, bz, eth
m81	4-Methoxy-2-methylaniline	$CH_3OC_6H_3(CH_3)NH_2$	137.18	13^2, 330	1.065	1.5647^{20}	13–14	248–249	>110	s alc
m82	3-Methoxy-3-methyl-1-butanol	$CH_3OC(CH_3)_2CH_2CH_2OH$	118.18	1^3, 2198	0.926	1.4280^{20}		173–175	71	

m83	2-Methoxy-1-methyl-ethyl cyanoacetate	$NCCH_2CO_2CH(CH_3)CH_2OCH_3$	157.17		1.030	1.4310^{20}		105^{2mm}	62	
m84	2-Methoxy-4-methyl-phenol	$CH_3OC_6H_3(CH_3)OH$	138.17	6, 878	1.092	1.5372^{20}	5	222	99	
m85	5-Methoxy-2-methyl-4-nitroaniline	$CH_3OC_6H_3(CH_3)(NO_2)NH_2$	182.18	13^3, 1575			168–170			
m86	1-Methoxy-2-methyl-propylene oxide	$(CH_3)_2C$—$CH(OCH_3)$ \ / O	102.13	17^3, 1035	0.904	1.3929^{20}		94	6	
m87	1-Methoxynaphthalene	$C_{10}H_7OCH_3$	158.20	6, 606	1.090	1.6220^{20}		135^{12mm}	>110	
m88	2-Methoxynaphthalene	$C_{10}H_7OCH_3$	158.20	6, 640			73–75	274		s bz, eth, CS_2
m89	2-Methoxy-4-nitro-aniline	$CH_3OC_6H_3(NO_2)NH_2$	168.15	13, 390			140–142			
m90	2-Methoxy-5-nitro-aniline	$CH_3OC_6H_3(NO_2)NH_2$	168.15	13, 389			117–119			s alc, hot bz, HOAc
m91	4-Methoxy-2-nitro-aniline	$CH_3OC_6H_3(NO_2)NH_2$	168.15	13, 521			123–126			sl s aq; s alc, eth
m92	2-Methoxynitro-benzene	$CH_3OC_6H_4NO_2$	153.14	6, 217	1.2527^{20}_{4}	1.5161^{20}	10.5	277	>110	0.17 aq; s alc, eth
m93	4-Methoxy-3-nitro-benzoic acid	$CH_3OC_6H_3(NO_2)CO_2H$	197.15	10, 181			192–194			
m94	2-Methoxy-5-nitro-pyridine	$CH_3O(C_5H_3N)NO_2$	154.13	21^3, 33			108–109			
m95	4-Methoxy-2-nitro-toluene	$CH_3OC_6H_3(NO_2)CH_3$	167.16	6, 411	1.207	1.5525^{20}	17	267	>110	
m96	4-Methoxyphenethyl-amine	$CH_3OC_6H_3CH_2CH_2NH_2$	151.21	13, 626	1.033	1.5379^{20}		140^{20mm}	>110	
m97	2-Methoxyphenol	$CH_3OC_6H_4OH$	124.14	6, 768	1.112(lg)	1.5429	28	205	82	1.5 aq; misc alc, eth
m98	3-Methoxyphenol	$CH_3OC_6H_4OH$	124.14	6, 813	1.131	1.5510^{20}	< −17.5	115^{5mm}	>110	misc alc, eth; sl s aq
m99	4-Methoxyphenol	$CH_3OC_6H_4OH$	124.14	6, 843			55–57	243	>110	v s bz; s alk
m100	3-(4-Methoxy-phenoxy)-1,2-propanediol	$CH_3OC_6H_4OCH_2CH(OH)CH_2OH$	198.22	6^3, 4411			76–80			
m101	4-Methoxyphenyl-acetic acid	$CH_3OC_6H_4CH_2CO_2H$	166.18	10, 190			86–88	140^{3mm}		1 aq; v s alc; s eth
m102	2-Methoxyphenyl-acetone	$CH_3OC_6H_4CH_2OCH_3$	164.20	8^3, 397	1.054	1.5250^{20}		130^{10mm}	>110	s alc, eth
m103	2-(Methoxyphenyl)-acetonitrile	$CH_3OC_6H_4CH_2CN$	147.18	10, 188			65–68	143^{15mm}		s hot bz

(Continued)

TABLE 2.20 Physical Constants of Organic Compounds (*Continued*)

No.	Name	Formula	Formula weight	Beilstein reference	Density, g/mL	Refractive index	Melting point, °C	Boiling point, °C	Flash point, °C	Solubility in 100 parts solvent
m104	4-(Methoxyphenyl)-acetonitrile	$CH_3OC_6H_4CH_2CN$	147.18	10, 191	1.085	1.5300^{20}		286–287	>110	
m105	1-Methoxy-2-propanol	$CH_3OCH_2CH(OH)CH_3$	90.12	1^2, 536	0.919^{20}_{20}	1.4021^{21}	−97	120.1	33	misc aq, acet, bz, eth
m106	2-Methoxypropene	$CH_3C(OCH_3){=}CH_2$	72.11	1, 435	0.735	1.3820^{20}		34–36	−29	
m107	*trans*-1-Methoxy-4-(1-propenyl)benzene	$CH_3OC_6H_4CH{=}CHCH_3$	148.21	6, 566	0.9883^{20}_{4}	1.5615^{20}	21.4	237	90	misc chl, eth; 50 alc; s bz, EtOAc
m108	2-Methoxy-4-propenyl-phenol	$CH_3OC_6H_3(OH)CH{=}CHCH_3$	164.20	6, 955	1.087^{20}_{4}	1.5748^{20}	−10	266	>112	misc alc, eth; sl s aq
m109	2-Methoxy-4-(2-propenyl)phenol	$CH_3OC_6H_3(OH)CH_2CH{=}CH_2$	164.20	6, 961	1.0664^{20}_{4}	1.5408^{20}	−9.2	255	>112	misc alc, chl, eth; s HOAc, alk; i aq
m110	3-Methoxypropionitrile	$CH_3OCH_2CH_2CN$	85.11	3^1, 113	0.937	1.4030^{20}		165	61	
m111	4-Methoxypropio-phenone	$CH_3OC_6H_4COCH_2CH_3$	164.20	8, 103	1.071	1.5465^{20}	27–29	274	61	
m112	3-Methoxypropylamine	$CH_3O(CH_2)_3NH_2$	89.14	4^3, 739	0.874	1.4175^{20}		118^{733mm}	22	
m113	2-Methoxypyridine	$CH_3O(C_5H_4N)$	109.13	21, 44	1.038	1.5029^{29}		142	32	misc aq
m114	6-Methoxy-1,2,3,4-tetrahydro-naphthalene		162.23	6^2, 537	1.033	1.5402^{20}		90^{1mm}	>110	
m115	6-Methoxy-1-tetralone		176.22	9^2, 889			77–79	171^{11mm}		
m116	2-Methoxytoluene	$CH_3OC_6H_4CH_3$	122.17	6, 352	0.9851^{15}_{15}	1.5161^{20}		170–172	51	i aq; v s alc, eth
m117	3-Methoxytoluene	$CH_3OC_6H_4CH_3$	122.17	6, 376	0.9697^{25}_{25}	1.5131^{20}		175–176	54	s alc, bz, eth; i aq
m118	4-Methoxytoluene	$CH_3OC_6H_4CH_3$	122.17	6, 392	0.969^{25}_{25}	1.5112^{20}		174	53	s alc, eth; i aq
m119	Methoxytrimethyl-silane	$CH_3OSi(CH_3)_3$	104.23	4^3, 1856	0.7560^{20}_{4}	1.3678^{20}		57–58	−30	
m120	*N*-Methylacetamide	$CH_3CONHCH_3$	73.10	4, 58	0.9460^{35}	1.4253^{35}	30.6	206	108	s aq
m121	4′-Methylacetanilide	$CH_3OCONHC_6H_4CH_3$	149.19	12, 920			150	307		
m122	Methyl acetate	$CH_3CO_2CH_3$	74.08	2, 224	0.9342^{20}_{4}	1.3619^{20}	−98	57	−10 (CC)	24 aq; misc alc, eth
m123	Methyl acetoacetate	$CH_3COCH_2CO_2CH_3$	116.12	3, 632	1.0757^{20}	1.4186^{20}	27.5	171.7	77	50 aq; misc alc
m124	4′-Methylaceto-phenone	$CH_3C_6H_4COCH_3$	134.18	7, 307	1.0051	1.5328^{20}	22–24	226	92	i aq; v s alc, eth
m125	Methyl 4-acetoxy-benzoate	$CH_3CO_2C_6H_4CO_2CH_3$	194.19	10, 159			82–84			
m126	Methyl acrylate	$H_2C{=}CHCO_2CH_3$	86.09	2, 399	0.9541^{20}_{4}	1.4040^{20}	−76.5	80.2	−3 (CC)	6 aq; s alc, eth
m127	Methylamine	CH_3NH_2	31.06	4, 32	0.699^{-11}_{4}		−93.5	−6.3	0	959 mL aq; 10.5 bz
m128	1-(Methylamino)-anthraquinone		237.26	14, 179			170–172			

m129	Methyl 2-aminobenzoate	$H_2NC_6H_4CO_2CH_3$	151.17	14, 317	1.168^{19}_{4}	1.5820^{20}	24	256	104	sl s aq; v s alc, eth
m130	Methyl 3-aminocrotonate	$CH_3C(NH_2){=}CHCO_2CH_3$	115.13	3, 632			81–83			
m131	2-(Methylamino)ethanol	$CH_3NHCH_2CH_2OH$	75.11	4, 276	0.937^{20}	1.4387^{20}		159	72	misc aq, alc, eth
m132	4-Methylaminophenol sulfate	$(CH_3NC_6H_4OH)_2 \cdot H_2SO_4$	344.39	13, 441			260 dec			4 aq; sl s alc; i eth
m133	Methyl 2-(aminosulfonyl)benzoate	$H_2HSO_2C_6H_4CO_2CH_3$	215.23	11, 377			126–128			
m134	*N*-Methylaniline	$C_6H_5NHCH_3$	107.16	12, 135	0.989^{20}_{4}	1.5684^{20}	−57	196	78	sl s aq; s alc, eth
m135	*N*-Methylanilinium trifluoroacetate	$C_6H_5NHCH_3 \cdot HO_2CCF_3$	221.18				65–66			
m136	2-Methylanthraquinone		222.24	7, 809			170–173			v s bz; s alc, eth
m137	Methylarsonic acid	$CH_3AsO(OH)_2$	139.96	4, 613			161			v s aq; s alc
m138	4-Methylbenzaldehyde	$CH_3C_6H_4CHO$	120.15	7, 297	1.0194^{17}_{4}	1.5447^{20}		205	80	misc alc, eth; sl s aq
m139	Methyl benzenesulfonate	$C_6H_5SO_2OCH_3$	172.20	11^2, 20	1.2889^{0}_{4}	1.5151^{20}	−4	154^{20mm}		v s alc, chl, eth
m140	2-Methylbenzimidazole		132.17	23, 145			176–177			s alk, hot aq; sl s alc
m141	Methyl benzoate	$C_6H_5CO_2CH_3$	136.15	9, 109	1.0933^{15}_{4}	1.5205^{15}	−15	199.5	83	0.2 aq; misc alc, eth
m142	2-Methylbenzoic acid	$CH_3C_6H_4CO_2H$	136.15	9, 462	1.062		103.7	258–259		sl s aq; v s alc
m143	3-Methylbenzoic acid	$CH_3C_6H_4CO_2H$	136.15	9, 475	1.054		111–113	263		0.09 aq; v s alc
m144	4-Methylbenzoic acid	$CH_3C_6H_4CO_2H$	136.15	9, 483			180	274–275		v s alc, eth
m145	4-Methylbenzophenone	$CH_3C_6H_4COC_6H_5$	196.25	7, 440			57	326		v s bz, eth
m146	2-Methylbenzothiazole		149.22	27, 46	1.173	1.6170^{20}	12–14	238	102	s alc, HOAc; i aq
m147	2-Methylbenzoxazole		133.15	27, 46	1.121	1.5497^{20}	8–10	178	75	
m148	α-Methylbenzyl acetate	$CH_3CO_2CH(CH_3)C_6H_5$	164.20	6, 476	1.028	1.4945^{20}		95^{12mm}	91	
m149	α-Methylbenzyl alcohol	$C_6H_5CH(CH_3)OH$	122.17	6, 475	1.0191^{13}_{4}	1.5265^{20}	20	204^{745mm}	85	v s alc; s bz, chl
m150	2-Methylbenzyl alcohol	$CH_3C_6H_4CH_2OH$	122.17	6, 484		1.5408^{20}	33–36	110^{14mm}	104	5 aq; 5 alc; s eth
m151	(±)-α-Methylbenzylamine	$C_6H_5CH(CH_3)NH_2$	121.18	12, 1094	0.940	1.5260^{20}		185	79	4.2 aq; misc alc, eth
m152	4-Methylbenzylamine	$CH_3C_6H_4CH_2NH_2$	121.18	12, 1141	0.952	1.5340^{20}	12–13	195	75	

(*Continued*)

TABLE 2.20 Physical Constants of Organic Compounds (*Continued*)

No.	Name	Formula	Formula weight	Beilstein reference	Density, g/mL	Refractive index	Melting point, °C	Boiling point, °C	Flash point, °C	Solubility in 100 parts solvent
m153	Methylbis(trimethylsilyloxy)vinyl ether	$CH_3Si[OSi(CH_3)_2]CH{=}CH_2$	148.55	4^4, 4184	0.864	1.3970^{20}		$48^{8.8mm}$	51	
m154	Methyl bromoacetate	$BrCH_2CO_2CH_3$	152.98	2, 213	1.616	1.4586^{20}		52^{15mm}	62	s alc
m155	(±)-Methyl 2-bromobutyrate	$CH_3CH_2CH(Br)CO_2CH_3$	181.04	2, 282	1.573	1.452^{20}		138^{50mm}	68	
m156	Methyl 2-bromopropionate	$CH_3CH(Br)CO_2CH_3$	167.01	2, 253	1.497	1.5420^{20}		51^{19mm}	51	s alc
m157	2-Methyl-1,3-butadiene	$H_2C{=}C(CH_3)CH{=}CH_2$	68.12	1, 252	0.681^{20}_{4}	1.4216^{20}	−146.0	34.1	−53	misc alc, eth
m158	2-Methylbutane	$CH_3CH_2CH(CH_3)_2$	72.15	1, 134	0.6197^{20}	1.3537^{20}	−159.9	27.8	−56	0.005 aq; misc alc
m159	2-Methyl-1-butenethiol	$CH_3CH_2CH(CH_3)CH_2SH$	104.22	1^2, 421	0.848	1.4465^{20}		117	19	s alc, eth; i aq
m160	2-Methyl-2-butanethiol	$CH_3CH_2C(CH_3)_2SH$	104.22	1^1, 196	0.842	1.4385^{20}	−103.9	99.1	−1	s alc, eth; i aq
m161	2-Methyl-1-butanol	$CH_3CH_2CH(CH_3)CH_2OH$	88.15	1, 388	0.816^{20}_{4}	1.4100^{20}	<−70	128	43	3 aq; misc alc, eth
m162	2-Methyl-2-butanol	$CH_3CH_2C(CH_3)_2OH$	88.15	1, 388	0.8096^{20}	1.4050^{20}	−9.0	102.0	21	11 aq; misc alc, bz, chl, eth
m163	3-Methyl-1-butanol	$(CH_3)_2CHCH_2CH_2OH$	88.15	1, 392	0.8129^{15}_{4}	1.4085^{15}	−117	131	45	2 aq; misc alc, bz, chl, eth, PE, HOAc
m164	3-Methyl-2-butanol	$(CH_3)_2CHCH(OH)CH_3$	88.15	1, 391	0.8179^{20}	1.4091^{20}		112.9	38	2.8 aq; misc alc, eth
m165	3-Methyl-2-butanone	$(CH_3)_2CHCOCH_3$	86.13	1, 682	0.802^{20}_{4}	1.3880^{20}	−92	94.3	6	misc alc, eth
m165a	2-Methyl-1-butene	$C_2H_5C(CH_3){=}CH_2$	70.14	1, 211	0.650	1.3780^{20}	−137.6	31	< −34	
m166	2-Methyl-2-butene	$CH_3CH{=}C(CH_3)_2$	70.14	1, 211	0.6620^{20}_{4}	1.3878^{20}	−133.8	38.6	−45	misc alc, eth; i aq
m167	3-Methyl-1-butene	$(CH_3)_2CHCH{=}CH_2$	70.14	1, 213	0.6272^{20}_{4}	1.3638^{20}	−168	20	−56	misc alc, eth
m168	*cis*-2-Methyl-2-butenoic acid	$CH_3CH{=}C(CH_3)CO_2H$	100.12	2, 428	0.983^{47}_{4}	1.4437^{47}	45	185		s alc, eth; v s hot aq
m169	*trans*-2-Methyl-2-butenoic acid	$CH_3CH{=}C(CH_3)CO_2H$	100.12	2, 430	0.969	1.4342^{81}	64	198		s alc, eth; v s hot aq
m170	3-Methyl-2-butenoic acid	$(CH_3)_2C{=}CHCO_2H$	100.12	2, 432	1.006^{24}		69	194–195		s aq, alc, eth
m171	2-Methyl-3-buten-2-ol	$(CH_3)_2C(OH)CH{=}CH_2$	86.13	1, 444	0.824	1.4170^{20}	2.6	98–99	13	
m172	3-Methyl-2-buten-1-ol	$(CH_3)_2C{=}CHCH_2OH$	86.13	1, 444	0.848	1.4440^{20}		140	43	
m173	3-Methyl-3-buten-1-ol	$H_2C{=}C(CH_3)CH_2CH_2OH$	86.13		0.853	1.4337^{20}			36	
m174	2-Methyl-1-buten-3-yne	$H_2C{=}C(CH_3)C{\equiv}CH$	66.10	1^1, 126	0.695	1.4140^{20}	−113	32	−6	
m175	*N*-Methylbutylamine	$CH_3CH_2CH_2CH_2NCH_3$	87.17	4, 157	0.736	1.3995^{20}	−75	91	1	
m176	1-Methylbutylamine	$CH_3CH_2CH_2CH(CH_3)NH_2$	87.17	4, 177	0.7384^{20}_{4}	1.4029^{20}		91	35	misc aq, alc, eth
m177	3-Methylbutyl 3-methylbutyrate	$(CH_3)_2CHCH_2CH_2O_2CCH_2$-$CH(CH_3)_2$	172.27	2, 312	0.8541^{25}	1.4100^{25}		190.4	84	misc alc, eth
m178	3-Methyl-1-butyne	$(CH_3)_2CHC{\equiv}CH$	68.12	1, 251	0.666^{20}_{4}	1.3740^{20}	−89.8	26.4		misc alc, eth
m179	2-Methyl-3-butyne-2-ol	$(CH_3)_2C(OH)C{\equiv}CH$	84.12	1^1, 235	0.8672^{20}_{20}	1.4209^{20}	2.6	104	25	misc aq, acet, bz

m180	2-Methylbutyraldehyde	$CH_3CH_2CH(CH_3)CHO$	86.13	1[1], 352	0.804	1.3919^{20}		90–92	4	
m181	3-Methylbutyraldehyde	$(CH_3)_2CHCH_2CHO$	86.13	1, 684	0.785^{20}_{20}	1.3882^{20}	−51	92–93	19	misc alc, eth; sl s aq
m182	Methyl butyrate	$CH_3CH_2CH_2CO_2CH_3$	102.13	2, 270	0.898^{20}_{4}	1.3860^{20}	−85.8	103	11	1.4 aq; misc alc, eth
m183	2-Methylbutyric acid	$CH_3CH_2CH(CH_3)CO_2H$	102.13	2, 305	1.4055^{20}			176.5	73	
m184	3-Methylbutyric acid	$(CH_3)_2CHCH_2CO_2H$	102.13	2, 309	0.9308^{20}_{4}	1.4033^{20}	−29.3	176.5	70	4 aq; s alc, chl, eth
m185	3-Methylbutyronitrile	$(CH_3)_2CHCH_2CN$	83.13	2[2], 278	0.7925^{19}_{4}	1.3927^{20}	−101	129		misc alc, eth
m186	3-Methylbutyryl chloride	$(CH_3)_2CHCH_2COCl$	120.58	2, 315	0.985^{20}_{4}	1.4161^{20}		115–117	18	dec aq, alc; s eth
m187	Methyl carbamate	$H_2NCO_2CH_3$	75.07	3, 21	1.136^{56}_{4}		56–58	177		220 aq; 73 alc; s eth
m188	Methyl chloroacetate	$ClCH_2CO_2CH_3$	108.52	2, 197	1.238^{20}_{20}	1.4220^{20}	−32	130	51	i aq; misc alc, eth
m189	Methyl 2-chloroaceto-acetate	$CH_3COCH(Cl)CO_2CH_3$	150.56		1.236	1.4465^{20}	−32.7	137	71	
m190	Methyl 4-chloroaceto-acetate	$ClCH_2COCH_2CO_2CH_3$	150.56	3[2], 426	1.305	1.4564^{20}		85^{4mm}	102	
m191	Methyl 3-chloro-benzoate	$ClC_6H_4CO_2CH_3$	170.60	9, 338	1.227	1.4923^{20}	21	101^{12mm}	104	
m192	Methyl-4-chloro-benzoate	$ClC_6H_4CO_2CH_3$	170.60	9, 340	1.382^{20}		42–44		106	s alc
m193	Methyl 4-chloro-butyrate	$ClCH_2CH_2CH_2CO_2CH_3$	136.58	2, 278	1.1268^{14}	1.4321^{20}		175–176	59	v s eth; s alc, acet
m194	Methyl chloroformate	$ClCO_2CH_3$	94.50	3, 9	1.223^{20}_{4}	1.3865^{20}		70–72	17	misc alc, bz, chl, eth
m195	Methyl 3-(chloro-formyl)propionate	$CH_3O_2CCH_2CH_2COCl$	150.56	2[2], 553	1.223	1.4402^{20}		65^{3mm}	73	
m196	Methyl 2-chloro-propionate	$CH_3CH(Cl)CO_2CH_3$	122.55	2, 248	1.075	1.4193^{20}		132–133	38	s alc
m197	2-Methylcinnam-aldehyde	$C_6H_5CH{=}C(CH_3)CHO$	146.19	7, 369	1.0407^{17}_{4}	1.6045^{20}		149^{27mm}	79	
m198	Methyl *trans*-cinnamate	$C_6H_5CH{=}CHCO_2CH_3$	162.19	9, 581			36–38	262	>110	
m199	6-Methylcoumarin		160.17	17, 337			75–76	303^{725mm}		
m200	Methyl crotonate	$CH_3CH{=}CHCO_2CH_3$	100.12	2, 410	0.9444^{20}_{4}	1.4242^{20}		121	4	v s alc, eth; i aq
m201	Methyl cyanoacetate	$NCCH_2CO_2CH_3$	99.09	2, 584	1.1225^{25}	1.4166^{25}	−22.5	201	>110	misc alc, eth
m202	Methylcyclohexane	$C_6H_{11}CH_3$	98.19	5, 29	0.7694^{20}	1.4221^{20}	−126.6	100.9	−4	
m203	Methyl cyclohexane-carboxylate	$C_6H_{11}CO_2CH_3$	142.20	9, 8	0.9954^{16}_{4}	1.4430^{20}		183	60	i aq; s alc, eth
m204	4-Methyl-1,2-cyclo-hexanedicarboxylic anhydride		168.19		1.162	1.4774^{20}			>110	

(*Continued*)

TABLE 2.20 Physical Constants of Organic Compounds (*Continued*)

No.	Name	Formula	Formula weight	Beilstein reference	Density, g/mL	Refractive index	Melting point, °C	Boiling point, °C	Flash point, °C	Solubility in 100 parts solvent
m205	1-Methylcyclohexanol	$CH_3C_6H_{10}OH$	114.19	6, 11	0.9251^{25}	1.4587^{25}	25	155	67	i aq; b bz, chl
m206	*cis*-2-Methylcyclohexanol	$CH_3C_6H_{10}OH$	114.19	6^2, 17	0.9360^{20}_{4}	1.4640^{30}	7	165	58	misc alc, eth
m207	*trans*-2-Methylcyclohexanol	$CH_3C_6H_{10}OH$	114.19	6, 11	0.9247^{20}_{4}	1.4616^{20}	−2	167.5	65	misc alc; s eth
m208	*cis*-3-Methylcyclohexanol	$CH_3C_6H_{10}OH$	114.19	6, 12	0.9155^{20}	1.4572^{20}	−6	168	62	misc alc, eth
m209	*trans*-3-Methylcyclohexanol	$CH_3C_6H_{10}OH$	114.19	6, 12	0.9214^{20}	1.4580^{20}	−0.5	167	62	
m210	*cis*-4-Methylcyclohexanol	$CH_3C_6H_{10}OH$	114.19	6, 14	0.9170^{20}	1.4614^{20}	−9.2	173	70	misc alc, eth
m211	*trans*-4-Methylcyclohexanol	$CH_3C_6H_{10}OH$	114.19	6, 14	0.9118^{21}_{4}	1.4559^{20}		174	70	misc alc; s eth
m212	2-Methylcyclohexanone	$CH_3C_6H_9(=O)$	112.17	7, 14	0.925^{20}_{4}	1.4478^{20}		162	46 (CC)	i aq; s alc, eth
m213	3-Methylcyclohexanone	$CH_3C_6H_9(=O)$	112.17	7, 15	0.9155^{20}_{4}	1.4460^{20}		169	51	i aq; s alc, eth
m214	4-Methylcyclohexanone	$CH_3C_6H_9(=O)$	112.17	7, 18	0.916^{20}_{4}	1.4455^{20}		171	40	i aq; s alc, eth
m215	1-Methyl-1-cyclohexene		96.17	5, 66	0.809^{20}_{4}	1.4502^{20}	−121	111	−3	i aq; s alc, eth
m216	4-Methyl-1-cyclohexene		96.17	5, 67	0.799	1.4412^{20}	−115.5	102	−1	i aq; s alc, eth
m217	6-Methyl-3-cyclohexene-1-methanol		126.20		0.954	1.4830^{20}				
m218	*N*-Methylcyclohexylamine	$C_6H_{11}NHCH_3$	113.20	12, 6	0.868	1.4560^{20}		149	??	
m219	3-Methylcyclohexylamine	$CH_3C_6H_{10}NH_2$	113.20	12, 10	0.855	1.4525^{20}		150^{730mm}	22	
m220	4-Methylcyclohexylamine	$CH_3C_6H_{10}NH_2$	113.20	12, 12	0.955	1.4531^{20}		151–154	26	
m221	Methylcyclopentadiene dimer		160.26	5^4, 1435	0.941	1.4976^{20}	−51	200	26	
m222	Methylcyclopentane	$C_5H_9CH_3$	84.16	5, 27	0.7487^{20}	1.4097^{20}	−142.4	71.8	−23	0.013 aq
m223	3-Methyl-1,2-cyclopentanedione		112.13	7^1, 310			105–107			
m224	2-Methylcyclopentanone		98.15	7^2, 13	0.9200^{20}_{4}	1.4347^{20}	−76	139	26	s aq; v s alc, eth

m225	Methyl cyclopropanecarboxylate	$C_3H_5CO_2CH_3$	100.12	9^1, 3	0.985	1.4181^{20}		119	17	
m226	Methyl decanoate	$CH_3(CH_2)_8CO_2CH_3$	186.30	2, 356	0.873	1.4255^{20}	−18	223	94	i aq; misc alc, eth
m227	Methyl dichloroacetate	$Cl_2CHCO_2CH_3$	142.97	2, 203	1.3808^{19}	1.4421^{20}	−52	143	80	i aq; s alc
m228	Methyl 2,2-dichloro-1-methylcyclopropanecarboxylate		183.03		1.245	1.4639^{20}		74^{8mm}	74	
m229	Methyl 2,3-dichloropropionate	$ClCH_2CH(Cl)CO_2CH_3$	157.00	2^1, 111	1.3282^{20}_4	1.4447^{20}		92^{50mm}	42	s alc
m230	*N*-Methyldiethanolamine	$CH_3N(CH_2CH_2OH)_2$	119.16	4, 284	1.0377^{20}	1.4685^{20}		248	126	misc aq, alc
m231	Methyl 3,4-dimethoxybenzoate	$(CH_3O)_2C_6H_3CO_2CH_3$	196.20	10, 396			59–62	283		
m232	Methyl 3,5-dimethoxybenzoate	$(CH_3O)_2C_6H_3CO_2CH_3$	196.20	10, 405			43	298	>110	
m233	Methyl 3-(dimethylamino)propionate	$(CH_3)_2NCH_2CH_2CO_2CH_3$	131.18	4, 403	0.917	1.4184^{20}		154	51	
m234	Methyl 2,5-dimethyl-3-furoate		154.17	18, 398	1.037	1.4750^{20}		198	80	
m235	Methyl 2,2-dimethylpropionate	$(CH_3)_3CCO_2CH_3$	116.16	2^1, 139	0.873	1.3880^{20}		101–103	−1	misc alc, eth; sl s aq
m236	*N*-Methyldioctylamine	$(C_8H_{17})_2NCH_3$	255.49	4^3, 381	1.066	1.4424^{20}	−30.1	165^{15mm}	>110	
m237	4-Methyl-1,3-dioxane		102.13	19^4, 49	0.976	1.4150^{20}	−45	114	22	
m238	*N*-Methyldiphenylamine	$(C_6H_5)_2NCH_3$	183.26	12, 180	1.048^{20}_4	1.6193^{20}	−7.6	135^{6mm}		i aq; s alc, eth
m239	Methyl diphenylglycolate	$(C_6H_5)_2C(OH)CO_2CH_3$	242.27	10, 344			74–76	187^{13mm}		
m240	3-Methyl-1,1-diphenylurea	$(C_6H_5)_2NCONHCH_3$	226.28	12,2, 852			172–174			
m241	Methyleneaminoacetonitrile	$CH_2{=}NCH_2CN$	68.08	Merck: 11, 5976			129			s hot aq, alc; sl s bz
m242	*N,N′*-Methylenebisacrylamide	$H_2C{=}CHC({=}O)NHCH_2$-$NHC({=}O)CH{=}CH_2$	154.17				>300			
m243	2,2′-Methylenebis-(4-chlorophenol)	$CH_2[C_6H_3(Cl)OH]_2$	269.13	6,3, 5408			168–172			100 EtOH; 100 eth; s PE
m244	4,4′-Methylenebis-(2,6-di-*tert*-butylphenol	$CH_2\{C_6H_2[C(CH_3)_3]_2OH\}_2$	424.67	6^4, 6811			156–158	289^{40mm}		

(*Continued*)

TABLE 2.20 Physical Constants of Organic Compounds (*Continued*)

No.	Name	Formula	Formula weight	Beilstein reference	Density, g/mL	Refractive index	Melting point, °C	Boiling point, °C	Flash point, °C	Solubility in 100 parts solvent
m245	4,4′-Methylenebis-(*N,N*-dimethyl-aniline)	$CH_2[C_6H_4N(CH_3)_2]_2$	254.38	13, 239			88–89			
m246	1,1′-Methylenebis(3-methylpiperidine)	$CH_2[CH_3C_5H_9N]_2$	210.37		0.887	1.4734^{20}		160^{50mm}	>110	
m247	4,4′-Methylenebis-(phenylisocyanate)	$CH_2(C_6H_4NCO)_2$	250.26	13,*3*, 461	1.180		42–44	200^{5mm}	>110	
m248	Methylene blue		373.90	27, 393			190 dec			4 aq; 1.3 alc; s chl
m249	4,4′-Methylenedi-aniline	$CH_2(C_6H_4NH_2)_2$	198.26	13, 238			89–91	399	221	v s alc, bz, eth; sl s aq
m250	3,4-Methylenedioxy-benzaldehyde		150.13	19, 115			37	264	>110	0.2 aq; v s alc, eth
m251	1,2-Methylenedioxy-benzene		122.12	19, 20	1.064	1.5398		173	55	
m252	3,4-Methylenedioxy-6-propylbenzyldi-ethyleneglycol butyl ether		338.45	19^3, 779	1.059	1.498		180^{1mm}	171	misc alc, bz, geons
m253	Methylenesuccinic acid	$H_2C{=}C(CO_2H)CH_2CO_2H$	130.10	2, 760	1.573		167			8.2 aq; 20 alc; v sl s bz, chl, eth, PE
m254	*N*-Methylethylene-diamine	$CH_3NHCH_2CH_2NH_2$	74.13	4^1, 415	0.841	1.4395^{20}		114–116	42	
m255	*N*-Methylformamide	$HC({=}O)NHCH_3$	59.07	4, 58	0.9988^{25}	1.4300^{25}	−4	199.5	98	misc aq
m256	*N*-Methylformanilide	$C_6H_5N(CH_3)CHO$	135.17	12, 234	1.095	1.5610^{20}	8–13	244	126	
m257	Methyl formate	HCO_2CH_3	60.05	2, 18	0.9815^{15}	1.3465^{15}	−99	31.7	−19	30 aq; misc alc
m258	5-Methylfuraldehyde		110.11	17, 289	1.1072^{18}_4	1.5263^{20}		187	72	s aq; v s alc; misc eth
m259	2-Methylfuran		82.10	17, 36	0.915^{20}_4	1.4332^{20}	−88	63–66	−22	0.3 aq
m259a	Methyl 2-furoate		126.11	18, 274	1.179^{20}	1.4879^{20}		181	73	s alc, eth; sl s aq
m260	Methylgermanium tribromide	CH_3GeBr_3	327.35		2.6337^{20}_4	1.5770^{20}		168		
m261	*N*-Methylglucamine		195.22	Merck: 12, 6154			128–129			100 aq^{25}; 1.2 alc^{70}
m262	Methyl-α-D-gluco-pyranoside		194.18	31, 179	1.46^{30}_4		168	$200^{0.2mm}$		63 aq; 1.6 alc; i eth
m263	(±)-2-Methylglutaro-nitrile	$NCCH_2CH_2CH(CH_3)CN$	108.14	2, 656	0.950	1.4340^{20}	−45	269–271	126	
m264	*N*-Methylglycine	$CH_3NHCH_2CO_2H$	89.09	4, 345			208 dec			42 aq; sl s alc
m265	Methyl glycolate	$HOCH_2CO_2CH_3$	90.08	3, 236	1.168^{18}_4	1.4170^{20}	74	151	67	s aq; misc alc, eth

m266	Methyl heptanoate	$CH_3(CH_2)_5CO_2CH_3$	144.22	2, 339	0.8815^{20}_{4}	1.4115^{20}	−55.8	173.5	52	s alc, eth; sl s aq
m267	5-Methyl-2-heptanol	$(CH_3)_2CH(CH_2)_3CH(OH)CH_3$	130.23	1, 421	0.803	1.4240^{20}		172	67	
m268	5-Methyl-3-heptanone	$C_2H_5CH(CH_3)CH_2COC_2H_5$	128.22	1^1, 363	0.823	1.4142^{20}		157–162	43	
m269	6-Methyl-5-hepten-2-one	$(CH_3)_2C{=}CHCH_2CH_2COCH_3$	126.20	1^3, 3010	0.855^{16}_{4}	1.4392^{20}	−67	73^{18mm}	50	misc alc, eth
m269a	Methyl hexadecanoate	$CH_3(CH_2)_{14}CO_2CH_3$	270.46	2, 372	0.852	1.4512^{20}	32–34	196^{15mm}	>110	s alc, chl, eth
m270	Methyl hexanoate	$CH_3(CH_2)_4CO_2CH_3$	130.19	2, 323	0.9038^{0}_{4}	1.4038^{23}	−71	151	45	v s alc, eth
m271	5-Methyl-2-hexanone	$(CH_3)_2CHCH_2CH_2COCH_3$	114.19	1^2, 756	0.888^{20}_{4}	1.4062^{20}	−73.9	144	3641	0.5 aq; misc alc, eth
m272	1-Methylhexylamine	$CH_3(CH_2)_4CH(NH_2)CH_3$	115.22	4, 194	0.7665^{18}	1.4175^{20}		144	54	sl s aq; s alc, eth
m273	1-Methylhydantoin		114.10	24, 244			157	subl		s aq, alc; 3 eth
m274	Methylhydrazine	CH_3NHNH_2	46.07	4^2, 957	0.866	1.4225^{20}	−52.4	87.5	21	misc aq, alc; s PE
m275	Methyl hydrazinocarboxylate	$H_2NNHCO_2CH_3$	90.08	3^1, 46			70–73	108^{12mm}		
m276	Methyl hydrogen glutarate	$HO_2CCH_2CH_2CH_2CO_2CH_3$	146.14	2^2, 565	1.169	1.4381^{20}		151^{10mm}	>110	
m277	Methyl hydrogen hexanedioate	$HO_2C(CH_2)_4CO_2H$	160.17	2, 652	1.081	1.4401^{20}	8–9	162^{10mm}	>110	s alc
m278	Methyl hydrogen succinate	$HO_2CCH_2CH_2CO_2H$	132.12	2, 608			56–59	151^{20mm}		v s aq, alc, eth
m279	Methyl hydroperoxide	CH_3OOH	48.04	1^2, 270	1.997^{15}_{4}	1.3642^{15}		38^{65mm}		misc aq, alc, eth; s bz
m280	Methylhydroquinone	$CH_3C_6H_3{-}1,4\text{-}(OH)_2$	124.14	6, 874			128–130			
m281	Methyl 4-hydroxybenzoate	$HOC_6H_4CO_2CH_3$	152.15	10, 158			126–128	270 dec		v s alc, eth, acet; 0.25 aq
m282	Methyl 2-hydroxyisobutyrate	$(CH_3)_2C(OH)CO_2CH_3$	118.13	3^2, 223	1.023	1.4112^{20}		127	42	v s aq, alc
m283	Methyl 4-hydroxyphenylacetate	$HOC_6H_4CH_2CO_2CH_3$	166.18	10, 191			57–60	163^{5mm}		
m284	2-Methylimidazole		82.11	23, 46	1.030	1.4960^{20}	−60	198	92	misc aq
m285	2-Methylimidazole		82.11	23, 65			142–143	268		
m286	4-Methylimidazole		82.11	23, 69			53–56	263	>110	
m287	2-Methyl-1*H*-indole		131.18	20, 311	1.07^{20}_{4}		58–60	273		v s alc, eth; s hot aq
m288	2-Methylindoline		133.19	20, 279	1.023	1.5681^{20}		229	93	
m289	*N*-Methylisatoic anhydride		177.16	27, 265			165 dec			
m290	Methyl isobutyrate	$(CH_3)_2CHCO_2CH_3$	102.13	2, 290	0.891^{20}	1.3840^{20}	−84.7	92.5	3	misc alc, eth; sl s aq
m291	Methyl isocyanate	CH_3NCO	57.05	4, 77	0.967	1.3695^{20}	−45	39	−6	s aq
m292	Methyl isodehydracetate		182.18	18, 410			68–70	167^{14mm}		

(*Continued*)

TABLE 2.20 Physical Constants of Organic Compounds (*Continued*)

No.	Name	Formula	Formula weight	Beilstein reference	Density, g/mL	Refractive index	Melting point, °C	Boiling point, °C	Flash point, °C	Solubility in 100 parts solvent
m293	*N*-Methylisopropylamine	$(CH_3)_2CHNHCH_3$	73.14	4^1, 153	0.702	1.3840^{20}	50–53		−31	
m294	Methyl isothiocyanate	CH_3NCS	73.12	4, 77	1.069	1.5258^{37}	35	118	32	v s alc, eth; sl s aq
m295	5-Methylisoxazole		83.09	27, 16	1.018	1.4386^{20}		122	30	
m296	Methyl lactate	$CH_3CH(OH)CO_2CH_3$	104.10	3, 280	1.088^{20}_{4}	1.4131^{20}		144–145	49	s aq (dec), alc, eth
m297	Methyl mandelate	$C_6H_5CH(OH)CO_2CH_3$	166.18	10, 202	1.1756^{20}		54–56	135^{12mm}	>110	s aq, alc, bz, chl
m298	Methyl mercaptoacetate	$HSCH_2CO_2CH_3$	106.14		1.187	1.4657^{20}		43^{10mm}	30	s alc, eth
m299	Methyl 3-mercaptopropionate	$HSCH_2CH_2CO_2CH_3$	120.17	3^2, 214	1.085	1.4660^{20}		55^{14mm}	60	
m300	Methyl methacrylate	$H_2C{=}C(CH_3)CO_2CH_3$	100.12	2^2, 398	0.9433^{20}	1.4140^{20}	−48	100	10	1,6 aq; s ketones, esters, CCl_4
m301	Methyl methanesulfonate	$CH_3SO_2OCH_3$	110.13	4, 4	1.2943^{20}_{4}	1.4138^{20}		202–203	104	20 aq; 100 DMF
m302	Methyl methoxyacetate	$CH_3OCH_2CO_2CH_3$	104.11	3, 236	1.0511^{20}_{4}	1.3964^{20}		130	35	v s alc, eth; sl s aq
m303	Methyl 4-methoxyacetoacetate	$CH_3OCH_2COCH_2CO_2CH_3$	146.14	3^4, 1939	1.129	1.4316^{20}		$89^{8.5mm}$	89	
m304	Methyl 2-methoxybenzoate	$CH_3OC_6H_4CO_2CH_3$	166.18	10, 71	1.157	1.5335^{20}		248	>110	
m305	Methyl 4-methoxybenzoate	$CH_3OC_6H_4CO_2CH_3$	166.18	10, 159			51	245	>110	
m306	Methyl 4-methoxyphenylacetate	$CH_3OC_6H_4CH_2CO_2CH_3$	180.20	10, 191	1.135	1.5165^{20}		158^{19mm}	36	
m307	Methyl 4-methoxypropionate	$CH_3OCH_2CH_2CO_2CH_3$	118.13	3, 297	1.009	1.4020		142–143	47	
m308	1-Methyl-4-(methylamino)piperidine		128.22		0.882	1.4672^{20}			55	
m309	Methyl 2-methylbenzoate	$CH_3C_6H_4CO_2CH_3$	150.18	9, 463	1.073	1.5190^{20}		207–208	82	
m310	Methyl 3-methylbenzoate	$CH_3C_6H_4CO_2CH_3$	150.18	9, 475	1.063	1.5160^{20}		113^{27mm}	95	
m311	Methyl 4-methylbenzoate	$CH_3C_6H_4CO_2CH_3$	150.18	9, 484			33–36	104^{15mm}	90	
m312	Methyl 2-methylbutyrate	$C_2H_5CH(CH_3)CO_2CH_3$	116.16	2, 304	0.885	1.3931^{20}		115	32	sl s aq; misc alc, eth
m313	2-Methyl-6-methylene-2-octanol	$C_2H_5C({=}CH_2)(CH_2)_3C(CH_3)_2OH$	156.27		0.784	1.4431^{20}		84^{10mm}	76	

m314	Methyl 2-methyl-3-furancarboxylate		140.14		1.116	1.4730^{20}		75^{20mm}	63	
m315	Methyl *S*-methylthiomethyl sulfoxide	$CH_3S(=O)CH_2SCH_3$	124.22		1.191	1.5487^{20}		$95^{2.5mm}$	>110	
m316	Methyl 3-(methylthio)propionate	$CH_3SCH_2CH_2CO_2CH_3$	134.20		1.077	1.4650^{20}		75^{13mm}	72	
m317	4-Methylmorpholine		101.15	27, 6	0.920	1.4349^{20}	−66	116	23	s aq, alc, eth
m318	1-Methylnaphthalene	$C_{10}H_7CH_3$	142.20	5, 566	1.0202^{20}	1.6170^{20}	−30.4	245	82	v s alc, eth
m319	2-Methylnaphthalene	$C_{10}H_7CH_3$	142.20	5, 567	1.029^{20}_{4}	1.6026^{40}	34.4	241	97	v s alc, eth
m320	Methyl 1-naphthaleneacetate	$C_{10}H_7CH_2CO_2CH_3$	200.24	9^3, 3206	1.142	1.5961^{20}		162^{5mm}	>110	
m321	2-Methyl-1,4-naphthoquinone		172.18	7^2, 656			105–107			1.4 alc; 10 bz; s chl
m322	Methyl 1-naphthyl ketone	$C_{10}H_7COCH_3$	170.21	7, 401	1.1336^{0}_{4}	1.6284^{20}	11	302	>110	s alc, eth; i aq
m323	Methyl 2-naphthyl ketone	$C_{10}H_7COCH_3$	170.21	7, 402			53–55	301	>110	sl s alc; s CS_2
m324	Methyl nitrate	CH_3ONO_2	77.04	1, 284	1.2075^{20}_{4}	1.3748^{20}	−83	64 expl		sl s aq; s alc, eth
m325	Methyl nitrite	CH_3ONO	61.04	1, 284	0.991(lq)			−17.3		s alc, eth
m326	*N*-Methyl-4-nitroaniline	$O_2NC_6H_4NHCH_3$	152.15	12, 714			152–154			
m327	2-Methyl-3-nitroaniline	$CH_3C_6H_3(NO_2)NH_2$	152.15	12, 848			88–90	305		
m328	2-Methyl-4-nitroaniline	$CH_3C_6H_3(NO_2)NH_2$	152.15	12, 846	1.1586^{140}_{4}		131–133			v s alc; s bz
m329	2-Methyl-5-nitroaniline	$CH_3C_6H_3(NO_2)NH_2$	152.15	12, 844			104–107			s alc, acet, eth
m330	4-Methyl-2-nitroaniline	$CH_3C_6H_3(NO_2)NH_2$	152.15	12, 1000			115–116			v s alc; s eth
m331	Methyl 2-nitrobenzoate	$O_2NC_6H_4CO_2CH_3$	181.15	9, 372	1.280	1.5340^{20}	−13	$106^{0.1mm}$	>110	s alc, eth
m332	Methyl 3-nitrobenzoate	$O_2NC_6H_4CO_2CH_3$	181.15	9, 378			78–80	279		
m333	Methyl 4-nitrobenzoate	$O_2NC_6H_4CO_2CH_3$	181.15	9, 390			94–96			

(*Continued*)

TABLE 2.20 Physical Constants of Organic Compounds (*Continued*)

No.	Name	Formula	Formula weight	Beilstein reference	Density, g/mL	Refractive index	Melting point, °C	Boiling point, °C	Flash point, °C	Solubility in 100 parts solvent
m334	2-Methyl-3-nitrobenzoic acid	$CH_3C_6H_3(NO_2)CO_2H$	181.15	9, 471			182–184			
m335	3-Methyl-4-nitrobenzoic acid	$CH_3C_6H_3(NO_2)CO_2H$	181.15	9, 481			216–218			
m336	4-Methyl-3-nitrobenzoic acid	$CH_3C_6H_3(NO_2)CO_2H$	181.15	9, 502			187–190			
m337	5-Methyl-2-nitrobenzoic acid	$CH_3C_6H_3(NO_2)CO_2H$	181.15	9, 482			134–136			
m338	2-Methyl-5-nitroimidazole		127.10	23^1, 23			252–254			
m339	3-Methyl-4-nitrophenol	$CH_3C_6H_3(NO_2)OH$	153.14	6, 386			127–129			
m340	4-Methyl-2-nitrophenol	$CH_3C_6H_3(NO_2)OH$	153.14	6, 412	1.240_4^{40}	1.574^{40}	32–35	125^{22mm}	108	v s alc, eth
m341	2-Methyl-2-nitro-1-propanol	$O_2NC(CH_3)_2CH_2OH$	119.12	1, 378			86–89	95^{10mm}		350 aq
m342	2-Methyl-2-nitropropyl methacrylate	$H_2C{=}C(CH_3)CO_2CH_2$-$C(CH_3)_2NO_2$	187.20	2^3, 1288	1.087	1.4500^{20}		102^{4mm}	>110	
m343	*N*-Methyl-*N*-nitroso-4-toluenesulfonamide	$CH_3C_6H_4SO_2N(CH_3)NO$	214.24	11^1, 29			62			
m344	Methyl 2-nonynoate	$CH_3(CH_2)_5C{\equiv}CCO_2CH_3$	168.24	2, 490	0.915	1.4484^{20}		121^{20mm}	100	
m345	Methyl-5-norbornene-2,3-dicarboxylic anhydride		178.19	17^2, 461	1.232	1.5060^{20}			>110	
m346	Methyl octadecanoate	$CH_3(CH_2)_{16}CO_2CH_3$	298.51	2, 379			38	215^{15mm}	>110	s alc, eth
m347	Methyl *cis*-9-octadecenoate	$CH_3(CH_2)_7CH{=}CH$-$(CH_2)_7CO_2CH_3$	296.50	2, 467	0.839^{20}	1.4521^{20}	−19.9	168^{2mm}	>110	misc abs alc, eth
m348	7-Methyl-1,6-octadiene	$(CH_3)_2C{=}CH(CH_2)_3CH{=}CH_2$	124.23	1^4, 1049	0.753	1.4360^{20}		143–144	26	
m349	Methyl octanoate	$CH_3(CH_2)_6CO_2CH_3$	158.24	2, 348	0.8775_4^{20}	1.4160^{25}	−40	192.9	72	v s alc, eth; i aq
m350	Methyl 2-octynoate	$CH_3(CH_2)_4C{\equiv}CCO_2CH_3$	154.21	2, 487	0.920	1.4460^{20}		217–220	88	
m351	3-Methyl-2-oxazolidinone		101.11		1.170	1.4541^{20}	15	90^{1mm}	>110	
m352	2-Methyl-2-oxazoline		85.11	27, 13	1.005	1.4340^{20}		110	20	
m353	3-Methyl-3-oxetanemethanol		102.13	17^3, 1128	1.024	1.4460^{20}		80^{40mm}	98	
m354	Methyl 2-oxocyclopentanecarboxylate	$(O{=})C_5H_7CO_2CH_3$	142.16	10, 597	1.145	1.4560^{20}		105^{19mm}	>110	

m355	Methyl 2-oxo-propionate	$CH_3C({=}O)CO_2CH_3$	102.09	3, 616	1.130	1.4065^{20}		134–137	39	misc alc, eth; sl s aq
m356	*trans*-2-Methyl-1,3-pentadiene	$CH_3CH{=}CHC(CH_3){=}CH_2$	82.15	1, 255	0.718	1.4469^{20}		75–76	−12	
m357	2-Methylpentane	$CH_3CH_2CH_2CH(CH_3)_2$	86.18	1, 148	0.6532^{20}	1.3725^{20}	−154	60.3	< −29	
m358	3-Methylpentane	$(CH_3CH_2)_2CHCH_3$	86.18	1, 149	0.6643^{20}	1.3765^{20}	−163	63	< −7	
m359	2-Methyl-1,5-pentane-diamine	$H_2N(CH_2)_3CH(CH_3)CH_2NH_2$	116.21	4, 270	0.860	1.4590^{20}	80			
m360	2-Methyl-2,4-pentanediol	$(CH_3)_2C(OH)CH_2CH(OH)CH_3$	118.18	1, 486	0.9216^{20}_{4}	1.4270^{20}	−50	198	102	misc aq
m361	4-Methylpentanenitrile	$(CH_3)_2CHCH_2CH_2CN$	97.16	2, 329	0.8035^{20}_{4}	1.4061^{20}	−51.1	156.5	45	s alc; misc eth
m362	Methyl pentanoate	$CH_3(CH_2)_3CO_2CH_3$	116.16	2, 301	0.875	1.3962^{20}		128	22	sl s aq; misc alc, eth
m363	2-Methylpentanoic acid	$CH_3CH_2CH_2CH(CH_3)CO_2H$	116.16	2^2, 288	0.9242^{20}_{20}	1.4135^{20}	−85	196.4	107	1.3 aq
m364	2-Methyl-1-pentanol	$CH_3CH_2CH_2CH(CH_3)CH_2OH$	102.18	1, 409	0.8262^{20}	1.4180^{20}		148	54	s alc, eth
m365	3-Methyl-3-pentanol	$(CH_3CH_2)_2C(CH_3)OH$	102.18	1, 411	0.8281^{20}	1.4186^{20}	−23.6	123	46	misc alc, eth; sl s aq
m366	4-Methyl-2-pentanol	$(CH_3)_2CHCH_2CH(OH)CH_3$	102.18	1, 410	0.8080^{20}	1.4112^{20}	−90	132	41	1.6 aq
m367	4-Methyl-2-pentanone	$(CH_3)_2CHCH_2COCH_3$	100.16	1, 691	0.7978^{20}	1.3958^{20}	−84	116.5	18	1.7 aq; misc alc, bz, eth
m368	2-Methyl-2-pentenal	$CH_3CH_2CH{=}C(CH_3)CHO$	98.15	1^4, 3471	0.861	1.4503^{20}		138	31	s alc
m369	4-Methyl-2-pentenoic acid	$(CH_3)_2CHCH{=}CHCO_2H$	114.14	2^2, 406	0.9529	1.4489	35	115^{20mm}	46	i aq; v s alc
m370	4-Methyl-3-penten-2-one	$(CH_3)_2C{=}CHCOCH_3$	98.15	1, 736	0.8653^{20}	1.4440^{20}	−59	129.5	31	3.1 aq
m370a	4-Methyl-2-pentyl acetate	$(CH_3)_2CHCH_2CH(CH_3)O_2CCH_3$	144.21		0.8805^{25}	1.3980^{20}		147.5	45	
m371	1-Methylpentylamine	$CH_3(CH_2)_3CH(NH_2)CH_3$	101.19	4, 190	0.767^{20}_{4}		−19	116–118	13	s aq, alc, PE
m372	3-Methyl-1-pentyn-3-ol	$CH_3CH_2C(CH_3)(OH)C{\equiv}CH$	98.15	1^2, 506	0.8688^{20}_{4}	1.4318^{20}	−30.6	122	26	13 aq, misc bz, acet PE, EtOAc; s eth
m373	4-Methylphenetole	$CH_3C_6H_4OCH_2CH_3$	136.19	6, 393	0.945	1.5044^{20}		189–191	70	
m374	*N*-(4-Methylphenyl)-acetamide	$CH_3C_6H_4NHCOCH_3$	149.19	12, 920	1.212^{15}		150–153	307		s alc, EtOAc, HOAc
m375	Methyl phenylacetate	$C_6H_5CH_2CO_2CH_3$	150.18	9, 434	1.044	1.5075^{20}		218	90	i aq; misc alc, eth
m376	2-Methyl-1-phenyl-2-propanol	$CH_6H_5CH_2C(CH_3)_2OH$	150.22	6, 523	0.974	1.5140^{20}	25–26	96^{18mm}	81	
m377	1-Methyl-3-phenyl-propylamine	$C_6H_5CH_2CH_2CH(CH_3)NH_2$	149.24	12, 1165	0.922	1.5123^{20}		222	97	

(*Continued*)

TABLE 2.20 Physical Constants of Organic Compounds (*Continued*)

No.	Name	Formula	Formula weight	Beilstein reference	Density, g/mL	Refractive index	Melting point, °C	Boiling point, °C	Flash point, °C	Solubility in 100 parts solvent
m378	3-Methyl-1-phenyl-2-pyrazolin-5-one		174.20	24, 20			129–130	287^{265mm}		
m379	Methyl phenyl sulfide	$C_6H_5SCH_3$	124.21	6, 297	1.058	1.5882^{20}	−15	188	57	i aq; s alc
m380	*N*-Methyl-*N*-phenyl-urethane	$C_6H_5N(CH_3)CO_2CH_2CH_3$	179.22	12, 417	1.074	1.5149^{20}		243–244	>110	
m381	*N*-Methylpiperazine		100.17		0.903	1.4655^{20}		138	42	v s aq, alc, eth
m382	2-Methylpiperazine		100.17	23, 17			65–67	155.6	65	78 aq; 37 acet; 32 bz
m383	*N*-Methylpiperidine	$C_5H_{10}NCH_3$	99.19	20, 19	0.816	1.4378^{20}		106–107	3	v s aq; misc alc, eth
m384	2-Methylpiperidine	$CH_3C_5H_9N$	99.19	20, 95	0.844	1.4459^{20}	−5	119	8	v s aq; misc alc, eth
m385	3-Methylpiperidine	$CH_3C_5H_9N$	99.19	20, 100	0.845	1.4470^{20}		126	17	v s aq
m386	4-Methylpiperidine	$CH_3C_5H_9N$	99.19	20, 101	0.838	1.4458^{20}		124	7	v s aq
m387	1-Methyl-3-piperdine-methanol		129.20	21^2, 8	1.013	1.4772^{20}		140–145	94	
m388	1-Methyl-4-piperidone		113.16	21^2, 215	0.920	1.4614^{20}			60	
m389	2-Methylpropan-aldehyde	$(CH_3)_2CHCHO$	72.11	1, 671	0.7891^{20}	1.3727^{20}	−65	64.1	−40	9 aq; misc alc, bz, chl, eth
m390	2-Methylpropane	$(CH_3)_3CH$	58.12	1, 124		1.3810^{-25}	−138	−11.7	−87	13 mL aq; 1320 mL alc; 2890 mL eth
m391	*N*-Methyl-1,3-propane-diamine	$H_2NCH_2CH_2CH_2NHCH_3$	88.15	4^1, 419	0.844	1.4468^{20}		139–141	35	
m392	2-Methyl-1,2-propane-diamine	$(CH_3)_2C(NH_2)CH_2NH_2$	88.15	4, 266	0.841	1.4410^{20}			23	
m393	2-Methyl-1,3-propane-diol	$HOCH_2CH(CH_3)CH_2OH$	90.12	1, 480	1.015	1.4450^{20}	−91	125^{20mm}	>110	
m394	1-Methyl-1-propane-thiol	$CH_3CH_2CH(SH)CH_3$	90.19	1, 373	0.8246_4^{25}	1.4338^{25}	−165	84–85	21	sl s aq; v s alc, eth
m395	2-Methyl-1-propane-thiol	$(CH_3)_2CHCH_2SH$	90.19	1, 378	0.8357^{20}	1.4396^{20}	−79	88.5	−9	v s alc, eth
m396	2-Methyl-2-propane-thiol	$(CH_3)_3CSH$	90.19	1, 383	0.7943_4^{25}	1.4198^{25}	1.1	64.1	−4	i aq
m397	2-Methyl-1-propanol	$(CH_3)_2CHCH_2OH$	74.12	1, 373	0.8016^{20}	1.3958^{20}	−108	108	28	10 aq; misc alc, eth
m398	2-Methyl-2-propanol	$(CH_3)_3COH$	74.12	1, 379	0.7888^{20}	1.3877^{20}	25.8	82.4	11	misc aq, alc, eth
m399	2-Methylpropene	$(CH_3)_2C{=}CH_2$	56.11	1, 207	0.6266_4^{mp}		−140	−6.9		v s alc, eth
m400	2-Methyl-2-propen-1-ol	$H_2C{=}C(CH_3)CH_2OH$	72.11	1, 443	0.857	1.4260^{20}		113–115	33	
m401	Methyl propionate	$CH_3CH_2CO_2CH_3$	85.11	2, 239	0.915_4^{20}	1.3770^{20}	−88	79.7	6	6 aq; misc alc, eth

m402	Methyl propionyl-acetate	$C_2H_5COCH_2CO_2CH_3$	130.15	3^3, 1212	1.037	1.4220^{20}		74^{5mm}	71	
m403	4′-Methylpropio-phenone	$CH_3C_6H_4COCH_2CH_3$	148.21	7, 317	0.993	1.5280^{20}	7.2	238–239	96	
m404	Methyl propyl ether	$CH_3CH_2CH_2OCH_3$	74.12	1, 354	0.738^{20}			39.1		
m405	2-Methyl-2-propyl-1,3-propanediol	$CH_3CH_2CH_2C(CH_3)(CH_2OH)_2$	132.20	1^1, 254		58–60	232	>110		sl s aq; misc alc, eth
m406	Methyl propyl sulfide	$CH_3SCH_2CH_2CH_3$	90.18	1^3, 1432	0.8424^{20}	1.4442^{20}	−113.0	95.5		s aq
m407	Methyl 2-propynyl ether	$CH_3OCH_2C{\equiv}CH$	70.09	1, 4541	0.830	1.3961^{20}		62	−18	
m408	2-Methylpyrazine		94.12	23, 94	1.030	1.5042^{20}	−29	135	50	v s aq, alc, eth
m409	2-Methylpyridine	$CH_3C_5H_4N$	93.13	20, 234	0.9443^{20}	1.4957^{20}	−66.7	129	39	misc aq; s alc, eth
m410	3-Methylpyridine	$CH_3C_5H_4N$	93.13	20, 239	0.9566^{20}	1.5040^{20}	−18.3	144	36	misc aq, alc, eth
m411	4-Methylpyridine	$CH_3C_5H_4N$	93.13	20, 240	0.9548^{20}	1.5037^{20}	3.8	145	57	misc aq, alc, eth
m412	Methyl 3-pyridine-carboxylate	$(C_5H_4N)CO_2CH_3$	137.14	22, 39			39	209		s aq, alc, bz
m413	Methyl 4-pyridine-carboxylate	$(C_5H_4N)CO_2CH_3$	137.14	22, 46	1.001	1.5122^{20}	8.5	207–209	82	
m414	1-Methyl-2-pyridone		109.13	21, 268	1.112	1.5690^{20}	30–32	250^{740mm}	>110	
m415	Methyl 3-pyridyl-carbamate		152.15	22^3, 4076			121–123			
m416	2-[3-(6-Methyl-2-pyridyl)propoxy]-ethanol		195.26		1.052	1.5150^{20}			>110	
m417	*N*-Methylpyrrole		81.12	20, 163	0.914	1.4875^{20}	−57	112–113	15	i aq; misc alc, eth
m418	*N*-Methylpyrrolidine		85.15	20, 4	0.819^{20}_{4}	1.4247^{20}		80–81	−21	misc aq, eth
m419	*N*-Methyl-2-pyrrolidinone		99.13	21, 237	1.0279^{25}	1.4680^{25}	−24.4	202	96	misc aq, alc, bz, eth
m420	2-Methylquinoline		143.19	20, 387	1.058	1.6108^{20}	−2	248	79	i aq; s chl, eth
m421	4-Methylquinoline		143.19	20, 395	1.0826^{20}_{4}	1.6200^{20}	9–10	263	>110	misc alc, bz, eth
m422	6-Methylquinoline		143.19	20, 397	1.063	1.6140^{20}		259	>110	
m423	2-Methylquinozaline		144.18	23^1, 44	1.118	1.6156^{20}	180	245–247	107	misc aq
m424	Methyl salicylate	$HOC_6H_4CO_2CH_3$	152.15	10, 70	1.1831^{20}	1.5360^{20}	−8	223	96	0.7 aq; misc alc, HOAc; s chl, eth
m425	α-Methylstyrene	$C_6H_5C(CH_3){=}CH_2$	118.18	5, 484	0.909	1.5375^{20}	−24	165.5	45	
m426	4-Methylstyrene	$CH_3C_6H_4CH{=}CH_2$	118.18	5, 485	0.897	1.5412^{20}		170–175	45	
m427	*mono*-Methyl succinate	$HO_2CCH_2CH_{162}CO_2CH_3$	132.12	2, 608			56–59	151^{20mm}		
m428	Methyl tetradecanoate	$CH_3(CH_2)_{12}CO_2CH_3$	242.40	2^2, 326	0.855	1.4362^{20}	18.4	323	>110	misc alc, bz, eth

(*Continued*)

TABLE 2.20 Physical Constants of Organic Compounds (*Continued*)

No.	Name	Formula	Formula weight	Beilstein reference	Density, g/mL	Refractive index	Melting point, °C	Boiling point, °C	Flash point, °C	Solubility in 100 parts solvent
m429	2-Methyltetrahydrofuran		86.13	17, 12	0.8552^{20}	1.4056^{20}		78	−11	
m430	3-Methyltetrahydropyran		100.16	17^3, 77	0.863	1.4204^{20}		109^{733mm}	6	
m431	3-Methyltetrahydrothiophene-1,1-dioxide		134.20		1.191	1.4772^{20}		276	>110	
m432	4-Methylthiazole		99.16	27.16	1.090	1.5257^{20}		134	32	
m433	4-Methyl-5-thiazoleethanol		143.21	27,*3*, 1754	1.196	1.5508^{20}		135^{7mm}	>110	
m434	2-Methyl-2-thiazoline		101.17	27, 13	1.067	1.5200^{20}	−101	145	37	
m435	(Methylthio)acetonitrile	CH_3SCH_2CN	87.14		1.039	1.4826^{20}		63^{15mm}	67	
m436	3-(Methylthio)aniline	$CH_3SC_6H_4NH_2$	139.22	13^1, 141	1.130	1.6423^{20}		165^{16mm}	>110	
m437	4-(Methylthio)benzaldehyde	$CH_3SC_6H_4CHO$	152.22	8^1, 533	1.144	1.6452^{20}		90^{1mm}	>110	
m438	2-(Methylthio)benzothiazole		181.28	27, 109			43–46		>110	
m439	3-(Methylthio)-2-butanone	$CH_3CH(SCH_3)COCH_3$	118.20	1^4, 3993	0.975	1.4710^{20}		$50–54^{20mm}$	44	
m440	Methyl thiocyanate	CH_3SCN	73.12	3, 175	1.068^{20}	1.4680^{20}	−5	133	38	i aq; misc alc, eth
m441	2-Methylthiophene		98.17	17, 37	1.0193^{20}	1.5199^{20}	−63	113	7	
m442	3-Methylthiophene		98.17	17, 38	1.0218^{20}	1.5180^{20}	−69	115.4	11	i aq; misc alc, eth
m443	5-Methyl-2-thiophenecarboxaldehyde		126.18	17^1, 151	1.170	1.5860^{20}		114^{25mm}	82	
m444	*N*-Methyl-2-thiourea	$CH_3NHC(=S)NH_2$	90.15	4, 70			119–121			v s aq, alc
m445	*N*-Methyl-*o*-toluamide	$CH_3C_6H_4CONHCH_3$	149.19	9, 465	1.158^{15}		69–71			
m446	*N*-Methyl-*p*-toluenesulfonamide	$CH_3C_6H_4SO_2NHCH_3$	185.25	11, 105			76–79			
m447	Methyl *p*-toluenesulfonate	$CH_3C_6H_4SO_2OCH_3$	186.23	11, 99	1.234		27.5	145^{5mm}	>110	
m448	Methyltriacetoxysilane	$CH_3Si(O_2CCH_3)_3$	220.26	4^3, 1896	1.175^{20}_{4}	1.408^{20}	40–45	88^{3mm}	85	
m449	Methyl trichloroacetate	$Cl_3CCO_2CH_3$	177.42	2, 208	1.488	1.4558^{20}		153	72	
m450	Methyltrichlorosilane	CH_3SiCl_3	149.48	4^3, 1896	1.273	1.4110^{20}		66	−15	
m451	Methyltriethoxysilane	$CH_3Si(OCH_3)_3$	178.30	4, 629	0.895	1.3840^{20}		141–143	23	
m452	Methyl trifluoroacetate	$F_3CCO_2CH_3$	128.05	2^3, 427	1.273	1.2907^{20}		43	−7	

m453	Methyl trifluoromethanesulfonate	$F_3CSO_2OCH_3$	164.10	3^4, 34	1.450	1.3244^{20}		94–99	38	
m454	Methyl 3,4,5-trihydroxybenzoate	$(HO)_3C_6H_2CO_2CH_3$	184.15	10, 483			201–203			
m455	Methyltrimethoxysilane	$CH_3Si(OCH_3)_3$	136.22	4^4, 4203	0.955	1.3703^{20}		102	11	
m456	Methyl trimethylacetate	$(CH_3)_3CCO_2CH_3$	116.16	2, 320	0.873	1.3900^{20}		101	6	
m457	*N*-Methyl-*N*-(trimethylsilyl)trifluoroacetamide	$F_3CC(=O)N(CH_3)Si(CH_3)_3$	199.25		1.075	1.3802^{20}		132	25	
m458	(Methyl)triphenylphosphonium bromide	$[CH_3P(C_6H_5)_3]^+ Br^-$	357.24	16, 760			230–234			
m459	2-Methylundecanal	$CH_3(CH_2)_8CH(CH_3)CHO$	184.32		0.830^{15}_4	1.4321^{20}		171	93	s alc, eth
m460	Methyl urea	$CH_3NHCONH_2$	74.08	4, 64	1.204		101–102			v s aq, alc; i eth
m461	*N*-Methyl–*N*-vinylacetamide	$CH_3CON(CH_3)CH{=}CH_2$	99.13	4^3, 442	0.959	1.4829^{20}		70^{25mm}	58	
m462	Methyl vinyl ether	$CH_3OCH{=}CH_2$	58.08	1^3, 1857	0.7511^{20}_4	1.3947	−123	5.5	−56	0.8 aq; v s alc
m463	Morpholine		87.12	27, 5	1.0005^{20}	1.4548^{20}	−4.9	128	375	misc aq, alc, bz, eth
m464	4-Morpholinepropionitrile		140.19	27^3, 337	1.037	1.4715^{20}	21	121^{2mm}		
m465	*N*-Morpholino-1-cyclohexene		167.25		0.995	1.5128^{20}		120^{10mm}	68	
m466	3-(*N*-Morpholino)-1,2-propanediol		161.20		1.157		37–38	191^{30mm}	>110	
m467	Myrcene	$(CH_3)_2C{=}CHCH_2CH_2\text{-}C(=CH_2)CH{=}CH_2$	136.24	1, 264	0.8013^{20}	1.4709^{20}		167	39	s alc, chl, eth, HOAc
n1	1-Naphthaldehyde	$C_{10}H_7CHO$	156.18	7, 400	1.150^{20}_4	1.6520^{20}	1–2	161^{15mm}	>110	s alc, eth
n2	Naphthalene	$C_{10}H_8$	128.17	5, 531	1.162^{20}_4	1.5821^{100}	80	217.7	79	0.3 aq; 7 alc; 33 bz; 50 chl
n3	1-Naphthalenecarboxylic acid	$C_{10}H_7CO_2H$	172.18	9, 647			160–162	300		sl s aq; v s hot alc, eth
n4	1,5-Naphthalenediamine	$C_{10}H_6(NH_2)_2$	158.20	13, 203			185–187			s hot aq, hot alc
n5	1,8-Naphthalenediamine	$C_{10}H_6(NH_2)_2$	158.20	13, 204	1.1265^{99}_4	1.6828^{99}	66.5	205^{12mm}		sl s aq; s alc, eth

(*Continued*)

TABLE 2.20 Physical Constants of Organic Compounds (*Continued*)

No.	Name	Formula	Formula weight	Beilstein reference	Density, g/mL	Refractive index	Melting point, °C	Boiling point, °C	Flash point, °C	Solubility in 100 parts solvent
n6	1-Naphthalenesulfonic acid	$C_{10}H_7SO_3H$	208.24	11, 155			90 dehydrates			v s aq, alc; sl s eth
n7	2-Naphthalenesulfonic acid	$C_{10}H_7SO_3H$	208.24	11, 171			124 dehydrates			v s aq, alc
n8	1,8-Naphthalic anhydride		198.18	17, 521			268			sl s HOAc
n9	1-Naphthol	$C_{10}H_7OH$	144.17	6, 596	1.0954^{99}_{4}	1.6206^{99}	96	288		v s alc, bz, chl, eth
n10	2-Naphthol	$C_{10}H_7OH$	144.17	6, 627	1.217^{4}		123	285	161	0.1 aq; 125 alc; 6 chl; 77 eth; s alk
n11	1,4-Naphthoquinone		158.16	7, 724	1.422		126			s bz, chl, eth, hot alc
n12	(2-Naphthoxy)acetic acid	$C_{10}H_7OCH_2CO_2H$	202.21	6, 645			155–157			
n13	2-(1-Naphthyl)-acetamide	$C_{10}H_7CH_2ONH_2$	185.23	9, 666			182			i aq; s bz, CS_2
n14	1-Naphthyl acetate	$C_{10}H_7O_2CCH_3$	186.21	6, 608			43–46		>110	s alc, eth
n15	1-Naphthylacetic acid	$C_{10}H_7CH_2CO_2H$	186.21	9, 666			135	dec		3.3 alc; v s chl, eth
n16	1-Naphthylacetonitrile	$C_{10}H_7CH_2CN$	167.21	9, 667		1.6192^{20}	33–35	194^{18mm}	>110	s alc
n17	1-Naphthylamine	$C_{10}H_7NH_2$	143.18	12, 1212	1.1232^{25}_{25}	1.6703	50	301	157	0.2 aq; v s alc, eth
n18	1-Naphthyl isocyanate	$C_{10}H_7NCO$	169.19	12, 1244	1.177	1.6344^{20}	4	267	>110	
n19	Nicotine		162.24	23, 117	1.0097^{20}_{4}	1.5882^{20}	−79	123^{17mm}	101	misc aq; v s alc, eth, PE
n20	Nitrilotriacetic acid	$N(CH_2CO_2H)_3$	191.14	4, 369			242 dec			0.1 aq; s hot alc
n21	3′-Nitroacetophenone	$O_2NC_6H_4COCH_3$	165.15	7, 288			76–78	202		s alc, eth
n22	4′-Nitroacetophenone	$O_2NC_6H_4COCH_3$	165.15	7, 288			78–80	202		s alc
n23	2-Nitroaniline	$O_2NC_6H_4CH_2$	138.13	12, 687	1.442^{15}		71	284		s hot aq, alc, chl
n24	3-Nitroaniline	$O_2NC_6H_4CH_2$	138.13	12, 698	1.43		114	306		0.1 aq; 5 alc; 6 eth
n25	4-Nitroaniline	$O_2NC_6H_4CH_2$	138.13	12, 711	1.437^{14}		147	332	165	4 alc; 3.3 eth; s bz
n26	3-Nitrobenzaldehyde	$O_2NC_6H_4CHO$	151.12	7, 250	1.2792^{20}_{4}		58	164^{23mm}		s alc, chl, eth
n27	4-Nitrobenzaldehyde	$O_2NC_6H_4CHO$	151.12	7, 256	1.496		106–107			s alc, bz, HOAc
n28	2-Nitrobenzamide	$O_2NC_6H_4CONH_2$	166.12	9, 373	1.462^{32}		174–178	317		s hot aq, hot alc, eth
n29	3-Nitrobenzamide	$O_2NC_6H_4CONH_2$	166.12	9, 381			140–143			
n30	Nitrobenzene	$C_6H_5NO_2$	123.11	5, 233	1.2051^{15}_{4}	1.5546^{15}	5.8	210.8	88	v s alc, bz, eth
n31	3-Nitrobenzene-1,2-dicarboxylic acid	$O_2NC_6H_3(CO_2H)_2$	211.13	9, 823			216 dec			2 aq; v s hot alc
n32	5-Nitrobenzene-1,3-dicarboxylic acid	$O_2NC_6H_3(CO_2H)_2$	211.13	9, 840			260			0.15 aq; v s alc, eth

n33	2-Nitrobenzenesulfonyl chloride	$O_2NC_6H_4SO_2Cl$	221.62	11, 67			65–67			s eth; d hot aq, alc
n34	5-Nitrobenzimidazole		163.14	23, 135			207–209			s alc, acid
n35	2-Nitrobenzoic acid	$O_2NC_6H_4CO_2H$	167.12	9, 370	1.58		146–148			0.7 aq; 33 alc; 22 eth
n36	3-Nitrobenzoic acid	$O_2NC_6H_4CO_2H$	167.12	9, 376	1.494		140–142			0.3 aq; 33 alc; 40 acet
n37	4-Nitrobenzoic acid	$O_2NC_6H_4CO_2H$	167.12	9, 389	1.58		242.8			9 alc; 2 eth; 5 acet
n38	4-Nitrobenzonitrile	$O_2NC_6H_4CN$	148.12	9, 397			146–149			s HOAC; sl s aq, alc
n39	3-Nitrobenzoyl chloride	$O_2NC_6H_4COCl$	185.57	9, 381			32–35	275–278	>110	dec aq, alc; v s eth
n40	4-Nitrobenzoyl chloride	$O_2NC_6H_4COCl$	185.57	9, 394			75	205^{105mm}		dec aq, alc; s eth
n41	2-Nitrobenzyl alcohol	$O_2NC_6H_4CH_2OH$	153.14	6, 447			70–72	270		
n42	3-Nitrobenzyl alcohol	$O_2NC_6H_4CH_2OH$	153.14	6, 449			30–32	180^{3mm}	>110	s aq, alc, eth
n43	4-Nitrobenzyl alcohol	$O_2NC_6H_4CH_2OH$	153.14	6, 450			92–94	185^{12mm}		v s alc, eth; sl s aq
n44	4-Nitrobenzyl bromide	$O_2NC_6H_4CH_2Br$	216.04	5, 334			98–100			2 alc; v s eth
n45	4-Nitrobenzyl chloride	$O_2NC_6H_4CH_2Cl$	171.58	5, 329			70–73			8 alc; s eth
n46	2-Nitrobiphenyl	$O_2NC_6H_4C_6H_5$	199.21	5, 582	1.44_4^{25}	1.613^{25}	36.7	325	179	s alc, acet, CCl_4
n47	4-Nitrobiphenyl	$O_2NC_6H_4C_6H_5$	199.21	5, 583			112–114	340		sl s alc; s chl, eth
n48	1-Nitrobutane	$CH_3CH_2CH_2CH_2NO_2$	103.18	1, 123	0.975_{20}^{20}	1.4112	−81.3	152.8	47	sl s aq; misc alc, eth
n49	3-Nitro-2-butanol	$CH_3CH(NO_2)CH(OH)CH_3$	119.12	1, 373	1.1296_4^{25}	1.4414^{20}		92^{10mm}	91	
n50	3-Nitrocinnamic acid	$O_2NC_6H_4CH{=}CHCO_2H$	193.16	Merck: 12, 6692			200–201			1 alc
n51	2-Nitrodiphenylamine	$O_2NC_6H_4NHC_6H_5$	214.22	12, 690			76			i aq; s alc
n52	Nitroethane	$CH_3CH_2NO_2$	75.07	1, 99	1.0528_{20}^{20}	1.3920^{20}	−90	114	28	4.5 aq; misc alc, eth; s alk, chl
n53	5-Nitro-2-furaldehyde semicarbazone		198.14	17^3, 4467			242–244			s alk, chl, alk; 0.2 alc
n54	1-nitroguanidine	$O_2NNHC({=}NH)NH_2$	104.07	3, 126			dec >225			0.4 aq; sl s MeOH
n55	5-Nitro-1*H*-indazole		163.14	23, 129			207–209			s alc, bz, eth, acet
n56	Nitromethane	CH_3NO_2	61.04	1, 74	1.1322_4^{25}	1.3795^{25}	−28.4	101.2	35	11 aq; s alc, eth
n57	1-Nitronaphthalene	$C_{10}H_7NO_2$	173.17	5, 553	1.223		59–60	304		s alc; v s chl, eth
n58	3-Nitro-2-pentanol	$CH_3CH_2CH(NO_2)CH(OH)CH_3$	133.15	1, 385	1.0818_4^{25}	1.4430^{20}		100^{10mm}	90	
n59	2-Nitrophenol	$O_2NC_6H_4OH$	139.11	6, 213	1.495		45	216		s alc, bz, eth, alk
n60	4-Nitrophenol	$O_2NC_6H_4OH$	139.11	6, 226	1.270_4^{120}		113–114	279		s aq; v s alc, chl, eth
n61	4-Nitrophenyl acetate	$O_2NC_6H_4O_2CCH_3$	181.15	6, 233			77–79			s aq; v s alc, bz, eth
n62	2-Nitrophenylacetic acid	$O_2NC_6H_4CH_2CO_2H$	181.15	9, 454			139–142			s hot aq, alc
n63	4-Nitrophenylacetic acid	$O_2NC_6H_4CH_2CO_2H$	181.15	9, 455			153–155			s alc, bz, eth; sl s aq

(Continued)

TABLE 2.20 Physical Constants of Organic Compounds (*Continued*)

No.	Name	Formula	Formula weight	Beilstein reference	Density, g/mL	Refractive index	Melting point, °C	Boiling point, °C	Flash point, °C	Solubility in 100 parts solvent
n64	4-Nitrophenylaceto-nitrile	$O_2NC_6H_4CH_2CN$	162.15	9, 456			115–117			s alc, eth; i aq
n65	2-Nitro-1,4-phenylene-diamine	$O_2NC_6H_3(NH_2)_2$	153.14	13, 120			137–140			
n66	4-Nitro-1,2-phenylene-diamine	$O_2NC_6H_3(NH_2)_2$	153.14	13, 29			199–201			sl s aq; s HCl
n67	4-Nitrophenyl-hydrazine	$O_2NC_6H_4NHNH_2$	153.14	15, 468			156 dec			s alc, chl, eth, hot bz
n68	2-Nitrophenyl phenyl ether	$O_2NC_6H_4OC_6H_5$	215.21	6^2, 222	1.2539^{20}	1.575^{20}	< −20	184^{8mm}		s alc, eth
n69	4-Nitrophenyl phenyl ether	$O_2NC_6H_4OC_6H_5$	215.21	6, 232			53–56	320	>110	s bz, eth
n70	3-Nitro-1,2-phthalic acid	$O_2NC_6H_3(CO_2H)_2$	211.13	9, 823			213–216 dec			
n71	4-Nitro-1,2-phthalic acid	$O_2NC_6H_3(CO_2H)_2$	211.13	9, 828			170–172			
n72	3-Nitrophthalic anhydride		193.11	17, 486			163–165			sl s aq, bz
n73	1-Nitropropane	$CH_3CH_2CH_2NO_2$	89.09	1, 115	1.0009^{20}	1.4016^{20}	−108	131.1	36	1.4 aq; misc org solv
n74	2-Nitropropane	$(CH_3)_2CHNO_2$	89.09	1, 116	0.9821^{20}	1.3949^{20}	−91.3	120.3	24	1.7 aq; misc org solv
n75	2-Nitro-1-propanol	$CH_3CH(NO_2)CH_2OH$	105.09	1, 358	1.1841^{25}_{4}	$1,4379^{20}$		99^{10mm}	100	s aq, alc, eth
n76	4-Nitropyridine-*N*-oxide	$O_2NC_5H_4N(\rightarrow O)$	140.10	20^3, 2528			159–162			
n77	Nitrosobenzene	C_6H_5NO	107.11	6, 230			67–69	59^{18mm}		
n78	*N*-Nitrosodimethyl-amine	$(CH_3)_2NNO$	74.08	8, 84	1.0048^{20}_{4}	1.4368^{20}		151	61	v s aq, alc, eth
n79	4-Nitrosodiphenyl-amine	$C_6H_5NC_6H_4NO$	198.22	Merck: 12, 6737			144–145			v s alc, bz, chl, eth
n80	1-Nitroso-2-naphthol	$C_{10}H_6(NO)OH$	173.16	7, 712			109–110			3 alc; s bz, eth, alk; 0.1 aq
n81	1-Nitroso-2-naphthol-3,6-disulfonic acid disodium salt hydrate		377.26	11^2, 190			>300			2.5 aq; sl s alc

n82	4-Nitrosophenol	HOC_6H_4NO	123.11	7, 622			126	dec 144		s aq; v s alc, eth; explodes on contact with conc acid, alk, or fire
n83	2-Nitrotoluene	$CH_3C_6H_4NO_2$	137.14	5, 318	1.1622^{19}_{15}	1.5472^{20}	−10	222	106	s alc, bz
n84	3-Nitrotoluene	$CH_3C_6H_4NO_2$	137.14	5, 321	1.1581^{20}_{4}	1.5459^{20}	15.5	231.9	101	misc alc, eth; s bz
n85	4-Nitrotoluene	$CH_3C_6H_4NO_2$	137.14	5, 323	1.392		52	238	106	s alc, bz, chl, eth
n86	2-Nitro-α,α,α-trifluorotoluene	$CF_3C_6H_4NO_2$	191.11	5^2, 251			31–32	105^{20mm}	95	v s alc, bz
n87	3-Nitro-α,α,α-trifluorotoluene	$CF_3C_6H_4NO_2$	191.11	5, 327	1.436^{16}_{4}	1.4715^{20}	−2.4	200–205	87	s alc, eth
n88	5-Nitrouracil		157.09	24, 320			>300			
n89	Nonadecane	$CH_3(CH_2)_{17}CH_3$	268.51	1, 174	0.7776^{32}_{4}	1.4335^{38}	32	330	168	s eth; sl s alc
n90	Nonane	$CH_3(CH_2)_7CH_3$	128.26	1, 165	0.7176^{20}_{4}	1.4054^{20}	−53.5	150.8	31	s abs alc, eth
n91	1,9-Nonanediamine	$H_2N(CH_2)_9NH_2$	158.29	4, 272			37–38	258	>110	
n92	Nonanedinitrile	$NC(CH_2)_7CN$	150.23	2, 709	0.929	1.4460^{20}		176^{11mm}	>110	v s alc, bz, eth
n93	1,9-Nonanedioic acid	$HO_2C(CH_2)_7CO_2H$	188.22	2, 707	1.029^{20}_{4}		106.5	286^{100mm}		0.24 aq; v s alc; 3 eth
n94	1,9-Nonanediol	$HO(CH_2)_7OH$	160.26	1, 493			47–49	177^{15mm}	>110	
n95	Nonanenitrile	$CH_3(CH_2)_7CN$	139.24	2, 354	0.851^{15}_{4}	1.4260^{20}	−34.2	224.0	81	s alc, eth
n96	Nonanoic acid	$CH_3(CH_2)_7CO_2H$	158.24	2, 352	0.906^{20}_{4}	1.4330^{20}	12.5	254.5	100	s alc, chl, eth
n97	γ-Nonanoic lactone		156.23	17, 245	0.976	1.4475^{20}		122^{6mm}	>110	
n98	1-Nonanol	$CH_3(CH_2)_8OH$	144.26	1, 423	0.8279^{20}_{4}	1.4338^{20}	−5.5	215	75	0.6 aq; misc alc, eth
n99	2-Nonanone	$CH_3(CH_2)_6COCH_3$	142.24	1, 709	0.832	1.4210^{20}	−21	192^{743mm}	64	
n100	3-Nonanone	$CH_3(CH_2)_5COCH_2CH_3$	142.24	1, 709	0.821	1.4204^{20}		187–188	67	
n101	5-Nonanone	$(CH_3CH_2CH_2CH_2)_2CO$	142.24	1, 710	0.806^{20}_{4}	1.4190^{20}	−50	186–187	60	misc alc, eth
n102	Nonanoyl chloride	$CH_3(CH_2)_7COCl$	176.69	2, 353	0.946^{15}_{4}	1.4377^{20}	−60.5	215.4	95	dec aq, alc; s eth
n103	3-Nonen-2-one	$CH_3(CH_2)_4CH{=}CHCOCH_3$	140.23	1^3, 3017	0.848	1.4484^{20}		85^{12mm}	81	
n104	Nonyl aldehyde	$CH_3(CH_2)_7CHO$	142.24	1, 708	0.827^{19}_{19}	1.4240^{20}		185	63	
n105	Nonylamine	$CH_3(CH_2)_8NH_2$	143.27	4, 198	0.782	1.4330^{20}		201	62	sl s aq; s alc, eth
n106	Nopol		166.26	6^3, 396	0.973	1.4930^{20}		230–240	98	
n107	Norbornane		96.17	5^1, 45			82–84			s alc
n108	2-Norbornanone		110.16	7, 57			94–96	168–172	33	
n109	*exo*-2-Norbornyl formate		140.18	6,3, 219	1.048	1.4622^{20}		67^{16mm}	53	
n110	(+)-Norephedrine	$C_6H_5CH(OH)CH(CH_3)NH_2$	151.21	13^2, 371			51–54		>110	
o1	*cis,cis*-9,12-Octadecadienoic acid	$CH_3(CH_2)_4CH{=}CHCH_2$-$CH{=}CH(CH_2)_7CO_2H$	280.44	2, 496	0.9025^{20}_{4}	1.4699^{20}	−5	230^{16mm}		v s eth; 10 PE; s abs alc

(Continued)

TABLE 2.20 Physical Constants of Organic Compounds (*Continued*)

No.	Name	Formula	Formula weight	Beilstein reference	Density, g/mL	Refractive index	Melting point, °C	Boiling point, °C	Flash point, °C	Solubility in 100 parts solvent
o2	Octadecanamide	$CH_3(CH_2)_{16}CONH_2$	283.50	2, 383			102–104	251^{12mm}		s hot alc, hot eth
o3	Octadecane	$CH_3(CH_2)_{16}CH_3$	254.50	1, 173	0.7767_4^{28}	1.4367^{28}	28.2	316.3	165	s acet, eth; sl s alc
o4	1-Octadecanethiol	$CH_3(CH_2)_{17}SH$	286.57	1,*3*, 1838		1.4648	31–35	360	185	s eth; sl s alc
o5	Octadecanoic acid	$CH_3(CH_2)_{16}CO_2H$	284.48	2, 377	0.847^{70}	1.4299^{80}	69	383		4.9 alc; 20 bz; 50 chl; 3.9 acet; 16.6 CCl_4; s toluene, pentyl acetate
o6	1-Octadecanol	$CH_3(CH_2)_{17}OH$	270.50	1, 431	0.8123_4^{58}	1.4388^{20}	59.6	203^{10mm}		s alc, eth
o7	9,12,15-Octadecatri-enoic acid	$CH_3(CH_2CH{=}CH_2)_3CH_2$-$(CH_2)_6CO_2H$	278.44	2, 499	0.914_4^{18}	1.4800^{20}		230^{17mm}	>110	s alc, bz, eth
o8	1-Octadecene	$CH_3(CH_2)_{15}CH{=}CH_2$	252.49	1, 226	0.791_4^{18}	1.4439^{20}	17.7	314.9	148	s hot acet
o9	9-Octadecen-1-amine	$CH_3(CH_2)_7CH{=}CH(CH_2)_8NH_2$	267.50		0.813	1.4596^{20}			154	
o10	*cis*-9-Octadecenoic acid	$CH_3(CH_2)_7CH{=}CH(CH_2)_7CO_2H$	282.47	2, 463	0.8936_4^{20}	1.4581^{20}	13.4	360	189	s alc, bz, chl, eth
o11	*trans*-9-Octadecenoic acid	$CH_3(CH_2)_7CH{=}CH(CH_2)_7CO_2H$	282.47	2^2, 441	0.851^{79}	1.4308^{99}	44-45	288^{100mm}		s bz, chl, eth
o12	*cis*-9-Octadecen-1-ol	$CH_3(CH_2)_7CH{=}CH(CH_2)_8OH$	268.49	1, 453	0.850_4^{20}	1.4610^{20}	13–19	195^{8mm}	>110	s alc, eth; i aq
o13	9-Octadecenoyl chloride	$CH_3(CH_2)_7CH{=}CH{-}(CH_2)_7COCl$	300.92	2, 469	0.912	1.4630^{20}		180^{3mm}	>110	
o14	Octadecyl acrylate	$H_2C{=}CHCO_2(CH_2)_{17}CH_3$	324.55	2^4, 1468	0.800		32–34		>110	
o15	Octadecylamine	$CH_3(CH_2)_{17}NH_2$	269.52	4, 196	0.777^{27}		55–57	232^{32mm}	>110	s alc, bz, eth
o16	Octadecyl isocyanate	$CH_3(CH_2)_{17}NCO$	299.51	4^3, 439	0.847	1.4501^{20}	15–16	173^{5mm}	148	
o17	Octadecyltrichloro-silane	$CH_3(CH_2)_{17}SiCl_3$	387.94		0.984	1.4602^{20}		223^{10mm}	89	
o18	Octadecyl vinyl ether	$CH_3(CH_2)_{17}OCH{=}CH_2$	296.54		0.821_4^{30}	1.4440^{30}	28	187^{5mm}	177	
o19	1,7-Octadiene	$H_2C{=}CH(CH_2)_4CH{=}CH_2$	110.20		0.746	1.4220^{20}		114–121	9	
o20	1*H*,1*H*,5*H*-Octafluoro-1-pentanol	$HCF_2CF_2CF_2CF_2CH_2OH$	232.07	1^4, 1648	1.6647^{20}	1.3178^{20}		140–141	75	
o21	Octamethylcyclotetra-siloxane	$[\text{-}(CH_3)_2SiO\text{-}]_4$	296.62	4^3, 1885	0.956	1.3958^{20}	17–18	176	60	
o22	Octamethyltrisiloxane	$[(CH_3)_3SiO]_2Si(CH_3)_2$	236.54	4^3, 1879	0.8200^{20}	1.3848^{20}	*ca.* −80	153	29	s bz, PE; sl s alc
o23	Octane	$CH_3(CH_2)_6CH_3$	114.23	1, 159	0.7028_4^{20}	1.3974^{20}	−56.8	125.7	22	s eth; sl s alc

o24	1,8-Octanediamine	$H_2N(CH_2)_8NH_2$	144.26	4, 271			50–52	225–226	165	
o25	1,8-Octanedioic acid	$HO_2C(CH_2)_6CO_2H$	174.20	2, 691			140–144	230^{15mm}		0.16 aq; 0.6 eth; s alc
o26	1,2-Octanediol	$CH_3(CH_2)_5CH(OH)CH_2OH$	146.23	1^3, 2217			36–38	132^{10mm}	>110	
o27	1,8-Octanediol	$HO(CH_2)_8OH$	146.23	1, 490			59–61	172^{20mm}		v s alc; sl s aq, eth
o28	Octanenitrile	$CH_3(CH_2)_6CN$	125.22	2, 349	0.8135^{20}	1.4202^{20}	−45.6	198	73	s eth; sl s alc
o29	1-Octanethiol	$CH_3(CH_2)_7SH$	146.30	1^3, 1710	0.843	1.4525^{20}	−49.2	199.0	68	s alc
o30	Octanoic acid	$CH_3(CH_2)_6CO_2H$	144.21	2, 347	0.9088_4^{20}	1.4279^{20}	16.6	239	>110	0.07 aq; v s alc, chl, eth, PE
o31	γ-Octanoic lactone		142.20	17, 244	0.981	1.4440^{20}		234	>110	
o32	1-Octanol	$CH_3(CH_2)_7OH$	130.23	1, 418	0.8258_4^{20}	1.4290^{20}	−15.5	195	81	0.06 aq; misc alc, chl, eth
o33	(±)-2-Octanol	$CH_3(CH_2)_5CH(OH)CH_3$	130.23	1, 419	0.8193_4^{20}	1.4202^{20}	−31.6	175	71	0.1 aq; misc, alc, eth
o34	3-Octanol	$CH_3(CH_2)_4CH(OH)CH_2CH_3$	130.23	1^1, 208	0.819	1.4260^{20}		174–176	65	
o35	4-Octanol	$CH_3(CH_2)_3CH(OH)CH_2CH_2CH_3$	130.23		0.8192^{20}	1.425^{20}		176.6	71	
o36	2-Octanone	$CH_3(CH_2)_5COCH_3$	128.22	1, 704	0.819_4^{20}	1.4150^{20}	−16	173	52	i aq; misc alc, eth
o37	3-Octanone	$CH_3(CH_2)_4COCH_2CH_3$	128.22	1, 706	0.8220_4^{20}	1.4150^{20}		167–168	46	i aq; misc alc, eth
o38	4-Octanone	$CH_3(CH_2)_3COCH_2CH_2CH_3$	128.22	1, 706	0.809	1.4139^{20}		164	45	
o39	Octanoyl chloride	$CH_3(CH_2)_6COCl$	162.66	2, 348	0.955	1.4350^{20}	< −70	195	80	dec aq, alc; s eth
o40	1-Octene	$CH_3(CH_2)_5CH{=}CH_2$	112.22	1, 221	0.7149_4^{20}	1.4087^{20}	−102	121	21	i aq; misc alc, eth
o41	2-Octen-1-ylsuccinic anhydride		210.27		1.000	1.4694^{20}	8–12	168^{10mm}	>110	
o42	Octyl acetate	$CH_3CO_2(CH_2)_7CH_3$	172.27	2, 134	0.868	1.4185^{20}		211	88	sl s aq; misc alc
o43	Octyl aldehyde	$CH_3(CH_2)_6CHO$	128.22	1, 704	0.821_4^{20}	1.4183^{20}	12–15	171	51	sl s aq; misc alc
o44	Octylamine	$CH_3(CH_2)_7NH_2$	129.25	4, 196	0.782	1.4290^{20}	−5/−1	175–177	62	i aq; s alc, eth
o45	Octyl cyanoacetate	$NCCH_2CO_2(CH_2)_7CH_3$	197.28		0.934	1.4490^{20}		$95^{0.11mm}$	>110	
o46	Octyl gallate	3,4,5-$(HO)_3C_6H_2CO_2(CH_2)_7CH_3$	282.34	10^3, 2079			101–104			
o47	1-Octyl-2-pyrrolidine		197.32		0.920	1.4650^{20}	−25	172^{15mm}	>110	
o48	Octyltrichlorosilane	$CH_3(CH_2)_7SiCl_3$	247.67	4^3, 1907	1.070^{20}	1.4473^{20}		226^{730mm}	96	
o49	1-Octyne	$CH_3(CH_2)_5C{\equiv}CH$	110.19	1, 258	0.7457^{20}	1.4159^{20}	−79.3	126.2	17	i aq; s alc, eth
o50	1-Octyn-3-ol	$CH_3(CH_2)_4CH(OH)C{\equiv}CH$	126.20	1^3, 1996	0.864	1.4410^{20}		83^{19mm}	63	
o51	L-(+)-Ornithine	$H_2N(CH_2)_3CH(NH_2)CO_2H$	132.16	4, 420			140			v s aq, alc; sl s eth
o52	Oxalic acid	HO_2CCO_2H	90.04	2, 502	1.90_4^{17}		190 dec			14 aq^{20}; 40 alc; 1.3 eth
o53	Oxalic acid dihydrate	$HO_2CCO_2H \cdot 2H_2O$	126.07	2, 502	1.653_4^{19}		$-2H_2O$, 102			14 aq; 40 alc; 1 eth

(Continued)

TABLE 2.20 Physical Constants of Organic Compounds (*Continued*)

No.	Name	Formula	Formula weight	Beilstein reference	Density, g/mL	Refractive index	Melting point, °C	Boiling point, °C	Flash point, °C	Solubility in 100 parts solvent
o54	Oxalyl bromide	BrC(=O)C(=O)Br	215.84	2^1, 236		1.5220^{20}	−19	103^{720mm}	none	
o55	Oxalyl chloride	ClC(=O)C(=O)Cl	126.93	2, 542	1.455	1.4290^{20}	−10	64	none	s eth; viol dec aq, alc
o56	Oxalyl dihydrazide	$H_2NNHC(=O)C(=O)NHNH_2$	118.10	2, 559			240 dec			s hot aq; sl s alc, eth
o57	Oxamic hydrazide	$H_2NC(=O)C(=O)NHNH_2$	103.08	2, 559			218 dec			s alk; sl s aq; i eth
o58	Oxamide	$H_2NC(=O)C(=O)NH_2$	88.07	2, 545	1.667^{20}_{4}		dec 350			sl s hot aq, alc
o59	2-Oxazolidone		87.08	27, 135			86–89	220^{48mm}		
o60	2-Oxobutyric acid	$CH_3CH_2C(=O)CO_2H$	102.09	3, 629	1.200^{17}_{4}	1.3972^{20}	32–34	82^{16mm}	81	v s aq, alc; v sl s eth
o61	2-Oxohexamethylene-imine		113.16	21^2, 216	1.02^{75}_{4}		69.2	270	125	84 aq; v s alc, eth, chlorinated HC's
o62	5-Oxohexanonitrile	$CH_3CO(CH_2)_3CN$	111.14	3^3, 1234	0.975	1.4328^{20}		240	107	
o63	4-Oxopentanoic acid	$CH_3COCH_2CH_2CO_2H$	116.12	3, 671	1.1447^{25}_{4}	1.4396^{20}	33–35	246	137	v s aq, alc, bz, eth
o64	2-Oxopropionaldehyde	$CH_3C(=O)CHO$	72.06	1, 762	1.0455^{24}	1.4209^{20}		72	none	s aq, alc
o65	2-Oxopropionic acid	$CH_3C(=O)CO_2H$	88.06	3, 608	1.267^{15}_{4}	1.4315^{20}	11.8	165 dec	82	misc aq, alc, eth
o66	2-Oxo-1-pyrrolidine-propionitrile		138.17		1.120	1.4880^{20}		$140^{0.3mm}$	>110	
o66a	2,2′-Oxybis[2-methyl]-propane	$(CH_3)_3COC(CH_3)_3$	130.23		0.7658	1.3949^{20}		107		dec acids
o67	2,2′-Oxydiacetic acid	$HO_2CCH_2OCH_2CO_2H$	134.09	3, 234			142–145	dec		v s aq, alc; sl s eth
o68	4,4′-Oxydianiline	$H_2NC_6H_4OC_6H_4NH_2$	200.24	13, 441			190–192		218	
o69	3,3′-Oxydipropio-nitrile	$NCCH_2CH_2OCH_2CH_2CN$	124.14		1.043	1.4405^{20}		$112^{0.5mm}$	>110	
p1	Paraformaldehyde	$(CH_2O)_x$		1, 566			165 dec		71	s(slow) aq; s alk; i alc, eth
p2	Paraldehyde	$[-HC(CH_3)O-]_3$	132.16	19, 385	0.9984^{15}	1.4049^{20}	12.6	124		11 aq; misc alc, chl
p3	Parathion	$(C_2H_5O)_2P(=S)C_6H_4NO_2$	291.27		1.26^{25}_{4}	1.5370^{25}	6	375		v s alc, bz, eth
p4	Pentabromophenol	C_6Br_5OH	488.62	6, 206			223–226			sl s alc, eth
p5	Pentachloroacetone	$Cl_2CHC(=O)CCl_3$	230.31	1, 690	1, 656	1.4967^{20}	21 (anhyd)	192	none	i aq; v s acet
p6	Pentachlorobenzene	C_6HCl_5	250.34	5, 205	1.8342^{16}		82–85	275–277	none	v s bz, chl, eth
p7	Pentachloroethane	Cl_2CHCCl_3	202.30	1, 87	1.6712^{25}_{4}	1.5030^{20}	−29.0	160	none	0.05 aq; misc alc, eth
p8	Pentachloronitro-benzene	$C_6Cl_5(NO_2)$	295.34	5, 247	1.718^{25}_{4}		140–143			s bz, chl
p9	Pentachlorophenol	C_6Cl_5OH	266.34	6, 194	1.978^{22}_{4}		190–191	310		v s alc; s bz; 148 eth
p10	Pentachloropyridine	C_5Cl_5N	251.33	20, 232			124–126			
p11	Pentadecane	$CH_3(CH_2)_{13}CH_3$	212.42	1, 172	0.7684^{20}_{4}	1.4319^{20}	9.9	270	132	v s alc, eth
p12	Pentadecanenitrile	$CH_3(CH_2)_{13}CN$	223.40	2^1, 163	0.825	1.4420^{20}	20–23	322	>110	
p13	8-Pentadecanone	$[CH_3(CH_2)_7]_2C=O$	226.40	1, 717			41–43	178	>110	s alc

p14	3-Pentadecylphenol	$CH_3(CH_2)_{14}C_6H_4OH$	304.52				50–53	195^{1mm}	>110	
p15	1,2-Pentadiene	$CH_3CH_2CH{=}C{=}CH_2$	68.12	1, 251	0.6926^{20}_{4}	1.4209^{20}	−137.3	44.9		
p16	*cis*-1,3-Pentadiene	$CH_3CH{=}CHCH{=}CH_2$	68.12	1, 251	0.6910^{10}	1.4363^{20}	−140.8	44.1	−28	
p17	*trans*-1,3-Pentadiene	$CH_3CH{=}CHCH{=}CH_2$	68.12	1, 251	0.6760^{20}	1.4301^{20}	−87.5	42.0	−28	
p18	1,4-Pentadiene	$H_2C{=}CHCH_2CH{=}CH_2$	68.12	1, 251	0.6608^{22}_{4}	1.3888^{20}	−148.3	26.0	4	
p19	Pentaerythritol	$C(CH_2OH)_4$	136.15	1, 528	1.38^{25}_{4}	1.548	260			6 aq; v sl s alc; i eth
p20	Pentaerythritol diacrylate monostrearate	$CH_3(CH_2)_{16}CO_2CH_2$-$C(CH_2O_2CCH{=}CH_2)_2$-CH_2OH	510.72		1.018		29–31		>110	
p21	Pentaerythritol triacrylate	$(H_2C{=}CHCO_2CH_2)_3CCH_2OH$	298.30		1.180	1.4864^{20}			>110	
p22	Pentaerythrityl tetranitrate	$C(CH_2ONO_2)_4$	316.15	1^2, 602	1.1773^{20}_{4}		140	explodes on shock		s acet; sl s eth, alc
p23	Pentaethylenehexamine	$H_2N(CH_2CH_2NH)_4CH_2CH_2NH_2$	232.38	4^4, 1245	0.950	1.5096^{20}			>110	
p24	Pentamethylbenzene	$C_6H(CH_3)_5$	148.25	5, 443	0.917^{20}_{4}	1.527^{20}	54.4	231	91	v s alc, bz
p25	1,2,3,4,5-Pentamethylcyclopentadiene		136.24		0.870	1.4733^{20}		58^{13mm}	44	
p26	*N,N,N′,N′,N″*-Pentamethyldiethylenetriamine	$[(CH_3)_2NCH_2CH_2]_2NCH_3$	173.30	4,4, 1245	0.830	1.4420^{20}	−20	198	53	
p27	1,5-Pentamethylenetetrazole		138.17	26^2, 213			59–61	194^{12mm}		
p28	Pentanal	$CH_3CH_2CH_2CH_2CHO$	86.13	1, 676	0.8095^{20}_{4}	1.3942^{20}	−92	103	12	1.4 aq; misc alc, eth
p29	Pentane	$CH_3CH_2CH_2CH_2CH_3$	72.15	1, 130	0.6262^{20}_{4}	1.3575^{20}	−129.7	36.0	−49	misc alc, eth
p30	1,5-Pentanediamine	$H_2N(CH_2)_5NH_2$	102.18	4, 266	0.873^{25}_{4}	1.4591^{20}	−129.7	178–180	62	s aq, alc; sl s eth
p31	1,2-Pentanediol	$CH_3CH_2CH_2CH(OH)CH_2OH$	104.15	1^2, 548	0.971	1.4397^{20}		206	104	
p32	1,5-Pentanediol	$HO(CH_2)_5OH$	104.15	1, 481	0.9941^{20}	1.4494^{20}	−18	239	129	s aq, alc; sl s eth
p33	2,3-Pentanedione	$CH_3CH_2C({=}O)C({=}O)CH_3$	100.11	1, 776	0.957	1.4068^{20}	−52	110–112	19	
p34	2,4-Pentanedione	$CH_3COCH_2COCH_3$	100.11	1, 777	0.9721^{25}	1.4510^{20}	−23.1	138	34	17 aq; misc alc, eth
p35	Pentanenitrile	$CH_3CH_2CH_2CH_2CN$	83.13	2, 301	0.8035^{15}_{4}	1.3991^{15}	−92	141.3	40	i aq; s alc, eth
p36	1-Pentanesulfonic acid, sodium salt	$CH_3(CH_2)_4SO_3^-\ Na^+$	174.19	4^3, 23			>300			4 aq

(*Continued*)

TABLE 2.20 Physical Constants of Organic Compounds (*Continued*)

No.	Name	Formula	Formula weight	Beilstein reference	Density, g/mL	Refractive index	Melting point, °C	Boiling point, °C	Flash point, °C	Solubility in 100 parts solvent
p37	1-Pentanethiol	$CH_3(CH_2)_4SH$	104.22	1, 384	0.840	1.4460^{20}	−75.7	126.6	18	i aq; misc alc, eth
p38	Pentanoic acid	$CH_3(CH_2)_3CO_2H$	102.13	2, 299	0.9390_4^{20}	1.4080^{20}	−33.7	186	96	2.4 aq; v s alc, eth
p39	1-Pentanol	$CH_3(CH_2)_4OH$	88.15	1, 383	0.8146_4^{20}	1.4100^{20}	−79	137.5	33	2.7 aq^{22}; misc alc, eth
p40	2-Pentanol	$CH_3CH_2CH_2CH(OH)CH_3$	88.15	1, 384	0.8098_4^{20}	1.4054^{20}	−73	119.3	34	16.6 aq^{20}; misc alc, eth
p41	3-Pentanol	$CH_3CH_2CH(OH)CH_2CH_3$	88.15	1, 385	0.8150_4^{25}	1.4077^{25}	−69	116	41	5.5 aq^{20}; s alc, eth
p42	2-Pentanone	$CH_3CH_2CH_2COCH_3$	86.13	1, 676	0.8095^{20}	1.3900^{20}	−76.8	102	7	misc acet, bz, eth, PE
p43	3-Pentanone	$CH_3CH_2COCH_2CH_3$	86.13	1, 679	0.8143^{20}	1.3920^{20}	−39.0	102.0	13	3.4 aq
p44	Pentanophenone	$C_6H_5CO(CH_2)_3CH_3$	162.23	7, 327	0.988	1.5143^{20}		107^{5mm}	102	s alc, eth
p45	Pentanoyl chloride	$CH_3(CH_2)_3COCl$	120.58	2, 301	1.016	1.4216^{20}		125–127	32	
p46	1,4,7,10,13-Pentaoxa-cyclopentadecane	$[-CH_2CH_2O-]_5$	220.27		1.109	1.4650^{20}		$135^{0.2mm}$	>110	
p47	2,5,8,11,14-Pentaoxa-pentadecane	$CH_3(OCH_2CH_3)_4OCH_3$	222.28	1^3, 2107	1.0087_4^{20}	1.4330^{20}	−27	275–276	140	s aq; misc hydrocarbon solvents
p48	1-Pentene	$CH_3CH_2CH_2CH{=}CH_2$	70.14	1, 210	0.6429_4^{20}	1.3714^{20}	−165	30.1	−18	misc alc, bz, eth
p49	*cis*-2-Pentene	$CH_3CH_2CH{=}CHCH_3$	70.14	1, 210	0.6503_4^{20}	1.3813^{20}	−151	37.0	−20	misc alc, eth
p50	*trans*-2-Pentene	$CH_3CH_2CH{=}CHCH_3$	70.14	1, 210	0.6482_4^{20}	1.3792^{20}	−140	36.3	−45	misc alc, eth
p51	*cis*-2-Pentenenitrile	$CH_3CH_2CH{=}CHCN$	81.12	2^2, 400	0.820	1.4269^{20}		128	23	
p52	*trans*-3-Pentenenitrile	$CH_3CH{=}CHCH_2CN$	81.12	2, 427	0.837	1.4221^{20}		144–147	40	
p53	Pentyl acetate	$CH_3(CH_2)_4O_2CCH_3$	130.19	2, 131	0.8753^{20}	1.4020^{20}	−70.8	149.2	16	0.17 aq; misc alc, eth
p54	Pentylamine	$CH_3(CH_2)_4NH_2$	87.16	4, 175	0.7544^{20}	1.448^{20}	−55	104	−1	v s aq; misc eth; s alc
p55	Pentylbenzene	$CH_3(CH_2)_4C_6H_5$	148.25	5, 434	0.8594_4^{20}	1.4885^{20}	−78.3	202.2	65	s alc, misc bz, eth
p56	2-Pentylcinnam-aldehyde	$C_6H_5CH{=}C[(CH_2)_4CH_3]CHO$	202.30	7^2, 310	0.970	1.5571^{20}		290	>110	
p57	4-*tert*-Pentylphenol	$CH_3CH_2C(CH_3)_2C_6H_4OH$	164.25	6, 548	0.962_4^{20}		93	262		s alc, eth
p58	1-Pentyne	$CH_3CH_2CH_2C{\equiv}CH$	68.11	1, 250	0.6901_4^{20}	1.3852^{20}	−106	40.2	−34	v s alc; misc eth
p59	Perfluoro-1-octane-sulfonyl fluoride	$CF_3(CF_2)_7SO_2F$	502.12	2^4, 996	1.824	1.3010^{20}		154–155	none	
p60	Peroxyacetic acid	$CH_3C({=}O)CO_2H$	76.05	2, 169	1.226_4^{15}	1.3876^{20}	−0.2	110	41	v s aq, alc, eth
p61	Petroleum ether	Principally pentanes and hexanes		Merck: 12, 7329	0.640	1.3630^{20}		35–60	−49	misc bz, chl, eth, CCl_4; s glacial HOAc
p62	Phenanthrene		178.23	5, 667	1.063		100	340		1.6 alc; 50 bz; 30 eth
p63	1,10-Phenanthroline		180.21	23, 227			114–117			0.3 aq; 1.4 bz; s alc, acet
p64	Phenethylisobutyrate	$(CH_3)_2CHCO_2CH_2CH_2C_6H_5$	192.26	6^2, 451	0.988	1.4880^{20}		250	108	
p65	Phenol	C_6H_5OH	94.11	6, 110	1.0576_4^{41}	1.5418^{41}	41	182	79	6.7 aq; 8.2 bz; v s alc, chl, eth, alk

p66	Phenolphthalein		318.33	18, 143	1.299		261–263			8.2 alc; 1 eth
p67	Phenothiazine		199.28	27, 63			185.1	371		v s bz; s eth; sl s alc
p68	Phenoxyacetic acid	$C_6H_5OCH_2CO_2H$	152.15	6, 161			98–100	285 sl dec		1.3 aq; v s alc, bz, HOAc, CS_2, eth
p69	Phenoxyacetyl chloride	$C_6H_5OCH_2COCl$	170.60	6, 162	1.235	1.5340^{20}		225–226	108	dec aq, alc; s eth
p70	4-Phenoxyaniline	$C_6H_5OC_6H_4NH_2$	185.23	13, 438			84	189^{14mm}		s hot aq; v s alc, eth
p71	2-Phenoxybutyric acid	$CH_3CH_2CH(OC_6H_5)CO_2H$	180.20	6, 163			79–83	258		sl s aq
p72	2-Phenoxyethanol	$C_6H_5OCH_2CH_2OH$	138.17	6, 146	1.102^{22}_{4}	1.5370^{20}	14	245.2	>110	s aq; v s alc, eth
p73	1-Phenoxy-2-propanol	$C_6H_5OCH_2CH(OH)CH_3$	152.19	6^1, 85	1.063^{25}_{4}	1.523^{20}	13–18	240	135	
p74	2-Phenoxypropionic acid	$CH_3CH(OC_6H_5)CO_2H$	166.18	6, 163			116–119	265		s alc; sl s aq
p75	3-Phenoxypropyl bromide	$C_6H_5O(CH_2)_3Br$	215.10	6, 142	1.365	1.5460^{20}		134^{14mm}	96	
p76	3-Phenoxytoluene	$C_6H_5OC_6H_4CH_3$	184.24	6, 377	1.051	1.5727^{20}		271–273	>110	
p77	Phenylacetaldehyde	$C_6H_5CH_2CHO$	120.15	7, 292	1.027^{25}_{25}	1.5290^{20}	33–34	195	86	sl s aq; s alc, eth
p78	Phenylacetaldehyde dimethyl acetal	$C_6H_5CH_2CH(OCH_3)_2$	166.22	7, 293	1.004	1.4930^{20}		221	83	
p79	Phenylacetaldehyde ethylene acetal		164.21	19^4, 220	1.100	1.5220^{20}		120^{12mm}	107	
p80	Phenyl acetate	$C_6H_5O_2CCH_3$	136.15	6, 152	1.073	1.5030^{20}		196	76	misc alc, eth, chl
p81	Phenylacetic acid	$C_6H_5CH_2CO_2H$	136.15	9, 431	1.091^{77}_{4}		76.5	265.5		s hot aq, alc, eth
p82	Phenylacetonitrile	$C_6H_5CH_2CN$	117.15	9, 441	1.0214	1.5233^{20}	−23.8	233.5	101	i aq; misc alc, eth
p83	Phenylacetyl chloride	$C_6H_5CH_2COCl$	154.60	9, 436	1.169	1.5325^{20}		95^{12mm}	102	dec aq, alc
p84	Phenylacetylene	$C_6H_5C{\equiv}CH$	102.14	5, 511	0.9300	1.5470^{20}	−44.9	142.4	31	misc alc, eth
p85	Phenylacetylurea	$C_6H_5CH_2CONHCONH_2$	178.19	Merck: 12, 7343			212–216			sl s alc, bz, chl, eth
p86	(±)-3-Phenylalanine	$C_6H_5CH_2CH(NH_2)CO_2H$	165.19	14, 495			271–273			1.4 aq
p87	Phenyl 4-amino-salicylate	$H_2NC_6H_3$-2-$(OH)CO_2C_6H_5$	229.24	Merck; 12, 7426			153			0.7 mg aq
p88	4-Phenylazoaniline	$C_6H_5N{=}NC_6H_4NH_2$	197.24	16^1, 310			123–126	>360		v s alc, bz, chl, eth
p89	Phenylazoformic acid 2-phenylhydrazide	$C_6H_5N{=}NCONHNHC_6H_5$	240.27	16, 24			156–159 dec			
p90	4-Phenylazophenol	$C_6H_5N{=}NC_6H_4OH$	198.23	16, 96			150–152	230^{20mm}		v s alc, eth
p91	2-Phenylbenzimidazole		194.24	23, 230			293–296			s abs alc; sl s bz, chl
p92	Phenyl benzoate	$C_6H_5CO_2C_6H_5$	198.22	9, 116	1.235		69–72	298–299		v s hot alc; sl s eth
p93	*N*-Phenylbenzylamine	$C_6H_5CH_2NHC_6H_5$	183.25	12, 1023	1.061		35–38	306–307	>110	s alc, chl, eth
p94	*trans*-4-Phenyl-3-buten-2-one	$C_6H_5CH{=}CHCOCH_3$	146.19	7, 364	1.0097^{45}_{4}	1.5836^{45}	41.5	260–262	65	v s alc, bz, chl, eth

(*Continued*)

TABLE 2.20 Physical Constants of Organic Compounds (*Continued*)

No.	Name	Formula	Formula weight	Beilstein reference	Density, g/mL	Refractive index	Melting point, °C	Boiling point, °C	Flash point, °C	Solubility in 100 parts solvent
p95	2-Phenyl-3-butyn-2-ol	$CH_3C(OH)(C_6H_5)C{\equiv}CH$	146.19	6^2, 559			47–49	217–218	96	0.8 aq; s alc, bz, acet
p96	3-Phenylbutyraldehyde	$CH_3CH(C_6H_5)CH_2CHO$	148.21	7^1, 168	0.997	1.5179^{20}		94^{16mm}	96	
p97	2-Phenylbutyric acid	$CH_3CH_2CH(C_6H_5)CO_2H$	164.20	9^2, 356	1.055	1.5160^{20}	42–44	270–2	>110	s bz, eth
p98	2-Phenylbutyronitrile	$CH_3CH_2CH(C_6H_5)CN$	145.21	9, 541	0.974	1.5086^{20}		114^{15mm}	105	
p99	Phenyl chloroformate	$C_6H_5O_2CCl$	156.57	6, 159	1.248	1.5107^{20}		71^{9mm}	75	
p100	Phenyl dichloro-phosphate	$C_6H_5OP(O)Cl_2$	210.98	6, 179	1.412	1.5230^{20}		241–243	>110	
p101	*N*-Phenyldiethanol-amine	$C_6H_5N(CH_2CH_2OH)_2$	181.24	12, 183	1.120^{60}_{20}		56–80	350 sl dec		5 aq; v s alc; 29 eth; 25 bz
p102	4-Phenyl-1,3-dioxane		164.21	19^1, 616	1.111	1.5300^{20}		250–251	>110	
p103	2-Phenyl-1,3-dioxolane		150.18		1.106	1.5260^{20}		$80^{0.3mm}$	98	
p104	1,2-Phenylenediamine	C_6H_4-1,2-$(NH_2)_2$	108.14	13, 6			103	257		v s alc, chl, eth; sl s aq
p105	1,3-Phenylenediamine	C_6H_4-1,3-$(NH_2)_2$	108.14	13, 33	1.139^{15}		63.5	285		s aq, alc, acet, chl
p106	1,4-Phenylenediamine	C_6H_4-1,4-$(NH_2)_2$	108.14	13, 61			146	267	156	1 aq; s alc, chl, eth
p107	1,4-Phenylene diiso-cyanate	C_6H_4-1,4-$(NCO)_2$	160.13	13, 105			97–98	260	>110	
p108	1-Phenyl-1,2-ethanediol	$C_6H_5CH(OH)CH_2OH$	138.17	6, 907			66–68	272–274		v s aq, alc, bz, eth, chl, HOAc
p109	1-Phenylethanol	$CH_3CH(OH)(C_6H_5)$	122.17	6, 475	1.0130^{20}	1.5270^{20}	20	204	85	2.3 aq
p110	2-Phenylethanol	$C_6H_5CH_2CH_2OH$	122.17	6, 478	1.023^{25}_{25}	1.5317^{20}	−27	221	102	2 aq; misc alc, eth
p111	2-Phenylethyl acetate	$CH_3CO_2CH_2CH_2C_6H_5$	164.20	9, 510	0.984	1.4985^{20}		238–239	101	2 aq; misc alc, eth
p112	2-Phenylethylamine	$C_6H_5CH_2CH_2NH_2$	212.18	12, 1096	0.9640^{25}_{4}	1.5290^{25}	<0	197.5	90	80 aq^{15}; s alc; i eth
p113	1-Phenylethyl propionate	$C_2H_5CO_2CH(CH_3)C_6H_5$	178.23	5^3, 1680	1.007	1.4895^{20}		92^{5mm}	94	
p114	(±)-2-Phenylglycine	$C_6H_5CH(NH_2)CO_2H$	151.17	14, 460			subl 255			s org solvents, alk
p115	1-Phenylheptane	$C_6H_5(CH_2)_6CH_3$	176.30	5, 451	0.860	1.4850^{20}		233	95	
p116	1-Phenylhexane	$C_6H_5(CH_2)_5CH_3$	162.28	5^2, 337	0.861	1.4860^{20}	−61	226	83	misc eth
p117	Phenylhydrazine	$C_6H_5NHNH_2$	108.14	15^2, 44	1.0978^{20}_{4}	1.6080^{20}	19.5	243	88	misc alc, bz, chl, eth
p118	Phenyl 1-hydroxy-2-naphthoate	$HOC_{10}H_6CO_2C_6H_5$	264.28	10, 332			94–96			
p119	Phenyl 3-hydroxy-2-naphthoate	$C_{10}H_6(OH)CO_2C_6H_5$	264.28	10, 335			129–132	261^{160mm}		
p120	2-Phenylimidazole		144.18	23, 182			144–147			
p121	2-Phenyl-2-imidazoline		146.19	23, 154			94–99			
p122	2-Phenyl-1,3-indandione		222.28	7, 808			148–150			

p123	2-Phenylindole		193.25	20, 467			188–190	250^{10mm}		
p124	Phenyl isocyanate	C_6H_5NCO	119.12	12, 437	1.0956^{20}_{4}	1.5350^{20}	−30	162–163	55	dec aq, alc; s eth
p125	Phenyl isothiocyanate	C_6H_5NCS	135.19	12, 453	1.1288^{25}_{4}	1.6497^{20}	−21	221	87	i aq; s alc, eth
p126	*N*-Phenylmaleimide		173.17	21, 400			85–87	163^{12mm}		s alc, chl, eth
p127	Phenylmalonic acid	$C_6H_5CH(CO_2H)_2$	180.16				153 dec			
p128	Phenylmercury(II) acetate	$C_6H_5HgO_2CCH_3$	336.74	Merck: 12, 7453			150–152			0.17 aq; s alc, bz, acet
p129	Phenylmercury(II) chloride	C_6H_5HgCl	313.15	Merck: 12, 7454			250–252			s bz, eth, pyr
p130	Phenylmercury(II) hydroxide	C_6H_5HgOH	294.70	16, 952			190 dec			
p131	*N*-Phenylmorpholine		163.22	27, 6	1.058^{270}		51–54	268	>110	1.0 aq; v s hot alc
p132	*N*-Phenyl-1-naphthylamine	$C_{10}H_7NHC_6H_5$	219.29	12, 1224			60–62	226^{15mm}		s alc, bz, chl, eth
p133	*N*-Phenyl-2-naphthylamine	$C_{10}H_7NHC_6H_5$	219.29	12, 1275			107–109	395		
p134	2-Phenyl-2-oxazoline		147.18	27, 47	1.118	1.5670^{20}	12	$75^{0.3mm}$		
p135	2-Phenylphenol	$C_6H_5C_6H_4OH$	170.21	6^2, 623	1.213		57–59	282	123	s alc, chl, eth, alk
p136	4-Phenylphenol	$C_6H_5C_6H_4OH$	170.21	6, 674			165–167	321	165	s alc, chl, eth, alk
p137	*N*-Phenyl-1,4-phenylenediamine	$C_6H_5NHC_6H_4NH_2$	184.24	13, 76			73–75			
p138	Phenylphosphinic acid	$C_6H_5PH(O)OH$	142.09	16, 791			85–87			
p139	Phenylphosphonic acid	$C_6H_5P(O)(OH)_2$	158.09	16, 803			163–166			
p140	Phenylphosphonic dichloride	$C_6H_5P(O)Cl_2$	194.99	16, 804	1.375	1.5600^{20}	3	258	>110	
p141	*N*-Phenylpiperazine		162.24	23^3, 49	1.0621^{20}_{4}	1.5875^{20}		286	>110	i aq; misc alc
p142	1-Phenylpiperidine		161.25	20, 22	1.001	1.5620^{20}	3–4	257–258	106	
p143	2-Phenyl-1,2-propanediol	$CH_3C(C_6H_5)(OH)CH_2OH$	152.19	6, 930			44–45	162^{26mm}	>110	
p144	3-Phenyl-1-propanethiol	$C_6H_5CH_2CH_2CH_2SH$	152.26	6^1, 253	1.010	1.5494^{20}		109^{10mm}	90	
p145	1-Phenyl-1-propanol	$C_6H_5CH(OH)CH_2CH_3$	136.19	6, 502	0.9915^{25}_{4}	1.5200^{20}		219	90	misc alc, bz
p146	3-Phenyl-1-propanol	$C_6H_5CH_2CH_2CH_2OH$	136.19	6, 503	1.008	1.5257^{20}	−18	235	109	s aq; misc alc, eth
p147	1-Phenyl-2-propanone	$C_6H_5CH_2COCH_3$	134.18	7^2, 233	1.0157^{20}_{4}	1.5160^{20}	27	100^{13mm}	84	v s alc, eth; misc bz
p148	2-Phenylpropionaldehyde	$CH_3CH(C_6H_5)CHO$	134.18	7, 305	1.009^{20}_{4}	1.5175^{20}		202–205	76	i aq; s alc

(Continued)

TABLE 2.20 Physical Constants of Organic Compounds (*Continued*)

No.	Name	Formula	Formula weight	Beilstein reference	Density, g/mL	Refractive index	Melting point, °C	Boiling point, °C	Flash point, °C	Solubility in 100 parts solvent
p149	3-Phenylpropionaldehyde	$C_6H_5CH_2CH_2CHO$	134.18	7, 304	1.019	1.5230^{20}		98^{12mm}	95	
p150	3-Phenylpropionic acid	$C_6H_5CH_2CH_2CO_2H$	150.18	9, 508	1.047^{100}_{4}		47–49	280	>110	0.6 aq; s bz, alc, chl, eth, HOAc, PE
p151	1-Phenyl-3-pyrazolidinone		162.19	24, 2			121–123			10 hot aq; s hot alc, alk, acid
p152	2-Phenylpyridine	$C_6H_5C_5H_4N$	155.20	20, 424	1.086	1.6332^{20}		268–270	>110	s alc, eth
p153	2-Phenyl-4-quinolinecarboxylic acid		249.27	22, 103			214–215			0.8 alc; 1 eth; 0.3 chl
p154	Phenyl salicylate	$C_6H_5(OH)CO_2C_6H_5$	214.22	10, 76	1.25		44–46	173^{12mm}	>110	17 alc; 66 bz; s acet, chl, eth; 0.015 aq
p155	Phenylsuccinic acid	$HO_2CCH_2CH(C_6H_5)CO_2H$	194.19	9, 865			167–169	$-H_2O$, >168		s hot aq, alc, eth
p156	(Phenylthio)acetic acid	$C_6H_5SCH_2CO_2H$	168.21	6, 313			64–66			
p157	*S*-Phenyl thioisobutyrate	$(CH_3)_2CHC(=O)SC_6H_5$	152.22	6,4, 1524	1.056	1.5460^{20}		129^{10mm}	>110	
p158	1-Phenyl-2-thiourea	$C_6H_5NHC(S)NH_2$	152.22	12, 388	1.3		154			0.25 aq; s alc, alk
p159	Phenyltrichlorosilane	$C_6H_5SiCl_3$	211.55	16, 911	1.329^{20}	1.5230^{20}		201	91	
p160	Phenyltriethoxysilane	$C_6H_5Si(OC_2H_5)_3$	240.38	16, 911	0.996	1.4604^{20}		113^{10mm}	42	
p161	Phenyltrimethoxysilane	$C_6H_5Si(OCH_3)_3$	198.30	16^4, 1556	1.062	1.4680^{20}		233	99	
p162	Phenyltrimethylammonium bromide	$[C_6H_5N(CH_3)_3]^+ Br^-$	216.13	12, 158			215 dec			v s aq; s hot alc
p163	Phenyltrimethylammonium chloride	$[C_6H_5N(CH_3)_3]^+ Cl^-$	171.67	12, 158			237 subl			s aq; v s alc; sl s eth
p164	Phenyltrimethylammonium iodide	$[C_6H_5N(CH_3)_3]^+ I^-$	263.12	12, 159			227 subl			s aq, alc; sl s acet
p165	Phenyltrimethylammonium tribromide	$[C_6H_5N(CH_3)_3]^+ Br_3^-$	375.95	12, 159			114–116			
p166	Phenyltrimethylsilane	$C_6H_5Si(CH_3)_3$	150.30	16^1, 525	0.873	1.4907^{20}		168–170	44	
p167	Phenylurea	$C_6H_5NHCONH_2$	136.15	12, 346	1.302		145–147	238		s hot aq, hot alc, eth

p168	1,2-Phthalic acid	C_6H_4-1,2-$(CO_2H)_2$	166.13	9, 791	1.593^{20}_{4}		230 rapid heating			0.6 aq; ;10 alc; 0.5 eth; v sl s chl
p169	Phthalic anhydride		148.12	17, 469	1.53		131–134	295	151	0.6 aq(dec); s alc
p170	Phthalide		134.13	17, 310	1.164^{99}_{4}		72–74	290		s alc
p171	Phthalimide		147.13	21, 458			234–236			v s alk; v sl s bz, PE
p172	1,2-Phthaloyl dichloride	C_6H_5-1,2-$(COCl)_2$	203.02	9, 805	1.409^{20}	1.5684^{20}	15–16	280–282	>110	dec by aq, alc; s eth
p173	Phthalylsulfathioazole		403.44	Merck: 12, 7533			272–277			s alk; sl s alc; i chl
p174	Picric acid	2,4,6-$(O_2N)_3C_6H_2OH$	229.11	6, 265	1.763^{20}_{4}		122–123	explodes >300		1.3 aq; 8.2 alc; 10 bz; 2.9 chl; 1.6 eth
p175	(+)-α-Pinene		136.24	5, 146	0.8591^{20}_{4}	1.4650^{20}	−62	156	35	misc alc, eth
p176	(−)-β-Pinene		136.24	5, 154	0.8590^{20}	1.4780^{20}	−61	166	38	
p177	α-Pinene oxide		152.24	5, 152	0.964	1.4690^{20}		103^{50mm}	65	
p178	Piperazine		86.14	23, 4		1.446^{113}	108–110	145–146	109	v s aq; 50 alc; i eth
p179	1,4-Piperazinebis-(ethanesulfonic acid)		302.37	Merck: 12, 7633			>300			
p180	Piperidine		85.15	20, 6	0.8622^{20}_{4}	1.4525^{20}	−13	106	4	misc aq; s alc, bz, chl
p181	1-Piperidinecarbo-nitrile		110.16	20, 56	0.951	1.4705^{20}		102^{10mm}	97	
p182	*N*-Piperidineethanol		129.20	20, 25	0.8732^{25}_{25}	1.4804^{20}		199–202	68	misc aq; s alc
p183	2-Piperidineethanol		129.20	21, 2	1.010^{17}		38–40	234	102	v s aq, alc, eth
p184	1-Piperidinepropionic acid		157.21	20^{3}, 1049			105–110	$108^{0.5mm}$		
p185	Piperidinepropionitrile		138.21		0.933	1.4695^{20}		111^{16mm}	102	
p186	2-(2-Piperidineethyl)-pyridine		190.29		0.985	1.5260^{20}		150^{17mm}	>110	
p187	L-Proline		115.13	22, 2			228 dec			
p188	Propane	$CH_3CH_2CH_3$	44.10	1, 103	0.584^{-42}	1.340^{-42}	−188	−42.1	−104	volumes per 100 vols solvent: 6.5 aq; 790 alc; 926 eth; 1300 chl; 1450 bz
p189	1,2-Propanediamine	$CH_3CH(NH_2)CH_2NH_2$	74.13	4, 257	0.878^{15}	1.4460^{20}		119–120	33	misc aq, bz; s alc, eth
p190	1,3-Propanediamine	$H_2NCH_2CH_2CH_2NH_2$	74.13	4, 261	0.884^{25}_{4}	1.4575^{20}	−12	140	48	misc alc, eth; s aq
p191	1,2-Propanediol	$CH_3CH(OH)CH_2OH$	76.10	1, 472	1.0364^{20}_{4}	1.4331^{20}	−60	188	107	misc aq, acet, chl; s alc, eth
p192	1,3-Propanediol	$HOCH_2CH_2CH_2OH$	76.10	1, 475	1.0538^{20}	1.4396^{20}	−27	214	79	misc aq, alc

(*Continued*)

TABLE 2.20 Physical Constants of Organic Compounds (*Continued*)

No.	Name	Formula	Formula weight	Beilstein reference	Density, g/mL	Refractive index	Melting point, °C	Boiling point, °C	Flash point, °C	Solubility in 100 parts solvent
p193	1,3-Propanediol bis-(4-aminobenzoate)	$CH_2(CH_2CO_2CC_6H_4NH_2)_2$	314.34	14^3, 1034	1.140		124–127			
p194	1,2-Propanediol dibenzoate	$C_6H_5CO_2CH_2CH(CH_3)$-$O_2CC_6H_5$	284.31	9, 129	1.160	1.5450^{20}	−3	232^{12mm}	>110	
p195	1,3-Propanedithiol	$HSCH_2CH_2CH_2SH$	108.23	1, 476	1.0772^{20}_{4}	1.5405^{20}	−79	172.9	58	misc alc, bz, eth, chl
p196	1-Propanesulfonyl chloride	$CH_3CH_2CH_2SO_2Cl$	142.60	4, 8	1.2864^{15}_{4}	1.4542^{20}		66^{8mm}	80	dec hot aq, hot alc
p197	1,3-Propane sultone		122.14	19^3, 4	1.392		31–33	180^{30mm}	>110	
p198	1-Propanethiol	$CH_3CH_2CH_2SH$	76.16	1, 359	0.836^{25}_{4}	1.4380^{20}	−113	67–68	−20	s alc, eth
p199	2-Propanethiol	$CH_3CH(SH)CH_3$	76.16	1, 367	0.809^{25}_{4}	1.4255^{20}	−131	52.6	−34	misc alc, eth; sl s aq
p200	1,2,3-Propanetriol tris(acetate)	$H_3CCO_2CH(CH_2O_2CCH_3)_3$	218.21	2, 147	1.1580^{20}	1.4302^{20}	−78	259	138	7.2 aq; misc alc, bz, chl, eth
p201	1-Propanol	$CH_3CH_2CH_2OH$	60.10	1, 350	0.8037^{20}	1.3840^{20}	−127	97.2	23	misc aq, alc, eth
p202	2-Propanol	$(CH_3)_2CHOH$	60.10	1, 360	0.7855^{20}_{4}	1.3772^{20}	−89.5	82.4	12	misc aq, alc, chl, eth
p203	2-Propenal	$H_2C{=}CHCHO$	56.07	1, 725	0.841^{20}_{4}	1.4017^{20}	−88	52.6	−18	21 aq; s alc, eth
p204	Propene	$H_2C{=}CHCH_3$	42.08	1, 196	0.610^{-48}_{4}	1.3567^{-40}	−185.2	−47.7	−108	vols in 100 vols solvent: 45 aq; 1200 alc; 500 acet
p205	2-Propene-1-thiol	$H_2C{=}CHCH_2SH$	74.15	1, 440	0.925^{23}_{4}	1.4765^{20}		67–68	21	misc alc, eth
p206	*trans*-1,2,3-Propene-tricarboxylic acid		174.11	2, 849			190 dec			50 aq^{25}; 50 88% alc^{12}; sl s eth
p207	1-Propen-2-yl acetate	$H_2C{=}C(O_2CCH_3)CH_3$	100.12		0.909	1.4000^{20}		97	18	
p208	4-(1-Propenyloxy-methyl)-1,3-dioxo-lan-2-one		158.16		1.100	1.4610^{20}		251–252	>110	
p209	2-Propenylphenol	$CH_3CH{=}CHC_6H_4OH$	134.18	6^1, 279	1.044	1.5780^{20}		230–231	90	
p210	*β*-Propiolactone		72.06	17^1, 130	1.1460^{20}_{4}	1.4131^{20}	−33.4	162	70	37 aq(hyd); misc alc (reacts); bz, eth, acet
p211	Propionaldehyde	CH_3CH_2CHO	58.08	1, 629	0.8071^{20}_{4}	1.3636^{20}	−81	48	−30	30 aq; misc alc, eth
p212	Propionamide	$CH_3CH_2CONH_2$	73.10	2, 243	0.9597^{80}_{4}	1.4160^{110}	79	222.2		v s aq, alc, chl, eth
p213	Propionic acid	$CH_3CH_2CO_2H$	74.09	2, 234	0.9934^{20}_{4}	1.3809^{20}	−20.5	141.1	52	misc aq; s alc, chl, eth
p214	Propionic anhydride	$[CH_3CH_2C({=}O)]_2O$	130.14	2, 242	1.0110^{20}	1.4037^{20}	−45	170	63	dec aq; s alc, chl, eth
p215	Propionitrile	CH_3CH_2CN	55.08	2, 245	0.7818^{20}_{4}	1.3658^{20}	−92.8	97.2	2	10 aq; misc alc, eth
p216	Propionyl chloride	CH_3CH_2COCl	92.53	2, 243	1.065^{20}_{4}	1.4051^{20}	−94	80	11	dec by aq, alc
p217	Propiophenone	$C_6H_5COCH_2CH_3$	134.18	7^2, 231	1.0105^{20}_{4}	1.5258^{20}	21	218.0	87	misc bz, eth, abs alc
p218	2-Propoxyethanol	$CH_3CH_2CH_2OCH_2CH_2OH$	104.15	1, 468	0.913	1.4130^{20}	−75	150–153	48	
p219	2-(2-Propoxyethyl)-pyridine	$C_5H_4NCH_2CH_2OCH_2CH_2CH_3$	165.24		0.954	1.4880^{20}			95	

p220	1-Propoxy-2-propanol	$C_5H_4NCH_2CH_2OCH_2CH(OH)CH_3$	118.18	1^2, 536	0.885	1.4110^{20}		140–160	48	
p221	Propoxytrimethylsilane	$CH_3CH_2CH_2OSi(CH_3)_3$	132.28	4,*4*, 3994	0.768^{20}_{4}	1.3840^{20}		100^{735mm}	−2	
p222	Propyl acetate	$CH_3CH_2CH_2O_2CCH_3$	102.13	2, 129	0.8878	1.3844^{20}	−93	101.6	13	2.3 aq; misc alc, eth
p223	Propylamine	$CH_3CH_2CH_2NH_2$	59.11	4, 136	0.7173^{20}	1.3872^{20}	−83	42.2	−37	misc aq, alc, eth
p224	2-(Propylamino)-ethanol	$C_3H_7NHCH_2CH_2OH$	103.17	4, 282	0.900	1.4415^{20}		182^{746mm}	78	
p225	Propylbenzene	$CH_3CH_2CH_2C_6H_5$	120.20	5, 390	0.8621^{20}_{4}	1.4912^{20}	−99.2	159.2	47	s alc, eth
p226	Propyl benzoate	$C_6H_5CO_2CH_2CH_2CH_3$	164.20	9, 112	1.032^{20}	1.5010^{20}	−51.6	230	98	i aq; s alc, eth
p227	Propyl butyrate	$CH_3CH_2CH_2CO_2CH_2CH_2CH_3$	130.19	2, 271	0.879^{15}_{4}	1.4000^{20}	−95	143	38	sl s aq; misc alc, eth
p228	Propyl chloroformate	$ClCO_2CH_2CH_2CH_3$	122.55	3, 11	1.090	1.4034^{20}		105–106	28	misc bz, chl, eth
p229	Propylcyclohexane	$CH_3CH_2CH_2C_6H_{11}$	126.24	5^2, 23	0.7929^{20}_{4}	1.4370^{20}	−94.9	156.7	35	s bz, eth
p230	Propylene carbonate		102.09	19^3, 1564	1.2041^{20}	1.4210^{20}	−48.8	242	135	v s aq, alc, bz, eth
p231	Propyleneimine	CH_3CH-CH_2 NH	57.09	20, 3	0.8017^{25}	1.4084^{25}		66.0	−15	misc aq, alc, PE
p232	1,2-Propylene oxide	CH_3CH-CH_2 O	58.08	17, 6	0.8287^{20}	1.3660^{20}	−112	34	−35 (CC)	41 aq; misc alc, eth
p233	Propylene sulfide	CH_3CH-CH_2 S	74.15	17^2, 15	0.946	1.4760^{20}		72–75	10	
p234	Propyl formate	$CH_3CH_2CH_2O_2CH$	88.10	2, 21	0.9058^{20}	1.3779^{20}	−92.9	80.9	−3	2 aq; misc alc, eth
p235	Propyl 4-hydroxy-benzoate	$HOC_6H_4CO_2CH_2CH_2CH_3$	180.20	10, 160			95–98			0.05 aq; v s alc, eth
p236	Propyl isocyanate	$CH_3CH_2CH_2NCO$	85.11	4^1, 366	0.908	1.3940^{20}		83–84	0	
p237	Propyl lactate	$CH_3CH(OH)CO_2CH_2CH_2CH_3$	132.16	3, 265	0.996^{20}	1.4167^{25}		86^{40mm}		s aq, alc, eth
p238	Propyl nitrate	$CH_3CH_2CH_2ONO_2$	105.09	1, 355	1.0538^{20}_{4}	1.3976^{20}	−100	110.1	23 (may explode on heating)	s alc, eth
p239	2-Propylpentanoic acid	$(CH_3CH_2CH_2)_2CHCO_2H$	144.21	2, 350	0.921	1.4250^{20}		220	111	
p240	2-Propylphenol	$CH_3CH_2CH_2C_6H_4OH$	136.19	6, 499	1.015^{20}	1.5279^{20}		224–226	93	s alc, eth
p241	Propylphosphonic dichloride	$CH_3CH_2CH_2P(O)Cl_2$	160.97	4, 596	1.290	1.4643^{20}		90^{50mm}	>110	
p242	Propyltrichlorosilane	$CH_3CH_2CH_2SiCl_3$	177.53	4, 630	1.1851^{20}_{4}	1.429^{20}		123–124	2	
p243	1-Propyl-4-piperidone		141.22		0.936	1.4600^{20}		56^{1mm}	75	
p244	Propyl propionate	$CH_3CH_2CO_2CH_2CH_2CH_3$	116.16	2, 240	0.883^{20}	1.3935^{20}	−76	122.5	19	0.5 aq; 103 alc; 83 eth

(Continued)

TABLE 2.20 Physical Constants of Organic Compounds (*Continued*)

No.	Name	Formula	Formula weight	Beilstein reference	Density, g/mL	Refractive index	Melting point, °C	Boiling point, °C	Flash point, °C	Solubility in 100 parts solvent
p245	Propyl 3,4,5-trihydroxybenzoate	$(HO)_3C_6H_2CO_2CH_2CH_2CH_3$	212.20	Merck: 12, 8044			150			0.35 aq; 1 alc; 83 eth
p246	Propyne	$CH_3C{\equiv}CH$	40.06	1, 246	0.691^{-20}_{4}	1.3725^{-20}	−102.8	−23.2		v s alc; 3000 mL eth
p247	2-Propynyl benzenesulfonate	$C_6H_5SO_3CH_2C{\equiv}CH$	196.23	11^3, 37	1.243	1.5250^{20}	−30	142^{2mm}	100	
p248	2-Propynoic acid	$HC{\equiv}CCO_2H$	70.05	2, 477	1.138^{20}_{4}	1.4320^{20}	9	102^{200mm}	58	s aq, alc, eth
p249	2-Propyn-1-ol	$HC{\equiv}CH_2OH$	56.06	1, 454	0.9478^{20}	1.4320^{20}	−51.8	114	36	misc aq, alc, bz, chl
p250	(+)-Pulegone		152.24	7, 87	0.9346^{15}_{4}	1.4870^{20}		224	85	misc alc, chl, eth
p251	Pyrazine		80.09	23, 91	1.031^{61}_{4}	1.4953^{61}	55	115	55	v s aq, alc, eth
p252	Pyrazinecarbonitrile		105.10	25^3, 777	1.174	1.5340^{20}		87^{6mm}	96	
p253	Pyrazinecarboxylic acid		124.10	25, 125			225 dec			sl s hot aq; 0.008 abs alc; i bz, chl, eth
p254	Pyrazole		68.08	23, 39		1.4203	68	187		s aq, alc, bz, eth
p255	Pyrene		202.26	5, 693	1.271^{23}		151	404		s org solvents
p256	Pyridazine		80.09	23, 89	1.1035^{25}_{4}	1.5230^{23}	−8	208	85	misc aq, bz; v s alc, eth
p257	Pyridine	C_5H_5N	79.10	20, 181	0.9827^{25}_{4}	1.5067^{25}	−41.6	115.2	20	misc aq, alc, eth
p258	Pyridine-d_5	C_5D_5N	84.14	20^3, 2305	1.050	1.5092^{20}		114.4	20	
p259	2-Pyridinealdoxime	(C_5H_4N)-2-CH=NOH	122.13	21^1, 288			110–112			
p260	4-Pyridinealdoxime	(C_5H_4N)-4-CH=NOH	122.13				130–133			
p261	2-Pyridinecarboxaldehyde	(C_5H_4N)-2-CHO	107.11	21^1, 287	1.126	1.5370^{20}		181	54	
p262	3-Pyridinecarboxaldehyde	(C_5H_4N)-3-CHO	107.11	21^1, 288	1.135	1.5493^{20}		97^{15mm}	60	
p263	4-Pyridinecarboxaldehyde	(C_5H_4N)-4-CHO	107.11	21, 287	1.122	1.5440^{20}		78^{12mm}	54	s aq, eth
p264	3-Pyridinecarboxamide	(C_5H_4N)-3-$CONH_2$	122.13	22, 40	1.400	1.466	130–133			100 aq; 66 alc
p265	2-Pyridinecarboxylic acid	(C_5H_4N)-2-CO_2H	123.11	22, 33			134–136	sublimes		s aq, alc, bz; v s HOAc
p266	3-Pyridinecarboxylic acid	(C_5H_4N)-3-CO_2H	123.11	22, 38	1.473		236.6	sublimes		1.4 aq; s alk; v s hot aq, hot alc
p267	4-Pyridinecarboxylic acid	(C_5H_4N)-4-CO_2H	123.11	22, 45			319	260^{15mm}		0.52 aq; i alc, bz, eth
p268	2,3-Pyridinedicarboxylic acid	(C_5H_4N)-2,3-$(CO_2H)_2$	167.12	22, 150			188–190 dec			0.56 aq; s alk
p269	2,5-Pyridinedicarboxylic acid	(C_5H_4N)-2,5-$(CO_2H)_2$	167.12	22, 153			256 dec			s hot acid
p270	2,6-Pyridinedicarboxylic acid	(C_5H_4N)-2,6-$(CO_2H)_2$	167.12	22, 154			248–250 dec			sl s aq; v sl s alc

p271	Pyridine-*N*-oxide	C_5H_5NO	95.10	20^2, 131			61–65	270		
p272	Pyridinium *p*-toluene-sulfonate	$C_5H_5NH^+ \ ^-O_3SC_6H_4CH_3$	251.31	20^2, 129			117–119			
p273	2-Pyridylcarbinol	(C_5H_4N)-2-CH_2OH	109.13	21^1, 203	1.131	1.5420^{20}		113^{16mm}	>110	v s aq, alc, eth
p274	3-Pyridylcarbinol	(C_5H_4N)-3-CH_2OH	109.13	21, 50	1.124	1.5445^{20}		154^{28mm}	>110	v s aq, eth
p275	3-(3-Pyridyl)-1-propanol	(C_5H_4N)-3-$CH_2CH_2CH_2OH$	137.18	21^3, 549	1.063	1.5300^{20}		133^{3mm}	>110	
p276	3-(4-Pyridyl)-1-propanol	(C_5H_4N)-4-$CH_2CH_2CH_2OH$	137.18	21^4, 550	1.061		35–39	289	>110	
p277	Pyrimidine		80.09	23, 89	1.016	1.5040^{20}	22	124	31	misc aq; s alc, eth
p278	2,4(1*H*,3*H*)-Pyrimidinedione		112.09	24, 312			335			0.3 aq; s alk
p279	Pyrrole		67.09	20, 159	0.9691^{20}_{4}	1.5085^{20}	−23.4	130	39	4.5 aq; v s alc, eth
p280	Pyrrolidine		71.12	20, 4	0.8586^{20}	1.4431^{20}	−58	86.5	3	misc aq; s alc, chl, eth
p281	1-Pyrrolidinebutyro-nitrile		138.21		0.926	1.4605^{20}		115^{18mm}	99	
p282	1-Pyrrolidinecarbo-dithioic acid, am-monium salt		164.29				153–155			
p283	1-Pyrrolidinecarbo-nitrile		96.13		0.954	1.4690^{20}		$77^{1.8mm}$	107	
p284	1-Pyrrolidino-1-cyclohexene		151.25		0.940	1.5225^{20}		115^{15mm}	39	
p285	2-Pyrrolidinone		85.11	21, 236	1.116^{25}_{4}	1.4806^{25}	25	251	129	misc aq, alc, bz, chl, eth, EtOAc
p286	3-(*N*-Pyrrolidino)-1,2-propanediol		145.20	20^1, 4			46–48	158^{30mm}	>110	
q1	Quinhydrone		218.20	7, 617	1.401^{20}_{4}		171–173			s hot aq, alc, eth
q2	Quinine		324.44	23, 511		1.625	173–175			125 alc; 1.2 bz; 83 chl
q3	Quinoline		129.16	20, 339	1.095^{20}_{4}	1.6273^{20}	−15	237	101	0.6 aq; misc alc, eth
q4	Quinoxaline		130.15	23, 176	1.334^{48}_{4}	1.6231^{48}	29–32	220–223	98	v s aq, alc, bz, eth
q5	2-Quinoxalinol		146.15	24, 147			271–272			
r1	D-Raffinose penta-hydrate		594.52	31, 462			80–82	dec 118		14 aq; 10 MeOH
r2	Resorcinol	C_6H_4-1,3-$(OH)_2$	110.11	6, 796	1.272		110–112	280		111 aq; 111 alc; v s eth
r3	Resorcinol 1,3-diacetate	C_6H_4-1,3-$(O_2CCH_3)_2$	194.19	6, 816	1.178	1.5030^{20}		146^{12mm}	>110	

(*Continued*)

TABLE 2.20 Physical Constants of Organic Compounds (*Continued*)

No.	Name	Formula	Formula weight	Beilstein reference	Density, g/mL	Refractive index	Melting point, °C	Boiling point, °C	Flash point, °C	Solubility in 100 parts solvent
r4	Resorcinol monoacetate	$CH_3CO_2C_6H_4$-3-(OH)	152.15	6, 816	1.223	1.5370^{20}		*ca* 283	>110	i aq; misc alc, bz, chl, acet; s alk OH's
r5	Resorcinol monobenzoate	$C_6H_5CO_2C_6H_4$-3-(OH)	214.20					133–135		
r6	Rhodamine B		479.02	19, 345			210–211 dec			v s aq, alc
r7	Rhodanine		133.19	27, 242	0.868		167–170 may explode on rapid heating			v s hot aq, alc, eth
r8	Riboflavin		376.37	Merck: 12, 8367			dec 278–282			v s alk(dec); i acet, bz, eth; sl s pentyl acetate, cyclohexanol
r9	D-Ribose		150.13	1, 859			88–92			s aq; sl s alc
s1	Saccharin		183.19	27, 168	0.828		228–230			0.34 aq; 3 alc; 8 acet
s2	Safrole		162.19	19, 39	1.095^{20}	1.5370^{20}	11.2	232–234	97	v s alc; misc chl, eth
s3	Semicarbazide hydrochloride	$H_2NNHCONH_2 \cdot HCl$	111.53	3, 98			175–177 dec			v s aq, alc; i eth
s4	L-Serine	$HOCH_2CH(NH_2)CO_2H$	105.09	4, 505			222 dec			s aq; v sl s alc, eth
s5	D-Sorbitol		182.17	1, 533	1.472^{-5}		98–100 if hydrated; 111 anhyd			83 aq; s hot alc, acet
s6	L-Sorbose		180.16	1, 927	1.65^{15}		163–165			55 aq; v sl s alc
s7	Squalane	$[(CH_3)_2CH(CH_3)_2CH(CH_3)$-$(CH_2)_3CH(CH_3)CH_2CH_2$-$]_2$	422.83	1^1, 72	0.8115^{15}	1.4530^{15}	−38	350	218	s bz, chl, eth, PE
s8	Squalene	$CH_3[C(CH_3){=}CHCH_2CH_2]_5$-$C(CH_3){=}C(CH_3)_2$	470.73	1^1, 130	0.8584^{20}_{4}	1.4965^{20}	−75	285^{25mm}	200	v s eth, acet, PE
s9	*trans*-Stilbene	$C_6H_5CH{=}CHC_6H_5$	180.25	5, 630	0.970		122–124	307		v s bz, eth
s10	(−)-Strychnine		334.42	27^2, 723	1.36^{20}_{4}		284–286	270^{5mm}		0.66 alc; 20 chl; 0.55 bz; 0.15 mg aq
s11	Styrene	$C_6H_5CH{=}CH_2$	104.15	5, 474	0.9060^{20}	1.5463^{20}	−31	145	31	s alc, acet, eth, CS_2
s12	Styrene oxide	$H_2C{-}CHC_6H_5$ (epoxide, bridged by O)	120.15	17, 49	1.054	1.5338^{20}	−37	194	79	
s13	Succinamic acid	$H_2NCOCH_2CH_2CO_2H$	117.10	2, 614			153–156			s aq; sl s alc; i eth

s14	Succinamide	$H_2NCOCH_2CH_2CONH_2$	116.12	2, 614			265 dec			0.45 aq; i alc, eth
s15	Succinic acid	$HO_2CCH_2CH_2CO_2H$	118.09	2, 601	1.552		188	235 dec		7.7 aq; 5.4 alc; 2.8 acet; 0.88 eth; i bz
s16	Succinic anhydride		100.07	17, 407			119.6	261		s alc, chl; v sl s eth
s17	Succinimide		99.09	21, 369	1.41		123–125	285–290		33 aq; 4 alc; i eth
s18	Succinonitrile	$NCCH_2CH_2CN$	80.09	2, 615	0.9864^{60}	1.4173^{60}	54.5	266	132	*see* b456
s19	Succinyl chloride	$ClCOCH_2CH_2COCl$	154.98	2, 613	1.395^{15}_{4}	1.473^{15}	16–17	190	76	dec by aq, alc; s bz
s20	Sucrose		342.30	31, 424	1.587^{25}_{4}		185–187			200 aq; 0.59 alc
s21	Sulfadiazine		250.28	Merck: 12, 9071			252–256			sl s aq, alc, acet; v s dil mineral acids, alk
s22	Sulfamethazine		278.34	Merck: 12, 9083			198–201			0.15 aq; s alk
s23	Sulfamic acid	HSO_3NH_2	97.09	Merck: 12, 9090	2.15		205 dec			15 aq; sl s alc, acet; s bases
s24	Sulfanilamide	$H_2NC_6H_4SO_2NH_2$	172.21	14, 698			164–166			0.76 aq; 2.7 alc; 20 acet; s acid, alk
s25	Sulfanic acid	$4\text{-}(H_2N)\text{-}C_6H_4SO_3H$	173.19	14, 695			d 288			1.45 aq; sl s hot MeOH
s26	Sulfoacetic acid	$HCO_2CH_2SO_3H$	140.11	4, 21			84–86	245 dec		s aq, alc; i eth, chl
s27	2-Sulfobenzoic acid cyclic anhydride		184.17	19, 110				186^{18mm}		s bz, chl, eth; i aq
s28	4,4′-Sulfonylbis(2,6-dibromophenol)	$[2,6\text{-}(Br)_2{-}C_6H_2OH]_2SO_2$	565.88	6, 865			303–306			
s29	4,4′-Sulfonylbis(methyl benzoate)	$(CH_3O_2CC_6H_4)_2SO_2$	334.35	10^2, 109			195–196			
s30	4,4′-Sulfonyldiphenol	$(HOC_6H_4)_2SO_2$	250.27	6, 861	1.3663^{15}		245–247			s alc, eth, acet; i aq
s31	5-Sulfosalicylic acid	$HO_3SC_6H_3(OH)CO_2H$	254.21	11, 411			120 anhyd			v s aq, alc; s eth
t1	D-(−)-Tartaric acid		150.09	3, 520	1.7598^{20}_{4}		172–174			139 aq^{20}; 59 MeOH; 33 EtOH; s glyc; 0.4 eth
t2	L-(+)-Tartaric acid		150.09	3, 481	1.7598^{20}_{4}		168–170			139 aq^{20}; 59 MeOH; 33 EtOH; s glyc; 0.4 eth
t3	*meso*-Tartaric acid monohydrate	$HO_2CCH(OH)CH(OH)\text{-}CO_2H \cdot H_2O$	168.11	3, 528	1.666^{20}_{4}; 1.737 also		140; also 159–160			125 aq^{20}

(*Continued*)

TABLE 2.20 Physical Constants of Organic Compounds (*Continued*)

No.	Name	Formula	Formula weight	Beilstein reference	Density, g/mL	Refractive index	Melting point, °C	Boiling point, °C	Flash point, °C	Solubility in 100 parts solvent
t4	DL-Tartaric acid monohydrate	$HO_2CCH(OH)CH(OH)$-$CO_2H \cdot H_2O$	168.11	3, 522	1.697_4^{20}		210–212			20.6 aq^{20}; 5 alc^{25}; 1 eth
t5	Tartrazine		534.37	25, 252						v s aq
t6	Terephthaldicarbox-aldehyde	C_6H_4-1,4-$(CHO)_2$	134.13	7, 675			115–116	245–248		
t7	*m*-Terphenyl	$C_6H_5—C_6H_4—C_6H_5$	230.31	5, 695	1.195		87	363		
t8	*o*-Terphenyl	$C_6H_5—C_6H_4—C_6H_5$	230.31	5^2, 611	1.16		56.2	332	>110	
t9	*p*-Terphenyl	$C_6H_5—C_6H_4—C_6H_5$	230.31	5, 695	1.213		210	376	>110	
t10	α-Terpinene		136.24	5, 126	0.8375_4^{20}	1.4775^{20}		174	46	misc alc, eth
t11	γ-Terpinene		136.24	5, 128	0.853_4^{15}	1.4754^{16}		183	51	
t12	Terpinen-4-ol		154.25	6, 55	0.9338_4^{20}	1.4820^{20}	36.4	90^{6mm}	79	v s alc, eth
t13	α-Terpineol		154.25	6, 57	0.9337^{20}	1.4813^{20}	40.5	220	90	
t14	1,2,4,5-Tetrabromo-benzene	$C_6H_2Br_4$	393.72	5, 214			180–182			
t15	3,4,5,6-Tetrabromo-cresol	$CH_3C_6Br_4(OH)$	423.75	6, 362			205–208			s alc, eth, alk
t16	1,1,2,2,-Tetrabromo-ethane	$Br_2CHCHBr_2$	345.67	1, 94	2.9655^{20}	1.6358^{20}	0	243.5	none	misc alc, chl, eth, HOAc
t17	Tetrabromophthalic anhydride		463.72	17, 485			274–276			sl s bz; i aq, alc
t18	α,α,α′,α′-Tetrabromo-*o*-xylene	C_6H_4-1,2-$(CHBr_2)_2$	421.77	5, 367			114–116			v s chl
t19	α,α,α′,α′-Tetrabromo-*m*-xylene	C_6H_4-1,3-$(CHBr_2)_2$	421.77	5, 375			105–108			
t20	α,α,α′,α′-Tetrabromo-*p*-xylene	C_6H_4-1,4-$(CHBr_2)_2$	421.77	5, 386			254–256			
t21	Tetrabutylammonium bromide	$(C_4H_9)_4N^+ Br^-$	322.38	4^2, 634			102–104			
t22	Tetrabutylammonium chloride	$(C_4H_9)_4N^+ Cl^-$	277.92	4^3, 292			73–75			
t23	Tetrabutylammonium hydrogen sulfate	$(C_4H_9)_4N^+ HSO_4^-$	339.54				171–173			
t24	Tetrabutylammonium iodide	$(C_4H_9)_4N^+ I^-$	369.38	4, 157			145–147			sl s aq; s alc, eth
t25	Tetrabutylammonium tetrafluoroborate	$(C_4H_9)_4N^+ BF_4^-$	329.28	4^3, 293			160–162			
t26	Tetrabutylammonium tribromide	$(C_4H_9)_4N^+ Br_3^-$	482.20	4^4, 557			74–76			

t27	*N,N,N′,N′*-Tetrabutyl-1,6-hexanediamine	$\{\text{-}(CH_2)_3N[(CH_2)_3]_2\}_2$	340.64		0.820	1.4510^{20}		83^{2mm}	57	
t28	Tetrabutyl ortho-silicate	$Si[O(CH_2)_3CH_3]_4$	320.55	1^2, 398	0.899^{20}_{4}	1.4131^{20}		275	78	
t29	Tetrabutyl phos-phonium bromide	$[CH_3(CH_2)_3]_4PBr$	339.35				100–103			
t30	Tetrabutyltin	$(C_4H_9)_4Sn$	347.15		1.057	1.4742^{20}	−97	145^{10mm}	107	
t31	1,1,3,3,-Tetrachloro-acetone	$Cl_2CHC({=}O)CHCl_2$	195.86	1, 656	1.624^{15}_{4}	1.497^{18}		182^{745mm}	none	v s acet, chl
t32	1,2,3,4-Tetrachloro-benzene	$C_6H_2Cl_4$	215.89	5, 204			46–47	254	>110	v s eth; sl s alc
t33	1,2,4,5-Tetrachloro-benzene	$C_6H_2Cl_4$	215.89	5, 205	1.858^{22}		139–142	240–246	>110	s bz, chl, eth
t34	Tetrachloro-1,2-benzoquinone	$C_6Cl_4\text{-}1,2\text{-}({=}O)_2$	245.88	7, 602			127–129			
t35	Tetrachloro-1,4-benzo-quinone	$C_6Cl_4\text{-}1,4\text{-}({=}O)_2$	245.88	7, 636			290 dec			s eth; sl s chl; i aq
t36	Tetrachloro-1,2-difluoroethane	$Cl_2CFCFCl_2$	203.83		1.6447^{25}	1.4130^{25}	26.0	92.8		0.012 aq
t36a	1,1,1,2-Tetrachloro-ethane	$ClCH_2CCl_3$	167.85	1, 86	1.5406^{20}	1.4821^{20}	−70.2	130.5	47	
t37	1,1,2,2-Tetrachloro-ethane	$Cl_2CHCHCl_2$	167.85	1, 86	1.5866^{25}_{4}	1.4910^{25}	−44	147	62	0.3 aq; misc alc, chl, eth, PE
t38	Tetrachloroethylene	$Cl_2C{=}CCl_2$	165.83	1, 187	1.6230^{20}_{4}	1.5057^{20}	−22	121	45	misc alc, chl, eth
t39	2,3,5,6-Tetrachloro-nitrobenzene	$HC_6Cl_4NO_2$	260.89	5, 247	1.744^{25}_{4}		98–101	304		s alc, bz, chl
t40	Tetrachlorophthalic anhydride		285.90	17, 484			254–258	371		dec hot aq; sl s eth
t41	Tetracosane	$CH_3(CH_2)_{22}CH_3$	338.66	1, 175	0.7786^{51}	1.4283^{70}	51	391	>110	9.4 chl; s eth
t42	Tetradecafluorohexane	$CF_3(CF_2)_4CF_3$	338.05	1^3, 388	1.669	1.2515^{20}	−4	58–60	none	
t43	Tetradecane	$CH_3(CH_2)_{12}CH_3$	198.40	1, 171	0.7627^{20}_{4}	1.4290^{20}	5.5	253.6	99	v s alc, eth
t44	Tetradecanoic acid	$CH_3(CH_2)_{12}CO_2H$	228.38	2, 365	0.8525^{70}_{4}	1.4273^{70}	54	250^{100mm}		v s bz, chl, eth; s alc
t45	1-Tetradecanol	$CH_3(CH_2)_{13}OH$	214.39	1, 428	0.8151^{50}	1.4358^{50}	39.5	289	>110	s eth; sl s alc
t46	Tetradecanoyl chloride	$CH_3(CH_2)_{12}COCl$	246.82	2, 368	0.908	1.4490^{20}	−1	168^{15mm}	>110	dec aq, alc; s eth
t47	1-Tetradecene	$CH_3(CH_2)_{11}CH{=}CH_2$	196.38	1, 226	0.775^{15}_{4}	1.4360^{20}	−12.9	251.2	115	v s alc, eth
t48	Tetraethoxysilane	$(CH_3CH_2O)_4Si$	208.33	1, 334	0.934^{20}_{4}	1.383^{20}	−77	168	46	dec aq; s alc
t49	Tetraethylammonium bromide	$(CH_3CH_2)_4N^+\ Br^-$	210.16	4, 104	1.397^{20}_{4}		285 dec			v s aq, alc, acet, chl

(Continued)

TABLE 2.20 Physical Constants of Organic Compounds (*Continued*)

No.	Name	Formula	Formula weight	Beilstein reference	Density, g/mL	Refractive index	Melting point, °C	Boiling point, °C	Flash point, °C	Solubility in 100 parts solvent
t50	Tetraethylammonium chloride	$(CH_3CH_2)_4N^+ Cl^-$	165.71	4, 104	1.0801^{21}_{4}					141 aq; s alc; 8.2 chl
t51	Tetra(ethylene glycol)	$(HOCH_2CH_2OCH_2CH_2)_2O$	194.23	1, 468	1.125^{20}_{20}	1.4577^{20}	−6	328	182	misc aq, alc, bz, eth
t52	Tetra(ethylene glycol) diacrylate	$(H_2C{=}CHCO_2CH_2CH_2O\text{-}CH_2CH_2)_2O$	302.33		1.110	1.4650^{20}			>110	
t53	Tetra(ethylene glycol) diethyl ether	$C_2H_5(OCH_2CH_2)_4OC_2H_5$	250.34	1^3, 2107	0.970	1.4324^{20}		159^{11mm}	>110	s aq
t54	Tetra(ethylene glycol) dimethacrylate	$[H_2C{=}C(CH_3)CO_2CH_2CH_2\text{-}OCH_2CH_2]_2O$	330.37	2^4, 1531	1.080	1.4630^{20}		220	>110	
t55	Tetra(ethylene glycol) dimethyl ether	$CH_3(OCH_2CH_2)_4OCH_3$	222.28	1^3, 2107	1.0087^{20}_{4}	1.4330^{20}	−30	275–276	140	s aq
t56	Tetraethylene-pentamine	$(H_2NCH_2CH_2NHCH_2CH_2)_2NH$	189.31	4,3, 543	0.999^{20}_{20}	1.5055^{20}	−40	340	185	misc aq, alc, eth
t57	*N,N,N′,N′*-Tetraethyl-ethylenediamine	$(C_2H_5)_2NCH_2CH_2N(C_2H_5)_2$	172.32	4, 251	0.808	1.4343^{20}		189–192	58	
t58	Tetraethylgermanium	$(C_2H_5)_4Ge$	188.84	4, 631	0.998	1.4420^{20}	−90	165.5	35	s alc, eth; i aq
t59	Tetraethyllead	$(C_2H_5)_4Pb$	323.45	4, 639	1.653^{20}_{4}	1.5190^{20}	−136	85^{15mm}	72	s bz; misc eth
t60	Tetraethylsilane	$(C_2H_5)_4Si$	144.34	4, 625	0.7658^{20}	1.4268^{20}	−82	154.7	26	i aq
t61	*N,N,N′,N′*-Tetraethyl-sulfamide	$(C_2H_5)_2NSO_2N(C_2H_5)_2$	208.33	4, 129	1.030	1.4480^{20}		249–251	>110	
t62	Tetraethylthiuram disulfide	$[(C_2H_5)_2NC({=}S)S^-]_2$	296.54	4, 122	1.30		71–72			3.8 alc; 7.1 eth; s bz, acet, chl; 0.02 aq
t63	Tetraethyltin	$(C_2H_5)_4Sn$	234.94	4, 632	1.199^{20}	1.4730^{20}	−112	181	53	i aq; s eth
t64	1,1,1,2-Tetrafluoro-ethane	FCH_2CF_3	102.03	1,4, 123			−26.5			
t65	Tetrafluoroethylene	$F_2C{=}CF_2$	100.02	1^3, 638	1.151^{-40}		−142.5	−76		i aq
t66	2,2,3,3-Tetrafluoro-1-propanol	$HCF_2CF_2CH_2OH$	132.06	1^4, 1438	1.4853^{20}_{4}	1.3197^{20}	−15	109–110	43	
t67	1,2,3,6-Tetrahydro-benzaldehyde	C_6H_9CHO	110.16	7^1, 48	0.940	1.4745^{20}		163–164	57	
t68	1,2,3,4-Tetrahydro-carbazole		171.24	20, 416			118–120	325–330		
t69	Tetrahydrofuran		72.11	17, 10	0.8892^{20}_{4}	1.4052^{20}	−108.5	65	−14	misc aq, alc, eth, PE
t70	2,5-Tetrahydrofuran-dimethanol		132.16		1.1542^{25}_{4}	1.4766^{25}	<−50	265		misc aq, alc, bz, chl; s eth
t71	Tetrahydro-2-furan-methanol		102.13	17^2, 106	1.0524^{20}	1.4520^{20}	<−80	178	75	misc aq, alc, bz, chl, eth, acet
t72	Tetrahydro-2-furan-methylamine		101.15	18^2, 415	0.980	1.4560^{20}		154^{744mm}	45	

t73	Tetrahydrofurfuryl acetate		144.17	17^2, 107	1.061	1.4370^{20}		196	84	
t74	Tetrahydrofurfuryl acrylate		156.18	17^3, 1104	1.064	1.4600^{20}		87^{9mm}	>110	
t75	Tetrahydrofurfuryl chloride		120.58	17^3, 61	1.110	1.4550^{20}		150–151	47	
t76	Tetrahydrofurfuryl methacrylate		170.21	17^3, 1105	1.044	1.4580^{20}		$52^{0.4mm}$	90	
t77	2(3)-(Tetrahydrofuryl-oxy)tetrahydropyran		186.25		1.030	1.4610^{20}			97	
t78	1,2,3,4-Tetrahydro-isoquinoline		133.19	20, 275	1.064	1.5668^{20}	−30	232–233	98	
t79	Tetrahydrolinalool	$(CH_3)_2CHCH_2CH_2CH_2$-$C(CH_3)(OH)CH_2CH_3$	158.29	1, 426	0.826	1.4340^{20}	76	73^{6mm}	76	
t80	1,2,3,4-Tetrahydro-naphthalene	$C_{10}H_{12}$	132.21	5, 491	0.9702^{20}_4	1.5414^{20}	−35.8	207.6	77	misc alc, bz, chl, eth, acet, PE
t81	*cis*-1,2,3,6-Tetrahydro-phthalic anhydride		152.15	17, 462			97–103		157	
t82	*cis*-1,2,3,6-Tetrahydro-phthalimide		151.17				129–133			
t83	Tetrahydropyran		86.14	17, 12	0.8814^{20}_4	1.4200^{20}	−45	88	−155	misc aq, alc, eth
t84	Tetrahydropyran-2-methanol		116.16		1.0254^{20}	1.4580^{20}	−70	187	93	misc aq, alc, bz, eth
t85	3,4,5,6-Tetrahydro-pyrimidinethiol		116.19	24, 5			210–212			
t86	1,2,3,4-Tetrahydro-quinoline		133.19	20, 262	1.061	1.5940^{20}	15–16	249	100	s aq; misc alc, eth
t87	Tetrahydrothiophene		88.17	17^1, 5	0.9987^{20}	1.5040^{20}	−96	121	12	misc alc, eth; i aq
t88	2,2′,4,4′-Tetrahydroxy-benzophenone	$[(HO)_2C_6H_3]_2C{=}O$	246.22	8, 496			200–203			
t89	Tetrakis(dimethyl-amino)ethylene	$[(CH_3)_2N]_2C{=}C[N(CH_3)_2]_2$	200.23	4^4, 167	0.861	1.4800^{20}		$59^{0.9mm}$	53	
t90	*N,N,N′,N′*-Tetrakis(2-hydroxypropyl)-ethylenediamine	$[CH_3CH(OH)CH_2]_2NCH_2$-$CH_2N[CH_2CH(OH)CH_3]_2$	292.40	4^4, 1685	1.013	1.4812^{20}		$181^{0.8mm}$	>110	
t91	1,1,8,8-Tetramethoxy-octane	$(CH_3O)_2CH(CH_2)_6CH(OCH_3)_2$	234.34		0.949	1.4300^{20}		130^{5mm}	52	
t92	1,1,3,3-Tetramethoxy-propane	$[(CH_3O)_2CH]_2CH_2$	164.20		0.997	1.4081^{20}		183	54	

(*Continued*)

TABLE 2.20 Physical Constants of Organic Compounds (*Continued*)

No.	Name	Formula	Formula weight	Beilstein reference	Density, g/mL	Refractive index	Melting point, °C	Boiling point, °C	Flash point, °C	Solubility in 100 parts solvent
t93	Tetramethyl- ammonium bromide	$(CH_3)_4N^+ Br^-$	154.06	4, 51	1.56		>300			55 aq
t94	Tetramethyl- ammonium chloride	$(CH_3)_4N^+ Cl^-$	109.60	4, 51	1.169^{20}_{4}		>300			s aq, hot alc
t95	Tetramethyl- ammonium iodide	$(CH_3)_4N^+ I^-$	201.06	4, 51	1.829		>300			sl s aq; v s abs alc
t96	*N,N*-3,5-Tetramethyl- aniline	$(CH_3)_2C_6H_3N(CH_3)_2$	149.24	12, 1131	0.913	1.5443^{20}		226–228	90	
t97	1,2,3,4-Tetramethyl- benzene	C_6H_2-1,2,3,4-$(CH_3)_4$	134.22	5, 430	0.905^{20}_{4}	1.5187^{20}	−6.2	205.0	68	misc alc, eth
t98	1,2,3,5-Tetramethyl- benzene	C_6H_2-1,2,3,5-$(CH_3)_4$	134.22	5, 430	0.8906^{20}_{4}	1.5134^{20}	−23.7	198.0	63	s alc; v s eth
t99	1,2,4,5-Tetramethyl- benzene	C_6H_2-1,2,4,5-$(CH_3)_4$	134.22	5, 431	0.838^{81}_{4}		79.3	196.8	73	v s alc, bz, eth
t100	2,2,3,3-Tetramethyl- butane	$(CH_3)_3CC(CH_3)_3$	114.23	1, 165	0.8242^{20}		−100.7	106.5	4	
t101	*N,N,N′,N′*-Tetra- methyl-1,3-butane- diamine	$(CH_3)_2NCH(CH_3)CH_2$- $CH_2N(CH_3)_2$	144.26	4^3, 570	0.787	1.4318^{20}		165	40	
t102	*N,N,N′,N′*-Tetra- methyl-1,4-butane- diamine	$(CH_3)_2N(CH_2)_4N(CH_3)_2$	144.26	4, 265	0.786^{20}	1.4280^{20}		169	46	s aq, alc, eth
t103	1,1,3,3-Tetramethyl- butylamine	$(CH_3)_3CCH_2C(CH_3)_2NH_2$	129.25	4, 198	0.805	1.4240^{20}		137–143	32	s alc, eth, PE; i aq
t104	1,3,5,7-Tetramethyl- cyclotetrasiloxane	$[-SiH(CH_3)O-]_4$	240.51	4^4, 4099	0.9912^{20}_{4}	1.3870^{20}	−69	134–135		
t105	*N,N,N′,N′*-Tetra- methyldiamino- methane	$(CH_3)_2NCH_2N(CH_3)_2$	102.18	4, 54	0.749	1.4005^{20}		85	−12	
t106	1,1,3,3-Tetramethyl- disiloxane	$[(CH_3)_2CH]_2O$	134.33	4^4, 3991	0.757^{20}_{4}	1.3700^{20}		70–71	−10	
t107	Tetramethylene sulfone		120.17	17^1, 5	1.2606^{30}_{4}	1.4820^{30}	27.6	285	177	misc aq, acet, toluene; s octanes, olifines, naphthenes
t108	*N,N,N′,N′*-Tetra- methylethylene- diamine	$(CH_3)_2NCH_2CH_2N(CH_3)_2$	116.21	4, 250	0.770	1.4179^{20}	−55	120–122	10	

t109	Tetramethylgermanium	$(CH_3)_4Ge$	132.73	4,2, 1008	0.978	1.3890^{20}	−88	43.4	−37	
t110	1,1,3,3-Tetramethylguanadine	$[(CH_3)_2N]_2C{=}NH$	115.18	4^1, 335	0.918	1.4692^{20}		163	60	
t111	*N,N,N′,N′*-Tetramethyl-1,6-hexanediamine	$[(CH_3)_2N(CH_2)_3\text{-}]_2$	172.32	4^1, 423	0.806	1.4359^{20}		209–210	73	
t112	Tetramethyl lead	$(CH_3)_4Pb$	267.33	4, 639	1.995^{20}_4		−27.5	110	38	misc alc, eth
t113	*N,N,N′,N′*-Tetramethylmethanediamine	$(CH_3)_2NCH_2N(CH_3)_2$	102.18	4, 54	0.749	1.4005^{20}		85	−12	
t114	2,6,10,14-Tetramethylpentadecane	$[(CH_3)_2CH(CH_3)_3\text{-}CH(CH_3)CH_2]_2CH_2$	268.53	Merck: 12, 7932	0.7827^{20}_4	1.4385^{20}	−100	296	>110	s bz, chl, eth, PE
t115	2,2,6,6-Tetramethylpiperidinyl-1-oxy (free radical)		156.25				36–40		67	
t116	*N,N,N′,N′*-Tetramethyl-1,3-propanediamine	$(CH_3)_2N(CH_2)_3N(CH_3)_2$	130.24	4, 262	0.779	1.4234^{20}		145–146	31	
t117	Tetramethylpyrazine		136.20	23, 99			84–86	190		
t118	Tetramethylsilane	$(CH_3)_4Si$	88.23	4, 625	0.6411^{20}_4	1.3580^{20}	−99.5	26.5	−27	v s alc, eth
t119	1,1,3,3-Tetramethyl-2-thiourea	$(CH_3)_2NC({=}S)N(CH_3)_2$	132.23	4^1, 336			75–77	245		
t120	Tetramethylthiuram disulfide	$[(CH_3)_2NCS_2\text{-}]_2$	240.43	4, 76	1.29		155–156			0.002 alc, 0.002 eth; 0.012 acet; 0.025 bz; s chl
t121	Tetramethyltin	$(CH_3)_4Sn$	178.83	4, 631	1.3149^{25}	1.5201	−54	74–75	−12	
t122	1,1,3,3-Tetramethylurea	$(CH_3)_2NC({=}O)N(CH_3)_2$	116.16	4, 74	0.9687^{20}_4	1.4493^{25}	−0.6	176–177	77	misc aq, common org solvents
t123	Tetranitromethane	$C(NO_2)_4$	196.03	1, 80	1.6229^{25}_4	1.4358^{25}	13.8	126	>110	v s alc, eth, alk
t124	1,4,7,10-Tetraoxacyclododecane (12-Crown-4)		176.21		1.089	1.4630^{20}	16	$70^{0.5mm}$	>110	
t125	2,4,8,10-Tetraoxaspiro[5.5]undecane		160.17	19, 436			52–55	$83^{1.5mm}$	108	
t126	Tetraphenylboron sodium	$(C_6H_5)_4B^-\ Na^+$	342.23	Merck: 12, 8839			>300			v s aq, acet; s chl

(Continued)

TABLE 2.20 Physical Constants of Organic Compounds (*Continued*)

No.	Name	Formula	Formula weight	Beilstein reference	Density, g/mL	Refractive index	Melting point, °C	Boiling point, °C	Flash point, °C	Solubility in 100 parts solvent
t127	1,1,4,4-Tetraphenyl-1,3-butadiene	$(C_6H_5)_2C{=}CHCH{=}C(C_6H_5)_2$	358.49	5, 750			207–209			
t128	Tetraphenyltin	$(C_6H_5)_4Sn$	427.11		1.490^{0}		224–227	>420	110	
t129	Tetrapropoxysilane	$(C_3H_7O)_4Si$	264.4	1, 355	0.916^{2}_{4}	1.401^{20}		94^{5mm}	95	
t130	Tetrapropylammonium bromide	$(CH_3CH_2CH_2)_4N^+ Br^-$	266.27	4^1, 364			270 dec			s aq
t131	1*H*-Tetrazole		70.06	26, 346			157–158			s aq, alc, acet
t132	2-Thenoyltrifluoro-acetone		222.18				40–44	98^{8mm}		
t133	Theobromine		180.17	26, 457			357	sublimes 290–295		100 aq; 0.045 alc; s alk; i bz, chl, eth
t134	Theophylline		180.17	26, 455			274–275			0.83 aq; 1.25 alc; 0.9 chl; s hot aq, alk, dil acids
t135	Thiamine HCl		337.27	Merck: 12, 9430			dec 260			100 aq; 1 alc; 5.5 glyc
t136	Thiazole		85.13	27, 15	1.200	1.5390^{20}		117–118	22	s alc, eth; sl s aq
t137	N^2-(2-Thiazolyl)-sulfanilamide		255.32	27^3, 4623			202			0.06 aq; 0.52 alc; s acet, dil mineral acids, alkalis
t138	Thioacetamide	$CH_3C({=}S)NH_2$	75.13	2, 232			112–114			16 aq; 16 alc; sl s eth
t139	Thiobenzoic acid	$C_6H_5C({=}O)SH$	138.19	9, 419	1.174	1.6050^{20}	15–18	122^{30mm}	>110	misc eth; v s alc; i aq
t140	4,4′-Thiobis(2-*tert*-butyl-6-methyl-phenol)		358.54	6^4, 6043			163–165	316^{40mm}	240	
t141	Thiocarbanilide	$C_6H_5NHC({=}S)NHC_6H_5$	228.32	12, 394	1.32^{24}		152–155			v s alc, eth
t142	*p*-Thiocresol	$HSC_6H_4CH_3$	124.21	6, 416			42–44	195	68	s alc, eth; i aq
t143	2,2′-Thiodiacetic acid	$(HO_2CCH_2)_2S$	150.15	3, 253			128–131			s aq, alc
t144	2,2′-Thiodiethanol	$(HOCH_2CH_2)_2S$	122.19	1, 470	1.1824^{20}_{4}	1.5203^{20}	−10.2	282	160	misc aq, alc; sl s eth
t145	4,4′-Thiodiphenol	$(HOC_6H_4)_2S$	218.27	6, 860			154–156			
t146	3,3′-Thiodipropionic acid	$(HO_2CCH_2CH_2)_2S$	178.21				131–134			3.7 aq; v s hot aq, alc, acet
t147	Thiolacetic acid	$CH_3C({=}O)SH$	76.12	2, 230	1.065	1.4630	< −17	88–91	11	s aq; v s alc
t148	*N*-Thionylaniline	$C_6H_5N{=}SO$	139.18	12, 578	1.236	1.6270^{20}		200	84	
t149	Thionyl bromide	$SOBr_2$	207.88	Merck: 12, 9484	2.683	1.6750^{20}	−52	138		misc bz, chl, CCl_4; hyd by aq

t150	Thionyl chloride	$SOCl_2$	118.97	Merck: 12, 9485	1.635	1.517^{20}	−101	76	none	misc bz, chl, CCl_4; hyd by aq
t151	Thiophene	C_4H_4S	84.14	17, 29	1.0573^{25}_{4}	1.5257^{25}	−39.4	84	−1	misc alc, eth; i aq
t152	2-Thiopheneacetic acid	$(C_4H_3S)CH_2CO_2H$	142.18	18, 293			63–67	160^{22mm}		
t153	2-Thiophenecarbonyl chloride	$(C_4H_3S)COCl$	146.60	18, 290	1.371	1.5900^{20}		206–208	90	
t154	2-Thiophenecarbox-aldehyde	$(C_4H_3S)CHO$	112.15	17, 285	1.200	1.5900^{20}		198	77	s eth
t155	2-Thiophenecarboxylic acid	$(C_4H_3S)CO_2H$	128.15	18, 289			127–130	260		s aq, chl; v s alc, eth
t156	Thiophenol	C_6H_5SH	110.18	6, 294	1.073	1.5880^{20}	−14.9	169	50	v s alc; misc bz, eth
t157	Thiophenoxyacetic acid	$C_6H_5SCH_2CO_2H$	168.21	6, 313			64–66			
t158	Thiophosphoryl chloride	$PSCl_3$	169.40		1.668	1.5550^{20}	−36 (β) −40 (α)	125	none	s bz, chl, CCl_4, CS_2
t159	Thiopropionic acid	$CH_3CH_2C(=O)SH$	90.14	2, 264	1.014	1.4640^{20}		108–110	11	
t160	3-Thiosemicarbazide	$H_2NC(=S)NHNH_2$	91.14	3, 195			182–184			s aq, alc
t161	Thiourea	$H_2NC(=S)NH_2$	76.12	3, 180	1.405		176–178			9 aq; s alc; sl s eth
t162	Thioxanthen-9-one		212.27	17, 357			212–213	373^{715mm}		v s bz, chl, hot HOAc
t162a	Thymol		150.22	6, 532	0.9699^{25}_{4}	1.5227^{20}	51.5	233	102	0.1 aq; 100 alc; 140 eth; s HOAc, alk OH
t163	Titanium(IV) ethoxide	$Ti(OC_2H_5)_4$	228.15	1, 335	1.088	1.5043^{20}		152^{10mm}	28	
t164	Titanium(IV) iso-propoxide	$Ti[OCH(CH_3)_2]_4$	284.26	1^{2}, 382	0.963	1.4660^{20}	18–20	220	22	s bz, chl, eth
t165	Titanium(IV) propox-ide	$Ti(OCH_2CH_2CH_3)_4$	284.26	1^{3}, 1423	1.033	1.4986^{20}		170^{3mm}	42	
t166	Toluene	$C_6H_5CH_3$	92.14	5, 280	0.8660^{20}_{4}	1.4960^{20}	−94.9	110.6	4	misc alc, chl, eth, acet, HOAc; 0.067 aq
t167	2,4-Toluenediamine	$CH_3C_6H_3$-2,4-$(NH_2)_2$	122.17	13, 124			99	292		s hot aq, alc, eth
t168	2,5-Toluenediamine	$CH_3C_6H_3$-2,5-$(NH_2)_2$	122.17	13, 144			64	273–274		v s aq, alc, eth
t169	2,6-Toluenediamine	$CH_3C_6H_3$-2,6-$(NH_2)_2$	122.17	13, 148			104–106			s aq, alc
t170	3,4-Toluenediamine	$CH_3C_6H_3$-3,4-$(NH_2)_2$	122.17	13, 148			91–93	156^{18mm}		v s aq
t171	Toluene-2,4-diiso-cyanate	$CH_3C_6H_3$-2,4-$(NCO)_2$	174.16	13, 138	1.2244^{20}_{4}	1.5689^{20}	20–21	251	132	dec aq, alc; misc acet, bz, eth
t172	*p*-Toluenesulfinic acid	$CH_3C_6H_4SO_2H$	156.21	11, 9			85			v s alc, eth; sl s aq
t173	*o*-Toluenesulfonamide	$CH_3C_6H_4SO_2NH_2$	171.22	11, 86			156–158			
t174	*p*-Toluenesulfon-amide	$CH_3C_6H_4SO_2NH_2$	171.22	11, 104			138–140			0.2 aq; 3.6 alc

(Continued)

TABLE 2.20 Physical Constants of Organic Compounds (*Continued*)

No.	Name	Formula	Formula weight	Beilstein reference	Density, g/mL	Refractive index	Melting point, °C	Boiling point, °C	Flash point, °C	Solubility in 100 parts solvent
t175	*p*-Toluenesulfonyl-hydrazide	$CH_3C_6H_4SO_2NHNH_2$	186.23	11[2], 66			110 dec			
t176	*p*-Toluenesulfonic acid	$CH_3C_6H_4SO_3H$	172.20	11, 97			107 anhyd	140^{20mm}		67 aq; s alc, eth
t177	*p*-Toluenesulfonyl chloride	$CH_3C_6H_4SO_2Cl$	190.65	11, 103			67–69	134^{10mm}		v s alc, bz, eth; i aq
t178	*p*-Toluenesulfonyl fluoride	$CH_3C_6H_4SO_2F$	174.19	11[2], 54			41–42	112^{16mm}	105	
t179	*p*-Toluenesulfonyl isocyanate	$CH_3C_6H_4SO_2NCO$	197.21			1.4355^{20}		144^{10mm}	>110	
t180	*m*-Toluidine	$CH_3C_6H_4NH_2$	107.16	12, 853	0.989^{20}_4	1.5680^{20}	−31	203	85 (CC)	misc alc, eth
t181	*o*-Toluidine	$CH_3C_6H_4NH_2$	107.16	12, 772	0.998^{20}	1.5720^{20}	−16.3	200	85	1.7 aq; s alc, eth
t182	*p*-Toluidine	$CH_3C_6H_4NH_2$	107.16	12, 880	0.9619^{20}	1.5532^{59}	43.8	200	87	7.4 aq; v s alc, eth
t183	*m*-Tolunitrile	$CH_3C_6H_4CN$	117.15	9, 477	0.976^{15}	1.5256^{20}	−23	210	86	0.09 aq; v s alc, eth
t184	*o*-Tolunitrile	$CH_3C_6H_4CN$	117.15	9, 466	0.989	1.5279^{20}	−13	205	84	i aq; misc alc, eth
t185	*p*-Tolunitrile	$CH_3C_6H_4CN$	117.15	9, 489	0.9785^{30}_4		29.5	217	85	i aq; v s alc, eth
t186	2-(*p*-Toluoyl)benzoic acid	$CH_3C_6H_4COC_6H_4CO_2H$	240.26	10, 759			137–139			v s alc, bz, eth, acet
t187	*m*-Toluoyl chloride	$CH_3C_6H_4COCl$	154.60	9, 477	1.173	1.5485^{20}		86^{5mm}	76	
t188	*o*-Toluoyl chloride	$CH_3C_6H_4COCl$	154.60	9, 464	1.185	1.5549^{20}		90^{12mm}	76	
t189	*p*-Toluoyl chloride	$CH_3C_6H_4COCl$	154.60	9, 484	1.169	1.5530^{20}	−2	225–227	82	
t190	*p*-Tolyl acetate	$CH_3CO_2C_6H_4CH_3$	150.18	6, 397	1.048	1.5010^{20}		210–211	90	
t191	1-(*o*-Tolyl)biguanide	$CH_3C_6H_4NHC({=}NH)NH{-}C({=}NH)NH_2$	191.24	12[3], 1873			143–145		>110	
t192	*m*-Tolyl isocyanate	$CH_3C_6H_4NCO$	133.15	12, 864	1.033	1.5305^{20}		76^{12mm}	65	s alc, eth; i aq
t193	1,2,4-Triacetoxy-benzene	$C_6H_3(O_2CCH_3)_3$	252.22	6, 1089			98–100			
t194	Triacetoxyvinylsilane	$(CH_3CO_2)_3SiCH{=}CH_2$	232.26		1.167	1.4220^{20}		128^{25mm}	76	
t195	Triallylamine	$(H_2C{=}CHCH_2)_3N$	137.23	4, 208	0.790	1.4510^{20}	150–151		30	
t196	Triallyl-1,3,5-triazine-2,4,6(1*H*,3*H*,5*H*)-trione		249.27		1.159	1.5129^{20}		152^{4mm}	>110	
t197	1*H*-1,2,4-Triazole		69.07	26, 13			119–121	260		s aq, alc
t198	Tribenzylamine	$(C_6H_5CH_2)_3N$	287.41	12, 1038	0.991^{95}_4		91–94		65	s hot alc, eth
t199	Tribromoacetaldehyde	Br_3CCHO	280.76	1, 626	2.665	1.5850^{20}		174	65	s aq, alc, chl, eth
t200	Tribromoacetic acid	Br_3CCO_2H	296.76	2, 220			130–133	245		s aq, alc, eth
t201	2,4,6-Tribromoaniline	$Br_3C_6H_2NH_2$	329.83	12, 663	2.35		120–122	300		s hot alc, chl, eth
t202	2,2,2-Tribromoethanol	Br_3CCH_2OH	282.77	1[2], 338			73–79	93^{10mm}		2 aq; s alc, bz, eth
t203	1,1,2-Tribromo-ethylene	$BrCH{=}CBr_2$	264.74	1, 191	1.708^{21}	1.6247^{25}		162.5		

t204	Tribromomethane	$CHBr_3$	252.77	1, 68	2.9000^{15}	1.6005^{15}	8.1	149.6	83	0.3 aq; misc eth, MeOH
t205	2,4,6-Tribromophenol	$Br_3C_6H_2OH$	330.82	6, 203	2.55		87–89	290^{746mm}		s alc, chl, eth; i aq
t206	1,2,3-Tribromopropane	$BrCH_2CH(Br)CH_2Br$	280.78	1, 112	2.390	1.584^{18}	16.5	220	93	s alc, eth
t207	Tributoxyborane	$(C_4H_9O)_3B$	230.16	1^2, 398	0.8567^{20}	1.4092^{20}	<-70	234	93	hyd aq
t208	Tributylamine	$(C_4H_9)_3N$	185.36	4, 157	0.7784	1.4280^{20}	−70	216	86	v s alc, eth; s acet
t209	Tributylborane	$(C_4H_9)_3B$	182.16	4^2, 1022	0.747			109^{20mm}	−36	i aq; s most org solv
t210	2,4,6-Tri-*tert*-butyl-phenol	$[(CH_3)_3C]_3C_6H_2OH$	262.44		0.864^{27}_{4}		129–132	277		
t211	Tributyl phosphate	$(C_4H_9O)_3P(O)$	266.32	1^2, 397	0.9727^{25}	1.4226^{25}	−79	289	146	0.04 aq; misc org solv
t212	Tributyl phosphite	$(C_4H_9O)_3P$	250.32	1^1, 187	0.925^{20}_{4}	1.4326^{20}		125^{7mm}	91	misc alc, bz, eth, PE
t213	Tributyltin chloride	$(C_4H_9)_3SnCl$	325.49	4^3, 1926	1.200	1.4905^{20}		173^{25mm}	>110	
t214	Tributyltin ethoxide	$(C_4H_9)_3SnOC_2H_5$	335.10		1.098	1.4672^{20}		$92^{0.1mm}$	40	
t215	Tributyltin hydride	$(C_4H_9)_3SnH$	291.05	4^4, 4312	1.082	1.4730^{20}		$80^{0.4mm}$	40	
t216	Tributyltin methoxide	$(C_4H_9)_3SnOCH_3$	321.07	4^4, 4331	1.115	1.4720^{20}		$97^{0.06mm}$	98	
t217	Trichloroacetamide	Cl_3CCONH_2	162.40	2, 211			141–143	238–240		
t218	Trichloroacetaldehyde	Cl_3CCHO	147.40	Merck: 12, 9755	1.510^{20}_{4}	1.4557^{20}	−57.5	97.8		dec aq, alc; s eth
t219	Trichloroacetic acid	Cl_3CCO_2H	163.39	2, 206	1.629^{61}_{4}	1.6200^{20}	57.5	196.5	>110	120 aq; v s alc, eth
t220	Trichloroacetic anhydride	$(Cl_3CCO)_2O$	308.75	2, 210	1.690	1.4838^{20}		141^{60mm}	none	
t221	1,1,3-Trichloroacetone	$ClCH_2COCHCl_2$	161.42	1, 655	1.508	1.4892^{20}	13–15	172	79	
t222	Trichloroacetonitrile	Cl_3CCN	144.39	2, 212	1.4403^{25}_{4}	1.4409^{20}	−42	86	none	
t223	2,2′,4′-Trichloro-acetophenone	$Cl_2C_6H_3COCH_2Cl$	223.49	7, 283			52–55	135^{4mm}	>110	
t224	Trichloroacetyl chloride	Cl_3CCOCl	181.83	2, 210	1.629	1.4689^{20}	−146	118		
t225	2,4,5-Trichloroaniline	$Cl_3C_6H_2NH_2$	196.46	12, 627			93–95	270		s alc
t226	2,4,6-Trichloroaniline	$Cl_3C_6H_2NH_2$	196.46	12, 627			73–75	262		s alc, eth
t227	1,2,3-Trichlorobenzene	$C_6H_3Cl_3$	181.45	5, 203	1.69	1.5776^{20}	53–55	218–220	126	v s bz, CS_2; sl s alc
t228	1,2,4-Trichlorobenzene	$C_6H_3Cl_3$	181.45	5, 204	1.454^{20}	1.5707^{20}	17	213–214	110	misc bz, eth, PE
t229	1,3,5-Trichlorobenzene	$C_6H_3Cl_3$	181.45	5, 204	1.66	1.5662^{19}	63.5	208	107	v s bz, eth, PE
t230	Trichloro-3-chloro-propylsilane	$Cl(CH_2)_3SiCl_3$	211.98		1.350	1.4666^{20}		181–183		
t231	1,1,1-Trichloroethane	CH_3CCl_3	133.41	1, 85	1.3390^{20}	1.4379^{20}	−30.4	74	−1	s acet, bz, eth
t232	1,1,2-Trichloroethane	$ClCH_2CHCl_2$	133.41	1, 85	1.4397^{20}	1.4714^{20}	−37	114	32	misc alc, eth
t233	2,2,2-Trichloroethanol	Cl_3CCH_2OH	149.40	1, 338	1.557	1.4900^{20}	18	151–153		8 aq; misc alc, eth
t234	2,2,2-Trichloroethyl chloroformate	$ClCO_2CH_2CCl_3$	211.86		1.539	1.4703^{20}		171–172		

(*Continued*)

TABLE 2.20 Physical Constants of Organic Compounds (*Continued*)

No.	Name	Formula	Formula weight	Beilstein reference	Density, g/mL	Refractive index	Melting point, °C	Boiling point, °C	Flash point, °C	Solubility in 100 parts solvent
t235	Trichloroethylene	$ClCH{=}CCl_2$	131.39	1, 187	1.4642^{20}	1.4773^{20}	−84.8	87	32	0.1 aq; misc alc, chl, eth
t236	Trichloroethylsilane	$C_2H_5SiCl_3$	163.51	4, 630	1.2373^{20}	1.4256^{20}	−105.6	100.5	22	
t237	Trichlorofluoromethane	Cl_3CF	137.37	Merck: 12, 9770	1.485^{21}	1.384^{20}	−111	23.8		0.14 aq; s alc, eth
t238	α,α,2-Trichloro-6-fluorotoluene	$ClC_6H_3(F)CHCl_2$	213.47	5^3, 701	1.446	1.5506^{20}		228–230	>110	
t239	Trichloroisocyanuric acid		232.41	25, 256			249–251			
t240	Trichloromethanesulfenyl chloride	Cl_3CSCl	185.89	3, 135	1.700^{20}_{4}	1.5436^{20}		146–148		
t241	1,1,1-Trichloro-2-methyl-2-propanol	$(CH_3)_2C(OH)CCl_3$	177.46	1, 382			99 anhyd	167		s alc, bz, chl, eth
t242	Trichloromethylsilane	CH_3SiCl_3	149.48	4^3, 1896	1.273^{20}_{4}	1.4108^{20}	−90	66	−9	
t243	1,2,4-Trichloro-5-nitrobenzene	$Cl_3C_6H_2NO_2$	226.45	5, 246	1.790^{20}		49–55	288	>110	v s bz, eth
t244	2,4,5-Trichlorophenol	$Cl_3C_6H_2OH$	197.45	6^2, 180			67–69	253		615 acet; 163 bz; 525 eth; 615 MeOH; i aq
t245	2,4,6-Trichlorophenol	$Cl_3C_6H_2OH$	197.45	6, 190	1.4901^{75}_{4}		69	246	none	525 acet; 113 bz; 354 eth; 525 MeOH; i aq
t246	(2,4,5-Trichlorophenoxy)acetic acid	$Cl_3C_6H_2OCH_2CO_2H$	255.49	6^3, 702			154–158			s alc; v sl s aq
t247	1,2,3-Trichloropropane	$ClCH_2CH(Cl)CH_2Cl$	147.43	1, 106	1.3889^{20}	1.4854^{20}	−14.7	157	71	misc alc, eth; i aq
t248	2,4,6-Trichloropyrimidine		183.43	23, 90		1.5700^{20}	23–25	>110		
t249	Trichlorosilane	$HSiCl_3$	135.45	Merck: 12, 9776	1.342	1.4000^{20}	−127	31–32	−13	dec aq; s bz, chl
t250	4-(Trichlorosilyl)butyronitrile	$Cl_3Si(CH_2)_3CN$	202.54	4,*4*, 4272	1.300	1.4630^{20}		237–238	92	
t251	α,α,α-Trichlorotoluene	$C_6H_5CCl_3$	195.48	5, 300	1.3723^{20}	1.5580^{20}	−5	219–223	127	s alc, bz, eth
t252	α,2,4-Trichlorotoluene	$Cl_2C_6H_3CH_2Cl$	195.48	5^4, 819	1.407	1.5760^{20}	−2.6	248	>110	
t253	α,2,6-Trichlorotoluene	$Cl_2C_6H_3CH_2Cl$	195.48			1.5761^{20}	36–39	119^{14mm}	>110	v s alc, eth
t254	α,3,4-Trichlorotoluene	$Cl_2C_6H_3CH_2Cl$	195.48	5, 300	1.411	1.5766^{20}		124^{14mm}	>110	
t255	2,4,6-Trichloro-1,3,5-triazine		184.41	26, 35			146–148	190		i aq; s alc
t256	1,1,1-Trichlorotrifluoroethane	Cl_3CCF_3	187.38		1.579	1.3699^{20}	13–14	46		

t257	1,1,2-Trichlorotrifluoroethane	$Cl_2CFCClF_2$	187.38	1^3, 157	1.5635^{25}	1.3557^{25}	−35	47.7		0.017 aq
t258	Trichlorovinylsilane	$H_2C{=}CHSiCl_3$	161.49		1.270	1.4360^{20}	−95	90	10	
t259	Tricyclo[5.2.1.0^{2,6}]-decane		136.24	5, 164			77–79	193	40	
t260	Tricyclo[5.2.1.0^{2,6}]-decan-8-one		150.22	7^2, 133	1.063	1.5025^{20}		132^{30mm}		
t261	Tridecane	$CH_3(CH_2)_{11}CH_3$	184.37	1, 171	0.7563^{20}_{4}	1.4256^{20}	−5 to −4	235	70	v s alc, eth
t262	Tridecanoic acid	$CH_3(CH_2)_{11}CO_2H$	214.35	2, 364			41–42	236^{100mm}	>110	v s alc, eth; i aq
t263	2-Tridecanone	$CH_3(CH_2)_{10}COCH_3$	198.35	1, 715	0.822	1.4350^{20}	29–31	134^{10mm}	>110	
t264	7-Tridecanone	$[CH_3(CH_2)_5]_2CO$	198.35	1, 715	0.825		30–32	264	>110	
t265	1-Tridecene	$CH_3(CH_2)_{10}CH{=}CH_2$	182.35	1, 225	0.7658^{20}	1.4340^{20}	−13	232.8	79	s alc; v s eth
t266	Triethanolamine	$(HOCH_2CH_2)_3N$	149.19	4, 285	1.1242^{20}_{4}	1.4853^{20}	20.5	335.4	179	misc aq, alc, acet; 4.5 bz; 1.6 eth; s chl
t267	3,4,5-Triethoxybenzoic acid	$(C_2H_5O)_3C_6H_2CO_2H$	254.29	10, 481			110–112			
t268	Triethoxyborane	$(C_2H_5O)_3B$	145.99	1, 335	0.864	1.3740^{20}		117–118	11	dec aq
t269	Triethoxysilane	$(C_2H_5O)_3SiH$	164.28	1, 334	0.890	1.3770^{20}		134–135	26	
t270	3-(Triethoxysilyl)-propionitrile	$(C_2H_5O)_3SiCH_2CH_2CN$	217.34	4^4, 4271	0.979	1.4140^{20}		224	100	
t271	3-(Triethoxysilyl)-propyl isocyanate	$(C_2H_5O)_3Si(CH_2)_3NCO$	247.37		0.999	1.4200^{20}		283	77	
t272	Triethoxyvinylsilane	$(C_2H_5O)_3SiCH{=}CH_2$	190.32		0.903^{20}_{4}	1.3978^{20}		160–161	34	
t273	Triethylaluminum	$(C_2H_5)_3Al$	114.17	4, 643	0.832^{25}		−50	194	−18	dec aq, air
t274	Triethylamine	$(C_2H_5)_3N$	101.19	4, 99	0.7275^{20}	1.4010^{20}	−114.7	88.8	−7	5.5 aq; misc alc, eth; s acet, EtOAc
t275	Triethylantimony	$(C_2H_5)_3Sb$	208.94	4, 618	1.324^{16}	1.42	−29	159.5		
t276	Triethylarsine	$(C_2H_5)_3As$	162.11	4, 602	1.150^{20}_{4}			140^{736mm}		i aq; misc alc, eth
t277	Triethylborane	$(C_2H_5)_3B$	98.00	4, 641	0.6961^{23}	1.3970^{20}	−02.9	95		i aq; dec by air
t278	Triethyl citrate	$HOC(CO_2C_2H_5)(CH_2CO_2C_2H_5)_2$	276.29	3, 568	1.137	1.4420^{20}		127^{1mm}	>110	
t279	Triethylenediamine		112.18	23^3, 484			158–160		62	45 aq; 13 acet; 77 alc; 51 bz
t280	Tri(ethylene glycol)	$(HOCH_2CH_2OCH_2\text{-})_2$	150.17	1, 468	1.1274^{15}	1.4550^{20}	−7	285	177	misc aq, alc, bz
t281	Tri(ethylene glycol) dimethacrylate	$[H_2C{=}C(CH_3)CO_2CH_2\text{-}CH_2OCH_2\text{-}]_2$	286.33	2^4, 1531	1.092	1.4605^{20}		172^{5mm}	>110	
t282	Tri(ethylene glycol) dimethyl ether	$(CH_3OCH_2CH_2OCH_2\text{-})_2$	178.23	Merck: 12, 9820	0.990^{20}_{4}	1.4224^{20}	−45	216	111	misc aq, hydrocarbon solvents

(*Continued*)

TABLE 2.20 Physical Constants of Organic Compounds (*Continued*)

No.	Name	Formula	Formula weight	Beilstein reference	Density, g/mL	Refractive index	Melting point, °C	Boiling point, °C	Flash point, °C	Solubility in 100 parts solvent
t283	Tri(ethylene glycol) divinyl ether	$H_2C{=}CH(OCH_2H_2)_3OCH{=}CH_2$	202.25	1^3, 2106	0.990	1.4530^{20}		126^{18mm}	>110	
t284	Tri(ethylene glycol) monomethyl ether	$CH_3(OCH_2CH_2)_3OH$	164.20	1^3, 2105	1.026	1.4399^{20}		122^{10mm}	>110	
t285	Triethylenetetramine	$(H_2NCH_2CH_2NHCH_2\text{-})_2$	146.24	4, 255	0.982	1.4971	12	266	143	
t286	Triethylgallium	$(C_2H_5)_3Ga$	156.91		1.0576^{30}		−82.3	142.6		
t287	1,3,5-Triethylhexa-hydro-1,3,5-triazine		171.20	26, 2	0.894	1.4595^{20}		207–208	80	
t288	Triethylindium	$(C_2H_5)_3In$	202.01		1.260^{20}	1.538^{20}	−32	144		
t289	Triethyl orthoacetate	$CH_3C(OC_2H_5)_3$	162.23	2, 129	0.8847^{25}_{4}	1.3950^{25}		142	36	misc alc, chl, eth
t290	Triethyl orthoformate	$HC(OC_2H_5)_3$	148.20	2, 20	0.891^{20}_{4}	1.3910^{20}	−76	146	30	dec aq; s alc, eth
t291	Triethyl ortho-propionate	$CH_3CH_2C(OC_2H_5)_3$	176.26	2, 240	0.876	1.3995^{20}		155–160	60	v s alc, eth
t292	Triethyl phosphate	$(C_2H_5O)_3P(O)$	182.16	1, 332	1.0695^{20}	1.4058^{20}	−56	215	115	s aq(dec), alc, eth
t293	Triethylphosphine	$(C_2H_5)_3P$	118.16	4, 582	0.800^{15}_{4}	1.4563^{20}	−88	128–129	−17	i aq; misc alc, eth; pyrophoric
t294	Triethyl phosphite	$(C_2H_5O)_3P$	166.16	1, 330	0.969^{20}_{4}	1.4130^{20}		156	54	i aq(hyd); misc alc, acet, bz, eth, PE
t295	Triethyl phosphono-acetate	$(CH_3CH_2O)_2P(O)CH_2CO_2C_2H_5$	224.19	4^1, 573	1.130	1.4310^{20}		145^{9mm}	>110	
t296	Triethyl phosphono-formate	$(CH_3CH_2O)_2P(O)CO_2C_2H_5$	212.17	3^2, 103	1.110	1.4320^{20}		135^{12mm}	>110	
t297	Triethylsilane	$(C_2H_5)_3SiH$	116.28	4, 625	0.731^{20}_{4}	1.412^{20}		107–108	−3	i aq; misc alc, eth
t298	Triethyl thiophosphate	$(C_2H_5O)_3P(S)$	198.22	1, 333	1.082	1.4480^{20}		100^{16mm}	107	
t299	2,2,2-Trifluoro-acetamide	CF_3CONH_2	113.04	2^2, 186			70–75	162.5		
t300	Trifluoroacetic acid	CF_3CO_2H	114.02	2^2, 186	1.4890^{20}	1.2850^{20}	−15.3	73		misc aq
t301	Trifluoroacetic anhydride	$[CF_3C(O)]_2O$	210.03	2^2, 186	1.487	<1.300	−65	39–40		
t302	1,1,1-Trifluoroacetone	$CF_3C(O)CH_3$	112.05	1^2, 717	1.252	<1.30		22	−30	
t303	1,3,5-Trifluorobenzene	$C_6H_3F_3$	132.09		1.277	1.4150^{20}	−5.5	75–76	−7	
t304	α,α,α-Trifluoro-*m*-cresol	$CF_3C_6H_4OH$	162.11	6^1, 187	1.333	1.4588^{20}	−1.8	178–179	73	
t305	2,2,2-Trifluoroethanol	CF_3CH_2OH	100.04	1^3, 1342	1.3842^{20}_{4}	1.2907^{20}	−43.5	74	29	
t306	2,2,2-Trifluoroethyl trifluoroacetate	$CF_3CH_2O_2CCF_3$	196.05	2^3, 427	1.4725^{18}_{4}	1.2812^{18}	−65.5	55	0	
t307	Trifluoromethane	HCF_3	70.01	1, 59	1.52^{-100}		−160	−84		75 mL aq; 500 mL alc
t308	Trifluoromethane-sulfonic acid	CF_3SO_3H	150.07	3^4, 34	1.695^{25}	1.3250^{25}	34	162	none	v s aq; misc eth

t309	Trifluoromethane-sulfonic anhydride	$(CF_3SO_2)_2O$	282.13	3^4, 35	1.677	1.3212^{20}		84	none	dec aq, alc
t310	3-(Trifluoromethyl)-aniline	$CF_3C_6H_4NH_2$	161.13	12, 870	1.290	1.4800^{20}	5–6	187	85	
t311	α,α,α-Trifluorotoluene	$C_6H_5CF_3$	146.11	5, 290	1.1886^{20}	1.4145^{20}	−29	102	12	
t312	Trihexyl *O*-acetyl-citrate	$CH_3CO_2C[CO_2(CH_2)_5CH_3]$-$[CH_2CO_2(CH_2)_5CH_3]_2$	486.65		1.005	1.4470^{20}			>110	
t313	Trihexylamine	$[CH_3(CH_2)_5]_3N$	269.52	4, 188	0.794	1.4415^{20}		163–265	>110	v s alc, eth; i aq
t314	Trihexyl *O*-butyl-citrate	$C_3H_7CO_2C[CO_2(CH_2)_5CH_3]$-$[CH_2CO_2(CH_2)_5CH_3]_2$	514.71		0.993	1.4480^{20}	−55		>110	
t315	Trihexylchlorosilane	$[CH_3(CH_2)_5]_3SiCl$	319.12		0.871^{20}_4	1.456^{20}		155^{5mm}		
t316	Trihexylsilane	$[CH_3(CH_2)_5]_3SiH$	284.60	4^4, 3915	0.799	1.448^{20}		161	>110	
t317	1,2,3-Trihydroxy-benzene	$C_6H_3(OH)_3$	126.11	6, 1071	1.45		133	309		59 aq; 77 alc; 62 eth
t318	1,3,5-Trihydroxy-benzene	$C_6H_3(OH)_3$	126.11	6, 1092			218–221			1 aq; 10 alc; s eth
t319	3,4,5-Trihydroxy-benzoic acid	$(HO)_3C_6H_2CO_2H$	170.12	10, 470			258–265			1.1 aq; 17 alc; 1 eth; 20 acet; i bz, chl, PE
t320	2,3,4-Trihydroxy-benzophenone	$(HO)_3C_6H_2COC_6H_5$	230.22	8, 417			140–142			
t321	1,2,6-Trihydroxy-hexane	$HO(CH_2)_4CH(OH)CH_2OH$	134.18	1,*4*, 2784	1.109	1.4760^{20}		178^{5mm}	79	
t322	Triisobutylaluminum	$[(CH_3)_2CHCH_2]_3Al$	198.33	4, 643	0.786	1.4494^{20}	4–6	86^{10mm}	−18	pyrophoric
t323	Triisobutylamine	$[(CH_3)_2CHCH_2]_3N$	185.36	4, 166	0.766	1.4230^{20}		192–193	57	
t324	Triisodecyl phosphite	$[(CH_3)_2CH(CH_2)_7O]_3P$	502.80		0.884	1.4600^{20}	<0	166	235	
t325	Triisopropanolamine	$[CH_3CH(OH)CH_2]_3N$	191.27	4^3, 762	0.9996^{50}_{20}		48–52	305.4	152	v s aq
t326	Triisopropoxyborane	$[(CH_3)_2CHO]_3B$	188.08	1, 363	0.815	1.3764^{20}		139–141	10	
t327	1,3,5-Triisopropyl-benzene	$C_6H_3[CH(CH_3)_2]_3$	204.36	5, 458	0.845	1.4880^{20}		232–236	86	
t328	Triisopropyl ortho-formate	$CH[OCH(CH_3)_2]_3$	190.29	2^3, 39	0.854	1.3970^{20}		66^{18mm}	42	
t329	Triisopropyl phosphite	$[(CH_3)_2CHO]_3P$	208.24	1, 363	0.914^{20}_4	1.4110^{20}		64^{11mm}	67	i aq(sl hyd)
t330	Triisopropylsilane	$[(CH_3)_2CH]_3SiH$	158.36	4^3, 1851	0.773	1.4344^{20}		86^{35mm}	37	
t331	3,4,5-Trimethoxy-benzaldehyde	$(CH_3O)_3C_6H_2CHO$	196.20	8, 391			73–75	165^{10mm}		
t332	1,2,3-Trimethoxy-benzene	$C_6H_3(OCH_3)_3$	168.19	6, 1081	1.112		43–45	241	>110	

(Continued)

TABLE 2.20 Physical Constants of Organic Compounds (*Continued*)

No.	Name	Formula	Formula weight	Beilstein reference	Density, g/mL	Refractive index	Melting point, °C	Boiling point, °C	Flash point, °C	Solubility in 100 parts solvent
t333	1,2,4-Trimethoxybenzene	$C_6H_3(OCH_3)_3$	168.19	6, 1088	1.126	1.5330^{20}		247	>110	
t334	1,3,5-Trimethoxybenzene	$C_6H_3(OCH_3)_3$	168.19	6, 1101			51–53	255	85	
t335	3,4,5-Trimethoxybenzoic acid	$(CH_3O)_3C_6H_2CO_2H$	212.20	10, 481			168–171	227^{10mm}		v s alc, eth; s chl
t336	3,4,5-Trimethoxybenzoyl chloride	$(CH_3O)_3C_6H_2COCl$	230.65	10, 487			81–84	185^{18mm}		
t337	3,4,5-Trimethoxybenzyl alcohol	$(CH_3O)_3C_6H_2CH_2OH$	198.22	6, 1159	1.233	1.5439^{20}		228^{25mm}	>110	
t338	Trimethoxyborane	$(CH_3O)_3B$	103.91	1, 287	0.920^{23}_{4}	1.3568^{20}	−34	67–68	−13	hyd aq; misc alc, eth
t339	Trimethoxyboroxine	$[-OB(OCH_3)-]_3$	173.53		1.195	1.3996^{20}	10	130	10	
t340	1,1,2-Trimethoxyethane	$CH_3OCH_2CH(OCH_3)_2$	120.15	1^3, 3183	0.932	1.3921^{20}		59^{56mm}	23	
t341	1,1,3-Trimethoxypropane	$CH_3OCH_2CH_2CH(OCH_3)_2$	134.18	1, 820	0.942	1.4004^{20}		46^{17mm}	40	
t342	1,1,3-Trimethoxypropylsilane	$CH_3OCH_2CH_2CH_2Si(OCH_3)_3$	164.28		0.932	1.3900^{20}		142	40	
t343	Trimethoxysilane	$(CH_3O)_3SiH$	122.20	1^2, 274	0.960	1.3579^{20}	−115	81	−4	
t344	3-(Trimethoxysilyl)propylamine	$H_2N(CH_2)_3Si(OCH_3)_3$	179.29		1.027	1.4240^{20}		92^{15mm}	83	
t345	*N*-[3-(Trimethylsilyl)propyl]aniline	$C_6H_5NH(CH_2)_3Si(OCH_3)_3$	255.39		1.070	1.5550^{20}		310	>110	
t346	*N*[1]-[3-(Trimethoxysilyl)propyl]ethylenediamine	$(CH_3O)_3Si(CH_2)_3NHCH_2CH_2NH_2$	224.36		1.019	1.4450^{20}		146^{15mm}	>110	
t347	3-(Trimethoxysilyl)propyl methacrylate	$(CH_3O)_3Si(CH_2)_3O_2CC(CH_3){=}CH_2$	248.35		1.045^{20}_{4}	1.4310^{20}		190	92	
t348	[3-(Trimethoxysilyl)propyl]urea	$(CH_3O)_3Si(CH_2)_3NHCONH_2$	222.32		1.150	1.4600^{20}		217–250	98	
t349	Trimethylacetic acid	$(CH_3)_3CCO_2H$	102.13	2, 319	0.889		33–35	163–164	63	
t350	Trimethylacetic anhydride	$[(CH_3)_3CCO]_2O$	186.25	2, 320	0.918	1.4090^{20}		193	57	
t351	Trimethylacetyl chloride	$(CH_3)_3CCOCl$	120.58	2, 320	0.979	1.4120^{20}		105–106	8	

t352	Trimethylaluminum	$(CH_3)_3Al$	72.09	4, 643	0.752^{20}	1.432^{12}	15	125–126	−18	s alk; v sl s alc
t354	Trimethylamine	$(CH_3)_3N$	59.11	4, 43	0.656	1.3631^{0}	−117	2.9	−7	41 aq; misc alc; s bz, chl, eth
t355	2,4,6-Trimethylaniline	$(CH_3)_3C_6H_2NH_2$	135.21	12, 1160	0.963	1.5510^{20}		233	96	
t356	1,3,3-Trimethyl-6-azabicyclo[3.2.1]octane		153.27		0.902	1.4716^{20}		194	75	
t357	1,2,3-Trimethylbenzene	$C_6H_3(CH_3)_3$	120.20	5, 399	0.8944^{20}_{4}	1.5139^{20}	−25.4	176.1	48	i aq; s alc, eth
t358	1,2,4-Trimethylbenzene	$C_6H_3(CH_3)_3$	120.20	5, 400	0.8756^{20}_{4}	1.5048^{20}	−43.9	169	48	s alc, bz, eth
t359	1,3,5-Trimethylbenzene	$C_6H_3(CH_3)_3$	120.20	5, 406	0.8637^{20}_{4}	1.4994^{20}	−44.7	165	44	misc alc, bz, eth
t360	Trimethyl 1,2,4-benzenetri-carboxylate	$C_6H_3(CO_2CH_3)_3$	252.22	9^1, 429	1.261	1.5214^{20}	38–40	194^{12mm}	>110	
t361	2,2,3-Trimethylbutane	$(CH_3)_2CHC(CH_3)_3$	100.20	1^2, 121	0.6901^{20}_{4}	1.3890^{20}	−24.9	80.9	−6	s alc, eth
t362	2,3,3-Trimethyl-2-butanol	$(CH_3)_3CC(CH_3)_2OH$	116.20	1^2, 447	0.8380^{25}_{4}	1.4233^{22}	15–17	130.5		misc alc, eth
t363	1,2,4-Trimethylcyclohexane	$C_6H_9(CH_3)_3$	126.24	5, 42	0.786	1.4330^{20}		141–143	18	
t364	3,5,5-Trimethylcyclohex-2-ene-1-one		138.2	7, 65	0.918	1.4720^{20}	−8.1	215	80	1.2 aq
t365	2,6,6-Trimethyl-2-cyclohexene-1,4-dione		152.19	7^4, 2032		1.4910^{20}	26–28	94^{11mm}	96	
t366	Trimethyl-1,6-diisocyanatohexane	$OCNCH_2CH_2C(CH_3)CH_2$-$C(CH_3)CH_2CNO$	210.28		1.012	1.4620^{20}		149	>110	
t367	2,2,6-Trimethyl-4*H*-1,3-dioxin-4-one		142.16	19^3, 1604	1.088	1.4620^{20}	12–13	67^{2mm}	86	
t368	4,4′-Trimethylenebis-(1-methylpiperidine)		238.42		0.896	1.4820^{20}	13	215^{50mm}	>110	
t369	4,4′-Trimethylene-dipiperidine		210.37				65–58			
t370	3,5,5-Trimethylhexanal	$(CH_3)_3CCH_2CH(CH_3)CH_2CHO$	142.24	1^3, 2894	0.817	1.4215^{20}		$68^{2.4mm}$	46	
t370a	3,5,5-Trimethylhexane	$(CH_3)_2CHCH_2CH(CH_3)CH(CH_3)_2$	128.26		0.7218^{20}	1.4051^{20}	−128	131		
t371	3,5,5-Trimethyl-1-hexanol	$(CH_3)_3CCH_2CH(CH_3)$-CH_2CH_2OH	144.25	1^3, 1755	0.8236^{20}_{4}	1.4300^{25}	< −70	193–194	80	s alc, eth
t372	3,5,5-Trimethylhexanoyl chloride	$(CH_3)_3CCH_2CH(CH_3)$-CH_2COCl	176.89	2^3, 834	0.930	1.4360^{20}		188–190	140	

(*Continued*)

TABLE 2.20 Physical Constants of Organic Compounds (*Continued*)

No.	Name	Formula	Formula weight	Beilstein reference	Density, g/mL	Refractive index	Melting point, °C	Boiling point, °C	Flash point, °C	Solubility in 100 parts solvent
t374	Trimethylhydroquinone	$(CH_3)_3C_6H(OH)_2$	152.19	6, 931			172–174			s aq; v s alc, bz, eth
t375	1,3,3-Trimethyl-2-norbornanol		154.25	6, 70	0.9641^{20}_{4}		39–45	201	73	s alc, eth
t376	1,3,3-Trimethyl-2-norbornanone		152.24	7, 96	0.948^{18}	1.4635^{18}	5	192–194	52	v s alc, eth
t377	Trimethyl orthoacetate	$CH_3C(OCH_3)_3$	120.15	2^2, 128	0.9428^{25}_{4}	1.3859^{25}		107–109	16	v s alc, eth
t378	Trimethyl orthoformate	$HC(OCH_3)_3$	106.12	2, 19	0.9676^{20}_{4}	1.3790^{20}		100.6	15	
t379	2,4,4-Trimethyl-2-oxazoline		113.16		0.887	1.4213^{20}		112–113	12	
t380	2,2,3-Trimethylpentane	$(CH_3)_3CCH(CH_3)CH_2CH_3$	114.23	1^1, 62	0.7160^{20}_{4}	1.4030^{20}	−112.3	110	<21	s eth; sl s alc
t381	2,2,4-Trimethylpentane	$(CH_3)_2CHCH_2C(CH_3)_3$	114.23	1^2, 127	0.6919^{20}_{4}	1.3915^{20}	−107.4	99.2	−12	s bz, chl, eth
t382	2,3,4-Trimethylpentane	$(CH_3)_2CH[CH(CH_3)]_2CHCH_3$	114.23	1^3, 500	0.7190^{20}_{4}	1.4042^{20}	−109.2	113–114	5	s alc, org solv
t383	2,2,4-Trimethyl-1,3-pentanediol	$(CH_3)_2CHCH(OH)C(CH_3)_2CH_2OH$	146.22	1^3, 2225	0.928^{55}_{15}	1.4513^{15}	52–56	232	113	1.8 aq; 75 alc; 22 bz; 25 acet
t384	2,4,4-Trimethyl-1-pentene	$(CH_3)_3CCH_2C(CH_3){=}CH_2$	112.22	1^3, 849	0.7150^{20}_{4}	1.4112^{20}	−93	101–102	−6	
t385	2,3,5-Trimethylphenol	$(CH_3)_3C_6H_2OH$	136.19	6, 518			92–95	230–231		
t386	2,3,6-Trimethylphenol	$(CH_3)_3C_6H_2OH$	136.19				62–64			
t387	2,4,6-Trimethylphenol	$(CH_3)_3C_6H_2OH$	136.19	6, 518			71–74	220		
t388	2,4,6-Trimethyl-1,3-phenylenediamine	$(CH_3)_3C_6H(NH_2)_2$	152.23	13^1, 190			88–91			
t389	Trimethyl phosphate	$(CH_3O)_3P(O)$	140.08	1, 286	1.197^{20}	1.3967^{20}	−46	197	107	100 aq; s alc
t390	Trimethyl phosphite	$(CH_3O)_3P$	124.08	1, 285	1.046^{20}_{4}	1.4080^{20}	−78	111–112	27	dec aq; misc alc, acet, bz, PE
t391	Trimethyl phosphonoacetate	$(CH_3O)_2P(O)CH_2CO_2CH_3$	182.11		1.125	1.4370^{20}		$118^{0.85mm}$	>110	
t392	1,2,4-Trimethylpiperazine		128.22		0.851^{25}_{25}	1.4480^{25}	−50	151^{746mm}		s aq, alc, acet, bz
t393	2,4,6-Trimethylpyridine	$C_5H_2N(CH_3)_3$	121.18	20, 250	0.9166^{22}_{4}	1.4959^{25}	−46	171	57	3.5 aq; misc eth; s alc, bz, chl
t394	*N*-(Trimethylsilyl)-	$CH_3CONHSi(CH_3)_3$	131.25				46–49	186	57	

t395	Trimethylsilyl acetate	$CH_3CO_2Si(CH_3)_3$	132.24	4[3], 1857	0.882	1.3880[20]	−32	108	4	
t396	*N*-(Trimethylsilyl)-imidazole		140.26		0.956	1.4751[20]		94[14mm]	5	
t397	Trimethylsilyl methacrylate	$H_2C{=}C(CH_3)CO_2Si(CH_3)_3$	158.28		0.890	1.4150[20]		51[20mm]	32	
t398	Trimethylsilyl trifluoromethane sulfonate	$CF_3SO_3Si(CH_3)_3$	222.26		1.228	1.3600[20]		77[80mm]	25	
t399	Trimethylsulfonium iodide	$[(CH_3)_3S]I$	204.07					215–220 sublime		
t400	Trimethylsulfoxonium iodide	$[(CH_3)_3S(O)]I$	220.07				169 dec			
t400a	1,7,7-Trimethyltricyclo[2.2.1.0^{2,6}]-heptane		136.24	5, 164	0.8668[80]	1.4296[80]	67.5	152.5		
t401	Trimethylvinylsilane	$(CH_3)_3SiCH{=}CH_2$	100.24		0.649	1.3920[20]		55	< −34	
t402	2,4,6-Trinitroaniline	$(O_2N)_3C_6H_2NH_2$	228.12	12, 763	1.762[14]		188–190	explodes		s hot acet; sl s alc
t403	1,2,4-Trinitrobenzene	$C_6H_3(NO_2)_3$	213.11	5, 271	1.73[16]		61–62	explodes		5.5 alc; 7.1 eth; i aq
t404	1,3,5-Trinitrobenzene	$C_6H_3(NO_2)_3$	213.11	5, 271	1.688^{20}_{4}		122.5	explodes		0.035 aq; 1.9 alc; 1.5 eth; 6.2 bz
t405	2,4,6-Trinitrotoluene	$(O_2N)_3C_6H_2CH_3$	227.13	5, 347	1.654^{20}_{4}		80.1	explodes		1.5 alc; 4 eth; s bz, acet; 0.01 aq
t406	Trioctylamine	$[(CH_3(CH_2)_7]_3N$	353.68	4, 196	0.809	1.4485[20]		365–367	>110	
t407	1,3,5-Trioxane		90.08	19, 381	1.170[65]		60.2	115	45	17.2 aq[18]; v s alc, bz, eth, EtOAc
t408	4,7,10-Trioxa-1,13-tridecanediamine	$O[CH_2CH_2O(CH_2)_3NH_2]_2$	220.31	4,*4*, 1625	1.005	1.4640[20]		148[4mm]	>110	
t409	Tripentaerythritol	$(HOCH_2)_3CCH_2OCH_2$-$C(CH_2OH)_2CH_2OCH_2$-	372.41				225 dec			
t410	Triphenylamine	$(C_6H_5)_3N$	245.33	12, 181	0.774^{0}_{0}		125–127	347–348		
t411	Triphenylantimony	$(C_6H_5)_3Sb$	353.07	16, 891	1.4343[25]		52–54	377	>110	v s bz, eth; sl s alc
t412	Triphenylarsine	$(C_6H_5)_3As$	306.24	16, 828	1.2225[48]	1.6139[48]	60–62	233[14mm]		v s bz, eth; s alc
t413	1,3,5-Triphenylbenzene	$(C_6H_5)_3C_6H_3$	306.41	5, 737	1.205		172–174	460		v s bz; s abs alc, eth
t414	Triphenylborane	$(C_6H_5)_3B$	242.13	16[2], 636			145	203[15mm]		
t415	Triphenylmethane	$(C_6H_5)_3CH$	244.34	5, 698	1.0134^{99}_{4}		92–94	360		v s hot alc, eth; 49 chl; 7 bz; s PE

(*Continued*)

TABLE 2.20 Physical Constants of Organic Compounds (*Continued*)

No.	Name	Formula	Formula weight	Beilstein reference	Density, g/mL	Refractive index	Melting point, °C	Boiling point, °C	Flash point, °C	Solubility in 100 parts solvent
t416	Triphenylmethanol	$(C_6H_5)_3COH$	260.34	6, 713	1.199_4^0		160–163	360		v s alc, bz, eth; i aq
t417	Triphenylmethyl bromide	$(C_6H_5)_3CBr$	323.24	5, 704			152–154	230^{15mm}		
t418	Triphenylmethyl chloride	$(C_6H_5)_3CCl$	278.78	5, 700			110–112	235^{20mm}		
t419	Triphenyl phosphate	$(C_6H_5O)_3P(O)$	326.29	6, 179			50–52	244^{10mm}	223	misc alc; s bz, acet, chl, eth; i aq
t420	Triphenylphosphine	$(C_6H_5)_3P$	262.29	16, 759	1.075_4^{81}		79–81	377	181	v s eth; s bz, chl, HOAc; sl s alc; i aq
t421	Triphenylphosphine oxide	$(C_6H_5)_3P(O)$	278.29	16, 783			156–158			
t422	Triphenyl phosphite	$(C_6H_5O)_3P$	310.29	6, 177	1.184	1.5903^{20}	22–24	360	218	s alc, bz, chl, eth
t423	Triphenylsilane	$(C_6H_5)_3SiH$	260.41	16^2, 605			42–44	152^{2mm}	76	
t424	Triphenyltin acetate	$CH_3CO_2Sn(C_6H_5)_3$	409.06	16^4, 1606			124–126			s eth; sl s alc, bz
t425	Triphenyltin chloride	$(C_6H_5)_3SnCl$	385.46	16, 914			108 dec	$240^{13.5mm}$		
t426	Triphenyltin hydroxide	$(C_6H_5)_3SnOH$	367.02	16, 914			124–126			
t427	Tripropoxyborane	$(CH_3CH_2CH_2O)_3B$	188.08	1^2, 369	0.8576_4^{20}	1.3948^{20}		175–177	32	v s alc; misc eth
t428	Tripropylaluminum	$(CH_3CH_2CH_2)_3Al$	156.25	4, 643	0.823		−107	84^{2mm}	−18	
t429	Tripropylamine	$(CH_3CH_2CH_2)_3N$	143.27	4, 139	0.753	1.4160^{20}	−93.5	155–158	36	s aq, alc, eth
t430	Tripropylene glycol	$H(OCH_2CH_2CH_2)_3OH$	192.26		1.021	1.442^{25}		273	141	s aq
t431	Tripropylene glycol butyl ether	$HO(CH_2CH_2CH_2O)_3(CH_2)_3CH_3$	248.4		0.932	1.430^{20}		276	135	
t432	Tripropylene glycol monomethyl ether	$HO(CH_2CH_2CH_2O)_3CH_3$	206.29	1^4, 2475	0.967	1.428^{25}	−42	242.4	127	misc aq, alc, eth
t433	Tripropyl orthoformate	$HC(OCH_2CH_2CH_3)_3$	190.28	2, 21	0.8805_4^{20}	1.4072^{20}		108^{40mm}	72	
t434	Tris(2-aminoethyl)-amine	$(H_2NCH_2CH_2)_3N$	146.24	4, 256	0.977	1.4970^{20}		114^{15mm}	>110	
t435	Tris(2-butoxyethyl) phosphate	$(C_4H_9OCH_2CH_2O)_3P(O)$	398.48		1.006	1.4359^{20}		228^{4mm}	110	
t436	Tris(2-chloroethyl) phosphate	$(ClCH_2CH_2O)_3P(O)$	285.49	1^2, 337	1.390	1.4721^{20}		330	232	
t437	Tris(2-chloroethyl) phosphite	$(ClCH_2CH_2O)_3P$	269.49		1.353_4^{20}	1.4863^{20}		115^{2mm}	190	misc alc, bz, eth
t438	Tris(2-ethylhexyl) phosphate	$[C_4H_9CH(C_2H_5)CH_2O]_3P(O)$	434.65	1^3, 1734	0.924	1.4437^{20}		215^{4mm}	>110	i aq

t439	Tris(hydroxymethyl)-aminomethane	$(HOCH_2)_3CNH_2$	121.14	4, 303			171–172	220^{10mm}		
t440	1,1,1-Tris(hydroxy-methyl)ethane	$CH_3C(CH_2OH)_3$	120.15	1, 520			200–203			
t441	*N*-[Tris(hydroxy-methyl)methyl]-glycine	$(HOCH_2)_3CNHCH_2CO_2H$	179.17	Merck: 12, 9783			187			satd aq^0 is 0.8*M*
t442	Tris(hydroxymethyl)-nitromethane	$(HOCH_2)_3CNO_2$	151.12	1, 520			214 pure 175 tech			220 aq; v s alc; sl s bz
t443	Tris[2-(2-methoxy-ethoxy)ethyl]amine	$(CH_3OCH_2CH_2OCH_2CH_2)_3N$	323.43		1.011	1.4486^{20}			>110	
t444	Tris(2-methoxy-ethoxy)-vinylsilane	$H_2C{=}CHSi(OCH_2CH_2{-}OCH_3)_3$	280.39	4^4, 4257	1.034^{25}_4	1.427^{25}		284–286	>110	
t445	Tris(2-methoxyethyl) borate	$(CH_3OCH_2CH_2O)_3B$	236.08	1^3, 2118	1.010	1.4150^{20}		135^{15mm}	87	
t446	Tris(2-methylallyl)-amine	$[H_2C{=}C(CH_3)CH_2]_3N$	179.31	4^3, 462	0.794	1.4575^{20}		85^{15mm}	53	
t447	Tris(2,2,2-trifluoro-ethyl) phosphite	$(CF_3CH_2O)_3P$	328.07	1^4, 1371	1.487	1.3245^{20}		131^{743mm}	>110	
t448	Tris[3-(trimethoxy-silyl)propyl] isocyanurate		615.86		1.170	1.4610^{20}		250	102	
t449	Tris(trimethylsilyl) borate	$[(CH_3)_3SiO]_3B$	278.38	4^3, 1861	0.831	1.3861^{20}		186	42	
t450	1,3,5-Trithiane		138.27	19, 382			216–218			s bz; sl s alc, eth
t451	Trithiocarbonic acid	$(HS)_2CS$	110.21	3, 221	1.483^{20}_4	1.8225^{20}	−26.9	57.8		dec aq, alc; sl s eth
t452	Tri-*o*-tolyl phosphate	$(CH_3C_6H_4O)_3P(O)$	368.37	Merck: 12, 9893	1.1955^{20}	1.5575^{20}	11	410	225	sl s aq, alc; s eth
t453	1,2,4-Trivinylcyclo-hexane	$(H_2C{=}CH)_3C_6H_9$	162.28		0.836	1.4780^{20}		88^{20mm}	68	
t454	L-(−)-Tryptophan		204.23	22, 546			280–285 dec			1.14 aq^{25}; s hot alc, alk; i eth, chl
t455	L-Tyrosine	$(HO)C_6H_4CH_2CH(NH_2)CO_2H$	181.19	14, 605	1.456		342–344			0.045 aq; 0.01 alc; s alk; i eth
u1	Undecanal	$CH_3(CH_2)_9CHO$	170.30	1, 712	0.825	1.4322^{20}	−4	115^{5mm}	96	i aq; s alc, eth
u2	Undecane	$CH_3(CH_2)_9CH_3$	156.31	1, 170	0.7402^{20}_4	1.4173^{20}	−25.6	196	60	i aq; misc alc, eth

(*Continued*)

TABLE 2.20 Physical Constants of Organic Compounds (*Continued*)

No.	Name	Formula	Formula weight	Beilstein reference	Density, g/mL	Refractive index	Melting point, °C	Boiling point, °C	Flash point, °C	Solubility in 100 parts solvent
u3	Undecanenitrile	$CH_3(CH_2)_9CN$	167.30	2, 358	0.823	1.4330^{20}		253	>110	
u4	Undecanoic acid	$CH_3(CH_2)_9CO_2H$	186.30	2, 358	0.8907	1.4294^{45}	28.5	228^{160mm}	>110	s alc, chl, eth; i aq
u5	Undecanoic γ-lactone		184.28	17, 247	0.949	1.4500^{20}		166^{13mm}	>110	
u6	Undecanoic δ-lactone		184.28	17^3, 4257	0.969	1.4590^{20}		$155^{10.5mm}$	>110	
u7	1-Undecanol	$CH_3(CH_2)_{10}OH$	172.31	1, 427	0.8324	1.4402^{20}	11	242.8	>110	
u8	2-Undecanol	$CH_3(CH_2)_8CH(OH)CH_3$	172.31	1, 427	0.828	1.4370^{20}	2–3	131^{28mm}	88	
u9	2-Undecanone	$CH_3(CH_2)_8COCH_3$	170.30	1, 173	0.829	1.4300^{20}	11–13	231–232	88 (CC)	s alc, bz, chl, eth, acet; i aq
u10	3-Undecanone	$CH_3(CH_2)_7COCH_2CH_3$	170.30	1, 713	0.827	1.4291^{20}	12–13	225–229	89	
u11	6-Undecanone	$CH_3(CH_2)_4CO(CH_2)_4CH_3$	170.30	1, 174	0.831	1.4280^{20}	14.6	228	88	i aq; v s alc, eth
u12	10-Undecenal	$H_2C{=}CH(CH_2)_8CHO$	168.28	1,*3*, 3029	0.810	1.4427^{20}			92	
u12a	1-Undecene	$H_2C{=}CH(CH_2)_8CH_3$	154.30	1, 225	0.7503^{20}	1.4261^{20}	−49	193	71	
u13	10-Undecenoic acid	$H_2C{=}CH(CH_2)_8CO_2H$	184.28	2, 458	0.907^{24}_{4}	1.4493^{20}	24.5	137^{2mm}	148	s alc, chl, eth; i aq
u14	10-Undecen-1-ol	$H_2C{=}CH(CH_2)_9OH$	170.30	1, 452	0.850^{15}	1.4500^{20}	−2	245	93	
u15	10-Undecenoyl chloride	$H_2C{=}CH(CH_2)_8COCl$	202.73	2, 459	0.944	1.4540^{20}		122^{10mm}	93	
u16	Urea	$(H_2N)_2CO$	60.06	3, 42	1.335		133–135	dec >mp		100 aq; 20 alc
u17	Uric acid		168.11	26, 513	1.893^{20}		>300 dec			s alk; i aq, alc, eth
u18	Uridine		244.20	31, 23			166–167			s aq; hot alc, pyr
v1	Valeric anhydride	$[CH_3(CH_2)_3CO]_2O$	186.25	2, 301	0.942	1.4210^{20}	−57	112^{16mm}	101	
v2	γ-Valerolactone		100.12	17, 235	1.057	1.4330^{20}	−31	207–208	81	
v3	δ-Valerolactone		100.12	17, 235	1.079	1.4580^{20}		$60^{0.5mm}$	100	
v4	L-Valine	$(CH_3)_2CHCH(NH)CO_2H$	117.15	4, 427	1.230		>315 subl			8.8 aq; v sl s alc, eth
v5	Vinyl acetate	$H_2C{=}CHO_2CCH_3$	86.09	2^1, 63	0.932^{20}_{4}	1.3954^{20}	−93	72–73	−8	2 aq; misc alc, eth
v6	Vinyl benzoate	$C_6H_5CO_2CH{=}CH_2$	148.16	9^1, 65	1.070	1.5290^{20}		96^{20mm}	82	
v7	4-Vinylbenzyl chloride	$H_2C{=}CHC_6H_4CH_2Cl$	152.62		1.083	1.5740^{20}		229	104	
v8	Vinylcyclohexane	$C_6H_{11}CH{=}CH_2$	110.20	5^1, 35	0.805	1.4463^{20}		126–127	20	
v9	4-Vinyl-1-cyclohexene		108.18	5^1, 63	0.803^{20}_{4}	1.4640^{20}	−101	127	20	
v10	2-Vinyl-1,3-dioxolane		100.12		1.001	1.4300^{20}		115–116	14	
v11	*N*-Vinylformamide	$HCONHCH{=}CH_2$	71.08		1.014	1.4940^{20}	−16	210	102	
v12	1-Vinylimidazole		94.12	23^4, 569	1.039	1.5308^{20}		79^{13mm}	81	
v13	5-Vinyl-2-norbornene		120.20		0.841	1.4802^{20}	−80	141	27	
v14	Vinyl propionate	$CH_3CH_2CO_2CH{=}CH_2$	100.12	2^3, 532	0.919	1.4030^{20}	−80	94–95	6	
v15	2-Vinylpyridine	$(C_5H_4N)CH{=}CH_2$	105.14	20, 256	0.975	1.5490^{20}		158–159	46	v s alc, chl, eth
v16	4-Vinylpyridine	$(C_5H_4N)CH{=}CH_2$	105.14	20^2, 170	0.975	1.5500^{20}		65^{15mm}	51	sl s hot aq, hot alc

v17	*N*-Vinyl-2-pyrrolidinone		111.14		1.040	1.5120^{20}		93^{13mm}	93	
v18	Vinyltrimethoxysilane	$H_2C{=}CHSi(OCH_3)_3$	148.24		0.968	1.3920^{20}		123	22	
x1	Xanthene		182.22	17, 73			101	310–312		s bz, eth; sl s alc, aq
x2	Xanthen-9-carboxylic acid		226.23	18^2, 279			217 dec			s hot alc, eth
x3	9-Xanthenone		196.21	17, 354			174–176	350^{730mm}		0.5 alc; v s chl
x4	*m*-Xylene	$C_6H_4(CH_3)_2$	106.17	5, 370	0.8642^{20}	1.4972^{20}	−47.9	139	27	misc alc, eth; 0.02 aq
x5	*o*-Xylene	$C_6H_4(CH_3)_2$	106.17	5, 362	0.8808^{20}_{4}	1.5054^{20}	−25.2	144–145	32	misc alc, eth; 0.017 aq
x6	*p*-Xylene	$C_6H_4(CH_3)_2$	106.17	5, 382	0.8611^{20}_{4}	1.4958^{20}	13	138	27	v s eth; s alc; 0.02 aq
x7	Xylitol	$HOCH_2(CHOH)_3CH_2OH$	152.15	1, 531	1.52		95–97			64 aq; 1.2 EtOH; 6.0 MeOH
x8	D-(+)-Xylose		150.13	31, 47	1.535^{0}		156–158			117 aq; s hot alc, pyr
x9	*m*-Xylylenediamine	$C_6H_4(CH_2NH_2)_2$	136.20	13, 186	1.032	1.5709^{20}	>110			

TABLE 2.21 Melting Points of Derivatives of Organic Compounds

(a) Derivatives of Alcohols

	3,5-Dinitro-benzoate $\theta_{C,m}$/°C		3,5-Dinitro-benzoate $\theta_{C,m}$/°C
Methanol	109	2-Methylpropan-2-ol	142
Ethanol	94	Pentan-1-ol	46
Propan-1-ol	75	Hexan-1-ol	61
Propan-2-ol	122	Phenylmethanol	113
Butan-1-ol	64	Cyclohexanol	113
2-Methylpropan-1-ol	88	Ethane-1,2-diol (glycol)	169*
Butan-2-ol	76		

(b) Derivatives of Phenols

	3,5-Dinitro benzoate $\theta_{C,m}$/°C	4-Methyl-benzene-sulphonate $\theta_{C,m}$/°C		3,5-Dinitro-benzoate $\theta_{C,m}$/°C	4-Methyl-benzene sulphonate $\theta_{C,m}$/°C
Phenol	146	96	Benzene-1,2-diol	152*	—
2-Methylphenol	138	55	Benzene-1,3-diol	201*	81*
3-Methylphenol	165	51	Benzene-1,4-diol	317*	159*
4-Methylphenol	189	70	2-Nitrophenol	155	83
Naphthalen-1-ol	217	88	3-Nitrophenol	159	113
Naphthalen-2-ol	210	125	4-Nitrophenol	188	97

(c) Derivatives of Aldehydes and Ketones

	2,4-Dinitro-Phenyl-hydrazone $\theta_{C,m}$/°C		2,4-Dinitro-Phenyl-hydrazone $\theta_{C,m}$/°C
Methanal	166	Propanone	126
Ethanal	168	Butanone	116
Propanal	155	Pentan-3-one	156
Butanal	126	Pentan-2-one	144
Benzaldehyde	237	Heptan-4-one	75
2-Hydroxybenzaldehyde	252 dec.	Phenylethanone	250
Ethanedial	327	Diphenylmethanone	239
Trichloroethanal	131	Cyclohexanone	162

(d) Derivatives of Amines

	Ethanoyl derivative $\theta_{C,m}$/°C	Benzoyl derivative $\theta_{C,m}$/°C	4-Methyl-benzene sulphonyl derivative $\theta_{C,m}$/°C
Methylamine	28	80	75
Ethylamine	205*	69	62
Propylamine	47	85	52
Butylamine	229‡	70	65
(Phenylmethyl) amine	60	105	116
Phenylamine	114	163	103
Cyclohexylamine	104	147	87
2-Methylphenylamine	112	143	110
3-Methylphenylamine	66	125	114
4-Methylphenylamine	152	158	118
Dimethylamine	116‡	42	87
Diethylamine	186‡	42	60
Diphenylamine	103	180	142

* Disubstituted derivative.
‡ Boiling temperature.

TABLE 2.22 Melting Points of n-Paraffins

Number of carbon atoms	Melting point °C	Melting point °F
1	−182	−296
2	−183	−297
3	−188	−306
4	−138	−216
5	−130	−202
6	−95	−139
7	−91	−132
8	−57	−71
9	−54	−65
10	−30	−22
11	−26	−15
12	−10	14
13	−5	23
14	6	43
15	10	50
16	18	64
17	22	72
18	28	82
19	32	90
20	36	97
30	66	151
40	82	180
50	92	198
60	99	210

TABLE 2.23 Boiling Point and Density of Alkyl Halides

	Chloride		Bromide		Iodide	
Name	B.p., °C	Density at 20°C	B.p., °C	Density at. 20°C	B.p., °C	Density at 20°C
Methyl	−24		5		43	2.279
Ethyl	12.5		38	1.440	72	1.933
n-Propyl	47	.890	71	1.335	102	1.747
n-Butyl	78.5	.884	102	1.276	130	1.617
n-Pentyl	108	.883	130	1.223	157	1.517
n-Hexyl	134	.882	156	1.173	180	1.441
n-Heptyl	160	.880	180		204	1.401
n-Octyl	185	.879	202		225.5	
Isopropyl	36.5	.859	60	1.310	89.5	1.705
Isobutyl	69	.875	91	1.261	120	1.605
see-Butyl	68	.871	91	1.258	119	1.595
tert-Butyl	51	.840	73	1.222	100*d*	
Cyclohexyl	142.5	1.000	165			
Vinyl(Haloethene)	−14		16		56	
Allyl (3-Halopropene)	45	.938	71	1.398	103	
Crotyl (1-Halo-2-butene)	84				132	

(Continued)

TABLE 2.23 Boiling Point and Density of Alkyl Halides (*Continued*)

	Chloride		Bromide		Iodide	
Name	B.p., °C	Density at 20°C	B.p., °C	Density at 20°C	B.p., °C	Density at 20°C
Methylvinylcarbinyl (3-Halo-1-butene)	64					
Propargyl (3-Halopropyne)	65		90	1.520	115	
Benzyl	179	1.102	201		93^{10}	
α-Phenylethyl	92^{15}		85^{10}			
β-Phenylethyl	92^{20}		92^{11}		127^{19}	
Diphenylmethyl	173^{19}		184^{20}			
Triphenylmethyl	310		230^{15}			
Dihalomethane	40	1.336	99	2.49	180*d*	3.325
Trihalomethane	61	1.489	151	2.89	*subl.*	4.008
Tetrahalomethane	77	1.595	189.5	3.42	*subl.*	4.32
1,1-Dihaloethane	57	1.174	110	2.056	179	2.84
1,2-Dihaloethane	84	1.257	132	2.180	*d*	2.13
Trihaloethylene	87		164	2.708		
Tetrahaloethylene	121				*subl.*	
Benzal halide	205		140^{20}			
Benzotrihalide	221	1.38				

TABLE 2.24 Properties of Carboxylic Acids

Name	Formula	M.p., °C	B.p., °C	Solub., g/100 g H_2O
Formic	$HCOOH$	8	100.5	∞
Acetic	CH_3COOH	16.6	118	∞
Propionic	CH_3CH_2COOH	–22	141	∞
Butyric	$CH_3(CH_2)_2COOH$	–6	164	∞
Valeric	$CH_3(CH_2)_3COOH$	–34	187	3.7
Caproic	$CH_3(CH_2)_4COOH$	–3	205	1.0
Caprylic	$CH_3(CH_2)_6COOH$	16	239	0.7
Capric	$CH_3(CH_2)_8COOH$	31	269	0.2
Lauric	$CH_3(CH_2)_{10}COOH$	44	225^{100}	i.
Myristic	$CH_3(CH_2)_{12}COOH$	54	251^{100}	i.
Palmitic	$CH_3(CH_2)_{14}COOH$	63	269^{100}	i.
Stearic	$CH_3(CH_2)_{16}COOH$	70	287^{100}	i.
Oleic	*cis*-9-Octadecenoic	16	223^{10}	i.
Linoleic	*cis,cis*-9,12-Octadecadienoic	–5	230^{16}	i.
Linolenic	*cis,cis,cis*-9,12,15-Octadecatrienoic	–11	232^{17}	i.
Cyclohexanecarboxylic	*cyclo*-$C_6H_{11}COOH$	31	233	0.20
Phenylacetic	$C_6H_5CH_2COOH$	77	266	1.66
Benzoic	C_6H_5COOH	122	250	0.34
o-Toluic	*o*-$CH_3C_6H_4COOH$	106	359	0.12
m-Toluic	*m*-$CH_3C_6H_4COOH$	112	263	0.10
p-Toluic	*p*-$CH_3C_6H_4COOH$	180	275	0.03
o-Chlorobenzoic	*o*-ClC_6H_4COOH	141		0.22
m-Chlorobenzoic	*m*-ClC_6H_4COOH	154		0.04
p-Chlorobenzoic	*p*-ClC_6H_4COOH	242		0.009
o-Bromobenzoic	*o*-BrC_6H_4COOH	148		0.18
m-Bromobenzoic	*m*-BrC_6H_4COOH	156		0.04

TABLE 2.24 Properties of Carboxylic Acids (*Continued*)

Name	Formula	M.p., °C	B.p., °C	Solub., g/100 g H_2O
p-Bromobenzoic	*p*-BrC_6H_4COOH	254		0.006
o-Nitrobenzoic	*o*-$O_2NC_6H_4COOH$	147		0.75
m-Nitrobenzoic	*m*-$O_2NC_6H_4COOH$	141		0.34
p-Nitrobenzoic	*p*-$O_2NC_6H_4COOH$	242		0.03
Phthalic	*o*-$C_6H_4(COOH)_2$	231		0.70
Isophthalic	*m*-$C_6H_4(COOH)_2$	348		0.01
Terephthalic	*p*-$C_6H_4(COOH)_2$	300 *subl.*		0.002
Salicylic	*o*-HOC_6H_4COOH	159		0.22
p-Hydroxybenzoic	*p*-HOC_6H_4COOH	213		0.65
Anthranilic	*o*-$H_2NC_6H_4COOH$	146		0.52
m-Aminobenzoic	*m*-$H_2NC_6H_4COOH$	179		0.77
p-Aminobenzoic	*p*-$H_2NC_6H_4COOH$	187		0.3
o-Methoxybenzoic	*o*-$CH_3OC_6H_4COOH$	101		0.5
m-Methoxybenzoic	*m*-$CH_3OC_6H_4COOH$	110		
p-Methoxybenzoic (Anisic)	*p*-$CH_3OC_6H_4COOH$	184		0.04

TABLE 2.25 The Structure, Melting Point, and Boiling Points of Polycyclic Aromatic Hydrocarbons

Structure	IUPAC nomeclature (synonyms)	Molecular weight	Melting point (°C)	Boiling point $(°C)^{760}$
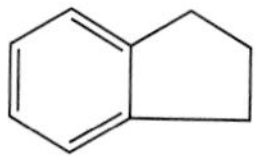	Indan Hydrindene 2,3-Dihydroindene	118.18	–51	178
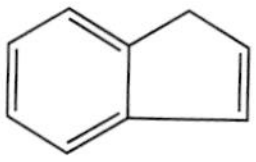	Indene Indonaphthene	116.16	–2	183
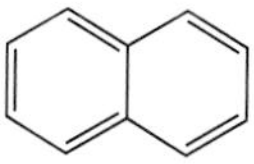	Naphthalene Tar Camphor White Tar Moth Flakes	128.19	81	218
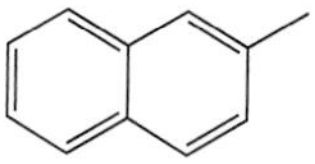	2-Methylnaphthalene β-Methylnaphthalene	142.20	35	241
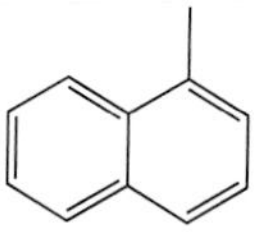	1-Methylnaphthalene α-Methylnaphthalene	142.20	–22	245
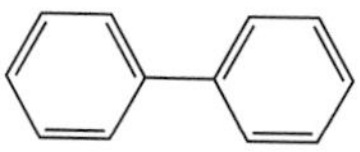	Biphenyl Diphenyl Phenylbenzene Bibenzene	154.21	71	255
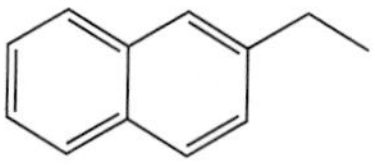	2-Ethylnaphthalene β-Ethylnaphthalene	156.23	–7	258

TABLE 2.25 The Structure, Melting Point, and Boiling Points of Polycyclic Aromatic Hydrocarbons (*Continued*)

Structure	IUPAC nomeclature (synonyms)	Molecular weight	Melting point (°C)	Boiling point (°C)^{760}
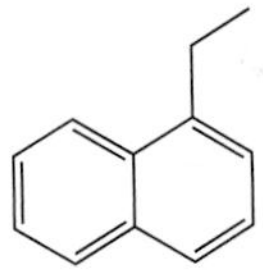	1-Ethylnaphthalene	156.23	–14	259
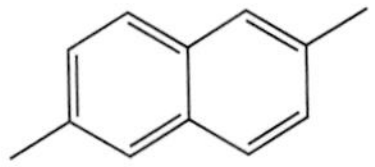	2,6-Dimethylnaphthalene	156.23	110	262
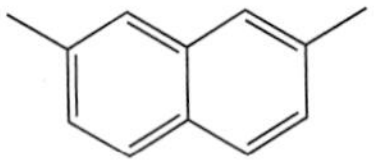	2,7-Dimethylnaphthalene	156.23	97	262
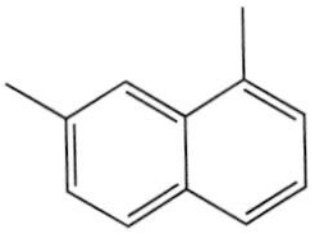	1,7-Dimethylnaphthalene	156.23		263
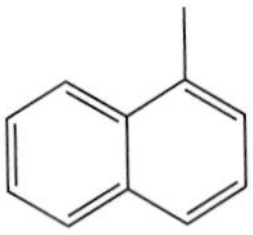	1,3-Dimethylnaphthalene	156.23		265
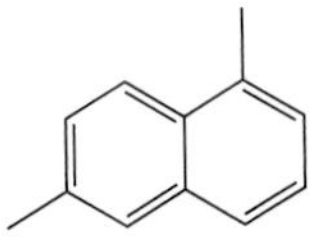	1,6-Dimethylnaphthalene	156.23		266
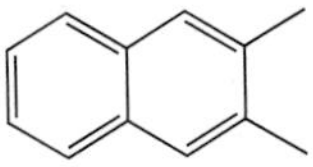	2,3-Dimethylnaphthalene Guaiene	156.23	105	268
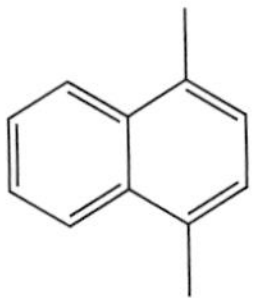	1,4-Dimethylnaphthalene α-Dimethylnaphthalene	156.23	8	268
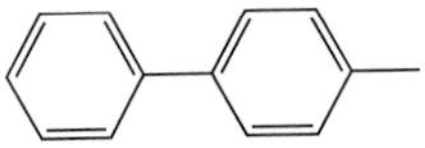	4-Methylbiphenyl	168.24	50	268
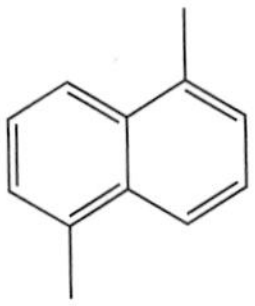	1,5-Dimethylnaphthalene	156.23	80	269

TABLE 2.25 The Structure, Melting Point, and Boiling Points of Polycyclic Aromatic Hydrocarbons (*Continued*)

Structure	IUPAC nomeclature (synonyms)	Molecular weight	Melting point (°C)	Boiling point (°C)^{760}
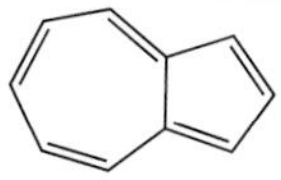	Azulene	128.19	100	270 d
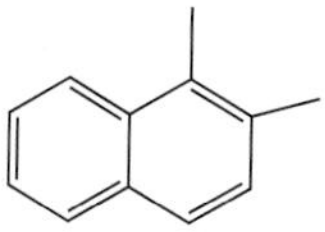	1,2-Dimethylnaphthalene	156.23	–4	271
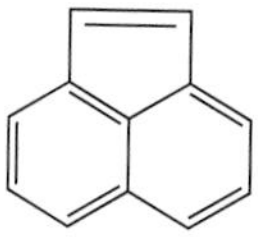	Acenaphthylene	152.21	93	–270 d
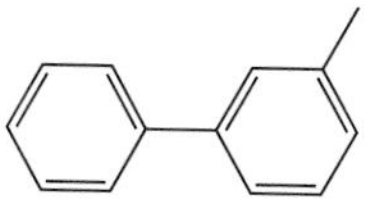	3-Methylbiphenyl	168.24	5	273
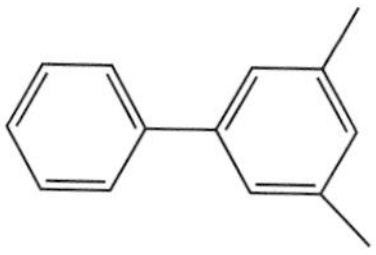	3,5-Dimethylbiphenyl	182.27		275
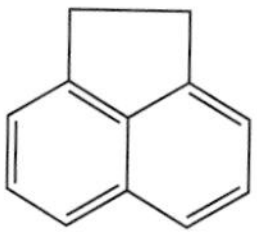	Acenaphthene Naphthyleneethylene	154.21	96	279
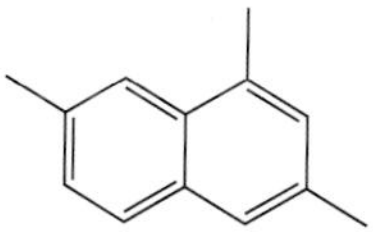	1,3,7-Trimethylnaphthalene	170.25	14	280
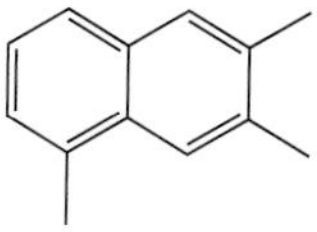	2,3,5-Trimethylnaphthalene	170.25	25	285
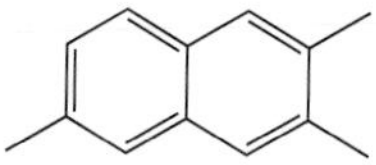	2,3,6-Trimethylnaphthalene	170.25	101	286
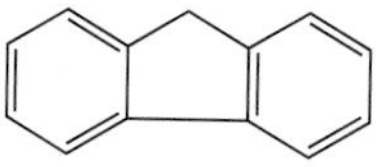	Fluorene 2,3-Benzindene Diphenylenemethane	166.23	117	294
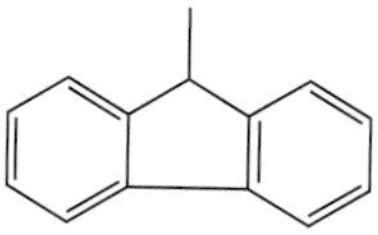	9-Methylfluorene	180.25	47	

(*Continued*)

TABLE 2.25 The Structure, Melting Point, and Boiling Points of Polycyclic Aromatic Hydrocarbons (*Continued*)

Structure	IUPAC nomeclature (synonyms)	Molecular weight	Melting point (°C)	Boiling point (°C)^{760}
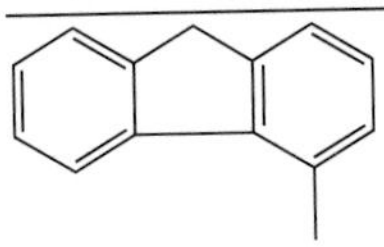	4-Methylfluorene	180.25		
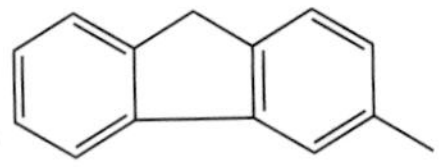	3-Methylfluorene	180.25	85	316
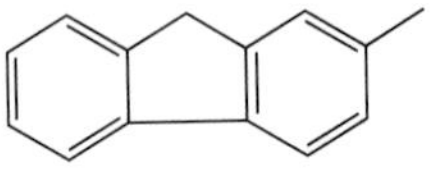	2-Methylfluorene	180.25	104	318
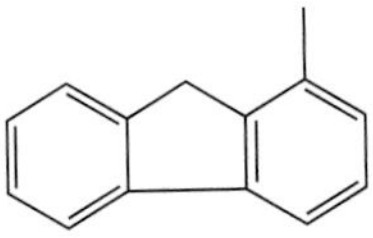	1-Methylfluorene	180.25		–318
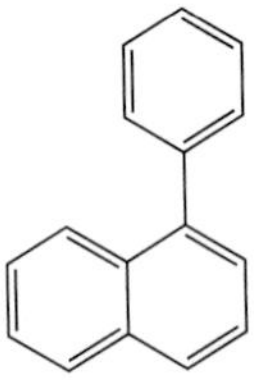	1-Phenylnaphthalene α-Phenylnaphthalene	204.28	–45	334
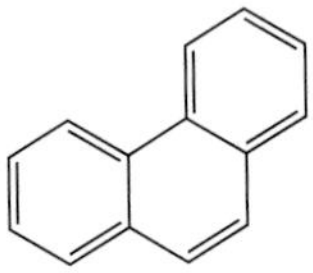	Phenanthrene o-Diphenyleneethylene	178.24	101	338
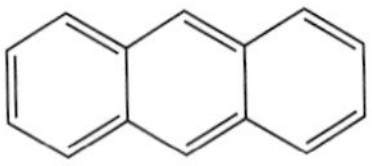	Anthracene	178.24	216	340
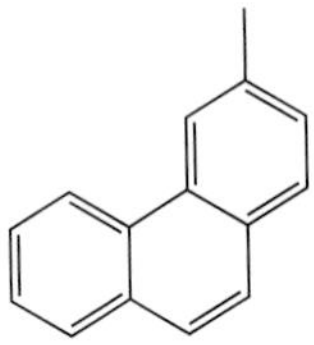	3-Methylphenanthrene	192.26	65	352
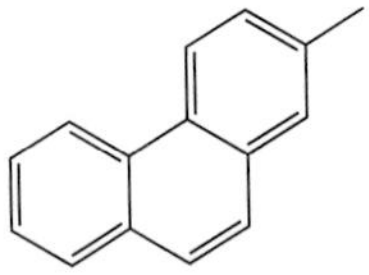	2-Methylphenanthrene	192.26		355

TABLE 2.25 The Structure, Melting Point, and Boiling Points of Polycyclic Aromatic Hydrocarbons (*Continued*)

Structure	IUPAC nomeclature (synonyms)	Molecular weight	Melting point (°C)	Boiling point (°C)[760]
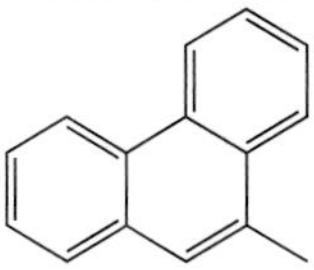	9-Methylphenanthrene	192.26	92	355
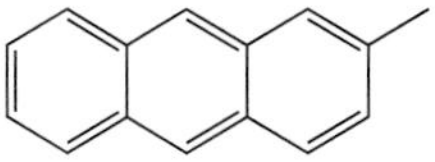	2-Methylanthracene	192.26	209	359 sub
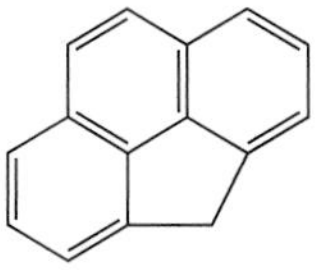	4,5-Methylenephenanthrene 4H-Cyclopenteno[def]phenanthrene 4H-Cyclopenta[def]phenanthrene 4,5-Phenanthrylenemethane	190.24	116	359
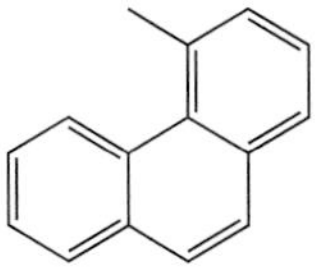	4-Methylphenanthrene	192.26		
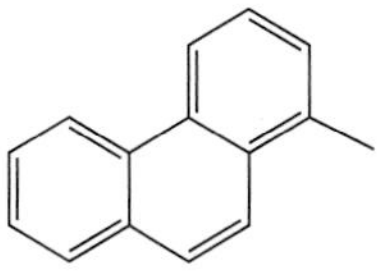	1-Methylphenanthrene	192.26	123	359
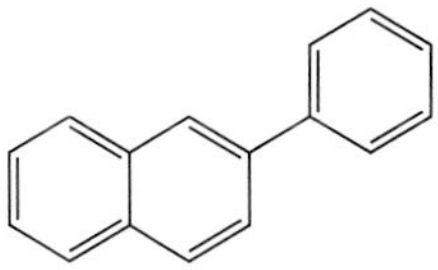	2-Phenylnaphthalene β-Phenylnaphthalene	204.28	104	360
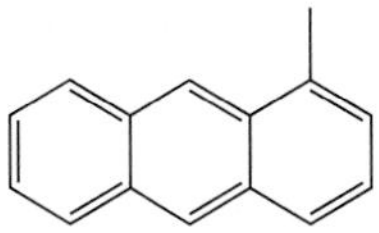	1-Methylanthracene	192.26	86	363
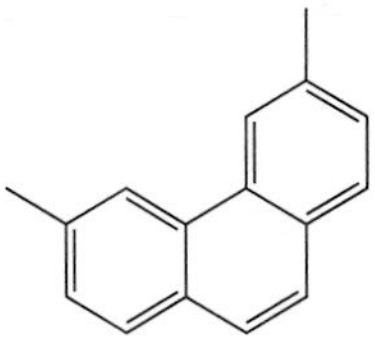	3,6-Dimethylphenanthrene	206.29		363
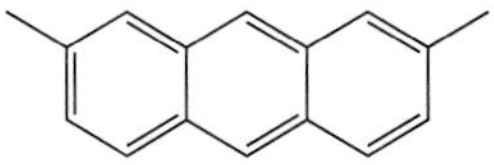	2,7-Dimethylanthracene	206.29	241	–370

(*Continued*)

TABLE 2.25 The Structure, Melting Point, and Boiling Points of Polycyclic Aromatic Hydrocarbons (*Continued*)

Structure	IUPAC nomeclature (synonyms)	Molecular weight	Melting point (°C)	Boiling point (°C)[760]
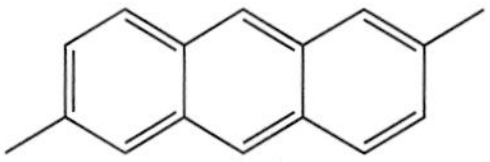	2,6-Dimethylanthracene	206.29	250	–370
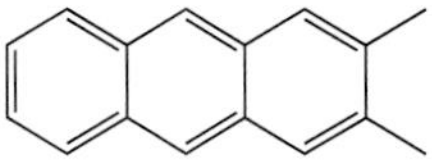	2,3-Dimethylanthracene	206.29	252	
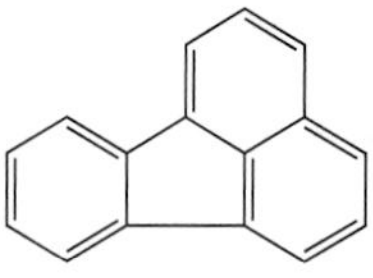	Fluoranthene Idryl 1,2-Benzacenaphthene Benzo[jk]fluorine Benz[a]acenaphthylene	202.26	111	383
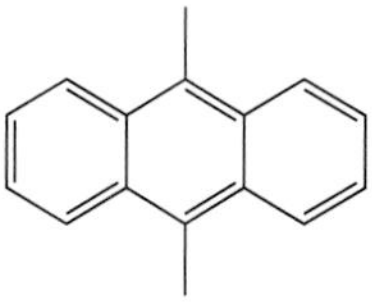	9,10-Dimethylanthracene	206.29	183	
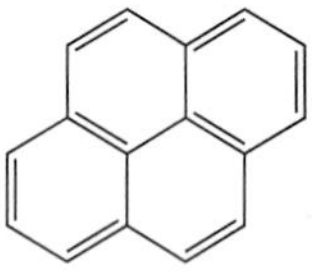	Pyrene Benzo[def]phenanthrene	202.26	156	393
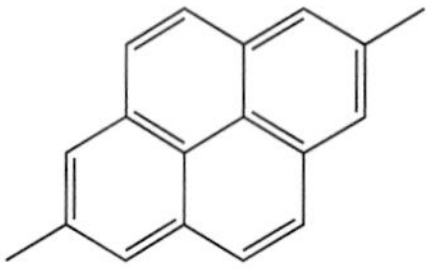	2,7-Dimethylpyrene	230.32		396
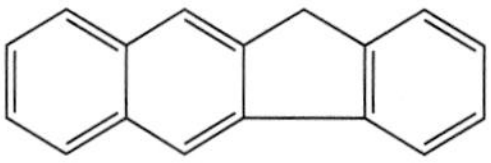	Benzo[b]fluorene 11 H-Benzo[b]fluorene 2,3-Benzofluorene Isonaphthofluorene	216.29	209	402
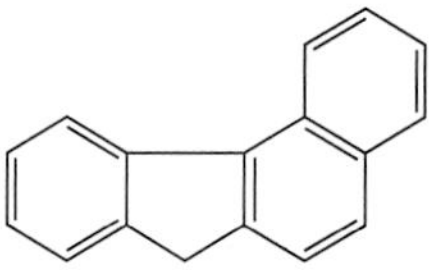	Benzo[c]fluorene 7H-Benzo[c]fluorene 3,4-Benzofluorene	216.29		406
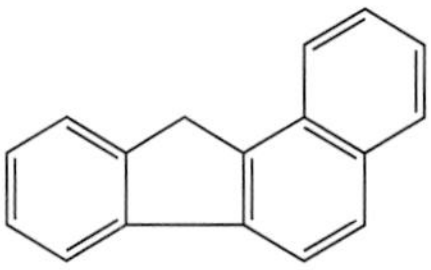	Benzo[a]fluorene 11 H-Benzo[a]fluorene 1,2-Benzofluorene Chrysofluorene	216.29	190	407
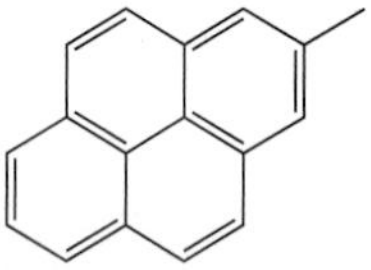	2-Methylpyrene 4-Methylpyren	216.29		410

TABLE 2.25 The Structure, Melting Point, and Boiling Points of Polycyclic Aromatic Hydrocarbons (*Continued*)

Structure	IUPAC nomeclature (synonyms)	Molecular weight	Melting point (°C)	Boiling point $(°C)^{760}$
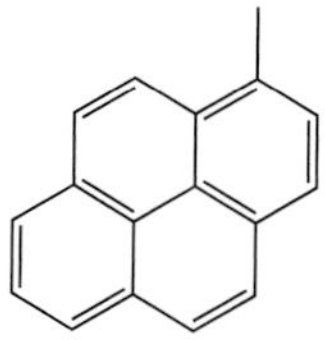	1-Methylpyrene 3-Methylpyren	216.29		410
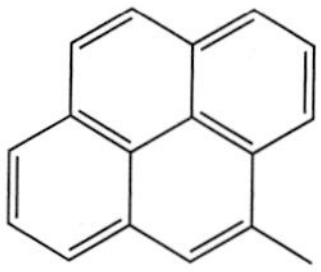	4-Methylpyrene 1-Methylpyren	216.29		410
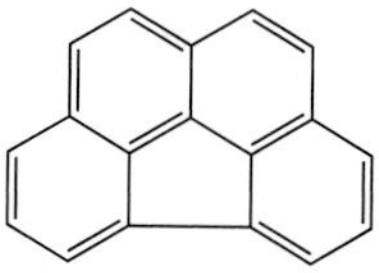	Benzo[ghi]fluoranthene	226.28		432
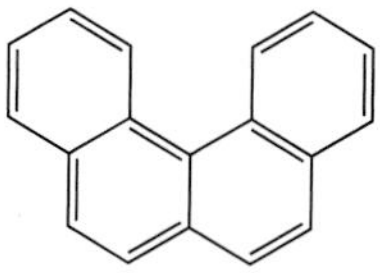	Benzo[c]phenanthrene 3,4-Benzophenanthrene	238.30	68	
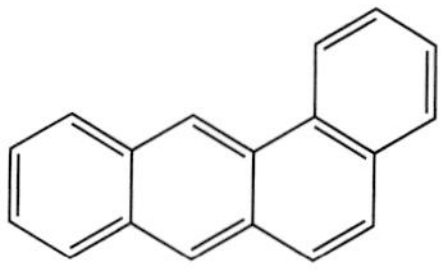	Benz[a]anthracene 1,2-Benzanthracene Tetraphene 2,3-Benzophenanthrene Naphthanthracene	228.30	162	435 sub
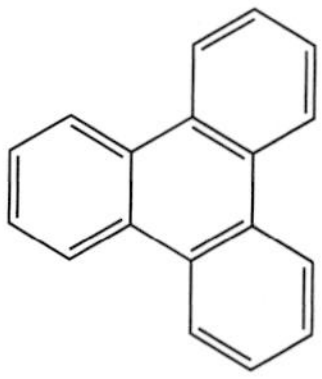	Triphenylene 9,10-Benzophenanthrene lsochrysene	228.30	199	439
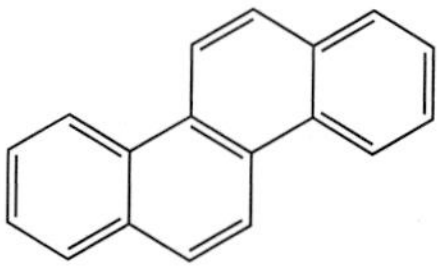	Chrysene 1,2-Benzophenanthrene Benzo[a]phenanthrene	228.30	256	441
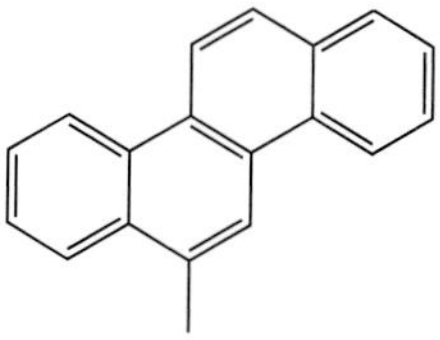	6-Methylchrysene	242.32		

(*Continued*)

TABLE 2.25 The Structure, Melting Point, and Boiling Points of Polycyclic Aromatic Hydrocarbons (*Continued*)

Structure	IUPAC nomeclature (synonyms)	Molecular weight	Melting point (°C)	Boiling point (°C)^{760}
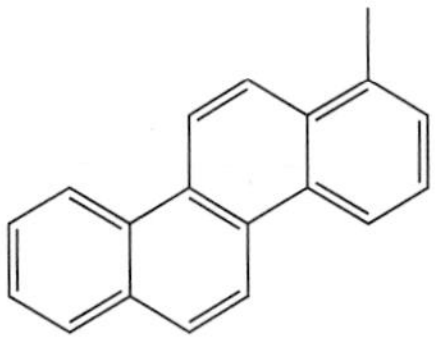	1-Methylchrysene	242.32	257	
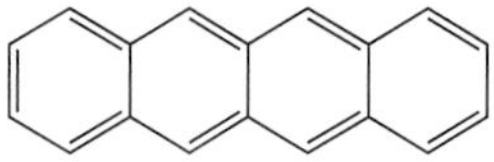	Naphthacene Benz[b]anthracene 2,3-Benzanthracene Tetracene	228.30	257	450 sub
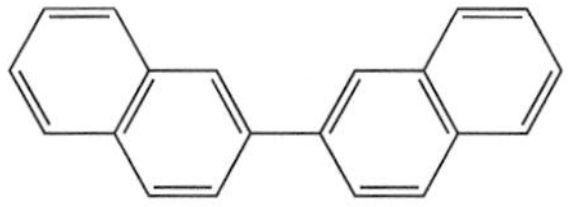	2,2′-Dinaphthyl 2,2′-Binaphthyl *β,β′*-Binaphthyl *β,β′*-Dinaphthyl	254.34	188	452^{753} sub
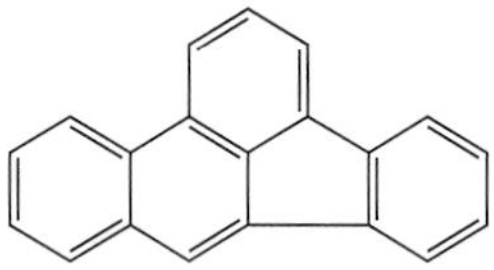	Benzo[b]fluoranthene 2,3-Benzofluoranthene 3,4-Benzofluoranthene Benz[e]acephenanthrylene	252.32	168	481
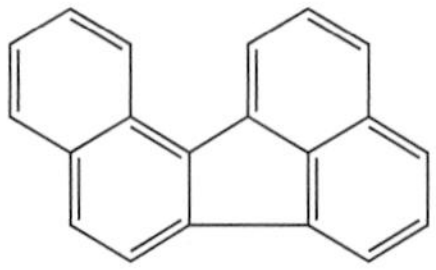	Benzo[j]fluoranthene 7,8-Benzofluoranthene 10,11-Benzofluoranthene	252.32	166	~480
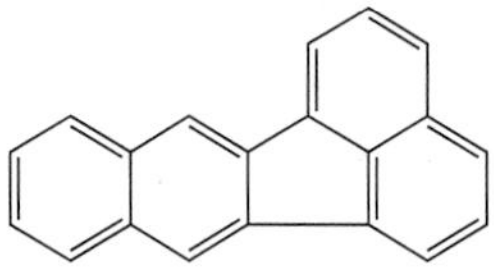	Benzo[k]fluoranthene 8,9-benzofluoranthene 11,12-Benzofluoranthene	252.32	217	481
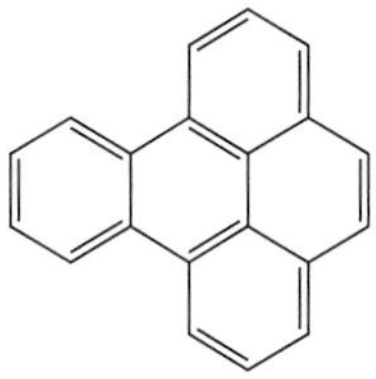	Benzo[e]pyrene 4,5-Benzpyrene 1,2-Benzopyrene	252.32	179	493
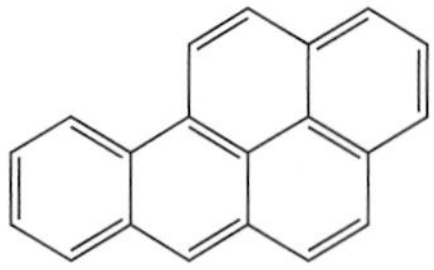	Benzo[a]pyrene 1,2-Benzpyrene 3,4-Benzopyrene Benzo[def]chrysene	252.32	177	496

TABLE 2.25 The Structure, Melting Point, and Boiling Points of Polycyclic Aromatic Hydrocarbons (*Continued*)

Structure	IUPAC nomeclature (synonyms)	Molecular weight	Melting point (°C)	Boiling point $(°C)^{760}$
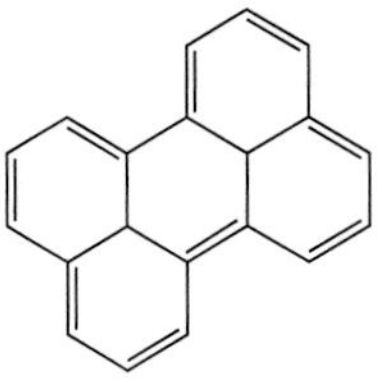	Perylene peri-Dinaphthalene	252.32	278	
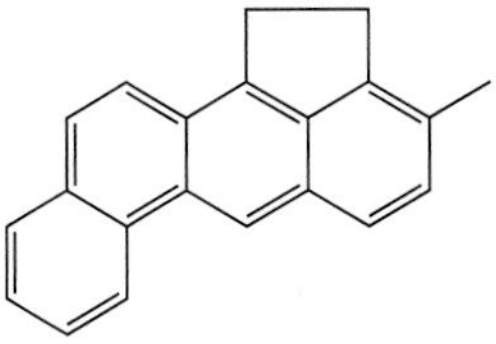	3-Methylcholanthrene 20-Methylcholanthrene	268.38	180	
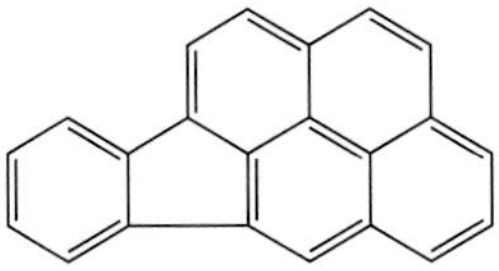	Indeno[1,2,3-cd]pyrene o-Phenylenepyrene	276.34		
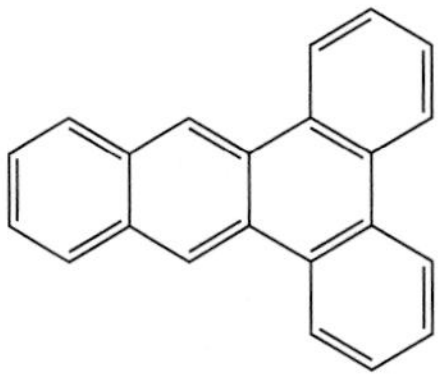	Dibenz[a,c]anthracene 1,2:3,4-Dibenzanthracene Naphtho-2′,3′,:9,10-phenanthrene	278.36	205	
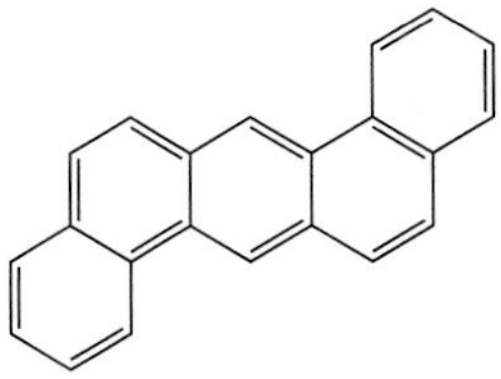	Dibenz[a,h]anthracene 1,2:5,6-Dibenzanthracene	278.36	270	
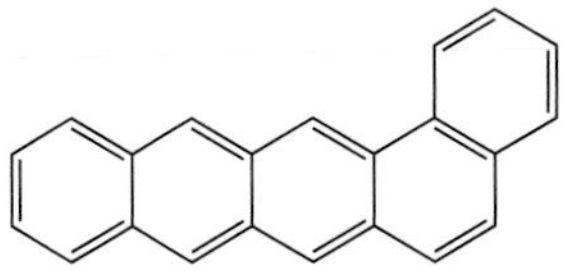	Dibenz[a,i]anthracene 1,2:6,7-Dibenzanthracene 1,2-Benzonaphthacene Isopentaphene	278.36	264	
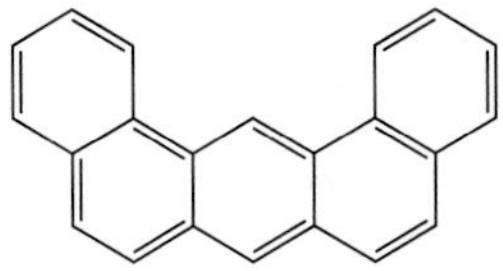	Dibenz[a,j]anthracene 1,2:7,8-Dibenzanthracene α,α'-Dibenzanthracene Dinaphthanthracene	278.36	198	

(*Continued*)

TABLE 2.25 The Structure, Melting Point, and Boiling Points of Polycyclic Aromatic Hydrocarbons (*Continued*)

Structure	IUPAC nomeclature (synonyms)	Molecular weight	Melting point (°C)	Boiling point (°C)[760]
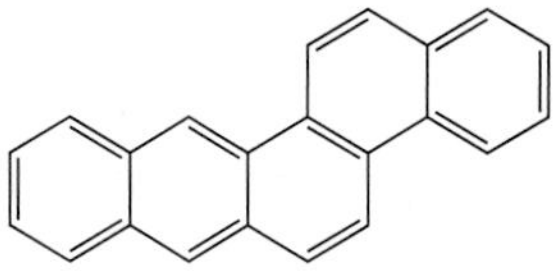	Benzo[b]chrysene 1,2:6,7-Dibenzophenanthrene 3,4-Benzotetraphene Naphtho-2′,1′:1,2-anthracene	278.36	294	
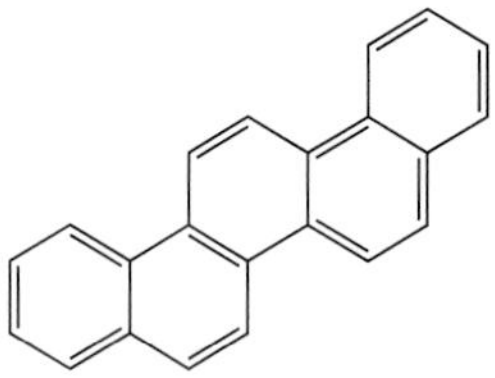	Picene Dibenzo[α;i]phenanthrene 3,4-Benzochrysene 1,2:7,8-Dibenzophenanthrene	278.36	368	519
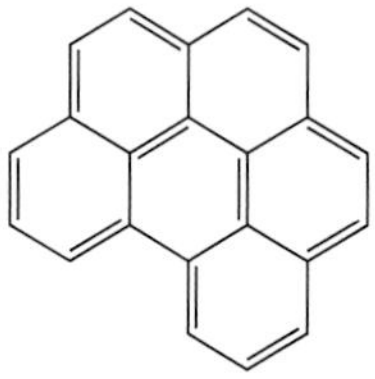	Benzo[ghi]perylene 1,12-Benzoperylene	276.34	278	
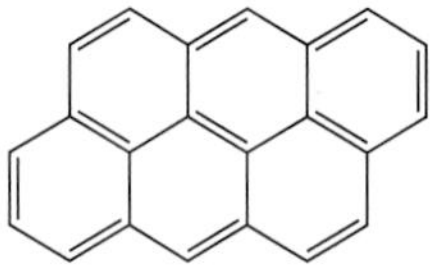	Anthanthrene Dibenzo[def, mno]chrysene	276.34		
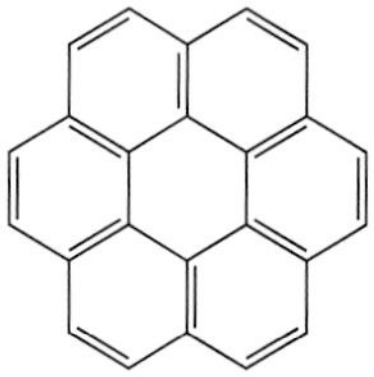	Coronene Hexabenzobenzene	300.36	439 cor	525?
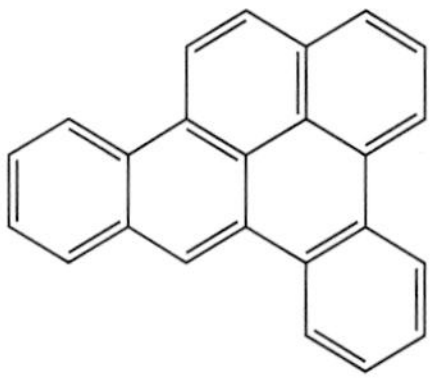	Dibenzo[a,e]pyrene	302.38	234	

*Key: d = decomposes; sub = sublimes.

TABLE 2.26 Properties of Naturally Occurring Amino Acids

Name	Three-letter code	One-letter code	Side chains (–*R*) *R*-CH(NH$_2$)COOH	Mol weight	pK_a	ΔH_{ion} kJ · mol^{-1}	Volume Å^3	ASA$_{mc}$ Å^2	ASA$_{sc}^{npl}$ Å^2	ASA$_{sc}^{pol}$ Å^2
Alanine	Ala	A	-CH$_3$	71.08			88.6	46	67	
Arginine	Arg	R	-(CH$_2$)$_3$-CNH(=NH)NH$_3$	156.20	12	44.9	173.4	45	89	107
Asparagine	Asn	N	-CH$_2$-CONH$_2$	114.11			117.7	45	44	69
Aspartic acid	Asp	D	-CH$_2$-COOH	115.09	4.5	4.6	111.1	45	48	58
Cystein	Cys	C	-CH$_2$-SH	103.14	9.1–9.5	36.0	108.5	36	35	69
Glutamine	Gln	Q	-(CH$_2$)$_2$-CONH$_2$	128.14			143.9	45	53	91
Glutamic acid	Glu	E	-(CH$_2$)$_2$-COOH	129.12	4.6	1.6	138.4	45	61	77
Glycine	Gly	G	-H	57.06			60.1	85		
Histidine	His	H	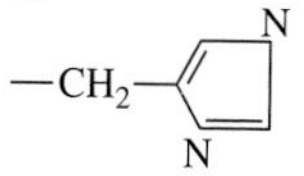	137.15	6.2	43.6	153.2	43	102	49
Isoleucine	Ile	I	-CH(CH$_3$)-C$_2$H$_5$	113.17			166.7	42	140	
Leucine	Leu	L	-CH(CH$_3$)$_2$-CH$_2$	113.17			166.7	43	137	
Lysine	Lys	K	-(CH$_2$)$_4$-NH$_2$	128.18	10.4	53.6	168.6	44	119	48
Methionine	Met	M	-(CH$_2$)$_2$-S-CH$_3$	131.21			162.9	44	117	43
Phenylalanine	Phe	F	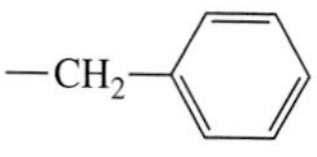	147.18			189.9	43	175	
Proline	Pro	P	*	97.12			122.7	38	105	
Serine	Ser	S	-CH$_2$-OH	87.08			89.0	42	44	36
Threonine	Thr	T	-CH$_2$-(CH$_3$)-OH	101.11			116.1	44	74	28
Tryptophane	Trp	W	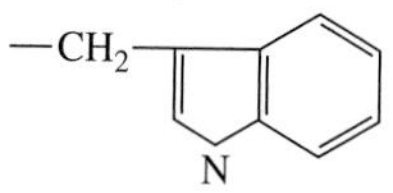	186.21			277.8	42	190	27
Tyrosine	Tyr	Y	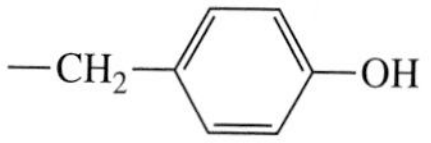	163.18	9.7	25.1	193.6	42	144	43
Valine	Val	V	-CH-(CH$_3$)$_2$	99.14			140	43	117	
			α-amino		6.8–7.9					
			α-carboxyl		3.5–4.3					

[a]Enthalpies of ionization of side chains at 25°C, ΔH_{ion}, are from [20]; van der Waals volume from [21]; ASA$_{mc}$, surface area of the backbone, ASA$_{sc}^{npl}$, nonpolar surface area of the side chains, and ASA$_{sc}^{pol}$, polar surface area of the side chains are taken [17].

TABLE 2.27 Hildebrand Solubility Parameters of Organic Liquids

Solvent	δ (Mpa$^{1/2}$)	H-bonding tendency[b]	Solvent	δ (Mpa$^{1/2}$)	H-bonding tendency[b]
Acetaldehyde	21.1	m	Ethyl chloride	18.8	m
Acetic acid	20.7	s	Ethylenediamine	25.2	s
Acetone	20.2	m	Ethylene dichloride	20.0	p
Acetonitrile	24.3	p	Ethylene glycol	29.9	s
Acetyl chloride	19.4	m	Ethylene glycol dimethylether	17.6	m
N-Acetylpiperidine	22.9	s			
Acrylic acid	24.5	s	Ethylene oxide	22.7	m
Allyl acetate	18.8	m	Ethyl formate	19.2	m
Allyl alcohol	24.1	s	Ethyl methacrylate	17.0	m
Ammonia	33.3	s	Formic acid	24.7	s
Benzene	18.8	p	Furan	19.2	m
Bromobenzene	20.2	p	Heptane	15.1	p
1,3-Butadiene	14.5	p	Hexane	14.9	p
Butane	13.9	p	1-Hexene	15.1	p
1,3-Butanediol	23.7	s	Hydrazine	37.0	s
1-Butanol	23.3	s	Hydrogen	6.1	p
2-Butanol	22.1	s	Isobutanol	21.5	s
tert-Butanol	21.7	s	Isobutyl acetate	17.0	m
Butyl acetate	17.4	m	Isobutylene	13.7	p
Butyl amine	17.8	s	Isoprene	15.1	p
Butyl ether	16.0	m	Isopropanol	23.5	s
Butyl lactate	19.2	m	Isopropyl acetate	17.2	m
Carbon disulfide	20.4	p	Methane	11.0	p
Chloroacetonitrile	25.8	p	Methanol	29.6	s
Chlorobenzene	19.4	p	Methyl acetate	19.6	m
Chloroethane	18.8	m	Methyl acrylate	18.2	m
Chloromethane	19.8	m	Methyl butyl ketone	17.0	m
Cyclohexane	16.8	p	Methyl ethyl ketone	19.0	m
Cyclohexanol	23.3	s	Methyl formate	20.9	m
Cyclopentane	17.8	p	Methyl isopropyl ketone	17.4	m
Decalin	18.0	p	Methyl methacrylate	18.0	m
Decane	13.5	p	Nitrobenzene	20.5	p
Diamyl ether	14.9	m	Nitroethane	22.7	p
Dibenzyl ether	19.2	m	Octane	15.6	p
Dibutyl amine	16.6	s	Pentane	14.3	p
Dibutyl fumarate	18.4	m	Propane	13.1	p
Dibutyl phenyl phosphate	17.8	m	1-Propanol	24.3	s
Dibutyl phthalate	19.0	m	2-Propanol	23.5	s
Diethylamine	16.4	s	Pyridine	21.9	s
Diethlene glycol	24.8	s	Quinoline	22.1	s
Diethyl ether	15.1	m	Silicon tetrachloride	15.1	p
Diisopropyl ether	14.1	m	Styrene	19.0	p
Diisopropyl ketone	16.4	m	Succinic anhydride	31.5	s
N,N-Dimethylformamide	24.8	m	Tetra chloromethane	17.6	p
Dimethyl sulfone	29.7	m	Tetrahydrofuran	18.6	m
Dimethylsulfoxide	24.5	m	Toluene	18.2	p
1,4-Dioxane	20.5	m	1,1,2-Trichloroethane	19.6	p
Ethane	12.3	p	Trichloromethane	19.0	p
Ethanol	26.0	s	Water	47.9	s
Ethyl acetate	18.6	m	Xylene	18.0	p
Ethylamine	20.5	s			
Ethylbenzene	18.0	p			

[b] p denotes poor; m, moderate; s, strong.

TABLE 2.28 Hansen Solubility Parameters of Organic Liquids

Solvent	V (cm^3/mol)	Solubility parameter (MPa$^{1/2}$)			
		δ_d	δ_p	δ_h	δ_t
Acetic acid	57.1	14.5	8.0	13.5	21.3
Acetone	74.0	15.5	10.4	7.0	20.1
Acetonitrile	52.6	15.3	18.0	6.1	24.6
Acetyl chloride	71.0	15.8	10.6	3.9	19.4
Benzene	29.4	18.4	0.0	2.0	18.6
Benzaldehyde	101.5	19.4	7.4	5.3	21.5
Benzyl chloride	115.0	18.8	7.2	2.7	20.3
Bromoform	87.5	21.5	4.1	6.1	22.7
N-Butane	101.4	14.1	0.0	0.0	14.1
Butyronitrile	27.0	15.3	12.5	5.1	20.5
Carbon tetrachloride	97.1	17.8	0.0	0.6	17.8
Carbon disulfide	60.0	20.5	0.0	0.6	20.5
Chlorobenzene	102.1	19.0	4.3	2.0	19.6
Chloroform	80.7	17.8	3.1	5.7	19.0
Cyclohexanol	106.0	17.4	4.1	13.5	22.5
Cyclohexylamine	115.2	17.4	3.1	6.5	18.8
N-Decane	195.9	15.8	0.0	0.0	15.8
Diacetone alcohol	124.2	15.8	8.2	4.8	20.9
o-Dichlorobenzene	112.8	19.2	6.3	3.3	20.5
Diethyl carbonate	121.0	16.6	3.1	6.1	18.0
Diethyl ketone	106.4	15.8	7.6	4.7	18.2
Dimethyl phthalate	163.0	18.6	4.8	4.9	22.1
Dimethyl sulfoxide	71.3	18.4	16.4	10.2	26.6
Ethanol	58.5	15.8	8.8	19.4	26.6
Ethyl acetate	98.5	15.8	5.3	7.2	18.2
Ethyl bromide	76.9	16.6	8.0	5.1	19.0
Ethyl formate	80.2	15.5	8.4	8.4	19.6
Ethylene carbonate	66.0	19.4	21.7	5.1	29.5
Ethylene dichloride	79.4	19.0	7.4	4.1	20.9
Formic acid	37.8	14.3	11.9	16.6	25.0
Furan	72.5	17.8	1.8	5.3	18.6
Methanol	40.7	15.1	12.3	22.3	29.7
Methyl acetate	79.7	15.5	7.2	7.6	18.8
Methyl chloride	55.4	15.3	6.1	3.9	17.0
Methylene dichloride	63.9	18.2	6.3	6.1	20.3
Nitrobenzene	102.7	20.1	8.6	4.1	22.1
Nitroethane	71.5	16.0	15.5	4.5	22.7
Nitromethane	54.3	15.8	18.8	5.1	25.0
1-Octanol	157.7	17.0	3.3	11.9	20.9
2-Octanol	159.1	16.2	4.9	11.0	20.3
Phenol	87.5	18.0	5.9	14.9	24.1
1-Propanol	75.2	16.0	6.8	17.4	24.6
2-Propanol	76.8	15.8	6.1	16.4	23.5
Quinoline	118.0	19.4	7.0	7.6	22.1
Styrene	115.6	18.6	1.0	4.1	19.0
Tetrahydrofuran	81.7	16.8	5.7	8.0	19.4
Toluene	106.8	18.0	1.4	2.0	18.2
Trimethyl phosphate	99.9	16.8	16.0	10.2	25.4
Water	18.0	15.5	16.0	42.4	47.9

TABLE 2.29 Group Contributions to the Solubility Parameter

	F_i				F_i		
Group	(1)	(2)	(3)	Group	(1)	(2)	(3)
—Br	340	258	300	—C≡N	410	355	480
—Cl	250–270	205	230				
—F	…	41	80	—C(=O)—NH_2	…	…	600
—H	80–100	…	…				
—I	425	…	…	NH_2—C(=O)—O—	…	…	725
—NO_2	440	…	…	—CO—	275	263	335
—ONO_2	440	…	…	—COO—	310	327	250
—O—	70	115	125	—COOH	…	…	319
—OH	…	226	369	—CO_3—	…	…	375
—PO_4	500	…	…	—C≡C—	222	…	…
—S—	225	209	225	CH≡C—	285	…	…
—SH	315	…	…				
>C<	−93	32	0	—C(=O)—O—C(=O)—	…	567	375
—CH=	19	84	40	—C=C—C=C—	20–30	23	…
—CF_2—	150	115	…				
—CF_3	274	156	…	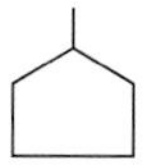	105–115	21	…
—CH<	28	86	68	—C_6H_4	658	705	673
—CH=	111	122	109	—C_6H_5	735	683	741
—CH_2—	133	131	137				
CH_2<	190	127	…	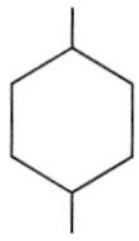	95–105	−23	…
—CH_3	214	148	205	—$C_{10}H_7$	1146	…	…

[a] Adapted from D. W. Van Krevelen, *Properties of Polymers*, 2nd ed. (Elsevier, Amsterdam, 1976), p. 134. The references referred to for the F_i values are (1) P.A. Small, J. Appl. Chem. **3**, 71 (1953); (2) K. L. Hoy, J. Paint Technol. **42**, 76 (1970); (3) D. W. Van Krevelen, *Properties of Polymers*, 2nd ed. (Elsevier, Amsterdam, 1976), p. 134.

2.3 VISCOSITY AND SURFACE TENSION

The *dynamic viscosity*, or coefficient of viscosity, η of a Newtonian fluid is defined as the force per unit area necessary to maintain a unit velocity gradient at right angles to the direction of flow between two parallel planes a unit distance apart. The SI unit is pascal-second or netwon-second per meter squared [$N \cdot s \cdot m^{-2}$]. The c.g.s. unit of viscosity is the poise [P]; 1 cP≡1 $mN \cdot s \cdot m^{-2}$.

Kinematic viscosity v is the ratio of the dynamic viscosity to the density of a fluid. The SI unit is meter squared per second [$m^2 \cdot s^{-1}$]. The c.g.s. units are called stokes [$cm^2 \cdot s^{-1}$]; poises = stokes × density.

Fluidity ϕ is the reciprocal of the dynamic viscosity.

The primary reference liquid for viscosity measurements is water. The absolute viscosity of water at 20°C is 1.0019 (±0.0003) mN · s · m^{-2} (or centipoise), as determined by Swindells, Coe, and Godfrey, *J. Research Natl. Bur. Standards* **48:**1 (1952). The relative viscosity of water, $\eta/\eta_{20°}$, is 0.8885 at 25°C, 0.7960 at 30°C, and 0.6518 at 40°C. Values at temperatures between 15 and 60°C are best represented by Cragoe's equation:

$$\log \frac{\eta}{\eta_{20°}} = \frac{1.2348(20-t) - 0.001467(t-20)^2}{t+96}$$

The *Reynolds number* for flow in a tube is defined by $d\bar{v}\rho/\eta$, where d is the diameter of the tube, $\bar{v}$ is the average velocity of the fluid along the tube, ρ is the density of the fluid, and η is its dynamic viscosity. At flow velocities corresponding with values of the Reynolds number of greater than 2000, turbulence is encountered.

The surface tension of a liquid, γ, is the force per unit length on the surface that opposes the expansion of the surface area. In the literature the surface tensions are expressed in dyn · cm^{-1}; 1 dyn · cm^{-1} = 1 mN · m^{-1} in the SI system. For the large majority of compounds the dependence of the surface tension on the temperature can be given as

$$\gamma = a - bt$$

where a and b are constants and t is the temperature in degrees Celsius. The values of a and b given in Tables 2.30 can be used to calculate the values of surface tension for the particular compound within its liquid range. For example, the least-squares constants for acetic anhydride (liquid from −73 to 140°C) are 35.52 and 0.1436, respectively. At 20°C, $\gamma = 35.52 - 0.1436(20) =$ 32.64 dyn · cm^{-1}.

TABLE 2.30 Viscosity and Surface Tension of Organic Compounds

For the majority of substances the dependence of the surface tension γ on the temperature can be given as:

$$\gamma = a - bt$$

where a and b are constants and t is the temperature in degrees Celsius. In the SI system the surface tensions are expressed in $mN \cdot m^{-1}$ (= $dyn \cdot cm^{-1}$).

A compilation of some 2200 liquid compounds has been prepared by J. J. Jasper, *J. Phys. Chem. Reference Data* **1**:841 (1972).

The SI unit of viscosity is pascal-second (Pa · s) or Newton-second per meter squared ($N \cdot s \cdot m^{-2}$). Values tabulated are $mN \cdot s \cdot m^{-2}$ (= centipoise, cP). The temperature in degrees Celsius at which the viscosity of a substance was measured is shown in parentheses after the value.

Substance	Surface tension, $mN \cdot m^{-1}$		Liquid range, °C	Viscosity, $mN \cdot s \cdot m^{-2}$
	a	*b*		
Acetaldehyde	23.90	0.1360	−123 to 21	0.2797(0), 0.2557(10), 0.22(20)
Acetaldoxime	34.23	0.1134	12(β) or 46.5(α) to 114.5	
Acetamide	47.66	0.1021	81 to 222	1.63(94), 1.32(105), 1.06(120)
Acetanilide	46.21	0.0912	114 to 304	2.22(120), 1.90(130)
Acetic acid	29.58	0.0994	16.7 to 118	1.056(25), 0.786(50), 0.424(110)
Acetic anhydride	35.52	0.1436	−73 to 139	1.241(0), 0.907(20), 0.699(40)
Acetone	26.26	0.112	−94 to 56	0.395(0), 0.306(25), 0.256(50)
Acetonitrile	29.58	0.1178	−44 to 81.6	0.397(10), 0.329(30), 0.2753(50)
Acetophenone	41.92	0.1154	20 to 202	1.511(30), 1.192(45), 0.634(100)
Acetyl chloride	26.7(15)		−113 to 51	0.368(25), 0.294(50)
Acrylic acid	28.1(30)		14 to 141	
Acrylonitrile	29.58	0.1178	−83.5 to 77.3	
Allyl acetate	28.73	0.1186	up to 104	
Allyl alcohol	27.53	0.0902	−129 to 97	1.218(25), 0.759(50), 0.553(70)
Allylamine	27.49	0.1287	−88 to 55	
Allyl isothiocyanate	36.76	0.1074	−80 to 152	
2-Aminoethanol	51.11	0.1117	10.3 to 171	
Aniline	44.83	0.1085	−6 to 186	3.847(25), 2.029(50), 1.247(75)
Benzaldehyde	40.72	0.1090	−26 to 179	
Benzamide	47.26	0.0705	129 to 290	
Benzene	28.88(20)	27.56(30)	5.5 to 80	0.649(20), 0.566(30), 0.395(60)
Benzenesulfonyl chloride	45.48	0.1117	14.5 to 251	
Benzenethiol	41.41	0.1202	−14.9 to 169	
Benzonitrile	41.69	0.1159	−12.7 to 191	1.447(15), 1.111(30), 0.883(50)
Benzophenone	46.31	0.1128	48 to 305	
Benzoyl bromide	45.85	0.1397	−24 to 219	
Benzoyl chloride	41.34	0.1084	−1 to 197	
Benzyl alcohol	38.25	0.1381	−15.2 to 205	5.474(25), 2.760(50), 1.618(75)
Benzylamine	42.33	0.1213	10 to 180	1.624(25), 1.080(50), 0.769(75)
Benzyl benzoate	48.07	0.1065	21 to 323	8.454(25)
Benzyl chloride	39.92	0.1227	−43 to 179	

TABLE 2.30 Viscosity and Surface Tension of Organic Compounds (*Continued*)

Substance	Surface tension, mN · m^{-1} *a*	*b*	Liquid range, °C	Viscosity, mN · s · m^{-2}
Benzyl ethyl ether	32.82(20)	29.97(40)	up to 186	
Biphenyl	41.52	0.0931	69 to 256	
Bis(2-ethoxyethyl) ether	29.74	0.1176	−45 to 188	
Bis(2-hydroxyethyl) ether	46.97	0.0880	−10.4 to 246	
Bis(2-methoxyethyl) ether	32.47	0.1164	−68 to 162	
Bromobenzene	38.14	0.1160	−30.6 to 156	1.196(15), 0.985(30), 0.385(1423)
1-Bromobutane	28.71	0.1126	−112.4 to 101.6	0.633(20), 0.606(25), 0.471(50)
(±)-2-Bromobutane	27.48	0.1107	−112.7 to 91.4	
Bromochloromethane	33.32(20)		−88 to 68	
Bromocyclohexane	36.13	0.1117	up to 165.8	
1-Bromodecane	31.26	0.0856	−30 to 240	
Bromodichloromethane	35.11	0.1294	−55 to 87	
1-Bromododecane	32.58	0.0882	−11 to bp	
Bromoethane	26.52	0.1159	−119 to 38.2	0.477(10), 0.374(25)
Bromoform	48.14	0.1308	8 to 149	
1-Bromoheptane	30.74	0.0982	−58 to 180	
1-Bromohexadecane	33.37	0.0861	17.8 to 336	
1-Bromohexane	29.81	0.0967	−85 to 158	
Bromomethane	26.52	0.1159	−94 to 3.56	
1-Bromo-3-methylbutane	28.10	0.0996	−112 to 119.7	
1-Bromo-2-methylpropane	26.96	0.1059	−119 to 91.5	
1-Bromonaphthalene	46.44	0.1018	−1.8 to 281	
1-Bromononane	31.36	0.0894	ca. −55 to 201	
1-Bromooctane	31.00	0.0928	−55 to 201	
1-Bromopentane	29.51	0.1049	−88 to 129.6	
p-Bromophenol	48.88	0.1070	64 to 238	
1-Bromopropane	28.30	0.1218	−110.1 to 71	0.539(15), 0.459(30), 0.338(70)
2-Bromopropane	26.21	0.1183	−89 to 59.5	0.536(15), 0.437(30), 0.359(50)
3-Bromopropene	29.45	0.1257	−119 to 70	0.620(0), 0.471(25), 0.373(50)
1-Bromotetradecane	32.93	0.0878	6 to >178	
o-Bromotoluene	36.62	0.0998	−26 to 181	
p-Bromotoluene	36.40	0.0997	28.5 to 184	
1-Bromoundecane	31.94	0.0861	−9 to >138	
Butanal	26.67	0.0925	−99 to 74.8	
Butane	14.87	0.1206	−138.3 to −0.5	
1,3-Butanediol	37.8(25)		<−50 to 207.5	
2,3-Butanediol	36(25)		25 to 182	
Butanenitrile			−112 to 117.6	0.553(25), 0.418(50), 0.330(75)
Butanesulfonyl chloride	37.33	0.0977		
1-Butanethiol	28.07	0.1142	−116 to 98.5	
Butanoic acid	28.35	0.0920	−6 to 163.5	1.540(20), 0.980(40), 0.323(60)
Butanoic anhydride	28.93(20)	28.44(25)	−66 to 199.5	
1-Butanol	27.18	0.0898	−89.5 to 117.7	5.185(0), 2.948(20), 1.782(40)
(±)-2-Butanol	23.47(20)	22.62(30)	−114.7 to 99.5	3.907(20), 1.332(50), 0.698(75)
2-Butanone	26.77	0.1122	−86.7 to 79.6	0.428(20), 0.349(40), 0.249(75)
1-Butene	15.19	0.1323	−185 to −6.5	
2-Butene	16.11	0.1289	−106 to 0.9	
3-Butenenitrile	31.40	0.1085	−87 to 119	
2-Butoxyethanol	28.18	0.0816	−75 to 168	

(*Continued*)

TABLE 2.30 Viscosity and Surface Tension of Organic Compounds (*Continued*)

Substance	Surface tension, mN · m⁻¹ a	b	Liquid range, °C	Viscosity, mN · s · m⁻²
2-(2-Butoxyethoxy)ethanol	30.0(25)		−68.1 to 230.4	
Butyl acetate	27.55	0.1068	−77 to 126	0.734(20), 0.688(25), 0.500(50)
(±)-*sec*-Butyl acetate	23.33(22)	21.24(42)	−99 to 112	0.676(25), 0.493(50), 0.370(75)
tert-Butyl acetate	24.69	0.1102	up to 98	
Butylamine	26.24	0.1122	−50 to 77	0.830(0), 0.574(25), 0.409(50)
sec-Butylamine	23.75	0.1057	−104 to 63	0.770(0), 0.571(25), 0.367(50)
tert-Butylamine	19.44	0.1028	−66 to 44	
Butylbenzene	31.28	0.1025	−88 to 183	1.035(20), 0.683(50), 0.515(75)
sec-Butylbenzene	30.48	0.0979	−82.7 to 173	
tert-Butylbenzene	30.10	0.0985	−58.1 to 168.5	
Butyl butanoate	27.65	0.0965	−91.5 to 166	
Butyl ethyl ether	22.75	0.1049	−124 to 92	
Butyl formate	27.08	0.1026	−91.5 to 106	0.940(0), 0.691(20), 0.472(50)
Butyl methyl ether	22.17	0.1057	−115.5 to 70	
Butyl nitrate	30.35	0.1126	up to 133	
Butyl propanoate	27.37	0.0993	−89 to 146.8	
4-*tert*-Butylpyridine	35.48	0.0951	ca. −44 to 197	
Butyl stearate	33.0(25)	32.7(30)	26 to 343	
Butyl vinyl ether	21.99(20)		−92 to 94.2	
Carbon disulfide	35.29	0.1484	−111.6 to 46.5	0.429(0), 0.363(20), 0.352(25)
Carbon tetrachloride	29.49	0.1224	−23 to 76.7	1.321(0), 0.908(25), 0.656(50)
D-(+)-Carvone	36.54	0.0920	<15 to 230	
Chloroacetic acid	43.27	0.1117	61 to 189	3.15(50), 1.92(75)
o-Chloroaniline	43.41	0.0904	−14 to 208.8	3.316(25), 1.913(50), 1.248(75)
p-Chloroaniline	48.69	0.1099	72.5 to 232	
Chlorobenzene	35.97	0.1191	−45.3 to 131.7	0.799(20), 0.631(40), 0.512(60)
1-Chlorobutane	25.97	0.1117	−123.1 to 78.4	0.556(0), 0.422(25), 0.329(50)
2-Chlorobutane	24.40	0.1118	−131.3 to 68.2	0.439(15)
Chlorocyclohexane	33.90	0.1101	−44 to 142	
1-Chlorododecane	31.56	0.0904	−9 to 116	
1-Chloro-2,3-epoxypropane	39.76	0.1360	−57.2 to 116.1	1.03(25)
Chloroethane	21.18(5)	20.58(10)	−139 to 12.3	0.416(−25), 0.319(0), 0.279(10)
2-Chloroethanol	38.9(20)		−67.5 to 128.6	3.913(15)
Chloroform	29.91	0.1295	−63.6 to 61.1	0.706(0), 0.596(15), 0.514(30)
1-Chloroheptane	28.94	0.0961	−69 to 161	
1-Chlorohexane	28.32	0.1038		
1-Chloro-3-methylbutane	25.51	0.1076	−104 to 99	
1-Chloro-2-methylpropane	24.40	0.1099	−130.3 to 68.9	0.462(20), 0.373(40)
2-Chloro-2-methylpropane	20.06(15)	18.35(30)	−26 to 50.8	0.543(15)
1-Chloronaphthalene	44.12	0.1035	−2.3 to 259	2.940(25)
o-Chloronitrobenzene	48.10	0.1171	33 to 246	
m-Chloronitrobenzene	49.71	0.1417	44 to 236	
p-Chloronitrobenzene	45.84	0.1046	84 to 242	
1-Chlorooctane	29.64	0.0961	−58 to 182	
1-Chloropentane	27.09	0.1076	−99 to 108	0.580(20)
o-Chlorophenol	42.5	0.1122	9.8 to 175	3.589(25), 1.835(50), 1.131(75)
m-Chlorophenol	43.7	0.1009	33 to 214	11.55(25), 4.725(45), 4.041(50)
p-Chlorophenol	46.0	0.1049	43 to 220	4.99(50)
1-Chloropropane	24.41	0.1246	−122.8 to 47	0.436(0), 0.372(15), 0.318(30)

TABLE 2.30 Viscosity and Surface Tension of Organic Compounds (*Continued*)

Substance	Surface tension, mN · m^{-1} *a*	*b*	Liquid range, °C	Viscosity, mN · s · m^{-2}
2-Chloropropane	21.37	0.0883	−117 to 36	0.401(0), 0.335(15), 0.299(30)
3-Chloro-1-propene	25.50	0.0946	−134.5 to 45	0.347(15)
o-Chlorotoluene			−35.6 to 159	1.267(25), 0.883(50), 0.662(75)
m-Chlorotoluene			−47.8 to 161.8	0.964(25), 0.710(50), 0.547(75)
p-Chlorotoluene	34.93	0.1082	7.5 to 162.4	0.837(25), 0.621(50), 0.483(75)
Chlorotrimethylsilane	19.51	0.0875	−40 to 57	
o-Cresol	39.43	0.1011	30 to 191	3.035(50), 1.562(75), 0.961(100)
m-Cresol	38.00	0.0924	12 to 202	12.9(25), 4.417(50), 2.093(75)
p-Cresol	38.58	0.0962	34.8 to 202	5.607(45)
Cycloheptanol	35.02	0.0923	2 to 185	
Cyclohexane	27.62	0.1188	6.6 to 80.7	0.980(20), 0.912(25), 0.650(50)
Cyclohexanol	35.33	0.0966	25.4 to 161	57.5(25), 41.07(30), 12.3(50)
Cyclohexanone	37.67	0.1242	−31 to 155.7	2.453(15), 1.803(30), 1.321(50)
Cyclohexene	29.23	0.1223	−103.5 to 83	0.882(0), 0.625(25), 0.467(50)
Cyclohexylamine	34.19	0.1188	−18 to 134	1.079(25), 0.692(50), 0.485(75)
Cyclooctane	32.02	0.1090	14.8 to 151.1	
Cyclopentane	25.53	0.1462	−94 to 50	0.555(0), 0.413(25), 0.321(50)
Cyclopentanol	35.04	0.1011	−19 to 140	0.439(20)
Cyclopentanone	35.55	0.1100	−51 to 130.6	
Cyclopentene	25.94	0.1495	−135.1 to 44.2	
cis-Decahydronaphthalene	32.18(20)	31.01(30)	−43 to 195.8	3.042(25), 1.875(50), 1.271(75)
trans-Decahydronaphthalene	29.89(20)	28.87(30)	−30.4 to 187.3	1.948(25), 1.289(50), 0.917(75)
Decamethylcyclopentasiloxane	19.56	0.0565	−38 to >101	
Decamethyltetrasiloxane	86.20(25)		−68 to 194	1.28(20)
Decane	25.67	0.0920	−29.7 to 174.1	1.277(0), 0.838(25), 0.598(50)
1-Decanol	30.34	0.0732	6.9 to 232	10.9(25), 4.590(50)
1-Decene	25.84	0.0919	−66 to 170.6	0.805(20)
Dibenzylamine	43.27	0.1086	−26 to 300	
Dibenzyl ether	38.2(35)		2 to 298	3.711(25)
p-Dibromobenzene	41.84	0.1007	87.3 to 220	
1,4-Dibromobutane	48.24	0.1190	−20 to 198	
1,2-Dibromoethane	42.85	0.1320	10 to 131.7	1.721(20), 1.286(40), 0.648(100)
1,2-Dibromopropane	36.81	0.1155	−55.5 to 142	1.5(25)
Dibromotetrafluoroethane	18.9(20)	18.1(25)	−110.5 to 47	0.72(25)
Dibutylamine	26.50	0.0952	−62 to 159.6	0.918(25), 0.619(50), 0.449(75)
Dibutyl decanedioate			−10 to 345	9.03(25)
Dibutyl ether	24.78	0.0934	−95 to 140	0.637(25), 0.466(50), 0.356(75)
Dibutyl maleate	32.46	0.0865	< −80 to 281	5.62(20), 4.76(25)
Dibutyl *o*-phthalate	33.40(20)		−35 to 340	19.91(20), 11.17(35), 7.85(45)
Dichloroacetic acid	37.8	0.0927	9 to 194	3.23(50), 1.92(75)
o-Dichlorobenzene	35.55(30)		−17 to 180.4	1.324(25), 0.962(50), 0.739(75)
m-Dichlorobenzene	38.30	0.1147	−24.8 to 173.1	1.044(25), 0.783(50), 0.628(75)
p-Dichlorobenzene	34.66	0.0879	53 to 174.1	0.839(55), 0.668(79)
1,4-Dichlorobutane	37.79	0.1174	−38 to 163	
1,1-Dichloroethane	27.03	0.1186	−97 to 57.3	0.505(15), 0.464(25), 0.362(50)
1,2-Dichloroethane	35.43	0.1428	−35.7 to 83.5	1.125(0), 0.779(25), 0.576(50)
1,1-Dichloroethylene			−122.6 to 31.6	0.442(0), 0.358(20)
cis-1,2-Dichloroethylene	28(20)		−80.1 to 60	0.785(−25), 0.575(0), 0.444(25)
trans-1,2-Dichloroethylene	25(20)		−49.8 to 48.7	0.522(−25), 0.398(0), 0.317(25)
2,2′-Dichloroethyl ether	40.57	0.1306	up to 178.5	2.41(20), 2.065(25)

(*Continued*)

TABLE 2.30 Viscosity and Surface Tension of Organic Compounds (*Continued*)

Substance	Surface tension, mN · m^{-1}		Liquid range, °C	Viscosity, mN · s · m^{-2}
	a	*b*		
Dichloromethane	30.41	0.1284	−95 to 40	0.533(0), 0.449(15), 0.393(30)
2,4-Dichlorophenol	46.59	0.1221	42 to 210	
1,2-Dichloropropane	31.42	0.1240	−100 to 96	0.865(20), 0.700(25)
1,3-Dichloropropane	36.40	0.1233	−99.5 to 122	
2,2-Dichloropropane	23.60(20)	22.53(30)	−35 to 69	0.769(15), 0.619(30)
α,α-Dichlorotoluene	41.26	0.1035	−16 to 205	
Diethanolamine			28 to 269	368(30), 109.5(50), 28.7(75)
1,1-Diethoxyethane	23.46	0.1030	−100 to 102.2	
1,2-Diethoxyethane			−74 to 121.4	0.65(20)
Dimethoxymethane	23.87	0.1291	up to 88	
Diethylamine	22.71	0.1143	−50 to 55.5	
N,N-Diethylaniline	36.59	0.1040	−38 to 217	3.838(0), 1.15(50), 0.750(75)
Diethyl carbonate	28.62	0.1100	−43 to 126	0.868(15), 0.748(25)
Diethyl decanedioate	34.68	0.0959		
Diethyl ether	18.92	0.0908	−116 to 34.6	0.283(0), 0.224(25)
Diethyl ethyl phosphonate	30.63	0.0975	up to 198	1.627(15), 0.969(45), 0.743(65)
Di(2-ethylhexyl) *o*-phthalate			−50 to 384	33.67(35), 21.40(45)
Diethyl maleate	34.67	0.1039	−8.8 to 225.3	3.57(20), 3.14(25)
Diethyl 1,3-propanedioate (malonate)	33.91	0.1042	−49.9 to 199.3	2.15(20), 1.94(25)
Diethyl oxalate	34.32	0.1119	−40.6 to 185.4	2.311(15), 1.618(30)
Diethyl *o*-phthalate	38.47	0.0963	−40 to 295	9.18(35), 6.41(45)
Diethyl succinate	33.97	0.1041	−21 to 217.7	
Diethyl sulfate	35.47	0.0976	−25 to 208	
Diethyl sulfide	27.33	0.1106	−104 to 92.1	0.558(0), 0.422(25)
1,2-Dihydroxybenzene	47.6	0.0849	104 to 245.5	
1,3-Dihydroxybenzene	54.8	0.0717	110 to 276	
Diiodomethane	70.21	0.1613	6 to 181	
Diisobutylamine	24.00	0.0912	−77 to 139	
Diisopentyl ether	24.76	0.0871	up to 172.5	1.40(11), 1.012(20)
Diisopropylamine	21.03	0.1077	−61 to 83.5	0.393(25), 0.300(50), 0.237(75)
Diisopropyl ether	19.89	0.1048	−87 to 68	0.379(25)
1,2-Dimethoxybenzene	34.4	0.0642	22.5 to 206	3.281(25), 2.184(40)
1,1-Dimethoxyethane	23.90	0.1159	−113 to 64.5	
1,2-Dimethoxyethane	48.0(25)		−68 to 85	0.670(−10), 0.530(10), 0.455(25)
Dimethoxymethane	23.59	0.1199	−104.8 to 42	0.340(15), 0.325(20)
N,N-Dimethylacetamide	32.40(30)	29.50(50)	−20 to 165.5	1.956(25), 1.279(50), 0.896(75)
Dimethylamine	29.50	0.1265	−92 to 6.9	0.300(−25), 0.232(0)
N,N-Dimethylaniline	38.14	0.1049	2.5 to 194	1.300(25), 0.911(50), 0.675(75)
2,4-Dimethylaniline	39.34	0.0996	−14 to 214	
2,2-Dimethylbutane	18.29	0.0990	−100 to 49.7	0.351(25), 0.330(30)
2,3-Dimethylbutane	19.38	0.1000	−128 to 58	0.361(25), 0.342(30)
2,3-Dimethyl-1-butanol	26.22	0.0992	−14 to 118	
Dimethyl carbonate	31.94	0.1343	0.5 to 91	
1,1-Dimethylcyclopentane	23.78	0.1016	−70 to 87.5	
Dimethyl ether	14.97	0.1478	−141 to −24.9	
N,N-Dimethylformamide	36.76(20)	34.40(40)	−60 to 153	1.176(0), 0.794(25), 0.624(50)
2,4-Dimethylheptane	23.21	0.0929	< −100 to 133	
2,5-Dimethylheptane	23.21	0.0929	< −100 to 136	

TABLE 2.30 Viscosity and Surface Tension of Organic Compounds (*Continued*)

Substance	Surface tension, mN · m^{-1} a	Surface tension, mN · m^{-1} b	Liquid range, °C	Viscosity, mN · s · m^{-2}
2,6-Dimethylheptane	22.17	0.0887	−103 to 135	
Dimethyl hexanedioate	38.26	0.1138	8 to >112	14(20)
Dimethyl maleate	40.73	0.1220	−19 to 202	3.54(20), 3.21(25)
Dimethyl malonate	39.72	0.1208	−62 to 181	
2,2-Dimethylpentane	19.94	0.0957	−124 to 79	
2,3-Dimethylpentane	21.96	0.0995	up to 90	0.406(20)
2,4-Dimethylpentane	20.09	0.0972	−120 to 80.4	0.361(20)
3,3-Dimethylpentane	21.59	0.0996	−135 to 86	
2,4-Dimethylphenol	34.57	0.0869	24.5 to 211	
2,5-Dimethylphenol	36.72	0.0850	74.5 to 211.5	1.55(80)
3,4-Dimethylphenol	35.75	0.0910	61 to 227	3.00(80)
3,5-Dimethylphenol	34.09	0.0807	64 to 222	2.42(80)
Dimethyl *o*-phthalate			5.5 to 284	14.4(25), 5.309(50), 2.824(75)
2,2-Dimethylpropane	12.05(20)	10.98(30)	−16.6 to 9.5	0.328(0), 0.303(5)
Dimethyl succinate	39.00	0.1191	19 to 196.4	
Dimethyl sulfate	41.26	0.1163	−31.8 to 188	
Dimethyl sulfide	26.07	0.0805	−98 to 37	0.356(0), 0.289(20), 0.265(36)
Dimethyl sulfite	36.48	0.1253	up to 127	0.715(30), 0.436(80)
Dimethyl sulfoxide	43.54(20)	42.41(30)	18.5 to 189	2.47(20), 1.192(55), 0.849(80)
1,4-Dioxane	36.23	0.1391	11.8 to 101.2	1.439(15), 1.087(30), 0.787(50)
Dipentyl ether	26.66	0.0925	−69 to 190	1.188(15), 0.922(30)
Dipentyl *o*-phthalate	32.56	0.0739		17.03(35), 11.51(45)
Dipentyl sulfide	29.55	0.0876		
Dipentylamine	45.36	0.1017	53 to 302	4.66 (55), 1.04(130)
Diphenyl ether	28.70	0.0780	27 to 258	2.130(50), 1.407(75), 1.023(100)
1,2-Dipropoxyethane	25.03	0.0972		
Dipropoxymethane	25.17	0.0953		
Dipropylamine	24.86	0.1022	−63 to 109	0.517(25), 0.377(50), 0.288(75)
Dipropyl carbonate	28.94	0.1015	up to 168	
Dipropylene glycol butyl ether	28.2(25)		up to >103	4.23(25)
Dipropylene glycol ethyl ether	27.7(25)			3.11(25)
Dipropylene glycol isopropyl ether	25.9(25)		up to 80	386(25)
Dipropylene glycol methyl ether	28.8(25)		−117 to 188	3.1(25)
Dipropyl ether	22.60	0.1047	−126 to 89.6	0.542(0), 0.396(25), 0.304(50)
Dodecane	27.12	0.0884	−10 to 216	2.277(0), 1.378(25), 0.930(50)
1-Dodecanol	31.25	0.0748	24 to 259	
Epichlorohydrin	39.76	0.1360	−26 to 117	1.20(25)
1,2-Epoxybutane	23.9(20)		−150 to 63	0.419(15), 0.358(30)
1,2-Ethanediamine	44.77	0.1398	11 to 117.3	1.54(20), 1.226(30)
1,2-Ethanediol	50.21	0.0890	−12.6 to 197.3	26.09(15), 13.55(30)
Ethanesulfonic acid	45.74	0.0824	−17 to >123	
Ethanesulfonyl chloride	43.43	0.1177	up to 177	
Ethanethiol	25.06	0.0793	−148 to 35	0.364(0), 0.287(25)
Ethanol	24.05	0.0832	−114 to 78	1.786(0), 1.074(25), 0.694(50)
Ethanolamine	51.11	0.1117	10.5 to 171	21.1(25), 8.560(50), 3.935(75)
Ethoxybenzene (phenetol)	35.17	0.1104	−29.5 to 170	1.364(15), 1.197(25), 0.817(50)
2-Ethoxyethanol	30.59	0.0897	−70 to 135	2.04(20), 1.85(25)
Ethyl acetate	26.29	0.1161	−84 to 77	0.578(0), 0.423(25), 0.325(50)

(*Continued*)

TABLE 2.30 Viscosity and Surface Tension of Organic Compounds (*Continued*)

Substance	Surface tension, mN · m^{-1}		Liquid range, °C	Viscosity, mN · s · m^{-2}
	a	*b*		
Ethyl acetoacetate	34.42	0.1015	−45 to 181	1.419(20), 1.508(25)
Ethylamine	22.63	0.1372	−81 to 16.6	
N-Ethylaniline	39.00	0.1070	−63.5 to 203	2.047(25), 1.231(50), 0.825(75)
Ethylbenzene	31.48	0.1094	−95 to 136	1360.631(25), 0.482(50), 0.380(75)
Ethyl benzoate	37.16	0.1059	−35 to 212	2.407(15), 1.751(30)
Ethyl butanoate	26.55	0.1045	−98 to 121	0.771(15), 0.613(25)
2-Ethylbutanoic acid	26.3(20)		−14 to 194	3.3(20)
2-Ethyl-1-butanol	25.06(15)	24.32(25)	< −15 to 146	8.021(15), 5.892(25)
Ethyl carbamate			50 to 184	0.916(105), 0.715(120)
Ethyl chloroacetate	34.18	0.1177	−21 to 144	
Ethyl chloroformate	28.90	0.1084	−81 to 93	
Ethyl *trans*-cinnamate	39.99	0.1045	10 to 271	8.7(20)
Ethyl crotonate	29.31	0.1066	up to 138	
Ethyl cyanoacetate	38.80	0.1092	−22 to 206	3.256(15), 2.148(30)
Ethylcyclohexane	27.78	0.1054	−111 to 132	1.139(0), 0.784(25), 0.579(50)
Ethyl dichloroacetate	34.89	0.1158	up to 155	
Ethyl dodecanoate	30.05	0.0863	−10 to 271	
Ethylene carbonate			36 to 248	1.85(40)
Ethylenediamine	44.77	0.1398	11 to 117	1.540(18)
Ethylene glycol	50.21	0.0890	up to 198	26.09(15), 13.35(30), 6.554(50)
Ethyleneimine	7.9(20)		−78 to 56	0.418(25)
Ethylene oxide	27.66	0.1664	−111 to 10.6	0.3(0)
Ethyl formate	26.47	0.1315	−80 to 54	0.419(15), 0.358(30), 0.300(50)
Ethyl fumarate	33.90	0.1056	68 to >148	
Ethylhexadecanoate	32.86	0.0859	22 to >191	
Ethyl hexanoate	27.73	0.0960	up to 168	
2-Ethyl-1-hexanol	30.0(22)		−70 to 185	6.271(25), 2.631(50), 1.360(75)
Ethyl isobutanoate	25.33	0.1046	−88 to 110	
Ethyl isothiocyanate	38.69	0.1326	−6 to 132	
Ethyl lactate	30.72	0.0983	−26 to 155	2.44(25)
Ethyl 3-methylbutanoate	25.79	0.1006	−99 to 135	
Ethyl methyl ether	18.56	0.1317	−113 to 7.4	
Ethyl methyl sulfide	27.63	0.1286	−106 to 67	0.373(20), 0.354(25)
Ethyl nitrate	30.81	0.1345	−95 to 88	
3-Ethylpentane	22.52	0.1032	−119 to 93.5	
Ethyl pentanoate	27.15	0.0999	−91 to 145	0.847(20)
Ethyl propanoate	26.72	0.1168	−74 to 99	0.564(15), 0.473(30), 0.380(50)
Ethyl propyl ether	21.92	0.1054	−79 to 63	0.401(0), 0.323(20), 0.225(60)
Ethyl salicylate	31.00	0.1091	2 to 234	1.772(45)
Ethyl thiocyanate	37.28	0.1226	up to 145	
o-Ethyltoluene	32.33	0.1060	−81 to 165	
p-Ethyltoluene	30.98	0.1075	−62 to 162	
Ethyl trichloroacetate	32.97	0.1073	up to 168	
Fluorobenzene	29.67	0.1204	−42 to 85	0.620(15), 0.517(30), 0.423(50)
1-Fluorohexane	23.41	0.1001	−103 to 93	
1-Fluoropentane	22.81	0.1315	−120 to 63	
o-Fluorotoluene			−62 to 115	0.680(20), 0.601(30)
m-Fluorotoluene	32.31	0.1257	−87 to 115	0.608(20), 0.534(30)
p-Fluorotoluene	30.44	0.1109	−56 to 117	0.622(20), 0.522(30)

TABLE 2.30 Viscosity and Surface Tension of Organic Compounds (*Continued*)

Substance	Surface tension, $mN \cdot m^{-1}$ *a*	*b*	Liquid range, °C	Viscosity, $mN \cdot s \cdot m^{-2}$
Formamide	59.13	0.0842	2.6 to 220	4.320(15), 2.296(30), 1.833(50)
Formanilide	44.30	0.0875	47 to 271	1.65(120)
Formic acid	39.87	0.1098	8 to 101	1.966(15), 1.607(25), 1.030(50)
Furan	24.10(20)	23.38(25)	−86 to 31	0.380(20), 0.361(25)
2-Furancarboxaldehyde	46.41	0.1327	−36.5 to 162	2.501(0), 1.587(25), 1.143(50)
2-Furanmethanol	ca. 38(20)		−31 to 171	4.62(25)
Glycerol	63.14(17)	62.5(25)	18 to 290	934(25), 152(50), 39.8(75)
Glycerol tris(acetate)	37.88	0.081		
Glycerol tris(nitrate)	55.74	0.2504	13 to >160	36.0(20), 13.6(40)
Glycerol tris(oleate)	36.03	0.0699	−5 to >233	
Glycerol tris(palmitate)	32.26	0.0672	65 to 320	
Glycerol tris(sterate)	32.73	0.0685		
Heptanal	28.64	0.0920	−43 to 153	0.977(15)
Heptane	22.10	0.0980	−91 to 98	0.523(0), 0.416(20), 0.341(40)
Heptanoic acid	29.88	0.0848	−8 to 222	3.84(25), 2.282(50), 1.488(75)
1-Heptanol			−34 to 176	8.53(15), 5.810(25), 2.603(50)
2-Heptanol			up to 159	3.955(25), 1.799(50), 0.987(75)
3-Heptanol			−70 to 157	1.957(50), 0.976(75), 0.584(100)
4-Heptanol				4.207(25), 1.695(50), 0.882(75)
2-Heptanone	28.76	0.1056	−35 to 151	0.854(15), 0.686(30), 0.407(50)
4-Heptanone	28.11	0.1060	−32 to 143.7	0.736(20)
1-Heptene	22.28	0.0991	−120 to 93.6	0.441(0), 0.340(25), 0.273(50)
Heptylamine	25.96	0.0783	−23 to 156	1.314(25), 0.865(50), 0.600(75)
Hexadecane	29.18	0.0854	18.2 to 286.8	3.032(25), 1.879(50), 1.260(75)
1,5-Hexadiene	20.93	0.1028	−140.7 to 59.5	0.275(20), 0.244(36)
Hexafluorobenzene	22.6(20)		5.1 to 80.3	2.789(25), 1.730(50), 1.151(75)
Hexamethyldisiloxane	17.01	0.0763	−67 to 101	
Hexamethylphosphoramide	33.8(20)		7 to 232	3.47(20)
Hexane	20.44	0.1022	−95.4 to 68.7	0.405(0), 0.313(20), 0.271(40)
Hexanenitrile	29.64	0.0907	−80 to 163.6	1.041(15), 0.830(30), 0.650(50)
Hexanoic acid	28.05(20)	27.55(25)	−3 to 205	3.525(15), 2.511(30)
1-Hexanol	27.81	0.0801	−44.6 to 157.5	6.203(15), 3.872(30), 2.271(50)
2-Hexanone	28.18	0.1092	−55.5 to 127.6	0.584(25), 0.429(50), 0.329(75)
1-Hexene	20.47	0.1027	−140 to 63.5	0.326(0), 0.252(25), 0.202(50)
Hexyl acetate	28.44	0.0970	−81 to 171	
4-Hydroxy-4-methyl-2-pentanone	31.0(20)		−44 to 168	6.621(0), 2.798(25), 1.829(50)
Iodobenzene	41.52	0.1123	−31 to 188	1.554(25), 1.117(50), 0.854(75)
1-Iodobutane	30.82	0.1031	−103.5 to 131	
2-Iodobutane	30.32	0.1056	−104 to 120	
Iodoethane	31.67	0.1286	−111 to 72.4	0.617(15), 0.540(30), 0.444(50)
1-Iodoheptane	32.18	0.0887	−48 to 204	
1-Iodohexadecane	34.49	0.0880	23 to >207	
1-Iodohexane	31.63	0.0845	up to 180	
Iodomethane	33.42	0.1234	−66.5 to 42.5	0.594(0), 0.500(20), 0.424(40)
1-Iodo-2-methylpropane	30.26	0.1072	−93.5 to 121	0.875(20), 0.697(40)
1-Iodooctane	32.51	0.0915	−46 to 226	
1-Iodopentane	31.41	0.1014	−85 to 155	
1-Iodopropane	31.64	0.1136	−101 to 102.6	0.837(15), 0.670(30), 0.541(50)

(*Continued*)

TABLE 2.30 Viscosity and Surface Tension of Organic Compounds (*Continued*)

Substance	Surface tension, $mN \cdot m^{-1}$		Liquid range, °C	Viscosity, $mN \cdot s \cdot m^{-2}$
	a	*b*		
2-Iodopropane	29.35	0.1107	−90 to 89.5	0.732(15), 0.620(30), 0.506(50)
p-Iodotoluene	39.23	0.0965	up to 211	
α-Ionone	34.10	0.0949	>124	
β-Ionone	35.36	0.0950	>128	
Isobutanenitrile	24.93(20)	23.84(30)	−71.5 to 104	0.551(15), 0.456(30)
Isobutyl acetate	25.59	0.1013	−99 to 116.5	0.676(25), 0.493(50), 0.370(75)
Isobutylamine	24.48	0.1092	−86.6 to 68	0.770(0), 0.571(25), 0.367(50)
Isobutylbenzene	29.39	0.0961	−51.5 to 172.8	
Isobutyl formate	26.14	0.1122	−95.5 to 98.4	0.680(20)
Isobutyl propanoate	30.92	0.1270	−71 to 137	
Isopentyl acetate	26.75	0.0989	−78.5 to 142	0.872(20), 0.790(25)
Isophorone			−8.1 to 215.2	4.201(0), 2.329(25), 1.415(50)
Isopropyl acetate	24.44	0.1072	−73 to 89	0.559(20)
Isopropylamine	19.91	0.0972	−95 to 31.7	0.454(0), 0.325(25)
Isopropylbenzene	30.32	0.1054	−96 to 154	1.075(0), 0.737(25), 0.547(50)
Isopropyl formate	24.56	0.1147		0.512(20)
Lactonitrile	38.31	0.0960	−40 to >103	2.01(30)
D-Limonene	29.50	0.0929	−96.5 to 178	
(±)-Mandelonitrile	45.90	0.0988	−10 to 170	
Methacrylic acid	26.5(25)		16 to 163	1.32(20)
Methacrylonitrile	24.4(20)		−35.8 to 90.3	0.392(20)
Methanesulfonic acid	52.28	0.0893	20 to >167	
Methanethiol	28.09	0.1696	−123 to 6.0	
Methanol	24.00	0.0773	−97.7 to 64.7	0.793(0), 0.676(10), 0.544(25)
o-Methoxybenzaldehyde	45.34	0.1105	37 to 238	
p-Methoxybenzaldehyde	44.69	0.1047	−1 to 248	
Methoxybenzene	38.11	0.1204	−37.5 to 153.8	1.152(15), 1.056(25), 0.747(50)
2-Methoxyethanol	33.30	0.0984	−85.1 to 124	1.71(20), 1.60(25)
2-(2-Methoxyethoxy)ethanol	34.8(25)	29.9(75)	−50 to 194	3.48(25), 1.61(60)
1-Methoxy-2-nitrobenzene	48.62	0.1185	10.5 to 277	
o-Methoxyphenol	41.2	0.0943	28 to 205	
p-Methoxytoluene	36.20	0.1071	up to 174	
N-Methylacetamide	33.67(30)	30.62(50)	30.6 to 206	3.88(30), 2.54(45)
Methyl acetate	27.95	0.1289	−98 to 57	0.477(0), 0.364(25), 0.284(50)
Methyl acetoacetate	34.98	0.0944	27.5 to 171.7	
Methyl acrylate			−76.5 to 80.2	1.398(20)
Methylamine	22.87	0.1488	−93.5 to −6.3	0.319(−25)
N-Methylaniline	39.32	0.0970	−57 to 196	2.042(25), 1.222(50), 0.825(75)
o-Methylaniline				3.823(25), 1.936(50), 1.198(75)
m-Methylaniline				3.306(25), 1.679(50), 1.014(75)
Methyl benzoate	40.10	0.1171	−15 to 199.5	2.298(15), 0.206(20), 1.673(30)
2-Methyl-1,2-butadiene				0.266(0.3), 0.233(20)
2-Methylbutane	17.20	0.1103	up to 30	0.376(−25), 0.277(0), 0.214(25)
Methyl butanoate	27.48	0.1145	−85.8 to 103	0.580(20), 0.459(40), 0.406(50)
3-Methylbutanoic acid	27.28	0.0886	−29.3 to 176.5	2.731(15), 2.411(20)
2-Methyl-1-butanol	21.5(25)		< −70 to 128	5.50(20), 4.453(25), 1.963(50)
2-Methyl-2-butanol	24.18	0.0748	−9.0 to 102.0	5.48(15), 2.81(30)
3-Methyl-1-butanol	25.76	0.0820	−117 to 131	4.81(15), 2.96(30), 1.842(50)
3-Methyl-2-butanol	23.0(25)		up to 112.9	3.51(25)

TABLE 2.30 Viscosity and Surface Tension of Organic Compounds (*Continued*)

Substance	Surface tension, mN · m^{-1} *a*	Surface tension, mN · m^{-1} *b*	Liquid range, °C	Viscosity, mN · s · m^{-2}
2-Methyl-1-butene	18.81	0.1148	−137.6 to 31	
2-Methyl-2-butene	19.70	0.1271	−133.8 to 38.6	
3-Methyl-1-butene	16.42	0.1031	−168 to 20	
2-Methylbutyl acetate	26.75	0.0989	−99 to 117	0.872(20)
3-Methylbutyronitrile	27.58	0.0827	−101 to 129	
Methyl chloroacetate	37.90	0.1304	−32 to 130	
Methyl cyanoacetate	41.32	0.1074	−22.5 to 201	3.824(50), 3.398(55), 2.687(65)
Methylcyclohexane	26.11	0.1130	−126.6 to 100.9	0.679(25), 0.501(50), 0.390(75)
cis-2-Methylcyclohexanol	32.45	0.0770 (mixed isomers)	7 to 165	18.08(25), 13.60(30)
trans-2-Methylcyclohexanol			−2 to 167.5	37.13(25), 25.14(30)
cis-3-Methylcyclohexanol	29.08	0.0629 (mixed isomers)	−6 to 168	19.7(25), 17.23(30)
trans-3-Methylcyclohexanol	28.80(30)		−0.5 to 167	25.62(16), 15.60(30)
cis-4-Methylcyclohexanol	29.07	0.0690 (mixed isomers)	−9.2 to 173	
2-Methylcyclohexanone	34.06	0.1027	up to 162	
3-Methylcyclohexanone	33.06	0.0925	up to 169	
4-Methylcyclohexanone	32.83	0.0935	up to 171	
Methylcyclopentane	24.63	0.1163	−142.2 to 71.8	0.653(0), 0.478(25), 0.364(50)
Methyl decanoate	30.33	0.0912	−18 to 223	
Methyl dichloroacetate	37.00	0.1219	−52 to 143	
Methyl dodecanoate	31.37	0.0893	4.8 to 262	
N-Methylformamide	37.96(30)	35.02(50)	−4 to 199.5	1.678(25), 1.155(50), 0.824(75)
Methyl formate	28.29	0.1572	−99 to 31.7	0.424(0), 0.360(15), 0.325(25)
Methyl heptanoate	28.95	0.0987	−55.8 to 173.5	
4-Methyl-3-heptanol			−123 to 170	1.085(25), 0.702(50), 0.497(75)
5-Methyl-3-heptanol			−91 to 172	1.178(25), 0.762(50), 0.536(75)
Methyl hexadecanoate (palmitate)	31.50	0.0775	32 to >196	
2-Methylhexane	21.22	0.0966	−118 to 90	0.378(20)
3-Methylhexane	21.73	0.0970	−119 to 92	0.372(20), 0.350(25)
Methyl hexanoate	28.47	0.1045	−71 to 151	
Methyl isobutanoate	25.99	0.1131	−84.7 to 92.5	0.672(0), 0.523(20), 0.419(40)
1-Methyl-4-isopropylbenzene (*p*-cymene)	28.83	0.0877		3.402(20)
Methyl methacrylate	28-29(30)		−48 to 100	0.632(20)
1-Methylnaphthalene	39.96	0.0934	−30.4 to 245	
Methyl octadecanoate	32.20	0.0775	38 to >215	
2-Methyloctane	23.76	0.0940	−80.3 to 143.2	
4-Methyloctane	24.22	0.0940	−113 to 142	
Methyl octanoate	29.93	0.1002	−40 to 192.9	
Methyl oleate	31.3(25)	25.4(100)	−19.9 to >218	4.88(20)
2-Methylpentane	19.37	0.0997	−154 to 60.3	0.372(0), 0.286(25), 0.226(50)
3-Methylpentane	20.26	0.1060	−163 to 63	0.395(0), 0.307(25), 0.292(30)

(*Continued*)

TABLE 2.30 Viscosity and Surface Tension of Organic Compounds (*Continued*)

Substance	Surface tension, $mN \cdot m^{-1}$		Liquid range, °C	Viscosity, $mN \cdot s \cdot m^{-2}$
	a	*b*		
4-Methylpentanenitrile	28.89	0.0917	−51.1 to 156.5	0.980(20), 0.843(30)
Methyl pentanoate	27.85	0.1044	up to 128	0.713(20)
2-Methyl-1-pentanol	26.98	0.0819	up to 148	
3-Methyl-1-pentanol	26.92	0.0789	up to 153	
4-Methyl-1-pentanol	25.93	0.0743	up to 152	
2-Methyl-2-pentanol	25.07	0.0861	−103 to 121	
3-Methyl-2-pentanol	27.14	0.0919	up to 134	
4-Methyl-2-pentanol	24.67	0.0821	−90 to 122	4.074(25)
2-Methyl-3-pentanol	26.43	0.0914	up to 126	
3-Methyl-3-pentanol	25.48	0.0888	−23.6 to 123	
4-Methyl-2-pentanone	23.64(20)	19.62(60)	−84 to 116.5	0.585(20), 0.522(30), 0.406(50)
Methyl phenyl sulfide	42.81	0.1238	−15 to 188	
N-Methyl propanamide	31.29(20)	29.12(50)	−43 to >146	6.06(20), 4.58(30), 3.56(40)
2-Methylpropanenitrile			−72 to 108	0.551(15), 0.456(30)
Methyl propanoate	27.58	0.1258	−88 to 80	0.581(0), 0.431(25), 0.333(50)
2-Methylpropanoic acid	25.55(20)	25.13(25)	−47 to 154	1.857(0), 1.226(25), 0.863(50)
2-Methyl-1-propanol	24.53	0.0795	−108 to 108	4.70(15), 2.876(30)
2-Methyl-2-propanol	20.02(15)	19.10(30)	25.8 to 82.4	1.421(50), 0.678(75)
2-Methylpropene	14.84	0.1319	−140 to −6.9	
1-Methylpropyl acetate	25.72	0.1054		
2-Methyl-1-propylamine	24.48	0.1092	−87 to 68	21.7(25)
2-Methylpropyl formate	26.14	0.1122	−96 to 98	0.680(20)
2-Methylpyridine	36.11	0.1243	−66.7 to 129	0.805(20), 0.710(30)
3-Methylpyridine	37.35	0.1153	−18.3 to 144	
4-Methylpyridine	37.71	0.1141	3.8 to 145	
N-Methyl-2-pyrrolidinone			−24.4 to 202	1.666(25)
Methyl salicylate	42.15	0.1174	−8 to 223	1.102(75), 0.815(100)
Methyl tetradecanoate	31.00	0.0800	18.4 to 323	
2-Methyltetrahydrofuran			< −75 to 78	0.777(−20), 0.601(0), 0.536(10)
Methyl thiocyanate	40.66	0.1305	−5 to 133	64.3(0)
Morpholine	37.63(20)	36.24(30)	−4.9 to 128	2.53(15), 1.79(30), 1.247(50)
Naphthalene			80 to 217.7	0.967(80), 0.780(100)
p-Nitroaniline	60.62	0.0923	147 to 332	
Nitrobenzene	48.62	0.1185	5.8 to 210.8	2.165(15), 1.863(25), 1.262(50)
Nitroethane	35.27	0.1255	−90 to 114	0.940(0), 0.688(25), 0.526(50)
Nitromethane	40.72	0.1678	−28.4 to 101.2	0.692(15), 0.596(30), 0.481(50)
1-Nitro-2-methoxybenzene	48.62	0.1185	95 to 273	
o-Nitrophenol	47.35	0.1174	45 to 216	2.343(45)
1-Nitropropane	32.62	0.1009	−108 to 131.1	0.798(25), 0.589(50), 0.460(75)
2-Nitropropane	32.18	0.1158	−91.3 to 120.3	0.750(25)
o-Nitrotoluene	44.10	0.1174	−10 to 222	2.37(20), 1.63(40)
m-Nitrotoluene	43.54	0.1118	15.5 to 231.9	0.233(20), 1.60(40)
p-Nitrotoluene	42.26	0.0974	52 to 238	1.20(60)
Nonane	24.72	0.0935	−53.5 to 150.8	0.964(0), 0.666(25), 0.488(50)
Nonanoic acid			12.5 to 254.5	7.011(25), 3.712(50), 2.234(75)
1-Nonanol	29.79	0.0789	−5.5 to 215	14.3(20), 9.123(25), 4.032(50)
5-Nonanone	28.72	0.0975	−50 to 187	1.199(25), 0.834(50), 0.619(75)
1-Nonene	24.90	0.0938	−81 to 146	0.620(20), 0.586(25)
Octadecane	29.98	0.0843	28.1 to 316.3	2.487(50), 1.609(75), 1.132(100)

TABLE 2.30 Viscosity and Surface Tension of Organic Compounds (*Continued*)

Substance	Surface tension, $mN \cdot m^{-1}$		Liquid range, °C	Viscosity, $mN \cdot s \cdot m^{-2}$
	a	*b*		
Octamethylcyclotetrasiloxane	20.19	0.0811	17 to 176	2.20(20)
Octane	23.52	0.0951	−56.8 to 125.7	0.546(20), 0.433(40), 0.355(60)
Octanenitrile	29.61	0.0802	−45.6 to 205	1.811(15), 1.356(30)
Octanoic acid	29.21(20)	28.7(25)	16.6 to 239	5.020(25), 2.656(50), 1.654(75)
1-Octanol	29.09	0.0795	−15.5 to 195	10.64(15), 6.125(30), 3.232(50)
2-Octanol	27.96	0.0820	−31.6 to 180	
1-Octene	23.68	0.0958	−102 to 121	0.470(20), 0.447(25)
Oleic acid	32.80(20)	27.94(90)	13.4 to 360	38.80(20), 27.64(25)
4-Oxopentanoic acid	41.69	0.0763	33 to 246	
Paraldehyde	28.28	0.1062	12.6 to 124	1.079(25), 0.692(50), 0.485(75)
Parathion	39.2(25)		6 to 375	15.30(25)
Pentachloroethane	37.09	0.1178	−29.9 to 160	2.741(15), 2.070(30), 1.491(50)
Pentadecane	28.78	0.0857	9.9 to 270	2.814(22)
Pentanal	27.96	0.1010	−92 to 103	
Pentane	18.25	0.1121	−129.7 to 36.0	0.351(−25), 0.274(0), 0.224(25)
1,5-Pentanediol	43.2(20)		−18 to 239	128(20)
2,4-Pentanedione	33.28	0.1144	−23.1 to 138	0.6(20)
Pentanenitrile	27.44(20)	26.33(30)	−92 to 141.3	0.779(15), 0.637(30)
Pentanoic acid	28.90	0.0887	−33.7 to 186	2.359(15), 1.774(30), 0.979(70)
1-Pentanol	27.54	0.0874	−79 to 137.5	4.650(15), 3.619(25), 1.820(50)
2-Pentanol	25.96	0.1004	−73 to 119.3	5.130(15), 2.780(30), 1.447(50)
3-Pentanol	24.60(20)	23.76(30)	−69 to 116	7.337(15), 3.306(30), 1.473(50)
2-Pentanone	24.89	0.0655	−76.8 to 102	0.641(0), 0.473(25), 0.362(50)
3-Pentanone	27.36	0.1047	−39.0 to 102	0.592(0), 0.444(25), 0.345(50)
1-Pentene	18.20	0.1099	−165 to 30.1	0.313(−25), 0.241(0), 0.195(25)
cis-2-Pentene	19.71	0.1172	−151 to 37.0	
trans-2-Pentene	18.90	0.0997	−140 to 36.3	
Pentyl acetate	27.66	0.0994	−70.8 to 149.2	0.924(20), 0.862(25)
Pentylamine	24.4(13)		−55 to 104	1.030(0), 0.702(25), 0.493(50)
Phenol	43.54	0.1069	41 to 182	3.437(50), 1.784(75), 1.099(100)
2-Phenylacetamide	46.26	0.0788	157 to bp	
Phenyl acetate			<45 to 196	1.799(45)
Phenylacetonitrile	44.57	0.1155	−23.8 to 233.5	1.93(25)
1-Phenylethanol	42.88	0.1038	20 to 204	
Phenylhydrazine	48.14	0.1292	19.5 to 243	13.0(25), 4.553(50), 1.850(75)
Phenyl isothiocyanate	42.73	0.1086	−30 to 163	
Phenyl salicylate	45.20	0.0976	44 to >173	
(±)-α-Pinene	28.35	0.0944	−64 to 156	1.61(25)
L-β-Pinene	28.26	0.0934	−61 to 166	1.70(20), 1.41(25)
Piperidine	31.79	0.1153	−11 to 106	1.573(25), 0.958(50), 0.649(75)
1,2-Propanediol (see propylene glycol)				
1,3-Propanediol	47.43	0.0903	−27 to 214	56.0(20), 18.0(40)
Propanenitrile (propionitrile)	29.63	0.1153	−92.8 to 97.2	0.294(25), 0.240(50), 0.202(75)
1-Propanethiol	27.38	0.1272	−113 to 68	0.503(0), 0.385(25)
2-Propanethiol	24.26	0.1174	−131 to 52.6	0.477(0), 0.357(25), 0.280(50)
Propanoic acid	28.68	0.0993	−20.5 to 141.1	1.030(25), 0.749(50), 0.569(75)
Propanoic anhydride	30.30(20)	29.70(25)	−45 to 170	1.144(20), 1.061(25)
1-Propanol	25.26	0.0777	−127 to 97.2	2.522(15), 1.722(30), 1.107(50)

(*Continued*)

TABLE 2.30 Viscosity and Surface Tension of Organic Compounds (*Continued*)

Substance	Surface tension, mN · m⁻¹		Liquid range, °C	Viscosity, mN · s · m⁻²
	a	*b*		
2-Propanol	22.90	0.0789	−89.5 to 82.4	2.859(15), 1.765(30), 1.028(50)
2-Propen-1-ol (allyl alcohol)	27.53	0.0902	−129 to 98	1.363(20), 0.914(40)
Propionaldehyde (propanal)			−81 to 48	0.357(15), 0.321(25)
Propionamide	39.05	0.0909	79 to 222.2	
Propyl acetate	26.60	0.1120	−93 to 101.6	0.768(0), 0.544(25), 0.406(50)
Propylamine	24.86	0.1243	−83 to 42.2	0.376(25)
Propylbenzene	31.13	0.1075	−99.2 to 159.2	
Propyl benzoate	36.55	0.1069	−51.6 to 98	
Propyl butanoate	27.06	0.1000	−95 to 143	0.831(20)
1,2-Propylene glycol			−60 to 188	40.4(0), 11.3(25), 4.770(50)
Propyleneimine			up to 66	0.491(25)
1,2-Propylene oxide			−112 to 34	0.327(20), 0.28(25)
Propyl formate	26.77	0.1119	−92.9 to 80.9	0.669(0), 0.574(20), 0.417(40)
Propyl isobutanoate	25.83	0.1015	up to 135	0.831(20)
Propyl nitrate	29.67	0.1237	−100 to 110.1	
Propyl pentanoate	27.72	0.0984	−75.9 to 122.5	1.053(20)
Propyl propanoate	26.85	0.1059	−76 to 122.5	0.673(20)
Propyne	14.51	0.1482	−102.8 to −23.2	
2-Propyn-1-ol	38.59	0.1270	−51.8 to 114	1.68(20)
Pyridazine	50.55	0.1036	−8 to 208	
Pyridine	39.82	0.1306	−41.6 to 115.2	1.361(0), 0.879(25), 0.637(50)
Pyrimidine	32.85	0.1010	22 to 124	
Pyrrole	39.81	0.1100	−23.4 to 130	2.085(0), 1.225(25), 0.828(50)
Pyrrolidine	31.48	0.0900	−58 to 86.5	1.071(0), 0.704(25), 0.512(50)
2-Pyrrolidone			25 to 251	13.3(25)
Quinoline	45.25	0.1063	−15 to 237	3.337(25), 1.892(50), 1.201(75)
Salicylaldehyde	45.38	0.1242	−7 to 197	2.90(20), 1.71(30), 1.669(45)
Squalane			−38 to 350	6.08(20)
Squalene			−75 to >285	12(25)
Stearic acid			67 to >184	11.6(70)
Styrene	32.0(20)	30.98(30)	−31 to 145	1.050(0), 0.696(25), 0.507(50)
Succinonitrile	53.26	0.1079	54.5 to 266	2.591(60), 2.008(75)
1,1,2,2-Tetrabromoethane	52.37	0.1463	0 to 243.5	13.50(11), 9.797(20)
1,1,2,2-Tetrachlorodifluoroethane	26.13	0.1133	26.0 to 92.8	1.21(25), 1.208(30)
1,1,2,2-Tetrachloroethane	38.75	0.1268	−70.2 to 130.5	1.844(15), 1.456(30)
Tetrachloroethylene	32.86(15)	31.27(30)	−22 to 121	1.932(15), 0.798(30), 0.654(53)
Tetradecane	28.30	0.0869	5.5 to 253.6	2.128(25), 1.376(50), 0.953(75)
Tetradecanoic acid	33.90	0.0932	54 to >250	
1-Tetradecanol	32.72	0.0703	39.5 to 289	
Tetraethylene glycol	45(25)		−6 to 328	44.9(25)
Tetraethyl lead	30.50	0.0969	−136 to >85	
Tetraethylsilane	25.22	0.1079	−82 to 154.7	
Tetraethyl silicate	23.63	0.0979	−82.5 to 169	
Tetrahydrofuran	26.5(25)		−108.5 to 65	0.605(0), 0.460(25), 0.359(50)
2,5-Tetrahydrofurandimethanol			< −50 to 265	225(25)
Tetrahydro-2-furanmethanol	39.96	0.1008	< −80 to 178	6.24(20)
1,2,3,4-Tetrahydronaphthalene	35.55	0.0954	−35.8 to 207.6	2.202(20), 2.003(25)
Tetrahydropyran			−45 to 88	0.826(20), 0.764(25)

TABLE 2.30 Viscosity and Surface Tension of Organic Compounds (*Continued*)

Substance	Surface tension, $mN \cdot m^{-1}$ a	b	Liquid range, °C	Viscosity, $mN \cdot s \cdot m^{-2}$
Tetrahydropyran-2-methanol	34.1(25)		−70 to 187	11.0(20)
Tetrahydrothiophene-1,1-dioxide (sulfolane)	35.5(30)		27.6 to 287.3	9.87(30), 6.280(50), 3.818(75)
Tetrahydrothiophene oxide				52(30), 19(80)
Thiacyclohexane	36.06(20)	33.74(40)		
Thiacyclopentane	38.44	0.1342		1.042(20), 0.971(25)
2,2′-Thiodiethanol	53.8(20)		−10.2 to 282	65.2(20)
Thiophene	34.00	0.1328	−39.4 to 84	0.871(0), 0.662(20), 0.353(82)
Thymol	33.95	0.0821	49 to 232	
Toluene	30.90	0.1189	−94.9 to 110.6	0.623(15), 0.523(30), 0.424(50)
p-Toluenesulfonyl chloride	42.41	0.0903	67 to >134	
o-Toluidine	42.87	0.1094	−16.5 to 200	5.195(15), 4.39(20)
m-Toluidine	40.33	0.0979	−31 to 203	4.418(15), 2.741(30)
p-Toluidine	39.58	0.0957	43.8 to 200	1.945(45), 1.557(60)
m-Tolunitrile	38.85	0.1013	−23 to 210	
p-Tolunitrile	39.79	0.1100	29.5 to 85	
Tribenzylamine	42.41	0.0953	91-94 to bp	
Tribromomethane	48.14	0.1308	8.1 to 149.6	2.152(15), 1.741(30), 1.367(50)
1,2,3-Tribromopropane	47.99	0.1267	16.5 to 220	
Tributylamine	26.47	0.0831	−70 to 216	1.35(25)
Tributyl borate	26.2(20)	25.8(25)	< −70 to 234	1.776(20), 1.601(25)
Tributyl phosphite	27.57	0.0865	up to >125	1.9(25)
Tributyl phosphate	28.71	0.0666	−79 to 289	11.1(15), 3.39(25)
Trichloroacetaldehyde	27.66	0.1197	−57.5 to 97.8	
Trichloroacetic acid	35.4	0.0895	57.5 to 196.5	
1,1,1-Trichloroethane	28.28	0.1242	−30.4 to 74	0.903(15), 0.725(30), 0.578(50)
1,1,2-Trichloroethane	37.40	0.1351	−37 to 114	0.119(20), 0.110(25)
Trichloroethylene	29.5(20)	28.8(25)	−84.8 to 87	0.703(0), 0.545(25), 0.444(50)
Trichlorofluoromethane	18(25)		−111 to 23.8	0.740(−25), 0.539(0)
2,4,6-Trichlorophenol	43.13	0.0955	69 to 246	
1,2,3-Trichloropropane	37.8(20)	37.05(25)	−14.7 to 157	
Trichlorosilane	20.43	0.1076	−127 to 32	0.332(20), 0.316(25)
α,α,α-Trichlorotoluene			−5 to 223	3.07(10), 2.55(17)
1,1,2-Trichloro-1,2,2-trifluoroethane	17.75(20)	16.56(30)	−35 to 47.7	0.711(20), 0.627(30)
Tridecane	27.73	0.0872	−5 to 235	2.909(0), 1.724(25), 1.129(50)
1-Tridecene	28.01	0.0884	−13 to 232.8	
Triethanolamine			20.5 to 335.4	609(25), 114(50), 31.5(75)
Triethylamine	22.70	0.0992	−114.7 to 88.8	0.455(0), 0.347(25), 0.273(50)
Triethylene glycol	47.33	0.0880	−7 to 285	49.0(20), 8.5(60)
Triethyl phosphate	31.81	0.0928	−56 to 215	1.684(40), 1.376(55)
Triethyl phosphite	25.73	0.0878	up to 156	0.72(25)
Trifluoroacetic acid	15.64	0.1844	−15.3 to 73	0.926(20), 0.808(25), 0.571(50)
2,2,2-Trifluoroethanol	20.6(33)		−43.5 to 74	1.996(20)
Trimethylamine	16.24	0.1133	−117 to 2.9	0.321(−33.5)
1,2,3-Trimethylbenzene	30.91	0.1040	−25.4 to 176.1	
1,2,4-Trimethylbenzene	31.76	0.1025	−43.9 to 169	0.894(15), 0.730(30)
1,3,5-Trimethylbenzene	29.79	0.0897	−44.7 to 165	1.154(20)
2,2,3-Trimethylbutane	20.70	0.0973	−24.9 to 80.9	0.579(20)

(*Continued*)

TABLE 2.30 Viscosity and Surface Tension of Organic Compounds (*Continued*)

Substance	Surface tension, $mN \cdot m^{-1}$		Liquid range, °C	Viscosity, $mN \cdot s \cdot m^{-2}$
	a	*b*		
cis,cis-1,3,5-Trimethylcyclohexane				0.632(20), 0.558(30)
trans-1,3,5-Trimethylcyclohexane			−107.4 to 140.5	0.714(20), 0.624(30)
Trimethylene sulfide	36.3(20)	35.0(30)	−73.2 to 95	0.638(20), 0.607(25)
3,5,5-Trimethyl-1-hexanol			< −70 to 194	11.06(25)
2,2,3-Trimethylpentane	22.46	0.0895	−112.3 to 110	0.598(20)
2,2,4-Trimethylpentane	20.55	0.0888	−107.4 to 99.2	0.502(20)
Trimethyl phosphite	27.18(20)	24.88(40)	−78 to 112	0.61(20)
2,4,6-Trimethylpyridine			−46 to 171	1.498(20)
Triphenylamine	46.2	0.0955	125 to 348	
Triphenyl phosphite			22 to 360	6.95(45)
Tripropylamine	24.58	0.0878	−93.5 to 158	
Tripropylene glycol	34(25)		up to 273	56.1(25)
Tripropylene glycol butyl ether	28.8(25)		up to 276	6.58(25)
Tripropylene glycol ethyl ether	28.2(25)			5.17(25)
Tripropylene glycol isopropyl ether	27.4(25)			7.7(25)
Tripropylene glycol methyl ether	30.0(25)		−42 to 242.4	5.96(25)
Tris(*m*-tolyl) phosphite				37.55(15), 9.132(45), 5.075(65)
Tris(*p*-tolyl) phosphite				35.52(15), 8.794(45), 5.017(65)
Tri-*o*-tolyl phosphate	40.9(20)		11 to 410	38.8(35), 16.8(55)
Undecane	26.26	0.0901	−25.6 to 196	1.707(0), 1.098(25), 0.761(50)
Vinyl acetate	23.95(20)	22.54(30)	−93 to 73	0.421(20)
o-Xylene	32.51	0.1101	−25.2 to 145	1.084(0), 0.760(25), 0.561(50)
m-Xylene	31.23	0.1104	−47.9 to 139	0.795(0), 0.581(25), 0.445(50)
p-Xylene	30.69	0.1074	13 to 138	0.603(25), 0.457(50), 0.359(75)

TABLE 2.31 Viscosity of Aqueous Glycerol Solutions

% Weight glycerol	Grams per liter	Relative density 25°/25°C	Viscosity, mN · s · m^{-2}		
			20°C	25°C	30°C
100	1261	1.262 01	1 495	942	622
99	1246	1.259 45	1 194	772	509
98	1231	1.256 85	971	627	423
97	1216	1.254 25	802	521	353
96	1201	1.251 65	659	434	296
95	1186	1.249 10	543.5	365	248
80	966.8	1.209 25	61.8	45.72	34.81
50	563.2	1.127 20	6.032	5.024	4.233
25	265.0	1.061 15	2.089	1.805	1.586
10	102.2	1.023 70	1.307	1.149	1.021

TABLE 2.32 Viscosity of Aqueous Sucrose Solutions

% Weight sucrose	Grams per liter	Relative density 20°/4°C	Viscosity, mN · s · m^{-2}		
			15°C	20°C	25°C
75	1034	1.379 0	4 039	2 328	1 405
70	943.0	1.347 2	746.9	481.6	321.6
65	855.6	1.316 3	211.3	147.2	105.4
60	771.9	1.286 5	79.49	58.49	40.03
50	614.8	1.299 6	19.53	15.43	12.40
40	470.6	1.176 4	7.463	6.617	5.164
30	338.1	1.127 0	3.757	3.187	2.735

2.4 REFRACTION AND REFRACTIVE INDEX

The refractive index n is the ratio of the velocity of light in a particular substance to the velocity of light in vacuum. Values reported refer to the ratio of the velocity in air to that in the substance saturated with air. Usually the yellow sodium doublet lines are used; they have a weighted mean of 589.26 nm and are symbolized by D. When only a single refractive index is available, approximate values over a small temperature range may be calculated using a mean value of 0.000 45 per degree for dn/dt, and remembering that n_D decreases with an increase in temperature. If a transition point lies within the temperature range, extrapolation is not reliable.

The *specific refraction* r_D is given by the Lorentz and Lorenz equation,

$$R_D = \frac{n_D^2 - 1}{n_D^2 + 2} \cdot \frac{1}{\rho}$$

where ρ is the density at the same temperature as the refractive index, and is independent of temperature and pressure. The molar refraction is equal to the specific refraction multiplied by the molecular weight. It is a more or less additive property of the groups or elements comprising the compound. A set of atomic refractions is given in Table 1.12; an extensive discussion will be found in Bauer, Fajans, and Lewin, in *Physical Methods of Organic Chemistry*, 3d ed., A. Weissberger (ed.), vol. 1, part II, chap. 28, Wiley-Interscience, New York, 1960.

The empirical Eykman equation

$$\frac{n_D^2 - 1}{n_D + 0.4} \cdot \frac{1}{\rho} = \text{constant}$$

offers a more accurate means for checking the accuracy of experimental densities and refractive indices, and for calculating one from the other, than does the Lorentz and Lorenz equation.

The refractive index of moist air can be calculated from the expression

$$(n-1)\times 10^6 = \frac{103.49}{T}p_1 + \frac{177.4}{T}p_2 + \frac{86.26}{T}\left(1+\frac{5748}{T}\right)p_3$$

where p_1 is the partial pressure of dry air (in mmHg), p_2 is the partial pressure of carbon dioxide (in mmHg), p_3 is the partial pressure of water vapor (in mmHg), and T is the temperature (in kelvins).

Example: 1-Propynyl acetate has $n_D = 1.4187$ and density = 0.9982 at 20°C; the molecular weight is 98.102. From the Lorentz and Lorenz equation,

$$r_D = \frac{(1.4187)^2+1}{(1.4187)^2+2}\cdot\frac{1}{0.9982} = 0.2528$$

The molar refraction is

$$Mr_D = (98.102)(0.2528) = 24.80$$

From the atomic and group refractions in Table 5.19, the molar refraction is computed as follows:

6 H	6.600
5 C	12.090
1 C≡C	2.398
1 O(ether)	1.643
1 O(carbonyl)	2.211
	$Mr_D = 24.942$

TABLE 2.33 Atomic and Group Refractions

Group	Mr_D	Group	Mr_D
H	1.100	N (primary aliphatic amine)	2.322
C	2.418	N (*sec*-aliphatic amine)	2.499
Double bond (C═C)	1.733	N (*tert*-aliphatic amine)	2.840
Triple bond (C≡C)	2.398	N (primary aromatic amine)	3.21
Phenyl (C_6H_5)	25.463	N (*sec*-aromatic amine)	3.59
Naphthyl ($C_{10}H_7$)	43.00	N (*tert*-aromatic amine)	4.36
O (carbonyl) (C═O)	2.211	N (primary amide)	2.65
O (hydroxyl) (O—H)	1.525	N (*sec* amide)	2.27
O (ether, ester) (C—O—)	1.643	N (*tert* amide)	2.71
F (one fluoride)	0.95	N (imidine)	3.776
(polyfluorides)	1.1	N (oximido)	3.901
Cl	5.967	N (carbimido)	4.10
Br	8.865	N (hydrazone)	3.46
I	13.900	N (hydroxylamine)	2.48
S (thiocarbonyl) (C═S)	7.97	N (hydrazine)	2.47
S (thiol) (S—H)	7.69	N (aliphatic cyanide) (C≡N)	3.05
S (dithia) (—S—S—)	8.11	N (aromatic cyanide)	3.79
Se (alkyl selenides)	11.17	N (aliphatic oxime)	3.93
3-membered ring	0.71	NO (nitroso)	5.91
4-membered ring	0.48	NO (nitrosoamine)	5.37
		NO_2 (alkyl nitrate)	7.59
		(alkyl nitrite)	7.44
		(aliphatic nitro)	6.72
		(aromatic nitro)	7.30
		(nitramine)	7.51

TABLE 2.34 Refractive Indices of Organic Compounds

Substance	Formula	Density, g/ml	Refractive index
Acenaphthene	$C_{12}H_{10}$	1.220	1.6048/98.8°
Acetaldehyde	C_2H_4O	0.788/16°	1.3316
Acetamide	C_2H_5ON	1.159	1.4274/78°
Acetanilide	C_8H_9ON	1.21/4°	
Acetic acid	$C_2H_4O_2$	1.0492	1.3718
Acetic anhydride	$C_4H_6O_3$	1.0850/15°	1.3904
Acetone	C_3H_6O	0.787/25°	1.3620/15°
Acetonitrile	C_2H_3N	0.7828	1.3460
Acetophenone	C_8H_8O	1.0329/15°	1.5342/19°
Acetyl chloride	C_2H_3OCl	1.1051	1.3898
Acetylene	C_2H_2	0.61/–80°	
Adipic acid	$C_6H_{10}O_4$	1.366	
Alloxan + $_4H_2O$	$C_4H_{10}O_8N_2$		
Allyl alcohol	C_3H_6O	0.8573/15°	1.4135
p-Aminobenzoic acid	$C_7H_7O_2N$		
2-Aminopyridine	$C_5H_6N_2$		
n-Amyl alcohol	$C_5H_{12}O$	0.8154	1.414/13°
act-Amyl alcohol	$C_5H_{12}O$	0.816	
sec-Amyl alcohol	$C_5H_{12}O$	0.8103	1.4053
tert-Amyl alcohol	$C_5H_{12}O$	0.809	1.4045
Aniline	C_6H_7N	1.026/15°	1.5863
Aniline hydrochloride	C_6H_8NCl	1.222/4°	
Anisole	C_7H_8O	0.9925/25°	1.5150/22°
Anthracene	$C_{14}H_{10}$	1.243	
Anthraquinone	$C_{14}H_8O$	1.419/4°	
Azobenzene	$C_{12}H_{10}N_2$		
Benzaldehyde	C_7H_6O	1.0504/15°	1.5463/17.6°
Benzene	C_6H_6	0.8790	1.5011
Benzoic acid	$C_7H_6O_2$	1.2656/15°	1.5397/15°
Benzoic anhydride	$C_{14}H_{10}O_3$	1.1989/15°	1.5767/15°
Benzoin	$C_{14}H_{12}O_2$		
Benzonitrile	C_7H_5N	1.0093/15°	1.5289
Benzophenone (*a*)	$C_{13}H_{10}O$	1.085/50°	
Benzoquinone	$C_6H_4O_2$		
Benzoyl chloride	C_7H_5OCl	1.212	1.5537
Benzoyl peroxide	$C_{14}H_{10}O_4$		
Benzyl alcohol	C_7H_8O	1.049/15°	1.5396
Benzyl benzoate	$C_{14}H_{12}O_2$	1.114/18°	1.5681/21°
Benzyl chloride	C_7H_7Cl	1.0983	1.5415/15°
Benzyl cinnamate	$C_{16}H_{14}O_2$		
Borneol (DL)	$C_{10}H_{18}O$	1.01	
a-Bromonaphthalene	$C_{10}H_7Br$	1.4888/16.5°	1.6601/16.5°
Bromobenzene	C_6H_5Br	1.4978/15°	1.5625/15°
Bromoform	$CHBr_3$	2.900/15°	1.6005/15°
n-Butane	C_4H_{10}	0.5788 (at sat. pressure)	
n-Butyl alcohol	$C_4H_{10}O$	0.8098	1.3993
iso-Butyl alcohol	$C_4H_{10}O$	0.8169	1.3968/17.5°
sec-Butyl alcohol	$C_4H_{10}O$	0.808	1.3949/25°
tert-Butyl alcohol	$C_4H_{10}O$	0.7887	1.3878
n-Butyl chloride	C_4H_9Cl	0.9074/0	1.4015
n-Butyric acid	$C_4H_8O_2$	0.9587	1.3991
iso-Butyric acid	$C_4H_8O_2$	0.950	

(Continued)

TABLE 2.34 Refractive Indices of Organic Compounds (*Continued*)

Substance	Formula	Density, g/ml	Refractive index
Camphene (DL)	$C_{10}H_{16}$	0.879	1.4402/80°
Camphor(D)	$C_{10}H_{16}O$	0.992/10°	
Carbitol (Diethyleneglycol-monomethylether)	$C_6H_{14}O_3$	0.9902	
Carbon disulphide	CS_2	1.2927/0°	1.6276
Carbon tetrabromide	CBr_4	2.9109/99.5°	
Carbon tetrachloride	CCl_4	1.6320/0°	1.4607
Cellosolve (Glycolmonoethylether)	$C_4H_{10}O_2$	0.9311	
Chloral hydrate	$C_2H_3O_2Cl_3$	1.9081	
Chloroacetic acid	$C_2H_3O_2Cl$	1.39/75°	1.4297/65°
Chlorobenzene	C_6H_5Cl	1.066	1.5248
Chloroform	$CHCl_3$	1.4985/15°	1.4467
Cholesterol	$C_{27}H_{46}O$	1.067	
Cineol (Eucalyptol)	$C_{10}H_{18}O$	0.9267	1.4584/18°
Cinnamic acid (trans)	$C_9H_8O_2$	1.247	
Cinnamyl alcohol	$C_9H_{10}O$	1.0440	1.5819
Citric acid	$C_6H_8O_7$	1.542/18°	
o-Cresol	C_7H_8O	1.051	1.5372/40°
m-Cresol	C_7H_8O	1.035	1.5406
p-Cresol	C_7H_8O	1.035	1.5316
Cumene	C_9H_{12}	0.8615	1.4909
Cyclohexane	C_6H_{12}	0.7786	1.4262
Cyclohexanol	$C_6H_{12}O$	0.9624	1.4656/22°
Cyclohexanone	$C_6H_{10}O$	0.9478	1.4507
Cyclohexene	C_6H_{10}	0.8108	1.4467
p-Cymene	$C_{10}H_{14}$	0.8766	1.5006
cis-Decalin	$C_{10}H_{18}$	0.8963	1.4811
trans-Decalin	$C_{10}H_{18}$	0.8703/18°	1.4697/18°
Dibenzyl	$C_{14}H_{14}$	0.995	
n-Dibutyl phthalate	$C_{16}H_{22}O_4$	1.0465	
Diethylamine	$C_4H_{11}N$	0.7108/18°	1.3873/18°
Difluorodichloro-methane (Freon 12)	$CC_{12}F_2$		
Difluoromonochloro-methane (Freon 22)	$CHClF_2$		
Dimethylamine	C_2H_7N	0.6804/0°	1.350/17°
Dimethylaniline	$C_8H_{11}N$	0.9557	1.5582
Dioxane	$C_4H_8O_2$	1.0338	1.4224
Diphenyl	$C_{12}H_{10}$	1.180/0°	1.5852/79°
Diphenylamine	$C_{12}H_{11}N$	1.159	
Epichlorhydrin	C_3H_5OCl	1.180	1.4420/11.6°
Ethane	C_2H_6		
Ethanolamine	C_2H_7ON	1.022	1.4539
di-Ethanolamine	$C_4H_{11}O_2N$	1.0966	1.4776
tri-Ethanolamine	$C_6H_{15}O_3N$	1.1242	1.4852
Ether (diethyl)	$C_4H_{10}O$	0.714/20°	1.3538
Ethyl acetate	$C_4H_8O_2$	0.9245	1.3701/25°
Ethyl acetoacetate	$C_6H_{10}O_3$	1.0282	1.4209/16°
Ethyl alcohol	C_2H_6O	0.7893	1.3610/20.5°
Ethylamine	C_2H_7N	0.7057/0°	

TABLE 2.34 Refractive Indices of Organic Compounds (*Continued*)

Substance	Formula	Density, g/ml	Refractive index
Ethylbenzene	C_8H_{10}	0.8669	1.4959
Ethyl benzoate	$C_9H_{10}O_2$	1.0509/15°	1.5068/17.3°
Ethyl bromide	C_2H_5Br	1.4555	1.4239
Ethyl chloride	C_2H_5Cl	0.9214/0°	
Ethylene	C_2H_4		
Ethylenediamine	$C_2H_8N_2$	0.902/15°	1.4540/26.1°
Ethylene dibromide	$C_2H_4Br_2$	2.1785	1.5379
Ethylene dichloride	$C_2H_4Cl_2$	1.2521	1.4443
Ethylene glycol	$C_2H_6O_2$	1.1155	1.4274
Ethylene oxide	C_2H_4O	0.877/7°	1.3597/7°
Ethyl formate	$C_3H_6O_2$	0.9168	1.3598
Ethyl iodide	C_2H_5I	1.9133/30°	1.5168/15°
Ethyl mercaptan	C_2H_6S	0.8315/25°	1.4351
Ethyl nitrate	$C_2H_5O_3N$	1.109	1.3853
Ethyl nitrite	$C_2H_5O_2N$	0.900/15°	
Ethyl oxalate	$C_6H_{10}O_4$	1.0785	1.4101
Ethyl salicylate	$C_9H_{10}O_3$	1.131	1.5226
Ethyl sulphate	$C_4H_{10}O_4S$	1.180/18°	1.4010/18°
Eugenol	$C_{10}H_{12}O_2$	1.0620/25°	1.5439/19°
Fluorescein	$C_{20}H_{12}O_5$		
Fluorobenzene	C_6H_5F	1.0236	1.4677
Formaldehyde	CH_2O	0.815/–20°	
Formamide	CH_3ON	1.1334	1.4472
Formic acid	CH_2O_2	1.220	1.3714
Fructose	$C_6H_{12}O_6$	1.598	
Fumaric acid	$C_4H_4O_4$	1.635	
Furfural	$C_5H_4O_2$	1.1594	1.5261
Furfuryl alcohol	$C_5H_6O_2$	1.1282/23°	1.4852
Furan	C_4H_4O	0.9644/0°	1.4216
Glucose	$C_6H_{12}O_6$	1.544/25°	
Glycerol	$C_3H_8O_3$	1.2604/17.5°	1.4730
Glyceryl trioleate	$C_{57}H_{104}O_6$	0.8992/50°	1.4561/60°
Glyceryl tripalmitate	$C_{51}H_{98}O_6$	0.8752/70°	1.4381/80°
Glyceryl tristearate	$C_{57}H_{110}O_6$	0.8559/90°	1.4385/80°
Glycine	$C_2H_5O_2N$		
Guaiacol	$C_7H_8O_2$	1.1287/21.4°	
n-Heptane	C_7H_{16}	0.6838	1.3877
Hexachlorotethane	C_2Cl_6	2.091	
Hexamine	$C_6H_{12}N_4$		
n-Hexane	C_6H_{14}	0.6594	1.3749
Hippuric acid	$C_9H_9O_3N$	1.371	
Hydroquinone	$C_6H_6O_2$	1.358	
Indene	C_9H_8	0.996	1.5766
Iodoform	CHI_3	4.008	
Isobutane	C_4H_{10}	0.5572 (at sat. press.)	
Isopentane	C_5H_{12}	0.6192	1.3538
isoprene	C_5H_8	0.6806	1.4194
Isooctane	C_8H_{18}	0.6919	1.3915
Isoquinoline	C_9H_7N	1.099	1.6223/25°
Lactic acid	$C_3H_6O_3$	1.2485	1.4414
Lactose + H_2O	$C_{12}H_{24}O_1$	1.525	
Maleic acid	$C_4H_4O_4$	1.5920	

(*Continued*)

TABLE 2.34 Refractive Indices of Organic Compounds (*Continued*)

Substance	Formula	Density, g/ml	Refractive index
Maleic anhydride	$C_4H_2O_3$	0.934	
Malonic acid	$C_3H_4O_4$	1.631/15°	
Maltose + H_2O	$C_{12}H_{24}O_1$	1.540	
Menthol (L)	$C_{10}H_{20}O$	0.903/15°	
Mesitylene	C_9H_{12}	0.8652	1.4994
Metaldehyde	$(C_2H_4O)_n$		
Methane	CH_4		
Methyl acetate	$C_3H_6O_2$	0.9280	1.3593/20°
Methyl alcohol	CH_4O	0.7910	1.3276/25°
Methylamine	CH_5N	0.699/–10.8°	
Methylaniline	C_7H_9N	0.9891	1.5702/21.2°
Methyl anthranilate	$C_8H_9O_2N$	1.1682/18.6°	
Methyl benzoate	$C_8H_8O_2$	1.0937/15°	1.5205/15°
Methyl bromide	CH_3Br	1.732/0°	
Methyl carbonate	$C_3H_6O_3$	1.0694	1.3687
Methyl chloride	CH_3Cl	0.991/–25°	
Methylene bromide	CH_2Br_2	2.8098/15°	
Methylene chloride	CH_2Cl_2	1.3348/15°	1.4237
Methyl ethyl ketone	C_4H_8O	0.8054	1.3814/15°
Methyl formate	$C_2H_4O_2$	0.9867/15°	1.344
Methyl iodide	CH_3I	2.251/30°	1.5293/21°
Methyl methacrylate	$C_5H_8O_2$	0.936	1.413
Methyl sulphate	$C_2H_6O_4S$	1.3348/15°	1.3874
Methyl salicylate	$C_8H_8O_3$	1.1787/25°	1.538/18.1°
Monofluorotrichlo-romethane (Freon 11)	CCl_3F	1.494/17°	
Morpholine	C_4H_9ON	0.9994	1.4545
Naphthalene	$C_{10}H_8$	1.14	1.5822/100°
α-Naphthol	$C_{10}H_8O$	1.099/99°	1.6206/98.7°
β-Naphthol	$C_{10}H_8O$	1.272	
α-Naphthylamine	$C_{10}H_9N$	1.1196/25°	1.6703/51°
β-Naphthylamine	$C_{10}H_9N$	1.0614/98°	1.6493/98°
Nicotine (L)	$C_{10}H_{14}N_2$	1.0097	1.5280
Nitrobenzene	$C_6H_5O_2N$	1.1732/25°	1.5530
Nitroethane	$C_2H_5O_2N$	1.050	1.3916
Nitromethane	CH_3O_2N	1.137	1.3818
1-Nitropropane	$C_3H_7O_2N$	1.001	1.4015
2-Nitropropane	$C_3H_7O_2N$	0.990	1.3941
n-Octane	C_8H_{18}	0.7025	1.3974
n-Octyl alcohol	$C_8H_{18}O$	0.8270	1.4292
Oleic acid	$C_{18}H_{34}O_2$	0.898	1.4582
Oxalic acid	$C_2H_2O_4$		
Palmitic acid	$C_{16}H_{32}O_2$	0.8527/62°	1.4339/60°
Paraformaldehyde	$(CH_2O)n$		
Paraldehyde	$C_6H_{12}O_3$	0.9943	1.4049
n-Pentane	C_5H_{12}	0.6262	1.3575
Phosgene	$COCl_2$		
Phenanthrene	$C_{14}H_{10}$	1.17	1.6567/129°
Phenol	C_6H_6O	1.073	1.5245/40.6°
Phthalic acid	$C_8H_6O_4$	1.593	
Phthalic anhydride	$C_8H_4O_3$	1.527/4°	
Phthalimide	$C_8H_5O_2N$		

TABLE 2.34 Refractive Indices of Organic Compounds (*Continued*)

Substance	Formula	Density, g/ml	Refractive index
α-Picoline	C_6H_7N	0.9443	1.5010
β-Picoline	C_6H_7N	0.9566	1.5068
γ-Picoline	C_6H_7N	0.9548	1.5058
Picric acid	$C_6H_3O_7N_3$	1.763	
Picryl chloride	$C_6H_2O_6N_3Cl$	1.797	
Pinene (Turpentine)	$C_{10}H_{16}$	0.861	1.4685/15°
Piperidine	$C_5H_{11}N$	0.8606	1.4530
Propane	C_3H_8		
n-Propyl acetate	$C_5H_{10}O_2$	0.887	1.3844
n-Propyl alcohol	C_3H_8O	0.8035	1.3850
iso-Propyl alcohol	C_3H_8O	0.7855	1.3776
Propylene	C_3H_6	0.5139 (at sat. press.)	
Pyridine	C_5H_5N	0.9831	1.5102
Pyrocatechol	$C_6H_6O_2$	1.344	
Pyrogallol	$C_6H_6O_3$		
Quinhydrone	$C_{12}H_{10}O_4$	1.401	
Quinoline	C_9H_7N	1.095	1.6269
Resorcinol	$C_6H_6O_2$	1.285/15°	
Salicylic acid	$C_7H_6O_3$	1.443	
Stearic acid	$C_{18}H_{36}O_2$	0.9408	1.4335/70°
Styrene	C_8H_8	0.9060	1.5469
Succinic acid	$C_4H_6O_4$	1.564/15°	
Succinic anhydride	$C_4H_4O_3$	1.234	
Sucrose	$C_{12}H_{22}O_{11}$	1.588/15°	
Sylvan (2-Methylfuran)	C_5H_6O	0.916	
Tartaric acid (*meso*-)	$C_4H_6O_6$	1.666	
Tartaric acid (racemic) + H_2O	$C_4H_8O_7$	1.697	
Tartaric acid (D)	$C_4H_6O_6$	1.7598	
Tartaric acid (L)	$C_4H_6O_6$	1.7598	
Tetralin	$C_{10}H_{12}$		1.5453/17°
Thiophen	C_4H_4S	1.0644	1.5287
Thiourea	CH_4N_2S	1.405	
Thymol	$C_{10}H_{14}O$	0.969	
Toluene	C_7H_8	0.8670	1.4969
o-Toluidine	C_7H_9N	1.0035	1.5688
m-Toluidine	C_7H_9N	0.987/25°	1.5686
p-Toluidine	C_7H_9N	0.961/50°	1.5532/59.1°
Trichloroethylene	C_2HCl_3	1.4597/15°	1.4782
Tri-*o*-cresyl phosphate	$C_{21}H_{21}O_4P$		
Tri-*p*-cresyl phosphate	$C_{21}H_{21}O_4P$		
Triethylamine	$C_6H_{15}N$	0.7495/0°	1.4003
Trimethylamine	C_3H_9N	0.6709/0°	
Trinitrotoluene	$C_7H_5O_6N_3$	1.654	
Triphenylmethane	$C_{19}H_{16}$		
Urea	CH_4ON_2	1.335	
Uric acid	$C_5H_4O_3N_4$	1.893	
n-Valeric acid	$C_5H_{10}O_2$	0.942	1.4086
iso-Valeric acid	$C_5H_{10}O_2$	0.937/15°	1.4018/22.4°
Vanillin	$C_8H_8O_3$		
o-Xylene	C_8H_{10}	0.8802	1.5054
m-Xylene	C_8H_{10}	0.8642	1.4972
p-Xylene	C_8H_{10}	0.8611	1.4958

(*Continued*)

TABLE 2.35 Solvents Having the Same Refractive Index and the Same Density at 25°C

Solvent 1	Solvent 2	Refractive index		Density, g/mL	
		1	2	1	2
Acetone	Ethanol	1.357	1.359	0.788	0.786
Ethyl formate	Methyl acetate	1.358	1.360	0.916	0.935
Ethanol	Propionitrile	1.359	1.363	0.786	0.777
2,2-Dimethylbutane	2-Methylpentane	1.366	1.369	0.644	0.649
2-Methylpentane	Hexane	1.369	1.372	0.649	0.655
Isopropyl acetate	2-Chloropropane	1.375	1.376	0.868	0.865
3-Butanone	Butyraldehyde	1.377	1.378	0.801	0.799
Butyraldehyde	Butyronitrile	1.378	1.382	0.799	0.786
Dipropyl ether	Butyl ethyl ether	1.379	1.380	0.753	0.746
Propyl acetate	Ethyl propionate	1.382	1.382	0.883	0.888
Propyl acetate	1-Chloropropane	1.382	1.386	0.883	0.890
Butyronitrile	2-Methyl-2-propanol	1.382	1.385	0.786	0.781
Ethyl propionate	1-Chloropropane	1.382	1.386	0.888	0.890
1-Propanol	2-Pentanone	1.383	1.387	0.806	0.804
Isobutyl formate	1-Chloropropane	1.383	1.386	0.881	0.890
1-Chloropropane	Butyl formate	1.386	1.387	0.890	0.888
Butyl formate	Methyl butyrate	1.387	1.391	0.888	0.875
Methyl butyrate	2-Chlorobutane	1.392	1.395	0.875	0.868
Butyl acetate	2-Chlorobutane	1.392	1.395	0.877	0.868
4-Methyl-2-pentanone	Pentanonitrile	1.394	1.395	0.797	0.795
4-Methyl-2-pentanone	1-Butanol	1.394	1.397	0.797	0.812
2-Methyl-1-propanol	Pentanonitrile	1.394	1.395	0.798	0.795
2-Methyl-1-propanol	2-Hexanone	1.394	1.395	0.798	0.810
2-Butanol	2,4-Dimethyl-3-pentanone	1.395	1.399	0.803	0.805
2-Hexanone	1-Butanol	1.395	1.397	0.810	0.812
Pentanonitrile	2,4-Dimethyl-3-pentanone	1.395	1.399	0.795	0.805
2-Chlorobutane	Isobutyl butyrate	1.395	1.399	0.868	0.860
Butyric acid	2-Methoxyethanol	1.396	1.400	0.955	0.960
1-Butanol	3-Methyl-2-pentanone	1.397	1.398	0.812	0.808
1-Chloro-2-methylpropane	Isobutyl butyrate	1.397	1.399	0.872	0.860
1-Chloro-2-methylpropane	Pentyl acetate	1.397	1.400	0.872	0.871
Methyl methacrylate	3-Methyl-2-pentanone	1.398	1.398	0.795	0.808
Triethylamine	2,2,3-Trimethylpentane	1.399	1.401	0.723	0.712
Butylamine	Dodecane	1.399	1.400	0.736	0.746
Isobutyl butyrate	1-Chlorobutane	1.399	1.401	0.860	0.875
1-Nitropropane	Propionic anhydride	1.399	1.400	0.995	1.007
Pentyl acetate	1-Chlorobutane	1.400	1.400	0.871	0.881
Pentyl acetate	Tetrahydrofuran	1.400	1.404	0.871	0.885
Dodecane	Dipropylamine	1.400	1.400	0.746	0.736
1-Chlorobutane	Tetrahydrofuran	1.401	1.404	0.871	0.885
Isopentanoic acid	2-Ethoxyethanol	1.402	1.405	0.923	0.926
Dipropylamine	Cyclopentane	1.403	1.404	0.736	0.740
2-Pentanol	4-Heptanone	1.404	1.405	0.804	0.813
3-Methyl-1-butanol	Hexanonitrile	1.404	1.405	0.805	0.801
3-Methyl-1-butanol	4-Heptanone	1.404	1.405	0.805	0.813
Hexanonitrile	4-Heptanone	1.405	1.405	0.801	0.813
Hexanonitrile	1-Pentanol	1.405	1.408	0.801	0.810
Hexanonitrile	2-Methyl-1-butanol	1.405	1.409	0.801	0.815
4-Heptanone	1-Pentanol	1.405	1.408	0.813	0.810

TABLE 2.35 Solvents Having the Same Refractive Index and the Same Density at 25°C (*Continued*)

Solvent 1	Solvent 2	Refractive index		Density, g/mL	
		1	2	1	2
2-Ethoxyethanol	Pentanoic acid	1.405	1.406	0.926	0.936
2-Heptanone	1-Pentanol	1.406	1.408	0.811	0.810
2-Heptanone	2-Methyl-1-butanol	1.406	1.409	0.811	0.815
2-Heptanone	Dipentyl ether	1.406	1.410	0.811	0.799
2-Pentanol	3-Isopropyl-2-pentanone	1.407	1.409	0.804	0.808
1-Pentanol	Dipentyl ether	1.408	1.410	0.810	0.799
2-Methyl-1-butanol	Dipentyl ether	1.409	1.410	0.815	0.799
Isopentyl isopentanoate	Allyl alcohol	1.410	1.411	0.853	0.847
Dipentyl ether	2-Octanone	1.410	1.414	0.799	0.814
2,4-Dimethyldioxane	3-Chloropentene	1.412	1.413	0.935	0.932
2,4-Dimethyldioxane	Hexanoic acid	1.412	1.415	0.935	0.923
Diethyl malonate	Ethyl cyanoacetate	1.412	1.415	1.051	1.056
3-Chloropentene	Octanoic acid	1.413	1.415	0.932	0.923
2-Octanone	1-Hexanol	1.414	1.416	0.814	0.814
2-Octanone	Octanonitrile	1.414	1.418	0.814	0.810
3-Octanone	3-Methyl-2-heptanone	1.414	1.416	0.830	0.818
3-Methyl-2-heptanone	1-Hexanol	1.415	1.416	0.818	0.814
3-Methyl-2-heptanone	Octanonitrile	1.415	1.418	0.818	0.810
1-Hexanol	Octanonitrile	1.416	1.418	0.814	0.810
Dibutylamine	Allylamine	1.416	1.419	0.756	0.758
Allylamine	Methylcyclohexane	1.419	1.421	0.758	0.765
Butyrolactone	1,3-Propanediol	1.434	1.438	1.051	1.049
Butyrolactone	Diethyl maleate	1.434	1.438	1.051	1.064
2-Chloromethyl-2-propanol	Diethyl maleate	1.436	1.438	1.059	1.064
N-Methylmorpholine	Dibutyl decanedioate	1.436	1.440	0.924	0.932
1,3-Propanediol	Diethyl maleate	1.438	1.438	1.049	1.064
Methyl salicylate	Diethyl sulfide	1.438	1.442	0.836	0.831
Methyl salicylate	1-Butanethiol	1.438	1.442	0.836	0.837
1-Chlorodecane	Mesityl oxide	1.441	1.442	0.862	0.850
Diethylene glycol	Formamide	1.445	1.446	1.128	1.129
Diethylene glycol	Ethylene glycol diglycidyl ether	1.445	1.447	1.128	1.134
Formamide	Ethylene glycol diglycidyl ether	1.446	1.447	1.129	1.134
2-Methylmorpholine	Cyclohexanone	1.446	1.448	0.951	0.943
2-Methylmorpholine	1-Amino-2-propanol	1.446	1.448	0.951	0.961
Dipropylene glycol monoethyl ether	Tetrahydrofurfuryl alcohol	1.446	1.450	1.043	1.050
1-Amino-2-methyl-2-pentanol	2-Butylcyclohexanone	1.449	1.453	0.904	0.901
2-Propylcyclohexanone	4-Methylcyclohexanol	1.452	1.454	0.923	0.908
Carbon tetrachloride	4,5-Dichloro-1,3-dioxolane-2-one	1.459	1.461	1.584	1.591
N-Butyldiethanolamine	Cyclohexanol	1.461	1.465	0.965	0.968
D-α-Pinene	*trans*-Decahydronaphthalene	1.464	1.468	0.855	0.867
Propylbenzene	*p*-Xylene	1.490	1.493	0.858	0.857
Propylbenzene	Toluene	1.490	1.494	0.858	0.860

(*Continued*)

TABLE 2.35 Solvents Having the Same Refractive Index and the Same Density at 25°C (*Continued*)

Solvent 1	Solvent 2	Refractive index 1	Refractive index 2	Density, g/mL 1	Density, g/mL 2
Phenyl 1-hydroxyphenyl ether	1,3-Dimorpholyl-2-propanol	1.491	1.493	1.081	1.094
Phenetole	Pyridine	1.505	1.507	0.961	0.978
2-Furanmethanol	Thiophene	1.524	1.526	1.057	1.059
m-Cresol	Benzaldehyde	1.542	1.544	1.037	1.041

2.5 VAPOR PRESSURE AND BOILING POINT

The *vapor pressure* is the pressure exerted by a pure component at equilibrium at any temperature when both liquid and vapor phases exist and thus extends from a minimum at the triple point temperature to a maximum at the critical temperature, the critical pressure the and is the most important of the basic thermodynamic properties affecting liquids and vapors.

Except at very high total pressures (above about 10 MPa), there is no effect of total pressure on vapor pressure. If such an effect is present, a correction can be applied. The pressure exerted above a solid-vapor mixture may also be called vapor pressure but is normally only available as experimental data for common compounds that sublime.

Numerous mathematical formulas relating the temperature and pressure of the gas phase in equilibrium with the condensed phase have been proposed. The Antoine equation (Eq. 1) gives good correlation with experimental values. Equation 2 is simpler and is often suitable over restricted temperature ranges. In these equations, and the derived differential coefficients for use in the Haggenmacher and Clausius-Clapeyron equations, the p term is the vapor pressure of the compound in pounds per square inch (psi), the t term is the temperature in degrees Celsius, and the T term is the absolute temperature in kelvins (t°C + 273.15).

Eq.	Vapor-pressure equation	dp/dT	$-[d(\ln p)/d(1/T)]$
1	$\log p = A - \dfrac{B}{t + C}$	$\dfrac{2.303pB}{(t + C)^2}$	$\dfrac{2.303BT^2}{(t + C)^2}$
2	$\log p = A - \dfrac{B}{T}$	$\dfrac{2.303pB}{T^2}$	$2.303B$
3	$\log p = A - \dfrac{B}{T} - C \log T$	$p\left(\dfrac{2.303B}{T^2} - \dfrac{C}{T}\right)$	$2.303B - CT$

Equations 1 and 2 are easily rearranged to calculate the temperature of the normal boiling point:

$$t = \frac{B}{A - \log p} - C$$

$$T = \frac{B}{A - \log P}$$

The constants in the Antoine equation may be estimated by selecting three widely spaced data points and substituting in the following equations in sequence:

$$\left(\frac{y_3 - y_2}{y_2 - y_1}\right)\left(\frac{t_2 - t_1}{t_3 - t_2}\right) = 1 - \left(\frac{t_3 - t_1}{t_3 + C}\right)$$

$$B = \left(\frac{y_3 - y_1}{t_3 - t_1}\right)(t_1 + C)(t_3 + C)$$

$$A = y_2 + \left(\frac{B}{t_2 + C}\right)$$

In these equations, $y_i = \log p_i$.

TABLE 2.36 Vapor Pressures of Various Organic Compounds

Substance	Eq.	Range, °C	*A*	*B*	*C*
Acenaphthene	1	147–187	7.728 19	2 534.234	245.576
	2	147–288	8.033	2834.99	
Acetaldehyde	1	liq	8.005 52	1 600.017	291.809
Acetic acid	1	liq	7.387 82	1 533.313	222.309
Acetic anhydride	1	liq	7.149 48	1 444.718	199.817
Acetone	1	liq	7.117 14	1 210.595	229.664
Acetonitrile	1	liq	7.119 88	1 314.4	230
Acetophenone	2	30–100	9.135 2	2 878.8	
Acetyl bromide	1	liq	5.197 02	545.784	150.396
Acetyl chloride	1	liq	6.948 87	1 115.954	223.554
Acetylene	1	–130 to –83	9.140 2	1 232.6	280.9
	1	–82 to –72	7.099 9	711.0	253.4
Acetyl iodide	1	liq	4.181 44	355.452	108.160
Acrylic acid	1	20–70	8.538 67	2305.843	266.547
Acrylonitrile	1	–20 to 140	7.038 55	1 232.53	222.47
Allyl isothiocyanate	1	10–50	5.126 58	791.434	154.019
m-Aminobenzotrifluoride	1	0–96	7.651 86	1 940.6	218.0
		96–300	7.170 30	1 650.21	193.58
p-Aminophenol	1	130–185	–3.357 50	699.157	–331.343
Aniline	1	102–185	7.320 10	1 731.515	206.049
Anthracene	2	100–160	8.91	3 761	
	1	176–380	7.674 01	2 819.63	247.02
9,10-Anthracenedione	2	224–286	12.305	5 747.9	
	2	285–370	8.002	3 341.94	
Benzene	1	–12 to 3	9.106 4	1 885.9	244.2
	1	8–103	6.905 65	1 211.033	220.790
Benzenethiol	1	52–198	6.990 19	1 529.454	203.048
Benzoic acid	2	60–110	9.033	3 333.3	
Benzonitrile	1	liq	6.746 31	1 436.72	181.0
Benzophenone	1	48–202	7.349 66	2 331.4	195.0
	1	200–306	7.162 94	2 051.855	173.074
Benzotrifluoride	1	–20 to 180	7.007 08	1 331.30	220.58
Benzoyl chloride	2	140–200	7.924 5	2 372.1	
Benzyl acetate	1	46–156	8.457 05	2 623.206	259.067
Benzyl alcohol	1	122–205	7.198 17	1 632.593	172.790

(Continued)

TABLE 2.36 Vapor Pressures of Various Organic Compounds (*Continued*)

Substance	Eq.	Range, °C	*A*	*B*	*C*
Biphenyl	1	69–271	7.245 41	1 998.725	202.733
2-(2-Biphenylyloxy)ethanol	1	240–300	8.005 87	2 776.761	206.914
Bromobenzene	1	56–154	6.860 64	1 438.817	205.441
2-Bromobenzyl cyanide	1	85–152	5.044 59	734.821	59.273
1-Bromobutane	1	−78 to 23	5.281 38	685.001	160.880
Bromochloromethane	1	16–68	6.496 06	942.267	192.587
Bromochlorodifluoromethane	1	−95 to 10	6.839 98	935.632	240.330
2-Bromo-2-chloro-1,1,1-trifluoroethane	1	−51 to 55	6.945 02	1 127.856	227.341
Bromocyclohexane	1	68–260	6.979 80	1 572.19	217.38
p-Bromodiphenyl ether	1	25–190	7.009 3	1 902.7	153.3
	1	190–400	6.681 43	1 683.84	132.90
Bromoethane	1	28–75	6.988 6	1 121.9	234.7
Bromoethene	1	−88 to 16	6.997 4	1 009.9	251.6
2-Bromoethylbenzene	1	127–217	7.800	2 235.4	238.7
4-Bromoethylbenzene	1	liq	6.982 09	1 632.60	193
2-Bromo-2-methylpropane	1	0–72.8	7.395 9	1 512.7	262.2
1-Bromonaphthalene	1	liq	7.003 50	1 927.05	186.0
o-Bromostyrene	1	liq	6.910 38	1 631.2	195
p-Bromostyrene	1		7.228 38	1 743.67	218.0
4-Bromotoluene	1	85–280	7.007 62	1 612.35	206.36
2-Bromovinylbenzene	1	110–129	0.564 97	82.913	−191.71
4-Bromovinylbenzene	1	119–147	12.504 2	7 349.00	559.02
1,2-Butadiene	1	−69 to −34	7.398 22	1 219.877	259.776
	1	−26 to 30	6.993 83	1 041.117	242.274
1,3-Butadiene	1	−80 to −62	7.035 55	998.106	245.233
	1	−58 to 15	6.849 99	930.546	238.854
n-Butane	1	−77 to 19	6.808 96	935.86	238.73
1-Butanethiol	1	−2 to 123	6.927 54	1 281.018	218.100
2-Butanethiol	1	−13 to 110	6.886 98	1 229.904	222.021
1-Butanol	1	15–131	7.476 80	1 362.39	178.77
2-Butanol	1	25–120	7.474 31	1 314.19	186.55
2-Butanone	1	43–88	7.063 56	1 261.34	221.97
1-Butene	1	−82 to 13	6.792 90	908.80	238.54
2-Butene *cis*	1	−73 to 23	6.884 68	967.32	237.87
trans	1	−76 to 20	6.883 37	967.50	240.84
Butyl acetate	1	60–126	7.127 12	1 430.418	210.745
n-Butylamine trimethylboron	1	0–99	8.465 21	1 980.98	193.60
n-Butylbenzene	1	62–213	6.983 17	1 577.965	201.378
sec-Butylbenzene	1	87–174	6.942 19	1 533.95	204.39
t-Butylbenzene	1	84–170	6.922 55	1 505.987	203.490
n-Butyl borate	1	117–218	7.406 87	1 905.035	186.134
n-Butyl-*t*-butyl ether	1	83–124	6.955 56	1 348.702	206.303
Butyl carbitol	1	50–153	7.741 14	2 056.904	195.655
Butyl cellosolve	1	93–170	6.956 59	1 399.903	172.154
sec-Butyl chloroacetate	1	30–172	7.933 38	2 103.30	249.29
n-Butylcyclohexane	1	60–211	6.910 30	1 538.518	200.833
sec-Butylcyclohexane	1	91–180	6.890 96	1 530.70	202.373
t-Butylcyclohexane	1	84–173	6.856 80	1 501.724	206.108
n-Butylcyclopentane	1	41–185	6.899 35	1 457.08	205.99
n-Butyl formate	1	29–112	7.693 6	1 698.7	247.4
sec-Butyl formate	1	30–100	6.493	972.9	176.0
n-Butyl-α-hydroxyisobutyrate	1	112–185	8.421 7	2 617.32	287.09

TABLE 2.36 Vapor Pressures of Various Organic Compounds (*Continued*)

Substance	Eq.	Range, °C	*A*	*B*	*C*
1-*n*-Butylnaphthalene	1	25–170	7.434 47	2 227.7	202.2
	1	170–345	7.081 4	1 971.5	180
2-*n*-Butylnaphthalene	1	25–170	7.438 08	2 242.2	202.3
	1	170–345	7.084 8	1 984.3	180
n-Butyl nitrate	1	0–70	8.054 27	1 992.83	254.30
1-Butyl pentafluoropropionate	1	82–116	6.651 00	1 108.02	177.04
2-*sec*-Butylphenol	1	179–240	6.951 93	1 593.74	163.79
2-*t*-Butylphenol	1	135–225	7.217 56	1 822.81	196.23
4-*t*-Butylphenol	1	198–252	7.000 38	1 627.51	155.24
Butyl phenyl ether	1	119–210	7.299 7	1 882.70	215.82
n-Butyl propionate	1	32–93	9.484 89	2 852.58	296.98
n-Butyl trifluoroacetate	1	71–104	8.567 94	2 305.22	301.06
1-Butyl trimethylsilyl ether	1	71–124	7.763 00	1 884.68	261.31
1-Butyne	1	−68 to 27	6.981 98	988.75	233.01
2-Butyne	1	−51 to −34	7.037 91	896.91	199.06
	1	−31 to 47	7.073 38	1 101.71	235.81
n-Butyraldehyde	1	31–74	6.385 44	913.59	185.48
Butyric acid	1	90–163	7.739 9	1 764.7	199.9
Camphor	2	0–180	8.799	2 797.39	
	1	178–232	6.106	1 043.6	116.4
Capric acid	1	153–187	6.255 3	1 106.3	57.96
Caproic acid	1	98–179	6.924 9	1 340.8	126.6
Capronitrile	1	92–164	7.123 1	1 597.2	212.8
Caprylic acid	1	130–206	7.770 64	1 933.05	159.36
Carbazole	1	253–358	7.086 3	2 179.4	163.5
Carbitol	1	40–151	7.640 81	1 801.31	183.97
Chloroacetic acid	1	104–190	7.550 16	1 723.365	179.98
4-Chloroacetophenone	1	122–212	7.084 57	1 693.63	190.95
Chloroacetyl chloride	1	28–107	7.149 77	1 340.79	208.70
N-Chloroaniline	1	61–125	3.037 67	171.35	−14.99
2-Chloroaniline	1	20–108	7.562 65	1 998.6	220.0
	1	108–300	7.192 40	1 762.74	200.0
3-Chloroaniline	1	15–125	7.559 39	2 073.75	215
	1	125–310	7.236 03	1 857.75	196.64
o-Chloroanisole	1	115–186	7.121 36	1 655.80	188.77
Chlorobenzene	1	62–131.7	6.978 08	1 431.05	217.55
o-Chlorobenzotrichloride	1	30–150	7.504 30	2 228.07	220.0
	1	150–350	7.117 94	1 951.37	196.27
1-Chloro-4-bromobenzene	2	23–63	11.629	3 643.30	
1-Chlorobutane	1	−17 to 78.6	6.836 94	1 173.79	218.13
2-Chlorobutane	1	0–40	6.799 23	1 149.12	224.68
1-Chlorodecane	1	86–225.9	6.939 86	1 639.06	177.94
1-Chlorododecane	1	116–246	6.834 08	1 654.82	155.09
Chloroethane	1	−56 to 12.2	6.986 47	1 030.01	238.61
2-Chloroethylbenzene	1		6.981 69	1 556.0	201.0
3-Chloroethylbenzene	1		6.990 82	1 577.3	200
4-Chloroethylbenzene	1		6.983 09	1 577.0	200
Chloroethylene	1	−65 to −13	6.891 17	905.01	239.48
Chloroform	1	−35 to 61	6.493 4	929.44	196.03
1-Chloroheptane	1	34–160	6.916 70	1 453.96	199.83
1-Chlorohexadecane	1	166–327	7.282 03	2 152.61	162.73
1-Chlorohexane	1	15–136	7.051 36	1 461.72	215.57
Chlorohexylisocyanate	1	90–180	7.740 95	2 340.50	241.90

(*Continued*)

TABLE 2.36 Vapor Pressures of Various Organic Compounds (*Continued*)

Substance	Eq.	Range, °C	*A*	*B*	*C*
Chloromethane	1	−75 to −5	7.093 49	948.58	249.34
Chloromethoxytrichlorosilane	1	0–50	7.312 92	1 545.71	226.10
2-Chloro-2-methylpropane	1	22–47	4.896	334.99	114.0
1-Chlorononane	1	69–205	7.046 54	1 655.57	192.26
1-Chlorooctane	1	54–184	7.051 52	1 600.24	200.28
Chloropentafluorobenzene	1	36–140	7.068 83	1 389.19	213.75
p-Chlorophenetole	1	122–212	7.084 57	1 693.63	190.95
2-Chlorophenol	1	80–200	6.877 31	1 471.61	193.17
β-Chloro-β-phenylethyl alcohol	1	166–259	6.917 33	1 635.63	145.87
1-Chlorophenylisocyanate	1	50–160	12.265 9	6 532.55	499.59
m-Chlorophenylisocyanate	1	71–158	6.797 29	1 512.43	180.90
Chloroprene	1	20–60	6.161 50	783.45	179.7
1-Chloropropane	1	−25 to 47	6.926 48	1 110.19	227.94
2-Chloropropane	1	0–30	7.771	1 582	288
3-Chloro-1-propene	1	13–44	5.297 16	418.375	128.168
2-Chloropropionitrile	1	0–84	7.329 73	1 732.55	211.79
	1	84–240	7.200 85	1 657.25	205.3
γ-Chloropropyltrichlorosilane	1	87–179	7.156 4	1 679.07	210.38
1-Chlorotetradecane	1	142–296.8	7.200 7	2 018.9	170.6
o-Chlorotoluene	1	0–65	7.367 97	1 735.8	230.0
	1	65–220	6.947 63	1 497.2	209.0
1-Chloro-2,4,6-trinitrobenzene	1	200–270	3.080 9	184.93	−117.9
1-Chloroundecane	1	101–245	6.967 6	1 709.4	172.9
o-Chlorovinylbenzene	1	98–155	6.956 6	1 602.2	204.5
p-Chlorovinylbenzene	1	100–127	9.969 1	4 093.5	392.4
2-Chlorovinyldichloroarsine *cis*	1	68–109	5.487 9	785.09	115.61
trans	1	50–150	6.814 0	1 465.07	178.53
3-Chlorovinyldichloroarsine	1	66–110	2.810 5	97.17	−27.51
o-Cresol	1	120–191	6.911 7	1 435.50	165.16
m-Cresol	1	150–201	7.508 0	1 856.36	199.07
p-Cresol	1	128–202	7.035 08	1 511.08	161.85
Cyanic acid	1	−76 to −6	7.568 59	1 251.86	243.79
Cyclobutane	1	−60 to 12	6.916 31	1 054.54	241.37
Cyclobutanone	1	−24 to 25	6.116 68	933.95	183.19
Cyclobutene	1	−77 to 2	7.305 7	1 166.0	261.06
Cycloheptane	1	68–159	6.853 95	1 331.57	216.35
1,3,5-Cycloheptatriene	1	0–65	6.974 33	1 376.84	220.75
Cyclohexane	1	20–81	6.841 30	1 201.53	222.65
Cyclohexanethiol	1	84–203	6.886 73	1 476.70	209.83
Cyclohexanol	1	94–161	6.255 3	912.87	109.13
Cyclohexene	1		6.886 17	1 229.973	224.10
Cyclohexyl acetate	1	95–172	7.975 86	2 167.99	252.30
Cyclohexylamine	1	61–128	6.689 54	1 229.42	188.80
1-Cyclohexylamino-2-propanol	1	150–238	7.011 56	1 655.02	162.59
Cyclohexylpentafluoropropionate	1	82–155	7.725 5	1 844.73	224.89
Cyclohexyltrifluoroacetate	1	72–147	7.802 35	1 954.66	249.33
Cyclohexyltrimethylsilyl ether	1	91–168	8.090 52	2 276.62	267.94
Cyclooctane	1	97–194	6.861 87	1 437.79	210.02
1,3,5,7-Cyclooctatetraene	1	0–75	7.006 69	1 472.11	215.84
Cyclopentane	1	−40 to 72	6.886 76	1 124.162	231.36
Cyclopentanethiol	1	81–173	6.914 97	1 388.63	212.05
Cyclopentanone	1	0–26	2.902 47	162.90	63.22
Cyclopentene	1		6.920 66	1 121.818	223.45

TABLE 2.36 Vapor Pressures of Various Organic Compounds (*Continued*)

Substance	Eq.	Range, °C	*A*	*B*	*C*
Cyclopentyl-1-thiaethane	1	83–199	6.940 83	1 480.70	208.47
Cyclopropane	1	−90 to −32	6.887 88	856.01	246.50
o-Cymene	1	81–180	7.266 10	1 768.45	224.95
m-Cymene	1	79–176	7.123 74	1 644.95	212.76
p-Cymene	1	107–178	7.050 74	1 608.91	208.72
Decahydronaphthalene *cis*	1	68–228	6.875 29	1 594.460	203.39
trans	1	61–219	6.856 81	1 564.683	206.26
Decane	1	58–203	6.943 65	1 495.17	193.86
1-Decanethiol	1	109–271	6.998 1	1 713.6	177.0
1-Decanol	1	25–52	11.560	4 055	273.2
	1	103–230	6.922 44	1 472.01	133.98
1-Decene	1	54–199	6.934 77	1 484.98	195.707
Decylbenzene	1	203–298	7.035 96	1 903.98	160.33
Decylcyclohexane	1	197–298	7.019 37	1 899.33	161.35
Decylcyclopentane	1	182–279	6.999 12	1 822.05	163.05
Deuterodiborane	1	−155 to −94	6.480 83	545.20	244.73
Diacetone alcohol	1	28–115	8.502 42	2 400.56	263.79
1,3-Diacetylbenzene	1	50–145	0.056 24	64.188	−196.97
1,4-Diacetylbenzene	1	116–157	2.803 71	177.25	−46.43
Diacetylene	1	−78 to 0	4.990 79	356.36	143.22
Diallyl sulfide	1	10–40	4.829 30	643.18	142.34
4,4′-Diaminodiphenylmethane	1	198–272	3.172 31	210.49	−137.41
Diamyl ether	1	105–187	7.067 10	1 604.77	196.58
Dibenzyl ketone	2	285–325	8.257	3 244.42	
1,2-Dibromobenzene	1	20–117	7.501 28	2 093.7	230
	1	117–300	7.102 65	1 825.77	207.0
Dibromodichloroethane	1	25–130	5.197 53	763.44	110.81
Dibromodifluoromethane	1	−26 to 23	7.152 22	1 181.612	253.85
1,2-Dibromoethane	1	52–131	6.721 48	1 280.82	201.75
1,2-Dibromoethylene *cis*	1	26–78	7.038 74	1 349.84	209.26
trans	1	4–71	4.581 11	393.641	103.56
1,2-Dibromopropane	1	0–50	7.303 98	1 644.4	232.0
	1	50–250	6.891 05	1 419.60	212.0
1,3-Dibromopropane	1	0–71	7.549 84	1 890.56	240.0
	1	71–275	7.198 74	1 678.26	222.0
Di-*n*-butyl ether	1	89–140	6.796 3	1 297.29	191.03
Di-*t*-butyl ether	1	4–109	6.932 9	1 348.53	233.79
Di-*n*-butyl phthalate	1	126–202	6.639 80	1 744.20	113.69
Di-*n*-butyl sebacate	1	128–208	7.587 66	2 364.89	147.54
Di-*n*-butyl sulfide	1	10–40	6.769 3	1 208.80	217.51
1,2-Dichlorobenzene	1	131–181	7.143 78	1 704.49	219.42
1,3-Dichlorobenzene	1	91–173	7.040 1	1 607.05	213.38
1,4-Dichlorobenzene	1	95–174	7.020 8	1 590.9	210.2
Dichlorobenzotrichloride	1	20–167	7.439 54	2 190.0	200
	1	167–340	6.985 24	1 868.91	172.00
Dichlorobenzyl chloride	1	20–138	7.504 57	2 125.9	213.8
	1	138–350	7.147 35	1 881.38	192.93
1,1-Dichloroethane	1	−39 to 18	6.977 0	1 174.02	229.06
1,2-Dichloroethane	1	−31 to 99	7.025 3	1 271.3	222.9
1,1-Dichloroethylene	1	−28 to 32	6.972 2	1 099.4	237.2
1,2-Dichloroethylene *cis*	1	0–84	7.022 3	1 205.4	230.6
trans	1	−38 to 85	6.965 1	1 141.9	231.9
2,2′-Dichloroethyl sulfide	1	15–76	8.587 41	2 588.23	246.06

(*Continued*)

TABLE 2.36 Vapor Pressures of Various Organic Compounds (*Continued*)

Substance	Eq.	Range, °C	*A*	*B*	*C*
1,2-Dichloroethyltrichlorosilane	1	102–181	7.826	2 144.9	253.1
Dichloromethane	1	−40 to 40	7.409 2	1 325.9	252.6
2-(2,4-Dichlorophenoxy)-ethanol	1	212–286	7.240 09	2 004.31	157.25
3,4-Dichlorophenylisocyanate	1	60–190	8.679 3	3 312.3	333.9
1,2-Dichloropropane	1	45–96	6.980 7	1 308.1	222.8
3,4-Dichlorotoluene	1	0–105	7.343 94	1 882.5	215.0
	1	105–330	6.979 25	1 655.44	195.0
Diethanolamine	1	194–241	8.138 8	2 327.9	174.4
1,1-Diethoxyethane	1	0–70	6.757 63	1 191.60	203.12
Diethoxymethane	1	0–75	6.908 41	1 229.52	217.01
Diethylaluminum chloride	1	44–125	8.229 70	2 484.53	255.45
Diethylamine	1	31–61	5.801 6	583.30	144.1
N,N-Diethylaniline	1	50–218	7.466 0	1 993.57	218.5
1,2-Diethylbenzene	1	liq	6.987 80	1 576.940	200.51
1,3-Diethylbenzene	1	liq	7.003 60	1 575.310	200.96
1,4-Diethylbenzene	1	liq	6.998 20	1 588.310	201.97
Diethyldichlorosilane	1	48–128	6.862 9	1 346.3	207.7
Diethyl disulfide	1	15–61	7.349 89	1 695.00	227.29
	1	61–230	6.975 07	1 485.970	208.96
Diethylene glycol	1	130–243	7.636 7	1 939.4	162.7
Diethyl ether	1	−61 to 20	6.920 32	1 064.07	228.80
Diethyl ethylphosphate	1	76–134	4.101 6	315.17	15.50
N,N-Diethylformamide	1	30–90	6.395 4	1 203.8	165.6
Diethyl ketone	1		6.857 91	1 216.3	204
3,3-Diethylpentane	1	63–147	6.896 03	1 453.48	215.83
3,5-Diethylphenol	1	114–248	7.651 3	2 228	218.5
Diethylpropylphosphonate	1	87–134	4.558 1	446.50	26.17
Diethyl sulfide	1	0–150	6.928 36	1 257.83	218.66
1,2-*bis*-Difluoroamino-4-methyl-pentane	1	−20 to 20	8.009 11	1 944.92	245.44
Difluoromethane	1	−82 to −32	7.138 9	821.7	244.7
1,2-Dihydroxybenzene	1	118–246	7.577	2 054	187
1,3-Dihydroxybenzene	1	151–276	7.889	2 231	169
1,2-Diiodoethylene *cis*	1	29–152	5.522	797.8	106.4
trans	1	77–130	6.093 1	1 197.0	172.3
Diisoamyl sulfide	1	10–80	−1.959 8	390.61	−219.33
p-Diisopropylbenzene	1	120–211	6.993 3	1 663.88	194.41
Diisopropyl ether	1	23–67	6.849 5	1 139.34	218.7
2,4-Diisopropylphenol	1	122–255	6.714	1 506	138
1,2-Dimethoxyethane	1	0–60	6.718 9	1 050.5	209.2
N,N-Dimethylacetamide	1	30–90	9.720 9	3 273.8	334.5
Dimethylamine	1	−72 to 6.9	7.082 12	960.242	221.67
bis-Dimethylaminoborane	1	−25 to 62.5	5.584 52	774.371	170.64
N-Dimethylaminodiborane	1	−38 to 14	8.340 1	1 917.35	302.73
bis-Dimethylaminodifluorosilane	1	24–88	5.952	748.7	146.9
N,N-Dimethylaniline	1	71–197	7.367 7	1 857.08	220.36
Dimethyl beryllium	1	100–180	19.089 9	11 535.45	496.64
1,4-Dimethyl-bicyclo(2,2,1)-heptane	1	56–119	6.761 96	1 342.66	213.53
2,3-Dimethyl-bicyclo(2,2,1)-heptane *trans*	1	72–138	6.868 15	1 420.32	212.94
2,3-Dimethyl-1,3-butadiene	1	0–68.5	7.119 7	1 299.69	238.09
2,2-Dimethylbutane	1	−42 to 73	6.754 83	1 081.176	229.34

TABLE 2.36 Vapor Pressures of Various Organic Compounds (*Continued*)

Substance		Eq.	Range, °C	*A*	*B*	*C*
2,3-Dimethylbutane		1	−35 to 81	6.809 83	1 127.187	228.90
2,3-Dimethyl-2-butanethiol		1	56–167	6.839 56	1 354.24	215.96
2,3-Dimethyl-1-butene		1	−36 to 78	6.862 36	1 134.675	229.37
2,3-Dimethyl-2-butene		1	−21 to 97	6.950 58	1 215.428	225.44
3,3-Dimethyl-1-butene		1	−47 to 64	6.677 51	1 010.516	224.91
Dimethyl cadmium		1	−2 to 23	6.490 55	1 126.36	201.07
1,1-Dimethylcyclohexane		1	10–147	6.798 21	1 321.705	217.85
1,2-Dimethylcyclohexane	*cis*	1	18–158	6.837 46	1 367.311	215.84
	trans	1	13–151	6.833 08	1 353.881	219.13
1,3-Dimethylcyclohexane	*cis*	1	11–147	6.838 83	1 338.473	218.07
	trans	1	15–152	6.834 55	1 343.687	215.39
1,4-Dimethylcyclohexane	*cis*	1	15–152	6.832 87	1 345.613	216.15
	trans	1	10–147	6.817 73	1 330.437	218.58
1,1-Dimethylcyclopentane		1	−12 to 113	6.817 24	1 219.474	221.95
1,2-Dimethylcyclopentane	*cis*	1	−3 to 125	6.850 08	1 269.140	220.21
	trans	1	−9 to 117	6.844 22	1 242.748	221.69
1,3-Dimethylcyclopentane	*cis*	1	−10–116	6.837 15	1 237.456	222.01
	trans	1	−9 to 117	6.838 17	1 240.023	221.62
Dimethyldichlorosilane		1	28–72	7.062 1	1 280.29	235.65
1,2-Dimethyldisilane		1	−46 to 0	4.024 3	255.4	129.2
Dimethyl ether		1	−71 to −25	6.976 03	889.264	241.96
N,N-Dimethylformamide		1	30–90	6.928 0	1 400.87	196.43
2,2-Dimethylhexane		1		6.837 15	1 273.59	215.07
2,3-Dimethylhexane		1		6.870 04	1 315.50	214.16
2,4-Dimethylhexane		1		6.853 05	1 287.88	214.79
2,5-Dimethylhexane		1		6.859 84	1 287.27	214.41
3,3-Dimethylhexane		1		6.851 21	1 307.88	217.44
3,4-Dimethylhexane		1		6.879 86	1 330.04	214.86
1,1-Dimethylhydrazine		1	−35 to 20	7.408 13	1 305.91	225.53
1,2-Dimethylhydrazine		1	1–25	5.611 9	633.59	143.17
N,N-Dimethylhydroxylamine		1	17–90	7.565 8	1 415.96	201.93
O,N-Dimethylhydroxylamine		1	−45 to 42.2	7.405 4	1 245.58	233.06
Dimethylmalononitrile		1	49–140	7.035 5	1 546.99	202.00
1,3-Dimethylnaphthalene		1	20–148	7.634 7	2 295.4	232.4
		1	148–310	7.269 8	2 076.0	210
1,4-Dimethylnaphthalene		1	20–148	7.634 7	2 345.8	232.6
(same for 1,6- and 1,7-)		1	148–310	7.269 8	2 076.0	210
1,8-Dimethylnaphthalene		1	25–150	7.407 89	2 123.2	201.2
		1	150–320	7.056 4	1 879	180
2,3-Dimethylnaphthalene		1	20–155	7.403 96	2 111.9	201.1
		1	155–315	7.052 7	1 869	180
2,6-Dimethylnaphthalene		1	20–150	7.396 8	2 080.3	200.8
		1	150–310	7.046 0	1 841	180
2,7-Dimethylnaphthalene		1	25–150	7.398 75	2 085.9	200.9
		1	150–310	7.047 8	1 846	180
2,2-Dimethylpentane		1	−19 to 103	6.814 80	1 190.033	223.30
2,3-Dimethylpentane		1	−10 to 115	6.853 82	1 238.017	221.82
2,4-Dimethylpentane		1	−17 to 105	6.826 21	1 192.04	225.32
3,3-Dimethylpentane		1	−14 to 112	6.826 67	1 228.663	225.32
2,4-Dimethyl-3-pentanone		1	48–125	6.968 53	1 382.84	213.06
Dimethyl-*o*-phthalate		1	82–151	4.522 32	700.31	51.42
2,2-Dimethylpropane		1	−14 to 29	6.604 27	883.42	227.78
2,2-Dimethyl-1-propanol		1	55–115	7.875 3	1 604.7	208.2

(*Continued*)

TABLE 2.36 Vapor Pressures of Various Organic Compounds (*Continued*)

Substance	Eq.	Range, °C	*A*	*B*	*C*
2,5-Dimethylpyrrole	1	100–199	7.203 06	1 509.60	181.76
2,4-Dimethylquinoline	1	185–269	7.025 4	1 830.29	174.44
2,6-Dimethylquinoline	1	188–267	6.931 12	1 748.73	166.37
Dimethyl sulfide	1	−22 to 20	7.150 9	1 195.58	242.68
3,3-Dimethyl-2-thiabutane	1	liq	6.847 09	1 259.648	218.69
2,2-Dimethyl-3-thiapentane	1	liq	6.850 86	1 323.24	212.89
2,4-Dimethyl-3-thiapentane	1	liq	6.871 18	1 327.12	212.55
2,3-Dimethylthiophene	1	50–205	6.924 9	1 430.0	212
2,4-Dimethylthiophene	1	50–205	6.993 9	1 450.7	212.0
2,5-Dimethylthiophene	1	47–200	6.961 1	1 427.7	213.2
3,4-Dimethylthiophene	1	54–205	6.996 1	1 467.1	211.5
1,3-Dinitrobenzene	1	252–292	4.337	229.2	−137
2,4-Dinitrotoluene	1	200–299	5.798	1 118	61.8
2,6-Dinitrotoluene	1	150–260	4.372	380	−43.6
3,5-Dinitrotoluene	1	220–270	1.556	30.59	−302
1,4-Dioxane	1	20–105	7.431 55	1 554.68	240.34
Dipentene	1	21–170	7.111 6	1 613.42	207.8
2,2′-Diphenol	1	171–325	8.193 5	3 067.6	253.1
Diphenyldichlorosilane	1	192–281	6.999 03	1 918.20	161.41
Diphenyl ether	1	204–271	7.011 04	1 799.71	177.74
Diphenylmethane	1	217–282	6.291	1 261	105
Di-*n*-propyl ether	1	26–89	6.947 6	1 256.5	219.0
Disilanyl chloride	1	−46 to 18	7.104 8	1 211.8	245.2
2,3-Dithiabutane	1	6–135	6.977 92	1 346.342	218.86
5,6-Dithiadecane	1	101–263	6.963 8	1 684.1	181.3
3,4-Dithiahexane	1	40–182	6.975 07	1 485.970	208.96
4,5-Dithiaoctane	1	72–226	6.975 29	1 603.793	195.85
Dodecane	1	91–247	6.997 95	1 639.27	181.84
1-Dodecanethiol	1		7.024 4	1 817.8	164.1
Dodecanoic acid	1	106–176	7.860 8	2 159.1	143.2
1-Dodecanol	1	138–214	7.539 86	2 003.29	168.13
1-Dodecene	1	89–244	6.976 07	1 621.11	182.45
Durenol	1	108–249	7.758	2 432	250
Eicosane	1	198–379	7.152 2	2 032.7	132.1
1-Eicosanethiol	1		7.114	2 125	119
1-Eicosene	1	liq	7.135 1	2 043.0	137.9
Ethane	1	−142 to −75	6.829 15	663.72	256.68
Ethanethiol	1	−49 to 56	6.952 06	1 084.531	231.39
Ethanol	1	−2 to 100	8.321 09	1 718.10	237.52
Ethanolamine	1	65–171	7.456 8	1 577.67	173.37
Ethyl acetate	1	15–76	7.101 79	1 244.95	217.88
m-Ethylacetophenone	1	19–143	3.767 2	708.05	182.6
p-Ethylacetophenone	1	21–94	4.274 6	629.34	120.9
Ethylamine	1	−20 to 90	7.054 13	987.31	220.0
N-Ethylaniline	1	50–207	7.422 8	1 903.4	214.3
Ethylbenzene	1	26–164	6.957 19	1 424.255	213.21
2-Ethyl-1-butene	1	−28 to 88	6.997 12	1 218.352	231.30
Ethyl butyl ether	1	38–92	6.944 4	1 256.4	216.9
Ethyl chloroacetate	1	25–146	6.967	1 355.9	188.2
p-Ethylchlorobenzene	1	109–184	6.951 1	1 557.1	198.1
Ethylcyclohexane	1	20–160	6.867 28	1 382.466	214.99
Ethylcyclopentane	1	−0.1 to 129	6.887 09	1 298.599	220.68
Ethylene	1	−153 to −91	6.744 19	594.99	256.16

TABLE 2.36 Vapor Pressures of Various Organic Compounds (*Continued*)

Substance	Eq.	Range, °C	*A*	*B*	*C*
Ethylene glycol	1	50–200	8.090 8	2 088.9	203.5
Ethylene glycol monoethyl ether	1	63–134	7.874 6	1 843.5	234.2
Ethylene glycol monomethyl ether	1	56–124	7.849 8	1 793.9	236.9
Ethylene oxide	1	−49 to 12	7.128 43	1 054.54	237.76
Ethyl formate	1	4–54	7.009 0	1 123.94	218.2
3-Ethylhexane	1		6.890 98	1 327.88	212.60
2-Ethyl-1-hexanol	1	74–184	6.914 7	1 339.7	147.8
2-Ethyl-2-hexenal	1	54–175	6.861 3	1 457.4	190.6
Ethyl iodoacetate	1	29–89	4.073 7	374.64	54.8
Ethyl isothiocyanate	1	10–50	7.106 0	1 567.5	234.2
Ethyl methyl ether	1	5–7.7	5.518	434.5	158
Ethyl methyl ketone	1		6.974 21	1 209.6	216
3-Ethyl-5-methylphenol	1	195–247	7.040 83	1 615.44	152.6
2-Ethyl-4-methyl-1-pentanol	1	70–176	6.582 6	1 134.6	129.2
Ethyl nitrate	1	0–60	7.163 7	1 338.8	224.9
3-Ethylpentane	1	−7 to 119	6.875 64	1 251.827	219.89
2-Ethylphenol	1	86–208	7.800 3	2 140.4	227
3-Ethylphenol	1	97–218	7.468	1 856	187
4-Ethylphenol	1	101–218	8.291	2 423	229
Ethyl phenyl ether	1	117–181	7.021 38	1 508.39	194.49
Ethyl *n*-propanoate	1	34–98	6.994 9	1 260.6	207.4
Ethyl *n*-propyl ether	1	20–63	6.985 1	1 188.5	226.4
Ethyl *n*-propyl ketone	1	75–133	7.000 82	1 365.79	208.01
m-Ethylstyrene	1		7.039 28	1 614.0	198
p-Ethylstyrene	1		6.900 71	1 570.9	198
Ethyl trichloroacetate	1	44–95	7.725 4	1 927.0	233.7
Ethyl trichlorosilane	1	28–96	6.606	1 118	201
Ethyl triexthoxysilane	1	64–153	6.886 8	1 377.9	183.0
Ethyl vinyldichlorosilane	1	45–122	6.859	1 331	210.8
Fenchyl alcohol	1	59–200	5.693	797.6	84.6
Fluoranthene	1	197–384	6.373	1 756	118
Fluorene	1	161–300	7.761 8	2 637.1	243.2
Fluorobenzene	1	−18 to 84	7.187 0	1 381.8	235.6
m-Fluorobenzotrifluoride	1	40–137	7.006 59	1 304.35	215.67
bis-(Fluorocarbonyl)-peroxide	1	−47 to −7	9.608 4	2 247.64	319.83
p-Fluorotoluene	1	68–155	6.994 26	1 374.055	217.40
Formaldehyde	1	−109 to −22	7.195 8	970.6	244.1
Formic acid	1	37–101	7.581 8	1 699.2	260.7
Formyl fluoride	1	−95 to −61	5.270	362	175
Furan	1	2–61	6.975 27	1 060.87	227.74
2-Furfuraldehyde	1	56–161	6.575 9	1 198.7	162.8
Glycerol	1	183–260	6.165	1 036	28
Glyceryl-1,3-diacetate	1	100–190	6.407 3	1 092.0	119.3
Guaiacol	1	82–205	6.161	1 051	116
Hemellitenol	1	123–248	6.972	1 563	134
Heptadecane	1	161–337	7.014 3	1 865.1	149.20
1-Heptadecene	1		7.008 67	1 868.9	152.50
Heptane	1	−2 to 124	6.896 77	1 264.90	216.54
1-Heptanethiol	1	58–206	6.952 49	1 525.311	197.70
Heptanoic acid	1	112–150	5.287 4	665.54	42.07
1-Heptanol	1	60–176	6.647 67	1 140.64	126.56
1-Heptene	1	−6 to 118	6.901 87	1 258.345	219.30
Hexadecane	1	149–321	7.028 67	1 830.51	154.45

(*Continued*)

TABLE 2.36 Vapor Pressures of Various Organic Compounds (*Continued*)

Substance	Eq.	Range, °C	*A*	*B*	*C*
1-Hexadecanethiol	1		7.075	1 990	140
1-Hexadecanol	1	50–103	7.281 7	1 909.7	128.1
	1	145–190	6.158 6	1 380.0	91
1-Hexadecene	1		7.040 11	1 840.52	157.57
1,5-Hexadiene	1	0–59	6.574 1	1 013.5	214.8
Hexafluoroacetone	1	−79 to −27	6.650 2	725.90	219.9
Hexafluorobenzene	1	5–114	7.032 95	1 227.98	215.49
Hexafluorodisiloxane	1	−39 to −23	7.471 2	1 169.3	278.1
Hexafluoroethane	1	−93 to −78	6.793 35	657.06	246.2
Hexahydroindane *cis*	1	77–168	6.868 22	1 497.33	207.67
trans	1	71–161	6.861 19	1 475.70	209.66
Hexamethyldisiloxane	1	36–138	6.773 79	1 202.03	208.25
Hexane	1	−25 to 92	6.876 01	1 171.17	224.41
1-Hexanethiol	1	40–181	6.946 64	1 454.004	204.95
1-Hexanol	1	35–157	7.860 45	1 761.26	196.66
2-Hexanol	1	25–142	7.261 0	1 371.7	173.2
3-Hexanol	1	25–138	7.689	1 670.0	211.8
1-Hexene	1	16–64	6.857 70	1 148.62	225.35
3-Hexyne	1	−20 to 24	5.895	863.3	194
Hydroquinone	1	159–286	8.137	2 461	183
3-Hydroxy-3-methyl-2-butanone	1	45–146	7.340 9	1 653.6	227.5
Iodobenzene	1	20–188	7.011 9	1 640.1	208.8
Iodoethane	1	30–60	6.959	1 232	229
Isoamyl acetate	1	41–95	7.436	1 606.6	216
Isobutylbenzene	1	86–174	6.935 56	1 530.05	204.59
Isobutyl borate	1	99–200	7.197	1 745.8	193
Isobutyl cellosolve	1	71–159	7.694 8	1 825.9	219.6
Isobutylcyclohexane	1	85–172	6.867 97	1 493.10	203.16
Isobutyl nitrate	1	0–70	8.164 3	2 022.7	262.4
Isobutyraldehyde	1	13–63	6.735 1	1 053.2	209.1
Isobutyric acid	1	58–152	4.894	382.6	38
Isocaproic acid	1	96–133	6.258	1 038.6	130
Isopropylbenzene	1	39–181	6.936 66	1 460.793	207.78
Isopropyl borate	1	65–139	8.070	2 120	269
o-Isopropylbromobenzene	1	132–210	6.717 8	1 462.7	170.9
Isopropyl caprate	1	90–178	9.959	4 013.9	326.5
Isopropyl caprylate	1	65–146	8.032 2	2 213.6	220.9
Isopropyl cellosolve	1	67–140	7.500 0	1 639.2	213.3
Isopropyl chloroacetate	1	35–153	8.382	2 328	275
Isopropylcyclohexane	1	71–155	6.873 14	1 453.20	209.44
Isopropylcyclopentane	1	47–127	6.887 36	1 380.12	218.05
Isopropyl laurate	1	117–196	8.532 6	2 951.6	240.7
Isopropyl myristate	1	140–193	10.418 0	4 866.48	314.17
Isopropyl nitrate	1	0–70	7.266 6	1 434.4	255.2
Isopropyl palmitate	1	160–197	10.916 4	5 572.0	364.8
o-Isopropylphenol	1	97–215	8.167	2 343	229
p-Isopropylphenol	1	108–228	8.666	2 810	258
Isopropyl phenyl ether	1	72–175	6.517 6	1 238.0	163.0
Isopropyl stearate	1	182–207	0.079 3	10.41	−221
Isopseudocumenol	1	106–233	5.602	768	49
Isoquinoline	1	167–244	6.912 2	1 723.4	184.3
Isovaleric acid	1	86–104	3.946 55	255.41	11.3
Ketene	1	−88 to −49	7.615	1 036	269

TABLE 2.36 Vapor Pressures of Various Organic Compounds (*Continued*)

Substance	Eq.	Range, °C	*A*	*B*	*C*
Lauric acid	1	106–176	7.860 8	2 159.1	143.2
Lepidine	1	199–266	7.271 2	1 946.14	177.64
2,3-Lutidine	1	155–162	7.447 8	1 832.6	240.1
2,4-Lutidine	1	150–160	7.339 0	1 733.4	230.4
2,5-Lutidine	1	85–157	7.081 0	1 539.6	209.6
2,6-Lutidine	1	79–144	7.056 7	1 470.2	208.0
3,4-Lutidine	1	172–180	7.362 0	1 840.1	231.5
3,5-Lutidine	1	163–173	7.333 1	1 783.6	228.7
Mesitol	1	94–221	6.659	1 392	148
Mesityl oxide	1	14–130	6.635 8	1 186.1	186.0
Methacrylonitrile	1		6.980 2	1 274.96	220.7
Methane c	1	−195 to −183	7.193 09	451.64	268.49
liq	1	−181 to −152	6.695 61	405.42	267.78
Methanol	1	−14 to 65	7.897 50	1 474.08	229.13
	1	64–110	7.973 28	1 515.14	232.85
Methoxybenzene	1	110–164	7.052 69	1 489.99	203.57
N-Methylacetamide	1	40–90	2.631 1	121.7	−9.3
Methyl acetate	1	1–56	7.065 2	1 157.63	219.73
Methylal	1	0–35	6.872 2	1 049.2	220.6
Methylamine	1	−83 to −6	7.336 9	1 011.5	233.3
N-Methylaniline	1	50–200	7.081 9	1 631.3	192.4
Methyl benzoate	1	111–199	7.273	1 847	221
Methyl borate	1	31–68	7.646 0	1 491.5	245.5
Methyl boric anhydride	1	0–55	8.004 1	1 726.1	257.9
2-Methyl-1,3-butadiene	1	−52 to −24	7.011 87	1 126.159	238.88
	1	−19 to 55	6.885 64	1 071.578	233.51
3-Methyl-1,2-butadiene	1	−45 to −20	7.151 95	1 194.537	239.47
	1	−20 to 62	6.943 50	1 103.901	230.89
2-Methylbutane	1	−57 to 49	6.833 15	1 040.73	235.45
2-Methyl-1-butanethiol	1	liq	6.913 85	1 347.317	215.07
3-Methyl-1-butanethiol	1	liq	6.914 91	1 342.509	214.45
2-Methyl-2-butanethiol	1	liq	6.828 37	1 254.885	218.76
2-Methyl-1-butanol	1	34–129	7.067 30	1 195.26	156.83
3-Methyl-1-butanol	1	25–153	7.258 21	1 314.36	169.36
2-Methyl-2-butanol	1	25–102	6.519 3	863.4	135.3
3-Methyl-2-butanol	1	25–111	6.942 1	1 090.9	157.2
2-Methyl-1-butene	1	−53 to 52	6.846 37	1 039.69	236.65
3-Methyl-1-butene	1	−63 to 41	6.824 55	1 012.37	236.65
2-Methyl-2-butene	1	−48 to 60	6.966 59	1 124.33	236.63
Methyl butyl ether	1	23–69	6.887 1	1 162.1	219.9
3-Methyl-1-butyne	1	−55 to 47	6.884 80	1 014.81	227.11
2-Methyl-3-butyn-2-ol	1	21–106	6.657 5	976.5	154.1
Methyl *n*-butyrate	1		6.972 11	1 272.73	208.5
Methyl caprate	1	107–188	7.190 0	1 783.8	181.6
Methyl caproate	1	44–105	7.409 3	1 672.74	218.98
Methyl caprylate	1	100–146	6.916 5	1 496.3	176.5
Methyl carbitol	1	112–193	7.424	1 751	192
Methyl cellosolve acetate	1	70–144	7.125 1	1 447.0	196.1
Methyl chloroacetate	1	45–130	7.004 4	1 306.3	187.3
Methylcyclohexane	1	−3 to 127	6.823 00	1 270.763	221.42
Methylcyclopentane	1	−24 to 96	6.862 83	1 186.059	226.04
Methyldichlorosilane	1	1–41	7.027 8	1 167.8	240.7
1-Methyl-2-ethylbenzene	1	48–194	7.003 14	1 535.374	207.30

(*Continued*)

TABLE 2.36 Vapor Pressures of Various Organic Compounds (*Continued*)

Substance		Eq.	Range, °C	*A*	*B*	*C*
1-Methyl-3-ethylbenzene		1	46–190	7.015 82	1 529.184	208.51
1-Methyl-4-ethylbenzene		1	46–191	6.998 02	1 527.113	208.92
1-Methyl-1-ethylcyclopentane		1	43–122	6.859 20	1 347.602	217.21
1-Methyl-2-ethylcyclopentane	*cis*	1	49–129	6.905 88	1 388.412	216.89
2-Methyl-3-ethylpentane		1		6.867 31	1 318.12	215.31
3-Methyl-3-ethylpentane		1		6.867 31	1 347	219.68
3-Methyl-5-ethylphenol		1	111–233	7.958	2 236	208
2-Methyl-5-ethylpyridine		1	52–177	5.050	517	59
N-Methylformamide		1	96–200	7.497 4	1 849.4	201.1
Methyl formate		1	21–32	3.027	3.02	−11.9
2-Methylheptane		1	42–119	6.917 35	1 337.47	213.69
3-Methylheptane		1	43–120	6.899 44	1 331.53	212.41
4-Methylheptane		1		6.900 65	1 327.66	212.57
2-Methylhexane		1	−9 to 115	6.873 18	1 236.026	219.55
3-Methylhexane		1	−8 to 117	6.867 64	1 240.196	219.22
Methylhydrazine		1	2–25	6.576 2	1 007.5	181.4
N-Methylhydroxylamine		1	40–65	7.045 6	1 223.3	172.1
O-Methylhydroxylamine		1	−63 to 48	7.363 9	1 225.3	225.2
Methyl isobutyl ketone		1	22–116	6.672 7	1 168.4	191.9
1-Methyl-2-isopropylbenzene		1	liq	6.940 4	1 548.05	203.15
1-Methyl-3-isopropylbenzene		1	liq	6.940 5	1 539.05	203.93
1-Methyl-4-isopropylbenzene		1	liq	6.923 7	1 537.06	203.05
3-Methylisoquinoline		1	176–225	6.969 2	1 717.3	166.9
Methyl isothiocyanate		1	10–50	2.896 8	103.6	45.4
Methyl laurate		1	158–212	6.767 1	1 589.72	140.5
Methyl linolate		1	166–206	6.111 1	1 660.1	118.8
Methyl methacrylate		1	39–89	8.409 2	2 050.5	274.4
Methyl myristate		1	166–238	7.622 3	2 283.93	184.8
1-Methylnaphthalene		1	108–278	7.035 92	1 826.948	195.00
2-Methylnaphthalene		1	105–274	7.068 50	1 840.268	198.40
Methyl oleate		1	166–205	7.544 1	2 656.9	200.7
Methyl palmitate		1	148–202	9.594 4	4 146.43	297.76
2-Methylpentane		1	−32 to 83	6.839 10	1 135.410	226.57
3-Methylpentane		1	−30 to 87	6.848 87	1 152.368	227.13
2-Methyl-2-pentanethiol		1	56–165	6.858 5	1 343.79	212.8
2-Methyl-1-pentanol		1	25–150	7.520 1	1 564.7	189.2
2-Methyl-4-pentanol		1	25–133	8.467 1	2 174.9	257.8
2-Methyl-1-pentene		1	−30 to 85	6.850 30	1 138.516	224.70
3-Methyl-1-pentene		1	−38 to 77	6.755 23	1 086.316	226.20
4-Methyl-1-pentene		1	−38 to 77	6.835 29	1 121.302	229.68
2-Methyl-2-pentene		1	−26 to 90	6.923 67	1 183.837	225.51
3-Methyl-2-pentene	*cis*	1	−26 to 91	6.910 73	1 186.402	226.70
	trans	1	−23 to 94	6.926 34	1 194.527	224.83
4-Methyl-2-pentene	*cis*	1	−35 to 79	6.841 29	1 120.707	226.59
	trans	1	−33 to 81	6.880 30	1 142.874	227.14
Methyl phenyl ether		1	110–164	7.052 69	1 489.99	203.57
2-Methylpiperidine		1	51–158	6.818 59	1 274.61	205.40
2-Methylpropane		1	−87 to 7	6.910 48	946.35	246.68
2-Methyl-1-propanethiol		1	−10 to 113	6.887 46	1 237.282	220.31
2-Methyl-2-propanethiol		1	1–88	6.787 81	1 115.565	221.31
2-Methyl-1-propanol		1	20–115	7.327 05	1 248.48	172.92
2-Methyl-2-propanol		1	26–83	9.170 6	2 206.4	267.9
2-Methylpropene		1	−82 to 12	6.684 66	866.25	234.64

TABLE 2.36 Vapor Pressures of Various Organic Compounds (*Continued*)

Substance	Eq.	Range, °C	*A*	*B*	*C*
N-Methylpropionamide	1	30–90	−0.9103	119.4	−148.0
Methyl propionate	1	21–79	6.942 4	1 170.2	208.8
2-Methyl-2-propylamine	1	19–75	6.783 2	993.33	210.50
Methyl propyl ether	1	0–39	6.118 6	708.69	179.9
2-Methylpyridine	1	80–168	7.032 4	1 415.73	211.63
3-Methylpyridine	1	74–185	7.050 21	1 481.78	211.25
4-Methylpyridine	1	75–186	7.041 77	1 480.68	210.50
1-Methylpyrrole	1	49–149	7.085 0	1 368.66	212.80
6-Methylquinoline	1	187–266	6.927 2	1 746.08	166.46
7-Methylquinoline	1	238–258	7.597 7	2 229.4	214.9
Methyl salicylate	1	79–220	7.083 3	1 712.8	187.1
Methyl stearate	1	204–240	2.357 0	68.92	−156.5
o-Methylstyrene	1	32–112	7.212 9	1 664.08	214.59
	1	75–255	6.884 61	1 485.41	200.0
m-Methylstyrene	1	10–72	7.275 34	1 695.4	220.0
	1	72–250	6.879 28	1 471.44	200.0
p-Methylstyrene	1	68–170	7.011 2	1 535.1	200.7
α-Methylstyrene	1		6.923 66	1 486.88	202.4
β-Methylstyrene	1		6.923 39	1 499.80	201.0
Methyl sulfoxide	1	20–50	7.763 7	2 048.7	231.6
3-Methyl-2-thiabutane	1	−13 to 109	6.901 96	1 232.170	221.67
2-Methylthiacyclopentane	1	liq	6.944 12	1 409.503	214.41
3-Methylthiacyclopentane	1	67–179	6.949 1	1 431.8	213.6
2-Methyl-3-thiapentane	1	liq	6.891 30	1 293.05	215.04
Methyl-2-thiazole	1	80–128	7.042 1	1 407.05	209.33
2-Methylthiophene	1	9–138	6.938 97	1 326.48	214.31
3-Methylthiophene	1	11–141	6.986 11	1 363.83	216.78
Methyl trichlorosilane	1	13–64	7.088 2	1 289.2	239.9
2-Methyl-5-vinylpyridine	1	69–183	6.156	1 023	129
Morpholine	1	0–44	7.718 13	1 745.8	235.0
	1	44–170	7.160 30	1 447.70	210.0
Naphthalene c	1	86–250	7.010 65	1 733.71	201.86
liq	1	125–218	6.818 1	1 585.86	184.82
1-Naphthol	1	141–282	7.284 21	2 077.56	184.0
2-Naphthol	1	144–288	7.347 14	2 135.00	183.0
Nicotine	1	134–246	6.789	1 650	176
o-Nitroaniline	2	150–260	8.868 4	3 336.50	
m-Nitroaniline	2	170–260	8.818 8	3 440.9	
p-Nitroaniline	2	190–260	9.559 5	4 039.73	
Nitrobenzene	1	134–211	7.115 6	1 746.6	201.8
m-Nitrobenzotrifluoride	1	10–105	7.653 15	2 006.1	220.0
	1	104–280	7.180 25	1 710.60	195.12
Nitromethane	1	56–136	7.281 66	1 446.94	227.60
1-Nitropropane	1	59–131	7.114 6	1 467.45	215.23
o-Nitrotoluene	1	129–222	5.851	946	96
p-Nitrotoluene	1	148–233	6.994 8	1 720.39	184.9
Nonadecane	1	184–366	7.015 3	1 932.8	137.6
1-Nonadecene	1	liq	7.115 1	1 997.4	142.7
Nonafluorocyclopentane	1	17–75	6.945 3	1 051.7	220.1
Nonane	1	39–179	6.938 93	1 431.82	202.01
1-Nonanethiol	1	93–251	6.983 9	1 655.6	183.7
Nonanoic acid	1	137–177	3.235 9	143.97	−75.6
1-Nonanol	1	94–214	7.827 8	1 953.8	181.9

(*Continued*)

TABLE 2.36 Vapor Pressures of Various Organic Compounds (*Continued*)

Substance	Eq.	Range, °C	*A*	*B*	*C*
1-Nonene	1	35–175	6.954 30	1 436.20	205.69
Octadecane	1	172–352	7.002 2	1 894.3	143.30
1-Octadecanethiol	1	liq	7.096	2 061	129
1-Octadecanol	1	120–218	6.461 6	1 599	90
1-Octadecene	1		7.060 65	1 997.4	147.50
Octane	1	19–152	6.918 68	1 351.99	209.15
1-Octanethiol	1	76–229	6.969 09	1 593.0	190.61
1-Octanol	1	0–80	12.070 1	4 506.8	319.9
	1	70–195	6.837 90	1 310.62	136.05
2-Octanol	1	72–180	6.388 8	1 060.4	122.5
3-Octanol	1	76–176	5.221 5	560.3	64.7
4-Octanol	1	71–176	5.739 6	760.5	89.5
1-Octene	1	15–147	6.934 95	1 355.46	213.05
5-Oxyhydrindene	1	120–251	9.213 7	3 665.8	326.4
Pentachloroethane	1	25–162	6.740	1 378	197
Pentadecane	1	136–304	7.023 59	1 789.95	161.38
1-Pentadecene	1		7.022 91	1 788.58	163.347
1,2-Pentadiene	1	−42 to −26	7.259 90	1 250.293	241.96
	1	−21 to 67	6.918 20	1 104.991	228.85
1,3-Pentadiene *cis*	1	−43 to −22	7.193 87	1 223.602	240.62
	1	−18 to 66	6.910 89	1 101.923	229.37
trans	1	−45 to −20	7.102 12	1 185.389	239.41
	1	−18 to 64	6.913 17	1 103.840	231.72
1,4-Pentadiene	1	−57 to −37	7.174 01	1 155.378	244.30
	1	−33 to 47	6.835 43	1 017.995	231.46
2,3-Pentadiene	1	−39 to −18	7.202 53	1 231.768	237.56
	1	−14 to 70	6.962 16	1 126.837	227.84
Pentafluorobenzene	1	49–94	7.036 65	1 254.07	216.02
Pentafluorochloroacetone	1	−40 to 32	6.848 4	925.3	225.4
Pentafluorochlorethane	1	−95 to −39	6.833 34	802.97	242.27
Pentafluorophenol	1	105–155	7.066 0	1 379.15	183.91
2,2,3,3,3-Pentafluoropropanol	1	0–23	6.308 7	830.56	153.8
Pentafluorotoluene	1	39–138	7.084 78	1 392.20	213.67
bis-Pentamethyldisilanoxydisilane	1	169–201	8.556 64	3 051.316	258.85
bis-Pentamethyldisilanyl ether	1	88–183	8.161 44	2 575.250	273.32
Pentane	1	−50 to 58	6.852 96	1 064.84	233.01
Pentanenitrile	1	69–141	7.104 9	1 519.4	218.4
1-Pentanethiol	1	19–153	6.933 11	1 369.479	211.31
Pentanoic acid	1	72–174	5.412	591	60
1-Pentanol	1	37–138	7.177 58	1 314.56	168.11
2-Pentanol	1	25–120	7.275 75	1 271.92	170.37
3-Pentanol	1	21–116	7.414 93	1 354.42	183.41
2-Pentanone	1	56–111	7.021 93	1 313.85	215.01
3-Pentanone	1	56–111	7.025 29	1 310.28	214.19
1-Pentene	1	−55 to 51	6.844 24	1 044.01	233.50
2-Pentene *cis*	1	−49 to 58	6.843 08	1 052.44	228.69
trans	1	−49 to 58	6.899 83	1 080.76	232.57
1-Pentyne	1	−44 to 61	6.967 34	1 092.52	227.18
2-Pentyne	1	−33 to 78	7.046 14	1 189.87	229.60
Perdeuterobenzene	1	10–82	6.892 35	1 198.39	219.43
Perdeuterocyclohexane	1	10–80	6.837 86	1 190.38	222.40
Perfluorobutane	1	−39 to −4	7.035 1	990.27	240.4
Perfluorobutene	1	−28 to 20	9.222	2 401.6	382

TABLE 2.36 Vapor Pressures of Various Organic Compounds (*Continued*)

Substance	Eq.	Range, °C	*A*	*B*	*C*
Perfluorocyclobutane	1	−32 to 0	6.815 29	862.49	225.19
Perfluorocyclohexane	1	19–65	6.04	597	136
Perfluorocyclopentane	1	17–56	7.039 6	1 069.3	234.6
Perfluoroheptane	1	−2 to 106	6.937 72	1 181.14	208.66
Perfluorohexane	1	30–57	6.875 2	1 080.8	213.4
Perfluoromethylcyclohexane	1	33–111	6.824 06	1 133.76	211.22
Perfluorooctane	1	37–105	5.902 5	1 225.93	198.99
Perfluoropentane	1	9–65	7.017 9	1 072.9	230.0
Perfluoropiperidine	1	29–81	6.853 4	1 059.95	217.2
Perfluoropropane	1	−79 to −36	6.919 4	825.8	241.2
Perfluoropropene	1	−41 to 20	7.355	1 012.1	257
Phenanthrene	1	176–379	7.260 82	2 379.04	203.76
Phenol	1	107–182	7.133 0	1 516.79	174.95
β-Phenylethyl acetate	1	149–233	6.834 3	1 555.2	160.8
α-Phenylethyl alcohol	1	82–190	1.508	91	−263
o-Phenylethylphenol	1	169–250	4.506 0	516.8	−32.1
p-Phenylethylphenol	1	174–251	4.304 1	459.3	−52.4
Phenylisocyanate	1	10–80	−0.708 0	106.4	−146.6
4-Phenylphenol	1	177–308	8.657 5	3 022.8	216.1
Phosgene	1	−68 to 68	6.842 97	941.25	230
Phthalic anhydride	2	160–285	8.022	2 868.5	
α-Pinene	1	19–156	6.852 5	1 446.4	208.0
β-Pinene	1	19–166	6.898 4	1 511.7	210.2
Piperidine	1	42–144	6.855 69	1 238.80	205.43
Propadiene	1	−99 to −16	5.713 7	458.06	196.07
Propane	1	−108 to −25	6.803 38	804.00	247.04
1-Propanethiol	1	−25 to 91	6.928 46	1 183.307	224.62
2-Propanethiol	1	−37 to 75	6.877 34	1 113.895	226.16
1-Propanol	1	2–120	7.847 67	1 499.21	204.64
2-Propanol	1	0–101	8.117 78	1 580.92	219.61
2-Propen-1-ol	1	21–97	11.187 0	4 068.5	392.7
Propionic acid	1	56–139.5	6.403	950.2	130.3
Propionic anhydride	1	67–167	5.819 5	810.3	108.7
Propionitrile	1	−84 to 22	5.278 2	665.52	159.10
Propiophenone	1	132–201	7.370	1 894	205
Propyl acetate	1	39–101	7.016 15	1 282.28	208.60
1-Propylamine	1	23–77	6.926 51	1 044.05	210.84
2-Propylamine	1	4–61	6.890 25	985.69	214.07
n-Propylbenzene	1	43–188	6.951 42	1 491.297	207.14
n-Propyl borate	1	85–179	7.399 8	1 741	206
n-Propyl caprate	1	97–186	8.701 22	2 945.99	253.63
n-Propyl caproate	1	43–120	8.667 1	2 556.0	262.9
n-Propyl caprylate	1	70–153	8.516 7	2 599.5	246.2
n-Propyl cellosolve	1	77–149	7.146 4	1 440.6	187.7
n-Propylcyclohexane	1	40–186	6.886 46	1 460.800	207.94
n-Propylcyclopentane	1	21–158	6.903 92	1 384.386	213.16
Propylene	1	−112 to −32	6.778 11	770.85	245.51
1,2-Propylene oxide	1	−35 to 130	7.064 92	1 113.6	232
n-Propyl formate	1	26–82	6.848	1 127	203
n-Propyl laurate	1	124–205	8.068 9	2 692.4	222.5
n-Propyl myristate	1	147–200	9.216 8	3 744.68	272.87
n-Propyl nitrate	1	0–70	6.954 9	1 294.4	206.7
n-Propyl palmitate	1	166–204	14.129 2	9 759.2	539.7

(*Continued*)

TABLE 2.36 Vapor Pressures of Various Organic Compounds (*Continued*)

Substance	Eq.	Range, °C	*A*	*B*	*C*
o-(*n*-Propyl)phenol	1	104–222	9.215	3 254	292
p-(*n*-Propyl)phenol	1	0–234	8.329 6	2 661	254
n-Propyl phenyl ether	1	101–190	7.734 3	2 146.2	252.3
Propyne	1	−90 to −6	6.784 85	803.73	229.08
Pseudocumenol	1	107–232	6.915	1 547	152
Pyrene	1	200–395	5.618 4	1 122.0	15.2
Pyridine	1	67–153	7.041 15	1 373.80	214.98
Pyrogallol	1	177–309	6.092	1 031	12
Pyrrole	1	66–166	7.294 70	1 501.56	210.42
Quinaldine	1	178–248	7.179 00	1 857.84	184.50
Quinoline	1	164–238	6.817 59	1 668.73	186.26
Spiropentane	1	3–71	6.917 00	1 090.08	231.10
Styrene	1	32–82	7.140 16	1 574.51	224.09
Terpenyl acetate	1	37–150	6.443 46	1 377.27	143.85
α-Terpineol	1	84–217	8.141 2	2 479.4	253.7
Terpinolene	1	40–179	7.169	1 706	211
Tetrabutyl tin	1	100–300	6.545	1 649	148
1,1,2,2-Tetrachloro-1,2-difluoro-ethane	1	10–91.5	10.995	4 437.1	455.2
1,1,1,2-Tetrachloroethane	1	59–130	6.898 75	1 365.88	209.74
1,1,2,2-Tetrachloroethane	1	25–130	6.631 7	1 228.1	179.9
Tetrachloroethylene	1	37–120	6.976 83	1 386.92	217.53
Tetrachloromethane	1		6.879 26	1 212.021	226.41
Tetradecane	1	122–286	7.013 00	1 740.88	167.72
1-Tetradecanethiol	1		7.048 5	1 909.2	151.9
1-Tetradecanol	1	130–264	6.674 1	1 204.5	54.0
1-Tetradecene	1	119–283	7.030 65	1 754.09	171.52
1,2,3,4-Tetrafluorobenzene	1	6–50	7.084 6	1 339.23	223.49
1,2,3,5-Tetrafluorobenzene	1	6–50	6.986 17	1 245.20	218.35
Tetrafluoroethylene	1	−131 to −65	6.896 59	683.84	245.93
Tetrafluoromethane	1		6.972 31	540.50	260.10
Tetrahydrofuran	1	23–100	6.995 15	1 202.29	226.25
Tetraiodothiophene	1	−65 to 24	5.585 44	871.25	175.59
Tetralin	1	94–206	7.070 55	1 741.30	208.26
1,2,3,4-Tetramethylbenzene	1	80–217	7.059 4	1 690.54	199.48
1,2,3,5-Tetramethylbenzene	1	75–228	7.077 9	1 675.43	201.14
1,2,4,5-Tetramethylbenzene	1	74–227	7.080 0	1 672.43	201.43
2,2,3,3-Tetramethylbutane	1	0–65	6.876 65	1 329.93	226.36
Tetramethyl lead	1	0–60	6.937 7	1 335.3	219.1
2,2,3,3-Tetramethylpentane	1	57–141	6.830 60	1 398.67	213.84
2,2,3,4-Tetramethylpentane	1	52–134	6.834 18	1 375.59	214.94
2,2,4,4-Tetramethylpentane	1	43–123	6.796 20	1 324.59	216.02
Tetramethylsilane	1	−64 to 21	6.822 39	1 033.72	235.62
2-Thiabutane	1	−26 to 90	6.938 49	1 182.562	224.78
Thiacyclobutane	1	−5 to 120	7.016 67	1 321.331	224.51
Thiacyclohexane	1	29–170	6.905 18	1 422.47	211.72
Thiacyclopentane	1	14–148	6.995 40	1 401.939	219.61
Thiacyclopropane	1	−35 to 77	7.037 25	1 194.37	232.42
3-Thiaheptane	1	33–172	6.941 02	1 421.32	205.81
4-Thiaheptane	1	32–170	6.935 77	1 413.44	205.73
2-Thiahexane	1	17–150	6.945 83	1 363.808	212.07
3-Thiahexane	1	14–144	6.933 80	1 341.57	212.51

TABLE 2.36 Vapor Pressures of Various Organic Compounds (*Continued*)

Substance	Eq.	Range, °C	*A*	*B*	*C*
2-Thiapentane	1	−4 to 120	6.955 45	1 284.32	219.66
3-Thiapentane	1	−13 to 109	6.928 36	1 257.833	218.66
2-Thiapropane	1	−47 to 58	6.948 79	1 090.755	230.80
Thiazole	1	63–118	7.142 01	1 425.35	216.26
Thiophene	1	−12 to 108	6.959 26	1 246.02	221.35
Toluene	1	6–137	6.954 64	1 344.800	219.48
o-Toluidine	1	118–200	7.082 03	1 627.72	187.13
m-Toluidine	1	122–203	7.093 67	1 631.43	183.91
p-Toluidine	1		7.260 22	1 758.55	201.0
m-Tolyl pentafluoropropionate	1	98–174	7.427 20	1 707.59	201.70
p-Tolyl pentafluoropropionate	1	99–176	8.078 6	2 223.8	252.1
m-Tolyl trifluoroacetate	1	91–166	7.681 0	1 874.84	223.48
p-Tolyl trifluoroacetate	1	92–169	7.913 8	2 055.41	238.99
Tribromomethane	1	30–101	6.821 8	1 376.7	201.0
1,2,3-Tribromopropane	1	128–205	7.037 2	1 735.32	195.42
Trichloroacetic acid	1	112–198	7.273 0	1 594.3	165.4
Trichloroacetonitrile	1	17–83	7.183 5	1 368.3	232.5
Trichloroacetyl chloride	1	32–119	6.990 75	1 390.47	220.11
1,1,1-Trichloroethane	1	−6 to 17	8.643 4	2 136.6	302.8
1,1,2-Trichloroethane	1	50–114	6.951 85	1 314.41	209.20
Trichloroethylene	1	18–86	6.518 3	1 018.6	192.7
Trichlorofluoromethane	1		6.884 28	1 043.004	236.88
Trichlorosilane	1	2–32	6.773 9	1 009.0	227.2
bis-Trichlorosilylethane	1	91–160	7.835 11	2 241.769	249.84
1,1,1-Trichloro-2,2,2-trifluoro-ethane	1	14–36	4.437 3	204.1	83.9
1,1,2-Trichloro-1,2,2-trifluoro-ethane	1	−25 to 83	6.880 3	1 099.9	227.5
Tridecane	1	107–267	7.007 56	1 690.67	174.22
1-Tridecene	1	105–264	6.981 02	1 672.00	174.95
Triethanolamine	1	252–305	10.067 5	4 542.78	297.76
Triethyl aluminum	1	57–126	11.646 1	4 466.59	322.87
Triethylamine	1	50–95	5.858 8	695.7	144.8
Triethyl borate	1	29–109	7.511 1	1 641.7	236.3
Triethylsilanol	1	24–140	7.793 7	1 756.1	202.4
Trifluoroacetic acid	1	12–72	8.389	1 895	273
Trifluoroacetic anhydride	1	−2 to 39	6.135 8	1 026.1	202.0
Trifluoroacetonitrile	1	−132 to −68	7.127 6	773.82	249.9
1,3,5-Trifluorobenzene	1	6–50	6.919 8	1 197.13	219.12
Trifluorochloroethylene	1	−67 to −11	6.896 16	848.33	293.64
1,1,1-Trifluoroethane	1	−110 to −48	6.903 78	788.20	243.23
2,2,2-Trifluoroethanol	1	−0.5 to 25	6.788 2	978.13	173.06
Trifluoromethane	1	−128 to −82	7.088 6	705.33	249.78
bis-(Trifluoromethyl)-acetoxyphos-phine	1	0–40	7.391 31	1 426.254	220.37
2,2,2-Trifluoro-1-methylbenzene	1	55–139	6.970 45	1 306.35	217.38
bis-(Trifluoromethyl)-chlorophos-phine	1	−80 to 0	7.661 06	1 386.652	267.14
Trifluoromethylhypofluorite	1	145–189	6.950 6	650.1	−18.4
bis-(Trifluoromethyl)-iodophos-phine	1	0–47	6.901 39	1 180.723	222.95
Triisobutylene	1	56–179	7.002 1	1 613.47	212.5

(*Continued*)

TABLE 2.36 Vapor Pressures of Various Organic Compounds (*Continued*)

Substance	Eq.	Range, °C	*A*	*B*	*C*
Trimethyl aluminum	1	64–127	7.570 29	1 734.72	242.78
Trimethylamine	1	−80 to 3	6.857 55	955.94	237.52
1,2,3-Trimethylbenzene	1	57–205	7.040 82	1 593.958	207.08
1,2,4-Trimethylbenzene	1	52–198	7.043 83	1 573.257	208.56
1,3,5-Trimethylbenzene	1	49–193	7.074 36	1 569.622	209.58
2,2,3-Trimethylbutane	1	−19 to 106	6.792 30	1 200.563	226.05
Trimethylchlorosilane	1	2–55	7.055 8	1 245.5	240.7
1,1,3-Trimethylcyclohexane	1	55–137	6.839 51	1 394.88	215.73
1,1,2-Trimethylcyclopentane	1	36–115	6.822 38	1 309.81	218.58
1,1,3-Trimethylcyclopentane	1	29–106	6.809 31	1 275.92	219.89
1,2,4-Trimethylcyclopentane					
cis, cis, trans	1	39–118	6.857 38	1 335.69	219.16
cis, trans, cis	1	33–110	6.851 3	1 307.10	219.92
1,3,5-Trimethyl-2-ethylbenzene	1	88–210	6.790 8	1 505.8	174.7
1,4,5-Trimethyl-2-ethylbenzene	1	87–132	3.029 3	116.4	−34.6
2,2,5-Trimethylhexane	1	46–125	6.837 75	1 325.54	210.91
2,4,4-Trimethylhexane	1	51–131	6.856 54	1 371.81	214.40
Trimethylhydrazine	1	−16 to 14	7.106 80	1 189.88	222.06
O,N,N-Trimethylhydroxylamine	1	−79 to 23	6.765 8	979.55	222.2
2,2,3-Trimethylpentane	1		6.825 46	1 294.88	218.42
2,2,4-Trimethylpentane	1	24–100	6.811 89	1 257.84	220.74
2,3,3-Trimethylpentane	1		6.843 53	1 328.05	220.38
2,3,4-Trimethylpentane	1	36–114	6.853 96	1 315.08	217.53
2,4,4-Trimethyl-1-pentene	1	−3 to 128	6.834 57	1 273.416	220.62
2,4,4-Trimethyl-2-pentene	1	2–131	6.859 22	1 272.717	214.99
2,3,5-Trimethylphenol	1	186–247	7.080 12	1 685.90	166.14
Trimethylsilanol	1	18–85	8.126 6	1 657.6	219.2
2,4,5-Trimethylstyrene	1	79–216	7.331 5	1 880.7	205.7
2,4,6-Trimethylstyrene	1	90–208	7.089 1	1 702.61	195.93
1,2,4-Trinitrobenzene	1	250–300	3.194	87	−199
1,3,5-Trinitrobenzene	1	202–312	5.534 5	993.6	11.2
2,4,6-Trinitrobenzene	1	249–342	9.621 1	4 987.9	329.9
2,4,6-Trinitrotoluene	1	230–250	7.671 52	2 669.4	205.6
α-Trioxane	1	56–114	7.818 6	1 783.3	247.1
Trivinylarsine	1	22–66	7.894 1	2 115.6	293.9
Trivinyl bismuth	1	20–74	7.237 2	1 667.0	215.1
Trivinylphosphine	1	16–61	7.928 4	2 102.0	301.3
Trivinylstibine	1	20–70	8.322 1	2 446.3	303.8
Undecane	1	75–226	6.972 20	1 569.57	187.70
1-Undecanethiol	1		7.012 2	1 767.4	170.4
1-Undecene	1	72–222	6.966 77	1 563.21	189.87
Urethane	1		7.421 64	1 758.21	205.0
Vinyl acetate	1	22–72	7.210 1	1 296.13	226.66
o-Xylene	1	32–172	6.998 91	1 474.679	213.69
m-Xylene	1	28–166	7.009 08	1 462.266	215.11
p-Xylene	1	27–166	6.990 52	1 453.430	215.31
2,3-Xylenol	1	149–218	7.053 97	1 617.57	170.74
2,4-Xylenol	1	144–212	7.055 39	1 587.46	169.34
2,5-Xylenol	1	144–212	7.051 56	1 592.70	170.74
2,6-Xylenol	1	145–204	7.070 70	1 628.32	187.60
3,4-Xylenol	1	172–229	7.079 19	1 621.45	159.26
3,5-Xylenol	1	155–223	7.130 76	1 639.86	164.16

TABLE 2.37 Boiling Points of Common Organic Compounds at Selected Pressures

Compound		Pressure, mm Hg										Melting Point, °C
		1	5	10	20	40	60	100	200	400	760	
Name	Formula	Temperature, °C										
Acenaphthalene	$C_{12}H_{10}$		114.8	131.2	148.7	168.2	181.2	197.5	222.1	250.0	277.5	95
Acetal	$C_6H_{14}O_2$	−23.0	−2.3	+8.0	19.6	31.9	39.8	50.1	66.3	84.0	102.2	
Acetaldehyde	C_2H_4O	−81.5	−65.1	−56.8	−47.8	−37.8	−31.4	−22.6	−10.0	+4.9	20.2	−123.5
Acetamide	C_2H_5NO	65.0	92.0	105.0	120.0	135.8	145.8	158.0	178.3	200.0	222.0	81
Acetanilide	C_8H_9NO	114.0	146.6	162.0	180.0	199.6	211.8	227.2	250.5	227.0	303.8	113.5
Acetic acid	$C_2H_4O_2$	−17.2	+6.3	17.5	29.9	43.0	51.7	63.0	80.0	99.0	118.1	16.7
anhydride	$C_4H_6O_3$	1.7	24.8	36.0	48.3	62.1	70.8	82.2	100.0	119.8	139.6	−73
Acetone	C_3H_6O	−59.4	−40.5	−31.1	−20.8	−9.4	−2.0	+7.7	22.7	39.5	56.5	−94.6
Acetonitrile	C_2H_3N	−47.0	−26.6	−16.3	−5.0	+7.7	15.9	27.0	43.7	62.5	81.8	−41
Acetophenone	C_8H_8O	37.1	64.0	78.0	92.4	109.4	119.8	133.6	154.2	178.0	202.4	20.5
Acetyl chloride	C_2H_3OCl	−50.0	−35.0	−27.6	−19.6	−10.4	−4.5	+3.2	16.1	32.0	50.8	−112.0
Acetylene	C_2H_2	−142.9	−133.0	−128.2	−122.8	−116.7	−112.8	−107.9	−100.3	−92.0	−84.0	−81.5
Acridine	$C_{13}H_9N$	129.4	165.8	184.0	203.5	224.2	238.7	256.0	284.0	314.3	346.0	110.5
Acrolein (2-propenal)	C_3H_4O	−64.5	−46.0	−36.7	−26.3	−15.0	−7.5	+2.5	17.5	34.5	52.5	−87.7
Acrylic acid	$C_3H_4O_2$	+3.5	27.3	39.0	52.0	66.2	75.0	86.1	103.3	122.0	141.0	14
Adipic acid	$C_6H_{10}O_4$	159.5	191.0	205.5	222.0	240.5	251.0	265.0	287.8	312.5	337.5	152
Allene (propadiene)	C_3H_4	−120.6	−108.0	−101.0	−93.4	−85.2	−78.8	−72.5	−61.3	−48.5	−35.0	−136
Allyl alcohol (propen-1-ol-3)	C_3H_6O	−20.0	+0.2	10.5	21.7	33.4	40.3	50.0	64.5	80.2	96.6	−129
chloride (3-chloropropene)	C_3H_5Cl	−70.0	−52.0	−42.9	−32.8	−21.2	−14.1	−4.5	10.4	27.5	44.6	−136.4
isopropyl ether	$C_6H_{12}O$	−43.7	−23.1	−12.9	−1.8	+10.9	18.7	29.0	44.3	61.7	79.5	
isothiocyanate	C_4H_5NS	−2.0	+25.3	38.3	52.1	67.4	76.2	89.5	108.0	129.8	150.7	−80
n-propyl ether	$C_6H_{12}O$	−39.0	−18.2	−7.9	+3.7	16.4	25.0	35.8	52.6	71.4	90.5	
4-Allylveratrole	$C_{11}H_{14}O_2$	85.0	113.9	127.0	142.8	158.3	169.6	183.7	204.0	226.2	248.0	
iso-Amyl acetate	$C_7H_{14}O_2$	0.0	+23.7	35.2	47.8	62.1	71.0	83.2	101.3	121.5	142.0	
n-Amyl alcohol	$C_5H_{12}O$	+13.6	34.7	44.9	55.8	68.0	75.5	85.8	102.0	119.8	137.8	
iso-Amyl alcohol	$C_5H_{12}O$	+10.0	30.9	40.8	51.7	63.4	71.0	80.7	95.8	113.7	130.6	−117.2
sec-Amyl alcohol (2-pentanol)	$C_5H_{12}O$	+1.5	22.1	32.2	42.6	54.1	61.5	70.7	85.7	102.3	119.7	
tert-Amyl alcohol	$C_5H_{12}O$	−12.9	+7.2	17.2	27.9	38.8	46.0	55.3	69.7	85.7	101.7	−11.9
sec-Amylbenzene	$C_{11}H_{16}$	29.0	55.8	69.2	83.8	100.0	110.4	124.1	145.2	168.0	193.0	
iso-Amyl benzoate	$C_{12}H_{16}O_2$	72.0	104.5	121.6	139.7	158.3	171.4	186.8	210.2	235.8	262.0	
bromide (1-bromo-3-methylbutane)	$C_5H_{11}Br$	−20.4	+2.1	13.6	26.1	39.8	48.7	60.4	78.7	99.4	120.4	

(*Continued*)

TABLE 2.37 Boiling Points of Common Organic Compounds at Selected Pressures (*Continued*)

Compound		Pressure, mm Hg										Melting Point, °C
		1	5	10	20	40	60	100	200	400	760	
Name	Formula	Temperature, °C										
n-butyrate	$C_9H_{18}O_2$	21.2	47.1	59.9	74.0	90.0	99.8	113.1	133.2	155.3	178.6	
formate	$C_6H_{12}O_2$	−17.5	+5.4	17.1	30.0	44.0	53.3	65.4	83.2	102.7	123.3	
iodide (1-iodo-3-methylbutane)	$C_5H_{11}I$	−2.5	+21.9	34.1	47.6	62.3	71.9	84.4	103.8	125.8	148.2	
isobutyrate	$C_9H_{18}O_2$	14.8	40.1	52.8	66.6	81.8	91.7	104.4	124.2	146.0	168.8	
Amyl isopropionate	$C_8H_{16}O_2$	+8.5	33.7	46.3	60.0	75.5	85.2	97.6	117.3	138.4	160.2	
iso-Amyl isovalerate	$C_{10}H_{20}O_2$	27.0	54.4	68.6	83.8	100.6	110.3	125.1	146.1	169.5	194.0	
n-Amyl levulinate	$C_{10}H_{18}O_3$	81.3	110.0	124.0	139.7	155.8	165.2	180.5	203.1	227.4	253.2	
iso-Amyl levulinate	$C_{10}H_{18}O_3$	75.6	104.0	118.8	134.4	151.7	162.6	177.0	198.1	222.7	247.9	
nitrate	$C_5H_{11}NO_3$	+5.2	28.8	40.3	53.5	67.6	76.3	88.6	106.7	126.5	147.5	
4-*tert*-Amylphenol	$C_{11}H_{16}O$		109.8	125.5	142.3	160.3	172.6	189.0	213.0	239.5	266.0	93
Anethole	$C_{10}H_{12}O$	62.6	91.6	106.0	121.8	139.3	149.8	164.2	186.1	210.5	235.3	22.5
Angelonitrile	C_5H_7N	−8.0	+15.0	28.0	41.0	55.8	65.2	77.5	96.3	117.7	140.0	
Aniline	C_6H_7N	34.8	57.9	69.4	82.0	96.7	106.0	119.9	140.1	161.9	184.4	−6.2
2-Anilinoethanol	$C_8H_{11}NO$	104.0	134.3	149.6	165.7	183.7	194.0	209.5	230.6	254.5	279.6	
Anisaldehyde	$C_8H_8O_2$	73.2	102.6	117.8	133.5	150.5	161.7	176.7	199.0	223.0	248.0	2.5
o-Anisidine (2-methoxyaniline)	C_7H_9NO	61.0	88.0	101.7	116.1	132.0	142.1	155.2	175.3	197.3	218.5	5.2
Anthracene	$C_{14}H_{10}$	145.0	173.5	187.2	201.9	217.5	231.8	250.0	279.0	310.2	342.0	217.5
Anthraquinone	$C_{14}H_8O_2$	190.0	219.4	234.2	248.3	264.3	273.3	285.0	314.6	346.2	379.9	286
Azelaic acid	$C_9H_{16}O_4$	178.3	210.4	225.5	242.4	260.0	271.8	286.5	309.6	332.8	356.5	106.5
Azelaldehyde	$C_9H_{18}O$	33.3	58.4	71.6	85.0	100.2	110.0	123.0	142.1	163.4	185.0	
Azobenzene	$C_{12}H_{10}N_2$	103.5	135.7	151.5	168.3	187.9	199.8	216.0	240.0	266.1	293.0	68
Benzal chloride (α,α-Dichlorotoluene)	$C_7H_6Cl_2$	35.5	64.0	78.7	94.3	112.1	123.4	138.3	160.7	187.0	214.0	−16.1
Benzaldehyde	C_7H_6O	26.2	50.1	62.0	75.0	90.1	99.6	112.5	131.7	154.1	179.0	−26
Benzanthrone	$C_{17}H_{10}O$	225.0	274.5	297.2	322.5	350.0	368.8	390.0	426.5			174
Benzene	C_6H_6	−36.7	−19.6	−11.5	−2.6	+7.6	15.4	26.1	42.2	60.6	80.1	+5.5
Benzenesulfonylchloride	$C_6H_5ClO_2S$	65.9	96.5	112.0	129.0	147.7	158.2	174.5	198.0	224.0	251.5	14.5
Benzil	$C_{14}H_{10}O_2$	128.4	165.2	183.0	202.8	224.5	238.2	255.8	283.5	314.3	347.0	95
Benzoic acid	$C_7H_6O_2$	96.0	119.5	132.1	146.7	162.6	172.8	186.2	205.8	227.0	249.2	121.7
anhydride	$C_{14}H_{10}O_3$	143.8	180.0	198.0	218.0	239.8	252.7	270.4	299.1	328.8	360.0	42
Benzoin	$C_{14}H_{12}O_2$	135.6	170.2	188.1	207.0	227.9	241.7	258.0	284.4	313.5	343.0	132
Benzonitrile	C_7H_5N	28.2	55.3	69.2	83.4	99.6	109.8	123.5	144.1	166.7	190.6	−12.9

Benzophenone	$C_{13}H_{10}O$	108.2	141.7	157.6	175.8	195.7	208.2	224.4	249.8	276.8	305.4	48.5
Benzotrichloride (*α,α,α*-Trichlorotoluene)	$C_7H_5Cl_3$	45.8	73.7	87.6	102.7	119.8	130.0	144.3	165.6	189.2	213.5	−21.2
Benzotrifluoride (*α,α,α*-Trifluorotoluene)	$C_7H_5F_3$	−32.0	−10.3	−0.4	12.2	25.7	34.0	45.3	62.5	82.0	102.2	−29.3
Benzoyl bromide	C_7H_5BrO	47.0	75.4	89.8	105.4	122.6	133.4	147.7	169.2	193.7	218.5	0
chloride	C_7H_5ClO	32.1	59.1	73.0	87.6	103.8	114.7	128.0	149.5	172.8	197.2	−0.5
nitrile	C_8H_5NO	44.5	71.7	85.5	100.2	116.6	127.0	141.0	161.3	185.0	208.0	33.5
Benzyl acetate	$C_9H_{10}O_2$	45.0	73.4	87.6	102.3	119.6	129.8	144.0	165.5	189.0	213.5	−51.5
alcohol	C_7H_8O	58.0	80.8	92.6	105.8	119.8	129.3	141.7	160.0	183.0	204.7	−15.3
Benzylamine	C_7H_9N	29.0	54.8	67.7	81.8	97.3	107.3	120.0	140.0	161.3	184.5	
Benzyl bromide (*α*-bromotoluene)	C_7H_7Br	32.2	59.6	73.4	88.3	104.8	115.6	129.8	150.8	175.2	198.5	−4
chloride (*α*-chlorotoluene)	C_7H_7Cl	22.0	47.8	60.8	75.0	90.7	100.5	114.2	134.0	155.8	179.4	−39
cinnamate	$C_{16}H_{14}O_2$	173.8	206.3	221.5	239.3	255.8	267.0	281.5	303.8	326.7	350.0	39
Benzyldichlorosilane	$C_7H_8Cl_2Si$	45.3	70.2	83.2	96.7	111.8	121.3	133.5	152.0	173.0	194.3	
Benzyl ethyl ether	$C_9H_{12}O$	26.0	52.0	65.0	79.6	95.4	105.5	118.9	139.6	161.5	185.0	
phenyl ether	$C_{13}H_{12}O$	95.4	127.7	144.0	160.7	180.1	192.6	209.2	233.2	259.8	287.0	
isothiocyanate	C_8H_7NS	79.5	107.8	121.8	137.0	153.0	163.8	177.7	198.0	220.4	243.0	
Biphenyl	$C_{12}H_{10}$	70.6	101.8	117.0	134.2	152.5	165.2	180.7	204.2	229.4	254.9	69.5
1-Biphenyloxy-2,3-epoxypropane	$C_{15}H_{14}O_2$	135.5	169.9	187.2	205.8	226.3	239.7	255.0	280.4	309.8	340.0	
d-Bornyl acetate	$C_{12}H_{20}O_2$	46.9	75.7	90.2	106.0	123.7	135.7	149.8	172.0	197.5	223.0	29
Bornyl *n*-butyrate	$C_{14}H_{24}O_2$	74.0	103.4	118.0	133.8	150.7	161.8	176.4	198.0	222.2	247.0	
formate	$C_{11}H_{18}O_2$	47.0	74.8	89.3	104.0	121.2	131.7	145.8	166.4	190.2	214.0	
isobutyrate	$C_{14}H_{24}O_2$	70.0	99.8	114.0	130.0	147.2	157.6	172.2	194.2	218.2	243.0	
propionate	$C_{13}H_{22}O_2$	64.6	93.7	108.0	123.7	140.4	151.2	165.7	187.5	211.2	235.0	
Brassidic acid	$C_{22}H_{42}O_2$	209.6	241.7	256.0	272.9	290.0	301.5	316.2	336.8	359.6	382.5	61.5
Bromoacetic acid	$C_2H_3BrO_2$	54.7	81.6	94.1	108.2	124.0	133.8	146.3	165.8	186.7	208.0	49.5
4-Bromoanisole	C_7H_7BrO	48.8	77.8	91.9	107.8	125.0	136.0	150.1	172.7	197.5	223.0	12.5
Bromobenzene	C_6H_5Br	+2.9	27.8	40.0	53.8	68.6	78.1	90.8	110.1	132.3	156.2	−30.7
4-Bromobiphenyl	$C_{12}H_9Br$	98.0	133.7	150.6	169.8	190.8	204.5	221.8	248.2	277.7	310.0	90.5
1-Bromo-2-butanol	C_4H_9BrO	23.7	45.4	55.8	67.2	79.5	87.0	97.6	112.1	128.3	145.0	
1-Bromo-2-butanone	C_4H_7BrO	+6.2	30.0	41.8	54.2	68.2	77.3	89.2	107.0	126.3	147.0	
cis-1-Bromo-1-butene	C_4H_7Br	−44.0	−23.2	−12.8	−1.4	+11.5	19.8	30.8	47.8	66.8	86.2	
trans-1-Bromo-butene	C_4H_7Br	−38.4	−17.0	−6.4	+5.4	18.4	27.2	38.1	55.7	75.0	94.7	−100.3
2-Bromo-1-butene	C_4H_7Br	−47.3	−27.0	−16.8	−5.3	+7.2	15.4	26.3	42.8	61.9	81.0	−133.4
cis-2-Bromo-2-butene	C_4H_7Br	−39.0	−17.9	−7.2	+4.6	17.7	26.2	37.5	54.5	74.0	93.9	−111.2
trans-2-Bromo-2-butene	C_4H_7Br	−45.0	−24.1	−13.8	−2.4	+10.5	18.7	29.9	46.5	66.0	85.5	−114.6
1,4-Bromochlorobenzene	C_6H_4BrCl	32.0	59.5	72.7	87.8	103.8	114.8	128.0	149.5	172.6	196.9	
1-Bromo-1-chloroethane	C_2H_4BrCl	−36.0	−18.0	−9.4	0.0	+10.4	17.0	28.0	44.7	63.4	82.7	16.6

(*Continued*)

TABLE 2.37 Boiling Points of Common Organic Compounds at Selected Pressures (*Continued*)

Compound		Pressure, mm Hg										Melting Point, °C
		1	5	10	20	40	60	100	200	400	760	
Name	Formula	Temperature, °C										
1-Bromo-2-chloroethane	C_2H_4BrCl	−28.8	−7.0	+4.1	16.0	29.7	38.0	49.5	66.8	86.0	106.7	−16.6
2-Bromo-4,6-dichlorophenol	$C_6H_3BrCl_2O$	84.0	115.6	130.8	147.7	165.8	177.6	193.2	216.5	242.0	268.0	68
1-Bromo-4-ethyl benzene	C_8H_9Br	30.4	42.5	74.0	90.2	108.5	121.0	135.5	156.5	182.0	206.0	−45.0
(2-Bromoethyl)-benzene	C_8H_9Br	48.0	76.2	90.5	105.8	123.2	133.8	148.2	169.8	194.0	219.0	
2-Bromoethyl 2-chloroethyl ether	C_4H_8BrClO	36.5	63.2	76.3	90.8	106.6	116.4	129.8	150.0	172.3	195.8	
(2-Bromoethyl)-cyclohexane	$C_8H_{15}Br$	38.7	66.6	80.5	95.8	113.0	123.7	138.0	160.0	186.2	213.0	
1-Bromoethylene	C_2H_3Br	−95.4	−77.8	−68.8	−58.8	−48.1	−41.2	−31.9	−17.2	−1.1	+15.8	−138
Bromoform (tribromomethane)	$CHBr_3$		22.0	34.0	48.0	63.6	73.4	85.9	106.1	127.9	150.5	8.5
1-Bromonaphthalene	$C_{10}H_7Br$	84.2	117.5	133.6	150.2	170.2	183.5	198.8	224.2	252.0	281.1	5.5
2-Bromo-4-phenylphenol	$C_{12}H_9BrO$	100.0	135.4	152.3	171.8	193.8	207.0	224.5	251.0	280.2	311.0	95
3-Bromopyridine	C_5H_4BrN	16.8	42.0	55.2	69.1	84.1	94.1	107.8	127.7	150.0	173.4	
2-Bromotoluene	C_7H_7Br	24.4	49.7	62.3	76.0	91.0	100.0	112.0	133.6	157.3	181.8	−28
3-Bromotuluene	C_7H_7Br	14.8	50.8	64.0	78.1	93.9	104.1	117.8	138.0	160.0	183.7	39.8
4-Bromotoluene	C_7H_7Br	10.3	47.5	61.1	75.2	91.8	102.3	116.4	137.4	160.2	184.5	28.5
3-Bromo-2,4,6-trichlorophenol	$C_6H_2BrCl_3O$	112.4	146.2	163.2	181.8	200.5	213.0	229.3	253.0	278.0	305.8	
2-Bromo-1,4-xylene	C_8H_9Br	37.5	65.0	78.8	94.0	110.6	121.6	135.7	156.4	181.0	206.7	+9.5
1,2-Butadiene (methyl allene)	C_4H_6	−89.5	−72.7	−64.2	−54.9	−44.3	−37.5	−28.3	−14.2	+1.8	18.5	
1,3-Butadiene	C_4H_6	−102.8	−87.6	−79.7	−71.0	−61.3	−55.1	−46.8	−33.9	−19.3	−4.5	−108.9
n-Butane	C_4H_{10}	−101.5	−85.7	−77.8	−68.9	−59.1	−52.8	−44.2	−31.2	−16.3	−0.5	−135
iso-Butane (2-methylpropane)	C_4H_{10}	−109.2	−94.1	−86.4	−77.9	−68.4	−62.4	−54.1	−41.5	−27.1	−11.7	−145
1,3-Butanediol	$C_4H_{10}O_2$	22.2	67.5	85.3	100.0	117.4	127.5	141.2	161.0	183.8	206.5	77
1,2,3-Butanetriol	$C_4H_{10}O_3$	102.0	132.0	146.0	161.0	178.0	188.0	202.5	222.0	243.5	264.0	
1-Butene	C_4H_8	−104.8	−89.4	−81.6	−73.0	−63.4	−57.2	−48.9	−36.2	−21.7	−6.3	−130
cis-2-Butene	C_4H_8	−96.4	−81.1	−73.4	−64.6	−54.7	−48.4	−39.8	−26.8	−12.0	+3.7	−138.9
trans-2-Butene	C_4H_8	−99.4	−84.0	−76.3	−67.5	−57.6	−51.3	−42.7	−29.7	−14.8	+0.9	−105.4
3-Butenenitrile	C_4H_5N	−19.6	+2.9	14.1	26.6	40.0	48.8	60.2	78.0	98.0	119.0	
iso-Butyl acetate	$C_6H_{12}O_2$	−21.2	+1.4	12.8	25.5	39.2	48.0	59.7	77.6	97.5	118.0	−98.9
n-Butyl acrylate	$C_7H_{12}O_2$	−0.5	+23.5	35.5	48.6	63.4	72.6	85.1	104.0	125.2	147.2	−64.6
alcohol	$C_4H_{10}O$	−1.2	+20.0	30.2	41.5	53.4	60.3	70.1	84.3	100.8	117.5	−79.9
iso-Butyl alcohol	$C_4H_{10}O$	−9.0	+11.6	21.7	32.4	44.1	51.7	61.5	75.9	91.4	108.0	−108
sec-Butyl alcohol	$C_4H_{10}O$	−12.2	+7.2	16.9	27.3	38.1	45.2	54.1	67.9	83.9	99.5	−114.7

tert-Butyl alcohol	$C_4H_{10}O$	−20.4	−3.0	+5.5	14.3	24.5	31.0	39.8	52.7	68.0	82.9	25.3
iso-Butyl amine	$C_4H_{11}N$	−50.0	−31.0	−21.0	−10.3	+1.3	8.8	18.8	32.0	50.7	68.6	−85.0
n-Butylbenzene	$C_{10}H_{14}$	22.7	48.8	62.0	76.3	92.4	102.6	116.2	136.9	159.2	183.1	−88.0
iso-Butylbenzene	$C_{10}H_{14}$	14.1	40.5	53.7	67.8	83.3	93.3	107.0	127.2	149.6	172.8	−51.5
sec-Butylbenzene	$C_{10}H_{14}$	18.6	44.2	57.0	70.6	86.2	96.0	109.5	128.8	150.3	173.5	−75.5
tert-Butylbenzene	$C_{10}H_{14}$	13.0	39.0	51.7	65.6	80.8	90.6	103.8	123.7	145.8	168.5	−58
iso-Butyl benzoate	$C_{11}H_{14}O_2$	64.0	93.6	108.6	124.2	141.8	152.0	166.4	188.2	212.8	237.0	
n-Butyl bromide (1-bromobutane)	C_4H_9Br	−33.0	−11.2	−0.3	+11.6	24.8	33.4	44.7	62.0	81.7	101.6	−112.4
iso-Butyl *n*-butyrate	$C_8H_{16}O_2$	+4.6	30.0	42.2	56.1	71.7	81.3	94.0	113.9	135.7	156.9	
carbamate	$C_5H_{11}NO_2$		83.7	96.4	110.1	125.3	134.6	147.2	165.7	186.0	206.5	65
Butyl carbitol (diethylene glycol butyl ether)	$C_8H_{18}O_3$	70.0	95.7	107.8	120.5	135.5	146.0	159.8	181.2	205.0	231.2	
n-Butyl chloride (1-chlorobutane)	C_4H_9Cl	−49.0	−28.9	−18.6	−7.4	+5.0	13.0	24.0	40.0	58.8	77.8	−123.1
iso-Butyl chloride	C_4H_9Cl	−53.8	−34.3	−24.5	−13.8	−1.9	+5.9	16.0	32.0	50.0	68.9	−131.2
sec-Butyl chloride (2-Chlorobutane)	C_4H_9Cl	−60.2	−39.8	−29.2	−17.7	−5.0	+3.4	14.2	31.5	50.0	68.0	−131.3
tert-Butyl chloride	C_4H_9Cl					−19.0	−11.4	−1.0	+14.6	32.6	51.0	−26.5
sec-Butyl chloroacetate	$C_6H_{11}ClO_2$	17.0	41.8	54.6	68.2	83.6	93.0	105.5	124.1	146.0	167.8	
2-*tert*-Butyl-4-cresol	$C_{11}H_{16}O$	70.0	98.0	112.0	127.2	143.9	153.7	167.0	187.8	210.0	232.6	
4-*tert*-Butyl-2-cresol	$C_{11}H_{16}O$	74.3	103.7	118.0	134.0	150.8	161.7	176.2	197.8	221.8	247.0	
iso-Butyl dichloroacetate	$C_6H_{10}Cl_2O_2$	28.6	54.3	67.5	81.4	96.7	106.6	119.8	139.2	160.0	183.0	
2,3-Butylene glycol (2,3-butanediol)	$C_4H_{10}O_2$	44.0	68.4	80.3	93.4	107.8	116.3	127.8	145.6	164.0	182.0	22.5
2-Butyl-2-ethylbutane-1,3-diol	$C_{10}H_{12}O_2$	94.1	122.6	136.8	151.2	167.8	178.0	191.9	212.0	233.5	255.0	
2-*tert*-Butyl-4-ethylphenol	$C_{12}H_{15}O$	76.3	106.2	121.0	137.0	154.0	165.4	179.0	200.3	223.8	247.8	
n-Butyl formate	$C_5H_{10}O_2$	−26.4	−4.7	+6.1	18.0	31.6	39.8	51.0	67.9	86.2	106.0	
iso-Butyl formate	$C_5H_{10}O_2$	−32.7	−11.4	−0.8	+11.0	24.1	32.4	43.4	60.0	79.0	98.2	−95.3
sec-Butyl formate	$C_5H_{10}O_2$	−34.4	−13.3	−3.1	+8.4	21.3	29.6	40.2	56.8	75.2	93.6	
sec-Butyl glycolate	$C_6H_{12}O_3$	28.3	53.6	66.0	79.8	94.2	104.0	116.4	135.5	155.6	177.5	
iso-Butyl iodide (1-iodo-2-methylpropane)	C_4H_9I	−17.0	+5.0	17.0	29.8	42.8	51.8	63.5	81.0	100.3	120.4	−90.7
isobutyrate	$C_8H_{16}O_2$	+4.1	28.0	39.9	52.4	67.2	75.9	88.0	106.3	126.3	147.5	−80.7
isovalerate	$C_9H_{18}O_2$	16.0	41.2	53.8	67.7	82.7	92.4	105.2	124.8	146.4	168.7	
levulinate	$C_9H_{16}O_3$	65.0	92.1	105.9	120.2	136.2	147.0	160.2	181.8	205.5	229.9	
naphthylketone (1-isovaleronaphthone)	$C_{15}H_{16}O$	136.0	167.9	184.0	201.6	219.7	231.5	246.7	269.7	294.0	320.0	
2-*sec*-Butylphenol	$C_{10}H_{14}O$	57.4	86.0	100.8	116.1	133.4	143.9	157.3	179.7	203.8	228.0	
2-*tert*-Butylphenol	$C_{10}H_{14}O$	56.6	84.2	98.1	113.0	129.2	140.0	153.5	173.8	196.3	219.5	
4-*iso*-Butylphenol	$C_{10}H_{14}O$	72.1	100.9	115.5	130.3	147.2	157.0	171.2	192.1	214.7	237.0	
4-*sec*-Butylphenol	$C_{10}H_{14}O$	71.4	100.5	114.8	130.3	147.8	157.9	172.4	194.3	217.6	242.1	

(Continued)

TABLE 2.37 Boiling Points of Common Organic Compounds at Selected Pressures (*Continued*)

Compound		Pressure, mm Hg										Melting Point, °C
		1	5	10	20	40	60	100	200	400	760	
Name	Formula	Temperature, °C										
4-*tert*-Butylphenol	$C_{10}H_{14}O$	70.0	99.2	114.0	129.5	146.0	156.0	170.2	191.5	214.0	238.0	99
2-(4-*tert*-Butylphenoxy)ethyl acetate	$C_{14}H_{20}O_3$	118.0	150.0	165.8	183.3	201.5	212.8	228.0	250.3	277.6	304.4	
4-*tert*-Butylphenyl dichlorophosphate	$C_{10}H_{13}Cl_2O_2P$	96.0	129.6	146.0	164.0	184.3	197.2	214.3	240.0	268.2	299.0	
tert-Butyl phenyl ketone (pivalophenone)	$C_{11}H_{14}O$	57.8	85.7	99.0	114.3	130.4	140.8	154.0	175.0	197.7	220.0	
iso-Butyl propionate	$C_7H_{14}O_2$	−2.3	+20.9	32.3	44.8	58.5	67.6	79.5	97.0	116.4	136.8	−71
4-*tert*-Butyl-2,5-xylenol	$C_{12}H_{18}O$	88.2	119.8	135.0	151.0	169.8	180.3	195.0	217.5	241.3	265.3	
4-*tert*-Butyl-2,6-xylenol	$C_{12}H_{18}O$	74.0	103.9	119.0	135.0	152.2	163.6	176.0	196.0	217.8	239.8	
6-*tert*-Butyl-2,4-xylenol	$C_{12}H_{18}O$	70.3	100.2	115.0	131.0	148.5	158.2	172.0	192.3	214.2	236.5	
6-*tert*-Butyl-3,4-xylenol	$C_{12}H_{18}O$	83.9	113.6	127.0	143.0	159.7	170.0	184.0	204.5	226.7	249.5	
Butyric acid	$C_4H_8O_2$	25.5	49.8	61.5	74.0	88.0	96.5	108.0	125.5	144.5	163.5	−74
iso-Butyric acid	$C_4H_8O_2$	14.7	39.3	51.2	64.0	77.8	86.3	98.0	115.8	134.5	154.5	−47
Butyronitrile	C_4H_7N	−20.0	+2.1	13.4	25.7	38.4	47.3	59.0	76.7	96.8	117.5	
iso-Valerophenone	$C_{11}H_{14}O$	58.3	87.0	101.4	116.8	133.8	144.6	158.0	180.1	204.2	228.0	
Camphene	$C_{10}H_{16}$			47.2	60.4	75.7	85.0	97.9	117.5	138.7	160.5	50
Campholenic acid	$C_{10}H_{16}O_2$	97.6	125.7	139.8	153.9	170.0	180.0	193.7	212.7	234.0	256.0	
d-Camphor	$C_{10}H_{16}O$	41.5	68.6	82.3	97.5	114.0	124.0	138.0	157.9	182.0	209.2	178.5
Camphylamine	$C_{10}H_{19}N$	45.3	74.0	83.7	97.6	112.5	122.0	134.6	153.0	173.8	195.0	
Capraldehyde	$C_{10}H_{20}O$	51.9	78.8	92.0	106.3	122.2	132.0	145.3	164.8	186.3	208.5	
Capric acid	$C_{10}H_{20}O_2$	125.0	142.0	152.2	165.0	179.9	189.8	200.0	217.1	240.3	268.4	31.5
n-Caproic acid	$C_6H_{12}O_2$	71.4	89.5	99.5	111.8	125.0	133.3	144.0	160.8	181.0	202.0	−1.5
iso-Caproic acid	$C_6H_{12}O_2$	66.2	83.0	94.0	107.0	120.4	129.6	141.4	158.3	181.0	207.7	−35
iso-Caprolactone	$C_6H_{10}O_2$	38.3	66.4	80.3	95.7	112.3	123.2	137.2	157.8	182.1	207.0	
Capronitrile	$C_6H_{11}N$	9.2	34.6	47.5	61.7	76.9	86.8	99.8	119.7	141.0	163.7	
Capryl alcohol (2-octanol)	$C_8H_{18}O$	32.8	57.6	70.0	83.3	98.0	107.4	119.8	138.0	157.5	178.5	−38.6
Caprylaldehyde	$C_8H_{16}O$	73.4	92.0	101.2	110.2	120.0	126.0	133.9	145.4	156.5	168.5	
Caprylic acid (octanoic acid)	$C_8H_{16}O_2$	92.3	114.1	124.0	136.4	150.6	160.0	172.2	190.3	213.9	237.5	16
Caprylonitrile	$C_8H_{15}N$	43.0	67.6	80.4	94.6	110.6	121.2	134.8	155.2	179.5	204.5	
Carbazole	$C_{12}H_9N$						248.2	265.0	292.5	323.0	354.8	244.8
Carbon dioxide	CO_2	−134.3	−124.4	−119.5	−114.4	−108.6	−104.8	−100.2	−93.0	−85.7	−78.2	−57.5
disulfide	CS_2	−73.8	−54.3	−44.7	−34.3	−22.5	−15.3	−5.1	+10.4	28.0	46.5	−110.8
monoxide	CO	−222.0	−217.2	−215.0	−212.8	−210.0	−208.1	−205.7	−201.3	−196.3	−191.3	−205.0

oxyselenide (carbonyl selenide)	COSe	−117.1	−102.3	−95.0	−86.3	−76.4	−70.2	−61.7	−49.8	−35.6	−21.9	
oxysulfide (carbonyl sulfide)	COS	−132.4	−119.8	−113.3	−106.0	−98.3	−93.0	−85.9	−75.0	−62.7	−49.9	−138.8
tetrabromide	CBr_4					96.3	106.3	119.7	139.7	163.5	189.5	90.1
tetrachloride	CCl_4	−50.0	−30.0	−19.6	−8.2	+4.3	12.3	23.0	38.3	57.8	76.7	−22.6
tetrafluoride	CF_4	−184.6	−174.1	−169.3	−164.3	−158.8	−155.4	−150.7	−143.6	−135.5	−127.7	−183.7
Carvacrol	$C_{10}H_{14}O$	70.0	98.4	113.2	127.9	145.2	155.3	169.7	191.2	213.8	237.0	+0.5
Carvone	$C_{10}H_{14}O$	57.4	86.1	100.4	116.1	133.0	143.8	157.3	179.6	203.5	227.5	
Chavibetol	$C_{10}H_{12}O_2$	83.6	113.3	127.0	143.2	159.8	170.7	185.5	206.8	229.8	254.0	
Chloral (trichloroacetaldehyde)	C_2HCl_3O	−37.8	−16.0	−5.0	+7.2	20.2	29.1	40.2	57.8	77.5	97.7	−57
hydrate (trichloroacetaldehyde hydrate)	$C_2H_3Cl_3O_2$	−9.8	+10.0	19.5	29.2	39.7	46.2	55.0	68.0	82.1	96.2	51.7
Chloranil	$C_6Cl_4O_2$	70.7	89.3	97.8	106.4	116.1	122.0	129.5	140.3	151.3	162.6	290
Chloroacetic acid	$C_2H_3ClO_2$	43.0	68.3	81.0	94.2	109.2	118.3	130.7	149.0	169.0	189.5	61.2
anhydride	$C_4H_4Cl_2O_3$	67.2	94.1	108.0	122.4	138.2	148.0	159.8	177.8	197.0	217.0	46
2-Chloroaniline	C_6H_6ClN	46.3	72.3	84.8	99.2	115.6	125.7	139.5	160.0	183.7	208.8	0
3-Chloroaniline	C_6H_6ClN	63.5	89.8	102.0	116.7	133.6	144.1	158.0	179.5	203.5	228.5	−10.4
4-Chloroaniline	C_6H_5Cl	59.3	87.9	102.1	117.8	135.0	145.8	159.9	182.3	206.6	230.5	70.5
Chlorobenzene	C_6H_5Cl	−13.0	+10.6	22.2	35.3	49.7	58.3	70.7	89.4	10.0	132.2	−45.2
2-Chlorobenzotrichloride (2-α,α,α-tetrachlorotoluene)	$C_7H_4Cl_4$	69.0	101.8	117.9	135.8	155.0	167.8	185.0	208.0	233.0	262.1	28.7
2-Chlorobenzotrifluoride (2-chloro-α,α,α-trifluorotoluene)	$C_7H_4ClF_3$	0.0	24.7	37.1	50.6	65.9	75.4	88.3	108.3	130.0	152.2	−6.0
2-Chlorobiphenyl	$C_{12}H_9Cl$	89.3	109.8	134.7	151.2	169.9	182.1	197.0	219.6	243.8	267.5	34
4-Chlorobiphenyl	$C_{12}H_9Cl$	96.4	129.8	146.0	164.0	183.8	196.0	212.5	237.8	264.5	292.9	75.5
α-Chlorocrotonic acid	$C_4H_5ClO_2$	70.0	95.6	108.0	121.2	135.6	144.4	155.9	173.8	193.2	212.0	
Chlorodifluoromethane	$CHClF_2$	−122.8	−110.2	−103.7	−96.5	−88.6	−83.4	−76.4	−65.8	−53.6	−40.8	−160
Chlorodimethylphenylsilane	$C_8H_{11}ClSi$	29.8	56.7	70.0	84.7	101.2	111.5	124.7	145.5	168.6	193.5	
1-Chloro-2-ethoxybenzene	C_8H_9ClO	45.8	72.8	86.5	101.5	117.8	127.8	141.8	162.0	185.0	208.0	
2-(2-Chloroethoxy) ethanol	$C_4H_9ClO_2$	53.0	78.3	90.7	104.1	118.4	127.5	139.5	157.2	176.5	196.0	
bis-2-Chloroethyl acetacetal	$C_6H_{12}Cl_2O_2$	56.2	83.7	97.6	112.2	127.8	138.0	150.7	169.8	190.5	212.6	
1-Chloro-2-ethylbenzene	C_8H_9Cl	17.2	43.0	56.1	70.3	86.2	96.4	110.0	130.2	152.2	177.6	−80.2
1-Chloro-3-ethylbenzene	C_8H_9Cl	18.6	45.2	58.1	73.0	89.2	99.6	113.6	133.8	156.7	181.1	−53.3
1-Chloro-4-ethylbenzene	C_8H_9Cl	19.2	46.4	60.0	75.5	91.8	102.0	116.0	137.0	159.8	184.3	−62.6
2-Chloroethyl chloroacetate	$C_4H_6Cl_2O_2$	46.0	72.1	86.0	100.0	116.0	126.2	140.0	159.8	182.2	205	
2-Chloroethyl 2-chloroisopropyl ether	$C_5H_{10}Cl_2O$	24.7	50.1	63.0	77.2	92.4	102.2	115.8	135.7	156.5	180.0	
2-Chloroethyl 2-chloropropyl ether	$C_5H_{10}Cl_2O$	29.8	56.5	70.0	84.8	101.5	111.8	125.6	146.3	169.8	194.1	
2-Chloroethyl α-methylbenzyl ether	$C_{10}H_{13}ClO$	62.3	91.4	106.0	121.8	139.6	150.0	164.8	186.3	210.8	235.0	
Chloroform (trichloromethane)	$CHCl_3$	−58.0	−39.1	−29.7	−19.0	−7.1	+0.5	10.4	25.9	42.7	61.3	−63.5

(*Continued*)

TABLE 2.37 Boiling Points of Common Organic Compounds at Selected Pressures (*Continued*)

Compound		Pressure, mm Hg										Melting Point, °C
		1	5	10	20	40	60	100	200	400	760	
Name	Formula	Temperature, °C										
1-Chloronaphthalene	$C_{10}H_7Cl$	80.6	104.8	118.6	134.4	153.2	165.6	180.4	204.2	230.8	259.3	−20
4-Chlorophenethyl alcohol	C_8H_9ClO	84.0	114.3	129.0	145.0	162.0	173.5	188.1	210.0	234.5	259.3	
2-Chlorophenol	C_6H_5ClO	12.1	38.2	51.2	65.9	82.0	92.0	106.0	126.4	149.8	174.5	7
3-Chlorophenol	C_6H_5ClO	44.2	72.0	86.1	101.7	118.0	129.4	143.0	164.8	188.7	214.0	32.5
4-Chlorophenol	C_6H_5ClO	49.8	78.2	92.2	108.1	125.0	136.1	150.0	172.0	196.0	220.0	42
2-Chloro-3-phenylphenol	$C_{12}H_9ClO$	118.0	152.2	169.7	186.7	207.4	219.6	237.0	261.3	289.4	317.5	+6
2-Chloro-6-phenylphenol	$C_{12}H_9ClO$	119.8	153.7	170.7	189.8	208.2	220.0	237.1	261.6	289.5	317.0	
Chloropicrin (trichloronitromethane)	CCl_3NO_2	−25.5	−3.3	+7.8	20.0	33.8	42.3	53.8	71.8	91.8	111.9	−64
1-Chloropropene	C_3H_5Cl	−81.3	−63.4	−54.1	−44.0	−32.7	−25.1	−15.1	+1.3	18.0	37.0	−99.0
2-Chloropyridine	C_5H_4ClN	13.3	38.8	51.7	65.8	81.7	91.6	104.6	125.0	147.7	170.2	
3-Chlorostyrene	C_8H_7Cl	25.3	51.3	65.2	80.0	96.5	107.2	121.2	142.2	165.7	190.0	
4-Chlorostyrene	C_8H_7Cl	28.0	54.5	67.5	82.0	98.0	108.5	122.0	143.5	166.0	191.0	−15.0
1-Chlorotetradecane	$C_{14}H_{29}Cl$	98.5	131.8	148.2	166.2	187.0	199.8	215.5	240.3	267.5	296.0	+0.9
2-Chlorotoluene	C_7H_7Cl	+5.4	30.6	43.2	56.9	72.0	81.8	94.7	115.0	137.1	159.3	
3-Chlorotoluene	C_7H_7Cl	+4.8	30.3	43.2	57.4	73.0	83.2	96.3	116.6	139.7	162.3	
4-Chlorotoluene	C_7H_7Cl	+5.5	31.0	43.8	57.8	73.5	83.3	96.6	117.1	139.8	162.3	+7.3
Chlorotriethylsilane	$C_6H_{15}ClSi$	−4.9	+19.8	32.0	45.5	60.2	69.5	82.3	101.6	123.6	146.3	
1-Chloro-1,2,2-trifluoroethylene	C_2ClF_3	−116.0	−102.5	−95.9	−88.2	−79.7	−74.1	−66.7	−55.0	−41.7	−27.9	−157.5
Chlorotrifluoromethane	$CClF_3$	−149.5	−139.2	−134.1	−128.5	−121.9	−117.3	−111.7	−102.5	−92.7	−81.2	
Chlorotrimethylsilane	C_3H_9ClSi	−62.8	−43.6	−34.0	−23.2	−11.4	−4.0	+6.0	21.9	39.4	57.9	
trans-Cinnamic acid	$C_9H_8O_2$	127.5	157.8	173.0	189.5	207.1	217.8	232.4	253.3	276.7	300.0	133
Cinnamyl alcohol	$C_9H_{10}O$	72.6	102.5	117.8	133.7	151.0	162.0	177.8	199.8	224.6	250.0	33
Cinnamylaldehyde	C_9H_8O	76.1	105.8	120.0	135.7	152.2	163.7	177.7	199.3	222.4	246.0	−7.5
Citraconic anhydride	$C_5H_4O_3$	47.1	74.8	88.9	103.8	120.3	131.3	145.4	165.8	189.8	213.5	
cis-α-Citral	$C_{10}H_{16}O$	61.7	90.0	103.9	119.4	135.9	146.3	160.0	181.8	205.0	228.0	
d-Citronellal	$C_{10}H_{18}O$	44.0	71.4	84.8	99.8	116.1	126.2	140.1	160.0	183.8	206.5	
Citronellic acid	$C_{10}H_{18}O_2$	99.5	127.3	141.4	155.6	171.9	182.1	195.4	214.5	236.6	257.0	
Citronellol	$C_{10}H_{20}O$	66.4	93.6	107.0	121.5	137.2	147.2	159.8	179.8	201.0	221.5	
Citronellyl acetate	$C_{12}H_{22}O_2$	74.7	100.2	113.0	126.0	140.5	149.7	161.0	178.8	197.8	217.0	
Coumarin	$C_9H_6O_2$	106.0	137.8	153.4	170.0	189.0	200.5	216.5	240.0	264.7	291.0	70
o-Cresol (2-cresol; 3-methylphenol)	C_7H_8O	38.2	64.0	76.7	90.5	105.8	115.5	127.4	146.7	168.4	190.8	30.8

m-Cresol (3-cresol; 3-methylphenol)	C_7H_8O	52.0	76.0	87.8	101.4	116.0	125.8	138.0	157.3	179.0	202.8	10.9
p-Cresol (4-cresol; 4-methylphenol)	C_7H_8O	53.0	76.5	88.6	102.3	117.7	127.0	140.0	157.3	179.4	201.8	35.5
cis-Crotonic acid	$C_4H_6O_2$	33.5	57.4	69.0	82.0	96.0	104.5	116.3	133.9	152.2	171.9	15.5
trans-Crotonic acid	$C_4H_6O_2$			80.0	93.0	107.8	116.7	128.0	146.0	165.5	185.0	72
cis-Crotononitrile	C_4H_5N	−29.0	−7.1	+4.0	16.4	30.0	38.5	50.1	68.0	88.0	108.0	
trans-Crotononitrile	C_4H_5N	−19.5	+3.5	15.0	27.8	41.8	50.9	62.8	81.1	101.5	122.8	
Cumene	C_9H_{12}	+2.9	26.8	38.3	51.5	66.1	75.4	88.1	107.3	129.2	152.4	−96.0
4-Cumidene	$C_9H_{13}N$	60.0	88.2	102.2	117.8	134.2	145.0	158.0	180.0	203.2	227.0	
Cuminal	$C_{10}H_{12}O$	58.0	87.3	102.0	117.9	135.2	146.0	160.0	182.8	206.7	232.0	
Cuminyl alcohol	$C_{10}H_{14}O$	74.2	103.7	118.0	133.8	150.3	161.7	176.2	197.9	221.7	246.6	
2-Cyano-2-*n*-butyl acetate	$C_7H_{11}NO_2$	42.0	68.7	82.0	96.2	111.8	121.5	133.8	152.2	173.4	195.2	
Cyanogen	C_2N_2	−95.8	−83.2	−76.8	−70.1	−62.7	−57.9	−51.8	−42.6	−33.0	−21.0	−34.4
bromide	CBrN	−35.7	−13.3	−10.0	−1.0	+8.6	14.7	22.6	33.8	46.0	61.5	58
chloride	CClN	−76.7	−61.4	−53.8	−46.1	−37.5	−32.1	−24.9	−14.1	−2.3	+13.1	−6.5
iodide	CIN	25.2	47.2	57.7	68.6	80.3	88.0	97.6	111.5	126.1	141.1	
Cyclobutane	C_4H_8	−92.0	−76.0	−67.9	−58.7	−48.4	−41.8	−32.8	−18.9	−3.4	+12.9	−50
Cyclobutene	C_4H_6	−99.1	−83.4	−75.4	−66.6	−56.4	−50.0	−41.2	−27.8	−12.2	+2.4	
Cyclohexane	C_6H_{12}	−45.3	−25.4	−15.9	−5.0	+6.7	14.7	25.5	42.0	60.8	80.7	+6.6
Cyclohexaneethanol	$C_8H_{16}O$	50.4	77.2	90.0	104.0	119.8	129.8	142.7	161.7	183.5	205.4	
Cyclohexanol	$C_6H_{12}O$	21.0	44.0	56.0	68.8	83.0	91.8	103.7	121.7	141.4	161.0	23.9
Cyclohexanone	$C_6H_{10}O$	+1.4	26.4	38.7	52.5	67.8	77.5	90.4	110.3	132.5	155.6	−45.0
2-Cyclohexyl-4,6-dinitrophenol	$C_{12}H_{14}N_2O_5$	132.8	161.8	175.9	191.2	206.7	216.0	229.0	248.7	269.8	291.5	
Cyclopentane	C_5H_{10}	−68.0	−49.6	−40.4	−30.1	−18.6	−11.3	−1.3	+13.8	31.0	49.3	−93.7
Cyclopropane	C_3H_6	−116.8	−104.2	−97.5	−90.3	−82.3	−77.0	−70.0	−59.1	−46.9	−33.5	−126.6
Cymene	$C_{10}H_{14}$	17.3	43.9	57.0	71.1	87.0	97.2	110.8	131.4	153.5	177.2	−68.2
cis-Decalin	$C_{10}H_{18}$	22.5	50.1	64.2	79.8	97.2	108.0	123.2	145.4	169.9	194.6	−43.3
trans-Decalin	$C_{10}H_{18}$	−0.8	+30.6	47.2	65.3	85.7	98.4	114.6	136.2	160.1	186.7	−30.7
Decane	$C_{10}H_{22}$	16.5	42.3	55.7	69.8	85.5	95.5	108.6	128.4	150.6	174.1	−29.7
Decan-2-one	$C_{10}H_{20}O$	44.2	71.9	85.8	100.7	117.1	127.8	142.0	163.2	186.7	211.0	+3.5
1-Decene	$C_{10}H_{20}$	14.7	40.3	53.7	67.8	83.3	93.5	106.5	126.7	149.2	172.0	
Decyl alcohol	$C_{10}H_{22}O$	69.5	97.3	111.3	125.8	142.1	152.0	165.8	186.2	208.8	231.0	+7
Decyltrimethylsilane	$C_{13}H_{30}Si$	67.4	96.4	111.0	126.5	144.0	154.3	169.5	191.0	215.5	240.0	
Dehydroacetic acid	$C_8H_8O_4$	91.7	122.0	137.3	153.0	171.0	181.5	197.5	219.5	244.5	269.0	
Desoxybenzoin	$C_{14}H_{12}O$	123.3	156.2	173.5	192.0	212.0	224.5	241.3	265.2	293.0	321.0	60
Diacetamide	$C_4H_7NO_2$	70.0	95.0	108.0	122.6	138.2	148.0	160.6	180.8	202.0	223.0	78.5
Diacetylene (1,3-butadiyne)	C_4H_2	−82.5	−68.0	−61.2	−53.8	−45.9	−41.0	−34.0	−20.9	−6.1	+9.7	−34.9
Diallyldichlorosilane	$C_6H_{10}Cl_2Si$	+9.5	34.8	47.4	61.3	76.4	86.3	99.7	119.4	142.0	165.3	

(Continued)

TABLE 2.37 Boiling Points of Common Organic Compounds at Selected Pressures (*Continued*)

Compound		Pressure, mm Hg										Melting Point, °C
		1	5	10	20	40	60	100	200	400	760	
Name	Formula	Temperature, °C										
Dialyl sulfide	$C_6H_{10}S$	−9.5	14.4	26.6	39.7	54.2	63.7	75.8	94.8	116.1	138.6	−83
Diisoamyl ether	$C_{10}H_{22}O$	18.6	44.3	57.0	70.7	86.3	96.0	109.6	129.0	150.3	173.4	
oxalate	$C_{12}H_{22}O_4$	85.4	116.0	131.4	147.7	165.7	177.0	192.2	215.0	240.0	265.0	
sulfide	$C_{10}H_{22}S$	43.0	73.0	87.6	102.7	120.0	130.6	145.3	166.4	191.0	216.0	
Dibenzylamine	$C_{14}H_{15}N$	118.3	149.8	165.6	182.2	200.2	212.2	227.3	249.8	274.3	300.0	−26
Dibenzyl ketone (1,3-diphenyl-2-propanone)	$C_{15}H_{14}O$	125.5	159.8	177.6	195.7	216.6	229.4	246.6	272.3	301.7	330.5	34.5
1,4-Dibromobenzene	$C_6H_4Br_2$	61.0	79.3	87.7	103.6	120.8	131.6	146.5	168.5	192.5	218.6	87.5
1,2-Dibromobutane	$C_4H_8Br_2$	7.5	33.2	46.1	60.0	76.0	86.0	99.8	120.2	143.5	166.3	−64.5
dl-2,3-Dibromobutane	$C_4H_8Br_2$	+5.0	30.0	41.6	56.4	72.0	82.0	95.3	115.7	138.0	160.5	
meso-2,3-Dibromobutane	$C_4H_8Br_2$	+1.5	26.6	39.3	53.2	68.0	78.0	91.7	111.8	134.2	157.3	−34.5
1,2-Dibromodecane	$C_{10}H_{20}Br_2$	95.7	123.6	137.3	151.0	167.4	177.5	190.2	209.6	229.8	250.4	
Di(2-bromoethyl) ether	$C_4H_8Br_2O$	47.7	75.3	88.5	103.6	119.8	130.0	144.0	165.0	188.0	212.5	
α,β-Dibromomaleie Anhydride	$C_4H_2Br_2O_3$	50.0	78.0	92.0	106.7	123.5	133.8	147.7	168.0	192.0	215.0	
1,2-Dibromo-2-methylpropane	$C_4H_8Br_2$	−28.8	−3.0	+10.5	25.7	42.3	53.7	68.8	92.1	119.8	149.0	−70.3
1,3-Dibromo-2-methylpropane	$C_4H_8Br_2$	14.0	40.0	53.0	67.5	83.5	93.7	107.4	117.8	150.6	174.6	
1,2-Dibromopentane	$C_5H_{10}Br_2$	19.8	45.4	58.0	72.0	87.4	97.4	110.1	130.2	151.8	175.0	
1,2-Dibromopropane	$C_3H_6Br_2$	−7.0	+17.3	29.4	42.3	57.2	66.4	78.7	97.8	118.5	141.6	−5.5
1,3-Dibromopropane	$C_3H_6Br_2$	+9.7	35.4	48.0	62.1	77.8	87.8	101.3	121.7	144.1	167.5	−34.4
2,3-Dibromopropene	$C_3H_4Br_2$	−6.0	+17.9	30.0	43.2	57.8	67.0	79.5	98.0	119.5	141.2	
2,3-Dibromo-1-propanol	$C_3H_6Br_2O$	57.0	84.5	98.2	113.5	129.8	140.0	153.0	173.8	196.0	219.0	
Diisobutylamine	$C_8H_{19}N$	−5.1	+18.4	30.6	43.7	57.8	67.0	79.2	97.6	118.0	139.5	−70
2,6-Ditert-butyl-4-cresol	$C_{15}H_{24}O$	85.8	116.2	131.0	147.0	164.1	175.2	190.0	212.8	237.6	262.5	
4,6-Ditert-butyl-2-cresol	$C_{15}H_{24}O$	86.2	117.3	132.4	149.0	167.4	179.0	194.0	217.5	243.4	269.3	
4,6-Ditert-butyl-3-cresol	$C_{15}H_{24}O$	103.7	135.2	150.0	167.0	185.3	196.1	211.0	233.0	257.1	282.0	
2,6-Ditert-butyl-4-ethylphenol	$C_{16}H_{26}O$	89.1	121.4	137.0	154.0	172.1	183.9	198.0	220.0	244.0	268.6	
4,6-Ditert-butyl-3-ethylphenol	$C_{16}H_{26}O$	111.5	142.6	157.4	174.0	192.3	204.4	218.0	241.7	264.6	290.0	
Diisobutyl oxalate	$C_{10}H_{18}O_4$	63.2	91.2	105.3	120.3	137.5	147.8	161.8	183.5	205.8	229.5	
2,4-Ditert-butylphenol	$C_{14}H_{22}O$	84.5	115.4	130.0	146.0	164.3	175.8	190.0	212.5	237.0	260.8	
Dibutyl phthalate	$C_{16}H_{22}O_4$	148.2	182.1	198.2	216.2	235.8	247.8	263.7	287.0	313.5	340.0	
sulfide	$C_8H_{18}S$	+21.7	51.8	66.4	80.5	96.0	105.8	118.6	138.0	159.0	182.0	−79.7

Diisobutyl *d*-tartrate	$C_{12}H_{22}O_6$	117.8	151.8	169.0	188.0	208.5	221.6	239.5	264.7	294.0	324.0	73.5
Dicarvaryl-mono-(6-chloro-2-xenyl) phosphate	$C_{32}H_{34}ClO_4P$	204.2	234.5	249.3	264.5	280.5	290.7	304.9	323.8	342.0	361.0	
Dicarvacryl-2-tolyl phosphate	$C_{27}H_{33}O_4P$	180.2	209.3	221.8	237.0	251.5	260.3	272.5	290.0	309.8	330.0	
Dichloroacetic acid	$C_2H_2Cl_2O_2$	44.0	69.8	82.6	96.3	111.8	121.5	134.0	152.3	173.7	194.4	9.7
1,2-Dichlorobenzene	$C_6H_4Cl_2$	20.0	46.0	59.1	73.4	89.4	99.5	112.9	133.4	155.8	179.0	−17.6
1,3-Dichlorobenzene	$C_6H_4Cl_2$	12.1	39.0	52.0	66.2	82.0	92.2	105.0	125.9	149.0	173.0	−24.2
1,4-Dichlorobenzene	$C_6H_4Cl_2$			54.8	69.2	84.8	95.2	108.4	128.3	150.2	173.9	53.0
1,2-Dichlorobutane	$C_4H_8Cl_2$	−23.6	−0.3	+11.5	24.5	37.7	47.8	60.2	79.7	100.8	123.5	
2,3-Dichlorobutane	$C_4H_8Cl_2$	−25.2	−3.0	+8.5	21.2	35.0	43.9	56.0	74.0	94.2	116.0	−80.4
1,2-Dichloro-1,2-difluoroethylene	$C_2Cl_2F_2$	−82.0	−65.6	−57.3	−48.3	−38.2	−31.8	−23.0	−10.0	+5.0	20.9	−112
Dichlorodifluoromethane	CCl_2F_2	−118.5	−104.6	−97.8	−90.1	−81.6	−76.1	−68.6	−57.0	−43.9	−29.8	
Dichlorodiphenyl silane	$C_{12}H_{10}Cl_2Si$	109.6	142.4	158.0	176.0	195.5	207.5	223.8	248.0	275.5	304.0	
Dichlorodiisopropyl ether	$C_6H_{12}Cl_2O$	29.6	55.2	68.2	82.2	97.3	106.9	119.7	139.0	159.8	182.7	
Di(2-chloroethoxy) methane	$C_5H_{10}Cl_2O_2$	53.0	80.4	94.0	109.5	125.5	135.8	149.6	170.0	192.0	215.0	
Dichloroethoxymethylsilane	$C_8H_8Cl_2OSi$	−33.8	−12.1	−1.3	+11.3	24.4	32.6	44.1	61.0	80.3	100.6	
1,2-Dichloro-3-ethylbenzene	$C_8H_8Cl_2$	46.0	75.0	90.0	105.9	123.8	135.0	149.8	172.0	197.0	222.1	−40.8
1,2-Dichloro-4-ethylbenzene	$C_8H_8Cl_2$	47.0	77.2	92.3	109.6	127.5	139.0	153.3	176.0	201.7	226.6	−76.4
1,4-Dichloro-2-ethylbenzene	$C_8H_8Cl_2$	38.5	68.0	83.2	99.8	118.0	129.0	144.0	166.2	191.5	216.3	−61.2
cis-1,2-Dichloroethylene	$C_2H_2Cl_2$	−58.4	−39.2	−29.9	−19.4	−7.9	−0.5	+9.5	24.6	41.0	59.0	−80.5
trans-1,2-Dichloro ethylene	$C_2H_2Cl_2$	−65.4	−47.2	−38.0	−28.0	−17.0	−10.0	−0.2	+14.3	30.8	47.8	−50.0
Di(2-chloroethyl) ether	$C_4H_8Cl_2O$	23.5	49.3	62.0	76.0	91.5	101.5	114.5	134.0	155.4	178.5	
Dichlorofluoromethane	$CHCl_2F$	−91.3	−75.5	−67.5	−58.6	−48.8	−42.6	−33.9	−20.9	−6.2	+8.9	−135
1,5-Dichlorohexamethyltrisiloxane	$C_6H_{18}Cl_2O_2Si_3$	26.0	52.0	65.1	79.0	94.8	105.0	118.2	138.3	160.2	184.0	−53.0
Dichloromethylphenylsilane	$C_7H_8Cl_2Si$	35.7	63.5	77.4	92.4	109.5	120.0	134.2	155.5	180.2	205.5	
1,1-Dichloro-2-methylpropane	$C_4H_8Cl_2$	−31.0	−8.4	+2.6	14.6	28.2	37.0	48.2	65.8	85.4	106.0	
1,2-Dichloro-2-methylpropane	$C_4H_8Cl_2$	−25.8	−4.2	+6.7	18.7	32.0	40.2	51.7	68.9	87.8	108.0	
1,3-Dichloro-2-methylpropane	$C_4H_8Cl_2$	−3.0	+20.6	32.0	44.8	58.6	67.5	78.8	96.1	115.4	135.0	
2,4-Dichlorophenol	$C_6H_4Cl_2O$	53.0	80.0	92.8	107.7	123.4	133.5	146.0	165.2	187.5	210.0	45.0
2,6-Dichlorophenol	$C_6H_4Cl_2O$	59.5	87.6	101.0	115.5	131.6	141.8	154.6	175.5	197.7	220.0	
α,α-Dichlorophenylacetonitrile	$C_8H_5Cl_2N$	56.0	84.0	98.1	113.8	130.0	141.0	154.5	176.2	199.5	223.5	
Dichlorophenylarsine	$C_6H_5AsCl_2$	61.8	100.0	116.0	133.1	151.0	163.2	178.9	202.8	228.8	256.5	
1,2-Dichloropropane	$C_3H_6Cl_2$	−38.5	−17.0	−6.1	+6.0	19.4	28.0	39.4	57.0	76.0	96.8	
2,3-Dichlorostyrene	$C_8H_6Cl_2$	61.0	90.1	104.6	120.5	137.8	149.0	163.5	185.7	210.0	235.0	
2,4-Dichlorostyrene	$C_8H_6Cl_2$	53.5	82.2	97.4	111.8	129.2	140.0	153.8	176.0	200.0	225.0	
2,5-Dichlorostyrene	$C_8H_6Cl_2$	55.5	83.9	98.2	114.0	131.0	142.0	155.8	178.0	202.5	227.0	
2,6-Dichlorostyrene	$C_8H_6Cl_2$	47.8	75.7	90.0	105.5	122.4	133.3	147.6	169.0	193.5	217.0	

(*Continued*)

TABLE 2.37 Boiling Points of Common Organic Compounds at Selected Pressures (*Continued*)

Compound		Pressure, mm Hg										Melting Point, °C
		1	5	10	20	40	60	100	200	400	760	
Name	Formula	Temperature, °C										
3,4-Dichlorostyrene	$C_8H_6Cl_2$	57.2	86.0	100.4	116.2	133.7	144.6	158.2	181.5	205.7	230.0	
3,5-Dichlorostyrene	$C_8H_6Cl_2$	53.5	82.2	97.4	111.8	129.2	140.0	153.8	176.0	200.0	225.0	
1,2-Dichlorotetraethylbenzene	$C_{14}H_{20}Cl_2$	105.6	138.7	155.0	172.5	192.2	204.8	220.7	245.6	272.8	302.0	
1,4-Dichlorotetraethylbenzene	$C_{14}H_{20}Cl_2$	91.7	126.1	143.8	162.0	183.2	195.8	212.0	238.5	265.8	296.5	
1,2-Dichloro-1,1,2,2-tetrafluoroethane	$C_2Cl_2F_4$	−95.4	−80.0	−72.3	−63.5	−53.7	−47.5	−39.1	−26.3	−12.0	+3.5	−94
Dichloro-4-tolysilane	$C_7H_8Cl_2Si$	46.2	71.7	84.2	97.8	113.2	122.6	135.5	153.5	175.2	196.3	
3,4-Dichloro-α,α,α-trifluorotoluene	$C_7H_3Cl_2F_3$	11.0	38.3	52.2	67.3	84.0	95.0	109.2	129.0	150.5	172.8	−12.1
Dicyclopentadiene	$C_{10}H_8$		34.1	47.6	62.0	77.9	88.0	101.7	121.8	144.2	166.6	32.9
Diethoxydimethylsilane	$C_6H_{16}O_2Si$	−19.1	+2.4	13.3	25.3	38.0	46.3	57.6	74.2	93.2	113.5	
Diethoxydiphenylsilane	$C_{16}H_{20}O_2Si$	111.5	142.8	157.6	174.3	193.2	205.0	220.0	243.8	259.7	296.0	
Diethyl adipate	$C_{10}H_{18}O_4$	74.0	106.6	123.0	138.3	154.6	165.8	179.0	198.2	219.1	240.0	−21
Diethylamine	$C_4H_{11}N$			−33.0	−22.6	−11.3	−40.	+6.0	21.0	38.0	55.5	38.9
N-Diethylaniline	$C_{10}H_{15}N$	49.7	78.0	91.9	107.2	123.6	133.8	147.3	168.2	192.4	215.5	−34.4
Diethyl arsanilate	$C_{10}H_{16}AsNO_3$	38.0	62.6	74.8	88.0	102.6	111.8	123.8	141.9	161.0	181.0	
1,2-Diethylbenzene	$C_{10}H_{14}$	22.3	48.7	62.0	76.4	92.5	102.6	116.2	136.7	159.0	183.5	−31.4
1,3-Diethylbenzene	$C_{10}H_{14}$	20.7	46.8	59.9	74.5	90.4	100.7	114.4	134.8	156.9	181.1	−83.9
1,4-Diethylbenzene	$C_{10}H_{14}$	20.7	47.1	60.3	74.7	91.1	101.3	115.3	136.1	159.0	183.8	−43.2
Diethyl carbonate	$C_5H_{10}O_3$	−10.1	+12.3	23.8	36.0	49.5	57.9	69.7	86.5	105.8	125.8	−43
cis-Diethyl citraconate	$C_9H_{14}O_4$	59.8	88.3	103.0	118.2	135.7	146.2	160.0	182.3	206.5	230.3	
Diethyl dioxosuccinate	$C_8H_{10}O_5$	70.0	98.0	112.0	126.8	143.8	153.7	167.7	188.0	210.8	233.5	
Diethylene glycol	$C_4H_{10}O_3$	91.8	120.0	133.8	148.0	164.3	174.0	187.5	207.0	226.5	244.8	
Diethyleneglycol-*bis*-chloroacetate	$C_8H_{12}Cl_2O_5$	148.3	180.0	195.8	212.0	229.0	239.5	252.0	271.5	291.8	313.0	
Diethylene glycol dimethyl ether												
Di(2-methoxyethyl) ether	$C_6H_{14}O_3$	13.0	37.6	50.0	63.0	77.5	86.8	99.5	118.0	138.5	159.8	
glycol ethyl ether	$C_6H_{14}O_3$	45.3	72.0	85.8	100.3	116.7	126.8	140.3	159.0	180.3	201.9	
Diethyl ether	$C_4H_{10}O$	−74.3	−56.9	−48.1	−38.5	27.7	−21.8	−11.5	+2.2	17.9	34.6	−116.3
ethylmalonate	$C_9H_{16}O_4$	50.8	77.8	91.6	106.0	122.4	132.4	146.0	166.0	188.7	211.5	
fumarate	$C_8H_{12}O_4$	53.2	81.2	95.3	110.2	126.7	137.7	151.1	172.2	195.8	218.5	+0.6
glutarate	$C_9H_{16}O_4$	65.6	94.7	109.7	125.4	142.8	153.2	167.8	189.5	212.8	237.0	
Diethylhexadecylamine	$C_{20}H_{43}N$	139.8	175.8	194.0	213.5	235.0	248.5	265.5	292.8	324.6	355.0	

Diethyl itaconate	$C_9H_{14}O_4$	51.3	80.2	95.2	111.0	128.2	139.3	154.3	177.5	203.1	227.9	
ketone (3-pentanone)	$C_5H_{10}O$	−12.7	+7.5	17.2	27.9	39.4	46.7	56.2	70.6	86.3	102.7	−42
malate	$C_8H_{14}O_5$	80.7	110.4	125.3	141.2	157.8	169.0	183.9	205.3	229.5	253.4	
maleate	$C_8H_{12}O_4$	57.3	85.6	100.0	115.3	131.8	142.4	156.0	177.8	201.7	225.0	
malonate	$C_7H_{12}O_4$	40.0	67.5	81.3	95.9	113.3	123.0	136.2	155.5	176.8	198.9	−49.8
mesaconate	$C_9H_{14}O_4$	62.8	91.0	105.3	120.3	137.3	147.9	161.6	183.2	205.8	229.0	
oxalate	$C_6H_{10}O_4$	47.4	71.8	83.8	96.8	110.6	119.7	130.8	147.9	166.2	185.7	−40.6
phthalate	$C_{12}H_{14}O_4$	108.8	140.7	156.0	173.6	192.1	204.1	219.5	243.0	267.5	294.0	
sebacate	$C_{14}H_{26}O_4$	125.3	156.2	172.1	189.8	207.5	218.4	234.4	255.8	280.3	305.5	1.3
2,5-Diethylstyrene	$C_{12}H_{16}$	49.7	78.4	92.6	108.5	125.8	136.8	151.0	173.2	198.0	223.0	
Diethyl succinate	$C_8H_{14}O_4$	54.6	83.0	96.6	111.7	127.8	138.2	151.1	171.7	193.8	216.5	−20.8
isosuccinate	$C_8H_{14}O_4$	39.8	66.7	80.0	94.7	111.0	121.4	134.8	155.1	177.7	201.3	
sulfate	$C_4H_{10}O_4S$	47.0	74.0	87.7	102.1	118.0	128.6	142.5	162.5	185.5	209.5	−25.0
sulfide	$C_4H_{10}S$	−39.6	−18.6	−8.0	+3.5	16.1	24.2	35.0	51.3	69.7	88.0	−99.5
sulfite	$C_4H_{10}O_3S$	10.0	34.2	46.4	59.7	74.2	83.8	96.3	115.8	137.0	159.0	
d-Diethyl tartrate	$C_8H_{14}O_6$	102.0	133.0	148.0	164.2	182.3	194.0	208.5	230.4	254.8	280.0	17
dl-Diethyl tartrate	$C_8H_{14}O_6$	100.0	131.7	147.2	163.8	181.7	193.2	208.0	230.0	254.3	280.0	
3,5-Diethyltoluene	$C_{11}H_{16}$	34.0	61.5	75.3	90.2	107.0	117.7	131.7	152.4	176.5	200.7	
Diethylzinc	$C_4H_{10}Zn$	−22.4	0.0	+11.7	24.2	38.0	47.2	59.1	77.0	97.3	118.0	−28
1-Dihydrocarvone	$C_{10}H_{16}O$	46.6	75.5	90.0	106.0	123.7	134.7	149.7	171.8	197.0	223.0	
Dihydrocitronellol	$C_{10}H_{22}O$	68.0	91.7	103.0	115.0	127.6	136.7	145.9	160.2	176.8	193.5	
1,4-Dihydroxyanthraquinone	$C_{14}H_8O_4$	196.7	239.8	259.8	282.0	307.4	323.3	344.5	377.8	413.0	450.0	194
Dimethylacetylene (2-butyne)	C_4H_6	−73.0	−57.9	−50.5	−42.5	−33.9	−27.8	−18.8	−5.0	+10.6	27.2	−32.5
Dimethylamine	C_2H_7N	−87.7	−72.2	−64.6	−56.0	−46.7	−40.7	−32.6	−20.4	−7.1	+7.4	−96
N,N-Dimethylaniline	$C_8H_{11}N$	29.5	56.3	70.0	84.8	101.6	111.9	125.8	146.5	169.2	193.1	+2.5
Dimethyl arsanilate	$C_8H_{12}AsNO_3$	15.0	39.6	51.8	65.0	79.7	88.6	101.0	119.8	140.3	160.5	
Di(α-methylbenzyl) ether	$C_{16}H_{18}O$	96.7	128.3	144.0	160.3	179.6	191.5	206.8	229.7	254.8	281.9	
2,2-Dimethylbutane	C_6H_{14}	−69.3	−50.7	−41.5	−31.1	−19.5	−12.1	−2.0	+13.4	31.0	49.7	−99.8
2,3-Dimethylbutane	C_6H_{14}	−63.6	−44.5	−34.9	−24.1	−12.4	−4.9	+5.4	21.1	39.0	58.0	−128.2
Dimethyl citraconate	$C_7H_{10}O_4$	50.8	78.2	91.8	106.5	122.6	132.7	145.8	165.8	188.0	210.5	
1,1-Dimethylcyclohexane	C_8H_{16}	−24.4	−1.4	+10.3	23.0	37.3	45.7	57.9	76.2	97.2	119.5	−34
cis-1,2-Dimethylcyclohexane	C_8H_{16}	−15.9	+7.3	18.4	31.1	45.3	54.4	66.8	85.6	107.0	129.7	−50.0
trans-1,2-Dimethylcyclohexane	C_8H_{16}	−21.1	+1.7	13.0	25.6	39.7	48.7	61.0	79.6	100.9	123.4	−80.0
trans-1,3-Dimethylcyclohexane	C_8H_{16}	−19.4	+3.4	14.9	27.4	41.4	50.4	62.5	81.0	102.1	124.4	−92.0
cis-1,3-Dimethylcyclohexane	C_8H_{16}	−22.7	0.0	+11.2	23.6	37.5	46.4	58.5	76.9	97.8	120.1	−76.2
cis-1,4-Dimethylcyclohexane	C_8H_{16}	−20.0	+3.2	14.5	27.1	41.1	50.1	62.3	80.8	101.9	124.3	−87.4
trans-1,4-Dimethylcyclohexane	C_8H_{16}	−24.3	−1.7	+10.1	22.6	36.5	45.4	57.6	76.0	97.0	119.3	−36.9
Dimethyl ether	C_2H_6O	−115.7	−101.1	−93.3	−85.2	−76.2	−70.4	−62.7	−50.9	−37.8	−23.7	−138.5

(*Continued*)

TABLE 2.37 Boiling Points of Common Organic Compounds at Selected Pressures (*Continued*)

Compound		Pressure, mm Hg										Melting Point, °C
		1	5	10	20	40	60	100	200	400	760	
Name	Formula	Temperature, °C										
2,2-Dimethylhexane	C_8H_{18}	−29.7	−7.9	+3.1	15.0	28.2	36.7	48.2	65.7	85.6	106.8	
2,3-Dimethylhexane	C_8H_{18}	−23.0	−1.1	+9.9	22.1	35.6	44.2	56.0	73.8	94.1	115.6	
2,4-Dimethylhexane	C_8H_{18}	−26.9	−5.3	+5.2	17.2	30.5	39.0	50.6	68.1	88.2	109.4	
2,5-Dimethylhexane	C_8H_{18}	−26.7	−5.5	+5.3	17.2	30.4	38.9	50.5	68.0	87.9	109.1	−90.7
3,3-Dimethylhexane	C_8H_{18}	−25.8	−4.4	+6.1	18.2	31.7	40.4	52.5	70.0	90.4	112.0	
3,4-Dimethylhexane	C_8H_{18}	−22.1	+0.2	11.3	23.5	37.1	45.8	57.7	75.6	96.0	117.7	
Dimethyl itaconate	$C_7H_{10}O_4$	69.3	94.0	106.6	119.7	133.7	142.6	153.7	171.0	189.8	208.0	38
1-Dimethyl malate	$C_6H_{10}O_5$	75.4	104.0	118.3	133.8	150.1	160.4	175.1	196.3	219.5	242.6	
Dimethyl maleate	$C_6H_8O_4$	45.7	73.0	86.4	101.3	117.2	127.1	140.4	160.0	182.2	205.0	
malonate	$C_5H_8O_4$	35.0	59.8	72.0	85.0	100.0	109.7	121.9	140.0	159.8	180.7	−62
trans-Dimethyl mesaconate	$C_7H_{10}O_4$	46.8	74.0	87.8	102.1	118.0	127.8	141.5	161.0	183.5	206.0	
2,7-Dimethyloctane	$C_{10}H_{22}$	+6.3	30.5	42.3	55.8	71.2	80.8	93.9	114.0	136.0	159.7	−52.8
Dimethyl oxalate	$C_4H_6O_4$	20.0	44.0	56.0	69.4	83.6	92.8	104.8	123.3	143.3	163.3	
2,2-Dimethylpentane	C_7H_{16}	−49.0	−28.7	−18.7	−7.5	+5.0	13.0	23.9	40.3	59.2	79.2	−123.7
2,3-Dimethylpentane	C_7H_{16}	−42.0	−20.8	−10.3	+1.1	13.9	22.1	33.3	50.1	69.4	89.8	−135
2,4-Dimethylpentane	C_7H_{16}	−48.0	−27.4	−17.1	−5.9	+6.5	14.5	25.4	41.8	60.6	80.5	−119.5
3,3-Dimethylpentane	C_7H_{16}	−45.9	−25.0	−14.4	−2.9	+9.9	18.1	29.3	46.2	65.5	86.1	−135.0
2,3-Dimethylphenol (2,3-xylenol)	$C_8H_{10}O$	56.0	83.8	97.6	112.0	129.2	139.5	152.2	173.0	196.0	218.0	75
2,4-Dimethylphenol (2,4-xylenol)	$C_8H_{10}O$	51.8	78.0	91.3	105.0	121.5	131.0	143.0	161.5	184.2	211.5	25.5
2,5-Dimethylphenol (2,5-xylenol)	$C_8H_{10}O$	51.8	78.0	91.3	105.0	121.5	131.0	143.0	161.5	184.2	211.5	74.5
3,4-Dimethylphenol (3,4-xylenol)	$C_8H_{10}O$	66.2	93.8	107.7	122.0	138.0	148.0	161.0	181.5	203.6	225.2	62.5
3,5-Dimethylphenol (3,5-xylenol)	$C_8H_{10}O$	62.0	89.2	102.4	117.0	133.3	143.5	156.0	176.2	197.8	219.5	68
Dimethylphenylsilane	$C_8H_{12}Si$	+5.3	30.3	42.6	56.2	71.4	81.3	94.2	114.2	136.4	159.3	
Dimethyl phthalate	$C_{10}H_{10}O_4$	100.3	131.8	147.6	164.0	182.8	194.0	210.0	232.7	257.8	283.7	
3,5-Dimethyl-1,2-pyrone	$C_7H_8O_2$	78.6	107.6	122.0	136.4	152.7	163.8	177.5	198.0	221.0	245.0	51.5
4,6-Dimethylresorcinol	$C_8H_{10}O_2$	49.0	76.8	90.7	105.8	122.5	133.2	147.3	167.8	192.0	215.0	
Dimethyl sebacate	$C_{12}H_{22}O_4$	104.0	139.8	156.2	175.8	196.0	208.0	222.6	245.0	269.6	293.5	38
2,4-Dimethylstyrene	$C_{10}H_{12}$	34.2	61.9	75.8	90.8	107.7	118.0	132.3	153.2	177.5	202.0	
2,5-Dimethylstyrene	$C_{10}H_{12}$	29.0	55.9	69.0	84.0	100.2	110.7	124.7	145.6	168.7	193.0	
α,α-Dimethylsuccinic anhydride	$C_6H_8O_3$	61.4	88.1	102.0	116.3	132.6	142.4	155.3	175.8	197.5	219.5	
Dimethyl sulfide	C_2H_6S	−75.6	−58.0	−49.2	−39.4	−28.4	−21.9	−12.0	+2.6	18.7	36.0	−83.2

d-Dimethyl tartrate	$C_6H_{10}O_6$	102.1	133.2	148.2	164.3	182.4	193.8	208.8	230.5	255.0	280.0	61.5
dl-Dimethyl tartrate	$C_6H_{10}O_6$	100.4	131.8	147.5	164.0	182.4	193.8	209.5	232.3	257.4	282.0	89
N,N-Dimethyl-2-toluidine	$C_9H_{13}N$	28.8	54.1	66.2	80.2	95.0	105.2	118.1	138.3	161.5	184.8	−61
N,N-Dimethyl-4-toluidine	$C_9H_{13}N$	50.1	74.3	86.7	100.0	116.3	126.4	140.3	161.6	185.4	209.5	
Di(nitrosomethyl) amine	$C_2H_5N_3O_2$	+3.2	27.8	40.0	53.7	68.2	77.7	90.3	110.0	131.3	153.0	
Diosphenol	$C_{10}H_{16}O_2$	66.7	95.4	109.0	124.0	141.2	151.3	165.6	186.2	209.5	232.0	
1,4-Dioxane	$C_4H_8O_2$	−35.8	−12.8	−1.2	+12.0	25.2	33.8	45.1	62.3	81.8	101.1	10
Dipentene	$C_{10}H_{16}$	14.0	40.4	53.8	68.2	84.3	94.6	108.3	128.2	150.5	174.6	
Diphenylamine	$C_{12}H_{11}N$	108.3	141.7	157.0	175.2	194.3	206.9	222.8	247.5	274.1	302.0	52.9
Diphenyl carbinol (benzhydrol)	$C_{13}H_{12}O$	110.0	145.0	162.0	180.9	200.0	212.0	227.5	250.0	275.6	301.0	68.5
chlorophosphate	$C_{12}H_{10}ClPO_3$	121.5	160.5	182.0	203.8	227.9	244.2	265.0	299.5	337.2	378.0	
disulfide	$C_{12}H_{10}S_2$	131.6	164.0	180.0	197.0	214.8	226.2	241.3	262.6	285.8	310.0	61
1,2-Diphenylethane (dibenzyl)	$C_{14}H_{14}$	86.8	119.8	136.0	153.7	173.7	186.0	202.8	227.8	255.0	284.0	51.5
Diphenyl ether	$C_{12}H_{10}O$	66.1	97.8	114.0	130.8	150.0	162.0	178.8	203.3	230.7	258.5	27
1,1-Diphenylethylene	$C_{14}H_{12}$	87.4	119.6	135.0	151.8	170.8	183.4	198.6	222.8	249.8	277.0	
trans-Diphenylethylene	$C_{14}H_{12}$	113.2	145.8	161.0	179.8	199.0	211.5	227.4	251.7	278.3	306.5	124
1,1-Diphenylhydrazine	$C_{12}H_{12}N_2$	126.0	159.3	176.1	194.0	213.5	225.9	242.5	267.2	294.0	322.2	44
Diphenylmethane	$C_{13}H_{12}$	76.0	107.4	122.8	139.8	157.8	170.2	186.3	210.7	237.5	264.5	26.5
Diphenyl sulfide	$C_{12}H_{10}S$	96.1	129.0	145.0	162.0	182.8	194.8	211.8	236.8	263.9	292.5	
Diphenyl-2-tolyl thiophosphate	$C_{18}H_{17}O_3PS$	159.7	179.6	201.6	215.5	230.6	240.4	252.5	270.3	290.0	310.0	
1,2-Dipropoxyethane	$C_8H_{18}O_2$	−38.8	−10.3	+5.0	22.3	42.3	55.8	74.2	103.8	140.0	180.0	
1,2-Diisopropylbenzene	$C_{12}H_{18}$	40.0	67.8	81.8	96.8	114.0	124.3	138.7	159.8	184.3	209.0	
1,3-Diisopropylbenzene	$C_{12}H_{18}$	34.7	62.3	76.0	91.2	107.9	118.2	132.3	153.7	177.6	202.0	−105
Dipropylene glycol	$C_6H_{14}O_3$	73.8	102.1	116.2	131.3	147.4	156.5	169.9	189.9	210.5	231.8	
Dipropyleneglycol monobutyl ether	$C_{10}H_{22}O_3$	64.7	92.0	106.0	120.4	136.3	146.3	159.8	180.0	203.8	227.0	
isopropyl ether	$C_9H_{20}O_3$	46.0	72.8	86.2	100.8	117.0	126.8	140.3	160.0	183.1	205.6	
Di-*n*-propyl ether	$C_6H_{14}O$	−43.3	−22.3	−11.8	0.0	+13.2	21.6	33.0	50.3	69.5	89.5	−122
Diisopropyl ether	$C_6H_{14}O$	−57.0	−37.4	−27.4	−16.7	−4.5	+3.4	13.7	30.0	48.2	67.5	−60
Di-*n*-propyl ketone (4-heptanone)	$C_7H_{14}O$	23.0	44.4	55.0	66.2	78.1	85.8	96.0	111.2	127.3	143.7	−32.6
Di-*n*-propyl oxalate	$C_8H_{14}O_4$	53.4	80.2	93.9	108.6	124.6	134.8	148.1	168.0	190.3	213.5	
Diisopropyl oxalate	$C_8H_{14}O_4$	43.2	69.0	81.9	95.6	110.5	120.0	132.6	151.2	171.8	193.5	
Di-*n*-propyl succinate	$C_{10}H_{18}O_4$	77.5	107.6	122.2	138.0	154.8	166.0	180.3	202.5	226.5	250.8	
Di-*n*-propyl *d*-tartrate	$C_{10}H_{18}O_6$	115.6	147.7	163.5	180.4	199.7	211.7	227.0	250.1	275.6	303.0	
Diisopropyl *d*-tartrate	$C_{10}H_{18}O_6$	103.7	133.7	148.2	164.0	181.8	192.6	207.3	228.2	251.8	275.0	
Divinyl acetylene (1,5-hexadiene-3-yne)	C_6H_6	−45.1	−24.4	−14.0	−2.8	+10.0	18.1	29.5	46.0	64.4	84.0	
1,3-Divinylbenzene	$C_{10}H_{10}$	32.7	60.0	73.8	88.7	105.5	116.0	130.0	151.4	175.2	199.5	−66.9
Docosanae	$C_{22}H_{46}$	157.8	195.4	213.0	233.5	254.5	268.3	286.0	314.2	343.5	376.0	44.5

(*Continued*)

TABLE 2.37 Boiling Points of Common Organic Compounds at Selected Pressures (*Continued*)

Compound		Pressure, mm Hg										Melting Point, °C
		1	5	10	20	40	60	100	200	400	760	
Name	Formula	Temperature, °C										
n-Dodeccane	$C_{12}H_{26}$	47.8	75.8	90.0	104.6	121.7	132.1	146.2	167.2	191.0	216.2	−9.6
1-Dodecene	$C_{12}H_{24}$	47.2	74.0	87.8	102.4	118.6	128.5	142.3	162.2	185.5	208.0	−31.5
n-Dodecyl alcohol	$C_{12}H_{26}O$	91.0	120.2	134.7	150.0	167.2	177.8	192.0	213.0	235.7	259.0	24
Dodecylamine	$C_{12}H_{27}N$	82.8	111.8	127.8	141.6	157.4	168.0	182.1	203.0	225.0	248.0	
Dodecyltrimethylsilane	$C_{15}H_{34}Si$	91.2	122.1	137.7	153.8	172.1	184.2	199.5	222.0	248.0	273.0	
Elaidic acid	$C_{18}H_{34}O_2$	171.3	206.7	223.5	242.3	260.8	273.0	288.0	312.4	337.0	362.0	51.5
Epichlorohydrin	C_3H_5ClO	−16.5	+5.6	16.6	29.0	42.0	50.6	62.0	79.3	98.0	117.9	−25.6
1,2-Epoxy-2-methylpropane	C_4H_8O	−69.0	−50.0	−40.3	−29.5	−17.3	−9.7	+1.2	17.5	36.0	55.5	
Erucic acid	$C_{22}H_{42}O_2$	206.7	239.7	254.5	270.6	289.1	300.2	314.4	336.5	358.8	381.5	33.5
Estragole (*p*-methoxy allyl benzene)	$C_{10}H_{12}O$	52.6	80.0	93.7	108.4	124.6	135.2	148.5	168.7	192.0	215.0	
Ethane	C_2H_6	−159.5	−148.5	−142.9	−136.7	−129.8	−125.4	−119.3	−110.2	−99.7	−88.6	−183.2
Ethoxydimethylphenylsilane	$C_{10}H_{16}OSi$	36.3	63.1	76.2	91.0	107.2	127.5	131.4	151.5	175.0	199.5	
Ethoxytrimethylsilane	$C_5H_{14}OSi$	−50.9	−31.0	−20.7	−9.8	+3.7	11.5	22.1	38.1	56.3	75.7	
Ethoxytriphenylsilane	$C_{20}H_{20}OSi$	167.0	198.2	213.5	230.0	247.0	258.3	273.5	295.0	319.5	344.0	
Ethyl acetate	$C_4H_8O_2$	−43.4	−23.5	−13.5	−3.0	+9.1	16.6	27.0	42.0	59.3	77.1	−82.4
acetoacetate	$C_6H_{10}O_3$	28.5	54.0	67.3	81.1	96.2	106.0	118.5	138.0	158.2	180.8	−45
Ethylacetylene (1-butyne)	C_4H_6	−92.5	−76.7	−68.7	−59.9	−50.5	−43.4	−34.9	−21.6	−6.9	+8.7	−130
Ethyl acrylate	$C_5H_8O_2$	−29.5	−8.7	+2.0	13.0	26.0	33.5	44.5	61.5	80.0	99.5	−71.2
α-Ethylacrylic acid	$C_5H_8O_2$	47.0	70.7	82.0	94.4	108.1	116.7	127.5	144.0	160.7	179.2	
α-Ethylacrylonitrile	C_5H_7N	−29.0	−6.4	+5.0	17.7	31.8	40.6	53.0	71.6	92.2	114.0	
Ethyl alcohol (ethanol)	C_2H_6O	−31.3	−12.0	−2.3	+8.0	19.0	26.0	34.9	48.4	63.5	78.4	−112
Ethylamine	C_2H_7N	−82.3	−66.4	−58.3	−48.6	−39.8	−33.4	−25.1	−12.3	+2.0	16.6	−80.6
4-Ethylaniline	$C_8H_{11}N$	52.0	80.0	93.8	109.0	125.7	136.0	149.8	170.6	194.2	217.4	−4
N-Ethylaniline	$C_8H_{11}N$	38.5	66.4	80.6	96.0	113.2	123.6	137.3	156.9	180.8	204.0	−63.5
2-Ethylanisole	$C_9H_{12}O$	29.7	55.9	69.0	83.1	98.9	109.0	122.3	142.1	164.2	187.1	
3-Ethylanisole	$C_9H_{12}O$	33.7	60.3	73.9	88.5	104.8	115.5	129.2	149.7	172.8	196.5	
4-Ethylanisole	$C_9H_{12}O$	33.5	60.2	73.9	88.5	104.7	115.4	128.4	149.2	172.3	196.5	
Ethylbenzene	C_8H_{10}	−9.8	+13.9	25.9	38.6	52.8	61.8	74.1	92.7	113.8	136.2	−94.9
Ethyl benzoate	$C_9H_{10}O_2$	44.0	72.0	86.0	101.4	118.2	129.0	143.2	164.8	188.4	213.4	−34.6
benzoylacetate	$C_{11}H_{12}O_3$	107.6	136.4	150.3	166.8	181.8	191.9	205.0	223.8	244.7	265.0	
bromide	C_2H_5Br	−74.3	−56.4	−47.5	−37.8	−26.7	−19.5	−10.0	+4.5	21.0	38.4	−117.8

α-bromoisobutyrate	$C_6H_{11}BrO_2$	10.6	35.8	48.0	61.8	77.0	86.7	99.8	119.7	141.2	163.6	
n-butyrate	$C_6H_{12}O_2$	−18.4	+4.0	15.3	27.8	41.5	50.1	62.0	79.8	100.0	121.0	−93.3
isobutyrate	$C_6H_{12}O_2$	−24.3	−2.4	+8.4	20.6	33.8	42.3	53.5	71.0	90.0	110.0	−88.2
Ethylcamphoronic anhydride	$C_{11}H_{16}O_5$	118.2	149.8	165.0	181.8	199.8	211.5	226.6	248.5	272.8	298.0	
Ethyl isocaproate	$C_8H_{16}O_2$	11.0	35.8	48.0	61.7	76.3	85.8	98.4	117.8	139.2	160.4	
carbamate	$C_3H_7NO_2$		65.8	77.8	91.0	105.6	114.8	126.2	144.2	164.0	184.0	49
carbanilate	$C_9H_{11}NO_2$	107.8	131.8	143.7	155.5	168.8	177.3	187.9	203.8	220.0	237.0	52.5
Ethylcetylamine	$C_{18}H_{39}N$	133.2	168.2	186.0	205.5	226.5	239.8	256.8	283.3	313.0	342.0	
Ethyl chloride	C_2H_5Cl	−89.8	−73.9	−65.8	−56.8	−47.0	−40.6	−32.0	−18.6	−3.9	+12.3	−139
chloroacetate	$C_4H_7ClO_2$	+1.0	25.4	37.5	50.4	65.2	74.0	86.0	103.8	123.8	144.2	−26
chloroglyoxylate	$C_4H_5ClO_3$	−5.1	+18.0	29.9	42.0	56.0	65.2	76.6	94.5	114.7	135.0	
α-chloropropionate	$C_5H_9ClO_2$	+6.6	30.2	41.9	54.3	68.2	77.3	89.3	107.2	126.2	146.5	
trans-cinnamate	$C_{11}H_{12}O_2$	87.6	108.5	134.0	150.3	169.2	181.2	196.0	219.3	245.0	271.0	12
3-Ethylcumene	$C_{11}H_{16}$	28.3	55.5	68.8	83.6	99.9	110.2	124.3	145.4	168.2	193.0	
4-Ethylcumene	$C_{11}H_{16}$	31.5	58.4	72.0	86.7	103.3	113.8	127.2	148.3	171.8	195.8	
Ethyl cyanoacetate	$C_5H_7NO_2$	67.8	93.5	106.0	119.8	133.8	142.1	152.8	169.8	187.8	206.0	
Ethylcyclohexane	C_8H_{16}	−14.5	+9.2	20.6	33.4	47.6	56.7	69.0	87.8	109.1	131.8	−111.3
Ethylcyclopentane	C_7H_{14}	−32.2	−10.8	−0.1	+11.7	25.0	33.4	45.0	62.4	82.3	103.4	−138.6
Ethyl dichloroacetate	$C_4H_6Cl_2O_2$	9.6	34.0	46.3	59.5	74.0	83.6	96.1	115.2	135.9	156.5	
N,*N*-diethyloxamate	$C_8H_{15}NO_3$	76.0	106.3	121.7	137.7	154.4	166.0	180.3	202.8	226.5	252.0	
N-Ethyldiphenylamine	$C_{14}H_{15}N$	98.3	130.2	146.0	162.8	182.0	193.7	209.8	233.0	258.8	286.0	
Ethylene	C_2H_4	−168.3	−158.3	−153.2	−147.6	−141.3	−137.3	−131.8	−123.4	−113.9	−103.7	−169
Ethylene-*bis*-(chloroacetate)	$C_6H_8Cl_2O_4$	112.0	142.4	158.0	173.5	191.0	201.8	215.0	237.3	259.5	283.5	
Ethylene chlorohydrin (2-chloroethanol)	C_2H_5ClO	−4.0	+19.0	30.3	42.5	56.0	64.1	75.0	91.8	110.0	128.8	−69
diamine (1,2-ethanediamine)	$C_2H_8N_2$	−11.0	+10.5	21.5	33.0	45.8	53.8	62.5	81.0	99.0	117.2	8.5
dibromide (1,2-dibromethane)	$C_2H_4Br_2$	−27.0	+4.7	18.6	32.7	48.0	57.9	70.4	89.8	110.1	131.5	10
dichloride (1,2-dichloroethane)	$C_2H_4Cl_2$	−44.5	−24.0	−13.6	−2.4	+10.0	18.1	29.4	45.7	64.0	82.4	−35.3
glycol (1,2-ethanediol)	$C_2H_6O_2$	53.0	79.7	92.1	105.8	120.0	129.5	141.8	158.5	178.5	197.3	−15.6
glycol diethyl ether (1,2-diethoxyethane)	$C_6H_{14}O_2$	−33.5	−10.2	+1.6	14.7	29.7	39.0	51.8	71.8	94.1	119.5	
glycol dimethyl ether (1,2-dimethoxyethane)	$C_4H_{10}O_2$	−48.0	−26.2	−15.3	−3.0	+10.7	19.7	31.8	50.0	70.8	93.0	
glycol monomethyl ether (2-methoxyethanol)	$C_3H_8O_2$	−13.5	+10.2	22.0	34.3	47.8	56.4	68.0	85.3	104.3	124.4	
oxide	C_2H_4O	−89.7	−73.8	−65.7	−56.6	−46.9	−40.7	−32.1	−19.5	−4.9	+10.7	−111.3
Ethyl α-ethylacetoacetate	$C_8H_{14}O_3$	40.5	67.3	80.2	94.6	110.3	120.6	133.8	153.2	175.6	198.0	
fluoride	C_2H_5F	−117.0	−103.8	−97.7	−90.0	−81.8	−76.4	−69.3	−58.0	−45.5	−32.0	
formate	$C_3H_6O_2$	−60.5	−42.2	−33.0	−22.7	−11.5	−4.3	−5.4	20.2	37.1	54.3	−79

(*Continued*)

TABLE 2.37 Boiling Points of Common Organic Compounds at Selected Pressures (*Continued*)

Compound: Name	Formula	1	5	10	20	40	60	100	200	400	760	Melting Point, °C
		Pressure, mm Hg: Temperature, °C										
2-furoate	$C_7H_8O_3$	37.6	63.8	77.1	91.5	107.5	117.5	130.4	150.1	172.5	195.0	34
glycolate	$C_4H_8O_3$	14.3	38.8	50.5	63.9	78.1	87.6	99.8	117.8	138.0	158.2	
3-Ethylhexane	C_8H_{18}	−20.0	+2.1	12.8	25.0	38.5	47.1	58.9	76.7	97.0	118.5	
2-Ethylhexyl acrylate	$C_{11}H_{20}O_2$	50.0	77.7	91.8	106.3	123.7	134.0	147.9	168.2	192.2	216.0	
Ethylidene chloride (1,1-dichloroethane)	$C_2H_4Cl_2$	−60.7	−41.9	−32.3	−21.9	−10.2	−2.9	+7.2	22.4	39.8	57.4	−96.7
fluoride (1,1-difluoroethane)	$C_2H_4F_2$	−112.5	−98.4	−91.7	−84.1	−75.8	−70.4	−63.2	−52.0	−39.5	−26.5	−117
Ethyl iodide	C_2H_5I	−54.4	−34.3	−24.3	−13.1	−0.9	+7.2	18.0	34.1	52.3	72.4	−105
Ethyl *l*-leucinate	$C_8H_{17}NO_2$	27.8	57.3	72.1	88.0	106.0	117.8	131.8	149.8	167.3	184.0	
Ethyl levulinate	$C_7H_{12}O_3$	47.3	74.0	87.3	101.8	117.7	127.6	141.3	160.2	183.0	206.2	
Ethyl mercaptan (ethanethiol)	C_2H_6S	−76.7	−59.1	−50.2	−40.7	−29.8	−22.4	−13.0	+1.5	17.7	35.0	−121
Ethyl methylcarbamate	$C_4H_9NO_2$	26.5	51.0	63.2	76.1	91.0	100.0	112.0	130.0	149.8	170.0	
Ethyl methyl ether	C_3H_8O	−91.0	−75.6	−67.8	−59.1	−49.4	−43.3	−34.8	−22.0	−7.8	+7.5	
1-Ethylnaphthalene	$C_{12}H_{12}$	70.0	101.4	116.8	133.8	152.0	164.1	180.0	204.6	230.8	258.1	−27
Ethyl α-naphthyl ketone (1-propionaphthone)	$C_{13}H_{12}O$	124.0	155.5	171.0	188.1	206.9	218.2	233.5	255.5	280.2	306.0	
Ethyl 3-nitrobenzoate	$C_9H_9NO_4$	108.1	140.2	155.0	173.6	192.6	205.0	220.3	244.6	270.6	298.0	47
3-Ethylpentane	C_7H_{16}	−37.8	−17.0	−6.8	+4.7	17.5	25.7	36.9	53.8	73.0	93.5	−118.6
4-Ethylphenetole	$C_{10}H_{14}O$	48.5	75.7	89.5	103.8	119.8	129.8	143.5	163.2	185.7	208.0	
2-Ethylphenol	$C_8H_{10}O$	46.2	73.4	87.0	101.5	117.9	127.9	141.8	161.6	184.5	207.5	−45
3-Ethylphenol	$C_8H_{10}O$	60.0	86.8	100.2	114.5	130.0	139.8	152.0	171.8	193.3	214.0	−4
4-Ethylphenol	$C_8H_{10}O$	59.3	86.5	100.2	115.0	131.3	141.7	154.2	175.0	197.4	219.0	46.5
Ethyl phenyl ether (phenetole)	$C_8H_{10}O$	18.1	43.7	56.4	70.3	86.6	95.4	108.4	127.9	149.8	172.0	−30.2
Ethyl propionate	$C_5H_{10}O_2$	−28.0	−7.2	+3.4	14.3	27.2	35.1	45.2	61.7	79.8	99.1	−72.6
Ethyl propyl ether	$C_5H_{12}O$	−64.3	−45.0	−35.0	−24.0	−12.0	−4.0	+6.8	23.3	41.6	61.7	
Ethyl salicylate	$C_9H_{10}O_3$	61.2	90.0	104.2	119.3	136.7	147.6	161.5	183.7	207.0	231.5	1.3
3-Ethylstyrene	$C_{10}H_{12}$	28.3	55.0	68.3	82.8	99.2	109.6	123.2	144.0	167.2	191.5	
4-Ethylstyrene	$C_{10}H_{12}$	26.0	52.7	66.3	80.8	97.3	107.6	121.5	142.0	165.0	189.0	
Ethylisothiocyanate	C_3H_5NS	13.2	+10.6	22.8	36.1	50.8	59.8	71.9	90.0	110.1	131.0	−5.9
2-Ethyltoluene	C_9H_{12}	9.4	34.8	47.6	61.2	76.4	86.0	99.0	119.0	141.4	165.1	
3-Ethyltoluene	C_9H_{12}	7.2	32.3	44.7	58.2	73.3	82.9	95.9	115.5	137.8	161.3	−95.5
4-Ethyltoluene	C_9H_{12}	7.6	32.7	44.9	58.5	73.6	83.2	96.3	116.1	136.4	162.0	

Ethyl trichloroacetate	$C_4H_5Cl_3O_2$	20.7	45.5	57.7	70.6	85.5	94.4	107.4	125.8	146.0	167.0	
Ethyltrimethylsilane	$C_5H_{14}Si$	−60.6	−41.4	−31.8	−21.0	−9.0	−1.2	+9.2	25.0	42.8	62.0	
Ethyltrimethyltin	$C_5H_{14}Sn$	−30.0	−7.6	+3.8	16.1	30.0	38.4	50.0	67.3	87.6	108.8	
Ethyl isovalerate	$C_7H_{14}O_2$	−6.1	+17.0	28.7	41.3	55.2	64.0	75.9	93.8	114.0	134.3	−99.3
2-Ethyl-1,4-xylene	$C_{10}H_{14}$	25.7	52.0	65.6	79.8	96.0	106.2	120.0	140.2	163.1	186.9	
4-Ethyl-1,3-xylene	$C_{10}H_{14}$	26.3	53.0	66.4	80.6	97.2	107.4	121.2	141.8	164.4	188.4	
5-Ethyl-1,3-xylene	$C_{10}H_{14}$	22.1	48.8	62.1	76.5	92.6	103.0	116.5	137.4	159.6	183.7	
Eugenol	$C_{10}H_{12}O_2$	78.4	108.1	123.0	138.7	155.8	167.3	182.2	204.7	228.3	253.5	
iso-Eugenol	$C_{10}H_{12}O_2$	86.3	117.0	132.4	149.0	167.0	178.2	194.0	217.2	242.3	267.5	−10
Eugenyl acetate	$C_{12}H_{14}O_3$	101.6	132.3	148.0	164.2	183.0	194.0	209.7	232.5	257.4	282.0	295
Fencholic acid	$C_{10}H_{16}O_2$	101.7	128.7	142.3	155.8	171.8	181.5	194.0	215.0	237.8	264.1	19
d-Fenchone	$C_{10}H_{16}O$	28.0	54.7	68.3	83.0	99.5	109.8	123.6	144.0	166.8	191.0	5
dl-Fenchyl alcohol	$C_{10}H_{18}O$	45.8	70.3	82.1	95.6	110.8	120.2	132.3	150.0	173.2	201.0	35
Fluorene	$C_{13}H_{10}$		129.3	146.0	164.2	185.2	197.8	214.7	240.3	268.6	295.0	113
Fluorobenzene	C_6H_5F	−43.4	−22.8	−12.4	−1.2	+11.5	19.6	30.4	47.2	65.7	84.7	−42.1
2-Fluorotoluene	C_7H_7F	−24.2	−2.2	+8.9	21.4	34.7	43.7	55.3	73.0	92.8	114.0	−80
3-Fluorotoluene	C_7H_7F	−22.4	−0.3	+11.0	23.4	37.0	45.8	57.5	75.4	95.4	116.0	−110.8
4-Fluorotoluene	C_7H_7F	−21.8	+0.3	11.8	24.0	37.8	46.5	58.1	76.0	96.1	117.0	
Formaldehyde	CH_2O			−88.0	−79.6	−70.6	−65.0	−57.3	−46.0	−33.0	−19.5	−92
Formamide	CH_3NO	70.5	96.3	109.5	122.5	137.5	147.0	157.5	175.5	193.5	210.5	
Formic acid	CH_2O_2	−20.0	−5.0	+2.1	10.3	24.0	32.4	43.8	61.4	80.3	100.6	8.2
trans-Fumaryl chloride	$C_4H_2Cl_2O_2$	+15.0	38.5	51.8	65.0	79.5	89.0	101.0	120.0	140.0	160.0	
Furfural (2-furaldehyde)	$C_5H_4O_2$	18.5	42.6	54.8	67.8	82.1	91.5	103.4	121.8	141.8	161.8	
Furfuryl alcohol	$C_5H_6O_2$	31.8	56.0	68.0	81.0	95.7	104.0	115.9	133.1	151.8	170.0	
Geraniol	$C_{10}H_{18}O$	69.2	96.8	110.0	125.6	141.8	151.5	165.3	185.6	207.8	230.0	
Geranyl acetate	$C_{12}H_{20}O_2$	73.5	102.7	117.9	133.0	150.0	160.3	175.2	196.3	219.8	243.3	
Geranyl *n*-butyrate	$C_{14}H_{24}O_2$	96.8	125.2	139.0	153.8	170.1	180.2	193.8	214.0	235.0	257.4	
Geranyl isobutyrate	$C_{14}H_{24}O_2$	90.9	119.6	133.0	147.9	164.0	174.0	187.7	207.6	228.5	251.0	
Geranyl formate	$C_{11}H_{18}O_2$	61.8	90.3	104.3	119.8	136.2	147.2	160.7	182.6	205.8	230.0	
Glutaric acid	$C_5H_8O_4$	155.5	183.8	196.0	210.5	226.3	235.5	247.0	265.0	283.5	303.0	97.5
Glutaric anhydride	$C_5H_6O_3$	100.8	133.3	149.5	166.0	185.5	196.2	212.5	236.5	261.0	287.0	
Glutaronitrile	$C_5H_6N_2$	91.3	123.7	140.0	156.5	174.6	189.5	205.5	230.0	257.3	286.2	
Glutaryl chloride	$C_5H_6Cl_2O_2$	56.1	84.0	97.8	112.3	128.3	139.1	151.8	172.4	195.3	217.0	
Glycerol	$C_3H_8O_3$	125.5	153.8	167.2	182.2	198.0	208.0	220.1	240.0	263.0	290.0	17.9
Glycerol dichlorohydrin (1,3-dichloro-2-propanol)	$C_3H_6Cl_2O$	28.0	52.2	64.7	78.0	93.0	102.0	114.8	133.3	153.5	174.3	
Glycol diacetate	$C_6H_{10}O_4$	38.3	64.1	77.1	90.8	106.1	115.8	128.0	147.8	168.3	190.5	−31

(*Continued*)

TABLE 2.37 Boiling Points of Common Organic Compounds at Selected Pressures (*Continued*)

Compound		Pressure, mm Hg										Melting Point, °C
		1	5	10	20	40	60	100	200	400	760	
Name	Formula	Temperature, °C										
Glycolide (1,4-dioxane-2,6-dione)	$C_4H_4O_4$		103.0	116.6	132.0	148.6	158.2	173.2	194.0	217.0	240.0	97
Guaicol (2-methoxyphenol)	$C_7H_8O_2$	52.4	79.1	92.0	106.0	121.6	131.0	144.0	162.7	184.1	205.0	28.3
Heneicosane	$C_{21}H_{44}$	152.6	188.0	205.4	223.2	243.4	255.3	272.0	296.5	323.8	350.5	40.4
Heptacosane	$C_{27}H_{56}$	211.7	248.6	266.8	284.6	305.7	318.3	333.5	359.4	385.0	410.6	59.5
Heptadecane	$C_{17}H_{36}$	115.0	145.2	160.0	177.7	195.8	207.3	223.0	247.8	274.5	303.0	22.5
Heptaldehyde (enanthaldehyde)	$C_7H_{14}O$	12.0	32.7	43.0	54.0	66.3	74.0	84.0	102.0	125.5	155.0	−42
n-Heptane	C_7H_{16}	−34.0	−12.7	−2.1	+9.5	22.3	30.6	41.8	58.7	78.0	98.4	−90.6
Heptanoic acid (enanthic acid)	$C_7H_{14}O_2$	78.0	101.3	113.2	125.6	139.5	148.5	160.0	179.5	199.6	221.5	−10
1-Heptanol	$C_7H_{16}O$	42.4	64.3	74.7	85.8	99.8	108.0	119.5	136.6	155.6	175.8	34.6
Heptanoyl chloride (enanthyl chloride)	$C_7H_{13}ClO$	34.2	54.6	64.6	75.0	86.4	93.5	102.7	116.3	130.7	145.0	
2-Heptene	C_7H_{14}	−35.8	−14.1	−3.5	+8.3	21.5	30.0	41.3	58.6	78.1	98.5	
Heptylbenzene	$C_{13}H_{20}$	64.0	94.6	110.0	126.0	144.0	154.8	170.2	193.3	217.8	244.0	
Heptyl cyanide (enanthonitrile)	$C_7H_{13}N$	21.0	47.8	61.6	76.3	92.6	103.0	116.8	137.7	160.0	184.6	
Hexachlorobenzene	C_6Cl_6	114.4	149.3	166.4	185.7	206.0	219.0	235.5	258.5	283.5	309.4	230
Hexachloroethane	C_2Cl_6	32.7	49.8	73.5	87.6	102.3	112.0	124.2	143.1	163.8	185.6	186.6
Hexacosane	$C_{26}H_{54}$	204.0	240.0	257.4	275.8	295.2	307.8	323.2	348.4	374.6	399.8	56.6
Hexadecane	$C_{16}H_{34}$	105.3	135.2	149.8	164.7	181.3	193.2	208.5	231.7	258.3	287.5	18.5
1-Hexadecene	$C_{16}H_{32}$	101.6	131.7	146.2	162.0	178.8	190.8	205.3	226.8	250.0	274.0	4
n-Hexadecyl alcohol (cetyl alcohol)	$C_{16}H_{34}O$	122.7	158.3	177.8	197.8	219.8	234.3	251.7	280.2	312.7	344.0	49.3
n-Hexadecylamine (cetylamine)	$C_{16}H_{35}N$	123.6	157.8	176.0	195.7	215.7	228.8	245.8	272.2	300.4	330.0	
Hexaethylbenzene	$C_{18}H_{30}$		134.3	150.3	168.0	187.7	199.7	216.0	241.7	268.5	298.3	130
n-Hexane	C_6H_{14}	−53.9	−34.5	−25.0	−14.1	−2.3	+5.4	15.8	31.6	49.6	68.7	−95.3
1-Hexanol	$C_6H_{14}O$	24.4	47.2	58.2	70.3	83.7	92.0	102.8	119.6	138.0	157.0	−51.6
2-Hexanol	$C_6H_{14}O$	14.6	34.8	45.0	55.9	67.9	76.0	87.3	103.7	121.8	139.9	
3-Hexanol	$C_6H_{14}O$	+2.5	25.7	36.7	49.0	62.2	70.7	81.8	98.3	117.0	135.5	
1-Hexene	C_6H_{12}	−57.5	−38.0	−28.1	−17.2	−5.0	+2.8	13.0	29.0	46.8	66.0	−98.5
n-Hexyl levulinate	$C_{11}H_{20}O_3$	90.0	120.0	134.7	150.2	167.8	179.0	193.6	215.7	241.0	266.8	
n-Hexyl phenyl ketone (enanthophenone)	$C_{13}H_{18}O$	100.0	130.3	145.5	161.0	178.9	189.8	204.2	225.0	248.3	271.3	
Hydrocinnamic acid	$C_9H_{10}O_2$	102.2	133.5	148.7	165.0	183.3	194.0	209.0	230.8	255.0	279.8	48.5
Hydrogen cyanide (hydrocyanic acid)	CHN	−71.0	−55.3	−47.7	−39.7	−30.9	−25.1	−17.8	−5.3	+10.2	25.9	−13.2

Hydroquinone	$C_6H_6O_2$	132.4	153.3	163.5	174.6	192.0	203.0	216.5	238.0	262.5	286.2	170.3
4-Hydroxybenzaldehyde	$C_7H_6O_2$	121.2	153.2	169.7	186.8	206.0	217.5	233.5	256.8	282.6	310.0	115.5
α-Hydroxyisobutyric acid	$C_4H_8O_3$	73.5	98.5	110.5	123.8	138.0	146.4	157.7	175.2	193.8	212.0	79
α-Hydroxybutyronitrile	C_5H_9NO	41.0	65.8	77.8	90.7	104.8	113.9	125.0	142.0	159.8	178.8	
4-Hydroxy-3-methyl-2-butanone	$C_5H_{10}O_2$	44.6	69.3	81.0	94.0	108.2	117.4	129.0	146.5	165.5	185.0	
4-Hydroxy-4-methyl-2-pentanone	$C_6H_{12}O_2$	22.0	46.7	58.8	72.0	86.7	96.0	108.2	126.8	147.5	167.9	−47
3-Hydroxypropionitrile	C_3H_5NO	58.7	87.8	102.0	117.9	134.1	144.7	157.7	178.0	200.0	221.0	
Indene	C_9H_8	16.4	44.3	58.5	73.9	90.7	100.8	114.7	135.6	157.8	181.6	−2
Iodobenzene	C_6H_5I	24.1	50.6	64.0	78.3	94.4	105.0	118.3	139.8	163.9	188.6	−28.5
Iodononane	$C_9H_{19}I$	70.0	96.2	109.0	123.0	138.1	147.7	159.8	179.0	199.3	219.5	
2-Iodotoluene	C_7H_7I	37.2	65.9	79.8	95.6	112.4	123.8	138.1	160.0	185.7	211.0	
α-Ionone	$C_{13}H_{20}O$	79.5	108.8	123.0	139.0	155.6	166.3	181.2	202.5	225.2	250.0	
Isoprene	C_5H_8	−79.8	−62.3	−53.3	−43.5	−32.6	−25.4	−16.0	−1.2	+15.4	32.6	−146.7
Lauraldehyde	$C_{12}H_{24}O$	77.7	108.4	123.7	140.2	157.8	168.7	184.5	207.8	231.8	257.0	44.5
Lauric acid	$C_{12}H_{24}O_2$	121.0	150.6	166.0	183.6	201.4	212.7	227.5	249.8	273.8	299.2	48
Levulinaldehyde	$C_5H_8O_2$	28.1	54.9	68.0	82.7	98.3	108.4	121.8	142.0	164.0	187.0	
Levulinic acid	$C_5H_8O_3$	102.0	128.1	141.8	154.1	169.5	178.0	190.2	208.3	227.4	245.8	33.5
d-Limonene	$C_{10}H_{16}$	14.0	40.4	53.8	68.2	84.3	94.6	108.3	128.5	151.4	175.0	−96.9
Linalyl acetate	$C_{12}H_{20}O_2$	55.4	82.5	96.0	111.4	127.7	138.1	151.8	173.3	196.2	220.0	
Maleic anhydride	$C_4H_2O_3$	44.0	63.4	78.7	95.0	111.8	122.0	135.8	155.9	179.5	202.0	58
Menthane	$C_{10}H_{20}$	+9.7	35.7	48.3	62.7	78.3	88.6	102.1	122.7	146.0	169.5	
1-Menthol	$C_{10}H_{20}O$	56.0	83.2	96.0	110.3	126.1	136.1	149.4	168.3	190.2	212.0	42.5
Menthyl acetate	$C_{12}H_{22}O_2$	57.4	85.8	100.0	115.4	132.1	143.2	156.7	178.8	202.8	227.0	
benzoate	$C_{17}H_{24}O_2$	123.2	154.2	170.0	186.3	204.3	215.8	230.4	253.2	277.1	301.0	54.5
formate	$C_{11}H_{20}O_2$	47.3	75.8	90.0	105.8	123.0	133.8	148.0	169.8	194.2	219.0	
Mesityl oxide	$C_6H_{10}O$	−8.7	+14.1	26.0	37.9	51.7	60.4	72.1	90.0	109.8	130.0	−59
Methacrylic acid	$C_4H_6O_2$	25.5	48.5	60.0	72.7	86.4	95.3	106.6	123.9	142.5	161.0	15
Methacrylonitrile	C_4H_5N	−44.5	−23.3	−12.5	−0.6	+12.8	21.5	32.8	50.0	70.3	90.3	
Methane	CH_4	−205.9	−119.0	−195.5	−191.8	−187.7	−185.1	−181.4	−175.5	−168.8	−161.5	−182.5
Methanethiol	CH_4S	−90.7	−75.3	−67.5	−58.8	−49.2	−43.1	−34.8	−22.1	−7.9	+6.8	−121
Methoxyacetic acid	$C_3H_6O_3$	52.5	79.3	92.0	106.5	122.0	131.8	144.5	163.5	184.2	204.0	
N-Methylacetanilide	$C_9H_{11}NO$		103.8	118.6	135.1	152.2	164.2	179.8	202.3	227.4	253.0	102
Methyl acetate	$C_3H_6O_2$	−57.2	−38.6	−29.3	−19.1	−7.9	−0.5	+9.4	24.0	40.0	57.8	−98.7
acetylene (propyne)	C_3H_4	−111.0	−97.5	−90.5	−82.9	−74.3	−68.8	−61.3	−49.8	−37.2	−23.3	−102.7
acrylate	$C_4H_6O_2$	−43.7	−23.6	−13.5	−2.7	+9.2	17.3	28.0	43.9	61.8	80.2	
alcohol (methanol)	CH_4O	−44.0	−25.3	−16.2	−6.0	+5.0	12.1	21.2	34.8	49.9	64.7	−97.8
Methylamine	CH_5N	−95.8	−81.3	−73.8	−65.9	−56.9	−51.3	−43.7	−32.4	−19.7	−6.3	−93.5

(Continued)

TABLE 2.37 Boiling Points of Common Organic Compounds at Selected Pressures (*Continued*)

Compound		Pressure, mm Hg										Melting Point, °C
		1	5	10	20	40	60	100	200	400	760	
Name	Formula	Temperature, °C										
N-Methylaniline	C_7H_9N	36.0	62.8	76.2	90.5	106.0	115.8	129.8	149.3	172.0	195.5	−57
Methyl anthranilate	$C_8H_9NO_2$	77.6	109.0	124.2	141.5	159.7	172.0	187.8	212.4	238.5	266.5	24
benzoate	$C_8H_8O_2$	39.0	64.4	77.3	91.8	107.8	117.4	130.8	151.4	174.7	199.5	−12.5
2-Methylbenzothiazole	C_8H_7NS	70.0	97.5	111.2	125.5	141.2	150.4	163.9	183.2	204.5	225.5	15.4
α-Methylbenzyl alcohol	$C_8H_{10}O$	49.0	75.2	88.0	102.1	117.8	127.4	140.3	159.0	180.7	204.0	
Methyl bromide	CH_3Br	−96.3	−80.6	−72.8	−64.0	−54.2	−48.0	−39.4	−26.5	−11.9	+3.6	−93
2-Methyl-l-butene	C_5H_{10}	−89.1	−72.8	−64.3	−54.8	−44.1	−37.3	−28.0	−13.8	+2.5	20.2	−135
2-Methyl-2-butene	C_5H_{10}	−75.4	−57.0	−47.9	−37.9	−26.7	−19.4	−9.9	+4.9	21.6	38.5	−133
Methyl isobutyl carbinol (2-methyl-4-pentanol)	$C_6H_{14}O$	−0.3	+22.1	33.3	45.4	58.2	67.0	78.0	94.9	113.5	131.7	
n-butyl ketone (2-hexanone)	$C_6H_{12}O$	+7.7	28.8	38.8	50.0	62.0	69.8	79.8	94.3	111.0	127.5	−56.9
isobutyl ketone (4-methyl-2-pentanone)	$C_6H_{12}O$	−1.4	+19.7	30.0	40.8	52.8	60.4	70.4	85.6	102.0	119.0	−84.7
n-butyrate	$C_5H_{10}O_2$	−26.8	−5.5	+5.0	16.7	29.6	37.4	48.0	64.3	83.1	102.3	
isobutyrate	$C_5H_{10}O_2$	−34.1	−13.0	−2.9	+8.4	21.0	28.9	39.6	55.7	73.6	92.6	−84.7
caprate	$C_{11}H_{22}O_2$	63.7	93.5	108.0	123.0	139.0	148.6	161.5	181.6	202.9	224.0	−18
caproate	$C_7H_{14}O_2$	+5.0	30.0	42.0	55.4	70.0	79.7	91.4	109.8	129.8	150	
caprylate	$C_9H_{18}O_2$	34.2	61.7	74.9	89.0	105.3	115.3	128.0	148.1	170.0	193.0	−40
chloride	CH_3Cl		−99.5	−92.4	−84.8	−76.0	−70.4	−63.0	−51.2	−38.0	−24.0	−97.7
chloroacetate	$C_3H_5ClO_2$	−2.9	19.0	30.0	41.5	54.5	63.0	73.5	90.5	109.5	130.3	−31.9
cinnamate	$C_{10}H_{10}O_2$	77.4	108.1	123.0	140.0	157.9	170.0	185.8	209.6	235.0	263.0	33.4
α-Methylcinnamic acid	$C_{10}H_{10}O_2$	125.7	155.0	169.8	185.2	201.8	212.0	224.8	245.0	266.8	288.0	
Methylcyclohexane	C_7H_{14}	−35.9	−14.0	−3.2	+8.7	22.0	30.5	42.1	59.6	79.6	100.9	−126.4
Methylcyclopentane	C_8H_{12}	−53.7	−33.8	−23.7	−12.8	−0.6	+7.2	17.9	34.0	52.3	71.8	−142.4
Methylcyclopropane	C_4H_8	−96.0	−80.6	−72.8	−64.0	−54.2	−48.0	−39.3	−26.0	−11.3	+4.5	
Methyl *n*-decyl ketone (*n*-dodecan-2-one)	$C_{12}H_{24}O$	77.1	106.0	120.4	136.0	152.4	163.8	177.5	199.0	222.5	246.5	
dichloroacetate	$C_3H_4Cl_2O_2$	3.2	26.7	38.1	50.7	64.7	73.6	85.4	103.2	122.6	143.0	
N-Methyldiphenylamine	$C_{13}H_{13}N$	103.5	134.0	149.7	165.8	184.0	195.4	210.1	232.8	257.0	282.0	−7.6
Methyl *n*-dodecyl ketone (2-tetradecanone)	$C_{14}H_{28}O$	99.3	130.0	145.5	161.3	179.8	191.4	206.0	228.2	253.3	278.0	
Methylene bromide (dibromomethane)	CH_2Br_2	−35.1	−13.2	−2.4	+9.7	23.3	31.6	42.3	58.5	79.0	98.6	−52.8
chloride (dichloromethane)	CH_2Cl_2	−70.0	−52.1	−43.3	−33.4	−22.3	−15.7	−6.3	+8.0	24.1	40.7	−96.7

Methyl ethyl ketone (2-butanone)	C_4H_8O	−48.3	−28.0	−17.7	−6.5	+6.0	14.0	25.0	41.6	60.0	79.6	−85.9
2-Methyl-3-ethylpentane	C_8H_{18}	−24.0	−1.8	+9.5	21.7	35.2	43.9	55.7	73.6	94.0	115.6	−114.5
3-Methyl-3-ethylpentane	C_8H_{18}	−23.9	−1.4	+9.9	22.3	36.2	45.0	57.1	75.3	96.2	118.3	−90
Methyl fluoride	CH_3F	−147.3	−137.0	−131.6	−125.9	−119.1	−115.0	−109.0	−99.9	−89.5	−78.2	
formate	$C_2H_4O_2$	−74.2	−57.0	−48.6	−39.2	−28.7	−21.9	−12.9	+0.8	16.0	32.0	−99.8
α-Methylglutaric anhydride	$C_6H_8O_3$	93.8	125.4	141.8	157.7	177.5	189.9	205.0	229.1	255.5	282.5	
Methyl glycolate	$C_3H_6O_3$	+9.6	33.7	45.3	58.1	72.3	81.8	93.7	111.8	131.7	151.5	
2-Methylheptadecane	$C_{18}H_{38}$	119.8	152.0	168.7	186.0	204.8	216.3	231.5	254.5	279.8	306.5	
2-Methylheptane	C_8H_{18}	−21.0	+1.3	12.3	24.4	37.9	46.6	58.3	76.0	96.2	117.6	−109.5
3-Methylheptane	C_8H_{18}	−19.8	+2.6	13.3	25.4	38.9	47.6	59.4	77.1	97.4	118.9	−120.8
4-Methylheptane	C_8H_{18}	−20.4	+1.5	12.4	24.5	38.0	46.6	58.3	76.1	96.3	117.7	−121.1
2-Methyl-2-heptene	C_8H_{16}	−16.1	+6.7	17.8	30.4	44.0	52.8	64.6	82.3	102.2	122.5	
6-Methyl-3-hepten-2-ol	$C_8H_{16}O$	41.6	65.0	76.7	89.3	102.7	111.5	122.6	139.5	156.6	175.5	
6-Methyl-5-hepten-2-ol	$C_8H_{16}O$	41.9	66.0	77.8	90.4	104.0	112.8	123.8	140.0	156.6	174.3	
2-Methylhexane	C_7H_{16}	−40.4	−19.5	−9.1	+2.3	14.9	23.0	34.1	50.8	69.8	90.0	−118.2
3-Methylhexane	C_7H_{16}	−39.0	−18.1	−7.8	+3.6	16.4	24.5	35.6	52.4	71.6	91.9	
Methyl iodide	CH_3I		−55.0	−45.8	−35.6	−24.2	−16.9	−7.0	+8.0	25.3	42.4	−64.4
laurate	$C_{13}H_{26}O_2$	87.8	117.9	133.2	149.0	166.0	176.8	190.8				5
levulinate	$C_6H_{10}O_3$	39.8	66.4	79.7	93.7	109.5	119.3	133.0	153.4	175.8	197.7	
methacrylate	$C_5H_8O_2$	−30.5	−10.0	+1.0	11.0	25.5	34.5	47.0	63.0	82.0	101.0	
myristate	$C_{15}H_{30}O_2$	115.0	145.7	160.8	177.8	195.8	207.5	222.6	245.3	269.8	295.8	18.5
α-naphthyl ketone (1-acetonaphthone)	$C_{12}H_{10}O$	115.6	146.3	161.5	178.4	196.8	208.6	223.8	246.7	270.5	295.5	
β-naphthyl ketone (2-acetonaphthone)	$C_{12}H_{10}O$	120.2	152.3	168.5	185.7	203.8	214.7	229.8	251.6	275.8	301.0	55.5
n-nonyl ketone (undecan-2-one)	$C_{11}H_{22}O$	68.2	95.5	108.9	123.1	139.0	148.6	161.0	181.2	202.3	224.0	15
palmitate	$C_{17}H_{34}O_2$	134.3	166.8	184.3	202.0							30
n-pentadecyl ketone (2-heptdecanone)	$C_{17}H_{34}O$	129.6	161.6	178.0	196.4	214.3	226.7	242.0	265.8	291.7	319.5	
2-Methylpentane	C_6H_{14}	−60.9	−41.7	−32.1	−21.4	−9.7	−1.9	+8.1	24.1	41.6	60.3	−154
3-Methylpentane	C_6H_{14}	−59.0	−39.8	−30.1	−19.4	−7.3	+0.1	10.5	26.5	44.2	63.3	−118
2-Methyl-1-pentanol	$C_6H_{14}O$	15.4	38.0	49.6	61.6	74.7	83.4	94.2	111.3	129.8	147.9	
2-Methyl-2-pentanol	$C_6H_{14}O$	−4.5	+16.8	27.6	38.8	51.3	58.8	69.2	85.0	102.6	121.2	−103
Methyl *n*-pentyl ketone (2-heptanone)	$C_7H_{14}O$	19.3	43.6	55.5	67.7	81.2	89.8	100.0	116.1	133.2	150.2	
phenyl ether (anisole)	C_7H_8O	+5.4	30.0	42.2	55.8	70.7	80.1	93.0	112.3	133.8	155.5	−37.3
2-Methylpropene	C_4H_8	−105.1	−96.5	−81.9	−73.4	−63.8	−57.7	−49.3	−36.7	−22.2	−6.9	−140.3
Methyl propionate	$C_4H_8O_2$	−42.0	−21.5	−11.8	−1.0	+11.0	18.7	29.0	44.2	61.8	79.8	−87.5
4-Methylpropiophenone	$C_{10}H_{12}O$	59.6	89.3	103.8	120.2	138.0	149.3	164.2	187.4	212.7	238.5	
2-Methylpropionyl bromide	C_4H_7BrO	13.5	38.4	50.6	64.1	79.4	88.8	101.6	120.5	141.7	163.0	
Methyl propyl ether	$C_4H_{10}O$	−72.2	−54.3	−45.4	−35.4	−24.3	−17.4	−8.1	+6.0	22.5	39.1	

(*Continued*)

TABLE 2.37 Boiling Points of Common Organic Compounds at Selected Pressures (*Continued*)

Compound		Pressure, mm Hg										Melting Point, °C
		1	5	10	20	40	60	100	200	400	760	
Name	Formula	Temperature, °C										
n-propyl ketone (2-pentanone)	$C_5H_{10}O$	−12.0	+8.0	17.9	28.5	39.8	47.3	56.8	71.0	86.8	103.3	−77.8
isopropyl ketone (3-methyl-2-butanone)	$C_5H_{10}O$	−19.9	−1.0	+8.3	18.3	29.6	36.2	45.5	59.0	73.8	88.9	−92
2-Methylquinoline	$C_{10}H_9N$	75.3	104.0	119.0	134.0	150.8	161.7	176.2	197.8	211.7	246.5	−1
Methyl salicylate	$C_8H_8O_3$	54.0	81.6	95.3	110.0	126.2	136.7	150.0	172.6	197.5	223.2	−8.3
α-Methyl styrene	C_9H_{10}	7.4	34.0	47.1	61.8	77.8	88.3	102.2	121.8	143.0	165.4	−23.2
4-Methyl styrene	C_9H_{10}	16.0	42.0	55.1	69.2	85.0	95.0	108.6	128.7	151.2	175.0	
Methyl *n*-tetradecyl ketone												
(2-hexadecanone)	$C_{16}H_{32}O$	109.8	151.5	167.3	184.6	203.7	215.0	230.5	254.4	279.8	307.0	
thiocyanate	C_2H_3NS	−14.0	+9.8	21.6	34.5	49.0	58.1	70.4	89.8	110.8	132.9	−51
isothiocyanate	C_2H_3NS	−34.7	−8.3	+5.4	20.4	38.2	47.5	59.3	77.5	97.8	119.0	35.5
undecyl ketone (2-tridecanone)	$C_{13}H_{26}O$	86.8	117.0	131.8	147.8	165.7	176.6	191.5	214.0	238.3	262.5	28.5
isovalerate	$C_6H_{12}O_2$	−19.2	+2.9	14.0	26.4	39.8	48.2	59.8	77.3	96.7	116.7	
Monovinylacetylene (butenyne)	C_4H_4	−93.2	−77.7	−70.0	−61.3	−51.7	−45.3	−37.1	−24.1	−10.1	+5.3	
Myrcene	$C_{10}H_{16}$	14.5	40.0	53.2	67.0	82.6	92.6	106.0	126.0	148.3	171.5	
Myristaldehyde	$C_{14}H_{28}O$	99.0	132.0	148.3	166.2	186.0	198.3	214.5	240.4	267.9	297.8	23.5
Myristic acid (tetradecanoic acid)	$C_{14}H_{28}O_2$	142.0	174.1	190.8	207.6	223.5	237.2	250.5	272.3	294.6	318.0	57.5
Napthalene	$C_{10}H_8$	52.6	74.2	85.8	101.7	119.3	130.2	145.5	167.7	193.2	217.9	80.2
1-Naphthoic acid	$C_{11}H_8O_2$	156.0	184.0	196.8	211.2	225.0	234.5	245.8	263.5	281.4	300.0	160.5
2-Naphthoic acid	$C_{11}H_8O_2$	160.8	189.7	202.8	216.9	231.5	241.3	252.7	270.3	289.5	308.5	184
1-Naphthol	$C_{10}H_8O$	94.0	125.5	142.0	158.0	177.8	190.0	206.0	229.6	255.8	282.5	96
2-Naphthol	$C_{10}H_8O$		128.6	145.5	161.8	181.7	193.7	209.8	234.0	260.6	288.0	122.5
1-Naphthylamine	$C_{10}H_9N$	104.3	137.7	153.8	171.6	191.5	203.8	220.0	244.9	272.2	300.8	50
2-Naphthylamine	$C_{10}H_9N$	108.0	141.6	157.6	175.8	195.7	208.1	224.3	249.7	277.4	306.1	111.5
Nicotine	$C_{10}H_{14}N_2$	61.8	91.8	107.2	123.7	142.1	154.7	169.5	193.8	219.8	247.3	
2-Nitroaniline	$C_6H_6N_2O_2$	104.0	135.7	150.4	167.7	186.0	197.8	213.0	236.3	260.0	284.5	71.5
3-Nitroaniline	$C_6H_6N_2O_2$	119.3	151.5	167.8	185.5	204.2	216.5	232.1	255.3	280.2	305.7	114
4-Nitroaniline	$C_6H_6N_2O_2$	142.4	177.6	194.4	213.2	234.2	245.9	261.8	284.5	310.2	336.0	146.5
2-Nitrobenzaldehyde	$C_7H_5NO_3$	85.8	117.7	133.4	150.0	168.8	180.7	196.2	220.0	246.8	273.5	40.9
3-Nitrobenzaldehyde	$C_7H_5NO_3$	96.2	127.4	142.8	159.0	177.7	189.5	204.3	227.4	252.1	278.3	58
Nitrobenzene	$C_6H_5NO_2$	44.4	71.6	84.9	99.3	115.4	125.8	139.9	161.2	185.8	210.6	+5.7
Nitroethane	$C_2H_5NO_2$	−21.0	+1.5	12.5	24.8	38.0	46.5	57.8	74.8	94.0	114.0	−90

Nitroglycerin	$C_3H_5N_3O_9$	127	167	188	210	235	251					11
Nitromethane	CH_3NO_2	−29.0	−7.9	+2.8	14.1	27.5	35.5	46.6	63.5	82.0	101.2	−29
2-Nitrophenol	$C_6H_5NO_3$	49.3	76.8	90.4	105.8	122.1	132.6	146.4	167.6	191.0	214.5	45
2-Nitrophenyl acetate	$C_8H_7NO_4$	100.0	128.0	142.0	155.8	172.8	181.7	194.1	213.0	233.5	253.0	
1-Nitropropane	$C_3H_7NO_2$	−9.6	+13.5	25.3	37.9	51.8	60.5	72.3	90.2	110.6	131.6	−108
2-Nitropropane	$C_3H_7NO_2$	−18.8	4.1	15.8	28.2	41.8	50.3	62.0	80.0	99.8	120.3	−93
2-Nitrotoluene	$C_7H_7NO_2$	50.0	79.1	93.8	109.6	126.3	137.6	151.5	173.7	197.7	222.3	−4.1
3-Nitrotoluene	$C_7H_7NO_2$	50.2	1.0	96.0	112.8	130.7	142.5	156.9	180.3	206.8	231.9	15.5
4-Nitrotoluene	$C_7H_7NO_2$	53.7	85.0	100.5	117.7	136.0	147.9	163.0	186.7	212.5	238.3	51.9
4-Nitro-1,3-xylene (4-nitro-*m*-xylene)	$C_8H_9NO_2$	65.6	95.0	109.8	125.8	143.3	153.8	168.5	191.7	217.5	244.0	+2
Nonacosane	$C_{29}H_{60}$	234.2	260.8	286.4	303.6	323.2	334.8	350.0	373.2	397.2	421.8	63.8
Nona lecane	$C_{19}H_{40}$	133.3	166.3	183.5	200.8	220.0	232.8	248.0	271.8	299.8	330.0	32
n-Nonane	C_9H_{20}	+1.4	25.8	38.0	51.2	66.0	75.5	88.1	107.5	128.2	150.8	−53.7
1-Nonanol	$C_9H_{20}O$	59.5	86.1	99.7	113.8	129.0	139.0	151.3	170.5	192.1	213.5	−5
2-Nonanone	$C_9H_{18}O$	32.1	59.0	72.3	87.2	103.4	113.8	127.4	148.2	171.2	195.0	−19
Octacosane	$C_{28}H_{58}$	226.5	260.3	277.4	295.4	314.2	326.8	341.8	364.8	388.9	412.5	61.6
Octadecane	$C_{18}H_{38}$	119.6	152.1	169.6	187.5	207.4	219.7	236.0	260.6	288.0	317.0	28
n-Octane	C_8H_{18}	−14.0	+8.3	19.2	31.5	45.1	53.8	65.7	83.6	104.0	125.6	−56.8
n-Octanol (1-octanol)	$C_8H_{18}O$	54.0	76.5	88.3	101.0	115.2	123.8	135.2	152.0	173.8	195.2	−15.4
2-Octanone	$C_8H_{18}O$	23.6	48.4	60.9	74.3	89.8	90.0	111.7	130.4	151.0	172.9	−16
n-Octyl acrylate	$C_{11}H_{20}O_2$	58.5	87.7	102.0	117.8	135.6	145.6	159.1	180.2	204.0	227.0	
iodide (1-Iodooctane)	$C_8H_{17}I$	45.8	74.8	90.0	105.9	123.8	135.4	150.0	173.3	199.3	225.5	−45.9
Oleic acid	$C_{18}H_{34}O_2$	176.5	208.5	223.0	240.0	257.2	269.8	286.0	309.8	334.7	360.0	14
Palmitaldehyde	$C_{16}H_{32}O$	121.6	154.6	171.8	190.0	210.0	222.6	239.5	264.1	292.3	321.0	34
Palmitic acid	$C_{16}H_{32}O_2$	153.6	188.1	205.8	223.8	244.4	256.0	271.5	298.7	326.0	353.8	64.0
Palmitonitrile	$C_{16}H_{31}N$	134.3	168.3	185.8	204.2	223.8	236.6	251.5	277.1	304.5	332.0	31
Pelargonic acid	$C_9H_{18}O_2$	108.2	126.0	137.4	149.8	163.7	172.3	184.4	203.1	227.5	253.5	12.5
Pentachlorobenzene	C_6HCl_5	98.6	129.7	144.3	160.0	178.5	190.1	205.5	227.0	251.6	276.0	85.5
Pentachloroethane	C_2HCl_5	+1.0	27.2	39.8	53.9	69.9	80.0	93.5	114.0	137.2	160.5	−22
Pentachloroethylbenzene	$C_6H_5Cl_5$	96.2	130.0	148.0	166.0	186.2	199.0	216.0	241.8	269.3	299.0	
Pentachlorophenol	C_6HCl_5O				192.2	211.2	223.4	239.6	261.8	285.0	309.3	188.5
Pentacosane	$C_{25}H_{52}$	194.2	230.0	248.2	266.1	285.6	298.4	314.0	339.0	365.4	390.3	53.3
Pentadecane	$C_{15}H_{32}$	91.6	121.0	135.4	150.2	167.7	178.4	194.0	216.1	242.8	270.5	10
1,3-Pentadiene	C_5H_8	−71.8	−53.8	−45.0	−34.8	−23.4	−16.5	−6.7	+8.0	24.7	42.1	
1,4-Pentadiene	C_5H_8	−83.5	−66.2	−57.1	−47.7	−37.0	−30.0	−20.6	−6.7	+8.3	26.1	
Pentaethylbenzene	$C_{16}H_{26}$	86.0	120.0	135.8	152.4	171.9	184.2	200.0	224.1	250.2	277.0	
Pentaethylchlorobenzene	$C_{16}H_{25}Cl$	90.0	183.8	140.7	158.1	178.2	191.0	208.0	230.3	257.2	285.0	

(*Continued*)

TABLE 2.37 Boiling Points of Common Organic Compounds at Selected Pressures (*Continued*)

Compound: Name	Formula	1	5	10	20	40	60	100	200	400	760	Melting Point, °C
		Pressure, mm Hg / Temperature, °C										
n-Pentane	C_5H_{12}	−76.6	−62.5	−50.1	−40.2	−29.2	−22.2	−12.6	+1.9	18.5	36.1	−129.7
iso-Pentane (2-methylbutane)	C_5H_{12}	−82.9	−65.8	−57.0	−47.3	−36.5	−29.6	−20.2	−5.9	+10.5	27.8	−159.7
neo-Pentane (2,2-dimethylpropane)	C_5H_{12}	−102.0	−85.4	−76.7	−67.2	−56.1	−49.0	−39.1	−23.7	−7.1	+9.5	−16.6
2,3,4-Pentanetriol	$C_5H_{12}O_3$	155.0	159.3	204.5	220.5	239.6	249.8	263.5	284.5	307.0	327.2	
1-Pentene	C_5H_{10}	−80.4	−63.3	−54.5	−46.0	−34.1	−27.1	−17.7	−3.4	+12.8	30.1	
α-Phellandrene	$C_{10}H_{16}$	20.0	45.7	58.0	72.1	87.8	97.6	110.6	130.6	152.0	175.0	
Phenanthrene	$C_{14}H_{10}$	118.2	154.3	173.0	193.7	215.8	229.9	249.0	277.1	308.0	340.2	99.5
Phenethyl alcohol (phenyl cellosolve)	$C_8H_{10}O_2$	58.2	85.9	100.0	114.8	130.5	141.2	154.0	175.0	197.5	219.5	
2-Phenetidine	$C_8H_{11}NO$	67.0	94.7	108.6	123.7	139.9	149.8	163.5	184.0	207.0	228.0	
Phenol	C_6H_6O	40.1	62.5	73.8	86.0	100.1	108.4	121.4	139.0	160.0	181.9	40.6
2-Phenoxyethanol	$C_8H_{10}O_2$	78.0	196.6	121.2	136.0	152.2	163.2	176.5	197.6	221.0	245.3	11.6
2-Phenoxyethyl acetate	$C_{10}H_{12}O_3$	82.6	143.5	128.0	144.5	162.3	174.0	189.2	211.3	235.0	259.7	−6.7
Phenyl acetate	$C_8H_8O_2$	38.2	64.8	78.0	92.3	108.1	118.1	131.6	151.2	173.5	195.9	
Phenylacetic acid	$C_8H_8O_2$	97.0	127.0	141.3	156.0	173.6	184.5	198.2	219.5	243.0	265.5	76.5
Phenylacetonitrile	C_8H_7N	60.0	89.0	103.5	119.4	136.3	147.7	161.8	184.2	208.5	233.5	−23.8
Phenylacetyl chloride	C_8H_7ClO	48.0	75.3	89.0	103.6	119.8	129.8	143.5	163.8	186.0	210.0	
Phenyl benzoate	$C_{13}H_{10}O_2$	106.8	141.5	157.8	177.0	197.6	210.8	277.8	254.0	283.5	314.0	70.5
4-Phenyl-3-buten-2-one	$C_{10}H_{10}O$	81.7	112.2	127.4	143.8	161.3	172.6	187.8	211.0	235.4	261.0	41.5
Phenyl isocyanate	C_7H_5NO	10.6	36.0	48.5	62.5	77.7	87.7	100.6	120.8	142.7	165.6	
isocyanide	C_7H_5N	12.0	37.0	49.7	63.4	78.3	88.0	101.0	120.8	142.3	165.0	
Phenylcyclohexane	$C_{12}H_{16}$	67.5	96.5	111.3	126.4	144.0	154.2	169.3	191.3	214.6	240.0	+75
Phenyl dichlorophosphate	$C_6H_5Cl_2O_2P$	66.7	95.9	110.0	125.9	143.4	153.6	168.0	189.8	213.0	239.5	
m-Phenylene diamine (1,3-phenylenediamine)	$C_6H_8N_2$	99.8	131.2	147.0	163.8	182.5	194.0	209.9	233.0	259.0	285.5	62.8
Phenylglyoxal	$C_8H_6O_2$		75.0	87.8	100.7	115.5	124.2	136.2	153.8	173.5	193.5	73
Phenylhydrazine	$C_6H_8N_2$	75.8	101.6	115.8	131.5	148.2	158.7	173.5	195.4	218.2	243.5	19.5
N-Phenyliminodiethanol	$C_{10}H_{15}NO_2$	145.0	170.2	195.8	213.4	233.0	245.3	260.6	284.5	311.3	337.8	
1-Phenyl-1,3-pentanedione	$C_{11}H_{12}O_2$	98.0	128.5	144.0	159.9	178.0	189.8	204.5	226.7	251.2	276.5	
2-Phenylphenol	$C_{12}H_{10}O$	100.0	131.6	146.2	163.3	180.3	192.2	205.9	227.9	251.8	275.0	56.5
4-Phenylphenol	$C_{12}H_{10}O$			176.2	193.8	213.0	225.3	240.9	263.2	285.5	308.0	164.5
3-Phenyl-1-propanol	$C_9H_{12}O$	74.7	102.4	116.0	131.2	147.4	156.8	170.3	191.2	212.8	235.0	

Phenyl isothiocyanate	C_7H_5NS	47.2	75.6	89.8	115.5	122.5	133.3	147.7	169.6	194.0	218.5	−21.0
Phorone	$C_9H_{14}O$	42.0	63.3	81.5	95.6	111.3	121.4	134.0	153.5	175.3	197.2	28
iso-Phorone	$C_9H_{14}O$	38.0	66.7	81.2	96.8	114.5	125.6	140.6	163.3	188.7	215.2	
Phosgene (carbonyl chloride)	CCl_2O	−92.9	−77.0	−69.3	−60.3	−50.3	−44.0	−35.6	−22.3	−7.6	+8.3	−104
Phthalic anhydride	$C_8H_4O_3$	96.5	124.3	134.0	151.7	172.0	185.3	202.3	228.0	256.8	284.5	130.8
Phthalide	$C_8H_6O_2$	95.5	127.7	144.0	161.3	181.0	193.5	210.0	234.5	261.8	290.0	73
Phthaloyl chloride	$C_8H_4Cl_2O_2$	86.3	118.3	134.2	151.0	170.0	182.2	197.8	222.0	248.3	275.8	88.5
2-Picoline	C_6H_7N	−11.1	+12.6	24.4	37.4	51.2	59.9	71.4	89.0	108.4	128.8	−70
Pimelic acid	$C_7H_{12}O_4$	163.4	196.2	212.0	229.3	247.0	258.2	272.0	294.5	318.5	342.1	103
α-Pinene	$C_{10}H_{16}$	−1.0	+24.6	37.3	51.4	66.8	76.8	90.1	110.2	132.3	155.0	−55
β-Pinene	$C_{10}H_{16}$	+4.2	30.0	42.3	58.1	71.5	81.2	94.0	114.1	136.1	158.3	
Piperidine	$C_5H_{11}N$		−7.0	+3.9	15.8	29.2	37.7	49.0	66.2	85.7	106.0	−9
Piperonal	$C_8H_6O_3$	87.0	117.4	132.0	148.0	165.7	177.0	191.7	214.3	238.5	263.0	37
Propane	C_3H_8	−128.9	−115.4	−108.5	−100.9	−92.4	−87.0	−79.6	−68.4	−55.6	−42.1	−187.1
Propenylbenzene	C_9H_{10}	17.5	43.8	57.0	71.5	87.7	97.8	111.7	132.0	154.7	179.0	−30.1
Propionamide	C_3H_7NO	65.0	91.0	105.0	119.0	134.8	144.3	156.0	174.2	194.0	213.0	79
Propionic acid	$C_3H_6O_2$	4.6	28.0	39.7	52.0	65.8	74.1	85.8	102.5	122.0	141.1	−22
anhydride	$C_6H_{10}O_3$	20.6	45.3	57.7	70.4	85.6	94.5	107.2	127.8	146.0	167.0	−45
Propionitrile	C_3H_5N	−35.0	−13.6	−3.0	+8.8	22.0	30.1	41.4	58.2	77.7	97.1	−91.9
Propiophenone	$C_9H_{10}O$	50.0	77.9	92.2	107.6	124.3	135.0	149.3	170.2	194.2	218.0	21
n-Propyl acetate	$C_5H_{10}O_2$	−26.7	−5.4	+5.0	16.0	28.8	37.0	47.8	64.0	82.0	101.8	−92.5
iso-Propyl acetate	$C_5H_{10}O_2$	−38.3	−17.4	−7.2	+4.2	17.0	25.1	35.7	51.7	69.8	89.0	
n-Propyl alcohol (1-propanol)	C_3H_8O	−15.0	+5.0	14.7	25.3	36.4	43.5	52.8	66.8	82.0	97.8	−127
iso-Propyl alcohol (2-propanol)	C_3H_8O	−26.1	−7.0	+2.4	12.7	23.8	30.5	39.5	53.0	67.8	82.5	−85.8
n-Propylamine	C_3H_9N	−64.4	−46.3	−37.2	−27.1	−16.0	−9.0	+0.5	15.0	31.5	48.5	−83
Propylbenzene	C_9H_{12}	6.3	31.3	43.4	56.8	71.6	81.1	94.0	113.5	135.7	159.2	−99.5
Propyl benzoate	$C_{10}H_{12}O_2$	54.6	83.8	98.0	114.3	131.8	143.3	157.4	180.1	205.2	231.0	−51.6
n-Propyl bromide (1-bromopropane)	C_3H_7Br	−53.0	−33.4	−23.3	−12.4	−0.3	+7.5	18.0	34.0	52.0	71.0	−109.9
iso-Propyl bromide (2-bromopropane)	C_3H_7Br	−61.8	−42.5	−32.8	−22.0	−10.1	−2.5	+8.0	23.8	41.5	60.0	−89.0
n-Propyl n-butyrate	$C_7H_{14}O_2$	−1.6	+22.1	34.0	47.0	61.5	70.3	82.6	101.0	121.7	142.7	−95.2
isobutyrate	$C_7H_{14}O_2$	−6.2	+16.8	28.3	40.6	54.3	63.0	73.9	91.8	112.0	133.9	
iso-Propyl isobutyrate	$C_7H_{14}O_2$	−16.3	+5.8	17.0	29.0	42.4	51.4	62.3	80.2	100.0	120.5	
Propyl carbamate	$C_4H_9NO_2$	52.4	77.6	90.0	103.2	117.7	126.5	138.3	155.8	175.8	195.0	
n-Propyl chloride (1-chloropropane)	C_3H_7Cl	−68.3	−50.0	−41.0	−31.0	−19.5	−12.1	−2.5	+12.2	29.4	46.4	−112.8
iso-Propyl chloride (2-chloropropane)	C_3H_7Cl	−78.8	−61.1	−52.0	−42.0	−31.0	−23.5	−13.7	+1.3	18.1	36.5	−117
iso-Propyl chloroacetate	$C_5H_9ClO_2$	+3.8	28.1	40.2	53.9	68.7	78.0	90.3	108.8	128.0	148.6	
Propyl chloroglyoxylate	$C_5H_7ClO_3$	9.7	32.3	43.5	55.6	68.8	77.2	88.0	104.7	123.0	150.0	

(Continued)

TABLE 2.37 Boiling Points of Common Organic Compounds at Selected Pressures (*Continued*)

Compound		Pressure, mm Hg										Melting Point, °C
		1	5	10	20	40	60	100	200	400	760	
Name	Formula	Temperature, °C										
Propylene	C_3H_6	−131.9	−120.7	−112.1	−104.7	−96.5	−91.3	−84.1	−73.3	−60.9	−47.7	−185
Propylene glycol (1,2-Propanediol)	$C_3H_8O_2$	45.5	70.8	83.2	96.4	111.2	119.9	132.0	149.7	168.1	188.2	
Propylene oxide	C_3H_6O	−75.0	−57.8	−49.0	−39.3	−28.4	−21.3	−12.0	+2.1	17.8	34.5	−112.1
n-Propyl formate	$C_4H_8O_2$	−43.0	−22.7	−12.6	−1.7	+10.8	18.8	29.5	45.3	62.6	81.3	−92.9
iso-Propyl formate	$C_4H_8O_2$	−52.0	−32.7	−22.7	−12.1	−0.2	+7.5	17.8	33.6	50.5	68.3	
4,4′-iso-Propylidenebisphenol	$C_{15}H_{16}O_2$	193.0	224.2	240.8	255.5	273.0	282.9	297.0	317.5	339.0	360.5	
n-Propyl iodide (1-iodopropane)	C_3H_7I	−36.0	−13.5	−2.4	+10.0	23.6	32.1	43.8	61.8	81.8	102.5	−98.8
iso-Propyl iodide (2-iodopropane)	C_3H_7I	−43.3	−22.1	−11.7	0.0	+13.2	21.6	32.8	50.0	69.5	89.5	−90
n-Propyl levulinate	$C_8H_{14}O_3$	59.7	86.3	99.9	114.0	130.1	140.6	154.0	175.6	198.0	221.2	
iso-Propyl levulinate	$C_8H_{14}O_3$	48.0	74.5	88.0	102.4	118.1	127.8	141.8	161.6	185.2	208.2	
Propyl mercaptan (1-propanethiol)	C_3H_8S	−56.0	−36.3	−26.3	−15.4	−3.2	+4.6	15.3	31.5	49.2	67.4	−112
2-iso-Propylnaphthalene	$C_{13}H_{14}$	76.0	107.9	123.4	140.3	159.0	171.4	187.6	211.8	238.5	266.0	
iso-Propyl β-naphthyl ketone (2-isobutyronaphthone)	$C_{14}H_{14}O$	133.2	165.4	181.0	197.7	215.6	227.0	242.3	264.0	288.2	313.0	
2-iso-Propylphenol	$C_9H_{12}O$	56.6	83.8	97.0	111.7	127.5	137.7	150.3	170.1	192.6	214.5	15.5
3-iso-Propylphenol	$C_9H_{12}O$	62.0	90.3	104.1	119.8	136.2	146.6	160.2	182.0	205.0	228.0	26
4-iso-Propylphenol	$C_9H_{12}O$	67.0	94.7	108.0	123.4	139.8	149.7	163.3	184.0	206.1	228.2	61
Propyl propionate	$C_6H_{12}O_2$	−14.2	+8.0	19.4	31.6	45.0	53.8	65.2	82.7	102.0	122.4	−76
4-iso-Propylstyrene	$C_{11}H_{14}$	34.7	62.3	76.0	91.2	108.0	118.4	132.8	153.9	178.0	202.5	
Propyl isovalerate	$C_8H_{16}O_2$	+8.0	32.8	45.1	58.0	72.8	82.3	95.0	113.9	135.0	155.9	
Pulegone	$C_{10}H_{16}O$	58.3	82.5	94.0	106.8	121.7	130.2	143.1	162.5	189.8	221.0	
Pyridine	C_5H_5N	−18.9	+2.5	13.2	24.8	38.0	46.8	57.8	75.0	95.6	115.4	−42
Pyrocatechol	$C_6H_6O_2$		104.0	118.3	134.0	150.6	161.7	176.0	197.7	221.5	245.5	105
Pyrocaltechol diacetate (1,2-phenylene diacetate)	$C_{10}H_{10}O_4$	98.0	129.8	145.7	161.8	179.8	191.6	206.5	228.7	253.3	278.0	
Pyrogallol	$C_6H_6O_3$		151.7	167.7	185.3	204.2	216.3	232.0	255.3	281.5	309.0	133
Pyrotartaric anhydride	$C_5H_6O_3$	69.7	99.7	114.2	130.0	147.8	158.6	173.8	196.1	221.0	247.4	
Pyruvic acid	$C_3H_4O_3$	21.4	45.8	57.9	70.8	85.3	94.1	106.5	124.7	144.7	165.0	13.
Quinoline	C_9H_7N	59.7	89.6	103.8	119.8	136.7	148.1	163.2	186.2	212.3	237.7	−15.
iso-Quinoline	C_9H_7N	63.5	92.7	107.8	123.7	141.6	152.0	167.6	190.0	214.5	240.5	24.
Resorcinol	$C_6H_6O_2$	108.4	138.0	152.1	168.0	185.3	195.8	209.8	230.8	253.4	276.5	110.

Safrole	$C_{10}H_{10}O_2$	63.8	93.0	107.6	123.0	140.1	150.3	165.1	186.2	210.0	233.0	11
Salicylaldehyde	$C_7H_6O_2$	33.0	60.1	73.8	88.7	105.2	115.7	129.4	150.0	173.7	196.5	−7
Salicylic acid	$C_7H_6O_3$	113.7	136.0	146.2	156.8	172.2	182.0	193.4	210.0	230.5	256.0	159
Sebacic acid	$C_{10}H_{18}O_4$	183.0	215.7	232.0	250.0	268.2	279.8	294.5	313.2	332.8	352.3	134.
Selenophene	C_4H_4Se	−39.0	−16.0	−4.0	+9.1	24.1	33.8	47.0	66.7	89.8	114.3	
Skatole	C_9H_9N	95.0	124.2	139.6	154.3	171.9	183.6	197.4	218.8	242.5	266.2	95
Stearaldehyde	$C_{18}H_{36}O$	140.0	174.6	192.1	210.6	230.8	244.2	260.0	285.0	313.8	342.5	63.5
Stearic acid	$C_{18}H_{36}O_2$	173.7	209.0	225.0	243.4	263.3	275.5	291.0	316.5	343.0	370.0	69.3
Stearyl alcohol (1-octadecanol)	$C_{18}H_{36}O$	150.3	185.6	202.0	220.0	240.4	252.7	269.4	293.5	320.3	349.5	58.5
Styrene	C_8H_8	−7.0	+18.0	30.8	44.6	59.8	69.5	82.0	101.3	122.5	145.2	−30.6
Styrene dibromide [(1,2-dibromoethyl) benzene]	$C_8H_8Br_2$	86.0	115.6	129.8	145.2	161.8	172.2	186.3	207.8	230.0	245.0	
Suberic acid	$C_8H_{14}O_4$	172.8	205.5	219.5	238.2	254.6	265.4	279.8	300.5	322.8	345.5	142
Succinic anhydride	$C_4H_4O_3$	92.0	115.0	128.2	145.3	163.0	174.0	189.0	212.0	237.0	261.0	119.6
Succinimide	$C_4H_5NO_2$	115.0	143.2	157.0	174.0	192.0	203.0	217.4	240.0	263.5	287.5	125.5
Succinyl chloride	$C_4H_4Cl_2O_2$	39.0	65.0	78.0	91.8	107.5	117.2	130.0	149.3	170.0	192.5	17
α-Terpineol	$C_{10}H_{18}O$	52.8	80.4	94.3	109.8	126.0	136.3	150.1	171.2	194.3	217.5	35
Terpenoline	$C_{10}H_{16}$	32.3	58.0	70.6	84.8	100.0	109.8	122.7	142.0	163.5	185.0	
1,1,1,2-Tetrabromoethane	$C_2H_2Br_4$	58.0	83.3	95.7	108.5	123.2	132.0	144.0	161.5	181.0	200.0	
1,1,2,2-Tetrabromoethane	$C_2H_2Br_4$	65.0	95.5	110.0	126.0	144.0	155.1	170.0	192.5	217.5	243.5	
Tetraisobutylene	$C_{16}H_{32}$	63.8	93.7	108.5	124.5	142.2	152.6	167.5	190.0	214.6	240.0	
Tetracosane	$C_{24}H_{50}$	183.8	219.6	237.6	255.3	276.3	288.4	305.2	330.5	358.0	386.4	51.1
1,2,3,4-Tetrachlorobenzene	$C_6H_2Cl_4$	68.5	99.6	114.7	131.2	149.2	160.0	175.7	198.0	225.5	254.0	46.5
1,2,3,5-Tetrachlorobenzene	$C_6H_2Cl_4$	58.2	89.0	104.1	121.6	140.0	152.0	168.0	193.7	220.0	246.0	54.5
1,2,4,5-Tetrachlorobenzene	$C_6H_2Cl_4$					146.0	157.7	173.5	196.0	220.5	245.0	139
1,1,2,2-Tetrachloro-1,2-difluoroethane	$C_2Cl_4F_2$	−37.5	−16.0	−5.0	+6.7	19.8	28.1	33.6	55.0	73.1	92.0	2
1,1,1,2-Tetrachloroethane	$C_2H_2Cl_4$	−16.3	+7.4	19.3	32.1	46.7	56.0	68.0	87.2	108.2	130.5	−6
1,1,2,2-Tetrachloroethane	$C_2H_2Cl_4$	−3.8	+20.7	33.0	46.2	60.8	70.0	83.2	102.2	124.0	145.9	
1,2,3,5-Tetrachloro-4-ethylbenzene	$C_8H_6Cl_4$	77.0	110.0	126.0	143.7	162.1	175.0	191.6	215.3	243.0	270.0	−3
Tetrachloroethylene	C_2Cl_4	−20.6	+2.4	13.8	26.3	40.1	49.2	61.3	79.8	100.0	120.8	−1
2,3,4,6-Tetrachlorophenol	$C_6H_2Cl_4O$	100.0	130.3	145.5	161.0	179.1	190.0	205.2	227.2	250.4	275.0	69
3,4,5,6-Tetrachloro-1,2-xylene	$C_8H_6Cl_4$	94.4	125.0	140.3	156.0	174.2	185.8	200.5	223.0	248.3	273.5	
Tetradecane	$C_{14}H_{30}$	76.4	106.0	120.7	135.6	152.7	164.0	178.5	201.8	226.8	252.5	5
Tetradecylamine	$C_{14}H_{31}N$	102.6	135.8	152.0	170.0	189.0	200.2	215.7	239.8	264.6	291.2	
Tetradecyltrimethylsilane	$C_{17}H_{38}Si$	120.0	150.7	166.2	183.5	201.5	213.3	227.8	250.0	275.0	300.0	
Tetraethoxysilane	$C_8H_{20}O_4Si$	16.0	40.3	52.6	65.8	81.1	90.7	103.6	123.5	146.2	168.5	
1,2,3,4-Tetraethylbenzene	$C_{14}H_{22}$	65.7	96.2	111.6	127.7	145.8	156.7	172.4	196.0	221.4	248.0	11

(*Continued*)

TABLE 2.37 Boiling Points of Common Organic Compounds at Selected Pressures (*Continued*)

Compound		Pressure, mm Hg										Melting Point, °C
		1	5	10	20	40	60	100	200	400	760	
Name	Formula	Temperature, °C										
Tetraethylene glycol	$C_8H_{18}O_5$	153.9	183.7	197.1	212.3	228.0	237.8	250.0	268.4	288.0	307.8	
Tetraethylene glycol chlorohydrin	$C_8H_{17}ClO_4$	110.1	141.8	156.1	172.6	190.0	200.5	214.7	236.5	258.2	281.5	
Tetraethyllead	$C_8H_{20}Pb$	38.4	63.6	74.8	88.0	102.4	111.7	123.8	142.0	161.8	183.0	−136
Tetraethylsilane	$C_8H_{20}Si$	−1.0	+23.9	36.3	50.0	65.3	74.8	88.0	108.0	130.2	153.0	
Tetralin	$C_{10}H_{12}$	38.0	65.3	79.0	93.8	110.4	121.3	135.3	157.2	181.8	207.2	−31
1,2,3,4-Tetramethylbenzene	$C_{10}H_{14}$	42.6	68.7	81.8	95.8	111.5	121.8	135.7	155.7	180.0	204.4	−6
1,2,3,5-Tetramethylbenzene	$C_{10}H_{14}$	40.6	65.8	77.8	91.0	105.8	115.4	128.3	149.9	173.7	197.9	−24
1,2,4,5-Tetramethylbenzene	$C_{10}H_{14}$	45.0	65.0	74.6	88.0	104.2	114.8	128.1	149.5	172.1	195.9	79
2,2,3,3-Tetramethylbutane	C_8H_{18}	−17.4	+3.2	13.5	24.6	36.8	44.5	54.8	70.2	87.4	106.3	−102
Tetramethylene dibromide (1,4-dibromobutane)	$C_4H_8Br_2$	32.0	58.8	72.4	87.6	104.0	115.1	128.7	149.8	173.8	197.5	−20
Tetramethyllead	$C_4H_{12}Pb$	−29.0	−6.8	+4.4	16.6	30.3	39.2	50.8	68.8	89.0	110.0	−27
Tetramethyltin	$C_4H_{12}Sn$	−51.3	−31.0	−20.6	−9.3	+3.5	11.7	22.8	39.8	58.5	78.0	
Tetrapropylene glycol monoisopropyl ether	$C_{15}H_{32}O_{15}$	116.6	147.8	163.0	179.8	197.7	209.0	223.3	245.0	268.3	292.7	
Thioacetic acid (mercaptoacetic acid)	$C_2H_4O_2S$	60.0	87.7	101.5	115.8	131.8	142.0	154.0				−16.5
Thiodiglycol (2,2′-thiodiethanol)	$C_4H_{10}O_2S$	42.0	96.0	128.0	165.0	210.0	240.5	285				
Thiophene	C_4H_4S	−40.7	−20.8	−10.9	0.0	+12.5	20.1	30.5	46.5	64.7	84.4	−38.3
Thiophenol (benzenethiol)	C_6H_6S	18.6	43.7	56.0	69.7	84.2	93.9	106.6	125.8	146.7	168.0	
α-Thujone	$C_{10}H_{16}O$	38.3	65.7	79.3	93.7	110.0	120.2	134.0	154.2	177.8	201.0	
Thymol	$C_{10}H_{14}O$	64.3	92.8	107.4	122.6	139.8	149.8	164.1	185.5	209.6	231.8	
Tiglaldehyde	C_5H_8O	−25.0	−1.6	+10.0	23.2	37.0	45.8	57.7	75.4	95.5	116.8	51
Tiglic acid	$C_5H_8O_2$	52.0	77.8	90.2	103.8	119.0	127.8	140.5	158.0	179.2	198.5	64
Tiglonitrile	C_5H_7N	−25.5	−2.4	+9.2	22.1	36.7	46.0	58.2	77.8	99.7	122.0	
Toluene	C_7H_8	−26.7	−4.4	+6.4	18.4	31.8	40.3	51.9	69.5	89.5	110.6	−95
Toluene-2,4-diamine	$C_7H_{10}N_2$	106.5	137.2	151.7	167.9	185.7	196.2	211.5	232.8	256.0	280.0	99
2-Toluic nitrile (2-tolunitrile)	C_8H_7N	36.7	64.0	77.9	93.0	110.0	120.8	135.0	156.0	180.0	205.2	−13
4-Toluic nitrile (4-tolunitrile)	C_8H_7N	42.5	71.3	85.8	101.7	109.5	130.0	145.2	167.3	193.0	217.6	29
2-Toluidine	C_7H_9N	44.0	69.3	81.4	95.1	110.0	119.8	133.0	153.0	176.2	199.7	−16
3-Toluidine	C_7H_9N	41.0	68.0	82.0	96.7	113.5	123.8	136.7	157.6	180.6	203.3	−31
4-Toluidine	C_7H_9N	42.0	68.2	81.8	95.8	111.5	121.5	133.7	154.0	176.9	200.4	44
2-Tolyl isocyanide	C_8H_7N	25.2	51.0	64.0	78.2	94.0	104.0	117.7	137.8	159.9	183.5	

4-Tolylhydrazine	$C_7H_{10}N_2$	82.4	110.0	123.8	138.6	154.1	165.0	178.0	198.0	219.5	242.0	65.5
Tribromoacetaldehyde	C_2HBr_3O	18.5	45.0	58.0	72.1	87.8	97.5	110.2	130.0	151.6	174.0	
1,1,2-Tribromobutane	$C_4H_7Br_3$	45.0	73.5	87.8	103.2	120.2	131.6	146.0	167.8	192.0	216.2	
1,2,2-Tribromobutane	$C_4H_7Br_3$	41.0	69.0	83.2	98.6	116.0	127.0	141.8	163.5	188.0	213.8	
2,2,3-Tribromobutane	$C_4H_7Br_3$	38.2	66.0	79.8	94.6	111.8	122.2	136.3	157.8	182.2	206.5	
1,1,2-Tribromoethane	$C_2H_3Br_3$	32.6	58.0	70.6	84.2	100.0	110.0	123.5	143.5	165.4	188.4	−26
1,2,3-Tribromopropane	$C_3H_5Br_3$	47.5	75.8	90.0	105.8	122.8	134.0	148.0	170.0	195.0	220.0	16.5
Triisobutylamine	$C_{12}H_{27}N$	32.3	57.4	69.8	83.0	97.8	107.3	119.7	138.0	157.8	179.0	−22
Triisobutylene	$C_{12}H_{24}$	18.0	44.0	56.5	70.0	86.7	96.7	110.0	130.2	153.0	179.0	
2,4,6-Tritertbutylphenol	$C_{18}H_{30}O$	95.2	126.1	142.0	158.0	177.4	188.0	203.0	226.2	250.6	276.3	
Trichloroacetic acid	$C_2HCl_3O_2$	51.0	76.0	88.2	101.8	116.3	125.9	137.8	155.4	175.2	195.6	57
Trichloroacetic anhydride	$C_4Cl_6O_3$	56.2	85.3	99.6	114.3	131.2	141.8	155.2	176.2	199.8	223.0	
Trichloroacetyl bromide	C_2BrCl_3O	−7.4	+16.7	29.3	42.1	57.2	66.7	79.5	98.4	120.2	143.0	
2,4,6-Trichloroaniline	$C_6H_4Cl_3N$	134.0	157.8	170.0	182.6	195.8	204.5	214.6	229.8	246.4	262.0	78
1,2,3-Trichlorobenzene	$C_6H_3Cl_3$	40.0	70.0	85.6	101.8	119.8	131.5	146.0	168.2	193.5	218.5	52.5
1,2,4-Trichlorobenzene	$C_6H_3Cl_3$	38.4	67.3	81.7	97.2	114.8	125.7	140.0	162.0	187.7	213.0	17
1,3,5-Trichlorobenzene	$C_6H_3Cl_3$		63.8	78.0	93.7	110.8	121.8	136.0	157.7	183.0	208.4	63.5
1,2,3-Trichlorobutane	$C_4H_7Cl_3$	+0.5	27.2	40.0	55.0	71.5	82.0	96.2	118.0	143.0	169.0	
1,1,1-Trichloroethane	$C_2H_3Cl_3$	−52.0	−32.0	−21.9	−10.8	+1.6	9.5	20.0	36.2	54.6	74.1	−30.6
1,1,2-Trichloroethane	$C_2H_3Cl_3$	−24.0	−2.0	+8.3	21.6	35.2	44.0	55.7	73.3	93.0	113.9	−36.7
Trichloroethylene	C_2HCl_3	−43.8	−22.8	−12.4	−1.0	+11.9	20.0	31.4	48.0	67.0	86.7	−73
Trichlorofluoromethane	CCl_3F	−84.3	−67.6	−59.0	−49.7	−39.0	−32.3	−23.0	−9.1	+6.8	23.7	
2,4,5-Trichlorophenol	$C_6H_3Cl_3O$	72.0	102.1	117.3	134.0	151.5	162.5	178.0	201.5	226.5	251.8	62
2,4,6-Trichlorophenol	$C_6H_3Cl_3O$	76.5	105.9	120.2	135.8	152.2	163.5	177.8	199.0	222.5	246.0	68.
Tri-2-chlorophenylthiophosphate	$C_{18}H_{12}Cl_3O_3PS$	188.2	217.2	231.2	246.7	261.7	271.5	283.8	302.8	322.0	341.3	
1,1,1-Trichloropropane	$C_3H_5Cl_3$	−28.8	−7.0	+4.2	16.2	29.9	38.3	50.0	67.7	87.5	108.2	−77.
1,2,3-Trichloropropane	$C_3H_5Cl_3$	+9.0	33.7	46.0	59.3	74.0	83.6	96.1	115.6	137.0	158.0	−14.
1,1,2-Trichloro-1,2,2-trifluoroethane	$C_2Cl_3F_3$	−68.0	−49.4	−40.3	−30.0	−18.5	−11.2	−1.7	+13.5	30.2	47.6	−35
Tricosane	$C_{23}H_{48}$	170.0	206.3	223.0	242.0	261.3	273.8	289.8	313.5	339.8	366.5	47.
Tridecane	$C_{13}H_{28}$	59.4	98.3	104.0	120.2	137.7	148.2	162.5	185.0	209.4	234.0	−6.2
Tridecanoic acid	$C_{13}H_{26}O_2$	137.8	166.3	181.0	195.8	212.4	222.0	236.0	255.2	276.5	299.0	41
Triethoxymethylsilane	$C_7H_{18}O_3Si$	−1.5	+22.8	34.6	47.2	61.7	70.4	82.7	101.0	121.8	143.5	
Triethoxyphenylsilane	$C_{12}H_{20}O_3Si$	71.0	98.8	112.6	127.2	143.5	153.2	167.5	188.0	210.5	233.5	
1,2,4-Triethylbenzene	$C_{12}H_{18}$	46.0	74.2	88.5	104.0	121.7	132.2	146.8	168.3	193.7	218.0	
1,3,4-Triethylbenzene	$C_{12}H_{18}$	47.9	76.0	90.2	105.8	122.6	133.4	147.7	168.3	193.2	217.5	

(Continued)

TABLE 2.37 Boiling Points of Common Organic Compounds at Selected Pressures (*Continued*)

Compound		Pressure, mm Hg										Melting Point, °C
		1	5	10	20	40	60	100	200	400	760	
Name	Formula	Temperature, °C										
Triethylborine	$C_6H_{15}B$			−148.0	−140.6	−131.4	−125.2	−116.0	−101.0	−81.0	−56.2	
Triethyl camphoronate	$C_{15}H_{26}O_6$		150.2	166.0	183.6	201.8	213.5	228.6	250.8	276.0	301.0	135
citrate	$C_{12}H_{20}O_7$	107.0	138.7	144.0	171.1	190.4	202.5	217.8	242.2	267.5	294.0	
Triethyleneglycol	$C_6H_{14}O_4$	114.0	144.0	158.1	174.0	191.3	201.5	214.6	235.2	256.6	278.3	
Triethylheptylsilane	$C_{13}H_{30}Si$	70.0	99.8	114.6	130.3	148.0	158.2	174.0	196.0	221.0	247.0	
Triethyloctylsilane	$C_{14}H_{32}Si$	73.7	104.8	120.6	137.7	155.7	168.0	184.3	208.0	235.0	262.0	
Triethyl orthoformate	$C_7H_{16}O_3$	+5.5	29.2	40.5	53.4	67.5	76.0	88.0	106.0	125.7	146.0	
phosphate	$C_6H_{15}O_4P$	39.6	67.8	82.1	97.8	115.7	126.3	141.6	163.7	187.0	211.0	
Triethylthallium	$C_6H_{15}Tl$	+9.3	37.6	51.7	67.7	85.4	95.7	112.1	136.0	163.5	192.1	−63.0
Trifluorophenylsilane	$C_6H_5F_3Si$	−31.0	−9.7	+0.8	12.3	25.4	33.2	44.2	60.1	78.7	98.3	
Trimethallyl phosphate	$C_{12}H_{21}PO_4$	93.7	131.0	149.8	169.8	192.0	207.0	225.7	255.0	288.5	324.0	
2,3,5-Trimethylacetophenone	$C_{11}H_{14}O$	79.0	108.0	122.3	137.5	154.2	165.7	179.7	201.3	224.3	247.5	
Trimethylamine	C_3H_9N	−97.1	−81.7	−73.8	−65.0	−55.2	−48.8	−40.3	−27.0	−12.5	+2.9	−117
2,4,5-Trimethylaniline	$C_9H_{13}N$	68.4	95.9	109.0	123.7	139.8	149.5	162.0	182.3	203.7	234.5	67
1,2,3-Trimethylbenzene	C_9H_{12}	16.8	42.9	55.9	69.9	85.4	95.3	108.8	129.0	152.0	176.1	−25
1,2,4-Trimethylbenzene	C_9H_{12}	13.6	38.3	50.7	64.5	79.8	89.5	102.8	122.7	145.4	169.2	−44
1,3,5-Trimethylbenzene	C_9H_{12}	9.6	34.7	47.4	61.0	76.1	85.8	98.9	118.6	141.0	164.7	−44
2,2,3-Trimethylbutane	C_7H_{16}			−18.8	−7.5	+5.2	13.3	24.4	41.2	60.4	80.9	−25.
Trimethyl citrate	$C_9H_{14}O_7$	106.2	146.2	160.4	177.2	194.2	205.5	219.6	241.3	264.2	287.0	78.
Trimethyleneglycol (1,3-propandiol)	$C_3H_8O_2$	59.4	87.2	100.6	115.5	131.0	141.1	153.4	172.8	193.8	214.2	
1,2,4-Trimethyl-5-ethylbenzene	$C_{11}H_{16}$	43.7	71.2	84.6	99.7	106.0	126.3	140.3	160.3	184.5	208.1	
1,3,5-Trimethyl-2-ethylbenzene	$C_{11}H_{16}$	38.8	67.0	80.5	96.0	113.2	123.8	137.9	158.4	183.5	208.0	
2,2,3-Trimethylpentane	C_8H_{18}	−29.0	−7.1	+3.9	16.0	29.5	38.1	49.9	67.8	88.2	109.8	−112.
2,2,4-Trimethylpentane	C_8H_{18}	−36.5	−15.0	−4.3	+7.5	20.7	29.1	40.7	58.1	78.0	99.2	−107.
2,3,3-Trimethylpentane	C_8H_{18}	−25.8	−3.9	+6.9	19.2	33.0	41.8	53.8	72.0	92.7	114.8	−101.
2,3,4-Trimethylpentane	C_8H_{18}	−26.3	−4.1	+7.1	19.3	32.9	41.6	53.4	71.3	91.8	113.5	−109.
2,2,4-Trimethyl-3-pentanone	$C_8H_{16}O$	14.7	36.0	46.4	57.6	69.8	77.3	87.6	102.2	118.4	135.0	
Trimethyl phosphate	$C_3H_9O_4P$	26.0	53.7	67.8	83.0	100.0	110.0	124.0	145.0	167.8	192.7	
2,4,5-Trimethylstryene	$C_{11}H_{14}$	48.1	77.0	91.6	107.1	124.2	135.5	149.8	171.8	196.1	221.2	
2,4,6-Trimethylsytrene	$C_{11}H_{14}$	37.5	65.7	79.7	94.8	111.8	122.3	136.8	157.8	182.3	207.0	
Trimethylsuccinic anhydride	$C_7H_{10}O_3$	53.5	82.6	97.4	113.8	131.0	142.2	156.5	179.8	205.5	231.0	

Triphenylmethane	$C_{19}H_{16}$	169.7	188.4	197.0	206.8	215.5	221.2	228.4	239.7	249.8	259.2	93.4
Triphenylphosphate	$C_{18}H_{15}O_4P$	193.5	230.4	249.8	269.7	290.3	305.2	322.5	349.8	379.2	413.5	49.4
Tripropyleneglycol	$C_9H_{20}O_4$	96.0	125.7	140.5	155.8	173.7	184.6	199.0	220.2	244.3	267.2	
Tripropyleneglycol monobutyl ether	$C_{13}H_{28}O_4$	101.5	131.6	147.0	161.8	179.8	190.2	204.4	224.4	247.0	269.5	
Tripropyleneglycol monoisopropyl ether	$C_{12}H_{26}O_4$	82.4	112.4	127.3	143.7	161.4	173.2	187.8	209.7	232.8	256.6	
Tritolyl phosphate	$C_{21}H_{21}O_4P$	154.6	184.2	198.0	213.2	229.7	239.8	252.2	271.8	292.7	313.0	
Undecane	$C_{11}H_{24}$	32.7	59.7	73.9	85.6	104.4	115.2	128.1	149.3	171.9	195.8	−25.6
Undecanoic acid	$C_{11}H_{22}O_2$	101.4	133.1	149.0	166.0	185.6	197.2	212.5	237.8	262.8	290.0	29.5
10-Undecenoic acid	$C_{11}H_{20}O_2$	114.0	142.8	156.3	172.0	188.7	199.5	213.5	232.8	254.0	275.0	24.5
Undecan-2-ol	$C_{11}H_{24}O$	71.1	99.0	112.8	127.5	143.7	153.7	167.2	187.7	209.8	232.0	
n-Valeric acid	$C_5H_{10}O_2$	42.2	67.7	79.8	93.1	107.8	116.6	128.3	146.0	165.0	184.4	−34.5
iso-Valeric acid	$C_5H_{10}O_2$	34.5	59.6	71.3	84.0	98.0	107.3	118.9	136.2	155.2	175.1	−37.6
γ-Valerolactone	$C_5H_8O_2$	37.5	65.8	79.8	95.2	101.9	122.4	136.5	157.7	182.3	207.5	
Valeronitrile	C_5H_9N	−6.0	+18.1	30.0	43.3	57.8	66.9	78.6	97.7	118.7	140.8	
Vanillin	$C_8H_8O_3$	107.0	138.4	154.0	170.5	188.7	199.8	214.5	237.3	260.0	285.0	81.5
Vinyl acetate	$C_4H_6O_2$	−48.0	−28.0	−18.0	−7.0	+5.3	13.0	23.3	38.4	55.5	72.5	
2-Vinylanisole	$C_9H_{10}O$	41.9	68.0	81.0	94.7	110.0	119.8	132.3	151.0	172.1	194.0	
3-Vinylanisole	$C_9H_{10}O$	43.4	69.9	83.0	97.2	112.5	122.3	135.3	154.0	175.8	197.5	
4-Vinylanisole	$C_9H_{10}O$	45.2	72.0	85.7	100.0	116.0	126.1	139.7	159.0	182.0	204.5	
Vinyl chloride (1-chloroethylene)	C_2H_3Cl	−105.6	−90.8	−83.7	−75.7	−66.8	−61.1	−53.2	−41.3	−28.0	−13.8	−153.7
cyanide (acrylonitrile)	C_3H_3N	−51.0	−30.7	−20.3	−9.0	+3.8	11.8	22.8	38.7	58.3	78.5	−82
fluoride (1-fluoroethylene)	C_2H_3F	−149.3	−138.0	−132.2	−125.4	−118.0	−113.0	−106.2	−95.4	−84.0	−72.2	−160.5
Vinylidene chloride (1,1-dichloroethene)	$C_2H_2Cl_2$	−77.2	−60.0	−51.2	−41.7	−31.1	−24.0	−15.0	−1.0	+14.8	31.7	−122.5
4-Vinylphenetole	$C_{10}H_{12}O$	64.0	91.7	105.6	120.3	136.3	146.4	159.8	180.0	202.8	225.0	
2-Xenyl dichlorophosphate	$C_{12}H_9Cl_2PO$	138.2	171.1	187.0	205.0	223.8	236.0	251.5	275.3	301.5	328.5	
2,4-Xyaldehyde	$C_9H_{10}O$	59.0	85.9	99.0	114.0	129.7	139.8	152.2	172.3	194.1	215.5	75
2-Xylene (2-xylene)	C_8H_{10}	−3.8	+20.2	32.1	45.1	59.5	68.8	81.3	100.2	121.7	144.4	−25.2
3-Xylene (3-xylene)	C_8H_{10}	−6.9	+16.8	28.3	41.1	55.3	64.4	76.8	95.5	116.7	139.1	−47.9
4-Xylene (4-xylene)	C_8H_{10}	−8.1	+15.5	27.3	40.1	54.4	63.5	75.9	94.6	115.9	138.3	+13.3
2,4-Xylidine	$C_8H_{11}N$	52.6	79.8	93.0	107.6	123.8	133.7	146.8	166.4	188.3	211.5	
2,6-Xylidine	$C_8H_{11}N$	44.0	72.6	87.0	102.7	120.2	131.5	146.0	168.0	193.7	217.9	

TABLE 2.38 Organic Solvents Arranged by Boiling Points

Name	BP, °C	Name	BP, °C
Ethylene oxide	10.6	1-Propanol	97.2
Chloroethane	12.3	Heptane	98.4
Furan	31.4	1-Chloro-3-methylbutane	99
Methyl formate	31.5	Ethyl propionate	99.1
Diethyl ether	34.6	2-Butanol	99.6
Propylene oxide	34.5	Formic acid	100.8
Pentane	36.1	Methylcyclohexane	100.9
Bromoethane	38.4	1,4-Dioxane	101.2
Dichloromethane	39.8	Nitromethane	101.2
Dimethoxymethane	42.3	Propyl acetate	101.5
Carbon disulfide	46.3	2-Pentanone	101.7
1-Isopropoxy-2-propanol	47.9	3-Pentanone	102.0
Ethyl formate	54.2	2-Methyl-2-butanol	102.0
Acetone	56.2	1,1-Diethoxyethane	102.7
Methyl acetate	56.3	Butyl formate	106.6
1,1-Dichloroethane	57.3	2-Methyl-1-propanol	107.9
Dichloroethylene	60.6	Toluene	110.6
Chloroform	61.2	*sec*-Butyl acetate	112.3
Methanol	64.7	1,1,2-Trichloroethane	113.5
Tetrahydrofuran	66.0	Nitroethane	114.1
Diisopropyl ether	68.0	Pyridine	115.2
Hexane	68.7	3-Pentanol	115.6
1-Chloro-2-methylpropane	68.9	4-Methyl-2-pentanone	115.7
1,1,1-Trichloroethane	74.0	1-Chloro-2,3-epoxypropane	116.1
1,3-Dioxolane	74–75		
Carbon tetrachloride	76.7	1-Butanol	117.7
Ethyl acetate	77.1	Acetic acid	117.9
1-Chlorobutane	77.9	Isobutyl acetate	118.0
Ethanol	78.3	2-Pentanol	119.3
2-Butanone	79.6	1-Bromo-3-methylbutane	119.7
2-Methyltetrahydrofuran	80.0	1-Methoxy-2-propanol	120.1
Benzene	80.1	2-Nitropropane	120.3
Cyclohexane	80.7	Tetrachloroethylene	121.1
Propyl formate	80.9	Ethyl butyrate	121.6
Acetonitrile	81.6	3-Hexanone	123
2-Propanol	82.4	2,4-Dimethyl-3-pentanone	124
1,1,-Dimethylethanol	82.4	2-Methoxyethanol	124.6
Cyclohexene	83.0	Octane	125.7
Diisopropylamine	83.5	Butyl acetate	126.1
1,2-Dichloroethane	83.7	Diethyl carbonate	126.8
Thiophene	84.2	2-Hexanone	127.2
Trichloroethylene	87.2	1-Chloro-2-propanol	127.4
Isopropyl acetate	88.2	2-Chloroethanol	128.6
1-Bromo-2-methylpropane	91.5	3-Methyl-1-penten-2-one	129.5
2,5-Dimethylfuran	93–94	1-Nitropropane	131.2
Ethyl chloroformate	94	Chlorobenzene	131.7
Allyl alcohol	96.6	1,2-Dibromoethane	131.7
1,2-Dichloropropane	96.8	4-Methyl-2-pentanol	131.7

TABLE 2.38 Organic Solvents Arranged by Boiling Points (*Continued*)

Name	BP, °C	Name	BP, °C
3-Methyl-1-butanol	132.0	Phenol	181.8
Cyclohexylamine	134.8	2-Ethyl-1-hexanol	184.3
2-Ethoxyethanol	134.8	Aniline	184.4
Ethylbenzene	136.2	Benzyl ethyl ether	185.0
1-Pentanol	138	Diethyl oxalate	185.4
p-Xylene	138.4	1,2-Propanediol	188
m-Xylene	139.1	Bis(2-ethoxyethyl) ether	188.4
Acetic anhydride	140.0	Dimethyl sulfoxide	189.0
2,4-Pentanedione	140.6	1,2-Ethanediol diacetate	190.2
Isopentyl acetate	142	Benzonitrile	191.0
Dibutyl ether	142.4	2,5-Hexanedione	191.4
4-Heptanone	143.7	2-(2-Methoxyethoxy)-ethanol	194.1
o-Xylene	144.4		
2-Methoxyethyl acetate	144.5	*N,N*-Dimethylaniline	194.2
1,1,2,2-Tetrachloroethane	146.3	1-Octanol	195.2
3-Heptanone	147.8	1,2-Ethanediol	197.3
Tribromomethane	149.6	Diethyl malonate	199.3
Nonane	150.8	Methyl benzoate	199.5
2-Heptanone	151	*o*-Toluidine	200.4
Isopropylbenzene	152.4	*p*-Toluidine	200.6
N,N-Dimethylformamide	153.0	2-(2-Ethoxyethoxy)-ethanol	202
Methoxybenzene	153.8		
Ethyl lactate	154.5	Acetophenone	202.1
Cyclohexanone	155.7	1,2-Dibutoxyethane	203.6
Bromobenzene	156.2	1-Phenylethanol	203.9
1,2,3-Trichloropropane	156.9	*m*-Toluidine	203.4
1-Hexanol	157.5	Benzyl alcohol	205.5
Propylbenzene	159.2	Camphor	207
Cyclohexanol	161.1	1,3-Butanediol	207.5
Bis(2-methoxyethyl)ether	160	1,2,3,4-Tetrahydro-naphthalene	207.6
Isopentyl propionate	160.2		
2-Heptanol	160.4	γ-Valerolactone	207–208
Pentachloroethane	160.5	*o*-Chloroaniline	208.8
2-Furaldehyde	161.8	Nitrobenzene	210.8
2,6-Dimethyl-4-heptanone	168.1	Ethyl benzoate	212.4
4-Hydroxy-4-methyl-2-pentanone	169.2	3,5,5-Trimethylcyclo-hex-2-en-1-one	215.2
2-Furanmethanol	170.0	Naphthalene	217.7
Ethoxybenzene	170	2-(2-Ethoxyethoxy)ethyl acetate	218.5
2-Butoxyethanol	170.2		
Diisopentyl ether	173.4	Acetamide	221.2
Decane	174.2	Methyl salicylate	223.0
1,3-Dichloro-2-propanol	174.3	Diethyl maleate	225.3
Cyclohexyl acetate	174–175	1,4-Butanediol	230
1-Heptanol	175.8	Propyl benzoate	231.2
Furfuryl acetate	175–177	1-Decanol	230.2
1,3,3-Trimethyl-2-oxabicyclo-[2.2.2]octane	177.4	Phenylacetonitrile	233.5
		Quinoline	237
		Tributyl borate	238.5
4-Isopropyl-1-methylbenzene	177.1	Propylene carbonate	240
		2-Phenoxyethanol	240
Isopentyl butyrate	178.6	Bis(2-hydroxyethyl) ether	245
Bis(2-chloroethyl) ether	178.8	Dibutyl oxalate	245.5
2-Octanol	179	Butyl benzoate	250
1,2-Dichlorobenzene	180.4	1,2,3-Propanetriol triacetate	258–259
Ethyl acetoacetate	180.8		

(*Continued*)

TABLE 2.38 Organic Solvents Arranged by Boiling Points (*Continued*)

Name	BP, °C	Name	BP, °C
1-Chloronaphthalene	259.3	2,2′-(Ethylenedioxy)-bisethanol	285
Isopentyl benzoate	262	Glycerol	290
Bis[2-(methoxyethoxy)-ethyl]ether	275.3	Diethyl *o*-phthalate	295
1-Methoxy-2-nitrobenzene	277	Benzyl benzoate	323.5
Isopentyl salicylate	277–278	Dibutyl *o*-phthalate	340.0
1-Bromonaphthalene	281.1	Dibutyl decanedioate	344–345
Dimethyl *o*-phthalate	283.7		

TABLE 2.39 Boiling Points of n-Paraffins

Carbon number	Boiling point, °C	Boiling point, °F
5	36	97
6	69	156
7	98	209
8	126	258
9	151	303
10	174	345
11	196	385
12	216	421
13	235	456
14	253	488
15	271	519
16	287	548
17	302	576
18	317	602
19	331	627
20	344	651
21	356	674
22	369	696
23	380	716
24	391	736
25	402	755
26	412	774
27	422	792
28	432	809
29	441	825
30	450	841
31	459	858
32	468	874
33	476	889
34	483	901
35	491	916
36	498	928
37	505	941
38	512	958
39	518	964
40	525	977
41	531	988
42	537	999
43	543	1009
44	548	1018

2.6 FLAMMABILITY PROPERTIES

The *flash point* of a substance is the lowest temperature at which the substance gives off sufficient vapor to form an ignitable mixture with air near its surface or within a vessel. The *fire point* is the temperature at which the flame becomes self-sustained and the burning continues. At the flash point, the flame does not need to be sustained. The fire point is usually a few degrees above the flash point. ASTM test methods include procedures using a closed cup (ASTM D-56, ASTM D-93, and ASTM D-3828), which is preferred, and an open cup (ASTM D-92, ASTM D-I310). When several values are available, the lowest temperature is usually taken in order to assure safe operation of the process.

The *ignition temperature* (or *ignition point*) is the minimum temperature required to initiate self-sustained combustion of a substance (solid, liquid, or gaseous) and independent of external ignition sources or heat.

Flash points, lower and upper flammability limits, and auto-ignition temperatures are the three properties that are used to indicate safe operating limits of temperature when processing organic materials. Prediction methods are somewhat erratic, but, together with comparisons with reliable experimental values for families or similar compounds, they are valuable in setting a conservative value for each of the properties.

The upper and lower flammability limits are the boundary-line mixtures of vapor or gas with air, which, if ignited, will just propagate flame and are given in terms of percent by volume of gas or vapor in the air. Each of these limits also has a temperature at which the flammability limits are reached. The temperature corresponding to the lower-limit partial vapor pressure should equal the flash point. The temperature corresponding to the upper-limit partial vapor pressure is somewhat above the lower limit and is usually considerably below the auto-ignition temperature. Flammability limits are calculated at one atmosphere total pressure and are normally considered synonymous with explosive limits. Limits in oxygen rather than air are sometimes measured and available. Limits are generally reported at 298°K and 1 atmosphere. If the temperature or the pressure is increased, the lower limit will decrease while the upper limit will increase, giving a wider range of compositions over which flame will propagate.

The auto-ignition temperature is the minimum temperature for a substance to initiate self-combustion in air in the absence of a spark or flame. The temperature is no lower than and is generally considerably higher than the temperature corresponding to the upper flammability limit. Large differences can occur in reported values determined by different procedures. The lowest reasonable value should be accepted in order to assure safety. Values are also sometimes given in oxygen rather than in air.

One simple method of estimating auto-ignition temperatures is to compare values for a compound with other members of its homologous series on a plot vs. carbon number as the temperature decreases and carbon number increases.

TABLE 2.40 Boiling Points, Flash Points, and Ignition Temperatures of Organic Compounds

Compound	Boiling point °F (°C)	Flash point, °F (°C)	Ignition point, °F (°C)
Acetal $CH_3CH(OC_2H_5)_2$ (Acetaldehydediethylacetal)	215 (102)	−5 (−21)	446 (230)
Acetaldehyde CH_3CHO (Acetic aldehyde) (Ethanal)	70 (21)	−38 (−39)	347 (175)
Acetaldehydediethylacetal	See Acetal.		
Acetaldel	See Aldol.		
Acetanilide $CH_3CONHC_6H_5$	582 (306)	337 (169) (oc)	985 ± 10 (530)
Acetic Acid, Glacial CH_3COOH	245 (118)	103 (39)	867 (463)
Acetic Acid, Isopropyl Ester	See Isopropyl Acetate.		
Acetic Acid, Methyl Ester	See Methyl Acetate.		
Acetic Acid, n-Propyl Ester	See Propyl Acetate.		
Acetic Aldehyde	See Acetaldehyde		
Acetic Anhydride $(CH_3CO)_2O$ (Ethanoic anhydride)	284 (140)	120 (49)	600 (316)
Acetic Ester	See Ethyl Acetate.		
Acetic Ether	See Ethyl Acetate.		
Acetoacetanilide $CH_3COCH_2CONHC_6H_5$		365 (185)	
o-Acetoacet Anisidide $CH_3COCH_2CONHC_6H_4OCH_3$		325 (168)	
Acetoacetic Acid, Ethyl Ester	See Ethyl acetoacetate.		
Acetoethylamide	See N-Ethylacetamide.		
Acetone CH_3COCH_3 (Dimethyl Ketone) (2-Propanone)	133 (56)	−4 (−20)	869 (465)
Acetone Cyanohydrin $(CH_3)_2C(OH)CN$ (2-Hydroxy2-Methyl Propionitrile)	248 (120) Decomposes	165 (74)	1270 (688)
Acetonitrile CH_3CN (Methyl Cyanide)	179 (82)	42 (6)	975 (524)
Acetonyl Acetone $(CH_2COCH_3)_2$ (2,5-Hexanedione)	378 (192)	174 (79)	920 (499)
Acetophenone $C_6H_5COCH_3$ (Phenyl Methyl Ketone)	396 (202)	170 (77)	1058 (570)
p-Acetotoluidide $CH_3CONHC_6H_4CH_3$	583 (306)	334 (168)	
Acetyl Acetone	See 2,4-Pentanedione.		
Acetyl Chloride CH_3COCl (Ethanoyl Chloride)	124 (51)	40 (4)	734 (390)

TABLE 2.40 Boiling Points, Flash Points, and Ignition Temperatures of Organic Compounds (*Continued*)

Compound	Boiling point °F (°C)	Flash point, °F (°C)	Ignition point, °F (°C)
Acetylene	−118	Gas	581
CH:CH	(−83)		(305)
(Ethine)			
(Ethyne)			
N-Acetyl Ethanolamine	304–308	355	860
CH_3C:$ONHCH_2CH_2OH$	(151–153)	(179)	(460)
(N-(2-Hydroxyethyl)	@10 mm	(oc)	
acetamide)	Decomposes		
N-Acetyl Morpholine	Decomposes	235	
$CH_3CONCH_2CH_2OCH_2CH$:		(113)	
Acetyl Oxide		See Acetic Anhydride.	
Acetylphenol		See Phenyl Acetate.	
Acrolein	125	−15	428
CH_2:CHCHO	(52)	(−26)	(220)
(Acrylic Aldehyde)			Unstable
Acrylic Acid (Glacial)	287	122	820
$CH_2CHCOOH$	(142)	(50)	(438)
Acrylic Aldehyde		See Acrolein.	
Acrylonitrile	171	32	898
CH_2:CHCN	(77)	(0)	(481)
(Vinyl Cyanide)			
(Propenenitrile)			
Adipic Acid	509	385	788
$HOOC(CH_2)_4COOH$	(265)	(196)	(420)
	@100 mm		
Adipic Ketone		See Cyclopentanone.	
Adiponitrile	563	200	
$NC(CH_2)_4CN$	(295)	(93)	
Alcohol		See Ethyl Alcohol, Methyl Alcohol.	
Aldol	174–176	150	482
$CH_3CH(OH)CH_2CHO$	(79–80)	(66)	(250)
(3-Hydroxybutanal)	@12 mm		
(β-Hydroxybuteraldehyde)	Decomposes @176 (80)		
Allyl Acetate	219	72	705
CH_3COCH_2CH:CH_2	(104)	(22)	(374)
Allyl Alcohol	206	70	713
CH_2:$CHCH_2OH$	(97)	(21)	(378)
Allylamine	128	−20	705
CH_2:$CHCH_2NH_2$	(53)	(−29)	(374)
(2-Propenylamine)			
Allyl Bromide	160	30	563
CH_2:$CHCH_2Br$	(71)	(−1)	(295)
(3-Bromopropene)			
Allyl Caproate	367–370	150	
$CH_3(CH_2)_4COOCH_2CH$:Cl	(186–188)	(66)	
(Allyl Hexanoate)			
(2-Propenyl Hexanoate)			

(*Continued*)

TABLE 2.40 Boiling Points, Flash Points, and Ignition Temperatures of Organic Compounds (*Continued*)

Compound	Boiling point °F (°C)	Flash point, °F (°C)	Ignition point, °F (°C)
Allyl Chloride	113	−25	737
CH_2:$CHCH_2Cl$	(45)	(−32)	(485)
(3-Chloropropene)			
Allyl Chlorocarbonate		See Allyl Chloroformate.	
Allyl Chloroformate	223–237	88	
CH_2:$CHCH_2OCOCl$	(106–114)	(31)	
(Allyl Chlorocarbonate)			
Allylene		See Propyne.	
Allyl Ether	203	20	
$(CH_2$:$CHCH_2)_2O$	(95)	(−7)	
(Diallyl Ether)			
Allylidene Diacetate	225	180	
CH_2:$CHCH(OCOCH_3)_2$	(107)	(82)	
	@50 mm		
Allyl Isothiocyanate		See Mustard Oil.	
Allylpropenyl		See 1,4-Hexadiene.	
Allyl Trichloride		See 1,2,3-Trichloropropane.	
Allyl Vinyl Ether		See Vinyl Allyl Ether.	
Alpha Methyl Pyridine		2-Picoline.	
Aminobenzene		See Aniline.	
2-Aminobiphenyl		See 2-Biphenylamine.	
1-Aminobutane		See Butylamine.	
2-Amino-1-Butanol	352	165	
$CH_3CH_2CHNH_2CH_2OH$	(178)	(74)	
1-Amino-4-Ethoxybenzene		See p-Phenetidine.	
β-Aminoethyl Alcohol		See Ethanolamine.	
Amyl Acetate	300	60	680
$CH_3COOC_5H_{11}$	(149)	(16)	(360)
(1-Pentanol Acetate)		70	
Comm.		(21)	
sec-Amyl Acetate	249	89	
$CH_3COOCH(CH_3)(CH_2)_2CH_3$	(121)	(32)	
(2-Pentanol Acetate)			
Amyl Alcohol	280	91	572
$CH_3(CH_2)_3CH_2OH$	(138)	(33)	(300)
(1-Pentanol)			
sec-Amyl Alcohol	245	94	650
$CH_3CH_2CH_2CH(OH)CH_3$	(118)	(34)	(343)
(Diethyl Carbinol)			
Amylamine	210	30	2.2 22
$C_5H_{11}NH_2$	(99)	(−1)	
(Pentylamine)			
sec-Amylamine	198	20	
$CH_3(CH_2)_2CH(CH_3)NH_2$	(92)	(−7)	
(2-Aminopentane)			
(Methylpropylcarbinylamine)			
p-tert-Amylaniline	498–504	215	
$(C_2H_5)(CH_2)_2CC_6H_4NH_2$	(259–262)	(102)	
Amylbenzene	365	150	
$C_6H_5C_5H_{11}$	(185)	(66)	
(Phenylpentane)		(oc)	

TABLE 2.40 Boiling Points, Flash Points, and Ignition Temperatures of Organic Compounds (*Continued*)

Compound	Boiling point °F (°C)	Flash point, °F (°C)	Ignition point, °F (°C)
Amyl Bromide $CH_3CH_2CH_2CH_2CH_2Br$ (1-Bromopentane)	128–9 (53–54) @746 mm	90 (32)	
Amyl Butyrate $C_5H_{11}OOCC_3H_7$	365 (185)	135 (57)	
Amyl Carbinol		See Hexyl Alcohol.	
Amyl Chloride $CH_3(CH_2)_3CH_2Cl$ (1-Chloropentane)	223 (106)	55 (13)	500 (260)
tert-Amyl Chloride $CH_3CH_2CCl(CH_3)CH_3$	187 (86)		653 (345)
Amyl Chlorides (Mixed) $C_5H_{11}Cl$	185–228 (85–109)	38 (3)	
Amylcyclohexane $C_5H_{11}C_6H_{11}$	395 (202)		462 (239)
Amylene		See 1-Pentene.	
β-Amylene-cis $C_2H_5CH{:}CHCH_3$ (2-Pentene-cis)	99 (37)	<–4 (<–20)	
β-Amylene-trans $C_2H_5CH{:}CHCH_3$ (2-Pentene-trans)	97 (36)	<–4 (<–20)	
Amylene Chloride		See 1,5-Dichloropentane.	
Amyl Ether $C_5H_{11}OC_5H_{11}$ (Diamyl Ether) (Pentyloxypentane)	374 (190)	135 (57)	338 (170)
Amyl Formate $HCOCC_5H_{11}$	267 (131)	79 (26)	
Amyl Lactate $C_2H_5OCOOCH_2$- $CH(CH_3)C_2H_5$	237–239 (114–115) @36 mm	175 (79)	
Amyl Laurate $C_{11}H_{23}COOC_5H_{11}$	554–626 (290–330)	300 (149)	
Amyl Maleate $(CHCOOC_5H_{11})_2$	518–599 (270–315)	270 (132)	
Amyl Mercaptan $C_5H_{11}SH$ (1-Pentanethiol)	260 (127)	65 (18)	
Amyl Mercaptans (Mixed) $CH_3(CH_2)_4SH$	176–257 (80–125)	65 (18)	
Amyl Naphthalene $C_{10}H_7C_5H_{11}$	550 (288)	255 (124)	
Amyl Nitrate $CH_3(CH_2)_4NO_2$	306–315 (153–157)	118 (48)	
Amyl Nitrite $CH_3(CH_2)_4NO_2$	220 (104)	410 (210)	
Amyl Oleate $C_{17}H_{33}COOC_5H_{11}$	392–464 (200–240) @20 mm	366 (186)	
Amyl Oxalate $(COOC_5H_{11})_2$ (Diamyl Oxalate)	464–523 (240–273)	245 (118)	

(*Continued*)

TABLE 2.40 Boiling Points, Flash Points, and Ignition Temperatures of Organic Compounds (*Continued*)

Compound	Boiling point °F (°C)	Flash point, °F (°C)	Ignition point, °F (°C)
o-Amyl Phenol $C_5H_{11}C_6H_4OH$	455–482 (235–250)	219 (104)	
p-tert-Amyl Phenol		See Pentaphen.	
p-sec-Amylphenol $C_5H_{11}C_6H_4OH$	482–516 (250–269)	270 (132)	
2-(p-tert-Amylphenoxy) Ethanol $C_5H_{11}C_6H_4OCH_2CH_2OH$	567–590 (297–310)	280 (138)	
2-(p-tert-Amylphenoxy) Ethyl Laurate $C_{11}H_{23}COO(CH_2)_2OC_6H_4C_5H_{11}$	464–500 (240–260) @6 mm	410 (210)	
p-tert-Amylphenyl Acetate $CH_3COOC_6H_4C_5H_{11}$	507–511 (264–266)	240 (116)	
p-tert-Amylphenyl Butyl Ether $C_5H_{11}C_6H_4OC_4H_9$	540–550 (282–288)	275 (135)	
Amyl Phenyl Ether $CH_3(CH_2)_4OC_6H_5$ (Amoxybenzene)	421–444 (216–229)	185 (85)	
p-tert-Amylphenyl Methyl Ether $C_5H_{11}C_6H_4OCH_3$	462–469 (239–243)	210 (99)	
Amyl Phthalate		See Diamyl Phthalate.	
Amyl Propionate $C_2H_5COO(CH_2)_4CH_3$ (Pentyl Propionate)	275–347 (135–175)	106 (41)	712 (378)
Amyl Salicylate $HOC_6H_4COOC_5H_{11}$	512 (267)	270 (132)	
Amyl Stearate $CH_3(CH_2)_{16}COOC_5H_{11}$	680 (360)	365 (185)	
Amyl Sulfides, (Mixed) $C_5H_{11}S$	338–356 (170–180)	185 (85)	
Amyl Tolene $C_5H_{11}C_6H_4CH_3$	400–415 (204–213)	180 (82)	
Amyl Xylyl Ether $C_5H_{11}OC_6H_3(CH_3)_2$	480–500 (249–260)	205 (96)	
Aniline $C_6H_5NH_2$ (Aminobenzene) (Phenylamine)	364 (184)	158 (70)	1139 (615)
Aniline Hydrochloride $C_6H_5NH_2HCl$	473 (245)	380 (193)	
2-Anilinoethanol $C_6H_5NHCH_2CH_2OH$ (β-Anilinoethanol Ethoxyaniline) (β-Hydroxyethylaniline)	547 (286)	305 (152)	
β-Anilinoethanol Ethoxyaniline		See 2-Anilinoethanol.	
o-Anisaldehyde		See o-Methoxy Benzaldehyde.	
o-Anisidine $H_2NC_6H_4OCH_3$ (2-Methoxyaniline)	435 (224)	244 (118)	

TABLE 2.40 Boiling Points, Flash Points, and Ignition Temperatures of Organic Compounds (*Continued*)

Compound	Boiling point °F (°C)	Flash point, °F (°C)	Ignition point, °F (°C)
Anisole $C_6H_5OCH_3$ (Methoxybenzene) (Methyl Phenyl Ether)	309 (154)	125 (52)	887 (475)
Anol		See Cyclohexanol.	
Anthracene $(C_6H_4CH)_2$	644 (340)	250 (121)	1004 (540)
Anthraquinone $C_6H_4(CO)_2C_6H_4$	716 (380)	365 (185)	
Asphalt (Petroleum Pitch)	>700 (>371)	400+ (204+)	905 (485)
Aziridine		See Ethyleneimine.	
Azobisisobutyronitrile $N{:}CC(CH_3)_2N{:}NC(CH_3)_2C{:}N$	Decomposes	147 (64)	
Benzaldehyde C_6H_5CHO (Benzenecarbonal)	355 (179)	145 (63)	377 (192)
Benzedrine $C_6H_5CH_2CH(CH_3)NH_2$ (1-Phenyl Isopropyl Amine)	392 (200)	<212 (<100)	
Benzene C_6H_6 (Benzol)	176 (80)	12 (−11)	928 (498)
Benzine		See Petroleum Ether.	
Benzocyclobutene	306 (152)	95 (35)	477 (247)
Benzoic Acid C_6H_5COOH	482 (250)	250 (121)	1058 (570)
Benzol		See Benzene.	
p-Benzoquinone $C_6H_4O_2$ (Quinone)	Sublimes	100–200 (38–93)	1040 (560)
Benzotrichloride $C_6H_5CCl_3$ (Toluene, α,α,α-Trichloro) (Phenyl Chloroform)	429 (221)	260 (127)	412 (211)
Benzotrifluoride $C_6H_5CF_3$	216 (102)	54 (12)	
Benzoyl Chloride C_6H_5COCl (Benzene Carbonyl Chloride)	387 (197)	162 (72)	
Benzyl Acetate $CH_3COOCH_2C_6H_5$	417 (214)	195 (90)	860 (460)
Benzyl Alcohol $C_6H_5CH_2OH$ (Phenyl Carbinol)	403 (206)	200 (93)	817 (436)
Benzyl Benzoate $C_6H_5COOCH_2C_6H_5$	614 (323)	298 (148)	896 (480)
Benzyl Butyl Phthalate $C_4H_9COOC_6H_4COOCH_2C_6H_5$ (Butyl Benzyl Phthalate)	698 (370)	390 (199)	

(*Continued*)

TABLE 2.40 Boiling Points, Flash Points, and Ignition Temperatures of Organic Compounds (*Continued*)

Compound	Boiling point °F (°C)	Flash point, °F (°C)	Ignition point, °F (°C)
Benzyl Carbinol		See Phenethyl Alcohol.	
Benzyl Chloride	354	153	1085
$C_6H_5CH_2Cl$	(179)	(67)	(585)
(α-Chlorotoluene)			
Benzyl Cyanide	452	235	
$C_6H_5CH_2CN$	(233.5)	(113)	
(Phenyl Acetonitrile)			
(α-Tolunitrile)			
N-Benzyldiethylamine	405–420	170	
$C_6H_5CH_2N(C_2H_5)_2$	(207–216)	(77)	
Benzyl Ether		See Dibenzyl Ether.	
Benzyl Mercaptan	383	158	
$C_6H_5CH_2SH$	(195)	(70)	
(α-Toluenethiol)			
Benzyl Sallcilate	406	>212	
$OHC_6H_4COOCH_2C_6H_5$	(208)	(>100)	
(Salycilic Acid Benzyl Ester)			
Bicyclohexyl	462	165	473
$[CH_2(CH_2)_4CH]_2$	(239)	(74)	(245)
(Dicyclohexyl)			
Biphenyl	489	235	1004
$C_6H_5C_6H_5$	(254)	(113)	(540)
(Diphenyl)			
(Phenylbenzene)			
2-Biphenylamine	570	842	
$NH_2C_6H_4C_6H_5$	(299)	(450)	
(2-Aminobiphenyl)			
Bromobenzene	313	124	1049
C_6H_5Br	(156)	(51)	(565)
(Phenyl Bromide)			
1-Bromo Butane		See Butyl Bromide.	
4-Bromodiphenyl	592	291	
$C_6H_5C_6H_4Br$	(311)	(144)	
Bromoethane		See Ethyl Bromide.	
Bromomethane		See Methyl Bromide.	
1-Bromopentane		See Amyl Bromide.	
3-Bromopropene		See Allyl Bromide.	
o-Bromotoluene	359	174	
$BrC_6H_4CH_3$	(182)	(79)	
p-Bromotoluene	363	185	
$BrC_6H_4CH_3$	(184)	(85)	
1,3-Butadiene	24		788
CH_2:CHCH:CH_2	(–4)	Gas	(420)
Butadiene Monoxide	151	<–58	
CH_2:CHCHOCH_2	(66)	(<–50)	
(Vinylethylene Oxide)			
Butanal		See Butyraldehyde.	
Butanal Oxime		See Butyraldoxime.	
Butane	31	–76	550
$CH_3CH_2CH_2CH_3$	(–1)	(–60)	(287)
1,3-Butanediamine	289–302	125	
$NH_2CH_2CH_2CHNH_2CH_3$	(143–150)	(52)	

TABLE 2.40 Boiling Points, Flash Points, and Ignition Temperatures of Organic Compounds (*Continued*)

Compound	Boiling point °F (°C)	Flash point, °F (°C)	Ignition point, °F (°C)
1,2-Butanediol $CH_3CH_2CHOHCH_2OH$ (1,2-Dihydroxybutane) (Ethylethylene Glycol)	381 (194)	104 (40)	
1,3-Butanediol	See β-Butylene Glycol.		
1,4-Butanediol $HOCH_2CH_2CH_2CH_2OH$	442 (228)	250 (121)	
2,3-Butanediol $CH_3CHOHCHOHCH_3$	363 (184)	756 (402)	
2,3-Butanedione $CH_3COCOCH_3$ (Diocetyl)	190 (88)	80 (27)	
1-Butanethiol $CH_3CH_2CH_2CH_2SH$ (Butyl Mercaptan)	208 (98)	35 (2)	
2-Butanethiol C_4H_9SH (sec-Butyl Mercaptan)	185 (85)	−10 (−23)	
1-Butanol	See Butyl Alcohol.		
2-Butanol	See sec-Butyl Alcohol.		
2-Butanone	See Methyl Ethyl Ketone.		
2-Butenal	See Crotonaldehyde.		
1-Butene $CH_3CH_2CH:CH_2$ (α-Butylene)	21 (−6)		725 (385)
2-Butene-cis $CH_3CH:CHCH_3$	38.7 (4)		617 (325)
2-Butene-trans $CH_3CH:CHCH_3$ (β-Butylene)	−34 (1)		615 (324)
Butenediol $HOCH_2CH:CHCH_2OH$ (2-Butene-1,4-Diol)	286–300 (141–149) @20 mm	263 (128)	
2-Butene-1,4-Diol	See Butenediol.		
2-Butene Nitrile	See Crotononitrile.		
Butoxybenzene	See Butyl Phenyl Ether.		
1-Butoxybutane	See Dibutyl Ether.		
2,β-Butoxyethoxyethyl Chloride $C_4H_9CH_2CH_2OCH_2CH_2Cl$	392–437 (200–225)	190 (88)	
1-(Butoxyethoxy)2Propanol $CH_3CH(OH)CH_2OC_2H_4OC_2H_4C_2H_5$	445 (229)	250 (121)	509 (265)
β-Butoxyethyl Salicylate $OCH_6H_4COOCH_2CH_2OC_4$	367–378 (186–192)	315 (157)	
N-Butyl Acetamide $CH_3CONHC_4H_9$	455–464 (235–240)	240 (116)	
N-Butylacetanilide $CH_3(CH_2)_3N(C_6H_5)COCH_3$	531–538 (277–281)	286 (141)	
Butyl Acetate $CH_3COOC_4H_9$ (Butylethanoate)	260 (127)	72 (22)	797 (425)

(*Continued*)

TABLE 2.40 Boiling Points, Flash Points, and Ignition Temperatures of Organic Compounds (*Continued*)

Compound	Boiling point °F (°C)	Flash point, °F (°C)	Ignition point, °F (°C)
sec-Butyl Acetate $CH_3COOCH(CH_3)C_2H_5$	234 (112)	88 (31)	
Butyl Acetoacetate $CH_3COCH_2COO(CH_2)_3CH_3$	417 (214)	185 (85)	
Butyl Acetyl Ricinoleate $C_{17}H_{32}(OCOCH_3)$-$(COOC_4H_9)$	428 (220)	230 (110)	725 (385)
Butyl Acrylate $CH_2{:}CHCOOC_4H_9$	260 (127) Polymerizes	84 (29)	559 (292)
Butyl Alcohol $CH_3(CH_2)_2CH_2OH$ (1-Butanol) (Propylcarbinol) (Propyl Methanol)	243 (117)	98 (37)	650 (343)
sec-Butyl Alcohol $CH_3CH_2CHOHCH_3$ (2-Butanol) (Methyl Ethyl Carbinol)	201 (94)	75 (24)	761 (405)
tert-Butyl Alcohol $(CH_3)_2COHCH_3$ (2-Methyl-2-Propanol) (Trimethyl Carbinol)	181 (83)	52 (11)	892 (478)
Butylamine $C_4H_9NH_2$ (1-Amino Butane)	172 (78)	10 (−12)	594 (312)
sec-Butylamine $CH_3CH_2CH(NH_2)CH_3$	145 (63)	16 (−9)	
tert-Butylamine $(CH_3)_3C{:}NH_2$	113 (45)		716 (380)
Butylamine Oleate $C_{17}H_{33}COONH_3C_4H_9$		150 (66)	
tert-Butylaminoethyl Methacrylate $(CH_3)_3CNHC_2H_4OOCC(CH_3){:}CH_2$	200–221 (93–105)	205 (96)	
N-Butylaniline $C_6H_5NHC_4H_9$	465 (241)	225 (107)	
Butylbenzene $C_6H_5C_4H_9$	356 (180)	160 (71)	770 (410)
sec-Butylbenzene $C_6H_5CH(CH_3)C_2H_5$	344 (173)	126 (52)	784 (418)
tert-Butylbenzene $C_6H_5C(CH_3)_3$	336 (169)	140 (60)	842 (450)
Butyl Benzoate $C_6H_5COOC_4H_9$	482 (250)	225 (107)	
2-Butylbiphenyl $C_6H_5C_6H_4C_4H_9$	−554 (−290)	>212 (>100)	806 (430)
Butyl Bromide $CH_3(CH_2)_2CH_2Br$ (1-Bromo Butane)	215 (102)	65 (18)	509 (265)
Butyl Butyrate $CH_3(CH_2)_2COOC_4H_9$	305 (152)	128 (53)	

TABLE 2.40 Boiling Points, Flash Points, and Ignition Temperatures of Organic Compounds (*Continued*)

Compound	Boiling point °F (°C)	Flash point, °F (°C)	Ignition point, °F (°C)
Butylcarbamic Acid, Ethyl Ester		See N-Butylurethane.	
tert-Butyl Carbinol	237	98	
$(CH_3)_3CCH_2OH$	(114)	(37)	
(2,2-Dimethyl-1-Propanol)			
Butyl Carbitol		See Diethylene Glycol Monobutyl Ether.	
4-tert-Butyl Catechol	545	266	
$(OH)_2C_6H_3C(CH_3)_3$	(285)	(130)	
Butyl Chloride	170	15	464
C_4H_9Cl	(77)	(−9)	(240)
(1-Chlorobutane)			
sec-Butyl Chloride	155	<32	
$CH_3CHClC_2H_5$	(68)	(<0)	
(2-Chlorobutane)			
tert-Butyl Chloride	124	<32	
$(CH_3)_3CCl$	(51)	(<0)	
(2-Chloro-2-Methyl-Propane)			
4-tert-Butyl-2-Chlorophenol	453–484	225	
$ClC_6H_3(OH)C(CH_3)_3$	(234–251)	(107)	
tert-Butyl-m-Cresol	451–469	116	
$C_6H_3(C_4H_9)(CH_3)OH$	(233–243)	(47)	
p-tert-Butyl-o-Cresol	278–280	244	
$(OH)C_6H_3CH_3C(CH_3)_3$	(137–138)	(118)	
Butylcyclohexane	352–356		475
$C_4H_9C_6H_{11}$	(178–180)		(246)
(1-Cyclohexylbutane)			
sec-Butylcyclohexane	351		531
$CH_3CH_2CH(CH_3)C_6H_{11}$	(177)		(277)
(2-Cyclohexylbutane)			
tert-Butylcyclohexane	333–336		648
$(CH_3)_3CC_6H_{11}$	(167–169)		(342)
N-Butylcyclohexylamine	409	200	
$C_6H_{11}NH(C_4H_9)$	(209)	(93)	
Butylcyclopentane	314	480	
$C_4H_9C_5H_9$	(157)	(250)	
Butyl Ether		See Dibutyl Ether.	
Butylethylacetaldehyde		See 2-Ethylhexanal.	
Butyl Ethylene		See 1-Hexene.	
Butyl Ethyl Ether		See Ethyl Butyl Ether.	
Butyl Formate	225	64	612
$HCOOC_4H_9$	(107)	(18)	(322)
(Butyl Methanoate)			
(Formic Acid, Butyl Ester)			
Butyl Glycolate	~356	142	
$CH_2OHCOOC_4H_9$	(~180)	(61)	
tert-Butyl Hydroperoxide		<80	
$(CH_3)_3COOH$		(<27)	
n-Butyl Isocyanate	235	66	
$CH_3(CH_2)_3NCO$	(113)	(19)	
(Butyl Isocyanate)			
Butyl Isovalerate	302	127	
$C_4H_9OOCCH_2CH(CH_3)_2$	(150)	(53)	

(*Continued*)

TABLE 2.40 Boiling Points, Flash Points, and Ignition Temperatures of Organic Compounds (*Continued*)

Compound	Boiling point °F (°C)	Flash point, °F (°C)	Ignition point, °F (°C)
Butyl Lactate $CH_3CH(OH)COOC_4H_9$	320 (160)	160 (71)	720 (382)
Butyl Mercaptan		See 1-Butanethiol.	
tert-Butyl Mercaptan		See 2-Methyl-2-Propanethiol.	
Butyl Methacrylate $CH_2{:}C(CH_3)COO(CH_2)_3CH_3$	325 (163)	126 (52)	
Butyl Methanoate		See Butyl Formate.	
N-Butyl Monoethanolamine $C_4H_9NHC_2H_4OH$	378 (192)	170 (77)	
Butyl Naphthalene $C_4H_9C_{10}H_7$		680 (360)	
Butyl Nitrate $CH_3(CH_2)_3ONO_2$	277 (136)	97 (36)	
2-Butyloctanol $C_6H_{13}CH(C_4H_9)CH_2OH$	486 (252)	230 (110)	
Butyl Oleate $C_{17}H_{33}COOC_4H_9$	440.6–442.4 (227–228) @15 mm	356 (180)	
Butyl Oxalate $(COOC_4H_9)_2$ (Butyl Ethanedioate)	472 (244)	265 (129) (oc)	
tert-Butyl Peracetate diluted with 25% of benzene $CH_3CO(O_2)C(CH_3)_3$	Explodes on heating.	<80 (<27)	
tert-Butyl Perbenzoate $C_6H_5COOOC(CH_3)_3$	Explodes on heating.	>190 (>88)	
tert-Butyl Peroxypivalate diluted with 25% of mineral spirits $(CH_3)_3COOCOC(CH_3)_3$	Explodes on heating.	>155 (>68)	
β-(p-tert-Butyl Phenoxy) Ethanol $(CH_3)_3CC_6H_4OCH_2CH_2OH$	293-313 (145-156)	248 (120)	
β-(p-tert-Butylphenoxy) Ethyl Acetate $(CH_3)_3CC_6H_6OCH_2CH_2OCOCH_3$	579–585 (304–307)	324 (162)	
Butyl Phenyl Ether $CH_3(CH_2)_3OC_6H_5$ (Butoxybenzene)	410 (210)	180 (82)	
4-tert-Butyl-2-Phenylphenol $C_6H_5C_6H_3OHC(CH_3)_3$	385–388 (196–198)	320 (160)	
Butyl Propionate $C_2H_5COOC_4H_9$	295 (146)	90 (32)	799 (426)
Butyl Ricinoleate $C_{18}H_{33}O_3C_4H_9$	790 (421)	230 (110)	
Butyl Sebacate $[(CH_2)_4COOC_4H_9]_2$	653 (345)	353 (178)	
Butyl Stearate $C_{17}H_{35}COOC_4H_9$	650 (343)	320 (160)	671 (355)
tert-Butylstyrene	426 (219)	177 (81)	

TABLE 2.40 Boiling Points, Flash Points, and Ignition Temperatures of Organic Compounds (*Continued*)

Compound	Boiling point °F (°C)	Flash point, °F (°C)	Ignition point, °F (°C)
tert-Butyl Tetralin $C_4H_9C_{10}H_{11}$		680 (360)	
Butyl Trichlorosilane $CH_3(CH_2)_3SiCl_3$	300 (149)	130 (54)	
N-Butylurethane $CH_3(CH_2)_3NHCOOC_2H_5$ (Butylcarbamic Acid, Ethyl Ester) (Ethyl Butylcarbamate)	396–397 (202–203)	197 (92)	
Butyl Vinyl Ether		See Vinyl Butyl Ether.	
2-Butyne $CH_3C⋮CCH_3$ (Crotonylene)	81 (27)	−4 (<−20)	
Butyraldehyde $CH_3(CH_2)_2CHO$ (Butanal) (Butyric Aldehyde)	169 (76)	−8 (−22)	425 (218)
Butyraldol $C_8H_{16}O_2$	280 (138) @ 50 mm	165 (74)	
Butyraldoxime C_4H_8NOH (Butanal Oxime)	306 (152)	136 (58)	
Butyric Acid $CH_3(CH_2)_2COOH$	327 (164)	161 (72)	830 (443)
Butyric Acid, Ethyl Ester		See Ethyl Butyrate.	
Butyric Aldehyde		See Butyraldehyde.	
Butyric Anhydride $[CH_3(CH_2)_2CO]_2O$	388 (196)	180 (54)	535 (279)
Butyric Ester		See Ethyl Butyrate.	
Butyrolactone $CH_2CH_2CH_2COO$ (ring)	399 (204)	209 (98)	
Butyrone		See 4-Heptanone.	
Butyronitrile $CH_3CH_2CH_2CN$	243 (117)	76 (24)	935 (501)
Caproic Acid $(CH_3)(CH_2)_4COOH$ (Hexanoic Acid)	400 (204)	215 (102)	716 (380)
Carbolic Acid		See Phenol.	
Carbon Bisulfide		See Carbon Disulfide.	
Carbon Disulfide CS_2 (Carbon Bisulfide)	115 (46)	−22 (−30)	194 (90)
Cetane		See Hexadecane.	
Chloroacetic Acid $CH_2ClCOOH$	372 (189)	259 (126)	>932 (>500)
Chloroacetophenone $C_6H_5COCH_2Cl$ (Phenacyl Chloride)	477 (247)	244 (118)	
2-Chloro-4,6-di-tert-Amylphenol $(C_5H_{11})_2C_6H_2ClOH$	320–354 (160–179) @ 22 mm	250 (121)	
Chloro-4-tert-Amylphenol $C_5H_{11}C_6H_3ClOH$	487–509 (253–265)	225 (107)	

(*Continued*)

TABLE 2.40 Boiling Points, Flash Points, and Ignition Temperatures of Organic Compounds (*Continued*)

Compound	Boiling point °F (°C)	Flash point, °F (°C)	Ignition point, °F (°C)
2-Chloro-4-tert-Amyl-Phenyl Methyl Ether $C_5H_{11}C_6H_3ClOCH_3$	518–529 (270–276)	230 (110)	
p-Chlorobenzaldehyde ClC_6H_4CHO	417 (214)	190 (88)	
Chlorobenzene C_6H_5Cl (Chlorobenzol) (Monochlorobenzene) (Phenyl Chloride)	270 (132)	82 (28)	1099 (593)
Chlorobenzol		See Chlorobenzene.	
o-Chlorobenzotrifluoride $ClC_6H_4CF_3$ (o-Chloro-α,α,α-trifluorotoluene)	306 (152)	138 (59)	
Chlorobutadiene		See 2-Chloro-1,3-Butadiene.	
2-Chloro-1,3-Butadiene $CH_2{:}CCl{:}CH{:}CH_2$ (Chlorobutadiene) (Chloroprene)	138 (59)	–4 (–20)	
1-Chlorobutane		See Butyl Chloride.	
2-Chlorobutene-2 $CH_3CCl{:}CHCH_3$	143–159 (62–71)	–3 (–19)	
Chlorodinitrobenzene		See Dinitrochlorobenzene.	
Chloroethane		See Ethyl Chloride.	
2-Chloroethanol CH_2ClCH_2OH (2-Chloroethyl Alcohol) (Ethylene Chlorohydrin)	264–266 (129–130)	140 (60)	797 (425)
2-Chloroethyl Acetate $CH_3COOCH_2CH_2Cl$	291 (144)	151 (66)	
2-Chloroethyl Alcohol		See 2-Chloroethanol.	
Chloro-4-Ethylbenzene $C_2H_5C_6H_4Cl$	364 (184)	147 (64)	
Chloroethylene		See Vinyl Chloride.	
2-Chloroethyl Vinyl Ether		See Vinyl 2-Chloroethyl Ether.	
2-Chloroethyl-2-Xenyl Ether $C_6H_5C_6H_4OCH_2CH_2Cl$	613 (323)	320 (160)	
1-Chlorohexane $CH_3(CH_2)_4CH_2Cl$ (Hexyl Chloride)	270 (132)	95 (35)	
Chloroisopropyl Alcohol		See 1-Chloro-2-Propanol.	
Chloromethane		See Methyl Chloride.	
1-Chloro-2-Methyl Propane		See Isobutyl Chloride.	
1-Chloronaphthalene $C_{10}H_7Cl$	505 (263)	250 (121)	>1036 (>558)
2-Chloro-5-Nitrobenzotrifluoride $C_6H_3CF_3(2\text{-}Cl, 5\text{-}NO_2)$ (2-Chloro-α,α,α-Trifluoro-5-Nitrotoluene)	446 (230)	275 (135)	
1-Chloro-1-Nitroethane $C_2H_4NO_2Cl$	344 (173)	133 (56)	

TABLE 2.40 Boiling Points, Flash Points, and Ignition Temperatures of Organic Compounds (*Continued*)

Compound	Boiling point °F (°C)	Flash point, °F (°C)	Ignition point, °F (°C)
1-Chloro-1-Nitropropane $CHNO_2ClC_2H_5$	285 (141)	144 (62)	
2-Chloro-2-Nitropropane $CH_3CNO_2ClCH_3$	273 (134)	135 (57)	
1-Chloropentane	See Amyl Chloride.		
β-Chlorophenetole $C_6H_5OCH_2CH_2Cl$ (β-Phenoxyethyl Chloride)	306–311 (152–155)	225 (107)	
o-Chlorophenol ClC_6H_4OH	347 (175)	147 (64)	
p-Chlorophenol C_6H_4OHCl	428 (220)	250 (121)	
2-Chloro-4-Phenylphenol $C_6H_5C_6H_3ClOH$	613 (323)	345 (174)	
Chloroprene	See 2-Chloro-1,3-Butadiene.		
1-Chloropropane	See Propyl Chloride.		
2-Chloropropane	See Isopropyl Chloride.		
2-Chloro-1-Propanol $CH_3CHClCH_2OH$ (β-Chloropropyl Alcohol) (Propylene Chlorohydrin	271–273 (133–134)	125 (52)	
1-Chloro-2-Propanol $CH_2ClCHOHCH_3$ (Chloroisopropyl Alcohol) (sec-Propylene Chlorohydrin)	261 (127)	125 (52)	
1-Chloro-1-Propene	See 1-Chloropropylene.		
3-Chloropropene	See Allyl Chloride.		
α-Chloropropionic Acid $CH_3CHClCOOH$	352–374 (178–190)	225 (107)	932 (500)
3-Chloropropionitrile $ClCH_2CH_2CN$	348.8 (176)	168 (76)	
2-Chloropropionyl Chloride	Decomposes 230 (110)	88 (31)	
β-Chloropropyl Alcohol	See 2-Chloro-1-Propanol.		
1-Chloropropylene $CH_3CH:CHCl$ (1-Chloro-1-Propene)	95–97 (35–36)	<21 (<–6)	
2-Chloropropylene $CH_3CCl:CH_2$ (β-Chloropropylene) (2-Chloropropene)	73 (23)	<–4 (<–20)	
2-Chloropropylene Oxide	See Epichlorohydrin.		
γ-Chloropropylene Oxide	See Epichlorohydrin.		
Chlorotoluene $C_6H_4ClCH_3$ (Tolyl Chloride)	320 (160)	126 (52)	
α-Chlorotoluene	See Benzyl Chloride.		
Chlorotrifluoroethylene	See Trifluorochloroethylene.		
2-Chloro-α,α,α-Trifluoro-5-Nitrotoluene	See 2-Chloro-5-Nitrobenzotrifluoride.		

(Continued)

TABLE 2.40 Boiling Points, Flash Points, and Ignition Temperatures of Organic Compounds (*Continued*)

Compound	Boiling point °F (°C)	Flash point, °F (°C)	Ignition point, °F (°C)
o-Chloro-α,α,α-Trifluorotoluene		See o-Chlorobenzotrifluoride.	
Coal Oil		See Fuel Oil No. 1.	
Coal Tar Light Oil		<80 (<27)	
Coal Tar Pitch		405 (207)	
Creosote Oil	382–752 (194–400)	165 (74)	637 (336)
o-Cresol $CH_3C_6H_4OH$ (Cresylic Acid) (o-Hydroxytoluene) (o-Methyl Phenol)	376 (191)	178 (81)	1110 (599)
p-Cresyl Acetate $CH_3C_6H_4OCOCH_3$ (P-Tolyl Acetate)		195 (91)	
Cresyl Diphenyl Phosphate $(C_6H_5O)_2[(CH_3)_2C_6H_4O]$-$PO_4$	734 450 (390) (232)		
Cresylic Acid		See o-Cresol.	
Crotonaldehyde CH_3CH:$CHCHO$ (2-Butenal) (Crotonic Aldehyde) (Propylene Aldehyde)	216 55 (102) (13)	450 (232)	
Crotonic Acid CH_3CH:$CHCOOH$	372 (189)	190 (88)	745 (396)
Crotononitrile CH_3CH:$CHCN$ (2-Butenenitrile)	230–240.8 (110–116)	<212 (<100)	
Crotonyl Alcohol CH_3CH:$CHCH_2OH$ (2-Buten-1-ol) (Crotyl Alcohol)	250 (121)	81 (27)	660 (349)
1-Crotyl Bromide CH_3CH:$CHCH_2Br$ (1-Bromo-2-Butene)			
1-Crotyl Chloride CH_3CH:$CHCH_2Cl$ (1-Chloro-2-Butene)			
Cumene $C_6H_5CH(CH_3)_2$ (Cumol) (2-Phenyl Propane) (Isopropyl Benzene)	306 (152)	96 (36)	795 (424)
Cumene Hydroperoxide $C_6H_5C(CH_3)_2OOH$	Explodes on heating.	175 (79)	
Cyanamide NH_2CN	500 (260) Decomposes	286 (141)	
2-Cyanoethyl Acrylate $CH_2CHCOOCH_2CH_2CN$	Polymerizes	255 (124)	
N-(2-Cyanoethyl) Cyclohexylamine $C_6H_{11}NHC_2H_4CN$		255 (124)	

TABLE 2.40 Boiling Points, Flash Points, and Ignition Temperatures of Organic Compounds (*Continued*)

Compound	Boiling point °F (°C)	Flash point, °F (°C)	Ignition point, °F (°C)
Cyclamen Aldehyde $(CH_3)_2CHC_6H_4CH(CH_3)CH_2CHO$ (Methyl Para-Isopropyl Phenyl Propyl Aldehyde)		190 (88)	
Cyclobutane C_4H_8 (Tetramethylene)	55 (13)		
1,5,9-Cyclododecatriene $C_{12}H_{18}$	448 (231)	160 (71)	
Cycloheptane $CH_2(CH_2)_5CH_2$	246 (119)	<70 (<21)	
Cyclohexane C_6H_{12} (Hexahydrobenzene) (Hexamethylene)	179 (82)	–4 (–20)	473 (245)
1,4-Cyclohexane Dimethanol $C_8H_{16}O_2$	525 (274)	332 (167)	600 (316)
Cyclohexanethiol $C_6H_{11}SH$ (Cyclohexylmercaptan)	315–319 (157–159)	110 (43)	
Cyclohexanol $C_6H_{11}OH$ (Anol) (Hexolin) (Hydralin)	322 (161)	154 (68)	572 (300)
Cyclohexanone $C_6H_{10}O$ (Pimelic Ketone)	313 (156)	111 (44)	788 (420)
Cyclohexene $CH_2CH_2CH_2CH_2CH{:}CH$	181 (83)	<20 (<–7)	471 (244)
3-Cyclohexene-1-Carboxaldehyde		See 1,2,3,6-Tetrahydrobenzaldehyde.	
Cyclohexenone C_6H_8O	313 (156)	93 (34)	
Cyclohexyl Acetate $CH_3CO_2C_6H_{11}$ (Hexolin Acetate)	350 (177)	136 (58)	635 (335)
Cyclohexylamine $C_6H_{11}NH_2$ (Aminocyclohexane) (Hexahydroaniline)	274 (134)	88 (31)	560 (293)
Cyclohexylbenzene $C_6H_5C_6H_{11}$ (Phenylcyclohexone)	459 (237)	210 (99)	
Cyclohexyl Chloride $CH_2(CH_2)_4CHCl$ (Chlorocyclohexane)	288 (142)	90 (32)	
Cyclohexylcyclohexanol $C_6H_{11}C_6H_{10}OH$	304–313 (151–156)	270 (132)	
Cyclohexyl Formate $CH_2(CH_2)_4HCOOCH$	324 (162)	124 (51)	

(*Continued*)

TABLE 2.40 Boiling Points, Flash Points, and Ignition Temperatures of Organic Compounds (*Continued*)

Compound	Boiling point °F (°C)	Flash point, °F (°C)	Ignition point, °F (°C)
Cyclohexylmethane	See Methylcyclohexane.		
o-Cyclohexylphenol $C_6H_{11}C_6H_4OH$	298 (148) @10 mm	273 (134)	
Cyclohexyltrichlorosilane $C_6H_{11}SiCl_3$	406 (208)	196 (91)	
1,5-Cyclooctadiene C_8H_{10}	304 (151)	95 (35)	
Cyclopentane C_5H_{10}	121 (49)	<20 (<–7)	682 (361)
Cyclopentene $CH{:}CHCH_2CH_2CH_2$ (ring)	111 (44)	–20 (–29)	743 (395)
Cyclopentanol $CH_2(CH_2)_3CHOH$ (ring)	286 (141)	124 (51)	
Cyclopentanone $OCCH_2CH_2CH_2CH_2$ (ring) (Adipic Ketone)	267 (131)	79 (26)	
Cyclopropane $(CH_2)_3$ (Trimethylene)	–29 (–34)		928 (498)
p-Cymene $CH_3C_6H_4CH(CH_3)_2$ Tech. (4-Isopropyl-1-Methyl Benzene)	349 (176)	117 (47) 127 (53)	817 (436) 833 (445)
Decahydronaphthalene $C_{10}H_{18}$ (Decalin)	382 (194)	136 (58)	482 (250)
Decahydronaphthalene-trans $C_{10}H_{18}$	369 (187)	129 (54)	491 (255)
Decalin	See Decahydronaphthalene.		
Decane $CH_3(CH_2)_8CH_3$	345 (174)	115 (46)	410 (210)
Decanol $CH_3(CH_2)_8CH_2OH$ (Decyl Alcohol)	444.2 (229)	180 (82)	550 (288)
1-Decene $CH_3(CH_2)_7CH{:}CH_2$	342 (172)	<131 (<55)	455 (235)
Decyl Acrylate $CH_3(CN_2)_9OCOCH{:}CH_2$	316 (158) @50 mm	441 (227)	
Decyl Alcohol	See Decanol.		
Decylamine $CH_3(CH_2)_9NH_2$ (1-Aminodecane)	429 (221)	210 (99)	
Decylbenzene $C_{10}H_{21}C_6H_5$	491–536 (255–280)	225 (107)	
tert-Decylmercaptan $C_{10}H_{21}SH$	410–424 (210–218)	190 (88)	
Decylnaphthalene $C_{10}H_{21}C_{10}H_7$	635–680 (335–360)	350 (177)	
Decyl Nitrate $CH_3(CH_2)_9ONO_2$	261 (127) @11 mm	235 (113)	

TABLE 2.40 Boiling Points, Flash Points, and Ignition Temperatures of Organic Compounds (*Continued*)

Compound	Boiling point °F (°C)	Flash point, °F (°C)	Ignition point, °F (°C)
Diacetone Alcohol $CH_3COCH_2C(CH_3)_2OH$	328 (164)	148	1118
Diacetyl	See 2,3-Butanedione.		
Diallyl Ether	See Allyl Ether.		
Diallyl Phthalate $C_6H_4(CO_2C_3H_5)_2$	554 (290)	330 (166)	
1,3-Diaminobutane	See 1,3-Butanediamine.		
1,3-Diamino-2-Propanol $NH_2CH_2CHOHCH_2NH_2$	266 (130)	270 (132)	
1,3-Diaminopropane	See 1,3-Propanediamine.		
Diamylamine $(C_5H_{11})_2NH$	356 (180)	124 (51)	
Diamylbenzene $(C_5H_{11})_2C_6H_4$	491–536 (255–280)	225 (107)	
Diamylbiphenyl $C_5H_{11}(C_6H_4)_2C_5H_{11}$ (Diaminodiphenyl)	687–759 (364–404)	340 (171)	
Di-tert-Amylcyclohexanol $(C_5H_{11})_2C_6H_9OH$	554–572 (290–300)	270 (132)	
Diamyidlphenyl	See Diamylbiphenyl.		
Diamylene $C_{10}H_{20}$	302 (150)	118 (48)	
Diamyl Ether	See Amyl Ether.		
Diamyl Maleate $(CHCOOC_5H_{11})_2$	505–572 (263–300)	270 (132)	
Diamyl Naphthalene $C_{10}H_6(C_5H_{11})_2$	624 (329)	315 (159)	
2,4-Diamylphenol $(C_5H_{11})_2C_6H_3OH$	527 (275)	260 (127)	
Di-tert-Amylphenoxy Ethanol $C_6H_3(C_5H_{11})_2OC_2H_4OH$	615 (324)	300 (149)	
Diamyl Phthalate $C_6H_4(COOC_5H_{11})_2$ (Amyl Phthalate)	475–490 (246–254) @ 50 mm	245 (118)	
Diamyl Sulfide $(C_5H_{11})_2S$	338–356 (170–180)	185 (85)	
o-Dianisldine $[NH_2(OCH_3)C_6H_3]_2$ (o-Dimethoxybenzidine		403 (206)	
Dibenzyl Ether $(C_6H_5CH_2)_2O$ (Benzyl Ether)	568 (298)	275 (135)	
Dibutoxy Ethyl Phthalate $C_6H_4(COOC_2H_4OC_4H_9)_2$	437 (225)	407 (208) (oc)	
Dibutoxymethane $CH_2(OC_4H_9)_2$	330–370 (166–188)	140 (60)	
Dibutoxy Tetraglycol $(C_4H_9OC_2H_4OC_2H_4)_2O$ (Tetraethylene Glycol Dibutyl Ether)	635 (335)	305 (152)	
N,N-Dibutylacetamide $CH_3CON(C_4H_9)_2$	469–482 (243–250)	225 (107)	

(*Continued*)

TABLE 2.40 Boiling Points, Flash Points, and Ignition Temperatures of Organic Compounds (*Continued*)

Compound	Boiling point °F (°C)	Flash point, °F (°C)	Ignition point, °F (°C)
Dibutylamine	322	117	
$(C_4H_9)_2NH$	(161)	(47)	
Di-sec-Butylamine	270–275	75	
$[C_2H_5(CH_3)CH]_2NH$	(132–135)	(24)	
Dibutylaminoethanol	432	200	
$(C_4H_9)_2NC_2H_4OH$	(222)	(93)	
1-Dibutylamino-2-Propanol		See Dibutylisopropanolamine.	
N,N-Dibutylanlline	505–527	230	
$C_6H_5N(CH_2CH_2CH_2CH_3)_2$	(263–275)	(110)	
Di-tert-Butyl-p-Cresol	495–511	261	
$C_6H_2(C_4H_9)_2(CH_3)OH$	(257–266)	(127)	
Dibutyl Ether	286	77	382
$(C_4H_9)_2O$	(141)	(25)	(194)
(1-Butoxybutane)			
(Butyl Ether)			
2,5-Di-tert-Butylhydroquinone		420	790
$[C(CH_3)_3]_2C_6H_2(OH)_2$		(216)	(421)
(DTBHQ)			
Dibutyl Isophthalate		322	
$C_6H_4(CO_2C_4H_9)_2$		(161)	
N,N[1]-Di-sec-Butyl-p-		270	625
Phenylenediamine		(132)	(329)
$C_6H_4[\text{-NHCH}(CH_3)\text{-}$			
$CH_2CH_3]_2$			
Dibutylisopropanolamine	444	205	
$CH_3CHOHCH_2N(C_4H_9)_2$	(229)	(96)	
Dibutyl Maleate	Decomposes	285	
$(\text{-CHCO}_2C_4H_9)_2$		(141)	
Dibutyl Oxalate	472	220	
$C_4H_9OOCCOOC_4H_9$	(244)	(104)	
Di-tert-Butyl Peroxide	231	65	
$(CH_3)_3COOC(CH_3)_3$	(111)	(18)	
Dibutyl Phthalate	644	315	757
$C_6H_4(CO_2C_4H_9)_2$	(340)	(157)	(402)
(Dibutyl-o-Phthatate)			
n-Dibutyl Tartrate	650	195	544
$(COOC_4H_9)_2(CHOH)_2$	(343)	(91)	(284)
(Dibutyl-d-2,3-Dihydroxybutanedioate)			
N,N-Dibutyltoluene-sulfonamide	392	330	
$CH_3C_6H_4SO_3N(C_4H_9)_2$	(200) @10 mm	(166)	
Dicaproate		See Triethylene Glycol.	
Dicapryl Phthalate	441–453	395	
$C_6H_4[COOCH(CH_3)C_6H_{13}]_2$	(227–234) @4.5 mm	(202)	
Dichloroacetyl Chloride	225–226	151	
$CHCl_2COCl$	(107–108)	(66)	
(Dichloroethanoyl Chloride)			
3,4-Dichloroaniline	522	331	
$NH_2C_6H_3Cl_2$	(272)	(166)	
o-Dichlorobenzene	356	151	1198
$C_6H_4Cl_2$	(180)	(66)	(648)
(o-Dichlorobenzol)			

TABLE 2.40 Boiling Points, Flash Points, and Ignition Temperatures of Organic Compounds (*Continued*)

Compound	Boiling point °F (°C)	Flash point, °F (°C)	Ignition point, °F (°C)
p-Dichlorobenzene	345	150	
$C_6H_4Cl_2$	(174)	(66)	
2,3-Dichlorobutadiene-1,3	212	50	694
CH_2:C(Cl)C(Cl):CH_2	(100)	(10)	(368)
1,2-Dichlorobutane		527	
$CH_3CH_2CHClCH_2Cl$		(275)	
1,4-Dichlorobutane	311	126	
$CH_2ClCH_2CH_2CH_2Cl$	(155)	(52)	
2,3-Dichlorobutane	241–253	194	
$CH_3CHClCHClCH_3$	(116–123)	(90)	
1,3-Dichloro-2-Butene	262	80	
CH_2ClCH:$CClCH_3$	(128)	(27)	
3,4-Dichlorobutene-1	316	113	
$CH_2ClCHClCHCH_2$	(158)	(45)	
1,3-Dichlorobutene-2	258	80	
CH_2ClCH:$CClCH_3$	(126)	(27)	
Dichlorodimethylsilane		See Dimethyldichlorosilane.	
1,1-Dichloroethane		See Ethylidene Dichloride.	
1,2-Dichloroethane		See Ethylene Dichloride.	
Dichloroethanoyl Chloride		See Dichloroacetyl Chloride.	
1,1-Dichloroethylene		See Vinylidene Chloride.	
Dichloroisopropyl Ether	369	185	
$ClCH_2CH(CH_3)OCH(CH_3)CH_2Cl$	(187)	(85)	
[Bis (β-Chloroisopropyl) Ether]			
2,2-Dichloro Isopropyl Ether	369	185	
$[ClCH_2CH(CH_3)]_2O$	(187)	(85)	
[Bis(2-Chloro-1-Mothylethyl) Ether]			
Dichloromethane		See Methylene Chloride.	
1,1-Dichloro-1-Nitro Ethane	255	168	
$CH_3CCl_2NO_2$	(124)	(76)	
1,1-Dichloro-1-Nitro Propane	289	151	
$C_2H_5CCl_2NO_2$	(143)	(66)	
1,5-Dichloropentane	352–358	>80	
$CH_2Cl(CH_2)_3CH_2Cl$	(178–181)	(>27)	
(Amylene Chloride)			
(Pentamethylene Dichloride)			
2,4-Dichlorophenol	410	237	
$Cl_2C_6H_3OH$	(210)	(114)	
1,2-Dichloropropane		See Propylene Dichloride.	
1,3-Dichloro-2-Propanol	346	165	
$CH_2ClCHOHCH_2Cl$	(174)	(74)	
1,3-Dichloropropene	219	95	
CHCl:$CHCH_2Cl$	(104)	(35)	
2,3-Dichloropropene	201	59	
CH_2CClCH_2Cl	(94)	(15)	
α, β-Dichlorostyrene		225	
C_6H_5CCl:CHCl		(107)	
Dicyclohexyl		See Bicyclohexyl.	
Dicyclohexylamine	496	>210	
$(C_6H_{11})_2NH$	(258)	(>99)	

(*Continued*)

TABLE 2.40 Boiling Points, Flash Points, and Ignition Temperatures of Organic Compounds (*Continued*)

Compound	Boiling point °F (°C)	Flash point, °F (°C)	Ignition point, °F (°C)
Dicyclopentadiene $C_{10}H_{12}$	342 (172)	90 (32)	937 (503)
Didecyl Ether $(C_{10}H_{21})_2O$ (Decyl Ether)		419 (215)	
Diesel Fuel Oil No. 1-D		100 Min. (38)	
Diesel Fuel Oil No. 2-D		125 Min. (52)	
Diesel Fuel Oil No. 4-D		130 Min. (54)	
Diethanolomine $(HOCH_2CH_2)_2NH$	514 (268)	342 (172)	1224 (662)
1,2-Diethoxyethane		See Diethyl Glycol.	
Diethylacetaldehyde		See 2-Ethylbutyraldehyde.	
Diethylacetic Acid		See 2-Ethylbutyric Acid.	
N,N-Diethyl-acetoacetamide $CH_3COCH_2CON(C_2H_5)_2$	Decomposes	250 (121)	
Diethyl Acetoacetate $CH_3COC(C_2H_5)_2COOC_2H_5$	412–424 (211–218) Decomposes	170 (77)	
Diethylamine $(C_2H_5)_2NH$	134 (57)	–9 (–23)	594 (312)
2-Diethyl (Amino) Ethanol		See N,N-Diethylethanolamine.	
2-(Diethylamino) Ethyl Acrylate CH_2:$CHCOOCH_2CH_2$- $HN(CH_3CH_2)_2$	Decomposes	195 (91)	
3-(Diethylamino)-Propylamine $(C_2H_5)_2NCH_2CH_2CH_2NH_2$ (N,N-Diethyl-1,3-Propanediamine)	337 (169)	138 (59)	
N,N-Diethylaniline $C_6H_5N(C_2H_5)_2$ (Phenyldiethylamine)	421 (216)	185 (85)	1166 (630)
o-Diethyl Benzene $C_6H_4(C_2H_5)_2$	362 (183)	135 (57)	743 (395)
m-Diethyl Benzene $C_6H_4(C_2H_5)_2$	358 (181)	133 (56)	842 (450)
p-Diethyl Benzene $C_6H_4(C_2H_5)_2$	358 (181)	132 (55)	806 (430)
N,N-Diethyl-1,3-Butanediamine $C_2H_5NHCH_2CH_2CHN(C_2H_5)CH_3$ [1,3-Bis(ethylamino) Buiane]	354–365 (179–185)	115 (46)	
D1-2-Ethylbutyl Phthalate $C_6H_4[COOCH_2CH(C_2H_5)_2]_2$	662 350	381 (194)	
Diethyl Carbamyl Chloride $(C_2H_5)_2NCOCl$	369–374 (187–190)	325–342 (163–172)	
Diethyl Carbinol		See sec-Amyl Alcohol.	
Diethyl Carbonate $(C_2H_5)_2CO_3$ (Ethyl Carbonate)	259 (126)	77 (25)	

TABLE 2.40 Boiling Points, Flash Points, and Ignition Temperatures of Organic Compounds (*Continued*)

Compound	Boiling point °F (°C)	Flash point, °F (°C)	Ignition point, °F (°C)
Diethylcyclohexane $C_{10}H_{20}$	344 (173)	120 (49)	464 (240)
1,3-Diethyl-1,3-Diphenyl Urea $[(C_2H_5)(C_6H_5)N]_2CO$	620 (327)	302 (150)	
Diethylene Diamine	299 (150)	144 (62)	
Diethylene Dioxide		See p-Dioxane.	
Diethylene Glycol $O(CH_2CH_2OH)_2$ (2,2-Dihydroxyethyl Ether)	472 (244)	255 (124)	435 (224)
Diethylene Glycol Methyl Ether $CH_3OC_2H_4OC_2H_4OH$ (2-(2-Methoxyethoxy) Ethanol)	379 (193)	205 (96)	465 (240)
Diethylene Glycol Methyl Ether Acetate $CH_3COOC_2H_4OC_2H_4OCH_3$	410 (210)	180 (82)	
Diethylene Glycol Monobutyl Ether $C_4H_9OCH_2CH_2OCH_2CH_2OH$	448 (231)	172 (78)	400 (204)
Diethylene Glycol Monoethyl Ether Acetate $C_4H_9O(CH_2)_2O(CH_2)_2OOCCH_3$	476 (247)	240 (116)	570 (298.9)
Diethylene Glycol Monoethyl Ether $CH_2OHCH_2OCH_2\text{-}CH_2OC_2H_5$	396 (202)	201 (94)	400 (204)
Diethylene Glycol Monoethyl Ether Acetate $C_2H_5O(CH_2)_2O(CH_2)_2OOCCH_3$	424 (218)	225 (107)	680 (360)
Diethylene Glycol Monoisobutyl Ether $(CH_3)_2CHCH_2O(CH_2)_2O(CH_2)_2OH$	422–437 (217–225)	222 (106)	452–485 (233–252)
Diethylene Glycol Monomethyl Ether $CH_3O(CH_2)O(CH_2)_2OH$	381 (194)	205 (96)	
Diethylene Glycol Mono-Methyl Ether Formal $CH_2(CH_3OCH_2CH_2OCH_2CH_2O)_2$	581 (305)	310 (154)	
Diethylene Glycol Phthalate $C_6H_4[COO(CH_2)_2OC_2H_5]_2$		343 (173)	
Diethylene Oxide		See Tetrahydrofuran.	
Diethylene Triamine $NH_2CH_2CH_2NHCH_2CH_2NH_2$	404 (207)	208 (98)	676 (358)
N,N-Diethylethanolamine $(C_2H_5)_2NC_2H_4OH$ (2-(Diethylamino) Ethanol)	324 (162)	140 (60)	608 (320)
Diethyl Ether		See Ethyl Ether.	
N,N-Diethylethylene-diamine $(C_2H_5)_2NC_2H_4NH_2$	293 (145)	115 (46)	
Diethyl Fumarate $C_2H_5OCOCH{:}CHCOOC_2H_5$	442 (217)	220 (104)	
Diethyl Glycol $(C_2H_5OCH_2)_2$ (1,2-Diethoxyethane)	252 (122)	95 (35)	401 (205)

(*Continued*)

TABLE 2.40 Boiling Points, Flash Points, and Ignition Temperatures of Organic Compounds (*Continued*)

Compound	Boiling point °F (°C)	Flash point, °F (°C)	Ignition point, °F (°C)
Diethyl Ketone	217	55	842
$C_2H_5COC_2H_5$	(103)	(13)	(450)
(3-Pentanone)			
N,N-Diethyllauramide	331–351	>150	
$C_{11}H_{23}CON(C_2H_5)_2$	(166–177)	(>66)	
	@2 mm		
Diethyl Maleate	438	250	662
$(-CHCO_2C_2H_3)_2$	(226)	(121)	(350)
Diethyl Malonate	390	200	
$CH_2(COOC_2H_3)_2$	(199)	(93)	
(Ethyl Malonate)			
Diethyl Oxide		See Ethyl Ether.	
3,3-Diethylpentane	295	554	
$CH_3CH_2C(C_2H_5)_2CH_2CH_3$	(146)	(290)	
Diethyl Phthalate	565	322	855
$C_6H_4(COOC_2H_5)_2$	(296)	(161)	(457)
p-Diethyl Phthalate		See Diethyl Terephthalate.	
N,N-Diethylstearamide	246–401	375	
$C_{17}H_{35}CON(C_2H_5)_2$	(119–205)	(191)	
	@1 mm		
Diethyl Succinate	421	195	
$(CH_2COOCH_2CH_3)_2$	(216)	(90)	
Diethyl Sulfate	Decomposes,	220	817
$(C_2H_5)_2SO_4$	giving	(104)	(436)
(Ethyl Sulfate)	Ethyl Ether		
Diethyl Tartrate	536	200	
$CHOHCOO(C_2H_5)_2$	(280)	(93)	
Diethyl Terephthalate	576	243	
$C_6H_4(COOC_2H_5)_2$	(302)	(117)	
(p-Diethyl Phthalate)			
3,9-Diethyl-6-tridecanol		See Heptadecanol.	
Diglycol Chlortormate	256–261	295	
$O:(CH_2CH_2OCOCl)_2$	(124–127)	(146)	
	@5 mm		
Diglycol Chlorohydrin	387	225	
$HOCH_2CH_2OCH_2CH_2Cl$	(197)	(107)	
Diglycol Diacetate	482	255	
$(CH_3COOCH_2CH_2)_2O$	(250)	(124)	
Diglycol Dilevulleate		340	
$(CH_2CH_2OOC-$		(171)	
$(CH_2)_2COCH_3)_2:O$			
Diglycol Laurate	559–617	290	
$C_{16}H_{32}O_4$	(293–325)	(143)	
Dihexyl		See Dodecane.	
Dihexylamine	451–469	220	
$[CH_3(CH_2)_5]_2NH$	(233–243)	(104)	
Dihexyl Ether		See Hexyl Ether.	
Dihydropyran	186	0	
$CH_2CH_2CH_2:CHCHO$	(86)	(−18)	
o-Dihydroxybenione	473	260	
$C_6H_4(OH)_2$	(245)	(127)	
(Pyrocalechol)			

TABLE 2.40 Boiling Points, Flash Points, and Ignition Temperatures of Organic Compounds (*Continued*)

Compound	Boiling point °F (°C)	Flash point, °F (°C)	Ignition point, °F (°C)
p-Dihydroxybenione	547	329	959
$C_6H_4(OH)_2$	(286)	(165)	(515)
(Hydroquinone)			
1,2-Dihydroxybenione		See 1,2-Butanediol.	
2,2-Dihydroxyethyl Ether		See Diethylene Glycol.	
2,5-Dihydroxyhexane		See 2,5-Hexanediol.	
Diisobutylamine	273–286	85	
$[(CH_3)_2CHCH_2]_2NH$	(134–141)	(29)	
[Bis(β-Methylpropyl) Amine]			
Diisobutyl Carbinol	353	165	
$[(CH_3)_2CHCH_2]_2CHOH$	(178)	(74)	
(Nonyl Alcohol)			
Diisobutylene		See 2,4,4-Trimethyl-1-Pentene.	
Diisobutylene	214	23	736
$(CH_3)_3CCH_2C(CH_3){:}CH_2$	(101)	(−5)	(391)
(2,4,4-Trimethy-H_2-Pentane)			
Diisobutyl Ketone	335	120	745
$[(CH_3)_2CHCH_2]_2CO$	(168)	(49)	(396)
(2,6-Dimethyl-4 Heptanone)			
(Isovalerone)			
Diisobutyl Phthalate	321	365	810
$C_6H_4[COOCH_2OH(CH_3)_2]_2$	(327)	(185)	(432)
Diisodecyl Adipoia	660	225	
$C_{10}H_{21}O_2C(CH_2)_2CO_2\text{-}C_{10}H_{21}$	(349)	(107)	
Diisodecyl Phthalate	182	450	755
$C_6H_4(COOC_{10}H_{21})_2$	(250)	(232)	(402)
Diisooctyl Phthalate	398	450	
$(C_8H_{17}COO)_2C_2H_4$	(370)	(232)	
Diisopropanolamine	480	260	705
$[CH_3CH(OH)\text{-}CH_2]_2NH$	(249)	(127)	(374)
Diisopropyl		See 2,3-Dimethylbutane.	
Diisopropylamine	183	30	600
$[(CH_3)_2CH]_2NH$	(84)	(−1)	(316)
Diisopropyl Benzene	401	170	840
$[(CH_3)_2CH]_2C_6H_4$	(205)	(77)	(449)
N,N-Diisopropyl-ethanolamine	376	175	
$[(CH_3)_2CH]_2NC_2H_4OH$	(191)	(79)	
Diisopropyl Ether		See Isopropyl Ether.	
Diisopropyl Maleate	444	220	
$(CH_3)_2CHOCOCH{:}$	(229)	(104)	
$CHCOOCH(CH_3)_2$			
Diisopropylmethanol		See 2,4-Dimethyl-3-Pentanol.	
Diisopropyl Peroxydicarbonate	Explodes		
$(CH_3)_2CHOCOOCOOCH(CH_3)_2$	on heating.		
Diketene	261	93	
$CH_2{:}CCH_2C(O)O$ (ring closed between C and O)	(127)	(34)	
(Vinylaceto-β-Lactone)			
2,5-Dimethoxyaniline	518	302	735
$NH_2C_6H_3(OCH_3)_2$	(270)	(150)	(391)
2,5-Dimethoxy Chlorobenzene	460–467	243	
$C_8H_9ClO_2$	(238–242)	(117)	
1,2-Dimethoxyethane		See Ethylene Glycol Dimethyl Ether.	

(*Continued*)

TABLE 2.40 Boiling Points, Flash Points, and Ignition Temperatures of Organic Compounds (*Continued*)

Compound	Boiling point °F (°C)	Flash point, °F (°C)	Ignition point, °F (°C)
Dimethoxyethyl Phthalate $C_6H_4(COOCH_2CH_2OCH_3)_2$ (Bis(2-methoxyethyl) Phthalate)	644 (340)	410 (210)	750 (399)
Dimethoxymethane	See Methylal.		
Dimethoxy Tetraglycol $CH_3OCH_2(CH_2OCH_2)_3CH_2OCH_3$ (Tetraethylene Glycol Dimethyl Ether)	528 (276)	285 (141)	
Dimethylacetamide $(CH_3)_2NC:OCH_3$ (DMAC)	330 (165)	158 (70)	914 (490)
Dimethylamine $(CH_3)_2NH$	45 (7)	Gos	752 (400)
1,2-Dimethylbenzene	See o–Xylene.		
1,3-Dimethylbenzene	See m-Xylene.		
1,4-Dimethylbenzene	See p-Xylene.		
Dimethylbenzylcarbinyl Acetate $C_6H_5CH_2C(CH_3)_2OOCCH_3$ (alpha, alpha-Dimethyl-phenethyl Acelate)		205 (96)	
2,2-Dimethylbutane $(CH_3)_3CCH_2CH_3$ (Neohexane)	122 (50)	–54 (–48)	761 (405)
2,3-Dimethylbutane $(CH_3)_2CHCH(CH_3)_2$ (Diisopropyl)	136 (58)	–20 (–29)	761 (405)
1,3-Dimethylbutanol	See Methyl Isobutyl Carbinol.		
2,3-Dimethyl-1-Butene $CH_3CH(CH_3)C(CH_3):CH_2$	133 (56)	<–4 (<–20)	680 (360)
2,3-Dimethyl-2-Butene $CH_3C(CH_3):C(CH_3)_2$	163 (73)	<–4 (<–20)	753 (401)
1,3-Dimethylbutyl Acetate $CH_3COOCH(CH_3)CH_2CH(CH_3)_2$	284–297 (140–147)	113 (45)	
1,3-Dimethylbutylamine $CH_3CHNH_2(CH_2)CH(CH_3)_2$ (2-Amino-4-Methylpeniane)	223–228 (106–109)	55 (13)	
Dimethyl Carbinol	See Isopropyl Alcohol.		
Dimethyl Carbonate	See Methyl Carbonate.		
Dimethyl Chloracetal $ClCH_2CH(OCH_3)_2$	259–270 (126–132)	111 (44)	450 (232)
Dimethylcyanamide $(CH_3)_2NCN$	320 (160)	160 (71)	
1,2-Dimethylcyclohexane $(CH_3)_2C_6H_{10}$	260 (127)		579 (304)
1,3-Dimethylcyclohexane $(CH_3)_2C_6H_{10}$ (Hexahydroxylene)	~256 (124)	~50 (10)	583 (306)
1,4-Dimethylcyclohexane $(CH_3)_2C_6H_{10}$ (Hexahydroxylol)	248 (120)	52 (11)	579 (304)
1,4-Dimethylcyclohexane-cis $C_6H_{10}(CH_3)_2$	255 (124)	61 (16)	

TABLE 2.40 Boiling Points, Flash Points, and Ignition Temperatures of Organic Compounds (*Continued*)

Compound	Boiling point °F (°C)	Flash point, °F (°C)	Ignition point, °F (°C)
1,4-Dimethylcyclohexane-trans $C_6H_{10}(CH_3)_2$	246 (119)	51 (11)	
Dimethyl Decalin $C_{10}H_{16}(CH_2)_2$	455 (235)	184 (84)	455 (235)
Dimethyldichlorosilane $(CH_3)_2SiCl_2$ (Dichlorodimethylsilane)	158 (70)	<70 (<21)	
Dimethyldioxane $CH_3CHCH_2OCH_2(CH_3)CHO$ (ring)	243 (117)	75 (24)	
1,3-Dimethyl-1-3-Diphenylcyclobutane $(C_6H_5CCH_3)_2(CH_2)_2$	585–588 (307–309)	289 (143)	
Dimethylene Oxide		See Ethylene Oxide.	
Dimethyl Ether		See Methyl Ether.	
Dimethyl Ethyl Carbinol		See 2-Methyl-2-Butanol.	
2,4-Dimethyl-3-Ethylpentane $CH_3CH(CH_3)CH(CH_2H_5)CH(CH_3)_2$ (3-Ethyl-2,4-Dimethylpentane)	279 (137)	734 (390)	
N,N-Dimethylformamide $HCON(CH_3)_2$	307 (153)	136 (58)	833 (445)
2,5-Dimethylfuran $OC(CH_3){:}CHCH{:}C(CH_3)$ (ring)	200 (93)	45 (7)	
Dimethyl Glycol Phthalate $C_6H_4[COO(CH_2)_2OCH_3]_2$	446 (230)	369 (187)	
3,3-Dimethylheptane $CH_3(CH_2)_3C(CH_3)_2CH_2CH_3$	279 (137)	617 (325)	
2,6-Dimethyl-4-Heptanone		See Diisobutyl Ketone.	
2,3-Dimethylhexane $CH_3CH(CH_3)CH(CH_3)C_2H_5CH_3$	237 (114)	45 (7)	820 (438)
2,4-Dimethylhexane $CH_3CH(CH_3)CH(CH_3)C_2H_5CH_3$	229 (109)	50 (10)	
Dimethyl Hexynol $C_4H_9CCH_3(OH)C{\vdots}CH$ (3,5-Dimethyl-1-Hexyn-3-ol)	302 (150)	135 (57)	
1,1-Dimethylhydrazine $(CH_3)_2NNH_2$ (Dimethylhydrazine, Unsymmetrical)	145 (63)	5 (−15)	480 (249)
Dimethylisophthalate $CH_3OOCC_6H_4COOCH_3$		280 (138)	
N,N-Dimethyliso-propanolamine $(CH_3)_2NCH_2CH(OH)CH_3$	257 (125)	95 (35)	
Dimethyl Ketone		See Acetone.	
Dimethyl Maleate $(-CHCOOCH_3)_2$	393 (201)	235 (113)	
2,6-Dimethylmorpholine $CH(CH_3)CH_2OCH_2CH(CH_3)NH$ (ring)	296 (147)	112 (44)	
2,3-Dimethyloctane $CH_3(CH_2)_4CH(CH_3)CH(CH_3)CH_3$	327 (164)	<131 (<55)	437 (225)
3,4-Dimethyloctane $C_3H_7CH(CH_3)CH(CH_3)C_3H_7$	324 (162)	<131 (<55)	

(*Continued*)

TABLE 2.40 Boiling Points, Flash Points, and Ignition Temperatures of Organic Compounds (*Continued*)

Compound	Boiling point °F (°C)	Flash point, °F (°C)	Ignition point, °F (°C)
2,3-Dimethylpentaldehyde $CH_3CH_2CH(CH_3)CH(CH_3)CHO$	293 (145)	94 (34)	
2,3-Dimethylpentane $CH_3CH(CH_3)CH(CH_3)CH_2CH_3$	194 (90)	<20 (<−7)	635 (335)
2,4-Dimethylpentane $(CH_3)_2CHCH_2CH(CH_3)_2$	177 (81)	10 (−12)	
2,4-Dimethyl-3-Pentanol $(CH_3)_2CHCHOHCH(CH_3)_2$ (Diisopropylmethanol)	284 (140)	120 (49)	
Dimethyl Phthalate $C_6H_4(COOCH_3)_2$	540 (282)	295 (146)	915 (490)
Dimethylpiperazine-cis $C_6H_{14}N_2$	329 (165)	155 (68)	
2,2-Dimethylpropane $(CH_3)_4C$ (Neopentane)	49 (9)		842 (450)
2,2-Dimethyl-1-Propanol		See tert-Butyl Carbinol.	
2,5-Dimethylpyrazine $CH_3C{:}CHN{:}C(CH_3)CH{:}N$	311 (155)	147 (64)	
Dimethyl Sebacate $[-(CH_2)_4COOCH_3]_2$ (Methyl Sebacate)	565 (296)	293 (145)	
Dimethyl Sulfate $(CH_3)_2SO_4$ (Methyl Sulfate)	370 (188)	182 (83)	370 (188)
Dimethyl Sulfide $(CH_3)_2S$	99 (37)	<0 (<−18)	403 (206)
Dimethyl Sulfoxide $(CH_3)_2SO$	372 (189)	203 (95) (oc)	419 (215)
Dimethyl Terephthalate $C_6H_4(COOCH_3)_2$ (Dimethyl-1,4-Benzene Dicarboxylate) (DMT)	543 (284)	308 (153)	965 (518)
2,4-Dinitroaniline $(NO_2)_2C_6H_3NH_2$		435 (224)	
1,2-Dinitro Benzol $C_6H_4(NO_2)_2$ (o-Dinitrobenzene)	604 (318)	302 (150)	
Dinitrochlorobenzene $C_6H_3Cl(NO_2)_2$ (Chlorodinitrobenzene)	599 (315)	382 (194)	
2,4-Dinitrotoluene $(NO_2)_2C_6H_3CH_3$	572 (300)	404 (207)	
Dioctyl Adipate $[-(CH_2)_2COOCH_2$- $CH(C_2H_5)C_4$-$H_9]_2$ [Bis(2-Ethylhexyl) Adipate] [Di(2-Ethylhexyl) Adipate]	680 (360)	402 (206)	710 (377)
Dioctyl Azelate $(CH_2)_7[COOCH_2CH(C_2H_5)C_4H_9]_2$ (Bis(2-Ethylhexyl) Azelate) (Di(2-Ethylhexyl) Azelate)	709 (376)	440 (227)	705 (374)

TABLE 2.40 Boiling Points, Flash Points, and Ignition Temperatures of Organic Compounds (*Continued*)

Compound	Boiling point °F (°C)	Flash point, °F (°C)	Ignition point, °F (°C)
Dioctyl Ether	558	>212	401
$(C_8H_{17})_2O$	(292)	(>100)	(205)
(Octyl Ether)			
Dioctyl Phthalate		420	735
$C_6H_4[CO_2CH_2$-		(215)	(390)
$CH(C_2H_5)C_4H_9]_2$			
[Di(2-Ethylhexyl) Phthalate]			
[Bis(2-Ethylhexyl) Phthalate]			
p-Dioxane	214	54	356
$OCH_2CH_2OCH_2CH_2$	(101)	(12)	(180)
(Diethylene Dioxide)			
Dioxolane	165	35	
$OCH_2CH_2OCH_2$	(74)	(2)	
Dipe ntene	339	113	458
$C_{10}H_{16}$	(170)	(45)	(237)
(Cinene)			
(Limonene)			
Diphenyl		See Biphenyl.	
Diphenylamine	575	307	1173
$(C_6H_5)_2NH$	(302)	(153)	(634)
(Phenylaniline)			
1,1-Diphenylbutane	561	>212	851
$(C_6H_5)_2CHC_3H_7$	(294)	(>100)	(455)
1,3-Diphenyl-2-buten-1-one		See Dypnone.	
Diphenyldichlorosllane	581	288	
$(C_6H_5)_2SiCl_2$	(305)	(142)	
Diphenyldodecyl Phosphite		425	
$(C_6H_5O)_2POC_{10}H_{21}$		(218)	
1,1-Diphenylethane (uns)	546	>212	824
$(C_6H_5)_2CHCH_3$	(286)	(>100)	(440)
1,2-Diphenylethane (sym)	544	264	896
$C_6H_5CH_2CH_2C_6H_5$	(284)	(129)	(480)
Diphenyl Ether		See Diphenyl Oxide.	
Diphenylmethane	508	266	905
$(C_6H_5)_2CH_2$	(264)	(130)	(485)
(Ditane)			
Diphenyl Oxide	496	239	1144
$(C_6H_5)_2O$	(258)	(115)	(618)
(Diphenyl Ether)			
1,1-Diphenylpentane	586	>212	824
$(C_6H_5)_2CHC_4H_9$	(308)	(>100)	(440)
1,1-Diphenylpropane	541	>212	860
$CH_3CH_2CH(C_6H_5)_2$	(283)	(>100)	(460)
Diphenyl Phthalate	761	435	
$C_6H_4(COOC_6H_5)_2$	(405)	(224)	
Dipropylamine	229	63	570
$(C_3H_7)_2NH$	(109)	(17)	(299)
Dipropylene Glycol	449	250	
$(CH_3CHOHCH_2)_2O$	(232)	(121)	
Dipropylene Glycol Methyl Ether	408	186	
	(209)	(86)	
$CH_3OC_3H_6OC_3H_6OH$			

(*Continued*)

TABLE 2.40 Boiling Points, Flash Points, and Ignition Temperatures of Organic Compounds (*Continued*)

Compound	Boiling point °F (°C)	Flash point, °F (°C)	Ignition point, °F (°C)
Dipropyl Ether		See n-Propyl Ether.	
Dipropyl Ketone		See 4-Heptanone.	
Ditane		See Diphenylmethane.	
Ditridecyl Phthalate	547	470	
$C_6H_4(COOC_{13}H_{27})_2$	@5 mm (286)	(243)	
Divinyl Acetylene	183	<–4	
$(⋮CCH{:}CH_2)_2$	(84)	(<–20)	
(1,5-Hexadien-3-yne)			
Divinylbenzene	392	169	
$C_6H_4(CH{:}CH_2)_2$	(200)	(76)	
Divinyl Ether	83	<–22	680
$(CH_2{:}CH)_2O$	(28)	(<–30)	(360)
(Ethenylaxyethene)			
(Vinyl Ether)			
Dodecane	421	165	397
$CH_3(CH_2)_{10}CH_3$	(216)	(74)	(203)
(Dihexyl)			
1-Dodecanethiol	289	262	
$CH_3(CH_2)_{11}SH$	(143)	(128)	
(Dodecyl Mercaptan)	@15 mm		
(Lauryl Mercaptan)			
1-Dodecanol	491	260	527
$CH_3(CH_2)_{11}OH$	(255)	(127)	(275)
(Louryl Alcohol)			
Dodecyl Bromide		See Lauryl Bromide.	
Dodecylene (α)	406	<212	491
$C_{16}H_{21}CH{:}CH_2$	(208)	(<100)	(255)
(1-Dodecane)			
Dodecyl Mercaptan		See 1-Dodecanethiol.	
tert-Dodecyl Mercaptan	428–451	205	
$C_{12}H_{25}SH$	(220–233)	(96)	
4-Dodecyloxy-2-Hydroxy-		498	715
Benzophenone		(254)	(379)
$C_{25}H_{34}O_3$			
Dodecyl Phenol	597–633	325	
$C_{12}H_{25}C_6H_4OH$	(314–334)	(163)	
		(oc)	
Dypnone	475	350	
$C_6H_5COCH{:}C(CH_3)C_6H_5$	(246)	(177)	
(1,3-Diphenyl-2-Buten-1-one)	@50 mm		
Eicosane	651	>212	450
$C_{20}H_{42}$	(344)	(>100)	(232)
Epichlorohydrin	239	88	772
CH_2CHOCH_2Cl	(115)	(31)	(411)
(2-Chloropropylene Oxide)			
(γ-Chloropropylene Oxide)			
1,2-Epoxyethane		See Ethylene Oxide.	
Erythrene		See 1,3-Butadiene.	
Ethanal		See Acetaldehyde.	
Ethane	–128		882
CH_3CH_3	(–89)		(472)

TABLE 2.40 Boiling Points, Flash Points, and Ignition Temperatures of Organic Compounds (*Continued*)

Compound	Boiling point °F (°C)	Flash point, °F (°C)	Ignition point, °F (°C)
1,2-Ethanediol		See Ethylene Glycol.	
1,2-Ethanediol Diformate	345	200	
$HCOOCH_2CH_2OOCH$	(174)	(93)	
(Ethylene Formate)			
(Ethylene Glycol Diformate)			
(Glycol Diformate)			
Ethanethiol		See Ethyl Mercaptan.	
Ethanoic Acid		See Acetic Acid.	
Ethanoic Anhydride		See Acetic Anhydride.	
Ethanol		See Ethyl Alcohol.	
Ethanolamine	342	186	770
$NH_2CH_2CH_2OH$	(172)	(86)	(410)
(2-Amino Ethanol)			
(β-Aminoethyl Alcohol)			
Ethanoyl Chloride		See Acetyl Chloride.	
Ethene		See Ethylene.	
Ethenyl Ethanoate		See Vinyl Acetate.	
Ethenyloxyethene		See Divinyl Ether.	
Ether		See Ethyl Ether.	
Ethine		See Acetylene.	
Ethoxyacetylene	124	<20	
$C_2H_5OC{:}CH$	(51)	(<–7)	
Ethoxybenzene	342	145	
$C_6H_5OC_2H_5$	(172)	(63)	
(Ethyl Phenyl Ether) (Phenetole)			
2-Ethoxy-3,4-Dihydro-2-Pyran	289	111	
$C_7H_{12}O_2$	(143)	(44)	
2-Ethoxy Ethanol		See Ethylene Glycol Monoethyl Ether.	
2-Ethoxyethyl Acetate	313	117	716
$CH_3COOCH_2CH_2OC_2H_5$	(156)	(47)	(380)
(Ethyl Glycol Acetate)			
3-Ethoxypropanal	275	100	
$C_2H_5OC_2H_4CHO$	(135)	(38)	
(3-Ethoxypropionaldehyde)			
1-Ethoxypropane		See Ethyl Propyl Ether.	
3-Ethoxypropionaldehyde	275	100	
$C_2H_5OCH_2CH_2CHO$	(135)	(38)	
3-Ethoxypropionic Acid	426	225	
$C_2H_5OCH_2CH_2COOH$	(219)	(107)	
Ethoxytriglycol	492	275	
$C_2H_5O(C_2H_4O)_3H$	(256)	(135)	
(Triethylene Glycol, Ethyl Ether)			
Ethyl Abietale	662	352	
$C_{19}H_{29}COOC_2H_5$	(350)	(178)	
N-Ethylacetamide	401	230	
$CH_3CONHC_2H_5$	(205)	(110)	
(Acetoethylamide)			
N-Ethyl Acetanilide	400	126	
$CH_3CON(C_2H_5)(C_6H_5)$	(204)	(52)	
Ethyl Acetate	171	24	800
$CH_3COOC_2H_5$	(77)	(–4)	(426)
(Acetic Ester)			
(Acetic Ether)			
(Ethyl Ethanoate)			

(*Continued*)

TABLE 2.40 Boiling Points, Flash Points, and Ignition Temperatures of Organic Compounds (*Continued*)

Compound	Boiling point °F (°C)	Flash point, °F (°C)	Ignition point, °F (°C)
Ethyl Acetoacetate $C_2H_5CO_2CH_2COCH_3$ (Acetoacetic Acid, Ethyl Ester) (Ethyl 3-Oxobutanoate)	356 (180)	135 (57)	563 (295)
Ethyl Acetyl Glycolate $CH_3COOCH_2COOC_2H_5$ (Ethyl Glycolate Acetate)	−365 (−185)	180 (82)	
Ethyl Acrylate $CH_2{:}CHCOOC_2H_5$	211 (99)	50 (10)	702 (372)
Ethyl Alcohol C_2H_5OH (Grain Alcohol, Ethanol)	173 (78)	55 (13)	685 (363)
Ethylamine $C_2H_5NH_2$ 70% aqueous solution (Aminoethane)	62 (17)	<0 (<−18)	725 (385)
Ethyl Amino Ethanol $C_2H_5NHC_2H_4OH$ [2-(Ethylamino)ethanol]	322 (161)	160 (71)	
Ethylaniline $C_2H_5NH(C_6H_5)$	401 (205)	185 (85)	
Ethylbenzene $C_2H_5C_6H_5$ (Ethylbenzol) (Phenylethane)	277 (136)	70 (21)	810 (432)
Ethyl Benzoate $C_6H_5COOC_2H_5$	414 (212)	190 (88)	914 (490)
Ethylbenzol		See Ethylbenzene.	
Ethyl Bromide C_2H_5Br (Bromoethane)	100 (38)	None	952 (511)
Ethyl Bromoacetate $BrCH_2COOC_2H_5$	318 (159)	118 (48)	
2-Ethylbutanol		See 2-Ethylbutyraldehyde.	
Ethyl Butanoate		See Ethyl Butyrate.	
2-Ethyl-1-Butanol		See 2-Ethylbutyl Alcohol.	
2-Ethyl-1-Butene $(C_2H_5)_2C{:}CH_2$	144 (62)	<−4 (<−20)	599 (315)
3-(2-Ethylbutoxy) Propionic Acid $CH_3CH_2CH(C_2H_5)CH_2{-}OCH_2CH_2COOH$	392 (200) @100 mm	280 (138)	
2-Ethylbutyl Acetate $CH_3COOCH_2CH(C_2H_5)_2$	324 (162)	130 (54)	
2-Ethylbutyl Acrylate $CH_2{:}CHCOOCH_2CH{-}$ $(C_2H_5)C_2H_5$	180 (82) @10 mm	125 (52)	
2-Ethylbutyl Alcohol $(C_2H_5)_2CHCH_2OH$ (2-Ethyl-1-Butanol)	301 (149)	135 (57) (oc)	
Ethylbutylamine $CH_3CH_2CH_2CH_2{-}NHCH_3CH_2$	232 (111)	64 (18)	
Ethyl Butylcarbamate		See N-Butylurethane.	
Ethyl Butyl Carbonate $(C_2H_5)(C_4H_9)CO_3$	275 (135)	122 (50)	

TABLE 2.40 Boiling Points, Flash Points, and Ignition Temperatures of Organic Compounds (*Continued*)

Compound	Boiling point °F (°C)	Flash point, °F (°C)	Ignition point, °F (°C)
Ethyl Butyl Ether $C_2H_5OC_4H_9$ (Butyl Ethyl Ether)	198 (92)	40 (4)	
2-Ethyl Butyl Glycol $(C_2H_5)_2CHCH_2OC_2H_4OH$ [2-(2-Ethylbutoxy)ethanol]	386 (197)	180 (82)	
Ethyl Butyl Ketone $C_2H_5CO(CH_2)_3CH_3$ (3-Heptanone)	299 (148)	115 (46)	
2-Ethyl-2-Butyl-1,3-Propanediol $HOCH_2C(C_2H_5)(C_4H_9)$- CH_2OH	352 (178) @50 mm	280 (138)	
2-Ethylbutyraldehyde $(C_2H_5)_2CHCHO$ (Diethyl Acetaldehyde) (2-Ethylbutanal)	242 (117)	70 (21)	
Ethyl Butyrate $CH_3CH_2CH_2COOC_2H_5$ (Butyric Acid, Ethyl Ester) (Butyric Ester) (Ethyl Butanoate)	248 (120)	75 (24)	865 (463)
2-Ethylbutyric Acid $(C_2H_5)_2CHCOOH$ (Diethyl Acetic Acid)	380 (193)	210 (99)	752 (400)
2-Ethylcaproaldehyde		See 2-Ethylhexanal.	
Ethyl Caproate $C_5H_{11}COOC_2H_5$ (Ethyl Hexoate) (Ethyl Hexanoate)	333 (167)	120 (49)	
Ethyl Caprylate $CH_3(CH_2)_6COOC_2H_5$ (Ethyl Octoate)	405–408 (207–209)	175 (79)	
Ethyl Octanoate		See Diethyl Carbonate.	
Ethyl Chloride C_2H_5Cl (Chloroethane) (Hydrochloric Ether) (Muriatic Ether)	54 (12)	−58 (−50)	966 (519)
Ethyl Chloroacetate $ClCH_2COOC_2H_5$	295 (146)	147 (64)	
Ethyl Chlorocarbonate		See Ethyl Chloroformate.	
Ethyl Chloroformate $ClCOOC_2H_5$ (Ethyl Chlorocarbonate) (Ethyl Chloromethanoate)	201 (94)	61 (16)	932 (500)
Ethyl Chloromethanoate		See Ethyl Chloroformate.	
Ethyl Crotonate $CH_3CH{:}CHCOOC_2H_5$	282 (139)	36 (2)	
Ethyl Cyanoacetate $CH_2CNCOOC_2H_5$	401–408 (205–209)	230 (110)	
Ethylcyclobutane $C_2H_5C_4H_7$	160 (71)	<4 (<−16)	410 (210)

(*Continued*)

TABLE 2.40 Boiling Points, Flash Points, and Ignition Temperatures of Organic Compounds (*Continued*)

Compound	Boiling point °F (°C)	Flash point, °F (°C)	Ignition point, °F (°C)
Ethylcyclohexane	269	95	460
$C_2H_5C_6H_{11}$	(132)	(35)	(238)
N-Ethylcyclohexylamine		86	
$C_6H_{11}NHC_2H_5$		(30)	
Ethylcyclopentane	218	<70	500
$C_2H_5C_5H_9$	(103)	(<21)	(260)
Ethyl Decanoate	469	>212	
$C_9H_{19}COOC_2H_5$	(243)	(>100)	
(Ethyl Caprate)			
N-Ethyldiethanolamine	487	280	
$C_2H_5N(C_2H_4OH)_2$	(253)	(138)	
Ethyl Dimethyl Methane		See Isopentane.	
Ethylene	−155		842
$H_2C{:}CH_2$	(−104)		(450)
(Ethene)			
Ethylene Acetate		See Glycol Diacetate.	
Ethylene Carbonate	351	290	
OCH_2CH_2OCO (ring)	(177)	(143)	
	@100 mm		
Ethylene Chlorohydrin		See 2-Chloroethanol.	
Ethylene Cyanohydrin	445	265	
$CH_2(OH)CH_2CN$	(229)	(129)	
(Hydracrylonitrile)	Decomposes		
Ethylenediamine	241	104	725
$H_2NCH_2CH_2NH_2$	(116)	(40)	(385)
Anydrous 76%	239–252	150	
	(115–122)		(66)
Ethylene Dichloride	183	56	775
CH_2ClCH_2Cl	(84)	(13)	(413)
(1,2-Dichloroethone)			
2,2-Ethylenedioxydiethanol		See Triethylene Glycol.	
Ethylene Formate		See 1,2-Ethanediol Diformate.	
Ethylene Glycol	387	232	748
HOC_2H_4OH	(197)	(111)	(398)
(1,2-Ethanediol)			
(Glycol)			
Ethylene Glycol n-Butyl Ether	340	150	
$HOCH_2CH_2OC_4H_9$	(171)	(66)	
Ethylene Glycol Diacetate		See Glycol Diacetate.	
Ethylene Glycol Dibutyl Ether	399	185	
$C_4H_9OC_2H_4OC_4H_9$	(204)	(85)	
Ethylene Glycol Diethyl Ether	251	95	406
$C_2H_5OCH_2CH_2OC_2H_5$	(122)	(35)	
Ethylene Glycol Diformate		See 1,2-Ethanediol Diformate.	
Ethylene Glycol Dimethyl Ether	174	29	395
	(79)	(−2)	(202)
$CH_3O(CH_2)_2OCH_3$	@630 mm		
(1,2-Dimethoxyethane)			
Ethylene Glycol Ethylbutyl Ether	386	180	
	(197)	(85)	
$(C_2H_5)_2CHCH_2OCH_2CH_2OH$			

TABLE 2.40 Boiling Points, Flash Points, and Ignition Temperatures of Organic Compounds (*Continued*)

Compound	Boiling point °F (°C)	Flash point, °F (°C)	Ignition point, °F (°C)
Ethylene Glycol Ethylhexyl Ether $C_4H_9CH(C_2H_5)CH_2OCH_2CH_2OH$	442 (228)	230 (110)	
Ethylene Glycol Isopropyl Ether $(CH_3)_2CHOCH_2CH_2OH$	289 (143)	92 (33)	
Ethylene Glycol Monoacetate $CH_2OHCH_2OOCCH_3$ (Glycol Monoacetate)	357 (181)	215 (102)	
Ethylene Glycol Monoacrylate $CH_2{:}CHCOOC_2H_4CH$ (2-Hydroxyethylacrylate)	410 (210)	220 (104) (oc)	
Ethylene Glycol Monobenzyl Ether $C_6H_5CH_2OCH_2CH_2OH$	493 (256)	265 (129)	665 (352)
Ethylene Glycol Monobutyl Ether $C_4H_9O(CH_2)_2(OH)$ (2-Butoxyethanol)	340 (171)	143 (62)	460 (238)
Ethylene Glycol Monobutyl Ether Acetate $C_4H_9O(CH_2)_2OOCCH_3$	377 (192)	160 (71)	645 (340)
Ethylene Glycol Monoethyl Ether $HOCH_2CH_2OC_2H_5$ (2-Ethoxyethanol)	275 (135)	110 (43)	455 (235)
Ethylene Glycol Monoethyl Ether Acetate $CH_3COOCH_2CH_2OC_2H_5$ (Cellosolve Acetate)	313 (156)	124 (52)	715 (379)
Ethylene Glycol Monoisobutyl Ether $(CH_3)_2CHCH_2OCH_2CH_2OH$	316–323 (158–162)	136 (58)	540 (282)
Ethylene Glycol Monomethyl Ether $CH_3OCH_2CH_2OH$ (2-Methoxyethanol)	255 (124)	102 (39)	545 (285)
Ethylene Glycol Monomethyl Ether Acetal $CH_3CH(OCH_2CH_2OCH_3)_2$	405 (207)	200 (93)	
Ethylene Glycol Monomethyl Ether Acetate $CH_3O(CH_2)_2OOCCH_3$	293 (145)	120 (49)	740 (392)
Ethylene Glycol Monomethyl Ether Formal $CH_2(OCH_2CH_2OCH_3)_2$	394 (201)	155 (68)	
Ethylene Glycol Phenyl Ether $C_6H_5OC_2H_4OH$ (2-Phenoxyethanol)	473 (245)	260 (127)	

(*Continued*)

TABLE 2.40 Boiling Points, Flash Points, and Ignition Temperatures of Organic Compounds (*Continued*)

Compound	Boiling point °F (°C)	Flash point, °F (°C)	Ignition point, °F (°C)
Ethylene Oxide	51	−20	1058
CH_2OCH_2 (ring)	(11)		with No Air
(Dimethylene Oxide)			
(1,2-Epoxyethane)			
(Oxirane)			
Ethylenimine	132	12	608
$NHCH_2CH_2$ (ring)	(56)	(−11)	(320)
(Aziridine)			
Ethyl Ethanoate		See Ethyl Acetate.	
N-Ethylethanolomine	322	160	
$C_2H_5NHC_2H_4OH$	(161)	(71)	
Ethyl Ether	95	−49	356
$C_2H_5OC_2H_5$	(35)	(−45)	(180)
(Diethyl Ether)			
(Diethyl Oxide)			
(Ether)			
(Ethyl Oxide)			
Ethylethylene Glycol		See 1,2-Butanediol.	
Ethyl Fluoride			
C_2H_5F	−36		
(1-Fluoroethane)	(−38)		
Ethyl Formate	130	−4	851
$HCO_2C_2H_5$	(54)	(−20)	(455)
(Ethyl Methanoate)			
(Formic Acid, Ethyl Ester)			
Ethyl Formate (ortho)	291	86	
$(C_2H_5O)_3CH$	(144)	(30)	
(Triethyl Orthoformate)			
Ethyl Glycol Acetate		See 2-Ethoxyethyl Acetate.	
2-Ethylhexaldehyde		See 2-Ethylhexanal.	
2-Ethylhexanal	325	112	375
$C_4H_9CH(C_2H_5)CHO$	(163)	(44)	(190)
(Butylethylacelaldehyde)			
(2-Ethylcaproaldehyde)			
(2-Ethylhexaldehyde)			
2-Ethyl-1,3-Hexanediol	472	260	680
$C_3H_7CH(OH)CH(C_2H_5)CH_2OH$	(244)	(127)	(360)
2-Ethylhexanoic Acid	440	245	700
$C_4H_9CH(C_2H_5)COOH$	(227)	(118)	(371)
(2-Ethyl Hexoic Acid)			
2-Ethylhexanol	359	164	448
$C_4H_9CH(C_2H_5)CH_2OH$	(182)	(73)	(231)
(2-Ethylhexyl Alcohol)			
(Octyl Alcohol)			
2-Ethylhexenyl		See 2-Ethyl-3-Propylacrolein.	
2-Ethylhexoic Acid		See 2-Ethylhexanoic Acid.	
2-Ethylhexyl Acetate	390	160	515
$CH_3COOCH_2CH(C_2H_5)C_4H_9$	(199)	(71)	(268)
(Octyl Acetate)			
2-Ethylhexyl Acrylate	266	180	485
$CH{:}CHCOOCH_2CH—$	(130)	(82)	(252)
$(C_2H_5)C_4H_9$	@50 mm		

TABLE 2.40 Boiling Points, Flash Points, and Ignition Temperatures of Organic Compounds (*Continued*)

Compound	Boiling point °F (°C)	Flash point, °F (°C)	Ignition point, °F (°C)
2-Ethylhexylamine $C_4H_9CH(C_2H_5)CH_2NH_2$	337 (169)	140 (60)	
N-2-(Ethylhexyl) Anlline $C_6H_5NHCH_2CH(C_2H_5)C_4H_9$	379 (193) @50 mm	325 (163)	
2-Ethylhexyl Chloride $C_4H_9CH(C_2H_5)CH_2Cl$	343 (173)	140 (60)	
N-(2-Ethylhexyl)cyclohexylamine $C_6H_{11}NH[CH_2CH—(C_2H_5)C_4H_9]$	342 (172) @50 mm	265 (129)	
2-Ethylhexyl Ether $[C_4H_9CH(C_2H_5)CH_2]_2O$	517 (269)	235 (113)	
1,1-Ethylidene Dichloride CH_3CHCl_2 (1,1-Dichloroethane)	135–138 (57–59)	2 (–17)	
1,2-Ethylidene Dichloride $ClCH_2CH_2Cl$	183 (84)	55 (13)	824 (440)
Ethyl Isobutyrate $(CH_3)_2CHCOOC_2H_5$	230 (110)	<70 (<21)	
2-Ethylisohexanol $(CH_3)_2CHCH_2CH(C_2H_5)CH_2OH$ (2-Ethyl Isohexyl Alcohol) (2-Ethyl-4-Methyl Pentanol)	343–358 (173–181)	158 (70)	600 (316)
Ethyl Lactate $CH_3CHOHCOOC_2H_5$	309 (154)	115 (46)	752 (400)
Tech.		131 (55)	
Ethyl Malonate		See Diethyl Malonate.	
Ethyl Mercaptan C_2H_5SH (Ethanethiol) (Ethyl Sulfhydrate)	9 (35)	<0 (<–18)	572 (300)
Ethyl Methacrylate $CH_2:C(CH_3)COOC_2H_5$ (Ethyl Methyl Acrylate)	239–248 (115–120)	68 (20)	
Ethyl Methanoate		See Ethyl Formate.	
Ethyl Methyl Acrylate		See Ethyl Methacrylate.	
Ethyl Methyl Ether		See Methyl Ethyl Ether.	
7-Ethyl-2-Methyl-4-Hendecanol $C_4H_9CH(C_2H_5)C_2H_4$-$CHOHCH_2CH(CH_3)_2$	507 (264)	285 (141)	
Ethyl Methyl Ketone		See Methyl Ethyl Ketone	
4-Ethylmorpholine $CH_2CH_2OC_2H_4NCH_2CH_3$	280 (138)	90 (32)	
1-Ethylnaphthalene $C_{10}H_7C_2H_5$	496 (258)		896 (480)
Ethyl Nitrate $CH_3CH_2ONO_2$ (Nitric Ether)	190 (88)	50 (10)	
Ethyl Nitrite C_2H_5ONO (Nitrous Ether)	63 (17)	–31 (–35)	194 (90)

(*Continued*)

TABLE 2.40 Boiling Points, Flash Points, and Ignition Temperatures of Organic Compounds (*Continued*)

Compound	Boiling point °F (°C)	Flash point, °F (°C)	Ignition point, °F (°C)
3-Ethyloctane $C_5H_{11}CH(C_2H_5)C_2H_5$	333 (167)		446 (230)
4-Ethyloctane $C_4H_9CH(C_2H_5)C_3H_7$	328 (164)		445 (229)
Ethyl Oxalate $(COOC_2H_5)_2$ (Oxalic Ether) (Diethyl Oxalate)	367 (186)	168 (76)	
Ethyl Oxide		See Ethyl Ether.	
p-Ethylphenol $HOC_6H_4C_2H_5$	426 (219)	219 (104)	
Ethyl Phenylacetate $C_6H_5CH_2COOC_2H_5$	529 (276)	210 (99)	
Ethyl Phenyl Ether		See Ethoxybenzene.	
Ethyl Phenyl Ketone $C_2H_5COC_6H_5$ (Propiophenone)	425 (218)	210 (99)	
Ethyl Phthalyl Ethyl Glycolate $C_2H_5OCOC_6H_4OCO$- $CH_2OCOC_2H_5$	608 (320)	365 (185)	
Ethyl Propenyl Ether $CH_3CH{:}CHOCH_2CH_3$	158 (70)	>19 (>–7)	
Ethyl Proplonate $C_2H_5COOC_2H_5$	210 (99)	54 (12)	824 (440)
2-Ethyl-3-Propylacrolein $C_3H_7CH{:}C(C_2H_5)CHO$ (2-Ethylhexenal)	347 (175)	155 (68)	
2-Ethyl-3-Propylacrylic Acid $C_3H_7CH{:}C(C_2H_5)COOH$	450 (232)	330 (166)	
Ethyl Propyl Ether $C_2H_5OC_3H_7$ (1-Ethoxypropane)	147 (64)	<–4 (<–20)	
m-Ethyltoluene $CH_3C_6H_4C_2H_5$ (1-Methyl-3-Ethylbenzene)	322 (161)		896 (480)
o-Ethyltoluene $CH_3C_6H_4C_2H_5$ (1-Methyl-2-Ethylbenzene)	329 (165)		824 (440)
p-Ethyltoluene $CH_3C_6H_4C_2H_5$ (1-Methyl-4-Ethylbenzene)	324 (162)		887 (475)
Ethyl p-Toluene Sulfonamide $C_7H_7SO_2NHC_2H_5$	208 (98) @745 mm	260 (127)	
Ethyl p-Toluene Sulfonate $C_7H_7SO_3C_2H_5$	345 (174)	316 (158)	
Ethyl Vinyl Ether		See Vinyl Ethyl Ether.	
Ethyne		See Acetylene.	
Fluorobenzene C_6H_5F	185 (85)	5 (–15)	
Formal		See Methylal.	
Formalin		See Formaldehyde.	

TABLE 2.40 Boiling Points, Flash Points, and Ignition Temperatures of Organic Compounds (*Continued*)

Compound	Boiling point °F (°C)	Flash point, °F (°C)	Ignition point, °F (°C)
Formaldehyde	−3	Gas	795
HCHO	(−19)	185	(424)
37% Methanol-free	214	(85)	
	(101)		
37%, 15% Methanol		122	
(Formalin)		(50)	
(Methylene Oxide)			
Formamide	410	310	
$HCONH_2$	(210)	(154)	
	Decomposes		
Formic Acid	213	156	1004
HCOOH	(101)	(69)	(539)
90% Solution		122	813
		(50)	(434)
Formic Acid, Butyl Ester		See Butyl Formate.	
Formic Acid, Ethyl Ester		See Ethyl Formate.	
Formic Acid, Methyl Ester		See Methyl Formate.	
Fuel Oil No. 1	304–574	100–162	410
(Kerosene)	(151–301)	(38–72)	(210)
(Range Oil)			
Fuel Oil No. 2		126–204	494
		(52–96)	(257)
Fuel Oil No. 4		142–240	505
		(61–116)	(263)
Fuel Oil No. 5			
Light		156–336	
Heavy		(69–169)	
		160–250	
		(71–121)	
Fuel Oil No. 6		150–270	765
		(66–132)	(407)
2-Furaldehyde		See Furfural.	
Furan	88	<32	
CH:CHCH:CHO	(31)	(<0)	
(Furfuran)			
Furfural	322	140	600
OCH:CHCH:CHCHO	(161)	(60)	(316)
(2-Furaldehyde)			
(Furfuraldehyde)			
(Furol)			
Furfuraldehyde		See Furfural.	
Furfuran		See Furan.	
Furfuryl Acetate	356–367	185	
$OCH:CHCH:CCH_2OOCCH_3$	(180–186)	(85)	
Furfuryl Alcohol	340	167	915
$OCH:CHCH:CCH_2OH$	(171)	(75)	(491)
		(oc)	
Furfurylamine	295	99	
$C_4H_3OCH_2NH_2$	(146)	(37)	
Furol		See Furfural.	
Fusel Oil		See Isoamyl Alcohol.	
Gas Oil	500–700	150+	640
	(260–371)	(66+)	(338)

(*Continued*)

TABLE 2.40 Boiling Points, Flash Points, and Ignition Temperatures of Organic Compounds (*Continued*)

Compound	Boiling point °F (°C)	Flash point, °F (°C)	Ignition point, °F (°C)
Gasoline	100–400	–45	
C_5H_{12} to C_9H_{20}	(38–204)	(–43)	
56–60 Octane		–45	536
73 Octane		(–43)	(280)
92 Octane		–36	853
100 Octane		(–38)	(456)
Gasoline			
100–130 (Aviation Grade)		–50 (–46)	824 (440)
Gasoline			
115–145 (Aviation Grade)		–50 (–46)	880 (471)
Gasoline (Casinghead)		0 (–18)	
Glycerine	340	390	698
$HOCH_2CHOHCH_2OH$	(171)	(199)	(370)
(Glycerol)			
α,β-Glycerine Dichiorohydrin	360	200	
$CH_2ClCHClCH_2OH$	(182)	(93)	
Glycerol		See Glycerine.	
Glyceryl Triacetate	496	280	812
$(C_3H_5)(OOCCH_3)_3$	(258)	(138)	(433)
(Triacelin)			
Glyceryl Tributyrate	597	356	765
$C_3H_5(OOCC_3H_7)_3$	(314)	(180)	(407)
(Tributyrin)			
(Butyrin)			
(Glycerol Tributyrate)			
Glyceryl Trinitrate		See Nitroglycerine.	
Glyceryl Tripropionate	540	332	790
$(C_2H_5COO)_3C_3H_5$	(282)	(167)	(421)
(Tripropionin)			
Glycidyl Acrylate	135	141	779
CH_2:CHCOOCH$_2$CHCH$_2$O (ring)	(57) @2 mm	(61)	(415)
Glycol		See Ethylene Glycol.	
Glycol Diacetate	375	191	900
$(CH_2OOCCH_3)_2$	(191)	(88)	(482)
(Ethylene Acetate)			
(Ethylene Glycol Diaceate)			
Glycol Dichloride		See Ethylene Dichloride.	
Glycol Diformate		See 1,2-Ethanediol Diformate.	
Glycol Dimercaptoacetate	280	396	
$(HSCH_2C:OOCH_2\text{—})_2$	(138)	(202)	
(GDMA)	1.2 mm		
Glycol Monoacetate		See Ethylene Glycol Monoacetate.	
Grain Alcohol		See Ethyl Alcohol.	
Hendecane	384	149	
$CH_3(CH_2)_9CH_3$	(196)	(65)	
(Undecane)			
Heptadecanol	588	310	
$C_4H_9CH(C_2H_5)C_2H_4$– $CH(OH)C_2H_4CH(C_2H_5)_2$	(309)	(154)	
(3,9-Diethyl-6-Tridecanol)			

TABLE 2.40 Boiling Points, Flash Points, and Ignition Temperatures of Organic Compounds (*Continued*)

Compound	Boiling point °F (°C)	Flash point, °F (°C)	Ignition point, °F (°C)
Heptane	209	25	399
$CH_3(CH_2)_5CH_3$	(98)	(−4)	(204)
2-Heptanol	320	160	
$CH_3(CH_2)_4CH(OH)CH_3$	(160)	(71)	
3-Heptanol	313	140	
$CH_3CH_2CH(OH)C_4H_9$	(156)	(60)	
3-Heptanone		See Ethyl Butyl Ketone.	
4-Heptanone	290	120	
$(C_3H_7)_2CO$	(143)	(49)	
(Butyrone)			
(Dipropyl Ketone)			
1-Heptene		See Heptylene.	
3-Heptene (mixed cis and trans)	203	21	
$C_3H_7CH{:}CHC_2C_5$	(95)	(−6)	
(3-Heptylene)			
Heptylamine	311	130	
$CH_3(CH_2)_6NH_2$	(155)	(54)	
(1-Aminoheptane)			
Heptylene	201	<32	500
$C_5H_{11}CH{:}CH_2$	(94)	(<0)	(260)
(1-Heptene)			
Heptylene-2-trans	208	<32	
$C_4H_9CH{:}CHCH_3$	(98)	(<0)	
(2-Heptene-trans)			
Hexachlorobutadiene			1130
$CCl_2{:}CClCCl{:}CCl_2$			(610)
Hexachloro Diphenyl Oxide			1148
$(C_6H_2Cl_3)_2O$			(620)
[Bis(Trichlorophenyl) Ether]			
Hexadecane	549	>212	396
$CH_3(CH_2)_{14}CH_3$	(287)	(>100)	(202)
(Cetane)			
tert-Hexadecanethiol	298–307	(265)	
$C_{16}H_{33}SH$	(148–153)	(129)	
(Hexadecyl-tert-Mercaptan)	@11 mm		
Hexadecylene-1	525	>212	464
$CH_3(CH_2)_{13}CH{:}CH_2$	(274)	(>100)	(240)
(1-Hexadecene)			
Hexadecyltrichiorosilane	516	295	
$C_{16}H_{33}SiCl_3$	(269)	(146)	
2,4-Hexadienal	339	154	
$CH_3CH{:}CHCH{:}CHC(O)H$	(171)	(68)	
1,4-Hexadiene	151	−6	
$CH_3CH{:}CHCH_2CH{:}CH_2$	(66)	(−21)	
(Allylpropenyl)			
Hexanal	268	90	
$CH_3(CH_2)_4CHO$	(131)	(32)	
(Caproaldehyde)			
(Hexaldehyde)			
Hexane	156	−7	437
$CH_3(CH_2)_4CH_3$	(69)	(−22)	(225)
(Hexyl Hydride)			

(*Continued*)

TABLE 2.40 Boiling Points, Flash Points, and Ignition Temperatures of Organic Compounds (*Continued*)

Compound	Boiling point °F (°C)	Flash point, °F (°C)	Ignition point, °F (°C)
1,2-Hexanediol		See Hexylene Glycol.	
2,5-Hexanediol $CH_3CH(OH)CH_2—CH_2CH(OH)CH_3$ (2,5-Dihydroxyhexane)	429 (221)	230 (110)	
2,5-Hexanedione		See Acetonyl Acetone.	
1,2,6-Hexanetriol $HOCH_2CH(OH)—(CH_2)_3CH_2OH$	352 (178) @5 mm	375 (191)	
Hexanoic Acid		See Caproic Acid.	
1-Hexanol		See Hexyl Alcohol.	
2-Hexanone		See Methyl Butyl Ketone.	
3-Hexanone $C_2H_5COC_3H_7$ (Ethyl n-Propyl Ketone)	253 (123)	95 (35)	
1-Hexene $CH_2:CH(CH_2)_3CH_3$ (Butyl Ethylene)	146 (63)	<20 (<–7)	487 (253)
2-Hexene-cis $C_3H_7CH:CHCH_3$	156 (69)	<–4 (<–20)	
3-Hexenol-cis $CH_3CH_2CH:CHCH_2CH_2OH$ (3-Hexen-l-ol) (Leaf Alcohol)	313 (156)	130 (54)	
Hexyl Acetate $(CH_3)_2CH(CH_2)_3OOCCH_3$ (Methylamyl Acetate)	285 (141)	113 (45)	
Hexyl Alcohol $CH_3(CH_2)_4CH_2OH$ (Amyl Carbinol) (1-Hexanol)	311 (155)	145 (63)	
sec-Hexyl Alcohol $C_4H_9CH(OH)CH_3$ (2-Hexanol)	284 (140)	136 (58)	
Hexylamine $CH_3(CH_2)_5NH_2$	269 (132)	85 (29)	
Hexyl Chloride		See 1-Chlorohexane.	
Hexyl Cinnamic Aldehyde $C_6H_{13}C(CHO):CHC_6H_5$ (Hexyl Cinnamaldehyde)	486 (252)	>212 (>100)	
Hexylene Glycol $CH_2OHCHOH(CH_2)_3CH_3$ (1,2-Hexanediol)	385 (196)	215 (102)	
Hexyl Ether $C_6H_{13}OC_6H_{13}$ (Dihexyl Ether)	440 (227)	170 (77)	365 (185)
Hexyl Methacrylate $C_6H_{13}OOCC(CH_3):CH_2$	388–464 (198–240)	180 (82)	
Hydracrylonitrile		See Ethylene Cyanohydrin.	
Hydralin		See Cyclohexanol.	
Hydroquinone $C_6H_4(OH)_2$ (Quinol) (Hydroquinol)	547 (286)	329 (165)	960 (516)

TABLE 2.40 Boiling Points, Flash Points, and Ignition Temperatures of Organic Compounds (*Continued*)

Compound	Boiling point °F (°C)	Flash point, °F (°C)	Ignition point, °F (°C)
Hydroquinone Di-(β-Hydroxyethyl) Ether $C_6H_4(-OCH_2CH_2OH)_2$	365–392 @ 0.3 mm (185–200)	435 (224)	875 (468)
Hydroquinone Monomethyl Ether $CH_3OC_6H_4OH$ (4-Methoxy Phenol) (Para-Hydroxyanisole)	475 (246)	270 (132)	790 (421)
o-Hydroxybenzaldehyde		See Salicylaldehyde.	
3-Hydroxybutanal		See Aldol.	
β-Hydroxybutyraldehyde		See Aldol.	
Hydroxycitronellal $(CH_3)_2C(OH)(CH_2)_3$— $CH(CH_3)CH_2CHO$ (Citronellal Hydrate) (3,7-Dimethyl-7-Hydroxyoctanal)	201–205 (94–96) @1 mm	>212 (>100)	
N-(2-Hydroxyethyl)-acetamide		See N-Acetyl Ethanolamine.	
2-Hydroxyethyl Acrylate (HEA)	410 (210)	214 (101)	1.8 @100°C
β-Hydroxyethylaniline		See 2-Anilinoethanol.	
N-(2-Hydroxyethyl) Cyclohexylamine $C_6H_{11}NH_2$ $CH_2OHCH_2NHCH_2CH_2NH_2$		249 (121)	
4-(2-Hydroxyethyl) Morpholine $C_2H_4OC_2H_4NC_2H_4OH$	437 (225)	210 (99)	
1-(2-Hydroxyethyl) Piperazine $HOCH_2CH_2—NCH_2CH_2NHCH_2CH_2$	475 (246)	255 (124)	
n-(2-Hydroxyethyl) Propylenediamine $CH_3CH(NHC_2H_4OH)CH_2NH_2$	465 (241)	260 (127)	
4-Hydroxy-4-Methyl-2-Pentanone		See Diacetone Alcohol.	
2-Hydroxy-2-methylpropionitrile		See Acetone Cyanohydrin.	
Hydroxypropyl Acrylate		See Propylene Glycol Monoacrylate.	
o-Hydroxytoluene		See o-Cresol.	
Ionone Alpha (α-Ionone) $C(CH_3)_2CH_2CH_2CH{:}C(CH_3)$— $CHCH{:}CHC(CH_3){:}O$ (α-Cyclocitrylideneacetone) [4-(2,6,6-Trimethyl-2-Cyclohexen-1-yl)-3-Buten-2-one]	259–262 (126–128) @12 mm	>212 (>100)	
Ionone Beta (β-Ionone) $C(CH_3)_2CH_2CH_2CH_2$– $C(CH_3){:}CCHCHC(CH_3){:}O$ (β-Cyclocitrylidene-acetone) [4-(2,6,6-Trimethyl-1-Cyclohexen-1-yl)-3-Buten-2-one]	284 (140) @18 mm	>212 (>100)	

(*Continued*)

TABLE 2.40 Boiling Points, Flash Points, and Ignition Temperatures of Organic Compounds (*Continued*)

Compound	Boiling point °F (°C)	Flash point, °F (°C)	Ignition point, °F (°C)
Isoamyl Acetate	290	77	680
$CH_3COOCH_2CH_2CH(CH_3)_2$	(143)	(25)	(360)
(Banana Oil)			
(3-Methyl-1-Butanol Acetate)			
(2-Methyl Butyl Ethanoate)			
Isoamyl Alcohol	270	109	662
$(CH_3)_2CHCH_2CH_2OH$	(132)	(43)	(350)
(Isobutyl Carbinol)			
(Fusel Oil)			
(3-Methyl-1-Butanol)			
tert-Isoamyl Alcohol		See 2-Methyl-2-Butanol.	
Isoamyl Butyrate	352	138	
$C_3H_7CO_2(CH_2)_2CH(CH_3)_2$	(178)	(59)	
(Isopentyl Butyrate)			
Isoamyl Chloride	212	<70	
$(CH_3)_2CHCH_2CH_2Cl$	(100)	(<21)	
(1-Chloro-3-Methylbutane)			
Isobornyl Acetate	428–435	190	
$C_{10}H_{17}OOCCH_3$	(220–224)	(88)	
Isobutane	11		860
$(CH_3)_3CH$	(−12)		(460)
(2-Methylpropane)			
Isobutyl Acetate	244	64	790
$CH_3COOCH_2CH(CH_3)_2$	(118)	(18)	(421)
(β-Methyl Propyl Ethanoate)			
Isobutyl Acrylate	142–145	86	800
$(CH_3)_2CHCH_2OOCCH{:}CH_2$	(61–63)	(30)	(427)
	@15 mm		
Isobutyl Alcohol	225	82	780
$(CH_3)_2CHCH_2OH$	(107)	(28)	(415)
(Isopropyl Carbinol)			
(2-Methyl-1-Propanol)			
Isobutylamine	150	15	712
$(CH_3)_2CHCH_2NH_2$	(66)	(−9)	(378)
Isobutylbenzene	343	131	802
$(CH_3)_2CHCH_2C_6H_5$	(173)	(55)	(427)
Isobutyl Butyrate	315	122	
$C_3H_7CO_2CH_2(CH_3)_2$	(157)	(50)	
Isobutyl Carbinol		See Isoamyl Alcohol.	
Isobutyl Chloride	156	<70	
$(CH_3)_2CHCH_2Cl$	(69)	(<21)	
(1-Chloro-3-Methyl-propane)			
Isobutylcyclohexane	336		525
$(CH_3)_2CHCH_2C_6H_{11}$	(169)		(274)
Isobutylene		See 2-Methylpropene.	
Isobutyl Formate	208	<70	608
$HCOOCH_2CH(CH_3)_2$	(98)	(<21)	(320)
Isobutyl Heptyl Ketone	412–426	195	770
$(CH_3)_2CHCH_2COCH_2$—	(211–219)	(91)	(410)
$CH(CH_3)CH_2CH(CH_3)_2$			
(2,6,8-Trimethyl-4-Non-anone)			
Isobutyl Isobutyrate	291–304	101	810
$(CH_3)_2CHCOOCH_2$—$CH(CH_3)_2$	(144–151)	(38)	(432)

TABLE 2.40 Boiling Points, Flash Points, and Ignition Temperatures of Organic Compounds (*Continued*)

Compound	Boiling point °F (°C)	Flash point, °F (°C)	Ignition point, °F (°C)
Isobutyl Phenylacetate $(CH_3)_2CHCH_2OOCCH_2C_6H_5$	477 (247)	>212 (>100)	
Isobutyl Phosphate $PO_4(CH_2CH(CH_3)_2)_3$ (Triisobutyl Phosphate)	302 (150) @20 mm	275 (135)	
Isobutyl Vinyl Ether		See Vinyl Isobutyl Ether.	
Isobutyraldehyde $(CH_3)_2CHCHO$ (2-Methylpropanal)	142 (61)	−1 (−18)	385 (196)
Isobutyric Acid $(CH_3)_2CHCOOH$	306 (152)	132 (56)	900 (481)
Isobutyric Anhydride $[(CH_3)_2CHCO]_2O$	360 (182)	139 (59)	625 (329)
Isobutyronitrile $(CH_3)_2CHCN$ (2-Methylpropanenitrile) (Isopropylcyanide)	214–216 (101–102)	47 (8)	900 (482)
Isodecaldehyde $C_9H_{19}CO$	387 (197)	185 (85)	
Isodecane $C_7H_{15}CH(CH_3)_2$ (2-Methylnonane)	333 (167)		410 (210)
Isodecanoic Acid $C_9H_{19}COOH$	489 (254)	300 (149)	
Isoevgenol $(CH_3CHCH)C_6H_3OHOCH_3$ (1-Hydroxy-2 Methoxy-4-Propenylbanzene)	514 (268)	>212 (>100)	
Isoheptane $(CH_3)_2CHC_4H_9$ (2-Methylhexane) (Ethylisobutylmelhane)	194 (90)	<0 (−18)	
tert-Isohexyl Alcohol $C_2H_5(CH_3)C(OH)C_2H_5$ (3-Methyl-3-Pentanol)	252 (122)	115 (46)	
Isooctane $(CH_3)_2CHCH_4C(CH_3)_3$ (2,2,4-Trimethylpentane)	210 (99)	40 (4.5)	784 (418)
Isooctyl Alcohol $C_7H_{15}CH_2OH$ (Isooctanol)	83–91 (182–195)	180 (82)	
Isooctyl Nitrate $C_8H_{17}NO_3$	106–109 (41–43) @1 mm	205 (96)	
Isooctyl Vinyl Ether		See Vinyl Isooctyl Ether.	
Isopentaldehyde $(CH_3)_2CHCH_2CHO$	250 (121)	48 (9)	
Isopentane $(CH_3)_2CHCH_2CH_3$ (2-Methylbutane) (Ethyl Dimethyl Methane)	82 (28)	<–60 (<–51)	788 (420)

(*Continued*)

TABLE 2.40 Boiling Points, Flash Points, and Ignition Temperatures of Organic Compounds (*Continued*)

Compound	Boiling point °F (°C)	Flash point, °F (°C)	Ignition point, °F (°C)
Isopentanoic Acid	361		781
$(CH_3)_2CHCH_2COOH$	(183)		(416)
(Isovaleric Acid)			
Isophorone	419	184	860
$COCHC(CH_3)CH_2C(CH_3)_2CH_2$	(215)	(84)	(460)
Isophthaloyl Chloride	529	356	
$C_6H_4(COCl)_2$	(276)	(180)	
(m-Phthalyl Dichloride)			
Isoprene	93	−65	743
$CH_2{:}C(CH_3)CH{:}CH_2$	(34)	(−54)	(395)
(2-Methyl-1,3-Butadiene)			
Isopropanol		See Isopropyl Alcohol.	
Isopropenyl Acetate	207	60	808
$CH_3COOC(CH_3){:}CH_2$	(97)	(16)	(431)
(1-Methylvinyl Acetate)			
Isopropenyl Acetylene	92	<19	
$CH_2{:}C(CH_3)C{:}CH$	(33)	(<−7)	
2-Isopropoxypropane		See Isopropyl Ether.	
3-Isopropoxyproplonitrile	149	155	
$(CH_3)_2CHOCH_2CH_2CN$	(65)	(68)	
	@10 mm		
Isopropyl Acetate	194	35	860
$(CH_3)_2CHOOCCH_3$	(90)	(2)	(460)
Isopropyl Alcohol	181	53	750
$(CH_3)_2CHOH$	(83)	(12)	(399)
(Isopropanol)			
(Dimethyl Carbinol)		57	
(2-Propanol)			
87.9% iso		(14)	
Isopropylamine	89	−35	756
$(CH_3)_2CHNH_2$	(32)	(−37)	(402)
Isopropylbenzene		See Cumene.	
Isopropyl Benzoate	426	210	
$C_6H_5COOCH(CH_3)_2$	(219)	(99)	
Isopropyl Bicyclohexyl	530–541	255	446
$C_{15}H_{28}$	(277–283)	(124)	(230)
2-Isopropylbiphenyl	518	285	815
$C_{15}H_{16}$	(270)	(141)	(435)
Isopropyl Carbinol		See Isobutyl Alcohol.	
Isopropyl Chloride	95	−26	1100
$(CH_3)_2CHCl$	(35)	(−32)	(593)
(2-Chloropropane)			
Isopropylcyclohexane	310		541
$(CH_3)_2CHC_6H_{11}$	(154.5)		(283)
(Hexahydrocumene)			
(Normanthane)			
Isopropylcyclohexylamine		93	
$C_6H_{11}NHCHC_2H_6$		(34)	
Isopropyl Ether	156	−18	830
$(CH_3)_2CHOCH(CH_3)_2$	(69)	(−28)	(443)
(2-Isopropoxypropane)			
(Diisopropyl Ether)			

TABLE 2.40 Boiling Points, Flash Points, and Ignition Temperatures of Organic Compounds (*Continued*)

Compound	Boiling point °F (°C)	Flash point, °F (°C)	Ignition point, °F (°C)
Isopropylethylene		See 3-Methyl-1-Butene.	
Isopropyl Formate HCOOCH$(CH_3)_2$ (Isopropyl Methanoate)	153 (67)	22 (−6)	905 (485)
4-Isopropylheptane $C_3H_7CH(C_3H_7)C_3H_7$ (m-Dihydroxybenzene)	155 (68)		491 (255)
Isopropyl-2-Hydroxypropanoate		See Isopropyl Lactate.	
Isopropyl Lactate $CH_3CHOHCCOCH(CH_3)_2$ (Isopropyl-2-Hydroxypropionate)	331–334 (166–168)	130 (54)	
Isopropyl Methanoate		See Isopropyl Formate.	
4-Isopropyl-1-Methyl Benzene		See p-Cymene.	
Isopropyl Vinyl Ether		See Vinyl Isopropyl Ether.	
Isovalerone		See Diisobutyl Ketone.	
Jet Fuel Jet A and Jet A-1	400–550 (204–288)	110–150 (43–66)	
Jet Fuel Jet B		−10 to +30 (−23 to −1)	
Jet Fuel JP-4		−10 to +30 (−23 to −1)	464 (240)
Jet Fuel JP-5		95–145 (35–63)	475 (246)
Jet Fuel JP-6	250 (121)	100 (38)	446 (230)
Kerosene		See Fuel Oil No. 1.	
Lactonitrile $CH_3CH(OH)CN$	361 (183)	171 (77)	
Lanolin (Wool Grease)		460 (238)	833 (445)
Lard Oil (Commercial or Animal)		395 (202)	833 (445)
No. 1		440 (227)	
Lard Oil (Pure)		500 (260)	
No. 2		419 (215)	
Mineral		404 (207)	
Lauryl Alcohol		See 1-Dodecanol.	
Lauryl Bromide $CH_3(CH_2)_{10}CH_2Br$ (Dodecyl Bromide)	356 (180) @45 mm	291 (144)	
Lauryl Mercaptan		See 1-Dodecanethiol.	
Linalool $(CH_3)_2C{:}CHCH_2CH_2C(CH_3)$— OHCA:$CH_2$ (3,7-Dimethyl-1,6-Octadiene-3-01)	383–390 (195-199)	160 (71)	
Linseed Oil	600+ (316+)	432 (222)	650 (343)

(*Continued*)

TABLE 2.40 Boiling Points, Flash Points, and Ignition Temperatures of Organic Compounds (*Continued*)

Compound	Boiling point °F (°C)	Flash point, °F (°C)	Ignition point, °F (°C)
Lubricating Oil (Paraffin Oil, includes Motor Oil)	680 (360)	300–450 (149–232)	500–700 (260–371)
Lubricating Oil, Spindle (Spindle Oil)		169 (76)	478 (248)
Lubricating Oil, Turbine (Turbine Oil)		400 (204)	700 (371)
Lynalyl Acetate $(CH_3)_2C{:}CHCH_2CH_2$—$C($—$OOCCH_3)CH{:}CH_2$ (Bergamol)	226–230 (108–110)	185 (85)	
Maleic Anhydride $(COCH)_2O$	396 (202)	215 (102)	890 (477)
Marsh Gas		See Methane.	
2-Mercaptoethanol $HSCH_2CH_2OH$	315 (157)	165 (74)	
Mesitylene		See 1,3,5-Trimethylbenzene.	
Mesityl Oxide $(CH_3)_2CCHCOCH_3$	266 (130)	87 (31)	652 (344)
Metaldehyde $(C_2H_4O)_4$	subl. 233–240 (112–116)	97 (36)	
α-Methacrolein		See 2-Methylpropenal.	
Methacrylic Acid $CH_2{:}C(CH_3)COOH$	316 (158)	171 (77)	154 (68)
Methacrylonitrile C_4H_5N	194 (90)	34 (1.1)	
Methallyl Alcohol $CH_2C(CH_3)CH_2OH$	237 (114)	92 (33)	
Methallyl Chloride $CH_2C(CH_3)CH_2Cl$	162 (72)	11 (−12)	
Methane CH_4 (Marsh Gas)	−259 (−162)		999 (537)
Methanol		See Methyl Alcohol.	
Methanethiol		See Methyl Mercaptan.	
o-Methoxybenzaldehyde $CH_3OC_6H_4CHO$ (o-Anisaldehyde)	275 (135)	104 (40)	
Methoxybenzene		See Anisole.	
3-Methoxybutanol $CH_3CH(OCH_3)CH_2CH_2OH$	322 (161)	165 (74)	
3-Methoxybutyl Acetate $CH_3OCH(CH_3)CH_2CH_2OOCCH_3$ (Butoxyl)	275-343 (135–173)	170 (77)	
3-Methoxybutyraldehyde $CH_3CH(OCH_3)CH_2CHO$ (Aldol Ether)	262 (128)	140 (60)	
2-Methoxyethanol		See Ethylene Glycol Monomethyl Ether.	

TABLE 2.40 Boiling Points, Flash Points, and Ignition Temperatures of Organic Compounds (*Continued*)

Compound	Boiling point °F (°C)	Flash point, °F (°C)	Ignition point, °F (°C)
2-Methoxyethyl Acrylate	142	180	
$C_2H_3COOC_2H_4OCH_3$	(61)	(82)	
	@17 mm		
Methoxy Ethyl Phthalate	376–412	275	
(Methox)	(191–211)	(135)	
3-Methoxypropionitrile	320	149	
$CH_3OC_2H_4CN$	(160)	(65)	
3-Methoxypropylamine	241	90	
$CH_3OC_3H_6NH_2$	(116)	(32)	
Methoxy Triglycol	480	245	
$CH_3O(C_2H_4O)_3H$	(249)	(118)	
(Triethylene Glycol, Methyl Ether)			
Methoxytriglycol Acetate	266	260	
$CH_3COO(C_2H_4O)_3CH_3$	(130)	(127)	
Methyl Abietate	680–689	356	
$C_{19}H_{29}COOCH_3$	(360–365)	(180)	
(Abalyn)	Decomposes		
Methyl Acetate	140	14	850
CH_3COOCH_3	(60)	(–10)	3.1
(Acetic Acid Methyl Ester)		(454)	16
(Methyl Acetic Ester)			
Methyl Acetic Ester		See Methyl Acetate.	
Methyl Acetoacetate	338	170	536
$CH_3CO_2CH_2COCH_3$	(170)	(77)	(280)
P-Methyl Acetophenone	439	205	
$CH_3C_6H_4COCH_3$	(226)	(96)	
(Methyl-p-Tolyl Ketone)			
(p-Acetotoluene)			
Methylacetylene		See Propyne.	
Methyl Acrylate	176	27	875
CH_2:$CHCOOCH_3$	(80)	(–3)	(468)
Methylal	111	–26	459
$CH_3OCH_2OCH_3$	(44)	(–32)	(237)
(Dimethoxymethane)			
(Formal)			
Methyl Alcohol	147	52	867
CH_3OH	(64)	(11)	(464)
(Methanol)			
(Wood Alcohol)			
Methylamine	21	806	4
CH_3NH_2	(–6)	(430)	
2-(Methylamino) Ethanol		See N-Methylethanolamine.	
Methylamyl Acetate		See Hexyl Acetate.	
Methylamyl Alcohol		See Methyl Isobutyl Carbinol.	
Methyl Amyl Ketone	302	102	740
$CH_3CO(CH_2)_4CH_3$	(150)	(39)	(393)
2-Heptanone			
2-Methylaniline		See o-Toluidine.	
4-Methylaniline		See p-Toluidine.	
Methyl Anthranilate	275	>212	
$H_2NC_6H_4CO_2CH_3$	@15 mm	(>100)	
(Methyl-ortho-Amino Benzoate)	(135)		
(Nevoli Oil, Artificial)			

(*Continued*)

TABLE 2.40 Boiling Points, Flash Points, and Ignition Temperatures of Organic Compounds (*Continued*)

Compound	Boiling point °F (°C)	Flash point, °F (°C)	Ignition point, °F (°C)
Methylbenzene		See Toluene	
Methyl Benzoate $C_6H_5COOCH_3$ (Niobe Oil)	302 (150)	181 (83)	
α-Methylbenzyl Alcohol		See Phenyl Methyl Carbinol.	
α-Methylbenzylamine $C_6H_5CH(CH_3)NH_2$	371 (188)	175 (79)	
α-Methylbenzyl Dimethyl Amine $C_6H_5CH(CH_3)N(CH_3)_2$	384 (196)	175 (79)	
α-Methylbenzyl Ether $C_6H_5CH(CH_3)OCH(CH_3)C_6H_5$	548 (287)	275 (135)	
2-Methylbiphenyl $C_6H_5C_6H_4CH_3$	492 (255)	280 (137)	936 (502)
Methyl Borate $B(OCH_3)_3$ (Trimethyl Borate)	156 (69)	<80 (<27)	
Methyl Bromide CH_3Br (Bromomethane)	38.4 (4)	999 (537)	
2-Methyl-1,3-Butadiene		See Isoprene.	
2-Methylbutane		See Isopentane.	
3-Methyl-2-Butanethiol $C_5H_{11}SH$ (Sec-Isoamyl Mercaptan)	230 (110)	37 (3)	
2-Methyl-1-Butanol $CH_3CH_2CH(CH_3)CH_2OH$	262 (128)	122 (50)	725 (385)
2-Methyl-2-Butanol $CH_3CH_2(CH_3)_2COH$ (tert-Isoamyl Alcohol) (Dimethyl Ethyl Carbinol)	215 (102)	67 (19)	819 (437)
3-Methyl-1-Butanol		See Isoamyl Alcohol.	
3-Methyl-1-Butanol Acetate		See Isoamyl Acetate.	
2-Methyl-1-Butene $CH_2{:}C(CH_3)CH_2CH_3$	88 (31)	<20 (<–7)	
2-Methyl-2-Butene $(CH_3)_2C{:}CCHCH_3$ (Trimethylethylene)	101 (38)	<20 (<–7)	
3-Methyl-1-Butene $(CH_3)_2CHCH{:}CH_2$ (Isopropylethylene)	68 (20)	<20 (<–7)	689 (365)
N-Methylbutylamine $CH_3CH_2CH_2CH_2NHCH_3$	196 (91)	55 (13)	
2-Methyl Butyl Ethanoate		See Isoamyl Acetate.	
Methyl Butyl Ketone $CH_3CO(CH_2)_3CH_3$ (2-Hexanone)	262 (128)	77 (25)	795 (423)
3-Methyl Butynol $(CH_3)_2C(OH)C{:}CH$	218 (103)	77 (25)	
2-Methylbutyraldehyde $CH_3CH_2CH(CH_3)CHO$	198–199 (92–93)	49 (9)	
Methyl Butyrate $CH_3OOCCH_2CH_2CH_3$	215 (102)	57 (14)	

TABLE 2.40 Boiling Points, Flash Points, and Ignition Temperatures of Organic Compounds (*Continued*)

Compound	Boiling point °F (°C)	Flash point, °F (°C)	Ignition point, °F (°C)
Methyl Carbonate	192	66	
$CO(OCH_3)_2$	(89)	(19)	
(Dimethyl Carbonate)		(oc)	
Methyl Cellosolve Acetate	292	~111	
$CH_3COOC_2H_4OCH_3$	(144)	(~44)	
(2-Methoxyethyl Acetate)			
Methyl Chloride	−11	−50	1170
CH_3Cl	(−24)		(632)
(Chloromethane)			
Methyl Chloroacetate	266	135	
$CH_2ClCOOCH_3$	(130)	(57)	
(Methyl Chloroethanoate)			
Methyl Chloroethanoate		See Methyl Chloroacetate.	
Methyl-p-Cresol		140	
$CH_3C_6H_4OCH_3$		(60)	
(p-Methylanisole)			
Methyl Cyanide		See Acetonitrile.	
Methylcyclohexane	214	25	482
$CH_2(CH_2)_4CHCH_3$	(101)	(−4)	(250)
(Cyclohexylmethane)			
(Hexahydrotoluene)			
2-Methylcyclohexanol	329	149	565
$C_7H_{13}OH$	(165)	(65)	(296)
3-Methylcyclohexonol		158	563
$CH_3C_6H_{10}OH$		(70)	(295)
4-Methylcyclohexanol	343	158	563
$C_7H_{13}OH$	(173)	(70)	(295)
Methylcyclohexanone	325	118	
$C_7H_{12}O$	(163)	(48)	
4-Methylcyclohexene	217	30	
$CH{:}CHCH_2CH(CH_3)CH_2CH_2$	(103)	(−1)	
Methylcyclohexyl Acetate	351–381	147	
$C_9H_{16}O_2$	(177–194)	(64)	
Methyl Cyclopentadiene	163	120	833
C_6H_8	(73)	(49)	(445)
Methylcyclopentane	161	<20	496
C_6H_{12}	(72)	(<−7)	(258)
2-Methyldecane	374		437
$CH_3(CH_2)_7CH(CH_3)_2$	(190)		(225)
Methyldichlorosilane	106	15	>600
CH_3HSiCl_2	(41)	(−9)	(316)
N-Methyldiethanolamine	464	260	
$CH_3N(C_2H_4OH)_2$	(240)	(127)	
1-Methyl-3,5-Diethyl-benzene	394		851
$(CH_3)C_6H_3(C_2H_5)_2$	(201)		(455)
(3,5-Diethyltoluene)			
Methyl Dihydroabietate	689–698	361	
$C_{19}H_{31}COOCH_3$	(365–370)	(183)	
Methylene Chloride	104		1033
CH_2Cl_2	(40)	None	(556)
(Dichloromethane)			

(*Continued*)

TABLE 2.40 Boiling Points, Flash Points, and Ignition Temperatures of Organic Compounds (*Continued*)

Compound	Boiling point °F (°C)	Flash point, °F (°C)	Ignition point, °F (°C)
Methylenedianiline	748–750	428	
$H_2NC_6H_4CH_2C_6H_4NH_2$	(398–399)		
(MDA)	@78 mm		
(p,p′-DiaminodiPhenylmethane)		(220)	
Methylene Dlisocyanate		185	
$CH_2(NCO)_2$		(85)	
Methylene Oxide		See Formaldehyde.	
N-Methylethanolamine			
$CH_3NHCH_2CH_2OH$	319	165	
(2-(Methylamino) Ethanol)	(159)	(74)	
Methyl Ether	−11	Gas	662
$(CH_3)_2O$	(−24)		(350)
(Dimethyl Ether)			
(Methyl Oxide)			
Methyl Ethyl Carbinol		See sec-Butyl Alcohol.	
2-Methyl-2-Ethyl-	244	74	
1,3-Dioxolane	(118)	(23)	
$(CH_3)(C_2H_5)COCH_2CH_2O$			
Methyl Ethylene Glycol		See Propylene Glycol.	
Methyl Ethyl Ether	51	−35	374
$CH_3OC_2H_5$	(11)	(−37)	(190)
(Ethyl Methyl Ether)			
2-Methyl-4-Ethylhexane	273	<70	536
$(CH_3)_2CHCH_2CH(C_2H_5)_2$	(134)	(<21)	(280)
(4-Ethyl-2-Methylhexane)			
3-Methyl-4-Ethylhexane	284	75	
$C_2H_5CH(CH_3)CH(C_2H_5)_2$	(140)	(24)	
(3-Ethyl-4-Methylhexane)			
Methyl Ethyl Ketone	176	16	759
$C_2H_5COCH_3$	(80)	(−9)	(404)
(2-Butanone)			
(Ethyl Methyl Ketone)			
Methyl Ethyl Ketoxime	306–307	156–170	
$CH_3C(C_2H_5)$:HOH	(152–153)	(69–77)	
2-Methyl-3-Ethylpentane	241	<70	860
$(CH_3)_2CHCH(C_2H_5)_2$	(116)	(<21)	(460)
(3-Ethyl-2-Methylpentane)			
2-Methyl-5-Ethyl-piperidine	326	126	
$NHCH(CH_3)CH_2CH_2CH(C_2H_5)CH_2$	(163)	(52)	
2-Methyl-5-Ethylpyridine	353	155	
$N:C(CH_3)CH:CHC(C_2H_5):CH$	(178)	(68)	
Methyl Formate	90	−2	840
CH_3OOCH	(32)	(−19)	(449)
(Formic Acid, Methyl Ether)			
2-Methylfuran	144–147	−22	
$C_4H_3OCH_3$	(62–64)	(−30)	
(Sylvan)			
Methyl Glycol Acetate		111	
$CH_2OHCHOHCH_2CO_1CH_3$		(44)	
(Propylene Glycol Acetate)			

TABLE 2.40 Boiling Points, Flash Points, and Ignition Temperatures of Organic Compounds (*Continued*)

Compound	Boiling point °F (°C)	Flash point, °F (°C)	Ignition point, °F (°C)
Methyl Heptolocyl Ketone	329	255	
$C_{17}H_{35}COCH_3$	(165)	(124)	
	@3 mm		
Methylheptenone	343–345	135	
$(CH_3)_2C{:}CH(CH_2)_2COCH_3$	(173–174)	(57)	
(6-Methyl-5-Hepten-2-one)			
Methyl Heptine Carbonate		190	
$CH_3(CH_2)_4C{:}CCOOCH_3$		(88)	
(Methyl 2-Octynoate)			
Methyl Heptyl Ketone	361–383	140	680
$C_7H_{15}COCH_4$	(183–195)	(60)	(360)
(5-Methyl-2-Octanone)			
2-Methylhexane	194	<0	536
$(CH_3)_2CH(CH_2)_3CH_3$	(90)	(<–18)	(280)
3-Methylhexane	198	25	536
$CH_3CH_2CH(CH_3)CH_2CH_2CH_3$	(92)	(–4)	(280)
Methyl Hexyl Ketone	344	125	
$CH_3COC_6H_{13}$	(173.5)	(52)	
(2-Octanone)			
(Octanone)			
Methyl-3-Hydroxybutyrate	347	180	
$CH_3CHOHCH_2COOCH_3$	(175)	(82)	
Methyl Ionone	291	>212	
$C_{14}H_{22}O$	(144)	(>100)	
(Irone)	@16 mm		
Methyl Isoamyl Ketone	294	96	375
$CH_3COCH_2CH_2CH(CH_3)_2$	(146)	(36)	(191)
Methyl Isobutyl Carbinol	266–271	106	
$CH_3CHOHCH_2CHCH_3CH_3$	(130–133)	(41)	
(1,3-Dimethylbutanol)			
(4-Methyl-2-Pentanol)			
(Methylamyl Alcohol)			
Methylisobutylcarbinol Acetate		See 4-Methyl-2-Pentanol Acetate.	
Methyl Isobutyl Ketone	244	64	840
$CH_3COCH_2CH(CH_3)_2$	(118)	(18)	(448)
(Hexone)			
(4-Methyl-2-Pentanone)			
Methyl Isopropenyl Ketone	208		
$CH_2COC{:}CH_2(CH_3)$	(98)		
Methyl Isocyanate	102	19	994
CH_3NCO	(39)	(–7)	(534)
(Methyl Carbonimide)			
Methyl Iso Eugenol	504–507	>212	
$CH_3CH{:}CHC_6H_3(OCH_3)_2$	(262–264)	(>100)	
(Propenyl Guaiacol)			
Methyl Lactate	293	121 725	@2.2 mm
$CH_3CHOHCOOCH_3$	(145)	(49) (385)	
	@2.2 mm		212
			(100)
Methyl Mercaptan	42.4		
CH_3SH	(6)		
(Methanethiol)			

(*Continued*)

TABLE 2.40 Boiling Points, Flash Points, and Ignition Temperatures of Organic Compounds (*Continued*)

Compound	Boiling point °F (°C)	Flash point, °F (°C)	Ignition point, °F (°C)
β-Methyl Mercapto-propionaldehyde $CH_3SC_2H_4CHO$ (3-(Methylthio) Propionalde-hyde)	~329 (~165)	142 (61)	491 (255)
Methyl Methacrylate $CH_2{:}C(CH_3)COOCH_3$	212 (100)	50 (10)	
Methyl Methanoate		See Methyl Formate.	
4-Methylmorpholine $C_2H_4OC_2H_4NCH_3$	239 (115)	75 (24)	
1-Methylnaphthalene $C_{10}H_7CH_3$	472 (244)		984 (529)
Methyl Nonyl Ketone $C_9H_{19}COCH_3$	433 (223)	192 (89)	
Methyl Oxide		See Methyl Ether.	
Methyl Pentadecyl Ketone $C_{15}H_{31}COCH_3$	313 (156) @3 mm	248 (120)	
2-Methyl-1,3-Pentadiene $CH_2{:}C(CH_3)CH{:}CHCH_3$	169 (76)	<–4 (<–20)	
4-Methyl-1,3-Pentadiene $CH_2{:}CHCH_2{:}C(CH_3)_2$	168 (76)	–30 (–34)	
Methylpentaldehyde $CH_3CH_2CH_2C(CH_3)HCHO$ (Methyl Pentanal)	243 (117)	68 (20)	
Methyl Pentanal		See Methylpentaldehyde.	
2-Methylpentane $(CH_3)_2CH(CH_2)_2CH_3$ (Isohexane)	140 (60)	<20 (<–7)	583 (306)
3-Methylpentane $CH_3CH_2CH(CH_3)CH_2CH_3$	146 (63)	<20 (<–7)	532 (278)
2-Methyl-1,3-Pentanediol $CH_3CH_2CH(OH)CH(CH_3)CH_2OH$	419 (215)	230 (110)	
2-Methyl-2,4-Pentanediol $(CH_3)_2C(OH)CH_2CH(OH)CH_3$	385 (196)	205 (96)	
2-Methylpentanoic Acid $C_3H_7CH(CH_3)COOH$	381 (194)	225 (107)	712 (378)
2-Methyl-1-Pentanol $CH_3(CH_2)_2CH(CH_3)CH_2OH$	298 (148)	129 (54)	590 (310)
4-Methyl-2-Pentanol		See Methyl Isobutyl Carbinol.	
4-Methyl-2-Pentanol Acetate $CH_3COOCH(CH_3)CH_2CH(CH_3)_2$ (Methylisobutylcarbinol Acetate)	295 (146)	110 (43)	660 (349)
4-Methyl-2-Pentanone		See Methyl Isobutyl Ketone.	
2-Methyl-1-Pentene $CH_2{:}C(CH_3)CH_2CH_2CH_3$	143 (62)	<20 (<–7)	572 (300)
4-Methyl-1-Pentene $CH_2{:}CHCH_2CH(CH_3)_2$	129 (54)	<20 (<–7)	572 (300)

TABLE 2.40 Boiling Points, Flash Points, and Ignition Temperatures of Organic Compounds (*Continued*)

Compound	Boiling point °F (°C)	Flash point, °F (°C)	Ignition point, °F (°C)
2-Methyl-2-Pentene	153	<20	
$(CH_3)_2C:CHCH_2CH_3$	(67)	(<–7)	
4-Methyl-2-Pentene	133–137	<20	
$CH_3CH:CHCH(CH_3)_2$	(56–58)	(<–7)	
3-Methyl-1-Pentynol	250	101	
$(C_2H_5)(CH_3)C(OH)C:CH$	(121)	(38)	
o-Methyl Phenol		See o-Cresol.	
Methyl Phenylacetate	424	195	
$C_6H_5CH_2COOCH_3$	(218)	(91)	
Methylphenyl Carbinol	399	200	
$C_6H_5CH(CH_3)OH$	(204)	(93)	
(α-Methylbenzyl Alcohol)			
(Styralyl Alcohol)			
(sec-Phenethyl Alcohol)			
Methyl Phenyl Carbinyl Acetate		195	
$C_6H_5CH(CH_3)OOCH_3$		(91)	
(α-Methyl-Benzyl Acetate)			
(Styrolyl Acetate)			
(sec-Phenylethyl Acetate)			
(Phenyl Methylcarbinyl Acetate)			
Methyl Phenyl Ether		See Anisole.	
Methyl Phthalyl Ethyl Glycolate	590	380	
$CH_3COOC_6H_4COO–$	(310)	(193)	
$CH_2COOC_2H_5$			
1-Methyl Piperazine	280	108	
$CH_3NCH_2CH_2NHCH_2CH_2$	(138)	(42)	
2-Methylpropanal		See Isobutyraldehyde.	
2-Methylpropane		See Isobutane.	
2-Methyl-2-Propanethiol	149–153	<–20	
$(CH_3)_3CSH$	(65–67)	(<–29)	
(tert-Butyl Mercaptan)			
2-Methyl Propanol-1		See Isobutyl Alcohol.	
2-Methyl-2-Propanol		See tert-Butyl Alcohol.	
2-Methylpropenal	154	35	
$CH_2:C(CH_3)CHO$	(68)	(2)	
(Methacrolein)			
(α-Methyl Acrolein)			
2-Methylpropene	20		869
$CH_2:C(CH_3)CH_3$	(–7)		(465)
(γ-Butylene)			
(Isobutylene)			
Methyl Propionate	176	28	876
$CH_3COOCH_2CH_3$	(80)	(–2)	(469)
Methyl Propyl Acetylene	185	<14	
$CH_3C_2H_4ClCCH_3$	(85)	(<–10)	
(2-Hexyne)			
Methyl Propyl Carbinol	247	105	
$CH_3CHOHC_3H_7$	(119)	(41)	
(2-Pentanol)			

(*Continued*)

TABLE 2.40 Boiling Points, Flash Points, and Ignition Temperatures of Organic Compounds (*Continued*)

Compound	Boiling point °F (°C)	Flash point, °F (°C)	Ignition point, °F (°C)
Methylpropylcarbinylumine		See sec-Amylamine.	
Methyl n-Propyl Ether $CH_3OC_3H_7$	102 (39)	<–4 (<–20)	
Methyl Propyl Ketone $CH_3COC_3H_7$ (2-Pentanone)	216 (102)	45 (7)	846 (452)
2-Methylpyrazine N:C(CH$_3$)CH:NCH:CH (ring)		122 (50)	
2-Methyl Pyridine		See 2-Picoline.	
Methylpyrrole N(CH$_3$)CH:CHCH:CH (ring)	234 (112)	61 (16)	
Methylpyrrolidine $CH_3NC_4H_5$	180 (82)	7 (–14)	
1-Methyl-2-Pyrrolidone CH$_3$NCOCH$_2$CH$_2$CH$_2$ (ring) (N-Methyl-2-Pyrrolidone)	396 (202)	204 (96)	655 (346)
Methyl Salicylate $HOC_6H_4COOCH_3$ (Oil of Wintergreen) (Gaultheria Oil) (Betula Oil) (Sweet-Birch Oil)	432 (222)	205 (96)	850 (454)
Methyl Stearate $C_{17}H_{35}COOCH_3$	421 (216)	307 (153)	
α-Methylstyrene **1-Methylethenyl Benzene** **1-Methyl-1-phenylethene**	329–331 (165–166)	129 (54)	1066 (574)
Methyl Sulfate		See Dimethyl Sulfate.	
2-Methyltetrahydrofuran $C_4H_7OCH_3$	176 (80)	12 (–11)	
Methyl Toluene Sulfonate $CH_3C_6H_4SO_3CH_3$	315 (157) @8 mm	306 (152)	
Methyltrichlorosilane CH_3SiCl_3 (Methyl Silico Chloroform) (Trichloromethylsilane)	151 (66)	15 (–9)	>760 (>404)
Methyl Undecyl Ketone $C_{11}H_{23}COCH_3$ (2-Tridecanone)	248 (120)	225 (107)	
1-Methylvinyl Acetate		See Isopropenyl Acetate.	
Methyl Vinyl Ether		See Vinyl Methyl Ether.	
Methyl Vinyl Ketone $CH_3COCH:CH_2$	177 (81)	20 (–7)	915 (491)
Mineral Wax		See Wax, Ozocerite.	
Morpholine OC$_2$H$_4$NHCH$_2$CH$_2$ (ring)	262 (128)	98 (37)	555 (290)
Mustard Oil $C_3H_5N:C:S$ (Allyl Isothiocyanate)	304 (151)	115 (46)	

TABLE 2.40 Boiling Points, Flash Points, and Ignition Temperatures of Organic Compounds (*Continued*)

Compound	Boiling point °F (°C)	Flash point, °F (°C)	Ignition point, °F (°C)
Naphtha, Coal		107 (42)	531 (277)
Naphtha, Petroleum		See Petroleum Ether.	
Naphtha V.M. & P., 50° Flash (10)	240–290 (116–143)	50 (10)	450 (232)
Naphtha V.M. & P., High Flash	280–350 (138–177)	85 (29)	450 (232)
Naphtha V.M. & P., Regular	212–320 (100–160)	28 (−2)	450 (232)
Naphthalene $C_{10}H_8$	424 (218)	174 (79)	979 (526)
β-Naphthol $C_{10}H_7OH$ (β-Hydroxy Naphthalene) (2-Naphthol)	545 (285)	307 (153)	
1-Naphthylamine $C_{10}H_7NH_2$	572 (300)	315 (157)	
Nechexane		See 2,2-Dimethylbutane.	
Neopentone		See 2,2-Dimethylpropane.	
Neopentyl Glycol $HOCH_2C(CH_3)_2CH_2OH$ (2,2-Dimethyl 1,3 Propanediol)	410 (210)	265 (129)	750 (399)
Nicoline $C_{10}H_{14}N_2$	475 (246)		471 (244)
Niobe Oil		See Methyl Benzoate.	
Nitric Ether		See Ethyl Nitrate.	
p-Nitroaniline $NO_2C_6H_4NH_2$	637 (336)	390 (199)	
Nitrobenzene $C_6H_5NO_2$ (Nitrobenzol) (Oil of Mirbane)	412 (211)	190 (88)	900 (482)
1,3-Nitrobenzotrifluoride $C_6H_4NO_2CF_3$ α,μ,α-Trifluoronitrotoluene	397 (203)	217 (103)	
Nitrobenzol		See Nitrobenzene.	
Nitrobiphenyl $C_6H_5C_6H_4NO_2$	626 (330)	290 (143)	
p-Nitrochlorobenzene $C_6H_4ClNO_2$ (1-Chloro-4-Nitrobenzene)	468 (242)	261 (127)	
Nitrocyclohexane $CH_2(CH_2)_4CHNO_2$	403 (206) Decomposes	190 (88)	
Nitroethane $C_2H_5NO_2$	237 (114)	82 (28)	778 (414)
Nitroglycerine $C_3H_5(NO_3)_3$ (Glyceryl Trinitrate)	502 (261) Explodes	Explodes	518 (270)
Nitromethane CH_3NO_2	214 (101)	95 (35)	785 (418)
1-Nitronaphthalene $C_{10}H_7NO_2$	579 (304)	327 (164)	

(*Continued*)

TABLE 2.40 Boiling Points, Flash Points, and Ignition Temperatures of Organic Compounds (*Continued*)

Compound	Boiling point °F (°C)	Flash point, °F (°C)	Ignition point, °F (°C)
1-Nitropropane $CH_3CH_2CH_2NO_2$	268 (131)	96 (36)	789 (421)
2-Nitropropane $CH_3CH(NO_2)CH_3$ (sec-Nitropropane)	248 (120)	75 (24)	802 (428)
sec-Nitropropane		See 2-Nitropropane.	
m-Nitrotoluene $C_6H_4CH_3NO_2$	450 (232)	223 (106)	
o-Nitrotoluene $C_6H_4CH_3NO_2$	432 (222)	223 (106)	
p-Nitrotoluene $HO_2C_6H_4CH_3$	461 (238)	223 (106)	
2-Nitro-p-toludine $CH_3C_6H_3(NH_2)NO_2$		315 (157)	
Nitrous Ether		See Ethyl Nitrite.	
Nonadecane $CH_3(CH_2)_{17}CH_3$	628 (331)	>212 (>100)	446 (230)
Nonane C_9H_{20}	303 (151)	88 (31)	401 (205)
Nonane (iso) $C_6H_{13}CH(CH_3)_2$ (2-Methyloctane)	290 (143)		428 (220)
Nonane $C_5H_{11}CH(CH_3)C_2H_5$ (3-Methyloctane)	291 (144)		428 (220)
Nonane $C_4H_9CH(CH_3)C_3H_7$ (4-Methyloctane)	288 (142)		437 (225)
Nonene C_9H_{18} (Nonylene)	270–290 (132–143)	78 (26)	
Nonyl Acetate $CH_2COOC_9H_{19}$	378 (192)	155 (68)	
Nonyl Alcohol		See Diisobutyl Carbinol.	
Nonylbenzene $C_9H_{19}C_6H_5$	468–486 (242–252)	210 (99)	
tert-Nonyl Mercaptan $C_9H_{19}SH$	370–385 (188–196)	154 (68)	
Nonylnaphthalene $C_9H_{19}C_{10}H_7$	626–653 (330–345)	<200 (<93)	
Nonylphenol $C_6H_4(C_9H_{19})OH$	559–567 (293–297)	285 (141)	
2,5-Norbornadiene C_7H_8 (NBD)	193 (89)	–6 (–21)	
Octadecane $C_{18}H_{38}$	603 (317)	>212 (>100)	441 (227)
Octadecylene α $CH_3(CH_2)_{15}CH:CH_2$ (1-Octadecene)	599 (315)	>212 (>100)	482 (250)
Octadecyltrichlorosilane $C_{18}H_{37}SiCl_3$ (Trichlorooctadecylsilane)	716 (380)	193 (89)	

TABLE 2.40 Boiling Points, Flash Points, and Ignition Temperatures of Organic Compounds (*Continued*)

Compound	Boiling point °F (°C)	Flash point, °F (°C)	Ignition point, °F (°C)
Octadecyl Vinyl Ether		See Vinyl Octodecyl Ether.	
Octane	258	56	403
$CH_3(CH_2)_6CH_3$	(126)	(13)	(206)
1-Octanethiol	390	156	
$C_8H_{17}SH$	(199)	(69)	
(n-Octyl Mercapian)			
1-Octanol		See Octyl Alcohol.	
2-Octanol	363	190	
$CH_3CHOH(CH_2)_5CH_3$	(184)	(88)	
1-Octene	250	70	446
$CH_2{:}C_7H_{14}$	(121)	(21)	(230)
Octyl Acetate		See 2-Ethylhexyl Acetate.	
Octyl Alcohol	381	178	
$CH_3(CH_2)_6CH_2OH$	(194)	(81)	
(1-Octanol)			
Octylamine	338	140	
$CH_3(CH_2)_6CH_2NH_2$	(170)	(60)	
(1-Aminooctane)			
tert-Octylamine	284	91	
$(CH_3)_3CCH_2C(CH_3)_2NH_2$	(140)	(33)	
(1,1,3,3-Tetramethyl-butylamine)			
Octyl Chloride	359	158	
$CH_3(CH_2)_7Cl$	(182)	(70)	
Octylene Glycol	475	230	635
$(CH_3(CH_2)_2CHOH)_2$	(246)	(110)	(335)
tert-Octyl Mercaptan	318–329	115	
$C_8H_{17}SH$	(159–165)	(46) (oc)	
p-Octylphenyl Salicylate		420	780
$C_{21}H_{26}O_3$		(216)	(416)
Oil of Mirbane		See Nitrobenzene.	
Oil of Wintergreen		See Methyl Salicylate.	
Oleic Acid	547	372	685
$C_8H_{17}CH{:}CH(CH_2)_7COOH$	(286)	(189)	(363)
(Red Oil)			
Distilled		364 (184)	
Oxalic Ether		See Ethyl Oxalate.	
Oxirane		See Ethylene Oxide.	
Paraffin Oil		444	
(See also Lubricating Oil)		(229)	
Paraformaldehyde		158	572
$HO(CH_2O)_nH$		(70)	(300)
Paraldehyde	255	96	460
$(CH_3CHO)_3$	(124)	(36)	(238)
1,2,3,4,5-Pentamethyl Benzene	449	200	800 est
$C_6H(CH_3)_5$	(232)	(93)	(427)
(Pentamethylbenzene)			

(*Continued*)

TABLE 2.40 Boiling Points, Flash Points, and Ignition Temperatures of Organic Compounds (*Continued*)

Compound	Boiling point °F (°C)	Flash point, °F (°C)	Ignition point, °F (°C)
Pentamethylene Dichloride		See 1,5-Dichloropentane.	
Pentamethylene Glycol		See 1,5-Pentanediol.	
Pentamethylene Oxide $O(CH_2)_4CH_2$ (Tetrahydropyran)	178 (81)	−4 (−20)	
Pentanal		See Valeraldehyde.	
Pentane $CH_3(CH_2)_3CH_3$	97 (36)	<−40 (<−40)	500 (260)
1,5-Pentanediol $HO(CH_2)_5OH$ (Pentamethylene Glycol)	468 (242)	265 (129)	635 (335)
2,4-Pentanedione $CH_3COCH_2COCH_3$ (Acetyl Acetone)	284 (140)	93 (34)	644 (340)
Pentanoic Acid C_4H_9COOH (Valeric Acid)	366 (186)	205 (96)	752 (400)
1-Pentanol		See Amyl Alcohol.	
2-Pentanol		See Methyl Propyl Carbinol.	
3-Pentanol $CH_3CH_2CH(OH)CH_2CH_3$ (tert-n-Amyl Alcohol)	241 (116)	105 (41)	815 (435)
1-Pentanol Acetate		See Amyl Acetate.	
2-Pentanol Acetate		See sec-Amyl Acetate.	
2-Pentanone		See Methyl Propyl Ketone.	
3-Pentanone		See Diethyl Ketone.	
Pentaphen $C_5H_{11}C_6H_4OH$ (p-tert-Amyl Phenol)	482 (250)	232 (111)	
1-Pentene $CH_3(CH_2)_2CH{:}CH_2$ (Amylene)	86 (30)	0 (−18)	527 (275)
1-Pentene-cis		See β-Amylene-cis.	
2-Pentene-trans		See β-Amylene-trans.	
Pentylamine		See Amylamine.	
Pentyloxypentane		See Amyl Ether.	
Pentyl Propionate		See Amyl Propionate.	
1-Pentyne $HC_1CC_3H_7$ (n-Propyl Acetylene)	104 (40)	<−4 (<−20)	
Perchloroethylene $Cl_2C{=}CCl_2$ (Tetrachloroethylene)	250 (121)	None	None
Perhydrophenanthrene $C_{14}H_{24}$ (Tetradecahydro Phenanthrene)	187–192 (86–89)		475 (246)
Petroleum, Crude Oil		20–90 (−7 to 32)	

TABLE 2.40 Boiling Points, Flash Points, and Ignition Temperatures of Organic Compounds (*Continued*)

Compound	Boiling point °F (°C)	Flash point, °F (°C)	Ignition point, °F (°C)
Petroleum Ether (Benzine) (Petroleum Naphtha)	95–140 (35–60)	<0 (<–18)	550 (288)
Petroleum Pitch		See Asphalt.	
β-Pheliandrene CH_2:CCH:CHCH[CH(CH_3)$_2$]CH_2CH_2 (p-Mentha-1(7), 2-Diene)	340 (171)	120 (49)	
Phenanthrene $(C_6H_4CH)_2$ (Phenanthrin)	644 (340)	340 (171)	
Phenethyl Alcohol $C_6H_5CH_2CH_2OH$ (Benzyl Carbinol) (Phenylethyl Alcohol)	430 (221)	205 (96)	
o-Phenetidine $H_2NC_6H_4OC_2H_5$ (2-Ethoxyaniline) (o-Amino-Phenetole)	442–446 (228–230)	239 (115)	
p-Phenetidine $C_2H_5OC_0H_4NH_2$ (1-Amino-4-Ethoxy-benzene) (p-Aminophenetole)	378–484 (192–251)	241 (116)	
Phenetole		See Ethoxybenzene.	
Phenol C_6H_5OH (Carbolic Acid)	358 (181)	175 (79)	1319 (715)
2-Phenoxyethanol		See Ethylene Glycol, Phenyl Ether.	
Phenoxy Ethyl Alcohol $C_6H_5O(CH_2)_2OH$ (2-Phenoxyethanol) (Phenyl Cellosolve)	468 (242)	250 (121)	
N-(2-Phenoxyethyl) Anlline $C_6H_5O(CH_2)_3NHC_6H_5$	396 (202)	338 (170)	
β-Phenoxyethyl Chloride		See β-Chlorophenetole.	
Phenylacetaldehyde $C_6H_5CH_2CHO$ (α-Toluic Aldehyde)	383 (195)	160 (71)	
Phenyl Acetate $CH_3COOC_6H_5$ (Acetylphenol)	384 (196)	176 (80)	
Phenylocetic Acid $C_6H_5CH_2COOH$ (α-Toluic Acid)	504 (262)	>212 (>100)	
Phenylamine		See Aniline.	
N-Phenylaniline		See Diphenylamine.	
Phenylbenzene		See Biphenyl.	
Phenyl Bromide		See Bromobenzene.	
Phenyl Carbinol		See Benzyl Alcohol.	
Phenyl Chloride		See Chlorobenzene.	
Phenyicyclohexane		See Cyclohexylbenzene.	
Phenyl Didecyl Phosphite $(C_6H_5O)P(OC_{10}H_{21})_2$		425 (218)	

(*Continued*)

TABLE 2.40 Boiling Points, Flash Points, and Ignition Temperatures of Organic Compounds (*Continued*)

Compound	Boiling point °F (°C)	Flash point, °F (°C)	Ignition point, °F (°C)
N-Phenyldiethanolamine	376	385	730
$C_6H_5N(C_2H_4OH)_2$	(191)	(196)	(387)
Phenyidiethylamine		See N,N-Diethylaniline.	
o-Phenylenediamine	513	313	
$NH_2C_6H_4NH_2$	(267)	(156)	
(1,2-Diaminobenzene)			
Phenylethane		See Ethylbenzene.	
N-Phenylethanolamine	545	305	
$C_6H_5NHC_2H_4OH$	(285)	(152)	
Phenylethyl Acetate (β)	435	230	
$C_6H_5CH_2CH_2OOCCH_3$	(224)	(110)	
Phenylethyl Alcohol		See Phenethyl Alcohol.	
Phenylethylene		See Styrene.	
N-Phenyl-N-Ethyl-	514	270	685
ethanolamine	(268)	(132)	(362)
$C_6H_5N(C_2H_5)C_2H_4OH$	@740 mm	(oc)	
Phenylhydrazine	Decomposes	190	
$C_6H_5NHNH_2$		(88)	
Phenylmethane		See Toluene.	
Phenylmethyl Ethanol Amine	378	280	
$C_6H_5N(CH_3)C_2H_4OH$	(192)	(138)	
(2-(N-Methylaniline)-	@100 mm		
Ethanol)			
Phenyl Methyl Ketone		See Acetophenone.	
4-Phenylmorpheline	518	220	
$C_6H_5NC_2H_4OCH_2CH_2$ (ring)	(270)	(104)	
		(oc)	
Phenylpentane		See Amylbenzene.	
o-Phenylphenol	547	255	986
$C_6H_5C_6H_4OH$	(286)	(124)	(530)
Phenylpropane		See Propylbenzene.	
2-Phenylpropane		See Cumene.	
Phenylpropyl Alcohol	426	212	
$C_6H_5(CH_2)_3OH$	(219)	(100)	
(Hydrocinnamic Alcohol)			
(3-Phenyl-l-propanol)			
(Phenylethyl Carbinol)			
Phenyl Propyl Aldehyde		205	
$C_6H_5CH_2CH_2CHO$		(96)	
(3-Phenylpropionaldehyde)			
(Hydrocinnamic Aldehyde)			
Phenyl Toluene o	500	>212	923
$C_6H_5C_6H_4CH_3$	(260)	(>100)	(495)
(2-Methylbiphenyl)			
Phorone	388	185	
$(CH_3)_2CCHCOCHC(CH_3)_2$	(198)	(85)	
Phosphine	−126		212
PH_3	(−88)		(100)
Phthalic Acid	552	334	
$C_6H_4(COOH)_2$	(289)	(168)	
Phthalic Anhydride	543	305	1058
$C_6H_4(CO)_2O$	(284)	(152)	(570)

TABLE 2.40 Boiling Points, Flash Points, and Ignition Temperatures of Organic Compounds (*Continued*)

Compound	Boiling point °F (°C)	Flash point, °F (°C)	Ignition point, °F (°C)
m-Phthalyl Dichloride		See Isophthaloyl Chloride.	
2-Picoline	262	102	1000
$CH_3C_5H_4N$	(128)	(39)	(538)
(2-Methylpyridine)		(oc)	
4-Picoline	292	134	
$CH_3C_5H_4N$	(144)	(57)	
(4-Methylpyridine)			
Pinane	336		523
$C_{10}H_{18}$	(151)		(273)
α-Pinene	312	91	491
$C_{10}H_{16}$	(156)	(33)	(255)
Pine Oil	367–439	172	
Steam Distilled	(186–226)	(78)	
		138	
		(59)	
Pine Pitch	490	285	
	(254)	(141)	
Pine Tar	208	130	671
	(98)	(54)	(355)
Pine Tar Oil		144	
(Wood Tar Oil)		(62)	
Piperazine	294	178	
$HNCH_2CH_2NHCH_2CH_2$ (ring)	(146)	(81)	
		(oc)	
Piperidine	223	61	
$(CH_2)_5NH$	(106)	(16)	
(Hexahydropyridine)			
Polyamyl Naphthalene	667–747	360	
Mixture of Polymers	(353–397)	(182)	
Polyethylene Glycols		360–550	
$OH(C_2H_5O)_nC_2H_4OH$		(182–287)	
Polyoxyethylene Lauryl Ether		>200	
$C_{12}H_{25}O(OCH_2CH_2)_nOH$		(>93)	
Polypropylene Glycols	Decomposes	365	
$OH(C_3H_6O)_nC_3H_4OH$		(185)	
Polyvinyl Alcohol Mixture of		175	
Polymers		(79)	
Potasslum Xanthate	392	205	
$KS_2C\text{-}OC_2H_5$	(200)	(96)	
	Decomposes		
Propanal	120	−22	405
CH_3CH_2CHO	(49)	(−30)	(207)
(Propionaldehyde)			
Propane	−44		842
$CH_3CH_2CH_3$	(−42)		(450)
1,3-Propanediamine	276	75	
$NH_2CH_2CH_2CH_2NH_2$	(136)	(24)	
(1,3-Diaminopropane)			
(Trimethylenediamine)			
1,2-Propanediol			See Propylene Glycol.
1,3-Propanediol			See Trimethylene Glycol.
1-Propanol			See Propyl Alcohol.

(*Continued*)

TABLE 2.40 Boiling Points, Flash Points, and Ignition Temperatures of Organic Compounds (*Continued*)

Compound	Boiling point °F (°C)	Flash point, °F (°C)	Ignition point, °F (°C)
2-Propanol		See Isopropyl Alcohol.	
2-Propanone		See Acetone.	
Propanoyl Chloride		See Propionyl Chloride.	
Propargyl Alcohol	239	97	
HC_1CCH_2OH	(115)	(36)	
(2-Propyn-1-ol)			
Propargyl Bromide	192	50	615
HC_1CCH_2Br	(89)	(10)	(324)
(3-Bromopropyne)			
Propene		See Propylene.	
2-Propenylamine		See Allylamine.	
Propenyl Ethyl Ether	158	<20	
$CH_3CH{:}CHOCH_2CH_3$	(70)	(<−7)	
β-Propiolactone	311	165	
$C_3H_4O_2$	(155)	(74)	
Propionaldehyde		See Propanal.	
Propionic Acid	297	126	870
CH_3CH_2COOH	(147)	(52)	(465)
Propionic Anhydride	336	145	545
$(CH_3CH_2CO)_2O$	(169)	(63)	(285)
Propionic Nitrile	207	36	
CH_3CH_2CN	(97)	(2)	
(Propionitrile)			
Propionic Chloride	176	54	
CH_3CH_2COCl	(80)	(12)	
(Propanoyl Chloride)			
Propyl Acetate	215	55	842
$C_3H_7OOCCH_3$	(102)	(13)	(450)
(Acetic Acid, n-Propyl Ester)			
Propyl Alcohol	207	74	775
$CH_3CH_2CH_2OH$	(97)	(23)	(412)
(1-Propanol)			
Propylamine	120	−35	604
$CH_3(CH_2)_2NH_2$	(49)	(−37)	(318)
Propylbenzene	319	86	842
$C_3H_7C_6H_5$	(159)	(30)	(450)
(Phenylpropane)			
2-Propylbiphenyl	~536	>212	833
$C_6H_5C_6H_4C_3H_7$	(~280)	(>100)	(445)
n-Propyl Bromide	160		914
C_3H_7Br	(71)		(490)
(1-Bromopropane)			
n-Propyl Butyrate	290	99	
$C_3H_7COOC_3H_7$	(143)	(37)	
Propyl Carbinol		See Butyl Alcohol.	
Propyl Chloride	115	<0	968
C_3H_7Cl	(46)	(<−18)	(520)
Propyl Chlorothiolformate	311	145	
C_3H_7SCOCl	(155)	(63)	
Propylcyclohexane	313–315		478
$H_7C_3C_6H_{11}$	(156–157)		(248)
Propylcyclopentane	269		516
$C_3H_7C_5H_9$	(131)		(269)
(1-Cyclopentylpropane)			

TABLE 2.40 Boiling Points, Flash Points, and Ignition Temperatures of Organic Compounds (*Continued*)

Compound	Boiling point °F (°C)	Flash point, °F (°C)	Ignition point, °F (°C)
Propylene $CH_2:CHCH_3$ (Propene)	−53 (−47)	Gas	851 (455)
Propylene Aldehyde		See Crotonaldehyde.	
Propylene Carbonate $OCH_2CH_2CH_2OCO$ (ring)	468 (242)	275 (135)	
Propylene Chlorohydrin		See 2-Chloro-1-Propanol.	
sec-Propylene Chlorohydrin		See 1-Chloro-2-Propanol.	
Propylenedlamine $CH_3CH(NH_2)CH_2NH_2$	246 (119)	92 (33) (oc)	780 (416)
Propylene Dichloride $CH_3CHClCH_2Cl$ (1,2-Dichloropropane)	205 (96)	60 (16)	1035 (557)
Propylene Glycol $CH_3CHOHCH_2OH$ (Methyl Ethylene Glycol) (1,2-Propanediol)	370 (188)	210 (99)	700 (371)
Propylene Glycol Acetate		See Methyl Glycol Acetate.	
Propylene Glycol Isopropyl Ether	283 (140)	110 (43)	
Propylene Glycol Methyl Ether $CH_3OCH_2CHOHCH_3$ (1-Methoxy-2-propanol)	248 (120)	90 (32)	
Propylene Glycol Methyl Ether Acetate (99% Pure)	295 (146)	108 (42)	
Propylene Glycol Monoacylate $CH_2:CHCOO(C_3H_6)OH$ (Hydroxypropyl Acrylate)	410 (210)	207 (97)	
Propylene Oxide OCH_2CHCH_3 (ring)	94 (35)	−35 (−37)	840 (449)
n-Propyl Ether $(C_3H_7)_2O$ (Dipropyl Ether)	194 (90)	70 (21)	370 (188)
Propyl Formate $HCOOC_3H_7$	178 (81)	27 (−3)	851 (455)
Propyl Methanol		See Butyl Alcohol.	
Propyl Nitrate $CH_3CH_2CH_2NO_3$	231 (111)	68 (20)	347 (175)
Propyl Proplonate $CH_3CH_2COOCH_2CH_2CH_3$	245 (118)	175 (79)	
Propyltrichlorosilane $(C_3H_7)SiCl_3$	254 (123.5)	98 (37)	
Propyne CH_3C_1CH (Allylene) (Methylacetylene)	−10 (−23)		
Pseudocumene		See 1,2,4-Trimethylbenzene.	
Pyridine CH < $(CHCH)_2$ >N	239 (115)	68 (20)	900 (482)

(*Continued*)

TABLE 2.40 Boiling Points, Flash Points, and Ignition Temperatures of Organic Compounds (*Continued*)

Compound	Boiling point °F (°C)	Flash point, °F (°C)	Ignition point, °F (°C)
Pyrrole $(CHCH)_2NH$ (Azole)	268 (131)	102 (39)	
Pyrrolidine $NHCH_2CH_2CH_2CH_2$ (Tetrahydropyrrole)	186–189 (86–87)	37 (3)	
2-Pyrrolidine $NHCOCH_2CH_2CH_2$	473 (245)	265 (129)	
Quinoline $C_6H_4N{:}CHCH{:}CH$	460 (238)		896 (480)
Range Oil	See Fuel Oil No. 1.		
Rape Seed Oil (Colza Oil)		325 (163)	836 (447)
Resorcinol $C_6H_4(OH)_2$ (Dihydroxybenzol)	531 (277)	261 (127)	1126 (608)
Rhodinol $CH_2{:}C(CH_3)(CH_2)_3CH$— $(CH_3)(CH_2)_2OH$	237–239 (114–115) @12 mm	>212 (>100)	
Rosin Oil	>680 (>360)	266 (130)	648 (342)
Salicylaldehyde HOC_6H_4CHO (o-Hydroxybenzaldehyde)	384 (196)	172 (78)	
Salicylic Acid HOC_6H_4COOH	Sublimes @169 (76)	315 (157)	1004 (540)
Safrole $C_3H_5C_6H_3O_2CH_2$ (4-allyl-1,2-Mathylenedioxy- benzene)	451 (233)	212 (100)	
Santatol $C_{15}H_{24}O$ (Arheol)	~575 (~300)	>212 (>100)	
Sesame Oil		491 (255)	
Soy Bean Oil		540 (282)	833 (445)
Sperm Oil No. 1		428 (220)	586 (308)
No. 2		460 (238)	
Stearic Acid $CH_3(CH_2)_{16}COOH$	726 (386)	385 (196)	743 (395)
Steryl Alcohol $CH_3(CH_2)_{17}OH$ (1-Ocladecanol)	410 (210) @15 mm		842 (450)
Styrene $C_6H_5CH{:}CH_2$ (Cinnamene) (Phenylethylene) (Vinyl Benzene)	295 (146)	88 (31)	914 (490)

TABLE 2.40 Boiling Points, Flash Points, and Ignition Temperatures of Organic Compounds (*Continued*)

Compound	Boiling point °F (°C)	Flash point, °F (°C)	Ignition point, °F (°C)
Styrene Oxide $C_6H_5CHOCH_2$		165 (74)	929 (498)
Succinonitrile $NCCH_2CH_2CN$ (Ethylene Dicyanide)	509–513 (265–267)	270 (132)	
Sulfolane $CH_2(CH_2)_3SO_2$ (Tetrahydrothiophene-1,1-Dioxide) (Tetramethylune Sulfone)	545 (285)	350 (177)	
Tartaric Acid (d, 1) $(CHOHCO_2H)_2$		410 (210) (oc)	797 (425)
Terephthalle Acid $C_6H_4(COOH)_2$ (para-Phthalic Acid) (Benzene-para-Dicarboxylic Acid)	Sublimes above 572 (300)	500 (260)	925 (496)
Terephthaloyl Chloride $C_6H_4(COCl)_2$ (Terephthalyl Dichloride) (p-Phthalyl Dichloride) (1,4-Benzenedicarbonyl Chloride)	498 (259)	356 (180)	
o-Terphenyl $(C_6H_5)_2C_6H_4$	630 (332)	325 (163)	
m-Terphenyl $(C_6H_5)_2C_6H_4$	685 (363)	375 (191)	
Terpineol $C_{10}H_{17}OH$ (Terpilenol)	417–435 (214–224)	195 (91)	
Terpinyl Acetate $C_{10}H_{17}OOCCH_3$	428 (220)	200 (93)	
Tetraamylbenzene $(C_5H_{11})_4C_6H_2$	608–662 (320–350)	295 (146)	
1,1,2,2-Tetrabromoethane $CHBr_2CHBr_2$ **(Acetylene Tetrabromide)**	275 (135)		635 (335)
1,2,4,5-Tetrachlorobenzene $C_6H_{12}Cl_4$	472 (245)	311 (155)	
Tetradecane $CH_3(CH_2)_{12}CH_3$	487 (253)	212 (100)	392 (200)
Tetradecanol $C_{14}H_{29}OH$	507 (264)	285 (141) (oc)	
1-Tetradecene $CH_2{:}CH(CH_2)_{11}CH_3$	493 (256)	230 (110)	455 (235)
tert-Tetradecyl Mercaptan $C_{14}H_{29}SH$	496–532 (258–278)	250 (121)	
Tetraethoxypropane $(C_2H_5O)_4C_3H_4$	621 (327)	190 (88)	
Tetra (2-Ethylbutyl) Silicate $[C_2H_5CH(C_2H_5)CH_2O]_4Si$	460 (238) @50 mm	335 (168)	
Tetraethylene Glycol $HOCH_2(CH_2OCH_2)_3CH_2OH$	Decomposes	360 (182)	

(*Continued*)

TABLE 2.40 Boiling Points, Flash Points, and Ignition Temperatures of Organic Compounds (*Continued*)

Compound	Boiling point °F (°C)	Flash point, °F (°C)	Ignition point, °F (°C)
Tetraethylene Glycol, Dimethyl Ether		See Dimethoxy Tetraglycol.	
Tetraethylene Pentamine $H_2N(C_2H_4NH)_3C_2H_4NH_2$	631 (333)	325 (163)	610 (321)
Tetra (2-Ethylhexyl) Silicate $[C_4H_9CH(C_2H_5)CH_2O]_4Si$		390 (199)	
Tetrafluoroethylene $F_2C:CF_2$ (TFE) (Perfluoroethylene)	−105 (−76)		392 (200)
1,2,3,6-Tetrahydrobenzaldehyde $CH_2CH:CHCH_2CH_2CHCHO$ (3-Cyclohexene-1-Carboxaldehyde)	328 (164)	135 (57)	
endo-Tetrahydrodicyclopentadiene $C_{10}H_{16}$ (Tricyclodecane)	379 (193)		523 (273)
Tetrahydrofuran $OCH_2CH_2CH_2CH_2$ (Diethylene Oxide) (Tetramethylene Oxide)	151 (66)	6 (−14)	610 (321)
Tetrahydrofurfuryl Alcohol $C_4H_7OCH_2OH$	352 (178) @743 mm	167 (75)	540 (282)
Tetrahydrofurfuryl Oleale $C_4H_7OCH_2OOCC_{17}H_{33}$	392–545 (200–285) @16 mm	390 (199)	
Tetrahydronaphthalene $C_6H_2(CH_3)_2C_2H_4$ (Tetralin)	405 (207)	160 (71)	725 (385)
Tetrahydropyran		See Pentamethylene Oxide.	
Tetrahydropyran-2-Methanol $OCH_2CH_2CH_2CH_2CHCH_2OH$	368 (187)	200 (93)	
Tetrahydropyrrole		See Pyrrolidine.	
Tetralin		See Tetrahydronophthalene.	
1,1,3,3-Tetramethoxy-propane $[(CH_3O)_2CH]_2CH_2$	361 (183)	170 (77)	
1,2,3,4-Tetramethylbenzene 95% $C_6H_2(CH_3)_4$ (Prohnitene)	399–401 (204–205)	166 (74)	800 est. (427)
1,2,3,5-Tetramethylbenzene 85.5% $C_6H_2(CH_3)_4$ (Isodurene)	387–389 (197–198)	160 (71)	800 est. (427)
1,2,4,5-Tetramethylbenzene 95% $C_6H_2(CH_3)_4$ (Durene)	385 (196)	130 (54)	
Tetramethylene		See Cyclobutane	
Tetramethyleneglycol $CH_2OH(CH_2)_2CH_2OH$	230 (110)	734 (390)	
Tetramethylene Oxide		See Tetrahydrofuran.	
Tetramethyl Lead, Compounds $Pb(CH_3)_4$		100 (38)	

TABLE 2.40 Boiling Points, Flash Points, and Ignition Temperatures of Organic Compounds (*Continued*)

Compound	Boiling point °F (°C)	Flash point, °F (°C)	Ignition point, °F (°C)
2,2,3,3-Tetramethyl Pentane $(CH_3)_3CC(CH_3)_2CH_2CH_3$	273 (134)	<70 (<21)	806 (430)
2,2,3,4-Tetramethyl-pentane $(CH_3)_3CCH(CH_3)CH(CH_3)_2$	270 (132)	<70 (<21)	
	172	<70	
Thialdine $SCH(CH_3)SCH(CH_3)NHCHCH_3$	Decomposes	200 (93)	
2,2-Thiodiethanol $(HOCH_2CH_2)_2S$ (Thiodiethylene Glycol)	540 (282)	320 (160)	
Thiodiethylene Glycol		See 2,2-Thiodiethanol.	
Thiodiglycol $(CH_2CH_2OH)_2S$ (Thiodiethylene Glycol) (Beta-bis-Hydroxyethyl Sulfide) (Dihydroxyethyl Sulfide)	541 (283)	320 (160)	568 (298)
Thiophene $SCH{:}CHCH{:}CH$	184 (84)	30 (−1)	
1,4-Thioxane $O(CH_2CH_2)_2S$ (1,4-Oxathiane)	300 (149)	108 (42)	
Toluene $C_6H_5CH_3$ (Methylbenzene) (Phenylmethane) (Toluol)	231 (111)	40 (4)	896 (480)
Toluene-2,4-Diisocyanate $CH_3C_6H_3(NCO)_2$	484 (251)	260 (127)	
p-Toluenesulfonic Acid $C_6H_4(SO_3H)(CH_3)$	295 (140) @ 20 mm	363 (184)	
Toluhydroquinone $C_6H_3(OH)_2CH_3$ (Methylhydroquinone)	545 (285)	342 (172)	875 (468)
o-Toluidine $CH_3C_6H_4NH_2$ (2-Methylaniline)	392 (200)	185 (85)	900 (482)
p-Toluidine $CH_3C_6H_4NH_2$ (4-Mothylaniline)	392 (200)	188 (87)	900 (482)
Toluol		See Toluene.	
m-Tolydiethanolamine $(HOC_2H_4)_2NC_6H_4CH_3$ (MTDEA)	400 (204)	740 (393)	0.6
2,4-Tolylene Diisocyanate		See Toluene-2,4-Diisocyanate.	
o-Tolyl Phosphate		See Tri-o-Cresyl Phosphate.	
o-Tolyl p-Toluene Sulfonate $C_{14}H_{14}O_3S$		363 (184)	
Transformer Oil (Tronsil Oil)		295 (146)	
Triacetin		See Glyceryl Triacetate.	

(*Continued*)

TABLE 2.40 Boiling Points, Flash Points, and Ignition Temperatures of Organic Compounds (*Continued*)

Compound	Boiling point °F (°C)	Flash point, °F (°C)	Ignition point, °F (°C)
Triamylamine $(C_5H_{11})_3N$	453 (234)	215 (102)	
Triamylbenzene $(C_5H_{11})_3C_6H_3$	575 (302)	270 (132)	
Tributylamine $(C_4H_9)_3N$	417 (214)	187 (86)	
Tri-n-Butyl Borate $B(OC_4H_9)_3$	446 (230)	200 (93)	
Tributyl Citrate $C_3H_4(OH)(COOC_4H_9)_3$	450 (232)	315 (157)	695 (368)
Tributyl Phosphate $(C_4H_9)_3PO_4$	560 (293)	295 (146)	
Tributylphosphine $(C_4H_9)_3P$	473 (245)		392 (200)
Tributyl Phosphite $(C_4H_9)_3PO_3$	244–250 (118–121) @7 mm	248 (120)	
1,2,4-Trichlorobenzene $C_6H_3CI_3$	415 (213)	222 (105)	1060 (571)
1,1,1-Trichloroethane CH_3CCI_3 (Methyl Chloroform)	165 (74)		
Trichloroethylene $CIHC{:}CCI_2$	188 (87)		788 (420)
1,2,3-Trichloropropane $CH_2CICHCICH_2CI$ (Allyl Trichloride) (Glyceryl Trichlorohydrin)	313 (156)	160 (71)	
Trichlorosllane $HSiCI_3$	89 (32)	7 (−14)	
Tri-o-Cresyl Phosphate $(CH_3C_6H_4)_3PO_4$ (o-Tolyl Phosphate)	770 (410) Decomposes	437 (225)	725 (385)
Tridecanol $CH_3(CH_2)_{12}OH$	525 (274)	250 (121)	
2-Tridecanone		See Methyl Undecyl Ketone.	
Tridecyl Acrylate $CH_2{:}CHCOOC_{13}H_{27}$	302 (150) @10 mm	270 (132)	
Tridecyl Alcohol $C_{12}H_{25}CH_2OH$ (Tridecanol)	485–503 (252–262)	180 (82)	
Tridecyl Phosphite $(C_{10}H_{21}O)_3P$	356 (180) @0.1 mm	455 (235)	
Triethanolamine $(CH_2OHCH_2)_3N$ (2,2′,2″-Nitrilotriethanol)	650 (343)	354 (179)	
1,1,3-Triethoxyhexane $CH(OC_2H_5)_2CH_2CH$-$(OC_2H_5)C_8H_7$	271 (133) @50 mm Decomposes @760 mm	210 (99)	
Triethylamine $(C_2H_5)_3N$	193 (89)	16 (−7)	480 (249)

TABLE 2.40 Boiling Points, Flash Points, and Ignition Temperatures of Organic Compounds (*Continued*)

Compound	Boiling point °F (°C)	Flash point, °F (°C)	Ignition point, °F (°C)
1,2,4-Triethylbenzene	423	181	
$(C_2H_5)_3C_6H_3$	(217)	(83)	
Triethyl Cltrate	561	303	
$HOC(CH_2CO_2C_2H_5)CO_2H_2H_5$	(294)	(151)	
Triethylene Glycol	546	350	700
$HOCH_2(CH_2OCH_2)_2CH_2OH$	(286)	(177)	(371)
(Dicaproate)			
Triethylene Glycol Diacetate	572	345	
$CH_3COO(CH_2CH_2O)_3COCH_3$	(300)	(174)	
(TDAC)			
Triethylene Glycol, Dimethyl Ether	421	232	
$CH_3(OCH_2)_3OCH_3$	(216)	(111)	
Triethylene Glycol, Ethyl Ether		See Ethoxytriglycol.	
Triethylene Glycol, Methyl Ether		See Methoxy Triglycol.	
Triethyleneglycol Monobutyl Ether	270	290	
$C_4H_9O(C_2H_4O)_3H$	(132)	(143)	
Triethylenetetramine	532	275	640
$N_2NCH_2(CH_2NHCH_2)_2CH_2NH_2$	(278)	(135)	(338)
Triethyl Phosphate	408–424	240	850
$(C_2H_5)_3PO_4$	(209–218)	(115)	(454)
(Ethyl Phosphate)			
Trifluorochloroethylene	−18		
CF_2:CFCI (R-1113)	(−28)		
(Chlorotrifluoroethylene)			
Triglycol Dichloride	466	250	
$ClCH_2(CH_3OCH_2)_2CH_2Cl$	(241)	(121)	
Trihexyl Phosphite	275–286	320	
$(C_6H_{13})_3PO_3$	(135–141) @2 mm	(160)	
Triisopropanolamine	584	320	608
$[(CH_3)_2COH]_3N$	(307)	(160)	(320)
(1,1′,1″-Nitrolotri-2-propanol)			
Triisopropylbenzene	495	207	
$C_6H_3(CH_3CHCH_3)_3$	(237)	(97)	
Triisopropyl Borate	288	82	
$(C_3H_7O)_3B$	(142)	(28)	
Triiauryl Trithiophosphite		398	
$[CH_3(CH_2)_{11}S]_3P$		(203)	
Trimethylamine	38		374
$(CH_3)_3N$	(3)		(190)
1,2,3-Trimethylbenzene	349	111	878
$C_6H_3(CH_3)_3$	(176)	(44)	(470)
(Hemellitol)			
1,2,4-Trimethylbenzene	329	112	932
$C_6H_3(CH_3)_3$	(165)	(44)	(500)
(Pseudocumene)			
1,3,5-Trimethylbenzene	328	122	1039
$C_6H_3(CH_3)_3$	(164)	(50)	(559)
(Mesitylene)			
Trimethyl Borate		See Methyl Borate.	
2,2,3-Trimethylbutane	178	<32	774
$(CH_3)_3C(CH_3)CHCH_3$	(81)	(<0)	(412)
(Triptane—an isomer of Heptane)			

(*Continued*)

TABLE 2.40 Boiling Points, Flash Points, and Ignition Temperatures of Organic Compounds (*Continued*)

Compound	Boiling point °F (°C)	Flash point, °F (°C)	Ignition point, °F (°C)
2,3,3-Trimethyl-1-Butene $(CH_3)_3CC(CH_3):CH_2$ (Heplylene)	172 (78)	<32 (<0)	707 (375)
Trimethyl Carbinol		See tert-Butyl Alcohol.	
Trimethylchlorosiiane $(CH_3)_3SiCI$	135 (57)	−18 (−28)	
1,3,5-Trimethylcyclohexane $(CH_3)_3C_6H_9$ (Hexahydromesitylene)	283 (139)		597 (314)
Trimethylcyclohexanol $CH(OH)CH_2C(CH_3)_2CH_2CH(CH_3)CH_2$ (ring)	388 (198)	165 (74)	
3,3,5-Trimethyl-1-Cyclohexanol $CH_2CH(CH_3)CH_2C(CH_3)_2CH_2CHOH$ (ring)	388 (198)	190 (88)	
Trimethylene		See Cyclopropane.	
Trimethylenediamine		See 1,3-Propanediamine.	
Trimethylene Glycol $HO(CH_2)_3OH$ (1,3-Propanediol)	417 (214)		752 (400)
Trimethylethylene		See 2-methyl-2-Butene.	
2,5,5-Trimethylheptane $C_2H_5C(CH_3)_2(CH_2)_2CH(CH_3)_2$	304 (151)	<131 (<55)	527 (275)
2,2,5-Trimethylhexane $(CH_3)_3C(CH_2)_2CH(CH_3)_2$	255 (124)	55 (13) (oc)	
3,5,5-Trimethylhexanol $CH_3C(CH_3)_2CH_2CH(CH_3)CH_2$-$CH_2OH$	381 (194)	200 (93)	
2,4,8-Trimethyl-6-Nonanol $C_4H_9CH(OH)C_7H_{15}$ (2,6,8-Trimethyl-4-nonanol)	491 (255)	199 (93)	
2,6,8-Trimethyl-4-Nonanol $(CH_3)_2CHCH_2CH(OH)CH_2$-$CH(CH_3)CH_2CH(CH_3)_2$	438 (226)	200 (93)	
2,6,8-Trimethyl-4-Nonanone $(CH_3)_2CHCH_2CH(CH_3)CH_2$-$COCH_2CH(CH_3)_2$	425 (218)	195 (91)	
2,2,4-Trimethylpentane $(CH_3)_3CCH_2CH(CH_3)_2$	211 (99)	10 (−12)	779 (415)
2,3,3-Trimethylpentane $CH_3CH_2C(CH_3)_2CH(CH_3)_2$	239 (115)	<70 (<21)	797 (425)
2,2,4-Trimethyl-1,3-Pentanediol $(CH_3)_2CHCH(OH)C(CH_3)_2$-$CH_2OH$	419–455 (215–235)	235 (113)	655 (346)
2,2,4-Trimethyl pentanediol Diisobutyrate $C_{16}H_{30}O_4$	536 (280)	250 (121)	795 (424)
2,2,4-Trimethyl 1,3-Pentanediol Isobutyrate $(CH_3)_2CHCH(OH)C(CH_3)_2$-$CH_2OOCCH(CH_3)_2$	356–360 125 mm (180–182)	248 (120)	740 (393)
2,2,4-Trimethylpentanediol Isobutyrate Benzoate $C_{19}H_{28}O_4$	167 (75) @10 mm	325 (163)	

TABLE 2.40 Boiling Points, Flash Points, and Ignition Temperatures of Organic Compounds (*Continued*)

Compound	Boiling point °F (°C)	Flash point, °F (°C)	Ignition point, °F (°C)
2,3,4-Trimethyl-1-Pentene $H_2C{:}C(CH_3)CH(CH_3)CH(CH_3)_2$	214 (101)	<70 (<21)	495 (257)
2,4,4-Trimethyl-1-Pentene $CH_2{:}C(CH_3)CH_2C(CH_3)_3$ (Diisobutylene)	214 (101)	23 (−5)	736 (391)
2,4,4-Trimethyl-2-Pentene $CH_3CH{:}C(CH_3)C(CH_3)_3$	221 (105)	35 (2) (oc)	581 (305)
3,4,4-Trimethyl-2-Pentene $(CH_3)_3CC(CH_3){:}CHCH_3$	234 (112)	<70 (<21)	617 (325)
Trimethyl Phosphite $(CH_3O)_3P$	232–234 (111–112)	130 (54)	
Trioctyl Phosphite $(C_8H_{17}O)_3P$ [Tris (2-Ethylhexyl) Phosphite]	212 (100) @0.01 mm	340 (171)	
Trioxane $OCH_2OCH_2OCH_2$ (ring)	239 (115) Sublimes	113 (45)	777 (414)
Triphenylmethane $(C_6H_5)_3CH$	678 (359)	>212 (>100)	
Triphenyl Phosphate $(C_6H_5)_3PO_4$	750 (399)	428 (220)	
Triphenylphosphine		See Triphenylphosphorus.	
Triphenyl Phosphite $(C_6H_5O)_3PO_3$	311–320 (155–160) @0.1 mm	425 (218)	
Triphenylphosphorus $(C_6H_5)_3P$ (Triphenylphosphine)	711 (377)	356 (180)	
Tripropylamine $(CH_3CH_2CH_2)_3N$	313 (156)	105 (41)	
Tripropylene C_9H_{18} (Propylene Trimer)	271–288 (133–142)	75 (24)	
Tripropylene Glycol $H(OC_3H_6)_3OH$	514 (268)	285 (141)	
Tripropylene Glycol Methyl Ether $HO(C_3H_6O)_2C_3H_6OCH_3$	470 (243)	250 (121)	
Tris (2-Ethylhexyl) Phosphite		See Trioctyl Phosphite.	
Tung Oil (China Wood Oil)		552 (289)	855 (457)
Turkey Red Oil		476 (247)	833 (445)
Turpentine	300 (149)	95 (35)	488 (253)
Undecane		See Hendecane.	
2-Undecanol $C_4H_9CH(C_2H_5)C_2H_4$-$CH(OH)CH_3$	437 (225)	235 (113)	
Valeraldehyde $CH_3(CH_2)_3CHO$ (Pentanal)	217 (103)	54 (12)	432 (222)

(*Continued*)

TABLE 2.40 Boiling Points, Flash Points, and Ignition Temperatures of Organic Compounds (*Continued*)

Compound	Boiling point °F (°C)	Flash point, °F (°C)	Ignition point, °F (°C)
Valeric Acid		See Pentanoic Acid.	
Vinyl Acetate	161	18	756
CH$_2$:CHOOCCH$_3$	(72)	(−8)	(402)
(Ethenyl Ethanoate)			
Vinylaceto-β-Lactone		See Diketene.	
Vinyl Acetylene	41		
CH$_2$:CHC:CH	(5)		
(1-Buten-3-yne)			
Vinyl Allyl Ether	153	<68	
CH$_2$:CHOCH$_2$CH$_2$O(CH$_2$)$_3$CH$_3$	(67)	(<20)	
(Allyl Vinyl Ether)			
Vinylbenzene		See Styrene.	
Vinylbenzylchloride	444	220	
ClCH$_2$H$_6$H$_4$CH:CH$_2$	(229)	(104)	
Vinyl Bromide	60	None	986
	(15.8)		(530)
Vinyl Butyl Ether	202	15	437
CH$_2$:CHOCH$_4$H$_9$	(94)	(−9)	(255)
(Butyl Vinyl Ether)			
Vinyl Butyrate	242	68	
CH$_2$:CHOCOC$_3$H$_7$	(117)	(20)	
Vinyl 2-Chloroethyl Ether	228	80	
CH$_2$:CHOCH$_2$CH$_2$Cl	(109)	(27)	
(2-Chloroethyl Vinyl Ether)			
Vinyl Chloride	7	−108.4	882
CH$_2$CHCl	(−14)	(−78)	(472)
(Chloroethylene)			
Vinyl Crotonate	273	78	
CH$_2$:CHOCOCH:CHCH$_3$	(134)	(26)	
Vinyl Cyanide		See Acrylonitrile.	
4-Vinyl Cyclohexene	266	61	517
C$_8$H$_{12}$	(130)	(16)	(269)
Vinyl Ether		See Divinyl Ether.	
Vinyl Ethyl Alcohol	233	100	
CH$_2$:CH(CH$_2$)$_2$OH	(112)	(38)	
(3-Buten-1-ol)			
Vinyl Ethyl Ether	96	<−50	395
CH$_2$:CHOC$_2$H$_5$	(36)	(<−46)	(202)
(Ethyl Vinyl Ether)			
Vinyl 2-Ethylhexoate	365	165	
CH$_2$:CHOCOCH(C$_2$H$_5$)C$_4$H$_9$	(185)	(74)	
Vinyl 2-Ethylhexyl Ether	352	135	395
C$_{10}$H$_{20}$O	(178)	(57)	(202)
(2-Ethylhexyl Vinyl Ether)			
2-Vinyl-5-Ethylpyridine	248	200	
N:C(CH:CH$_2$)CH:CHC(C$_2$H$_5$):CH	(120)	(93)	
	@50 mm		
Vinyl Fluoride	−97.5		
CH$_2$:CHF	(−72)		
Vinylidene Chloride	89	−19	1058
CH$_2$:CCl$_2$	(32)	(−28)	(570)
(1,1-Dichloroethylene)			
Vinylidene Fluoride	−122.3		
CH$_2$:CF$_2$	(−86)		

TABLE 2.40 Boiling Points, Flash Points, and Ignition Temperatures of Organic Compounds (*Continued*)

Compound	Boiling point °F (°C)	Flash point, °F (°C)	Ignition point, °F (°C)
Vinyl Isobutyl Ether CH_2:$CHOCH_2CH(CH_3)CH_3$ (Isobutyl Vinyl Ether)	182 (83)	15 (−9)	
Vinyl Isooctyl Ether CH_2:$CHO(CH_2)_5CH(CH_3)_2$ (Isooctyl Vinyl Ether)	347 (175)	140 (60)	
Vinyl Isopropyl Ether CH_2:$CHOCH(CH_3)_2$ (Isopropyl Vinyl Ether)	133 (56)	−26 (−32)	522 (272)
Vinyl 2-Methoxyethyl Ether CH_2:$CHOC_2H_4OCH_3$ (1-Methoxy-2-Vinyloxyethane)	228 (109)	64 (18)	
Vinyl Methyl Ether CH_2:$CHOCH_3$ (Methyl Vinyl Ether)	43 (6)		549 (287)
Vinyl Octadecyl Ether CH_2:$CHO(CH_2)_{17}CH_3$ (Octadecyl Vinyl Ether)	297–369 (147–187) @5 mm	350 (177)	
Vinyl Propionate CH_2:$CHOCOC_2H_5$	203 (95)	34 (1)	
1-Vinylpyrrolidone CH_2:$CHNCOCH_2CH_2CH_2$ (Vinyl-2-Pyrrolidone)	205 (96) @14 mm	209 (98)	
Vinyl-2-Pyrrolidone	See 1-Vinylpyrrolidone.		
Vinyl Trichlorosilane CH_2:$CHSiCl_3$	195 (91)	70 (21)	
Wax, Microcrystalline		>400 (>204)	
Wax, Ozocerite (Mineral Wax)		236 (113)	
Wax, Paraffin	>700 (>371)	390 (199)	473 (245)
White Tar	See Naphthalene.		
Wood Alcohol	See Methyl Alcohol.		
Wood Tar Oil	See Pine Tar Oil.		
Wool Grease	See Lanolin.		
m-Xylene $C_6H_4(CH_3)_2$ (1,3-Dimethylbenzene)	282 (139)	81 (27)	982 (527)
o-Xylene $C_6H_4(CH_3)_2$ (1,2-Dimethylbenzene) (o-Xylol)	292 (144)	90 (32)	867 (463)
p-Xylene $C_6H_4(CH_3)_2$ (1,4-Dimethylbenzene)	281 (138)	81 (27)	984 (528)
o-Xylidine $C_6H_3(CH_3)_2NH_2$ (o-Dimethylaniline)	435 (224)	206 (97)	
o-Xylol	See o-Xylene.		

TABLE 2.41 Properties of Combustible Mixtures in Air

The *autoignition temperature* is the minimum temperature required for self-sustained combustion in the absence of an external ignition source. The value depends on specified test conditions. The *flammable (explosive) limits* specify the range of concentration of the vapor in air (in percent by volume) for which a flame can propagate. Below the lower flammable limit, the gas mixture is too lean to burn; above the flammable limit, the mixture is too rich.

Substance	Autoignition temperature, °C	Flammable (explosive) limits, percent by volume of fuel (25°C, 760 mm)	
		Lower	Upper
Acetaldehyde	175	4.0	60
Acetanilide	540		
Acetic acid, glacial	463	4.0	19.9
Acetic anhydride	316	2.7	10.3
Acetone	465	2.5	12.8
Acetonitrile	524	3.0	16.0
Acetophenone	570		
Acetylacetone	340		
Acetylene	305	3.0	65
Acetyl chloride	390		
Acrolein	220	2.8	31.0
Acrylic acid (2-propenoic acid)	438	2.4	8.0
Acrylonitrile	481	3.0	17.0
Adiponitrile	550	2	5
Allyl acetate	374		
Allyl alcohol	378	2.5	18.0
Allylamine	374	2.2	22
Ammonia, anhydrous	651	16	25
Aniline	615	1.3	11
Asphalt	485		
Benzaldehyde	192		
Benzene	498	1.2	7.8
Benzoyl peroxide	80		
Benzyl acetate	460		
Benzyl alcohol	436		
Benzyl benzoate	480		
Benzyl chloride	585	1.1	
Bis(2-aminoethyl)amine	399		
Bis(2-chloroethyl) ether	369	2.7	
Biscyclohexyl	245	0.7	5.1
Bis(2-hydroethyl) ether	229		
Bromobenzene	565		
1-Bromobutane	265	2.6	6.6
Bromoethane	511	6.8	8.0
Bromomethane	537	10	16.0
1-Bromopropane	490		
3-Bromopropene	295	4.4	7.3
1,3-Butadiene	420	2.0	11.5
Butanal (butyraldehyde)	218	1.9	12.5
Butane	287	1.9	8.5
1,3-Butanediol	395		
2,3-Butanediol	402		
Butanenitrile	501	1.65	
Butanoic acid (butyric acid)	443	2.0	10.0
Butanoic anhydride (butyric anhydride)	279	0.9	5.8

TABLE 2.41 Properties of Combustible Mixtures in Air (*Continued*)

Substance	Autoignition temperature, °C	Flammable (explosive) limits, percent by volume of fuel (25°C, 760 mm)	
		Lower	Upper
1-Butanol	343	1.4	11.2
2-Butanol	415	1.7	11
2-Butanone	404	1.4	11.4
trans-2-Butenal (crotonaldehyde)	232	2.1	15.9
1-Butene	384	1.6	9.3
cis-2-Butene	324	1.7	
trans-2-Butene	324	1.8	9.7
1-Butene oxide		1.5	18.3
3-Buten-1-ol		4.7	34
2-Butoxyethanol	238	4	13
2-(2-Butoxyethoxy)ethyl acetate	299		
Butyl acetate	425	1.7	7.6
sec-Butyl acetate		1.7	9.8
Butylamine	312	1.7	9.8
tert-Butylamine	380	1.7	8.9
Butylbenzene	410	0.8	5.8
sec-Butylbenzene	418	0.8	6.9
tert-Butylbenzene	450	0.7	5.7
Butyl formate	322	1.7	8.2
Butyl methyl ketone	423	1	8
Butyl 2-methyl-2-propenoate	294	2	8
Butyl propanoate	427		
Butyl stearate	355		
Butyl vinyl ether	255		
2-Butyne		1.4	
Camphor	466	0.6	3.5
Carbon disulfide	90	1.3	50.0
Carbon monoxide	609	12.5	74.2
Carbonyl sulfide		12	28.5
Chlorobenzene	593	1.3	9.6
1-Chloro-1,3-butadiene		4.0	20.0
1-Chlorobutane	240	1.8	10.1
2-Chloro-2-butene		2.3	9.3
1-Chloro-2,3-epoxypropane	411	4	21
1-Chloro-1,1-difluoroethane		6.2	17.9
1-Chloro-2,4-dinitrobenzene		2.0	22
1-Chloro-2,3-epoxypropane	411	3.8	21
Chloroethane	519	3.8	15.4
2-Chloroethanol	425	4.9	15.9
Chloromethane	632	8.1	17.4
1-Chloro-3-methylbutane		1.5	7.4
1-Chloro-2-methylpropane		2.0	8.8
3-Chloro-2-methyl-1-propene		2.3	9.3
1-Chloronaphthalene	>588		
1-Chloropentane	260	1.6	8.6
1-Chloropropane	520	2.6	11.1
2-Chloropropane	593	2.8	10.7
1-Chloro-1-propene		4.5	16
2-Chloro-1-propene		4.5	16
3-Chloro-1-propene	485	2.9	11.1
Chlorotrifluoroethylene		24	40.3
m-Cresol	558	1.1	

(*Continued*)

TABLE 2.41 Properties of Combustible Mixtures in Air (*Continued*)

Substance	Autoignition temperature, °C	Flammable (explosive) limits, percent by volume of fuel (25°C, 760 mm)	
		Lower	Upper
o-Cresol	599	1.4	
p-Cresol	558	1.1	
Cumene	424	0.9	6.5
Cyanogen		6.6	32
Cyclobutane		1.8	
Cyclohexane	245	1.3	8
Cyclohexanol	300	1	9
Cyclohexanone	420	1.1	9.4
Cyclohexene	244	1.2	
Cyclohexyl acetate	334		
Cyclohexylamine	293	1	9
Cyclopentane	361	1.5	
Cyclopentene	395		
Cyclopropane	500	2.4	10.4
p-Cymene	436	0.7	5.6
trans-Decahydronaphthalene	255	0.7	5.4
Decane	210	0.8	5.4
Decene	235		
Diborane(6)	38 to 52	0.8	88
Dibutylamine		1.1	6
Dibutyl decanedioate (dibutyl sebacate)	365	0.44	
Dibutyl ether	194	1.5	7.6
Dibutyl *o*-phthalate	402	0.5	
1,2-Dichlorobenzene	648	2.2	9.2
1,1-Dichloroethane	458	5.4	11.4
1,2-Dichloroethane	413	6.2	16
1,1-Dichloroethylene	570	6.5	15.5
cis-1,2-Dichloroethylene	460	3	15
trans-1,2-Dichloroethylene	460	6	13
Dichloromethane	556	13	23
1,2-Dichloropropane	557	3.4	14.5
Diethanolamine [2,2′-iminobis(ethanol)]	662	2	13
1,1-Diethoxyethane (acetal)	230	1.6	10.4
Diethylamine	312	1.8	10.1
Diethylene glycol [bis(2-hydroxyethyl) ether]	224	2	17
Diethylene glycol dibutyl ether	310		
Diethylene glycol monoethyl ether acetate	425		
Diethylene glycol monomethyl ether	240	1.4	22.7
Diethylenetriamine	358	2	6.7
Diethyl ether	180	1.9	36.0
3,3-Diethylpentane	290	0.7	5.7
Diethyl peroxide		2.3	15.9
Diethyl sulfate	436		
1,1-Difluoroethylene		5.5	21.3
1,3-Dihydroxybenzene (resorcinol)	664		
1,4-Dihydroxybenzene	516		
Diisopropylamine	316	1.1	7.1
Diisopropyl ether	443	1.4	7.9
Dimethoxymethane	237	2.2	13.8
N,N-Dimethylacetamide	490	2.0	11.5
Dimethylamine (anhydrous)	400	2.8	14.4
N,N-Dimethylaniline	371		

TABLE 2.41 Properties of Combustible Mixtures in Air (*Continued*)

Substance	Autoignition temperature, °C	Flammable (explosive) limits, percent by volume of fuel (25°C, 760 mm)	
		Lower	Upper
2,3-Dimethylaniline		1.0	
2,2-Dimethylbutane	405	1.2	7.0
2,3-Dimethylbutane	405	1.2	7.0
3,3-Dimethyl-2-butanone	423	1	8
cis-1,2-Dimethylcyclohexane	304		
trans-1,2-Dimethylcyclohexane	304		
Dimethyl ether	350	3.4	27.0
N,N-Dimethylformamide	445	2.2	15.2
2,6-Dimethyl-4-heptanol		0.8	6.1
2,6-Dimethyl-4-heptanone	396	0.8	6.2
2,3-Dimethylhexane	438		
1,1-Dimethylhydrazine	249	2	95
2,3-Dimethylpentane	335	1.1	6.7
Dimethyl 1,2-phthalate	490	0.9	
2,2-Dimethylpropane	450	1.4	7.5
Dimethyl sulfate	188		
Dimethyl sulfide	206	2.2	19.7
Dimethyl sulfoxide	215	2.6	42
1,4-Dioxane	180	2.0	22
Dipentene	237		
Dipentyl ether	170		
Diphenylamine	634		
Diphenyl ether	618	0.8	1.5
Dipropylamine	299		
Dipropyl ether	188	1.3	7.0
Divinyl ether	360	1.7	27.0
Dodecane	203	0.6	
1-Dodecanol	275		
1,2-Epoxybutane	439	1.7	19
Ethane	515	3.0	12.5
1,2-Ethanediamine	385	2.5	12.0
1,2-Ethanediol	398	3.2	22
Ethanethiol	299	2.8	18.2
Ethanol	363	3.3	19
Ethanolamine	410	3.0	23.5
2-Ethoxyethanol	235	3	18
2-Ethoxyethyl acetate	379	2	8
1-Ethoxypropane		1.7	9.0
Ethyl acetate	426	2	11.5
Ethyl acetoacetate	295	1.4	9.5
Ethyl acrylate	372	1.4	14
Ethylamine	385	3.5	14.0
Ethylbenzene	432	0.8	6.7
Ethyl benzoate	490		
Ethyl butanoate	463		
2-Ethylbutanoic acid	463		
Ethyl chloroformate	500		
Ethylcyclobutane	210	1.2	7.7
Ethylcyclohexane	238	0.9	6.6
Ethylene	490	2.7	36.0
Ethylene glycol diacetate	482	1.6	8.4

(Continued)

TABLE 2.41 Properties of Combustible Mixtures in Air (*Continued*)

Substance	Autoignition temperature, °C	Flammable (explosive) limits, percent by volume of fuel (25°C, 760 mm)	
		Lower	Upper
Ethylene glycol dimethyl ether	202		
Ethylene glycol ethyl ether acetate	379	2	8
Ethylene glycol monobutyl ether	238	4	13
Ethylene glycol methyl ether acetate	392	2	12
Ethylene glycol monoethyl ether	235	3	18
Ethyleneimine	320	3.3	54.8
Ethylene oxide	429	3.0	100
Ethyl formate	455	2.8	16.0
2-Ethylhexanal	197		
2-Ethyl-1,3-hexanediol	360		
2-Ethyl-1-hexanol	231	0.88	9.7
2-Ethylhexyl acetate	268	0.76	8.14
Ethyl lactate	400	1.5	
Ethyl methyl ether		2.0	10.0
3-Ethyl-2-methylpentane	460		
Ethyl nitrate	85 explodes	3.8	
Ethyl nitrite	90 explodes	3.0	50.0
Ethyl propanoate	440	1.9	11
Ethyl vinyl ether	202	1.7	28
Formaldehyde	430	7.0	73.0
Formic acid, 90%	434	18	57
2-Furaldehyde (furfural)	316	2.1	19.3
Furan		2.3	14.3
Furfuryl alcohol	491	1.8	16.3
Gasoline, 50-100 octane	280 to 456	1.4	7.6
Glycerol	370	3	19
Heptane	204	1.05	6.7
2-Heptanone (methyl pentyl ketone)	393	1.1	7.9
4-Heptanone (diisobutyl ketone)	396	0.8	7.1
1-Heptene	260		
1,1,2,3,4,4-Hexachlorobutadiene	610		
Hexane	225	1.1	7.5
1,6-Hexanedioic acid	420		
Hexanoic acid	380		
2-Hexanone	423	1	8
1-Hexene	253		
Hydrazine	23 to 270	4.7	100
Hydrogen	400	4.1	74.2
Hydrogen cyanide, 96%	538	5.6	40.0
Hydrogen sulfide	260	4	46
N-Hydroxyethyl-1,2-ethanediamine	368		
1-Hydroxy-2-methylbenzene	599	1.4	
1-Hydroxy-3-methylbenzene	559	1.1	
1-Hydroxy-4-methylbenzene (see *p*-cresol)			
4-Hydroxy-4-methyl-2-pentanone	643	1.8	6.9
Isobutanal	196	1.6	10.6
Isobutyl acetate	421	1	10.5
Isobutylamine	378	2	12
Isobutylbenzene	427	0.8	6.0
Isobutyl isobutyrate	432	0.96	7.59
Isopentane	420	1.4	7.6
Isopentyl acetate	360	1.0	7.5

TABLE 2.41 Properties of Combustible Mixtures in Air (*Continued*)

Substance	Autoignition temperature, °C	Flammable (explosive) limits, percent by volume of fuel (25°C, 760 mm)	
		Lower	Upper
Isoprene	220	2	9
Isopropyl acetate	460	1.8	8
Isopropyl alcohol	399	2.5	12.7
Isopropylamine	402	2.3	10.4
Isopropylbenzene (cumene)	424	0.8	6.5
Isopropyl formate	485		
4-Isopropyl-1-methylbenzene	436		
Kerosene	210	0.7	5.0
Maleic anhydride	477	1.4	7.1
Methacrylic acid	68	1.6	8.8
Methacrylonitrile		2	6.8
Methane	650	5.3	15.0
Methanethiol		3.9	21.8
Methanol	464	6.0	36
Methoxybenzene (anisole)	475		
2-Methoxyethanol	285	1.8	14
2-Methoxyethyl acetate	392	1.5	12.3
Methyl acetate	454	3.1	16
Methyl acetoacetate	280		
Methyl acetylacetate	280		
Methyl acrylate	468	2.8	25
Methylamine	430	4.9	20.7
2-Methylbutane		1.4	7.6
2-Methyl-1-butanol	385	1.4	9.0
2-Methyl-2-butanol	437	1.2	9.0
3-Methyl-1-butanol	350	1.2	9.0
3-Methylbutyl acetate	360	1.0	7.5
2-Methyl-2-butene	275	1.6	8.7
3-Methyl-1-butene	365	1.5	9.1
2-Methyl-1-buten-3-one		1.8	9.0
Methyl chloroformate	504		
Methylcyclohexane	250	1.2	6.7
cis-2-Methylcyclohexanol	296		
trans-2-Methylcyclohexanol	296		
cis-4-Methylcyclohexanol	295		
trans-4-Methylcyclohexanol	295		
Methylcyclopentane	258	1.0	8.35
Methyl formate	449	4.5	23
2-Methylhexane	280	1.0	6.0
3-Methylhexane	280		
5-Methyl-2-hexanone	191	1.0	8.2
Methylhydrazine	196	2.5	97.±2
Methyl isobutyl ketone (MIBK)	448	1	8
2-Methyllactonitrile	688		
Methyl methacrylate		1.7	8.2
1-Methyl-4-(1-methylethenyl)-cyclohexene (dipentene)	237		
1-Methylnaphthalene	529		
2-Methylpentane	264	1.0	7.0
3-Methylpentane	278	1.2	7.0
2-Methyl-2,4-pentanediol	306	1	9
2-Methyl-1-pentanol	310	1.1	9.65
4-Methyl-2-pentanol		1.0	5.5

(*Continued*)

TABLE 2.41 Properties of Combustible Mixtures in Air (*Continued*)

Substance	Autoignition temperature, °C	Flammable (explosive) limits, percent by volume of fuel (25°C, 760 mm)	
		Lower	Upper
4-Methyl-2-pentanone	452	2	8.0
4-Methyl-3-penten-2-one	344	1.4	7.2
2-Methylpropanal	223	1.6	10.6
2-Methyl-1-propanamine	378	2	12
2-Methylpropane	460	1.8	8.4
2-Methylpropanenitrile	482		
Methyl propanoate	469	2.5	13
2-Methylpropanoic acid	481	2.0	9.2
2-Methyl-1-propanol	415	1.7	10.6
2-Methyl-2-propanol (*t*-butyl alcohol)	478	2.4	8.0
2-Methyl-1-propene	465	1.8	9.6
2-Methylpropyl acetate	421	1.3	10.5
2-Methylpropyl formate	320	1.7	8
2-Methylpyridine	538		
N-Methyl-2-pyrrolidone	346	1	10
Methyl salicylate	454		
α-Methylstyrene	574	1.9	6.1
Methyl vinyl ether		2.6	39
Morpholine	290	1	11
Naphtha, coal tar	277		
Naphthalene	526	0.9	5.9
Neoprene		4.0	20
Nicotine	244	0.75	4.0
Nitrobenzene	482	1.8	9
2-Nitrobiphenyl	179		
Nitroethane	414	3.4	17
Nitroglycerine	270		
Nitromethane	418	7.3	22
1-Nitropropane	421	2.2	
2-Nitropropane	428	2.6	11
Nonane	205	0.8	2.9
Octadecanoic acid (stearic acid)	395		
cis-9-Octadecenoic acid (oleic acid)	362		
Octane	206	1.0	6.5
1-Octene	230		
Paraldehyde	238	1.3	
Pentaborane(9)		0.42	
Pentanamine		2.2	22
Pentane	260	1.5	7.8
1,5-Pentanediol	335		
Pentanoic acid	400		
1-Pentanol	300	1.2	10.0
2-Pentanol	343		
3-Pentanol	435	1.2	9.0
2-Pentanone (methyl propyl ketone)	452	1.5	8.2
3-Pentanone (diethyl ketone)	450	1.6	
1-Pentene	275	1.5	8.7
Pentyl acetate	360	1.1	7.5
Pentylamine		2.2	22
Petroleum ether (solvent naphtha)	288	1.1	5.9
Phenol	715	1.8	8.6

TABLE 2.41 Properties of Combustible Mixtures in Air (*Continued*)

Substance	Autoignition temperature, °C	Flammable (explosive) limits, percent by volume of fuel (25°C, 760 mm)	
		Lower	Upper
Phosphorus, red	260		
Phosphorus, white	30		
Phosphorus pentasulfide	142		
o-Phthalic anhydride	570	1.7	10.4
Picric acid	300 (explodes)		
α-Pinene	275		
β-Pinene	275		
Piperidine		1	10
1-Propanal	207	2.6	17
1-Propanamine (propylamine)	318	2.0	10.4
Propane	450	2.1	9.5
1,2-Propanediol	371	2.6	12.5
1,3-Propanediol	400		
Propanenitrile	512	3.1	14
1,2,3-Propanetriol (glycerol)	370	3	19
1,2,3-Propanetriol triacetate (triacetin)	433	1.0	
Propanoic acid	465	2.9	12.1
Propanoic anhydride	285	1.3	9.5
1-Propanol	412	2.2	13.7
2-Propanol	399	2.0	12.7
Propene	460	2.4	10.1
Propyl acetate	450	1.7	8
Propylbenzene	450	0.8	6.0
Propyl formate	455		
Propyl nitrate	175	2	100
Propyne		1.7	
Pyridine	482	1.8	12.4
Quinoline	480		
Sodium	115 (dry air)		
Styrene	490	0.9	6.8
Sulfur (di-) dichloride	233		
1,1,2,2-Tetrabromoethane	335		
Tetrabromoethylene	335		
1,1,1,2-Tetrachloroethane		5	12
1,1,2,2-Tetrachloroethane		20	54
Tetrahydrofuran	321	2	11.8
Tetrahydrofurfuryl alcohol	282	1.5	9.7
1,2,3,4-Tetrahydronaphthalene	385	0.8	5.0
2,2,3,3-Tetramethylpentane	430	0.8	4.9
2,2-Thiodiethanol	298		
Titanium, powder	250		
Toluene	480	1.1	7.1
Toluene diisocyanate		0.9	9.5
o-Toluidine (also *p*-)	482		
Tributylamine		1	5
1,1,1-Trichloroethane	537	7.5	12.5
1,1,2-Trichloroethane	460	6	28
Trichloroethylene	420	8	10.5
(Trichloromethyl)benzene	211		

(*Continued*)

TABLE 2.41 Properties of Combustible Mixtures in Air (*Continued*)

Substance	Autoignition temperature, °C	Flammable (explosive) limits, percent by volume of fuel (25°C, 760 mm)	
		Lower	Upper
Trichloromethylsilane	>404	7.6	>20
1,2,3-Trichloropropane		3.2	12.6
Trichlorosilane	104		
1,1,2-Trichloro-1,2,2-trifluoroethane (Freon 113)	680		
Tri-*o*-cresyl phosphate	385		
Triethanolamine		1	10
Triethylamine	249	1.2	8.0
Triethylene glycol	371	0.9	9.2
Triethyl phosphate	454		
Trimethylamine	190	2.0	11.6
1,2,3-Trimethylbenzene (hemimellitene)	470	0.8	6.6
1,2,4-Trimethylbenzene (pseudocumene)	500	0.9	6.4
1,3,5-Trimethylbenzene	559	1	5
2,2,3-Trimethylbutane	412		
1,1,3-Trimethyl-3-cyclohexen-5-one	462	0.8	3.8
3,5,5-Trimethylcyclohex-2-ene-1-one	460	0.8	3.8
2,2,3-Trimethylpentane	346		
2,2,4-Trimethylpentane	418	1.1	6.0
2,3,3-Trimethylpentane	425		
Trioxane	414	3.6	28.7
Tri-*o*-tolyl phosphate	385		
Turpentine		0.8	
Vinyl acetate	402	2.6	13.4
Vinyl bromide	530	9	15
Vinyl butanoate		1.4	8.8
Vinyl chloride	472	3.6	33.0
4-Vinyl-1-cyclohexene	269		
Vinyl fluoride		2.6	21.7
Vinylidene	573	5.6	16.0
m-Xylene	527	1.1	7.0
o-Xylene	463	0.9	6.7
p-Xylene	528	1.1	7.0

2.7 AZEOTROPIC MIXTURES

An azeotrope is liquid mixture of two or more components that boils at a temperature either higher or lower than the boiling point of any of the individual components. In industrial situation, if the components of a solution are very close in boiling point and cannot be separated by conventional distillation, a substance can be added that forms an azeotrope with one component, modifying its boiling point and making it separable by distillation.

TABLE 2.42 Binary Azeotropic (Constant-Boiling) Mixtures

A. Binary azeotropes containing water

System	BP of azeotrope, °C	Composition, wt %	
		Water	Other component
Inorganic acids			
Hydrogen bromide	126	52.5	47.5
Hydrogen chloride	108.58	79.78	20.22
Hydrogen fluoride	111.35	64.4	35.6
Hydrogen iodide	127	43	57
Hydrogen peroxide	zeotrope		
Nitric acid	120.7	32.6	67.4
Perchloric acid	203	28.4	71.6
Organic acids			
Formic acid	107.2	22.6	77.4
Acetic acid	zeotrope		
Propionic acid	99.9	82.3	17.7
Isobutyric acid	99.3	79	21
Butyric acid	99.4	81.6	18.4
Pentanoic acid	99.8	89	11
Isopentanoic acid	99.5	81.6	18.4
Perfluorobutyric acid	97	71	29
Crotonic acid	99.9	97.8	2.2
Alcohols			
Ethanol	78.17	4	96
Allyl alcohol	88.9	27.7	72.3
1-Propanol	71.7	71.7	28.3
2-Propanol	80.3	12.6	87.4
1-Butanol	92.7	42.5	57.5
2-Butanol	87.0	26.8	73.2
2-Methyl-2-propanol	79.9	11.7	88.3
1-Pentanol	95.8	54.4	45.6
2-Pentanol	91.7	36.5	63.5
3-Pentanol	91.7	36.0	64.0
2,2-Dimethyl-2-propanol	87.35	27.5	72.5
1-Hexanol	97.8	67.2	32.8
1-Octanol	99.4	90	10
Cyclopentanol	96.25	58	42
1-Heptanol	98.7	83	17
Phenol	99.52	90.8	9.2
2-Methoxyphenol	99.5	87.5	12.5
1-Phenylphenol	99.95	98.75	1.25
Benzyl alcohol	99.9	91	9
2,3-Dimethyl-2,3-butanediol	zeotrope		
Furfuryl alcohol	98.5	80	20

TABLE 2.42 Binary Azeotropic (Constant-Boiling) Mixtures (*Continued*)

System	BP of azeotrope, °C	Composition, wt %	
		Water	Other component
	Aldehydes		
Propionaldehyde	47.5	2	98
Butyraldehyde	68	6	94
Pentanal	83	19	81
Paraldehyde	90	28.5	71.5
Furaldehyde	97.5	65	35
	Amines		
N-Methylbutylamine	82.7	15	85
Furfurylamine	99	74	26
Piperidine	92.8	35	65
Pyridine	93.6	41.3	58.7
2-Methylpyridine	93.5	48	52
3-Methylpyridine	97	60	40
4-Methylpyridine	97.35	62.8	37.2
2,6-Dimethylpyridine	96.02	51.8	48.2
Dibutylamine	97	50.5	49.5
Dihexylamine	99.8	92.8	7.2
Triallylamine	95	38	62
Tributylamine	99.65	79.7	20.3
Aniline	98.6	80.8	19.2
N-Ethylaniline	99.2	83.9	16.1
1-Methyl-2-(2-pyridyl)pyrrolidine	99.85	97.5	2.5
	Halogenated hydrocarbons		
Chloroform	56.1	2.8	97.2
Carbon tetrachloride	42.6	2.8	97.2
Trichloroethylene	73.4	17	83
Tetrachloroethylene	88.5	17.2	82.8
1,2-Dichloroethane	72	8.3	91.7
1-Chloropropane	44	2.2	97.8
1,2-Dichloropropane	78	12	88
Chlorobenzene	90.2	28.4	71.6
	Esters		
Ethyl formate	52.6	5	95
Isopropyl formate	65.0	3	97
Propyl formate	71.6	2.3	97.7
Isobutyl formate	80.4	7.8	92.2
Butyl formate	83.8	14.5	85.5
Isopentyl formate	90.2	21	79
Pentyl formate	91.6	28.4	71.6
Benzyl formate	99.2	80	20
Ethyl acetate	70.38	8.47	91.53
Allyl acetate	83	14.7	85.3

TABLE 2.42 Binary Azeotropic (Constant-Boiling) Mixtures (*Continued*)

System	BP of azeotrope, °C	Composition, wt %	
		Water	Other component
	Esters (*continued*)		
Isopropyl acetate	76.6	10.6	89.4
Propyl acetate	82.4	14	86
Isobutyl acetate	87.4	16.5	83.5
Butyl acetate	90.2	28.7	71.3
Isopentyl acetate	93.8	36.3	63.7
Pentyl acetate	95.2	41	59
Hexyl acetate	97.4	61	39
Phenyl acetate	98.9	75.1	24.9
Benzyl acetate	99.6	87.5	12.5
Methyl propionate	71.4	3.9	96.1
Ethyl propionate	81.2	10	90
Isopropyl propionate	85.2	19.9	80.1
Propyl propionate	88.9	23	77
Isobutyl propionate	92.75	52.2	47.8
Isopentyl propionate	96.55	48.5	51.5
Methyl butyrate	82.7	11.5	88.5
Ethyl butyrate	87.9	21.5	78.5
Propyl butyrate	94.1	36.4	63.6
Isobutyl butyrate	96.3	46	54
Butyl butyrate	97.2	53	47
Isopentyl butyrate	98.05	63.5	36.5
Methyl isobutyrate	77.7	6.8	93.2
Ethyl isobutyrate	85.2	15.2	84.8
Propyl isobutyrate	92.2	30.8	69.2
Isobutyl isobutyrate	95.5	39.4	60.6
Isopentyl isobutyrate	97.4	56.0	44.0
Methyl isopentanoate	87.2	19.2	80.8
Ethyl isopentanoate	92.2	30.2	69.8
Propyl isopentanoate	96.2	45.2	54.8
Isobutyl isopentanoate	97.4	55.8	44.2
Isopentyl isopentanoate	98.8	74.1	25.9
Ethyl pentanoate	94.5	40	60
Ethyl hexanoate	97.2	54	46
Methyl benzoate	99.08	79.2	20.8
Ethyl benzoate	99.4	84.0	16.0
Propyl benzoate	99.7	90.9	9.1
Butyl benzoate	99.9	94	6
Isopentyl benzoate	99.9	95.6	4.4
Ethyl phenylacetate	99.7	91.3	8.7
Methyl cinnamate	99.9	95.5	4.5
Methyl phthalate	99.95	97.5	2.5
Diethyl *o*-phthalate	99.98	98.0	2.0
Ethyl chloroacetate	95.2	45.1	54.9
Butyl chloroacetate	98.12	75.5	24.5
Methyl acrylate	71	7.2	92.8
Isobutyl carbonate	98.6	74	26
Ethyl crotonate	93.5	38	62
Methyl lactate	99	80	20

(Continued)

TABLE 2.42 Binary Azeotropic (Constant-Boiling) Mixtures (*Continued*)

System	BP of azeotrope, °C	Composition, wt %	
		Water	Other component
1,2-Ethanediol diacetate	99.7	84.6	15.4
Ethyl nitrate	74.35	22	78
Propyl nitrate	84.8	20	80
Isobutyl nitrate	89.0	25	75
Methyl sulfate	98.6	73	27
	Ethers		
Ethyl vinyl ether	34.6	1.5	98.5
Diethyl ether	34.2	1.3	98.7
Ethyl propyl ether	59.5	4	96
Diisopropyl ether	62.2	4.5	95.5
Butyl ethyl ether	76.6	11.9	88.1
Diisobutyl ether	88.6	23	77
Dibutyl ether	92.9	33	67
Diisopentyl ether	97.4	54	46
1,1-Diethoxyethane	82.6	14.5	85.5
Diphenyl ether	99.33	96.75	3.25
Methoxybenzene	95.5	40.5	59.5
	Hydrocarbons		
Pentane	34.6	1.4	98.6
Hexane	61.6	5.6	94.4
Heptane	79.2	12.9	87.1
2,2,4-Trimethylpentane	78.8	11.1	88.9
Nonane	94.8	82	18
Undecane	98.85	96.0	4.0
Dodecane	99.45	98	2
Acrolein	52.4	2.6	97.4
Cyclohexene	70.8	8.93	91.07
Cyclohexane	69.5	8.4	91.6
1-Octene	88.0	28.7	71.3
Benzene	69.25	8.83	91.17
Toluene	84.1	13.5	86.5
Ethylbenzene	92.0	33.0	67.0
m-Xylene	92	35.8	64.2
Isopropylbenzene	95	43.8	56.2
Naphthalene	98.8	84	16
	Ketones		
Acetone	zeotrope		
2-Butanone	73.5	11	89
2-Pentanone	83.3	19.5	80.5
Cyclopentanone	94.6	42.4	57.6
4-Methyl-2-pentanone	87.9	24.3	75.7

TABLE 2.42 Binary Azeotropic (Constant-Boiling) Mixtures (*Continued*)

System	BP of azeotrope, °C	Composition, wt %	
		Water	Other component
	Ketones (*continued*)		
2-Heptanone	95	48	52
3-Heptanone	94.6	42.2	57.8
4-Heptanone	94.3	40.5	59.5
4-Hydroxy-4-methyl-2-pentanone	98.8	87.3	12.7
4-Methyl-3-penten-2-one	91.8	34.8	65.2
	Nitriles		
Acetonitrile	76.5	16.3	83.7
Isobutyronitrile	82.5	23	177
Butyronitrile	88.7	32.5	67.5
Acrylonitrile	70.6	14.3	85.7
	Miscellaneous		
Hydrazine	120	32.3	67.7
Acetamide	zeotrope		
Nitromethane	83.59	23.6	76.4
Nitroethane	87.22	28.5	71.5
2,5-Dimethylfuran	77.0	11.7	88.3
Trioxane	91.4	30	70
Carbon disulfide	42.6	2.8	97.2

B. Binary azeotropes containing organic acids

System	BP of azeotrope, °C	Composition, wt %	
		Acid	Other componen
	Formic acid		
2-Methylbutane	27.2	4	96
Pentane	34.2	20	80
Hexane	60.6	28	72
Methylcyclopentane	63.3	29	71
Cyclohexane	70.7	70	30
Methylcyclohexane	80.2	46.5	53.5
Heptane	78.2	56.5	43.5
Octane	90.5	63	37
Benzene	71.05	31	69
Toluene	85.8	50	50
o-Xylene	95.5	74	26
m-Xylene	92.8	71.8	28.2
Styrene	97.8	73	27

(*Continued*)

TABLE 2.42 Binary Azeotropic (Constant-Boiling) Mixtures (*Continued*)

System	BP of azeotrope, °C	Composition, wt %	
		Acid	Other component
	Formic acid (*continued*)		
Iodomethane	42.1	6	94
Chloroform	59.15	15	85
Carbon tetrachloride	66.65	18.5	81.5
Trichloroethylene	74.1	25	75
Tetrachloroethylene	88.2	50	50
Bromoethane	38.2	3	97
1,2-Dibromoethane	94.7	51.5	48.5
1,2-Dichloroethane	77.4	14	86
1-Bromopropane	64.7	27	73
2-Bromopropane	56.0	14	86
1-Chloropropane	45.6	8	92
2-Chloropropane	34.7	1.5	98.5
1-Chloro-2-methylpropane	63.0	19	81
Bromobenzene	98.1	68	32
Chlorobenzene	93.7	59	41
Fluorobenzene	73.0	27	73
o-Chlorotoluene	100.2	83	17
Pyridine	127.43	61.4	38.6
2-Methylpyridine	158.0	25	75
2-Pentanone	105.3	32	68
3-Pentanone	105.4	33	67
Nitromethane	97.07	45.5	54.5
Diethyl sulfide	82.2	35	65
Diisopropyl sulfide	93.5	62	38
Dipropyl sulfide	98.0	83	17
Carbon disulfide	42.55	17	83
	Acetic acid		
Hexane	68.3	6.0	94.0
Heptane	91.7	23	67
Octane	105.7	53.7	46.3
Nonane	112.9	69	31
Decane	116.75	79.5	20.5
Undecane	117.9	95	5
Cyclohexane	78.8	9.6	90.4
Methylcyclohexane	96.3	31	69
Benzene	80.05	2.0	98.0
Toluene	100.6	28.1	71.9
o-Xylene	116.6	78	22
m-Xylene	115.35	72.5	27.5
p-Xylene	115.25	72	28
Ethylbenzene	114.65	66	34
Styrene	116.8	85.7	14.3
Isopropylbenzene	116.0	84	16
Triethylamine	163	67	33
Nitromethane	101.2	96	4

TABLE 2.42 Binary Azeotropic (Constant-Boiling) Mixtures (*Continued*)

System	BP of azeotrope, °C	Composition, wt %	
		Acid	Other component
	Acetic acid (*continued*)		
Nitroethane	112.4	30	70
Pyridine	138.1	51.1	48.9
2-Methylpyridine	144.1	40.4	59.6
3-Methylpyridine	152.5	30.4	69.6
4-Methylpyridine	154.3	30.3	69.7
2,6-Dimethylpyridine	148.1	22.9	77.1
Carbon tetrachloride	76	98.46	1.54
Trichloroethylene	86.5	96.2	3.8
Tetrachloroethylene	107.4	61.5	38.5
1,2-Dibromoethane	114.4	55	45
2-Iodopropane	88.3	9	91
1-Bromobutane	97.6	18	82
1-Bromo-2-methylpropane	90.2	12	88
Chlorobenzene	114.7	58.5	41.5
Trichloronitromethane	107.65	80.5	19.5
1,4-Dioxane	119.5	77	23
Diisopropyl sulfide	111.5	48	52
	Propionic acid		
Heptane	97.8	2	98
Octane	120.9	21.5	78.5
Nonane	134.3	54.0	46.0
Decane	139.8	80.5	19.5
o-Xylene	135.4	43	57
p-Xylene	132.5	34	66
1,3,5-Trimethylbenzene	139.3	77	23
Isopropylbenzene	139.0	65	35
Propylbenzene	139.5	75	25
Camphene	138.0	65	35
α-Pinene	136.4	58.5	41.5
Methoxybenzene	140.8	96	4
Pyridine	148.6	67.2	32.8
2-Methylpyridine	154.5	55.0	45.0
1,2-Dibromoethane	127.8	17.5	82.5
1-Iodo-2-methylpropane	119.5	9	91
Chlorobenzene	128.9	18	82
Dipropyl sulfide	136.5	45	55
	Butyric acid		
Undecane	162.4	84.4	15.5
o-Xylene	143.0	10	90
m-Xylene	138.5	6	94
p-Xylene	137.8	5.5	94.5
Ethylbenzene	135.8	4	96

(*Continued*)

TABLE 2.42 Binary Azeotropic (Constant-Boiling) Mixtures (*Continued*)

System	BP of azeotrope, °C	Composition, wt %	
		Acid	Other component
	Butyric acid (*continued*)		
Styrene	143.5	15	85
1,2,4-Trimethylbenzene	159.5	45	55
1,3,5-Trimethylbenzene	158.0	38	62
Isopropylbenzene	149.5	20	80
Propylbenzene	154.5	28	72
Butylbenzene	162.5	75	25
Naphthalene	zeotrope		
Indene	163.7	84	16
Camphene	152.3	2.8	97.2
Methoxybenzene	152.9	12	88
Pyridine	163.2	92.0	8.0
2-Furaldehyde	159.4	42.5	57.5
1,2-Dibromoethane	131.1	3.5	96.5
1-Iodobutane	129.8	2.5	97.5
Chlorobenzene	131.75	2.8	97.2
1,4-Dichlorobenzene	162.0	57	43
o-Bromotoluene	163.0	72	28
m-Bromotoluene	163.6	79.5	20.5
p-Bromotoluene	161.5	75	25
α-Chlorotoluene	160.8	65	35
Ethyl bromoacetate	157.4	84	16
Propyl chloroacetate	160.5	40	60
	Isobutyric acid		
2,7-Dimethyloctane	148.6	48	52
o-Xylene	141.0	22	78
m-Xylene	139.9	15	85
p-Xylene	136.4	13	87
Styrene	142.0	27	73
1,2,4-Trimethylbenzene	152.3	63	37
Isopropylbenzene	146.8	35	65
Propylbenzene	149.3	49	51
Camphene	148.1	45	55
D-Limonene	152.5	78	22
Methoxybenzene	149.0	42	58
Ethyl bromoacetate	153.0	40	60
Ethyl 2-oxopropionate	153.0	60	40
1,2-Dibromoethane	130.5	6.5	93.5
1-Iodobutane	128.8	7	93
1-Bromohexane	148.0	35	65
Bromobenzene	148.6	35	65
Chlorobenzene	131.5	8	92
o-Bromotoluene	153.9	85	15
α-Chlorotoluene	153.5	80	20
Diisopentyl ether	154.2	93	7
Ethyl bromoacetate	153.0	40	60

TABLE 2.42 Binary Azeotropic (Constant-Boiling) Mixtures (*Continued*)

C. Binary azeotropes containing alchohols

System	BP of azeotrope, °C	Composition, wt %	
		Alcohol	Other component
	Methanol		
Pentane	30.9	7	93
Cyclopentane	38.8	14	86
Cyclohexane	53.9	36.4	63.6
Methylcyclohexane	59.2	54	46
Heptane	59.1	51.5	48.5
Octane	62.8	67.5	32.5
Nonane	64.1	83.4	16.6
Benzene	57.5	39.1	60.9
Fluorobenzene	59.7	32	68
Toluene	63.5	72.5	27.5
Bromomethane	3.55	99.55	0.45
Iodomethane	37.8	95.5	4.5
Bromodichloromethane	63.8	60	40
Chloroform	53.4	87.4	12.6
Carbon tetrachloride	55.7	79.44	20.56
Bromoethane	34.9	5.3	94.7
1,2-Dichloroethane	61.0	32	68
Trichloroethylene	59.3	38	62
1-Bromopropane	54.5	21	79
2-Bromopropane	48.6	15.0	85.0
1-Chloropropane	40.5	9.5	90.5
2-Chloropropane	33.4	6	94
2-Iodopropane	61.0	38	62
1-Chlorobutane	57.0	27	73
Isobutyl formate	64.6	95	5
Methyl acetate	53.5	19	81
Methyl acrylate	62.5	54	46
Methyl nitrate	52.5	73	27
Acetone	55.5	12.1	87.9
1,4-Dioxane	zeotrope		
Dipropyl ether	63.8	72	28
Methyl *tert*-butyl ether	51.3	14.3	85.7
Diethyl sulfide	61.2	62	38
Carbon disulfide	39.8	71	29
Thiophene	59.7	16.4	83.6
Nitromethane	64.4	9.1	90.9
	Ethanol		
Pentane	34.3	5	95
Cyclopentane	44.7	7.5	92.5
Hexane	58.7	21	79
Cyclohexane	64.8	29.2	70.8
Heptane	70.9	49	51

(*Continued*)

TABLE 2.42 Binary Azeotropic (Constant-Boiling) Mixtures (*Continued*)

System	BP of azeotrope, °C	Composition, wt %	
		Alcohol	Other component
	Ethanol (*continued*)		
Octane	77.0	78	22
Benzene	67.9	31.7	68.3
Fluorobenzene	70.0	75	25
Toluene	76.7	68	32
Bromodichloromethane	75.5	72	28
Iodomethane	41.2	96.8	3.2
Chloroform	59.3	93	7
Trichloronitromethane	77.5	34	66
Carbon tetrachloride	65.0	84.2	15.8
1,2-Dichloroethane	70.5	37	63
3-Chloro-1-propene	44	5	95
1-Bromopropane	62.8	20.5	79.5
2-Bromopropane	55.6	10.5	89.5
1-Chloropropane	45.0	6	94
2-Chloropropane	35.6	2.8	97.2
1-Iodopropane	75.4	44	56
2-Iodopropane	71.5	27	73
1-Bromobutane	75.0	43	57
1-Chlorobutane	65.7	20.3	79.7
2-Butanone	74.8	40	60
1,1-Diethoxyethane	78.0	76	24
Dipropyl ether	74.5	44	56
Acetronitrile	72.5	44	56
Acrylonitrile	70.8	41	59
Nitromethane	76.1	29	71
Carbon disulfide	42.6	91	9
Diethyl sulfide	72.6	56	44
	1-Propanol		
Hexane	65.7	4	96
Cyclohexane	74.7	18.5	81.5
Methylcyclohexane	87.0	34.7	65.3
Heptane	84.6	34.7	65.3
Octane	93.9	70	30
Benzene	77.1	16.9	83.1
Toluene	92.5	51.2	48.8
o-Xylene	zeotrope		
m-Xylene	97.1	94	6
p-Xylene	96.9	92.2	7.8
Styrene	97.0	8	92
Propyl formate	80.7	3	97
Butyl formate	95.5	64	36
Propyl acetate	94.7	51	49
Ethyl propionate	93.4	48	52
Methyl butyrate	94.4	49	51
Dipropyl ether	85.7	30	70

TABLE 2.42 Binary Azeotropic (Constant-Boiling) Mixtures (*Continued*)

System	BP of azeotrope, °C	Composition, wt %	
		Alcohol	Other component
	1-Propanol (*continued*)		
1,1-Diethoxyethane	92.4	37	63
1,4-Dioxane	95.3	55	45
Chloroform	zeotrope		
Carbon tetrachloride	73.4	92.1	7.9
Trichloronitromethane	94.1	58.5	41.5
Iodethane	70	93	7
1,2-Dichloroethane	80.7	19	81
Tetrachloroethylene	94.0	52	48
1-Bromopropane	69.7	9	91
1-Chlorobutane	74.8	18	82
Chlorobenzene	96.5	80	20
Fluorobenzene	80.2	18	82
Nitromethane	89.1	48.4	51.6
1-Nitropropane	97.0	8.8	91.2
Carbon disulfide	45.7	94.5	5.5
	2-Propanol		
Pentane	35.5	6	94
Hexane	62.7	23	77
Cyclohexane	69.4	32	68
Heptane	76.4	50.5	49.5
Octane	81.6	84	16
Benzene	71.7	33.7	66.3
Fluorobenzene	74.5	30	70
Toluene	80.6	69	31
Chloroform	60.8	4.2	95.8
Trichloronitromethane	81.9	35	65
Carbon tetrachloride	69.0	18	82
1,2-Dichloroethane	74.7	43.5	56.5
Iodoethane	67.1	15	85
3-Bromo-1-propene	66.5	20	80
1-Chloropropane	46.4	2.8	97.2
1-Bromopropane	66.8	20.5	79.5
2-Bromopropane	57.8	12	88
1-Iodopropane	79.8	42	58
2-Iodopropane	76.0	32	68
1-Chlorobutane	70.8	23	77
Ethyl acetate	75.3	25	75
Isopropyl acetate	81.3	60	40
Methyl propionate	76.4	37	63
Acrylonitrile	71.7	56	44
Butylamine	74.7	60	40
2-Butanone	77.5	32	68
1,1-Diethoxyethane	81.3	63	37
Ethyl propyl ether	62.0	10	90
Diisopropyl ether	66.2	14.1	85.9

(*Continued*)

TABLE 2.42 Binary Azeotropic (Constant-Boiling) Mixtures (*Continued*)

System	BP of azeotrope, °C	Composition, wt %	
		Alcohol	Other component
	1-Butanol		
Cyclohexane	79.8	9.5	90.5
Cyclohexene	82.0	5	95
Hexane	68.2	3.2	96.8
Methylcyclohexane	95.3	20	80
Heptane	93.9	18	82
Octane	108.5	45.2	54.8
Nonane	115.9	71.5	28.5
Toluene	105.5	27.8	72.2
o-Xylene	116.8	75	25
m-Xylene	116.5	71.5	28.5
p-Xylene	115.7	68	32
Ethylbenzene	115.9	65.1	34.9
Butyl formate	105.8	23.6	76.4
Isopentyl formate	115.9	69	31
Butyl acetate	117.2	47	53
Isobutyl acetate	114.5	50	50
Ethyl butyrate	115.7	64	36
Ethyl isobutyrate	109.2	17	83
Methyl isopentanoate	113.5	40	60
Ethyl borate	113.0	52	48
Ethyl carbonate	116.5	63	37
Isobutyl nitrate	112.8	45	55
Dibutyl ether	117.8	82.5	17.5
Diisobutyl ether	113.5	48	52
1,1-Diethoxyethane	101.0	13	87
Carbon tetrachloride	76.6	97.6	2.4
Tetrachloroethylene	110.0	68	32
2-Bromo-2-methylpropane	90.2	7	93
2-Iodo-2-methylpropane	110.5	30	70
Chlorobenzene	115.3	56	44
Paraldehyde	115.8	52	48
Hexaldehyde	116.8	77.1	22.9
Ethylenediamine	124.7	35.7	64.3
Pyridine	118.6	69	31
1-Nitropropane	115.3	32.2	67.8
Butyronitrile	113.0	50	50
Diisopropyl sulfide	112.0	45	55
	2-Methyl-2-propanol		
Cyclohexene	80.5	14.2	85.8
Cyclohexane	78.3	14	86
Methylcyclopentane	71.0	5	95
Hexane	68.3	2.5	97.5
Methylcyclohexane	92.6	32	68
Heptane	90.8	27	73
2,5-Dimethylhexane	98.7	42	58
1,3-Dimethylcyclohexane	102.2	56	44
2,2,4-Trimethylpentane	92.0	27	73
Benzene	79.3	7.4	92.6
Chlorobenzene	107.1	63	37
Fluorobenzene	84.0	9	91

TABLE 2.42 Binary Azeotropic (Constant-Boiling) Mixtures (*Continued*)

System	BP of azeotrope, °C	Composition, wt %	
		Alcohol	Other component
	2-Methyl-2-propanol (*continued*)		
Toluene	101.2	45	55
Ethylbenzene	107.2	80	20
p-Xylene	107.1	88.6	11.4
Butyl formate	103.0	40	60
Isobutyl formate	97.4	12	88
Propyl acetate	101.0	17	83
Isobutyl acetate	107.6	92	8
Methyl butyrate	101.3	25	75
Ethyl isobutyrate	105.5	52	48
Methyl chloroacetate	107.6	12	88
Dipropyl ether	89.5	10	90
Isobutyl vinyl ether	82.7	6.2	93.8
1,1-Diethoxyethane	98.2	20	80
2-Pentanone	101.8	19	81
3-Pentanone	101.7	20	80
1,2-Dichloroethane	83.5	6.5	93.5
1-Bromobutane	95.0	21	79
1-Chlorobutane	77.7	4	96
2-Bromo-2-methylpropane	88.8	12	88
2-Iodo-2-methylpropane	104.0	36	64
1-Nitropropane	105.3	15.2	84.8
Isobutyl nitrate	105.6	36	64
Diisopropyl sulfide	105.8	73	27
	3-Methyl-1-butanol		
Heptane	97.7	7	93
Octane	117.0	30	70
Toluene	109.7	10	90
Ethylbenzene	125.7	49	51
Isopropylbenzene	131.6	94	6
Camphene	130.9	24	76
Bromobenzene	131.7	85	15
o-Fluorotoluene	112.1	14.0	86.0
Butyl acetate	125.9	16.5	83.5
Paraldehyde	123.5	22.0	78.0
Dibutyl ether	129.8	65	35
	Cyclohexanol		
o-Xylene	143.0	14	86
m-Xylene	138.9	5	95
Propylbenzene	153.8	40	60
Indene	160.0	75	25
Camphene	151.9	41	59
Cineole	160.6	92	8

(*Continued*)

TABLE 2.42 Binary Azeotropic (Constant-Boiling) Mixtures (*Continued*)

System	BP of azeotrope, °C	Composition, wt % Alcohol	Composition, wt % Other component
	Allyl alcohol		
Methylcyclohexane	85.0	42	58
Hexane	65.5	4.5	95.5
Cyclohexane	74.0	58	42
2,5-Dimethylhexane	89.3	50	50
Octane	93.4	68	32
Benzene	76.75	17.36	82.64
Toluene	92.4	50	50
Propyl acetate	94.2	53	47
Methyl butyrate	93.8	55	45
1,2-Dichloroethane	79.9	18	82
3-Iodo-1-propene	89.4	28	72
Chlorobenzene	96.2	85	15
Diethyl sulfide	85.1	45	55
	Phenol		
2,7-Dimethyloctane	159.5	6	94
Decane	168.0	35	65
Tridecane	180.6	83.1	16.9
Butylbenzene	175.0	46	54
1,2,4-Trimethylbenzene	166.0	25	75
1,3,5-Trimethylbenzene	163.5	21	79
Indene	177.8	47	53
Camphene	156.1	22	78
Benzaldehyde	175.6	51.0	49.0
1-Octanol	195.4	13	87
2-Octanol	184.5	50	50
Dipentyl ether	180.2	78	22
Diisopentyl ether	172.2	15	85
2-Methylpyridine	185.5	75.4	24.6
3-Methylpyridine	188.9	71.2	29.8
4-Methylpyridine	190.0	67.5	32.5
2,4-Dimethylpyridine	193.4	57.0	43.0
2,6-Dimethylpyridine	185.5	72.5	27.5
2,4,6-Trimethylpyridine	195.2	52.3	47.7
Aniline	185.8	41.9	58.1
Ethylene diacetate	195.5	39.2	60.8
Iodobenzene	177.7	53	47
	Benzyl alcohol		
Naphthalene	204.1	60	40
D-Limonene	176.4	11	89
1,3,5-Triethylbenzene	203.2	57	43
o-Cresol	zeotrope		
m-Cresol	207.1	61	39

TABLE 2.42 Binary Azeotropic (Constant-Boiling) Mixtures (*Continued*)

System	BP of azeotrope, °C	Composition, wt %	
		Alcohol	Other component
	Benzyl alcohol (*continued*)		
p-Cresol	206.8	62	38
N-Methylaniline	195.8	30	70
N,*N*-Dimethylaniline	193.9	6.5	93.5
N-Ethylaniline	202.8	50	50
N,*N*-Diethylaniline	204.2	72	28
Iodobenzene	187.8	12	88
Nitrobenzene	204.0	58	42
o-Bromotoluene	181.3	7	93
Borneol	205.1	85.8	14.2
	2-Ethoxyethanol		
Methylcyclohexane	98.6	15	85
Heptane	96.5	14	86
Octane	116.0	38	62
Toluene	110.2	10.8	89.2
Ethylbenzene	127.8	48	52
p-Xylene	128.6	50	50
Styrene	130.0	55	45
Propylbenzene	134.6	80	20
Isopropylbenzene	133.2	67	33
Camphene	131.0	65	35
Propyl butyrate	133.5	72	28
	2-Butoxyethanol		
Dipentene	164.0	53	47
1,3,5-Trimethylbenzene	162.0	32	68
Butylbenzene	169.6	73.4	26.6
Camphene	154.5	30	70
o-Cresol	191.6	15	85
Phenetole	167.1	52	48
Cineole	168.9	58.5	41.5
Benzaldehyde	171.0	91	9
Diisobutyl sulfide	163.8	42	58
	1,2-Ethanediol		
Heptane	97.9	3	97
Decane	161.0	23	77
Tridecane	188.0	55	45
Toluene	110.1	2.3	97.7
Styrene	139.5	16.5	83.5
Stilbene	196.8	87	13
m-Xylene	135.1	6.55	93.45
p-Xylene	134.5	6.4	93.6
1,3,5-Trimethylbenzene	156	13	87
Propylbenzene	152	19	81

(*Continued*)

TABLE 2.42 Binary Azeotropic (Constant-Boiling) Mixtures (*Continued*)

System	BP of azeotrope, °C	Composition, wt %	
		Alcohol	Other component
1,2-Ethanediol (*continued*)			
Isopropylbenzene	147.0	18	82
Naphthalene	183.9	51	49
1-Methylnaphthalene	190.3	60.0	40.0
2-Methylnaphthalene	189.1	57.2	42.8
Anthracene	197	98.3	1.7
Indene	168.4	26	74
Acenaphthene	194.65	74.2	25.8
Fluorene	196.0	82	18
Camphene	152.5	20	80
Camphor	186.2	40	60
Biphenyl	192.3	66.5	33.5
Diphenylmethane	193.3	68.5	31.5
Benzyl alcohol	193.1	56	44
2-Phenylethanol	194.4	69	31
o-Cresol	189.6	27	73
m-Cresol	195.2	60	40
3,4-Dimethylphenol	197.2	89	11
Menthol	188.6	51.5	48.5
Ethyl benzoate	186.1	46.5	53.5
o-Bromotoluene	166.8	25	75
Dibutyl ether	139.5	6.4	93.6
Methoxybenzene	150.5	10.5	89.5
Diphenyl ether	193.1	60	40
Benzyl phenyl ether	195.5	87	13
Acetophenone	185.7	52	48
2,4-Dimethylaniline	188.6	47	53
N,N-Dimethylaniline	175.9	33.5	66.5
m-Toluidine	188.6	42	58
2,4,6-Trimethylpyridine	170.5	9.7	90.3
Quinoline	196.4	79.5	20.5
Tetrachloroethylene	119.1	94	6
1,2-Dibromoethane	129.8	4	96
Chlorobenzene	130.1	94.4	5.6
α-Chlorotoluene	167.0	30	70
Nitrobenzene	185.9	59	41
o-Nitrotoluene	188.5	48.5	51.5
1,2-Ethanediol monoacetate			
Indene	180.0	20	80
1-Octanol	189.5	71	29
Phenol	197.5	65	35
o-Cresol	199.5	51	49
m-Cresol	206.5	31	69
p-Cresol	206.0	33	67
Dipentyl ether	180.8	42	58
Diisopentyl ether	170.2	28	72
m-Bromotoluene	182.0	32	68

TABLE 2.42 Binary Azeotropic (Constant-Boiling) Mixtures (*Continued*)

D. Binary azeotropes containing ketones

System	BP of azeotrope, °C	Composition, wt %	
		Ketone	Other component
Acetone			
Cyclopentane	41.0	36	64
Pentane	32.5	20	80
Cyclohexane	53.0	67.5	32.5
Hexane	49.8	59	41
Heptane	55.9	89.5	10.5
Diethylamine	51.4	38.2	61.8
Methyl acetate	55.8	48.3	51.7
Diisopropyl ether	54.2	61	39
Chloroform	64.4	78.1	21.9
Carbon tetrachloride	56.1	11.5	88.5
Carbon disulfide	39.3	67	33
Ethylene sulfide	51.5	57	43
2-Butanone			
Cyclohexane	71.8	40	60
Hexane	64.2	28.6	71.4
Heptane	77.0	70	30
2,5-Dimethylhexane	79.0	95	5
Benzene	78.33	44	56
2-Methyl-2-propanol	78.7	69	31
Butylamine	74.0	35	65
Ethyl acetate	77.1	11.8	88.2
Methyl propionate	79.0	60	40
Butyl nitrite	76.7	30	70
1-Chlorobutane	77.0	38	62
Fluorobenzene	79.3	75	25

E. Miscellaneous binary azeotropes

System	BP of azeotrope, °C	Composition, wt %	
		Solvent	Other component
Solvent: acetamide			
Dipentene	169.2	18	82
Biphenyl	213.0	50.5	49.5
Diphenylmethane	215.2	56.5	43.5
1,2-Diphenylethane	218.2	68	32
o-Xylene	142.6	11	89

(*Continued*)

TABLE 2.42 Binary Azeotropic (Constant-Boiling) Mixtures (*Continued*)

System	BP of azeotrope, °C	Composition, wt %	
		Solvent	Other component
Solvent: acetamide (*continued*)			
m-Xylene	138.4	10	90
p-Xylene	137.8	8	92
Styrene	144	12	88
4-Isopropyl-1-methylbenzene	170.5	19	81
Naphthalene	199.6	27	73
1-Methylnaphthalene	209.8	43.8	56.2
2-Methylnaphthalene	208.3	40	60
Indene	177.2	17.5	82.5
Acenaphthene	217.1	64.2	35.8
Camphene	155.5	12	88
Camphor	199.8	23	77
Benzaldehyde	178.6	6.5	93.5
3,4-Dimethylphenol	221.1	96	4
2-Methoxy-4-(2-propenyl)phenol	220.8	88	12
N-Methylaniline	193.8	14	86
N-Ethylaniline	199.0	18	82
N,*N*-Diethylaniline	198.1	24	76
Diphenyl ether	214.6	52	48
Safrole	208.8	32	68
Tetrachloroethylene	120.5	97.4	2.6
Solvent: aniline			
Nonane	149.2	13.5	86.5
Decane	167.3	36	64
Undecane	175.3	57.5	42.5
Dodecane	180.4	71.5	28.5
Tridecane	182.9	86.2	13.8
Tetradecane	183.9	95.2	4.8
Butylbenzene	177.8	46	54
1,2,4-Trimethylbenzene	168.6	13.5	86.5
1,3,5-Trimethylbenzene	164.3	12.0	88.0
Indene	179.8	41.5	58.5
1-Octanol	183.9	83	17
o-Cresol	191.3	8	92
Dipentyl ether	177.5	55	45
Diisopentyl ether	169.3	28	72
Hexachloroethane	176.8	66	34
Solvent: pyridine			
Heptane	95.6	25.3	74.7
Octane	109.5	56.1	43.9
Nonane	115.1	89.9	10.1
Toluene	110.1	22.2	77.8
Phenol	183.1	13.1	86.9
Piperidine	106.1	8	92

TABLE 2.42 Binary Azeotropic (Constant-Boiling) Mixtures (*Continued*)

System	BP of azeotrope, °C	Composition, wt %	
		Solvent	Other component
	Solvent: thiophene		
Methylcyclopentane	71.5	14	86
Cyclohexane	77.9	41.2	58.8
Hexane	68.5	11.2	88.8
Heptane	83.1	83.2	16.8
2,3-Dimethylpentane	80.9	64	36
2,4-Dimethylpentane	76.6	42.7	57.3
	Solvent: benzene		
Methylcyclopentane	71.7	16	84
Cyclohexene	78.9	64.7	35.3
Cyclohexane	77.6	51.9	48.1
Hexane	68.5	4.7	95.3
Heptane	80.1	99.3	0.7
2,2-Dimethylpentane	75.9	46.3	53.7
2,3-Dimethylpentane	79.4	78.8	21.2
2,4-Dimethylpentane	75.2	48.3	51.7
2,2,4-Trimethylpentane	80.1	97.7	2.3
	Solvent: bis(2-hydroxyethyl) ether		
Biphenyl	232.7	48	52
Diphenylmethane	236.0	52	48
1,3,5-Trimethylbenzene	210.0	22	78
Naphthalene	212.6	22	78
1-Methylnaphthalene	277.0	45	55
2-Methylnaphthalene	225.5	39	61
Acenaphthene	239.6	62	38
Fluorene	243.0	80	20
Benzyl acetate	214.9	7	93
Bornyl acetate	223.0	18	82
Ethyl fumarate	217.1	10	90
Dimethyl *o*-phthalate	245.4	96.3	3.7
Methyl salicylate	220.6	15	85
2-Hydroxy-1-isopropyl-4-methylbenzene	232.3	13	87
1,2-Dihydroxybenzene	259.5	46	54
Safrole	225.5	33	67
Isosafrole	233.5	46	54
Benzyl phenyl ether	241.5	80	20
Nitrobenzene	210.0	10	90
m-Nitrotoluene	224.2	25	75
o-Nitrophenol	216.0	10.5	89.5
Quinoline	233.6	29	71
p-Dibromobenzene	212.9	13	87

(*Continued*)

TABLE 2.43 Ternary Azeotropic Mixtures

A. Ternary azeotropes containing water and alcohols

System	BP of azeotrope, °C	Composition, wt %		
		Water	Alcohol	Other component
		Methanol		
Chloroform	52.3	1.3	8.2	90.5
2-Methyl-1,3-butadiene	30.2	0.6	5.4	94.0
Methyl chloroacetate	67.9	6.3	81.2	13.5
		Ethanol		
Acetonitrile	72.9	1	55	44
Acrylonitrile	69.5	8.7	20.3	71.0
Benzene	64.9	7.4	18.5	74.1
Butylamine	81.8	7.5	42.5	50.0
Butyl methyl ether	62	6.3	8.6	85.1
Carbon disulfide	41.3	1.6	5.0	93.4
Carbon tetrachloride	62	4.5	10.0	85.5
Chloroform	55.3	2.3	3.5	94.2
Crotonaldehyde	78.0	4.8	87.9	7.3
Cyclohexane	62.6	4.8	19.7	75.5
1,2-Dichloroethane	66.7	5	17	78
1,1-Diethoxyethane	77.8	11.4	27.6	61.0
Diethoxymethane	73.2	12.1	18.4	69.5
Ethyl acetate	70.2	9.0	8.4	82.6
Heptane	68.8	6.1	33.0	60.9
Hexane	56.0	3	12	85
Toluene	74.4	12	37	51
Trichloroethylene	67.0	5.5	16.1	78.4
Triethylamine	74.7	9	13	78
		1-Propanol		
Benzene	67	7.6	10.1	82.3
Carbon tetrachloride	65.4	5	11	84
Cyclohexane	66.6	8.5	10.0	81.5
1,1-Dipropoxyethane	87.6	27.4	51.6	21.0
Dipropoxymethane	86.4	8.0	44.8	47.2
Dipropyl ether	74.8	11.7	20.2	68.1
3-Pentanone	81.2	20	20	60
Propyl acetate	82.5	17.0	10.0	73.0
Propyl formate	70.8	13	5	82
Tetrachloroethylene	81.2	12.5	20.7	66.8
		2-Propanol		
Benzene	66.5	7.5	18.7	73.8
Butylamine	83	12.5	40.5	47.0

TABLE 2.43 Ternary Azeotropic Mixtures (*Continued*)

A. Ternary azeotropes containing water and alcohols

System	BP of azeotrope, °C	Composition, wt %		
		Water	Alcohol	Other component
2-Propanol (*continued*)				
Cyclohexane	64.3	7.5	18.5	74.0
Toluene	76.3	13.1	38.2	48.7
Trichloroethylene	69.4	7	20	73
1-Butanol				
Butyl acetate	89.4	37.3	27.4	35.3
Butyl formate	83.6	21.3	10.0	68.7
Dibutyl ether	90.6	29.9	34.6	35.5
Heptane	78.1	41.4	7.6	51.0
Hexane	61.5	19.2	2.9	77.9
Nonane	90.0	69.9	18.3	11.8
Octane	86.1	60.0	14.6	25.4
2-Butanol				
Carbon tetrachloride	65	4.05	4.95	91.00
Cyclohexane	69.7	8.9	10.8	80.3
Isooctane	76.3	9	19	72
2-Methyl-1-propanol				
Isobutyl acetate	86.8	30.4	23.1	46.5
Isobutyl formate	80.2	17.3	6.7	76.0
Toluene	81.3	17.9	16.4	65.7
2-Methyl-2-propanol				
Benzene	67.3	8.1	21.4	70.5
Carbon tetrachloride	64.7	3.1	11.9	85.0
Cyclohexane	65.0	8	21	71
3-Methyl-1-butanol				
Isopentyl acetate	93.6	44.8	31.2	24.0
Isopentyl formate	89.8	32.4	19.6	48.0
Allyl alcohol				
Benzene	68.2	8.6	9.2	82.2
Carbon tetrachloride	65.2	5	11	84
Cyclohexane	66.2	8	11	81
Hexane	59.7	8.5	5.1	86.4

TABLE 2.43 Ternary Azeotropic Mixtures (*Continued*)

B. Other ternary azeotropes					
System	BP of azeotrope, °C	Composition, wt%	System	BP of azeotrope, °C	Composition, wt%
Water	32.5	0.4	Water	80.7	17.4
Acetone		7.6	Nitromethane		58.3
2-Methyl-1,3-butadiene		92.0	Nonane		24.3
Water	66	8.2	Water	77.4	12.4
Acetonitrile		23.3	Nitromethane		44.3
Benzene		68.5	Octane		43.3
Water	67	6.4	Water	33.1	2.1
Acetonitrile		20.5	Nitromethane		6.5
Trichloroethylene		73.1	Pentane		91.4
Water	68.6	3.5	Water	82.8	20.6
Acetonitrile		9.6	Nitromethane		73.3
Triethylamine		86.9	Undecane		6.1
Water	63.6	5	Water	93.5	40.5
2-Butanone		35	Pyridine		54.5
Cyclohexane		60	Dodecane		5.0
Water	55.0	4	Water	93.1	38.5
Butyraldehyde		21	Pyridine		51.0
Hexane		75	Undecane		10.5
Water	107.6	21.3	Water	92.3	35.5
Formic acid		76.3	Pyridine		45.5
Isopentanoic acid		2.4	Decane		19.0
Water	107.0	15.5	Water	107.6	19.5
Formic acid		66.8	Formic acid		75.9
Isobutyric acid		17.7	Butyric acid		4.6
Water	71.4	7.9	Water	107.2	18.6
Nitromethane		29.7	Formic acid		71.9
Heptane		62.4	Propionic acid		9.5

Water	105	11.0
Hydrogen bromide		10.4
Chlorobenzene		78.6
Water	96.9	20.2
Hydrogen chloride		5.3
Chlorobenzene		74.5
Water	107.3	64.8
Hydrogen chloride		15.8
Phenol		19.4
Water	116.1	54
Hydrogen fluoride		10
Fluorosilic acid		36
Water	75.1	11.5
Nitroethane		75.1
Heptane		64.0
Water	59.5	8.4
Nitroethane		9.3
Hexane		82.3
Water	82.4	19.1
Nitromethane		68.1
Decane		12.8
Water	90.5	30.5
Pyridine		37.0
Nonane		32.5
Water	86.7	22.4
Pyridine		25.5
Octane		52.0
Water	78.6	14.0
Pyridine		15.5
Heptane		70.5
Acetic acid	134.4	23
Pyridine		55
Acetic anhydride		22
Acetic acid	134.1	31.4
Pyridine		38.2
Decane		30.4
Acetic acid	129.1	13.5
Pyridine		25.2
Ethylbenzene		61.3
Acetic acid	98.5	3.4
Pyridine		10.6
Heptane		86.0
Acetic acid	128.0	20.7
Pyridine		29.4
Nonane		49.9
Acetic acid	115.7	10.4
Pyridine		20.1
Octane		69.5
Water	83.1	21.5
Nitromethane		75.3
Dodecane		3.2
Acetic acid	129.2	10.2
Pyridine		22.5
p-Xylene		67.3
Acetic acid	163.0	75.0
2,6-Dimethylpyridine		13.8
Undecane		11.2
Acetic acid	147.0	12.6
2,6-Dimethylpyridine		74.3
Decane		13.1
Acetic acid	141.3	19.9
2-Methylpyridine		46.8
Decane		33.3
Acetic acid	135.0	12.8
2-Methylpyridine		38.4
Nonane		48.8
Acetic acid	121.3	3.6
2-Methylpyridine		24.8

TABLE 2.43 Ternary Azeotropic Mixtures (*Continued*)

B. Other ternary azeotropes

System	BP of azeotrope, °C	Composition, wt%
Octane		71.6
Acetic acid	77.2	7.6
Benzene		34.4
Cyclohexane		58.0
Acetic acid	132	15
2-Methyl-1-butanol		54
Isopentyl acetate		31
Propionic acid	149.3	29.5
2-Methylpyridine		32.0
Decane		38.5
Acetic acid	132.2	17.7
Pyridine		30.5
o-Xylene		51.8
Methanol	47.4	14.6
Methyl acetate		36.8
Hexane		48.6
Ethanol	63.2	10.4
Acetone		24.3
Chloroform		65.3
Ethanol	70.1	8
Acetonitrile		34
Triethylamine		58
Ethanol	64.7	29.6
Benzene		12.8
Cyclohexane		57.6
Ethanol	57.3	9.5
Chloroform		56.1

System	BP of azeotrope, °C	Composition, wt%
Hexane		34.4
1-Propanol	73.8	15.5
Benzene		30.4
Cyclohexane		54.2
2-Propanol	69.1	31.1
Benzene		15.0
Cyclohexane		53.9
1-Butanol	77.4	4
Benzene		48
Cyclohexane		48
1-Butanol	108.7	11.9
Pyridine		20.7
Toluene		76.4
Propionic acid	140.1	16.5
2-Methylpyridine		21.5
Nonane		42.0
Propionic acid	123.7	4.5
2-Methylpyridine		10.5
Octane		85.0
Propionic acid	153.4	43.0
2-Methylpyridine		40.0
Undecane		17.0
Propionic acid	147.1	55.5
Pyridine		26.4
Undecane		18.1
Methanol	57.5	23

Acetone		30	3-Methylpyridine		16.4
Chloroform		47	1,2-Ethanediol	188.6	29.5
Methanol	47	14.6	Phenol		54.8
Acetone		30.8	2,4,6-Trimethylpyridine		15.7
Hexane		59.6	Acetone	60.8	3.6
Methanol	53.7	17.4	Chloroform		68.8
Acetone		5.8	Hexane		27.6
Methyl acetate		76.8	Acetone	49.7	51.1
Methanol	50.8	17.8	Methyl acetate		5.6
Methyl acetate		48.6	Hexane		43.3
Cyclohexane		33.6	Chloroform	62.0	79.7
1,2-Ethanediol	185.0	8.7	Ethyl formate		5.3
Phenol		74.6	2-Bromopropane		15.7
2,6-Dimethylpyridine		16.7	1,4-Dioxane	101.8	44.3
1,2-Ethanediol	185.1	5.9	2-Methyl-1-propanol		26.7
Phenol		79.1	Toluene		29.0
2-Methylpyridine		15.0			
1,2,-Ethanediol	186.4	15.9			
Phenol		67.7			

2.8 FREEZING MIXTURES

A freezing mixture a mixture of substances (such as salt and ice) to obtain a temperature below the freezing point of the solvent (such as water).

TABLE 2.44 Compositions of Aqueous Antifreeze Solutions

Freezing point of ethyl alcohol-water mixtures*

Specific gravity 20°/4°C. (68°F.)	% alcohol by weight	% alcohol by volume	Freezing point	
			°C.	°F.
0.99363	2.5	3.13	−1.0	30.2
0.98971	4.8	6.00	−2.0	28.4
0.98658	6.8	8.47	−3.0	26.6
0.98006	11.3	14.0	−5.0	23.0
0.97670	13.8	17.0	−6.1	21.0
0.97336	16.4	20.2	−7.5	18.5
0.97194	17.5	21.5	−8.7	16.3
0.97024	18.8	23.1	−9.4	15.1
0.96823	20.3	24.8	−10.6	12.9
0.96578	22.1	27.0	−12.2	10.0
0.96283	24.2	29.5	−14.0	6.8
0.95914	26.7	32.4	−16.0	3.2
0.95400	29.9	36.1	−18.9	−2.0
0.94715	33.8	40.5	−23.6	−10.5
0.93720	39.0	46.3	−28.7	−19.7
0.92193	46.3	53.8	−33.9	−29.0
0.90008	56.1	63.6	−41.0	−41.8
0.86311	71.9	78.2	−51.3	−60.3

Freezing point of methyl (wood) alcohol-water mixtures*

Specific gravity 15.6°C. (60°F.)	% alcohol by weight	% alcohol by volume	Freezing point	
			°C.	°F.
0.993	3.9	5	−2.2	28
0.986	8.1	10	−5.0	23
0.980	12.2	15	−8.3	17
0.974	16.4	20	−11.7	11
0.968	20.6	25	−15.6	4
0.963	24.9	30	−20.0	−4
0.956	29.2	35	−25.0	−13
0.949	33.6	40	−30.0	−22
0.942	38.0	45	−35.6	−32

*Values are for pure alcohol. Since some commercial antifreezes contain small amounts of water, slightly higher volume concentrations than those given in the table may be required. Antifreezes also contain corrosion inhibitors and other additives to make them function properly as cooling liquids. These affect freezing point slightly and specific gravity to a greater degree.

TABLE 2.44 Compositions of Aqueous Antifreeze Solutions (*Continued*)

Freezing point of Prestone-water mixtures†

% Prestone		Specific gravity	Freezing point	
By weight	By volume	15°/15C. (59°F.)	°C.	°F.
10	9.2	1.013	−3.6	25.6
15	13.8	1.019	−5.6	22.0
20	18.3	1.026	−7.9	17.8
25	23.0	1.033	−10.7	12.8
30	28.0	1.040	−14.0	6.8
40	37.8	1.053	−22.3	−8.2
50	47.8	1.067	−33.8	−28.8
60	58.1	1.079	−49.3	−56.7

Freezing point of ethyl alcohol-water mixtures

Specific gravity 15.6°C. (60°F.)	% alcohol by volume	Freezing point °C.	Freezing point °F.
0.990	5	−1.7	29
0.984	10	−3.3	26
0.978	15	−6.1	21
0.972	20	−8.3	17
0.964	25	−11.1	12
0.955	30	−14.4	6
0.945	35	−17.8	0
0.933	40	−18.3	−1
0.922	45	−18.9	−2
0.910	50	−20.0	−4
0.899	55	−21.7	−7
0.887	60	−23.3	−10
0.875	65	−24.4	−12
0.864	70	−26.7	−16
0.852	75	−32.2	−26
0.840	80	−41.7	−43

†Eveready Prestone marketed for antifreeze purposes, is 97% ethylene glycol containing fractional percentages of soluble and insoluble ingredients to prevent foaming, creepage and water corrosion in automobile cooling systems.

TABLE 2.44 Compositions of Aqueous Antifreeze Solutions (*Continued*)

Freezing point of propylene glycol-water mixtures*

Specific gravity 15.6°C. (60°F.)	% glycol by volume	Freezing point °C.	Freezing point °F.
1.004	5	−1.1	30
1.006	10	−2.2	28
1.012	15	−3.9	25
1.017	20	−6.7	20
1.020	25	−8.9	16
1.024	30	−12.8	9
1.028	35	−16.1	3
1.032	40	−20.6	−5
1.037	45	−26.7	−16
1.040	50	−33.3	−28

Freezing point of glycerol-water mixtures†

% Glycerol by weight	Specific gravity 15°/15°C. (59°F.)	Specific gravity 20°/20°C. (68°F.)	Freezing point °C.	Freezing point °F.
10	1.02415	1.02395	−1.6	29.1
20	1.04935	1.04880	−4.8	23.4
30	1.07560	1.07470	−9.5	14.9
40	1.10255	1.10135	−15.5	4.3
50	1.12985	1.12845	−22.0	−7.4
60	1.15770	1.15605	−33.6	−28.5
70	1.18540	1.18355	−37.8	−36.0
80	1.21290	1.21090	−19.2	−2.3
90	1.23950	1.23755	−1.6	29.1
100	1.26557	1.26362	17.0	62.6

*Values are for pure alcohol. Since some commercial antifreezes contain small amounts of water, slightly higher volume concentrations than those given in the table may be required. Antifreezes also contain corrosion inhibitors and other additives to make them function properly as cooling liquids. These affect freezing point slightly and specific gravity to a greater degree.

†The values are those reported by Bosart and Snoddy (*Jour. Ind. Eng. Chem.*, **19,** 506 (1927), and Lane (*Jour. Ind. Eng. Chem.*, **17,** 924 (1925)) but modified by adding 2°F to all temperatures below 0°F.

TABLE 2.44 Compositions of Aqueous Antifreeze Solutions (*Continued*)

Freezing point of magnesium chloride brines

% $MgCl_2$ by weight	Spec. grav. 15.6°C. (60°F.)	Freezing point °C.	Freezing point °F.	% $MgCl_2$ by weight	Spec. grav. 15.6°C. (60°F.)	Freezing point °C.	Freezing point °F.
5	1.043	−3.11	26.4	18	1.161	−22.1	−7.7
6	1.051	−3.89	25.0	19	1.170	−25.6	−12.2
7	1.060	−4.72	23.5	20	1.180	−27.4	−17.3
8	1.069	−5.67	21.8	21	1.190	−30.6	−23.0
9	1.078	−6.67	20.0	22	1.200	−32.8	−27.0
10	1.086	−7.83	17.9	23	1.210	−28.9	−20.0
11	1.096	−9.05	15.7	24	1.220	−25.6	−14.0
12	1.105	−10.5	13.1	25	1.230	−23.3	−10.0
13	1.114	−12.1	10.3	26	1.241	−21.1	−6.0
14	1.123	−13.7	7.3	27	1.251	−19.4	−3.0
15	1.132	−15.6	4.0	28	1.262	−18.3	−1.0
16	1.142	−17.6	0.4	29	1.273	−17.2	+1.0
17	1.151	−19.7	−3.5	30	1.283	−16.7	2.0

Freezing point of sodium chloride brines

% NaCl by weight	Spec. grav. 15°C. (59°F.)	Freezing point °C.	Freezing point °F.	% NaCl by weight	Spec. grav. 15°C. (59°F.)	Freezing point °C.	Freezing point °F.
0	1.000	0.00	32.0	15	1.112	−10.88	12.4
1	1.007	−0.58	31.0	16	1.119	−11.90	10.6
2	1.014	−1.13	30.0	17	1.127	−12.93	8.7
3	1.021	−1.72	28.9	18	1.135	−14.03	6.7
4	1.028	−2.35	27.8	19	1.143	−15.21	4.6
5	1.036	−2.97	26.7	20	1.152	−16.46	2.4
6	1.043	−3.63	25.5	21	1.159	−17.78	+0.0
7	1.051	−4.32	24.2	22	1.168	−19.19	−2.5
8	1.059	−5.03	22.9	23	1.176	−20.69	−5.2
9	1.067	−5.77	21.6	23.3 (*E*)	1.179	−21.13	−6.0
10	1.074	−6.54	20.2	24	1.184	−17.0*	+1.4*
11	1.082	−7.34	18.8	25	1.193	−10.4*	13.3*
12	1.089	−8.17	17.3	26	1.201	−2.3*	27.9*
13	1.097	−9.03	15.7	26.3	1.203	0.0*	32.0*
14	1.104	−9.94	14.1				

*Saturation temperatures of sodium chloride dihydrate; at these temperatures $NaCl \cdot 2H_2O$ separates leaving the brine of the eutectic composition (*E*).

Propylene glycol, a satisfactory antifreeze with the advantage of being nontoxic, can be combined with glycerol, also an efficient nontoxic antifreeze, to give a mixture that can be tested for freezing point with an ethylene glycol (Prestone) hydrometer. A mixture of 70% propylene glycol and 30% glycerol (% by weight of water-free materials), when diluted, can be tested on the standard instrument used for ethylene glycol solutions.

2.9 BOND LENGTHS AND STRENGTHS

Distances between centers of bonded atoms are called *bond lengths*, or *bond distances*. Bond lengths vary depending on many factors, but in general, they are very consistent. Of course the bond orders affect bond length, but bond lengths of the same order for the same pair of atoms in various molecules are very consistent.

The *bond order* is the number of electron pairs shared between two atoms in the formation of the bond. Bond order for C═C and O═O is 2. The amount of energy required to break a bond is called *bond dissociation energy* or simply *bond energy*. Since bond lengths are consistent, bond energies of similar bonds are also consistent.

Bonds between the same type of atom are *covalent bonds*, and bonds between atoms when their electronegativity differs slightly are also predominant covalent in character. Theoretically, even ionic bonds have some covalent character. Thus, the boundary between ionic and covalent bonds is not a clear line of demarcation.

For covalent bonds, bond energies and bond lengths depend on many factors: electron afinities, sizes of atoms involved in the bond, differences in their electronegativity, and the overall structure of the molecule. There is a general trend in that *the shorter the bond length, the higher the bond energy* but there is no formula to show this relationship, because of the widespread variation in bond character.

TABLE 2.45 Bond Lengths between Carbon and Other Elements

Bond type	Bond Length, µm
Carbon-carbon	
Single bond	
Paraffinic: —C—C—	154.1(3)
In presence of —C═C— or of aromatic ring	153(1)
In presence of —C═O bond	151.6(5)
In presence of two carbon-oxygen bonds	149(1)
In presence of two carbon-carbon double bonds	142.6(5)
Aryl-C═O	147(2)
In presence of one carbon-carbon triple bond: —C—C≡C—	146.0(3)
In presence of one carbon-nitrogen triple bond: —C—C≡N	146.6(5)
In compounds with tendency to dipole formation, e.g., C═C—C═O	144(1)
In aromatic compounds	139.5(5)
In presence of carbon-carbon double and triple bounds: —C═C—C≡C—	142.6(5)
In presence of two carbon-carbon triple bounds: —C≡C—C═C—	137.3(4)
Double bond	
Single: —C═C—	133.7(6)
Conjugated with a carbon-carbon double bond: —C═C—C═C—	133.6(5)
Conjugated with a carbon-oxygen double bond: —C═C—C═O	136(1)
Cumulative: —C═C═C—or —C═C═O	130.9(5)
Triple bond	
Simple: —C≡C—	120.4(2)
Conjugated: —C≡C—C═C—, —C≡C—C═O, or —C≡C—aryl	120.6(4)

Bond type	Bond length, pm			
	Carbon-halogen			
	Fluorine	Chlorine	Bromine	Iodine
Paraffinic: R—X	137.9(5)	176.7(2)	193.8(5)	213.9(1)
Olenfinic: —C═C—X	133.3(5)	171.9(5)	189(1)	209.2(5)
Aromatic: Ar-X	132.8(5)	170(1)	185(1)	205(1)
Acetylenic: —C≡C—X	(127)	163.5(5)	179.5(10)	199(2)

TABLE 2.45 Bond Lengths between Carbon and Other Elements (*Continued*)

Bond type	Bond Length, μm
Carbon-carbon	
Paraffinic	
In methane (in CD_4, 109.2)	109.4
In monosubstituted carbon: H—C—Y	109.6(5)
In disubstituted carbon: H—C(X)(Y)—	107.3(5)
In trisubstituted carbon: H—C(X)(Z)—Y	107.0(7)
Olefinic	
Simple: H—C═C—	108.3(5)
Cumulative carbon-carbon double bonds: H—C═C═C—	107(1)
Cumulative carbon-carbon-oxygen double bonds: H—C—C═C═O	108(1)
Aromatic	108.4(5)
Acetylenic (in C_2H_2, 105.9)	105.5(5)
In small rings	108.1(5)
In presence of a carbon triple bond: H—C≡C—	111.5(4)
Carbon-nitrogen	
Single bond	
Paraffinic:	
3-covalent nitrogen: RNH_2, R_2NH, R_3N	147.2(5)
4-covalent nitrogen: RNH_3^+, $R_3N\text{-}BX_3$	147.9(5)
In —C—N═	147.5(10)
In aromatic compounds	143(1)
In conjugated heterocyclic systems (partial double bond)	135.3(5)
In —N—C═O (partial double bond)	132.2(5)
Double bond: —C═N—	132
Triple bond (in CN radical, 117.74): —C≡N	115.7(5)
Carbon-oxygen	
Single bond	
Paraffinic and saturated heterocyclic: —C—O—	142.6(5)
Strained, as in epoxides: —C—C— (bridged by O)	143.5(5)
In aromatic compounds, as Ar-OH	136(1)
Longer bond in carboxylic acids and esters (HCOOH, 131.2)	135.8(5)
In conjugated heterocyclics, as furan	137.1(16)
Double bond	
In CO^+	111.5
In CO	112.8
In CO_2^+	117.7
In HCO	119.8(8)
In carbonyls	114.5(10)
In aldehydes and ketones	121.5(5)
In acyl halides: R—CO—X	117.1(4)
Shorter bond in carboxylic acids and esters	123.3(5)
In zwitterion forms	126(1)

TABLE 2.45 Bond Lengths between Carbon and Other Elements (*Continued*)

Bond type	Bond Length, μm
Carbon-oxygen	
In O=C=	116.0(1)
In isocyanates: RN=C=O	117(1)
In conjugated systems, as in partial triple bond: O=C—C=C	121.5(5)
In 1,4-quinones	115(2)
In metal acetylacetonates	128(2)
In calcite: $CaCO_3$	129(1)
Carbon-selenium	
Single bond	
Paraffinic: —C—Se—	198(2)
In presence of fluorine, as in perfluorocompounds: —CF—Se—	195(2)
Double bond	
In Se=C=, as SeCS and SeCO	170.9(3)
In CSe radical	167
Carbon-silicon	
Alkyl substituent: H_3C—Si or H_2C—Si	187.0(5)
Aryl substituent: aryl—Si	184.3(5)
Electronegative substituent: R—Si—X	185.4(5)
Carbon-sulfur	
Single bond	
Paraffinic: —C—S—	181.7(5)
In presence of fluorine, as in perfluoro compounds: —CF—S—	183.5(1)
In heterocyclic systems: partial double bonds	171.8(5)
Double bonds	
In S=C; thiophene, S=CR_2	171(1)
In sulfoxides and sulfones	180(1)
In presence of second carbon-carbon double bond: S=C—C=C—	155.5(1)
In SC radical [in CS_2^+, 155.4(5)]	153.49(2)

Bond type	Bond length, pm	Bond type	Bond length, pm
Other elements and carbon			
C-Al	224(4)	C-Cr	192(4)
C-As	198(1)	C-Fe	184(2)
C-B	156(1)	C-Ge	
C-Be	193	Alkyl	193(3)
C-Bi	230	Aryl	194.5(5)
C-Co	183(2)		
C-Hg	207(1)	C-Sn	
in $Hg(CN)_2$	199(2)	Alkyl	214.3(5)
C-In	216(4)	Electronegative substituent	218(2)
C-Mo	208(4)		
C-Ni	210.7(5)	C-Te	190.4
C-Pb (alkyl)	230(1)	C-Tl	270.5(5)
C-Pd	227(4)	C-W	206
C-Sb (paraffinic)	220.2(16)		

TABLE 2.46 Bond Dissociation Energies

Bond	ΔHf_{298}, kJ/mol	Bond	ΔHf_{298}, kJ/mol
Carbon (*continued*)		Carbon (*continued*)	
$(CH_3)_2C—CH_3$	335	$C_6H_5CH_2—N(CH_3)_2$	255(4)
$(CH_3)_2C—C(CH_3)_2$	282.4	$CH_3—(N{=}NCH_3)$	219.7
$CH_3—C_6H_5$	389	$C_2H_5—(N{=}NC_2H_5)$	209.2
$CH_3—CH_2C_6H_5$	301	$(CH_3)_3C—N{=}NC(CH_3)_3$	182.0
$(CH_3)_3C—C(C_6H_5)_3$	63	Aryl—$CH_2N{=}NCH_2$—aryl	157
CH_3—allyl	301	$CF_3—(N{=}NCF_3)$	231.0
CH_3—vinyl	121	$H_2C{=}NH$	644(21)
$CH_3—C{\equiv}CH$	490	$HC{\equiv}N$	937
$CH_2{=}CH—CH{=}CH_2$	418	$CH_3—NO$	174.9(38)
$HC{\equiv}C—C{\equiv}CH$	628	$C_2H_5—NO$	175.7(54)
$H_2C{=}CH_2$	682	$C_3H_7—NO$	167.8(75)
$HC{\equiv}CH$	962	$(CH_3)_2CH—NO$	171.5(54)
$CH_3—CN$	506(21)	n-$C_4H_9—NO$	215.5(42)
$CH_3—CH_2CN$	305(8)	$C_6H_5—NO$	215.5(42)
$CH_3—CH(CH_3)CN$	331(8)	$Cl_3C—NO$	134
$CH_3—C(C_6H_5)CN(CH_3)$	251	$F_3C—NO$	130
$CH_3CH_2—CH_2CN$	321.8(71)	$C_6F_5—NO$	211.3(42)
$NC—CN$	603(21)	$NC—NO$	121(13)
$C_6H_5—C_6H_5$	418	$CH_3—NO_2$	247(13)
$CH_3—CF_3$	423.4(46)	$C_2H_5—NO_2$	259
$CH_2F—CH_2F$	368(8)	$C—O$	1076.5(4)
$CF_3—CF_3$	406(13)	$CH_3—OCH_3$	335
$CF_2{=}CF_2$	318(13)	$CH_3—OC_6H_5$	381
$CF_3—CN$	501	$CH_3—OCH_2C_6H_5$	280
$CH_3—CHO$	314	$C_2H_5—OC_6H_5$	213
$CH_3—CO$	342.7	$C_6H_5CH_2—OCOCH_3$	285
$CH_3CO—CF_3$	308.8	$C_6H_5CH_2—OCOC_6H_5$	289
$CH_3CO—COCH_3$	280(8)	$CH_3CO—OCH_3$	406
$C_6H_5CO—COC_6H_5$	277.8	$CH_3—OSOCH_3$	280
Aryl—CH_2COCH_2—aryl	273.6	$CH_2{=}CHCH_2—OSOCH_3$	209
$C_6H_5CH_2—COOH$	284.9	$C_6H_5CH_2—OSOCH_3$	222
$(C_6H_5CH_2)_2CH—COOH$	248.5	$C{=}O$	749
$C—Cl$	397(29)	$H_2C{=}O$	732
$C—F$	536(21)	$OC{=}O$	532.2(4)
$C—H$	337.2(8)	$SC{=}O$	628
$C—I$	209(21)	$C{\equiv}O$	1075
$C—N$	770(4)	$C—P$	513(8)
$CF_3—NF_2$	272(13)	$C—S$	699(8)
$CH_3—NH_2$	331(13)	$CH_3—SH$	305(13)
$C_6H_5CH_2—NH_2$	301(4)	$CH_3—SC_6H_5$	285(8)
$CH_3—NHC_6H_5$	285	$CH_3—SCH_2C_6H_5$	247(8)
$CH_3—N(CH_3)C_6H_5$	272	$OC—S$	310.4
$C_6H_5CH_2—NHCH_3$	289(4)	$C—Se$	582(96)

2.10 DIPOLE MOMENTS AND DIELECTRIC CONSTANTS

The permanent dipole moment of an isolated molecule depends on the magnitude of the charge and on the distance separating the positive and negative charges. It is defined as

$$\mu = \left(\sum_i q_i r_i \right)$$

where the summation extends over all charges (electrons and nuclei) in the molecule. The numerical values of the dipole moment, expressed in the c.g.s. system of units, are in debye units, D, where 1 D = 10^{-18} esu of charge × centimeters. The conversion factor to SI units is

$$1\ \text{D} = 3.335\ 64 \times 10^{-30}\ \text{C} \cdot \text{m [coulomb-meter]}$$

Tables 2.49 contain a selected group of compounds for which the dipole moment is given. An extensive collection of dipole moments (approximately 7000 entries) is contained in A. L. McClellan, *Tables of Experimental Dipole Moments*, W. H. Freeman, San Francisco, 1963. A critical survey of 500 compounds in the gas phase is given by Nelson, Lide, and Maryott, NSRDS-NBS 10, Washington, D.C., 1967.

If two oppositely charged plates exist in a vacuum, there is a certain force of attraction between them, as stated by Coulomb's law:

$$F = \frac{1}{4\pi\varepsilon_0} \cdot \frac{q_1 q_2}{\varepsilon r^2}$$

where ***F*** is the force, in newtons, acting on each of the charges q_1 and q_2, r is the distance between the charges, ε is the dielectric constant of the medium between the plates, and ε_0 is the permittivity of free space. q_1, q_2 are expressed in coulombs and r in meters. If another substance, such as a solvent, is in the space separating these charges (or ions in a solution), their attraction for each other is less. The dielectric constant is a measure of the relative effect a solvent has on the force with which two oppositely charged plates attract each other. The dielectric constant is a unitless number.

Dielectric constants for a selected group of inorganic and organic compounds are included in Tables 2.49 and 1.52. An extensive list has been compiled by Maryott and Smith, *National Bureau Standards Circular 514*, Washington, D.C., 1951.

For gases the values of the dielectric constant can be adjusted to somewhat different conditions of temperature and pressure by means of the equation

$$\frac{(\varepsilon - 1)_{t,p}}{(\varepsilon - 1)_{20°,1\ \text{atm}}} = \frac{p}{760[1 + 0.003\ 411(t - 20)]}$$

where p is the pressure (in mmHg) and t is the temperature (in °C). The errors associated with this equation probably do not exceed 0.02% for gases between 10 and 30°C and for pressures between 700 and 800 mm. The dielectric constants of selected gases will be found in Table 1.52.

TABLE 2.47 Bond Dipole Moments

	Moment, D*	
Group	Aromatic C—X	Aliphatic C—X
$C—CH_3$	0.37	0.0
$C—C_2H_5$	0.37	0.0
$C—C(CH_3)_3$	0.5	0.0
$C—CH{=}CH_2$	<0.4	0.6
$C—C{\equiv}CH$	0.7	0.9
C—F	1.47	1.79

TABLE 2.48 Group Dipole Moments

Group	Moment, D*	
	Aromatic C—X	Aliphatic C—X
C—Cl	1.59	1.87
C—Br	1.57	1.82
C—I	1.40	1.65
C—CH_2F	1.77	
C—CF_3	2.54	2.32
C—CH_2Cl	1.85	1.95
C—$CHCl_2$	2.04	1.94
C—CCl_3	2.11	1.57
C—CH_2Br	1.86	1.96
C—C≡N	4.05	3.4
C—NC	3.5	3.5
C—CH_2CN	1.86	2.0
C—C=O	2.65	2.4
C—CHO	2.96	2.49
C—COOH	1.64	1.63
C—CO—CH_3	2.96	2.49
C—CO—OCH_3	1.83	1.75
C—CO—OC_2H_5	1.9	1.8
C—OH	1.6	1.7
C—OCH_3	1.28	1.28
C—OCF_3	2.36	
C—$OCOCH_3$	1.69	
C—OC_6H_5	1.16	1.16
C—CH_2OH	1.58	1.68
C—NH_2	1.53	1.46
C—$NHCH_3$	1.71	
C—$N(CH_3)_2$	1.58	0.86
C—$NHCOCH_3$	3.69	
C—$N(C_6H_5)_2$	(0.3)	−0.3
C—NCO	2.32	2.8
C—N_3	1.44	
C—NO	3.09	
C—NO_2	4.01	2.70
C—CH_2NO_2	3.3	3.4
C—SH	1.22	1.55
C—SCH_3	1.34	1.40
C—SCF_3	2.50	
C—SCN	3.59	3.6
C—NCS	2.9	3.3
C—SC_6H_5	1.51	1.5
C—SF_5	3.4	
C—$SOCF_3$	3.88	
$(C—)_2SO_2$	5.05	4.53
$(C—)_2SO_2CH_3$	4.73	
$(C—)_2SO_2CF_3$	4.32	
C—SeH	1.08	
C—$SeCH_3$	1.31	1.32
C—$Si(CH_3)_3$	0.44	0.4

*To convert debye units D into coulomb-meters, multiply by 3.33564×10^{-30}.

TABLE 2.49 Dielectric Constant (Permittivity) and Dipole Moment of Organic Compounds

The temperature in degrees Celsius at which the dielectric constant and dipole moment were measured is shown in this table in parentheses after the value. In some cases, the dipole moment was determined with the substance dissolved in a solvent, and the solvent used is also shown in parentheses after the temperature.

The dielectric constant (permittivity) tabulated is the relative dielectric constant, which is the ratio of the actual electric displacement to the electric field strength when an external field is applied to the substance, which is the ratio of the actual dielectric constant to the dielectric constant of a vacuum. The table gives the static dielectric constant ϵ, measured in static fields or at relatively low frequencies where no relaxation effects occur.

The dipole moment is given in debye units D. The conversion factor to SI units is I D = 3.33564×10^{-30}C · m.

Alternative names for entries are listed in Table 2.20 at the bottom of each double page.

List of Abbreviations

B, benzene	*g, gas*
C, CCl_4	*Hx, hexane*
cHex, cyclohexane	*lq, liquid*
D, 1,4-dioxane	

Substance	Dielectric constant, ϵ	Dipole moment, D
Acetaldehyde	21.8 (10), 21.0 (18)	2.75
Acetaldehyde oxime	4.70 (25)	0.830 (20, lq), 0.90 (25, B)
Acetamide	67.6 (91)	3.76
Acetanilide		3.65 (25, B)
Acetic acid	6.20 (20)	1.70
Acetic anhydride	23.3 (0), 22.45 (20)	2.8
Acetone	21.0 (20), 20.7 (25), 17.6 (56)	2.88
Acetonitrile	36.64 (20), 26.6 (82)	3.924
Acetophenone	17.44 (25), 8.64 (202)	3.02
(±)-*erythro*-2-Acetoxy-2-bromobutane	7.268 (25)	
(±)-*threo*-2-Acetoxy-2-bromobutane	7.414 (25)	
Acetyl bromide	16.2 (20)	2.43 (20, B)
Acetyl chloride	16.9 (2), 15.8 (22)	2.72
Acetylene	2.484 (−77)	
Acrylonitrile	33.0 (20)	3.87
Allene	2.025 (−4)	
Allylamine		1.2
Allyl alcohol	19.7 (20)	1.61
Allyl isocyanate	15.15 (15)	
Allyl isothiocyanate	17.2 (18)	3.2 (20, B)
Allyl nitrite	9.12 (25)	
2-Aminoethanol	31.94 (20), 37.72 (25)	2.59 (25, D)
2-(2-Aminoethylamino)ethanol	21.81 (20)	
N-(2-Aminoethyl)-1,2-ethanediamine	12.62 (20)	1.9
Aniline	7.06 (20), 5.93 (70)	1.13
Benzaldehyde	19.7 (0), 17.85 (20)	3.0
Benzaldehyde oxime (mp 30)	3.8 (20)	1.2 (25, B)
(mp 128)		1.5 (25, B)
Benzamide		3.42 (25, B)
Benzene	2.292(15), 2.283 (20), 2.274 (25)	0
Benzeneacetonitrile	17.87 (26)	3.5
Benzenesulfonyl chloride	28.90 (50)	4.50 (20, B)
Benzenethiol	4.38 (25), 4.26 (30)	1.13 (25, lq), 1.19 (20, B)
Benzonitrile	25.9 (20), 24.0 (40)	4.18
Benzophenone	14.60 (18), 11.4 (50)	3.09 (50, lq), 2.98 (25, B)

TABLE 2.49 Dielectric Constant (Permittivity) and Dipole Moment of Organic Compounds (*Continued*)

Substance	Dielectric constant, ϵ	Dipole moment, D
Benzoyl bromide	21.33 (20), 20.74 (25)	3.40 (20, B)
Benzoyl chloride	29.0 (0), 23 (23)	3.16 (25, B)
Benzoyl fluoride	22.7 (20)	
Benzyl acetate	5.1 (21), 5.34 (930)	1.80 (25, B)
Benzyl alcohol	13.0 (20), 11.92 (30), 9.5 (70)	1.71
Benzylamine	5.5 (1), 5.18 (20)	1.15 (20, lq), 1.38 (25, B)
Benzyl benzoate	5.26 (30)	2.06 (30, B)
Benzyl chloride	7.0 (13), 6.85 (25)	1.83 (20, B)
Benzylethylamine	4.3 (20)	
Benzyl ethyl ether	3.90 (25)	
Benzyl formate	6.34 (30)	
N-Benzylmethylamine	4.4 (19)	
Biphenyl	2.53 (75)	0
Bis(2-aminoethyl)amine	12.62 (20)	
Bis(2-chloroethyl) ether	21.20 (20)	2.6
Bis(3-chloropropyl) ether	10.10 (20)	
Bis(2-ethoxyethyl) ether		1.92 (25, B)
Bis(2-hydroxyethyl) ether	31.69 (20)	2.31 (20, B)
Bis(2-hydroxyethyl)sulfide	28.61 (20)	
Bis(2-hydroxypropyl) ether	20.38 (20)	
Bis(2-methoxyethyl) ether	7.23 (25)	
(±)-Bornyl acetate	4.6 (21)	1.89 (22)
3-Bromoaniline	13.0 (20)	2.67 (20, B)
4-Bromoaniline	7.06 (30)	2.88 (25, B)
2-Bromoanisole	8.96 (30)	
4-Bromoanisole	7.40 (30)	
Bromobenzene	5.45 (20), 5.40 (25)	1.70
1-Bromobutane	7.88 (−10), 7.32 (10), 7.07 (20)	2.08
(±)-2-Bromobutane	8.64 (25)	2.23
2-Bromobutanoic acid	7.2 (20)	
cis-2-Bromo-2-butene	5.38 (20)	
trans-2-Bromo-2-butene	6.76 (20)	
1-Bromo-2-chlorobenzene	6.80 (20)	2.15 (20, B)
1-Bromo-3-chlorobenzene	4.58 (20)	1.52 (22, B)
1-Bromo-4-chlorobenzene		0.1 (25, B)
1-Bromo-2-chloroethane	7.41 (10)	1.09
cis-1-Bromo-2-chloroethene	7.31 (17)	
trans-1-Bromo-2-chloroethene	2.50 (17)	
Bromochlorodifluoromethane	3.92 (−150)	
Bromochloromethane	7.79	1.66 (25, B)
3-Bromo-1-chloro-2-methylpropane	8.90 (30)	
Bromocyclohexane	11 (−65), 8.003(30)	1.08 (25, lq), 2.3 (25, B)
1-Bromodecane	4.75 (1), 4.44 (25)	2.08 (20, lq), 1.90 (25, lq)
Bromodichloromethane		1.31 (25, B)
1-Bromododecane	4.07 (25)	2.01 (25, lq), 1.89 (25, B)
Bromoethane	13.6 (−60), 9.39 (20), 9.01 (25)	2.03 (g), 2.04 (20, lq)
1-Bromo-2-ethoxypentane	6.45 (25)	2.32 (25, B)
2-Bromo-3-ethoxypentane	6.40 (25)	2.07 (25, B)
3-Bromo-2-ethoxypentane	8.24 (25)	2.15 (25, B)
1-Bromo-2-ethylbenzene	5.55 (25)	
1-Bromo-3-ethylbenzene	5.56 (25)	
1-Bromo-4-ethylbenzene	5.42 (25)	

(*Continued*)

TABLE 2.49 Dielectric Constant (Permittivity) and Dipole Moment of Organic Compounds (*Continued*)

Substance	Dielectric constant, ϵ	Dipole moment, D
Bromoethylene	5.63 (5), 4.78 (25)	1.42
1-Bromo-2-fluorobenzene	4.72 (25)	
1-Bromo-3-fluorobenzene	4.85 (25)	
1-Bromo-4-fluorobenzene	2.60 (25)	
Bromoform	4.39 (20)	1.00, 0.92 (25, lq)
1-Bromoheptane	5.33 (25), 4.48 (90)	2.17, 2.02 (20, lq)
2-Bromoheptane	6.46 (22)	2.08 (20, B)
3-Bromoheptane	6.93 (22)	2.06 (20, B)
4-Bromoheptane	6.81 (22)	2.06 (20, B)
1-Bromohexadecane	3.71 (25)	1.98 (20, lq), 1.96 (25, C)
1-Bromohexane	6.30 (1), 5.82 (25)	2.06 (20, lq)
Bromomethane	9.82 (0), 9.71 (3), 1.0068 (100, g)	1.82
(Bromomethyl)benzene	6.658 (20)	
1-Bromo-3-methylbutane	8.04 (−56), 6.33 (18)	1.95 (20, B)
2-Bromo-2-methylbutane	9.21 (25)	
2-Bromo-3-methylbutanoic acid	6.5 (20)	
1-Bromo-2-methylpropane	10.98 (20), 7.2 (25)	1.92 (25, lq), 1.99 (20, B)
2-Bromo-2-methylpropane	10.98 (20)	
1-Bromonaphthalene	5.83 (25), 5.12 (20)	1.29 (25, lq)
3-Bromonitrobenzene	20.2 (55)	
1-Bromononane	5.42 (−20), 4.74 (25)	1.95 (25, lq)
1-Bromooctane	6.35 (−50)	1.99 (20, lq), 1.88 (25, lq)
1-Bromopentadecane	3.9 (20)	
1-Bromopentane	9.9 (−90), 6.32 (25)	2.20
3-Bromopentane	8.37 (25)	
1-Bromopropane	8.09 (20)	2.18
2-Bromopropane	9.46 (20)	2.21
2-Bromopropanoic acid	11.0 (21)	
3-Bromopropene	7.0 (20)	1.9
2-Bromopyridine	23.18 (25)	
1-Bromotetradecane	3.84 (25)	1.92 (20, lq), 1.83 (25, lq)
o-Bromotoluene	4.64 (20), 4.28 (58)	1.45 (20, B)
m-Bromotoluene	5.566 (20), 5.36 (58)	1.77 (20, B)
p-Bromotoluene	5.503 (20), 5.49 (58)	1.95 (20, B)
Bromotrichloromethane	2.40 (20)	
Bromotrifluoromethane	3.73 (−150)	0.65
1-Bromoundecane	4.73 (−9)	
1,3-Butadiene	2.050 (−8)	0.403
Butanal	13.45 (25)	2.72
Butane	1.7697 (22)	0
1,2-Butanediol	22.4 (25)	
1,3-Butanediol	28.8 (25)	
1,4-Butanediol	33 (15), 31.9 (25), 30 (38)	4.07
1,3-Butanediol dinitrate	18.85 (20)	
2,3-Butanediol dinitrate	28.85 (20)	
1,3-Butanedione	4.04 (25)	
Butanenitrile	24.83 (20)	4.07
Butanesulfonyl chloride		3.94 (25, D)
1,2,3,4-Butanetetrol	28.2 (120)	
1-Butanethiol	5.20 (15), 5.07 (25), 4.59 (50)	1.54 (25, lq or B)
2-Butanethiol	5.645 (15)	
Butanoic acid	2.97 (20)	1.65 (30, B)

TABLE 2.49 Dielectric Constant (Permittivity) and Dipole Moment of Organic Compounds (*Continued*)

Substance	Dielectric constant, ϵ	Dipole moment, D
Butanoic anhydride	12.8 (20)	
1-Butanol	17.84 (20), 8.2 (118)	1.66
(±)-2-Butanol	17.26 (20), 16.6 (25)	1.66 (30, B)
2-Butanone	18.56 (20), 15.3 (60)	2.78
2-Butanone oxime	3.4 (20)	
trans-2-Butenal		3.67
1-Butene	2.2195 (−53), 1.0032 (20, g)	0.438
cis-2-Butene	1.960 (23)	0.253
trans-2-Butene		0
3-Butenenitrile	28.1 (20)	4.53
2-Butoxyethanol	9.43 (25)	2.08 (25, B)
Butoxyethyne	6.62 (25)	2.05 (25, lq)
N-Butylacetamide	104.0 (20)	
N-*sec*-Butylacetamide	100.0 (100)	
Butyl acetate	6.85 (−73), 5.07 (20)	1.86 (22, B)
sec-Butyl acetate	5.135 (20)	1.9
tert-Butyl acetate	5.672 (20)	1.91 (25, B)
tert-Butylacetic acid	2.85 (23)	
Butyl acrylate	5.25 (28)	
Butylamine	4.71 (20)	1.00
sec-Butylamine	4.4 (21)	1.28 (25, B)
tert-Butylamine		1.29 (25, B)
Butylbenzene	2.36 (20)	0
sec-Butylbenzene	2.36 (20)	0
tert-Butylbenzene	2.36 (20)	0.83
Butyl butanoate	4.39 (25)	
Butyl ethyl ether		1.24
Butyl formate	6.10 (30), 2.43 (80)	2.08 (26, lq), 2.03 (25, B)
Butyl isocyanate	12.29 (20)	
Butyl methyl ether		1.25 (25, B)
2-*tert*-Butyl-4-methylphenol		1.31 (20, B)
Butyl nitrate	13.10 (20)	2.99 (20, B)
tert-Butyl nitrite	11.47 (25)	
Butyl oleate	4.00 (25)	
N-Butylpropanamide	100.6 (25)	
Butyl propanoate	4.838 (20)	1.79 (23, B)
4-*tert*-Butylpyridine		2.87 (25, C)
Butylsilane	2.537 (20)	
Butyl stearate	3.11 (30)	1.88 (24, B)
Butyl trichloroacetate	7.480 (20)	
Butyl vinyl ether		1.25 (25, Hx)
4-Butyrolactone	39.0 (20)	4.27
Camphor	11.35 (20)	2.91 (20, B), 3.10 (25, B)
Carbon disulfide	3.0 (−112), 2.64 (20)	0
Carbon tetrachloride	2.24 (20), 2.228 (25)	0
Carbon tetrafluoride	1.0006 (25, g)	0
D-(+)-Carvone	11 (22)	2.8 (15, B)
Chloroacetic acid	20 (20), 12.35 (65)	2.31 (30, B)
o-Chloroaniline	13.40 (20)	1.78 (20, B)
m-Chloroaniline	13.3 (20)	2.68 (20, B)
p-Chloroaniline		2.99 (25, B)
Chlorobenzene	5.69 (20), 4.2 (120)	1.69

(*Continued*)

TABLE 2.49 Dielectric Constant (Permittivity) and Dipole Moment of Organic Compounds (*Continued*)

Substance	Dielectric constant, ϵ	Dipole moment, D
2-Chloro-1,3-butadiene	4.914 (20)	
1-Chlorobutane	9.07 (−30), 7.276 (20)	2.05 (g), 2.0 (20, B)
2-Chlorobutane	8.564 (20), 7.09 (30)	2.04 (g), 2.1 (20, B)
Chlorocyclohexane	10.9 (−47), 7.951 (30)	2.2 (25, B)
Chlorodifluoromethane	6.11 (24)	1.42 (g)
2-Chloro-*N,N*-dimethylacetamide	39.2 (25)	
1-Chlorododecane	4.2 (20)	2.11 (25, lq), 1.94 (20, B)
1-Chloro-2,3-epoxypropane	25.6 (1), 22.6 (22)	1.8 (25, C)
Chloroethane	1.013 (19, g), 9.45 (20)	2.05
2-Chloroethanol	25.80 (20), 13 (132)	1.78
(2-Chloro)ethylbenzene	4.36 (25)	
(3-Chloro)ethylbenzene	5.18 (25)	
(4-Chloro)ethylbenzene	5.16 (25)	
2-Chlorofluorobenzene	6.10 (25)	
3-Chlorofluorobenzene	4.96 (25)	
4-Chlorofluorobenzene	3.34 (25)	
Chloroform	4.807 (25), 4.31 (50)	1.04
1-Chloroheptane	5.52 (20)	1.86 (22, B)
2-Chloroheptane	6.52 (22)	2.05 (22, B)
3-Chloroheptane	6.70 (22)	2.06 (22, B)
4-Chloroheptane	6.54 (22)	2.06 (22, B)
1-Chlorohexane	6.104 (20)	1.94 (20, B)
6-Chloro-1-hexanol	21.6 (−31)	
1-Chloro-2-isocyanatoethane	29.1 (15)	
Chloromethane	1.0069 (g), 12.6 (−20), 10.0 (22)	1.892
1-Chloro-3-methylbutane	7.63 (−70), 6.05 (20)	1.94 (20, B)
2-Chloro-2-methylbutane	12.31 (−50)	
4-Chloromethyl-1,3-dioxolan-2-one	97.5 (40)	
Chloromethyl methyl ether		1.88 (C)
(Chloromethyl)oxirane	22.6 (20)	1.8
1-Chloro-2-methylpropane	7.87 (−38), 7.027 (20)	2.00
2-Chloro-2-methylpropane	10.95 (0), 9.66 (20)	2.13
1-Chloronaphthalene	5.04 (25)	1.33 (25, lq), 1.52 (25, B)
o-Chloronitrobenzene	37.7 (50), 32 (80)	4.64
m-Chloronitrobenzene	20.9 (50), 18 (80)	3.73
p-Chloronitrobenzene	8.09 (120)	2.83
2-Chloro-2-nitropropane	31.9 (−23)	
4-Chloro-3-nitrotoluene	28.07 (28)	
1-Chlorooctane	5.05 (25)	2.14 (25, lq)
Chloropentafluoroethane		0.52
1-Chloropentane	6.654 (20)	2.16
o-Chlorophenol	7.40 (21), 6.31 (25)	2.19
m-Chlorophenol	6.255 (20)	2.19 (25, B)
p-Chlorophenol	11.18 (41)	2.11
1-Chloropropane	8.59 (20)	2.05
2-Chloropropane	9.82 (20)	2.17
3-Chloro-1,2-propanediol	31.0 (20)	
3-Chloro-1,2-propanediol dinitrate	17.50 (20)	
3-Chloro-1-propanol	36.0 (−58)	
1-Chloro-2-propanol	59.0 (−120)	
1-Chloro-2-propanone	30 (19)	2.22 (g), 2.37 (20, Hx)
2-Chloro-1-propene	8.92 (26)	1.647

TABLE 2.49 Dielectric Constant (Permittivity) and Dipole Moment of Organic Compounds (*Continued*)

Substance	Dielectric constant, ϵ	Dipole moment, D
3-Chloro-1-propene	8.2 (20)	1.94
2-Chloropyridine	27.32 (20)	
4-Chlorothiophenol	3.59 (65)	
o-Chlorotoluene	4.72 (20), 4.2 (55)	1.56
m-Chlorotoluene	5.76 (20), 5.0 (60)	1.77 (20, lq), 1.8 (22, B)
p-Chlorotoluene	6.25 (20), 5.6 (55)	2.21
Chlorotrifluoromethane	1.0013 (29, g), 3.01 (−150)	0.50
2-Chloro-1-trifluoromethyl-5-nitrobenzene	9.8 (30)	
4-Chloro-1-trifluoromethyl-3-nitrobenzene	12.8 (30)	
3-Chloro-1,1,1-trifluoropropane	7.32 (22)	
Chlorotrimethylsilane		2.09 (20, B)
Cineole	4.57 (25)	
Cinnamaldehyde	17 (20), 16.9 (24)	3.74
o-Cresol	6.76 (25)	1.45 (25, B)
m-Cresol	12.44 (25)	1.61 (25, B)
p-Cresol	13.05 (25)	1.54 (20, B)
Crotonic acid		2.13 (30, B)
Cyanoacetic acid	33.4 (4)	
Cyanoacetylene	72.3 (19)	3.724
2-Cyanopyridine	93.77 (30)	
3-Cyanopyridine	20.54 (50)	
4-Cyanopyridine	5.23 (80)	
Cyclobutanone	14.27 (25)	2.89
Cycloheptane	2.078 (30)	
Cycloheptanone	13.16 (25)	
1,3-Cyclohexadiene	2.68 (−89)	0.38 (20, B)
1,4-Cyclohexadiene	2.211 (23)	
Cyclohexane	2.05 (15), 2.02 (25)	0
Cyclohexanecarboxylic acid	2.6 (31)	
1,4-Cyclohexanedione	15.0 (25), 4.40 (78)	1.41
Cyclohexanethiol	5.420 (25)	
Cyclohexanol	16.40 (20), 15.0 (25), 7.24 (100)	1.86 (25, C)
Cyclohexanone	20 (−40), 16.1 (20)	2.87
Cyclohexanone oxime	3.04 (89)	0.83 (25, B)
Cyclohexene	2.6 (−105), 2.218 (20)	0.332
Cyclohexylamine	4.55 (20)	1.22 (20, lq), 1.26 (20, B)
Cyclohexylbenzene		0
Cyclohexylmethanol	9.7 (60), 8.1 (80)	1.68 (20, B)
Cyclohexyl nitrite	9.33 (25)	
o-Cyclohexylphenol	3.97 (55)	
p-Cyclohexylphenol	4.42 (131)	
Cyclooctane	2.116 (22)	0
cis-Cyclooctene	2.306 (23)	
Cyclopentane	1.9687 (20)	0
Cyclopentanecarbonitrile	22.68 (20)	
Cyclopentanol	25 (−20), 18.5 (10)	1.72 (25, C)
Cyclopentanone	16 (−51), 13.58 (25)	3.30
Cyclopentene	2.083 (22)	0.20
p-Cymene	2.243 (20), 2.23 (25)	0
cis-Decahydronaphthalene	2.22 (20)	0

(*Continued*)

TABLE 2.49 Dielectric Constant (Permittivity) and Dipole Moment of Organic Compounds (*Continued*)

Substance	Dielectric constant, ϵ	Dipole moment, D
trans-Decahydronaphthalene	2.18 (20)	0
Decamethylcyclopentasiloxane	2.5 (20)	
Decamethyltetrasiloxane	2.4 (20)	0.79 (25, lq)
Decane	1.991 (20), 1.844 (130)	0
1-Decanol	8.1 (20)	1.71 (20, B), 1.62 (25, B)
1-Decene	2.14 (20)	0
meso-2,3-Diacetoxybutane	6.644 (25)	
Diallyl sulfide	4.9 (20)	1.33 (25, B)
Dibenzofuran	3.0 (100)	0.88 (25, B)
Dibenzylamine	3.6 (20)	0.97 (20, lq), 1.02 (20, B)
Dibenzyl decanedioate	4.6 (25)	
Dibenzyl ether	3.82 (20)	1.39 (21, B)
o-Dibromobenzene	7.86 (20)	2.13 (20, B)
m-Dibromobenzene	4.21 (20)	1.5 (20, B)
p-Dibromobenzene	2.57 (95)	0
1,2-Dibromobutane	4.74 (20)	
1,3-Dibromobutane	9.14 (20)	
1,4-Dibromobutane	8.68 (30)	2.16 (20, lq), 2.06 (20, B)
2,3-Dibromobutane	6.36 (20), 5.75 (25)	2.20
meso-2,3-Dibromobutane	6.245 (25)	
(±)-2,3-Dibromobutane	5.758 (25)	
1,2-Dibromodichloromethane	2.54 (25)	
1,2-Dibromodifluoromethane	2.94 (0)	0.66
1,2-Dibromoethane	4.96 (20), 4.78 (25), 4.09 (131)	1.11
cis-1,2-Dibromoethylene	7.08 (25)	
trans-1,2-Dibromoethylene	2.88 (25)	
Dibromomethane	7.77 (10)	1.43
cis-1,2-Dibromoethylene	7.7 (0), 7.08 (25)	1.35 (B)
trans-1,2-Dibromoethylene	2.9 (0), 2.88 (25)	0
1,2-Dibromoheptane	3.8 (25)	1.78 (25, D)
2,3-Dibromoheptane	5.1 (25)	2.15 (25, B)
3,4-Dibromoheptane	4.7 (25)	2.15 (25, B)
meso-3,4-Dibromohexane	4.67 (25)	
(±)-3,4-Dibromohexane	6.732 (25)	
1,6-Dibromohexane	8.52 (25)	
Dibromomethane	7.77 (10), 6.7 (40)	1.43
1,2-Dibromo-2-methylpropane	4.1 (20)	
1,2-Dibromopentane	4.39 (25)	
(±)-*erythro*-2,3-Dibromopentane	5.43 (25)	
(±)-*threo*-2,3-Dibromopentane	6.507 (25)	
1,4-Dibromopentane	9.05 (20)	
1,5-Dibromopentane	9.14 (30)	
1,2-Dibromopropane	4.60 (10), 4.3 (20)	1.13
1,3-Dibromopropane	9.48 (20)	
Dibromotetrafluoroethane	2.34 (25)	
Dibutylamine	2.78 (20)	1.06 (20, lq), 1.05 (20, B)
Dibutyl decanedioate	4.54 (20)	2.64 (25, B)
Dibutyl ether	3.08 (20)	1.18
Dibutyl maleate		2.70 (25, B)
Dibutyl *o*-phthalate	6.58 (20), 6.436 (30), 5.99 (45)	2.97 (20, lq), 2.85 (30, B)
Dibutyl sulfide	4.29 (25)	1.6
Dichloroacetic acid	8.33 (20), 7.8 (61)	

TABLE 2.49 Dielectric Constant (Permittivity) and Dipole Moment of Organic Compounds (*Continued*)

Substance	Dielectric constant, ϵ	Dipole moment, D
Dichloroacetic anhydride	15.8 (25)	
1,1,-Dichloroacetone	14.6 (20)	
o-Dichlorobenzene	10.12 (20), 9.93 (25), 7.10 (90)	2.50
m-Dichlorobenzene	5.02 (20), 5.04 (25), 4.22 (90)	1.72
p-Dichlorobenzene	2.394 (55)	0
1,2-Dichlorobutane	7.74 (25)	
1,4-Dichlorobutane	9.30 (35)	2.22
Dichlorodifluoromethane	3.50 (−150), 2.13 (29)	0.51
4-Chloro-1,3-dioxalan-2-one	62.0 (40)	
4,5-Dichloro-1,3-dioxalan-2-one	31.8 (40)	
1,1-Dichloroethane	10.10 (20)	2.06
1,2-Dichloroethane	12.7 (−10), 10.42 (20)	1.48
1,1-Dichloroethylene	4.60 (20), 4.60 (25)	1.34
cis-1,2-Dichloroethylene	9.20 (25)	1.90
trans-1,2-Dichloroethylene	2.14 (20)	0
2,2′-Dichloroethyl ether	21.2 (20)	2.61 (20, B)
Dichlorofluoromethane	5.34 (28)	1.29 (g)
1,6-Dichlorohexane	8.60 (35)	
Dichloromethane	9.14 (20), 8.93 (25), 1.0065 (100, g)	1.60
1,3-Dichloroisopropyl nitrate	13.28 (20)	
(Dichloromethyl)benzene	6.9 (20)	2.1
Dichloromethyl isocyanate	7.36 (15)	
1,2-Dichloro-2-methylpropane	7.15 (23)	
2,4-Dichloro-1-nitrobenzene	13.06 (28)	
1,1-Dichloro-1-nitroethane	16.3 (30)	
1,2-Dichloropentane	6.89 (20)	
1,5-Dichloropentane	9.92 (25)	
2,4-Dichlorophenol		1.60 (25, B)
1,2-Dichloropropane	8.37 (20), 8.93 (26), 7.90 (35)	1.87 (25, B)
1,3-Dichloropropane	10.27 (30)	2.08
2,2-Dichloropropane	11.37 (20)	2.62
1,1-Dichloro-2-propanone	14 (20)	
1,2-Dichlorotetrafluoroethane	2.48 (0), 2.26 (25)	0.53
2,4-Dichlorotoluene	5.68 (28)	1.7
2,6-Dichlorotoluene	3.36 (28)	
3,4-Dichlorotoluene	9.39 (28)	3.0
Diethanolamine	25.75 (20)	2.84 (25, B)
1,1-Diethoxyethane	3.80 (25)	1.08
1,2-Diethoxyethane	3.90 (20)	1.99 (20, B), 1.65 (25, B)
Diethoxymethane	2.527 (20)	
N,N-Diethylacetamide	32.1 (20)	
N,N-Diethylacetoacetamide	40.8 (25)	
Diethylamine	3.680 (20)	0.92
N,N-Diethylaniline	5.5 (19)	1.40 (20, lq), 1.80 (20, B)
Diethyl carbonate	2.82 (24)	1.10
N,N-Diethyl-*N′,N′*-dimethylurea	17.89 (25)	
Diethyl decanedioate	5.0 (30)	2.38 (20, lq), 2.52 (20, B)
Diethylene glycol	3.182 (20)	2.3
Diethylene glycol diethyl ether	5.70	
Diethyl ether	4.267 (20), 3.97 (40)	1.15
Diethyl ethyl phosphonate	11.00 (15), 9.86 (45)	2.95 (32, lq), 2.91 (20, C)
N,N-Diethylformamide	29.6 (20)	

(*Continued*)

TABLE 2.49 Dielectric Constant (Permittivity) and Dipole Moment of Organic Compounds (*Continued*)

Substance	Dielectric constant, ϵ	Dipole moment, D
Diethyl fumarate	6.56 (23)	2.40 (20, B)
Diethyl glutarate	6.7 (30)	2.46 (30, lq)
Diethyl glycol	31.82 (20)	
Di(2-ethylhexyl) *o*-phthalate	5.3 (20), 4.91 (35), 4.77 (45)	2.8
Diethyl maleate	8.58 (23), 7.56 (25)	2.56 (25, B)
Diethyl methanephosphate	13.405 (40)	
Diethyl 1,3-propanedioate (malonate)	8.03 (25), 7.55 (31)	2.49 (20, lq), 2.54 (25, B)
Diethyl nonanedioate	5.13 (30)	
Diethyl oxalate	8.266 (20)	2.49 (20, D)
Diethyl *o*-phthalate	7.34 (35), 7.13 (45)	2.8 (25, B)
Diethylsilane	2.544 (20)	
Diethyl succinate	6.098 (20)	2.3
Diethyl sulfate	29.2 (20)	4.46 (25, D)
Diethyl sulfide	5.72 (25), 5.24 (50)	1.54
Diethyl sulfite	15.6 (20), 14 (50)	
Diethylzinc	2.55 (20)	0.62 (25, B)
o-Difluorobenzene	13.38 (28)	2.46
m-Difluorobenzene	5.01 (28)	1.51
1,1-Difluoroethane		2.27
Difluoromethane	53.74 (−121)	1.978
2,3-Dihydropyran	5.136 (35)	
1,2-Dihydroxybenzene	17.57 (115)	2.60 (25, B)
1,3-Dihydroxybenzene	13.55 (120)	2.09 (44, B)
1,4-Dihydroxybenzene		1.4 (44, B)
1,2-Diiodobenzene	5.7 (20), 5.41 (50)	1.70 (20, B)
1,3-Diiodobenzene	4.3 (25), 4.11 (50)	1.22 (20, B)
1,4-Diodobenzene	2.88 (120)	0.19 (20, B)
cis-1,2-Diiodoethylene	4.46 (72)	0.71 (B)
trans-1,2-Diiodoethylene	3.19 (77)	0
Diiodomethane	5.316 (25)	1.08 (25, B)
Diisobutylamine	2.7 (22)	1.10 (25, B)
1,6-Diisocyanatohexane	14.41 (15)	
Diisopentylamine	2.5 (18)	1.48 (30, B)
Diisopentyl ether	2.82 (20)	0.98 (20, lq), 1.23 (25, B)
Diisopropylamine		1.26 (25, B)
Diisopropyl ether	3.88 (25), 3.805 (30)	1.13
1,2-Dimethoxybenzene	4.45 (20), 4.09 (25)	1.32 (25, B)
Dimethoxydimethylsilane	3.663 (25)	
1,2-Dimethoxyethane	7.60 (10), 7.30 (23.5)	1.71 (25, B)
Dimethoxymethane	2.644 (20)	0.74
N,N-Dimethylacetamide	38.85 (21), 37.78 (25)	3.80
2-Dimethylamino-2-methyl-1-propanol	12.36 (25)	
Dimethylamine	6.32 (0), 5.26 (25)	1.01
N,N-Dimethylaniline	4.90 (25), 4.4 (70)	1.68
2,4-Dimethylaniline	4.9 (20)	1.40 (25, B)
2,3-Dimethyl-1,3-butadiene	2.102 (20)	
N,N-Dimethylbutanamide	29.7 (20)	
2,2-Dimethylbutane	1.869 (20)	0
2,3-Dimethylbutane	1.889 (20)	0
3,3-Dimethyl-2-butanone	12.73 (20)	

TABLE 2.49 Dielectric Constant (Permittivity) and Dipole Moment of Organic Compounds (*Continued*)

Substance	Dielectric constant, ϵ	Dipole moment, D
2,2-Dimethyl-1-butanol	10.5 (20)	
Dimethyl carbonate	3.087 (25)	0.90
cis-1,2-Dimethylcyclohexane	2.06 (25)	0
trans-1,2-Dimethylcyclohexane	2.04 (25)	0
1,1-Dimethylcyclopentane		0
Dimethyl disulfide	9.6 (25)	1.8
Dimethyl ether	6.18 (−15), 5.02 (25), 2.97 (110)	1.30
N,N-Dimethylformamide	38.25 (20), 36.71 (25)	3.82 (25, B)
2,4-Dimethylheptane	1.9 (20)	0
2,5-Dimethylheptane	1.9 (20)	0
2,6-Dimethylheptane	2 (20)	0
2,6-Dimethyl-4-heptanone	9.91 (20)	2.66 (25, C)
2,2-Dimethylhexane	1.95 (20)	0
2,5-Dimethylhexane	1.96 (21)	0
3,3-Dimethylhexane	1.96 (20)	0
3,4-Dimethylhexane	1.98 (19)	0
Dimethyl hexanedioate	6.84 (20)	2.28 (20, B)
1,3-Dimethylimidazolidin-2-one	37.60 (25)	
Dimethyl maleate		2.48 (25, C)
Dimethyl malonate	9.82 (20)	2.41 (20, B)
Dimethyl methanephosphate	22.3 (20)	
N,N-Dimethyl methanesulfonamide	80.4 (50)	
1,2-Dimethylnaphthalene	2.61 (25)	0
1,6-Dimethylnaphthalene	2.73 (20)	0
4,4-Dimethyloxazolidine-2-one	39.2 (60)	
N,N-Dimethylpentanamide	26.4 (20)	
2,2-Dimethylpentane	1.915 (20)	0
2,3-Dimethylpentane	1.929 (20)	0
2,4-Dimethylpentane	1.902 (20)	0
3,3-Dimethylpentane	1.942 (20)	0
Dimethyl pentanedioate	7.87 (20)	
2,4-Dimethyl-3-pentanone		2.7
2,3-Dimethylphenol	4.81 (70)	
2,4-Dimethylphenol	5.06 (30)	1.48 (20, B), 1.98 (60, B)
2,5-Dimethylphenol	5.36 (65)	1.43 (20, B), 1.52 (60, B)
2,6-Dimethylphenol	4.90 (40)	1.4
3,4-Dimethylphenol	9.02 (60)	1.77 (20, B)
3,5-Dimethylphenol	9.06 (50)	1.76 (20, B)
Dimethyl *o*-phthalate	8.66 (20), 8.25 (25), 8.11 (45)	2.8 (25, B)
2,2-Dimethylpropanal	9.051 (20)	2.66
N,N-Dimethylpropanamide	34.6 (20)	
2,2-Dimethylpropanamide	20.13 (25)	
2,2-Dimethylpropane	1.769 (23), 1.678 (98)	0
2,2-Dimethylpropane nitrile	21.1 (20)	3.95
N,N-Dimethylpropanamide	33.1	
2,2-Dimethyl-1-propanol	8.35 (60)	
2,5-Dimethylpyrazine	2.436 (20)	0
2,6-Dimethylpyrazine	2.653 (35)	
2,4-Dimethylpyridine	9.60 (20)	2.3
2,6-Dimethylpyridine	7.33 (20)	1.7
2,6-Dimethylpyridine-1-oxide	46.11 (25)	
2,3-Dimethylquinoxaline	2.3 (25)	0

(*Continued*)

TABLE 2.49 Dielectric Constant (Permittivity) and Dipole Moment of Organic Compounds (*Continued*)

Substance	Dielectric constant, ϵ	Dipole moment, D
Dimethyl succinate	7.19 (20)	2.09 (20, B)
Dimethyl sulfate	55.0 (25)	4.31 (25, D)
Dimethyl sulfide	6.70 (21)	1.554
Dimethyl sulfite	22.5 (23)	2.93 (20, B)
Dimethyl sulfone	47.39 (110)	
Dimethyl sulfoxide	47.24 (20), 41.9 (55)	3.96 (25, B)
cis-2,5-Dimethyltetrahydrofuran	5.03 (23)	
N,N-Dimethylthioformamide	47.5 (25)	
N,N-Dimethyl-*o*-toluidine	3.4 (20)	0.88 (25, B)
N,N-Dimethyl-*p*-toluidine	3.9(20)	1.29 (25, B)
m-Dinitrobenzene	22.9 (92)	
2,2-Dinitropropane	42.4 (52)	
Dinonyl hexanedioate		2.53 (25, B)
Dinonyl *o*-phthalate	4.65 (35), 4.52 (45)	
Dioctyl decanedioate	4.0 (27)	
Dioctyl *o*-phthalate	5.1 (25)	3.06 (25, C)
1,4-Dioxane	2.219 (20), 2.21 (25)	0
1,3-Dioxolane		1.19
1,3-Dioxolan-2-one	89.78 (40)	
Dipentene	2.38 (25)	
Dipentyl ether	2.80 (25)	0.98 (20, lq), 1.24 (25, B)
Dipentyl *o*-phthalate	5.79 (35), 5.62 (45)	2.71 (20, lq)
Dipentyl sulfide	3.83 (25)	1.59 (25, B)
Dipentylamine	3.3 (52)	1.31 (20, C), 1.01 (25, B)
1,2-Diphenylethane	2.4 (110)	0 (110, lq), 0.45 (25, B)
Diphenyl ether	3.73 (10), 3.63 (30)	1.3
Diphenylmethane	2.7 (18), 2.57 (26)	0.26 (30, lq), 0.3 (25, B)
Dipropylamine	2.923 (20)	1.01 (20, lq), 1.03 (20, B)
Dipropyl ether	3.38 (24)	1.21
N,N-Dipropylformamaide	23.5 (20)	
Dipropyl sulfone	32.62 (30)	
Dipropyl sulfoxide	30.37 (30)	
Divinyl ether	3.94 (15)	0.78
Dodecamethylcyclohexasiloxane	2.6 (20)	
Dodecamethylpentasiloxane	2.5 (20)	
Dodecane	2.05 (−10), 2.01 (20)	0
1-Dodecanol	5.15 (20), 6.5 (25)	1.52 (20, B)
1-Dodecene	2.15 (20)	0
6-Dodecyne	2.17 (25)	
1,2-Epoxybutane		2.01 (20, B)
Erythritol	28 (128)	
Ethane	1.936 (−178), 1.0015 (0)	0
1,2-Ethanediamine	16.8 (18), 13.82 (20)	1.96
1,2-Ethanediol	41.4 (20), 37.7 (25)	2.28
1,2-Ethanediol diacetate	7.7 (17)	2.34 (30, B)
1,2-Ethanediol dinitrate	28.26 (20)	
1,2-Ethanediol monoacetate	12.95 (30)	
1,2-Ethanedithiol	7.26 (20)	
Ethanesulfonyl chloride		3.89 (25, B)
Ethanethiol	6.9 (15), 6.667 (25)	1.58
Ethanol	25.3 (20), 20.21 (55)	1.69
Ethanolamine	31.94 (20)	

TABLE 2.49 Dielectric Constant (Permittivity) and Dipole Moment of Organic Compounds (*Continued*)

Substance	Dielectric constant, ϵ	Dipole moment, D
Ethoxyacetylene	8.05 (25)	
4-Ethoxyaniline	7.43 (25)	
Ethoxybenzene (phenetol)	4.216 (20)	1.45
2-Ethoxyethanol	13.38 (25)	2.24 (30, B)
2-Ethoxyethyl acetate	7.567 (30)	2.25 (30, B)
1-Ethoxy-2-methylbutane	3.96 (20)	
1-Ethoxynaphthalene	3.3 (19)	
1-Ethoxypentane	3.6 (23)	
α-Ethoxytoluene	3.9 (20)	
Ethoxytrimethylsilane	3.013 (25)	
N-Ethylacetamide	135.0 (20)	
Ethyl acetate	6.081 (20), 5.30 (77)	1.78
Ethyl acetoacetate	14.0 (20)	3.22 (18, B, keto form) 2.04 (−80, CS_2, enol form)
Ethyl acrylate	6.05 (30)	2.0
Ethylamine	8.7 (0), 6.94 (10)	1.22
N-Ethylaniline	5.87 (20)	
4-Ethylaniline	4.84 (25)	
Ethylbenzene	2.446 (20)	0.59
Ethyl benzoate	6.20 (20)	2.00
Ethyl 2-bromoacetate	8.75 (30)	
Ethyl α-bromobutanoate	8 (20)	2.40 (25, B)
Ethyl 2-bromo-2-methylpropanoate	8.55 (30)	
Ethyl 2-bromopropanoate	9.4 (20), 8.57 (30)	
N-Ethylbutanamide	107.0 (25)	
Ethyl butanoate	5.18 (28)	1.74 (22, B)
2-Ethylbutanoic acid	2.72 (23)	
2-Ethyl-1-butanol	6.19 (90)	
Ethyl *tert*-butyl ether	7.07 (25)	
Ethyl carbamate	14.2 (50), 14.14 (55)	2.59 (30, D)
Ethyl chloroacetate	11.4 (21)	2.65 (25, B)
Ethyl chlorocarbonate	9.736 (36)	
Ethyl *cis*-3-chlorocrotonate	7.67 (76)	
Ethyl *trans*-3-chlorocrotonate	4.70 (54)	
Ethyl chloroformate	11 (20)	2.56 (35, B)
Ethyl 2-chloropropanoate	11.95 (30)	
Ethyl 3-chloropropanoate	10.19 (30)	
Ethyl *trans*-cinnamate	6.1 (18), 5.83 (20)	1.86 (20, B)
Ethyl crotonate	5.4 (20)	1.95 (24, B)
Ethyl cyanoacetate	31.62 (−10), 26.9 (20)	2.2
Ethylcyclobutane	1.965 (20)	
Ethylcyclohexane	2.054 (20)	0
Ethylcyclopropane	1.933 (20)	
Ethyl dichloroacetate	12 (2), 10 (22)	2.63 (25, B)
Ethyl dodecanoate	3.4 (20), 2.7 (143)	1.3 (20, lq)
Ethylene	1.001 44 (0, g), 1.483 (−3)	0
Ethylene carbonate	89.78 (40), 69.4 (91)	4.87 (25, B)
Ethylenediamine	13.82 (20)	1.98
Ethylene dinitrate	28.3 (20)	3.58 (25, B)
2,2′-(Ethylenedioxy)diethanol	23.69 (20)	5.58 (lq)
Ethylene glycol	41.4 (20), 37.7 (25)	2.28
Ethylene glycol diacetate	7.7 (17)	

(*Continued*)

TABLE 2.49 Dielectric Constant (Permittivity) and Dipole Moment of Organic Compounds (*Continued*)

Substance	Dielectric constant, ϵ	Dipole moment, D
Ethyleneimine	18.3 (25)	1.90
Ethylene oxide	14 (−1), 12.42 (20)	1.89
Ethylene sulfite	39.6 (25)	
N-Ethylformamide	102.7 (25)	
Ethyl formate	8.57 (15), 7.16 (25)	1.94
Ethyl fumarate	6.5 (23)	
Ethyl furan-2-carboxylate	9.02 (20)	
Ethylhexadecanoate	3.2 (20), 2.71 (104)	1.2 (lq)
3-Ethylhexane	1.96 (20)	0
2-Ethyl-1,2-hexanediol	18.73 (20)	
Ethyl hexanoate	4.45 (20)	1.80 (20, B)
2-Ethyl-1-hexanol	7.58 (25), 4.41 (90)	1.74 (25, B)
2-Ethylhexyl acetate		1.8
Ethyl 2-iodopropanoate	8.6 (20)	
Ethyl isocyanate	19.7 (20)	
Ethyl isopentyl ether	3.96 (20)	
Ethyl isothiocyanate	19.6 (20)	3.67 (20, B)
Ethyl lactate	15.4 (30)	2.4 (20, B)
Ethyl maleate	8.6 (23)	
Ethyl methacrylate	5.68 (30)	
Ethyl 3-methylbutanoate	4.71 (20)	
Ethyl-*N*-methyl carbamate	21.10 (25)	
Ethyl methyl carbonate	2.985 (20)	
Ethyl methyl ether		1.17
3-Ethyl-2-methylpentane	1.99 (18)	0
Ethyl nitrate	19.7 (20)	2.93 (20, B)
Ethyl 9-octadecanoate	3.2 (25)	1.83 (20, lq)
3-Ethyloxazolidine-2-one	66.8 (25)	
4-Ethyloxazolidine-2-one	42.6 (25)	
Ethyl 4-oxopentanoate	12 (21)	
3-Ethylpentane	1.942 (20)	0
Ethyl pentanoate	4.71 (18)	1.76 (28, B)
3-Ethyl-3-pentanol	3.158 (20)	
Ethyl pentyl ether	3.6 (23)	1.2 (20, B)
Ethyl phenylacetate	5.3 (21)	1.82 (30)
Ethyl phenyl sulfide		4.08 (25, B)
N-Ethyl propanamide	126.8 (25)	
Ethyl propanoate	5.76 (20)	1.75 (22, B)
Ethyl propyl ether		1.16 (25, B)
2-Ethylpyridine	8.33 (20)	
4-Ethylpyridine	10.98 (20)	
Ethyl salicylate	7.99 (30)	2.85 (25, B)
Ethyl stearate	2.98 (40), 2.69 (100)	1.65 (40, lq)
Ethyl thiocyanate	29.3 (21)	3.33 (20, B)
p-Ethyltoluene	2.24 (25)	0
Ethyl trichloroacetate	8.428 (20)	2.56 (25, B)
Ethyltrimethylsilazine	2.275 (30)	
Ethyl vinyl ether		1.26 (20, B)
Fluorobenzene	5.465 (20), 5.42 (25), 4.7 (60)	1.60
4-Fluorobenzene sulfonylchloride	12.65 (40)	
2-Fluoroiodobenzene	8.22 (25)	
3-Fluoroiodobenzene	4.62 (25)	

TABLE 2.49 Dielectric Constant (Permittivity) and Dipole Moment of Organic Compounds (*Continued*)

Substance	Dielectric constant, ϵ	Dipole moment, D
4-Fluoroiodobenzene	3.12 (25)	
Fluoromethane	51.0 (−142)	1.858
2-Fluoro-2-methylbutane	5.89 (20)	1.92 (25, B)
1-Fluoropentane	3.93 (20)	1.85 (25, B)
o-Fluorotoluene	4.23 (25), 4.22 (30), 3.9 (60)	1.37
m-Fluorotoluene	5.41 (25), 4.9 (60)	1.82
p-Fluorotoluene	5.88 (25), 5.86 (30), 5.3 (60)	2.00
Formamide	111.0 (20), 103.5 (40)	3.73
Formanilide		3.37 (25, C)
Formic acid	58.5 (15), 57.0 (21), 51.1 (25)	1.41
2-Furaldehyde	42.1 (20), 34.9 (50)	3.63 (25, B)
Furan	2.88 (4)	0.66
2-Furfuryl acetate	5.85 (20)	
Furfuryl alcohol	16.85 (25)	1.92 (25, lq)
Glycerol	46.5 (20), 42.5 (25)	2.68 (25, D)
Glycerol tris(acetate)	7.2 (20)	2.73 (25, B)
Glycerol tris(nitrate)	19.25 (20)	3.38 (25, B)
Glycerol tris(oleate)	3.2 (26)	3.11 (23, B)
Glycerol tris(palmitate)	2.9 (65)	2.80 (23, B)
Glycerol tris(sterate)	2.8 (70)	2.86 (23, B)
1,6-Heptadiene	2.161 (20)	
Heptacosafluorotributylamine	2.15 (20)	
2,2,3,3,4,4,4-Heptafluoro-1-butanol	14.4 (25)	
Heptanal	9.1 (20)	2.26 (40, lq), 2.58 (22, B)
Heptane	1.921 (20), 1.85 (70)	0
1-Heptanethiol	4.194 (20)	
Heptanoic acid	3.04 (15), 2.6 (71)	
1-Heptanol	11.75 (20)	1.73 (20, B)
(±)-2-Heptanol	9.72 (21)	1.73 (20, B)
(±)-3-Heptanol	7.07 (23)	1.73 (20, B)
4-Heptanol	6.18 (23)	1.72 (20, B)
2-Heptanone	11.95 (20), 8.27 (100)	2.61 (22, B)
3-Heptanone	12.7 (20)	2.81 (22, B)
4-Heptanone	12.60 (20), 9.46 (80)	2.74 (20, B)
1-Heptene	2.09 (20)	0
Heptylamine	3.81 (20)	
Hexachloroacetone	3.93 (19)	
Hexachloro-1,3-butadiene	2.55 (20)	
Hexadecamethylcyclooctasiloxane	2.7 (20)	
Hexadecane	2.046 (30)	0
1-Hexadecanol	3.8 (50)	1.67 (25, B)
1,5-Hexadiene	2.125 (26)	
2,4-Hexadiene	2.207 (25)	0.31 (25, B)
cis,cis-2,4-Hexadiene	2.163 (24)	
trans,trans-2,4-Hexadiene	2.123 (24)	
Hexafluoroacetone	2.104 (−71)	
Hexafluorobenzene	2.029 (25)	0
1,1,1,3,3,3-Hexafluoro-2-propanol	16.70 (20)	
Hexamethyldisiloxane	2.2 (20)	0.37 (25, lq)
Hexamethylphosphorotriamide	31.3 (20)	5.5, 4.31 (25, lq)
Hexane	1.904 (15), 1.890 (20)	0
Hexanedinitrile	32.45 (25)	3.8 (25, B)

(*Continued*)

TABLE 2.49 Dielectric Constant (Permittivity) and Dipole Moment of Organic Compounds (*Continued*)

Substance	Dielectric constant, ϵ	Dipole moment, D
Hexanenitrile	17.26 (25)	
1-Hexanethiol	4.436 (20)	
1,2,6-Hexanetriol	31.5 (12)	
Hexanoic acid	2.600 (25)	1.13 (25, lq)
1-Hexanol	13.03 (20), 8.5 (75)	1.55 (20, B)
(±)-2-Hexanol	11.06 (25)	
3-Hexanol	9.66 (25)	
2-Hexanone	14.6 (15), 14.56 (20)	2.68 (22, B)
1-Hexene	2.051 (20)	0
cis-2-Hexene		0
trans-2-Hexene	1.978 (22)	0
cis-3-Hexene	2.069 (23)	0
trans-3-Hexene	1.954 (20)	0
Hexyl acetate	4.42 (20)	
Hexylamine	4.08 (20)	
1-Hexyne	2.621 (23)	0.83
2-Hydroxyacetophenone	21.33 (25)	
2-Hydroxybutanoic acid	37.7 (23)	
3-Hydroxybutanoic acid	31.5 (23)	
N-(2-Hydroxyethyl)acetamide	96.6 (25)	
4-Hydroxy-4-methyl-2-pentanone	18.2 (25)	3.24 (20, B)
3-Hydroxypropanoic acid	30.0 (23)	
Iodobenzene	4.59 (20)	1.70
1-Iodobutane	6.27 (20), 4.52 (130)	2.10
2-Iodobutane	7.873 (20)	2.12
1-Iodododecane	3.9 (20)	1.87 (20, C)
Iodoethane	10.2 (−50), 7.82 (20)	1.91
1-Iodoheptane	4.92 (22)	1.86 (22, B)
3-Iodoheptane	6.39 (22)	1.95 (22, B)
1-Iodohexadecane	3.5 (20)	
1-Iodohexane	5.37 (20)	1.94 (20, C)
Iodomethane	6.97 (20)	1.62
1-Iodo-3-methylbutane	5.6 (19)	1.85 (20, B)
2-Iodo-2-methylbutane	8.19 (20)	2.20 (20, B)
1-Iodo-2-methylpropane	6.47 (20)	1.89 (20, B)
2-Iodo-2-methylpropane	6.65 (10)	
1-Iodooctane	4.6 (25)	1.80 (25, lq), 1.90 (20, C)
2-Iodooctane	5.8 (20)	2.07 (20, C)
1-Iodopentane	5.78 (20)	1.90 (20, B)
3-Iodopentane	7.432 (20)	
1-Iodopropane	7.07 (20)	2.03
2-Iodopropane	8.19 (25)	2.01 (20, B)
3-Iodopropene	6.1 (19)	
p-Iodotoluene	4.4 (35)	1.72 (22, B)
α-Ionone	11 (18)	
β-Ionone	12 (20)	
Iron pentacarbonyl	2.602 (20)	
Isobutanenitrile	20.4 (24)	3.61 (25, B)
Isobutene	2.1225 (15)	0.503
N-Isobutylacetamide	111.0 (20)	
Isobutyl acetate	5.068 (20)	1.87 (22, B)
Isobutylamine	4.43 (21)	1.27 (25, B)

TABLE 2.49 Dielectric Constant (Permittivity) and Dipole Moment of Organic Compounds (*Continued*)

Substance	Dielectric constant, ϵ	Dipole moment, D
Isobutylbenzene	2.319 (20), 2.298 (30)	0.31 (20, lq)
Isobutyl butanoate	4.1 (20)	1.9
Isobutyl chlorocarbonate	9.1 (20)	
Isobutyl formate	6.41 (20)	1.89 (20, B)
Isobutyl isocyanate	11.64 (20)	
Isobutyl nitrate	2.7 (20)	
Isobutyl pentanoate	3.8 (19)	
Isobutylsilane	2.497 (20)	
Isobutyl trichloroacetate	7.667 (20)	
Isobutyl vinyl ether	3.34 (20)	
Isobutyronitrile	20.4 (24)	3.61 (25, B)
Isopentyl acetate	4.72 (20), 4.63 (30)	1.84 (22, B), 1.76 (30, lq)
Isopentyl butanoate	4.0 (20)	
Isopentyl pentanoate	3.6 (19)	1.8 (28, B)
Isopentyl propanoate	4.2 (20)	
Isopropyl acetate		1.86 (22, B)
Isopropylamine	5.627 (20)	1.19
Isopropylbenzene	2.38 (20)	0.79
Isopropyl carborane	45.0 (20)	
N-Isopropylformamide	65.7 (25)	
1-Isopropyl-4-methylbenzene	2.24 (20)	0
Isopropyl nitrite	13.92 (−13)	
Isoquinoline	11.0 (25)	2.73
Lactic acid	22 (17)	
Lactonitrile	38 (20)	
D-Limonene	2.4 (20), 2.37 (25)	1.57 (25, B)
(±)-Limonene	2.3 (20)	0.63 (25, B)
Maleic anhydride	52.75 (53)	
(±)-Mandelonitrile	17.8 (23)	
D-Mannitol	24.6 (170)	
Menthol		1.55 (20, B)
Methacrylic acid		1.65
Methacrylonitrile		3.69
Methane	1.676 (−182), 1.000 94 (0)	0
Methanesulfonyl chloride	34.0 (20)	
Methanethiol		1.52 (g)
Methanol	41.8 (−20), 33.0 (20)	1.70
2-Methoxyaniline	5.230 (30)	
3-Methoxyaniline	8.76 (25)	
4-Methoxyaniline	7.85 (60)	
o-Methoxybenzaldehyde		4.34 (20, B)
p-Methoxybenzaldehyde	22.3 (22), 22.0 (30), 10.4 (248)	3.26 (35, B)
Methoxybenzene	4.30 (21), 3.9 (70)	1.38
2-Methoxyethanol	17.2 (25), 16.0 (30)	2.36
N-(2-Methoxyethyl)acetamide	80.7 (25)	
2-Methoxyethyl acetate	8.25 (20)	2.13 (30, B)
1-Methoxy-2-nitrobenzene	45.75 (20)	4.83
o-Methoxyphenol	11.95 (25)	
m-Methoxyphenol	11.59 (25)	
p-Methoxyphenol	11.05 (60)	
2-Methoxy-4-(2-propenyl)phenol		2.46 (25, B)
o-Methoxytoluene	3.5 (20)	

(*Continued*)

TABLE 2.49 Dielectric Constant (Permittivity) and Dipole Moment of Organic Compounds (*Continued*)

Substance	Dielectric constant, ϵ	Dipole moment, D
m-Methoxytoluene	3.5 (20)	
p-Methoxytoluene	4.0 (20)	
Methoxytrimethylsilane	3.248 (25)	
N-Methylacetamide	178.9 (30), 138.6 (60)	4.39 (20, D)
Methyl acetate	7.07 (15), 7.03 (20), 6.68 (25)	1.72
Methyl acrylate	7.03 (30)	1.77 (25, B)
Methylamine	16.7 (−58), 11.4 (−10), 10.0 (18)	1.31
Methyl 2-aminobenzoate	21.9 (25)	
N-Methylaniline	5.96 (20)	1.67 (25, B)
2-Methylaniline	6.138 (25)	
3-Methylaniline	5.816 (25)	
4-Methylaniline	5.058 (25)	
N-Methylbenzenesulfonamide	67.1 (30)	
Methyl benzoate	6.64 (30)	1.86 (25, B)
2-Methyl-1,2-butadiene	2.1 (25)	0.15
2-Methyl-1,3-butadiene	2.098 (20)	0.25
2-Methylbutane	1.871 (0), 1.845 (20)	0.13
2-Methyl-2-butanethiol	5.083 (20)	
Methyl butanoate	5.6 (20), 5.48 (29)	1.72 (22, B)
3-Methylbutanoic acid	2.64 (20)	0.63 (25)
2-Methyl-1-butanol	15.63 (25)	1.9
2-Methyl-2-butanol	5.78 (25)	1.72 (20, B)
3-Methyl-1-butanol	15.63 (20), 14.7 (25), 5.82 (130)	1.82 (25, B)
3-Methyl-2-butanol	12.1 (25)	
3-Methyl-2-butanone	10.37 (20)	
2-Methyl-1-butene	2.180 (20)	0.52 (20, lq)
2-Methyl-2-butene	1.979 (23)	0.11 (25, lq), 0.34 (25, B)
3-Methyl-1-butene	1.0028 (100, g)	0.320
2-Methyl-1-butene-2-one	10.39 (30)	
2-Methylbutyl acetate	4.63 (30)	1.82 (22)
3-Methylbutyl 3-methylbutanoate	4.39 (15)	
3-Methylbutyronitrile	18 (220)	3.62 (25, C)
Methyl carbamate	18.48 (55)	
Methyl chloroacetate	12.0 (20)	
N-Methyl-2-chloroacetamide	92.3 (50)	
Methyl 4-chlorobutanoate	9.51 (30)	
Methyl crotonate	6.664 (20)	
Methyl cyanoacetate	29.3 (20), 19.23 (50), 17.57 (65)	
Methylcyclohexane	2.024 (20)	0
2-Methylcyclohexanol		1.95 (25, B)
cis-3-Methylcyclohexanol	16.05 (20)	1.91
trans-3-Methylcyclohexanol	8.05 (20)	1.75
4-Methylcyclohexanol		1.9 (25, B)
2-Methylcyclohexanone	16 (−15), 14.0 (20)	2.98 (25, B)
3-Methylcyclohexanone	18 (−80), 12.4 (20)	3.06 (25, B)
4-Methylcyclohexanone	15 (−41), 12.35 (20)	3.07 (25, B)
Methylcyclopentane	1.985 (20)	0
1-Methylcyclopentanol	7.11 (37)	
Methyl decanoate		1.65 (20, Hx)
Methyl dodecanoate		1.70 (20, Hx)
N-Methylformamide	200.1 (15), 189.0 (20), 182.4 (25)	3.83
Methyl formate	9.20 (15), 8.5 (20)	1.77

TABLE 2.49 Dielectric Constant (Permittivity) and Dipole Moment of Organic Compounds (*Continued*)

Substance	Dielectric constant, ϵ	Dipole moment, D
2-Methylfuran	2.76 (20)	0.65
Methyl furan-2-carboxylate	11.01 (20)	
(mono)Methyl glutarate	8.37 (20)	
2-Methylheptane	1.95 (20)	0
2-Methyl-2-heptanol	3.38 (−7), 2.46 (25)	
2-Methyl-3-heptanol	3.37 (20), 3.75 (60)	1.63 (20, B)
2-Methyl-4-heptanol	3.30 (20), 3.65 (60)	
3-Methyl-3-heptanol	3.74 (20), 2.89 (60)	
3-Methyl-4-heptanol	9.1 (−20), 7.4 (20)	
4-Methyl-3-heptanol	5.25 (20), 4.62 (55)	
4-Methyl-4-heptanol	2.87 (20), 3.27 (60)	
2-Methylhexane	1.922 (20)	0
3-Methylhexane	1.920 (20)	0
Methyl hexanoate	4.615 (20)	1.70 (20, Hx)
2-Methyl-2-hexanol	3.257 (24)	
3-Methyl-2-hexanol	4.990 (24)	
3-Methyl-3-hexanol	3.248 (25)	
5-Methyl-2-hexanone	13.53 (20)	
Methyl isobutanoate		1.98 (20, B)
Methylisocyanate	21.75 (16)	2.8
Methyl methacrylate	6.32 (30)	1.68 (25, B)
N-Methyl methanesulfonamide	104.4 (25)	
Methyl *o*-methoxybenzene	7.7 (21)	
Methyl *p*-methoxybenzoate	4.3 (33)	
N-Methyl-2-methylbutanamide	123.0 (34)	
N-Methyl-3-methylbutanamide	114.0 (26)	
Methyl 3-(methylthio)propanoate	8.66 (30)	
1-Methylnaphthalene	2.92 (20)	0
Methyl nitrate	23.9 (20)	
Methyl nitrite	20.77 (−73)	
Methyl *o*-nitrobenzoate	28 (25)	3.67 (30, B)
2-Methyloctane	1.97 (20)	0
3-Methyloctane		0
4-Methyloctane	1.97 (20)	0
Methyl oleate	3.211 (20)	
2-Methyl-1,3-pentadiene	2.422 (25)	
3-Methyl-1,3-pentadiene	2.426 (25)	
4-Methyl-1,3-pentadiene	2.599 (20)	
N-Methylpentanamide	131.0 (13)	
2-Methylpentane	1.886 (20)	0
3-Methylpentane	1.886 (20)	0
2-Methyl-2,4-pentanediol	23.4 (20)	2.9
4-Methylpentanenitrile	17.5 (22)	3.53 (25, B)
Methyl pentanoate	4.992 (20)	1.62 (22, B)
3-Methyl-1-pentanol	15.2 (25)	
3-Methyl-3-pentanol	4.322 (20)	
4-Methyl-2-pentanone	15.6 (0), 15.1 (20), 11.78 (40)	
4-Methylpentenenitrile	17.5 (22)	3.5
4-Methyl-3-penten-2-one	15.6 (0)	2.8
1-Methyl-1-phenylhydrazine	7.3 (19)	1.84 (15, B)
Methyl phenyl sulfide		1.38 (20, B)
Methyl phenyl sulfone	37.9 (100)	

(*Continued*)

TABLE 2.49 Dielectric Constant (Permittivity) and Dipole Moment of Organic Compounds (*Continued*)

Substance	Dielectric constant, ϵ	Dipole moment, D
2-Methylpropanal		2.6
N-Methylpropanamide	170.0 (20), 151 (40)	3.59
2-Methyl-1-propanamine	4.43 (21)	1.3
2-Methylpropane	1.752 (25)	0.132
2-Methylpropanenitrile	24.42 (20)	4.29
2-Methyl-1-propanethiol	4.961 (25)	
2-Methyl-2-propanethiol	5.475 (20)	1.66
Methyl propanoate	6.200 (20)	1.70 (22, B)
2-Methylpropanoic acid	2.58 (20)	1.08 (25, lq)
2-Methylpropanoic anhydride	13.6 (19)	
2-Methyl-1-propanol	26 (−34), 17.93 (20)	1.64
2-Methyl-2-propanol	12.47 (25), 10.9 (30), 8.49 (50)	1.67 (22, B)
2-Methylpropene		0.50
2-Methyl-2-propenenitrile		3.69
2-Methylpropenoic acid		1.6
2-Methylpropyl acetate	5.07 (20)	1.87 (22, B)
2-Methyl-1-propylamine	4.43 (21)	1.27 (27)
(2-Methylpropyl)benzene	2.32 (20)	0
2-Methylpropyl formate	6.41 (20)	1.88 (22)
2-Methylpyridine	10.18 (20)	1.85
3-Methylpyridine	11.10 (30)	2.41 (25, B)
4-Methylpyridine	12.2 (20)	2.70
2-Methylpyridine-1-oxide	36.4 (50)	
3-Methylpyridine-1-oxide	28.26 (45)	
N-Methylpyrrolidine	32.2 (25)	
N-Methyl-2-pyrrolidinone	32.55 (20), 32.2 (25)	4.09 (30, B)
Methyl salicylate	9.41 (30), 8.80 (41)	2.47 (25, B)
3-Methyl sulfolane	29.4 (25)	
Methyl tetradecanoate		1.62 (25, B)
2-Methyltetrahydrofuran	6.97 (25)	
Methyl tetrahydrothiophene-2-carboxylate	7.30 (20)	
Methyl thiocyanate	4.3 (19)	3.34 (20, B)
2-Methylthiophene		0.674
3-Methylthiophene		0.95
Methyl thiophene-2-carboxylate	8.81 (20)	
Methyl trifluoromethyl sulfone	32.0 (20)	
Morpholine	7.42 (25)	1.55
β-Myrcene	2.3 (25)	
Naphthalene	2.54 (90)	0
1-Naphthonitrile	16 (70)	
2-Naphthonitrile	17 (70)	
o-Nitroaniline	47.3 (80), 34.5 (90)	4.28 (20, B)
m-Nitroaniline	35.6 (125)	
p-Nitroaniline	78.5 (155), 56.3 (160)	6.3 (25, B)
o-Nitroanisole	45.75 (20)	4.83
m-Nitroanisole	25.7 (45)	
p-Nitroanisole	26.95 (65)	
Nitrobenzene	35.6 (20), 34.82 (25), 24.9 (90)	4.22
m-Nitrobenzyl alcohol	22 (20)	
2-Nitrobiphenyl		3.83 (20, B)
Nitroethane	29.11 (15), 28.06 (30), 27.4 (35)	3.23

TABLE 2.49 Dielectric Constant (Permittivity) and Dipole Moment of Organic Compounds (*Continued*)

Substance	Dielectric constant, ϵ	Dipole moment, D
2-Nitro-ethylbenzene	21.9 (0)	
Nitromethane	37.27 (20), 35.87 (30), 35.1 (35)	3.46
1-Nitro-2-methoxybenzene		4.83
o-Nitrophenol	16.50 (50)	3.14 (25, B)
m-Nitrophenol	35.45 (100)	
p-Nitrophenol	42.20 (120)	
1-Nitropropane	24.70 (15), 23.24 (30), 22.7 (35)	3.66
2-Nitropropane	26.74 (15), 25.52 (30)	3.73
N-Nitrosodimethylamine	53 (20)	4.01 (20, B)
o-Nitrotoluene	26.36 (20), 22.0 (58)	3.72 (20, B)
m-Nitrotoluene	24.95 (30), 22 (58)	4.20 (20, B)
p-Nitrotoluene	22.2 (58)	4.47 (25, B)
Nonane	1.972 (20), 1.85 (110)	0
Nonanoic acid	2.48 (22)	0.8
1-Nonanol		1.72 (20, B)
1-Nonene	2.18 (20)	0
(*trans, trans*)-9,12-Octadecadienoic acid	2.70 (70), 2.60 (120)	1.40 (18, Hx)
Octamethylcyclotetrasiloxane	2.4 (20)	0.42 (25, lq), 0.67 (25, B)
Octamethyltrisiloxane	2.3 (20)	0.64 (25, lq)
Octane	1.948 (20), 1.83 (110)	0
Octanenitrile	13.90 (20)	
Octanoic acid	2.85 (15), 2.45 (20)	1.15 (25, lq)
1-Octanol	11.3 (10), 10.30 (20)	1.72 (20, B)
2-Octanol	8.13 (20), 6.52 (40)	1.65 (20, B)
2-Octanone	9.51 (20), 7.42 (100)	2.72 (15, B)
1-Octene	2.113 (20)	0
cis-2-Octene	2.06 (25)	0
trans-2-Octene	2.00 (25)	0
Oleic acid	2.34 (20)	1.2
Oxalyl chloride	3.470 (21)	0.93 (20, B)
Palmitic acid	2.3 (70)	
Paraldehyde	13.9 (25)	1.43
Parathion		4.98 (25, B)
Pentachloroethane	3.73 (20), 3.716 (25)	0.92
2,3,4,5,6-Pentachlorotoluene	4.8 (20)	
Pentadecane		0
cis-1,3-Pentadiene	2.32 (25)	0.50 (25, B)
1,4-Pentadiene	2.054 (24)	
Pentanal	10.1 (17), 10.00 (20)	2.59 (20, B)
Pentane	2.011 (−90), 1.837 (20)	0
1,2-Pentanediol	17.31 (24)	
1,4-Pentanediol	26.74 (23)	
1,5-Pentanediol	26.2 (20)	2.45 (20, D)
2,3-Pentanediol	17.37 (24)	
2,4-Pentanediol	24.69 (21)	
2,4-Pentanedione	26.52 (30)	3.03
Pentanenitrile	20.04 (20)	4.12, 3.57 (25, B)
1-Pentanethiol	4.85 (20), 4.55 (25), 4.23 (50)	1.54 (25, lq)
Pentanoic acid	2.66 (21)	1.61 (20, D)
1-Pentanol	16.9 (20), 15.13 (25)	1.71 (20, B)
2-Pentanol	13.71 (25)	1.66 (22, B)

(*Continued*)

TABLE 2.49 Dielectric Constant (Permittivity) and Dipole Moment of Organic Compounds (*Continued*)

Substance	Dielectric constant, ϵ	Dipole moment, D
3-Pentanol	13.35 (25)	1.64 (22, B)
2-Pentanone	15.45 (20), 11.73 (80)	2.72 (22, B)
3-Pentanone	19.4 (−20), 17.00 (20)	2.72 (20, B)
2-Pentanone oxime	3.3 (25)	
1-Pentene	2.011 (20)	0.5
cis-2-Pentene		0
trans-2-Pentene		0
Pentyl acetate	4.79 (20)	1.75
Pentylamine	4.27 (20)	1.55 (30, B)
Pentyl formate	5.7 (19)	1.90
Pentyl nitrate	9.0 (18)	
Pentyl nitrite	7.21 (25)	
tert-Pentyl nitrite	10.88 (25)	
Phenanthrene	2.8 (20)	0
Phenol	12.40 (30), 9.78 (60)	1.224
Phenoxyacetylene	4.76 (25)	1.42 (25, lq)
Phenyl acetate	5.40 (25)	1.54 (22, B)
Phenylacetic acid	3.47 (80)	
Phenylacetonitrile	17.87 (26), 8.5 (234)	3.47 (27, B)
Phenylacetylene	2.98 (20)	0.72 (20, B)
1-Phenylethanol	8.77 (20), 7.6 (90)	1.51 (20, B)
2-Phenylethanol	12.31 (20)	
Phenylhydrazine	7.15 (20)	1.67 (25, B)
Phenyl isocyanate	8.94 (20)	
Phenyl isothiocyanate	10 (20)	
1-Phenylpropene	2.7 (20)	
2-Phenylpropene	2.3 (20)	
3-Phenylpropene	2.6 (20)	
Phenyl salicylate	6.3 (50)	
Phosgene	4.7 (0), 4.3 (22)	
Phthalide	36 (75)	
(±)-α-Pinene	2.64 (25), 2.26 (30)	0.60 (25, B)
L-β-Pinene	2.76 (20)	
Piperidine	4.33 (20)	1.19 (25, B)
Propanal	18.5 (17)	2.52
Propane	1.668 (20)	0.084
1,2-Propanediamine	10.2	
1,3-Propanediamine	9.55	1.96 (25, B)
1,2-Propanediol	32.0 (20), 27.5 (30)	2.27 (25, D)
1,3-Propanediol	35.1 (20)	2.52 (25, D)
1,2-Propanediol dinitrate	26.80 (20)	
1,3-Propanediol dinitrate	18.97 (20)	
1,2-Propanedithiol	7.24 (20)	
1,3-Propanedithiol	8.11 (30)	
Propanenitrile	29.7 (20)	4.05
1-Propanethiol	5.94 (15), 1.55 (25)	1.68
2-Propanethiol	5.95 (25)	1.61
1,2,3-Propanetriol 1-acetate	38.57 (−31), 7.11 (20)	
Propanoic acid	3.30 (10), 3.44 (25)	1.76
Propanoic anhydride	18.30 (20)	
1-Propanol	20.8 (20), 20.33 (25)	1.55
2-Propanol	20.18 (20), 18.3 (25), 16.2 (40)	1.58

TABLE 2.49 Dielectric Constant (Permittivity) and Dipole Moment of Organic Compounds (*Continued*)

Substance	Dielectric constant, ϵ	Dipole moment, D
2-Propenal		3.12
Propene	2.137 (−53), 1.88 (20), 1.44 (90)	0.366
Propenenitrile	33.0 (20)	3.87
2-Propen-1-ol	21.6 (15), 19.7 (20)	1.60
Propionaldehyde (propanal)	18.5 (17)	2.75
Propionamide		3.4 (30, B)
Propyl acetate	5.62 (20)	1.86 (25, B)
N-Propylacetamide	117.8 (25)	
Propylamine	5.31 (20), 5.08 (26)	1.17
Propylbenzene	2.37 (20), 2.351 (30)	0
Propyl benzoate	5.78 (30)	
Propyl butanoate	4.3 (20)	
Propyl carbamate	12.06 (65)	
Propylene carbonate	66.14 (20)	4.9
Propyleneimine		1.77 (*cis*), 1.60 (*trans*)
1,2-Propylene oxide		2.00
Propyl formate	7.72 (19), 6.92 (30)	1.91 (22, B)
Propyl nitrate	14 (18)	3.01 (20, B)
Propyl nitrite	12.35 (−23)	
Propyl pentanoate	4 (19)	
N-Propylpropanamide	118.1 (25)	
Propyl propanoate	5.25 (20)	1.79 (22, B)
Propyl trichloroacetate	8.32 (25)	
Propyne	3.218 (−27)	0.784
2-Propyn-1-ol	20.8 (20)	1.13
Pulegone	9.5 (20)	2.00 (25, B)
Pyridazine		4.22
Pyrazine	2.80 (50)	0
Pyridine	13.26 (20), 12.3 (25), 9.4 (116)	2.215
Pyridine-1-oxide	35.94 (70)	
Pyrimidine		2.33
1*H*-Pyrrole	8.00 (20), 8.13 (25)	1.74
Pyrrolidine	8.30 (20)	1.58 (20, B)
2-Pyrrolidone		3.55 (25, B)
Quinoline	9.16 (20), 9.00 (25)	2.29
Safrole	3.1 (21)	
Salicylaldehyde	18.35 (20)	2.86 (20, B)
D-Sorbitol	35.5 (80)	
Squalane	1.911 (100)	0
Squalene		0.68 (25, B)
Stearic acid	2.29 (70), 2.26 (100)	1.76 (25, D)
Styrene	2.47 (20), 2.43 (25), 2.32 (75)	0.13 (25, lq)
Succinonitrile	62.6 (25), 56.5 (57), 54 (68)	3.68 (30, toluene)
α-Terpinene	2.45 (25)	
Terpinolene	2.29 (25)	
1,1,2,2-Tetrabromoethane	8.6 (3), 7.0 (22), 6.72 (30)	1.41
1,1,2,2-Tetrachlorodifluoroethane	2.52 (35)	
1,1,1,2-Tetrachloroethane	9.22 (−66)	
1,1,2,2-Tetrachloroethane	8.50 (20)	1.32
Tetrachloroethylene	2.30 (25), 2.268 (30)	0
1,1,3,4-Tetrachlorohexafluorobutane	2.86 (20)	

(*Continued*)

TABLE 2.49 Dielectric Constant (Permittivity) and Dipole Moment of Organic Compounds (*Continued*)

Substance	Dielectric constant, ϵ	Dipole moment, D
Tetradecafluorohexane	1.76 (25)	
Tetradecamethylhexasiloxane	2.5 (20)	1.58 (20, lq)
Tetradecane		0
Tetradecanoic acid		0.76 (25, B)
1-Tetradecanol	4.72 (38), 4.40 (48)	1.69 (25, C)
Tetraethylene glycol	20.44 (20)	5.84 (20, lq)
Tetraethyl lead		0.3 (20, B)
Tetraethylsilane	2.09 (20)	0
Tetraethyl silicate	4.1 (20)	1.72 (32, B)
Tetrafluoromethane	1.685 (−147)	
2,2,3,3-Tetrafluoro-1-propanol	21.03 (25)	
Tetrahydrofuran	11.6 (−70), 7.52 (22)	1.75 (25, B)
Tetrahydro-2-furanmethanol	13.61 (23), 13.48 (30)	2.12 (35, lq)
2-Tetrahydrofurfuryl acetate	9.65 (20)	
1,2,3,4-Tetrahydronaphthalene	2.77 (25)	0
1,2,3,4-Tetrahydro-2-naphthol	11.7 (20), 6.7 (90)	
Tetrahydropyran	5.66 (20), 5.61 (25)	1.74
Tetrahydrothiophene		1.9
Tetrahydrothiophene-1,1-dioxide (sulfolane)	43.26 (30)	4.81 (25, B)
Tetrahydrothiophene-*S*-oxide	42.96 (25), 42.5 (30)	
Tetrakis(methylthio)methane	2.818 (70)	
Tetramethoxymethane	2.40 (20)	
Tetramethyl germanium	1.817 (24)	
1,1,3,3-Tetramethylguanidine	11.5 (25)	
Tetramethylsilane	1.921 (20)	0
Tetramethyl silicate	6.0 (20)	
1,1,2,2-Tetramethylurea	23.10 (20)	3.47 (25, B)
Tetranitromethane	2.317 (25)	0
Tetrathiomethylmethane	2.82 (70)	
Thiacyclopentane		1.90 (25, B)
Thioacetic acid	14.30 (25)	
Thiophene	2.74 (20), 2.57 (25)	0.55
Thymol		1.55 (25, B)
Toluene	2.385 (20), 2.364 (30)	0.375
o-Toluidine	6.34 (18), 6.14 (25), 5.71 (58)	1.60 (25, B)
m-Toluidine	5.95 (18), 5.82 (25), 5.45 (58)	1.45 (25, B)
p-Toluidine	5.06 (60)	1.52 (25, B)
m-Tolunitrile		4.21 (22, B)
p-Tolunitrile		4.47 (20, B)
Tribenzylamine		0.65 (20, B)
2,2,2-Tribromoacetaldehyde	7.6 (20)	1.70 (20, C)
Tribromochloromethane	2.60 (60)	
Tribromofluoromethane	3.00 (20)	
Tribromomethane	4.404 (10), 4.39 (20)	0.99
Tribromonitromethane	9.03 (25)	
1,2,3-Tribromopropane	6.45 (20), 6.00 (30)	1.59 (25, B)
Tributylamine	2.34 (20)	0.78 (25, B)
Tributyl borate	2.23 (20)	0.78 (25, C)
Tributyl phosphate	8.34 (20), 7.96 (30)	3.07 (25, B)
Tributyl phosphite		1.92 (20, C)
Trichloroacetaldehyde	7.6 (−40), 6.9 (20), 6.8 (25)	1.96 (25, B)

TABLE 2.49 Dielectric Constant (Permittivity) and Dipole Moment of Organic Compounds (*Continued*)

Substance	Dielectric constant, ϵ	Dipole moment, D
Trichloroacetic acid	4.34 (60)	1.1 (25, B, dimer)
Trichloroacetic anhydride	5.0 (25)	
Trichloroacetonitrile	7.85 (19)	1.93 (19, lq)
4,4,4-Trichlorobutanal	10.0 (18)	
1,2,2-Trichloro-1,1-difluoroethane	4.01 (30)	
1,1,1-Trichloroethane	7.1 (7), 7.24 (20)	1.755
1,1,2-Trichloroethane	7.19 (25)	1.45
Trichloroethylene	3.42 (16), 3.39 (28)	0.77 (30, lq), 0.95 (30, B)
Trichloroethylsilane		2.0
Trichlorofluoromethane	3.00 (25), 2.28 (29)	0.45
(Trichloromethyl)benzene	6.9 (21)	2.0
Trichloromethylsilane		1.87 (25, B)
Trichloronitromethane	7.32 (25)	
2,4,6-Trichlorophenol		1.88 (25, D)
1,2,3-Trichloropropane	7.5 (20)	1.61
Trichlorosilane		0.86
α,α,α-Trichlorotoluene	6.9 (21)	2.17 (20, B)
1,1,2-Trichloro-1,2,2-trifluoroethane	2.41 (25)	
Tridecane	2.02 (20)	0
1-Tridecene	2.14 (20)	0
Triethanolamine	29.36 (25)	3.57 (25, B)
Triethoxymethane	4.779 (20)	
Triethylaluminum	2.9 (20)	
Triethylamine	2.418 (20)	0.66
Triethylborane	1.874 (20)	
Triethylene glycol	23.69 (20)	5.58 (20, lq)
Triethylenetetramine	10.76 (20)	
Triethyl orthovanadate	3.333 (25)	
Triethyl phosphate	13.43 (15), 13.20 (25), 10.93 (65)	3.08 (25, B)
Triethylphosphine oxide	35.5 (50)	
Triethylphosphine sulfide	39.0 (98)	
Triethyl phosphite	5.0	1.82 (25, D)
Trifluoroacetic acid	8.42 (20), 5.76 (50)	2.28
Trifluoroacetic anhydride	2.7 (25)	
1,1,1-Trifluoroethane		2.347
2,2,2-Trifluoroethanol	27.68 (20)	2.03 (25, cHex)
Trifluoromethane	5.2 (26)	1.651
(Trifluoromethyl)benzene	9.22 (25)	2.86
1-Trifluoromethyl-3-nitrobenzene	17.0 (30)	
α,α,α-Trifluorotoluene	9.2 (30), 8.1 (60)	
Trimethoxymethylsilane	4.9 (25)	
Trimethylamine	2.44 (25)	0.612
1,2,3-Trimethylbenzene	2.66 (20), 2.609 (30)	0
1,2,4-Trimethylbenzene	2.38 (20), 2.36 (30)	0
1,3,5-Trimethylbenzene	2.28 (20)	0
Trimethyl borate	2.276 (20)	0.82 (25, C)
2,2,3-Trimethylbutane	1.930 (20)	0
Trimethylchlorosilane	10.21 (0)	
Trimethylene sulfide		1.85
2,2,5-Trimethylhexane		0
2,3,5-Trimethylhexane		0
2,2,3-Trimethylpentane	1.962 (20)	0

(*Continued*)

TABLE 2.49 Dielectric Constant (Permittivity) and Dipole Moment of Organic Compounds (*Continued*)

Substance	Dielectric constant, ϵ	Dipole moment, D
2,2,4-Trimethylpentane	1.940 (20)	0
2,3,3-Trimethylpentane	1.98 (20)	0
2,3,4-Trimethylpentane	1.97 (20)	0
Trimethyl phosphate	20.6 (20)	3.2
Trimethylphosphine sulfide		71.6 (20)
Trimethyl phosphite		1.83 (20, C)
2,4,6-Trimethylpyridine	7.807 (25)	1.95 (25, B)
2,4,6-Trinitrophenol	4.0 (21)	
1,3,5-Trioxane	15.55 (65)	2.08
Triphenyl phosphite	3.67 (45), 3.57 (65)	2.04 (25, B)
Tris(4-ethylphenyl) phosphite	3.74 (15), 3.61 (45)	2.08 (25, B)
Tris(2-methylphenyl) phosphate	6.7 (25)	2.9
Tris(3-methylphenyl) phosphate		3.0
Tris(4-methylphenyl) phosphate		3.2
Tris(*m*-tolyl) phosphite	3.67 (15), 3.53 (45)	1.62 (25, B)
Tris(*p*-tolyl) phosphite	3.88 (15), 3.74 (45)	1.77 (25, B)
Tri-*o*-tolyl phosphate	6.92 (40)	2.84 (40, C)
Undecane	2.00 (20), 1.84 (150)	0
2-Undecanone		2.71 (15, B)
1-Undecene	2.14 (20)	0
Urea		4.59 (25, D)
Vinyl acetate		1.79 (25, B)
Vinyl chloride	6.26 (17)	1.45
Vinyl isocyanate	10.62 (25)	
2-Vinylpyridine	9.126 (20)	
4-Vinylpyridine	10.50 (20)	
o-Xylene	2.562 (20), 2.54 (30)	0.62
m-Xylene	2.359 (20), 2.35 (30)	0.33 (20, lq), 0.37 (20, B)
p-Xylene	2.273 (20), 2.22 (50)	0
Xylitol	40.0 (20)	

2.11 IONIZATION ENERGY

The ionization energy or ionization potential is the energy necessary to remove an electron from the neutral atom. It is a minimum for the alkali metals that have a single electron outside a closed shell. It generally increases across a row on the periodic maximum for the noble gases that have closed shells. For example, sodium requires only 496 kJ/mol or 5.14 eV/atom to ionize it while neon, the noble gas immediately preceding it in the periodic table, requires 2081 kJ/mol or 21.56 eV/atom. The ionization energy is one of the primary energy considerations used in quantifying chemical bonds.

The electron affinity is a measure of the energy change when an electron is added to a neutral atom to form a negative ion. For example, when a neutral chlorine atom in the gaseous form picks up an electron to form a Cl^- ion, it releases energy of 349 kJ/mol or 3.6 eV/atom. It is said to have an electron affinity of -349 kJ/mol and this large number indicates that it forms a stable negative ion. Small numbers indicate that a less stable negative ion is formed. Group VIA and VIIA in the periodic table have the largest electron affinities.

Note: 1 kJ/mol = .010364 eV/atom

TABLE 2.50 Ionization Energy of Molecular and Radical Species

This table gives the first ionization potential in MJ · mol^{-1} and in electron volts. Also listed is the enthalpy of formation of the ion at 25°C (298 K).

Species	Ionization energy		$\Delta_f H$ (ion)
	In MJ · mol^{-1}	In electron volts	in kJ · mol^{-1}
Acenaphthene	0.741	7.68	896
Acenaphthylene	0.793	8.22(4)	1053
Acetaldehyde	0.98696(7)	10.2290(7)	821
Acetamide	0.931(3)	9.65(3)	693
Acetic acid	1.029(2)	10.66(2)	596
Acetic anhydride	0.965	10.0	398
Acetone	0.9364	9.705	719
Acetonitrile	1.1766(5)	12.194(5)	1252
Acetophenone	0.896(3)	9.29(3)	810
Acetyl chloride	1.047(5)	10.85(5)	804
Acetyl fluoride	1.111(2)	11.51(2)	667
Acetylene	1.1000(2)	11.400(2)	1328
Allene	0.935(1)	9.69(1)	1126
Allyl alcohol	0.933(5)	9.67(5)	808
Allylamine	0.845	8.76	891
3-Amino-I-propanol	0.87	9.0	651
Aniline	0.7449(2)	7.720(2)	832
Anthracene	0.719(3)	7.45(3)	949
Azoxybenzene	0.78	8.1	1123
Azulene	0.715(2)	7.41(2)	1004
Benzaldehyde	0.916(2)	9.49(2)	878
Benzamide	0.912	9.45	811
Benzene	0.89212(2)	9.2459(2)	975
Benzenethiol	0.801(2)	8.30(2)	913
Benzoic acid	0.914	9.47	620
Benzonitrile	0.928	9.62	1146
Benzophenone	0.873(5)	9.05(5)	923
p-Benzoquinone	0.969(2)	10.04(18)	847
Benzoyl chloride	0.920	9.54	816
Benzyl alcohol	0.82	8.5	720
Benzylamine	0.834(5)	8.64(5)	917
Biphenyl	0.767(2)	7.95(2)	950
Bromoacetylene	0.995(2)	10.31(2)	1242
Beomobenzene	0.866(2)	8.98(2)	971
Bromochlorodifluoromethane	1.141	11.83	702
Bromochloromethane	1.039(1)	10.77(1)	1085
Bromodichloromethane	1.02	10.6	973
Bromethane	0.992	10.28	930
Bromethylene	0.946(2)	9.80(2)	1025
Bromomethane	1.0171(3)	10.541(3)	979
1-Bromonaphthalene	0.781	8.09	956
Bromopentafluorobenzene	0.923(2)	9.57(2)	212
1-Bromopropane	0.982(1)	10.18(1)	898
2-Bromopropane	0.972(1)	10.07(1)	874
3-Bromopropene	0.972(1)	10.07(1)	1018
p-Bromotoluene	0.837(1)	8.67(1)	908
Bromotrichloromethane	1.02	10.6	980
Bromotrifluoromethane	1.10	11.4	451
1,2-Butadiene	0.871	9.03	1034

TABLE 2.50 Ionization Energy of Molecular and Radical Species (*Continued*)

Species	Ionization energy In MJ · mol^{-1}	Ionization energy In electron volts	$\Delta_f H$ (ion) in kJ · mol^{-1}
1,3-Butadiene	0.8750	9.069	985
Butanal	0.949(2)	9.84(2)	742
Butanenitrile	1.08	11.2	1110
2-Butanone	0.918(4)	9.51(4)	677
trans-2-Butenal	0.939(1)	9.73(1)	835
1-Butene	0.924(2)	9.58(2)	924
cis-2-Butene	0.8788(8)	9.108(8)	871
trans-2-Butene	0.8780(8)	9.100(8)	866
1-Buten-3-yne	0.924(2)	9.58(2)	1230
Butyl acetate	0.965	10.0	479
sec-Butyl acetate	0.955	9.90	453
Butyl ethyl ether	0.903	9.36	610
Butylbenzene	0.838(1)	8.69(1)	826
sec-Butylbenzene	0.837(1)	8.68(1)	820
tert-Butylbenzene	0.834(2)	8.64(2)	812
Butylcyclohexane	0.908	9.41	695
Butylcyclopentane	0.960(3)	9.95(3)	793
p-tert-Butylphenol	0.75	7.8	552
p-tert-Butyltoluene	0.799	8.28	745
1-Butyne	0.9821(5)	10.178(5)	1147
2-Butyne	0.9226(5)	9.562(5)	1068
Camphor	0.845(3)	8.76(3)	577
Caprolactam	0.875(2)	9.07(2)	629
Carbazole	0.730(3)	7.57(3)	961
Carbon	1.0865	11.260	1803
Carbon (C_2)	1.188	12.31	2000
Carbon dioxide	1.3289(2)	13.773(2)	935
Carbon monoxide	1.35217	14.0139	1242
Carbon oxyselenide	1.000(1)	10.36(1)	929
Carbon oxysulfide	1.07812(15)	11.1736(15)	936
Carbon sulfide	0.97149(19)	10.0685(20)	1089
Carbon sulfide (CS)	1.093(1)	11.33(1)	1368
Carbonyl fluoride	1.257	13.03	617
Carbonyltrihydroboron (BH_3CO)	1.075(2)	11.14(2)	962
Chloroacetaldehyde	1.011(3)	10.48(3)	815
Chloroacetic acid	0.984	10.2	597
Chloroacetyl chloride	1.06	11.0	815
Chloroacetylene	1.021(2)	10.58(2)	1276
m-Chloroaniline	0.781(10)	8.09(10)	835
o-Chloroaniline	0.820	8.50	883
p-Chloroaniline	0.789	8.18	844
Chlorobenzene	0.874(2)	9.06(2)	929
Chlorodibromomethane	0.1022(1)	10.59(1)	1030
1-Chloro-1,1-difluoroethane	1.156(1)	11.98(1)	626
1-Chloro-2,2-difluoroethylene	0.946(4)	9.80(4)	628
Chlorodifluoromethane	1.18	12.2	693
Chloroethane	1.058(2)	10.97(2)	946
2-Chloroethanol	1.015	10.52	756
Chloroethylene	0.964(2)	9.99(2)	985
Chlorofluoromethane	1.130(1)	11.71(1)	870
Chloromethane	1.083(1)	11.22(1)	1001
Chloromethylene	0.949	9.84	1247

TABLE 2.50 Ionization Energy of Molecular and Radical Species (*Continued*)

Species	Ionization energy		$\Delta_f H$ (ion)
	In MJ · mol^{-1}	In electron volts	in kJ · mol^{-1}
Chloromethylidine (CCl)	0.86(2)	8.9(2)	1244
1-Chloronaphthalene	0.784	8.13	906
m-Chloronitrobenzene	0.957(10)	9.92(10)	995
p-Chloronitrobenzene	0.961(10)	9.96(10)	999
Chloropentafluorobenzene	0.938(2)	9.72(2)	126
Chloropentafluoroethane	1.22	12.6	99
m-Chlorophenol	0.835	8.65	680
p-Chlorophenol	0.834	8.69	692
1-Chloropropane	1.044(3)	10.82(3)	912
2-Chloropropane	1.040(2)	10.78(2)	895
3-Chloropropene	0.96	9.9	950
m-Chlorotoluene	0.852(2)	8.83(2)	869
o-Chlorotoluene	0.852(2)	8.83(2)	869
p-Chlorotoluene	0.838(2)	8.69(2)	855
Chlorotrifluoroethylene	0.947	9.81(3)	373
Chlorotrifluoromethane	1.195	12.39	485
Chrysene	0.732	7.59(2)	1016
Coronene	0.703	7.29	1026
m-Cresol	0.800	8.29	668
o-Cresol	0.785	8.14	660
p-Cresol	0.784	8.13	659
cis-Crotonic acid	0.973	10.08	625
trans-Crotonic acid	0.96	9.9	604
Cumene	0.842	8.73(1)	847
Cyanamide	1.00	10.4	1137
Cyanate (NCO)	1.135(1)	11.76(1)	1290
Cyanide (CN)	1.360	14.09	1795
Cyanoacetylene	1.123(1)	11.64(1)	1475
Cyanogen	1.290(1)	13.37(1)	1597
Cyanogen chloride	1.191(1)	12.34(1)	1329
Cyanogen fluoride	1.285(1)	13.32(1)	1323
Cyclobutane	0.957(5)	9.92(5)	986
Cyclobutanone	0.9025	9.354	815
Cyclobutene	0.910	9.43	1067
Cycloheptane	0.962	9.97	844
Cyclohexane	0.951(3)	9.86(3)	828
Cyclohexanol	0.941	9.75	651
Cyclohexanone	0.882(1)	9.14(1)	656
Cyclohexene	0.8631(10)	8.945(10)	859
Cyclohexylamine	0.832(23)	8.62(24)	727
Cyclohexylcyclohexane	0.908	9.41	690
Cyclooctane	0.942	9.76	817
Cyclopropane	0.951	9.86	1005
Cyclopropanecarbonitrile	0.989	10.25	1173
Cyclopropanone	0.88(1)	9.1(1)	895
Cyclopropene	0.930	9.67(1)	1209
Cyclopropylamine	0.84	8.7	916
Cyclopropylbenzene	0.806	8.35	956
cis-Decahydronaphthalene	0.893	9.26	724
trans-Decahydronaphthalene	0.892	9.24	710
Decane	0.931	9.65	682
1-Decene	0.909(1)	9.42(1)	786

(*Continued*)

TABLE 2.50 Ionization Energy of Molecular and Radical Species (*Continued*)

Species	Ionization energy In MJ · mol^{-1}	Ionization energy In electron volts	$\Delta_f H$ (ion) in kJ · mol^{-1}
Diazomethane	0.8683(1)	8.999(1)	1098
1,4-Dibromobutane	0.979	10.15	879
1,2-Dibromoethane	1.001	10.37	963
Dibromofluoromethane	1.069(3)	11.07(3)	687
Dibromomethane	1.013(2)	10.50(2)	1013
1,2-Dibromopropane	0.975	10.1	903
1,3-Dibromopropane	0.990	10.26	919
1,2-Dibromotetrafluoroethane	1.07	11.1	280
Dibutyl ether	0.910	9.43	575
Di-*sec*-butyl ether	0.879	9.11	511
Di-*tert*-butyl ether	0.850	8.81	486
Dibutyl sulfide	0.79	8.2	624
Di-*tert*-butyl sulfide	0.77	8.0	583
Dibutylamine	0.742(3)	7.69(3)	586
Dichloroacetyl chloride	1.06	11.0	819
Dichloroacetylene	0.974	10.09	1183
m-Dichlorobenzene	0.879(1)	9.11(1)	907
o-Dichlorobenzene	0.876(1)	9.08(1)	909
p-Dichlorobenzene	0.856(1)	8.89(1)	882
Dichlorodifluoromethane	1.134(4)	11.75(4)	656
Dichlorodimethylsilane	1.03	10.7	576
1,1-Dichloroethane	1.067	11.06	937
1,2-Dichloroethane	1.065	11.04	931
1,1-Dichloroethylene	0.945(4)	9.79(4)	947
cis-1,2-Dichloroethylene	0.932(1)	9.66(1)	936
trans-1,2-Dichloroethylene	0.931(2)	9.65(2)	935
Dichlorofluoromethane	1.11	11.5	829
Dichloromethane	1.092(1)	11.32(1)	996
Dichloromethylene	1.000	10.36	1163
1,2-Dichloropropane	1.049(5)	10.87(5)	886
1,3-Dichloropropane	1.047(5)	10.85(5)	888
1,2-Dichlorotetrafluoroethane	1.18	12.2	252
Dicyclopropyl ketone	0.88	9.1	1041
1,1-Diethoxyethane	0.944	9.78	490
Diethyl oxalate	0.95	9.8	205
m-Diethylbenzene	0.819(1)	8.49(1)	798
o-Diethylbenzene	0.821	8.51	804
p-Diethylbenzene	0.810	8.40	790
Diethylene glycol dimethyl ether	0.96	9.8	448
m-Difluorobenzene	0.900(1)	9.33(1)	591
o-Difluorobenzene	0.895(1)	9.28(1)	602
p-Difluorobenzene	0.882(1)	9.14(1)	575
1,1-Difluoroethane	1.145(3)	11.87(3)	643
1,1-Difluoroethylene	0.993(1)	10.29(1)	650
cis-1,2-Difluoroethylene	0.987	10.23	690
Difluoromethane	1.226	12.71	774
Difluoromethylene	1.102(1)	11.42(1)	897
2,5-Dihydrothiophene	0.81	8.4	898
Diiodomethane	0.913(2)	9.46(2)	1030
Diisobutyl sulfide	0.807(5)	8.36(5)	627
Diisobutylamine	0.754	7.81	574
Diisopropyl ether	0.888(5)	9.20(5)	569
Diisopropyl sulfide	0.833(5)	8.63(5)	630

TABLE 2.50 Ionization Energy of Molecular and Radical Species (*Continued*)

Species	Ionization energy		$\Delta_f H$ (ion)
	In $MJ \cdot mol^{-1}$	In electron volts	in $kJ \cdot mol^{-1}$
Diisopropylamine	0.746(3)	7.73(3)	602
Diketene	0.93(2)	9.6(2)	736
Dimethoxymethane	0.92	9.5	569
Dimethyl disulfide	0.71	7.4(3)	690
Dimethyl ether	0.9673(23)	10.025(25)	783
Dimethyl oxalate	0.965	10.0	287
***o*-Dimethyl phthalate**	0.930(7)	9.64(7)	277
Dimethyl sulfide	0.838(1)	8.69(1)	801
Dimethyl sulfoxide	0.878	9.01	718
Dimethylamine	0.794(8)	8.23(8)	776
***N,N*-Dimethylaniline**	0.687(2)	7.12(2)	787
2,2-Dimethylbutane	0.971	10.06	787
2,3-Dimethylbutane	0.967	10.02	791
3,3-Dimethyl-2-butanone	0.879(2)	9.11(2)	589
2,3-Dimethyl-1-butene	0.875(1)	9.07(1)	812
2,3-Dimethyl-2-butene	0.798(1)	8.27(1)	729
3,3-Dimethyl-1-butyne	0.946(5)	9.80(5)	1050
1,1-Dimethylcyclohexane	0.909	9.42	728
cis-1,2-Dimethylcyclohexane	<0.944	<9.78	772
cis-1,3-Dimethylcyclohexane	<0.963	<9.98	778
cis-1,4-Dimethylcyclohexane	<0.958	<9.93	782
trans-1,2-Dimethylcyclohexane	0.908	9.41	728
trans-1,3-Dimethylcyclohexane	0.920	9.53	743
trans-1,4-Dimethylcyclohexane	0.922	9.56	738
cis-1,2-Dimethylcyclopentane	0.957(5)	9.92(5)	828
trans-1,2-Dimethylcyclopentane	0.960(5)	9.95(5)	823
N,N-Dimethylformamide	0.881(2)	9.13(2)	689
2,6-Dimethyl-4-heptanone	0.872(3)	9.04(3)	515
1,1-Dimethylhydrazine	0.702(4)	7.28(4)	786
2,4-Dimethyl-3-pentanone	0.864(1)	8.95(1)	552
2,3-Dimethylpyridine	0.854(2)	8.85(2)	922
2,4-Dimethylpyridine	0.854(3)	8.85(3)	918
2,5-Dimethylpyridine	0.849(5)	8.80(5)	916
2,6-Dimethylpyridine	0.847(3)	8.86(3)	913
3,4-Dimethylpyridine	0.883	9.15	953
3,5-Dimethylpyridine	0.893	9.25	965
N,N-Dimethyl-*o*-toluidine	0.714(2)	7.40(2)	814
1,3-Dioxane	0.95	9.8	607
1,4-Dioxane	0.887(1)	9.19(1)	571
1,3-Dioxolane	0.96	9.9	658
Diphenyl ether	0.781(3)	8.09(3)	766
Diphenylacetylene	0.762(2)	7.90(2)	1164
Diphenylamine	0.691(4)	7.16(4)	908
1,2-Diphenylethane	0.84(1)	8.7(1)	983
Diphenylmethane	0.825(3)	8.55(3)	963
Dipropyl ether	0.894(5)	9.27(5)	602
Dipropyl sulfide	0.801(2)	8.30(2)	676
Dipropylamine	0.746(3)	7.73(3)	641
Divinyl ether	0.84	8.7	827
5,7-Dodecadiyne	0.837	8.67	1079
Dodecafluorocyclohexane	1.27	13.2	−1095
Epichlorohydrin	0.98	10.2	875

(*Continued*)

TABLE 2.50 Ionization Energy of Molecular and Radical Species (*Continued*)

Species	Ionization energy In $MJ \cdot mol^{-1}$	Ionization energy In electron volts	$\Delta_f H$ (ion) in $kJ \cdot mol^{-1}$
Ethylene glycol	0.980	10.16	593
Ethylene oxide	1.0195(10)	10.566(10)	967
Ethyleneimine	0.89(1)	9.2(1)	1014
p-Ethylphenol	0.756	7.84	613
Ethynyl (HC≡C)	1.13	11.7	1694
Fluoranthene	0.768(4)	7.95(4)	1057
Fluorene	0.761(3)	7.89(3)	950
Fluoroacetylene	1.086	11.26	1195
Fluorobenzene	0.8877(5)	9.200(5)	772
Fluoroethane	1.12	11.6	856
Fluoroethylene	1.0000(15)	10.363(15)	861
Fluoromethane	1.203(2)	12.47(2)	956
Fluoromethylene	1.012	10.49	1121
Fluoromethylidene (CF)	0.879(1)	9.11(1)	1134
p-Fluoronitrobenzene	0.955	9.90	826
1-Fluoropropane	1.09	11.3	806
2-Fluoropropane	1.069(2)	11.08(2)	776
3-Fluoropropene	0.975	10.11	821
m-Fluorotoluene	0.860(1)	8.91(1)	709
o-Fluorotoluene	0.860(1)	8.91(1)	709
p-Fluorotoluene	0.848(1)	8.79(1)	701
Formaldehyde	1.0492(2)	10.874(2)	940
Formamide	0.980(6)	10.16(6)	796
Formic acid	1.093(1)	11.33(1)	715
Fulminic acid (HCNO)	1.045	10.83	1263
Fulvene	0.807	8.36	1031
Fumaric acid	1.03	10.7	355
Furan	0.8571(3)	8.883(3)	822
Glyoxal	0.975	10.1	763
1-Heptanal	0.931(2)	9.65(2)	668
Heptane	0.957(5)	9.92(5)	770
1-Heptanol	0.949(3)	9.84(3)	614
2-Heptanol	0.936(3)	9.70(3)	580
3-Heptanol	0.934(3)	9.68(3)	578
4-Heptanol	0.927(3)	9.61(3)	572
2-Heptanone	0.897(1)	9.30(1)	596
1-Heptene	0.911	9.44	849
2-Heptene	0.853(2)	8.84(2)	782
3-Heptene	0.861	8.92	790
Hexachlorobenzene	0.866	8.98	822
Hexachloroethane	1.07	11.1	920
1,5-Hexadiene	0.896(5)	9.29(5)	980
Hexafluoroacetone	1.104	11.44	−294
Hexafluorobenzene	0.9558	9.906	10
Hexafluoroethane	1.29	13.4	−50
Hexafluoropropene	1.023(3)	10.60(3)	−103
Hexamethylbenzene	0.757	7.85	670
1-Hexanal	0.933(5)	9.67(5)	686
Hexane	0.977	10.13	810
Hexanoic acid	0.976	10.12	463
1-Hexanol	0.954(3)	9.89(3)	639

TABLE 2.50 Ionization Energy of Molecular and Radical Species (*Continued*)

Species	Ionization energy		$\Delta_f H$ (ion)
	In MJ · mol^{-1}	In electron volts	in kJ · mol^{-1}
2-Hexanol	0.946(3)	9.80(3)	611
3-Hexanol	0.929(3)	9.63(3)	599
2-Hexanone	0.902(2)	9.35(2)	626
3-Hexanone	0.880(2)	9.12(2)	600
1-Hexene	0.911(4)	9.44(4)	869
cis-2-Hexene	0.865(1)	8.97(1)	818
trans-2-Hexene	0.865(1)	8.97(1)	814
Hexylamine	0.833(5)	8.63(5)	699
1-Hexyne	0.960	9.95(5)	1081
Hydrogen cyanide (HCN)	1.312(1)	13.60(1)	1447
Hydrogen isocyanide (HNC)	1.21(1)	12.5(1)	1407
p-Hydroquinone	0.767(3)	7.95(3)	504
Imidazole	0.850(1)	8.81(1)	997
Indane	0.90	9.3	864
Indene	0.785(1)	8.14(1)	949
Iodobenzene	0.8380	8.685	1003
Iodoethane	0.9018	9.346	893
1-Iodohexane	0.8857	9.179	794
Iodomethane	0.9203	9.538	936
1-Iodopropane	0.8943	9.269	862
2-Iodopropane	0.8853	9.175	844
Isobutylbenzene	0.838(1)	8.68(1)	816
Isocyanic acid	1.120(3)	11.61(3)	1016
Isophthalic acid	0.963(20)	9.98(20)	268
Isopropylcyclohexane	0.900	9.33	704
Isoquinoline	0.8239(3)	8.539(3)	1032
Isoxazole	0.958(5)	9.93(5)	1038
Ketene	0.927(2)	9.61(2)	880
Maleic anhydride	1.04	10.8	645
Mesityl oxide	0.876(3)	9.08(3)	692
Methacrylic acid	0.979	10.15	611
Methane	1.207	12.51	1133
Methanethiol	9.108(5)	9.440(5)	888
Methanol	1.047(1)	10.85(1)	845
Methoxy	0.83	8.6	845
Methoxybenzene (Anisole)	0.792(2)	8.21(2)	724
2-Methoxyethanol	0.93	9.6	562
Methyl	0.949(1)	9.84(1)	1095
Methyl acetate	0.991(2)	10.27(2)	581
Methyl acrylate	0.96	9.9	611
Methyl azide	0.947(2)	9.81(2)	1227
Methyl benzoate	0.899(3)	9.32(3)	611
Methyl chloroacetate	0.99	10.3	575
Methyl 2,2-dimethylpropanoate	0.955(4)	9.90(4)	466
Methyl formate	1.0435(5)	10.815(5)	688
Methyl pentanoate	1.00(2)	10.4(2)	532
Methyl pentyl ether	0.933	9.67	657
Methyl vinyl ether	0.862(2)	8.93(2)	761
Methylacrylonitrile	0.998	10.34	1127
Methylamine	0.865(2)	8.97(2)	843
2-Methylaniline	0.718(2)	7.44(2)	772

(*Continued*)

TABLE 2.50 Ionization Energy of Molecular and Radical Species (*Continued*)

Species	Ionization energy In MJ · mol^{-1}	In electron volts	$\Delta_f H$ (ion) in kJ · mol^{-1}
3-Methylaniline	0.724(2)	7.50(2)	778
4-Methylaniline	0.698(2)	7.24(2)	753
N-Methylaniline	0.707(2)	7.33(2)	791
Methylcyclohexane	0.930	9.64	775
1-Methylcyclohexanol	0.95(2)	9.8(2)	586
Methylcyclopentane	0.950(3)	9.85(3)	845
Methylcyclopropane	0.913	9.46	936
2-Methyldecane	0.934	9.68	685
Methylene	1.0031(3)	10.396(3)	1386
N-Methylformamide	0.945	9.79	756
2-Methylheptane	0.949	9.84	734
5-Methyl-2-hexanone	0.895(1)	9.28(1)	586
Methylhydrazine	0.740(2)	7.67(2)	835
Methylidyne	1.027(1)	10.64(1)	1622
Methylisocyanate	1.030(2)	10.67(2)	900
1-Methyl-4-isopropylbenzene (*p*-Cymene)	0.800	8.29	771
1-Methylnaphthalene	0.757	7.85	870
2-Methylnaphthalene	0.75	7.8	866
Methyloxirane	0.986(2)	10.22(2)	892
2-Methylpentane	0.976	10.12	802
3-Methylpentane	0.973	10.08	801
2-Methyl-3-pentanone	0.878(1)	9.10(1)	592
3-Methyl-2-pentanone	0.889(1)	9.21(1)	600
4-Methyl-2-pentanone	0.897(1)	9.30(1)	609
2-Methyl-1-pentene	0.876(1)	9.08(1)	817
2-Methyl-2-pentene	0.828	8.58	761
4-Methyl-1-pentene	0.912(1)	9.45(1)	862
4-Methyl-*cis*-2-pentene	0.866(1)	8.98(1)	809
4-Methyl-*trans*-2-pentene	0.865(1)	8.97(1)	804
2-Methylpropanal	0.9364(5)	9.705(5)	721
2-Methylpropanenitrile	1.09	11.3	1115
2-Methylpropenal	0.951	9.86	834
2-Methylpropene (Isobutene)	0.8915(3)	9.239(3)	875
2-Methylpyridine	0.870(3)	9.02(3)	970
3-Methylpyridine	0.872(3)	9.04(3)	979
4-Methylpyridine	0.872(3)	9.04(3)	976
Methylsilane	1.03	10.7	1003
m-Methylstyrene	0.786(2)	8.15(2)	908
o-Methylstyrene	0.888(2)	9.20(2)	908
p-Methylstyrene	0.78(1)	8.1(1)	895
Methyltrichlorosilane	1.096(3)	11.36(3)	548
Naphthalene	0.785(1)	8.14(1)	936
1-Naphthol	0.749(3)	7.76(3)	719
2-Naphthol	0.757(5)	7.85(5)	727
Nickel carbonyl	0.798(4)	8.27(4)	200
m-Nitroaniline	0.802(2)	8.31(2)	865
o-Nitroaniline	0.798(1)	8.27(1)	861
p-Nitroaniline	0.804(1)	8.34(1)	850
Nitrobenzene	0.951(2)	9.86(2)	1019
Nitroethane	1.050(5)	10.88(5)	948

TABLE 2.50 Ionization Energy of Molecular and Radical Species (*Continued*)

Species	Ionization energy		$\Delta_f H$ (ion)
	In MJ · mol^{-1}	In electron volts	in kJ · mol^{-1}
Nitromethane	1.063(4)	11.02(4)	988
m-Nitrophenol	0.86	9.0	755
o-Nitrophenol	0.88	9.1	782
p-Nitrophenol	0.88	9.1	761
1-Nitropropane	1.043(3)	10.81(3)	919
2-Nitropropane	1.033(5)	10.71(5)	894
m-Nitrotoluene	0.15(2)	9.48(2)	944
o-Nitrotoluene	0.912(4)	9.45(4)	966
p-Nitrotoluene	0.91	9.4	936
Nonane	0.938	9.72	710
2-Nonanone	0.884	9.16	545
5-Nonanone	0.875	9.07	530
Octafluoronaphthalene	0.854	8.85	−368
Octafluoropropane	1.291	13.38	−491
Octafluorotoluene	0.96	9.9	−233
Octane	0.948	9.82	739
1-Octene	0.910(1)	9.43(1)	829
1-Octyne	0.960(2)	9.95(2)	1040
2-Octyne	0.898(1)	9.31(1)	961
3-Octyne	0.890(1)	9.22(1)	952
4-Octyne	0.888(1)	9.20(1)	946
Oxazole	0.93	9.6	910
Oxetane	0.9328(5)	9.668(5)	853
2-Oxetanone	0.936(1)	9.70(1)	653
Oxomethyl (HCO)	0.782(5)	8.10(5)	826
Pentafluorobenzene	0.929	9.63	122
Pentafluorophenol	0.888(2)	9.20(2)	−71
2,3,4,5,6-Pentafluorotoluene	0.91	9.4	64
Pentanchloroethane	1.06	11.0	919
Pentylamine	0.837	8.67	728
Perylene	0.666(1)	6.90(1)	975
Phenanthrene	0.758(2)	7.86(2)	963
Phenetole	0.784(2)	8.13(2)	683
Phenol	0.817	8.47	721
Phenylacetic acid	0.797	8.26	479
m-Phenylenediamine	0.689	7.14	777
o-Phenylenediamine	0.69	7.2	787
p-Phenylenediamine	0.663(5)	6.87(5)	759
Phthalic anhydride	0.96	10.0	593
α-Pinene	0.779	8.07	808
Propanal	0.9603(5)	9.953(5)	773
Propanamide	0.92	9.5	720
Propane	1.057(5)	10.95(5)	952
Propanenitrile	1.142(2)	11.84(2)	1194
1-Propanethiol	0.8872(5)	9.195(5)	819
2-Propanethiol	0.882	9.14	806
Propanoic acid	1.0155(3)	10.525(3)	568
1-Propanol	0.986(3)	10.22(3)	731
2-Propanol	0.976(8)	10.12(8)	704

(*Continued*)

TABLE 2.50 Ionization Energy of Molecular and Radical Species (*Continued*)

Species	Ionization energy		$\Delta_f H$ (ion)
	In $MJ \cdot mol^{-1}$	In electron volts	in $kJ \cdot mol^{-1}$
Propenal	0.975(6)	10.103(6)	900
Propene	0.939(2)	9.73(2)	959
Propenenitrile	1.053(1)	10.91(1)	1237
Propenoic acid	1.023	10.60	701
1-Propylamine	0.847(2)	8.78(2)	777
2-Propylamine	0.841(3)	8.72(3)	758
Propylbenzene	0.841(1)	8.72(1)	849
Propylcyclohexane	0.913	9.46	720
Propylcyclopentane	0.965(4)	10.00(4)	817
Propyleneimine	0.87	9.0	960
Propynal	1.04	10.8	1155
Propyne	1.000(1)	10.36(1)	1186
2-Propyn-1-ol	1.014	10.51	1060
Pyrene	0.715	7.41	933
Pyridazine	0.834	8.64	1112
Pyrimidine	0.891	9.23	1087
Pyrrole	0.7920(5)	8.208(5)	900
2-Pyrrolidone	0.89	9.2	674
Quinoline	0.832(1)	8.62(1)	1041
cis-Stilbene	0.753(2)	7.80(2)	1005
trans-Stilbene	0.743(3)	7.70(3)	977
Styrene	0.813(6)	8.43(6)	961
Succinic anhydride	1.02	10.6	500
Succinonitrile	1.158(24)	12.10(25)	1377
Terephthalic acid	0.951(20)	9.86(20)	232
m-Terphenyl	0.773(1)	8.01(1)	1057
o-Terphenyl	0.77	8.0	1056
p-Terphenyl	0.751(1)	7.78(1)	1035
Tetrabromomethane	0.995(2)	10.31(2)	1079
Tetrachloro-1,2-difluoroethane	1.09	11.3	563
1,1,1,2-Tetrachloroethane	1.07	11.1	920
1,1,2,2-Tetrachloroethane	1.121	11.62	971
Tetrachloroethylene	0.899	9.32	887
Tetrachloromethane	1.107(1)	11.47(1)	1011
Tetraethylsilane	0.86	8.9	595
1,2,3,4-Tetrafluorobenzene	0.920(1)	9.53(1)	284
1,2,3,5-Tetrafluorobenzene	0.920(1)	9.53(1)	263
1,2,4,5-Tetrafluorobenzene	0.902(1)	9.35(1)	254
Tetrafluoroethylene	0.976(2)	10.12(2)	315
Tetrahydrofurane	0.908(2)	9.41(2)	724
1,2,3,4-Tetrahydronaphthalene	0.817	8.47	842
1,2,4,5-Tetramethylbenzene	0.776(1)	8.04(1)	730
2,2,3,3-Tetramethylbutane	0.95	9.8	720
Thiacyclobutane	0.838	8.69	899
Thiophene	0.856(4)	8.87(4)	971
p-Tolualdehyde	0.900(5)	9.33(5)	825
Toluene	0.851(1)	8.82(1)	901
m-Toluic acid	0.910(20)	9.43(20)	579
o-Toluic acid	0.88	9.1	558
p-Toluic acid	0.891(20)	9.23(20)	560

TABLE 2.50 Ionization Energy of Molecular and Radical Species (*Continued*)

Species	Ionization energy		$\Delta_f H$ (ion)
	In MJ · mol^{-1}	In electron volts	in kJ · mol^{-1}
m-Tolunitrile	0.901	9.34	1085
o-Tolunitrile	0.905	9.38	1085
p-Tolunitrile	0.899	9.32	1083
Tribromomethane	1.011(2)	10.48(2)	1035
Tributylamine	0.71	7.4	492
Trichloroacetyl chloride	1.06	11.0	827
1,2,4-Trichlorobenzene	0.872	9.04	880
1,3,5-Trichlorobenzene	0.899(2)	9.32(2)	899
1,1,1-Trichloroethane	1.06	11.0	917
1,1,2-Trichloroethane	1.06	11.0	911
Trichloroethylene	0.914(1)	9.47(1)	895
Trichlorofluoromethane	1.136(2)	11.77(2)	868
Trichloromethane	1.097(2)	11.37(2)	992
Trichloromethylbenzene	0.926	9.60	914
1,1,2-Trichlorotrifluoroethane	1.157(2)	11.99(2)	429
Triethanolamine	0.76	7.9	206
Triethylamine	0.724	7.50	631
Trifluoroacetic acid	1.106	11.46	75
Trifluoroacetonitrile	1.337	13.86	838
1,1,1-Trifluoro-2-bromo-2-chloroethane	1.06	11.0	362
1,1,1-Trifluoroethane	1.24(1)	12.9(1)	496
Trifluoroethylene	0.978	10.14	489
Trifluoroiodomethane	0.987	10.23	397
Trifluoromethane	1.337	13.86	643
Trifluoromethyl (CF_3)	0.86	8.9	399
Trifluoromethylbenzene	0.9345(4)	9.685(4)	335
3,3,3-Trifluoropropene	1.05	10.9	437
Triiodomethane	0.893(2)	9.25(2)	1010
Trimethylamine	0.755462	7.82960	731
1,2,3-Trimethylbenzene	0.812(2)	8.42(2)	803
1,2,4-Trimethylbenzene	0.798(1)	8.27(1)	784
1,3,5-Trimethylbenzene	0.811(1)	8.41(1)	796
Trimethylborate	0.96	10.0	65
Trimethylchlorosilane	0.979	10.15	624
3,5,5-Trimethylcyclohex-2-en-1-one	0.875	9.07	670
2,2,4-Trimethylpentane	0.951	9.86	713
2,2,4-Trimethyl-3-pentanone	0.849(1)	8.80(1)	511
2,4,6-Trimethylpyridine	0.88(1)	8.9(1)	580
Trioxane	0.99	10.3	528
Undecane	0.922	9.56	650
Urea	0.94	9.7	690
Vinyl acetate	0.887	9.19	572
m-Xylene	0.826(1)	8.56(1)	843
o-Xylene	0.826(1)	8.56(1)	844
p-Xylene	0.814(1)	8.44(1)	832
2,3-Xylenol	0.797	8.26	640
2,4-Xylenol	0.77	8.0	609
2,6-Xylenol	0.777(2)	8.05(2)	615
3,4-Xylenol	0.781	8.09	624

TABLE 2.51 Thermal Conductivities of Gases as a Function of Temperature

The coefficient *k*, expressed in $J \cdot sec^{-1} \cdot cm^{-1} \cdot K^{-1}$, is the quantity of heat in joules, transmitted per second through a sample one centimeter in thickness and one square centimeter in area when the temperature difference between the two sides is one degree kelvin (or Celsius). The tabulated values are in microjoules.

	Temperature, °C										
Substance	−40	−20	0	20	40	60	80	100	120	140	160
Acetone		80	95	107	124	140	156	173	190	207	
Acetaldehyde				109	126	142	159	176	195		
Acetonitrile						112	124	137	151	166	
Acetylene	118^{-75}		184	205	224	248	269	290			
Air			242	256	270	284	299	311	324	336	342^{149}
Ammonia	164^{-60}		218	238	259	280	301	321			
Argon			166	176	186	196	206	211			
Benzene						126	146	165	184	205	266
Boron trifluoride				186						241	
Bromine			42	45	50	54	59				
Bromomethane					82	94	104	117			
1-Butanamine			$135^{6.5}$					176^{110}			
Butane			135	154	174	193	213	233			
Carbon dioxide			144	160	176	192	207	215			
Carbon disulfide			67	76	85						
Carbon monoxide			228	245	262	278					
Carbon tetrachloride			59	64	70	75	80	86			109^{184}
Chlorine	64	72	79	85	93	100					
Chlorodifluorimethane		103	110	116	122						
Chloroethane			90	105	120	134	151	167	186	204	
Chloroform					75	84	91	99	107	116	
Chloromethane			84	105	117	130	142	155			
Cyclohexane			77	99	120	141	163				
Cyclopropane											
2-Methyl-2-propanol								225			
Neon	410	433	454	476	497	518	537	556			
Nitric oxide	205	221	238	254	269	285	301	317			
Nitrogen	211	226	241	256	270	282	295	307	320	333	385^{227}
Nitromethane									139	155	

Nitrous oxide	121	137	152	168	184						
Octafluorocyclobutane				120					190		
Oxygen	211	228	245	261	278	294	311	328			
Pentane			130					218			
Propane	116	132	151	171	192	215	238	262	330	353	379
2-Propanol				151[31]						250[127]	
Sulfur dioxide			83		163			106			
Sulfur hexafluoride				126					201	275[227]	338[327]
Tetrafluoromethane				235					235		
Thiophene								152[110]			
1,1,2-Trichlorotrifluoroethane				87					133		
Triethylamine								195	216	239	
Water		142	159	175	191	207	224	241	257		
Xenon	36[-73]			54					72	89[227]	104[327]
Deuterium	1150	1222	1297	1372	1448	1523					
Deuterium oxide									263		358[220]
Dibromomethane									74[110]		
Dichlorodifluoromethane		81	84	92	100			138			194[200]
1,1-Dichloroethane			69	81	93	105	117	129	144		
1,2-Dichloroethane								127	140		
Dichlorofluoromethane		91	94	97	100						
Dichloromethane			93					161			
1,2-Dichlorotetrafluoroethane				99					153		211[227]
Diethylamine			118			179	199	218	243	268	
Diethyl ether			113	135	157	178	200	222	244	269	351[213]
1,4-Dioxane								167	187	207	
Ethane	137	159	182	204	228	257	288	316	344		
Ethanol			126	141	155			209			
Ethene					230[49]						
Ethyl acetate					115	133	151	170	191	211	234
Ethylamine			136	153	169	206					
Ethylene	137	158	178	220	241	262	282				
Ethylene oxide								193	256	279	
Ethyl formate			79	100	121	142	164	186	206	226	
Ethyl nitrate								159	178	197	
Fluorine	212	230	247	264	278	294	309	325			
Helium	1276	1343	1423	1481	1540	1598	1661	1720	1778		
Heptane			100	115	130			174			

(*Continued*)

TABLE 2.51 Thermal Conductivities of Gases as a Function of Temperature (*Continued*)

Substance	Temperature, °C										
	−40	−20	0	20	40	60	80	100	120	140	160
Hexane			109				178	201	224	247	271
Hydrogen	1494	1607	1724	1828	1925	2025					
Hydrogen bromide	64	70	77	84	90	97	104				
Hydrogen chloride	107	117	128	138	148				191		240[227]
Hydrogen cyanide		99	110	121	132	143					
Hydrogen sulfide		116	129	143	156	169					
Iodomethane			46	53	60	68	75	82	89		
Krypton		79	85		95			110			
Methane	257	280	307	334	361	387	416	445			
Methanol						174	197	221	241	263	284
Methyl acetate			67			150[70]		177	195	215	237
2-Methylbutane			122					215			421
2-Methylpropane			141	156	176	196		233[93]	271		

TABLE 2.52 Thermal Conductivity of Various Substances

All values of thermal conductivity, k, are in millijoules $cm^{-1} \cdot s^{-1} \cdot K^{-1}$.

	Thermal conductivity in $mJ \cdot cm^{-1} \cdot s^{-1} \cdot K^{-1}$						
Substance	−25°C	0°C	20°C	25°C	50°C	75°C	100°C
Acetaldehyde			1.900				
Acetic acid				1.58	1.53	1.49	1.44
Acetic anhydride			2.209				
Acetone	1.987^{-80}	1.69	1.61		1.51^{40}		
Acetonitrile	2.08	1.98		1.88	1.78	1.68	
Allyl alcohol				1.80^{30}			
Aniline			1.77^{17}				
Argon	1.259^{-189}						
Benzaldehyde				1.51	1.41	1.31	1.21
Benzene				1.411	1.329	1.247	
Bromobenzene			1.113				
Bromoethane			1.029				
1-Bromo-2-methylpropane		1.163^{12}					
1-Bromopentane			0.983				
Bromopropane		1.075^{12}					
Butanoic acid		1.506^{12}					
1-Butanol		1.538		1.54	1.49		
2-Butanone	1.58	1.51		1.45	1.39	1.33	
Butyl acetate			1.368				
2-Butyne	1.37	1.29		1.21			
Carbon disulfide		1.54		1.49			
Carbon tetrachloride	1.100^{-20}	1.071	1.029		0.974		
Chlorobenzene	1.36	1.31		1.27	1.22	1.17	1.12
Chloroethane	1.45	1.32		1.19	1.06	0.93	
Chloroform	1.27	1.22		1.17	1.12	1.07	1.02
(Chloromethyl)oxirane	1.42	1.37		1.31	1.25	1.19	1.14
1-Chloro-2-methylpropane		1.163^{12}					
1-Chloropentane		1.184^{12}					
Chloropropane		1.184^{12}					
4-Chlorotoluene			1.297				
m-Cresol			1.498			1.452^{80}	
Cyclohexane			1.243	1.23	1.17	1.11	
Cyclohexene	1.42	1.36		1.30	1.24	1.18	
Cyclohexanol				1.34	1.31		
Cyclopentane	1.40	1.33		1.26			
Cyclopentene	1.43	1.36		1.29			
Decane	1.44	1.38		1.32	1.26	1.19	1.13
1-Decanol				1.62	1.56	1.50	1.45
Dibromomethane	1.20	1.14		1.08	1.03	0.97	
Dibutyl phthalate	1.44	1.40		1.36	1.33	1.29	1.25
1,2-Dichloroethane		1.264					
Dichlorofluoromethane	0.134						
Dichloromethane	1.590^{-20}	1.564	1.477				
Diethyl ether	1.50	1.40		1.30	1.20	1.10	1.00
Diisopropyl ether			1.096				
2,3-Dimethylbutane				1.038^{32}	0.996		
N,N-Dimethylformamide				1.84	1.78	1.71	1.65
Dimethyl phthalate		1.501		1.473	1.443	1.409	1.373

(Continued)

TABLE 2.52 Liquid Thermal Conductivity of Various Substances (*Continued*)

	Thermal conductivity in $mJ \cdot cm^{-1} \cdot s^{-1} \cdot K^{-1}$						
Substance	−25°C	0°C	20°C	25°C	50°C	75°C	100°C
1,4-Dioxane				1.59	1.47	1.35	1.23
Diphenyl ether					1.39	1.35	1.31
Dodecane		1.57		1.52	1.46	1.40	1.35
1-Dodecanol				1.46	1.42	1.39	1.35
Ethanol		1.76		1.69	1.62		
Ethanolamine				2.99	2.86	2.74	2.61
Ethoxybenzene			1.497				
Ethyl acetate	1.62	1.53		1.44	1.35	1.26	
Ethylbenzene				1.30	1.24	1.18	1.12
Ethylene glycol		2.56		2.56	2.56	2.56	2.56
Ethyl formate		1.581^{12}					
Furan	1.42	1.34		1.26			
Glycerol				2.92	2.95	2.97	3.00
Heptane	1.378	1.303	1.259	1.228	1.152	1.077	
1-Heptanol		1.66		1.59	1.53	1.47	1.41
Hexadecane				1.40	1.35	1.30	1.25
Hexane	1.37	1.28	1.218	1.20	1.11	1.92	0.93
1-Hexanol	1.59	1.54		1.50	1.45	1.41	1.37
2-Hexanone	1.51	1.45		1.39	1.33	1.27	1.21
1-Hexene	1.37	1.29		1.21	1.13		
Hydrochloric acid, 38%			4.402^{32}				
Hydrogen	1.180^{-253}						
Iodobenzene	1.063^{-20}		1.276			0.937^{80}	
Iodoethane				1.109^{30}			
1-Iodo-2-methylpropane		0.870^{12}					
1-Iodopentane		0.849^{12}					
Iodopropane		0.920^{12}					
Isopentyl acetate			1.297				
Isopropylbenzene				1.28	1.20	1.12	1.07
Mercury	72.5	77.7		82.5	86.8	90.7	94.3
Methanol	2.14	2.07	2.021	2.00	1.93		
Methoxybenzene	1.70	1.63		1.56	1.50	1.43	1.36
Methyl acetate	1.74	1.64		1.53	1.43	1.33	1.22
Methyl butanoate			1.402				
3-Methylbutanoic acid		1.305					
3-Methyl-1-butanol				1.477^{30}			
Methylcyclohexane				1.276^{30}			
Methylcyclopentane				1.209	1.151^{38}		
N-Methylformamide				2.03	2.01	1.99	1.96
1-Methyl-4-isopropylbenzene	1.32	1.27		1.22	1.17	1.12	1.07
2-Methylpentane				1.084^{32}	1.033		
Methyl pentanoate		1.318^{12}					
4-Methylpentanoic acid		1.427^{12}					
4-Methyl-3-pentene-2-one	1.70	1.63		1.56	1.49	1.42	1.34
2-Methyl-1-propanol		1.423^{12}					
2-Methyl-2-propanol				1.159^{38}		1.067^{77}	
Nitrobenzene			1.510				
Nitromethane				2.151^{30}			
Nonane	1.44	1.38		1.31	1.24	1.151^{80}	1.11

TABLE 2.52 Liquid Thermal Conductivity of Various Substances (*Continued*)

	Thermal conductivity in $mJ \cdot cm^{-1} \cdot s^{-1} \cdot K^{-1}$						
Substance	−25°C	0°C	20°C	25°C	50°C	75°C	100°C
1-Nonanol		1.66		1.61	1.55	1.49	1.43
Octadecane					1.46	1.42	1.37
Octane	1.43	1.35		1.28	1.20	1.13	1.06
1-Octanol		1.68	1.657	1.61	1.54	1.47	1.41
Palmitic acid						1.598	
Pentachloroethane			1.251				
Pentane	1.32	1.22	1.138	1.13	1.03	0.95	0.87
Pentanoic acid		1.360^{12}					
1-Pentanol		1.57		1.53	1.49	1.45	
1-Pentene	1.31	1.24		1.16			
Pentyl acetate			1.289				
Phenol					1.56	1.53	1.51
Phenylhydrazine				1.724			
1,2-Propanediol		2.02		2.00	1.99	1.98	1.97
Propanoic acid		1.728^{12}					
1-Propanol	1.62	1.58		1.54	1.49	1.45	1.41
2-Propanol	1.46	1.41		1.35	1.29	1.24	1.18
1,2-Propylene glycol		2.008					
Propyl formate		1.494^{12}					
Pyridine		1.69		1.65	1.61	1.58	
Silicon tetrachloride				0.99	0.96		
Sodium							753.1^{300}
Sodium chloride (aq, satd)	5.732						
Stearic acid						1.598	
Styrene	1.48	1.42		1.37	1.31	1.26	1.20
Sulfuric acid, 90%				3.540^{32}			
1,1,2,2-Tetrachloroethane		1.138					
Tetrachloroethylene	1.17		1.10	1.04	0.97		
Tetrachloromethane	1.04		0.99	0.93	0.88		
Tetradecane				1.36	1.31	1.26	1.21
1-Tetradecanol					1.67	1.62	1.57
Tetrahydrofuran	1.32	1.26		1.20	1.14		
Thiophene				1.99	1.95	1.91	1.86
Toluene	1.590^{-80}	1.386	1.347	1.311	1.236	1.161	
1,1,1-Trichloroethane	1.06		1.01	0.96			
Trichloroethylene	1.359^{-60}	1.24		1.160	1.08	1.00	
Trichloromethane	1.27	1.22		1.17	1.12	1.07	
Tridecane				1.37	1.32	1.27	1.22
Triethylamine	1.464^{-80}		1.209		1.113^{44}		
Trimethylamine	1.43	1.33					
1,3,5-Trimethylbenzene	1.47	1.41		1.36	1.30	1.24	1.18
2,2,4-Trimethylpentane				0.966^{38}		0.841^{77}	
Undecane				1.40	1.35	1.29	1.23
Water		5.610	5.983	6.071	6.435	6.668	6.791
m-Xylene				1.30	1.24	1.18	1.13
o-Xylene				1.31	1.26	1.20	1.14
p-Xylene				1.30	1.24	1.18	1.12

2.13 ENTHALPIES AND GIBBS ENERGIES OF FORMATION, ENTROPIES, AND HEAT CAPACITIES (CHANGE OF STATE)

The tables in this section contain values of the enthalpy and Gibbs energy of formation, entropy, and heat capacity at 298.15 K (25°C). No values are given in these tables for metal alloys or other solid solutions, for fused salts, or for substances of undefined chemical composition.

The physical state of each substance is indicated in the column headed "State" as crystalline solid (c), liquid (lq), or gaseous (g). Solutions in water are listed as aqueous (aq).

The values of the thermodynamic properties of the pure substances given in these tables are, for the substances in their standard states, defined as follows: For a pure solid or liquid, the standard state is the substance in the condensed phase under a pressure of 1 atm (101, 325 Pa). For a gas, the standard state is the hypothetical ideal gas at unit fugacity, in which state the enthalpy is that of the real gas at the same temperature and at zero pressure.

The values of $\Delta_f H°$ and $\Delta_f G°$ that are given in the tables represent the change in the appropriate thermodynamic quantity when one mole of the substance in its standard state is formed, isothermally at the indicated temperature, from the elements, each in its appropriate standard reference state. The standard reference state at 25°C for each element has been chosen to be the standard state that is thermodynamically stable at 25°C and 1 atm pressure. The standard reference states are indicated in the tables by the fact that the values of $\Delta_f H°$ and $\Delta_f G°$ are exactly zero.

The values of $S°$ represent the virtual or "thermal" entropy of the substance in the standard state at 298.15 K (25°C), omitting contributions from nuclear spins. Isotope mixing effects are also excluded except in the case of the ^{1}H—^{2}H system.

Solutions in water are designated as aqueous, and the concentration of the solution is expressed in terms of the number of moles of solvent associated with 1 mol of the solute. If no concentration is indicated, the solution is assumed to be dilute. The standard state for a solute in aqueous solution is taken as the hypothetical ideal solution of unit molality (indicated as std. state or ss). In this state the partial molal enthalpy and the heat capacity of the solute are the same as in the infinitely dilute real solution.

For some tables the uncertainty of entries is indicated within parentheses immediately following the value; viz., an entry 34.5(4) implies 34.5 ± 0.4 and an entry 34.5(12) implies 34.5 ± 1.2.

References: D. D. Wagman, et al., *The NBS Tables of Chemical Thermodynamic Properties*, in *J. Phys. Chem. Ref. Data*, **11: 2,** 1982; M. W. Chase, et al., *JANAF Thermochemical Tables*, 3rd ed., American Chemical Society and the American Institute of Physics, 1986 (supplements to JANAF appear in *J. Phys. Chem. Ref. Data*); Thermodynamic Research Center, *TRC Thermodynamic Tables*, Texas A&M University, College Station, Texas; I. Barin and O. Knacke, *Thermochemical Properties of Inorganic Substances*, Springer-Verlag, Berlin, 1973; J. B. Pedley, R. D. Naylor, and S. P. Kirby, *Thermochemical Data of Organic Compounds*, 2nd ed., Chapman and Hall, London, 1986; V. Majer and V. Svoboda, *Enthalpies of Vaporization of Organic Compounds*, International Union of Pure and Applied Chemistry, Chemical Data Series No. 32, Blackwell, Oxford, 1985.

2.13.1 THERMODYNAMIC RELATIONS

Enthalpy of Formation. Once standard enthalpies are assigned to the elements, it is possible to determine standard enthalpies for compounds. For the reaction:

$$\text{C(graphite)} + O_2(g) \rightarrow CO_2(g) \qquad \Delta H° = -393.51 \text{ kJ} \tag{6.1}$$

Since the elements are in their standard states, the enthalpy change for the reaction is equal to the standard enthalpy of CO_2 less the standard enthalpies of C and O_2, which are zero in each instance. Thus,

$$\Delta_f H° = -393.51 - 0 - 0 = -393.51 \text{ kJ} \tag{6.2}$$

Tables of enthalpies, such as Tables 2.53 and 1.56, can be used to determine the enthalpy for any reaction at 1 atm and 298.15 K involving the elements and any of the compounds appearing in the tables.

The solution of 1 mole of HCl gas in a large amount of water (infinitely dilute real solution) is represented by:

$$HCl(g) + \text{inf } H_2O \rightarrow H^+(aq) + Cl^-(aq) \tag{6.3}$$

The heat evolved in the reaction is $\Delta H° = -74.84$ kJ. With the value of $\Delta_f H°$ from Table 2.53, one has for the reaction:

$$\Delta_f H° = \Delta_f H°[H^+ (aq)] + \Delta_f H°[Cl^-(aq)] - \Delta_f H°[HCl(g)]$$

for the standard enthalpy of formation of the pair of ions H^+ and Cl^- in aqueous solution (standard state, $m = 1$). To obtain the $\Delta_f H°$ values for individual ions, the enthalpy of formation of H^+(aq) is arbitrarily assigned the value zero at 298.15 K. Thus, from Eq. (6.4):

$$\Delta_f H°[Cl^-(aq)] = -74.84 + (-92.31) = -167.15 \text{ kJ}$$

With similar data from Tables 2.53 and 1.56, the enthalpies of formation of other ions can be determined. Thus, from the $\Delta_f H°$[KCl(aq, std. state, $m = 1$ or aq, ss)] of −419.53 kJ and the foregoing value for $\Delta_f H°[Cl^-(aq, ss)]$:

$$\begin{aligned}\Delta_f H°[K^+(aq, ss)] &= \Delta_f H°[KCl(aq, ss)] - \Delta_f H°[Cl^-(aq,ss)]\\ &= -419.53 - (-167.15) = -252.38 \text{ kJ}\end{aligned}$$

Enthalpy of Vaporization (or Sublimation) When the pressure of the vapor in equilibrium with a liquid reaches 1 atm, the liquid boils and is completely converted to vapor on absorption of the enthalpy of vaporization ΔHv at the normal boiling point T_b. A rough empirical relationship between the normal boiling point and the enthalpy of vaporization (*Trouton's rule*) is:

$$\frac{\Delta Hv}{T_b} = 88 \text{ J} \cdot \text{mol}^{-1} \cdot \text{K}^{-1}$$

It is best applied to nonpolar liquids which form unassociated vapors.

To a first approximation, the enthalpy of sublimation ΔHs at constant temperature is:

$$\Delta Hs = \Delta Hm + \Delta Hv$$

where ΔHm is the enthalpy of melting.

The *Clapeyron* equation expresses the dynamic equilibrium existing between the vapor and the condensed phase of a pure substance:

$$\frac{dP}{dT} = \frac{\Delta Hv}{T\Delta V}$$

where ΔV is the volume increment between the vapor phase and the condensed phase. If the condensed phase is solid, the enthalpy increment is that of sublimation.

Substitution of $V = RT/P$ into the foregoing equation and rearranging gives the *Clausius-Clapeyron* equation,

$$\frac{dP}{p\,dT} = \frac{\Delta Hv}{RT^2}$$

or

$$\Delta Hv = -R\frac{d(\ln P)}{1/T}$$

which may be used for calculating the enthalpy of vaporization of any compound provided its boiling point at any pressure is known. If an Antoine equation is available, differentiation and insertion into the foregoing equation gives:

$$\Delta Hv = \frac{4.5757 T^2 B}{(T + C - 273.15)^2}$$

Inclusion of a compressibility factor into the foregoing equation, as suggested by the *Haggenmacher* equation improves the estimate of ΔHv:

$$\Delta Hv = \frac{RT^2}{P}\left(\frac{dP}{dT}\right)\left(1 - \frac{T_c^3 P}{T^3 P_c}\right)^{1/2}$$

where T_c and P_c are critical constants (Table 2.55). Although critical constants may be unknown, the compressibility factor is very nearly constant for all compounds belonging to the same family, and an estimate can be deduced from a related compound whose critical constants are available.

Heat Capacity (or Specific Heat) The temperature dependence of the heat capacity is complex. If the temperature range is restricted, the heat capacity of any phase may be represented adequately by an expression such as:

$$C_p = a + bT + cT^2$$

in which a, b, and c are empirical constants. These constants may be evaluated by taking three pieces of data: $(T_1, C_{p,1})$, $(T_2, C_{p,2})$, and $(T_3, C_{p,1})$, and substituting in the following expressions:

$$\frac{C_{p,1}}{(T_1 - T_2)(T_1 - T_3)} + \frac{C_{p,2}}{(T_2 - T_1)(T_2 - T_3)} + \frac{C_{p,3}}{(T_3 - T_2)(T_3 - T_1)} = c$$

$$\frac{C_{p,1} - C_{p,2}}{T_1 - T_2} - [(T_1 + T_2)c] = b$$

$$(C_{p,1} - bT_1) - cT_1^2 = a$$

Smoothed data presented at rounded temperatures, such as are available in Tables 2.54 and 1.57, plus the C_p° values at 298 K listed in Table 2.53, are especially suitable for substitution in the foregoing parabolic equations. The use of such a parabolic fit is appropriate for interpolation, but data extrapolated outside the original temperature range should not be sought.

Enthalpy of a System The enthalpy increment of a system over the interval of temperature from T_1 to T_2, under the constraint of constant pressure, is given by the expression:

$$H_2 - H_1 = \int_{T_1}^{T_2} C_p\, dT$$

The enthalpy over a temperature range that includes phase transitions, melting, and vaporization, is represented by:

$$H_2 - H_1 = \int_{T_1}^{T_2} C_p(\text{c,II})\, dT + \Delta Ht + \int_{T_1}^{T_m} C_p(c,\text{I})dT + \Delta Hm$$
$$+ \int_{T_m}^{T_b} C_p(\text{lq})\, dT + \Delta Hv + \int_{T_b}^{T_2} C_p(g)\, dT$$

Integration of heat capacities, as expressed by Eq. (6.13), leads to:

$$\Delta H = a(T_2 - T_1) + \frac{b\left(T_2^2 - T_1^2\right)}{2} + \frac{c\left(T_2^3 - T_1^3\right)}{3}$$

Entropy In the physical change of state,

$$\Delta Sm = \frac{\Delta Hm}{T_m}$$

is the entropy of melting (or fusion),

$$\Delta Sv = \frac{\Delta Hv}{T_b}$$

is the entropy of vaporization, and

$$\Delta Ss = \frac{\Delta Hs}{Ts}$$

is the entropy of sublimation

A general expression for the entropy of a system, involving any phase transitions, is

$$S_2 - S_1 = \int_{T_1}^{T_t} \frac{C_p(c,\mathrm{II})\,dT}{T} + \frac{\Delta Ht}{T} + \int_{T_b}^{T_m} \frac{C_p(c,\mathrm{I})\,dT}{T} + \frac{\Delta Hm}{T}$$
$$+ \int_{T_m}^{T_b} \frac{C_p(\mathrm{1q})\,dT}{T} + \frac{\Delta Hv}{T} + \int_{T_b}^{T_m} \frac{C_p(\mathrm{g})\,dT}{T}$$

If C_p is independent of temperature,

$$\Delta S = C_p(\ln\ T_2 - \ln T_1) = 2.303\ C_p \log \frac{T_2}{T_1}$$

If the heat capacities change with temperature, an empirical equation may be inserted in before integration. Usually the integration is performed graphically from a plot of either C_p/T versus T or C_p versus ln T.

TABLE 2.53 Enthalpies and Gibbs Energies of Formation, Entropies, and Heat Capacities of Organic Compounds

Substance	Physical state	$\Delta_f H°$ kJ · mol^{-1}	$\Delta_f G°$ kJ · mol^{-1}	$S°$ J · deg^{-1}·mol^{-1}	$C_p°$ J · deg^{-1} · mol^{-1}
Acenaphthene	c	70.34		188.9	190.4
Acenaphthylene	c	186.7			166.4
Acetaldehyde	lq	−192.2	−127.6	160.4	89.0
	g	−166.1	−133.0	263.8	55.3
Acetaldoxime	c	−77.9			
	lq	−81.6			
Acetamide	c	−317.0		115.0	91.3
Acetamidoguanidine nitrate	c	−494.0			
1-Acetamido-2-nitroguanidine	c	−193.6			
5-Acetamidotetrazole	c	−5.0			
Acetanilide	c	−210.6			
Acetic acid	lq	−484.4	−390.2	159.9	123.6
	g	−432.2	−374.2	283.5	63.4
ionized; std. state, $m = 1$	aq	−486.34	−369.65	86.7	−6.3
Acetic anhydride	lq	−624.4	−489.14	268.8	168.2[30]

(Continued)

TABLE 2.53 Enthalpies and Gibbs Energies of Formation, Entropies, and Heat Capacities of Organic Compounds (*Continued*)

Substance	Physical state	$\Delta_f H°$ kJ · mol^{-1}	$\Delta_f G°$ kJ · mol^{-1}	$S°$ J · deg^{-1}·mol^{-1}	$C_p°$ J · deg^{-1} · mol^{-1}
Acetone	lq	−248.4	−152.7	198.8	126.3
	g	−217.1	−152.7	295.3	74.5
Acetonitrile	lq	31.4	86.5	149.7	91.5
	g	74.0	91.9	243.4	52.2
Acetophenone	lq	−142.5	−17.0	249.6	204.6
Acetyl bromide	lq	−223.5			
Acetyl chloride	lq	−272.9	−208.2	201.0	117.0
	g	−242.8	−205.8	295.1	67.8
Acetylene	g	227.4	209.0	201.0	44.1
Acetylene-d_2	g	221.5	205.9	208.9	49.3
Acetylenedicarboxylic acid	c	−578.2			
Acetyl fluoride	g	−442.1			
1-Acetylimidazole	c	−574.0			
Acetyl iodide	lq	−163.5			
Acridine	c	179.4			
Adamantane	c	−194.1			
Adenine	c	96.0	299.6	151.1	147.0
(+)-Alanine	c	−561.2	−369.4	132.3	
(−)-Alanine	c	−604.0	−370.5	129.3	
(±)-Alanine	c	−563.6	−372.3	132.3	
β-Alanine	c	−558.0			
(±)-*N*-Alanylglycine	c	−777.8	−489.9	213.5	
(−)-Alanylglycine	c	−827.0	−533.0	195.2	
Allene	g	190.5			
Alloxan monohydrate	c	−1000.7	−762.3	186.7	
Allylamine	lq	−10.0			
Allyl *tert*-butyl sulfide	lq	−91.0			
Allyl ethyl sulfone	lq	−406.0			
Allyl methyl sulfone	lq	−385.1			
Allyl trichloroacetate	lq	−395.3			
Allyl (*see* Propene)					
Aminetrimethylboron	c	−284.1	−79.3	218.0	
3-Aminoacetophenone	c	−173.3			
4-Aminoacetophenone	c	−182.1			
2-Aminoacridine	c	166.4			
9-Aminoacridine	c	159.2			
2-Aminobenzoic acid	c	−400.9			
3-Aminobenzoic acid	c	−411.6			
4-Aminobenzoic acid	c	−412.9			
2-Aminobiphenyl	c	112.2			
4-Aminobiphenyl	c	81.2			
4-Aminobutanoic acid	c	−581.0			
2-Aminoethanesulfonic acid	c	−785.9	−562.3	154.1	140.7
ionized; std. state, $m = 1$	aq	−719.8	−509.8	200.1	
2-Aminoethanol	lq				195.5
2-Aminohexanoic acid (norleucine)	c	−639.1			
4-Aminohexanoic acid	c	−646.2			
5-Aminohexanoic acid	c	−643.3			
6-Aminohexanoic acid	c	−639.1			
(−)-2-Amino-3-hydroxy-butanoic acid	c	−759.5			

TABLE 2.53 Enthalpies and Gibbs Energies of Formation, Entropies, and Heat Capacities of Organic Compounds (*Continued*)

Substance	Physical state	$\Delta_f H°$ kJ · mol^{-1}	$\Delta_f G°$ kJ · mol^{-1}	$S°$ J · deg^{-1}·mol^{-1}	$C_p°$ J · deg^{-1} · mol^{-1}
2-Amino-2-(hydroxymethyl)-1,1-propanediol	c	717.8			
3-Aminonitroguanidine	c	22.1			
5-Aminopentanoic acid	c	−604.1			
5-Aminotetrazole	c	−207.8			
3-Amino-1,2,4-triazole	c	76.8			
Aniline	lq	31.3	149.2	191.4	191.9
	g	87.5	−7.0	317.9	107.9
Anthracene	c	129.2	286.0	207.6	210.5
9,10-Anthraquinone	c	−207.5			
D-(−)-Arabinose [also (+)-]	c	−1057.9			
(+)-Arginine	c	−623.5	−240.5	250.8	232.0
L-(+)-Ascorbic acid	c	−1164.6			
L-(+)-Asparagine	c	−789.4	−530.6	174.6	
L-(+)-Aspartic acid	c	−973.3	−730.7	170.2	
cis-Azobenzene	c	310.2			
trans-Azobenzene	c	365.2			
Azoisopropane	g	35.8			
Azomethane	g	148.8	239.7	289.9	78.0
Azomethane-d_6	g	119.3	218.3	305.7	90.6
Azopropane	g	51.5			
Azulene	g	289.1	353.4	338.1	128.5
Barbituric acid	c	−637.2			
Benzaldehyde	lq	−87.0	9.4		172.0
Benzamide	c	−202.6			
Benzanilide	c	−93.4			
1,2-Benzanthracene	c	170.9			
2,3-Benzanthracene	c	160.4	359.2	215.5	
1,2-Benzanthracene-9,10-dione	c	−231.9			
Benzene	lq	49.0	124.4	173.4	136.0
	g	82.6	129.7	269.2	82.4
Benzeneboronic acid	c	−720.1			
1,2-Benzenediamine	c	−0.3			
1,3-Benzenediamine	c	−7.8			
1,4-Benzenediamine	c	3.1			
1,3-Benzenedicarboxylic acid	c	803.0			
1,4-Benzenedicarboxylic acid	c	816.1			
1,2,4,5-Benzenetetra-carboxylic acid	c	1571.0			
Benzenethiol (thiophenol)	lq	63.7	134.0	222.8	173.2
	g	111.3	147.6	336.9	104.9
1,2,3-Benzenetricarboxylic acid	c	−1160.0			
1,2,4-Benzenetricarboxylic acid	c	−1179.0			
1,3,5-Benzenetricarboxylic acid	c	−1190.0			
1,2,3-Benzenetriol	c	−551.1			
1,2,4-Benzenetriol	c	−563.8			
1,3,5-Benzenetriol	c	−584.6			
p-Benzidine	c	70.7			
Benzil	c	−153.9			
Benzoic acid	c	−385.2	−245.3	167.6	146.8
Benzoic anhydride	c	−415.4			

(*Continued*)

TABLE 2.53 Enthalpies and Gibbs Energies of Formation, Entropies, and Heat Capacities of Organic Compounds (*Continued*)

Substance	Physical state	$\Delta_f H°$ kJ · mol^{-1}	$\Delta_f G°$ kJ · mol^{-1}	$S°$ J · deg^{-1}·mol^{-1}	$C_p°$ J · deg^{-1} · mol^{-1}
Benzonitrile	lq	163.2		209.1	165.2
	g	215.8	260.8	321.0	109.1
Benzo[*def*]phenanthrene	c	125.5	269.5	224.8	236.0
Benzophenone	c	−34.5	140.2	245.2	224.8
Benzo[*f*]quinoline	c	150.6			
Benzo[*h*]quinoline	c	149.7			
1,4-Benzoquinone	c	−185.7	−83.6	162.8	129.0
Benzo[*b*]thiophene	c	100.6			
1,2,3-Benzotriazole	c	250.0			
Benzotrifluoride	lq	−636.7			
Benzoyl bromide	lq	−107.3			
Benzoyl chloride	lq	−158.0			
Benzoylformic acid	c	−482.4			
N-Benzoylglycine	c	−609.8	−369.57	239.3	
Benzoyl iodide	lq	−53.5			
3,4-Benzphenanthrene	c	184.9			
Benzylamine	lq	34.2			
Benzyl alcohol	lq	−160.7	−27.5	216.7	218.0
Benzyl bromide	lq	16.0			
Benzyl chloride	lq	−32.6			182.4
N-Benzyldiphenylamine	c	184.7			
Benzyl ethyl sulfide	lq	−4.9			
Benzyl iodide	lq	57.3			
Benzyl methyl ketone	lq	−151.9			
Benzyl methyl sulfide	lq	26.2			
Bicyclo[1.1.0]butane	g	217.1			
Bicyclo[2.2.1]hepta-2,5-dione	lq	213.0			
Bicyclo[2.2.1]heptane	c	−95.1			
Bicyclo[4.1.0]heptane	lq	−36.7			
Bicyclo[2.2.1]heptene	lq	90.0	203.9		130.0
Bicyclo[3.1.0]hexane	g	38.6			
Bicyclohexyl	lq	−273.7			
Bicyclo[2.2.2]octane	c	−146.9			
Bicyclo[4.2.0]octane	g	−26.2			
Bicyclo[5.1.0]octane	g	−16.6			
Bicyclo[2.2.2]oct-2-ene	g	−23.3			
Bicyclopropyl	g	129.3			
Biphenyl	c	99.4	254.2	209.4	198.4
2-Biphenylcarboxylic acid	c	−349.0			
(1,1′-Biphenyl)-4,4′-diamine	c	70.7			
Biphenylene	c	334.0			
Bis(2-chloroethyl) ether	lq				220.9
Bis(dimethylthiocarbonyl) disulfide	c	41.6			
Bis(2-hydroxyethyl) ether	lq	−1621.0		441.0	135.1
	g	−571.1			
Bromoacetone	g	−181.0			
Bromoacetylene	g			253.7	55.7
Bromobenzene	lq	60.9	126.0	219.2	154.3
4-Bromobenzoic acid	c	−378.3			

TABLE 2.53 Enthalpies and Gibbs Energies of Formation, Entropies, and Heat Capacities of Organic Compounds (*Continued*)

Substance	Physical state	$\Delta_f H°$ kJ · mol^{-1}	$\Delta_f G°$ kJ · mol^{-1}	$S°$ J · deg^{-1}·mol^{-1}	$C_p°$ J · deg^{-1} · mol^{-1}
1-Bromobutane	lq	−143.8	−12.9	369.8	109.3
2-Bromobutane	lq	−154.8	−19.25		
	g	−120.3	−25.8	370.3	110.8
Bromochlorodifluoromethane	g	−471.5	−448.4	318.5	74.6
1-Bromo-2-chloroethane	lq				130.1[27]
Bromochlorofluoromethane	g	−295.0	−278.6	304.3	63.2
Bromochloromethane	lq				52.7
	g	−50.2	−39.3	287.6	
1-Bromo-2-chloro-1,1,2-trifluoroethane	g	−644.8			
2-Bromo-2-chloro-1,1,1-trifluoroethane	g	−690.4			
1-Bromodecane	lq	−344.7			
Bromodichlorofluoromethane	g	−269.5	−246.8	330.6	80.0
Bromodichloromethane	g	−58.6	−42.5	316.4	67.4
Bromodifluoromethane	g	−424.9	−447.3	295.1	58.7
Bromoethane	lq	−90.5	−25.8	198.7	100.8
	g	−61.9	−23.9	286.7	64.5
Bromoethylene (vinyl bromide)	lq				107.7[15]
	g	79.2	81.7	275.8	55.4
Bromofluoromethane	g	−252.7	−241.5	276.3	49.2
1-Bromoheptane	lq	−218.4			
1-Bromohexane	lq	−194.2		453.0	203.5
Bromoiodomethane	g	50.2	39.2	307.5	
Bromomethane	lq				78.7[7]
	g	−35.4	−26.3	246.4	42.5
2-Bromo-2-methylpropane	lq	−163.8			151.0
	g	−132.4	−28.2	332.0	116.5
1-Bromooctane	lq	−245.1			
Bromopentafluoroethane	g	−1064.4			
1-Bromopentane	lq	−170.2			132.2
	g	−129.0	−5.7	408.8	
1-Bromopropane	lq	−121.8			86.4
	g	−87.0	−22.5	330.9	
2-Bromopropane	lq	−130.5			132.2
	g	−99.4	−27.2	316.2	89.4
cis-1-Bromopropene	g	40.8			
3-Bromopropene	g	45.2			
N-Bromosuccinimide	c	−335.9			
α-Bromotoluene	lq	23.4			
Bromotrichloromethane	g	−41.1	−12.4	332.8	85.3
Bromotrifluoroethane	g	−694.5			
Bromotrifluoromethane	g	−648.3	−622.6	297.8(5)	69.3
Bromotrimethylsilane	lq	−325.9			
Bromotrinitromethane	g	80.3			
Brucine	c	−496.2			
1,2-Butadiene	g	162.3	199.5	293.0	80.1
1,3-Butadiene	lq	88.5		199.0	123.6
	g	110.0	150.7	278.7	79.5
1,3-Butadiyne	g	472.8	444.0	250.0	73.6
Butanal	lq	−239.2			163.7

(*Continued*)

TABLE 2.53 Enthalpies and Gibbs Energies of Formation, Entropies, and Heat Capacities of Organic Compounds (*Continued*)

Substance	Physical state	$\Delta_f H°$ kJ · mol^{-1}	$\Delta_f G°$ kJ · mol^{-1}	$S°$ J · deg^{-1}·mol^{-1}	$C_p°$ J · deg^{-1} · mol^{-1}
	g	−204.9	−114.8	243.7	103.4
Butanamide	lq	−346.9			
Butane	lq				$104.5^{-0.5}$
	g	−125.6	−17.2	310.1	97.5
1,2-Butanediamine	lq	−120.2			
(±)-1,2-Butanediol	lq	−523.6			
1,3-Butanediol	lq	−501.0			227.2^{30}
1,4-Butanediol	lq	−503.3		223.4	200.1
2,3-Butanediol	lq	−541.5			213.0
Butanedinitrile	c	139.7			
	lq				160.5^{62}
2,3-Butanedione	lq	−365.8			
1,4-Butanedithiol	lq	−105.7			
Butanenitrile	lq	−5.8			159^{67}
	g	33.6	108.7	325.4	97.0
1-Butanethiol	lq	−124.7	4.1	276.0	171.2
2-Butanethiol	lq	−131.0	−0.17	271.4	
Butanoic acid	lq	−533.8	−377.7	222.2	178.6
Butanoic anhydride	lq				283.7
1-Butanol	lq	−327.3	−162.5	225.8	177.0
	g	−275.0	−150.8	362.8	122.6
(±)-2-Butanol	lq	−342.6	−177.0	214.9	196.9
	g	−292.9	−167.6	359.5	113.3
2-Butanone	lq	−273.3	−151.4	239.1	158.9
	g	−238.5		339.9	101.7
Butanophenone	lq	−188.9			
trans-2-Butenal	lq	−138.7			95.4
cis-Butenedinitrile	c	268.2			
1-Butene	lq	−20.8		227.0	118.0
	g	0.1	71.3	305.6	85.7
cis-2-Butene	lq	−29.8		219.9	127.0
	g	−7.1	65.9	300.8	78.9
trans-2-Butene	g	−11.4	63.0	296.5	87.8
cis-2-Butenenitrile	lq	95.1			
trans-2-Butenenitrile	lq	95.1			
3-Butenenitrile	g	159.7	193.4	298.4	82.1
cis-2-Butenoic acid	lq	−347.0			
trans-2-Butenoic acid	c	−430.5			
cis-2-Butenedioic acid	c	−788.7			
trans-2-Butenedioic acid	c	−811.1			
1-Buten-3-yne	g	304.6	306.0	279.4	73.2
2-Butoxyethanol	lq				281.0
N-Butylacetamide	lq	−380.8			
Butyl acetate	lq	−529.2			227.8
Butylamine	lq	−127.7			179.2
	g	−92.0	49.2	363.3	118.6
sec-Butylamine	lq	−137.5			
	g	−104.6	40.7	351.3	117.2
tert-Butylamine	g	−150.6			192.1
	g	−121.0	28.9	337.9	120.0
Butylbenzene	lq	63.2			243.4
	g	−13.1	144.7	439.5	416.3

TABLE 2.53 Enthalpies and Gibbs Energies of Formation, Entropies, and Heat Capacities of Organic Compounds (*Continued*)

Substance	Physical state	$\Delta_f H°$ kJ · mol^{-1}	$\Delta_f G°$ kJ · mol^{-1}	$S°$ J · deg^{-1}·mol^{-1}	$C_p°$ J · deg^{-1} · mol^{-1}
sec-Butylbenzene	lq	−66.4			
tert-Butylbenzene	lq	−70.7			238.0
sec-Butyl butanoate	lq	−492.6			
Butyl chloroacetate	lq	−538.4			
Butyl 2-chlorobutanoate	lq	−655.2			
Butyl 3-chlorobutanoate	lq	−610.9			
Butyl 4-chlorobutanoate	lq	−618.0			
Butyl 2-chloropropanoate	lq	−572.0			
Butyl 3-chloropropanoate	lq	−558.2			
Butyl crotonate	lq	−467.8			
Butylcyclohexane	lq	−263.1		345.0	271.0
	g	−213.4	56.4	458.5	207.1
Butylcyclopentane	g	−168.3	61.4	456.2	177.5
Butyl dichloroacetate	lq	−550.2			
Butyl ethyl ether	lq				159.0
Butyl ethyl sulfide (3-thiaheptane)	g	−125.2	32.0	453.0	162.0
tert-Butyl ethyl sulfide	lq	−187.3			
Butyl formate	lq				200.2
tert-Butyl hydroperoxide	lq	−293.6			
Butyllithium	lq	−132.2			
Butyl methyl ether	lq	−290.6		295.3	192.7
tert-Butyl methyl ether	lq	−313.6		265.3	187.5
Butyl methyl sulfide (2-thiahexane)	lq	−142.8	17.1	307.5	200.9
tert-Butyl methyl sulfide	lq	−156.9		276.1	199.9
Butyl methyl sulfone	lq	−535.8			
tert-Butyl methyl sulfone	c	−556.0			
cis-Butyl 9-octadecanoate	lq	−816.9			
tert-Butyl peroxide	lq	−380.9			
Butyl trichloroacetate	lq	−545.8			
Butylurea	c	−419.5			
Butyl vinyl ether	lq	−218.8			232.0
1-Butyne	g	165.2	202.1	290.8	81.4
2-Butyne	g	145.7	185.4	283.3	78.0
2-Butynedinitrile	g	529.2			
2-Butynedioic acid	c	−577.4			
3-Butynoic acid	c	−241.8			
γ-Butyrolactone	lq	−420.9			141.4
(+)-Camphor	c	−319.4			271.2
ϵ-Caprolactam	c	−329.4			
9*H*-Carbazole	c	101.7			
Carbonyl bromide	g	−96.2	−110.9	309.1	61.8
Carbonyl chloride	g	−219.1	−204.9	283.5	57.7
Carbonyl chloride fluoride	g			276.7	52.4
Carbonyl fluoride	g	−639.8			46.8
Chloroacetamide	c	−338.5			
Chloroacetic acid	c	−510.5			
Chloroacetyl chloride	lq	−283.7			
Chloroacetylene	g			242.0	54.3
2-Chlorobenzaldehyde	lq	−118.4			

(*Continued*)

TABLE 2.53 Enthalpies and Gibbs Energies of Formation, Entropies, and Heat Capacities of Organic Compounds (*Continued*)

Substance	Physical state	$\Delta_f H°$ kJ · mol^{-1}	$\Delta_f G°$ kJ · mol^{-1}	$S°$ J · deg^{-1}·mol^{-1}	$C_p°$ J · deg^{-1} · mol^{-1}
3-Chlorobenzaldehyde	lq	−126.0			
4-Chlorobenzaldehyde	c	−146.4			
Chlorobenzene	lq	11.0	89.2	209.2	150.2
2-Chlorobenzoic acid	c	−404.5			
3-Chlorobenzoic acid	c	−423.3			
4-Chlorobenzoic acid	c	−428.9			163.2
Chloro-1,4-benzoquinone	c	−220.6			
1-Chlorobutane	lq	−188.1			175.0
	g	−154.6	−38.8	358.1	107.6
(±)-2-Chlorobutane	lq	−192.8			
	g	−161.2	−53.5	359.6	108.5
2-Chlorobutanoic acid	lq	−575.5			
3-Chlorobutanoic acid	lq	−556.3			
4-Chlorobutanoic acid	lq	−566.3			
Chlorocyclohexane	lq	−207.2			
1-Chloro-1,1-difluoroethane	lq				130.5^{21}
	g			307.2	82.5
1-Chloro-2,2-difluoroethylene	g	−315.5	−289.1	303.0	72.1
2-Chloro-1,1-difluoroethylene	g	−331.4	−305.0	302.4	
Chlorodifluoromethane	lq				93.0^{-41}
	g	−482.6	−450.0	281.0	55.9
2-Chloro-1,4-dihydroxybenzene	c	−382.81			
Chlorodimethylsilane	lq	−79.8			
1-Chloro-2,3-epoxypropane	lq	−148.5			125.1
1-Chloroethane	lq	−136.8	−59.3	190.8	104.3
	g	−112.1	−60.5	275.8	62.6
2-Chloroethanol	lq	−295.4			
1-Chloro-2-ethylbenzene	lq	−54.1			
1-Chloro-4-ethylbenzene	lq	−51.7			
Chloroethylene (vinyl chloride)	lq				89.4
	g	37.3	53.6	263.9	53.7
2-Chloroethyl ethyl ether	g	−301.3			
2-Chloroethyl vinyl ether	g	−170.1			
Chloroethyne	g	213.0	197.0	241.9	54.3
1-Chloro-1-fluoroethane	g	−313.4			
2-Chlorohexane	lq	−246.1			
Chlorofluoromethane	g	−290.8	−265.5	264.3	47.0
Chlorohydroquinone	c	−382.8			
Chloroiodomethane	g	12.6	15.4	296.1	
Chloromethane	lq				75.6^{-24}
	g	−81.9	−58.5	234.6	40.8
1-Chloro-3-methylbutane	lq	−216.0			175.1
	g	−179.7			
2-Chloro-2-methylbutane	g	−202.2			
2-Chloro-3-methylbutane	g	−185.1			
1-Chloro-2-methylpropane	lq	−191.1			158.6
	g	−159.4	−49.7	355.0	108.5
2-Chloro-2-methylpropane	lq	−211.2			172.8
	g	−182.2	−64.1	322.2	114.2
1-Chloronaphthalene	lq	54.6			212.6
2-Chloronaphthalene	c	55.2			
1-Chlorooctane	lq	−291.3			198.5

TABLE 2.53 Enthalpies and Gibbs Energies of Formation, Entropies, and Heat Capacities of Organic Compounds (*Continued*)

Substance	Physical state	$\Delta_f H°$ kJ · mol^{-1}	$\Delta_f G°$ kJ · mol^{-1}	$S°$ J · deg^{-1}·mol^{-1}	$C_p°$ J · deg^{-1} · mol^{-1}
Chloropentafluoroacetone	g	−1121.0			
Chloropentafluoroethane	lq				184.2
	g	−1188.8			
1-Chloropentane	lq	−213.2			
	g	−175.0	−37.4	397.0	130.5
3-Chlorophenol	c	−206.4			
4-Chlorophenol	c	−197.9			
1-Chloropropane	lq	−160.6			132.2
	g	−131.9	−50.7	319.1	84.6
2-Chloropropane	lq	−172.1			
	g	−144.9	−62.5	304.2	87.3
2-Chloro-1,3-propanediol	lq	−517.5			
3-Chloro-1,2-propanediol	lq	−525.3			
2-Chloropropanoic acid	lq	−522.5			131.6
3-Chloropropanoic acid	c	−549.3			
2-Chloro-1-propene	g	−21.0			
3-Chloro-1-propene (allyl chloride)	lq				125.1
	g	−0.63	43.6	306.7	75.4
N-Chlorosuccinimide	c	−358.1			
α-Chlorotoluene	lq	−32.6			
o-Chlorotoluene	lq				166.8
2-Chloro-1,1,1-trifluoroethane	g			326.4	154.6
Chlorotrifluoroethylene	g	−505.5	−523.8	322.1	83.9
Chlorotrifluoromethane	g	−707.8	−667.4	285.4	66.9
Chlorotrimethylsilane	lq	−384.1			
Chlorotrinitromethane	lq	−27.1			
	g	18.4			
Chrysene	c	145.3			
(−)-Cinchonidine	c	29.7			
Cinchonine	c	31.0			
cis-Cinnamic acid	c	−315.0			
trans-Cinnamic acid	c	−338.5			
Cinnamic anhydride	c	−347.7			
Citric acid	c	−1543.9	−1236.4	166.2	
Codeine monohydrate	c	−632.6			
Creatine	c	−537.2			
o-Cresol	c	−204.6		165.4	154.6
	lq				233.6^{40}
	g	−128.6	37.1	357.6	130.3
m-Cresol	lq	−194.0		212.6	224.9
	g	−132.3	−40.5	356.8	122.5
p-Cresol	c	−199.3		167.3	150.2
	lq				221.0^{40}
	g	−125.4	−30.9	347.6	124.5
Cuban	c	541.3			
Cyanamide	c	58.8			
Cyanide (CN)	g	437.6	407.5	202.6	29.2
Cyanogen	g	306.7	297.2	241.9	56.9
Cyanogen bromide	g	140.5	165.3	248.3	46.9
Cyanogen chloride	g	138.0	131.0	236.2	45.0

(*Continued*)

TABLE 2.53 Enthalpies and Gibbs Energies of Formation, Entropies, and Heat Capacities of Organic Compounds (*Continued*)

Substance	Physical state	$\Delta_f H°$ kJ · mol^{-1}	$\Delta_f G°$ kJ · mol^{-1}	$S°$ J · deg^{-1}·mol^{-1}	$C_p°$ J · deg^{-1} · mol^{-1}
Cyanogen fluoride	g	−639.8		224.7	41.8
Cyanogen iodide	c	166.2	185.0	96.2	
	g	205.5	196.6	256.8	48.3
Cyclobutane	g	27.7	110.0	265.4	72.2
Cyclobutanecarbonitrile	lq	103.0			
Cyclobutene	g	156.7	174.7	263.5	67.1
Cyclobutylamine	g	41.2			
Cyclododecane	c	−306.6			
1,3-Cycloheptadiene	g	94.3			
Cycloheptane	lq	−156.6	54.1	242.6	123.1
Cycloheptanone	lq	−299.4			
1,3,5-Cycloheptatriene	lq	142.2	243.1	214.6	162.8
Cycloheptene	g	−9.2			
Cyclohexane	lq	−156.4	26.7	204.4	154.9
	g	−123.4	31.8	298.3	106.3
cis-Cyclohexane-1,2-dicarboxylic acid	c	−961.1			
trans-Cyclohexane-1,2-dicarboxylic acid	c	−970.7			
Cyclohexanethiol	lq	−140.7		255.6	192.6
	g	−96.1			
Cyclohexanol	lq	−348.1	−133.3	199.6	208.2
Cyclohexanone	lq	−271.2		255.6	182.2
	g	−226.1	−90.8	322.2	109.7
Cyclohexene	lq	−38.5	101.6	214.6	148.3
1-Cyclohexenylmethanol	lq	−382.4			
Cyclohexylamine	lq	−147.7			
Cyclohexylbenzene	lq	−76.6			261.3
Cyclohexylcyclohexane	lq	−329.3			
Cyclooctane	lq	−167.7			
Cyclooctanone	lq	−326.0			
1,3,5,7-Cyclooctatetraene	lq	254.5	358.6	220.3	184.0
Cyclooctene	lq	−74.0			
1,3-Cyclopentadiene	g	134.3	179.3	267.8	
Cyclopentane	lq	−105.1	36.4	204.3	128.9
	g	−76.4	38.6	292.9	83.0
cis-1,2-Cyclopentanediol	c	−484.9			
trans-1,2-Cyclopentanediol	c	−489.9			
Cyclopentanethiol	lq	−89.5	46.8	256.9	165.2
Cyclopentanol	lq	−300.1	−127.8	206.3	184.1
Cyclopentanone	lq	−235.7			154.5
Cyclopentene	lq	4.4	108.5	201.3	122.4
	g	34.0	110.8	291.8	75.1
1-Cyclopentenylmethanol	lq	34.3			
Cyclopentylamine	lq	−95.1		241.0	181.2
Cyclopropane	g	53.3	104.4	237.4	55.6
Cyclopropanecarbonitrile	g	182.8			
Cyclopropene	g	277.1	286.3	223.3	
Cyclopropylamine	lq	45.8		187.7	147.1
	g	77.0			
Cyclopropylbenzene	lq	100.3			
(−)-Cysteine	c	−534.1			

TABLE 2.53 Enthalpies and Gibbs Energies of Formation, Entropies, and Heat Capacities of Organic Compounds (*Continued*)

Substance	Physical state	$\Delta_f H°$ kJ · mol^{-1}	$\Delta_f G°$ kJ · mol^{-1}	$S°$ J · deg^{-1}·mol^{-1}	$C_p°$ J · deg^{-1} · mol^{-1}
(−)-Cystine	c	−1032.7			
Cytosine	c	−221.3		132.6	
Decafluorobutane	lq				127.2^{20}
cis-Decahydronaphthalene	lq	−219.4	68.9	265.0	232.0
trans-Decahydronaphthalene	lq	−230.6	57.7	265.0	228.5
Decanal	g	−330.9	−66.5	578.6	239.7
Decane	lq	−300.9	17.5	425.5	314.4
Decanedioic acid	c	−1082.8			
1,10-Decanediol	c	−693.5			
1-Decanenitrile	lq	−158.4			
1-Decanethiol	lq	−276.5		476.1	350.4
	g	−211.5	61.4	610.1	255.6
Decanoic acid	c	−713.7			
1-Decanol	lq	−478.1	−132.2	430.5	370.6
1-Decene	lq	−173.8	105.0	425.0	300.8
1-Decyne	g	41.2	252.2	524.5	219.7
Deoxybenzoin	c	−71.0			
Diacetamide	c	−489.0			
Diacetyl peroxide	lq	−535.3			
1,2-Diallyl phthalate	lq	−550.6			
2,2′-Diaminodiethylamine	lq				254^{40}
2,6-Diaminopyridine	c	−6.5			
Diazomethane	g	192.5	217.8	242.8	52.5
Dibenz[*de,kl*]anthracene	c	182.8			
1,2-Dibenzoylethane	c	−255.6			
trans-1,2-Dibenzoylethylene	c	−114.7	109.8	319.2	
Dibenzoylmethane	c	−223.5			
Dibenzoyl peroxide	c	−369.6			
Dibenzyl	c	44.1	260.0	269.4	255.2
Dibenzyl sulfide	c	99.0			
Dibenzyl sulfone	c	−282.6			
1,2-Dibromobutane	g	−91.5	−13.1	408.8	127.1
1,3-Dibromobutane	lq	−148.0			
1,4-Dibromobutane	g	−87.8			
2,3-Dibromobutane	g	−102.0			
Dibromochlorofluoromethane	g	−231.8	−223.4	342.8	82.4
Dibromochloromethane	g	−20.9	−18.8	327.7	69.2
1,2-Dibromo-1-chloro-1,2,2-trifluoroethane	lq	−691.7			
	g	−656.6			
1,2-Dibromocycloheptane	lq	−157.6			
1,2-Dibromocyclohexane	lq	−162.8			
1,2-Dibromocyclooctane	lq	−173.3			
Dibromodifluoroethane	g	−36.9		327.7	80.8
Dibromodichloromethane	g	−29.3	−19.5	347.8	87.1
Dibromodifluoromethane	g	−429.7	−419.1	325.3	77.0
1,1-Dibromoethane	lq	−66.2			
1,2-Dibromoethane	lq	−79.2	−20.9	223.3	136.0
	g	−37.5			
cis-1,2-Dibromoethylene	g			313.3	68.8
trans-1,2-Dibromoethylene	g			313.5	70.3
Dibromofluoromethane	g	−223.4	−221.1	316.8	65.1

(*Continued*)

TABLE 2.53 Enthalpies and Gibbs Energies of Formation, Entropies, and Heat Capacities of Organic Compounds (*Continued*)

Substance	Physical state	$\Delta_f H°$ kJ · mol⁻¹	$\Delta_f G°$ kJ · mol⁻¹	$S°$ J · deg⁻¹·mol⁻¹	$C_p°$ J · deg⁻¹ · mol⁻¹
Dibromomethane	lq				105.3
	g	−14.8	−16.2	293.2	54.7
1,3-Dibromo-2-methylpropane	g	−137.6			
1,3-Dibromotetrafluoroethane	lq	−817.7			
	g	−789.1			
1,2-Dibromopropane	lq				160.0
	g	−71.5	−17.7	376.1	102.8
1,2-Dibromotetrafluoroethane	lq				180.3
Dibutoxymethane	lq	−549.4			
Dibutylamine	lq	−206.0			292.9
Dibutyl disulfide	g	−160.6	53.9	572.8	231.1
Di-*tert*-butyl disulfide	lq	−255.2			
Dibutyl ether	lq	−377.9			278.2
	g	−332.8	−88.5	500.4	204.0
Di-*sec*-butyl ether	lq	−401.5			
	g	−360.9			
Di-*tert*-butyl ether	lq	−399.6			276.1
	g	−362.0			
Dibutylmercury	lq	−97.9			
Dibutyl peroxide	lq	−380.7			
Dibutyl 1,2-phthalate	c	−842.6			498.0
Dibutyl sulfate	lq	−904.6			
Dibutyl sulfide	lq	−220.7	32.2	405.1	284.3
Di-*tert*-butyl sulfide	lq	−232.4			
Dibutyl sulfite	lq	−693.1			
Dibutyl sulfone	c	−610.2			
Dichloroacetic acid	lq	−496.3			
ionized	aq	−507.1			
Dichloroacetyl chloride	lq	−280.4			
1,2-Dichlorobenzene	lq	−17.5			162.4
	g	30.2	82.7	341.5	113.5
1,3-Dichlorobenzene	lq	−20.7			171
	g	25.7	78.6	343.5	113.8
1,4-Dichlorobenzene	c	−42.3			
	lq			175.4	147.8
	g	22.5	77.2	336.7	113.9
Dichlorodifluoromethane	lq				117.2
	g	−477.4	−439.4	300.8	72.3
1,3-Dichlorobutane	g	−195.0			
1,4-Dichlorobutane	g	−183.4			
Dichlorodimethylsilane	g	−461.1		335.4	101.1
Dichlorodiphenylsilane	lq	−278.2			
1,1-Dichloroethane	lq	−158.4			126.3
	g	−127.7	−73.8	305.1	76.2
1,2-Dichloroethane	lq	−167.4			128.4
	g	−126.4	−73.9	308.4	78.7
1,1-Dichloroethylene	lq	−23.9			111.3
	g	2.8	25.4	289.1	67.0
cis-1,2-Dichloroethylene	g	4.6	24.4	289.5	65.1
trans-1,2-Dichloroethylene	lq	−23.1			116.8
	g	5.0	28.6	289.9	66.7
Dichlorofluoromethane	g	−283.0	−253.0	293.1	61.0

TABLE 2.53 Enthalpies and Gibbs Energies of Formation, Entropies, and Heat Capacities of Organic Compounds (*Continued*)

Substance	Physical state	$\Delta_f H°$ kJ · mol^{-1}	$\Delta_f G°$ kJ · mol^{-1}	$S°$ J · deg^{-1}·mol^{-1}	$C_p°$ J · deg^{-1} · mol^{-1}
1,1-Dichloro-1-fluoroethane	g			320.2	88.7
1,1-Dichlorofluoroethylene	g			313.9	76.5
1,1-Dichlorofluoromethane	lq				112.6
Dichloromethane	lq	−124.2		177.8	101.2
	g	−95.4	−68.9	270.3	51.0
Dichloropentadienyliron	c	141.0			
1,2-Dichloropropane	lq	−198.8			
	g	−162.8	−83.1	354.8	98.2
1,3-Dichloropropane	g	−159.2	−82.6	367.2	99.6
2,2-Dichloropropane	g	−173.2	−84.6	326.0	105.9
1,3-Dichloro-2-propanol	lq	−385.4			
2,3-Dichloro-1-propanol	lq	−381.3			
2,3-Dichloropropene	lq	−73.3			
1,2-Dichlorotetrafluoromethane	lq				164.2
	g	−916.3			
2,2-Dichlorotetrafluoroethane	lq	−960.2			111.7
2,2-Dichloro-1,1,1-trifluoroethane	g			352.8	102.5
Dicyanoacetylene	lq	500.4			
Dicyanobenzene	c	275.4			
1,4-Dicyanobutane	lq	85.1			128.7
1,4-Dicyano-2-butyne	c	366.5			
Dicyanodiamide	c	22.6	179.5	129.3	118.8
Dicyclopentadiene	c	116.7			
Diethanolamine	c	−493.8			
	lq				233.5[30]
1,1-Diethoxyethane	lq	−491.4			238.0
1,2-Diethoxyethane	lq	−451.4			259.4
Diethoxymethane	lq	−450.4			
1,3-Diethoxypropane	lq	−482.1			
2,2-Diethoxypropane	lq	−538.5			
Diethylamine	lq	−103.7			169.2
	g	−72.2	72.1	352.2	115.7
Diethylamine hydrochloride	c	−358.6			
Diethylbarbituric acid (veronal)	c	−747.7			
1,2-Diethylbenzene	g	−19.0	141.1	434.3	182.6
1,3-Diethylbenzene	g	−21.8	136.7	439.3	176.9
1,4-Diethylbenzene	g	−22.3	137.9	434.0	176.2
Diethyl carbonate	lq	−681.5			212.4
cis-1,2-Diethylcyclopropane	lq	−79.9			
trans-1,2-Diethylcyclopropane	lq	83.3			
Diethyl disulfide	lq	−120.0	9.5	269.3	171.4
	g	−79.4	22.3	414.5	141.3
Diethylenediamine	c	−13.4	240.2	85.8	
Diethylene glycol	lq	−628.5			244.8
	g	−571.1		441.0	135.1
Diethylene glycol dibutyl ether	lq				452[20]
Diethylene glycol diethyl ether	lq				341.4[15]
Diethylene glycol dimethyl ether	lq				274.1

(*Continued*)

TABLE 2.53 Enthalpies and Gibbs Energies of Formation, Entropies, and Heat Capacities of Organic Compounds (*Continued*)

Substance	Physical state	$\Delta_f H°$ kJ · mol^{-1}	$\Delta_f G°$ kJ · mol^{-1}	$S°$ J · deg^{-1}·mol^{-1}	$C_p°$ J · deg^{-1} · mol^{-1}
Diethylene glycol monoethyl ether	lq				301.0
Diethylene glycol monomethyl ether	lq				271.1
Diethyl ether	lq	−279.5	−116.7	172.4	172.6
	g	−252.1	−122.3	342.7	119.5
Di-2-ethylhexyl phthalate	lq				704.7
Diethyl malonate	lq	−805.5			260.7
Diethylmercury	lq	30.1			182.8
Diethyl oxalate	lq	−805.5			
3,3-Diethylpentane	lq	−275.4			278.2
Diethyl peroxide	lq	−223.3			
Diethyl 1,2-phthalate	lq	−776.6		425.1	366.1
Diethyl selenide	lq	−96.2			
Diethyl sulfate	lq	−813.2			
Diethyl sulfide	lq	−119.4		269.3	171.4
	g	−83.6	17.8	368.0	117.0
Diethyl sulfite	lq	−600.7			
Diethyl sulfone	c	−515.5			
Diethyl sulfoxide	lq	−268.0			
N,N-Diethylurea	c	−372.2			
Diethylzinc	lq	16.7			
1,2-Difluorobenzene	lq	−330.0		222.6	159.0
	g	−293.8	−242.0	321.9	106.5
1,3-Difluorobenzene	lq	−343.9		223.8	159.1
	g	−309.2	−257.0	320.4	106.3
1,4-Difluorobenzene	lq	−342.3			157.5
	g	−306.7	−252.8	315.6	106.9
2,2′-Difluorobiphenyl	c	−295.9			
4,4′-Difluorobiphenyl	c	−296.5			
1,1-Difluoroethane	lq				118.4
	g	−497.0	−443.0	282.4	67.8
1,1-Difluoroethylene	g	−335.0	−321.5	266.2	60.1
Difluoromethane	g	−452.2	−425.4	246.6	42.9
9,10-Dihydroanthracene	c	66.4			
1,2-Dihydronaphthalene	lq	71.5			
1,4-Dihydronaphthalene	lq	84.2			
Dihydro-2*H*-pyran	lq	−157.4			
5,12-Dihydrotetracene	c	106.4			
2,3-Dihydrothiophene	lq	52.9			
	g	90.7	133.5	303.5	79.8
2,5-Dihydrothiophene	g	86.9	131.6	297.1	83.3
2,5-Dihydrothiophene-1,1-dioxide	c	318.9			
2′,4-Dihydroxyacetophenone	c	−573.6			
1,2-Dihydroxybenzene (pyrocatechol)	c	−354.1	−210.0	150.2	132.2
1,3-Dihydroxybenzene	c	−368.0	−209.2	147.7	131.0
1,4-Dihydroxybenzene (*p*-hydroquinone)	c	−364.5	−207.0	140.2	136.0
Dihydroxymalonic acid	c	−1216.3			

TABLE 2.53 Enthalpies and Gibbs Energies of Formation, Entropies, and Heat Capacities of Organic Compounds (*Continued*)

Substance	Physical state	$\Delta_f H°$ kJ · mol^{-1}	$\Delta_f G°$ kJ · mol^{-1}	$S°$ J · deg^{-1}·mol^{-1}	$C_p°$ J · deg^{-1} · mol^{-1}
2,4-Dihydroxy-5-methyl-pyrimidine	c	−468.2			
2,4-Dihydroxy-6-methyl-pyrimidine	c	−456.9			
Diiodoacetylene	g			313.1	70.3
1,2-Diiodobenzene	c	172.4			
1,3-Diiodobenzene	c	187.0			
1,4-Diiodobenzene	lq	−30.0			
	c	160.7			
1,2-Diiodoethane	g	75.0	78.5	348.5	82.3
Diiodomethane	lq	66.9	90.4	174.1	134.0
	g	119.5	95.8	309.7	57.7
1,2-Diiodopropane	g	35.6			
1,3-Diiodopropane	lq	−9.0			
Diisobutylamine	lq	−218.5			
Diisopentyl ether	lq				379^{100}
Diisopropylamine	lq	−178.5			
Diisopropyl ether	lq	−351.5			216.8
	g	−319.2	−121.9	390.2	158.3
Diisopropylmercury	lq	−13.0			
Diisopropyl sulfide	lq	−181.6		313.0	232.0
	g	−142.1	27.1	415.5	169.2
Diketene	lq	−233.1			
1,2-Dimethoxybenzene	lq	−290.4			
1,1-Dimethoxybutane	lq	−468.1			
2,2-Dimethoxybutane	lq	−485.1			
1,1-Dimethoxyethane	lq	−420.2			
1,2-Dimethoxyethane	lq	−376.7			193.3
Dimethoxymethane	lq	−377.8		244.0	161.3
1,1-Dimethoxypentane	lq	−494.6			
2,2-Dimethoxypentane	lq	−509.2			
1,1-Dimethoxypropane	lq	−443.3			
2,2-Dimethoxypropane	lq	−459.0			
1,1-Dimethoxy-2-methyl-propane	lq	−476.2			
N,N-Dimethylacetamide	lq	−278.3			175.6
Dimethylamine	lq	−43.9	70.0	182.3	137.7
	g	−18.5	68.5	273.0	70.7
4-(Dimethylamino)benz-aldehyde	c	−137.6			
Dimethylaminomethanol	lq	−253.6			
N,N-Dimethylaminotri-methylsilane	lq	−279.5			
N,N-Dimethylaniline	lq	47.7			214.6^{29}
2,6-Dimethylaniline	lq				238.9
2,3-Dimethylbenzoic acid	c	−450.4			
2,4-Dimethylbenzoic acid	c	−458.5			
2,5-Dimethylbenzoic acid	c	−456.1			
2,6-Dimethylbenzoic acid	c	−440.7			
3,4-Dimethylbenzoic acid	c	−468.8			
3,5-Dimethylbenzoic acid	c	−466.4			
3,3′-Dimethylbiphenyl	lq	20.0			

(*Continued*)

TABLE 2.53 Enthalpies and Gibbs Energies of Formation, Entropies, and Heat Capacities of Organic Compounds (*Continued*)

Substance	Physical state	$\Delta_f H°$ kJ · mol^{-1}	$\Delta_f G°$ kJ · mol^{-1}	$S°$ J · deg^{-1}·mol^{-1}	$C_p°$ J · deg^{-1} · mol^{-1}
2,2-Dimethylbutane	lq	−213.8		272.5	191.9
	g	−186.1	−9.2	358.2	141.9
2,3-Dimethylbutane	lq	−207.4		287.8	189.7
	g	−178.3	−4.1	365.8	140.5
3,3-Dimethyl-2-butanone	lq	−328.6			
2,3-Dimethyl-1-butene		−62.6	79.0	365.6	143.5
2,3-Dimethyl-2-butene	lq	−101.4		270.2	174.7
	g	−68.2	76.1	364.6	123.6
3,3-Dimethyl-1-butene	g	−60.5	98.2	343.8	126.5
2,3-Dimethyl-2-butenoic acid	c	−455.6			
Dimethylcadmium	lq	63.6	139.3	201.9	132.0
1,1-Dimethylcyclohexane	lq	−218.7	26.5	267.2	209.2
	g	−180.9	35.2	365.0	154.4
cis-1,2-Dimethylcyclohexane	lq	−211.8		274.1	210.2
	g	−172.1	41.2	374.5	165.5
trans-1,2-Dimethylcyclohexane	lq	−218.2		273.2	209.4
	g	−180.0	34.5	370.9	159.0
cis-1,3-Dimethylcyclohexane	lq	−222.9		272.6	209.4
	g	−184.6	29.8	370.5	157.3
trans-1,3-Dimethylcyclohexane	lq	−215.7		276.3	212.8
	g	−176.5	36.3	376.2	157.3
cis-1,4-Dimethylcyclohexane	lq	−215.6		271.1	212.1
	g	−176.6	38.0	370.5	157.3
trans-1,4-Dimethylcyclohexane	lq	−222.4		268.0	210.2
	g	−184.5	31.7	364.8	157.7
1,1-Dimethylcyclopentane	g	−138.2	39.0	359.3	133.3
cis-1,2-Dimethylcyclopentane	lq	−165.3		269.2	
	g	−129.5	45.7	366.1	134.14
trans-1,2-Dimethylcyclopentane	g	−136.6	38.4	366.8	134.5
cis-1,3-Dimethylcyclopentane	g	−135.9	39.2	366.8	134.5
trans-1,3-Dimethylcyclopentane	g	−133.6	41.5	366.8	134.5
1,1-Dimethylcyclopropane	lq	−33.3			
cis-1,2-Dimethylcyclopropane	lq	−26.3			
trans-1,2-Dimethylcyclopropane	lq	−30.7			
cis-2,4-Dimethyl-1,3-dioxane	lq	−465.2			
4,5-Dimethyl-1,3-dioxane	lq	−451.6			
5,5-Dimethyl-1,3-dioxane	lq	−461.3			
4,4′-Dimethyldiphenylamine	c	−11.72			
Dimethyl disulfide	lq	−62.6	7.0	235.4	146.1
Dimethyl ether	g	−184.1	−112.6	266.4	64.4
N,N-Dimethylformamide	lq	−239.3			150.6
Dimethyl fumarate	lq	−729.3			
Dimethylglyoxime	c	−199.7			
2,2-Dimethylheptane	lq	−288.2			
2,6-Dimethyl-4-heptanone	lq	−408.5			297.3
2,2-Dimethylhexane	lq	−261.9	3.0	331.9	
2,3-Dimethylhexane	lq	−252.6	9.1	342.7	
2,4-Dimethylhexane	lq	−257.0	3.7	345.7	
2,5-Dimethylhexane	lq	−260.4	2.5	338.7	249.2
3,3-Dimethylhexane	lq	−257.5	5.2	339.4	246.6
3,4-Dimethylhexane	lq	−251.8	8.5	347.2	

TABLE 2.53 Enthalpies and Gibbs Energies of Formation, Entropies, and Heat Capacities of Organic Compounds (*Continued*)

Substance	Physical State	$\Delta_f H°$ kJ · mol^{-1}	$\Delta_f G°$ kJ · mol^{-1}	$S°$ J · deg^{-1}·mol^{-1}	$C_p°$ J · deg^{-1} · mol^{-1}
Dimethyl hexanedioate	lq	−886.6			
cis-2,2-Dimethyl-3-hexene	lq	−126.4			
trans-2,2-Dimethyl-3-hexene	lq	−144.9			
cis-2,5-Dimethyl-3-hexene	lq	−151.0			
trans-2,5-Dimethyl-3-hexene	lq	−159.2			
5,5-Dimethylhydantoin	c	−533.3			
1,1-Dimethylhydrazine	lq	48.9	206.7	198.0	164.1
1,2-Dimethylhydrazine	lq	52.7	212.6	199.2	171.0
3,5-Dimethylisoxazole	lq	−63.2			
Dimethyl maleate	lq	−703.8			263.2
Dimethylmaleic anhydride	c	−581.6			
Dimethyl malonate	lq	−795.8			
Dimethylmercury	lq	59.8	140.3	209.0	
	g	94.4	146.1	306.0	83.3
6,6-Dimethyl-2-methylene-bicyclo[3.1.1]heptane	lq	−7.7			
Dimethyl oxalate	lq	−756.3			
2,2-Dimethylpentane	lq	−238.3		300.3	221.1
	g	−205.9	0.1	392.9	166.0
2,3-Dimethylpentane	lq	−233.1			218.3
	g	−198.9	0.7	414.0	166.0
2,4-Dimethylpentane	lq	−234.6		303.2	224.2
	g	−201.7	3.1	396.6	166.0
3,3-Dimethylpentane	lq	−234.2			
	g	−201.2	2.6	399.7	166.0
Dimethyl pentanedioate	lq	−205.9			
2,4-Dimethyl-3-pentanone	lq	−352.9		318.0	233.7
	g	−311.5			
2,4-Dimethyl-1-pentene	g	−83.8			
4,4-Dimethyl-1-pentene	g	−81.6			
2,4-Dimethyl-2-pentene	g	−88.7			
cis-4,4-Dimethyl-2-pentene	g	−72.6			
trans-4,4-Dimethyl-2-pentene	g	−88.8			
2,7-Dimethylphenanthrene	c	36.4			
4,5-Dimethylphenanthrene	c	89.0			
9,10-Dimethylphenanthrene	c	47.7			
2,3-Dimethylphenol	c	−241.2			206.9
2,4-Dimethylphenol	lq	−228.7			
2,5-Dimethylphenol	c	−246.6			
2,6-Dimethylphenol	c	−237.4			
3,4-Dimethylphenol	c	−242.3			
3,5-Dimethylphenol	c	−244.4			
Dimethyl 1,2-phthalate	lq	−678			303.1
Dimethyl 1,3-phthalate	c	−730.0			
Dimethyl 1,4-phthalate	c	−732.6			261.1
2,2-Dimethylpropane	lq				163.9[6]
	g	−168.0	−1.5	306.4	121.6
2,2-Dimethylpropanenitrile	lq	−39.8		232.0	179.4
2,2-Dimethyl-1,3-propanediol	c	−551.2			
2,2-Dimethylpropanoic acid	lq	−564.4			
2,2-Dimethylpropanoic anhydride	lq	−779.9			

(*Continued*)

TABLE 2.53 Enthalpies and Gibbs Energies of Formation, Entropies, and Heat Capacities of Organic Compounds (*Continued*)

Substance	Physical State	$\Delta_f H°$ kJ · mol^{-1}	$\Delta_f G°$ kJ · mol^{-1}	$S°$ J · deg^{-1}·mol^{-1}	$C_p°$ J · deg^{-1} · mol^{-1}
2,2-Dimethyl-1-propanol	lq	−399.4			
2,3-Dimethylpyridine	lq	19.4		243.7	189.5
2,4-Dimethylpyridine	lq	16.2		248.5	184.8
2,5-Dimethylpyridine	lq	18.7		248.8	184.7
2,6-Dimethylpyridine	lq	12.7		249.2	185.2
3,4-Dimethylpyridine	lq	18.3		240.7	191.8
3,5-Dimethylpyridine	lq	22.5		241.7	184.5
Dimethyl succinate	lq	−835.1			
2,2-Dimethylsuccinic acid	c	−987.8			
meso-2,3-Dimethylsuccinic acid	c	−977.5			
Dimethyl sulfate	lq	−735.5			
Dimethyl sulfide	lq	−65.4			118.1
	g	−37.5	7.0	285.9	74.1
Dimethyl sulfite	lq	−523.6			
Dimethyl sulfone	c	−450.1	−302.5	142.0	
	lq	−373.1	−272		
	g			310.6	100.0
Dimethyl sulfoxide	lq	−204.2	−99.2	188.3	153.0
1,5-Dimethyltetrazole	c	188.7			
2,2-Dimethylthiacyclopropane	lq	−24.2			
5,5-Dimethyl-4-thia-1-hexene	lq	−90.7			
N,N-Dimethylurea	c	−319.1			
N,N'-Dimethylurea	c	−312.1			
Dimethylzinc	lq	23.4		201.6	129.2
2,3-Dinitroaniline	c	−11.7			
2,4-Dinitroaniline	c	−67.8			
2,5-Dinitroaniline	c	−44.4			
2,6-Dinitroaniline	c	−50.6			
3,4-Dinitroaniline	c	−32.6			
3,5-Dinitroaniline	c	−38.9			
2,4-Dinitroanisole	c	−186.6			
2,6-Dinitroanisole	c	−189.1			
1,2-Dinitrobenzene	c	−1.8	211.5	216.3	
1,3-Dinitrobenzene	c	−27.4	184.6	220.9	
1,4-Dinitrobenzene	c	−38.7			
1,1-Dinitroethane	lq	−148.2			
1,2-Dinitroethane	lq	−165.2			
Dinitromethane	lq	−104.9			
	g	−58.9			
1,5-Dinitronaphthalene	c	30.5			
2,4-Dinitro-1-naphthol	c	−181.4			
2,4-Dinitrophenol	c	−232.6			
2,6-Dinitrophenol	c	−210.0			
1,1-Dinitropropane	lq	−163.2			
1,3-Dinitropropane	lq	−207.1			
2,2-Dinitropropane	lq	−181.2			
2,4-Dinitroresorcinol	c	−415.5			
2,4-Dinitrotoluene	c	−71.6			
2,6-Dinitrotoluene	c	−51.0			
1,3-Dioxane	lq	−379.7			143.9
1,4-Dioxane	lq	−353.9	−188.1	270.2	153.6

TABLE 2.53 Enthalpies and Gibbs Energies of Formation, Entropies, and Heat Capacities of Organic Compounds (*Continued*)

Substance	Physical State	$\Delta_f H°$ kJ · mol^{-1}	$\Delta_f G°$ kJ · mol^{-1}	$S°$ J · deg^{-1}·mol^{-1}	$C_p°$ J · deg^{-1} · mol^{-1}
	g	−315.8	−180.8	299.8	94.1
1,3-Dioxolane	lq	−333.5			118.0
	g	−298.0			
1,3-Dioxolan-2-one	c	−581.6			133.9[50]
1,3-Dioxol-2-one	lq	−459.9			
Dipentene	lq	−50.8			249.4
Dipentyl ether	lq				250
N,N-Diphenylacetamide	c	−43.1			
Diphenylacetylene	c	312.4			225.9
Diphenylamine	c	130.6			
Diphenylboron bromide	lq	−16.1			
cis,cis-1,4-Diphenylbutadiene	c	198.8			
trans,trans-1,4-Diphenylbutadiene	c	178.8			
Diphenylbutadiyne	c	518.4			
1,4-Diphenylbutane	c	−9.9			
1,4-Diphenyl-1,4-butanedione	c	−256.2	7.8	324.7	
1,4-Diphenyl-2-butene-1,4-dione	c	−114.7	111.5	319.2	
Diphenyl carbonate	c	−401.2	−175.9	278.4	
Diphenyl disulfide	c	−148.5			
Diphenyl disulfone	c	−643.2			
Diphenyleneimine	c	126.8			
1,1-Diphenylethane	lq	48.7	245.1	335.9	
1,2-Diphenylethane	lq	51.5	67.2	270.3	
Diphenylethanedione	c	−154.0			
Diphenyl ether	c	−32.1		233.9	216.6
	lq	−14.9	144.2	291.3	268.6
1,1-Diphenylethylene	lq	172.4			
Diphenylethyne	c	312.4			
6,6-Diphenylfulvene	c	197.4			
1,2-Diphenylhydrazine	c	221.3			
Diphenylmercury	c	279.5			
Diphenylmethane	c	71.7		239.3	
	lq	89.7	276.9		233.1
1,3-Diphenyl-2-propanone	c	−84.0			
Diphenyl sulfide	lq	163.4			
Diphenyl sulfone	c	−225.0			
Diphenyl sulfoxide	c	9.7			
1,3-Diphenylurea	c	−122.6			
Dipropylamine	lq	−156.1			253.0[75]
Dipropyl disulfide	lq	−171.3	19.1	373.6	
Dipropyl ether	lq	−328.8		323.9	221.6
	g	−292.9	−105.6	422.5	158.3
Dipropylmercury	lq	−20.9			
Dipropyl sulfate	lq	−859.0			
Dipropyl sulfide	lq	−171.5			
	g	−125.3	33.2	448.4	161.2
Dipropyl sulfite	lq	−646.8			
Dipropyl sulfone	lq	−548.2			
Dipropyl sulfoxide	lq	−329.4			
2,2'-Dipyridyl ketone	c	−19.7			
1,3-Dithiane	g	−10.0	72.4	333.5	110.4

(*Continued*)

TABLE 2.53 Enthalpies and Gibbs Energies of Formation, Entropies, and Heat Capacities of Organic Compounds (*Continued*)

Substance	Physical State	$\Delta_f H°$ kJ · mol^{-1}	$\Delta_f G°$ kJ · mol^{-1}	$S°$ J · deg^{-1}·mol^{-1}	$C_p°$ J · deg^{-1} · mol^{-1}
1,2-Dithiolane	g	0.0	47.7	313.5	86.5
1,3-Dithiolane	g	10.0	54.7	323.3	84.7
Divinyl ether	lq	−39.8			
	g	−13.6			
Divinyl sulfone	lq	−207.4			
Docosanoic acid	c	−983.0			
cis-13-Docosenic acid	c	−866.0			
trans-13-Docosenic acid	c	−960.7			
Dodecane	lq	−350.9	28.1	490.6	376.0
	g	−289.7	50.0	622.5	280.3
Dodecanedioic acid	c	−1130.0			
Dodecanoic acid	c	−774.6			
	lq	−737.9			404.3
1-Dodecanol	lq	−528.5			438.1
1-Dodecene	lq	−226.2		484.8	360.7
	g	−165.4	137.9	618.3	269.6
1-Dodecyne	g	−0.04	268.6	602.4	265.4
Dulcitol	c	−1346.8			
1,2-Epoxybutane	lq	−168.9		230.9	147.0
Ergosterol	c	−789.9			
Ethane	g	−84.0	−32.0	229.1	52.5
Ethane-d_6	g	−107.4	−47.3	244.5	64.6
1,2-Ethanediamine	lq	−63.0		209.2	172.6
1,2-Ethanediol	lq	−455.3	−323.2	163.2	149.3
	g	−392.2	−304.5	303.8	82.7
Ethanedithioamide	c	−20.8			
Ethanedioyl dichloride	lq	−367.6			
1,2-Ethanedithiol	lq	−54.4			
Ethanethiol	lq	−73.6	−5.5	207.0	117.9
	g	−46.1	−4.8	296.1	72.7
Ethanol	lq	−277.6	−174.8	161.0	112.3
	g	−234.8	−167.9	281.6	65.6
Ethene (*see* Ethylene)					
Ethoxybenzene	lq	−152.6			228.5
2-Ethoxyethyl acetate	lq				376.0
2-Ethoxyethanol	lq				210.8
Ethyl acetate	lq	−479.3	−332.7	257.7	170.7
	g	−443.6	−327.4	362.8	113.6
Ethylamine	lq				130.0
	g	−47.4	36.3	283.8	71.5
Ethyl 4-aminobenzoate	c	−418.0			
N-Ethylaniline	lq	4.0	188.7	239.3	
Ethylbenzene	lq	−12.3			183.2
	g	29.9	130.6	360.5	
Ethyl benzoate	lq				246.0
2-Ethylbenzoic acid	c	−441.3			
3-Ethylbenzoic acid	c	−445.8			
4-Ethylbenzoic acid	c	−460.7			
2-Ethyl-1-butene	g	−56.0	80.0	376.6	133.6
Ethyl *trans*-2-butenoate (ethyl crotonate)	lq	−420.1			228.0
Ethyl carbamate	c	−520.5			
Ethyl 4-chlorobutanoate	lq	−566.5			

TABLE 2.53 Enthalpies and Gibbs Energies of Formation, Entropies, and Heat Capacities of Organic Compounds (*Continued*)

Substance	Physical State	$\Delta_f H°$ kJ · mol^{-1}	$\Delta_f G°$ kJ · mol^{-1}	$S°$ J · deg^{-1}·mol^{-1}	$C_p°$ J · deg^{-1} · mol^{-1}
Ethyl chloroformate	lq	−505.1			
Ethylcyclobutane	g	−27.5			
Ethylcyclohexane	lq	−211.9	29.1	280.9	211.8
	g	−171.7	39.3	382.6	158.8
1-Ethylcyclohexene	lq	−106.7			
Ethylcyclopentane	lq	−163.4	37.3	279.9	185.8
1-Ethylcyclopentene	g	−19.7			
Ethylcyclopropane	lq	−24.8			
Ethyl diethylcarbamate	lq	−592.3			
Ethyl 2,2-dimethylpropanoate	lq	−577.2			
	g	−536.0			
Ethylene	g	52.5	68.4	219.3	42.9
Ethylene-d_4	g	38.2	59.2	230.5	51.9
Ethylene carbonate	c	−581.5			133.9
Ethylenediaminetetra-acetic acid	c	−1759.4			
Ethylenediammonium chloride	c	−513.4			
2,2′-(Ethylenedioxy)bis-ethanol	lq	−804.2			
Ethylene glycol dibutyl ether	lq				350[20]
Ethylene glycol diethyl ether	lq	−451.4			259.4
Ethylene glycol dimethyl ether	lq	−376.6			193.3
Ethyleneimine	lq	91.9			
	g	126.5(9)	178.0	250.6	52.6
Ethylene oxide	lq	−78.0	−11.8	153.9	88.0
	g	−52.6(6)	−13.1	242.4	47.9
Ethyl formate	lq				149.3
2-Ethylhexanal	lq	−342.5			
3-Ethylhexane	lq	−250.4			
	g	−210.7			
2-Ethyl-1-hexanol	lq	−432.8		347.0	317.5
Ethyl hydroperoxide	g	198.9			
Ethylidenecyclohexane	lq	−103.5			
Ethylidenecyclopentane	lq	−56.7			
Ethyl isocyanide	lq	108.4			
Ethyl isopropyl sulfide	lq	−156.1			
Ethyl lactate	lq				254
Ethyllithium	c	−58.6			
Ethylmercury bromide	c	−107.5			
Ethylmercury chloride	c	−141.1			
Ethylmercury iodide	c	−65.7			
1-Ethyl-2-methylbenzene	g	1.3	131.1	399.2	157.9
2-Ethyl-3-methyl-1-butene	g	−79.5			
Ethyl 2-methylbutanoate	lq	−566.8			
Ethyl 3-methylbutanoate	lq	−570.9			
Ethyl methyl ether	g	−216.4	−117.7	309.2	93.3
3-Ethyl-2-methylpentane	lq	−249.6			
	g	−211.0	21.3	441.1	
3-Ethyl-3-methylpentane	lq	−252.8			
	g	−214.8	19.9	433.0	
3-Ethyl-2-methyl-1-pentene	g	−100.3			

(*Continued*)

TABLE 2.53 Enthalpies and Gibbs Energies of Formation, Entropies, and Heat Capacities of Organic Compounds (*Continued*)

Substance	Physical state	$\Delta_f H°$ kJ · mol^{-1}	$\Delta_f G°$ kJ · mol^{-1}	$S°$ J · deg^{-1}·mol^{-1}	$C_p°$ J · deg^{-1} · mol^{-1}
Ethyl methyl sulfide	lq	−91.6		239.1	144.6
	g	−59.6	11.4	333.1	95.1
Ethyl nitrate	g	−154.1	−36.9	348.3	97.4
Ethyl nitrite	g	−104.2		103.5	99.2
1-Ethyl-2-nitrobenzene	lq	−48.7			
1-Ethyl-4-nitrobenzene	lq	−55.4			
Ethyl 3-oxobutanoate	lq				248.0
3-Ethylpentane	lq	−224.9		314.5	219.6
	g	−189.6	11.0	411.5	166.0
Ethyl pentanoate	lq	−553.0			
2-Ethylphenol	lq		−208.8		
3-Ethylphenol	lq	−214.3			
4-Ethylphenol	c	−224.4			206.9
Ethylphosphonic acid	c	−1051.4			
Ethylphosphonic dichloride	lq	−613.4			
Ethyl propanoate	lq	−502.7			196.1
	g	−463.3	−323.7		
Ethyl propyl ether	g	−272.2		295.0	197.2
Ethyl propyl sulfide	lq	−144.8		309.5	198.4
	g	−104.7	23.6	414.1	139.3
2-Ethylpyridine	lq	7.4			
S-Ethyl thioacetate	lq	−268.2			
2-Ethyltoluene	g	1.3	131.1	399.2	157.9
3-Ethyltoluene	g	−1.8	126.4	404.2	152.2
4-Ethyltoluene	g	−3.2	85.3	398.9	151.5
N-Ethylurea	c	−357.8			
Ethyl β-vinylacrylate	lq	−338.1			
Ethyl vinyl ether	lq	−167.4			
	g	−140.8			
Ethynylbenzene	g	327.3	361.8	321.7	114.9
Ethynylsilane	g			269.4	72.6
Fluoranthene	c	189.9	345.6	230.5	230.2
Fluoroacetamide	c	−496.6			
Fluoroacetic acid	c	−688.3			
Fluoroacetylene	g			269.4	72.6
Fluorobenzene	lq	−150.6		205.9	146.4
	g	−116.0	−69.0	302.6	94.4
2-Fluorobenzoic acid	c	−567.6			
3-Fluorobenzoic acid	c	−582.0			
4-Fluorobenzoic acid	c	−585.7			
Fluoroethane	g	−263.2	−211.0	264.5	58.6
2-Fluoroethanol	lq	−465.7			
Fluoroethylene	g	−138.8			
Fluoromethane	g	−237.8	−213.8	222.8	37.5
1-Fluoropropane	g	−285.9	−200.3	304.2	82.6
2-Fluoropropane	g	−293.5	−204.2	292.1	82.0
Fluorosyltrifluoromethane	g	−766.0	−707.0	322.4	79.4
4-Fluorotoluene	lq	−186.9	−79.8	237.1	171.2
Fluorotribromomethane	g	−190.4	−193.1	345.8	
Fluorotrinitromethane	lq	−220.9			
Formaldehyde	g	−108.6	−102.5	218.8	35.4

TABLE 2.53 Enthalpies and Gibbs Energies of Formation, Entropies, and Heat Capacities of Organic Compounds (*Continued*)

Substance	Physical state	$\Delta_f H°$ kJ · mol^{-1}	$\Delta_f G°$ kJ · mol^{-1}	$S°$ J · deg^{-1}·mol^{-1}	$C_p°$ J · deg^{-1} · mol^{-1}
Formamide	lq	−254.0			107.6
	g	−193.9	−141.0	248.6	45.4
Formanilide	c	−151.5			
Formic acid	lq	−424.7	−361.4	129.0	99.5
	g	−378.7	−351.0	248.7	45.2
Formyl fluoride	g	−376.6	−368.1	246.5(8)	40.0
D-(−)-Fructose	c	−1265.6			
D-(+)-Fucose	c	−1099.1			
Fullerene-C_{60}	c	2327.0	2302.0	426.0	520.0
Fumaric acid	c	−811.7	−655.6	168.0	142.0
Fumaronitrile	c	268.2			
Furan	lq	−62.3		177.0	114.8
	g	−34.9	0.88	267.2	65.4
2-Furancarboxaldehyde	lq	−201.6			163.2
2-Furancarboxylic acid	c	−498.4			
2-Furanmethanol	lq	−276.2	−154.2	215.5	204.0
Furfuryl alcohol	lq	−276.2			204.0
Furylacrylic acid	c	−459.0			
Furylethylene	lq	−10.5			
D-(+)-Galactose	c	−1286.3	−918.8	205.4	
D-Gluconic acid	c	−1587.0			
D-(+)-Glucose	c	−1273.3	−910.4	212.1	
D-(−)-Glutamic acid	c	−1009.7	−727.5	191.2	
L-(+)-Glutamic acid	c	−1005.2	−731.3	188.2	
L-Glutamine	c	−826.4			
Glutaric acid	c	−960.0			
Glyceraldehyde	lq	−598.0			
Glycerol	lq	−668.5	−477.0	206.3	218.9
Glyceryl 1-acetate	lq	−909.1			
Glyceryl 1-benzoate	c	−777.3			
Glyceryl 2-benzoate	c	−772.8			
Glyceryl 1,3-diacetate	lq	−1120.7			
Glyceryl 1-dodecanoate	c	−1160.9			
Glyceryl 2-dodecanoate	c	−1152.6			
Glyceryl 1-hexadecanoate	c	−1281.5			
Glyceryl 1-hexanoate	c	−1109.0			
Glyceryl 2-hexanoate	c	−1095.8			
Glyceryl 1-octadecanoate	c	−1324.8			
Glyceryl 1-tetradecanoate	c	−1222.6			
Glyceryl triacetate	lq	−1330.8			
Glyceryl trinitrate	lq	−370.9			
Glyceryl tris(dodecanoate)	c	−2046.0			
Glyceryl tris(tetradecanoate)	c	−2176.0			
Glycine	c	−528.5	−368.6	103.5	99.2
ionized; std. state	aq	−469.8	−315.0	111.0	
$^{+}H_3NCH_2COOH$; std. state	aq	−517.9	−384.2	190.2	
Glycylglycine	c	−747.7	−490.6	190.0	
Glyoxal	g	−212.0			
Glyoxime	c	−90.5			
Glyoxylic acid	c	−835.5			
Guanidine	c	−56.0			

(*Continued*)

TABLE 2.53 Enthalpies and Gibbs Energies of Formation, Entropies, and Heat Capacities of Organic Compounds (*Continued*)

Substance	Physical state	$\Delta_f H°$ kJ · mol^{-1}	$\Delta_f G°$ kJ · mol^{-1}	$S°$ J · deg^{-1}·mol^{-1}	$C_p°$ J · deg^{-1} · mol^{-1}
Guanidine carbonate	c	−971.9	−557.4	295.4	258.9
Guanidine nitrate	c	−387.0			
Guanidine sulfate	c	−1205.0			
Guanine	c	−183.9	47.4	160.3	
Guanylurea nitrate	c	−427.2			
L-Gulonic acid-γ-lactone	c	−1219.6			
Heptadecane	g	−393.9	82.1	817.3	394.7
Heptadecanoic acid	c	−924.4			475.7
1-Heptadecene	g	−268.4	179.9	813.1	383.9
Heptanal	lq	−311.5	−100.6	335.4	230.1
	g	−264.0	−86.7	461.7	
Heptane	lq	−224.2			224.9
	g	−187.7	8.0	427.9	166.0
Heptanedioic acid	c	−1009.4			
Heptanenitrile	lq	−82.8			
1-Heptanethiol	g	−150.0	36.2	493.3	186.9
Heptanoic acid	lq	−610.2			265.4
1-Heptanol	lq	−403.3	−142.3	320.1	272.1
	g	−336.4	−120.9	480.3	178.7
2-Heptanone	lq				232.6
1-Heptene	lq	−97.9		327.6	211.8
	g	−62.3	95.8	423.6	155.2
cis-2-Heptene	lq	−105.1			
trans-2-Heptene	lq	−109.5			
cis-3-Heptene	lq	−104.3			
trans-3-Heptene	lq	−109.3			
1-Heptyne	g	103.0	226.7	407.7	151.1
Hexabromoethane	g			441.9	139.3
Hexachlorobenzene	c	−127.6	1.1	260.2	201.3
	g	−35.5	44.2	441.2	173.2
Hexachloroethane	c	−202.8		237.3	198.2
	g	−143.6	−54.9	398.7	136.7
Hexadecafluoroethylcyclohexane	lq	−3420.0			
Hexadecafluoroheptane	lq	−3420.8	−3093.0	561.8	419.0
Hexadecane	lq	−456.1			501.6
	g	−374.8	83.7	778.3	371.8
Hexadecanoic acid	c	−891.5	−316.1	452.4	460.7
1-Hexadecanol	c	−686.7	−98.7	451.9	422.0
	lq	−635.4	−96.6	606.7	
1-Hexadecene	lq	−328.7		587.9	488.9
	g	−248.5	171.5	774.1	361.0
1,5-Hexadiene	lq	54.1			
2,4-Hexadienoic acid	c	−390.8			
1,5-Hexadiyne	lq	384.2			
Hexafluoroacetone	g	−1249.3			
Hexafluoroacetylacetone	c	−2286.7			
Hexafluorobenzene	lq	−991.3		280.8	156.6
	g	−955.4	−79.4	383.2	
Hexafluoroethane	g	−1344.2	−1255.8	332.3	106.7
cis-Hexahydroindane	g	−127.2			

(*Continued*)

TABLE 2.53 Enthalpies and Gibbs Energies of Formation, Entropies, and Heat Capacities of Organic Compounds (*Continued*)

Substance	Physical state	$\Delta_f H°$ kJ · mol^{-1}	$\Delta_f G°$ kJ · mol^{-1}	$S°$ J · deg^{-1}·mol^{-1}	$C_p°$ J · deg^{-1} · mol^{-1}
trans-Hexahydroindane	g	−131.4			
Hexamethylbenzene	c	−162.4	117.4	306.3	245.6
1,1,1,3,3,3-Hexamethyldisilazane	lq	−518.0			
Hexamethyldisiloxane	lq	−814.6	−541.8	433.8	311.4
	g	−777.7	−534.5	535.0	238.5
Hexamethylenetetramine	c	125.5	434.8	163.4	
Hexamethylphosphoric triamide	lq				321
Hexanal	g	−248.4	−100.1	422.9	148.2
Hexanamide	c	−423.0			
	lq	−397.0			
Hexane	lq	−198.8	−3.8	296.1	195.6
	g	−167.1(8)	−0.25	388.4	143.1
1,6-Hexanedioic acid	lq	−985.4	−207.3		232.2
1,2-Hexandediol	lq	−577.1			
1,6-Hexanediol	c	−569.9			
Hexanedinitrile	lq	85.1			128.7
1-Hexanethiol	g	−129.9	27.8	454.3	164.1
Hexanoic acid	lq	−583.9			225.0
1-Hexanol	lq	−377.5	−152.3	287.4	240.4
	g	−317.6	−135.6	441.4	155.6
2-Hexanol	lq	−392.9			
3-Hexanol	lq	−392.4			286.2
2-Hexanone	lq	−322.0			213.3
3-Hexanone	lq	−320.2		305.3	216.9
1-Hexene	lq	−74.1	83.6	295.1	183.3
	g	−43.5	84.45	384.6	132.3
cis-2-Hexene	lq	−83.9			
	g	−52.3	76.2	386.5	125.7
trans-2-Hexene	lq	−85.5			
	g	−53.9	76.4	380.6	132.4
cis-3-Hexene	lq	−79.0			
	g	−47.6	83.0	379.6	123.6
trans-3-Hexene	lq	−86.1			
Hexyl acetate	lq				282.8
	g	−54.4	77.6	374.8	132.8
1-Hexyne	g	123.6	218.6	368.7	128.2
(−)-Histidine	c	−466.7			
Hydantoin	c	−448.5			
Hydrazine	lq	50.6	149.2	121.2	98.9
Hydrazinecarbothioamide	c	24.7			
Hydrazobenzene	c	221.3			
Hydroxyacetic acid	c	−663.6			
2′-Hydroxyacetophenone	c	−357.7			
3′-Hydroxyacetophenone	c	370.7			
4′-Hydroxyacetophenone	c	−364.4			
2-Hydroxybenzaldehyde	lq	−279.9			
2-Hydroxybenzaldoxime	c	−183.7			
2-Hydroxybenzoic acid	c	−589.9	−421.3	178.2	159.1
3-Hydroxybenzoic acid	c	−584.9	−417.3	177.0	157.3
4-Hydroxybenzoic acid	c	−584.5	−416.5	175.7	155.1

(Continued)

TABLE 2.53 Enthalpies and Gibbs Energies of Formation, Entropies, and Heat Capacities of Organic Compounds (*Continued*)

Substance	Physical state	$\Delta_f H°$ kJ · mol^{-1}	$\Delta_f G°$ kJ · mol^{-1}	$S°$ J · deg^{-1}·mol^{-1}	$C_p°$ J · deg^{-1} · mol^{-1}
(±)-2-Hydroxybutanoic acid	lq	−679.1			
2-Hydroxy-2,4,6-cyclohepta-trienone	c	−239.2			
2-Hydroxyisobutanoic acid	c	−744.3			
2-Hydroxy-1-isopropyl-4-methylbenzene	c	−309.6			
3-Hydroxy-4-methoxybenz-aldehyde	c	−453.6			
4-Hydroxy-4-methyl-2-pentanone	lq				221.3
2-Hydroxymethyl-1,3-propane-diol	c	−744.6			
3-Hydroxy-2-naphthalene-carboxylic acid	c	−547.7			
5-Hydroxy-1-pentanal	lq	−479.9			
trans-(−)-4-Hydroxyproline	c	−661.1			
(*S*)-2-Hydroxypropanoic acid	c	−694.0			
2-Hydroxypropanonitrile	lq	−138.9	34.3		
2-Hydroxypyridine	c	−166.3			
3-Hydroxypyridine	c	−132.0			
4-Hydroxypyridine	c	−144.6			
8-Hydroxyquinoline	c	−81.2			
(−)-2-Hydroxysuccinic acid	c	−1103.7	−884.7		
(±)-2-Hydroxysuccinic acid	c	−1105.7			
Hypoxanthene	c	−110.8	76.9	145.6	134.5
Icosane	g	−455.8	117.3	934.1	463.3
Icosanoic acid	c	−1011.9			545.1
Icosene	g	−330.2	205.1	929.9	452.5
Imidazole	c	49.8			
Iminodiacetic acid	c	−932.6			
Indane	lq	11.5	150.8	56.0	190.3
1*H*-Indazole	c	151.9			
Indene	lq	110.6	217.6	215.3	186.9
1*H*-Indole	c	86.7			
Indole-2,3-dione	c	−268.2			
Iodoacetone	g	−130.5			
Iodobenzene	lq	117.1		205.4	158.7
	g	164.9	187.8	334.1	100.8
2-Iodobenzoic acid	c	−302.3			
3-Iodobenzoic acid	c	−316.9			
4-Iodobenzoic acid	c	−316.1			
Iodocyclohexane	lq	−97.2			
Iodoethane	lq	−40.0	14.7	211.7	115.1
	g	−8.1	19.2	306.0	66.9
Iodoethylene	g			285.0	57.9
Iodomethane	g	14.4	15.6	254.1	44.1
2-Iodo-2-methylpropane	lq	−107.5			162.3
	g	−72.0	23.6	342.2	118.3
1-Iodonaphthalene	lq	161.5			
2-Iodonaphthalene	c	144.3			
2-Iodophenol	c	−95.8			
3-Iodophenol	c	−94.5			

TABLE 2.53 Enthalpies and Gibbs Energies of Formation, Entropies, and Heat Capacities of Organic Compounds (*Continued*)

Substance	Physical state	$\Delta_f H^\circ$ kJ · mol^{-1}	$\Delta_f G^\circ$ kJ · mol^{-1}	S° J · deg^{-1}·mol^{-1}	C_p° J · deg^{-1} · mol^{-1}
4-Iodophenol	c	−95.4			
1-Iodopropane	lq	−66.0			126.8
	g	−30.0			
2-Iodopropane	lq	−74.8			91.0
	g	−40.3	20.1	324.5	90.1
3-Iodopropanoic acid	c	−460.0			
3-Iodo-1-propene	g	91.5			
α-Iodotoluene	lq	57.7			
3-Iodotoluene	lq	79.1			
4-Iodotoluene	lq	67.4			
Isobutanenitrile	g	25.4	103.6	313.3	96.4
Isobutylamine	lq	−132.6			183.2
Isobutylbenzene	lq	−69.8			
Isobutyl trichloroacetate	lq	−553.4			
Isocyanomethane	g	163.5	165.7	246.9	52.9
(−)-Isoleucine	c	−637.9	−347.2	208.0	188.3
(±)-Isoleucine	c	−635.3			
Isoxazole	g	78.6			
Isopropenyl acetate	lq	−386.4			
Isopropyl acetate	lq	−518.9			199.4
Isopropylamine	lq	−112.3		218.3	163.8
	g	−83.7	32.2	312.2	97.5
Isopropylbenzene	lq	−41.1	124.3	279.8	210.7
	g	4.0	137.0	388.6	151.7
1-Isopropyl-2-methylbenzene	lq	−73.3			
1-Isopropyl-3-methylbenzene	lq	−78.6			
1-Isopropyl-4-methylbenzene	lq	−78.0	119.1	306.6	
Isopropyl methyl ether	lq	−278.8		253.8	161.9
	g	−252.0	−120.9	332.3	111.1
2-Isopropyl-5-methylphenol	c	−309.7			
Isopropyl methyl sulfide	lq	−105.7		263.1	172.4
	g	−90.5	13.4	359.3	117.2
Isopropyl nitrate	g	−191.0	−40.7	373.2	120.7
2-Isopropylphenol	lq	−233.7			
3-Isopropylphenol	lq	−252.5			
4-Isopropylphenol	lq	−265.9			
Isopropyl thioacetate	lq	−298.2			
Isopropyl trichloroacetate	lq	−536.0			
Isoquinoline	c	144.5			
	lq				196.8
Ketene	g	−47.5	−48.3	247.6	51.8
(+)-Lactic acid	c	−694.1	−522.9	142.3	
(±)-Lactic acid	lq	−674.5	−518.2	192.1	
β-Lactose	c	−2236.7	−1567.0	386.2	
(+)-Leucine	c	−637.3	−347.2	208.0	
(−)-Leucine	c	−637.4	−346.3	211.8	201.0
(+)-Limonene	lq	−54.5			249.0
(±)-Lysine	c	−678.6			
Malic acid	c	−789.4	−625.1	160.8	137.0
Maleic anhydride	c	−469.8			
(R)-Malic acid	c	−1105.7			
(S)-Malic acid	c	−1103.6			

(*Continued*)

TABLE 2.53 Enthalpies and Gibbs Energies of Formation, Entropies, and Heat Capacities of Organic Compounds (*Continued*)

Substance	Physical state	$\Delta_f H°$ kJ · mol^{-1}	$\Delta_f G°$ kJ · mol^{-1}	$S°$ J · deg^{-1}·mol^{-1}	$C_p°$ J · deg^{-1} · mol^{-1}
Malonamide	c	−546.0			
Malonic acid	c	−891.0			
Malonodiamide	c	−546.1			
Malononitrile	c	186.6			
D-(+)-Maltose	c	−2220.9	−1726.3		
(±)-Mandelic acid	c	−579.4			
(+)-Mannitol	c	−1337.1	−942.2	238.5	
D-(+)-Mannose	c	−1263.0			
2-Mercaptopropanoic acid	lq	−468.2	−343.9	228.9	
Methane	g	−74.6	−50.5	186.3	35.7
Methane-d_4	g	−88.2	−59.5	198.9	40.3
Methanethiol	lq	−46.7	−7.7	169.2	90.5
	g	−22.9	−9.9	255.1	50.3
Methanol	lq	−239.1	−166.6	126.8	81.2
	g	−201.0	−162.3	239.9	44.1
(−)-Methionine	c	−577.5	−505.8	231.5	
2-Methoxybenzaldehyde	c	−266.5			
3-Methoxybenzaldehyde	lq	−276.1			
4-Methoxybenzaldehyde	lq	−267.2			
Methoxybenzene	lq	−114.8			199.0
	g	−67.9			
2-Methoxybenzoic acid	c	−538.5			
3-Methoxybenzoic acid	c	−553.5			
4-Methoxybenzoic acid	c	−561.7			
2-Methoxyethanol	lq				171.1
2-Methyoxyethyl acetate	lq				310.0
2-Methoxytetrahydropyran	lq	−442.3			
5-Methoxytetrazole	c	69.1			
1-Methoxy-2,4,6-trinitrobenzene	c	−157.5			
Methyl (CH_3)	g	145.7	147.9	194.2	38.7
Methyl acetate	lq	−445.8			141.9
	g	−413.3		324.4	86.0
Methyl acrylate	lq	−362.2	−243.2	239.5	158.8
	g	−333.0	−237.6		
Methylamine	lq	−47.2	35.7	150.2	102.1
	g	−22.5	32.7	242.9	50.1
N-Methylaniline	lq	32.2			207.1
o-Methylaniline	lq	−6.3			209.6
	g	56.4	167.6	351.0	130.2
m-Methylaniline	lq	−8.1			227.0
	g	54.6	165.4	352.5	125.5
p-Methylaniline	lq	−23.5			
	g	55.3	167.7	347.0	126.2
Methyl benzoate	lq	−343.5			221.3
2-Methylbenzoic acid	c	−416.5			
	lq				174.9
3-Methylbenzoic acid	c	−426.1			
	lq				163.6
4-Methylbenzoic acid	c	−429.2			
	lq				169.0

TABLE 2.53 Enthalpies and Gibbs Energies of Formation, Entropies, and Heat Capacities of Organic Compounds (*Continued*)

Substance	Physical state	$\Delta_f H°$ kJ · mol^{-1}	$\Delta_f G°$ kJ · mol^{-1}	$S°$ J · deg^{-1}·mol^{-1}	$C_p°$ J · deg^{-1} · mol^{-1}
2-Methylbenzoic anhydride	c	−533.5			
4-Methylbenzoic anhydride	c	−520.9			
1-Methylbicyclo[4.1.0]heptane	lq	−59.9			
1-Methylbicyclo[3.1.0]hexane	lq	−33.2			
2-Methylbiphenyl	lq	108.0			
3-Methylbiphenyl	lq	85.4			
4-Methylbiphenyl	c	55.2			
2-Methyl-1,3-butadiene	lq	48.2		229.3	152.6
	g	75.5	145.9	315.6	104.6
3-Methyl-1,2-butadiene	g	129.7	198.6	319.7	105.4
2-Methylbutane	lq	−178.4		260.4	164.8
	g	−154.0	−14.8	343.6	118.8
2-Methyl-2-butanethiol	lq	−162.8		290.1	198.1
	g	−127.1	9.2	386.9	143.5
3-Methyl-1-butanethiol	g	−114.9			
3-Methyl-2-butanethiol	lq	−158.8			
2-Methylbutanoic acid	lq	−554.4			
3-Methylbutanoic acid	lq	−561.6			197.1
2-Methyl-1-butanol	lq	−356.6			220.1
3-Methyl-1-butanol	lq	−356.4			210.0
2-Methyl-2-butanol	lq	−379.5	−175.3	229.3	247.1
(±)-3-Methyl-2-butanol	lq	−366.6			232.2
3-Methyl-2-butanone	lq	−299.5		268.5	179.9
	g	−262.5			
2-Methyl-1-butene	lq	−61.1		254.0	157.2
	g	−35.3	65.6	339.5	110.0
3-Methyl-1-butene	lq	−51.5		253.3	156.1
	g	−27.6	74.8	333.5	118.6
2-Methyl-2-butene	lq	−68.6		251.0	152.8
	g	−41.8	59.7	338.6	105.0
trans-2-Methyl-2-butenedioic acid [also *cis*]	c	−824.4			
cis-2-Methyl-2-butenoic acid	c	−455.6			
trans-2-Methyl-2-butenoic acid	c	−490.8			
3-Methylbutyl acetate	lq				248.5
3-Methyl-1-butyne	g	136.4	205.5	319.0	104.7
Methyl *trans*-2-butenoate	lq	−382.8			
Methylcyclobutane	lq	−44.5			
Methylcyclobutanecarboxylic acid	lq	−395.0			
Methylcyclohexane	lq	−190.1	20.3	247.9	184.9
	g	−154.7	27.3	343.3	135.0
cis-2-Methylcyclohexanol	lq	−390.2			200[17]
trans-2-Methylcyclohexanol	lq	−415.8			200[17]
cis-3-Methylcyclohexanol	lq	−416.1			292[17]
trans-3-Methylcyclohexanol	lq	−394.4			202[17]
cis-4-Methylcyclohexanol	lq	−413.2			202[17]
trans-4-Methylcyclohexanol	lq	−433.3			202[17]
2-Methylcyclohexene	lq	−81.2			
Methylcyclopentane	lq	−138.0	31.5	247.9	158.7
	g	−106.2	35.8	339.9	109.8

(*Continued*)

TABLE 2.53 Enthalpies and Gibbs Energies of Formation, Entropies, and Heat Capacities of Organic Compounds (*Continued*)

Substance	Physical state	$\Delta_f H°$ kJ · mol^{-1}	$\Delta_f G°$ kJ · mol^{-1}	$S°$ J · deg^{-1}·mol^{-1}	$C_p°$ J · deg^{-1} · mol^{-1}
1-Methylcyclopentanol	lq	−343.3			
2-Methylcyclopentanone	lq	−265.3			
1-Methylcyclopentene	g	−3.8	102.1	326.4	100.8
3-Methylcyclopentene	g	7.4	115.0	330.5	100.0
4-Methylcyclopentene	g	14.6	121.6	328.9	100.0
1-Methylcyclopropene	lq	1.7			
	g	243.6			
Methylenecyclobutane	g	121.6			
Methylenebutanedioic acid	c	−841.1			
Methylenecyclohexane	lq	−61.3			
Methylenecyclohexene	lq	−12.7			
Methylenecyclopropane	g	200.5			
Methyl decanoate	lq	−640.4			
Methyl 2,2-dimethylpropanoate	lq	−530.0			257.9
2-Methyl-1,3-dioxane	lq	−436.4			
4-Methyl-1,3-dioxane	c	416.1			
N-Methyldiphenylamine	lq	120.5			
4-Methyldiphenylamine	c	49.0			
Methyl dodecanoate	lq	−693.0			
Methylene (CH_2)	g	390.4	372.9	194.9	33.8
Methylenebutanedioic acid	c	−841.1			
Methylenecyclohexane	lq	−61.3			
2-Methylenecyclohexanol	lq	−277.6			
3-Methylenecyclohexene	lq	−12.7			
2-Methylenecyclopentanol	lq	46.9			
Methylenecyclopropane	g	200.5			
Methylenesuccinic acid	c	−841.2			
Methylene sulfate	c	−688.7			
N-Methylformamide	lq				123.8
Methyl formate	lq	−386.1			119.1
	g	−357.4	−297.2	285.3	64.4
Methyl 2-furancarboxylate	lq	−450.0			
2-Methyl-2,5-furandione	lq	−504.5			
α-Methyl-(+)-glucoside	c	−1233.4			
N-Methylglycine	c	−513.3			
Methylglyoxal	g	−27.1			
Methylglyoxime	c	−126.8			
2-Methylheptane	lq	−255.0		356.4	252.0
	g	−215.4	12.8	452.5	
3-Methylheptane	lq	−252.3		362.6	250.2
	g	−212.5	13.7	461.6	
4-Methylheptane	lq	−251.6			251.1
	g	−212.0	16.7	453.3	
Methyl heptanoate	lq	−567.1			285.1
2-Methylhexane	lq	−229.5		323.3	222.9
	g	−194.6	3.2	420.0	166.0
3-Methylhexane	lq	−226.4			214.2
	g	−192.3	4.6	424.1	166.0
Methyl hexanoate	lq	−540.2			
5-Methyl-1-hexene	g	−65.7			
cis-3-Methyl-3-hexene	g	−79.4			

TABLE 2.53 Enthalpies and Gibbs Energies of Formation, Entropies, and Heat Capacities of Organic Compounds (*Continued*)

Substance	Physical state	$\Delta_f H°$ kJ · mol^{-1}	$\Delta_f G°$ kJ · mol^{-1}	$S°$ J · deg^{-1}·mol^{-1}	$C_p°$ J · deg^{-1} · mol^{-1}
trans-3-Methyl-3-hexene	g	−76.8			
Methylhydrazine	lq	54.2	179.9	165.9	134.9
	g	94.7	186.9	278.7	71.1
2-Methyl-1*H*-indole	c	60.7			
3-Methyl-1*H*-indole	c	68.2			
Methyl isocyanate	lq	−92.0			
Methyl isocyanide	g	163.5	165.7	246.8	52.9
1-Methyl-4-isopropylbenzene	lq	−78.0			236.4
Methyl isopropyl sulfide	g	−90.4	13.4	359.3	117.2
Methyl isothiocyanate	c	79.4			
	g	131.0	144.4	252.3	65.5
5-Methylisoxazole	lq	−5.6			
Methylmercury bromide	c	−86.2			
Methylmercury chloride	c	−116.3			
Methylmercury iodide	c	−43.5			
Methyl 2-methylbutanoate	lq	−534.3			
Methyl 3-methylbutanoate	lq	−538.9			
7-Methyl-3-methylene-1,6-octadiene	lq	14.5			
(*R*)-1-Methyl-4-(1-methylethenyl)cyclohexene	lq	−54.5			249[20]
1-Methylnaphthalene	lq	56.3	189.4	254.8	224.4
2-Methylnaphthalene	c	44.9	192.6	220.0	196.0
	g	106.7	216.2	380.0	159.8
Methyl nitrate	lq	−156.3	−43.5	217.2	157.3
	g	−124.4	−39.3	318.5	76.5
Methyl nitrite	g	−66.1	1.0	284.3	63.2
Methyl nitroacetate	lq	−464.0			
2-Methyl-5-nitroaniline	c	−91.3			
4-Methyl-3-nitroaniline	c	−71.7			
1-Methyl-2-nitrobenzene	lq	−9.7			
1-Methyl-3-nitrobenzene	lq	−31.5			
1-Methyl-4-nitrobenzene	c	−48.1			
2-Methyl-2-nitropropane	c	−229.8			
2-Methyl-2-nitro-1,3-propanediol	c	−575.3			
2-Methyl-2-nitro-1-propanol	c	−410.0			
2-Methylnonane	lq	−309.8		420.1	313.3
5-Methylnonane	lq	−307.9		423.8	314.4
Methyl phenylcarbamate	c	−186.7			
Methyl *cis*-9-octadecanoate	lq	−734.5			
Methyl octanoate	lq	−590.3			
2-Methyl-2-oxazoline	g	−130.5			
2-Methylpentane	lq	−204.6		290.6	193.7
	g	−174.8	−5.0	380.5	144.2
3-Methylpentane	lq	−202.4		292.5	190.7
	g	−172.1	2.1	379.8	143.1
2-Methyl-2,4-pentanediol	lq				236.0
Methyl pentanoate	lq	−514.2			229.3
2-Methyl-1-pentanol	lq				248.0
2-Methyl-3-pentanol	lq	−396.4			

(*Continued*)

TABLE 2.53 Enthalpies and Gibbs Energies of Formation, Entropies, and Heat Capacities of Organic Compounds (*Continued*)

Substance	Physical state	$\Delta_f H°$ kJ · mol^{-1}	$\Delta_f G°$ kJ · mol^{-1}	$S°$ J · deg^{-1}·mol^{-1}	$C_p°$ J · deg^{-1} · mol^{-1}
3-Methyl-2-pentanol	lq				275.9
3-Methyl-3-pentanol	lq				293.4
4-Methyl-2-pentanol	lq	−394.7			273.0
2-Methyl-3-pentanone	lq	−325.9			
4-Methyl-2-pentanone	lq				213.3
2-Methyl-1-pentene	g	−59.4	77.6	382.2	135.6
2-Methyl-2-pentene	g	−66.9	71.2	378.4	126.6
3-Methyl-1-pentene	g	−49.5	86.4	376.8	142.4
cis-3-Methyl-2-pentene	g	−62.3	73.2	378.4	126.6
trans-3-Methyl-2-pentene	g	−63.1	71.3	381.8	126.6
4-Methyl-1-pentene	g	−51.3	90.0	367.7	126.5
cis-4-Methyl-2-pentene	g	−57.5	82.1	373.3	133.6
trans-4-Methyl-2-pentene	g	−61.5	79.6	368.3	141.4
Methyl 2-methylpropenoate	lq				191.2
4-Methyl-3-penten-2-one	lq				212.5
Methyl pentyl sulfide	g	122.9	35.1	450.7	163.7
3-Methyl-1-phenyl-1-butanone	lq	−220.2			
Methyl phenyl sulfide	lq	43.0			
Methyl phenyl sulfone	c	−345.4			
Methylphosphonic acid	c	−1054			
(±)-2-Methylpiperidine	lq	−124.9			
2-Methylpropanal	lq	−247.4			
	g	−215.8			
N-Methylpropanamide	lq				179
2-Methylpropanamine	lq	−132.6			183.2
2-Methylpropane	g	−134.2	−20.9	294.6	130.5^{-12}
2-Methyl-1,2-propanediamine	lq	−133.9			
2-Methyl-1,2-propanediol	lq	−539.7			
2-Methylpropanenitrile	lq	−13.8			
2-Methyl-1-propanethiol	g	−97.3	5.6	362.9	118.3
2-Methyl-2-propanethiol	g	−109.6	0.7	338.0	121.0
2-Methylpropanoic acid	lq				173
2-Methyl-1-propanol	lq	−334.7		214.7	181.2
	g	−283.9	−167.35	359.0	111.3
2-Methyl-2-propanol	lq	−359.2		193.3	219.8
	g	−312.5	−177.7	326.7	113.6
2-Methylpropene	g	−16.9	58.1	293.6	89.1
2-Methylpropenoic acid	lq				161.1
1-Methyl-2-propylbenzene	lq	−72.5			
1-Methyl-3-propylbenzene	lq	−76.2			
1-Methyl-4-propylbenzene	lq	−75.1			
(2-Methylpropyl)benzene	lq	−69.8			240.6
Methyl propyl ether	lq	−266.0		262.9	165.4
	g	−238.2	−109.9	349.5	112.5
Methyl propyl sulfide	g	−82.3	18.4	371.7	117.4
2-Methylpyridine	lq	56.7	166.5	217.9	158.4
	g	99.2	177.1	325.0	100.0
3-Methylpyridine	lq	61.9	214.0	216.3	158.7
	g	106.4	184.3	325.0	99.6
4-Methylpyridine	lq	59.2		209.1	159.0
1-Methyl-1*H*-pyrrole	lq	62.4			

TABLE 2.53 Enthalpies and Gibbs Energies of Formation, Entropies, and Heat Capacities of Organic Compounds (*Continued*)

Substance	Physical state	$\Delta_f H°$ kJ · mol^{-1}	$\Delta_f G°$ kJ · mol^{-1}	$S°$ J · deg^{-1}·mol^{-1}	$C_p°$ J · deg^{-1} · mol^{-1}
2-Methyl-1*H*-pyrrole	lq	23.3			
3-Methyl-1*H*-pyrrole	lq	20.5			
N-Methylpyrrolidone	lq	−262.2			307.8
2-Methylquinoline	c	164.4			
Methyl salicylate	lq	−531.8			249.0
Methylsilane	g			256.5	65.9
α-Methylstyrene	g	113.0	208.5	383.7	145.2
cis-(β)-Methylstyrene	g	121.3	216.9	383.7	145.2
trans-(β)-Methylstyrene	g	117.2	213.7	380.3	146.0
Methylsuccinic acid	c	−958.2			
Methylsuccinic anhydride	lq	−617.6			
Methyl tetradecanoate	lq	−743.9			
2-Methylthiacyclopentane	g	−63.3			
4-Methylthiazole	lq	68.0			
Methylthiirane	g	45.8			
2-Methylthiophene	lq	44.6			149.8
	g	83.5	122.9	320.6	95.4
3-Methylthiophene	lq	43.1			
	g	82.6	121.8	321.3	94.9
Methyl *p*-tolyl sulfone	c	−372.8			
5-Methyluracil	c	−462.8			
Methylurea	c	−332.8			
Morphine monohydrate	c	−711.7			
Morpholine	lq				164.8
Murexide	c	−1212.1			
Naphthalene	c	77.9	201.6	167.4	165.7
	g	150.6	224.1	333.1	131.9
1-Naphthaleneacetic acid	c	−359.2			
2-Naphthaleneacetic acid	c	−371.9			
1-Naphthoic acid	c	333.5			
2-Naphthoic acid	c	−346.1			
1-Naphthol	c	−121.0			166.9
2-Naphthol	lq	−124.2			
1,4-Naphthoquinone	c	−183.4			
1-Naphthyl acetate	c	−288.2			
2-Naphthyl acetate	c	−304.3			
1-Naphthylamine	c	67.8			
2-Naphthylamine	c	59.7			
Nicotine	lq	39.3			
Nitrilotriacetic acid	c	−1311.9	−1307.5		
Nitroacetone	lq	−278.6			
2-Nitroaniline	c	−26.1	178.2	176.2	166.0
3-Nitroaniline	c	−38.3	174.1	176.2	158.8
4-Nitroaniline	c	−42.0	151.0	176.2	167.0
Nitrobenzene	lq	12.5	146.2	224.3	185.8
2-Nitrobenzoic acid	c	−378.5	−196.4	208.4	
3-Nitrobenzoic acid	c	−394.7	−220.5	205.0	
4-Nitrobenzoic acid	c	−392.2	−222.0	210.0	181.2
3-Nitrobiphenyl	c	65.1			
4-Nitrobiphenyl	c	40.5			
1-Nitrobutane	g	−143.9	10.1	394.5	124.9

(*Continued*)

TABLE 2.53 Enthalpies and Gibbs Energies of Formation, Entropies, and Heat Capacities of Organic Compounds (*Continued*)

Substance	Physical state	$\Delta_f H°$ kJ · mol^{-1}	$\Delta_f G°$ kJ · mol^{-1}	$S°$ J · deg^{-1}·mol^{-1}	$C_p°$ J · deg^{-1} · mol^{-1}
2-Nitrobutane	g	−163.6	−6.2	383.3	123.5
3-Nitro-2-butanol	lq	−390.0			
N-Nitrodiethylamine	lq	−106.2			
2-Nitrodiphenylamine	c	64.4			
Nitroethane	lq	−143.9			134.4
	g	−102.3	−4.9	315.4	78.2
2-Nitroethanol	lq	−350.7			
2-Nitrofuran	c	−104.1			
5-Nitrofurancarboxylic acid	c	−516.8			
1-Nitroguanidine	c	−92.4			
Nitromethane	lq	−113.1	−14.4	171.8	106.6
	g	−74.3	−6.8	275.0	57.3
(Nitromethyl)benzene	lq	−22.8			
1-Nitronaphthalene	c	42.6			
1-Nitroso-2-naphthol	c	−50.5			
2-Nitroso-1-naphthol	c	−61.8			
4-Nitroso-1-naphthol	c	−107.8			
1-Nitropropane	lq	−167.2			175.3
	g	−123.8			
2-Nitropropane	lq	−180.3			170.3
	g	−139.0			
1-Nitro-2-propanone	c	−294.7			
4-Nitrosodiphenylamine	c	213.0			
β-Nitrostyrene	c	30.5			
4-Nitrotoluene	c	−48.1			172.3
Nonadecane	g	−435.1	108.9	895.2	440.4
1-Nonadecene	g	−309.6	196.7	891.0	429.7
1-Nonanal	g	−310.3	−74.9	539.6	216.8
Nonane	lq	−274.7			284.4
	g	−228.2	24.8	505.7	211.7
1-Nonanethiol	g	−190.8	53.0	571.2	232.7
Nonanoic acid	lq	−659.7			362.4
1-Nonanol	g	−376.3	−110.5	558.6	224.3
2-Nonanone	lq	−397.2			
5-Nonanone	lq	−398.2		401.4	303.6
1-Nonene	g	−103.5	112.7	501.5	201.0
Norleucine	c	−639.1			
Octadecane	c	−567.4		480.2	485.6
	g	−414.6	100.5	856.2	417.6
Octadecanoic acid	c	−947.7			501.5
1,8-Octadecanoic acid	c	−1038.1			
1-Octadecene	g	−289.0	188.3	852.0	406.8
cis-9-Octadecenoic acid	lq	−743.5			577.0^{50}
trans-9-Octadecenoic acid	c	−910.9			
1,7-Octadiyne	lq	334.4			
Octafluorocyclobutane	lq				209.8^{-6}
	g	−1542.6	−1398.8	400.4	156.2
Octafluoropropane	g	−1783.1			
Octafluorotoluene	lq	−1311.1		355.5	262.3
1-Octanal	g	−289.6	−83.3	500.7	194.0
Octanamide	c	−473.2			

TABLE 2.53 Enthalpies and Gibbs Energies of Formation, Entropies, and Heat Capacities of Organic Compounds (*Continued*)

Substance	Physical state	$\Delta_f H°$ kJ · mol^{-1}	$\Delta_f G°$ kJ · mol^{-1}	$S°$ J · deg^{-1}·mol^{-1}	$C_p°$ J · deg^{-1} · mol^{-1}
Octane	lq	−250.1			254.6
	g	−208.6	16.4	466.7	188.9
1-Octanenitrile	lq	−107.3			
1-Octanethiol	g	−44.9	44.6	582.2	209.8
Octanoic acid	lq	−636.0			297.9
1-Octanol	lq	−426.5	−143.1	377.4	305.1
2-Octanol	lq				330.1
2-Octanone	lq	−384.5	−140.3	373.8	273.3
1-Octene	lq	−121.8			241.0
	g	−81.4	104.2	462.5	178.1
cis-2-Octene	lq	−135.7			239.0
trans-2-Octene	lq	−135.7			239.0
1-Octyne	g	82.4	235.4	496.6	174.0
(±)-Ornithine	c	−652.7			
Oxalic acid	c	−821.7	−697.9	109.8	91.0
Oxalic acid dihydrate	c	−1492.0			
Oxaloyl dichloride	lq	−367.6			
Oxaloyl dihydrazide	c	−295.2			
Oxamic acid	c	−661.2			
Oxamide	c	−504.4	−342.7	118.0	
Oxazole	g	−5.5			
2-Oxetanone	lq	−329.9		175.3	122.1
Oxindole	c	−172.4			
2-Oxohexamethyleneimine	c	−329.4	−95.1	168.6	156.8
Oxomethyl (HCO)	g	43.1	28.0	224.7	34.6
2-Oxo-1,5-pentanedioic acid	c	−1026.2			
4-Oxopentanoic acid	c	−697.1			
2-Oxopropanoic acid	lq	−584.5	−463.4	179.5	
8-Oxypurine	c	−64.4			
Papaverine	c	−502.3			
Paraformaldehyde	c	−177.6			
Paraldehyde	lq	−687.0			
Pentachloroethane	lq	−187.6			173.8
	g	−142.0	−70.3	381.5	118.1
Pentachlorofluoroethane	g	−317.2	−234.0	391.8	
Pentachlorophenol	c	−292.4	−144.1	251.9	202.0
Pentacyclo[4.2.0.0^{2,5}.0^{3,8}.0^{4,7}]-octane	c	541.8			
Pentadecane	g	−352.8	75.2	739.4	349.0
Pentadecanoic acid	c	−861.7			443.3
1-Pentadecene	g	−227.2	163.1	735.2	338.2
1-Pentadecyne	g	−61.8	293.9	719.3	33.41
1,2-Pentadiene	g	140.7	210.4	333.5	105.4
cis-1,3-Pentadiene	g	81.5	145.8	324.3	94.6
trans-1,3-Pentadiene	g	76.5	146.73	319.7	103.3
1,4-Pentadiene	g	105.7	170.3	333.5	105.0
2,3-Pentadiene	g	133.1	205.9	324.7	101.3
Pentaerythritol	c	−920.6	−613.8	198.1	190.4
Pentaerythritol tetranitrate	c	−538.6			
Pentafluorobenzoic acid	c	−1239.6			
Pentafluoroethane	g	−1104.6	−1029.3	333.7	95.7

(*Continued*)

TABLE 2.53 Enthalpies and Gibbs Energies of Formation, Entropies, and Heat Capacities of Organic Compounds (*Continued*)

Substance	Physical state	$\Delta_f H°$ kJ · mol^{-1}	$\Delta_f G°$ kJ · mol^{-1}	$S°$ J · deg^{-1}·mol^{-1}	$C_p°$ J · deg^{-1} · mol^{-1}
Pentafluorophenol	c	−1024.1			
2,3,4,5,6-Pentafluorotoluene	lq	−883.8		306.4	225.8
Pentamethylbenzene	c	−133.6			
	g	−74.5	123.3	443.9	216.5
Pentamethylbenzoic acid	c	−536.1			
Pentanal	g	−228.5	−108.3	383.0	125.4
Pentanamide	c	−379.5			
1-Pentanamine	lq				218.0
Pentane	lq	−173.5	−9.3	262.7	167.2
	g	−146.9	−8.4	349.0	120.2
1,5-Pentanediol	lq	−531.5			321.3
2,4-Pentanedione	lq	−423.8			208.2
	g	−380.6		397.9	120.1
1,5-Pentanedithiol	g	−71.0			
Pentanenitrile	lq	−33.1			180
1-Pentanethiol	lq	−151.3			
Pentanoic acid	lq	−559.4		259.8	210.3
	g	−491.9	−357.2	439.8	
1-Pentanol	lq	−351.6			208.1
	g	−294.7	−146.0	402.5	133.1
2-Pentanol	lq	−365.2			
	g	−311.0			
3-Pentanol	lq	−368.9			239.7
	g	−311.4	−158.2	382.0	
2-Pentanone	lq	−297.3			184.1
	g	−259.0	−137.1	376.2	121.0
3-Pentanone	lq	−296.5		266.0	190.9
1-Pentene	lq	−46.0		262.6	154.0
	g	−21.2	79.1	345.8	109.6
cis-2-Pentene	lq	−53.7		258.6	151.7
	g	−27.6	71.8	346.3	101.8
trans-2-Pentene	lq	−58.2		256.5	157.0
	g	−31.9	69.9	340.4	108.5
cis-2-Pentenenitrile	lq	71.8			
trans-2-Pentenenitrile	lq	74.9			
trans-3-Pentenenitrile	lq	80.9			
2-Pentenoic acid	lq	−446.4			
3-Pentenoic acid	lq	−434.8			
4-Pentenoic acid	lq	−430.6			
cis-3-Penten-1-yne	lq	226.5			
trans-3-Penten-1-yne	lq	228.2			
Pentyl acetate	lq				261.0
1-Pentyne	g	144.4	210.3	329.8	106.7
2-Pentyne	g	128.9	194.2	331.8	98.7
Perfluoropiperidine	lq	−2020.5	−1768.5	393.4	296.8
Perylene	c	182.8			
α-Phellandrene	lq	41.3			
Phenanthrene	c	116.2	268.3	215.1	220.6
9,10-Phenanthrenedione	c	−154.7			
Phenazine	c	237.0			
Phenol	c	−165.1	−50.4	144.0	127.4

TABLE 2.53 Enthalpies and Gibbs Energies of Formation, Entropies, and Heat Capacities of Organic Compounds (*Continued*)

Substance	Physical state	$\Delta_f H°$ kJ · mol^{-1}	$\Delta_f G°$ kJ · mol^{-1}	$S°$ J · deg^{-1}·mol^{-1}	$C_p°$ J · deg^{-1} · mol^{-1}
	lq				199.8[41]
	g	−96.4	−32.9	315.6	103.6
Phenoxyacetic acid	c	−513.8			
Phenyl acetate	lq	−334.9			
Phenylacetic acid	c	−398.7			
Phenylacetylene	g	327.3	363.5	321.7	114.9
(±)-3-Phenyl-2-alanine	c	−466.9	−211.7	213.6	203.0
Phenyl benzoate	c	−241.0			
Phenylboron dichloride	lq	−299.4			
1-Phenylcyclohexene	lq	−16.8			
Phenylcyclopropane	lq	100.3			
N-Phenyldiacetimide	c	−362.5			
1,3-Phenylenediamine	c	−7.8		154.5	159.6
Phenyl formate	lq	−268.7			
N-Phenylglycine	c	−402.5			
(±)-2-Phenylglycine	c	−431.8			
Phenylhydrazine	lq	141.0			217.0
Phenyl 2-hydroxybenzoate	c	−436.6			
Phenylmethanethiol	lq	43.5			
Phenylmethyl acetate	lq				148.5
N-Phenyl-2-naphthylamine	c	159.8			
1-Phenyl-1-propanone	lq	−167.2			
1-Phenyl-2-propanone	lq	−151.9			
1-Phenylpyrrole	c	154.3			
2-Phenylpyrrole	c	139.2			
Phenylsuccinic acid	c	−841.0			
S-Phenyl thioacetate	lq	−122.0			
Phenyl vinyl ether	lq	−26.2			
Phosgene	g	−220.9	−206.8	283.8	57.7
Phthalamide	c	−433.1			
1,2-Phthalic acid	c	−782.0	−591.6	207.9	188.3
1,3-Phthalic acid	c	−803.0			
1,4-Phthalic acid	c	−816.1			
Phthalic anhydride	c	−460.1	−331.0	180.0	160.0
Phthalonitrile	c	280.6			
Picric acid	c	−214.4			
α-Pinene	lq	−16.4			
β-Pinene	lq	−7.7			
Piperazine	c	−45.6	240.2	85.8	
2,5-Piperazinedione	c	−446.5			
Piperidine	lq	−86.4		210.0	179.9
2-Piperidone	c	−306.6	−112.1	164.9	(lq 307.8)
L-Proline	c	515.2			
Propadiene	g	190.5	202.4	243.9	59.0
Propanal	lq	−215.3			137.2
	g	−185.6	−130.5	304.5	80.7
Propanamide	c	−338.2			
Propane	lq				98.3[−43]
	g	−103.8	−23.4	270.2	73.6
Propanediamide	c	−546.1			
(±)-1,2-Propanediamine	lq	−97.8			

(*Continued*)

TABLE 2.53 Enthalpies and Gibbs Energies of Formation, Entropies, and Heat Capacities of Organic Compounds (*Continued*)

Substance	Physical state	$\Delta_f H°$ kJ · mol^{-1}	$\Delta_f G°$ kJ · mol^{-1}	$S°$ J · deg^{-1}·mol^{-1}	$C_p°$ J · deg^{-1} · mol^{-1}
1,2-Propanediol	lq	−485.7			190.8
1,3-Propanediol	lq	−464.9			
1,2-Propanedione	lq	−309.1			
Propanedinitrile	lq	186.4			
1,2-Propanedithiol	lq	−79.4			
1,3-Propanedithiol	lq	−79.4			
Propanenitrile	lq	15.5	89.2	189.3	119.3
1-Propanethiol	lq	−99.9		242.5	144.6
	g	−67.9	2.2	336.4	94.8
2-Propanethiol	lq	−105.0		233.5	145.3
	g	−76.2	−2.6	324.3	96.0
1,2,3-Propanetriol tris(acetate)	lq	−1330.8		458.3	384.7
Propanoic acid	lq	−510.7	−383.5	191.0	152.8
Propanoic anhydride	lq	−679.1	−475.6		235.0
1-Propanol	lq	−302.6	−170.6	193.6	143.7
	g	−255.1	−161.8	322.7	85.6
2-Propanol	lq	−318.1	−180.3	181.1	155.0
	g	−272.6	−173.4	309.2	89.3
2-Propenal	g	−85.8	−64.6		
Propene	g	20.0	62.8	266.6	64.3
trans-1-Propene-1,2-dicarboxylic acid	c	−824.4			
2-Propenenitrile	lq	147.1			108.8
	g	180.6	195.4	274.1	63.8
cis-1,2,3-Propenetricarboxylic acid	c	−1224.7			
trans-1,2,3-Propenetricarboxylic acid	c	−1233.0			
2-Propenoic acid	lq	−383.8			145.7
	g	−336.5	−286.3	315.2	77.8
2-Propen-1-ol	lq	−171.8			138.9
	g	−124.5	−71.3	307.6	76.0
2-Propenyl acetate	lq	−386.2			184.1
cis-1-Propenylbenzene	g	121.3	216.9	383.7	145.2
trans-1-Propenylbenzene	g	117.2	213.7	380.3	146.0
2-Propenylbenzene	lq	88.0			
Propyl acetate	lq				196.2
Propylamine	lq	−101.5			162.5
	g	−70.2	39.8	325.1	91.2
Propylbenzene	lq	−38.3		287.8	214.7
	g	7.9	137.2	400.7	152.3
Propylcarbamate	c	−552.6			
Propylchloroacetate	lq	−515.6			
Propylchlorocarbonate	g	−492.7			
Propylcyclohexane	lq	−237.4		311.9	242.0
	g	−192.5	47.3	419.5	184.2
Propylcyclopentane	lq	−188.8		310.8	216.8
	g	−147.1	52.6	417.3	154.6
Propylene carbonate	lq	−613.2			218.6
Propylene oxide	lq	−123.0		196.5	120.4
	g	−94.7	−25.8	286.9	72.6

TABLE 2.53 Enthalpies and Gibbs Energies of Formation, Entropies, and Heat Capacities of Organic Compounds (*Continued*)

Substance	Physical state	$\Delta_f H°$ kJ · mol^{-1}	$\Delta_f G°$ kJ · mol^{-1}	$S°$ J · deg^{-1}·mol^{-1}	$C_p°$ J · deg^{-1} · mol^{-1}
Propyl formate	lq	−500.3			171.4
Propyl nitrate	g	−173.9	−27.3	385.4	121.3
S-Propyl thioacetate	lq	−294.1			
Propyl trichloroacetate	lq	−513.0			
Propyl vinyl ether	lq	−190.9			
2-Propynyl-1-amine	lq	205.7			
Propyne	g	184.9	194.4	248.1	60.7
2-Propynoic acid	lq	−193.2			
1*H*-Purine	c	169.4			
Pyrazine	c	139.8			
1*H*-Pyrazole	c	116.0			
	lq	105.4			
Pyrene	c	125.5		224.9	229.7
Pyridazine	lq	224.8			
Pyridine	lq	100.2	181.3	177.9	132.7
	g	140.4	190.2	282.8	78.1
3-Pyridinecarbonitrile	c	193.4			
3-Pyridinecarboxylic acid	c	−344.9			
Pyrimidine	lq	145.9			
1*H*-Pyrrole	lq	63.1		156.4	127.7
Pyrrole-2-carboxaldehyde	c	−106.4			
Pyrrole-2-carboldoxime	c	12.1			
Pyrrolidine	lq	−41.0		204.1	156.6
	g	−3.6	114.7	309.5	81.1
(±)-2-Pyrrolidinecarboxylic acid	c	−524.2			
2-Pyrrolidone	c	−286.2			164.4
Quinhydrone	c	−82.8	−323.0	325.9	277.0
Quinidine	c	−160.3			
Quinine	c	−155.2			
Quinoline	lq	141.2	275.7	217.2	194.9
Raffinose	c	−3184			
L-(+)-Rhamnose	c	−1073.2			
D-(−)-Ribose	c	−1047.2			
Salicylaldehyde	lq	−279.9			222[18]
Salicylaldoxime	c	−183.7			
Salicylic acid	c	−589.5	−418.1	178.2	
Semicarbazide std. state	aq	−166.9	−40.6	297.9	
(−)-Serine	c	−732.7			
(±)-Serine	c	−739.0			
L-(−)-Sorbose	c	−1271.5	−908.4	220.9	
5,5′-Spirobis(1,3-dioxane)	c	−702.1			
Spiro[2.2]pentane	lq	157.5		193.7	134.5
	g	185.2	265.3	282.2	88.1
cis-Stilbene	lq	183.3			
trans-Stilbene	c	136.9	317.6	251.0	
(−)-Strychnine	c	−171.5			
Styrene	lq	103.8	202.4	237.6	182.0
	g	147.9	213.8	345.1	122.1
Succinic acid	c	−940.5	−747.4	167.3	153.1
Succinic acid monoamide	c	−581.2			

(*Continued*)

TABLE 2.53 Enthalpies and Gibbs Energies of Formation, Entropies, and Heat Capacities of Organic Compounds (*Continued*)

Substance	Physical state	$\Delta_f H°$ kJ · mol^{-1}	$\Delta_f G°$ kJ · mol^{-1}	$S°$ J · deg^{-1}·mol^{-1}	$C_p°$ J · deg^{-1} · mol^{-1}
Succinic anhydride	c	−608.6			
Succinimide	c	−459.0			
Succinonitrile	lq	139.7		191.6	145.6
(+)-Sucrose	c	−2226.1	−1544.7	360.2	
(±)-Tartaric acid	c	−1290.8			
(−)-Tartaric acid	c	−1282.4			
meso-Tartaric acid	c	−1279.9			
α-Terpinene	g	−20.5			
1,1,2,2,-Tetrabromoethane	lq				165.7
Tetrabromoethylene	g			387.1	102.7
Tetrabromomethane	c	29.4	47.7	212.5	144.3
	g	83.9	67.0	358.1	91.2
Tetrabutyltin	lq	−304.6			
Tetracene	c	158.8			
Tetrachloro-1,4-benzoquinone	c	−288.7			
1,1,2,2,-Tetrachloro-1,2-difluoroethane	lq				178.6
	g	−489.9	−407.1	382.8	123.4
1,1,1,2-Tetrachloroethane	lq				153.8
	g	−149.4	−80.3	355.9	102.7
1,1,2,2,-Tetrachloroethane	lq	−195.0	−95.0	246.9	162.3
	g	−149.2	−85.6	362.7	100.8
Tetrachloroethylene	lq	−50.6			143.4
	g	−10.9	3.0	266.9	
Tetrachloromethane	lq	−128.2	−62.6	216.2	130.7
	g	−95.7	−53.6	309.9	83.4
1,1,1,3-Tetrachloropropane	lq	−207.8			
1,2,2,3-Tetrachloropropane	lq	−251.8			
1,1,2,2-Tetracyanocyclopropane	c	590			
Tetracyanoethylene	c	623.8			
Tetracyanomethane	c	611.6			
Tetradecane	g	−332.1	66.9	700.4	326.1
Tetradecanoic acid	c	−833.5			432.0
1-Tetradecanol	c	−629.6			388.0
1-Tetradecene	g	−206.5	154.8	696.2	315.3
Tetraethylene glycol	lq	−981.6			428.8
Tetraethylgermanium	lq	−210.5			
Tetraethyllead	lq	52.7	336.4	464.6	307.4
Tetraethylsilane	lq				298.1
Tetraethyltin	lq	−95.8			
1,1,1,2-Tetrafluoroethane	g	−895.8	−826.2	316.2	86.3
Tetrafluoroethylene	g	−658.9	−623.7	300.0	80.5
Tetrafluoromethane	g	−933.6	−888.3	261.6	61.0
2,2,3,3-Tetrafluoro-1-propanol	g	−1061.3			
Tetrahydrofuran	lq	−216.2		204.3	124.0
	g	−184.2		302.4	76.3
Tetrahydro-2-furanmethanol	lq	−435.6			181.2
1,2,3,4-Tetrahydronaphthalene	lq	−29.2			217
5,6,7,8-Tetrahydro-1-naphthol	c	−285.3			

TABLE 2.53 Enthalpies and Gibbs Energies of Formation, Entropies, and Heat Capacities of Organic Compounds (*Continued*)

Substance	Physical state	$\Delta_f H°$ kJ · mol^{-1}	$\Delta_f G°$ kJ · mol^{-1}	$S°$ J · deg^{-1}·mol^{-1}	$C_p°$ J · deg^{-1} · mol^{-1}
Tetrahydro-2*H*-pyran	lq	−258.3			156.5
Tetrahydro-2*H*-pyran-2-one	lq	−436.7			
1,2,3,6-Tetrahydropyridine	lq	33.5			
Tetrahydrothiophene	lq	−72.9			
	g	−34.1	−45.8	309.6	92.5
Tetrahydrothiophene-1,1-dioxide	lq				180[20]
Tetraiodoethylene	c	305.0			
Tetraiodomethane	g	474.0	217.1	391.9	95.9
Tetramethylammonium bromide	c	−251.0			
Tetramethylammonium chloride	c	−276.4			
Tetramethylammonium iodide	c	−203.4			
1,2,3,4-Tetramethylbenzene	lq	−90.2	106.7	290.6	
1,2,3,5-Tetramethylbenzene	lq	−96.4	98.7	416.5	240.7
1,2,4,5-Tetramethylbenzene	c	−119.9	101.3	245.6	215.1
2,3,5,6-Tetramethylbenzoic acid	c	−506.1			
2,2,3,3-Tetramethylbutane	c	−269.0		273.7	239.2
	g	−225.6	22.0	389.4	192.5
1,1,2,2-Tetramethylcyclopropane	lq	−119.7			
Tetramethyllead	lq	97.9	262.8	320.1	
	g	135.9	270.7	420.5	144.0
2,2,3,3-Tetramethylpentane	lq	−278.3			271.5
2,2,3,4-Tetramethylpentane	lq	−277.7			
2,2,4,4-Tetramethylpentane	lq	−280.0			266.3
2,3,3,4-Tetramethylpentane	lq	−277.9			
Tetramethylsilane	lq	−264.0			204.1
	g	−239.1	−100.0	359.1	143.9
Tetramethylsuccinic acid	c	−1012.5			
Tetramethylthiacyclopropane	c	−83.0			
Tetramethyltin	g	−18.8			
Tetranitromethane	lq	38.4			
1,1,1,2-Tetraphenylethane	c	223.0			
1,1,2,2-Tetraphenylethane	c	216.0			
Tetraphenylethylene	c	311.5			
Tetraphenylhydrazine	c	457.9			
Tetraphenylmethane	c	247.1	574.0		
Tetraphenyltin	c	412.1			
Tetrapropylgermanium	g	−229.7			
Tetrapropyltin	lq	−211.3			
1,2,3,4-(1*H*)-Tetrazole	c	237.0			
Theobromine	c	−361.5			
2-Thiaadamantane	c	−143.5			
Thiacyclobutane	g	60.6	107.1	285.0	68.3
Thiacycloheptane	g	−61.3	84.1	361.9	124.6
Thiacyclohexane	lq	−106.3		218.2	163.3
	g	−63.5	53.1	323.0	109.7
Thiacyclopentane	g	−33.8	46.0	309.4	90.9
Thiacyclopropane	g	82.2	96.9	255.3	53.7
Thianthrene	c	−182.5			
Thiirane	g	82.0	96.8	255.2	53.3

(*Continued*)

TABLE 2.53 Enthalpies and Gibbs Energies of Formation, Entropies, and Heat Capacities of Organic Compounds (*Continued*)

Substance	Physical state	$\Delta_f H°$ kJ · mol^{-1}	$\Delta_f G°$ kJ · mol^{-1}	$S°$ J · deg^{-1}·mol^{-1}	$C_p°$ J · deg^{-1} · mol^{-1}
Thiirene	g	300.0	275.8	255.3	54.7
Thioacetamide	c	−71.7			
Thioacetic acid	lq	−216.9			
	g	−175.1	−154.0	313.2	80.9
1,2-Thiocresol	lq	44.2			
Thiohydantoic acid	c	−554.8			
Thiohydantoin	c	−249.0			
2-Thiolactic acid	lq	−468.4			
Thiophene	lq	80.2	121.2	181.2	123.8
	g	115.0	126.8	278.9	72.9
Thiophenol	lq	64.1	134.0	222.8	173.2
	g	111.6	147.6	336.9	104.9
Thiosemicarbazide	c	25.1			
Thiourea	c	−89.1	21.8	115.9	
	g	22.9			
(−)-Threonine	c	−807.2			
(±)-Threonine	c	−758.8			
Thymine	c	−462.8			150.8
Thymol	c	−309.7			
Toluene	lq	12.4	113.8	221.0	157.0
	g	50.4	122.0	320.7	103.6
1*H*-1,2,4-Triazol-3-amine	c	76.8			
2,4,6-Triamino-1,3,5-triazine	c	−72.4	184.5	149.1	
2-Triazoethanol	lq	94.6			
Tribenzylamine	c	140.6			
Tribromoacetaldehyde	lq	−130.3			
Tribromochloromethane	g	12.6	9.1	357.8	89.4
Tribromofluoromethane	g	−190.0	−193.1	345.9	84.4
Tribromomethane	lq	−28.5	8.0	220.9	130.7
	g	23.8	−5.0	330.9	71.2
Tributoxyborane	lq	−1199.6			
Tributylamine	lq	−281.6			
Tributyl phosphate	lq	−1456			
Tributylphosphine oxide	c	−460			
Trichloroacetaldehyde	lq	−234.5			151.0
2,2,2-Trichloroacetamide	c	−358.2			
Trichloroacetic acid	c	−503.3			
ionized	aq	−517.6			
Trichloroacetonitrile	g			336.6	96.1
Trichloroacetyl chloride	lq	−280.8			
Trichlorobenzoquinone	c	−269.9			
1,1,1-Trichloroethane	lq	−177.4		227.4	144.3
	g	−144.6	−76.2	323.1	93.3
1,1,2-Trichloroethane	lq	−191.5		232.6	150.9
	g	−151.2	−77.5	337.1	89.0
Trichloroethylene	lq	−43.6			124.4
	g	−9.0	19.9	324.8	80.3
Trichlorofluoromethane	lq	−301.3	−236.8	255.4	121.6
	g	−268.3	−249.3	309.7	78.0
Trichloromethane	lq	−134.5	73.7	201.7	114.2
	g	−102.7	−76.0	295.7	65.7

TABLE 2.53 Enthalpies and Gibbs Energies of Formation, Entropies, and Heat Capacities of Organic Compounds (*Continued*)

Substance	Physical state	$\Delta_f H°$ kJ · mol^{-1}	$\Delta_f G°$ kJ · mol^{-1}	$S°$ J · deg^{-1}·mol^{-1}	$C_p°$ J · deg^{-1} · mol^{-1}
1,2,2-Trichloropropane	g	−185.8	−97.8	382.9	112.2
1,2,3-Trichloropropane	lq	−230.6			183.6
	g	−182.9			
1,2,3-Trichloropropene	lq	−101.8			
1,1,2-Trichlorotrifluoroethane	lq	−805.8			170.1
1,1,1-Tricyanoethane	c	351.0			
Tricyanoethylene	c	439.3			
Tridecane	g	−311.5	58.5	661.5	303.2
Tridecanoic acid	c	−806.6			
1-Tridecene	g	−186.0	146.3	657.3	292.4
Triethanolamine	c	−664.2			389.0
Triethoxyborane	lq	−1047.4			
Triethoxymethane	lq	−687.3			
Triethylaluminum	lq	−236.8			
Triethylamine	lq	−127.7			219.9
	g	−92.8	110.3	405.4	160.9
Triethylaminoborane	lq	−198.6			
Triethyl arsenite	lq	−706.7			
Triethylarsine	lq	13.0			
Triethylbismuthine	lq	169.9			
Triethylborane	lq	−194.6	9.4	336.7	241.2
	g	−157.7	16.1	437.8	
Triethylenediamine	c	−14.2	239.7	157.6	
Triethylene glycol	lq	−804.2			
Triethyl phosphate	lq	−1243			
Triethylphosphine	lq	−89.1			
Triethyl phosphite	lq	−861.5			
Triethylstibine	lq	5.0			
Triethylsuccinic acid	c	−1066.5			
Triethyl thiophosphate	lq	−972.8			
Trifluoroacetic acid	lq	−1069.9			
Trifluoroacetonitrile	g	−497.9	−461.9	298.1	77.9
1,1,1-Trifluoroethane	g	−744.6	−678.3	279.9	78.2
1,1,2-Trifluoroethane	g	−730.7			
2,2,2-Trifluoroethanol	lq	−932.4			
Trifluoroethylene	g	−490.4	−469.5	292.6	69.2
Trifluoroiodoethane	g	−644.5			
Trifluoroiodomethane	g	−587.8	−572.0	307.5	70.9
Trifluoromethane	g	−695.4	−658.9	259.6	51.1
(Trifluoromethyl)benzene	g	−599.1	−511.3	372.6	130.4
1,1,1-Trifluoro-2,4-pentanedione	lq	−1040.2			
3,3,3-Trifluoropropene	g	−614.2			
Trihexylamine	lq	−433.0			
(±)-Trihydroxyglutaric acid	c	−1490			
2,4,6-Trihydroxypryimidine	c	−634.7			
Triiodomethane	g	251.0	178.0	356.2	75.1
Triisopropyl phosphite	lq	−980.3			
Trimethoxyborane	g	−899.1			
Trimethoxyethane	lq	−612.0			
Trimethoxymethane	lq	−570.0			

(*Continued*)

TABLE 2.53 Enthalpies and Gibbs Energies of Formation, Entropies, and Heat Capacities of Organic Compounds (*Continued*)

Substance	Physical state	$\Delta_f H°$ kJ · mol^{-1}	$\Delta_f G°$ kJ · mol^{-1}	$S°$ J · deg^{-1}·mol^{-1}	$C_p°$ J · deg^{-1} · mol^{-1}
Trimethylacetic acid	lq	−564.4			
Trimethylacetic anhydride	lq	−779.9			
2′,4′,5′-Trimethylacetophenone	lq	−252.3			
2′,4′,6′-Trimethylaceto-phenone	lq	−267.4			
Trimethylaluminum	lq	−136.4	−9.9	209.4	155.6
Trimethylamine	lq	−45.7		208.5	137.9
	g	−23.7	98.9	287.1	91.8
std. state	aq	−76.0	93.0	133.5	
Trimethylamine-aluminum chloride adduct	c	−879.1			
Trimethylamine-borane	c	−142.5	70.7	187.0	
Trimethylammonium ion, std. state	aq	−112.9	37.2	196.7	
Trimethyl arsenite	lq	−590.8			
Trimethylarsine	g	11.7			
1,2,3-Trimethylbenzene	lq	−58.5	107.5	267.8	216.4
1,2,4-Trimethylbenzene	lq	−61.8	102.3	284.2	215.0
1,3,5-Trimethylbenzene	lq	−63.4	103.9	273.6	209.3
2,3,4-Trimethylbenzoic acid	c	−486.6			
2,3,5-Trimethylbenzoic acid	c	−488.7			
2,3,6-Trimethylbenzoic acid	c	−475.7			
2,4,5-Trimethylbenzoic acid	c	−495.7			
2,4,6-Trimethylbenzoic acid	c	−477.9			
3,4,5-Trimethylbenzoic acid	c	−500.9			
2,6,6-Trimethylbicyclo-[3.1.1]-2-heptene	lq	16.4			
Trimethylbismuthine	g	192.9			
Trimethylborane	g	−124.3	−35.9	314.7	88.5
2,2,3-Trimethylbutane	g	−204.5	4.3	383.3	164.6
2,2,3-Trimethylbutane	lq	−236.5		292.2	213.5
2,3,3-Trimethyl-1-butene	lq	−117.7			
Trimethylchlorosilane	lq	−382.8	−246.4	278.2	
	g	−352.8	−243.5	369.1	
cis,cis-1,3,5-Trimethyl-cyclohexane	g	−215.4	33.9	390.4	179.6
1,1,2-Trimethylcyclopropane	lq	−96.2			
Trimethylene oxide (Oxetane)	lq	−110.8			
	g	−80.5	−9.8	273.9	
Trimethylgallium	g	−46.9			
2,3,5-Trimethylhexane	lq	−284.0			
Trimethylindium	g	170.7			
2,2,3-Trimethylpentane	lq	−256.9	9.3	327.6	188.9
	g	−220.0	17.1	425.2	
2,2,4-Trimethylpentane	lq	−259.2	6.9	328.0	239.1
	g	−224.0	13.7	423.2	
2,3,3-Trimethylpentane	lq	−253.5	10.6	334.4	245.6
	g	−216.3	18.9	431.5	
2,3,4-Trimethylpentane	lq	−255.0	10.7	329.3	247.3
2,2,4-Trimethyl-3-pentanone	lq	−381.6			
2,4,4-Trimethyl-1-pentene	lq	−145.9	86.4	306.3	

TABLE 2.53 Enthalpies and Gibbs Energies of Formation, Entropies, and Heat Capacities of Organic Compounds (*Continued*)

Substance	Physical state	$\Delta_f H°$ kJ · mol^{-1}	$\Delta_f G°$ kJ · mol^{-1}	$S°$ J · deg^{-1}·mol^{-1}	$C_p°$ J · deg^{-1} · mol^{-1}
2,4,4-Trimethyl-2-pentene	lq	−142.4	88.0	311.7	
Trimethylphosphine	lq	−122.2			
Trimethylphosphine oxide	c	−477.8			
Trimethyl phosphite	lq	−741.0			
Trimethylsilane	g			331.0	117.9
Trimethylsilanol	lq	−545.0			
Trimethylstibine	g	32.2			
Trimethylsuccinic acid	c	−1000.8			
Trimethylsuccinic anhydride	c	−688.3			
Trimethylthiacyclopropane	lq	−60.5			
Trimethyltin bromide	lq	−185.4			
Trimethyltin chloride	lq	−213.0			
Trimethylurea	c	−330.5			
Trinitroacetonitrile	lq	183.7			
2,4,6-Trinitroanisole	c	−157.3			
1,3,5-Trinitrobenzene	c	−37.2			
1,1,1-Trinitroethane	lq	−96.9			
Trinitroglycerol	lq	−370.9			
Trinitromethane	lq	−32.8			
	g	−0.2			
2,4,6-Trinitrophenetole	c	−204.6			
2,4,6-Trinitrophenol	c	−214.3			
2,4,6-Trinitrophenylhydrazine	c	36.8			
2,4,6-Trinitrotoluene	c	−65.5			
2,4,6-Trinitro-1,3-xylene	c	−102.5			
Trioctylamine	lq	−584.9			
1,3,6-Trioxacyclooctane	lq	−515.9			
1,3,5-Trioxane	c	−522.5		133.0	114.4
Triphenylamine	c	234.7	504.2		
Triphenylarsine	c	310.0			
Triphenylbismuthine	c	469.0			
Triphenylborane	c	48.5			
Triphenylene	c	151.8	329.2	254.7	
1,1,1-Triphenylethane	c	157.2			
1,1,2-Triphenylethane	c	130.2			
Triphenylethylene	c	233.5	514.6		
2,4,6-Triphenylimidazole	c	272			
Triphenylmethane	c	171.2	412.5	312.1	295.0
Triphenylmethanol	c	−3.4	272.8	329.3	
Triphenyl phosphate	c	−757			
Triphenylphosphine	c	232.2			
Triphenylphosphine oxide	c	−60.3			
Triphenylstibine	c	329.3			
Tripropoxyborane	lq	−1127.2			
Tripropylamine	lq	−207.2			
Tripropynylamine	lq	814.2			
Tris(acetylacetonato)-chromium	c	−1533.0			
Tris(diethylamino)phosphine	lq	−289.5			
1,1,1-Tris(hydroxymethyl)-ethane	c	−744.6			

(*Continued*)

TABLE 2.53 Enthalpies and Gibbs Energies of Formation, Entropies, and Heat Capacities of Organic Compounds (*Continued*)

Substance	Physical state	$\Delta_f H°$ kJ · mol^{-1}	$\Delta_f G°$ kJ · mol^{-1}	$S°$ J · deg^{-1}·mol^{-1}	$C_p°$ J · deg^{-1} · mol^{-1}
Tris(hydroxymethyl)nitro-methane	c	−735.6			
Tris(isopropoxy)borane	lq	−293.3			
Tris(trimethylsilyl)amine	c	−725.1			
(−)-Tryptophane	c	−415.3	−119.4	251.0	238.2
(−)-Tyrosine	c	−685.1	−385.7	214.0	216.4
Undecane	lq	−327.2	22.8	458.1	344.9
Undecanoic acid	c	−735.9			
1-Undecanol	lq	−504.8			
1-Undecene	g	−144.8	129.5	579.4	246.7
10-Undecenoic acid	c	−577			
Uracil	c	−429.4			120.5
Urea	c	−333.1	−196.8	104.6	93.1
	g	−245.8			
Urea nitrate	c	−564.0			
Urea oxalate	c	−1528.4			
5-Ureidohydantoin	c	−718.0	−434.0	195.1	
Uric acid	c	−618.8	−358.8	173.2	166.1
(±)-Valine	c	−628.9	−359.0	178.9	168.8
Valylphenylalanine	c	−767.8			
Vinyl acetate	g	−314.4			
Vinylbenzene	lq	103.8			
Vinylcyclohexane	lq	−88.7			
4-Vinylcyclohexene	lq	26.8			
Vinylcyclopentane	lq	−34.8			
Vinylcyclopropane	lq	122.5			
2-Vinylpyridine	lq	157.1			
Xanthine	c	−379.6	−165.9	161.1	151.3
Xanthone	c	−191.5			
1,2-Xylene	lq	−24.4	110.3	246.5	186.1
	g	19.1	122.1	352.8	133.3
1,3-Xylene	lq	−25.4	107.7	252.2	183.3
	g	17.3	118.9	357.7	127.6
1,4-Xylene	lq	−24.4	110.1	247.4	181.5
	g	18.0	121.1	352.4	126.9
Xylitol	c	−1118.5			
D-(+)-Xylose	c	−1057.8			

TABLE 2.54 Heat of Fusion, Vaporization, Sublimation, and Specific Heat at Various Temperatures of Organic Compounds

Abbreviations Used in the Table

ΔHm, enthalpy of melting (at the melting point) in kJ · mol^{-1}
ΔHv, enthalpy of vaporization (at the boiling point) in kJ · mol^{-1}
ΔHs, enthalpy of sublimation (or vaporization at 298 K) in kJ · mol^{-1}
C_p, specific heat (at temperature specified on the Kelvin scale) for the physical state in existence (or specified: c, lq, g) at that temperature in J · K^{-1} · mol^{-1}
ΔHt, enthalpy of transition (at temperature specified, superscript, measured in degrees Celsius) in kJ · mol^{-1}

				C_p			
Substance	ΔHm	ΔHv	ΔHs	400 K	600 K	800 K	1000 K
Acenaphthene	21.54	54.73	86.2				
Acenaphthylene			73.0				
Acetaldehyde	3.24	25.8	25.5	66.3(g)	85.9	101.3	112.5
Acetamide	15.71	56.1	78.7				
Acetanilide		64.7	80.8				
Acetic acid	11.54	23.7	23.4	79.7	106.2	125.5	139.3
Acetic anhydride	10.5	38.2	48.3	129.1	174.1	204.6	226.4
Acetone	5.69	29.1	31.0	92.1	122.8	144.9	162.0
Acetonitrile, $\Delta Ht = 0.22^{-56}$	8.17	29.8	32.9	61.2	76.8	89.0	98.3
Acetophenone		38.8	55.9				
Acetyl bromide			33.1				
Acetyl chloride			30.1	78.9	97.0	110.0	119.7
Acetylene	3.8	17.0	21.3	50.1	58.1	63.5	68.0
Acetylene-d_2				54.8	61.9	67.4	71.8
Acetylenedicarbonitrile			28.8	94.8	106.2	114.1	119.8
Acetyl fluoride			25.1				
Acetyl iodide			38.5				
Acrylic acid	11.16	44.1	54.3	96.0	123.4	142.0	155.3
Acrylonitrile	6.23	32.6	33.5	76.8	96.7	110.6	120.8
Adamantane			59.7				
Adenine			108.8				
α-Alanine			138.1				
Allyl *tert*-butyl sulfide			44.4				
Allyl ethyl sulfone			83.7				
Allyl ethyl sulfoxide			71.6				
Allyl methyl sulfone			79.5				
Allyl trichloroacetate			52.3				
3-Aminoacetophenone	12.1						
4-Aminoacetophenone	15.9						
2-Aminobenzoic acid	20.5		104.9				
3-Aminobenzoic acid	21.8		128.0				
4-Aminobenzoic acid	20.9		116.1				
2-Aminoethanol	20.5	50.9					
Aniline	10.56	42.4	55.8	143.0	192.8	225.1	230.9
Anthracene	28.83	56.5	101.5				
9,10-Anthraquinone		88.5	112.1				
cis-Azobenzene	22.04		92.9				
trans-Azobenzene	22.6	93.8					
Azobutane			49.3				
Azomethane				93.9	123.1	145.7	162.6
Azomethane-d_6				110.7	142.8	165.2	180.6

(*Continued*)

TABLE 2.54 Heat of Fusion, Vaporization, Sublimation, and Specific Heat at Various Temperatures of Organic Compounds (*Continued*)

				C_p			
Substance	Δ*Hm*	Δ*Hv*	Δ*Hs*	400 K	600 K	800 K	1000 K
Azoisopropane			36.0				
Azopropane			39.9				
trans-Azoxybenzene	17.93						
Azulene	12.1	55.5	76.8	176.4	248.2	295.4	327.4
Benzaldehyde	9.32	42.5	49.8				
Benzamide	18.49						
1,2-Benzanthracene			123.0				
2,3-Benzanthracene			126				
1,2-Benzanthracene-9,10-dione			82.8				
Benzene	9.95	30.7	33.8	113.5(g)	160.1	190.5	211.4
Benzeneacetic acid	14.49						
1,3-Benzenedicarboxylic acid			106.7				
1,4-Benzenedicarboxylic acid			98.3				
Benzenethiol	11.48	39.9	47.6				
Benzil	23.54						
Benzoic acid	18.06	50.6	91.1	138.4	196.7	234.9	260.7
Benzoic anhydride	17.2		96.4				
Benzonitrile	10.88	45.9	52.5	140.8	187.4	217.9	238.8
Benzo[*def*]phenanthrene	17.1		100.2				
Benzophenone	18.19		94.1				
1,4-Benzoquinone	18.53		62.8				
Benzo[*f*]quinoline			83.1				
Benzo[*h*]quinoline			80.8				
Benzo[*b*]thiophene, Δ*Ht* = $3.0^{-11.6}$	11.8						
Benzotrifluoride			37.6				
Benzoyl bromide			58.6				
Benzoyl chloride			54.8				
Benzoyl iodide			61.9				
4-Benzphenanthrene			106.3				
Benzyl acetate		49.4					
Benzyl alcohol	8.97	50.5	60.3				
Benzylamine			60.2				
Benzyl benzoate		53.6	77.8				
Benzyl bromide			47.3				
Benzyl chloride			51.5				
Benzyl ethyl sulfide			56.9				
Benzyl iodide			47.3				
Benzyl mercaptan			56.6				
Benzyl methyl ketone			49.0				
Benzyl methyl sulfide			53.6				
Bicyclo[1.1.0]butane			23.4				
Bicyclo[2.2.1]hepta-2,5-dione		32.9					
Bicyclo[2.2.1]heptane			40.2				
Bicyclo[4.1.0]heptane			38.0				
Bicyclo[2.2.1]-2-heptene			38.8				
Bicyclo[3.1.0]hexane			32.8				
Bicyclohexyl			58.0				
Bicyclo[2.2.2]octane			48.0				
Bicyclo[4.2.0]octane			42.0				
Bicyclo[5.1.0]octane			43.5				

TABLE 2.54 Heat of Fusion, Vaporization, Sublimation, and Specific Heat at Various Temperatures of Organic Compounds (*Continued*)

				C_p			
Substance	ΔHm	ΔHv	ΔHs	400 K	600 K	800 K	1000 K
Bicyclo[2.2.2]-2-octene			43.8				
Bicyclopropyl			33.5				
Biphenyl	18.6	45.6	81.8	221.0	307.7	363.7	401.7
Biphenylene			84.3				
Bis(2-butoxyethyl) ether		55.9					
Bis(2-chloroethyl) ether	8.66	45.2					
Bis(2-ethoxyethyl) ether		49.0					
Bis(2-ethoxymethyl) ether		36.2	44.7				
Bis(2-hydroxyethyl) ether		52.3	57.3				
Bis(2-methoxyethyl) ether		43.1					
Bromobenzene	10.62	37.9	44.5	127.4	171.5	199.9	219.2
4-Bromobenzoic acid			87.9				
1-Bromobutane	6.69	32.5	36.7	136.6	180.0	211.2	234.4
(±)-2-Bromobutane	6.89	30.8	34.4	138.1	214.7	238.2	
1-Bromo-2-chloroethane		33.7	38.2				
Bromochloromethane		30.0	32.8				
1-Bromo-3-chloropropane		37.6	44.1				
1-Bromo-2-chloro-1,1,2-trifluoroethane		28.3	30.1				
Bromochloro-2,2,2-trifluoroethane		28.1	29.8				
1-Bromododecane		74.8					
Bromoethane	5.86	27.0	28.0	79.2	102.8	119.6	132.2
Bromoethylene	5.12	23.4	18.2	66.6	83.0	94.1	102.3
1-Bromoheptane			50.6			74.8	
1-Bromohexadecane			94.4				
1-Bromohexane			45.9				
Bromomethane, $\Delta Ht = 0.47^{-99.4}$	5.98	23.9	22.8	50.0	62.7	72.2	79.5
1-Bromo-2-methylpropane		31.3	34.8				
2-Bromo-2-methylpropane	1.97	29.2	31.8	146.1	190.7	220.3	241.6
$\Delta Ht = 5.7^{-64.5}$							
$\Delta Ht = 1.0^{-41.6}$							
1-Bromonaphthalene	15.16	39.3	52.5				
1-Bromooctane			55.8				
1-Bromopentane	11.46	35.0	41.3	165.6	219.0	257.5	286.0
1-Bromopropane	6.53	29.8	32.0	107.5	140.8	164.9	182.8
2-Bromopropane		28.3	30.2	110.2	144.0	167.7	185.2
3-Bromopropene		30.2	32.7				
Bromotrichloromethane	2.54						
Bromotrifluoromethane				79.3	91.3	97.5	100.9
Bromotrimethylsilane			32.6				
1,2-Butadiene	7.0	24.0	23.2	98.4	128.5	150.7	167.4
1,3-Butadiene	7.98	22.5	20.9	101.2	154.1	169.5	
1,3-Butadiyne				84.4	96.8	105.1	111.3
Butanal	11.09	31.5	34.5	126.4	165.7	195.0	216.3
Butanamide	17.6		85.9				
Butane, $\Delta Ht = 2.1^{-165.6}$	4.66	22.4	21.0	123.9	168.6	201.8	226.9
1,2-Butanediamine			46.3				
Butanedinitrile	3.7	48.5	70.0				
1,3-Butanediol		58.5	67.8				
1,4-Butanediol			76.6				
2,3-Butanediol			59.2				

(*Continued*)

TABLE 2.54 Heat of Fusion, Vaporization, Sublimation, and Specific Heat at Various Temperatures of Organic Compounds (*Continued*)

				C_p			
Substance	Δ*Hm*	Δ*Hv*	Δ*Hs*	400 K	600 K	800 K	1000 K
2,3-Butanedione			38.7				
1,4-Butanedithiol			55.1				
Butanenitrile	5.02	33.7	39.3	118.8	155.1	181.9	201.8
meso-1,2,3,4-Butanetetrol			135.1				
1,4-Butanedithiol			49.7				
1-Butanethiol	10.46	32.2	36.6	146.2	194.7	233.0	263.4
2-Butanethiol	6.5	30.6	34.0	148.0	194.2	227.2	251.1
1,2,4-Butanetriol		58.6					
Butanoic acid	11.08	41.8	40.5				
Butanoic anhydride		50.0					
1-Butanol	9.28	43.3	52.3	137.2	183.7	218.0	243.8
2-Butanol		40.8	49.7	141.0	187.1	220.4	245.3
2-Butanone	8.44	31.3	34.8	124.7	163.6	192.8	214.8
trans-2-Butenal			34.5				
1-Butene	3.9	22.1	20.2	109.0	147.1	174.9	195.9
cis-2-Butene	7.58	23.3	22.2	101.8	141.4	171.0	193.1
trans-2-Butene	9.8	22.7	21.4	108.9	145.6	184.9	194.9
cis-2-Butenedinitrile			72.0				
cis-2-Butenedioic acid			110.0				
trans-2-Butenedioic acid			136.3				
cis-2-Butene-1,4-diol		66.1					
trans-2-Butene-1,4-diol		69.0					
cis-2-Butenenitrile			38.9				
trans-2-Butenenitrile			40.0				
3-Butenenitrile			40.0				
cis-2-Butenoic acid	12.57						
trans-2-Butenoic acid	12.98						
cis-2-Buten-1-ol		46.4					
1-Buten-3-yne				89.0	111.6	127.2	138.7
2-Butoxyethanol			56.6				
1-*tert*-Butoxy-2-ethoxyethane			50.9				
2-(2-Butoxyethoxy)ethanol		28.0					
2-Butoxyethyl acetate			59.5				
1-*tert*-Butoxy-2-methoxyethane		38.5	47.8				
N-Butylacetamide			76.1				
Butyl acetate		36.3	43.9				
tert-Butyl acetate		33.1	38.0				
Butylamine		31.8	35.7	148.3	197.9	234.4	261.7
sec-Butylamine		29.9	32.8	148.1	199.0	236.1	261.7
tert-Butylamine	0.88	28.3	29.6	152.6	204.5	240.5	266.9
Butylbenzene	11.22	38.9	51.4	229.1	314.6	373.9	416.3
sec-Butylbenzene	9.83	38.0	48.0				
tert-Butylbenzene	8.39	37.6	47.7				
sec-Butyl butanoate			47.3				
Butyl chloroacetate			51.0				
Butyl 2-chlorobutanoate			52.7				
Butyl 3-chlorobutanoate			53.1				
Butyl 4-chlorobutanoate			54.4				
Butyl 2-chloropropanoate			54.4				
Butyl 3-chlorobutanoate			55.4				

TABLE 2.54 Heat of Fusion, Vaporization, Sublimation, and Specific Heat at Various Temperatures of Organic Compounds (*Continued*)

				C_p			
Substance	ΔHm	ΔHv	ΔHs	400 K	600 K	800 K	1000 K
Butyl crotonate			51.9				
sec-Butyl crotonate			49.4				
Butylcyclohexane	14.16	38.5	49.4	276.1	289.5	469.9	525.9
Butylcyclopentane	11.3	36.2	45.9	241.7	336.3	407.3	480.3
N-Butyldiacetimide			64.4				
Butyl dichloroacetate			52.3				
Butylethylamine		34.0	40.2				
Butyl ethyl ether		31.6	36.3				
Butyl ethyl sulfide	12.4	37.0	44.5	202.4	271.8	325.3	367.2
tert-Butyl ethyl sulfide	7.1	33.5	39.3				
Butyl formate		36.6	41.1				
tert-Butyl hydroperoxide			47.7				
Butylisopropylamine		34.5	42.1				
Butyllithium			107.1				
Butyl methyl ether		29.6	32.4				
sec-Butyl methyl ether		28.1	30.2				
tert-Butyl methyl ether		27.9	29.8				
Butyl methyl sulfide	12.5	34.5	40.5	174.6	233.0	278.4	314.1
tert-Butyl methyl sulfide	8.4	31.5	35.8				
Butyl methyl sulfone			76.2				
tert-Butyl methyl sulfone			82.4				
Butyl octadecanoate	56.90						
tert-Butyl peroxide			31.8				
Butyl propyl ether		33.7	40.2				
Butyl thiolacetate			48.1				
Butyl trichloroacetate			53.6				
Butyl vinyl ether		31.6	36.2				
1-Butyne	6.0	24.5	23.3	99.9	129.0	150.4	166.7
2-Butyne	9.23	26.5	26.6	94.6	124.2	147.0	164.4
2-Butynedinitrile			28.8				
4-Butyrolactone	9.57	52.2					
Butyrophenone			60.7				
(+)-Camphor	6.84	59.5					
9*H*-Carbazole	26.9		84.5				
Chloroacetic acid	12.28		75.3				
Chloroacetyl chloride			38.9				
2-Chloroaniline	11.88	44.4	56.8				
2-Chlorobenzaldehyde			53.1				
Chlorobenzene	9.61	35.2	41.0	128.1	172.2	200.4	219.6
2-Chlorobenzoic acid	25.73		79.5				
3-Chlorobenzoic acid			82.0				
4-Chlorobenzoic acid			87.9				
Chloro-1,4-benzoquinone			69.0				
1-Chlorobutane		30.4	33.5	135.1	179.0	210.5	234.0
2-Chlorobutane		29.2	31.5	136.1	180.7	212.7	236.8
Chlorocyclohexane			43.5				
1-Chloro-1,1-difluoroethane	2.69	22.4					
Chlorodifluoromethane	4.12	20.2		65.4	78.9	87.2	92.4
2-Chloro-1,4-dihydroxybenzene			69.0				
Chlorodimethylsilane		26.2					

(*Continued*)

TABLE 2.54 Heat of Fusion, Vaporization, Sublimation, and Specific Heat at Various Temperatures of Organic Compounds (*Continued*)

Substance	ΔHm	ΔHv	ΔHs	C_p 400 K	600 K	800 K	1000 K
Chlorodiphenylsilane			69.5				
1-Chloro-2,3-epoxypropane		33.1	40.6				
Chloroethane	4.45	24.7		77.6	101.6	118.8	131.7
2-Chloroethanol		41.4					
1-Chloro-2-ethylbenzene			47.3				
1-Chloro-4-ethylbenzene			48.1				
Chloroethylene	4.75	20.8		65.0	82.1	93.5	101.9
2-Chloroethyl vinyl ether		38.2					
Chloroethyne				60.2	66.8	71.0	74.3
1-Chloroheptane			47.7				
1-Chlorohexane		35.7	42.8				
Chlorohydroquinone			69.0				
Chloromethane	6.43	21.4	18.9	48.2	61.3	71.3	78.9
1-Chloro-2-methylbenzene	8.37	37.5					
1-Chloro-3-methylbenzene	10.46						
1-Chloro-4-methylbenzene		38.7					
1-Chloro-3-methylbutane		32.0	36.2				
1-Chloro-2-methylpropane		29.2	31.7	136.1	180.7	212.7	236.8
2-Chloro-2-methylpropane	2.09	27.6	29.0	142.3	184.9	215.5	238.5
$\Delta Ht = 1.7^{-90.1}$							
$\Delta Ht = 5.8^{-53.6}$							
1-Chloronaphthalene	12.90	52.1	65.3				
2-Chloronaphthalene			82.0				
1-Chloro-3-nitrobenzene	19.37						
1-Chloro-4-nitrobenzene	20.77						
1-Chlorooctane			52.4				
Chloropentafluoroacetone			25.3				
Chloropentafluorobenzene		34.8	41.1				
Chloropentafluoroethane	1.88	19.4					
1-Chloropentane		33.2	38.2	164.2	218.0	256.8	285.6
2-Chloropentane		31.8	36.0				
2-Chlorophenol	12.52						
3-Chlorophenol	14.91		53.1				
4-Chlorophenol	14.07		51.9				
1-Chloropropane	5.54	27.2	28.4	106.1	139.9	164.2	182.4
2-Chloropropane	7.39	26.3	26.9	108.7	143.1	167.1	184.8
3-Chloro-1-propene		29.0	28.2	92.6	111.0	137.8	151.9
Chlorotrifluoroethylene	5.6	20.8					
Chlorotrifluoromethane		15.8		77.5	90.3	96.9	100.5
Chlorotrimethylsilane		27.6	30.1				
Chlorotrinitromethane			45.4				
Chrysene	26.15		124.5				
Coronene	19.2						
1,2-Cresol	13.94	45.2	76.0	166.3	220.8	257.5	287.9
1,3-Cresol	9.41	47.4	61.7	162.1	218.7	256.4	286.6
1,4-Cresol	11.89	47.5	73.9	161.7	218.0	255.7	286.5
Cubane			80.3				
Cyanamide	8.76	68.6					
Cyanogen	8.1	23.3	19.7	61.9(g)	68.2	72.9	76.4
Cyclobutane, $\Delta Ht = 5.8^{-126.8}$	1.1	24.2	23.5	100.0	145.4	177.5	200.7

TABLE 2.54 Heat of Fusion, Vaporization, Sublimation, and Specific Heat at Various Temperatures of Organic Compounds (*Continued*)

				C_p			
Substance	ΔHm	ΔHv	ΔHs	400 K	600 K	800 K	1000 K
Cyclobutanecarbonitrile		36.9	44.3				
Cyclobutanenitrile			40.0				
Cyclobutene				90.3	126.8	151.7	169.6
Cyclobutylamine			35.6				
Cyclododecane			76.4				
Cycloheptane	1.88	33.2	38.5	175.0	261.2	322.3	365.7
$\Delta Ht = 5.0^{-138.4}$							
$\Delta Ht = 0.3^{-75.0}$							
$\Delta Ht = 0.5^{-60.8}$							
Cycloheptanone			51.9				
1,3,5-Cycloheptatriene	1.2	38.7		155.4	209.5	245.1	270.2
$\Delta Ht = 2.4^{-119.2}$							
Cyclohexane	2.63	30.0	33.0	149.9	225.2	279.3	317.2
$\Delta Ht = 6.7^{-87}$							
Cyclohexanecarbonitrile			51.9				
Cyclohexanethiol		37.1	44.6				
Cyclohexanol	1.76	45.5	62.0	172.1	248.1	302.0	339.5
$\Delta Ht = 8.2^{-9.7}$							
Cyclohexanone		40.3	45.1	150.6	221.3	272.0	305.4
Cyclohexene	3.29	30.5	33.5	144.9	206.9	248.9	278.7
$\Delta Ht = 4.3^{-134.4}$							
1-Cyclohexenecarbonitrile			53.5				
Cyclohexylamine		36.1	43.7				
Cyclohexylbenzene	15.30		59.9				
Cyclohexylcyclohexane		51.9	58.0				
cis, cis-1,5-Cyclooctadiene			43.4				
Cyclooctane	2.41	35.9	43.3	200.1	297.1	365.3	414.3
$\Delta Ht = 6.3^{-106.7}$							
$\Delta Ht = 0.5^{-89.4}$							
Cyclooctanone			54.4				
1,3,5,7-Cyclooctatetraene	11.3	36.4	43.1	160.9	220.8	260.4	288.2
Cyclooctene			47.0				
Cyclopentadiene			28.4				
Cyclopentane	0.61	27.3	28.5	118.7	178.1	220.1	250.4
$\Delta Ht = 4.8^{-150.8}$							
$\Delta Ht = 0.3^{-135.1}$							
Cyclopentanecarbonitrile			43.4				
1-Cyclopentenecarbonitrile			45.0				
Cyclopentanethiol	7.8	35.3	41.4	144.5	203.6	245.2	275.5
Cyclopentanol			57.6				
Cyclopentanone		36.4	42.7				
Cyclopentene	3.36		28.1	104.9	155.6	191.5	217.3
$\Delta Ht = 0.5^{-186.1}$							
Cyclopentylamine	8.31		40.2				
Cyclopropane	5.44	20.1	16.9	76.6	109.4	140.5	148.1
Cyclopropanecarbonitrile		35.6	41.9				
Cyclopropylamine	13.18		31.3				
Cyclopropylbenzene			50.2				
Cyclopropyl methyl ketone		34.1	38.4				
Decafluorobutane		22.9					

(*Continued*)

TABLE 2.54 Heat of Fusion, Vaporization, Sublimation, and Specific Heat at Various Temperatures of Organic Compounds (*Continued*)

				C_p			
Substance	ΔHm	ΔHv	ΔHs	400 K	600 K	800 K	1000 K
cis-Decahydronaphthalene	9.49	41.0	50.2	237.0	352.0	432.5	489.5
$\Delta Ht = 2.1^{-57.1}$							
trans-Decahydronaphthalene	14.41	40.2	43.5	237.6	352.3	432.6	489.2
Decanal				300.4	400.4	472.8	525.9
Decane	28.78	38.8	51.4	298.1	403.2	480.8	536.4
Decanedioic acid	40.8		160.7				
Decanenitrile			66.8				
1-Decanethiol	31.0	46.4	65.5	320.6	429.4	510.9	573.1
Decanoic acid	28.02		118.8				
1-Decanol	37.7	49.8	81.5	187.2	418.2	495.9	553.3
1-Decene	21.10	38.7	50.4	283.6	381.9	453.0	505.9
$\Delta Ht = 8.0^{-74.8}$							
1-Decyne				274.6	363.8	428.5	476.6
Deoxybenzoin			93.3				
Dibenz[*de,kl*]anthracene			125.5				
Dibenzoyl peroxide	31.4		102.5				
Dibenzyl ether		20.2					
Dibenzyl sulfide			93.3				
Dibenzyl sulfone			125.5				
1,2-Dibromobutane			50.3	153.9	195.4	224.3	244.8
1,4-Dibromobutane			53.1				
2,3-Dibromobutane			37.7				
1,2-Dibromo-1-chloro-1,1,2-trifluoroethane		31.2	35.0				
1,2-Dibromocycloheptane			52.0				
1,2-Dibromocyclohexane			50.5				
1,2-Dibromocyclooctane			54.6				
1,2-Dibromoethane	10.84	34.8	41.7	99.7	122.3	137.8	149.8
1,2-Dibromoheptane			54.4				
Dibromomethane		32.9	37.0	63.0	74.8	82.5	88.0
1,2-Dibromopropane	8.94	35.6	41.7	124.4	157.4	179.5	195.6
1,3-Dibromopropane	13.6		47.5				
1,2-Dibromotetrafluoroethane	7.04	27.0	28.4				
1,2-Dibutoxyethane		47.8	58.8				
Dibutoxymethane			48.1				
Dibutylamine		38.4	49.5				
N,N-Dibutyl-1-butanamine		46.9					
Dibutyl decanedioate		92.9					
Dibutyl disulfide		46.9	64.5	286.1	376.5	442.8	493.1
Di-*tert*-butyl disulfide			54.3				
Dibutyl ether		36.5	45.0	254.3	340.1	403.8	451.3
Di-*sec*-butyl ether		34.1	40.8				
Di-*tert*-butyl ether		32.2	37.6				
Dibutylmercury			63.5				
Di-*tert*-butyl peroxide			31.8				
Dibutyl 1,2-phthalate		79.2	91.6				
Dibutyl sulfate			75.9				
Dibutyl sulfide	19.4	41.3	53.0	259.8	348.6	420.8	475.8
Di-*tert*-butyl sulfide		33.3	43.8				
Dibutyl sulfite			67.8				
Dibutyl sulfone			100.4				

TABLE 2.54 Heat of Fusion, Vaporization, Sublimation, and Specific Heat at Various Temperatures of Organic Compounds (*Continued*)

Substance	ΔHm	ΔHv	ΔHs	C_p 400 K	600 K	800 K	1000 K
Dichloroacetyl chloride			39.3				
1,2-Dichlorobenzene	12.93	39.7	50.2	142.8	184.4	210.4	227.7
1,3-Dichlorobenzene	12.64	38.6	48.6	143.0	184.5	210.4	227.7
1,4-Dichlorobenzene	17.15	38.8	49.0	143.3	184.8	210.7	227.9
2,6-Dichlorobenzoquinone			69.9				
2,2′-Dichlorobiphenyl			96.2				
4,4′-Dichlorobiphenyl			103.8				
1,2-Dichlorobutane		33.9	39.6				
1,4-Dichlorobutane			46.4				
Dichlorodifluoromethane	4.14	20.1		82.4	93.6	99.1	100.0
Dichlorodimethylsilane			34.3				
Dichlorodiphenylsilane			69.5				
1,1-Dichloroethane	8.84	28.9	30.6	91.4	113.7	128.8	139.8
1,2-Dichloroethane	8.83	32.0	35.2	92.1	112.6	127.2	138.1
1,1-Dichloroethylene	6.51	26.1	26.5	78.7	93.9	103.4	110.0
cis-1,2-Dichloroethylene	7.20	30.2	31.0	77.0	93.0	102.9	109.8
trans-1,2-Dichloroethylene	11.98	28.9	29.3	77.7	93.2	102.9	109.8
2,2-Dichloroethyl ether		38.4					
Dichlorofluoromethane		25.2		70.2	82.4	89.6	94.2
1,2-Dichlorohexafluoropropane		26.3	26.9				
1,2-Dichlorohexane			48.2				
Dichloromethane	6.00	28.1	28.8	59.6	72.4	80.8	86.8
1,2-Dichloro-4-methylbenzene	10.68						
1,2-Dichloropentane		36.5	43.9				
1,5-Dichloropentane			50.7				
(±)-1,2-Dichloropropane	6.40	31.8	36.0	119.7	152.6	175.6	192.8
1,3-Dichloropropane		35.2	40.8	120.0	151.5	173.9	190.4
2,2-Dichloropropane		29.3	32.6	127.9	159.2	179.9	194.8
1,3-Dichloro-2-propanol			66.9				
1,2-Dichlorotetrafluoroethane	6.32	23.3					
Dicyanoacetylene			28.8				
Dicyclopentadienyliron			73.6				
Dicyclopropyl ketone			53.7				
Diethanolamine	25.10	65.2					
1,1-Diethoxyethane		36.3	43.2				
1,2-Diethoxyethane		36.3	43.2				
Diethoxymethane		31.3	35.7				
1,3-Diethoxypropane		37.2	45.9				
2,2-Diethoxypropane			31.8				
Diethylamine		29.1	31.3	143.9	197.2	235.0	263.2
1,2-Diethylbenzene	16.8	39.4	52.8	234.4	316.6	374.6	416.3
1,3-Diethylbenzene	11.0	39.4	52.5	230.2	314.6	379.7	415.8
1,4-Diethylbenzene	10.6	39.4	52.5	228.8	313.1	372.5	414.9
Diethyl carbonate		36.2	43.6				
Diethyl disulfide	9.4	37.6	45.2	171.1	218.6	251.8	276.0
Diethylene glycol diethyl ether	13.60	49.0	58.4				
Diethylene glycol dimethyl ether		36.2	44.7				
Diethylene glycol monoethyl ether		47.5					
Diethylene glycol monomethyl ether		46.6					
Diethyl ether	7.27	26.5	27.1	138.1	183.8	218.7	244.8

(*Continued*)

TABLE 2.54 Heat of Fusion, Vaporization, Sublimation, and Specific Heat at Various Temperatures of Organic Compounds (*Continued*)

				C_p			
Substance	Δ*Hm*	Δ*Hv*	Δ*Hs*	400 K	600 K	800 K	1000 K
Diethyl malonate		54.8					
Diethyl oxalate		42.0	63.5				
Diethyl peroxide			30.5				
3,3-Diethylpentane	10.09	34.6	42.0				
Diethyl 1,2-phthalate			88.3				
Diethyl sulfide	11.90	31.8	35.8	145.0	192.9	229.7	258.5
Diethyl sulfite			48.5				
Diethyl sulfone			86.2				
Diethyl sulfoxide			62.3				
Diethylzinc			40.2				
1,2-Difluorobenzene	11.1	32.2	36.2	137.1	181.3	209.7	229.0
1,3-Difluorobenzene	8.58	31.1	34.6	137.0	180.5	207.8	225.6
1,4-Difluorobenzene		31.8	35.5	137.4	180.1	207.8	225.7
2,2′-Difluorobiphenyl			95.0				
4,4′-Difluorobiphenyl			91.2				
1,1-Difluoroethane		21.6	19.1	83.4	107.5	124.3	136.3
1,1-Difluoroethylene				71.8	89.2	100.2	107.7
Difluoromethane				51.1	65.8	76.2	83.7
9,10-Dihydroanthracene			93.3				
Dihydro-2*H*-pyran			32.2				
5,12-Dihydrotetracene			115.9				
2,3-Dihydrothiophene		33.2	37.7				
2,5-Dihydrothiophene		34.8	40.0				
2,4-Dihydrothiophene-1,1-dioxide			62.8				
1,4-Dihydroxybenzene	27.11		99.2				
1,2-Diiodobenzene			64.9				
1,2-Diiodoethane			65.7	96.0	116.8	131.3	141.6
Diiodomethane	44.80	42.5	51.0	65.9	76.9	83.9	89.1
Diisobutylamine			39.3				
Diisobutyl ether		34.0	40.9				
Diisobutyl sulfide			48.7				
Diisopropylamine		30.4	34.6				
Diisopropyl ether	11.03	29.1	32.1	196.2	262.0	311.3	348.0
Diisopropylmercury			53.6				
Diisopropyl sulfide	10.4	33.8	39.6	211.9	277.1	322.7	356.6
Diketene		36.8	42.9				
1,2-Dimethoxybenzene	16.04	48.2	66.9				
1,1-Dimethoxyethane			30.5				
1,2-Dimethoxyethane	12.60	32.4	36.4				
Dimethoxymethane	8.33		35.1				
2,2-Dimethoxypropane			29.4				
N,N-Dimethylacetamide	10.42	43.4	50.2				
Dimethylamine	5.94	26.4	25.0	87.4	118.9	142.0	159.8
Dimethylaminomethanol			50.2				
N,N-Dimethylaminotrimethylsilane			31.8				
N,N-Dimethylaniline			52.8				
1,4-Dimethylbicyclo[2.2.1]heptane		33.3	38.9				
2,3-Dimethylbicyclo[2.2.1]-2-heptene		34.9	42.2				
2,2-Dimethylbutane	0.58	26.3	27.7	182.8	251.0	298.7	333.5

TABLE 2.54 Heat of Fusion, Vaporization, Sublimation, and Specific Heat at Various Temperatures of Organic Compounds (*Continued*)

Substance	ΔHm	ΔHv	ΔHs	C_p 400 K	C_p 600 K	C_p 800 K	C_p 1000 K
$\Delta Ht = 5.4^{-147.3}$							
$\Delta Ht = 0.3^{-132.3}$							
2,3-Dimethylbutane	0.80	27.4	29.1	181.2	247.7	314.6	331.0
$\Delta Ht = 6.5^{-137.1}$							
2,2-Dimethyl-1-butanol		42.6	56.1				
2,3-Dimethyl-1-butanol		47.3					
3,3-Dimethyl-1-butanol		46.4					
2,3-Dimethyl-2-butanol		40.4	51.0				
(±)-3,3-Dimethyl-2-butanol		43.9					
3,3-Dimethyl-2-butanone		33.4	37.9				
2,3-Dimethyl-1-butene		27.4	29.2	178.2	231.8	272.0	302.1
3.3-Dimethyl-1-butene	1.1	25.7	27.1	162.8	223.4	266.1	297.1
$\Delta Ht = 4.3^{-148.3}$							
2,3-Dimethyl-2-butene	5.46	29.6	32.5	156.8	216.7	262.7	297.7
$\Delta Ht = 3.5^{-76.3}$							
Di(3-methylbutyl) ether		35.2					
Dimethylcadmium			38.0				
1,1-Dimethylcyclohexane	2.06	32.5	37.9	212.1	310.0	379.5	427.6
$\Delta Ht = 6.0^{-120.0}$							
cis-1,2-Dimethylcyclohexane	1.64	33.5	39.7	213.8	309.6	377.0	424.3
$\Delta Ht = 8.3^{-100.6}$							
trans-1,2-Dimethylcyclohexane	10.49	33.0	38.4	217.2	312.1	378.7	425.5
cis-1,3-Dimethylcyclohexane	10.82	32.9	38.3	214.2	310.5	378.7	426.8
trans-1,3-Dimethylcyclohexane	9.86	33.4	39.2	213.8	308.8	375.7	423.0
cis-1,4-Dimethylcyclohexane	9.31	33.3	39.0	213.8	308.8	375.7	423.0
trans-1,4-Dimethylcyclohexane	12.33	32.6	37.9	215.9	312.1	378.9	425.7
1,1-Dimethylcyclopentane	1.1	30.3	33.8	182.2	262.6	318.7	359.1
$\Delta Ht = 6.5^{-126.4}$							
cis-1,2-Dimethylcyclopentane	1.7	31.7	35.7	182.7	262.4	317.9	358.0
$\Delta Ht = 6.7^{-131.7}$							
trans-1,2-Dimethylcyclopentane	7.2	30.9	34.6	182.9	262.2	317.3	357.4
cis-1,3-Dimethylcyclopentane	7.4	30.4	34.2	182.9	262.2	317.3	357.4
trans-1,3-Dimethylcyclopentane	7.3	30.8	34.5	182.9	262.2	317.3	357.4
cis-2,4-Dimethyl-1,3-dioxane			39.9				
4,5-Dimethyl-1,3-dioxane			42.5				
5,5-Dimethyl-1,3-dioxane			41.3				
Dimethyl disulfide	9.19	33.8	37.9	110.3	137.4	157.6	172.8
Dimethyl ether	4.94	21.5	18.5	79.6	105.3	125.7	141.4
N,N-Dimethylformamide	16.15	38.4	46.9				
Dimethylglyoxime			97.1				
2,2-Dimethylheptane	8.90						
2,6-Dimethyl-4-heptanone		39.9	50.9				
2,2-Dimethylhexane	6.78	32.1	37.3				
2,3-Dimethylhexane		33.2	38.8				
2,4-Dimethylhexane		32.5	37.8				
2,5-Dimethylhexane	12.95	32.5	37.9				
3,3-Dimethylhexane	6.98	32.3	37.5				
3,4-Dimethylhexane		33.2	39.0				
cis-2,2-Dimethyl-3-hexene			37.2				

(*Continued*)

TABLE 2.54 Heat of Fusion, Vaporization, Sublimation, and Specific Heat at Various Temperatures of Organic Compounds (*Continued*)

				C_p			
Substance	ΔHm	ΔHv	ΔHs	400 K	600 K	800 K	1000 K
trans-2,2-Dimethyl-3-hexene			37.3				
1,1-Dimethylhydrazine	10.1	32.6	35.0				
1,2-Dimethylhydrazine		35.2	39.3				
3,5-Dimethylisoxazole			45.2				
Dimethyl maleate	14.7		44.3				
Dimethylmercury			34.6				
6,6-Dimethyl-2-methylene-bicyclo[3.1.1]heptane		40.2	46.4				
2,4-Dimethyloctane		36.5	47.1				
Dimethyl oxalate	21.07		47.4				
3,3-Dimethyloxetane		30.9	33.9				
2,2-Dimethylpentane	5.86	29.2	32.4	211.0	285.9	340.7	381.6
2,3-Dimethylpentane		30.5	34.3	211.0	285.9	340.7	381.6
2,4-Dimethylpentane	6.69	29.6	32.9	211.0	285.9	340.7	381.6
3,3-Dimethylpentane	7.07	29.6	33.0	211.0	285.9	340.7	381.6
2,2-Dimethyl-3-pentanone		36.1	42.3				
2,4-Dimethyl-3-pentanone	11.18	34.6	41.5				
2,4-Dimethyl-1-pentene			33.2				
4,4-Dimethyl-1-pentene			29.0				
2,4-Dimethyl-2-pentene			34.4				
cis-4,4-Dimethyl-2-pentene			32.7				
trans-4,4-Dimethyl-2-pentene			32.7				
2,7-Dimethylphenanthrene			106.7				
4,5-Dimethylphenanthrene			104.6				
9,10-Dimethylphenanthrene			119.5				
2,3-Dimethylphenol	21.02		84.0				
2,4-Dimethylphenol		47.1	65.0				
2,5-Dimethylphenol	23.38	46.9	85.0				
2,6-Dimethylphenol	18.90	44.5	75.3				
3,4-Dimethylphenol	18.13	49.7	85.0				
3,5-Dimethylphenol	18.00	49.3	82.0				
Dimethyl 1,2-phthalate	162.7						
2,2-Dimethylpropane	3.10	22.7	21.8	157.1	218.5	254.3	283.7
$\Delta Ht = 2.6^{-133.1}$							
2,2-Dimethylpropanenitrile		32.4	37.3				
2,2-Dimethyl-1-propanol		9.6					
2,3-Dimethylpyridine		39.1	47.7				
2,4-Dimethylpyridine		38.5	47.5				
2,5-Dimethylpyridine			47.8				
2,6-Dimethylpyridine	10.04	37.5	45.4				
3,4-Dimethylpyridine		40.0	50.5				
3,5-Dimethylpyridine		39.5	49.5				
Dimethyl sulfate			48.5				
Dimethyl sulfide	7.99	27.0	27.7	88.4	113.0	132.2	147.2
Dimethyl sulfite			40.2				
Dimethyl sulfone			77.0				
Dimethyl sulfoxide	14.37	43.1	52.9				
2,2-Dimethylthiacyclopropane			35.8				
Dimethylzinc			29.5				
Dinitromethane			46.0				

TABLE 2.54 Heat of Fusion, Vaporization, Sublimation, and Specific Heat at Various Temperatures of Organic Compounds (*Continued*)

				C_p			
Substance	ΔHm	ΔHv	ΔHs	400 K	600 K	800 K	1000 K
2,4-Dinitrophenol			104.6				
2,6-Dinitrophenol			112.1				
1,1-Dinitropropane			62.5				
1,3-Dioxane		34.4	39.1				
1,4-Dioxane	12.85	34.2	38.6	126.5	181.8	218.2	243.3
$\Delta Ht = 2.4^{-0.3}$							
1,3-Dioxolane	27.48		35.6				
Diphenylamine	17.86		89.1				
Diphenyl carbonate	23.4		90.0				
Diphenyl disulfide			95.0				
Diphenyl disulfone			161.9				
Diphenylenimine			84.5				
1,2-Diphenylethane		51.5	91.4				
1,1-Diphenylethylene			73.2				
Diphenyl ether	17.22	48.2	67.0				
6,6-Diphenylfulvene			104.6				
Diphenylmercury			112.8				
Diphenylmethane	18.2		67.5				
1,3-Diphenyl-2-propanone			89.1				
Diphenyl sulfide			67.8				
Diphenyl sulfone			106.3				
Diphenyl sulfoxide			97.1				
1,2-Dipropoxyethane			50.6				
Dipropylamine		33.5	40.0				
Dipropyl disulfide	13.8	41.9	54.1	186.2	298.3	350.2	390.0
Dipropyl ether	8.83	31.3	35.7	196.2	262.0	311.3	348.0
Dipropylmercury			55.2				
Dipropyl sulfate			66.9				
Dipropyl sulfide	12.1	36.6	44.2	201.7	272.5	328.2	372.6
Dipropyl sulfite			58.6				
Dipropyl sulfone			79.9				
Dipropyl sulfoxide			74.5				
Divinyl ether			26.2				
Divinyl sulfone			56.5				
Dodecane	36.55	44.5	61.5	356.2	481.3	572.2	656.5
Dodecanedioic acid			153.1				
Dodecanenitrile			76.1				
Dodecanoic acid	36.64		132.6				
Dodecanol	31.4	63.5	92.0				
1-Dodecene	17.42	44.0	60.8	341.8	460.0	545.6	608.8
$\Delta Ht = 4.6^{-60.2}$							
1,2-Epoxybutane		30.3					
1,2-Epoxypropane		21.6					
Ergosterol			118.4				
Ethane	2.86	14.7	5.2	65.5	89.3	108.0	122.6
Ethane-d_6				81.7	108.5	127.4	140.5
1,2-Ethanediamine	22.58	38.0	45.0				
1,2-Ethanediol	11.23	50.5	67.8	113.2	136.9	166.9	
1,2-Ethanediol diacetate		45.5	61.4				
1,2-Ethanedithiol		37.9	44.7				

(Continued)

TABLE 2.54 Heat of Fusion, Vaporization, Sublimation, and Specific Heat at Various Temperatures of Organic Compounds (*Continued*)

Substance	ΔHm	ΔHv	ΔHs	C_p 400 K	C_p 600 K	C_p 800 K	C_p 1000 K
Ethanethiol	4.98	26.8	27.3	88.2	113.9	133.2	148.0
Ethanol	5.02	38.6	42.3	81.2	107.7	127.2	141.9
Ethanolamine	20.50	49.8					
Ethoxybenzene		40.7	51.0				
2-Ethoxyethanol		39.2	48.2				
2-(2-Ethoxyethoxy)ethanol		47.5					
2-(2-Ethoxyethoxy)ethyl acetate		91.2					
2-Ethoxyethyl acetate			52.7				
1-Ethoxy-2-methoxyethane		34.3	39.8				
N-Ethylacetamide			64.9				
Ethyl acetate	10.48	31.9	35.6	137.4	182.6	213.4	234.5
Ethyl acrylate		34.7					
Ethylamine		28.0	26.6	90.6	119.6	141.8	158.5
N-Ethylaniline			52.3				
Ethylbenzene	9.18	35.6	42.2	170.5	236.1	281.0	312.8
2-Ethylbenzoic acid			100.7				
3-Ethylbenzoic acid			99.1				
4-Ethylbenzoic acid			97.5				
2-Ethyl-1-butanol		43.2	63.2				
Ethyl butanoate		35.5	42.7				
2-Ethylbutanoic acid		51.2					
2-Ethyl-1-butene		28.8	31.1	170.3	228.0	269.5	300.8
Ethyl *trans*-2-butenoate			44.4				
Ethyl chloroacetate		40.4	49.5				
Ethyl 4-chlorobutanoate			52.7				
Ethyl chloroformate			42.3				
Ethyl *trans*-cinnamate		58.6					
Ethyl crotonate			44.3				
Ethyl cyanoacetate		64.4					
Ethylcyclobutane		28.7	31.2				
Ethylcyclohexane	8.33	34.0	40.6	215.9	310.0	377.0	423.8
1-Ethylcyclohexene			43.3				
Ethylcyclopentane	6.9	32.0	36.4	183.6	258.2	314.7	356.3
1-Ethylcyclopentene		38.5					
Ethyl dichloroacetate			50.6				
Ethyl 2,2-dimethylpropanoate		34.5	41.2				
Ethylene	3.35	13.5		53.1	70.7	83.8	93.9
Ethylene-d_4				63.9	82.3	95.6	104.9
Ethylene carbonate	13.19	50.1	73.2				
2,2′-(Ethylenedioxy)bis(ethanol)		71.4	79.1				
Ethylene glycol (*see* 1,2-Ethanediol)							
Ethylene glycol diacetate			61.4				
Ethylene oxide	5.2	25.5	24.8	62.6	86.3	102.9	114.9
Ethylenimine		30.3	34.6	70.4	98.6	117.7	131.6
N-Ethylformamide			58.4				
Ethyl formate	9.20	29.9	32.0				
2-Ethylhexanal			49.0				
2-Ethylhexane		33.6	39.6				
Ethyl hexanoate			51.7				
2-Ethylhexanoic acid		56.0	75.6				

TABLE 2.54 Heat of Fusion, Vaporization, Sublimation, and Specific Heat at Various Temperatures of Organic Compounds (*Continued*)

Substance	ΔHm	ΔHv	ΔHs	C_p 400 K	600 K	800 K	1000 K
2-Ethyl-1-hexanol		45.2					
2-Ethylhexyl acetate		43.5	48.1				
2-Ethyl hydroperoxide			43.1				
Ethylidenecyclohexane			42.0				
Ethylidenecyclopentane		18.1					
Ethyl isocyanide			33.5				
Ethyl isopentanoate	8.7	43.9					
Ethyl isopentyl ether		33.0	39.0				
Ethylisopropylamine		29.9	33.1				
Ethyl isopropyl ether		28.2	30.1				
Ethyl isopropyl sulfide	8.7	32.7	37.8				
Ethyl lactate		46.4	49.4				
Ethyllithium			116.7				
Ethylmercury bromide			76.6				
Ethylmercury chloride			76.1				
Ethylmercury iodide			79.5				
1-Ethyl-2-methylbenzene	10.0	38.9	47.7	202.9	275.3	326.8	363.6
1-Ethyl-3-methylbenzene	7.6	38.5	46.9	198.7	273.6	325.5	363.2
1-Ethyl-4-methylbenzene	13.4	38.4	46.6	197.5	272.0	324.7	362.2
Ethyl 2-methylbutanoate			44.4				
Ethyl 3-methylbutanoate		37.0	43.9				
2-Ethyl-3-methyl-1-butene			34.5				
1-Ethyl-1-methylcyclopentane		33.2	38.9				
Ethyl methyl ether		26.7		109.1	144.7	172.3	193.2
3-Ethyl-2-methylpentane	11.34	32.9	38.5				
3-Ethyl-3-methylpentane	10.84	32.8	38.0				
3-Ethyl-2-methyl-1-pentene			37.5				
Ethyl 2-methylpropanoate		33.7	39.8				
Ethyl methyl sulfide	9.8	29.5	31.9	116.4	152.3	179.6	200.6
Ethyl nitrate	8.5	33.1	36.3	120.2	155.1	178.7	195.4
1-Ethyl-2-nitrobenzene			59.8				
1-Ethyl-4-nitrobenzene			62.8				
3-Ethylpentane	9.55	31.1	35.2	211.0	285.9	340.7	381.6
Ethyl pentanoate		37.0	47.0				
Ethyl pentyl ether		34.4	41.0				
2-Ethylphenol			63.6				
3-Ethylphenol			68.2				
4-Ethylphenol			80.3				
Ethylphosphonic acid			50.6				
Ethylphosphonic dichloride			42.7				
Ethyl propanoate		33.9	39.2				
Ethyl propyl ether		28.9	31.4				
Ethyl propyl sulfide	10.6	34.2	40.0	173.3	232.7	279.0	315.6
Ethyl trichloroacetate			51.0				
S-Ethyl thiolacetate	34.4	40.0					
Ethyl 2-vinylacrylate			48.5				
Ethyl vinyl ether		26.2	26.6				
Fluoranthrene	18.87		99.2				
9*H*-Fluorene	19.58						
Fluorobenzene	11.31	31.2	34.6	125.5	171.0	200.1	220.0

(*Continued*)

TABLE 2.54 Heat of Fusion, Vaporization, Sublimation, and Specific Heat at Various Temperatures of Organic Compounds (*Continued*)

Substance	Δ*Hm*	Δ*Hv*	Δ*Hs*	C_p 400 K	C_p 600 K	C_p 800 K	C_p 1000 K
4-Fluorobenzoic acid			91.2				
Fluoroethane				74.1	98.6	116.4	129.7
Fluoromethane		16.7		44.2	57.9	68.8	77.2
1-Fluorooctane		40.4	49.7				
1-Fluoropropane				102.7	137.3	162.7	181.5
2-Fluoropropane				103.5	138.7	163.8	182.2
2-Fluorotoluene		35.4					
4-Fluorotoluene	9.4	34.1	39.4	152.4	207.9	245.2	271.3
Fluorotrichloromethane		25.0					
Fluorotrinitromethane			34.7				
Formaldehyde		23.3		39.2(g)	48.2	55.9	62.0
Formamide	6.69		60.2				
Formic acid	12.7	22.7	20.1	53.8	67.0	76.8	83.5
Formyl fluoride		21.7		46.4	56.2	63.1	67.9
Fumaric acid			136.0				
Fumaronitrile			72.0				
Furan, Δ*Ht* = $2.1^{-123.2}$	3.80	27.1	27.5	88.7	122.6	164.9	158.5
2-Furancarboxaldehyde	14.35	43.2	50.6				
2-Furancarboxylic acid			108.5				
Furanmethanol	13.13	53.6	64.4				
Glutaric acid	20.9						
Glycerol	18.28	61.0	85.8				
Glyceryl triacetate			85.7				
Glyceryl tributanoate			107.1				
Glyceryl trinitrate	21.87		100.0				
Heptadecane, Δ*Ht* = $11.0^{11.1}$	40.5	52.9	86.0	501.4	676.8	803.7	897.9
Heptadecanoic acid	58.8						
1-Heptadecene	31.4	51.8	85.0	486.9	655.5	777.1	866.9
1-Heptanal	23.6		47.7	213.4	283.3	333.9	371.1
Heptane	14.16	31.8	36.6	211.0	285.9	340.7	381.6
1-Heptanenitrile			51.9				
1-Heptanethiol	25.4	39.8	50.6	233.5	312.1	372.0	418.4
Heptanoic acid			74.0				
1-Heptanol	13.2	48.1	66.8	224.4	300.9	357.0	392.5
2-Heptanol		49.8					
3-Heptanol		42.5					
2-Heptanone		38.3	47.2				
4-Heptanone		36.2					
1-Heptene, Δ*Ht* = 0.3^{-136}	12.66	31.1	35.5	196.5	264.6	314.1	351.0
trans-2-Heptene	11.72						
Heptylamine			50.0				
Heptyl methyl ether			46.9				
Hexachlorobenzene	23.85		92.6	201.2	233.4	250.9	260.8
Hexachloroethane, Δ*Ht* = $8.0^{71.3}$	9.8	45.9	59.0	151.5	166.6	173.6	177.3
Hexadecafluoroethylcyclohexane			38.5				
Hexadecafluoroheptane			36.4				
Hexadecane	51.8	51.2	81.4	472.3	687.7	757.4	846.0
Hexadecanoic acid	42.04		154.4				
1-Hexadecanol, Δ*Ht* = 16.6^{34}	34.29		169.5	485.7	652.7	773.6	863.2
1-Hexadecene	30.2	50.4	80.3	457.9	616.4	731.82	815.0

TABLE 2.54 Heat of Fusion, Vaporization, Sublimation, and Specific Heat at Various Temperatures of Organic Compounds (*Continued*)

Substance	ΔHm	ΔHv	ΔHs	C_p 400 K	C_p 600 K	C_p 800 K	C_p 1000 K
Hexadienoic acid	13.6						
Hexafluoroacetone		19.8	21.3				
Hexafluoroacetylacetone		27.1	30.6				
Hexafluorobenzene	11.58	31.7	35.7	183.6	219.9	241.1	253.7
Hexafluoroethane, $\Delta Ht = 3.7^{-169.2}$	2.7	16.2		125.6	149.0	160.7	166.8
cis-Hexahydroindane			57.5				
trans-Hexahydroindane			56.1				
Hexamethylbenzene	20.6	48.2	74.7	310.4	406.4	474.9	525.3
$\Delta Ht = 1.1^{-156.7}$							
$\Delta Ht = 1.8^{110.7}$							
1,1,1,3,3,3-Hexamethyldisilazane			41.4				
Hexamethyldisiloxane			37.2				
Hexamethylphosphoric triamide	14.28						
Hexanal				184.2	243.9	287.4	319.7
Hexanamide	25.1		98.7				
Hexane	13.08	28.9	31.6	181.9	246.8	294.4	330.1
1,6-Hexanedioic acid	34.85		129.3				
1,6-Hexanediol	25.5		83.3				
Hexanenitrile		38.0	47.9				
1-Hexanethiol	18.0(1)	37.2	45.8	204.5	273.1	325.1	366.7
Hexanoic acid	15.40	71.1	72.2				
1-Hexanol	15.40	44.5	61.6	195.3	261.8	310.7	346.9
2-Hexanol		41.0	58.5				
3-Hexanol	44.3	46.0					
2-Hexanone	14.90	36.4	43.1				
3-Hexanone	13.49	35.4	42.5				
1-Hexene	9.35	28.3	30.6	167.5	225.5	267.9	299.3
cis-2-Hexene	8.86	29.1	32.2	161.5	221.8	165.3	297.9
trans-2-Hexene	8.26	28.9	31.6	166.1	223.4	266.1	297.9
cis-3-Hexene	8.25	28.7	31.4	161.1	222.6	265.7	297.9
trans-3-Hexene	11.08	28.9	31.7	168.2	225.5	267.4	298.7
Hexylamine		36.5	45.1				
Hexyl methyl ether		34.9	42.1				
1-Hexyne				158.5	207.5	243.3	270.1
Hydrazine	12.7	45.3					
2-Hydroxybenzaldehyde		38.2					
2-Hydroxybenzoic acid			95.1				
2-Hydroxy-2,4,6-cycloheptatrienone			83.7				
2-Hydroxy-1-isopropyl-4-methylbenzene			91.2				
4-Hydroxy-4-methyl-2-pentanone		28.5	47.7				
3-Hydroxypropanonitrile		56.1					
2-Hydroxypyridine			86.6				
3-Hydroxypyridine			88.3				
4-Hydroxypyridine			103.8				
8-Hydroxyquinoline			108.8				
Icosane	69.88	57.5	100.8	588.5	794.0	942.6	1052.7
Icosanoic acid	72.0		199.6				
1-Icosene	34.3	55.9	99.8	574.0	772.7	916.0	1021.7
Indane		39.6	48.8				
Indene			52.9				

(*Continued*)

TABLE 2.54 Heat of Fusion, Vaporization, Sublimation, and Specific Heat at Various Temperatures of Organic Compounds (*Continued*)

				C_p			
Substance	Δ*Hm*	Δ*Hv*	Δ*Hs*	400 K	600 K	800 K	1000 K
Indole			69.9				
Iodobenzene	9.76	39.5	47.7	130.1	173.3	201.1	220.1
Iodobenzoic acid			87.9				
1-Iodobutane		34.7	40.6				
2-Iodobutane		33.3	38.5				
Iodocyclohexane			47.3				
Iodoethane		29.4	31.9	80.3	103.1	119.9	132.4
1-Iodohexane			49.8				
Iodomethane		27.3	28.0	51.6	63.9	73.1	80.2
1-Iodo-2-methylpropane		33.5	38.8				
2-Iodo-2-methylpropane	14.5	31.4	35.4	148.8	191.7	221.1	242.3
1-Iodonaphthalene			72.4				
2-Iodonaphthalene			90.8				
1-Iodopentane			45.3				
1-Iodopropane		32.1	36.2	109.9	142.7	166.5	184.2
2-Iodopropane		30.7	34.1	111.2	144.7	168.2	185.5
3-Iodo-1-propene			38.1				
2-Iodotoluene (also 3-, 4-)			54.4				
Isobutanonitrile		32.4	37.2	119.5	156.4	183.0	202.5
Isobutyl acetate		35.9					
Isobutylamine		30.6	33.9				
Isobutylbenzene	12.51	37.8	47.9				
Isobutylcyclohexane			47.6				
Isobutyl dichloroacetate			52.3				
Isobutyl formate		33.6					
Isobutyl isobutanoate		38.2	46.4				
Isobutyl isopropyl ether		31.6	36.6				
Isobutyl methyl ether		28.0	30.1				
Isobutyl propyl ether		28.3	30.3				
Isobutyl trichloroacetate			53.1				
Isobutyl vinyl ether		30.7	34.6				
2-Isopropoxyethanol		40.4	50.1				
Isopropyl acetate		32.9	37.2				
Isopropylamine	7.33	27.8	28.4				
Isopropylbenzene	7.79	37.5	45.1	200.8	277.0	328.9	365.3
Isopropylcyclohexane			44.0				
Isopropylcyclopentane		33.6	39.4				
Isopropylmethylamine		28.7	30.9				
1-Isopropyl-2-methylbenzene	10.0	38.4	50.6				
1-Isopropyl-3-methylbenzene	13.7	38.1	50.0				
1-Isopropyl-4-methylbenzene	9.7	38.2	50.2				
Isopropyl methyl ether		26.1	26.4	138.0	184.8	220.4	247.2
2-Isopropyl-5-methylphenol			91.2				
Isopropyl methyl sulfide	9.4	30.7	34.2	145.1	192.5	229.9	260.6
Isopropyl nitrate		34.9	38.8	150.5	195.9	226.5	247.9
Isopropylpropylamine		32.1	37.2				
Isopropyl propyl sulfide		35.1	41.8				
Isopropyl trichloroacetate			51.9				
Isoquinoline	7.45	49.0	60.3				
Ketene			20.4	59.5	70.7	78.7	86.4

TABLE 2.54 Heat of Fusion, Vaporization, Sublimation, and Specific Heat at Various Temperatures of Organic Compounds (*Continued*)

Substance	ΔHm	ΔHv	ΔHs	C_p 400 K	600 K	800 K	1000 K
(−)-Leucine			150.6				
(+)-Limonene			48.1				
Maleic acid			110.0				
Maleic anhydride			71.5				
Malononitrile			79.1				
D-Mannitol	22.6						
Methacrylonitrile		31.8					
Methane	0.94	8.2		40.5	52.2	62.9	71.8
Methane-d_4				48.6	63.4	74.8	83.0
Methanethiol, $\Delta Ht = 0.22^{-135.6}$	5.91	24.6	23.8	58.7	73.5	85.0	94.1
Methanol, $\Delta Ht = 0.6^{-115.8}$	3.18	35.2	37.4	51.4	67.0	79.7	89.5
4-Methoxybenzaldehyde		56.8	64.5				
Methoxybenzene		39.0	46.9				
2-Methoxybenzoic acid			104.7				
3-Methoxybenzoic acid			107.4				
4-Methoxybenzoic acid			109.8				
3-Methoxy-1-butanol		50.8					
2-Methoxyethanol		37.5	45.2				
2-(2-Methoxyethoxy)ethanol		46.6					
2-Methoxyethyl acetate		43.9	50.3				
2-Methoxy-1-propoxyethane		36.3	43.7				
2-Methoxytetrahydropyran			42.7				
1-Methoxy-2,4,6-trinitrobenzene			133.1				
N-Methylacetamide	9.72	59.4					
Methyl acetate		30.3	32.3				
Methyl acetoacetate		36.0					
Methyl acrylate		33.1	29.2				
Methylamine	6.13	25.6	24.4	60.2	78.9	93.9	105.7
4-Methylaniline	18.22						
Methyl benzoate	9.74	43.2	55.6				
2-Methylbenzoic acid	20.17						
3-Methylbenzoic acid	15.72						
4-Methylbenzoic acid	22.73						
1-Methylbicyclo[4.1.0]heptane			39.2				
1-Methylbicyclo[3.1.0]hexane		31.1	34.8				
2-Methyl-1,3-butadiene	4.79	25.9	26.8	133.1	173.2	200.8	221.3
3-Methyl-1,3-butadiene		27.2	28.0	129.7	168.6	197.5	219.2
2-Methylbutane	5.15	24.7	24.9	152.7	208.7	249.8	280.8
3-Methylbutanenitrile		35.1	41.7				
2-Methylbutanethiol		33.8	39.5				
3-Methyl-1-butanethiol	7.5		39.4				
2-Methyl-2-butanethiol $\Delta Ht = 8.0^{-114.0}$	0.6	31.4	35.7	179.0	236.7	279.4	308.8
Methyl butanoate		33.8	39.3				
2-Methylbutanoic acid			46.9				
3-Methylbutanoic acid	7.32	43.2	57.5				
2-Methyl-1-butanol		45.2	55.2				
3-Methyl-1-butanol		44.1	55.6				
2-Methyl-2-butanol, $\Delta Ht = 2.0^{-127.2}$	4.45	39.0	50.1				
3-Methyl-2-butanol		41.8	53.0				

(*Continued*)

TABLE 2.54 Heat of Fusion, Vaporization, Sublimation, and Specific Heat at Various Temperatures of Organic Compounds (*Continued*)

				C_p			
Substance	Δ*Hm*	Δ*Hv*	Δ*Hs*	400 K	600 K	800 K	1000 K
3-Methyl-2-butanone		32.4	36.8				
2-Methyl-1-butene	7.9	25.5	25.9	138.9	187.1	222.4	248.7
3-Methyl-1-butene	5.4	24.1	23.8	147.5	192.1	225.3	250.3
2-Methyl-2-butene	7.6	26.3	27.1	133.6	181.7	217.8	245.0
Methyl 2-butenoate			41.0				
3-Methyl-1-butyne		26.2	25.8	130.1	169.9	198.3	219.2
2-Methylbutyl acetate		37.5					
Methyl chloroacetate		39.2	46.7				
Methyl cyanoacetate		48.2	61.7				
Methyl cyclobutanecarboxylate		37.1	44.7				
Methylcyclohexane	6.75	31.3	35.4	185.6	269.7	329.5	371.5
1-Methylcyclohexanol		79.0	80				
cis-2-Methylcyclohexanol		48.5	63.2				
trans-2-Methylcyclohexanol		53.0	63.2				
cis-3-Methylcyclohexanol			65.3				
trans-3-Methylcyclohexanol			65.3				
cis-4-Methylcyclohexanol			65.7				
trans-4-Methylcyclohexanol			66.1				
1-Methylcyclohexene			37.9				
Methylcyclopentane	6.93	29.1	31.6	151.1	219.4	267.8	303.1
1-Methyl-1-cyclopentene			32.6	136.0	195.8	238.5	269.0
3-Methyl-1-cyclopentene			31.0	136.4	197.1	239.3	269.9
4-Methyl-1-cyclopentene			32.2	136.4	196.7	238.4	269.5
Methyl cyclopropanecarboxylate		35.3	41.3				
2-Methyldecane		40.3	54.3				
4-Methyldecane		40.7	53.8				
Methyl decanoate			66.7				
Methyl dichloroacetate		39.3	47.7				
Methyldichlorosilane			28.0				
Methyl 2,2-dimethylpropanoate		33.4	38.8				
2-Methyl-1,3-dioxane			38.6				
4-Methyl-1,3-dioxane			39.2				
4-Methyl-1,3-dioxolan-2-one	9.62						
Methyl dodecanoate			77.2				
N-Methylethanediamine		37.6	45.2				
1-Methylethyl acetate		32.9	37.3				
1-Methylethyl thiolacetate		35.7	42.3				
N-Methylformamide			56.2				
Methyl formate	7.45	27.9	28.4	81.6	105.4	121.8	133.9
Methyl 2-furancarboxylate			45.2				
Methylglyoxal			38.1				
2-Methylheptane	11.88	33.3	39.7				
3-Methylheptane	11.38	33.7	39.8				
4-Methylheptane	10.84	33.4	39.7				
Methyl heptanoate			51.6				
2-Methylhexane	8.87	30.6	34.9	211.0	285.9	340.7	381.6
3-Methylhexane		30.9	35.1	212.0	285.9	340.7	381.6
Methyl hexanoate		38.6	48.0				
5-Methyl-1-hexene			34.3				
cis-3-Methyl-3-hexene			36.5				

TABLE 2.54 Heat of Fusion, Vaporization, Sublimation, and Specific Heat at Various Temperatures of Organic Compounds (*Continued*)

Substance	ΔHm	ΔHv	ΔHs	C_p 400 K	C_p 600 K	C_p 800 K	C_p 1000 K
trans-3-Methyl-3-hexene			35.9				
Methylhydrazine	10.4	36.1	40.4				
Methyl isobutanoate		32.6	37.3				
Methyl isocyanide			30.8				
1-Methyl-4-isopropylbenzene	9.60	38.2					
3-Methylisoxazole			41.0				
5-Methylisoxazole			41.0				
Methylmercury bromide			67.8				
Methylmercury chloride			64.4				
Methylmercury iodide			65.3				
Methyl methacrylate		36.0	60.7				
Methyl 2-methylbutanoate			41.8				
Methyl-3-methylbutanoate			41.0				
1-Methylnaphthalene	6.94	45.5		212.3	292.0	345.1	381.6
$\Delta Ht = 5.0^{-32.4}$							
2-Methylnaphthalene	11.97	46.0	61.7	211.2	290.0	343.2	381.2
$\Delta Ht = 5.6^{15.4}$							
Methyl nitrate	8.2	31.6	32.1	91.5	115.2	131.7	143.1
Methyl nitrite		20.9	22.6	76.3	97.7	112.8	123.5
1-Methyl-4-nitrobenzene			79.1				
2-Methylnonane		38.2	49.6				
3-Methylnonane		38.3	49.7				
5-Methylnonane		38.1	49.3				
2-Methyloctane	18.00						
Methyl octanoate			56.4				
Methyl oxirane		27.4	27.9				
2-Methylpentane	6.27	27.8	29.9	184.1	211.7	296.2	331.4
3-Methylpentane	5.30	28.1	30.3	181.9	246.9	294.6	330.1
2-Methyl-2,4-pentanediol		57.3					
3-Methylpentanenitrile		35.1	41.6				
Methyl pentanoate		35.4	43.1				
2-Methylpentanoic acid		52.1	57.5				
2-Methyl-1-pentanol		50.2	55.7				
2-Methyl-2-pentanol		39.6	54.8				
2-Methyl-3-pentanol		41.8	54.4				
3-Methyl-1-pentanol		46.3	62.3				
3-Methyl-2-pentanol		43.4	56.9				
4-Methyl-1-pentanol		44.5	60.5				
4-Methyl-2-pentanol		44.2	50.6				
3-Methyl-3-pentanol		41.8					
2-Methyl-3-pentanone		33.8	39.8				
3-Methyl-2-pentanone		34.2	40.5				
4-Methyl-2-pentanone		34.5	40.6				
2-Methyl-1-pentene		28.1	30.5	170.7	227.6	269.5	300.4
3-Methyl-1-pentene		26.9	28.7	177.8	232.6	272.8	302.5
4-Methyl-1-pentene		27.1	28.7	162.8	221.3	264.0	296.2
2-Methyl-2-pentene		29.0	31.6	163.2	222.6	245.2	297.5
cis-3-Methyl-2-pentene		28.8	31.2	163.2	222.6	265.3	297.5
trans-3-Methyl-2-pentene		29.3	31.5	163.2	222.6	265.3	297.5
cis-4-Methyl-2-pentene		27.6	29.5	167.6	226.4	267.8	299.2

(*Continued*)

TABLE 2.54 Heat of Fusion, Vaporization, Sublimation, and Specific Heat at Various Temperatures of Organic Compounds (*Continued*)

Substance	ΔHm	ΔHv	ΔHs	C_p 400 K	600 K	800 K	1000 K
trans-4-Methyl-2-pentene		28.0	30.0	171.1	229.3	269.9	300.4
4-Methyl-3-penten-2-one		36.1		214.0			
Methyl pentyl ether		32.0	36.9				
Methyl pentyl sulfide		37.4	45.2	203.6	272.2	324.6	366.0
3-Methyl-1-phenyl-1-butanone			59.5				
2-Methyl-1-phenylpropane	12.5	37.8	49.5				
Methyl phenyl sulfide			54.3				
Methyl phenyl sulfone			92.0				
Methylphosphonic acid			48.1				
2-Methylpiperidine			40.5				
2-Methylpropanal			31.5				
2-Methylpropane	4.66	21.3	19.3	124.6	169.5	202.9	227.6
2-Methylpropanenitrile		32.4	37.1				
2-Methyl-1-propanethiol	5.0	31.0	34.6	147.7	193.6	225.0	247.6
2-Methyl-2-propanethiol	2.5	28.5	30.8	151.2	199.2	232.3	256.2
$\Delta Ht = 4.1^{-121.6}$							
$\Delta Ht = 0.7^{-116.2}$							
$\Delta Ht = 1.0^{-73.8}$							
Methyl propanoate		32.2	35.9				
2-Methylpropanoic acid	5.02		35.3				
2-Methyl-1-propanol	6.32	41.8	50.8				
2-Methyl-2-propanol	6.79	39.1	46.7	142.9	189.8	222.9	247.5
$\Delta Ht = 0.8^{13}$							
2-Methylpropene	5.93	22.1	20.6	111.2	147.7	175.1	196.0
Methyl propyl ether		26.8	27.6	138.1	183.8	218.7	244.8
Methyl propyl sulfide	9.9	32.1	36.2	144.9	191.9	227.8	255.8
2-Methylpyridine	9.72	36.2	42.5	133.6	186.4	222.6	243.3
3-Methylpyridine	14.18	37.4	44.4	133.1	186.1	222.3	247.8
4-Methylpyridine	11.57	37.5	44.6				
1-Methyl-1*H*-pyrrole			40.8				
Methyl salicylate		46.7					
α-Methylstyrene				187.4	254.0	300.4	333.9
cis-β-Methylstyrene				187.4	254.0	300.4	333.9
trans-β-Methylstyrene				189.1	256.1	301.3	334.7
Methyl tetradecanoate			37.0				
2-Methylthiacyclopentane		36.4	41.8				
4-Methylthiazole		37.6	43.8				
2-Methylthiophene	9.20	33.9	38.9	123.1	165.6	194.3	214.6
3-Methylthiophene	10.53	34.2	39.4	122.9	164.6	192.3	211.7
Methyl trichloroacetate			48.3				
Methyl tridecanoate			82.7				
Methyl undecanoate			71.4				
5-Methyluracil			134.1				
Morpholine		37.1	44.0				
Naphthalene	18.98	43.2	72.6	180.1(g)	251.5	297.3	329.2
1-Naphthalenecarboxylic acid			110.4				
2-Naphthalenecarboxylic acid			113.6				
1-Naphthol	23.33		91.2				
2-Naphthol	17.51		94.2				
1,4-Naphthoquinone			72.4				

TABLE 2.54 Heat of Fusion, Vaporization, Sublimation, and Specific Heat at Various Temperatures of Organic Compounds (*Continued*)

				C_p			
Substance	ΔHm	ΔHv	ΔHs	400 K	600 K	800 K	1000 K
1-Naphthylamine			90.0				
2-Naphthylamine			88.3				
2-Nitroaniline	16.11		90.0				
3-Nitroaniline	23.68		96.7				
4-Nitroaniline	21.1		109				
Nitrobenzene	11.59	40.8	55.0				
1-Nitrobutane		38.9	48.6	157.5	210.1	247.0	273.6
2-Nitrobutane		36.8	43.8	157.4	211.1	248.7	276.0
Nitroethane	9.85	38.0	41.6	99.0	131.6	154.0	170.2
Nitromethane	9.70	34.0	38.3	70.3	91.7	106.9	117.9
(Nitromethyl)benzene			53.6				
2-Nitrophenol	17.44						
3-Nitrophenol	19.2						
4-Nitrophenol	18.25						
1-Nitronaphthalene			107.1				
1-Nitropropane		38.5	43.4	128.5	171.0	200.7	222.0
2-Nitropropane		36.8	41.3	129.2	172.3	201.8	222.8
2-Nitroso-1-naphthol			56.5				
4-Nitroso-1-naphthol			87.4				
1-Nitroso-2-naphthol			86.6				
2-Nitrotoluene		16.5	47.2				
3-Nitrotoluene		15.0	49.9				
4-Nitrotoluene	16.81	15.5	50.2				
Nonadecane, $\Delta Ht = 13.8^{22.8}$	45.82	56.0	95.8	559.4	754.9	896.3	1000.8
1-Nonadecene	33.5	54.6	94.9	545.0	733.7	869.7	969.9
1-Nonal			72.3	271.1	361.5	426.4	474.5
Nonane, $\Delta Ht = 6.3^{-56.0}$	15.47	36.9	46.4	269.0	364.1	433.3	484.9
1-Nonanethiol	33.5	44.4		291.6	390.3	464.6	521.5
Nonanoic acid	20.28		82.4				
1-Nonanol		54.4	76.9	282.4	379.1	449.6	501.7
2-Nonanone			56.4				
5-Nonanone	24.93		53.3				
1-Nonene	18.08	36.3	45.5	254.6	342.8	406.8	454.0
cis-Octadecafluorodecahydronaphthalene		35.6	45.2				
trans-Octadecafluorodecahydronaphthalene		35.8	45.4				
Octadecafluoropropylcyclohexane		24.5	43.1				
Octadecafluorooctane		33.4	41.1				
Octadecane	61.39	54.5	152.8	530.4	715.8	850.0	949.4
Octadecanedioic acid	56.6						
Octadecanoic acid	56.59		166.5				
Octadecanol			113.4				
1-Octadecene	32.6	53.3	90.0	516.0	694.5	823.4	918.4
cis-9-Octadecenoic acid		64.7					
Octafluorocyclobutane	2.77	23.2		186.1	225.3	245.4	257.3
Octafluorotoluene	11.58						
Octamethylcyclotetrasiloxane		45.6					
Octanal				242.3	322.2	380.3	422.6
Octanamide			110.5				
Octane	20.65	34.4	41.5	240.0	325.0	387.0	433.5
1,8-Octanedioic acid			143.1				

(*Continued*)

TABLE 2.54 Heat of Fusion, Vaporization, Sublimation, and Specific Heat at Various Temperatures of Organic Compounds (*Continued*)

				C_p			
Substance	ΔHm	ΔHv	ΔHs	400 K	600 K	800 K	1000 K
Octanenitrile		41.3	56.8				
1-Octanethiol	24.3	42.3		262.6	351.3	418.3	469.9
Octanoic acid	21.36	58.5	81.7				
1-Octanol	42.30	46.9	71.0	253.4	340.0	403.3	450.1
(±)-2-Octanol		44.4					
(±)-3-Octanol		36.5					
4-Octanol		40.5					
2-Octanone	24.42						
1-Octene	15.57	34.1	40.4	225.6	303.7	360.5	402.5
1-Octyne		35.8	42.3	216.5	285.7	336.0	410.9
2-Octyne		37.3	44.5				
3-Octyne		36.9	43.9				
4-Octyne		36.0	42.7				
Oxalic acid			98.0				
Oxaloyl chloride			31.8				
Oxamide			113.0				
Oxetane		28.7	29.9				
2-Oxetanone			47.0				
2-Oxohexamethyleneimine	16.2	54.8	83.3				
4-Oxopentanoic acid	9.22						
1,1′-Oxybis(2-ethoxy)ethane			58.4				
2,2′-Oxybis(ethanol)		52.3	57.3				
Paraldehyde			41.4				
Pentachloroethane	11.34	36.9	45.6	133.7	152.1	162.0	168.1
Pentachlorofluoroethane	1.9						
Pentachlorophenol			67.4				
Pentacyclo-[4.2.0.0^{2,5}.0^{3,8}.0^{4,7}]octane			80.3				
Pentadecane, $\Delta Ht = 9.2^{-2.25}$	34.8	49.5	76.1	443.3	598.6	711.1	794.5
Pentadecanoic acid	50.2		162.7				
1-Pentadecene	28.9	48.7	75.1	428.9	577.3	684.5	763.6
1,2-Pentadiene		27.6	28.7	131.4	170.7	199.6	220.9
cis-1,3-Pentadiene		27.6	28.3	123.4	166.9	196.7	218.4
trans-1,3-Pentadiene		27.0	27.8	130.5	171.1	199.6	220.1
1,4-Pentadiene	6.14	25.2	25.7	131.0	170.2	220.5	
2,3-Pentadiene		28.2	29.5	125.1	164.9	195.0	217.6
Pentaerythritol		92	143.9				
Pentaerythritol tetranitrate			151.9				
Pentafluorobenzene	10.85	32.2	36.3				
Pentafluorobenzoic acid			91.6				
Pentafluoroethane				113.8	137.8	151.1	158.9
Pentafluorophenol	12.85		67.4				
2,3,4,5,6-Pentafluorotoluene	12.99	34.8	41.1				
Pentamethylbenzene $\Delta Ht = 2.0^{23.7}$	12.3	45.1	60.8	272.0	360.2	423.8	470.0
2,2,4,6,6-Pentamethylheptane			49.0				
Pentanal			38.8	155.2	205.0	241.4	267.8
Pentanamide			89.3				
Pentane	8.42	25.8	26.4	152.8	207.7	248.1	278.5
1,5-Pentanediol		60.7					

TABLE 2.54 Heat of Fusion, Vaporization, Sublimation, and Specific Heat at Various Temperatures of Organic Compounds (*Continued*)

				C_p			
Substance	Δ*Hm*	Δ*Hv*	Δ*Hs*	400 K	600 K	800 K	1000 K
1,5-Pentanedithiol			59.3				
2,4-Pentanedione		34.3	41.8				
Pentanenitrile	4.73	36.1	43.6				
1-Pentanethiol	17.5	34.9	41.2	175.4	234.0	279.4	315.1
Pentanoic acid	14.16	44.1	62.4				
1-Pentanol	9.83	44.4	57.0	166.3	222.8	264.4	295.4
2-Pentanol		41.4	54.2				
3-Pentanol		43.5	54.0				
2-Pentanone	10.63	33.4	38.4	152.4	202.2	239.0	266.1
3-Pentanone	11.59	33.5	38.5				
1-Pentene	5.81	25.2	25.5	138.5	186.4	221.5	247.7
cis-2-Pentene	7.12	26.1	26.9	132.1	182.5	218.8	245.9
trans-2-Pentene	8.36	26.1	26.8	136.7	184.2	219.5	246.1
cis-2-Pentenenitrile		36.4	43.2				
trans-2-Pentenenitrile		37.8	44.9				
trans-3-Pentenenitrile		37.1	44.8				
Pentyl acetate		41.0					
Pentylamine		34.0	40.1				
Pentylcyclohexane			53.9				
Pentyl propyl ether		35.0	42.8				
1-Pentyne		27.7	28.4	130.1	169.0	197.1	218.4
2-Pentyne		29.3	30.8	122.2	161.9	192.1	215.1
Perylene	31.75						
α-Phellandrene			50.6				
Phenanthrene	16.46	55.7	75.5				
9,10-Phenanthrenedione			91.6				
Phenazine			99.9				
Phenol	11.29	45.7	57.8	135.8	182.2	211.8	232.2
Phenyl acetate			54.8				
Phenylacetonitrile		52.9					
Phenylacetylene			41.8	150.4	200.9	233.4	255.9
(−)-3-Phenyl-1-alanine			155.2				
α-Phenylbenzeneacetic acid	31.27						
Phenyl benzoate			99.0				
Phenylboron dichloride			33.9				
Phenylcyclopropane			50.2				
N-Phenyldiacetimide			90.0				
Phenyl formate			52.9				
Phenylhydrazine	16.43		61.7				
1-Phenyl-1-propanone			58.5				
1-Phenyl-2-propanone			49.0				
Phenyl salicylate			92.1				
Phenyl vinyl ether			49.9				
Phthalamide			57.3				
1,3-Phthalic acid			106.7				
1,4-Phthalic acid			98.3				
Phthalic anhydride			88.7				
Phthalonitrile			86.9				
Piperidine	14.85	31.7	39.3				
Propadiene		18.6		72.0	92.1	106.4	117.2

(*Continued*)

TABLE 2.54 Heat of Fusion, Vaporization, Sublimation, and Specific Heat at Various Temperatures of Organic Compounds (*Continued*)

				C_p			
Substance	ΔHm	ΔHv	ΔHs	400 K	600 K	800 K	1000 K
Propanal		28.3	29.6	96.6	126.4	148.3	164.0
Propanamide	17.6		85.9				
Propane	3.53	19.0	14.8	94.0	128.7	154.8	174.6
1,2-Propanediamine			44.2				
1,3-Propanediamine		40.9	50.2				
Propanedinitrile			79.1				
1,2-Propanediol		54.1	58.0				
1,3-Propanediol		57.9	37.1				
1,2-Propanedione			38.1				
1,2-Propanedithiol			49.7				
Propanenitrile, $\Delta Ht = 1.7^{-96.2}$	5.05	31.8	36.0	88.6	114.7	134.5	149.4
1-Propanethiol, $\Delta Ht = 4.0^{-131.1}$	5.5	29.5	31.9	116.6	153.6	182.4	205.1
2-Propanethiol	5.7	27.9	29.5	118.6	154.9	181.0	200.5
1,2,3-Propanetriol triacetate		57.8	85.7				
1,2,3-Propanetriol trinitrate	21.9						
Propanoic acid	10.66	32.3	32.1				
Propanoic anhydride		41.7	52.6				
1-Propanol	5.20	41.4	47.4	108.2	144.6	171.7	192.2
2-Propanol	5.37	39.9	45.4	112.0	149.6	176.3	195.9
Propanolactone			47.0				
2-Propenal		28.3	31.3				
Propene	3.00	18.4	14.2	80.5	108.0	128.7	144.4
2-Propenenitrile	6.23						
Propenoic acid	11.16						
2-Propen-1-ol		40.0	47.3	95.4	126.0	147.6	163.4
cis-1-Propenylbenzene				187.4	254.0	300.4	333.9
2-Propoxyethanol		41.4	52.1				
Propyl acetate		33.9	39.7				
1-Propylamine	10.97	29.6	31.3	119.3	159.0	188.0	210.1
Propylbenzene	9.27	38.2	46.2	200.1	275.6	327.6	364.7
Propyl benzoate		49.8	51.9				
Propyl carbamate			81.2				
Propyl chloroacetate			48.5				
Propylcyclohexane	10.37	36.1	45.1	247.3	350.6	423.4	474.5
Propylcyclopentane	10.0	34.7	41.1	212.7	297.2	361.0	407.9
Propylene oxide	6.5	27.4	28.3	92.7	125.8	149.3	166.5
Propyl formate		33.6	37.5				
Propyl nitrate		35.9	40.6	149.8	194.5	225.4	247.2
Propyl propanoate		35.5	43.5				
Propyl trichloroacetate			53.1				
Propyl vinyl ether			29.3				
Propyne		22.1		72.5	91.2	105.2	115.9
2-Propyn-1-ol		42.1					
Pyrazine			56.3				
Pyrene	17.11						
Pyridazine			53.5				
Pyridine	8.28	35.1	40.2	106.4	149.5	177.8	197.4
Pyrimidine		49.8	50.0				
1*H*-Pyrrole	7.91	38.8	45.1				
Pyrrolidine, $\Delta Ht = 0.5^{-66}$	8.58	33.0	37.6	114.4	168.7	206.5	233.6

TABLE 2.54 Heat of Fusion, Vaporization, Sublimation, and Specific Heat at Various Temperatures of Organic Compounds (*Continued*)

				C_p			
Substance	ΔHm	ΔHv	ΔHs	400 K	600 K	800 K	1000 K
Quinoline	10.66	49.7	53.9				
Salicylic acid			95.1				
5,5′-Spirobis(1,3-dioxane)			72.8				
Spiro[2.2]pentane	5.8	26.8	27.5	119.5	167.8	200.5	223.9
cis-Stilbene			69.0				
trans-Stilbene	27.4		99.2				
Styrene	11.0	38.7	43.9	160.3	218.2	256.9	284.2
Succinic acid	32.95		117.5				
Succinic anhydride	20.41						
Succinonitrile	3.92						
p-Terphenyl	35.5						
1,1,2,2-Tetrabromoethane		48.7	70.0				
Tetrabromomethane		45.1	110	97.1	102.6	106.7	105.9
Tetrabutyltin			19.8				
Tetracene			125.5				
Tetrachloro-1,4-benzoquinone			98.7				
1,1,2,2-Tetrachloro-1,2-difluoroethane	3.70	35.0					
1,1,1,2-Tetrachloro-2,2-fluorooctane	3.99						
1,1,1,2-Tetrachloroethane				118.7	139.2	151.6	159.7
1,1,2,2-Tetrachloroethane		37.6	45.7	116.7	137.7	150.0	158.0
Tetrachloroethylene	10.56	34.7	39.7	105.0	116.6	122.6	125.8
Tetrachloromethane	3.28	29.8	32.4	91.7	99.7	103.1	104.8
$\Delta Ht = 4.6^{-47.9}$							
Tetracyanoethylene			81.2				
Tetracyanomethane			61.1				
Tetradecane	45.6	47.6	71.3	414.3	559.5	664.8	743.1
Tetradecanenitrile			85.3				
Tetradecanoic acid	45.38		139.8				
1-Tetradecanol	49.0		102.2				
1-Tetradecene	27.6	46.9	70.2	399.8	538.2	638.2	712.1
Tetraethylene glycol		62.6	98.7				
Tetraethylgermanium			44.8				
Tetraethyllead			56.9				
Tetraethylsilane	13.01						
Tetraethyltin			51.0				
1,1,1,2-Tetrafluoroethane				104.2	128.7	143.1	152.1
Tetrafluoroethylene	7.7	16.8		91.9	106.8	115.5	120.8
Tetrafluoromethane	0.7	12.6		72.4	86.8	94.5	98.8
$\Delta Ht = 1.5^{-196.9}$							
Tetrahydrofuran	8.54	29.8	32.0				
Tetrahydrofuran-2,5-dimethanol		63.6					
Tetrahydrofuran-2-methanol		45.2	51.6				
1,2,3,4-Tetrahydronaphthalene	12.45	43.9	55.2				
Tetrahydropyran		31.2	34.6				
Tetrahydropyran-2-methanol		44.4					
Tetrahydrothiophene		34.7	39.4				
Tetrahydrothiophene-1,1-dioxide	1.43						
Tetraiodomethane				100.4	104.4	105.9	106.7
Tetramethoxysilane		194.6					
1,2,3,4-Tetramethylbenzene	11.2	45.0	57.2	237.7	316.7	374.1	416.2

(*Continued*)

TABLE 2.54 Heat of Fusion, Vaporization, Sublimation, and Specific Heat at Various Temperatures of Organic Compounds (*Continued*)

				C_p			
Substance	ΔHm	ΔHv	ΔHs	400 K	600 K	800 K	1000 K
1,2,3,5-Tetramethylbenzene	10.7	43.8	53.7	233.3	313.0	371.5	414.3
1,2,4,5-Tetramethylbenzene	21.0	45.5	53.4	232.2	311.2	369.9	413.0
2,2,3,3-Tetramethylbutane	7.54	31.4	42.9				
$\Delta Ht = 2.0^{-120.7}$							
Tetramethylene sulfone	1.4	61.5					
Tetramethyllead			38.1				
2,2,3,3-Tetramethylpentane	2.33						
2,2,3,4-Tetramethylpentane	0.50						
2,2,4,4-Tetramethylpentane	9.75	32.5	38.5				
2,3,3,4-Tetramethylpentane	9.00						
Tetramethylsilane	6.88						
Tetramethyltin			33.1				
1,1,3,3-Tetramethylurea	14.10	45.6					
Tetranitromethane		40.7	49.9				
Tetraphenylmethane			150.6				
Tetraphenyltin			66.3				
Tetrapropylgermanium			61.5				
Tetrapropyltin			66.9				
1,2,3,4-(1*H*)-Tetrazole			97.5				
Thiacyclobutane		32.3	36.0				
Thiacycloheptane			47.3	175.7	272.0	330.5	368.2
Thiacyclohexane	2.5	36.0	42.6	149.4	219.1	267.8	302.7
$\Delta Ht = 1.1^{-71.8}$							
$\Delta Ht = 7.8^{-33.1}$							
Thiacyclopentane	7.4	34.7	39.5	121.1	167.5	199.4	222.3
Thiacyclopropane		29.2	30.3	69.2	92.0	107.2	118.0
Thioacetamide			83.3				
Thioacetic acid			37.2	93.1	111.8	127.2	136.5
1,2-Thiocresol			51.5				
2,2′-Thiodiethanol		66.8					
Thiophene, $\Delta Ht = 0.6^{-101.6}$	5.09	31.5	34.7	96.3	129.5	150.7	165.4
Thiophenol	11.5	39.9	47.6	137.1	184.6	215.9	237.6
Thymol	17.27						
Toluene	6.85	33.2	38.0	140.1	197.5	236.9	264.9
o-Toluidine		44.6	56.7				
m-Toluidine	3.89	44.9	57.3				
p-Toluidine	18.22	44.3					
Triacetamide			60.4				
2,4,6-Triamino-1,3,5-triazine			124.3				
Tribromomethane		39.7	46.1	78.7	88.0	93.3	96.7
Tributoxyborane		56.1	52.3				
Tributyl phosphate		61.4	72.0				
Trichloroacetic acid	5.88						
Trichloroacetonitrile		34.1					
Trichloroacetyl chloride			41.0				
1,3,5-Trichlorobenzene	18.2						
Trichlorobenzoquinone			88.7				
1,1,1-Trichloroethane	2.73	29.9	32.5	107.6	128.4	141.1	149.8
$\Delta Ht = 7.5^{-49.0}$							
1,1,2-Trichloroethane	11.54	34.8	40.2	104.7	126.1	139.2	148.2

TABLE 2.54 Heat of Fusion, Vaporization, Sublimation, and Specific Heat at Various Temperatures of Organic Compounds (*Continued*)

				C_p			
Substance	ΔHm	ΔHv	ΔHs	400 K	600 K	800 K	1000 K
Trichloroethylene		31.4	34.5	91.2	104.9	112.7	117.8
Trichloromethane	8.8	29.2	31.3	74.3	85.3	91.5	95.5
Trichloromethylsilane	8.94						
1,2,3-Trichloropropane	8.9	37.1		31.7	38.9	43.8	47.3
1,1,1-Trichlorotrifluoroethane		26.9	28.1				
1,1,2-Trichlorotrifluoroethane	2.47	27.0	28.4				
1,1,1-Trichloro-3,3,3-trifluoropropane		32.2	36.8				
Tricyanoethylene			81.2				
Tridecane, $\Delta Ht = 7.7^{-18.2}$	28.50	45.7	66.4	385.2	520.4	618.5	691.2
Tridecanenitrile			85.3				
Tridecanoic acid	43.1		146.4				
1-Tridecene	22.83	45.0	65.3	370.8	499.1	592.0	660.2
Triethanolamine	27.2	67.5					
Triethoxyborane			43.9				
Triethoxymethane			46.0				
Triethylaluminum			73.2				
Triethylamine		31.0	34.8	203.8	276.6	328.7	367.4
Triethylaminoborane			60.7				
Triethylarsine			43.1				
Triethyl arsenite			50.6				
Triethylbismuthine			46.0				
Triethylborane			36.8				
Triethylenediamine	6.1		61.9				
$\Delta Ht = 9.6^{79.8}$							
Triethylene glycol		71.4	79.1				
Triethylphosphine			39.8				
Triethyl phosphate			57.3				
Triethyl phosphite			41.8				
Triethylstibine			43.5				
Trifluoroacetic acid		33.3	38.5				
ΔH(dimer dissoc) $= 58.8^{100}$							
Trifluoroacetonitrile	5.0						
1,1,1-Trifluoro-2-bromo-2-chloroethane		28.1	29.6				
1,1,1-Trifluoroethane	6.19	19.2		95.2	118.7	133.8	144.1
2,2,2-Trifluoroethanol		40.0					
Trifluoroethylene				81.1	97.5	107.5	113.9
Trifluoromethane	4.1	16.7		61.1	76.0	85.1	91.0
(Trifluoromethyl)benzene	13.46	32.6	37.6	169.8	226.8	262.6	286.4
Triiodomethane	16.3		69.9	82.0	90.0	94.7	97.8
Triisopropylborane			41.8				
Triisopropyl phosphite			46.0				
Trimethoxyborane			34.7				
1,1,1-Trimethoxyethane			39.2				
Trimethoxymethane			38.1				
2′,4′,5′-Trimethylacetophenone			63.2				
2′,4′,6′-Trimethylacetophenone			62.3				
Trimethylaluminum			63.2				
Trimethylamine	6.55	22.9	21.7	117.5	160.4	190.9	213.3
Trimethyl arsenite			42.3				
Trimethylarsine			28.9				

(*Continued*)

TABLE 2.54 Heat of Fusion, Vaporization, Sublimation, and Specific Heat at Various Temperatures of Organic Compounds (*Continued*)

				C_p			
Substance	ΔHm	ΔHv	ΔHs	400 K	600 K	800 K	1000 K
1,2,3-Trimethylbenzene	8.37	40.0	49.1	196.2	267.8	320.9	359.4
$\Delta Ht = 0.7^{-54.5}$							
$\Delta Ht = 1.3^{-42.9}$							
1,2,4-Trimethylbenzene		39.3	47.9	196.5	269.0	321.9	360.2
1,3,5-Trimethylbenzene	9.51	39.0	47.5	194.2	268.1	321.5	360.1
2,6,6-Trimethylbicyclo[3.1.1]-2-heptene			44.8				
Trimethylbismuthine			34.7				
Trimethylborane			20.2				
2,2,3-Trimethylbutane	2.20	28.9	32.0	212.7	291.3	346.1	386.3
$\Delta Ht = 2.5^{-151.8}$							
2,3,3-Trimethyl-1-butene			32.2				
cis,cis-1,3,5-Trimethylcyclohexane				242.9	351.2	427.6	482.0
Trimethylene oxide		28.7	29.9				
Trimethylene sulfide	8.3	32.3	36.0	91.6	127.4	152.3	170.2
$\Delta Ht = 0.7^{-96.5}$							
Trimethylgallium			38.1				
2,2,5-Trimethylhexane	6.2	33.7	40.2				
2,3,5-Trimethylhexane	10.00	34.4	41.4				
Trimethylindium			48.5				
2,4,7-Trimethyloctane		38.2	49.9				
2,2,3-Trimethylpentane	8.62	31.9	36.9				
2,2,4-Trimethylpentane	9.04	30.8	35.1				
2,3,3-Trimethylpentane	0.86	32.1	37.3				
$\Delta Ht = 7.7^{-109.0}$							
2,3,4-Trimethylpentane	9.27	32.4	37.7				
2,2,4-Trimethyl-1,3-pentanediol	8.6	55.7					
2,2,4-Trimethyl-3-pentanone		35.6	43.3				
2,4,4-Trimethyl-1-pentene		31.4	35.8				
2,4,4-Trimethyl-2-pentene		32.6	37.5				
Trimethylphosphine			28.0				
Trimethylphosphine oxide			50.2				
Trimethyl phosphate			36.8				
2,3,6-Trimethylpyridine		40.0	50.6				
2,4,6-Trimethylpyridine	9.53	39.9	50.3				
Trimethylsilanol			45.6				
Trimethylstibine			31.4				
Trimethylsuccinic anhydride			74.1				
Trimethylthiacyclopropane			39.3				
Trimethyltin bromide			47.3				
2,4,6-Trinitroanisole			133.1				
1,3,5-Trinitrobenzene	16.7		99.6				
Trinitromethane		32.6	46.7				
2,4,6-Trinitrophenetole			120.5				
2,4,6-Trinitrotoluene			104.7				
1,3,6-Trioxacycloactane			48.8				
1,3,5-Trioxane	15.11		56.6				
Triphenylarsine			99.3				
Triphenylbismuthine			110.9				
Triphenylborane			81.6				
Triphenylene			118.0				

TABLE 2.54 Heat of Fusion, Vaporization, Sublimation, and Specific Heat at Various Temperatures of Organic Compounds (*Continued*)

				C_p			
Substance	ΔHm	ΔHv	ΔHs	400 K	600 K	800 K	1000 K
Triphenylmethane			100.0				
Triphenylphosphine			96				
Triphenylstibine			106.3				
Tripropoxyborane			49.4				
Tris(diethylamino)phosphine			60.7				
Tris(trimethylsilyl)amine			54.4				
Tropolone			83.7				
Undecane	22.32	41.5	56.4	327.1	442.7	525.9	588.3
$\Delta Ht = 6.9^{-36.6}$							
Undecanenitrile			71.1				
Undecanoic acid	25.9		121.3				
1-Undecene, $\Delta Ht = 9.2^{-55.8}$	16.99	40.9	55.4	312.7	421.1	499.3	557.3
Uracil			126.5				
Urea	15.1	87.9					
(−)-Valine			162.8				
Vinyl acetate		34.4	34.8				
Vinyl benzene			39.6				
Vinylcyclohexane			39.7				
4-Vinyl-1-cyclohexene		33.5	38.3				
1,2-Xylene	13.61	36.2	43.4	171.7	234.2	278.8	311.1
1,3-Xylene	11.55	35.7	42.7	167.5	232.2	277.9	310.6
1,4-Xylene	16.81	35.7	42.4	166.1	230.8	276.7	309.7

2.14 CRITICAL PROPERTIES

Critical temperature (T_c), critical pressure(P_c), and critical volume (V_c) represent three widely used pure component constants. These critical constants are very important properties in chemical engineering field because almost all other thermo chemical properties are predictable from boiling point and critical constants with using corresponding state theory. Therefore, precise prediction of critical constants is very necessary.

2.14.1 Critical Temperature

The critical temperature of a compound is the temperature above which a liquid phase cannot be formed no matter what the pressure on the system. The critical temperature is important in determining the phase boundaries of any compound and is a required input parameter for most phase equilibrium thermal property or volumetric property calculations using analytic equations of state or the theorem of corresponding states. Critical temperatures are predicted by various empirical methods according to the type of compound or mixture being considered.

2.14.2 Critical Pressure

The critical pressure of a compound is the vapor pressure of that compound at the critical temperature. Below the critical temperature, any compound above its vapor pressure will be a liquid.

2.14.3 Critical Volume

The critical volume of a compound is the volume occupied by a specified mass of a compound at its critical temperature and critical pressure.

2.14.4 Critical Compressibility Factor

The critical compressibility factor of a compound is used as a characterization parameter in corresponding states methods to predict volumetric and thermal properties. The factor varies from approximately 0.23 for water to 0.26–0.28 for most hydrocarbons to above 0.30 for light gases.

TABLE 2.55 Critical Properties

Substance	T_c, °C	P_c, atm	P_c, MPa	V_c, $cm^3 \cdot mol^{-1}$	ρ_c, $g \cdot cm^{-3}$
Acetaldehyde	193	55	5.57	154	0.286
Acetic acid	319.56	57.1	5.786	171.3	0.351
Acetic anhydride	333	39.5	4.0	290	0.352
Acetone	235.0	46.4	4.700	209	0.278
Acetonitrile	272.4	47.7	4.85	173	0.237
Acetophenone	436.4	38	3.85	386	0.311
Acetyl chloride	235	58	5.88	204	0.325
Acetylene	35.2	60.6	6.14	113	0.231
Acrylic acid	342	56	5.67	210	0.343
Acrylonitrile	263	45	4.56	210	0.253
Allene	120	54.0	5.47	162	0.247
Allyl alcohol	272.0	56.4	5.71	203	0.286
2-Aminoethanol	341	44	4.46	196	0.312
Aniline	426	49.5	4.89	287	0.324
Anthracene	610	28.6	2.90	554	0.333
Benzaldehyde	422	45.9	4.65	324	0.327
Benzene	288.90	48.31	4.895	255	0.306
Benzoic acid	479	41.55	4.21	341	0.358
Benzonitrile	426.3	41.55	4.21	339	0.304
Benzyl alcohol	422	42.4	4.3	334	0.324
Biphenyl	516	38.0	3.85	502	0.307
Bromobenzene	397	44.6	4.52	324	0.485
Bromochlorodifluoromethane	158.8	41.98	4.254	246	0.672
Bromoethane	230.8	61.5	6.23	215	0.507
Bromomethane	173.4	85	8.61	156	0.609
Bromopentafluorobenzene	397	44.6	4.52		
1-Bromopropane	−1.8				0.462
2-Bromopropane	−14.2				0.462
Bromotrifluoromethane	67.1	39.2	3.97	200	0.76
1,2-Butadiene	170.6	44.4	4.50	219	0.247
1,3-Butadiene	152	42.7	4.33	221	0.245
Butanal	264.1	42.6	4.32	258	0.279
Butane	151.97	37.34	3.784	255	0.228
Butanenitrile	312.3	38.3	3.88	285	0.242
Butanoic acid	351	39.8	4.03	290	0.304
1-Butanol	289.9	43.56	4.414	275	0.270
2-Butanol	263.1	41.47	4.202	269	0.276
2-Butanone	263.63	41.52	4.207	267	0.270
1-Butene	146.5	39.7	4.02	240	0.234
cis-2-Butene	147.5	40.5	4.10	238	0.240
trans-2-Butene	147.5	40.5	4.10	238	0.236
3-Butenenitrile	312.3	38.3	3.88	265	0.253
1-Buten-3-yne	182	49	4.96	202	0.258
Butyl acetate	306.7	31	3.14	400	0.290
1-Butylamine	258.8	41.9	4.25	277	0.264
sec-Butylamine	241.2	41.4	4.20	278	0.263
tert-Butylamine	210.8	37.9	3.84	292	0.250

TABLE 2.55 Critical Properties (*Continued*)

Substance	T_c, °C	P_c, atm	P_c, MPa	V_c, $cm^3 \cdot mol^{-1}$	ρ_c, $g \cdot cm^{-3}$
Butylbenzene	387.4	28.5	2.89	497	0.270
sec-Butylbenzene	391	29.1	2.94	510	0.263
tert-Butylbenzene	387	29.3	2.97	490	0.273
Butyl benzoate	450	26	2.63	561	0.318
Butyl butanoate	338				0.292
Butylcyclohexane	394	31.1	3.15	534	0.63
sec-Butylcyclohexane	396	26.4	2.67		
tert-Butylcyclohexane	385.9	26.3	2.66		
Butylcyclopentane	357.9				
Butyl ethyl ether	257.9	30	3.04	390	0.262
2-Butylhexadecafluoro-tetrahydrofuran	227.1	15.86	1.607	588	0.707
Butylisopropylamine	290.5				
tert-Butyl methyl sulfide	296.7				
1-Butyne	190.6	46.5	4.71	220	0.246
2-Butyne	215.5	50.2	5.09	221	0.246
4-Butyrolactone	436				
Carbon tetrachloride	283.3	45.0	4.56	276	0.558
Carbon tetrafluoride	−45.7	36.9	3.74	140	0.629
Chlorobenzene	359.3	44.6	4.52	308	0.365
1-Chlorobutane	268.9	36.4	3.69	312	0.297
2-Chlorobutane	247.5	39	3.95	305	0.303
1-Chloro-1,1-difluoroethane	137.1	40.7	4.12	231	0.435
2-Chloro-1,1-difluoroethylene	127.5	44.0	4.46	197	0.499
Chlorodifluoromethane	96.1	49.1	4.98	165	0.525
1-Chloro-2,3-epoxypropane	351				
Chloroethane	187.3	52.0	5.27	199	0.324
Chloroform	263.3	54.0	5.47	239	0.504
1-Chlorohexane	321.5				
Chloromethane	143.1	65.9	6.679	139	0.353
2-Chloro-2-methylpropane	234	39	3.95	295	0.314
Chloropentafluoroacetone	137.6	28.4	2.88		
Chloropentafluorobenzene	297.9	31.8	3.22		
Chloropentafluoroethane	80.1	31.9	3.229	252	0.613
1-Chloropentane	295.4				
1-Chloropropane	230	45.2	4.58	254	0.309
2-Chloropropane	212	46.6	4.72	230	0.341
3-Chloropropene	241	47	4.76	234	0.336
Chlorotrifluoromethane	29	38.98	3.946	180	0.579
Chlorotrifluorosilane	35.4	34.2	3.47		
Chlorotrimethylsilane	224.7	31.6	3.20		
1,2-Cresol	424.5	49.4	5.01	282	0.384
1,3-Cresol	432.7	45.0	4.56	309	0.346
1,4-Cresol	431.5	50.8	5.15	277	0.391
Cyanogen	126.7	62.2	6.30	145	0.360
Cyclobutane	186.8	49.2	4.99	210	0.267
Cycloheptane	316	36.7	3.72	390	0.252
Cyclohexane	280.4	40.2	4.07	308	0.273
trans-Cyclohexanedimethanol	451	34.85	3.531		
Cyclohexanethiol	390.9				
Cyclohexanol	376.9	42.0	4.26	327	0.306
Cyclohexanone	379.9	39.5	4.0	312	0.315
Cyclohexene	287.33	42.9	4.35	292	0.281

(*Continued*)

TABLE 2.55 Critical Properties (*Continued*)

Substance	T_c, °C	P_c, atm	P_c, MPa	V_c, $cm^3 \cdot mol^{-1}$	ρ_c, $g \cdot cm^{-3}$
Cyclohexylamine	341.5				
Cyclopentane	238.6	44.49	4.508	260	0.27
Cyclopentanethiol	360.4				
Cyclopentanone	353	53	5.37	268	0.314
Cyclopentene	232.9				
1-Cyclopentylheptane	406	19.2	1.94	649	0.260
1-Cyclopentylpentadecane	506.9	10.1	1.02	1096	0.256
Cyclopropane	124.7	54.2	5.49	170	0.248
p-Cymene	379	2.80	2.84	492	0.273
Decafluorobutane	113.3	22.93	2.323	378	0.629
cis-Decahydronaphthalene	429.2	31.6	3.20	480	0.288
trans-Decahydronaphthalene	414.0	31	3.14	480	0.288
Decane	344.6	20.8	2.11	624	0.228
Decanenitrile	348.8	32.1	3.25		
1-Decanol	413.9	22	2.23	600	0.264
1-Decene	343.3	21.89	2.218	585	0.240
Dibutyl sulfide	380				
Decylcyclohexane	477	13.4	1.36		
Decylcyclopentane	450	15.0	1.52		
Diallyl sulfide	380				
1,2-Dibromo-2-chlorotrifluoro-ethane	287.6				
Dibromodifluoromethane	198.3	40.8	4.13	249	0.843
1,2-Dibromoethane	309.9	71.1	7.2	242	0.776
Dibromomethane	310	71	7.19		
1,2-Dibromotetrafluoroethane	214.7	33.49	3.393	329	0.790
Dibutylamine	334.4	30.7	3.11	517	0.250
Dibutyl ether	311.0	29.7	3.01	500	0.260
Dibutyl sulfide	377	24.7	2.50	537	0.272
1,2-Dichlorobenzene	424.2	40.5	4.10	360	0.408
1,3-Dichlorobenzene	411	38	3.85	359	0.408
1,4-Dichlorobenzene	412	39	3.95	372	0.395
Dichlorodifluoromethane	111.80	40.82	4.136	217	0.558
1,1-Dichloroethane	250	50.0	5.07	236	0.419
Dichlorodifluorosilane	95.8	34.5	3.50		
1,2-Dichloroethane	288	53	5.4	225	0.440
1,1-Dichloroethylene	222	51.3	5.20	218	0.445
cis-1,2-Dichloroethylene	271.1			224	0.433
trans-1,2-Dichloroethylene	234.4	54.4	5.51	224	0.433
Dichlorofluoromethane	178.43	51.1	5.18	196	0.522
1,2-Dichlorohexafluoropropane	172.9				
Dichloromethane	237	60.2	6.10	193	0.440
1,2-Dichloropropane	304	44	4.49	226	0.500
Dichlorosilane	176	46.1	4.67		
1,1-Dichlorotetrafluoroethane	145.5	32.6	3.30	294	0.582
1,2-Dichlorotetrafluoroethane	145.63	32.1	3.252	297	0.582
Dideuterium oxide (D_2O)	371.0	215.7	21.86		0.363
Diethanolamine	442.0	32.3	3.27	349	0.301
1,1-Diethoxyethane (Acetal)	254				
Diethylamine	226.84	37.3	3.758	301	0.243
1,4-Diethylbenzene	384.8	27.7	2.81	480	0.280
Diethyl disulfide	368.9				
Diethylene glycol	408	46	4.66	316	0.336
Diethyl ether	193.59	35.9	3.638	280	0.265

TABLE 2.55 Critical Properties (*Continued*)

Substance	T_c, °C	P_c, atm	P_c, MPa	V_c, $cm^3 \cdot mol^{-1}$	ρ_c, $g \cdot cm^{-3}$
3,3-Diethyl-2-methylpentane	366.8	25.0	2.53	501	0.284
3,3-Diethylpentane	337	26.4	2.67		
Diethyl sulfide	284	39.1	3.96	318	0.284
Difluoroamine (HNF_2)	130	93	9.42		
1,2-Difluorobenzene	284.2			300	0.381
cis-Difluorodiazine	−1	70	7.09		
trans-Difluorodiazine	−13	55	5.57		
1,1-Difluoroethane	113.6	44.4	4.50	181	0.365
1,1-Difluoroethylene	29.8	44.0	4.46	154	0.417
Dihexyl ether	384	18	1.82	720	0.259
Diisopropyl sulfide	391				
Diisopropyl ether	227.17	27.9	2.832	386	0.265
1,2-Dimethoxyethane	263	38.2	3.87	271	0.333
Dimethoxymethane	242.1	44.2	4.48		
N,N-Dimethylacetamide	364	38.7	3.92		
Dimethylamine	164.07	52.7	5.340	187	0.241
N,N-Dimethylaniline	414	35.8	3.63		
2,2-Dimethylbutane	215.7	30.49	3.090	359	0.240
2,3-Dimethylbutane	499.9	30.90	3.131	358	0.241
3,3-Dimethyl-2-butanone	289.8				
2,3-Dimethyl-1-butene	228	32.0	3.24	343	0.245
3,3-Dimethyl-1-butene	217	32.1	3.25	340	0.248
2,3-Dimethyl-2-butene	250.9	33.2	3.36	351	0.240
1,1-Dimethylcyclohexane	318	29.3	2.97	416	0.378
cis-1,2-Dimethylcyclohexane	333.0	29.0	2.94	460	0.244
trans-1,2-Dimethylcyclohexane	323.0	29.3	2.97	460	0.244
cis-1,3-Dimethylcyclohexane	317.9	29.3	2.97	450	0.249
trans-1,3-Dimethylcyclohexane	325	29.3	2.97	460	0.244
cis-1,4-Dimethylcyclohexane	325.0	29.0	2.94	460	0.244
trans-1,4-Dimethylcyclohexane	317.0	29.0	2.94	459	0.249
1,1-Dimethylcyclopentane	274	34.0	3.44	360	0.273
cis-1,2-Dimethylcyclopentane	291.7	34.0	3.44	368	0.267
trans-1,2-Dimethylcyclopentane	277.2	34.0	3.44	362	0.271
cis-1,3-Dimethylcyclopentane	318.9				
Dimethyl disulfide	59.5				
Dimethyl ether	126.9	53.0	5.37	190	0.242
N,N-Dimethylformamide	376.5	51.5	5.22	262	0.279
2,2-Dimethylheptane	303.7	23.19	2.350	519	0.247
2,2-Dimethylhexane	276.8	25.0	2.529	478	0.239
2,3-Dimethylhexane	290.4	25.94	2.628	468	0.244
2,4-Dimethylhexane	280.5	25.22	2.556	472	0.242
2,5-Dimethylhexane	277.0	24.54	2.487	482	0.237
3,3-Dimethylhexane	289.0	26.19	2.654	443	0.258
3,4-Dimethylhexane	295.8	26.57	2.692	466	0.245
1,1-Dimethylhydrazine	250	53.6	5.43	230	0.261
2,4-Dimethyl-3-isopentane	341.3	23.1	2.34	521	0.273
2,3-Dimethyloctane	340.1	21.6	2.19	567	0.251
2,4-Dimethyloctane	326.3	21.1	2.14	566	0.251

(*Continued*)

TABLE 2.55 Critical Properties (*Continued*)

Substance	T_c, °C	P_c, atm	P_c, MPa	V_c, $cm^3 \cdot mol^{-1}$	ρ_c, $g \cdot cm^{-3}$
2,5-Dimethyloctane	330	21.2	2.15	569	0.250
2,6-Dimethyloctane	330	21.1	2.15	576	0.247
2,7-Dimethyloctane	329.8	20.7	2.10	590	0.241
3,3-Dimethyloctane	339	21.9	2.22	557	0.255
3,4-Dimethyloctane	341	22.1	2.24	551	0.258
3,5-Dimethyloctane	333.2	21.6	2.19	555	0.256
3,6-Dimethyloctane	335.2	21.6	2.19	562	0.253
4,5-Dimethyloctane	333.8	21.8	2.21	548	0.260
4,5-Dimethyloctane	339.1	22.1	2.24	546	0.261
Dimethyl oxalate	355	39.2	3.97		
2,2-Dimethylpentane	247.4	27.4	2.773	416	0.241
2,3-Dimethylpentane	264.3	28.70	2.908	393	0.255
2,4-Dimethylpentane	246.7	27.01	2.737	418	0.240
3,3-Dimethylpentane	263.3	29.07	2.946	414	0.242
2,3-Dimethylphenol	449.7	48	4.86	470	0.26
2,4-Dimethylphenol	434.5	43	4.36	509	0.24
2,5-Dimethylphenol	433.8	48	4.86	470	0.26
2,6-Dimethylphenol	427.9	42	4.26	509	0.24
3,4-Dimethylphenol	456.7	49	4.96	552	0.27
3,5-Dimethylphenol	442.5	36	3.65	611	0.25
2,2-Dimethylpropane	160.7	31.55	3.197	307	0.238
2,2-Dimethyl-1-propanol	276	39	3.95	319	
2,3-Dimethylpyridine	382.3				
2,4-Dimethylpyridine	373.9				
2,5-Dimethylpyridine	371				
2,6-Dimethylpyridine	350.7			316	0.339
3,4-Dimethylpyridine	410.7				
3,5-Dimethylpyridine	394.1				
Dimethyl sulfide	229.9	54.6	5.53	201	0.309
N,N-Dimethyl-1,2-toluidine	395	30.8	3.12		
1,4-Dioxane	314	51.5	5.21	238	0.370
Diphenyl ether	493.7	31	3.14		
Diphenylmethane	494	29.4	2.98		
Dipropylamine	282.7	35.8	3.63	407	0.249
Dipropyl ether	257.5	29.91	3.028		
Docosafluorodecane	269	14.3	1.45		
Dodecafluorocyclohexane	184.1	24	2.43		
Dodecafluorocyclohexene	188.7				
Dodecafluoro-1-hexene	181.3				
Dodecafluoropentane	149	20.1	2.03		
Dodecane	385	18.0	1.82	754	0.226
1-Dodecanol	405.9	19	1.92	718	0.260
1-Dodecene	384.5	18.3	1.85		
Dodecylbenzene	501	15.6	1.58	1000	0.246
Dodecylcyclopentane	477	12.8	1.30		
Ethane	32.3	48.2	4.90	148	0.203
1,2-Ethanediamine	319.8	62.1	6.29	206	0.292
1,2-Ethanediol	445	76	7.7	186	0.334
Ethanethiol	225.5	54.2	5.49	207	0.300
Ethanol	240.9	60.57	6.137	167	0.276
Ethoxybenzene	374.0	33.8	3.42		
Ethyl acetate	250.2	38.31	3.882	286	0.308

TABLE 2.55 Critical Properties (*Continued*)

Substance	T_c, °C	P_c, atm	P_c, MPa	V_c, $cm^3 \cdot mol^{-1}$	ρ_c, $g \cdot cm^{-3}$
Ethyl acetoacetate	400				
Ethyl acrylate	279	37.0	3.75	320	0.313
Ethylamine	183	55.5	5.62	182	0.248
Ethylbenzene	344.00	35.61	3.609	374	0.284
Ethyl benzoate	424	32	3.24	451	0.111
Ethylbutanoate	293	30.2	3.06	421	0.28
2-Ethyl-1-butanol	145.7				
Ethyl crotonate	326				
Ethylcyclohexane	336	29.9	3.03	450	0.249
Ethylcyclopentane	296.4	33.5	3.39	375	0.262
3-Ethyl-2,2-dimethylhexane	338.6	22.8	2.31	526	0.271
4-Ethyl-2,2-dimethylhexane	321.5	21.9	2.22	539	0.264
3-Ethyl-2,3-dimethylhexane	353.7	23.9	2.42	516	0.276
4-Ethyl-2,3-dimethylhexane	344.2	23.1	2.34	524	0.271
3-Ethyl-2,4-dimethylhexane	343.0	23.1	2.34	522	0.273
4-Ethyl-2,4-dimethylhexane	347.8	24.4	2.47	524	0.271
3-Ethyl-2,5-dimethylhexane	330.4	22.1	2.24	537	0.265
3-Ethyl-3,4-dimethylhexane	351.4	23.9	2.42	511	0.278
Ethylene	9.3	49.7	5.036	129	0.218
Ethylene glycol dimethyl ether	263	38.2	3.87	271	0.333
Ethylene glycol ethyl ether acetate	334.2	31.25	3.166	443	0.298
Ethylene glycol monobutyl ether	360.8			424	0.279
Ethylene oxide	196	71.0	7.275	140	0.314
Ethyl formate	235.4	46.8	4.74	229	0.323
3-Ethylhexane	292.4	25.74	2.608	455	0.251
2-Ethyl-1-hexanol	367.5	27.2	2.76	494	0.264
Ethyl isopentanoate	315				
Ethyl isopropyl ether	217.2				
2-Ethyl-1-methylbenzene	378	30.0	3.04	460	0.26
3-Ethyl-1-methylbenzene	364	28.0	2.84	490	0.24
4-Ethyl-1-methylbenzene	367	29.0	2.94	470	0.26
Ethyl 3-methylbutanoate	314.9				
1-Ethyl-1-methylcyclopentane	319	29.5	2.99		
Ethyl methyl ether	164.8	43.4	4.40	221	0.272
3-Ethyl-2-methylheptane	337.8	22.0	2.23	544	0.262
4-Ethyl-2-methylheptane	328.7	21.6	2.19	545	0.261
5-Ethyl-2-methylheptane	333.6	21.6	2.19	555	0.256
3-Ethyl-3-methylheptane	347.0	22.8	2.31	532	0.267
4-Ethyl-3-methylheptane	341.2	22.5	2.28	530	0.269
5-Ethyl-3-methylheptane	333.5	22.0	2.23	541	0.263
3-Ethyl-4-methylheptane	342.4	22.5	2.28	533	0.267
4-Ethyl-4-methylheptane	342.4	22.8	2.31	525	0.271
Ethyl methyl ketone	262.4	41.0	4.154	267	0.270
3-Ethyl-2-methylpentane	294.0	26.65	2.700	443	0.258
3-Ethyl-3-methylpentane	303.5	27.71	2.808	455	0.351
Ethyl 2-methylpropanoate	280	30	3.04	410	0.28
Ethyl methyl sulfide	260	42	4.26		
2-Ethylnaphthalene	502	31.0	3.14	521	0.300
Ethyl nonanoate	401				

(*Continued*)

TABLE 2.55 Critical Properties (*Continued*)

Substance	T_c, °C	P_c, atm	P_c, MPa	V_c, $cm^3 \cdot mol^{-1}$	ρ_c, $g \cdot cm^{-3}$
3-Ethyloctane	340	21.6	2.19	561	0.241
4-Ethyloctane	337	21.5	2.18	552	0.258
Ethyl octanoate	386				
3-Ethylpentane	267.6	28.53	2.891	416	0.241
1,2-Ethylphenol	429.9				
1,3-Ethylphenol	443.3				
1,4-Ethylphenol	443.3				
Ethyl propanoate	272.9	33.18	3.362	345	0.296
Ethyl propyl ether	227.1	32.1	3.25	244	0.361
m-Ethyltoluene	364.0	28.1	2.837	490	0.245
o-Ethyltoluene	378.0	30.1	3.04	460	0.261
p-Ethyltoluene	367	29.0	2.94	479	0.256
3-Ethyl-2,2,3-trimethylpentane	372.9	25.4	2.57	503	0.283
3-Ethyl-2,2,4-trimethylpentane	342.2	23.4	2.37	518	0.275
3-Ethyl-2,3,4-trimethylpentane	369.2	25.1	2.54	506	0.281
Ethyl vinyl ether	202	40.17	4.07	260	0.277
Fluorobenzene	286.94	44.91	4.551	357	0.269
Fluoroethane	102.2	49.6	5.03	169	0.284
Fluoromethane	44.7	58.0	5.88	124	0.274
4-Fluorotoluene	316.4				
Formaldehyde	135	65	6.6	105	0.286
Formic acid	315				
2-Furaldehyde	397	58.1	5.89		
Furan	217.1	54.3	5.50	218	0.312
Glycerol	453	66	6.69	255	0.361
Heptadecane	460	13.0	1.32	1006	0.140
1-Heptadecanol	736	14.0	1.42	960	0.267
Heptane	267.1	27.0	2.74	428	0.232
1-Heptanol	359.5	30.18	3.058	435	0.267
2-Heptanol	335.2	29.81	3.021	432	0.269
3-Heptanol	332.3				
2-Heptanone	338.4	33.91	3.436	421	0.271
1-Heptene	264.2	28.83	2.921	402	0.246
Heptylcyclopentane	406	19.2	1.945		
Hexadecafluoroheptane	201.7	16.0	1.62	664	0.584
Hexadecane	444	14	1.42	930	0.243
1-Hexadecene	444	13.2	1.34	933	0.241
Hexadecylcyclopentane	518	9.6	0.97		
1,5-Hexadiene	234	34	3.44	328	0.250
Hexafluoroacetone	84.1	29.0	2.94	329	0.505
Hexafluorobenzene	243.6	32.30	3.273	335	0.505
Hexafluoroethane	19.7			224	0.617
Hexamethylbenzene	494			600	0.271
Hexane	234.5	29.85	3.025	368	0.233
Hexanenitrile	360.7	32.57	3.30		
Hexanoic acid	389	31.6	3.20		
1-Hexanol	337.2	33.72	3.417	381	0.268
2-Hexanol	312.8	32.67	3.310		

TABLE 2.55 Critical Properties (*Continued*)

Substance	T_c, °C	P_c, atm	P_c, MPa	V_c, $cm^3 \cdot mol^{-1}$	ρ_c, $g \cdot cm^{-3}$
3-Hexanol	309.3	33.2	3.36		
2-Hexanone	313.9	32.8	3.32		
3-Hexanone	309.7	32.76	3.320		
1-Hexene	231.0	31.64	3.206	348	0.242
cis-2-Hexene	245	32.4	3.28	351	0.240
trans-2-Hexene	243	32.3	3.27	351	0.240
cis-3-Hexene	244	32.4	3.28	350	0.240
trans-3-Hexene	246.8	32.1	3.25	350	0.240
Hexylcyclopentane	387.0	21.1	2.14		
Icosafluorononane	251	15.4	1.56		
Icosane	494	10.3	1.04	1190	0.237
1-Icosanol	497	12.0	1.22		
Indane	411.8	39.0	3.95	381	0.310
Iodine	546	115	11.7	155	0.164
Iodobenzene	448	44.6	4.52	351	0.581
Iodoethane	281.0				
Iodomethane	255	65	6.59	190	0.75
1-Iodopropane	323				
Isobutyl acetate	288	31.2	3.16	414	0.281
Isobutylamine	246	40.2	4.07	284	0.258
Isobutylbenzene	377	30.1	3.05	480	0.280
Isobutyl bromide	294.1				
Isobutyl butanoate	338				
Isobutylcyclohexane	386	30.8	3.12		
Isobutyl formate	278	38.3	3.88	350	0.29
Isobutyl isobutanoate	329				
Isobutyl 3-methylbutanoate	348				
Isobutyl propanoate	319				
Isopentyl acetate	326				
Isopentyl butanoate	346				
Isopentyl propanoate	338				
Isopropyl acetate	258				
Isopropylamine	198.7	44.8	4.54	221	0.267
Isopropylbenzene	357.9	31.67	3.209	429	0.281
Isopropylcycloheptane	334.5				
Isopropylcyclohexane	367	28	2.84		
Isopropylcyclopentane	328	29.6	3.00		
4-Isopropylheptane	334.5	22.0	2.23	537	0.265
Isopropylmethylamine	217.6				
2-Isopropyl-1-methylbenzene	397	28.6	2.90		
3-Isopropyl-1-methylbenzene	393	29.0	2.94		
4-Isopropyl-1-methylbenzene	380	27.9	2.83		
3-Isopropyl-2-methylhexane	359.3	22.6	2.29	529	0.269
Isopropyl methyl sulfide	276.4				
Isoquinoline	530	50.3	5.10	374	0.345
Isoxazole	278.9				
Ketene	380	64	6.5	145	0.290
Methane	−82.60	45.44	4.604	99.0	0.162
Methanethiol	196.8	71.4	7.23	145	0.332
Methanol	239.4	79.78	8.084	118	0.272
Methoxybenzene	372.5	41.9	4.25		0.321
Methyl acetamide	417				
Methyl acetate	233.40	46.9	4.75	228	0.325

(*Continued*)

TABLE 2.55 Critical Properties (*Continued*)

Substance	T_c, °C	P_c, atm	P_c, MPa	V_c, $cm^3 \cdot mol^{-1}$	ρ_c, $g \cdot cm^{-3}$
Methyl acrylate	263	42	4.26	265	0.325
Methylamine	157.6	75.14	7.614	140	0.222
N-Methylaniline	428	51.3	5.20	373	0.287
Methyl benzoate	438	36	3.65	396	0.344
2-Methyl-1,3-butadiene	211	38.0	3.85	276	0.247
3-Methyl-1,3-butadiene	223	40.6	4.11	267	0.255
2-Methylbutane	187.3	33.4	3.38	306	0.236
2-Methyl-1-butanethiol	318.8				
2-Methyl-2-butanethiol	297.0				
Methyl butanoate	281.3	34.3	3.475	340	0.300
3-Methylbutanoic acid	356	33.6	3.40		
2-Methyl-1-butanol	302.3	38.9	3.94	322	0.274
3-Methyl-1-butanol	304.1	38.8	3.93	329	0.268
2-Methyl-2-butanol	270.6	36.6	3.71	319	0.276
3-Methyl-2-butanol	283.0	38.2	3.87		
3-Methyl-2-butanone	280.3	38.0	3.85	310	0.278
2-Methyl-1-butene	196.9	34.0	3.445	294	0.239
3-Methyl-1-butene	191.6	34.7	3.52	300	0.234
2-Methyl-2-butene	207.9	34.0	3.445	318	0.221
Methylcyclohexane	299.1	34.26	3.471	368	0.267
Methylcyclopentane	259.58	37.35	3.784	319	0.264
Methyl dodecanoate	439			758	0.283
N-Methylethylamine	223.5	36.6	3.71	243	0.243
Methyl formate	214.1	59.20	5.998	172	0.349
2-Methylfuran	254	46.6	4.72	247	0.333
2-Methylheptane	286.6	24.52	2.484	488	0.234
3-Methylheptane	290.6	25.13	2.546	464	0.246
4-Methylheptane	288.7	25.09	2.542	476	0.240
2-Methylhexane	257.3	26.98	2.734	421	0.238
3-Methylhexane	262.2	27.77	2.814	404	0.248
Methylhydrazine	294	79.3	8.035	271	0.170
Methyl 2-hydroxybenzoate	436				
Methyl isobutanoate	267.7	33.9	3.43	339	0.301
Methyl isocyanate	218	55	5.57		
1-Methylnaphthalene	499	35.5	3.60	445	0.320
2-Methylnaphthalene	488	34.6	3.51	462	0.308
2-Methyloctane	313.9	22.80	2.310		
2-Methylpentane	224.6	29.91	3.031	367	0.235
3-Methylpentane	231.4	30.85	3.126	367	0.235
2-Methyl-2,4-pentanediol	405	33.9	3.43		
Methyl pentanoate	294				
2-Methyl-2-pentanol	286.4				
2-Methyl-3-pentanol	302.9	34.1	3.46		
3-Methyl-3-pentanol	302.5	34.7	3.52		
4-Methyl-1-pentanol	330.4				
4-Methyl-2-pentanol	301.3	42.4	4.30	380	0.269
3-Methyl-2-pentanone	298.8				
4-Methyl-2-pentanone	298	32.3	3.27	371	0.270
2-Methyl-2-pentene	245	32.4	3.28	351	0.240
cis-3-Methyl-2-pentene	245	32.4	3.28	351	0.240
trans-3-Methyl-2-pentene	248	32.3	3.27	350	0.240
cis-4-Methyl-2-pentene	217	30	3.04	360	0.234
trans-4-Methyl-2-pentene	220	30	3.04	360	0.234

TABLE 2.55 Critical Properties (*Continued*)

Substance	T_c, °C	P_c, atm	P_c, MPa	V_c, $cm^3 \cdot mol^{-1}$	ρ_c, $g \cdot cm^{-3}$
2-Methylpropanal	240	41	4.15	274	0.263
2-Methyl-1-propanamine	246	40.2	4.07	278	0.263
N-Methylpropanamide	412				
2-Methylpropane	134.70	35.83	3.630	263	0.221
2-Methyl-1-propanethiol	286.4				
2-Methyl-2-propanethiol	257.0				
Methyl propanoate	257.5	39.5	4.00	282	0.312
2-Methylpropanoic acid	332	36.5	3.7	292	0.302
2-Methyl-1-propanol	274.6	42.39	4.295	273	0.272
2-Methyl-2-propanol	233.1	39.20	3.972	275	0.270
2-Methylpropene	144.73	39.48	4.000	239	0.235
2-Methylpropyl acetate	288	31.2	3.16	414	0.281
Methyl propyl ether	203.2				
Methyl propyl sulfide	301.0				
2-Methylpyridine	347.9	45.4	4.60	292	0.319
3-Methylpyridine	371.9	44.2	4.48	288	0.323
4-Methylpyridine	373	46.4	4.70	292	0.319
1-Methyl-2-pyrrolidinone	448.7			311	0.319
1-Methylstyrene	381	33.6	3.40	397	0.298
2-Methyltetrahydrofuran	264	37.1	3.76	267	0.322
2-Methylthiophene	333.1	47.9	4.85	275	0.356
3-Methylthiophene	337.7	48.9	4.95	275	0.356
Methyl vinyl ether	163	47	4.76	205	0.283
Morpholine	345	54	54.7	253	0.344
Naphthalene	475.3	39.98	4.051	407	0.31
Nitrobenzene	459				
Nitroethane	284	37	3.75		
Nitromethane	315	57.9	5.87	173	0.352
1-Nitropropane	402.0				
2-Nitropropane	344.8				
Nonadecane	483	11.0	1.12	1130	0.238
Nonane	321.5	22.6	2.29	555	0.231
Nonanoic acid	438	23.7	2.40		
1-Nonanol	404			546	0.264
1-Nonene	319	23.1	2.34	580	0.218
Nonylbenzene	468	18.7	1.89	790	0.259
Nonylcyclopentane	437.4	16.3	1.65		
Octadecafluorooctane	229	16.4	1.66		
Octadecane	472.3	12.73	1.29	1070	0.238
1-Octadecanol	474	14	1.42		
1-Octadecene	466	11.2	1.13		
Octafluorocyclobutane	115.31	27.48	2.784	325	0.616
Octafluoronaphthalene	399.9				
Octafluoropropane	72.7	26.5	2.69	299	0.628
Octamethylcyclotetrasiloxane	313	13.2	1.33	970	0.306
Octane	295.6	24.6	2.49	492	0.232
Octanenitrile	401.3	28.1	2.85		
Octanoic acid	422	26.1	2.64		
1-Octanol	379.4	27.41	2.777	490	0.266
2-Octanol	356.5	27.18	2.754	494	0.278
1-Octene	293.6	26.40	2.675	464	0.242
cis-2-Octene	307	27.3	2.77		
Octylcyclopentane	421	17.7	1.79		

(*Continued*)

TABLE 2.55 Critical Properties (*Continued*)

Substance	T_c, °C	P_c, atm	P_c, MPa	V_c, $cm^3 \cdot mol^{-1}$	ρ_c, $g \cdot cm^{-3}$
Pentachloroethane	373.0				
Pentadecane	433.9	15	1.52	880	0.241
1-Pentadecene	431	14.4	1.46		
Pentadecylcyclopentane	507	10.1	1.02		
1,2-Pentadiene	230	40.2	4.07	276	0.248
cis-1,3-Pentadiene	223	39.4	3.99	275	0.248
1,4-Pentadiene	205	37.4	3.79	276	0.248
Pentafluorobenzene	258.9	34.7	3.52		
2,3,4,5,6-Pentafluorotoluene	275.5				
2,2,3,3,4-Pentamethyl-pentane	370.7	25.5	2.58	508	0.280
2,2,3,4,4-Pentamethyl-pentane	354.2	23.7	2.40	521	0.273
Nonadecane	483	11.0	1.12	1130	0.238
Nonane	321.5	22.6	2.29	555	0.231
Nonanoic acid	438	23.7	2.40		
1-Nonanol	404			546	0.264
1-Nonene	319	23.1	2.34	580	0.218
Nonylbenzene	468	18.7	1.89	790	0.259
Nonylcyclopentane	437.4	16.3	1.65		
Octadecafluorooctane	229	16.4	1.66		
Octadecane	472.3	12.73	1.29	1070	0.238
1-Octadecanol	474	14	1.42		
1-Octadecene	466	11.2	1.13		
Octafluorocyclobutane	115.31	27.48	2.784	325	0.616
Octafluoronaphthalene	399.9				
Octafluoropropane	72.7	26.5	2.69	299	0.628
Octamethylcyclotetrasiloxane	313	13.2	1.33	970	0.306
Octane	295.6	24.6	2.49	492	0.232
Octanenitrile	401.3	28.1	2.85		
Octanoic acid	422	26.1	2.64		
1-Octanol	379.4	27.41	2.777	490	0.266
2-Octanol	356.5	27.18	2.754	494	0.278
1-Octene	293.6	26.40	2.675	464	0.242
cis-2-Octene	307	27.3	2.77		
Octylcyclopentane	421	17.7	1.79		
Osmium tetroxide	132	170	17.2		
Oxygen	−118.56	49.77	5.043	73.4	0.436
Oxygen difluoride	−58.0	48.9	4.95	97.7	0.553
Ozone	−12.10	53.8	5.45	88.9	0.540

Substance	T_c, °C	P_c, atm	P_c, MPa	V_c, $cm^3 \cdot mol^{-1}$	ρ_c, $g \cdot cm^{-3}$
Pentachloroethane	373.0				
Pentadecane	433.9	15	1.52	880	0.241
1-Pentadecene	431	14.4	1.46		
Pentadecylcyclopentane	507	10.1	1.02		
1,2-Pentadiene	230	40.2	4.07	276	0.248
cis-1,3-Pentadiene	223	39.4	3.99	275	0.248
1,4-Pentadiene	205	37.4	3.79	276	0.248
Pentafluorobenzene	258.9	34.7	3.52		
2,3,4,5,6-Pentafluorotoluene	275.5				
2,2,3,3,4-Pentamethyl-pentane	370.7	25.5	2.58	508	0.280
2,2,3,4,4-Pentamethyl-pentane	354.2	23.7	2.40	521	0.273

TABLE 2.55 Critical Properties (*Continued*)

Substance	T_c, °C	P_c, atm	P_c, MPa	V_c, $cm^3 \cdot mol^{-1}$	ρ_c, $g \cdot cm^{-3}$
3-Pentanol	286.5				
2-Pentanone	287.93	36.46	3.694	301	0.286
3-Pentanone	288.31	36.9	3.729	336	0.256
1-Pentene	191.63	34.81	3.527	293	0.239
cis-2-Pentene	202	36.4	3.69		
trans-2-Pentene	198	34.7	3.52	304	0.231
Pentyl acetate	332				
Pentylbenzene	406.8	25.7	2.60	550	0.269
Pentyl formate	303				
1-Pentyne	220.3	40	4.05	278	0.245
Perchloryl fluoride	95.3	53.0	5.37	161	0.637
Phenanthrene	596			554	0.322
Phenol	421.1	60.5	6.13	229	0.41
1-Phenylhexadecane	535	12.7	1.29	1200	0.252
1-Phenylpentadecane	526.9	13.3	1.35	1140	0.253
1-Phenyltetradecane	519	14.0	1.42	1110	0.247
Phthalic anhydride	537	47	4.76	368	0.402
Piperidine	321.0	48.8	4.94	288	0.296
Propadiene	120	54.0	5.47	162	0.247
Propanal	231.3	52.0	5.27	204	0.285
Propane	96.68	41.92	4.248	200	0.217
1,2-Propanediol	352	60	6.08	237	0.321
1,3-Propanediol	385	59	5.98	241	0.316
Propanenitrile	288.2	42.0	4.26	230	0.240
1-Propanethiol	262.5				
2-Propanethiol	244.2				
Propanoic acid	331	44.7	4.53	222	0.32
1-Propanol	263.7	51.01	5.169	218.5	0.275
2-Propanol	235.2	47.02	4.764	220	0.273
2-Propenal	233	51	5.17	197	0.285
Propene	91.9	45.6	4.62	181	0.233
2-Propen-1-ol	272.0			208	0.279
Propyl acetate	276.6	33.2	3.36	345	0.296
Propylamine	223.9	46.6	4.72	233	0.254
Propylbenzene	365.20	31.58	3.200	440	0.273
Propyl butanoate	327				
Propylcyclopentane	358.7	29.6	3.00	425	0.264
Propylcyclohexane	336.7	27.7	2.81		
Propylene oxide	209.1	48.6	4.92	186	0.312
Propyl formate	264.9	40.1	4.06	285	0.309
Propyl 2-methylpropanoate	316				
Propyl 3-methylpropanoate	336				
Propyl propanoate	305				
Propyne	129.3	55.5	5.62	164	0.245
Pyridine	346.9	55.96	5.67	243	0.325
Pyrrole	366.6	62.6	6.34	200	0.335
Pyrrolidine	295.1	55.2	5.59	238	0.300
Quinoline	509	48.0	4.86	437	0.300
Spiro[2.2]pentane	233.3				
Styrene	363.8	36.3	3.68	347	0.300

(*Continued*)

TABLE 2.55 Critical Properties (*Continued*)

Substance	T_c, °C	P_c, atm	P_c, MPa	V_c, $cm^3 \cdot mol^{-1}$	ρ_c, $g \cdot cm^{-3}$
1,2-Terphenyl	617.9	38.5	3.90	755	0.305
1,3-Terphenyl	651.7	34.6	3.51	768	0.300
1,4-Terphenyl	652.9	32.8	3.32	762	0.302
1,1,2,2-Tetrachlorodifluoro-ethane	278	34	3.44	371	0.549
1,1,2,2-Tetrachloroethane	388.00				
Tetrachloroethylene	347.1	44.3	4.49	290	0.572
Tetrachloromethane	283.5	44.57	4.516	276	0.557
Tetradecafluoro-1-heptene	205.1				
Tetradecafluorohexane	174.5	18.8	1.90		
Tetradecafluoromethylcyclohexane	213.7	23	2.33		
Tetradecane	420.9	16	1.62	830	0.239
1-Tetradecene	416	15.4	1.56		
Tetradecylcyclopentane	499	11.1	1.12		
Tetraethylsilane	330.6	25.68	2.602		
Tetrafluoroethylene	33.4	38.9	3.91	175	0.58
Tetrafluorohydrazine	33.3	37	3.75		
Tetrafluoromethane	−45.5	36.9	3.74	140	0.629
Tetrahydrofuran	267.0	51.22	5.19	224	0.322
1,2,3,4-Tetrahydronaphthalene	447	36.0	3.65	408	0.324
Tetrahydropyran	299.1	47.1	4.77	263	0.328
Tetrahydrothiophene	358.9				
1,2,4,5-Tetramethylbenzene	402	29	2.94	480	0.280
2,2,3,3-Tetramethylbutane	294.7	28.3	2.87	461	0.248
2,2,3,3-Tetramethylhexane	350.0	24.8	2.51	573	0.248
2,2,3,4-Tetramethylhexane	347.3	23.4	2.37	525	0.271
2,2,3,5-Tetramethylhexane	328.2	22.4	2.27	540	0.263
2,2,4,4-Tetramethylhexane	337.1	22.2	2.25	535	0.266
2,2,4,5-Tetramethylhexane	325.4	21.9	2.22	544	0.262
2,2,5,5-Tetramethylhexane	308.4	21.6	2.19	573	0.248
2,3,3,4-Tetramethylhexane	360.0	24.5	2.48	514	0.277
2,3,3,5-Tetramethylhexane	337.0	22.9	2.32	531	0.268
2,3,4,4-Tetramethylhexane	353.5	23.9	2.42	518	0.275
2,3,4,5-Tetramethylhexane	340.1	23.1	2.34	530	0.269
3,3,4,4-Tetramethylhexane	373.6	25.4	2.57	506	0.281
2,2,3,3-Tetramethylpentane	334.6	27.05	2.741		
2,2,3,4-Tetramethylpentane	319.6	25.68	2.602		
2,2,4,4-Tetramethylpentane	301.6	24.52	2.485		
2,3,3,4-Tetramethylpentane	334.6	26.80	2.716		
Tetramethylsilane	175.49	27.84	2.821	362	0.244
Thiacyclopentane	358.8				
2-Thiapropane	230.0	54.6	5.53	201	0.309
Thiophene	306.3	56.16	5.69	219	0.385
Thiophenol	416.4				
Thymol	425				
Toluene	318.60	40.54	4.108	316	0.292
1,2-Toluidine	434	43.1	4.37	343	0.312
1,3-Toluidine	434	42.2	4.28	343	0.312
1,4-Toluidine	433	45.2	4.58		
Toluonitrile	450				
Tributoxyborane	472	19.6	1.99	863	0.267

TABLE 2.55 Critical Properties (*Continued*)

Substance	T_c, °C	P_c, atm	P_c, MPa	V_c, $cm^3 \cdot mol^{-1}$	ρ_c, $g \cdot cm^{-3}$
Tributylamine	365.3	18	1.82		
1,1,1-Trichloroethane	272	42.4	4.30		
1,1,2-Trichloroethane	329	41	4.15	294	0.454
Trichloroethylene	271.1	49.5	5.02	256	0.513
Trichlorofluoromethane	198.1	43.5	4.41	248	0.554
Trichlorofluorosilane	165.4	35.3	3.57		
Trichloromethane	263.3	54.0	5.47	239	0.500
Trichloromethylsilane	244	32.4	3.28	348	0.430
1,2,3-Trichloropropane	378	39	3.95	348	0.424
1,2,2-Trichlorotrifluoroethane	214.2	33.7	3.42	325	0.576
Tridecane	402	16.6	1.68	780	0.236
1-Tridecene	401	16.8	1.70		
Tridecylcyclopentane	488	11.9	1.21		
Triethanolamine	514.3	24.2	2.45		
Triethylamine	262.5	29.92	3.032	389	0.26
Trifluoroacetic acid	218.2	32.15	3.258	204	0.559
Trifluoroamine oxide (NOF_3)	29.5			169	0.593
1,1,1-Trifluoroethane	73.2	37.1	3.76	194	0.434
Trifluoromethane	25.8	47.7	4.83	133	0.525
(Trifluoromethyl)benzene	286.8				
Trimethylamine	159.64	40.34	4.087	254	0.233
1,2,3-Trimethylbenzene	391.4	34.09	3.454	430	0.280
1,2,4-Trimethylbenzene	376.0	31.90	3.232	430	0.280
1,3,5-Trimethylbenzene	364.2	30.86	3.127	433	0.278
2,2,3-Trimethylbutane	258.1	29.15	2.954	398	0.252
2,2,3-Trimethyl-1-butene	260	28.6	2.90	400	0.245
1,1,2-Trimethylcyclopentane	306.4	29.0	2.94		
1,1,3-Trimethylcyclopentane	296.4	27.9	2.83		
cis,trans,cis-1,2,4-Trimethylcyclopentane	298	27.7	2.81		
cis,cis,trans-1,2,4-Trimethylcyclopentane	306	28.4	2.88		
2,2,3-Trimethylheptane	338.6	22.4	2.27	546	0.261
2,2,4-Trimethylheptane	321.4	21.4	2.17	552	0.258
2,2,5-Trimethylheptane	325.0	21.4	2.17	559	0.256
2,2,6-Trimethylheptane	320.3	21.0	2.13	573	0.248
2,3,3-Trimethylheptane	344.4	22.9	2.32	538	0.265
2,3,4-Trimethylheptane	340.6	22.6	2.29	538	0.265
2,3,5-Trimethylheptane	339.7	22.1	2.24	547	0.260
2,3,6-Trimethylheptane	331.0	21.6	2.19	560	0.254
2,4,4-Trimethylheptane	327.2	21.9	2.22	541	0.263
2,4,5-Trimethylheptane	333.8	22.1	2.24	544	0.262
2,4,6-Trimethylheptane	317.2	21.2	2.15	560	0.254
2,5,5-Trimethylheptane	329.8	21.9	2.22	550	0.259
3,3,4-Trimethylheptane	349.4	23.4	2.37	526	0.271
3,3,5-Trimethylheptane	336.5	22.9	2.32	579	0.246
3,4,4-Trimethylheptane	347.8	23.4	2.37	524	0.271
3,4,5-Trimethylheptane	339.7	22.1	2.24	547	0.261
2,2,3-Trimethylhexane	315	24.6	2.49		
2,2,4-Trimethylhexane	300.6	23.4	2.37		

(*Continued*)

TABLE 2.55 Critical Properties (*Continued*)

Substance	T_c, °C	P_c, atm	P_c, MPa	V_c, $cm^3 \cdot mol^{-1}$	ρ_c, $g \cdot cm^{-3}$
2,2,5-Trimethylhexane	295	23.0	2.33	519	0.247
2,4,7-Trimethyloctane	335.7				
2,2,3-Trimethylpentane	290.4	26.94	2.730	436	0.262
2,2,4-Trimethylpentane	270.9	25.34	2.568	468	0.244
2,3,3-Trimethylpentane	300.5	27.83	2.820	455	0.251
2,3,4-Trimethylpentane	293.4	26.94	2.730	461	0.248
2,2,4-Trimethyl-1,3-pentanediol	398	25.6	2.59	364.6	0.4010
2,3,6-Trimethylpyridine	381.4				
2,4,6-Trimethylpyridine	379.9				
2,4,6-Trimethyl-1,3,5-trioxane	290				
1*H*-Undecafluoropentane	170.8				
Undecane	365.7	19.4	1.97	657	0.238
1-Undecene	364	19.7	2.00		0.240
Vinyl acetate	228.4	22.4	2.27	265	0.325
Vinyl chloride	156.6	55.3	5.60	169	0.370
Vinyl fluoride	54.7	51.7	5.24	114	0.320
Vinyl formate	202	57	5.78	210	0.343
1,2-Xylene	357.2	36.83	3.732	370	0.288
1,3-Xylene	343.9	34.95	3.541	375	0.282
1,4-Xylene	343.1	34.65	3.511	379	0.280

TABLE 2.56 Lydersen's Critical Property Increments

	Δ_T	Δ_p	Δ_v
Nonring Increments			
$-CH_3$	0.020	0.227	55
$-\overset{\vert}{C}H_2$	0.020	0.227	55
$-\underset{\vert}{\overset{\vert}{C}}H$	0.012	0.210	51
$-\underset{\vert}{\overset{\vert}{C}}-$	0.00	0.210	41
$=CH_2$	0.018	0.198	45
$=\overset{\vert}{C}H$	0.018	0.198	45
$=\overset{\vert}{C}H-$	0.0	0.198	36
$=C=$	0.0	0.198	36
$\equiv CH$	0.005	0.153	(36)
$\equiv C-$	0.005	0.153	(36)
Ring Increments			
$-CH_2-$	0.013	0.184	44.5
$-\underset{\vert}{\overset{\vert}{C}}H$	0.012	0.192	46
$-\underset{\vert}{\overset{\vert}{C}}-$	(−0.007)	(0.154)	(31)
$=\overset{\vert}{C}H$	0.011	0.154	37
$=\overset{\vert}{C}H-$	0.011	0.154	36
$=C=$	0.011	0.154	36
Halogen Increments			
$-F$	0.018	0.224	18
$-Cl$	0.017	0.320	49
$-Br$	0.010	(0.50)	(70)
$-I$	0.012	(0.83)	(95)
Oxygen Increments			
$-OH$ (alcohols)	0.082	0.06	(18)
$-OH$ (phenols)	0.031	(−0.02)	(3)
$-O-$ (nonring)	0.021	0.16	20
$-O-$ (ring)	(0.014)	(0.12)	(8)
$-\overset{\vert}{C}=O$ (nonring)	0.040	0.29	60
$-\overset{\vert}{C}=O$ (ring)	(0.033)	(0.2)	(50)
$H\overset{\vert}{C}=O$ (aldehyde)	0.048	0.33	73
$-COOH$ (acid)	0.085	(0.4)	80
$-COO-$ (ester)	0.047	0.47	80
$=O$ (except for combinations above)	(0.02)	(0.12)	(11)
Nitrogen Increments			
$-NH_2$	0.031	0.095	28
$-\overset{\vert}{N}H$ (nonring)	0.031	0.135	(37)

TABLE 2.56 Lydersen's Critical Property Increments (*Continued*)

	Δ_T	Δ_p	Δ_v
Nitrogen Increments (*continued*)			
—NH (ring) (with one more bond)	(0.024)	(0.09)	(27)
—NH— (nonring) (with one more bond)	0.014	0.17	(42)
—N— (ring) (with one more bond)	(0.007)	(0.13)	(32)
—CN	(0.060)	(0.36)	(80)
—NO_2	(0.055)	(0.42)	(78)
Sulfur Increments			
—SH	0.015	0.27	55
—S— (nonring)	0.015	0.27	55
—S— (ring)	(0.008)	(0.24)	(45)
=S	(0.003)	(0.24)	(47)
Miscellaneous			
—Si— (with two more bonds)	0.03	(0.54)	
—B— (with one more bond)	(0.03)		
Nonring:			

†There are no increments for hydrogen. All bonds shown as free are connected with atoms other than hydrogen. Values in parentheses are based upon too few experimental values to be reliable. From vapor-pressure measurements and a calculational technique similar to Fishtine [6], it has been suggested that the C—H ring increment (C also bonded to three ring atoms) common to two condensed saturated rings be given the value of $\Delta_T = 0.064$.

TABLE 2.57 Vetere Group Contribution to Estimate Critical Volume

Group	ΔV_i	Group	ΔV_i
Nonring:		—C=O (nonring) (with one more bond)	1.765
In linear chain:			
CH_3, CH_2, CH, C	3.360	—C=O (ring) (with one more bond)	1.500
In side chain			
CH_3, CH_2, CH, C	2.888	—HC=O (aldehyde) (with one more bond)	2.333
=CH_2, =CH, =C— (CH and C with one more bond)	2.940		
=C=	2.908	—COOH	1.652
≡CH, ≡C—	2.648	—COO—	1.607
Ring:			
CH_2, CH, C	2.813	—NH_2 (with one more bond)	2.184
=CH, =C— (CH and C with one more bond)	2.538	—NH (nonring) (with one more bond)	2.333
		—NH (ring) (with one more bond)	1.736
F	0.770		
Cl	1.237	—N— (nonring) (with one more bond)	1.793
Br	0.899		
I	0.702	—N— (ring) (with one more bond)	1.883
		—CN	2.784
—OH (alcohols)	0.704	—NO_2	1.559
—OH (phenols)	1.553		
—O— (nonring)	1.075	—SH	1.537
—O— (ring)	0.790	—S— (nonring)	0.591
—O— (epoxy)	—0.252	—S— (ring)	0.911

TABLE 2.58 Van der Waalls' Constants for Gases

The van der Waals' equation of state for a real gas is:

$$\left(P+\frac{n^2a}{V^2}\right)(V-nb)=nRT \qquad \text{for } n \text{ moles}$$

where P is the pressure. V the volume (in liters per mole = 0.001 m^3 per mole in the SI system), T the temperature (in degrees Kelvin), n the amount of substance (in moles), and R the gas constant. To use the values of a and b in the table, P must be expressed in the same units as in the gas constant. Thus, the pressure of a standard atmosphere may be expressed in the SI system as follows:

$$1 \text{ atm} = 101{,}325 \text{ N}\cdot\text{m}^{-2} = 101{,}325 \text{ Pa} = 1.01325 \text{ bar}$$

The appropriate value for the gas constant is:

$$0.083\,144\,1 \text{ L}\cdot\text{bar}\cdot\text{K}^{-1}\cdot\text{mol}^{-1} \quad \text{or} \quad 0.082\,056 \text{ L}\cdot\text{atm}\cdot\text{K}^{-1}\cdot\text{mol}^{-1}$$

The van der Waals' constants are related to the critical temperature and pressure, T_c and P_c, in Table 2.55 by:

$$a=\frac{27\,R^2T_c^2}{64\,P_c} \quad \text{and} \quad b=\frac{RT_c}{8\,P_c}$$

Substance	a, L$^2\cdot$bar$\cdot$mol^{-2}	b, L$\cdot$mol^{-1}
Acetaldehyde	11.37	0.08695
Acetic acid	17.71	0.1065
Acetic anhydride	26.8	0.157
Acetone	16.02	0.1124
Acetonitrile	17.89	0.1169
Acetyl chloride	12.80	0.08979
Acetylene	4.516	0.05218
Acrylic acid	19.45	0.1127
Acrylonitrile	18.37	0.1222
Allene	8.235	0.07467
Allyl alcohol	15.17	0.1036
Aluminum trichloride	42.63	0.2450
2-Aminoethanol	7.616	0.0431
Ammonia	4.225	0.03713
Ammonium chloride	2.380	0.00734
Aniline	29.14	0.1486
Antimony tribromide	42.08	0.1658
Argon	1.355	0.03201
Arsenic trichloride	17.23	0.1039
Arsine	6.327	0.06048
Benzaldehyde	30.30	0.1553
Benzene	18.82	0.1193
Benzonitrile	33.89	0.1727
Benzyl alcohol	34.7	0.173
Biphenyl	47.16	0.2130
Bismuth trichloride	33.89	0.1025
Boron trichloride	15.60	0.1222
Boron trifluoride	3.98	0.05443
Bromine (Br_2)	9.75	0.0591
Bromobenzene	28.96	0.1541
Bromochlorodifluoromethane	12.79	0.1055
Bromoethane	11.89	0.08406
Bromomethane	6.753	0.05390
Bromotrifluoromethane	8.502	0.0891

(*Continued*)

TABLE 2.58 Van der Waalls' Constants for Gases (*Continued*)

Substance	a, $L^2 \cdot bar \cdot mol^{-2}$	b, $L \cdot mol^{-1}$
1,2-Butadiene	12.76	0.1025
1,3-Butadiene	12.17	0.1020
Butanal	19.48	0.1292
Butane	13.93	0.1168
Butanenitrile	25.76	0.1568
Butanoic acid	28.18	0.1609
1-Butanol	20.90	0.1323
2-Butanol	20.94	0.1326
2-Butanone	19.97	0.1326
1-Butene	12.76	0.1084
cis-2-Butene	12.58	0.1066
trans-2-Butene	12.58	0.1066
3-Butenenitrile	25.76	0.1568
Butyl acetate	31.22	0.1919
1-Butylamine	19.41	0.1301
sec-Butylamine	18.37	0.1273
tert-Butylamine	17.78	0.1310
Butylbenzene	44.071	0.2378
sec-Butylbenzene	43.74	0.2347
tert-Butylbenzene	42.77	0.2310
Butyl benzoate	57.97	0.2857
Butylcyclohexane	41.19	0.2201
sec-Butylcyclohexane	48.89	0.2604
tert-Butylcyclohexane	48.34	0.2614
Butyl ethyl ether	27.05	0.1815
2-Butylhexadecafluorotetrahydrofuran	45.41	0.3235
1-Butyne	13.31	0.1023
2-Butyne	13.68	0.0998
Carbon dioxide	3.658	0.04284
Carbon disulfide	11.25	0.07262
Carbon monoxide	1.472	0.03948
Carbon oxysulfide (COS)	6.975	0.06628
Carbon tetrachloride	20.01	0.1281
Carbon tetrafluoride	4.029	0.06319
Carbonyl chloride	10.65	0.08340
Carbonyl sulfide	3.933	0.05817
Chlorine	6.343	0.05422
Chlorine pentafluoride	9.581	0.08214
Chlorobenzene	25.80	0.1454
1-Chlorobutane	23.22	0.1527
2-Chlorobutane	20.01	0.1370
1-Chloro-1,1-difluoroethane	11.91	0.1035
2-Chloro-1,1-difluoroethylene	10.49	0.09335
Chloroethane	11.7	0.090
Chloroform	15.34	0.1019
Chloromethane	7.566	0.06477
2-Chloro-2-methylpropane	18.98	0.1334
Chloropentafluoroacetone	17.08	0.1482
Chloropentafluorobenzene	29.53	0.1843
Chloropentafluoroethane	11.27	0.1137
1-Chloropropane	16.11	0.1141
2-Chloropropane	14.53	0.1068
Chlorotrifluoromethane	6.873	0.08110

TABLE 2.58 Van der Waalls' Constants for Gases (*Continued*)

Substance	a, $L^2 \cdot bar \cdot mol^{-2}$	b, $L \cdot mol^{-1}$
Chlorotrifluorosilane	7.994	0.09240
Chlorotrimethylsilane	22.58	0.1617
m-Cresol	31.86	0.1609
o-Cresol	28.33	0.1447
p-Cresol	28.11	0.1422
Cyanogen	7.803	0.06952
Cyclobutane	12.39	0.0960
Cycloheptane	27.20	0.1645
Cyclohexane	21.95	0.1413
Cyclohexanol	28.93	0.1586
Cyclohexanone	31.1	0.170
Cyclohexene	75.04	0.1339
Cyclopentane	16.94	0.1180
Cyclopentanone	75.84	0.1211
Cyclopentene	15.61	0.1097
Cyclopropane	8.293	0.07420
p-Cymene	43.65	0.2386
Decane	52.88	0.3051
Decanenitrile	34.71	0.1988
1-Decanol	57.45	0.2971
1-Decene	49.96	0.2888
Deuterium (normal)	0.2583	0.02397
Deuterium oxide	5.584	0.03090
Diborane (B_2H_6)	6.048	0.07437
Dibromodifluoromethane	15.69	0.1186
1,2-Dibromoethane	13.98	0.08664
1,2-Dibromotetrafluoroethane	20.45	0.1494
Dibutylamine	34.61	0.2030
Dibutyl ether	33.06	0.2017
Dibutyl sulfide	49.3	0.2702
1,2-Dichlorobenzene	34.59	0.1767
1,3-Dichlorobenzene	35.44	0.1846
1,4-Dichlorobenzene	34.64	0.1802
Dichlorodifluoromethane	10.45	0.09672
Dichlorodifluorosilane	11.34	0.1095
1,1-Dichloroethane	15.73	0.1072
1,2-Dichloroethane	17.0	0.108
1,1-Dichloroethylene	13.74	0.09893
trans-1,2-Dichloroethylene	13.63	0.09573
Dichlorofluoromethane	11.48	0.09060
Dichloromethane	12.44	0.08689
1,2-Dichloropropane	21.62	0.1335
Dichlorosilane	12.59	0.09992
1,1-Dichlorotetrafluoroethane	15.49	0.1318
1,2-Dichlorotetrafluoroethane	15.72	0.1338
Dideuterium oxide	5.535	0.03062
Diethanolamine	45.61	0.2273
Diethylamine	19.40	0.1383
1,4-Diethylbenzene	45.03	0.2439
Diethylene glycol	29.02	0.1519
Diethyl ether	17.46	0.1333
3,3-Diethylhexane	47.69	0.2707
3,4-Diethylhexane	47.93	0.2760

(*Continued*)

TABLE 2.58 Van der Waalls' Constants for Gases (*Continued*)

Substance	a, $L^2 \cdot bar \cdot mol^{-2}$	b, $L \cdot mol^{-1}$
3,3-Diethyl-2-methylpentane	47.20	0.2629
3,3-Diethylpentane	40.64	0.2374
Diethyl sulfide	22.85	0.1462
Difluoroamine	5.028	0.04446
cis-Difluorodiazine	3.043	0.03987
trans-Difluorodiazine	3.539	0.04851
1,1-Difluoroethane	9.691	0.08931
1,1-Difluoroethylene	6.000	0.07058
Difluoromethane	6.184	0.06268
Dihexyl ether	69.17	0.3752
Dihydrogen disulfide	16.15	0.1006
Diisopropyl ether	25.26	0.1836
Dimethoxyethane	21.65	0.1439
Dimethoxymethane	17.28	0.1195
N,N-Dimethoxyacetamide	30.19	0.1689
Dimethylamine	10.44	0.08510
N,N-Dimethylaniline	37.92	0.1967
2,2-Dimethylbutane	22.55	0.1644
2,3-Dimethylbutane	23.29	0.1660
2,3-Dimethyl-1-butene	22.59	0.2566
3,3-Dimethyl-1-butene	21.55	0.1567
2,3-Dimethyl-2-butene	23.83	0.1621
1,1-Dimethylcyclohexane	34.30	0.2068
cis-1,2-Dimethylcyclohexane	36.44	0.2143
trans-1,2-Dimethylcyclohexane	34.89	0.2086
cis-1,3-Dimethylcyclohexane	34.30	0.2068
trans-1,3-Dimethylcyclohexane	35.11	0.2093
cis-1,4-Dimethylcyclohexane	35.47	0.2114
trans-1,4-Dimethylcyclohexane	34.54	0.2086
1,1-Dimethylcyclopentane	25.37	0.1653
cis-1,2-Dimethylcyclopentane	27.04	0.1706
trans-1,2-Dimethylcyclopentane	25.67	0.1663
Dimethyl ether	8.690	0.07742
N,N-Dimethylformamide	23.57	0.1293
2,2-Dimethylheptane	41.29	0.2551
2,2-Dimethylhexane	34.87	0.2260
2,3-Dimethylhexane	35.24	0.2228
2,4-Dimethylhexane	34.97	0.2251
2,5-Dimethylhexane	35.49	0.2299
3,3-Dimethylhexane	34.72	0.2201
3,4-Dimethylhexane	35.06	0.2196
1,1-Dimethylhydrazine	14.69	0.1001
2,4-Dimethyl-3-isopentane	47.05	0.2729
Dimethyl oxalate	28.97	0.1644
2,2-Dimethylpentane	28.49	0.1951
2,3-Dimethylpentane	28.96	0.1921
2,4-Dimethylpentane	28.79	0.1974
3,3-Dimethylpentane	28.48	0.1892
2,3-Dimethylphenol	31.35	0.1545
2,4-Dimethylphenol	33.49	0.1687
2,5-Dimethylphenol	29.99	0.1512
2,6-Dimethylphenol	33.64	0.1710
3,4-Dimethylphenol	31.32	0.1529

TABLE 2.58 Van der Waalls' Constants for Gases (*Continued*)

Substance	a, $L^2 \cdot bar \cdot mol^{-2}$	b, $L \cdot mol^{-1}$
3,5-Dimethylphenol	40.92	0.2037
2,2-Dimethylpropane	17.17	0.1410
2,3-Dimethylpropane	23.13	0.1669
2,2-Dimethyl-1-propanol	22.25	0.1444
Dimethyl sulfide	13.34	0.09453
N,N-Dimethyl-1,2-toluidine	41.71	0.2225
1,4-Dioxane	19.29	0.1171
Diphenyl ether	54.61	0.2538
Diphenylmethane	60.46	0.2798
Dipropylamine	24.82	0.1591
Dipropyl ether	27.12	0.1821
Dodecafluorocyclohexane	25.09	0.1955
Dodecafluoropentane	25.58	0.2161
Dodecane	69.14	0.3741
1-Dodecanol	72.69	0.3598
1-Dodecene	68.17	0.3694
Ethane	5.570	0.06499
1,2-Ethanediamine	16.30	0.09796
Ethanethiol	13.23	0.09447
Ethanol	12.56	0.08710
Ethoxybenzene	35.70	0.1996
Ethyl acetate	20.57	0.1401
Ethyl acrylate	23.70	0.1530
Ethylamine	10.79	0.08433
Ethylbenzene	30.86	0.1782
Ethyl benzoate	43.73	0.2236
Ethyl butanoate	30.53	0.1922
Ethylcyclohexane	35.70	0.2089
Ethylcyclopentane	27.90	0.1746
3-Ethyl-2,2-dimethylhexane	47.24	0.2752
4-Ethyl-2,2-dimethylhexane	46.45	0.2784
3-Ethyl-2,3-dimethylhexane	47.35	0.2692
4-Ethyl-2,3-dimethylhexane	47.49	0.2742
3-Ethyl-2,4-dimethylhexane	47.31	0.2736
4-Ethyl-2,4-dimethylhexane	45.52	0.2613
3-Ethyl-2,5-dimethylhexane	47.42	0.2800
3-Ethyl-3,4-dimethylhexane	47.00	0.2682
Ethylene	4.612	0.05821
Ethylene glycol dimethyl ether	21.65	0.1439
Ethylene glycol ethyl ether acetate	33.97	0.05594
Ethylene oxide	8.922	0.06779
Ethyl formate	15.91	0.1115
3-Ethylhexane	35.76	0.2253
Ethyl mercaptan	11.24	0.08098
2-Ethyl-1-methylbenzene	40.66	0.2226
3-Ethyl-1-methylbenzene	41.67	0.2331
4-Ethyl-1-methylbenzene	40.63	0.2262
1-Ethyl-1-methylcyclopentane	34.18	0.2058
Ethyl methyl ether	12.70	0.1034
3-Ethyl-2-methylheptane	48.81	0.2847
Ethyl methyl ketone	20.13	0.1340
3-Ethyl-2-methylpentane	34.74	0.2183
3-Ethyl-2-methylpentane	34.53	0.2134

(*Continued*)

TABLE 2.58 Van der Waalls' Constants for Gases (*Continued*)

Substance	a, L^2 · bar · mol^{-2}	b, L · mol^{-1}
Ethyl 2-methylpropanoate	29.05	0.1872
Ethyl methyl sulfide	19.45	0.1300
3-Ethylpentane	29.49	0.1944
Ethyl phenyl ether	35.16	0.1963
Ethyl propanoate	25.86	0.1688
Ethyl propyl ether	22.45	0.1600
m-Ethyltoluene	41.73	0.2334
o-Ethyltoluene	40.67	0.2226
p-Ethyltoluene	40.63	0.2262
Ethyl vinyl ether	16.17	0.1213
Fluorine	1.171	0.02896
Fluorobenzene	20.10	0.1279
Fluoroethane	8.170	0.07758
Fluoroethylene	5.984	0.06504
Fluoromethane	5.009	0.05617
Formaldehyde	7.356	0.06425
Furan	12.74	0.0926
2-Furaldehyde (furfural)	22.23	0.1182
Germanium tetrachloride	23.12	0.1489
Germanium tetrahydride	5.743	0.06555
Glycerol	22.98	0.07037
Hafnium tetrachloride	26.01	0.1282
Helium (equilibrium)	0.0346	0.02356
Heptane	30.89	0.2038
1-Heptanol	37.22	0.2097
2-Heptanol	35.72	0.2093
2-Heptanone	31.78	0.1850
1-Heptene	28.82	0.09400
Hexadecafluoroheptane	40.58	0.3046
1,5-Hexadiene	21.79	0.1532
Hexafluoraoacetone	12.66	0.1264
Hexafluorobenzene	26.63	0.1641
Hexane	24.97	0.1753
Hexanenitrile	35.50	0.1996
Hexanoic acid	39.94	0.2150
1-Hexanol	31.35	0.1829
2-Hexanol	30.25	0.1840
3-Hexanol	29.44	0.1803
2-Hexanone	30.27	0.1837
3-Hexanone	29.84	0.1824
1-Hexene	23.12	0.1634
cis-2-Hexene	23.86	0.1641
trans-2-Hexene	23.75	0.1640
cis-3-Hexene	23.77	0.1638
trans-3-Hexene	24.25	0.1663
Hexylcyclopentane	59.38	0.3206
Hydrazine	8.46	0.0462
Hydrogen (normal)	0.2484	0.02651
Hydrogen bromide	4.500	0.04415
Hydrogen chloride	3.700	0.04061
Hydrogen cyanide	11.29	0.08806
Hydrogen deuteride	0.2527	0.02516
Hydrogen fluoride	9.565	0.0739

TABLE 2.58 Van der Waalls' Constants for Gases (*Continued*)

Substance	a, $L^2 \cdot bar \cdot mol^{-2}$	b, $L \cdot mol^{-1}$
Hydrogen iodide	6.309	0.05303
Hydrogen selenide	5.523	0.0479
Hydrogen sulfide	4.544	0.04339
Indane	34.63	0.1802
Iodobenzene	33.54	0.1658
Iodomethane	12.34	0.08327
Isobutyl acetate	29.05	0.1845
Isobutylamine	19.30	0.1325
Isobutylbenzene	40.40	0.2215
Isobutylcyclohexane	40.39	0.2195
Isobutyl formate	22.82	0.1476
Isopropylamine	14.30	0.1080
Isopropylbenzene	36.20	0.2044
Isopropylcyclohexane	42.06	0.2342
Isopropylcyclopentane	35.11	0.2082
4-Isopropylheptane	48.28	0.2832
2-Isopropyl-1-methylbenzene	45.14	0.2401
3-Isopropyl-1-methylbenzene	44.00	0.2354
4-Isopropyl-1-methylbenzene	43.94	0.2398
3-Isopropyl-2-methylhexane	50.93	0.2870
Ketene	19.1	0.1044
Krypton	2.325	0.0396
Mercury	5.193	0.01057
Methane	2.300	0.04301
Methanethiol	8.911	0.06756
Methanol	9.472	0.06584
Methoxybenzoate	28.60	0.1579
Methyl acetate	15.75	0.1108
Methyl acrylate	19.67	0.1308
Methylamine	7.106	0.05879
2-Methyl-1,3-butadiene	17.74	0.1307
3-Methyl-1,3-butadiene	17.46	0.1245
2-Methylbutane	18.29	0.1415
Methyl butanoate	25.83	0.1661
3-Methylbutanoic acid	33.94	0.1923
2-Methyl-1-butanol	24.51	0.1518
3-Methyl-1-butanol	24.72	0.1526
2-Methyl-2-butanol	23.24	0.1523
3-Methyl-2-butanol	23.30	0.1493
3-Methyl-2-butanone	23.20	0.1494
2-Methyl-1-butene	16.9	0.129
3-Methyl-1-butene	18.08	0.1405
2-Methyl-2-butene	17.26	0.1279
Methylcyclohexane	27.51	0.1713
Methylcyclopentane	21.87	0.1463
N-Methylethylamine	19.39	0.1391
Methyl formate	11.54	0.08406
2-Methylfuran	14.67	0.1160
2-Methylheptane	36.78	0.2342
3-Methylheptane	36.40	0.2301
4-Methylheptane	36.21	0.2297
2-Methylhexane	30.01	0.2016
3-Methylhexane	29.70	0.1977

(*Continued*)

TABLE 2.58 Van der Waalls' Constants for Gases (*Continued*)

Substance	a, $L^2 \cdot bar \cdot mol^{-2}$	b, $L \cdot mol^{-1}$
Methylhydrazine	11.67	0.07334
Methyl isobutanoate	24.87	0.1639
Methyl isocyanate	12.6	0.09161
1-Methyl-2-isopropylbenzene	42.7	0.234
1-Methyl-4-isopropylbenzene	45.27	0.2478
Methyl 2-methylpropanoate	24.50	0.163 7
2-Methyloctane	43.50	0.2641
2-Methylpentane	23.83	0.1707
3-Methylpentane	23.75	0.1677
2-Methyl-2,4-pentanediol	39.05	0.2054
Methyl pentanoate	29.39	0.1847
2-Methyl-3-pentanol	27.96	0.1730
3-Methyl-3-pentanol	27.45	0.1699
4-Methyl-2-pentanol	22.38	0.1388
4-Methyl-2-pentanone	29.08	0.1815
2-Methyl-2-pentene	23.86	0.1641
cis-3-Methyl-2-pentene	23.86	0.1641
trans-3-Methyl-2-pentene	24.60	0.1656
cis-4-Methyl-2-pentene	23.03	0.1675
trans-4-Methyl-2-pentene	23.32	0.1685
2-Methylpropanal	18.49	0.1285
2-Methyl-1-propanamine	19.30	0.1325
2-Methylpropane (isobutane)	13.36	0.1168
Methyl propanoate	20.51	0.1377
2-Methylpropanoic acid	28.9	0.170
2-Methyl-1-propanol	20.35	0.1324
2-Methyl-2-propanol	18.81	0.1324
2-Methylpropene	12.73	0.1086
2-Methylpropyl acetate	29.05	0.1845
2-Methylpropyl formate	22.54	0.1476
2-Methylpyridine	24.45	0.1403
3-Methylpyridine	27.08	0.1496
4-Methylpyridine	25.89	0.1428
1-Methylstyrene	36.69	0.1999
2-Methyltetrahydrofuran	22.37	0.1484
2-Methylthiophene	22.10	0.1299
3-Methylthiophene	21.98	0.1282
Methyl vinyl ether	11.65	0.09520
Morpholine	20.36	0.1174
Naphthalene	40.32	0.1920
Neon	0.208	0.01709
Niobium pentafluoride	25.22	0.1220
Nitric oxide (NO)	1.46	0.0289
Nitroethane	24.13	0.1544
Nitrogen-14	15.18	0.1288
Nitrogen chloride difluoride	6.447	0.06089
Nitrogen dioxide (NO_2)	5.36	0.0443
Nitrogen trifluoride	3.58	0.05364
Nitrous oxide (N_2O)	3.852	0.04435
Nitromethane	17.18	0.1041
Nitrosyl chloride	6.191	0.05014
Nonane	45.11	0.2702
1-Nonanol	50.00	0.2634

TABLE 2.58 Van der Waalls' Constants for Gases (*Continued*)

Substance	*a*, $L^2 \cdot bar \cdot mol^{-2}$	*b*, $L \cdot mol^{-1}$
1-Nonene	43.68	0.2629
Octadecafluorooctane	44.27	0.3143
Octafluorocyclobutane	15.81	0.1450
Octafluoropropane	12.96	0.1338
Octamethylcyclotetrasiloxane	75.30	0.4579
Octane	37.86	0.2370
1-Octanol	44.71	0.2371
2-Octanol	41.98	0.2376
1-Octene	35.01	0.2227
cis-2-Octene	35.42	0.2176
Osmium tetraoxide	2.79	0.2447
Oxygen	1.382	0.03186
Oxygen difluoride	2.726	0.04516
Ozone	3.570	0.04977
Pentadecane	95.91	0.4834
1-Pentadecene	99.00	0.5011
1,2-Pentadiene	18.13	0.1284
cis-1,3-Pentadiene	17.98	0.1292
1,4-Pentadiene	17.58	0.1311
Pentafluorobenzene	23.45	0.1571
2,2,3,3,4-Pentamethylpentane	46.85	0.2593
2,2,3,4,4-Pentamethylpentane	47.82	0.2716
Pentanal	25.21	0.1622
Pentane	19.13	0.1449
Pentanenitrile	34.16	0.1772
Pentanoic acid	33.68	0.1867
1-Pentanol	25.81	0.1572
2-Pentanol	24.89	0.1585
2-Pentanone	24.85	0.1578
3-Pentanone	24.65	0.1565
1-Pentene	17.86	0.1370
cis-2-Pentene	17.83	0.1338
trans-2-Pentene	18.30	0.1391
Pentylbenzene	51.85	0.2718
Pentyl formate	27.97	0.1730
1-Pentyne	17.53	0.1266
Perchloryl fluoride (ClO_3F)	7.371	0.07130
Phenol	22.93	0.1177
Phosgene	10.65	0.08340
Phosphine	4.693	0.05155
Phosphonium chloride	4.111	0.04545
Phosphorus	53.6	0.157
Phosphorus chloride difluoride	8.47	0.0833
Phosphorus dichloride fluoride	12.50	0.0962
Phosphorus trifluoride	4.954	0.06510
Phosphoryl chloride difluoride	11.90	0.1001
Phosphoryl trifluoride	8.26	0.0849
Piperidine	20.84	0.1250
Propadiene	8.23	0.0747
Propanal	14.08	0.0995
Propane	9.385	0.09044
1,2-Propanediol	18.74	0.1068
1,3-Propanediol	21.11	0.1143

(*Continued*)

TABLE 2.58 Van der Waalls' Constants for Gases (*Continued*)

Substance	a, $L^2 \cdot bar \cdot mol^{-2}$	b, $L \cdot mol^{-1}$
Propanenitrile	21.57	0.1369
Propanoic acid	23.49	0.1386
1-Propanol	16.26	0.1080
2-Propanol	15.82	0.1109
2-Propenal	14.44	0.1017
Propene	8.411	0.08211
Propyl acetate	26.23	0.1700
Propylamine	15.26	0.1095
Propylbenzene	37.14	0.2073
Propylcyclopentane	38.80	0.2189
Propylcyclohexane	38.59	0.2255
Propylene oxide	13.78	0.1019
Propyl formate	20.79	0.1377
Propyne	8.40	0.0744
Pyridine	19.77	0.1136
Pyrrole	18.82	0.1049
Pyrrolidine	16.84	0.1056
Quinoline	36.70	0.1672
Radon	6.601	0.06239
Selenium	33.4	0.0675
Silicon chloride trifluoride	7.95	0.0921
Silicon tetrachloride	20.96	0.1470
Silicon tetrafluoride	5.259	0.072361
Silicon tetrahydride (silane)	4.30	0.0579
Styrene	32.15	0.1799
Sulfur (S)	24.3	0.0660
Sulfur dioxide	6.714	0.05636
Sulfur hexafluoride (SF_6)	7.857	0.08786
Sulfur trioxide	8.57	0.0622
1,1,2,2-Tetrachlorodifluoroethane	25.74	0.1665
Tetrachloroethylene	24.98	0.1435
Tetrachloromethane	20.01	0.1281
Tetradecafluorohexane	30.75	0.2448
Tetradecafluoromethylcyclohexane	29.66	0.2171
1-Tetradecanol	89.91	0.4289
Tetraethylsilane	40.85	0.2411
Tetrafluoroethylene	6.954	0.08085
Tetrafluorohydrazine (N_2F_4)	7.426	0.08564
Tetrafluoromethane	4.040	0.06325
Tetrahydrofuran	16.39	0.1082
Tetrahydropyran	20.02	0.1247
1,2,4,5-Tetramethylbenzene	45.8	0.2422
2,2,3,3-Tetramethylbutane	32.76	0.2056
2,2,3,3-Tetramethylhexane	45.11	0.2580
2,2,3,4-Tetramethylhexane	47.36	0.2721
2,2,3,5-Tetramethylhexane	46.45	0.2753
2,2,4,4-Tetramethylhexane	48.26	0.2819
2,2,4,5-Tetramethylhexane	47.05	0.2802
2,2,5,5-Tetramethylhexane	45.03	0.2760
2,3,3,4-Tetramethylhexane	47.13	0.2653
2,3,3,5-Tetramethylhexane	46.79	0.2733
2,3,4,4-Tetramethylhexane	47.32	0.2691
2,3,4,5-Tetramethylhexane	46.86	0.2723

TABLE 2.58 Van der Walls' Constants for Gases (*Continued*)

Substance	a, $L^2 \cdot bar \cdot mol^{-2}$	b, $L \cdot mol^{-1}$
3,3,4,4-Tetramethylhexane	47.46	0.2615
2,2,3,3-Tetramethylpentane	39.29	0.2304
2,2,3,4-Tetramethylpentane	39.37	0.2367
2,2,4,4-Tetramethylpentane	38.76	0.2403
2,3,3,4-Tetramethylpentane	39.65	0.2325
Tetramethylsilane	20.81	0.1653
Thiophene	17.21	0.1058
Tin(IV) chloride	27.25	0.1641
Titanium(IV) chloride	25.47	0.1423
Toluene	24.89	0.1499
1,2-Toluidine	33.36	0.1681
1,3-Toluidine	34.06	0.1717
1,4-Toluidine	31.74	0.1602
Tributoxyborane	81.34	0.3891
Tributylamine	65.31	0.3645
1,1,1-Trichloroethane	20.14	0.1317
1,1,2-Trichloroethane	25.47	0.1508
Trichloroethylene	17.21	0.1127
Trichlorofluoromethane	14.68	0.1111
Trichlorofluorosilane	15.67	0.1277
Trichloromethane	15.34	0.1019
Trichloromethylsilane	23.77	0.1638
1,2,3-Trichloropropane	31.29	0.1713
1,1,2-Trichlorotrifluoroethane	20.25	0.1481
1,2,2-Trichlorotrifluoroethane	20.25	0.1481
Tridecane	79.09	0.4176
1-Tridecanol	81.20	0.3942
1-Tridecene	77.93	0.4121
Tridecylcyclopentane	139.6	0.6536
Triethanolamine	32.14	0.3340
Triethylamine	27.59	0.1836
Trifluoroacetic acid	21.61	0.1567
1,1,1-Trifluoroethane	9.302	0.09572
Trifluoromethane	5.378	0.06403
Trimethylamine	13.37	0.1101
1,2,3-Trimethylbenzene	37.28	0.1999
1,2,4-Trimethylbenzene	38.03	0.2088
1,3,5-Trimethylbenzene	37.87	0.2118
2,2,3-Trimethylbutane	27.86	0.1869
2,2,3-Trimethyl-1-butene	28.57	0.1910
1,1,2-Trimethylcyclopentane	33.31	0.2048
1,1,3-Trimethylcyclopentane	33.42	0.2091
2,2,3-Trimethylheptane	48.07	0.2801
2,2,4-Trimethylheptane	47.49	0.2847
2,3,4-Trimethylheptane	47.96	0.2785
3,3,4-Trimethylheptane	47.68	0.2730
2,2,3-Trimethylhexane	40.5	0.2452
2,2,4-Trimethylhexane	40.50	0.2516
2,2,5-Trimethylhexane	40.38	0.2533
2,2,3-Trimethylpentane	33.92	0.2145
2,2,4-Trimethylpentane	33.61	0.2202
2,3,3-Trimethylpentane	34.03	0.2114
2,3,4-Trimethylpentane	34.28	0.2157

(*Continued*)

TABLE 2.58 Van der Waalls' Constants for Gases (*Continued*)

Substance	a, $L^2 \cdot bar \cdot mol^{-2}$	b, $L \cdot mol^{-1}$
2,2,4-Trimethyl-1,3-pentanediol	19.96	0.2692
Tungsten(VI) fluoride (WF_6)	13.25	0.1063
Undecane	60.88	0.3396
1-Undecene	59.17	0.3310
Uranium(VI) fluoride (UF_6)	16.01	0.1128
Vinyl acetate	32.31	0.2296
Vinyl chloride	9.62	0.07975
Vinyl fluoride	5.98	0.06502
Vinyl formate	11.38	0.08541
Xenon	4.192	0.05156
Xenon difluoride	12.46	0.7037
Xenon tetrafluoride	15.52	0.09035
m-Xylene	31.41	0.1814
o-Xylene	31.06	0.1756
p-Xylene	31.54	0.1824
Water	5.537	0.03052
Zirconium(IV) chloride	30.59	0.1401

2.15 EQUILIBRIUM CONSTANTS

The equilibrium constant, K, relates to a chemical reaction at equilibrium. It can be calculated if the equilibrium concentration of each reactant and product in a reaction at equilibrium is known.

There are several types of equilibrium constants. *Each is constant at a constant temperature.*

TABLE 2.59 *pK*, Values of Organic Materials in Water at 25°C

Ionic strength μ is zero unless otherwise indicated. Protonated cations are designated by (+ 1), (+ 2), etc., after the pK_a value; neutral species by (0), if not obvious; and negatively charged acids by (–1), (–2), etc.

Substance	pK_1	pK_2	pK_3	pK_4
Abietic acid	7.62			
Acetamide	−0.37(+1)			
Acetamidine	1.60(+1)			
N-(2-Acetamido)-2-aminoethane-sulfonic acid (20°C)	6.88			
2-Acetamidobenzoic acid	3.63			
3-Acetamidobenzoic acid	4.07			
4-Acetamidobenzoic acid	4.28			
2-(Acetamido)butanoic acid	3.716			
N-(2-Acetamido)iminodiacetic acid (20°C)	6.62			
3-Acetamidopyridine	4.37(+1)			
Acetanilide	0.4(+1)	13.39(0)$^{40°C}$		
Acetic acid	4.756			
Acetic acid-*d* (in D_2O)	5.32			

TABLE 2.59 *pK*, Values of Organic Materials in Water at 25°C (*Continued*)

Substance	pK_1	pK_2	pK_3	pK_4
Acetoacetic acid (18°C)	3.58			
Acetohydrazine	3.24(+1)			
Acetone oxime	12.2			
2-Acetoxybenzoic acid (acetylsalicyclic acid)	3.48			
3-Acetoxybenzoic acid	4.00			
4-Acetoxybenzoic acid	4.38			
Acetylacetic acid (18°C)	3.58			
N-Acetyl-α-alanine	3.715			
N-Acetyl-β-alanine	4.455			
2-Acetylaminobutanoic acid	3.72			
3-Acetylaminopropionic acid	4.445			
2-Acetylbenzoic acid	4.13			
3-Acetylbenzoic acid	3.83			
4-Acetylbenzoic acid	3.70			
2-Acetylcyclohexanone	14.1			
N-Acetylcysteine (30°C)	9.52			
Acetylenedicarboxylic acid	1.75	4.40		
N-Acetylglycine	3.670			
N-Acetylguanidine	8.23(+1)			
N-α-Acetyl-L-histidine	7.08			
Acetylhydroxamic acid (20°C)	9.40			
N-Acetyl-2-mercaptoethylamine	9.92(SH)			
4-Acetyl-β-mercaptoisoleucine (30°C)	10.30			
2-Acetyl-1-naphthol (30°C)	13.40			
N-Acetylpenicillamine (30°C)	9.90			
2-Acetylphenol	9.19			
4-Acetylphenol	8.05			
2-Acetylpyridine	2.643(+1)			
3-Acetylpyridine	3.256(+1)			
4-Acetylpyridine	3.505(+1)			
Aconitine	8.11(+1)			
Acridine	5.60(+1)			
Acrylic acid	4.26			
Adenine	4.17(+1)	9.75(0)		
Adeninedeoxyriboside-5′-phosphoric acid	—	4.4	6.4	
Adenine-*N*-oxide	2.69(+1)	8.49(0)		
Adenosine	3.5(+1)	12.34(0)		
Adenosine-5′-diphosphoric acid	—	4.2(−1)	7.20(−2)	
Adenosine-2′-phosphoric acid	3.81(+1)	6.17(0)		
Adenosine-3′-phosphoric acid	3.65(0)	5.88(−1)		
Adenosine-5′-phosphoric acid	3.74(0)	6.05(−1)	13.06(−2)	
Adenosine-5′-triphosphoric acid	—	4.00(−1)	6.48(−2)	
Adipamic acid (adipic acid monoamide)	4.629			
Adipic acid	4.418	5.412		
α-Alanine	2.34(+1)	9.69(0)		
β-Alanine	3.55(+1)	10.238(0)		
α-Alanine, methyl ester (μ = 0.10)	7.743(+1)			

(*Continued*)

TABLE 2.59 *pK*, Values of Organic Materials in Water at 25°C (*Continued*)

Substance	pK_1	pK_2	pK_3	pK_4
β-Alanine, methyl ester (μ = 0.10)	9.170(+1)			
N-D-Alanyl-α-D-alanine (μ = 0.1)	3.32(+1)	8.13(0)		
N-L-Alanyl-α-L-alanine (μ = 0.1)	3.32(+1)	8.13(0)		
N-L-Alanyl-α-D-alanine	3.12(+1)	8.30(0)		
N-α-Alanylglycine	3.11(+1)	8.11(0)		
Alanylglycylglycine	3.190(+1)	8.15(0)		
β-Alanylhistidine	2.64	6.86	9.40	
Albumin (bovine serum (μ = 0.15)	10–10.3			
2-Aldoxime pyridine	3.42(+1)	10.22(0)		
Alizarin Black SN	5.79	12.8		
Alizarin-3-sulfonic acid	5.54	11.01		
Allantoin	8.96			
Allothreonine	2.108(+1)	9.096(0)		
Alloxanic acid	6.64			
Allylacetic acid	4.68			
Allylamine	9.69(+1)			
5-Allylbarbituric acid	4.78(+1)			
5-Allyl-5-(-methylbutyl)barbituric acid	8.08			
2-Allylphenol	10.28			
1-Allylpiperidine	9.65(+1)			
2-Allylpropionic acid	4.72			
3-Amidotetrazoline	3.95(+1)			
2-Aminoacetamide	7.95(+1)			
Aminoacetonitrile	5.34(+1)			
9-Aminoacridine (20°C)	9.95(+1)			
4-Aminoantipyrine	4.94(+1)			
2-Aminobenzenesulfonic acid	2.459(0)			
3-Aminobenzenesulfonic acid	3.738(0)			
4-Aminobenzenesulfonic acid	3.227(0)			
2-Aminobenzoic acid	2.09(+1)	4.79(0)		
3-Aminobenzoic acid	3.07(+1)	4.79(0)		
4-Aminobenzoic acid	2.41(+1)	4.85(0)		
2-Aminobenzoic acid, methyl ester	2.36(+1)			
3-Aminobenzoic acid, methyl ester	3.58(+1)			
4-Aminobenzoic acid, methyl ester	2.45(+1)			
3-Aminobenzonitrile	2.75(+1)			
4-Aminobenzonitrile	1.74(+1)			
4-Aminobenzophenone	2.15(+1)			
2-Aminobenzothiazole (20°C)	4.48(+1)			
2-Aminobenzoylhydrazide	1.85	3.47	12.80	
2-Aminobiphenyl	3.78(+1)			
3-Aminobiphenyl	4.18(+1)			
4-Aminobiphenyl	4.27(+1)			
4-Amino-3-bromomethylpyridine	7.47(+1)			
4-Amino-3-bromopyridine (20°C)	7.04(+1)			
2-Aminobutanoic acid	2.286(+1)	9.830(0)		
3-Aminobutanoic acid	—	10.14(0)		
4-Aminobutanoic acid	4.031(+1)	10.556(0)		
2-Aminobutanoic acid, methyl ester (μ = 0.1)	7.640(+1)			

TABLE 2.59 *pK*, Values of Organic Materials in Water at 25°C (*Continued*)

Substance	pK_1	pK_2	pK_3	pK_4
4-Aminobutanoic acid, methyl ester (μ = 0.1)	9.838(+1)			
D-(+)-2-Amino-1-butanol	9.52(+1)			
3-Amino-*N*-butyl-3-methyl-2-butanone oxime	9.09(+1)			
4-Aminobutylphosphonic acid	2.55	7.55	10.9	
2-Amino-*N*-carbamoylbutanoic acid	3.886(+1)			
4-Amino-*N*-carbamoylbutanoic acid	4.683(+1)			
2-Amino-*N*-carbamoyl-2-methyl-propanoic acid	4.463			
1-Amino-1-cycloheptanecarboxylic acid	2.59(+1)	10.46(0)		
1-Amino-1-cyclohexanecarboxylic acid	2.65(+1)	10.03(0)		
2-Amino-1-cyclohexanecarboxylic acid	3.56(+1)	10.21(0)		
1-Aminocyclopentane	10.65(+1)			
1-Aminocyclopropane	9.10(+1)			
10-Aminodecylphosphonic acid	——	8.0	11.25	
10-Aminodecylsulfonic acid	2.65(+1)			
1-Amino-2-di(aminomethyl)butane	3.58(+3)	8.59(+2)	9.66(+1)	
2-Amino-*N*,*N*-dihydroxyethyl-2-hydroxyl-1,3-propanediol	6.484(+1)			
2-Amino-*N*,*N*-dimethylbenzoic acid	1.63(+1)	8.42(0)		
4-Amino-2,5-dimethylphenol	5.28(+1)	10.40(0)		
4-Amino-3,5-dimethylpyridine (20°C)	9.54(+1)			
12-Aminododecanoic acid	4.648(+1)			
2-Aminoethane-1-phosphoric acid	5.838	10.64		
1-Aminoethanesulfonic acid	−0.33	9.06		
2-Aminoethanesulfonic acid	1.5	9.061		
2-Aminoethanethiol (cysteamine) (μ = 0.01)	8.23(+1)			
2-Aminoethanol (ethanolamine)	9.50(+1)			
2-[2-(2-Aminoethyl)amino-ethyl]pyridine	3.50	6.59	9.51	
2-Amino-2-ethyl-1-butanol	9.82(+1)			
3-(2-Aminoethyl)indole	——	10.2		
3-Amino-*N*-ethyl-3-methyl-2-butanone oxime	9.23(+1)			
N-(2-Aminoethyl)morpholine	4.06(+2)	9.15(+1)		
p-(2-Aminoethyl)phenol	9.3	10.9		
2-Aminoethylphosphonic acid	2.45(+1)	7.0(0)	10.8(−1)	
N-(2-Aminoethyl)piperidine (30°C)	6.38	9.89		
2-(2-Aminoethyl)pyridine (μ = 0.5)	4.24(+2)	9.78(+1)		
4-Amino-3-ethylpyridine (20°C)	9.51(+1)			
N-(2-Aminoethyl)pyrrolidine (30°C)	6.56(+2)	9.74(+1)		

(*Continued*)

TABLE 2.59 *pK*, Values of Organic Materials in Water at 25°C (*Continued*)

Substance	pK_1	pK_2	pK_3	pK_4
2-Aminofluorine	10.34(+1)			
2-Amino-D-β-glucose (μ = 0.05)	2.20(+1)	9.08(0)		
2-Amino-*N*-glycylbutanoic acid	3.155(+1)	8.331(0)		
7-Aminoheptanoic acid	4.502			
2-Aminohexanoic acid	2.335(+1)	9.834(0)		
6-Aminohexanoic acid	4.373(+1)	10.804(0)		
C-Amino-*C*-hydrazinocarbonyl-methane	2.38(+2)	7.69(+1)		
2-Amino-3-hydroxybenzoic acid	2.5(+1)	5.192(0)	10.118(OH)	
L-2-Amino-3-hydroxybutanoic acid (threonine)	2.088(+1)	9.100(0)		
DL-2-Amino-4-hydroxybutanoic acid (μ = 0.1)	2.265(+1)	9.257(0)		
DL-4-Amino-3-hydroxybutanoic acid (μ = 0.1)	3.834(+1)	9.487(0)		
2-Amino-2′-hydroxydiethyl sulfide	9.27(+1)			
4-Amino-2-hydroxypyrimidine (cytosine)	4.58(+1)	12.15(0)		
3-Amino-*N*-isopropyl-3-methyl-2-butanone oxime	9.09(+1)			
4-Amino-3-isopropylpyridine (20°C)	9.54(+1)			
1-Aminoisoquinoline (20°C, μ = 0.01)	7.62(+1)			
3-Aminoisoquinoline (20°C, μ = 0.005)	5.05(+1)			
4-Aminoisoxazolidine-3-one	7.4(+1)			
Aminomalonic acid	3.32(+1)	9.83(0)		
DL-2-Amino-4-mercaptobutanoic acid	2.22(+1)	8.87(0)	10.86(SH)	
2-Amino-3-mercapto-3-Methylbutanoic acid	1.8(+1)	7.9(0)	10.5(SH)	
2-Amino-6-methoxybenzothiazole	4.50(+1)			
3-Amino-4-methylbenzenesulfonic acid	3.633			
4-Amino-3-methylbenzenesulfonic acid	3.125			
2-Amino-4-methylbenzothiazole	4.7(+1)			
1-Amino-3-methylbutane	10.64(+1)			
3-Amino-3-methyl-2-butanone oxime	9.09(+1)			
3-Amino-*N*-methyl-3-methyl-2-butanone oxime	9.23(+1)			
2-Amino-3-methylpentanoic acid	2.320(+1)	9.758(0)		
3-Aminomethyl-6-methylpyridine (30°C)	8.70(+1)			
Aminomethylphosphonic acid	2.35	5.9	10.8	
2-Amino-2-methyl-1,3-propanediol	8.801			
2-Amino-2-methyl-1-propanol	9.694(+1)			
2-Amino-2-methylpropanoic acid	2.357(+1)	10.205(0)		
(2-Aminomethyl(pyridine (μ = 0.5)	2.31(+2)	8.79(+1)		

TABLE 2.59 *pK*, Values of Organic Materials in Water at 25°C (*Continued*)

Substance	pK_1	pK_2	pK_3	pK_4
2-Amino-3-methylpyridine	7.24(+1)			
4-Amino-3-methylpyridine	9.43(+1)			
2-Amino-4-methylpyridine	7.48(+1)			
2-Amino-5-methylpyridine	7.22(+1)			
2-Amino-6-methylpyridine	7.41(+1)			
2-Amino-4-methylpyrimidine (20°C)	4.11(+1)			
Aminomethylsulfonic acid	5.57(+1)			
N-Aminomorpholine	4.19(+1)			
4-Amino-1-naphthalenesulfonic acid	2.81			
1-Amino-2-naphthalenesulfonic acid	1.71			
1-Amino-3-naphthalenesulfonic acid	3.20			
1-Amino-5-naphthalenesulfonic acid	3.69			
1-Amino-6-naphthalenesulfonic acid	3.80			
1-Amino-7-naphthalenesulfonic acid	3.66			
1-Amino-8-naphthalenesulfonic acid	5.03			
2-Amino-1-naphthalenesulfonic acid	2.35			
2-Amino-4-naphthalenesulfonic acid	3.79			
2-Amino-6-naphthalenesulfonic acid	3.79	8.94		
2-Amino-8-naphthalenesulfonic acid	3.89			
3-Amino-1-naphthoic acid	2.61	4.39		
4-Amino-2-naphthoic acid	2.89	4.46		
8-Amino-2-naphthol	4.20(+1)			
DL-2-Aminopentanoic acid (DL-norvaline)	2.318(+1)	9.808		
3-Aminopentanoic acid	4.02(+1)	10.399(0)		
4-Aminopentanoic acid	3.97(+1)	10.46(0)		
5-Aminopentanoic acid	4.20(+1)	9.758(0)		
5-Aminopentanoic acid, ethyl ester	10.151			
2-Aminophenol	9.28	9.72		
3-Aminophenol	9.83	9.87		
4-Aminophenol	8.50	10.30		
4-Aminophenylacetic acid (20°C)	3.60	5.26		
2-Aminophenylarsonic acid	ca 2	3.77	8.66	
3-Aminophenylarsonic acid	ca 2	4.02	8.92	
4-Aminophenylarsonic acid	ca 2	4.02	8.62	
3-Aminophenylboric acid	4.46	8.81		
4-Aminophenylboric acid	3.71	9.17		
4-Aminophenyl (4-chlorophenyl) sulfone	1.38			
2-Aminophenylphosphonic acid	—	4.10	7.29	
3-Aminophenylphosphonic acid	—	—	7.16	

(*Continued*)

TABLE 2.59 *pK*, Values of Organic Materials in Water at 25°C (*Continued*)

Substance	pK_1	pK_2	pK_3	pK_4
4-Aminophenylphosphonic acid	—	—	7.53	
1-Amino-1,2,3-propanetricarboxylic acid (μ = 2.2)	2.10(+1)	3.60(0)	4.60(−1)	9.82(−2)
3-Aminopropanoic acid	3.551(+1)	10.235(0)		
1-Amino-1-propanol	9.96(+1)			
DL-2-Amino-1-propanol	9.469(+1)			
3-Amino-1-propanol	9.96(+1)			
3-Aminopropene	9.691(+1)			
3-Amino-*N*-propyl-3-methyl-2-butanone oxime	9.09(+1)			
2-Aminopropylsulfonic acid	—	9.15		
2-Aminopyridine	6.71(+1)			
3-Aminopyridine	6.03(+1)			
4-Aminopyridine	9.114(+1)			
2-Aminopyridine-1-oxide	2.58(+1)			
3-Aminopyridine-1-oxide	1.47(+1)			
4-Aminopyridine-1-oxide	3.54(+1)			
8-Aminoquinaldine	4.86(+1)			
2-Aminoquinoline (20°C, μ = 0.01)	7.34(+1)			
3-Aminoquinoline (20°C, μ = 0.01)	4.95(+1)			
4-Aminoquinoline (20°C, μ = 0.01)	9.17(+1)			
5-Aminoquinoline (20°C, μ = 0.01)	5.46(+1)			
6-Aminoquinoline (20°C, μ = 0.01)	5.63(+1)			
8-Aminoquinoline (20°C, μ = 0.01)	3.99(+1)			
4-Aminosalicyclic acid	1.991(+1)	3.917(0)	13.74	
5-Aminosalicyclic acid	2.74(+1)	5.84(0)		
2-Amino-3-sulfopropanoic acid	1.89(+1)	8.70(0)		
4-Amino-2,3,5,6-tetramethylpyridine (20°C)	10.58(+1)			
5-Amino-1,2,3,4-tetrazole (20°C)	1.76	6.07		
2-Aminothiazole (20°C)	5.36(+1)			
1-Amino-3-thiobutane (30°C)	9.18(+1)			
5-Amino-3-thio-1-pentanol (30°C)	9.12(+1)			
2-Aminothiophenol	<2(+1)	7.90(0)		
2-Amino-4,4,4-trifluorobutanoic acid		8.171(0)		
3-Amino-4,4,4-trifluorobutanoic acid		5.831(0)		
3-Amino-2,4,6-trinitroluene		9.5(+1)		
Angiotensin II	10.37			
Anhydroplatynecine	9.40			
Aniline	4.60(+1)			
2-Anilinoethylsulfonic acid	3.80(+1)			
3-Anilinoethylsulfonic acid	4.85(+1)			
Anthracene-1-carboxylic acid	3.68			
Anthracene-2-carboxylic acid	4.18			
Anthracene-9-carboxylic acid	3.65			

TABLE 2.59 *pK*, Values of Organic Materials in Water at 25°C (*Continued*)

Substance	pK_1	pK_2	pK_3	pK_4
Anthraquinone-1-carboxylic acid (20°C)	3.37			
Anthraquinone-2-carboxylic acid (20°C)	3.42			
9,10-Anthraquinone monoxime	9.78			
9,10-Anthraquinone-1-sulfonic acid	0.27			
9,10-Anthraquinone-2-sulfonic acid	0.38			
Antipyrine	1.45(+1)			
Apomorphine (15°C)		8.92		
D-(−)-Arabinose	12.34			
L-(+)-Arginine	2.17	9.04(+1)	12.47(−1)	
Arsenazo III [pK_5 10.5(−4); pK_6 12.0(−5)]		1.2	2.7	7.9(−3)
Arsenoacetic acid		4.67	7.68	
Arsenoacrylic acid		4.23	8.60	
Arsenobutanoic acid		4.92	7.64	
2-Arsenocrotonic acid		4.61	8.75	
3-Arsenocrotonic acid		4.03	8.81	
Arsenopentanoic acid		4.89	7.75	
L-(+)-Ascorbic acid (vitamin C)	4.17	11.57		
L-(+)-Asparagine	2.01(0)	8.80(+1)		
L-Asparaginylglycine		4.53	9.07	
D-Aspartic acid	1.89(0)	3.65	9.60	
Aspartic diamide (μ = 0.2)	7.00			
Aspartylaspartic acid		3.40	4.70	8.26
α-Aspartylhistidine (38°C, μ = 0.1)		3.02	6.82	7.98
β-Aspartylhistidine (38°C, μ = 0.1)		2.95	6.93	8.72
N-Aspartyl-*p*-tyrosine (μ = 0.01)		3.57	8.92	10.23(OH)
Aspidospermine	7.65			
Atropine (17°C)	4.35(+1)			
1-Azacycloheptane	11.11(+1)			
1-Azacyclooctane	11.1(+1)			
Azetidine	11.29(+1)			
Aziridine	8.04(+1)			
Barbituric acid		8.372(0)		
m-Benzbetaine	3.217(+1)			
p-Benzbetaine	3.245(+1)			
Benzenearsonic acid (22°C)		8.48(−1)		
Benzene-1-arsonic acid-4-carboxylic acid		4.22 (COOH)	5.59	
Benzeneboronic acid	13.7			
Benzene-1-carboxylic acid-2-phosphoric acid		3.78	9.17	
Benzene-1-carboxylic acid-3-phosphoric acid		4.03	7.03	
Benzene-1-carboxylic acid-4-phosphoric acid	1.50	3.95	6.89	
Benzenediazine	11.08(+1)			
1,3-Benzenedicarboxylic acid (isophthalic acid)	3.62(0)	4.60(−1)		

(*Continued*)

TABLE 2.59 *pK*, Values of Organic Materials in Water at 25°C (*Continued*)

Substance	pK_1	pK_2	pK_3	pK_4
1,4-Benzenedicarboxylic acid (terephthalic acid)	3.54(0)	4.46(−1)		
1,3-Benzenedicarboxylic acid mononitrile	3.60(0)			
1,4-Benzenedicarboxylic acid mononitrile	3.55(0)			
Benzenehexacarboxylic acid (pK_5 6.32; pK_6 7.49)	0.68	2.21	3.52	5.09
Benzenepentacarboxylic acid (pK_5 6.46)	1.80	2.73	3.96	5.25
Benzenesulfinic acid	1.50			
Benzenesulfonic acid	2.554			
1,2,3,4-Benzenetetracarboxylic acid	2.05	3.25	4.73	6.21
1,2,3,5-Benzenetetracarboxylic acid	2.38	3.51	4.44	5.81
1,2,4,5-Benzenetetracarboxylic acid	1.92	2.87	4.49	5.63
1,2,3-Benzenetricarboxylic acid	2.88	4.75	7.13	
1,2,4-Benzenetricarboxylic acid	2.52	3.84	5.20	
1,3,5-Benzenetricarboxylic acid	2.12	4.10	5.18	
Benzil-α-dioxime	12.0			
Benzilic acid	3.09			
Benzimidazole	5.53(+1)	12.3(0)		
Benzohydroxamic acid (20°C)	8.89(0)			
Benzoic acid	4.204			
5,6-Benzoquinoline (20°C)	5.00(+1)			
7,8-Benzoquinoline (20°C)	4.15(+1)			
1,4-Benzoquinone monoxime	6.20			
Benzosulfonic acid	0.70			
1,2,3-Benzotriazole	8.38(+1)			
1-Benzoylacetone	8.23			
Benzoylamine	9.34(+1)			
2-Benzoylbenzoic acid	3.54			
Benzoylglutamic acid	3.49	4.99		
N-Benzoyglycine (hippuric acid)	3.65			
Benzoylhydrazine	3.03(+2)	12.45(+1)		
Benzoylpyruvic acid	6.40	12.10		
3-Benzoyl-1,1,1-trifluoroacetone	6.35			
Benzylamine	9.35(+1)			
Benzylamine-4-carboxylic acid	3.59	9.64		
2-Benzyl-2-phenylsuccinic acid (20°C)	3.69	6.47		
2-Benzylpyridine	5.13(+1)			
4-Benzylpyridine-1-oxide	−1.018(+1)			
1-Benzylpyrrolidine	9.51(+1)			
2-Benzylpyrrolidine	10.31(+1)			
Benzylsuccinic acid (20°C)	4.11	5.65		
3-(Benzylthio)propanoic acid	4.463			
Berberine (18°C)	11.73(+1)			
Betaine	1.832(+1)			
Biguanide	2.96(+2)	11.51(+1)		
2,2′-Biimidazolyl (μ = 0.3)	5.01(+1)			
2-Biphenylcarboxylic acid	3.46			
(1,1′-Biphenyl)-4,4′-diamine	3.63(+2)	4.70(+1)		
Bis(2-aminoethyl) ether (30°C)	8.62(+2)	9.59(+1)		

TABLE 2.59 *pK*, Values of Organic Materials in Water at 25°C (*Continued*)

Substance	pK_1	pK_2	pK_3	pK_4
N,N'-Bis(2-aminoethyl)-ethylenediamine (20°C)	3.32(+4)	6.67(+3)	9.20(+2)	9.92(+1)
N,N-Bis(2-hydroxyethyl)-2-aminoethane sulfonic acid (BES) (20°C)	7.15			
N,N-Bis(2-hydroxyethyl)glycine (bicine) (20°C)	8.35			
Bis(2-hydroxyethyl)iminotris (hydroxymethyl)methane (bis-tris)	6.46(+1)			
1,3-Bis[tris(hydroxymethyl)methylamino]propane (20°C)	6.80(+1)			
Bromoacetic acid	2.902			
2-Bromoaniline	2.53(+1)			
3-Bromoaniline	3.53(+1)			
4-Bromoaniline	3.88(+1)			
2-Bromobenzoic acid	2.85			
3-Bromobenzoic acid	3.810			
4-Bromobenzoic acid	3.99			
2-Bromobutanoic acid (35°C)	2.939			
erythro-2-Bromo-3-chlorosuccinic acid (19°C, μ = 0.1)	1.4	2.6		
threo-2-Bromo-chlorosuccinic acid (19°C, μ = 0.1)	1.5	2.8		
trans-2-Bromocinnamic acid	4.41			
3-Bromo-4-(dimethylamino)pyridine (20°C)	6.52(+1)			
2-Bromo-4,6-dinitroaniline	−6.94(+1)			
3-Bromo-2-hydroxymethylbenzoic acid (20°C)	3.28			
6-Bromo-2-hydroxymethylbenzoic acid (20°C)	2.25			
7-Bromo-8-hydroxyquinoline-5-sulfonic acid	2.51	6.70		
3-Bromomandelic acid	3.13			
3-Bromo-4-methylaminopyridine (20°C)	7.49(+1)			
(2-Bromomethyl)butanoic acid	3.92			
Bromomethylphosphonic acid	1.14	6.52		
2-Bromo-6-nitrobenzoic acid	1.37			
2-Bromophenol	8.452			
3-Bromophenol	9.031			
4-Bromophenol	9.34			
2-(2'-Bromophenoxy)acetic acid	3.12			
2-(3'-Bromophenoxy)acetic acid	3.09			
2-(4'-Bromophenoxy)acetic acid	3.13			
2-Bromo-2-phenylacetic acid	2.21			
2-(Bromophenyl) acetic acid	4.054			
4-(Bromophenyl)acetic acid	4.188			
4-Bromophenylarsonic acid	3.25	8.19		
4-Bromophenylphosphinic acid (17°C)	2.1			
2-Bromophenylphosphonic acid	1.64	7.00		

(*Continued*)

TABLE 2.59 *pK*, Values of Organic Materials in Water at 25°C (*Continued*)

Substance	pK_1	pK_2	pK_3	pK_4
3-Bromophenylphosphonic acid	1.45	6.69		
4-Bromophenylphosphonic acid	1.60	6.83		
3-Bromophenylselenic acid	4.43			
4-Bromophenylselenic acid	4.50			
2-Bromopropanoic acid	2.971			
3-Bromopropanoic acid	3.992			
Bromopropynoic acid	1.855			
2-Bromopyridine	0.71(+1)			
3-Bromopyridine	2.85(+1)			
4-Bromopyridine	3.71(+1)			
3-Bromoquinoline	2.69(+1)			
Bromosuccinic acid	2.55	4.41		
2-Bromo-*p*-tolylphosphonic acid	1.81	7.15		
Brucine (15°C)	2.50(+2)	8.16(+1)		
2-Butanamine (*sec*-butylamine)	10.56(+1)			
1,2-Butanediamine	6.399(+2)	9.388(+1)		
1,4-Butanediamine	9.35(+2)	10.82(+1)		
2,3-Butanediamine	6.91(+2)	10.00(+1)		
1,2,3,4-Butanetetracarboxylic acid	3.43	4.58	5.85	7.16
cis-2-Butenoic acid (isocrotonic acid)	4.44			
trans-2-Butenoic acid (*trans*-crotonic acid) (35°C)	4.676			
3-Butenoic acid (vinylacetic acid)	4.68			
3-Butoxybenzoic acid (20°C)	4.25			
Butylamine	10.64(+1)			
tert-Butylamine	10.685(+1)			
4-*tert*-Butylaniline	3.78(+1)			
N-*tert*-Butylaniline	7.10(+1)			
Butylarsonic acid (18°C)	4.23	8.91		
2-*tert*-Butylbenzoic acid	3.57			
3-*tert*-Butylbenzoic acid	4.199			
4-*tert*-Butylbenzoic acid	4.389			
N-Butylethylenediamine	7.53(+2)	10.30(+1)		
N-Butylglycine	2.35(+1)	10.25(0)		
tert-Butylhydroperoxide	12.80			
1-(*tert*-Butyl)-2-hydroxybenzene	10.62			
1-(*tert*-Butyl)-3-hydroxybenzene	10.119			
1-(*tert*-Butyl)-4-hydroxybenzene	10.23			
Butylmethylamine	10.90(+1)			
2-Butyl-1-methyl-2-pyrroline	11.84(+1)			
4-*tert*-Butylphenylactic acid	4.417			
Butylphosphinic acid	3.41			
tert-Butylphosphinic acid	4.24			
tert-Butylphosphonic acid	2.79	8.88		
1-Butylpiperidine (μ = 0.02)	10.43(+1)			
2-*tert*-Butylpyridine	5.76(+1)			
3-*tert*-Butylpyridine	5.82(+1)			
4-*tert*-Butylpyridine	5.99(+1)			
2-*tert*-Butylthiazole (μ = 0.1)	3.00(+1)			
4-*tert*-Butylthiazole (μ = 0.1)	3.04(+1)			
2-Butyn-1,4-dioic acid	1.75	4.40		
2-Butynoic acid (tetrolic acid)	2.620			

TABLE 2.59 *pK*, Values of Organic Materials in Water at 25°C (*Continued*)

Substance	pK_1	pK_2	pK_3	pK_4
Butyric acid	4.817			
4-Butyrobetaine (20°C)	3.94(+1)			
Caffeine (40°C)	10.4			
Calcein (pK_5 >12)	<4	5.4	9.0	10.5
Calmagite	8.14	12.35		
D-Camphoric acid	4.57	5.10		
Canaline	2.40	3.70	9.20	
Canavanine	2.50(+2)	6.60(+1)	9.25(0)	
N-Carbamoylacetic acid	3.64			
N-Carbamoyl-α-D-alanine	3.89(+1)			
N-Carbamoyl-β-alanine	4.99(+1)			
DL-*N*-Carbamoylalanine	3.892(+1)			
N-Carbamoylglycine	3.876			
2-Carbamoylpyridine (20°C)	2.10(+1)			
3-Carbamoylpyridine	3.328(+1)			
4-Carbamoylpyridine (20°C)	3.61(+1)			
β-Carboxymethylaminopropanoic acid	3.61(+1)	9.46(0)		
Chloroacetic acid	2.867			
N-(2′-Chloroacetyl)glycine	3.38(0)			
cis-3-Chloroacrylic acid (18°C, μ = 0.1)	3.32			
trans-3-chloroacrylic acid (18°C, μ = 0.1)	3.65			
2-Chloroaniline	2.64(+1)			
3-Chloroaniline	3.52(+1)			
4-Chloroaniline	3.99(+1)			
2-Chlorobenzoic acid	2.877			
3-Chlorobenzoic acid	3.83			
4-Chlorobenzoic acid	3.986			
2-Chlorobutanoic acid	2.86			
3-Chlorobutanoic acid	4.05			
4-Chlorobutanoic acid	4.50			
2-Chloro-3-butenoic acid	2.54			
3-Chlorobutylarsonic acid (18°C)	3.95	8.85		
trans-2′-Chlorocinnamic acid	4.234			
trans-3′-Chlorocinnamic acid	4.294			
trans-4′-Chlorocinnamic acid	4.413			
2-Chlorocrotonic acid	3.14			
3-Chlorocrotonic acid	3.84			
Chlorodifluoroacetic acid	0.46			
1-Chloro-1,2-dihydroxybenzene	8.522			
1-Chloro-2,6-dimethyl-4-hydroxybenzene	9.549			
4-Chloro-2,6-dinitrophenol	2.97			
2-Chloroethylarsonic acid	3.68	8.37		
3-Chlorohexyl-1-arsonic acid (18°C)	3.51	8.31		
2-Chloro-3-hydroxybutanoic acid	2.59			
3-Chloro-2-(hydroxymethyl)benzoic acid (20°C)	3.27			

(*Continued*)

TABLE 2.59 *pK*, Values of Organic Materials in Water at 25°C (*Continued*)

Substance	pK_1	pK_2	pK_3	pK_4
6-Chloro-2-(hydroxy-methyl)benzoic acid (20°C)	2.26			
7-Chloro-8-hydroxyquinoline-5-sulfonic acid	2.92	6.80		
2-Chloroisocrotonic acid	2.80			
3-Chloroisocrotonic acid	4.02			
3-Chlorolactic acid	3.12			
3-Chloromandelic acid	3.237			
3-Chloro-4-methoxyphenyl-phosphonic acid	2.25	6.7		
3-Chloro-4-methylaniline	4.05(+1)			
4-Chloro-*N*-methylaniline	3.9(+1)			
4-Chloro-3-methylphenol	9.549			
Chloromethylphosphonic acid	1.40	6.30		
2-Chloro-2-methylpropanoic acid	2.975			
2-Chloro-6-nitroaniline	−2.41(+1)			
4-Chloro-2-nitroaniline	−1.10(+1)			
2-Chloro-3-nitrobenzoic acid	2.02			
2-Chloro-4-nitrobenzoic acid	1.96			
2-Chloro-5-nitrobenzoic acid	2.17			
2-Chloro-6-nitrobenzoic acid	1.342			
4-Chloro-2-nitrophenol	6.48			
2-Chlorophenol	8.55			
3-Chlorophenol	9.10			
4-Chlorophenol	9.43			
(4-Chloro-3-nitrophenoxy)acetic acid	2.959			
2-Chloro-4-nitrophenylphosphonic acid	1.12	6.14		
3-Chloropentyl-1-arsonic acid (18°C)	3.71	8.77		
2-Chlorophenoxyacetic acid	3.05			
3-Chlorophenoxyacetic acid	3.07			
4-Chlorophenoxyacetic acid	3.10			
4-Chlorophenoxy-2-methylacetic acid	3.26			
2-Chlorophenylacetic acid	4.066			
3-Chlorophenylacetic acid	4.140			
4-Chlorophenylacetic acid	4.190			
2-Chlorophenylalanine	2.23(+1)	8.94(0)		
3-Chlorophenylalanine	2.17(+1)	8.91(0)		
DL-4-Chlorophenylalanine	2.08(+1)	8.96(0)		
4-Chlorophenylarsonic acid	3.33	8.25		
2-Chlorophenylphosphonic acid	1.63	6.98		
3-Chlorophenylphosphonic acid	1.55	6.65		
4-Chlorophenylphosphonic acid	1.66	6.75		
3-(2′-Chlorophenyl)propanoic acid	4.577			
3-(3′-Chlorophenyl)propanoic acid	4.585			
3-(4′-Chlorophenyl)propanoic acid	4.607			
3-Chlorophenylselenic acid	4.47			
4-Chlorophenylselenic acid	4.48			
4-Chloro-1,2-phthalic acid	1.60			

TABLE 2.59 *pK*, Values of Organic Materials in Water at 25°C (*Continued*)

Substance	pK_1	pK_2	pK_3	pK_4
2-Chloropropanoic acid	2.84			
3-Chloropropanoic acid	3.992			
2-Chloropropylarsonic acid (18°C)	3.76	8.39		
3-Chloropropylarsonic acid (18°C)	3.63	8.53		
Chloropropynoic acid	1.854			
2-Chloropyridine	0.49(+1)			
3-Chloropyridine	2.84(+1)			
4-Chloropyridine	3.83(+1)			
7-Chlorotetracycline	3.30(+1)	7.44	9.27	
4-Chloro-2-(2′-thiazolylazo)phenol	7.09			
4-Chlorothiophenol	5.9			
N-Chloro-*p*-toluenesulfonamide	4.54(+1)			
3-Chloro-*o*-toluidine	2.49(+1)			
4-Chloro-*o*-toluidine	3.385(+1)			
5-Chloro-*o*-toluidine	3.85(+1)			
6-Chloro-*o*-toludine	3.62(+1)			
Chrome Azurol S	2.45	4.86	11.47	
Chrome Dark Blue	7.56	9.3	12.4	
Cinchonine	5.85(+2)	9.92(+1)		
cis-Cinnamic acid	3.879			
trans-Cinnamic acid	4.438			
Citraconic acid	2.29(0)	6.15(−1)		
Citric acid	3.128	4.761	6.396	
L-(+)-Citrulline	2.43(+1)	9.41(0)		
Cocaine	8.41(+1)			
Codeine	7.95(+1)			
Colchicine	1.65(+1)			
Coniine (μ = 0.5)	11.24(+1)			
Creatine (40°C)	3.28(+1)			
Creatinine	3.57(+1)			
o-Cresol	10.26			
m-Cresol	10.00			
p-Cresol	10.26			
Cumene hydroperoxide	12.60			
Cupreine	7.63(+1)			
Cyanamide	10.27			
Cyanoacetic acid	2.460			
Cyanoacetohydrazide	2.34(+2)	11.17(+1)		
2-Cyanobenzoic acid	3.14			
3-Cyanobenzoic acid	3.60			
4-Cyanobenzoic acid	3.55			
4-Cyanobutanoic acid	4.44			
trans-1-Cyanocyclohexane-2-carboxylic acid	3.865			
4-Cyano-2,6-dimethylphenol	8.27			
4-Cyano-3,5-dimethylphenol	8.21			
2-Cyanoethylamine	7.7(+1)			
N-(2-Cyano)ethylnorcodeine	5.68(+1)			
Cyanomethylamine	5.34(+1)			
2-Cyano-2-methyl-2-phenylacetic acid	2.290			
1-Cyanomethylpiperidine	4.55(+1)			
2-Cyano-2-methylpropanoic acid	2.422			

(*Continued*)

TABLE 2.59 *pK*, Values of Organic Materials in Water at 25°C (*Continued*)

Substance	pK_1	pK_2	pK_3	pK_4
3-Cyanophenol	8.61			
o-Cyanophenoxyacetic acid	2.98			
m-Cyanophenoxyacetic acid	3.03			
p-Cyanophenoxyacetic acid	2.93			
2-Cyanopropanoic acid	2.37			
3-Cyanopropanoic acid	3.99			
2-Cyanopyridine	−0.26(+1)			
3-Cyanopyridine	1.45(+1)			
4-Cyanopyridine	1.90(+1)			
Cyanuric acid	6.78			
Cyclobutanecarboxylic acid	4.785			
1,1-Cyclobutanedicarboxylic acid	3.13	5.88		
cis-1,2-Cyclobutanedicarboxylic acid	3.90	5.89		
trans-1,2-Cyclobutanedicarboxylic acid	3.79	5.61		
cis-1,3-Cyclobutanedicarboxylic acid	4.04	5.31		
trans-1,3-Cyclobutanedicarboxylic acid	3.81	5.28		
Cyclohexanecarboxylic acid	4.90			
1,1-Cyclohexanediacetic acid	3.49	6.96		
cis-1,2-Cyclohexanediacetic acid (20°C)	4.42	5.45		
trans-1,2-Cyclohexanediacetic acid (20°C)	4.38	5.42		
cis-1,2-Cyclohexanediamine	6.43(+2)	9.93(+1)		
trans-1,2-Cyclohexanediamine	6.34(+2)	9.74(+1)		
1,1-Cyclohexanedicarboxylic acid	3.45	4.11		
cis-1,2-Cyclohexanedicarboxylic acid (20°C)	4.34	6.76		
trans-1,2-Cyclohexanedicarboxylic acid (20°C)	4.18	5.93		
cis-1,3-Cyclohexanedicarboxylic acid (16°C)	4.10	5.46		
trans-1,3-Cyclohexanedicarboxylic acid (19°C)	4.31	5.73		
trans-1,4-Cyclohexanedicarboxylic acid (16°C)	4.18	5.42		
1,3-Cyclohexanedione	5.26			
cis,cis-1,3,5-Cyclohexanetriamine	6.9(+3)	8.7(+2)	10.4(+1)	
Cyclohexanonimine	9.15			
cis-4-Cyclohexene-1,2-dicarboxylic acid (20°C)	3.89	6.79		
trans-4-Cyclohexene-1,2-dicarboxylic acid (20°C)	3.95	5.81		
Cyclohexylacetic acid	4.51			
Cyclohexylamine	10.64(+1)			
2-(Cyclohexylamino)ethanesulfonic acid (CHES) (20°C)	9.55			
3-Cyclohexylamino-1-propanesulfonic acid (CAPS) (20°C)	10.40			
4-Cyclohexylbutanoic acid	4.95			

TABLE 2.59 *pK*, Values of Organic Materials in Water at 25°C (*Continued*)

Substance	pK_1	pK_2	pK_3	pK_4
Cyclohexylcyanoacetic acid	2.367			
1,2-Cyclohexylenedinitriloacetic acid (μ = 0.1)	2.4	3.5	6.16	12.35
3-Cyclohexylpropanoic acid	4.91			
2-Cyclohexylpyrrolidine	10.76(+1)			
2-Cyclohexyl-2-pyrroline	7.91(+1)			
Cyclohexylthioacetic acid	3.488			
Cyclopentanecarboxylic acid	4.905			
cis-Cyclopentane-1-carboxylic acid-2-acetic acid	4.40	5.79		
trans-Cyclopentane-1-carboxylic acid-2-acetic acid	4.39	5.67		
Cyclopentane-1,2-diamine-*N*,*N*′,*N*′-tetraacetic acid (μ = 0.1)	—	—	—	10.20
Cyclopentane-1,1-dicarboxylic acid	3.23	4.08		
cis-Cyclopentane-1,2-dicarboxylic acid	4.43	6.67		
trans-Cyclopentane-1,2-dicarboxylic acid	3.96	5.85		
cis-Cyclopentane-1,3-dicarboxylic acid	4.26	5.51		
trans-Cyclopentane-1,3-dicarboxylic acid	4.32	5.42		
Cyclopentylamine	10.65(+1)			
1,1-Cyclopentyldiacetic acid	3.80	6.77		
cis-Cyclopentyl-1,2-diacetic acid	4.42	5.42		
trans-Cyclopentyl-1,2-diacetic acid	4.43	5.43		
Cyclopropanecarboxylic acid	4.827			
Cyclopropane-1,1-dicarboxylic acid	1.82	5.43		
cis-Cyclopropane-1,2-dicarboxylic acid	3.33	6.47		
trans-Cyclopropane-1,2-dicarboxylic acid	3.65	5.13		
Cyclopropylamine	9.10(+1)			
5-Cyclopropyl-1,2,3,4-tetrazole	4.90(+1)			
L-Cysteic acid (3-sulfo-L-alanine)	1.89(+1)	8.7(0)		
L-(+)-Cysteine	1.96	8.18	10.29(SH)	
L-(+)-Cysteine, ethyl ester	6.69 (NH_3^+)	9.17(SH)		
L-(+)-Cysteine, methyl ester	6.56 (NH_3^+)	8.99(SH)		
L-Cysteinyl-L-asparagine	2.97	7.09	8.47	
L-Cystine (35°C)	1.6(+2)	2.1(+1)	8.02(0)	8.71(−1)
Cystinylglycylglycine (35°C)	3.12	3.21	6.01	6.87
Cytidine	4.08(+1)	12.24(0)		
Cytidine-2′-phosphoric acid	0.8(+1)	4.36(0)	6.17(−1)	
Cytidine-3′-phosphoric acid	0.80(+1)	4.31(0)	6.04(−1)	13.2(sugar)
Cytidine-5′-phosphoric acid	—	4.39(0)	6.62(−1)	
Cytosine	4.58(+1)	12.15(0)		
Decanedioic acid (sebacic acid)	4.59	5.59		
Dehydroascorbic acid (20°C)	3.21	7.92	10.3	
2′-Deoxyadenosine (μ = 0.1)	3.8(+1)			

(*Continued*)

TABLE 2.59 *pK*, Values of Organic Materials in Water at 25°C (*Continued*)

Substance	pK_1	pK_2	pK_3	pK_4
Deoxycholic acid	6.58			
2-Deoxyglucose	12.52			
2-Deoxyguanosine (μ = 0.1)	2.5(+1)			
5-Desoxypyridoxal (μ = 0)	4.17(+1)	8.14(OH)		
1,1-Diacetic acid semicarbazide (30°C, μ = 0.1)	2.96	4.04		
Diacetylacetone	7.42			
Diallylamine (μ = 0.02)	9.29(+1)			
5,5-Diallybarbituric acid	7.78(0)			
1,3-Diamino-2-aminomethylpropane	6.44(+3)	8.56(+2)	10.38(+1)	
3,5-Diaminobenzoic acid	5.30			
1,3-Diamino-*N*,*N*′-bis-(2-aminoethyl)propane (μ = 0.5)	6.01(+4)	7.26(+3)	9.49(+2)	10.23(+1)
2,4-Diaminobutanoic acid (20°C)	1.85(+2)	8.24(+1)	10.40(0)	
2,2′-Diaminodiethyl sulfide (30°C)	8.84(+2)	9.64(+1)		
1,8-Diamino-3,6-dithiooctane (30°C)	8.43(+2)	9.31(+1)		
2,7-Diaminooctanedioic acid (20°C, μ = 0.1)	1.84(+2)	2.64(+1)	9.23(0)	9.89(−1)
1,8-Diamino-3,6-octanedione (30°C)	8.60(+2)	9.57(+1)		
1,8-Diamino-3-oxa-6-thiooctane	8.54(+2)	9.46(+1)		
2,3-Diaminopropanoic acid (μ = 0.1)	1.33(+2)	6.674(+1)	9.623(0)	
2,3-Diaminopropanoic acid, methyl ester (μ = 0.1)	4.412(+1)	8.250(0)		
1,3-Diamino-2-propanol (20°C)	7.93(+2)	9.69(+1)		
2,5-Diaminopyridine (20°C)	2.13(+2)	6.48(+1)		
1,4-Diazabicyclo[2.2.2]octane	2.90(+2)	8.60(+1)		
Dibenzylamine	8.52(+1)			
Dibenzylsuccinic acid (20°C)	3.96	6.66		
Dibromoacetic acid	1.39			
3,5-Dibromoaniline	2.35(+1)			
3,5-Dibromophenol	8.056			
2,2-Dibromopropanoic acid	1.48			
2,3-Dibromopropanoic acid	2.33			
rac-2,3-Dibromosuccinic acid (20°C)	1.43	2.24		
meso-2,3-Dibromosuccinic acid (20°C)	1.51	2.71		
3,5-Dibromo-*p*-L-tyrosine	2.17(+1)	6.45(0)	7.60(−1)	
Dibutylamine	11.25(+1)			
Di-*sec*-butylamine	10.91(+1)			
2,6-Di-*tert*-butylpyridine	3.58(+1)			
rac-2,3-Di-*tert*-butylsuccinic acid (μ = 0.1)	3.58	10.2		
1,12-Dicarboxydodecaborane	9.07	10.23		
Dichloroacetic acid	1.26			
Dichloroacetylacetic acid	2.11			
3,5-Dichloroaniline	2.37(+1)			
1,3-Dichloro-2,5-dihydroxybenzene (μ = 0.65)	7.30	9.99		

TABLE 2.59 *pK*, Values of Organic Materials in Water at 25°C (*Continued*)

Substance	pK_1	pK_2	pK_3	pK_4
2,5-Dichloro-3,6-dihydroxy-*p*-benzoquinone	1.09	2.42		
Dichloromethylphosphonic acid	1.14	5.61		
2,4-Dichloro-6-nitroaniline	−3.00(+1)			
2,5-Dichloro-4-nitroaniline	−1.74(+1)			
2,6-Dichloro-4-nitroaniline	−3.31(+1)			
2,3-Dichlorophenol	7.44			
2,4-Dichlorophenol	7.85			
2,6-Dichlorophenol	6.78			
3,4-Dichlorophenol	8.630			
3,5-Dichlorophenol	8.179			
2,4-Dichlorophenoxyacetic acid (2,4-D)	2.64			
4,6-Dichlorophenoxy-2-methylacetic acid	3.13			
3,6-Dichlorophthalic acid	1.46			
2,2-Dichloropropanoic acid	2.06			
2,3-Dichloropropanoic acid	2.85			
rac-2,3-Dichlorosuccinic acid (20°C)	1.43	2.81		
meso-2,3-Dichlorosuccinic acid	1.49	2.97		
3,5-Dichloro-*p*-tyrosine	2.12	6.47	7.62	
2-Dicyanoethylamine	5.14(+1)			
2,2-Dicyanopropanoic acid	−2.8			
Dicyclohexylamine	11.25(+1)			
Dicyclopentylamine	10.93(+1)			
Didodecylamine	10.99(+1)			
Diethanolamine	8.88(+1)			
Di(ethoxyethyl)amine	8.47(+1)			
3,5-Diethoxyphenol	9.370			
3-(Diethoxyphosphinyl)benzoic acid	3.65			
4-(Diethoxyphosphinyl)benzoic acid	3.60			
3-(Diethoxyphosphinyl)phenol	8.66			
4-(Diethoxyphosphinyl)phenol	8.28			
Diethylamine	10.8(+1)			
2-(Diethylamino)ethyl-4-aminobenzoate	8.85(+1)			
α-(Diethylamino)toluene	9.44(+1)			
N,N-Diethylaniline	6.56(+1)			
5,5-Diethylbarbituric acid (veronal)	8.020(0)			
N,N-Diethylbenzylamine	9.48(+1)			
Diethylbiguanide (30°C)	2.53(+1)	11.68(0)		
Diethylenetriamine	4.42(+3)	9.21(+2)	10.02(+1)	
Diethylenetriaminepentaacetic acid (pK_5, 10.58)	1.80(0)	2.55(−1)	4.33(−2)	8.60(−3)
N,N-Diethylethylenediamine	7.70(+2)	10.46(+1)		
2,2-Diethylglutaric acid	3.62	7.12		
N,N-Diethylglycine	2.04(+1)	10.47(0)		
Diethylglycolic acid (18°C)	3.804			
Diethylmalonic acid	2.151	7.417		
Diethylmethylamine	10.43(+1)			
rac-2,3-Diethylsuccinic acid	3.63	6.46		

(*Continued*)

TABLE 2.59 *pK*, Values of Organic Materials in Water at 25°C (*Continued*)

Substance	pK_1	pK_2	pK_3	pK_4
meso-2,3-Diethylsuccinic acid	3.54	6.59		
N,N-Diethyl-*o*-toluidine	7.18(+1)			
Difluoroacetic acid	1.33			
3,3-Difluoroacrylic acid	3.17			
Diglycolic acid	2.96			
Diguanidine	12.8			
Dihexylamine	11.0(+1)			
Dihydroarecaidine	9.70			
Dihydroarecaidine, methyl ester	8.39			
Dihydrocodeine	8.75(+1)			
Dihydroergonovine	7.38(+1)			
α-Dihydrolysergic acid	3.57	8.45		
γ-Dihydrolysergic acid	3.60	8.71		
α-Dihydrolysergol	8.30			
β-Dihydrolysergol	8.23			
Dihydromorphine	9.35			
3,4-Dihydroxyalanine	2.32(+1)	8.68(0)	9.87(−1)	
1,2-Dihydroxyanthraquinone-3-sulfonic acid (alizarin-3-sulfonic acid)	—	5.54(−1)	11.01(−2)	
3,4-Dihydroxybenzaldehyde	7.55			
1,2-Dihydroxybenzene (pyrocatechol) (μ = 0.1)	9.356(0)	12.98(−1)		
1,3-Dihydroxybenzene (resorcinol)	9.44(0)	12.32(−1)		
1,4-Dihydroxybenzene (hydroquinone)	9.91(0)	12.04(−1)		
4,5-Dihydroxybenzene-1,3-disulfonic acid	—	—	7.66(−2)	12.6(−3)
2,3-Dihydroxybenzoic acid (30°C)	2.98	10.14		
2,4-Dihydroxybenzoic acid (β-resorcyclic acid)	3.29	8.98		
2,5-Dihydroxybenzoic acid	2.97	10.50		
2,6-Dihydroxybenzoic acid	1.30			
3,4-Dihydroxybenzoic acid	4.48	8.67	11.74	
3,5-Dihydroxybenzoic acid	4.04			
2,5-Dihydroxy-*p*-benzoquinone	2.71	5.18		
3,4-Dihydroxy-3-cyclobutene-1,2-dione	0.541	3.480		
2,3-Dihydroxy-2-cyclopenten-1-one (20°C)	4.72			
1,4-Dihydroxy-2,6-dinitrobenzene	4.42	9.14		
Di(2,2′-hydroxyethyl)amine	8.8(+1)			
N,N-Di(2-hydroxyethyl)glycine	8.333			
Dihydroxymaleic acid	1.10			
Dihydroxymalic acid	1.92			
1,3-Dihydroxy-2-methylbenzene (μ = 0.65)	10.05	11.64		
2,2-Di(hydroxymethyl)-3-hydroxypropanoic acid	4.460			
2,4-Dihydroxy-5-methylpyrimidine	9.90			
2,4-Dihydroxy-6-methylpyrimidine	9.52			
1,4-Dihydroxynaphthalene (26°C, μ = 0.65)	9.37	10.93		
1,2-Dihydroxy-3-nitrobenzene	6.68			

TABLE 2.59 *pK*, Values of Organic Materials in Water at 25°C (*Continued*)

Substance	pK_1	pK_2	pK_3	pK_4
1,2-Dihydroxy-4-nitrobenzene (μ = 0.1)	6.701			
2,4-Dihydroxy-1-phenylazobenzene (μ = 0.1)	11.98			
2,4-Dihydroxyoxazolidine	6.11(+1)			
2,4-Dihydroxypteridine	<1.3	7.92		
2,6-Dihydroxypurine	7.53(0)	11.84(−1)		
2,4-Dihydroxypyridine (20°C)	1.37(+1)	6.45(0)	13(−1)	
Dihydroxytartaric acid	1.95	4.00		
1,4-Dihydroxy-2,3,5,6-tetramethylbenzene (μ = 0.65)	11.25	12.70		
3,5-Diiodoaniline	2.37(+1)			
2,5-Diiodohistamine	2.31(+2)	8.20(+1)	10.11(0)	
2,5-Diiodohistidine (μ = 0.1)	2.72	8.18	9.76	
3,5-Diiodophenol	8.103			
3,5-Diiodotyrosine	2.117(+1)	6.479(0)	7.821(−1)	
Diisopropylmalonic acid	2.124	8.848		
Dilactic acid	2.955			
threo-1,4-Dimercapto-2,3-butanediol	8.9			
meso-2,3-Dimercaptosuccinic acid	2.71	3.48	8.89(SH)	10.79(SH)
3,5-Dimethoxyaniline	3.86(+1)			
2,6-Dimethoxybenzoic acid	3.44			
1,10-Dimethoxy-3,8-dimethyl-4,7-phenanthroline	7.21			
Di(2-methoxyethyl)amine	9.51(+1)			
3,5-Dimethoxyphenol	9.345			
(3,4-Dimethoxy)phenylacetic acid	4.333			
Dimethylamine	10.77(+1)			
4-Dimethylaminobenzaldehyde	1.647(+1)			
N,N-Dimethylaminocyclohexane	10.72(+1)			
4-Dimethylamino-2,3-dimethyl-1-phenyl-3-pyrazolin-5-one	4.18(+1)			
4-Dimethylamino-3,5-dimethylpyridine (20°C)	8.15(+1)			
2-(Dimethylamino)ethanol	9.26(+1)			
2-[2-(Dimethylamino)ethyl]pyridine	3.46(+2)	8.75(+1)		
3-(Dimethylaminoethyl)pyridine	4.30(+2)	8.86(+1)		
4-(Dimethylaminoethyl)pyridine	4.66(+2)	8.70(+1)		
4-(Dimethylamino)-3-ethylpyridine (20°C)	8.66(+1)			
4-(Dimethylamino)-3-isopropylpyridine (20°C)	8.27(+1)			
2-(Dimethylaminomethyl)pyridine	2.58(+2)	8.12(+1)		
3-(Dimethylaminomethyl)pyridine	3.17(+2)	8.00(+1)		
4-(Dimethylaminomethyl)pyridine	3.39(+2)	7.66(+1)		
4-(Dimethylamino)-3-methylpyridine (20°C)	8.68(+1)			
4-(Dimethylaminophenyl)phosphonic acid	2.0(+1)	4.2	7.35	
3-(Dimethylamino)propanoic acid	9.85(+1)			
4-(Dimethylamino)pyridine (20°C)	6.09(+1)			

(*Continued*)

TABLE 2.59 *pK*, Values of Organic Materials in Water at 25°C (*Continued*)

Substance	pK_1	pK_2	pK_3	pK_4
N,N-Dimethylaniline	5.15(+1)			
2,3-Dimethylaniline	4.70(+1)			
2,4-Dimethylaniline	4.89(+1)			
2,5-Dimethylaniline	4.53(+1)			
2,6-Dimethylaniline	3.95(+1)			
3,4-Dimethylaniline	5.17(+1)			
3,5-Dimethylaniline	4.765(+1)			
N,N-Dimethylaniline-4-phosphonic acid (17°C)	2.0(+1)	4.2	7.39	
Dimethylarsinic acid (cacodylic acid)	1.67	6.273		
1,3-Dimethylbarbituric acid	4.68(+1)			
2,3-Dimethylbenzoic acid	3.771			
2,4-Dimethylbenzoic acid	4.217			
2,5-Dimethylbenzoic acid	3.990			
2,6-Dimethylbenzoic acid	3.362			
3,4-Dimethylbenzoic	4.41			
3,5-Dimethylbenzoic acid	4.302			
N,N-Dimethylbenzylamine	9.02(+1)			
Dimethylbiguanide	2.77(+1)	11.52		
2,2-Dimethylbutanoic acid (18°C)	5.03			
Dimethylchlorotetracycline (μ = 0.01)	3.30(+1)			
2,6-Dimethyl-4-cyanophenol	8.27			
3,5-Dimethyl-4-cyanophenol	8.21			
5,5-Dimethyl-1,3-cyclohexanedione	5.15			
cis-3,3-Dimethyl-1,2-cyclopropane-dicarboxylic acid	2.34	8.31		
trans-3,3-Dimethyl-1,2-cyclopro-panedicarboxylic acid	3.92	5.32		
3,5-Dimethyl-4-(dimethylamino)-pyridine (20°C)	8.12(+1)			
2,2-Dimethyl-1,3-dioxane-4,6-dione	5.1			
1,1-Dimethylethanethiol (μ = 0.1)	11.22			
N,N-Dimethylethylenediamine-*N,N*-diacetic acid	6.63	9.53		
N,N'-Dimethylethylenediamine-*N,N'*-diacetic acid	7.40	10.16		
N,N-Dimethylethylenediamine-*N,N'*-diacetic acid	5.99	9.97		
N,N-Dimethylglycine	2.146(+1)	9.940(0)		
Dimethylglycolic acid (18°C)	4.04			
N,N-Dimethylglycylglycine	3.11(+1)	8.09(0)		
Dimethylglyoxime	10.60			
5,5-Dimethyl-2,4-hexanedione	10.01			
5,5-Dimethylhydantoin	9.19			
2,4-Dimethyl-8-hydroxyquinoline	6.20(+1)	10.60(0)		
3,4-Dimethyl-8-hydroxyquinoline	5.80(+1)	10.05(0)		
2,4-Dimethyl-8-hydroxyquinoline-7-sulfonic acid	3.20 (NH^+)	10.14(OH)		
Dimethylhydroxytetracycline	7.5	9.4		
2,4-Dimethylimidazole	8.38(+1)			

TABLE 2.59 *pK*, Values of Organic Materials in Water at 25°C (*Continued*)

Substance	pK_1	pK_2	pK_3	pK_4
Dimethylmalic acid	3.17	6.06		
2,2-Dimethylmalonic acid	3.17	6.06		
3,5-Dimethyl-4-(methylamino) pyridine (20°C)	9.96(+1)			
2,3-Dimethylnaphthalene-1-carboxylic acid	3.33			
2,6-Dimethyl-4-nitrophenol	7.190			
3,5-Dimethyl-4-nitrophenol	8.245			
α,α-Dimethyloxaloacetic acid	1.77	4.62		
3,3-Dimethylpentanedioic acid	3.70	6.34		
2,2-Dimethylpentanoic acid	4.969			
4,4-Dimethylpentanoic acid (18°C)	4.79			
2,3-Dimethylphenol	10.50			
2,4-Dimethylphenol	10.58			
2,5-Dimethylphenol	10.22			
2,6-Dimethylphenol	10.59			
3,4-Dimethylphenol	10.32			
3,5-Dimethylphenol	10.15			
2,6-Dimethylphenoxyacetic acid	3.356			
Dimethylphenylsilylacetic acid	5.27			
N,N'-Dimethylpiperazine	4.630(+2)	8.539(+1)		
1,2-Dimethylpiperidine	10.22			
cis-2,6-Dimethylpiperidine	11.07(+1)			
2,2-Dimethylpropanoic acid (pivalic acid)	5.031			
2,2'-Dimethylpropylphosphonic acid	2.84	8.65		
2,4-Dimethylpyridine (2,4-lutidine)	6.74(+1)			
2,5-Dimethylpyridine (2,5-lutidine)	6.43(+1)			
2,6-Dimethylpyridine (2,6-lutidine)	6.71(+1)			
3,4-Dimethylpyridine (3,4-lutidine)	6.47(+1)			
3,5-Dimethylpyridine (3,5-lutidine)	6.09(+1)			
2,4-Dimethylpyridine-1-oxide	1.627(+1)			
2,5-Dimethylpyridine-1-oxide	1.208(+1)			
2,6-Dimethylpyridine-1-oxide	1.366(+1)			
3,4-Dimethylpyridine-1-oxide	1.493(+1)			
3,5-Dimethylpyridine-1-oxide	1.181(+1)			
2,3-Dimethylquinoline	4.94(+1)			
2,6-Dimethylquinoline	5.46(+1)			
meso-2,2-Dimethylsuccinic acid	3.77	5.936		
rac-2,2-Dimethylsuccinic acid	3.93	6.20		
D-2,3-Dimethylsuccinic acid	3.82	5.93		
meso-2,3-Dimethylsuccinic acid	3.67	5.30		
rac-2,3-Dimethylsuccinic acid	3.94	6.20		
2,4-Dimethylthiazole (μ = 0.1)	3.98			
2,5-Dimethylthiazole (μ = 0.1)	3.91			
4,5-Dimethylthiazole (μ = 0.1)	3.73			
N,N-Dimethyl-*o*-toluidine	5.86(+1)			
N,N-Dimethyl-*p*-toluidine	7.24(+1)			
2,4-Dinitroaniline	−4.25(+1)			
2,6-Dinitroaniline	−5.23(+1)			
3,5-Dinitroaniline	0.229(+1)			
2,3-Dinitrobenzoic acid	1.85			

(*Continued*)

TABLE 2.59 *pK*, Values of Organic Materials in Water at 25°C (*Continued*)

Substance	pK_1	pK_2	pK_3	pK_4
2,4-Dinitrobenzoic acid	1.43			
2,5-Dinitrobenzoic acid	1.62			
2,6-Dinitrobenzoic acid	1.14			
3,4-Dinitrobenzoic acid	2.82			
3,5-Dinitrobenzoic acid	2.85			
1,1-Dinitrobutane (20°C)	5.90			
1,1-Dinitrodecane	3.60			
1,1-Dinitroethane (20°C)	5.21			
Dinitromethane (20°C)	3.60			
1,1-Dinitropentane	5.337			
2,4-Dinitrophenol	4.08			
2,5-Dinitrophenol	5.216			
2,6-Dinitrophenol	3.713			
3,4-Dinitrophenol	5.424			
3,5-Dinitrophenol	6.732			
2,4-Dinitrophenylacetic acid	3.50			
1,1-Dinitropropane (20°C)	5.5			
2,6-Dioxo-1,2,3,6-tetrahydro-4-pyrimidinecarboxylic acid (orotic acid)	1.8(+1)	9.55(0)		
Diphenylacetic acid	3.939			
Diphenylamine	0.9(+1)			
2,2-Diphenylglutaric acid (20°C)	3.91	5.38		
1,3-Diphenylguanidine	10.12			
2,2-Diphenylheptanedioic acid (20°C)	4.28	5.39		
2,2-Diphenylhexanedioic acid (20°C)	4.17	5.40		
3,3-Diphenylhexanedioic acid	4.22	5.19		
Diphenylhydroxyacetic acid (35°C)	3.05			
Diphenylketimine	6.82			
2,2-Diphenylnonanedioic acid (20°C)	4.33	5.38		
meso-2,2-Diphenylsuccinic acid	3.48			
rac-2,2-Diphenylsuccinic acid	3.58			
2,2-Diphenylsuccinic acid, 1-methyl ester (20°C)	4.47			
2,2-Diphenylsuccinic acid, 4-methyl ester (20°C)	3.900			
Diphenylthiocarbazone	4.50	15		
Dipropylamine	10.91(+1)			
Dipropylenetriamine	7.72(+3)	9.56(+2)	10.65(+1)	
2,2-Dipropylglutaric acid	3.688	7.31		
Dipropylmalonic acid	2.04	7.51		
2,2′-Dipyridyl	−0.52(+2)	4.352(+1)		
2,3′-Dipyridyl (20°C)	1.52(+2)	4.42(+1)		
2,4′-Dipyridyl (20°C)	1.19(+2)	4.77(+1)		
3,3′-Dipyridyl (20°C, μ = 0.2)	3.0(+2)	4.60(+1)		
3.4′-Dipyridyl (20°C, μ = 0.2)	3.0(+2)	4.85(+1)		
4,4′-Dipyridyl	3.17(+2)	4.82(+1)		
Dithiodiacetic acid (18°C)	3.075	4.201		
1,4-Dithioerythritol	9.5			

TABLE 2.59 *pK*, Values of Organic Materials in Water at 25°C (*Continued*)

Substance	pK_1	pK_2	pK_3	pK_4
Dithiooxamide (rubeanic acid)	10.89			
Dulcitol	13.46			
Ecgonine	10.91			
Emetine	7.36(+1)	8.23(0)		
Epinephrine enantiomorph	9.39(+1)			
Epinephrine, pseudo	9.53(+1)			
Ergometrinine	7.32(+1)			
Ergonovine	6.73(+1)			
Eriochrome Black T	6.3	11.55		
1,2-Ethanediamine	6.85(+2)	9.92(+1)		
Ethane-1,2-diamino-*N,N′*-dimethyl-*N,N′*-diacetic acid (20°C)	6.047(0)	10.068(−1)		
1,2-Ethanedithiol	8.96	10.54		
Ethanethiol (μ = 0.015)	10.61			
Ethoxyacetic acid (18°C)	3.65			
2-Ethoxyaniline (*o*-phenetidine)	4.47(+1)			
3-Ethoxyaniline	4.17(+1)			
4-Ethoxyaniline	5.25(+1)			
2-Ethoxybenzoic acid (20°C)	4.21			
3-Ethoxybenzoic acid (20°C)	4.17			
4-Ethoxybenzoic acid (20°C)	4.80			
Ethoxycarbonylethylamine	9.13(+1)			
2-Ethoxyethanethiol	9.38			
2-Ethoxyethylamine	6.26(+1)			
2-Ethoxyphenol	10.109			
3-Ethoxyphenol	9.655			
(4-Ethoxyphenyl)phosphonic acid	2.06	7.28		
4-Ethoxypyridine	6.67(+1)			
Ethyl acetoacetate	10.68			
3-Ethylacrylic acid	4.695			
N-Ethylalanine	2.22(+1)	10.22(0)		
Ethylamine	10.63(+1)			
(3-Ethylamino)phenylphosphonic acid	1.1(+1)	4.90(0)	7.24(−1)	
N-Ethylaniline	5.11(+1)			
2-Ethylaniline	4.42(+1)			
3-Ethylaniline	4.70(+1)			
4-Ethylaniline	5.00(+1)			
Ethylarsonic acid (18°C)	3.89	8.35		
Ethylbarbituric acid	3.69(+1)			
2-Ethylbenzimidazole (μ = 0.16)	6.27(+1)			
2-Ethylbenzoic acid	3.79			
4-Ethylbenzoic acid	4.35			
Ethylbiguanide	2.09(+1)	11.47(0)		
2-Ethylbutanoic acid (20°C)	4.710			
S-Ethyl-L-cysteine (μ = 0.1)	2.03(+1)	8.60(0)		
Ethylenebiguanide (30°C)	1.74	2.88	11.34	11.76
Ethylenebis(thioacetic acid) (18°C)	3.382(0)	4.352(−1)		
Ethylenediamine-*N,N′*-diacetic acid	6.42	9.46		
Ethylenediamine-*N,N*-dimethyl-*N′,N′*-diacetic acid	6.047	10.068		

(*Continued*)

TABLE 2.59 *pK*, Values of Organic Materials in Water at 25°C (*Continued*)

Substance	pK_1	pK_2	pK_3	pK_4
Ethylenediamine-*N*,*N*-dipropanoic acid (30°C)	6.87	9.60		
Ethylenediamine-*N*,*N*,*N'*,*N'*-tetra-acetic acid (μ = 0.1)	1.99	2.67	6.16	10.26
Ethylenediamine-*N*,*N*,*N'*,*N'*-tetra-propanoic acid (30°C)	3.00	3.43	6.77	9.60
Ethylene glycol	14.22			
Ethyleneimine	8.04(+1)			
cis-Ethylene oxide dicarboxylic acid	1.93	3.92		
trans-Ethylene oxide dicarboxylic acid	1.93	3.25		
N-Ethylethylenediamine	7.63(+2)	10.56(+1)		
N-Ethylglycine (μ = 0.1)	2.34(+1)	10.23(0)		
3-Ethylglutaric acid	4.28	5.33		
Ethyl hydroperoxide	11.80			
Ethyl hydrogen malonate	3.55			
3-Ethyl-2-hydroxypyridine	5.00(+1)			
Ethylmalonic acid	2.90(0)	5.55(−1)		
N-Ethyl mercaptoacetamide	8.14(SH)			
Ethyl 2-mercaptoacetate	7.95(SH)			
Ethyl 3-mercaptopropanoate	9.48(SH)			
3-Ethyl-4-(methylamino)pyridine (20°C)	9.90(+1)			
5-Ethyl-5-(1-methylbutyl)barbituric acid	8.11(0)			
Ethyl methyl ketoxime	12.45			
Ethylmethylmalonic acid	2.86(0)	6.41(−1)		
1-Ethyl-2-methylpiperidine	10.66(+1)			
3-Ethyl-6-methylpyridine (20°C)	6.51(+1)			
3-Ethyl-4-methylpyridine-1-oxide	−1.534(+1)			
5-Ethyl-2-methylpyridine-1-oxide	−1.288(+1)			
1-Ethyl-2-methyl-2-pyrroline	11.84(+1)			
Ethylmorphine (15°C)	8.08			
Ethyl nitroacetate	5.85			
3-Ethylpentane-2,4-dione	11.34			
2-Ethylpentanoic acid (18°C)	4.71			
5-Ethyl-5-pentylbarbituric acid	7.960			
2-Ethylphenol	10.2			
3-Ethylphenol	10.07			
4-Ethylphenol	10.0			
4-Ethylphenylacetic acid	4.373			
5-Ethyl-5-phenylbarbituric acid	7.445			
Ethylphosphinic acid	3.29			
Ethylphosphonic acid	2.43	8.05		
1-Ethylpiperidine (μ = 0.01)	10.45(+1)			
2,2-Ethylpropylglutaric acid	3.511			
Ethylpropylmalonic acid	3.14	7.43		
2-Ethylpyridine	5.89(+1)			
3-Ethylpyridine (20°C)	5.80(+1)			
4-Ethylpyridine	5.87(+1)			
Ethyl 3-pyridinecarboxylate	3.35(+1)			

TABLE 2.59 *pK*, Values of Organic Materials in Water at 25°C (*Continued*)

Substance	pK_1	pK_2	pK_3	pK_4
Ethyl 4-pyridinecarboxylate	3.45(+1)			
2-Ethylpyridine-1-oxide	−1.19(+1)			
3-Ethylpyridine-1-oxide	−0.965(+1)			
Ethylpyrrolidine	10.43(+1)			
2-Ethyl-2-pyrroline	7.87(+1)			
Ethylsuccinic acid	4.08(0)			
S-Ethylthioacetic acid	5.06			
N-Ethyl-*o*-toluidine	4.92(+1)			
N-Ethylveratramine	7.40(+1)			
β-Eucaine	9.35(+1)			
Fluoroacetic acid	2.586			
2-Fluoroacrylic acid	2.55			
2-Fluoroaniline	3.20(+1)			
3-Fluoroaniline	3.58(+1)			
4-Fluoroaniline	4.65(+1)			
2-Fluorobenzoic acid	3.27			
3-Fluorobenzoic acid	3.865			
4-Fluorobenzoic acid	4.14			
Fluoromandelic acid	4.244			
2-Fluorophenol	8.73			
3-Fluorophenol	9.29			
4-Fluorophenol	9.89			
2-Fluorophenoxyacetic acid	3.08			
3-Fluorophenoxyacetic acid	3.08			
4-Fluorophenoxyacetic acid	3.13			
4-Fluorophenylacetic acid	4.25			
2′-Fluorophenylalanine	2.14(+1)	9.01(0)		
3′-Fluorophenylalanine	2.10(+1)	8.98(0)		
4-Fluorophenylalanine	2.13(+1)	9.05(0)		
2-Fluorophenylphosphonic acid	1.64	6.80		
3-Fluorophenylselenic acid	4.34			
4-Fluorophenylselenic acid	4.50			
2-Fluoropyridine	−0.44(+1)			
3-Fluoropyridine	2.97(+1)			
5-Fluorouracil	8.00(0)	ca 13(−1)		
Folic acid (pteroylglutamic acid)	8.26			
Formic acid	3.751			
N-Formylglycine	3.43			
2-Formyl-3-hydroxypyridine (20°C)	3.40(+1)	6.95(OH)		
4-Formyl-3-hydroxypyridine	4.05(+1)	6.77(OH)		
2-Formyl-3-methoxypyridine (20°C)	3.89(+1)	12.95		
Formyl-3-methoxypyridine (20°C)	4.45(+1)	11.7		
D-(−)-Fructose	12.03			
Fumaric acid	3.10	4.60		
2-Furancarboxylic acid (2-furoic acid)	3.164			
D-(+)-Galactose	12.35			
Galactose-1-phosphoric acid	1.00	6.17		
Glucoascorbic acid	4.26	11.58		
D-Gluconic acid	3.86			

(*Continued*)

TABLE 2.59 *pK*, Values of Organic Materials in Water at 25°C (*Continued*)

Substance	pK_1	pK_2	pK_3	pK_4
α-D-Glucose-1-phosphate	1.11(0)	6.504(−1)		
trans-Glutaconic acid	3.77	5.08		
D-(−)-Glutamic acid	2.162(+1)	4.272(0)	9.358(−1)	
L-Glutamic acid	2.19(+1)	4.25(0)	9.67(−1)	
Glutamic acid, 1-ethyl ester	3.85(+1)	7.84(0)		
Glutamic acid, 5-ethyl ester	2.15(+1)	9.19(0)		
L-Glutamine (μ = 0.2)	2.17(+1)	9.13(0)		
Glutaric acid	3.77	6.08		
Glutaric acid monoamide	4.600(0)			
Glutarimide	11.43			
Glutathione	2.12(+1)	3.53(0)	8.66	9.12
DL-Glyceric acid	3.64			
Glycerol	14.15			
Glyceryl-1-phosphoric acid	—	6.656(−1)		
Glyceryl-2-phosphoric acid	1.335(0)	6.650(−1)		
Glycine	2.341(+1)	9.60(0)		
Glycine amide	8.03(+1)			
Glycine, ethyl ester	7.66(+1)			
Glycine hydroxamic acid	7.10	9.10		
Glycine, methyl ester	7.59(+1)			
Glycine-*O*-phenylphosphorylserine	2.96	8.07		
Glycolic acid	3.831			
N-Glycl-α-alanine	3.15(+1)	8.33(0)		
Glycylalanylalanine	3.38(+1)	8.10(0)		
N-Glycylasparagine	2.942			
Glycyclaspartic acid	2.81(+1)	4.45(0)	8.60(−1)	
Glycyl-DL-glutamine (18°C)	2.88(+1)	8.33(0)		
N-Glycylglycine	3.126(+1)	8.252(0)		
Glycylglycylcysteine (35°C)	2.71	2.71	7.94	7.94
Glycylglycylglycine	3.225(+1)	8.090(0)		
Glycyl-L-histidine (μ = 0.16)	6.79	8.20		
Glycylisoleucine	8.00			
N-Glycyl-L-leucine	3.180(+1)	8.327(0)		
Glycyl-*O*-phosphorylserine	2.90	6.02	8.43	
L-Glycylproline (μ = 0.1)	2.81(+1)	8.65(0)		
N-Glycylsarcosine (μ = 0.1)	2.98(+1)	8.55(0)		
N-Glycylserine	2.98(+1)	8.38(0)		
Glycylserylglycine	3.32	7.99		
Glycyltyrosine	2.93	8.45	10.49	
Glycylvaline	3.15	8.18		
Glyoxaline	7.03(+1)			
Glyoxylic acid	3.30(0)			
Guanidineacetic acid	2.82(+1)			
Guanine	3.3(+1)	9.2	12.3	
Guanine deoxyriboside-3′-phosphoric acid	—	2.9	6.4	9.7
Guanosine	1.9(+1)	9.25(0)	12.33(OH)	
Guanosine-5′-diphosphoric acid (μ = 0.1; pK_5 9.6)	—	—	2.9	6.3
Guanosine-3′-phosphoric acid	0.7	2.3	5.92	9.38
Guanosine-5′-phosphoric acid (μ = 0.1)	—	2.4	6.1	9.4

TABLE 2.59 *pK*, Values of Organic Materials in Water at 25°C (*Continued*)

Substance	pK_1	pK_2	pK_3	pK_4
Guanosine-5′-triphosphoric acid [μ = 0.1; pK_5 7.10(−3); pK_6 9.3(−4)]	—	—	—	3.0(−2)
Guanylurea	1.80	8.20		
Harmine (20°C)	7.61(+1)			
Heptafluorobutanoic acid	0.17			
4,4,5,5,6,6,6-Heptafluorohexanoic acid	4.18			
4,4,5,5,6,6,6-Heptafluoro-2-hexenoic acid	3.23			
Heptanedioic acid (pimelic acid)	4.484	5.424		
2,4-Heptanedione	8.43(keto); 9.15(enol)			
Heptanoic acid	4.893			
Heroin	7.6(+1)			
2,4-Hexadienoic acid (sorbic acid)	4.77			
1,1,1,3,3,3-Hexafluoro-2,2-propanediol	8.801			
1,1,1,3,3,3-Hexafluoro-2-propanol	9.42			
Hexahydroazepine	11.07			
Hexamethyldisilazine	7.55			
1,2,3,8,9,10-Hexamethyl-4,7-phenanthroline (20°C)	7.26			
1,6-Hexanediamine	9.830(+2)	10.930(+1)		
1,6-Hexanedioic acid	4.418	5.412		
2,4-Hexanedione	8.49 (enol); 9.32 (keto)			
2,2′,4,4′,6,6′-Hexanitrodiphenylamine	5.42(+1)			
Hexanoic acid (20°C)	4.849			
trans-2-Hexenoic acid	4.74			
trans-3-Hexenoic acid	4.72			
3-Hexen-4-oic acid	4.58			
4-Hexen-5-oic acid	4.74			
Hexylamine	10.64(+1)			
Hexylarsonic acid	4.16	9.19		
Hexylphosphonic acid	2.6	7.9		
DL-Histidine	1.82(+2)	6.00(+1)	9.16(0)	
Histidine amide (μ = 0.2)	5.78(+2)	7.64(+1)		
Histidine, methyl ester (μ = 0.1)	5.01(+2)	7.23(+1)		
Histidylglycine	2.40(+2)	5.80(+1)	7.82(0)	
Histidylhistidine (μ = 0.16)	5.40(+2)	6.80(+1)	7.95(0)	
DI-Homatropine	9.7(+1)			
DI-Homocysteine	2.222(+1)	8.87	10.86	
Homocysteine (μ = 0.1)	1.593(+2)	2.523(+1)	8.676(0)	9.413(−1)
Hydantoin	9.12			
Hydrastine	6.23(+1)			
Hydrazine-*N*,*N*-diacetic acid	<0.1	2.8	3.8	
Hydrazine-*N*′-*N*′-diacetic acid	2.40	3.12	7.32	
4-Hydrazinocarbonylpyridine (20°C)	1.82	3.52	10.79	
N-Hydroxyacetamide	9.40			

(*Continued*)

TABLE 2.59 *pK*, Values of Organic Materials in Water at 25°C (*Continued*)

Substance	pK_1	pK_2	pK_3	pK_4
2′-Hydroxyacetophenone	9.90			
3′-Hydroxyacetophenone	9.19			
4′-Hydroxyacetophenone	8.05			
1-Hydroxyacridine (15°C)	5.72			
2-Hydroxyacridine (15°C)	5.62			
3-Hydroxyacridine (15°C)	5.30			
α-Hydroxyasparagine	2.28(+1)	7.20(0)		
β-Hydroxyasparagine	2.09(+1)	8.29(0)		
Hydroxyaspartic acid	1.91(+1)	3.51(0)	9.11(−1)	
2-Hydroxybenzaldehyde (salicylaldehyde)	8.34			
3-Hydroxybenzaldehyde	9.00			
4-Hydroxybenzaldehyde	7.620			
2-Hydroxybenzaldehyde oxime	1.37(+1)	9.18	12.11	
2-Hydroxybenzamide	8.36			
2-Hydroxybenzenemethanol (2-hydroxybenzyl alcohol)	9.92			
3-Hydroxybenzenemethanol	9.83			
4-Hydroxybenzenemethanol	9.82			
4-Hydroxybenzenesulfonic acid	—	9.055(−1)		
2-Hydroxybenzohydroxamic acid	5.19			
2-Hydroxybenzoic acid (salicyclic acid)	2.98	12.38		
3-Hydroxybenzoic acid	4.076	9.85		
4-Hydroxybenzoic acid	4.582	9.23		
4-Hydroxybenzonitrile	7.95			
2-Hydroxy-5-bromobenzoic acid	2.61			
2-Hydroxybutanoic acid (30°C)	3.65			
L-3-Hydroxybutanoic acid (30°C)	4.41			
4-Hydroxybutanoic acid (30°C)	4.71			
2-Hydroxy-5-chlorobenzoic acid	2.63			
trans-2′-Hydroxycinnamic acid	4.614			
trans-3′-Hydroxycinnamic acid	4.40			
10-Hydroxycodeine	7.12			
cis-2-Hydroxycyclohexane-1-carboxylic acid	4.796			
trans-2-Hydroxycyclohexane-1-carboxylic acid	4.682			
cis-3-Hydroxycyclohexane-1-carboxylic acid	4.602			
trans-3-Hydroxycyclohexane-1-carboxylic acid	4.815			
cis-4-Hydroxycyclohexane-1-carboxylic acid	4.836			
trans-4-Hydroxycyclohexane-1-carboxylic acid	4.687			
1-Hydroxy-2,4-dihydroxymethylbenzene	9.79			
N-(Hydroxyethyl)biguanide	2.8(+2)	11.53(+1)		
N-(2-Hydroxyethyl)ethylenediamine	7.21(+2)	10.12(+1)		
N′-(2-Hydroxyethyl)ethylenediamine-*N*,*N*,*N*′-triacetic acid	2.39	5.37	9.93	

TABLE 2.59 *pK*, Values of Organic Materials in Water at 25°C (*Continued*)

Substance	pK_1	pK_2	pK_3	pK_4
N-(2-Hydroxyethyl)iminodiacetic acid (μ = 0.1)	2.2	8.65		
N-(2-Hydroxyethyl)piperazine-*N*′-ethansulfonic acid (20°C)	7.55			
4′-(2-Hydroxyethyl)-1′-piperazine-propanesulfonic acid (20°C)	8.00			
2-Hydroxyethyltrimethylamine	8.94(+1)			
L-β-Hydroxyglutamic acid	2.09	4.18	9.20	
1-Hydroxy-4-hydroxymethylbenzene	9.84			
5-Hydroxy-2-(hydroxymethyl)-4*H*-pyran-4-one	7.90	8.03		
3-Hydroxy-2-hydroxymethylpyridine (20°C, μ = 0.2)	5.00(+1)	9.07(OH)		
3-Hydroxy-4-hydroxymethylpyridine (20°C, μ = 0.2)	5.00(+1)	8.95(OH)		
8-Hydroxy-7-iodoquinoline-5-sulfonic acid	2.51(0)	7.417(−1)		
Hydroxylysine (38°C, μ = 0.1)	2.13(+2)	8.62(+1)	9.67(0)	
2-Hydroxy-3-methoxybenzaldehyde	7.912			
3-Hydroxy-4-methoxybenzaldehyde (isovanillin)	8.889			
4-Hydroxy-3-methoxybenzaldehyde (vanillin)	7.396			
4-Hydroxy-3-methoxybenzoic acid	4.355			
1-Hydroxy-2-methoxybenzylamine	8.70(+1)	10.52(0)		
2-Hydroxy-1-methoxybenzylamine	8.89(+1)	10.52(0)		
3-Hydroxy-2-methoxybenzylamine	8.94(+1)	10.42(0)		
2-Hydroxymethyl-2-benzeneacetic acid	4.12			
(2-Hydroxy-5-methylbenzene)-methanol	10.15			
2-Hydroxy-3-methylbenzoic acid	2.99			
2-Hydroxy-4-methylbenzoic acid	3.17			
2-Hydroxy-5-methylbenzoic acid	4.08			
2-Hydroxy-6-methylbenzoic acid	3.32			
2-Hydroxy-2-methylbutanoic acid (18°C)	3.991			
3-Hydroxy-2-methylbutanoic acid (18°C)	4.648			
4-Hydroxy-4-methylpentanoic acid (18°C)	4.873			
1-Hydroxymethylphenol	9.95			
Hydroxymethylphosphoric acid	1.91	7.15		
2-Hydroxy-2-methylpropanoic acid (μ = 0.1)	3.717			
2-Hydroxy-4-methylpyridine	4.529(+1)			
8-Hydroxy-2-methylquinoline	5.55(+1)	10.31(0)		
8-Hydroxy-4-methylquinoline	5.56(+1)	10.00(0)		
8-Hydroxy-2-methylquinoline-5-sulfonic acid	4.80(0)	9.30(−1)		

(*Continued*)

TABLE 2.59 *pK*, Values of Organic Materials in Water at 25°C (*Continued*)

Substance	pK_1	pK_2	pK_3	pK_4
8-Hydroxy-4-methylquinoline-7-sulfonic acid	4.78(0)	10.01(−1)		
8-Hydroxy-6-methylquinoline-5-sulfonic acid	4.20(0)	8.7(−1)		
2-Hydroxy-1-naphthoic acid (20°C)	3.29	9.68		
2-Hydroxy-2-nitrobenzoic acid	2.23			
2-Hydroxy-3-nitrobenzoic acid	1.87			
2-Hydroxy-5-nitrobenzoic acid	2.12			
2-Hydroxy-6-nitrobenzoic acid	2.24			
2-Hydroxy-4-nitrophenylphosphonic acid	1.22	5.39		
8-Hydroxy-7-nitroquinoline-5-sulfonic acid	1.94(0)	5.750(−1)		
3-Hydroxy-4-nitrotoluene (μ = 0.1)	7.41			
4-Hydroxypentanoic acid (18°C)	4.686			
4-Hydroxy-3-pentenoic acid	4.30			
3-Hydroxyphenazine (15°C)	2.67			
4-Hydroxyphenylarsonic acid	3.89	8.37 (phenol)	10.05	
3-Hydroxyphenylboric acid	8.55	10.84		
2-Hydroxy-2-phenylpropanoic acid	3.532			
2-(2-Hydroxyphenyl)pyridine (20°C)	4.19(+1)	10.64		
trans-4-Hydroxyproline	1.818(+1)	9.662(0)		
Hydroxypropanedioic acid (tartronic acid)	2.37	4.74		
2-Hydroxypropanoic acid	3.858			
1-Hydroxy-2-propylbenzene	10.50			
4-Hydroxypteridine	1.3(+1)	7.89(0)		
2-Hydroxypyridine	1.25(+1)	11.62(0)		
3-Hydroxypyridine	4.80(+1)	8.72(0)		
4-Hydroxypyridine	3.23(+1)	11.09(0)		
2-Hydroxypyridine-*N*-oxide	−0.62(+1)	5.97(0)		
2-Hydroxypyrimidine	2.24(+1)	9.17(0)		
4-Hydroxypyrimidine	1.85(+1)	8.59(0)		
8-Hydroxyquinazoline	3.41(+1)	8.65(0)		
2-Hydroxyquinoline (20°C)	−0.31(+1)	11.74		
3-Hydroxyquinoline (20°C)	4.30(+1)	8.06(0)		
4-Hydroxyquinoline (20°C)	2.27(+1)	11.25(0)		
5-Hydroxyquinoline (20°C)	5.20(+1)	8.54(0)		
6-Hydroxyquinoline (20°C)	5.17(+1)	8.88(0)		
7-Hydroxyquinoline (20°C)	5.48(+1)	8.85(0)		
8-Hydroxyquinoline (20°C)	4.91(+1)	9.81(0)		
8-Hydroxyquinoline-5-sulfonic acid	4.092(+1)	8.776(0)		
DL-Hydroxysuccinic acid (malic acid)	3.458	5.097		
L-Hydroxysuccinic acid	3.40	5.05		
Hydroxytetracycline	3.27(+1)	7.32(0)	9.11(−1)	
5-Hydroxy-1,2,3,4-tetrazole	3.32			
4-Hydroxy-3-(2′-thiazolyazo)toluene	8.36			

TABLE 2.59 *pK*, Values of Organic Materials in Water at 25°C (*Continued*)

Substance	pK_1	pK_2	pK_3	pK_4
2-Hydroxytoluene	10.33			
3-Hydroxytoluene	10.10			
4-Hydroxytoluene	10.276			
4-Hydroxy-α,α,α-trifluorotoluene	8.675			
1-Hydroxy-2,4,6-trihydroxymethyl-benzene	9.56			
Hydroxyuracil	8.64			
Hydroxyvaline	2.55(+1)	9.77(0)		
Hyoscyamine	9.68(+1)			
Hypoxanthene	1.79(+1)	8.91(0)	12.07(−1)	
Hypoxanthine	5.3			
Imidazole	6.993(+1)	10.58(0)		
Imidazolidinetrione (parabanic acid)	6.10			
4-(4-Imidazolyl)butanoic acid (μ = 0.1)	4.26(+1)	7.26(0)		
2-(4-Imidazolyl)ethylamine	5.784(+2)	9.756(+1)		
3-(4-Imidazolyl)propanoic acid (μ = 0.16)	3.96(+1)	7.57(0)		
3,3′-Iminobispropanoic acid	4.11(0)	9.61(−1)		
3,3′-Iminobispropylamine (30°C)	8.02(+2)	9.70(+1)	10.70(0)	
2,2′-Iminodiacetic acid (diglycine) (30°C, μ = 0.1)	2.54(0)	9.12(−1)		
4-Indanol	10.32			
Indole-3-acetic acid	4.75			
Inosine	ca 1.5(+1)	8.96(0)	12.36	
Inosine-5′-phosphoric acid	1.54(0)	6.66(−1)		
Inosine-5′-triphosphoric acid [pK_5 7.68(−4)]	——	——	2.2(−2)	6.92(−3)
Iodoacetic acid	3.175			
2-Iodoaniline	2.54(+1)			
3-Iodoaniline	3.58(+1)			
4-Iodoaniline	3.82(+1)			
2-Iodobenzoic acid	2.86			
3-Iodobenzoic acid	3.86			
4-Iodobenzoic acid	4.00			
5-Iodohistamine	4.06(+1) (imidazole)	9.20(+1) (NH_3^+)	11.88(0) (imino)	
7-Iodo-8-hydroxyquinoline-5-sulfonic acid	2.514	7.417		
Iodomandelic acid	3.264			
Iodomethylphosphoric acid	1.30	6.72		
2-Iodophenol	8.464			
3-Iodophenol	8.879			
4-Iodophenol	9.200			
2-Iodophenoxyacetic acid	3.17			
3-Iodophenoxyacetic acid	3.13			
4-Iodophenoxyacetic acid	3.16			
2-Iodophenylacetic acid	4.038			
3-Iodophenylacetic acid	4.159			
4-Iodophenylacetic acid	4.178			

(*Continued*)

TABLE 2.59 *pK*, Values of Organic Materials in Water at 25°C (*Continued*)

Substance	pK_1	pK_2	pK_3	pK_4
2-Iodophenylphosphoric acid	1.74	7.06		
2-Iodopropanoic acid	3.11			
3-Iodopropanoic acid	4.08			
2-Iodopyridine	1.82(+1)			
3-Iodopyridine	3.25(+1)			
4-Iodopyridine (20°C)	4.02(+1)			
Isoasparagine	2.97(+1)	8.02(0)		
Isobutylacetic acid (18°C)	4.79			
Isobutylamine	10.41(+1)			
Isochlorotetracycline	3.1(+1)	6.7(0)	8.3(−1)	
Isocreatine	2.84(+1)			
Isogluatamine	3.81(+1)	7.88(0)		
Isohistamine (μ = 0.1)	6.036(+2)	9.274(+1)		
L-Isoleucine	2.35(+1)	9.68(0)		
Isolysergic acid	3.33(0)	8.46(NH)		
Isopilocarpine (15°C)	7.18(+1)			
2-(Isopropoxy)benzoic acid (20°C)	4.24			
3-(Isopropoxy)benzoic acid (20°C)	4.15			
4-(Isopropoxy)benzoic acid (20°C)	4.68			
Isopropylamine	10.64(+1)			
N-Isopropylaniline	5.50(+1)			
5-Isopropylbarbituric acid	4.907(+1)			
2-Isopropylbenzene acid	3.64			
4-Isopropylbenzene acid	4.36			
N-Isopropylglycine (μ = 0.1)	2.36(+1)	10.06(0)		
Isopropylmalonic acid	2.94	5.88		
Isopropylmalonic acid mononitrile	2.401			
3-Isopropyl-4-(methylamino)pyridine (20°C)	9.96(+1)			
3-Isopropylpentanedioic acid	4.30	5.51		
4-Isopropylphenylacetic acid	4.391			
Isopropylphosphinic acid	3.56			
Isopropylphosphonic acid	2.66	8.44		
2-Isopropylpyridine	5.83(+1)			
3-Isopropylpyridine (20°C)	5.72(+1)			
4-Isopropylpyridine	6.02(+1)			
DL-Isoproterenol	8.64(+1)			
Isoquinoline	5.40(+1)			
Isoretronecanol	10.83			
L-Isoserine (μ = 0.16)	2.72(+1)	9.25(0)		
Isothiocyanatoacetic acid	6.62			
L-(+)-Lactic acid	3.858			
L-Leucine	2.33(+1)	9.60(0)		
Leucine amide	7.80(+1)			
Leucine, ethyl ester (μ = 0.1)	7.57(+1)			
L-Leucyl-L-asparagine	3.00(+1)	8.12(0)		
L-Leucyl-L-glutamine	2.99(+1)	8.11(0)		
DL-Leucylglycine	3.25(+1)	8.28(0)		
Leucylisoserine (20°C)	3.188(+1)	8.207(0)		
D-Leucyl-L-tyrosine	3.12(+1)	8.38(0)	10.35(−1)	
L-Leucyl-L-tyrosine	3.46(+1)	7.84(0)	10.09(−1)	
Lysergic acid	3.44(+1)	7.68(0)		

TABLE 2.59 *pK*, Values of Organic Materials in Water at 25°C (*Continued*)

Substance	pK_1	pK_2	pK_3	pK_4
L-(+)-Lysine	2.18(+2)	8.94(+1)	10.53(0)	
Lysine, methyl ester (μ = 0.1)	6.965(+1)	10.251(0)		
L-Lysyl-L-alanine	3.22(+1)	7.62(0)	10.70(−1)	
L-Lysyl-D-alanine	3.00(+1)	7.74(0)	10.63(−1)	
Lysylglutamic acid	2.93(+2)	4.47(+1)	7.75(0)	10.50(+1)
L-Lysyl-L-lysine (μ = 0.1)	3.01(+2)	7.53(+1)	10.05(0)	10.01(−1)
L-Lysyl-D-lysine (μ = 0.1)	2.85(+2)	7.53(+1)	9.92(0)	10.89(−1)
L-Lysyl-L-lysyl-L-lysine (μ = 0.1)	3.08(+2)	7.34(+1)	9.80(0)	10.54(−1)
L-Lysyl-D-lysyl-L-lysine (μ = 0.1)	2.91(+2)	7.29(+1)	9.79(0)	10.54(−1)
L-Lysyl-D-lysyl-lysine (μ = 0.1)	2.94(+2)	7.15(+1)	9.60(0)	10.38(−1)
α-D-Lyxose	12.11			
Maleic acid	1.910	6.33		
Malonamic acid	3.641(0)			
Malonic acid	2.826	5.696		
Malonitrile (cyanoacetic acid)	2.460			
Mandelic acid	3.411			
D-(+)-Mannose	12.08			
Mercaptoacetic acid (thioglycolic acid)	3.60(0)	10.56(SH)		
2-Mercaptobenzoic acid (20°C)	4.05(0)			
2-Mercaptobutanoic acid	3.53(0)			
Mercaptodiacetic acid	3.32	4.29		
2-Mercaptoethanesulfonic acid (20°C)		9.5(−1)		
2-Mercaptoethanol	9.88			
2-Mercaptoethylamine	8.27(+1)	10.53(0)		
2-Mercaptohistidine	1.84(+1)	8.47(0)	11.4(SH)	
Mercapto-*S*-phenylacetic acid (μ = 0.1)	3.9			
2-Mercaptopropane (μ = 0.1)	10.86			
3-Mercapto-1,2-propanediol (μ = 0.5)	9.43			
2-Mercaptopropanoic acid	4.32(0)	10.20(SH)		
3-Mercaptopropanoic acid	—	10.84(SH)		
2-Mercaptopyridine (20°C)	−1.07(+1)	10.00(0)		
3-Mercaptopyridine (20°C)	2.26(+1)	7.03(0)		
4-Mercaptopyridine (20°C)	1.43(+1)	8.86(0)		
2-Mercaptoquinoline (20°C)	−1.44(+1)	10.21(0)		
3-Mercaptoquinoline (20°C)	2.33(+1)	6.13(0)		
4-Mercaptoquinoline (20°C)	0.77(+1)	8.83(0)		
Mercaptosuccinic acid	3.30(0)	4.94(−1)	10.94(SH)	
Mesitylenic acid	4.32			
Mesoxaldialdehyde	3.60			
Methacrylic acid	4.66			
Methanethiol	10.70			
DL-Methionine	2.28(+1)	9.21(0)		
2-(*N*-Methoxyacetamido)pyridine	2.01(+1)			
3-(*N*-Methoxyacetamido)pyridine	3.52(+1)			
4-(*N*-Methoxyacetamido)pyridine	4.62(+1)			
Methoxyacetic acid	3.570			
3-Methoxy-D-α-alanine	2.037(+1)	9.176(0)		

(*Continued*)

TABLE 2.59 *pK*, Values of Organic Materials in Water at 25°C (*Continued*)

Substance	pK_1	pK_2	pK_3	pK_4
2-Methoxyaniline	4.53(+1)			
3-Methoxyaniline	4.20(+1)			
4-Methoxyaniline	5.36(+1)			
2-Methoxybenzoic acid	4.09			
3-Methoxybenzoic acid	4.08			
4-Methoxybenzoic acid	4.49			
N,N-Methoxybenzylamine	9.68(+1)			
2-Methoxycarbonylaniline	2.23(+1)			
3-Methoxycarbonylaniline	3.64(+1)			
4-Methoxycarbonylaniline	2.38(+1)			
Methoxycarbonylmethylamine	7.66(+1)			
2-Methoxycarbonylpyridine	2.21(+1)			
3-Methoxycarbonylpyridine	3.13(+1)			
4-Methoxycarbonylpyridine	3.26(+1)			
trans-2-Methoxycinnamic acid	4.462			
trans-3-Methoxycinnamic acid	4.376			
trans-4-Methoxycinnamic acid	4.539			
2-Methoxyethylamine	9.45(+1)			
2-Methoxy-4-nitrophenylphosphonic acid	1.53	6.96		
2-Methoxyphenol	9.99			
3-Methoxyphenol	9.652			
4-Methoxyphenol	10.20			
(2′-Methoxy)phenoxyacetic acid	3.231			
(3′-Methoxy)phenoxyacetic acid	3.141			
(4′-Methoxy)phenoxyacetic acid	3.213			
4′-Methoxyphenylacetic acid	4.358			
(4-Methoxyphenyl)phosphinic acid (17°C)	2.35			
(2-Methoxyphenyl)phosphonic acid	2.16	7.77		
(4-Methoxyphenyl)phosphonic acid (17°C)	2.4	7.15		
3-(2′-Methoxyphenyl)propanoic acid	4.804			
3-(3′-Methoxyphenyl)propanoic acid	4.654			
3-(4′-Methoxyphenyl)propanoic acid	4.689			
3-Methoxyphenylselenic acid	4.65			
4-Methoxyphenylselenic acid	5.05			
2-Methoxy-4-(2-propenyl)phenol	10.0			
2-Methoxypyridine	3.06(+1)			
3-Methoxypyridine	4.91(+1)			
4-Methoxypyridine	6.47(+1)			
4-Methoxy-2-(2′-thiazoylazo)phenol	7.83			
2-Methylacrylic acid (18°C)	4.66			
N-Methylalanine	2.22(+1)	10.19(0)		
O-Methylallothreonine (μ = 0.1)	1.92(+1)	8.90(0)		
Methylamine	10.62(+1)			
2-(*N*-Methylamino)benzoic acid	1.93(+1)	5.34(0)		
3-(*N*-Methylamino)benzoic acid	—	5.10(0)		
4-(*N*-Methylamino)benzoic acid	—	5.05		

TABLE 2.59 *pK*, Values of Organic Materials in Water at 25°C (*Continued*)

Substance	pK_1	pK_2	pK_3	pK_4
Methylaminodiacetic acid (20°C)	2.146	10.088		
2-(Methylamino)ethanol	9.88(+1)			
2-(2-Methylaminoethyl)pyridine (30°C)	3.58(+2)	9.65(+1)		
2-(Methylaminomethyl)6-methylpyridine (μ = 0.5)	3.03(+2)	9.15(+1)		
2-(Methylaminomethyl)pyridine (30°C)	2.92(+2)	8.82(+1)		
4-Methylamino-3-methylpyridine (20°C)	9.83(+1)			
(3-Methylamino)phenylphosphonic acid	1.1(+1)	4.72(+1)	7.30(−1)	
(4-Methylamino)phenylphosphonic acid	—	—	7.85(−1)	
3-(Methylamino)pyridine (30°C)	8.70(+1)			
4-(Methylamino)pyridine (20°C)	9.65(+1)			
4-(Methylamino)-2,3,5,6-tetramethylpyridine (20°C)	10.06(+1)			
N-Methylaniline	4.85(+1)			
Methylarsonic acid (18°C)	3.41	8.18		
1-Methylbarbituric acid	4.35(+1)			
5-Methylbarbituric acid	3.386(+1)			
2-(*N*-Methylbenzamido)pyridine	1.44(+1)			
3-(*N*-Methylbenzamido)pyridine	3.66(+1)			
4-(*N*-Methylbenzamido)pyridine	4.68(+1)			
2-Methylbenzimidazole (μ = 0.16)	6.29(+1)			
2-Methylbenzoic acid (*o*-toluic acid)	3.90			
3-Methylbenzoic acid	4.269			
4-Methylbenzoic acid	4.362			
N-Methyl-1-benzoylecgonine	8.65			
Methylbiguanidine	3.00(+2)	11.44(+1)		
2-Methyl-2-butanethiol	11.35			
2-Methylbutanoic acid	4.761			
3-Methylbutanoic acid (20°C)	4.767			
(*E*)-2-Methyl-2-butendioic acid (mesaconic acid)	3.09	4.75		
3-Methyl-2-butenoic acid	5.12			
(*E*)-2-Methyl-2-butenoic acid (tiglic acid)	4.96			
(*Z*)-2-Methyl-2-butenoic acid (angelic acid)	4.30			
4-Methylcarboxylphenol	8.47			
(*E*)-2-Methylcinnamic acid	4.500			
(*E*)-3-Methylcinnamic acid	4.442			
(*E*)-4-Methylcinnamic acid	4.564			
1-Methylcyclohexane-1-carboxylic acid	5.13			
cis-2-Methylcyclohexane-1-carboxylic acid	5.03			
trans-2-Methylcyclohexane-1-carboxylic acid	5.73			
cis-3-Methylcyclohexane-1-carboxylic acid	4.88			

(*Continued*)

TABLE 2.59 *pK*, Values of Organic Materials in Water at 25°C (*Continued*)

Substance	pK_1	pK_2	pK_3	pK_4
trans-3-Methylcyclohexane-1-carboxylic acid	5.02			
cis-4-Methylcyclohexane-1-carboxylic acid	5.04			
trans-4-Methylcyclohexane-1-carboxylic acid	4.89			
2-Methylcyclohexyl-1,1-diacetic acid	3.53	6.89		
3-Methylcyclohexyl-1,1-diacetic acid	3.49	6.08		
4-Methylcyclohexyl-1,1,1-diacetic acid	3.49	6.10		
3-Methylcyclopentyl-1,1-diacetic acid	3.79	6.74		
S-Methyl-L-cysteine	8.97			
N-Methylcytidine	3.88			
5-Methylcytidine	4.21			
N-Methyl-2′-deoxycytidine	3.97			
5-Methyl-2′-deoxycytidine	4.33			
2-Methyl-3,5-dinitrobenzoic acid	2.97			
5-Methyldipropylenetriamine (30°C)	6.32(+3)	9.19(+2)	10.33(+1)	
2,2′-Methylenebis(4-chlorophenol)	7.6	11.5		
2,2′-Methylenebis(4,6-dichlorophenol)	5.6	10.56		
Methylenebis(thioacetic acid (18°C)	3.310	4.345		
3,3′-(Methylenedithio)dialanine	2.200(+1)	8.16(0)		
Methylenesuccinic acid	3.85	5.45		
N-Methylethylamine	4.23(+1)			
N-Methylethylenediamine	6.86(+1)	10.15(+1)		
α-Methylglucoside	13.71			
3-Methylglutaric acid	4.24	5.41		
N-Methylglycine (sarcosine)	2.12(+1)	10.20(0)		
5-Methyl-2,4-heptanedione	8.52(enol); 9.10(keto)			
5-Methyl-2,4-hexanedione	8.66(enol); 9.31(keto)			
5-Methyl-4-hexenoic acid	4.80			
3-Methylhistamine	5.80(+1)	9.90(0)		
1-Methylhistidine	1.69	6.48	8.85	
2-Methylhistidine (18°C)	1.7	7.2	9.5	
2-Methyl-8-hydroxyquinoline (μ = 0.005)	4.58(+1)	11.71(0)		
4-Methyl-8-hydroxyquinoline	4.67(+1)	11.62(0)		
1-Methylimidazole	7.06(+1)			
4-Methylimidazole	7.55(+1)			
N-Methyliminodiacetic acid	2.15	10.09		
S-Methylisothiourea	9.83(+1)			
O-Methylisourea	9.72(+1)			
Methylmalonic acid	3.07	5.87		
2-(*N*-Methylmethanesulfonamido)pyridine	1.73(+1)			

TABLE 2.59 *pK*, Values of Organic Materials in Water at 25°C (*Continued*)

Substance	pK_1	pK_2	pK_3	pK_4
3-(*N*-Methylmethanesulfonamido)pyridine	3.94(+1)			
4-(*N*-Methylmethanesulfonamido)pyridine	5.14(+1)			
2-Methyl-6-methylaminopyridine (20°C)	3.17(+1)	8.84(0)		
3-Methyl-4-methylaminopyridine (20°C)	—	9.84(0)		
4-Methyl-2,2′-(4-methylpyridyl)pyridine	5.32(+1)			
N-Methylmorpholine	7.13(+1)			
2-Methyl-1-naphthoic acid	3.11			
N-Methyl-1-naphthylamine	3.70(+1)			
2-Methyl-4-nitrobenzoic acid	1.86			
2-Methyl-6-nitrobenzoic acid	1.87			
1-Methyl-2-nitroterephthalic acid	3.11			
4-Methyl-2-nitroterephthalic acid	1.82			
3-Methylpentanedioic acid	4.25	5.41		
3-Methylpentane-2,4-dione	10.87			
2-Methylpentanoic acid	4.782			
3-Methylpentanoic acid	4.766			
4-Methylpentanoic acid	4.845			
cis-3-Methyl-2-pentenoic acid	5.15			
trans-3-Methyl-2-pentenoic acid	5.13			
4-Methyl-2-pentenoic acid	4.70			
4-Methyl-3-pentenoic acid	4.60			
6-Methyl-1,10-phenanthroline	5.11(+1)			
(2-Methylphenoxy)acetic acid	3.227			
(3-Methylphenoxy)acetic acid	3.203			
(4-Methylphenoxy)acetic acid	3.215			
(2-Methylphenyl)acetic acid (18°C)	4.35			
(4-Methylphenyl)acetic acid	4.370			
5-Methyl-5-phenylbarbituric acid	8.011(0)			
3-(2-Methylphenyl)propanoic acid	4.66			
3-(3-Methylphenyl)propanoic acid	4.677			
3-(4-Methylphenyl)propanoic acid	4.684			
1-Methyl-2-phenylpyrrolidine	8.80			
5-Methyl-1-phenyl-1,2,3-triazole-4-carboxylic acid	3.73			
Methylphosphinic acid	3.08			
Methylphosphonic acid	2.38	7.74		
3-Methyl-*o*-phthalic acid	3.18			
4-Methyl-*o*-phthalic acid	3.89			
N-Methylpiperazine (μ = 0.1)	4.94(+2)	9.09(+1)		
2-Methylpiperazine	5.62(+2)	9.60(+1)		
N-Methylpiperidine	10.19(+1)			
2-Methylpiperidine	10.95(+1)			
3-Methylpiperidine	11.07(+1)			
4-Methylpiperidine (μ = 0.5)	11.23(+1)			
2-Methyl-1,2-propanediamine	6.178(+2)	9.420(+1)		
2-Methyl-2-propanethiol	11.2			
2-Methylpropanoic acid	4.853			

(*Continued*)

TABLE 2.59 *pK*, Values of Organic Materials in Water at 25°C (*Continued*)

Substance	pK_1	pK_2	pK_3	pK_4
2-Methyl-2-propylamine	10.682(+1)			
2-Methyl-2-propylglutaric acid	3.626			
2-Methylpyridine	5.96(+1)			
3-Methylpyridine	5.68(+1)			
4-Methylpyridine	6.00(+1)			
Methyl 4-pyridinecarboxylate	3.26(+1)			
6-Methylpyridine-2-carboxylic acid	5.83			
2-Methylpyridine-1-oxide	1.029(+1)			
3-Methylpyridine-1-oxide	10.921(+1)			
4-Methylpyridine-1-oxide	1.258(+1)			
O-Methylpyridoxal (μ = 0.16)	4.74			
Methyl-2-pyridyl ketoxime	9.97			
1-Methyl-2-(3-pyridyl)pyrrolidine	3.41	7.94		
1-Methylpyrrolidine	10.46(+1)			
1-Methyl-3-pyrroline	9.88(+1)			
5-Methylquinoline	4.62(+1)			
Methylsuccinic acid	4.13	5.64		
Methylsulfonylacetic acid	2.36			
3-Methylsulfonylaniline	2.68(+1)			
4-Methylsulfonylaniline	1.48(+1)			
3-Methylsulfonylbenzoic acid	3.52			
4-Methylsulfonylbenzoic acid	3.64			
4-Methylsulfonyl-3,5-dimethyl-phenol	8.13			
3-Methylsulfonylphenol	9.33			
4-Methylsulfonylphenol	7.83			
1-Methyl-1,2,3,4-tetrahydro-3-pyridinecarboxylic acid (arecaidine; isoguvacine)	9.07			
5-Methyl-1,2,3,4-tetrazole	3.32			
2-Methylthiazole (μ = 0.1)	3.40(+1)			
4-Methylthiazole (μ = 0.1)	3.16(+1)			
5-Methylthiazole (μ = 0.1)	3.03(+1)			
Methylthioacetic acid	3.72			
4-Methylthioaniline	4.40(+1)			
2-Methylthioethylamine (30°C)	9.18(+1)			
Methylthioglycolic acid	7.68			
3-(*S*-Methylthio)phenol	9.53			
4-(*S*-Methylthio)phenol	9.53			
2-Methylthiopyridine (20°C)	3.59(+1)			
3-Methylthiopyridine (20°C)	4.42(+1)			
4-Methylthiopyridine (20°C)	5.94(+1)			
5-Methylthio-1,2,3,4-tetrazole	4.00(+1)			
O-Methylthreonine	2.02(+1)	9.00(0)		
O-Methyltyrosine	2.21(+1)	9.35(0)		
1-Methylxanthine	7.70	12.0		
3-Methylxanthine	8.10	11.3		
7-Methylxanthine	8.33	ca 13		
9-Methylxanthine	6.25			
Morphine (20°C)	7.87(+1)	9.85(0)		
Morpholine	8.492(+1)			
2-(*N*-Morpholino)ethanesulfonic acid (MES) (20°C)	6.15			

TABLE 2.59 *pK*, Values of Organic Materials in Water at 25°C (*Continued*)

Substance	pK_1	pK_2	pK_3	pK_4
3-(*N*-Morpholino)-2-hydroxypropanesulfonic acid (37°C)	6.75			
3-(*N*-Morpholino)propanesulfonic acid (20°C)	7.20			
Murexide	0.0	9.20	10.50	
Myosmine	5.26			
1-Naphthalenecarboxylic acid (1-naphthoic acid)	3.695			
2-Naphthalenecarboxylic acid	4.161			
1-Naphthol (20°C)	9.30			
2-Naphthol (20°C)	9.57			
Naphthoquinone monoxime	8.01			
1-Naphthylacetic acid	4.236			
2-Naphthylacetic acid	4.256			
1-Naphthylamine	3.92(+1)			
2-Naphthylamine	4.11(+1)			
1-Naphthylarsonic acid	3.66	8.66		
1-Naphthysulfonic acid	0.57			
Narceine (15°C)	3.5(+1)	9.3		
Narcotine	6.18(+1)			
Nicotine	3.15(+1)	7.87(0)		
Nicotyrine	4.76(+1)			
Nitrilotriacetic acid (NTA) (20°C)	1.65	2.94	10.33	
Nitroacetic acid	1.68			
2-Nitroaniline	−0.28(+1)			
3-Nitroaniline	2.46(+1)			
4-Nitroaniline	1.01(+1)			
2-Nitrobenzene-1,4-dicarboxylic acid	1.73			
3-Nitrobenzene-1,2-dicarboxylic acid	1.88			
4-Nitrobenzene-1,2-dicarboxylic acid	2.11			
2-Nitrobenzoic acid	2.18			
3-Nitrobenzoic acid	3.46			
4-Nitrobenzoic acid	3.441			
trans-2-Nitrocinnamic acid	4.15			
trans-3-Nitrocinnamic acid	4.12			
trans-4-Nitrocinnamic acid	4.05			
Nitroethane	8.57			
2-Nitrohydroquinone	7.63	10.06		
N-Nitroiminodiacetic acid	2.21	3.33		
3-Nitromesitol	8.984			
Nitromethane	10.12			
1-Nitro-6,7-phenanthroline (μ = 0.2)	3.23(+1)			
5-Nitro-1,10-phenanthroline	3.232(+1)			
6-Nitro-1,10-phenanthroline	3.23(+1)			
2-Nitrophenol	7.222			
3-Nitrophenol	8.360			
4-Nitrophenol	7.150			
(2-Nitrophenoxy)acetic acid	2.896			

(*Continued*)

TABLE 2.59 *pK*, Values of Organic Materials in Water at 25°C (*Continued*)

Substance	pK_1	pK_2	pK_3	pK_4
(3-Nitrophenoxy)acetic acid	2.951			
(4-Nitrophenoxy)acetic acid	2.893			
2-Nitrophenylacetic acid	4.00			
3-Nitrophenylacetic acid	3.97			
4-Nitrophenylacetic acid	3.85			
2-Nitrophenylarsonic acid	3.37	8.54		
3-Nitrophenylarsonic acid	3.41	7.80		
4-Nitrophenylarsonic acid	2.90	7.80		
7-(4-Nitrophenylazo)-8-hydroxy-5-quinolinesulfonic acid	3.14(0)	7.495(−1)		
3-Nitrophenylphosphonic acid	1.30	6.27		
4-Nitrophenylphosphonic acid	1.24	6.23		
3-(2′-Nitrophenyl)propanoic acid	4.504			
3-(4′-Nitrophenyl)propanoic acid	4.473			
3-Nitrophenylselenic acid	4.07			
4-Nitrophenylselenic acid	4.00			
1-Nitropropane	8.98			
2-Nitropropane	7.675			
2-Nitropropanoic acid	3.79			
2-Nitropyridine (μ = 0.02)	−2.06(+1)			
3-Nitropyridine (μ = 0.02)	0.79(+1)			
4-Nitropyridine (μ = 0.02)	1.23(+1)			
N-Nitrosoiminodiacetic acid	2.28	3.38		
4-Nitrosophenol	6.48			
Nitrourea	4.15(+1)			
1,9-Nonanedioic acid (azelaic acid)	4.53	5.40		
Nonanoic acid (pelargonic acid)	4.95			
DL-Norleucine	2.335(+1)	9.834(0)		
Novocaine	8.85(+1)			
2,2,3,3,4,4,5,5-Octafluoropentanoic acid	2.65			
1,8-Octanedioic acid (suberic acid)	4.512	5.404		
Octanoic acid (caprylic acid)	4.895			
Octopine-DD	1.35	2.30	8.68	11.25
Octopine-LD	1.40	2.30	8.72	11.34
Octylamine	10.65(+1)			
L-(+)-Ornithine	1.94(+2)	8.65(+1)	10.76(0)	
Oxalic acid	1.271	4.272		
3,6-Oxaoctanedioic acid (μ = 1.0)	3.055	3.676		
Oxoacetic acid	3.46			
2-Oxabutanedioic acid (oxaloacetic acid)	2.56	4.37		
2-Oxobutanoic acid	2.50			
5-Oxohexanoic acid (5-ketohexanoic acid) (18°C)	4.662			
3-Oxo-1,5-pentanedioic acid	3.10			
4-Oxopentanoic acid (levulinic acid)	4.59			
2-Oxopropanoic acid (pyruvic acid)	2.49			
Oxytetracycline	3.10(+1)	7.26	9.11	
Papaverine	5.90(+1)			

TABLE 2.59 *pK*, Values of Organic Materials in Water at 25°C (*Continued*)

Substance	pK_1	pK_2	pK_3	pK_4
Pentamethylenebis(thioacetic acid) (18°C)	3.485	4.413		
3,3-Pentamethylenepentanedioic acid	3.49	6.96		
1,5-Pentanediamine	10.05(+2)	10.916(+1)		
2,4-Pentanedione	8.24(enol); 8.95(keto)			
1-Pentanoic acid (valeric acid)	4.842			
2-Pentenoic acid	4.70			
3-Pentenoic acid	4.52			
4-Pentenoic acid	4.677			
Pentylarsonic acid	4.14	9.07		
N-Pentylveratramine	7.28(+1)			
Perhydrodiphenic acid (20°C)	4.96	6.68		
Perlolidine (18°C)	4.01	11.39		
Peroxyacetic acid	8.20			
1,7-Phenanthroline	4.30(+1)			
1,10-Phenanthroline	4.857(+1)			
6,7-Phenanthroline	4.857(+1)			
Phenazine	1.2(+1)			
Phenethylthioacetic acid	3.795			
Phenol	9.99			
Phenol-3-phosphoric acid	1.78	7.03	10.2	
Phenol-4-phosphoric acid	1.99	7.25	9.9	
Phenolphthalein	9.4			
3-Phenolsulfonic acid	——	9.05(−1)		
Phenosulsulfonephthalein	7.9			
Phenoxyactic acid	3.171			
2-Phenoxybenzoic acid	3.53			
3-Phenoxybenzoic acid	3.95			
4-Phenoxybenzoic acid	4.52			
5-Phenoxy-1,2,3,4-tetrazole	3.49(+1)			
Phenylacetic acid	4.312			
L-3-Phenyl-α-alanine	1.83(+1)	9.12(0)		
3-Phenyl-α-alanine, methyl ester	7.05(+1)			
Phenylalanylarginine (μ = 0.01)	2.66(+1)	7.57(0)	12.40(−1)	
Phenylalanylglycine (μ = 0.01)	3.10(+1)	7.71(0)		
7-Phenylazo-8-hydroxy-5-quinolinesulfonic acid	3.41(0)	7.850(−1)		
5-Phenylbarbituric acid	2.544(+1)			
2-Phenyl-2-benzylsuccinic acid	3.69	6.47		
1-Phenylbiguanide	2.13(+2)	10.76(+1)		
4-Phenylbutanoic acid	4.757			
Phenylbutazone	4.5(+1)			
2-Phenylenediamine	<2(+2)	4.47(+1)		
3-Phenylenediamine	2.65(+2)	4.88(+1)		
4-Phenylenediamine	3.29(+2)	6.08(+1)		
2-Phenylethylamine	9.83(+1)			
β-Phenylethylboronic acid	10.0			
DL-α-Phenylglycine	1.83(+1)	4.39(0)		
Phenylguanidine	10.77(+1)			

(*Continued*)

TABLE 2.59 *pK*, Values of Organic Materials in Water at 25°C (*Continued*)

Substance	pK_1	pK_2	pK_3	pK_4
Phenylhydrazine	5.20(+1)			
2-Phenyl-3-hydroxypropanoic acid	3.53			
3-Phenyl-3-hydroxypropanoic acid	4.40			
Phenyliminodiacetic acid (20°C)	2.40	4.98		
Phenylmalonic acid	2.58	5.03		
Phenylmethanethiol	10.70			
2-Phenyl-2-phenethylsuccinic acid (20°C)	3.74	6.52		
2-Phenylphenol	9.55			
3-Phenylphenol	9.63			
4-Phenylphenol	9.55			
Phenylphosphinic acid (17°C)	2.1			
Phenylphosphonic acid	1.83	7.07		
O-Phenylphosphorylserine	2.13(+1)	8.79		
O-Phenylphosphorylserylglycine	3.18(+1)	6.95(0)		
O-Phenylphosphoryl-L-seryl-L-leucine	3.16(+1)	7.12(0)		
N-Phenylpiperazine (μ = 0.1)	8.71(+1)			
2-Phenylpropanoic acid	4.38			
3-Phenylpropanoic acid (35°C)	4.664			
3-Phenyl-1-propylamine	10.39(+1)			
Phenylpropynoic acid (35°C)	2.269			
Phenylselenic acid	4.79			
Phenylselenoacetic acid (μ = 0.1)	3.75			
β-Phenylserine (μ = 0.16)	8.79(0)			
Phenylsuccinic acid (20°C)	3.78	5.55		
Phenylsulfenylacetic acid	2.66			
Phenylsulfonylacetic acid	2.44			
5-Phenyl-1,2,3,4-tetrazole	4.38(+1)			
1-Phenyl-1,2,3-triazole-4-carboxylic acid	2.88			
1-Phenyl-1,2,3-triazole-4,5-dicarboxylic acid	2.13	4.93		
Phosphoramidic acid	3.08	8.63		
O-Phosphorylethanolamine	5.838(+1)	10.638(0)		
O-Phosphorylserylglycine	3.13	5.41	8.01	
O-Phosphoryl-L-seryl-L-leucine	3.11	5.47	8.26	
Phosphoserine	2.08	5.65	9.74	
Phthalamide	3.79(0)			
Phthalazine	3.47(+1)			
o-Phthalic acid	2.950	5.408		
Phthalimide	9.90(0)			
Physostigmine	1.76(+1)	7.88(0)		
Picric acid (2,4,6-trinitrophenol) (18°C)	0.419			
Pilocarpine	1.3(+1)	6.85(0)		
Piperazine	5.333(+2)	9.781(+1)		
1,4-Piperazinebis(ethanesulfonic acid) (20°C)	6.80			
Piperazine-2-carboxylic acid	1.5	5.41	9.53	
Piperdine	11.123(+1)			
2-Piperidinecarboxylic acid	2.12(+1)	10.75(0)		
3-Piperidinecarboxylic acid	3.35(+1)	10.64(0)		

TABLE 2.59 *pK*, Values of Organic Materials in Water at 25°C (*Continued*)

Substance	pK_1	pK_2	pK_3	pK_4
4-Piperidinecarboxylic acid	3.73(+1)	10.72(0)		
1-(2-Piperidinyl)-2-propanone (15°C)	9.45			
Piperine (15°C)	1.98(+1)			
Proline	1.99(+1)	10.96(0)		
1,2-Propanediamine	6.607(+2)	9.702(+1)		
1,3-Propanediamine	8.49(+2)	10.47(+1)		
1-Propanethiol	10.86			
1,2,3-Propanetriamine	3.72(+3)	7.95(+2)	9.59(+1)	
1,2,3-Propanetricarboxylic acid	3.67	4.87	6.38	
Propanoic acid	4.874			
Propenoic acid	4.247			
N-Propionyglycine	3.718(0)			
2-Propoxybenzoic acid (20°C)	4.24			
3-Propoxybenzoic acid (20°C)	4.20			
4-Propoxybenzoic acid (20°C)	4.78			
N-Propylalanine	2.21(+1)	10.19(0)		
Propylamine	10.568(+1)			
Propylarsonic acid (18°C)	4.21	9.09		
Propylenimine	8.18(+1)			
N-Propylglycine (μ = 0.1)	2.38(+1)	10.03(0)		
L-Propylglycine	3.19(+1)	8.97(0)		
Propylmalonic acid	2.97	5.84		
Propylphosphinic acid	3.46			
Propylphosphonic acid	2.49	8.18		
2-Propylpyridine	6.30(+1)			
N-Propylveratramine	7.20(+1)			
2-Propynoic acid	1.887			
Pseudoecgonine	9.70			
Pseudoisocyanine (μ = 0.2)	4.59(+2)			
Pseudotropine	9.86(+1)			
Pteroylglutamic acid	8.26			
Purine	2.52(+1)	8.92(0)		
Pyrazine	0.6(+1)			
Pyrazinecarboxamide	0.5(+1)			
Pyrazole	2.61(+1)			
Pyridazine	2.33(+1)			
Pyridine	5.17(+1)			
Pyridine-d_5	5.83(+1)			
2-Pyridinealdoxime	3.56(+1)	10.17(0)		
3-Pyridinealdoxime	4.07(+1)	10.39(0)		
4-Pyridinealdoxime	4.73(+1)	10.03(0)		
2-Pyridinecarbaldehyde	3.84(+1)			
3-Pyridinecarbaldehyde	3.80(+1)			
4-Pyridinecarbaldehyde	4.74(+1)			
3-Pyridinecarbamide (nicotinamide)	3.33(+1)			
3-Pyridinecarbonitrile	1.35(+1)			
Pyridine-2-carboxylic acid (picolinic acid)	1.01(+1)	5.29(0)		
Pyridine-3-carboxylic acid (nicotinic acid)	2.07(+1)	4.75(0)		

(*Continued*)

TABLE 2.59 *pK*, Values of Organic Materials in Water at 25°C (*Continued*)

Substance	pK_1	pK_2	pK_3	pK_4
Pyridine-4-carboxylic acid (isonicotinic acid)	1.84(+1)	4.86(0)		
Pyridine-2,3-dicarboxylic acid	2.36(+1)	7.08(0)		
Pyridine-2,4-dicarboxylic acid	2.23(+1)	7.02(0)		
Pyridine-2,6-dicarboxylic acid	2.16(+1)	6.92(0)		
Pyridine-1-oxide	0.688(+1)			
Pyridoxal	4.20(+1)	8.66(ring OH)		
Pyridoxal-5-phosphate (μ = 0.15)	<2.5	4.14	6.20	8.69
Pyridoxamine (μ = 0.1)	3.37(+2)	8.01(+1)	10.13(ring OH)	
Pyridoxamine-5-phosphate (μ = 0.15; pK_5 10.92)	2.5	3.69	5.76	8.61
Pyridoxine (vitamin B_6) (18°C)	5.00(+1)	8.96(ring OH)		
3-(2′-Pyridyl)alanine	1.37(+2)	4.02(+1)	9.22(0)	
3-(3′-Pyridyl)alanine	1.77(+2)	4.64(+1)	9.10(0)	
2-(2′-Pyridyl)benzimidazole (μ = 0.16)	5.58(+1)			
2-(2′-Pyridyl)imidazole (μ = 0.005)	8.98(+1)			
4-(2′-Pyridyl)imidazole (μ = 0.1)	5.49(+1)			
Pyrimidine	1.30(+1)			
2,4(1*H*,3*H*)-Pyrimidinedione (uracil)	0.6(+1)	9.46(0)		
2,4,5,6(1*H*,3*H*)-Pyrimidinetetrone-5-oxime	4.57(0)			
Pyrocatecholsulfonephthaleine	7.82	9.76	11.73	
Pyroxilidine	11.11(+1)			
Pyrrole-1-carboxylic acid	4.45			
Pyrrole-2-carboxylic acid	4.45			
Pyrrole-3-carboxylic acid	4.453			
Pyrrolidine	11.305(+1)			
Pyrrolidine-2-carboxylic acid (proline)	1.952(+1)	10.640(0)		
2-[2-(*N*-Pyrrolidinyl)ethyl]pyridine	3.60(+2)	9.39(+1)		
3-[2-(*N*-Pyrrolidinyl)ethyl]pyridine	4.28(+2)	9.28(+1)		
4-[2-(*N*-Pyrrolidinyl)ethyl]pyridine	4.65(+2)	9.27(+1)		
2-(1-Pyrrolidinylmethyl)pyridine	2.54(+1)	8.56(+1)		
3-(1-Pyrrolidinylmethyl)pyridine	3.14(+2)	8.36(+1)		
4-(1-Pyrrolidinylmethyl)pyridine	3.38(+2)	8.16(+1)		
3-Pyrroline	−0.27(+1)			
Quinidine	4.0(+1)	8.54(0)		
Quinine	4.11(+1)	8.52(0)		
Quinoline	4.80(+1)			
Quinoxaline	0.72(+1)			
D-Raffinose	12.74			
Riboflavin (vitamin B_2) (μ = 0.01)	ca −0.2	9.69		
α-D-Ribofuranose	12.11			
D-Ribose-5′-phosphonic acid	—	6.70(−1)	13.05(−2)	

TABLE 2.59 *pK*, Values of Organic Materials in Water at 25°C (*Continued*)

Substance	pK_1	pK_2	pK_3	pK_4
D-Saccharic acid	5.00(0)			
Saccharin (*o*-benzoic sulfimide)	2.32			
Sarcosine	2.12(+1)	10.20(0)		
Sarcosine amide	8.35(+1)			
Sarcosine dimethylamide	8.86(+1)			
Sarcosine methylamide	8.28(+1)			
Sarcosylglycine (μ = 0.16)	3.15(+1)	8.56(0)		
Sarcosylleucine	3.15(+1)	8.67(0)		
Sarcosylsarcosine	2.92(+1)	9.15(0)		
Sarcosylserine	3.17(+1)	8.63(0)		
3-Selenosemicarbazide (μ = 0.1)	0.8(+1)			
Semicarbazide (μ = 0.1)	3.53(+1)			
L-Serine	2.21(+1)	9.15(0)	13.6	
Serine, methyl ester (μ = 0.1)	7.03(+1)			
Serylglycine (μ = 0.15)	2.10(+1)	7.33(0)		
L-Seryl-L-leucine	3.08(+1)	7.45(0)		
Solanine	7.34(+1)			
D-Sorbitol (17.5°C)	13.60			
L-(−)-Sorbose (18°C)	11.55			
Sparteine	4.49(+1)	11.76(0)		
Spinaceamine (μ = 0.1)	4.895(+2)	8.90(+1)		
Spinacine	1.649(+2)	4.936(+1)	8.663(0)	
L-Strychnine (15°C)	2.50	8.20		
Succinamic acid (succinic acid monoamide)	4.39(0)			
Succinic acid	4.207	5.635		
DL-Succinimide	9.623			
β-(4′-Sulfaminophenyl)alanine	1.99(+1)	8.64(0)	10.26(−1)	
3-Sulfamylbenzoic acid	3.54			
4-Sulfamylbenzoic acid	3.47			
4-Sulfamylphenylphosphoric acid	1.42	6.38	10.0	
Sulfanilamide	10.43(+1)			
Sulfoacetic acid	—	4.0		
3-Sulfobenzoic acid	—	3.78		
4-Sulfobenzoic acid	—	3.72		
3-Sulfophenol	0.39	9.07		
4-Sulfophenol	0.58	8.70		
2-Sulfopropanoic acid	1.99			
5-Sulfosalicyclic acid	2.49	12.00		
Sylvic acid	7.62			
D-Tartaric acid	3.036	4.366		
meso-Tartaric acid	3.22	4.81		
Tetracycline (μ = 0.005)	3.30(+1)	7.68	9.69	
Tetradehydroyohimbine	10.59(+1)			
Tetraethylenepentamine [μ = 0.1; pK_5 9.67(+1)]	2.98(+5)	4.72(+4)	8.08(+3)	9.10(+2)
1,4,5,6-Tetrahydro-1,2-dimethylpyridine	11.38(+1)			
1,4,5,6-Tetrahydro-2-methylpyridine	9.53(+1)			
cis-Tetrahydronaphthalene-2,3-dicarboxylic acid (20°C)	3.98	6.47		

(*Continued*)

TABLE 2.59 *pK*, Values of Organic Materials in Water at 25°C (*Continued*)

Substance	pK_1	pK_2	pK_3	pK_4
trans-Tetrahydronaphthalene-2,3-dicarboxylic acid (20°C)	4.00	5.70		
5,6,7,8-Tetrahydro-1-naphthol	10.28			
5,6,7,8-Tetrahydro-2-naphthol	10.48			
Tetrahydroserpentine	10.55(+1)			
2,3,5,6-Tetramethylbenzoic acid	3.415			
Tetramethylenebis(thioacetic acid) (18°C)	3.463	4.423		
Tetramethylenediamine	9.22(+2)	10.75(+1)		
N,N,N',N'-Tetramethylethylenediamine	2.20(+2)	6.35(+1)		
2,3,5,6-Tetramethyl-4-methylaminopyridine	0.07(+1)			
2,2,6,6-Tetramethylpiperidine (μ = 0.5)	1.24(+1)			
2,3,5,6-Tetramethylpyridine (20°C)	7.90(+1)			
Tetramethylsuccinic acid	3.50	7.28		
1,2,3,4-Tetrazole	4.90			
Thebaine	7.95(+1)			
2-Thenoyltrifluoroacetone	5.70(0)			
Theobromine	0.68(+1)	7.89		
Theophylline	<1(+1)	8.80		
Thiazoline	2.53(+1)			
Thioacetic acid	3.33			
o-Thiocresol	6.64			
m-Thiocresol	6.58			
p-Thiocresol	6.52			
Thiocyanatoacetic acid	2.58			
2,2'-Thiodiacetic acid	3.32	4.29		
4,4'-Thiodibutanoic acid (18°C)	4.351	5.275		
3,3'-Thiodipropanoic acid (18°C)	4.085	5.075		
3-Thio-*S*-methylcarbazide (μ = 0.1)	7.563(+1)			
1-Thionylcarboxylic acid	3.53			
2-Thionylcarboxylic acid	4.10			
2-Thiophenecarboxylic acid (30°C)	3.529			
3-Thiophenecarboxylic acid (3-thenoic acid)	4.10			
Thiophenol	6.50			
3-Thiosemicarbazide (μ = 0.1)	1.5(+1)			
3-Thiosemicarbazide-1,1-diacetic acid (30°C)	2.94	4.07		
Thiourea	2.03(+1)			
Thorin	3.7	8.3	11.8	
Thymidine	9.79	12.85		
p-Toluenesulfinic acid	1.7			
Toluhydroquinone	10.03	11.62		
o-Toluidine	4.45(+1)			
m-Toluidine	4.71(+1)			
p-Toluidine	5.08(+1)			
o-Tolylacetic acid (18°C)	4.36			
p-Tolylacetic acid (18°C)	4.36			
o-Tolylarsonic acid	3.82	8.85		

TABLE 2.59 *pK*, Values of Organic Materials in Water at 25°C (*Continued*)

Substance	pK_1	pK_2	pK_3	pK_4
m-Tolylarsonic acid	3.82	8.60		
p-Tolylarsonic acid	3.70	8.68		
o-Tolylphosphonic acid	2.10	7.68		
m-Tolylphosphonic acid	1.88	7.44		
p-Tolylphosphonic acid	1.84	7.33		
3-Tolylselenic acid	4.80			
4-Tolylselenic acid	4.88			
Triacetylmethane	5.81			
Triallylamine	8.31(+1)			
1,3,5-Triazine-2,4,6-triol	7.20	11.10		
1*H*-1,2,3-Triazole	——	9.26		
1*H*-1,2,4-Triazole	2.386(+1)	9.972		
1,2,3-Triazole-4-carboxylic acid	3.22	8.73		
1,2,3-Triazole-4,5-dicarboxylic acid	1.86	5.90	9.30	
1,2,4-Triazolidine-3,5-dione (urazole)	5.80			
Tribomoacetic acid	−0.147			
2,4,6-Tribromobenzoic acid	1.41			
Trichloroacetic acid	0.52			
Trichloroacrylic acid	1.15			
3,3,3-Trichlorolactic acid	2.34			
Trichloromethylphosphonic acid	1.63	4.81		
2,4,5-Trichlorophenol	7.37			
3,4,5-Trichlorophenol	7.839			
Tricine (20°C)	8.15			
Triethanolamine	7.76(+1)			
Triethylamine	10.72(+1)			
Triethylenediamine	4.18(+2)	8.19(+1)		
Triethylenetetramine (20°C)	3.32(+4)	6.67(+3)	9.20(+2)	9.92(+1)
Triethylsuccinic acid	2.74			
Trifluoroacetic acid	0.50			
Trifluoroacrylic acid	1.79			
4,4,4-Trifluoro-2-aminobutanoic acid	1.600(+1)	8.169(0)		
4,4,4-Trifluoro-3-aminobutanoic acid	2.756(+1)	5.822(0)		
4,4,4-Trifluorobutanoic acid	4.16			
α,α,α-Trifluoro-*m*-cresol	8.950			
4,4,4-Trifluorocrotonic acid	3.15			
5,5,5-Trifluoroleucine	2.045(+1)	8.942(0)		
3-(Trifluoromethyl)aniline	3.5(+1)			
4-(Trifluoromethyl)aniline	2.6(+1)			
3-Trifluoromethylphenol	8.950			
5-Trifluoromethyl-1,2,3,4-tetrazole	1.70			
6,6,6-Trifluoronorleucine	2.164(+1)	9.463(0)		
5,5,5-Trifluoronorvaline	2.042(+1)	8.916(0)		
5,5,5-Trifluoropentanoic acid	4.50			
3,3,3-Trifluoropropanoic acid	3.06			
4,4,4-Trifluorothreonine	1.554(+1)	7.822(0)		
4,4,4-Trifluorovaline	1.537(+1)	8.098(0)		

(*Continued*)

TABLE 2.59 *pK*, Values of Organic Materials in Water at 25°C (*Continued*)

Substance	pK_1	pK_2	pK_3	pK_4
1,2,3-Trihydroxybenzene (pyrogallol)	9.03(0)	11.63(−1)		
1,3,5-Trihydroxybenzene (phloroglucinol)	8.45(0)	8.88(−1)		
2,4,6-Trihydroxybenzoic acid	1.68(0)			
3,4,5-Trihydroxybenzoic acid	4.19(0)	8.85(−1)		
3,4,5-Trihydroxycyclohex-1-ene-1-carboxylic acid [D-(−)-shikimic acid]	4.15			
2,4,6-Tri(hydroxymethyl)phenol	9.56			
Triisobutylamine	10.42(+1)			
Trimethylamine	9.80(+1)			
3-(Trimethylamino)phenol	8.06			
4-(Trimethylamino)phenol	8.35			
2,4,6-Trimethylaniline	4.38(+1)			
2,4,6-Trimethylbenzoic acid	3.448			
Trimethylenebis(thioacetic acid) (18°C)	3.435	5.383		
2,3,4-Trimethylphenol	10.59			
2,4,5-Trimethylphenol	10.57			
2,4,6-Trimethylphenol	10.88			
3,4,5-Trimethylphenol	10.25			
2,3,6-Trimethylpyridine (μ = 0.5)	7.60(+1)			
2,4,6-Trimethylpyridine	7.43(+1)			
2,4,6-Trimethylpyridine-1-oxide	1.990(+1)			
3-(Trimethylsilyl)benzoic acid	4.089			
4-(Trimethylsilyl)benzoic acid	4.192			
2,4,5-Trimethylthiazole (μ = 0.1)	4.55			
2,4,6-Trinitroaniline (picramide)	−10.23(+1)			
2,4,6-Trinitrobenzene acid	0.654			
2,2,2-Trinitroethanol	2.36			
Trinitromethane (20°C)	0.17			
Triphenylacetic acid	3.96			
Tripropylamine	10.66(+1)			
Tris(2-hydroxyethyl)amine	7.762(+1)			
Tri(hydroxymethyl)aminomethane (TRIS)	8.08(+1)			
2-[Tris(hydroxymethyl)methyl amino]-1-ethanesulfonic acid (TES)	7.50			
3-[Tris(hydroxymethyl)methyl amino]-1-propanesulfonic acid (TAPS) (20°C)	8.4			
N-[Tris(hydroxymethyl)methyl]-glycine (tricine)	2.023(+1)	8.135		
Tris(trimethylsilyl)amine	4.70(+1)			
Trithiocarbonic acid (20°C)	2.64			
Tropacocaine (15°C)	9.88(+1)			
3-Tropanol (tropine)	10.33(+1)			
Trypsin (μ = 0.1)	6.25			
L-Tryptophan	2.38(+1)	9.39(0)		
DL-Tyrosine	2.18(+1)	9.11(0)	10.6(OH)	

TABLE 2.59 *pK*, Values of Organic Materials in Water at 25°C (*Continued*)

Substance	pK_1	pK_2	pK_3	pK_4
Tyrosine amide	7.48	9.89		
Tyrosine, ethyl ester	7.33	9.80		
Tyrosylarginine (μ = 0.01)	2.65(+1)	7.39(0)	9.36(−1)	11.62(−2)
Tyrosyltyrosine	3.52(+1)	7.68(0)	9.80(−1)	10.26(−2)
α-Ureidobutanoic acid	3.886(0)			
γ-Ureidobutanoic acid	4.683(0)			
β-Ureidopropanoic acid	4.487(0)			
Uric acid	5.40	5.53		
Uridine	9.30			
Uridine-5′-diphosphoric acid	7.16			
Uridine-5′-phosphoric acid (5′-uridylic acid)	6.63			
Uridine-5′-triphosphoric acid	7.58			
DL-Valine	2.32(+1)	9.61(0)		
L-Valine	2.296(+1)	9.79(0)		
Valine amide (μ = 0.2)	8.00			
L-Valine, methyl ester	7.49(+1)			
L-Valylglycine	3.23(+1)	8.00(0)		
Vetramine	7.49(+1)			
Veratrine	8.85(+1)			
Vinylmethylamine	9.69(+1)			
2-Vinylpyridine	4.98(+1)			
4-Vinylpyridine	5.62(+1)			
Vitamin B_{12}	7.64(+1)			
Xanthine (40°C)	0.68(+1)			
Xanthosine	<2.5(+1)	5.67(0)	12.00(−1)	
Xylenol Orange [pK_5 10.46(−4); pK_6 12.28(−5)]	—	2.58(−1)	3.23(−2)	6.37(−3)
D-(+)-Xylose	12.15(0)			
Zincon	—	4	7.85	15

TABLE 2.60 Selected Equilibrium Constants in Aqueous Solution at Various Temperatures

Abbreviations Used in the Table

(+1), protonated cation	*(−2), doubly ionized anion*
(0), neutral molecule	*pK_{auto}, negative logarithm (base 10) of autoprotolysis constant*
(−1), singly ionized anion	*pK_{sp}, negative logarithm (base 10) of solubility product*

Substance	Temperature, °C									
	0	5	10	15	20	25	30	35	40	50
Acetic acid (0)	4.780	4.770	4.762	4.758	4.757	4.756	4.757	4.762	4.769	4.787
DL-*N*-Acetylalanine (+1)		3.699	3.699	3.703	3.708	3.715	3.725	3.733	3.745	3.774
β-Acetylaminopropionic (+1)		4.479	4.465	4.465	4.449	4.445	4.444	4.443	4.445	4.457
N-Acetylglycine (+1)		3.682	3.676	3.673	3.667	3.670	3.673	3.678	3.685	3.706
α-Alanine										
(+1)	2.42		2.39		2.35	2.34	2.33	2.33	2.33	2.33
(0)	10.59		10.29		10.01	9.87	9.74	9.62	9.49	9.26
2-Aminobenzenesulfonic acid (0), pK_2	2.633	2.591	2.556	2.521	2.448	2.459	2.431	2.404	2.380	2.338
3-Aminobenzenesulfonic acid (0), pK_2	4.075	4.002	3.932	3.865	3.799	3.738	3.679	3.622	3.567	3.464
4-Aminobenzenesulfonic acid (0), pK_2	3.521	3.457	3.398	3.338	3.283	3.227	3.176	3.126	3.079	2.989
3-Aminobenzoic acid (0)					4.90	4.79	4.75		4.68	4.60
4-Aminobenzoic acid (0)					4.95	4.85	4.90		4.95	5.10
2-Aminobutyric acid										
(+1)			2.334			2.286		2.289$^{37.5°C}$		2.297
(0)			10.530			9.380		9.518$^{37.5°C}$		9.234
4-Aminobutyric acid										
(+1)			4.057	4.046	4.038	4.031	4.027	4.025	4.027	4.032
(0)			11.026	10.867	10.706	10.556	10.409	10.269	10.114	9.874
2-Aminoethylsulfonic acid (0)			9.452	9.316	9.186	9.061	8.940	8.824	8.712	9.499
2-Amino-3-methylpentanoic acid										
(+1)	2.365$^{1°C}$		2.338$^{12.5°C}$			2.320		2.317$^{37.5°C}$		2.332
(0)	10.460$^{1°C}$		10.100$^{12.5°C}$			9.758		9.439$^{37.5°C}$		9.157

2-Amino-2-methyl-1,3-propanediol	9.612	9.433	9.266	9.104	8.951	8.801	8.659	8.519	8.385	8.132
2-Amino-2-methylpropionic acid										
(+1)	2.419$^{1°C}$		2.380$^{12.5°C}$			2.357		2.351$^{37.5°C}$		2.356
(0)	10.960$^{1°C}$		10.580$^{12.5°C}$			10.205		9.872$^{37.5°C}$		9.561
2-Aminopentanoic acid										
(+1)	2.376$^{1°C}$		2.347			2.318			2.309	2.313
(0)	10.508$^{1°C}$			10.154$^{12.5°C}$		9.808		9.490$^{37.5°C}$		9.198
3-Aminopropionic acid										
(+1)	3.656	3.627		3.583		3.551		3.524	3.517	
(0)	11.000	10.830		10.526		10.235		9.963	9.842	
4-Aminopyridine (+1)	9.873	9.704	9.549	9.398	9.252	9.114	8.978	8.846	8.717	8.477
Ammonium ion (+1)	10.081	9.904	9.731	9.564	9.400	9.245	9.093	8.947	8.805	8.539
Arginine										
(+1)	1.914	1.885	1.870	1.849	1.837	1.823	1.814	1.801	1.800	1.787
(0)	9.718	9.563	9.407	9.270	9.123	8.994	8.859	8.739	8.614	8.385
Barbituric acid										
(+1)				3.969	3.980	4.02	4.00	4.008	4.017	4.032
(0)				8.493	8.435	8.372	8.302	8.227	8.147	7.974
Benzoic acid (0)		4.231	4.220	4.215	4.206	4.204	4.203	4.207	4.219	4.223
Boric acid (0)	9.508	9.439	9.380	9.327	9.280	9.236	9.197	9.161	9.132	9.080
Bromoacetic acid (0)				2.875	2.887	2.902	2.918	2.936		
3-Bromobenzoic acid (0)				3.818	3.813	3.810	3.808	3.810	3.813	
4-Bromobenzoic acid (0)				4.011	4.005	3.99	4.001	4.001	4.003	
Bromopropynoic acid (0)			1.786	1.814	1.839	1.855	1.879	1.900	1.919	
3-*tert*-Butylbenzoic acid (0)				4.266	4.231	4.199	4.170	4.143	4.119	
4-*tert*-Butylbenzoic acid (0)				4.463	4.425	4.389	4.354	4.320	4.287	
2-Butynoic acid (0)			2.618	2.626	2.611	2.620	2.618	2.621	2.631	
Butyric acid (0)	4.806	4.804	4.803	4.805	4.810	4.817	4.827	4.840	4.854	4.885
DL-*N*-Carbamoylalanine (+1)		3.898	3.894	3.891	3.890	3.892	3.896	3.902	3.908	3.931
N-Carbamoylglycine (+1)		3.911	3.900	3.889	3.879	3.876	3.874	3.873	3.875	3.888
Carbon dioxide + water										
(0)	6.577	6.517	6.465	6.429	6.382	6.352	6.327	6.309	6.296	6.285
(−1)	10.627	10.558	10.499	10.431	10.377	10.329	10.290	10.250	10.220	10.172
Chloroacetic acid (0)				2.845	2.856	2.867	2.883	2.900		
3-Chlorobenzoic acid (0)				3.838	3.831	3.83	3.825	3.826	3.829	

(Continued)

TABLE 2.60 Selected Equilibrium Constants in Aqueous Solution at Various Temperatures (*Continued*)

Substance	Temperature, °C									
	0	5	10	15	20	25	30	35	40	50
4-Chlorobenzoic acid (0)				4.000	3.991	3.986	3.981	3.980	3.981	
Chloropropynoic acid (0)			1.766	1.796	1.820	1.845	1.864	1.879	1.893	
Citric acid										
(0)	3.220	3.200	3.176	3.160	3.142	3.128	3.116	3.109	3.099	3.095
(−1)	4.837	4.813	4.797	4.782	4.769	4.761	4.755	4.751	4.750	4.757
(−2)	6.393	6.386	6.383	6.384	6.388	6.396	6.406	6.423	6.439	6.484
Cyanoacetic acid (0)		2.445	2.447	2.452	2.460	2.460	2.482	2.496	2.511	
2-Cyano-2-methylpropionic acid										
(0)		2.342	2.360	2.379	2.400	2.422	2.446	2.471	2.498	
5,5-Diethylbarbituric acid (0)	8.40	8.30	8.22	8.169	8.094	8.020	7.948	7.877	7.808	7.673
Diethylmalonic acid										
(0)			2.129	2.136	2.144	2.151	2.160	2.172	2.187	
(−1)			7.400	7.401	7.408	7.417	7.428	7.441	7.457	
2,3-Dimethylbenzoic acid (0)				3.663	3.687	3.771	3.726	3.762	3.788	
2,4-Dimethylbenzoic acid (0)				4.154	4.187	4.217	4.244	4.268	4.290	
2,5-Dimethylbenzoic acid (0)				3.911	3.954	3.990	4.020	4.045	4.065	
2,6-Dimethylbenzoic acid (0)				3.234	3.304	3.362	3.409	3.445	3.472	
3,5-Dimethylbenzoic acid (0)				4.292	4.299	4.302	4.304	4.306	4.306	
N,*N*′-Dimethylethyleneamine-*N*,*N*′-diacetic acid										
(0)	6.294		6.169		6.047		5.926		5.803	
(−1)	10.446		10.268		10.068		9.882		9.684	
N,*N*-Dimethylglycine (0)		10.34		10.14		9.94		9.76		
3,5-Dinitrobenzoic acid (0)			2.60		2.73		2.85		2.96	3.07
2-Ethylbutyric acid (0)	4.623		4.664		4.710	4.751	4.758		4.812	4.869
5-Ethyl-5-phenylbarbituric acid (0)				7.592	7.517	7.445	7.377	7.311	7.248	7.130
Fluoroacetic acid (0)				2.555	2.571	2.586	2.604	2.624		
Formic acid (0)	3.786	3.772	3.762	3.757	3.753	3.751	3.752	3.758	3.766	3.782
2-Furancarboxylic acid (0)						3.164	3.200	3.216	3.239	
Glucose-1-phosphate (0)		6.506	6.500	6.499	6.500	6.504	6.510	6.519	6.531	6.561
Glycerol-1-phosphoric acid (−1)		6.642	6.641	6.643	6.648	6.656	6.666	6.679	6.695	6.733
Glycerol-2-phosphoric acid										
(0)		1.223	1.245	1.271	1.301	1.335	1.372	1.413	1.457	1.554

(−1)		6.657	6.650	6.646	6.646	6.650	6.657	6.666	6.679	6.712
Glycine										
(+1)			2.397	2.380	2.36	2.351	2.34	2.33	2.327	2.32
(0)		10.34	10.193	10.044	9.91	9.780	9.65	9.53	9.412	9.19
Glycolic acid (0)	3.875		$3.844^{12.5°C}$			3.831		$3.833^{37.5°C}$		3.849
Glycylasparagine (+1)		2.968	2.958	2.952	2.943	2.942	2.942	2.944	2.947	2.959
N-Glycylglycine (+1)	3.201					3.126				3.159
			$8.594^{12.5°C}$			8.252		$7.948^{37.5°C}$		7.668
Hexanoic acid (0)	4.840		4.839		4.849		4.865		4.890	4.920
Hydrogen cyanide (0)			9.63	9.49	9.36	9.21	9.11	8.99	8.88	
Hydrogen peroxide (0)	12.23			11.86	11.75	11.65	11.55	11.45		11.21
Hydrogen sulfide										
(0)		7.33	7.24	7.13	7.05	6.97	6.90	6.82	6.79	6.69
(−1)		13.5		13.2		12.90	12.75	12.6		
4-Hydroxybenzoic acid (0)				4.596	4.586	4.582	4.577	4.576	4.578	
Hydroxylamine (0)				6.186	6.063	5.948		5.730		
2-Hydroxy-1-naphthoic acid										
(0)					3.29		3.24		3.19	3.26
(−1)					9.68		9.65		9.61	9.58
4-Hydroxyproline										
(+1)	$1.900^{1°C}$		$1.850^{12.5°C}$			1.818		$1.798^{37.5°C}$		1.796
(0)	$10.274^{1°C}$		$9.958^{12.5°C}$			9.662		$9.394^{37.5°C}$		9.138
2-Hydroxypropionic acid (0)	3.880	3.873	3.868	3.861	3.857	3.858	3.861	3.867	3.873	3.895
DL-2-Hydroxysuccinic acid										
(0)	3.537	3.520	3.494	3.482	3.472	3.458	3.452	3.446	3.444	3.445
(−1)	5.119	5.108	5.098	5.096	5.096	5.097	5.099	5.104	5.117	5.149
Hypobromous acid (0)				8.83		8.60		8.47	$8.37^{45°C}$	
Hypochlorous acid (0)	7.82	7.75	7.69	7.63	7.58	7.54	7.50	7.46		7.05
Imidazole (+1)	7.581	7.467	7.334	7.216	7.103	6.993	6.887	6.784	6.685	6.497
Iodoacetic acid (0)				3.143	3.158	3.175	3.193	3.213		
DL-Isoleucine										
(+1)	2.365		$2.338^{12.5°C}$			2.318		$2.317^{37.5°C}$		2.332
(0)	10.460		$10.100^{12.5°C}$			9.758		$9.439^{37.5°C}$		9.157
Isopropylmalonic acid, mononitrile (0)		2.299	2.320	2.343	2.365	2.401	2.427	2.452	2.481	
Lactic acid (0)	3.880	3.873	3.868	3.862	3.857	3.858	3.861	3.867	3.873	3.895
Lead sulfate, pK_{sp}	8.01			7.87		7.80		7.73		7.63
DL-Leucine										
(+1)	$2.383^{1°C}$		$2.348^{12.5°C}$			2.328		$2.327^{37.5°C}$		2.333
(0)	$10.458^{1°C}$		$10.095^{1.5°C}$			9.744		$9.434^{37.5°C}$		9.142

(*Continued*)

TABLE 2.60 Selected Equilibrium Constants in Aqueous Solution at Various Temperatures (*Continued*)

Substance	Temperature, °C									
	0	5	10	15	20	25	30	35	40	50
Malonic acid (−1)	5.670	5.665	5.667	5.673	5.683	5.696	5.710	5.730	5.753	5.803
Mannose (0)			12.45			12.08			11.81	
Mercury(I) chloride, pK_{sp}			18.65	18.48	18.27	17.88		16.79		
Methanol (solvent), pK_{auto}		17.12		16.84		16.71	16.65	16.53		
Methylamine (+1)	11.496		11.130		10.787	10.62	10.466		10.161	9.876
Methylaminodiacetic acid										
(0)	2.138		2.142		2.146		2.150		2.154	
(−1)	10.474		10.287		10.088		9.920		9.763	
3-Methylbenzoic acid (0)				4.303	4.285	4.269	4.256	4.244	4.235	
4-Methylbenzoic acid (0)				4.390	4.376	4.362	4.349	4.336	4.322	
3-Methylbutyric acid (0)	4.726		4.742		4.767		4.794		4.831	4.871
4-Methylpentanoic acid (0)	4.827		4.827		4.837		4.853		4.879	4.908
5-Methyl-5-phenylbarbituric acid										
(0)				8.104	8.057	8.011	7.966	7.922	7.879	7.797
2-Methylpropionic acid (0)	4.825		4.827		4.840	4.853	4.886		4.918	4.955
2-Methyl-2-propylamine (+1)		11.439	11.240	11.048	10.862	10.682	10.511	10.341		
Nitric acid (0)	−1.65					−1.38				−1.20
Nitrilotriacetic acid										
(0)	1.69		1.65		1.65		1.66		1.67	
(−1)	2.95		2.95		2.94		2.96		2.98	
(−2)	10.59		10.45		10.33		10.23			
4-Nitrobenzoic acid (0)				3.448	3.444	3.441	3.441	3.442	3.445	
Nitrous acid (0)				3.244	3.177	3.138		3.100		
DL-Norleucine										
(+1)	2.394		$2.356^{12.5°C}$			2.335		$2.324^{37.5°C}$		2.328
(0)	10.564		$10.190^{12.5°C}$			9.834		$9.513^{37.5°C}$		9.224
Oxalic acid (−1)	4.210	4.216	4.227	4.240	4.254	4.272	4.295	4.318	4.349	4.409
2,4-Pentanedione (0)	9.07					8.95			8.90	
Pentanoic acid (0)	4.823		4.763		4.835	4.842	4.851		4.861	4.906
Phenylalanine (0)			9.75			9.31			8.96	
Phosphoric acid (0)	2.056	2.073	2.088	2.107	2.127	2.148	2.171	2.196	2.224	2.277
(−1)	7.313	7.282	7.254	7.231	7.213	7.198	7.189	7.185	7.181	7.183

o-Phthalic acid										
(0)	2.925	2.927	2.931	2.937	2.943	2.950	2.958	2.967	2.978	3.001
(−1)	5.432	5.418	5.410	5.405	5.405	5.408	5.416	5.427	5.442	5.485
Piperidine (+1)	11.963	11.786	11.613	11.443	11.280	11.123	10.974	10.818	10.670	10.384
Proline										
(+1)	2.011		1.964$^{12.5°C}$			1.952		1.950$^{37.5°C}$		1.958
(0)	11.296		10.972$^{12.5°C}$			10.640		10.342$^{37.5°C}$		10.064
Propenoic acid (0)				4.267	4.250	4.247	4.249	4.267	4.301	
N-Propionylglycine (+1)		3.728	3.723	3.718	3.716	3.718	3.721	3.725	3.731	3.750
Propynoic acid (0)			1.791	1.829	1.867	1.887	1.940	1.932	1.963	
Pyrrolidine (+1)	12.17	11.98	11.81	11.63	11.43	11.30	11.15	10.99	10.84	11.56
Serine										
(+1)	2.296$^{1°C}$		2.232$^{12.5°C}$			2.186		2.154$^{37.5°C}$		2.132
(0)	9.880$^{1°C}$		9.542$^{12.5°C}$			9.208		8.904$^{37.5°C}$		8.628
Silver bromide, pK_{sp}		13.33		12.83	12.57	12.30	12.07	11.83	11.61	11.19
Silver chloride, pK_{sp}		10.595		10.152		9.749		9.381	9.21	8.88
Succinic acid										
(0)	4.285	4.263	4.245	4.232	4.218	4.207	4.198	4.191	4.188	4.186
(−1)	5.674	5.660	5.649	5.642	5.639	5.635	6.541	5.647	5.654	5.680
Sulfuric acid (−1)	1.778	1.812$^{4.3°C}$		1.894		1.987	2.05	2.095	2.17	2.246
Sulfurous acid (0)	1.63		1.74			1.89		1.98		2.12
D-Tartaric acid										
(0)	3.118	3.095	3.075	3.057	3.044	3.036	3.025	3.019	3.018	3.021
(−1)	4.426	4.407	4.391	4.381	4.372	4.366	4.365	4.367	4.372	4.391
2,3,5,6-Tetramethylbenzoic acid (0)				3.310	3.367	3.415	3.453	3.483	3.505	
Threonine										
(+1)	2.200$^{1°C}$		2.132$^{12.5°C}$			2.088		2.070$^{37.5°C}$		2.055
(0)	9.748$^{1°C}$		9.420$^{12.5°C}$			9.100		8.812$^{37.5°C}$		8.548
o-Toluidine (0)				4.58	4.495	4.45	4.345	4.28	4.20	
1,2,4-Triazole										
(+1)				2.451	2.418	2.386	2.327			
(0)				10.205	10.083	9.972	9.768			
3,4,5-Trihydroxybenzoic acid (0)					4.19		4.30		4.38	4.53
Tris(2-hydroxyethyl)amine (+1)	8.290	8.173	8.067	7.963	7.861	7.762	7.666	7.570	7.477	7.299
2,4,6-Trimethylbenzoic (0)				3.325	3.391	3.448	3.498	3.541	3.577	
3-Trimethylsilylbenzoic acid (0)				4.142	4.116	4.089	4.060	4.029	3.996	
4-Trimethylsilylbenzoic acid (0)				4.270	4.230	4.192	4.155	4.119	4.084	
β-Ureidopropionic acid (0)		4.514	4.505	4.497	4.490	4.487	4.486	4.486	4.488	4.500
DL-Valine										
(+1)	2.320		2.297$^{12.5°C}$			2.296		2.292$^{37.5°C}$		2.310
(0)	10.413		10.064$^{12.5°C}$			9.719		9.405$^{37.5°C}$		9.124

TABLE 2.61 *pK*, Values for Proton-Transfer Reactions in Non-aqueous Solvents

Acid	Methanol	Ethanol	Other Solvents
Acetic acid	9.52	10.32	11.4[a], 9.75[d]
p-Aminobenzoic acid	10.25		
Ammonium ion	10.7		6.40[b]
Anilinium ion	6.0	5.70	
Benzoic acid		10.72	10.0[a]
Bromocresol purple	11.3	11.5	
Bromocresol green	9.8	10.65	
Bromophenol blue	8.9	9.5	
Bromothymol blue	12.4	13.2	
Di-*n*-butylammonium ion			10.3[a]
o-Chloroanilinium ion	3.4		
Cyanoacetic acid		7.49	
2,5-Dichloroanilinium ion			9.48[b]
Dimethylaminoazobenzene		5.2	6.32[b]
N,N'-Dimethylanilinium ion		4.37	
Formic acid		9.15	
Hydrobromic acid			5.5[c]
Hydrochloric acid			8.55[b], 8.9[c]
Methyl orange	3.8	3.4	
Methyl red (acid range)	4.1	3.55	
(alkaline range)	9.2	10.45	
Methyl yellow	3.4	3.55	
Neutral red	8.2	8.2	
o-Nitrobenzoic acid	7.6		
m-Nitrobenzoic acid	8.3		
p-Nitrobenzoic acid	8.4		
Perchloric acid			4.87[b]
Phenol	14.0		
Phenol red	12.8	13.4	
Phthalic acid, pK_2	11.65		11.5[d], 6.10[d](pK_1)
Picric acid	3.8	3.8	8.9[c]
Pyridinium ion			6.1[b]
Salicylic acid	8.7	7.9	
Stearic acid	10.0		
Succinic acid, pK_2	11.4		
Sulfuric acid, pK_1			7.24[b,c]
Tartaric acid, pK_2	9.9		
Thymol blue (alkaline range)	14.0	15.2	
(acid range)	4.7	5.35	
Thymolbenzein (acid range)	3.5		
(alkaline range)	13.1		
p-Toluenesulfonic acid			8.44[b]
p-Toluidinium ion		6.24	
Tribenzylammonium ion			5.40[b]
Tropeoline 00	2.2		
Urea (protonated cation)			6.96[b]
Veronal	12.6		

[a] Dimethylsulfoxide. [b] Glacial acetic acid. [c] Acetonitrile. [d] Acetone + 10% water.

2.16 INDICATORS

An acid-base indicator is a conjugate acid-base pair of which the acid form and the base form are of different colors. These indicators are used to show the relative acidity or alkalinity of the test material.

Acid-base indicators are dyes that are themselves weak acids and bases. The conjugate acid-base forms of the dye are of different colors. An indicator does not change color from pure acid to pure alkaline at specific hydrogen ion concentration, but, rather, color change occurs over a range of hydrogen ion concentrations. This range is termed the *color change interval* and is expressed as a pH range. The chemical structures of the dyes are often complex but can be represented chemically by the symbol HIn. The acid-base indicator reaction is represented as:

$$HIn + H_2O \quad H_3O^+ + In \tag{1}$$

TABLE 2.62 Acid-Base Indicators

Indicator	pH range		Color	
	Minimum	Maximum	Acid	Alkaline
Brilliant cresyl blue	0.0	1.0	red-orange	blue
Methyl violet	0.0	1.6	yellow	blue
Crystal violet	0.0	1.8	yellow	blue
Ethyl violet	0.0	2.4	yellow	blue
Methyl Violet 6B	0.1	1.5	yellow	blue
Cresyl red	0.2	1.8	red	yellow
2-(p-Dimethylaminophenylazo) pyridine	0.2	1.8	yellow	blue
Malachite green	0.2	1.8	yellow	blue-green
Methyl green	0.2	1.8	yellow	blue
Cresol red (o-Cresolsulfonephthalein)	1.0	2.0	red	yellow
Quinaldine red	1.0	2.2	colorless	red
p-Methyl red	1.0	3.0	red	yellow
Metanil yellow	1.2	2.3	red	yellow
Pentamethoxy red	1.2	2.3	red-violet	colorless
Metanil yellow	1.2	2.4	red	yellow
p-Phenylazodiphenylamine	1.2	2.6	red	yellow
Thymol blue (Thymolsulfonephthalein)	1.2	2.8	red	yellow
m-Cresol purple	1.2	2.8	red	yellow
p-Xylenol blue	1.2	2.8	red	yellow
Benzopurpurin 4B	1.2	3.8	violet	red
Tropeolin OO	1.3	3.2	red	yellow
Orange IV	1.4	2.8	red	yellow
4-o-Tolylazo-o-toluidine	1.4	2.8	orange	yellow
Methyl violet 6B	1.5	3.2	blue	violet
Phloxine B	2.1	4.1	colorless	pink
Erythrosine, disodium salt	2.2	3.6	orange	red
Benzopupurine 4B	2.2	4.2	violet	red
N,N-dimethyl-p-(m-tolylazo) aniline	2.6	4.8	red	yellow
2,4-Dinitrophenol	2.8	4.0	colorless	yellow
N,N-Dimethyl-p-phenylazoaniline	2.8	4.4	red	yellow
Methyl yellow	2.9	4.0	red	yellow
Bromophenol blue	3.0	4.6	yellow	blue-violet
Tetrabromophenol blue	3.0	4.6	yellow	blue
Direct purple	3.0	4.6	blue-purple	red

(Continued)

TABLE 2.62 Acid-Base Indicators (*Continued*)

Indicator	pH range		Color	
	Minimum	Maximum	Acid	Alkaline
Congo red	3.1	4.9	blue	red
Methyl orange	3.1	4.4	red	yellow
Bromochlorophenol blue	3.2	4.8	yellow	blue
Ethyl orange	3.4	4.8	red	yellow
p-Ethoxychrysoidine	3.5	5.5	red	yellow
Alizarin sodium sulfonate	3.7	5.2	yellow	violet
α-Naphthyl red	3.7	5.7	red	yellow
Bromocresol green	3.8	5.4	yellow	blue
Resazurin	3.8	6.4	orange	violet
Bromophenol green	4.0	5.6	yellow	blue
2,5-Dinitrophenol	4.0	5.8	colorless	yellow
Methyl red	4.2	6.2	red	yellow
2-(*p*-Dimethylaminophenylazo) pyridine	4.4	5.6	red	yellow
Lacmoid	4.4	6.2	red	blue
Azolitmin	4.5	8.3	red	blue
Litmus	4.5	8.3	red	blue
Alizarin red S	4.6	6.0	yellow	red
Chlorophenol red	4.8	6.4	yellow	red
Cochineal	4.8	6.2	red	violet
Propyl red	4.8	6.6	red	yellow
Hematoxylin	5.0	6.0	red	blue
Bromocresol purple	5.2	6.8	yellow	violet
Bromophenol red	5.2	7.0	yellow	red
Chlorophenol red	5.4	6.8	yellow	red
p-Nitrophenol	5.6	6.6	colorless	yellow
Alizarin	5.6	7.2	yellow	red
Bromothymol blue	6.0	7.6	yellow	blue
Indo-oxine	6.0	8.0	red	blue
Bromophenol blue	6.2	7.6	yellow	blue
m-Dinitrobenzoylene urea	6.4	8.0	colorless	yellow
Phenol red (Phenolsulfonephthalein)	6.4	8.0	yellow	red
Rosolic acid	6.4	8.0	yellow	red
Brilliant yellow	6.6	7.9	yellow	orange
Quinoline blue	6.6	8.6	colorless	blue
Neutral red	6.8	8.0	red	orange
Phenol red	6.8	8.4	yellow	yellow
m-Nitrophenol	6.8	8.6	colorless	yellow
Cresol red (o-Cresolsulfonephthalein)	7.0	8.8	yellow	red
α-Naphtholphthalein	7.3	8.8	yellow	blue
Curcumin	7.4	8.6	yellow	red
m-Cresol purple (m-Cresolsulfonephthalein)	7.4	9.0	yellow	violet
Tropeolin OOO	7.6	8.9	yellow	rose-red
2,6-Divanillydenecyclohexanone	7.8	9.4	yellow	red
Thymol blue (Thymolsulfonephthalein)	8.0	9.6	yellow	purple
p-Xylenol blue	8.0	9.6	yellow	blue
Turmeric	8.0	10.0	yellow	orange
Phenolphthalein	8.0	10.0	colorless	red
o-Cresolphthalein	8.2	9.8	colorless	red
p-Naphtholphthalein	8.2	10.0	colorless	pink
Ethyl bis(2,4-dimethylphenyl acetate)	8.4	9.6	colorless	blue

TABLE 2.62 Acid-Base Indicators (*Continued*)

Indicator	pH range		Color	
	Minimum	Maximum	Acid	Alkaline
Ethyl bis(2,4-dinitrophenyl acetate)	8.4	9.6	colorless	blue
α-Naphtholbenzein	8.5	9.8	yellow	green
Thymolphthalein	9.4	10.6	colorless	blue
Nile blue A	10.0	11.0	blue	purple
Alizarin yllow CG	10.0	12.0	yellow	lilac
Alizarin yellow R	10.2	12.0	yellow	orange red
Salicyl yellow	10.0	12.0	yellow	orange-brown
Diazo violet	10.1	12.0	yellow	violet
Nile blue	10.1	11.1	blue	red
Curcumin	10.2	11.8	yellow	red
Malachite green hydrochloride	10.2	12.5	green-blue	colorless
Methyl blue	10.6	13.4	blue	pale violet
Brilliant cresyl blue	10.8	12.0	blue	yellow
Alizarin	11.0	12.4	red	purple
Nitramine	11.0	13.0	colorless	orange brown
Poirier's blue	11.0	13.0	blue	violet-pink
Tropeolin O	11.0	13.0	yellow	orange
Indigo carmine	11.4	13.0	blue	yellow
Sodium indigosulfonate	11.4	13.0	blue	yellow
Orange G	11.5	14.0	yellow	pink
2,4,6-Trinitrotoluene	11.7	12.8	colorless	orange
1,3,5-Trinitrobenzene	12.0	14.0	colorless	orange
2,4,6-Trinitrobenzoic acid	12.0	13.4	blue	violet-pink
Clayton yellow	12.2	13.2	yellow	amber

TABLE 2.63 Mixed Indicators

Mixed indicators give sharp color changes and are especially useful in titrating to a given titration exponent (p*I*).

The information given in this table is from the two-volume work *Volumetric Analysis* by Kolthoff and Stenger, published by Interscience Publishers, Inc., New York, 1942 and 1947, and reproduced with their permission.

Composition of Indicator Solution	p*I*	Color: Acid	Color: Alkaline	Notes
1 part 0.1% methyl yellow in alc. * 1 part 0.1% methylene blue in alc.	3.25	Blue-violet	Green	Still green at pH 3.4, blue-violet at 3.2†
1 part 0.14% xylene cyanol FF in alc. * 1 part 0.1% methyl orange in aq.	3.8	Violet	Green	Color is gray at pH 3.8
1 part 0.1% methyl orange in aq. * 1 part 0.25% indigo carmine in aq.	4.1	Violet	Green	Good indicator, especially in artificial light
1 part 0.1% methyl orange in aq. 1 part 0.1% aniline blue in aq.	4.3	Violet	Green	
1 part 0.1% bromcresol green sodium salt in aq. 1 part 0.02% methyl orange in aq.	4.3	Orange	Blue-green	Yellow at pH 3.5, greenish yellow at 4.0, weakly green at 4.3
3 parts 0.1% bromcresol green in alc. 1 part 0.2% methyl red in alc.	5.1	Wine-red	Green	Very sharp color change†
1 part 0.2% methyl red in alc. * 1 part 0.1% methylene blue in alc.	5.4	Red-violet	Green	Color is red-violet at pH 5.2, a dirty blue at 5.4, and a dirty green at 5.6
1 part 0.1% chlorphenol red sodium salt in aq. 1 part 0.1% aniline blue in water	5.8	Green	Violet	Pale violet at pH 5.8
1 part 0.1% bromcresol green sodium salt in aq. 1 part 0.1% chlorphenol red sodium salt in aq.	6.1	Yellow-green	Blue-violet	Blue-green at pH 5.4, blue at 5.8, blue with a touch of violet at 6.0, blue-violet at 6.2
1 part 0.1% bromcresol purple sodium salt in aq. 1 part 0.1% bromthymol blue sodium salt in aq.	6.7	Yellow	Violet-blue	Yellow-violet at pH 6.2, violet at 6.6, blue-violet at 6.8
2 parts 0.1% bromthymol blue sodium salt in aq. 1 part 0.1% azolitmin in aq.	6.9	Violet	Blue	

1 part 0.1% neutral red in alc. 1 part 0.1% methylene blue in alc.	*	7.0	Violet-blue	Green	Violet blue at pH 7.0†
1 part 0.1% neutral red in alc. 1 part 0.1% bromthymol blue in alc.		7.2	Rose	Green	Dirty green at pH 7.4, pale rose at 7.2, clear rose at 7.0
2 parts 0.1% cyanine in 50% alc. 1 part 0.1% phenol red in 50% alc.		7.3	Yellow	Violet	Orange at pH 7.2, beautiful violet at 7.4, color fades on standing
1 part 0.1% bromthymol blue sodium salt in aq. 1 part 0.1% phenol red sodium salt in aq.		7.5	Yellow	Violet	Dirty green at pH 7.2, pale violet at 7.4, strong violet at 7.6†
1 part 0.1% cresol red sodium salt in aq. 3 parts 0.1% thymol blue sodium salt in aq.		8.3	Yellow	Violet	Rose at pH 8.2, distinctly violet at 8.4†
2 parts 0.1% α-naphtholphthalein in alc. 1 part 0.1% cresol red in alc.		8.3	Pale rose	Violet	Pale violet at pH 8.2, strong violet at 8.4
1 part 0.1% α-naphtholphthalein in alc. 3 parts 0.1% phenolphthalein in alc.		8.9	Pale rose	Violet	Pale green at pH 8.6, violet at 9.0
1 part 0.1% phenolphthalein in alc. 2 parts 0.1% methyl green in alc.	*	8.9	Green	Violet	Pale blue at pH 8.8, violet at 9.0
1 part 0.1% thymol blue in 50% alc. 3 parts 0.1% phenolphthalein in 50% alc.		9.0	Yellow	Violet	From yellow thru green to violet†
1 part 0.1% phenolphthalein in alc. 1 part 0.1% thymolphthalein in alc.		9.9	Colorless	Violet	Rose at pH 9.6, violet at 10; sharp color change
1 part 0.1% phenolphthalein in alc. 2 parts 0.2% Nile blue in alc.		10.0	Blue	Red	Violet at pH 10†
2 parts 0.1% thymolphthalein in alc. 1 part 0.1% alizarin yellow in alc.		10.2	Yellow	Violet	Sharp color change
2 parts 0.2% Nile blue in aq. 1 part 0.1% alizarin yellow in alc.		10.8	Green	Red-brown	

* Store in a dark bottle. † Excellent indicator.

TABLE 2.64 Fluorescent Indicators

Name	pH range	Color change acid to base	Indicator solution
Benzoflavine	−0.3 to 1.7	Yellow to green	1
3,6-Dihydroxyphthalimide	0 to 2.4	Blue to green	1
	6.0 to 8.0	Green to yellow/green	
Eosin (tetrabromofluorescein)	0 to 3.0	Non-fl to green	4, 1%
4-Ethoxyacridone	1.2 to 3.2	Green to blue	1
3,6-Tetramethyldiaminoxanthone	1.2 to 3.4	Green to blue	1
Esculin	1.5 to 2.0	Weak blue to strong blue	
Anthranilic acid	1.5 to 3.0	Non-fl to light blue	2 (50% ethanol)
	4.5 to 6.0	Light blue to dark blue	
	12.5 to 14	Dark blue to non-fl	
3-Amino-1-naphthoic acid	1.5 to 3.0	Non-fl to green	2 (as sulfate in 50% ethanol)
	4.0 to 6.0	Green to blue	
	11.6 to 13.0	Blue to non-fl	
1-Naphthylamino-6-sulfonamide (also the 1-, 7-)	1.9 to 3.9	Non-fl to green	3
	9.6 to 13.0	Green to non-fl	
2-Naphthylamino-6-sulfonamide (also the 2-, 8-)	1.9 to 3.9	Non-fl to dark blue	3
	9.6 to 13.0	Dark blue to non-fl	
1-Naphthylamino-5-sulfonamide	2.0 to 4.0	Non-fl to yellow/orange	3
	9.5 to 13.0	Yellow/orange to non-fl	
1-Naphthoic acid	2.5 to 3.5	Non-fl to blue	4
Salicylic acid	2.5 to 4.0	Non-fl to dark blue	4 (0.5%)
Phloxin BA extra (tetrachlorotetrabromofluorescein)	2.5 to 4.0	Non-fl to dark blue	2
Erythrosin B (tetraiodofluorescein)	2.5 to 4.0	Non-fl to light green	4 (0.2%)
2-Naphthylamine	2.8 to 4.4	Non-fl to violet	1
Magdala red	3.0 to 4.0	Non-fl to purple	
p-Aminophenylbenzenesulfonamide	3.0 to 4.0	Non-fl to light blue	3
2-Hydroxy-3-naphthoic acid	3.0 to 6.8	Blue to green	4 (0.1%)
Chromotropic acid	3.1 to 4.4	Non-fl to light blue	4 (5%)
1-Naphthionic acid	3 to 4	Non-fl to blue	4
	10 to 12	Blue to yellow-green	
1-Naphthylamine	3.4 to 4.8	Non-fl to blue	1
5-Aminosalicylic acid	3.1 to 4.4	Non-fl to light green	1 (0.2% fresh)
Quinine	3.0 to 5.0	Blue to weak violet	1 (0.1%)
	9.5 to 10.0	Weak violet to non-fl	
o-Methoxybenzaldehyde	3.1 to 4.4	Non-fl to green	4 (0.2%)
o-Phenylenediamine	3.1 to 4.4	Green to non-fl	5
p-Phenylenediamine	3.1 to 4.4	Non-fl to orange/yellow	5
Morin (2′,4′,3,5,7-pentahydroxyflavone)	3.1 to 4.4	Non-fl to green	6 (0.2%)
	8 to 9.8	Green to yellow/green	
Thioflavine S	3.1 to 4.4	Dark blue to light blue	6 (0.2%)
Fluorescein	4.0 to 4.5	Pink/green to green	4 (1%)
Dichlorofluorescein	4.0 to 6.6	Blue green to green	1
β-Methylesculetin	4.0 to 6.2	Non-fl to blue	1
	9.0 to 10.0	Blue to light green	
Quininic acid	4.0 to 5.0	Yellow to blue	6 (satd)
β-Naphthoquinoline	4.4 to 6.3	Blue to non-fl	3
Resorufin (7-oxyphenoxazone)	4.4 to 6.4	Yellow to orange	

TABLE 2.64 Fluorescent Indicators (*Continued*)

Name	pH range	Color change acid to base	Indicator solution
Acridine	5.2 to 6.6	Green to violet	2
3,6-Dihydroxyxanthone	5.4 to 7.6	Non-fl to blue/violet	1
5,7-Dihydroxy-4-methylcoumarin	5.5 to 5.8	Light blue to dark blue	
3,6-Dihydroxyphthalic acid dinitrile	5.8 to 8.2	Blue to green	1
1,4-Dihydroxybenzenedisulfonic acid	6 to 7	Non-fl to light blue	4 (0.1%)
Luminol	6 to 7	Non-fl to blue	
2-Naphthol-6-sulfonic acid	5–7 to 8–9	Non-fl to blue	4
Quinoline	6.2 to 7.2	Blue to non-fl	6 (satd)
1-Naphthol-5-sulfonic acid	6.5 to 7.5	Non-fl to green	6 (satd)
Umbelliferone	6.5 to 8.0	Non-fl to blue	
Magnesium-8-hydroxyquinolinate	6.5 to 7.5	Non-fl to yellow	6 (0.1% in 0.01 *M* HCl)
Orcinaurine	6.5 to 8.0	Non-fl to green	6 (0.03%)
Diazo brilliant yellow	6.5 to 7.5	Non-fl to blue	
Coumaric acid	7.2 to 9.0	Non-fl to green	1
β-Methylumbelliferone	>7.0	Non-fl to blue	2 (0.3%)
Harmine	7.2 to 8.9	Blue to yellow	
2-Naphthol-6,8-disulfonic acid	7.5 to 9.1	Blue to light blue	4
Salicylaldehyde semicarbazone	7.6 to 8.0	Yellow to blue	2
1-Naphthol-2-sulfonic acid	8.0 to 9.0	Dark blue to light blue	4
Salicylaldehyde acetylhydrazone	8.3	Non-fl to green/blue	2
Salicylaldehyde thiosemicarbazone	8.4	Non-fl to blue/green	2
1-Naphthol-4-sulfonic acid	8.2	Dark blue to light blue	4
Naphthol AS	8.2 to 10.3	Non-fl to yellow/green	4
2-Naphthol	8.5 to 9.5	Non-fl to blue	2
Acridine orange	8.4 to 10.4	Non-fl to yellow/green	1
Orcinsulfonephthalein	8.6 to 10.0	Non-fl to yellow	
2-Naphthol-3,6-disulfonic acid	9.0 to 9.5	Dark blue to light blue	4
Ethoxyphenylnaphthostilbazonium chloride	9 to 11	Green to non-fl	1
o-Hydroxyphenylbenzothiazole	9.3	Non-fl to blue green	2
o-Hydroxyphenylbenzoxazole	9.3	Non-fl to blue/violet	2
o-Hydroxyphenylbenzimidazole	9.9	Non-fl to blue/violet	2
Coumarin	9.5 to 10.5	Non-fl to light green	
6,7-Dimethoxyisoquinoline-1-carboxylic acid	9.5 to 11.0	Yellow to blue	0.1% in glycerine/ethanol/water in 2:2:18 ratio
1-Naphthylamino-4-sulfonamide	9.5 to 13.0	Dark blue to white/blue	3

Indicator solutions: 1, 1% solution in ethanol; 2, 0.1% solution in ethanol; 3, 0.05% solution in 90% ethanol; 4, sodium or potassium salt in distilled water, 5; 0.2% solution in 70% ethanol; 6, distilled water.

TABLE 2.65 Selected List of Oxidation-Reduction Indicators

Name	Reduction Potential (30°C) in Volts at pH = 0	Reduction Potential (30°C) in Volts at pH = 7	Suitable pH Range	Color Change Upon Oxidation
Bis(5-bromo-1,10-phenanthroline) ruthenium(II) dinitrate	1.41*			Red to faint blue
Tris(5-nitro-1,10-phenanthroline) iron(II) sulfate	1.25*			Red to faint blue
Iron(II)-2,2′,2″-tripyridine sulfate	1.25*			Pink to faint blue
Tris(4,7-diphenyl-1,10-phenanthroline) iron(II) disulfate	1.13 (4.6 *M* H_2SO_4)* 0.87 (1.0 *M* H_2SO_4)*			Red to faint blue
o,m′-Diphenylaminedicarboxylic acid	1.12			Colorless to blue-violet
Setopaline	1.06 (*trans*)†			Yellow to orange
p-Nitrodiphenylamine	1.06			Colorless to violet
Tris(1,10-phenanthroline)-iron(II) sulfate	1.06 (1.00 *M* H_2SO_4)* 1.00 (3.0 *M* H_2SO_4)* 0.89 (6.0 *M* H_2SO_4)*			Red to faint blue
Setoglaucine O	1.01 (*trans*)†			Yellow-green to yellow-red
Xylene cyanole FF	1.00 (*trans*)†			Yellow-green to pink
Erioglaucine A	1.00 (*trans*)†			Green-yellow to bluish red
Eriogreen	0.99 (*trans*)†			Green-yellow to orange
Tris(2,2′-bipyridine)-iron(II) hydrochloride	0.97*			Red to faint blue
2-Carboxydiphenylamine [*N*-phenylanthranilic acid]	0.94			Colorless to pink
Benzidine dihydrochloride	0.92			Colorless to blue
o-Toluidine	0.87			Colorless to blue
Bis(1,10-phenanthroline)-osmium(II) perchlorate	0.859 (0.1 *M* H_2SO_4)			Green to pink
Diphenylamine-4-sulfonate (Na salt)	0.85			Colorless to violet
3,3′-Dimethoxybenzidine dihydrochloride [*o*-dianisidine]	0.85			Colorless to red
Ferrocyphen	0.81			Yellow to violet
4′-Ethoxy-2,4-diaminoazobenzene	0.76			Red to pale yellow
N,N-Diphenylbenzidine	0.76			Colorless to violet

Diphenylamine	0.76			Colorless to violet
N,N-Dimethyl-*p*-phenylenediamine	0.76			Colorless to red
Variamine blue B hydrochloride	0.712‡	0.310	1.5–6.3	Colorless to blue
N-Phenyl-1,2,4-benzenetriamine	0.70			Colorless to red
Bindschedler's green	0.680‡	0.224	2–9.5	
2,6-Dichloroindophenol (Na salt)	0.668‡	0.217	6.3–11.4	Colorless to blue
2,6-Dibromophenolindophenol	0.668‡	0.216	7.0–12.3	Colorless to blue
Brilliant cresyl blue [3-amino-9-dimethyl-amino-10-methylphenoxyazine chloride]	0.583	0.047	0–11	Colorless to blue
Iron(II)-tetrapyridine chloride	0.59			Red to faint blue
Thionine [Lauth's violet]	0.563‡	0.064	1–13	Colorless to violet
Starch (soluble potato, I_3^- present)	0.54			Colorless to blue
Gallocyanine (25°C)		0.021		Colorless to violet-blue
Methylene blue	0.532‡	0.011	1–13	Colorless to blue
Nile blue A [aminonaphthodiethylamino-phenoxazine sulfate]	0.406‡	−0.119	1.4–12.3	Colorless to blue
Indigo-5,5′,7,7′-tetrasulfonic acid (Na salt)	0.365‡	−0.046	<9	Colorless to blue
Indigo-5,5′,7-trisulfonic acid (Na salt)	0.332‡	−0.081	<9	Colorless to blue
Indigo-5,5′-disulfonic acid (Na salt)	0.291‡	−0.125	<9	Colorless to blue
Phenosafranine	0.280‡	−0.252	1–11	Colorless to violet-blue
Indigo-5-monosulfonic acid (Na salt)	0.262‡	−0.157	<9	Colorless to blue
Safranine T	0.24‡	−0.289	1–12	Colorless to violet-blue
Bis(dimethylglyoximato)-iron(II) chloride	0.155		6–10	Red to colorless
Induline scarlet	0.047‡	−0.299	3–8.6	Colorless to red
Neutral red		−0.323	2–11	Colorless to red-violet

* Transition point is at higher potential than the tabulated formal potential because the molar absorptivity of the reduced form is very much greater than that of the oxidized form.

† *Trans* = first noticeable color transition; often 60 mV less than E°

‡ Values of E° are obtained by extrapolation from measurements in weakly acid or weakly alkaline systems.

TABLE 2.66 Indicators for Approximate pH Determination

No. 1. Dissolve 60 mg methyl yellow, 40 mg methyl red, 80 mg bromthymol blue, 100 mg thymol blue and 20 mg phenolphthalein in 100 ml of ethanol and add enough 0.1N NaOH to produce a yellow color.
No. 2. Dissolve 18.5 mg methyl red, 60 mg bromthymol blue and 64 mg phenolphthalein in 100 ml of 50% ethanol and add enough 0.1N NaOH to produce a green color.

	Color			Color	
pH	No. 1	No. 2	pH	No. 1	No. 2
1	cherry-red	red	7	yellowish-green	greenish-yellow
2	rose	red	8	green	green
3	red-orange	red	9	bluish-green	greenish-blue
4	orange-red	deeper red	10	blue	violet
5	orange	orange-red	11	—	reddish-violet
6	yellow	orange-yellow			

TABLE 2.67 Oxidation-Reduction Indicators

Common name	Reference	Transition potential, volts (*N* hydrogen electrode = 0.000)	Color: Reduced form	Color: Oxidized form
p-ethoxychrysoidine	1	0.76	red	yellow
diphenylamine	2	0.776	colorless	purple
diphenylbenzidine	3	0.776	colorless	purple
diphenylamine-sulfonic acid or barium salt	4	0.84	colorless	purple
naphthidine	5	—	colorless	red
dimethylferroin	6	0.97	red	yellowish-green
eriogreen B	7	0.99	yellow	orange
erioglaucin A	7	1.0	yellowish-green	red
xylene cyanole FF	11	1.0		
2,2′-dipyridyl ferrous ion	6	1.03	red	colorless
N-phenylanthranilic acid	8	1.08	colorless	pink
methylferroin	6	1.08	red	pale-blue
ferroin (*o*-phenanthrolineferrous ion)	9	1.12	red	pale-blue
chloroferroin	6	1.17	red	pale-blue
nitroferroin	6	1.31	red	pale greenish-blue
α-naphtolflavone	10	—	pale straw	brownish-orange

2.17 ELECTRODE POTENTIALS

The potential of a polarographic or voltammetric *indicator electrode* at the point, on the rising part of a polarographic or voltammetric wave, where the difference between the total current and the *residual current* is equal to one-half of the *limiting current*. The quarter-wave potential, the three-quarter-wave potential, etc., may be similarly defined.

TABLE 2.68 Half-Wave Potentials (vs. Saturated Calomel Electrode) of Organic Compounds at 25°C. The solvent system in this table are listed below:

A, acetonitrile and a perchlorate salt such as $LiClO_4$ or a tetraalkyl ammonium salt
B, acetic acid and an alkali acetate, often plus a tetraalkyl ammonium iodide
C, 0.05 to 0.175M tetraalkyl ammonium halide and 75% 1,4-dioxane
D, buffer plus 50% ethanol (EtOH)

Abbreviations Used in the Table

Bu, butyl — *Me, methyl*
Et, ethyl — *MeOH, methanol*
EtOH, ethanol — *PrOh, propanol*
M, molar

Compound	Solvent system	$E_{1/2}$
	Unsaturated aliphatic hydrocarbons	
Acrylonitrile	C but 30% EtOH	−1.94
Allene	C	−2.29
1,3-Butadiene	A	−2.03
	C	−2.59
1,3-Butadiyne	C	−1.89
1-Buten-2-yne	C	−2.40
1,4-Cyclohexadiene	A	−1.6
Cyclohexene	A	−1.89
1,3,5,7-Cyclooctatetraene	B	−1.42
	C	−1.51
Diethyl fumarate	B, pH 4.0	−0.84
Diethyl maleate	B, pH 4.0	−0.95
2,3-Dimethyl-1,3-butadiene	A	−1.83
Dimethylfulvene	C	−1.89
Diphenylacetylene	C	−2.20
1,1-Diphenylethylene	B	−1.52
	C	−2.19
Ethyl methacrylate	0.1 *N* LiCl+25% EtOH	−1.9
2-Methyl-1,3-butadiene	A	−1.84
2-Methyl-1-butene	A	−1.97
1-Piperidino-4-cyano-4-phenyl-1,3-butadiene	$LiClO_4$ in dimethylformamide	−0.16
trans-Stilbene	B	−1.51
Tetrakis(dimethylamino)ethylene	A	−0.75

(*Continued*)

TABLE 2.68 Half-Wave Potentials (vs. Saturated Calomel Electrode) of Organic Compounds at 25°C (*Continued*)

Compound	Solvent system	$E_{1/2}$
	Aromatic hydrocarbons	
Acenaphthene	A	−0.95
	B	−1.36
	C	−2.58
Anthracene	A	−0.84
	B	−1.20
	C	−1.94
Azulene	A	−0.71
	C	−1.66, −2.26, −2.56
	Aromatic hydrocarbons (*continued*)	
1,2-Benzanthracene	C	−2.03, −2.54
2,3-Benzanthracene	A	−0.54, −1.20
Benzene	A	−2.08
1,2-Benzo[*a*]pyrene	A	−0.76
Biphenyl	A	−1.48
	B	−1.91
	C	−2.70
Chrysene	A	−1.22
1,2,5,6-Dibenzanthracene	A	−1.00, −1.26
1,2-Dihydronaphthalene	C	−2.57
9,10-Dimethylanthracene	A	−0.65
2,3-Dimethylnaphthalene	A	−1.08, −1.34
9,10-Diphenylanthracene	A	−0.92
Fluorene	A	−1.25
	B	−1.65
	C	−2.65
Hexamethylbenzene	A	−1.16
	B	−1.52
Indan	A	−1.59, −2.02
Indene	A	−1.23
	C	−2.81
1-Methylnaphthalene	A	−1.24
	B	−1.53
	C	−2.46
2-Methylnaphthalene	A	−1.22
	B	−1.55
	C	−2.46
Naphthalene	A	−1.34
	B	−1.72
Pentamethylbenzene	A	−1.28
	B	−1.62
Phenanthrene	A	−1.23
	B	−1.68
	C	−2.46, −2.71
Phenylacetylene	C	−2.37
Pyrene	A	−1.06, −1.24
trans-Stilbene	B	−1.51
	C	−2.26

(*Continued*)

TABLE 2.68 Half-Wave Potentials (vs. Saturated Calomel Electrode) of Organic Compounds at 25°C (*Continued*)

Compound	Solvent system	$E_{1/2}$
	Aromatic hydrocarbons (*continued*)	
Styrene	C	−2.35
1,2,3,5-Tetramethylbenzene	A	−1.50, −1.99
1,2,4,5-Tetramethylbenzene	A	−1.29
Tetraphenylethylene	C	−2.05
1,4,5,8-Tetraphenylnaphthalene	A	−1.39
Toluene	A	−1.98
1,2,3-Trimethylbenzene	A	−1.58
1,2,4-Trimethylbenzene	A	−1.41
1,3,5-Trimethylbenzene	A	−1.50
	B	−1.90
Triphenylene	A	−1.46, −1.55
Triphenylmethane	C	−1.01, −1.68, −1.96
o-Xylene	A	−1.58, −2.04
m-Xylene	A	−1.58
p-Xylene	A	−1.56
	Aldehydes	
Acetaldehyde	B, pH 6.8–13	−1.89
Benzaldehyde	McIlvaine buffer, pH 2.2	−0.96, −1.32
Bromoacetaldehyde	pH 8.5	−0.40
	pH 9.8	−1.58, −1.82
Chloroacetaldehyde	Ammonia buffer, pH 8.4	−1.06, −1.66
Cinnamaldehyde	Buffer + EtOH, pH 6.0	−0.9, −1.5, −1.7
Crotonaldehyde	B, pH 1.3–2.0	−0.92
	Ammonia buffer, pH 8.0	−1.30
Dichloroacetaldehyde	Ammonia buffer, pH 8.4	−1.03, −1.67
3,7-Dimethyl-2,6-octadienal	0.1 *M* Et_4NI	−1.56, −2.22
Formaldehyde	0.05 *M* KOH+0.1 *M* KCl, pH 12.7	−1.59
2-Furaldehyde	pH 1–8	−0.86–0.07 pH
	pH 10	−1.43
Glucose	Phosphate buffer, pH 7	−1.55
Glyceraldehyde	Britton-Robinson buffer, pH 5.0	−1.47
	Britton-Robinson buffer, pH 8.0	−1.55
Glycolaldehyde	0.1 *M* KOH, pH 13	−1.70
Glyoxal	B, pH 3.4	−1.41
4-Hydroxybenzaldehyde	Britton-Robinson buffer, pH 1.8	−1.16
	Britton-Robinson buffer, pH 6.8	−1.45
4-Hydroxy-2-methoxybenzaldehyde	McIlvaine buffer, pH 2.2	−1.05
	McIlvaine buffer, pH 5.0	−1.16, −1.36
	McIlvaine buffer, pH 8.0	−1.47
o-Methoxybenzaldehyde	Britton-Robinson buffer, pH 1.8	−1.02
	Britton-Robinson buffer, pH 6.8	−1.49
p-Methoxybenzaldehyde	Britton-Robinson buffer, pH 1.8	−1.17
	Britton-Robinson buffer, pH 6.8	−1.48
Methyl glyoxal	A, pH 4.5	−0.83
m-Nitrobenzaldehyde	Buffer+10% EtOH, pH 2.0	−0.28, −1.20

(*Continued*)

TABLE 2.68 Half-Wave Potentials (vs. Saturated Calomel Electrode) of Organic Compounds at 25°C (*Continued*)

Compound	Solvent system	$E_{1/2}$
	Aldehydes (*continued*)	
Phthalaldehyde	Buffer, pH 3.1	−0.64, −1.07
	Buffer, pH 7.3	−0.89, −1.29
2-Propenal (acrolein)	pH 4.5	−1.36
	pH 9.0	−1.1
Propionaldehyde	0.1 *M* LiOH, pH 13	−1.93
Pyrrole-2-carbaldehyde	0.1 *M* HCl+50% EtOH	−1.25
Salicylaldehyde	McIlvaine buffer, pH 2.2	−0.99, −1.23
	McIlvaine buffer, pH 5.0	−1.20, −1.30
	McIlvaine buffer, pH 8.0	−1.32
Trichloroacetaldehyde	Ammonia buffer, pH 8.4	−1.35, −1.66
	0.1 *M* KCl+50% EtOH	−1.55
	Ketones	
Acetone	B, pH 9.3	−1.52
	C	−2.46
Acetophenone	D+McIlvaine buffer, pH 4.9	−1.33
	D+McIlvaine buffer, pH 7.2	−1.58
	D+McIlvaine buffer, pH 1.3	−1.08
7*H*-Benz[*de*]anthracen-7-one	0.1 *N* H_2SO_4+75% MeOH	−0.96
Benzil	D+McIlvaine buffer, pH 1.3	−0.27
	D+McIlvaine buffer, pH 4.9	−0.50
Benzoin	D+McIlvaine buffer, pH 1.3	−0.90
	D+McIlvaine buffer, pH 8.6	−1.49
Benzophenone	D+McIlvaine buffer, pH 1.3	−0.94
	D+McIlvaine buffer, pH 8.6	−1.36
Benzoylacetone	Buffer, pH 2.6	−1.60
	Buffer, pH 5.3 and pH 7.6	−1.68
	Buffer, pH 9.7	−1.72
Bromoacetone	0.1 *M* LiCl	−0.29
2,3-Butanedione	0.1 *M* HCl	−0.84
3-Buten-2-one	0.1 *M* KCl	−1.42
Butyrophenone	0.1 *M* NH_4Cl+50% EtOH	−1.55
D-Carvone	0.1 *M* Et_4NI+80% EtOH	−1.71
Chloroacetone	0.1 *M* LiCl	−1.18
Coumarin	McIlvaine buffer, pH 2.0	−0.95
	McIlvaine buffer, pH 5.0	−1.11, −1.44
Cyclohexanone	C	−2.45
cis-Dibenzoylethylene	D, pH 1	−0.30
	D, pH 11	−0.62, −1.65
trans-Dibenzoylethylene	D, pH 1	−0.12
	D, pH 11	−0.57, −1.52
Dibenzoylmethane	D, pH 1.3	−0.59
	D, pH 11.3	−1.30, −1.62
9,10-Dihydro-9-oxoanthracene	D, pH 2.0	−0.93
1,5-Diphenyl-1,5-pentanedione	A	−2.10
1,5-Diphenylthiocarbazone	D, pH 7.0	−0.6
Flavanone	Acetate buffer+Me_4NOH+50% 2-PrOH, pH 6.1	−1.30
	Acetate buffer+Me_4NOH+50% 2-PrOH, pH 9.6	−1.51

TABLE 2.68 Half-Wave Potentials (vs. Saturated Calomel Electrode) of Organic Compounds at 25°C (*Continued*)

Compound	Solvent system	$E_{1/2}$
	Ketones (*continued*)	
Fluorescein	Acetate buffer, pH 2.0	−0.50
	Phthalate buffer, pH 5.0	−0.65
	Borate buffer, pH 10.1	−1.18, −1.44
Fructose	0.02 *M* LiCl	−1.76
Girard derivatives of aliphatic ketones	pH 8.2	−1.52
o-Hydroxyacetophenone	D, pH 5	−1.36
p-Hydroxyacetophenone	D, pH 5	−1.46
1,2,3-Indantrione (ninhydrin)	Britton-Robinson buffer, pH 2.5	−0.67, −0.83
	Britton-Robinson buffer, pH 4.5	−0.73, −1.01
	Britton-Robinson buffer, pH 6.8	−0.10, −0.90, −1.20
	Britton-Robinson buffer, pH 9.2	−1.35
α-Ionone	C	−1.59, −2.08
Isatin	Phosphate buffer+citrate buffer, pH 2.9	−0.3, −0.5
	Phosphate buffer+citrate buffer, pH 4.3	−0.3, −0.5, −0.8
	Phosphate buffer+citrate buffer, pH 5.4	−0.8
4-Methyl-3,5-heptadien-2-one	A	−0.64
4-Methyl-2,6-heptanedione	A	−1.28
4-Methyl-3-penten-2-one	D+McIlvaine buffer, pH 1.3	−1.01
	D+McIlvaine buffer, pH 11.3	−1.60
4-Phenyl-3-buten-2-one	D, pH 1.3	−0.72
	D, pH 8.6	−1.27
Phthalide	0.1 *M* Bu_4NI+50% dioxane	−0.20
Phthalimide	pH 4.2	−1.1, −1.5
	pH 9.7	−1.2, −1.4
Pulegone	C	−1.74
Quinalizarin	Phosphate buffer+1% EtOH, pH 8.0	−0.56
Testosterone	D+Britton-Robinson buffer, pH 2.6	−1.20
	D+Britton-Robinson buffer, pH 5.8	−1.40
	D+Britton-Robinson buffer, pH 8.8	−1.53, −1.79
	Quinones	
Anthraquinone	Acetate buffer+40% dioxane, pH 5.6	−0.51
	Phosphate buffer+40% dioxane, pH 7.9	−0.71
o-Benzoquinone	Britton-Robinson buffer, pH 7.0	+0.20
	Britton-Robinson buffer, pH 9.0	+0.08
2,3-Dimethylnaphthoquinone	D, pH 5.4	−0.22
1,2-Naphthoquinone	Phosphate buffer, pH 5.0	−0.03
	Phosphate buffer, pH 7.0	−0.13
1,4-Naphthoquinone	Britton-Robinson buffer, pH 7.0	−0.07
	Britton-Robinson buffer, pH 9.0	−0.19

(*Continued*)

TABLE 2.68 Half-Wave Potentials (vs. Saturated Calomel Electrode) of Organic Compounds at 25°C (*Continued*)

Compound	Solvent system	$E_{1/2}$
	Acids	
Acetic acid	A	−2.3
Acrylic acid	pH 5.6	−0.85
Adenosine-5′-phosphoric acid	$HClO_4$+$KClO_4$, pH 2.2	−1.13
4-Aminobenzenesulfonic acid	0.05 *M* Me_4NI	−1.58
3-Aminobenzoic acid	pH 5.6	−0.67
Anthranilic acid	pH 5.6	−0.67
Ascorbic acid	Britton-Robinson buffer, pH 3.4	+0.17
	Britton-Robinson buffer, pH 7.0	−0.06
Barbituric acid	Borate buffer, pH 9.3	−0.04
Benzoic acid	A	−2.1
Benzoylformic acid	Britton-Robinson buffer, pH 2.2	−0.48
	Britton-Robinson buffer, pH 5.5	−0.85, −1.26
	Britton-Robinson buffer, pH 7.2	−0.98, −1.25
	Britton-Robinson buffer, pH 9.2	−1.25
Bromoacetic acid	pH 1.1	−0.54
2-Bromopropionic acid	pH 2.0	−0.39
Crotonic acid	C	−1.94
Dibromoacetic acid	pH 1.1	−0.03, −0.59
Dichloroacetic acid	pH 8.2	−1.57
5,5-Diethylbarbituric acid	Borate buffer, pH 9.3	0.00
Flavanol	D, pH 5.6	−1.25
	D, pH 7.7	−1.40
Folic acid	Britton-Robinson buffer, pH 4.6	−0.73
Formic acid	0.1 *M* KCl	−1.66
Fumaric acid	HCl+KCl, pH 2.6	−0.83
	Acetate buffer, pH 4.0	−0.93
	Acetate buffer, pH 5.9	−1.20
2,4-Hexadienedioic acid	Acetate buffer, pH 4.5	−0.97
Iodoacetic acid	pH 1	−0.16
Maleic acid	Britton-Robinson buffer, pH 2.0	−0.70
	Britton-Robinson buffer, pH 4.0	−0.97
	Britton-Robinson buffer, pH 6.0	−1.11, −1.30
	Britton-Robinson buffer, pH 10.0	−1.51
Mercaptoacetic acid	B, pH 6.8	−0.38
Methacrylic acid	D+0.1 *M* LiCl	−1.69
Nitrobenzoic acids	Buffer+10% EtOH, pH 2.0	−0.2, −0.7
Oxalic acid	B, pH 5.4–6.1	−1.80
2-Oxo-1,5-pentanedioic acid	HCl+KCl, pH 1.8	−0.59
	Ammonia buffer, pH 8.2	−1.30
2-Oxopropionic acid	Britton-Robinson buffer, pH 5.6	−1.17
	Britton-Robinson buffer, pH 6.8	−1.22, −1.53
	Britton-Robinson buffer, pH 9.7	−1.51
Phenolphthalein	Phthalate buffer, pH 2.5	−0.67
	Phthalate buffer, pH 4.7	−0.80
	D, pH 9.6	−0.98, −1.35
Picric acid	pH 4.2	−0.34
	pH 11.7	−0.36, −0.56, −0.96

TABLE 2.68 Half-Wave Potentials (vs. Saturated Calomel Electrode) of Organic Compounds at 25°C (*Continued*)

Compound	Solvent system	$E_{1/2}$
	Acids (*continued*)	
1,2,3-Propenetricarboxylic acid	pH 7.0	−2.1
Trichloroacetic acid	Ammonia buffer, pH 8.2	−0.84, −1.57
	Phosphate buffer, pH 10.4	−0.9, −1.6
3,4,5-Trihydroxybenzoic acid	Phosphate buffer, pH 2.9	+0.50
	Phosphate buffer, pH 8.8	+0.1
p-Aminophenol	Britton-Robinson buffer, pH 6.3	+0.14
	Britton-Robinson buffer, pH 8.6	−0.04
	Britton-Robinson buffer, pH 12.0	−0.16
o-Chlorophenol	pH 5.6	−0.63
m-Chlorophenol	pH 5.6	−0.73
p-Chlorophenol	pH 5.6	−0.65
o-Cresol	pH 5.6	−0.56
m-Cresol	pH 5.6	−0.61
p-Cresol	pH 5.6	−0.54
1,2-Dihydroxybenzene	pH 5.6	−0.35
1,3-Dihydroxybenzene	pH 5.6	−0.61
1,4-Dihydroxybenzene	pH 5.6	−0.23
o-Methoxyphenol	pH 5.6	−0.46
m-Methoxyphenol	pH 5.6	−0.62
p-Methoxyphenol	pH 5.6	−0.41
1-Naphthol	A	−0.74
2-Naphthol	A	−0.82
1,2,3-Trihydroxybenzene	Britton-Robinson buffer, pH 3.1	+0.35
	Britton-Robinson buffer, pH 6.5	+0.10
	Britton-Robinson buffer, pH 9.5	−0.10
	Halogen compounds	
Bromobenzene	A	−1.98
	C	−2.32
1-Bromobutane	C	−2.27
Bromoethane	C	−2.08
Bromomethane	C	−1.63
1-Bromonaphthalene (also 2-bromonaphthalene)	A	−1.55, −1.60
3-Bromo-1-propene	C	−1.29
p-Bromotoluene	A	−1.72
Carbon tetrachloride	C	−0.78, −1.71
Chlorobenzene	A	−2.07
Chloroform	C	−1.63
Chloromethane	C	−2.23
3-Chloro-1-propene	C	−1.91
α-Chlorotoluene	C	−1.81
p-Chlorotoluene	A	−1.76
N-Chloro-*p*-toluenesulfonamide	0.5 *M* K_2SO_4	−0.13
9,10-Dibromoanthracene	A	−1.15, −1.47
p-Dibromobenzene	C	−2.10
1,2-Dibromobutane	D+1% Na_2SO_3	−1.45

(Continued)

TABLE 2.68 Half-Wave Potentials (vs. Saturated Calomel Electrode) of Organic Compounds at 25°C (*Continued*)

Compound	Solvent system	$E_{1/2}$
	Halogen compounds (*continued*)	
Dibromoethane	C	−1.48
meso-2,3-Dibromosuccinic acid	Acetate buffer, pH 4.0	−0.23, −0.89
Dichlorobenzenes	C	−2.5
Dichloromethane	C	−1.60
Diiodomethane	C	−1.12, −1.53
Hexabromobenzene	C	−0.8, −1.5
Hexachlorobenzene	C	−1.4, −1.7
Iodobenzene	A	−1.72
Iodoethane	C	−1.67
Iodomethane	A	−2.12
	C	−1.63
Tetrabromomethane	C	−0.3, −0.75, −1.49
Tetraidomethane	C	−0.45, −1.05, −1.46
Tribromomethane	C	−0.64, −1.47
α,α,α-Trichlorotoluene	C	−0.68, −1.65, −2.00
	Nitro and nitroso compounds	
1,2-Dinitrobenzene	Phthalate buffer, pH 2.5	−0.12, −0.32, −1.26
	Borate buffer, pH 9.2	−0.38, −0.74
1,3-Dinitrobenzene	Phthalate buffer, pH 2.5	−0.17, −0.29
	Borate buffer, pH 9.2	−0.46, −0.68
1,4-Dinitrobenzene	Phthalate buffer, pH 2.5	−0.12, −0.33
	Borate buffer, pH 9.2	−0.35, −0.80
Methyl nitrobenzoates	Buffer+10% EtOH, pH 2.0	−0.20 to −0.25 −0.68 to −0.74
p-Nitroacetophenone	Britton-Robinson buffer, pH 2.2	−0.16, −0.61, −1.09
	Britton-Robinson buffer, pH 10.0	−0.51, −1.40, −1.73
o-Nitroaniline	0.03 *M* LiCl+0.02 *M* benzoic acid in EtOH	−0.88
m-Nitroaniline	Britton-Robinson buffer, pH 4.3	−0.3, −0.8
	Britton-Robinson buffer, pH 7.2	−0.5
	Britton-Robinson buffer, pH 9.2	−0.7
p-Nitroaniline	pH 2.0	−0.36
	Acetate buffer, pH 4.6	−0.5
o-Nitroanisole	Buffer+10% EtOH, pH 2.0	−0.29, −0.58
p-Nitroanisole	Buffer+10% EtOH, pH 2.0	−0.35, −0.64
1-Nitroanthraquinone	Britton-Robinson buffer, pH 7.0	−0.16
Nitrobenzene	HCl+KCl+8% EtOH, pH 0.5	−0.16, −0.76
	Phthalate buffer, pH 2.5	−0.30
	Borate buffer, pH 9.2	−0.70
Nitrocresols	Britton-Robinson buffer, pH 2.2	−0.2 to −0.3
	Britton-Robinson buffer, pH 4.5	−0.4 to −0.5
	Britton-Robinson buffer, pH 8.0	−0.6
Nitroethane	Britton-Robinson buffer+30% MeOH, pH 1.8	−0.7
	Britton-Robinson buffer+30% MeOH, pH 4.6	−0.8

TABLE 2.68 Half-Wave Potentials (vs. Saturated Calomel Electrode) of Organic Compounds at 25°C (*Continued*)

Compound	Solvent system	$E_{1/2}$
	Nitro and nitroso compounds (*continued*)	
2-Nitrohydroquinone	Phosphate buffer+citrate buffer, pH 2.1	−0.2
	Phosphate buffer+citrate buffer, pH 5.2	−0.4
	Phosphate buffer+citrate buffer, pH 8.0	−0.5
Nitromethane	Britton-Robinson buffer+30% MeOH, pH 1.8	−0.8
	Britton-Robinson buffer+30% MeOH, pH 4.6	−0.85
o-Nitrophenol	Britton-Robinson buffer+10% EtOH, pH 2.0	−0.23
	Britton-Robinson buffer+10% EtOH, pH 4.0	−0.4
	Britton-Robinson buffer+10% EtOH, pH 8.0	−0.65
	Britton-Robinson buffer+10% EtOH, pH 10.0	−0.80
m-Nitrophenol	Britton-Robinson buffer+10% EtOH, pH 2.0	−0.37
	Britton-Robinson buffer+10% EtOH, pH 4.0	−0.40
	Britton-Robinson buffer+10% EtOH, pH 8.0	−0.64
	Britton-Robinson buffer+10% EtOH, pH 10.0	−0.76
p-Nitrophenol	Britton-Robinson buffer+10% EtOH, pH 2.0	−0.35
	Britton-Robinson buffer+10% EtOH, pH 4.0	−0.50
	Britton-Robinson buffer+10% EtOH, pH 8.0	−0.82
1-Nitropropane	Britton-Robinson buffer+30% MeOH, pH 1.8	−0.73
	Britton-Robinson buffer+30% MeOH, pH 8.6	−0.88
	Britton-Robinson buffer+30% MeOH, pH 8.0	−0.95
2-Nitropropane	McIlvaine buffer, pH 2.1	−0.53
	McIlvaine buffer, pH 5.1	−0.81
Nitrosobenzene	McIlvaine buffer, pH 6.0	−0.03
	McIlvaine buffer, pH 8.0	−0.14
1-Nitroso-2-naphthol	D+buffer, pH 4.0	+0.02
	D+buffer, pH 7.0	−0.20
	D+buffer, pH 9.0	−0.31
N-Nitrosophenylhydroxylamine	pH 2.0	−0.84
o-Nitrotoluene	Phthalate buffer, pH 2.5	−0.35, −0.66
	Phthalate buffer, pH 7.4	−0.60, −1.06

(*Continued*)

TABLE 2.68 Half-Wave Potentials (vs. Saturated Calomel Electrode) of Organic Compounds at 25°C (*Continued*)

Compound	Solvent system	$E_{1/2}$
	Nitro and nitroso compounds (*continued*)	
m-Nitrotoluene (also *p*-nitrotoluene)	Phthalate buffer, pH 2.5	−0.30, −0.53
	Phthalate buffer, pH 7.4	−0.58, −1.06
Tetranitromethane	pH 12.0	−0.41
1,3,5-Trinitrobenzene	Phthalate buffer, pH 4.1	−0.20, −0.29, −0.34
	Borate buffer, pH 9.2	−0.34, −0.48, −0.65
	Heterocyclic compounds containing nitrogen	
Acridine	D, pH 8.3	−0.80, −1.45
Cinchonine	B, pH 3	−0.90
2-Furanmethanol	Britton-Robinson buffer, pH 2.0	−0.96
	Britton-Robinson buffer, pH 5.8	−1.38, −1.70
2-Hydroxyphenazine	Britton-Robinson buffer, pH 4.0	−0.24
8-Hydroxyquinoline	B, pH 5.0	−1.12
	Phosphate buffer, pH 8.0	−1.18, −1.71
3-Methylpyridine	D+0.1 *M* LiCl	−1.76
4-Methylpyridine	D+0.1 *M* LiCl	−1.87
Phenazine	Phosphate buffer+citrate buffer, pH 7.0	−0.36
Pyridine	Phosphate buffer+citrate buffer, pH 7.0	−1.75
Pyridine-2-carboxylic acid	B, pH 4.1	−1.10
	B, pH 9.3	−1.48, −1.94
Pyridine-3-carboxylic acid	0.1 *M* HCl	−1.08
Pyridine-4-carboxylic acid	Britton-Robinson buffer, pH 6.1	−1.14
	pH 9.0	−1.39, −1.68
Pyrimidine	Citrate buffer, pH 3.6	−0.92, −1.24
	Ammonia buffer, pH 9.2	−1.54
Quinoline-8-carboxylic acid	pH 9	−1.11
Quinoxaline	Phosphate buffer+citrate buffer, pH 7.0	−0.66, −1.52
	Azo, hydrazine, hydroxylamine, and oxime compounds	
Azobenzene	D, pH 4.0	−0.20
	D, pH 7.0	−0.50
Azoxybenzene	Buffer+20% EtOH, pH 6.3	−0.30
Benzoin 1-oxime	Buffer, pH 2.0	−0.88
	Buffer, pH 5.6	−1.08
	Buffer, pH 8.2	−1.67
Benzoylhydrazine	0.13 *M* NaOH, pH 13.0	−0.30
Dimethylglyoxime	Ammonia buffer, pH 9.6	−1.63
Hydrazine	Britton-Robinson buffer, pH 9.3	−0.09
Hydroxylamine	Britton-Robinson buffer, pH 4.6	−1.42
	Britton-Robinson buffer, pH 9.2	−1.65

TABLE 2.68 Half-Wave Potentials (vs. Saturated Calomel Electrode) of Organic Compounds at 25°C (*Continued*)

Compound	Solvent system	$E_{1/2}$
	Azo, hydrazine, hydroxylamine, and oxime compounds (*continued*)	
Oxamide	Acetate buffer	−1.55
Phenylhydrazine	McIlvaine buffer, pH 2	+0.19
	0.13 *M* NaOH, pH 13.0	−0.36
Phenylhydroxylamine	McIlvaine buffer+10% EtOH, pH 2	−0.68
	McIlvaine buffer+10 EtOH, pH 4–10	−0.33 0.061 pH
Salicylaldoxime	Phosphate buffer, pH 5.4	−1.02
Thiosemicarbazide	Borate buffer, pH 9.3	−0.26
Thiourea	0.1 *M* sulfuric acid	+0.02
	Indicators and dyestuffs	
Brilliant Green	HCl+KCl, pH 2.0	−0.2, −0.5
Indigo carmine	pH 2.5	−0.24
Indigo disulfonate	pH 7.0	−0.37
Malachite Green G	HCl+KCl, pH 2.0	−0.2, −0.5
Metanil yellow	Phosphate buffer+1% EtOH, pH 7.0	−0.51
Methylene blue	Britton-Robinson buffer, pH 4.9	−0.15
	Britton-Robinson buffer, pH 9.2	−0.30
Methylene green	Phosphate buffer+1% EtOH, pH 7.0	−0.12
Methyl orange	Phosphate buffer+1% EtOH, pH 7.0	−0.51
Morin	D, pH 7.6	−1.7
Neutral red	Britton-Robinson buffer, pH 2.0	−0.21
	Britton-Robinson buffer, pH 7.0	−0.57
	Peroxide	
Ethyl peroxide	0.02 *M* HCl	−0.2

2.18 ELECTRICAL CONDUCTIVITY

TABLE 2.69 Electrical Conductivity of Various Pure Liquids

Liquid	Temp. °C	mhos/cm or $ohm^{-1} \cdot cm^{-1}$	Liquid	Temp. °C	mhos/cm or $ohm^{-1} \cdot cm^{-1}$
Acetaldehyde	15	1.7×10^{-6}	Epichlorohydrin	25	3.4×10^{-8}
Acetamide	100	$<4.3 \times 10^{-5}$	Ethyl acetate	25	$<1 \times 10^{-9}$
Acetic acid	0	5×10^{-9}	Ethyl acetoacetate	25	4×10^{-8}
	25	1.12×10^{-8}	Ethyl alcohol	25	1.35×10^{-9}
Acetic anhydride	0	1×10^{-6}	Ethylamine	0	4×10^{-7}
	25	4.8×10^{-7}	Ethyl benzoate	25	$<1 \times 10^{-9}$
Acetone	18	2×10^{-8}	Ethyl bromide	25	$<2 \times 10^{-8}$
	25	6×10^{-8}	Ethylene bromide	19	$<2 \times 10^{-10}$
Acetonitrile	20	7×10^{-6}	Ethylene chloride	25	3×10^{-8}
Acetophenone	25	6×10^{-9}	Ethyl ether	25	$<4 \times 10^{-13}$
Acetyl bromide	25	2.4×10^{-6}	Ethylidene chloride	25	$<1.7 \times 10^{-8}$
Acetyl chloride	25	4×10^{-7}	Ethyl iodide	25	$<2 \times 10^{-8}$
Alizarin	233	1.45×10^{-6}(?)	Ethyl isothiocyanate	25	1.26×10^{-7}
Allyl alcohol	25	7×10^{-6}	Ethyl nitrate	25	5.3×10^{-7}
Ammonia	−79	1.3×10^{-7}	Ethyl thiocyanate	25	1.2×10^{-6}
Aniline	25	2.4×10^{-8}	Eugenol	25	$<1.7 \times 10^{-8}$
Anthracene	230	3×10^{-10}			
Arsenic tribromide	35	1.5×10^{-6}	Formamide	25	4×10^{-6}
Arsenic trichloride	25	1.2×10^{-6}	Formic acid	18	5.6×10^{-5}
				25	6.4×10^{-5}
Benzaldehyde	25	1.5×10^{-7}	Furfural	25	1.5×10^{-6}
Benzene	. . .	7.6×10^{-8}			
Benzoic acid	125	3×10^{-9}	Gallium	30	36,800
Benzonitrile	25	5×10^{-8}	Glycerol	25	6.4×10^{-8}
Benzyl alcohol	25	1.8×10^{-6}	Glycol	25	3×10^{-7}
Benzylamine	25	$<1.7 \times 10^{-8}$	Guaiacol	25	2.8×10^{-7}
Benzyl benzoate	25	$<1 \times 10^{-9}$			
Bromine	17.2	1.3×10^{-13}	Heptane	. . .	$<1 \times 10^{-13}$
Bromobenzene	25	$<2 \times 10^{-11}$	Hexane	18	$<1 \times 10^{-18}$
Bromoform	25	$<2 \times 10^{-8}$	Hydrogen bromide	−80	8×10^{-9}
iso-Butyl alcohol	25	8×10^{-8}	Hydrogen chloride	−96	1×10^{-8}
			Hydrogen cyanide	0	3.3×10^{-6}
			Hydrogen iodide	B.P.	2×10^{-7}
Capronitrile	25	3.7×10^{-6}	Hydrogen sulfide	B.P.	1×10^{-11}
Carbon disulfide	1	7.8×10^{-18}			
Carbon tetrachloride	18	4×10^{-18}	Iodine	110	1.3×10^{-10}
Chlorine	−70	$<1 \times 10^{-16}$			
Chloroacetic acid	60	1.4×10^{-6}	Kerosene	25	$<1.7 \times 10^{-8}$
m-Chloroaniline	25	5×10^{-8}			
Chloroform	25	$<2 \times 10^{-8}$	Mercury	0	10,629.6
Chlorohydrin	25	5×10^{-7}	Methyl acetate	25	3.4×10^{-6}
m-Cresol	25	$<1.7 \times 10^{-8}$	Methyl alcohol	18	4.4×10^{-7}
Cyanogen	. . .	$<7 \times 10^{-9}$	Methyl ethyl ketone	25	1×10^{-7}
Cymene	25	$<2 \times 10^{-8}$	Methyl iodide	25	$<2 \times 10^{-8}$
			Methyl nitrate	25	4.5×10^{-6}
Dichloroacetic acid	25	7×10^{-8}	Methyl thiocyanate	25	1.5×10^{-6}
Dichlorohydrin	25	1.2×10^{-5}			
Diethylamine	−33.5	2.2×10^{-9}	Naphthalene	82	4×10^{-10}
Diethyl carbonate	25	1.7×10^{-8}	Nitrobenzene	0	5×10^{-9}
Diethyl oxalate	25	7.6×10^{-7}	Nitromethane	18	6×10^{-7}
Diethyl sulfate	25	2.6×10^{-7}	*o*- or *m*-Nitrotoluene	25	$<2 \times 10^{-7}$
Dimethyl sulfate	0	1.6×10^{-7}	Nonane	25	$<1.7 \times 10^{-8}$

TABLE 2.69 Electrical Conductivity of Various Pure Liquids (*Continued*)

Liquid	Temp. °C	mhos/cm or $ohm^{-1} \cdot cm^{-1}$	Liquid	Temp. °C	mhos/cm or $ohm^{-1} \cdot cm^{-1}$
Oleic acid	15	$<2 \times 10^{-10}$	Salicylaldehyde	25	1.6×10^{-7}
			Stearic acid	80	$<4 \times 10^{-13}$
Pentane	19.5	$<2 \times 10^{-10}$	Sulfonyl chloride, $SOCl_2$	25	2×10^{-6}
Petroleum	...	3×10^{-13}			
Phenetole	25	$<1.7 \times 10^{-8}$	Sulfur	115	1×10^{-12}
Phenol	25	$<1.7 \times 10^{-8}$		130	5×10^{-12}
Phenyl isothiocyanate	25	1.4×10^{-6}		440	1.2×10^{-7}
Phosgene	25	7×10^{-9}	Sulfur dioxide	35	1.5×10^{-8}
Phosphorus	25	4×10^{-7}	Sulfuric acid	25	1×10^{-2}
Phosphorus oxychloride	25	2.2×10^{-6}	Sulfuryl chloride, SO_2Cl_2	25	3×10^{-8}
Pinene	23	$<2 \times 10^{-10}$			
Piperidine	25	$<2 \times 10^{-7}$	Toluene	...	$<1 \times 10^{-14}$
Propionaldehyde	25	8.5×10^{-7}	*o*-Toluidine	25	$<2 \times 10^{-6}$
Propionic acid	25	$<1 \times 10^{-9}$	*p*-Toluidine	100	6.2×10^{-8}
Propionitrile	25	$<1 \times 10^{-7}$	Trichloroacetic acid	25	3×10^{-9}
n-Propyl alcohol	18	5×10^{-8}	Trimethylamine	−33.5	2.2×10^{-10}
	25	2×10^{-8}	Turpentine	...	2×10^{-13}
iso-Propyl alcohol	25	3.5×10^{-6}			
n-Propyl bromide	25	$<2 \times 10^{-8}$	*iso*-Valeric acid	80	$<4 \times 10^{-13}$
Pyridine	18	5.3×10^{-8}	Water	18	4×10^{-8}
Quinoline	25	2.2×10^{-8}	Xylene	...	$<1 \times 10^{-15}$

TABLE 2.70 Limiting Equivalent Ionic Conductances in Aqueous Solutions

	Temperature, °C		
Ion	0	18	25
Fluoroacetate⁻			44.4
Fluorobenzoate⁻			33
Formate⁻		47	54.6
Fumarate(2−)			61.8
Glutarate(2−)			52.6
Hydrogenoxalate (1−)			40.2
Iodoacetate⁻			40.6
Lactate(1−)			38.8
Malate(2−)			58.8
Malonate(1−)			63.5
3-Methylbutanoate⁻			32.7
Methylsulfonate⁻			48.8
Naphthylacetate⁻			28.4
1,8-Octanedioate(2−)			36
Octylsulfonate⁻			29
Oxalate(2−)			74.11
Phenylacetate⁻			30.6
m-Phthalate(2−)			54.7
o-Phthalate(2−)			52.3
Picrate⁻			30.37
Propanoate⁻			35.8
Propylsulfonate⁻			37.1
Salicylate⁻			36
Succinate(2−)			58.8
Tartrate(2−)		55	59.6
Trichloroacetate⁻			36.6
Trimethylacetate⁻			31.9

TABLE 2.71 Properties of Organic Semiconductors

Substance	Formula	Resistivity, ohm-cm	Band Gap: Conductivity, eV	Band Gap: Photo Conduct, eV
POLYACENES				
Anthracene	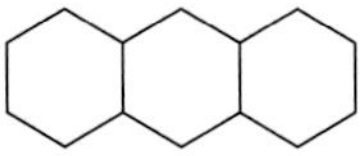	300	0.83	—
Tetracene	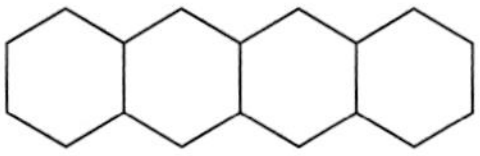	10	0.85	3.6
Pyrene	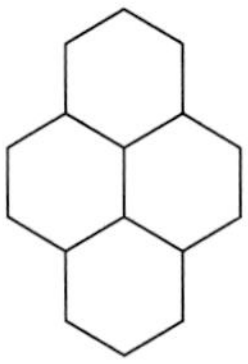	300	1.01	3.2
Perylene	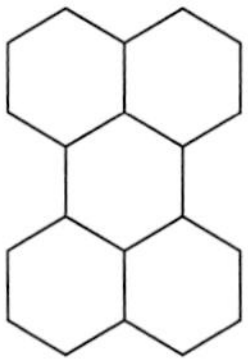	10	0.98	—
Chrysene	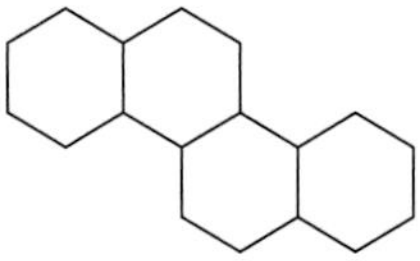	100	1.10	3.2
Coronene	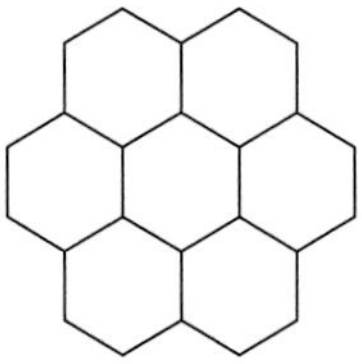	0.2	1.15	—
Pyranthrene	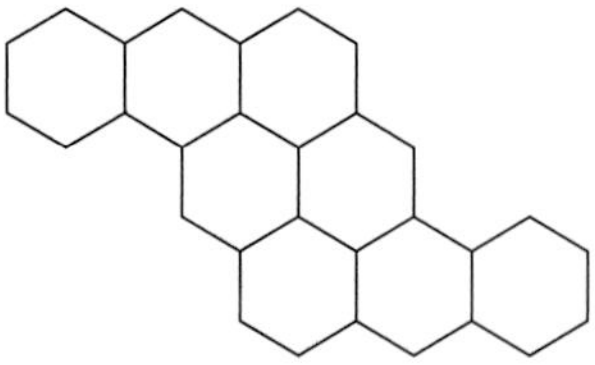	10^7	0.54	0.85

TABLE 2.71 Properties of Organic Semiconductors (*Continued*)

Substance	Formula	Resistivity, ohm-cm	Band Gap: Conductivity, eV	Band Gap: Photo Conduct, eV
POLYACENES WITH QUINONOID ATTACHEMENTS				
Violanthrone	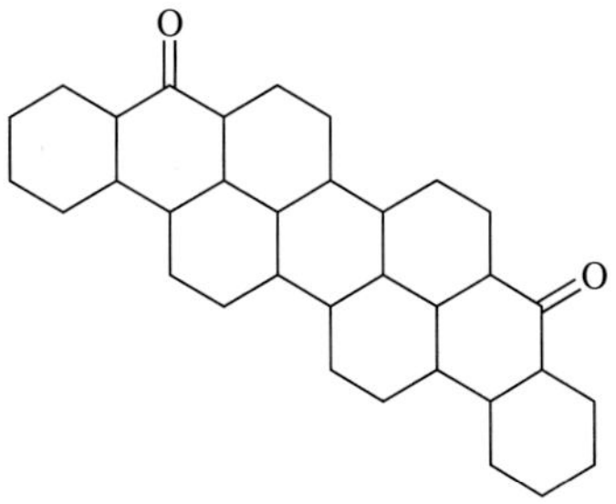	1000	0.39	0.84
Pyranthrone	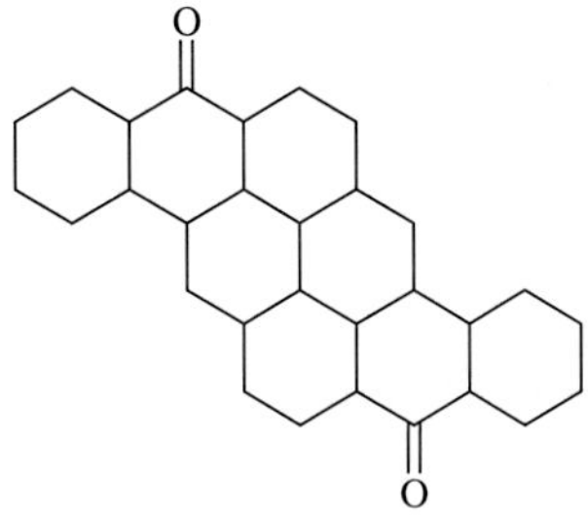	10^6	0.54	1.14
AZO-AROMATIC COMPOUNDS				
Indanthrone black	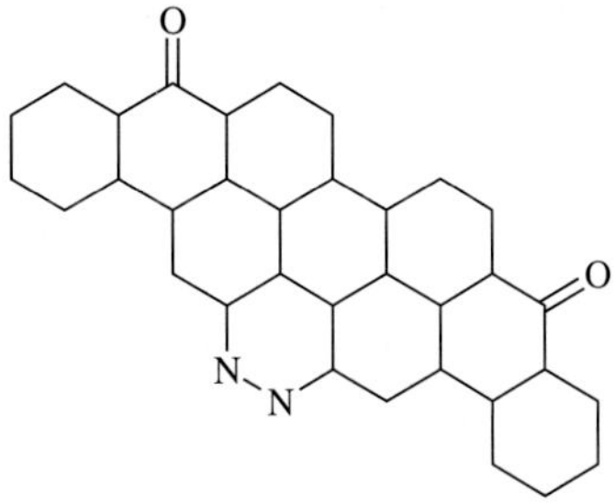	300	0.28	—
1,9,4,10-Anthradipyrimidine	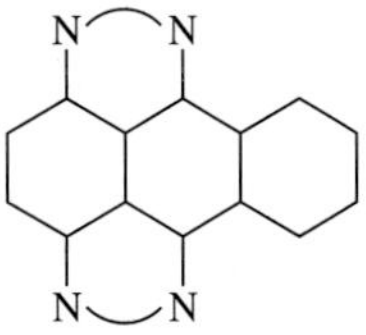	1000	1.61	—

(*Continued*)

TABLE 2.71 Properties of Organic Semiconductors (*Continued*)

Substance	Formula	Resis-tivity, ohm-cm	Band Gap: Conduc-tivity, eV	Band Gap: Photo Conduct, eV
PHTHALOCYANINES	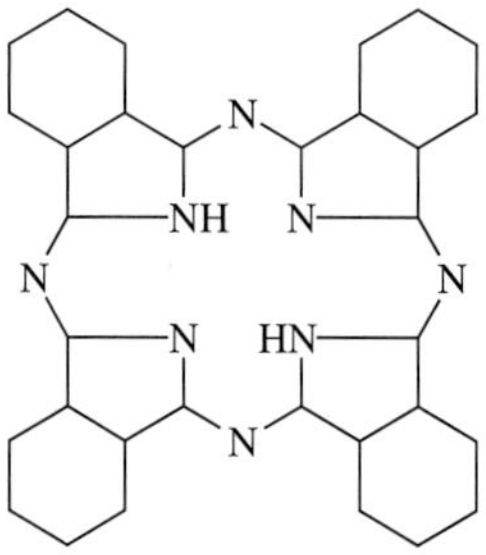	10^4	1.2	1.56
FREE RADICALS α,α-Diphenyl β-pieryl hydrazyl	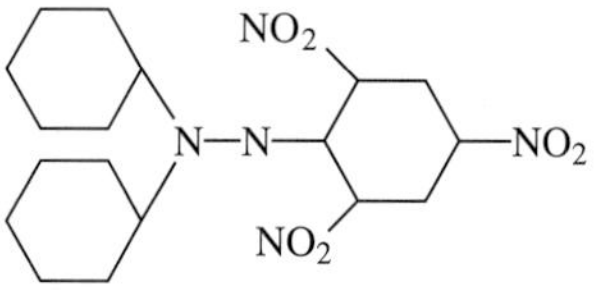	10^6	0.74	-

2.19 LINEAR FREE ENERGY RELATIONSHIPS

Many equilibrium and rate processes can be systematized when the influence of each substituent on the reactivity of substrates is assigned a characteristic constant σ and the reaction parameter ρ is known or can be calculated. The Hammett equation

$$\log \frac{K}{K^\circ} = \sigma\rho$$

describes the behavior of many *meta*- and *para*-substituted aromatic species. In this equation K° is the acid dissociation constant of the reference in aqueous solution at 25°C and K is the corresponding constant for the substituted acid. Separate sigma values are defined by this reaction for *meta* and *para* substituents and provide a measure of the total electronic influence (polar, inductive, and resonance effects) in the absence of conjugation effects. Sigma constants are not valid of substituents *ortho* to the reaction center because of anomalous (mainly steric) effects. The inductive effect is transmitted about equally to the *meta* and *para* positions. Consequently, σ_m is an approximate measure of the size of the inductive effect of a given substituent and $\sigma_p - \sigma_m$ is an approximate measure of a substituent's resonance effect. Values of Hammett sigma constants are listed in Table 2.72.

Taft sigma values σ^* perform a similar function with respect to aliphatic and alicyclic systems. Values of σ^* are listed in Table 2.72.

The reaction parameter ρ depends upon the reaction series but not upon the substituents employed. Values of the reaction parameter for some aromatic and aliphatic system are given in Tables 2.73 and 2.74.

Since substituent effects in aliphatic systems and in *meta* positions in aromatic systems are essentially inductive in character, σ^* and σ_m values are often related by the expression.

$\sigma_m = 0.217\ \sigma^* - 0.106$. Substituent effects fall off with increasing distance from the reaction center; generally a factor of 0.36 corresponds to the interposition of a —CH_2— group, which enables σ^* values to be estimated for R—CH_2— groups not otherwise available.

Two modified sigma constants have been formulated for situations in which the substituent enters into resonance with the reaction center in an electron-demanding transition state (σ^+) or for an electron-rich transition state (σ^-).σ^- constants give better correlations in reactions involving phenols, anilines, and pyridines and in nucleophilic substitutions. Values of some modified sigma constants are given in Table 2.75.

TABLE 2.72 Hammett and Taft Substituent Constants

Substituent	Hammett constants σ_m	Hammett constants σ_p	Taft constant σ^*
$—AsO_3H^-$	−0.09	−0.02	0.06
$—B(OH)_2$	0.01	0.45	
—Br	0.39	0.23	2.84
$—CH_2Br$			1.00
m-BrC_6H_4—		0.09	
p-BrC_6H_4—		0.08	
$—CH_3$	−0.07	−0.17	0.0
$—CH_2CH_3$	−0.07	−0.15	−0.10
$—CH_2CH_2CH_3$	−0.05	−0.15	−0.12
$—CH(CH_3)_2$	−0.07	−0.15	−0.19
$—CH_2CH_2CH_2CH_3$	−0.07	−0.16	−0.13
$—CH_2CH(CH_3)_2$	−0.07	−0.12	−0.13
$—CH(CH_3)CH_2CH_3$		−0.12	−0.19
$—C(CH_3)_3$	−0.10	−0.20	−0.30
$—CH_2CH_2CH_2CH_2CH_3$			−0.25
$—CH_2CH_2CH(CH_3)_2$			−0.17
$—CH_2C(CH_3)_3$		−0.23	−0.12
$—CH_2CH_2CH_2CH_2CH_2CH_2CH_3$			−0.37
Cyclopropyl—	−0.07	−0.21	
Cyclohexyl—			−0.15
—3,4-$(CH_2)_2$ (fused)		−0.26	
—3,4-$(CH_2)_3$— (fused ring)		−0.48	
—3,4-$(CH)_4$— (fused ring)	0.06	0.04	
$—CH{=}CH_2$	0.02		0.56
$—CH{=}C(CH_3)_2$			0.19
$—CH{=}CHCH_3$, *trans*			0.36
$—CH_2—CH{=}CH_2$			0.0
$—CH{=}CHC_6H_5$	0.14	−0.05	0.41
$—C{\equiv}CH$	0.21	0.23	2.18
$—C{\equiv}CC_6H_5$	0.14	0.16	1.35
$—CH_2—C{\equiv}CH$			0.81
$—C_6H_5$	0.06	−0.01	0.60
p-$CH_3C_6H_4$—		−0.5	
Naphthyl— (both 1- and 2-)			0.75
$—CH_2C_6H_5$		0.46	0.22
$—CH_2CH_2—C_6H_5$			−0.06
$—CH(CH_3)C_6H_5$			0.37
$—CH(C_6H_5)_2$			0.41
$—CH_2—C_{10}H_7$			0.44
2-Furoyl—			0.25
3-Indolyl—			−0.06
2-Thienyl—			1.31

(Continued)

TABLE 2.72 Hammett and Taft Substituent Constants (*Continued*)

Substituent	Hammett constants σ_m	Hammett constants σ_p	Taft constant σ^*
2-Thienylmethylene—			0.31
—CHO	0.36	0.22	
—$COCH_3$	0.38	0.50	1.65
—$COCH_2CH_2$		0.48	
—$COCH(CH_3)_2$		0.47	
—$COC(CH_3)_3$		0.32	
—$COCF_3$	0.65		3.7
—COC_6H_5	0.34	0.46	2.2
—$CONH_2$	0.28	0.36	1.68
—$CONHC_6H_5$			1.56
—CH_2COCH_3			0.60
—CH_2CONH_2			0.31
—$CH_2CH_2CONH_2$			0.19
—$CH_2CH_2CH_2CONH_2$			0.12
—$CH_2CONHC_6H_5$			0.0
—COO^-	−0.1	0.0	−1.06
—COOH	0.36	0.43	2.08
—CO—OCH_3	0.32	0.39	2.00
—CO—OCH_2CH_3	0.37	0.45	2.12
—CH_2CO—OCH_3			1.06
—CH_2CO—OCH_2CH_3			0.82
—CH_2COO			−0.06
—CH_2CH_2COOH	−0.03	−0.07	
—Cl	0.37	0.23	2.96
—CCl_3	0.47		2.65
—$CHCl_2$			1.94
—CH_2Cl	0.12	0.18	1.05
—CH_2CH_2Cl			0.38
—CH_2CCl_3			0.75
—$CH_2CH_2CCl_3$			0.25
—CH=CCl_2			1.00
—CH_2CH=CCl_2			0.19
p-ClC_6H_4—		0.08	
—F	0.34	0.06	3.21
—CF_3	0.43	0.54	2.61
—CHF_2			2.05
—CH_2F			1.10
—CH_2CF_3			0.90
—$CH_2CF_2CF_2CF_3$			0.87
—C_6F_5	−0.12	−0.03	
—$Ge(CH_3)_3$		0.0	
—$Ge(CH_2CH_3)_3$		0.0	
—H	0.00	0.00	0.49
—I	0.35	0.28	2.46
—CH_2I			0.85
—IO_2	0.70	0.76	
—N_2^+	1.76	1.91	
—N_3 (azide)	0.33	0.08	2.62
—NH_2	−0.16	−0.66	0.62
—NH_3^+	1.13	1.70	3.76
—CH_2—NH_2			0.50
—CH_2—NH_3^+			2.24
—NH—CH_3	−0.30	−0.84	

TABLE 2.72 Hammett and Taft Substituent Constants (*Continued*)

Substituent	Hammett constants σ_m	Hammett constants σ_p	Taft constant σ^*
$—NH—C_2H_5$	−0.24	−0.61	
$—NH—C_4H_9$	−0.34	−0.51	
$—NH(CH_3)_2^+$			4.36
$—NH_2—CH_3^+$	0.96		3.74
$—NH_2—C_2H_5^+$	0.96		3.74
$—N(CH_3)_3^+$	0.88	0.82	4.55
$—N(CH_3)_2$	−0.2	−0.83	0.32
$—CH_2—N(CH_3)_3^+$			1.90
$—N(CF_3)_2$	0.45	0.53	
p-$H_2N—C_6H_5—$		−0.30	
$—NH—CO—CH_3$	0.21	0.00	1.40
$—NH—CO—C_2H_5$			1.56
$—NH—CO—C_6H_5$	0.22	0.08	1.68
$—NH—CHO$	0.25		1.62
$—NH—CO—NH_2$	0.18		1.31
$—NH—OH$	−0.04	−0.34	
$—NH—CO—OC_2H_5$	0.33		1.99
$—CH_2—NH—CO—CH_3$			0.43
$—NH—SO_2—C_6H_5$			1.99
$—NH—NH_2$	−0.02	−0.55	
$—CN$	0.56	0.66	3.30
$—CH_2—CN$	0.17	0.01	1.30
$—NO$		0.12	
$—NO_2$	0.71	0.78	4.0
$—CH_2—NO_2$			1.40
$—CH_2—CH_2—NO_2$			0.50
$—CH{=}CHNO_2$	0.33	0.26	
m-$O_2N—C_6H_4$		0.18	
p-$O_2N—C_6H_4$		0.24	
$(NO_2)_3C_6H_2—$ (picryl)	0.43	0.41	
$—N(CO—CH_3)(CO—C_6H_5)$			1.37
$—N(CO—CH_3)$(naphthyl)			1.65
$—O^-$	−0.71	−0.52	
$—OH$	0.12	−0.37	1.34
$—O—CH_3$	0.12	−0.27	1.81
$—O—C_2H_5$	0.10	−0.24	1.68
$—O—C_3H_7$	0.00	−0.25	1.68
$—O—CH(CH_3)_2$	0.05	−0.45	1.62
$—O—C_4H_9$	−0.05	−0.32	1.68
—O—cyclopentyl			1.62
—O—cyclohexyl	0.29		1.81
$—O—CH_2—$cyclohexyl	0.18		1.31
$—O—C_6H_5$	0.25	−0.32	2.43
$—O—CH_2—C_6H_5$		−0.42	
$—OCF_3$	0.40	0.35	
$3,4$-$O—CH_2—O—$		−0.27	
$3,4$-$O—(CH_2—)_2O—$		−0.12	
$—O—CO—CH_3$	0.39	0.31	
$—ONO_2$			3.86
$—O—N{=}C(CH_3)_2$			1.81
$—ONH_3^+$			2.92
$—CH_2—O^-$			0.27

(*Continued*)

TABLE 2.72 Hammett and Taft Substituent Constants (*Continued*)

Substituent	Hammett constants σ_m	Hammett constants σ_p	Taft constant σ^*
$—CH_2—OH$	0.08	0.08	0.31
$—CH_2—O—CH_3$			0.52
$—CH(OH)—CH_3$			0.12
$—CH(OH)—C_6H_5$			0.50
p-$HO—C_6H_4—$		−0.24	
p-$CH_3O—C_6H_4—$		−0.10	
$—CH_2—CH(OH)—CH_3$			−0.06
$—CH_2—C(OH)(CH_3)_2$			−0.25
$—P(CH_3)_2$	0.1	0.05	
$—P(CH_3)_3^+$	0.8	0.9	
$—P(CF_3)_2$	0.6	0.7	
$—PO_3H^-$	0.2	0.26	
$—PO(OC_2H_5)_2$	0.55	0.60	
$—SH$	0.25	0.15	1.68
$—SCH_3$	0.15	0.00	1.56
$—S(CH_3)_2^+$	1.0	0.9	
$—SCH_2CH_3$	0.23	0.03	1.56
$—SCH_2CH_2CH_3$			1.49
$—SCH_2CH_2CH_2CH_3$			1.44
—S—cyclohexyl			1.93
$—SC_6H_5$	0.30		1.87
$—SC(C_6H_5)_3$			0.69
$—SCH_2C_6H_5$			1.56
$—SCH_2CH_2C_6H_5$			1.44
$—CH_2SH$	0.03		0.62
$—CH_2SCH_2C_6H_5$			0.37
$—SCF_3$	0.40	0.50	
$—SCN$	0.63	0.52	3.43
$—S—CO—CH_3$	0.39	0.44	
$—S—CONH_2$	0.34		2.07
$—SO—CH_3$	0.52	0.49	
$—SO—C_6H_5$			3.24
$—CH_2—SO—CH_3$			1.33
$—SO_2—CH_3$	0.60	0.68	3.68
$—SO_2—CH_2CH_3$			3.74
$—SO_2—CH_2CH_2CH_3$			3.68
$—SO_2—C_6H_5$	0.67		3.55
$—SO_2—CF_3$	0.79	0.93	
$—SO_2—NH_2$	0.46	0.57	
$—CH_2—SO_2—CH_3$			1.38
$—SO_3^-$	0.05	0.09	0.81
$—SO_3H$		0.50	
$—SeCH_3$	0.1	0.0	
—Se—cyclohexyl			2.37
$—SeCN$	0.67	0.66	3.61
$—Si(CH_3)_3$	−0.04	−0.07	−0.81
$—Si(CH_2CH_3)_3$		0.0	
$—Si(CH_3)_2C_6H_5$			−0.87
$—Si(CH_3)_2—O—Si(CH_3)_3$			−0.81
$—CH_2Si(CH_3)_3$	−0.16	−0.22	−0.25
$—CH_2CH_2Si(CH_3)_3$			−0.25
$—Sn(CH_3)_3$		0.0	
$—Sn(CH_2CH_3)_3$		0.0	

TABLE 2.73 pK_a° and Rho Values for Hammett Equation

Acid	pK_a°	ρ
Arenearsonic acids		
pK_1	3.54	1.05
pK_2	8.49	0.87
Areneboronic acids (in aqueous 25% ethanol)	9.70	2.15
Arenephosphonic acids		
pK_1	1.84	0.76
pK_2	6.97	0.95
α-Aryladoximes	10.70	0.86
Benzeneseleninic acids	4.78	1.03
Benzenesulfonamides (20°C)	10.00	1.06
Benzenesulfonanilides (20°C)		
$X—C_6H_4—SO_2—NH—C_6H_5$	8.31	1.16
$C_6H_5—SO_2—NH—C_6H_4—X$	8.31	1.74
Benzoic acids	4.21	1.00
Cinnamic acids	4.45	0.47
Phenols	9.92	2.23
Phenylacetic acids	4.30	0.49
Phenylpropiolic acids (in aqueous 35% dioxane)	3.24	0.81
Phenylpropionic acids	4.45	0.21
Phenyltrifluoromethylcarbinols	11.90	1.01
Pyridine-1-oxides	0.94	2.09
2-Pyridones	11.65	4.28
4-Pyridones	11.12	4.28
Pyrroles	17.00	4.28
5-Substituted pyrrole-2carboxylic acids	2.82	1.40
Thiobenzoic acids	2.61	1.0
Thiophenols	6.50	2.2
Trifluoroacetophenone hydrates	10.00	1.11
5-Substituted topolones	6.42	3.10
Protonated cations of		
Acetophenones	−6.0	2.6
Anilines	4.60	2.90
C-Aryl-*N*-dibutylamidines (in aqueous 50% ethanol)	11.14	1.41
N,N-Dimethylanilines	5.07	3.46
Isoquinolines	5.32	5.90
1-Naphthylamines	3.85	2.81
2-Naphthylamines	4.29	2.81
Pyridines	5.18	5.90
Quinolines	4.88	5.90

TABLE 2.74 pK_a° and Rho Values for Taft Equation

Acid	pK°_a	ρ
RCOOH	4.66	1.62
RCH_2COOH	4.76	0.67
RC≡C—COOH	2.39	1.89
H_2C═C(R)—COOH	4.39	0.64
$(CH_3)_2C$═C(R)—COOH	4.65	0.47
cis-C_6H_5—CH═C(R)—COOH	3.77	0.63
trans-C_6H_5—CH═C(R)—COOH	4.61	0.47
R—CO—CH_2—COOH	4.12	0.43
HON═C(R)—COOH	4.84	0.34
RCH_2OH	15.9	1.42
$RCH(OH)_2$	14.4	1.42
R_1CO—NHR_2	22.0	3.1*
CH_3CO—C(R)═C(OH)CH_3	9.25	1.78
CH_3CO—CH(R)—CO—OC_2H_5	12.59	3.44
R—CO—NHOH	9.48	0.98
R_1R_2C═NOH (R_1, R_2 not acyl groups)	12.35	1.18
(R)(CH_3CO)C═NOH	9.00	0.94
$RC(NO_2)_2H$	5.24	3.60
RSH	10.22	3.50
RCH_2SH	10.54	1.47
R—CO—SH	3.52	1.62
Protonated cations of		
RNH_2	10.15	3.14
R_1R_2NH	10.59	3.23
$R_1R_2R_3N$	9.61	3.30
R_1R_2PH	3.59	2.61
$R_1R_2R_3P$	7.85	2.67

*σ^* for R_1CO and R_2.

TABLE 2.75 Special Hammett Sigma Constants

Substituent	σ_m^+	σ_p^+	σ_p^-
—CH_3	−0.07	−0.31	−0.17
—$C(CH_3)_3$	−0.06	−0.26	
—C_6H_5	0.11	−0.18	
—CF_3	0.52	0.61	0.74
—F	0.35	−0.07	0.02
—Cl	0.40	0.11	0.23
—Br	0.41	0.15	0.26
—I	0.36	0.14	
—CN	0.56	0.66	0.88
—CHO			1.13
—$CONH_2$			0.63
—$COCH_3$			0.85
—COOH	0.32	0.42	0.73
—CO—OCH_3	0.37	0.49	0.66
—CO—OCH_2CH_3	0.37	0.48	0.68
—N_2^+			3.2
—NH_2	0.16	−1.3	−0.66
—$N(CH_3)_2$		−1.7	
—$N(CH_3)_3^+$	0.36	0.41	
—NH—CO—CH_3		−0.60	
—NO_2	0.67	0.79	1.25
—OH		−0.92	
—O^-			−0.81
—OCH_3	0.05	−0.78	−0.27
—SF_5			0.70
—SCF_3			0.57
—SO_2CH_3			1.05
—SO_2CF_3			1.36

2.20 POLYMERS

Polymers are mixtures of macromolecules with similar structures and molecular weights that exhibit some average characteristic properties. In some polymers long segments of linear polymer chains are oriented in a regular manner with respect to one another. Such polymers have many of the physical characteristics of crystals and are said to be *crystalline*. Polymers that have polar functional groups show a considerable tendency to be crystalline. Orientation is aided by alignment of dipoles on different chains. Van der Waals' interactions between long hydrocarbon chains may provide sufficient total attractive energy to account for a high degree of regularity within the polymers.

Irregularities such as branch points, comonomer units, and cross-links lead to *amorphous* polymers. They do not have true melting points but instead have glass transition temperatures at which the rigid and glasslike material becomes a viscous liquid as the temperature is raised.

Elastomers. Elastomers is a generic name for polymers that exhibit rubberlike elasticity. Elastomers are soft yet sufficiently elastic that they can be stretched several hundred percent under tension. When the stretching force is removed, they retract rapidly and recover their original dimensions.

Polymers that soften or melt and then solidify and regain their original properties on cooling are called *thermoplastic*. A thermoplastic polymer is usually a single strand of linear polymer with few if any cross-links.

Thermosetting Polymers. Polymers that soften or melt on warming and then become infusible solids are called *thermosetting*. The term implies that thermal decomposition has not taken place.

Thermosetting plastics contain a cross-linked polymer network that extends through the finished article, making it stable to heat and insoluble in organic solvents. Many molded plastics are shaped while molten and are then heated further to become rigid solids of desired shapes.

Synthetic Rubbers. Synthetic rubbers are polymers with rubberlike characteristics that are prepared from dienes or olefins. Rubbers with special properties can also be prepared from other polymers, such as polyacrylates, fluorinated hydrocarbons, and polyurethanes.

Structural Differences. Polymers exhibit structural differences. A *linear* polymer consists of long segments of single strands that are oriented in a regular manner with respect to one another. *Branched* polymers have substituents attached to the repeating units that extend the polymer laterally. When these units participate in chain propagation and link together chains, a *cross-linked* polymer is formed. A *ladder* polymer results when repeating units have a tetravalent structure such that a polymer consists of two backbone chains regularly cross-linked at short intervals.

Generally polymers involve bonding of the most substituted carbon of one monomeric unit to the least substituted carbon atom of the adjacent unit in a *head-to-tail* arrangement. Substituents appear on alternate carbon atoms. *Tacticity* refers to the configuration of substituents relative to the backbone axis. In an *isotactic* arrangement, substituents are on the same plane of the backbone axis; that is, the configuration at each chiral center is identical.

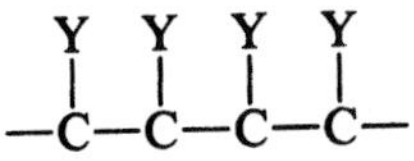

In a *syndiotactic* arrangement, the substituents are in an ordered alternating sequence, appearing alternately on one side and then on the other side of the chain, thus

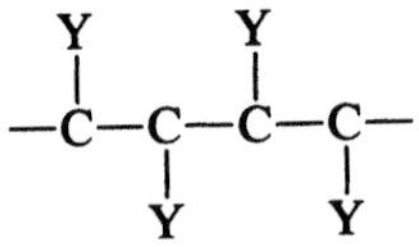

In an *atactic* arrangement, substituents are in an unordered sequence along the polymer chains.

Copolymerization. Copolymerization occurs when a mixture of two or more monomer types polymerizes so that each kind of monomer enters the polymer chain. The fundamental structure resulting from copolymerization depends on the nature of the monomers and the relative rates of monomer reactions with the growing polymer chain. A tendency toward alternation of monomer units is common.

—X—Y—X—Y—X—Y—

Random copolymerization is rather unusual. Sometimes a monomer which does not easily form a homopolymer will readily add to a reactive group at the end of a growing polymer chain. In turn, that monomer tends to make the other monomer much more reactive.

In *graft copolymers* the chain backbone is composed of one kind of monomer and the branches are made up of another kind of monomer.

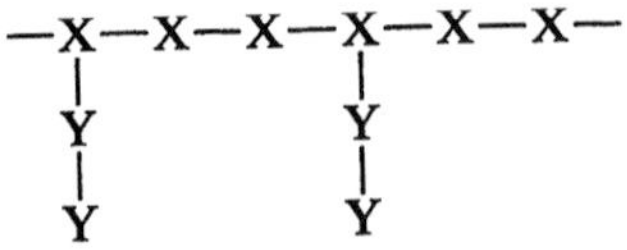

The structure of a *block copolymer* consists of a homopolymer attached to chains of another homopolymer.

—XXXX—YYY—XXXX—YYY—

Configurations around any double bond give rise to cis and trans stereoisomerism.

2.20.1 Additives

Antioxidants

Antioxidants markedly retard the rate of autoxidation throughout the useful life of the polymer. Chain-terminating antioxidants have a reactive —NH or —OH functional group and include compounds such as secondary aryl amines or hindered phenols. They function by transfer of hydrogen to free radicals, principally to peroxy radicals. Butylated hydroxytoluene is a widely used example.

Peroxide-decomposing antioxidants destroy hydroperoxides, the sources of free radicals in polymers. Phosphites and thioesters such as tris(nonylphenyl) phosphite, distearyl pentaerythritol diphosphite, and dialkyl thiodipropionates are examples of peroxide-decomposing antioxidants.

Antistatic Agents

External antistatic agents are usually quaternary ammonium salts of fatty acids and ethoxylated glycerol esters of fatty acids that are applied to the plastic surface. Internal antistatic agents are compounded into plastics during processing. Carbon blacks provide a conductive path through the bulk of the plastic. Other types of internal agents must bloom to the surface after compounding in order to be active. These latter materials are ethoxylated fatty amines and ethoxylated glycerol esters of fatty acids, which often must be individually selected to match chemically each plastic type.

Antistatic agents require ambient moisture to function. Consequently their effectiveness is dependent on the relative humidity. They provide a broad range of protection at 50% relative humidity. Much below 20% relative humidity, only materials which provide a conductive path through the bulk of the plastic to ground (such as carbon black) will reduce electrostatic charging.

Chain-Transfer Agents

Chain-transfer agents are used to regulate the molecular weight of polymers. These agents react with the developing polymer and interrupt the growth of a particular chain. The products, however, are free radicals that are capable of adding to monomers and initiating the formation of new chains. The overall effect is to reduce the average molecular weight of the polymer without reducing the rate of polymerization. Branching may occur as a result of chain transfer between a growing but rather short chain with another and longer polymer chain. Branching may also occur if the radical end of a growing chain abstracts a hydrogen from a carbon atom four or five carbons removed from the end. Thiols are commonly used as chain-transfer agents.

Coupling Agents

Coupling agents are molecular bridges between the interface of an inorganic surface (or filler) and an organic polymer matrix. Titanium-derived coupling agents interact with the free protons at the inorganic interface to form organic monomolecular layers on the inorganic surface. The titanate-coupling-agent molecule has six functions:

$$\overset{1}{(RO)_m}{-}Ti{-}(\overset{2}{O}{-}\overset{3}{Y}{-}\overset{4}{R^2}{-}\overset{5\,6}{Z})_n$$

where

Type	m	n
Monoalkoxy	1	3
Coordinate	4	2
Chelate	1	2

Function 1 is the attachment of the hydrolyzable portion of the molecule to the surface of the inorganic (or proton-bearing) species.
Function 2 is the ability of the titanate molecule to transesterify.
Function 3 affects performance as determined by the chemistry of alkylate, carboxyl, sulfonyl, phenolic, phosphate, pyrophosphate, and phosphite groups.
Function 4 provides van der Waals' entanglement via long carbon chains.
Function 5 provides thermoset reactivity via functional groups such as methacrylates and amines.
Function 6 permits the presence of two or three pendent organic groups. This allows all functionality to be controlled to the first-, second-, or third-degree levels.
Silane coupling agents are represented by the formula

$$Z\text{—}R\text{—}SiY_3$$

where Y represents a hydrolyzable group (typically alkoxy); Z is a functional organic group, such as amino, methacryloxy, epoxy; and R typically is a small aliphatic linkage that serves to attach the functional organic group to silicon in a stable fashion. Bonding to surface hydroxy groups of inorganic compounds is accomplished by the $—SiY_3$ portion, either by direct bonding of this group or more commonly via its hydrolysis product $—Si(OH)_3$. Subsequent reaction of the functional organic group with the organic matrix completes the coupling reaction and establishes a covalent chemical bond from the organic phase through the silane coupling agent to the inorganic phase.

Flame Retardants

Flame retardants are thought to function via several mechanisms, dependent upon the class of flame retardant used. Halogenated flame retardants are thought to function principally in the vapor phase either as a diluent and heat sink or as a free-radical trap that stops or slows flame propagation. Phosphorus compounds are thought to function in the solid phase by forming a glaze or coating over the substrate that prevents the heat and mass transfer necessary for sustained combustion. With some additives, as the temperature is increased, the flame retardant acts as a solvent for the polymer, causing it to melt at lower temperatures and flow away from the ignition source.

Mineral hydrates, such as alumina trihydrate and magnesium sulfate heptahydrate, are used in highly filled thermoset resins.

Foaming Agents (Chemical Blowing Agents)

Foaming agents are added to polymers during processing to form minute gas cells throughout the product. Physical foaming agents include liquids and gases. Compressed nitrogen is often used in injection molding. Common liquid foaming agents are short-chain aliphatic hydrocarbons in the C_5 to C_7 range and their chlorinated or fluorinated analogs.

The chemical foaming agent used varies with the temperature employed during processing. At relatively low temperatures (15 to 200°C), the foaming agent is often 4,4′-oxybis-(benzenesulfonylhydrazide) or *p*-toluenesulfonylhydrazide. In the midrange (160 to 232°C), either sodium hydrogen carbonate or 1,1′ azobisformamide is used. For the high range (200 to 285°C), there are *p*-toluenesulfonyl semicarbazide, 5-phenyltetrazole and analogs, and trihydrazinotriazine.

Inhibitors

Inhibitors slow or stop polymerization by reacting with the initiator or the growing polymer chain. The free radical formed from an inhibitor must be sufficiently unreactive that it does not function as a chain-transfer agent and begin another growing chain. Benzoquinone is a typical free-radical chain inhibitor. The resonance-stabilized free radical usually dimerizes or disproportionates to produce inert products and end the chain process.

Lubricants

Materials such as fatty acids are added to reduce the surface tension and improve the handling qualities of plastic films.

Plasticizers

Plasticizers are relatively nonvolatile liquids which are blended with polymers to alter their properties by intrusion between polymer chains. Diisooctyl phthalate is a common plasticizer. A plasticizer must be compatible with the polymer to avoid bleeding out over long periods of time. Products containing plasticizers tend to be more flexible and workable.

Ultraviolet Stabilizers

2-Hydroxybenzophenones represent the largest and most versatile class of ultraviolet stabilizers that are used to protect materials from the degradative effects of ultraviolet radiation. They function by absorbing ultraviolet radiation and by quenching electronically excited states.

Hindered amines, such as 4-(2,2,6,6-tetramethylpiperidinyl) decanedioate, serve as radical scavengers and will protect thin films under conditions in which ultraviolet absorbers are ineffective. Metal salts of nickel, such as dibutyldithiocarbamate, are used in polyolefins to quench singlet oxygen or electronically excited states of other species in the polymer. Zinc salts function as peroxide decomposers.

Vulcanization and Curing

Originally, vulcanization implied heating natural rubber with sulfur, but the term is now also employed for curing polymers. When sulfur is employed, sulfide and disulfide cross-links form between polymer chains. This provides sufficient rigidity to prevent *plastic flow*. Plastic flow is a process in which coiled polymers slip past each other under an external deforming force; when the force is released, the polymer chains do not completely return to their original positions.

Organic peroxides are used extensively for the curing of unsaturated polyester resins and the polymerization of monomers having vinyl unsaturation. The —O—O— bond is split into free radicals which can initiate polymerization or cross-linking of various monomers or polymers.

Plastics

Homopolymer. Acetal homopolymers are prepared from formaldehyde and consist of high-molecular-weight linear polymers of formaldehyde.

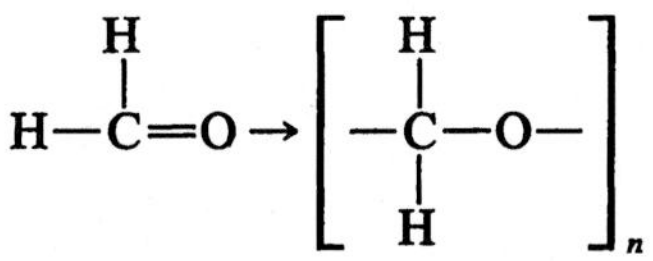

The good mechanical properties of this homopolymer result from the ability of the oxymethylene chains to pack together into a highly ordered crystalline configuration as the polymers change from the molten to the solid state.

Key properties include high melt point, strength and rigidity, good frictional properties, and resistance to fatigue. Higher molecular weight increases toughness but reduces melt flow.

Copolymer.   Acetal copolymers are prepared by copolymerization of 1,3,5-trioxane with small amounts of a comonomer. Carbon-carbon bonds are distributed randomly in the polymer chain. These carbon-carbon bonds help to stabilize the polymer against thermal, oxidative, and acidic attack.

Acrylics

Poly(methyl Methacrylate).   The monomer used for poly(methyl methacrylate), 2-hydroxy-2-methylpropanenitrile, is prepared by the following reaction:

$$CH_3-\underset{\underset{O}{\|}}{C}-CH_3 + HCN \rightarrow CH_3-\underset{\underset{CN}{|}}{\overset{\overset{OH}{|}}{C}}-CH_3$$

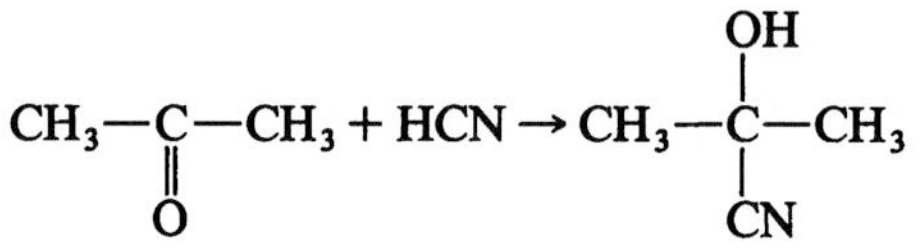

2-Hydroxy-2-methylpropanenitrile is then reacted with methanol (or other alcohol) to yield methacrylate ester. Free-radical polymerization is initiated by peroxide or azo catalysts and produce poly(methyl methacrylate) resins having the following formula:

$$\left[-CH_2-\underset{\underset{COOCH_3}{|}}{\overset{\overset{CH_3}{|}}{C}}-\right]_n$$

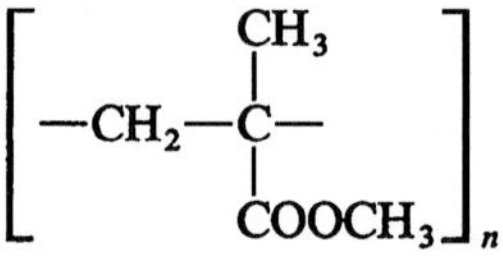

Key properties are improved resistance to heat, light, and weathering. This polymer is unaffected by most detergents, cleaning agents, and solutions of inorganic acids, alkalies, and aliphatic hydrocarbons. Poly(methyl methacrylate) has light transmittance of 92% with a haze of 1 to 3% and its clarity is equal to glass.

Poly(methyl Acrylate).   The monomer used for preparing poly(methyl acrylate) is produced by the oxidation of propylene. The resin is made by free-radical polymerization initiated by peroxide or azo catalysts and has the following formula:

$$\left[-CH_2-\underset{\underset{COOCH_3}{|}}{CH}-\right]_n$$

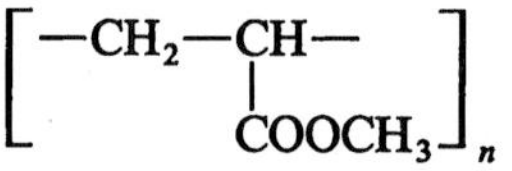

Resins vary from soft, elastic, film-forming materials to hard plastics.

Poly(acrylic Acid) and Poly(methacrylic Acid).   Glacial acrylic acid and glacial methacrylic acid can be polymerized to produce water-soluble polymers having the following structures:

$$\left[-CH_2-\underset{\underset{COOH}{|}}{CH}-\right]_n \qquad \left[-CH_2-\underset{\underset{COOH}{|}}{\overset{\overset{CH_3}{|}}{C}}-\right]_n$$

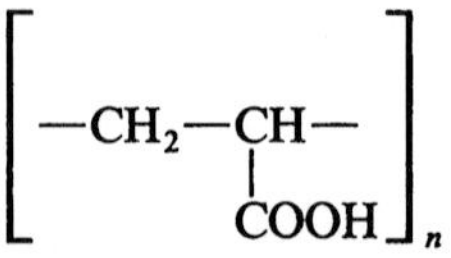

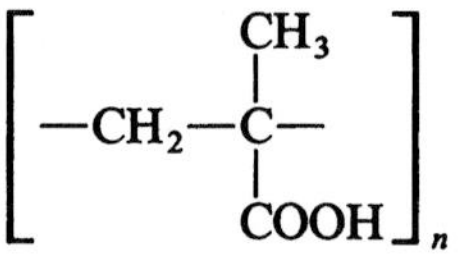

These monomers provide a means for introducing carboxyl groups into copolymers. In copolymers these acids can improve adhesion properties, improve freeze-thaw and mechanical stability of polymer dispersions, provide stability in alkalies (including ammonia), increase resistance to attack by oils, and provide reactive centers for cross-linking by divalent metal ions, diamines, or epoxides.

Functional Group Methacrylate Monomers. Hydroxyethyl methacrylate and dimethylaminoethyl methacrylate produce polymers having the following formulas:

$$\left[-CH_2-\underset{\displaystyle COOCH_2CH_2OH}{\overset{\displaystyle CH_3}{\underset{|}{\overset{|}{C}}}}- \right]_n \qquad \left[-CH_2-\underset{\displaystyle COOCH_2CH_2N(CH_3)_2}{\overset{\displaystyle CH_3}{\underset{|}{\overset{|}{C}}}}- \right]_n$$

The use of hydroxyethyl (also hydroxypropyl) methacrylate as a monomer permits the introduction of reactive hydroxyl groups into the copolymers. This offers the possibility for subsequent cross-linking with an HO-reactive difunctional agent (diisocyanate, diepoxide, or melamine-formaldehyde resin). Hydroxyl groups promote adhesion to polar substrates.

Use of dimethylaminoethyl (also *tert*-butylaminoethyl) methacrylate as a monomer permits the introduction of pendent amino groups which can serve as sites for secondary cross-linking, provide a way to make the copolymer acid-soluble, and provide anchoring sites for dyes and pigments.

Poly(acrylonitrile). Poly(acrylonitrile) polymers have the following formula:

$$\left[-CH_2-\underset{\displaystyle CN}{\underset{|}{CH}}- \right]_n$$

Alkyds

Alkyds are formulated from polyester resins, cross-linking monomers, and fillers of mineral or glass. The unsaturated polyester resins used for thermosetting alkyds are the reaction products of polyfunctional organic alcohols (glycols) and dibasic organic acids.

Key properties of alkyds are dimensional stability, colorability, and arc track resistance. Chemical resistance is generally poor.

Alloys

Polymer alloys are physical mixtures of structurally different homopolymers or copolymers. The mixture is held together by secondary intermolecular forces such as dipole interaction, hydrogen bonding, or van der Waals' forces.

Homogeneous alloys have a single glass transition temperature which is determined by the ratio of the components. The physical properties of these alloys are averages based on the composition of the alloy.

Heterogeneous alloys can be formed when graft or block copolymers are combined with a compatible polymer. Alloys of incompatible polymers can be formed if an interfacial agent can be found.

Allyls

Diallyl Phthalate (and Diallyl 1,3-Phthalate). These allyl polymers are prepared from

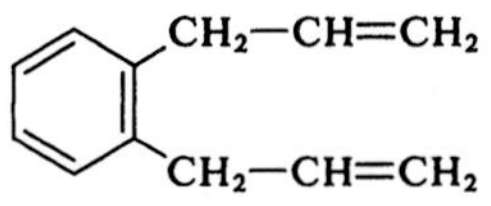

These resulting polymers are solid, linear, internally cyclized, thermoplastic structures containing unreacted allylic groups spaced at regular intervals along the polymer chain.

Molding compounds with mineral, glass, or synthetic fiber filling exhibit good electrical properties under high humidity and high temperature conditions, stable low-loss factors, high surface and volume resistivity, and high arc and track resistance.

Cellulosics

10.3.6.1 Cellulose Triacetate. Cellulose triacetate is prepared according to the following reaction:

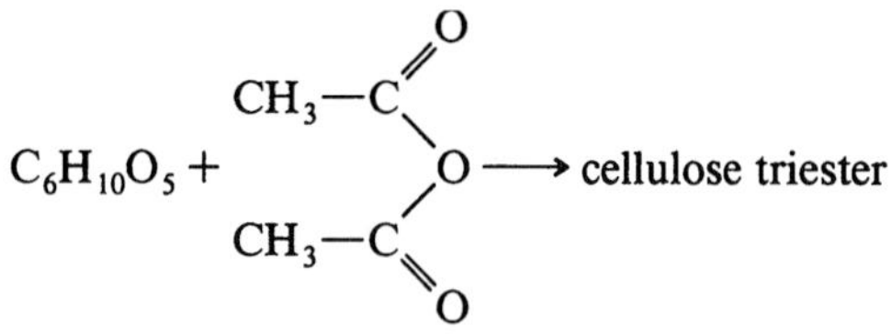

Because cellulose triacetate has a high softening temperature, it must be processed in solution. A mixture of dichloromethane and methanol is a common solvent.

Cellulose triacetate sheeting and film have good gauge uniformity and good optical clarity. Cellulose triacetate products have good dimensional stability and resistance to water and have good folding endurance and burst strength. It is highly resistant to solvents such as acetone. Cellulose triacetate products have good heat resistance and a high dielectric constant.

Cellulose Acetate, Propionate, and Butyrate. Cellulose acetate is prepared by hydrolyzing the triester to remove some of the acetyl groups; the plastic-grade resin contains 38 to 40% acetyl. The propionate and butyrate esters are made by substituting propionic acid and its anhydride (or butyric acid and its anhydride) for some of the acetic acid and acetic anhydride. Plastic grades of cellulose-acetate-propionate resin contain 39 to 47% propionyl and 2 to 9% acetyl; cellulose-acetate-butyrate resins contain 26 to 39% butyryl and 12 to 15% acetyl.

These cellulose esters form tough, strong, stiff, hard plastics with almost unlimited color possibilities. Articles made from these plastics have a high gloss and are suitable for use in contact with food.

Cellulose Nitrate. Cellulose nitrate is prepared according to the following reaction:

$$C_6H_{10}O_5 + HNO_3 \rightarrow [—C_6H_7O_2(OH)(ONO_2)_2—]_n$$

The nitrogen content for plastics is usually about 11%, for lacquers and cement base it is 12%, and for explosives it is 13%. The standard plasticizer added is camphor.

Key properties of cellulose nitrate are good dimensional stability, low water absorption, and toughness. Its disadvantages are its flammability and lack of stability to heat and sunlight.

Ethyl Cellulose. Ethyl cellulose is prepared by reacting cellulose with caustic to form caustic cellulose, which is then reacted with chloroethane to form ethyl cellulose. Plastic-grade material contains 44 to 48% ethoxyl.

Although not as resistant as cellulose esters to acids, it is much more resistant to bases. An outstanding feature is its toughness at low temperatures.

Rayon. Viscose rayon is obtained by reacting the hydroxy groups of cellulose with carbon disulfide in the presence of alkali to give xanthates. When this solution is poured (spun) into an acid medium, the reaction is reserved and the cellulose is regenerated (coagulated).

Epoxy

Epoxy resin is prepared by the following condensation reaction:

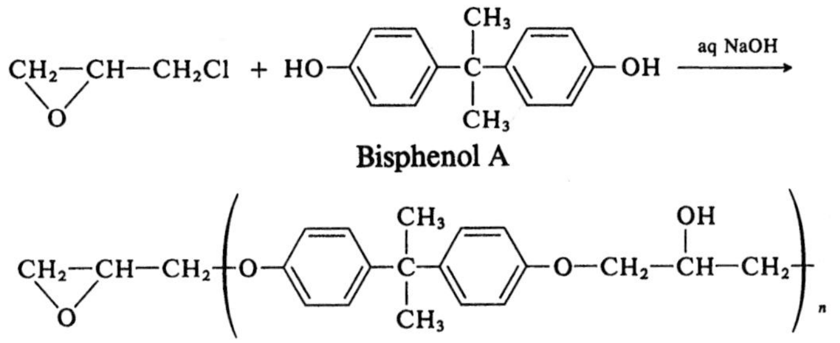

The condensation leaves epoxy end groups that are then reacted in a separate step with nucleophilic compounds (alcohols, acids, or amines). For use as an adhesive, the epoxy resin and the curing resin (usually an aliphatic polyamine) are packaged separately and mixed together immediately before use.

Epoxy novolac resins are produced by glycidation of the low-molecular-weight reaction products of phenol (or cresol) with formaldehyde. Highly cross-linked systems are formed that have superior performance at elevated temperatures.

Fluorocarbon

10.3.8.1 Poly(tetrafluoroethylene). Poly(tetrafluoroethylene) is prepared from tetrafluoroethylene and consists of repeating units in a predominantly linear chain:

$$F_2C{=}CF_2 \rightarrow [{-}CF_2{-}CF_2{-}]_n$$

Tetrafluoroethylene polymer has the lowest coefficient of friction of any solid. It has remarkable chemical resistance and a very low brittleness temperature (−100°C). Its dielectric constant and loss factor are low and stable across a broad temperature and frequency range. Its impact strength is high.

Fluorinated Ethylene-Propylene Resin. Polymer molecules of fluorinated ethylene-propylene consist of predominantly linear chains with this structure:

$$\left[{-}CF_2{-}CF_2{-}CF_2{-}\underset{\displaystyle CF_3}{\underset{|}{CF}}{-}\right]_n$$

Key properties are its flexibility, translucency, and resistance to all known chemicals except molten alkali metals, elemental fluorine and fluorine precursors at elevated temperatures, and concentrated perchloric acid. It withstands temperatures from −270° to 250°C and may be sterilized repeatedly by all known chemical and thermal methods.

Perfluoroalkoxy Resin. Perfluoroalkoxy resin has the following formula:

$$\left[{-}CF_2{-}CF_2{-}\underset{\displaystyle R}{\underset{|}{\underset{\displaystyle O}{\underset{|}{CF}}}}{-}CF_2{-}CF_2{-}\right]_n \qquad \text{where R is } {-}C_nF_{2n+1}$$

It resembles polytetrafluoroethylene and fluorinated ethylene propylene in its chemical resistance, electrical properties, and coefficient of friction. Its strength, hardness, and wear resistance are about equal to the former plastic and superior to that of the latter at temperatures above 150°C.

Poly(vinylidene Fluoride). Poly(vinylidene fluoride) consists of linear chains in which the predominant repeating unit is

$$[-CH_2-CF_2-]_n$$

It has good weathering resistance and does not support combustion. It is resistant to most chemicals and solvents and has greater strength, wear resistance, and creep resistance than the preceding three fluorocarbon resins.

Poly(1-Chloro-1,2,2-Trifluoroethylene). Poly(1-chloro-1,2,2-trifluoroethylene consists of linear chains in which the predominant repeating unit is

$$\left[-CF_2-\underset{\displaystyle Cl}{\underset{|}{C}}F-\right]_n$$

It possesses outstanding barrier properties to gases, especially water vapor. It is surpassed only by the fully fluorinated polymers in chemical resistance. A few solvents dissolve it at temperatures above 100°C, and it is swollen by a number of solvents, especially chlorinated solvents. It is harder and stronger than perfluorinated polymers, and its impact strength is lower.

Ethylene-Chlorotrifluoroethylene Copolymer. Ethylene-chlorotrifluoroethylene copolymer consists of linear chains in which the predominant 1:1 alternating copolymer is

$$\left[-CH_2-CH_2-CF_2-\underset{\displaystyle Cl}{\underset{|}{C}}F-\right]_n$$

This copolymer has useful properties from cryogenic temperatures to 180°C. Its dielectric constant is low and stable over a broad temperature and frequency range .

Ethylene-Tetrafluoroethylene Copolymer. Ethylene-tetrafluoroethylene copolymer consists of linear chains in which the repeating unit is

$$[-CH_2-CH_2-CF_2-CF_2-]_n$$

Its properties resemble those of ethylene-chlorotrifluoroethylene copolymer.

Poly(vinyl Fluoride). Poly(vinyl fluoride) consists of linear chains in which the repeating unit is

$$[-CH_2-CHF-]_n$$

It is used only as a film, and it has good resistance to abrasion and resists staining. It also has outstanding weathering resistance and maintains useful properties from –100 to 150°C.

Nitrile Resins

The principal monomer of nitrile resins is acrylonitrile (see "Polyacrylonitrile"), which constitutes about 70% by weight of the polymer and provides the polymer with good gas barrier and chemical resistance properties. The remainder of the polymer is 20 to 30% methylacrylate (or styrene), with 0 to 10% butadiene to serve as an impact-modifying termonomer.

Melamine Formaldehyde

The monomer used for preparing melamine formaldehyde is formed as follows:

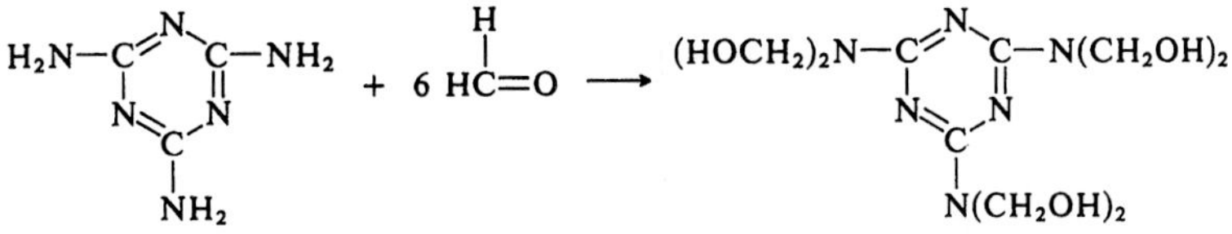

Hexamethylolmelamine

Hexamethylolmelamine can further condense in the presence of an acid catalyst; ether linkages can also form (see "Urea Formaldehyde"). A wide variety of resins can be obtained by careful selection of pH, reaction temperature, reactant ratio, amino monomer, and extent of condensation. Liquid coating resins are prepared by reacting methanol or butanol with the initial methylolated products. These can be used to produce hard, solvent-resistant coatings by heating with a variety of hydroxy, carboxyl, and amide functional polymers to produce a cross-linked film.

Phenolics

Phenol-Formaldehyde Resin. Phenol-formaldehyde resin is prepared as follows:

$$C_6H_5OH + H_2C{=}O \rightarrow [\,{-}C_6H_2(OH)CH_2{-}\,]_n$$

One-Stage Resins. The ratio of formaldehyde to phenol is high enough to allow the thermosetting process to take place without the addition of other sources of cross-links.

Two-Stage Resins. The ratio of formaldehyde to phenol is low enough to prevent the thermosetting reaction from occurring during manufacture of the resin. At this point the resin is termed *novolac* resin. Subsequently, hexamethylenetetramine is incorporated into the material to act as a source of chemical cross-links during the molding operation (and conversion to the thermoset or cured state).

Polyamides

Nylon 6, 11, and 12. This class of polymers is polymerized by addition reactions of ring compounds that contain both acid and amine groups on the monomer.

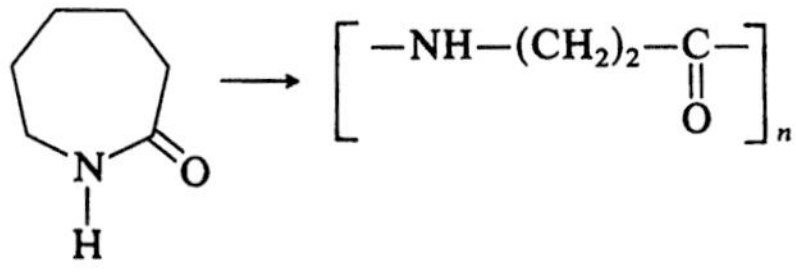

Nylon 6 is polymerized from 2-oxohexamethyleneimine (6 carbons); nylon 11 and 12 are made this way from 11- and 12-carbon rings, respectively.

10.3.12.2 Nylon 6/6, 6/9, and 6/12. As illustrated below, nylon 6/6 is polymerized from 1,6-hexanedioic acid (six carbons) and 1,6-hexanediamine (six carbons).

$$HOOC{-}(CH_2)_4{-}COOH + H_2N{-}CH_2{-}(CH_2)_4{-}CH_2{-}NH_2 \rightarrow$$

1,6-Hexanedioic acid **1,6-Hexanediamine**

$$\left[{-}NH{-}(CH_2)_6{-}NH{-}\underset{\displaystyle O}{\overset{\|}{C}}{-}(CH_2)_4{-}\underset{\displaystyle O}{\overset{\|}{C}}{-}\right]_n$$

Poly(hexamethylene 1,6-hexanediamide)

Other nylons are made this way from direct combinations of monomers to produce types 6/9, 6/10, and 6/12.

Nylon 6 and 6/6 possess the maximum stiffness, strength, and heat resistance of all the types of nylon. Type 6/6 has a higher melt temperature, whereas type 6 has a higher impact resistance and better processibility. At a sacrifice in stiffness and heat resistance, the higher analogs of nylon are useful primarily for improved chemical resistance in certain environments (acids, bases, and zinc chloride solutions) and for lower moisture absorption.

Aromatic nylons, $[—NH—C_6H_4—CO—]_n$ (also called aramids), have specialty uses because of their improved clarity.

Poly(amide-imide)

Poly(amide-imide) is the condensation polymer of 1,2,4-benzenetricarboxylic anhydride and various aromatic diamines and has the general structure:

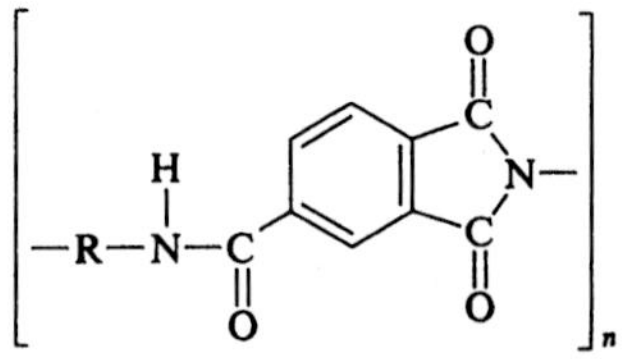

It is characterized by high strength and good impact resistance, and retains its physical properties at temperatures up to 260°C. Its radiation (gamma) resistance is good.

Polycarbonate

Polycarbonate is a polyester in which dihydric (or polyhydric) phenols are joined through carbonate linkages. The general-purpose type of polycarbonate is based on 2,2-bis(4′-hydroxybenzene)propane (bisphenol A) and has the general structure:

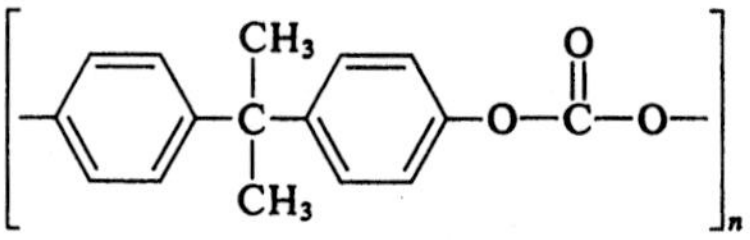

Polycarbonates are the toughest of all thermoplastics. They are window-clear, amazingly strong and rigid, autoclavable, and nontoxic. They have a brittleness temperature of −135°C.

Polyester

Poly(butylene Terephthalate). Poly(butylene terephthalate) is prepared in a condensation reaction between dimethyl terephthalate and 1,4-butanediol and its repeating unit has the general structure

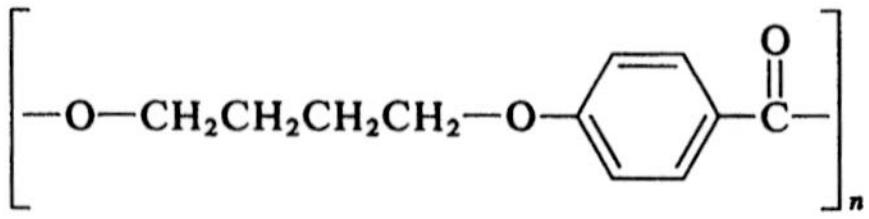

This thermoplastic shows good tensile strength, toughness, low water absorption, and good frictional properties, plus good chemical resistance and electrical properties.

Poly(ethylene Terephthalate). Poly(ethylene terephthalate) is prepared by the reaction of either terephthalic acid or dimethyl terephthalate with ethylene glycol, and its repeating unit has the general structure.

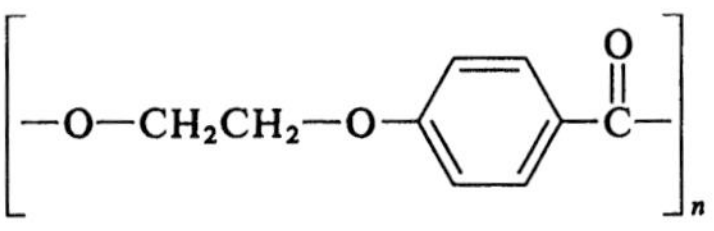

The resin has the ability to be oriented by a drawing process and crystallized to yield a high-strength product.

Unsaturated Polyesters. Unsaturated polyesters are produced by reaction between two types of dibasic acids, one of which is unsaturated, and an alcohol to produce an ester. Double bonds in the body of the unsaturated dibasic acid are obtained by using maleic anhydride or fumaric acid.

PCTA Copolyester. Poly(1,4-cyclohexanedimethylene terephthalic acid) (PCTA) copolyester is a polymer of cyclohexanedimethanol and terephthalic acid, with another acid substituted for a portion of the terephthalic acid otherwise required. It has the following formula:

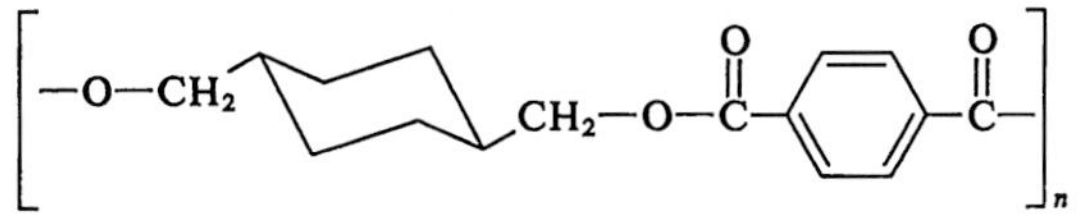

Polyimides. Polyimides have the following formula:

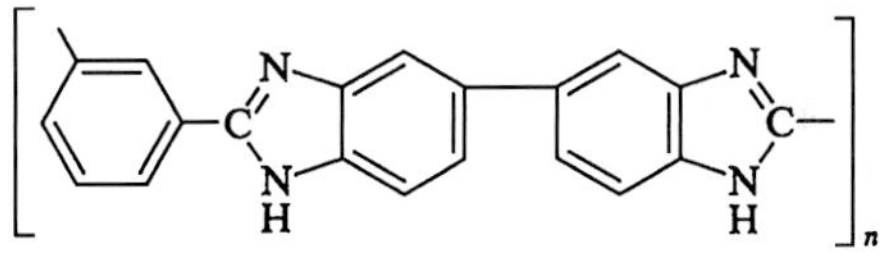

They are used as high-temperature structural adhesives since they become rubbery rather than melt at about 300°C.

Poly(methylpentene)

Poly(methylpentene) is obtained by a Ziegler-type catalytic polymerization of 4-methyl-1-pentene.

Its key properties are its excellent transparency, rigidity, and chemical resistance, plus its resistance to impact and to high temperatures. It withstands repeated autoclaving, even at 150°C.

Polyolefins

10.3.17.1 Polyethylene. Polymerization of ethylene results in an essentially straight-chain high-molecular-weight hydrocarbon.

$$CH_2{=}CH_2 \rightarrow [-CH_2-CH_2-]_n$$

Branching occurs to some extent and can be controlled. Minimum branching results in a "high-density" polyethylene because of its closely packed molecular chains. More branching gives a less compact solid known as "low-density" polyethylene.

A key property is its chemical inertness. Strong oxidizing agents eventually cause some oxidation, and some solvents cause softening or swelling, but there is no known solvent for polyethylene at room temperature. The brittleness temperature is −100°C for both types. Polyethylene has good low-temperature toughness, low water absorption, and good flexibility at subzero temperatures.

Polypropylene. The polymerization of propylene results in a polymer with the following structure:

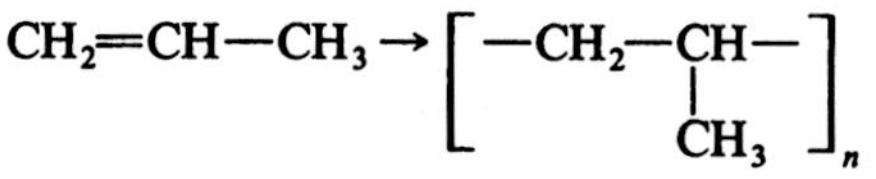

The desired form in homopolymers is the isotactic arrangement (at least 93% is required to give the desired properties). Copolymers have a random arrangement. In block copolymers a secondary reactor is used where active polymer chains can further polymerize to produce segments that use ethylene monomer.

Polypropylene is translucent and autoclavable and has no known solvent at room temperature. It is slightly more susceptible to strong oxidizing agents than polyethylene.

Polybutylene. Polybutylene is composed of linear chains having an isotactic arrangement of ethyl side groups along the chain backbone.

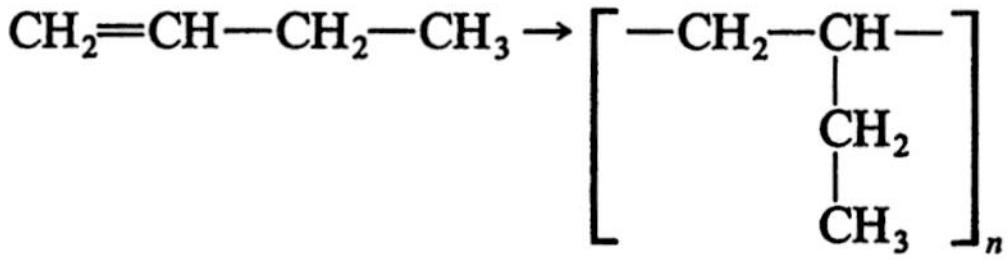

It has a helical conformation in the stable crystalline form.

Polybutylene exhibits high tear, impact, and puncture resistance. It also has low creep, excellent chemical resistance, and abrasion resistance with coilability.

Ionomer. Ionomer is the generic name for polymers based on sodium or zinc salts of ethylene-methacrylic acid copolymers in which interchain ionic bonding, occurring randomly between the long-chain polymer molecules, produces solid-state properties.

The abrasion resistance of ionomers is outstanding, and ionomer films exhibit optical clarity. In composite structures ionomers serve as a heat-seal layer.

Poly(phenylene Sulfide)

Poly(phenylene sulfide) has the following formula:

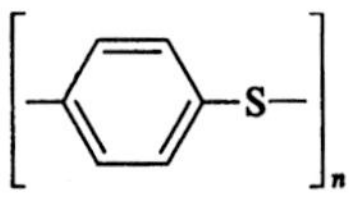

The recurring *para*-substituted benzene rings and sulfur atoms form a symmetrical rigid backbone.

The high degree of crystallization and the thermal stability of the bond between the benzene ring and sulfur are the two properties responsible for the polymer's high melting point, thermal stability, inherent flame retardance, and good chemical resistance. There are no known solvents of poly (phenylene sulfide) that can function below 205°C.

Polyurethane

10.3.19.1 Foams. Polyurethane foams are prepared by the polymerization of polyols with isocyanates.

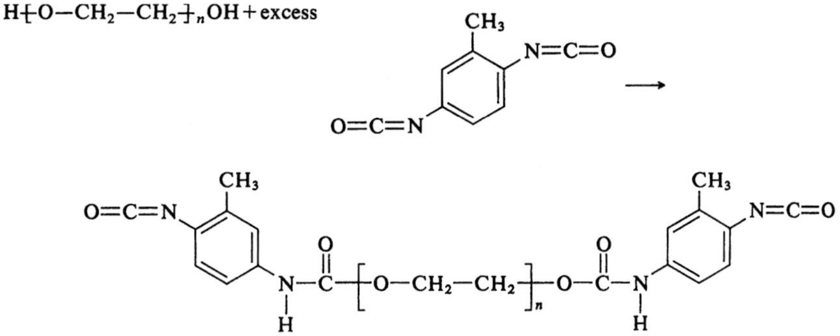

Commonly used isocyanates are toluene diisocyanate, methylene diphenyl isocyanate, and polymeric isocyanates. Polyols used are macroglycols based on either polyester or polyether. The former [poly(ethylene phthalate) or poly(ethylene 1,6-hexanedioate)] have hydroxyl groups that are free to react with the isocyanate. Most flexible foam is made form 80/20 toluene diisocyanate (which refers to the ratio of 2,4-toluene diisocyanate to 2,6-toluene diisocyanate). High-resilience foam contains about 80% 80/20 toluene diisocyanate and 20% poly(methylene diphenyl isocyanate), while semiflexible foam is almost always 100% poly(methylene diphenyl isocyanate). Much of the latter reacts by trimerization to form isocyanurate rings.

Flexible foams are used in mattresses, cushions, and safety applications. Rigid and semiflexible foams are used in structural applications and to encapsulate sensitive components to protect them against shock, vibration, and moisture. Foam coatings are tough, hard, flexible, and chemically resistant.

Elastomeric Fiber. Elastomeric fibers are prepared by the polymerization of polymeric polyols with diisocyanates.

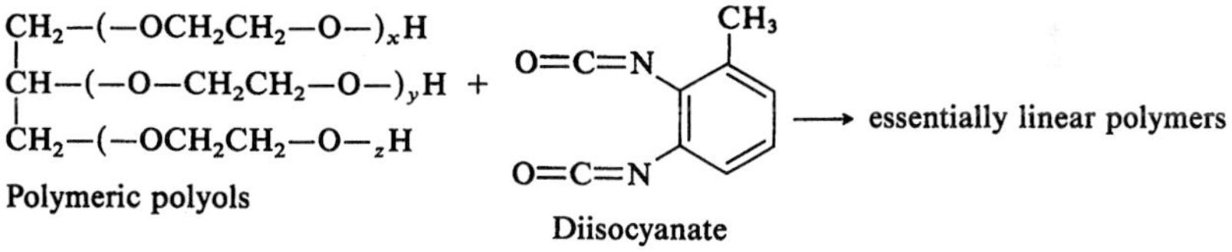

The structure of elastomeric fibers is similar to that illustrated for polyurethane foams.

Silicones

Silicones are formed in the following multistage reaction :

$$R_2SiCl_2 + 2H_2O \rightarrow R_2Si(OH)_2 + 2HCl$$
$$\downarrow$$
$$[\text{—Si(R)}_2\text{—O—}]_n$$

The silanols formed above are unstable and under dehydration. On polycondensation, they give polysiloxanes (or silicones) which are characterized by their three-dimensional branched-chain structure. Various organic groups introduced within the polysiloxane chain impart certain characteristics and properties to these resins.

Methyl groups impart water repellency, surface hardness, and noncombustibility.

Phenyl groups impart resistance to temperature variations, flexibility under heat, resistance to abrasion, and compatibility with organic products.

Vinyl groups strengthen the rigidity of the molecular structure by creating easier cross-linkage of molecules.

Methoxy and alkoxy groups facilitate cross-linking at low temperatures.

Oils and gums are nonhighly branched- or straight-chain polymers whose viscosity increases with the degree of polycondensation.

Styrenics

Polystyrene Polystyrene has the following formula:

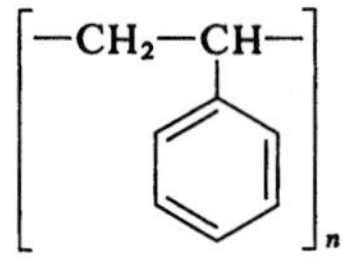

Polystyrene is rigid with excellent dimensional stability, has good chemical resistance to aqueous solutions, and is an extremely clear material.

Impact polystyrene contains polybutadiene added to reduce brittleness. The polybutadiene is usually dispersed as a discrete phase in a continuous polystyrene matrix. Polystyrene can be grafted onto rubber particles, which assures good adhesion between the phases.

Acrylonitrile-Butadiene-Styrene (ABS) Copolymers. This basic three-monomer system can be tailored to yield resins with a variety of properties. Acrylonitrile contributes heat resistance, high strength, and chemical resistance. Butadiene contributes impact strength, toughness, and retention of low-temperature properties. Styrene contributes gloss, processibility, and rigidity. ABS polymers are composed of discrete polybutadiene particles grafted with the styrene-acrylonitrile copolymer; these are dispersed in the continuous matrix of the copolymer.

Styrene-Acrylonitrile (SAN) Copolymers. SAN resins are random, amorphous copolymers whose properties vary with molecular weight and copolymer composition. An increase in molecular weight or in acrylonitrile content generally enhances the physical properties of the copolymer but at some loss in case of processing and with a slight increase in polymer color.

SAN resins are rigid, hard, transparent thermoplastics which process easily and have good dimensional stability—a combination of properties unique in transparent polymers.

Sulfones

Below are the formulas for three polysulfones.

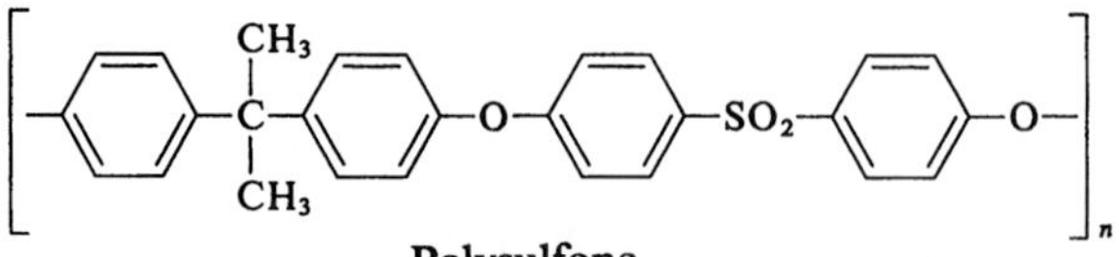

Polysulfone

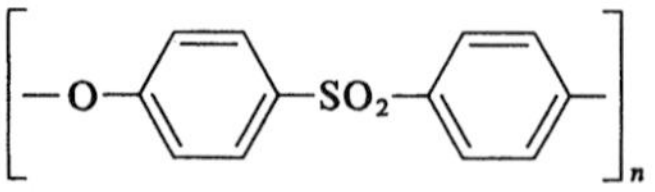

Poly(ester sulfone)

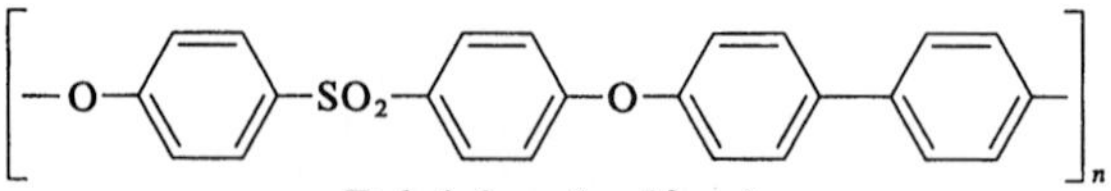

Poly(phenyl sulfone)

The isopropylidene linkage imparts chemical resistance, the ether linkage imparts temperature resistance, and the sulfone linkage imparts impact strength. The brittleness temperature of polysulfones is −100°C. Polysulfones are clear, strong, nontoxic, and virtually unbreakable. They do not hydrolyze during autoclaving and are resistant to acids, bases, aqueous solutions, aliphatic hydrocarbons, and alcohols.

Thermoplastic Elastomers

Polyolefins. In these thermoplastic elastomers the hard component is a crystalline polyolefin, such as polyethylene or polypropylene, and the soft portion is composed of ethylene-propylene rubber. Attractive forces between the rubber and resin phases serve as labile cross-links. Some contain a chemically cross-linked rubber phase that imparts a higher degree of elasticity.

Styrene-Butadiene-Styrene Block Copolymers. Styrene blocks associate into domains that form hard regions. The midblock, which is normally butadiene, ethylene-butene, or isoprene blocks, forms the soft domains. Polystyrene domains serve as cross-links.

Polyurethanes. The hard portion of polyurethane consists of a chain extender and polyisocyanate. The soft component is composed of polyol segments.

Polyesters. The hard portion consists of copolyester, and the soft portion is composed of polyol segments.

Vinyl

Poly(vinyl Chloride) (PVC). Polymerization of vinyl chloride results in the formation of a polymer with the following formula:

$$CH_2{=}CHCl \rightarrow \left[-CH_2-\underset{\displaystyle Cl}{\underset{|}{CH}}- \right]_n$$

When blended with phthalate ester plasticizers, PVC becomes soft and pliable.

Its key properties are good resistance to oils and a very low permeability to most gases.

Poly(vinyl Acetate) Poly(vinyl acetate) has the following formula:

$$\left[-CH_2-\underset{\displaystyle O-CO-CH_3}{\underset{|}{CH}}- \right]_n$$

Poly(vinyl acetate) is used in latex water paints because of its weathering, quick-drying, recoatability, and self-priming properties. It is also used in hot-melt and solution adhesives.

Poly(vinyl Alcohol) Poly(vinyl alcohol) has the following formula:

$$\left[-CH_2-\underset{\displaystyle OH}{\underset{|}{CH}}- \right]_n$$

It is used in adhesives, paper coating and sizing, and textile warp size and finishing applications.

Poly(vinyl Butyral) Poly(vinyl butyral) is prepared according to the following reaction:

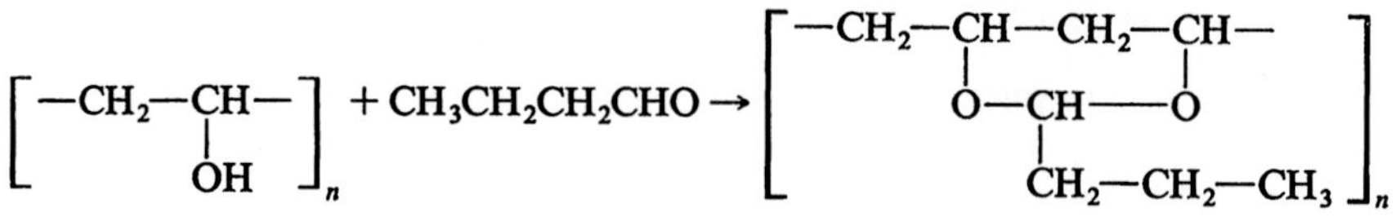

$$\left[-CH_2-\underset{\underset{OH}{|}}{CH}- \right]_n + CH_3CH_2CH_2CHO \rightarrow \left[\begin{array}{c} -CH_2-\underset{\underset{O-CH-O}{|}}{CH}-CH_2-CH- \\ \underset{\underset{CH_2-CH_2-CH_3}{|}}{} \end{array} \right]_n$$

Its key characteristics are its excellent optical and adhesive properties. It is used as the interlayer film for safety glass.

Poly(vinylidene Chloride)   Poly(vinylidene chloride) is prepared according to the following reaction:

$$CH_2{=}CCl_2 + CH_2{=}CHCl \rightarrow [-CH_2-CCl_2-CH_2-CHCl-]_n$$
Random copolymer

Urea Formaldehyde

The reaction of urea with formaldehyde yields the following products, which are used as monomers in the preparation of urea formaldehyde resin.

$$H_2N-CO-NH_2 + H_2CO \rightarrow H_2N-CO-NH-CH_2OH$$
$$+ HOCH_2-NH-CO-NH-CH_2OH$$

The reaction conditions can be varied so that only one of those monomers is formed. 1-Hydroxy-methylurea and 1,3-bis(hydroxymethyl)urea condense in the presence of an acid catalyst to produce urea formaldehyde resins. A wide variety of resins can be obtained by careful selection of the pH, reaction temperature, reactant ratio, amino monomer, and degree of polymerization. If the reaction is carried far enough, an infusible polymer network is produced.

Liquid coating resins are prepared by reacting methanol or butanol with the initial hydroxy-methylureas. Ether exchange reactions between the amino resin and the reactive sites on the polymer produce a cross-linked film.

2.20.3   Rubber

Gutta Percha

Gutta percha is a natural polymer of isoprene (3-methyl-1,3-butadiene) in which the configuration around each double bond is *trans*. It is hard and horny and has the following formula:

$$\left[\begin{array}{c} \underset{\underset{C}{||}}{CH_3} \\ CH_2{}^{\diagup}{}^{\diagdown}{}_{CH}{}^{\diagup}{}^{CH_2}{}^{\diagdown} \end{array} \right]_n$$

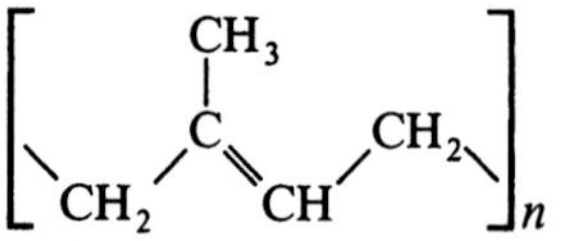

Natural Rubber

Natural rubber is a polymer of isoprene in which the configuration around each double bond is *cis* (or *Z*):

$$\left[\begin{array}{c} H_3C{}^{\diagdown}{}_{C=CH}{}^{\diagdown} \\ -CH_2{}^{\diagup} \qquad CH_2- \end{array} \right]_n$$

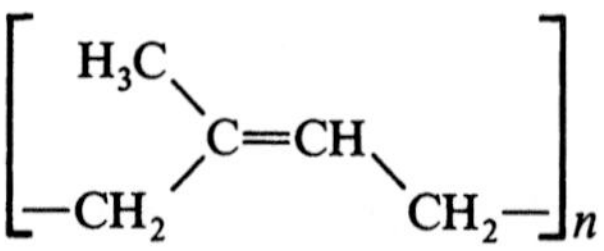

Its principal advantages are high resilience and good abrasion resistance.

Chlorosulfonated Polyethylene

Chlorosulfonated polyethylene is prepared as follows:

$$[-CH_2-CH_2-]_n + HSO_3Cl \rightarrow \left[-CH_2-\underset{\displaystyle SO_3H}{\underset{|}{CH}}-\right]_n + HCl$$

Cross-linking, which can occur as a result of side reactions, causes an appreciable gel content in the final product.

The polymer can be vulcanized to give a rubber with very good chemical (solvent) resistance, excellent resistance to aging and weathering, and good color retention in sunlight.

Epichlorohydrin

Epichlorohydrin is a product of covulcanization of epichlorohydrin (epoxy) polymers with rubbers, especially *cis*-polybutadiene.

Its advantages include impermeability to air, excellent adhesion to metal, and good resistance to oils, weathering, and low temperature.

Nitrile Rubber (NBR, GRN, Buna N)

Nitrile rubber can be prepared as follows:

$$\underset{\text{2 parts}}{CH_2{=}CH-CH{=}CH_2} + \underset{\text{1 part}}{CH_2{=}CH-CN} \rightarrow$$

$$\left[-CH_2-CH{=}CH-CH_2-CH_2-\underset{\displaystyle CN}{\underset{|}{CH}}-CH_2-CH{=}CH-CH_2-\right]_n$$

Nitrile rubber is also known as nitrile-butadiene rubber (NBR), government rubber nitrile (GRN), and Buna N.

It possesses resistance to oils up to 120°C and excellent abrasion resistance and adhesion to metal.

Polyacrylate

Polyacrylate has the following formula:

$$\left[-CH_2-\underset{\displaystyle CN}{\underset{|}{CH}}-\right]_n$$

It possesses oil and heat resistance to 175°C and excellent resistance to ozone.

cis-Polybutadiene Rubber (BR)

cis-Polybutadiene is prepared by polymerization of butadiene by mostly, 1,4-addition.

$$CH_2{=}CH-CH{=}CH_2 \rightarrow [-CH_2-CH{=}CH-CH_2-]_n$$

The polybutadiene produced is in the *Z* (or *cis*) configuration.

cis-Polybutadiene has good abrasion resistance, is useful at low temperature, and has excellent adhesion to metal.

Polychloroprene (Neoprene)

Polychloroprene is prepared as follows:

$$CH_2=CH-\underset{\underset{Cl}{|}}{C}=CH_2 \rightarrow [-CH_2-CH=C(Cl)-CH_2-]_n$$

It has very good weathering characteristics, is resistant to ozne and to oil, and is heat-resistant to 100°C.

Ethylene-Propylene-Diene Rubber (EPDM)

Ethylene-propylene-diene rubber is polymerized from 60 parts ethylene, 40 parts propylene, and a small amount of nonconjugated diene. The nonconjugated diene permits sulfur vulcanization of the polymer instead of using peroxide.

It is a very lightweight rubber and has very good weathering and electrical properties, excellent adhesion, and excellent ozone resistance.

Polyisobutylene (Butyl Rubber)

Polyisobutylene is prepared as follows:

$$\underset{\text{98 parts}}{H_3C-\underset{\underset{CH_3}{|}}{C}=CH_2} + \underset{\text{2 parts}}{CH_2=\underset{\underset{CH_3}{|}}{C}-CH=CH_2} \rightarrow$$

$$\left[\left(-\underset{\underset{CH_3}{|}}{\overset{\overset{CH_3}{|}}{C}}-CH_2- \right)_n -CH_2-\underset{\underset{CH_3}{|}}{C}=CH-CH_2- \right]$$

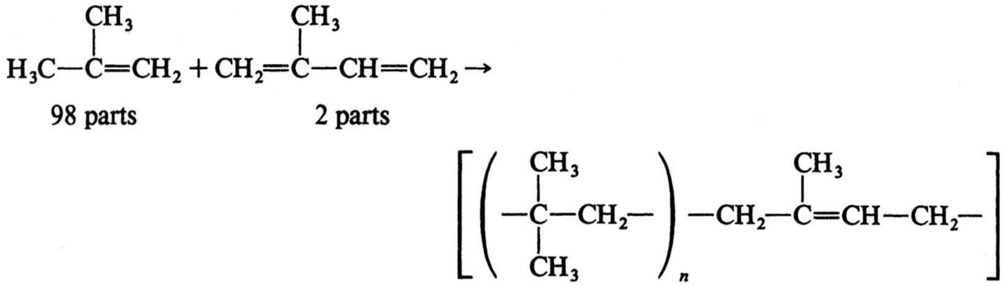

It possesses excellent ozone resistance, very good weathering and electrical properties, and good heat resistance.

(Z)-Polyisoprene (Synthetic Natural Rubber)

Polymerization of isoprene by 1,4-addition produces polyisoprene that has a *cis* (or Z) configuration.

$$\left[\underset{-CH_2}{\overset{H_3C}{\diagdown}} C = C \underset{CH_2-}{\overset{H}{\diagup}} \right]_n$$

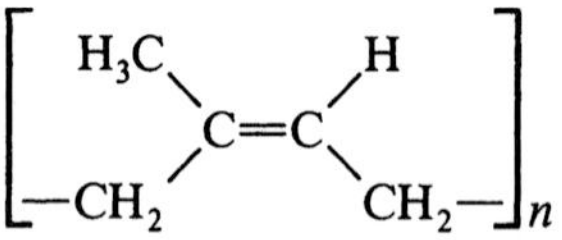

Polysulfide Rubbers

Polysulfide rubbers are prepared as follows:

$$Cl-R-Cl + Na-S-S-S-S-Na \rightarrow HS[-R-S-S-S-S-]_nR-SH$$

where R can be

$$-CH_2CH_2-,\ -CH_2CH_2-O-CH_2CH_2-,$$

or

$$-CH_2CH_2-O-CH_2-O-CH_2CH_2-.$$

Polysulfide rubbers posses excellent resistance to weathering and oils and have very good electrical properties.

Poly(vinyl Chloride) (PVC)

Poly(vinyl chloride) has the following structures:

$$\left[\begin{array}{c} -CH_2-CH- \\ | \\ Cl \end{array} \right]_n$$

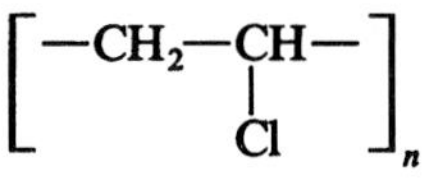

PVC polymer plus special plasticizers are used to produce flexible tubing which has good chemical resistance.

Silicone Rubbers

Silicone rubbers are prepared as follows:

$$Cl-\underset{\underset{CH_3}{|}}{\overset{\overset{CH_3}{|}}{Si}}-Cl \xrightarrow{H_2O} HO-\underset{\underset{CH_3}{|}}{\overset{\overset{CH_3}{|}}{Si}}-OH \xrightarrow{polymerize} \left[-\underset{\underset{CH_3}{|}}{\overset{\overset{CH_3}{|}}{Si}}-O- \right]_n$$

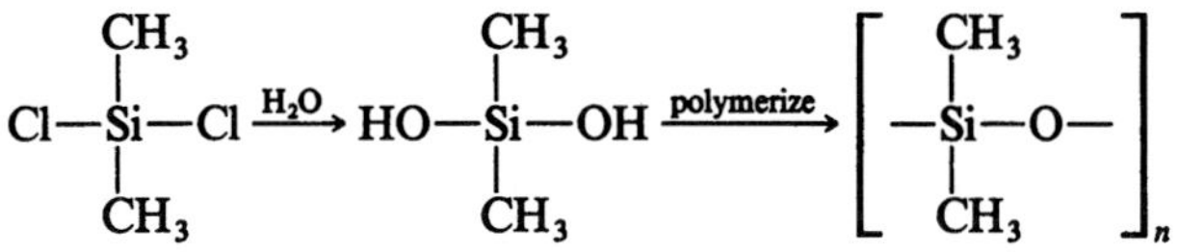

Other groups may replace the methyl groups.
Silicone rubbers have excellent ozone and weathering resistance, good electrical properties, and good adhesion to metal.

Styrene-Butadiene Rubber (GRS, SBR, Buna S)

Styrene-butadiene rubber is prepared from the free-radical copolymerization of one part by weight of styrene and three parts by weight of 1,3-butadiene. The butadiene is incorporated by both 1,4-addition (80%) and 1,2-addition (20%). The configuration around the double bond of the 1,4-adduct is about 80% *trans*. The product is a random copolymer with these general features:

| *trans*-1,4-Adduct | 1,2-Adduct | *trans*-1,4-Adduct | Styrene | *cis*-1,4-Adduct |

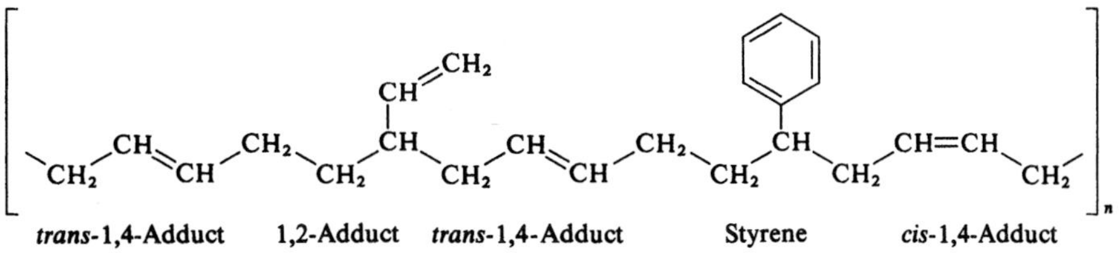

Styrene-butadiene rubber (SBR) is also known as government rubber styrene (GRS) and Buna S.

Urethane

See Table 2.79

TABLE 2.76 Names and Structures of Polymers

Common name	Acronym, alternate name	Class	Structure of repeat unit
Amylose		Polysaccharide	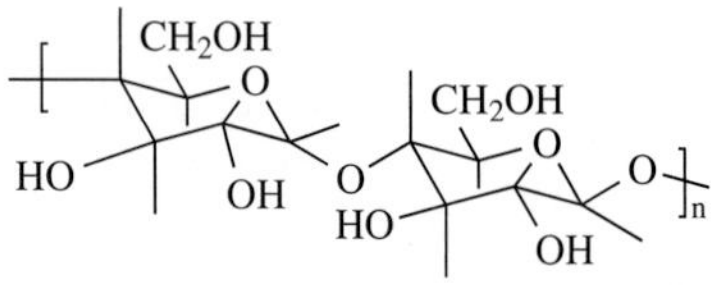
Cellulose	Rayon Cellophane Regenerated cellulose	Polysaccharide	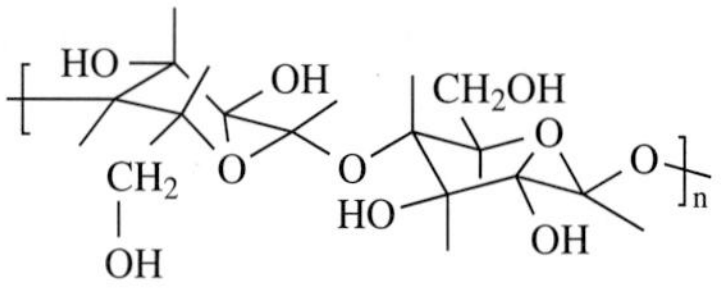
Cellulose acetate	CA	Cellulose ester	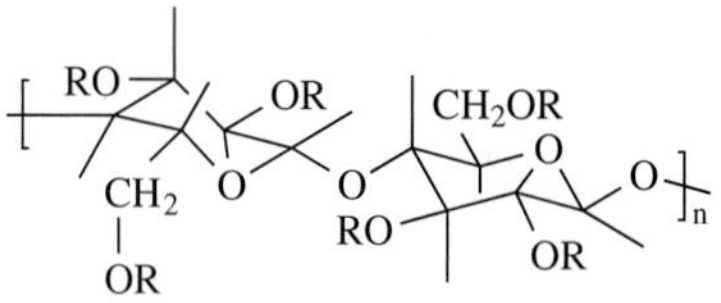 $R = -\overset{O}{\overset{\parallel}{C}}-CH_3$
Cellulose nitrate	CN	Cellulose ester	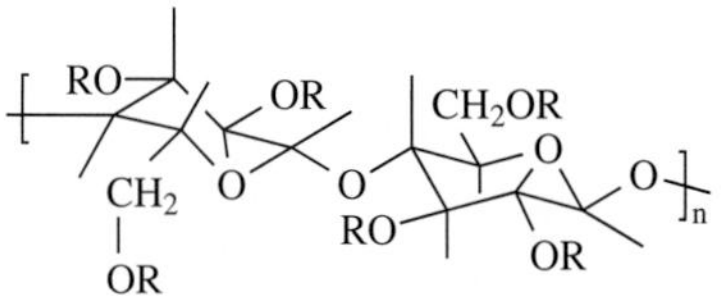 $R = -NO_2$
Hydroxypropylcellulose	HPC	Cellulose ester	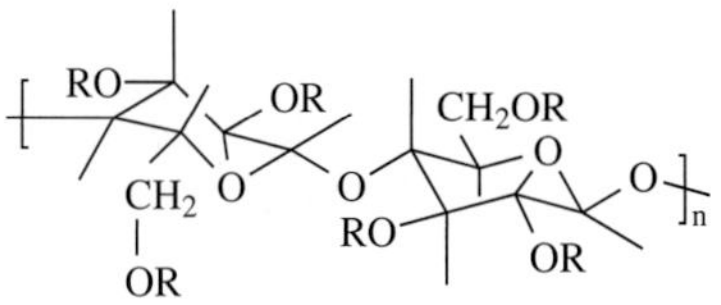 $R = -(CH_2)_3-OH$
Ladder polymer	Double-strand polymer		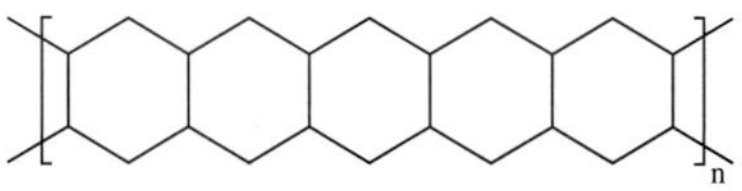
Phenol-formaldehyde	Bakelite	Phenolic polymer	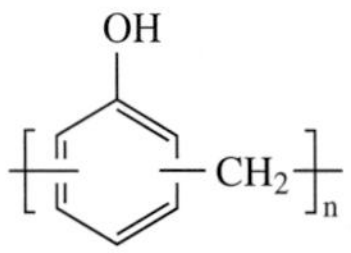

TABLE 2.76 Names and Structures of Polymers (*Continued*)

Common name	Acronym, alternate name	Class	Structure of repeat unit
Polyacetal		Polyacetal	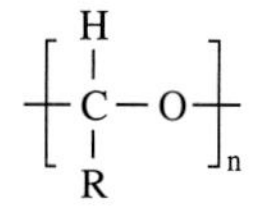
Polyacetylene		Polyalkyne	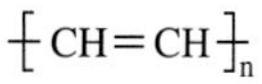
Polyacrylamide		Vinyl polymer	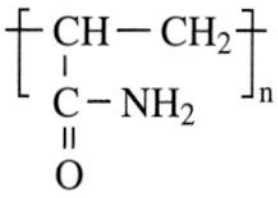
Poly(acrylic acid)		Vinyl polymer	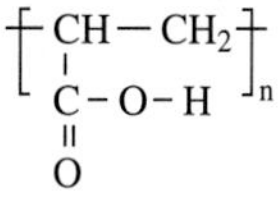
Polyacrylonitrile	PAN	Vinyl polymer	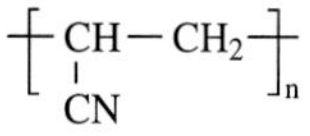
Poly(L-alanine)		Polypeptide	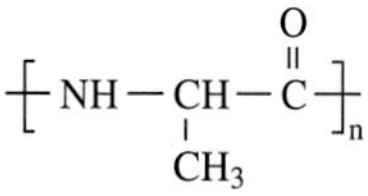
Polyamide	Nylon	Polyamide	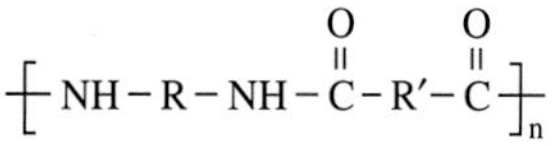
Polyaniline		Polyamine	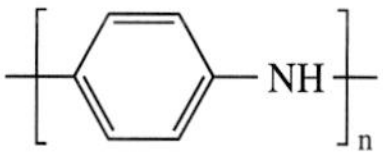
Polybenzimidazole	PBI	Polyhetero-aromatic	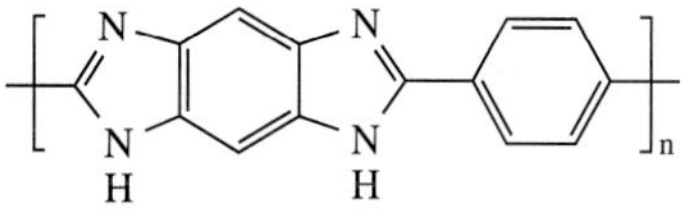
Polybenzobisoxazole	PBO	Polyhetero-aromatic	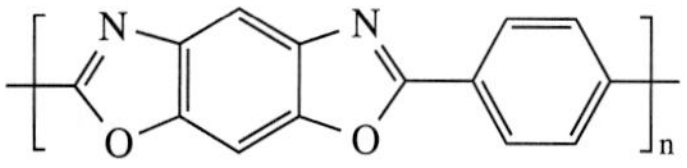
Polybenzobisthiazole	PBT	Polyhetero-aromatic	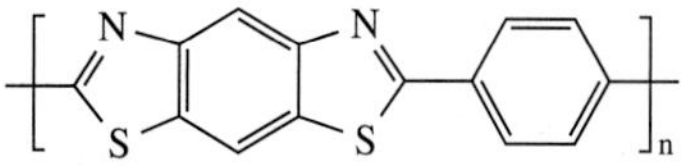

(*Continued*)

TABLE 2.76 Names and Structures of Polymers (*Continued*)

Common name	Acronym, alternate name	Class	Structure of repeat unit
Poly(γ-benzyl-L-glutamate)	PBLG	Polypeptide	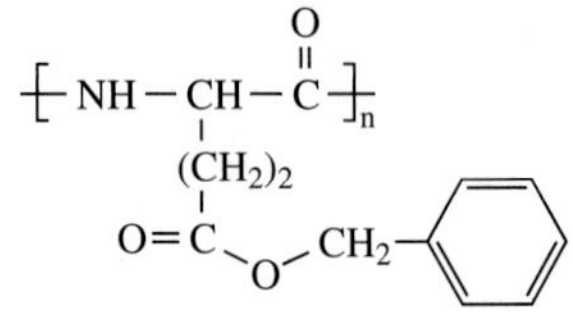
1,2-Polybutadiene	PBD	Diene polymer	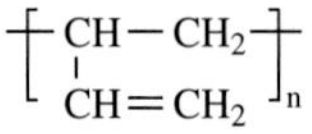
cis-1,4-Polybutadiene	PBD	Diene polymer	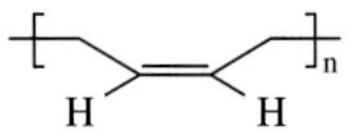
trans-1,4-Polybutadiene	PBD	Diene polymer	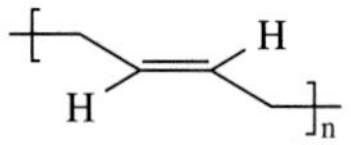
Poly(butene-1)	PB-1	Poly(α-olefin)	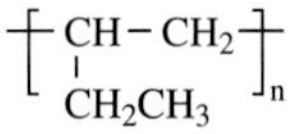
Polybutylene-terephthalate	PBT	Polyester	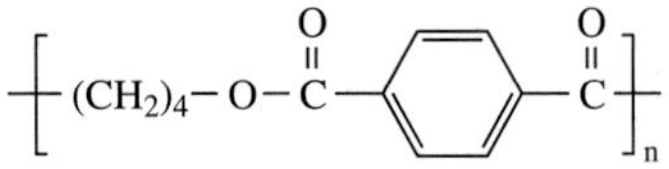
Poly(ϵ-caprolactam)	Nylon-6	Polyamide	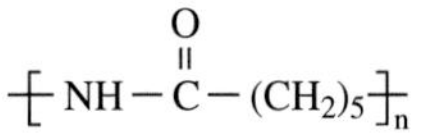
Poly(ϵ-caprolactone)		Polyester	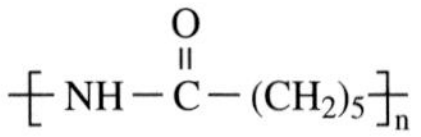
Polycarbonate	PC	Polyester	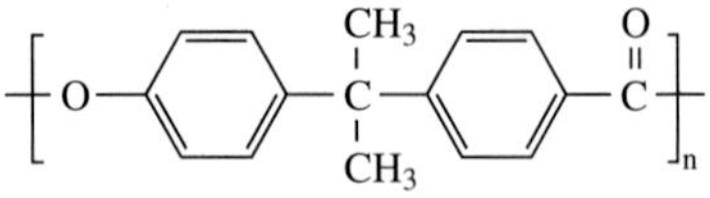
cis, trans-1,4-Polychloroprene	Neoprene	Diene polymer	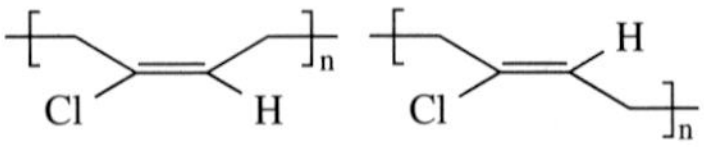 *cis* *trans*
Polychlorotrifluoro ethylene	PCTFE	Vinyl polymer	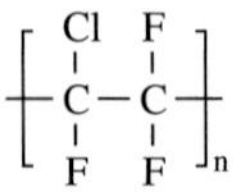

TABLE 2.76 Names and Structures of Polymers (*Continued*)

Common name	Acronym, alternate name	Class	Structure of repeat unit
Polydiethylsiloxane	PDES	Polysiloxane	$-[Si(CH_2CH_3)_2-O]_n-$
Polydimethylsiloxane	PDMS	Polysiloxane	$-[Si(CH_3)_2-O]_n-$
Polydiphenylsiloxane	PDPS	Polysiloxane	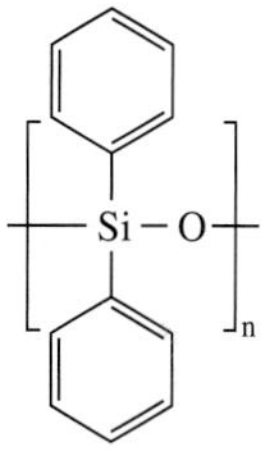
Polyester		Polyester	$-[O-R-O-C(=O)-R'-C(=O)]_n-$
Polyetheretherketone	PEEK	Polyketone	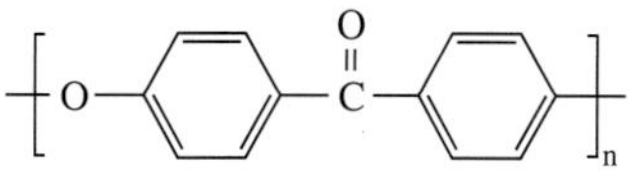
Polyethylene	PE	Polyolefin	$-[CH_2-CH_2]_n-$
Poly(ethylene imine)		Polyamine	$-[CH_2-CH_2-NH]_n-$
Poly(ethylene oxide) [Poly(ethylene glycol)]	PEO (PEG)	Polyether	$-[CH_2-CH_2-O]_n-$
Polyethylene-terephthalate	PET	Polyester	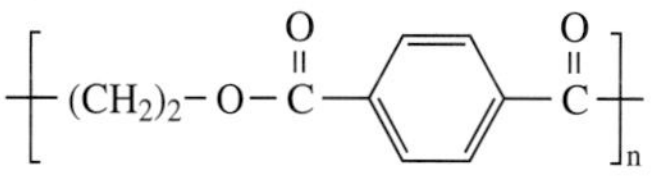
Polyglycine		Polypeptide	$-[NH-CH_2-C(=O)]_n-$
Poly(hexamethylene adipamide)	Nylon-66	Polyamide	$-[NH-(CH_2)_6-NH-C(=O)-(CH_2)_4-C(=O)]_n-$
Polyhydroxybutyrate	PHB	Polyester	$-[O-CH(CH_3)-CH_2-C(=O)]_n-$

(*Continued*)

TABLE 2.76 Names and Structures of Polymers (*Continued*)

Common name	Acronym, alternate name	Class	Structure of repeat unit
Polyimide	PI	Polyimide	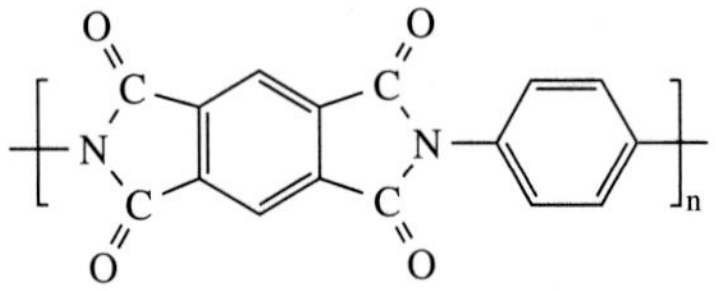
Poly(imino-1,3-phenylene iminoisophthaloyl) (Nomex)		Polyaramide	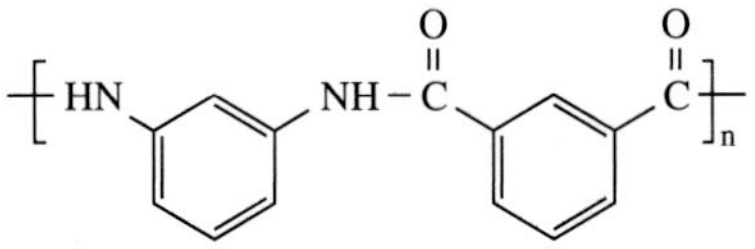
Poly(imino-1,4-phenylene iminoterephthaloyl) (Kevlar)		Polyaramide	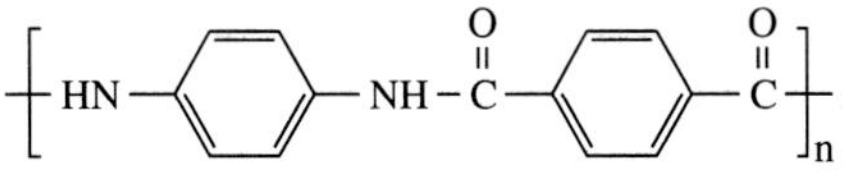
Polyisobutylene	Butyl rubber	Vinylidene polymer	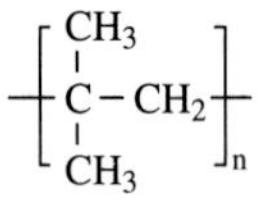
Polyisocyanate	PIC	Polyamide	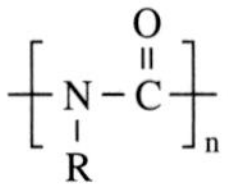
Polyisocyanide		Polyisocyanide	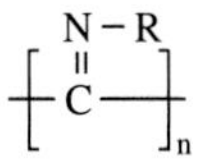
cis-1,4-Polyisoprene	*cis*-PIP, Natural rubber	Diene polymer	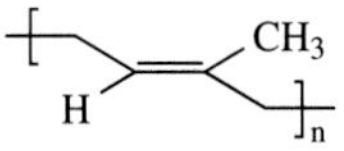
tran-1,4-Polyisoprene	*trans*-PIP, Gutta percha	Diene polymer	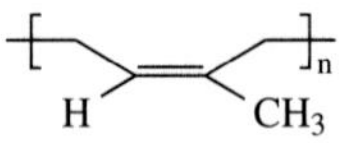
Polylactam		Polyamide	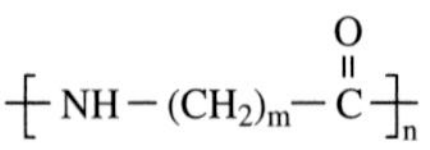
Polylactone		Polyester	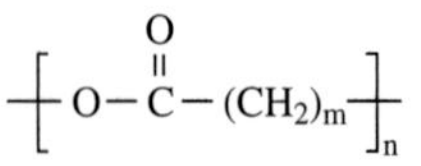
Poly(*p*-methyl styrene)		Vinyl polymer	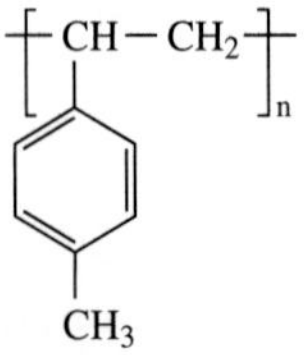

TABLE 2.76 Names and Structures of Polymers (*Continued*)

Common name	Acronym, alternate name	Class	Structure of repeat unit
Poly(methyl acrylate)	PMA	Vinyl polymer	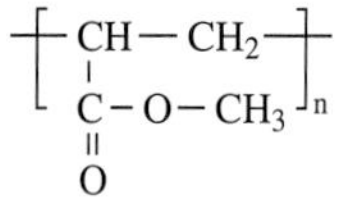 $\left[\begin{array}{c}\text{CH}-\text{CH}_2\\ \mid\\ \text{C}-\text{O}-\text{CH}_3\\ \parallel\\ \text{O}\end{array}\right]_n$
Poly(methyl methacrylate)	PMMA	Vinylidene polymer	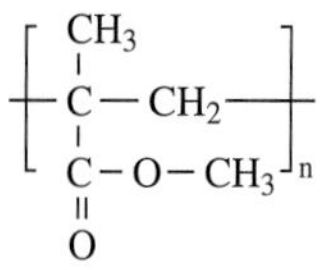 $\left[\begin{array}{c}\text{CH}_3\\ \mid\\ \text{C}-\text{CH}_2\\ \mid\\ \text{C}-\text{O}-\text{CH}_3\\ \parallel\\ \text{O}\end{array}\right]_n$
Poly(α-methyl styrene)		Vinylidene polymer	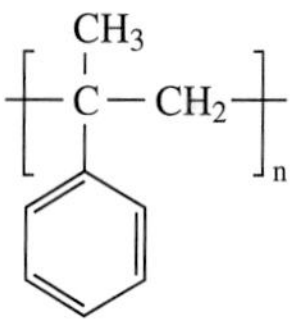 $\left[\begin{array}{c}\text{CH}_3\\ \mid\\ \text{C}-\text{CH}_2\\ \mid\\ \text{C}_6\text{H}_5\end{array}\right]_n$
Poly(methylene oxide)	PMO	Polyether	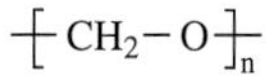 $\left[\text{CH}_2-\text{O}\right]_n$
Polymethylphenyl-siloxane	PMPS	Polysiloxane	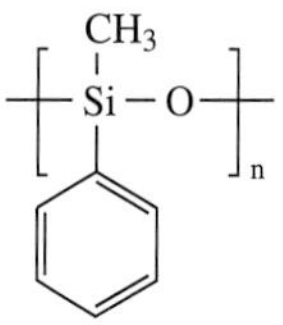 $\left[\begin{array}{c}\text{CH}_3\\ \mid\\ \text{Si}-\text{O}\\ \mid\\ \text{C}_6\text{H}_5\end{array}\right]_n$
Polynitrile		Polyimine	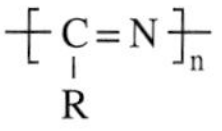 $\left[\begin{array}{c}\text{C}=\text{N}\\ \mid\\ \text{R}\end{array}\right]_n$
Polynucleotide		Polynucleotide	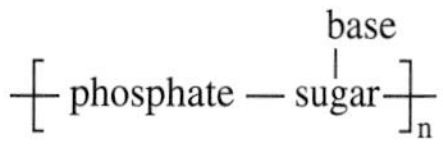 $\left[\begin{array}{c}\text{base}\\ \mid\\ \text{phosphate}-\text{sugar}\end{array}\right]_n$
Poly(n-pentene-2)		Poly(α-olefin)	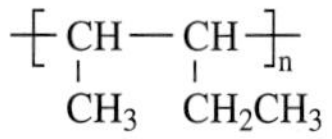 $\left[\begin{array}{c}\text{CH}-\text{CH}\\ \mid\qquad\mid\\ \text{CH}_3\quad\text{CH}_2\text{CH}_3\end{array}\right]_n$
Poly(n-pentene-1)		Poly(α-olefin)	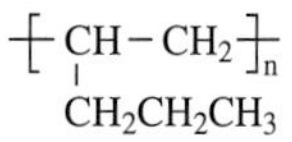 $\left[\begin{array}{c}\text{CH}-\text{CH}_2\\ \mid\\ \text{CH}_2\text{CH}_2\text{CH}_3\end{array}\right]_n$
Polypeptides [Poly(α-amino acid)]		Polypeptide	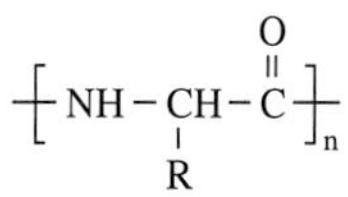 $\left[\begin{array}{c}\qquad\qquad\text{O}\\ \qquad\qquad\parallel\\ \text{NH}-\text{CH}-\text{C}\\ \mid\\ \text{R}\end{array}\right]_n$
Poly(p-phenylene oxide)	PPO	Polyether	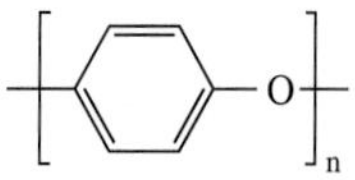 $\left[\text{C}_6\text{H}_4-\text{O}\right]_n$

(*Continued*)

TABLE 2.76 Names and Structures of Polymers (*Continued*)

Common name	Acronym, alternate name	Class	Structure of repeat unit
Poly(*p*-phenylene sulfide)	PPS	Polysulfide	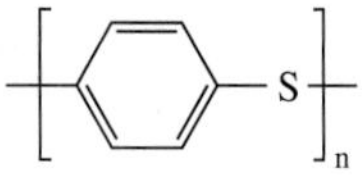
Poly(*p*-phenylene vinylene)		Polyaromatic	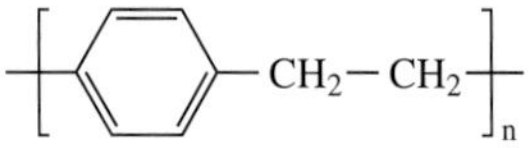
Poly(*p*-phenylene)	PP	Polyaromatic	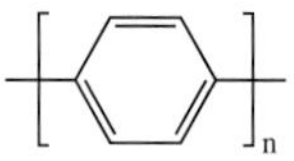
Polyphosphate		Inorganic polymer	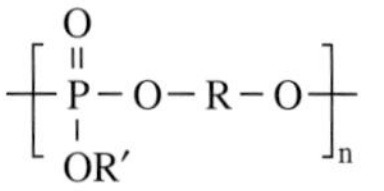
Polyphosphazene		Inorganic polymer	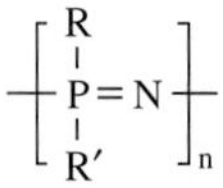
Polyphosphonate		Inorganic polymer	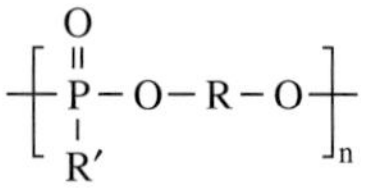
Polypropylene	PP	Poly(α-olefin)	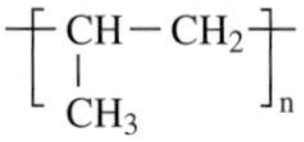
Poly(propylene oxide)	PPO	Polyether	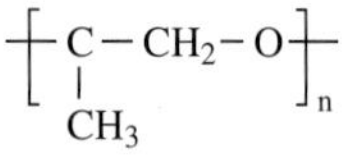
Poly(pyromellitimide-1,4-diphenyl ether) (Kapton)		Polyimide	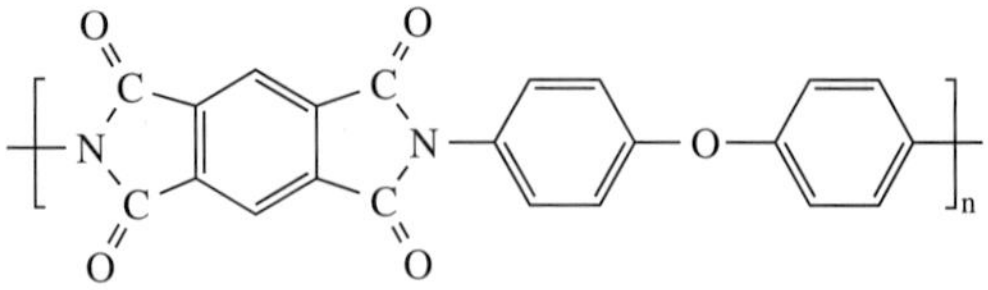
Polypyrrole		Polyhetero-cyclic	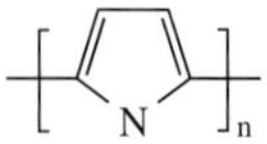
Polysilane		Inorganic polymer	

TABLE 2.76   Names and Structures of Polymers (*Continued*)

Common name	Acronym, alternate name	Class	Structure of repeat unit
Polyailazane		Inorganic polymer	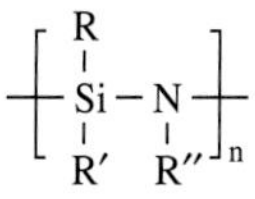 $\left[\begin{array}{c} R \\ | \\ Si-N \\ | \quad | \\ R' \quad R'' \end{array}\right]_n$
Polysiloxane	Silicones	Inorganic polymer	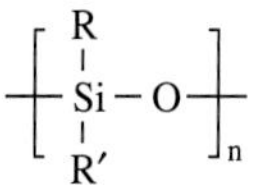 $\left[\begin{array}{c} R \\ | \\ Si-O \\ | \\ R' \end{array}\right]_n$
Polystyrene	PS Styrofoam	Vinyl polymer	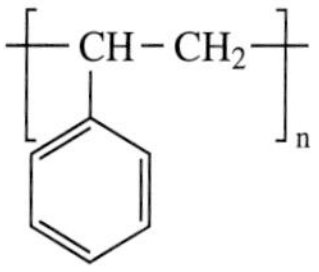 $\left[\!\!\begin{array}{c} CH-CH_2 \\ | \\ C_6H_5 \end{array}\!\!\right]_n$
Polysulfide	Thiokol	Polysulfide	$\left[R-S_m\right]_n$
Polysulfur		Polysulfur	$\left[S\right]_{8n}$
Polytetrafluoroethylene (Teflon)	PTFE	Poly(α-olefin)	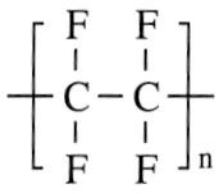 $\left[\begin{array}{cc} F & F \\ | & | \\ C-C \\ | & | \\ F & F \end{array}\right]_n$
Poly(tetramethylene oxide)	PTMO	Polyether	$\left[CH_2-CH_2-CH_2-CH_2-O\right]_n$
Polythiophene		Polyhetero-cyclic	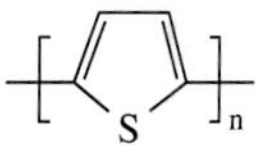 $\left[\!\!\begin{array}{c} \\ S \end{array}\!\!\right]_n$
Polyurea		Polyurea	$\left[NH-R-NH-\overset{\overset{\textstyle O}{\|}}{C}-NH-R'-NH-\overset{\overset{\textstyle O}{\|}}{C}\right]_n$
Polyurethane	Adiprene	Polyurethane	$\left[O-R-O-\overset{\overset{\textstyle O}{\|}}{C}-NH-R'-NH-\overset{\overset{\textstyle O}{\|}}{C}\right]_n$
Poly(L-valine)		Polypeptide	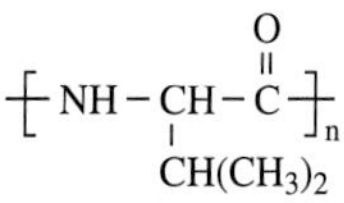 $\left[\begin{array}{c} NH-CH-\overset{\overset{\textstyle O}{\|}}{C} \\ | \\ CH(CH_3)_2 \end{array}\right]_n$
Poly(vinyl acetate)	PVAc	Vinyl polymer	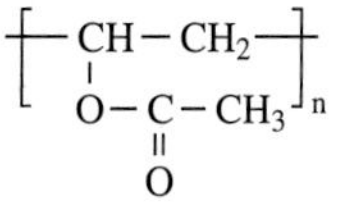 $\left[\begin{array}{c} CH-CH_2 \\ | \\ O-\underset{\underset{\textstyle O}{\|}}{C}-CH_3 \end{array}\right]_n$

(*Continued*)

TABLE 2.76 Names and Structures of Polymers (*Continued*)

Common name	Acronym, alternate name	Class	Structure of repeat unit
Poly(vinyl alcohol)	PVA	Vinyl polymer	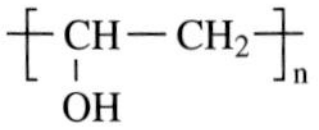
Poly(vinyl chloride)	PVC	Vinyl polymer	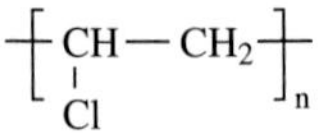
Poly(vinyl fluoride)	PVF	Vinyl polymer	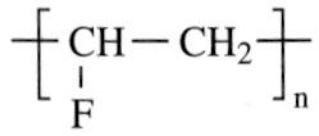
Poly(2-vinyl pyridine)	PVP	Vinyl polymer	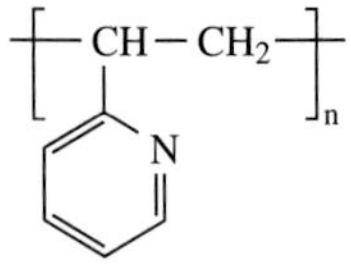
Poly(N-vinyl pyrrolidone)		Vinyl polymer	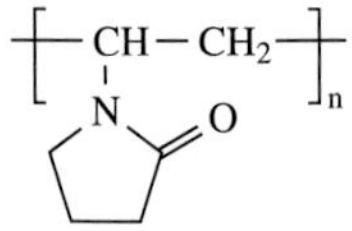
Poly(vinylidene chloride)	PVDC Saran	Vinylidene polymer	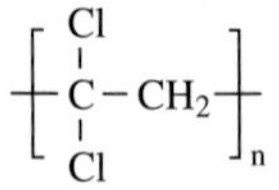
Poly(vinylidiene fluoride)	PVDF	Vinylidiene polymer	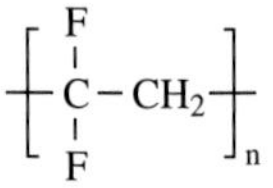
Vinyl polymer		Vinyl polymer	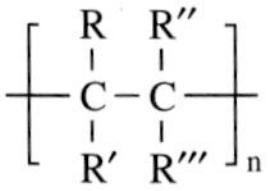

TABLE 2.77 Plastics

- Acetals
- Acrylics
  - Poly(methyl methacrylate) (PMMA)
  - Poly(acrylonitrile)
- Alkyds
- Alloys
  - Acrylic-poly(vinyl chloride) alloy
  - Acrylonitrile-butadiene-styrene-poly(vinyl chloride) alloy (ABS-PVC)
  - Acrylonitrile-butadiene-styrene-polycarbonate alloy (ABS-PC)
- Allyls
  - Allyl-diglycol-carbonate polymer
  - Diallyl phthalate (DAP) polymer
- Cellulosics
  - Cellulose acetate resin
  - Cellulose-acetate-propionate resin
  - Cellulose-acetate-butyrate resin
  - Cellulose nitrate resin
  - Ethyl cellulose resin
  - Rayon
- Chlorinated polyether
- Epoxy
- Fluorocarbons
  - Poly(tetrafluoroethylene) (PTFE)
  - Poly(chlorotrifluoroethylene) (PCTFE)
  - Perfluoroalkoxy (PFA) resin
  - Fluorinated ethylene-propylene (FEP) resin
- Poly(methylpentene)
- Polyolefins (PO)
  - Low-density polyethylene (LDPE)
  - High-density polyethylene (HDPE)
  - Ultrahigh-molecular-weight polyethylene (UHMWPE)
  - Polypropylene (PP)
  - Polybutylene (PB)
  - Polyallomers
- Poly(phenylene oxide)
- Poly(phenylene sulfide) (PPS)
- Polyurethanes
- Silicones
- Styrenics
  - Polystyrene (PS)
  - Acrylonitrile-butadiene-styrene (ABS) copolymer
  - Styrene-acrylonitrile (SAN) copolymer
  - Styrene-butadiene copolymer
- Sulfones
  - Polysulfone (PSF)
- Fluorocarbons (*continued*)
  - Poly(vinylidene fluoride) (PVDF)
  - Ethylene-chlorotrifluoroethylene copolymer
  - Ethylene-tetrafluoroethylene copolymer
  - Poly(vinyl fluoride) (PVF)
- Melamine formaldehyde
- Melamine phenolic
- Nitrile resins
- Phenolics
- Polyamides
  - Nylon 6
  - Nylon 6/6
  - Nylon 6/9
  - Nylon 6/12
  - Nylon 11
  - Nylon 12
  - Aromatic nylons
- Poly(amide-imide)
- Poly(aryl ether)
- Polycarbonate (PC)
- Polyesters
  - Poly(butylenes terephthalate) (PBT) [also called polytetramethylene terephthalate (PTMT)]
  - Poly(ethylene terephthalate) (PET)
  - Unsaturated polyesters (SMC, BMC)
    - Butadiene-maleic acid copolymer (BMC)
    - Styrene-maleic acid copolymer (SMC)
- Polyimide
- Sulfones (*continued*)
  - Poly(ether sulfone)
  - Poly(phenyl sulfone)
- Thermoplastic elastomers
  - Polyolefin
  - Polyester
  - Block copolymers
    - Styrene-butadiene block copolymer
    - Styrene-isoprene block copolymer
    - Styrene-ethylene block copolymer
    - Styrene-butylene block copolymer
- Urea formaldehyde
- Vinyls
  - Poly(vinyl chloride) (PVC)
  - Poly(vinyl acetate) (PVAC)
  - Poly(vinylidene chloride)
  - Poly(vinyl butyrate) (PVB)
  - Poly(vinyl formal)
  - Poly(vinyl alcohol) (PVAL)

TABLE 2.78 Properties of Commercial Plastics

Properties	Acetal				
	Homopolymer	Copolymer	20% glass-reinforced homopolymer	25% glass-reinforced copolymer	21% poly(tetrafluoroethylene)-filled homopolymer
Physical					
Melting temperature, °C					
Crystalline	175	175	181	175	181
Amorphous					
Specific gravity	1.42	1.41	1.56	1.61	1.54
Water absorption (24 h), %	0.25–0.40	0.22	0.25	0.29	0.20
Dielectric strength, $KV \cdot mm^{-1}$	19.7	19.7	19.3	22.8	15.7
Electrical					
Volume (dc) resistivity, ohm-cm	10^{15}	10^{15}	5×10^{14}		3×10^{16}
Dielectric constant (60 Hz)	3.7	3.7	3.9		3.1
Dielectric constant (10^6 Hz)	3.7	3.7	3.9		3.1
Dissipation (power) factor (60 Hz)					
Dissipation factor (10^6 Hz)	0.005	0.005	0.005		0.005
Mechanical					
Compressive modulus, $10^3\ lb \cdot in^{-2}$	670	450			
Compressive strength, rupture or 1% yield, $10^3\ lb \cdot in^{-2}$	5.29	16 (10% yield)	18 (10% yield)	17 (10% yield)	13 (10% yield)
Elongation at break, %	25–75	40–75	7	3	15–22
Flexural modulus at 23°C, $10^3\ lb \cdot in^2$	380–430	375	730	1100	340–350
Flexural strength, rupture or yield, $10^3\ lb \cdot in^{-2}$	14	13	15	28	
Hardness, Rockwell (or Shore)	M94	M78	M90	M79	M78
Impact strength (Izod) at 23°C, $J \cdot m^{-1}$	69–123	53–80	43	96	37–64
Tensile modulus, $10^3\ lb \cdot in^{-2}$	520	410	1000	1250	

Tensile strength at break, 10^3 lb · in^{-2}	10	10	8.5	18.5	7.6
Tensile yield strength, 10^3 lb · in^{-2}	9.5–12	8.5			6.9–7.6
<u>Thermal</u>					
Burning rate, mm · min^{-1}	27.9				
Coefficient of linear thermal expansion, 10^{-6}°C	100	85	36–81		75
Deflection temperature under flexural load (264 lb · in^{-2}), °C	124	110	157	163	100
Maximum recommended service temperature, °C	84				
Specific heat, cal · g^{-1}	0.35				
Thermal conductivity, W · m^{-1} · K^{-1}	0.23	0.23			

(Continued)

TABLE 2.78 Properties of Commercial Plastics (*Continued*)

Properties	Acrylic					Alloy	
	Poly(methyl methacrylate)	Cast sheet	Impact-modified	Heat-resistant	Alkyd, molded	Acrylic poly(vinyl chloride) alloy	Acrylonitrile-butadiene-styrene-poly(vinyl chloride) alloy
Physical							
Melting temperature, °C							
Crystalline	90–105	90–105	80–100	100–125		105	
Amorphous							
Specific gravity	1.17–1.20	1.18–1.20	1.11–1.18	1.16–1.19	2.22–2.24		
Water absorption (24 h), %	0.1–0.4	0.2–0.4	0.2–0.8	0.2–0.3		0.06	
Dielectric strength, $KV \cdot mm^{-1}$	15.7–19.9	17.7–21.7	15.0–19.9	15.7–19.9		>15.7	19.7
Electrical							
Volume (dc) resistivity, ohm-cm	$>10^{14}$	$>10^{14}$					
Dielectric constant (60 Hz)	3.3–4.5	3.5–4.5			3.8–5.0		
Dielectric constant (10^6 Hz)		3.0–3.5			3.6–4.7		
Dissipation (power) factor (60 Hz)		0.04–0.06			0.012–0.026		
Dissipation factor (10^6 Hz)		0.02–0.03			0.01–0.016		
Mechanical							
Compressive modulus, 10^4 lb · in^{-2}	370–460	390–475	240–370	350–460		330–400	
Compressive strength, rupture or 1% yield, 10^3 lb · in^{-2}	12–18	11–19	4–14	17	16–20	8.4	
Elongation at break, %	2–10	2–7	20–70	3–5		100	
Flexural modulus at 23°C, 10^3 lb · in^{-2}	420–460	390–475	200–380	460–500		330–400	340
Flexural strength, rupture or yield, 10^3 lb · in^{-2}	13–19	12–17	7–13	12–16		10.7	9.6
Hardness, Rockwell (or Shore)	M85–M105	M80–M100	R105–R120	M95–M105	E76	R99–R105	R100
Impact strength (Izod) at 23°C, J · m^{-1}	16–27	16–21	43–133	16–21	27–240	800	560

Tensile modulus, 10^3 lb · in^{-2}	380–450	350–450	200–400	350–460		330–335	330
Tensile strength at break, 10^3 lb · in^{-2}	7–11	8–11	5–9	10	4.5–6.5	6.5	5.8
Tensile yield strength, 10^3 lb · in^{-2}					10–13		
Thermal							
Burning rate, mm · min^{-1}		0.5–2.2			Self-extinguishing		
Coefficient of linear thermal expansion, 10^{-6}°C	50–90	50–90	50–80	50–60	40–55		46
Deflection temperature under flexural load (264 lb · in^{-2}), °C	74–99	71–102	74–95	88–104	177–204	71	
Maximum recommended service temperature, °C		60–71			220		
Specific heat, cal · g^{-1}	0.36	0.35					
Thermal conductivity, W · m^{-1}, K^{-1}	0.17–0.25	0.17–0.25	0.17–0.21	0.19			

(*Continued*)

TABLE 2.78 Properties of Commercial Plastics (*Continued*)

Properties	Alloy	Allyl			Cellulosic		
	Polycarbonate acrylonitrile-butadiene-styrene alloy	Allyl-diglycol-carbonate polymer	Diallyl phthalate molding		Cellulose acetate		Cellulose-acetate-butyrate resin
			Glass-filled	Mineral-filled	Sheet	Molding	Sheet
Physical							
Melting temperature, °C							
Crystalline		Thermoset	Thermoset	Thermoset	230	230	140
Amorphous	150						
Specific gravity	1.12–1.20	1.3–1.4	1.7–2.0	1.65–1.85	1.27–1.34	1.29–1.34	1.15–1.22
Water absorption (24 h), %	0.21–0.24	0.2	0.12–0.35	0.2–0.5	2–7	1.7–6.5	0.9–2.2
Dielectric strength, $kV \cdot mm^{-1}$	17.7	15.0	15.7–17.7	15.7–17.7	11–24	9–24	9–18
Electrical							
Volume (dc) resistivity, ohm-cm					10^{10}–10^{13}	10^{10}–10^{13}	10^{10}–10^{12}
Dielectric constant (60 Hz)					3.4–7.4	3.5–7.5	3.7–4.3
Dielectric constant (10^6 Hz)					3.2–7.0	3.2–7.0	3.3–3.8
Dissipation (power) factor (60 Hz)					0.01–0.06	0.01–0.06	0.01–0.04
Dissipation factor (10^6 Hz)					0.01–0.06	0.01–0.10	0.01–0.04
Mechanical							
Compressive modulus, $10^3\ lb \cdot in^{-2}$		300					
Compressive strength, rupture or 1% yield, $10^3\ lb \cdot in^{-2}$	11	21–23	25–35	20–32	22–33	25–36	
Elongation at break, %	10–15		3–5	3–5	17–40	6–40	50–100
Flexural modulus at 23°C, $10^3\ lb \cdot in^{-2}$	300–400	250–330	1200–1500	1000–1400			740–1300

Flexural strength, rupture or yield, 10^3 lb · in^{-2}	13.0–13.7	6–13	9–20	8.5–11	6–10	2–16	4–9
Hardness, Rockwell (or Shore)	R117	M95–M100	E80–E87	E61	R85–R120	R100–R123	R50–R95
Impact strength (Izod) at 23°C, J · m^{-1}	560	11–21	21–800	16–43	107–454	53–214	133–288
Tensile modulus, 10^3 lb · in^{-2}	370–380	300	1400–2200	1200–2200			200–250
Tensile strength at break, 10^3 lb · in^{-2}	7.0–7.3	5–6	6–11	5–8	4.5–8.0	1.9–9.0	2.6–6.9
Tensile yield strength, 10^3 lb · in^{-2}	8.5				2.2–7.4	4.1–7.6	
Thermal							
Burning rate, mm · min^{-1}						1.3–3.8	1.3–3.8
Coefficient of linear thermal expansion, 10^{-6}°C	63–67	5.4–9.6	0.68–2.4	2.8	100–150	80–180	110–170
Deflection temperature under flexural load (264 lb · in^{-2}), °C	104–116	60–88	165–288+	160–288	44–91	51–98	49–58
Maximum recommended service temperature, °C							
Specific heat, cal · g^{-1}					0.3–0.4	0.3–0.42	0.3–0.4
Thermal conductivity, W · m^{-1} · K^{-1}	0.25–0.38	0.20–0.21	0.21–0.63	0.30–1.04	0.17–0.34	0.17–0.34	0.17–0.34

(*Continued*)

TABLE 2.78 Properties of Commercial Plastics (*Continued*)

Properties	Cellulosic					Epoxy	
						Bisphenol	
	Cellulose-acetate butyrate resin, molding	Cellulose-acetate propionate resin, molding	Ethyl cellulose	Cellulose nitrate	Chlorinated polyether	Glass-fiber-reinforced	Mineral-filled
Physical							
Melting temperature, °C							
Crystalline	140	190	135		125	Thermoset	Thermoset
Amorphous							
Specific gravity	1.15–1.22	1.17–1.24	1.09–1.17	1.35–1.40	1.4	1.6–2.0	1.6–2.1
Water absorption (24 h), %	0.9–2.2	1.2–2.8	0.8–1.8			0.04–0.20	0.03–0.20
Dielectric strength, kV · mm^{-1}	9–13	12–17.7	13.8–19.7			9.8–15.7	9.8–15.7
Electrical							
Volume (dc) resistivity, ohm-cm	10^{10}–10^{12}			10^{10}			
Dielectric constant (60 Hz)	3.5–6.4			7.0–7.5			
Dielectric constant (10^6 Hz)	3.2–6.2		3.01	6.6			
Dissipation (power) factor (60 Hz)	0.01–0.04						
Dissipation factor (10^6 Hz)	0.01–0.04						
Mechanical							
Compressive modulus, 10^3 lb · in^{-2}						3000	
Compressive strength, rupture or 1% yield, 10^3 lb · in^{-2}	2.1–7.5	2.4–7.0		2.1–8.0		18,000–40,000	18,000–40,000
Elongation at break, %	40–88	29–100	5–40	40–45	600–800	4	
Flexural modulus at 23°C, 10^3 lb · in^{-2}	90–300	120–350				2–4.5	

Flexural strength, rupture or yield, 10^3 lb · in^{-2}	1.8–9.3	2.9–11.4	4–12	9–11	5	8–30	6–18
Hardness, Rockwell (or Shore)	R31–R116	R10–R122	R50–R115	R95–R115	R100	M100–M112	M100–M112
Impact strength (Izod) at 23°C, J · m^{-1}	53–582	27 to no break	21	267–374	21	16–533	16–22
Tensile modulus, 10^3 lb · in^{-2}	50–200	60–215		190–220		3	
Tensile strength at break, 10^3 lb · in^{-2}	2.6–6.9	2.0–7.8	2–8	7–8	1.5–1.8	5–20	4–10
Tensile yield strength, 10^3 lb · in^{-2}							
<u>Thermal</u>							
Burning rate, mm · min^{-1}	1.3–3.8				Self-extinguishing		
Coefficient of linear thermal expansion, 10^{-6}°C	110–170	110–170	100–200	80–120	6.6	11–50	20–60
Deflection temperature under flexural load (264 lb · in^{-2}), °C	44–94	44–109	45–88	60–71	185	107–260	107–260
Maximum recommended service temperature, °C					255		
Specific heat, cal · g^{-1}	0.3–0.4			0.31–0.41			
Thermal conductivity, W · m^{-1} · K^{-1}	0.17–0.30	0.17–0.30	0.16–0.30	0.23		0.17–0.42	0.17–1.48

(Continued)

TABLE 2.78 Properties of Commercial Plastics (*Continued*)

	Epoxy			Fluorocarbon			
	Casting resin		Novolac resin	Poly(tetrafluoroethylene)		Poly(chloro-trifluoro-ethylene)	Perfluoroalkoxy
Properties	Unfilled	Flexible	Mineral-filled	Granular	Glass-fiber-reinforced		
Physical							
Melting temperature, °C							
Crystalline	Thermoset	Thermoset	Thermoset	327	327	220	310
Amorphous							
Specific gravity	1.11–1.40	1.05–1.35	1.7–2.1	2.14–2.20	2.2–2.3	2.1–2.2	2.12–2.17
Water absorption (24 h), %	0.08–0.15	0.27–0.50	0.05–0.2	0.01		0.03	
Dielectric strength, $kV \cdot mm^{-1}$	11.8–19.7	9.3–15.8	11.8–13.8	18.9	12.6	19.7–23	19.7
Electrical							
Volume (dc) resistivity, ohm-cm	10^{12}–10^{17}			10^{18}		10^{18}	
Dielectric constant (60 Hz)	3.5–5.0			2.1		2.3–2.7	
Dielectric constant (10^6 Hz)	3.5–5.0			2.1		2.3–2.5	
Dissipation (power) factor (60 Hz)				0.0002		0.001	
Dissipation factor (10^6 Hz)				0.0002		0.005	
Mechanical							
Compressive modulus, 10^3 lb · in^{-2}				60			
Compressive strength, rupture or 1% yield, 10^3 lb · in^{-2}	15–25	1–14	30	1.7		4.6–7.4	
Elongation at break, %	3–6	20–70	2–4	200–400	200–300	80–250	300
Flexural modulus at 23°C, 10^3 lb · in^{-2}			2000	80	235	120	

Flexural strength, rupture or yield, 10^{-3} lb · in^{-2}	13–21	1–13	16–20		2	7.4–9.3	
Hardness, Rockwell (or Shore)	M80–M110			(D50–D55)	(D60–D70)	R75–R95	(D64)
Impact strength (Izod) at 23°C, J · m^{-1}	10.7–53	187–267	21	160	144	133–160	No break
Tensile modulus, 10^3 lb · in^{-2}	350	1–350		58–80		150–300	
Tensile strength at break, 10^3 lb · in^{-2}	4–13	2–10	6–12	2–5	2–2.7	4.5–6	4–4.3
Tensile yield strength, 10^3 lb · in^{-2}			30				
Thermal							
Burning rate, mm · min^{-1}				Self-extinguishing	Self-extinguishing	Self-extinguishing	
Coefficient of linear thermal expansion, 10^{-6}°C	45–65	20–100	22–30	100	77–100	70	
Deflection temperature under flexural load (264 lb · in^{-2}), °C	46–288	23–121	149–260	121 (66 lb · in^{-2})		126 (66 lb · in^{-2})	74 (66 lb · in^{-2})
Maximum recommended service temperature, °C				260		200	
Specific heat, cal · g^{-1}				0.25		0.22	
Thermal conductivity, W · m^{-1} · K^{-1}	0.17–0.21			0.25	0.34–0.40	0.19–0.22	0.25

(Continued)

TABLE 2.78 Properties of Commercial Plastics (*Continued*)

Properties	Fluorocarbon					Melamine formaldehyde	
	Fluorinated ethylene-propylene resin	Poly(vinylidene fluoride)	Ethylene-tetrafluoroethylene copolymer		Ethylene-chlorotrifluoro-ethylene copolymer	Cellulose-filled	Glass-fiber-reinforced
			Unfilled	Glass-fiber-reinforced			
Physical							
Melting temperature, °C							
Crystalline	275	156	270	270	245	Thermoset	Thermoset
Amorphous							
Specific gravity	2.14–2.17	1.75–1.78	1.7	1.8	1.68	1.47–1.52	1.5–2.0
Water absorption (24 h), %	<0.01	0.04–0.06	0.03	0.02	0.01	0.1–0.8	0.09–1.3
Dielectric strength, $kV \cdot mm^{-1}$	20–24	10	16	17	19	11–16	5–15
Electrical							
Volume (dc) resistivity, ohm-cm							
Dielectric constant (60 Hz)	2.1	8–9	2.6		2.6		
Dielectric constant (10^6 Hz)	2.1	8–9	2.6		2.6		
Dissipation (power) factor (60 Hz)		High					
Dissipation factor (10^6 Hz)		High					
Mechanical							
Compressive modulus, 10^3 lb · in^{-2}		120	120	1200	240		
Compressive strength, rupture or 1% yield, 10^3 lb · in^{-2}	2.2	8.7–10	7.1	10		33–45	20–35
Elongation at break, %	250–330	25–500	100–400	8	200–300	0.6–1.0	0.6
Flexural modulus at 23°C, 10^3 lb · in^{-2}	80–95	200	200	950	240	1100	
Flexural strength, rupture or yield, 10^3 lb · in^{-2}		8.6–11	5.5	10.7	7	9–16	14–23

Hardness, Rockwell (or Shore)	(D60–D65)	(D80)	R50 (D75)	R74	R95	M115–M125	M115
Impact strength (Izod) at 23°C, $J \cdot m^{-1}$	No break	192–214	No break	480	No break	11–21	32–961
Tensile modulus, 10^3 lb $\cdot$ in^{-2}	50	120	120	1200	240	1.1–1.4	1.6–2.4
Tensile strength at break, 10^3 lb $\cdot$ in^{-2}	2.7–3.1	5.5–7.4	6.5	12	7	5–13	5–10.5
Tensile yield strength, 10^3 lb $\cdot$ in^{-2}							
Thermal							
Burning rate, mm $\cdot$ min^{-1}	Not combustible	Not combustible	Not combustible	Not combustible	Not combustible	Self-extinguishing	Self-extinguishing
Coefficient of linear thermal expansion, 10^{-6}°C	83–105	85	59	10–32	80	40–45	15–28
Deflection temperature under flexural load (264 lb $\cdot$ in^{-2}), °C	70 (66 lb $\cdot$ in^{-2})	80–90	71	210	77	177–199	190–204
Maximum recommended service temperature, °C	205	150				210	
Specific heat, cal $\cdot$ g^{-1}	0.28						
Thermal conductivity, $W \cdot m^{-1} \cdot K^{-1}$	0.25	0.19–0.24	0.24		0.16	0.27–0.41	0.41–0.49

(Continued)

TABLE 2.78 Properties of Commercial Plastics (*Continued*)

Properties	Melamine phenolic, woodflour- and cellulose-filled	Nitrile	Phenolic				
			Unfilled	Woodflour-filled	Glass-fiber-reinforced	Cellulose-filled	Mineral-filled
Physical							
Melting temperature, °C							
Crystalline	Thermoset		Thermoset	Thermoset	Thermoset	Thermoset	Thermoset
Amorphous		95					
Specific gravity	1.5–1.7	1.15	1.24–1.32	1.37–1.46	1.69–2.0	1.38–1.42	1.42–1.84
Water absorption (24 h), %	0.3–0.65	0.28	0.1–0.36	0.3–1.2	0.03–1.2	0.5–0.9	0.1–0.3
Dielectric strength, $kV \cdot mm^{-1}$	8.7–12.8	8.7–9.5	9.8–15.8	10.2–15.8	5.5–15.8	11.8–15	7.9–13.8
Electrical							
Volume (dc) resistivity, ohm-cm		1.9×10^{15}	1×10^{12} to 7×10^{12}				
Dielectric constant (60 Hz)			6.5–7.5				
Dielectric constant (10^6 Hz)			4.0–5.5				
Dissipation (power) factor (60 Hz)			0.10–0.15				
Dissipation factor (10^6 Hz)			0.04–0.05				
Mechanical							
Compressive modulus, $10^3\ lb \cdot in^{-2}$							
Compressive strength, rupture or 1% yield, $10^3\ lb \cdot in^{-2}$	26–30	12	18–32	25–31	26–70	22–31	22.5–34.6
Elongation at break, %	0.4–0.8	3–4	1.5–2.0	0.4–0.8	0.2	1–2	0.1–0.5
Flexural modulus at 23°C, $10^3\ lb \cdot in^{-2}$	1000–2000	500–590	700–1500	1000–1200	2000–33,000	900–1300	1000–2000
Flexural strength, rupture or yield, $10^3\ lb \cdot in^{-2}$	8–10	14	11–17	7–14	15–60	5.5–11	11–14

Hardness, Rockwell (or Shore)	E95–E100	M72–M76	M93–M120	M100–M115	E54–E101	M95–115	E88
Impact strength (Izod) at 23°C, $J \cdot m^{-1}$	11–21	80–256	13–21	11–32	27–960	21–59	14–19
Tensile modulus, 10^3 lb $\cdot$ in^{-2}	800–1700	510–580	700–1500	800–1700	1900–3300		2400
Tensile strength at break, 10^3 lb $\cdot$ in^{-2}	6–8	9	6–9	5–9	7–18	3.5–6.5	6–9.7
Tensile yield strength, 10^3 lb $\cdot$ in^{-2}			12–15				
Thermal							
Burning rate, mm $\cdot$ min^{-1}			Self-extinguishing				
Coefficient of linear thermal expansion, 10^{-6}°C	10–40	66	68	30–45	8–21	20–31	19–26
Deflection temperature under flexural load (264 lb $\cdot$ in^{-2}), °C	140–154	73	74–80	149–188	177–316	149–177	320–246
Maximum recommended service temperature, °C							
Specific heat, cal $\cdot$ g^{-1}							
Thermal conductivity, $W \cdot m^{-1} \cdot K^{-1}$	0.17–0.30	0.26	0.15	0.17–0.34	0.34–0.59	0.25–0.38	0.42–0.57

(Continued)

TABLE 2.78 Properties of Commercial Plastics (*Continued*)

	Polyamide						
	Nylon 6				Nylon 6/6		
Properties	Molding and extrusion	30–35% glass-fiber-reinforced	High-impact copolymer	Molding	33% glass-fiber-reinforced	Molybdenum disulfide-filled	Nylon 6/6-nylon 6 copolymer
Physical							
Melting temperature, °C							
Crystalline	216	216	216	265	265	265	240
Amorphous							
Specific gravity	1.12–1.14	1.35–1.42	1.08–1.17	1.13–1.15	1.38	1.15–1.17	1.08–1.14
Water absorption (24 h), %	2.9	1.2	1.3–1.5	1.0–1.3	1.0	0.8–1.1	1.5–2.0
Dielectric strength, $kV \cdot mm^{-1}$	15.8	15.8	22	24		14	15.8
Electrical							
Volume (dc) resistivity, ohm-cm	10^{12}			10^{12}–10^{15}			10^{10}
Dielectric constant (60 Hz)	9.8			4.0			16
Dielectric constant (10^6 Hz)	3.7			3.6			4
Dissipation (power) factor (60 Hz)	0.14			0.01–0.02			0.4
Dissipation factor (10^6 Hz)	0.12			0.02–0.03			0.1
Mechanical							
Compressive modulus, $10^3\ lb \cdot in^{-2}$	250						
Compressive strength, rupture or 1% yield, $10^3\ lb \cdot in^{-2}$	13–16	19		15 (yield)	24.9	12.5	
Elongation at break, %	30–100	3–6	150–270	60	3	15	40
Flexural modulus at 23°C, $10^3\ lb \cdot in^{-2}$	390	1500	110–320	420	1300	450	150–410
Flexural strength, rupture or yield, $10^3\ lb \cdot in^{-2}$	14	33	5–12	17	41	17	

Hardness, Rockwell (or Shore)	R119	M101	R81–R110	R120	M100	R119	R119
Impact strength (Izod) at 23°C, $J \cdot m^{-1}$	32–53	160	96 to no break	43–53	117	240	37
Tensile modulus, 10^3 lb · in^{-2}	380	1450				550	150–410
Tensile strength at break, 10^3 lb · in^{-2}	11.8	25	7.5–11	12	28	13.7	7.4–12.4
Tensile yield strength, 10^3 lb · in^{-2}	8			8			
Thermal							
Burning rate, mm · min^{-1}	Self-extinguishing	Self-extinguishing	Self-extinguishing	Self-extinguishing	Self-extinguishing	Self-extinguishing	Self-extinguishing
Coefficient of linear thermal expansion, 10^{-6}°C	80–90	20–30	30–40	80	15–20	54	
Deflection temperature under flexural load (264 lb · in^{-2}), °C	68–85	210	45–54	75	249	127	77
Maximum recommended service temperature, °C	107			135			
Specific heat, cal · g^{-1}	0.4			0.4			
Thermal conductivity, $W \cdot m^{-1} \cdot K^{-1}$	0.24	0.24		0.24	0.22		

(Continued)

TABLE 2.78 Properties of Commercial Plastics (*Continued*)

Properties	Polyamide						Poly(amide-imide), unfilled
	Nylon 6/9, molding and extrusion	Nylon 6/12		Nylon 11, molding and extrusion	Nylon 12, molding and extrusion	Aromatic nylon (aramid), molded and unfilled	
		Molding	30–35% glass-fiber-reinforced				
Physical							
Melting temperature, °C							
Crystalline	205	217	217	194	179	275	
Amorphous							275
Specific gravity	1.08–1.10	1.06–1.08	1.31–1.38	1.03–1.05	1.01–1.02	1.30	1.40
Water absorption (24 h), %	0.5	0.4	0.2	0.3	0.25	0.6	0.28
Dielectric strength, $kV \cdot mm^{-1}$	24	16	21	17	18	31	24
Electrical							
Volume (dc) resistivity, ohm-cm		10^{15}			10^{14}		
Dielectric constant (60 Hz)		4.0			3.8		
Dielectric constant (10^6 Hz)		3.5			3.0		
Dissipation (power) factor (60 Hz)		0.02			0.07		
Dissipation factor (10^6 Hz)		0.02			0.04		
Mechanical							
Compressive modulus, $10^3\ lb \cdot in^{-2}$				180		290	413
Compressive strength, rupture or 1% yield, $10^3\ lb \cdot in^{-2}$		2.4			7.5	30	40
Elongation at break, %	1125	150	4	300	300	5	12–18
Flexural modulus at 23°C, $10^3\ lb \cdot in^{-2}$	290	290	1120	150	165	640	664
Flexural strength, rupture or yield, $10^3\ lb \cdot in^{-2}$					1.5	25.8	30

Hardness, Rockwell (or Shore)	R111	R114	E40–E50	R108	R106–R109	E90	E78
Impact strength (Izod) at 23°C, $J \cdot m^{-1}$	59	53	139	96	107–300	75	133
Tensile modulus, 10^3 lb · in^{-2}	275	290	1200	185	180		730
Tensile strength at break, 10^3 lb · in^{-2}	8.5	8.8	24	8	8–9	17.5	26.9
Tensile yield strength, 10^3 lb · in^{-2}		8.8					
Thermal							
Burning rate, mm · min^{-1}				Self-extinguishing			
Coefficient of linear thermal expansion, 10^{-6}°C		90		55–100	67–100	40	36
Deflection temperature under flexural load (264 lb · in^{-2}), °C	57–60	82	93–218	54	54	260	274
Maximum recommended service temperature, °C				100–120			260
Specific heat, cal · g^{-1}		0.4		0.58			
Thermal conductivity, $W \cdot m^{-1} \cdot K^{-1}$		0.22		0.34	0.22	0.22	0.25

(*Continued*)

TABLE 2.78 Properties of Commercial Plastics (*Continued*)

Properties	Poly(aryl ether), unfilled	Polycarbonate		Thermoplastic polyester			
				Poly(butylene terephthalate)		Poly(ethylene terephthalate)	
		Low viscosity	30% glass-fiber reinforced	Unfilled	30% glass-fiber-reinforced	Unfilled	30% glass-fiber-reinforced
Physical							
Melting temperature, °C							
Crystalline				232–267	232–267	245	245
Amorphous	160	140	150				
Specific gravity	1.14	1.2	1.4	1.31–1.38	1.52	1.34–1.39	1.27
Water absorption (24 h), %	0.25	0.15	0.14	0.08–0.09	0.06–0.08	0.1–0.2	0.05
Dielectric strength, $kV \cdot mm^{-1}$	17	15	19	16–22	18–22		22
Electrical							
Volume (dc) resistivity, ohm-cm		2×10^{16}	$>10^{16}$		10^{16}	10^{16}	
Dielectric constant (60 Hz)		3.17	3.35				
Dielectric constant (10^6 Hz)		2.96	3.31			3.25	
Dissipation (power) factor (60 Hz)		0.0009	0.011				
Dissipation factor (10^6 Hz)		0.010	0.007				
Mechanical							
Compressive modulus, $10^3\ lb \cdot in^{-2}$		350	1300				
Compressive strength, rupture or 1% yield, $10^3\ lb \cdot in^{-2}$		12.5	18	8.6–14.5	18–23.5	11–15	25
Elongation at break, %	80	110	3–5	50–300	2–4	50–300	3
Flexural modulus at 23°C, $10^3\ lb \cdot in^{-2}$	300	340	1100	330–400	1100–1200	35–450	1440
Flexural strength, rupture or yield, $10^3\ lb \cdot in^{-2}$	11	13.5	23	12–16.7	26–29	14–18	33.5

Hardness, Rockwell (or Shore)	R117	M70	M92	M68–M78	M90	M94–M101	M100
Impact strength (Izod) at 23°C, $J \cdot m^{-1}$	427	14	107	43–53	69–85	13–32	101
Tensile modulus, 10^3 lb · in^{-2}	320	345	1250	280	1300	400–600	1440
Tensile strength at break, 10^3 lb · in^{-2}	7.5	9.5	19	8.2	17–19	8.5–10.5	23
Tensile yield strength, 10^3 lb · in^{-2}		9.0					
Thermal							
Burning rate, mm · min^{-1}		Self-extinguishing	Self-extinguishing				
Coefficient of linear thermal expansion, 10^{-6}°C	65	68	22	60–95	25	65	29
Deflection temperature under flexural load (264 lb · in^{-2}), °C	149	138–145	146	50–85	220	38–41	224
Maximum recommended service temperature, °C		143					
Specific heat, cal · g^{-1}		0.3				0.27	
Thermal conductivity, $W \cdot m^{-1} \cdot K^{-1}$	0.30	0.20	0.22	0.18–0.30	0.30	0.15	

(*Continued*)

TABLE 2.78 Properties of Commercial Plastics (*Continued*)

Properties	Thermoplastic polyester		Thermosetting and alkyd polyester				
	Aromatic polyester		Unsaturated polyester		Alkyd molding compounds		
	Extrusion-transparent	Injection molding	Styrene-maleic acid copolymer, low-shrink	Butadiene-maleic acid copolymer	Putty, mineral-filled	Glass-fiber-reinforced	Polyimide, unfilled
Physical							
Melting temperature, °C							
Crystalline			Thermoset	Thermoset	Thermoset	Thermoset	
Amorphous	81						310–365
Specific gravity		1.39					1.36–1.43
Water absorption (24 h), %		0.01					0.24
Dielectric strength, $kV \cdot mm^{-1}$		14					22
Electrical							
Volume (dc) resistivity, ohm-cm							$>10^{16}$
Dielectric constant (60 Hz)							
Dielectric constant (10^6 Hz)							3–4
Dissipation (power) factor (60 Hz)							
Dissipation factor (10^6 Hz)							
Mechanical							
Compressive modulus, $10^3 lb \cdot in^{-2}$					2000–3000		
Compressive strength, rupture or 1% yield, $10^3 lb \cdot in^{-2}$		10	15–30	14–30	12–38	15–36	30–40
Elongation at break, %	225	7–10	3–5				8–10
Flexural modulus at 23°C, $10^3 lb \cdot in^{-2}$	290	700	1000–2500		2000	2000	450–500
Flexural strength, rupture or yield, $10^3 lb \cdot in^{-2}$	10.6	12	9–35	16–24	6–17	8.5–26	19–28.8
Hardness, Rockwell (or Shore)	R105		40–70 (Barcol)	50–60 (Barcol)	E98	E95	E52–E99

Impact strength (Izod) at 23°C, $J \cdot m^{-1}$	101		133–800	214–694	16–27	27–854	80
Tensile modulus, 10^3 lb · in^{-2}		300	1000–2500	1500–2500	500–3000		300
Tensile strength at break, 10^3 lb · in^{-2}	6	11	4.5–20	5–10	3–9	4–9.5	10.5–17.1
Tensile yield strength, 10^3 lb · in^{-2}	7						12.5
Thermal							
Burning rate, mm · min^{-1}							
Coefficient of linear thermal expansion, 10^{-6}°C		29	6–30		20–50	15–33	45–56
Deflection temperature under flexural load (264 lb · in^{-2}), °C	63	282	190–260	160–177	177–260	204–260	277–360
Maximum recommended service temperature, °C							
Specific heat, cal · g^{-1}							0.27
Thermal conductivity, $W \cdot m^{-1} \cdot K^{-1}$		0.29		0.76–0.93	0.51–0.89	0.6–0.89	0.10–0.11

(Continued)

TABLE 2.78 Properties of Commercial Plastics (*Continued*)

Properties	Poly(methyl pentene), unfilled	Polyolefin					
		Polyethylene					
		Low-density	Medium-density	High-density	Ultra high-molecular-weight	Glass-fiber-reinforced, high-density	Ethylene-vinyl acetate copolymer
Physical							
Melting temperature, °C							
Crystalline	230–240						65–90
Amorphous		95–130	120–140	120–140	125–135	120–140	
Specific gravity	0.84	0.910–0.925	0.926–0.94	0.941–0.965	0.94	1.28	0.92–0.95
Water absorption (24 h), %	0.01	<0.01	<0.01	<0.01	<0.01	0.02	0.05–0.13
Dielectric strength, $kV \cdot mm^{-1}$		18–39	18–39	18–39	28	20	24–30
Electrical							
Volume (dc) resistivity, ohm-cm		$>10^{15}$	$>10^{15}$	$<10^{15}$			
Dielectric constant (60 Hz)		2.3	2.3	2.3			
Dielectric constant (10^6 Hz)		2.3	2.3	2.3			
Dissipation (power) factor (60 Hz)		<0.0005	<0.0005	<0.0005			
Dissipation factor (10^6 Hz)		<0.0005	<0.0005	<0.0005			
Mechanical							
Compressive modulus, $10^3\ lb \cdot in^{-2}$	114–171						
Compressive strength, rupture or 1% yield, $10^3\ lb \cdot in^{-2}$	5–6.6			2.7–3.6		7	
Elongation at break, %	10–50	90–800	50–600	20–130	450–525	1.5	550–900
Flexural modulus at 23°C, $10^3\ lb \cdot in^{-2}$	110–260	8–60	60–115	100–260	130–140	800	1–20
Flexural strength, rupture or yield, $10^3\ lb \cdot in^{-2}$	4–6.5					11	

Hardness, Rockwell (or Shore)	L67–L74	(D40–D51)	(D50–D60)	R30–R50	R50	R75	
Impact strength (Izod) at 23°C, $J \cdot m^{-1}$	16–64	No break	27–854	27–1068	No break	59	No break
Tensile modulus, 10^3 lb · in^{-2}	160–280	14–38	25–55	60–180			20–120
Tensile strength at break, 10^3 lb · in^{-2}	3.5–4	0.6–2.3	1.2–3.5	3.1–5.5	5.6	9	1.4–2.8
Tensile yield, strength, 10^3 lb · in^{-2}		0.8–1.2	1.0–2.2	3–4	3.1–4.0		
Thermal							
Burning rate, mm · min^{-1}		1.0	1.0	1.0			
Coefficient of linear thermal expansion, 10^{-6}°C	117	100–200	140–160	110–130	130	48	160–200
Deflection temperature under flexural load (264 lb · in^{-2}), °C	41	32–41	41–49	43–54	43–49	121	34
Maximum recommended service temperature, °C	175	70	93	200			
Specific heat, cal · g^{-1}		0.55	0.55	0.46–0.55			
Thermal conductivity, $W \cdot m^{-1} \cdot K^{-1}$	0.17	0.34	0.34–0.42	0.46–0.51		0.46	

(Continued)

TABLE 2.78 Properties of Commercial Plastics (*Continued*)

Properties	Polyolefin					Poly(phenylene sulfide)	
		Polypropylene					
	Polybutylene extrusion	Homopolymer	Copolymer	Impact copolymer	Polyallomer	Injection molding	40% glass-fiber-reinforced
Physical							
Melting temperature, °C							
Crystalline	126	168	160–168		120–135	290	290
Amorphous							
Specific gravity	0.91–0.925	0.90–0.91	0.89–0.905	0.90	0.90	1.3	1.6
Water absorption (24 h),%	0.01–0.02	0.01–0.03	0.03	<0.03	<0.01	<0.02	0.05
Dielectric strength, $kV \cdot mm^{-1}$	18	24	24	24	31	15	18
Electrical							
Volume (dc) resistivity, ohm-cm		10^{17}	10^{17}	10^{17}			
Dielectric constant (60 Hz)		2.2–2.6	2.3				
Dielectric constant (10^6 Hz)		2.2–2.6	2.3	2.3			
Dissipation (power) factor (60 Hz)		<0.0005	0.0001–0.0005				
Dissipation factor (10^6 Hz)		0.0005–0.002	0.0001–0.0002	0.0003			
Mechanical							
Compressive modulus, $10^3\ lb \cdot in^{-2}$	31	150–300					
Compressive strength, rupture or 1% yield, $10^3\ lb \cdot in^{-2}$		5.5–8.0	3.5–8.0			16	21
Elongation at break, %	300–380	100–600	200–700	8–20	400–500	1–2	1
Flexural modulus at 23°C, $10^3\ lb \cdot in^{-2}$	45–50	170–250	130–200	130–190	70–110	550	1700
Flexural strength, rupture or yield, $10^3\ lb \cdot in^{-2}$	2–2.3	6–8	5–7			14	29

Hardness, Rockwell (or Shore)		R80–R102	R50–R96	R40–R90	R50–R85	R123	R123
Impact strength (Izod) at 23°C, $J \cdot m^{-1}$	No break	21–53	53–1068	80–900	91–203	<27	75
Tensile modulus, 10^3 lb · in^{-2}	30–40	165–225	100–170			480	1100
Tensile strength at break, 10^3 lb · in^{-2}	3.8–4.4	4.5–6	4–5.5		3–3.8	9.5	19.5
Tensile yield strength, 10^3 lb · in^{-2}	1.7–2.5	4.5–5.4	3.5–4.3	2.5–3.1	3–3.4		
Thermal							
Burning rate, mm · min^{-1}							
Coefficient of linear thermal expansion, 10^{-6}°C	128–150	81–100	68–95	60–90	83–100	49	22
Deflection temperature under flexural load (264 lb · in^{-2}), °C	54–60	48–57	45–57	90–105 (66 lb · in^{-2})	51–56	135	249
Maximum recommended service temperature, °C		160	240	140–160			
Specific heat, cal · g^{-1}		0.44–0.46	0.45–0.50	0.45–0.50			
Thermal conductivity, $W \cdot m^{-1} \cdot K^{-1}$	0.22	0.12	0.15–0.17	0.12–0.17	0.09–0.17	0.29	0.29

(Continued)

TABLE 2.78 Properties of Commercial Plastics (*Continued*)

Properties	Polyurethane			Silicone			Styrenic
	Casting resin		Thermoplastic elastomer	Cast resin, flexible	Mineral- and/or glass-filled	Epoxy molding and encapsulating compound	Polystyrene
	Liquid	Unsaturated					Crystal
Physical							
Melting temperature, °C							
Crystalline	Thermoset	Thermoset		Thermoset	Thermoset	Thermoset	
Amorphous			120–160				85–105
Specific gravity	1.1–1.5	1.05	1.05–1.25	0.99–1.5	1.8–1.94	1.84	1.04–1.05
Water absorption (24 h), %	0.02–1.5	0.1–0.2	0.7–0.9				0.03–0.10
Dielectric strength, $kV \cdot mm^{-1}$	12–20		13–25	22	8–15	10	24
Electrical							
Volume (dc) resistivity, ohm-cm	10^{11}–10^{15}		10^{11}–10^{13}	10^{14}–10^{15}			$>10^{16}$
Dielectric constant (60 Hz)	4.0–7.5		5.4–7.6	2.7–4.2			
Dielectric constant (10^6 Hz)							2.5
Dissipation (power) factor (60 Hz)							
Dissipation factor (10^6 Hz)							
Mechanical							
Compressive modulus, 10^3 lb · in^{-2}	10–100		4–9				
Compressive strength, rupture or 1% yield, 10^3 lb · in^{-2}	20		20		10–16	28	11.5–16
Elongation at break, %	100–1000	3–6	100–1100	100–700			1–2
Flexural modulus at 23°C, 10^3 lb · in^{-2}	10–100	610	10–350		1000–2500		380–450

Flexural strength, rupture or yield, 10^3 lb · in^{-2}	0.7–4.5	19	0.7–9		9–14	17	8–14
Hardness, Rockwell (or Shore)			(A65–D80)	(A15–A65)	M80–M90		M60–M75
Impact strength (Izod) at 23°C, J · m^{-1}	1334 to flexible	21	No break		13–427	16	13–21
Tensile modulus, 10^3 lb · in^{-2}	10–100		10–350				350–485
Tensile strength at break, 10^3 lb · in^{-2}	0.175–10	10–11	1.5–8.4	0.35–1.0	4–6.5	6–8	5.3–7.9
Tensile yield strength, 10^3 lb · in^{-2}							
Thermal							
Burning rate, mm · min^{-1}					0–78		
Coefficient of linear thermal expansion, 10^{-6}°C	100–200		100–200	300–800	20–50	30	70–80
Deflection temperature under flexural load (264 lb · in^{-2}), °C	Varies over wide range	87–93	Varies over wide range		260	74–100	
Maximum recommended service temperature, °C					371		93
Specific heat, cal · g^{-1}	0.43		0.43				0.3
Thermal conductivity, W · m^{-1} · K^{-1}	0.21		0.07–0.31	0.15–0.31	0.30	0.68	0.09–0.13

(Continued)

TABLE 2.78 Properties of Commercial Plastics (*Continued*)

Properties	Styrenic						
	Polystyrene	Acrylonitrile-butadiene-styrene copolymer					
			Molding				
	Heat-resistant	Extrusion	Heat-resistant	High-impact	Flame-retarded	Platable	20% glass-reinforced
Physical							
Melting temperature, °C							
Crystalline							
Amorphous	110–125	88–120	110–125	100–110	110–125	100–110	
Specific gravity	1.05–1.09	1.02–1.06	1.05–1.08	1.01–1.04	1.16–1.21	1.06–1.07	1.22
Water absorption (24 h), %	0.03–0.12	0.20–0.45	0.20–0.45	0.20–0.45	0.2–0.6		
Dielectric strength, kV · mm^{-1}	20	14–20	14–20	14–20	14–20	16–22	18
Electrical							
Volume (dc) resistivity, ohm-cm							
Dielectric constant (60 Hz)				2.4–5.0			
Dielectric constant (10^6 Hz)				2.4–3.8			
Dissipation (power) factor (60 Hz)				0.003–0.008			
Dissipation factor (10^6 Hz)				0.007–0.015			
Mechanical							
Compressive modulus, 10^3 lb · in^{-2}		150–390	190–440	140–300	130–310		
Compressive strength, rupture or 1% yield, 10^3 lb · in^{-2}	11.5–16	5.2–10	7.2–10	4.5–8	6.5–7.5		14
Elongation at break, %	2–60	20–100	3–20	5–70	5–25		
Flexural modulus at 23°C, 10^3 lb · in^{-2}	340–470	130–420	300–400	250–350	300–400	340–390	710

Flexural strength, rupture or yield, 10^3 lb · in^{-2}	8.9–14	4–14	10–13	8–11	9–14	10.5–11.5	15.5
Hardness, Rockwell (or Shore)	L80–L108	R75–R115	R100–R115	R85–R105	R100–R120	R103–R109	M85
Impact strength (Izod) at 23°C, J · m^{-1}	21–181	133–640	107–347	347–400	160–640	267–283	64
Tensile modulus, 10^3 lb · in^{-2}	320–460	130–380	300–350	230–330	320–400	330–380	740
Tensile strength at break, 10^3 lb · in^{-2}	5–7.8	2.5–8.0	6–7.5	4.8–6.3	5–8	6–6.4	11
Tensile yield strength, 10^3 lb · in^{-2}			5.5–7	4–5.5	4–6		
Thermal							
Burning rate, mm · min^{-1}		1.3		1.3			
Coefficient of linear thermal expansion, 10^{-6}°C	60–70	60–130	60–93	95–110	65–95	47–53	21
Deflection temperature under flexural load (264 lb · in^{-2}), °C	93–120	77–104 annealed	104–116 annealed	96–102 annealed	90–107 annealed	96–102 annealed	99
Maximum recommended service temperature, °C				110			
Specific heat, cal g^{-1}				0.3–0.4			
Thermal conductivity, W · m^{-1} · K^{-1}			0.19–0.34				

(Continued)

TABLE 2.78 Properties of Commercial Plastics (*Continued*)

Properties	Styrenic			Sulfone			
	Styrene-acrylonitrile copolymer		Styrene-butadiene copolymer, high-impact	Polysulfone		Poly(ether sulfone)	Poly(phenyl sulfone)
	Unfilled	20% glass-fiber-reinforced		Unfilled	20% glass-fiber-reinforced		
Physical							
Melting temperature, °C							
Crystalline							
Amorphous	115–125	115–125	90–110	200	200	230	220
Specific gravity	1.07–1.08	1.22	1.03–1.06	1.24	1.46	1.37	1.29
Water absorption (24 h), %	0.2–0.3	0.15–0.20	0.05–0.10	0.22	0.23	0.43	1.1–1.3 (saturated)
Dielectric strength, $kV \cdot mm^{-1}$	16–20	20	18	17	17	17	16
Electrical							
Volume (dc) resistivity, ohm-cm				10^{15}			
Dielectric constant (60 Hz)				3.14	3.7		
Dielectric constant (10^6 Hz)				3.26	3.7		
Dissipation (power) factor (60 Hz)				0.004	0.002		
Dissipation factor (10^6 Hz)				0.008	0.009		
Mechanical							
Compressive modulus, $10^3\ lb \cdot in^{-2}$	530			370			
Compressive strength, rupture of 1% yield, $10^3\ lb \cdot in^{-2}$	14–17	19	4–9	13.9	22		
Elongation at break, %	1–4	1–2	13–50	50–100	2	30–80	60
Flexural modulus at 23°C, $10^3\ lb \cdot in^{-2}$	550	100–1100	280–450	390	1000	375	330
Flexural strength, rupture or yield, $10^3\ lb \cdot in^{-2}$	14–17	20	5.3–9.4	15.4	23	18.7	12.4

Hardness, Rockwell (or Shore)	M80–M90	R122	M10–M68	M69, R120	M123	M88	
Impact strength (Izod) at 23°C, $J \cdot m^{-1}$	19–27	53	32–192	64	59	85	640
Tensile modulus, 10^3 lb · in^{-2}	400–560	1150–1200	280–465	360	1200	350	310
Tensile strength at break, 10^3 lb · in^{-2}	9–12	15.8–18	3.2–4.9		17		
Tensile yield strength, 10^3 lb · in^{-2}			2.9–4.9	10.2		12.2	10.4
Thermal							
Burning rate, mm · min^{-1}							
Coefficient of linear thermal expansion, 10^{-6}°C	36–38	38–40	70–101	52–56	25	55	31
Deflection temperature under flexural load (264 lb · in^{-2}), °C	88–104	99	74–93	174	182	203	204
Maximum recommended service temperature, °C				149			
Specific heat, cal · g^{-1}							
Thermal conductivity, $W \cdot m^{-1} \cdot K^{-1}$	0.12	0.26–0.28	0.12–0.21	0.12	0.38	0.14–0.19	

(Continued)

TABLE 2.78 Properties of Commercial Plastics (*Continued*)

Properties	Thermoplastic elastomers				Urea formaldehyde, alpha-cellulose filled	Vinyl	
						Poly(vinyl chloride) and poly(vinyl acetate)	
	Polyolefin	Polyester	Block copolymers of styrene and butadiene or styrene and isoprene	Block copolymers of styrene and ethylene or styrene and butylene		Rigid	Flexible and unfilled
Physical							
Melting temperature, °C							
Crystalline		168–206			Thermoset		
Amorphous						75–105	75–105
Specific gravity	0.88–0.90	1.17–1.25	0.9–1.2	0.9–1.2	1.47–1.52	1.30–1.58	1.16–1.35
Water absorption (24 h), %	0.01		0.19–0.39		0.4–0.8	0.04–0.4	0.15–0.75
Dielectric strength, $kV \cdot mm^{-1}$	24–26		16–21		12–16	14–20	12–16
Electrical							
Volume (dc) resistivity, ohm-cm					0.5–5.0	10^{12}–10^{15}	10^{11}–10^{14}
Dielectric constant (60 Hz)					7.7–9.5	3.2–4.0	5.0–9.0
Dielectric constant (10^6 Hz)					6.7–8.0	3.0–4.0	3.0–4.0
Dissipation (power) factor (60 Hz)					0.036–0.043	0.01–0.02	0.03–0.05
Dissipation factor (10^6 Hz)					0.025–0.035	0.006–0.02	0.06–0.1
Mechanical							
Compressive modulus, 10^3 lb · in^{-2}			3.6–120				
Compressive strength, rupture or 1% yield, 10^3 lb · in^{-2}					25–45	8–13	0.9–1.7
Elongation at break, %	150–300	350–450	500–1350	600–800	<1	40–80	200–450
Flexural modulus at 23°C, 10^3 lb · in^{-2}	1.5–2.0	7–75	4–150	4–100	1300–1600	300–500	

Flexural strength, rupture or yield, 10^3 lb · in^{-2}					10–18	10–16	
Hardness, Rockwell (or Shore)	(A65–A92)	(D40–D72)	(A40–A90)	(A50–A90)	M110–M120	(D65–D95)	(A50–A100)
Impact strength (Izod) at 23°C, J · m^{-1}	No break	208 to no break	No break	No break	13–21	21–1068	Varies over wide range
Tensile modulus, 10^3 lb · in^{-2}		1.1–2.5	0.8–50		1000–1500	350–600	
Tensile strength at break, 10^3 lb · in^{-2}	0.65–2.0	3.7–5.7	0.6–3.0	1–3	5.5–13	6–75	1.5–3.5
Tensile yield strength, 10^3 lb · in^{-2}							
Thermal							
Burning rate, mm · min^{-1}					Self-extinguishing	Self-extinguishing	Slow to self-extinguishing
Coefficient of linear thermal expansion, 10^{-6}°C	130–170		130–137		22–36	50–100	70–250
Deflection temperature under flexural load (264 lb · in^{-2}), °C			<0–49		127–143	60–77	
Maximum recommended service temperature, °C					77	70–74	80–105
Specific heat, cal · g^{-1}					0.6	0.2–0.28	0.36–0.5
Thermal conductivity, W · m^{-1} · K^{-1}	0.19–0.21		0.15		0.30–0.42	0.15–0.21	0.13–0.17

(Continued)

TABLE 2.78 Properties of Commercial Plastics (*Continued*)

Properties	Vinyl						
	Poly(vinyl chloride) and poly(vinyl acetate)						
	Flexible and filled	Poly(vinyl chloride), 15% glass-fiber-reinforced	Poly(vinylidene chloride)	Poly(vinyl formal)	Chlorinated poly(vinyl chloride)	Poly(vinyl butyral), flexible	
Physical							
Melting temperature, °C							
Crystalline			210				
Amorphous	75–105	75–105		105	110	49	
Specific gravity	1.3–1.7	1.54	1.65–1.72	1.2–1.4	1.49–1.56	1.05	
Water absorption (24 h),%	0.5–1.0	0.01	0.1	0.5–3.0	0.02–0.15	1.0–2.0	
Dielectric strength, $kV \cdot mm^{-1}$	9.8–12	24–31	16–24	19		14	
Electrical							
Volume (dc) resistivity, ohm-cm			$10^{14}–10^{16}$				
Dielectric constant (60 Hz)			4.5–6.0				
Dielectric constant (10^6 Hz)							
Dissipation (power) factor (60 Hz)							
Dissipation factor (10^6 Hz)							
Mechanical							
Compressive modulus, $10^3\ lb \cdot in^{-2}$					335–600		
Compressive strength, rupture or 1% yield, $10^3\ lb \cdot in^{-2}$	1.0–1.8	9	2–2.7		9–22		
Elongation at break, %	200–400	2–3	50–250	5–20	4–65	150–450	
Flexural modulus at 23°C, $10^3\ lb \cdot in^{-2}$		750			380–450		

Flexural strength, rupture or yield, 10^3 lb · in^{-2}		13.5	4.2–6.2	17–18	14.5–17	
Hardness, Rockwell (or Shore)	(A50–A100)	R118	M50–M65	M85	R117–R122	A10–A100
Impact strength (Izod) at 23°C, J · m^{-1}	Varies over wide range	53	16–53	43–75	53–299	Varies over wide range
Tensile modulus, 10^3 lb · in^{-2}		870	50–80	350–600	360–475	
Tensile strength at break, 10^3 lb · in^{-2}	1–3.5	9.5	3–5	10–12	7.5–9	0.5–3.0
Tensile yield strength, 10^3 lb · in^{-2}						
Thermal						
Burning rate, mm · min^{-1}			Self-extinguishing			Slow
Coefficient of linear thermal expansion, 10^{-6}°C			190	64	68–78	
Deflection temperature under flexural load (264 lb · in^{-2}), °C		68	54–71	71–77	94–112	
Maximum recommended service temperature, °C			100			
Specific heat, cal · g^{-1}			0.32			
Thermal conductivity, W · m^{-1} · K^{-1}	0.13–0.17		0.13	0.16	0.14	

(*Continued*)

TABLE 2.79 Properties of Natural and Synthetic Rubbers

Rubber	Specific gravity	Durometer hardness (or Shore)	Ultimate elongation % (23°C)	Tensile strength, lb · in^{-2} (23°C)	Service temperature, °C	
					Minimum	Maximum
Gutta percha (hard rubber)	1.2–1.95	(65–95)	3–8	4000–10,000		104
Natural rubber (NR)	0.93	20–100	750–850	3000–4500	−56	82
Chlorosulfonated polyethylene	1.10	50–95	100–500	500–3000	−54	121
Epichlorohydrin	1.27	60–90	100–400	1000–2500	−46	121
Fluoroelastomers	1.4–1.95	60–90	100–350	2000–3000	−40	232
Isobutene-isoprene rubber (IIR) [also known as government rubber I(GR-I)]	0.91	(40–70)	750–950	2300–3000		121
Nitrile rubber (butadiene-acrylonitrile rubber) (also known as Buna N and NBR)	1.00	30–100	100–600	500–4000	−54	121
Polyacrylate	1.10	40–100	100–400	1000–2200	−18	149
Polybutadiene rubber (BR)	0.93	30–100	100–700	2500–3000	−62	79–100
Polychloroprene (neoprene)	1.23	20–90	800–1000	2000–3500	−54	121
Poly(ethylene-propylene-diene) (EPDM)	0.85	30–100	100–300	1000–3000	−40	149
Polyisobutylene (butyl rubber)	0.92	30–100	100–700	1000–3000	−54	100
Polyisoprene	0.94	20–100	100–750	2000–3000	−54	79–82
Polysulfide (Thiokol ST)	1.34	20–80	100–400	700–1250	−54	82–100
Poly(vinyl chloride) (Koroseal)	1.32	(80–90)		2400–3000		71
Silicone, high-temperature				700–800		316
Silicone	0.98	20–95	50–800	500–1500	−84	232
Styrene-butadiene rubber (SBR) (also known as Buna S)	0.94	40–100	400–600	1600–3700	−60	107
Urethane	0.85	62–95	100–700	1000–8000	−54	100

TABLE 2.80 Density of Polymers Listed by Trade Name

Common or trade name	ρ(g/cm^3)
Acetate Rayon	1.32
Acrylic	1.16
Acrylonitrile-styrene copolymer	1.075–1.10
Acrylonitrile-styrene-butadiene copolymer (ABS)	1.04–1.07
Aniline-formaldehyde	1.22–1.25
Benzylcellulose	1.22
Bisphenol-*A* polycarbonate (BPAPC)	1.20
Butyl rubber	0.92
Cellulose I	1.582–1.630
Cellulose II	1.583–1.62
Cellulose III	1.61
Cellulose IV	1.61
Cellulose acetate	1.28–1.32
Cellulose acetate-butyrate	1.14–1.22
Cellulose formate fiber	1.45
Cellulose nitrate	1.35–1.40
Cellulose propionate	1.18–1.24
Cellulose triacetate	1.28–1.33
Cellulose tributyrate	1.16
Chlorinated polyether	1.40
Cotton	1.50–1.54
Cotton, acetylated	1.43
Ethylcellulose	1.09–1.17
Ethylene-propylene copolymer (EPM)	0.86
Glass	3.54
Glass and asbestos	2.5
Kevlar	1.44
Lignocellulose	1.45
Maleic anhydride-styrene copolymer	1.286
Melamine-formaldehyde	1.16
Methyl polyvinyl ketone	1.12
Methylcellulose	1.362
Nomex	1.38
Nylon 6	1.12–1.24
Nylon 66	1.13–1.15, 1.22–1.25
Nylon-610	1.156
Nylon-12	1.02–1.034
Rubber, butyl	0.92
Rubber (unvulcanized)	0.91
Rubber (hard) (Ebonite)	1.11–1.17
Rubber, chlorinated (Neoprene) (CR), unvulcanized	1.23
Rubber, chlorinated (Neoprene) (CR), vulcanized	1.32–1.42
Rubber, fluorinated silicone	1.0
Rubber, silicone	0.80
Rubber, silicone (vulcanized)	1.3–2.3
Rubber, styrene-butadiene (SBR), (unvulcanized)	0.93–0.94
Rubber, styrene-butadiene (SBR), (vulcanized)	0.961
Silk	1.25–1.35
Toluene-sulfonamide-formaldehyde	1.21–1.35
Urea-formaldehyde	1.16
Urea-thiourea-formaldehyde	1.477
Viscose Rayon	1.5
Wool	1.28–1.33

TABLE 2.81 Density of Polymers Listed by Chemical Name

Chemical name	$\rho(g/cm^3)$
Poly-	
acetylaldehyde	1.07
acrolein	1.322
acrylic acid	1.22
acrylonitrile (PAN)	1.01–1.17, 1.20
acrylonitrile-vinyl acetate	1.14
amide-6 (PA-6)	1.12–1.24
amide-66 (PA-66)	1.13–1.15, 1.22–1.25
amide-610 (PA-610)	1.156
amide-12 (PA-12)	1.02–1.034
aryl ether ether ketone (PEEK)	1.20
arylate	1.21
bisphenol carbonate (BPAPC)	1.20
butadiene-1,2, isotactic	0.96
butadiene-1,2, syndiotactic	0.96
butadiene-1,4-*cis*	1.01
butadiene-1,4-*trans*	0.93–0.97, 1.01
1-butene	0.85
butene	0.91–0.92
butyl acrylate	1.08
sec.-butyl acrylate	1.05
butylene	0.60
tert.-butyl methacrylate	1.03
-*n*-butyl methacrylate	1.055
sec.-butyl methacrylate	1.04
tert.-butylstyrene	0.957
caprolactam, nylon	0.985
carbonate (PC)	1.14–1.2
chlorobutadiene	1.25
chloroprene (Neoprene rubber) (CR), unvulcanized	1.23
chloroprene (Neoprene rubber) (CR), vulcanized	1.32–1.42
chlorotrifluoroethylene	2.03
dichlorostyrene	1.38
2,2-dimethylpropyl acrylate	1.04
dimethylsiloxane	0.970
dodecyl methacrylate	0.93
1-ethylpropyl acrylate	1.04
etheretherketone (PEEK)	1.27
ethyl acrylate	1.095, 1.12
ethyl methacrylate	1.11, 1.12
ethylbutadiene	0.891
ethylene	0.870, 0.910–0.965
ethylene (amorphous)	0.85
ethylene (crystalline)	0.99
ethylene (high density: HDPE)	0.941–0.965
ethylene (linear low density: LLDPE)	0.918–0.935
ethylene (low density: LDPE)	0.910–0.925
ethylene (medium density: MDPE)	0.926–0.940
ethylene glycol	1.0951
ethylene glycol fumarate	1.385
ethylene glycol isophthalate, cryst.	1.358

TABLE 2.81 Density of Polymers Listed by Chemical Name (*Continued*)

Chemical name	ρ(g/cm^3)
ethylene glycol phthalate	1.352
ethylene glycol waxes	1.15–1.20
ethylene isophthalate	1.34
ethylene phthalate	1.34
ethylene terephthalate (PETP)	1.33–1.42
formaldehyde	1.425
–*n*-hexyl methacrylate	1.01
imide	1.43
isobutene	0.917
isobutyl methacrylate	1.02–1.04
isobutylene	0.87–0.93
isoprene (1,4–)	0.900–0.913
–*N*-isopropylacrylamide	1.070–1.118
isopropyl acrylate	1.08
isopropyl methacrylate	1.04
methacrylonitrile	1.10
methyl acrylate	1.07–1.223
methyl methacrylate (PMMA)	1.16–1.20
4-methyl-1-pentene	0.84
myrcene	0.895
oxymethylene (POM)	1.41–1.435
phenylene oxide	1.00–1.06
polysulfide (Thiokol A)	1.60
polysulfide (Thiokol B)	1.65
propyl methacrylate	1.06–1.08
propylene (PP)	0.85–0.92
propylene, amorphous	0.87
propylene, head-to-head	0.878
propylene, isotactic	0.90–0.92
propylene, isotactic (crystalline)	0.92–0.939
propylene, syndiotactic (crystalline)	0.93
propylene oxide	1.00
styrene (PS)	1.04–1.09
styrene, crystalline	1.08–1.111
styrene-butadiene thermoplastic elastomer	0.93–1.10
sulfone	1.24
tetrafluoroethylene (PTFE)	2.28–2.344
trifluorochloroethylene	2.11–2.13
vinyl acetate (PVAC)	1.08–1.25
vinyl alcohol (PVA)	1.21–1.31
vinyl butyral	1.07–1.20
vinyl chloride	1.37–1.44
vinyl chloride-co-methyl acrylate	1.34
vinyl chloride, flexible	1.25–1.35
vinyl chloride, rigid	1.35–1.55
vinyl chloride acrylonitrile (60/40)	1.28
vinylethylene	0.889
vinyl formal	1.2–1.4
vinyl pyrrolidone (PVP)	1.25
vinyl-vinylidene chloride	1.70
vinylcarbazole	1.20
vinylidene chloride (PVDC)	1.65–1.875
vinylidene fluoride (PVDF)	1.75–1.78
vinylisobutyl ether	0.91–0.92
–*m*-xylene adipamide	1.22

TABLE 2.82 Density of Polymers at Various Temperatures

Temperature (deg C)	0	20	40	60	80	100	120	140	160	180	200	220	240	260	280	300	320	340	360	380
Natural rubber, unvulcanized	0.9283	0.9162																		
Natural rubber, cured	0.9211	0.9093																		
Polyamide, Nylon 6												1.176	1.165	1.154	1.143					
Polyamide, Nylon 6,6													0.963							
														1.100	1.086	1.071				
Poly(butene-1), isotactic								0.797	0.786	0.776	0.765	0.755	0.745							
Poly(*n*-butyl methacrylate)			1.045	1.032	1.018	1.004	0.990	0.975	0.961	0.947	0.933									
	*g*1.063[a]	1.057	1.043	1.030	1.017	1.005	0.993													
Poly(*e*-caprolactone)						1.037	1.023	1.010												
Polycarbonate, (with Bisphenol A)			*g*1.192	*g*1.186	*g*1.180	*g*1.174	*g*1.167	*g*1.161	1.150	1.136	1.123	1.109	1.095	1.081	1.067	1.053	1.039	1.025		
Poly(cyclohexyl methacrylate)		*g*1.101	*g*1.095	*g*1.090	*g*1.084		1.066	1.054	1.041	1.028	1.015									
Poly (2,6-dimethylphenylene ether)			*g*1.061	*g*1.057	*g*1.052	*g*1.048	*g*1.043	*g*1.039	*g*1.035	*g*1.030	*g*1.026	1.012	0.997	0.983	0.968	0.953	0.939			
Poly(dimethyl siloxane)		0.9742	0.9566	0.9393	0.9222	0.9053	0.8887	0.8722	0.8560	0.8400	0.8242									
			0.9566	0.9389																
Polyetheretherketone																		1.113	1.098	1.084
Polyethylene, branched							0.801	0.790	0.780	0.769	0.759	0.749								
						0.785	0.774	0.763	0.752											
Polyethylene, linear						0.7847	0.7735	0.7624	0.7514											
						0.789	0.778	0.766	0.753											
Poly(ethylene terephthalate)															1.172	1.156	1.140	1.125		
Poly(ethyl methacrylate)	*g*1.131	*g*1.125	*g*1.119	*g*1.113	1.103															
Polyisobutylene	0.9297	0.9195	0.9093	0.8992	0.8891	0.8791	0.8691	0.8592												
Poly(methyl methacrylate)			*g*1.181	*g*1.177	*g*1.171	*g*1.166	1.153	1.139	1.126	1.112	1.097	1.082	1.067	1.052						
		*g*1.184	*g*1.179	*g*1.174	*g*1.168		1.148	1.136	1.123											
							1.153	1.141	1.129	1.117	1.106	1.094								
	*g*1.175	*g*1.170	*g*1.165	*g*1.160	*g*1.155	*g*1.150	1.140	1.128												
Poly(methyl methacrylate), isotactic		*g*1.220		1.204	1.189	1.174	1.160	1.146	1.132	1.119										
Poly(*o*-methyl styrene)			*g*1.016	1.011	*g*1.006			0.9881	0.9777	0.9674	0.9571									
Polyoxyethylene				1.063	1.048	1.033	1.018	1.004	0.990	0.976										
Polyoxymethylene									1.167	1.151										
Polypropylene, atactic					0.827	0.816	0.802													

Polypropylene, isotactic										0.764	0.754	0.744	0.734	0.724	0.714	0.705			
										0.763	0.753	0.743							
Polystyrene				1.0260	1.0142	1.0025	0.9909	0.9795	0.9681	0.9569									
		*g*1.044	*g*1.040	*g*1.035			1.0125	1.0021	0.9919	0.9818	0.9717								
			*g*1.040	*g*1.034	1.026	1.016	1.005	0.994	0.984	0.973	0.961	0.950	0.939	0.928	0.916	0.905	0.893		
Polysulfone, (with Bisphenol *A*)			*g*1.232	*g*1.226	*g*1.221	*g*1.216	*g*1.211	*g*1.206	*g*1.201	*g*1.195	1.183	1.170	1.157	1.144	1.130	1.117	1.104	1.091	1.078
Polytetrafluoroethylene																		1.548	1.504
Polytetrahydrofuran			0.944	0.931	0.919	0.907	0.895												
Poly(vinyl acetate)	*g*1.196	*g*1.189	1.1783	1.1615	1.1449	1.1285													
Poly(vinyl chloride)				1.352	1.338	1.322													
Poly(vinyl methyl ether)		1.0580	1.0436	1.0294	1.0152	1.0011	0.9871												

[a] *g* = glass.

TABLE 2.83 Surface Tension (Liquid Phase) of Polymers

Polymer	MW	γ_{LV} at 20°C (mN/m)	$-d\gamma/dT$ [mN/(mK)]
Poly(oxyhexafluoropropylene)	∞	18.4 (25°C)	0.059 (M_n ~ 7000)
Poly[heptadecafluorodecyl)methylsiloxane]	M_n ~ 19600	18.5 (25°C)	···
Poly(dimethylsiloxane)	∞	21.3 (20°C)	0.048 (10^6 cS)
Poly[methyl(trifluoropropyl)siloxane]	∞	24.4 (25°C)	···
Poly(tetrafluoroethylene)	∞	25.6	0.053 (M_n = 1038)
Poly(oxyisobutylene)	M ~ 30000	27.5	0.066
Poly(vinyl octanoate)	···	28.7	0.061
Polypropylene, atactic	Melt index ~ 1000	29.4	0.056
Paraffin wax	···	30.0 (20°C)	~0.06
Poly(1,2-butadiene)	M_n ~ 1000	30.4 (25°C)	···
Poly(*t*-butyl methacrylate)	M_v ~ 6000	30.5	0.059
Poly(oxypropylene)	M_n ~ 4100	30.7 (25°C)	0.073
Poly(*i*-butyl methacrylate)	M_v ~ 35000	30.9	0.060
Poly(chlorotrifluoroethylene)	M_n ~ 1280	30.9	0.067
Poly(vinyl hexadecanoate)	···	30.9	0.066
Poly(*n*-butyl methacrylate)	M_v ~ 37000	31.2	0.059
Poly(oxytetramethylene)	M_n ~ 32000	31.8	0.060
Poly(methoxyethylene)	M_n ~ 46500	31.8	0.075
Poly(*n*-butyl acrylate)	M ~ 32000	33.7	0.070
Polyethylene, branched	M_n ~ 7000	34.3	0.060
Poly(isobutylene)	∞	35.6 (24°C)	0.064 (M_n ~ 2700)
Polyethylene, linear	M_w ~ 67000	35.7	0.057
Poly(oxydecamethylene)	···	36.1	0.068
Poly(vinyl acetate)	M_w ~ 120000	36.5	0.066
Poly(2-methylstyrene)	M_n ~ 3000	38.7	0.058
Poly(oxydodecamethyleneoxyisophthaloyl)	···	40.0	0.070
Polystyrene	M_v ~ 44000	40.7	0.072
Poly(methyl acrylate)	M_n ~ 25000	41.0	0.070
Poly(methyl methacrylate)	M_v ~ 3000	41.1	0.076
Poly(epichlorohydrin)	M_n ~ 1500	43.2 (25°C)	···
Polychloroprene	M_v ~ 30000	43.6	0.086
Poly(oxyethyleneoxyterephthaloyl)	M_n ~ 16000	44.5	0.064
Poly(oxyethylene)	∞	45.0 (24°C)	0.076 (M_n ~ 6000)
Poly(hexamethylene adipamide)	M_n ~ 17000	46.4	0.064
Poly(oxyisophthaloyloxypropylene)	···	49.3	0.083

TABLE 2.84 Interfacial Tension (Liquid Phase) of Polymers

Polymer pair	γ_{12} at 20°C (mN/m)	$-d\gamma/dT$ [mN/(mK)]
Polychloroprene/polystyrene	0.5 (140°C)	. . .
Polychloroprene/poly(*n*-butyl methacrylate)	1.6 (140°C)	. . .
Poly(methyl methacrylate)/poly (*t*-butyl methacrylate)	3.0	0.005
Poly(methyl methacrylate)/polystyrene	3.2	0.013
Poly(dimethylsiloxane)/polypropylene	3.2	0.002
Poly(methyl methacrylate)/poly(*n*-butyl methacrylate)	3.4	0.012
Poly(dimethylsiloxane)/poly(*t*-butyl methacrylate)	3.6	0.003
Polybutadiene/poly(dimethylsiloxane)	4.0	0.009
Poly(methyl acrylate)/poly(*n*-butyl acrylate)	4.0	0.008
Poly(dimethylsiloxane)/poly(isobutylene)	4.0	0.016
Poly(n-butyl methacrylate)/poly(vinyl acetate)	4.2	0.011
Poly(dimethylsiloxane)/poly(*n*-butyl methacrylate)	4.2	0.004
Polystyrene/poly(vinyl acetate)	4.2	0.004
Polyethylene/polystyrene	4.4 (200°C)	. . .
Poly(oxyethylene)/poly(oxtetramethylene)	4.5	0.005
Polychloroprene/Polyethylene, branched	4.6	0.008
Polyethylene, linear/poly(*n*-butyl acrylate)	5.0	0.014
Polyethylene, branched/poly(oxytetramethylene)	5.0	0.007
Poly(dimethylsiloxane)/polyethylene, branched	5.3	0.002
Poly(oxytetramethylene)/poly(vinyl acetate)	5.5	0.008
Polyethylene, branched/poly(*i*-butyl methacrylate)	5.5	0.010
Polyethylene, branched/poly(oxydodecamethyleneoxyisophthaloyl)	5.9	0.011
Polyethylene, branched/poly(*t*-butyl methacrylate)	5.9	0.016
Poly(dimenthylsiloxane)/polystyrene	6.1	~0
Poly(dimethylsiloxane)/poly(oxytetramethylene)	6.4	0.001
Poly(dimethylsiloxane)/polychloroprene	7.1	0.005
Polyethylene, linear/poly(*n*-butyl methacrylate)	7.1	0.015
Polyethylene, linear/polystyrene	8.3	0.020
Poly(dimentylsiloxane)/poly(vinyl acetate)	8.4	0.008
Poly(isobutylene)/poly(vinyl acetate)	9.9	0.020
Polyethylene, linear/poly(methyl acrylate)	10.6	0.018
Polyethylene/poly(caprolactam)	10.7 (250°C)	. . .
Poly(dimethylsiloxane)/poly(oxyethylene)	10.9	0.008
Polyethylene, branched/poly(oxyethylene)	11.6	0.016
Polyethylene, linear/poly(methyl methacrylate)	11.9	0.018
Polyethylene, linear/poly(vinyl acetate)	14.5	0.027
Polyethylene, linear/poly(hexamethylene adipamide)	14.9	0.018
Polyethylene, branched/poly(oxyisophthaloyloxpropylene)	15.4	0.030

TABLE 2.85 Thermal Expansion Coefficients of Polymers

Temperature (deg C)	0	20	40	60	80	100	120	140	160	180	200	220	240	260	280	300	320	340	360	380
Natural Rubber, unvulcanized	6.6	6.6																		
Natural Rubber, cured	6.5	6.4																		
		6.7																		
Polyamide, Nylon 6													4.7	4.7	4.7					
Polyamide, Nylon 6,6														6.6	6.6					
															6.8					
Poly(butene-1), isotactic								6.7	6.7	6.7	6.7	6.7	6.7							
Poly(*n*-butyl methacrylate)			6.2	6.5	6.8	7.0	7.2	7.3	7.4	7.4	7.4									
	*g*3.8[a]	6.4	6.4	6.3	6.2	6.1	6.1													
Poly(*e*-caprolactone)							6.4	6.3												
Polycarbonate, (with Bisphenol *A*)			*g*2.6	*g*2.6	*g*2.6	*g*2.6	*g*2.6	*g*2.6	5.8	5.9	6.1	6.2	6.3	6.4	6.6	6.7	6.8	6.9		
Poly(cyclohexyl methacrylate)		*g*2.4	*g*2.5	*g*2.5	*g*2.5		5.9	6.0	6.2	6.3	6.4									
Poly(2,6-dimethylphenylene ether)			*g*2.1	*g*2.1	*g*2.1	*g*2.1	*g*2.1	*g*2.1	*g*2.1	*g*2.1	*g*2.1	7.1	7.3	7.4	7.6	7.7	7.8			
Poly(dimethyl siloxane)		9.06	9.11	9.17	9.23	9.29	9.35	9.41	9.47	9.53	9.59									
		9.0																		
			9.4	9.2																
Polyetheretherketone																		6.7	6.7	6.7
Polyethylene, branched							6.7	6.7	6.7	6.7	6.7	6.7								
								7.5	7.2	6.9										
Polyethylene, linear								7.14	7.18	7.24	7.32									
										7.0	7.0									
									7.6	7.9										
Poly(ethylene terephthalate)															6.8	6.8	6.8	6.8		
Poly(ethyl methacrylate)	*g*2.7	*g*2.7	*g*2.7	*g*2.7	6.0															
Polyisobutylene	5.51	5.54	5.58	5.61	5.65	5.68	5.72	5.75												
Poly(methyl methacrylate)			*g*1.8	*g*2.1	*g*2.4	*g*2.7	5.5	5.8	6.1	6.4	6.7	7.0	7.2	7.5						
		*g*1.8	*g*2.2	*g*2.5	*g*2.9		5.4	5.7	6.0											
							5.2	5.2	5.2	5.2	5.2	5.2								
	*g*2.1	*g*2.1	*g*2.1	*g*2.1	*g*2.1	*g*2.1	5.2	5.2												

Poly(methyl methacrylate), isotactic		*g*2.2		6.4	6.3	6.2	6.1	5.9	5.8	5.7									
Poly(*o*-methyl styrene)			*g*2.6	*g*2.6	*g*2.6			5.3	5.3	5.3									
Polyoxyethylene						7.1	7.1	7.1	7.1	7.1	7.1	7.1							
Polyoxymethylene										7.9									
											6.8	6.8							
Polypropylene, atactic					6.1	7.7	9.3												
Polypropylene, isotactic										6.6	6.6	6.6							
										6.7	6.7	6.7	6.7	6.7	6.7	6.7			
										6.7	6.7								
Polystyrene						5.78	5.79	5.81	5.82	5.84	5.85	5.87							
		*g*2.0	*g*2.3	*g*2.5			5.1	5.1	5.1	5.1	5.1								
			*g*2.9	*g*2.9	5.0	5.2	5.3	5.4	5.6	5.7	5.8	5.9	6.0	6.2	6.3	6.4	6.5		
Polysulfone, (with Bisphenol *A*)			*g*2.1	*g*2.1	*g*2.1	*g*2.1	*g*2.1	*g*2.1	*g*2.2	*g*2.2	5.5	5.6	5.7	5.8	5.8	5.9	6.0	6.1	6.1
Polytetrafluoroethylene																		14.4	14.7
Polytetrahydrofuran					6.7	6.7	6.7	6.7	6.7										
Poly(vinyl acetate)	*g*2.8	*g*2.8	7.13	7.17	7.20	7.23													
Poly(vinyl chloride)						4.7	5.5	6.2											
Poly(vinyl methyl ether)			6.87	6.92	6.96	7.01	7.06												

[a] *g* = glass

TABLE 2.86 Heat Capacities of Polymers

Polymer	Abbre-viations	Molecular[a] weight g/mol	T_g (K)	Temp. (K)	C_p[b] kJ/kg·K	C_p[b] J/mol·K	ΔC_p[c] J/mol·K
			1. Main-chain carbon polymers				
			Poly(acrylics)				
Poly(*iso*-butyl acrylate)	PiBA	128.17	249	220	1.2156	155.80	36.60
				240	1.3365	171.30	
				300	1.8108	232.09	
				500	2.3388	299.77	
Poly(*n*-butyl acrylate)	PnBA	128.17	218	80	0.5598	71.75	45.40
				180	1.0632	136.27	
				300	1.8201	233.28	
				440	2.1803	279.45	
Poly(ethyl acrylate)	PEA	100.12	249	90	0.5792	57.99	45.60
				200	1.0301	103.13	
				300	1.7867	178.88	
				500	2.2189	222.16	
Poly(methyl acrylate)	PMA	86.09	279	100	0.6154	52.98	42.30
				200	0.9816	84.51	
				300	1.765	151.99	
				500	2.143	184.49	
			Poly(dienes)				
1,4-Poly(butadiene)	PBD	54.09					
cis-			171	50	0.3694	19.98	29.10
				150	0.8967	48.50	
				300	1.960	106.00	
				350	2.214	114.90	
trans-			180	50	0.3465	18.74	28.20
				150	0.9057	48.99	
				300	NA	NA	
				500	2.616	141.50	
Poly(1-butene)	PB	56.11	249	100	0.6733	37.78	23.06
				200	1.2190	68.40	
				300	2.086	117.02	
				600	3.071	172.31	
Poly(1-butenylene)	PBUT	55.10					
cis-			171	30	0.2140	11.79	28.91
				130	0.7775	42.838	
				300	1.924	106.03	
				450	2.409	132.73	
trans-			190	30	0.1761	9.704	26.48
				130	0.7898	43.516	
				300	1.924	106.03	
				450	2.409	132.73	
			Poly(alkenes)				
Poly(ethylene)	PE	14.03	252	100	0.674	9.45 (c)	10.1
				200	1.110	15.57	
				300	1.555	21.81 (s)	
					2.202	30.89 (m)	
				600	3.127	43.87	
Poly(1-hexene)	PHE	84.16	223	100	0.7020	59.08 (a)	25.1
				200	1.3319	112.09	
				250	1.903	160.18 (a)	
				290	2.079	174.98 (a)	

TABLE 2.86 Heat Capacities of Polymers (*Continued*)

Polymer	Abbre-viations	Molecular[a] weight g/mol	T_g (K)	Temp. (K)	C_p[b] kJ/kg·K	C_p[b] J/mol·K	ΔC_p[c] J/mol·K
Poly(isobutene)	PiB	56.11	200	50	0.2440	13.69 (a)	22.29
				150	0.8660	48.59	
				300	1.962	110.09 (a)	
				380	2.311	129.66	
Poly(2-methylbutadiene)	PMBD	68.12					
cis-			200	50	0.3573	24.34	30.87 (a)
				150	0.9025	61.48	
				300	1.911	130.20	
				360	2.216	144.80	
Poly(4-methyl-1-pentene)	P4MPE	84.16	303	80	0.5610	47.21	33.7 (a)
				180	1.090	91.75	
				250	1.4449	121.60	
				300	1.728	145.40	
Poly(1-pentene)	PPE	70.14	233	200	1.253	87.90	27.03 (a)
				220	1.338	93.82	
				300	2.058	144.34	
				470	2.770	194.32	
Poly(propylene)	PP	42.08	260	100	0.6238	26.25 (c)	17.37
				200	1.132	47.63 (c)	
				300	1.622	68.24 (s)	
					2.099	88.34 (m)	
				600	3.178	133.73 (a)	
Poly(methacrylics)							
Poly(*n*-butyl methacrylate)	PnBMA	142.20	293	80	0.5472	77.81	29.70
				200	1.1557	164.34	
				300	1.8524	263.41	
				450	2.3673	336.63	
Poly(*i*-butyl methacrylate)	PiBMA	142.20	326	230	1.2229	173.90	39.00
				300	1.5710	223.40	
				350	2.0190	287.10	
				400	2.1127	300.43	
Poly(ethyl methacrylate)	PEMA	114.15	338	80	0.5155	58.84	31.70
				300	1.4666	167.42	
				350	1.9489	222.47	
				380	2.0462	233.57	
Poly(hexyl methacrylate)	PHMA	170.25	268	270	1.8264	310.77	—
				300	1.9091	324.83	
				420	2.2396	381.06	
Poly(methacrylic acid)	PMAA	86.09	—	100	0.5248	45.18	—
				200	0.9456	81.41	
				300	1.307	112.50	
Poly(methacrylamide)	PMAM	85.11	—	100	0.5904	50.25	—
				200	1.032	87.81	
				300	1.395	118.70	
Poly(methyl methacrylate)	PMMA	100.12	378	100	0.5742	57.49	33.5
				300	1.3755	137.72	
				400	2.0766	207.91	
				550	2.4323	243.52	

(*Continued*)

TABLE 2.86 Heat Capacities of Polymers (*Continued*)

Polymer	Abbre-viations	Molecular[a] weight g/mol	T_g (K)	Temp. (K)	$C_p{}^b$ kJ/kg·K	$C_p{}^b$ J/mol·K	$\Delta C_p{}^c$ J/mol·K
			Poly(styrenes)				
Poly(styrene)	PS	104.15	373	100	0.4548	47.37 (g)	30.7 (a)
				300	1.2230	127.38	
					1.2730	132.58	
				400	1.9322	201.24	
				600	2.4417	254.30	
—,α-methyl	PαMS	118.18	441	100	0.4712	55.69	25.3
				300	1.2752	150.70 (g)	
				460	2.1868	258.44	
				490	2.3331	275.72	
—, *p*-bromo-	PBS	183.05	410	300	0.79650	145.800	31.9
				350	0.92349	169.045	
				420	1.2651	231.582	
				550	1.4641	267.995	
—, *p*-chloro-	PCS	138.60	406	300	1.0229	141.780	31.1
				350	1.19848	166.110	
				410	1.6331	226.345	
				550	1.9134	265.195	
—, *p*-fluoro-	PFS	122.14	384	130	0.47611	58.152	33.3
				200	0.62048	75.786	
				300	0.93079	113.687	
				380	1.2672	154.773	
—, *p*-iodo-	PIS	230.05	424	300	0.67607	155.53	37.9
				400	0.89102	204.980	
				430	1.1145	256.41	
				550	1.2570	289.17	
—, *p*-methyl-	PMS	118.18	380	300	1.2743	150.600	34.6
				350	1.4917	176.290	
				390	1.9449	229.846	
				500	2.2766	269.05	
			Poly(vinyl halides) and Poly(vinyl nitriles)				
Poly(acrylonitrile)	PAN	53.06	378	100	0.5695	30.22	—
				200	0.9286	49.27	
				300	1.297	68.83	
				370	1.624	86.16	
Poly(chlorotrifluoroethylene)	PC3FE	116.47	325	80	0.2787	32.46	—
				200	0.6257	72.87	
				300	0.85945	100.10	
				320	0.90667	105.60	
Poly(tetrafluoroethylene)	PTFE	50.01	240	100	0.3873	19.37	7.82
				200	0.6893	34.47	
				300	0.9016	45.09 (s)	
					1.028	51.42 (m)	
				700	1.454	72.69	
Poly(trifluoroethylene)	P3FE	82.02	304	100	0.4049	33.21	21.00
				200	0.7128	58.46	
				300	1.078	88.40	
Poly(vinyl chloride)	PVC	62.50	354	100	0.4291	26.82 (g)	19.37(a)
				300	0.9496	59.35 (g)	
				360	1.457	91.08	
				380	1.569	98.05	

TABLE 2.86 Heat Capacities of Polymers (*Continued*)

Polymer	Abbre-viations	Molecular[a] weight g/mol	T_g (K)	Temp. (K)	C_p[b] kJ/kg·K	C_p[b] J/mol·K	ΔC_p[c] J/mol·K
Poly(vinylidene chloride)	PVC2	96.95	255	100	0.3745	36.31	70.26
				200	0.5932	57.51	
				250	0.7115	68.98	
				300	NA	NA	
Poly(vinylidene fluoride)	PVF2	64.03	233	100	0.4435	28.40	22.80
				150	0.6185	39.60	
				230	0.8918	57.10	
				250	0.7856	50.30	
				300	NA	NA	
Poly(vinyl fluoride)	PVF	46.04	314	100	0.5204	23.96	17.80(a)
				200	0.8692	40.02	
				300	1.301	59.91	
				310	1.353	62.29	
			Others				
Poly(*p*-phenylene)	PPP	76.10	—	80	0.3708	28.22 (sc)	—
				150	0.58135	44.241 (sc)	
				250	0.92926	70.717 (sc)	
				300	1.117	85.040 (sc)	
Poly(vinyl acetate)	PVAc	86.09	304	80	0.3230	27.81	53.7
				300	1.183	101.86	
				320	1.8409	158.48	
				370	1.898	163.37	
Poly(vinyl alcohol)	PVA	44.05	358	60	0.2674	11.78	—
				150	0.7187	31.66	
				250	1.185	52.21	
				300	1.546	68.11	
Poly(vinyl benzoate)	PVBZ	148.16	347	190	0.71808	106.39	69.5
				300	1.1025	163.35	
				400	1.8390	272.47	
				500	2.0333	301.25	
Poly(*p*-xylylene)	PPX	104.15	286	220	0.91445	95.241 (sc)	37.6(a)
				250	1.0576	110.149 (sc)	
				300	1.3022	135.622 (sc)	
				410	1.8686	194.619 (sc)	
			2. Main-chain heteroatom polymers				
			Poly(amides)				
Poly(iminoadipoy-liminododecamethylene)	Nylon 612	310.48	319	230	1.2296	381.78	214.8(a)
				300	1.5926	494.48	
				400	2.4842	771.30	
				600	3.1596	980.986	
Poly(imioadipoy-liminohexamethylene)	Nylon 66	226.32	323	230	1.1139	252.10	145.0(a)
				300	1.4638	331.30	
				400	2.3794	538.50	
				600	2.793	632.1	
Poly(iminohexamethylene-iminoazelaoyl)	Nylon 69	268.40	331	230	1.1980	321.53	—
				300	1.5204	408.080	
				400	2.3840	639.874	
				600	3.0720	824.534	
Poly(iminohexamethylene-iminosebacoyl)	Nylon 610	282.43	323	230	1.2069	340.870	—
				300	1.5644	441.820	
				400	2.3975	677.125	
				600	3.1041	876.685	

(*Continued*)

TABLE 2.86 Heat Capacities of Polymers (*Continued*)

Polymer	Abbre-viations	Molecular[a] weight g/mol	T_g (K)	Temp. (K)	C_p[b] kJ/kg·K	C_p[b] J/mol·K	ΔC_p[c] J/mol·K
Poly(imino-(1-oxohexamethylene))	Nylon 6	113.16	313	70	0.4400	49.78	93.6(a)
				300	1.5023	170.00	
				400	2.5186	285.00	
				600	2.7881	315.50	
Poly(imino-1-oxododecamethylene)	Nylon 12	197.32	314	230	1.2874	254.020	—
				300	1.6952	334.49	
				400	2.4709	487.565	
				600	3.2786	646.945	
Poly(imino-1-oxoundecamethylene)	Nylon 11	183.30	316	230	1.2996	238.21	—
				300	1.7507	320.91	
				400	2.4567	450.314	
				600	3.2449	594.794	
Poly(methacrylamide)	PMAM	85.11	—	100	0.5904	50.25	—
				200	1.032	87.81	
				250	1.214	103.30	
				300	1.395	118.70	
Poly(amino acids)							
Poly(*L*-alanine)	PALA	71.08	—	230	1.102	78.33	—
				300	1.315	93.47	
				350	1.498	106.5	
				390	1.622	115.3	
Poly(*L*-asparagine)	PASN	114.10	—	230	0.958	109.3	—
				300	1.218	139.0	
				350	1.397	159.4	
				390	1.537	175.4	
Polyglycine	PGLY	57.05	—	230	0.929	53.00	—
				300	1.170	66.75	
				350	1.356	77.36	
				390	1.516	86.49	
Poly(*L*-methionine)	PMET	131.19	—	220	0.936	122.8	—
				300	1.347	176.7	
				350	1.595	209.3	
				390	1.768	232.0	
Poly(*L*-phenylalanine)	PPHE	147.18	—	220	0.830	122.1	—
				300	1.153	169.7	
				350	1.382	203.4	
				390	1.548	227.8	
Poly(*L*-serine)	PSER	87.08	—	220	0.959	83.50	—
				300	1.297	112.9	
				350	1.541	134.2	
				390	1.747	152.1	
Poly(*L*-valine)	PVAL	99.13	—	230	1.213	120.2	—
				300	1.455	144.2	
				350	1.647	163.3	
				390	1.802	178.6	
Poly(esters)							
Poly(butylene adipate)	PBAD	200.24	199	80	0.54302	108.734	140.046
				150	0.87449	175.107	
				300	1.9706	394.595	
				450	2.2147	443.470	

TABLE 2.86 Heat Capacities of Polymers (*Continued*)

Polymer	Abbre-viations	Molecular[a] weight g/mol	T_g (K)	Temp. (K)	C_p[b] kJ/kg·K	C_p[b] J/mol·K	ΔC_p[c] J/mol·K
Poly(butylene terephthalate)	PBT	220.23	248	150	0.61075	134.505(sc)	106.77
			320	200	0.82262	181.166	77.812
				300	1.6134	355.311	
				400	1.8187	400.532	
				570	2.1678	477.407	
Poly(ethylene terephthalate)	PET	192.16	342	100	0.4393	84.42	77.8 (a)
				300	1.172	225.2	
				400	1.8203	349.80	
				600	2.1136	406.15	
Poly(tridecanolactone)	PTDL	212.34	237	185	0.95	202	—
				260	1.45	308	
				300	1.79	380	
				395	2.15	457	
Poly(trimethylene adipate)	PTMA	186.21	—	300	NA	NA	—
				310	1.8710	348.401	
				330	1.9137	356.341	
				360	1.9776	368.252	
Poly(trimethylene succinate)	PTMS	158.15	—	300	NA	NA	—
				310	1.8401	291.014	
				330	1.8721	296.074	
				360	1.9201	303.664	
Poly(γ-butyrolactone)	PBL	86.09	214	100	0.6012	51.760	57.4
				210	1.024	88.170	
				300	1.810	155.858(m)	
				350	1.870	161.031(m)	
Poly(ϵ-caprolactone)	PCL	114.15	209	100	0.62322	71.140	59.5
				200	1.0243	116.923	
				300	1.4229	162.42	
					1.8138	207.04 (s)	
				350	1.9415	221.62 (m)	
Poly(glycolide)	PGL	58.04	318	100	0.5250	30.470	44.4
				300	1.127	65.42	
				400	1.999	116.039 (m)	
				550	2.098	121.75 (m)	
Poly(β-propiolactone)	PPL	72.07	249	100	0.5568	40.130	50.4
				240	1.044	75.220	
				300	1.878	135.354 (m)	
				400	2.081	149.994 (m)	
Poly(ethylene oxalate)	PEOL	116.07	306	100	0.49910	57.930	56.23
				300	1.1175	129.705	
				320	1.6395	190.295 (m)	
				360	1.7012	197.456 (m)	
Poly(ethylene sebacate)	PES	228.29	245	120	0.66292	151.338 (s)	154.059
				200	0.95269	217.490 (sc)	
				300	1.9245	439.34 (m)	
				410	2.1923	500.500 (m)	
			Poly(oxides)				
Poly(oxy-2,6-dimethyl-1,4-phenylene)	PPO	120.15	482	80	0.4418	53.08	31.9 (a)
				300	1.2459	149.70	
				500	2.1232	255.10	
				570	2.2555	271.00	

(*Continued*)

TABLE 2.86 Heat Capacities of Polymers (*Continued*)

Polymer	Abbre-viations	Molecular[a] weight g/mol	T_g (K)	Temp. (K)	C_p[b] kJ/kg·K	C_p[b] J/mol·K	ΔC_p[c] J/mol·K
Poly(oxyethylene)	POE	44.05	206	100	0.6114	26.93 (s)	38.96
				200	0.9507	41.88 (s)	
				300	1.257	55.36 (s)	
					1.995	87.89 (m)	
				450	2.223	97.91	
Polyoxymethylene	POM	30.03	190	100	0.5554	16.68 (s)	27.47
				150	0.7266	21.82 (s)	
				300	1.283	38.52 (s)	
					1.920	57.67 (m)	
				600	2.292	68.83	
Poly(oxy-1,4-phenylene)	POPh	92.10	358	300	1.185	109.10 (s)	21.4 (a)
				350	1.367	125.90 (s)	
				400	1.694	156.00 (m)	
				600	2.003	184.50 (m)	
Poly(oxypropylene)	POPP	58.08	198	80	0.537	31.21 (s)	32.15
				180	1.014	58.89 (s)	
				300	1.915	111.23 (m)	
				370	2.105	122.27 (m)	
Poly(oxytetramethylene)	PO4M	72.11	189	80	0.5465	39.41 (s)	46.49
				180	1.033	74.52 (s)	
				300	1.985	143.15 (m)	
				340	2.081	150.04 (m)	
Poly(oxytrimethylene)	PO3M	58.08	195	80	0.5095	29.59 (s)	50.73
				180	0.9464	54.97 (s)	
				300	1.373	79.73 (s)	
					2.055	119.34 (m)	
				330	2.107	122.37	
			Others				
Poly(diethyl siloxane)	PDES	102.21	135	50	0.38820	39.678 (sc)	30.189
				100	0.73995	75.630 (sc)	
				300	1.6184	165.417 (m)	
				360	1.7525	179.125 (m)	
Poly(dimethyl itaconate)	PDMI	158.16	377	110	0.59700	94.419 (a)	54.23
				300	1.3183	208.507 (a)	
				400	1.9282	304.968 (m)	
				450	2.0009	316.463 (m)	
Poly(dimethyl siloxane)	PDMS	74.15	146	50	0.3672	27.23	27.7 (a)
				100	0.7131	52.88	
				300	1.591	118.0	
				340	1.657	122.9	
Poly(4-hydroxybenzoic acid)	PHBA	120.11	434	170	0.58914	70.762	34
				300	1.0207	122.60	
				400	1.3662	164.091	
				434	1.4686	176.399	
Poly(4,4′-isopropylidene diphenylenecarbonate)	PC	254.27	418	100	0.43143	109.70 (s)	48.5
				300	1.207	306.8 (s)	
				450	1.9570	497.60 (m)	
				560	2.207	561.3 (m)	

TABLE 2.86 Heat Capacities of Polymers (*Continued*)

Polymer	Abbre-viations	Molecular[a] weight g/mol	T_g (K)	Temp. (K)	C_p[b] kJ/kg·K	C_p[b] J/mol·K	ΔC_p[c] J/mol·K
Poly(oxy-1,4-phenylene-oxy-1,4-phenylene-carbonyl-1,4-phenylene)	PEEK	288.30	419	300	NA	NA	78.1
				419	1.789	515.8	
				500	1.928	555.9	
				750	2.358	679.8	
Poly(oxy-1,4-phenylene-sulphonyl-1,4-phenylene-oxy-1,4-phenylene-(1-methylidene)-1,4-phenylene)	PBISP	442.54	458.5	200	0.75870	335.754	102.482
				300	1.1161	493.934	
				500	1.9436	860.132	
				540	2.0251	896.19	
Poly(1,4-phenylene sulphonyl)	PAS	140.16	492.6	150	0.597	83.7	—
				300	1.009	141.4	
				500	1.571	220.2	
				620	1.642	230.1	
Poly(1-propene sulphone)	P1PS	106.14		10	0.01580	1.677	—
				30	1.165	123.7	
Trigonal selenium	SEt	78.96	303.4	100	0.2304	18.19 (s)	13.29
				300	0.318	25.11	
				400	0.3338	26.36 (s)	
					0.4777	37.72 (m)	
				600	0.4343	34.29	

[a]This is the molecular weight of the repeat unit of the polymer.
[b]Except the data for PTDL and P1PS, C_p data reported in the unit of kJ/kg·K were converted from the C_p data which were directly cited from the literature, using the molecular weight of the repeat unit.
[c]Specific heat increment at T_g.

TABLE 2.87 Thermal Conductivity of Polymers

Polymer	Temperature (K)	k (W/m K)
Polyamides		
Polylauryllactam (nylon-12)		0.25
		0.19
Polycaprolactam (nylon-6)		
Moldings	293	0.24
Crystalline	303	0.43
Amorphous	303	0.36
Melt	523	0.210
Poly(hexamethylene adipamide) (nylon-6,6)		
Moldings	293	0.24
Crystalline	303	0.43
Amorphous	303	0.36
Melt	523	0.15
Poly(hexamethylene dodecanediamide) (nylon-6, 12)		0.22
Poly(hexamethylene sebacamide) (nylon-6, 10)		0.22
Polyundecanolactam (nylon-11)		0.23
Polycarbonates, polyesters, polyethers, and polyketones		
Polyacetal		0.23
		0.3
Polyaryletherketone	293	0.30
Poly(butylene terephthalate) (PBT)	293	0.29
		0.16
Polycarbonate (Biphenol A)	293	0.20
Temperature dependence	300–573	
	150–400	
Poly(dially carbonate)		0.21
Poly(2,6-dimethyl-1,4-phenylene ether)		0.12
Polyester		
Cast, rigid		0.17
Chlorinated		0.33
Polyetheresteramide	303	0.24–0.34
	353	0.20–0.26
Polyetheretherketone (PEEK)		0.25
Poly(ethylene terephthalate) (PET)	293	0.15
Temperature dependence	200–350	
Poly(oxymethylene)	293	0.292
	293	0.44
Temperature dependence	100–400	
Poly(phenylene oxide)		
Molding grade		0.23
Epoxides		
Epoxy resin		
Casting grade	293	0.19
Temperature dependence	300–500	0.19–0.34
Halogenated olefin polymers		
Polychlorotrifluoroethylene	293	0.29
	311–460	0.146–0.248
Poly(ethylene-tetrafluoroethylene) copolymer		0.238
Polytetrafluoroethylene	293	0.25
	298	0.25
	345	0.34
Low-temperature dependence	5–20.8	

TABLE 2.87 Thermal Conductivity of Polymers (*Continued*)

Polymer	Temperature (K)	k (W/m K)
Poly(tetrafluoroethylene-hexafluoropropylene) copolymer (Teflon EEP)		0.202
Poly(vinyl chloride)		
Rigid	293	0.21
Flexible	293	0.17
Chlorinated	293	0.14
Temperature dependence	103	0.129
	273	0.158
	373	0.165
Poly(vinylidene chloride)	293	0.13
Poly(vinylidene fluoride)	293	0.13
	298–433	0.17–0.19
Hydrocarbon polymers		
Polybutene		0.22
Polybutadiene		
Extrusion grade	293	0.22
Poly(butadiene-styrene) copolymer (SBR)		
23.5% styrene content		
Pure gum vulcanizate		0.190–0.250
Carbon black vulcanizate		0.300
Polychloroprene (neoprene)		
Unvulcanized	293	0.19
Pure gum vulcanizate		0.192
Carbon black vulcanizate		0.210
Poly(1,3-cyclopentylenevinylene) [poly(2-norbornene)]		0.29
Polyethylene		
Low density		0.33
Medium density		0.42
High density		0.52
Temperature dependence	20–573	
Molecular-weight dependence		
Poly(ethylene-propylene) copolymer		0.355
Polyisobutylene		0.13
Polyisoprene (natural rubber)		
Unvulcanized		0.13
Pure gum vulcanizate		0.15
Carbon black vulcanizate		0.28
Poly(4-methyl-1-pentene)		0.167
Polypropylene	293	0.12
		0.2
Temperature dependence		
Polystyrene	273	0.105
	373	0.128
	473	0.13
	573	0.14
	673	0.160
Poly(*p*-xylylene) (PPX)		12
Polyimides		
Polyetherimide		0.07
Polyimide		
Thermoplastic	293	0.11
Thermoset		0.23–0.50
Temperature dependence	300–500	

(*Continued*)

TABLE 2.87 Thermal Conductivity of Polymers (*Continued*)

Polymer	Temperature (K)	k (W/m K)
Phenolic resins		
Poly(phenol-formaldehyde) resin		
Casting grade		0.15
Molding grade		0.25
Poly(phenol-furfural) resin		
Molding grade	293	0.25
Polysaccharides		
Cellulose		
Cotton		0.071
Rayon		0.054–0.07
Sulfite pulp, wet		0.8
Sulfite pulp, dry		0.067
Laminated Kraft paper		0.13
Alkali cellulose		0.046–0.067
Different papers	303–333	0.029–0.17
Cellulose acetate	293	0.20
Cellulose acetate butyrate	293	0.33
Cellulose nitriate		0.23
Cellulose propionate		0.20
Ethylcellulose		0.21
Polysiloxanes		
Poly(dimethylsiloxane)	230	0.25
	290	0.22
	340	0.20
	410	0.17
Poly(methylphenylsiloxane)		
9.5% phenyl, $d = 1110$ kg/m^3	273	0.158
	323	0.150
	373	0.144
48% phenyl, $d = 1070$ kg/m^3	273	0.143
	323	0.136
	373	0.127
62% phenyl, $d = 1110$ kg/m^3	273	0.141
	323	0.137
	373	0.132
Polysulfide and polysulfones		
Polyarylsulfone		0.18
Polyethersulfone		0.18
Poly(phenylene sulfide)	293	0.29
	240–310	0.288
Poly(phenylene sulfone)		0.18
Udel polysulfone		0.26
Polyurethanes		
Polyurethane		
Casting resin	293	0.21
Elastomer	293	0.31
Vinyl Polymers		
Polyacrylonitrile	293	0.26
Poly(acrylonitrile-butadiene)copolymer (NBR)		
35% acrylonitrile	333	0.251
	413	0.184

TABLE 2.87 Thermal Conductivity of Polymers (*Continued*)

Polymer	Temperature (K)	*k* (W/m K)
Poly(acrylonitrile-butadiene-styrene) copolymer (ABS)		
Injection molding grade		0.33
Poly(acrylonitrile-styrene) copolymer	293	0.18
Poly(*i*-butyl methacrylate)		
At 0.82 atm		0.13
Poly(*n*-butyl methacrylate)		
At 0.82 atm		0.45
Poly(butyl methacrylate-triethylene glycol dimethacrylate) copolymer		0.15
Poly(chloroethylene-vinyl acetate) copolymer	293	0.134
	325	0.146
	375	0.218
Poly(dially phthalate)		0.21
Poly(ethyl acrylate)	310.9	0.213
	422.1	0.230
	533.2	0.213
Poly(ethyl methacrylate)		
At 0.82 atm	273	0.175
Poly(ethylene vinyl acetate)		0.34
Poly(methyl methacrylate)	293	0.21
Poly(methyl methacrylate-acrylonitrile) copolymer		0.18
Poly(methyl methacrylate-styrene) copolymer		0.21–0.21
Poly(vinyl acetate)		0.159
Poly(vinyl acetate-vinyl chloride) copolymer		0.167
Poly(vinyl alcohol)		0.2
Poly(*N*-vinyl carbozole)	293	0.126
	443	0.168
Poly(vinyl fluoride)	243	0.14
	333	0.17
Poly(vinyl formal) Molding grade	293	0.27

TABLE 2.88 Thermal Conductivity of Foamed Polymers

Name	k (W/m K)
Poly(acrylonitrile-butadiene) copolymer	
$d = 160$–400 kg/m^3	0.036–0.043
Cellulose acetate	
$d = 96$–128 kg/m^3	0.045–0.46
Polychloroprene (Neoprene)	
$d = 112$ kg/m^3	0.040
$d = 192$ kg/m^3	0.065
Poly(dimethylsiloxane)	
Sheet, $d = 160$ kg/m^3	0.086
Epoxy	
$d = 32$–48 kg/m^3	0.016–0.022
$d = 80$–128 kg/m^3	0.035–0.040
Polythylene	
Extruded plank	
$d = 35$ kg/m^3	0.053
$d = 64$ kg/m^3	0.058
$d = 96$ kg/m^3	0.058
$d = 144$ kg/m^3	0.058
Sheet, extruded, $d = 43$ kg/m^3	0.040–0.049
Sheet, crosslinked, $d = 26$–38 kg/m^3	0.036–0.040
Polyisocyanurate	
$d = 24$–56 kg/m^3	0.012–0.02
Polyisoprene (natural rubber)	
$d = 56$ kg/m^3	0.036
$d = 320$ kg/m^3	0.043
Phenolic resin	
$d = 32$–64 kg/m^3	0.029–0.032
$d = 112$–160 kg/m^3	0.035–0.040
Polypropylene	
$d = 64$–96 kg/m^3	0.039
Polystyrene	
$d = 16$ kg/m^3	0.040
$d = 32$ kg/m^3	0.036
$d = 64$ kg/m^3	0.033
$d = 96$ kg/m^3	0.036
$d = 160$ kg/m^3	0.039
Poly(styrene-butadiene) copolymer (SBR)	
$d = 72$ kg/m^3	0.030
Poly(urea-formaldehyde) resin	
$d = 13$–19 kg/m^3	0.026–0.030
Polyurethane	
Air blown, $d = 20$–70 kg/m^3	
At 0°C	0.033
At 20°C	0.036
At 70°C	0.040
CO_2 blown, $d = 64$ kg/m^3, at 20°C	0.016
20% closed cells, at 20°C	0.033
90% closed cells, at 20°C	0.016
500 μm cell size, at 20°C	0.024
100 μm cell size, at 20°C	0.016
Poly(vinyl chloride)	
$d = 56$ kg/m^3	0.035
$d = 112$ kg/m^3	0.040

TABLE 2.89 Thermal Conductivity of Polymers with Fillers

Name	k (W/m K)
Polyacetal	
5–20% polytetrafluoroethylene (PTFE)	0.20
Poly(acrylonitrile-butadiene-styrene) copolymer (ABS)	
20% glass fiber	0.20
Polyaryletherketone	
40% glass fiber	0.44
Poly(butylene terephthalate) (PBT)	
30% glass fiber	0.29
	0.21
40–45% glass fiber	0.42
Polycarbonate	
10% glass fiber	0.22
30% glass fiber	0.32
Polychloroprene (Neoprene)	
33% carbon black	0.210
Poly(dially phthalate)	
Glass fiber	0.21–0.62
Epoxy resin	
50% aluminum	1.7–3.4
25% Al_2O_3	0.35–0.52
50% Al_2O_3	0.52–0.69
75% Al_2O_3	1.4–1.7
30% mica	0.24
50% mica	0.39
Silica	0.42–0.84
Polyetheretherketone (PEEK)	
30% glass fiber	0.21
30% carbon fiber	0.21
Polyethylene	
30% glass fiber	0.36–0.46
Poly(ethylene terephthalate) (PET)	
30% glass fiber	0.29
45% glass fiber	0.31
30% graphite fiber	0.71
40% polyacrylonitrile (PAN) carbon fiber	0.72
Polyimide	
Thermoplastic, 15% graphite	0.87
Thermoplastic, 40% graphite	1.73
Thermoset, 50% glass fiber	0.41
Polyisoprene (natural rubber)	
33% carbon black	0.28
Poly(melamine-formaldehyde) resin	
Asbestor	0.544–0.73
Cellulose fiber	0.27–0.42
Glass fiber	0.42–0.48
Macerated fabric	0.443
Wood flour/cellulose	0.17–0.48
Poly(melamine-phenolic) resin	
Cellulose fiber	0.17–0.29
Wood flour	0.17–0.29
Nylon-6 (polycaprolactam)	
30–35% glass fiber	0.24–0.28
Nylon-6,6 [poly(hexamethylene adipamide)]	
30–33% glass fiber	0.21–0.49
40% glass fiber and mineral	0.46
30% graphite or polyacrylonitrile (PAN) carbon fiber	1.0
Nylon-6,12 [poly(hexamethylenedodecanediamide)]	
30–35% glass fiber	0.427
Poly(phenylene oxide)	
30% glass fiber	0.16
Poly(phenylene sulfide)	
40% glass fiber	0.288
30% carbon fiber	0.28–0.75
Polypropylene	
40% talc	0.32
40% $CaCO_3$	0.29
40% glass fiber	0.37
Polystyrene	
20% glass fiber	0.25
Poly(styrene-acrylonitrile) copolymer	
20% glass fiber	0.28
Poly(styrene-butadiene) copolymer (SBR)	
33% carbon black	0.300
Polytetrafluoroethylene	
25% glass fiber	0.33–0.41
Poly(urea-formaldehyde) resin	
33% α-cellulose	0.423

TABLE 2.90 Resistance of Selected Polymers and Rubber to Various Chemicals at 20°C

The information in this table is intended to be used only as a general guide. The chemical resistance classifications are E = excellent (30 days of exposure causes no damage), G = good (some damage after 30 days), F = fair (exposure may cause crazing, softening, swelling, or loss of strength), N = not recommended (immediate damage may occur).

	Chemical												
	Acids, dilute or weak	Acids, strong and concentrated	Alcohols, aliphatic	Aldehydes	Alkalies, concentrated	Esters	Ethers	Glycols	Hydrocarbons, aliphatic	Hydrocarbons, aromatic	Hydrocarbons, halogenated	Ketones	Oxidizing agents, strong
Polymers													
Acetals	F	N	F	N	N	N	N	G	N	N	N	N	N
Acrylics: poly(methyl methacrylate)	G	N	E	—	N	N	E	E	G	N	N	N	N
Allyls: diallyl phthalate	G	—	—	—	N	—	—	—	E	G	G	N	—
Cellulosics: cellulose-acetate-butyrate and cellulose-acetate-propionate polymers	F	N	N	N	N	N	N	G	F	N	N	N	—
Fluorocarbons	E	E	E	E	E	E	E	E	E	E	E	E	E
Polyamides	N	N	G	E	E	G	—	G	G	F	F	G	N
Polycarbonates	G	N	G	F	N	N	N	G	N	N	N	N	N
Polyesters	G	G	N	—	N	N	F	G	G	F	F	N	F
Poly(methyl pentene)	E	E	G	G	E	G	N	E	F	G	N	F	F
Low-density polyethylene	E	E	E	G	E	G	N	E	F	F	N	G	F
High-density polyethylene	E	E	E	E	E	G	N	E	G	G	N	G	F
Polybutadiene	G	F	E	—	—	—	—	—	—	E	E	E	—
Polypropylene and polyallomer	E	E	E	E	E	G	N	E	G	F	N	G	F
Polystyrene	N	N	E	—	N	N	—	E	N	N	N	N	N
Styrene-acrylonitrile copolymers	—	—	N	—	N	—	—	F	N	—	—	—	—
Styrene-acrylonitrile-butadiene copolymers	—	N	G	—	G	N	—	—	F	N	N	N	—
Sulfones: polysulfone	G	N	F	F	E	N	F	G	F	N	N	N	G
Vinyls: poly(vinyl chloride)	E	G	E	G	G	N	F	F	G	N	N	N	G
Rubbers													
Natural rubber	—	—	E	—	—	N	N	E	N	N	N	N	—
Nitrile rubber	—	—	E	—	—	N	G	E	E	N	N	N	—
Polychloroprene	—	—	E	—	—	N	F	E	F	N	N	N	—
Polyisobutylene	—	—	E	—	—	F	F	E	N	N	N	N	—
Polysulfide rubbers: Thiokol	—	—	E	—	—	E	E	E	E	F	N	N	—
Styrene-butadiene rubber	—	—	E	—	—	N	N	E	N	N	N	N	—

TABLE 2.91 Gas Permeability Constants (10^{10} P) at 25°C for Polymers and Rubber

The gas permeability constant P is

$$P = \frac{\text{amount of permeant}}{(\text{area}) \times (\text{time}) \times (\text{driving forced across the film})}$$

The gas permeability constant is the amount of gas expressed in cubic centimeters passed in 1 s through a 1–cm^2 area of film when the pressure across a film thickness of 1 cm is 1 cmHg and the temperature is 25°C. All tabulated values are multiplied by 10^{10} and are in units of seconds^{-1} $(\text{centimeters of Hg})^{-1}$. Other temperatures are indicated by exponents and are expressed in degrees Celsius.

	Gas						
Polymer or rubber	He	N_2	H_2	O_2	CO_2	H_2O	Other
Cellulose (cellophane)	0.005^{20}	0.003 2	0.006 5	0.002 1	0.004 7	1900	0.006^{45} (H_2S); 0.001 7 (SO_2)
Cellulose acetate	13.6^{20}	0.28^{30}	3.5^{20}	0.78^{30}	22.7^{30}	5500	3.5^{30} (H_2S); 17^{0} (ethylene oxide); 6.8^{60} (bromomethane)
Cellulose nitrate	6.9	0.12	2.0^{20}	1.95	2.12	6290	57.1 (NH_3); 1.76 (SO_2)
Ethyl cellulose	400^{30}	8.4^{30}	87^{20}	26.5^{30}	41.0^{30}	12000^{20}	705 (NH_3); 204 (SO_2); 420^{0} (ethylene oxide)
Gutta percha		2.17	14.4	6.16	35.4	510	
Natural rubber		9.43	52.0	23.3	15.3	2290	15.7 (CO); 30.1 (CH_4); 1.68 (C_3H_8); 98.9 (C_2H_2); 550 ($CH_3C{\equiv}CH$); 3.59 (SF_6)
Nylon 6	0.53^{20}	$0.009\ 5^{30}$		0.038^{30}	0.10^{30}	177	0.33^{30} (H_2S); 1.2^{20} (NH_3); 0.84^{60} (CH_3Br)
Nylon 11	1.95^{30}		1.78^{30}		1.00^{40}		0.344^{30} (Ne); 0.189^{40} (Ar); 13.6^{50} (propyne)
Poly(acrylonitrile)				0.000 2	0.000 8	300	
Acrylonitrile-styrene copolymer (66:34)				0.048	0.21	2000	
Poly(1,3-butadiene)		6.42	41.9	19.0	138.0	5070	
Poly (*cis*-1,4-butadiene)	32.6	19.2					19.2 (Ne); 41.0 (Ar)
Butadiene-acrylonitrile copolymer (80:20)	12.2	1.06	15.9	3.85	30.8		24.8 (C_2H_2); 7.7 (propyne)

(*Continued*)

TABLE 2.91 Gas Permeability Constants (10^{10} P) at 25°C for Polymers and Rubber (*Continued*)

Polymer or rubber	Gas						
	He	N_2	H_2	O_2	CO_2	H_2O	Other
Butadiene-styrene copolymer (80 : 20)	13.4	1.71					5.01 (Ne); 4.49 (Ar)
Butadiene-styrene copolymer (92 : 8)	22.9	5.11					9.70 (Ne); 12.7 (Ar)
Polychloroprene		1.2	13.6	4.0	25.8		3.79 (Ar); 3.27 (CH_4)
Polyethylene, low-density	4.9	0.969	12.0^{30}	2.88	12.6	90	2.88 (CH_4); 6.81 (C_2H_6); 9.43 (C_3H_8); 1.48 (CO); 49^{0} (ethylene oxide); 14.4 (propene); 42.2 (propyne); 0.170 (SF_6); 472^{60} (CH_3Br)
Polyethylene, high-density	1.14	0.143	3.0^{20}	0.403	0.36	12.0	0.388 (CH_4); 0.590 (C_2H_6); 0.537 (C_3H_8); 0.008 3 (SF_6); 1.69 (Ar); 4.01 (propene)
Poly(ethylene terephthalate)							
Crystalline	1.32	0.006 5	3.70^{20}	0.035	0.17	130	0.003 2 (CH_4); 0.08^{60} (CH_3Br)
Amorphous	3.28	0.013		0.059	0.30		0.009 (CH_4)
Poly(ethyl methacrylate)	6.82	0.220		1.15	5.00	3200	2.98 (Ne); 0.565 (Ar); 0.370 (Kr); 3.83 (H_2S); 0.000 001 65 (SF_6)
Isobutene-isoprene copolymer (98 : 2)	8.38	0.324	7.20	1.30	5.16	110^{38}	13.6^{50} (C_3H_8)
Isoprene-acrylonitrile copolymer (76 : 24)	7.77	0.181	7.41	0.852	4.32		
Isoprene-methacrylonitrile copolymer (76 : 24)		0.596	13.6	2.34	14.1		
Methacrylonitrile-styrene-butadiene copolymer (88 : 7 : 5)				0.004 8	0.014	600	
Poly(methylpentene)	101	7.83	136	32.0	92.6		
Polypropylene	38^{20}	0.44^{30}	41^{20}	2.3^{30}	9.2^{30}	51	0.33^{20} (H_2S); 9.2^{20} (NH_3)
Silicone rubber, 10% filler	233^{0}	227^{0}	464^{0}	489^{0}	3240	$43{,}000^{35}$	191^{0} (Ne); 550^{0} (Ar); 1020^{0} (Kr); 2550^{0} (Xe); 19000^{0} (butane)
Polystyrene	18.7	0.788	23.3	2.63	10.5	1200	
Poly(tetrafluoroethylene)		1.4	9.8	4.2	11.7		15.7 (NO_2); 37.5 (N_2O_4)
Poly(trifluoroethylene)	6.8^{20}	0.003	0.94^{20}	0.025^{40}	0.048^{40}	0.29	1.2^{0} (ethylene oxide); 4.6^{60} (CH_3Br)
Poly(vinyl acetate)	12.6^{30}		89^{30}	0.50^{30}			2.64^{30} (Ne); 0.19^{30} (Ar); 0.078^{30} (Kr); 0.050^{30} (CH_4)
Poly(vinyl alcohol)	0.001^{30}	$<0.001^{14}$	0.009	0.008 9	0.001^{23}		0.007 (H_2S); 0.002^{0} (ethylene oxide)
Poly(vinyl chloride)	2.05	0.011 8	1.70	0.045 3	0.157	275	3.92 (Ne); 0.011 5 (Ar); 0.028 6 (CH_4)
Poly(vinylidene chloride)	0.31^{34}	$0.000\ 94^{30}$		$0.005\ 3^{30}$	0.03^{30}	0.5	0.03^{30} (H_2S); 0.008^{60} (CH_3Br)

TABLE 2.92 Vapor Permeability Constants (10^{10} P) at 35°C for Polymers

Polymer	Vapor				
	Benzene	Hexane	Carbon tetrachloride	Ethanol	Ethyl acetate
Cellulose	1.4	0.912	0.836	85.8	13.4
Cellulose acetate	512	2.80	3.74	2980	3595
Poly(acrylonitrile)	2.61	1.59	1.47	0	1.34
Polyethylene, low-density	5300	2910	3810	55.9	513
Polystyrene	10,600		6820	0	soluble
Poly(vinyl alcohol)	3.58	2.34	1.61	32.7	2.53

TABLE 2.93 Hildebrand Solubility Parameters of Polymers

Polymer	δ (MPa$^{1/2}$)	T (°C)	Method
Cellulose	32.02		
Cellulose diacetate	23.22		Calc.
Cellulose nitrate (11.83% N)	21.44		Calc.
Epoxy resin	22.3		
Natural rubber	16.2		
	17.09		
Poly(4-acetoxystyrene)	22.7	25	Visc.
Poly(acrylic acid)			
—, butyl ester	18.0	35	
	18.52		Swelling
—, methyl ester	20.77		Swelling
	20.7		Swelling
Poly(acrylonitrile)	26.09	25	Calc.
Poly(butadiene)	16.2	75	IPGC
	17.15		Calc.
Poly(butadiene-co-acrylonitrile)			
BUNA N (72/55)	18.93	25	Calc.
(61/39)	20.5	75	IPGC
Poly(butadiene-co-styrene)			
BUNA S (85/15)	17.41		Calc.
	17.39		Obs.
Poly(butadiene-co-vinylpyridine)			
(75/25)	19.13		
Poly(chloroprene)	18.42	25	
	19.19		Calc.
	17.6		Swelling
Poly(dimethyl siloxane)	14.9	30	Calc.
Poly(ethylene)	16.6		Calc.
Poly(ethylene)	16.4		Calc.
	16.2		Obs.
Poly(ethylene-co-vinyl-acetate)	18.6	25	IPGC
	17.0	75	IPGC
Poly(*tetra*-fluoroethylene)	12.7		Calc.
Poly(heptamethylene *p,p′*-bibenzoate)	19.50	25	Visc.
Poly(4-hydroxystyrene)	23.9	25	Visc.
Poly(isobutene)	16.06	35	Av.
	16.47		Swelling
	16.06	25	
Poly(isobutene-co-isoprene) butyl rubber	16.47		
Poly(isoprene)			
1,4-*cis*	15.18	25	Calc.
	16.68	25	
	16.57	35	
	20.46	35	Swelling
	16.6		Swelling
	16.68	25	Calc.
Poly(methacrylic acid)			
—, isobutyl ester	14.7	140	IPGC
—, ethyl ester	18.31		Swelling
—, methyl ester	18.58	25	
Poly(methacrylonitrile)	21.9		Calc.
Poly(methylene)	14.3	20	Extrap.
poly(α-methyl styrene)	18.75	30	Visc.

TABLE 2.93 Hildebrand Solubility Parameters of Polymers (*Continued*)

Polymer	δ (MPa$^{1/2}$)	T (°C)	Method
Poly(σ-methylstyrene-co-acrylonitrile)	16.4	180	IPGC
Poly(oxyethylene)	20.2	25	IPGC
Poly(propylene)	18.8	25	
Poly(styrene)	18.72	35	
Poly(styrene-co-*n*-butyl-methacrylate)	15.1	140	IPGC
Poly(thioethylene)	19.19		Swelling
Poly(vinyl acetate)	19.62	25	Calc.
Poly(vinyl alcohol)	25.78		
Poly(vinyl chloride)	19.28		Calc.
	19.8		Obs.
Poly(vinyl chloride), chlorinated	19.0	25	Visc.
Poly(vinyl propionate)	18.01	35	

TABLE 2.94 Hansen Solubility Parameters of Polymers

Polymer (trade name, supplier)	Solubility parameter (MPa$^{1/2}$)			
	δ_d	δ_p	δ_h	δ_t
Acrylonitrile-butadiene elastomer (Hycar 1052, BF Goodrich)	18.6	8.8	4.2	21.0
Alcohol soluble resin (Pentalyn 255, Hercules)	17.5	9.3	14.3	24.4
Alcohol soluble resin (Pentalyn 830, Hercules)	20.5	5.8	10.9	23.5
Alkyd, long oil (66% oil length, Plexal P65, Polyplex)	20.42	3.44	4.56	21.20
Alkyd, short oil (Coconut oil 34% phthalic anhydride; Plexal C34)	18.50	9.21	4.91	21.24
Blocked isocyanate (Phenol, Suprasec F5100, ICI)	20.19	13.16	13.07	27.42
Cellulose acetate (Cellidore A, Bayer)	18.60	12.73	11.01	25.08
Cellulose nitrate (1/2 s; H-23, Hagedon)	15.41	14.73	8.84	23.08
Epoxy (Epikote 1001, Shell)	20.36	12.03	11.48	26.29
Ester gum (Ester gum BL, Hercules)	19.64	4.73	7.77	21.65
Furfuryl alcohol resin (Durez 14383, Hooker Chemical)	21.16	13.56	12.81	28.21
Hexamethoxymethyl melamine (Cymel 300 American Cyanimid)	20.36	8.53	10.64	24.51
Isoprene elastomer (Cariflex IR 305, Shell)	16.57	1.41	–0.82	16.65
Methacrylonitrile/methacrylic acid copolymer	17.39	14.32	12.28	25.78
Nylon 66	18.62	5.11	12.28	22.87
Nylon 66 (Zytel, DuPont)	18.62	0.00	14.12	23.37
Petroleum hydrocarbon resin (Piceopale 110, Penn. Ind. Chem.)	17.55	11.19	3.60	17.96

(*Continued*)

TABLE 2.94 Hansen Solubility Parameters of Polymers (*Continued*)

Polymer (trade name, supplier)	Solubility parameter ($MPa^{1/2}$)			
	δ_d	δ_p	δ_h	δ_t
Phenolic resin				
(Resole, Phenodur 373 U Chemische Werke Albert)	19.74	11.62	14.59	27.15
Phenolic resin, pure				
(Super Beckacite 1001, Reichhold)	23.26	6.55	8.35	25.57
Poly(4-acetoxy,α-acetoxy styrene)	17.80	10.23	7.37	21.89
Poly(4-acetoxystyrene)	17.80	9.00	8.39	21.69
Poly (acrylonitrile)	18.21	16.16	6.75	25.27
Polyamid, thermoplastic				
(Versamid 930, General Mills)	17.43	−1.92	14.89	23.02
Poly(*p*-benzamide)	18.0	11.9	7.9	23.0
cis-Poly(butadiene)elastomer				
(Bunahuls CB10, Chemische Werke Huels)	17.53	2.25	3.42	18.00
Poly(isobutylene)				
(Lutonal IC/123, BASF)	14.53	2.52	4.66	15.47
Poly(ethyl methacrylate)				
(Lucite 2042, DuPont)	17.60	9.66	3.97	20.46
Poly(ethylene terephthalate)	19.44	3.48	8.59	21.54
Poly(4-hydroxystyrene)	17.60	10.03	13.71	24.55
Poly(methacrylic acid)	17.39	12.48	15.96	26.80
Poly(methacrylonitrile)	18.00	15.96	7.98	25.37
Poly(methyl methacrylate)				
Poly(sulfone), Bisphenol A				
(Polystyrene LG, BASF)	21.28	5.75	4.30	22.47
Poly(sulfone), Bisphenol A				
(Udel)	19.03	0.00	6.96	20.26
Poly(vinyl acetate)				
(Mowilith 50, Hoechst)	20.93	11.27	9.66	25.66
Poly(vinyl butyral)				
(Butvar B76, Shawinigan)	18.60	4.36	13.03	23.12
Poly(vinyl chloride)				
(Vipla KR $K = 50$, Montecatini)	18.23	7.53	8.35	21.42
Poly(vinyl chloride)	18.72	10.03	3.07	21.46
Poly(vinyl chloride)	18.82	10.03	3.07	21.54
Saturated polyester				
(Desmophen 850, Bayer)	21.54	14.94	12.28	28.95
Styrene-butadiene (SBR) raw elastomer				
(Polysar 5630, Polymer Corp.)	17.55	3.36	2.70	18.07
Terpene resin				
(Piccolyte S-1000, Penn. Ind. Chem.)	16.47	0.37	2.84	16.72
Urea-formaldehyde resin				
(Plastopal H, BASF)	20.81	8.29	12.71	25.74
Vinylidene cyanide/4-acetoxy,α-acetoxy styrene copolymer	21.48	11.25	7.16	21.89
Vinylidene cyanide/4-chloro-styrene copolymer	16.98	12.07	8.18	22.38
(Rohm and Haas)	18.64	10.52	7.51	22.69
Poly(styrene)				

TABLE 2.95 Refractive Indices of Polymers

Polymer name	Refractive index (20°C, 68°F)	Polymer name	Refractive index (20°C, 68°F)
Acetal homopolymer	1.48	Polyethylene (medium Density)	1.52
Acrylics	1.49–1.52	Polyethylene (high density)	1.54
Ally diglycol carbonate	1.50	Polyethylene dimethacrylate	1.51
Cellulose acetate	1.46–1.50	Poly(ethylene terephthalate)	1.57–1.58
Cellulose acetate butyrate	1.46–1.49	Poly(methyl-α–chloroacrylate)	1.52
Cellulose ester	1.47–1.50	Poly(methyl methacrylate)	1.49
Cellulose nitrate	1.49–1.51	Polypropylene	1.49
Cellulose propionate	1.46–1.49	Poly(propyl methacrylate)	1.48
Chlorotrifluoroethylene (CTFE)	1.42	Polystyrene	1.57–1.60
Diallyl isophthalate	1.57	Polysulfone	1.63
Epoxies	1.55–1.65	Poly(tetrafluoroethylene) (PTFE)	1.35
Ethyl cellulose	1.47	Poly(trifluorochloroethylene)	1.43
Fluorinated ethylene-propylene	1.34	Poly(trifluoroethylene)	1.35–1.37
Methylpentene polymer	1.485	Poly(vinyl alcohol)	1.49–1.53
Nylon	1.52–1.53	Poly(vinyl acetal)	1.48
Phenol formaldehyde	1.50–1.70	Poly(vinyl acetate)	1.46–1.47
Phenoxy polymer	1.60	Poly(vinyl butyral)	1.49
Polyacetal	1.48	Poly(vinyl chloride)	1.52–1.55
Polyallomer	1.49	Poly(vinyl cyclohexene dioxide)	1.53
Polyallyl methacrylate	1.52	Poly(vinyl formal)	1.60
Polyamide nylon 6/6	1.53	Poly(vinyl naphthalene)	1.68
Polyamide nylon 11	1.52	Poly(vinylidene chloride)	1.60–1.63
Polybutylene	1.50	Poly(vinylidene fluoride)	1.42
Polycarbornate	1.57–1.59	Silicone polymer	1.43
Poly(cyclohexyl methacrylate)	1.51	Styrene acrylonitrile copolymer	1.56–1.57
Poly(diallyl phthalate)	1.57	Styrene butadiene thermoplastic	1.52–1.55
Polyester	1.53–1.58	Styrene methacrylate copolymer	1.53
Poly(ester-styrene)	1.54–1.57	Urea formaldehyde	1.54–1.58
Polyethylene (low density)	1.51	Urethane	1.50–1.60

2.21 FATS, OILS, AND WAXES

Fats, oils, and waxes belong to the group of naturally occurring organic materials called lipids. Lipids are those constituents of plants or animals that are insoluble in water but soluble in other organic solvents.

The fats and oils of vegetable and animal origin belong to the class of triglycerides, i.e., fatty acid tri-esters of glycerol. The component fatty acid (acyl) radicals can be saturated or unsaturated. Their chain lengths, degrees of unsaturation, and relative positions in the molecule determine the character of the fat or fatty oil. Thus a triglyceride of the (saturated) plamitic or stearic acids (i.e., solid fatty acids with sixteen and eighteen carbon atoms respectively) will be a solid. Oleic acid is liquid at room temperature; it is an unsaturated fatty acid with eighteen carbon atoms and one double bond. It occurs in olive oil, also in peanut and sesame oils. Linseed oil contains linoleic and linolenic acids (in addition to oleic, plamitic, and stearic acids). These acids are still more unsaturated in character; there are two double bonds in the molecule of linoleic acid, and three in that of linolenic acid.

Waxes are usually the plastic substances deposited by insects or obtained from plants. Waxes are esters of various fatty acids with higher, usually monohydric alcohols. The wax of pharmacy is principally yellow wax (beeswax), the material of which honeycomb is made. It consists chiefly of cerotic acid and myricin and is used in making ointments, cerates, etc. Other waxes include petroleum wax that is a mixture of paraffin hydrocarbons that melts above room temperature.

TABLE 2.96 Physical Properties of Fats and Oils

Fat or oil	Solidification point, °C	Specific gravity (15°C/15°C)	Refractive index	Acid value	Saponification value	Iodine value
			Animal origin			
Butterfat	20–23	$0.91^{40°C}_{15°C}$	$1.45^{40°C}$	0.5–35	210–230	26–38
Chicken fat	21–27	0.924		1.2	193–205	66–72
Cod-liver oil	−3	0.92–0.93	$1.481^{25°C}$	5.6	171–189	137–166
Deer fat		0.96–0.97		0.8–5.3	195–200	26–36
Dolphin	−3 to +5	0.91–0.93		2–12	203 (body); 290 (jaw)	127 (body); 33 (jaw)
Goat butter		$0.91–0.94^{38°C}_{38°C}$			233–236	25–37
Goose fat	22–24	0.92–0.93		0.6	191–193	58–67
Herring oil		0.92–0.94	$1.4610^{60°C}$	1.8–44	170–194	102–149
Horse fat	20–45	0.92–0.93		0–2.4	195–200	75–86
Human fat	15	0.903	1.460		193–200	57–73
Lard oil	−2 to +4	0.913–0.915	1.462	0.1–2.5	193–198	63–79
Lard oil, fatty tissue	27–30	0.93–0.94	1.462	0.5–0.8	195–203	47–67
Menhaden oil	−5	0.92–0.93	$1.465^{60°C}$	3–12	189–193	148–185
Neat's-foot oil	−2 to +10	0.91–0.92	$1.464^{25°C}$	0.1–0.6	193–199	58–75
Porpoise, body oil	−16	0.926		1.2	203	127
Rabbit fat	17–23	0.93–0.94		1.4–7.2	199–203	70–100
Sardine oil	20–22	0.92–0.93	$1.466^{60°C}$	4–25	188–196	130–152
Seal	3	0.915–0.926		1.9–40	188–196	130–152
Shark		0.916–0.919			157–164	115–139
Sperm oil	15.5	0.878–0.884		13	120–137	80–84
Tallow, beef	31–38	0.895		0.25	196–200	35–42
Tallow, mutton	32–41	0.937–0.953	$1.457^{40°C}$	2–14	195–196	48–61
Whale oil	−2 to 0	0.917–0.924	$1.460^{60°C}$	1.9	160–202	90–146
			Plant origin			
Acorn	−10	0.916			199	100
Almond	−20 to −15	0.914–0.921		0.5–3.5	183–208	93–103
Babassu oil	22–26	$0.893^{60°C}$	$1.443^{60°C}$		247	16
Beechnut oil	−17	0.922			191–196	97–111

Castor oil	−18 to −17	0.960–0.967	1.477	0.1–0.8	175–183	84
Chaulmoogra oil, USP	<−25	$0.950^{25°C}$			196–213	98–110
Chinese vegetable tallow	24–34	0.918–0.922		2.4	179–206	23–41
Cocoa butter	21.5–23	0.964–0.974	$1.457^{40°C}$	1.1–1.9	193–195	33–42
Coconut oil	14–22	0.926	$1.449^{40°C}$	2.5–10	153–262	6–10
Corn (maize) oil	−20 to −10	0.921–0.928	$1.473^{40°C}$	1.4–2.0	187–193	111–128
Cottonseed oil	−13 to +12	$0.918^{25°C}_{25°C}$	$1.474^{40°C}$	0.6–0.9	194–196	103–111
Hazelnut oil	−18 to −17	0.917			191–197	87
Hemp-seed oil	−28 to −15	0.928–0.934		0.45	190–195	145–162
Linseed oil	−27 to −19	0.930–0.938	$1.478^{25°C}$	1–3.5	188–195	175–202
Mustard, black, oil	16	0.918–0.921	$1.475^{40°C}$	5.7–7.3	173–175	99–110
Neem oil	−3	0.917	$1.462^{40°C}$		195	71
Niger-seed oil		0.925	$1.471^{40°C}$		190	129
Oiticica oil		$0.974^{25°C}$				140–180
Olive oil	−6	0.914–0.918	$1.468^{40°C}$	0.3–1.0	185–196	79–88
Palm oil	35–42	0.915	$1.458^{40°C}$	10	200–205	49–59
Palm kernel oil	24	0.918–0.925	$1.457^{40°C}$	0.3–0.6	220–231	26–32
Peanut oil	3	0.917–0.926	$1.469^{40°C}$	0.8	186–194	88–98
Perilla oil		0.930–0.937	$1.481^{25°C}$		188–194	185–206
Pistachio-nut oil	−10 to −5	0.913–0.919			191	83–87
Poppy-seed oil	−18 to −16	0.924–0.926	$1.469^{40°C}$	2.5	193–195	128–141
Pumpkin-seed oil	−15	0.923–0.925			188–193	121–130
Rapeseed oil	−10	0.913–0.917	$1.471^{40°C}$	0.36–1.0	168–179	94–105
Safflower oil	−18 to −13	0.925–0.928	$1.462^{60°C}$	0.6	188–203	122–141
Sesame oil	−6 to −4	$0.919^{25°C}_{25°C}$	$1.465^{40°C}$	9.8	188–193	103–117
Soybean oil	−16 to −10	0.924–0.927	$1.473^{40°C}$	0.3–1.8	189–194	122–134
Sunflower-seed oil	−17	0.924–0.926	$1.469^{40°C}$	11.2	188–193	129–136
Tung oil	−2.5	0.94–0.95	$1.517^{25°C}$	2	190–197	163–171
White-mustard-seed oil	−16 to −8	0.912–0.916		5.4	171–174	94–98
Wheat-germ oil						125

TABLE 2.97 Physical Properties of Waxes

Wax	Melting point, °C	Specific gravity (15°C/15°C)	Refractive index	Acid value	Saponification value	Iodine value
Bamboo leaf	79–80	$0.961^{25°C}$		14–15	43–44	7.8
Bayberry (myrtle)	47–49	0.99	$1.436^{80°C}$	3–4	205–212	4–9.5
Beeswax, ordinary	62–66	0.95–0.97	$1.44–1.48^{40°C}$	17–21	88–100	8–11
Beeswax, East Indian	61–67	0.95–0.97	$1.44^{40°C}$	5–10.5	87–117	4–10.5
Beeswax, white, USP	61–69	0.95–0.98	$1.45–1.47^{65°C}$	17–24	90–96	7–11
Candelilla	73–77	0.98–0.99	$1.45–1.46^{85°C}$	19–24	55–64	14–20
Cape berry	40–45	1.01	$1.45^{45°C}$	2.5–4.0	211–215	0.5–2.5
Caranda	80–85	0.99–1.00		5.0–9.5	64–79	8–9
Carnauba, No. 1 yellow	86–88	0.99–1.00		1.5–2.5	75–86	
Carnauba, No. 3, crude	86–90	0.99–1.01		3.0–8.5	75–89	
Carnauba, No. 3, refined	86–89	0.96–0.97	$1.47^{40°C}$	3.0–5.0	76–85	7–13.5
Castor oil, hydrogenated	83–88	$0.98–0.99^{20°C}$		1.0–5.0	177–181	2.5–8.5
Chinese insect	80–85	0.95–0.97	$1.46^{40°C}$	2–9	78–93	1.0–2.5
Cotton	68–71	0.96		32	71	25
Cranberry	207–218	0.97–0.98		42–59	131–134	44–53
Esparto	75–79	0.985–0.995		22–27	58–73	7–15
Flax	61–70	0.91–0–0.99		17–48	37–102	22–29
Japan	49–56	0.97–1.00		4–15	210–235	4–15
Jojoba	11–12	$0.86–0.90^{25°C}$	$1.465^{25°C}$	0.2–0.6	92–95	82–88
Microcrystalline, amber	64–91	0.91–0.94	$1.42–1.45^{80°C}$	0	0	0
Microcrystalline, white	71–89	0.93–0.94	$1.441^{80°C}$	0	0	0
Montan, crude	76–86	$1.01–1.02^{25°C}$		22–31	59–92	14–18
Montan, refined	77–84	1.02–1.04		23–45	72–115	10–14
Ouricury	86–89	0.99–1.01		12–19	88–96	6.9–7.8
Ozokerite	56–82	0.90–1.00		0	0	4–8
Palm	74–86	0.99–1.05		5–11	64–104	9–17
Paraffin, American	49–63	0.896–0.925	$1.44–1.48^{80°C}$	0	0	0
Shellac	79–82	0.97–0.98		12–24	64–83	6–9
Sisal hemp	74–81	1.007–1.010		16–19	56–58	28–29
Spermaceti	41–49	0.905–0.960		0.5–3.0	121–135	2.5–8.5
Sugarcane, refined	76–82	0.96–0.98	$1.51^{25°C}$	8–23	55–70	13–29
Wool	38–40	0.97	$1.48^{40°C}$	6–22	82–130	15–47

2.22 PETROLEUM PRODUCTS

Petroleum is an extremely complex naturally occurring mixture of hydrocarbon compounds, usually with minor amounts of nitrogen-, oxygen-, and sulfur-containing compounds as well as trace amounts of metal-containing compounds. Petroleum products are, for example, fuels and lubricants that are manufactured from petroleum as well as other products of industrial interest. Petrochemicals are also manufactured from petroleum.

TABLE 2.98 Physical Properties of Petroleum Products

	Molecular weight	Specific gravity	Boiling point, °F	Ignition temperature, °F	Flash point, °F	Flammability limits in air, % v/v
Benzene	78.1	0.879	176.2	1040	12	1.35–6.65
n-Butane	58.1	0.601	31.1	761	−76	1.86–8.41
iso-Butane	58.1		10.9	864	−117	1.80–8.44
n-Butene	56.1	0.595	21.2	829	Gas	1.98–9.65
iso-Butene	56.1		19.6	869	Gas	1.8–9.0
Diesel fuel	170–198	0.875			100–130	
Ethane	30.1	0.572	−127.5	959	Gas	3.0–12.5
Ethylene	28.0		−154.7	914	Gas	2.8–28.6
Fuel oil No. 1		0.875	304–574	410	100–162	0.7–5.0
Fuel oil No. 2		0.920		494	126–204	
Fuel oil No.4	198.0	0.959		505	142–240	
Fuel oil No.5		0.960			156–336	
Fuel oil No. 6		0.960			150	
Gasoline	113.0	0.720	100–400	536	−45	1.4–7.6
n-Hexane	86.2	0.659	155.7	437	−7	1.25–7.0
n-Heptane	100.2	0.668	419.0	419	25	1.00–6.00
						0.7–5.00
Kerosene	154.0	0.800	304–574	410	100–162	5.0–15.0
Methane	16.0	0.553	−258.7	900–1170	Gas	0.90–5.90
Naphthalene	128.2		424.4	959	174	
Neohexane	86.2	0.649	121.5	797	−54	1.19–7.58
Neopentane	72.1		49.1	841	Gas	1.38–7.11
n-Octane	114.2	0.707	258.3	428	56	0.95–32
iso-Octane	114.2	0.702	243.9	837	10	0.79–5.94
n-Pentane	72.1	0.626	97.0	500	−40	1.40–7.80
iso-Pentane	72.1	0.621	82.2	788	−60	1.31–9.16
n-pentene	70.1	0.641	86.0	569	—	1.65–7.70
Propane	44.1		−43.8	842	Gas	2.1–10.1
Propylene	42.2		−53.9	856	Gas	2.00–11.1
Toluene	92.1	0.867	321.1	992	40	1.27–6.75
Xylene	106.2	0.861	281.1	867	63	1.00–6.00

SECTION 3
SPECTROSCOPY

SECTION 3
SPECTROSCOPY

3.1 INFRARED ABSORPTION SPECTROSCOPY

Infrared (IR) absorption spectroscopy is a common technique that is used to identify the major functional groups in a compound. The identification of these groups depends upon the amount of infrared radiation absorbed and the particular frequency (measured in cm^{-1}, wave-numbers) at which these groups absorb. Thus, infrared absorption spectroscopy is the measurement of the wavelength and intensity of the absorption of mid-infrared light by a sample. Mid-infrared light (2.5 – 50 μm, 4000 – 200 cm^{-1}) is energetic enough to excite molecular vibrations to higher energy levels. The wavelength of many infrared absorption bands are characteristic of specific types of chemical bonds, and infrared spectroscopy finds its greatest utility for qualitative analysis of organic and organometallic molecules. Infrared spectroscopy is used to confirm the identity of a particular compound and as a tool to help determine the structure of a molecule.

Significant for the identification of the source of an absorption band are *intensity* (weak, medium or strong), *shape* (broad or sharp), and *position* (cm^{-1}) in the spectrum. Characteristic examples are provided in the table below to assist the user in becoming familiar with the intensity and shape absorption bands for representative absorptions.

TABLE 3.1 Absorption Frequencies of Single Bonds to Hydrogen

Abbreviations Used in the Table

m, moderately strong — *var, of variable strength*
m–s, moderate to strong — *w, weak*
s, strong — *w–m, weak to moderately strong*

Group	Band, cm^{-1}	Remarks
	Saturated C—H	
H —C—H H	2975–2950 (s) 2885–2865 (w)	Two or three bands usually; asymmetrical and symmetrical CH stretching, respectively. In presence of double bond adjacent to CH_3 group symmetrical band splits into two.
	1450–1260 (m)	Sensitive to adjacent negative substituents
H —C— acyclic H	ca 2930 (s) 2870–2840 (w)	Frequency increased in strained systems. Symmetrical band splits into two bands when double bond adjacent.
	1480–1440 (m)	Scissoring mode
	ca 720 (w)	Rocking mode
	Alkane residues attached to carbon	
Cyclopropane	ca 3050 (w)	CH stretching
	540–500	
	470–460 (s)	Aliphatic cyclopropanes
Cyclobutanes	580–490 (s)	Alkyl derivatives: 550–530 cm^{-1}
Cyclopentanes	595–490 (s)	Alkyl derivatives: 585–530 cm^{-1}

(Continued)

TABLE 3.1 Absorption Frequencies of Single Bonds to Hydrogen (*Continued*)

Group	Band, cm^{-1}	Remarks
	Alkane residues attached to carbon (*continued*)	
$>C(CH_3)_2$	ca 1380 (m) 1175–1165 (m) 1150–1130 (m)	A roughly symmetrical doublet If no H on central carbon, then one band at ca 1190 cm^{-1}
$—C(CH_3)_3$	1395–1385 (m) 1365 (s)	Split into two bands
Aryl-CH_3 Aryl-C_2H_5 Aryl-C_3H_7 (or C_4H_9)	390–260 (m) 565–540 (m–s) 585–565 (m)	 Two bands
$—(CH_2)_n—$ $n = 1$ $n = 2$ $n = 3$ $n \geq 4$	 785–770 (w–m) 745–735 (w–m) 735–725 (w–m) 725–720 (w–m)	 Rocking vibrations
	Alkane residues attached to miscellaneous atoms	
Epoxide C—H 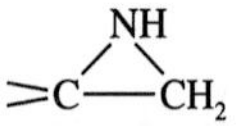	ca 3050 (m–s) ca 3050 (m–s)	
$—CH_2—$halogen	ca 3050 (m–s) 1435–1385 (m) 1300–1240 (s)	Halogens except fluorine
—CHO	2900–2800 (w) 2775–2700 (w) 1420–1370 (m)	
$—CO—CH_3$	3100–2900 (w) 1450–1400 (s) 1360–1355 (s)	
$—O—CH_3$ ethers	2835–2810 (s) 1470–1430 (m–s) ca 1030 (w–m)	 Two bands
$—O—C(CH_3)_3$	1200–1155 (s)	
$—O—CH_2—O—$	2790–2770 (m)	
$—O—CH_2—$ esters	1475–1460 (m–s) 1470–1435 (m–s)	 Acyclic esters. Frequency increased ca 30 cm^{-1} for cyclic and small ring systems.

TABLE 3.1 Absorption Frequencies of Single Bonds to Hydrogen (*Continued*)

Group	Band, cm^{-1}	Remarks
	Alkane residues attached to miscellaneous atoms (*continued*)	
$—O—CO—CH_3$	1450–1400 (s) 1385–1365 (s) 1360–1355 (s)	Acetate esters The high intensity of these bands often dominates this region of the spectrum.
$—CH_2—\overset{\vert}{C}{=}C{<}$	1445–1430 (m)	
$—CH_2—SO_2—$	ca 1250 (m)	
$P—CH_3$ $Se—CH_3$ $B—CH_3$ $Si—CH_3$ $Sn—CH_3$ $Pb—CH_3$ $As—CH_3$ $Ge—CH_3$ $Sb—CH_3$ $Bi—CH_3$ $—CH_2—$(Cd, Hg, Zn, Sn)	1320–1280 (s) ca 1280 (m) 1460–1405 (m) 1320–1280 (m) 1265–1250 (m–s) 1200–1180 (m) 1170–1155 (m) 1265–1240 (m) 1240–1230 (m) 1215–1195 (m) 1165–1145 (m) 1430–1415 (m)	
$N—CH_3$ and $N—CH_2—$ $N—CH_2—CH_2—N$ $N—CH_3$ Amine · HCl Amino acid · HCl Amides $N—CH_2—$ amides	2820–2780 (s) 1440–1390 (m) 1480–1450 (s) 1475–1395 (m) 1490–1480 (m) 1420–1405 (s) ca 1440 (m)	 Ethylenediamine complexes Ethylenediamine complexes
$S—CH_3$	2990–2955 (m–s) 2900–2865 (m–s) 1440–1415 (m) 1325–1290 (m) 1030–960 (m) 710–685 (w–m)	
$S—CH_2—$	2950–2930 (m) 2880–2845 (m) 1440–1415 (m) 1270–1220 (s)	
$—C{\equiv}CH$	ca 3300 (s) 700–600	Sharp Bending
${>}C{=}C{<}^{H}$	3040–3010 (m)	

(*Continued*)

TABLE 3.1 Absorption Frequencies of Single Bonds to Hydrogen (*Continued*)

Group	Band, cm^{-1}	Remarks
	Alkane residues attached to miscellaneous atoms (*continued*)	
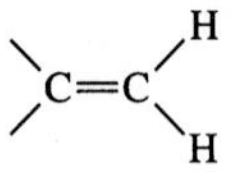	3095–3075 (m) 2985–2970 (m)	CH stretching sometimes obscured by much stronger bands of saturated CH groups
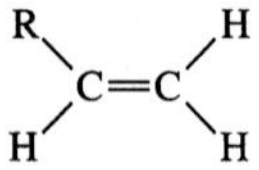	995–980 (s) 940–900 (s) ca 635 (s) 485–445 (m–s)	
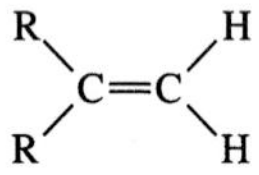	895–885 (s) 560–530 (s) 470–435 (m)	
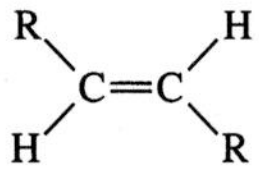	980–955 (s) 455–370 (m–s)	
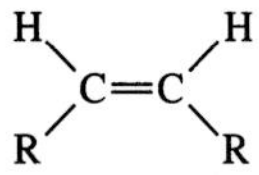	730–655 (m) 670–455 (s)	
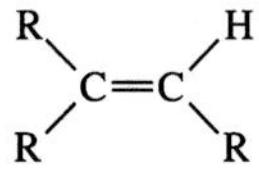	850–790 (m) 570–515 (s) 525–470 (s)	
—O—CH=CH_2	965–960 (s) 945–940 (m) 820–810 (s)	
—S—CH=CH_2	ca 965 (s) ca 860 (s)	
—CO—CH=CH_2	995–980 (s) 965–955 (m)	
—CO—OCH=CH_2	950–935 (s) 870–850 (s)	
—CO—C=CH_2	ca 930 (s)	
—CO—OC=CH_2	880–865	
—O—CH=CH— *trans*	940–920 (s)	
—CO—CH=CH— *trans*	ca 990 (s)	
	Hydroxyl group O—H compounds	
Primary aliphatic alcohols	3640–3630 (s) 1350–1260 (s) 1085–1030 (s)	Only in very dilute solutions in nonpolar solvents OH bending Also broad band at 700–600 cm^{-1}

TABLE 3.1 Absorption Frequencies of Single Bonds to Hydrogen (*Continued*)

Group	Band, cm^{-1}	Remarks
(Hydroxyl group O—H compounds) (*Continued*)		
Secondary aliphatic alcohols	3625–3620 (s) 1350–1260 (s) 1125–1085 (s)	See comments under primary aliphatic alcohols Also for α-unsaturated and cyclic tertiary aliphatic alcohols
Tertiary aliphatic alcohols	3620–3610 (s) 1410–1310 (s) 1205–1125 (s)	See comments under primary aliphatic alcohols
Aryl—OH	ca 3610 (s) 1410–1310 (s) 1260–1180 (s) 1085–1030 (s)	See comments under primary aliphatic alcohols Also for unsaturated secondary aliphatic alcohols
Carboxylic acids	3300–2500 (w–m) 995–915 (s)	Broad Broad diffuse band
Enol form of β-diketones	2700–2500 (var)	Broad
Free oximes	3600–3570 (w–m)	Shoulder
Free hydroperoxides	3560–3530 (m)	
Peroxy acids	ca 3280 (m)	
Phosphorus acids	2700–2560 (m)	Broad
Water in solution	3710	When solution is damp
Intermolecular H bond Dimeric Polymeric	 3600–3500 3400–3200 (s)	 Rather sharp. Absorptions arising from H bond with polar solvents also appear in this region. Broad
Intramolecular H bond Polyvalent alcohols Chelation	 3600–3500 (s) 3200–2500	 Sharper than dimeric band above Broad and occasionally weak; the lower the frequency, the stronger the intramolecular bond
Water of crystallation (solid state spectra)	3600–3100 (w)	Usually a weak band at 1640–1615 cm^{-1} also. Water in trace amounts in KBr disks shows a broad band at 3450 cm^{-1}.

(*Continued*)

TABLE 3.1 Absorption Frequencies of Single Bonds to Hydrogen (*Continued*)

Group	Band, cm^{-1}	Remarks
	Amine, imine, ammonium, and amide N — H	
Primary amines	3550–3300 (m)	Two bands in this range
Aliphatic	1650–1560 (m)	
	1090–1020 (w–m)	With α-carbon branching at 795 cm^{-1} and strong
	850–810 (w–m)	
	495–445 (m–s)	Broad
	ca 290 (s)	Broad
Aromatic	1350–1260 (s)	Also for secondary aryl amines
	445–345	
Amino acids	3100–3030 (m)	Values for solid states; broad bands also (but not always) near 2500 and 200 cm^{-1}
	2800–2400 (m)	Number of sharp bands; dilute solution
	1625–1560 (m)	
	1550–1550 (m)	
Amino salts	3550–3100 (m)	Values for solid state
	ca 3380	Dilute solutions
	ca 3280	
Secondary amines	3550–3400 (w)	Only one band, whereas primary amines show two bands
	1580–1490 (w)	Often too weak to be noticed
	1190–1170 (m)	
	1145–1130 (m)	
	455–405 (w–m)	
Salts	ca 2500	Sharp; broad values for solid state
	ca 2400	Sharp; broad values for solid state
	1620–1560 (m–s)	
Tertiary amines		
$R_1R_2R_3NH^+$	2700–2250	Group of relatively sharp bands; broad bands in solid state
Ammonium ion	3300–3030 (s)	Group of bands
	1430–1390 (s)	
Imines =N=H	3350–3310 (w)	Aliphatic
	3490 (s)	Aryl
	3490 (s)	Pyrroles, indoles; band sharp
Imine salts	2700–2330 (m–s)	Dilute solutions
	2200–1800 (m)	One or more bands; useful to distinguish from protonated tertiary amines
Primary amide —$CONH_2$	ca 3500 (m)	Lowered ca 150 cm^{-1} in the solid state and on H bonding; often several bands 3200–3050 cm^{-1}
	ca 3400 (m)	
Secondary amide —CONH—	3460–3400 (m)	Two bands; lowered on H bonding and in solid state. Only one band with lactams
	3100–3070 (w)	Extra band with bonded and solid-state samples

TABLE 3.1 Absorption Frequencies of Single Bonds to Hydrogen (*Continued*)

Group	Band, cm^{-1}	Remarks
	Amine, imine, ammonium, and Miscellaneous R — H	
—S—H	2600–2550 (w)	Weaker than OH and less affected by H bonding
P—H	2440–2350 (m)	Sharp
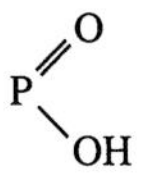	2700–2560 (m)	Associated OH
R—D	100/137 times the corresponding RH frequency	Useful when assigning RH bands; deuteration leads to a known shift to lower frequency

TABLE 3.2 Absorption Frequencies of Triple Bonds

Abbreviations Used in the Table

m, moderately strong — *var, of variable strength*
m–s, moderate to strong — *w–m, weak to moderately strong*
s, strong

Group	Band, cm^{-1}	Remarks
Alkynes		
Terminal	3300 (s)	CH stretching
	2140–2100 (w–m)*	C≡C stretching
	1375–1225 (w–m)	
	695–575 (m–s)	Two bands if molecule has axial symmetry
	ca 630 (s)	Alkyl monosubstituted
Nonterminal	2260–2150 (var)*	Symmetrical or nearly symmetrical substitution makes the C≡C stretching frequency inactive. When more than one C≡C linkage is present, and sometimes when there is only one, there are frequently more absorption bands in this region than there are triple bonds to account for them.
R_1—C≡C—R_2	540–465 (m)	The longer the chain, the lower the frequency
Aryl—C≡C—	ca 550 (m) ca 350 (var)	
—C≡C—halogen (Cl, Br, I)	185–160 (var)	

(*Continued*)

TABLE 3.2 Absorption Frequencies of Triple Bonds (*Continued*)

Group	Band, cm^{-1}	Remarks
Nitriles —C≡N	2260–2200 (var)	Stronger and toward the lower end of the range when conjugated; occasionally very weak or absent
Aliphatic	580–555 (m–s) 560–525 (m–s) 390–350 (s)	
Aromatic	580–540 (s) 430–380 (m)	
Isonitriles $R—\overset{+}{N}≡\overset{-}{C}$ or R—N=C:	2175–2150 (s) 2150–2115 (s) 1595	Very sensitive to changes in substituents Not found for nitriles
Cyanamides $>N—C≡N \rightleftharpoons >\overset{+}{N}=C=\overset{-}{N}$	2225–2210 (s)	
Thiocyanates R—S—C≡N	2175–2140 (s) 404–400 (s) ca 600 (m–s)	Aryl thiocyanates at the upper end of the range, alkyl at the lower end Aliphatic derivatives
Nitrile *N*-oxides —C≡N→O	2305–2285 (s) 1395–1365 (s)	Aryl derivatives
Diazonium salts $R—\overset{+}{N}≡N$	2300–2230 (m–s)	
Selenocyanates R—Se—C≡N	ca 2160 (m–s) 545–520 ca 390 ca 350	

*Conjugation with olefinic or acetylenic groups lowers the frequency and raises the intensity. Conjugation with carbonyl groups usually has little effect on the position of absorption.

TABLE 3.3 Absorption Frequencies of Cumulated Double Bonds

Abbreviations Used in the Table

m–s, moderate to strong *vs, very strong*
s, strong *w, weak*

Group	Band, cm^{-1}	Remarks
Carbon dioxide O=C=O	2349 (s)	Appears in many spectra as a result of inequalities in path length
Isocyanates —N=C=O	2275–2250 (vs)	Position unaffected by conjugation

TABLE 3.3 Absorption Frequencies of Cumulated Double Bonds (*Continued*)

Group	Band, cm^{-1}	Remarks
Isoselenocyanates —N=C=Se	2200–2000 (s) 675–605	Broad; usually two bands
Azides $—N_3$ or $—N{=}\overset{+}{N}{=}\overset{-}{N}$	2140–2030 (s) 1340–1180 (w)	 Not observed for azides
—N=C=N—	2155–2130 (s)	Split into unsymmetrical doublet by conjugation with aryl groups: 2145–2125 (vs) and 2115–2105 (vs)
Isothiocyanates —N=C=S	2140–1990 (vs) 649–600 (m–s) 565–510 (m–s) 470–440 (m–s)	Broad; usually a doublet
Ketenes >C=C=O	ca 2150 (s)	
Ketenimines C=C=N—	2050–2000 (s)	
Allenes >C=C=C<	2000–1915 (m–s)	Two bands when terminal allene or when bonded to electron-attracting groups
Thionylamines —N=S=O	1300–1230 (s) 1180–1110 (s)	
Diazoalkanes $R_2C{=}\overset{+}{N}{=}\overset{-}{N}$ $—CH{=}\overset{+}{N}{=}\overset{-}{N}$	2030–2000 (s) 2050–2035 (s)	
Diazoketones $—CO—CH{=}\overset{+}{N}{=}\overset{-}{N}$	2100–2080 2075–2050	Monosubstituted Disubstituted

3.1.1 Intensities of Carbonyl Bands

Acids generally absorb more strongly than esters, and esters more strongly than ketones or aldehydes. Amide absorption is usually similar in intensity to that of ketones but is subject to much greaer variations.

3.1.2 Position of Carbonyl Absorption

The general trends of structural variation on the position of C═O stretching frequencies may be summarized as follows:

1. The more electronegative the group X in the system R—CO—X—, the higher is the frequency.
2. α, β Unsaturation causes a lowering of frequency of 15 to 40 cm^{-1}, except in amides, where little shift is observed and that usually to higher frequency.
3. Further conjugation has relatively little effect.
4. Ring strain in cyclic compounds causes a relatively large shift to higher frequency. This phenomenon provides a remarkably reliable test of ring size, distinguishing clearly between four-, five-, and larger-membered-ring ketones, lactones, and lactams. Six-ring and larger ketones, lactones, and lactams show the normal frequency found for the open-chain compounds.
5. Hydrogen bonding to a carbonyl group causes a shift to lower frequency of 40 to 60 cm^{-1}. Acids, amides, enolized β-keto carbonyl systems, and *o*-hydroxyphenol and *o*-aminophenyl carbonyl compounds show this effect. All carbonyl compounds tend to give slightly lower values for the carbonyl stretching frequency in the solid state compared with the value for dilute solutions.
6. Where more than one of the structural influences on a particular carbonyl group is operating, the net effect is usually close to additive.

TABLE 3.4 Absorption Frequencies of Carbonyl Bands

All bands quoted are strong.

Groups	Band, cm^{-1}	Remarks
Acid anhydrides		
—CO—O—CO—		
Saturated	1850–1800 1790–1740	Two bands usually separated by about 60 cm^{-1}. The higher-frequency band is more intense in acyclic anhydrides, and the lower-frequency band is more intense in cyclic anhydrides.
Aryl and α,β-unsaturated	1830–1780 1790–1710	
Saturated five-ring	1870–1820 1800–1750	
All classes	1300–1050	One or two strong bands due to CO stretching
Acid chlorides —COCl		
Saturated	1815–1790	Acid fluorides higher, bromides and iodides lower
Aryl and α,β-unsaturated	1790–1750	
Acid peroxide		
CO—O—O—CO—		
Saturated	1820–1810 1800–1780	
Aryl and α,β-unsaturated	1805–1780 1785–1755	
Esters and lactones		
—CO—O—		

TABLE 3.4 Absorption Frequencies of Carbonyl Bands (*Continued*)

Groups	Band, cm^{-1}	Remarks
Saturated	1750–1735	
Aryl and α,β-unsaturated	1730–1715	
Aryl and vinyl esters		
C═C—O—CO—alkyl	1800–1750	The C═C stretching band also shifts to higher frequency.
Esters with electronegative α substituents; e.g.,		
>CCl—CO—O—	1770–1745	
α-Keto esters	1755–1740	
Six-ring and larger lactones	Similar values to the corresponding open-chain esters	
Five-ring lactone	1780–1760	
α,β-Unsaturated five-ring lactone	1770–1740	When α-CH is present, there are two bands, the relative intensity depending on the solvent.
β,γ-Unsaturated five-ring lactone, vinyl ester type	ca 1800	
Four-ring lactone	ca 1820	
β-Keto ester in H bonding enol form	ca 1650	Keto from normal; chelate-type H bond causes shift to lower frequency than the normal ester. The C═C band is strong and is usually near 1630 cm^{-1}.
All classes	1300–1050	Usually two strong bands due to CO stretching.
Aldehydes —CHO		
(See also Table 3.44 for C—H.) All values given below are lowered in liquid-film or solid-state spectra by about 10–20 cm^{-1}. Vapor-phase spectra have values raised about 20 cm^{-1}.		
Saturated	1740–1720	
Aryl	1715–1695	*o*-Hydroxy or amino groups shift this value to 1655–1625 cm^{-1} because of intramolecular H bonding.
α,β-Unsaturated	1705–1680	
$\alpha,\beta,\gamma,\delta$-Unsaturated	1680–1660	
β-Ketoaldehyde in enol form	1670–1645	Lowering caused by chelate-type H bonding.
Ketones >C═O		
All values given below are lowered in liquid-film or solid-state spectra by about 10–20 cm^{-1}. Vapor-phase spectra have values raised about 20 cm^{-1}.		

(*Continued*)

TABLE 3.4 Absorption Frequencies of Carbonyl Bands (*Continued*)

Groups	Band, cm^{-1}	Remarks
Ketones **>C=O** (*continued*)		
Saturated	1725–1705	
Aryl	1700–1680	
α,β-Unsaturated	1685–1665	
$\alpha,\beta,\alpha',\beta'$-Unsaturated and diaryl	1670–1660	
Cyclopropyl	1705–1685	
Six-ring ketones and larger	Similar values to the corresponding open-chain ketones	
Five-ring ketones	1750–1740	α,β Unsaturation, $\alpha,\beta,\alpha',\beta'$ unsaturation, etc., have a similar effect on these values as on those of open-chain ketones.
Four-ring ketones	ca 1780	
α-Halo ketones	1745–1725	Affected by conformation; highest values are obtained when both halogens are in the same plane as the C=O.
α,α'-Dihalo ketones	1765–1745	
1,2-Diketones, *syn-trans*-open chains	1730–1710	Antisymmetrical stretching frequency of both C=O's. The symmetrical stretching is inactive in the infrared but active in the Raman.
syn-cis-1,2-Diketones, six-ring	1760 and 1730	
syn-cis-1,2-Diketones, five ring	1775 and 1760	
o-Amino-aryl or *o*-hydroxy-aryl ketones	1655–1635	Low because of intramolecular H bonding. Other substituents and steric hindrance affect the position of the band.
Quinones	1690–1660	C=C band is strong and is usually near 1600 cm^{-1}.
Extended quinones	1655–1635	
Tropone	1650	Near 1600 cm^{-1} when lowered by H bonding as in tropolones
Carboxylic acids **$-CO_2H$**		
All types	3000–2500	OH stretching; a characteristic group of small bands due to combination bands
Saturated	1725–1700	The monomer is near 1760 cm^{-1}, but is rarely observed. Occasionally both bands, the free monomer, and the H-bonded dimer can be seen in solution spectra. Ether solvents give one band near 1730 cm^{-1}.
α,β-Unsaturated	1715–1690	
Aryl	1700–1680	
α-Halo-	1740–1720	
Carboxylate ions **$-CO_2^-$**		
Most types	1610–1550 1420–1300	Antisymmetrical and symmetrical stretching, respectively
Amides—CO—N< (See also Table 7.49 for NH stretching and bending.)		

TABLE 3.4 Absorption Frequencies of Carbonyl Bands (*Continued*)

Groups	Band, cm^{-1}	Remarks
Amides —CO—N< (*continued*)		
Primary —$CONH_2$		
In solution	ca 1690	Amide I; C=O stretching
Solid state	ca 1650	
In solution	ca 1600	Amide II: mostly NH bending
Solid state	ca 1640	Amide I is generally more intense than amide II. (In the solid state, amides I and II may overlap.)
Secondary —CONH—		
In solution	1700–1670	Amide I
Solid state	1680–1630	
In solution	1550–1510	Amide II; found in open-chain amides only
Solid state	1570–1515	Amide I is generally more intense than amide II.
Tertiary	1670–1630	Since H bonding is absent, solid and solution spectra are much the same.
Lactams		
Six-ring and larger rings	ca 1670	
Five-ring	ca 1700	Shifted to higher frequency when the N atom is in a bridged system
Four-ring	ca 1745	
R—CO—N—C=C		Shifted +15 cm^{-1} by the additional double bond
C=C—CO—N		Shifted by up to +15 cm^{-1} by the additional double bond. This is an unusual effect by α,β unsaturation. It is said to be due to the inductive effect of the C=C on the well-conjugated CO—N system, the usual conjugation effect being less important in such a system.
Imides —CO—N—CO—		
Cyclic six-ring	ca 1710 and ca 1700	Shift of +15 cm^{-1} with α,β unsaturation
Cyclic five-ring	ca 1770 and ca 1700	
Ureas N—CO—N		
RNHCONHR	ca 1660	
Six-ring	ca 1640	
Five-ring	ca 1720	
Urethanes R—O—CO—N	1740–1690	Also shows amide II band when nonsubstituted on N
Thioesters and Acids RCO—S—R′		
RCOSH	ca 1720	α,β-Unsaturated or aryl acid or ester shifted about −25 cm^{-1}
RCOS—alkyl	ca 1690	
RCOS—aryl	ca 1710	

TABLE 3.5 Absorption Frequencies of Other Double Bonds

Abbreviations Used in the Table

m, moderately strong — *vs, very strong*
m–s, moderate to strong — *w, weak*
var, of variable strength

Group	Band, cm^{-1}	Remarks
	Alkenes $>C{=}C<$	
Nonconjugated	1680–1620 (w–m)	May be very weak if symmetrically substituted
Conjugated with aromatic ring	1640–1610 (m)	More intense than with unconjugated double bonds
Internal (ring)	3060–2995 (m)	Highest frequencies for smallest ring
Carbons: $n = 3$	ca 1665 (w–m)	
$n = 4$	ca 1565 (w–m)	
$n = 5$	ca 1610 (w–m)	
	1370–1340 (s)	Characteristic
$n \geq 6$	1650–1645 (w–m)	
Exocyclic $C{=}C(CH_2)_n$ $n = 2$	1780–1730 (m)	
$n = 3$	ca 1680 (m)	
$n \geq 4$	1655–1650 (m)	
Fulvene 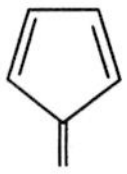	1645–1630 (m)	
	1370–1340 (s)	
	790–765 (s)	
Dienes, trienes, etc.	1650 (s) and 1600 (s)	Lower-frequency band usually more intense and may hide or overlap the higher-frequency band
α,β-Unsaturated carbonyl compounds	1640–1590 (m)	Usually much weaker than the C=O band
Enol esters, enol ethers, and enamines	1700–1650 (s)	
	Imines, oximes, and amidines $>C{=}N{-}$	
Imines and oximes		
Aliphatic	1690–1640 (w)	
α,β-Unsaturated and aromatic	1650–1620 (m)	
Conjugated cyclic systems	1660–1480 (var)	
	960–930 (s)	NO stretching of oximes
Imino ethers —O—C=N—	1690–1640 (var)	Usually a strong doublet

TABLE 3.5 Absorption Frequencies of Other Double Bonds (*Continued*)

Group	Band, cm^{-1}	Remarks
	Imines, oximes, and amidines >C=N— (*Continued*)	
Imino thioethers —S—C=N=	1640–1605 (var)	
Imine oxides >C=$\overset{+}{N}$—$\bar{O}$	1620–1550 (s)	
Amidines >N—C=N—	1685–1580 (var)	
Benzamidines Aryl—C=N—N	1630–1590	
Guanidine >N—C(—N)=N—	1725–1625 (s)	
Azines >C=N—N=C<	1670–1600	
Hydrazoketones —CO—C=N—N	1600–1530 (vs)	
	Azo compounds —N=N—	
Azo —N=N—		
Aliphatic	ca 1575 (var)	Very weak or inactive
Aromatic		
cis	ca 1510 (w)	
trans	1440–1410 (w)	
Azoxy —$\overset{+}{N}$(O^-)=N—		
Aliphatic	1590–1495 (m–s) 1345–1285 (m–s)	
Aromatic	1480–1450 (m–s) 1340–1315 (m–s)	
Azothio —N=$\overset{+}{N}$—$\bar{S}$—	1465–1445 (w) 1070–1055 (w)	
	Nitro compounds N=O	
Nitro C—NO_2		
Aliphatic	ca 1560 (s) 1385–1350 (s)	The two bands are due to asymmetrical and symmetrical stretching of the N=O bond. Electron-withdrawing substituents adjacent to nitro group increase the frequency of the asymmetrical band and decrease that of the symmetrical frequency.

(*Continued*)

TABLE 3.5 Absorption Frequencies of Other Double Bonds (*Continued*)

Group	Band, cm^{-1}	Remarks
	Nitro compounds N=O (*Continued*)	
Nitro C—NO_2 (*continued*)		
Aromatic	1570–1485 (s) 1380–1320 (s)	See above remark; also bulky orthosubstituents shift band to higher frequencies. Strong H bonding shifts frequency to lower end of range.
	865–835 (s)	Strong and sometimes at ca 750 cm^{-1}
	580–520 (var)	
α,β-Unsaturated Nitroalkenes	1530–1510 (s) 1360–1335 (s)	
Nitrates —O—NO_2	1650–1625 (vs) 1285–1275 (vs) 870–855 (vs) 760–755 (w–m) 710–695 (w–m)	
Nitramines >N—NO_2	1630–1550 (s) 1300–1250 (s)	
Nitrates —O—N=O	1680–1610 (vs)	Two bands
	815–750 (s)	*Trans* form
	850–810 (s)	*Cis* form
	690–615 (s)	
Thionitrites —S—N=O	730–685 (m–s)	
Nitroso ⩾C—N=O	1600–1500 (s)	
N—$\overset{+}{N}$=$\overset{-}{O}$		
Aliphatic	1530–1495 (m–s)	
Aromatic	1480–1450 (m–s) 1335–1315 (m–s)	
Nitrogen oxides N→O		
Pyridine	1320–1230 (m–s) 1190–1150 (m–s)	
Pyrazine	1380–1280 (m–s) 1040–990 (m–s) ca 850 (m)	Affected by ring substituents

TABLE 3.6 Absorption Frequencies of Aromatic Bands

Abbreviations Used in the Table

m, moderately strong	*var, of variable strength*
m–s, moderate to strong	*w–m, weak to moderately strong*
s, srong	

Group	Band, cm^{-1}	Remarks
Aromatic rings	ca 1600 (m)	
	ca 1580 (m)	Stronger when ring is further conjugated
	ca 1470 (m)	When substituent on ring is electron acceptor
	ca 1510 (m)	When substituent on ring is electron donor
Five adjacent H	900–860 (w–m)	
	770–730 (s)	
	720–680 (s)	
	625–605 (w–m)	
	ca 550 (w–m)	Substituents: C=C, C≡C, C≡N
1,2-Substitution	770–735 (s)	
	555–495 (w–m)	
	470–415 (m–s)	
1,3-Substitution	810–750 (s)	
	560–505 (m)	
	460–415 (m–s)	490–460 cm^{-1} when substituents are electron-accepting groups
1,4-Substitution	860–800 (s)	
	650–615 (w–m)	
	520–440 (m–s)	520–490 cm^{-1} when substituents are electron-donating groups
1,2,3-Trisubstitution	800–760 (s)	
	720–685 (s)	
	570–535 (s)	
	ca 485	
1,2,4-Trisubstitution	900–885 (m)	
	780–760 (s)	
	475–425 (m–s)	
1,3,5-Trisubstitution	950–925 (var)	
	865–810 (s)	
	730–680 (m–s)	
	535–495 (s)	
	470–450 (w–m)	
Pentasubstitution	900–860 (m–s)	
	580–535 (s)	
Hexasubstitution	415–385 (m–s)	

TABLE 3.7 Absorption Frequencies of Miscellaneous Bands

Abbreviations Used in the Table

m, moderately strong	*vs, very strong*
m–s, moderate to strong	*w, weak*
s, strong	*w–m, weak to moderately strong*
var, of variable strength	

Group	Band, cm^{-1}	Remarks
	Ethers	
Saturated aliphatic ⩾C—O—C⩽	1150–1060 (vs) 1140–900 (s)	Two peaks may be observed for branched chain, usually 1140–1110 cm^{-1}. Usually 930–900 cm^{-1}; may be absent for symmetric ethers
Alkyl–aryl =C—O—C⩽ |	1270–1230 (vs) 1120–1020 (s)	=CO stretching CO stretching
Vinyl	1225–1200 (s)	Usually about 1205 cm^{-1}
Diaryl =C—O—C= | |	1200–1120 (s) 1100–1050 (s)	
Cyclic	1270–1030 (s)	
Epoxides 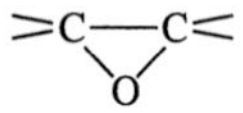	1260–1240 (m–s) 880–805 (m) 950–860 (var) 865–785 (m) 770–750 (m)	 Monosubstituted *Trans* form *Cis* form Trisubstituted
Ketals and acetals	1190–1140 (s) 1195–1125 (s) 1100–1000 (s) 1060–1035 (s)	 Strongest band Sometimes obscured
Phthalanes	915–895 (s)	
Aromatic methylenedioxy	1265–1235 (s)	
	Peroxides	
—O—O—	900–830 (w) 1150–1030 (m–s) ca 1000 (m)	 Alkyl Aryl

TABLE 3.7 Absorption Frequencies of Miscellaneous Bands (*Continued*)

Group	Band, cm^{-1}	Remarks
	Sulfur compounds	
Thiols		
—S—H	2600–2450 (w)	
—CO—SH	840–830 (m)	
—CS—SH	ca 860 (s)	Broad
Thiocarbonyl		
>C=S	1200–1050 (s)	Behaves generally in manner similar to carbonyl band
>N—C(—)=S	1570–1395 1420–1260 1140–940	
—S—C(—)=S	ca 580 (s)	
Sulfoxides		
>S=O	1075–1040 (vs) 730–690 (var) 395–360 (var)	Halogen or oxygen atom bonded to sulfur increases the frequency.
Sulfones		
$>SO_2$	1360–1290 (vs) 1170–1120 (vs) 610–545 (m–s) 525–495 (m–s)	Halogen or oxygen atom bonded to sulfur increases the frequency.
Sulfonamides		
$—SO_2—N<$	1380–1330 (vs) 1170–1140 (vs) 950–860 (m) 715–700 (w–m)	
Sulfonates		
$—SO_2—O—$	1420–1330 (s) 1200–1145 (s)	May appear as doublet
Thiosulfonates		
$—SO_2—S—$	ca 1340 (vs)	
Sulfates $—O—SO_2—O—$	1415–1380 (s) 1200–1185 (s)	Electronegative substituents increase frequencies.
Primary alkyl salts	1315–1220 (s) 1140–1075 (m)	Strongly influenced by metal ion
Secondary alkyl salts	1270–1210 (vs) 1075–1050 (s)	Doublet; both bands strongly influenced by metal ion

(*Continued*)

TABLE 3.7 Absorption Frequencies of Miscellaneous Bands (*Continued*)

Group	Band, cm^{-1}	Remarks
	Sulfur compounds (*Continued*)	
Stretching frequencies of C—S and S—S bonds		
—S—CH_3	710–685 (w–m)	
—S—CH_2—	660–630 (w–m)	
—S—CH<	630–600 (w–m)	
—S—C≤	600–570 (w–m)	
—S—aryl	1110–1070 (m) 710–685 (w–m)	
R—S—S—R	705–570 (w) 520–500 (w)	
Aryl—S—S—aryl	500–430 (w–m)	
Polysulfides	500–470 (w–m)	
CH_2—S—CH_2—	695–655 (w–m)	CSC stretching
$(R—S)_2C{=}O$	880–825 (s) 570–560 (var)	
—CO—S—	1035–935 (s)	
—CS—S	ca 580 (s)	
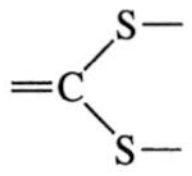	1050–900 (m–s) 980–850 (m–s) 900–800 (m–s)	Monoionic Ionic 1,1-dithiolates
	Phosphorus compounds	
P—H	2455–2265 (m) 1150–965 (w–m)	Sharp. Phosphines lie in the region 2285–2265 cm^{-1}.
—PH_2	1100–1085 (m) 1065–1040 (w–m) 940–910 (m)	
P—alkyl	795–650 (m–s)	
P—aryl	1130–1090 (s) 750–680 (s)	
P—O—alkyl	1050–970 (s)	Broad
P—O—aryl	1240–1190 (s)	
P—O—P	970–910	Broad
P=O	1350–1150 (s)	May appear as doublet
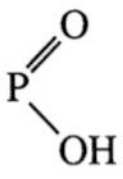	2725–2520 (w–m) 2350–2080 (w–m) 1740–1600 (w–m) 1335 (s) 1090–910 (s) 540–450 (w–m)	H-bonded; broad Broad; may be doublet for aryl acids P=O stretching

TABLE 3.7 Absorption Frequencies of Miscellaneous Bands (*Continued*)

Group	Band, cm^{-1}	Remarks
	Phosphorus compounds (*Continued*)	
P═S	865–655 (m–s) 595–530 (var)	
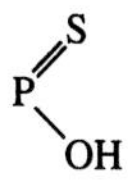	3100–3000 (w) 2360–2200 (w) 935–910 (s) 810–750 (m–s) 655585 (var)	 PO stretching P═S stretching P═S stretching
	Silicon compounds	
Si—H	2250–2100 (s) 985–800	 SiH_3 has two bands.
Si—C	860–760	Accompanied by CH_2 rocking
Si—C⋸	1280–1250 (s)	Sharp
Si—C_2H_5	1250–1220 (m) 1020–1000 (m) 970–945 (m)	
Si—Aryl	1125–1090 (vs)	Splits into two bands when two aryl groups are attached to one silicon atom, but has only one band when three aryl groups attached
⋹Si—OH	870–820	OH deformation band
⋹Si—O—Si⋸	1100–1000	
⋹Si—N—Si⋸	940–870 (s)	
⋹Si—Cl	550–470 (s) 250–150	
>$SiCl_2$	595–535 (s) 540–460 (m)	
—$SiCl_3$	625–570 (s) 535–450 (m)	
	Boron compounds	
Boranes >BH or —BH_2	 2640–2450 (m–s) 2640–2570 (m–s) 2535–2485 (m–s) 2380–2315 (s) 2285–2265 (s) 2140–2080 (w–m) 2580–2450 (m)	 Free H in BH Free H in BH_2 plus second band In complexes; second band for BH_2 Bridged H Borazoles and borazines

(*Continued*)

TABLE 3.7 Absorption Frequencies of Miscellaneous Bands (*Continued*)

Group	Band, cm^{-1}	Remarks
	Boron compounds (*Continued*)	
BH_4^-	2310–2195 (s)	Two bands
B—N	1550–1330 750–635	Borazines and borazoles
B—O	1390–1310 (s) 1280–1200	BO stretching Metal orthoborates
B—Cl B—Br	1090–890 (s)	Plus other bands at lower frequencies for BX_2 and BX_3
B—F	1500–840 (var)	Isotope splitting present
XBF_2	1500–1410 (s) 1300–1200 (s)	
X_2BF	1360–1300 (s)	
BF_3 complexes	1260–1125 (s) 1030–800 (s)	Band splitting may be added to isotopic splittings.
BF_4^-	ca 1030 (vs)	
	Halogen compounds	
C—F Aliphatic, mono-F Aliphatic, di-F Aliphatic, poly-F Aromatic	 1110–1000 (vs) 780–680 (s) 1250–1050 (vs) 1360–1090 (vs) 1270–1100 (m) 680–520 (m–s) 420–375 (var) 340–240 (s)	 Two bands Number of bands
—CF_3 Aliphatic Aromatic	 1350–1120 (vs) 780–680 (s) 680–590 (s) 600–540 (s) 555–505 (s) 1330–1310 (m–s) 600–580 (s)	
C—Cl Primary alkanes	 730–720 (s) 685–680 (s) 660–650 (s)	

TABLE 3.7 Absorption Frequencies of Miscellaneous Bands (*Continued*)

Group	Band, cm^{-1}	Remarks
	Halogen compounds (*Continued*)	
C—Cl (*continued*)		
Secondary alkanes	ca 760 (m)	
	675–655 (m–s)	
	615–605 (s)	
Tertiary alkanes	635–610 (m–s)	
	580–560 (m–s)	
Poly-Cl	800–700 (vs)	
Aryl:		
1,2-	1060–1035 (m)	
1,3-	1080–1075 (m)	
1,4-	1100–1090 (m)	
Chloroformates	ca 690 (s)	
	485–470 (s)	
Axial Cl	730–580 (s)	
Equatorial Cl	780–740 (s)	
C—Br		
Primary alkanes	645–635 (s)	
	565–555 (s)	
	440–430 (var)	
Secondary alkanes	620–605 (s)	
	590–575 (m–w)	
	540–530 (s)	
Tertiary alkanes	600–595 (m–s)	
	525–505 (s)	
Axial	690–550 (s)	
Equatorial	750–685 (s)	
Aryl:		
1,2-	1045–1025 (m)	
1,3-; 1,4-	1075–1065 (m)	
Other bands	400–260 (s)	
	325–175 (m–s)	
	290–225 (m–s)	
C—I		
Primary alkanes	600–585 (s)	
	515–500 (s)	
Secondary alkanes	ca 575 (s)	
	550–520 (s)	
	490–480 (s)	
Tertiary alkanes	580–560 (s)	
	510–485 (m)	
	485–465 (s)	
Aromatic	1060–1055 (m–s)	
	310–160 (s)	
	265–185	
Axial	ca 640 (s)	
Equatorial	ca 655 (s)	

(*Continued*)

TABLE 3.7 Absorption Frequencies of Miscellaneous Bands (*Continued*)

Group	Band, cm^{-1}	Remarks
	Inorganic ions (*Continued*)	
Ammonium	3300–3030	Several bands, all strong
Cyanate	2220–2130 (s)	
Cyanide	2200–2000	
Carbonate	1450–1410	
Hydrogen sulfate	1190–1160 (s) 1180–1000 (s) 880–840 (m)	
Nitrate	1410–1350 (vs) 860–800 (m)	
Nitrite	1275–1230 (s) 835–800 (m)	 Shoulder
Phosphate	1100–1000	
Sulfate	1130–1080 (s)	
Thiocyanate	ca 2050 (s)	

TABLE 3.8 Absorption Frequencies in the Near Infrared

Values in parentheses are molar absorptivity.

Class	Band, cm^{-1}	Remarks
Acetylenes	9800–9430 6580–6400 (1.0)	 Overtone of ≡CH stretching
Alcohols (nonhydrogen-bonded)	7140–7010 (2.0)	Overtone of OH stretching
Aldehydes		
Aliphatic	4640–4520 (0.5)	Combination of C=O and CH stretchings
Aromatic	ca 8000 ca 4525 ca 4445	
Formate	4775–4630 (1.0)	

TABLE 3.8 Absorption Frequencies of the Near Infrared (*Continued*)

Class	Band, cm^{-1}	Remarks
Alkanes		
$—CH_3$	9000–8350 (0.02)	
	5850–5660 (0.1)	
	4510–4280 (0.3)	
$—CH_2—$	9170–8475 (0.02)	
	5830–6640 (0.1)	
	4420–4070 (0.25)	
⋝CH	8550–8130	All bands very weak
	7000–6800	
	5650–5560	
Cyclopropane	6160–6060	
	4500–4400	
Alkenes		
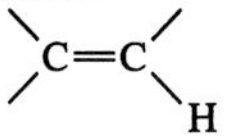	6850–6370 (1.0)	
$>C{=}CH_2$ and $—CH{=}CH_2$	7580–7300 (0.02)	
	6140–5980 (0.2)	
	4760–4700 (1.2)	
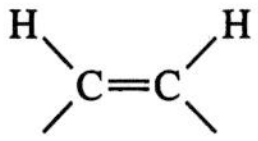	4760–4660 (0.15)	*Trans* isomers have no unique bands.
$—O—CH{=}CH_2$	6250–6040 (0.3)	
$—CO—CH{=}CH_2$	7580–7410 (0.02)	
	6190–5990 (0.3)	
	4820–4750 (0.2–0.5)	
Amides		
Primary	7400–6540 (0.7)	Two bands; overtone of NH stretch
	5160–5060 (3.0)	Second overtone of C=O stretch;
	5040–4990 (0.5)	second overtone of NH deforma-
	4960–4880 (0.5)	tion; combination of C=O and NH
Secondary	7330–7140 (0.5)	Overtone of NH stretch
	5050–4960 (0.4)	Combination of NH stretch and NH bending
Amines, aliphatic		
Primary	9710–9350	Second overtone of NH stretch
	6670–6450 (0.5)	Two bands; overtone of NH stretch
	5075–4900 (0.7)	Two bands; combination of NH stretch and NH bending
Secondary	9800–9350	Second overtone of NH stretch
	6580–6410 (0.5)	Overtone of NH stretch
Amines, aromatic		
Primary	9950–9520 (0.4)	
	7040–6850 (0.2)	
	6760–6580 (1.4)	
	5140–5040 (1.5)	
Secondary	10 000–9710	
	6800–6580 (0.5)	

(*Continued*)

TABLE 3.8 Absorption Frequencies of the Near Infrared (*Continued*)

Class	Band, cm^{-1}	Remarks
Aryl-H	7660–7330 (0.1) 6170–5880 (0.1)	Overtone of CH stretch
Carbonyl	5200–5100	
Carboxylic acids	7000–6800	
Epoxide (terminal)	6135–5960 (0.2) 4665–4520 (1.2)	Cyclopropane bands in same region
Glycols	7140–7040	
Hydroperoxides		
Aliphatic	6940–6750 (2.0) 4960–4880 (0.8)	
Aromatic	7040–6760 (1.0) 4950–4850 (1.3)	Two bands
Imides	9900–9620 6540–6370	
Nitriles	5350–5200 (0.1)	
Oximes	7140–7050	
Phosphines	5350–5260 (0.2)	
Phenols		
Nonbonded	7140–6800 (3.0) 5000–4950	
Intramolecularly bonded	7000–6700	
Thiols	5100–4950 (0.05)	

TABLE 3.9 Infrared Transmitting Materials

Material	Wavelength range, μm	Wavenumber range, cm^{-1}	Refractive index at 2 μm
NaCl, rock salt	0.25–17	40 000–590	1.52
KBr, potassium bromide	0.25–25	40 000–400	1.53
KCl, potassium chloride	0.30–20	33 000–500	1.5
AgCl, silver chloride*	0.40–23	25 000–435	2.0
AgBr, silver bromide*	0.50–35	20 000–286	2.2
CaF_2, calcium fluoride (Irtran-3)	0.15–9	66 700–1 110	1.40
BaF_2, barium fluoride	0.20–11.5	50 000–870	1.46
MgO, magnesium oxide (Irtran-5)	0.39–9.4	25 600–1 060	1.71

TABLE 3.9 Infrared Transmitting Materials (*Continued*)

Material	Wavelength range, μm	Wavenumber range, cm^{-1}	Refractive index at 2 μm
CsBr, cesium bromide	1–37	10 000–270	1.67
CsI, cesium iodide	1–50	10 000–200	1.74
TlBr-TlI, thallium bromide-iodide (KRS-5)*	0.50–35	20 000–286	2.37
ZnS, zinc sulfide (Irtran-2)	0.57–14.7	17 500–680	2.26
ZnSe, zinc selenide* (vacuum deposited) (Irtran-4)	1–18	10 000–556	2.45
CdTe, cadmium telluride (Irtran-6)	2–28	5 000–360	2.67
Al_2O_3, sapphire*	0.20–6.5	50 000–1538	1.76
SiO_2, fused quartz	0.16–3.7	62 500–2 700	
Ge, germanium*	0.50–16.7	20 000–600	4.0
Si, silicon*	0.20–6.2	50 000–1 613	3.5
Polyethylene	16–300	625–33	1.54

* Usual for internal reflection work.

TABLE 3.10 Infrared Transmission Characteristics of Selected Solvents

Transmission below 80%, obtained with a 0.10-mm cell path, is shown as shaded area.

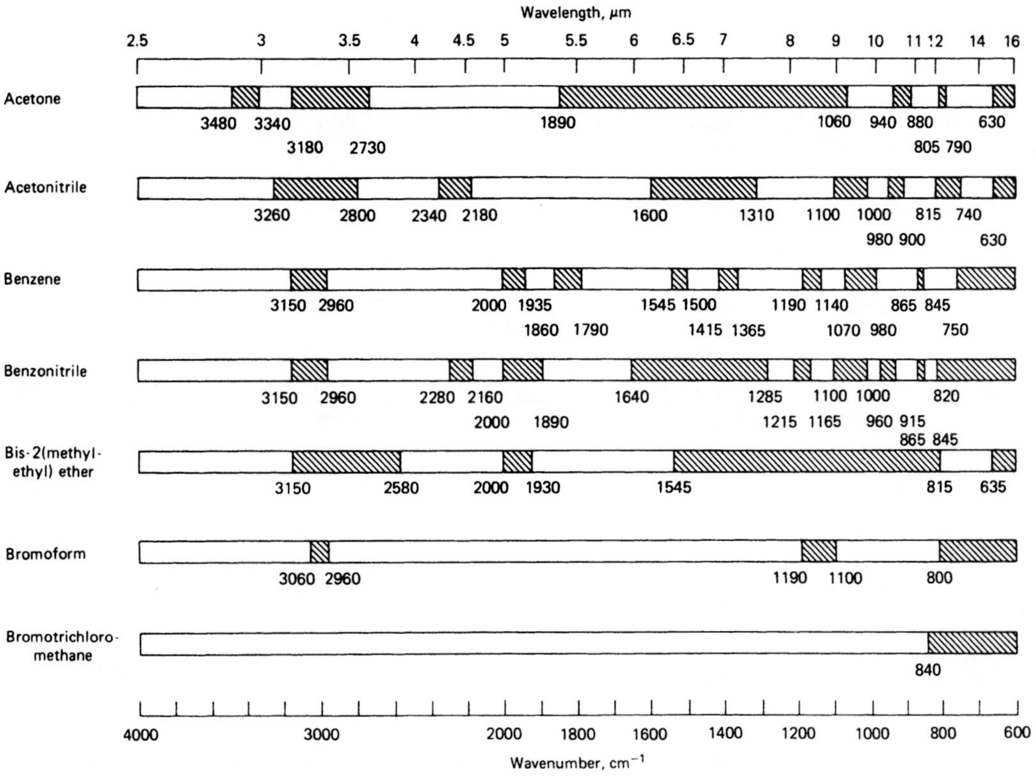

(*Continued*)

TABLE 3.10 Infrared Transmission Characteristics of Selected Solvents (*Continued*)

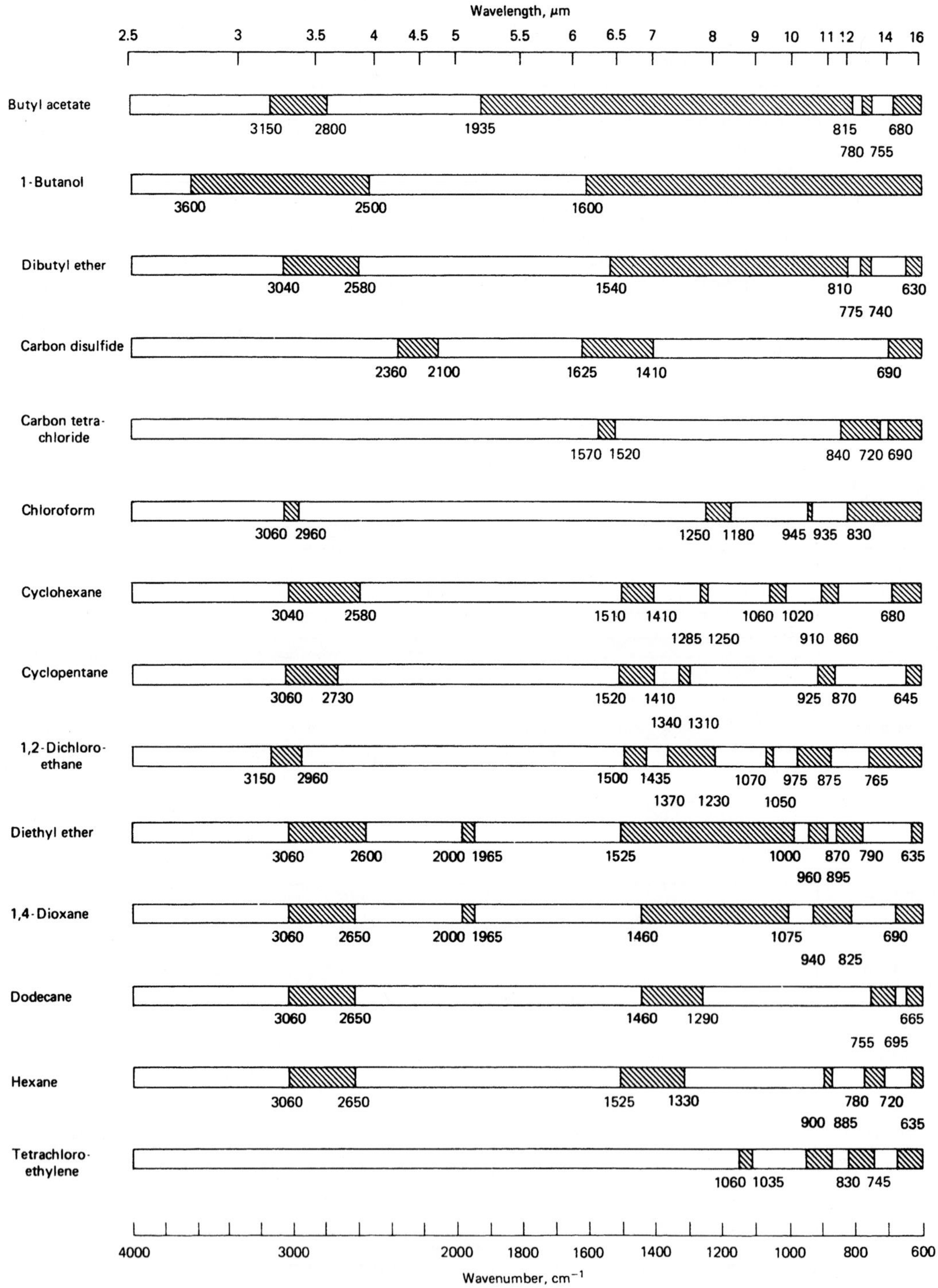

TABLE 3.10 Infrared Transmission Characteristics of Selected Solvents (*Continued*)

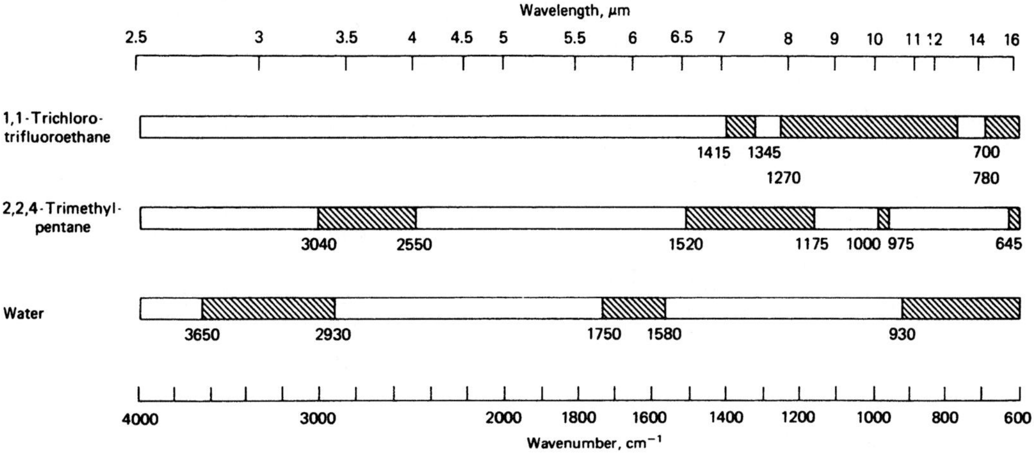

TABLE 3.11 Values of Absorbance for Percent Absorption

To convert percent absorption (% *A*) to absorbance, find the present absorption to the nearest whole digit in the left-hand column; read across to the column located under the tenth of a percent desired, and read the value of absorbance. The value of absorbance corresponding to 26.8% absorption is thus 0.1355.

% *A*	.0	.1	.2	.3	.4	.5	.6	.7	.8	.9
0.0	.0000	.0004	.0009	.0013	.0017	.0022	.0026	.0031	.0035	.0039
1.0	.0044	.0048	.0052	.0057	.0061	.0066	.0070	.0074	.0079	.0083
2.0	.0088	.0092	.0097	.0101	.0106	.0110	.0114	.0119	.0123	.0128
3.0	.0132	.0137	.0141	.0146	.0150	.0155	.0159	.0164	.0168	.0173
4.0	.0177	.0182	.0186	.0191	.0195	.0200	.0205	.0209	.0214	.0218
5.0	.0223	.0227	.0232	.0236	.0241	.0246	.0250	.0255	.0259	.0264
6.0	.0269	.0273	.0278	.0283	.0287	.0292	.0297	.0301	.0306	.0311
7.0	.0315	.0320	.0325	.0329	.0334	.0339	.0343	.0348	.0353	.0357
8.0	.0362	.0367	.0372	.0376	.0381	.0386	.0391	.0395	.0400	.0405
9.0	.0410	.0414	.0419	.0424	.0429	.0434	.0438	.0443	.0448	.0453
10.0	.0458	.0462	.0467	.0472	.0477	.0482	.0487	.0491	.0496	.0501
11.0	.0506	.0511	.0516	.0521	.0526	.0531	.0535	.0540	.0545	.0550
12.0	.0555	.0560	.0565	.0570	.0575	.0580	.0585	.0590	.0595	.0600
13.0	.0605	.0610	.0615	.0620	.0625	.0630	.0635	.0640	.0645	.0650
14.0	.0655	.0660	.0665	.0670	.0675	.0680	.0685	.0691	.0696	.0701
15.0	.0706	.0711	.0716	.0721	.0726	.0731	.0737	.0742	.0747	.0752
16.0	.0757	.0762	.0768	.0773	.0778	.0783	.0788	.0794	.0799	.0804
17.0	.0809	.0814	.0820	.0825	.0830	.0835	.0841	.0846	.0851	.0857
18.0	.0862	.0867	.0872	.0878	.0883	.0888	.0894	.0899	.0904	.0910
19.0	.0915	.0921	.0926	.0931	.0937	.0942	.0947	.0953	.0958	.0964
20.0	.0969	.0975	.0980	.0985	.0991	.0996	.1002	.1007	.1013	.1018
21.0	.1024	.1029	.1035	.1040	.1046	.1051	.1057	.1062	.1068	.1073
22.0	.1079	.1085	.1090	.1096	.1101	.1107	.1113	.1118	.1124	.1129
23.0	.1135	.1141	.1146	.1152	.1158	.1163	.1169	.1175	.1180	.1186

TABLE 3.11 Values of Absorbance for Percent Absorption (*Continued*)

%A	.0	.1	.2	.3	.4	.5	.6	.7	.8	.9
24.0	.1192	.1198	.1203	.1209	.1215	.1221	.1226	.1232	.1238	.1244
25.0	.1249	.1255	.1261	.1267	.1273	.1278	.1284	.1290	.1296	.1302
26.0	.1308	.1314	.1319	.1325	.1331	.1337	.1343	.1349	.1355	.1361
27.0	.1367	.1373	.1379	.1385	.1391	.1397	.1403	.1409	.1415	.1421
28.0	.1427	.1433	.1439	.1445	.1451	.1457	.1463	.1469	.1475	.1481
29.0	.1487	.1494	.1500	.1506	.1512	.1518	.1524	.1530	.1537	.1543
30.0	.1549	.1555	.1561	.1568	.1574	.1580	.1586	.1593	.1599	.1605
31.0	.1612	.1618	.1624	.1630	.1637	.1643	.1649	.1656	.1662	.1669
32.0	.1675	.1681	.1688	.1694	.1701	.1707	.1713	.1720	.1726	.1733
33.0	.1739	.1746	.1752	.1759	.1765	.1772	.1778	.1785	.1791	.1798
34.0	.1805	.1811	.1818	.1824	.1831	.1838	.1844	.1851	.1858	.1864
35.0	.1871	.1878	.1884	.1891	.1898	.1904	.1911	.1918	.1925	.1931
36.0	.1938	.1945	.1952	.1959	.1965	.1972	.1979	.1986	.1993	.2000
37.0	.2007	.2013	.2020	.2027	.2034	.2041	.2048	.2055	.2062	.2069
38.0	.2076	.2083	.2090	.2097	.2104	.2111	.2118	.2125	.2132	.2140
39.0	.2147	.2154	.2161	.2168	.2175	.2182	.2190	.2197	.2204	.2211
40.0	.2218	.2226	.2233	.2240	.2248	.2255	.2262	.2269	.2277	.2284
41.0	.2291	.2299	.2306	.2314	.2321	.2328	.2336	.2343	.2351	.2358
42.0	.2366	.2373	.2381	.2388	.2396	.2403	.2411	.2418	.2426	.2434
43.0	.2441	.2449	.2457	.2464	.2472	.2480	.2487	.2495	.2503	.2510
44.0	.2518	.2526	.2534	.2541	.2549	.2557	.2565	.2573	.2581	.2588
45.0	.2596	.2604	.2612	.2620	.2628	.2636	.2644	.2652	.2660	.2668
46.0	.2676	.2684	.2692	.2700	.2708	.2716	.2725	.2733	.2741	.2749
47.0	.2757	.2765	.2774	.2782	.2790	.2798	.2807	.2815	.2823	.2832
48.0	.2840	.2848	.2857	.2865	.2874	.2882	.2890	.2899	.2907	.2916
49.0	.2924	.2933	.2941	.2950	.2958	.2967	.2976	.2984	.2993	.3002
50.0	.3010	.3019	.3028	.3036	.3045	.3054	.3063	.3072	.3080	.3089
51.0	.3098	.3107	.3116	.3125	.3134	.3143	.3152	.3161	.3170	.3179
52.0	.3188	.3197	.3206	.3215	.3224	.3233	.3242	.3251	.3261	.3270
53.0	.3279	.3288	.3298	.3307	.3316	.3325	.3335	.3344	.3354	.3363
54.0	.3372	.3382	.3391	.3401	.3410	.3420	.3429	.3439	.3449	.3458
55.0	.3468	.3478	.3487	.3497	.3507	.3516	.3526	.3536	.3546	.3556
56.0	.3565	.3575	.3585	.3595	.3605	.3615	.3625	.3635	.3645	.3655
57.0	.3665	.3675	.3686	.3696	.3706	.3716	.3726	.3737	.3747	.3757
58.0	.3768	.3778	.3788	.3799	.3809	.3820	.3830	.3840	.3851	.3862
59.0	.3872	.3883	.3893	.3904	.3915	.3925	.3936	.3947	.3958	.3969
60.0	.3979	.3990	.4001	.4012	.4023	.4034	.4045	.4056	.4067	.4078
61.0	.4089	.4101	.4112	.4123	.4134	.4145	.4157	.4168	.4179	.4191
62.0	.4202	.4214	.4225	.4237	.4248	.4260	.4271	.4283	.4295	.4306
63.0	.4318	.4330	.4342	.4353	.4365	.4377	.4389	.4401	.4413	.4425
64.0	.4437	.4449	.4461	.4473	.4485	.4498	.4510	.4522	.4535	.4547
65.0	.4559	.4572	.4584	.4597	.4609	.4622	.4634	.4647	.4660	.4672
66.0	.4685	.4698	.4711	.4724	.4737	.4750	.4763	.4776	.4789	.4802
67.0	.4815	.4828	.4841	.4855	.4868	.4881	.4895	.4908	.4921	.4935
68.0	.4948	.4962	.4976	.4989	.5003	.5017	.5031	.5045	.5058	.5072
69.0	.5086	.5100	.5114	.5129	.5143	.5157	.5171	.5186	.5200	.5214
70.0	.5229	.5243	.5258	.5272	.5287	.5302	.5317	.5331	.5346	.5361
71.0	.5376	.5391	.5406	.5421	.5436	.5452	.5467	.5482	.5498	.5513

TABLE 3.11 Values of Absorbance for Percent Absorption (*Continued*)

%A	.0	.1	.2	.3	.4	.5	.6	.7	.8	.9
72.0	.5528	.5544	.5560	.5575	.5591	.5607	.5622	.5638	.5654	.5670
73.0	.5686	.5702	.5719	.5735	.5751	.5768	.5784	.5800	.5817	.5834
74.0	.5850	.5867	.5884	.5901	.5918	.5935	.5952	.5969	.5986	.6003
75.0	.6021	.6038	.6055	.6073	.6091	.6108	.6126	.6144	.6162	.6180
76.0	.6198	.6216	.6234	.6253	.6271	.6289	.6308	.6326	.6345	.6364
77.0	.6383	.6402	.6421	.6440	.6459	.6478	.6498	.6517	.6536	.6556
78.0	.6576	.6596	.6615	.6635	.6655	.6676	.6696	.6716	.6737	.6757
79.0	.6778	.6799	.6819	.6840	.6861	.6882	.6904	.6925	.6946	.6968
80.0	.6990	.7011	.7033	.7055	.7077	.7100	.7122	.7144	.7167	.7190
81.0	.7212	.7235	.7258	.7282	.7305	.7328	.7352	.7375	.7399	.7423
82.0	.7447	.7471	.7496	.7520	.7545	.7570	.7595	.7620	.7645	.7670
83.0	.7696	.7721	.7747	.7773	.7799	.7825	.7852	.7878	.7905	.7932
84.0	.7959	.7986	.8013	.8041	.8069	.8097	.8125	.8153	.8182	.8210
85.0	.8239	.8268	.8297	.8327	.8356	.8386	.8416	.8447	.8477	.8508
86.0	.8539	.8570	.8601	.8633	.8665	.8697	.8729	.8761	.8794	.8827
87.0	.8861	.8894	.8928	.8962	.8996	.9031	.9066	.9101	.9136	.9172
88.0	.9208	.9245	.9281	.9318	.9355	.9393	.9431	.9469	.9508	.9547
89.0	.9586	.9626	.9666	.9706	.9747	.9788	.9830	.9872	.9914	.9957

TABLE 3.12 Transmittance-Absorbance Conversion Table

This table gives absorbance values to four significant figures corresponding to % transmittance values, which are given to three significant figures. The values of % transmittance are given in the left-hand column and in the top row. For example, 8.4% transmittance corresponds to an absorbance of 1.076.

Interpolation is facilitated and accuracy is maximized if the % transmittance is between 1 and 10, by multiplying its value by 10, finding the absorbance corresponding to the result, and adding 1. For example, to find the absorbance corresponding to 8.45% transmittance, note that 84.5% transmittance corresponds to an absorbance of 0.0731, so that 8.45% transmittance corresponds to an absorbance of 1.0731. For % transmittance values between 0.1 and 1, multiply by 100, find the absorbance corresponding to the result, and add 2.

Conversely, to find the % transmittance corresponding to an absorbance between 1 and 2, subtract 1 from the absorbance, find the % transmittance corresponding to the result, and divide by 10. For example, an absorbance of 1.219 can best be converted to % transmittance by noting that an absorbance of 0.219 would correspond to 60.4% transmittance; dividing this by 10 gives the desired value, 6.04% transmittance. For absorbance values between 2 and 3, subtract 2 from the absorbance, find the % transmittance corresponding to the result, and divide by 100.

% Trans-mittance	0.0	0.1	0.2	0.3	0.4	0.5	0.6	0.7	0.8	0.9
0	......	3.000	2.699	2.523	2.398	2.301	2.222	2.155	2.097	2.046
1	2.000	1.959	1.921	1.886	1.854	1.824	1.796	1.770	1.745	1.721
2	1.699	1.678	1.658	1.638	1.620	1.602	1.585	1.569	1.553	1.538
3	1.523	1.509	1.495	1.481	1.469	1.456	1.444	1.432	1.420	1.409
4	1.398	1.387	1.377	1.367	1.357	1.347	1.337	1.328	1.319	1.310
5	1.301	1.292	1.284	1.276	1.268	1.260	1.252	1.244	1.237	1.229
6	1.222	1.215	1.208	1.201	1.194	1.187	1.180	1.174	1.167	1.161
7	1.155	1.149	1.143	1.137	1.131	1.125	1.119	1.114	1.108	1.102
8	1.097	1.092	1.086	1.081	1.076	1.071	1.066	1.060	1.056	1.051
9	1.046	1.041	1.036	1.032	1.027	1.022	1.018	1.013	1.009	1.004
10	1.000	0.9957	0.9914	0.9872	0.9830	0.9788	0.9747	0.9706	0.9666	0.9626

(*Continued*)

TABLE 3.12 Transmittance-Absorbance Conversion Table (*Continued*)

% Trans-mittance	0.0	0.1	0.2	0.3	0.4	0.5	0.6	0.7	0.8	0.9
11	0.9586	0.9547	0.9508	0.9469	0.9431	0.9393	0.9355	0.9318	0.9281	0.9245
12	0.9208	0.9172	0.9136	0.9101	0.9066	0.9031	0.8996	0.8962	0.8928	0.8894
13	0.8861	0.8827	0.8794	0.8761	0.8729	0.8697	0.8665	0.8633	0.8601	0.8570
14	0.8539	0.8508	0.8477	0.8447	0.8416	0.8386	0.8356	0.8327	0.8297	0.8268
15	0.8239	0.8210	0.8182	0.8153	0.8125	0.8097	0.8069	0.8041	0.8013	0.7986
16	0.7959	0.7932	0.7905	0.7878	0.7852	0.7825	0.7799	0.7773	0.7747	0.7721
17	0.7696	0.7670	0.7645	0.7620	0.7595	0.7570	0.7545	0.7520	0.7496	0.7471
18	0.7447	0.7423	0.7399	0.7375	0.7352	0.7328	0.7305	0.7282	0.7258	0.7235
19	0.7212	0.7190	0.7167	0.7144	0.7122	0.7100	0.7077	0.7055	0.7033	0.7011
20	0.6990	0.6968	0.6946	0.6925	0.6904	0.6882	0.6861	0.6840	0.6819	0.6799
21	0.6778	0.6757	0.6737	0.6716	0.6696	0.6676	0.6655	0.6635	0.6615	0.6596
22	0.6576	0.6556	0.6536	0.6517	0.6498	0.6478	0.6459	0.6440	0.6421	0.6402
23	0.6383	0.6364	0.6345	0.6326	0.6308	0.6289	0.6271	0.6253	0.6234	0.6216
24	0.6198	0.6180	0.6162	0.6144	0.6126	0.6108	0.6091	0.6073	0.6055	0.6038
25	0.6021	0.6003	0.5986	0.5969	0.5952	0.5935	0.5918	0.5901	0.5884	0.5867
26	0.5850	0.5834	0.5817	0.5800	0.5784	0.5766	0.5751	0.5735	0.5719	0.5702
27	0.5686	0.5670	0.5654	0.5638	0.5622	0.5607	0.5591	0.5575	0.5560	0.5544
28	0.5528	0.5513	0.5498	0.5482	0.5467	0.5452	0.5436	0.5421	0.5406	0.5391
29	0.5376	0.5361	0.5346	0.5331	0.5317	0.5302	0.5287	0.5272	0.5258	0.5243
30	0.5229	0.5214	0.5200	0.5186	0.5171	0.5157	0.5143	0.5129	0.5114	0.5100
31	0.5086	0.5072	0.5058	0.5045	0.5031	0.5017	0.5003	0.4989	0.4976	0.4962
32	0.4949	0.4935	0.4921	0.4908	0.4895	0.4881	0.4868	0.4855	0.4841	0.4828
33	0.4815	0.4802	0.4789	0.4776	0.4763	0.4750	0.4737	0.4724	0.4711	0.4698
34	0.4685	0.4672	0.4660	0.4647	0.4634	0.4622	0.4609	0.4597	0.4584	0.4572
35	0.4559	0.4547	0.4535	0.4522	0.4510	0.4498	0.4486	0.4473	0.4461	0.4449
36	0.4437	0.4425	0.4413	0.4401	0.4389	0.4377	0.4365	0.4353	0.4342	0.4330
37	0.4318	0.4306	0.4295	0.4283	0.4271	0.4260	0.4248	0.4237	0.4225	0.4214
38	0.4202	0.4191	0.4179	0.4168	0.4157	0.4145	0.4134	0.4123	0.4112	0.4101
39	0.4089	0.4078	0.4067	0.4056	0.4045	0.4034	0.4023	0.4012	0.4001	0.3989
40	0.3979	0.3969	0.3958	0.3947	0.3936	0.3925	0.3915	0.3904	0.3893	0.3883
41	0.3872	0.3862	0.3851	0.3840	0.3830	0.3820	0.3809	0.3799	0.3788	0.3778
42	0.3768	0.3757	0.3747	0.3737	0.3726	0.3716	0.3706	0.3696	0.3686	0.3675
43	0.3665	0.3655	0.3645	0.3635	0.3625	0.3615	0.3605	0.3595	0.3585	0.3575
44	0.3565	0.3556	0.3546	0.3536	0.3526	0.3516	0.3507	0.3497	0.3487	0.3478
45	0.3468	0.3458	0.3449	0.3439	0.3429	0.3420	0.3410	0.3401	0.3391	0.3382
46	0.3372	0.3363	0.3354	0.3344	0.3335	0.3325	0.3316	0.3307	0.3298	0.3288
47	0.3279	0.3270	0.3261	0.3251	0.3242	0.3233	0.3224	0.3215	0.3206	0.3197
48	0.3188	0.3179	0.3170	0.3161	0.3152	0.3143	0.3134	0.3125	0.3116	0.3107
49	0.3098	0.3089	0.3080	0.3072	0.3063	0.3054	0.3045	0.3036	0.3028	0.3019
50	0.3010	0.3002	0.2993	0.2984	0.2976	0.2967	0.2958	0.2950	0.2941	0.2933
51	0.2924	0.2916	0.2907	0.2899	0.2890	0.2882	0.2874	0.2865	0.2857	0.2848
52	0.2840	0.2832	0.2823	0.2815	0.2807	0.2798	0.2790	0.2782	0.2774	0.2765
53	0.2757	0.2749	0.2741	0.2733	0.2725	0.2716	0.2708	0.2700	0.2692	0.2684
54	0.2676	0.2668	0.2660	0.2652	0.2644	0.2636	0.2628	0.2620	0.2612	0.2604
55	0.2596	0.2588	0.2581	0.2573	0.2565	0.2557	0.2549	0.2541	0.2534	0.2526

TABLE 3.12 Transmittance-Absorbance Conversion Table (*Continued*)

% Trans-mittance	0.0	0.1	0.2	0.3	0.4	0.5	0.6	0.7	0.8	0.9
56	0.2518	0.2510	0.2503	0.2495	0.2487	0.2480	0.2472	0.2464	0.2457	0.2449
57	0.2441	0.2434	0.2426	0.2418	0.2411	0.2403	0.2396	0.2388	0.2381	0.2373
58	0.2366	0.2358	0.2351	0.2343	0.2336	0.2328	0.2321	0.2314	0.2306	0.2299
59	0.2291	0.2284	0.2277	0.2269	0.2262	0.2255	0.2248	0.2240	0.2233	0.2226
60	0.2218	0.2211	0.2204	0.2197	0.2190	0.2182	0.2175	0.2168	0.2161	0.2154
61	0.2147	0.2140	0.2132	0.2125	0.2118	0.2111	0.2104	0.2097	0.2090	0.2083
62	0.2076	0.2069	0.2062	0.2055	0.2048	0.2041	0.2034	0.2027	0.2020	0.2013
63	0.2007	0.2000	0.1993	0.1986	0.1979	0.1972	0.1965	0.1959	0.1952	0.1945
64	0.1938	0.1931	0.1925	0.1918	0.1911	0.1904	0.1898	0.1891	0.1884	0.1878
65	0.1871	0.1864	0.1858	0.1851	0.1844	0.1838	0.1831	0.1824	0.1818	0.1811
66	0.1805	0.1798	0.1791	0.1785	0.1778	0.1772	0.1765	0.1759	0.1752	0.1746
67	0.1739	0.1733	0.1726	0.1720	0.1713	0.1707	0.1701	0.1694	0.1688	0.1681
68	0.1675	0.1669	0.1662	0.1656	0.1649	0.1643	0.1637	0.1630	0.1624	0.1618
69	0.1612	0.1605	0.1599	0.1593	0.1586	0.1580	0.1574	0.1568	0.1561	0.1555
70	0.1549	0.1543	0.1537	0.1530	0.1524	0.1518	0.1512	0.1506	0.1500	0.1494
71	0.1487	0.1481	0.1475	0.1469	0.1463	0.1457	0.1451	0.1445	0.1439	0.1433
72	0.1427	0.1421	0.1415	0.1409	0.1403	0.1397	0.1391	0.1385	0.1379	0.1373
73	0.1367	0.1361	0.1355	0.1349	0.1343	0.1337	0.1331	0.1325	0.1319	0.1314
74	0.1308	0.1302	0.1296	0.1290	0.1284	0.1278	0.1273	0.1267	0.1261	0.1255
75	0.1249	0.1244	0.1238	0.1232	0.1226	0.1221	0.1215	0.1209	0.1203	0.1198
76	0.1192	0.1186	0.1180	0.1175	0.1169	0.1163	0.1158	0.1152	0.1146	0.1141
77	0.1135	0.1129	0.1124	0.1118	0.1113	0.1107	0.1101	0.1096	0.1090	0.1085
78	0.1079	0.1073	0.1068	0.1062	0.1057	0.1051	0.1046	0.1040	0.1035	0.1029
79	0.1024	0.1018	0.1013	0.1007	0.1002	0.0996	0.0991	0.0985	0.0980	0.0975
80	0.0969	0.0964	0.0958	0.0953	0.0947	0.0942	0.0937	0.0931	0.0926	0.0921
81	0.0915	0.0910	0.0904	0.0899	0.0894	0.0888	0.0883	0.0878	0.0872	0.0867
82	0.0862	0.0857	0.0851	0.0846	0.0841	0.0835	0.0830	0.0825	0.0820	0.0814
83	0.0809	0.0804	0.0799	0.0794	0.0788	0.0783	0.0778	0.0773	0.0768	0.0762
84	0.0757	0.0752	0.0747	0.0742	0.0737	0.0731	0.0726	0.0721	0.0716	0.0711
85	0.0706	0.0701	0.0696	0.0691	0.0685	0.0680	0.0675	0.0670	0.0665	0.0660
86	0.0655	0.0650	0.0645	0.0640	0.0635	0.0630	0.0625	0.0620	0.0615	0.0610
87	0.0605	0.0600	0.0595	0.0590	0.0585	0.0580	0.0575	0.0570	0.0565	0.0560
88	0.0555	0.0550	0.0545	0.0540	0.0535	0.0531	0.0526	0.0521	0.0516	0.0511
89	0.0506	0.0501	0.0496	0.0491	0.0487	0.0482	0.0477	0.0472	0.0467	0.0462
90	0.0458	0.0453	0.0448	0.0443	0.0438	0.0434	0.0429	0.0424	0.0419	0.0414
91	0.0410	0.0405	0.0400	0.0395	0.0391	0.0386	0.0381	0.0376	0.0372	0.0367
92	0.0362	0.0357	0.0353	0.0348	0.0343	0.0339	0.0334	0.0329	0.0325	0.0320
93	0.0315	0.0311	0.0306	0.0301	0.0297	0.0292	0.0287	0.0283	0.0278	0.0273
94	0.0269	0.0264	0.0259	0.0255	0.0250	0.0246	0.0241	0.0237	0.0232	0.0227
95	0.0223	0.0218	0.0214	0.0209	0.0205	0.0200	0.0195	0.0191	0.0186	0.0182
96	0.0177	0.0173	0.0168	0.0164	0.0159	0.0155	0.0150	0.0146	0.0141	0.0137
97	0.0132	0.0128	0.0123	0.0119	0.0114	0.0110	0.0106	0.0101	0.0097	0.0092
98	0.0088	0.0083	0.0079	0.0074	0.0070	0.0066	0.0061	0.0057	0.0052	0.0048
99	0.0044	0.0039	0.0035	0.0031	0.0026	0.0022	0.0017	0.0013	0.0009	0.0004

TABLE 3.13 Wavenumber/Wavelength Conversion Table

This table is based on the conversion: wavenumber (in cm^{-1}) = 10,000/wavelength (in μm). For example, 15.4 μm is equal to 649 cm^{-1}.

Wavelength (μm)	0	0.1	0.2	0.3	0.4	0.5	0.6	0.7	0.8	0.9 cm^{-1}
1.0	10000	9091	8333	7692	7143	6667	6250	5882	5556	5263
2.0	5000	4762	4545	4348	4167	4000	3846	3704	3571	3448
3.0	3333	3226	3125	3030	2941	2857	2778	2703	2632	2564
4.0	2500	2439	2381	2326	2273	2222	2174	2128	2083	2041
5.0	2000	1961	1923	1887	1852	1818	1786	1754	1724	1695
6.0	1667	1639	1613	1587	1563	1538	1515	1493	1471	1449
7.0	1429	1408	1389	1370	1351	1333	1316	1299	1282	1266
8.0	1250	1235	1220	1205	1190	1176	1163	1149	1136	1124
9.0	1111	1099	1087	1075	1064	1053	1042	1031	1020	1010
10.0	1000	990	980	971	962	952	943	935	926	917
11.0	909	901	893	885	877	870	862	855	847	840
12.0	833	826	820	813	806	800	794	787	781	775
13.0	769	763	758	752	746	741	735	730	725	719
14.0	714	709	704	699	694	690	685	680	676	671
15.0	667	662	658	654	649	645	641	637	633	629
16.0	625	621	617	613	610	606	602	599	595	592
17.0	588	585	581	578	575	571	568	565	562	559
18.0	556	552	549	546	543	541	538	535	532	529
19.0	526	524	521	518	515	513	510	508	505	503
20.0	500	498	495	493	490	488	485	483	481	478
21.0	476	474	472	469	467	465	463	461	459	457
22.0	455	452	450	448	446	444	442	441	439	437
23.0	435	433	431	429	427	426	424	422	420	418
24.0	417	415	413	412	410	408	407	405	403	402
25.0	400	398	397	395	394	392	391	389	388	386
26.0	385	383	382	380	379	377	376	375	373	372
27.0	370	369	368	366	365	364	362	361	360	358
28.0	357	356	355	353	352	351	350	348	347	346
29.0	345	344	342	341	340	339	338	337	336	334
30.0	333	332	331	330	329	328	327	326	325	324
31.0	323	322	321	319	318	317	316	315	314	313
32.0	313	312	311	310	309	308	307	306	305	304
33.0	303	302	301	300	299	299	298	297	296	295
34.0	294	293	292	292	291	290	289	288	287	287
35.0	286	285	284	283	282	282	281	280	279	279
36.0	278	277	276	275	274	274	273	272	272	271
37.0	270	270	269	268	267	267	266	265	265	264
38.0	263	262	262	261	260	260	259	258	258	257
39.0	256	256	255	254	254	253	253	252	251	251
40.0	250									

3.2 RAMAN SPECTROSCOPY

Raman spectroscopy is the measurement of the wavelength and intensity of inelastically scattered light from molecules. The Raman scattered light occurs at wavelengths that are shifted from the incident light by the energies of molecular vibrations.

The mechanism of Raman scattering is different from that of infrared absorption but Raman and IR spectra provide complementary information for the identification of organic functionalities. Raman spectra arise from the absorption of monochromatic light by a sample before it is emitted as scattered light. As in infrared spectra, Raman spectra are recorded in wavenumbers. Frequently a Raman spectrum will reveal something that was missed in the infrared spectrum. This is because a bond that has no dipole moment (i.e., it is electrically symmetrical) will appear in the Raman spectrum but will not appear in the infrared spectrum. Typical applications for Raman spectroscopy are in structure determination, multicomponent qualitative analysis, and quantitative analysis.

The Raman scattering transition moment is:

$$\mathrm{R} = < \mathrm{X}_i \mid \mathrm{a} \mid \mathrm{X}_j >$$

where X_i and X_j are the initial and final states, respectively, and a is the polarizability of the molecule:

$$a = a_o + (r - r_e)(\mathrm{da/dr}) + \cdots \text{ higher terms}$$

where r is the distance between atoms and a_o is the polarizability at the equilibrium bond length, r_e. Polarizability can be defined as the ease of which an electron cloud can be distorted by an external electric field. Since a_o is a constant and $< X_i \mid X_j > = 0$, R simplifies to:

$$R = < X_i \mid (r - r_e)(\mathrm{da/dr}) \mid X_j >$$

The result is that there must be a change in polarizability during the vibration for that vibration to inelastically scatter radiation.

The polarizability depends on how tightly the electrons are bound to the nuclei. In the symmetric stretch the strength of electron binding is different between the minimum and maximum internuclear distances. Therefore the polarizability changes during the vibration and this vibrational mode scatters Raman light (the vibration is Raman active). In the asymmetric stretch the electrons are more easily polarized in the bond that expands but are less easily polarized in the bond that compresses. There is no overall change in polarizability and the asymmetric stretch is Raman inactive.

Raman line intensities are proportional to:

$$\nu \cdot \sigma(\nu) \cdot I \cdot \exp(-E_i/kT) \cdot C$$

where ν is the frequency of the incident radiation, $\sigma(\nu)$ is the Raman cross section (typically 10^{-29} cm^2), I is the radiation intensity, $\exp(-E_i/kT)$ is the Boltzmann factor for state i, and C is the analyte concentration.

TABLE 3.14 Raman Frequencies of Single Bonds to Hydrogen and Carbon

Abbreviations Used in the Table

m, moderately strong	*vw, very weak*
m–s, moderate to strong	*w, weak*
m–vs, moderate to very strong	*w–m, weak to moderately strong*
s, strong	*w–vs, weak to very strong*
vs, very strong	

Group	Band, cm^{-1}	Remarks
	Saturated C—H and C—C	
$—CH_3$	2969–2967 (s)	
	2884–2883 (s)	
	ca 1205 (s)	In aryl compounds
	1150–1135	In unbranched alkyls
	1060–1056	In unbranched alkyls
	975–835 (s)	Terminal rocking of methyl group
	280–220	$CH_2—CH_3$ torsion
$—CH_2—$	2949–2912 (s)	
	2861–2849 (s)	
	1473–1443 (m–vs)	Intensity proportional to number of CH_2 groups
	1305–1295 (s)	
	1140–1070 (m)	Often two bands; see above
	888–837 (w)	
	425–150	
	500–490	Substituent on aromatic ring
$—CH(CH_3)_2$	1350–1330 (m)	
	835–750 (s)	If attached to C=C bond, 870–800 cm^{-1}. If attached to aryl ring, 740 cm^{-1}
$—C(CH_3)_3$	1265–1240 (m)	Not seen in *tert*-butyl bromide
	1220–1200 (m)	Not seen in *tert*-butyl bromide
	760–685 (vs)	If attached to C=C or aromatic ring, 760–720 cm^{-1}
Internal tertiary carbon atom	855–805 (w)	
	455–410	
Internal quaternary carbon atom	710–680 (vs)	
	490–470	
Two adjacent tertiary carbon atoms	730–920	Often a band at 530–524 cm^{-1} indicates presence of adjacent tertiary and quaternary carbon atoms.
	770–725	

TABLE 3.14 Raman Frequencies of Single Bonds to Hydrogen and Carbon (*Continued*)

Group	Band, cm^{-1}	Remarks
	Saturated C—H and C—C (*Continued*)	
Dialkyl substitution at α-carbon atom	800–700 (m–s) 680–650 (vs) 605–550	
Cyclopropane	3101–3090 3038–3019 1210–1180 (s)	Shifts to 1200 cm^{-1} for monoalkyl or 1,2-dialkyl substitution and to 1320 cm^{-1} for *gem*-1,1-dialkyl substitution
Cyclobutane	1001–960 (vs)	Shifts to 933 cm^{-1} for monoalkyl, to 887 cm^{-1} for *cis*-1,3-dialkyl, and to 891 cm^{-1} plus 855 cm^{-1} (doublet) for *trans*-1,3,-dialkyl substitution
Cyclopentane	900–800 (s)	
Cyclohexane	825–815 (vs) 810–795 (vs)	Boat configuration Chair configuration
Cycloheptane	ca 733	
Cyclooctane	ca 703	
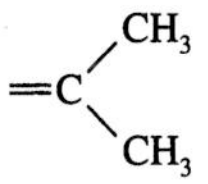	1392–1377 450–400 (vw) 270–250 (m)	
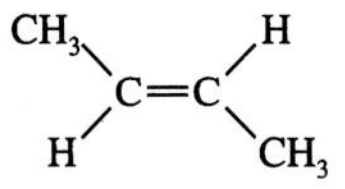	1380–1379 492–455 (vw) 220–200 (m)	
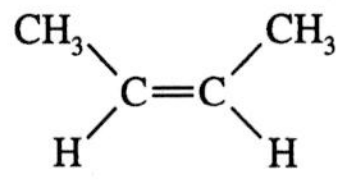	1372–1368 970–952 (m) 592–545 (vw) 420–400 (m) 310–290 (m)	
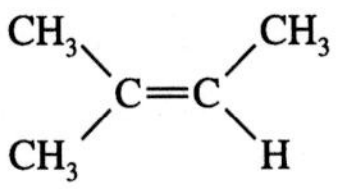	1385–1375 522–488 (w)	

(*Continued*)

TABLE 3.14 Raman Frequencies of Single Bonds to Hydrogen and Carbon (*Continued*)

Group	Band, cm^{-1}	Remarks
	Saturated C—H and C—C (*Continued*)	
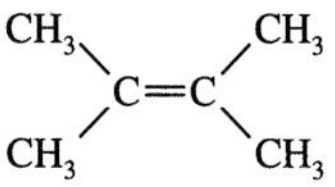	1392–1386 690–678 (m–s) 510–485 (m) 424–388 (w)	
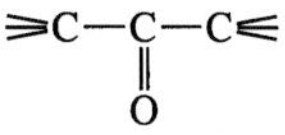	1170–1100 (w–m) 600–580 (m–s)	
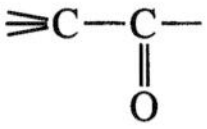	1120–1090 (m–vs) 600–510 (w–m)	Tertiary or quaternary carbon adjacent to carbonyl group lowers the frequency 300 cm^{-1}
$—CH_2—CO—$	1420–1410 (s)	
—CHO	2850–2810 (m) 2720–2695 (vs)	Often appears as a shoulder
	Unsaturated C—H	
—C≡C—H	3340–3270 (w–m)	Alkyl substituents at higher frequencies; unsaturated or aryl substituents at lower frequencies
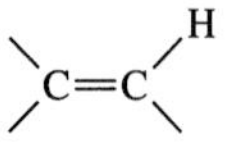	3040–2995 (m)	
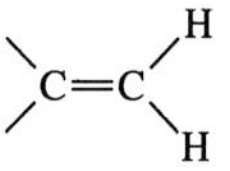	3095–3050 (m) 2990–2983 (s)	Asymmetric $=CH_2$ stretch Symmetric $=CH_2$ stretch
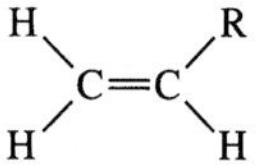	1419–1415 (m) 1309–1288 (m)	Plus =CH and =CH stretching bands
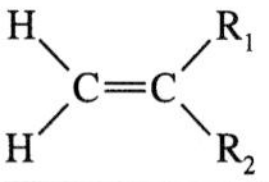	1413–1399 (m) 909–885 (m) 711–684 (w)	Plus $=CH_2$ stretching bands
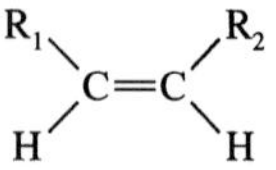	1270–1251 (m)	Plus =CH stretching band

TABLE 3.14 Raman Frequencies of Single Bonds to Hydrogen and Carbon (*Continued*)

Group	Band, cm^{-1}	Remarks
	Saturated C—H (*Continued*)	
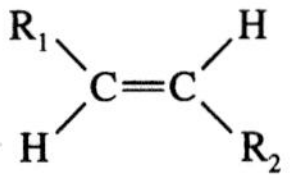	1314–1290 (m)	Plus =CH stretching band
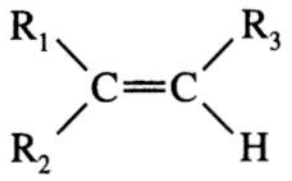	1360–1322 (w) 830–800 (vw)	Plus =CH stretching band
	Hydroxy O—H	
Free —OH Intermolecularly bonded Aromatic —OH	3650–3250 (w) 3400–3300 (w) ca 3160 (s)	
—OH	1460–1320 (w) 1276–1205 (w–m) 1260 (w–m)	Common to all OH substituents Primary Secondary
C—C—OH primary	1070–1050 (m–s) 1030–960 (m–s) 480–430 (w–m)	CCO stretching CCO deformation
C—C—OH Secondary Tertiary	 1135–1120 (m–s) 825–815 (vs) 500–490 (w–m) 1210–1200 (m–s) 755–730 (vs) 360–350 (w–m)	
—CO—O—H	1305–1270	CO stretching
	N—H and C—N bonds	
Amine >N—H Associated Nonbonded Salts —NH_2	 3400–3250 (s) 3550–3250 (s) 2986–2974 1650–1590 (w–vs)	 Primary amines show two bands Often obscured by intense CH stretching bands Bending

(*Continued*)

TABLE 3.14 Raman Frequencies of Single Bonds to Hydrogen and Carbon (*Continued*)

Group	Band, cm^{-1}	Remarks
	N—H and C—N Bonds (*Continued*)	
Amides		
Primary	3540–3500 (w) 3400–3380 (w) 1310–1250 (s)	Both bands lowered ca 150 cm^{-1} in solid state and H bonding Interaction of NH bending and CN stretching; lowered 50 cm^{-1} in nonbonded state
	1150–1095 (m)	Rocking of NH_2
Secondary	3491–3404 (m–s)	Two bands; lowered in frequency on H bonding and in solid state
	1190–1130 (m) 931–865 (m–s) 430–395 (w–m)	
—CO—N	607–555 (m)	O═CN bending
C—N(—C)—C	1070–1045 (m)	Stretching
$\geqslant$C—N$\leqslant$		
Primary carbon	1090–1060 (m)	CN stretching
Secondary α carbon	1140–1035 (m)	Two bands but often obscured. Strong band at 800 cm^{-1}
Tertiary α carbon	1240–1020 (m)	Two bands; Strong band also at 745 cm^{-1}

TABLE 3.15 Raman Frequencies of Triple Bonds

Abbreviations Used in the Table

m, moderately strong — *s–vs, strong to very strong*
m–s, moderate to strong — *vs, very strong*
s, strong

Group	Band, cm^{-1}	Remarks
R—C≡CH	2160–2100 (vs) 650–600 (m) 356–335 (s)	Monoalkyl substituted; C≡C stretch C≡CH deformation C≡C—C bending of monoalkyls
R_1—C≡C—R_2	2300–2190 (vs)	C≡C stretching of disubstituted alkyls; sometimes two bands
—C≡C—C≡C—	2264–2251 (vs)	
—C≡N	2260–2240 (vs)	Unsaturated nonaryl substituents lower the frequency and enhance the intensity
	2234–2200 (vs)	Lowered ca 30 cm^{-1} with aryl and conjugated aliphatics
	840–800 (s–vs)	CCCN symmetrical stretching
	385–350 (m–s)	
	200–160 (vs)	Aliphatic nitriles

TABLE 3.15 Raman Frequencies of Triple Bonds (*Continued*)

Group	Band, cm^{-1}	Remarks
H—C≡N	2094 (vs)	
Azides $—\overset{-}{N}—\overset{+}{N}≡N$	2170–2080 (s) 1258–1206 (s)	Asymmetric NNN stretching Symmetric NNN stretching; HN_3 at 1300 cm^{-1}
Diazonium salts $R—\overset{+}{N}≡N$	2300–2240 (s)	
Isonitriles $—\overset{+}{N}≡\overset{-}{C}$	2146–2134 2124–2109	Stretching of aliphatics Stretching of aromatics
Thiocyanates —S—C≡N	2260–2240 (vs) 650–600 (s)	Stretching of C≡N Stretching of SC

TABLE 3.16 Raman Frequencies of Cumulated Double Bonds

Abbreviations Used in the Table

s, strong *vw, very weak*
vs, very strong *w, weak*

Group	Band, cm^{-1}	Remarks
Allenes C=C=C	2000–1960 (s) 1080–1060 (vs) 356	Pseudo-asymmetric stretching Symmetric stretching C=C=C bending
Carbodiimides (cyanamides) —N=C=N—	2140–2125 (s) 2150–2100 (vs) 1460 1150–1140 (vs)	Asymmetric stretching of aliphatics Asymmetric stretching of aromatics; two bands Symmetrical stretching of aliphatics Symmetric stretching of aryls
Cumulenes (trienes) C=C=C=C	2080–2030 (vs) 878	
Isocyanates —N=C=O	2300–2250 (vw) 1450–1400 (s)	Asymmetric stretching Symmetric stretching
Isothiocyanates —N=C=S	2220–2100 690–650	Two bands Alkyl derivatives

(*Continued*)

TABLE 3.16 Raman Frequencies of Cumulated Double Bonds (*Continued*)

Group	Band, cm^{-1}	Remarks
Ketenes		
C=C=O	2060–2040 (vs)	Pseudo-asymmetric stretching
	1130 (s)	Pseudo-symmetric stretching
	1374 (s)	Alkyl derivatives
	1120 (s)	Aryl derivatives
Sulfinylamines		
R—N=S=O	1306–1214 (w)	Asymmetric stretching
	1155–989 (s)	Symmetric stretching

TABLE 3.17 Raman Frequencies of Carbonyl Bands

Abbreviations Used in the Table

m, moderately strong *s–vs, strong to very strong*
m–s, moderate to strong *vs, very strong*
s, strong *w, weak*

Group	Band, cm^{-1}	Remarks
Acid anhydrides		
—CO—O—CO—		
Saturated	1850–1780 (m)	
	1771–1770 (m)	
Conjugated, noncyclic	1775	
	1720	
Acid fluorides —CO—F		
Alkyl	1840–1835	
Aryl	1812–1800	
Acid chlorides —CO—Cl		
Alkyl	1810–1770 (s)	
Aryl	1774	
	1731	
Acid bromides —CO—Br		
Alkyl	1812–1788	
Aryl	1775–1754	
Acid iodides —CO—I		
Alkyl	ca 1806	
Aryl	ca 1752	
Lactones	1850–1730 (s)	

TABLE 3.17 Raman Frequencies of Carbonyl Bands (*Continued*)

Group	Band, cm^{-1}	Remarks
Esters		
Saturated	1741–1725	Alkyl branching on carbon adjacent to C═O lowers frequency by 5–15 cm^{-1}
Aryl and α,β-unsaturated	1727–1714	
Diesters		
Oxalates	1763–1761	
Phthalates	1738–1728	
C≡C—CO—O—	1716–1708	
Carbamates	1694–1688	
Aldehydes	1740–1720 (s–vs)	
Ketones		
Saturated	1725–1700 (vs)	
Aryl	1700–1650 (m)	
Alicyclic		
$n = 4$	1782 (m)	
$n = 5$	1744 (m)	
$n \geq 6$	1725–1699 (m)	
Carboxylic acids		
Mono-	1686–1625 (s)	These α-substituents increase the frequency: F, Cl, Br, OH
Poly-	1782–1645	Solid state; often two bands
	1750–1710	In solution; very broad band
Amino acids	1743–1729	
Carboxylate ions	1690–1550 (w)	
	1440–1340 (vs)	
Amino acid anion	1743–1729	Often masked by water deformation band near 1630 cm^{-1}
	1600–1570 (w)	
Amides (see also Table 7.30)		
Primary		
Associated	1686–1576 (m–s)	
	1650–1620 (m)	
Nonbonded	1715–1675 (m)	
	1620–1585 (m)	
Secondary		
Associated	1680–1630 (w)	Both *cis* and *trans* forms
	1570–1510 (w)	*Trans* form
	1490–1440	*Cis* form
Nonbonded	1700–1650	Both *cis* and *trans* forms
	1550–1500	*Trans* form (no *cis* band)
Tertiary	1670–1630 (m)	
Lactams	1750–1700 (m)	

TABLE 3.18 Raman Frequencies of Other Double Bonds

Abbreviations Used in the Table

m, moderately strong	*vs, very strong*
m–s, moderate to strong	*w, weak*
s, strong	*s–vs, strong to very strong*
w–m, weak to moderately strong	

Group	Band, cm^{-1}	Remarks
	Alkenes >C=C<	
>C=C<	1680–1576 (m–s)	General range
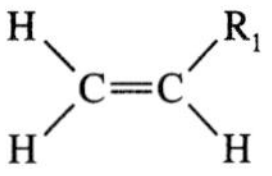	1648–1638 (vs)	C=C stretching
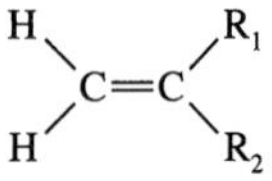	ca 1650 (vs) 270–252 (w)	C=C stretching C=C—C skeletal deformation
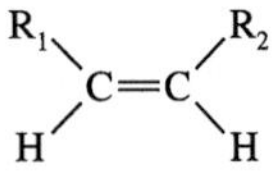	ca 1660 (vs) 970–952 (w)	C=C stretching Asymmetric CC stretching
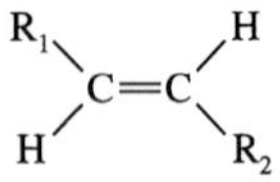	1676–1665 (s)	C—C stretching
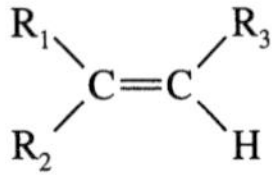	1678–1664 (vs) 522–488 (w)	C=C stretching C=C—C skeletal deformation
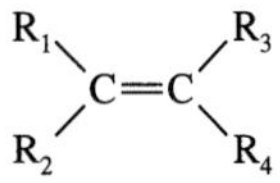	1680–1665 (s) 690–678 (m–s) 510–485 (m) 424–388 (w)	C=C stretching Symmetrical CC stretching Skeletal deformation Skeletal deformation

Haloalkene	X = fluorine	X = chlorine	X = bromine	X-iodine
		>C=C< stretch of haloalkanes		
H_2C=CHX	1654	1603–1601	1596–1593	1581
HXC=CHX				
cis	1712	1590–1587	1587–1583	1543
trans	1694	1578–1576	1582–1581	1537
$H_2C=CX_2$	1728	1616–1611	1593	
X_2C=CHX	1792	1589–1582	1552	
$X_2C=CX_2$	1872	1577–1571	1547	1465 (solid)

TABLE 3.18 Raman Frequencies of Other Double Bonds (*Continued*)

Group	Band, cm^{-1}	Remarks
	>C=N— bonds	
Aldimines (azomethines) 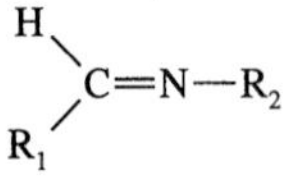	1673–1639 1405–1400 (s)	Dialkyl substituents at higher frequency; diaryl substituents at lower end of range
Aldoximines and ketoximes >C=N—OH	1680–1617 (vs) 1335–1330 (w)	
Azines >C=N—N=C<	1625–1608 (s)	
Hydrazones 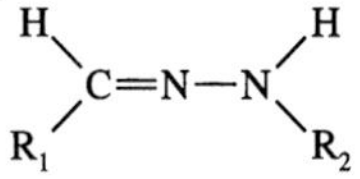	1660–1610 (s–vs)	
Imido ethers 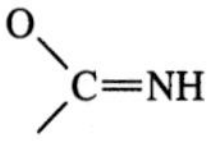	1658–1648	NH stretching at 3360–3327 cm^{-1}
Semicarbazones and thiosemicarbazones 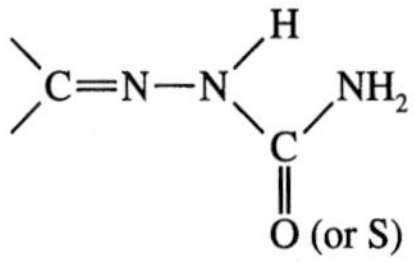	1665–1642 (vs) 1620–1610 (vs)	Aliphatic. Thiosemicarbazones fall in lower end of range Aromatic derivatives
	Azo compounds —N=N—	
—N=N—	1580–1570 (vs) 1442–1380 (vs) 1060–1030 (vs)	Nonconjugated Conjugated to aromatic ring CN stretching in aryl compounds
	Nitro compounds N=O	
Alkyl nitrites	1660–1620 (s)	N=O stretching
Alkyl nitrates	1635–1622 (w–m) 1285–1260 (vs) 610–562 (m)	Asymmetric NO_2 stretching Symmetric NO_2 stretching NO_2 deformation

(*Continued*)

TABLE 3.18 Raman Frequencies of Other Double Bonds (*Continued*)

Group	Band, cm^{-1}	Remarks
Nitroalkanes	1560–1548 (m–s)	Sensitive to substitutes attached
Primary	1395–1370 (s)	to CNO_2 group
	915–898 (m–s)	
	894–873 (m–s)	
	618–609 (w)	
	640–615 (w)	Shoulder
	494–472 (w–m)	Broad; useful to distinguish from secondary nitroalkanes
Secondary	1553–1547 (m)	
	1375–1360 (s)	
	908–868 (m)	
	863–847 (s)	
	625–613 (m)	
	560–516 (s)	Sharp band
Tertiary	1543–1533 (m)	
	1355–1345 (s)	
Nitrogen oxides		
$\gg\overset{+}{N} \rightarrow \overset{-}{O}$	1612–1602 (s)	
	1252 (m)	
	1049–1017 (s)	
	835 (s)	
	541 (w)	
	469 (w)	

TABLE 3.19 Raman Frequencies of Aromatic Compounds

Abbreviations Used in the Table

m, moderately strong
m–s, moderate to strong
m–vs, moderate to very strong
s, strong
s–vs, strong to very strong
var, of variable strength
vs, very strong
w, weak
w–m, weak to moderately strong

Group	Band, cm^{-1}	Remarks
	Common features	
Aromatic compounds	3070–3020 (s)	CH stretching
	1630–1570 (m–s)	C—C stretching
	Substitution patterns of the benzene ring	
Monosubstituted	1180–1170 (w–m)	
	1035–1015 (s)	
	1010–990 (vs)	Characteristic feature; found also with 1,3- and 1,3,5-substitutions
	630–605 (w)	

TABLE 3.19 Raman Frequencies of Aromatic Compounds (*Continued*)

Group	Band, cm^{-1}	Remarks
	Substitution patterns of the benzene ring (*Continued*)	
1,2-Disubstituted	1230–1215 (m)	
	1060–1020 (s)	Characteristic feature
	740–715 (m)	Lowered 60 cm^{-1} for halogen substituents
1,3-Disubstituted	1010–990 (vs)	Characteristic feature
	750–640 (s)	
1,4-Disubstituted	1230–1200 (s–vs)	
	1180–1150 (m)	
	830–750 (vs)	Lower frequency with Cl substituents
	650–630 (m–w)	
Isolated hydrogen	1379 (s–vs)	
	1290–1200 (s)	
	745–670 (m–vs)	
	580–480 (s)	Characteristic feature
1,2,3-Trisubstituted	1100–1050 (m)	
	670–500 (vs)	The lighter the mass of the substituent, the higher the frequency
	490–430 (w)	
1,2,4-Trisubstituted	750–650 (vs)	Lighter mass at higher frequencies
	580–540 (var)	
	500–450 (var)	
1,3,5-Trisubstituted	1010–990 (vs)	
Completely substituted	1296 (s)	
	550 (vs)	
	450 (m)	
	361 (m)	
	Other aromatic compounds	
Naphthalenes	1390–1370	Ring breathing
	1026–1012	α or β substituents
	767–762	β substituents
	535–512	α substituents
	519–512	β substituents
Disubstituted naphalenes	773–737 (s)	1,2-; 1,3-; 2,3-; 2,6-; 2,7-
	726–705 (s)	1,3-; 1,4-(two bands); 1,6-; 1,7-(two bands)
	690–634 (s)	1,2-; 1,4-(two bands); 1,5-; 1,8-(two bands)
	608	1,3-
	575–569	1,2-; 1,3-; 1,6-
	544–537	1,2-; 1,7-; 1,8-
Anthracenes	1415–1385	Ring breathing

TABLE 3.20 Raman Frequencies of Sulfur Compounds

Abbreviations Used in the Table

m, moderately strong	*s–vs, strong to very strong*
m–s, moderate to strong	*vs, very strong*
s, strong	*w–m, weak to moderately srong*

Group	Band, cm^{-1}	Remarks
—S—H	2590–2560 (s)	SH stretching for both aliphatic and aromatic
>C=S	1065–1050 (m) 735–690 (vs)	 Solid state
>S=O In $(RO_2)_2SO$ In $(R_2N)_2SO$ In R_2SO SOF_2 $SOCl_2$ $SOBr_2$	 1209–1198 1108 1070–1010 (w–m) 1308 1233 1121	 One or two bands Broad
—SO_2—	1330–1260 (m–s) 1155–1110 (s) 610–540 (m) 512–485 (m)	Asymmetric SO_2 stretching Symmetric SO_2 stretching Scissoring mode of aryls Scissoring mode of alkyls
—SO_2—N<	ca 1322 (m) 1163–1138 (s) 524–510 (s)	Asymmetric SO_2 stretching Symmetric SO_2 stretching Scissoring mode
—SO_2—O	1363–1338 (w–m) 1192–1165 (vs) 589–517 (w–m)	SO_2 stretching. Aryl substituents occur at higher range Scissoring (two bands). Aryl substituents occur at higher range of frequencies
—SO_2—S—	1334–1305 (m–s) 1128–1126 (s) 559–553 (m–s)	
X—SO_2—X	1412–1361 (w–m) (F) (Cl) 1263–1168 (s) (F) (Cl) 596–531 (s)	
—O—SO_2—O—	1388–1372 (s) 1196–1188 (vs)	
—O—C(=S)—S—	670–620 (vs) 480–450 (vs)	C=S stretching CS stretching
⩾C—SH	920 (m) 850–820 (m)	C—SH deformation of aryls

TABLE 3.20 Raman Frequencies of Sulfur Compounds (*Continued*)

Group	Band, cm^{-1}	Remarks
⩾C—S—	752 (vs), 731 (vs)	With vinyl group attached
	742–722 (m–s)	With CH_3 attached
	698 (w), 678 (s)	With allyl group attached
	693–639 (s)	Ethyl or longer alkyl chain
	651–610 (s–vs)	Isopropyl group attached
	589–585 (vs)	*tert*-Butyl group attached
$(CH_2)_n$ S (ring)		
$n = 2$	1112	
$n = 4$	688	
$n = 5$	659	
⩾C—(S—S)$_n$—C⩽	715–620 (vs)	Two bands; CS stretching
	525–510 (vs)	Two bands; SS stretching
Didi-*n*-alkyl disulfides	576 (s)	CS stretching
Di-*tert*-butyl disulfide	543 (m)	SS stretching
Trisulfides	510–480 (s)	SS stretching

TABLE 3.21 Raman Frequencies of Ethers

Abbreviations Used in the Table

m, moderately strong *var, of variable strength*
s, strong *vs, very strong*

Group	Band, cm^{-1}	Remarks
⩾C—O—C⩽		
Aliphatic	1200–1070 (m)	Asymmetrical COC stretching. Symmetrical substitution gives higher frequencies
	930–830 (s)	Symmetrical COC stretching
	800–700 (s)	Braching at α carbon gives higher frequencies.
	550–400	
Aromatic	1310–1210 (m)	
	1050–1010 (m)	
⩾C—O—C(—)—O—C⩽	1145–1129 (m)	
	900–800 (vs)	
	537–370 (s)	
	396–295	
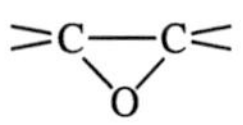	1280–1240 (s)	Ring breathing
—O—O—	800–770 (var)	
$(CH_2)_n$ O (ring)		
$n = 3$	1040–1010 (s)	
$n = 4$	920–900 (s)	
$n = 5$	820–800 (s)	

TABLE 3.22 Raman Frequencies of Halogen Compounds

Abbreviations Used in the Table

m–s, moderate strong *var, of variable strength*
s, strong *vs, very strong*

Group	Band, cm^{-1}	Remarks
C—F	1400–870	Correlations of limited applicability because of vibrational coupling with stretching
C—Cl	350–290 (s)	CCCl bending; general
Primary	660–650 (vs)	
Secondary	760–605 (s)	May be one to four bands
Tertiary	620–540 (var)	May be one to three bands
=C—Cl	844–564 438–396 381–170	
$=CCl_2$	601–441 300–235	
C—Br	690–490 (s) 305–258 (m–s)	Often several bands; primary at higher range of frequencies. Tertiary has very strong band at ca 520 cm^{-1}
=C—Br	745–565 356–318 240–115	
$=CBr_2$	467–265 185–145	
C—I	663–595 309 154–85	
=C—I	ca 180	Solid state
$=CI_2$	ca 265 ca 105	Solid state Solid state

TABLE 3.23 Raman Frequencies of Miscellaneous Compounds

Abbreviations Used in the Table

m–s, moderately strong *vs, very strong*
s, strong *vvs, very very strong*

Group	Band, cm^{-1}	Remarks
C—As	570–550 (vs)	CAs stretching
	240–220 (vs)	CAsC deformation
C—Pb	480–420 (s)	CPb stretching
C—Hg	570–510 (vvs)	CHg stretching
C—Si	1300–1200 (s)	CSi stretching
C—Sn	600–450 (s)	CSn stretching
P—H	2350–2240 (m)	PH stretching
	Heterocyclic rings	
Trimethylene oxide	1029	
Trimethylene imine	1026	
Tetrahydrofuran	914	
Pyrrolidine	899	
1,3-Dioxolane	939	
1,4-Dioxane	834	
Piperidine	815	
Tetrahydropyran	818	
Morpholine	832	
Piperazine	836	
Furan	1515–1460	2-Substituted
	1140	
Pyrazole	1040–990	
Pyrrole	1420–1360 (vs)	
	1144	
Thiophene	1410 (s)	
	1365 (s)	
	1085 (vs)	
	1035 (s)	
	832 (vs)	
	610 (s)	
Pyridine	1030 (vs)	
	990 (vs)	

TABLE 3.24 Principal Argon-Ion Laser Plasma Lines

Wavelength, nm	Wavenumber, cm^{-1}	Relative intensity	Shift relative to 488.0 nm, cm^{-1}	Shift relative to 514.5 nm, cm^{-1}
487.9860	20 486.67	5000	0	
488.9033	20 448.23	200	38.4	
490.4753	20 382.70	130	104.0	
493.3206	20 265.13	970	221.5	
496.5073	20 135.07	960	351.6	
497.2157	20 106.39	330	380.3	
500.9334	19 957.16	1500	529.5	

(*Continued*)

TABLE 3.24 Principal Argon-Ion Laser Plasma Lines (*Continued*)

Wavelength, nm	Wavenumber, cm^{-1}	Relative intensity	Shift relative to 488.0 nm, cm^{-1}	Shift relative to 514.5 nm, cm^{-1}
501.7160	19 926.03	620	560.6	
506.2036	19 749.39	1400	737.3	
514.1790	19 443.06	360	1043.6	
514.5319	19 429.73	1000	1056.9	0
516.5774	19 352.79	38	1133.9	76.9
517.6233	19 313.69	41	1173.0	116.0
521.6816	19 163.44	20	1323.2	266.3
528.6895	18 909.43	150	1577.2	520.3
539.7522	18 521.87	18	1964.8	907.9
545.4307	18 329.04	19	2157.6	1100.7
555.8703	17 984.81	30	2501.9	1444.9
560.6734	17 830.75	48	2655.9	1599.0
565.0705	17 692.00	29	2794.7	1737.7
565.4450	17 680.28	27	2806.4	1749.4
569.1650	17 564.73	27	2921.9	1865.0
577.2326	17 319.24	69	3167.4	2110.5
581.2746	17 198.80	49	3287.9	2230.9
598.5920	16 701.24	23	3785.4	2728.5
610.3546	16 379.38	91	4107.3	3050.4
611.4929	16 348.90	1750	4137.8	3080.8
612.3368	16 326.36	100	4160.3	3103.4
613.8660	16 285.69	97	4201.0	3144.0
617.2290	16 196.96	1400	4289.7	3232.8
624.3125	16 013.19	590	4473.5	3416.5
639.9215	15 622.60	160	4864.1	3807.1
641.6308	15 580.98	50	4905.7	3848.8

3.3 ULTRAVIOLET SPECTROSCOPY

Ultraviolet spectroscopy involves the excitation of an electron in its ground state level to a higher energy level. This is accomplished by irradiating a sample with ultraviolet light (electromagnetic radiation with wavelengths in the range of 200 nanometers (nm) to 400 nm). The wavelength of maximum absorption (λ_{max}) can be calculated by using Woodward's Rules.

λ_{max} has a specific degree of absorbance associated with it. The absorbance at a particular wavelength is dependent upon the intensity or molar absorbtivity, ε, of the incident light. The molar absorbtivity is related to the absorbance:

$$\varepsilon = \log (I_0/I)/c.l$$

where I_0 is the initial light intensity, I is the final light intensity, c is the concentration of sample in moles per liter, l is the path length of sample tube in centimeters.

Beer's Law relates the absorbance A to I_0 and I ($A = \log [I_0/I]$). Hence the equation for molar absorbtivity is:

$$\varepsilon = A/c.l$$

where A is the absorbance at λ_{max}.

Molecules with two or more isolated chromophores (absorbing groups) absorb light of nearly the same wavelength as does a molecule containing only a single chromophore of a particular type. The intensity of the absorption is proportional to the number of that type of chromophore present in the molecule.

The solvent chosen must dissolve the sample, yet be relatively transparent in the spectral region of interest. In order to avoid poor resolution and difficulties in spectrum interpretation, a solvent should not be employed for measurements that are near the wavelength of or are shorter than the wavelength of its ultraviolet cutoff, that is, the wavelength at which absorbance for the solvent alone approaches one absorbance unit.

Appreciable interaction between chromophores does not occur unless they are linked directly to each other, or forced into close proximity as a result of molecular stereochemical configuration. Interposition of a single methylene group, or *meta* orientation about an aromatic ring, is sufficient to insulate chromophores almost completely from each other. Certain combinations of functional groups afford chromophoric systems that give rise to characteristic absorption bands.

Sets of empirical rules, often referred to as Woodward's Rules or the Woodward-Fieser Rules, enable the absorption maxima of dienes and enones and dienones to be predicted. To the respective base values (absorption wavelength of parent compound) are added the increments for the structural features or substituent groups present. When necessary, a solvent correction is also applied.

Ring substitution on the benzene ring affords shifts to longer wavelengths and intensification of the spectrum. With electron-withdrawing substituents, practically no change in the maximum position is observed. The spectra of heteroaromatics are related to their isocyclic analogs, but only in the crudest way. As with benzene, the magnitude of substituent shifts can be estimated, but tautomeric possibilities may invalidate the empirical method.

When electronically complementary groups are situated *para* to each other in disubstituted benzenes, there is a more pronounced shift to a longer wavelength than would be expected from the additive effect due to the extension of the chromophore from the electron-donating group through the ring to the electron-withdrawing group. When the *para* groups are not complementary, or when the groups are situated *ortho* or *meta* to each other, disubstituted benzenes show a more or less additive effect of the two substituents on the wavelength maximum.

TABLE 3.25 Electronic Absorption Bands for Representative Chromophores

Chromophore	System	λ_{max}	ϵ_{max}
Acetylide	—C≡C—	175–180	6 000
Aldehyde	—CHO	210	strong
		280–300	11–18
Amine	$—NH_2$	195	2 800
Azido	>C=N—	190	5 000
Azo	—N=N—	285–400	3–25
Bromide	—Br	208	300
Carbonyl	>C=O	195	1 000
		270–285	18–30
Carboxyl	—COOH	200–210	50–70
Disulfide	—S—S—	194	5 500
		255	400
Ester	—COOR	205	50
Ether	—O—	185	1 000
Ethylene	—C=C—	190	8 000
Iodide	—I	260	400
Nitrate	$—ONO_2$	270 (shoulder)	12
Nitrile	—C≡N	160	——
Nitrite	—ONO	220–230	1 000–2 000
		300–400	10
Nitro	$—NO_2$	210	strong
Nitroso	—NO	302	100

(Continued)

TABLE 3.25 Electronic Absorption Bands for Representative Chromophores (*Continued*)

Chromophore	System	λ_{max}	ϵ_{max}
Oxime	—NOH	190	5 000
Sulfone	$—SO_2—$	180	——
Sulfoxide	>S=O	210	1 500
Thiocarbonyl	>C=S	205	strong
Thioether	—S—	194	4 600
		215	1 600
Thiol	—SH	195	1 400
	$—(C=C)_2—$ (acyclic)	210–230	21 000
	$—(C=C)_3—$	260	35 000
	$—(C=C)_4—$	300	52 000
	$—(C=C)_5—$	330	118 000
	$—(C=C)_2—$ (alicyclic)	230–260	3 000–8 000
	C=C—C≡C	219	6 500
	C=C—C=N	220	23 000
	C=C—C=O	210–250	10 000–20 000
		300–350	weak
	$C=C—NO_2$	229	9 500
Benzene		184	46 700
		204	6 900
		255	170
Diphenyl		246	20 000
Naphthalene		222	112 000
		275	5 600
		312	175
Anthracene		252	199 000
		375	7 900
Phenanthrene		251	66 000
		292	14 000
Naphthacene		272	180 000
		473	12 500
Pentacene		310	300 000
		585	12 000
Pyridine		174	80 000
		195	6 000
		257	1 700
Quinoline		227	37 000
		270	3 600
		314	2 750
Isoquinoline		218	80 000
		266	4 000
		317	3 500

TABLE 3.26 Ultraviolet Cutoffs of Spectrograde Solvents

Solvent	Wavelength, nm	Solvent	Wavelength, nm
Acetic acid	260	Hexadecane	200
Acetone	330	Hexane	210
Acetonitrile	190	Isobutyl alcohol	230
Benzene	280	Methanol	210
1-Butanol	210	2-Methoxyethanol	210
2-Butanol	260	Methylcyclohexane	210
Butyl acetate	254	Methylene chloride	235
Carbon disulfide	380	Methyl ethyl ketone	330
Carbon tetrachloride	265	Methyl isobutyl ketone	335
1-Chlorobutane	220	2-Methyl-1-propanol	230
Chloroform (stabilized with ethanol)	245	*N*-Methylpyrrolidone	285
Cyclohexane	210	Nitromethane	380
1,2-Dichloroethane	226	Pentane	210
Diethyl ether	218	Pentyl acetate	212
1,2-Dimethoxyethane	240	1-Propanol	210
N,N-Dimethylacetamide	268	2-Propanol	210
N,N-Dimethylformamide	270	Pyridine	330
Dimethylsulfoxide	265	Tetrachloroethylene (stabilized with thymol)	290
1,4-Dioxane	215	Tetrahydrofuran	220
Ethanol	210	Toluene	286
2-Ethoxyethanol	210	1,1,2-Trichloro-1,2,2-trifluoroethane	231
Ethyl acetate	255	2,2,4-Trimethylpentane	215
Ethylene chloride	228	*o*-Xylene	290
Glycerol	207	Water	191
Heptane	197		

TABLE 3.27 Absorption Wavelength of Dienes

Heteroannular and acyclic dienes usually display molar absorptivities in the 8000 to 20,000 range, whereas homoannular dienes are in the 5000 to 8000 range.

Poor correlations are obtained for cross-conjugated polyene systems such as

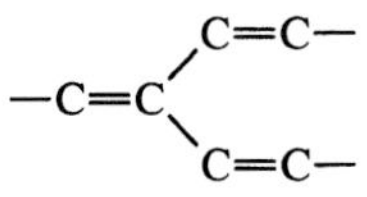

The correlations presented here are sometimes referred to as Woodward's rules or the Woodward-Fieser rules.

Base value for heteroannular or open chain diene, nm	214
Base value for homoannular diene, nm	253
Increment (in nm) for	
Double bond extending conjugation	30
Alkyl substituent or ring residue	5
Exocyclic double bond	5
Polar groupings:	
-*O*-acyl	0
-*O*-alkyl	6
-*S*-alkyl	30
-Cl, -Br	5
-*N*(alkyl)$_2$	60
Solvent correction (see Table 7.13)	———
Calculated wavelength =	total

TABLE 3.28 Absorption Wavelength of Enones and Dienones

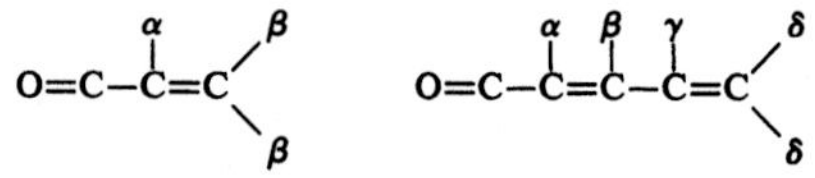

Base values, nm	
Acyclic α,β-unsaturated ketones	215
Acyclic α,β-unsaturated aldehyde	210
Six-membered cyclic α,β-unsaturated ketones	215
Five-membered cyclic α,β-unsaturated ketones	214
α,β-Unsaturated carboxylic acids and esters	195
Increments (in nm) for	
Double bond extending conjugation:	
Heteroannular	30
Homoannular	69
Alkyl group or ring residue:	
α	10
β	12
γ, δ	18
Polar groups:	
—OH	
α	35
β	30
γ	50
—O—CO—CH_3 and —O—CO—C_6H_5: α, β, γ, δ	6
—OCH_3	
α	35
β	30
γ	17
δ	31
—S—alkyl, β	85
—Cl	
α	15
β	12
—Br	
α	25
β	30
—N(alkyl)$_2$, β	95
Exocyclic double bond	5
Solvent correction (see Table 7.13)	
Calculated wavelength =	total

TABLE 3.29 Solvent Correction for Ultraviolet-Visible Spectroscopy

Solvent	Correction, nm
Chloroform	+1
Cyclohexane	
Diethyl ether	+11
1,4-Dioxane	+5
Ethanol	0
Hexane	+11
Methanol	0
Water	−8

TABLE 3.30 Primary Bands of Substituted Benzene and Heteroaromatics

In methanol.

Base value: 203.5 nm

Substituent	Wavelength shift, nm	Substituent	Wavelength shift, nm
—CH_3	3.0	—COOH	25.5
—CH=CH_2	44.5	—COO^-	20.5
—C≡CH	44	—CN	20.5
—C_6H_5	48	—NH_2	26.5
—F	0	—NH_3^+	−0.5
—Cl	6.0	—$N(CH_3)_2$	47.0
—Br	6.5	—NH—CO—CH_3	38.5
—I	3.5	—NO_2	57
—OH	7.0	—SH	32
—O^-	31.5	—SO—C_6H_5	28
—OCH_3	13.5	—SO_2CH_3	13
—OC_6H_5	51.5	—SO_2NH_2	14.0
—CHO	46.0	—CH=CH—C_6H_5	
—CO—CH_3	42.0	*cis*	79
—CO—C_6H_5	48	*trans*	92.0
		—CH=CH—COOH, *trans*	69.5

Heteroaromatic	Base value, nm	Heteroaromatic	Base value, nm
Furan	200	Pyridine	257
Pyrazine	257	Pyrimidine	ca 235
Pyrazole	214	Pyrrole	209
Pyridazine	ca 240	Thiophene	231

TABLE 3.31 Wavelength Calculation of the Principal Band of Substituted Benzene Derivatives

In ethanol.

Base value of parent chromophore, nm	
C_6H_5COOH or C_6H_5COO—alkyl	230
C_6H_5—CO—alkyl (or aryl)	246
C_6H_5CHO	250
Increment (in nm) for each substituent on phenyl ring	
—Alkyl or ring residue	
o-, *m*-	3
p-	10
—OH and —O— alkyl	
o-, *m*-	7
p-	25
—O^-	
o-	11
m-	20
p-	78*

(Continued)

TABLE 3.31 Wavelength Calculation of the Principal Band of Substituted Benzene Derivatives (*Continued*)

—Cl	
o-, *m*-	0
p-	10
—Br	
o-, *m*-	2
p-	15
—NH_2	
o-, *m*-	13
p-	58
—NHCO—CH_3	
o-, *m*-	20
p-	45
—$NHCH_3$	
p-	73
—$N(CH_3)_2$	
o-, *m*-	20
p-	85

*Value may be decreased markedly by steric hindrance to coplanarity.

3.4 FLUORESCENCE SPECTROSCOPY

Fluorescence spectroscopy is a measure of the optical emission from atoms that have been excited to higher energy levels by absorption of electromagnetic radiation. The main advantage of fluorescence detection compared to absorption measurements is the greater sensitivity achievable because the fluorescence signal has a very low background. The resonant excitation provides selective excitation of the analyte to avoid interferences. Fluorescence spectroscopy is useful to study the electronic structure of atoms and to make quantitative measurements. Analytical applications include flames and plasmas diagnostics, and enhanced sensitivity in atomic analysis. Because of the differences in the nature of the energy-level structure between atoms and molecules, discussion of laser-induced fluorescence from molecules is found in a separate document.

Analysis of solutions or solids requires that the analyte atoms be desolvated, vaporized, and atomized at a relatively low temperature in a heat pipe, flame, or graphite furnace. A hollow-cathode lamp or laser provides the resonant excitation to promote the atoms to higher energy levels. The atomic fluorescence is dispersed and detected by monochromators and photomultiplier tubes, similar to atomic-emission spectroscopy instrumentation.

TABLE 3.32 Fluorescene Spectroscopy of Some Organic Compounds

Compound	Solvent	pH	Excitation wavelength, nm	Emission wavelength nm
Acenaphthene	Pentane		291	341
Acridine	CF_3COOH		358	475
Adenine	Water	1	280	375
Adenosine	Water	1	285	395
Adenosine triphosphate	Water	1	285	395
Adrenalin			295	335
p-Aminobenzoic acid	Water	8	295	345
Aminopterin	Water	7	280, 370	460

TABLE 3.32 Fluorescene Spectroscopy of Some Organic Compounds (*Continued*)

Compound	Solvent	pH	Excitation wavelength, nm	Emission wavelength nm
1-Aminopyrene	CF_3COOH		330, 342	415
p-Aminosalicyclic acid	Water	11	300	405
Amobarbital	Water	14	265	410
Anilines	Water	7	280, 291	344, 361
Anthracene	Pentane		420	430
Anthranilic acid	Water	7	300	405
Azaindoles	Water	10	290, 299	310, 347
Benz[*c*]acridine	CF_3COOH		295, 380	480
Benz[*a*]anthracene	Pentane		284	382
1,2-Benzanthracene			280, 340	390, 410
Benzanthrone	CF_3COOH		370, 420	550
Benzo[*b*]chrysene	Pentane		283	398
11-*H*-Benzo[*a*]fluorene	Pentane		317	340
Benzoic acid	70% H_2SO_4		285	385
3,4-Benzopyrene	Benzene		365	390, 480
Benzo[*e*]pyrene	Pentane		329	389
Benzoquinoline	CF_3COOH		280	425
Benzoxanthane	Pentane		363	418
Bromolysergic acid diethyl amide	Water	1	315	460
Brucine	Water	7	305	500
Carbazole	*N,N*-Dimethyl formamide		291	359
Chlortetracycline			355	445
Chrysene	Pentane		250, 300, 310	260, 380
Cinchonine	Water	1	320	420
Coumarin	Ethanol		280	352
Dibenzo[*a,c*]anthracene	Pentane		280	381
Dibenzo[*b,k*]chrysene	Pentane		308	428
Dibenzo[*a,e*]pyrene	Pentane		370	401
3,4,8,9-Dibenzopyrene			370, 335, 390, 410	480, 510
5,12-Dihydronaphthacene	Pentane		282	340
1,4-Diphenylbutadiene	Pentane		328	370
Epinephrine	Water	7	295	335
Ethacridine	Water	2	370, 425	515
Fluoranthrene	Pentane		354	464
Fluorene	Pentane		300	321
Fluorescein	Water	7–11	490	515
Folic acid	Water	7	365	450
Gentisic acid	Water	7	315	440
Griseofulvin	Water	7	295, 335	450
Guanine	Water	1	285	365
Harmine	Water	1	300, 365	400
Hippuric acid	70% H_2SO_4		270	370
Homovanillic acid	Water	7	270	315
m-Hydroxybenzoic acid	Water	12	314	430
p-Hydroxycinnamic acid	Water	7	350	440
7-Hydroxycoumarin	Ethanol		325	441
5-Hydroxyindole	Water	1	290	355
5-Hydroxyindoleacetic acid	Water	7	300	355

(*Continued*)

TABLE 3.32 Fluorescene Spectroscopy of Some Organic Compounds (*Continued*)

Compound	Solvent	pH	Excitation wavelength, nm	Emission wavelength nm
3-Hydroxykynurenine	Water	11	365	460
p-Hydroxymandelic acid	Water	7	300	380
p-Hydroxyphenylacetic acid	Water	7	280	310
p-Hydroxyphenylpyruvic acid	Water	7	290	345
p-Hydroxyphenylserine	Water	1	290	320
5-Hydroxytryptophan	Water	7	295	340
Imipramine	Water	14	295	415
Indoleacetic acid	Water	8	285	360
Indoles	Water	7	269, 315	355
Indomethacin	Water	13	300	410
Kynurenic acid	Water	7	325	405
		11	325	440
Lysergic acid diethylamide	Water	1	325	445
Menadione	Ethanol		335	480
9-Methylanthracene	Pentane		382	410
3-Methylcholanthrene	Pentane		297	392
7-Methyldibenzopyrene	Pentane		460	467
2-Methylphenanthrene	Pentane		257	357
3-Methylphenanthrene	Pentane		292	368
1-Methylpyrene	Pentane		336	394
4-Methylpyrene	Pentane		338	386
Naphthacene			290, 310	480, 515
1-Naphthol	0.1 *M* NaOH 20% ethanol		365	480
2-Naphthol	0.1 *M* NaOH 20% ethanol		365	426
Oxytetracycline			390	520
Phenanthrene	Pentane		252	362
Phenylalanine	Water		215, 260	282
o-Phenylenepyrene	Pentane		360	506
Phenylephrine			270	305
Picene	Pentane		281	398
Procaine	Water	11	275	345
Pyrene	Pentane		330	382
Pyridoxal	Water	12	310	365
Quinacrine	Water	11	285	420
Quinidine	Water	1	350	450
Quinine	Water	1	250, 350	450
Reserpine	Water	1	300	375
Resorcinol	Water		265	315
Riboflavin	Water	7	270, 370, 445	520
Rutin	Water	1	430	520
Salicyclic acid	Water	11	310	435
Scoparone	Water	10	350, 365	430
Scopoletin	Water	10	365, 390	460
Serotonin	3 *M* HCl		295	550
Skatole	Water		290	370
Streptomycin	Water	13	366	445
p-Terphenyl	Pentane		284	338
Thiopental			315	530
Thymol	Water	7	265	300

TABLE 3.32 Fluorescene Spectroscopy of Some Organic Compounds (*Continued*)

Compound	Solvent	pH	Excitation wavelength, nm	Emission wavelength nm
Tocopherol	Hexane-ethanol		295	340
Tribenzo[*a,e,i*]pyrene	Pentane		384	448
Triphenylene	Pentane		288	357
Tryptamine	Water	7	290	360
Tryptophan	Water	11	285	365
Tyramine	Water	1	275	310
Tyrosine	Water	7	275	310
Uric acid	Water	1	325	370
Vitamin A	1-Butanol		340	490
Vitamin B_{12}	Water	7	275	305
Warfarin	Methanol		290, 342	385
Xanthine	Water	1	315	435
2,6-Xylenol			275	305
3,4-Xylenol			280	310
Yohimbine	Water	1	270	360
Zoxazolamine	Water	11	280	320

TABLE 3.33 Fluorescene Quantum Yield Values

Compound	Solvent	Q_F value vs. Q_F standard
	Q_F standard	
9-Aminoacridine	Water	0.99
Anthracene	Ethanol	0.30
POPOP*	Toluene	0.85
Quinine sulfate dihydrate	1*N* H_2SO_4	0.55
	Secondary standards	
Acridine orange hydrochloride	Ethanol	0.54 Quinine sulfate 0.58 Anthracene
1,8-ANS† (free acid)	Ethanol	0.38 Anthracene 0.39 POPOP
1,8-ANS (magnesium salt)	Ethanol	0.29 Anthracene 0.31 POPOP
Fluorescein	0.1*N* NaOH	0.91 Quinine sulfate 0.94 POPOP
Fluorescein, ethyl ester	0.1*N* NaOH	0.99 Quinine sulfate 0.99 POPOP
Rhodamine B	Ethanol	0.69 Quinine sulfate 0.70 Anthracene
2,6-TNS‡ (potassium salt)	Ethanol	0.48 Anthracene 0.51 POPOP

* POPOP, *p*-bis[2-(5-phenyloxazoyl)]benzene.
† ANS, anilino-8-naphthalene sulfonic acid.
‡ TNS, 2-*p*-toluidinylnaphthalene-6-sulfonate.

3.5 FLAME ATOMIC EMISSION, FLAME ATOMIC ABSORPTION, ELECTROTHERMAL (FURNACE) ATOMIC ABSORPTION, ARGON INDUCTION COUPLED PLASMA, AND PLASMA ATOMIC FLUORESCENCE

The tables of atomic emission and atomic absorption lines are presented in two parts. In Table 3.34 the data are arranged in alphabetic order by name of the element, whereas in Table 3.35 the sensitive lines of the elements are arranged in order of decreasing wavelengths.

The detection limits in the table correspond generally to the concentration of an element required to give a net signal equal to three times the standard deviation of the noise (background) in accordance with IUPAC recommendations. Detection limits can be confusing when steady-state techniques such as flame atomic emission or absorption, and plasma atomic emission or fluorescence, are compared with the electrothermal or furnace technique which uses the entire sample and detects an absolute amount of the analyte element. To compare the several methods on the basis of concentration, the furnace detection limits assume a 20-μL sample.

Data for the several flame methods assume an acetylene–nitrous oxide flame residing on a 5- or 10-cm slot burner. The sample is nebulized into a spray chamber placed immediately ahead of the burner. Detection limits are quite dependent on instrument and operating variables, particularly the detector, the fuel and oxidant gases, the slit width, and the method used for background correction and data smoothing.

3.5.1 Common Spectroscopic Relationships

Electromagnetic Radiation. Electromagnetic radiation travels in straight lines in a uniform medium, has a velocity of 299,792,500 $m \cdot s^{-1}$ in a vacuum, and possesses properties of both a wave motion and a particle (photon). *Wavelength* λ is the distance from crest to crest; *frequency* v is the number of waves passing a fixed point in a unit length of time. Wavelength and frequency are related by the relation

$$c = \lambda v$$

where c is the velocity of light (in a vacuum). In any material medium the speed of propagation is smaller than this and is given by the product nc, where n is the refractive index of the medium.

Radiation is absorbed or emitted only is discrete packets called photons and quanta:

$$E = hv$$

where E is the energy of the quantum and h is Planck's constant.

The relation between energy and mass is given by the *Einstein equation*:

$$\Delta E = \Delta mc^2$$

where ΔE is the energy release and Δm is the loss of mass. Strictly, the mass of a particle depends on its velocity, but here the masses are equated to their rest masses (at zero velocity).

The *Wien displacement law* states that the wavelength of maximum emission λ_m of a blackbody varies inversely with absolute temperature; the product $\lambda_m T$ remains constant. When λ_m is expressed in micrometers, the law becomes

$$\lambda_m T = 2898$$

In terms of σ_m, the wavenumber of maximum emission:

$$\sigma_m = 3.48T$$

Another useful version is $hv_m = 5kT$, where k is the Boltzmann constant.

Stefan's law states that the total energy J radiated by a blackbody per unit time and area (power per unit area) varies as the fourth power of the absolute temperature:

$$J = aT^{-4}$$

where a is a constant whose value is 5.67×10^{-8} $W \cdot m^{-2} \cdot K^{-4}$.

The relationship between the voltage of an X-ray tube (or other energy source), in volts, and the wavelength is given by the *Duane-Hunt equation*:

$$\lambda = \frac{hc}{eV} = \frac{12{,}398}{V}$$

where the wavelength is expressed in angstrom units.

Laws of Photometry. The time rate at which energy is transported in a beam of radiant energy is denoted by the symbol P_0 for the incident beam, and by P for the quantity remaining unabsorbed after passage through a sample or container. The ratio of radiant power transmitted by the sample to the radiant power incident on the sample is the *transmittance T*:

$$T = \frac{P}{P_0}$$

The logarithm (base 10) of the reciprocal of the transmittance is the *absorbance A*:

$$A = -\log T = \log\left(\frac{1}{T}\right)$$

When a beam of monochromatic light, previously rendered plane parallel, enters an absorbing medium at right angles to the plane-parallel surfaces of the medium, the rate of decrease in radiant power with the length of light path (cuvette interior) b, or with the concentration of absorbing material C (in grams per liter) will follow the exponential progression, often referred to as *Beer's law*:

$$T = 10^{-abC} \qquad \text{or} \qquad A = abC$$

where a is the absorptivity of the component of interest in the solution. When C is expressed in moles per liter,

$$T = 10^{-\epsilon bC} \qquad \text{or} \qquad A = \epsilon bC$$

where ε is the molar absorptivity.

The total fluorescence (or phosphorescence) intensity is proportional to the quanta of light absorbed, $P_0 - P$, and to the efficiency ϕ, which is the ratio of quanta absorbed to quanta emitted:

$$F = (P_0 - P)\phi = P_0\phi\,(1 - e^{-\epsilon bC})$$

When the terms εbC is not greater than 0.05 (or 0.01 in phosphorescence),

$$F = k\phi P_0 \epsilon bC$$

where the term k has been introduced to handle instrumental artifacts and the geometry factor because fluorescence (and phosphorescence) is emitted in all directions but is viewed only through a limited aperture.

The thickness of a transparent film or the path length of infrared absorption cells b, in centimeters, is given by

$$b = \frac{1}{2n_D}\left(\frac{n}{\bar{\nu}_1 - \bar{\nu}_2}\right)$$

where n is the number of fringes (peaks or troughs) between two wavenumbers $\bar{\nu}_1$ and $\bar{\nu}_2$, and n_D is the refractive index of the sample material (unity for the air path of an empty cuvette). If measurements are made in wavelength, as micrometers, the expression is

$$b = \frac{1}{2n_D}\left(\frac{n\lambda_1\lambda_2}{\lambda_2 - \lambda_1}\right)$$

Grating Equation. The light incident on each groove is diffracted or spread out over a range of angles, and in certain directions reinforcement or constructive interference occurs, as stated in the grating formula:

$$m\lambda = b(\sin i \pm \sin r)$$

where b is the distance between adjacent grooves, i is the angle of incidence, r is the angle of reflection (both angles relative to the grating normal), and m is the order number. A positive sign applies where incoming and emergent beams are on the same side of the grating normal.

The *blaze wavelength* is that wavelength for which the angle of reflectance from the groove face and the angle of reflection (usually the angle of incidence) from the grating are identical.

The *Bragg equation*

$$m\lambda = 2d \sin \theta$$

states the condition for reinforcement of reflection from a crystal lattice, where d is the distance between each set of atomic planes and θ is the angle of reflection.

Ionization of Metals in a Plasma. A loss in spectrochemical sensitivity results when a free metal atom is split into a positive ion and an electron:

$$\mathrm{M} = \mathrm{M}^+ + e^-$$

The degree of ionization α_i is defined as

$$\lambda = \frac{hc}{eV} = \frac{12{,}398}{V}$$

At equilibrium, when the ionization and recombination rates are balanced, the ionization constant K_i (in atm) is given by

$$K_i = \frac{[\mathrm{M}^+][e^-]}{[\mathrm{M}]} = \left(\frac{\alpha_i^2}{1 - \alpha_i^2}\right) P_{\Sigma \mathrm{M}}$$

where $P_{\Sigma \mathrm{M}}$ (in atm) is the total atom concentration of metal in all forms in the plasma.

The ionization constant can be calculated from the *Saha equation*:

$$\log K_i = -5040\frac{E_i}{T} + \frac{5}{2} \log T - 6.49 + \log \frac{g_{\mathrm{M}^+} g_{e^-}}{g_{\mathrm{M}}}$$

where E_i is the ionization potential of the metal in eV (Table 4.2), T is the absolute temperature of the plasma (in kelvins), and the g terms are the statistical weights of the ionized atom, the electron, and the neutral atom. For the alkali metals the final term is zero; for the alkaline earth metals, it is 0.6.

To suppress the ionization of a metal, another easily ionized metal (denoted a *deionizer* or *radiation buffer*) is added to the sample. To ensure that ionization is suppressed for the test element, the product $(K_i)_{\mathrm{M}}P_{\mathrm{M}}$ of the deionizer must exceed the similar product for the test element one hundredfold (for 1 percent residual ionization of the test element).

TABLE 3.34 Detection Limits in ng/mL

The detection limits in the table correspond generally to the concentration of analyte required to give a net signal equal to three times the standard deviation of the background in accordance with IUPAC recommendations.

Element	Wavelength, nm	Flame emission	Flame atomic absorption	Electrothermal atomic absorption	Argon ICP	Plasma atomic fluorescence
Aluminum	308.22		40		10	
	309.28		20	0.05	11	4
	394.40	3.6	45		36	
	396.15	7.5	30	0.01	20	5
Antimony	206.83				50	
	217.58		30		50	
	231.15	70			30	10
	259.81	200		0.08		0.1
Arsenic	189.04		160		35	
	193.76		120	1	50	
	197.20		240			
	228.81	455				
	234.90	250				10
Barium	455.36	3			0.9	
	493.41	4			1	
	553.55	1.5	9	0.04		2
Beryllium	234.86		1	0.05	0.4	
	313.04		2	0.003	1	
	313.11	100			1	0.2
Bismuth	223.06		18	0.35	30	
	227.66			2		
	306.77	60		0.5	30	2
Boron	182.59				8	
	249.77		700	15	3	60
(as BO_2)	518.00	50				
(as BO_2)	547.60	50				
Bromine	154.07				50	
Cadmium	214.44				1	
	226.50				0.6	
	228.80	6	1	0.008	228	
	326.11	3	0.5	0.014		0.001
Calcium	315.89				20	
	393.37				0.6	
	396.85				1.2	
	422.67	1.5	1	0.3		0.08
Carbon	193.09				44	
	247.86				1000	
Cerium	413.38				30	
	418.66				30	
	569.92	150				
Cesium	852.11	0.02	8	0.04		
	894.35	0.04	130			
Chlorine	134.72				50	
Chromium	267.72				3	
	283.58				20	
	284.98				30	
	357.87	6	2	0.05		0.4
	359.35	7				

(Continued)

TABLE 3.34 Detection Limits in ng/mL (*Continued*)

Element	Wavelength, nm	Flame emission	Flame atomic absorption	Electrothermal atomic absorption	Argon ICP	Plasma atomic fluorescence
Chromium	360.53	13				
(*cont.*)	425.44	3	6		66	
	427.48	4				
	428.97	5				
Cobalt	228.62				3	
	238.89				28	
	240.73	5	8	0.01	7	0.4
	345.35	30				
Copper	324.75	1.5	1	0.01	2	0.2
	327.40	3	2	0.02	4	
Dysprosium	353.17				3	
	340.78				6	
	404.60	30	50			300
	418.68		60			
	421.17		60			
Erbium	323.06				15	
	349.81				10	
	400.80	30	40	0.3		500
	408.77		40			
Europium	381.97				2	
	412.97				3	
	459.40	0.45	20	0.5		20
Gadolinium	335.05				10	
	368.41		4000			
	440.19	72	1000	8		800
Gallium	287.42		70			
	294.36		20		30	
	404.30	5	50			
	417.21	3	30	1	40	0.9
Germanium	209.43				50	
	219.87				100	
	265.12	400	40	7.5		50
Gold	242.80		10	0.5	5	
	267.60	500	8	0.5	10	0.3
Hafnium	263.87				10	
	277.34				10	
	307.29		2000			
Holmium	339.90				3	
	345.60				8	
	405.39	15	40	0.7		100
	410.38		30			
Indium	230.61				40	
	303.94	100	7	0.01		
	325.61	22	8			
	410.18	14	20			
	451.13	0.7	22		2	0.2
Iodine	178.38				20	
	183.00			3		
Iridium	208.88	400	500	0.5		
	212.68				20	
	224.27				20	

TABLE 3.34 Detection Limits in ng/mL (*Continued*)

Element	Wavelength, nm	Flame emission	Flame atomic absorption	Electrothermal atomic absorption	Argon ICP	Plasma atomic fluorescence
Iron	238.20				4	
	248.33		3	0.01		
	259.94				3	
	302.06	18	5			
	371.99	15	10			0.3
	385.99	12	21			
Lanthanum	379.48				15	
	392.76		8000			
	408.67				2	
	550.13	20				
	579.13	5	2000	0.5		
(as LaO)	441.82	100				
(as LaO)	560.25	300				
Lead	217.10		20	0.4		
	220.35				20	
	283.31	60	10	1		5
	368.35	30				
	405.78	20				
Lithium	460.29	0.06	30		50	
	610.36	0.001				
	670.78	0.003	0.3	1.5	5	0.4
Lutetium	261.54				1	
	307.76				6	
Magnesium	279.08				30	
	279.55				1.5	
	285.21	4.5	0.1	0.018	3.6	0.4
Manganese	256.37				2.7	
	257.61				0.5	
	259.37		60		3	
	260.57				6	
	279.48	1	1	0.05		0.4
	293.30				24	
	294.92				24	
	403.08	1.5	30			
Mercury	194.23				30	
	253.65	150	0.001	6	50	5
Molybdenum	202.03				5	
	203.84				8	
	281.62				1.2	
	313.26	220	30	0.06		12
	390.30	75	50			
Neodymium	292.45	200				
	401.23				10	
	430.36				30	
	492.45	150	600			2000
Nickel	231.60				6	
	232.00	8	4	0.5	10	
	341.48	15	2			
	352.45	8	2			2
Niobium	316.34				20	
	405.89	250	1000			1000

(*Continued*)

TABLE 3.34 Detection Limits in ng/mL (*Continued*)

Element	Wavelength, nm	Flame emission	Flame atomic absorption	Electrothermal atomic absorption	Argon ICP	Plasma atomic fluorescence
Osmium	225.58				20	
	228.23				40	
	263.71	2000	80			
	290.91		110			
Palladium	244.80	20	20	0.5		40
	340.46	25	80		40	
	363.47	50			60	
Phosphorus	178.28				50	
	213.62				50	
(as HPO)	524.90	100				
Platinum	214.42				20	
	265.95	2000	100	0.2	40	300
Potassium	404.41	1.3	100			
	404.72	2.6				
	766.49	0.15	1	0.004	200	0.6
	769.90	0.3	2			
Praseodymium	390.84				20	
	414.31				30	
	493.97	300				1000
Rhenium	197.31				8	
	345.19	690				
	346.05	200	200	10		
	346.47	275				
Rhodium	343.49	10	2	0.1	20	100
	369.24	20			30	
Rubidium	780.02	0.0065	0.3		500	3
	794.76	0.013				
Ruthenium	240.27				50	
	349.89	80	70	10	150	500
Samarium	442.43				10	
	476.03	30	500		100	
Scandium	255.24				21	
	357.24				1	
	361.38				1.5	
	391.18	21	20	6	120	10
	402.04	30				
	402.34	30				
Selenium	196.03		90	2.5	6	10
Silicon	251.61		80	0.5	10	50
	283.16				15	
Silver	328.07	2	0.9	0.001	4.5	0.1
	338.29	4			3	
Sodium	330.23	125		0.7	15	
	330.30	250				
	589.00	0.01	0.2	0.004	20	0.2
	589.59	0.02				
Strontium	407.78				1	
	421.55				0.5	
	460.73	0.1	2	0.01		0.3
Sulfur	180.73		10		70	
(as S_2)	394.00	1600				

TABLE 3.34 Detection Limits in ng/mL (*Continued*)

Element	Wavelength, nm	Flame emission	Flame atomic absorption	Electrothermal atomic absorption	Argon ICP	Plasma atomic fluorescence
Tantalum	240.06				20	
	271.47		800			
Tellurium	214.27	150	15	0.5		2
	238.58				60	
Terbium	350.92				10	
	384.87				40	
	431.89	150	600			500
Thallium	190.86				50	
	276.78		9	0.15		
	351.92				150	
	377.57	3		0.5		4
	535.05	1.5				
Thorium	283.73				30	
	401.91				30	
Thulium	313.13				3	
	371.79	4	10			100
	384.80				7	
Tin	189.99				15	
	224.60		110	1	30	
	284.00	100	200			10
	286.33		160	1.5		
Titanium	334.19	400				
	334.94				6	
	337.28				8	
	364.27	210	60	2.5		30
	365.35	180				
	399.86	150				
Tungsten	207.91				30	
	209.48				50	
	400.87	450	1000			2000
Uranium	358.49	100		30		
	385.96				70	
	409.01				140	
Vanadium	292.40				7.8	
	310.23				10	
	318.34	18				
	318.54	25	50	1		30
	437.92	15				
Ytterbium	328.94				1	
	369.42				2	
	398.80	0.45	5	0.1		10
Yttrium	360.07				3	
	362.09	40	50	10		50
	371.03				1	
	410.24	30	50			
Zinc	202.55				4	
	213.86	1000	0.8	0.005	2	0.0003
Zirconium	339.20				5	
	343.82				7	
	349.62				45	
	360.12	1000	350			

TABLE 3.35 Sensitive Lines of the Elements

In this table the sensitive lines of the elements are arranged in order of decreasing wavelengths. A Roman numeral II following an element designation indicates a line classified as being emitted by the singly ionized atom. In the column headed Sensitivity, the most sensitive line of the nonionized atom is indicated by U1, and other lines by U2, U3, and so on, in order of decreasing sensitivity. For the singly ionized atom the corresponding designations are V1, V2, V3, and so on.

Wavelength, nm	Element	Sensitivity	Wavelength, nm	Element	Sensitivity
894.35	Cs	U2	492.45	Nd	U1
852.11	Cs	U1	488.91	Re	U4
819.48	Na	U4	487.25	Sr	U3
818.33	Na	U3	483.21	Sr	U2
811.53	Ar	U2	482.59	Ra	U1
794.76	Rb	U2	481.95	Cl II	V4
780.02	Rb	U1	481.67	Br II	V3
769.90	K	U2	481.05	Zn	U3
766.49	K	U1	481.01	Cl II	V3
750.04	Ar	U4	479.45	Cl II	V2
706.72	Ar	U3	478.55	Br II	V2
696.53	Ar	U3	476.03	Sm	U1
690.24	F	U3	470.09	Br II	V1
685.60	F	U2	467.12	Xe	U2
670.78	Li	U1	462.43	Xe	U3
656.28	H	U2	460.73	Sr	U1
649.69	Ba II	V4	460.29	Li	U4
624.99	La	U3	459.40	Eu	U1
614.17	Ba II	V3	459.32	Cs	U4
610.36	Li	U2	455.54	Cs	U3
593.06	La	U4	455.40	Ba II	V1
589.59	Na	U2	451.13	In	U1
589.00	Na	U1	450.10	Xe	U4
587.76	He	U3	445.48	Ca	U2
587.09	Kr	U2	442.43	Sm II	V4
579.13	La	U1	440.85	V	U4
569.92	Ce	U1	440.19	Gd	U1
567.96	N II	V2	439.00	V	U3
567.60	N II	V4	437.49	Y II	V4
566.66	N II	V3	437.92	V	U1
557.02	Kr	U3	435.84	Hg	U3
553.55	Ba	U1	431.89	Tb	U1
550.13	La	U2	430.36	Nd II	V2
546.55	Ag	U4	430.21	W	U1
546.07	Hg	U2	429.67	Sm	U1
545.52	La	U3	428.97	Cr	U3
535.84	Hg	U3	427.48	Cr	U2
535.05	Tl	U1	425.43	Cr	U1
521.82	Cu	U3	422.67	Ca	U1
520.91	Ag	U3	421.56	Rb	U4
520.84	Cr	U8	421.55	Sr II	V1
520.60	Cr	U7	421.17	Dy	U2
515.32	Cu	U4	420.19	Rb	U3
498.18	Ti	U1	418.68	Dy	U2
496.23	Sr	U2	418.66	Ce II	V1
493.97	Pr	U1	417.21	Ga	U1
493.41	Ba II	V2	414.31	Pr II	V2

TABLE 3.35 Sensitive Lines of the Elements (*Continued*)

Wavelength, nm	Element	Sensitivity	Wavelength, nm	Element	Sensitivity
414.29	Y	U4	386.41	Mo	U2
413.38	Ce II	V1	385.99	Fe	U2
413.07	Ba II	V5	385.96	U II	V1
412.97	Eu II	V2	384.87	Tb II	V2
412.83	Y	U3	384.80	Tm II	V2
412.38	Nb	U4	383.83	Mg	U2
412.32	La II	V4	383.82	Mo	U2
411.00	N	U2	382.23	Mg	U3
410.38	Ho	U1	382.94	Mg	U4
410.24	Y	U1	381.97	Eu II	V1
410.18	In	U2	379.94	Ru	U3
410.09	Nb	U3	379.63	Mo	U1
409.99	N	U3	379.48	La II	V2
409.01	U II	V2	379.08	La II	V3
408.77	Er	U1	377.57	Tl	U3
408.67	La II	V1	377.43	Y II	V3
407.97	Nb	U2	374.83	Fe	U4
407.77	Sr II	V2	373.49	Fe	U2
407.74	Y	U2	372.80	Ru	U1
407.74	La II	V2	371.99	Fe	U1
407.43	W	U2	371.79	Tm	U1
405.89	Nb	U1	371.03	Y II	V1
405.78	Pb	U1	369.42	Yb II	V2
405.39	Ho	U2	369.24	Rh	U2
404.72	K	U4	368.41	Gd	U2
404.66	Hg	U5	368.35	Pb	U2
404.60	Dy	U1	365.48	Hg	U4
404.41	K	U3	365.35	Ti	U2
403.45	Mn	U3	365.01	Hg	U3
403.31	Mn	U2	364.28	Sc II	V3
403.30	Ga	U2	364.27	Sn	U3
403.08	Mn	U1	363.47	Pd	U2
402.37	Sc	U3	363.07	Sc II	V2
402.04	Sc	U3	362.09	Y	U2
401.91	Th II	V1	361.38	Sc II	V1
401.23	Nd II	V1	360.96	Pd	U2
400.87	W	U1	360.12	Zr	U1
400.80	Er	U1	360.07	Y II	V2
399.86	Cr	U1	360.05	Cr	U6
399.86	Ti	U1	359.62	Ru	U3
398.80	Yb	U1	359.34	Cr	U5
396.85	Ca II	V2	359.26	Sm II	V1
396.15	Al	U1	358.49	U	V1
394.91	La II	V2	357.87	Cr	U4
394.40	Al	U2	357.25	Zr II	V4
393.37	Ca II	V1	357.24	Sc II	V1
391.18	Sc	U1	356.83	Sn II	V1
390.84	Pr II	V1	355.31	Pd	U3
390.75	Sc	U2	354.77	Zr	U3
390.30	Mo	U1	353.17	Dy II	V1
389.18	Ba	V4	352.98	Co	U3
388.86	He	U2	352.94	Tl	U4
388.63	Fe	U5	352.69	Co	U4

(*Continued*)

TABLE 3.35 Sensitive Lines of the Elements (*Continued*)

Wavelength, nm	Element	Sensitivity	Wavelength, nm	Element	Sensitivity
352.45	Ni	U2	324.75	Cu	U1
351.96	Zr	U3	324.27	Pd	U4
351.92	Tl	U2	323.45	Cr	V3
351.69	Pd	U3	323.26	Li	U3
351.36	Ir	U2	323.06	Er II	V2
350.92	Tb II	V1	322.08	Ir	U1
350.63	Co	U3	318.54	V	U3
350.23	Co	U2	318.40	V	U2
349.89	Ru	U2	317.93	Ca II	V3
349.62	Zr II	V3	316.34	Nb II	V1
349.41	Er II	V1	315.89	Ca II	V4
348.11	Pd	U5	313.26	Mo	U2
347.40	Ni	U3	313.13	Tm II	V1
346.47	Re	U2	313.11	Be	U1
346.05	Re	U1	313.04	Be	U2
345.60	Ho II	V2	311.84	V II	V4
345.58	Co	U5	311.07	V II	V3
345.19	Re	U3	310.23	V II	V2
345.14	B II	V2	309.42	Nb II	V1
344.36	Co	U2	309.31	V II	V1
344.06	Fe	U2	309.27	Al	U3
343.82	Zr II	V2	308.22	Al	U4
343.67	Ru	U2	307.76	Lu II	V2
343.49	Rh	U1	307.29	Hf	U1
342.83	Ru	U4	306.77	Bi	U3
342.12	Pd	U3	306.47	Pt	U1
341.48	Ni	U3	303.94	In	U4
341.23	Co	U4	303.90	Ge	U2
340.78	Dy II	V2	303.41	Sn	U3
340.51	Co	U2	302.06	Fe	U3
340.46	Pd	U2	300.91	Sn	U4
339.90	Ho II	V1	294.91	Mn II	V4
339.20	Zr II	V1	294.44	W	U5
338.29	Ag	U2	294.36	Ga	U3
337.28	Ti II	V3	294.02	Ta	U3
336.12	Ti II	V2	293.30	Mn II	V4
335.05	Gd II	V1	292.98	Pt	U3
334.94	Ti II	V1	292.45	Nd	U2
334.50	Zn	U2	292.40	V II	V1
334.19	Ti	U4	290.91	Os	U2
332.11	Be	U3	289.80	Bi	U2
331.12	Ta	U3	289.10	Mo II	V4
330.03	Na	U6	288.16	Si	U1
330.26	Zn	U3	287.42	Ga	U4
330.23	Na	U5	287.15	Mo II	V3
328.94	Yb II	V1	286.33	Sn	U2
328.23	Zn	U5	286.04	As	U2
328.07	Ag	U1	285.21	Mg	U1
327.40	Cu	U2	284.82	Mo II	V2
326.95	Ge	U3	284.00	Sn	U1
326.23	Sn	U3	283.73	Th II	V1
326.11	Cd	U1	283.58	Cr II	V2
325.61	In	U3	283.31	Pb	U3

TABLE 3.35 Sensitive Lines of the Elements (*Continued*)

Wavelength, nm	Element	Sensitivity	Wavelength, nm	Element	Sensitivity
283.16	Si II	V1	239.56	Fe II	V2
283.03	Pt	U3	238.89	Co II	V2
281.62	Al II	V2	238.58	Te	U2
281.61	Mo II	V1	238.32	Te	U3
280.27	Mg II	V2	238.20	Fe II	V1
280.20	Pb	U4	234.90	As	U4
279.83	Mn	U3	234.86	Be	U1
279.55	Mg II	V1	232.00	Ni	U2
279.48	Mn	U3	231.60	Ni II	V1
279.08	Mg II	V2	231.15	Sb	U1
278.02	As	U1	230.61	In II	V1
277.34	Hf II	V1	228.81	As	U5
276.78	Tl	U4	228.80	Cd	U2
272.44	W	U4	228.71	Ni II	V1
271.90	Fe	U5	228.62	Co II	V1
271.47	Ta	U1	228.23	Os II	V2
270.65	Sn	U4	227.66	Bi	U3
267.72	Cr II	V1	227.02	Ni II	V2
267.60	Au	U2	226.50	Cd II	V2
266.92	Al II	V1	226.45	Ni II	V3
265.95	Pt	U1	225.58	Os II	V1
265.12	Ge	U1	225.39	Ni II	V4
265.05	Ba	U2	224.70	Cu II	V3
264.75	Ta	U2	224.64	Ag II	V3
263.87	Hf II	V1	224.60	Sn	U1
263.71	Os	U1	224.27	Ir II	V1
260.57	Mn II	V3	223.06	Bi	U1
259.94	Fe II	V1	220.35	Pb II	V1
259.81	Sb	U2	219.87	Ge II	V2
259.37	Mn	U2	219.23	Cu II	V2
257.61	Mn II	V1	217.58	Sb	U2
256.37	Mn II	V2	217.00	Pb II	V1
255.33	P	U3	214.44	Cd II	V1
255.24	Sc II	V3	214.42	Pt II	V1
253.65	Hg	U1	214.27	Te	U1
253.57	P	U1	213.86	Zn	U1
252.85	Si	U2	213.62	P	U1
252.29	Fe	U3	213.60	Cu II	V1
251.61	Si	U3	212.68	Ir II	V1
250.69	Si	U4	209.48	W II	V2
250.20	Zn II	V4	209.43	Ge II	V1
249.77	B	U1	208.88	Ir	U1
249.68	B	U2	207.91	W II	V1
248.33	Fe	U3	207.48	Se	U4
247.86	C	U2	206.83	Sb	U1
245.65	As	U4	206.28	Se	U3
243.78	Ag II	V2	206.19	Zn II	V2
242.80	Au	U1	203.99	Se	U1
241.05	Fe II	V4	203.84	Mo II	V3
240.73	Co	U1	202.55	Zn II	V1
240.49	Fe	V3	202.03	Mo II	V2
240.27	Ru	V1	197.31	Re II	V1
240.06	Ta II	V1	197.20	As	U3

(*Continued*)

TABLE 3.35 Sensitive Lines of the Elements (*Continued*)

Wavelength, nm	Element	Sensitivity	Wavelength, nm	Element	Sensitivity
196.03	Se	U2	183.00	I	U2
194.23	Hg II	V1	182.59	B II	V2
193.76	As	U1	180.73	S	U1
193.09	C	U1	178.38	I	U1
190.86	Tl II	V1	178.28	P	U1
189.99	Sn II	V1	154.07	Br II	V4
189.04	As	U2	134.72	Cl II	V1

3.6 NUCLEAR MAGNETIC RESONANCE SPECTROSCOPY

Nuclear magnetic resonance (NMR) spectroscopy is based on the principle that nuclei absorb radiation of slightly different frequency depending upon their local magnetic environments.

Certain atoms have a nuclear spin similar to the spin of an electron. The spinning of charged particles (the proton or protons in the nucleus bears a positive charge) generates a magnetic field. When an atom is placed in an external magnetic field, the magnetic field generated by the nucleus will be aligned with or against the external magnetic field. At some frequency of electromagnetic radiation, the nucleus will absorb energy and "flip" over so that it reverses its alignment with respect to the external magnetic field. This is known as the nuclear magnetic resonance (NMR) phenomenon. It is generally concerned with the nuclear magnetic resonance of hydrogen atoms and is therefore sometimes called proton magnetic resonance (PMR). It is also standard practice for the frequency of radiation to be kept constant while the strength of the external magnetic field is varied. At some value of the magnetic field strength, the energy required to flip the proton matches the energy of the radiation. Absorption will occur and a signal will be observed. The spectrum that results from all these absorptions is called an NMR spectrum. Absorptions that occur at relatively low field strengths are downfield relative to those that occur at higher field strengths. The field strength at which a proton will absorb energy is called the chemical shift (measured in parts per million, ppm or 6, relative to the absorbance of tetramethylsilane). The chemical shift of a proton depends upon the proton's electronic environment. Electron withdrawing atoms (or groups) that are nearby a proton will decrease the electron density about that proton; this is known as a deshielding effect. The proton's absorption will occur downfield from what is expected. Specifically, the proton will absorb at a smaller field strength than a proton experiencing no deshielding effects. Electron releasing atoms (or groups) that are nearby a proton will increase the proton's electron density; the proton is experiencing a shielding effect. The proton's absorbance will occur upfield (higher magnetic field strength) from what is expected.

The signal that arises from a proton's absorption may occur as a singlet, a doublet, a triplet, etc. The number of peaks in the signal depends upon the neighboring protons. Protons that are in identical electronic environments are equivalent protons; those that are in nonidentical electronic environments are nonequivalent protons. A proton that has n nonequivalent adjacent protons will have a signal with $n + 1$ peaks, called an $n + 1$ multiplet. This is the result of spin-spin splitting of the protons.

The differences in resonance frequencies are very small. For instance, the difference in resonance frequency for the protons in chloromethane and fluoromethane is 72 Hz. Since the incident radiation had a frequency of 60 MHz, this difference is about 1 part per million. This cannot be measured accurately; therefore, differences are measured as the difference between the resonant frequency of a reference compound and the substance to be analyzed. The most common reference is tetramethylsilane $(CH_3)_4Si$, TMS. Thus, when a compound is analyzed, the resonance of each individual proton is reported in terms of how far (in Hz) the proton is shifted from the protons of tetramethylsilane.

The shift from tetramethylsilane for a given proton depends upon the strength of the applied magnetic field. The protons in tetramethylsilane resonate at 0 ppm. Most protons in organic compounds will resonate at higher frequencies and the position of the absorbance gives valuable information about the molecular environment of a particular proton, leading to structural information about the compound under investigation.

The nucleus of carbon-13 is magnetic. This property enables detection of the nuclei of carbon-13 atoms by nuclear magnetic resonance. By detecting the location of carbon-13 atoms in carbon-based molecules, structural information about the molecules can also be produced. Other nuclei of different atoms can also be detected and structural information deduced.

TABLE 3.36 Nuclear Properties of the Elements

In the following table the magnetic moment μ is in multiples of the nuclear magneton $\mu_N(eh/4\pi Mc)$ with diamagnetic correction. The spin I is in multiples of $h/2\pi$, and the electric quadrupole moment Q is in multiples of 10^{-28} square meters. Nuclei with spin $^1/_2$ have no quadrupole moment. Sensitivity is for equal numbers of nuclei at constant field. NMR frequency at any magnetic field is the entry for column 5 multiplied by the value of the magnetic field in kilogauss. For example, in a magnetic field of 23.490 kG, protons will process at 4.2576 × 23.490 kG = 100.0 MHz. Radionuclides are denoted with an asterisk.

The data were extracted from M. Lederer and V. S. Shirley, *Table of Isotopes*, 7th ed., Wiley-Interscience, New York, 1978; A. H. Wapstra and G. Audi, "The 1983 Atomic Mass Evaluation," *Nucl. Phys.* **A432:**1–54 (1985); V. S. Shirley, ed., *Table of Radioactive Isotopes*, 8th ed., Wiley-Interscience, New York, 1986; and P. Raghavan, "Table of Nuclear Moments," *At. Data Nucl. Data Tables*, **42:**189 (1989).

Nuclide	Natural abundance, %	Spin I	Sensitivity at constant field relative to ^{1}H	NMR frequency for a 1-kG field, MHz	Magnetic moment μ/μ_N, J · T^{-1}	Electric quadrupole moment Q, 10^{-28} m^2
^{1}n	*	½	0.321 39	2.916 39	−1.913 043	
^{1}H	99.985	½	1.000 00	4.257 64	2.792 847	
^{2}H	0.015	1	0.009 65	0.653 57	0.857 438	0.002 860
^{3}H	*	½	1.213 54	4.541 37	2.978 963	
^{3}He	0.0001	½	0.442 12	3.243 52	−2.127 624	
^{6}Li	7.5	1	0.008 50	0.626 60	0.822 047	0.000 82
^{7}Li	92.5	3/2	0.293 55	1.654 78	3.256 427	−0.040 1
^{9}Be	100	3/2	0.013 89	0.598 6	−1.177 9	0.052 88
^{10}B	19.9	3	0.019 85	0.457 51	1.800 645	0.084 59
^{11}B	80.1	3/2	0.165 22	1.366 26	2.688 649	0.040 59
^{13}C	1.10	½	0.015 91	1.070 81	0.702 412	
^{14}N	99.634	1	0.001 01	0.307 76	0.403 761	0.020 2
^{15}N	0.366	½	0.001 04	0.431 72	−0.283 189	
^{17}O	0.038	5/2	0.029 10	0.577 41	−1.893 80	−0.025 58
^{19}F	100	½	0.834 00	4.007 65	2.628 867	
^{21}Ne	0.27	3/2	0.002 46	0.336 30	−0.661 797	0.101 55
^{22}Na	*	3	0.018 10	0.443 4	1.745	
^{23}Na	100	3/2	0.092 70	1.126 86	2.217 522	0.108 9
^{25}Mg	10.00	5/2	0.002 68	0.260 82	−0.855 46	0.199 4
^{27}Al	100	5/2	0.206 89	1.110 28	3.641 504	0.140 3
^{29}Si	4.67	½	0.007 86	0.846 53	−0.555 29	
^{31}P	100	½	0.066 52	1.725 10	1.131 60	
^{33}S	0.75	3/2	0.002 27	0.327 16	0.643 821	−0.067 8
^{35}S	*	3/2	0.008 50	0.508	1.00	0.045
^{35}Cl	75.77	3/2	0.004 72	0.417 64	0.821 874	−0.081 65
^{36}Cl	*	2	0.012 10	0.489 3	1.283 8	−0.016 8
^{37}Cl	24.23	3/2	0.002 72	0.347 64	0.684 124	−0.064 35
^{37}Ar	*	3/2	0.012 76	0.581 8	1.145	

(*Continued*)

TABLE 3.36 Nuclear Properties of the Elements (*Continued*)

Nuclide	Natural abundance, %	Spin I	Sensitivity at constant field relative to ^{1}H	NMR frequency for a 1-kG field, MHz	Magnetic moment μ/μ_N, $J \cdot T^{-1}$	Electric quadrupole moment Q, 10^{-28} m^2
^{39}K	93.258	3/2	0.000 51	0.198 93	0.391 466	0.060 1
^{40}K	0.0117	4	0.005 23	0.247 37	−1.298 099	−0.074 9
^{41}K	6.730	3/2	0.000 084	0.109 19	0.214 870	0.073 3
^{43}Ca	0.135	7/2	0.006 42	0.286 88	−1.317 26	−0.040 8
^{45}Sc	100	7/2	0.302 44	1.035 88	4.756 483	−0.22
^{47}Ti	7.3	5/2	0.002 10	0.240 40	−0.788 48	0.29
^{49}Ti	5.5	7/2	0.003 78	0.240 47	−1.104 17	0.24
^{50}V	0.250	6	0.055 71	0.425 04	3.345 689	0.21
^{51}V	99.750	7/2	0.383 60	1.121 30	5.148 706	−0.052
^{53}Cr	9.501	3/2	0.000 91	0.241 14	−0.474 54	−0.15
^{55}Mn	100	5/2	0.178 81	1.057 60	3.468 72	0.33
^{57}Fe	2.1	1/2	0.000 03	0.138 15	0.090 623	
^{59}Co	100	7/2	0.278 41	1.007 7	4.627	0.42
^{61}Ni	1.140	3/2	0.003 59	0.381 13	−0.750 02	0.162
^{63}Cu	69.17	3/2	0.093 42	1.129 79	2.223 29	−0.220
^{65}Cu	30.83	3/2	0.114 84	1.210 27	2.381 67	−0.204
^{67}Zn	4.1	5/2	0.002 87	0.266 93	0.875 479	0.150
^{69}Ga	60.108	3/2	0.069 71	1.024 75	2.016 59	0.170
^{71}Ga	39.892	3/2	0.143 00	1.302 04	2.562 27	0.100
^{73}Ge	7.73	9/2	0.001 41	0.148 97	−0.879 468	−0.173
^{75}As	100	3/2	0.025 36	0.731 48	1.439 475	0.314
^{77}Se	7.63	1/2	0.007 03	0.815 66	0.535 042	
^{79}Br	50.69	3/2	0.079 45	1.070 39	2.106 399	0.331
^{81}Br	49.31	3/2	0.099 51	1.153 81	2.270 562	0.276
^{83}Kr	11.5	9/2	0.001 90	0.164 42	−0.970 669	0.253
^{85}Rb	72.165	5/2	0.010 61	0.412 53	1.353 03	0.274
^{87}Rb	27.835	3/2	0.177 03	1.398 07	2.751 24	0.132
^{87}Sr	7.00	9/2	0.002 72	0.185 24	−1.093 603	0.335
^{89}Y	100	1/2	0.000 12	0.209 49	−0.137 415	
^{91}Zr	11.22	5/2	0.009 49	0.397 47	−1.303 62	−0.206
^{93}Nb	100	9/2	0.488 21	1.045 20	6.170 5	−0.32
^{95}Mo	15.92	5/2	0.003 27	0.278 74	−0.914 2	−0.022
^{97}Mo	9.55	5/2	0.003 49	0.284 62	−0.933 5	−0.255
^{99}Tc	*	9/2	0.381 74	0.963	5.684 7	−0.129
^{99}Ru	12.7	5/2	0.001 13	0.195 53	−0.641 3	0.079
^{101}Ru	17.0	5/2	0.001 59	0.219 2	−0.718 8	0.457
^{103}Rh	100	1/2	0.000 03	0.134 76	−0.088 40	
^{105}Pd	22.33	5/2	0.001 13	0.195 7	−0.642	0.660
^{107}Ag	51.839	1/2	0.000 066 9	0.173 30	−0.113 680	
^{109}Ag	48.161	1/2	0.000 101	0.199 24	−0.130 691	
^{111}Cd	12.80	1/2	0.009 66	0.906 89	−0.594 886	
^{113}Cd	12.22	1/2	0.011 06	0.948 68	−0.622 301	
^{113}In	4.3	9/2	0.351 21	0.936 52	5.528 9	0.799
^{115}In	95.7	9/2	0.353 48	0.938 54	5.540 8	0.81
^{115}Sn	0.34	1/2	0.035 61	1.400 74	−9.1 884	
^{117}Sn	7.68	1/2	0.046 05	1.526 06	−1.001 05	
^{119}Sn	8.59	1/2	0.052 73	1.596 56	−1.047 28	
^{121}Sb	57.36	5/2	0.163 02	1.025 49	3.363 4	−0.36
^{123}Sb	42.64	7/2	0.046 59	0.555 30	2.549 8	−0.49
^{123}Te	0.908	1/2	0.018 37	1.123 46	−0.736 948	

TABLE 3.36 Nuclear Properties of the Elements (*Continued*)

Nuclide	Natural abundance, %	Spin I	Sensitivity at constant field relative to ^{1}H	NMR frequency for a 1-kG field, MHz	Magnetic moment μ/μ_N, $J \cdot T^{-1}$	Electric quadrupole moment Q, 10^{-28} m^2
^{125}Te	7.139	½	0.032 20	1.354 51	−0.888 505	
^{127}I	100	5⁄2	0.095 40	0.857 76	2.813 327	−0.789
^{129}Xe	26.4	½	0.021 62	1.186 01	−0.777 976	
^{131}Xe	21.2	3⁄2	0.002 82	0.351 58	0.691 862	−0.12
^{133}Cs	100	7⁄2	0.048 38	0.562 32	2.582 025	−0.003 7
^{135}Ba	6.592	3⁄2	0.005 00	0.425 81	0.837 943	0.160
^{137}Ba	11.23	3⁄2	0.006 97	0.476 33	0.937 365	0.245
^{138}La	* 0.0902	5	0.094 04	0.566 14	3.713 646	0.45
^{139}La	99.9098	7⁄2	0.060 58	0.606 10	2.783 045	0.20
^{137}Ce	*	3⁄2	0.006 41	0.462	0.91	
^{139}Ce	*	3⁄2	0.006 41	0.462	0.91	
^{141}Ce	*	7⁄2	0.003 64	0.237	1.09	
^{141}Pr	100	5⁄2	0.334 83	1.303 55	4.275 4	−0.059
^{143}Nd	12.18	7⁄2	0.003 39	0.231 9	−1.065	−0.63
^{145}Nd	8.30	7⁄2	0.000 79	0.142 9	−0.656	−0.33
^{143}Pm	*	5⁄2	0.235 10	1.16	3.8	
^{147}Pm	*	7⁄2	0.049 40	0.57	2.6	0.70
^{147}Sm	15.0	7⁄2	0.001 52	0.177 47	−0.814 9	−0.26
^{149}Sm	13.8	7⁄2	0.000 85	0.146 31	−0.671 8	0.094
^{151}Eu	47.8	5⁄2	0.179 29	1.058 54	3.471 8	0.903
^{153}Eu	52.2	5⁄2	0.015 44	0.467 44	1.533 1	2.41
^{155}Gd	14.80	3⁄2	0.000 15	0.131 7	−0.259 1	1.27
^{157}Gd	15.65	3⁄2	0.000 33	0.1727	−0.339 9	1.35
^{159}Tb	100	3⁄2	0.069 45	1.023	2.014	1.432
^{161}Dy	18.9	5⁄2	0.000 48	1.465 3	−0.480 6	2.47
^{163}Dy	24.9	5⁄2	0.001 30	0.205 07	0.672 6	2.65
^{165}Ho	100	7⁄2	0.204 23	0.908 81	4.173	3.58
^{167}Er	22.95	7⁄2	0.000 507	0.122 81	−0.563 9	3.57
^{169}Tm	100	½	0.000 566	3.531	−0.231 6	
^{171}Yb	14.3	½	0.005 52	0.752 59	0.493 67	
^{173}Yb	16.12	5⁄2	0.001 35	0.207 301	−0.679 89	2.80
^{175}Lu	97.41	7⁄2	0.031 28	0.486 24	2.232 7	3.49
^{176}Lu	* 2.59	7	0.039 75	0.345 1	3.169	4.97
^{177}Hf	18.606	7⁄2	0.001 40	0.172 81	0.793 5	3.36
^{179}Hf	13.629	9⁄2	0.000 55	0.108 56	−0.640 9	3.79
^{180}Ta	0.012	9	0.102 51	0.404	4.77	
^{181}Ta	99.988	7⁄2	0.037 44	0.516 25	2.3705	3.17
^{183}W	14.3	½	0.000 08	0.179 56	0.117 785	
^{185}Re	37.40	5⁄2	0.138 70	0.971 7	3.1871	2.18
^{187}Re	* 62.60	5⁄2	0.143 00	0.981 7	3.219 7	2.07
^{187}Os	1.6	½	0.000 01	0.098 56	0.064 652	
^{189}Os	16.1	3⁄2	0.002 44	0.335 35	0.659 933	0.856
^{191}Ir	37.3	3⁄2	0.000 03	0.076 6	0.150 7	0.816
^{193}Ir	62.7	3⁄2	0.000 04	0.0832	0.163 7	0.751
^{195}Pt	33.8	½	0.010 39	0.929 20	0.609 52	
^{197}Au	100	3⁄2	0.000 03	0.074 06	0.145 746	0.547
^{199}Hg	16.87	½	0.005 94	0.771 21	0.505 885	
^{201}Hg	13.18	3⁄2	0.001 49	0.284 68	−0.560 226	0.386
^{203}Tl	29.524	½	0.195 981	2.473 10	1.622 258	
^{205}Tl	70.476	½	0.201 82	2.497 42	1.638 215	
^{207}Pb	22.1	½	0.009 55	0.903 38	0.592 58	

(*Continued*)

TABLE 3.36 Nuclear Properties of the Elements (*Continued*)

Nuclide	Natural abundance, %	Spin I	Sensitivity at constant field relative to ^1H	NMR frequency for a 1-kG field, MHz	Magnetic moment μ/μ_N, $J \cdot T^{-1}$	Electric quadrupole moment Q, 10^{-28} m^2
^{209}Bi	100	9/2	0.144 33	0.696 28	4.110 6	−0.50
^{229}Th	*	5/2	0.000 42	0.140	0.46	4.30
^{231}Pa	*	3/2	0.069 03	1.02	2.01	−1.72
^{235}U	* 0.7200	7/2	0.000 15	0.083	−0.38	4.936
^{237}Np	*	5/2	0.132 64	0.957	3.14	3.886
^{239}Pu	*	1/2	0.000 38	0.309	0.203	
^{243}Am	*	5/2	0.017 88	0.491	1.61	4.21

TABLE 3.37 Proton Chemical Shifts

Values are given on the officially approved δ scale; τ = 10.00 − δ

Abbreviations Used in the Table

R, alkyl group *Ar, aryl group*

Substituent group	Methyl protons	Methylene protons	Methine proton
HC—C—CH_2	0.95	1.20	1.55
HC—C—NR_2	1.05	1.45	1.70
HC—C—C=C	1.00	1.35	1.70
HC—C—C=O	1.05	1.55	1.95
HC—C—NRAr	1.10	1.50	1.80
HC—C—H(C=O)R	1.10	1.50	1.90
HC—C—(C=O)NR_2	1.10	1.50	1.80
HC—C—(C=O)Ar	1.15	1.55	1.90
HC—C—(C=O)OR	1.15	1.70	1.90
HC—C—Ar	1.15	1.55	1.80
HC—C—OH	1.20	1.50	1.75
HC—C—OR	1.20	1.50	1.75
HC—C—C≡CR	1.20	1.50	1.80
HC—C—C≡N	1.25	1.65	2.00
HC—C—SR	1.25	1.60	1.90
HC—C—OAr	1.30	1.55	2.00
HC—C—O(C=O)R	1.30	1.60	1.80
HC—C—SH	1.30	1.60	1.65
HC—C—(S=O)R and HC—C—SO_2R	1.35	1.70	
HC—C—NR_3^+	1.40	1.75	2.05
HC—C—O—N=O	1.40		
HC—C—O(C=O)CF_3	1.40	1.65	
HC—C—CL	1.55	1.80	1.95
HC—C—F	1.55	1.85	2.15
HC—C—NO_2	1.60	2.05	2.50
HC—C—O(C=O)Ar	1.65	1.75	1.85
HC—C—I	1.75	1.80	2.10
HC—C—Br	1.80	1.85	1.90
HC—CH_2	0.90	1.30	1.50
HC—C=C	1.60	2.05	
HC—C≡C	1.70	2.20	2.80

TABLE 3.37 Proton Chemical Shifts (*Continued*)

Substituent group	Methyl protons	Methylene protons	Methine proton
HC—(C=O)OR	2.00	2.25	2.50
HC—(C=O)NR_2	2.00	2.25	2.40
HC—SR	2.05	2.55	3.00
HC—O—O	2.10	2.30	2.55
HC—(C=O)R	2.10	2.35	2.65
HC—C≡N	2.15	2.45	2.90
HC—I	2.15	3.15	4.25
HC—CHO	2.20	2.40	
HC—Ar	2.25	2.45	2.85
HC—NR_2	2.25	2.40	2.80
HC—SSR	2.35	2.70	
HC—(C=O)Ar	2.40	2.70	3.40
HC—SAr	2.40		
HC—NRAr	2.60	3.10	3.60
HC—SO_2R and HC—(SO)R	2.60	3.05	
HC—Br	2.70	3.40	4.10
HC—NR_3^+	2.95	3.10	3.60
HC—NH(C=O)R	2.95	3.35	3.85
HC—SO_3R	2.95		
HC—Cl	3.05	3.45	4.05
HC—OH and HC—OR	3.20	3.40	3.60
HC—PAr_3	3.20	3.40	
HC—NH_2	3.50	3.75	4.05
HC—O(C=O)R	3.65	4.10	4.95
HC—OAr	3.80	4.00	4.60
HC—O(C=O)Ar	3.80	4.20	5.05
HC—O(C=O)CF_1	3.95	4.30	
HC—F	4.25	4.50	4.80
HC—NO_2	4.30	4.35	4.60
Cyclopropane		0.20	0.40
Cyclobutane		2.45	
Cyclopentane		1.65	
Cyclohexane		1.50	1.80
Cycloheptane		1.25	

Substituent group	Proton shift	Substituent group	Proton shift
HC≡CH	2.35	HO—C=O	10–12
HC≡CAr	2.90	HO—SO_2	11–12
HC≡C—C=C	2.75	HO—Ar	4.5–6.5
HAr	7.20	HO—R	0.5–4.5
HCO—O	8.1	HS—Ar	2.8–3.6
HCO—R	9.4–10.0	HS—R	1–2
HCO—Ar	9.7–10.5	HN—Ar	3–6
HO—N=C (oxime)	9–12	HN—R	0.5–5

(*Continued*)

TABLE 3.37 Proton Chemical Shifts (*Continued*)

Saturated heterocyclic ring systems

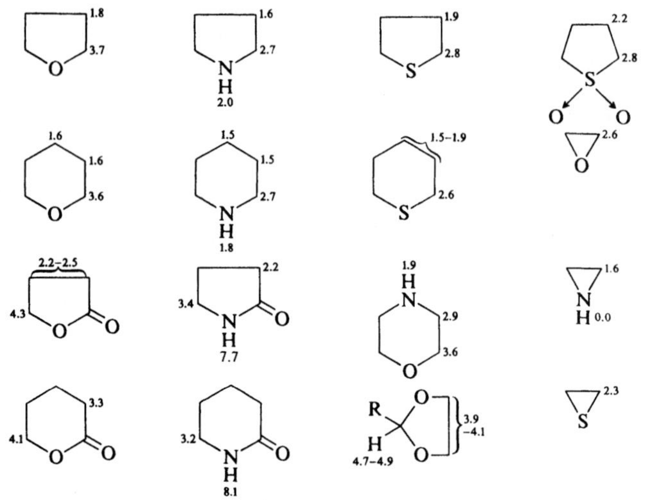

Unsaturated cyclic systems

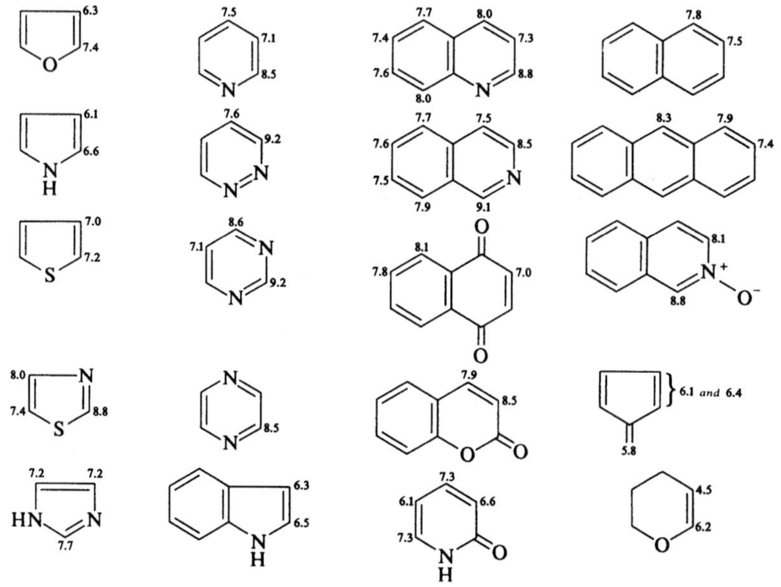

TABLE 3.38 Estimation of Chemical Shift for Protons of CH_2 and Methine Groups

$$\delta_{CH_2} = 0.23 + C_1 + C_2 \qquad \delta_{CH} = 0.23 + C_1 + C_2 + C_3$$

X*	C	X*	C	X*	C
—CH_3	0.5	—SR	1.6	—OR	2.4
—CF_3	1.1	—C≡C—Ar	1.7	—Cl	2.5
>C=C<	1.3	—CN	1.7	—OH	2.6
—C≡C—R	1.4	—CO—R	1.7	—N=C=S	2.9
—COOR	1.5	—I	1.8	—OCOR	3.1
—NR_2	1.6	—Ph	1.8	—OPh	3.2
—$CONR_2$	1.6	—Br	2.3		

*R, alkyl group; Ar, aryl group; Ph, phenyl group.

TABLE 3.39 Estimation of Chemical Shift of Proton Attached to a Double Bond

Positive Z values indicate a downfield shift, and an arrow indicates the point of attachment of the substituent group to the double bond.

$$\delta_{C=C\diagdown H} = 5.25 - Z_{gem} + Z_{cis} + Z_{trans}$$

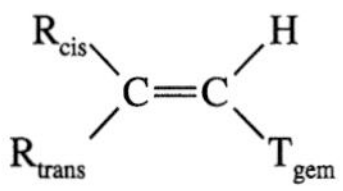

R	Z_{gem}, ppm	Z_{cis}, ppm	Z_{trans}, ppm
→H	0	0	0
→alkyl	0.45	−0.22	−0.28
→alkyl—ring (5- or 6-member)	0.69	−0.25	−0.28
→CH_2O—	0.64	−0.01	−0.02
→CH_2S—	0.71	−0.13	−0.22
→CH_2X (X: F, Cl, Br)	0.70	0.11	−0.04
→CH_2N<	0.58	−0.10	−0.08
→>C=C (isolated)	1.00	−0.09	−0.23
→>C=C (conjugated)	1.24	0.02	−0.05
→C≡N	0.27	0.75	0.55
→C≡C—	0.47	0.38	0.12
→>C=O (isolated)	1.10	1.12	0.87
→>C=O (conjugated)	1.06	0.91	0.74
→COOH (isolated)	0.97	1.41	0.71
→COOH (conjugated)	0.80	0.98	0.32
→COOR (isolated)	0.80	1.18	0.55
→COOR (conjugated)	0.78	1.01	0.46
→C(H)=O	1.02	0.95	1.17

(Continued)

TABLE 3.39 Estimation of Chemical Shift of Proton Attached to a Double Bond (*Continued*)

R	Z_{gem}, ppm	Z_{cis}, ppm	Z_{trans}, ppm
→C(=O)—N<	1.37	0.98	0.46
→C(=O)—Cl	1.11	1.46	1.01
→OR (R: aliphatic)	1.22	−1.07	−1.21
→OR (R: conjugated)	1.21	−0.60	−1.00
→OCOR	2.11	−0.35	−0.64
→CH_2—C(—)=O; →CH_2—C≡N	0.69	−0.08	−0.06
→CH_2—aromatic ring	1.05	−0.29	−0.32
→F	1.54	−0.40	−1.02
→Cl	1.08	0.18	0.13
→Br	1.07	0.45	0.55
→I	1.14	0.81	0.88
→N—R (R: aliphatic)	0.80	−1.26	−1.21
→N(—)—R (R: conjugated)	1.17	−0.53	−0.99
→N(—)—C(—)=O	2.08	−0.57	−0.72
→aromatic	1.38	0.36	−0.07
→CF_3	0.66	0.61	0.32
→aromatic (*o*-substituted)	1.65	0.19	0.09
→SR	1.11	−0.29	−0.13
→SO_2	1.55	1.16	0.93

TABLE 3.40 Chemical Shifts in Monosubstituted Benzene

$$\delta = 7.27 + \Delta_i$$

Substituent	Δ_{ortho}	Δ_{meta}	Δ_{para}
NO_2	0.94	0.18	0.39
CHO	0.58	0.20	0.26
COOH	0.80	0.16	0.25
$COOCH_3$	0.71	0.08	0.20
COCl	0.82	0.21	0.35
CCl_3	0.80	0.20	0.20
$COCH_3$	0.62	0.10	0.25
CN	0.26	0.18	0.30
$CONH_2$	0.65	0.20	0.22
$\overset{+}{N}H_3$	0.40	0.20	0.20
CH_2X*	0.0–0.1	0.0–0.1	0.0–0.1
CH_3	−0.16	−0.09	−0.17
CH_2CH_3	−0.15	−0.06	−0.18
$CH(CH_3)_2$	−0.14	−0.09	−0.18
$C(CH_3)_2$	−0.09	0.05	−0.23
F	−0.30	−0.02	−0.23
Cl	0.01	−0.06	−0.08

TABLE 3.40 Chemical Shifts in Monosubstituted Benzene (*Continued*)

Substituent	Δ_{ortho}	Δ_{meta}	Δ_{para}
Br	0.19	−0.12	−0.05
I	0.39	−0.25	−0.02
NH_2	−0.76	−0.25	−0.63
OCH_3	−0.46	−0.10	−0.41
OH	−0.49	−0.13	−0.20
OCOR	−0.20	0.10	−0.20
$NHCH_3$	−0.80	−0.30	−0.60
$N(CH_3)_2$	−0.60	−0.10	−0.62

*X = Cl, alkyl, OH, or NH_2.

TABLE 3.41 Proton Spin Coupling Constants

Structure	*J*, Hz	Structure	*J*, Hz
>C(H)H	12–15	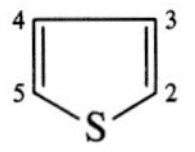 2–3	5–6
		3–4	3.5–5.0
		2–4	1.5
		2–5	3.4
>CH—CH< (free rotation)	6–8	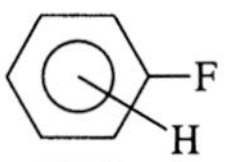 *o*	6–12
		m	4–8
>CH—OH (no exchange)	5	*p*	1.5–2.5
>CH—NH	4–8	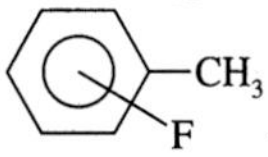 *o*	2.5
		m	1.5
>CH—SH	6–8	*p*	0
>CH—C(H)=O	1–3	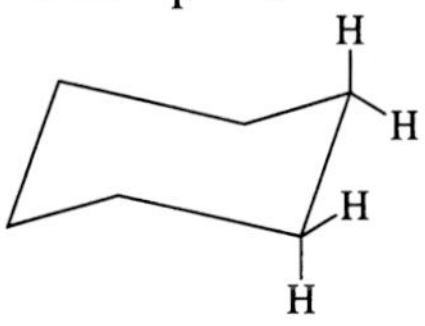 *a–a*	8–10
		a–e	2–3
		e–e	2–3
—N=C(H)H	8–16		
H_t, H_c >C=C< H_g, H *gem*	0–3	Cyclopentane *cis*	4–6
cis	6–14	*trans*	4–6
trans	11–18	Cyclobutane *cis*	8
		trans	8
H_c, H_t >C=C< CH, H_g *cis*	0.5–3	Cyclopropane *cis*	9–11
trans	0.5–3	*trans*	6–8
gem	4–10	*gem*	4–6
>C=CH—CH=C<	10–13	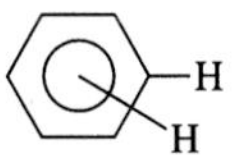 *o*	6–10
=CH—C(H)=O	6	*m*	1–3
		p	0–1
—CH_2—C≡C—CH<	0–3	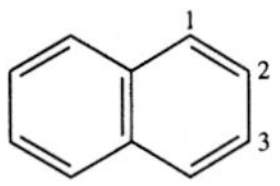 1–2	8–9
		2–3	6
>CH—C≡CH	0–3	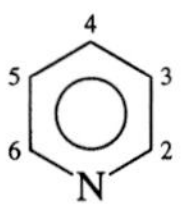 2–3	5–6
H—C=C—H (ring) 3-member	0–2	3–4	7–9
4-member	2–4	2–4	1–2
5-member	5–7	3–5	1–2
6-member	6–9	2–5	0–1
		2–6	0–1

(*Continued*)

TABLE 3.41 Proton Spin Coupling Constants (*Continued*)

Structure		*J*, Hz	Structure		*J*, Hz
	7-member	10–13	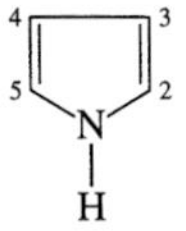	1–2	2–3
	*cis*	4–5		1–3	2–3
	trans	3		2–3	2–3
	gem	5–6		3–4	3–4
	*cis*	0		2–4	1–2
	trans	7		2–5	1–3
	gem	6	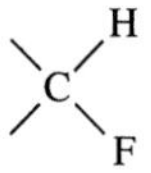		45–52
	*cis*	2			
	trans	6			
	gem	4			
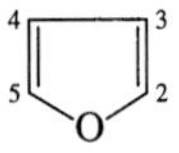	2–3	1.8	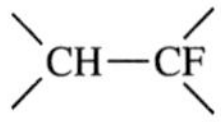	*gauche*	0–12
	3–4	3.5		*trans*	10–45
	2–4	0–1	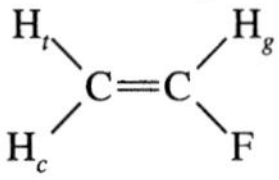	*gem*	72–90
	2–5	1–2		*cis*	−3–20
				trans	12–40
			HC≡CF		21
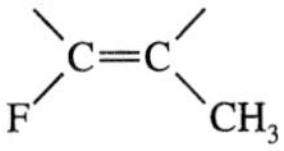		2–4	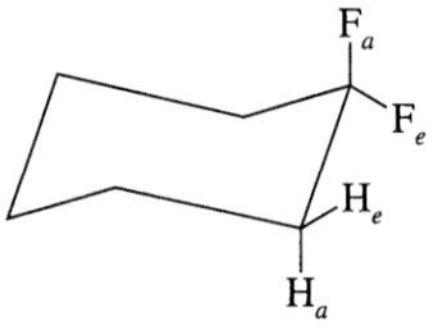	*a–a*	34
				a–e	12
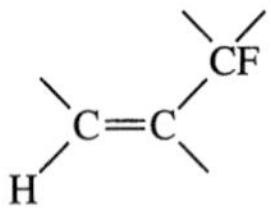		0–6		*e–e*, *e–e*	<5–8

TABLE 3.42 Proton Chemical Shifts of Reference Compounds

Relative to tetramethylsilane.

Compound	δ, ppm	Solvent(s)
Sodium acetate	1.90	D_2O
1,2-Dibromoethane	3.63	$CDCl_3$
1,1,2,2-Tetrachloroethane	5.95	$CDCl_3$; CCl_4
1,4-Benzoquinone	6.78	$CDCl_3$; CCl_4
1,4-Dichlorobenzene	7.23	CCl_4
1,3,5-Trinitrobenzene	9.21	DMSO-d_6*
	9.55	$CHCl_3$

*DMSO, dimethyl sulfoxide.

TABLE 3.43 Solvent Positions of Residual Protons in Incompletely Deuterated Solvents

Relative to tetramethylsilane.

Solvent	Group	δ, ppm
Acetic-d_3 acid-d_1	Methyl	2.05
	Hydroxyl	11.5*
Acetone-d_6	Methyl	2.057
Acetonitrile-d_3	Methyl	1.95
Benzene-d_6	Methine	6.78

TABLE 3.43 Solvent Positions of Residual Protons in Incompletely Deuterated Solvents (*Continued*)

Solvent	Group	δ, ppm
tert-Butanol-d_1 $(CH_3)_3COD$	Methyl	1.28
Chloroform-d_1	Methine	7.25
Cyclohexane-d_{12}	Methylene	1.40
Deuterium oxide	Hydroxyl	4.7*
Dimethyl-d_6-formamide-d_1	Methyl	2.75; 2.95
	Formyl	8.05
Dimethyl-d_6 sulfoxide	Methyl	2.51
	Absorbed water	3.3*
1,4-Dioxane-d_8	Methylene	3.55
Hexamethyl-d_{18}-phosphoramide	Methyl	2.60
Methanol-d_4	Methyl	3.35
	Hydroxyl	4.8*
Dichloromethane-d_2	Methylene	5.35
Pyridine-d_5	C-2 Methine	8.5
	C-3 Methine	7.0
	C-4 Methine	7.35
Toluene-d_8	Methyl	2.3
	Methine	7.2
Trifluoroacetic acid-d_1	Hydroxyl	11.3*

*These values may vary greatly, depending upon the solute and its concentration.

TABLE 3.44 Carbon-13 Chemical Shifts

Values given in ppm on the δ scale, relative to tetramethylsilane.

Substituent group	Primary carbon	Secondary carbon	Tertiary carbon	Quaternary carbon
Alkynes:				
C—C	5–30	25–45	23–58	28–50
C—O	45–60	42–71	62–78	73–86
C—N	13–45	44–58	50–70	60–75
C—S	10–30	22–42	55–67	53–62
C—halide (I to Cl)	3–25	3–40	34–58	35–75

Substituent group	δ, ppm	Substituent group	δ, ppm
Cyclopropane	−5–5	Aromatics:	
Cycloalkane C_4–C_{10}	5–25	Aryl-C	125–145
Mercaptanes	5–70	Aryl-P	119–128
Amines:		Aryl-N	128–138
R_2N—C	20–70	Aryl-O	133–152
Aryl—N	128–138	Azomethines	145–162
Sulfoxides, sulfones	35–55	Carbonates	159–162
Alcohols R—OH	45–87	Ureas	150–170
Ethers R—O—R	57–87	Anhydrides	150–175
Nitro $R—NO_2$	60–78	Amides	154–178
Alkynes:		Oximes	155–165
HC≡CR	63–73		
RC≡CR	72–95		

(*Continued*)

TABLE 3.44 Carbon-13 Chemical Shifts (*Continued*)

Substituent group	Primary carbon	Secondary carbon	Tertiary carbon	Quaternary carbon
Acetals, ketals	88–112	Esters:		
Thiocyanates R—SCN	96–118	Saturated		158–165
Alkenes:		α,β-Unsaturated		165–176
H_2C=	100–122	Isocyanides R—NC		162–175
R_2C=	110–150	Carboxylic acids:		
Heteroaromatics:		Nonconjugated		162–165
C=N	100–152	Conjugated		165–184
C_α	142–160	Salts (anion)		175–195
Cyanates R—OCN	105–120	Ketones:		
Isocyanates R—NCO	115–135	α-Halo		160–200
Isothiocyanates R—NCS	115–142	Nonconjugated		192–202
Nitriles, cyanides	117–124	α,β-Unsaturated		202–220
Thioureas	165–185	Imides		165–180
Aldehydes:		Acyl chlorides R—CO—Cl		165–183
α-Halo	170–190	Thioketones R—CS—R		190–202
Nonconjugated	182–192	Carbonyl $M(CO)_n$		190–218
Conjugated	192–208	Allenes =C=		197–205

Saturated heterocyclic ring systems

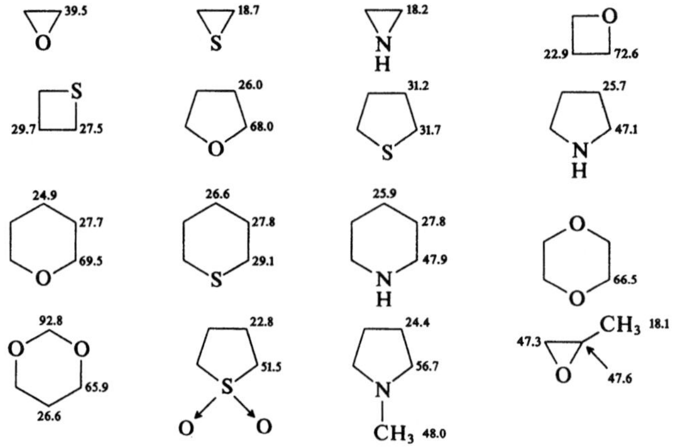

Unsaturated cyclic systems

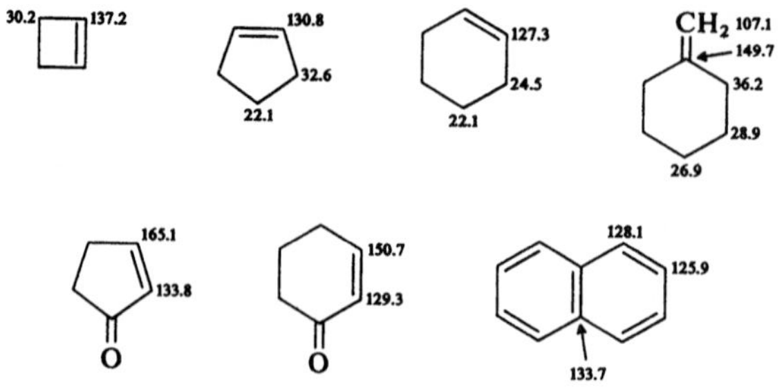

TABLE 3.44 Carbon-13 Chemical Shifts (*Continued*)

Unsaturated cyclic systems (*Continued*)

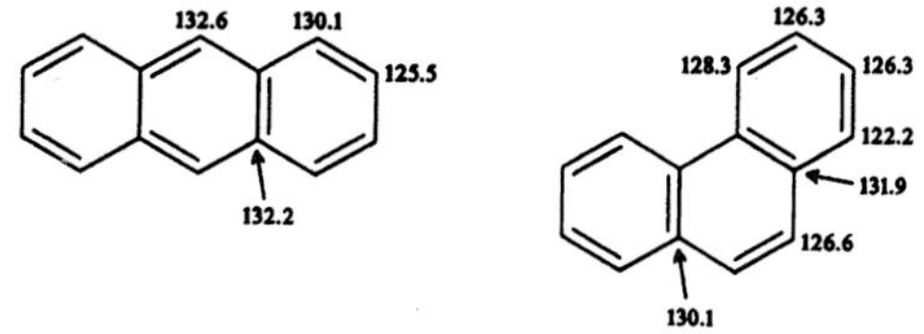

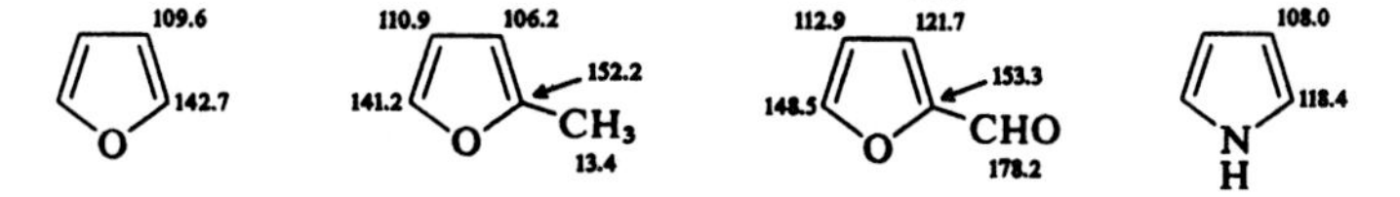

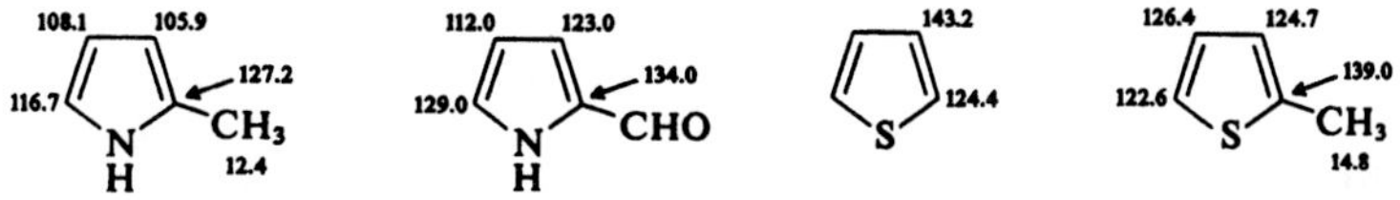

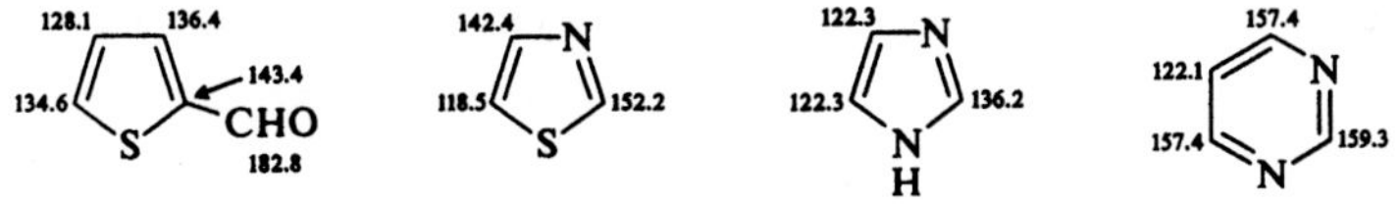

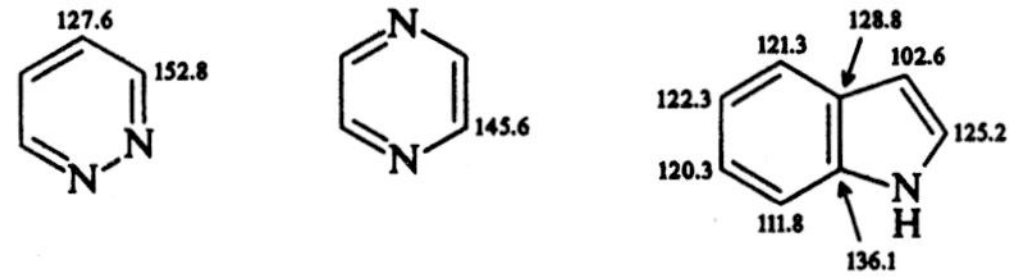

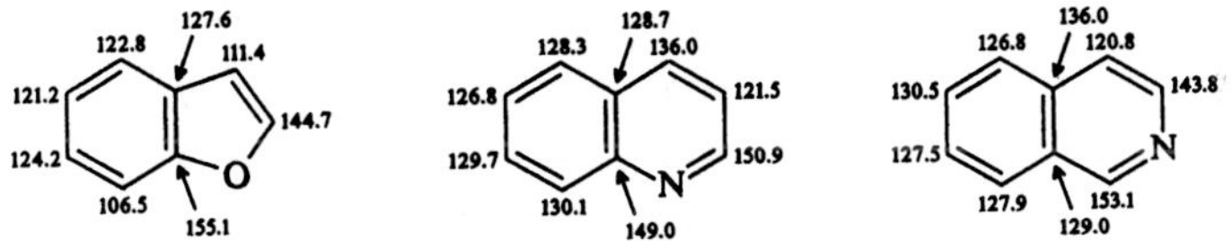

Saturated alicyclic ring systems

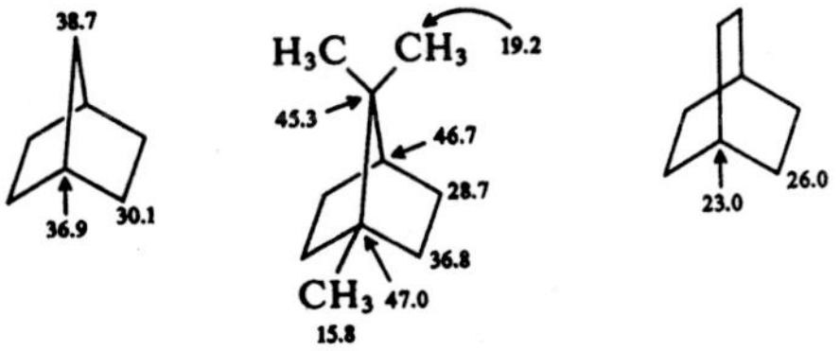

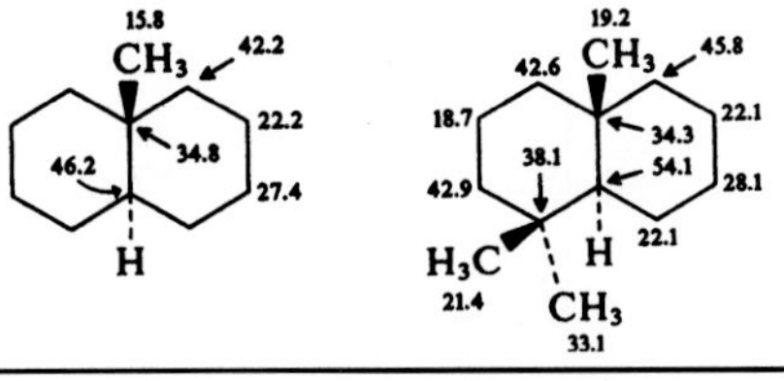

TABLE 3.45 Estimation of Chemical Shifts of Alkane Carbons

Relative to tetramethylsilane.

Positive terms indicate a downfield shift.

$$\delta_c = -2.6 + 9.1n_\alpha + 9.4n_\beta - 2.5n_\gamma + 0.3n_\delta + 0.1n_\epsilon \qquad \text{(plus any correction factors)}$$

where n_α is the number of carbons bonded directly to the ith carbon atom and n_β, n_γ, n_δ, and n_ϵ are the number of carbon atoms two, three, four, and five bonds removed. The constant is the chemical shift for methane.

Chain branching*	Correction factor	Chain branching*	Correction factor
1°(3°)	−1.1	4°(1°)	−1.5
1°(4°)	−3.4	2°(4°)	−7.2
2°(3°)	−2.5	3°(3°)	−9.5
3°(2°)	−3.7	4°(2°)	−8.4

* 1° signifies a CH_3— group; 2°, a —CH_2— group; 3°, a >CH— group; and 4°, a >C< group. 1° (3°) signifies a methyl group bound to a >CH— group, and so on.

Examples: For 3-methylpentane, CH_3—CH_2—CH(CH_3)—CH_2—CH_3,

$$\delta_{C\text{-}2} = -2.6 + 9.1(2) + 9.4(2) - 2.5 - 1(1)[2°(3°)] = 29.4$$

$$\delta_{C\text{-}3} = -2.6 + 9.1(3) + 9.4(2) + (2)[3°(2°)] = 36.2$$

TABLE 3.46 Effect of Substituent Groups on Alkyl Chemical Shifts

These increments are added to the shift value of the appropriate carbon atom as calculated from Table 3.45.

Straight: Y—CH_2—CH_2—CH_3 (α, β)

Branched: —CH_2—CH_2—CH(Y)—CH_2—CH_2— (γ, β, α, β, γ)

	α carbon		β carbon		
Substituent group Y*	Straight	Branched	Straight	Branched	γ carbon
—CO—OH	20.9	16	2.5	2	−2.2
—COO^- (anion)	24.4	20	4.1	3	−1.6
—CO—OR	20.5	17	2.5	2	−2
—CO—Cl	33	28		2	
—CO—NH_2	22	2.5			−0.5
—CHO	31		0		−2
—CO—R	30	24	1	1	−2
—OH	48.3	40.8	10.2	7.7	−5.8
—OR	58	51	8	5	−4
—O—CO—NH_2	51		8		
—O—CO—R	51	45	6	5	−3
—C—CO—Ar	53				
—F	68	63	9	6	−4
—Cl	31.2	32	10.5	10	−4.6
—Br	20.0	25	10.6	10	−3.1

TABLE 3.46 Effect of Substituent Groups on Alkyl Chemical Shifts (*Continued*)

	α carbon		β carbon		
Substituent group Y*	Straight	Branched	Straight	Branched	γ carbon
—I	−8	4	11.3	12	−1.0
—NH_2	29.3	24	11.3	10	−4.6
—NH_3^+	26	24	8	6	−5
—NHR	36.9	31	8.3	6	−3.5
—NR_2	42		6		−3
—NR_3^+	31		5		−7
—NO2	63	57	4	4	
—CN	4	1	3	3	−3
—SH	11	11	12	11	−6
—SR	20		7		−3
—CH=CH2	20		6		−0.5
—C6H5	23	17	9	7	−2
—C≡CH	4.5		5.5		−3.5

*R, alkyl group; Ar, aryl group.

TABLE 3.47 Estimation of Chemical Shifts of Carbon Attached to a Double Bond

The olefinic carbon chemical shift is calculated from the equation

$$\delta_c = 123.3 + 10.6n_\alpha + 7.2n_\beta - 7.9n_{\alpha'} - 1.8n_{\beta'} \qquad \text{(plus any steric correction terms)}$$

where n is the number of carbon atoms at the particular position, namely,

$$\begin{array}{cccc} \beta & \alpha & \alpha' & \beta' \\ \text{C} - & \text{C} = & \text{C} - & \text{C} \end{array}$$

Substituents on both sides of the double bond are considered separately. Additional vinyl carbons are treated as if they were alkyl carbons. The method is applicable to alicyclic alkenes; in small rings carbons are counted twice, i.e., from both sides of the double bond where applicable. The constant in the equation is the chemical shift for ethylene. The effect of other substituent groups is tabulated below.

Substituent group	β	α	α'	β'
—OR	2	29	−39	−1
—OH	6			−1
—O—CO—CH_3	−3	18	−27	4
—CO—CH_3		15	6	
—CHO		13.6	13.2	
—CO—OH		5.2	9.1	
—CO—OR		6	7	
—CN		−15.4	14.3	
—F		24.9	−34.3	
—Cl	−1	3.3	−5.4	2
—Br	0	−7.2	−0.7	2
—I		−37.4	7.7	
—C_6H_5		12	−11	

TABLE 3.47 Estimation of Chemical Shifts of Carbon Attached to a Double Bond (*Continued*)

Substituent pair		Steric correction term
α,α'	*trans*	0
α,α'	*cis*	−1.1
α,α	*gem*	−4.8
α',α'		+2.5
β,β		+2.3

TABLE 3.48 Carbon-13 Chemical Shifts in Substituted Benzenes

$$\delta_c = 128.5 + \Delta$$

Substituent group	Δ_{C-1}	Δ_{ortho}	Δ_{meta}	Δ_{para}
$-CH_3$	9.3	0.8	−0.1	−2.9
$-CH_2CH_3$	15.6	−0.4	0	−2.6
$-CH(CH_3)_2$	20.2	−2.5	0.1	−2.4
$-C(CH_3)_3$	22.4	−3.1	−0.1	−2.9
$-CH_2O-CO-CH_3$	7.7	0	0	0
$-C_6H_5$	13.1	−1.1	0.4	−1.2
$-CH{=}CH_2$	9.5	−2.0	0.2	−0.5
$-C{\equiv}CH$	−6.1	3.8	0.4	−0.2
$-CH_2OH$	12.3	−1.4	−1.4	−1.4
$-CO-OH$	2.1	1.5	0	5.1
$-COO^-$ (anion)	8	1	0	3
$-CO-OCH_3$	2.1	1.1	0.1	4.5
$-CO-CH_3$	9.1	0.1	0	4.2
$-CHO$	8.6	1.3	0.6	5.5
$-CO-Cl$	4.6	2.4	1	6.2
$-CO-CF_3$	−5.6	1.8	0.7	6.7
$-CO-C_6H_5$	9.4	1.7	−0.2	3.6
$-CN$	−15.4	3.6	0.6	3.9
$-OH$	26.9	−12.7	1.4	−7.3
$-OCH_3$	31.4	−14.0	1.0	−7.7
$-OC_6H_5$	29.2	−9.4	1.6	−5.1
$-O-CO-CH_3$	23.0	−6.4	1.3	−2.3
$-NH_2$	18.0	−13.3	0.9	−9.8
$-N(CH_3)_2$	22.4	−15.7	0.8	−11.5
$-N(C_6H_5)_2$	19	−4	1	−6
$-NHC_6H_5$	14.6	−10.7	0.7	−7.7
$-NH-CO-CH_3$	11.1	−9.9	0.2	−5.6
$-NO_2$	20.0	−4.8	0.9	5.8
$-F$	34.8	−12.9	1.4	−4.5
$-Cl$	6.2	0.4	1.3	−1.9
$-Br$	−5.5	3.4	1.7	−1.6
$-I$	−32.2	9.9	2.6	−1.4
$-CF_3$	−9.0	−2.2	0.3	3.2
$-NCO$	5.7	−3.6	1.2	−2.8
$-SH$	2.3	1.1	1.1	−3.1
$-SCH_3$	10.2	−1.8	0.4	−3.6
$-SO_2-NH_2$	15.3	−2.9	0.4	3.3
$-Si(CH_3)_3$	13.4	4.4	−1.1	−1.1

TABLE 3.49 Carbon-13 Chemical Shifts in Substituted Pyridines*

$$\delta_C(k) = C_k + \Delta_i$$

Substituent group	$C_2 = C_6 = 149.6$ Δ_{C-2} or Δ_{C-6}	Δ_{23}	Δ_{24}	Δ_{25}	Δ_{26}
$—CH_3$	9.1	−1.0	−0.1	−3.4	−0.1
$—CH_2CH_3$	14.0	−2.1	0.1	−3.1	0.2
$—CO—CH_3$	4.3	−2.8	0.7	3.0	−0.2
—CHO	3.5	−2.6	1.3	4.1	0.7
—OH	14.9	−17.2	0.4	−3.1	−6.8
$—OCH_3$	15.3	−13.1	2.1	−7.5	−2.2
$—NH_2$	11.3	−14.7	2.3	10.6	−0.9
$—NO_2$	8.0	−5.1	5.5	6.6	0.4
—CN	−15.8	−5.0	−1.7	3.6	1.9
—F	14.4	−14.7	5.1	−2.7	−1.7
—Cl	2.3	0.7	3.3	−1.2	0.6
—Br	−6.7	4.8	3.3	−0.5	1.4

Substituent group	Δ_{32}	$C_3 = C_5 = 124.2$ Δ_{C-3} or Δ_{C-5}	Δ_{34}	Δ_{35}	
$—CH_3$	1.3	9.0	0.2	−0.8	−2.3
$—CH_2CH_3$	0.3	15.0	−1.5	−0.3	−1.8
$—CO—CH_3$	0.5	−0.3	−3.7	−2.7	4.2
—CHO	2.4	7.9	0	0.6	5.4
—OH	−10.7	31.4	−12.2	1.3	−8.6
$—NH_2$	−11.9	21.5	−14.2	0.9	−10.8
—CN	3.6	−13.7	4.4	0.6	4.2
—Cl	−0.3	8.2	−0.2	0.7	−1.4
—Br	2.1	−2.6	2.9	1.2	−0.9
—I	7.1	−28.4	9.1	2.4	0.3

Substituent group	$\Delta_{42} = \Delta_{46}$	$\Delta_{43} = \Delta_{45}$	$C_4 = 136.2$ Δ_{C-4}
$—CH_3$	0.5	0.8	10.8
$—CH_2CH_3$	0	−0.3	15.9
$—CH{=}CH_2$	0.3	−2.9	8.6
$—CO—CH_3$	1.6	−2.6	6.8
—CHO	1.7	−0.6	5.5
$—NH_2$	0.9	−13.8	19.6
—CN	2.1	2.2	−15.7
—Br	3.0	3.4	−3.0

* May be used for disubstituted, polyheterocyclic, and polynuclear systems if deviations due to steric and mesomeric effects are allowed for.

TABLE 3.50 Carbon-13 Chemical Shifts Carbonyl Group

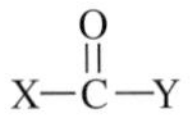

X	Y	δ_C	X	Y	δ_C
H—	—CH_3	199.7	CH_3—	—CH=CH_2	196.9
H—	—CCl_3	175.3	CH_3—	—C_6H_5	197.6
H—	—NH_2	165.5	CH_3—	—CH_2—CO—CH_3	201.9 (keto)
H—	—$N(CH_3)_2$	162.4			191.4 (enol)
H—	2-Furyl	153.3	CH_3—	—CH_2CHO	167.7
H—	2-Pyrrolyl	134.0	CH_3—	—C_6H_5—CH_3	196 (*m, p*)
H—	2-Thienyl	143.3			199 (*o*)
$(CH_3)_2$CH—	—OH	184.8	CH_3—	—2,6-$(CH_3)_2C_6H_5$	206
C_6H_5—	—OH	172.6	CH_3—	—OH	178
CF_3—	—OH	163.0	CH_3—	—O^- (anion)	181.5
CCl_3—	—OH	168.0	CH_3—	—OCH_3	170.7
$CH_3CH(NH_2)$—	—OH	176.5	CH_3—	—O—CH=CH_2	167.7
CF_3—	—OCH_2CH_3	158.1	CH_3—	—O—$CH(CH_3)_2$	170.3
H_2N—	—OCH_2CH_3	157.8	CH_3—	—O—CO—CH_3	167.3
2-Furyl	—OCH_3	159.1	CH_3—	—NH_2	172.7
$(CH_3)_2N$—	—C_6H_5	170.8	CH_3—	—$NHCH_3$	172
CH_2=$CHCH_2O$—CO—	—OCH_2CH=CH_2	157.6	CH_3—	—$N(CH_3)_2$	169.5
CH_3CH_2—	—CH_2CH_3	211.4	CH_3—	—Cl	169.6
CH_3—CH_2—	—O—CO—CH_2CH_3	170.3	CH_3—	—Br	165.6
CH_3—	—CH_3	205.8	CH_3—	—I	158.9
CH_3—	—CH_2CH_3	207			

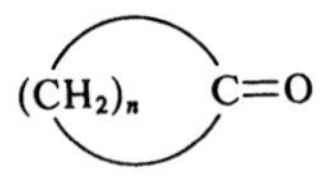

n	δ_C
3	207.9
4	218.2
5	211.3
6	211.4
7	216.0

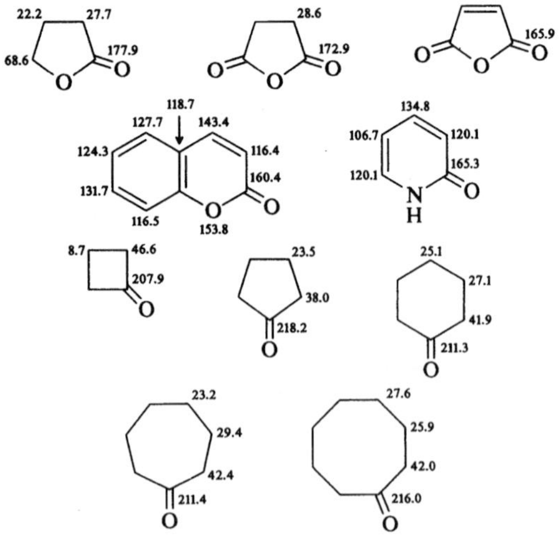

TABLE 3.51 One-Bond Carbon-Hydrogen Spin Coupling Constants

Structure	J_{CH}, Hz	Structure	J_{CH}, Hz
$H{-}CH_3$	125.0	$H{-}CH{=}O$; $CH_3{-}\underline{CH}{=}O$	172
$H{-}CH_2CH_3$	124.9	$H_2N{-}CH{=}O$	188.3
$CH_3{-}\underline{CH_2}{-}CH_3$	119.2	$(CH_3)_2N{-}\underline{CH}{=}O$	191
$H{-}C(\overline{CH_3})_2$	114.2	$H{-}COOH$	222
$H{-}CH_2CH_2OH$	126.9	$H{-}COO^-$ (anion)	195
$H{-}CH_2CH{=}CH_2$	122.4	$H{-}CO{-}OCH_3$	226
$H{-}CH_2C_6H_5$	129.4	$H{-}CO{-}F$	267
$H{-}CH_2C{\equiv}CH$	132.0	$CH_3CH_2{-}O{-}\underline{CHO}$	225.6
$H{-}CH_2CN$	136.1	$Cl_3{-}CHO$	207
$H{-}CH(CN)_2$	145.2	$H{-}C{\equiv}CH$	249
$H{-}CH_2{-}$halogen	149–152	$H{-}C{\equiv}CCH_3$	248
$H{-}CHF_2$	184.5	$H{-}C{\equiv}CC_6H_5$	251
$H{-}CHCl_2$	178.0	$H{-}C{\equiv}CCH_2OH$	241
$H{-}CH_2NH_2$	133.0	$H{-}CN$	269
$H{-}CH_2NH_3^+$	145.0	Cyclopropane	161
$H{-}CH_2OH$ (or $H{-}CH_2OR$)	140–141	Cyclobutane	136
$H{-}CH(OR)_2$	161–162	Cyclopentane	131
$H{-}C(OR)_3$	186	Cyclohexane	123
$H{-}C(OH)R_2$	143	Tetrahydrofuran 2,5	149
$H{-}CH_2NO_2$	146.0	3,4	133
$H{-}CH(NO_2)_2$	169.4	1,4-Dioxane	145
$H{-}CH_2COOH$	130.0	Benzene	159
$H{-}CH(COOH)_2$	132.0	Fluorobenzene 2,6	155
$H{-}CH{=}CH_2$	156.2	3,5	163
$H{-}C(CH_3){=}C(CH_3)_2$	148.4	4	161
$H{-}CH{=}C(tert\text{-}C_4H_9)_2$	152	Bromobenzene 2,6	171
$H{-}C(tert\text{-}C_4H_9){=}$ $C(tert\text{-}C_4H_9)_2$	143	3,5	164
		4	161
Methylenecycloalkane C_4–C_7	153–155	Benzonitrile 2,6	173
$H{-}CH{=}C{=}CH_2$	168	3,6	166
$H{-}C(C_6H_5){=}CH(C_6H_5)$		4	163
cis	155	Nitrobenzene 2,6	171
trans	151	3,5	167
Cyclopropene	220	4	163
$(H_t)(H_c)C{=}C(H_g)F$ *gem*	200	Mesitylene	154
cis	159	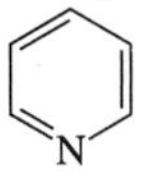 2,6	170
trans	162	3,5	163
$(H_t)(H_c)C{=}C(H_g)Cl$ *gem*	195	4	152
cis	163	2,4,6-Trimethylpyridine	158
trans	161		
$(H_t)(H_c)C{=}C(H_g)CHO$ *gem*	162	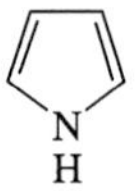 2,5	183
cis	157	3,4	170
trans	162		
$(H_t)(H_c)C{=}C(H_g)CN$ *gem*	177	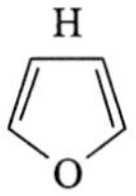 2,5	201
cis	163	3,4	175
trans	165		
$(H)(CH_3)C{=}N{-}OH$ *cis*	163	 2,5	185
trans	177	3,4	167
		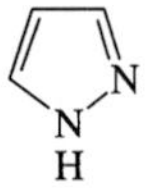 3,5	190
		4	178

(Continued)

TABLE 3.51 One-Bond Carbon-Hydrogen Spin Coupling Constants (*Continued*)

Structure		J_{CH}, Hz	Structure	J_{CH}, Hz
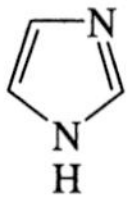	2 4	208 199	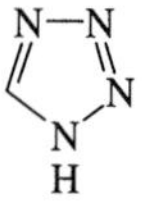	216
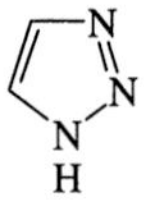		205		

TABLE 3.52 Two-Bond Carbon-Hydrogen Spin Coupling Constants

Structure	$^2J_{CH}$, Hz	Structure	$^2J_{CH}$, Hz
CH_3-CH_2-H	−4.5	$(CH_2)_n$ C=CH_2, $n = 4$	4.2
CCl_3-CH_2-H	5.9	$n = 5$	5.2
$ClCH_2-CH_2Cl$	−3.4	$n = 6$	5.5
$Cl_2CH-CHCl_2$	1.2	H(Cl)C=C(H)Cl *cis*	16.0
CH_3-CHO	26.7	*trans*	0.8
$CH_2=CH_2$	−2.4	$HC\equiv CH$	49.3
$(CH_3)_2C=O$	5.5	$C_6H_5O-C\equiv CH$	61.0
$CH_2=CH-CH=O$	26.9	$HC\equiv C-CHO$	33.2
$(C_2H_5)CH-CHO$	26.9	$ClCH_2-CHO$	32.5
$H_2NCH=CH-CHO$	6.0	$Cl_2CH-CHO$	35.3
$H_2NCH-CH-CHO$	20.0	Cl_3C-CHO	46.3
C_6H_6	1.0	$C_6H_5-C\equiv C\equiv CH_3$	10.8

TABLE 3.53 Carbon-Carbon Spin Coupling Constants

Structure*	J_{CC}, Hz	Structure	J_{CC}, Hz
H_3C-CH_3	35	C—CO—OR	59
H_3C-CHR_2	37	C—CN	52–57
H_3C-CH_2Ar	34	$C-C\equiv C$ $^2J_{CC} = 11.8$	67
H_3C-CH_2CN	33	$H_2C=CH_2$	68
$H_3C-CH_2-CH_2OH$		>C=C—CO—OH	70–71
C-1, C-2	38	>C=C—CN	71
C-2, C-3	34	>C=C—Ar	67–70
$H_3C-CH_2NH_2$	37	C_6H_6	57
C—C=O	38–40	$C_3H_5NO_2$	
C—C—C=O	36	1-2	55
C—C—Ar	43	2-3, 3-4	56
C—CO—O⁻ (anion)	52	$^3J_{2\text{-}5}$	7.6
C—CO—N	52		
C—CO—OH	57		

*R, alkyl group; Ar, aryl group.

TABLE 3.53 Carbon-Carbon Spin Coupling Constants (*Continued*)

Structure	J_{CC}, Hz	Structure	J_{CC}, Hz
C_6H_5I		Pyridine	
1-2	60	2-3	54
2-3	53	3-4	56
3-4	58	$^3J_{2\text{-}5}$	14
$^3J_{2\text{-}5}$	8.6	Furan	69
C_6H_5—OCH_3		Pyrrole	69
2-3	58	Thiophene	64
3-4	56	$H_2\underline{C}$=$\underline{C}$=$C(CH_3)_2$	100
$C_6H_5NH_2$		—$\overline{C}$≡$\overline{C}$—	170–176
1-2	61	**Structure**	$^2J_{CC}$, **Hz**
2-3	58		
3-4	57		
$^3J_{2\text{-}5}$	7.9	$\underline{C}H_3$—CO=$\underline{C}H_3$	16
$C_6H_5CH_3$	44	$\overline{C}H_3$—C≡$\overline{C}H$	11.8
		$\underline{C}H_3CH_2$—$\underline{C}N$	33

*R, alkyl group; Ar, aryl group.

TABLE 3.54 Carbon-Fluorine Spin Coupling Constants

Structure*	J_{CF}, Hz	Structure*	J_{CF}, Hz
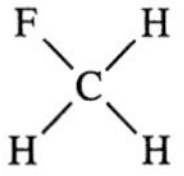	−158	*p*-F—C_6H_4—CF_3 *p*-F—C_6H_4—CO—CH_3 *p*-F—C_6H_4—NO_2	−252 −253 −257
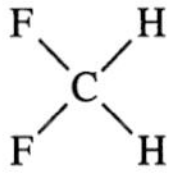	−235	F—C_6H_5 $^2J_{CF}$ = 21.0 $^3J_{CF}$ = 7.7 $^4J_{CF}$ = 3.4	−244
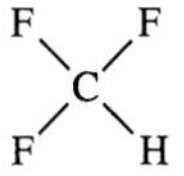	−274	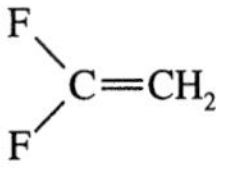	−287
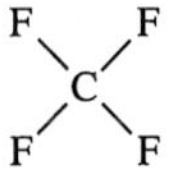	−259	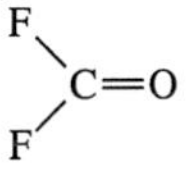	−308
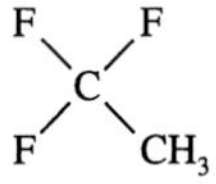	−271	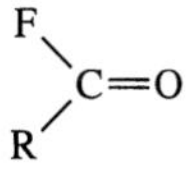	−353
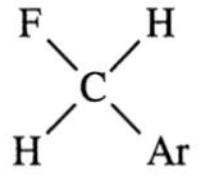	−165	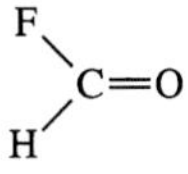	−369
F—CH_2CH_2— or F—CR_3 *p*-F—C_6H_4—OR *p*-F—C_6H_4—R	−167 −237 −241	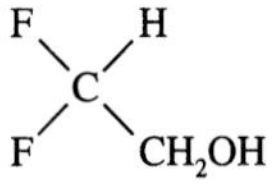	−241

(*Continued*)

TABLE 3.54 Carbon-Fluorine Spine Coupling Constants (*Continued*)

Structure*	J_{CF}, Hz	Structure*	J_{CF}, Hz
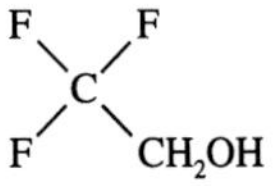	−278	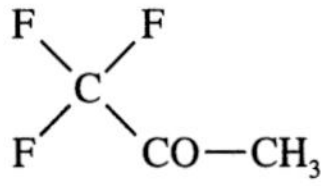	−289
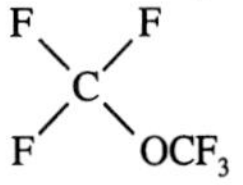	−265		

*Ar, aryl group; R, alkyl group.

TABLE 3.55 Carbon-13 Chemical Shifts of Deuterated Solvents

Relative to tetramethylsilane.

Solvent	Group	δ, ppm
Acetic-d_3 acid-d_1	Methyl	20.0
	Carbonyl	205.8
Acetone-d_6	Methyl	28.1
	Carbonyl	178.4
Acetonitrile-d_3	Methyl	1.3
	Carbonyl	117.7
Benzene-d_6		128.5
Carbon disulfide		193
Carbon tetrachloride		97
Chloroform-d_1		77
Cyclohexane-d_{12}		25.2
Dimethyl sulfoxide-d_6		39.5
1,4-Dioxane-d_6		67
Formic-d_1 acid-d_1	Carbonyl	165.5
Methanol-d_4		47–49
Methylene chloride-d_2		53.8
Nitromethane-d_3		57.3
Pyridine-d_5	C_3, C_5	123.5
	C_4	135.5
	C_2, C_6	149.9

TABLE 3.56 Carbon-13 Coupling Constants with Various Nuclei

Nuclei	Structure	^1J, Hz	^2J, Hz	^3J, Hz	^4J, Hz
^2H	$CDCl_3$	32			
	$CD_3—CO—CD_3$	20			
	$(CD_3)_2SO$	22			
	C_6D_6	26			
^7Li	CH_3Li	15			
^{11}B	$(C_6H_5)_4B^-$	49		3	

TABLE 3.56 Carbon-13 Coupling Constants with Various Nuclei (*Continued*)

Nuclei	Structure	^1J, Hz	^2J, Hz	^3J, Hz	^4J, Hz
^{14}N	$(CH_3)_4N^+$	10			
	CH_3NC	8			
^{29}Si	$(CH_3)_4Si$	52			
^{31}P	$(CH_3)_3P$	14			
	$(C_4H_9)_3P$	11	12	5	
	$(C_6H_5)_3P$	12	20	7	0
	$(CH_3)_4P^+$	56			
	$(C_4H_9)_4P^+$	48	4	15	
	$(C_6H_5)_4P^+$	88	11	13	3
	$R(RO)_2P{=}O$	142	5–7		
	$(C_4H_9O)_3P{=}O$		6	7	
^{77}Se	$(CH_3)_2Se$	62			
	$(CH_3)_3Se^+$	50			
^{113}Cd	$(CH_3)_2Cd$	513, 537			
^{119}Sn	$(CH_3)_4Sn$	340			
	$(CH_3)_3SnC_6H_5$	474	37	47	11
^{125}Te	$(CH_3)_2Te$	162			
^{199}Hg	$(CH_3)_2Hg$	687			
	$(C_6H_5)_2Hg$	1186	88	102	18
^{207}Pb	$(CH_3)_2Pb$	250			
	$(C_6H_5)_4Pb$	481	68	81	20

TABLE 3.57 Boron-11 Chemical Shifts

Values given in ppm on the δ scale, relative to $B(OCH_3)_3$.

Structure	δ, ppm	Structure	δ, ppm
R_3B	−67 to −68	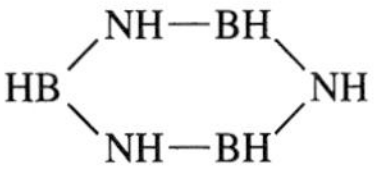	−12
Ar_3B	−43		
BF_3	24		
BCl_3	−12		
BBr_3	−6	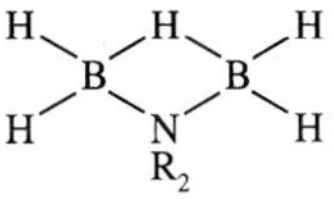	37
BI_3	41		
$B(OH)_3$	36		
$B(OR)_3$	0–1		
$B(NR_2)_3$	−13	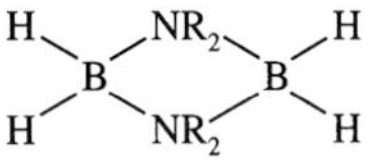	15
$C_6H_5BCl_2$	−36		
$C_6H_5B(OH)_2$	−14		
$C_6H_5B(OR)_2$	−10	$(CH_3)_2N{-}B(CH_3)_2$	62
$M(BH_4)$	55–61		
$B(BF_4)$	19–20		
Addition complexes		**Boranes**	
$R_2O \cdot BH_3$	18–19	B_2H_6	1
$R_3N \cdot BH_3$	25	B_4H_{10}	
$R_2NH \cdot BH_3$	33	(BH_2)	25
		(BH)	60
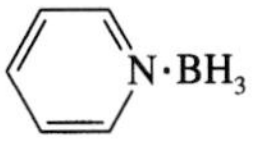	31		Base Apex

(*Continued*)

TABLE 3.57 Boron-11 Chemical Shifts (*Continued*)

Values given in ppm on the δ scale, relative to $B(OCH_3)_3$.

Structure	δ, ppm	Structure	δ, ppm
R_2O(or ROH)·BF_3	17–19		
R_2O(or ROH)·BCl_3	−7 to −8	B_5H_9	31 70
R_2O(or ROH)·BBr_3	23–24	B_5H_{11}	−16 50
R_2O(or ROH)·BI_3	74–82	$B_{10}H_{14}$	7 54
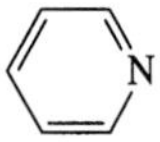 N·BBr_3	24		

TABLE 3.58 Nitrogen-15 (or Nitrogen-14) Chemical Shifts

Values given in ppm on the δ scale, relative to NH_3 *liquid.*

Substituent group	δ, ppm	Substituent group	δ, ppm
Aliphatic amines		Amides (*continued*)	
Primary	1–59	HCO—NH—Aryl	138–141
Secondary	7–81	RCO—NHR or RCO—NR_2	103–130
Tertiary	14–44	RCO—NH—Aryl	131–136
Cyclo, primary	29–44	Aryl—CO—H—Aryl	ca 126
Aryl amines	40–100	Guanidines	
Aryl hydrazines	40–100	Amino	30–60
Piperidines, decahydroquino-	30–82	Imino	166–207
lines		Thioureas	85–111
Amine cations		Thioamides	135–154
Primary	19–59	Cyanamides	
Secondary	40–74	R_2N—	−12 to −38
Tertiary	30–67	—CN	175–200
Quaternary	43–70	Carbodiimides	95–120
Enamines, tertiary type		Isocyanates	
Alkyl	29–82	Alkyl, primary	14–32
Cycloalkyl	55–104	Alkyl, secondary and tertiary	54–57
Aminophosphines	59–100	Aryl	ca 46
Amine *N*-oxides	95–122	Isothiocyanates	90–107
Ureas		Azides	52–80
Aliphatic	63–84		108–122
Aryl	105–108		240–260
Sulfonamides	79–164	Lactams	113–122
Amides		Hydrazones	
HCO—NHR		Amino	141–167
R = primary	100–115	Imino	319–327
R = secondary	104–148	Cyanates	155–182
R = tertiary	96–133	Nitrile *N*-oxides, fulminates	195–225

TABLE 3.58 Nitrogen-15 (or Nitrogen-14) Chemical Shifts (*Continued*)

Substituent group	δ, ppm	Substituent group	δ, ppm
Isonitriles		Oximes	340–380
Alkyl, primary	162–178	Nitramines	
Alkyl, secondary	191–199	Amine	252–280
Aryl	ca 180	$—NO_2$	328–355
Nitriles		Nitrates	310–353
Alkyl	235–241	*gem*-Polynitroalkanes	310–353
Aryl	258–268	Nitro	
Thiocyanates	265–280	Aryl	350–382
Diazonium		Alkyl	372–410
Internal	222–230	Hetero, unsaturated	354–367
Terminal	315–322	Azoxy	330–356
Diazo		Azo	504–570
Internal	226–303	Nitrosamines	222–250
Terminal	315–440		525–550
Nitrilium ions	123–150	Nitrites	555–582
Azinium ions	185–220	Thionitrites	720–790
Azine *N*-oxides	230–300	Nitroso	
Nitrones	270–285	Aliphatic amines, NO	535–560
Imides	170–178	Aryl	804–913
Imines	310–359		
Saturated cyclic systems			
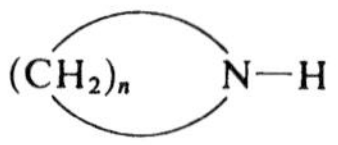		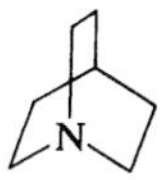	7.5 (in C_6H_6) 18.0 (in H_2O)
$n = 2$	−8.5	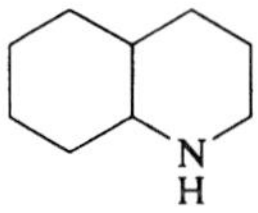	
$n = 3$	25.3		
$n = 4$	36.7		
$n = 5$	37.7		
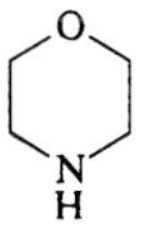	32.1	*cis*	42.4
		trans	52.9
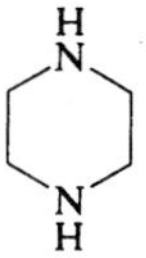	35.5		
Unsaturated cyclic systems			
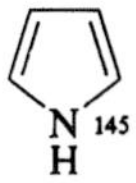	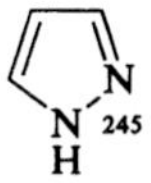	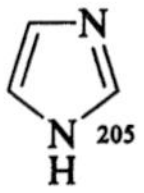	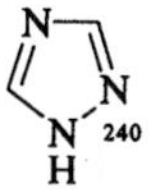
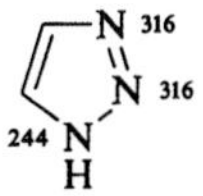	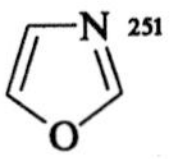	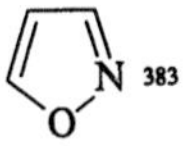	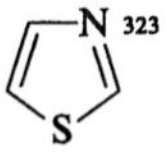

(*Continued*)

TABLE 3.58 Nitrogen-15 (or Nitrogen-14) Chemical Shifts (*Continued*)

Unsaturated cyclic systems (*contined*)

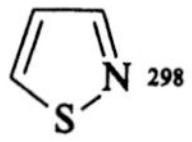

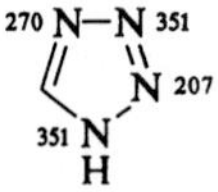

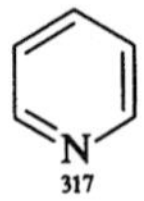

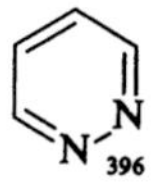

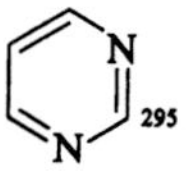

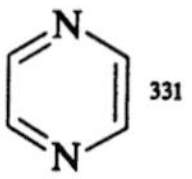

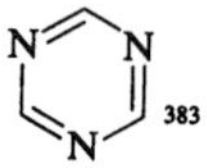

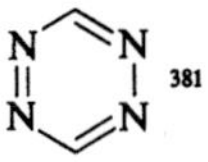

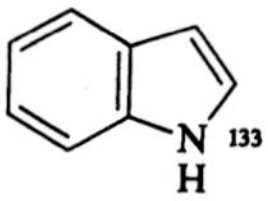

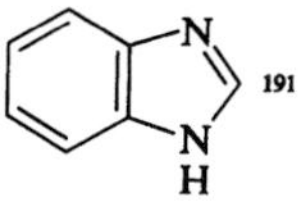

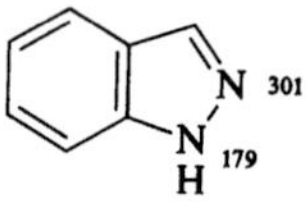

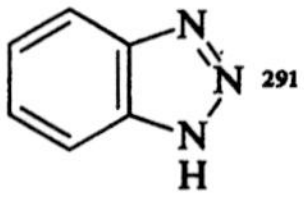

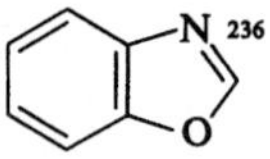

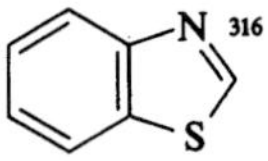

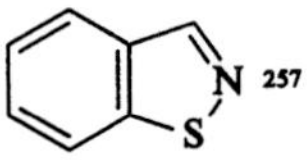

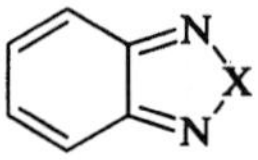

X	δ, ppm
O	517
S	331
Se	373

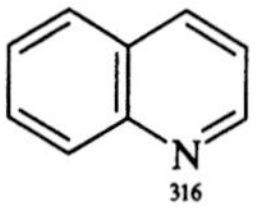

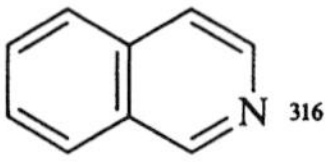

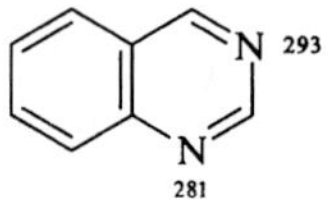

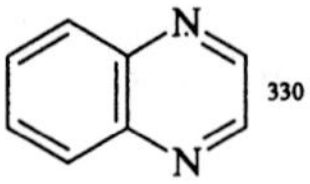

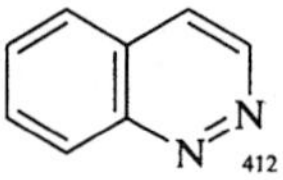

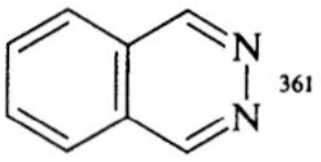

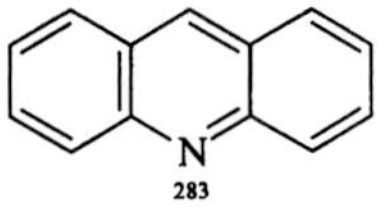

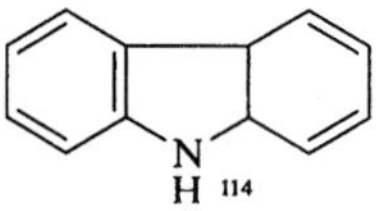

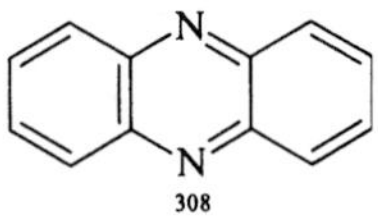

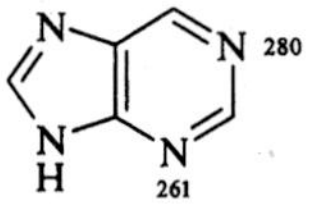

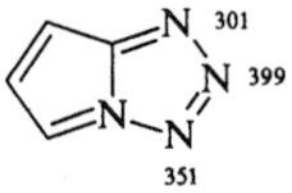

TABLE 3.59 Nitrogen-15 Chemical Shifts in Mono-substituted Pyridine

$$\delta = 317.3 + \Delta_i$$

Substituent	$\Delta_{C\text{-}2}$	$\Delta_{C\text{-}3}$	$\Delta_{C\text{-}4}$
$—CH_3$	−0.4	0.3	−8.0
$—CH_2CH_3$	−1.8		−6.6
$—CH(CH_3)_2$	−5.1		−5.9
$—C(CH_3)_3$	−2.5		−5.8
—CN	−0.9	−0.8	10.6
—CHO	10	11	29
$—CO—CH_3$	−9	15	11
$—CO—OCH_2CH_3$	11.8		−5
$—OCH_3$	−49	0	−23
—OH	−126	−2	−118
$—NO_2$	−23	1	22
$—NH_2$	−45	10	−46
—F	−42	−18	
—Cl	−4	4	−6
—Br	2	8	7

TABLE 3.60 Nitrogen-15 Chemical Shifts for Standards

Values given in ppm, relative to NH_3 liquid at 23°C.

Substance	δ, ppm	Conditions
Nitromethane (neat)	380.2	For organic solvents and acidic aqueous solutions
Potassium (or sodium) nitrate (saturated aqueous solution)	376.5	For neutral and basic aqueous solutions
$C(NO_2)_4$	331	For nitro compounds
$(CH_3)_2—CHO$ (neat)	103.8	For organic solvents and aqueous solutions
$(C_2H_5)_4N^+Cl^-$	64.4	Saturated aqueous solution
$(CH_3)_4N^+Cl^-$	43.5	Saturated aqueous solution
NH_4Cl	27.3	Saturated aqueous solution
NH_4NO_3	20.7	Saturated aqueous solution
NH_3	0.0	Liquid, 25°C
	−15.9	Vapor, 5 atm

TABLE 3.61 Nitrogen-15 to Hydrogen-1 Spin Coupling Constants

Structure	*J*, Hz	Structure	*J*, Hz
R—NH_2 and R_2NH	61–67	Aryl—$NHNH_2$	90
Aryl—NH_2	78	*p*-O_2N—aryl—$NHNH_2$	99
p-CH_3O—aryl—NH_2	79	Aryl—SO_2—NH_2	81
p-O_2N—aryl—NH_2	90–93	Aryl—SO_2—NHR	86
Amine salts (alkyl and aryl)	73–76	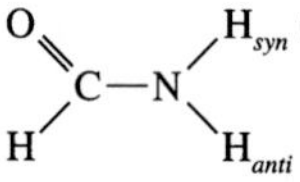 H_{syn} (to —CO—)	88
Aryl—NHOH	79	H_{anti}	92–93
Aryl—$NHCH_3$	87	$(R_3Si)_2NH$	67
Aryl—$NHCH_2F$	90	CF_3—S—NH_2	81
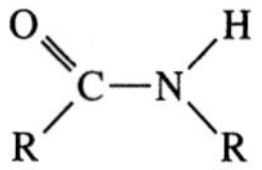	88–92	$(CF_3—S)_2NH$	99
Pyrrole	97	Pyridinium ion	90
HC≡NH^+	133–136	Quinolinium ion	96
⪾P—NH_2	82–90		

TABLE 3.62 Nitrogen-15 to Carbon-13 Spin Coupling Constants

Structure	*J*, Hz	Structure	*J*, Hz
Alkyl amines	4–4.5	Alkyl—NO_2	11
Cyclic alkyl amines	2–2.5	R—CN	18
Alkyl amines protonated	4–5	CH_3—$\overset{+}{N}$≡$\overset{-}{C}$	
Aryl amines	10–14	H_3C—N	10
Aryl amines protonated	9	—N≡C	9
CH_3CO—NH_2	14–15	Diaryl azoxy	
H_2N—CO—NH_2	20	*anti*	18
Aryl—NO_2	15	*syn*	13

TABLE 3.63 Nitrogen-15 to Fluorine-19 Spin Coupling Constants

Structure	*J*, Hz	Structure	*J*, Hz
NF_3	155	Pyridine	
F_4N_2	164	2-F	52
FNO_2	158	3-F	4
F_3NO	190	2,6-di-F	37
F_3C—O—NF_2	164–176	Pyridinium ion	
FCO—NF_2	221	2-F	23
$(NF_4)^+SbF_6^-$	323	3-F	3
$(NF_4)^+AsF_6^-$	328	Quinoline, 8-F	3
$(N_2F)^+AsF_6^-$	459	Aniline	
F_3C—NO_2	215	2-F	0
		3-F	0
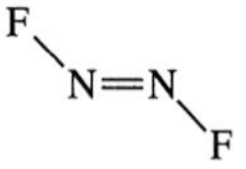 (^2J = 10)	190	4-F	1.5
		Anilinium ion	
		2-F	1.4
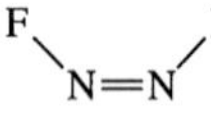 (^2J = 52)	203	3-F	0.2
		4-F	0

TABLE 3.64 Fluorine-19 Chemical Shifts

Values given in ppm on the δ scale, relative to CCl_3F.

Substituent group	δ, ppm	Substituent group	δ, ppm
$—SO_2—F$	−67 to −42 (aryl)(alkyl)	Cyclohexane-F	210 (axial) to 240 (equatorial)
$—CO—F$	−29 to −20		
$>N—CO—F$	−5		
$Aryl—CF_2Cl$	49		
$—CF_2I$	56	Perfluorocycloalkane	131–138
$—CF_2Br$	63	$>C\underline{F}—CF_3$	163–198
$R—CF_2Cl$	61–71	$>C\underline{F}(CF_3)_2$	180–191
$>C—CF_3$ and $aryl—CF_3$	56–73	$—CFH—$	198–231
$—CS—CF_3$	70	$—CFH_2$	235–244
$>CF—CF_3$	71–73	$F_2C{=}CF_2$	133
$—S—CF_3$	41	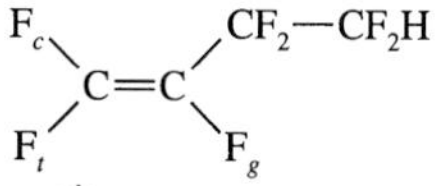	
$—S—CF_2—S—$	39		
$>P—CF_3$	46–66		
$>N—CF_3$	40–58		
$>N—CF_2—C$	85–127	*cis*	108
$—O—CF_2—R$	70–91	*trans*	92
$—O—CF_2—C\underline{F_3}$	70–91	*gem*	192
$—CH_2—CF_3$	76–77	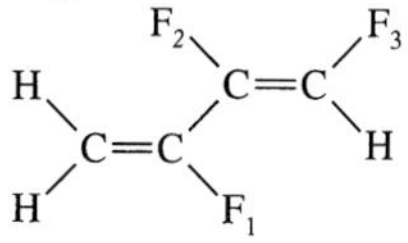	
$HO—CO—CF_3$	77		
$—CHF—C\underline{F_3}$	81		
$—CF_2—C\underline{F_3}$	78–88	F-1	126
$—CS—F$	81	F-2	155
$CF_3—C—N<$	84–96	F-3	162
$—CO—CF_2—C\underline{F_3}$	83	$Cl\underline{F}C{=}CH—CF_3$	61
$—CF_2—$	86–126	Cycloalkenes	
$—CF_2Br$	91	$=CF—C\underline{F_2}—C(CF_3 \text{ or } H)—$	101–113
$—C—CF_2—S—$	91–98	$—CF_2—C\underline{F_2}—C(CF_3 \text{ or } CH_3){=}$	110–114
$—CF{=}$	180–192	$—CF_2—C\underline{F_2}—CH{=}$	113–116
$—C\underline{F_2}—CF_3$	111	$—CF_2—C\underline{F_2}—CF{=}$	119–122
$—CO—CF_2—$	116–131	$Aryl—F$	113
$—C(halide)—CF_2—$	119–128	$C_{10}H_7—F$	
$—C\underline{F_2}—CF_3$	121–125	F-1	127
$—C\underline{F_2}—CF_2—$	121–129	F-2	114
$—CF_2—CH_2—$	122–133	$C_6H_5—C_6H_4—F$	
$—C\underline{F_2}—CHF_2$	128–132	F-2	117
$—C\underline{F_2}H$	136–143	F-3	113
 F_2	151–156	F-4	109
 F_2	147	C_6F_6	163
 F_2	96–133		
 F	159		

TABLE 3.65 Fluorine-19 Chemical Shifts for Standards

Substance	Formula	δ, ppm
Trichlorofluoromethane	$CFCl_3$	0.0
α,α,α-Trifluorotoluene	$C_6H_5CF_3$	63.8
Trifluoroacetic acid	CF_3COOH	76.5
Carbon tetrafluoride	CF_4	76.7
Fluorobenzene	C_6H_5F	113.1
Perfluorocyclobutane	C_4F_8	138.0

TABLE 3.66 Fluorine-19 to Fluorine-19 Spin Coupling Constants

Structure	J_{FF}, Hz
F_2C cycloalkane	
gem	212–260
Unsaturated compounds >C=C<	
gem	30–90
trans	115–130
cis	9–58
Aromatic compounds, monocyclic	
ortho	18–22
meta	0–7
para	12–15
Alkanes	
$C\underline{F}Cl_2—C\underline{F}_2—CFCl_2$	6
$C\underline{F}Cl_2—C\underline{F}_2—CCl_3$	5
$C\underline{F}_2Cl—C\underline{F}_2—CF_2Cl$	1
$C\underline{F}_3—C\underline{F}_2—CF_2Cl$ (or $—CF_3$)	<1
$CF_3—C\underline{F}_2—C\underline{F}_2Cl$	2
$C\underline{F}_3—CF_2—C\underline{F}_2Cl$	9
$C\underline{F}_3—CF_2—C\underline{F}_3$	7

TABLE 3.67 Silicon-29 Chemical Shifts

Values given in ppm on the δ scale relative to tetramethylsilane.

Substituent group X in $(CH_3)_{4-n}SiX_n$	*n*			
	1	2	3	4
—F	35	9	−52	−109
—Cl	30	32	13	−19
—Br	26	20	−18	−94
—I	9	−34	−18	−346
—H	−19	−42	−65	−93
$—C_2H_5$	2	5	7	8
$—C_6H_5$	−5	−9	−12	
$—CH{=}CH_2$	−7	−14	−21	−23
—Oalkyl	14–17	−3 to −6	−41 to −45	−79 to −83
—Oaryl	17	−6	−54	−101
—O—CO—alkyl	22	4	−43	−75
$—N(CH_3)_2$	6	−2	−18	−28

TABLE 3.67 Silicon-29 Chemical Shifts (*Continued*)

Structure	δ, ppm
Hydrides	
$H_3Si—$	−39 to −60
$—H_2Si—$	−5 to −37
$HSi\leqslant$	−2 to −39
Silicates	
Orthosilicate anions	−69 to −72
Silicon in end position	−77 to −81
Silicon in middle	−85 to −89
Branching silicons	−93 to −97
Cross-linked silicons	−107 to −120
Methyl siloxanes	
$(CH_3)_2Si—O—$ (end position)	6–8
$(CH_3)_2Si(O—)_2$ (middle)	−18 to −23
$CH_3Si(H)(O—)_2$ (middle)	−35 to −36
$CH_3Si(O—)_3$ (branching)	−65 to −66
$Si(O—)_4$ (cross-linked)	−105 to −110
Polysilanes	
$F_3Si—SiF_3$	−74
$Cl_3Si—SiCl_3$	−8
$(CH_3O)_3Si—Si(OCH_3)_3$	−53
$(CH_3)_3Si—Si(CH_3)_3$	−20
$(CH_3)_2Si[Si(CH_3)_3]_2$	−48
$HSi[Si(CH_3)_3]_3$	−117
$Si[Si(CH_3)_3]_4$	−135

TABLE 3.68 Phosphorus-31 Chemical Shifts

Values given in ppm on the δ scale, relative to 85% H_3PO_4.

Structure	Identical atoms attached directly to phosphorus	Non-identically substituted phosphorus		
		R = CH_3	R = C_2H_5	R = C_6H_5
P_4	461			
PR_3		62	20	6
PHR_2		99	56	41
PH_2R		164	128	122
PH_3	241			
PF_3	−97			
PRF_2			−168	−207
PCl_3	−220			
$PRCl_2$		−192	−196	−162
PR_2Cl		−94	−119	−81
PBr_3	−227			
$PRBr_2$		−184	−194	−152
PR_2Br		−91	−116	−71
PI_3	−178			
$P(CN)_3$	136			
$P(SiR_3)_3$		251		
$P(OR)_3$		−141	−139	−127
$P(OR)_2Cl$		−169	−165	−157
$P(OR)Cl_2$		−114	−177	−173
$P(SR)_3$		−125	−115	−132
$P(SR)_2Cl$		−188	−186	−183
$P(SR)Cl_2$		−206	−211	−204
$P(SR)_2Br$				−184

(*Continued*)

TABLE 3.68 Phosphorus-31 Chemical Shifts (*Continued*)

Structure	Identical atoms attached directly to phosphorus	Non-identically substituted phosphorus		
		R = CH_3	R = C_2H_5	R = C_6H_5
$P(SR)Br_2$		−204		
$P(NR_2)_3$		−123	−118	
$P(NR_2)Cl_2$		−166	−162	−151
$PR(NR_2)_2$		−86	−100	−100
$PR_2(NR_2)$		−39	−62	
$F_2P—PF_2$	−226			
$Cl_2P—PCl_2$	−155			
$I_2P—PI_2$	−170			
$PH_2^-K^+$	255			
$P(CF_3)_3$	3			
P_4O_6	−113			

Structure	Identical atoms attached directly to phosphorus	Non-identically substituted phosphorus		
		X = F	X = Cl	X = Br
$P(NCO)_3$	−97			
$P(NCO)_2X$		−128	−128	−127
$P(NCO)X_2$		−131	−166	
$P(NCS)_3$	−86			
$P(NCS)_2X$			−114	−112
$P(NCS)X_2$			−155	−153

Structure	Identical atoms attached directly to phosphorus	Non-identically substituted phosphorus		
		R = CH_3	R = C_2H_5	R = C_6H_5
$O{=}PR_3$		−36	−48	−25
$O{=}PHR_2$		−63		−23
$O{=}PF_3$	36			
$O{=}PRF_2$		−27	−29	−11
$O{=}PCl_3$	−2			
$O{=}PRCl_2$		−45	−53	−34
$O{=}PR_2Cl$		−65	−77	−43
$O{=}P(OR)_3$		−1	1	18
$O{=}P(OR)_2Cl$		−6	−3	6
$O{=}P(OR)Cl_2$		−6	−6	−2
$O{=}PH(OR)_2$		−19	−15	
$O{=}PR_2(OC_2H_5)$		−50	−52	−31
$O{=}PR(OC_2H_5)_2$		−30	−33	−17
$O{=}P(NR_2)_3$		−23	−24	−2
$O{=}PR_2(NR_2)$		−44		−26
$O{=}P(OR)_2NH_2$		−15	−12	−3
$O{=}P(OR)_2(NCS)$			19	29
$O{=}P(SR)_3$		−66	−61	−55
$O{=}PBr_3$	103			
$O{=}P(NCO)_3$	41			
$O{=}P(NCS)_3$	62			
$O{=}P(NH_2)_3$	−22			

TABLE 3.68 Phosphorus-31 Chemical Shifts (*Continued*)

Structure	Identical atoms attached directly to phosphorus	Structure	Identical atoms attached directly to phosphorus
PF_5	35	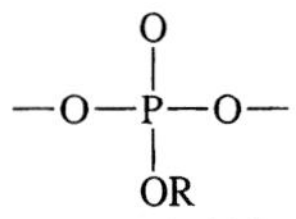	
$PF_6^-H^+$	144		
PBr_5	101		
$P(OC_2H_5)_5$	71		
PO_4^{3-}	−6	(middle group)	ca 18
$O{=}P[OSi(CH_3)_3]_3$	33	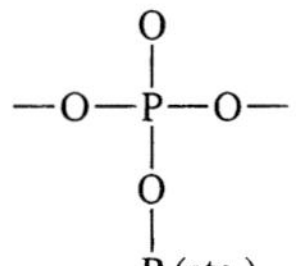	
$H_4P_2O_7$	11		
Phosphonates	−24 to −2		
Phosphonium cations			
Alkyl	−43 to −32		
Aryl	−35 to −18	P (etc.)	
$(O_3P{-}PO_3)^{4-}$	−9	(branch group)	ca 30
Polyphosphates			
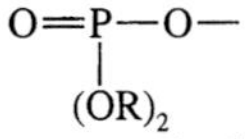 (end group)	ca 6		

Structure	Identical atoms attached directly to phosphorus	Non-identically substituted phosphorus		
		R = CH_3	R = C_2H_5	R = C_6H_5
$S{=}PR_3$		−59	−55	−43
$S{=}PCl_3$	−29			
$S{=}PRCl_2$		−80	−94	−75
$S{=}PR_2Cl$		−87	−109	−80
$S{=}PBr_3$	112			
$S{=}PRBr_2$		−21	−42	−20
$S{=}PR_2Br$		−64	−98	
$S{=}P(OR)_3$		−73	−68	−53
$S{=}P(OR)Cl_2$		−59	−56	−54
$S{=}P(OR)_2Cl$		−73	−68	−59
$S{=}PH(OR)_2$		−74	−69	−59
$S{=}P(SR)_3$		−98	−92	−92
$S{=}P(NH_2)_3$	−60			
$S{=}P(NR_2)_3$		−82	−78	
$Se{=}P(OR)_3$		−78	−71	−58
$Se{=}P(SR)_3$		−82	−76	
$P(OR)_5$			71	86
PRF_4		30	30	42
PR_2F_3		−9	−6	

TABLE 3.69 Phosphorus-31 Spin Coupling Constants

Substituent group	J_{PH}, Hz
$>$PH	180–225
—PH_2^-	134
RPH_2	160–210
$>$P—CH_3	1–6
$>$P—CH_2—	14
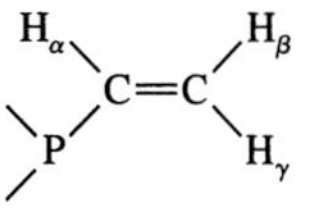	
α	12–22
β	30–40
γ	14–20
$(Halogen)_2$P—CH	16–20
$>$P—NH	10–28
$>$P—O—CH_3	11–15
$>$P—O—CH_2—R	6–10
$>$P—O—CHR_2	3–7
$>$P—SCH	5–20
$>$P—N—CH	8–25
$>$P—C—CH	0–4
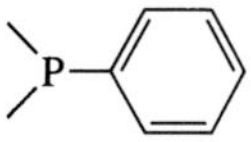	
ortho	7–10
meta	2–4
O=PHR_2	210–500
O—PH(S)R	490–540
O_2PHR	500–575
O_2PH(N)	560–630
O_2PH(S or Se)	630–655
O_3PH	630–760
S(or Se)=P—H	490–650
S(or Se)=PHR_2	420–454
O=P—CH_3	7–15
O=P—CH=C	15–30
O=P—CH—Aryl(or C=O)	15–30
$(Halogen)_2$P—N—CH	9–18
S=P—CH	11–15
$\geqslant$P—CH_3^+	12–17
$\geqslant$P—H^+	490–600

Substituent group	J_{PP}, Hz
$>$P—F	1320–1420 (1F) (3F)
RPF_2	1140–1290
R_2PF	1020–1110
RP(N)F	920–985 (alkyl) (aryl)
(—O)(—O)PF	1225–1305
(OCN)PF	1310
$>$N—P(F)$<$	1100–1200
$>$P—CF	60–90
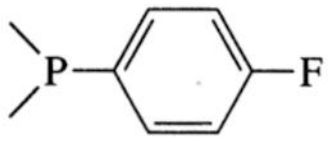	
ortho	0–60
meta	1–7
para	0–3

Substituent group		J_{PF}, Hz
—P(F) (trigonal bipyramidal)	axial	600–860
	equatorial	800–1000
O=P—CF		110–113
O=P—F		980–1190
<u>P</u>—O—P—<u>F</u>		2

Substituent group	J_{PB}, Hz
H_3B—P—N$<$	80

Substituent group	J_{PP}, Hz
$>$P—P$<$	220–400
O=P—P=O	330–500
S=P—P=S	15–500

TABLE 3.69 Phosphorus-31 Spin Coupling Constants (*Continued*)

Substituent group	J_{PP}, Hz	Substituent group	J_{PP}, Hz
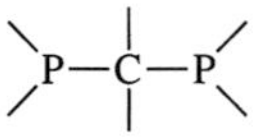	ca 70	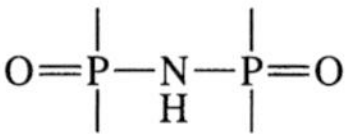	8–30
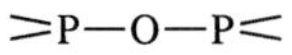	20–40	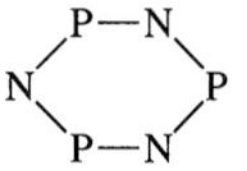	5–66
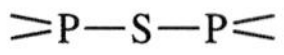	86–90		
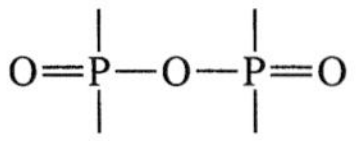	15–25	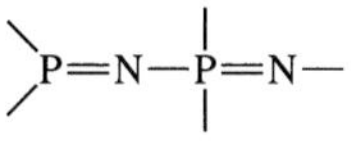	5–65

3.7 MASS SPECTROMETRY

3.7.1 Correlation of Mass Spectra with Molecular Structure

Molecular Identification. In the identification of a compound, the most important information is the molecular weight. The mass spectrometer is able to provide this information, often to four decimal places. One assumes that no ions heavier than the molecular ion form when using electron-impact ionization. The chemical ionization spectrum will often show a cluster around the nominal molecular weight.

Several relationships aid in deducing the empirical formula of the parent ion (and also molecular fragments). From the empirical formula hypothetical molecular structures can be proposed, using the entries in the formula indices of Beilstein and *Chemical Abstracts*.

Natural Isotopic Abundances. The relative abundances of natural isotopes produce peaks one or more mass units larger than the parent ion (Table 3.70(*a*)). For a compound $C_wH_xO_zN_y$, a formula allows one to calculate the percent of the heavy isotope contributions from a monoisotopic peak, P_M, to the P_{M+1} peak:

$$100\frac{P_{M+1}}{P_M} = 0.015x + 1.11w + 0.37y + 0.37z$$

Tables of abundance factors have been calculated for all combinations of C, H, N, and O up to mass 500 (J. H. Beynon and A. E. Williams, *Mass and Abundance Tables for Use in Mass Spectrometry,* Elsevier, Amsterdam, 1963).

Compounds that contain chlorine, bromine, sulfur, or silicon are usually apparent from prominent peaks at masses 2, 4, 6, and so on, units larger than the nominal mass of the parent of fragment ion. For example, when one chlorine atom is present, the $P + 2$ mass peak will be about one-third the intensity of the parent peak. When one bromine atom is present, the $P + 2$ mass peak will be about the same intensity as the parent peak. The abundance of heavy isotopes is treated in terms of the binominal expansion $(a + b)^m$, where a is the relative abundance of the light isotope, b is the relative abundance of the heavy isotope, and m is the number of atoms of the particular element present in the molecule. If two bromine atoms are present, the binominal expansion is

$$(a + b)^2 = a^2 + 2ab + b^2$$

Now substituting the percent abundance of each isotope (^{79}Br and ^{81}Br) into the expansion,

$$(0.505)^2 + 2(0.505)(0.495) + (0.495)^2$$

gives

$$0.255 + 0.500 + 0.250$$

which are the proportions of P:$(P + 2) : (P + 4)$, a triplet that is slightly distorted from a 1 : 2 : 1 pattern. When two elements with heavy isotopes are present, the binomial expansion $(a + b)^m(c + d)^n$ is used.

Sulfur-34 enhances the $P + 2$ peak by 4.2%; silicon-29 enhances the $P + 1$ peak by 4.7% and the $P + 2$ peak by 3.1%.

Exact Mass Differences. If the exact mass of the parent or fragment ions are ascertained with a high-resolution mass spectrometer, this relationship is often useful for combinations of C, H, N, and O (Table 3.70(b):

$$\frac{\text{Exact mass difference from nearest integral mass} + 0.0051z - 0.0031y}{0.0078} = \text{number of hydrogens}$$

One substitutes integral numbers (guesses) for z (oxygen) and y (nitrogen) until the divisor becomes an integral multiple of the numerator within 0.0002 mass unit.

For example, if the exact mass is 177.0426 for a compound containing only C, H, O, and N (Note the odd mass which indicates an odd number of nitrogen atoms), then

$$\frac{0.0426 + 0.0051z - 0.0031y}{0.0078} = 7 \text{ hydrigen atoms}$$

when $z = 3$ and $y = 1$. The empirical formula is $C_9H_7NO_3$ since

$$\frac{177 - 7(1) - 1(14) - 3(16)}{12} = 9 \text{ carbon atoms}$$

Number of Rings and Double Bonds. The total number of rings and double bonds can be determined from the empirical formula ($C_wH_xO_zN_y$) by the relationship

$$\frac{1}{2(2w - x + y + z)}$$

when covalent bonds comprise the molecular structure. Remember the total number for a benzene ring is four (one ring and three double bonds); a triple bond has two.

General Rules

1. If the nominal molecular weight of a compound containing only C, H, O, and N is even, so is the number of hydrogen atoms it contains.
2. If the nominal molecular weight is divisible by four, the number of hydrogen atoms is also divisible by four.
3. When the nominal molecular weight of a compound containing only C, H, O, and N is odd, the number of nitrogen atoms must be odd.

Metastable Peaks. If the mass spectrometer has a field-free region between the exit of the ion source and the entrance to the mass analyzer, metastable peaks m^* may appear as a weak, diffuse (often humped-shape) peak, usually at a nonintegral mass. The one-step decomposition process takes the general form:

$$\text{Original ion} \rightarrow \text{daughter ion} + \text{neutral fragment}$$

The relationship between the original ion and daughter ion is given by

$$m^* = \frac{(\text{mass of daughter ion})^2}{\text{mass of original ion}}$$

For example, a metastable peak appeared at 147.9 mass units in a mass spectrum with prominent peaks at 65, 91, 92, 107, 108, 155, 172, and 200 mass units. Try all possible combinations in the above expression. The fit is given by

$$147.9 = \frac{(172)^2}{200}$$

which provides this information:

$$200^1 \rightarrow 172^+ + 28$$

The probable neutral fragment lost is either $CH_2{=}CH_2$ or CO.

3.7.2 Mass Spectra and Structure

The mass spectrum is a fingerprint for each compound because no two molecules are fragmented and ionized in exactly the same manner on electron-impact ionization. In reporting mass spectra the data are normalized by assigning the most intense peak (denoted as base peak) a value of 100. Other peaks are reported as percentages of the base peak.

A very good general survey for interpreting mass spectral data is given by R. M. Silverstein, G. C. Bassler, and T. C. Morrill, *Spectrometric Identification of Organic Compounds*, 4th ed., Wiley, New York, 1981.

Initial Steps in Elucidation of a Mass Spectrum

1. Tabulate the prominent ion peaks, starting with the highest mass.
2. Usually only one bond is cleaved. In succeeding fragmentations a new bond is formed for each additional bond that is broken.
3. When fragmentation is accompanied by the formation of a new bond as well as by the breaking of an existing bond, a rearrangement process is involved. These will be even mass peaks when only C, H, and O are involved. The migrating atom is almost exclusively hydrogen; six-membered cyclic transition states are most important.
4. Tabulate the probable groups that (*a*) give rise to the prominent charged ion peaks and (*b*) list the neutral fragments.

General Rules for Fragmentation Patterns

1. Bond cleavage is more probable at branched carbon atoms: tertiary > secondary > primary. The positive charge tends to remain with the branched carbon.
2. Double bonds favor cleavage beta to the carbon (but see rule 6).
3. A strong parent peak often indicates a ring.
4. Saturated ring systems lose side chains at the alpha carbon. Upon fragmentation, two ring atoms are usually lost.
5. A heteroatom induces cleavage at the bond beta to it.
6. Compounds that contain a carbonyl group tend to break at this group; the positive charge remains with the carbonyl portion.

7. For linear alkanes, the initial fragment lost is an ethyl group (never a methyl group), followed by propyl, butyl, and so on. An intense peak at mass 43 suggests a chain longer than butane.
8. The presence of Cl, Br, S, and Si can be deduced from the unusual isotopic abundance patterns of these elements. These elements can be traced through the positively charged fragments until the pattern disappears or changes due to the loss of one of these atoms to a neutral fragment.
9. When unusual mass differences occur between some fragments ions, the pressure of F (mass difference 19), I (mass difference 127), or P (mass difference 31) should be suspected.

Characteristic Low-Mass Fragment Ions

Mass 30 = Primary amines
Masses 31, 45, 59 = Alcohol or ether
Masses 19 and 31 = Alcohol
Mass 66 = Monobasic carboxylic acid
Masses 77 and 91 = Benzene ring

Characteristic Low-Mass Neutral Fragments from the Molecular Ion

Mass 18 (H_2O) = From alcohols, aldehydes, ketones
Mass 19 (F) and 20 (HF) = Fluorides
Mass 27 (HCN) = Aromatic nitriles or nitrogen heterocycles
Mass 29 = Indicates either CHO or C_2H_5
Mass 30 = Indicates either CH_2O or NO
Mass 33 (HS) and 34 (H_2S) = Thiols
Mass 42 = CH_2CO via rearrangement from a methyl ketone or an aromatic acetate or an aryl-$NHCOCH_3$ group
Mass 43 = C_3H_7 or CH_3CO
Mass 45 = COOH or OC_2H_5

Table 3.71 is condensed, with permission, from the Catalog of Mass Spectral Data of the American Petroleum Institute Research Project 44. These, and other tables, should be consulted for further and more detailed information.

Included in the table are all compounds for which information was available through the C_7 compounds. The mass number for the five most important peaks for each compound are listed, followed in each case by the relative intensity in parentheses. The intensities in all cases are normalized to the *n*-butane 43 peak taken as 100. Another method for expressing relative intensities is to assign the base peak a value of 100 and express the relative intensities of the other peaks as a ratio to the base peak. Taking ethyl nitrate as an example, the tabulated values would be

Ethyl nitrate	91(0.01)(*P*)	46(100)	29(44.2)	30(30.5)	76(24.2)

The compounds are arranged in the table according to their molecular formulas. Each formula is arranged alphabetically, except that C is first if carbon occurs in the molecules, followed by H if it occurs. The formulas are then arranged alphabetically and according to increasing number of atoms of each kind, all C_4 compounds being listed before any C_5 compounds, and so on.

Nearly all these spectra have been recorded using 70-V electrons to bombard the sample molecules.

TABLE 3.70 Isotopic Abundances and Masses of Selecteded Elements

(*a*) Abundances of some polyisotopic elements, %

Element	Abundance	Element	Abundance	Element	Abundance
^{1}H	99.985	^{16}O	99.76	^{33}S	0.76
^{2}H	0.015	^{17}O	0.037	^{34}S	4.22
^{12}C	98.892	^{18}O	0.204	^{35}Cl	75.53
^{13}C	1.108	^{28}Si	92.18	^{37}Cl	24.47
^{14}N	99.63	^{29}Si	4.71	^{79}Br	50.52
^{15}N	0.37	^{30}Si	3.12	^{81}Br	49.48

(*b*) Selected isotope masses

Element	Mass	Element	Mass
^{1}H	1.0078	^{31}P	30.9738
^{12}C	12.0000	^{32}S	31.9721
^{14}N	14.0031	^{35}Cl	34.9689
^{16}O	15.9949	^{56}Fe	55.9349
^{19}F	18.9984	^{79}Br	78.9184
^{28}Si	27.9769	^{127}I	126.9047

TABLE 3.71 Table of Mass Spectra

		Mass numbers (and intensities) of:				
Molecular formula	Name	Parent peak	Base peak	Three next most intense peaks		
B_2H_6	Diborane	28(0.13)	26(54)	27(52)	24(48)	25(30)
$B_3H_6N_3$	Triborine triamine	81(21)	80(58)	79(37)	53(29)	52(22)
B_5H_9	Pentaborane	64(15)	59(30)	60(30)	62(24)	61(21)
$CBrClF_2$	Difluorochlorobromomethane	164(0.23)	85(86)	87(27)	129(17)	131(16)
CBr_2F_2	Difluorodibromomethane	208(1.7)	129(70)	131(68)	79(18)	31(18)
CCl_2F_2	Difluorodichloromethane	120(0.07)	85(33)	87(11)	50(3.9)	101(2.8)
CCl_3F	Fluorotrichloromethane	136(0.04)	101(54)	103(35)	66(7.0)	35(5.8)
CCl_4	Tetrachloromethane	152(0.0)	117(39)	119(37)	35(16)	47(16)
CF_3I	Trifluoroiodomethane	196(51)	196(51)	127(49)	69(40)	177(16)
CF_4	Tetrafluoromethane	88(0.0)	69(57)	50(6.8)	19(3.9)	31(2.8)
CHBrClF	Fluorochlorobromomethane	148(5.5)	67(120)	69(38)	31(13)	111(11)
$CHBrF_2$	Difluorobromomethane	130(13)	51(83)	31(18)	132(13)	79(13)
$CHCl_3$	Trichloromethane	118(1.3)	83(69)	85(44)	47(24)	35(13)
CHF_3	Trifluoromethane	70(0.25)	69(20)	51(18)	31(9.9)	50(2.9)
CHN	Hydrogen cyanide	27(92)	27(92)	26(15)	12(3.8)	28(1.6)
CH_2ClF	Fluorochloromethane	68(48)	68(48)	33(25)	70(15)	49(11)
CH_2Cl_2	Dichloromethane	84(41)	49(71)	86(26)	51(21)	47(13)
CH_2F_2	Difluoromethane	52(2.7)	33(26)	51(25)	31(7.3)	32(2.9)
CH_2O	Methanal (formaldehyde)	30(19)	29(21)	28(6.6)	14(0.94)	13(0.92)
CH_2O_2	Methanoic acid (formic)	46(72)	29(118)	45(56)	28(20)	17(20)
CH_3Cl	Chloromethane	50(66)	50(66)	15(54)	52(21)	49(6.6)

TABLE 3.71 Table of Mass Spectra (*Continued*)

Molecular formula	Name	Mass numbers (and intensities) of: Parent peak	Base peak	Three next most intense peaks		
CH_3F	Monofluoromethane	34(29)	15(31)	33(28)	14(5.3)	31(3.2)
CH_3I	Indomethane	142(78)	142(78)	127(29)	141(11)	15(10)
CH_3NO_2	Nitromethane	61(35)	30(65)	15(34)	46(23)	29(5.3)
CH_4	Methane	16(67)	16(67)	15(58)	14(11)	13(5.5)
CH_4O	Methanol	32(26)	31(38)	29(25)	28(2.4)	18(0.7)
CH_4S	Methanethiol	48(49)	47(65)	45(40)	46(9.5)	15(8.9)
CH_5N	Aminomethane (methylamine)	31(30)	30(53)	28(47)	29(8.7)	27(8.6)
CO	Carbon monoxide	28(78)	28(78)	12(3.7)	16(1.3)	29(0.9)
COS	Carbonyl sulfide	60(83)	60(83)	32(48)	28(6.9)	12(5.0)
CO_2	Carbon dioxide	44(76)	44(76)	28(5.0)	16(4.7)	12(1.9)
CS_2	Carbon disulfide	76(184)	76(184)	32(40)	44(33)	78(16)
C_2F_4	Tetrafluoroethene	100(20)	31(47)	81(34)	50(14)	12(3.6)
C_2F_6	Hexafluoroethane	138(0.14)	69(95)	119(39)	31(17)	50(9.6)
C_2F_6Hg	Hexafluorodimethylmercury	340(0.83)	69(111)	202(26)	271(22)	200(21)
C_2H_2	Ethyne	26(102)	26(102)	25(20)	24(5.7)	13(5.7)
C_2H_2ClN	Chloroethanenitrile	75(51)	75(51)	48(46)	40(23)	77(16)
$C_2H_2Cl_2$	*cis*-1,2-Dichloroethene	96(53)	61(72)	98(34)	63(23)	26(22)
$C_2H_2Cl_2$	*trans*-1,2,-Dichloroethene	96(49)	61(73)	98(32)	26(25)	63(23)
$C_2H_2Cl_4$	1,1,2,2-Tetrachloroethane	166(5.9)	83(95)	85(60)	95(11)	87(9.7)
$C_2H_2F_2$	1,1-Difluoroethene	64(32)	64(32)	45(21)	31(16)	33(13)
$C_2H_3Cl_3$	1,1,1-Trichloroethane	132(0.0)	97(37)	99(24)	61(19)	117(7.1)
$C_2H_3Cl_3$	1,1,2-Trichloroethane	132(3.9)	97(43)	83(41)	99(27)	85(26)
$C_2H_3F_3$	1,1,1-Trifluoroethane	84(0.94)	69(81)	65(31)	15(13)	45(10)
C_2H_3N	Ethanenitrile	41(89)	41(89)	40(46)	39(17)	38(10)
C_2H_4	Ethene (ethylene)	28(66)	28(66)	27(43)	26(41)	25(7.8)
C_2H_4BrCl	1-Chloro-2-bromoethane	142(7.9)	63(93)	27(82)	65(30)	26(24)
$C_2H_4Br_2$	1,2-Dibromoethane	186(1.6)	27(93)	107(72)	109(67)	26(23)
$C_2H_4Cl_2$	1,1-Dichloroethane	98(5.7)	63(89)	27(64)	65(28)	26(21)
$C_2H_4Cl_2$	1,2-Dichloroethane	98(1.7)	62(12)	27(11)	49(4.9)	64(3.9)
$C_2H_4N_2$	Diazoethane	56(16)	28(27)	27(25)	26(21)	41(5.2)
C_2H_4O	Ethanal (acetaldehyde)	44(30)	29(66)	43(18)	42(6.1)	26(6.1)
C_2H_4O	Ethylene oxide	44(30)	29(46)	15(30)	14(12)	43(7.1)
$C_2H_4O_2$	Ethanoic acid (acetic)	60(19)	43(37)	45(33)	15(21)	14(8.0)
$C_2H_4O_2$	Methyl formate	60(27)	31(96)	29(60)	32(33)	28(6.8)
C_2H_5Br	Bromoethane	108(35)	29(54)	27(48)	110(33)	26(16)
C_2H_5Cl	Chloroethane	64(36)	64(36)	28(32)	29(30)	27(27)
C_2H_5F	Fluoroethane	48(2.4)	47(24)	27(8.9)	33(8.2)	26(3.0)
C_2H_5N	Ethylenimine	43(31)	42(56)	28(44)	15(20)	41(11)
$C_2H_5NO_2$	Nitroethane	75(0.0)	29(85)	27(74)	30(19)	26(11)
$C_2H_5NO_3$	Ethyl nitrate	91(0.01)	46(95)	29(42)	30(29)	76(23)
C_2H_6	Ethane	30(26)	28(99)	27(33)	26(23)	29(21)
C_2H_6O	Ethanol	46(9.7)	31(63)	45(22)	29(14)	27(14)
C_2H_6O	Dimethyl ether	46(32)	45(71)	29(56)	15(41)	14(8.9)
$C_2H_6O_2$	Dimethyl peroxide	62(28)	29(47)	31(45)	15(16)	30(12)
C_2H_6S	2-Thiapropane	62(56)	47(69)	45(42)	46(29)	35(24)
C_2H_6S	Ethanethiol	62(44)	62(44)	29(43)	47(36)	27(35)
$C_2H_6S_2$	2,3-Dithiabutane	94(95)	94(95)	45(59)	79(56)	46(34)
$C_2H_6S_3$	2,3,4-Trithiapentane	126(54)	126(54)	45(32)	79(27)	47(19)
C_2H_7N	Aminoethane (ethylamine)	45(18)	30(96)	28(28)	44(19)	27(13)
C_2H_7N	*N*-Methylaminomethane	45(36)	44(71)	28(48)	15(14)	42(13)

TABLE 3.71 Table of Mass Spectra (*Continued*)

Molecular formula	Name	Parent peak	Base peak	Three next most intense peaks		
		Mass numbers (and intensities) of:				
$C_2H_8N_2$	1,2-Diaminoethane	60(2.7)	30(111)	18(14)	42(6.9)	43(5.9)
C_3F_6	Hexafluoropropene	150(16)	31(56)	69(44)	131(41)	100(20)
C_3F_8	Octafluoropropane	188(0.0)	69(171)	31(49)	169(42)	50(16)
C_3H_3N	Propenenitrile	53(55)	26(55)	52(41)	51(18)	27(10)
C_3H_4	Propadiene	40(72)	40(72)	39(69)	38(29)	37(23)
C_3H_4	Propyne (methylacetylene)	40(79)	40(79)	39(73)	38(29)	37(22)
C_3H_4ClN	3-Chloropropanenitrile	89(12)	49(68)	54(54)	51(29)	26(20)
C_3H_4O	Propenal (acrolein)	56(16)	27(25)	26(15)	28(13)	55(11)
C_3H_5Cl	1-Chloro-1-propene	76(30)	41(70)	39(43)	40(10)	78(9.6)
C_3H_5ClO	3-Chloro-1,2-epoxypropane	92(0.19)	57(55)	27(53)	29(40)	31(21)
$C_3H_5ClO_2$	Methyl chloroacetate	109(0.23)	59(56)	49(44)	15(43)	29(37)
$C_3H_5Cl_3$	1,2,3-Trichloropropane	146(0.71)	75(61)	110(22)	77(19)	61(18)
C_3H_5N	Propanenitrile	55(8.3)	28(83)	54(51)	26(17)	27(15)
C_3H_6	Cyclopropane	42(64)	42(64)	41(58)	39(44)	27(23)
C_3H_6	Propene	42(39)	41(58)	39(41)	27(22)	40(17)
$C_3H_6Cl_2$	1,1-Dichloropropane	112(0.0)	63(27)	41(25)	77(22)	62(19)
$C_3H_6Cl_2$	1,2-Dichloropropane	112(2.6)	63(51)	62(36)	27(29)	41(25)
C_3H_6O	1-Propen-3-ol (allyl alc.)	58(12)	57(43)	29(34)	31(26)	27(19)
C_3H_6O	Propanal	58(25)	29(66)	28(46)	27(38)	26(14)
C_3H_6O	Propanone (acetone)	58(24)	43(85)	15(26)	27(5.9)	42(5.9)
C_3H_6O	1,2-Epoxypropane	58(19)	28(44)	29(30)	27(28)	26(18)
$C_3H_6O_2$	1,3-Dioxolane	74(3.1)	73(52)	43(36)	44(30)	29(30)
$C_3H_6O_2$	Propanoic acid	74(27)	28(34)	29(28)	27(21)	45(19)
$C_3H_6O_2$	Ethyl formate	74(5.8)	31(82)	28(60)	29(54)	27(36)
$C_3H_6O_2$	Methyl acetate	74(22)	43(148)	29(16)	42(15)	59(8.4)
$C_3H_6O_3$	Methyl carbonate	90(3.3)	15(93)	45(54)	29(43)	31(34)
C_3H_7Br	1-Bromopropane	122(14)	43(94)	27(55)	41(47)	39(22)
C_3H_7Br	2-Bromopropane	122(11)	43(100)	27(50)	41(47)	39(24)
C_3H_7Cl	1-Chloropropane	78(3.6)	42(60)	29(27)	27(22)	41(14)
C_3H_7Cl	2-Chloropropane	78(14)	43(58)	27(20)	63(15)	41(13)
C_3H_7F	2-Fluoropropane	62(1.0)	47(84)	46(24)	61(12)	27(7.6)
C_3H_7N	2-Methylethylenimine	57(22)	28(76)	56(34)	30(24)	29(19)
C_3H_7N	*N*-Methylethylenimine	57(31)	42(94)	15(46)	28(25)	27(17)
C_3H_7NO	*N*,*N*-Dimethylformamide	73(54)	44(63)	42(29)	28(25)	15(24)
$C_3H_7NO_2$	1-Nitropropane	89(0.0)	43(68)	27(67)	41(58)	39(24)
$C_3H_7NO_2$	2-Nitropropane	89(0.0)	43(75)	41(55)	27(53)	39(23)
C_3H_8	Propane	44(25)	29(85)	28(50)	27(33)	43(19)
C_3H_8O	1-Propanol	60(7.2)	31(115)	27(18)	29(17)	59(10)
C_3H_8O	2-Propanol	60(0.45)	45(112)	43(19)	27(18)	29(11)
C_3H_8O	Methyl ethyl ether	60(24)	45(94)	29(46)	15(23)	27(19)
$C_3H_8O_2$	Dimethoxymethane	76(1.6)	45(117)	29(51)	75(51)	15(48)
$C_3H_8O_2$	2-Methoxy-1-ethanol	76(7.3)	45(122)	29(44)	15(38)	31(32)
C_3H_8S	2-Thiabutane	76(47)	61(73)	48(40)	47(30)	27(27)
C_3H_8S	1-Propanethiol	76(30)	47(43)	43(34)	27(34)	41(32)
C_3H_8S	2-Propanethiol	76(41)	43(65)	41(44)	27(41)	61(26)
C_3H_9N	1-Aminopropane	59(1.5)	30(20)	28(2.5)	27(1.3)	41(1.0)
C_3H_9N	Trimethylamine	59(37)	58(95)	42(44)	15(32)	30(17)
$C_3H_{12}B_3N_3$	*B*,*B*′,*B*″-Trimethylborazole	123(30)	108(102)	107(77)	67(38)	66(34)
C_4F_6	Hexafluorocyclobutene	162(21)	93(80)	31(51)	143(15)	74(6.9)
C_4F_6	Hexafluoro-1,3-butadiene	162(27)	93(90)	31(45)	74(10)	112(10)

(*Continued*)

TABLE 3.71 Table of Mass Spectra (*Continued*)

Molecular formula	Name	Mass numbers (and intensities) of: Parent peak	Base peak	Three next most intense peaks		
C_4F_6	Hexafluoro-2-butyne	162(18)	93(47)	143(38)	31(25)	69(20)
C_4F_8	Octafluorocyclobutane	200(0.12)	100(97)	131(84)	31(53)	69(24)
C_4F_8	Octafluoromethylpropene	200(14)	69(74)	181(54)	31(44)	93(22)
C_4F_8	Octafluoro-1-butene	200(11)	131(122)	31(86)	69(44)	93(16)
C_4F_{10}	Decafluorobutane	238(0.0)	69(178)	119(33)	31(22)	100(15)
$C_4HF_7O_2$	Heptafluorobutanoic acid	214(0.0)	45(26)	69(24)	119(17)	100(14)
C_4H_2	1,3-Butadiyne	50(133)	50(133)	49(57)	48(14)	25(12)
C_4H_4	1-Buten-3-yne	52(55)	52(55)	51(28)	50(23)	49(7.2)
C_4H_4O	Furan	68(36)	39(58)	38(9.7)	29(9.3)	40(6.7)
C_4H_4S	Thiophene	84(93)	84(93)	58(56)	45(49)	39(24)
$C_4H_4S_2$	2-Thiophenethiol	116(68)	116(68)	71(64)	45(31)	39(11)
C_4H_5N	3-Butenenitrile	67(27)	41(80)	39(36)	27(30)	40(20)
C_4H_5N	Pyrrole	67(67)	67(67)	39(46)	41(42)	40(36)
C_4H_6	1,2-Butadiene	54(65)	54(65)	27(35)	53(29)	39(28)
C_4H_6	1,3-Butadiene	54(46)	39(53)	27(36)	53(31)	28(24)
C_4H_6	1-Butyne	54(64)	54(64)	39(49)	53(27)	27(26)
C_4H_6	2-Butyne	54(93)	54(93)	27(42)	53(41)	39(24)
$C_4H_6Cl_2O_2$	Ethyl dichloroacetate	156(0.12)	29(192)	27(58)	83(23)	28(19)
$C_4H_6O_2$	2,3-Butanedione	86(13)	43(118)	15(40)	14(12)	42(8.6)
$C_4H_6O_2$	Methyl 2-propenoate	86(2.0)	55(98)	27(66)	15(27)	26(22)
$C_4H_7BrO_2$	2-Bromoethyl acetate	166(0.03)	43(158)	27(35)	106(31)	108(30)
C_4H_7Cl	2-Chloro-2-butene	90(27)	55(68)	27(21)	39(21)	29(18)
$C_4H_7ClO_2$	2-Chloroethyl acetate	122(0.0)	43(162)	73(43)	15(36)	27(29)
$C_4H_7ClO_2$	Ethyl chloroacetate	122(0.96)	29(130)	27(41)	77(37)	49(29)
C_4H_7N	2-Methylpropanenitrile	69(1.7)	42(79)	68(38)	28(26)	54(19)
C_4H_7N	*n*-Butanenitrile	69(0.15)	41(112)	29(70)	27(38)	28(11)
C_4H_8	Cyclobutane	56(41)	28(65)	41(58)	27(27)	26(15)
C_4H_8	2-Methylpropene	56(36)	41(85)	39(37)	28(18)	27(17)
C_4H_8	1-Butene	56(32)	41(87)	39(30)	27(26)	28(26)
C_4H_8	*cis*-2-Butene	56(36)	41(76)	39(27)	27(25)	28(24)
C_4H_8	*trans*-2-Butene	56(37)	41(80)	27(27)	39(26)	28(26)
$C_4H_8Cl_2$	1,2-Dichlorobutane	126(0.30)	41(39)	77(35)	27(20)	76(16)
$C_4H_8Cl_2$	1,4-Dichlorobutane	126(0.03)	55(87)	41(29)	27(24)	90(23)
$C_4H_8Cl_2$	*dl*-2,3-Dichlorobutane	126(0.95)	63(63)	62(58)	27(57)	55(29)
$C_4H_8Cl_2$	*meso*-2,3-Dichlorobutane	126(0.95)	63(64)	27(57)	62(54)	55(31)
$C_4H_8N_2$	Acetaldazine	84(23)	42(92)	15(47)	28(46)	69(38)
C_4H_8O	Butanal	72(19)	27(41)	29(38)	44(34)	43(32)
C_4H_8O	2-Butanone	72(17)	43(97)	29(24)	27(15)	57(6.0)
C_4H_8O	Ethyl ethenyl ether	72(27)	44(64)	43(56)	29(49)	27(43)
C_4H_8O	*cis*-2,3-Epoxybutane	72(3.6)	43(67)	44(39)	27(35)	29(33)
C_4H_8O	*trans*-2,3-Epoxybutane	72(3.5)	43(69)	44(35)	29(32)	27(31)
C_4H_8O	Tetrahydrofuran	72(22)	42(76)	41(39)	27(25)	71(20)
$C_4H_8O_2$	2-Methyl-1,3-dioxacyclopentane	88(0.33)	73(67)	43(48)	45(44)	29(34)
$C_4H_8O_2$	1,4-Dioxane	88(42)	28(138)	29(51)	58(33)	31(24)
$C_4H_8O_2$	2-Methylpropanoic acid	88(8.1)	43(77)	41(33)	27(26)	73(19)
$C_4H_8O_2$	*n*-Butanoic acid	88(1.0)	60(40)	73(12)	27(9.6)	41(9.1)
$C_4H_8O_2$	*n*-Propyl formate	88(0.41)	31(123)	42(89)	29(38)	27(36)
$C_4H_8O_2$	Ethyl acetate	88(7.1)	43(181)	29(46)	45(24)	27(24)
$C_4H_8O_2$	Methyl propanoate	88(23)	29(110)	57(83)	27(40)	59(27)
C_4H_8S	3-Methylthiacyclobutane	88(42)	46(101)	45(31)	39(24)	47(21)

TABLE 3.71 Table of Mass Spectra (*Continued*)

Molecular formula	Name	Parent peak	Base peak	Three next most intense peaks		
		Mass numbers (and intensities) of:				
C_4H_8S	Thiacyclopentane	88(44)	60(82)	45(29)	46(29)	47(22)
C_4H_9Br	1-Bromobutane	136(7.0)	57(86)	41(63)	29(50)	27(46)
C_4H_9Br	2-Bromobutane	136(0.72)	57(108)	41(65)	29(61)	27(36)
C_4H_9N	Pyrrolidine	71(24)	43(102)	28(38)	70(33)	42(20)
$C_4H_9NO_2$	*n*-Butyl nitrite	103(0.0)	27(55)	43(54)	41(50)	30(47)
C_4H_{10}	2-Methylpropane	58(3.2)	43(117)	41(45)	42(39)	27(33)
C_4H_{10}	*n*-Butane	58(12)	43(100)	29(44)	27(37)	28(33)
$C_4H_{10}Hg$	Diethylmercury	260(12)	29(188)	27(54)	28(21)	231(15)
$C_4H_{10}O$	2-Methyl-1-propanol	74(7.5)	43(84)	31(56)	42(48)	41(47)
$C_4H_{10}O$	2-Methyl-2-propanol	74(0.0)	59(92)	31(31)	41(19)	43(14)
$C_4H_{10}O$	1-Butanol	74(0.37)	31(52)	56(44)	41(31)	43(30)
$C_4H_{10}O$	2-Butanol	74(0.30)	45(116)	31(23)	59(22)	27(20)
$C_4H_{10}O$	Diethyl ether	74(22)	31(73)	59(34)	29(29)	45(28)
$C_4H_{10}O$	Methyl isopropyl ether	74(8.3)	59(126)	29(42)	43(37)	15(32)
$C_4H_{10}O_2$	1,1-Dimethoxyethane	90(0.06)	59(93)	29(52)	15(37)	31(37)
$C_4H_{10}O_2$	1,2-Dimethoxyethane	90(12)	45(177)	29(53)	15(50)	60(16)
$C_4H_{10}O_2$	2-Ethoxyethanol	90(0.49)	31(112)	29(57)	59(56)	27(31)
$C_4H_{10}O_2$	Diethyl peroxide	90(20)	29(116)	15(42)	45(34)	62(30)
$C_4H_{10}S$	3-Methyl-2-thiabutane	90(41)	41(49)	75(47)	43(41)	48(38)
$C_4H_{10}S$	2-Thiapentane	90(58)	61(126)	48(50)	41(43)	27(43)
$C_4H_{10}S$	3-Thiapentane	90(41)	75(59)	47(51)	27(39)	61(33)
$C_4H_{10}S$	2-Methyl-1-propanethiol	90(35)	41(60)	43(46)	56(34)	47(29)
$C_4H_{10}S$	2-Methyl-2-propanethiol	90(34)	41(68)	57(61)	29(44)	39(21)
$C_4H_{10}S$	1-Butanethiol	90(40)	56(74)	41(65)	27(42)	47(31)
$C_4H_{10}S$	2-Butanethiol	90(34)	41(56)	57(50)	61(46)	29(46)
$C_4H_{10}S_2$	2,3-Dithiahexane	122(37)	80(53)	43(36)	41(27)	27(25)
$C_4H_{10}S_2$	3,4-Dithiahexane	122(73)	29(82)	66(81)	27(57)	94(53)
$C_4H_{10}SO_3$	Ethyl sulfite	138(3.3)	29(131)	31(59)	45(42)	27(39)
$C_4H_{11}N$	*N*-Ethylaminoethane	73(17)	58(83)	30(81)	28(30)	27(24)
$C_4H_{11}N$	1-Amino-2-methylpropane	73(1.0)	30(22)	28(2.0)	41(1.2)	27(1.1)
$C_4H_{11}N$	2-Amino-2-methylpropane	73(0.25)	58(127)	41(26)	42(20)	15(18)
$C_4H_{11}N$	1-Aminobutane	73(12)	30(200)	28(23)	27(16)	18(12)
$C_4H_{11}N$	2-Aminobutane	73(1.2)	44(170)	18(25)	41(18)	58(18)
$C_4H_{12}Pb$	Tetramethyllead	268(0.14)	253(69)	223(59)	208(46)	251(36)
C_5F_{10}	Decafluorocyclopentane	250(0.62)	131(173)	100(41)	31(40)	69(28)
C_5F_{12}	Dodecafluoro-2-methylbutane	288(0.0)	69(277)	119(45)	131(23)	31(18)
C_5F_{12}	Dodecafluoropentane	288(0.08)	69(259)	119(76)	169(25)	31(24)
C_5HF_9	Nonafluorocyclopentane	232(0.07)	131(61)	113(49)	69(34)	31(19)
C_5H_5N	Pyridine	79(135)	79(135)	52(95)	51(48)	50(35)
C_5H_6	Cyclopentadiene	66(95)	66(95)	65(40)	39(35)	40(30)
C_5H_6	*trans*-2-Penten-4-yne	66(77)	66(77)	39(54)	65(38)	40(35)
$C_5H_6N_2$	2-Methylpyrazine	94(81)	94(81)	67(48)	26(33)	39(30)
$C_5H_6O_2$	Furfuryl alcohol	98(3.4)	98(3.4)	41(3.3)	39(3.3)	42(2.6)
C_5H_6S	2-Methylthiophene	98(68)	97(125)	45(26)	39(17)	53(11)
C_5H_6S	3-Methylthiophene	98(74)	97(138)	45(35)	39(14)	27(11)
C_5H_8	Methylenecyclobutane	68(38)	40(67)	67(48)	39(47)	53(21)
C_5H_8	Spiropentane	68(8.9)	67(58)	40(56)	39(52)	53(23)
C_5H_8	Cyclopentene	68(41)	67(99)	39(36)	53(23)	41(19)
C_5H_8	3-Methyl-1,2-butadiene	68(53)	68(53)	53(40)	39(28)	41(26)

(*Continued*)

TABLE 3.71 Table of Mass Spectra (*Continued*)

Molecular formula	Name	Mass numbers (and intensities) of: Parent peak	Base peak	Three next most intense peaks		
C_5H_8	2-Methyl-1,3-butadiene	68(40)	67(48)	53(41)	39(34)	27(23)
C_5H_8	1,2-Pentadiene	68(39)	68(39)	53(38)	39(37)	27(31)
C_5H_8	*cis*-1,3-Pentadiene	68(40)	67(53)	39(43)	53(38)	41(25)
C_5H_8	*trans*-1,3-Pentadiene	68(41)	67(52)	39(43)	53(39)	41(26)
C_5H_8	1,4-Pentadiene	68(40)	39(47)	67(35)	53(33)	41(30)
C_5H_8	2,3-Pentadiene	68(62)	68(62)	53(42)	39(36)	41(31)
C_5H_8	3-Methyl-1-butyne	68(8.5)	53(74)	67(45)	27(35)	39(21)
C_5H_8	1-Pentyne	68(8.7)	67(50)	40(44)	39(42)	27(34)
C_5H_8	2-Pentyne	68(67)	68(67)	53(61)	39(32)	27(27)
$C_5H_8N_2$	3,5-Dimethylpyrazole	96(47)	96(47)	95(37)	39(16)	54(12)
$C_5H_8O_2$	2,4-Pentanedione	100(22)	43(120)	85(33)	15(23)	27(11)
$C_5H_8O_2$	2-Propenyl acetate	100(0.16)	43(177)	41(30)	39(29)	15(28)
$C_5H_8O_2$	Methyl methacrylate	100(26)	41(78)	69(52)	39(31)	15(16)
$C_5H_9ClO_2$	Ethyl 3-chloropropanoate	136(0.70)	27(65)	29(62)	91(42)	63(37)
C_5H_{10}	*cis*-1,2-Dimethylcyclopropane	70(39)	55(77)	42(35)	39(32)	41(32)
C_5H_{10}	*trans*-1,2-Dimethylcyclopropane	70(42)	55(79)	42(34)	41(33)	39(30)
C_5H_{10}	Ethylcyclopropane	70(26)	42(93)	55(47)	41(39)	39(35)
C_5H_{10}	Cyclopentane	70(44)	42(148)	55(43)	41(43)	39(31)
C_5H_{10}	2-Methyl-1-butene	70(30)	55(97)	42(36)	39(34)	41(28)
C_5H_{10}	3-Methyl-1-butene	70(26)	55(102)	27(31)	42(28)	29(27)
C_5H_{10}	2-Methyl-2-butene	70(31)	55(88)	41(31)	39(28)	42(27)
C_5H_{10}	1-Pentene	70(27)	42(89)	55(53)	41(39)	39(31)
C_5H_{10}	*cis*-2-Pentene	70(30)	55(89)	42(41)	39(30)	29(26)
C_5H_{10}	*trans*-2-Pentene	70(31)	55(93)	42(41)	39(30)	41(28)
$C_5H_{10}O$	3-Methyl-1-butanal	86(3.0)	41(30)	43(26)	58(20)	29(20)
$C_5H_{10}O$	2-Pentanone	86(16)	43(106)	29(23)	27(23)	57(20)
$C_5H_{10}O$	3-Pentanone	86(15)	57(87)	29(87)	27(32)	28(9.4)
$C_5H_{10}O$	Ethyl-2-propenyl ether	86(6.2)	41(52)	29(48)	58(44)	57(42)
$C_5H_{10}O$	Ethyl isopropyl ether	86(21)	43(87)	44(69)	41(46)	27(45)
$C_5H_{10}O$	2-Methyltetrahydrofuran	86(8.9)	71(57)	43(55)	41(40)	27(27)
$C_5H_{10}O_2$	Tetrahydrofurfuryl alcohol	102(0.02)	71(8.9)	43(6.8)	41(4.8)	27(3.8)
$C_5H_{10}O_2$	2-Methoxyethyl ethenyl ether	102(3.0)	29(69)	45(58)	15(48)	58(45)
$C_5H_{10}O_2$	2,2-Dimethylpropanoic acid	102(2.0)	57(83)	41(38)	29(27)	39(12)
$C_5H_{10}O_2$	2-Methylbutanoic acid	102(0.32)	74(54)	57(34)	29(33)	41(28)
$C_5H_{10}O_2$	*n*-Butyl formate	102(0.27)	56(80)	41(48)	31(47)	29(42)
$C_5H_{10}O_2$	Isobutyl formate	102(0.27)	43(58)	56(48)	41(46)	31(38)
$C_5H_{10}O_2$	*sec*-Butyl formate	102(0.17)	45(99)	29(49)	27(32)	41(31)
$C_5H_{10}O_2$	*n*-Propyl acetate	102(0.07)	43(176)	61(34)	31(31)	27(26)
$C_5H_{10}O_2$	Isopropyl acetate	102(0.17)	43(155)	45(50)	27(22)	61(18)
$C_5H_{10}O_2$	Ethyl propanoate	102(10)	29(151)	57(97)	27(52)	28(24)
$C_5H_{10}O_2$	Methyl 2-methylpropanoate	102(8.9)	43(69)	71(23)	41(19)	59(17)
$C_5H_{10}O_2$	Methyl butanoate	102(1.0)	43(53)	74(37)	71(29)	27(23)
$C_5H_{10}O_3$	Ethyl carbonate	118(0.30)	29(114)	45(80)	31(60)	27(46)
$C_5H_{10}S$	2-Methylthiacyclopentane	102(37)	87(88)	41(30)	45(29)	59(18)
$C_5H_{10}S$	3-Methylthiacyclopentane	102(40)	60(45)	41(31)	45(25)	74(23)
$C_5H_{10}S$	Thiacyclohexane	102(43)	87(44)	68(33)	61(32)	41(28)
$C_5H_{10}S$	Cyclopentanethiol	102(19)	41(48)	69(47)	39(26)	67(18)
$C_5H_{11}N$	Piperidine	85(22)	84(43)	57(22)	56(22)	44(17)
$C_5H_{11}NO$	*N*-Methylmorpholine	101(4.4)	43(18)	42(8.6)	15(3.4)	71(2.9)
$C_5H_{11}NO_2$	3-Methylbutyl nitrite	117(0.0)	29(75)	41(68)	57(43)	30(42)

TABLE 3.71 Table of Mass Spectra (*Continued*)

Molecular formula	Name	Mass numbers (and intensities) of: Parent peak	Base peak	Three next most intense peaks		
C_5H_{12}	2,2-Dimethylpropane	72(0.01)	57(126)	41(52)	29(49)	27(20)
C_5H_{12}	2-Methylbutane	72(4.7)	43(74)	42(64)	41(49)	57(40)
C_5H_{12}	*n*-Pentane	72(10)	43(114)	42(66)	41(45)	27(39)
$C_5H_{12}O$	2-Methyl-1-butanol	88(0.18)	57(57)	29(55)	41(53)	56(50)
$C_5H_{12}O$	3-Methyl-1-butanol	88(0.02)	55(47)	42(42)	43(39)	41(38)
$C_5H_{12}O$	2-Methyl-2-butanol	88(0.0)	59(43)	55(37)	45(25)	73(22)
$C_5H_{12}O$	1-Pentanol	88(0.0)	42(41)	55(30)	41(25)	70(23)
$C_5H_{12}O$	Methyl *n*-butyl ether	88(3.1)	45(211)	56(36)	29(36)	27(28)
$C_5H_{12}O$	Methyl isobutyl ether	88(12)	45(186)	41(30)	29(30)	15(27)
$C_5H_{12}O$	Methyl *sec*-butyl ether	88(2.0)	52(142)	29(50)	27(27)	41(25)
$C_5H_{12}O$	Methyl *tert*-butyl ether	88(0.02)	73(119)	41(33)	43(32)	57(32)
$C_5H_{12}O$	Ethyl isopropyl ether	88(2.6)	45(143)	43(46)	73(40)	27(24)
$C_5H_{12}O_2$	Diethoxymethane	104(2.1)	31(104)	59(99)	29(62)	103(39)
$C_5H_{12}O_2$	1,1-Dimethoxypropane	104(0.05)	75(84)	73(62)	29(43)	45(37)
$C_5H_{12}S$	3,3-Dimethyl-2-thiabutane	104(30)	57(83)	41(62)	29(42)	39(16)
$C_5H_{12}S$	4-Methyl-2-thiapentane	104(37)	41(46)	56(38)	27(29)	39(23)
$C_5H_{12}S$	2-Methyl-3-thiapentane	104(82)	89(119)	62(79)	43(63)	61(58)
$C_5H_{12}S$	2-Thiahexane	104(38)	61(77)	56(50)	41(39)	27(33)
$C_5H_{12}S$	3-Thiahexane	104(30)	75(72)	27(53)	47(50)	62(33)
$C_5H_{12}S$	2,2-Dimethyl-1-propanethiol	104(31)	57(100)	41(55)	55(48)	29(42)
$C_5H_{12}S$	2-Methyl-1-butanethiol	104(28)	41(65)	29(44)	57(40)	70(40)
$C_5H_{12}S$	2-Methyl-2-butanethiol	104(18)	43(88)	71(54)	41(46)	55(34)
$C_5H_{12}S$	3-Methyl-2-butanethiol	104(23)	61(73)	43(55)	27(33)	55(28)
$C_5H_{12}S$	1-Pentanethiol	104(35)	42(91)	55(44)	41(39)	70(39)
$C_5H_{12}S$	2-Pentanethiol	104(28)	43(72)	61(52)	27(39)	55(38)
$C_5H_{12}S$	3-Pentanethiol	104(23)	43(56)	41(48)	75(29)	47(23)
$C_5H_{12}S_2$	4,4-Dimethyl-2,3-dithiapentane	136(12)	57(74)	41(38)	29(36)	80(13)
$C_5H_{12}S_2$	2-Methyl-3,4-dithiahexane	136(20)	94(49)	27(46)	43(39)	66(37)
$C_5H_{14}Pb$	Trimethylethyllead	282(0.64)	223(61)	253(52)	208(51)	221(33)
C_6F_6	Hexafluorobenzene	186(95)	186(95)	117(59)	31(58)	93(23)
C_6F_{12}	Dodecafluorocyclohexane	300(0.96)	131(138)	69(97)	100(40)	31(30)
C_6F_{14}	Tetradecafluoro-2-methylpentane	338(0.0)	69(317)	131(41)	119(36)	169(29)
C_6F_{14}	Tetradecafluorohexane	338(0.13)	69(268)	119(74)	169(51)	131(37)
C_6H_5Br	Bromobenzene	156(75)	77(98)	158(74)	51(41)	50(36)
C_6H_5Cl	Chlorobenzene	112(102)	112(102)	77(49)	114(33)	51(17)
$C_6H_5NO_2$	Nitrobenzene	123(39)	77(93)	51(55)	50(23)	30(15)
C_6H_6	Benzene	78(113)	78(113)	52(22)	77(20)	51(18)
C_6H_6	1,5-Hexadiyne	78(58)	39(65)	52(38)	51(32)	50(26)
C_6H_6	2,4-Hexadiyne	78(108)	78(108)	51(55)	52(38)	50(31)
C_6H_6S	Benzenethiol	110(68)	110(68)	66(26)	109(17)	51(15)
C_6H_7N	Aminobenzene (aniline)	93(19)	93(19)	66(6.5)	65(3.6)	39(3.5)
C_6H_7N	2-Methylpyridine	93(86)	93(86)	66(36)	39(28)	51(16)
C_6H_7NO	1-Methyl-2-pyridone	109(71)	109(71)	81(49)	39(34)	80(29)
C_6H_8	Methylcyclopentadiene	80(53)	79(87)	77(29)	39(19)	51(11)
C_6H_8	1,3-Cyclohexadiene	80(53)	79(92)	77(35)	39(21)	27(18)
C_6H_8O	2.5-Dimethylfuran	96(57)	43(65)	95(48)	53(37)	81(24)
C_6H_8S	2,3-Dimethylthiophene	112(44)	97(53)	111(44)	45(16)	27(9.4)
C_6H_8S	2,4-Dimethylthiophene	112(27)	111(36)	97(18)	45(9.4)	39(7.0)
C_6H_8S	2,5-Dimethylthiophene	112(67)	111(95)	97(59)	59(23)	45(19)
C_6H_8S	2-Ethylthiophene	112(27)	97(68)	45(16)	39(8.9)	27(5.4)

(*Continued*)

TABLE 3.71 Table of Mass Spectra (*Continued*)

Molecular formula	Name	Mass numbers (and intensities) of: Parent peak	Base peak	Three next most intense peaks		
C_6H_8S	3-Ethylthiophene	112(54)	97(147)	45(38)	39(20)	27(12)
C_6H_9N	2,5-Dimethylpyrrole	95(73)	94(127)	26(52)	80(22)	42(19)
C_6H_{10}	Isopropenylcyclopropane	82(20)	67(92)	41(47)	39(46)	27(22)
C_6H_{10}	1-Methylcyclopentene	82(26)	67(98)	39(21)	81(16)	41(16)
C_6H_{10}	Cyclohexene	82(33)	67(83)	54(64)	41(31)	39(30)
C_6H_{10}	2,3-Dimethyl-1,3-butadiene	82(41)	67(60)	39(55)	41(44)	54(22)
C_6H_{10}	2-Methyl-1,3-pentadiene	82(23)	67(48)	39(30)	41(26)	27(13)
C_6H_{10}	1,5-Hexadiene	82(1.3)	41(98)	67(80)	39(60)	54(52)
C_6H_{10}	3,3-Dimethyl-1-butyne	82(0.57)	67(101)	41(57)	39(31)	27(11)
C_6H_{10}	4-Methyl-1-pentyne	82(2.3)	67(82)	41(74)	43(64)	39(55)
C_6H_{10}	1-Hexyne	82(1.0)	67(131)	41(88)	27(85)	43(67)
C_6H_{10}	2-Hexyne	82(56)	67(58)	53(50)	27(39)	41(36)
C_6H_{10}	3-Hexyne	82(55)	67(59)	41(55)	39(37)	53(20)
$C_6H_{10}O$	Cyclohexanone	98(32)	55(102)	42(86)	41(35)	27(34)
$C_6H_{10}O$	4-Methyl-3-penten-2-one	98(40)	55(82)	83(82)	43(64)	29(38)
$C_6H_{10}O_2$	2,5-Hexanedione	114(4.0)	43(148)	15(25)	99(22)	14(14)
$C_6H_{10}O_3$	Propanoic anhydride	130(0.0)	57(190)	29(119)	27(62)	28(26)
$C_6H_{10}O_3$	Ethyl acetoacetate	130(8.3)	43(150)	29(52)	27(32)	15(27)
$C_6H_{11}N$	4-Methylpentanenitrile	97(0.13)	55(98)	41(51)	43(45)	27(39)
$C_6H_{11}N$	Hexanenitrile	97(0.54)	41(73)	54(49)	27(43)	55(40)
C_6H_{12}	1,1,2-Trimethylcyclopropane	84(38)	41(132)	69(81)	39(34)	27(24)
C_6H_{12}	1-Methyl-1-ethylcyclopropane	84(25)	41(78)	55(58)	69(53)	27(33)
C_6H_{12}	Isopropylcyclopropane	84(2.0)	56(114)	41(84)	39(30)	43(28)
C_6H_{12}	Ethylcyclobutane	84(3.8)	56(138)	41(89)	27(35)	55(34)
C_6H_{12}	Methylcyclopentane	84(18)	56(116)	41(74)	69(37)	42(33)
C_6H_{12}	Cyclohexane	84(58)	56(75)	41(44)	55(25)	42(21)
C_6H_{12}	2,3-Dimethyl-1-butene	84(27)	41(117)	69(96)	39(36)	27(24)
C_6H_{12}	3,3-Dimethyl-1-butene	84(23)	41(112)	69(107)	39(28)	27(26)
C_6H_{12}	2-Ethyl-1-butene	84(30)	41(74)	69(66)	55(56)	27(38)
C_6H_{12}	2,3-Dimethyl-2-butene	84(32)	41(108)	69(88)	39(35)	27(20)
C_6H_{12}	2-Methyl-1-pentene	84(29)	56(91)	41(73)	55(39)	39(36)
C_6H_{12}	3-Methyl-1-pentene	84(25)	55(85)	41(67)	69(60)	27(43)
C_6H_{12}	4-Methyl-1-pentene	84(12)	43(110)	41(80)	56(47)	27(37)
C_6H_{12}	2-Methyl-2-pentene	84(36)	41(120)	69(111)	39(35)	27(28)
C_6H_{12}	3-Methyl-*cis*-2-pentene	84(37)	41(104)	69(82)	55(46)	27(36)
C_6H_{12}	3-Methyl-*trans*-2-pentene	84(38)	41(102)	69(81)	55(47)	27(35)
C_6H_{12}	4-Methyl-*cis*-2-pentene	84(35)	41(122)	69(114)	39(35)	27(26)
C_6H_{12}	4-Methyl-*trans*-2-pentene	84(34)	41(123)	69(112)	39(34)	27(26)
C_6H_{12}	1-Hexene	84(20)	41(70)	56(60)	42(52)	27(48)
C_6H_{12}	*cis*-2-Hexene	84(27)	55(91)	42(51)	41(45)	27(45)
C_6H_{12}	*trans*-2-Hexene	84(32)	55(112)	42(54)	41(46)	27(41)
C_6H_{12}	*cis*-3-Hexene	84(28)	55(81)	41(62)	42(54)	27(32)
C_6H_{12}	*trans*-3-Hexene	84(32)	55(89)	41(72)	42(62)	27(35)
$C_6H_{12}N_2$	Acetone azine (ketazine)	112(31)	56(99)	15(31)	97(31)	39(26)
$C_6H_{12}O$	Cyclopentylmethanol	100(0.02)	41(35)	68(32)	69(31)	67(24)
$C_6H_{12}O$	4-Methyl-2-pentanone	100(12)	43(115)	58(37)	41(22)	57(22)
$C_6H_{12}O$	Ethenyl *n*-butyl ether	100(5.7)	29(80)	41(59)	56(45)	57(35)
$C_6H_{12}O$	Ethenyl isobutyl ether	100(5.8)	29(73)	41(65)	57(58)	56(40)
$C_6H_{12}O_2$	4-Hydroxy-4-methyl-2-pentanone	116(0.0)	43(149)	15(45)	58(32)	27(14)

TABLE 3.71 Table of Mass Spectra (*Continued*)

Molecular formula	Name	Mass numbers (and intensities) of: Parent peak	Base peak	Three next most intense peaks		
$C_6H_{12}O_2$	*n*-Butyl acetate	116(0.03)	43(172)	56(58)	41(30)	27(27)
$C_6H_{12}O_2$	*n*-Propyl propanoate	116(0.03)	57(147)	29(84)	27(57)	75(47)
$C_6H_{12}O_2$	Isopropyl proponoate	116(0.26)	57(116)	43(88)	29(54)	27(46)
$C_6H_{12}O_2$	Methyl 2,2-dimethylpropanoate	116(3.2)	57(85)	41(32)	29(24)	56(21)
$C_6H_{12}O_2$	Ethyl butanoate	116(2.2)	43(50)	71(45)	29(43)	27(31)
$C_6H_{12}O_3$	2,4,6-Trimethyl-1,3,5-trioxacyclo-hexane	132(0.12)	45(196)	43(107)	29(35)	89(23)
$C_6H_{12}S$	1-Cyclopentyl-1-thiaethane	116(31)	68(72)	41(64)	39(37)	67(37)
$C_6H_{12}S$	*cis*-2,5-Dimethylthiacyclopentane	116(32)	101(85)	59(34)	41(26)	74(24)
$C_6H_{12}S$	*trans*-2.5-Dimethylthiacyclopentane	116(32)	101(85)	59(34)	74(25)	41(25)
$C_6H_{12}S$	2-Methylthiacyclohexane	116(42)	101(81)	41(37)	27(32)	67(30)
$C_6H_{12}S$	3-Methylthiacyclohexane	116(41)	101(55)	41(47)	39(33)	45(28)
$C_6H_{12}S$	4-Methylthiacyclohexane	116(46)	116(46)	101(44)	41(40)	27(39)
$C_6H_{12}S$	Thiacycloheptane	116(60)	87(75)	41(66)	67(48)	47(46)
$C_6H_{12}S$	1-Methylcyclopentanethiol	116(20)	83(76)	55(58)	41(39)	67(33)
$C_6H_{12}S$	*cis*-2-Methylcyclopentanethiol	116(32)	55(55)	83(54)	60(48)	41(47)
$C_6H_{12}S$	*trans*-2-Methylcyclopentanethiol	116(28)	67(48)	55(46)	41(42)	83(40)
$C_6H_{12}S$	Cyclohexanethiol	116(21)	55(56)	41(45)	67(35)	83(32)
$C_6H_{13}N$	Cyclohexylamine	99(8.9)	56(92)	43(25)	28(13)	30(13)
$C_6H_{13}N$	3-Methylpiperidine	99(23)	44(49)	30(34)	28(27)	57(26)
$C_6H_{13}NO$	*N*-Ethylmorpholine	115(2.0)	42(9.8)	57(7.0)	100(5.2)	28(4.3)
C_6H_{14}	2,2-Dimethylbutane	86(0.04)	43(85)	57(82)	71(61)	41(51)
C_6H_{14}	2,3-Dimethylbutane	86(5.3)	43(157)	42(136)	41(49)	27(40)
C_6H_{14}	2-Methylpentane	86(4.4)	43(147)	42(78)	41(47)	27(40)
C_6H_{14}	3-Methylpentane	86(3.2)	57(105)	56(80)	41(67)	29(64)
C_6H_{14}	*n*-Hexane	86(12)	57(87)	43(71)	41(64)	29(55)
$C_6H_{14}N_2$	*cis*-2,5-Dimethylpiperazine	114(0.38)	58(10)	28(7.7)	30(4.7)	44(4.2)
$C_6H_{14}O$	2-Ethyl-1-butanol	102(0.0)	43(114)	70(40)	29(39)	27(38)
$C_6H_{14}O$	2-Methyl-1-pentanol	102(0.0)	42(110)	41(40)	29(34)	27(33)
$C_6H_{14}O$	3-Methyl-1-pentanol	102(0.0)	56(26)	41(20)	29(19)	55(18)
$C_6H_{14}O$	4-Methyl-2-pentanol	102(0.08)	45(111)	43(34)	41(17)	27(14)
$C_6H_{14}O$	1-Hexanol	102(0.0)	56(63)	43(52)	41(37)	55(36)
$C_6H_{14}O$	Ethyl *n*-butyl ether	102(3.8)	59(108)	31(87)	29(61)	27(42)
$C_6H_{14}O$	Ethyl *sec*-butyl ether	102(1.5)	45(150)	73(76)	29(51)	27(39)
$C_6H_{14}O$	Ethyl isobutyl ether	102(8.7)	59(124)	31(95)	29(53)	27(38)
$C_6H_{14}O$	Diisopropyl ether	102(1.4)	45(125)	43(66)	87(23)	27(19)
$C_6H_{14}O_2$	1,1-Diethoxyethane	118(0.0)	45(132)	73(69)	29(36)	27(27)
$C_6H_{14}O_2$	1,2-Diethoxyethane	118(1.2)	31(124)	59(88)	29(72)	45(53)
$C_6H_{14}O_3$	*bis*-(2-Methoxyethyl) ether	134(0.0)	59(140)	29(74)	58(57)	15(56)
$C_6H_{14}S$	2,2-Dimethyl-3-thiapentane	118(33)	57(147)	41(70)	29(54)	27(40)
$C_6H_{14}S$	2,4-Dimethyl-3-thiapentne	118(33)	43(94)	61(85)	41(48)	103(44)
$C_6H_{14}S$	2-Methyl-3-thiahexane	118(206)	43(540)	41(317)	42(301)	27(287)
$C_6H_{14}S$	4-Methyl-3-thiahexane	118(195)	89(585)	29(343)	27(296)	41(279)
$C_6H_{14}S$	5-Methyl-3-thiahexane	118(171)	75(520)	41(230)	47(224)	56(217)
$C_6H_{14}S$	3-Thiaheptane	118(35)	75(55)	29(33)	27(33)	62(28)
$C_6H_{14}S$	4-Thiaheptane	118(47)	43(86)	89(74)	41(57)	27(55)
$C_6H_{14}S$	2-Methyl-1-pentanethiol	118((19)	43(96)	41(51)	56(32)	27(31)
$C_6H_{14}S$	4-Methyl-1-pentanethiol	118(30)	56(142)	41(57)	43(57)	27(32)
$C_6H_{14}S$	4-Methyl-2-pentanethiol	118(6.3)	43(68)	69(61)	41(56)	84(42)
$C_6H_{14}S$	2-Methyl-3-pentanethiol	118(20)	41(64)	43(63)	75(50)	27(28)

(*Continued*)

TABLE 3.71 Table of Mass Spectra (*Continued*)

Molecular formula	Name	Mass numbers (and intensities) of: Parent peak	Base peak	Three next most intense peaks		
$C_6H_{14}S$	1-Hexanethiol	118(16)	56(66)	41(41)	27(40)	43(38)
$C_6H_{14}S_2$	2,5-Dimethyl-3,4-dithiahexane	150(31)	43(152)	108(41)	41(36)	27(30)
$C_6H_{14}S_2$	5-Methyl-3,4-dithiaheptane	150(14)	29(86)	94(66)	66(57)	27(41)
$C_6H_{14}S_2$	6-Methyl-3,4-dithiaheptane	150(4.9)	29(42)	66(40)	122(30)	94(29)
$C_6H_{14}S_2$	4,5-Dithiaoctane	150(44)	43(167)	27(65)	41(64)	108(35)
$C_6H_{15}N$	Triethylamine	101(21)	86(134)	30(46)	27(36)	58(35)
$C_6H_{15}N$	Di-*n*-propylamine	101(7.1)	30(89)	72(70)	44(36)	43(28)
$C_6H_{15}N$	Diisopropylamine	101(5.0)	44(171)	86(52)	58(24)	42(22)
$C_6H_{16}Pb$	Dimethyldiethyllead	296(0.98)	267(89)	223(83)	208(79)	221(44)
C_7F_{14}	Tetradecafluoromethylcyclohexane	350(0.0)	69(244)	131(107)	181(48)	100(38)
C_7F_{16}	Hexadecafluoroheptane	388(0.0)	69(330)	119(89)	169(68)	131(44)
C_7H_5N	Benzonitrile	103(246)	103(246)	76(80)	50(42)	51(24)
C_7H_7Br	1-Methyl-2-bromobenzene	170(48)	91(97)	172(46)	39(21)	63(20)
C_7H_7Br	1-Methyl-4-bromobenzene	170(46)	91(97)	172(45)	39(20)	65(19)
C_7H_7Cl	1-Methyl-2-chlorobenzene	126(44)	91(121)	63(20)	39(19)	89(18)
C_7H_7Cl	1-Methyl-3-chlorobenzene	126(51)	91(120)	63(19)	39(18)	128(16)
C_7H_7Cl	1-Methyl-4-chlorobenzene	126(44)	91(120)	125(19)	63(18)	39(17)
C_7H_7F	1-Methyl-3-fluorobenzene	110(79)	109(129)	83(17)	57(12)	39(12)
C_7H_7F	1-Methyl-4-fluorobenzene	110(73)	109(122)	83(16)	57(12)	39(9.3)
C_7H_8	Methylbenzene (toluene)	92(82)	91(108)	39(20)	65(14)	51(10)
C_7H_8S	1-Phenyl-1-thiaethane	124(76)	124(76)	109(34)	78(25)	91(19)
C_7H_9N	2,4-Dimethylpyridine	107(76)	107(76)	106(29)	79(16)	92(13)
$C_7H_{10}S$	2,3,4-Trimethylthiophene	126(50)	111(81)	125(47)	45(22)	39(18)
C_7H_{12}	Ethenylcyclopentane	96(13)	67(118)	39(44)	68(38)	54(35)
C_7H_{12}	Ethylidenecyclopentane	96(40)	67(180)	39(44)	41(30)	27(30)
C_7H_{12}	Bicyclo[2.2.1]heptane	96(12)	67(64)	68(50)	81(44)	54(30)
C_7H_{12}	3-Ethylcyclopentene	96(29)	67(193)	39(36)	41(35)	27(26)
C_7H_{12}	1-Methylcyclohexene	96(32)	81(83)	68(38)	67(37)	39(33)
C_7H_{12}	4-Methylcyclohexene	96(28)	81(84)	54(50)	39(44)	55(34)
C_7H_{12}	4-Methyl-2-hexyne	96(13)	81(71)	67(52)	41(48)	39(35)
C_7H_{12}	5-Methyl-2-hexyne	96(42)	43(49)	81(43)	27(39)	39(38)
C_7H_{12}	1-Heptyne	96(0.44)	41(75)	81(70)	29(65)	27(47)
C_7H_{14}	1,1,2,2,-Tetramethylcyclopropane	98(21)	55(92)	83(90)	41(69)	39(41)
C_7H_{14}	*cis*-1,2-Dimethylcyclopentane	98(19)	56(85)	70(77)	41(65)	55(65)
C_7H_{14}	*trans*-1,2-Dimethylcyclopentane	98(25)	56(93)	41(63)	55(61)	70(54)
C_7H_{14}	*cis*-1,3-Dimethylcyclopentane	98(12)	56(81)	70(78)	41(64)	55(59)
C_7H_{14}	*trans*-1,3-Dimethylcyclopentane	98(13)	56(81)	70(68)	41(63)	55(58)
C_7H_{14}	1,1-Dimethylcyclopentane	98(6.7)	56(81)	55(63)	69(56)	41(55)
C_7H_{14}	Ethylcyclopentane	98(14)	69(83)	41(78)	68(60)	55(46)
C_7H_{14}	Methylcyclohexane	98(41)	83(94)	55(78)	41(55)	42(34)
C_7H_{14}	Cycloheptane	98(37)	41(57)	55(54)	56(50)	42(49)
C_7H_{14}	2,3,3-Trimethyl-1-butene	98(20)	83(101)	55(83)	41(61)	39(33)
C_7H_{14}	3-Methyl-2-ethyl-1-butene	98(22)	41(71)	69(71)	55(62)	27(38)
C_7H_{14}	2,3-Dimethyl-1-pentene	98(13)	41(92)	69(86)	55(40)	39(35)
C_7H_{14}	2,4-Dimethyl-1-pentene	98(9.1)	56(117)	43(68)	41(61)	39(39)
C_7H_{14}	3,3-Dimethyl-1-pentene	98(9.4)	69(104)	41(85)	55(42)	27(36)
C_7H_{14}	3,4-Dimethyl-1-pentene	98(0.61)	56(75)	55(62)	43(55)	41(54)
C_7H_{14}	4,4-Dimethyl-1-pentene	98(2.6)	57(161)	41(86)	29(52)	55(49)
C_7H_{14}	3-Ethyl-1-pentene	98(19)	41(116)	69(91)	27(43)	39(37)
C_7H_{14}	2,3-Dimethyl-2-pentene	98(31)	83(80)	55(75)	41(63)	39(34)

TABLE 3.71 Table of Mass Spectra (*Continued*)

Molecular formula	Name	Mass numbers (and intensities) of: Parent peak	Base peak	Three next most intense peaks		
C_7H_{14}	2,4-Dimethyl-2-pentene	98(26)	83(97)	55(71)	41(52)	39(34)
C_7H_{14}	3,4-Dimethyl-*cis*-2-pentene	98(30)	83(87)	55(82)	41(52)	27(32)
C_7H_{14}	3,4-Dimethyl-*trans*-2-pentene	98(31)	83(89)	55(83)	41(52)	27(34)
C_7H_{14}	4,4-Dimethyl-*cis*-2-pentene	98(27)	83(96)	55(92)	41(62)	39(35)
C_7H_{14}	4,4-Dimethyl-*trans*-2-pentene	98(28)	83(105)	55(89)	41(58)	39(31)
C_7H_{14}	3-Ethyl-2-pentene	98(33)	41(86)	69(80)	55(74)	27(33)
C_7H_{14}	2-Methyl-1-hexene	98(4.6)	56(105)	41(54)	27(30)	39(27)
C_7H_{14}	3-Methyl-1-hexene	98(7.7)	55(76)	41(60)	69(57)	56(48)
C_7H_{14}	4-Methyl-1-hexene	98(4.9)	41(98)	57(94)	56(80)	29(70)
C_7H_{14}	5-Methyl-1-hexene	98(1.6)	56(91)	41(75)	55(47)	27(42)
C_7H_{14}	2-Methyl-2-hexene	98(28)	69(113)	41(99)	27(36)	39(33)
C_7H_{14}	3-Methyl-*cis*-2-hexene	98(30)	41(95)	69(90)	55(42)	27(36)
C_7H_{14}	4-Methyl-*trans*-2-hexene	98(23)	69(118)	41(106)	55(40)	39(35)
C_7H_{14}	5-Methyl-2-hexene	98(13)	56(90)	55(74)	43(71)	41(57)
C_7H_{14}	2-Methyl-*trans*-3-hexene	98(24)	69(86)	41(74)	55(62)	56(37)
C_7H_{14}	3-Methyl-*cis*-3-hexene	98(28)	69(98)	41(82)	39(33)	27(33)
C_7H_{14}	3-Methyl-*trans*-3-hexene	98(28)	69(97)	41(86)	55(63)	39(35)
C_7H_{14}	1-Heptene	98(15)	41(91)	56(79)	29(64)	55(54)
C_7H_{14}	*trans*-2-Heptene	98(27)	55(64)	56(59)	41(50)	27(35)
C_7H_{14}	*trans*-3-Heptene	98(27)	41(98)	56(65)	69(55)	55(47)
$C_7H_{14}O$	2,4-Dimethyl-3-pentanone	114(13)	43(226)	71(62)	27(49)	41(42)
$C_7H_{14}O_2$	*n*-Butyl propanoate	130(0.03)	57(152)	29(98)	56(54)	27(52)
$C_7H_{14}O_2$	Isobutyl propanoate	130(0.07)	57(187)	29(87)	56(27)	27(47)
$C_7H_{14}O_2$	*n*-Propyl *n*-butanoate	130(0.05)	43(96)	71(90)	27(54)	89(48)
$C_7H_{14}O_3$	*n*-Propyl carbonate	146(0.02)	43(171)	27(61)	63(55)	41(49)
$C_7H_{14}S$	*cis*-2-Methylcyclohexanethiol	130(28)	55(138)	97(70)	81(44)	41(44)
$C_7H_{15}N$	2,6-Dimethylpiperidine	113(5.3)	98(73)	44(43)	42(34)	28(26)
C_7H_{16}	2,2,3-Trimethylbutane	100(0.03)	57(110)	43(84)	56(67)	41(64)
C_7H_{16}	2,2-Dimethylpentane	100(0.06)	57(130)	43(95)	41(59)	56(52)
C_7H_{16}	2,3-Dimethylpentane	100(2.1)	43(94)	56(93)	57(67)	41(64)
C_7H_{16}	2,4-Dimethylpentane	100(1.6)	43(139)	57(93)	41(59)	56(50)
C_7H_{16}	3,3-Dimethylpentane	100(0.03)	43(166)	71(103)	27(38)	41(36)
C_7H_{16}	3-Ethylpentane	100(3.1)	43(175)	70(77)	70(77)	29(45)
C_7H_{16}	2-Methylhexane	100(5.9)	43(154)	42(59)	41(57)	85(49)
C_7H_{16}	3-Methylhexane	100(4.0)	43(110)	57(52)	71(52)	41(50)
C_7H_{16}	*n*-Heptane	100(17)	43(126)	41(65)	57(60)	29(58)
$C_7H_{16}O$	2-Heptanol	116(0.01)	45(131)	43(29)	27(25)	29(23)
$C_7H_{16}O$	3-Heptanol	116(0.01)	59(61)	69(41)	41(29)	31(25)
$C_7H_{16}O$	4-Heptanol	116(0.02)	55(102)	73(72)	43(45)	27(32)
$C_7H_{16}O$	*n*-Propyl *n*-butyl ether	116(3.7)	43(120)	57(102)	41(51)	29(49)
$C_7H_{16}O_2$	Di-*n*-propoxymethane	132(0.58)	43(194)	73(114)	27(45)	41(34)
$C_7H_{16}O_2$	Diisopropoxymethane	132(0.16)	43(133)	45(84)	73(71)	27(28)
$C_7H_{16}O_2$	1,1-Diethoxypropane	132(0.0)	59(138)	47(88)	87(84)	29(74)
$C_7H_{16}S$	2,2,4-Trimethyl-3-thiapentane	132(30)	57(149)	41(74)	29(35)	43(32)
$C_7H_{16}S$	2,4-Dimethyl-3-thiahexane	132(30)	61(94)	103(60)	41(51)	43(46)
$C_7H_{16}S$	2-Thiaoctane	132(34)	61(73)	56(53)	27(46)	41(44)
$C_7H_{16}S$	1-Heptanethiol	132(14)	41(48)	27(40)	56(39)	70(38)
$C_7H_{18}Pb$	Methyltriethyllead	310(0.84)	281(86)	208(76)	223(66)	237(60)
$C_7H_{18}Pb$	*n*-Butyltrimethyllead	310(0.14)	253(76)	223(75)	208(68)	295(52)
$C_7H_{18}Pb$	*sec*-Butyltrimethyllead	310(1.8)	253(94)	223(85)	208(74)	251(45)

(*Continued*)

TABLE 3.71 Table of Mass Spectra (*Continued*)

Molecular formula	Name	Mass numbers (and intensities) of: Parent peak	Base peak	Three next most intense peaks		
$C_7H_{18}Pb$	*tert*-Butyltrimethyllead	310(0.09)	252(95)	223(82)	208(65)	250(46)
C_8H_{10}	1,2-Dimethylbenzene	106(52)	91(91)	105(22)	39(15)	51(14)
C_8H_{10}	1,3-Dimethylbenzene	106(58)	91(93)	105(26)	39(17)	51(14)
C_8H_{10}	1,4-Dimethylbenzene	106(52)	91(85)	105(25)	51(13)	39(13)
C_8H_{10}	Ethylbenzene	106(45)	91(146)	51(19)	39(14)	65(12)
F_3N	Nitrogen trifluoride	71(10)	52(33)	33(13)	14(3.0)	19(2.7)
HCl	Hydrogen chloride	36(54)	36(54)	38(17)	35(9.2)	37(2.9)
H_2S	Hydrogen sulfide	34(75)	34(75)	32(33)	33(32)	1(4.1)
H_3P	Ammonia	17(32)	17(32)	16(26)	15(2.4)	14(0.7)
H_3N	Phosphine	34(59)	34(59)	33(20)	31(19)	32(7.5)
H_4N_2	Hydrazine	32(48)	32(48)	31(23)	29(19)	30(15)
NO	Nitric oxide	30(76)	30(76)	14(5.7)	15(1.8)	16(1.1)
NO_2	Nitrogen dioxide	46(6.6)	30(18)	16(4.0)	14(1.7)	47(0.02
N_2	Nitrogen	28(65)	28(65)	14(3.3)	29(0.47)	. . .
N_2O	Nitrous oxide	44(60)	44(60)	30(19)	14(7.8)	28(6.5)
O_2	Oxygen	32(54)	32(54)	16(2.7)	28(1.7)	34(0.22
O_2S	Sulfur dioxide	64(47)	64(47)	48(23)	32(4.9)	16(2.4)

Source: L. Meites, ed., *Handbook of Analytical Chemistry*, McGraw-Hill, New York, 1963. J. A. Dean, ed., Analytical Chemistry Handbook, McGraw-Hill, New York, 1995.

3.8 X-RAY METHODS

An X-ray tube operating at a voltage V (in keV) emits a continuous X-ray spectrum, the minimum wavelength of which is given by $\lambda_{min} = 12.398/V$ with the wavelength expressed in angstroms. For expressing the wavelength in kX units, divide by the factor 1.00202. Tables 3.72 and 3.73 are based on the K and L wavelength values as published by Y. Cauchois and H. Hulubei (*Tables de Constantes et Données Numériques,* I. *Longueurs d'Onde des Émissions X et des Discontinuités d'Absorption X,* Hermann, Paris, 1947) and by the International Union of Crystallography (*International Tables for X-Ray Crystallography,* Kynoch Press, Birmingham, England, 1962). Wavelength accuracy is only to about 1 in 25 000 except for the lines employed in X-ray diffraction work.

Use of energy-proportional detectors for X-rays creates a need for energy values of K and L absorption edges (Table 3.74) and emission series (Table 3.75). These values were obtained by a conversion to keV of tabulated experimental wavelength values and smoothed by a fit to Moseley's law. Although values are listed to 1 eV, chemical form may shift absorption edges and emission lines as much as 10 to 20 eV. S. Fine and C. F. Hendee [*Nucelonics,* **13**(3):36 (1955)] also give values for $K\beta_2$, $L\gamma_1$, and $L\beta_2$ lines.

The relative intensities of X-ray emission lines from targets varies for different elements. However, one can assume a ratio of $K\alpha_1/K\alpha_2 = 2$ for the commonly used targets. The ratio of $K\alpha_2/K\alpha_1$ from these targets varies from 6 to 3.5. The intensities of $K\beta_2$ radiations amount to about 1 percent of that of the corresponding $K\alpha_1$ radiation. In practical applications these ratios have to be corrected for differential absorption in the window of the tube and air path, the ratio of scattering factors for and differential absorption in the crystal, and for sensitivity characteristics of the detector. Generalizing, the intensities of radiations from the K and L series are as follows:

Emission line	$K\alpha_1$	$K\alpha_2$	$K\beta_1$	$K\beta_2$	$L\alpha_1$	$L\alpha_2$	$L\beta_1$	$L\beta_2$	$L\gamma_1$
Relative intensity	500	250	80–150	5	100	10	30	60	40

For angles at which the $K\alpha_1$, $K\alpha_2$ doublet is not resolved, a mean wavelength [$K\bar{\alpha} = (2K\alpha_1 + K\alpha_2)/3$] can be used.

Filters. The K spectra of the light metals, often used as target material in the production of X-rays for diffraction studies, contain three strong lines, α_1, α_2 and β_1, of which the α lines form a doublet with a narrow wavelength separation. The $K\beta$ radiation can be eliminated by using a thin foil filter, usually of the element of next lower atomic number to that of the target element: the $K\alpha$ lines are transmitted with a relatively small loss of intensity. Table 3.76, restricted to the K wavelengths of target elements in common use, lists the calculated thicknesses of β filters required to reduce the $K\beta_1/K\alpha_1$ integrated intensity ratio to $^1/_{100}$.

Interplanar Spacings. Diffractometer alignment procedures require the use of a well-prepared polycrystalline specimen. Two standard samples found to be suitable are silicon amd α-quartz (including Novaculite). The 2θ values of several of the most intense reflections for these materials are listed in Table 3.77 (*Tables of Interplanar Spacings d vs. Diffraction Angle 2θ for Selected Targets,* Picker Nuclear, White Plains, N.Y., 1966). To convert to d for $K\alpha$ or to d for $K\alpha_2$, multiply the tabulated d value (Table 3.77) for $K\alpha_1$ by the factor given below:

Element	$K\bar{\alpha}$	$K\alpha_2$
W	1.007 69	1.023 07
Ag	1.002 63	1.007 89
Mo	1.002 02	1.006 04
Cu	1.000 82	1.002 48
Ni	1.000 77	1.002 32
Co	1.000 72	1.002 16
Fe	1.000 67	1.002 04
Cr	1.000 57	1.001 70

Analyzing Crystals. The range of wavelengths usable with various analyzing crystals are governed by the d spacings of the crystal planes and by the geometric limits to which the goniometer can be rotated. The d value should be small enough to make the angle 2θ greater than approximately 10 or 15 deg, even at the shortest wavelength used: otherwise excessively long analyzing crystals would be needed to prevent the direct fluorescent beam from entering the detector. A small d value is also favorable for producing a large dispersion of the spectrum to give good separation of adjacent lines. On the other hand, a small d value imposes an upper limit to the range of wavelengths that can be analyzed. Actually the goniometer is limited mechanically to about 150 deg for a 2θ value. A final requirement is the reflection efficiency and minimization of higher-order reflections. Table 3.78 gives a list of crystals commonly used for X-ray spectroscopy.

The long-wavelength analyzers are prepared by dipping an optical flat into the film of the metal fatty acid about 50 times to produce a layer 180 molecules in thickness.

Lithium fluoride is the optimum crystal for all wavelengths less than 3 Å. Pentaerythritol (PET) and potassium hydrogen phthalate (KAP) are usually the crystals of choice for wavelengths from 3 to 20 Å. Two crystals suppress even-ordered reflections: silicon (111) and calcium fluoride (111).

Mass Absorption Coefficients. Radiation traversing a layer of substance is diminished in intensity by a constant fraction per centimeter thickness x of material. The emergent radiant power P, in terms of incident radiant power P_0, is given by

$$P = P_0 \exp(-\mu x)$$

which defines the total linear absorption coefficient μ. Since the reduction of intensity is determined by the quantity of matter traversed by the primary beam, the absorber thickness is best expressed on a mass basis, in g/cm^2. The mass absorption coefficient μ/ρ, expressed in units cm^2/g, where ρ is the density of the material, is approximately independent of the physical state of the material and, to a good approximation, is additive with respect to the elements composing a substance.

Table 3.79 contains values of μ/ρ for the common target elements employed in X-ray work. A more extensive set of mass absorption coefficients for *K*, *L*, and *M* emission lines within the wavelength range from 0.7 to 12 Å is contained in K. F. J. Heinrich's paper in T. D. McKinley, K. F. J. Heinrich, and D. B. Wittry (eds.), *The Electron Microprobe,* Wiley, New York, 1966, pp. 351-377. This article should be consulted to ascertain the probable accuracy of the values and for a compilation of coefficients and exponents employed in the computations.

TABLE 3.72 Wavelengths of X-Ray Emission Spectra in Angstroms

Atomic No.	Element	$K\alpha_2$	$K\alpha_1$	$K\beta_1$	$L\alpha_1$	$L\beta_1$
3	Li	240				
4	Be	113				
5	B	67				
6	C	44				
7	N	31.60				
8	O	23.71				
9	F	18.31				
10	Ne	14.616		14.464		
11	Na	11.909		11.617	407.6	
12	Mg	9.889		9.558	251.0	
13	Al	8.3392	8.3367	7.981	169.8	
14	Si	7.1277	7.1253	6.7681	123	
15	P	6.1549		5.8038		
16	S	5.3747	5.3720	5.0317		
17	Cl	4.7305	4.7276	4.4031		
18	Ar	4.1946	4.1916	3.8848		
19	K	3.7446	3.7412	3.4538	42.7	
20	Ca	3.3616	3.3583	3.0896	36.32	35.95
21	Sc	3.0345	3.0311	2.7795	31.33	31.01
22	Ti	2.75207	2.7484	2.5138	27.39	27.02
23	V	2.5073	2.5035	2.2843	24.26	23.85
24	Cr	2.29351	2.28962	2.08480	21.67	21.28
25	Mn	2.1057	2.1018	1.9102	19.45	19.12
26	Fe	1.93991	1.93597	1.75653	17.567	17.255
27	Co	1.79278	1.78892	1.62075	15.968	15.667
28	Ni	1.66169	1.65784	1.50010	14.566	14.279
29	Cu	1.54433	1.54051	1.39217	13.330	13.053
30	Zn	1.4389	1.4351	1.2952	12.257	11.985
31	Ga	1.3439	1.3400	1.20784	11.290	11.023
32	Ge	1.2580	1.2540	1.1289	10.435	10.174
33	As	1.1798	1.1758	1.0573	9.671	9.414
34	Se	1.1088	1.1047	0.9921	8.990	8.736
35	Br	1.0438	1.0397	0.9327	8.375	8.125

TABLE 3.72 Wavelengths of X-Ray Emission Spectra in Angstroms (*Continued*)

Atomic No.	Element	$K\alpha_2$	$K\alpha_1$	$K\beta_1$	$L\alpha_1$	$L\beta_1$
36	Kr	0.9841	0.9801	0.8785	7.822	7.574
37	Rb	0.9296	0.9255	0.8286	7.3181	7.076
38	Sr	0.8794	0.8752	0.7829	6.8625	6.6237
39	Y	0.8330	0.8279	0.7407	6.4485	6.2117
40	Zr	0.7901	0.7859	0.7017	6.0702	5.8358
41	Nb	0.7504	0.7462	0.6657	5.7240	5.4921
42	Mo	0.713543	0.70926	0.632253	5.4063	5.1768
43	Tc	0.6793	0.6749	0.6014	5.1126	4.8782
44	Ru	0.6474	0.6430	0.5725	4.8455	4.6204
45	Rh	0.6176	0.6132	0.5456	4.5973	4.3739
46	Pd	0.5898	0.5854	0.5205	4.3676	4.1460
47	Ag	0.563775	0.559363	0.49701	4.1541	3.9344
48	Cd	0.5394	0.5350	0.4751	3.9563	3.7381
49	In	0.5165	0.5121	0.4545	3.7719	3.5552
50	Sn	0.4950	0.4906	0.4352	3.5999	3.3848
51	Sb	0.4748	0.4703	0.4171	3.4392	3.2256
52	Te	0.4558	0.4513	0.4000	3.2891	3.0767
53	I	0.4378	0.4333	0.3839	3.1485	2.9373
54	Xe	0.4204	0.4160	0.3685	3.016	2.807
55	Cs	0.4048	0.4003	0.3543	2.9016	2.8920
56	Ba	0.3896	0.3851	0.3408	2.7752	2.5674
57	La	0.3753	0.3707	0.3280	2.6651	2.4583
58	Ce	0.3617	0.3571	0.3158	2.5612	2.3558
59	Pr	0.3487	0.3441	0.3042	2.4627	2.2584
60	Nd	0.3565	0.3318	0.2933	2.3701	2.1666
61	Pm	0.3249	0.3207	0.2821	2.282	2.0796
62	Sm	0.3137	0.3190	0.2731	2.1994	1.9976
63	Eu	0.3133	0.2985	0.2636	2.1206	1.9202
64	Gd	0.2932	0.2884	0.2544	2.0460	1.8462
65	Tb	0.2834	0.2788	0.2460	1.9755	1.7763
66	Dy	0.2743	0.2696	0.2376	1.9088	1.7100
67	Ho	0.2655	0.2608	0.2302	1.8447	1.6468
68	Er	0.2572	0.2525	0.2226	1.7843	1.5873
69	Tm	0.2491	0.2444	0.2153	1.7263	1.5299
70	Yb	0.2415	0.2368	0.2088	1.6719	1.4756
71	Lu	0.2341	0.2293	0.2021	1.6194	1.4235
72	Hf	0.2270	0.2222	0.1955	1.5696	1.3740
73	Ta	0.2203	0.2155	0.1901	1.5219	1.3270
74	W	0.213813	0.208992	0.184363	1.4764	1.2818
75	Re	0.2076	0.2028	0.1789	1.4329	1.2385
76	Os	0.2016	0.1968	0.1736	1.3911	1.1972
77	Ir	0.1959	0.1910	0.1685	1.3513	1.1578
78	Pt	0.1904	0.1855	0.1637	1.3130	1.1198
79	Au	0.1851	0.1802	0.1590	1.2764	1.0836
80	Hg	0.1799	0.1750	0.1544	1.2411	1.0486

(*Continued*)

TABLE 3.72 Wavelengths of X-Ray Emission Spectra in Angstroms (*Continued*)

Atomic No.	Element	$K\alpha_2$	$K\alpha_1$	$K\beta_1$	$L\alpha_1$	$L\beta_1$
81	Tl	0.1750	0.1701	0.1501	1.2074	1.0152
82	Pb	0.1703	0.1654	0.1460	1.1750	0.9822
83	Bi	0.1657	0.1608	0.1419	1.1439	0.9520
84	Po	0.1608	0.1559	0.1382	1.1138	0.9222
85	At	0.1570	0.1521	0.1343	1.0850	0.8936
86	Rn	0.1529	0.1479	0.1307	1.0572	0.8659
87	Fr	0.1489	0.1440	0.1272	1.0300	0.8400
88	Ra	0.1450	0.1401	0.1237	1.0047	0.8137
89	Ac	0.1414	0.1364	0.1205	0.9799	0.7890
90	Th	0.1378	0.1328	0.1174	0.9560	0.7652
91	Pa	0.1344	0.1294	0.1143	0.9328	0.7422
92	U	0.1310	0.1259	0.1114	0.9105	0.7200
93	Np	0.1278	0.1226	0.1085	0.8893	0.6984
94	Pu	0.1246	0.1195	0.1058	0.8682	0.6777
95	Am	0.1215	0.1165	0.1031	0.8481	0.6576
96	Cm	0.1186	0.1135	0.1005	0.8287	0.6388
97	Bk	0.1157	0.1107	0.0980	0.8098	0.6203
98	Cf	0.1130	0.1079	0.0956	0.7917	0.6023
99	Es	0.1103	0.1052	0.0933	0.7740	0.5850
100	Fm	0.1077	0.1026	0.0910	0.7570	0.5682

TABLE 3.73 Wavelengths of Absorption Edges in Angstroms

Atomic No.	Element	K	L_I	L_{II}	L_{III}
3	Li	226.5			
4	Be	110.68			
5	B	66.289			
6	C	43.68			
7	N	30.99			
8	O	23.32			
9	F	17.913			
10	Ne	14.183			
11	Na	11.478		400	
12	Mg	9.512	197.4	247.92	
13	Al	7.951	142.5	170	
14	Si	6.745	105.1	126.48	
15	P	5.787	81.0	96.84	
16	S	5.018	64.23	76.05	
17	Cl	4.397	52.08	61.37	62.93
18	Ar	3.871	43.19	50.39	50.60
19	K	3.436	36.35	42.02	42.17
20	Ca	3.070	31.07	35.20	35.49

TABLE 3.73 Wavelengths of Absorption Edges in Angstroms (*Continued*)

Atomic No.	Element	K	L_I	L_{II}	L_{III}
21	Sc	2.757	26.83	30.16	30.53
22	Ti	2.497	23.39	26.83	27.37
23	V	2.269	20.52	23.70	24.26
24	Cr	2.07012	16.7	17.9	20.7
25	Mn	1.896	16.27	18.90	19.40
26	Fe	1.74334	14.60	17.17	17.53
27	Co	1.60811	13.34	15.53	15.93
28	Ni	1.48802	12.27	14.13	14.58
29	Cu	1.38043	11.27	13.01	13.29
30	Zn	1.283	10.33	11.86	12.13
31	Ga	1.195	9.54	10.61	11.15
32	Ge	1.116	8.73	9.97	10.23
33	As	1.044	8.108	9.124	9.367
34	Se	0.9800	7.505	8.417	8.646
35	Br	0.9199	6.925	7.752	7.989
36	Kr	0.8655	6.456	7.165	7.395
37	Rb	0.8155	5.997	6.643	6.863
38	Sr	0.7697	5.582	6.172	6.387
39	Y	0.7276	5.233	5.756	5.962
40	Zr	0.6888	4.867	5.378	5.583
41	Nb	0.6529	4.581	5.025	5.223
42	Mo	0.61977	4.299	4.719	4.912
43	Tc	0.5888	4.064	4.427	4.629
44	Ru	0.5605	3.841	4.179	4.369
45	Rh	0.5338	3.626	3.942	4.130
46	Pd	0.5092	3.428	3.724	3.908
47	Ag	0.48582	3.254	3.514	3.698
48	Cd	0.4641	3.084	3.326	3.504
49	In	0.4439	2.926	3.147	3.324
50	Sn	0.4247	2.778	2.982	3.156
51	Sb	0.4066	2.639	2.830	3.000
52	Te	0.3897	2.510	2.687	2.855
53	I	0.3738	2.390	2.553	2.719
54	Xe	0.3585	2.274	2.429	2.592
55	Cs	0.3447	2.167	2.314	2.474
56	Ba	0.3314	2.068	2.204	2.363
57	La	0.3184	1.973	2.103	2.258
58	Ce	0.3065	1.891	2.009	2.164
59	Pr	0.2952	1.811	1.924	2.077
60	Nd	0.2845	1.735	1.843	1.995
61	Pm	0.2743	1.668	1.766	1.918
62	Sm	0.2646	1.598	1.702	1.845
63	Eu	0.2555	1.536	1.626	1.775
64	Gd	0.2468	1.477	1.561	1.709
65	Tb	0.2384	1.421	1.501	1.649

(*Continued*)

TABLE 3.73 Wavelengths of Absorption Edges in Angstroms (*Continued*)

Atomic No.	Element	K	L_{I}	L_{II}	L_{III}
66	Dy	0.2305	1.365	1.438	1.579
67	Ho	0.2229	1.319	1.390	1.535
68	Er	0.2157	1.269	1.339	1.483
69	Tm	0.2089	1.222	1.288	1.433
70	Yb	0.2022	1.181	1.243	1.386
71	Lu	0.1958	1.140	1.198	1.341
72	Hf	0.1898	1.099	1.154	1.297
73	Ta	0.1839	1.061	1.113	1.255
74	W	0.17837	1.025	1.074	1.215
75	Re	0.1731	0.9901	1.036	1.177
76	Os	0.1678	0.9557	1.001	1.140
77	Ir	0.1629	0.9243	0.9670	1.106
78	Pt	0.1582	0.8914	0.9348	1.072
79	Au	0.1534	0.8638	0.9028	1.040
80	Hg	0.1492	0.8353	0.8779	1.009
81	Tl	0.1447	0.8079	0.8436	0.9793
82	Pb	0.1408	0.7815	0.8155	0.9503
83	Bi	0.1371	0.7565	0.7891	0.9234
84	Po	0.1332	0.7322	0.7638	0.8970
85	At	0.1295	0.7092	0.7387	0.8720
86	Rn	0.1260	0.6868	0.7153	0.8479
87	Fr	0.1225	0.6654	0.6929	0.8248
88	Ra	0.1192	0.6446	0.6711	0.8027
89	Ac	0.1161	0.6248	0.6500	0.7813
90	Th	0.1129	0.6061	0.6301	0.7606
91	Pa	0.1101	0.5875	0.6106	0.7411
92	U	0.1068	0.5697	0.5919	0.7233
93	Np	0.1045	0.5531	0.5742	0.7042
94	Pu	0.1018	0.5366	0.5571	0.6867
95	Am	0.0992	0.5208	0.5404	0.6700
96	Cm	0.0967	0.5060	0.5246	0.6532
97	Bk	0.0943	0.4913	0.5093	0.6375
98	Cf	0.0920	0.4771	0.4945	0.6223
99	Es	0.0897	0.4636	0.4801	0.6076
100	Fm	0.0875	0.4506	0.4665	0.5935

TABLE 3.74 Critical X-Ray Absorption Energies in KeV

Atomic No.	Element	K	L_I	L_{II}	L_{III}
1	H	0.0136			
2	He	0.0246			
3	Li	0.0547			
4	Be	0.112			
5	B	0.187			
6	C	0.284			
7	N	0.400			
8	O	0.532			
9	F	0.692			
10	Ne	0.874	0.048	0.022	
11	Na	1.08	0.055	0.034	
12	Mg	1.30	0.0628	0.0502	
13	Al	1.559	0.0870	0.0720	
14	Si	1.838	0.118	0.0977	
15	P	2.142	0.153	0.128	
16	S	2.469	0.193	0.163	0.162
17	Cl	2.822	0.238	0.202	0.201
18	Ar	3.200	0.287	0.246	0.244
19	K	3.606	0.341	0.295	0.292
20	Ca	4.038	0.399	0.350	0.346
21	Sc	4.496	0.462	0.411	0.407
22	Ti	4.966	0.530	0.462	0.456
23	V	5.467	0.604	0.523	0.515
24	Cr	5.988	0.679	0.584	0.574
25	Mn	6.542	0.762	0.656	0.644
26	Fe	7.113	0.849	0.722	0.709
27	Co	7.713	0.929	0.798	0.783
28	Ni	8.337	1.02	0.877	0.858
29	Cu	8.982	1.10	0.954	0.935
30	Zn	9.662	1.20	1.05	1.02
31	Ga	10.39	1.30	1.17	1.14
32	Ge	11.10	1.42	1.24	1.21
33	As	11.87	1.529	1.358	1.32
34	Se	12.65	1.66	1.472	1.431
35	Br	13.48	1.791	1.599	1.552
36	Kr	14.32	1.92	1.729	1.674
37	Rb	15.197	2.064	1.863	1.803
38	Sr	16.101	2.212	2.004	1.937
39	Y	17.053	2.387	2.171	2.096
40	Zr	17.998	2.533	2.308	2.224
41	Nb	18.986	2.700	2.467	2.372
42	Mo	20.003	2.869	2.630	2.525
43	Tc	21.050	3.045	2.796	2.680

(*Continued*)

TABLE 3.74 Critical X-Ray Absorption Energies in KeV (*Continued*)

Atomic No.	Element	K	L_{I}	L_{II}	L_{III}
44	Ru	22.117	3.227	2.968	2.839
45	Rh	23.210	3.404	3.139	2.995
46	Pd	24.356	3.614	3.338	3.181
47	Ag	25.535	3.828	3.547	3.375
48	Cd	26.712	4.019	3.731	3.541
49	In	27.929	4.226	3.929	3.732
50	Sn	29.182	4.445	4.139	3.911
51	Sb	30.497	4.708	4.391	4.137
52	Te	31.817	4.953	4.621	4.347
53	I	33.164	5.187	4.855	4.559
54	Xe	34.551	5.448	5.103	4.783
55	Cs	35.974	5.706	5.360	5.014
56	Ba	37.432	5.995	5.629	5.250
57	La	38.923	6.264	5.902	5.490
58	Ce	40.43	6.556	6.169	5.728
59	Pr	41.99	6.837	6.446	5.968
60	Nd	43.57	7.134	6.728	6.215
61	Pm	45.19	7.431	7.022	6.462
62	Sm	46.85	7.742	7.316	6.720
63	Eu	48.51	8.059	7.624	6.984
64	Gd	50.23	8.383	7.942	7.251
65	Tb	52.00	8.713	8.258	7.520
66	Dy	53.77	9.053	8.587	7.795
67	Ho	55.61	9.395	8.918	8.074
68	Er	57.47	9.754	9.270	8.362
69	Tm	59.38	10.12	9.622	8.656
70	Yb	61.31	10.49	9.985	8.949
71	Lu	63.32	10.87	10.35	9.248
72	Hf	65.37	11.28	10.75	9.567
73	Ta	67.46	11.68	11.14	9.883
74	W	69.51	12.09	11.54	10.20
75	Re	71.67	12.52	11.96	10.53
76	Os	73.87	12.97	12.38	10.86
77	Ir	76.11	13.41	12.82	11.21
78	Pt	78.35	13.865	13.26	11.55
79	Au	80.67	14.351	13.731	11.92
80	Hg	83.08	14.838	14.205	12.278
81	Tl	85.52	15.344	14.695	12.65
82	Pb	87.95	15.861	15.200	13.03
83	Bi	90.54	16.386	15.709	13.42
84	Po	93.16	16.925	16.233	13.81
85	At	95.73	17.481	16.777	14.21

Table 3.74 Critical X-Ray Absorption Energies in KeV (*Continued*)

Atomic No.	Element	K	L_1	L_{11}	L_{111}
86	Rn	98.45	18.054	17.331	14.61
87	Fa	101.1	18.628	17.893	15.02
88	Ra	103.9	19.228	18.473	15.44
89	Ac	107.7	19.829	19.071	15.86
90	Th	109.8	20.452	19.673	16.278
91	Pa	112.4	21.096	20.295	16.720
92	U	115.0	21.757	20.944	17.163
93	Np	118.2	22.411	21.585	17.606
94	Pu	121.2	23.117	22.250	18.062
95	Am	124.3	23.795	22.935	18.524
96	Cm	127.2	24.502	23.629	18.992
97	Bk	131.3	25.231	24.344	19.466
98	Cf	133.6	26.010	25.070	19.954
99	Es	138.1	26.729	25.824	20.422
100	Fm	141.5	27.503	26.584	20.912

TABLE 3.75 X-Ray Emission Energies in KeV

Atomic No.	Element	$K\beta_1$	$K\alpha_1$	$L\beta_1$	$L\alpha_1$
3	Li		0.052		
4	Be		0.110		
5	B		0.185		
6	C		0.282		
7	N		0.392		
8	O		0.523		
9	F		0.677		
10	Ne		0.851		
11	Na	1.067	1.041		
12	Mg	1.297	1.254		
13	Al	1.553	1.487		
14	Si	1.832	1.740		
15	P	2.136	2.015		
16	S	2.464	2.308		
17	Cl	2.815	2.622		
18	Ar	3.192	2.957		
19	K	3.589	3.313		
20	Ca	4.012	3.691	0.344	0.341
21	Sc	4.460	4.090	0.399	0.395
22	Ti	4.931	4.510	0.458	0.452
23	V	5.427	4.952	0.519	0.512

(*Continued*)

TABLE 3.75 X-Ray Emission Energies in KeV (*Continued*)

Atomic No.	Element	$K\beta_1$	$K\alpha_1$	$L\beta_1$	$L\alpha_1$
24	Cr	5.946	5.414	0.581	0.571
25	Mn	6.490	5.898	0.647	0.636
26	Fe	7.057	6.403	0.717	0.704
27	Co	7.649	6.930	0.790	0.775
28	Ni	8.264	7.477	0.866	0.849
29	Cu	8.904	8.047	0.948	0.928
30	Zn	9.571	8.638	1.032	1.009
31	Ga	10.263	9.251	1.122	1.096
32	Ge	10.981	9.885	1.216	1.186
33	As	11.725	10.543	1.317	1.282
34	Se	12.495	11.221	1.419	1.379
35	Br	13.290	11.923	1.526	1.480
36	Kr	14.112	12.649	1.638	1.587
37	Rb	14.960	13.394	1.752	1.694
38	Sr	15.834	14.164	1.872	1.806
39	Y	16.736	14.957	1.996	1.922
40	Zr	17.666	15.774	2.124	2.042
41	Nb	18.621	16.614	2.257	2.166
42	Mo	19.607	17.478	2.395	2.293
43	Tc	20.612	18.370	2.538	2.424
44	Ru	21.655	19.278	2.683	2.558
45	Rh	22.721	20.214	2.834	2.696
46	Pd	23.816	21.175	2.990	2.838
47	Ag	24.942	22.162	3.151	2.984
48	Cd	26.093	23.172	3.316	3.133
49	In	27.274	24.207	3.487	3.287
50	Sn	28.483	25.270	3.662	3.444
51	Sb	29.723	26.357	3.843	3.605
52	Te	30.993	27.471	4.029	3.769
53	I	32.292	28.610	4.220	3.937
54	Xe	33.644	29.779	4.422	4.111
55	Cs	34.984	30.970	4.620	4.286
56	Ba	36.376	32.191	4.828	4.467
57	La	37.799	33.440	5.043	4.651
58	Ce	39.255	34.717	5.262	4.840
59	Pr	40.746	36.023	5.489	5.034
60	Nd	42.269	37.359	5.722	5.230
61	Pm	43.811	38.726	5.956	5.431
62	Sm	45.400	40.124	6.206	5.636
63	Eu	47.027	41.529	6.456	5.846
64	Gd	48.718	42.983	6.714	6.059
65	Tb	50.391	44.470	6.979	6.275

TABLE 3.75 X-Ray Emission Energies in KeV (*Continued*)

Atomic No.	Element	$K\beta_1$	$K\alpha_1$	$L\beta_1$	$L\alpha_1$
66	Dy	52.178	45.985	7.249	6.495
67	Ho	53.934	47.528	7.528	6.720
68	Er	55.690	49.099	7.810	6.948
69	Tm	57.487	50.730	8.103	7.181
70	Yb	59.352	52.360	8.401	7.414
71	Lu	61.282	54.063	8.708	7.654
72	Hf	63.209	55.757	9.021	7.898
73	Ta	65.210	57.524	9.341	8.145
74	W	67.233	59.310	9.670	8.396
75	Re	69.298	61.131	10.008	8.651
76	Os	71.404	62.991	10.354	8.910
77	Ir	73.549	64.886	10.706	9.173
78	Pt	75.736	66.820	11.069	9.441
79	Au	77.968	68.794	11.439	9.711
80	Hg	80.258	70.821	11.823	9.987
81	Tl	82.558	72.860	12.210	10.266
82	Pb	84.922	74.957	12.611	10.549
83	Bi	87.335	77.097	13.021	10.836
84	Po	89.809	79.296	13.441	11.128
85	At	92.319	81.525	13.873	11.424
86	Rn	94.877	83.800	14.316	11.724
87	Fr	97.483	86.119	14.770	12.029
88	Ra	100.136	88.485	15.233	12.338
89	Ac	102.846	90.894	15.712	12.650
90	Th	105.592	93.334	16.200	12.966
91	Pa	108.408	95.851	16.700	13.291
92	U	111.289	98.428	17.218	13.613
93	Np	114.181	101.005	17.740	13.945
94	Pu	117.146	103.653	18.278	14.279
95	Am	120.163	106.351	18.829	14.618
96	Cm	123.235	109.098	19.393	14.961
97	Bk	126.362	111.896	19.971	15.309
98	Cf	129.544	114.745	20.562	15.661
99	Es	132.781	117.646	21.166	16.018
100	Fm	136.075	120.598	21.785	16.379

TABLE 3.76 β Filters for Common Target Elements

Target Element	$K\bar{\alpha}$, Å	Excitation Voltage, keV	$K\beta_1\ K\alpha_1 = 1/100$			% Loss $K\alpha_1$
			Absorber	Thickness, mm	g/cm^2	
Ag	0.560834	25.52	Pd	0.062	0.074	60
Mo	0.71069	20.00	Zr	0.081	0.053	57
Cu	1.54178	8.981	Ni	0.015	0.013	45
Ni	1.65912	8.331	Co	0.013	0.011	42
Co	1.79021	7.709	Fe	0.012	0.009	39
Fe	1.93728	7.111	Mn	0.011	0.008	38
			MnO_2	0.026	0.013	45
Cr	2.29092	5.989	V	0.011	0.007	37
			V_2O_5	0.036	0.012	48
	$L\alpha_1$		$L\beta_1\ L\alpha_1 = 1/100$			% Loss $L\alpha_1$
W	1.4763	10.200	Cu	0.035		77

TABLE 3.77 Interplanar Spacing for K_a, Radiation, d versus 20

α-quartz (Including Novaculite)

hkl	100	101	110	102	200	112	202	211	203	301
d(Å)	4.260	3.343	2.458	2.282	2.128	1.817	1.672	1.541	1.375	1.372
W $K\alpha_1$: 2θ	2.81	3.58	4.87	5.25	5.63	6.59	7.17	7.78	8.72	8.74
Ag $K\alpha_1$: 2θ	7.53	9.60	13.07	14.08	15.10	17.71	19.26	20.91	23.47	23.52
Mo $K\alpha_1$: 2θ	9.55	12.18	16.59	17.88	19.19	22.51	24.49	26.61	29.89	29.96
Cu $K\alpha_1$: 2θ	20.83	26.64	36.52	39.45	42.44	50.16	54.86	59.98	68.14	68.31
Ni $K\alpha_1$: 2θ	22.44	28.71	39.42	42.60	45.85	54.28	59.44	65.08	74.15	74.34
Co $K\alpha_1$: 2θ	24.24	31.04	42.68	46.15	49.71	58.98	64.68	70.96	81.16	81.38
Fe $K\alpha_1$: 2θ	26.27	33.66	46.38	50.20	54.11	64.38	70.75	77.83	89.50	89.74
Cr $K\alpha_1$: 2θ	31.18	40.05	55.52	60.22	65.09	78.11	86.42	95.96	112.73	113.11

Silicon

hkl	111	220	311	400	331	422	511,333	440	531	620
d(Å)	3.1353	1.91997	1.63736	1.357630	1.24584	1.1085	1.0451	0.959986	0.917922	0.858637
W $K\alpha_1$: 2θ	3.82	6.24	7.32	8.83	9.62	10.82	11.48	12.50	13.07	13.98
Ag $K\alpha_1$: 2θ	10.24	16.75	19.67	23.78	25.95	29.23	31.04	33.88	35.48	38.02
Mo $K\alpha_1$: 2θ	12.99	21.29	25.02	30.28	33.08	37.32	39.67	43.36	45.45	48.79
Cu $K\alpha_1$: 2θ	28.44	47.30	56.12	69.13	76.38	88.03	94.96	106.71	114.10	127.55
Ni $K\alpha_1$: 2θ	30.66	51.16	60.83	75.26	83.42	96.80	104.96	119.42	129.12	149.76
Co $K\alpha_1$: 2θ	33.15	55.53	66.22	82.42	91.77	107.59	117.71	137.42	154.04	
Fe $K\alpha_1$: 2θ	35.97	60.55	72.48	90.96	101.97	121.67	135.70			
Cr $K\alpha_1$: 2θ	42.83	73.21	88.72	114.97	133.53					

TABLE 3.78 Analyzing Crystals for X-Ray Spectroscopy

Crystal	Reflecting Plane	2*d* Spacing, Å	Reflectivity
Quartz	$505\bar{2}$	1.624	Low
Aluminum	111	2.338	High
Topaz	303	2.712	Medium
Quartz	$20\bar{2}3$	2.750	Low
Lithium fluoride	220	2.848	High
Silicon	111	3.135	High
Quartz	112	3.636	Medium
Lithium fluoride	200	4.028	High
Sodium chloride	200	5.639	High
Calcium fluoride	111	6.32	High
Quartz	$10\bar{1}1$	6.686	High
Quartz	$10\bar{1}0$	8.50	Medium
Pentaerythritol (PET)	002	8.742	High
Ethylenediamine tartrate (EDT)	020	8.808	Medium
Ammonium dihydrogen phosphate (ADP)	110	10.648	Low
Gypsum	020	15.185	Medium
Mica	002	19.92	Low
Potassium hydrogen phthalate (KAP)	$10\bar{1}1$	26.4	Medium
Lead palmitate		45.6	
Strontium behenate		61.3	
Lead stearate		100.4	Medium

TABLE 3.79 Mass Absorption Coefficients for Kα_1 Lines and *W* $L\alpha_1$ Line

Emitter wavelength, Å / Absorber	Ag $K\alpha_1$ 0.559	Mo $K\alpha_1$ 0.709	Cu $K\alpha_1$ 1.541	Ni $K\alpha_1$ 1.658	Co $K\alpha_1$ 1.789	Fe $K\alpha_1$ 1.936	Cr $K\alpha_1$ 2.290	W $L\alpha_1$ 1.476
1 H	0.37	0.38	0.43	0.4	0.4	0.5	0.5	0.4
2 He	0.16	0.18	0.37	0.4	0.4	0.5	0.7	0.3
3 Li	0.18	0.22	0.50	0.6	0.7	0.9	1.5	0.4
4 Be	0.22	0.30	1.2	1.5	1.9	2.3	3.7	1.1
5 B	0.30	0.45	2.5	3.1	3.9	4.9	7.9	2.2
6 C	0.42	0.50	4.6	5.7	7.1	8.8	14.2	4.1
7 N	0.60	0.83	7.5	9.3	11.5	14.4	23.1	6.7
8 O	0.80	1.45	12.9	15.8	19.5	24.5	39.4	11.4
9 F	1.00	1.9	16.5	20.3	25.2	31.4	50.3	14.6
10 Ne	1.41	2.6	22.8	27.9	34.6	43.1	69.0	20.1
11 Na	1.75	3.5	30.3	37.2	45.9	57.2	91.4	26.8
12 Mg	2.27	4.6	39.5	48.4	59.8	74.6	119.1	34.9
13 Al	2.74	5.8	49.6	60.7	75.0	93.4	149.0	43.9
14 Si	3.44	7.3	61.4	75.2	92.8	115.5	183.8	54.4
15 P	4.20	8.8	74.7	91.4	112.9	140.5	223.6	66.2
16 S	5.15	10.6	89.2	109.2	134.7	167.4	266.1	79.1
17 Cl	5.86	12.4	104.8	128.2	158.1	196.6	312.4	92.8
18 Ar	6.40	14.5	121.4	148.5	183.0	227.3	360.7	107.6
19 K	8.0	16.7	139.8	171	211	262	415	124
20 Ca	9.7	18.9	158.6	194	239	296	469	141
21 Sc	10.5	21.8	180.5	221	272	337	534	160
22 Ti	11.8	25.3	203	247	304	378	597	180
23 V	13.3	27.7	228	278	342	424	77	202
24 Cr	15.7	31.0	254	311	382	474	88	226
25 Mn	17.4	34.5	282	344	423	63.5	101	250
26 Fe	19.9	38.1	311	380	57.6	71.4	113	276
27 Co	21.8	42.1	341	52.8	64.9	80.6	127	303
28 Ni	25.0	46.4	48.3	58.9	72.5	90.0	142	333
29 Cu	26.4	50.7	53.7	65.5	80.6	100.0	158	47.6
30 Zn	28.2	55.4	59.5	72.7	89.4	110.9	175	52.8
31 Ga	30.8	60.1	65.9	80.5	99.0	122.8	194	58.5
32 Ge	33.5	65.2	72.3	88.2	108.6	134.7	213	64.1
33 As	36.5	70.5	79.1	96.6	118.9	147	233	70.2
34 Se	38.5	76.0	86.1	105.1	129.4	161	254	76.4
35 Br	42.3	82.5	93.9	114.7	141.2	175	277	83.4
36 Kr	45.0	88.3	101.9	124.5	153.2	190	300	90.5
37 Rb	48	95	84	103	127	158	252	98
38 Sr	52	102	90	110	137	170	271	106

TABLE 3.79 Mass Absorption Coefficients for K_1 Lines and W $L\alpha_1$, Line (*Continued*)

Emitter wavelength, Å / Absorber	Ag $K\alpha_1$ 0.559	Mo $K\alpha_1$ 0.709	Cu $K\alpha_1$ 1.541	Ni $K\alpha_1$ 1.658	Co $K\alpha_1$ 1.789	Fe $K\alpha_1$ 1.936	Cr $K\alpha_1$ 2.290	W $L\alpha_1$ 1.476
39 Y	56	109	97	119	147	183	292	114
40 Zr	61	17	104	128	158	197	314	122
41 Nb	66	18	112	138	170	212	338	132
42 Mo	71	19	119	146	180	225	358	140
43 Tc	76	20	128	157	194	241	384	150
44 Ru	K 12	22	137	168	207	258	410	160
45 Rh	13	23	146	179	221	275	438	171
46 Pd	14	24	155	190	235	292	466	182
47 Ag	15	26	165	202	249	310	493	193
48 Cd	15	28	174	213	263	327	520	204
49 In	16	30	185	227	280	347	553	217
50 Sn	17	32	195	239	295	367	583	229
51 Sb	19	34	206	252	310	386	612	241
52 Te	19	36	216	265	326	405	644	253
53 I	21	37	230	281	346	431	684	269
54 Xe	22	39	239	293	361	448	710	280
55 Cs	24	42	332	404	495	612	822	295
56 Ba	25	44	349	425	522	645	622	311
57 La	26	46	365	444	545	673	647	325
58 Ce	28	48	383	466	571	603	216	341
59 Pr	29	51	401	487	597	453	229	356
60 Nd	31	54	420	510	534	473	241	373
61 Pm	32	56	440	535		164	254	392
62 Sm	33	59	456	473	417	173	268	406
63 Eu	35	61	L_I 405	354	148	182	282	423
64 Gd	36	64	424	370	156	191	296	
65 Tb	38	67	L_{II} 316	135	164	201	311	393 L_I
66 Dy	39	70	329	141	172	211	327	293 L_{II}
67 Ho	41	72	L_{III} 123	148	181	222	343	304
68 Er	43	75	129	156	189	233	360	316
69 Tm	45	79	135	163	199	244	377	120 L_{III}
70 Yb	46	82	141	171	208	256	395	126
71 Lu	48	84	148	179	218	267	414	132
72 Hf	51	88	155	187	228	280	433	138
73 Ta	52	91	162	196	238	293	453	144
74 W	55	95	169	204	249	306	473	151
75 Re	57	98	176	213	260	319	494	157
76 Os	59	102	184	223	271	333	515	164
77 Ir	61	106	192	232	283	347	538	171
78 Pt	64	109	200	242	295	362	560	179

(*Continued*)

TABLE 3.79 Mass Absorption Coefficients for K_1 Lines and W $L\alpha_1$, Line (*Continued*)

Emitter wavelength, Å / Absorber	Ag $K\alpha_1$ 0.559	Mo $K\alpha_1$ 0.709	Cu $K\alpha_1$ 1.541	Ni $K\alpha_1$ 1.658	Co $K\alpha_1$ 1.789	Fe $K\alpha_1$ 1.936	Cr $K\alpha_1$ 2.290	W $L\alpha_1$ 1.476
79 Au	67	113	209	252	307	377	584	186
80 Hg	69	117	218	263	321	394	609	194
81 Tl	72	121	227	275	334	411	635	203
82 Pb	74	125	236	286	348	428	662	211
83 Bi	78	129	247	298	363	446	690	220
84 Po		131	258	311	380	466	721	230
85 At			269	325	397	487	753	240
86 Rn	85		281	340	414	509	787	251
87 Fr		89	294	356	433	532	823	262
88 Ra	91		307	372	453	556	861	274
89 Ac			322	389	474	582	900	287
90 Th	97		337	408	497	610	944	301
91 Pa			353	427	520	639	988	315
92 U	104		372	450	548	673	898	332
93 Np			392	474	578	709	945	350
94 Pu		54	418	505	615	755	835	373

SECTION 4

GENERAL INFORMATION AND CONVERSION TABLES

SECTION 4
GENERAL INFORMATION AND CONVERSION TABLES

4.1 GENERAL INFORMATION

TABLE 4.1 SI Prefixes

Submultiple	Prefix	Symbol	Multiple	Prefix	Symbol
10^{-1}	deci	d	10	deka	da
10^{-2}	centi	c	10^{2}	hecto	h
10^{-3}	milli	m	10^{3}	kilo	k
10^{-6}	micro	μ	10^{6}	mega	M
10^{-9}	nano	n	10^{9}	giga	G
10^{-12}	pico	p	10^{12}	tera	T
10^{-15}	femto	f	10^{15}	peta	P
10^{-18}	atto	a	10^{18}	exa	E
10^{-21}	zepto	z	10^{21}	zetta	Z
10^{-24}	yocto	y	10^{24}	yotta	Y

Numerical (multiplying) prefixes

Number	Prefix	Number	Prefix	Number	Prefix
0.5	hemi	19	nonadeca	39	nonatriaconta
1	mono	20	icosa	40	tetraconta
1.5	sesqui	21	henicosa	41	hentetraconta
2	di (bis)*	22	docosa	42	dotetraconta
3	tri (tris)*	23	tricosa	43	tritetraconta
4	tetra (tetrakis)*	24	tetracosa	44	tetratetraconta
5	penta	25	pentacosa	45	pentatetraconta
6	hexa	26	hexacosa	46	hexatetraconta
7	hepta	27	heptacosa	47	heptatetraconta
8	octa	28	octacosa	48	octatetraconta
9	nona	29	nonacosa	49	nonatetraconta
10	deca	30	triaconta	50	pentaconta
11	undeca	31	hentriaconta	60	hexaconta
12	dodeca	32	dotriaconta	70	heptaconta
13	trideca	33	tritriaconta	80	octaconta
14	tetradeca	34	tetratriaconta	90	nonaconta
15	pentadeca	35	pentatriaconta	100	hecta
16	hexadeca	36	hexatriaconta	110	decahecta
17	heptadeca	37	heptatriaconta	120	icosahecta
18	octadeca	38	octatriaconta	130	triacontahecta

*In the case of complex entities such as organic ligands (particularly if they are substituted) the multiplying prefixes bis-, tris-, tetrakis-, pentakis-, . . . are used, i.e., -kis is added starting from tetra-. The modified entity is often placed within parentheses to avoid ambiguity.

TABLE 4.2 Greek Alphabet

Capital	Lower case	Name	Capital	Lower case	Name
Α	α	Alpha	Ν	ν	Nu
Β	β	Beta	Ξ	ξ	Xi
Γ	γ	Gamma	Ο	o	Omicron
Δ	δ	Delta	Π	π	Pi
Ε	ϵ	Epsilon	Ρ	ρ	Rho
Ζ	ζ	Zeta	Σ	σ	Sigma
Η	η	Eta	Τ	τ	Tau
Θ	θ	Theta	Υ	υ	Upsilon
Ι	ι	Iota	Φ	ϕ	Phi
Κ	κ	Kappa	Χ	χ	Chi
Λ	λ	Lambda	Ψ	ψ	Psi
Μ	μ	Mu	Ω	ω	Omega

4.2 PHYSICAL CONSTANTS AND CONVERSION FACTORS

TABLE 4.3 Physical Constants

A. Defined values			
Physical quantity	Name of SI unit	Symbol for SI unit	Definition
1. Base SI units			
Amount of substance	mole	mol	Amount of substance which contains as many specified entities as there are atoms of carbon-12 in exactly 0.012 kg of that nuclide. The elementary entities must be specified and may be atoms, molecules, ions, electrons, other particles, or specified groups of such particles.
Electric current	ampere	A	Magnitude of the current that, when flowing through each of two straight parallel conductors of infinite length, of negligible cross-section, separated by 1 meter in a vacuum, results in a force between the two wires of 2×10^{-7} newton per meter of length.
Length	meter	m	Distance light travels in a vacuum during 1/299 792 458 of a second.
Luminous intensity	candela	cd	Luminous intensity, in a given direction, of a source that emits monochromatic radiation of frequency 540×10^{12} hertz and that has a radiant intensity in that direction of 1/683 watt per steradian.
Mass	kilogram	kg	Mass of a cylinder of platinum-iridium alloy kept at Paris.
Temperature	kelvin	K	Defined as the fraction 1/273.16 of the thermodynamic temperature of the triple point of water.

TABLE 4.3 Physical Constants (*Continued*)

A. Defined values

Physical quantity	Name of SI unit	Symbol for SI unit	Definition
Time	second	s	Duration of 9 192 631 770 periods of the radiation corresponding to the transition between the two hyperfine levels of the ground state of the cesium-133 atom.
2. Supplementary SI units			
Plane angle	radian	rad	The plane angle between two radii of a circle which cut off on the circumference an arc equal in length to the radius.
Solid angle	steradian	sr	The solid angle which, having its vertex in the center of a sphere, cuts off an area of the surface of the sphere equal to that of a square with sides of length equal to the radius of the sphere.

B. Derived SI units

Physical quantity	Name of SI unit	Symbol for SI unit	Expression in terms of SI base units
Absorbed dose (of radiation)	gray	Gy	$J \cdot kg^{-1}$
Activity (radioactive)	becquerel	Bq	$s^{-1} = m^2 \cdot s^{-2}$
Capacitance (electric)	farad	F	$C \cdot V^{-1} = m^{-2} \cdot kg^{-1} \cdot s^4 \cdot A^2$
Charge (electric)	coulomb	C	$A \cdot s$
Conductance (electric)	siemens	S	$\Omega^{-1} = m^{-2} \cdot kg^{-1} \cdot s^3 \cdot A^2$
Dose equivalent (radiation)	sievert	Sv	$J \cdot kg^{-1} = m^2 \cdot s^{-2}$
Energy, work, heat	joule	J	$N \cdot m = m^2 \cdot kg \cdot s^{-2}$
Force	newton	N	$m \cdot kg \cdot s^{-2}$
Frequency	hertz	Hz	s^{-1}
Illuminance	lux	lx	$cd \cdot sr \cdot m^{-2}$
Inductance	henry	H	$V \cdot A^{-1} \cdot s = m^2 \cdot kg \cdot s^{-2} \cdot A^{-2}$
Luminous flux	lumen	Lm	$cd \cdot sr$
Magnetic flux	weber	Wb	$V \cdot s = m^2 \cdot kg \cdot s^{-2} \cdot A^{-1}$
Magnetic flux density	tesla	T	$V \cdot s \cdot m^{-2} = kg \cdot s^{-2} \cdot A^{-1}$
Potential, electric (electromotive force)	volt	V	$J \cdot C^{-1} = m^2 \cdot kg \cdot s^{-3} \cdot A^{-1}$
Power, radiant flux	watt	W	$J \cdot s^{-1} = m^2 \cdot kg \cdot s^{-3}$
Pressure, stress	pascal	Pa	$N \cdot m^{-2} = m^{-1} \cdot kg \cdot s^{-2}$
Resistance, electric	ohm	Ω	$V \cdot A^{-1} = m^2 \cdot kg \cdot s^{-3} \cdot A^{-2}$
Temperature, Celsius	degree Celsius	°C	°C = (K − 273.15)

C. Recommended consistent values of constants

Quantity	Symbol	Value*
Anomalous electron moment correction	$\mu_e - 1$	0.001 159 615(15)
Atomic mass constant	$m_u = 1$ u	1.660 540 2(10) × 10^{-27} kg
Avogadro constant	L, N_A	6.022 136 7(36) × 10^{23} mol^{-1}
Bohr magneton ($=eh/4\pi m_e$)	μ_B	9.274 015 4(31) × 10^{-24} $J \cdot T^{-1}$

(*Continued*)

TABLE 4.3 Physical Constants (*Continued*)

C. Recommended consistent values of constants

Quantity	Symbol	Value*
Bohr radius	a_0	$5.291\ 772\ 49(24) \times 10^{-11}$ m
Boltzmann constant	k	$1.380\ 658(12) \times 10^{-23}$ J · K^{-1}
Charge-to-mass ratio for electron	e/m_e	$1.758\ 805(5) \times 10^{-11}$ C · kg^{-1}
Compton wavelength of electron	λ_c	$2.426\ 309(4) \times 10^{-12}$ m
Compton wavelength of neutron	$\lambda_{c,n}$	$1.319\ 591(2) \times 10^{-15}$ m
Compton wavelength of proton	$\lambda_{c,p}$	$1.321\ 410(2) \times 10^{-15}$ m
Diamagnetic shielding factor, spherical water molecule	$1 + \sigma(H_2O)$	1.000 025 64(7)
Electron magnetic moment	μ_e	$9.284\ 770\ 1(31) \times 10^{-24}$ J · T^{-1}
Electron radius (classical)	r_e	$2.817\ 938(7) \times 10^{-15}$ m
Electron rest mass	m_e	$9.109\ 389\ 7(54) \times 10^{-31}$ kg
Elementary charge	e	$1.602\ 177\ 33(49) \times 10^{-19}$ C
Energy equivalents:		
1 electron mass		0.511 003 4(14) MeV
1 electronvolt	1 eV/k	$1.160\ 450(36) \times 10^4$ K
	1 eV/hc	$8.065\ 479(21) \times 10^3$ cm^{-1}
	1 eV/h	$2.417\ 970(6) \times 10^{14}$ Hz
1 neutron mass		939.573 1(27) MeV
1 proton mass		938.279 6(27) MeV
1 u		931.501 6(26) MeV
Faraday constant	F	96 485.309(29) C · mol^{-1}
Fine structure constant	α	0.007 297 353 08(33)
	α^{-1}	137.035 989 5(61)
First radiation constant	c_1	$3.741\ 774\ 9(22) \times 10^{-16}$ W · m^2
Gas constant	R	8.314 510(70) J · K^{-1} · mol^{-1}
g factor (Lande) for free electron	g_e	2.002 319 304 386(20)
Gravitational constant	G	$6.672\ 59(85) \times 10^{-11}$ m^3 · kg^{-1} · s^{-2}
Hartree energy	E_h	$4.359\ 748\ 2(26) \times 10^{-18}$ J
Josephson frequency-voltage ratio		$4.835\ 939(13) \times 10^{14}$ Hz · V^{-1}
Magnetic flux quantum	Φ_0	$2.067\ 851(5) \times 10^{-15}$ Wb
Magnetic moment of protons in water	μ_p/μ_B	$1.520\ 993\ 129(17) \times 10^{-3}$
Molar volume, ideal gas, $p = 1$ bar, $\theta = 0$°C		22.711 08(19) L · mol^{-1}
Neutron rest mass	m_n	$1.674\ 928\ 6(10) \times 10^{-27}$ kg
Nuclear magneton	μ_N	$5.050\ 786\ 6(17) \times 10^{-27}$ J · T^{-1}
Permeability of vacuum	μ_0	$4\pi \times 10^{-7}$ H · m^{-1} exactly
Permittivity of vacuum	ϵ_0	$8.854\ 187\ 816 \times 10^{-12}$ F · m^{-1}
	$\hbar = h/2\pi$	$1.054\ 572\ 66(63) \times 10^{-34}$ J · s
Planck constant	h	$6.626\ 0.75\ 5(40) \times 10^{-34}$ J · s
Proton magnetic moment	μ_p	$1.410\ 607\ 61(47) \times 10^{-26}$ J · T^{-1}
Proton magnetogyric ratio	γ_p	$2.675\ 221\ 28(81) \times 10^8$ s^{-1} · T^{-1}
Proton resonance frequency per field in H_2O	$\gamma_p'/2\pi$	42.576 375(13) MHz · T^{-1}
Proton rest mass	m_p	$1.672\ 623\ 1(10) \times 10^{-27}$ kg
Quantum-charge ratio	h/e	$4.135\ 701(11) \times 10^{-15}$ J · Hz^{-1} · C^{-1}
Quantum of circulation	h/m_e	$7.273\ 89(1) \times 10^{-4}$ J · s · kg^{-1}
Ratio, electron-to-proton magnetic moments	μ_e/μ_p	$6.582\ 106\ 88(7) \times 10^2$

TABLE 4.3 Physical Constants (*Continued*)

C. Recommended consistent values of constants

Quantity	Symbol	Value*
Rydberg constant	R_∞	$1.097\ 373\ 153\ 4(13) \times 10^7\ m^{-1}$
Second radiation constant	c_2	$1.438\ 769(12) \times 10^{-2}\ m \cdot K$
Speed of light in vacuum	c_0	$299\ 792\ 458\ m \cdot s^{-1}$ exactly
Standard acceleration of free fall	g_n	$9.806\ 65\ m \cdot s^{-2}$ exactly
Standard atmosphere	atm	101 325 Pa exactly
Stefan-Boltzmann constant	σ	$5.670\ 51(19) \times 10^{-8}\ W \cdot m^{-2} \cdot K^{-4}$
Thomson cross section	σ_e	$6.652\ 448(33) \times 10^{-29}\ m^2$
Wien displacement constant	b	$0.289\ 78(4)\ cm \cdot K$
Zeeman splitting constant	μ_B/hc	$4.668\ 58(4) \times 10^{-5}\ cm^{-1} \cdot G^{-1}$

D. Units in use together with SI units

Physical quantity	Name of unit	Symbol for unit	Value in SI units
Area	barn	b	10^{-28} m
Energy	electronvolt	eV ($e \times$ V)	$\approx 1.60218 \times 10^{-19}$ J
	megaelectronvolt[1]	MeV	
Length	ångström[2]	Å	10^{-10} m; 0.1 nm
Mass	tonne	t	10^3 kg; Mg
	unified atomic mass unit	u[= $m_a(^{12}C)/12$]	$\approx 1.66054 \times 10^{-27}$ kg
	dalton[3]	Da	
Plane angle	degree	°	(π/180) rad
	minute	′	(π/10 800) rad
	second	″	(π/648 000) rad
Pressure	bar[2]	bar	10^5 Pa = 10^5 N m^{-2}
Time	minute	min	60 s
	hour	h	3600 s
	day	d	86 400 s
Volume	liter (litre)	L, l	dm^3 = 10^{-3} m^3
	milliliter	mL, ml	cm^3 = 10^{-6} m^3

*The digits in parentheses following a numerical value represent the standard deviation of that value in terms of the final listed digits.

[1] The term million electronvolts is frequently used in place of megaelectronvolts.

[2] The ångström and bar are approved for temporary use with SI units; however, they should not be introduced if not used at present.

TABLE 4.4 Conversion Factors

Relations which are exact are indicated by an asterisk (*). Factors in parentheses are also exact. Other factors are within ±5 in the last significant figure.

To convert	Into	Multiply by
Abampere	ampere*	10
Abcoulomb	coulomb*	10
	statcoulomb	2.998×10^{10}
Abfarad	farad*	10^{9}
Abhenry	henry*	10^{-9}
Abmho	siemens*	10^{9}
Abvolt	volt	10^{-8}
Acre	hectare or square hectometer	0.404 685 64
	square chain (Gunter's)*	10
	square kilometer*	0.004 046 873
	square meter*	4046.873
	square mile*	(1/640)
	square rod*	160
	square yard*	4840
Acre (U.S. survey)	square meter	4046.873
Acre-foot	cubic foot*	4.3560×10^{4}
	cubic meter	1233.482
	gallon (U.S.)	3.259×10^{5}
Acre-inch	cubic foot*	3630
	cubic meter	102.7902
Ampere per square centimeter	ampere per square inch*	6.4516
Ampere-hour	coulomb*	3600
	faraday	0.037 31
Ampere-turn	gilbert	1.256 637
Ampere-turn per centimeter	ampere-turn per inch	2.540
Ångström	meter*	10^{-10}
	nanometer*	0.1
Apostilb	candela per square meter	0.318 309 9; $(1/\pi)$
	lambert*	10^{-4}
Are	acre	0.024 710 54
	square meter*	100
Assay ton	gram	29.1667
Astronomical unit	meter	$1.496\,00 \times 10^{-11}$
	light-year	$1.581\,284 \times 10^{-5}$
Atmosphere	bar*	1.013 25.0
	foot of water (at 4°C)	33.898 54
	inch of mercury (at 0°C)	29.921 26
	kilogram per square centimeter	1.033 227
	millimeter of mercury*	760
	millimeter of water (4°C)	$1.033\,227 \times 10^{4}$
	newton per square meter*	$1.013\,250 \times 10^{5}$
	pascal*	101 325.0
	pound per square inch	14.695 95
	ton per square inch	0.007 348
	torr*	760
Atomic mass unit	gram	1.6605×10^{-24}
Avogadro number	molecules per mole	$6.022\,137 \times 10^{23}$

TABLE 4.4 Conversion Factors (*Continued*)

To convert	Into	Multiply by
Bar	atmosphere	0.986 923
	dyne per square centimeter*	10^6
	kilogram per square centimeter	1.019 716
	millimeter of mercury	750.062
	millimeter of water (4°C)	$1.019\ 716 \times 10^4$
	newton per square meter	10^5
	pascal*	10^5
	pound per square inch	14.503 77
Barn	square meter*	10^{-28}
Barrel (British)	gallon (British)*	36
	liter	163.659
Barrel (petroleum)	gallon (British)	34.9723
	gallon (U.S.)*	42
	liter	158.987
Barrel (U.S. dry)	bushel (U.S.)	3.281 22
	cubic foot	4.083 33
	liter	115.6271
	quart (U.S. dry)	104.9990
Barrel (U.S. liquid)	gallon (U.S.)	31.5 (variable)
	liter	119.2405
Barye	dyne per square centimeter*	1
Becquerel	curie*	2.7×10^{-11}
Biot	ampere*	10
Board foot	cubic foot	(1/12)
	cubic meter	$2.359\ 737 \times 10^{-3}$
Bohr	meter	$5.291\ 77 \times 10^{-11}$
Bohr magneton	joule per tesla	$9.274\ 02 \times 10^{-24}$
Bolt (U.S. cloth)	foot*	120
	meter	36.576
Boltzmann constant	joule per degree	1.3806×10^{-23}
British thermal unit (Btu)	calorie	251.996
	cubic foot-atmosphere	0.367 717
	erg	1.0550×10^{10}
	foot-pound	778.169
	horsepower-hour (British)	$3.930\ 15 \times 10^{-4}$
	horsepower-hour (metric)	$3.984\ 66 \times 10^{-4}$
	joule (International table)	1055.056
	joule (thermochemical)	1054.350
	kilogram-calorie	0.2520
	kilogram-meter	107.5
	kilowatt-hour	$2.930\ 71 \times 10^{-4}$
	liter-atmosphere	10.4126
Btu per foot^3	kilocalorie per cubic meter	8.899 15
Btu (International table)/ft^3	joule per meter^3	$3.725\ 895 \times 10^4$
Btu (thermochemical)/ft^3	joule per meter^3	$3.723\ 402 \times 10^4$
Btu (International table)/hour	watt	0.293 071 1
Btu (thermochemical)/hour	watt	0.292 875 1
Btu (International table)/pound	joule per kilogram*	2.326×10^3
Btu (thermochemical)/pound	joule per kilogram	$2.324\ 444 \times 10^3$
Btu (thermochemical)/(ft^2 · h)	watt per meter^2	3.154 591
Btu (thermochemical)/minute	watt	17.572 50
Btu (thermochemical)/pound	joule per kilogram	$2.324\ 444 \times 10^3$
Btu per square foot	joule per square meter	$1.135\ 65 \times 10^4$
Bucket (British, dry)	gallon (British)*	4

(*Continued*)

TABLE 4.4 Conversion Factors (*Continued*)

To convert	Into	Multiply by
Bushel (British)	bushel (U.S.)	1.032 057
	cubic foot	1.284 35
	gallon (British)*	8
	gallon (U.S.)	9.607 60
	liter	36.3687
Bushel (U.S.)	barrel (U.S., dry)	0.304 765
	bushel (British)	0.968 939
	cubic foot	1.244 456
	cubic meter	0.035 239 07
	gallon (British)	7.751 51
	gallon (U.S.)	9.309 18
	liter	35.239 07
	peck (U.S.)*	4
	pint (U.S., dry)*	64
Cable length (international)	foot	607.611 55
	meter*	185.2
	mile (nautical)*	0.1
Cable length (U.S. or British)	foot*	720
	meter	219.456
	mile (nautical)	0.118 407
	mile (statute)	0.136 364
Caliber	inch*	0.01
	millimeter*	0.254
Calorie	Btu	0.003 968 320
	foot-pound	3.088 03
	foot-poundal	99.3543
	horsepower-hour (British)	$1.559\ 61 \times 10^{-6}$
	joule*	4.184
	kilowatt-hour	1.163×10^{-6}
	liter-atmosphere	0.041 320 5
Calorie (15°C)	joule	4.1858
Calorie (international)	joule	4.1868
Calorie per minute	foot-pound per second	0.051 467 1
	horsepower (British)	$9.357\ 65 \times 10^{-5}$
	watt*	0.069 78
Candela	Hefner unit	1.11
	lumen per steradian*	1
Candela per square centimeter	candela per square foot*	929.0304
	candela per square meter*	10^4
	lambert	3.141 593; (π)
Carat (metric)	gram*	0.2
Celsius temperature	Fahrenheit temperature	(9/5)°C + 32
	kelvin	°C − 273.15
Centigrade heat unit or chu	Btu*	1.8
	calorie	453.592
	joule	1899.10
Centimeter	foot	0.032 808 4
	inch	0.393 700 8
	mil	393.700 8
Centimeter of mercury (0°C)	pascal	1333.22
Centimeter of water (4°C)	pascal	98.063 8
Centimeter per second	foot per minute	1.986 50
	kilometer per hour*	0.036

TABLE 4.4 Conversion Factors (*Continued*)

To convert	Into	Multiply by
Centimeter per second (*continued*)	knot	0.019 438 4
	mile per hour	0.022 369 4
Centimeter per second squared	foot per second squared	0.032 808 4
	meter per second squared*	0.01
Centimeter-dyne	erg*	1
	joule*	10^{-7}
	meter-kilogram	1.020×10^{-8}
	pound-foot	7.376×10^{-8}
Centimeter-gram	erg*	980.665
	joule*	$9.806\ 65 \times 10^{-5}$
Centipoise	kilogram per (meter-second)*	0.001
	pascal-second*	0.001
	pound per (foot-second)	0.006 72
Chain (Ramsden's)	foot*	100
	meter*	30.48
Chain (Gunter's)	foot*	66
	meter*	20.1168
Circular inch	circular mil*	10^6
	square centimeter	5.067 075
	square inch	$(\pi/4)$
Circular millimeter	square millimeter	$(\pi/4)$
Circumference	degree*	360
	gon (grade)	400
	radian	(2π)
Cord	cord foot*	8
	cubic foot*	128
Coulomb	ampere-second*	1
Coulomb per square centimeter	coulomb per square inch*	6.4516
Cubic centimeter	cubic foot	$3.531\ 47 \times 10^{-5}$
	cubic inch	0.061 023 744
	dram (U.S., fluid)	0.270 512 2
	gallon (British)	$2.199\ 69 \times 10^{-4}$
	gallon (U.S.)	$2.641\ 72 \times 10^{-4}$
	liter*	0.001
	minim (U.S.)	16.230 73
	ounce (British, fluid)	0.035 195 1
	ounce (U.S., fluid)	0.033 814 02
	pint (British)	0.001 759 75
	pint (U.S., dry)	0.001 816 17
	pint (U.S., liquid)	0.002 113 376
Cubic centimeter-atmosphere	joule*	0.101 325
	watt-hour	$2.814\ 58 \times 10^{-5}$
Cubic centimeter per gram	cubic foot per pound	0.016 018 5
Cubic centimeter per second	cubic foot per minute	0.002 118 88
	liter per hour*	3.6
Cubic decimeter (dm^3)	liter*	1
Cubic foot	acre-foot	$2.295\ 68 \times 10^{-5}$
	board foot*	12
	cord*	(1/128)
	cord foot*	(1/16)
	cubic inch*	1728
	cubic meter*	0.028 316 846 592
	cubic yard	(1/27)

(Continued)

TABLE 4.4 Conversion Factors (*Continued*)

To convert	Into	Multiply by
Cubic foot (*continued*)	gallon (British)	6.228 835
	gallon (U.S.)	7.480 519
	liter	28.316 847
Cubic foot per hour	liter per minute	0.471 947
Cubic foot per pound	cubic meter per kilogram	0.062 428 0
Cubic foot-atmosphere	Btu	2.719 48
	calorie	685.298
	joule	2869.205
	kilogram-meter	292.577
	liter-atmosphere	28.3168
	watt-hour	0.797 001
Cubic inch	cubic foot	(1/1728)
	milliliter*	16.387 064
Cubic inch per minute	cubic centimeter per second	0.273 118
Cubic kilometer	cubic mile	0.239 913
Cubic meter per kilogram	cubic foot per pound	16.0185
Cubic yard	bushel (British)	21.0223
	bushel (U.S.)	21.6962
	cubic foot*	27
	cubic meter	0.764 554 86
	liter	764.555
Cubic yard per minute	cubic foot per second*	0.45
	gallon (British) per second	2.802 98
	gallon (U.S.) per second	3.366 23
	liter per second	12.742 58
Cubit	inch*	18
Cup (U.S.)	milliliter; centimeter^3	236.6
Cup (metric)	cubic centimeter*	200
Curie	becquerel*	3.7×10^{10}
Cycle per second	hertz*	1
Dalton	kilogram	$1.660\ 54 \times 10^{-27}$
	unified atomic mass*	1
Day (mean solar)	hour*	24
	minute*	1440
	second*	86 400
Debye	coulomb-meter	$3.335\ 64 \times 10^{-30}$
Decibel	neper	0.115 129 255
Degree (plane angle)	circumference	(1/366)
	gon (grade)	1.111 11
	minute (angle)*	60
	quadrant	(1/90)
	radian	(π/180)
	revolution	(1/360)
	second (angle)*	3600
Degree (angle) per foot	radian per meter	0.057 261 5
Degree (angle) per second	radian per second	0.017 453 3
Degree Celsius	degree Fahrenheit*	1.8
	degree Rankine*	1.8
	kelvin*	1
Degree Fahrenheit	degree Celsius	(5/9)
Degree Rankine	kelvin	(5/9)
Denier	tex	(1/9)
Dipole length (*e* cm)	coulomb-meter	$1.602\ 18 \times 10^{-21}$

TABLE 4.4 Conversion Factors (*Continued*)

To convert	Into	Multiply by
Drachm (British)	dram (apothecaries or troy)*	1
Drachm (British, fluid)	cubic centimeter	3.551 633
	dram (U.S., fluid)	0.960 760
	minim (British)	60
	ounce (British, fluid)	(1/8)
Dram (apothecaries or troy)	dram (weight)	2.194 285 7
	grain*	60
	gram*	3.887 934 6
	ounce (troy)*	(1/8)
	pennyweight*	2.5
	pound (troy)*	(1/96)
	scruple*	3
Dram (weight)	grain*	27.343 75
	gram	1.771 845 2
	ounce (weight)	(1/16)
	pound (weight)	(1/256)
Dram (U.S., fluid)	cubic centimeter	3.696 691 2
	gallon (U.S.)	(1/1024)
	gill (U.S.)	(1/32)
	milliliter	3.696 691 2
	minim (U.S.)*	60
	ounce (U.S., fluid)	(1/8)
	pint (U.S., fluid)	(1/128)
Dyne	kilogram (force)	$1.019\ 716 \times 10^{-6}$
	newton*	10^{-5}
	pound (force)	$2.248\ 09 \times 10^{-6}$
Dyne per centimeter	newton per meter*	0.001
Dyne per square centimeter	bar*	10^{-6}
	kilogram per square centimeter	$1.019\ 716 \times 10^{-6}$
	millimeter of mercury (0°C)	$7.500\ 617 \times 10^{-4}$
	millimeter of water (4°C)	0.010 197 16
	newton per square meter*	0.1
	pascal*	0.1
	pound per square inch (psi)	$1.450\ 38 \times 10^{-5}$
Dyne-centimeter	erg*	1
	foot-pound (force)	$7.375\ 62 \times 10^{-8}$
	foot-poundal	$2.373\ 04 \times 10^{-6}$
	joule*	10^{-7}
	kilogram-meter (force)	$1.019\ 716 \times 10^{-8}$
	newton-meter*	10^{-7}
Dyne-second/centimeter^2	poise*	1
	pascal-second*	0.1
Electron charge	coulomb	$1.602\ 18 \times 10^{-19}$
Electron charge-centimeter (*e* cm)	coulomb-meter	$1.602\ 18 \times 10^{-21}$
Electron charge-centimeter^2	coulomb-meter squared	$1.602\ 18 \times 10^{-23}$
Electron mass	atomic mass unit	0.000 548 6
	gram	9.1096×10^{-28}
Electronvolt	erg	$1.602\ 18 \times 10^{-12}$
	joule	$1.602\ 18 \times 10^{-19}$
	kilojoule per mole	96.4853
Ell	inch*	45
Em, pica	inch	0.167
	millimeter	4.217 52

(*Continued*)

TABLE 4.4 Conversion Factors (*Continued*)

To convert	Into	Multiply by
EMU[1] of capacitance	farad*	10^{9}
EMU of current	ampere*	10
EMU of electric potential	volt*	10^{-8}
EMU of inductance	henry*	10^{-9}
EMU of quantity (charge)	coulomb	10
EMU of resistance	ohm	10^{-9}
EMU of work	joule	10^{-7}
ESU[2] of capacitance	farad	$1.112\ 650 \times 10^{-12}$
ESU of current	ampere	$3.335\ 641 \times 10^{-10}$
ESU of electric potential	volt	299.792 5
ESU of inductance	henry	$8.987\ 552 \times 10^{11}$
ESU of quantity (charge)	coulomb	$3.335\ 556 \times 10^{-11}$
ESU of resistance	ohm	$8.987\ 552 \times 10^{11}$
ESU of work	joule	10^{-7}
Erg	dyne-centimeter*	1
	joule*	10^{-7}
	watt-hour	$2.777\ 78 \times 10^{-11}$
Erg per second	Btu	5.69×10^{-6}
	watt*	10^{-7}
Erg per ($cm^2 \times$ second)	watt per square meter*	0.001
Erg per gauss	ampere-centimeter squared*	10
	joule per tesla*	0.001
Fahrenheit scale	centigrade scale	(5/9)
Fahrenheit temperature (°F)	Celsius temperature (°C)	(°F − 32)(5/9)
Faraday (based on carbon-12)	coulomb	96 487.0
Faraday (chemical)	coulomb	96 495.7
Faraday (physical)	coulomb	96 521.9
Fathom	foot*	6
	meter	1.828 8
Fermi	meter*	10^{-15}
Foot	centimeter*	30.48
	inch*	12
	mile (nautical)	$1.645\ 788 \times 10^{-4}$
	mile (statute)	$1.893\ 939 \times 10^{-4}$
	yard	(1/3)
Foot of water (4°C)	atmosphere	0.029 499 8
	bar	0.029 499 8
	gram per square centimeter	30.48
	inch of mercury (0°C)	0.882 671
	pascal	2989.067
Foot per minute	centimeter per second*	0.508
	knot	0.009 874 73
	mile per hour	0.011 363 6
Foot-candle	lumen per square foot*	1
	lumen per square meter	10.7639
	lux	10.76391
Foot-lambert	candela per square centimeter	$3.426\ 26 \times 10^{-4}$
	candela per square foot	$(1/\pi)$
	lambert	0.001 076 39
	meter-lambert	10.7639

[1] EMU, the electromagnetic system of electrical units based on dynamics.
[2] ESU, the electrostatic system of electrical units based on static data.

TABLE 4.4 Conversion Factors (*Continued*)

To convert	Into	Multiply by
EMU[1] of capacitance	farad*	10^9
EMU of current	ampere*	10
EMU of electric potential	volt*	10^{-8}
EMU of inductance	henry*	10^{-9}
EMU of quantity (charge)	coulomb	10
EMU of resistance	ohm	10^{-9}
EMU of work	joule	10^{-7}
ESU[2] of capacitance	farad	$1.112\ 650 \times 10^{-12}$
ESU of current	ampere	$3.335\ 641 \times 10^{-10}$
ESU of electric potential	volt	299.792 5
ESU of inductance	henry	$8.987\ 552 \times 10^{11}$
ESU of quantity (charge)	coulomb	$3.335\ 556 \times 10^{-11}$
ESU of resistance	ohm	$8.987\ 552 \times 10^{11}$
ESU of work	joule	10^{-7}
Erg	dyne-centimeter*	1
	joule*	10^{-7}
	watt-hour	$2.777\ 78 \times 10^{-11}$
Erg per second	Btu	5.69×10^{-6}
	watt*	10^{-7}
Erg per ($cm^2 \times$ second)	watt per square meter*	0.001
Erg per gauss	ampere-centimeter squared*	10
	joule per tesla*	0.001
Fahrenheit scale	centigrade scale	(5/9)
Fahrenheit temperature (°F)	Celsius temperature (°C)	(°F − 32)(5/9)
Faraday (based on carbon-12)	coulomb	96 487.0
Faraday (chemical)	coulomb	96 495.7
Faraday (physical)	coulomb	96 521.9
Fathom	foot*	6
	meter	1.828 8
Fermi	meter*	10^{-15}
Foot	centimeter*	30.48
	inch*	12
	mile (nautical)	$1.645\ 788 \times 10^{-4}$
	mile (statute)	$1.893\ 939 \times 10^{-4}$
	yard	(1/3)
Foot of water (4°C)	atmosphere	0.029 499 8
	bar	0.029 499 8
	gram per square centimeter	30.48
	inch of mercury (0°C)	0.882 671
	pascal	2989.067
Foot per minute	centimeter per second*	0.508
	knot	0.009 874 73
	mile per hour	0.011 363 6
Foot-candle	lumen per square foot*	1
	lumen per square meter	10.7639
	lux	10.76391
Foot-lambert	candela per square centimeter	$3.426\ 26 \times 10^{-4}$
	candela per square foot	$(1/\pi)$
	lambert	0.001 076 39
	meter-lambert	10.7639

[1] EMU, the electromagnetic system of electrical units based on dynamics.
[2] ESU, the electrostatic system of electrical units based on static data.

(*Continued*)

TABLE 4.4 Conversion Factors (*Continued*)

To convert	Into	Multiply by
Foot-pound	Btu	0.001 285 07
	calorie	0.323 832
	foot-poundal	32.1740
	horsepower (British)	$5.050\ 51 \times 10^{-7}$
	joule	1.355 818
	kilogram-meter	0.138 255
	liter-atmosphere	0.013 380 9
	newton-meter	1.355 818
	watt-hour	$3.766\ 161 \times 10^{-4}$
Foot-pound per minute	horsepower (British)	$3.030\ 30 \times 10^{-5}$
	horsepower (metric)	$3.072\ 33 \times 10^{-5}$
	watt	0.022 597 0
Foot-poundal	Btu	$3.994\ 11 \times 10^{-5}$
	calorie	0.010 064 99
	foot-pound	0.031 081 0
	joule	0.042 140 11
	kilogram-meter	0.004 297 10
	liter-atmosphere	$4.158\ 91 \times 10^{-4}$
	watt-hour	$1.170\ 56 \times 10^{-5}$
Franklin	coulomb	$3.335\ 64 \times 10^{-10}$
Franklin per cm^3	coulomb per cubic meter	$3.335\ 64 \times 10^{-4}$
Franklin per cm^2	coulomb per square meter	$3.335\ 64 \times 10^{-6}$
Furlong	chain (Gunter's)*	10
	foot*	600
	meter*	201.168
	mile	(1/8)
Gallon (British, imperial)	bushel (British)	(1/8)
	cubic decimeter, liter*	4.546 90
	cubic foot	0.160 544
	gallon (U.S., fluid)	1.200 95
	gill (British)*	32
	liter	4.546 09
	ounce (British)*	160
	quart (British)*	4
Gallon (U.S.)	barrel (petroleum)	(1/42)
	cubic decimeter, liter	3.785 41
	cubic foot	0.133 680 56
	gallon (British)	0.832 674
	liter	3.785 41
	ounce (U.S., fluid)*	128
	quart (U.S., fluid)*	4
Gallon (U.S.) per minute	cubic foot per hour	8.020 83
	cubic meter per hour	0.227 125
	liter per minute	3.785 412
Gamma	microgram*	1
Gas constant	calorie per mole-degree	1.987
	joule per mole-degree	8.3143
	liter-atmosphere per mole-degree	0.082 057
Gauss	tesla*	10^{-4}
	weber per square meter*	10^{-4}
Gilbert	ampere-turn	0.795 775

TABLE 4.4 Conversion Factors (*Continued*)

To convert	Into	Multiply by
Gill (British)	cubic centimeter, mL	142.065
	cubic inch	8.669 36
	gallon (British)	(1/32)
	gill (U.S.)	1.200 95
	ounce (British, fluid)*	5
	pint (British)	(1/4)
Gill (U.S.)	cubic centimeter, mL	118.2941
	gallon (U.S.)	(1/32)
	liter	0.118 294 1
	ounce (U.S., fluid)*	4
	quart (U.S.)	(1/8)
Gon (grade)	circumference	(1/400)
	minute (angle)*	54
	radian	$(2\pi/400)$
Grade	radian	$(2\pi/400)$
Grain	carat (metric)*	0.323 994 55
	milligram*	64.798 91
	ounce (weight)	0.002 285 714 3
	ounce (troy)	(1/480)
	pennyweight	(1/24)
	pound	(1/7000)
	scruple	(1/20)
Gram	carat (metric)*	5
	dram	0.564 383 39
	grain	15.432 358
	ounce (weight)	0.035 273 962
	ounce (troy)	0.032 150 747
	pennyweight	0.643 014 93
	pound	0.002 204 622 6
	ton (metric)*	10^{-6}
Gram per (centimeter-second)	poise*	1
Gram per cubic centimeter	kilogram per liter*	1
	pound per cubic foot	62.4280
	pound per gallon (U.S.)	8.345 40
Gram per square meter	ounce per square foot	0.327 706
Gram per ton (long)	gram per ton (metric)	0.984 207
	gram per ton (short)	0.892 857
Gram (force)	dyne*	980.665
	newton*	0.009 806 65
Gram per square centimeter	pascal*	98.0665
Gram-centimeter	joule*	$9.806\,65 \times 10^{-5}$
Gram-square centimeter	pound-square foot	$2.373\,04 \times 10^{-6}$
Gray	joule per kilogram*	1
Hartree	electron volt	27.211 40
	hertz	$6.579\,683\,90 \times 10^{15}$
	joule	$4.359\,75 \times 10^{-18}$
Hectare	acre	2.471 054
	are*	100
	meter squared	10^4
Hefner unit	candela	0.9
Hemisphere	sphere*	0.5
	spherical right angle*	4
	steradian	(2π)

(*Continued*)

TABLE 4.4 Conversion Factors (*Continued*)

To convert	Into	Multiply by
Hertz	cycle per second*	1
Hogshead	gallon (U.S.)*	63
Horsepower (British)	Btu per hour	2544.43
	foot pound per hour*	1.98×10^6
	horsepower (metric)	1.013 87
	joule per second	745.700
	kilocalorie per hour	641.186
	kilogram-meter per second	76.0402
	watt	745.70
Horsepower (electric)	watt*	746
Horsepower-hour (British)	Btu	2544.43
	foot-pound*	1.98×10^6
	joule	$2.684\ 52 \times 10^6$
	kilocalorie	641.186
	kilogram-meter	$2.737\ 45 \times 10^5$
	watt-hour	745.7
Hour (mean solar)	day	(1/24)
	minute*	60
	second*	3600
	week	(1/168)
Hundredweight (long)	kilogram*	50.802 345 44
	pound*	112
	ton (long)	(1/20)
	ton (metric)	0.050 802 345
	ton (short)*	0.056
Hundredweight (short)	hundredweight (long)	0.892 857
Inch	centimeter*	2.54
	foot	(1/12)
	mil*	1000
Inch of mercury (0°C)	atmosphere	0.033 421 05
	inch of water (4°C)	13.5951
	millibar	33.863 88
	millimeter of water (4°C)	345.316
	pascal	3386.388
	pound per square inch, psi	0.491 1541
Inch of water (4°C)	inch of mercury (0°C)	0.073 5559
	millibar	2.490 89
	millimeter of mercury (0°C)	1.868 32
	pascal	249.089
	pound per square inch, psi	0.036 1273
Inch per minute	foot per hour*	5
	meter per hour*	1.524
	millimeter per second	0.423 333
Joule	Btu	$9.478\ 170 \times 10^{-4}$
	calorie*	0.2390
	centigrade heat unit, chu	5.265 65
	centimeter-dyne*	10^7
	cubic foot-atmosphere	0.000 348 529
	cubic foot-(pound per in^2)	0.005 121 959
	erg*	10^7
	foot-pound	0.737 562
	foot-poundal	23.7304
	horsepower-hour (British)	$3.725\ 06 \times 10^{-7}$
	liter-atmosphere	0.009 869 233

TABLE 4.4 Conversion Factors (*Continued*)

To convert	Into	Multiply by
Joule (*continued*)	newton-meter*	1
	watt-second*	1
Joule per centimeter	kilogram (force)	10.197 16
	newton*	100
	pound (force)	22.4809
Joule per gram	Btu per pound	0.429 923
	kilocalorie per kilogram	0.238 846
	watt-hour per pound	0.125 998
Joule per second	watt*	1
Kilogram (force)	dyne*	$9.806\,65 \times 10^5$
	newton*	9.806 65
	pound (force)	2.204 62
	poundal	70.9316
Kilometer	astronomical unit	$6.684\,59 \times 10^{-9}$
	mile (nautical)	0.539 956 80
	mile (statute)	0.621 371 192
Kilowatt	Btu per minute	56.8690
	foot-pound per second	737.562
	horsepower (British)	1.341 02
	horsepower (metric)	1.359 62
	joule per second*	1000
	kilocalorie per hour	859.845
Kilowatt-hour	Btu	3412.14
	horsepower-hour (British)	1.341 02
	joule*	3.6×10^6
	kilocalorie	859.845
Knot	foot per minute	101.2686
	kilometer per hour*	1.852
	mile (nautical) per hour*	1
	mile (statute) per hour	1.150 78
Lambda	decimeter cubed*	10^{-6}
	microliter*	1
Lambert	candela per square meter	$(1/\pi) \times 10^4$; 3183.099
	candela per square inch	2.053 61
	foot-lambert	929.030
Langley	joule per square meter*	4.184×10^4
League (nautical)	mile (nautical)*	3
League (statute)	mile (statute)*	3
Light-year	astronomical unit	$6.323\,97 \times 10^4$
	meter	$9.460\,73 \times 10^{15}$
Link	chain*	0.01
Liter	cubic decimeter (dm^3)*	1
	cubic foot	0.035 314 67
	gallon (British)	0.219 969
	gallon (U.S.)	0.264 172 1
	quart (British)	0.879 877
	quart (U.S.)	1.056 688
Liter per minute	cubic foot per hour	2.118 88
	gallon (British) per hour	13.198
	gallon (U.S.) per hour	15.8503
Liter-atmosphere	Btu	0.096 037 6
	calorie	24.2011
	cubic foot-atmosphere	0.035 314 7
	cubic foot-pound per in^2	0.518 983

(*Continued*)

TABLE 4.4 Conversion Factors (*Continued*)

To convert	Into	Multiply by
Liter-atmosphere (*continued*)	horsepower (British)	$3.774\ 42 \times 10^{-5}$
	horsepower (metric)	$3.826\ 77 \times 10^{-5}$
	joule*	101.325
	kilogram-meter	10.332 27
	watt-hour	0.028 145 8
Lumen per square centimeter	lux*	10^4
	phot*	1
Lumen per square meter	lumen per square foot	0.092 903 0
Lux	lumen per square meter*	1
Maxwell	weber*	10^{-8}
Meter	ångström*	10^{10}
	fathom	0.546 807
	foot	3.280 839 895
	inch	39.370 078 740
	mile (nautical)	$5.399\ 568 \times 10^{-4}$
	mile (statute)	$6.213\ 712 \times 10^{-4}$
Meter per second	foot per minute	196.850
	kilometer per hour*	3.6
	knot	1.943 844
	mile per hour	2.236 936
Meter-candle	lux*	1
Meter-lambert	candela per square meter	$(1/\pi)$
	foot-lambert	0.092 903 0
	lambert*	10^{-4}
Mho (ohm−1)	siemen*	1
Micron	meter	10^{-6}
Mil	inch*	0.001
	micrometer*	25.4
Mile (nautical)	foot	6076.115 49
	kilometer*	1.852
	mile (statute)	1.150 78
Mile (statute)	chain (Gunter's)*	80
	chain (Ramsden's)*	52.8
	foot*	5280
	furlong*	8
	kilometer*	1.609 344
	light-year	$1.701\ 11 \times 10^{-11}$
	link (Gunter's)*	8000
	link (Ramsden's)*	5280
	mile (nautical)	0.868 976
	rod*	320
Mile per gallon (British)	kilometer per liter	0.354 006
Mile per gallon (U.S.)	kilometer per liter	0.425 144
Mile per hour	foot per minute	88
	kilometer per hour*	1.609 344
	knot	0.868 976
Milliliter	cubic centimeter*	1
Millimeter of mercury (0°C)	atmosphere	(1/760)
	dyne per square centimeter	1333.224
	millimeter of water (4°C)	13.5951
	pascal	133.322
	pound per square inch (psi)	0.019 336 8
	torr*	1

TABLE 4.4 Conversion Factors (*Continued*)

To convert	Into	Multiply by
Millimeter of water (4°C)	atmosphere	0.009 678 41
	millibar*	0.098 066 5
	millimeter of mercury (0°C)	0.073 555 9
	pascal*	9.806 65
	pound per square inch	0.001 422 33
Minim (British)	milliliter	0.059 193 9
	minim (U.S.)	0.960 760
Minim (U.S.)	milliliter	0.061 611 5
Minute (plane angle)	circumference	$4.629\ 63 \times 10^{-5}$
	degree (angle)	(1/60)
	gon	(1/54)
	radian	(π/10,800)
Minute	hour	(1/60)
	second	60
Month (mean of 4-year period)	day	30.4375
	hour	730.5
	week	4.348 21
Nail (British)	inch*	2.25
Nanometer	ångström*	10
Neper	decibel	8.685 890
Nuclear magneton	joule per tesla	$5.050\ 79 \times 10^{-27}$
Neutron mass	atomic mass unit	1.008 66
	gram	1.6749×10^{-24}
Newton	dyne*	10^5
	kilogram (force)	0.101 971 6
	pound (force)	0.224 809
	poundal	7.233 01
Newton per square meter	*See* pascal	
Newton-meter	foot-pound	0.737 562
	joule*	1
	kilogram-meter	0.101 971 6
	watt-second*	1
Nit	candela per square meter*	1
Noggin (British)	gill (British)*	1
Nox	lux*	0.001
Oersted	ampere per meter (in practice)	(1000/4π); 79.577 47
Ohm (mean international)	ohm	1.000 49
Ohm (U.S. international)	ohm	1.000 495
Ohm per foot	ohm per meter	3.280 84
Ounce (avoirdupois)	dram*	16
	grain*	437.5
	gram*	28.349 5
	ounce (troy)	0.911 458 33
	pound	(1/16)
Ounce (troy)	grain*	480
	gram*	31.1035
	ounce (avoirdupois)	1.097 142 9
	pennyweight*	20
	pound (avoirdupois)	0.068 571 429
	scruple*	24
Ounce (British, fluid)	cubic centimeter	28.413 06
	gallon (British)	(1/160)
	milliliter	28.413 06

(*Continued*)

TABLE 4.4 Conversion Factors (*Continued*)

To convert	Into	Multiply by
Ounce (British, fluid) (*continued*)	minim (British)	480
	ounce (U.S., fluid)	0.960 760
	pint (British)	(1/20)
	quart (British)	(1/40)
Ounce (U.S., fluid)	cubic centimeter	29.573 530
	gallon (U.S.)	(1/128)
	milliliter	29.573 530
	pint (U.S., fluid)	(1/16)
	quart (U.S., fluid)	(1/32)
Ounce (avoirdupois) per cubic foot	kilogram per cubic meter	1.001 154
Ounce (avoirdupois) per gallon (U.S.)	gram per liter	7.489 15
Ounce (avoirdupois) per ton (long)	gram per ton (metric)	27.9018
	milligram per kilogram	27.9018
Ounce (avoirdupois) per ton (short)	gram per ton (metric)*	31.25
	milligram per kilogram*	31.25
Parsec	light-year	3.261 636
Part per million	milligram per kilogram*	1
	milliliter per cubic meter*	1
Pascal	atmosphere	$9.869\ 233 \times 10^{-6}$
	bar*	10^{-5}
	dyne per square centimeter*	10
	inch of mercury	$2.953\ 00 \times 10^{-4}$
	millimeter of mercury	$7.500\ 62 \times 10^{-3}$
	millimeter of water	0.101 972
	newton per square meter*	1
	pound per square inch	$1.450\ 377 \times 10^{-4}$
	poundal per square foot	0.671 969
Pascal-second	poise*	10
Peck (British)	gallon (British)*	2
Peck (U.S.)	bushel (U.S.)*	0.25
Pennyweight	grain*	24
	gram*	1.555 173 84
	ounce (troy)	(1/20)
	pound	0.003 428 571 4
Phot	lux*	10^4
Pica (printer's)	inch	0.167
	point*	12
Pint (British)	gallon (British)	(1/8)
	liter	0.568 261
	pint (U.S., fluid)	1.200 95
	quart (British)	0.5
Pint (U.S., dry)	bushel (U.S.)	(1/64)
	liter	0.550 610 5
	peck (U.S.)	(1/16)
	pint (British)	0.968 939
	quart (U.S., dry)	0.5
Pint (U.S., fluid)	gallon (U.S.)	(1/8)
	liter	0.473 176 5

TABLE 4.4 Conversion Factors (*Continued*)

To convert	Into	Multiply by
Pint (U.S., fluid)	pint (British)	0.832 674
(*continued*)	quart (U.S., fluid)*	0.5
Planck's constant	joule-second	$6.626\ 08 \times 10^{-34}$
Point (printer's, Didot)	millimeter	0.376 065 03
Point (printer's, U.S.)	millimeter*	0.351 459 8
Poise	dyne-second per square centimeter*	1
	pascal-second*	0.1
Polarizability volume ($4\pi\epsilon_0$ cm^3)	coulomb squared-(meter squared per joule)	$1.112\ 65 \times 10^{-16}$
Pole (British)	foot*	16.5
Pottle (British)	gallon (British)*	0.5
Pound	gram*	453.592 37
	ounce (weight)*	16
	ton (long)	$4.464\ 285\ 7 \times 10^{-4}$
	ton (short)	(1/2000)
Pound (troy)	grain	5760
	gram*	373.241 721 6
	ounce (troy)*	12
	pennyweight	240
	pound (weight)	0.822 857 14
	scruple*	288
Pound per cubic foot	kilogram per cubic meter	16.018 46
Pound per cubic inch	gram per cubic centimeter	27.679 905
	pound per cubic foot*	1728
Pound per foot	kilogram per meter	1.488 16
Pound per (foot-second)	pascal-second	1.488 16
Pound per gallon (U.S.)	gram per liter	119.8264
Pound per hour	kilogram per day	10.886 22
Pound per inch	kilogram per meter	17.857 97
Pound per minute	kilogram per hour	27.215 54
Pound per square foot	kilogram per square meter	4.882 43
Pound (force)	kilogram (force)	0.453 592
	newton	4.448 222
	poundal	32.1740
Pound per square inch	atmosphere	0.068 046 0
	bar	0.068 948 0
	inch of mercury (0°C)	2.036 02
	millimeter of mercury (0°C)	51.7149
	millimeter of water (4°C)	703.070
	pascal	6894.757
	pound per square foot	144
Pound-second per square inch	pascal-second	6894.76
Poundal	gram (force)	14.0981
	newton	0.138 255
	pound (force)	0.031 081 0
Poundal per square foot	pascal	1.488 164
Poundal-foot	newton-meter	0.042 140 1
Poundal-second per square foot	pascal-second	1.488 164
Proof (U.S.)	percent alcohol by volume*	0.5
Proton mass	atomic mass unit	1.007 28
	gram	1.6726×10^{-24}
Puncheon (British)	gallon (British)	70

(*Continued*)

TABLE 4.4 Conversion Factors (*Continued*)

To convert	Into	Multiply by
Quad	Btu	10^{15}
	joule	1.055×10^{18}
Quadrant	circumference*	0.25
	degree (angle)*	90
	gon (grade)*	100
	minute (angle)*	5400
	radian	(π/2)
Quadrupole area (*e* cm^2)	coulomb meter squared	$1.602\ 18 \times 10^{-23}$
Quart (British)	gallon (British)*	0.25
	liter	1.136 523
	ounce (British, fluid)*	40
	pint (British)*	2
	quart (U.S., fluid)	1.200 95
Quart (U.S., dry)	bushel (U.S.)	(1/32)
	cubic foot	0.038 889 25
	liter	1.101 221
	peck (U.S.)	(1/8)
	pint (U.S., dry)*	2
Quart (U.S., fluid)	gallon (U.S.)*	0.25
	liter	0.946 529
	ounce (U.S., fluid)*	32
	pint (U.S., fluid)	2
	quart (British)	0.832 674
Quartern (British, fluid)	gill (British)*	0.5
Quintal (metric)	kilogram*	100
Rad (absorbed dose)	gray*	0.01
	joule per kilogram*	0.01
Radian	circumference	(1/2π)
	degree (angle)	57.295 780
	minute (angle)	3437.75
	quadrant	(2/π)
	revolution	(1/2π)
Radian per centimeter	degree per millimeter	5.729 58
	degree per inch	145.531
Radian per second	revolution per minute	9.549 30
Radian per second squared	revolution per minute squared	572.958
Rankin (degree)	kelvin	(5/9)
Ream	quire*	20
	sheet	480 or 500
Register ton	cubic foot*	100
	cubic meter	2.831 685
Rem (dose equivalent)	sievert*	0.01
Revolution	degree (angle)	360
	gon*	400
	quadrant*	4
	radian	(2π)
Revolution per minute	degree (angle) per second*	6
	radian per second	0.104 720
Revolution per minute squared	radian per second squared	0.001 745 33
Revolution per second squared	radian per second squared	6.283 185
	revolution per minute squared	3600
Reyn	pascal-second	6894.76
	pound-second per square inch	1

TABLE 4.4 Conversion Factors (*Continued*)

To convert	Into	Multiply by
Rhe	per pascal-second*	10
Right angle	degree*	90
	radian	$(\pi/2)$
Rod (British, volume)	cubic foot*	1000
Rod (surveyer's measure)	chain (Gunter's)*	0.25
	foot*	16.5
	link (Gunter's)*	25
	meter*	5.0292
Roentgen	coulomb per kilogram	2.58×10^{-4}
Rood (British)	acre*	0.25
	square meter	1011.714 1
Rydberg	joule	$2.179\ 87 \times 10^{-18}$
Scruple	dram (troy)	(1/3)
	grain*	20
	gram*	1.295 978 2
	ounce (weight)	0.045 714 286
	ounce (troy)	(1/24)
	pennyweight	(10/12)
	pound	(1/350)
Second (plane angle)	degree	$2.777\ 78 \times 10^{-4}$
	minute	(1/60)
	radian	$(\pi/6.48 \times 10^5)$
Section	square mile*	1
Siemens	mho (ohm^{-1})*	1
Slug	geepound*	1
	kilogram	14.593 90
	pound	32.1740
Speed of light	centimeter per second	$2.997\ 924\ 58 \times 10^{10}$
Sphere	steradian	(4π)
Square centimeter	circular mil	$1.973\ 53 \times 10^5$
	circular millimeter	127.3240
	square inch	0.155 000 31
Square chain (Gunter's)	acre*	0.1
	square foot*	4356
	square meter	404.686
Square chain (Ramsden's)	square foot*	10^4
Square degree (angle)	steradian	$3.046\ 17 \times 10^{-4}$
Square foot	acre	$2.295\ 68 \times 10^{-5}$
	square centimeter	929.0304
	square meter	0.092 903 04
	square rod	0.003 673 09
Square inch	circular mil	$1.273\ 240 \times 10^6$
	circular millimeter	821.4432
	square centimeter	6.4516
Square kilometer	acre	247.1054
	hectare*	100
	square mile	0.386 102 16
Square link (Gunter's)	square foot*	0.4356
Square link (Ramsden's)	square foot*	1
Square meter	are*	0.01
	square foot	10.763 91
	square mile	$3.861\ 01 \times 10^{-7}$

(*Continued*)

TABLE 4.4 Conversion Factors (*Continued*)

To convert	Into	Multiply by
Square meter (*continued*)	square rod	0.039 536 9
	square yard	1.195 990
Square mile	acre*	640
	square kilometer	2.589 988 110
	township	(1/36)
Square rod	acre	(1/160)
	square foot	272.25
	square meter	25.292 853
Square yard	square foot*	9
	square inch*	1296
	square meter*	0.836 127 36
	square rod	0.033 057 85
Statampere	ampere	$3.335\ 641 \times 10^{-10}$
Statcoulomb	coulomb	$3.335\ 641 \times 10^{-10}$
Statfarad	farad	$1.112\ 650 \times 10^{-12}$
Stathenry	henry	$8.987\ 552 \times 10^{11}$
Statmho	siemens	$1.112\ 650 \times 10^{-12}$
Statohm	ohm	$8.987\ 552 \times 10^{11}$
Statvolt	volt	299.7925
Statweber	weber	299.7925
Steradian	sphere	$(1/4\pi)$
	spherical right angle	$(2/\pi)$
	square degree	3282.81
Stere	cubic meter*	1
Stilb	candela/cm^2	1
Stokes (kinematic viscosity)	square meter per second*	10^{-4}
Stone (British)	pound*	14
Svedberg	second*	10^{-13}
Tablespoon (metric)	cubic centimeter*; milliliter	14.79
Teaspoon (metric)	cubic centimeter*; milliliter	4.929
Tesla	weber per square meter*	1
Tex	denier*	9
	gram per kilometer*	1
Therm	Btu*	10^5
	joule*	$1.054\ 804 \times 10^8$
Ton (assay)	gram	29.166 67
Ton (long)	hundredweight (long)*	20
	hundredweight (short)*	22.4
	kilogram	1016.046 908 8
	pound*	2240
	ton (metric)	1.016 046 9
	ton (short)	1.12
Ton (metric)	hundredweight (long)	19.684 131
	hundredweight (short)	22.046 226
	kilogram*	1000
	pound	2204.6226
	ton (long)	0.984 206 53
	ton (short)*	1.102 311 3
Ton (short)	kilogram	907.184 74
	pound*	2000
Ton (force, long)	newton	1186.553
Ton (force, metric)	newton	9806.65

TABLE 4.4 Conversion Factors (*Continued*)

To convert	Into	Multiply by
Ton (force, short)	newton	8896.44
Ton (force, long)/ft^2	bar	1.072 518
	pascal	$1.072\ 518 \times 10^5$
Ton (force, metric)/m^2	bar	0.098 066 5
	pascal	9806.65
Ton (force, short)/ft^2	bar	0.957 605
	pascal	$9.576\ 05 \times 10^4$
Tonne (metric)	kilogram*	1000
Torr	atmosphere	(1/760)
	millibar	1.333 224
	millimeter of mercury* (0°C)	1
	pascal	133.322; (101 325/760)
Township (U.S.)	square kilometer	93.2396
	square mile*	36
Unified atomic mass unit	kilogram	$1.660\ 54 \times 10^{-27}$
Unit pole	weber	$1.256\ 637 \times 10^{-7}$
Volt (mean international)	volt	1.000 34
Volt (U.S. international)	volt	1.000 330
Volt-second	weber*	1
Watt	Btu per hour	3.412 14
	calorie per minute	14.3308
	erg per second*	10^7
	foot-pound per minute	44.2537
	horsepower (British)	0.001 341 02
	horsepower (metric)	0.001 359 62
	joule per second*	1
	kilogram-meter per second	0.101 972
Watt per square inch	watt per square meter	1550.003
Watt-hour	Btu	3.412 14
	calorie	859.845
	foot-pound	2655.22
	horsepower-hour (British)	0.001 341 02
	horsepower-hour (metric)	0.001 359 62
	joule*	3600
	liter-atmosphere	35.5292
Watt-second	joule*	1
Weber	maxwell*	10^8
Week	day*	7
	hour*	168
Wey (British, capacity)	bushel (British)	40 (variable)
Wey (British, mass)	pound	252 (variable)
X unit	meter	$1.002\ 02 \times 10^{-13}$
Yard	fathom*	0.5
	meter	0.9144
Year (mean of 4-years)	day	365.25
	week	52.178 87
Year (sidereal)	day (mean solar)	365.256 36

4.3 CONVERSION OF THERMOMETER SCALES

The following abbreviations are used: °F, degrees Fahrenheit; °C, degrees Celsius; K, degrees Kelvin; °Ré, degrees Reaumur; °R, degrees Rankine; °Z, degrees on any scale; (fp) "Z", the freezing point of water on the Z scale; and (bp) "Z", the boiling point of water on the Z scale. Reference: Dodds, *Chemical and Metallurigical Engineering* **38**:476 (1931).

$$\frac{^\circ\text{F}-32}{180}=\frac{^\circ\text{C}}{100}=\frac{^\circ\text{Ré}}{80}=\frac{\text{K}-273}{100}=\frac{^\circ\text{R}-492}{180}=\frac{^\circ\text{Z}-(\text{fp})\text{"Z"}}{(\text{bp})\text{"Z"}-(\text{fp})\text{"Z"}}$$

Examples

(1) To find the Fahrenheit temperature corresponding to –20°C:

$$\frac{^\circ\text{F}-32}{180}=\frac{^\circ\text{C}}{100} \quad \text{or} \quad \frac{^\circ\text{F}-32}{180}=\frac{-20}{100}$$

$$^\circ\text{F}-32=\frac{(-20)(180)}{100}=-36$$

$$^\circ\text{F}=-4$$

(2) To find the Reaumur temperature corresponding to 20°F:

$$\frac{^\circ\text{F}-32}{180}=\frac{^\circ\text{Ré}}{80}=\frac{20-32}{180}=\frac{^\circ\text{Ré}}{80}$$

i.e.,

$$20^\circ\text{F}=-5.33^\circ\text{Ré}$$

(3) To find the correct tempeature on a thermometer reading 80°C and that shows a reading of –0.30°C in a melting ice/water mixture and 99.0°C in steam at 760 mm pressure of mercury:

$$\frac{^\circ\text{C}}{100}=\frac{\text{Z}-(\text{fp})\text{"Z"}}{(\text{bp})\text{"Z"}-(\text{fp})\text{"Z"}}=\frac{80-(-0.30)}{99.0-(-0.30)}$$

i.e.,

$$^\circ\text{C}=80.87 \text{ (corrected)}$$

TABLE 4.5 Temperature Conversion

The column of figures in bold and which is headed "Reading in °F. or °C. to be converted" refers to the temperature either in degrees Fahrenheit or Celsius which it is desired to convert into the other scale. If converting from Fahrenheit degrees to Celsius degrees, the equivalent temperature will be found in the column headed "°C."; while if converting from degrees Celsius to degrees Fahrenheit, the equivalent temperature will be found in the column headed "°F." This arrangement is very similar to that of Sauveur and Boylston, copyrighted 1920, and is published with their permission.

°F.	Reading in °F. or °C. to be converted	°C.
....	−458	−272.22
....	−456	−271.11
....	−454	−270.00
....	−452	−268.89
....	−450	−267.78
....	−448	−266.67
....	−446	−265.56
....	−444	−264.44
....	−442	−263.33
....	−440	−262.22
....	−438	−261.11
....	−436	−260.00
....	−434	−258.89
....	−432	−257.78
....	−430	−256.67
....	−428	−255.56
....	−426	−254.44
....	−424	−253.33
....	−422	−252.22
....	−420	−251.11
....	−418	−250.00
....	−416	−248.89
....	−414	−247.78
....	−412	−246.67
....	−410	−245.56
....	−408	−244.44
....	−406	−243.33
....	−404	−242.22
....	−402	−241.11
....	−400	−240.00
....	−398	−238.89
....	−396	−237.78
....	−394	−236.67
....	−392	−235.56
....	−390	−234.44
....	−388	−233.33
....	−386	−232.22
....	−384	−231.11
....	−382	−230.00
....	−380	−228.89
....	−378	−227.78
....	−376	−226.67
....	−374	−225.56
....	−372	−224.44
....	−370	−223.33
....	−368	−222.22
....	−366	−221.11
....	−364	−220.00
....	−362	−218.89
....	−360	−217.78
....	−358	−216.67
....	−356	−215.56
....	−354	−214.44
....	−352	−213.33
....	−350	−212.22
....	−348	−211.11
....	−346	−210.00
....	−344	−208.89
....	−342	−207.78
....	−340	−206.67
....	−338	−205.56
....	−336	−204.44
....	−334	−203.33
....	−332	−202.22
....	−330	−201.11
....	−328	−200.00
....	−326	−198.89
....	−324	−197.78
....	−322	−196.67
....	−320	−195.56
....	−318	−194.44
....	−316	−193.33
....	−314	−192.22
....	−312	−191.11
....	−310	−190.00
....	−308	−188.89
....	−306	−187.78
....	−304	−186.67
....	−302	−185.56
....	−300	−184.44

(*Continued*)

TABLE 4.5 Temperature Conversion (*Continued*)

°F.	Reading in °F. or °C. to be converted	°C.	°F.	Reading in °F. or °C. to be converted	°C.
....	−298	−183.33	−342.4	−208	−133.33
....	−296	−182.22	−338.8	−206	−132.22
....	−294	−181.11	−335.2	−204	−131.11
....	−292	−180.00	−331.6	−202	−130.00
....	−290	−178.89	−328.0	−200	−128.89
....	−288	−177.78	−324.4	−198	−127.78
....	−286	−176.67	−320.8	−196	−126.67
....	−284	−175.56	−317.2	−194	−125.56
....	−282	−174.44	−313.6	−192	−124.44
....	−280	−173.33	−310.0	−190	−123.33
....	−278	−172.22	−306.4	−188	−122.22
....	−276	−171.11	−302.8	−186	−121.11
....	−274	−170.00	−299.2	−184	−120.00
−457.6	−272	−168.89	−295.6	−182	−118.89
−454.0	−270	−167.78	−292.0	−180	−117.78
−450.4	−268	−166.67	−288.4	−178	−116.67
−446.8	−266	−165.56	−284.8	−176	−115.56
−443.2	−264	−164.44	−281.2	−174	−114.44
−439.6	−262	−163.33	−277.6	−172	−113.33
−436.0	−260	−162.22	−274.0	−170	−112.22
−432.4	−258	−161.11	−270.4	−168	−111.11
−428.8	−256	−160.00	−266.8	−166	−110.00
−425.2	−254	−158.89	−263.2	−164	−108.89
−421.6	−252	−157.78	−259.6	−162	−107.78
−418.0	−250	−156.67	−256.0	−160	−106.67
−414.4	−248	−155.56	−252.4	−158	−105.56
−410.8	−246	−154.44	−248.8	−156	−104.44
−407.2	−244	−153.33	−245.2	−154	−103.33
−403.6	−242	−152.22	−241.6	−152	−102.22
−400.0	−240	−151.11	−238.0	−150	−101.11
−396.4	−238	−150.00	−234.4	−148	−100.00
−392.8	−236	−148.89	−230.8	−146	−98.89
−389.2	−234	−147.78	−227.2	−144	−97.78
−385.6	−232	−146.67	−223.6	−142	−96.67
−382.0	−230	−145.56	−220.0	−140	−95.56
−378.4	−228	−144.44	−216.4	−138	−94.44
−374.8	−226	−143.33	−212.8	−136	−93.33
−371.2	−224	−142.22	−209.2	−134	−92.22
−367.6	−222	−141.11	−205.6	−132	−91.11
−364.0	−220	−140.00	−202.0	−130	−90.00
−360.4	−218	−138.89	−198.4	−128	−88.89
−356.8	−216	−137.78	−194.8	−126	−87.78
−353.2	−214	−136.67	−191.2	−124	−86.67
−349.6	−212	−135.56	−187.6	−122	−85.56
−346.0	−210	−134.44	−184.0	−120	−84.44

TABLE 4.5 Temperature Conversion (*Continued*)

°F.	Reading in °F. or °C. to be converted	°C.
−180.4	−118	−83.33
−176.8	−116	−82.22
−173.2	−114	−81.11
−169.6	−112	−80.00
−166.0	−110	−78.89
−162.4	−108	−77.78
−158.8	−106	−76.67
−155.2	−104	−75.56
−151.6	−102	−74.44
−148.0	−100	−73.33
−144.4	−98	−72.22
−140.8	−96	−71.11
−137.2	−94	−70.00
−133.6	−92	−68.89
−130.0	−90	−67.78
−126.4	−88	−66.67
−122.8	−86	−65.56
−119.2	−84	−64.44
−115.6	−82	−63.33
−112.0	−80	−62.22
−108.4	−78	−61.11
−104.8	−76	−60.00
−101.2	−74	−58.89
−97.6	−72	−57.78
−94.0	−70	−56.67
−90.4	−68	−55.56
−86.8	−66	−54.44
−83.2	−64	−53.33
−79.6	−62	−52.22
−76.0	−60	−51.11
−72.4	−58	−50.00
−68.8	−56	−48.89
−65.2	−54	−47.78
−61.6	−52	−46.67
−58.0	−50	−45.56
−54.4	−48	−44.44
−50.8	−46	−43.33
−47.2	−44	−42.22
−43.6	−42	−41.11
−40.0	−40	−40.00
−36.4	−38	−38.89
−32.8	−36	−37.78
−29.2	−34	−36.67
−25.6	−32	−35.56
−22.0	−30	−34.44

°F.	Reading in °F. or °C. to be converted	°C.
−18.4	−28	−33.33
−14.8	−26	−32.22
−11.2	−24	−31.11
−7.6	−22	−30.00
−4.0	−20	−28.89
−0.4	−18	−27.78
+3.2	−16	−26.67
+6.8	−14	−25.56
+10.4	−12	−24.44
+14.0	−10	−23.33
+17.6	−8	−22.22
+19.4	−7	−21.67
+21.2	−6	−21.11
+23.0	−5	−20.56
+24.8	−4	−20.00
+26.6	−3	−19.44
+28.4	−2	−18.89
+30.2	−1	−18.33
+32.0	±0	−17.78
+33.8	+1	−17.22
+35.6	+2	−16.67
+37.4	+3	−16.11
+39.2	+4	−15.56
+41.0	+5	−15.00
+42.8	+6	−14.44
+44.6	+7	−13.89
+46.4	+8	−13.33
+48.2	+9	−12.78
+50.0	+10	−12.22
+51.8	+11	−11.67
+53.6	+12	−11.11
+55.4	+13	−10.56
+57.2	+14	−10.00
+59.0	+15	−9.44
+60.8	+16	−8.89
+62.6	+17	−8.33
+64.4	+18	−7.78
+66.2	+19	−7.22
+68.0	+20	−6.67
+69.8	+21	−6.11
+71.6	+22	−5.56
+73.4	+23	−5.00
+75.2	+24	−4.44
+77.0	+25	−3.89
+78.8	+26	−3.33

(*Continued*)

TABLE 4.5 Temperature Conversion (*Continued*)

°F.	Reading in °F. or °C. to be converted	°C.
+80.6	+27	−2.78
+82.4	+28	−2.22
+84.2	+29	−1.67
+86.0	+30	−1.11
+87.8	+31	−0.56
+89.6	+32	±0.00
+91.4	+33	+0.56
+93.2	+34	+1.11
+95.0	+35	+1.67
+96.8	+36	+2.22
+98.6	+37	+2.78
+100.4	+38	+3.33
+102.2	+39	+3.89
+104.0	+40	+4.44
+105.8	+41	+5.00
+107.6	+42	+5.56
+109.4	+43	+6.11
+111.2	+44	+6.67
+113.0	+45	+7.22
+114.8	+46	+7.78
+116.6	+47	+8.33
+118.4	+48	+8.89
+120.2	+49	+9.44
+122.0	+50	+10.00
+123.8	+51	+10.56
+125.6	+52	+11.11
+127.4	+53	+11.67
+129.2	+54	+12.22
+131.0	+55	+12.78
+132.8	+56	+13.33
+134.6	+57	+13.89
+136.4	+58	+14.44
+138.2	+59	+15.00
+140.0	+60	+15.56
+141.8	+61	+16.11
+143.6	+62	+16.67
+145.4	+63	+17.22
+147.2	+64	+17.78
+149.0	+65	+18.33
+150.8	+66	+18.89
+152.6	+67	+19.44
+154.4	+68	+20.00
+156.2	+69	+20.56
+158.0	+70	+21.11
+159.8	+71	+21.67

°F.	Reading in °F. or °C. to be converted	°C.
+161.6	+72	+22.22
+163.4	+73	+22.78
+165.2	+74	+23.33
+167.0	+75	+23.89
+168.8	+76	+24.44
+170.6	+77	+25.00
+172.4	+78	+25.56
+174.2	+79	+26.11
+176.0	+80	+26.67
+177.8	+81	+27.22
+179.6	+82	+27.78
+181.4	+83	+28.33
+183.2	+84	+28.89
+185.0	+85	+29.44
+186.8	+86	+30.00
+188.6	+87	+30.56
+190.4	+88	+31.11
+192.2	+89	+31.67
+194.0	+90	+32.22
+195.8	+91	+32.78
+197.6	+92	+33.33
+199.4	+93	+33.89
+201.2	+94	+34.44
+203.0	+95	+35.00
+204.8	+96	+35.56
+206.6	+97	+36.11
+208.4	+98	+36.67
+210.2	+99	+37.22
+212.0	+100	+37.78
+213.8	+101	+38.33
+215.6	+102	+38.89
+217.4	+103	+39.44
+219.2	+104	+40.00
+221.0	+105	+40.56
+222.8	+106	+41.11
+224.6	+107	+41.67
+226.4	+108	+42.22
+228.2	+109	+42.78
+230.0	+110	+43.33
+231.8	+111	+43.89
+233.6	+112	+44.44
+235.4	+113	+45.00
+237.2	+114	+45.56
+239.0	+115	+46.11
+240.8	+116	+46.67

TABLE 4.5 Temperature Conversion (*Continued*)

°F.	Reading in °F. or °C. to be converted	°C.	°F.	Reading in °F. or °C. to be converted	°C.
+242.6	+117	+47.22	+323.6	+162	+72.22
+244.4	+118	+47.78	+325.4	+163	+72.78
+246.2	+119	+48.33	+327.2	+164	+73.33
+248.0	+120	+48.89	+329.0	+165	+73.89
+249.8	+121	+49.44	+330.8	+166	+74.44
+251.6	+122	+50.00	+332.6	+167	+75.00
+253.4	+123	+50.56	+334.4	+168	+75.56
+255.2	+124	+51.11	+336.2	+169	+76.11
+257.0	+125	+51.67	+338.0	+170	+76.67
+258.8	+126	+52.22	+339.8	+171	+77.22
+260.6	+127	+52.78	+341.6	+172	+77.78
+262.4	+128	+53.33	+343.4	+173	+78.33
+264.2	+129	+53.89	+345.2	+174	+78.89
+266.0	+130	+54.44	+347.0	+175	+79.44
+267.8	+131	+55.00	+348.8	+176	+80.00
+269.6	+132	+55.56	+350.6	+177	+80.56
+271.4	+133	+56.11	+352.4	+178	+81.11
+273.2	+134	+56.67	+354.2	+179	+81.67
+275.0	+135	+57.22	+356.0	+180	+82.22
+276.8	+136	+57.78	+357.8	+181	+82.78
+278.6	+137	+58.33	+359.6	+182	+83.33
+280.4	+138	+58.89	+361.4	+183	+83.89
+282.2	+139	+59.44	+363.2	+184	+84.44
+284.0	+140	+60.00	+365.0	+185	+85.00
+285.8	+141	+60.56	+366.8	+186	+85.56
+287.6	+142	+61.11	+368.6	+187	+86.11
+289.4	+143	+61.67	+370.4	+188	+86.67
+291.2	+144	+62.22	+372.2	+189	+87.22
+293.0	+145	+62.78	+374.0	+190	+87.78
+294.8	+146	+63.33	+375.8	+191	+88.33
+296.6	+147	+63.89	+377.6	+192	+88.89
+298.4	+148	+64.44	+379.4	+193	+89.44
+300.2	+149	+65.00	+381.2	+194	+90.00
+302.0	+150	+65.56	+383.0	+195	+90.56
+303.8	+151	+66.11	+384.8	+196	+91.11
+305.6	+152	+66.67	+386.6	+197	+91.67
+307.4	+153	+67.22	+388.4	+198	+92.22
+309.2	+154	+67.78	+390.2	+199	+92.78
+311.0	+155	+68.33	+392.0	+200	+93.33
+312.8	+156	+68.89	+393.8	+201	+93.89
+314.6	+157	+69.44	+395.6	+202	+94.44
+316.4	+158	+70.00	+397.4	+203	+95.00
+318.2	+159	+70.56	+399.2	+204	+95.56
+320.0	+160	+71.11	+401.0	+205	+96.11
+321.8	+161	+71.67	+402.8	+206	+96.67

(*Continued*)

TABLE 4.5 Temperature Conversion (*Continued*)

°F.	Reading in °F. or °C. to be converted	°C.
+404.6	+207	+97.22
+406.4	+208	+97.78
+408.2	+209	+98.33
+410.0	+210	+98.89
+411.8	+211	+99.44
+413.6	+212	+100.00
+415.4	+213	+100.56
+417.2	+214	+101.11
+419.0	+215	+101.67
+420.8	+216	+102.22
+422.6	+217	+102.78
+424.4	+218	+103.33
+426.2	+219	+103.89
+428.0	+220	+104.44
+431.6	+222	+105.56
+435.2	+224	+106.67
+438.8	+226	+107.78
+442.4	+228	+108.89
+446.0	+230	+110.00
+449.6	+232	+111.11
+453.2	+234	+112.22
+456.8	+236	+113.33
+460.4	+238	+114.44
+464.0	+240	+115.56
+467.6	+242	+116.67
+471.2	+244	+117.78
+474.8	+246	+118.89
+478.4	+248	+120.00
+482.0	+250	+121.11
+485.6	+252	+122.22
+489.2	+254	+123.33
+492.8	+256	+124.44
+496.4	+258	+125.56
+500.0	+260	+126.67
+503.6	+262	+127.78
+507.2	+264	+128.89
+510.8	+266	+130.00
+514.4	+268	+131.11
+518.0	+270	+132.22
+521.6	+272	+133.33
+525.2	+274	+134.44
+528.8	+276	+135.56
+532.4	+278	+136.67
+536.0	+280	+137.78
+539.6	+282	+138.89

°F.	Reading in °F. or °C. to be converted	°C.
+543.2	+284	+140.00
+546.8	+286	+141.11
+550.4	+288	+142.22
+554.0	+290	+143.33
+557.6	+292	+144.44
+561.2	+294	+145.56
+564.8	+296	+146.67
+568.4	+298	+147.78
+572.0	+300	+148.89
+575.6	+302	+150.00
+579.2	+304	+151.11
+582.8	+306	+152.22
+586.4	+308	+153.33
+590.0	+310	+154.44
+593.6	+312	+155.56
+597.2	+314	+156.67
+600.8	+316	+157.78
+604.4	+318	+158.89
+608.0	+320	+160.00
+611.6	+322	+161.11
+615.2	+324	+162.22
+618.8	+326	+163.33
+622.4	+328	+164.44
+626.0	+330	+165.56
+629.6	+332	+166.67
+633.2	+334	+167.78
+636.8	+336	+168.89
+640.4	+338	+170.00
+644.0	+340	+171.11
+647.6	+342	+172.22
+651.2	+344	+173.33
+654.8	+346	+174.44
+658.4	+348	+175.56
+662.0	+350	+176.67
+665.6	+352	+177.78
+669.2	+354	+178.89
+672.8	+356	+180.00
+676.4	+358	+181.11
+680.0	+360	+182.22
+683.6	+362	+183.33
+687.2	+364	+184.44
+690.8	+366	+185.56
+694.4	+368	+186.67
+698.0	+370	+187.78
+701.6	+372	+188.89

TABLE 4.5 Temperature Conversion (*Continued*)

°F.	Reading in °F. or °C. to be converted	°C.	°F.	Reading in °F. or °C. to be converted	°C.
+705.2	+374	+190.00	+867.2	+464	+240.00
+708.8	+376	+191.11	+870.8	+466	+241.11
+712.4	+378	+192.22	+874.4	+468	+242.22
+716.0	+380	+193.33	+878.0	+470	+243.33
+719.6	+382	+194.44	+881.6	+472	+244.44
+723.2	+384	+195.56	+885.2	+474	+245.56
+726.8	+386	+196.67	+888.8	+476	+246.67
+730.4	+388	+197.78	+892.4	+478	+247.78
+734.0	+390	+198.89	+896.0	+480	+248.89
+737.6	+392	+200.00	+899.6	+482	+250.00
+741.2	+394	+201.11	+903.2	+484	+251.11
+744.8	+396	+202.22	+906.8	+486	+252.22
+748.4	+398	+203.33	+910.4	+488	+253.33
+752.0	+400	+204.44	+914.0	+490	+254.44
+755.6	+402	+205.56	+917.6	+492	+255.56
+759.2	+404	+206.67	+921.2	+494	+256.67
+762.8	+406	+207.78	+924.8	+496	+257.78
+766.4	+408	+208.89	+928.4	+498	+258.89
+770.0	+410	+210.00	+932.0	+500	+260.00
+773.6	+412	+211.11	+935.6	+502	+261.11
+777.2	+414	+212.22	+939.2	+504	+262.22
+780.8	+416	+213.33	+942.8	+506	+263.33
+784.4	+418	+214.44	+946.4	+508	+264.44
+788.0	+420	+215.56	+950.0	+510	+265.56
+791.6	+422	+216.67	+953.6	+512	+266.67
+795.2	+424	+217.78	+957.2	+514	+267.78
+798.8	+426	+218.89	+960.8	+516	+268.89
+802.4	+428	+220.00	+964.4	+518	+270.00
+806.0	+430	+221.11	+968.0	+520	+271.11
+809.6	+432	+222.22	+971.6	+522	+272.22
+813.2	+434	+223.33	+975.2	+524	+273.33
+816.8	+436	+224.44	+978.8	+526	+274.44
+820.4	+438	+225.56	+982.4	+528	+275.56
+824.0	+440	+226.67	+986.0	+530	+276.67
+827.6	+442	+227.78	+989.6	+532	+277.78
+831.2	+444	+228.89	+993.2	+534	+278.89
+834.8	+446	+230.00	+996.8	+536	+280.00
+838.4	+448	+231.11	+1000.4	+538	+281.11
+842.0	+450	+232.22	+1004.0	+540	+282.22
+845.6	+452	+233.33	+1007.6	+542	+283.33
+849.2	+454	+234.44	+1011.2	+544	+284.44
+852.8	+456	+235.56	+1014.8	+546	+285.56
+856.4	+458	+236.67	+1018.4	+548	+286.67
+860.0	+460	+237.78	+1022.0	+550	+287.78
+863.6	+462	+238.89	+1025.6	+552	+288.89

(*Continued*)

TABLE 4.5 Temperature Conversion (*Continued*)

°F.	Reading in °F. or °C. to be converted	°C.	°F.	Reading in °F. or °C. to be converted	°C.
+1029.2	+554	+290.00	+1191.2	+644	+340.00
+1032.8	+556	+291.11	+1194.8	+646	+341.11
+1036.4	+558	+292.22	+1198.4	+648	+342.22
+1040.0	+560	+293.33	+1202.0	+650	+343.33
+1043.6	+562	+294.44	+1205.6	+652	+344.44
+1047.2	+564	+295.56	+1209.2	+654	+345.56
+1050.8	+566	+296.67	+1212.8	+656	+346.67
+1054.4	+568	+297.78	+1216.4	+658	+347.78
+1058.0	+570	+298.89	+1220.0	+660	+348.89
+1061.6	+572	+300.00	+1223.6	+662	+350.00
+1065.2	+574	+301.11	+1227.2	+664	+351.11
+1068.8	+576	+302.22	+1230.8	+666	+352.22
+1072.4	+578	+303.33	+1234.4	+668	+353.33
+1076.0	+580	+304.44	+1238.0	+670	+354.44
+1079.6	+582	+305.56	+1241.6	+672	+355.56
+1083.2	+584	+306.67	+1245.2	+674	+356.67
+1086.8	+586	+307.78	+1248.8	+676	+357.78
+1090.4	+588	+308.89	+1252.4	+678	+358.89
+1094.0	+590	+310.00	+1256.0	+680	+360.00
+1097.6	+592	+311.11	+1259.6	+682	+361.11
+1101.2	+594	+312.22	+1263.2	+684	+362.22
+1104.8	+596	+313.33	+1266.8	+686	+363.33
+1108.4	+598	+314.44	+1270.4	+688	+364.44
+1112.0	+600	+315.56	+1274.0	+690	+365.56
+1115.6	+602	+316.67	+1277.6	+692	+366.67
+1119.2	+604	+317.78	+1281.2	+694	+367.78
+1122.8	+606	+318.89	+1284.8	+696	+368.89
+1126.4	+608	+320.00	+1288.4	+698	+370.00
+1130.0	+610	+321.11	+1292.0	+700	+371.11
+1133.6	+612	+322.22	+1295.6	+702	+372.22
+1137.2	+614	+323.33	+1299.2	+704	+373.33
+1140.8	+616	+324.44	+1302.8	+706	+374.44
+1144.4	+618	+325.56	+1306.4	+708	+375.56
+1148.0	+620	+326.67	+1310.0	+710	+376.67
+1151.6	+622	+327.78	+1313.6	+712	+377.78
+1155.2	+624	+328.89	+1317.2	+714	+378.89
+1158.8	+626	+330.00	+1320.8	+716	+380.00
+1162.4	+628	+331.11	+1324.4	+718	+381.11
+1166.0	+630	+332.22	+1328.0	+720	+382.22
+1169.6	+632	+333.33	+1331.6	+722	+383.33
+1173.2	+634	+334.44	+1335.2	+724	+384.44
+1176.8	+636	+335.56	+1338.8	+726	+385.56
+1180.4	+638	+336.67	+1342.4	+728	+386.67
+1184.0	+640	+337.78	+1346.0	+730	+387.78
+1187.6	+642	+338.89	+1349.6	+732	+388.89

TABLE 4.5 Temperature Conversion (*Continued*)

°F.	Reading in °F. or °C. to be converted	°C.	°F.	Reading in °F. or °C. to be converted	°C.
+1353.2	+734	+390.00	+1515.2	+824	+440.00
+1356.8	+736	+391.11	+1518.8	+826	+441.11
+1360.4	+738	+392.22	+1522.4	+828	+442.22
+1364.0	+740	+393.33	+1526.0	+830	+443.33
+1367.6	+742	+394.44	+1529.6	+832	+444.44
+1371.2	+744	+395.56	+1533.2	+834	+445.56
+1374.8	+746	+396.67	+1536.8	+836	+446.67
+1378.4	+748	+397.78	+1540.4	+838	+447.78
+1382.0	+750	+398.89	+1544.0	+840	+448.89
+1385.6	+752	+400.00	+1547.6	+842	+450.00
+1389.2	+754	+401.11	+1551.2	+844	+451.11
+1392.8	+756	+402.22	+1554.8	+846	+452.22
+1396.4	+758	+403.33	+1558.4	+848	+453.33
+1400.0	+760	+404.44	+1562.0	+850	+454.44
+1403.6	+762	+405.56	+1565.6	+852	+455.56
+1407.2	+764	+406.67	+1569.2	+854	+456.67
+1410.8	+766	+407.78	+1572.8	+856	+457.78
+1414.4	+768	+408.89	+1576.4	+858	+458.89
+1418.0	+770	+410.00	+1580.0	+860	+460.00
+1421.6	+772	+411.11	+1583.6	+862	+461.11
+1425.2	+774	+412.22	+1587.2	+864	+462.22
+1428.8	+776	+413.33	+1590.8	+866	+463.33
+1432.4	+778	+414.44	+1594.4	+868	+464.44
+1436.0	+780	+415.56	+1598.0	+870	+465.56
+1439.6	+782	+416.67	+1601.6	+872	+466.67
+1443.2	+784	+417.78	+1605.2	+874	+467.78
+1446.8	+786	+418.89	+1608.8	+876	+468.89
+1450.4	+788	+420.00	+1612.4	+878	+470.00
+1454.0	+790	+421.11	+1616.0	+880	+471.11
+1457.6	+792	+422.22	+1619.6	+882	+472.22
+1461.2	+794	+423.33	+1623.2	+884	+473.33
+1464.8	+796	+424.44	+1626.8	+886	+474.44
+1468.4	+798	+425.56	+1630.4	+888	+475.56
+1472.0	+800	+426.67	+1634.0	+890	+476.67
+1475.6	+802	+427.78	+1637.6	+892	+477.78
+1479.2	+804	+428.89	+1641.2	+894	+478.89
+1482.8	+806	+430.00	+1644.8	+896	+480.00
+1486.4	+808	+431.11	+1648.4	+898	+481.11
+1490.0	+810	+432.22	+1652.0	+900	+482.22
+1493.6	+812	+433.33	+1655.6	+902	+483.33
+1497.2	+814	+434.44	+1659.2	+904	+484.44
+1500.8	+816	+435.56	+1662.8	+906	+485.56
+1504.4	+818	+436.67	+1666.4	+908	+486.67
+1508.0	+820	+437.78	+1670.0	+910	+487.78
+1511.6	+822	+438.89	+1673.6	+912	+488.89

(*Continued*)

TABLE 4.5 Temperature Conversion (*Continued*)

°F.	Reading in °F. or °C. to be converted	°C.	°F.	Reading in °F. or °C. to be converted	°C.
+1677.2	+914	+490.00	+1868.0	+1020	+548.89
+1680.8	+916	+491.11	+1886.0	+1030	+554.44
+1684.4	+918	+492.22	+1904.0	+1040	+560.00
+1688.0	+920	+493.33	+1922.0	+1050	+565.56
+1691.6	+922	+494.44	+1940.0	+1060	+571.11
+1695.2	+924	+495.56	+1958.0	+1070	+576.67
+1698.8	+926	+496.67	+1976.0	+1080	+582.22
+1702.4	+928	+497.78	+1994.0	+1090	+587.78
+1706.0	+930	+498.89	+2012.0	+1100	+593.33
+1709.6	+932	+500.00	+2030.0	+1110	+598.89
+1713.2	+934	+501.11	+2048.0	+1120	+604.44
+1716.8	+936	+502.22	+2066.0	+1130	+610.00
+1720.4	+938	+503.33	+2084.0	+1140	+615.56
+1724.0	+940	+504.44	+2102.0	+1150	+621.11
+1727.6	+942	+505.56	+2120.0	+1160	+626.67
+1731.2	+944	+506.67	+2138.0	+1170	+632.22
+1734.8	+946	+507.78	+2156.0	+1180	+637.78
+1738.4	+948	+508.89	+2174.0	+1190	+643.33
+1742.0	+950	+510.00	+2192.0	+1200	+648.89
+1745.6	+952	+511.11	+2210.0	+1210	+654.44
+1749.2	+954	+512.22	+2228.0	+1220	+660.00
+1752.8	+956	+513.33	+2246.0	+1230	+665.56
+1756.4	+958	+514.44	+2264.0	+1240	+671.11
+1760.0	+960	+515.56	+2282.0	+1250	+676.67
+1763.6	+962	+516.67	+2300.0	+1260	+682.22
+1767.2	+964	+517.78	+2318.0	+1270	+687.78
+1770.8	+966	+518.89	+2336.0	+1280	+693.33
+1774.4	+968	+520.00	+2354.0	+1290	+698.89
+1778.0	+970	+521.11	+2372.0	+1300	+704.44
+1781.6	+972	+522.22	+2390.0	+1310	+710.00
+1785.2	+974	+523.33	+2408.0	+1320	+715.56
+1788.8	+976	+524.44	+2426.0	+1330	+721.11
+1792.4	+978	+525.56	+2444.0	+1340	+726.67
+1796.0	+980	+526.67	+2462.0	+1350	+732.22
+1799.6	+982	+527.78	+2480.0	+1360	+737.78
+1803.2	+984	+528.89	+2498.0	+1370	+743.33
+1806.8	+986	+530.00	+2516.0	+1380	+748.89
+1810.4	+988	+531.11	+2534.0	+1390	+754.44
+1814.0	+990	+532.22	+2552.0	+1400	+760.00
+1817.6	+992	+533.33	+2570.0	+1410	+765.56
+1821.2	+994	+534.44	+2588.0	+1420	+771.11
+1824.8	+996	+535.56	+2606.0	+1430	+776.67
+1828.4	+998	+536.67	+2624.0	+1440	+782.22
+1832.0	+1000	+537.78	+2642.0	+1450	+787.78
+1850.0	+1010	+543.33	+2660.0	+1460	+793.33

TABLE 4.5 Temperature Conversion (*Continued*)

°F.	Reading in °F. or °C. to be converted	°C.	°F.	Reading in °F. or °C. to be converted	°C.
+2678.0	+1470	+798.89	+3488.0	+1920	+1048.9
+2696.0	+1480	+804.44	+3506.0	+1930	+1054.4
+2714.0	+1490	+810.00	+3524.0	+1940	+1060.0
+2732.0	+1500	+815.56	+3542.0	+1950	+1065.6
+2750.0	+1510	+821.11	+3560.0	+1960	+1071.1
+2768.0	+1520	+826.67	+3578.0	+1970	+1076.7
+2786.0	+1530	+832.22	+3596.0	+1980	+1082.2
+2804.0	+1540	+837.78	+3614.0	+1990	+1087.8
+2822.0	+1550	+843.33	+3632.0	+2000	+1093.3
+2840.0	+1560	+848.89	+3650.0	+2010	+1098.9
+2858.0	+1570	+854.44	+3668.0	+2020	+1104.4
+2876.0	+1580	+860.00	+3686.0	+2030	+1110.0
+2894.0	+1590	+865.56	+3704.0	+2040	+1115.6
+2912.0	+1600	+871.11	+3722.0	+2050	+1121.1
+2930.0	+1610	+876.67	+3740.0	+2060	+1126.7
+2948.0	+1620	+882.22	+3758.0	+2070	+1132.2
+2966.0	+1630	+887.78	+3776.0	+2080	+1137.8
+2984.0	+1640	+893.33	+3794.0	+2090	+1143.3
+3002.0	+1650	+898.89	+3812.0	+2100	+1148.9
+3020.0	+1660	+904.44	+3830.0	+2110	+1154.4
+3038.0	+1670	+910.00	+3848.0	+2120	+1160.0
+3056.0	+1680	+915.56	+3866.0	+2130	+1165.6
+3074.0	+1690	+921.11	+3884.0	+2140	+1171.1
+3092.0	+1700	+926.67	+3902.0	+2150	+1176.7
+3110.0	+1710	+932.22	+3920.0	+2160	+1182.2
+3128.0	+1720	+937.78	+3938.0	+2170	+1187.8
+3146.0	+1730	+943.33	+3956.0	+2180	+1193.3
+3164.0	+1740	+948.89	+3974.0	+2190	+1198.9
+3182.0	+1750	+954.44	+3992.0	+2200	+1204.4
+3200.0	+1760	+960.00	+4010.0	+2210	+1210.0
+3218.0	+1770	+965.56	+4028.0	+2220	+1215.6
+3236.0	+1780	+971.11	+4046.0	+2230	+1221.1
+3254.0	+1790	+976.67	+4064.0	+2240	+1226.7
+3272.0	+1800	+982.22	+4082.0	+2250	+1232.2
+3290.0	+1810	+987.78	+4100.0	+2260	+1237.8
+3308.0	+1820	+993.33	+4118.0	+2270	+1243.3
+3326.0	+1830	+998.89	+4136.0	+2280	+1248.9
+3344.0	+1840	+1004.4	+4154.0	+2290	+1254.4
+3362.0	+1850	+1010.0	+4172.0	+2300	+1260.0
+3380.0	+1860	+1015.6	+4190.0	+2310	+1265.6
+3398.0	+1870	+1021.1	+4208.0	+2320	+1271.1
+3416.0	+1880	+1026.7	+4226.0	+2330	+1276.7
+3434.0	+1890	+1032.2	+4244.0	+2340	+1282.2
+3452.0	+1900	+1037.8	+4262.0	+2350	+1287.8
+3470.0	+1910	+1043.3	+4280.0	+2360	+1293.3

(*Continued*)

TABLE 4.5 Temperature Conversion (*Continued*)

°F.	Reading in °F. or °C. to be converted	°C.
+4298.0	+2370	+1298.9
+4316.0	+2380	+1304.4
+4334.0	+2390	+1310.0
+4352.0	+2400	+1315.6
+4370.0	+2410	+1321.1
+4388.0	+2420	+1326.7
+4406.0	+2430	+1332.2
+4424.0	+2440	+1337.8
+4442.0	+2450	+1343.3
+4460.0	+2460	+1348.9
+4478.0	+2470	+1354.4
+4496.0	+2480	+1360.0
+4514.0	+2490	+1365.6
+4532.0	+2500	+1371.1
+4550.0	+2510	+1376.7
+4568.0	+2520	+1382.2
+4586.0	+2530	+1387.8
+4604.0	+2540	+1393.3
+4622.0	+2550	+1398.9
+4640.0	+2560	+1404.4
+4658.0	+2570	+1410.0
+4676.0	+2580	+1415.6
+4694.0	+2590	+1421.1
+4712.0	+2600	+1426.7
+4730.0	+2610	+1432.2
+4748.0	+2620	+1437.8
+4766.0	+2630	+1443.3
+4784.0	+2640	+1448.9
+4802.0	+2650	+1454.4
+4820.0	+2660	+1460.0
+4838.0	+2670	+1465.6
+4856.0	+2680	+1471.1
+4874.0	+2690	+1476.7
+4892.0	+2700	+1482.2
+4910.0	+2710	+1487.8
+4928.0	+2720	+1493.3
+4946.0	+2730	+1498.9

°F.	Reading in °F. or °C. to be converted	°C.
+4964.0	+2740	+1504.4
+4982.0	+2750	+1510.0
+5000.0	+2760	+1515.6
+5018.0	+2770	+1521.1
+5036.0	+2780	+1526.7
+5054.0	+2790	+1532.2
+5072.0	+2800	+1537.8
+5090.0	+2810	+1543.3
+5108.0	+2820	+1548.9
+5126.0	+2830	+1554.4
+5144.0	+2840	+1560.0
+5162.0	+2850	+1565.6
+5180.0	+2860	+1571.1
+5198.0	+2870	+1576.7
+5216.0	+2880	+1582.2
+5234.0	+2890	+1587.8
+5252.0	+2900	+1593.3
+5270.0	+2910	+1598.9
+5288.0	+2920	+1604.4
+5306.0	+2930	+1610.0
+5324.0	+2940	+1615.6
+5342.0	+2950	+1621.1
+5360.0	+2960	+1626.7
+5378.0	+2970	+1632.2
+5396.0	+2980	+1637.8
+5414.0	+2990	+1643.3
+5432.0	+3000	+1648.9
+5450.0	+3010	+1654.4
+5468.0	+3020	+1660.0
+5486.0	+3030	+1665.6
+5504.0	+3040	+1671.1
+5522.0	+3050	+1676.7
+5540.0	+3060	+1682.2
+5558.0	+3070	+1687.8
+5576.0	+3080	+1693.3
+5594.0	+3090	+1698.9
+5612.0	+3100	+1704.4

4.4 DENSITY AND SPECIFIC GRAVITY

Alcoholometer. This hydrometer is used in determining the density of aqueous ethyl alcohol solutions; the reading in degrees is numerically the same as the percentage of alcohol by volume. The scale known as Tralle gives the percentage by volume. Wine and Must hydrometer relations are given below.

Ammoniameter. This hydrometer, employed in finding the density of aqueous ammonia solutions, has a scale graduated in equal divisions from 0° to 40°. To convert the reading to specific gravity multiply by 3 and subtract the resulting number from 1000.

Balling Hydrometer. See under Saccharometers.

Barkometer or Barktrometer. This hydrometer, which is used in determining the density of tanning liquors, has a scale from 0° to 80° Bk; the number to the right of the decimal point of a specific gravity reading is the corresponding Bk degree; thus, a specific gravity of 1.015 is 15° Bk.

Baumé Hydrometers. For liquids heavier than water: This hydrometer was originally based on the density of a 10% sodium chloride solution, which was given the value of 10°, and the density of pure water, which was given the value of 0°; the interval between these two values was divided into ten equal parts. Other reference points have been taken with the result that so much confusion exists that there are about 36 different scales in use, many of which are incorrect. In general a Baumé hydrometer should have inscribed on it the temperature at which it was calibrated and also the temperature of the water used in relating the density to a specific gravity. The following expression gives the relation between the specific gravity and several of the Baumé scales:

$$\text{Specific gravity} = \frac{m}{m - \text{Baumé}}$$

$m = 145$ at 60°/60°F (15.56°C) for the American Scale

$= 144$ for the old scale used in Holland

$= 146.3$ at 15°C for the Gerlach Scale

$= 144.3$ at 15°C for the Rational Scale generally used in Germany

For liquids lighter than water: Originally the density of a solution of 1 gram of sodium chloride in 9 grams of water at 12.5°C was given a value of 10°Bé. The scale between these points was divided into ten equal parts and these divisions were repeated throughout the scale giving a relation which could be expressed by the formula: Specific gravity = 145.88/(135.88 + Bé), which is approximately equal to 146/(136 + Bé). Other scales have since come into more general use such as that of the Bureau of Standards in which the specific gravity at 60°/60°F = 140/(130 + Bé) and that of the American Petroleum Institute (A.P.I. Scale) in which the specific gravity at 60°/60°F = 141.5/(131.5 + API°).

See also special table for conversion to density and Twaddell scale.

Beck's Hydrometer. This hydrometer is graduated to show a reading of 0° in pure water and a reading of 30° in a solution with a specific gravity of 0.850, with equal scale divisions above and below these two points.

Brix Hydrometer. See under Saccharometers.

Cartier's Hydrometer. This hydrometer shows a reading of 22° when immersed in a solution having a density of 22° Baumé but the scale divisions are smaller than on the Baumé hydrometer in the ratio of 16 Cartier to 15 Baumé.

Fatty Oil Hydrometer. The graduations on this hydrometer are in specific gravity within the range 0.908 to 0.938. The letters on the scale correspond to the specific gravity of the various common oils as follows: *R*, rape; *O*, olive; *A*, almond; *S*, sesame; *HL*, hoof oil; *HP*, hemp; *C*, cotton seed; *L*, linseed. See also Oleometer below.

Lactometers. These hydrometers are used in determining the density of milk. The various scales in common use are the following:

New York Board of Health has a scale graduated into 120 equal parts, 0° being equal to the specific gravity of water and 100° being equal to a specific gravity of 1.029.

Quevenne lactometer is graduated from 15° to 40° corresponding to specific gravities from 1.015 to 1.040.

Soxhlet lactometer has a scale from 25° to 35° corresponding to specific gravities from 1.025 to 1.035 respectively.

Oleometer. A hydrometer for determining the density of vegetable and sperm oils with a scale from 50° to 0° corresponding to specific gravities from 0.870 to 0.970. See also Fatty Oil Hydrometer above.

Saccharometers. These hydrometers are used in determining the density of sugar solutions. Solutions of the same concentration but of different carbohydrates have very nearly the same specific gravity and in general a concentration of 10 grams of carbohydrate per 100 mL of solution shows a specific gravity of 1.0386. Thus, the wt. of sugar in 1000 mL soln. is (a) for conc. <12g/100 mL: (wt. of 1000 mL soln. − 1000) ÷ 0.386; (b) for conc. >12g/100mL: (wt of 1000 mL soln. − 1000) ÷ 0.385.

Brix hydrometer is graduated so that the number of degrees is identical with the percentage by weight of cane sugar and is used at the temperature indicated on the hydrometer.

Balling's saccharometer is used in Europe and is practically identical with the Brix hydrometer.

Bates brewers' saccharometer which is used in determining the density of malt worts is graduated so that the divisions express pounds per barrel (32 gallons). The relation between degrees Bates (= b) and degrees Balling (= B) is shown by the following formula: $B = 260b/(360 + b)$.

See also below under Wine and Must Hydrometer.

Salinometer. This hydrometer, which is used in the pickling and meat packing plants, is graduated to show percentage of saturation of a sodium chloride solution. An aqueous solution is completely saturated when it contains 26.4% pure sodium chloride. The range from 0% to 26.4% is divided into 100 parts, each division therefore representing 1% of saturation. In another type of salinometer, the degrees correspond to percentages of sodium chloride expressed in grams of sodium chloride per 100 mL of water.

Sprayometer (Parrot and Stewart). This hydrometer which is used in determining the density of *lime sulfur* solutions has two scales; one scale is graduated from 0° to 38° Baumé and the other scale is from 1.000 to 1.350 specific gravity.

Tralle Hydrometer. See Alcoholometer above.

Twaddell Hydrometer. This hydrometer, which is used only for liquids heavier than water, has a scale such that when the reading is multiplied by 5 and added to 1000 the resulting number is the specific gravity with reference to water as 1000. To convert specific gravity at 60°/60°F to Twaddell degrees, take the decimal portion of the specific gravity value and multiply it by 200; thus a specific gravity of $1.032 = 0.032 \times 200 = 6.4°$ Tw. See also special table for conversion to density and Baumé scale.

Wine and Must Hydrometer. This instrument has three scales. One scale shows readings of 0° to 15° Brix for sugar (see Brix Hydrometer above); another scale from 0° to 15° Tralle is used for sweet wines to indicate the percentage of alcohol by volume; and a third scale from 0° to 20° Tralle is used for tart wines to indicate the percentage of alcohol by volume.

Conversion of Specific Gravity at 25°/25°C to Density at any Temperature from 0° to 40°C.* Liquids change volume with change in temperature, but the amount of this change, β (coefficient of cubical expansion), varies widely with different liquids, and to some extent for the same liquid at different temperatures.

The table below, which is calculated from the relationship:

$$F_{\beta t} = \frac{\text{density of water at } 25^\circ\text{C } (= 0.99705)}{1 - \beta(25 - t)}$$

may be used to find d^t, the density (weight of 1 mL) of a liquid at any temperature (t) between 0° and 40°C if the specific gravity at 25°/25°C (S) and the coefficient of cubical expansion (β) are known. Substitutions are made in the equations:

$$d^t = SF_{\beta_t} \tag{4.2}$$

$$S = \frac{d^t}{F_{\beta_t}} \tag{4.3}$$

Factors ($F\beta_t$)

Density t°*C = sp. gr.* 25°/25° × F_{β_t}

*β × 10³ \ °C.	0	5	10	15	20	25	30	35	40
1.3	1.0306	1.0237	1.0169	1.0102	1.0036	0.99705	0.99065	0.9843	0.9780
1.2	1.0279	1.0216	1.0154	1.0092	1.0031	0.99705	0.9911	0.9853	0.9794
1.1	1.0253	1.0195	1.0138	1.0082	1.0026	0.99705	0.9916	0.9963	0.9809
1.0	1.0227	1.0174	1.0123	1.0072	1.0021	0.99705	0.9921	0.9872	0.98234
0.9	1.0200	1.0153	1.0107	1.0060	1.0016	0.99705	0.99262	0.9882	0.9838
0.8	1.0174	1.0133	1.0092	1.0051	1.0011	0.99705	0.9931	0.98918	0.9851
0.7	1.0148	1.0113	1.0077	1.0041	1.0006	0.99705	0.9935	0.99015	0.98672
0.6	1.0122	1.0092	1.0061	1.0031	1.0001	0.99705	0.9941	0.9911	0.9882
0.5	1.0097	1.0072	1.0046	1.0021	0.99958	0.99705	0.9944	0.9921	0.9897
0.	1.0071	1.0051	1.0031	1.0011	0.99908	0.99705	0.9951	0.9931	0.9911

*β = coefficient of cubical expansion.

*Cf. Dreisbach, *Ind., Eng. Chem., Anal. Ed.* **12**:160 (1940).

Examples. All examples are based upon an assumed coefficient of cubical expansion, β, of 1.3×10^{-3}.

Example 1. To find the density of a liquid at 20°C, d^{20}, which has a specific gravity (S) of $1.2500\frac{25}{25}$:

From the table above F_{β_t} at 20°C = 1.0036.

$$d^{20} = d^t = SF_{\beta_t} = 1.2500 \times 1.0036 = 1.2545$$

Example 2. To find the density at 20°C (d^{20}) of a liquid which has a specific gravity of $1.2500\frac{17}{4}$:
Since the density of water at 4°C is equal to 1, specific gravity at 17°/4° = d^{17} = 1.2500.
Substitution in Equation 3 with F_{β_t} at 17°C, by interpolation from the table, equal to 1.00756, gives

$$\text{Sp. gr. } 25°/25° = S = 1.2500 \div 1.00756$$

Substitution of this value for S in Equation 2 with F_{β_t} at 20°C, from the table, equal to 1.0036, gives

$$d^{20} = d^t = (1.2500 \div 1.00756) \times 1.0036 = 1.2451$$

Example 3. To find the specific gravity at 20°/4°C of a liquid which has a specific gravity of $1.2500\frac{25}{4}$:

Since the density of water at 4°C is equal to 1, specific gravity 25°/4° = d^{25} = 1.2500; and, specific gravity 20°/4° = d^{20}.

Substitution in Equation 3, with d^t = 1.2500; and, with F_{β_t} at 25°C, from the table, equal to 0.99705, gives

$$\text{Sp. gr. } 25°/25° = S = 1.2500 \div 0.99705$$

Substitution of this value for S in Equation 2, with F_{β_t} at 20°C, from the table, equal to 1.0036, gives

$$\text{Sp. gr. } 20°/4° = d^{20} = (1.2500 \div 0.99705) \times 1.0036 = 1.2582$$

Example 4. To find the density at 25°C of a liquid which has a specific gravity of $1.2500\frac{15}{15}$:
Since the density of water at 15°C = 0.99910,

$$d^{15} = \text{sp. gr. } 15°/15° \times 0.99910 = 1.2500 \times 0.99910$$

Substitution in Equation 3, with F_{β_t} at 15°C, from the table, equal to 1.0102, gives

$$\text{Sp. gr. } 25°/25° = S = (1.2500 \times 0.99910) \div 1.0102$$

Substitution of this value for S in Equation 2, with F_{β_t} at 25°, from the table, equal to 0.99705, gives

$$d^{26} = d^t = (1.2500 \times 0.99910 \div 1.0102) \times 0.99705 = 1.2326$$

TABLE 4.6 Hydrometer Conversion

This table gives the relation between density (c.g.s.) and degrees on the Baumé and Twaddell scales. The Twaddell scale is never used for densities less than unity. See also Sec. 2.1.2.1, Hydrometers.

Density	Degrees Baumé (NIST* scale)	Degrees Baumé (A.P.I. †scale)
0.600	103.33	104.33
0.605	101.40	102.38
0.610	99.51	100.47
0.615	97.64	98.58
0.620	95.81	96.73
0.625	94.00	94.90
0.630	92.22	93.10
0.635	90.47	91.33
0.640	88.75	89.59
0.645	87.05	87.88
0.650	85.38	86.19
0.655	83.74	84.53
0.660	82.12	82.89
0.665	80.52	81.28
0.670	78.95	79.69
0.675	77.41	78.13
0.680	75.88	76.59
0.685	74.38	75.07
0.690	72.90	73.57
0.695	71.43	72.10
0.700	70.00	70.64
0.705	68.57	69.21
0.710	67.18	67.80
0.715	65.80	66.40
0.720	64.44	65.03
0.725	63.10	63.67
0.730	61.78	62.34
0.735	60.48	61.02
0.740	59.19	59.72
0.745	57.92	58.43
0.750	56.67	57.17
0.755	55.43	55.92
0.760	54.21	54.68
0.765	53.01	53.47
0.770	51.82	52.27
0.775	50.65	51.08
0.780	49.49	49.91
0.785	48.34	48.75
0.790	47.22	47.61
0.795	46.10	46.49
0.800	45.00	45.38
0.805	43.91	44.28
0.810	42.84	43.19
0.815	41.78	42.12
0.820	40.73	41.06
0.825	39.70	40.02
0.830	38.68	38.98
0.835	37.66	37.96
0.840	36.67	36.95
0.845	35.68	35.96
0.850	34.71	34.97
0.855	33.74	34.00
0.860	32.79	33.03
0.865	31.85	32.08
0.870	30.92	31.14
0.875	30.00	30.21
0.880	29.09	29.30
0.885	28.19	28.39
0.890	27.30	27.49
0.895	26.42	26.60
0.900	25.56	25.72
0.905	24.70	24.85
0.910	23.85	23.99
0.915	23.01	23.14
0.920	22.17	22.30
0.925	21.35	21.47
0.930	20.54	20.65
0.935	19.73	19.84
0.940	18.94	19.03
0.945	18.15	18.24
0.950	17.37	17.45
0.955	16.60	16.67
0.960	15.83	15.90
0.965	15.08	15.13
0.970	14.33	14.38
0.975	13.59	13.63
0.980	12.86	12.89
0.985	12.13	12.15
0.990	11.41	11.43
0.995	10.70	10.71
1.000	10.00	10.00

DENSITIES GREATER THAN UNITY

Density	Degrees Baumé (NIST* scale)	Degrees Baumé (A.P.I. †scale)
1.00	0.00	0
1.01	1.44	2
1.02	2.84	4

* NIST, National Institute for Science and Technology (formerly the National Bureau of Standards, U.S.).
† A.P.I is the American Petroleum Institute.

TABLE 4.6 Hydrometer Conversion (*Continued*)

Density	Degrees Baumé (NIST* scale)	Degrees Baumé (A.P.I. †scale)	Density	Degrees Baumé (NIST* scale)	Degrees Baumé (A.P.I. †scale)
1.03	4.22	6	1.52	49.60	104
1.04	5.58	8	1.53	50.23	106
1.05	6.91	10	1.54	50.84	108
1.06	8.21	12	1.55	51.45	110
1.07	9.49	14	1.56	52.05	112
1.08	10.78	16	1.57	52.64	114
1.09	11.97	18	1.58	53.23	116
1.10	13.18	20	1.59	53.80	118
1.11	14.37	22	1.60	54.38	120
1.12	15.54	24	1.61	54.94	122
1.13	16.68	26	1.62	55.49	124
1.14	17.81	28	1.63	56.04	126
1.15	18.91	30	1.64	56.58	128
1.16	20.00	32	1.65	57.12	130
1.17	21.07	34	1.66	57.65	132
1.18	22.12	36	1.67	58.17	134
1.19	23.15	38	1.68	58.69	136
1.20	24.17	40	1.69	59.20	138
1.21	25.16	42	1.70	59.71	140
1.22	26.15	44	1.71	60.20	142
1.23	27.11	46	1.72	60.70	144
1.24	28.06	48	1.73	61.18	146
1.25	29.00	50	1.74	61.67	148
1.26	29.92	52	1.75	62.14	150
1.27	30.83	54	1.76	62.61	152
1.28	31.72	56	1.77	63.08	154
1.29	32.60	58	1.78	63.54	156
1.30	33.46	60	1.79	63.99	158
1.31	34.31	62	1.80	64.44	160
1.32	35.15	64	1.81	64.89	162
1.33	35.98	66	1.82	65.31	164
1.34	36.79	68	1.83	65.77	166
1.35	37.59	70	1.84	66.20	168
1.36	38.38	72	1.85	66.62	170
1.37	39.16	74	1.86	67.04	172
1.38	39.93	76	1.87	67.46	174
1.39	40.68	78	1.88	67.87	176
1.40	41.43	80	1.89	68.28	178
1.41	42.16	82	1.90	68.68	180
1.42	42.89	84	1.91	69.08	182
1.43	43.60	86	1.92	69.48	184
1.44	44.31	88	1.93	69.87	186
1.45	45.00	90	1.94	70.26	188
1.46	45.68	92	1.95	70.64	190
1.47	46.36	94	1.96	71.02	192
1.48	47.03	96	1.97	71.40	194
1.49	47.68	98	1.98	71.77	196
1.50	48.33	100	1.99	72.14	198
1.51	48.97	102	2.00	72.50	200

* NIST, National Institute for Science and Technology (formerly the National Bureau of Standards, U.S.).

4.5 BAROMETRY AND BAROMETRIC CORRECTIONS

In principle, the mercurial barometer balances a column of pure mercury against the weight of the atmosphere. The height of the column above the level of the mercury in the reservoir can be measured and serves as a direct index of atmospheric pressure. The space above the mercury in a barometer tube should be a Torricellian vacuum, perfect except for the practically negligible vapor pressure of mercury. The perfection of the vacuum is indicated by the sharpness of the click noted when the barometer tube is inclined. A barometer should be in a vertical position, suspended rather than fastened to a wall, and in a good light but not exposed to direct sunlight or too near a source of heat. The standard conditions for barometric measurements are 0°C and gravity as at 45° latitude and sea level. There are numerous sources of error, but corrections for most of these are readily applied. Some of the corrections are very small, and their application may be questionable in view of the probably larger errors. The degree of consistency to be expected in careful measurements is about 0.13 mm with a 6.4-mm tube, increasing to 0.04 mm with a tube 12.7 mm in diameter.

In reading a barometer of the Fortin type (the usual laboratory instrument for precision measurements), the procedure should be as follows: (1) Observe and record the temperature as indicated by the thermometer attached to the barometer. The temperature correction is very important and may be affected by heat from the observer's body. (2) Set the mercury in the reservoir at zero level, so that the point of the pin above the mercury just touches the surface, making a barely noticeable dimple therein. Tap the tube at the top and verify the zero setting. (3) Bring the vernier down until the view at the light background is cut off at the highest point of the meniscus. Record the reading.

The corrections to be made on the reading are as follows: (1) Temperature, to correct for the difference in thermal expansion of the mercury and the brass (or glass) to which the scale is attached. This correction converts the reading into the value of 0°C. The brass scale table is applicable to the Fortin barometer. See Tables 4.8 (latitude-gravity correction), and Tables 4.9 (altitude-gravity correction), to compensate for differences in gravity, which would affect the height of the mercury column by variation in mass. If local gravity is unknown, an approximate correction may be made from the tables. Local values of gravity are often subject to irregularities which lead to errors even when the corrections here provided are made. It is, therefore, advisable to determine the local value of gravity, from which the correction can be effected in the following manner:

$$Bt = Br + \left(\frac{g_1 - g_0}{g_0} \right) \times Br$$

in which Bt and Br are the true and the observed heights of the barometer, respectively. g_0 is standard gravity (980 665 cm · s^{-2}), and g_1 is the local gravity. It may be noted that for most localities, g_1 is smaller than g_0, which makes the correction negative. These corrections compensate the reading to gravity at 45° latitude and sea level. (3) Correction for capillary depression of the level of the meniscus. This varies with the tube diameter and actual height of the meniscus in a particular case. Some barometers are calibrated to allow for an average value of the latter and approximating the correction. See table. (4) Correction for vapor pressure of mercury. This correction is usually negligible, being only 0.001 mm at 20°C and 0.006 mm at 40°C. This correction is added. See table of vapor pressure of mercury.

The corrections above do not apply to aneroid barometers. These instruments should be calibrated at regular intervals by checking them against a corrected mercurial barometer.

For records on weather maps, meteorologists customarily correct barometer readings to sea level, and some barometers may be calibrated accordingly. Such instruments are not suitable for laboratory use where true pressure under standard conditions is required. Scale corrections should be specified in the maker's instructions with the instrument, and are also indicated by the lack of correspondence between a gauge mark usually placed exactly 76.2 cm from the zero point and the 76.2-cm scale graduation.

TABLE 4.7 Barometer Temperature Correction—Metric Units

The values in the table below are to be subtracted from the observed readings to correct for the difference in the expansion of the mercury and the glass scale at different temperatures.

A. Glass scale

Temp. °C.	Observed barometer height in millimeters						
	700	730	740	750	760	770	800
	mm.	mm.	mm.	mm.	mm.	mm.	mm.
0	0.00	0.00	0.00	0.00	0.00	0.00	0.00
1	0.12	0.13	0.13	0.13	0.13	0.13	0.14
2	0.24	0.25	0.26	0.26	0.26	0.27	0.27
3	0.36	0.38	0.38	0.39	0.40	0.40	0.42
4	0.49	0.51	0.51	0.52	0.53	0.53	0.55
5	0.61	0.63	0.64	0.65	0.66	0.67	0.69
6	0.73	0.76	0.77	0.78	0.79	0.80	0.83
7	0.85	0.89	0.90	0.91	0.92	0.93	0.97
8	0.97	1.01	1.03	1.04	1.05	1.07	1.11
9	1.09	1.14	1.15	1.17	1.18	1.20	1.25
10	1.21	1.26	1.28	1.30	1.32	1.33	1.39
11	1.33	1.39	1.41	1.43	1.45	1.47	1.52
12	1.45	1.52	1.54	1.56	1.58	1.60	1.66
13	1.58	1.64	1.67	1.69	1.71	1.73	1.80
14	1.70	1.77	1.79	1.82	1.84	1.87	1.94
15	1.82	1.90	1.92	1.95	1.97	2.00	2.08
16	1.94	2.02	2.05	2.08	2.10	2.13	2.21
17	2.06	2.15	2.18	2.21	2.23	2.26	2.35
18	2.18	2.27	2.30	2.33	2.37	2.40	2.49
19	2.30	2.40	2.43	2.46	2.50	2.53	2.63
20	2.42	2.52	2.56	2.59	2.63	2.66	2.77
21	2.54	2.65	2.69	2.72	2.76	2.79	2.90
22	2.66	2.78	2.81	2.85	2.89	2.93	3.04
23	2.78	2.90	2.94	2.98	3.02	3.06	3.18
24	2.90	3.03	3.07	3.11	3.15	3.19	3.32
25	3.02	3.15	3.20	3.24	3.28	3.32	3.45
26	3.14	3.28	3.32	3.37	3.41	3.46	3.59
27	3.26	3.40	3.45	3.50	3.54	3.59	3.73
28	3.38	3.53	3.58	3.63	3.67	3.72	3.87
29	3.50	3.65	3.70	3.75	3.80	3.85	4.00
30	3.62	3.78	3.83	3.88	3.93	3.99	4.14
31	3.74	3.90	3.96	4.01	4.06	4.12	4.28
32	3.86	4.03	4.08	4.14	4.20	4.25	4.42
33	3.98	4.15	4.21	4.27	4.33	4.38	4.55
34	4.10	4.28	4.34	4.40	4.46	4.51	4.69
35	4.22	4.40	4.47	4.53	4.59	4.65	4.83

TABLE 4.7 Barometer Temperature Correction—Metric Units (*Continued*)

The values in the table below are to be subtracted from the observed readings to correct for the difference in the expansion of the mercury and the glass scale at different temperatures.

B. Brass scale

	Observed barometer height in millimeters						
	640	650	660	670	680	690	700
Temp. °C.	mm.	mm.	mm.	mm.	mm.	mm.	mm.
0	0.00	0.00	0.00	0.00	0.00	0.00	0.00
1	0.10	0.11	0.11	0.11	0.11	0.11	0.11
2	0.21	0.21	0.22	0.22	0.22	0.23	0.23
3	0.31	0.32	0.32	0.33	0.33	0.34	0.34
4	0.42	0.42	0.43	0.44	0.44	0.45	0.46
5	0.52	0.53	0.54	0.55	0.55	0.56	0.57
6	0.63	0.64	0.65	0.66	0.66	0.67	0.68
7	0.73	0.74	0.75	0.76	0.78	0.79	0.80
8	0.84	0.85	0.86	0.87	0.89	0.90	0.91
9	0.94	0.95	0.97	0.98	1.00	1.01	1.03
10	1.04	1.06	1.07	1.09	1.11	1.12	1.14
11	1.15	1.16	1.18	1.20	1.22	1.24	1.25
12	1.25	1.27	1.29	1.31	1.33	1.35	1.37
13	1.35	1.38	1.40	1.42	1.44	1.46	1.48
14	1.46	1.48	1.50	1.53	1.55	1.57	1.59
15	1.56	1.59	1.61	1.64	1.66	1.68	1.71
16	1.67	1.69	1.72	1.74	1.77	1.80	1.82
17	1.77	1.80	1.82	1.85	1.88	1.91	1.94
18	1.87	1.90	1.93	1.96	1.99	2.02	2.05
19	1.98	2.01	2.04	2.07	2.10	2.13	2.16
20	2.08	2.11	2.15	2.18	2.21	2.24	2.28
21	2.18	2.22	2.25	2.29	2.32	2.35	2.39
22	2.29	2.32	2.36	2.40	2.43	2.47	2.50
23	2.39	2.43	2.47	2.50	2.54	2.58	2.62
24	2.49	2.53	2.57	2.61	2.65	2.69	2.73
25	2.60	2.64	2.68	2.72	2.76	2.80	2.84
26	2.70	2.74	2.79	2.83	2.87	2.91	2.96
27	2.81	2.85	2.89	2.94	2.98	3.02	3.07
28	2.91	2.95	3.00	3.05	3.09	3.14	3.18
29	3.01	3.06	3.11	3.15	3.20	3.25	3.29
30	3.12	3.16	3.21	3.26	3.31	3.36	3.41
31	3.22	3.27	3.32	3.37	3.42	3.47	3.52
32	3.32	3.37	3.43	3.48	3.53	3.58	3.63
33	3.42	3.48	3.53	3.59	3.64	3.69	3.75
34	3.53	3.58	3.64	3.69	3.75	3.80	3.86
35	3.63	3.69	3.74	3.80	3.86	3.91	3.97

TABLE 4.7 Barometer Temperature Correction—Metric Units (*Continued*)

B. Brass scale (*continued*)

Observed barometer height in millimeters 710	720	730	740	750	760	770	780	Temp.
mm.	mm.	mm.	mm.	mm.	mm.	mm.	mm.	°C.
0.00	0.00	0.00	0.00	0.00	0.00	0.00	0.00	0
0.12	0.12	0.12	0.12	0.12	0.12	0.13	0.13	1
0.23	0.23	0.24	0.24	0.24	0.25	0.25	0.25	2
0.35	0.35	0.36	0.36	0.37	0.37	0.38	0.38	3
0.46	0.47	0.48	0.48	0.49	0.50	0.50	0.51	4
0.58	0.59	0.59	0.60	0.61	0.62	0.63	0.64	5
0.69	0.70	0.71	0.72	0.73	0.74	0.75	0.76	6
0.81	0.82	0.83	0.84	0.86	0.87	0.88	0.89	7
0.93	0.94	0.95	0.96	0.98	0.99	1.00	1.02	8
1.04	1.06	1.07	1.08	1.10	1.11	1.13	1.14	9
1.16	1.17	1.19	1.21	1.22	1.24	1.25	1.27	10
1.27	1.29	1.31	1.33	1.34	1.36	1.38	1.40	11
1.39	1.41	1.43	1.45	1.47	1.48	1.50	1.52	12
1.50	1.52	1.54	1.57	1.59	1.61	1.63	1.65	13
1.62	1.64	1.66	1.69	1.71	1.73	1.75	1.78	14
1.73	1.76	1.78	1.81	1.83	1.85	1.88	1.90	15
1.85	1.87	1.90	1.93	1.95	1.98	2.00	2.03	16
1.96	1.99	2.02	2.05	2.07	2.10	2.13	2.16	17
2.08	2.11	2.14	2.17	2.20	2.22	2.25	2.28	18
2.19	2.22	2.25	2.29	2.32	2.35	2.38	2.41	19
2.31	2.34	2.37	2.41	2.44	2.47	2.50	2.54	20
2.42	2.46	2.49	2.53	2.56	2.59	2.63	2.66	21
2.54	2.57	2.61	2.65	2.68	2.72	2.75	2.79	22
2.65	2.69	2.73	2.77	2.80	2.84	2.88	2.91	23
2.77	2.81	2.85	2.88	2.92	2.96	3.00	3.04	24
2.88	2.92	2.96	3.00	3.05	3.09	3.13	3.17	25
3.00	3.04	3.08	3.12	3.17	3.21	3.25	3.29	26
3.11	3.16	3.20	3.24	3.29	3.33	3.38	3.42	27
3.23	3.27	3.32	3.36	3.41	3.45	3.50	3.54	28
3.34	3.39	3.44	3.48	3.53	3.58	3.62	3.67	29
3.46	3.50	3.55	3.60	3.65	3.70	3.75	3.80	30
3.57	3.62	3.67	3.72	3.77	3.82	3.87	3.92	31
3.68	3.74	3.79	3.84	3.89	3.94	4.00	4.05	32
3.80	3.85	3.91	3.96	4.01	4.07	4.12	4.17	33
3.91	3.97	4.02	4.08	4.13	4.19	4.24	4.30	34
4.03	4.09	4.14	4.20	4.26	4.31	4.37	4.43	35

TABLE 4.7 Barometer Temperature Correction—Metric Units (*Continued*)

C. Correction of a barometer for capillarity (*Smithsonian Tables*)

Diameter of tube, millimeters	Height of meniscus in millimeters: 0.4	0.6	0.8	1.0	1.2	1.4	1.6	1.8
	Correction to be added in millimeters							
4	0.83	1.22	1.54	1.98	2.37	....	....	....
5	0.47	0.65	0.86	1.19	1.45	1.80	....	....
6	0.27	0.41	0.56	0.78	0.98	1.21	1.43	....
7	0.18	0.28	0.40	0.53	0.67	0.82	0.97	....
8	....	0.20	0.29	0.38	0.46	0.56	0.65	1.13
9	....	0.15	0.21	0.28	0.33	0.40	0.46	0.77
10	....	....	0.15	0.20	0.25	0.29	0.33	0.52
11	....	....	0.10	0.14	0.18	0.21	0.24	0.37
12	....	....	0.07	0.10	0.13	0.15	0.18	0.27
13	....	....	0.04	0.07	0.10	0.12	0.13	0.19
								0.14

TABLE 4.8 Barometric Latitude-Gravity—Metric Units

The values in the table below are to be subtracted from the barometric reading for latitudes from 0 to 45° inclusive, and are to be added from 46 to 90°.

Deg. Lat.	Barometer readings, millimeters: 680	700	720	740	760	780
	mm.	mm.	mm.	mm.	mm.	mm.
0	1.82	1.87	1.93	1.98	2.04	2.09
5	1.79	1.85	1.90	1.95	2.00	2.06
10	1.71	1.76	1.81	1.86	1.92	1.97
15	1.58	1.63	1.67	1.72	1.77	1.81
20	1.40	1.44	1.49	1.53	1.57	1.61
21	1.36	1.40	1.44	1.48	1.52	1.56
22	1.32	1.36	1.40	1.44	1.48	1.51
23	1.28	1.31	1.35	1.39	1.43	1.46
24	1.23	1.27	1.30	1.34	1.37	1.41
25	1.18	1.22	1.25	1.29	1.32	1.36
26	1.13	1.17	1.20	1.23	1.27	1.30
27	1.08	1.12	1.15	1.18	1.21	1.24
28	1.03	1.06	1.09	1.12	1.15	1.18
29	0.98	1.01	1.04	1.07	1.10	1.12
30	0.93	0.95	0.98	1.01	1.04	1.06
31	0.87	0.90	0.92	0.95	0.98	1.00
32	0.82	0.84	0.86	0.89	0.91	0.94
33	0.76	0.78	0.80	0.83	0.85	0.87
34	0.70	0.72	0.74	0.76	0.79	0.81
35	0.64	0.66	0.68	0.70	0.72	0.74
36	0.58	0.60	0.62	0.64	0.65	0.67
37	0.52	0.54	0.56	0.57	0.59	0.60
38	0.46	0.48	0.49	0.51	0.52	0.53

(*Continued*)

TABLE 4.8 Barometric Latitude-Graviy—Metric Units (*Continued*)

	Barometer readings, millimeters					
	680	700	720	740	760	780
Deg. Lat.	mm.	mm.	mm.	mm.	mm.	mm.
39	0.40	0.42	0.43	0.44	0.45	0.46
40	0.34	0.35	0.36	0.37	0.38	0.39
41	0.28	0.29	0.30	0.30	0.31	0.32
42	0.22	0.22	0.23	0.24	0.24	0.25
43	0.16	0.16	0.16	0.17	0.17	0.18
44	0.09	0.10	0.10	0.10	0.10	0.11
45	0.03	0.03	0.03	0.03	0.03	0.04
46	0.03	0.03	0.03	0.03	0.04	0.04
47	0.09	0.10	0.10	0.10	0.10	0.11
48	0.16	0.16	0.17	0.17	0.18	0.18
49	0.22	0.23	0.23	0.24	0.25	0.25
50	0.28	0.29	0.30	0.31	0.31	0.32
51	0.34	0.35	0.36	0.37	0.38	0.39
52	0.40	0.42	0.43	0.44	0.45	0.46
53	0.46	0.48	0.49	0.51	0.52	0.53
54	0.52	0.54	0.56	0.57	0.59	0.60
55	0.58	0.60	0.62	0.64	0.65	0.67
56	0.64	0.66	0.68	0.70	0.72	0.74
57	0.70	0.72	0.74	0.76	0.78	0.80
58	0.76	0.78	0.80	0.82	0.85	0.87
59	0.81	0.84	0.86	0.89	0.91	0.93
60	0.87	0.89	0.92	0.94	0.97	1.00
61	0.92	0.95	0.98	1.00	1.03	1.06
62	0.97	1.00	1.02	1.05	1.08	1.11
63	1.03	1.06	1.09	1.12	1.15	1.18
64	1.08	1.11	1.14	1.17	1.20	1.23
65	1.13	1.16	1.19	1.22	1.26	1.29
66	1.17	1.21	1.24	1.28	1.31	1.35
67	1.22	1.25	1.29	1.33	1.36	1.40
68	1.26	1.30	1.34	1.37	1.41	1.45
69	1.31	1.34	1.38	1.42	1.46	1.50
70	1.35	1.39	1.43	1.47	1.51	1.55
72	1.42	1.47	1.51	1.55	1.59	1.63
75	1.53	1.57	1.62	1.66	1.71	1.75
80	1.66	1.71	1.76	1.81	1.86	1.90
85	1.74	1.79	1.84	1.90	1.95	2.00
90	1.77	1.82	1.87	1.93	1.98	2.03

TABLE 4.9 Barometric Correction for Gravity—Metric Units

The values in Table 4.9 are to be subtracted from the readings taken on a mercurial barometer to correct for the decrease in gravity with increase in altitude.

Height above sealevel meters	Observed barometer height in millimeters								
	400	450	500	550	600	650	700	750	800
	mm.	mm.	mm.	mm.	mm.	mm.	mm.	mm.	mm.
100	....	....	....	....	....	....	0.02	0.02	0.02
200	....	....	....	....	....	....	0.04	0.05	0.05
300	....	....	....	....	....	....	0.07	0.07	0.07
400	....	....	....	....	....	....	0.09	0.10	0.10
500	....	....	....	....	....	....	0.11	0.12	0.13
600	....	....	....	....	....	0.12	0.13	0.14	....
700	....	....	....	....	....	0.14	0.15	0.16	....
800	....	....	....	....	....	0.16	0.18	0.19	....
900	....	....	....	....	....	0.18	0.20	0.22	....
1000	....	....	....	0.18	0.19	0.20	0.22	0.24	....
1100	....	....	....	0.19	0.21	0.22	0.24	....	....
1200	....	....	....	0.21	0.23	0.24	0.26	....	....
1300	....	....	....	0.22	0.24	0.26	0.29	....	....
1400	....	....	....	0.24	0.26	0.28	0.31	....	....
1500	....	....	0.24	0.26	0.28	0.30	0.33	....	....
1600	....	....	0.25	0.28	0.30	0.32	....	....	....
1700	....	....	0.27	0.30	0.32	0.34	....	....	....
1800	....	....	0.28	0.31	0.34	0.36	....	....	....
1900	....	....	0.30	0.33	0.36	0.39	....	....	....
2000	....	0.28	0.31	0.34	0.38	0.41	....	....	....
2100	....	0.30	0.33	0.36	0.40	....	....	....	....
2200	....	0.31	0.35	0.38	0.41	....	....	....	....
2300	....	0.32	0.36	0.40	0.43	....	....	....	....
2400	....	0.34	0.38	0.42	0.45	....	....	....	....
2500	0.31	0.35	0.39	0.43	0.47	....	....	....	....
2600	0.33	0.37	0.41	....	....	....	....	....	....
2800	0.35	0.40	0.44	....	....	....	....	....	....
3000	0.38	0.42	0.47	....	....	....	....	....	....
3200	0.40	0.46	....	....	....	....	....	....	....
3400	0.43	0.48	....	....	....	....	....	....	....

TABLE 4.10 Reduction of the Barometer to Sea Level-Metric Units

A barometer located at an elevation above sea level will show a reading lower than a barometer at sea level by an amount approximately 2.5 mm (0.1 in) for each 30.5 m (100 ft) of elevation. A closer approximation can be made by reference to the following tables, which take into account (1) the effect of altitude of the station at which the barometer is read, (2) the mean temperature of the air column extending from the station down to sea level, (3) the latitude of the station at which the barometer is read, and (4) the reading of the barometer corrected for its temperature, a correction which is applied only to mercurial barometers since the aneroid barometers are compensated for temperature effects.

Example. A barometer which has been corrected for its temperature reads 650 mm at a station whose altitude is 1350 m above sea level and at a latitude of 30°. The mean temperature (outdoor temperature) at the station is 20°C.

Table A (metric units) gives for these conditions a temperature-altitude factor of	135.2
The Latitude Factor Table gives for 135.2 at 30° lat. a correction of	+0.17
Therefore, the corrected value of the temperature-altitude factor is	135.37

Entering Table B (metric units), with a temperature-altitude factor of 135.37 and a barometric reading of 650 mm (corrected for temperature), the correction is found to be 109.6

Accordingly the barometric reading reduced to sea level is 650 + 109.6 = 759.6 mm.

Latitude Factor–English or Metric Units. For latitudes 0°–45° add the latitude factor, for 45°–90° subtract the latitude factor, from the values obtained in Table A.

Temp.—Alt. Factor From Table A	Latitude				
	0°	10°	20°	30°	45°
50	0.1	0.1	0.1	0.1	0.0
100	0.3	0.3	0.2	0.1	0.0
150	0.4	0.4	0.3	0.2	0.0
200	0.5	0.5	0.4	0.3	0.0
250	0.7	0.6	0.5	0.3	0.0
300	0.8	0.8	0.6	0.4	0.0
350	0.9	0.9	0.7	0.5	0.0
	90°	80°	70°	60°	45°

A. Values of the temperature-altitude factor for use in Table B.*

Altitude in Meters	Mean Temperature of Air Column in Centigrade Degrees										
	−16°	−8°	−4°	0°	6°	10°	14°	18°	20°	22°	26°
10	1.2	1.1	1.1	1.1	1.1	1.0	1.0	1.0	1.0	1.0	1.0
50	5.8	5.6	5.5	5.4	5.3	5.2	5.1	5.0	5.0	5.0	4.9
100	11.5	11.2	11.0	10.8	10.6	10.4	10.3	10.1	10.0	9.9	9.8
150	17.3	16.7	16.5	16.2	15.9	15.6	15.4	15.1	15.0	14.9	14.7
200	23.0	22.3	22.0	21.6	21.1	20.8	20.5	20.2	20.0	19.9	19.6
250	28.8	27.9	27.5	27.0	26.4	26.0	25.6	25.2	25.0	24.9	24.5
300	34.5	33.5	33.0	32.5	31.7	31.2	30.7	30.3	30.1	29.8	29.4
350	40.3	39.0	38.5	37.9	37.0	36.4	35.9	35.3	35.1	34.8	34.3
400	46.0	44.6	43.9	43.3	42.3	41.6	41.0	40.4	40.1	39.8	39.2
450	51.8	51.3	49.4	48.7	47.6	46.8	46.1	45.4	45.1	44.8	44.1
500	57.5	55.8	54.9	54.1	52.9	52.0	51.2	50.5	50.1	49.7	49.0
550	63.3	61.4	60.4	59.5	58.1	57.2	56.4	55.5	55.1	54.7	53.9
600	69.0	66.9	65.9	64.9	63.4	62.4	61.5	60.6	60.1	59.7	58.8
650	74.8	72.5	71.4	70.3	68.7	67.6	66.6	65.6	65.1	64.6	63.7

TABLE 4.10 Reduction of the Barometer to Sea Level—Metric Units (*Continued*)

Altitude in Meters	Mean Temperature of Air Column in Centigrade Degrees										
	− 16°	− 8°	− 4°	0°	6°	10°	14°	18°	20°	22°	26°
700	80.6	78.1	76.9	75.7	74.0	72.9	71.7	70.7	70.1	69.6	68.6
750	86.3	83.7	82.4	81.1	79.3	78.1	76.9	75.7	75.1	74.6	73.5
800	92.1	89.2	87.9	86.5	84.6	83.3	82.0	80.8	80.1	79.6	78.4
850	97.8	94.8	93.4	92.0	89.8	88.5	87.1	85.8	85.2	84.5	83.3
900	103.6	100.4	98.9	97.4	95.1	93.7	92.2	90.8	90.2	89.5	88.2
950	109.3	106.0	104.4	102.8	100.4	98.9	97.4	95.9	95.2	94.5	93.1
1000	115.1	111.5	109.8	108.2	105.7	104.1	102.5	100.9	100.2	99.4	98.0
1050	120.8	117.1	115.3	113.6	111.0	109.3	107.6	106.0	105.2	104.4	102.9
1100	126.6	122.7	120.8	119.0	116.3	114.5	112.7	111.0	110.2	109.4	107.8
1150	132.3	128.3	126.3	124.4	121.6	119.7	117.9	116.1	115.2	114.4	112.7
1200	138.1	133.8	131.8	129.8	126.8	124.9	123.0	121.1	120.2	119.3	117.6
1250	143.8	139.4	137.3	135.2	132.1	130.1	128.1	126.2	125.2	124.3	122.5
1300	149.6	145.0	142.8	140.6	137.4	135.3	133.2	131.2	130.2	129.3	127.4
1350	155.3	150.6	148.3	146.0	142.7	140.5	138.4	136.3	135.2	134.2	132.3
1400	161.1	156.2	153.8	151.4	148.0	145.7	143.5	141.3	140.2	139.2	137.2
1450	166.8	161.7	159.3	156.8	153.3	150.9	148.6	146.4	145.3	144.2	142.1
1500	172.6	167.3	164.8	162.3	158.5	156.1	153.7	151.4	150.3	149.1	147.0
1550	178.3	172.9	170.2	167.7	163.8	161.3	158.8	156.4	155.3	154.1	151.8
1600	184.1	178.5	175.7	173.1	169.1	166.5	164.0	161.5	160.3	159.1	156.7
1650	189.8	184.0	181.2	178.5	174.4	171.7	169.1	166.5	165.3	164.1	161.6
1700	195.6	189.6	186.7	183.9	179.7	176.9	174.2	171.6	170.3	169.0	166.5
1750	201.4	195.2	192.2	189.3	185.0	182.1	179.3	176.6	175.3	174.0	171.4
1800	207.1	200.8	197.7	194.7	190.2	187.3	184.5	181.7	180.3	179.0	176.3
1850	212.9	206.3	203.2	200.1	195.5	192.5	189.6	186.7	185.3	183.9	181.2
1900	218.6	211.9	208.7	205.5	200.8	197.7	194.7	191.8	190.3	188.9	186.1
1950	224.4	217.5	214.2	210.9	206.1	202.9	199.8	196.8	195.3	193.9	191.0
2000	230.1	223.0	219.7	216.3	211.4	208.1	204.9	201.9	200.3	198.8	195.0
2050	235.9	228.6	225.1	221.7	216.7	213.3	210.1	206.9	205.3	203.8	200.8
2100	241.6	234.2	230.6	227.1	221.9	218.5	215.2	211.9	210.4	208.8	205.7
2150	247.4	239.8	236.1	232.5	227.2	223.7	220.3	217.0	215.4	213.8	210.6
2200	253.1	245.4	241.6	237.9	232.5	228.9	225.4	222.0	220.4	218.7	215.5
2250	258.9	250.9	247.1	243.4	237.8	234.1	230.6	227.1	225.4	223.7	220.4
2300	264.6	256.5	252.6	248.8	243.1	239.3	235.7	232.1	230.4	228.7	225.3
2350	270.4	262.1	258.1	254.2	248.3	244.5	240.8	237.2	235.4	233.6	230.2
2400	276.1	267.7	263.6	259.6	253.6	249.7	245.9	242.2	240.4	238.6	235.1
2450	281.9	273.2	269.1	265.0	258.9	254.9	251.0	247.3	245.4	243.6	240.0
2500	287.6	278.8	274.5	270.4	264.2	260.1	256.2	252.3	250.4	248.5	244.9
2550	293.4	284.4	280.0	275.8	269.5	265.3	261.3	257.3	255.4	253.5	249.8
2600	299.1	290.0	285.5	281.2	274.8	270.5	266.4	262.4	260.4	258.5	254.7
2650	304.9	295.5	291.0	286.6	280.0	275.7	271.5	267.4	265.4	263.4	259.6
2700	310.6	301.1	296.5	292.0	285.3	280.9	276.6	272.5	270.4	268.4	264.5
2750	316.4	306.7	302.0	297.4	290.6	286.1	281.8	277.5	275.4	273.4	269.4
2800	322.1	312.3	307.5	302.8	295.9	291.3	286.9	282.6	280.4	278.3	274.3
2850	327.9	317.8	313.0	308.2	301.2	296.5	292.0	287.6	285.4	283.3	279.2
2900	333.6	323.4	318.4	313.6	306.4	301.7	297.1	292.6	290.4	288.3	284.1
2950	339.4	329.0	323.9	319.0	311.7	306.9	302.2	297.7	295.5	293.3	289.0
3000	345.1	334.5	329.4	324.4	317.0	312.1	307.4	302.7	300.5	298.2	293.8

* From *Smithsonian Meteorological Tables*, 3d ed., 1907.

TABLE 4.10 Reduction of the Barometer to Sea Level—Metric Units (*Continued*)

*B. Values in millimeters to be added.**

Temp. —Alt. Factor	Barometer Reading in Millimeters						
	790	770	750	730	710	690	670
1	0.9	0.9	0.9	0.8	0.8	0.8	
5	4.6	4.4	4.3	4.2	4.1	4.0	
10	9.1	8.9	8.7	8.5	8.2	8.0	
15	13.8	13.4	13.1	12.7	12.4	12.0	
20	18.4	17.9	17.5	17.0	16.5	16.1	
25		22.5	21.9	21.3	20.7	20.1	
30		27.1	26.4	25.7	25.0	24.2	
35		31.7	30.8	30.0	29.2	28.4	
40		36.3	35.3	34.4	33.5	32.5	31.6
45			39.9	38.8	37.8	36.7	35.6
	750	730	710	690	670	650	630
50	44.4	43.3	42.1	40.9	39.7		
55	49.0	47.7	46.4	45.1	43.8		
60	53.6	52.2	50.8	49.3	47.9		
65	58.3	56.7	55.2	53.6	52.1		
70		61.3	59.6	57.9	56.2		
75		65.8	64.0	62.2	60.4		
80		70.4	68.5	66.6	64.6	62.7	60.8
85		75.0	73.0	70.9	68.9	66.8	64.8
90			77.5	75.3	73.1	71.0	68.8
95			82.1	79.7	77.4	75.1	72.8
	710	690	670	650	630	610	
100	86.6	84.2	81.8	79.3	76.9		
105	91.2	88.7	86.1	83.5	81.0		
110	95.9	93.2	90.5	87.8	85.1		
115	100.5	97.7	94.8	92.0	89.2		
120		102.2	99.3	96.3	93.3		
125		106.8	103.7	100.6	97.5	94.4	
130		111.4	108.2	104.9	101.7	98.5	
135		116.0	112.7	109.3	105.9	102.6	
140		120.7	117.2	113.7	110.2	106.7	
145			121.7	118.1	114.5	110.8	
	670	650	630	610	590	570	
150	126.3	122.5	118.8	115.0			
155	130.9	127.0	123.1	119.2			
160	135.5	131.5	127.4	123.4			
165	140.2	136.0	131.8	127.6			
170		140.5	136.2	131.9	127.5	123.2	
175		145.1	140.6	136.2	131.7	127.2	
180		149.7	145.1	140.5	135.9	131.3	
185		154.3	149.5	144.8	140.0	135.3	
190		158.9	154.0	149.2	144.3	139.4	
195			158.6	153.5	148.5	143.5	

*From *Smithsonian Meteorological Tables*, 3d ed., 1907.

TABLE 4.10 Reduction of the Barometer to Sea Level—Metric Units (*Continued*)

*B. Values in millimeters to be added.**

Temp. —Alt. Factor	Barometer Reading in Millimeters					
	630	610	590	570	550	530
200	163.1	157.9	152.8	147.6		
205	167.7	162.4	157.1	151.7		
210	172.3	166.8	161.4	155.9		
215	176.9	171.3	165.7	160.1	154.5	148.9
220		175.8	170.1	164.3	158.5	152.8
225		180.4	174.5	168.5	162.6	156.7
230		184.9	178.9	172.8	166.7	160.7
235		189.5	183.3	177.1	170.9	164.7
240		194.1	187.8	181.4	175.0	168.7
245		198.8	192.3	185.7	179.2	172.7
	590	570	550	530	510	
250	196.8	190.1	183.4	176.8		
255	201.3	194.5	187.7	180.8		
260	205.9	198.9	191.9	185.0	178.0	
265	210.5	203.3	196.2	189.1	181.9	
270	215.1	207.8	200.5	193.2	185.9	
275	219.8	212.3	204.9	197.4	190.0	
280		216.8	209.2	201.6	194.0	
285		221.4	213.6	205.8	198.1	
290		225.9	218.0	210.1	202.1	
295		230.5	222.4	214.3	206.3	
	570	550	530	510	490	
300	235.1	226.9	218.6	210.4		
305	239.8	231.4	223.0	214.6	206.1	
310		235.9	227.3	218.7	210.1	
315		240.4	231.7	222.9	214.2	
320		245.0	236.1	227.2	218.3	
325		249.6	240.5	231.4	222.4	
330		254.2	244.9	235.7	226.5	
335		258.8	249.4	240.0	230.6	
340		263.5	253.9	244.4	234.8	
345			258.4	248.7	238.9	

*From *Smithsonian Meteorological Tables,* 3d ed., 1907.

TABLE 4.11 Pressure Conversion

psi	Inches H_2O at 4°C	Inches Hg at 0°C	mmH_2O at 4°C	mmHg at 0°C	atm	Pascals ($N \cdot m^{-2}$)
0.01	0.2768	0.0204	7.031	0.517	0.0007	68.95
0.02	0.5536	0.0407	14.06	1.034	0.0014	137.90
0.03	0.8304	0.0611	21.09	1.551	0.0020	206.8
0.04	1.107	0.0814	28.12	2.068	0.0027	275.8
0.05	1.384	0.1018	35.15	2.586	0.0034	344.7
0.06	1.661	0.1222	42.18	3.103	0.0041	413.7
0.07	1.938	0.1425	49.22	3.620	0.0048	482.6
0.08	2.214	0.1629	56.25	4.137	0.0054	551.6
0.09	2.491	0.1832	63.28	4.654	0.0061	620.5
0.10	2.768	0.2036	70.31	5.171	0.0068	689.5
0.20	5.536	0.4072	140.6	10.34	0.0136	1 379.9
0.30	8.304	0.6108	210.9	15.51	0.0204	2 068.5
0.40	11.07	0.8144	281.2	20.68	0.0272	2 758
0.50	13.84	1.018	351.5	25.86	0.0340	3 447
0.60	16.61	1.222	421.8	31.03	0.0408	4 137
0.70	19.38	1.425	492.2	36.20	0.0476	4 826
0.80	22.14	1.629	562.5	41.37	0.0544	5 516
0.90	24.91	1.832	632.8	46.54	0.0612	6 205
1.00	27.68	2.036	703.1	51.71	0.0689	6 895
2.00	55.36	4.072	1 072	103.4	0.1361	13 790
3.00	83.04	6.108	2 109	155.1	0.2041	20 684
4.00	110.7	8.144	2 812	206.8	0.2722	27 579
5.00	138.4	10.18	3 515	258.6	0.3402	34 474
6.00	166.1	12.22	4 218	310.3	0.4083	41 369
7.00	193.8	14.25	4 922	362.0	0.4763	48 263
8.00	221.4	16.29	5 625	413.7	0.5444	55 158
9.00	249.1	18.32	6 328	465.4	0.6124	62 053
10.0	276.8	20.36	7 031	517.1	0.6805	68 948
14.7	406.9	29.93	10 332	760.0	1.000	101 325
15.0	415.2	30.54	10 550	775.7	1.021	103 421
20.0	553.6	40.72	14 060	1 034	1.361	137 895
25.0	692.0	50.90	17 580	1 293	1.701	172 369
30.0	830.4	61.08	21 090	1 551	2.041	206 843
40.0	1 107	81.44	28 120	2 068	2.722	275 790
50.0	1 384	101.8	35 150	2 586	3.402	344 738
60.0	1 661	122.2	42 180	3 103	4.083	413 685
70.0	1 938	142.5	49 220	3 620	4.763	482 633
80.0	2 214	162.9	56 250	4 137	5.444	551 581
90.0	2 491	183.2	63 280	4 654	6.124	620 528
100.0	2 768	203.6	70 307	5 171	6.805	689 476
150.0	4 152	305.4		7 757	10.21	1 034 214
200.0	5 536	407.2		10 343	13.61	1 378 951
250.0	6 920	509.0			17.01	1 723 689
300.0	8 304	610.8			20.41	2 068 427
400.0					27.22	2 757 903
500.0					34.02	3 447 379

1 bar = 10^5 pascal.

TABLE 4.12 Conversion of Weighings in Air to Weighings in Vacuo

If the mass of a substance in air is m_f, its density ρ_m, the density of weights used in making the weighing ρ_w, and the density of air ρ_a, the true mass of the substance in vacuo, m_{vac}, is

$$m_{vac} = m_f + \rho_a m_f\left(\frac{1}{\rho_m} - \frac{1}{\rho_w}\right)$$

For most purposes it is sufficient to assume a density of 8.4 for brass weights, and a density of 0.0012 for air under ordinary conditions. The equation then becomes

$$m_{vac} = m_f + 0.0012 m_f\left(\frac{1}{\rho_m} - \frac{1}{8.4}\right)$$

The table which follows gives the values of k (buoyancy reduction factor), which is the correction necessary because of the buoyant effect of the air upon the object weighed; the table is computed for air with the density of 0.0012; m is the weight in grams of the object when weighted in air; weight of object reduced to "in vacuo" $= m + km/1000$.

	Buoyancy reduction factor, k			
Density of object weighed	Brass weights, density = 8.4	Pt or Pt-Ir weights, density = 21.5	Al or quartz weights, density = 2.7	Gold weights, density = 17
0.2	5.89	5.98	5.58	5.97
0.3	3.87	3.96	3.56	3.95
0.4	2.87	2.95	2.55	2.94
0.5	2.26	2.35	1.95	2.34
0.6	1.86	1.95	1.55	1.93
0.7	1.57	1.66	1.26	1.65
0.75	1.46	1.55	1.15	1.53
0.80	1.36	1.45	1.05	1.43
0.82	1.32	1.41	1.01	1.39
0.84	1.29	1.37	0.98	1.36
0.86	1.25	1.34	0.94	1.33
0.88	1.22	1.31	0.91	1.29
0.90	1.19	1.28	0.88	1.26
0.92	1.16	1.25	0.85	1.24
0.94	1.13	1.22	0.82	1.21
0.96	1.11	1.20	0.80	1.18
0.98	1.08	1.17	0.77	1.16
1.00	1.06	1.15	0.75	1.13
1.02	1.03	1.12	0.72	1.11
1.04	1.01	1.10	0.70	1.08
1.06	0.99	1.08	0.68	1.06
1.08	0.97	1.06	0.66	1.04
1.10	0.95	1.04	0.64	1.02
1.12	0.93	1.02	0.62	1.00
1.14	0.91	1.00	0.60	0.98
1.16	0.89	0.98	0.58	0.96
1.18	0.87	0.96	0.56	0.95
1.20	0.86	0.95	0.55	0.93
1.25	0.82	0.91	0.51	0.89
1.30	0.78	0.87	0.47	0.85

(Continued)

TABLE 4.12 Conversion of Weighings in Air to Weighings in Vacuo (*Continued*)

Density of object weighed	Buoyancy reduction factor, *k*			
	Brass weights, density = 8.4	Pt or Pt-Ir weights, density = 21.5	Al or quartz weights, density = 2.7	Gold weights, density = 17
1.35	0.75	0.83	0.44	0.82
1.40	0.71	0.80	0.40	0.79
1.50	0.66	0.74	0.35	0.73
1.6	0.61	0.69	0.30	0.68
1.7	0.56	0.65	0.25	0.64
1.8	0.52	0.61	0.21	0.60
1.9	0.49	0.58	0.18	0.56
2.0	0.46	0.54	0.15	0.53
2.2	0.40	0.49	0.09	0.48
2.4	0.36	0.44	0.05	0.43
2.6	0.32	0.41	0.01	0.39
2.8	0.29	0.37	−0.02	0.36
3.0	0.26	0.34	−0.05	0.33
3.5	0.20	0.29	−0.11	0.27
4	0.16	0.24	−0.15	0.23
5	0.10	0.18	−0.21	0.17
6	0.06	0.14	−0.25	0.13
7	0.03	0.12	−0.28	0.10
8	0.01	0.09	−0.30	0.08
9	−0.01	0.08	−0.32	0.06
10	−0.02	0.06	−0.33	0.05
12	−0.04	0.04	−0.35	0.03
14	−0.06	0.03	−0.37	0.02
16	−0.07	0.02	−0.38	0.00
18	−0.08	0.01	−0.39	0.00
20	−0.08	0.00	−0.39	−0.01
22	−0.09	0.00	−0.40	−0.02

TABLE 4.13 Factors for Reducing Gas Volumes to Normal (Standard) Temperature and Pressure (760 mmHg)

Examples: (*a*) 20 mL of dry gas at 22°C and 730 mm = 20 × 0.8888 = 17.78 mL at 0°C and 760 mm. (*b*) 20 mL of a gas over water at 22° and 730 mm = 20 × (factor corrected for aqueous tension; i.e., 730 – 19.8 or 710.2 mm) = 20 mL of dry gas at 22° and 710.2 mm = 20 × 0.86475 = 17.30 mL at 0°C and 760 mm. Mass in milligrams of 1 mL of gas at S.T.P.: acetylene, 1.173; carbon dioxide, 1.9769; hydrogen, 0.0899; nitric oxide (NO), 1.3402; nitrogen, 1.25057; oxygen, 1.42904.

Pressure mm of mercury	Temperature °C							
	10°	11°	12°	13°	14°	15°	16°	17°
670	0.8504	0.8474	0.8445	0.8415	0.8386	0.8357	0.8328	0.8299
672	0.8530	0.8500	0.8470	0.8440	0.8411	0.8382	0.8353	0.8324
674	0.8555	0.8525	0.8495	0.8465	0.8436	0.8407	0.8377	0.8349
676	0.8580	0.8550	0.8520	0.8490	0.8461	0.8431	0.8402	0.8373
678	0.8606	0.8576	0.8545	0.8516	0.8486	0.8456	0.8427	0.8398
680	0.8631	0.8601	0.8571	0.8541	0.8511	0.8481	0.8452	0.8423
682	0.8657	0.8626	0.8596	0.8566	0.8536	0.8506	0.8477	0.8448
684	0.8682	0.8651	0.8621	0.8591	0.8561	0.8531	0.8502	0.8472
686	0.8707	0.8677	0.8646	0.8616	0.8586	0.8556	0.8527	0.8497
688	0.8733	0.8702	0.8672	0.8641	0.8611	0.8581	0.8551	0.8522
690	0.8758	0.8727	0.8697	0.8666	0.8636	0.8606	0.8576	0.8547
692	0.8784	0.8753	0.8722	0.8691	0.8661	0.8631	0.8601	0.8572
694	0.8809	0.8778	0.8747	0.8717	0.8686	0.8656	0.8626	0.8596
696	0.8834	0.8803	0.8772	0.8742	0.8711	0.8681	0.8651	0.8621
698	0.8860	0.8828	0.8798	0.8767	0.8736	0.8706	0.8676	0.8646
700	0.8885	0.8854	0.8823	0.8792	0.8761	0.8731	0.8700	0.8671
702	0.8910	0.8879	0.8848	0.8817	0.8786	0.8756	0.8725	0.8695
704	0.8936	0.8904	0.8873	0.8842	0.8811	0.8781	0.8750	0.8720
706	0.8961	0.8930	0.8898	0.8867	0.8836	0.8806	0.8775	0.8745
708	0.8987	0.8955	0.8924	0.8892	0.8861	0.8831	0.8800	0.8770
710	0.9012	0.8980	0.8949	0.8917	0.8886	0.8856	0.8825	0.8794
712	0.9037	0.9006	0.8974	0.8943	0.8911	0.8880	0.8850	0.8819
714	0.9063	0.9031	0.8999	0.8968	0.8936	0.8905	0.8875	0.8844
716	0.9088	0.9056	0.9024	0.8993	0.8961	0.8930	0.8899	0.8869
718	0.9114	0.9081	0.9050	0.9018	0.8987	0.8955	0.8924	0.8894
720	0.9139	0.9107	0.9075	0.9043	0.9012	0.8980	0.8949	0.8918
722	0.9164	0.9132	0.9100	0.9068	0.9037	0.9005	0.8974	0.8943
724	0.9190	0.9157	0.9125	0.9093	0.9062	0.9030	0.8999	0.8968
726	0.9215	0.9183	0.9151	0.9118	0.9087	0.9055	0.9024	0.8993
728	0.9241	0.9208	0.9176	0.9144	0.9112	0.9080	0.9049	0.9017
730	0.9266	0.9233	0.9201	0.9169	0.9137	0.9105	0.9073	0.9042
732	0.9291	0.9259	0.9226	0.9194	0.9162	0.9130	0.9098	0.9067
734	0.9317	0.9284	0.9251	0.9219	0.9187	0.9155	0.9123	0.9092
736	0.9342	0.9309	0.9277	0.9244	0.9212	0.9180	0.9148	0.9117
738	0.9368	0.9334	0.9302	0.9269	0.9237	0.9205	0.9173	0.9141
740	0.9393	0.9360	0.9327	0.9294	0.9262	0.9230	0.9198	0.9166
742	0.9418	0.9385	0.9352	0.9319	0.9287	0.9255	0.9223	0.9191
744	0.9444	0.9410	0.9377	0.9345	0.9312	0.9280	0.9248	0.9216
746	0.9469	0.9436	0.9403	0.9370	0.9337	0.9305	0.9272	0.9240
748	0.9494	0.9461	0.9428	0.9395	0.9362	0.9329	0.9297	0.9265

(*Continued*)

TABLE 4.13 Factors for Reducing Gas Volumes to Normal (Standard) Temperature and Pressure (*Continued*)

Pressure mm of mercury	Temperature °C							
	10°	11°	12°	13°	14°	15°	16°	17°
750	0.9520	0.9486	0.9453	0.9420	0.9387	0.9354	0.9322	0.9290
752	0.9545	0.9511	0.9478	0.9445	0.9412	0.9379	0.9347	0.9315
754	0.9571	0.9537	0.9504	0.9470	0.9437	0.9404	0.9372	0.9339
756	0.9596	0.9562	0.9529	0.9495	0.9462	0.9429	0.9397	0.9364
758	0.9621	0.9587	0.9554	0.9520	0.9487	0.9454	0.9422	0.9389
760	0.9647	0.9613	0.9579	0.9546	0.9512	0.9479	0.9446	0.9414
762	0.9672	0.9638	0.9604	0.9571	0.9537	0.9504	0.9471	0.9439
764	0.9698	0.9663	0.9630	0.9596	0.9562	0.9529	0.9496	0.9463
766	0.9723	0.9689	0.9655	0.9620	0.9587	0.9554	0.9521	0.9488
768	0.9748	0.9714	0.9680	0.9646	0.9612	0.9579	0.9546	0.9513
770	0.9774	0.9739	0.9705	0.9671	0.9637	0.9604	0.9571	0.9538
772	0.9799	0.9764	0.9730	0.9696	0.9662	0.9629	0.9596	0.9562
774	0.9825	0.9790	0.9756	0.9721	0.9687	0.9654	0.9620	0.9587
776	0.9850	0.9815	0.9781	0.9746	0.9712	0.9679	0.9645	0.9612
778	0.9875	0.9840	0.9806	0.9772	0.9737	0.9704	0.9670	0.9637
780	0.9901	0.9866	0.9831	0.9797	0.9763	0.9729	0.9695	0.9662
782	0.9926	0.9891	0.9856	0.9822	0.9788	0.9754	0.9720	0.9686
784	0.9952	0.9916	0.9882	0.9847	0.9813	0.9778	0.9745	0.9711
786	0.9977	0.9942	0.9907	0.9872	0.9838	0.9803	0.9770	0.9736
788	1.0002	0.9967	0.9932	0.9897	0.9863	0.9828	0.9794	0.9761

Pressure mm of mercury	Temperature °C							
	18°	19°	20°	21°	22°	23°	24°	25°
670	0.8270	0.8242	0.8214	0.8186	0.8158	0.8131	0.8103	0.8076
672	0.8295	0.8267	0.8239	0.8211	0.8183	0.8155	0.8128	0.8100
674	0.8320	0.8291	0.8263	0.8235	0.8207	0.8179	0.8152	0.8124
676	0.8345	0.8316	0.8288	0.8259	0.8231	0.8204	0.8176	0.8149
678	0.8369	0.8341	0.8312	0.8284	0.8256	0.8228	0.8200	0.8173
680	0.8394	0.8365	0.8337	0.8308	0.8280	0.8252	0.8224	0.8197
682	0.8419	0.8390	0.8361	0.8333	0.8304	0.8276	0.8249	0.8221
684	0.8443	0.8414	0.8386	0.8357	0.8329	0.8301	0.8273	0.8245
686	0.8468	0.8439	0.8410	0.8382	0.8353	0.8325	0.8297	0.8269
688	0.8493	0.8464	0.8435	0.8406	0.8378	0.8349	0.8321	0.8293
690	0.8517	0.8488	0.8459	0.8430	0.8402	0.8373	0.8345	0.8317
692	0.8542	0.8513	0.8484	0.8455	0.8426	0.8398	0.8369	0.8341
694	0.8567	0.8537	0.8508	0.8479	0.8451	0.8422	0.8394	0.8366
696	0.8591	0.8562	0.8533	0.8504	0.8475	0.8446	0.8418	0.8390
698	0.8616	0.8587	0.8557	0.8528	0.8499	0.8471	0.8442	0.8414
700	0.8641	0.8611	0.8582	0.8553	0.8524	0.8495	0.8466	0.8438
702	0.8665	0.8636	0.8606	0.8577	0.8547	0.8519	0.8490	0.8462
704	0.8690	0.8660	0.8631	0.8602	0.8572	0.8543	0.8515	0.8486
706	0.8715	0.8685	0.8655	0.8626	0.8597	0.8568	0.8539	0.8510
708	0.8740	0.8710	0.8680	0.8650	0.8621	0.8592	0.8563	0.8534

TABLE 4.13 Factors for Reducing Gas Volumes to Normal (Standard) Temperature and Pressure (*Continued*)

Pressure mm of mercury	Temperature °C							
	18°	19°	20°	21°	22°	23°	24°	25°
710	0.8764	0.8734	0.8704	0.8675	0.8645	0.8616	0.8587	0.8558
712	0.8789	0.8759	0.8729	0.8699	0.8670	0.8640	0.8611	0.8582
714	0.8814	0.8783	0.8753	0.8724	0.8694	0.8665	0.8636	0.8607
716	0.8838	0.8808	0.8778	0.8748	0.8718	0.8689	0.8660	0.8631
718	0.8863	0.8833	0.8802	0.8773	0.8743	0.8713	0.8684	0.8655
720	0.8888	0.8857	0.8827	0.8797	0.8767	0.8738	0.8708	0.8679
722	0.8912	0.8882	0.8852	0.8821	0.8792	0.8762	0.8732	0.8703
724	0.8937	0.8906	0.8876	0.8846	0.8816	0.8786	0.8757	0.8727
726	0.8962	0.8931	0.8901	0.8870	0.8840	0.8810	0.8781	0.8751
728	0.8986	0.8956	0.8925	0.8895	0.8865	0.8835	0.8805	0.8775
730	0.9011	0.8980	0.8950	0.8919	0.8889	0.8859	0.8829	0.8799
732	0.9036	0.9005	0.8974	0.8944	0.8913	0.8883	0.8853	0.8824
734	0.9060	0.9029	0.8999	0.8968	0.8938	0.8907	0.8877	0.8848
736	0.9085	0.9054	0.9023	0.8992	0.8962	0.8932	0.8902	0.8872
738	0.9110	0.9079	0.9048	0.9017	0.8986	0.8956	0.8926	0.8896
740	0.9135	0.9103	0.9072	0.9041	0.9011	0.8980	0.8950	0.8920
742	0.9159	0.9128	0.9097	0.9066	0.9035	0.9005	0.8974	0.8944
744	0.9184	0.9153	0.9121	0.9090	0.9059	0.9029	0.8998	0.8968
746	0.9209	0.9177	0.9146	0.9115	0.9084	0.9053	0.9023	0.8992
748	0.9233	0.9202	0.9170	0.9139	0.9108	0.9077	0.9047	0.9016
750	0.9258	0.9226	0.9195	0.9164	0.9132	0.9102	0.9071	0.9041
752	0.9283	0.9251	0.9219	0.9188	0.9157	0.9126	0.9095	0.9065
754	0.9307	0.9276	0.9244	0.9212	0.9181	0.9150	0.9119	0.9089
756	0.9332	0.9300	0.9268	0.9237	0.9206	0.9174	0.9144	0.9113
758	0.9357	0.9325	0.9293	0.9261	0.9230	0.9199	0.9168	0.9137
760	0.9381	0.9349	0.9317	0.9286	0.9254	0.9223	0.9192	0.9161
762	0.9406	0.9374	0.9342	0.9310	0.9279	0.9247	0.9216	0.9185
764	0.9431	0.9399	0.9366	0.9335	0.9303	0.9272	0.9240	0.9209
766	0.9456	0.9423	0.9391	0.9359	0.9327	0.9296	0.9265	0.9233
768	0.9480	0.9448	0.9415	0.9383	0.9352	0.9320	0.9289	0.9258
770	0.9505	0.9472	0.9440	0.9408	0.9376	0.9344	0.9313	0.9282
772	0.9530	0.9497	0.9464	0.9432	0.9400	0.9369	0.9337	0.9306
774	0.9554	0.9522	0.9489	0.9457	0.9425	0.9393	0.9361	0.9330
776	0.9579	0.9546	0.9514	0.9481	0.9449	0.9417	0.9385	0.9354
778	0.9604	0.9571	0.9538	0.9506	0.9473	0.9441	0.9410	0.9378
780	0.9628	0.9595	0.9563	0.9530	0.9498	0.9466	0.9434	0.9402
782	0.9653	0.9620	0.9587	0.9555	0.9522	0.9490	0.9458	0.9426
784	0.9678	0.9645	0.9612	0.9579	0.9546	0.9514	0.9482	0.9450
786	0.9702	0.9669	0.9636	0.9603	0.9571	0.9538	0.9506	0.9474
788	0.9727	0.9694	0.9661	0.9628	0.9595	0.9563	0.9531	0.9499

(*Continued*)

TABLE 4.13 Factors for Reducing Gas Volumes to Normal (Standard) Temperature and Pressure (*Continued*)

Pressure mm of mercury	Temperature °C							
	26°	27°	28°	29°	30°	31°	32°	33°
670	0.8049	0.8022	0.7996	0.7969	0.7943	0.7917	0.7891	0.7865
672	0.8073	0.8046	0.8020	0.7993	0.7967	0.7940	0.7914	0.7889
674	0.8097	0.8070	0.8043	0.8017	0.7990	0.7964	0.7938	0.7912
676	0.8121	0.8094	0.8067	0.8041	0.8014	0.7988	0.7962	0.7936
678	0.8145	0.8118	0.8091	0.8064	0.8038	0.8011	0.7985	0.7959
680	0.8169	0.8142	0.8115	0.8088	0.8061	0.8035	0.8009	0.7982
682	0.8193	0.8166	0.8139	0.8112	0.8085	0.8059	0.8032	0.8006
684	0.8217	0.8190	0.8163	0.8136	0.8109	0.8082	0.8056	0.8029
686	0.8241	0.8214	0.8187	0.8160	0.8133	0.8106	0.8079	0.8053
688	0.8265	0.8238	0.8211	0.8183	0.8156	0.8129	0.8103	0.8076
690	0.8289	0.8262	0.8234	0.8207	0.8180	0.8153	0.8126	0.8100
692	0.8313	0.8286	0.8258	0.8231	0.8204	0.8177	0.8150	0.8123
694	0.8338	0.8310	0.8282	0.8255	0.8227	0.8200	0.8174	0.8147
696	0.8362	0.8334	0.8306	0.8278	0.8251	0.8224	0.8197	0.8170
698	0.8386	0.8358	0.8330	0.8302	0.8275	0.8248	0.8221	0.8194
700	0.8410	0.8382	0.8354	0.8326	0.8299	0.8271	0.8244	0.8217
702	0.8434	0.8406	0.8378	0.8350	0.8322	0.8295	0.8268	0.8241
704	0.8458	0.8429	0.8401	0.8374	0.8346	0.8319	0.8291	0.8264
706	0.8482	0.8453	0.8425	0.8397	0.8370	0.8342	0.8315	0.8288
708	0.8506	0.8477	0.8449	0.8421	0.8393	0.8366	0.8338	0.8311
710	0.8530	0.8501	0.8473	0.8445	0.8417	0.8389	0.8362	0.8335
712	0.8554	0.8525	0.8497	0.8469	0.8441	0.8413	0.8386	0.8358
714	0.8578	0.8549	0.8521	0.8493	0.8465	0.8437	0.8409	0.8382
716	0.8602	0.8573	0.8545	0.8516	0.8488	0.8460	0.8433	0.8405
718	0.8626	0.8597	0.8569	0.8540	0.8512	0.8484	0.8456	0.8429
720	0.8650	0.8621	0.8592	0.8564	0.8536	0.8508	0.8480	0.8452
722	0.8674	0.8645	0.8616	0.8588	0.8559	0.8531	0.8503	0.8475
724	0.8698	0.8669	0.8640	0.8612	0.8583	0.8555	0.8527	0.8499
726	0.8722	0.8693	0.8664	0.8635	0.8607	0.8579	0.8550	0.8522
728	0.8746	0.8717	0.8688	0.8659	0.8631	0.8602	0.8574	0.8546
730	0.8770	0.8741	0.8712	0.8683	0.8654	0.8626	0.8598	0.8569
732	0.8794	0.8765	0.8736	0.8707	0.8678	0.8649	0.8621	0.8593
734	0.8818	0.8789	0.8759	0.8730	0.8702	0.8673	0.8645	0.8616
736	0.8842	0.8813	0.8783	0.8754	0.8725	0.8697	0.8668	0.8640
738	0.8866	0.8837	0.8807	0.8778	0.8749	0.8720	0.8692	0.8663
740	0.8890	0.8861	0.8831	0.8802	0.8773	0.8744	0.8715	0.8687
742	0.8914	0.8884	0.8855	0.8826	0.8796	0.8768	0.8739	0.8710
744	0.8938	0.8908	0.8879	0.8849	0.8820	0.8791	0.8762	0.8734
746	0.8962	0.8932	0.8903	0.8873	0.8844	0.8815	0.8786	0.8757
748	0.8986	0.8956	0.8927	0.8897	0.8868	0.8838	0.8809	0.8781
750	0.9010	0.8980	0.8950	0.8921	0.8891	0.8862	0.8833	0.8804
752	0.9034	0.9004	0.8974	0.8945	0.8915	0.8886	0.8857	0.8828
754	0.9058	0.9028	0.8998	0.8968	0.8939	0.8909	0.8880	0.8851
756	0.9082	0.9052	0.9022	0.8992	0.8962	0.8933	0.8904	0.8875
758	0.9106	0.9076	0.9046	0.9016	0.8986	0.8957	0.8927	0.8898

TABLE 4.13 Factors for Reducing Gas Volumes to Normal (Standard) Temperature and Pressure (*Continued*)

Pressure mm of mercury	Temperature °C							
	26°	27°	28°	29°	30°	31°	32°	33°
760	0.9130	0.9100	0.9070	0.9040	0.9010	0.8980	0.8951	0.8922
762	0.9154	0.9124	0.9094	0.9064	0.9034	0.9004	0.8974	0.8945
764	0.9178	0.9148	0.9118	0.9087	0.9057	0.9028	0.8998	0.8969
766	0.9202	0.9172	0.9141	0.9111	0.9081	0.9051	0.9021	0.8992
768	0.9227	0.9196	0.9165	0.9135	0.9105	0.9075	0.9045	0.9015
770	0.9251	0.9220	0.9189	0.9159	0.9128	0.9098	0.9069	0.9039
772	0.9275	0.9244	0.9213	0.9182	0.9152	0.9122	0.9092	0.9062
774	0.9299	0.9268	0.9237	0.9206	0.9176	0.9146	0.9116	0.9086
776	0.9323	0.9292	0.9261	0.9230	0.9200	0.9169	0.9139	0.9109
778	0.9347	0.9316	0.9285	0.9254	0.9223	0.9193	0.9163	0.9133
780	0.9371	0.9340	0.9308	0.9278	0.9247	0.9217	0.9186	0.9156
782	0.9395	0.9363	0.9332	0.9301	0.9271	0.9240	0.9210	0.9180
784	0.9419	0.9387	0.9356	0.9325	0.9294	0.9264	0.9233	0.9203
786	0.9443	0.9411	0.9380	0.9349	0.9318	0.9287	0.9257	0.9227
788	0.9467	0.9435	0.9404	0.9373	0.9342	0.9311	0.9281	0.9250

Pressure mm of mercury	Temperature °C		
	34°	35°	36°
670	0.7839	0.7814	0.7789
672	0.7863	0.7837	0.7812
674	0.7886	0.7861	0.7835
676	0.7910	0.7884	0.7858
678	0.7933	0.7907	0.7882
680	0.7956	0.7931	0.7905
682	0.7980	0.7954	0.7928
684	0.8003	0.7977	0.7951
686	0.8027	0.8001	0.7975
688	0.8050	0.8024	0.7998
690	0.8073	0.8047	0.8021
692	0.8097	0.8071	0.8044
694	0.8120	0.8094	0.8068
696	0.8144	0.8117	0.8091
698	0.8167	0.8141	0.8114
700	0.8190	0.8164	0.8137
702	0.8214	0.8187	0.8161
704	0.8237	0.8211	0.8184
706	0.8261	0.8234	0.8207
708	0.8284	0.8257	0.8230
710	0.8307	0.8281	0.8254
712	0.8331	0.8304	0.8277
714	0.8354	0.8327	0.8300
716	0.8378	0.8350	0.8323
718	0.8401	0.8374	0.8347
720	0.8424	0.8397	0.8370
722	0.8448	0.8420	0.8393
724	0.8471	0.8444	0.8416
726	0.8495	0.8467	0.8440
728	0.8518	0.8490	0.8463

Pressure mm of mercury	Temperature °C		
	34°	35°	36°
730	0.8541	0.8514	0.8486
732	0.8565	0.8537	0.8509
734	0.8588	0.8560	0.8533
736	0.8612	0.8584	0.8556
738	0.8635	0.8607	0.8579
740	0.8658	0.8630	0.8602
742	0.8682	0.8654	0.8626
744	0.8705	0.8677	0.8649
746	0.8729	0.8700	0.8672
748	0.8752	0.8724	0.8695
750	0.8775	0.8747	0.8719
752	0.8799	0.8770	0.8742
754	0.8822	0.8794	0.8765
756	0.8846	0.8817	0.8788
758	0.8869	0.8840	0.8812
760	0.8892	0.8864	0.8835
762	0.8916	0.8887	0.8858
764	0.8939	0.8910	0.8881
766	0.8963	0.8934	0.8905
768	0.8986	0.8957	0.8928
770	0.9009	0.8980	0.8951
772	0.9033	0.9004	0.8974
774	0.9056	0.9027	0.8998
776	0.9080	0.9050	0.9021
778	0.9103	0.9074	0.9044
780	0.9127	0.9097	0.9067
782	0.9150	0.9120	0.9091
784	0.9173	0.9144	0.9114
786	0.9197	0.9167	0.9137
788	0.9220	0.9190	0.9160

4.6 VISCOSITY

Viscosity is the shear stress per unit area at any point in a confined fluid divided by the velocity gradient in the direction perpendicular to the direction of flow. If this ratio is constant with time at a given temperature and pressure for any species, the fluid is called a Newtonian fluid.

The *absolute viscosity* (μ) is the shear stress at a point divided by the velocity gradient at that point. The most common unit is the poise (1 kg/m sec) and the SI unit is the Pa.sec (1 kg/m sec). As many common fluids have viscosities in the hundredths of a poise the centipoise (cp) is often used. One centipoise is then equal to one mPa sec.

The *kinematic viscosity* (v) is ratio of the absolute viscosity to density at the same temperature and pressure. The most common unit corresponding to the poise is the stoke (1 cm^2/sec) and the SI unit is m^2/sec.

TABLE 4.14 Viscosity Conversion

Centistokes to Saybolt, Redwood, and Engler units.

Poise = cgs unit of absolute viscosity　　Centipoise = 0.01 poise
Stoke = cgs unit of kinematic viscosity　　Centistoke = 0.01 stoke
Centipoises = centistokes × density (at temperature under consideration)
Reyn (1 lb · s per sq in) = 69×10^5 centipoises

Cf. *Jour. Inst. Pet. Tech.*, Vol. 22, p. 21 (1936); *Reports of A. S. T. M. Committee D-2, 1936 and 1937.*

The values of Saybolt Universal Viscosity at 100°F and at 210°F are taken directly from the comprehensive *ASTM Viscosity Table, Special Technical Publication No. 43A* (1953) by permission of the publishers, American Society for Testing Materials, West Conshohocken, PA.

Centistokes	Saybolt Universal Viscosity at 100°F.	Saybolt Universal Viscosity at 130°F.	Saybolt Universal Viscosity at 210°F.	Redwood Seconds at 70°F.	Redwood Seconds at 140°F.	Redwood Seconds at 200°F.	Engler Degrees at all Temps.
2.0	32.62	32.68	32.85	30.2	31.0	31.2	1.14
3.0	36.03	36.10	36.28	32.7	33.5	33.7	1.22
4.0	39.14	39.22	39.41	35.3	36.0	36.3	1.31
5.0	42.35	42.43	42.65	37.9	38.5	38.9	1.40
6.0	45.56	45.65	45.88	40.5	41.0	41.5	1.48
7.0	48.77	48.86	49.11	43.2	43.7	44.2	1.56
8.0	52.09	52.19	52.45	46.0	46.4	46.9	1.65
9.0	55.50	55.61	55.89	48.9	49.1	49.7	1.75
10.0	58.91	59.02	59.32	51.7	52.0	52.6	1.84
11.0	62.43	62.55	62.86	54.8	55.0	55.6	1.93
12.0	66.04	66.17	66.50	57.9	58.1	58.8	2.02
14.0	73.57	73.71	74.09	64.4	64.6	65.3	2.22
16.0	81.30	81.46	81.87	71.0	71.4	72.2	2.43
18.0	89.44	89.61	90.06	77.9	78.5	79.4	2.64
20.0	97.77	97.96	98.45	85.0	85.8	86.9	2.87
22.0	106.4	106.6	107.1	92.4	93.3	94.5	3.10
24.0	115.0	115.2	115.8	99.9	100.9	102.2	3.34
26.0	123.7	123.9	124.5	107.5	108.6	110.0	3.58
28.0	132.5	132.8	133.4	115.3	116.5	118.0	3.82
30.0	141.3	141.6	142.3	123.1	124.4	126.0	4.07

TABLE 4.14 Viscosity Conversion (*Continued*)

Centistokes	Saybolt Universal Viscosity at			Redwood Seconds at			Engler Degrees at all Temps.
	100°F.	130°F.	210°F.	70°F.	140°F.	200°F.	
32.0	150.2	150.5	151.2	131.0	132.3	134.1	4.32
34.0	159.2	159.5	160.3	138.9	140.2	142.2	4.57
36.0	168.2	168.5	169.4	146.9	148.2	150.3	4.83
38.0	177.3	177.6	178.5	155.0	156.2	158.3	5.08
40.0	186.3	186.7	187.6	163.0	164.3	166.7	5.34
42.0	195.3	195.7	196.7	171.0	172.3	175.0	5.59
44.0	204.4	204.8	205.9	179.1	180.4	183.3	5.85
46.0	213.7	214.1	215.2	187.1	188.5	191.7	6.11
48.0	222.9	223.3	224.5	195.2	196.6	200.0	6.37
50.0	232.1	232.5	233.8	203.3	204.7	208.3	6.63
60.0	278.3	278.8	280.2	243.5	245.3	250.0	7.90
70.0	324.4	325.0	326.7	283.9	286.0	291.7	9.21
80.0	370.8	371.5	373.4	323.9	326.6	333.4	10.53
90.0	417.1	417.9	420.0	364.4	367.4	375.0	11.84
100.0*	463.5	464.4	466.7	404.9	408.2	416.7	13.16

*At higher values use the same ratio as above for 100 centistokes; *e.g.*, 102 centistokes = 102 × 4.635 Saybolt seconds at 100°F.

To obtain the Saybolt Universal viscosity equivalent to a kinematic viscosity determined at *t*°F., multiply the equivalent Saybolt Universal viscosity at 100°F. by 1 + (t – 100) 0.000064; *e.g.*, 10 centistokes at 210°F are equivalent to 58.91 × 1.0070, or 59.32 Saybolt Universal Viscosity at 210°F.

4.7 PHYSICAL CHEMISTRY EQUATIONS FOR GASES

A number of physical chemistry relationships, not enumerated in other sections (*see* Index), will be discussed in this section.

Boyle's law states that the volume of a given quantity of a gas varies inversely as the pressure, the temperature remaining constant. That is,

$$V = \frac{\text{constant}}{P} \quad \text{or} \quad PV = \text{constant}$$

A convenient form of the law, true strictly for ideal gases, is

$$P_1V_1 = P_2V_2$$

Charles' law, also known as *Gay-Lussac's law*, states that the volume of a given mass of gas varies directly as the absolute temperature if the pressure remains constant, that is,

$$\frac{V}{T} = \text{constant}$$

Combining the laws of Boyle and Charles into one expression gives

$$\frac{P_1V_1}{T_1} = \frac{P_2V_2}{T_2}$$

In terms of moles, *Avogadro's hypothesis* can be stated: The same volume is occupied by one mole of any gas at a given temperature and pressure. The number of molecules in one mole is known as the *Avogadro number constant* N_A.

The behavior of all gases that obey the laws of Boyle and Charles, and Avogadro's hypothesis, can be expressed by the ideal gas equation:

$$PV = nRT$$

where R is called the *gas constant* and n is the number of moles of gas. If pressure is written as force per unit area and the volume as area times length, then R has the dimensions of energy per degree per mole—8.314 J · K^{-1}· mol^{-1} or 1.987 cal · K^{-1} · mol^{-1}.

Dalton's law of partial pressures states that the total pressure exerted by a mixture of gases is equal to the sum of the pressures which each component would exert if placed separately into the container:

$$P_{\text{total}} = p_1 + p_2 + p_3 + \cdots$$

There are two ways to express the fraction which one gaseous component contributes to the total mixture: (1) the pressure fraction, p_i/P_{total}, and (2) the mole fraction, n_i/n_{total}.

4.7.1 Equations of State (PVT Relations for Real Gases)

1. *Virial equation* represents the experimental compressibility of a gas by an empirical equation of state:

$$PV = A_p + B_p P + C_p P^2 + \cdots$$

or

$$PV = A_v + B_v V + \frac{C_v}{V^2} + \cdots$$

where $A, B, C, \ldots$ are called the virial coefficients and are a function of the nature of the gas and the temperature.

2. *Van der Waals' equation:*

$$\left(P + \frac{an^2}{V^2}\right)(V - nb) = nRT$$

where the term an^2/V^2 is the correction for intermolecular attraction among the gas molecules and the nb term is the correction for the volume occupied by the gas molecules. The constants a and b must be fitted for each gas from experimental data; consequently the equation is semiempirical. The constants are related to the critical-point constants as follows:

$$a = 3P_c V^2$$

$$b = \frac{V_c}{3}$$

$$R = \frac{8P_c V_c}{3T_c}$$

Substitution into van der Waals' equation and rearrangement leads to only the terms P/P_c, V/V_c, and T/T_c, which are called the reduced variables P_R, V_R, and T_R. For 1 mole of gas,

$$\left(P_R + \frac{3}{V_R^2}\right)\left(V_R - \frac{1}{3}\right) = \frac{8}{3}T_R$$

3. *Berthelot's equation of state*, used by many thermodynamicists, is

$$PV = nRT\left[1 + \frac{9}{128}\frac{PT_c}{PT}\left(1 - 6\frac{T_c^2}{T^2}\right)\right]$$

This equation requires only knowledge of the critical temperature and pressure for its use and gives accurate results in the vicinity of room temperature for unassociated substances at moderate pressures.

4.7.2 Properties of Gas Molecules

Vapor Density. Substitution of the Antoine vapor-pressure equation for its equivalent log P in the ideal gas equation gives

$$\log \rho_{\text{vap}} = \log M - \log R - \log(t + 273.15) + A - \frac{B}{t + C}$$

where ρ_{vap} is the vapor density in g · mL^{-1} at t°C, M is the molecular weight, R is the gas constant, and A, B, and C are the constants of the Antoine equation for vapor pressure. Since this equation is based on the ideal gas law, it is accurate only at temperatures at which the vapor of any specific compound follows this law. This condition prevails at reduced temperatures (T_R) of about 0.5 K.

Velocities of Molecules. The mean square velocity of gas molecules is given by

$$\overline{u^2} = \frac{3kT}{m} = \frac{3RT}{M}$$

where k is Boltzmann's constant and m is the mass of the molecule.

The mean velocity is given by

$$\bar{u} = \left(\frac{8\overline{u^2}}{3\pi}\right)^{1/2}$$

Viscosity. On the assumption that molecules interact like hard spheres, the viscosity of a gas is

$$\eta = \left(\frac{5}{16\sigma^2}\right)\left(\frac{mkT}{\pi}\right)^{1/2}$$

where σ is the molecular diameter.

Mean Free Path. The mean free path of a gas molecule l and the mean time between collisions τ are given by

$$l = \frac{m}{\pi\rho\sigma^2\sqrt{2}}$$

$$\tau = \frac{1}{\bar{u}} = \frac{4\eta}{5P}$$

Graham's Law of Diffusion. The rates at which gases diffuse under the same conditions of temperature and pressure are inversely proportional to the square roots of their densities:

$$\frac{r_1}{r_2} = \left(\frac{\rho_2}{\rho_1}\right)^{1/2}$$

Since $\rho = MP/RT$ for an ideal gas, it follows that

$$\frac{r_1}{r_2} = \left(\frac{M_2}{M_1}\right)^{1/2}$$

Henry's Law. The solubility of a gas is directly proportional to the partial pressure exerted by the gas:

$$p_i = kx_i$$

Joule-Thompson Coefficient for Real Gases. This expresses the change in temperature with respect to change in pressure at constant enthalpy:

$$\mu_\pi = \left(\frac{\partial T}{\partial P}\right)_H$$

TABLE 4.15 Molar Equivalent of One Liter of Gas at Various Temperatures and Pressures

The values in this table, which give the number of moles in 1 liter of gas, are based on the properties of an "ideal" gas and were calculated by use of the formula:

$$\text{Moles/liter} = \frac{P}{760} \times \frac{273}{T} \times \frac{1}{22.40}$$

where P is the pressure in millimeters of mercury and T is the temperature in kelvins (= t°C + 273).
To convert to moles per cubic foot multiply the values in the table by 28.316.

Pressure mm of mercury	Temperature °C					
	10°	12°	14°	16°	18°	20°
655	0.03712	0.03686	0.03660	0.03634	0.03610	0.03585
660	3731	3714	3688	3662	3637	3612
665	3768	3742	3716	3690	3665	3640
670	3796	3770	3744	3718	3692	3667
675	3825	3798	3772	3745	3720	3695
680	0.03853	0.03826	0.03800	0.03773	0.03747	0.03694
685	3881	3854	3827	3801	3775	3749
690	3910	3882	3855	3829	3802	3776
695	3938	3910	3883	3856	3830	3804
700	3967	3939	3911	3884	3858	3831
702	0.03978	0.03950	0.03922	0.03895	0.03869	0.03842
704	3989	3961	3934	3906	3880	3853
706	4000	3972	3945	3917	3891	3864
708	4012	3984	3956	3929	3902	3875
710	4023	3995	3967	3940	3913	3886

TABLE 4.15 Molar Equivalent of One Liter of Gas at Various Temperatures and Pressures (*Continued*)

Pressure mm of mercury	Temperature °C					
	10°	12°	14°	16°	18°	20°
712	0.04035	0.04006	0.03978	0.03951	0.03924	0.03897
714	4046	4018	3989	3962	3935	3908
716	4057	4029	4001	3973	3946	3919
718	4068	4040	4012	3984	3957	3930
720	4080	4051	4023	3995	3968	3941
722	0.04091	0.04063	0.04034	0.04006	0.03979	0.03952
724	4103	4074	4045	4017	3990	3963
726	4114	4085	4057	4028	4001	3973
728	4125	4096	4068	4040	4012	3984
730	4136	4108	4079	4051	4023	3995
732	0.04148	0.04119	0.04090	0.04062	0.04034	0.04006
734	4159	4130	4101	4073	4045	4017
736	4171	4141	4112	4084	4056	4028
738	4182	4153	4124	4095	4067	4039
740	4193	4164	4135	4106	4078	4050
742	0.04204	0.04175	0.04146	0.04117	0.04089	0.04061
744	4216	4186	4157	4128	4100	4072
746	4227	4198	4168	4139	4111	4038
748	4239	4209	4179	4151	4122	4094
750	4250	4220	4191	4162	4133	4105
752	0.04261	0.04231	0.04202	0.04173	0.04144	0.04116
754	4273	4243	4213	4184	4155	4127
756	4284	4254	4224	4195	4166	4138
758	4295	4265	4235	4206	4177	4149
760	4307	4276	4247	4217	4188	4160
762	0.04318	0.04287	0.04258	0.04228	0.04199	0.04171
764	4329	4299	4269	4239	4210	4181
766	4341	4310	4280	4250	4221	4192
768	4352	4321	4291	4262	4232	4203
770	4363	4333	4302	4273	4243	4214
772	0.04375	0.04344	0.04314	0.04284	0.04254	0.04225
774	4386	4355	4325	4295	4265	4236
776	4397	4366	4336	4306	4276	4247
778	4409	4378	4347	4317	4287	4258
780	4420	4389	4358	4328	4298	4269

Pressure mm of mercury	Temperature °C					
	22°	24°	26°	28°	30°	32°
655	0.03561	0.03537	0.03515	0.03490	0.03467	0.03444
660	3588	3564	3541	3516	3493	3470
665	3614	3591	3568	3543	3520	3496
670	3642	3618	3595	3569	3546	3523
675	3669	3645	3622	3596	3572	3549

(*Continued*)

TABLE 4.15 Molar Equivalent of One Liter of Gas at Various Temperatures and Pressures (*Continued*)

Pressure mm of mercury	Temperature °C					
	22°	24°	26°	28°	30°	32°
680	0.03697	0.03672	0.03649	0.03623	0.03599	0.03575
685	3724	3699	3676	3649	3625	3602
690	3751	3726	3702	3676	3652	3628
695	3778	3753	3729	3703	3678	3654
700	3805	3780	3756	3729	3705	3680
702	0.03816	0.03790	0.03767	0.03740	0.03715	0.03691
704	3827	3801	3777	3750	3726	3701
706	3838	3812	3788	3761	3736	3712
708	3849	3823	3799	3772	3747	3722
710	3860	3834	3810	3783	3758	3733
712	0.03870	0.03844	0.03820	0.03793	0.03768	0.03744
714	3881	3855	3831	3804	3779	3754
716	3892	3866	3842	3815	3789	3765
718	3902	3877	3853	3825	3800	3775
720	3914	3888	3863	3836	3811	3786
722	0.03925	0.03898	0.03874	0.03847	0.03821	0.03796
724	3936	3909	3885	3857	3832	3807
726	3947	3920	3896	3868	3842	3817
728	3957	3931	3906	3878	3853	3828
730	3968	3941	3917	3889	3863	3838
732	0.03979	0.03952	0.03928	0.03900	0.03874	0.03849
734	3990	3963	3938	3910	3885	3859
736	4001	3974	3949	3921	3895	3870
738	4012	3985	3960	3932	3906	3880
740	4023	3995	3971	3942	3916	3891
742	0.04033	0.04006	0.03981	0.03953	0.03927	0.03901
744	4044	4017	3992	3964	3938	3912
746	4055	4028	4003	3974	3948	3922
748	4066	4039	4014	3985	3959	3933
750	4077	4049	4024	3996	3969	3943
752	0.04088	0.04060	0.04035	0.04006	0.03980	0.03954
754	4099	4071	4046	4017	3991	3964
756	4110	4082	4056	4028	4001	3975
758	4121	4093	4067	4038	4012	3985
760	4131	4103	4078	4049	4022	3996
762	0.04142	4114	4089	4060	4033	4006
764	4153	4125	4099	4070	4043	4017
766	4164	4136	4110	4081	4054	4027
768	4175	4147	4121	4092	4065	4038
770	4186	4158	4132	4102	4075	4048
772	0.04197	0.04168	0.04142	0.04113	0.04086	0.04059
774	4207	4179	4153	4124	4096	4070
776	4218	4190	4164	4134	4107	4080
778	4229	4201	4175	4145	4117	4091
780	4240	4211	4185	4155	4128	4101

TABLE 4.16 Corrections to Be Added to Molar Values to Convert to Molal

Temperature, °C	Aqueous solution $\Delta G°$ J·mol^{-1}	$\Delta H°$ J·mol^{-1}	$\Delta S°$ J·deg^{-1}·mol^{-1}	$\Delta C°_p$ J·deg^{-1}·mol^{-1}
0	0.4	−42.7	−0.17	55.2
10	0.8	58.1	0.21	45.6
20	4.2	148.1	0.50	38.9
30	10.9	230.5	0.79	35.1
40	20.1	313.4	1.09	33.0
50	32.2	397.9	1.34	32.6
60	46.8	482.4	1.59	32.2

4.8 COOLING

TABLE 4.17 Cooling Mixtures

The table below gives the lowest temperature that can be obtained from a mixture of the inorganic salt with finely shaved dry ice. With the organic substances, dry ice (−78°C) in small lumps can be added to the solvent until a slight excess of dry ice remains or liquid nitrogen (−196°C) can be poured into the solvent until a slush is formed that consists of the solid-liquid mixture at its melting point.

Substance	Quantity of substance, g	Quantity of water, mL	Temperature, °C
Ammonium nitrate	100	94	−4.0
Sodium nitrate	75	100	−5.3
Sodium thiosulfate 5-water	110	100	−8.0
Sodium chloride	36	100	−10.0
Sodium nitrate	50	100	−17.8
Sodium bromide	66	100	−28
Magnesium chloride	85	100	−34
Calcium chloride 6-water	100	81	−40.3
	100	70	−55

Substance	Temperature, °C	Substance	Temperature, °C
Ethylene glycol	−13	Acetone	−77
1,2-Dichlorobenzene	−17	Ethyl acetate	−84
Carbon tetrachloride	−22.9	2-Butanone	−87
Bromobenzene	−31	Hexane	−95
Methoxybenzene	−37	Methanol	−98
Bis(2-ethoxyethyl) ether	−44	Carbon disulfide	−112
Chlorobenzene	−45	Bromoethane	−119
N-Methylaniline	−57	Pentane	−130
p-Cymene	−68	2-Methylbutane	−160

TABLE 4.18 Molecular Lowering of the Melting or Freezing Point

Cryoscopic constants.

The cryoscopic constant K_f gives the depression of the melting point ΔT (in degrees Celsius) produced when 1 mol of solute is dissolved in 1000 g of a solvent. It is applicable only to dilute solutions for which the number of moles of solute is negligible in comparison with the number of moles of solvent. It is often used for molecular weight determinations.

$$M_2 = \frac{1000 w_2 K_f}{w_1 \Delta T}$$

where w_1 is the weight of the solvent and w_2 is the weight of the solute whose molecular weight is M_2.

Compound	K_f	Compound	K_f
Acetamide	4.04	Diphenylamine	8.60
Acetic acid	3.90	Diphenyl ether	7.88
Acetone	2.40	1,2-Ethanediamine	2.43
Ammonia	0.957	Ethoxybenzene	7.15
Aniline	5.87	Formamide	3.85
Antimony(III) chloride	17.95	Formic acid	2.77
Benzene	5.12	Glycerol	3.3 to 3.7
Benzonitrile	5.34	Hexamethylphosphoramide	6.93
Benzophenone	9.8		
Bicyclohexane	14.52	*N*-Methylacetamide	6.65
Biphenyl	8.0	2-Methyl-2-butanol	10.4
Borneol	35.8	Methylcyclohexane	14.13
Bornylamine	40.6	Methyl *cis*-9-octadecenoate	3.4
Butanedinitrile	18.26	2-Methyl-2-propanol	8.37
Camphene	31.08	Naphthalene	6.94
Camphoquinone	45.7	Nitrobenzene	6.852
D-(+)-Camphor	39.7	Octadecanoic acid	4.50
Carbon tetrachloride	29.8	2-Oxohexamethyleneimine	7.30
o-Cresol	5.60	Phenol	7.40
p-Cresol	6.96	Pyridine	4.75
Cyclohexane	20.0	Quinoline	1.95
Cyclohexanol	39.3	Succinonitrile	18.26
Cyclohexylcyclohexane	14.52	Sulfuric acid	1.86
Cyclopentadecanone	21.3	1,1,2,2-Tetrabromoethane	21.7
cis-Decahydronaphthalene	19.47	1,1,2,2-Tetrachloro-1,2-difluoroethane	37.7
trans-Decahydronaphthalene	20.81	Tetramethylene sulfone	64.1
Dibenz[*de,kl*]anthracene	25.7	*p*-Toluidine	5.372
Dibenzyl ether	6.27	Tribromomethane	14.4
1,2-Dibromoethane	12.5	1,3,3-Trimethyl-2-oxabicyclo-[2.2.2.]octane	6.7
Diethyl ether	1.79	Triphenylmethane	12.45
1,2-Dimethoxybenzene	6.38	Water	1.86
N,N-Dimethylacetamide	4.46	*p*-Xylene	4.3
2,2-Dimethyl-1-propanol	11.0		
Dimethyl sulfoxide	4.07		
1,4-Dioxane	4.63		

TABLE 4.19 Drying Agents

Drying agent	Most useful for	Residual water, mg H_2O per liter of dry air (25°C)	Grams water removed per gram of desiccant	Regeneration, °C
Al_2O_3	Hydrocarbons	0.002–0.005	0.2	175 (24 h)
$Ba(ClO_4)_2$[a]	Inert gas streams	0.6–0.8	0.17	140
BaO	Basic gases: hydrocarbons, aldehydes, alcohols	0.0007–0.003	0.12	1000
CaC_2[b]	Ethers		0.56	Impossible
$CaCl_2$[c]	Inert organics	0.1–0.2	0.15 (1 H_2O) 0.30 (2 H_2O)	250
CaH_2[d]	Hydrocarbons, ethers, amines, esters, higher alcohols	1×10^{-5}	0.85	Impossible
CaO	Ethers, esters, alcohols, amines	0.01–0.003	0.31	Difficult, 1000
$CaSO_4$	Most organic substances	0.005–0.07	0.07	225
Dow Desiccant 812[e]	Most materials	(5–200 ppm)		No
K_2CO_3	Most materials except acids and phenols		0.16	158
KOH	Amines	0.01–0.9		Impossible
$LiAlH_4$[f]	Hydrocarbons		1.9	Impossible
$Mg(ClO_4)_2$[a]	Gas streams	0.0005–0.002	0.24	250 (high vacuum)
MgO	All but acidic compounds	0.008	0.45	800
$MgSO_4$	Most organic compounds	1–12	0.15–0.75	Not feasible
Molecular sieves: 4X	Molecules with effective diameter >4Å	0.001	0.18	250
5X	Molecules with effective diameter >5Å	0.001	0.18	250
9.5% Na-Pb alloy[d]	Hydrocarbons, ethers	(For solvents only)	0.08	Impossible
Na_2SO_4	Ketones, acids, alkyl and aryl halides	12	1.25	150
P_2O_5	Gas streams; not suitable for alcohols, amines, ketones, or amines	2×10^{-5}	0.5	Not feasible
Silica gel	Most organic amines	0.002–0.07	0.2	200–350
Sulfuric acid	Air and inert gas streams	0.003–0.008	Indefinite	Not feasible

[a]May form explosive mixtures when contacting organic material. [b]Explosive C_2H_2 formed. [c]Slow in drying action.
[d]H_2 formed. [e]Used as column drying of organic liquids. [f]Strong reductant.

A saturated aqueous solution in contact with an excess of a definite solid phase at a given temperature will maintain constant humidity in an enclosed space. Table 4.20 gives a number of salts suitable for this purpose. The aqueous tension (vapor pressure, in millimeters of Hg) of a solution at a given temperature is found by multiplying the decimal fraction of the humidity by the aqueous tension at 100 percent humidity for the specific temperature. For example, the aqueous tension of a saturated solution of NaCl at 20°C is $0.757 \times 17.54 = 13.28$ mmHg and at 80°C it is $0.764 \times 355.1 = 271.3$ mmHg.

TABLE 4.20 Solutions for Maintaining Constant Humidity

Solid Phase	% Humidity at Specified Temperatures (°C)						
	10	20	25	30	40	60	80
$K_2Cr_2O_7$			98.0				
K_2SO_4	98	97	97	96	96	96	
KNO_3	95	93	92.5	91	88	82	
KCl	88	85.0	84.3	84	81.7	80.7	79.5
KBr		84	80.7		79.6	79.0	79.3
NaCl	76	75.7	75.3	74.9	74.7	74.9	76.4
$NaNO_3$			73.8	72.8	71.5	67.5	65.5
$NaNO_2$		66	65	63.0	61.5	59.3	58.9
$NaBr \cdot 2H_2O$		57.9	57.7		52.4	49.9	50.0
$Na_2Cr_2O_7 \cdot 2H_2O$	58	55	54		53.6	55.2	56.0
$Mg(NO_3)_2 \cdot 6H_2O$	57	55	52.9	52	49	43	
$K_2CO_3 \cdot 2H_2O$	47	44	42.8		42		
$MgCl_2 \cdot 6H_2O$	34	33	33.0	33	32	30	
$KF \cdot 2H_2O$				27.4	22.8	21.0	22.8
$KC_2H_3O_2 \cdot 1.5H_2O$	24	23	22.5	22	20		
$LiCl \cdot H_2O$	13	12	10.2	12	11	11	
KOH	13	9	8	7	6	5	
100% Humidity: Aqueous Tension (mm Hg)	9.21	17.54	23.76	31.82	55.32	149.4	355.1

TABLE 4.21 Concentration of Solutions of H_2SO_4, NaOH, and $CaCl_2$ Giving Specified Vapor Pressures and Percent Humidity at 25°C

Percent humidity	Aqueous tension, mmHg	H_2SO_4		NaOH		$CaCl_2$	
		Molality	Weight %	Molality	Weight %	Molality	Weight %
100	23.76	0.00	0.00	0.00	0.00	0.00	0.00
95	22.57	1.263	11.02	1.465	5.54	0.927	9.33
90	21.38	2.224	17.91	2.726	9.83	1.584	14.95
85	20.19	3.025	22.88	3.840	13.32	2.118	19.03
80	19.00	3.730	26.79	4.798	16.10	2.579	22.25
75	17.82	4.398	30.14	5.710	18.60	2.995	24.95
70	16.63	5.042	33.09	6.565	20.80	3.400	27.40
65	15.44	5.686	35.80	7.384	22.80	3.796	29.64
60	14.25	6.341	38.35	8.183	24.66	4.188	31.73
55	13.07	7.013	40.75	8.974	26.42	4.581	33.71
50	11.88	7.722	43.10	9.792	28.15	4.990	35.64
45	10.69	8.482	45.41	10.64	29.86	5.431	37.61
40	9.50	9.304	47.71	11.54	31.58	5.912	39.62
35	8.31	10.21	50.04	12.53	33.38	6.478	41.83
30	7.13	11.25	52.45	13.63	35.29	7.183	44.36
25	5.94	12.47	55.01	14.96	37.45		
20	4.75	13.94	57.76	16.67	40.00		
15	3.56	15.81	60.80	19.10	43.32		
10	2.38	18.48	64.45	23.05	47.97		
5	1.19	23.17	69.44				

Concentrations are expressed in percentage of anhydrous solute by weight.

TABLE 4.22 Relative Humidity from Wet and Dry Bulb Thermometer Readings

Dry bulb temperature, °C	Wet bulb depression, °C											
	0.5	1.0	1.5	2.0	2.5	3.0	3.5	4.0	4.5	5.0	5.5	6.0
	Relative humidity, %											
−10	83	67	51	35	19							
−5	88	76	64	52	41	29	18	7				
0	91	81	72	64	55	46	38	29	21	13	5	
2	91	84	76	68	60	52	44	37	29	22	14	7
4	92	85	78	71	63	57	49	43	36	29	22	16
6	93	86	79	73	66	60	54	48	41	35	29	24
8	93	87	81	75	69	63	57	51	46	40	35	29
10	94	88	82	77	71	66	60	55	50	44	39	34
12	94	89	83	78	73	68	63	58	53	48	43	39
14	95	90	85	79	75	70	65	60	56	51	47	42
16	95	90	85	81	76	71	67	63	58	54	50	46
18	95	91	86	82	77	73	69	65	61	57	53	49
20	96	91	87	83	78	74	70	66	63	59	55	51
22	96	92	87	83	80	76	72	68	64	61	57	54
24	96	92	88	84	80	77	73	69	66	62	59	56

(Continued)

TABLE 4.22 Relative Humidity from Wet and Dry Bulb Thermometer Readings (*Continued*)

Dry bulb temperature, °C	Wet bulb depression, °C											
	0.5	1.0	1.5	2.0	2.5	3.0	3.5	4.0	4.5	5.0	5.5	6.0
	Relative humidity, %											
26	96	92	88	85	81	78	74	71	67	64	61	58
28	96	93	89	85	82	78	75	72	69	65	62	59
30	96	93	89	86	83	79	76	73	70	67	64	61
35	97	94	90	87	84	81	78	75	72	69	67	64
40	97	94	91	88	85	82	80	77	74	72	69	67

Dry bulb temperature, °C	Wet bulb depression, °C											
	6.5	7.0	7.5	8.0	8.5	9.0	10.0	11.0	12.0	13.0	14.0	15.0
	Relative humidity, %											
4	9											
6	17	11	5									
8	24	19	14	8								
10	29	24	20	15	10	6						
12	34	29	25	21	16	12	5					
14	38	34	30	26	22	18	10					
16	42	38	34	30	26	23	15	8				
18	45	41	38	34	30	27	20	14	7			
20	48	44	41	37	34	31	24	18	12	6		
22	50	47	44	40	37	34	28	22	17	11	6	
24	53	49	46	43	40	37	31	26	20	15	10	5
26	54	51	49	46	43	40	34	29	24	19	14	10
28	56	53	51	48	45	42	37	32	27	22	18	13
30	58	55	52	50	47	44	39	35	30	25	21	17
32	60	57	54	51	49	46	41	37	32	28	24	20
34	61	58	56	53	51	48	43	39	35	30	26	23
36	62	59	57	54	52	50	45	41	37	33	29	25
38	63	61	58	56	54	51	47	43	39	35	31	27
40	64	62	59	57	54	53	48	44	40	36	33	29

TABLE 4.23 Relative Humidity from Dew Point Readings

Depression of dew point, °C	Dew point reading, °C: −10	0	10	20	30
	Relative humidity, %				
0.5	96	96	96	96	97
1.0	92	93	94	94	94
1.5	89	89	90	91	92
2.0	86	87	88	88	89
3.0	79	81	82	83	84
4.0	73	75	77	78	80
5.0	68	70	72	74	75
6.0	63	66	68	70	71
7.0	59	61	63	66	68
8.0	54	57	60	62	64
9.0	51	53	56	58	61
10.0	47	50	53	55	57
11.0	44	47	49	52	
12.0	41	44	47	49	
13.0	38	41	44	46	
14.0	35	38	41	44	
15.0	33	36	39	42	
16.0	31	34	37	39	
18.0	27	30	33	35	
20.0	24	26	29	32	
22.0	21	23	26		
24.0	18	21	23		
26.0	16	18	21		
28.0	14	16	19		
30.0	12	14	17		

TABLE 4.24 Mass of Water Vapor in Saturated Air

The values in the table are grams of water contained in a cubic meter (m^3) of saturated air at a total pressure 101 325 Pa (1 atm).

°C	$g \cdot m^{-3}$	°C	$g \cdot m^{-3}$	°C	$g \cdot m^{-3}$
−30	0.341	12	10.65	53	95.56
−29	0.375	13	11.35	54	100.0
−28	0.413	14	12.05	55	104.5
−27	0.456	15	12.80	56	109.1
−26	0.504	16	13.60	57	114.1
−25	0.554	17	14.45	58	119.2
−24	0.607	18	15.35	59	124.7
−23	0.667	19	16.30	60	130.2
−22	0.733	20	17.30	61	136.0
−21	0.804	21	18.35	62	142.1
−20	0.883	22	19.40	63	148.4
−19	0.968	23	20.55	64	154.9
−18	1.063	24	21.75	65	161.3
−17	1.164	25	23.05	66	167.9
−16	1.273	26	24.35	67	175.1
−15	1.375	27	25.75	68	182.6
−14	1.510	28	27.20	69	190.3
−13	1.650	29	28.75	70	198.2
−12	1.800	30	30.35	71	206.5
−11	1.965	31	32.05	72	215.1
−10	2.140	32	33.80	73	223.7
−9	2.331	33	35.60	74	233.0
−8	2.539	34	37.55	75	242.0
−7	2.761	35	39.55	76	251.2
−6	3.003	36	41.65	77	261.1
−5	3.250	37	43.90	78	271.6
−4	3.512	38	46.20	79	282.3
−3	3.810	39	48.60	80	293.4
−2	4.131	40	51.21	81	304.8
−1	4.473	41	53.86	82	316.6
0	4.849	42	56.61	83	328.7
1	5.199	43	59.51	84	341.2
2	5.569	44	62.53	85	353.6
3	5.947	45	65.52	86	366.2
4	6.35	46	68.61	87	379.9
5	6.80	47	72.00	88	394.1
6	7.25	48	75.56	89	408.6
7	7.75	49	79.24	90	423.5
8	8.25	50	83.05	91	439.0
9	8.80	51	87.04	92	454.8
10	9.40	52	91.22	93	471.2
11	10.00				

4.10 MOLECULAR WEIGHT

TABLE 4.25 Molecular Elevation of the Boiling Point

Ebullioscopic constants.

Molecular weights can be determined with the relation:

$$M = E_b \frac{1000\, w_2}{w_1\, \Delta T_b}$$

where ΔT_b is the elevation of the boiling point brought about by the addition of w_2 grams of solute to w_1 grams of solvent and E_b is the ebullioscopic constant. In the column headed "Barometric correction" is the number of degrees for each millimeter of difference between the barometric reading and 760 mmHg to be subtracted from E_b if the pressure is lower, or added if higher, than 760 mm. In general, the effect is within experimental error if the pressure is within 10 mm of 760 mm.

The ebullioscopic constant, a characteristic property of the solvent, may be calculated from the relation:

$$E_b = \frac{RT_b^2 M}{\Delta_{vap} H}$$

where R is the molar gas constant, M is the molar mass of the solvent, and $\Delta_{vap}H$ the molar enthalpy (heat) of vaporization of the solvent.

Compound	Barometric correction	E_b, °C kg · mol^{-1}
Acetic acid	0.0008	3.22
Acetic anhydride		3.79
Acetone	0.0004	1.80
Acetonitrile		1.44
Acetophenone		5.81
Aniline	0.0009	3.82
Benzaldehyde		4.24
Benzene	0.0007	2.64
Benzonitrile		4.02
Bromobenzene	0.0016	6.35
Bromoethane		1.73
1-Butanol		2.17
2-Butanone		2.28
cis-2-Butene-1,4-diol		2.73
D-(+)-Camphor	0.0015	4.91
Carbon disulfide	0.0006	2.42
Carbon tetrachloride	0.0013	5.26
Chlorobenzene	0.0011	4.36
1-Chlorobutane		3.13
Chloroethane		1.77
Chloroform	0.0009	3.80
Cyclohexane	0.0007	2.92
Cyclohexanol		3.5
Decane		6.10
1,2-Dibromomethane	0.0016	6.01
1,1-Dichloroethane		3.13
1,2-Dichloroethane		3.27

TABLE 4.25 Molecular Elevation of the Boiling Point (*Continued*)

Compound	Barometric correction	E_b, °C kg · mol^{-1}
Dichloromethane		2.42
Diethyl ether	0.0005	2.20
Diethyl sulfide		3.14
Dimethoxymethane		2.12
N,N-Dimethylacetamide		3.22
Dimethyl sulfide		1.85
Dimethyl sulfoxide		3.22
1,4-Dioxane		3.00
Ethanol	0.0003	1.22
Ethoxybenzene		4.90
Ethyl acetate	0.0007	2.82
Ethylene glycol		2.26
Formic acid		2.36
Glycerol		6.52
Heptane	0.0008	3.62
Hexane		2.90
2-Hydroxybenzaldehyde		5.87
Iodoethane		5.27
Iodomethane		4.31
4-Isopropyl-1-methylbenzene		5.92
Methanol	0.0002	0.86
Methoxybenzene		4.20
Methyl acetate	0.0005	2.21
N-Methylaniline		4.3
2-Methyl-2-butanol		2.64
3-Methyl-1-butanol		2.88
3-Methylbutyl acetate		4.83
N-Methylformamide		2.2
Methyl formate		1.66
2-Methyl-1-propanol		2.14
2-Methyl-2-propanol		1.99
Naphthalene	0.0014	5.94
Nitrobenzene		5.24
Nitroethane		2.46
Nitromethane		2.09
Octane		4.39
1-Octanol		5.06
Pentyl acetate		4.71
Phenol	0.0009	3.54
Piperidine		3.21
Propanoic acid		3.27
1-Propanol		1.66
2-Propanol		1.58
Propionitrile		1.97
Pyridine		2.83
Pyrrole		2.33
Pyrrolidine		2.32
Quinoline		5.62
Tetrachloroethylene		6.18
Tetrachloromethane		5.26
1,2,3,4-Tetrahydronaphthalene		5.58
Toluene	0.0008	3.40

TABLE 4.25 Molecular Elevation of the Boiling Point (*Continued*)

Compound	Barometric correction	E_b, °C kg · mol^{-1}
p-Toluidine		4.51
Trichloroethylene		4.52
Trichloromethane	0.0009	3.80
1,1,2-Trichloro-1,2,2-trifluoroethane		5.93
Triethylamine		3.57
Water	0.0001	0.512
o-Xylene		4.25

4.11 HEATING BATHS

TABLE 4.26 Substances That Can Be Used for Heating Baths

Medium	Melting point, °C	Boiling point, °C	Useful range, °C	Flash point, °C	Comments
Water	0	100	0–100	None	Ideal
Silicone oil	−50	—	30–250	315	Somewhat viscous at low temperature
Triethylene glycol	−7	285	0–250	165	Noncorrosive
Glycerol	18	290	−20 to 260	160	Water-soluble, nontoxic
Paraffin	50	—	60–300	199	Flammable
Dibutyl *o*-phthalate	−35	340	150–320	171	Generally used

4.12 SEPARATION METHODS

4.12.1 McReynolds' Constants

The *Kovats Retention indices* (R.I.) indicate where compounds will appear on a chromatogram with respect to unbranched alkanes injected with the sample. By definition, the R.I. for pentane is 500, for hexane is 600, for heptane is 700, and so on, regardless of the column used or the operating conditions, although the exact conditions and column must be specified, such as liquid loading, particular support used, and any pretreatment. For example, suppose that on a 20% squalane column at 100°C, the retention times for hexane, benzene, and octane are found to be 15, 16, and 25 min, respectively. On a graph of ln t_R' (naperian logarithm of the adjusted retention time) of the alkanes versus their retention indices, a R.I. of 653 for benzene is read off the graph. The number 653 for benzene means that it elutes halfway between hexane and heptane on a logarithmic time scale. If the experiment is repeated with a dinonyl phthalate column, the R.I for benzene is found to be 736 (lying between heptane and octane), which implies that dinonyl phthalate will retard benzene slightly more than squalane will; that is, dinonyl phthalate is slightly more polar than squalane by $\Delta I = 83$ units. The difference gives a measure of solute-solvent interaction due to all intermolecular forces other than London dispersion forces. The latter are the principal solute-solvent effects with squalane.

TABLE 4.27 Solvents of Chromatographic Interest

Solvent	Boiling point, °C	Solvent strength parameter		Viscosity, $mN \cdot s \cdot m^{-2}$ (20°C)	Refractive index (20°C)	UV cutoff, nm
		$e°$ (SiO_2)	$e°$ (Al_2O_3)			
Fluoroalkanes			−0.25		1.25	
Pentane	36	0.0	0.0	$0.24^{15°C}$	1.358	210
Hexane	69	0.0	0.0	0.31	1.375	210
2,2,4-Trimethylpentane	99		0.01	0.50	1.392	215
Decane	174		0.04	0.93	1.412	210
Cyclohexane	81	−0.05	0.04	0.98	1.426	210
Cyclopentane	49		0.05	0.44	1.407	210
Diisobutylene	101		0.06		1.411	
1-Pentene	30		0.08	$0.24^{0°C}$	1.371	
Carbon disulfide	46	0.14	0.15	0.36	1.626	380
Carbon tetrachloride	77	0.14	0.18	0.97	1.466	265
1-Chlorobutane	78		0.26	0.43	1.402	220
1-Chloropentane	98		0.26	0.58	1.412	225
o-Xylene	144		0.26	0.81	1.505	290
Diisopropyl ether	68		0.28	$0.38^{25°C}$	1.369	220
2-Chloropropane	35		0.29	0.33	1.378	225
Toluene	111		0.29	0.59	1.497	286
1-Chloropropane	47		0.30	0.35	1.389	225
Chlorobenzene	132		0.40	0.80	1.525	
Benzene	80	0.25	0.32	0.65	1.501	280
Bromoethane	38		0.37	0.40	1.424	
Diethyl ether	35	0.38	0.38	0.25	1.353	218
Diethyl sulfide	92		0.38	0.45	1.443	290
Chloroform	62	0.26	0.40	0.57	1.443	245
Dichloromethane	41		0.42	0.44	1.425	235
4-Methyl-2-pentanone	116		0.43	$0.42^{15°C}$	1.396	335
Tetrahydrofuran	66		0.45	0.55	1.407	220
1,2-Dichloroethane	84		0.49	0.80	1.445	228
2-Butanone	80		0.51	$0.42^{15°C}$	1.379	330
1-Nitropropane	131		0.53	$0.80^{25°C}$	1.402	380
Acetone	56	0.47	0.56	0.32	1.359	330

1,4-Dioxane	101	0.49	0.56	$1.44^{15°C}$	1.420	215
Ethyl acetate	77	0.38	0.58	0.45	1.372	255
Methyl acetate	56		0.60	$0.48^{15°C}$	1.362	260
1-Pentanol	138		0.61	4.1	1.410	210
Dimethyl sulfoxide	189		0.62	2.47	1.478	265
Aniline	184		0.62	4.40	1.586	
Diethylamine	56		0.63	0.33	1.386	275
Nitromethane	101		0.64	0.67	1.394	380
Acetonitrile	82	0.50	0.65	0.37	1.344	190
Pyridine	115		0.71	0.97	1.510	330
2-Butoxyethanol	170		0.74	$3.15^{25°C}$	1.420	220
1-Propanol	97		0.82	2.25	1.386	210
2-Propanol	82		0.82	2.50	1.377	210
Ethanol	78		0.88	1.20	1.361	210
Methanol	65		0.95	0.59	1.328	210
Ethylene glycol	198		1.11	21.8	1.432	210
Acetic acid	118		large	1.23	1.372	260
Water	100		large	1.00	1.333	191

TABLE 4.28 McReynolds' Constants for Stationary Phases in Gas Chromatography

Stationary phase	Chemical type	Similar stationary phases	Temp., °C		McReynolds' constants						USP code
			Min	Max	*x*'	*y*'	*z*'	*u*'	*s*'	Σ	
Boiling-point separation of broad molecular weight range of compounds; nonpolar phases											
Squalane	2,6,10,15,19,23-Hexa-methyltetracosane		20	150	0	0	0	0	0	0	
Paraffin oil					9	5	2	6	11	33	
Apiezon® L			50	300	32	22	15	32	42	143	
SPB-1	Poly(dimethylsiloxane)	SA-1, DB-1	−60	320	4	58	43	56	38	199	
SP™-2100	Poly(dimethylsiloxane)	DC-200, SE 30, UC W98, DC 200	0	350	17	57	45	67	43	229	G 9
OV-1	Methylsiloxane gum		100	350	16	55	44	65	42	227	G 2
OV-101	Methylsiloxane fluid		20	350	17	57	45	67	43	234	G 1
SPB-5	1% Vinyl, 5% phenyl methyl polysiloxane	SA-5, DB-5	−60	320	19	74	64	93	62	312	
SE-54	1% Vinyl, 5% phenyl methyl polysiloxane	PTE-5	50	300	19	74	64	93	62	312	G 36
SE-52	5% Phenyl methyl polysiloxane		50	300	32	72	65	98	67	334	G 27
OV-73	5.5% Phenyl methyl polysiloxane	SP-400	0	325	40	86	76	114	85	401	G 27
OV-3	Poly(dimethyldiphenyl-siloxane); 90%:10%		0	350	44	86	81	124	88	423	
Dexsil® 300	Carborane—methyl silicone		50	450	47	80	103	148	96	474	G 33
Dexsil® 400	Carborane—methyl-phenyl silicone		50	400	72	108	118	166	123	587	

OV-7	20% Phenyl methyl polysiloxane	DC 550	0	350	69	113	111	171	128	592	
SPB-20	20% Phenyl methyl polysiloxane	SPB-35, SPB-1701, DB-1301	<20	300	67	116	117	174	131	605	
Di-(2-ethylhexyl)-sebacate			−20	125	72	168	108	180	125	653	G 11
DC 550	25% Phenyl methyl polysiloxane		20	225	81	124	124	189	145	663	G 28
Unsaturated hydrocarbons and other compounds of intermediate polarity											
Diisodecyl phthalate			20	150	84	173	137	218	155	767	G 24
OV-11	35% Phenyl methyl polysiloxane		0	350	102	142	145	219	178	786	
OV-1701	Vinyl methyl polysiloxane	SPB-1701, SA-1701, DB-1701	0	250	67	170	152	228	171	789	
Poly-I 110				275	115	194	122	204	202	837	G 37
SP-2250	Poly(phenylmethylsiloxane); 50% phenyl	OV-17, DB-17	0	375	119	158	162	243	202	884	G 3
Dexsil® 410	Carborane—methylcyano ethyl silicone		50	400	72	286	174	249	171	952	
UCON® LB-550-X	Polyalkylene glycol		20	200	118	271	158	243	206	996	
UCON LB-1880-X	Polyalkylene glycol			200	123	275	161	249	212	1020	G 18
Poly-A 103				275	115	331	144	263	214	1072	G 10
OV-22	Poly(diphenyldimethylsiloxane); 65%:35%		0	350	160	188	191	283	253	1075	
Di(2-ethylhexyl) phthalate				150	135	254	213	320	235	1157	G 22
OV-25	Poly(diphenyldimethylsiloxane); 75%:25%		0	350	178	204	208	305	280	1175	G 17

(*Continued*)

TABLE 4.28 McReynolds' Constants for Stationary Phases in Gas Chromatography (*Continued*)

Stationary phase	Chemical type	Similar stationary phases	Temp., °C		McReynolds' constants						USP code
			Min	Max	*x'*	*y'*	*z'*	*u'*	*s'*	Σ	
			Moderately polar compounds								
DC QF-1			0	250	144	233	355	463	305	1500	
OV-210	50% Trifluoropropyl-methylpolysiloxane	SP-2401, DB-210	0	275	146	238	358	468	310	1520	G 6
OV-215	Poly(trifluoropropyl-methylsiloxane)		0	275	149	240	363	478	315	1545	
UCON-50-HB-2000	Polyalkylene glycol		0	200	202	394	253	392	341	1582	
Triton® X-100	Octylphenoxy poly-ethoxy ethanol		0	190	203	399	268	402	362	1634	
UCON 50-HB-5100	Polyglycol		0	200	214	418	278	421	375	1706	
XE-60	Poly(cyanoethylphenyl-methylsiloxane)		0	250	204	381	340	493	367	1785	G 26
OV-225	25% Cyanopropyl 25% phenyl methyl polysiloxane	DB-225, DB-23	0	265	228	369	338	492	386	1813	G 19
Ipegal CO-880	Nonylphenoxypoly-(ethyleneoxy)ethanol		100	200	259	461	311	482	426	1939	G 31
Triton® X-305	Octylphenoxy poly-ethoxy ethanol		200	250	262	467	314	488	430	1961	
			Polar compounds								
Hi-EFF-3BP	Neopentylglycol succinate		50	230	272	469	366	539	474	2120	G 21
Carbowax 20M-TPA	Polyethyleneglycol + terephthalic acid		60	250	321	367	368	573	520	2149	G 25

Supelcowax™ 10	Polyethyleneglycol + terephthalic acid	DB-WAX, SA-WAX	50	280	305	551	360	562	484	2262	
SP-1000	Polyethyleneglycol + terephthalic acid		60	220	304	552	359	549	498	2262	
Carbowax 20M	Polyethyleneglycol	SP-2300	25	275	322	536	368	572	510	2308	G 16
Nukol™		SP-1000, FFAP, OV-351			311	572	374	572	520	2349	
Carbowax 3350		Formerly Carbowax 4000	60	200	325	551	375	582	520	2353	G 15
OV-351	Polyethyleneglycol + nitroterephthalic acid	SP-1000	50	270	335	552	382	583	540	2392	
SP-2300	36% Cyanopropyl		25	275	316	495	446	637	530	2424	
Silar 5 CP	50% Cyanopropyl phenyl silicone	SP-2300	0	250	319	495	446	637	531	2428	G 7
FFAP			50	250	340	580	397	602	627	2546	G 35
Hi-EFF-10BP	Phenyldiethanolamine succinate		20	230	386	555	472	674	656	2744	G 21
Carbowax 1450		Formerly 1540	50	175	371	639	453	666	641	2770	G 14
SP-2380					402	629	520	744	623	2918	
SP-2310	55% Cyanopropyl	Silar 7 CP	25	275	440	637	605	840	670	3192	
SP-2330	68% Cyanopropyl	SP-2331, SH-60	25	275	490	725	630	913	778	3536	
Silar 9 CP	90% Cyanopropyl phenyl		50	250	489	725	631	913	778	3536	G 8
Hi-EFF-1BP	Diethyleneglycol succinate		20	200	499	751	593	840	860	3543	G 4
SP-2340	75% Cyanopropyl phenyl	OV-275, SH-80	<25	275	520	757	659	942	800	3678	
Silar 10 CP	100% Cyanopropyl silicone	SP-2340	25	275	523	757	659	942	801	3682	G 5
THEED	Amino alcohol		0	125	463	942	626	801	893	3725	
OV-275	Dicyanoallylsilicone		25	250	629	872	763	110	849	4219	
Absolute index values on squalane for reference compounds:				653	590	627	652		699		

Note: USP code is the United States Pharmacopeia designation.

Now the overall effects due to hydrogen bonding, dipole moment, acid-base properties, and molecular configuration can be expressed as

$$\sum \Delta I = ax' + by' + cz' + du' + es'$$

where $x' = \Delta I$ for benzene, $y' = \Delta I$ for 1-butanol, $z' = \Delta I$ for 2-pentanone, $u' = \Delta I$ for 1-nitropropane, and $s' = \Delta I$ for pyridine (or dioxane).

4.12.2 Chromatographic Behavior of Solutes

Retention Behavior. On a chromatogram the distance on the time axis from the point of sample injection to the peak of an eluted component is called the *uncorrected retention time* t_R. The corresponding retention volume is the product of retention time and flow rate, expressed as volume of mobile phase per unit time:

$$V_R = t_R F_c$$

The *average linear velocity* u of the mobile phase in terms of the column length L and the average linear velocity of eluent t_M (which is measured by the transit time of a nonretained solute) is

$$u = \frac{L}{t_M}$$

The *adjusted retention time* t'_R is given by

$$t'_R = t_R - t_M$$

When the mobile phase is a gas, a *compressibility factor* j must be applied to the adjusted retention volume to give the *net retention volume*:

$$V_N = jV'_R$$

The compressibility factor is expressed by

$$j = \frac{3}{2}\frac{[(P_i/P_o)^2 - 1]}{[(P_i/P_o)^3 - 1]}$$

where P_i is the carrier gas pressure at the column inlet and P_o that at the outlet.

Partition Ratio. The partition ratio is the additional time a solute band takes to elute, as compared with an unretained solute (for which $k' = 0$), divided by the elution time of an unretained band:

$$k' = \frac{t_R - t_M}{t_M} = \frac{V_R - V_M}{V_M}$$

Retention time may be expressed as

$$t_R = t_M(1 + k') = \frac{L}{u}(1 + k')$$

Relative Retention. The relative retention α of two solutes, where solute 1 elutes before solute 2, is given variously by

$$\alpha = \frac{k_2'}{k_1'} = \frac{V_{R,2}'}{V_{R,1}'} = \frac{t_{R,2}'}{t_{R,1}'}$$

The relative retention is dependent on (1) the nature of the stationary and mobile phases and (2) the column operating temperature.

Column Efficiency. Under ideal conditions the profile of a solute band resembles that given by a Gaussian distribution curve (Fig. 4.1). The efficiency of a chromatographic system is expressed by the effective plate number N_{eff}, defined from the chromatogram of a single band,

$$N_{\text{eff}} = \frac{L}{H} = 16\left(\frac{t_R'}{W_b}\right)^2 = 5.54\left(\frac{t_R'}{W_{1/2}}\right)^2$$

where L is the column length, H is the plate height, t_R' is the adjusted time for elution of the band center, W_b is the width at the base of the peak ($W_b = 4\sigma$) as determined from the intersections of tangents to the inflection points with the baseline, and $W_{1/2}$ is the width at half the peak height. Column efficiency, when expressed as the number of theoretical plates N_{theor} uses the uncorrected retention time in the foregoing expression. The two column efficiencies are related by

$$N_{\text{eff}} = N_{\text{theor}}\left(\frac{k'}{k'+1}\right)^2$$

Band Asymmetry. The peak asymmetry factor AF is often defined as the ratio of peak half-widths at 10% of peak height, that is, the ratio b/a, as shown in Fig. 4.2. When the asymmetry ratio lies outside the range 0.95–1.15 for a peak of $k' = 2$, the effective plate number should be calculated from the expression

$$N = \frac{41.7(t_R'/W_{0.1})}{(a/b)+1.25}$$

Resolution. The degree of separation or resolution, Rs, of two adjacent peaks is defined as the distance between band peaks (or centers) divided by the average bandwidth using W_b, as shown in Fig. 4.3.

$$\text{Rs} = \frac{t_{R,2} - t_{R,1}}{0.5(W_2 + W_1)}$$

For reasonable quantitative accuracy, peak maxima must be at least 4σ apart. If so, then Rs = 1.0, which corresponds approximately to a 3% overlap of peak areas. A value of Rs = 1.5 (for 6σ) represents essentially complete resolution with only 0.2% overlap of peak areas. These criteria pertain to roughly equal solute concentrations.

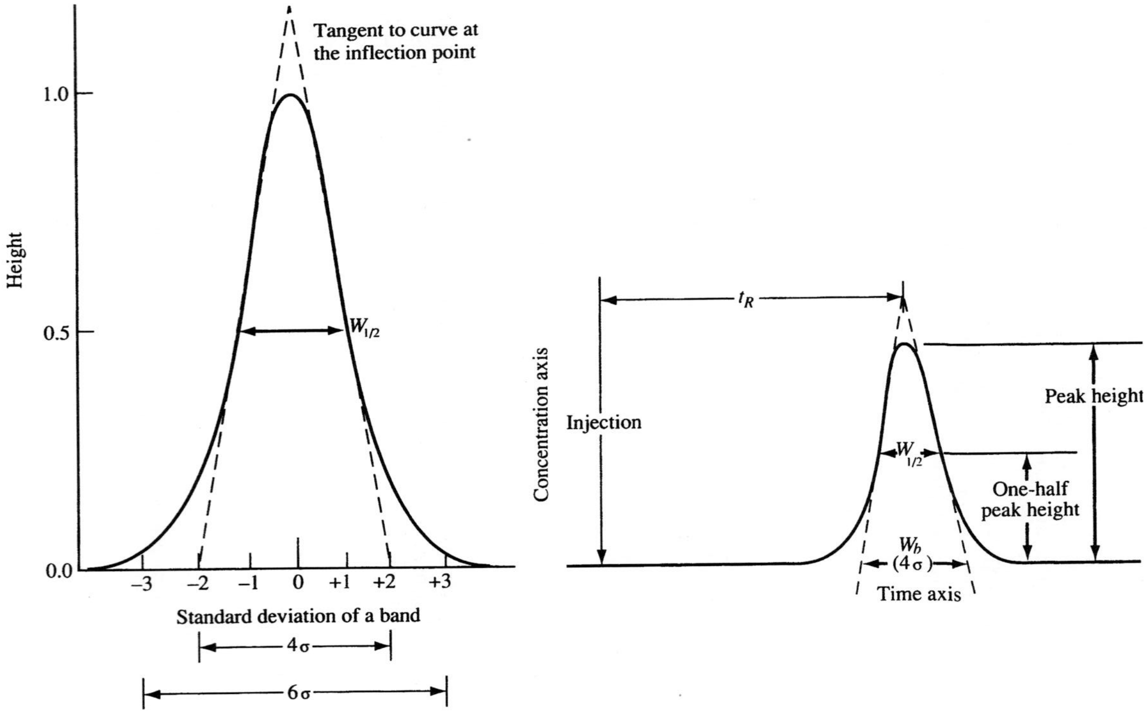

Profile of a solute band.

FIGURE 4.1

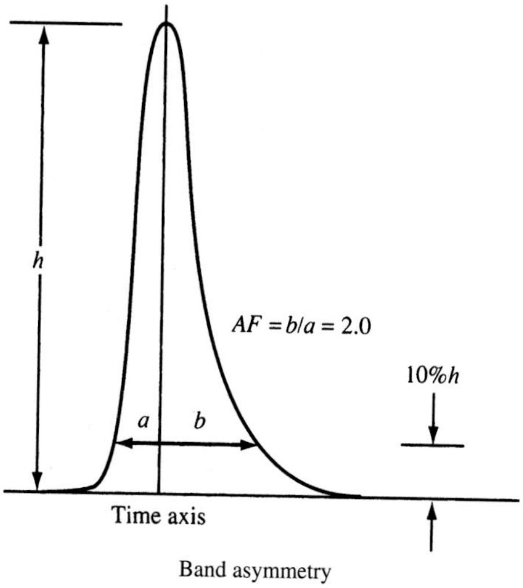

FIGURE 4.2

The fundamental resolution equation incorporates the terms involving the thermodynamics and kinetics of the chromatographic system:

$$\mathrm{Rs} = \frac{1}{4}\left(\frac{\alpha - 1}{\alpha}\right)\left(\frac{k'}{1+k'}\right)\left(\frac{L}{H}\right)^{1/2}$$

Three separate factors affect resolution: (1) a column selectivity factor that varies with α, (2) a capacity factor that varies with k' (taken usually as k_2), and (3) an efficiency factor that depends on the theoretical plate number.

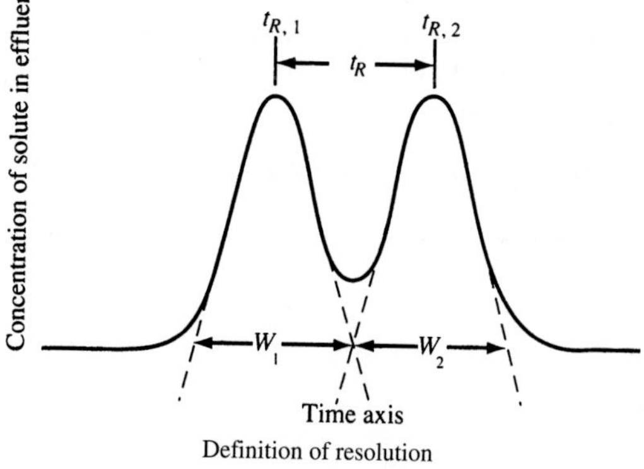

FIGURE 4.3

TABLE 4.29 Characteristics of Selected Supercritical Fluids

Fluid	Critical temperature, K (°C)	Critical pressure, atm (psi)
Ammonia	406 (133)	111.3 (1636)
Argon	151 (−122)	48.1 (707)
Benzene	562 (289)	48.3 (710)
Butane	425 (125)	37.5 (551)
Carbon dioxide	304 (31)	72.8 (1070)
Carbon disulfide	552 (279)	78.0 (1147)
Chlorotrifluoromethane	379 (106)	40 (588)
2,2-Dimethylpropane	434 (161)	31.6 (464)
Ethane	305 (32)	48.2 (706)
Fluoromethane	318 (45)	58.0 (853)
Heptane	540 (267)	27.0 (397)
Hexane	507 (234)	29.3 (431)
Hydrogen sulfide	373 (100)	88.2 (1296)
Krypton	209 (−64)	54.3 (798)
Methane	191 (−82)	45.4 (667)
Methanol	513 (240)	79.9 (1175)
2-Methylpropane	408 (65)	36.0 (529)
Nitrogen	126 (−147)	33.5 (492)
Nitrogen(I) oxide	310 (37)	71.5 (1051)
Pentane	470 (197)	33.3 (490)
Propane	470 (197)	41.9 (616)
Sulfur dioxide	431 (158)	77.8 (1144)
Sulfur hexafluoride	319 (46)	37.1 (545)
Trichloromethane	536 (263)	54.9 (807)
Trifluoromethane	299 (26)	47.7 (701)
Water	647 (374)	217.6 (3199)
Xenon	290 (17)	57.6 (847)

Time of Analysis. The retention time required to perform a separation is given by

$$t_R = 16\text{Rs}^2\left(\frac{\alpha}{\alpha-1}\right)^2\left[\frac{(1+k')^3}{(k')^2}\right]\left(\frac{H}{u}\right)$$

Now t_R is a minimum when $k' = 2$, that is, when $t_R = 3t_M$. There is little increase in analysis time when k' lies between 1 and 10. A twofold increase in the mobile-phase velocity roughly halves the analysis time (actually it is the ratio H/u which influences the analysis time). The ratio H/u can be obtained from the experimental plate height/velocity graph.

High-Performance Liquid Chromatography. Typical performances for various experimental conditions are given in Table 4.30. The data assume these reduced parameters: $h = 3$, $\nu = 4.5$. The *reduced plate height* is

$$h = \frac{H}{d_p} = \frac{L}{Nd_p}$$

The *reduced velocity* of the eluent is

$$\nu = \frac{ud_p}{D_M} = \frac{Ld_p}{t_M D_M}$$

TABLE 4.30 Typical Performances in HPLC for Various Conditions

Performances		Column parameters		
N	t_M, s	L, cm	d_p, μm	P, atm (psi)
2 500	30	2.3	3	18.4 (270)
2 500	30	3.7	5	18.4 (270)
2 500	30	7.5	10	18.4 (270)
5 000	30	4.5	3	74 (1088)
5 000	30	7.5	5	74 (1088)
5 000	30	15.0	10	74 (1088)
10 000	30	9.0	3	300 (4410)
10 000	30	15.0	5	300 (4410)
10 000	30	30.0	10	300 (4410)
10 000	30	9.0	3	300 (4410)
10 000	60	9.0	3	150 (2200)
10 000	90	9.0	3	100 (1470)
15 000	90	2.3	3	223 (3275)
15 000	120	2.3	3	167 (2459)
11 100	30	10.0	3	369 (5420)
11 100	37	10.0	3	300 (4410)
11 100	101	10.0	3	100 (1470)
27 800	231	25.0	3	300 (4410)

Assumed reduced parameters: $h = 3$, $v = 4.5$. These are optimum values from a graph of reduced plate height versus reduced linear velocity of the mobile phase.

In these expressions, d_p is the particle diameter of the stationary phase that constitutes one plate height. D_M is the diffusion coefficient of the solute in the mobile phase.

4.12.3 Ion-Exchange (Normal Pressure, Columnar)

Ion-exchange methods are based essentially on a reversible exchange of ions between an external liquid phase and an ionic solid phase. the solid phase consists of a polymeric matrix, insoluble, but permeable, which contains fixed charge groups and mobile counter ions of opposite charge. These counter ions can be exchanged for other ions in the external liquid phase. Enrichment of one or several of the components is obtained if selective exchange forces are operative. The method is limited to substances at least partially in ionized form.

Chemical Structure of Ion-Exchange Resins. An ion-exchange resin usually consists of polystyrene copolymerized with divinylbenzene to build up an inert three-dimensional, cross-linked matrix of hydrocarbon chains. Protruding from the polymer chains are the ion-exchange sites distributed statistically throughout the entire resin particle. The ionic sites are balanced by an equivalent number of mobile counter ions. The type and strength of the exchanger is determined by these active groups. Ion-exchangers are designated anionic or cationic, according to whether they have an affinity for negative or positive counter ions. Each main group is further subdivided into strongly or weakly ionized groups. A selection of commercially available ion-exchange resins is given in Table 4.31.

The cross-linking of a polystyrene resin is expressed as the proportion by weight percent of divinylbenzene in the reaction mixture; for example, "× 8" for 8 percent cross-linking. As the percentage is increased, the ionic groups come into effectively closer proximity, resulting in increased selectivity. Intermediate cross-linking, in the range of 4 to 8 percent, is usually used. An increase in cross-linking decreases the diffusion rate in the resin particles; the diffusion rate is the rate-controlling step in column operations. Decreasing the particle size reduces the time required for attaining equilibrium, but at the same time decreases the flow rate until it is prohibitively slow unless pressure is applied.

In most inorganic chromatography, resins of 100 to 200 mesh size are suitable; difficult separations may require 200 to 400 mesh resins. A flow rate of 1 $mL \cdot cm^{-2} \cdot min^{-1}$ is often satisfactory. With HPLC columns, the flow rate in long columns of fine adsorbent can be increased by applying pressure.

Macroreticular Resins. Macroreticular resins are an agglomerate of randomly packed microspheres which extend through the agglomerate in a continuous non-gel pore structure. The channels throughout the rigid pore structure render the bead centers accessible even in nonaqueous solvents, in which microreticular resins do not swell sufficiently. Because of their high porosity and large pore diameters, these resins can handle large organic molecules.

Microreticular Resins. Microreticular resins, by contrast, are elastic gels that, in the dry state, avidly absorb water and other polar solvents in which they are immersed. While taking up solvent, the gel structure expands until the retractile stresses of the distended polymer network balance the osmotic effect. In nonpolar solvents, little or no swelling occurs and diffusion is impaired.

Ion-Exchange Membranes. Ion-exchange membranes are extremely flexible, strong membranes, composed of analytical grade ion-exchange resin beads (90%) permanently enmeshed in a poly(tetrafluoroethylene) membrane (10%). The membranes offer an alternative to column and batch methods, and can be used in many of the same applications as traditional ion exchange resins. Three ion-exchange resin types have been incorporated into membranes: AG 1-X8, AG 50W-X8, and Chelex 100.

Functional Groups. *Sulfonate exchangers* contain the group SO_3^-, which is strongly acidic and completely dissociated whether in the H form or the cation form. These exchangers are used for cation exchange.

Carboxylate exchangers contain —COOH groups which have weak acidic properties and will only function as cation exchangers when the pH is sufficiently high (pH > 6) to permit complete dissociation of the —COOH site. Outside this range the ion exchanger can be used only at the cost of reduced capacity.

TABLE 4.31 Ion-Exchange Resins

Dowex is the trade name of Dow resins; X (followed by a numeral) is percent cross-linked. Mesh size (dry) are available in the range 50 to 100, 100 to 200, 200 to 400, and sometimes minus 400.

S-DVB is the acronym for styrene-divinylbenzene.

MP is the acronym for macroporous resin. Mesh size (dry) is available in the range 20 to 50, 100 to 200, and 200 to 400.

Bio-Rex is the trade name for certain resins sold by Bio-Rad Laboratories.

Amberlite and Duolite are trade names of Rohm & Haas resins.

Resin type and nominal percent cross-linkage	Minimum wet capacity, mequiv · mL^{-1}	Density (nominal), g · mL^{-1}	Comments
Anion exchange resins—gel type—strongly basic—quaternary ammonium functionality			
Dowex 1-X2	0.6	0.65	Strongly basic anion exchanger with S-DVB matrix for separation of small peptides, nucleotides, and large metal complexes. Molecular weight exclusion is <2700.
Dowex 1-X4	1.0	0.70	Strongly basic anion exchanger with S-DVB matrix for separation of organic acids, nucleotides, phosphoinositides, and other anions. Molecular weight exclusion is <1400.
Dowex 1-X8	1.2	0.75	Strongly basic anion exchanger with S-DVB matrix for separation of inorganic and organic anions with molecular weight exclusion <1000. 100–200 mesh is standard for analytical separations.
Dowex 2-X8	1.2	0.75	Strongly basic (but less basic than Dowex 1 type) anion exchanger with S-DVB matrix for deionization of carbohydrates and separation of sugars, sugar alcohols, and glycosides.
Amberlite IRA-400	1.4	1.11	8% cross-linkage. Used for systems essentially free of organic materials.
Amberlite IRA-402	1.3	1.07	Lower cross-linkage than IRA-400; better diffusion rate with large organic molecules.
Amberlite IRA-410	1.4	1.12	Dimethylethanolamine functionality and slightly lower basicity than IRA-400.
Amberlite IRA-458	1.2	1.08	Has an acrylic structure rather than S-DVB; hence more hydrophilic and resistant to organic fouling.
Anion exchange resin–gel type—intermediate basicity			
Bio-Rex 5	2.8	0.70	Intermediate basic anion exchanger with primarily tertiary amines on a polyalkyleneamine matrix for separation of organic acids.

(Continued)

TABLE 4.31 Ion-Exchange Resins (*Continued*)

Resin type and nominal percent cross-linkage	Minimum wet capacity, mequiv · mL^{-1}	Density (nominal), g · mL^{-1}	Comments
Anion exchange resins—gel type—weakly basic—polyamine functionality			
Dowex 4-X4	1.6	0.70	Weakly basic anion exchanger with tertiary amines on an acrylic matrix for the deionization of carbohydrates. Use at pH <7.
Amberlite IRA-68	1.6	1.06	Acrylic-DVB with unusually high capacity for large organic molecules.
Cation exchange resins—gel type—strongly acidic—sulfonic acid functionality			
Dowex 50W-X2	0.6	0.70	Strongly acidic cation exchanger with S-DVB matrix for separation of peptides, nucleotides, and cations. Molecular weight exclusion <2700.
Dowex 50W-X4	1.1	0.80	Strongly acidic cation exchanger with S-DVB matrix for separation of amino acids, nucleosides, and cations. Molecular weight exclusion is <1400.
Dowex 50W-X8	1.7	0.80	Strongly acidic cation exchanger with S-DVB matrix for separation of amino acids, metal cations, and cations. Molecular weight exclusion is <1000. 100–200 mesh is standard for analytical applications.
Dowex 50W-X12	2.1	0.85	Strongly acidic cation exchanger with S-DVB matrix used primarily for metal separations.
Dowex 50W-X16	2.4	0.85	Strongly acidic cation exchanger with S-DVB matrix and high cross linkage.
Amberlite IR-120	1.9	1.26	8% styrene-DVB type; high physical stability.
Amberlite IR-122	2.1	1.32	10% styrene-DVB type; high physical stability and high capacity.
Weakly acidic cation exchangers—gel type—carboxylic acid functionality			
Duolite C-433	4.5	1.19	Acrylic-DVB type; very high capacity. Used for metals removal and neutralization of alkaline solutions.
Bio-Rex 70	2.4	0.70	Weakly acidic cation exchanger with carboxylate groups on a macroreticular acrylic matrix for separation and fractionation of proteins, peptides, enzymes, and amines, particularly high molecular weight solutes. Does not denature proteins as do styrene-based resins.

TABLE 4.31 Ion-Exchange Resins (*Continued*)

Resin type and nominal percent cross-linkage	Minimum wet capacity, mequiv · mL^{-1}	Density (nominal), g · mL^{-1}	Comments
Selective ion exchange resins			
Duolite GT-73	1.3	1.30	Removal of Ag, Cd, Cu, Hg, and Pb.
Amberlite IRA-743A	0.6	1.05	Boron specific ion exchange resin.
Amberlite IRC-718	1.0	1.14	Removal of transition metals.
Chelex® 100	0.4	0.65	Weakly acidic chelating resin with S-DVB matrix for heavy metal concentration.
Anion exchanger—macroreticular type—strongly basic—quaternary ammonium functionality			
Amberlite IRA-910	1.1	1.09	Dimethylethanolamine styrene-DVB type which offers slightly less silica removal than Amberlite IRA resin, but offers improved regeneration efficiency.
Amberlite IRA-938	0.5	1.20	Pore size distribution between 2500 and 23000 nm; suitable for removal of high molecular weight organic materials.
Amberlite IRA-958	0.8		Acrylic-DVB; resistant to organic fouling.
AG MP-1	1.0	0.70	Strongly basic macroporous anion exchanger with S-DVB matrix for separation of some enzymes, radioactive anions, and other applications.
Cation exchange resin—macroreticular type—sulfonic acid functionality			
Amberlite 200	1.7	1.26	Styrene-DVB with 20% DVB by weight; superior physical stability and greater resistance to oxidation by factor of three over comparable gel type resin.
AG MP-50	1.5	0.80	Strongly acidic macroporous cation exchanger with S-DVB matrix for separation of radioactive cations and other applications.
Weak cation exchanger—macroreticular type—carboxylic acid or phenolic functionality			
Amberlite DP-1	2.5	1.17	Methacrylic acid-DVB; high resin capacity. Use pH >5.
Amberlite IRC-50	3.5	1.25	Methacrylic acid-DVB. Selectivity adsorbs organic gases such as antibiotics, alkaloids, peptides, and amino acids. Use pH >5.
Duolite C-464	3.0	1.13	Polyacrylic resin with high capacity and outstanding resistance to osmotic shock.

(*Continued*)

TABLE 4.31 Ion-Exchange Resins (*Continued*)

Resin type and nominal percent cross-linkage	Minimum wet capacity, mequiv · mL^{-1}	Density (nominal), g · mL^{-1}	Comments
Weak cation exchanger—macroreticular type—carboxylic acid or phenolic functionality (*continued*)			
Duolite A-7	2.2	1.12	Phenolic type resin. High porosity and hydrophilic matrix. pH range is 0 to 6.
Duolite A-368	1.7	1.04	Styrene-DVB; pH range is 0 to 9.
Amberlite IRA-35	1.1		Acrylic-DVB; pH range is 0 to 9.
Amberlite IRA-93	1.3	1.04	Styrene-DVB; pH range is 0 to 9. Excellent resistance to oxidation and organic fouling.
Liquid amines			
Amberlite LA-1			A secondary amine containing two highly branched aliphatic chains of M.W. 351 to 393. Solubility is 15 to 20 mg/mL in water. Used as 5 to 40% solutions in hydrocarbons.
Amberlite LA-2			A secondary amine of M.W. 353 to 395. Insoluble in water.
Microcrystalline exchanger			
AMP-1	4.0		Microcrystalline ammonium molybdophosphate with cation exchange capacity of 1.2 mequiv/g. Selectively adsorbs larger alkali metal ions from smaller alkali metal ions, particularly cesium.
Ion retardation resin			
AG 11 A8		0.70	Ion retardation resin containing paired anion (COO^-) and cation $(CH_3)_3N^+$ sites. Selectively retards ionic substances.

Source: J.A. Dean, ed., *Analytical Chemistry Handbook*, McGraw-Hill, New York, 1995.

Quaternary ammonium exchangers contain $—R_4N^+$ groups which are strongly basic and completely dissociated in the OH form and the anion form.

Tertiary amine exchangers possess $—R_3NH_2$ groups which have exchanging properties only in an acidic medium when a proton is bound to the nitrogen atom.

Aminodiacetate exchangers have the $—N(CH_2COOH)_2$ group which has an unusually high preference for copper, iron, and other heavy metal cations and, to a lesser extent, for alkaline earth cations. The resin selectivity for divalent over monovalent ions is approximately 5000 to 1. The resin functions as a chelating resin at pH 4 and above. At very low pH, the resin acts as an anion exchanger. This exchanger is the column packing often used for ligand exchange.

Ion-Exchange Equilibrium. Retention differences among cations with an anion exchanger, or among anions with a cation exchanger, are governed by the physical properties of the solvated ions. The stationary phase will show these preferences:

1. The ion of higher charge.
2. The ion with the smaller solvated radius. Energy is needed to strip away the solvation shell surrounding ions with large hydrated radii, even though their crystallographic ionic radii may be less than the average pore opening in the resin matrix.
3. The ion that has the greater polarizability (which determines the Van der Waals' attraction).

To accomplish any separation of two cations (or two anions) of the same net charge, the stationary phase must show a preference for one more than the other. No variation in the eluant concentration will improve the separation. However, if the exchange involves ions of different net charges, the separation factor does depend on the eluant concentration. The more dilute the counterion concentration in the eluant, the more selective the exchange becomes for polyvalent ions.

In the case of an ionized resin, initially in the H-form and in contact with a solution containing K^+ ions, an equilibrium exists:

$$\text{resin, } H^+ + K^+ \leftrightarrows \text{resin, } K^+ + H^+$$

which is characterized by the selectivity coefficient, $k_{K/H}$:

$$k_{K/H} = \frac{[K^+]_r[H^+]}{[H^+]_r[K^+]}$$

where the subscript r refers to the resin phase. Table 4.32 contains selectivity coefficients for cations and Table 4.33 for anions. Relative selectivities are of limited use for the prediction of the columnar exchange behavior of a cation because they do not take account of the influence of the aqueous phase. More specific information about the behavior to be expected from a cation in a column elution experiment is given by the equilibrium distribution coefficient K_d.

TABLE 4.32 Relative Selectivity of Various Counter Cations

Counterion	Relative selectivity for AG 50W-X8 resin	Counterion	Relative selectivity for AG 50W-X8 resin
H^+	1.0	Zn^{2+}	2.7
Li^+	0.86	Co^{2+}	2.8
Na^+	1.5	Cu^{2+}	2.9
NH_4^+	1.95	Cd^{2+}	2.95
K^+	2.5	Ni^{2+}	3.0
Rb^+	2.6	Ca^{2+}	3.9
Cs^+	2.7	Sr^{2+}	4.95
Cu^+	5.3	Hg^{2+}	7.2
Ag^+	7.6	Pb^{2+}	7.5
??	10.7	Ba^{2+}	8.7
Mn^{2+}	2.35	Ce^{3+}	22
Mg^{2+}	2.5	La^{3+}	22
Fe^{2+}	2.55		

TABLE 4.33 Relative Selectivity of Various Counter Anions

Counterion	Relative selectivity for Dowex 1-X8 resin	Relative selectivity for Dowex 2-X8 resin
OH^-	1.0	1.0
Benzenesulfonate$^-$	500	75
Salicylate$^-$	450	65
Citrate	220	23
I^-	175	17
Phenate$^-$	110	27
HSO_4^-	85	15
ClO_3^-	74	12
NO_3^-	65	8
Br^-	50	6
CN^-	28	3
HSO_3^-	27	3
BrO_3^-	27	3
NO_2^-	24	3
Cl^-	22	2.3
ClO_4^-	20	
SCN^-	8.0	
HCO_3^-	6.0	1.2
IO_3^-	5.5	0.5
$H_2PO_4^-$	5.0	0.5
Formate$^-$	4.6	0.5
Acetate$^-$	3.2	0.5
Propanoate$^-$	2.6	0.3
F^-	1.6	0.3

The partitioning of the potassium ion between the resin and solution phases is described by the concentration distribution ratio, D_c:

$$(D_c)_K = \frac{[K^+]_r}{[K^+]}$$

Combining the equations for the selectivity coefficient and for D_c:

$$(D_c)_K = k_{K/H}\frac{[H^+]_r}{[H^+]}$$

The foregoing equation reveals that essentially the concentration distribution ratio for trace concentrations of an exchanging ion is independent of the respective solution of that ion and that the uptake of each trace ion by the resin is directly proportional to its solution concentration. However, the concentration distribution ratios are inversely proportional to the solution concentration of the resin counterion.

To accomplish any separation of two cations (or two anions), one of these ions must be taken up by the resin in distinct preference to the other. This preference is expressed by the separation factor (or relative retention), $\alpha_{K/Na}$, using K^+ and Na^+ as the example:

$$\alpha_{K/Na} = \frac{(D_c)_K}{(D_c)_{Na}} = \frac{k_{K/H}}{k_{Na/H}} = K_{K/Na}$$

The more α deviates from unity for a given pair of ions, the easier it will be to separate them. If the selectivity coefficient is unfavorable for the separation of two ions of the same charge, no variation in the concentration of H^+ (the eluant) will improve the separation.

The situation is entirely different if the exchange involves ions of different net charges. Now the separation factor does depend on the eluant concentration. For example, the more dilute the counterion concentration in the eluant, the more selective the exchange becomes for the ion of higher charge.

In practice, it is more convenient to predict the behavior of an ion, for any chosen set of conditions, by employing a much simpler distribution coefficient, D_g, which is defined as the concentration of a solute in the resin phase divided by its concentration in the liquid phase, or:

$$D_g = \frac{\text{concentration of solute, resin phase}}{\text{concentration of solute, liquid phase}}$$

$$D_g = \frac{\%\text{ solute within exchanger}}{\%\text{ solute within solution}} \times \frac{\text{volume of solution}}{\text{mass of exchanger}}$$

D_g remains constant over a wide range of resin to liquid ratios. In a relatively short time, by simple equilibration of small known amounts of resin and solution followed by analysis of the phases, the distribution of solutes may be followed under many different sets of experimental conditions. Variables requiring investigation include the capacity and percent cross-linkage of resin, the type of resin itself, the temperature, and the concentration and pH of electrolyte in the equilibrating solution.

By comparing the ratio of the distribution coefficients for a pair of ions, a separation factor (or relative retention) is obtained for a specific experimental condition.

Instead of using D_g, separation data may be expressed in terms of a volume distribution coefficient D_v, which is defined as the amount of solution in the exchanger per cubic centimeter of resin bed divided by the amount per cubic centimeter in the liquid phase. The relation between D_g and D_v is given by:

$$D_v = D_g \rho$$

where ρ is the bed density of a column expressed in the units of mass of dry resin per cubic centimeter of column. The bed density can be determined by adding a known weight of dry resin to a graduated cylinder containing the eluting solution. After the resin has swelled to its maximum, a direct reading of the settled volume of resin is recorded.

Intelligent inspection of the relevant distribution coefficients will show whether a separation is feasible and what the most favorable eluant concentration is likely to be. In the columnar mode, an ion, even if not eluted, may move down the column a considerable distance and with the next eluant may appear in the eluate much earlier than indicated by the coefficient in the first eluant alone. A distribution coefficient value of 12 or lower is required to elute an ion completely from a column containing about 10 g of dry resin using 250 to 300 mL of eluant. A larger volume of eluant is required only when exceptionally strong tailing occurs. Ions may be eluted completely by 300 to 400 mL of eluant from a column of 10 g of dry resin at D_g values of around 20. The first traces of an element will appear in the eluate at around 300 mL when its D_g value is about 50 to 60.

Example Shaking 50 mL of 0.001 M cesium salt solution with 1.0 g of a strong cation exchanger in the H-form (with a capacity of 3.0 mequiv $\cdot$ g^{-1}) removes the following amount of cesium. The selectivity coefficient, $k_{Cs/H}$, is 2.56, thus:

$$\frac{[Cs^+]_r[H^+]}{[Cs^+][H^+]_r} = 2.56$$

The maximum amount of cesium which can enter the resin is 50 mL $\times$ 0.001 M = 0.050 equiv. The minimum value of $[H^+]_r$ = 3.00 − 0.05 = 2.95 mequiv, and the maximum value, assuming complete exchange of cesium ion for hydrogen ion, is 0.001 *M*. The minimum value of the distribution ratio is:

$$(D_c)_{Cs} = \frac{[Cs^+]_r}{[Cs^+]} = \frac{(2.56)(2.95)}{0.001} = 7550$$

$$\frac{\text{Amount of Cs, resin phase}}{\text{Amount of Cs, solution phase}} = \frac{(7550)(1.0\text{ g})}{50\text{ mL}} = 151$$

Thus, at equilibrium the 1.0 g of resin removed is:

$$\frac{100\% - x}{x} = 151$$

with all but 0.66% of cesium ions from solution. If the amount of resin were increased to 2.0 g, the amount of cesium remaining in solution would decrease to 0.33%, half the former value. However, if the depleted solution were decanted and placed in contact with 1 g of fresh resin, the amount of cesium remaining in solution would decrease to 0.004%. Two batch equilibrations would effectively remove the cesium from the solution.

4.13 GRAVIMETRIC ANALYSIS

TABLE 4.34 Gravimetric Factors

In the following table the elements are arranged in alphabetical order.

Example: To convert a given weight of Al_2O_3 to its equivalent of Al, multiply by the factor at the right, 0.52926; similarly to convert Al to Al_2O_3, multiply by the factor at the left, 1.8894.

Factor		Factor
	ALUMINUM **Al = 26.9815**	
0.74971	$Al \leftrightarrow Al_4C_3$	1.3341
0.058728	$Al \leftrightarrow Al(C_9H_6ON)_3$ (oxinate)	17.027
0.65829	$Al \leftrightarrow AlN$	1.5191
1.8894	$Al_2O_3 \leftrightarrow Al$	0.52926
1.4165	$Al_2O_3 \leftrightarrow Al_4C_3$	0.70596
0.38233	$Al_2O_3 \leftrightarrow AlCl_3$	2.6155
0.41804	$Al_2O_3 \leftrightarrow AlPO_4$	2.3921
0.29800	$Al_2O_3 \leftrightarrow Al_2(SO_4)_3$	3.3557
0.15300	$Al_2O_3 \leftrightarrow Al_2(SO_4)_3 \cdot 18H_2O$	6.5361
0.10746	$Al_2O_3 \leftrightarrow K_2SO_4 \cdot Al_2(SO_4)_3 \cdot 24H_2O$	9.3055
0.11246	$Al_2O_3 \leftrightarrow (NH_4)_2SO_4 \cdot Al_2(SO_4)_3 \cdot 24H_2O$	8.8922
4.5197	$AlPO_4 \leftrightarrow Al$	0.22125
1.3946	$CaF_2 \leftrightarrow AlF_3$	0.71704
0.58196	$P_2O_5 \leftrightarrow AlPO_4$	1.7183
	AMMONIUM **NH_4 = 18.03858**	
1.1013	$Ag \leftrightarrow NH_4Br$	0.90802
2.0166	$Ag \leftrightarrow NH_4Cl$	0.49590
0.74424	$Ag \leftrightarrow NH_4I$	1.3437
1.9171	$AgBr \leftrightarrow NH_4Br$	0.52161
2.6792	$AgCl \leftrightarrow NH_4Cl$	0.37323

TABLE 4.34 Gravimetric Factors (*Continued*)

Factor		Factor
1.6198	$AgI \leftrightarrow NH_4I$	0.61737
1.7663	$BaSO_4 \leftrightarrow (NH_4)_2SO_4$	0.56615
0.81583	$Br \leftrightarrow NH_4Br$	1.2257
1.9654	$Cl \leftrightarrow NH_4$	0.50881
0.66277	$Cl \leftrightarrow NH_4Cl$	1.5088
0.68162	$HCl \leftrightarrow NH_4Cl$	1.4671
0.87553	$I \leftrightarrow NH_4I$	1.1422
14.410	$MgNH_4PO_4 \cdot 6H_2O \leftrightarrow NH_3$	0.069398
13.604	$MgNH_4PO_4 \cdot 6H_2O \leftrightarrow NH_4$	0.073506
9.4249	$MgNH_4PO_4 \cdot 6H_2O \leftrightarrow (NH_4)_2O$	0.10610
0.82244	$N \leftrightarrow NH_3$	1.2159
0.77648	$N \leftrightarrow NH_4$	1.2879
0.26185	$N \leftrightarrow NH_4Cl$	3.8189
0.17499	$N \leftrightarrow NH_4NO_3$	5.7145
0.53793	$N \leftrightarrow (NH_4)_2O$	1.8590
0.21200	$N \leftrightarrow (NH_4)_2SO_4$	4.7169
0.94412	$NH_3 \leftrightarrow NH_4$	1.0592
0.35449	$NH_3 \leftrightarrow (NH_4)_2CO_3$	2.8210
0.21543	$NH_3 \leftrightarrow NH_4HCO_3$	4.6419
0.21277	$NH_3 \leftrightarrow NH_4NO_3$	4.6998
0.65407	$NH_3 \leftrightarrow (NH_4)_2O$	1.5289
0.48596	$NH_3 \leftrightarrow NH_4OH$	2.0578
0.25777	$NH_3 \leftrightarrow (NH_4)_2SO_4$	3.8794
3.1409	$NH_4Cl \leftrightarrow NH_3$	0.31838
2.9654	$NH_4Cl \leftrightarrow NH_4$	0.33723
2.0543	$NH_4Cl \leftrightarrow (NH_4)_2O$	0.48677
1.5263	$NH_4Cl \leftrightarrow NH_4OH$	0.65516
2.5020	$NH_4OH \leftrightarrow N$	0.39967
1.9428	$NH_4OH \leftrightarrow NH_4$	0.51472
13.032	$(NH_4)_2PtCl_6 \leftrightarrow NH_3$	0.076737
12.303	$(NH_4)_2PtCl_6 \leftrightarrow NH_4$	0.081279
4.1490	$(NH_4)_2PtCl_6 \leftrightarrow NH_4Cl$	0.24102
2.7728	$(NH_4)_2PtCl_6 \leftrightarrow NH_4NO_3$	0.36065
8.5235	$(NH_4)_2PtCl_6 \leftrightarrow (NH_4)_2O$	0.11732
6.3328	$(NH_4)_2PtCl_6 \leftrightarrow NH_4OH$	0.15791
3.3592	$(NH_4)_2PtCl_6 \leftrightarrow (NH_4)_2SO_4$	0.29769
1.3473	$(NH_4)_2SO_4 \leftrightarrow H_2SO_4$	0.74223
3.1710	$N_2O_5 \leftrightarrow NH_3$	0.31536
0.67470	$N_2O_5 \leftrightarrow NH_4NO_3$	1.4821
2.0740	$N_2O_5 \leftrightarrow (NH_4)_2O$	0.48215
5.7275	$Pt \leftrightarrow NH_3$	0.17460
5.4074	$Pt \leftrightarrow NH_4$	0.18493
1.8235	$Pt \leftrightarrow NH_4Cl$	0.54838
1.2187	$Pt \leftrightarrow NH_4NO_3$	0.82058
3.7462	$Pt \leftrightarrow (NH_4)_2O$	0.26694
2.7833	$Pt \leftrightarrow NH_4OH$	0.35928
1.4764	$Pt \leftrightarrow (NH_4)_2SO_4$	0.67733
2.3505	$SO_3 \leftrightarrow NH_3$	0.42545
0.60589	$SO_3 \leftrightarrow (NH_4)_2SO_4$	1.6505
	ANTIMONY **Sb = 121.760**	
0.36460	$Sb \leftrightarrow KSbO \cdot C_4H_4O_6 \cdot ½H_2O$	2.7428

(*Continued*)

TABLE 4.34 Gravimetric Factors (*Continued*)

Factor		Factor
0.83535	$Sb \leftrightarrow Sb_2O_4$	1.1971
0.75271	$Sb \leftrightarrow Sb_2O_5$	1.3285
0.43646	$Sb_2O_3 \leftrightarrow KSbO \cdot C_4H_4O_6 \cdot ½H_2O$	2.2912
0.90106	$Sb_2O_3 \leftrightarrow Sb_2O_5$	1.1098
0.72184	$Sb_2O_3 \leftrightarrow Sb_2S_5$	1.3853
0.46042	$Sb_2O_4 \leftrightarrow KSbO \cdot C_4H_4O_6 \cdot ½H_2O$	2.1719
1.2628	$Sb_2O_4 \leftrightarrow Sb$	0.79188
1.0549	$Sb_2O_4 \leftrightarrow Sb_2O_3$	0.94796
0.95053	$Sb_2O_4 \leftrightarrow Sb_2O_5$	1.0520
0.90523	$Sb_2O_4 \leftrightarrow Sb_2S_3$	1.1047
0.76147	$Sb_2O_4 \leftrightarrow Sb_2S_5$	1.3133
0.80110	$Sb_2O_5 \leftrightarrow Sb_2S_5$	1.2483
0.50862	$Sb_2S_3 \leftrightarrow KSbO \cdot C_4H_4O_6 \cdot ½H_2O$	1.9661
1.3950	$Sb_2S_3 \leftrightarrow Sb$	0.71683
1.1653	$Sb_2S_3 \leftrightarrow Sb_2O_3$	0.85812
1.0500	$Sb_2S_3 \leftrightarrow Sb_2O_5$	0.95234
1.6584	$Sb_2S_5 \leftrightarrow Sb$	0.60299
	ARSENIC	
	As = 74.9216	
1.3203	$As_2O_3 \leftrightarrow As$	0.75738
0.86079	$As_2O_3 \leftrightarrow As_2O_5$	1.1617
1.5339	$As_2O_5 \leftrightarrow As$	0.65195
1.6420	$As_2S_3 \leftrightarrow As$	0.60903
1.2436	$As_2S_3 \leftrightarrow As_2O_3$	0.80413
1.0705	$As_2S_3 \leftrightarrow As_2O_5$	0.93418
0.79324	$As_2S_3 \leftrightarrow As_2S_5$	1.2606
2.0699	$As_2S_5 \leftrightarrow As$	0.48311
1.5678	$As_2S_5 \leftrightarrow As_2O_3$	0.63787
1.3495	$As_2S_5 \leftrightarrow As_2O_5$	0.74103
4.6729	$BaSO_4 \leftrightarrow As$	0.21400
3.5392	$BaSO_4 \leftrightarrow As_2O_3$	0.28255
3.0465	$BaSO_4 \leftrightarrow As_2O_6$	0.32825
2.8482	$BaSO_4 \leftrightarrow AsO_3$	0.35110
2.5202	$BaSO_4 \leftrightarrow AsO_4$	0.39680
2.0719	$Mg_2As_2O_7 \leftrightarrow As$	0.48265
1.5692	$Mg_2As_2O_7 \leftrightarrow As_2O_3$	0.63726
1.3509	$Mg_2As_2O_7 \leftrightarrow As_2O_5$	0.74032
1.2629	$Mg_2As_2O_7 \leftrightarrow AsO_2$	0.79186
1.1174	$Mg_2As_2O_7 \leftrightarrow AsO_4$	0.89493
1.2619	$Mg_2As_2O_7 \leftrightarrow As_2S_3$	0.79249
2.5397	$MgNH_4AsO_4 \cdot ½H_2O \leftrightarrow As$	0.39374
1.9235	$MgNH_4AsO_4 \cdot ½H_2O \leftrightarrow As_2O_3$	0.51988
1.6558	$MgNH_4AsO_4 \cdot ½H_2O \leftrightarrow As_2O_5$	0.60395
1.5480	$MgNH_4AsO_4 \cdot ½H_2O \leftrightarrow AsO_3$	0.64600
1.3697	$MgNH_4AsO_4 \cdot ½H_2O \leftrightarrow AsO_4$	0.73008
	BARIUM	
	Ba = 137.34	
1.4369	$BaCO_3 \leftrightarrow Ba$	0.69592
0.94766	$BaCO_3 \leftrightarrow BaCl_2$	1.0552
0.76088	$BaCO_3 \leftrightarrow Ba(HCO_3)_2$	1.3143
1.2871	$BaCO_3 \leftrightarrow BaO$	0.77699
1.8446	$BaCrO_4 \leftrightarrow Ba$	0.54214

TABLE 4.34 Gravimetric Factors (*Continued*)

Factor		Factor
1.2165	$BaCrO_4 \leftrightarrow BaCl_2$	0.82205
1.2838	$BaCrO_4 \leftrightarrow BaCO_3$	0.77902
1.6521	$BaCrO_4 \leftrightarrow BaO$	0.60530
2.0345	$BaSiF_6 \leftrightarrow Ba$	0.49152
1.5936	$BaSiF_6 \leftrightarrow BaF_2$	0.62751
1.8222	$BaSiF_6 \leftrightarrow BaO$	0.54878
1.6994	$BaSO_4 \leftrightarrow Ba$	0.58843
1.1208	$BaSO_4 \leftrightarrow BaCl_2$	0.89224
0.95546	$BaSO_4 \leftrightarrow BaCl_2 \cdot 2H_2O$	1.0466
1.1827	$BaSO_4 \leftrightarrow BaCO_3$	0.84554
0.89308	$BaSO_4 \leftrightarrow Ba(NO_3)_2$	1.1197
1.5221	$BaSO_4 \leftrightarrow BaO$	0.65698
1.3783	$BaSO_4 \leftrightarrow BaO_2$	0.72554
1.3778	$BaSO_4 \leftrightarrow BaS$	0.72579
0.28701	$CO_2 \leftrightarrow BaO$	3.4842
0.22300	$CO_2 \leftrightarrow BaCO_3$	4.4842
	BERYLLIUM	
	Be = 9.0122	
8.8678	$BeCl_2 \leftrightarrow Be$	0.11277
2.7753	$BeO \leftrightarrow Be$	0.36033
0.31296	$BeO \leftrightarrow BeCl_2$	3.1953
0.14119	$BeO \leftrightarrow BeSO_4 \cdot 4H_2O$	7.0825
	BISMUTH	
	Bi = 208.980	
0.89699	$Bi \leftrightarrow Bi_2O_3$	1.1148
1.6648	$BiAsO_4 \leftrightarrow Bi$	0.60069
1.4933	$BiAsO_4 \leftrightarrow Bi_2O_4$	0.66968
0.48030	$Bi_2O_3 \leftrightarrow Bi(NO_3)_3 \cdot 5H_2O$	2.0820
0.81183	$Bi_2O_3 \leftrightarrow BiONO_3$	1.2318
1.2462	$BiOCl \leftrightarrow Bi$	0.80244
0.53689	$BiOCl \leftrightarrow Bi(NO_3)_3 \cdot 5H_2O$	1.8626
1.1178	$BiOCl \leftrightarrow Bi_2O_3$	0.89460
0.90748	$BiOCl \leftrightarrow BiONO_3$	1.1019
1.2301	$Bi_2S_3 \leftrightarrow Bi$	0.81291
1.1034	$Bi_2S_3 \leftrightarrow Bi_2O_3$	0.90627
	BORON	
	B = 10.81	
3.2199	$B_2O_3 \leftrightarrow B$	0.31057
0.81317	$B_2O_3 \leftrightarrow BO_2$	1.2298
0.59193	$B_2O_3 \leftrightarrow BO_3$	1.6894
0.89693	$B_2O_3 \leftrightarrow B_4O_7$	1.1149
0.56298	$B_2O_3 \leftrightarrow H_3BO_3$	1.7763
0.36510	$B_2O_3 \leftrightarrow Na_2B_4O_7 \cdot 10H_2O$	2.7389
6.4005	$B_6C \leftrightarrow C$	0.15624
11.646	$KBF_4 \leftrightarrow B$	0.085863
3.6171	$KBF_4 \leftrightarrow B_2O_3$	0.27647
2.0363	$KBF_4 \leftrightarrow H_3BO_3$	0.49108
1.3206	$KBF_4 \leftrightarrow Na_2B_4O_7 \cdot 10H_2O$	0.75723

(*Continued*)

TABLE 4.34 Gravimetric Factors (*Continued*)

Factor		Factor
	BROMINE **Br = 79.90**	
1.3499	$Ag \leftrightarrow Br$	0.74079
0.84333	$Ag \leftrightarrow BrO_3$	1.1858
1.3331	$Ag \leftrightarrow HBr$	0.75013
2.3499	$AgBr \leftrightarrow Br$	0.42555
1.4681	$AgBr \leftrightarrow BrO_3$	0.68117
2.3206	$AgBr \leftrightarrow HBr$	0.43091
0.55756	$Br \leftrightarrow AgCl$	1.7935
9.9892	$Br \leftrightarrow O$	0.10010
1.1858	$BrO_3 \leftrightarrow Ag$	0.84333
	CADMIUM **Cd = 112.40**	
0.61317	$Cd \leftrightarrow CdCl_2$	1.6309
0.47545	$Cd \leftrightarrow Cd(NO_3)_2$	2.1033
1.1423	$CdO \leftrightarrow Cd$	0.87539
0.70045	$CdO \leftrightarrow CdCl_2$	1.4276
0.54312	$CdO \leftrightarrow Cd(NO_3)_2$	1.8412
1.2852	$CdS \leftrightarrow Cd$	0.77807
0.78806	$CdS \leftrightarrow CdCl_2$	1.2689
0.61106	$CdS \leftrightarrow Cd(NO_3)_2$	1.6365
1.1251	$CdS \leftrightarrow CdO$	0.88883
0.69298	$CdS \leftrightarrow CdSO_4$	1.4430
1.8546	$CdSO_4 \leftrightarrow Cd$	0.53919
1.1372	$CdSO_4 \leftrightarrow CdCl_2$	0.87935
0.88177	$CdSO_4 \leftrightarrow Cd(NO_3)_2$	1.1341
1.6235	$CdSO_4 \leftrightarrow CdO$	0.61595
	CALCIUM **Ca = 40.08**	
3.2352	$BaSO_4 \leftrightarrow CaS$	0.30910
1.7144	$BaSO_4 \leftrightarrow CaSO_4$	0.58329
1.3556	$BaSO_4 \leftrightarrow CaSO_4 \cdot 2H_2O$	0.73766
0.36111	$Ca \leftrightarrow CaCl_2$	2.7692
0.51334	$Ca \leftrightarrow CaF_2$	1.9480
0.71471	$Ca \leftrightarrow CaO$	1.3992
2.4973	$CaCO_3 \leftrightarrow Ca$	0.40044
0.90179	$CaCO_3 \leftrightarrow CaCl_2$	1.1089
0.61742	$CaCO_3 \leftrightarrow Ca(HCO_3)_2$	1.6196
1.7848	$CaCO \leftrightarrow CaO$	0.56029
0.73520	$CaCO_3 \leftrightarrow CaSO_4$	1.3602
0.58134	$CaCO_3 \leftrightarrow CaSO_4 \cdot 2H_2O$	1.7202
1.3726	$CaCO_3 \leftrightarrow HCl$	0.72856
0.50526	$CaO \leftrightarrow CaCl_2$	1.9792
0.71825	$CaO \leftrightarrow CaF_2$	1.3923
0.34593	$CaO \leftrightarrow Ca(HCO_3)_2$	2.8907
0.75685	$CaO \leftrightarrow Ca(OH)_2$	1.3213
0.41192	$CaO \leftrightarrow CaSO_4$	2.4276
0.32572	$CaO \leftrightarrow CaSO_4 \cdot 2H_2O$	3.0701
2.5797	$Ca_3(PO_4)_2 \leftrightarrow Ca$	0.38765
1.8437	$Ca_3(PO_4)_2 \leftrightarrow CaO$	0.54239
0.75946	$Ca_3(PO_4)_2 \leftrightarrow CaSO_4$	1.3167

TABLE 4.34 Gravimetric Factors (*Continued*)

Factor		Factor
3.3967	$CaSO_4 \leftrightarrow Ca$	0.29440
1.2266	$CaSO_4 \leftrightarrow CaCl_2$	0.81526
1.3602	$CaSO_4 \leftrightarrow CaCO_3$	0.73520
1.7437	$CaSO_4 \leftrightarrow CaF_2$	0.57351
2.4276	$CaSO_4 \leftrightarrow CaO$	0.41192
1.7691	$Cl \leftrightarrow Ca$	0.56526
0.63885	$Cl \leftrightarrow CaCl_2$	1.5653
1.2644	$Cl \leftrightarrow CaO$	0.79089
0.78479	$CO_2 \leftrightarrow CaO$	1.2742
0.43970	$CO_2 \leftrightarrow CaCO_3$	2.2743
0.77989	$Mg_2As_2O_7 \leftrightarrow Ca_3(AsO_4)_2$	1.2822
0.71883	$MgO \leftrightarrow CaO$	1.3912
0.71755	$Mg_2P_2O_7 \leftrightarrow Ca_3(PO_4)_2$	1.3936
12.098	$(NH_4)_3PO_4 \cdot 12MoO_3 \leftrightarrow Ca_3(PO_4)_2$	0.082657
0.65824	$N_2O_5 \leftrightarrow Ca(NO_3)_2$	1.5192
0.45761	$P_2O_3 \leftrightarrow Ca_3(PO_4)_2$	2.1853
1.4277	$SO_3 \leftrightarrow CaO$	0.70044
0.58809	$SO_3 \leftrightarrow CaSO_4$	1.7004
0.46502	$SO_3 \leftrightarrow CaSO_4 \cdot 2H_2O$	2.1505
0.80523	$WO_3 \leftrightarrow CaWO_4$	1.2419
	CARBON	
	C = 12.011	
3.9913	$Ag \leftrightarrow HCN$	0.25054
1.6565	$Ag \leftrightarrow KCN$	0.60369
4.9541	$AgCN \leftrightarrow HCN$	0.20185
2.0561	$AgCN \leftrightarrow KCN$	0.48637
16.431	$BaCO_3 \leftrightarrow C$	0.060861
4.4842	$BaCO_3 \leftrightarrow CO_2$	0.22301
3.2887	$BaCO_3 \leftrightarrow CO_3$	0.30407
3.4842	$BaO \leftrightarrow CO_2$	0.28701
1.7421	$BaO \leftrightarrow CO_2$, bicarbonate	0.57402
0.19432	$CN \leftrightarrow AgCN$	5.1461
0.24120	$CN \leftrightarrow Ag$	4.1460
0.35000	$SCN \leftrightarrow AgSCN$	2.8572
0.47757	$SCN \leftrightarrow CuSCN$	2.0939
0.24885	$SCN \leftrightarrow BaSO_4$	4.0185
1.2742	$CaO \leftrightarrow CO_2$	0.78479
0.63712	$CaO \leftrightarrow CO_2$, bicarbonate	1.5696
0.33936	$CO_2 \leftrightarrow Ba(HCO_3)_2$	2.9467
3.6641	$CO_2 \leftrightarrow C$	0.27291
0.43970	$CO_2 \leftrightarrow CaCO_3$	2.2743
0.54297	$CO_2 \leftrightarrow Ca(HCO_3)_2$	1.8417
0.73341	$CO_2 \leftrightarrow CO_3$	1.3635
0.13507	$CO_2 \leftrightarrow Cs_2CO_3$	7.4033
0.22695	$CO_2 \leftrightarrow CsHCO_3$	4.4063
0.37986	$CO_2 \leftrightarrow FeCO_3$	2.6326
0.49483	$CO_2 \leftrightarrow Fe(HCO_3)_2$	2.0209
0.31843	$CO_2 \leftrightarrow K_2CO_3$	3.1404
0.43957	$CO_2 \leftrightarrow KHCO_3$	2.2749
0.46718	$CO_2 \leftrightarrow K_2O$	2.1405
0.59564	$CO_2 \leftrightarrow Li_2CO_3$	1.6789
0.64762	$CO_2 \leftrightarrow LiHCO_3$	1.5441

(*Continued*)

TABLE 4.34 Gravimetric Factors (*Continued*)

Factor		Factor
1.4730	$CO_2 \leftrightarrow Li_2O$	0.67887
0.52193	$CO_2 \leftrightarrow MgCO_3$	1.9159
0.60143	$CO_2 \leftrightarrow Mg(HCO_3)_2$	1.6627
1.0918	$CO_2 \leftrightarrow MgO$	0.91595
0.38286	$CO_2 \leftrightarrow MnCO_3$	2.6119
0.49737	$CO_2 \leftrightarrow Mn(HCO_3)_2$	2.0106
0.62041	$CO_2 \leftrightarrow MnO$	1.6118
0.41523	$CO_2 \leftrightarrow Na_2CO_3$	2.4083
0.52388	$CO_2 \leftrightarrow NaHCO_3$	1.9088
0.71008	$CO_2 \leftrightarrow Na_2O$	1.4083
0.45802	$CO_2 \leftrightarrow (NH_4)_2CO_3$	2.1833
0.55669	$CO_2 \leftrightarrow NH_4HCO_3$	1.7963
0.16471	$CO_2 \leftrightarrow PbCO_3$	6.0713
0.19055	$CO_2 \leftrightarrow Rb_2CO_3$	5.2477
0.30043	$CO_2 \leftrightarrow RbHCO_3$	3.3286
0.23542	$CO_2 \leftrightarrow Rb_2O$	4.2477
0.29811	$CO_2 \leftrightarrow SrCO_3$	3.3545
0.41984	$CO_2 \leftrightarrow Sr(HCO_3)_2$	2.3818
0.42474	$CO_2 \leftrightarrow SrO$	2.3545
	CERIUM **Ce = 140.12**	
0.36100	$Ce \leftrightarrow Ce(NO_3)_4$	2.7701
0.24746	$Ce \leftrightarrow Ce(NO_3)_4 \cdot 2NH_4NO_3 \cdot H_2O$	4.0411
0.81408	$Ce \leftrightarrow CeO_2$	1.2284
0.85377	$Ce \leftrightarrow Ce_2O_3$	1.1713
0.49302	$Ce \leftrightarrow Ce_2(SO_4)_3$	2.0283
1.0527	$Ce_2(C_2O_4)_3 \cdot 3H_2O \leftrightarrow Ce_2(SO_4)_3$	0.94998
2.1351	$Ce_2(C_2O_4)_3 \cdot 3H_2O \leftrightarrow Ce$	0.46835
0.44345	$CeO_2 \leftrightarrow Ce(NO_3)_4$	2.2551
0.30397	$CeO_2 \leftrightarrow Ce(NO_3)_4 \cdot 2NH_4NO_3 \cdot H_2O$	3.2898
0.42284	$Ce_2O_3 \leftrightarrow Ce(NO_3)_4$	2.3650
0.28984	$Ce_2O_3 \leftrightarrow Ce(NO_3)_4 \cdot 2NH_4NO_3 \cdot H_2O$	3.4502
0.95352	$Ce_2O_3 \leftrightarrow CeO_2$	1.0487
0.57746	$Ce_2O_3 \leftrightarrow Ce_2(SO_4)_3$	1.7317
	CESIUM **Cs = 137.905**	
0.85127	$AgCl \leftrightarrow CsCl$	1.1747
0.26675	$Cl \leftrightarrow Cs$	3.7489
0.21058	$Cl \leftrightarrow CsCl$	4.7488
0.78944	$Cs \leftrightarrow CsCl$	1.2667
0.57200	$Cs \leftrightarrow CsClO_4$	1.7483
0.81585	$Cs \leftrightarrow Cs_2CO_3$	1.2257
0.94326	$Cs \leftrightarrow Cs_2O$	1.0602
0.83693	$Cs_2O \leftrightarrow CsCl$	1.1948
0.77876	$Cs_2O \leftrightarrow Cs_2SO_4$	1.2841
2.5341	$Cs_2PtCl_6 \leftrightarrow Cs$	0.39461
2.0005	$Cs_2PtCl_6 \leftrightarrow CsCl$	0.49987
2.0675	$Cs_2PtCl_6 \leftrightarrow Cs_2CO_3$	0.48369
2.3903	$Cs_2PtCl_6 \leftrightarrow Cs_2O$	0.41835
1.3613	$Cs_2SO_4 \leftrightarrow Cs$	0.73457
1.0747	$Cs_2SO_4 \leftrightarrow CsCl$	0.93050
1.1106	$Cs_2SO_4 \leftrightarrow Cs_2CO_3$	0.90038
0.28410	$SO_3 \leftrightarrow Cs_2O$	3.5199

TABLE 4.34 Gravimetric Factors (*Continued*)

Factor		Factor
	CHLORINE Cl = 35.453	
3.0426	$Ag \leftrightarrow Cl$	0.32866
2.9585	$Ag \leftrightarrow HCl$	0.33801
4.0425	$AgCl \leftrightarrow Cl$	0.24737
3.9308	$AgCl \leftrightarrow HCl$	0.25440
3.5728	$BaCrO_4 \leftrightarrow Cl$	0.27990
0.56526	$Ca \leftrightarrow Cl$	1.7691
0.97235	$Cl \leftrightarrow HCl$	1.0284
0.58227	$ClO_3 \leftrightarrow AgCl$	1.7174
1.1193	$ClO_3 \leftrightarrow KCl$	0.89340
1.4279	$ClO_3 \leftrightarrow NaCl$	0.70033
0.69391	$ClO_4 \leftrightarrow AgCl$	1.4411
1.3339	$ClO_4 \leftrightarrow KCl$	0.74967
1.7017	$ClO_4 \leftrightarrow NaCl$	0.58766
1.1029	$K \leftrightarrow Cl$	0.90668
2.1029	$KCl \leftrightarrow Cl$	0.47553
0.19572	$Li \leftrightarrow Cl$	5.1092
0.34288	$Mg \leftrightarrow Cl$	2.9165
1.3429	$MgCl_2 \leftrightarrow Cl$	0.74467
1.2261	$MnO_2 \leftrightarrow Cl$	0.81560
0.64846	$Na \leftrightarrow Cl$	1.5421
1.6485	$NaCl \leftrightarrow Cl$	0.60663
0.50881	$NH_4 \leftrightarrow Cl$	1.9654
1.4671	$NH_4Cl \leftrightarrow HCl$	0.68162
1.8121	$(NH_4)_2SO_4 \leftrightarrow HCl$	0.55185
4.5580	$PbCrO_4 \leftrightarrow Cl$	0.21939
	CHROMIUM Cr = 51.996	
4.8721	$BaCrO_4 \leftrightarrow Cr$	0.20525
3.3335	$BaCrO_4 \leftrightarrow Cr_2O_3$	0.29998
2.5335	$BaCrO_4 \leftrightarrow CrO_3$	0.39472
2.1841	$BaCrO_4 \leftrightarrow CrO_4$	0.45786
0.70718	$BaCrO_4 \leftrightarrow Cr_2(SO_4)_3 \cdot 18H_2O$	1.4141
7.4935	$Cr_3C_2 \leftrightarrow Cr$	0.13345
1.9231	$CrO_3 \leftrightarrow Cr$	0.51999
1.4616	$Cr_2O_3 \leftrightarrow Cr$	0.68420
0.76000	$Cr_2O_3 \leftrightarrow CrO_3$	1.3158
0.65519	$Cr_2O_3 \leftrightarrow CrO_4$	1.5263
3.7349	$K_2CrO_4 \leftrightarrow Cr$	0.26774
1.9421	$K_2CrO_4 \leftrightarrow CrO_3$	0.51490
1.4710	$K_2Cr_2O_7 \leftrightarrow CrO_3$	0.67979
6.2155	$PbCrO_4 \leftrightarrow Cr$	0.16089
4.2527	$PbCrO_4 \leftrightarrow Cr_2O_3$	0.23515
3.2320	$PbCrO_4 \leftrightarrow CrO_3$	0.30941
2.7863	$PbCrO_4 \leftrightarrow CrO_4$	0.35890
0.90217	$PbCrO_4 \leftrightarrow Cr_2(SO_4)_3 \cdot 18H_2O$	1.1084
1.6642	$PbCrO_4 \leftrightarrow K_2CrO_4$	0.60090
2.1971	$PbCrO_4 \leftrightarrow K_2Cr_2O_7$	0.45515
	COBALT Co = 58.9332	
0.20249	$Co \leftrightarrow Co(NO_3)_2 \cdot 6H_2O$	4.9385
0.78648	$Co \leftrightarrow CoO$	1.2715

(*Continued*)

TABLE 4.34 Gravimetric Factors (*Continued*)

Factor		Factor
	COBALT (*continued*) **Co = 58.9332**	
0.20965	$Co \leftrightarrow CoSO_4 \cdot 7H_2O$	4.7698
7.6743	$K_3[Co(NO_2)_6] \leftrightarrow Co$	0.13030
6.0357	$K_3[Co(NO_2)_6] \leftrightarrow CoO$	0.16568
1.3620	$Co_3O_4 \leftrightarrow Co$	0.73422
1.0712	$Co_3O_4 \leftrightarrow CoO$	0.93355
2.4758	$Co_2P_2O_7 \leftrightarrow Co$	0.40391
1.9471	$Co_2P_2O_7 \leftrightarrow CoO$	0.51357
3.2233	$CoNH_4PO_4 \cdot H_2O \leftrightarrow Co$	0.31024
2.5351	$CoNH_4PO_4 \cdot H_2O \leftrightarrow CoO$	0.39447
2.6299	$CoSO_4 \leftrightarrow Co$	0.38024
2.0684	$CoSO_4 \leftrightarrow CoO$	0.48347
3.7514	$CoSO_4 \cdot 7H_2O \leftrightarrow CoO$	0.26657
7.0656	$(CoSO_4)_2 \cdot (K_2SO_4)_3 \leftrightarrow Co$	0.14153
5.5569	$(CoSO_4)_2 \cdot (K_2SO_4)_3 \leftrightarrow CoO$	0.17996
	COPPER **Cu = 63.544**	
0.25071	$Cu \leftrightarrow Cu_2C_2H_3O_2 \cdot (AsO_2)_3$	3.9887
0.79885	$Cu \leftrightarrow CuO$	1.2518
0.25449	$Cu \leftrightarrow CuSO_4 \cdot 5H_2O$	3.9295
1.9141	$CuSCN \leftrightarrow Cu$	0.52245
1.5291	$CuSCN \leftrightarrow CuO$	0.65400
0.31856	$CuO \leftrightarrow CuSO_4 \cdot 5H_2O$	3.1391
1.1259	$Cu_2O \leftrightarrow Cu$	0.88817
1.2523	$Cu_2S \leftrightarrow Cu$	0.79854
1.0004	$Cu_2S \leftrightarrow CuO$	0.99961
1.1122	$Cu_2S \leftrightarrow Cu_2O$	0.89908
0.31869	$Cu_2S \leftrightarrow CuSO_4 \cdot 5H_2O$	3.1379
0.91872	$Mg_2As_2O_7 \leftrightarrow Cu_2C_2H_3O_2(AsO_2)_3$	1.0885
	ERBIUM **Er = 167.26**	
1.1435	$Er_2O_3 \leftrightarrow Er$	0.87452
	FLUORINE **F = 18.9984**	
1.5936	$BaSiF_6 \leftrightarrow BaF_2$	0.62751
2.4513	$BaSiF_6 \leftrightarrow F$	0.40795
2.3277	$BaSiF_6 \leftrightarrow 6HF$	0.42960
1.9392	$BaSiF_6 \leftrightarrow H_2SiF_6$	0.51568
2.6847	$BaSiF_6 \leftrightarrow SiF_4$	0.37249
1.9666	$BaSiF_6 \leftrightarrow SiF_6$	0.50848
1.6256	$CaF_2 \leftrightarrow H_2SiF_6$	0.61516
1.6486	$CaF_2 \leftrightarrow SiF_6$	0.60658
3.5829	$CaSO_4 \leftrightarrow F$	0.27910
2.4024	$CaSO_4 \leftrightarrow HF$	0.29391
0.48666	$F \leftrightarrow CaF_2$	2.0548
0.51248	$HF \leftrightarrow CaF_2$	1.9513
1.2641	$H_2SiF_6 \leftrightarrow F$	0.79109
3.6011	$H_2SiF_6 \leftrightarrow 2HF$	0.27769

TABLE 4.34 Gravimetric Factors (*Continued*)

Factor		Factor
	FLUORINE (*continued*) **F = 18.9984**	
1.2004	$H_2SiF_6 \leftrightarrow 6HF$	0.83308
1.3844	$H_2SiF_6 \leftrightarrow SiF_4$	0.72233
1.0141	$H_2SiF_6 \leftrightarrow SiF_6$	0.98605
2.0556	$KF \cdot HF \leftrightarrow 2F$	0.48647
1.9520	$KF \cdot HF \leftrightarrow 2HF$	0.51228
0.67218	$KF \cdot HF \leftrightarrow 2KF$	1.4877
0.41489	$KF \cdot HF \leftrightarrow 2(KF \cdot 2H_2O)$	2.4103
1.9325	$K_2SiF_6 \leftrightarrow F$	0.51748
1.8351	$K_2SiF_6 \leftrightarrow 6HF$	0.54494
1.5288	$K_2SiF_6 \leftrightarrow H_2SiF_6$	0.65412
1.8957	$K_2SiF_6 \leftrightarrow 2KF$	0.52751
1.5504	$K_2SiF_6 \leftrightarrow SiF_6$	0.64500
1.9495	$NH_4F \leftrightarrow F$	0.51295
1.5013	$NH_4F \cdot HF \leftrightarrow 2F$	0.66611
1.4256	$NH_4F \cdot HF \leftrightarrow 2HF$	0.70145
0.49090	$NH_4F \cdot HF \leftrightarrow 2KF$	2.0371
0.30300	$NH_4F \cdot HF \leftrightarrow 2(KF \cdot 2H_2O)$	3.3003
1.5629	$(NH_4)_2SiF_6 \leftrightarrow F$	0.63985
1.4841	$(NH_4)_2SiF_6 \leftrightarrow 6HF$	0.67381
1.2364	$(NH_4)_2SiF_6 \leftrightarrow H_2SiF_6$	0.80881
2.4050	$(NH_4)_2SiF_6 \leftrightarrow 2NH_4F$	0.41580
1.2539	$(NH_4)_2SiF_6 \leftrightarrow SiF_6$	0.79753
2.2101	$NaF \leftrightarrow F$	0.45246
1.6498	$Na_2SiF_6 \leftrightarrow F$	0.60614
1.5666	$Na_2SiF_6 \leftrightarrow 6HF$	0.63831
1.3052	$Na_3SiF_6 \leftrightarrow H_2SiF_6$	0.76619
2.2394	$Na_2SiF_6 \leftrightarrow 2NaF$	0.44654
1.3236	$Na_2SiF_6 \leftrightarrow SiF_6$	0.75550
	GALLIUM **Ga = 69.72**	
1.3442	$Ga_2O_3 \leftrightarrow Ga$	0.74392
1.6898	$Ga_2S_3 \leftrightarrow Ga$	0.59178
	GERMANIUM **Ge = 72.59**	
1.4408	$GeO_2 \leftrightarrow Ge$	0.69404
3.6476	$K_2GeF_6 \leftrightarrow Ge$	0.27415
	GOLD **Au = 196.967**	
0.64936	$Au \leftrightarrow AuCl_3$	1.5400
0.47826	$Au \leftrightarrow HAuCl_4 \cdot 4H_2O$	2.0909
0.54995	$Au \leftrightarrow KAu(CN)_4 \cdot H_2O$	1.8183
	HYDROGEN **H = 1.0079**	
8.9365	$H_2O \leftrightarrow H$	0.11190
7.9364	$O \leftrightarrow H$	0.12600
0.35607	$HSCN \leftrightarrow AgSCN$	2.8084

(*Continued*)

TABLE 4.34 Gravimetric Factors (*Continued*)

Factor		Factor
	HYDROGEN (*continued*) **H = 1.0079**	
0.48586	HSCN ↔ CuSCN	2.0582
0.25317	HSCN ↔ $BaSO_4$	3.9499
	INDIUM **In = 114.82**	
1.2090	In_2O_3 ↔ In	0.82711
1.4189	In_2S_3 ↔ In	0.70476
	IODINE **I = 126.904**	
0.84333	Ag ↔ HI	1.1858
0.85004	Ag ↔ I	1.1764
1.1294	AgCl ↔ I	0.88543
1.8354	AgI ↔ HI	0.54483
1.8500	AgI ↔ I	0.54053
1.3423	AgI ↔ IO_3	0.74498
1.2298	AgI ↔ IO_4	0.81314
1.4066	AgI ↔ I_2O_5	0.71091
1.2836	AgI ↔ I_2O_7	0.77904
0.41592	Pd ↔ HI	2.4043
0.41921	Pd ↔ I	2.3854
1.4081	PdI_2 ↔ HI	0.71020
1.4192	PdI_2 ↔ I	0.70462
1.0297	PdI_2 ↔ IO_3	0.97113
0.94343	PdI_2 ↔ IO_4	1.0600
1.0791	PdI_2 ↔ I_2O_5	0.92671
0.98472	PdI_2 ↔ I_2O_7	1.0155
2.5899	TlI ↔ HI	0.38612
2.6105	TlI ↔ I	0.38307
1.8941	TlI ↔ IO_3	0.52797
1.7353	TlI ↔ IO_4	0.57627
1.9848	TlI ↔ I_2O_5	0.50383
1.8112	TlI ↔ I_2O_7	0.55211
	IRON **Fe = 55.845**	
2.2598	Ag ↔ $Fe_7(CN)_{18}$ (Prussian blue)	0.44252
0.54503	CN ↔ $Fe_7(CN)_{18}$	1.8347
0.61256	CO_2 ↔ FeO	1.6325
0.37986	CO_2 ↔ $FeCO_3$	2.6326
0.49483	CO_2 ↔ $Fe(HCO_3)_2$	2.0209
0.31396	Fe ↔ $Fe(HCO_3)_2$	3.1851
0.44061	Fe ↔ $FeCl_2$	2.2696
0.77730	Fe ↔ FeO	1.2865
0.69943	Fe ↔ Fe_2O_3	1.4297
0.72359	Fe ↔ Fe_3O_4	1.3820
0.36763	Fe ↔ $FeSO_4$	2.7201
0.20087	Fe ↔ $FeSO_4 \cdot 7H_2O$	4.9782
0.14242	Fe ↔ $FeSO_4 \cdot (NH_4)_2SO_4 \cdot 6H_2O$	7.0217
0.62011	FeO ↔ $FeCO_3$	1.6126

TABLE 4.34 Gravimetric Factors (*Continued*)

Factor		Factor
	IRON (*continued*) **Fe = 55.845**	
0.40390	$FeO \leftrightarrow Fe(HCO_3)_2$	2.4759
0.89982	$FeO \leftrightarrow Fe_2O_3$	1.1113
0.49223	$Fe_2O_3 \leftrightarrow FeCl_2$	2.0316
0.68915	$Fe_2O_3 \leftrightarrow FeCO_3$	1.4511
0.44887	$Fe_2O_3 \leftrightarrow Fe(HCO_3)_2$	2.2278
0.33422	$Fe_2O_3 \leftrightarrow Fe(HCO_3)_3$	2.9920
1.1113	$Fe_2O_3 \leftrightarrow FeO$	0.89982
1.0345	$Fe_2O_3 \leftrightarrow Fe_3O_4$	0.96662
0.52941	$Fe_2O_3 \leftrightarrow FePO_4$	1.8889
0.52561	$Fe_2O_3 \leftrightarrow FeSO_4$	1.9026
0.28719	$Fe_2O_3 \leftrightarrow FeSO_4 \cdot 7H_2O$	3.4820
0.20361	$Fe_2O_3 \leftrightarrow FeSO_4 \cdot (NH_4)_2SO_4 \cdot 6H_2O$	4.9113
0.39934	$Fe_2O_3 \leftrightarrow Fe_2(SO_4)_3$	2.5041
2.7006	$FePO_4 \leftrightarrow Fe$	0.37029
2.0992	$FePO_4 \leftrightarrow FeO$	0.47637
1.5741	$FeS \leftrightarrow Fe$	0.63527
1.2236	$FeS \leftrightarrow FeO$	0.81726
1.1010	$FeS \leftrightarrow Fe_2O_3$	0.90825
0.79699	$Mg_2As_2O_7 \leftrightarrow FeAsO_4$	1.2547
1.1144	$SO_3 \leftrightarrow FeO$	0.89738
0.52704	$SO_3 \leftrightarrow FeSO_4$	1.8974
	LANTHANUM **La = 138.91**	
1.1728	$La_2O_3 \leftrightarrow La$	0.85268
	LEAD **Pb = 207.2**	
0.77541	$Pb \leftrightarrow PbCO_3$	1.2896
0.80141	$Pb \leftrightarrow (PbCO_3)_2 \cdot Pb(OH)_2$	1.2478
0.85901	$Pb \leftrightarrow Pb(OH)_2$	1.1641
0.92831	$Pb \leftrightarrow PbO$	1.0772
1.3422	$PbCl_2 \leftrightarrow Pb$	0.74502
1.2460	$PbCl_2 \leftrightarrow PbO$	0.80255
1.5598	$PbCrO_4 \leftrightarrow Pb$	0.64110
0.85198	$PbCrO_4 \leftrightarrow Pb(C_2H_3O_2)_2 \cdot 3H_2O$	1.1737
1.2501	$PbCrO_4 \leftrightarrow (PbCO_3)_2 \cdot Pb(OH)_2$	0.79997
1.4480	$PbCrO_4 \leftrightarrow PbO$	0.69061
1.4142	$PbCrO_4 \leftrightarrow Pb_3O_4$	0.70711
1.0657	$PbCrO_4 \leftrightarrow PbSO_4$	0.93833
0.83529	$PbO \leftrightarrow PbCO_3$	1.1972
0.67388	$PbO \leftrightarrow Pb(NO_3)_2$	1.4839
0.93311	$PbO \leftrightarrow PbO_2$	1.0717
1.1544	$PbO_2 \leftrightarrow Pb$	0.86622
0.72219	$PbO_2 \leftrightarrow Pb(NO_3)_2$	1.3847
1.1547	$PbS \leftrightarrow Pb$	0.86600
1.0720	$PbS \leftrightarrow PbO$	0.93287
0.78895	$PbS \leftrightarrow PbSO_4$	1.2675
1.2993	$PbSO_4 \leftrightarrow BaSO_4$	0.76966
1.4636	$PbSO_4 \leftrightarrow Pb$	0.68323

(*Continued*)

TABLE 4.34 Gravimetric Factors (*Continued*)

Factor		Factor
	LEAD (*continued*) **Pb = 207.2**	
0.79944	$PbSO_4 \leftrightarrow Pb(C_2H_3O_2)_2 \cdot 3H_2O$	1.2509
1.1349	$PbSO_4 \leftrightarrow PbCO_3$	0.88112
1.1730	$PbSO_4 \leftrightarrow (PbCO_3)_2 \cdot Pb(OH)_2$	0.85254
0.91561	$PbSO_4 \leftrightarrow Pb(NO_3)_2$	1.0922
1.3587	$PbSO_4 \leftrightarrow PbO$	0.73599
1.2678	$PbSO_4 \leftrightarrow PbO_2$	0.78875
1.3270	$PbSO_4 \leftrightarrow Pb_3O_4$	0.75358
	LITHIUM **Li = 6.941**	
0.59562	$CO_2 \leftrightarrow Li_2CO_3$	1.6789
0.64759	$CO_2 \leftrightarrow LiHCO_3$	1.5442
1.4729	$CO_2 \leftrightarrow Li_2O$	0.67894
6.1086	$LiCl \leftrightarrow Li$	0.16369
2.8378	$LiCl \leftrightarrow Li_2O$	0.35239
5.3228	$Li_2CO_3 \leftrightarrow Li$	0.18787
0.87147	$Li_2CO_3 \leftrightarrow LiCl$	1.1475
0.54364	$Li_2CO_3 \leftrightarrow LiHCO_3$	1.8395
2.4730	$Li_2CO_3 \leftrightarrow Li_2O$	0.40436
4.5491	$LiHCO_3 \leftrightarrow Li_2O$	0.21983
3.7371	$LiF \leftrightarrow Li$	0.26759
2.1525	$Li_2O \leftrightarrow Li$	0.46457
0.27176	$Li_2O \leftrightarrow Li_2SO_4$	3.6798
5.5609	$Li_3PO_4 \leftrightarrow Li$	0.17983
0.91047	$Li_3PO_4 \leftrightarrow LiCl$	1.0983
1.0447	$Li_3PO_4 \leftrightarrow Li_2CO_3$	0.95717
0.56797	$Li_3PO_4 \leftrightarrow LiHCO_3$	1.7607
2.5837	$Li_3PO_4 \leftrightarrow Li_2O$	0.38704
0.70214	$Li_3PO_4 \leftrightarrow Li_2SO_4$	1.4242
0.60331	$Li_3PO_4 \leftrightarrow Li_2SO_4 \cdot H_2O$	1.6575
7.9153	$Li_2SO_4 \leftrightarrow Li$	0.12634
1.2967	$Li_2SO_4 \leftrightarrow LiCl$	0.77118
2.6797	$SO_3 \leftrightarrow Li_2O$	0.37317
0.72823	$SO_3 \leftrightarrow Li_2SO_4$	1.3732
	MAGNESIUM **Mg = 24.305**	
1.9390	$BaSO_4 \leftrightarrow MgSO_4$	0.51572
0.94693	$BaSO_4 \leftrightarrow MgSO_4 \cdot 7H_2O$	1.0560
6.5755	$Br \leftrightarrow Mg$	0.15208
0.86800	$Br \leftrightarrow MgBr_2$	1.1521
0.54691	$Br \leftrightarrow MgBr_2 \cdot 6H_2O$	1.8285
2.9173	$Cl \leftrightarrow Mg$	0.34278
0.74472	$Cl \leftrightarrow MgCl_2$	1.3429
0.25533	$Mg \leftrightarrow MgCl_2$	3.9165
0.28883	$Mg \leftrightarrow MgCO_3$	3.4683
10.4427	$I \leftrightarrow Mg$	0.095761
0.91261	$I \leftrightarrow MgI_2$	1.09576
0.34876	$Cl \leftrightarrow MgCl_2 \cdot 6H_2O$	2.8673
0.52193	$CO_2 \leftrightarrow MgCO_3$	1.9160

TABLE 4.34 Gravimetric Factors (*Continued*)

Factor		Factor
	MAGNESIUM (*continued*) **Mg = 24.305**	
1.0918	$CO_2 \leftrightarrow MgO$	0.91595
0.57616	$MgCO_3 \leftrightarrow Mg(HCO_3)_2$	1.7356
10.094	$MgNH_4PO_4 \cdot 6H_2O \leftrightarrow Mg$	0.099067
6.0879	$MgNH_4PO_4 \cdot 6H_2O \leftrightarrow MgO$	0.16426
1.6581	$MgO \leftrightarrow Mg$	0.60311
0.47807	$MgO \leftrightarrow MgCO_3$	2.0918
0.27544	$MgO \leftrightarrow Mg(HCO_3)_2$	3.6305
0.33489	$MgO \leftrightarrow MgSO_4$	2.9860
4.5784	$Mg_2P_2O_7 \leftrightarrow Mg$	0.21841
1.1687	$Mg_2P_2O_7 \leftrightarrow MgCl_2$	0.85562
0.54737	$Mg_2P_2O_7 \leftrightarrow MgCl_2 \cdot 6H_2O$	1.8269
0.40049	$Mg_2P_2O_7 \leftrightarrow MgCl_2 \cdot KCl \cdot 6H_2O$	2.4969
1.3198	$Mg_2P_2O_7 \leftrightarrow MgCO_3$	0.75770
0.76040	$Mg_2P_2O_7 \leftrightarrow Mg(HCO_3)_2$	1.3151
2.7607	$Mg_2P_2O_7 \leftrightarrow MgO$	0.36223
0.92452	$Mg_2P_2O_7 \leftrightarrow MgSO_4$	1.0816
0.45150	$Mg_2P_2O_7 \leftrightarrow MgSO_4 \cdot 7H_2O$	2.2149
4.9523	$MgSO_4 \leftrightarrow Mg$	0.20193
1.9864	$SO_3 \leftrightarrow MgO$	0.50343
0.6651	$SO_3 \leftrightarrow MgSO_4$	1.5034
0.38482	$SO_3 \leftrightarrow MgSO_4 \cdot 7H_2O$	3.0786
	MANGANESE **Mn = 54.9380**	
1.5457	$BaSO_4 \leftrightarrow MnSO_4$	0.64696
0.38286	$CO_2 \leftrightarrow MnCO_3$	2.6119
0.62041	$CO_2 \leftrightarrow MnO$	1.6118
0.47793	$Mn \leftrightarrow MnCO_3$	2.0924
0.77446	$Mn \leftrightarrow MnO$	1.2912
0.63193	$Mn \leftrightarrow MnO_2$	1.5825
0.69599	$Mn \leftrightarrow Mn_2O_3$	1.4368
0.76126	$MnCO_3 \leftrightarrow MnSO_4$	1.3136
1.5395	$Mn(HCO_3)_2 \leftrightarrow MnCO_3$	0.64955
0.61711	$MnO \leftrightarrow MnCO_3$	1.6205
0.40084	$MnO \leftrightarrow Mn(HCO_3)_2$	2.4947
0.89868	$MnO \leftrightarrow Mn_2O_3$	1.1127
0.46978	$MnO \leftrightarrow MnSO_4$	2.1286
1.3883	$Mn_3O_4 \leftrightarrow Mn$	0.72031
0.66351	$Mn_3O_4 \leftrightarrow MnCO_3$	1.5071
0.43098	$Mn_3O_4 \leftrightarrow Mn(HCO_3)_2$	2.3203
1.0752	$Mn_3O_4 \leftrightarrow MnO$	0.93008
0.96625	$Mn_3O_4 \leftrightarrow Mn_2O_3$	1.0349
0.87731	$Mn_3O_4 \leftrightarrow MnO_2$	1.1399
0.50510	$Mn_3O_4 \leftrightarrow MnSO_4$	1.9798
2.5831	$Mn_2P_2O_7 \leftrightarrow Mn$	0.38713
1.2345	$Mn_2P_2O_7 \leftrightarrow MnCO_3$	0.81002
2.0005	$Mn_2P_2O_7 \leftrightarrow MnO$	0.49987
1.6324	$Mn_2P_2O_7 \leftrightarrow MnO_2$	0.61261
0.93980	$Mn_2P_2O_7 \leftrightarrow MnSO_4$	1.0641
1.5836	$MnS \leftrightarrow Mn$	0.63146

(*Continued*)

TABLE 4.34 Gravimetric Factors (*Continued*)

Factor		Factor
	MANGANESE (*continued*) **Mn = 54.9380**	
0.75687	$MnS \leftrightarrow MnCO_3$	1.3212
1.2265	$MnS \leftrightarrow MnO$	0.81535
0.57617	$MnS \leftrightarrow MnSO_4$	1.7356
2.7486	$MnSO_4 \leftrightarrow Mn$	0.36383
1.1286	$SO_3 \leftrightarrow MnO$	0.88603
0.53021	$SO_3 \leftrightarrow MnSO_4$	1.8860
	MERCURY **Hg = 200.59**	
0.73882	$Hg \leftrightarrow HgCl_2$	1.3535
0.92613	$Hg \leftrightarrow HgO$	1.0798
0.86220	$Hg \leftrightarrow HgS$	1.1598
1.1767	$HgCl \leftrightarrow Hg$	0.84981
0.86939	$HgCl \leftrightarrow HgCl_2$	1.1502
0.89889	$HgCl \leftrightarrow HgNO_3$	1.1125
1.1316	$HgCl \leftrightarrow Hg_2O$	0.88371
1.0898	$HgCl \leftrightarrow HgO$	0.91760
1.0146	$HgCl \leftrightarrow HgS$	0.98564
0.98564	$HgS \leftrightarrow HgCl$	1.0146
0.85691	$HgS \leftrightarrow HgCl_2$	1.1670
0.92091	$HgS \leftrightarrow Hg(CN)_2$	1.0859
0.88598	$HgS \leftrightarrow HgNO_3$	1.1287
0.71673	$HgS \leftrightarrow Hg(NO_3)_2$	1.3952
0.67903	$HgS \leftrightarrow Hg(NO_3)_2 \cdot H_2O$	1.4727
1.1153	$HgS \leftrightarrow Hg_2O$	0.89658
1.0741	$HgS \leftrightarrow HgO$	0.93097
0.78426	$HgS \leftrightarrow HgSO_4$	1.2751
	MOLYBDENUM **Mo = 95.94**	
8.9876	$MoC \leftrightarrow C$	0.11126
1.5003	$MoO_3 \leftrightarrow Mo$	0.66653
0.73436	$MoO_3 \leftrightarrow (NH_4)_2MoO_4$	1.3617
2.0026	$MoS_3 \leftrightarrow Mo$	0.49935
1.3348	$MoS_4 \leftrightarrow MoO_3$	0.74918
0.98021	$MoS_3 \leftrightarrow (NH_4)_2MoO_4$	1.0202
1.0863	$(NH_4)_3PO_4 \cdot 12MoO_3 \leftrightarrow MoO_3$	0.92058
0.79771	$(NH_4)_3PO_4 \cdot 12MoO_3 \leftrightarrow (NH_4)_2MoO_4$	1.2536
3.8267	$PbMoO_4 \leftrightarrow Mo$	0.26132
2.5506	$PbMoO_4 \leftrightarrow MoO_3$	0.39207
1.8730	$PbMoO_4 \leftrightarrow (NH_4)_2MoO_4$	0.53390
	NEODYMIUM **Nd = 144.24**	
1.1664	$Nd_2O_3 \leftrightarrow Nd$	0.85735
	NICKEL **Ni = 58.71**	
0.20319	$Ni \leftrightarrow$ Ni dimethylglyoxime	4.9215
0.20188	$Ni \leftrightarrow Ni(NO_3)_2 \cdot 6H_2O$	4.9533

TABLE 4.34 Gravimetric Factors (*Continued*)

Factor		Factor
	NICKEL (*continued*) **Ni = 58.71**	
0.78585	Ni ↔ NiO	1.2725
0.20902	Ni ↔ $NiSO_4 \cdot 7H_2O$	4.7842
3.8675	Ni dimethylglyoxime ↔ NiO	0.25856
0.25690	NiO ↔ $Ni(NO_3)_2 \cdot 6H_2O$	3.8926
0.26598	NiO ↔ $NiSO_4 \cdot 7H_2O$	3.7597
2.6362	$NiSO_4$ ↔ Ni	0.37934
0.53220	$NiSO_4$ ↔ $Ni(NO_3)_2 \cdot 6H_2O$	1.8790
2.0716	$NiSO_4$ ↔ NiO	0.48271
0.55102	$NiSO_4$ ↔ $NiSO_4 \cdot 7H_2O$	1.8148
	NIOBIUM **Nb = 92.906**	
7.7351	Nb ↔ C	0.12928
8.7353	NbC ↔ C	0.11448
11.065	Nb_2O_5 ↔ 2C	0.090373
1.4305	Nb_2O_5 ↔ Nb	0.69904
	NITROGEN **N = 14.0067**	
3.2731	$AgNO_2$ ↔ HNO_2	0.30552
4.0488	$AgNO_2$ ↔ N_2O_3	0.24698
1.8722	KNO_3 ↔ N_2O_5	053412
0.22229	N ↔ HNO_3	4.4987
0.30446	N ↔ NO_2	3.2845
0.36855	N ↔ N_2O_3	2.7134
0.22590	N ↔ NO_3	4.4268
0.25936	N ↔ N_2O_5	3.8556
6.0680	$NaNO_3$ ↔ N	0.16480
1.5738	$NaNO_3$ ↔ N_2O_5	0.63539
0.47619	NO ↔ HNO_3	2.1000
0.65222	NO ↔ NO_2	1.5332
0.78951	NO ↔ N_2O_3	1.2666
0.48393	NO ↔ NO_3	2.0664
0.55561	NO ↔ N_2O_5	1.7998
0.27028	NH_3 ↔ HNO_3	3.6999
1.2159	NH_3 ↔ N	0.82244
0.31536	NH_3 ↔ N_2O_5	3.1710
0.27467	NH_3 ↔ NO_3	3.6407
0.84890	NH_4Cl ↔ HNO_3	1.1780
0.86270	NH_4Cl ↔ NO_3	1.1591
0.99050	NH_4Cl ↔ N_2O_5	1.0096
3.8189	NH_4Cl ↔ N	0.26185
3.5221	$(NH_4)_2PtCl_6$ ↔ HNO_3	0.28393
15.845	$(NH_4)_2PtCl_6$ ↔ N	0.063112
4.1096	$(NH_4)_2PtCl_6$ ↔ N_2O_6	0.24333
3.5794	$(NH_4)_2PtCl_6$ ↔ NO_3	0.27938
4.7169	$(NH_4)_2SO_4$ ↔ N	0.21200
1.2234	$(NH_4)_2SO_4$ ↔ N_2O_5	0.81739
1.5480	Pt ↔ HNO_3	0.64599
6.9640	Pt ↔ N	0.14360

(*Continued*)

TABLE 4.34 Gravimetric Factors (*Continued*)

Factor		Factor
	NITROGEN (*continued*) $N = 14.0067$	
1.5732	$Pt \leftrightarrow NO_3$	0.63566
1.8062	$Pt \leftrightarrow N_2O_5$	0.55364
0.63528	$SO_3 \leftrightarrow HNO_3$	1.5741
2.8579	$SO_3 \leftrightarrow N$	0.34990
0.74125	$SO_3 \leftrightarrow N_2O_5$	1.3491
	OSMIUM $Os = 190.2$	
1.3365	$OsO_4 \leftrightarrow Os$	0.74823
	PALLADIUM $Pd = 106.4$	
0.49873	$Pd \leftrightarrow PdCl_2 \cdot 2H_2O$	2.0051
0.46179	$Pd \leftrightarrow Pd(NO_3)_2$	2.1655
3.3854	$PdI_2 \leftrightarrow Pd$	0.29538
3.7342	$K_2PdCl_6 \leftrightarrow Pd$	0.26779
1.8624	$K_2PdCl_6 \leftrightarrow PdCl_2 \cdot 2H_2O$	0.53695
	PHOSPHORUS $P = 30.9738$	
13.514	$Ag_3PO_4 \leftrightarrow P$	0.073998
4.4075	$Ag_3PO_4 \leftrightarrow PO_4$	0.22689
5.8980	$Ag_3PO_4 \leftrightarrow P_2O_5$	0.16955
9.7730	$Ag_4P_2O_7 \leftrightarrow P$	0.10232
3.1874	$Ag_4P_2O_7 \leftrightarrow PO_4$	0.31374
4.2653	$Ag_4P_2O_7 \leftrightarrow P_2O_5$	0.23445
0.71833	$Al_2O_3 \leftrightarrow P_2O_5$	1.3921
1.2841	$AlPO_4 \leftrightarrow PO_4$	0.77877
1.7183	$AlPO_4 \leftrightarrow P_2O_5$	0.58196
2.1853	$Ca_3(PO_4)_2 \leftrightarrow P_2O_5$	0.45761
1.5881	$FePO_4 \leftrightarrow PO_4$	0.62970
2.1251	$FePO_4 \leftrightarrow P_2O_5$	0.47056
0.78392	$Mg_2P_2O_7 \leftrightarrow Na_2HPO_4$	1.2756
0.31073	$Mg_2P_2O_7 \leftrightarrow Na_2HPO_4 \cdot 12H_2O$	3.2182
0.53229	$Mg_2P_2O_7 \leftrightarrow NaNH_4HPO_4 \cdot 4H_2O$	1.8787
3.5929	$Mg_2P_2O_7 \leftrightarrow P$	0.27833
1.1718	$Mg_2P_2O_7 \leftrightarrow PO_4$	0.85340
1.5681	$Mg_2P_2O_7 \leftrightarrow P_2O_5$	0.63773
60.577	$(NH_4)_3PO_4 \cdot 12MoO_3 \leftrightarrow P$	0.016508
19.757	$(NH_4)_3PO_4 \cdot 12MoO_3 \leftrightarrow PO_4$	0.050616
26.438	$(NH_4)_3PO_4 \cdot 12MoO_3 \leftrightarrow P_2O_5$	0.037824
0.63773	$P_2O_5 \leftrightarrow Mg_2P_2O_7$	1.5681
0.49993	$P_2O_5 \leftrightarrow Na_2HPO_4$	2.0003
0.19816	$P_2O_5 \leftrightarrow Na_2HPO_4 \cdot 12H_2O$	5.0464
0.33946	$P_2O_5 \leftrightarrow NaNH_4HPO_4 \cdot 4H_2O$	2.9459
2.2913	$P_2O_5 \leftrightarrow P$	0.43644
58.057	$P_2O_5 \cdot 24MoO_3 \leftrightarrow P$	0.017225
18.935	$P_2O_5 \cdot 24MoO_3 \leftrightarrow PO_4$	0.052813
25.338	$P_2O_5 \cdot 24MoO_3 \leftrightarrow P_2O_5$	0.039466
11.526	$U_2P_2O_{11} \leftrightarrow P$	0.086762

TABLE 4.34 Gravimetric Factors (*Continued*)

Factor		Factor
	PHOSPHORUS (*continued*) **P = 30.9738**	
3.7590	$U_2P_2O_{11} \leftrightarrow PO_4$	0.26603
5.0303	$U_2P_2O_{11} \leftrightarrow P_2O_5$	0.19880
	PLATINUM **Pt = 195.09**	
0.93839	$K_2PtCl_6 \leftrightarrow H_2PtCl_6 \cdot 6H_2O$	1.0657
2.4912	$K_2PtCl_6 \leftrightarrow Pt$	0.40141
1.4426	$K_2PtCl_6 \leftrightarrow PtCl_4$	0.69320
1.1383	$K_2PtCl_6 \leftrightarrow PtCl_4 \cdot 5H_2O$	0.87854
2.2753	$(NH_4)_2PtCl_6 \leftrightarrow Pt$	0.43950
1.3176	$(NH_4)_2PtCl_6 \leftrightarrow PtCl_4$	0.75897
1.0885	$(NH_4)_2PtCl_6 \leftrightarrow PtCl_6$	0.91872
0.37668	$Pt \leftrightarrow H_2PtCl_6 \cdot 6H_2O$	2.6548
0.57907	$Pt \leftrightarrow PtCl_4$	1.7269
0.45691	$Pt \leftrightarrow PtCl_4 \cdot 5H_2O$	2.1886
	POTASSIUM **K = 39.098**	
0.90639	$Ag \leftrightarrow KBr$	1.1033
1.4469	$Ag \leftrightarrow KCl$	0.69116
0.88021	$Ag \leftrightarrow KClO_3$	1.1361
0.77856	$Ag \leftrightarrow KClO_4$	1.2844
1.6565	$Ag \leftrightarrow KCN$	0.60369
0.64978	$Ag \leftrightarrow KI$	1.5390
1.5779	$AgBr \leftrightarrow KBr$	0.63377
1.1244	$AgBr \leftrightarrow KBrO_3$	0.88939
1.9223	$AgCl \leftrightarrow KCl$	0.52020
1.1695	$AgCl \leftrightarrow KClO_3$	0.85508
1.0344	$AgCl \leftrightarrow KClO_4$	0.96672
2.0561	$AgCN \leftrightarrow KCN$	0.48637
1.4142	$AgI \leftrightarrow KI$	0.70712
1.0971	$AgI \leftrightarrow KIO_3$	0.91153
1.3045	$BaCrO_4 \leftrightarrow K_2CrO_4$	0.76659
1.7222	$BaCrO_4 \leftrightarrow K_2Cr_2O_7$	0.58065
1.7140	$BaSO_4 \leftrightarrow KHSO_4$	0.58342
2.1166	$BaSO_4 \leftrightarrow K_2S$	0.47245
1.3393	$BaSO_4 \leftrightarrow K_2SO_4$	0.74666
2.0436	$Br \leftrightarrow K$	0.48933
0.67145	$Br \leftrightarrow KBr$	1.4893
0.41473	$CaF_2 \leftrightarrow KF \cdot 2H_2O$	2.4112
0.72315	$CaSO_4 \leftrightarrow KF \cdot 2H_2O$	1.3828
0.90668	$Cl \leftrightarrow K$	1.1029
0.47553	$Cl \leftrightarrow KCl$	2.1029
0.28929	$Cl \leftrightarrow KClO_3$	3.4567
0.25589	$Cl \leftrightarrow KClO_4$	3.9080
0.75269	$Cl \leftrightarrow K_2O$	1.3286
0.46718	$CO_2 \leftrightarrow K_2O$	2.1405
0.31843	$CO_2 \leftrightarrow K_2CO_3$	3.1404
0.76441	$I \leftrightarrow KI$	1.3082
0.59299	$I \leftrightarrow KIO_3$	1.6864

(*Continued*)

TABLE 4.34 Gravimetric Factors (*Continued*)

Factor		Factor
	POTASSIUM (*continued*) **K = 39.098**	
0.31907	$K \leftrightarrow KClO_3$	3.1341
0.83016	$K \leftrightarrow K_2O$	1.2046
0.38673	$K \leftrightarrow KNO_3$	2.5858
3.0436	$KBr \leftrightarrow K$	0.32856
2.5267	$KBr \leftrightarrow K_2O$	0.39578
1.9067	$KCl \leftrightarrow K$	0.52447
1.0789	$KCl \leftrightarrow K_2CO_3$	0.92690
0.50685	$KCl \leftrightarrow K_2Cr_2O_7$	1.9730
0.74466	$KCl \leftrightarrow KHCO_3$	1.3429
0.73737	$KCl \leftrightarrow KNO_3$	1.3562
1.5829	$KCl \leftrightarrow K_2O$	0.63177
0.85563	$KCl \leftrightarrow K_2SO_4$	1.1687
1.6437	$KClO_3 \leftrightarrow KCl$	0.60836
3.5433	$KClO_4 \leftrightarrow K$	0.28222
1.8584	$KClO_4 \leftrightarrow KCl$	0.53811
2.9415	$KClO_4 \leftrightarrow K_2O$	0.33996
4.2456	$KI \leftrightarrow K$	0.23554
3.5245	$KI \leftrightarrow K_2O$	0.28373
0.38435	$K_2O \leftrightarrow KClO_3$	2.6018
0.68159	$K_2O \leftrightarrow K_2CO_3$	1.4672
0.32021	$K_2O \leftrightarrow K_2Cr_2O_7$	3.1229
0.47045	$K_2O \leftrightarrow KHCO_3$	2.1256
0.46584	$K_2O \leftrightarrow KNO_3$	2.1466
0.81194	$KOH \leftrightarrow K_2CO_3$	1.2316
1.1912	$KOH \leftrightarrow K_2O$	0.83946
6.2146	$K_2PtCl_6 \leftrightarrow K$	0.16091
3.5165	$K_2PtCl_6 \leftrightarrow K_2CO_3$	0.28438
3.2594	$K_2PtCl_6 \leftrightarrow KCl$	0.30680
2.4271	$K_2PtCl_6 \leftrightarrow KHCO_3$	0.41201
2.4034	$K_2PtCl_6 \leftrightarrow KNO_3$	0.41608
5.1592	$K_2PtCl_6 \leftrightarrow K_2O$	0.19383
2.7888	$K_2PtCl_6 \leftrightarrow K_2SO_4$	0.35857
0.51224	$K_2PtCl_6 \leftrightarrow K_2SO_4 \cdot Al_2(SO_4)_3 \cdot 24H_2O$	1.9522
0.48659	$K_2PtCl_6 \leftrightarrow K_2SO_4 \cdot Cr_2(SO_4)_3 \cdot 24H_2O$	2.0551
1.2609	$K_2SO_4 \leftrightarrow K_2CO_3$	0.79308
0.87031	$K_2SO_4 \leftrightarrow KHCO_3$	1.1490
0.63990	$K_2SO_4 \leftrightarrow KHSO_4$	1.5627
1.0238	$K_2SO_4 \leftrightarrow KNO_2$	0.97674
0.86179	$K_2SO_4 \leftrightarrow KNO_3$	1.1604
2.2285	$K_2SO_4 \leftrightarrow K$	0.44875
1.8499	$K_2SO_4 \leftrightarrow K_2O$	0.54056
1.5804	$K_2SO_4 \leftrightarrow K_2S$	0.63275
0.60582	$Mg_2As_2O_7 \leftrightarrow K_3AsO_4$	1.6506
0.71164	$Mg_2As_2O_7 \leftrightarrow K_2HAsO_4$	1.4052
0.40040	$Mn_2O_3 \leftrightarrow K_2MnO_4$	2.4975
0.49946	$Mn_2O_3 \leftrightarrow KMnO_4$	2.0022
0.44132	$MnS \leftrightarrow K_2MnO_4$	2.2659
0.55051	$MnS \leftrightarrow KMnO_4$	1.8165
0.13853	$N \leftrightarrow KNO_3$	7.2185
0.16844	$NH_3 \leftrightarrow KNO_3$	5.9368

TABLE 4.34 Gravimetric Factors (*Continued*)

Factor		Factor
	POTASSIUM (*continued*) **K = 39.098**	
0.29677	$NO \leftrightarrow KNO_3$	3.3697
0.44656	$N_2O_3 \leftrightarrow KNO_2$	2.2393
1.1466	$N_2O_5 \leftrightarrow K_2O$	0.87217
0.53412	$N_2O_5 \leftrightarrow KNO_3$	1.8722
2.4946	$Pt \leftrightarrow K$	0.40086
1.3084	$Pt \leftrightarrow KCl$	0.76431
2.0710	$Pt \leftrightarrow K_2O$	0.48287
0.38943	$SiO_2 \leftrightarrow K_2SiO_3$	2.5679
0.45941	$SO_3 \leftrightarrow K_2SO_4$	2.1767
	PRASEODYMIUM **Pr = 140.908**	
1.1703	$Pr_2O_3 \leftrightarrow Pr$	0.85449
	RHODIUM **Rh = 102.905**	
0.26758	$Rh \leftrightarrow Na_3RhCl_6$	3.7372
0.49178	$Rh \leftrightarrow RhCl_3$	2.0334
	RUBIDIUM **Rb = 85.468**	
1.6768	$AgCl \leftrightarrow Rb$	0.59636
1.1852	$AgCl \leftrightarrow RbCl$	0.84371
0.41480	$Cl \leftrightarrow Rb$	2.4108
0.29319	$Cl \leftrightarrow RbCl$	3.4107
0.70683	$Rb \leftrightarrow RbCl$	1.4148
0.74016	$Rb \leftrightarrow Rb_2CO_3$	1.3511
0.91441	$Rb \leftrightarrow Rb_2O$	1.0936
0.64023	$Rb \leftrightarrow Rb_2SO_4$	1.5620
1.0472	$RbCl \leftrightarrow Rb_2CO_3$	0.95497
0.90577	$RbCl \leftrightarrow Rb_2SO_4$	1.1040
2.1636	$RbClO_4 \leftrightarrow Rb$	0.46220
0.78828	$Rb_2CO_3 \leftrightarrow RbHCO_3$	1.2686
0.77299	$Rb_2O \leftrightarrow RbCl$	1.2937
0.70015	$Rb_2O \leftrightarrow Rb_2SO_4$	1.4283
3.3857	$Rb_2PtCl_6 \leftrightarrow Rb$	0.29536
2.3931	$Rb_2PtCl_6 \leftrightarrow RbCl$	0.41787
2.5060	$Rb_2PtCl_6 \leftrightarrow Rb_2CO_3$	0.39905
1.9754	$Rb_2PtCl_6 \leftrightarrow RbHCO_3$	0.50623
3.0959	$Rb_2PtCl_6 \leftrightarrow Rb_2O$	0.32301
1.1561	$Rb_2SO_4 \leftrightarrow Rb_2CO_3$	0.86498
0.91133	$Rb_2SO_4 \leftrightarrow RbHCO_3$	1.0973
	SELENIUM **Se = 78.96**	
0.61224	$Se \leftrightarrow H_2SeO_3$	1.6334
0.54466	$Se \leftrightarrow H_2SeO_4$	1.8360
0.71161	$Se \leftrightarrow SeO_2$	1.4053
0.62193	$Se \leftrightarrow SeO_3$	1.6079

(*Continued*)

TABLE 4.34 Gravimetric Factors (*Continued*)

Factor		Factor
	SILICON $Si = 28.086$	
2.6847	$BaSiF_6 \leftrightarrow SiF_4$	0.37249
4.6504	$BaSiF_6 \leftrightarrow SiO_2$	0.21503
2.1163	$K_2SiF_6 \leftrightarrow SiF_4$	0.47249
3.6661	$K_2SiF_6 \leftrightarrow SiO_2$	0.27277
3.3384	$SiC \leftrightarrow C$	0.29954
0.91111	$SiC \leftrightarrow CO_2$	1.0976
0.76933	$SiO_2 \leftrightarrow H_2SiO_3$	1.2998
2.1393	$SiO_2 \leftrightarrow Si$	0.46744
0.57730	$SiO_2 \leftrightarrow SiF_4$	1.7322
0.78972	$SiO_2 \leftrightarrow SiO_3$	1.2663
0.65250	$SiO_2 \leftrightarrow SiO_4$	1.5326
1.6651	$SiO_2 \leftrightarrow Si_2O$	0.60057
0.62514	$SiO_2 \leftrightarrow Si(OH)_4$	1.5997
	SILVER $Ag = 107.868$	
0.63501	$Ag \leftrightarrow AgNO_3$	1.5748
0.93096	$Ag \leftrightarrow Ag_2O$	1.0742
1.7408	$AgBr \leftrightarrow Ag$	0.57445
1.3286	$AgCl \leftrightarrow Ag$	0.75265
0.84371	$AgCl \leftrightarrow AgNO_3$	1.1852
1.2369	$AgCl \leftrightarrow Ag_2O$	0.80847
1.7935	$AgCl \leftrightarrow Br$	0.55756
1.2412	$AgCN \leftrightarrow Ag$	0.80566
2.1764	$Agl \leftrightarrow Ag$	0.45947
1.2935	$Ag_3PO_4 \leftrightarrow Ag$	0.77311
1.4031	$Ag_4P_2O_7 \leftrightarrow Ag$	0.71269
0.74079	$Br \leftrightarrow Ag$	1.3499
0.42555	$Br \leftrightarrow AgBr$	2.3499
0.32866	$Cl \leftrightarrow Ag$	3.0426
0.24737	$Cl \leftrightarrow AgCl$	4.0425
1.1764	$I \leftrightarrow Ag$	0.85004
0.54053	$I \leftrightarrow Agl$	1.8500
	SODIUM $Na = 22.9898$	
1.0483	$Ag \leftrightarrow NaBr$	0.95393
1.8457	$Ag \leftrightarrow NaCl$	0.54179
0.71966	$Ag \leftrightarrow Nal$	1.3895
1.8249	$AgBr \leftrightarrow NaBr$	0.54798
2.4523	$AgCl \leftrightarrow NaCl$	0.40778
1.5663	$Agl \leftrightarrow Nal$	0.63845
1.9440	$BaSO_4 \leftrightarrow NaHSO_4$	0.51440
1.6905	$BaSO_4 \leftrightarrow NaHSO_4 \cdot H_2O$	0.59156
2.9906	$BaSO_4 \leftrightarrow Na_2S$	0.33438
1.8518	$BaSO_4 \leftrightarrow Na_2SO_3$	0.54002
0.92564	$BaSO_4 \leftrightarrow Na_2SO_3 \cdot 7H_2O$	1.0803
1.6432	$BaSO_4 \leftrightarrow Na_2SO_4$	0.60857
0.72442	$BaSO_4 \leftrightarrow Na_2SO_4 \cdot 10H_2O$	1.3804

TABLE 4.34 Gravimetric Factors (*Continued*)

Factor		Factor
	SODIUM (*continued*) **Na = 22.9898**	
0.69198	$B_2O_3 \leftrightarrow Na_2B_4O_7$	1.4451
0.36510	$B_2O_3 \leftrightarrow Na_2B_4O_7 \cdot 10H_2O$	2.7389
3.4758	$Br \leftrightarrow Na$	0.28770
0.77657	$Br \leftrightarrow NaBr$	1.2877
2.5786	$Br \leftrightarrow Na_2O$	0.38781
0.94956	$CaCl_2 \leftrightarrow NaCl$	1.0531
0.94433	$CaCO_3 \leftrightarrow Na_2CO_3$	1.0590
0.92975	$CaF_2 \leftrightarrow NaF$	1.0756
0.52910	$CaO \leftrightarrow Na_2CO_3$	1.8900
1.2845	$CaSO_4 \leftrightarrow Na_2CO_3$	0.77854
1.5421	$Cl \leftrightarrow Na$	0.64846
0.60663	$Cl \leftrightarrow NaCl$	1.6485
1.1442	$Cl \leftrightarrow Na_2O$	0.87410
0.41520	$CO_2 \leftrightarrow Na_2CO_3$	2.4083
0.71008	$CO_2 \leftrightarrow Na_2O$	1.4083
1.2292	$H_3BO_3 \leftrightarrow Na_2B_4O_7$	0.81357
0.64853	$H_3BO_3 \leftrightarrow Na_2B_4O_7 \cdot 10H_2O$	1.5419
5.5198	$I \leftrightarrow Na$	0.18117
0.84662	$I \leftrightarrow NaI$	1.1812
4.0949	$I \leftrightarrow Na_2O$	0.24420
2.5029	$KBF_4 \leftrightarrow Na_2B_4O_7$	0.39954
1.3206	$KBF_4 \leftrightarrow Na_2B_4O_7 \cdot 10H_2O$	0.75724
0.91360	$Mg_2As_2O_7 \leftrightarrow Na_2HAsO_3$	1.0946
0.83497	$Mg_2As_2O_7 \leftrightarrow Na_2HAsO_4$	1.1976
0.81462	$MgCl_2 \leftrightarrow NaCl$	1.2276
0.67882	$Mg_2P_2O_7 \leftrightarrow Na_3PO_4$	1.4731
0.78392	$Mg_2P_2O_7 \leftrightarrow Na_2HPO_4$	1.2757
0.31073	$Mg_2P_2O_7 \leftrightarrow NaHPO_4 \cdot 12H_2O$	3.2182
0.53229	$Mg_2P_2O_7 \leftrightarrow NaNH_4 \cdot HPO_4 \cdot 4H_2O$	1.8787
0.49897	$Mg_2P_2O_7 \leftrightarrow Na_4P_2O_7 \cdot 10H_2O$	2.0041
4.4759	$NaBr \leftrightarrow Na$	0.22342
3.3205	$NaBr \leftrightarrow Na_2O$	0.30116
65.502	$NaOAc \cdot Mg(OAc)_2 \cdot UO_2(OAc)_2 \cdot 6\frac{1}{2}H_2O \leftrightarrow Na$	0.015267
14.635	Triple MgOAc $\leftrightarrow NaBr$	0.066331
28.416	Triple MgOAc $\leftrightarrow Na_2CO_3$	0.035192
25.768	Triple MgOAc $\leftrightarrow NaCl$	0.038809
17.926	Triple MgOAc $\leftrightarrow NaHCO_3$	0.055785
10.047	Triple MgOAc $\leftrightarrow NaI$	0.099535
37.650	Triple MgOAc $\leftrightarrow NaOH$	0.026560
48.594	Triple MgOAc $\leftrightarrow Na_2O$	0.020579
21.204	Triple MgOAc $\leftrightarrow Na_2SO_4$	0.047161
66.894	$NaOAc \cdot Zn(OAc)_2 \cdot UO_2(OAc)_2 \cdot 6H_2O \leftrightarrow Na$	0.014949
14.946	Triple ZnOAc $\leftrightarrow NaBr$	0.066909
29.020	Triple ZnOAc $\leftrightarrow Na_2CO_3$	0.034459
26.315	Triple ZnOAc $\leftrightarrow NaCl$	0.038002
18.307	Triple ZnOAc $\leftrightarrow NaHCO_3$	0.054624
10.260	Triple ZnOAc $\leftrightarrow NaI$	0.097464
38.451	Triple ZnOAc $\leftrightarrow NaOH$	0.026008
49.626	Triple ZnOAc $\leftrightarrow Na_2O$	0.020151
21.654	Triple ZnOAc $\leftrightarrow Na_2SO_4$	0.046180

(*Continued*)

TABLE 4.34 Gravimetric Factors (*Continued*)

Factor		Factor
	SODIUM (*continued*) **Na = 22.9898**	
2.5421	$NaCl \leftrightarrow Na$	0.39337
1.1028	$NaCl \leftrightarrow Na_2CO_3$	0.90678
0.69569	$NaCl \leftrightarrow NaHCO_3$	1.4374
0.82337	$NaCl \leftrightarrow Na_2HPO_4$	1.2145
1.8859	$NaCl \leftrightarrow Na_2O$	0.53025
0.82291	$NaCl \leftrightarrow Na_2SO_4$	1.2152
0.74267	$NaClO_3 \leftrightarrow AgCl$	1.3465
1.8213	$NaClO_3 \leftrightarrow NaCl$	0.54907
0.85432	$NaClO_4 \leftrightarrow AgCl$	1.1705
2.0950	$NaClO_4 \leftrightarrow NaCl$	0.47732
2.3051	$Na_2CO_3 \leftrightarrow Na$	0.43381
0.63084	$Na_2CO_3 \leftrightarrow NaHCO_3$	1.5852
1.7101	$Na_2CO_3 \leftrightarrow Na_2O$	0.58476
1.3250	$Na_2CO_3 \leftrightarrow NaOH$	0.75473
3.6541	$NaHCO_3 \leftrightarrow Na$	0.27367
2.7108	$NaHCO_3 \leftrightarrow Na_2O$	0.36889
6.5198	$NaI \leftrightarrow Na$	0.15338
4.8368	$NaI \leftrightarrow Na_2O$	0.20675
1.3480	$Na_2O \leftrightarrow Na$	0.74186
0.43659	$Na_2O \leftrightarrow Na_2HPO_4$	2.2905
0.36460	$Na_2O \leftrightarrow NaNO_3$	2.7427
0.77480	$Na_2O \leftrightarrow NaOH$	1.2907
0.93653	$Na_4P_2O_7 \leftrightarrow Na_2HPO_4$	1.0678
0.37122	$Na_4P_2O_7 \leftrightarrow Na_2HPO_4 \cdot 12H_2O$	2.6938
3.0892	$Na_2SO_4 \leftrightarrow Na$	0.32371
1.3401	$Na_2SO_4 \leftrightarrow Na_2CO_3$	0.74620
0.49640	$Na_2SO_4 \leftrightarrow Na_2CO_3 \cdot 10H_2O$	2.0145
2.2917	$Na_2SO_4 \leftrightarrow Na_2O$	0.43635
0.16480	$N \leftrightarrow NaNO_3$	6.0680
0.20038	$NH_3 \leftrightarrow NaNO_3$	4.9906
0.081461	$NH_3 \leftrightarrow NaNH_4HPO_4 \cdot 4H_2O$	12.276
0.35303	$NO \leftrightarrow NaNO_3$	2.8326
0.63539	$N_2O_5 \leftrightarrow NaNO_3$	1.5738
1.7427	$N_2O_5 \leftrightarrow Na_2O$	0.57383
0.49993	$P_2O_5 \leftrightarrow Na_2HPO_4$	2.0003
0.19816	$P_2O_5 \leftrightarrow Na_2HPO_4 \cdot 12H_2O$	5.0464
0.33946	$P_2O_5 \leftrightarrow NaNH_4HPO_4 \cdot H_2O$	2.9459
0.61564	$SO_2 \leftrightarrow NaHSO_3$	1.6243
0.50828	$SO_2 \leftrightarrow Na_2SO_3$	1.9674
0.25407	$SO_2 \leftrightarrow Na_2SO_3 \cdot 7H_2O$	3.9360
1.2918	$SO_2 \leftrightarrow Na_2O$	0.77414
0.56366	$SO_2 \leftrightarrow Na_2SO_4$	1.7741
	STRONTIUM **Sr = 87.62**	
0.29811	$CO_2 \leftrightarrow SrCO_8$	3.3545
0.77265	$SO_3 \leftrightarrow SrO$	1.2942
0.43588	$SO_3 \leftrightarrow SrSO_4$	2.2942
0.41402	$Sr \leftrightarrow Sr(NO_3)_2$	2.4153
1.6849	$SrCO_3 \leftrightarrow Sr$	0.59351

TABLE 4.34 Gravimetric Factors (*Continued*)

Factor		Factor
	STRONTIUM (*continued*) **Sr = 87.62**	
0.93124	$SrCO_3 \leftrightarrow SrCl_2$	1.0738
0.70424	$SrCO_3 \leftrightarrow Sr(HCO_3)_2$	1.4200
0.69759	$SrCO_3 \leftrightarrow Sr(NO_3)_2$	1.4335
1.1826	$SrO \leftrightarrow Sr$	0.84559
0.65363	$SrO \leftrightarrow SrCl_2$	1.5299
0.70189	$SrO \leftrightarrow SrCO_3$	1.4247
0.49430	$SrO \leftrightarrow Sr(HCO_3)_2$	2.0231
0.48963	$SrO \leftrightarrow Sr(NO_3)_2$	2.0424
2.0963	$SrSO_4 \leftrightarrow Sr$	0.47703
1.1586	$SrSO_4 \leftrightarrow SrCl_2$	0.86308
1.2442	$SrSO_4 \leftrightarrow SrCO_3$	0.80373
0.86793	$SrSO_4 \leftrightarrow Sr(NO_3)_2$	1.1522
1.7726	$SrSO_4 \leftrightarrow SrO$	0.56413
	SULFUR **S = 32.06**	
2.4064	$As_2S_3 \leftrightarrow H_2S$	0.41556
2.5577	$As_2S_3 \leftrightarrow S$	0.39097
3.8906	$BaSO_4 \leftrightarrow FeS_2$	0.25703
6.8486	$BaSO_4 \leftrightarrow H_2S$	0.14602
2.8436	$BaSO_4 \leftrightarrow H_2SO_3$	0.35166
2.3797	$BaSO_4 \leftrightarrow H_2SO_4$	0.42022
7.2792	$BaSO_4 \leftrightarrow S$	0.13738
3.6433	$BaSO_4 \leftrightarrow SO_2$	0.27448
2.9152	$BaSO_4 \leftrightarrow SO_3$	0.34302
2.4297	$BaSO_4 \leftrightarrow SO_4$	0.41158
4.2388	$CdS \leftrightarrow H_2S$	0.23591
4.5054	$CdS \leftrightarrow S$	0.22196
1.2250	$H_2SO_4 \leftrightarrow SO_3$	0.81631
1.6505	$(NH_4)_2SO_4 \leftrightarrow SO_3$	0.60589
1.3473	$(NH_4)_2SO_4 \leftrightarrow H_2SO_4$	0.74223
2.3492	$SO_3 \leftrightarrow H_2S$	0.42567
	TANTALUM **Ta = 180.948**	
0.81898	$Ta \leftrightarrow Ta_2O_5$	1.2210
0.50515	$Ta \leftrightarrow TaCl_5$	1.9796
16.065	$TaC \leftrightarrow C$	0.062246
1.0664	$TaC \leftrightarrow Ta$	0.93776
0.61680	$Ta_2O_5 \leftrightarrow TaCl_5$	1.6213
1.0376	$Ta_2O_5 \leftrightarrow Ta_2O_4$	0.96379
	TELLURIUM **Te = 127.60**	
0.65906	$Te \leftrightarrow H_2TeO_4$	1.5173
0.55565	$Te \leftrightarrow H_2TeO_4 \cdot 2H_2O$	1.7997
0.79950	$Te \leftrightarrow TeO_2$	1.2508
0.72665	$Te \leftrightarrow TeO_3$	1.3762
1.5645	$(TeO_2)_2SO_3 \leftrightarrow Te$	0.63918

(*Continued*)

TABLE 4.34 Gravimetric Factors (*Continued*)

Factor		Factor
	THALLIUM **Tl = 204.37**	
0.87198	$Tl \leftrightarrow Tl_2CO_3$	1.1468
0.85218	$Tl \leftrightarrow TlCl$	1.1735
0.61693	$Tl \leftrightarrow TlI$	1.6209
0.76724	$Tl \leftrightarrow TlNO_3$	1.3034
0.96232	$Tl \leftrightarrow Tl_2O$	1.0391
1.2838	$Tl_2CrO_4 \leftrightarrow Tl$	0.77895
1.4750	$TlHSO_4 \leftrightarrow Tl$	0.67798
1.9977	$Tl_2PtCl_6 \leftrightarrow Tl$	0.50057
1.7024	$Tl_2PtCl_6 \leftrightarrow TlCl$	0.58740
1.7420	$Tl_2PtCl_6 \leftrightarrow Tl_2CO_3$	0.57406
1.2325	$Tl_2PtCl_6 \leftrightarrow TlI$	0.81139
1.5327	$Tl_2PtCl_6 \leftrightarrow TlNO_3$	0.65243
1.9225	$Tl_2PtCl_6 \leftrightarrow Tl_2O$	0.52017
1.6176	$Tl_2PtCl_6 \leftrightarrow Tl_2SO_4$	0.61821
1.2350	$Tl_2SO_4 \leftrightarrow Tl$	0.80971
	THORIUM **Th = 232.038**	
1.1379	$ThO_2 \leftrightarrow Th$	0.87881
0.70627	$ThO_2 \leftrightarrow ThCl_4$	1.4159
0.44893	$ThO_2 \leftrightarrow Th(NO_3)_4 \cdot 6H_2O$	2.2275
	TIN **Sn = 118.69**	
0.62600	$Sn \leftrightarrow SnCl_2$	1.5974
0.52604	$Sn \leftrightarrow SnCl_2 \cdot 2H_2O$	1.9010
0.45562	$Sn \leftrightarrow SnCl_4$	2.1948
0.32297	$Sn \leftrightarrow SnCl_4 \cdot (NH_4Cl)_2$	3.0962
0.88121	$Sn \leftrightarrow SnO$	1.1348
0.78764	$Sn \leftrightarrow SnO_2$	1.2696
0.79478	$SnO_2 \leftrightarrow SnCl_2$	1.2582
0.66786	$SnO_2 \leftrightarrow SnCl_2 \cdot 2H_2O$	1.4973
0.57846	$SnO_2 \leftrightarrow SnCl_4$	1.7287
0.41005	$SnO_2 \leftrightarrow SnCl_4 \cdot (NH_4Cl)_2$	2.4387
1.1188	$SnO_2 \leftrightarrow SnO$	0.89382
	TITANIUM **Ti = 47.867**	
2.1059	$K_2TiF_6 \leftrightarrow F$	0.47485
3.0699	$K_2TiF_6 \leftrightarrow K$	0.32574
2.0660	$K_2TiF_6 \leftrightarrow 2KF$	0.48403
1.2752	$K_2TiF_6 \leftrightarrow 2(KF \cdot 2H_2O)$	0.78421
5.0150	$K_2TiF_6 \leftrightarrow Ti$	0.19940
3.0057	$K_2TiF_6 \leftrightarrow TiO_2$	0.33270
3.9853	$Ti \leftrightarrow C$	0.25092
4.9853	$TiC \leftrightarrow C$	0.20059
1.2509	$TiC \leftrightarrow Ti$	0.79940
1.6299	$TiF_4 \leftrightarrow F$	0.61354
1.6685	$TiO_2 \leftrightarrow Ti$	0.59934

TABLE 4.34 Gravimetric Factors (*Continued*)

Factor		Factor
	TUNGSTEN W = 183.85	
3.9348	$FeWO_4 \leftrightarrow Fe_3O_4$	0.25414
1.3099	$FeWO_4 \leftrightarrow WO_3$	0.76344
6.7515	$MgWO_4 \leftrightarrow MgO$	0.14812
1.1739	$MgWO_4 \leftrightarrow WO_3$	0.85189
4.2684	$MnWO_4 \leftrightarrow MnO$	0.23428
1.3060	$MnWO_4 \leftrightarrow WO_3$	0.76571
2.0387	$PbWO_4 \leftrightarrow PbO$	0.49051
2.4751	$PbWO_4 \leftrightarrow W$	0.40403
1.9626	$PbWO_4 \leftrightarrow WO_3$	0.50952
15.307	$W \leftrightarrow C$	0.065330
0.96837	$W \leftrightarrow W_2C$	1.0327
0.93868	$W \leftrightarrow WC$	1.0653
31.614	$W_2C \leftrightarrow C$	0.031632
16.307	$WC \leftrightarrow C$	0.061324
1.1741	$WO_2 \leftrightarrow W$	0.85175
4.1515	$WO_3 \leftrightarrow Fe$	0.24088
1.2611	$WO_3 \leftrightarrow W$	0.79297
	URANIUM U = 238.03	
1.1344	$UO_2 \leftrightarrow U$	0.88149
1.1792	$U_3O_8 \leftrightarrow U$	0.84800
1.0395	$U_3O_8 \leftrightarrow UO_2$	0.96200
0.55901	$U_3O_8 \leftrightarrow UO_2(NO_3)_2 \cdot 6H_2O$	1.7889
1.4998	$U_2P_2O_{11} \leftrightarrow U$	0.66675
1.3221	$U_2P_2O_{11} \leftrightarrow UO_2$	0.75639
	VANADIUM V = 50.941	
5.2413	$VC \leftrightarrow C$	0.19079
1.7852	$V_2O_5 \leftrightarrow V$	0.56017
0.79120	$V_2O_5 \leftrightarrow VO_4$	1.2639
	YTTERBIUM Yb = 173.04	
1.1387	$Yb_2O_3 \leftrightarrow Yb$	0.87820
	ZINC Zn = 65.38	
2.3955	$BaSO_4 \leftrightarrow ZnS$	0.41745
0.81171	$BaSO_4 \leftrightarrow ZnSO_4 \cdot 7H_2O$	1.2320
0.80338	$Zn \leftrightarrow ZnO$	1.2447
2.7288	$ZnNH_4PO_4 \leftrightarrow Zn$	0.36646
2.1922	$ZnNH_4PO_4 \leftrightarrow ZnO$	0.45616
0.59707	$ZnO \leftrightarrow ZnCl_2$	1.6748
0.64898	$ZnO \leftrightarrow ZnCO_3$	1.5409
0.28298	$ZnO \leftrightarrow ZnSO_4 \cdot 7H_2O$	3.5338
2.3304	$Zn_2P_2O_7 \leftrightarrow Zn$	0.42911
1.8722	$Zn_2P_2O_7 \leftrightarrow ZnO$	0.53413
1.4905	$ZnS \leftrightarrow Zn$	0.67091
1.1974	$ZnS \leftrightarrow ZnO$	0.83512
0.33885	$ZnS \leftrightarrow ZnSO_4 \cdot 7H_2O$	2.9511

(*Continued*)

TABLE 4.34 Gravimetric Factors (*Continued*)

Factor		Factor
	ZIRCONIUM **Zr = 91.22**	
2.4864	$K_2ZrF_6 \leftrightarrow F$	0.40219
2.4390	$K_2ZrF_6 \leftrightarrow 2KF$	0.41001
1.5054	$K_2ZrF_6 \leftrightarrow 2(KF \cdot 2H_2O)$	0.66427
3.1069	$K_2ZrF_6 \leftrightarrow Zr$	0.32187
2.3000	$K_2ZrF_6 \leftrightarrow ZrO_2$	0.43478
8.5946	$ZrC \leftrightarrow C$	0.11635
2.2004	$ZrF_4 \leftrightarrow F$	0.45447
1.3508	$ZrO_2 \leftrightarrow Zr$	0.74030
0.46470	$ZrO_2 \leftrightarrow ZrP_2O_7$	2.1519

TABLE 4.35 Elements Precipitated by General Analytical Reagents

This table includes the more common reagents used in gravimetric determinations. The lists of elements precipitated are not in all cases exhaustive. The usual solvent for a precipitating agent is indicated in parentheses after its name or formula. When the symbol of an element or radical is italicized, the element may be quantitatively determined by the use of the reagent in question.

Reagent	Conditions	Substances precipitated
Ammonia, NH_3 (aqueous)	After removal of acid sulfide group.	*Al*, Au, *Be*, Co, *Cr*, *Cu*, *Fe*, Ga, *In*, Ir, *La*, Nb, Ni, Os, P, *Pb*, *rare earths*, *Sc*, Si, *Sn*, Ta, *Th*, *Ti*, *U*, V, *Y*, *Zn*, *Zr*
Ammonium polysulfide, $(NH_4)_2S_x$ (aqueous)	After removal of acid sulfide and $(NH_4)_2S$ groups.	Co, Mn, Ni, Si, Tl, V, W, Zn
Anthranilic acid, $NH_2C_6H_4COOH$ (aqueous)	1% aqueous solution (pH 6); Cu separated from others at pH 2.9.	Ag, *Cd*, *Co*, *Cu*, Fe, *Hg*, *Mn*, *Ni*, *Pb*, *Zn*
α-Benzoin oxime, $C_6H_5CHOHC({=}NOH)C_6H_5$ (1–2% alcohol)	(a) Strongly acid medium. (b) Ammoniacal tartrate medium.	(a) Cr(VI), *Mo(VI)*, Nb, Pd(II), Ta(V), V(V), *W(VI)* (b) Above list
Benzidine, $H_2NC_6H_4C_6H_4NH_2$ (alcohol), 0.1*M* HCl		Cd, Fe(III), IO_3^-, *PO_4^{3-}*, SO_4^{2-}, *W(VI)*
N-Benzoylphenylhydroxylamine, $C_6H_5CO(C_6H_5)NOH$ (aqueous)	Similar to cupferron (*q.v.*). Cu, Fe(III), and Al complexes can be weighed as such; Ti compound must be ignited to the oxide.	See Cupferron
Cinchonine, $C_{19}H_{21}N_2OH$, 6*M* HCl		Ir, Mo, Pt, *W*
Cupferron, $C_6H_5N(NO)ONH_4$ (aqueous)	Group precipitant for several higher-charged metal ions from strongly acid solution. Precipitate ignited to metal oxide.	*Al*, *Bi*, *Cu*, *Fe*, *Ga*, La, Mo, *Nb*, Pd, rare earths, Sb, *Sn*, Ta, *Th*, *Ti*, Tl, *U*, *V*, W, *Zr*
1,2-Cyclohexanedionedioxime	More water soluble than dimethylglyoxime; less subject to coprecipitation with metal chelate.	See Dimethylglyoxime

TABLE 4.35 Elements Precipitated by General Analytical Reagents (*Continued*)

Reagent	Conditions	Substances precipitated
Diammonium hydrogen phosphate, $(NH_4)_2HPO_4$ (aqueous)	(a) Acid medium. (b) Ammoniacal medium containing citrate or tartrate.	(a) *Bi, Co,* Hf, In, Ti, *Zn, Zr* (b) Au, Ba, *Be,* Ca, Hg, In, La, *Mg, Mn,* Pb, rare earths, Sr, Th, U, *Zr*
Dimethylglyoxime, $[CH_3C(NOH)]_2$ (alcohol)	(a) Dilute HCl or H_2SO_4 medium. (b) Ammoniacal tartrate medium about pH 8. Weighed as such.	(a) Au, *Pd,* Se (b) *Ni* (and Co, Fe if present in large amounts)
Hydrazine, N_2H_4 (aqueous)		Ag, Au, *Cu, Hg,* Ir, *Os,* Pd, Pt, *Rh,* Ru, *Se, Te*
Hydrogen sulfide, H_2S	(a) 0.2–0.5*M* H^+. (b) Ammoniacal solution after removal of acid sulfide group.	(a) Ag, *As,* Au, Bi, Cd, *Cu, Ge, Hg,* In, *Ir, Mo,* Os, Pb, Pd, *Pt,* Re, *Rh,* Ru, Sb, Se, Sn, Te, Tl, V, W, Zn (b) Co, Fe, Ga, In, Mn, Ni, Tl, U, V, Zn
4-Hydroxyphenylarsonic acid, $C_6H_4(OH)AsO(OH)_2$ (aqueous)	Dilute acid solution.	Ce, *Fe, Sn, Th, Ti, Zr*
8-Hydroxyquinoline (oxine), C_9H_6NOH, (alcohol)	(a) HOAc–OAc^- buffer. (b) Ammoniacal solution.	(a) Ag, *Al,* Bi, *Cd, Co,* Cr, *Cu, Fe,* Ga, Hg, *In,* La, *Mn, Mo,* Nb, *Ni,* Pb, Pd, rare earths, Sb, Ta, Th, Ti, V, W, *Zn, Zr* (b) Same as in (a) except for Ag; in addition, Ba, *Be, Ca, Mg,* Sn, *Sr*
2-Mercaptobenzothiazole, $C_6H_4(SCN)SH$ (acetic acid solution)	Ammoniacal solution, except for Cu, when a dilute acid solution is used.	Ag, *Au, Bi, Cd, Cu,* Hg, *Ir, Pb, Pt, Rh,* Tl
Nitron (diphenylenedianilohydrotriazole), $C_{20}H_{16}N_4$, (5% acetic acid)	Dilute H_2SO_4 medium.	*B,* ClO_3^-, ClO_4^-, NO_3^-, ReO_4^-, *W*
1-Nitroso-2-naphthol, $C_{10}H_6(NO)OH$ (very dilute alkali)	Selective for Co; acid solution. Precipitate ignited to Co_3O_4.	Ag, Au, B, *Co,* Cr, *Cu, Fe,* Mo, Pd, Ti, V, W, Zr
Oxalic acid, $H_2C_2O_4$, (aqueous)	Dilute acid solution.	*Ag, Au,* Cu, *Hg, La,* Ni, *Pb, rare earths, Sc, Th,* U(IV), W, *Zr*
Phenylarsonic acid, $C_6H_5AsO(OH)_2$, (aqueous)	Selective precipitants for quadrivalent metals in acid solution. Metals weighed as dioxides.	*Bi,* Ce(IV), Fe, *Hf, Mg, Sn, Ta, Th,* Ti, U(IV), W, *Zr*
Phenylthiohydantoic acid, $C_6H_5N{=}C(NH_2)SCH_2COOH$ (aqueous or alcohol)		Bi, Cd, *Co,* Cu, Fe, Hg, Ni, Pb, Sb
Picrolonic acid, $C_{10}H_7O_5N_4H$ (aqueous)	Neutral solution.	*Ca,* Mg, *Pb, Th*
Propylarsonic acid, $C_3H_9AsO(OH)_2$ (aqueous)	Preferred for W; see Phenylarsonic acid.	
Pyridine plus thiocyanate	Dilute acid solution.	*Ag, Cd, Cu, Mn, Ni*
Quinaldic acid, C_9B_6NCOOH (aqueous)	Dilute acid solution.	Ag, *Cd,* Co, *Cu,* Fe, Hg, Mo, Ni, Pb, Pd, Pt(II), *U,* W, *Zn*
Salicylaldoxime, $C_7H_5(OH)NOH$ (alcohol)	Dilute acid solution.	Ag, *Bi,* Cd, Co, *Cu,* Fe, Hg, Mg, Mn, Ni, *Pb, Pd,* V, Zn
Silver nitrate, $AgNO_3$ (aqueous)	(a) Dilute HNO_3 solution. (b) Acetate buffer, pH 5–7.	(a) Br^-, Cl^-, I^-, SCN^- (b) *As(V),* CN^-, OCN^-, IO_3^-, *Mo(VI),* N_3^-, S^{2-}, *V(V)*

(*Continued*)

TABLE 4.35 Elements Precipitated by General Analytical Reagents (*Continued*)

Reagent	Conditions	Substances precipitated
Sodium tetraphenylborate, $NaB(C_6H_5)_4$ (aqueous)	Specific for K group of alkali metals from dilute HNO_3 or HOAc solution (pH 2), or pH 6.5 in presence of EDTA.	*Cs, K,* NH_4^+, *Rb*
Tannic acid (tannin), $C_{14}H_{10}O_9$ (aqueous)	Acts as negative colloid that is a flocculent for positively charged hydrous oxide sols. Noteworthy for W in acid solution, and for Ta (from Nb in acidic oxalate medium).	*Al, Be,* Cr, Ga, *Ge,* Nb, Sb, *Sn, Ta, Th, Ti, U,* V, *W, Zr*
Tartaric acid, HOOC(CHOH)$_2$COOH (aqueous)		*Ca, K, Mg, Sc, Sr, Ta*
Tetraphenylarsonium chloride, $(C_6H_5)_4AsCl$ (aqueous)	$(C_6H_5)_4AsTlCl_4$ and $(C_6H_5)_4AsReO_4$ weighed as such.	*Re, Tl*
Thioglycolic-β-aminonaphthalide, thionalide, $C_{10}H_7NHCOCH_2SH$ (alcohol)	(a) Acid solution.	(a) Ag, As, Au, Bi, *Cu, Hg, Os, Pb,* Pd, *Rh, Ru,* Sb, Sn, Tl
	(b) Carbonate medium containing tartrate.	(b) Au, Cd, Cu, Hg(II), Tl(I)
	(c) Carbonate medium containing tartrate and cyanide.	(c) Au, Bi, Pb, Sb, Sn, Tl
	(d) Strongly alkaline medium containing tartrate and cyanide.	(d) *Tl*

TABLE 4.36 Cleaning Solutions for Fritted Glassware

Material	Cleaning solution
Fatty materials	Carbon tetrachloride.
Organic matter	Hot concentrated sulfuric acid plus a few drops of sodium or potassium nitrate solution.
Albumen	Hot aqueous ammonia or hot hydrochloric acid.
Glucose	Hot mixed acid (sulfuric plus nitric acids).
Copper or iron oxides	Hot hydrochloric acid plus potassium chlorate.
Mercury residue	Not nitric acid.
Silver chloride	Aqueous ammonia or sodium thiosulfate.
Aluminous and siliceous residues	A 2% hydrofluoric acid solution followed by concentrated sulfuric acid; rinse immediately with distilled water followed by a few milliliters of acetone. Repeat rinsing until all trace of acid is removed.

TABLE 4.37 Common Fluxes

Flux	Melting point, °C	Types of crucible used for fusion	Type of substances decomposed
Na_2CO_3	851	Pt	For silicates, and silica-containing samples; alumina-containing samples; insoluble phosphates and sulfates
Na_2CO_3 plus an oxidizing agent such as KNO_3, $KClO_3$, or Na_2O_2		Pt (do not use with Na_2O_2) or Ni	For samples needing an oxidizing agent
NaOH or KOH	320–380	Au, Ag, Ni	For silicates, silicon carbide, certain minerals
Na_2O_2	Decomposes	Fe, Ni	For sulfides, acid-insoluble alloys of Fe, Ni, Cr, Mo, W, and Li; Pt alloys; Cr, Sn, Zn minerals
$K_2S_2O_7$	300	Pt or porcelain	Acid flux for insoluble oxides and oxide-containing samples
B_2O_3	577	Pt	For silicates and oxides when alkalis are to be determined
$CaCO_3$ plus NH_4Cl		Ni	For decomposing silicates in the determination of alkali element

TABLE 4.38 Membrane Filters

Filter pore size, μm	Maximum rigid particle to penetrate, μm	Filter pore size, μm	Maximum rigid particle to penetrate, μm
14	17	0.65	0.68
10	12	0.60	0.65
8	9.4	0.45	0.47
7	9.0	0.30	0.32
5	6.2	0.22	0.24
3	3.9	0.20	0.25
2	2.5	0.10	0.108
1.2	1.5	0.05	0.053
1.0	1.1	0.025	0.028
0.8	0.95		

TABLE 4.39 Porosities of Fritted Glassware

Porosity	Nominal maximum pore size, μm	Principal uses
Extra coarse	170–220	Filtration of very coarse materials. Gas dispersion, gas washing, and extractor beds. Support of other filter materials.
Coarse	40–60	Filtration of coarse materials. Gas dispersion, gas washing, gas absorption. Mercury filtration. For extraction apparatus.
Medium	10–15	Filtration of crystalline precipitates. Removal of "floaters" from distilled water.
Fine	4–5.5	Filtration of fine precipitates. As a mercury valve. In extraction apparatus.
Very fine	2–2.5	General bacteria filtrations.
Ultra fine	0.9–1.4	General bacteria filtrations.

TABLE 4.40 Tolerances for Analytical Weights

This table gives the individual and group tolerances established by the National Bureau of Standards (Washington, D.C.) for classes M, S, S-1, and P weights. Individual tolerances are "acceptance tolerances" for new weights. Group tolerances are defined by the National Bureau of Standards as follows: "The corrections of individual weights shall be such that no combination of weights that is intended to be used in a weighing shall differ from the sum of the nominal values by more than the amount listed under the group tolerances."

For class S-1 weights, two-thirds of the weights in a set must be within one-half of the individual tolerances given below. No group tolerances have been specified for class P weights. See *Natl. Bur. Standards Circ.* 547, sec. 1 (1954).

Denomination	Class M		Class S		Class S-1, individual tolerance, mg	Class P, individual tolerance, mg
	Individual tolerance, mg	Group tolerance, mg	Individual tolerance, mg	Group tolerance, mg		
100 g	0.50	None specified	0.25	None specified	1.0	2.0
50 g	0.25		0.12		0.60	1.2
30 g	0.15		0.074	0.154	0.45	0.90
20 g	0.10		0.074		0.35	0.70
10 g	0.050		0.074		0.25	0.50
5 g	0.034	0.065	0.054	0.105	0.18	0.36
3 g	0.034		0.054		0.15	0.14
2 g	0.034		0.054		0.13	0.26
1 g	0.034		0.054		0.10	0.20
500 mg	0.0054	0.0105	0.025	0.055	0.080	0.16
300 mg	0.0054		0.025		0.070	0.14
200 mg	0.0054		0.025		0.060	0.12
100 mg	0.0054		0.025		0.050	0.10
50 mg	0.0054	0.0105	0.014	0.034	0.042	0.085
30 mg	0.0054		0.014		0.038	0.076
20 mg	0.0054		0.014		0.035	0.070
10 mg	0.0054	0.0105	0.014	0.034	0.030	0.060
5 mg	0.0054		0.014		0.028	0.055
3 mg	0.0054		0.014		0.026	0.052
2 mg	0.0054		0.014		0.025	0.050
1 mg	0.0054		0.014		0.025	0.050
½ mg	0.0054		0.014		0.025	....

TABLE 4.41 Heating Temperatures, Composition of Weighing Forms, and Gravimetric Factors

The minimum temperature required for heating a pure precipitate to constant weight is frequently lower than that commonly recommended in gravimetric procedures. However, the higher temperature is very often still to be preferred in order to ensure that contaminating substances are expelled. The thermal stability ranges of various precipitates as deduced from thermograms are also tabulated. Where a stronger ignition is advisable, the safe upper limit can be ascertained.

Gravimetric factors are based on the 1993 International Atomic Weights. The factor Ag: 0.7526 given in the first line of the table indicates that the weight of precipitate obtained (AgCl) is to be multiplied by 0.7526 to calculate the corresponding weight of silver.

Element	Thermal stability range, °C	Final heating temperature, °C	Composition of weighing form	Gravimetric factors
Ag	70–600	130–150	AgCl	Ag: 0.7526
Al	>475	1200	Al_2O_3	Al: 0.5293
	>743	>743	$AlPO_4$	Al: 0.2212; Al_2O_3: 0.4180
	102–220	110	$Al(C_9H_6NO)_3$	Al: 0.0587; Al_2O_3: 0.1110
As	200–275	105–110	Al_2S_3	As: 0.6090; As_2O_3: 0.8041
		850	$Mg_2As_2O_7$	As: 0.4827; As_2O_3: 0.6373
		vacuum at 25	$MgNH_4AsO_4 \cdot 6H_2O$	As: 0.2589
Au	20–957	1060	Au	
Ba	780–1100	780	$BaSO_4$	Ba: 0.5884; BaO: 0.6570
	<60	<60	$BaCrO_4$	Ba: 0.5421; BaO: 0.6053
Be	>900	1000	BeO	Be: 0.3603
Bi		100	BiOCl	Bi: 0.8024; Bi_2O_3: 0.8946
		100	$Bi(C_{12}H_{10}NOS)_3$	Bi: 0.2387
	379–961	800	$BiPO_4$	Bi: 0.6875; Bi_2O_3: 0.7665
Br	70–946	130–150	AgBr	Br: 0.4256
Ca	478–635	475–525	$CaCO_3$	Ca: 0.4004; CaO: 0.5601
	838–1025	950–1000	CaO	Ca: 0.7147
		air-dried	$Ca(picrolonate)_2 \cdot 8H_2O$	Ca: 0.05642
Cd		>320	$CdSO_4$	Cd: 0.5392; CdO: 0.6159
		125	$Cd(C_{10}H_6NO_2)_2$	Cd: 0.2462
	218–420		CdS	Cd: 0.7781; CdO: 0.8888
Ce	>360	500–600	CeO_2	Ce: 0.8141
Cl	70–600	130–150	AgCl	Cl: 0.2474
Co	285–946	750–850	Co_3O_4	Co: 0.7342
		130	$Co(C_{10}H_6NO_2)_3 \cdot 2H_2O$	Co: 0.09639; CoO: 0.1226
		450–500	$CoSO_4$	Co: 0.3802
Cr		120	$PbCrO_4$	Cr: 0.1609
Cu		105–120	CuSCN	Cu: 0.5225; CuO: 0.6540
	<115	100–105	$Cu(C_7H_5NO_2)_2$	Cu: 0.1891
		105–115	$Cu(C_{13}H_{11}NO_2)$	Cu: 0.2201
		110–115	$Cu(C_{10}H_6NO_2) \cdot H_2O$	Cu: 0.1494
		105	$Cu(C_{12}H_{10}NOS)_2 \cdot H_2O$	Cu: 0.1237
F	66–538	130–140	PbClF	F: 0.07261
Fe	470–946	900	Fe_2O_3	Fe: 0.6994
Ga	408–946	900	Ga_2O_3	Ga: 0.7439
Hg		105	$Hg(C_{12}H_{10}NOS)_2$	Hg: 0.3169
I	60–900	130–150	AgI	I: 0.5405
In	345–1200	1200	In_2O_3	In: 0.8271
Ir			IrO_2	Ir: 0.8573
K	73–653	<653	$KClO_4$	K: 0.2822; K_2O: 0.3399
		<270	K_2PtCl_6	K: 0.1609; K_2O: 0.1938
			KIO_4	K: 0.1700
		120	$KB(C_6H_5)_4$	K: 0.1091

(Continued)

TABLE 4.41 Heating Temperatures, Composition of Weighing Forms, and Gravimetric Factors (*Continued*)

Element	Thermal stability range, °C	Final heating temperature, °C	Composition of weighing form	Gravimetric factors
Li		200	Li_2SO_4	Li: 0.1263; Li_2O: 0.2718
Mg		1050–1100	$Mg_2P_2O_7$	Mg: 0.2184; MgO: 0.3622
	88–300	155–160	$Mg(C_9H_6NO)_2$	Mg: 0.07775; MgO: 0.1289
Mn	>946	1000	Mn_3O_4	Mn: 0.7203
		1000	$Mn_2P_2O_7$	Mn: 0.3871; MnO: 0.4998
Mo		>505	$PbMoO_4$	Mo: 0.2613; MoO_3: 0.3291
		500–525	MoO_3	Mo: 0.6666
N (as NO_3^-)	20–242	105	Nitron nitrate	N: 0.3732; NO_3: 0.1652
Na	360–674	125	$NaMg(UO_2)_3(C_2H_3O_2)_9 \cdot 6.5\ H_2O$	Na: 0.01527; Na_2O: 0.02058
Nb	650–950	900	Nb_2O_3	Nb: 0.6990
Ni	79–172	110–120	$Ni(C_4H_7N_2O_2)_2$	Ni: 0.2032; NiO: 0.2586
Os		800 (in H_2)	Os metal	
P		>477	$Mg_2P_2O_7$	P: 0.2783; PO_4: 0.8536
	160–415	110	$(NH_4)_3[P(Mo_3O_{10})_4]$	P: 0.0165; P_2O_5: 0.0378
Pb	271–959	500–600	$PbSO_4$	Pb: 0.6832; PbO: 0.7359
		600	$PbMoO_4$	Pb: 0.5643; PbO: 0.6078
		120	$PbCrO_4$	Pb: 0.6411
	271–959	600–800	$PbSO_4$	Pb: 0.6832; PbO: 0.7359
		105	$Pb(C_{12}H_{10}NOS)_2$	Pb: 0.3240
Pd	45–171	110	$Pd(C_4H_7N_2O_2)_2$	Pd: 0.3162
Rb	70–674	<674	Rb_2PtCl_6	Rb: 0.2954; Rb_2O: 0.3230
Re		130	$(C_6H_5)_4AsReO_4$	Re: 0.2939
		110	Nitron perrhenate	Re: 0.3306
S		>780	$BaSO_4$	S: 0.1374; SO_3: 0.3430; SO_4: 0.4116
Sb		100	$Sb(C_{12}H_{10}NOS)_3$	Sb: 0.1581
SCN^-		130	AgSCN	SCN: 0.3500
		110–120	CuSCN	SCN: 0.4775
Se		120–130	Se metal	SeO_2: 1.4052
Si	358–946	>358	SiO_2	Si: 0.4675
Sn	>834	900	SnO_2	Sn: 0.7877
Sr		130–140	$Sr(NO_3)_2$	Sr: 0.4140
	100–300	100–300	$SrSO_4$	Sr: 0.4770; SrO: 0.5641
Te		105	Te metal	
Th	610–946	700–800	ThO_2	Th: 0.8788
		900	ThP_2O_7	Th: 0.5863
Ti	350–946	900	TiO_2	Ti: 0.5992
Tl(III)		100	$Tl(C_{12}H_{10}NOS)$	Tl: 0.4860
U		1000	U_3O_8	U: 0.8480; UO_2: 0.9620
V	581–946	700–800	V_2O_5	V: 0.5602
W	>674	800–900	WO_3	W: 0.7930
Zn	>1000	950–1000	ZnO	Zn: 0.8034
		1000	$Zn_2P_2O_7$	Zn: 0.4292; ZnO: 0.5342
		125	$Zn(C_{10}H_6NO_2)_2 \cdot H_2O$	Zn: 0.1529
Zr		>850	ZrP_2O_7	Zr: 0.3440; ZrO_2: 0.4647
		1200	ZrO_2	Zr: 0.7403

TABLE 4.42 Primary Standards for Aqueous Acid-Base Titrations

Standard	Formula weight	Preparation
		Basic substances for standardizing acidic solutions
$(HOCH_3)_3CNHH_2$	121.137	Tris(hydroxymethyl)aminomethane is available commercially as a primary standard. Dry at 100–103°C (<110°C). In titrations with a strong acid the equivalence point is at about pH 4.5–5. Equivalent weight is the formula weight. [J. H. Fossum, P. C. Markunas, and J. A. Riddick, *Anal. Chem.*, **23**:491 (1951).]
HgO	216.59	Dissolve 100 g pure $HgCl_2$ in 1 L H_2O, and add with stirring to 650 mL 1.5 *M* NaOH. Filter and wash with H_2O until washings are neutral to phenolphthalein. Dry to constant weight at or below 40°C, and store in a dark bottle. To 0.4 g HgO (≡ 40 mL 0.1*N* acid) add 10–15 g KBr plus 20–25 mL H_2O. Stir, excluding CO_2, until solution is complete. Titrate with acid to pH 5–8. Equivalent weight is one-half formula weight.
$Na_2B_4O_7 \cdot 10H_2O$	381.372	Recrystallize reagent-grade salt twice from water at temperatures below 55°C. Wash the crystals with H_2O, twice with ethanol, and twice with diethyl ether. Let stand in a hygrostat oversaturated $NaBr \cdot 2H_2O$ or saturated NaCl-sucrose solution. Use methyl red indicator. Equivalent weight is one-half the formula weight.
Na_2CO_3	105.989	Heat reagent-grade material for 1 hr at 255–265°C. Cool in an efficient desiccator. Titrate sample with acid to pH 4–5 (first green tint of bromocresol green), boil the solution to eliminate the carbon dioxide, cool, and again titrate to pH 4–5. Equivalent weight is one-half the formula weight.
NaCl	58.45	Accurately weigh about 6 g NaCl and dissolve in distilled water. Pass the solution through a well-rinsed cation exchange column (Dowex 50W) in the hydrogen form. The equivalent amount of HCl is washed from the column (in 10 column volumes) into a volumetric flask and made up to volume. Equivalent weight is the formula weight.
		Acidic substances for standardizing basic solutions
C_6H_5COOH	122.125	Pure benzoic acid is available from NIST (National Institute for Science and Technology). Dissolve 0.5 g in 20 mL of neutral ethanol (run a blank), excluding CO_2, add 20–50 mL, and titrate using phenolphthalein as indicator.
o-$C_6H_4(COOK)(COOH)$	204.22	Potassium hydrogen *o*-phthalate is available commercially as primary standard, also from NIST. Dry at <135°C. Dissolve in water, excluding CO_2, and titrate with phenolphthalein as indicator. For $Ba(OH)_2$ solution, perform the titration at an elevated temperature to prevent precipitation of Ba phthalate.
$KH(IO_3)_2$	389.915	Potassium hydrogen bis(iodate) is available commercially in a primary standard grade. Dry at 110°C. Dissolve a weighed amount of the salt in water, excluding CO_2, and titrate to pH 5–8. [I. M. Kolthoff and L. H. van Berk, *J. Am. Chem. Soc.*, **48**:2800(1926)].
NH_2SO_3H	97.09	Hydrogen amidosulfate (sulfamic acid) acts as a strong acid. Primary standard grade is available commercially. Since it does undergo slow hydrolysis, an acid end point (pH 4 to 6.5) should be chosen unless fresh reagent is available, then the end point can be in the range pH 4 to 9. [W. F. Wagner, J. A. Wuellner, and C. E. Feiler, *Anal. Chem.*, **24**:1491 (1952). M. J. Butler, G. F. Smith, and L. F. Audrieth, *Ind. Eng. Chem., Anal. Ed.*, **10**:690 (1938)].

TABLE 4.43 Titrimetric (Volumetric) Factors

Acids

The following factors are the equivalent of 1 mL of *normal acid*. Where the normality of the solution being used is other than normal, multiply the factors given in the table below by the normality of the solution employed.

The equivalents of the esters are based on the results of saponification.

The indicators methyl orange and phenolphthalein are indicated by the abbreviations MO and pH, respectively.

Substance	Formula	Grams
Ammonia	NH_3	0.017031
Ammonium	NH_4	0.018039
Ammonium chloride	NH_4Cl	0.053492
Ammonium hydroxide	NH_4OH	0.035046
Ammonium oleate	$C_{17}H_{33}CO_2NH_4$	0.29950
Ammonium oxide	$(NH_4)_2O$	0.026038
Amyl acetate	$CH_3CO_2C_5H_{11}$	0.13019
Barium carbonate (MO)	$BaCO_3$	0.09867
Barium hydroxide	$Ba(OH)_2$	0.085677
Barium oxide	BaO	0.07667
Bornyl acetate	$CH_3CO_2C_{10}H_{17}$	0.19629
Calcium carbonate (MO)	$CaCO_3$	0.05004
Calcium hydroxide	$Ca(OH)_2$	0.037047
Calcium oleate	$(C_{17}H_{33}CO_2)_2Ca$	0.30150
Calcium oxide	CaO	0.02804
Calcium stearate	$(C_{17}H_{35}CO_2)_2Ca$	0.30352
Casein (N 6.38)	....	0.089371
Ethyl acetate	$CH_3CO_2C_2H_5$	0.088107
Glue (N 5.60)	....	0.078445
Hydrochloric acid	HCl	0.036461
Magnesium carbonate (MO)	$MgCO_3$	0.04216
Magnesium oxide	MgO	0.02016
Menthyl acetate	$CH_3CO_2C_{10}H_{19}$	0.19831
Methyl acetate	$CH_3CO_2CH_3$	0.074080
Nicotine	$C_{10}H_{14}N_2$	0.16224
Nitrogen	N	0.014007
Potassium carbonate (MO)	K_2CO_3	0.06911
Potassium carbonate, acid (MO)	$KHCO_3$	0.10012
Potassium nitrate	KNO_3	0.10111
Potassium oleate	$C_{17}H_{33}CO_2K$	0.32057
Potassium oxide	K_2O	0.04710
Potassium stearate	$C_{17}K_{35}CO_2K$	0.32258
Protein (N 5.70)	....	0.079846
Protein (N 6.25)	....	0.087550
Sodium acetate	CH_3CO_2Na	0.082035
Sodium acetate	$CH_3CO_2Na \cdot 3H_2O$	0.13608
Sodium borate, tetra- (MO)	$Na_2B_4O_7$	0.10061
Sodium borate, tetra- (MO)	$Na_2B_4O_7 \cdot 10H_2O$	0.19069
Sodium carbonate (MO)	Na_2CO_3	0.052994
Sodium carbonate (MO)	$Na_2CO_3 \cdot H_2O$	0.062002
Sodium carbonate (MO)	$Na_2CO_3 \cdot 10H_2O$	0.14307
Sodium carbonate, acid (MO)	$NaHCO_3$	0.084007
Sodium hydroxide	$NaOH$	0.39997
Sodium oleate	$C_{17}H_{33}CO_2Na$	0.30445

TABLE 4.43 Titrimetric (Volumetric) Factors (*Continued*)

Acids (*Continued*)

Substance	Formula	Grams
Sodium oxalate	$Na_2C_2O_4$	0.067000
Sodium oxide	Na_2O	0.030990
Sodium phosphate (MO)	Na_2HPO_4	0.14196
Sodium phosphate (MO)	$Na_2PHO_4 \cdot 12H_2O$	0.35814
Sodium phosphate (MO)	Na_3PO_4	0.081970
Sodium phosphate (PH)	Na_3PO_4	0.16394
Sodium silicate	$Na_2Si_4O_9$	0.15111
Sodium stearate	$C_{17}H_{35}CO_2Na$	0.30647
Sodium sulfide (MO)	Na_2S	0.039022

Alkali

The following factors are the equivalent of the milliliter of *normal alkali.* Where the normality of the solution being used is other than normal, multiply the factors given in the table below by the normality of the solution employed.

The equivalents of the esters are based on the results of saponification.

The indicators methyl orange and phenolphthalein are indicated by the abbreviations MO and PH, respectively.

Substance	Formula	Grams
Abietic acid (PH)	$HC_{20}H_{29}O_2$	0.30246
Acetic acid (PH)	CH_3CO_2H	0.06005
Acetic anhydride (PH)	$(CH_3CO)_2O$	0.051045
Aluminum sulfate	$Al_2(SO_4)_3$	0.05702
Amyl acetate	$CH_3CO_2C_5H_{11}$	0.13019
Benzoic acid (PH)	$C_6H_5CO_2H$	0.12212
Borate tetra- (PH)	B_4O_7	0.03881
Boric acid (PH)	H_3BO_3	0.061833
Boric anhydride (PH)	B_2O_3	0.03486
Bornyl acetate	$CH_3CO_2C_{10}H_{17}$	0.19629
Butyric acid (PH)	$C_3H_7CO_2H$	0.088107
Calcium acetate	$(CH_3CO_2)_2Ca$	0.079085
Calcium oleate	$(C_{17}H_{33}CO_2)_2Ca$	0.30150
Calcium stearate	$(C_{17}H_{35}CO_2)_2Ca$	0.30352
Carbon dioxide (PH)	CO_2	0.022005
Chlorine	Cl	0.035453
Citric acid (PH)	$H_3C_6H_5O_7 \cdot H_2O$	0.070047
Ethyl acetate	$CH_3CO_2C_2H_5$	0.088107
Formaldehyde	HCHO	0.030026
Formic acid (PH)	HCO_2H	0.046026
Glycerol (sap. of acetyl)	$C_3H_5(OH)_3$	0.030698
Hydriodic acid	HI	0.12791
Hydrobromic acid	HBr	0.080917
Hydrochloric acid	HCl	0.036461
Lactic acid (PH)	$HC_3H_5O_3$	0.090079
Lead acetate	$(CH_3CO_2)_2Pb \cdot 3H_2O$	0.18966
Maleic acid (PH)	$(CHCO_2H)_2$	0.058037
Malic acid (PH)	$H_2C_4H_4O_5$	0.067045
Menthol (sap. of acetyl)	$C_{10}H_{19}OH$	0.15627

(*Continued*)

TABLE 4.43 Titrimetric (Volumetric) Factors (*Continued*)

Substance	Formula	Grams
Menthyl acetate	$CH_3CO_2C_{10}H_{19}$	0.19831
Methyl acetate	$CH_3CO_2CH_3$	0.074080
Nitrate	NO_3	0.062005
Nitric acid	HNO_3	0.063013
Nitrogen	N	0.014007
Nitrogen pentoxide	N_2O_5	0.054005
Oleic acid (PH)	$C_{17}H_{33}CO_2H$	0.28247
Oxalic acid (PH)	$(CO_2H)_2$	0.045018
Oxalic acid (PH)	$(CO_2H)_2 \cdot 2H_2O$	0.063033
Phosphoric acid (MO)	H_3PO_4	0.097995
Phosphoric acid (PH)	H_3PO_4	0.048998
Potassium carbonate, acid (MO)	$KHCO_3$	0.10012
Potassium oleate	$C_{17}K_{33}CO_2K$	0.32056
Potassium oxalate, acid (PH)	KHC_2O_4	0.12813
Potassium phthalate, acid (PH)	$HC_8H_4O_4K$	0.20423
Potassium stearate	$C_{17}H_{35}CO_2K$	0.32258
Sodium benzoate	$C_6H_5CO_2Na$	0.14411
Sodium borate, tetra- (PH)	$Na_2B_4O_7$	0.050305
Sodium borate, tetra- (PH)	$Na_2B_4O_7 \cdot 10H_2O$	0.095343
Sodium carbonate, acid (MO)	$NaHCO_3$	0.084007
Sodium oleate	$C_{17}H_{33}CO_2Na$	0.30445
Sodium salicylate	$C_6H_5OCO_2Na$	0.16011
Stearic acid (PH)	$C_{17}H_{35}CO_2H$	0.28449
Succinic acid (PH)	$(CH_2CO_2H)_2$	0.059045
Sulfate	SO_4	0.048031
Sulfur dioxide (PH)	SO_2	0.032031
Sulfur trioxide	SO_3	0.040031
Sulfuric acid	H_2SO_4	0.049039
Sulfurous acid (PH)	H_2SO_3	0.041039
Tartaric acid (PH)	$H_2C_4H_4O_6$	0.075044
Tartaric acid (PH)	$H_2C_4H_4O_6 \cdot H_2O$	0.084052

Iodine

The following factors are the equivalent of 1 mL of *normal iodine*. Where the normality of the solution being used is other than normal, multiply the factors given in the table below by the normality of the solution employed.

Substance	Formula	Grams
Acetone	$(CH_3)_2CO$	0.0096801
Ammonium chromate	$(NH_4)_2CrO_4$	0.050690
Antimony	Sb	0.06088
Antimony trioxide	Sb_2O_3	0.07287
Arsenic	As	0.037461
Arsenic pentoxide	As_2O_5	0.057460
Arsenic trioxide	As_2O_3	0.049460
Arsenite	AsO_3	0.061460
Bleaching powder	$CaOCl_2$	0.063493
Bromine	Br	0.079909
Chlorine	Cl	0.035453
Chromic oxide	Cr_2O_3	0.02533

TABLE 4.43 Titrimetric (Volumetric) Factors (*Continued*)

Iodine (*Continued*)

Substance	Formula	Grams
Chromium trioxide	CrO_3	0.033331
Copper	Cu	0.06354
Copper oxide	CuO	0.07954
Copper sulfate	$CuSO_4$	0.15960
Copper sulfate	$CuSO_4 \cdot 5H_2O$	0.24968
Ferric iron	Fe^{3+}	0.05585
Ferric oxide	Fe_2O_3	0.07985
Hydrogen sulfide	H_2S	0.017040
Iodine	I	0.126904
Lead chromate	$PbCrO_4$	0.10773
Lead dioxide	PbO_2	0.11959
Nitrous acid	HNO_2	0.023507
Oxygen	O	0.0079997
Potassium chlorate	$KClO_3$	0.020426
Potassium chromate	K_2CrO_4	0.064733
Potassium dichromate	$K_2Cr_2O_7$	0.049032
Potassium nitrite	KNO_2	0.042554
Potassium permanganate	$KMnO_4$	0.031608
Red lead	Pb_3O_4	0.34278
Sodium chromate	Na_2CrO_4	0.053991
Sodium dichromate	$Na_2Cr_2O_7$	0.043661
Sodium dichromate	$Na_2Cr_2O_7 \cdot 2H_2O$	0.049666
Sodium nitrite	$NaNO_2$	0.034498
Sodium sulfide	Na_2S	0.039022
Sodium sulfide	$Na_2S \cdot 9H_2O$	0.12009
Sodium sulfite	Na_2SO_3	0.063021
Sodium sulfite	$Na_2SO_3 \cdot 7H_2O$	0.12607
Sodium thiosulfate	$Na_2S_2O_3$	0.15811
Sulfur	S	0.016032
Sulfur dioxide	SO_2	0.032031
Sulfurous acid	H_2SO_3	0.041039
Tin	Sn	0.059345

Potassium dichromate

The following factors are the equivalent of 1 mL of *normal potassium dichromate.* Where the normality of the solution being used is other than normal, multiply the factors given in the table below by the normality of the solution employed.

Substance	Formula	Grams
Chromic oxide	Cr_2O_3	0.025332
Chromium trioxide	CrO_3	0.033331
Ferrous iron	Fe^{2+}	0.055847
Ferrous oxide	FeO	0.071846
Ferroso-ferric oxide	Fe_3O_4	0.077180
Ferrous sulfate	$FeSO_4$	0.15191
Ferrous sulfate	$FeSO_4 \cdot 7H_2O$	0.27802
Glycerol	$C_3H_5(OH)_3$	0.0065782
Lead chromate	$PbCrO_4$	0.10773
Zinc	Zn	0.032685

(*Continued*)

TABLE 4.43 Titrimetric (Volumetric) Factors (Continued)

Potassium permanganate

The following factors are the equivalent of 1 mL of *normal potassium permanganate*. Where the normality of the solution being used is other than normal, multiply the factors given in the table below by the normality of the solution employed.

Substance	Formula	Grams
Ammonium oxalate	$(NH_4)_2C_2O_4$	0.062049
Ammonium oxalate	$(NH_4)_2C_2O_4 \cdot H_2O$	0.071056
Ammonium peroxydisulfate	$(NH_4)_2S_2O_8$	0.11410
Antimony	Sb	0.060875
Barium peroxide	BaO_2	0.084669
Barium peroxide	$BaO_2 \cdot 8H_2O$	0.15673
Calcium carbonate	$CaCO_3$	0.050045
Calcium oxide	CaO	0.02804
Calcium peroxide	CaO_2	0.036039
Calcium sulfate	$CaSO_4$	0.068071
Calcium sulfate	$CaSO_4 \cdot 2H_2O$	0.086086
Ferric oxide	Fe_2O_3	0.079846
Ferroso-ferric oxide	Fe_3O_4	0.077180
Ferrous ammonium sulfate	$Fe(NH_4)_2(SO_4)_2 \cdot 6H_2O$	0.39214
Ferrous oxide	FeO	0.071846
Ferrous sulfate	$FeSO_4$	0.15191
Ferrous sulfate	$FeSO_4 \cdot 7H_2O$	0.27802
Formic acid	HCO_2H	0.023013
Hydrogen peroxide	H_2O_2	0.017007
Iodine	I	0.126904
Iron	Fe	0.055847
Manganese	Mn	0.010988
Manganese dioxide	MnO_2	0.043468
Manganous oxide (Volhard)	MnO	0.035469
Molybdenum trioxide titration from yellow ppt. after reduction	MoO_3	0.047979
Oxalic acid	$(CO_2H)_2$	0.045018
Oxalic acid	$(CO_2H)_2 \cdot 2H_2O$	0.063033
Phosphorus titration from yellow ppt. after reduction	P	0.0008604
Phosphorus pentoxide to titration from yellow ppt. after reduction	P_2O_5	0.0019715
Potassium dichromate	$K_2Cr_2O_7$	0.049032
Potassium nitrite	KNO_2	0.042552
Potassium persulfate	$K_2S_2O_8$	0.13516
Sodium nitrite	$NaNO_2$	0.034498
Sodium oxalate	$Na_2C_2O_4$	0.067000
Sodium persulfate	$Na_2S_2O_8$	0.11905
Tin	Sn	0.059345

TABLE 4.43 Titrimetric (Volumetric) Factors (*Continued*)

Silver nitrate

The following factors are the equivalent of *normal silver nitrate*. Where the normality of the solution being used is other than normal, multiply the factors given in the table below by the normality of the solution employed.

Substance	Formula	Grams
Ammonium bromide	NH_4Br	0.097948
Ammonium chloride	NH_4Cl	0.053492
Ammonium iodide	NH_4I	0.14494
Ammonium thiocyanate	NH_4SCN	0.076120
Barium chloride	$BaCl_2$	0.10412
Barium chloride	$BaCl_2 \cdot 2H_2O$	0.12214
Bromine	Br	0.079909
Cadmium chloride	$CdCl_2$	0.091653
Cadmium iodide	CdI_2	0.18310
Calcium chloride	$CaCl_2$	0.055493
Chlorine	Cl	0.035453
Ferric chloride	$FeCl_3$	0.054069
Ferrous chloride	$FeCl_2$	0.063377
Hydriodic acid	HI	0.12791
Hydrobromic acid	HBr	0.080917
Hydrochloric acid	HCl	0.036461
Iodine	I	0.126904
Lithium chloride	LiCl	0.042392
Lead chloride	$PbCl_2$	0.13905
Magnesium chloride	$MgCl_2$	0.047609
Magnesium chloride	$MgCl_2 \cdot 6H_2O$	0.10166
Potassium bromide	KBr	0.11901
Potassium chloride	KCl	0.074555
Potassium iodide	KI	0.16601
Potassium oxide	K_2O	0.047102
Potassium thiocyanate	KSCN	0.097184
Silver	Ag	0.10787
Silver iodide	AgI	0.23477
Silver nitrate	$AgNO_3$	0.16987
Sodium bromide	NaBr	0.10290
Sodium bromide	$NaBr \cdot 2H_2O$	0.13893
Sodium chloride	NaCl	0.058443
Sodium iodide	NaI	0.14989
Sodium iodide	$NaI \cdot 2H_2O$	0.18592
Sodium oxide	Na_2O	0.030990
Strontium chloride	$SrCl_2$	0.079263
Strontium chloride	$SrCl_2 \cdot 6H_2O$	0.13331
Zinc chloride	$ZnCl_2$	0.068138

(*Continued*)

TABLE 4.43 Titrimetric (Volumetric) Factors (*Continued*)

Sodium thiosulfate

The following factors are the equivalent of I_{mL} of *normal sodium thiosulfate*. Where the normality of the solution being used is other than normal, multiply the factors given in the table below by the normality of the solution employed.

Substance	Formula	Grams
Acetone	$(CH_3)_2CO$	0.0096801
Ammonium chromate	$(NH_4)_2CrO_4$	0.050690
Antimony	Sb	0.06088
Antimony trioxide	Sb_2O_3	0.07287
Bleaching powder	$CaOCl_2$	0.063493
Bromine	Br	0.079909
Chlorine	Cl	0.035453
Chromic oxide	Cr_2O_3	0.02533
Chromium trioxide	CrO_3	0.033331
Copper	Cu	0.06354
Copper oxide	CuO	0.07954
Copper sulfate	$CuSO_4$	0.15960
Copper sulfate	$CuSO_4 \cdot 5H_2O$	0.24968
Iodine	I	0.126904
Lead chromate	$PbCrO_4$	0.10773
Lead dioxide	PbO_2	0.11959
Nitrous acid	HNO_2	0.023507
Potassium chromate	K_2CrO_4	0.064733
Potassium dichromate	$K_2Cr_2O_7$	0.049032
Red lead	Pb_3O_4	0.34278
Sodium chromate	Na_2CrO_4	0.053991
Sodium dichromate	$Na_2Cr_2O_7$	0.043661
Sodium dichromate	$Na_2Cr_2O_7 \cdot 2H_2O$	0.049666
Sodium nitrite	$NaNO_2$	0.034498
Sodium thiosulfate	$Na_2S_2O_3$	0.15811
Sodium thiosulfate	$Na_2S_2O_3 \cdot 5H_2O$	0.24818
Sulfur	S	0.016032
Sulfur dioxide	SO_2	0.032031
Tin	Sn	0.059345

TABLE 4.44 Equations for the Redox Determinations of the Elements with Equivalent Weights

Al	$Al(C_9H_6NO)_3 + 3\ HCl = AlCl_3 + 3\ C_9H_7NO$ (8-hydroxyquinoline) $3\ C_9H_7NO + 6\ Br_2 = 3\ C_9H_5Br_2NO + 6\ HBr$ $Al/12 = 2.2485$; $Al_2O_3/24 = 4.2483$
As^0	$As + 5\ Ce(IV) + 4\ H_2O = H_3AsO_4 + 5\ Ce(III) + 5\ H^+$ $As/5 = 14.9843$
As(III)	$5\ H_3AsO_3 + 2\ KMnO_4 + 6\ HCl = 5\ H_3AsO_4 + 2\ MnCl_2 + 3\ H_2O$ $H_3AsO_3 + 2\ Ce(SO_4)_2 + H_2O = H_3AsO_4 + Ce_2(SO_4)_3 + H_2SO_4$ $As/2 = 37.4608$; $As_2O_3/4 = 49.460$ $3\ H_3AsO_3 + KBrO_3\ (+\ HCl) = 3\ H_3AsO_4 + KBr$ $H_3AsO_3 + I_2 + 2\ H_2O = H_3AsO_4 + 2\ I^- + 2\ H^+$ $As/2 = 37.4608$; $As_2O_3/4 = 49.460$
As(V)	$H_3AsO_4 + 2\ KI$ (excess) $+ 2\ HCl = H_3AsO_3 + I_2 + 2\ KCl + H_2O$ $I_2 + 2\ Na_2S_2O_3 = 2\ NaI + Na_2S_4O_6$ $As/2 = 37.4608$; $As_2O_3/4 = 49.460$
Ba	$BaCrO_4 + 6\ KI$ (excess) $+ 16\ HCl = 2\ BaCl_2 + 3\ I_2 + 6\ KCl + 2\ CrCl_3 + 8\ H_2O$ $I_2 + 2\ Na_2S_2O_3 = 2\ NaI + Na_2S_4O_6$ $Ba/3 = 45.78$ $BaCrO_4 + 3\ Fe^{2+}$ (excess) $+ 8\ H^+ = Ba^{2+} + Cr^{3+} + 3\ Fe^{3+} + 4\ H_2O$ Titrate excess Fe^{2+} with permanganate or dichromate; $Ba/3 = 45.78$
Br_2	$Br_2 + 2\ KI$ (excess) $= 2\ KBr + I_2$ $I_2 + 2\ Na_2S_2O_3 \rightarrow 2\ NaI = Na_2S_4O_6$ $Br_2/2 = 79.904$
Br^-	$Br^- + 3\ HClO = BrO_3^- + 3\ Cl^- + 3\ H^+$ $Br/6 = 13.317$
BrO_3^-	$BrO_3^- + 6\ I^-$ (excess) $+ 6\ H^+ = Br^- + 3\ I_2 + 3H_2O$ $I_2 + 2\ Na_2S_2O_3 = 2\ NaI + Na_2S_4O_6$ $KBrO_3/6 = 27.835$
CO	$5\ CO + I_2O_5 = 5\ CO_2 + I_2$ (at 125°C; adsorbed and measured colorimetrically) $5/2\ CO = 70.02$
$C_2O_4^{2-}$	Titrate as for CaC_2O_4
$C_2O_6^{2-}$	Acidify and titrate as for H_2O_2; $C_2O_6^{2-} + 2\ H^+ = H_2O_2 + CO_2$ $K_2C_2O_6/2 = 99.11$
Ca	$5\ CaC_2O_4 + 2\ KMnO_4 + 8\ H_2SO_4 = 5\ CaSO_4 + 10\ CO_2 + K_2SO_4 + 2\ MnSO_4 + 8\ H_2O$ $Ca/2 = 20.039$; $CaO/2 = 28.04$
Cd	$Cd(anthranilate)_2 + 4\ Br_2 = 2\ NH_2C_6H_2Br_2COOH + 4\ Br^-$ Titrate with $KBrO_3$—KBr until color of indigo changes to yellow. Add KI and back-titrate iodine liberated with thiosulfate. $Cd/8 = 14.05$
Ce	Oxidize Ce(III) to Ce(IV) with $(NH_4)_2S_2O_8$ plus Ag^+; destroy excess by boiling. $2\ Ce(SO_4)_2 + 2\ FeSO_4 = Ce_2(SO_4)_3 + Fe_2(SO_4)_3$ $Ce/1 = 140.12$; $Ce_2O_3/2 = 164.12$
Cl_2	Same as for Br_2; $Cl_2/2 = 35.453$
ClO^-	$ClO^- + 2\ I^- + 2\ H = Cl^- + I_2 + H_2O$ Titrate liberated I_2 with thiosulfate; $HClO/2 = 26.230$
ClO_2^-	$ClO_2^- + 4\ I^- + 4\ H^+ = Cl^- + 2\ I_2 + 2\ H_2O$ Titrate liberated I_2 with thiosulfate; $HClO/2 = 26.230$
ClO_3^-	$ClO_3^- + 6\ I^- + 6\ H_2O = Cl^- + 3\ I_2 + 3\ H_2O$ Titrated liberated I_2 with thiosulfate; $HClO_2/4 = 17.115$ $ClO_3^- + 3\ H_3AsO_3$ (excess; boil with strong HCl) $= Cl^- + 3\ H_3AsO_4$ Titrate excess H_3AsO_3 with bromate; $HClO_3/6 = 14.077$
Co	$Co(NH_3)_6^{2+} + Fe(CN)_6^{3-}$ [Citrate-NH_3 buffer] $= Co(NH_3)_6^{3+} + Fe(CN)_6^{4-}$ $Co/1 = 58.9332$

(*Continued*)

TABLE 4.44 Equations for the Redox Determinations of the Elements with Equivalent Weights (*Continued*)

	Precipitate Co anthranilate and treat as for cadmium; Co/8 = 7.3667
Cr	$Cr_2O_7^{2-} + 6\ Fe^{2+} + 14\ H^+ = 2\ Cr^{3+} + 6\ Fe^{3+} + 7\ H_2O$ Cr/3 = 17.332; $Cr_2O_3/6$ = 25.337
Cu	$2\ Cu^{2+} + 2\ I^- + 2SCN^- = 2CuSCN + I_2$ Titrate the liberated iodine with thiosulfate; Cu/1 = 63.546
	$4\ CuSCN + 7\ IO_3^- + 14\ H^+ + 7\ Cl^- = 4\ Cu^{2+} + 4\ SO_4^{2-} + 7\ ICl + 4\ HCN + 5\ H_2O$ Precipitate and wash CuSCN. Titrate with standard KIO_3 solution with 5 mL $CHCl_3$ until a definite I_2 color appears in the organic layer. Back-titrate the excess I_2 with standard thiosulfate solution. Cu/7 = 9.078; $KIO_3/4$ = 53.505
Fe(II)	$5\ Fe^{2+} + MnO_4^- + 8\ H^+ = 5\ Fe^{3+} + Mn^{2+} + 4\ H_2O$ Fe^{2+} + Ce(IV) = Fe^{3+} + Ce(III); use 1,10-phenanthroline iron(II) indicator. $6\ Fe^{2+} + Cr_2O_7^{2-} + 14\ H^+ = 6\ Fe^{3+} + 2\ Cr^{3+} + 7\ H_2O$; use diphenylamine sulfonate indicator. Fe/1 = 55.847; $Fe_2O_3/2$ = 79.845
Fe(III)	$Fe^{3+} + 4\ SCN^- = Fe(SCN)_4^-$; $Fe(SCN)_4^-$ + Ti(III) = Fe^{2+} + Ti(IV) + $4\ SCN^-$ Fe/1 = 55.847; $Fe_2O_3/2$ = 79.845
	$2\ Fe^{3+} + Zn = 2\ Fe^{2+} + Zn^{2+}$; then proceed by a method under Fe(II).
	$Fe^{3+} + Ag + Cl^- = Fe^{2+} + AgCl$; then proceed by a method under Fe(II).
	$2\ Fe^{3+} + SnCl_2$(slight excess) $+ 4\ Cl^- = 2\ Fe^{2+} + SnCl_6^{2-}$ $2\ HgCl_2 + SnCl_2 + 2\ Cl^- = Hg_2Cl_2 + SnCl_6^{2-}$ Pour above mixture into an H_3PO_4 plus $MnSO_4$ solution and titrate with $KMnO_4$ as under Fe(II). Fe/1 = 55.847; $Fe_2O_3/2$ = 79.845 $2\ Fe^{3+} + 2\ I^- = Fe^{2+} + I_2$ Titrate liberated iodine with thiosulfate; Fe/1 = 55.847; $Fe_2O_3/2$ = 79.845
I_2	$I_2 + 2\ S_2O_3^{2-} = 2\ I^- + S_4O_6^{2-}$ [titrate solution (pH $\circ$ 7.0) with thiosulfate until color is pale yellow. Add KI and starch and continue titration to disappearance of blue color. $I_2/2$ = 126.9045 $I_2 + H_3AsO_3 + H_2O = 2\ I^- + H_3AsO_4 + 2\ H^+$; use starch and KI as indicator. $I_2/2$ = 126.9045
I^-	$2\ I^- + Br_2$(excess) $= I_2 + 2Br^-$ Remove excess Br_2 formic acid and titrate I_2 with thiosulfate. $I_2/2$ = 126.9045
IO_3^-	$IO_3^- + 5\ I^-$(excess) $+ 6\ H^+ = 3\ I_2 + 3\ H_2O$; titrate I_2 with thiosulfate. $KIO_3/6$ = 35.67
IO_4^-	$IO_4^- + 7\ I^-$(excess) $+ 8\ H^+ = 4\ I_2 + 4\ H_2O$; use a neutral buffered solution. Titrate I_2 with thiosulfate. $KIO_4/2$ = 115.00
K	$K_2Na[Co(NO_2)_6]$; dissolve in H_2SO_4 and titrate with either $KMnO_4$ or Ce(IV). ca. K/5.5 but use an empirical factor.
Mg	$Mg(oxine)_2$; dissolve precipitate and use procedure for Al(8-hydroxyquinoline)$_3$. Mg/8 = 3.0381
Mn(II)	$2\ Mn^{2+} + 5\ BiO_3^- + 14\ H^+ = 2\ MnO_4^- + 5\ Bi^{3+} + 7\ H_2O$ $2\ MnO_4^- + 5\ AsO_3^{3-} + 6\ H^+ = 2\ Mn^{2+} + 5\ AsO_4^{3-} + 3\ H_2O$; Mn/5 = 10.9876
	$2\ Mn^{2+} + 5\ S_2O_8^{2-} + 8\ H_2O$ (Ag^+ catalyst) $= 2\ MnO_4^- + 10\ SO_4^{2-} + 16\ H^+$ Titrate the permanganate formed with iron(II) as under iron(II); Mn/5 = 10.9876
	$2\ Mn^{2+} + 5\ IO_4^- + 3\ H_2O = 2\ MnO_4^- + 5\ IO_3^- + 6\ H^+$ Slowly precipitate excess KIO_4 with $Hg(NO_3)_2$. Filter, add excess Fe^{2+} and titrate excess with standard $KMnO_4$ solution; Mn/5 = 10.9876
	$MnO_4^- + 4\ Mn^{2+} + 15\ H_2P_2O_7^{2-}$ [pH range 4 to 7] $= 5\ Mn(H_2P_2O_7)_3^{3-} + 4\ H_2O$ Use Pt—SCE indicator system; Mn/1 = 54.9380

TABLE 4.44 Equations for the Redox Determinations of the Elements with Equivalent Weights (*Continued*)

Mn(IV)	$MnO_2 + 2\ Fe^{2+}$(excesss standard) $+ 4\ H^+ = Mn^{2+} + 2\ Fe^{3+} + 2\ H_2O$ (use CO_2 atmosphere) $MnO_2 + H_2C_2O_4$(excess standard) $+ 2\ H^+ = Mn^{2+} + 2\ CO_2 + 2\ H_2O$ (use CO_2 atmosphere) In either of the above, titrate excess with $KMnO_4$. Mn/2 = 27.469; $MnO_2/2$ = 43.47
Mn(VI)	$MnO_4^{2-} + 2H_2C_2O_4 + 4\ H^+ = Mn^{2+} + 4\ CO_2 + 4\ H_2O$ Add excess oxalate and back-titrate with permanganate. Mn/4 = 13.7345
Mn(VII)	$2\ MnO_4^- + 5H_2C_2O_4\ 6\ H^+ = 2\ Mn^{2+} + 10\ CO_2 + 3\ H_2O$; Mn/5 = 10.9876
Mo	Mo(VI) + Zn = Mo(III) + Zn^{2+}; catch eluate in excess $Fe_2(SO_4)_3$ solution Mo(III) + 3 Fe^{3+} + 4 H_2O = MoO_4^{2+} + 3 Fe^{2+} 8 H^+; titrate Fe(II) with $KMnO_4$ Mo/3 = 31.98 Mo(VI) + Ag + Cl^- = Mo(V) + AgCl; pass through Ag reductor at 60–80°C. Mo(V) + Ce(IV) = Mo(VI) + Ce(III); Mo/l = 95.94
N_2H_4	$3\ N_2H_4 + 2\ BrO_3^-$(excess) $= 3\ N_2 + 2\ Br^- + 6\ H_2O$; add excess KI and titrate I_2 with thiosulfate. $N_2H_4/4$ = 8.01
NH_2OH	$NH_2OH + BrO_3^- = NO_3^- + Br^- + H^+ + H_2O$; proceed as above for N_2H_4. $NH_2OH/6$ = 5.505
HN_3	$2\ HN_3 + 2$ Ce(IV)(excess) $= 3\ N_2 + 2$ Ce(III) $+ 2\ H^+$; done under inert atmosphere. Add excess KI and titrate with thiosulfate. $HN_3/1$ = 43.03
NO_2^-	$5\ NO_2^- + 2\ MnO_4^-$(excess) $+ 6\ H^+ = 5\ NO_3^- + 2\ Mn^{2+} + 3\ H_2O$; determine excess $KMnO_4$ standard $Na_2C_2O_4$ solution. $NaNO_2/1$ = 69.00 $NO_2^- + 2$ Ce(IV)(excess) $+ H_2O = NO_3^- + 2$ Ce(III) $+ 2\ H^+$; warmed to 50°C. Add excess standard Fe(II) solution and back-titrate with standard Ce(IV) using erioglaucine indicator. $NaNO_2/1$ = 69.00
NO_3^-	NO_3^- + excess Fe^{2+} (Mo catalyst) + $4H^+$ = NO + Fe^{3+}. Add H_3PO_4 and back-titrate excess Fe(II) with $K_2Cr_2O_7$. $NaNO_3/3$ = 28.34
Nb(V)	Nb(V) + Zn = Nb(III) + Zn^{2+}; catch reduced solution under excess Fe(III). Nb(III) + 2 Fe^{3+} = Nb(V) + 2 Fe^{2+}; titrate Fe(II) with MnO_4 solution using 1,10-phenanthroline as indicator. Nb/2 = 46.453; Nb_2O_5 = 66.455
Ni	Precipitate Ni(anthranilate)$_2$ and proceed as under Cd. Ni/8 = 7.336
O_2	$O_2 + 2\ Mn^{2+} + 2\ OH^- = 2\ MnO_2 + 2\ H^+$; stoppered flask plus KI $MnO_2 + 2\ I^- + 4\ H^+ = Mn^{2+} + I_2\ 2H_2O$; titrate I_2 released with thiosulfate. $O_2/4$ = 7.007
O_3	$O_3 + 2\ I^- + H_2O = O_2 + I_2 + 2\ OH^-$; acidify and titrate with thiosulfate. $O_3/2$ = 24.00
H_2O_2	$5\ H_2O_2 + 2\ MnO_4^- + 6\ H^+ = 5\ O_2 + 2\ Mn^{2+} + 8\ H_2O$; $H_2O_2/2$ = 17.01 $H_2O_2 + 2$ Ce(IV) $+ 2\ H^+ = 2$ Ce(III) $+ 2\ H_2O$; use 1,10-phenanthroline indicator $H_2O_2/1$ = 34.02 $H_2O_2 + 2\ I^- + 2\ H^+ = I_2 + 2\ H_2O$; titrate I_2 with thiosulfate. $H_2O_2/2$ = 17.01 $H_2O_2 + 2$ Ti(III) $+ 2H^+ = 2$ Ti(IV) $+ 2H_2O$; end point is disappearance of the yellow color of peroxotitanic acid. $H_2O_2/2$ = 17.01
P	The yellow precipitate of $(NH_4)_3[P(Mo_3O_{10})_4]$ is dissolved in NH_4OH, then solution is strongly acidified with H_2SO_4. See molybdenum; 12 moles Mo per P. P/36 = 0.86038
HPH_2O_2	$HPH_2O_2 + 2\ I_2$(excess) $+ 2\ H_2O = H_3PO_4 + 4\ I^- + 4\ H^+$ (let stand 10 h) Make solution alkaline with $NaHCO_3$ and titrate excess I_2 with standard arsenite solution. $HPH_2O_2/4$ = 16.499
H_3PO_3	$H_3PO_3 + I_2$(excess) $+ H_2O = H_3PO_4 + 2\ I^- + 2\ H^+$ (use $CO_2/NaHCO_3$ buffer; let stand 40–60 min in stoppered flask). Titrate excess I_2 with standard arsenite solution. $H_3PO_3/2$ = 41.00

(*Continued*)

TABLE 4.44 Equations for the Redox Determinations of the Elements with Equivalent Weights (*Continued*)

Pb	Isolate Pb as $PbSO_4$, dissolve it in NaOAc and precipitate with $K_2Cr_2O_7$. Dissolve K_2CrO_4 in NaCl—HCl solution, add KI, and titrate I_2 with thiosulfate solution. $2\ PbCrO_4 + 6\ I^- + 16\ H^+ = 2\ Pb^{2+} + 2\ Cr^{3+} + 3\ I_2 + 8\ H_2O$ Pb/3 = 69.1; PbO/3 = 74.4
S^{2-}	$H_2S + I_2(\text{excess}) = S + 2\ I^- + 2\ H^+$ Back-titrate excess I_2 with standard thiosulfate solution. S/2 = 16.03; $H_2S/2$ = 17.04
	$H_2S + 4\ Br_2 + 4\ H_2O = SO_4^{2-} + 8\ Br^- + 10\ H^+$ Use excess KBr and standard $KBrO_3$ solution. Let stand until clear, add excess KI, and titrate with standard thiosulfate solution. $H_2S/8$ = 4.260; $SO_2/2$ = 32.03; SCN/6 = 9.681
SO_2, SO_3^{2-}	$SO_2 + I_2 + 2\ H_2O = SO_4^{2-} + 2\ I^- + 4\ H^+$ (Titrate excess I_2 with standard thiosulfate) $SO_2/2$ = 32.03
	$SO_2 + 4\ Br_2 + 2\ H_2O = SO_4^{2-} + 2\ Br^- + 4\ H^+$ (Titrate with standard $KBrO_3$—KBr solution until methyl orange is bleached.) $SO_2/2$ = 32.03
$S_2O_3^{2-}$	$2\ S_2O_3^{2-} + I_2 = S_4O_6^{2-} + 2\ I^-$ (Use starch indicator) $Na_2S_2O_3/1$ = 158.11
H_2SO_5	$SO_5^{2-} + H_3AsO_3 = SO_4^{2-} + H_3AsO_4$ $H_2SO_5/2$ = 57.04
$S_2O_8^{2-}$	$S_2O_8^{2-} + H_3AsO_3 + H_2O = 2\ SO_4^{2-} + H_3AsO_4 + 2\ H^+$ $H_2S_2O_8/2$ = 97.07
	$S_2O_8^{2-} + 2\ Fe^{2+} = 2\ SO_4^{2-} + 2\ Fe^{3+}$ $H_2S_2O_8/2$ = 97.07
Sb	$5\ Sb(III) + 2\ MnO_4^- + 16\ H^+ = 5\ Sb(V) + 2\ Mn^{2+} + 8\ H_2O$
	$3\ Sb(III) + BrO_3^- + 6\ H^+ = 3\ Sb(V) + Br^- + 3\ H_2O$
	$Sb(III) + I_2$ [tartrate buffer, pH >7] $= Sb(V) + 2\ I^-$
	$Sb(III) + 2\ Ce(IV) = Sb(V) + 2\ Ce(III)$ For all four methods: Sb/2 = 60.88; $Sb_2O_3/4$ = 72.88
SeO_3^{2-}	$5\ H_2SeO_3 + 2\ MnO_4^- + 6\ H^+ = 5\ H_2SeO_4 + 2\ Mn^{2+} + 3\ H_2O$ $Na_2SeO_3/2$ = 86.47
	$H_2SeO_3 + 4\ I^- + 4\ H^+ = Se + 2\ I_2 + 3\ H_2O$ (titrate I_2 with standard thiosulfate solution) $Na_2SeO_3/2$ = 86.47
	$H_2SeO_3 + 4\ S_2O_3^{2-} + 4\ H^+ = SeS_4O_6^{2-} + S_4O_6^{2-} + 3\ H_2O$ (add small excess of thiosulfate and back-titrate with standard iodine solution) $Na_2SeO_3/4$ = 47.23
SeO_4^{2-}	$SeO_4^{2-} + 2\ H^+ + 2\ Cl^- = SeO_3^{2-} + Cl_2 + H_2O$ (absorb Cl_2 in KI solution) $Cl_2 + 2\ I^- = 2\ Cl^- + I_2$ (titrate I_2 with standard thiosulfate) $Na_2SeO_4/2$ = 94.47
Sn(IV)	$SnCl_6^{2-} + Pb = Sn^{2+} + Pb^{2+} + 6\ Cl^-$ (in CO_2 atmosphere boil 40 min) $Sn^{2+} + I_2 + 6\ Cl^- = SnCl_6^{2-} + 2\ I^-$ (at 0–3°C) Sn/2 = 59.35; $SnO_2/2$ = 67.35
Sn(II)	$Sn(II) + 2\ Ce(IV) = Sn(IV) + 2\ Ce(III)$ Sn/2 = 59.35
Te(IV)	$3\ H_2TeO_3 + Cr_2O_7^{2-} + 8\ H^+ = 3\ H_2TeO_4 + 2\ Cr^{3+} + 4\ H_2O$ Te/2 = 63.80
Te(VI)	$H_2TeO_4 + 2\ Cl^- + 2\ H^+ = H_2TeO_3 + Cl_2 + H_2O$ (see SeO_4^{2-}) Te/2 = 63.80
Ti	$2\ Ti(IV) + Zn(\text{reductor}) = 2Ti(III) + Zn(II)$ $Ti(III) + Fe^{3+} = Ti(IV) + Fe^{2+}$ (in CO_2 atmosphere; use KSCN as indicator) Ti/1 = 47.88 or Ti(III) + Methylene blue = Ti(IV) + colorless leuco base (in CO_2 atmosphere) Ti/1 = 47.88
Tl	$2\ Tl^+ + MnO_4^- + 8\ H^+ = 2\ Tl^{3+} + Mn^{2+} + 4\ H_2O$ Tl/2 = 102.19
	$Tl^+ + 2\ Ce^{3+} = Tl^{3+} + 2\ Ce^{3+}$ (to a yellow color or use 1,10-phenanthroline) Tl/2 = 102.19
U	U(VI) + Zn = U(III) + U(IV) + Zn(II) [pass air through solution to oxidize U(III) to U(IV)] $5\ U^{4+} + 2\ MnO_4^- + 2\ H_2O = 5\ UO_2^{2+} + 2\ Mn^{2+} + 4\ H^+$ U/2 = 119.01; $U_3O_8/6$ = 140.35

TABLE 4.44 Equations for the Redox Determinations of the Elements with Equivalent Weights (*Continued*)

V	Oxidize V(IV) to V(V) with permanganate. Destroy excess with sodium azide and boiling. $VO_2^+ + Fe^{2+} + 2\,H^+ = VO^{2+} + Fe^{3+} + H_2O$ (diphenyaminesulfonic acid indicator) V/1 = 50.94
	Reduce V(V) with SO_2 and bubble CO_2 through boiling solution to remove excess SO_2. $5\,VO^{2+} + MnO_4^- + H_2O = 5\,VO_2^+ + Mn^{2+} + 2\,H^+$ V/1 = 50.94
	Reduce V(V) to V(II) with Zn; catch eluate in excess Fe^{3+}. $V^{2+} + 2\,Fe^{3+} + H_2O = VO^{2+} + 2\,Fe^{2+} + 2\,H^+$ Titrate VO^{2+}—Fe^{2+} mixture with permanganate to VO_2^+—Fe^{3+} V/3 = 16.98; $V_2O_5/6$ = 30.32
Zn	Dissolve precipitate of $Zn[Hg(SCN)_4]$ in 4*M* HCl in stoppered flask, add $CHCl_3$. $2\,SCN^- + 3\,IO_3^- + 2\,H^+ + CN^- = 2\,SO_4^{2-} + 3\,ICN + H_2O$ Zn/24 = 2.725
	$2\,Fe(CN)_6^{3-} + 2\,I^- + 3\,Zn^{2+} + 2\,K^{2+} = K_2Zn_3[Fe(CN)_6]_2 + I_2$ Remove I_2 as formed by standard thiosulfate solution. 3Zn/2 = 98.07 but empirical value of 99.07 is recommended.
	Precipitate Zn(anthranilate)$_2$; proceed as with Cd. Zn/8 = 8.174

Note: Additional procedural information plus interferences and general remarks will be found in J. A. Dean, ed., *Analytical Chemistry Handbook*, McGraw-Hill, New York, Second Edition, 2004.

TABLE 4.45 Standard Solutions for Precipitation Titrations

The list given below includes the substances that are most used and most useful for the standardization of solutions for precipitation titrations. Primary standard solutions are denoted by the letter (P) in Column 1.

Standard	Formula weight	Preparation
$AgNO_3$ (P)	169.89	Weigh the desired amount of ACS reagent grade* $AgNO_3$, dried at 105°C for 2 hr, and dissolve in double distilled water. Store in amber container and away from light. Check against NaCl.
$BaCl_2 \cdot 2H_2O$	244.28	Dissolve clear crystals of the salt in distilled water. Standardize against K_2SO_4 or Na_2SO_4.
$Hg(NO_3)_2 \cdot H_2O$	342.62	Dissolve the reagent grade salt in distilled water and dilute to desired volume. Standardize against NaCl.
KBr	119.01	The commercial reagent (ACS) may contain 0.2% chloride. Prepare an aqueous solution of approximately the desired concentration and standardize it against $AgNO_3$.
$K_4[Fe(CN)]_6 \cdot 3H_2O$	422.41	Dissolve the high-purity commercial salt in distilled water containing 0.2 g/L of Na_2CO_3. Kept in an amber container and away from direct sunlight, solutions are stable for a month or more. Standardize against zinc metal.
KSCN	97.18	Prepare aqueous solutions having the concentration desired. Standardize against $AgNO_3$ solution. Protect from direct sunlight.
K_2SO_4 (P)	174.26	Dissolve about 17.43 g, previously dried at 150°C and accurately weighed, in distilled water and dilute exactly to 1 L.
NaCl (P)	58.44	Dry at 130–150°C and weigh accurately, from a closed container, 5.844 g, dissolve in water, and dilute exactly to 1 L.
NaF (P)	41.99	Dry at 110°C and weigh the appropriate amount of ACS reagent. Dissolve in water and dilute exactly to 1 L.
Na_2SO_4 (P)	142.04	Weigh accurately 14.204 g, dried at 150°C, and dissolve in distilled water. Dilute to exactly 1 L.
$Th(NO_3)_4 \cdot 4H_2O$	552.12	Weigh the appropriate amount of crystals and dissolve in water. Standardize against NaF.

*Meets standards of purity (and impurity) set by the American Chemical Society.

TABLE 4.46 Indicators for Precipitation Titrations

Indicator	Preparation and use
	Specific reagents
$NH_4Fe(SO_4)_2 \cdot 12H_2O$	Use reagent (ACS)* grade salt, low in chloride. Dissolve 175 g in 100 mL 6 *M* HNO_3 which has been gently boiled for 10 min to expel nitrogen oxides. Dilute with 500 mL water. Use 2 mL per 100 mL of end-point volume.
K_2CrO_4	Use reagent (ACS)* grade salt, low in chloride. Prepare 0.1*M* aqueous solution (19.421 g/L). Use 2.5 mL per 100 mL of end-point volume.
Tetrahydroxy-1,4-benzoquinone (THQ)	Prepare fresh as required by dissolving 15 mg in 5 mL of water. Use 10 drops for each titration.
	Adsorption indicators
Bromophenol blue	Dissolve 0.1 g of the acid in 200 mL 95% ethanol.
2′,7′-Dichlorofluorescein	Dissolve 0.1 g of the acid in 100 mL 70% ethanol. Use 1 mL for 100 mL of initial solution.
Eosin, tetrabromofluorescein	See Dichlorofluorescein.
Fluorescein	Dissolve 0.4 g of the acid in 200 mL 70% ethanol. Use 10 drops.
Potassium rhodizonate, $C_4O_4(OK)_2$	Prepare fresh as required by dissolving 15 mg in 5 mL of water. Use 10 drops for each titration.
Rhodamine 6G	Dissolve 0.1 g in 200 mL 70% ethanol.
Sodium 3-alizarinsulfonate	Prepare a 0.2% aqueous solution. Use 5 drops per 120 mL end-point volume.
Thorin	Prepare a 0.025% aqueous solution. Use 5 drops.
	Protective colloids
Dextrin	Use 5 mL of 2% aqueous solution of chloride-free dextrin per 25 mL of 0.1*M* halide solution.
Polyethylene glycol 400	Prepare a 50% (v/v) aqueous solution of the surfactant. Use 5 drops per 100 mL end-point volume.

*Meets standards of purity (and impurity) set by the American Chemical Society.

TABLE 4.47 Properties and Applications of Selected Metal Ion Indicators

Indicator	Chemical name	Dissociation constants and colors of free indicator species	Colors of metal-indicator complexes	Applications
Calmagite 0.05 g/100 mL water; stable 1 year	1-(6-Hydroxy-*m*-tolylazo)-2-naphthol-4-sulfonic acid	H_2In^- (red); pK_2 = 8.1 HIn^{2-} (blue); pK_3 = 12.4 In^{3-} (orange)	Wine-red	Titrations performed with Eriochrome Black T as indicator may be carried out equally well with Calmagite
Eriochrome Black T 0.1 g/100 mL water; prepare fresh daily	1-(2-Hydroxy-1-naphthylazo)-6-nitro-2-naphthol-4-sulfonic acid	H_2In^- (red); pK_2 = 6.3 HIn^{2-} (blue); pK_3 = 11.5 In^{3-} (yellow-orange)	Wine-red	*Direct titration*: Ba, Ca, Cd, In, Mg, Mn, Pb, Sc, Sr, Tl, Zn, and lanthanides *Back titration*: Al, Ba, Bi, Ca, Co, Cr, Fe, Ga, Hg, Mn, Ni, Pb, Pd, Sc, Tl, V *Substitution titration*: Au, Ba, Ca, Cu, Hg, Pb, Pd, Sr
Murexide Suspend 0.5 g in water; use fresh supernatent liquid each day	5-[(Hexahydro-2,4,6-trioxo-5-pyrimidinyl)imino]-2,4,6(1*H*,3*H*,5*H*)-pyrimidinetrione monoammonium salt	H_4In^- (red-violet); pK_2 = 9.2 H_3In^{2-} (violet); pK_3 = 10.9 H_2In^{3-} (blue)	Red with Ca^{2+} Yellow with Co^{2+}, Ni^{2+}, and Cu^{2+}	*Direct titration*: Ca, Co, Cu, Ni *Back titration*: Ca, Cr, Ga *Substitution titration*: Ag, Au, Pd
PAN	1-(2-Pyridylazo)-2-naphthol	HIn (orange-red); pK_1 = 12.3 In^- (pink)	Red	*Direct titration*: Cd, Cu, In, Sc, Tl, Zn *Back titration*: Cu, Fe, Ga, Ni, Pb, Sc, Sn, Zn *Substitution titration*: Al, Ca, Co, Fe, Ga, Hg, In, Mg, Mn, Ni, Pb, V, Zn
Pyrocatechol Violet 0.1 g/100 mL; stable several weeks	Pyrocatecholsulfonephthalein	H_4In (red); pK_1 = 0.2 H_3In^- (yellow): pK_2 = 7.8 H_2In^{2-} (violet); pK_3 = 9.8 HIn^{3-} (red-purple); pK_4 = 11.7	Blue, except red with Th(IV)	*Direct titration*: Al, Bi, Cd, Co, Fe, Ga, Mg, Mn, Ni, Pb, Th, Zn *Back titration*: Al, Bi, Fe, Ga, In, Ni, Pd, Sn, Th, Ti
Salicylic acid	2-Hydroxybenzoic acid	H_2In; pK_1 = 2.98 HIn^-; pK_2 = 12.38	$FeSCN^{2+}$ at pH 3 is reddish-brown	Typical uses: Fe(III) titrated with EDTA to colorless iron-EDTA complex
Xylenol orange	3,3'-Bis[*N,N*-di(carboxyethyl)aminomethyl]-*o*-cresolsulfonephthalein	—COOH groups: pK_3 = 0.76; pK_4 = 1.15; pK_5 = 2.58; pK_6 = 3.23		Typical uses: Bi, Pb, Th

Source: J. A. Dean, ed., Analytical Chemistry Handbook, McGraw-Hill, New York, Second Edition, 2004.

TABLE 4.48 Variation of a_4 with pH

pH	$-\log \alpha_4$	pH	$-\log \alpha_4$
2.0	13.44	7.0	3.33
2.5	11.86	8.0	2.29
3.0	10.60	9.0	1.29
4.0	8.48	10.0	0.46
5.0	6.45	11.0	0.07
6.0	4.66	12.0	0.00

TABLE 4.49 Formation Constants of EDTA Complexes at 25°C, Ionic Strength Approaching Zero

Metal ion	$\log K_{MY}$	Metal ion	$\log K_{MY}$
Co(III)	36	V(IV)	18.0
V(III)	25.9	U(IV)	17.5
In	24.95	Ti(IV)	17.3
Fe(III)	24.23	Ce(III)	16.80
Th	23.2	Zn	16.4
Sc	23.1	Cd	16.4
Cr(III)	23	Co(II)	16.31
Bi	22.8	Al	16.13
Tl(III)	22.5	La	16.34
Sn(II)	22.1	Fe(II)	14.33
Ti(III)	21.3	Mn(II)	13.8
Hg(II)	21.80	Cr(II)	13.6
Ga	20.25	V(II)	12.7
Zr	19.40	Ca	11.0
Cu(II)	18.7	Be	9.3
Ni	18.56	Mg	8.64
Pd(II)	18.5	Sr	8.80
Pb(II)	18.3	Ba	7.78
V(V)	18.05	Ag	7.32

TABLE 4.50 Cumulative Formation Constants of Ammine Complexes at 20°C, Ionic Strength 0.1

Cation	$\log K_1$	$\log K_2$	$\log K_3$	$\log K_4$	$\log K_5$	$\log K_6$
Cadmium	2.65	4.75	6.19	7.12	6.80	5.14
Cobalt(II)	2.11	3.74	4.79	5.55	5.73	5.11
Cobalt(III)	6.7	14.0	20.1	25.7	30.8	35.2
Copper(I)	5.93	10.86				
Copper(II)	4.31	7.98	11.02	13.32	12.66	
Iron(II)	1.4	2.2				
Manganese(II)	0.8	1.3				
Mercury(II)	8.8	17.5	18.5	19.28		
Nickel	2.80	5.04	6.77	7.96	8.71	8.74
Platinum(II)						35.3
Silver(I)	3.24	7.05				
Zinc	2.37	4.81	7.31	9.46		

TABLE 4.51 Masking Agents for Various Elements

Element	Masking agent
Ag	Br^-, citrate, Cl^-, CN^-, I^-, NH_3, SCN^-, $S_2O_3^{2-}$, thiourea, thioglycolic acid, diethyldithiocarbamate, thiosemicarbazide, bis(2-hydroxyethyl)dithiocarbamate
Al	Acetate, acetylacetone, BF_4^-, citrate, $C_2O_4^{2-}$, EDTA, F^-, formate, 8-hydroxyquinoline-5-sulfonic acid, mannitol, 2,3-mercaptopropanol, OH^-, salicylate, sulfosalicylate, tartrate, triethanolamine, tiron
As	Citrate, 2,3-dimercaptopropanol, $NH_2OH \cdot HCl$, OH^-, S_2^{2-}, $S_2O_3^{2-}$, tartrate
Au	Br^-, CN^-, NH_3, SCN^-, $S_2O_3^{2-}$, thiourea
Ba	Citrate, cyclohexanediaminetetraacetic acid, *N,N*-dihydroxyethylglycine, EDTA, F^-, SO_4^{2-}, tartrate
Be	Acetylacetone, citrate, EDTA, F^-, sulfosalicylate, tartrate
Bi	Br^-, citrate, Cl^-, 2,3-dimercaptopropanol, dithizone, EDTA, I^-, OH^-, $Na_5P_3O_{10}$, SCN^-, tartrate, thiosulfate, thiourea, triethanolamine
Ca	BF_4^-, citrate, *N,N*-dihydroxyethylglycine, EDTA, F^-, polyphosphates, tartrate
Cd	Citrate, CN^-, 2,3-dimercaptopropanol, dimercaptosuccinic acid, dithizone, EDTA, glycine, I^-, malonate, NH_3, 1,10-phenanthroline, SCN^-, $S_2O_3^{2-}$, tartrate
Ce	Citrate, *N,N*-dihydroxyethylglycine, EDTA, F^-, PO_4^{3-}, reducing agents (ascorbic acid), tartrate, tiron
Co	Citrate, CN^-, diethyldithiocarbamate, 2,3-dimercaptopropanol, dimethylglyoxime, ethylenediamine, EDTA, F^-, glycine, H_2O_2, NH_3, NO_2^-, 1,10-phenanthroline, $Na_5P_3O_{10}$, SCN^-, $S_2O_3^{2-}$, tartrate
Cr	Acetate, (reduction with) ascorbic acid + KI, citrate, *N,N*-dihydroxyethylglycine, EDTA, F^-, formate, NaOH + H_2O_2, oxidation to CrO_4^{2-}, $Na_5P_3O_{10}$, sulfosalicylate, tartrate, triethylamine, tiron
Cu	Ascorbic acid + KI, citrate, CN^-, diethyldithiocarbamate, 2,3-dimercaptopropanol, ethylenediamine, EDTA, glycine, hexacyanocobalt(III)(3−), hydrazine, I^-, NaH_2PO_2, $NH_2OH \cdot HCl$, NH_3, NO_2^-, 1,10-phenanthroline, S^{2-}, SCN^- + SO_3^{2-}, $S_2O_3^{2-}$, sulfosalicylate, tartrate, thioglycolic acid, thiosemicarbazide, thiocarbohydrazide, thiourea
Fe	Acetylacetone, (reduction with) ascorbic acid, $C_2O_4^{2-}$, citrate, CN^-, 2,3-dimercaptopropanol, EDTA, F^-, NH_3, $NH_2OH \cdot HCl$, OH^-, oxine, 1,10-phenanthroline, 2,2'-bipyridyl, PO_4^{3-}, $P_2O_7^{4-}$, S^{2-}, SCN^-, $SnCl_2$, $S_2O_3^{2-}$, sulfamic acid, sulfosalicylate, tartrate, thioglycolic acid, thiourea, tiron, triethanolamine, trithiocarbonate
Ga	Citrate, Cl^-, EDTA, OH^-, oxalate, sulfosalicylate, tartrate
Ge	F^-, oxalate, tartrate
Hf	See Zr
Hg	Acetone, (reduction with) ascorbic acid, citrate, Cl^-, CN^-, 2,3-dimercaptopropan-1-ol, EDTA, formate, I^-, SCN^-, SO_3^{2-}, tartrate, thiosemicarbazide, thiourea, triethanolamine
In	Cl^-, EDTA, F^-, SCN^-, tartrate, thiourea, triethanolamine
Ir	Citrate, CN^-, SCN^-, tartrate, thiourea
La	Citrate, EDTA, F^-, oxalate, tartrate, tiron
Mg	Citrate, $C_2O_4^{2-}$, cyclohexane-1,2-diaminetetraacetic acid, *N,N*-dihydroxyethylglycine, EDTA, F^-, glycol, hexametaphosphate, OH^-, $P_2O_7^{4-}$, triethanolamine
Mn	Citrate, CN^-, $C_2O_4^{2-}$, 2,3-dimercaptopropanol, EDTA, F^-, $Na_5P_3O_{10}$, oxidation to MnO_4^-, $P_2O_7^{4-}$, reduction to Mn(II) with $NH_2OH \cdot HCl$ or hydrazine, sulfosalicylate, tartrate, triethanolamine, triphosphate, tiron
Mo	Acetylacetone, ascorbic acid, citrate, $C_2O_4^{2-}$, EDTA, F^-, H_2O_2, hydrazine, mannitol, $Na_5P_3O_{10}$, $NH_2OH \cdot HCl$, oxidation to molybdate, SCN^-, tartrate, tiron, triphosphate

(*Continued*)

TABLE 4.51 Masking Agents for Various Elements (*Continued*)

Element	Masking agent
Nb	Citrate, $C_2O_4^{2-}$, F^-, H_2O_2, OH^-, tartrate
Nd	EDTA
NH_4^+	HCHO
Ni	Citrate, CN^-, *N,N*-dihydroxyethylglycine, dimethylglyoxime, EDTA, F^-, glycine, malonate, $Na_5P_3O_{10}$, NH_3, 1,10-phenanthroline, SCN^-, sulfosalicylate, thioglycolic acid, triethanolamine, tartrate
Np	F^-
Os	CN^-, SCN^-, thiourea
Pa	H_2O_2
Pb	Acetate, $(C_6H_5)_4AsCl$, citrate, 2,3-dimercaptopropanol, EDTA, I^-, $Na_5P_3O_{10}$, SO_4^{2-}, $S_2O_3^{2-}$, tartrate, tiron, tetraphenylarsonium chloride, triethanolamine, thioglycolic acid
Pd	Acetylacetone, citrate, CN^-, EDTA, I^-, NH_3, NO_2^-, SCN^-, $S_2O_3^{2-}$, tartrate, triethanol amine
Pt	Citrate, CN^-, EDTA, I^-, NH_3, NO_2^-, SCN^-, $S_2O_3^{2-}$, tartrate, urea
Pu	Reduction to Pu(IV) with sulfamic acid
Rare earths	$C_2O_4^{2-}$, citrate, EDTA, F^-, tartrate
Re	Oxidation to perrhenate
Rh	Citrate, tartrate, thiourea
Ru	CN^-, thiourea
Sb	Citrate, 2,3-dimercaptopropanol, EDTA, F^-, I^-, OH^-, oxalate, S^{2-}, S_2^{2-}, $S_2O_3^{2-}$, tartrate, triethanolamine
Sc	Cyclohexane-1,2-diaminetetraacetic acid, F^-, tartrate
Se	Citrate, F^-, I^-, reducing agents, S^{2-}, SO_3^{2-}, tartrate
Sn	Citrate, $C_2O_3^{2-}$, 2,3-dimercaptopropanol, EDTA, F^-, I^-, OH^-, oxidation with bromine water, phosphate(3−), tartrate, triethanolamine, thioglycolic acid
Sr	Citrate, *N,N*-dihydroxyethylglycine, EDTA, F^-, SO_4^{2-}, tartrate
Ta	Citrate, F^-, H_2O_2, OH^-, oxalate, tartrate
Te	Citrate, F^-, I^-, reducing agents, S^{2-}, sulfite, tartrate
Th	Acetate, acetylacetone, citrate, EDTA, F^-, SO_4^{2-}, 4-sulfobenzenearsonic acid, sulfosalicylic acid, tartrate, triethanolamine
Ti	Ascorbic acid, citrate, F^-, gluconate, H_2O_2, mannitol, $Na_5P_3O_{10}$, OH^-, SO_4^{2-}, sulfosalicylic acid, tartrate, triethanolamine, tiron
Tl	Citrate, Cl^-, CN^-, EDTA, HCHO, hydrazine, $NH_2OH \cdot HCl$, oxalate, tartrate, triethanolamine
U	Citrate, $(NH_4)_2CO_3$, $C_2O_4^{2-}$, EDTA, F^-, H_2O_2, hydrazine + triethanolamine, phosphate(3−), tartrate
V	(Reduction with) ascorbic acid, hydrazine, or $NH_2OH \cdot HCl$, CN^-, EDTA, F^-, H_2O_2, mannitol, oxidation to vanadate, triethanolamine, tiron
W	Citrate, F^-, H_2O_2, hydrazine, $Na_5P_3O_{10}$, $NH_2OH \cdot HCl$, oxalate, SCN^-, tartrate, tiron, triphosphate, oxidation to tungstate(VI)
Y	Cyclohexane-1,2-diaminetetraacetic acid, F^-
Zn	Citrate, CN^-, *N,N*-dihydroxyethylglycine, 2,3-dimercaptopropanol, dithizone, EDTA, F^-, glycerol, glycol, hexacyanoferrate(II)(4−), $Na_5P_3O_{10}$, NH_3, OH^-, SCN^-, tartrate, triethanolamine
Zr	Arsenazo, carbonate, citrate, $C_2O_4^{2-}$, cyclohexane-1,2-diaminetetraacetic acid, EDTA, F^-, H_2O_2, PO_4^{3-}, $P_2O_7^{4-}$, pyrogallol, quinalizarinesulfonic acid, salicylate, SO_4^{2-} + H_2O_2, sulfosalicylate, tartrate, triethanolamine

TABLE 4.52 Masking Agents for Anions and Neutral Molecules

Anion or neutral molecule	Masking agent
Boric acid	F^-, glycol, mannitol, tartrate, and other hydroxy acids
Br^-	Hg(II)
Br_2	Phenol, sulfosalicylic acid
BrO_3^-	Reduction with arsenate(III), hydrazine, sulfite, or thiosulfate
Chromate(VI)	Reduction with arsenate(III), ascorbic acid, hydrazine, hydroxylamine, sulfite, or thiosulfate
Citrate	Ca(II)
Cl^-	Hg(II), Sb(III)
Cl_2	Sulfite
ClO_3^-	Thiosulfate
ClO_4^-	Hydrazine, sulfite
CN^-	HCHO, Hg(II), transition metal ions
EDTA	Cu(II)
F^-	Al(III), Be(II), boric acid, Fe(III), Th(IV), Ti(IV), Zr(IV)
$Fe(CN)_6^{3-}$	Arsenate(III), ascorbic acid, hydrazine, hydroxylamine, thiosulfate
Germanic acid	Glucose, glycerol, mannitol
I^-	Hg(II)
I_2	Thiosulfate
IO_3^-	Hydrazine, sulfite, thiosulfate
IO_4^-	Arsenate(III), hydrazine, molybdate(VI), sulfite, thiosulfate
MnO_4^-	Reduction with arsenate(III), ascorbic acid, azide, hydrazine, hydroxylamine, oxalic acid, sulfite, or thiosulfate
MoO_4^{2-}	Citrate, F^-, H_2O_2, oxalate, thiocyanate + Sn(II)
NO_2^-	Co(II), sulfamic acid, sulfanilic acid, urea
Oxalate	Molybdate(VI), permanganate
Phosphate	Fe(III), tartrate
S	CN^-, S^{2-}, sulfite
S^{2-}	Permanganate + sulfuric acid, sulfur
Sulfate	Cr(III) + heat
Sulfite	HCHO, Hg(II), permanganate + sulfuric acid
SO_5^{2-}	Ascorbic acid, hydroxylamine, thiosulfate
Se and its anions	Diaminobenzidine, sulfide, sulfite
Te	I^-
Tungstate	Citrate, tartrate
Vanadate	Tartrate

TABLE 4.53 Common Demasking Agents

Abbreviations: DPC, diphenylcarbazide; HDMG, dimethylglyoxime; PAN, 1-(2-pyridylazo)-2-naphthol.

Complexing agent	Ion demasked	Demasking agent	Application
CN^-	Ag^+	H^+	Precipitation of Ag
	Cd^{2+}	H^+	Free Cd^{2+}
		$HCHO + OH^-$	Detection of Cd (with DPC) in presence of Cu
	Cu^+	H^+	Precipitation of Cu
	Cu^{2+}	HgO	Determination of Cu
	Fe^{2+}	Hg^{2+}	Free Fe^{2+}
	Fe^{3+}	HgO	Determination of Fe
CN^- (*continued*)	HDMG	Pd^{2+}	Detection of CN^- (with Ni^{2+})
	Hg^{2+}	Pd^{2+}	Detection of Pd (with DPC)
	Ni^{2+}	HCHO	Detection of Ni (with HDMG)
		H^+	Free Ni^{2+}
		HgO	Determination of Ni
		Ag^+	Detection and determination of Ni (with HDMG) in presence of Co
		Ag^+, Hg^{2+}, Pb^{2+}	Detection of Ag, Hg, Pb (with HDMG)
	Pd^{2+}	H^+	Precipitation of Pd
		HgO	Determination of Pd
	Zn^{2+}	$Cl_3CCHO \cdot H_2O$	Titration of Zn with EDTA
		H^+	Free Zn
CO_3^{2-}	Cu^{2+}	H^+	Free Cu^{2+}
$C_2O_4^{2-}$	Al^{3+}	OH^-	Precipitation of $Al(OH)_3$
Cl^- (concentrated)	Ag^+	H_2O	Precipitation of AgCl
Ethylenediamine	Ag^+	SiO_2 (amorphous)	Differentiation of crystalline and amorphous SiO_2 (with CrO_4^{2-})
EDTA	Al^{3+}	F^-	Titration of Al
	Ba^{2+}	H^+	Precipitation of $BaSO_4$ (with SO_4^{2-})
	Co^{2+}	Ca^{2+}	Detection of Co (with diethyldithiocarbamate)
	Mg^{2+}	F^-	Titration of Mg, Mn
	Th(IV)	SO_4^{2-}	Titration of Th
	Ti(IV)	Mg^{2+}	Precipitation of Ti (with NH_3)
	Zn^{2+}	CN^-	Titration of Mg, Mn, Zn
	Many ions	KMO_4^-	Free ions
F^-	Al(III)	Be(II)	Precipitation of Al (with 8-hydroxylquinoline)
		OH^-	Precipitation of $Al(OH)_3$
	Fe(III)	OH^-	Precipitation of $Fe(OH)_3$
	Hf(IV)	Al(III) or Be(II)	Detection of Hg (with xylenol orange)
	Mo(VI)	H_3BO_3	Free molybdate
	Sn(IV)	H_3BO_3	Precipitation of Sn (with H_2S)
	U(VI)	Al(III)	Detection of U (with dibenzoylmethane)
	Zr(IV)	Al(III) or Be(II)	Detection of Zr (with xylenol orange)
		Ca(II)	Detection of Ca (with alizarin S)
		OH^-	Precipitation of $Zr(OH)_4$
H_2O_2	Hf(IV), Ti(IV), or Zr	Fe(III)	Free ions
NH_3	Ag^+	Br^-	Detection of Br^-
		H^+	Detection of Ag
		I^-	Detection of I and Br
		SiO_2 (amorphous)	Differentiation of crystalline and amorphous SiO_2 (with CrO_4^{2-})

TABLE 4.53 Common Demasking Agents (*Continued*)

Complexing agent	Ion demasked	Demasking agent	Application
NO_2^-	Co(III)	H^+	Free Co
PO_4^{3-}	Fe(III)	OH^-	Precipitation of $FePO_4$
	UO_2^{2-}	Al(III)	Detection of U (with dibenzoylmethane)
SCN^-	Fe(III)	OH^-	Precipitation of $Fe(OH)_3$
SO_4^{2-} (conc. H_2SO_4)	Ba^{2+}	H_2O	Precipitation of $BaSO_4$
$S_2O_3^{2-}$	Ag^+	H^+	Free Ag^+
	Cu^{2+}	OH^-	Detection of Cu (with PAN)
Tartrate	Al(III)	$H_2O_2 + Cu^{2+}$	Precipitation of $Al(OH)_3$

TABLE 4.54 Amino Acids pI and pKQ Values

This table lists the pK_a and pI (pH at the isoelectric point) values of α-amino acids commonly found in proteins along with their abbreviations. The dissociation constants refer to aqueous solutions at 25°C.

Name	Abbreviations		pK_a values			pI values
	3 Letter	1 Letter	—COOH	$—NH_3^+$	Other groups	
Alanine	Ala	A	2.34	9.69		6.00
Arginine	Arg	R	2.17	9.04	12.48	10.76
Asparagine	Asn	N	2.01	8.80		5.41
Aspartic acid	Asp	D	1.89	9.60	3.65	2.77
Cysteine	Cys	C	1.96	10.28	8.18	5.07
Glutamine	Gln	Q	2.17	9.13		5.65
Glutamic acid	Glu	E	2.19	9.67	4.25	3.22
Glycine	Gly	G	2.34	9.60		5.97
Histidine	His	H	1.82	9.17	6.00	7.59
Isoleucine	Ile	I	2.36	9.60		6.02
Leucine	Leu	L	2.36	9.60	5.98	
Lysine	Lys	K	2.18	8.98	10.53	9.74
Methionine	Met	M	2.28	9.21		5.74
Phenylalanine	Phe	F	1.83	9.13		5.48
Proline	Pro	P	1.99	10.60		6.30
Serine	Ser	S	2.21	9.15		5.68
Threonine	Thr	T	2.09	9.10		5.60
Tryptophan	Trp	W	2.83	9.39		5.89
Tyrosine	Tyr	Y	2.20	9.11	10.07	5.66
Valine	Val	V	2.32	9.62		5.96

Source: E. L. Smith, et al., *Principles of Biochemistry*, 7th ed., McGraw-Hill, New York, 1983; H. J. Hinz, ed., *Thermodynamic Data for Biochemistry and Biotechnology*, Springer-Verlag, Heidelberg, 1986.

TABLE 4.55 Tolerances of Volumetric Flasks

Capacity, mL	Tolerances,* ±mL		Capacity, mL	Tolerances,* ±mL	
	Class A	Class B		Class A	Class B
5	0.02	0.04	200	0.10	0.20
10	0.02	0.04	250	0.12	0.24
25	0.03	0.06	500	0.20	0.40
50	0.05	0.10	1000	0.30	0.60
100	0.08	0.16	2000	0.50	1.00

*Accuracy tolerances for volumetric flasks at 20°C are given by ASTM standard E288.

TABLE 4.56 Pipette Capacity Tolerances

Volumetric transfer pipets			Measuring and serological pipets	
	Tolerances,* ±mL			Tolerances,† ±mL
Capacity, mL	Class A	Class B	Capacity, mL	Class B
0.5	0.006	0.012	0.1	0.005
1	0.006	0.012	0.2	0.008
2	0.006	0.012	0.25	0.008
3	0.01	0.02	0.5	0.01
4	0.01	0.02	0.6	0.01
5	0.01	0.02	1	0.02
10	0.02	0.04	2	0.02
15	0.03	0.06	5	0.04
20	0.03	0.06	10	0.06
25	0.03	0.06	25	0.10
50	0.05	0.10		
100	0.08	0.16		

*Accuracy tolerances for volumetric transfer pipets are given by ASTM standard E969 and Federal Specification NNN-P-395.

†Accuracy tolerances for measuring pipets are given by Federal Specification NNN-P-350 and for serological pipets by Federal Specification NNN-P-375.

TABLE 4.57 Tolerances of Micropipets (Eppendorf)

Capacity, μL	Accuracy, %	Precision, %	Capacity, μL	Accuracy, %	Precision, %
10	1.2	0.4	100	0.5	0.2
40	0.6	0.2	250	0.5	0.15
50	0.5	0.2	500	0.5	0.15
60	0.5	0.2	600	0.5	0.15
70	0.5	0.2	900	0.5	0.15
80	0.5	0.2	1000	0.5	0.15

TABLE 4.58 Burette Accuracy Tolerances

		Accuracy, ±mL	
Capacity, mL	Subdivision, mL	Class A* and precision grade	Class B and standard grade
10	0.05	0.02	0.04
25	0.10	0.03	0.06
50	0.10	0.05	0.10
100	0.20	0.10	0.20

*Class A conforms to specifications in ASTM E694 for standard taper stopcocks and to ASTM E287 for Teflon or polytetrafluoroethylene stopcock plugs. The 10-mL size meets the requirements for ASTM D664.

TABLE 4.59 Factors for Simplified Computation of Volume

The volume is determined by weighing the water, having a temperature of t°C, contained or delivered by the apparatus at the same temperature. The weight of water, w grams, is obtained with brass weights in air having a density of 1.20 mg/mL.

For apparatus made of soft glass, the volume contained or delivered at 20°C is given by

$$v_{20} = wf_{20} \text{ mL}$$

where v_{20} is the volume at 20° and f_{20} is the factor (apparent specific volume) obtained from the table below for the temperature t at which the calibration is performed. The volume at any other temperature t' may then be obtained from

$$v' = v_{20}[1 + 0.00002(t' - 20)] \text{ mL}$$

For apparatus made of any other material, the volume contained or delivered at the temperature t is

$$v_t = wf_t \text{ mL}$$

where w is again the weight in air obtained with brass weights (in grams), and f_t is the factor given in the third column of the table for the temperature t. The volume at any temperature t' may then be obtained from

$$v'_t = v_t[1 + \beta(t' - t)]\text{mL}$$

where β is the cubical coefficient of thermal expansion of the material from which the apparatus is made. Approximate values of β for some frequently encountered materials are given in Table 4.60.

t, °C	f_{20}	f_t	t, °C	f_{20}	f_t
0	1.001 62	1.001 22	14	93	81
1	54	16	15	1.002 06	1.001 96
2	48	12	16	20	1.002 12
3	43	09	17	35	29
4	41	09	18	51	47
5	1.001 39	1.001 09	19	68	66
6	40	12	20	1.002 86	1.002 86
7	42	16	21	1.003 05	1.003 07
8	45	21	22	26	30
9	50	28	23	47	53
10	1.001 56	1.001 36	24	69	77
11	63	45	25	1.003 93	1.004 03
12	72	56	26	1.004 17	29
13	82	68	27	42	56

(*Continued*)

TABLE 4.59 Factors for Simplified Computation of Volume (*Continued*)

t, °C	f_{20}	f_t	t, °C	f_{20}	f_t
28	68	84	35	1.006 77	1.007 07
29	95	1.005 13	36	1.007 10	1.007 42
30	1.005 23	1.005 43	37	1.007 44	1.007 78
31	1.005 52	1.005 74	38	1.007 79	1.008 15
32	1.005 82	1.006 06	39	1.008 15	1.008 53
33	1.006 13	1.006 39	40	1.008 52	1.008 91
34	1.006 44	1.006 72			

TABLE 4.60 Cubical Coefficients of Thermal Expansion

This table lists values of β, the cubical coefficient of thermal expansion, taken from "Essentials of Quantitative Analysis," by Benedetti-Pichler, and from various other sources. The values of β represents the relative increases in volume for a change in temperature of 1°C at temperatures in the vicinity of 25°C, and is equal to 3α, where α is the linear coefficient of thermal expansion. Data are given for the types of glass from which volumetic apparatus is most commonly made, and also for some other materials which have been or may be used in the fabrication of apparatus employed in analytical work.

Material	β
Glasses	
Alkali-resistant, Corning 728	1.90×10^{-5}
Gerateglas, Schott G20	1.47
Kimble KG-33 (borosilicate)	0.96
N-51A ("Resistant")	1.47
R-6 (soft)	2.79
Pyrex, Corning 744	0.96
Vitreous silica	0.15
Vycor, Corning 790	0.24
Metals	
Brass	*ca.* 5.5
Copper	5.0
Gold	4.3
Monel metal	4.0
Platinum	2.7
Silver	5.7
Stainless steel	*ca.* 5.3
Tantalum	*ca.* 2.0
Tungsten	1.3
Plastics and other materials	
Hard rubber	24×10^{-5}
Polyethylene	45–90
Polystyrene	18–24
Porcelain	*ca.* 1.2
Teflon (polytetrafluoroethylene)	16.5

TABLE 4.61 General Solubility Rules for Inorganic Compounds

Nitrates	All nitrates are soluble.
Acetates	All acetates are soluble; silver acetate is moderately soluble.
Chlorides	All chlorides are soluble except AgCl, $PbCl_2$, and Hg_2Cl_2. $PbCl_2$ is soluble in hot water, slightly soluble in cold water.
Sulfates	All sulfates are soluble except barium and lead. Silver, mercury(I), and calcium are only slightly soluble.
Hydrogen sulfates	The hydrogen sulfates are more soluble than the sulfates.
Carbonates, phosphates, chromates, silicates	All carbonates, phosphates, chromates, and silicates are insoluble, except those of sodium, potassium, and ammonium. An exception is $MgCrO_4$ which is soluble.
Hydroxides	All hydroxides (except lithium, sodium, potassium, cesium, rubidium, and ammonia) are insoluble; $Ba(OH)_2$ is moderately soluble; $Ca(OH)_2$ and $Sr(OH)_2$ are slightly soluble.
Sulfides	All sulfides (except alkali metals, ammonium, magnesium, calcium, and barium) are insoluble. Aluminum and chromium sulfides are hydrolyzed and precipitate as hydroxides.
Sodium, potassium, ammonium	All sodium, potassium, and ammonium salts are soluble. Exceptions: $Na_4Sb_2O_7$, $K_2NaCo(NO_2)_6$, K_2PtCl_6, $(NH_4)_2PtCl_6$, and $(NH_4)_2NaCo(NO_2)_6$.
Silver	All silver salts are insoluble. Exceptions: $AgNO_3$ and $AgClO_4$; $AgC_2H_3O_2$ and Ag_2SO_4 are moderately soluble.

TABLE 4.62 Concentration of Commonly Used Acids and Bases

Freshly opened bottles of these reagents are generally of the concentrations indicated in the table. This may not be true of bottles long opened and this is especially true of ammonium hydroxide, which rapidly loses its strength. In preparing volumetric solutions, it is well to be on the safe side and take a little more than the calculated volume of the concentrated reagent, since it is much easier to dilute a concentrated solution than to strengthen one that is too weak.

A concentrated C.P. reagent usually comes to the laboratory in a bottle having a label which states its molecular weight w, its density (or its specific gravity) d, and its percentage assay p. When such a reagent is used to prepare an aqueous solution of desired molarity M, a convenient formula to employ is

$$V = \frac{100\ wM}{pd}$$

where V is the number of milliliters of concentrated reagent required for 1 liter of the dilute solution.

Example: Sulfuric acid has the molecular weight 98.08. If the concentrated acid assays 95.5% and has the specific gravity 1.84, the volume required for 1 liter of a 0.1 molar solution is

$$V = \frac{100 \times 95.08 \times 0.1}{95.5 \times 1.84} = 5.58\ \text{mL}$$

Reagent	Formula Weight	Density, g · mL^{-1} (20°C)	Weight % (approx)	Molarity	V, mL*
Acetic acid	60.05	1.05	99.8	17.45	57.3
Ammonium hydroxide	35.05	0.90	56.6	14.53	60.0
(as NH_3)	17.03		28.0		
Ethylenediamine	60.10	0.899	100	15.0	66.7
Formic acid	46.03	1.20	90.5	23.6	42.5
Hydrazine	32.05	1.011	95	30.0	33.3
Hydriodic acid	127.91	1.70	57	7.6	132
Hydrobromic acid	80.92	1.49	48	8.84	113
Hydrochloric acid	36.46	1.19	37.2	12.1	82.5

(*Continued*)

TABLE 4.62 Concentration of Commonly Used Acids and Bases (*Continued*)

Reagent	Formula Weight	Density, $g \cdot mL^{-1}$ (20°C)	Weight % (approx)	Molarity	V, mL*
Hydrofluoric acid	20.0	1.18	49.0	28.9	34.5
Nitric acid	63.01	1.42	70.4	15.9	63.0
Perchloric acid	100.47	1.67	70.5	11.7	85.5
Phosphoric acid	97.10	1.70	85.5	14.8	67.5
Pyridine	79.10	0.982	100	12.4	80.6
Potassium hydroxide (soln)	56.11	1.46	45	11.7	85.5
Sodium hydroxide (soln)	40.00	1.54	50.5	19.4	51.5
Sulfuric acid	98.08	1.84	96.0	18.0	55.8
Triethanolamine	149.19	1.124	100	7.53	132.7

*V, mL = volume in milliliters needed to prepare 1 liter of 1 molar solution.

TABLE 4.63 Standard Stock Solutions

Element	Procedure
Aluminum	Dissolve 1.000 g Al wire in minimum amount of 2 *M* HCl; dilute to volume.
Antimony	Dissolve 1.000 g Sb in (1) 10 ml HNO_3 plus 5 ml HCl, and dilute to volume when dissolution is complete; or (2) 18 ml HBr plus 2 ml liquid Br_2; when dissolution is complete add 10 ml $HClO_4$, heat in a well-ventilated hood while swirling until white fumes appear and continue for several minutes to expel all HBr, then cool and dilute to volume.
Arsenic	Dissolve 1.3203 g of As_2O_3 in 3 ml 8 *M* HCl and dilute to volume; or treat the oxide with 2 g NaOH and 20 ml water; after dissolution dilute to 200 ml, neutralize with HCl (pH meter), and dilute to volume.
Barium	(1) Dissolve 1.7787 g $BaCl_2 \cdot 2H_2O$ (fresh crystals) in water and dilute to volume. (2) Dissolve 1.516 g $BaCl_2$ (dried at 250°C for 2 hr) in water and dilute to volume. (3) Treat 1.4367 g $BaCO_3$ with 300 ml water, slowly add 10 ml of HCl and, after the CO_2 is released by swirling, dilute to volume.
Beryllium	(1) Dissolve 19.655 g $BeSO_4 \cdot 4H_2O$ in water, add 5 ml HCl (or HNO_3), and dilute to volume. (2) Dissolve 1.000 g Be in 25 ml 2 *M* HCl, then dilute to volume.
Bismuth	Dissolve 1.000 g Bi in 8 ml of 10 *M* HNO_3, boil gently to expel brown fumes, and dilute to volume.
Boron	Dissolve 5.720 g fresh crystals of H_3BO_3 and dilute to volume.
Bromine	Dissolve 1.489 g KBr (or 1.288 g NaBr) in water and dilute to volume.
Cadmium	(1) Dissolve 1.000 g Cd in 10 ml of 2 *M* HCl; dilute to volume. (2) Dissolve 2.282 g $3CdSO_4 \cdot 8H_2O$ in water; dilute to volume.
Calcium	Place 2.4973 g $CaCO_3$ in volumetric flask with 300 ml water, carefully add 10 ml HCl; after CO_2 is released by swirling, dilute to volume.
Cerium	(1) Dissolve 4.515 g $(NH_4)_4Ce(SO_4)_4 \cdot 2H_2O$ in 500 ml water to which 30 ml H_2SO_4 had been added, cool, and dilute to volume. Advisable to standardize against As_2O_3. (2) Dissolve 3.913 g $(NH_4)_2Ce(NO_3)_6$ in 10 ml H_2SO_4, stir 2 min, cautiously introduce 15 ml water and again stir 2 min. Repeat addition of water and stirring until all the salt has dissolved, then dilute to volume.
Cesium	Dissolve 1.267 g CsCl and dilute to volume. Standardize: Pipette 25 ml of final solution to Pt dish, add 1 drop H_2SO_4, evaporate to dryness, and heat to constant weight at $\not>$ 800°C. Cs (in μg/ml) = (40)(0.734)(wt of residue)
Chlorine	Dissolve 1.648 g NaCl and dilute to volume.
Chromium	(1) Dissolve 2.829 g $K_2Cr_2O_7$ in water and dilute to volume. (2) Dissolve 1.000 g Cr in 10 ml HCl, and dilute to volume.

*1000 μg/mL as the element in a final volume of 1 liter unless stated otherwise.

TABLE 4.63 Standard Stock Solutions (*Continued*)

Element	Procedure
Cobalt	Dissolve 1.000 g Co in 10 ml of 2 *M* HCl, and dilute to volume.
Copper	(1) Dissolve 3.929 g fresh crystals of $CuSO_4 \cdot 5H_2O$, and dilute to volume. (2) Dissolve 1.000 g Cu in 10 ml HCl plus 5 ml water to which HNO_3 (or 30%H_2O_2) is added dropwise until dissolution is complete. Boil to expel oxides of nitrogen and chlorine, then dilute to volume.
Dysprosium	Dissolve 1.1477 g Dy_2O_3 in 50 ml of 2 *M* HCl; dilute to volume.
Erbium	Dissolve 1.1436 g Er_2O_3 in 50 ml of 2 *M* HCl; dilute to volume.
Europium	Dissolve 1.1579 g Eu_2O_3 in 50 ml of 2 *M* HCl; dilute to volume.
Fluorine	Dissolve 2.210 g NaF in water and dilute to volume.
Gadolinium	Dissolve 1.152 g Gd_2O_3 in 50 ml of 2 *M* HCl; dilute to volume.
Gallium	Dissolve 1.000 g Ga in 50 ml of 2 *M* HCl; dilute to volume.
Germanium	Dissolve 1.4408 g GeO_2 with 50 g oxalic acid in 100 ml of water; dilute to volume.
Gold	Dissolve 1.000 g Au in 10 ml of hot HNO_3 by dropwise addition of HCl, boil to expel oxides of nitrogen and chlorine, and dilute to volume. Store in amber container away from light.
Hafnium	Transfer 1.000 g Hf to Pt dish, add 10 ml of 9 *M* H_2SO_4, and then slowly add HF dropwise until dissolution is complete. Dilute to volume with 10% H_2SO_4.
Holmium	Dissolve 1.1455 g Ho_2O_3 in 50 ml of 2 *M* HCl; dilute to volume.
Indium	Dissolve 1.000 g In in 50 ml of 2 *M* HCl; dilute to volume.
Iodine	Dissolve 1.308 g Kl in water and dilute to volume.
Iridium	(1) Dissolve 2.465 g Na_3IrCl_6 in water and dilute to volume. (2) Transfer 1.000 g Ir sponge to a glass tube, add 20 ml of HCl and 1 ml of $HClO_4$. Seal the tube and place in an oven at 300°C for 24 hr. Cool, break open the tube, transfer the solution to a volumetric flask, and dilute to volume. Observe all safety precautions in opening the glass tube.
Iron	Dissolve 1.000 g Fe wire in 20 ml of 5 *M* HCl; dilute to volume.
Lanthanum	Dissolve 1.1717 g La_2O_3 (dried at 110°C) in 50 ml of 5 *M* HCl, and dilute to volume.
Lead	(1) Dissolve 1.5985 g $Pb(NO_3)_2$ in water plus 10 ml HNO_3, and dilute to volume. (2) Dissolve 1.000 g Pb in 10 ml HNO_3, and dilute to volume.
Lithium	Dissolve a slurry of 5.3228 g Li_2CO_3 in 300 ml of water by addition of 15 ml HCl; after release of CO_2 by swirling, dilute to volume.
Lutetium	Dissolve 1.6079 g $LuCl_3$ in water and dilute to volume.
Magnesium	Dissolve 1.000 g Mg in 50 ml of 1 *M* HCl and dilute to volume.
Manganese	(1) Dissolve 1.000 g Mn in 10 ml HCl plus 1 ml HNO_3, and dilute to volume. (2) Dissolve 3.0764 g $MnSO_4 \cdot H_2O$ (dried at 105°C for 4 hr) in water and dilute to volume. (3) Dissolve 1.5824 g MnO_2 in 10 HCl in a good hood, evaporate to gentle dryness, dissolve residue in water and dilute to volume.
Mercury	Dissolve 1.000 g Hg in 10 ml of 5 *M* HNO_3 and dilute to volume.
Molybdenum	(1) Dissolve 2.0425 g $(NH_4)_2MoO_4$ in water and dilute to volume. (2) Dissolve 1.5003 g MoO_3 in 100 ml of 2 *M* ammonia, and dilute to volume.
Neodymium	Dissolve 1.7373 g $NdCl_3$ in 100 ml 1 *M* HCl and dilute to volume.
Nickel	Dissolve 1.000 g Ni in 10 ml hot HNO_3, cool, and dilute to volume.
Niobium	Transfer 1.000 g Nb (or 1.4305 g Nb_2O_5) to Pt dish, add 20 ml HF, and heat gently to complete dissolution. Cool, add 40 ml H_2SO_4, and evaporate to fumes of SO_3. Cool and dilute to volume with 8 *M* H_2SO_4.
Osmium	Dissolve 1.3360 g OsO_4 in water and dilute to 100 ml. Prepare only as needed as solution loses strength on standing unless Os is reduced by SO_2 and water is replaced by 100 ml 0.1 *M* HCl.
Palladium	Dissolve 1.000 g Pd in 10 ml of HNO_3 by dropwise addition of HCl to hot solution; dilute to volume.
Phosphorus	Dissolve 4.260 g $(NH_4)_2HPO_4$ in water and dilute to volume.
Platinum	Dissolve 1.000 g Pt in 40 ml of hot aqua regia, evaporate to incipient dryness, add 10 ml HCl and again evaporate to moist residue. Add 10 ml HCl and dilute to volume.

(*Continued*)

TABLE 4.63 Standard Stock Solutions (*Continued*)

Element	Procedure
Potassium	Dissolve 1.9067 g KCl (or 2.8415 g KNO_3) in water and dilute to volume.
Praseodymium	Dissolve 1.1703 g Pr_2O_3 in 50 ml of 2 *M* HCl; dilute to volume.
Rhenium	Dissolve 1.000 g Re in 10 ml of 8 *M* HNO_3 in an ice bath until initial reaction subsides, then dilute to volume.
Rhodium	Dissolve 1.000 g Rh by the sealed-tube method described under iridium.
Rubidium	Dissolve 1.4148 g RbCl in water. Standardize as described under cesium. Rb (in μg/ml) = (40)(0.320)(wt of residue).
Ruthenium	Dissolve 1.317 g RuO_2 in 15 ml of HCl; dilute to volume.
Samarium	Dissolve 1.1596 g Sm_2O_3 in 50 ml of 2 *M* HCl; dilute to volume.
Scandium	Dissolve 1.5338 g Sc_2O_3 in 50 ml of 2 *M* HCl; dilute to volume.
Selenium	Dissolve 1.4050 g SeO_2 in water and dilute to volume or dissolve 1.000 g Se in 5 ml of HNO_3, then dilute to volume.
Silicon	Fuse 2.1393 g SiO_2 with 4.60 g Na_2CO_3, maintaining melt for 15 min in Pt crucible. Cool, dissolve in warm water, and dilute to volume. Solution contains also 2000 μg/ml sodium.
Silver	(1) Dissolve 1.5748 g $AgNO_3$ in water and dilute to volume. (2) Dissolve 1.000 g Ag in 10 ml of HNO_3; dilute to volume. Store in amber glass container away from light.
Sodium	Dissolve 2.5421 g NaCl in water and dilute to volume.
Strontium	Dissolve a slurry of 1.6849 g $SrCO_3$ in 300 ml of water by careful addition of 10 ml of HCl; after release of CO_2 by swirling, dilute to volume.
Sulfur	Dissolve 4.122 g $(NH_4)_2SO_4$ in water and dilute to volume.
Tantalum	Transfer 1.000 g Ta (or 1.2210 g Ta_2O_5) to Pt dish, add 20 ml of HF, and heat gently to complete the dissolution. Cool, add 40 ml of H_2SO_4 and evaporate to heavy fumes of SO_3. Cool and dilute to volume with 50% H_2SO_4.
Tellurium	(1) Dissolve 1.2508 g TeO_2 in 10 ml of HCl; dilute to volume. (2) Dissolve 1.000 g Te in 10 ml of warm HCl with dropwise addition of HNO_3, then dilute to volume.
Terbium	Dissolve 1.6692 g of $TbCl_3$ in water, add 1 ml of HCl, and dilute to volume.
Thallium	Dissolve 1.3034 g $TlNO_3$ in water and dilute to volume.
Thorium	Dissolve 2.3794 g $Th(NO_3)_4 \cdot 4H_2O$ in water, add 5 ml HNO_3, and dilute to volume.
Thulium	Dissolve 1.142 g Tm_2O_3 in 50 ml of 2 *M* HCl; dilute to volume.
Tin	Dissolve 1.000 g Sn in 15 ml of warm HCl; dilute to volume.
Titanium	Dissolve 1.000 g Ti in 10 ml of H_2SO_4 with dropwise addition of HNO_3; dilute to volume with 5% H_2SO_4.
Tungsten	Dissolve 1.7941 g of $Na_2WO_4 \cdot 2H_2O$ in water and dilute to volume.
Uranium	Dissolve 2.1095 g $UO_2(NO_3)_2 \cdot 6H_2O$ (or 1.7734 g uranyl acetate dihydrate) in water and dilute to volume.
Vanadium	Dissolve 2.2963 g NH_4VO_3 in 100 ml of water plus 10 ml of HNO_3; dilute to volume.
Ytterbium	Dissolve 1.6147 g $YbCl_3$ in water and dilute to volume.
Yttrium	Dissolve 1.2692 g Y_2O_3 in 50 ml of 2 *M* HCl and dilute to volume.
Zinc	Dissolve 1.000 g Zn in 10 ml of HCl; dilute to volume.
Zirconium	Dissolve 3.533 g $ZrOCl_2 \cdot 8H_2O$ in 50 ml of 2 *M* HCl, and dilute to volume. Solution should be standardized.

TABLE 4.64 TLV Concentration Limits for Gases and Vapors

Exposure limits (threshold limit value or TLV) are those set by the Occupational Safety and Health Administration and represent conditions to which most workers can be exposed without adverse effects. The TLV value is expressed as a time weighted average airborne concentration over a normal 8-hour workday and 40-hour workweek.

Substance	Maximum allowable exposure		Toxicity
	ppm	$mg \cdot m^{-3}$	
Acetaldehyde	25	45	carcinogen
Acetic acid	10	25	
Acetic anhydride	5	21	
Acetone	750	1780	
Acetonitrile	40	67	
Acetophenone	10	49	
Acetylene			slightly narcotic
Acrolein	0.1	0.23	
Acrylic acid	2	5.9	
Acrylonitrile	2	4.3	
Acrylonitrile	20	45	
Allyl alcohol	2	4.8	
Allyl chloride	1	3	
Allyl glycidyl ether	5	22	
Ammonia	25	18	toxic
Aniline	2	7.6	carcinogen
Arsine	0.05	0.2	highly toxic
Benzene	10	32	carcinogen
Benzenethiol	0.5	2.3	
p-Benzoquinone	0.1		
Benzoyl chloride	0.5		
Benzoyl peroxide		5	
Benzyl acetate	10		
Benzyl chloride	1		carcinogen
Biphenyl	0.2		
Bis(2-aminoethyl)amine	1		
Bis(2-chloroethyl) ether	5	29	
Bis(2-chloromethyl) ether	0.001		carcinogen
Bis(2-ethylhexyl) phthalate		5	
Boron tribromide	1		
Boron trichloride			toxic
Boron trifluoride	1	3	highly toxic
Bromine	0.1	0.7	
Bromine pentafluoride	0.1		highly toxic
Bromine trifluoride			highly toxic
Bromochloromethane (Halon 1011)	200	1060	
Bromoethane	5	22	carcinogen
Bromoethylene	5	22	slightly toxic
Bromoform	0.5	5	
Bromomethane	5	19	highly toxic, carcinogen
1,3-Butadiene	2		slightly anesthetic, carcinogen
Butane	800	1900	slightly anesthetic
1-Butanethiol	0.5	1.8	

(Continued)

TABLE 4.64 TLV Concentration Limits for Gases and Vapors (*Continued*)

Substance	Maximum allowable exposure		Toxicity
	ppm	$mg \cdot m^{-3}$	
1-Butanol	50	152	
2-Butanol	100	303	
2-Butanone	200	590	
2-Butoxyethanol	25	121	
Butyl acetate	150	710	
sec-Butyl acetate	200	950	
tert-Butyl acetate	200	950	
Butyl acrylate	10		
tert-Butyl alcohol	100	300	
Butylamine	5	15	
tert-Butyl chromate (as CrO_3)		0.1	
Butyl glycidyl ether	50	270	
Butyl mercaptan	0.5	1.5	
p-tert-Butyltoluene	10		
(+)-Camphor	2	12	
Caprolactam	5		
Carbon dioxide	5000	9000	
Carbon disulfide	10	31	
Carbon monoxide	25	28	toxic
Carbon tetrachloride	10	65	
Carbonyl chloride	0.1		
Carbonyl fluoride	2		toxic
Chlordane		0.5	
Chlorine	0.5	1.5	highly toxic
Chlorine dioxide	0.1	0.3	
Chlorine trifluoride	0.1	0.4	highly toxic
Chloroacetaldehyde	1	3	
α-Chloroacetophenone	0.05	0.3	
Chloroacetyl chloride	0.05		
Chlorobenzene	10	46	
2-Chloro-1,3-butadiene	10		carcinogen
Chlorodifluoromethane (CFC 22)	1000	3540	
Chloroethane	100	264	low toxicity
2-Chloroethanol	1	3.3	
Chloroethylene (vinyl chloride)	5	13	toxic, carcinogen
Chloroform (trichloromethane)	10	49	
Chloromethane	50	103	toxic, carcinogen
1-Chloro-1-nitropropane	20	100	
Chloropentafluoroethane (CFC 115)	1000	6320	
3-Chloro-1-propene (allyl chloride)	1	3	carcinogen
o-Chlorotoluene	50	259	
Chlorotrifluoroethylene			toxic
Chromyl chloride (CrO_2Cl_2)	0.025		carcinogen
o-Cresol (also *m*-, *p*-)	5	22	
trans-Crotonaldehyde	2	5.7	
Cyanogen	10	20	highly toxic
Cyanogen chloride	0.3		
Cyclohexane	300	1030	
Cyclohexanol	50	206	
Cyclohexanone	25	100	

TABLE 4.64 TLV Concentration Limits for Gases and Vapors (*Continued*)

Substance	Maximum allowable exposure ppm	Maximum allowable exposure $mg \cdot m^{-3}$	Toxicity
Cyclohexene	300	1015	
Cyclohexylamine	10	41	
1,3-Cyclopentadiene	75		
Cyclopentane	600	1720	
Cyclopropane			anesthetic
2,4-D		10	
DDT		1	
Decaborane	0.05	0.3	
Diacetone alcohol	50	238	
2,2'-Diaminodiethylamine	1	4.2	
Diazomethane	0.2		carcinogen
Diborane	0.1	0.1	
Dibromodifluoromethane	100	860	
1,2-Dibromoethane			carcinogen
Dibutyl phthalate		5	
Dichloroacetylene	0.1		
o-Dichlorobenzene	25	150	
p-Dichlorobenzene	10	60	carcinogen
Dichlorodifluoromethane (Freon 12)	1000	4950	
1,1-Dichloroethane	100	405	
1,2-Dichloroethane	10	40	carcinogen
1,1-Dichloroethylene	5	20	carcinogen
cis-1,2-Dichloroethylene	200	793	
trans-1,2-Dichloroethylene	200	793	
Dichlorofluoromethane (Freon 21)	10	42	
Dichloromethane	50	174	carcinogen
1,1-Dichloro-1-nitroethane	10	60	
1,2-Dichloropropane	75	347	carcinogen
1,3-Dichloropropene	1		carcinogen
Dichlorosilane			highly toxic
1,2-Dichlorotetrafluoroethane (Freon 114)	1000	7000	
Dieldrin		0.25	
Diethanolamine	0.46		
Diethylamine	5	15	
Diethyl ether	400	1210	
Diglycidyl ether	0.5	2.8	
Diisobutyl ketone	25	150	
Diisopropylamine	5	20	
Diiopropyl ether	250	1040	
Dimethoxymethane	1000	3110	
N,N-Dimethylacetamide	10	35	
Dimethylamine	5	9.2	highly toxic
N,N-Dimethylaniline	5	25	
Dimethyl 1,2-dibromo-2,2-dichloroethylphosphate		3	
Dimethyl ether			slightly toxic, anesthetic
1-(1,1-Dimethylethyl)-4-methylbenzene	1	6.1	
N,N-Dimethylformamide	10	30	
2,6-Dimethyl-4-heptanone	25		
1,1-Dimethylhydrazine	0.5	1	carcinogen

(*Continued*)

TABLE 4.64 TLV Concentration Limits for Gases and Vapors (*Continued*)

Substance	Maximum allowable exposure		Toxicity
	ppm	$mg \cdot m^{-3}$	
Dimethyl phthalate		5	
2,2-Dimethylpropane			probably anesthetic
Dimethyl sulfate	0.1	0.5	carcinogen
Dinitrobenzene	0.15	1	
Dinitro-*o*-cresol		0.2	
Dinitrotoluene		1.5	
1,4-Dioxane	25	90	carcinogen
Diphenyl	0.2	1	
Diphenyl ether	1	7	
Dipropylene glycol methyl ether—skin	100	600	
Endrin—skin		0.1	
Epichlorohydrin	2	7.6	carcinogen
2,3-Epoxy-1-propanol (glycidol)	50	150	
1,2-Ethanediamine	10	25	
Ethanethiol	0.5		
Ethanol	1000	1880	
Ethanolamine	3	7.5	
2-Ethoxyethanol (Cellosolve)	5	18	
2-Ethoxyethyl acetate	5	27	
Ethyl acetate	400	1400	
Ethyl acrylate	5	20	
Ethylamine	5	9.2	highly toxic
Ethylbenzene	100	435	
Ethylene			anesthetic
Ethylene glycol	39		
Ethylene glycol dinitrate	0.2		
Ethyleneimine	0.05		carcinogen
Ethylene oxide	1		toxic, carcinogen
Ethyl formate	100	300	
Ethyl mercaptan	0.1	1	
Ethyl silicate	100	850	
Fluorine	1	2	highly toxic
Fluorotrichloromethane (Freon 11)	1000	5600	
Formaldehyde	0.3		carcinogen
Formamide	10	18	
Formic acid	5	9.4	
2-Furancarboxaldehyde (furfural)	2	7.9	
2-Furanmethanol	10	40	
Glycerol		10	
Heptachlor		0.5	
Heptane	400	1640	
2-Heptanone	50	233	
3-Heptanone	50	234	
Hexachloro-1,3-butadiene	0.02		carcinogen
Hexachlorocyclohexane (lindane)		0.5	
Hexachloroethane	1		carcinogen
Hexachloronaphthalene		0.2	
Hexamethylphosphoric triamide			carcinogen
Hexane	50	176	
2-Hexanone	5	20	

TABLE 4.64 TLV Concentration Limits for Gases and Vapors (*Continued*)

Substance	Maximum allowable exposure		Toxicity
	ppm	$mg \cdot m^{-3}$	
sec-Hexyl acetate	50	300	
Hexylene glycol	25		
Hydrazine	0.01	0.1	carcinogen
Hydrogen bromide	3	10	highly toxic
Hydrogen chloride	5	7	highly toxic
Hydrogen cyanide	4.7		highly toxic
Hydrogen fluoride	3	2	highly toxic
Hydrogen iodide			highly toxic
Hydrogen peroxide (90%)	1	1.4	
Hydrogen selenide	0.05	0.2	highly toxic
Hydrogen sulfide	10	15	highly toxic
4-Hydroxy-4-methyl-2-pentanone	50	238	
Indene	10		
Iodine	0.1	1	
Iodine pentafluoride			highly toxic
Iodomethane	2	12	
Isobutyl acetate	150	700	
Isobutyl alcohol	50	150	
Isopentyl acetate	100	525	
Isopentyl alcohol	100	360	
Isophorone	5	28	
Isopropyl acetate	250	1040	
Isopropylamine	5	12	
Isopropylbenzene (cumene)	50	246	
Isopropyl glycidyl ether	50	240	
Ketene	0.5	0.9	
Lindane		0.5	
Liquified petroleum gas	1000	1800	
Malathion		10	
Maleic anhydride	0.25	1	
Malononitrile	0.05	0.4	
Mesityl oxide	15	60	
Methacrylic acid	20	70	
Methanethiol	0.5		
Methanol	200	262	
2-Methoxyaniline (also 4-)	0.1		carcinogen
2-Methoxyethanol	5	16	
2-Methoxyethyl acetate	5	24	
Methyl acetate	200	610	
Methyl acetylene-propadiene (MAPP)	1000	1800	
Methyl acrylate	10	35	
Methylacrylonitrile	1		
Methylamine	5	6.4	highly toxic
o-Methylaniline (also *p*-)	2		carcinogen
m-Methylaniline	2		
N-Methylaniline	0.5	2.2	
3-Methyl-1-butanol	100	361	
Methyl *tert*-butyl ether	40		
Methylcyclohexane	400	1600	
1-Methylcyclohexanol	50	234	

(*Continued*)

TABLE 4.64 TLV Concentration Limits for Gases and Vapors (*Continued*)

Substance	Maximum allowable exposure		Toxicity
	ppm	$mg \cdot m^{-3}$	
cis-2-Methylcyclohexanol	50	234	
trans-2-Methylcyclohexanol	50	234	
cis-3-Methylcyclohexanol	50	234	
trans-3-Methylcyclohexanol	50	234	
cis-4-Methylcyclohexanol	50	234	
trans-4-Methylcyclohexanol	50	234	
Methyl formate	100	250	
5-Methyl-2-hexanone	50	234	
Methyl hydrazine	0.01		
Methyl isocyanate	0.02	0.05	
Methyl mercaptan	0.5	1	highly toxic
Methyl methacrylate	100	410	
Methyl oxirane	20		carcinogen
4-Methyl-2-pentanol	25	104	
4-Methyl-2-pentanone	50	205	
2-Methyl-2,4-pentanediol	25	121	
2-Methyl-1-propanol	50	152	
2-Methyl-2-propanol	100	303	
2-Methyl-2-propenenitrile	1	2.7	
o-Methylstyrene (also *m*-, *p*-)	50		
Morpholine	20	70	
Naphthalene	10	50	
Nickel carbonyl [$Ni(CO)_4$]	0.05	0.35	carcinogen
Nicotine		0.5	
Nitric acid	2	5	
Nitric oxide	25	30	highly toxic
Nitrobenzene	1	5	
p-Nitrochlorobenzene		1	
Nitroethane	100	310	
Nitrogen dioxide	3		highly toxic
Nitrogen trifluoride	10		
Nitrogen trioxide	10	29	highly toxic
Nitroglycerine	0.2	2	
Nitromethane	100	250	
1-Nitropropane	25	90	
2-Nitropropane	10	36	
Nitrosyl chloride			highly toxic
o-Nitrotoluene (also *m*-, *p*-)	2		
Nonane	200	1050	
Octachloronaphthalene		0.1	
Octane	300	1450	
Oxalic acid		1	
2-Oxetanone	0.05		carcinogen
Oxygen difluoride	0.05	0.1	
Ozone	0.1	0.2	
Parathion		0.1	
Pentaborane	0.005	0.01	
Pentachloronaphthalene		0.5	
Pentachlorophenol		0.5	
Pentanal	50		

TABLE 4.64 TLV Concentration Limits for Gases and Vapors (*Continued*)

Substance	Maximum allowable exposure		Toxicity
	ppm	$mg \cdot m^{-3}$	
Pentane	600	1770	
2-Pentanone	200	700	
3-Pentanone	200	700	
Pentyl acetate	100	530	
Perchloroethylene	100	670	
Perchloromethyl mercaptan	0.1	0.8	
Perchloryl fluoride	3	14	
Perfluoroacetone	0.1		
Phenol	5	19	
p-Phenylenediamine		0.1	
Phenylhydrazine	0.1		carcinogen
Phosgene	0.1	0.4	highly toxic
Phosphine	0.3	0.4	highly toxic
Phosphoric acid		1	
Phosphorus pentachloride		1	
Phosphorus pentafluoride			highly toxic
Phosphorus pentasulfide		1	
Phosphorus trichloride	0.5	3	
Phosphoryl chloride	0.1		
Phthalic anhydride	1	6	
Picric acid—skin		0.1	
Propane	1000	1800	low toxicity
Propanoic acid	10	30	
1-Propanol	200	500	
2-Propanol	400	980	
Propenal	0.1		
Propenenitrile	2		carcinogen
Propenoic acid	2		
Propyl acetate	200	835	
Propyleneimine	2	5	carcinogen
Propylene oxide	100	240	toxic
Propyl nitrate	25	110	
Propyne	1000	1650	
2-Propyn-1-ol	1	2.3	
Pyridine	5	15	
Quinone	0.1	0.4	
Selenium compounds (as Se)		0.2	
Selenium hexafluoride	0.05	0.4	
Silane	5	7	highly toxic
Silicon tetrafluoride			highly toxic
Stibine	0.1		
Stoddard solvent	100	575	
Strychnine		0.15	
Styrene	50	213	carcinogen
Sulfur dioxide	2		highly toxic
Sulfur hexafluoride	1000	6000	low toxicity
Sulfuric acid		1	
Sulfur monochloride	1	6	
Sulfur pentafluoride	0.01		
Sulfur tetrafluoride	0.1	0.4	

(*Continued*)

TABLE 4.64 TLV Concentration Limits for Gases and Vapors (*Continued*)

Substance	Maximum allowable exposure		Toxicity
	ppm	$mg \cdot m^{-3}$	
Sulfuryl fluoride	5	20	highly toxic
Tellurium hexafluoride	0.02	0.2	
Terphenyls	1	9	
1,1,2,2-Tetrabromoethane	1	14	
Tetrabromomethane	0.1		
1,1,1,2-Tetrachloro-2,2-difluoroethane	500	4170	
1,1,2,2-Tetrachloro-1,2-difluoroethane	500	4170	
1,1,2,2-Tetrachloroethane	1	6.9	carcinogen
Tetrachloroethylene	25	170	carcinogen
Tetrachloromethane	5	31	carcinogen
1,2,3,4-Tetrachloronaphthalene		2	
Tetraethyllead (as Pb)		0.100	
Tetrafluoromethane			low toxicity
Tetrahydrofuran	200	590	
Tetramethyllead (as Pb)		0.150	
Tetramethylsuccinonitrile	0.5	3	
Tetranitromethane	1	8	
Thionyl chloride	1		
Thiram		5	
Toluene	50	188	
Toluene-2,4-diisocyanate	0.02	0.14	
o-Toluidine (also *m*-, *p*-)	2	8.8	
Tribromomethane	0.5	5.2	
Tributyl phosphate	0.2	2.2	
1,2,4-Trichlorobenzene	5		
1,1,1-Trichloroethane	350	1910	
1,1,2-Trichloroethane	10	55	carcinogen
Trichloroethylene	50	270	carcinogen
Trichlorofluoromethane	1000	5600	
Trichloromethane	10	49	carcinogen
1,2,3-Trichloropropane	10	60	
1,1,2-Trichlorotrifluoroethane	1000		
Tri-*o*-cresol phosphate (also *m*-, *p*-)		0.1	
Triethanolamine	0.5		
Triethylamine	1		
Trifluorobromomethane (Freon 13B1)	1000	6100	
1,1,2-Trifluorotrichloroethane	1000	7600	
Triiodomethane	0.6		
Trimethylamine	5	12	highly toxic
1,2,3-Trimethylbenzene	25	123	
1,2,4-Trimethylbenzene (pseudocumene)	25	123	
1,3,5-Trimethylbenzene (mesitylene)	25	123	
Trinitrotoluene (TNT)		1.5	
Triphenyl phosphate		3	
Turpentine	100	560	
Vinyl acetate	10	35	carcinogen
Vinyl methyl ether			probably anesthetic
Warfarin		0.1	
o-Xylene (also *m*-, *p*-)	100	434	
2,3-Xylidine (also 2,4-, 2,5-, 2,6-, 3,4-, 3,5-)	0.5	2.5	

TABLE 4.65 Some Common Reactive and Incompatible Chemicals

Chemical	Keep out of contact with
Acetic acid	Chromium(VI) oxide, chlorosulfonic acid, ethylene glycol, ethyleneimine, hydroxyl compounds, nitric acid, oleum, perchloric acid, peroxides, permanganates, potasssium *tert*-butoxide, PCl_3
Acetylene	Bromine, chlorine, brass, copper and copper salts, fluorine, mercury and mercury salts, nitric acid, silver and silver salts, alkali hydrides, potassium metal
Alkali metals	Moisture, acetylene, metal halides, ammonium salts, oxygen and oxidizing agents, halogens, carbon tetrachloride, carbon, carbon dioxide, carbon disulfide, chloroform, chlorinated hydrocarbons, ethylene oxide, boric acid, sulfur, tellurium
Aluminum	Chlorinated hydrocarbons, halogens, steam
Ammonia, anhydrous	Mercury, halogens, hypochlorites, chlorites, chlorine(I) oxide, hydrofluoric acid (anhydrous), hydrogen peroxide, chromium(VI) oxide, nitrogen dioxide, chromyl(VI) chloride, sulfinyl chloride, magnesium perchlorate, peroxodisulfates, phosphorus pentoxide, acetaldehyde, ethylene oxide, acrolein, gold(III) chloride
Ammonium nitrate	Acids, metal powders, flammable liquids, chlorates, nitrites, sulfur, finely divided organic or combustible materials, perchlorates, urea
Ammonium perchlorate	Hot copper tubing, sugar, finely divided organic or combustible materials, potassium periodate and permanganate, powdered metals, carbon, sulfur
Aniline	Nitric acid, peroxides, oxidizing materials, acetic anhydride, chlorosulfonic acid, oleum, ozone
Benzoyl peroxide	Direct sunlight, sparks and open flames, shock and friction, acids, alcohols, amines, ethers, reducing agents, polymerization catalysts, metallic naphthenates
Bromine	Ammonia, carbides, dimethylformamide, fluorine, ozone, olefins, reducing materials including many metals, phosphine, silver azide
Calcium carbide	Moisture, selenium, silver nitrate, sodium peroxide, tin(II) chloride, potassium hydroxide plus chlorine, HCl gas, magnesium
Carbon, activated	Calcium hypochlorite, all oxidizing agents, unsaturated oils
Chlorates	Ammonium salts, acids, metal powders, sulfur, finely divided organic or combustible materials, cyanides, metal sulfides, manganese dioxide, sulfur dioxide, organic acids
Chlorine	Ammonia, acetylene, alcohols, alkanes, benzene, butadiene, carbon disulfide, dibutyl phthalate, ethers, fluorine, glycerol, hydrocarbons, hydrogen, sodium carbide, finely divided metals, metal acetylides and carbides, nitrogen compounds, nonmetals, nonmetal hydrides, phosphorus compounds, polychlorobiphenyl, silicones, steel, sulfides, synthetic rubber, turpentine
Chlorine dioxide	Ammonia, carbon monoxide, hydrogen, hydrogen sulfide, methane, mercury, nonmetals, phosphine, phosphorus pentachloride
Chlorites	Ammonia, organic matter, metals
Chloroform	Aluminum, magnesium, potassium, sodium, aluminum chloride, ethylene, powerful oxidants
Chlorosulfonic acid	Saturated and unsaturated acids, acid anhydrides, nitriles, acrolein, alcohols, ammonia, esters, HCl, HF, ketones, hydrogen peroxide, metal powders, nitric acid, organic materials, water
Chromic(VI) acid	Acetic acid, acetic anhydride, acetone, alcohols, alkali metals, ammonia, dimethylformamide, camphor, glycerol, hydrogen sulfide, phosphorus, pyridine, selenium, sulfur, turpentine, flammable liquids in general
Cobalt	Acetylene, hydrazinium nitrate, oxidants
Copper	Acetylene and alkynes, ammonium nitrate, azides, bromates, chlorates, iodates, chlorine, ethylene oxide, fluorine, peroxides, hydrogen sulfide, hydrazinium nitrate

TABLE 4.65 Some Common Reactive and Incompatible Chemicals (*Continued*)

Chemical	Keep out of contact with
Copper(II) sulfate	Hydroxylamine, magnesium
Cumene hydroperoxide	Acids (inorganic or organic)
Cyanides	Acids, water or steam, fluorine, magnesium, nitric acid and nitrates, nitrites
Cyclohexanol	Oxidants
Cyclohexanone	Hydrogen peroxide, nitric acid
Decaborane-14	Dimethyl sulfoxide, ethers, halocarbons
Diazomethane	Alkali metals, calcium sulfate
1,1-Dichloroethylene	Air, chlorotrifluoroethylene, ozone, perchloryl fluoride
Dimethylformamide	Halocarbons, inorganic and organic nitrates, bromine, chromium(VI) oxide, aluminum trimethyl, phosphorus trioxide
1,1-Dimethylhydrazine	Air, hydrogen peroxide, nitric acid, nitrous oxide
Dimethylsulfoxide	Acyl and aryl halides, boron compounds, bromomethane, nitrogen dioxide, magnesium perchlorate, periodic acid, silver difluoride, sodium hydride, sulfur trioxide
Dinitrobenzenes	Nitric acid
Dinitrotoluenes	Nitric acid
1,4-Dioxane	Silver perchlorate
Esters	Nitrates
Ethylamine	Cellulose, oxidizers
Ethers	Oxidizing materials, boron triiodide
Ethylene	Aluminum trichloride, carbon tetrachloride, chlorine, nitrogen oxides, tetrafluoroethylene
Ethylene oxide	Acids and bases, alcohols, air, 1,3-nitroaniline, aluminum chloride, aluminum oxide, ammonia, copper, iron chlorides and oxides, magnesium perchlorate, mercaptans, potassium, tin chlorides, alkane thiols
Ethyl ether	Liquid air, chlorine, chromium(VI) oxide, lithium aluminum hydride, ozone, perchloric acid, peroxides
Ethyl sulfate	Oxidizing materials, water
Flammable liquids	Ammonium nitrate, chromic acid, the halogens, hydrogen peroxide, nitric acid
Fluorine	Isolate from everything; only lead and nickel resist prolonged attack
Formamide	Iodine, pyridine, sulfur trioxide
Freon 113	Aluminum, barium, lithium, samarium, NaK alloy, titanium
Glycerol	Acetic anhydride, hypochlorites, chromium(VI) oxide, perchlorates, alkali peroxides, sodium hydride
Hydrazine	Alkali metals, ammonia, chlorine, chromates and dichromates, copper salts, fluorine, hydrogen peroxide, metallic oxides, nickel, nitric acid, liquid oxygen, zinc diethyl
Hydrides	Powerful oxidizing agents, moisture
Hydrocarbons	Halogens, chromium(VI) oxide, peroxides
Hydrogen	Halogens, lithium, oxidants, lead trifluoride
Hydrogen bromide	Fluorine, iron(III) oxide, ammonia, ozone
Hydrogen chloride	Acetic anhydride, aluminum, 2-aminoethanol, ammonia, chlorosulfonic acid, ethylenediamine, fluorine, metal acetylides and carbides, oleum, perchloric acid, potassium permanganate, sodium, sulfuric acid
Hydrogen fluoride	Acetic anhydride, 2-aminoethanol, ammonia, arsenic trioxide, chlorosulfonic acid, ethylenediamine, ethyleneimine, fluorine, HgO, oleum, phosphorus trioxide, propylene oxide, sodium, sulfuric acid, vinyl acetate
Hydrogen iodide	Fluorine, nitric acid, ozone, metals
Hydrogen peroxide	Copper, chromium, iron, most metals or their salts, alcohols, acetone, organic materials, flammable liquids, combustible materials
Hydrogen selenide	Hydrogen peroxide, nitric acid
Hydrogen sulfide	Fuming nitric acid, oxidizing gases, peroxides

TABLE 4.65 Some Common Reactive and Incompatible Chemicals (*Continued*)

Chemical	Keep out of contact with
Hydroquinone	Sodium hydroxide
Hydroxylamine	Barium oxide and peroxide, carbonyls, chlorine, copper(II) sulfate, dichromates, lead dioxide, phosphorus trichloride and pentachloride, permanganates, pyridine, sodium, zinc
Hypochlorites, salts of	Urea, amines, anthracene, carbon, carbon tetrachloride, ethanol, glycerol, mercaptans, organic sulfides, sulfur, thiols
Indium	Acetonitrile, nitrogen dioxide, mercury(II) bromide, sulfur
Iodine	Acetaldehyde, acetylene, aluminum, ammonia (aqueous or anhydrous), antimony, bromine pentafluoride, carbides, cesium oxide, chlorine, ethanol, fluorine, formamide, lithium, magnesium, phosphorus, pyridine, silver azide, sulfur trioxide
Iodine monochloride	Aluminum foil, organic matter, metal sulfides, phosphorus, potassium, rubber, sodium
Iodoform	Acetone, lithium, mercury(II) oxide, mercury(I) chloride, silver nitrate
Iodomethane	Silver chlorite, sodium
Iron disulfide	Water, powdered pyrites
Isothiourea	Acrylaldehyde, hydrogen peroxide, nitric acid
Ketones	Aldehydes, nitric acid, perchloric acid
Lactonitrile	Oxidizing materials
Lead	Ammonium nitrate, chlorine trifluoride, hydrogen peroxide, sodium azide and carbide, zirconium, oxidants
Lead(II) azide	Calcium stearate, copper, zinc, brass, carbon disulfide
Lead chromate	Iron hexacyanoferrate(4−)
Lead dioxide	Aluminum carbide, hydrogen peroxide, hydrogen sulfide, hydroxylamine, nitroalkanes, nitrogen compounds, nonmetal halides, peroxoformic acid, phosphorus, phosphorus trichloride, potassium, sulfur, sulfur dioxide, sulfides, tungsten, zirconium
Lead(II) oxide	Chlorinated rubber, chlorine, ethylene, fluorine, glycerol, metal acetylides, perchloric acid
Lead(II,IV) oxide	Same as for lead dioxide
Lithium hydride	Nitrous oxide, oxygen
Magnesium	Air, beryllium fluoride, ethylene oxide, halogens, halocarbons, HI, metal cyanides, metal oxides, metal oxosalts, methanol, oxidants, peroxides, sulfur, tellurium
Maleic anhydride	Alkali metals, amines, KOH, NaOH, pyridine
Manganese dioxide	Aluminum, hydrogen sulfide, oxidants, potassium azide, hydrogen peroxide, peroxosulfuric acid, sodium peroxide
Mercaptans	Powerful oxidizers
Mercury	Acetylenic compounds, chlorine, fulminic acid, ammonia, ethylene oxide, metals, methyl azide, oxidants, tetracarbonylnickel
Mercury(II) cyanide	Fluorine, hydrogen cyanide, magnesium, sodium nitrite
Mercury(I) nitrate	Phosphorus
Mercury(II) nitrate	Acetylene, aromatics, ethanol, hypophosphoric acid, phosphine, unsaturated organic compounds
Mercury(II) oxide	Chlorine, hydrazine hydrate, hydrogen peroxide, hypophosphorous acid, magnesium, phosphorus, sulfur, butadiene, hydrocarbons, methanethiol
Mesityl oxide	2-Aminoethanol, chlorosulfonic acid, nitric acid, ethylenediamine, sulfuric acid
Methanol	Beryllium dihydride, chloroform, oxidants, potassium *tert*-butoxide
Methylamine	Nitromethane
N-Methylformamide	Benzenesulfonyl chloride
Methyl isobutyl ketone	Potassium *tert*-butoxide

(*Continued*)

TABLE 4.65 Some Common Reactive and Incompatible Chemicals (*Continued*)

Chemical	Keep out of contact with
Methyl methacrylate	Air, benzoyl peroxide
4-Methylnitrobenzene	Sulfuric acid, tetranitromethane
2-Methylpyridine	Hydrogen peroxide, iron(II) sulfate, sulfuric acid
Methylsodium	4-Chloronitrobenzene
Molybdenum trioxide	Chlorine trifluoride, interhalogens, metals
Naphthalene	Chromium trioxide, dinitrogen pentaoxide
2-Naphthol	Antipyrine, camphor, phenol, iron(III) salts, menthol, oxidizing materials, permanganates, urethane
Neodymium	Phosphorus
Nickel	Aluminum, aluminum(III) chloride, ethylene, 1,4-dioxan, hydrogen, methanol, nonmetals, oxidants, sulfur compounds
Nickel carbonyl	Air, bromine, oxidizing materials
Niobium	Bromine trifluoride, chlorine, fluorine
Nitrates	Aluminum, BP, cyanides, esters, phosphorus, tin(II) chloride, sodium hypophosphite, thiocyanates
Nitric acid, fuming	Organic matter, nonmetals, most metals, ammonia, chlorosulfonic acid, chromium trioxide, cyanides, dichromates, hydrazines, hydrides, HCN, HI, hydrogen sulfide, sulfur dioxide, sulfur halides, sulfuric acid, flammable liquids and gases
Nitric oxide	Aluminum, BaO, boron, carbon disulfide, chromium, many chlorinated hydrocarbons, fluorine, hydrocarbons, ozone, phosphine, phosphorus, hydrazine, acetic anhydride, ammonia, chloroform, Fe, K, Mg, Mn, Na, sulfur
Nitrites	Organic nitrites in contact with ammonium salts, cyanides
Nitrobenzene	Nitric acid, nitrous oxide, silver perchlorate
Nitroethane	Hydroxides, hydrocarbons, metal oxides
Nitrogen trichloride	Ammonia, As, hydrogen sulfide, nitrogen dioxide, organic matter, ozone, phosphine, phosphorus, KCN, KOH, Se, dibutyl ether
Nitrogen dioxide	Cyclohexane, fluorine, formaldehyde, alcohols, nitrobenzene, petroleum, toluene
Nitrogen triiodide	Acids, bromine, chlorine, hydrogen sulfide, ozone
α-Nitroguanidine	Complex salts of mercury and silver
Nitromethane	Acids, alkylmetal halides, hydroxides, hydrocarbons, organic amines, formaldehyde, nitric acid, perchlorates
1-Nitropropane	*See* under Nitromethane; chlorosulfonic acid, oleum
Nitrosyl fluoride	Haloalkenes, metals, nonmetals
Nitrosyl perchlorate	Acetones, amines, diethyl ether, metal salts, organic materials
Nitrourea	Mercury(II) and silver salts
Nitrous acid	Phosphine, phosphorus trichloride, silver nitrate, semicarbazone
Nitryl chloride	Ammonia, sulfur trioxide, tin(IV) bromide and iodide
Oxalic acid	Furfuryl alcohol, silver, mercury, sodium chlorate, sodium chlorite, sodium hypochlorite
Oxygen	Acetaldehyde, acetone, alcohols, alkali metals, alkaline earth metals, Al-Ti alloys, ether, carbon disulfide, halocarbons, hydrocarbons, metal hydrides, 1,3,5-trioxane
Ozone	Alkenes, aromatic compounds, bromine, diethyl ether, ethylene, HBr, HI, nitric oxide, nitrogen dioxide, rubber, stibine
Palladium	Arsenic, carbon, ozonides, sulfur, sodium tetrahydridoborate
Paraformaldehyde	Liquid oxygen
Paraldehyde	Alkalies, HCN, iodides, nitric acid, oxidizers
Pentaborane-9	Dimethylsulfoxide
Pentacarbonyliron	Acetic acid, nitric oxide, transition metal halides, water, zinc

TABLE 4.65 Some Common Reactive and Incompatible Chemicals (*Continued*)

Chemical	Keep out of contact with
2-Pentanone	Bromine trifluoride
3-Pentanone	Hydrogen peroxide, nitric acid
Perchlorates	Carbonaceous materials, finely divided metals particularly magnesium and aluminum, sulfur, benzene, olefins, ethanol, sulfur, sulfuric acid
Perchloric acid	Acetic acid, acetic anhydride, alcohols, antimony compounds, azo pigments, bismuth and its alloys, methanol, carbonaceous materials, carbon tetrachloride, cellulose, dehydrating agents, diethyl ether, glycols and glycolethers, HCl, HI, hypophosphites, ketones, nitric acid, pyridine, steel, sulfoxides, sulfuric acid
Permanganates	All reducing agents, organic materials
Peroxides	Reducing agents, organic materials, thiocyanates
Peroxoacetic acid	Acetic anhydride, olefins, organic matter
Peroxobenzoic acid	Olefins, reducing materials
Peroxoformic acid	Metals and nonmetals, organic materials
Peroxosulfuric acid	Acetone, alcohols, aromatic compounds, catalysts
Phenol	Butadiene, peroxodisulfuric acid, peroxosulfuric acid, aluminum chloride plus nitrobenzene
Phenylhydrazine	Lead dioxide, oxidizers
Phosgene	Aluminum, alkali metals, 2-propanol
Phosphine	Air, boron trichloride, bromine, chlorine, nitric acid, nitrogen oxides, nitrous acid, oxygen, silver nitrate
Phosphorus pentachloride	Aluminum, chlorine, chlorine dioxide, chlorine trioxide, fluorine, magnesium oxide, nitrobenzene, diphosphorus trioxide, potassium, sodium, urea, water
Phosphorus pentafluoride	Water or steam
Phosphorus pentasulfide	Air, alcohols, water
Phosphorus pentoxide	Formic acid, HF, inorganic bases, metals, oxidants, water
Phosphorus, red	Organic materials
Phosphorus tribromide	Potassium, ruthenium tetroxide, sodium, water
Phosphorus trichloride	Acetic acid, aluminum, chromyl dichloride, dimethylsulfoxide, hydroxylamine, lead dioxide, nitric acid, nitrous acid, organic matter, potassium, sodium, water
Phosphorus, white	Air, oxidants of all types, halogens, metals
Phosphoryl chloride	Carbon disulfide, *N*,*N*-dimethylformamide, 2,5-dimethylpyrrole, 2,6-dimethylpyridine 1-oxide, dimethylsulfoxide, water, zinc
Phthalic acid	Nitric acid, sodium nitrite
Piperazine	Oxidizers
Platinum	Acetone, arsenic, hydrazine, lithium, proxosulfuric acid, phosphorus, selenium, tellurium
Potassium	*See* under Alkali metals
Potassium *tert*-butoxide	Organic compounds, sulfuric acid
Potassium hydride	Air, chlorine, acetic acid, acrolein, acrylonitrile, maleic anhydride, nitroparaffins, *N*-nitrosomethylurea, tetrahydrofuran, water
Potassium perchlorate	Aluminum plus magnesium, carbon, nickel plus titanium, reducing agents, sulfur, sulfuric acid
Potassium permanganate	Organic or readily oxidizable materials
Potassium sodium alloy	Air, carbon dioxide, carbon disulfide, halocarbons, metal oxides
2-Propyn-1-ol	Alkali metals, mercury(II) sulfate, oxidizing materials, phosphorus pentoxide, sulfuric acid
Pyridine	Chlorosulfonic acid, chromium trioxide, formamide, maleic anhydride, nitric acid, oleum, perchromates, silver perchlorate, sulfuric acid
Pyrrolidine	Oxidizing materials

(*Continued*)

TABLE 4.65 Some Common Reactive and Incompatible Chemicals (*Continued*)

Chemical	Keep out of contact with
Quinoline	Dinitrogen tetroxide, linseed oil, maleic anhydride, thionyl chloride
Salicylic acid	Iodine, iron salts, lead acetate
Silicon	Alkali carbonates, calcium, chlorine, cobalt(II) fluoride, manganese trifluoride, oxidants, silver fluoride, sodium-potassium alloy
Silver	Acetylene, ammonium compounds, ethyleneimine, hydrogen peroxide, oxalic acid, sulfuric acid, tartaric acid
Sodium	*See* under Alkali metals
Sodium peroxide	Glacial acetic acid, acetic anhydride, aniline, benzene, benzaldehyde, carbon disulfide, diethyl ether, ethanol or methanol, ethylene glycol, ethyl acetate, furfural, glycerol, metals, methyl acetate, organic matter
Sulfides	Acids, powerful oxidizers, moisture
Sulfur	Oxidizing materials, halogens
Sulfur dioxide	Halogens, metal oxides, polymeric tubing, potassium chlorate, sodium hydride
Sulfuric acid	Chlorates, metals, HCl, organic materials, perchlorates, permanganates, water
Sulfuryl dichloride	Alkalis, diethyl ether, dimethylsulfoxide, dinitrogen tetroxide, lead dioxide, phosphorus
Tellurium	Halogens, metals
Tetrahydrofuran	Tetrahydridoaluminates, KOH, NaOH
Tetranitroaniline	Reducing materials
Tetranitromethane	Aluminum, cotton, aromatic nitro compounds, hydrocarbons, cotton, toluene
Thiocyanates	Chlorates, nitric acid, peroxides
Thionyl chloride	Ammonia, dimethylsulfoxide, linseed oil, quinoline, sodium
Thiophene	Nitric acid
Thymol	Acetanilide, antipyrine, camphor, chlorohydrate, menthol, quinine sulfate, urethene
Tin(II) chloride	Boron trifluoride, ethylene oxide, hydrazine hydrate, nitrates, Na, K, hydrogen peroxide
Tin(IV) chloride	Alkyl nitrates, ethylene oxide, K, Na turpentine
Titanium	Aluminum, boron trifluoride, carbon dioxide, CuO, halocarbons, halogens, PbO, nitric acid, potassium chlorate, potassium nitrate, potassium permanganate, steam at high temperatures, water
Toluene	Sulfuric plus nitric acids, nitrogen dioxide, silver perchlorate, uranium hexafluoride
Toluidines	Nitric acid
2,4,6-Trinitrotoluene	Sodium dichromate, sulfuric acid
1,3,5-Trioxane	Oxidizing materials, acids
Urea	Sodium nitrite, phosphorus pentachloride
Vinylidene chloride	Chlorosulfonic acid, nitric acid, oleum

TABLE 4.66 Chemicals Recommended for Refrigerated Storage

A. Due to chemical decomposition or polymerization	
Acetaldehyde	Isoprene
Acrolein	Lecithin
Adenosinetriphosphoric acid	Mercaptoacetic acid
Bromacetaldehyde, diethyl acetal	Methyl acrylate
Bromosuccinimide	2-Methyl-1-butene
3-Buten-2-one	Methylenedi-1,4-phenylene diisocyanate
tert-Butyl hydroperoxide	4-Methyl-1-pentene
2-Chlorocyclohexanone	α-Methylstyrene
Cupferron	1-Naphthyl isocyanate
1,3-Cyclohexadiene	1-Pentene
1,3-Dihydroxy-2-propanone	Isopentyl acetate
Divinylbenzene	Pyruvic acid
Ethyl methacrylate, monomer	Styrene, stabilized
Glutathione	Tetramethylsilane
Glycidol	Thioacetamide
Histamine, base	Veratraldehyde
Hydrocinnamaldehyde	Vitamin E (and the acetate)

B. Due to flammability and high volatility	
Acetaldehyde	Iodomethane
Bromoethane	Isoprene
tert-Butylamine	Isopropylamine
Carbon disulfide	Methylal
1-Chloropropane	2-Methylbutane
3-Chloropropane	2-Methyl-2-butene
Cyclopentane	Methyl formate
Diethyl ether	Pentane
2,2-Dimethylbutane	Propylamine
Dimethyl sulfide	Propylene oxide
Furan	Trichlorosilane

TABLE 4.67 Chemicals Which Polymerize or Decompose on Extended Refrigeration

Formaldehyde	Sodium methoxide
Hydrogen peroxide	Sodium nitrate
Sodium chlorite [sodium chlorate (IV)]	Sodium peroxide
Sodium chromate(VI)	Strontium nitrate
Sodium dithionite	Urea
Sodium ethoxide	

4.15 SIEVES AND SCREENS

TABLE 4.68 U.S. Standard Sieves

Sieve no.	Sieve opening mm	Sieve opening inch	Sieve no.	Sieve opening mm	Sieve opening inch
	125	5.00	10	2.00	0.0787
	106	4.24	12	1.70	0.0661
	90	3.50	14	1.40	0.0555
	75	3.00	16	1.18	0.0469
	63	2.50	18	1.00	0.0394
	53	2.12	20	0.850	0.0331
	45	1.75	25	0.710	0.0278
	37.5	1.50	30	0.600	0.0234
	31.5	1.25	35	0.500	0.0197
	26.5	1.06	40	0.425	0.0165
	22.4	0.875	45	0.355	0.0139
	19.0	0.75	50	0.300	0.0117
	16.0	0.625	60	0.250	0.0098
	13.2	0.530	70	0.212	0.0083
	11.2	0.438	80	0.180	0.0070
	9.5	0.375	100	0.150	0.0059
	8.0	0.312	120	0.125	0.0049
	6.7	0.265	140	0.106	0.0041
3.5	5.60	0.223	170	0.090	0.0035
4	4.75	0.187	200	0.075	0.0029
5	4.00	0.157	230	0.063	0.0025
6	3.35	0.132	270	0.053	0.0021
7	2.80	0.111	325	0.045	0.0017
8	2.36	0.0937	400	0.038	0.0015

Specifications are from ASTM E.11-81/ISO 565. The sieve numbers are the approximate number of openings per linear inch.

4.16 THERMOMETRY

4.16.1 Temperature Measurement

The new international temperature scale, known as ITS-90, was adopted in September 1989. However, neither the definition of thermodynamic temperature nor the definition of the kelvin or the Celsius temperature scales has changed; it is the way in which we are to realize these definitions that has changed. The changes concern the recommended thermometers to be used in different regions of the temperature scale and the list of secondary standard fixed points. The changes in temperature determined using ITS-90 from the previous IPTS-68 are always less than 0.4 K, and almost always less than 0.2 K, over the range 0–300 K.

The ultimate definition of thermodynamic temperature is in terms of pV (pressure × volume) in a gas thermometer extrapolated to low pressure. The kelvin (K), the unit of thermodynamic temperature, is defined by specifying the temperature of one fixed point on the scale—the triple point of water which is defined to be 273.16 K. The Celsius temperature scale (°C) is defined by the equation

$$°C = K - 273.15$$

where the freezing point of water at 1 atm is 273.15 K.

TABLE 4.69 Fixed Points in the ITS-90

Fixed points	T, K	t, °C
Triple point of hydrogen	13.8033	−259.3467
Boiling point of hydrogen at 33 321.3 Pa	17.035	−256.115
Boiling point of hydrogen at 101 292 Pa	20.27	−252.88
Triple point of neon	24.5561	−248.5939
Triple point of oxygen	54.3584	−218.7916
Triple point of argon	83.8058	−189.3442
Triple point of mercury	234.3156	−38.8344
Triple point of water	273.16	0.01
Melting point of gallium	302.9146	29.7646
Freezing point of indium	429.7458	156.5985
Freezing point of tin	505.078	231.928
Freezing point of zinc	692.677	419.527
Freezing point of aluminum	933.473	660.323
Freezing point of silver	1234.93	961.78
Freezing point of gold	1337.33	1064.18
Freezing point of copper	1357.77	1084.62
Secondary reference points to extend the scale (IPTS-68):		
Freezing point of platinum	2042	1769
Freezing point of rhodium	2236	1963
Freezing point of iridium	2720	2447
Melting point of tungsten	3660	3387

The fixed points in the ITS-90 are given in Table 4.54. Platinum resistance thermometers are recommended for use between 14 K and 1235 K (the freezing point of silver), calibrated against the fixed points. Below 14 K either the vapor pressure of helium or a constant-volume gas thermometer is to be used. Above 1235 K radiometry is to be used in conjunction with the Planck radiation law,

$$L_\lambda = c_1 \lambda^{-5} \left(e^{c2/\lambda T} - 1\right)^{-1}$$

where L_λ is the spectral radiance at wavelength λ. The first radiation constant, c_1, is 3.741 83 × 10^{-16} W · m^2 and the second radiation constant, c_2, has a value of 0.014 388 m · K.

When a thermometer which has been standardized for total immersion is used with a part of the liquid column at a temperature below that of the bulb, the reading is low and a correction must be applied. The stem correction, in degrees Celsius, is given by

$$KL(t_o - t_m) = \text{degrees Celsius}$$

where K = constant, characteristic of the particular kind of glass and temperature (see Table 4.65)
L = length of exposed thermometer, °C (that is, the length not in contact with vapor or liquid being measured)
t_o = observed temperature on thermometer
t_m = mean temperature of exposed column (obtained by placing an auxiliary thermometer alongside with its bulb midpoint)

For thermometers containing organic liquids, it is sufficient to use the approximate value, K = 0.001. In such thermometers the value of K is practically independent of the kind of glass.

TABLE 4.70 Values of K for Stem Correction of Thermometers

Temperature, °C	Soft glass	Heat-resistant glass
0–150	0.000 158	0.000 165
200	0.000 159	0.000 167
250	0.000 161	0.000 170
300	0.000 164	0.000 174
350		0.000 178
400		0.000 183
450		0.000 188

4.17 THERMOCOUPLES

The thermocouple reference data in Tables 4.71 to 4.79 give the thermoelectric voltage in millivolts with the reference junction at 0°C. Note that the temperature for a given entry is obtained by adding the corresponding temperature in the top row to that in the left-hand column, regardless of whether the latter is positive or negative.

The noble metal thermocouples, Types B, R, and S, are all platinum or platinum-rhodium thermocouples and hence share many of the same characteristics. Metallic vapor diffusion at high temperatures can readily change the platinum wire calibration, hence platinum wires should only be used inside a nonmetallic sheath such as high-purity alumina.

Type B thermocouples (Table 4.72) offer distinct advantages of improved stability, increased mechanical strength, and higher possible operating temperatures. They have the unique advantage that the reference junction potential is almost immaterial, as long as it is between 0°C and 40°C. Type B is virtually useless below 50°C because it exhibits a double-value ambiguity from 0°C to 42°C.

Type E thermoelements (Table 4.73) are very useful down to about liquid hydrogen temperatures and may even be used down to liquid helium temperatures. They are the most useful of the commercially standardized thermocouple combinations for subzero temperature measurements because of their high Seebeck coefficient (58 μV/°C), low thermal conductivity, and corrosion resistance. They also have the largest Seebeck coefficient (voltage response per degree Celsius) above 0°C of any of the standardized thermocouples which makes them useful for detecting small temperature changes. They are recommended for use in the temperature range from –250 to 871°C in oxidizing or inert atmospheres. They should not be used in sulfurous, reducing, or alternately reducing and oxidizing atmospheres unless suitably protected with tubes. They should not be used in vacuum at high temperatures for extended periods of time.

Type J thermocouples (Table 4.74) are one of the most common types of industrial thermocouples because of the relatively high Seebeck coefficient and low cost. They are recommended for use in the temperature range from 0 to 760°C (but never above 760°C due to an abrupt magnetic transformation that can cause decalibration even when returned to lower temperatures). Use is permitted in vacuum and in oxidizing, reducing, or inert atmospheres, with the exception of sulfurous atmospheres above 500°C. For extended use above 500°C, heavy-gauge wires are recommended. They are not recommended for subzero temperatures. These thermocouples are subject to poor conformance characteristics because of impurities in the iron.

The Type K thermocouple (Table 4.75) is more resistant to oxidation at elevated temperatures than the Type E, J, or T thermocouple, and consequently finds wide application at temperatures above 500°C. It is recommended for continuous use at temperatures within the range –250 to 1260°C in inert or oxidizing atmospheres. It should not be used in sulfurous or reducing atmospheres, or in vacuum at high temperatures for extended times.

The Type N thermocouple (Table 4.76) is similar to Type K but it has been designed to minimize some of the instabilities in the conventional Chromel-Alumel combination. Changes in the alloy content have improved the order/disorder transformations occurring at 500°C and a higher silicon content of the positive element improves the oxidation resistance at elevated temperatures.

The Type R thermocouple (Table 4.77) was developed primarily to match a previous platinum-10% rhodium British wire which was later found to have 0.34% iron impurity in the rhodium. Comments on Type S also apply to Type R.

The Type S thermocouple (Table 4.78) is so stable that it remains the standard for determining temperatures between the antimony point (630.74°C) and the gold point (1064.43°C). The other fixed point used is that of silver. The Type S thermocouple can be used from −50°C continuously up to about 1400°C, and intermittently at temperatures up to the freezing point of platinum (1769°C). The thermocouple is most reliable when used in a clean oxidizing atmosphere, but may also be used in inert gaseous atmospheres or in a vacuum for short periods of time. It should not be used in reducing atmospheres, nor in those containing metallic vapor (such as lead or zinc), nonmetallic vapors (such as arsenic, phosphorus, or sulfur), or easily reduced oxides, unless suitably protected with nonmetallic protecting tubes.

The Type T thermocouple (Table 4.79) is popular for the temperature region below 0°C (but see under Type E). It can be used in vacuum, or in oxidizing, reducing, or inert atmospheres.

TABLE 4.71 Thermoelectric Values in Millivolts at Fixed Points for Various Thermocouples

Abbreviations Used in the Table

FP, freezing point — BP, boiling point
NBP, normal boiling point — TP, triple point

Fixed point	°C	Type B	Type E	Type J	Type K	Type N	Type R	Type S	Type T
Helium NPB	−268.934		−9.8331		−6.4569	−4.345			−6.2563
Hydrogen TP	−259.347*		−9.7927		−6.4393	−4.334			−6.2292
Hydrogen NBP	−252.88*		−9.7447		−6.4167	−4.321			−6.1977
Neon TP	−248.594*		−9.7046		−6.3966	−4.271			−6.1714
Neon NBP	−246.048		−9.6776		−6.3827	−4.300			−6.1536
Oxygen TP	−218.792*		−9.2499		−6.1446	−4.153			−5.8730
Nitrogen TP	−210.001		−9.0629	−8.0957	−6.0346	−4.083			−5.7533
Nitrogen NBP	−195.802		−8.7168	−7.7963	−5.8257	−3.947			−5.5356
Oxygen NBP	−182.962		−8.3608	−7.4807	−5.6051	−3.802			−5.3147
Carbon dioxide SP	−78.474		−4.2275	−3.7187	−2.8696	−1.939			−2.7407
Mercury TP	−38.834*		−2.1930	−1.4849		−0.985	−0.1830	−0.1895	−1.4349
Ice point	0.000	−0.000	0.000	0.000	0.000	0.000	0.000	0.000	0.000
Diphenyl ether TP	26.87	−0.0024	1.6091	1.3739	1.076	0.698	0.1517	0.1537	1.0679
Water BP	100.00	0.0332	6.3171	5.2677	4.0953	2.774	0.6472	0.6453	4.2773
Benzoic acid TP	122.37	0.0561	7.8468	6.4886	5.0160	3.446	0.8186	0.8129	5.3414
Indium FP	156.598*	0.1019	10.260	8.3743	6.0404	4.508	1.0956	1.0818	7.0364
Tin FP	231.928*	0.2474	15.809	12.552	9.4201	6.980	1.7561	1.7146	11.013
Bismuth FP	271.442	0.3477	18.821	14.743	11.029	8.336	2.1250	2.0640	13.219
Cadmium FP	321.108	0.4971	22.684	17.493	13.085	10.092	2.6072	2.5167	16.095
Lead FP	327.502	0.5182	23.186	17.846	13.351	10.322	2.6706	2.5759	16.473
Mercury BP	356.66	0.6197	25.489	19.456	14.571		2.9630	2.8483	18.218
Zinc FP	419.527*	0.8678	30.513	22.926	17.223		3.6113	3.4479	
Cu-Al eutectic FP	548.23	1.4951	40.901	30.109	22.696		5.0009	4.7140	
Antimony FP	630.74	1.9784	47.561	34.911	26.207		5.9331	5.5521	
Aluminum FP	660.37	2.1668	49.941	36.693	27.461		6.2759	5.8591	
Silver FP	961.93*	4.4908	73.495	55.669	39.779		10.003	9.1482	
Gold FP	1064.43*	5.4336		61.716	43.755		11.364	10.334	
Copper FP	1084.5	5.6263		62.880	44.520		11.635	10.570	
Nickel FP	1455	9.5766					16.811	15.034	
Cobalt FP	1494	10.025					17.360	15.504	
Palladium FP	1554	10.721					18.212	16.224	
Platinum FP	1772	13.262					21.103	18.694	

*Defining fixed points of the International Temperature Scale of 1990 (ITS-90). Except for the triple points, the assigned values of temperature are for equilibrium states at a pressure of one standard atmosphere (101 325 Pa).

TABLE 4.72 Type B Thermocouples: Platinum-30% Rhodium Alloy vs. Platinum-6% Rhodium Alloy

Thermoelectric voltage in millivolts; reference junction at 0°C.

°C	0	10	20	30	40	50	60	70	80	90
0	0.00	−0.0019	−0.0026	−0.0021	−0.0005	0.0023	0.0062	0.0112	0.0174	0.0248
100	0.0332	0.0427	0.0534	0.0652	0.0780	0.0920	0.1071	0.1232	0.1405	0.1588
200	0.1782	0.1987	0.2202	0.2428	0.2665	0.2912	0.3170	0.3438	0.3717	0.4006
300	0.4305	0.4615	0.4935	0.5266	0.5607	0.5958	0.6319	0.6690	0.7071	0.7462
400	0.7864	0.8275	0.8696	0.9127	0.9567	1.0018	1.0478	1.0948	1.1427	1.1916
500	1.2415	1.2923	1.3440	1.3967	1.4503	1.5048	1.5603	1.6166	1.6739	1.7321
600	1.7912	1.8512	1.9120	1.9738	2.0365	2.1000	2.1644	2.2296	2.2957	2.3627
700	2.4305	2.4991	2.5686	2.6390	2.7101	2.7821	2.8548	2.9284	3.0028	3.0780
800	3.1540	3.2308	3.3084	3.3867	3.4658	3.5457	3.6264	3.7078	3.7899	3.8729
900	3.9565	4.0409	4.1260	4.2119	4.2984	4.3857	4.4737	4.5624	4.6518	4.7419
1000	4.8326	4.9241	5.0162	5.1090	5.2025	5.2966	5.3914	5.4868	5.5829	5.6796
1100	5.7769	5.8749	5.9734	6.0726	6.1724	6.2728	6.3737	6.4753	6.5774	6.6801
1200	6.7833	6.8871	6.9914	7.0963	7.2017	7.3076	7.4140	7.5210	7.6284	7.7363
1300	7.8446	7.9534	8.0627	8.1724	8.2826	8.3932	8.5041	8.6155	8.7273	8.8394
1400	8.9519	9.0648	9.1780	9.2915	9.4053	9.5194	9.6338	9.7485	9.8634	9.9786
1500	10.0940	10.2097	10.3255	10.4415	10.5577	10.6740	10.7905	10.9071	11.0237	11.1405
1600	11.2574	11.3743	11.4913	11.6082	11.7252	11.8422	11.9591	12.0761	12.1929	12.3100
1700	12.4263	12.5429	12.6594	12.7757	12.8918	13.0078	13.1236	13.2391	13.3545	13.4696
1800	13.5845	13.6991	13.8135							

TABLE 4.73 Type E Thermocouples: Nickel-Chromium Alloy vs. Copper-Nickel Alloy

Thermoelectric voltage in millivolts; reference junction at 0°C.

°C	0	10	20	30	40	50	60	70	80	90
−200	−8.824	−9.063	−9.274	−9.455	−9.604	−9.719	−9.797	−9.835		
−100	−5.237	−5.680	−6.107	−6.516	−6.907	−7.279	−7.631	−7.963	−8.273	−8.561
−0	0.000	−0.581	−1.151	−1.709	−2.254	−2.787	−3.306	−3.811	−4.301	−4.777
0	0.000	0.591	1.192	1.801	2.419	3.047	3.683	4.394	4.983	5.646
100	6.317	6.996	7.683	8.377	9.078	9.787	10.501	11.222	11.949	12.681
200	13.419	14.161	14.909	15.661	16.417	17.178	17.942	18.710	19.481	20.256
300	21.033	21.814	22.597	23.383	24.171	24.961	25.754	26.549	27.345	28.143
400	28.943	29.744	30.546	31.350	32.155	32.960	33.767	34.574	35.382	36.190
500	36.999	37.808	38.617	39.426	40.236	41.045	41.853	42.662	43.470	44.278
600	45.085	45.891	46.697	47.502	48.306	49.109	49.911	50.713	51.513	52.312
700	53.110	53.907	54.703	55.498	56.291	57.083	57.873	58.663	59.451	60.237
800	61.022	61.806	62.588	63.368	64.147	64.924	65.700	66.473	67.245	68.015
900	68.783	69.549	70.313	71.075	71.835	72.593	73.350	74.104	74.857	75.608
1000	76.358									

TABLE 4.74 Type J Thermocouples: Iron vs. Copper-Nickel Alloy

Thermoelectric voltage in millivolts; reference junction at 0°C.

°C	0	10	20	30	40	50	60	70	80	90
−200	−7.890	−8.096								
−100	−4.632	−5.036	−5.426	−5.801	−6.159	−6.499	−6.821	−7.122	−7.402	−7.659
−0	0.000	−0.501	−0.995	−1.481	−1.960	−2.431	−2.892	−3.344	−3.785	−4.215
0	0.000	0.507	1.019	1.536	2.058	2.585	3.115	3.649	4.186	4.725
100	5.268	5.812	6.359	6.907	7.457	8.008	8.560	9.113	9.667	10.222
200	10.777	11.332	11.887	12.442	12.998	13.553	14.108	14.663	15.217	15.771
300	16.325	16.879	17.432	17.984	18.537	19.089	19.640	20.192	20.743	21.295
400	21.846	22.397	22.949	23.501	24.054	24.607	25.161	25.716	26.272	26.829
500	27.388	27.949	28.511	29.075	29.642	30.210	30.782	31.356	31.933	32.513
600	33.096	33.683	34.273	34.867	35.464	36.066	36.671	37.280	37.893	38.510
700	39.130	39.754	40.482	41.013	41.647	42.283	42.922			

TABLE 4.75 Type K Thermocouples: Nickel-Chromium Alloy vs. Nickel-Aluminum Alloy

Thermoelectric voltage in millivolts; reference junction at 0°C.

°C	0	10	20	30	40	50	60	70	80	90
−200	−5.891	−6.035	−6.158	−6.262	−6.344	−6.404	−6.441	−6.458		
−100	−3.553	−3.852	−4.138	−4.410	−4.669	−4.912	−5.141	−5.354	−5.550	−5.730
−0	0.000	−0.392	−0.777	−1.156	−1.517	−1.889	−2.243	−2.586	−2.920	−3.242
0	0.000	0.397	0.798	1.203	1.611	2.022	2.436	2.850	3.266	3.681
100	4.095	4.508	4.919	5.327	5.733	6.137	6.539	6.939	7.338	7.737
200	8.137	8.537	8.938	9.341	9.745	10.151	10.560	10.969	11.381	11.793
300	12.207	12.623	13.039	13.456	13.874	14.292	14.712	15.132	15.552	15.974
400	16.395	16.818	17.241	17.664	18.088	18.513	18.839	19.363	19.788	20.214
500	20.640	21.066	21.493	21.919	22.346	22.772	23.198	23.624	24.050	24.476
600	24.902	25.327	25.751	26.176	26.599	27.022	27.445	27.867	28.288	28.709
700	29.128	29.547	29.965	30.383	30.799	31.214	31.629	32.042	32.455	32.866
800	33.277	33.686	34.095	34.502	34.909	35.314	35.718	36.121	36.524	36.925
900	37.325	37.724	38.122	38.519	38.915	39.310	39.703	40.096	40.488	40.879
1000	41.269	41.657	42.045	42.432	42.817	43.202	43.585	43.968	44.349	44.729
1100	45.108	45.486	45.863	46.238	46.612	46.985	47.356	47.726	48.095	48.462
1200	48.828	49.129	49.555	49.916	50.276	50.633	50.990	51.344	51.697	52.049
1300	52.398	52.747	53.093	53.439	53.782	54.125	54.466	54.807		

TABLE 4.76 Type N Thermocouples: Nickel-14.2% Chromium-1.4% Silicon Alloy vs. Nickel-4.4% Silicon-0.1% Magnesium Alloy

Thermoelectric voltage in millivolts; reference junction at 0°C.

°C	0	10	20	30	40	50	60	70	80	90
−200	−3.990	−4.083	−4.162	−4.227	−4.277	−4.313	−4.336	−4.345		
−100	−2.407	−2.612	−2.807	−2.994	−3.170	−3.336	−3.491	−3.634	−3.766	−3.884
−0	0.000	−0.260	−0.518	−0.772	−1.023	−1.268	−1.509	−1.744	−1.972	−2.193
0	0.000	0.261	0.525	0.793	1.064	1.339	1.619	1.902	2.188	2.479
100	2.774	3.072	3.374	3.679	3.988	4.301	4.617	4.936	5.258	5.584
200	5.912	6.243	6.577	6.914	7.254	7.596	7.940	8.287	8.636	8.987
300	9.340	9.695	10.053	10.412	10.772	11.135	11.499	11.865	12.233	12.602
400	12.972	13.344	13.717	14.091	14.467	14.844	15.222	15.601	15.981	16.362
500	16.744	17.127	17.511	17.896	18.282	18.668	19.055	19.443	19.831	20.220
600	20.609	20.999	21.390	21.781	22.172	22.564	22.956	23.348	23.740	24.133
700	24.526	24.919	25.312	25.705	26.098	26.491	26.885	27.278	27.671	28.063
800	28.456	28.849	29.241	29.633	30.025	30.417	30.808	31.199	31.590	31.980
900	32.370	32.760	33.149	33.538	33.926	34.315	34.702	35.089	35.476	35.862
1000	36.248	36.633	37.018	37.402	37.786	38.169	38.552	38.934	39.315	39.696
1100	40.076	40.456	40.835	41.213	41.590	41.966	42.342	42.717	43.091	43.464
1200	43.836	44.207	44.577	44.947	45.315	45.682	46.048	46.413	46.777	47.140
1300	47.502									

TABLE 4.77 Type R Thermocouples: Platinum-13% Rhodium Alloy vs. Platinum

Thermoelectric voltage in millivolts; reference junction at 0°C.

°C	0	10	20	30	40	50	60	70	80	90
(Below zero)		−0.0515	−0.100	−0.1455	−0.1877	−0.2264				
0	0.0000	0.0543	0.1112	0.1706	0.2324	0.2965	0.3627	0.4310	0.5012	0.5733
100	0.6472	0.7228	0.8000	0.8788	0.9591	1.0407	1.1237	1.2080	1.2936	1.3803
200	1.4681	1.5571	1.6471	1.7381	1.8300	1.9229	2.0167	2.1113	2.2068	2.3030
300	2.4000	2.4978	2.5963	2.6954	2.7953	2.8957	2.9968	3.0985	3.2009	3.3037
400	3.4072	3.5112	3.6157	3.7208	3.8264	3.9325	4.0391	4.1463	4.2539	4.3620
500	4.4706	4.5796	4.6892	4.7992	4.9097	5.0206	5.1320	5.2439	5.3562	5.4690
600	5.5823	5.6960	5.8101	5.9246	6.0398	6.1554	6.2716	6.3883	6.5054	6.6230
700	6.7412	6.8598	6.9789	7.0984	7.2185	7.3390	7.4600	7.5815	7.7035	7.8259
800	7.9488	8.0722	8.1960	8.3203	8.4451	8.5703	8.6960	8.8222	8.9488	9.0758
900	9.2034	9.3313	9.4597	9.5886	9.7179	9.8477	9.9779	10.1086	10.2397	10.3712
1000	10.5032	10.6356	10.7684	10.9017	11.0354	11.1695	11.3041	11.4391	11.5745	11.7102
1100	11.8463	11.9827	12.1194	12.2565	12.3939	12.5315	12.6695	12.8077	12.9462	13.0849
1200	13.2239	13.3631	13.5025	13.6421	13.7818	13.9218	14.0619	14.2022	14.3426	14.4832
1300	14.6239	14.7647	14.9056	15.0465	15.1876	15.3287	15.4699	15.6110	15.7522	15.8935
1400	16.0347	16.1759	16.3172	16.4583	16.5995	16.7405	16.8816	17.0225	17.1634	17.3041
1500	17.4447	17.5852	17.7256	17.8659	18.0059	18.1458	18.2855	18.4251	18.5644	18.7035
1600	18.8424	18.9810	19.1194	19.2575	19.3953	19.5329	19.6702	19.8071	19.9437	20.0797
1700	20.2151	20.3497	20.4834	20.6161	20.7475	20.8777	21.0064			

TABLE 4.78 Type S Thermocouples: Platinum-10% Rhodium Alloy vs. Platinum

Thermoelectric voltage in millivolts; reference junction at 0°C.

°C	0	10	20	30	40	50	60	70	80	90
(Below zero)		−0.0527	−0.1028	−0.1501	−0.1944	−0.2357				
0	0.0000	0.0552	0.1128	0.1727	0.2347	0.2986	0.3646	0.4323	0.5017	0.5728
100	0.6453	0.7194	0.7948	0.8714	0.9495	1.0287	1.1089	1.1902	1.2726	1.3558
200	1.4400	1.5250	1.6109	1.6975	1.7849	1.8729	1.9617	2.0510	2.1410	2.2316
300	2.3227	2.4143	2.5065	2.5991	2.6922	2.7858	2.8798	2.9742	3.0690	3.1642
400	3.2597	3.3557	3.4519	3.5485	3.6455	3.7427	3.8403	3.9382	4.0364	4.1348
500	4.2336	4.3327	4.4320	4.5316	4.6316	4.7318	4.8323	4.9331	5.0342	5.1356
600	5.2373	5.3394	5.4417	5.5445	5.6477	5.7513	5.8553	5.9595	6.0641	6.1690
700	6.2743	6.3799	6.4858	6.5920	6.6986	6.8055	6.9127	7.0202	7.1281	7.2363
800	7.3449	7.4537	7.5629	7.6724	7.7823	7.8925	8.0030	8.1138	8.2250	8.3365
900	8.4483	8.5605	8.6730	8.7858	8.8989	9.0124	9.1262	9.2403	9.3548	9.4696
1000	9.5847	9.7002	9.8159	9.9320	10.0485	10.1652	10.2823	10.3997	10.5174	10.6354
1100	10.7536	10.8720	10.9907	11.1095	11.2286	11.3479	11.4674	11.5871	11.7069	11.8269
1200	11.9471	12.0674	12.1878	12.3084	12.4290	12.5498	12.6707	12.7917	12.9127	13.0338
1300	13.1550	13.2762	13.3975	13.5188	13.6401	13.7614	13.8828	14.0041	14.1254	14.2467
1400	14.3680	14.4892	14.6103	14.7314	14.8524	14.9734	15.9042	15.2150	15.3356	15.4561
1500	15.5765	15.6967	15.8168	15.9368	16.0566	16.1762	16.2956	16.4148	16.5338	16.6526
1600	16.7712	16.8895	17.0076	17.1255	17.2431	17.3604	17.4474	17.5942	17.7105	17.8264
1700	17.9417	18.0562	18.1698	18.2823	18.3937	18.5038	18.6124			

TABLE 4.79 Type T Thermocouples: Copper vs. Copper-Nickel Alloy

Thermoelectric voltage in millivolts; reference junction at 0°C.

°C	0	10	20	30	40	50	60	70	80	90
−200	−5.603	−5.753	−5.889	−6.007	−6.105	−6.181	−6.232	−6.258		
−100	−3.378	−3.656	−3.923	−4.177	−4.419	−4.648	−4.865	−5.069	−5.261	−5.439
−0	0.000	−0.383	−0.757	−1.121	−1.475	−1.819	−2.152	−2.475	−2.788	−3.089
0	0.000	0.391	0.789	1.196	1.611	2.035	2.467	2.908	3.357	3.813
100	4.277	4.749	5.227	5.712	6.204	6.702	7.207	7.718	8.235	8.757
200	9.286	9.820	10.360	10.905	11.456	12.011	12.572	13.137	13.707	14.281
300	14.860	15.443	16.030	16.621	17.217	17.816	18.420	19.027	19.638	20.252
400	20.869									

INDEX

INDEX